Third Edition

Developmental Mathematics

Basic Mathematics and Algebra

Third Edition

Developmental Mathematics

Basic Mathematics and Algebra

Margaret L. Lial
American River College

John Hornsby
University of New Orleans

Terry McGinnis

Stanley A. Salzman
American River College

Diana L. Hestwood
Minneapolis Community and Technical College

PEARSON

Boston Columbus Indianapolis New York San Francisco Upper Saddle River
Amsterdam Cape Town Dubai London Madrid Milan Munich Paris Montréal Toronto
Delhi Mexico City São Paulo Sydney Hong Kong Seoul Singapore Taipei Tokyo

Editorial Director	Christine Hoag
Editor in Chief	Maureen O'Connor
Executive Content Editor	Kari Heen
Content Editor	Christine Whitlock
Senior Content Editor	Lauren Morse
Assistant Editor	Rachel Haskell
Senior Managing Editor	Karen Wernholm
Senior Production Project Manager	Kathleen A. Manley
Digital Assets Manager	Marianne Groth
Supplements Production Coordinator	Kerri Consalvo
Media Producer	Stephanie Green
Software Development	Rebecca Williams, MathXL; Mary Durnwald, TestGen
Marketing Manager	Rachel Ross
Senior Author Support/Technology Specialist	Joe Vetere
Rights and Permissions Advisor	Cheryl Besenjak
Image Manager	Rachel Youdelman
Procurement Manager	Evelyn Beaton
Procurement Specialist	Debbie Rossi
Media Procurement Specialist	Ginny Michaud
Associate Director of Design	Andrea Nix
Senior Designer	Barbara Atkinson
Text Design, Production Coordination, Composition, and Illustrations	Cenveo Publisher Services/ Nesbitt Graphics, Inc.
Cover Image	*Under the Pines* © Lorraine Cota Manley

For permission to use copyrighted material, grateful acknowledgment is made to the copyright holders on page P-1, which is hereby made part of this copyright page.

Library of Congress Cataloging-in-Publication Data

Developmental mathematics : basic mathematics and algebra / Margaret L. Lial . . . [et al.].—3rd ed.

p. cm.

Includes bibliographical references.

ISBN: 978-0-321-85446-9

1. Arithmetic—Textbooks. 2. Algebra—Textbooks. 3. Problem solving—Textbooks. I. Lial, Margaret L.

QA107.2.D48 2013

513—dc23

2012013984

1 2 3 4 5 6 7 8 9 10—CRK—16 15 14 13 12

PEARSON

www.pearsonhighered.com

ISBN 13: 978-0-321-85446-9
ISBN 10: 0-321-85446-2

This book is dedicated to Margaret L. Lial

Always passionate about mathematics and teaching,

Always a valued colleague, a mentor, and a friend,

Always in our memory.

John Hornsby
Terry McGinnis
Stan Salzman
Diana Hestwood

Contents

*Available in MyMathLab®

Preface

In the third edition of *Developmental Mathematics: Basic Mathematics and Algebra*, we have addressed the diverse needs of today's students by creating a tightly coordinated text and technology package that includes integrated activities to help students improve their study skills, an attractive design, updated applications and graphs, helpful features, and careful explanations of concepts. We have also expanded the supplements and study aids. We have revamped the video series into a complete Lial Video Library with expanded video coverage and new, easier navigation. And we have added the new Lial MyWorkBook. We have also responded to the suggestions of users and reviewers and have added many new examples and exercises based on their feedback.

The text is designed for students who need to review basic mathematics and learn introductory algebra topics before moving on to an intermediate algebra course. This book begins with basic mathematics concepts, and then introduces geometry, statistics, and introductory algebra topics. Students will benefit from the text's student-oriented approach. Of particular interest to students and instructors will be the new guided solutions in margin problems and exercises and the new Concept Check exercises.

This text is part of a series that also includes the following books:

- *Essential Mathematics,* Fourth Edition, by Lial and Salzman
- *Basic College Mathematics,* Ninth Edition, by Lial, Salzman, and Hestwood
- *Prealgebra,* Fifth Edition, by Lial and Hestwood
- *Introductory Algebra,* Tenth Edition, by Lial, Hornsby, and McGinnis
- *Intermediate Algebra,* Tenth Edition, by Lial, Hornsby, and McGinnis
- *Introductory and Intermediate Algebra,* Fifth Edition, by Lial, Hornsby, and McGinnis
- *Prealgebra and Introductory Algebra,* Fourth Edition, by Lial, Hestwood, Hornsby, and McGinnis

WHAT'S NEW IN THIS EDITION

We are pleased to offer the following new textbook features and supplements.

▶ ***Engaging Chapter Openers*** The new Chapter Openers portray real-life situations that reflect the mathematical content and are relevant to students. Each opener also includes an expanded outline of the chapter contents. (See pp. 1, 315, and 761—Chapters 1, 5, and 11.)

▶ ***Guided Solutions*** Selected exercises in the margins and in the exercise sets, marked with a GS icon, step students through solutions as they learn new concepts or procedures. (See pp. 113 and 718, margin problems, p. 490, Exercises 11 and 12, and p. 722, Exercises 23 and 24.)

▶ ***Concept Check Exercises*** Each section exercise set now begins with a group of these exercises, designated **CONCEPT CHECK**. They are designed to facilitate students' mathematical thinking and conceptual understanding. Many of these emphasize vocabulary. (See pp. 299 and 721.) Additional Concept Check problems are sprinkled throughout the exercise sets and ask students to apply mathematical processes and concepts or to identify **What Went Wrong?** in incorrect solutions. (See pp. 77 and 675.)

▶ ***Vocabulary Tips*** (Chapters 1–8 and Appendix A) Many students at this level of mathematics do not have strong reading skills. The vocabulary tips included in the margins help them to learn the meaning of root words and prefixes commonly used in mathematics vocabulary (for example, *equ-, centi, tri-*), distinguish the mathematical meaning from the common usage of particular words (such as *volume, average*), and provide tips for remembering the difference between often-confused terms (such as LCM and LCD on p. 209).

- ***Helpful Teaching Tips*** All new **Teaching Tips**, located in the margins of the *Annotated Instructor's Edition,* provide helpful suggestions, emphasize common student trouble spots, and offer other pertinent information that instructors, especially those new to teaching this course, may find helpful. (See pp. 264 and 713.)

- ***Lial Video Library*** The Lial Video Library, available in MyMathLab and on the Video Resources DVD, provides students with a wealth of video resources to help them navigate the road to success. All video resources in the library include optional captions in English and Spanish. The Lial Video Library includes Section Lecture Videos, Solutions Clips, Quick Review Lectures, and Chapter Test Prep Videos. The Chapter Test Prep Videos are also available on YouTube (searchable using author name and book title), or by scanning the QR Code® on the inside back cover for easy access.

- ***MyWorkBook*** This new workbook provides Guided Examples and corresponding Now Try Exercises for each text objective. The extra practice exercises for every section of the text, with ample space for students to show their work, are correlated to Examples, Lecture Videos, and Exercise Solution Clips, to give students the help they need to successfully complete problems. Additionally, MyWorkBook lists the learning objectives and key vocabulary terms for every text section, along with vocabulary practice problems.

CONTENT CHANGES

The scope and sequence of topics in *Developmental Mathematics: Basic Mathematics and Algebra* has stood the test of time and rates highly with our reviewers. Therefore, the table of contents remains intact, making the transition to the new edition easier. Specific content changes include the following:

- We increased the emphasis on checking solutions and answers, as indicated by a new **CHECK** tag and ✓ in the exposition and examples.

- In preparation for Chapter 11 and the material on forms of linear equations, Section 10.5 provides a new example and corresponding exercises on solving a linear equation in two variables x and y for y.

- Section 11.4 on writing and graphing equations of lines provides increased development and coverage of the slope-intercept form, including two new examples.

- Sections 13.2–13.4 include three new examples on factoring trinomials.

- The following topics are among those that have been enhanced and/or expanded:

  Dividing real numbers involving zero (Section 9.6)
  Applying the properties of real numbers (Section 9.7)
  Solving applications involving consecutive integers and angle measures (Section 10.4)
  Using interval notation and solving linear inequalities (Section 10.6)
  Graphing linear equations in two variables using intercepts (Section 11.1)
  Finding numerical values of rational expressions and values for which rational expressions are undefined (Section 14.1)
  Solving systems of equations with decimal coefficients (Section 15.2)
  Solving quadratic equations by using the zero-factor property (Section 17.1)

HALLMARK FEATURES

We have enhanced the following popular features, each of which is designed to increase ease-of-use by students and/or instructors.

▶ ***Study Skills*** Thirteen carefully designed study skills activities provide opportunities for students to practice proven strategies for learning mathematics. Poor study skills and behaviors are major factors in low success rates in mathematics courses. Research shows that a few generic tips sprinkled here and there are not enough to help students change their study behaviors. Because students need specific instruction in study skills, we have contextualized them, integrating them into the text material. Topics include note taking, homework, study cards, math anxiety, test preparation, test taking, preparing for a final exam, and more. (See pp. 136–137, 176–177, and 178–179.) Most are located within the first few chapters so that students can use the skills throughout the course. (See the Contents for titles and locations.) The first activity, "Your Brain *Can* Learn Mathematics," explains how the brain actually learns and remembers so that students understand why the study skills will help them succeed in the course.

▶ ***Emphasis on Problem Solving*** A six-step method for solving applied problems with basic mathematics is introduced at the end of Chapter 1 (see p. 88), and this method is modified for solving applied problems with algebra in Chapter 10 (see p. 713). These steps are emphasized in boldface type and repeated in examples and exercises to reinforce the problem-solving process for students. (See pp. 418 and 716.) We also provide students with **Problem-Solving Hint** boxes that feature helpful problem-solving tips and strategies. (See pages 615 and 714.)

▶ ***Helpful Learning Objectives*** We begin each section with clearly stated, numbered objectives, and the included material is directly keyed to these objectives so that students and instructors know exactly what is covered in each section. (See pages 42 and 606.)

▶ ***Popular Cautions and Notes*** One of the most popular features of previous editions, we include information marked **CAUTION** and **Note** to warn students about common errors and emphasize important ideas throughout the exposition. The updated text design makes them easy to spot. (See pages 140, 141, 704, and 706.)

▶ ***Comprehensive Examples*** The new edition features a multitude of step-by-step, worked-out examples that include pedagogical color, helpful side comments, and special pointers. We give increased attention to checking example solutions—more checks, designated using a special **CHECK** tag and ✓, are included than in past editions. (See pages 690 and 1146–1147.)

▶ ***More Pointers*** Well received by both students and instructors in the previous edition, we incorporate more pointers in examples and discussions throughout this edition of the text. They provide students with important on-the-spot reminders and warnings about common pitfalls. (See pages 132 and 701.)

▶ ***Ample Margin Problems*** Margin problems, with answers immediately available at the bottom of the page, are found in every section of the text. (See pp. 318 and 608.) This expanded key feature allows students to immediately practice the material covered in the examples in preparation for the exercise sets. Many include new guided solutions.

▶ ***Updated Figures, Photos, and Hand-Drawn Graphs*** Today's students are more visually oriented than ever. As a result, we have made a concerted effort to include appealing mathematical figures, diagrams, tables, and graphs, including a "hand-drawn" style of graphs, whenever possible. (See pp. 778 and 1185–1188.) We have incorporated new depictions of well-known mathematicians, as well as new photos to accompany applications in examples and exercises. (See pp. 943 and 1085.)

- ***Optional Calculator Tips*** These tips, marked, offer helpful information and instruction for students using calculators in the course. (See pp. 268 and 609.)

- ***Relevant Real-Life Applications*** We include many new or updated applications from fields such as business, pop culture, sports, technology, and the health sciences that show the relevance of mathematics to daily life. (See pp. 613 and 723.)

- ***Extensive and Varied Exercise Sets*** The text contains a wealth of exercises to provide students with opportunities to practice, apply, connect, review, and extend the skills they are learning. Numerous illustrations, tables, graphs, and photos help students visualize the problems they are solving. Problem types include skill building, writing, and calculator exercises, as well as applications, matching, true/false, multiple-choice, and fill-in-the-blank problems. (See pp. 666–668, 725, and 813.)

  In the Annotated Instructor's Edition of the text, the writing exercises are marked with an icon so that instructors may assign these problems at their discretion. Exercises suitable for calculator work are marked in both the student and instructor editions with a calculator icon. Students can watch an instructor work through the complete solution for all exercises marked with a Play Button icon on the Videos on DVD or in MyMathLab.

- ***Flexible Relating Concepts Exercises*** These help students tie concepts together and develop higher level problem-solving skills as they compare and contrast ideas, identify and describe patterns, and extend concepts to new situations. (See pp. 584 and 1077.) These exercises, now located at the end of selected exercise sets, make great collaborative activities for pairs or small groups of students.

- ***Math in the Media*** Each of these one-page activities presents a relevant look at how mathematics is used in the media. Designed to help instructors answer the often-asked question, "When will I ever use this stuff?," these activities ask students to read and interpret data from newspaper articles, the Internet, and other familiar, real-world sources. (See pp. 63 and 454.) The activities are well-suited to collaborative work or they can be completed by individuals or used for open-ended class discussions.

- ***Step-by-Step Solutions to Selected Exercises*** Exercise numbers enclosed in a blue square, such as **11.**, indicate that a worked-out solution for the problem is included at the back of the text. These solutions are given for selected exercises that most commonly cause students difficulty. (See pp. S-1 through S-31.)

- ***Extensive Review Opportunities*** We conclude each chapter with the following important review components:

  A **Chapter Summary** that features a helpful list of **Key Terms,** organized by section, **New Symbols, Test Your Word Power** vocabulary quiz (with answers immediately following), and a **Quick Review** of each section's contents, complete with additional examples (See pp. 676–679.)

  A comprehensive set of **Chapter Review Exercises,** keyed to individual sections for easy student reference, as well as a set of **Mixed Review Exercises** that helps students further synthesize concepts (See pp. 756–758.)

  A **Chapter Test** that students can take under test conditions to see how well they have mastered the chapter material (See pp. 759–760.)

  Chapters 8 and 17 each conclude with a set of **Cumulative Review Exercises** (See pp. 601–604 and 1207–1212.)

STUDENT SUPPLEMENTS

Student's Solutions Manual

- By Jeffery A. Cole, Anoka-Ramsey Community College
- Provides detailed solutions to the odd-numbered section-level exercises and to all margin, Relating Concepts, Summary, Chapter Review, Chapter Test, and Cumulative Review Exercises

ISBNs: 0-321-85445-4, 978-0-321-85445-2

NEW MyWorkBook

- Provides Guided Examples and corresponding Now Try Exercises for each text objective
- Refers students to correlated Examples, Lecture Videos, and Exercise Solution Clips
- Includes extra practice exercises for every section of the text with ample space for students to show their work
- Lists the learning objectives and key vocabulary terms for every text section, along with vocabulary practice problems

ISBNs: 0-321-85465-9, 978-0-321-85465-0

NEW Lial Video Library

The Lial Video Library, available in MyMathLab and on the Video Resources DVD, provides students with a wealth of video resources to help them navigate the road to success! All video resources in the library include optional subtitles in English and Spanish. The Lial Video Library includes the following resources:

- **Section Lecture Videos** offer a new navigation menu that allows students to easily focus on the key examples and exercises that they need to review in each section.
- **Solutions Clips** show an instructor working through the complete solutions to selected exercises from the text. Exercises with a solution clip are marked in the text and e-book with a Play Button icon ▶.
- **Quick Review Lectures** provide a short summary lecture of each key concept from the Quick Reviews at the end of every chapter in the text.
- **The Chapter Test Prep Videos** let students watch instructors work through step-by-step solutions to all the Chapter Test exercises from the textbook. Chapter Test Prep videos are also available on YouTube™ (search using author name and book title) and in MyMathLab, or by scanning the QR Code® on the inside back cover for easy access.

INSTRUCTOR SUPPLEMENTS

Annotated Instructor's Edition

- Provides answers to all text exercises in color next to the corresponding problems
- Includes all **NEW** Teaching Tips located in the margins
- Icons identify writing and calculator exercises

ISBNs: 0-321-85472-1, 978-0-321-85472-8

Instructor's Solutions Manual (Download only)

- By Jeffery A. Cole, Anoka-Ramsey Community College
- Provides complete solutions to all exercises in the text
- Available for download at www.pearsonhighered.com

ISBNs: 0-321-85473-X, 978-0-321-85473-5

Instructor's Resource Manual with Tests and Mini-Lectures (Download only)

- Contains a test bank with two diagnostic pretests, six free-response and two multiple-choice test forms per chapter, and two final exams
- Contains a mini-lecture for each section of the text with objectives, key examples, and teaching tips
- Includes a correlation guide from the second to the third edition and phonetic spellings for all key terms in the text
- Includes resources to help both new and adjunct faculty with course preparation and classroom management, by offering helpful teaching tips correlated to the sections of the text
- Available for download at www.pearsonhighered.com

ISBNs: 0-321-85451-9, 978-0-321-85451-3

ADDITIONAL MEDIA SUPPLEMENTS

MyMathLab®

MyMathLab® Online Course (access code required)

MyMathLab from Pearson is the world's leading online resource in mathematics, integrating interactive homework, assessment, and media in a flexible, easy-to-use format. MyMathLab delivers **proven results** in helping individual students succeed. It provides **engaging experiences** that personalize, stimulate, and measure learning for each student. And, it comes from an **experienced partner** with educational expertise and an eye on the future.

To learn more about how MyMathLab combines proven learning applications with powerful assessment, visit **www.mymathlab.com** or contact your Pearson representative.

MyMathLab® Ready to Go Course (access code required)
These new Ready to Go courses provide students with all the same great MyMathLab features but make it easier for instructors to get started. Each course includes pre-assigned homework and quizzes to make creating a course even simpler. Ask your Pearson representative about the details for this particular course or to see a copy of this course.

MyMathLab® Plus/MyStatLab™ Plus
MyLabsPlus combines proven results and engaging experiences from MyMathLab® and MyStatLab™ with convenient management tools and a dedicated services team. Designed to support growing math and statistics programs, it includes additional features such as:

- **Batch Enrollment:** Schools can create the login name and password for every student and instructor, so everyone can be ready to start class on the first day. Automation of this process is also possible through integration with the school's Student Information System.
- **Login from a campus portal:** Students and instructors can link directly from their campus portal into MyLabsPlus courses. A Pearson service team works with each institution to create a single sign-on experience for instructors and students.
- **Advanced Reporting:** MyLabsPlus's advanced reporting allows instructors to review and analyze students' strengths and weaknesses by tracking their performance on tests, assignments, and tutorials. Administrators can review grades and assignments across all courses on the MyLabsPlus campus for a broad overview of program performance.
- **24/7 Support:** Students and instructors receive 24/7 support, 365 days a year, by email or online chat.

MyLabsPlus is available to qualified adopters. For more information, visit our website at *www.mylabsplus.com* or contact your Pearson representative.

MathXL®

MathXL® Online Course (access code required)
MathXL is the homework and assessment engine that runs MyMathLab. (MyMathLab is MathXL plus a learning management system.)

With MathXL, instructors can:

- Create, edit, and assign online homework and tests using algorithmically generated exercises correlated at the objective level to the textbook.
- Create and assign their own online exercises and import TestGen tests for added flexibility.
- Maintain records of all student work tracked in MathXL's online gradebook.

With MathXL, students can:

- Take chapter tests in MathXL and receive personalized study plans and/or personalized homework assignments based on their test results.
- Use the study plan and/or the homework to link directly to tutorial exercises for the objectives they need to study.
- Access supplemental animations and video clips directly from selected exercises.

MathXL is available to qualified adopters. For more information, visit our website at *www.mathxl.com*, or contact your Pearson representative.

TestGen®
TestGen (*www.pearsoned.com/testgen*) enables instructors to build, edit, print, and administer tests using a computerized bank of questions developed to cover all the objectives of the text. TestGen is algorithmically based, allowing instructors to create multiple but equivalent versions of the same question or test with the click of a button. Instructors can also modify test bank questions or add new questions. The software and testbank are available for download from Pearson Education's online catalog.

PowerPoint® Lecture Slides

- Present key concepts and definitions from the text
- Available for download at *www.pearsonhighered.com* or in MyMathLab

ACKNOWLEDGMENTS

The comments, criticisms, and suggestions of users, nonusers, instructors, and students have positively shaped this textbook over the years, and we are most grateful for the many responses we have received. The feedback gathered for this revision of the text was particularly helpful, and we especially wish to thank the following individuals who provided invaluable suggestions for this and the previous edition:

Mary Kay Abbey, *Montgomery College*
Carla Ainsworth, *Salt Lake Community College*
George Alexander, *University of Wisconsin College*
Randall Allbritton, *Daytona State College*
Sonya Armstrong, *West Virginia State University*
Jannette Avery, *Monroe Community College*
Linda Beattie, *Western New Mexico University*
Linda Beller, *Brevard Community College*
Solveig R. Bender, *William Rainey Harper College*
Carla J. Bissell, *University of Nebraska at Omaha*
Jean Bolyard, *Fairmont State University*
Vernon Bridges, *Durham Technical Community College*
Barbara Brown, *Anoka-Ramsey Community College*
Kim Brown, *Tarrant County College—Northeast Campus*
Hien Bui, *Hillsborough Community College*
Tim C. Caldwell, *Meridian Community College*
Russell Campbell, *Fairmont State University*
Ernie Chavez, *Gateway Community College*
John Close, *Salt Lake Community College*
Terry Joe Collins, *Hinds Community College*
Dawn Cox, *Cochise College*
Jane Cuellar, *Taft College*
Martha Daniels, *Central Oregon Community College*
Ky Davis, *Zane State College*
Julie Dewan, *Mohawk Valley Community College*
Bill Dunn, *Las Positas College*
Lucy Edwards, *Las Positas College*
Rob Farinelli, *Community College of Allegheny—Boyce Campus*
Scott Fallstrom, *Shoreline Community College*
Matthew Flacche, *Camden Community College*
Donna Foster, *Piedmont Technical College*
Randy Gallaher, *Lewis and Clark Community College*
Lourdes Gonzalez, *Miami-Dade College*
Mark Gollwitzer, *Greenville Technical College*
J. Lloyd Harris, *Gulf Coast Community College*
Terry Haynes, *Eastern Oklahoma State College*
Edith Hays, *Texas Woman's University*
Anthony Hearn, *Community College of Philadelphia*
Karen Heavin, *Morehead State University*
Lance Hemlow, *Raritan Valley Community College*
Elizabeth Heston, *Monroe Community College*
Joe Howe, *St. Charles County Community College*
Matthew Hudock, *St. Phillip's College*
Sharon Jackson, *Brookhaven College*
Rose Kaniper, *Burlington County College*
Rosemary Karr, *Collin College*
Amy Kauffman, *Ivy Tech Community College*
Harriet Kiser, *Georgia Highlands College*
Jeffrey Kroll, *Brazosport College*
Babara Krueger, *Cochise College*
Valerie Lazzara, *Palm Beach State College*
Christine Heinecke Lehmann, *Purdue University–North Central*
Douglas Lewis, *Yakima Valley Community College*
Sandy Lofstock, *California Lutheran University*
Lou Ann Mahaney, *Tarrant County College–Northeast Campus*
Valerie H. Maley, *Cape Fear Community College*
Susan McClory, *San Jose State University*
Gary McCracken, *Shelton State Community College*
Judy Mee, *Oklahoma City Community College*
Pam Miller, *Phoenix College*
Wayne Miller, *Lee College*
Jeffrey Mills, *Ohio State University*
Elizabeth Morrison, *Valencia Community College—West Campus*
Linda J. Murphy, *Northern Essex Community College*
Celia Nippert, *Western Oklahoma State College*
Elizabeth Olgilvie, *Horry-Georgetown Technical College*
Ted Panitz, *Cape Cod Community College*
Claire Peacock, *Chattanooga State Technical Community College*
Kathy Peay, *Sampson Community College*
Faith Peters, *Miami-Dade College*
Thea Philliou, *Santa Fe University of Art and Design*
Larry Potanski, *Pueblo Community College*
Manoj Raghunandanan, *Temple University*
Serban Raianu, *California State University—Dominguez Hills*
Janice Rech, *University of Nebraska at Omaha*
Janalyn Richards, *Idaho State University*
Jane Roads, *Moberly Area Community College*
Diann Robinson, *Ivy Tech State College–Lafayette*
Richard D. Rupp, *Del Mar College*
Rachael Schettenhelm, *Southern Connecticut State University*
Ellen Sawyer, *College of DuPage*
Lois Schuppig, *College of Mount St. Joseph*
Mary Lee Seitz, *Erie Community College—City Campus*
Julia Simms, *Southern Illinois University—Edwardsville*
Sounny Slitine, *Palo Alto College*
Dwight Smith, *Big Sandy Community and Technical College*
Lee Ann Spahr, *Durham Technical Community College*
Julia Speights, *Shelton State Community College*
Theresa Stalder, *University of Illinois—Chicago*
Carol Stewart, *Fairmont State University*
Kathryn Taylor, *Santa Ana College*
Sharon Testone, *Onondaga Community College*
Shae Thompson, *Montana State University*
Mike Tieleman-Ward, *Anoka Technical College*
Mark Tom, *College of the Sequoias*
Sven Trenholm, *North Country Community College*

Bettie A. Truitt, *Black Hawk College*
Cora S. West, *Florida State College at Jacksonville*
Cheryl Wilcox, *Diablo Valley College*
Johanna Windmueller, *Seminole State College of Florida*
Jackie Wing, *Angelina College*
Gabriel Yimesghen, *Community College of Philadelphia*
Kevin Yokoyama, *College of the Redwoods*
Karl Zilm, *Lewis and Clark Community College*

Our sincere thanks go to the dedicated individuals at Pearson who have worked hard to make this revision a success: Maureen O'Connor, Kathy Manley, Barbara Atkinson, Michelle Renda, Rachel Ross, Kari Heen, Christine Whitlock, Lauren Morse, Stephanie Green, and Rachel Haskell.

Abby Tanenbaum did an outstanding job helping us with manuscript preparation and researching applications, and prepared the index. We are truly grateful for her contributions to so many of our books over the years. Special thanks go to Jill Owens, an instructor at Diablo Valley College, who assisted with manuscript development and provided accuracy checking throughout this edition, and to Janis M. Cimperman, Associate Professor of Mathematics at St. Cloud State University, for her careful review of the new material in selected chapters.

We are also grateful to Marilyn Dwyer, Carol Merrigan, and Kathy Diamond of Cenveo/Nesbitt Graphics for their excellent production work; Bonnie Boehme and David Abel for supplying their copyediting expertise; Beth Anderson, for her fine photo research; Jeff Cole, for writing the Solutions Manuals; Perian Herring, for writing many of the Vocabulary Tips and accuracy checking the manuscript; Sam Tinsley, Sarah Sponholz, Janis Cimperman, Paul Lorczak, and Chris Heeren, for their accuracy checking; and Judy Martinez, for her excellent typing skills.

Linda Russell, who developed and wrote the Study Skills activities that appear throughout this text, also worked to make the text more readable for developmental-level students, wrote many of the vocabulary tips, and provided much-needed help during the production phase. Her many years of teaching and working with students at this level was invaluable.

The ultimate measure of this textbook's success is whether it helps students master basic skills, develop problem-solving techniques, and increase their confidence in learning and using mathematics. In order for us, as authors, to know what to keep and what to improve for the next edition, we need to hear from you, the instructor, and you, the student. Please tell us what you like and where you need additional help by sending an e-mail to *math@pearson.com*. We appreciate your feedback!

John Hornsby
Terry McGinnis
Stanley A. Salzman
Diana L. Hestwood

Developmental Mathematics

Basic Mathematics and Algebra

Third Edition

1 Whole Numbers

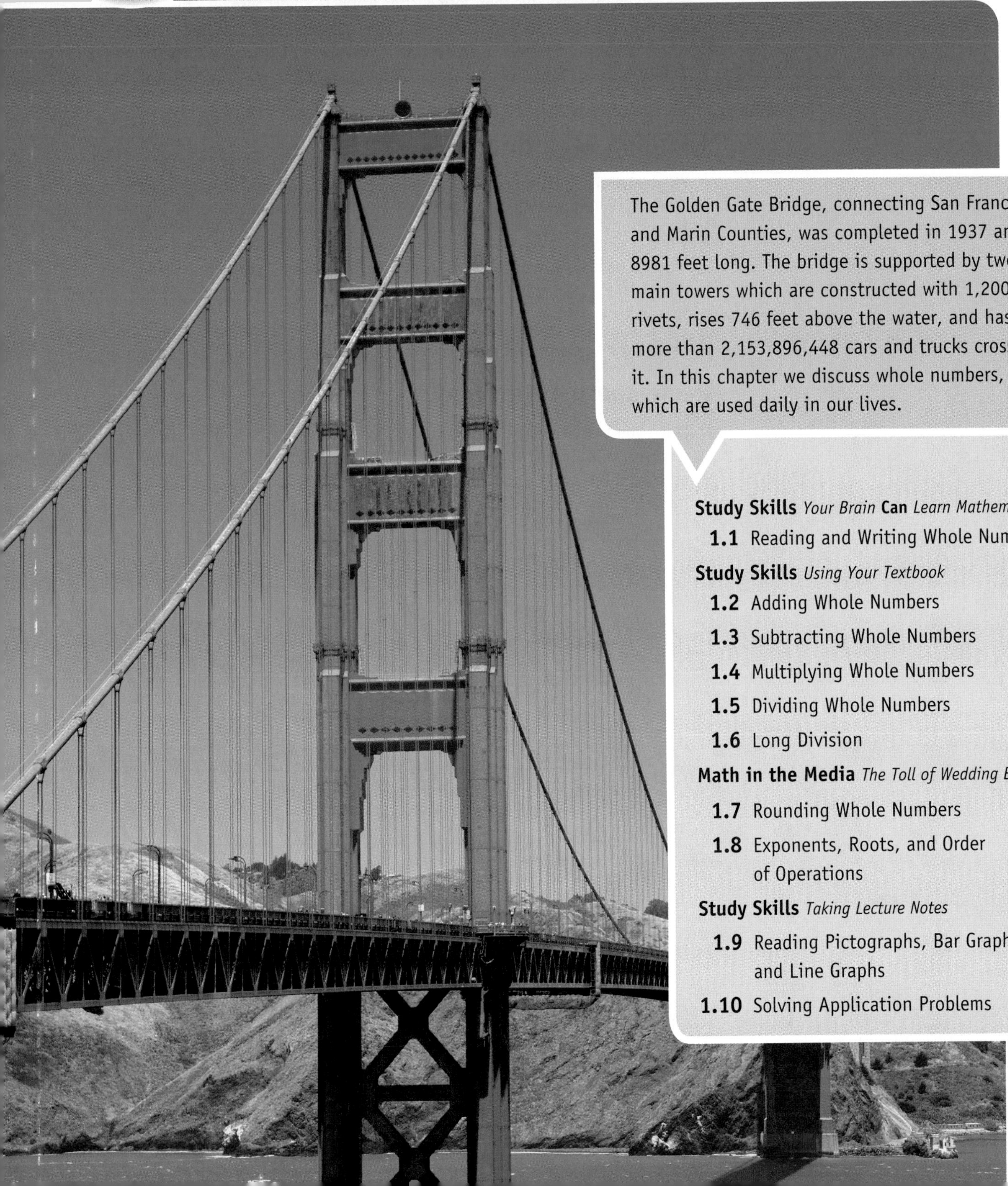

The Golden Gate Bridge, connecting San Francisco and Marin Counties, was completed in 1937 and is 8981 feet long. The bridge is supported by two main towers which are constructed with 1,200,000 rivets, rises 746 feet above the water, and has had more than 2,153,896,448 cars and trucks cross over it. In this chapter we discuss whole numbers, which are used daily in our lives.

Study Skills

YOUR BRAIN *CAN* LEARN MATHEMATICS

OBJECTIVES

1. **Describe how practice fosters dendrite growth.**
2. **Explain the effect of anxiety on the brain.**

Your brain knows how to learn, just as your lungs know how to breathe; however, there are important things you can do to maximize your brain's ability to do its work. This short introduction will help you choose effective strategies for learning mathematics. This is a simplified explanation of a complex process.

Your brain's outer layer is called the **neocortex,** which is where higher level thinking, language, reasoning, and purposeful behavior occur. The neocortex has about 100 billion (100,000,000,000) brain cells called **neurons.**

Learning Something New

- As you learn something new, threadlike branches grow out of each neuron. These branches are called **dendrites.**
- When the dendrite from one neuron grows close enough to the dendrite from another neuron, a connection is made. There is a small gap at the connection point called a **synapse.** One dendrite sends an electrical signal across the gap to another dendrite.
- ***Learning = growth and connecting of dendrites.***

Remembering New Skills

- When you practice a skill just once or twice, the connections between neurons are very weak. If you do not practice the skill again, the dendrites at the connection points wither and die back. You have forgotten the new skill!

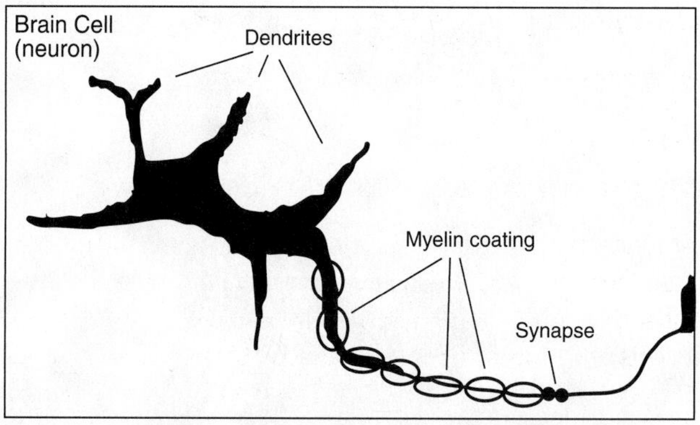

A neuron with several dendrites: one dendrite has developed a myelin coating through repeated practice.

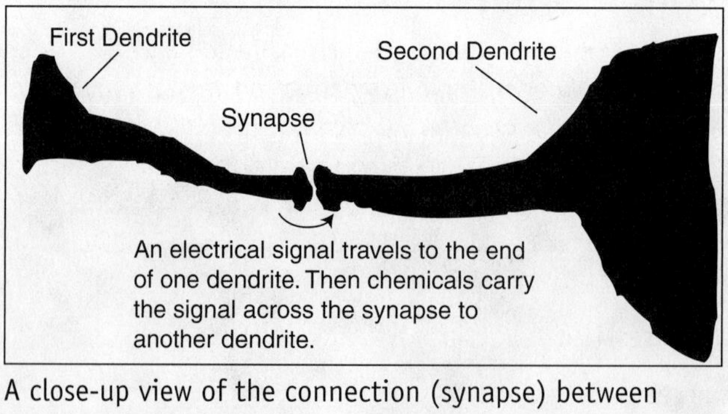

A close-up view of the connection (synapse) between two dendrites.

- If you practice a new skill many times, the dendrites for that skill become coated with a fatty protein called **myelin.** Each time one dendrite sends a signal to another dendrite, the myelin coating becomes thicker and smoother, allowing the signals to move faster and with less interference. Thinking can now occur more quickly and easily, and ***you will remember the skill for a long time*** because the dendrite connections are very strong.

Become An Effective Student

- You grow dendrites specifically for the thing you are studying. If you practice dividing fractions, you will grow specialized dendrites just for dividing fractions. If you ***watch other people*** solve fraction problems, ***you will grow dendrites for watching, not for solving.*** So, be sure you are actively learning and practicing.
- If you practice something the ***wrong*** way, you will develop strong dendrite connections for doing it the wrong way! So, as you study, check frequently that you are getting correct answers.
- As you study a new topic that is related to things you already know, you will grow new dendrites, but your brain will also send signals throughout the network of dendrites for the related topics. In this way, you build a complex **neural network** that allows you to apply concepts, see differences and similarities between ideas, and understand relationships between concepts.

In the first few chapters of this textbook you will find "brain friendly" activities that are designed to help you grow and develop your own reliable neural networks for mathematics. Since you must grow your own dendrites (no one can grow them for you), these activities show you how to

- develop new dendrites,
- strengthen existing ones, and
- encourage the myelin coating to become thicker so signals are sent with less effort.

When you incorporate the activities into your regular study routine, you will discover that you understand better, remember longer, and forget less.

Also remember that ***it does take time for dendrites to grow.*** Trying to cram in several new concepts and skills at the last minute is not possible. Your dendrites simply can't grow that quickly. You can't expect to develop huge muscles by lifting weights for just one evening before a body building competition! In the same way, practice the study techniques ***throughout the course*** to facilitate strong growth of dendrites.

When Anxiety Strikes

If you are under stress or feeling anxious, such as during a test, your body secretes **adrenaline** into your system. Adrenaline in the brain blocks connections between neurons. In other words, you can't think! If you've ever experienced "blanking out" on a test, you know what adrenaline does. You'll learn several solutions to that problem in later activities.

Start Your Course Right!

- Attend all class sessions (especially the first one).
- Gather the necessary supplies.
- Carefully read the syllabus for the course, and ask questions if you don't understand.

1.1 Reading and Writing Whole Numbers

OBJECTIVES

1. Identify whole numbers.
2. Give the place value of a digit.
3. Write a number in words or digits.
4. Read a table.

VOCABULARY TIP

Place value In our decimal number system, each place has a value of 10 times the place to its right.

1 Identify the place value of the 4 in each whole number.

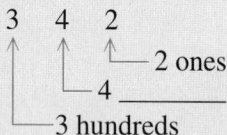

GS **(a)** 342

3 4 2
- 2 → 2 ones
- 4 → 4 ________
- 3 → 3 hundreds

(b) 714

(c) 479

Answers

1. **(a)** tens **(b)** ones **(c)** hundreds

Knowing how to read and write numbers is important in mathematics.

OBJECTIVE 1 Identify whole numbers. The **decimal system** of writing numbers uses the ten digits

0, 1, 2, 3, 4, 5, 6, 7, 8, 9

to write any number. These digits can be used to write **the whole numbers:**

0, 1, 2, 3, 4, 5, 6, 7, 8, 9, 10, 11, 12, 13...

The three dots indicate that the list goes on forever.

OBJECTIVE 2 Give the place value of a digit. Each digit in a whole number has a **place value,** depending on its position in the whole number. The following place value chart shows the names of the different places used most often and has the whole number 402,759,780 entered.

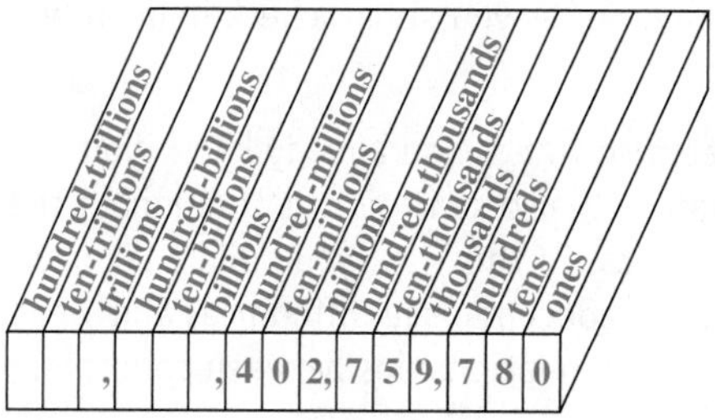

The United States is the leading consumer of coffee in the world. Each day we drink 402,759,780 cups of coffee. Each of the 7s in 402,759,780 represents a different amount because of its position, or *place value,* within the number. The *place value* of the 7 on the left is 7 hundred-thousands (700,000). The *place value* of the 7 on the right is 7 hundreds (700).

EXAMPLE 1 Identifying Place Values

Identify the place value of 8 in each whole number.

Each "8" has a different value.

(a) 28 — 8 ones **(b)** 85 — 8 tens **(c)** 869 — 8 hundreds

Notice that the value of 8 in each number is different, depending on its location (place) in the number.

◀ **Work Problem 1 at the Side.**

EXAMPLE 2 Identifying Place Values

Identify the place value of each digit in the number 725,283.

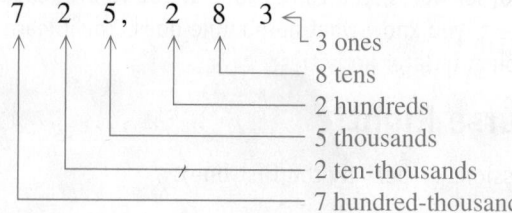

7 2 5, 2 8 3
- 3 → 3 ones
- 8 → 8 tens
- 2 → 2 hundreds
- 5 → 5 thousands
- 2 → 2 ten-thousands
- 7 → 7 hundred-thousands

Continued on Next Page

Notice the comma between the hundreds and thousands position in the number 725,283 in **Example 2.**

Work Problem 2 at the Side. ▶

Using Commas

Commas are used to separate each group of three digits, starting from the right. This makes numbers easier to read. (An exception: Commas are frequently omitted in four-digit numbers such as 9748 or 1329.) Each three-digit group is called a **period.** Some instructors prefer to just call them **groups.**

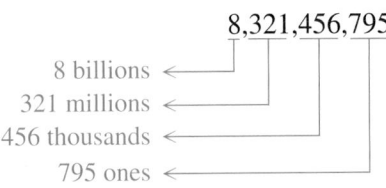

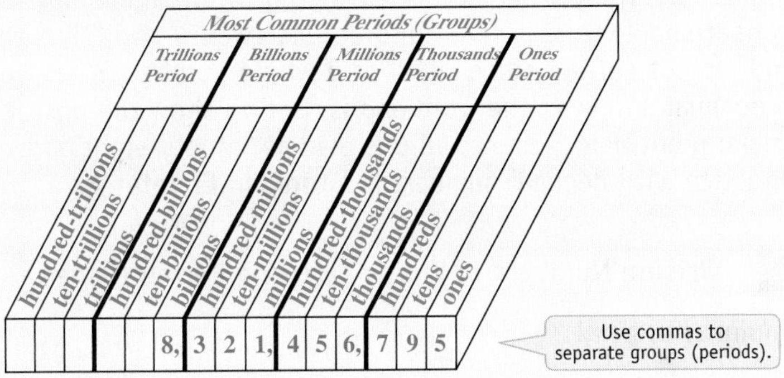

EXAMPLE 3 Knowing the Period or Group Names

Write the digits in each period of 8,321,456,795.

8,321,456,795

8 billions
321 millions
456 thousands
795 ones

Work Problem 3 at the Side. ▶

Use the following rule to read a number with more than three digits.

Writing Numbers in Words

Start at the left when writing a number in words or saying it aloud. Write or say the digit names in each period (group), followed by the name of the period, except for the period name "ones," which is *not* used.

OBJECTIVE 3 Write a number in words or digits. The following examples show how to write names for whole numbers.

EXAMPLE 4 Writing Numbers in Words

Write each number in words.

(a) 57

This number means 5 tens and 7 ones, or 50 ones and 7 ones. Write the number as

fifty-seven.

Continued on Next Page

2 Identify the place value of each digit.

GS **(a)** 14,218

8 ones
____ tens
____ hundreds
____ thousands
____ ten-thousands

(b) 460,329

3 In the number 3,251,609,328 identify the digits in each period (group).

(a) billions period

(b) millions period

(c) thousands period

(d) ones period

Answers

2. **(a)** 1 : ten-thousands
4 : thousands
2 : hundreds
1 : tens
8 : ones
(b) 4 : hundred-thousands
6 : ten-thousands
0 : thousands
3 : hundreds
2 : tens
9 : ones

3. **(a)** 3 **(b)** 251 **(c)** 609 **(d)** 328

4 Write each number in words.

(a) 18

(b) 36

(c) 418

(d) 902

5 Write each number in words.

GS **(a)** 3104

three __________,

one __________ four

GS **(b)** 95,372

ninety-five __________,
three hundred seventy-two

(c) 100,075,002

(d) 11,022,040,000

6 Rewrite each number using digits.

(a) one thousand, four hundred thirty-seven

(b) nine hundred seventy-one thousand, six

(c) eighty-two million, three hundred twenty-five

Answers

4. (a) eighteen
   (b) thirty-six
   (c) four hundred eighteen
   (d) nine hundred two
5. (a) three thousand, one hundred four
   (b) ninety-five thousand, three hundred seventy-two
   (c) one hundred million, seventy-five thousand, two
   (d) eleven billion, twenty-two million, forty thousand
6. (a) 1437 (b) 971,006 (c) 82,000,325

(b) 94

ninety-four

Remember: Start at the *left* to read a number.

(c) 874

eight hundred seventy-four

(d) 601

six hundred one

◀ **Work Problem 4 at the Side.**

CAUTION

The word *and* should never be used when writing whole numbers. You will often hear someone say "five hundred *and* twenty-two," but the use of "and" is not correct since "522" is a whole number. When you work with decimal numbers, the word *and* is used to show the position of the decimal point. For example, 98.6 is read as "ninety-eight *and* six tenths." Practice with decimal numbers is a topic in **Chapter 4**.

EXAMPLE 5 Writing Numbers in Words by Using Period Names

Write each number in words.

(a) 725,283

seven hundred twenty-five **thousand,** two hundred eighty-three

(seven hundred twenty-five: Number in period; thousand: Name of period; two hundred eighty-three: Number in period (not necessary to write "ones"))

(b) 7252

Careful: *Do not* use "and" when reading a whole number.

seven **thousand,** two hundred fifty-two

(thousand: Name of period; fifty-two: No period name needed)

(c) 111,356,075

one hundred eleven **million,** three hundred fifty-six **thousand,** seventy-five

The period name is not used for the ones period.

(d) 17,000,017,000

seventeen **billion,** seventeen **thousand**

◀ **Work Problem 5 at the Side.**

EXAMPLE 6 Writing Numbers in Digits

Rewrite each number using digits.

(a) six **thousand,** twenty-two

6022

With 4 digits or fewer, *no comma* is needed.

(b) two hundred fifty-six **thousand,** six hundred twelve

256,612

(c) nine **million,** five hundred fifty-nine

9,000,559

Zeros indicate there are no thousands.

◀ **Work Problem 6 at the Side.**

Calculator Tip

Does your calculator show a comma between each group of three digits? Probably not, but try entering a long number such as 34,629,075. Notice that there is no key with a comma on it, so you do not enter commas. A few calculators may show the position of the commas *above* the digits, like this

34'629'075

Most of the time you will have to write in the commas where needed.

OBJECTIVE 4 **Read a table.** A common way of showing number values is by using a **table.** Tables organize and display facts so that they are more easily understood. The following table shows some past facts and future predictions for the United States. These numbers give us a glimpse of what we can expect in the 21st century.

NUMBERS FOR THE 21ST CENTURY

Year	1990	2010	2020*
U.S. population	261 million	309 million	338 million
Household income	$42,936	$46,326	$55,735
Average yearly salary	$21,129	$28,834	$32,080

*Estimated figures

Source: *Family Circle* magazine; U.S. Census Bureau.

If you read from left to right along the row labeled "U.S. population," you find that the population in 1990 was 261 million, then the population in 2010 was 309 million, and the estimated population for 2020 is 338 million.

EXAMPLE 7 **Reading a Table**

Use the table to find each number, and write the number in words.

(a) The estimated household income in the year 2020

Read from left to right along the row labeled "Household income" until you reach the 2020 column and find $55,735.

Fifty-five thousand, seven hundred thirty-five dollars

(b) The average yearly salary in 1990

Read from left to right along the row labeled "Average yearly salary." In the 1990 column you find $21,129.

Remember: Use hyphens when necessary.

Twenty-one thousand, one hundred twenty-nine dollars

Work Problem 7 at the Side. ▶

Note

Notice in **Example 7** that hyphens are used when writing numbers in words. A hyphen is used when writing the numbers 21 through 99 (twenty-one through ninety-nine), except for numbers ending in zero (20, 30, 40, ... 90).

7 Use the table to find each number, and write the number in digits when given in words, or write the number in words when given in digits.

GS **(a)** The population in 2010

The U.S. population in the 2010 column is 309 million and is written in digits as 3__ __,000,000

(b) The estimated population in 2020

(c) Household income in 1990

(d) The estimated average yearly salary in 2020

Answers

7. **(a)** 0; 9; 309,000,000
(b) 338,000,000
(c) forty-two thousand, nine hundred thirty-six dollars
(d) thirty-two thousand, eighty dollars

1.1 Exercises

FOR EXTRA HELP

Download the MyDashBoard App

MyMathLab®

CONCEPT CHECK *Choose the letter of the correct response.*

1. Which is the digit in the hundreds place in the whole number 3065?
   A. 5 **B.** 3 **C.** 0 **D.** 6

2. Which is the digit in the ten-thousands place in the whole number 134,681?
   A. 6 **B.** 3 **C.** 8 **D.** 1

Write the digit for the given **place value** *in each whole number.* ***See Examples 1 and 2.***

3. 18,015
   ten-thousands
   hundreds

4. 86,332
   ten-thousands
   ones

5. 7,628,592,183
   millions
   thousands

6. 1,700,225,016
   billions
   millions

CONCEPT CHECK *Identify the correct period.*

7. Write the digits in the thousands period in the whole number 552,687,318.

8. Write the digits in the millions period in the whole number 947,321,876,528.

Write the digits for the given **period** *(group) in each whole number.* ***See Example 3.***

9. 3,561,435
   millions
   thousands
   ones

10. 100,258,100,006
    billions
    millions
    thousands
    ones

11. Do you think the fact that humans have four fingers and a thumb on each hand explains why we use a number system based on ten digits? Explain.

12. The decimal system uses ten digits. Fingers and toes are often referred to as digits. In your opinion, is there a relationship here? Explain.

CONCEPT CHECK *Answer* true *or* false *for each statement.*

13. The number 23,115 is written in words as twenty-three thousand and one hundred and fifteen.

14. The number 37,886 is written in words as thirty-seven thousand, eight hundred eighty-six.

Write each number in words. ***See Examples 4 and 5.***

15. 346,009
    GS three hundred forty six ______, ______

16. 218,033
    GS two hundred eighteen ______, thirty- ______

17. 25,756,665

18. 999,993,000

Write each number using digits. ***See Example 6.***

19. sixty-three thousand, one hundred sixty-three
    GS 6 __, 1 __ __

20. ninety-five thousand, one hundred eleven
    GS 9 __, __ __ 1

21. ten million, two hundred twenty-three

22. one hundred million, two hundred

Write the numbers from each sentence using digits. **See Example 6.**

23. There are three million, two hundred thousand parachute jumps in the United States each year. (*Source:* History Channel.)

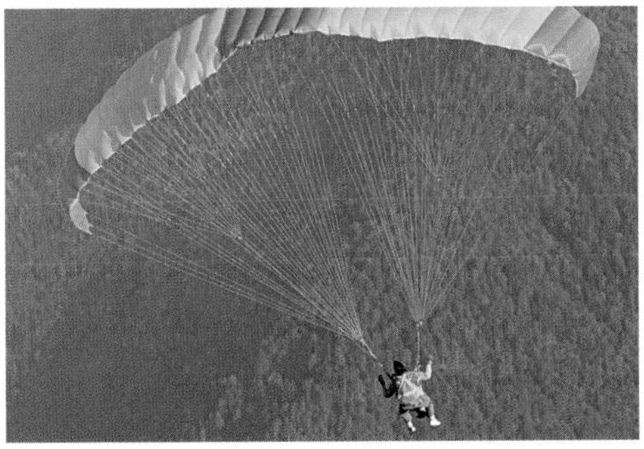

24. A full-grown caterpillar is 27,000 times its birth size. A 9-pound human baby growing at the same rate would weigh two hundred forty-three thousand pounds by college graduation. (*Source: Spirit Magazine.*)

25. The number of cans of Pepsi Cola sold each day is fifty million, fifty-one thousand, five hundred seven. (*Source:* Andy Rooney, *60 Minutes.*)

26. In the United States we drink one hundred forty-six billion, three hundred eighty-five million cups of coffee every year. (*Source: Spirit Magazine.*)

27. There are fifty-four million, seven hundred fifty thousand Hot Wheels sold each year. (*Source:* Andy Rooney, *60 Minutes.*)

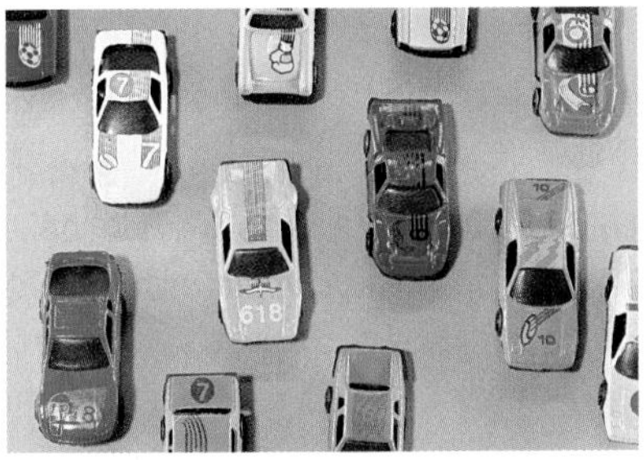

28. A middle-income family will typically spend two hundred twenty-one thousand dollars to raise a child to the age of eighteen. (*Source: Los Angeles Times.*)

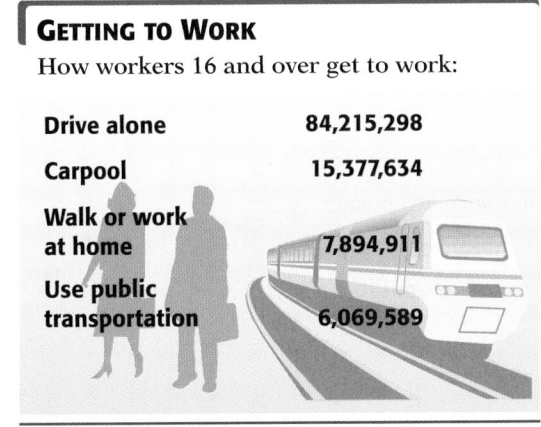

29. Rewrite eight hundred trillion, six hundred twenty-one million, twenty thousand, two hundred fifteen by using digits.

30. Rewrite 2,153,896,448, the number of vehicles that have crossed the Gloden Gate Bridge, in words.

The table at the right shows various ways people get to work. Use the table to answer Exercises 31–34. **See Example 7.**

31. Which method of transportation is least used? Write the number in words.

32. Which method of transportation is most used? Write the number in words.

33. Find the number of people who walk to work or work at home, and write it in words.

34. Find the number of people who carpool, and write it in words.

GETTING TO WORK

How workers 16 and over get to work:

Drive alone	**84,215,298**
Carpool	**15,377,634**
Walk or work at home	**7,894,911**
Use public transportation	**6,069,589**

Source: U.S. Census Bureau.

Study Skills
USING YOUR TEXTBOOK

OBJECTIVES

1. **Explain the meaning of text features such as section numbering, objectives, margin exercises.**
2. **Locate the Index, Answers, and Solutions sections.**

Be sure to read Your Brain* Can *Learn Mathematics before this activity. It is at the beginning of this chapter. You'll find out how your brain learns and remembers.

Your textbook can be very helpful. Find out what it has to offer. First, let's look at some general features that will help in all chapters.

Table of Contents

Look at page vii in the very front for the Table of Contents. Before you start Chapter 1, you should look at the Preface, which begins on page x. On page xiv is a list of Supplementary Resources for students. If you are interested in any of these, ask your instructor if they are available.

Section Numbering

Each chapter is divided into sections, and each section has a number, such as 1.3 or 3.5. Your instructor will use these numbers to assign readings and homework.

Chapter 3 ⟶ **3.5** ⟵ Section 5 within Chapter 3

Chapter Features

There are four features to pay special attention to as you work in your book.

▶ **Objectives.** Each section lists the objectives in the upper corner of the first page. The objectives are listed again as each one is introduced. An objective tells you *what you will be able to do after you complete the section*. An excellent way to check your learning is to go back to the list of objectives when you are finished with a section and ask yourself if you can do them all.

▶ **Margin Exercises.** The exercises in the blue shaded margins of the pages in your textbook give you immediate practice. **This is a perfect way to get your dendrites growing right away!** The answers are given at the bottom, so you can check yourself easily.

▶ **Cautions, Pointers, Notes, and Calculator Tips.**

- **Caution!** A yellow box is a comment about a common error that students make or a common trouble spot you will want to avoid.
- **Pointers** are little "clouds" next to worked examples. They point to specific places where common mistakes are made and give on-the-spot reminders.
- Look for the specially marked **Note** boxes. They contain hints, explanations, or interesting side comments about a topic.
- A small picture of a red calculator appears several places. In the main part of the chapter, the icon means that there is a **Calculator Tip,** which helps you learn more about using your calculator. A calculator in an Exercise section is a recommendation to use your calculator to work that exercise.
- **Vocabulary Tips** appear in the margins to help you learn the language of mathematics. Use them to create study cards for terminology that is new to you.

List a page number from Chapter 1 for each of these features:

A *Caution* appears on page _____.

A *Pointer* appears on page _____.

A *Note* appears on page _____.

A *Calculator Tip* appears on page _____.

End-of-Chapter Features

Go back to the Table of Contents again. What is listed at the end of each chapter?

- **Chapter Summary.** Turn to page 96 to find the Summary for Chapter 1. It lists the chapter's **Key Terms** (arranged in the order that they appear in the chapter) and **New Symbols** and/or **New Formulas.** Then, **Test Your Word Power** checks your understanding of the math vocabulary. Next is a **Quick Review section.** It lists each topic in the chapter and shows a worked example, with tips.
- **Review Exercises** Use these exercises as a way to check your understanding of all the concepts in the chapter. You can practice every type of problem. If you get stuck, the numbers inside the dark blue rectangles tell you which section of the chapter to go back to for more explanations. Make sure you do the **Mixed Review Exercises** to practice for tests.
- **Chapter Test.** Plan to take the test as a practice exam. That way you can be sure you really know how to work all types of problems without looking back at the chapter.
- **Cumulative Review (after Chapters 8 and 17)** These exercises help you maintain the skills you've learned in all previous chapters. Working on previous skills throughout the course will be a big help on the final exam.

How will you make good use of the features at the end of each chapter?

____

____

____

____

Answers

How do you find out if you've worked the exercises correctly? Your textbook provides many of the answers. Throughout each chapter you should work the sample problems in the **margins.** The answers for those are at the **bottom of each page** in the margin area.

For homework, you can find the answers to all of the **odd-numbered section exercises** in the **Answers to Selected Exercises** section near the end of your textbook. Also, *all* of the answers are given for the Chapter Review Exercises, Chapter Tests, and Cumulative Reviews. Check your textbook now, and find the page on which the Answers section begins.

Flag the Answers section with a sticky note or other device, so that you can turn to it quickly.

Solutions

The **Solutions** section near the end of the book shows how to solve some of the harder odd-numbered exercises step by step. In the exercise sets, look for the exercise numbers with a square of blue shading around them. These are the ones that have a solution in the back of the book.

Index

The **Index** is the last thing in your book. It lists all the topics, vocabulary, and concepts in alphabetical order. For example, look up the words below. There may be several subheadings listed under the main word, or several page numbers listed. Notice that the page number printed in bold type is where the word is introduced and defined. Write down the boldface page number for each word. Then go to that page and find the word.

Commutative property of multiplication is defined on page ____.
Factors of numbers are defined on page ____.
Rounding of mixed numbers is explained on page ____.

Why Are These Features Brain Friendly?

The textbook authors included text features that make it easier for you to understand the mathematics. **Your brain naturally seeks organization and predictability.** When you pay attention to the regular features of your textbook, you are allowing your brain to get familiar with all of the helpful tips, suggestions, and explanations that your book has to offer. You will make the best possible use of your textbook.

1.2 Adding Whole Numbers

OBJECTIVES

1. Add two single-digit numbers.
2. Add more than two numbers.
3. Add when regrouping (carrying) is not required.
4. Add with regrouping (carrying).
5. Use addition to solve application problems.
6. Check the answer in addition.

There are four baseballs at the left and two at the right. In all, there are six.

The process of finding the total is called **addition.** Here 4 and 2 were added to get 6. Addition is written with a + sign, so that

$$4 + 2 = 6.$$

OBJECTIVE 1 Add two single-digit numbers. In addition, the numbers being added are called **addends,** and the resulting answer is called the **sum** or **total.**

$$\begin{array}{rl} 4 & \leftarrow \text{Addend} \\ +\,2 & \leftarrow \text{Addend} \\ \hline 6 & \leftarrow \text{Sum (total)} \end{array}$$

Addition problems can also be written horizontally, as follows.

$$\underset{\text{Addend}}{\underset{\uparrow}{4}} \quad + \quad \underset{\text{Addend}}{\underset{\uparrow}{2}} \quad = \quad \underset{\text{Sum}}{\underset{\uparrow}{6}}$$

VOCABULARY TIP

Commutative property of addition You can remember the commutative property by thinking of the numbers "commuting," or changing places.

Commutative Property of Addition

By the **commutative property of addition,** changing the order of the addends in an addition problem does not change the sum.

For example, the sum of 4 + 2 is the same as the sum of 2 + 4. This allows the addition of the same numbers in a different order.

1 Add, and then change the order of numbers to write another addition problem.

GS **(a)** $2 + 6 =$ ____
$6 +$ ____ $=$ ____

(b) $9 + 5$

(c) $4 + 7$

(d) $6 + 9$

EXAMPLE 1 Adding Two Single-Digit Numbers

Add, and then change the order of numbers to write another addition problem.

(a) $5 + 3 = 8$ and $3 + 5 = 8$

(b) $7 + 8 = 15$ and $8 + 7 = 15$

(c) $8 + 3 = 11$ and $3 + 8 = 11$

(d) $8 + 8 = 16$

Changing the order in addition does not change the sum.

◀ **Work Problem 1 at the Side.**

Associative Property of Addition

By the **associative property of addition,** changing the grouping of the addends in an addition problem does not change the sum.

VOCABULARY TIP

Associative property of addition You can remember the associative property by thinking of two numbers associating with each other, and then one leaves to associate with another number.

For example, the sum of 3 + 5 + 6 may be found as follows.

$$(3 + 5) + 6 = 8 + 6 = 14 \qquad \text{Parentheses tell us to add } 3 + 5 \text{ first.}$$

Another way to add the same numbers is

Changing the grouping of addends does not change the sum.

$$3 + (5 + 6) = 3 + 11 = 14. \qquad \text{Parentheses tell us to add } 5 + 6 \text{ first.}$$

Either grouping gives a sum of 14 because of the associative property of addition.

Answers

1. **(a)** 8; $6 + 2 = 8$ **(b)** 14; $5 + 9 = 14$
(c) 11; $7 + 4 = 11$ **(d)** 15; $9 + 6 = 15$

OBJECTIVE ❷ **Add more than two numbers.** To add several numbers, first write them in a column. Add the first number to the second. Add this sum to the third number. Continue until all the numbers are used.

EXAMPLE 2 Adding More Than Two Numbers

Add 2, 5, 6, 1, and 4.

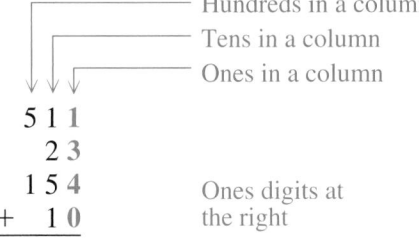

```
    2 ┐
    5 ┘── 2 + 5 = 7
   (6)──────────── 7 + 6 = 13
   (1)──────────────────── 13 + 1 = 14
 + (4)──────────────────────────── 14 + 4 = 18
 ─────
   18
```

Work Problem ❷ at the Side. ▶

Note

By the commutative and associative properties of addition, numbers may also be added starting at the bottom of a column. Adding from the top or adding from the bottom will give the same answer.

OBJECTIVE ❸ **Add when regrouping (carrying) is not required.** If numbers have two or more digits, you must arrange the numbers in columns so that the ones digits are in the same column, tens are in the same column, hundreds are in the same column, and so on. Next, you add column by column starting at the right.

EXAMPLE 3 Adding without Regrouping

Add $511 + 23 + 154 + 10$.

First line up the numbers in columns, with the ones column at the right.

```
          Hundreds in a column
          Tens in a column
          Ones in a column
    5 1 1
      2 3
    1 5 4     Ones digits at
 +    1 0     the right
 ────────
```

Now start at the right and add the ones digits. Add the tens digits next, and finally, the hundreds digits.

```
    5 1 1
      2 3
    1 5 4
 +    1 0   ◁ Always begin addition
 ────────     in the ones column.
    6 9 8
          Sum of ones
          Sum of tens
          Sum of hundreds
```

The sum of the four numbers is 698.

❷ Add each column of numbers.

(a)
```
   3
   8
   5
   4
 + 6
 ───
```

(b)
```
   5
   6
   3
   2
 + 4
 ───
```

(c)
```
   9
   6
   8
   7
 + 3
 ───
```

(d)
```
   3
   8
   6
   4
 + 8
 ───
```

Answers

2. **(a)** 26 **(b)** 20 **(c)** 33 **(d)** 29

3 Add.

(a)
$$\begin{array}{r} 26 \\ +\ 73 \\ \hline \end{array}$$

(b)
$$\begin{array}{r} 534 \\ +\ 265 \\ \hline \end{array}$$

(c)
$$\begin{array}{r} 42{,}305 \\ +\ 11{,}563 \\ \hline \end{array}$$

4 Add with regrouping

GS **(a)**
$$\begin{array}{r} ^{1}\ \\ 66 \\ +\ 27 \\ \hline \_3 \end{array}$$

(b)
$$\begin{array}{r} 58 \\ +\ 33 \\ \hline \end{array}$$

(c)
$$\begin{array}{r} 56 \\ +\ 37 \\ \hline \end{array}$$

(d)
$$\begin{array}{r} 34 \\ +\ 49 \\ \hline \end{array}$$

Answers

3. **(a)** 99 **(b)** 799 **(c)** 53,868

4. **(a)** 9; 93 **(b)** 91 **(c)** 93 **(d)** 83

◀ **Work Problem 3 at the Side.**

OBJECTIVE 4 Add with regrouping (carrying). If the sum of the digits in any column is more than 9, use **regrouping** (sometimes called **carrying**).

EXAMPLE 4 Adding with Regrouping

Add 47 and 29.
Add ones.

$$\begin{array}{r} 47 \\ +\ 29 \\ \hline \end{array}$$

7 ones and 9 ones = 16 ones

Regroup 16 ones as 1 ten and 6 ones. Write 6 ones in the ones column and write 1 ten in the tens column.

$$\begin{array}{r} ^{1}\ \\ 47 \\ +\ 29 \\ \hline 6 \end{array}$$

Write 1 ten in the tens column.
7 ones and 9 ones = 16 ones
Write 6 ones in the ones column.

Add the digits in the tens column, including the regrouped 1.

$$\begin{array}{r} ^{1}\ \\ 47 \\ +\ 29 \\ \hline 76 \end{array}$$

1 ten + 4 tens + 2 tens = 7 tens

◀ **Work Problem 4 at the Side.**

EXAMPLE 5 Adding with Regrouping

Add 324 + 7855 + 23 + 7 + 86.

Step 1 Add the digits in the ones column.

$$\begin{array}{r} ^{2}\ \\ 324 \\ 7855 \\ 23 \\ 7 \\ +\ 86 \\ \hline 5 \end{array}$$

Write 2 tens in the tens column.
Sum of the ones column is 25 ones.
Write 5 ones in the ones column.

Notice that 25 ones are regrouped as 2 tens and 5 ones.

Step 2 Add the digits in the tens column, including the regrouped 2.

$$\begin{array}{r} ^{1\,2}\ \\ 324 \\ 7855 \\ 23 \\ 7 \\ +\ 86 \\ \hline 95 \end{array}$$

Write 1 hundred in the hundreds column.
Sum of the tens column is 19 tens.
Write 9 in the tens column.

Notice that 19 tens are regrouped as 1 hundred and 9 tens.

Continued on Next Page

Step 3 Add the hundreds column, including the regrouped 1.

$$\begin{array}{r} {}^{1\,1\,2} \\ 324 \\ 7855 \\ 23 \\ 7 \\ +\ \ 86 \\ \hline 295 \end{array}$$

Write 1 thousand in the thousands column.

Notice that 12 hundreds are regrouped as 1 thousand and 2 hundreds.

Sum of the hundreds column is 12 hundreds.

Write 2 hundreds in the hundreds column.

Step 4 Add the thousands column, including the regrouped 1.

$$\begin{array}{r} {}^{1\,1\,2} \\ 324 \\ 7855 \\ 23 \\ 7 \\ +\ \ 86 \\ \hline 8295 \end{array}$$

Sum of the thousands column is 8.

Finally, $324 + 7855 + 23 + 7 + 86 = 8295$.

Work Problem 5 at the Side. ▶

Note

For additional speed, you can try to regroup mentally. Do not write the regrouped number, but just remember it as you move to the top of the next column. Try this method, and use it if it works for you.

Work Problem 6 at the Side. ▶

OBJECTIVE 5 Use addition to solve application problems. In **Section 1.10** we will describe how to solve application problems in more detail. The next two examples are application problems that require adding.

EXAMPLE 6 Applying Addition Skills

On this map of the Walt Disney World area in Florida, the distance in miles from one location to another is written alongside the road. Find the shortest route from Altamonte Springs to Clear Lake.

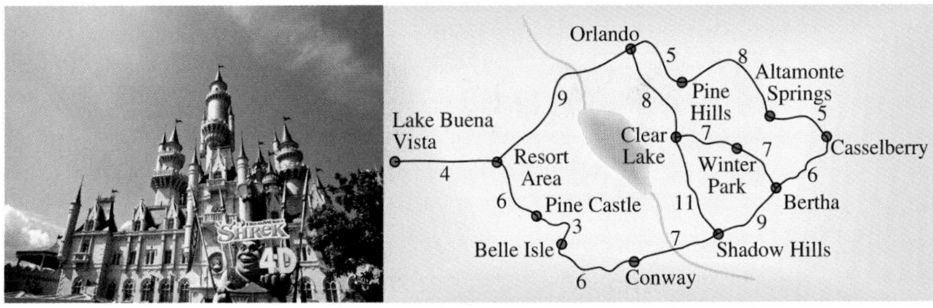

Continued on Next Page

5 Add by regrouping as necessary.

GS **(a)**
$$\begin{array}{r} {}^{1\ \ \ 2\ 1} \\ 1\,6\,2 \\ 4\,2\,7\,1 \\ 3\,7\,2 \\ +\ 8\,9\,7\,6 \\ \hline 1\,3,\_\ \_\ 1 \end{array}$$

(b)
$$\begin{array}{r} 7821 \\ 435 \\ 72 \\ 305 \\ +\ 1693 \\ \hline \end{array}$$

6 Add by regrouping mentally.

(a)
$$\begin{array}{r} 816 \\ 363 \\ 17 \\ 2 \\ 5 \\ +\ 7654 \\ \hline \end{array}$$

(b)
$$\begin{array}{r} 15{,}829 \\ 765 \\ 78 \\ 15 \\ 9 \\ 7 \\ +\ 13{,}179 \\ \hline \end{array}$$

Answers

5. **(a)** 7; 8; 13,781 **(b)** 10,326

6. **(a)** 8857 **(b)** 29,882

7 Use the map in **Example 6** to find the shortest route from Lake Buena Vista to Conway.

GS

$$\begin{array}{r} 4 \\ 6 \\ 3 \\ +\_ \\ \hline \_\_ \end{array}$$

8 The road is closed between Orlando and Clear Lake, so this route cannot be used. Use the map in **Example 6** to find the next shortest route from Orlando to Clear Lake.

Answers

7. 6; 19 miles

8.
5 Orlando to Pine Hills
8 Pine Hills to Altamonte Springs
5 Altamonte Springs to Casselberry
6 Casselberry to Bertha
7 Bertha to Winter Park
+ 7 Winter Park to Clear Lake
38 miles

Approach Add the mileage along various routes to determine the distances from Altamonte Springs to Clear Lake. Then select the shortest route.

Solution One way from Altamonte Springs to Clear Lake is through Orlando. Add the mileage numbers along this route.

8	Altamonte Springs to Pine Hills
5	Pine Hills to Orlando
+ 8	Orlando to Clear Lake
21 →	miles from Altamonte Springs to Clear Lake, going through Orlando

Remember: Shortest distance is the fewest total miles.

Another way is through Casselberry, Bertha, and Winter Park. Add the mileage numbers along this route.

5	Altamonte Springs to Casselberry
6	Casselberry to Bertha
7	Bertha to Winter Park
+ 7	Winter Park to Clear Lake
25 →	miles from Altamonte Springs to Clear Lake through Bertha and Winter Park

The shortest route from Altamonte Springs to Clear Lake is 21 miles through Orlando.

◀ Work Problem **7** at the Side.

EXAMPLE 7 Finding a Total Distance

Use the map in **Example 6** to find the total distance from Shadow Hills to Casselberry to Orlando and back to Shadow Hills.

Approach Add the mileage from Shadow Hills to Casselberry to Orlando and back to Shadow Hills to find the total distance.

Solution Use the numbers from the map.

9	Shadow Hills to Bertha
6	Bertha to Casselberry
5	Casselberry to Altamonte Springs
8	Altamonte Springs to Pine Hills
5	Pine Hills to Orlando
8	Orlando to Clear Lake
+ 11	Clear Lake to Shadow Hills
52 →	miles from Shadow Hills to Casselberry to Orlando and back to Shadow Hills

◀ Work Problem **8** at the Side.

EXAMPLE 8 Finding a Perimeter

Find the number of feet of hedges needed to enclose the Civil War Veterans Park shown.

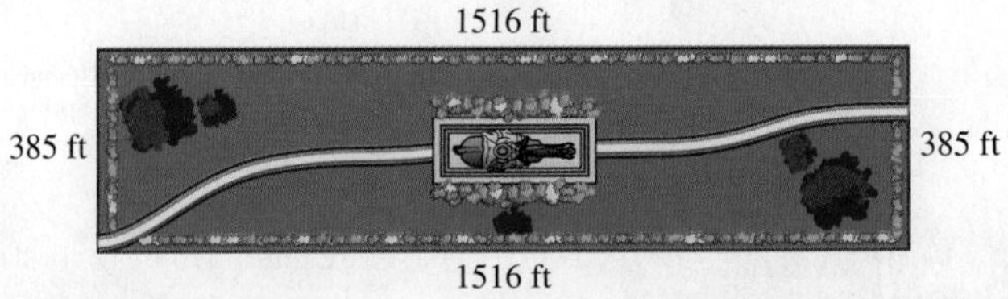

The short way to write feet is ft.

Approach Find the **perimeter,** or total distance around the park, by adding the lengths of all the sides.

Continued on Next Page

Solution Use the lengths shown.

$$\begin{array}{r} 1516 \\ 385 \\ 1516 \\ +\ 385 \\ \hline 3802 \text{ ft} \end{array}$$

The amount of hedge needed is 3802 ft, which is the perimeter of (distance around) the park.

Work Problem **9** at the Side. ▶

OBJECTIVE 6 Check the answer in addition. Checking the answer is an important part of problem solving. A common method for checking addition is to re-add from bottom to top. This is an application of the commutative and associative properties of addition.

EXAMPLE 9 Checking Addition

Check the following addition.

$$\begin{array}{r} \mathbf{1428} \\ 738 \\ 63 \\ 125 \\ 17 \\ +\ 485 \\ \hline \mathbf{1428} \end{array}$$

Add down. ↓ To check, add up. ↑

Adding down and adding up should give the same answer.

Add from bottom to top to check addition.

Here the answers agree, so the sum is probably correct.

EXAMPLE 10 Checking Addition

Check the following additions. Are they correct?

(a)

$$\begin{array}{rr} & \mathbf{1033} \\ 785 & 785 \\ 63 & 63 \\ +\ 185 & +\ 185 \\ \hline 1033 & \mathbf{1033} \end{array}$$

Correct, because both answers are the same.

To check, add up.

(b)

$$\begin{array}{rr} & \mathbf{2454} \\ 635 & 635 \\ 73 & 73 \\ 831 & 831 \\ +\ 915 & +\ 915 \\ \hline 2444 & \mathbf{2444} \end{array}$$

Error, because the answers are different.

To check, add up.

Avoid wrong answers by checking your work.

Re-add to find that the correct sum is 2454.

Work Problem **10** at the Side. ▶

9 Solve the problem. Find the number of feet of fencing needed to enclose the solar electricity generating project shown.

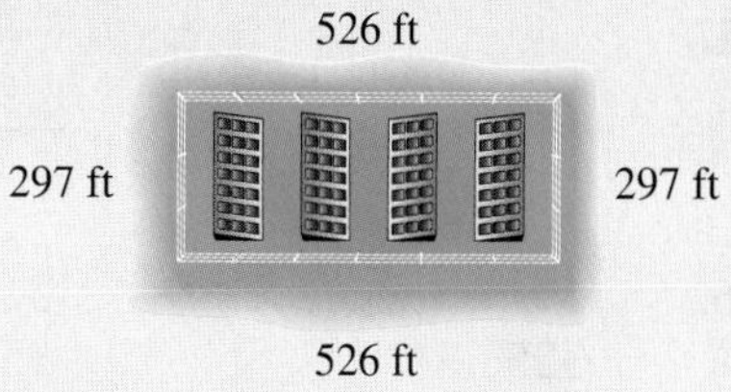

10 Check the following additions. If an answer is incorrect, find the correct answer.

(a) $\begin{array}{r} 63 \\ 4 \\ 9 \\ +\ 28 \\ \hline 104 \end{array}$

(b) $\begin{array}{r} 927 \\ 395 \\ 64 \\ +\ 251 \\ \hline 1637 \end{array}$

(c) $\begin{array}{r} 79 \\ 218 \\ 7 \\ +\ 639 \\ \hline 953 \end{array}$

(d) $\begin{array}{r} 21{,}892 \\ 11{,}746 \\ +\ 43{,}925 \\ \hline 79{,}563 \end{array}$

Answers

9. 1646 ft

10. **(a)** correct **(b)** correct
(c) incorrect; should be 943
(d) incorrect; should be 77,563

1.2 Exercises

FOR EXTRA HELP

Download the MyDashBoard App

MyMathLab®

Add. **See Examples 1–3.**

1. $\begin{array}{r} 43 \\ +\ 54 \\ \hline \end{array}$

2. $\begin{array}{r} 18 \\ +\ 11 \\ \hline \end{array}$

3. $\begin{array}{r} 56 \\ +\ 33 \\ \hline \end{array}$

4. $\begin{array}{r} 83 \\ +\ 15 \\ \hline \end{array}$

5. $\begin{array}{r} 317 \\ +\ 572 \\ \hline \end{array}$

6. $\begin{array}{r} 574 \\ +\ 325 \\ \hline \end{array}$

7. $\begin{array}{r} 318 \\ 151 \\ +\ 420 \\ \hline \end{array}$

8. $\begin{array}{r} 135 \\ 253 \\ +\ 410 \\ \hline \end{array}$

9. $\begin{array}{r} 6310 \\ 252 \\ +\ 1223 \\ \hline \end{array}$

10. $\begin{array}{r} 121 \\ 5705 \\ +\ 3163 \\ \hline \end{array}$

CONCEPT CHECK *Determine whether the following additions are* correct *or* incorrect.

11. 932 + 44 + 613 = 1589

12. 517 + 131 + 250 = 1098

13. 1251 + 4311 + 2114 = 7686

14. 3241 + 1513 + 2014 = 6768

Add. **See Examples 1–3.**

15. 12,142 + 43,201 + 23,103

16. 41,124 + 12,302 + 23,500

17. 3213 + 5715

18. 6344 + 1655

19. 38,204 + 21,020 $\begin{array}{r} 38,2\ 0\ 4 \\ +\ 21,0\ 2\ 0 \\ \hline 59,\_\ \_\ 4 \end{array}$

20. 63,251 + 36,305 $\begin{array}{r} 6\ 3,\ 2\ 5\ 1 \\ +\ 3\ 6,\ 3\ 0\ 5 \\ \hline 9\ \_,\ \_\ 5\ \_ \end{array}$

CONCEPT CHECK *Determine which answers are* correct *or* incorrect.

21. $\begin{array}{r} 87 \\ +\ 63 \\ \hline 150 \end{array}$

22. $\begin{array}{r} 19 \\ +\ 92 \\ \hline 101 \end{array}$

23. $\begin{array}{r} 86 \\ +\ 69 \\ \hline 155 \end{array}$

24. $\begin{array}{r} 37 \\ +\ 85 \\ \hline 132 \end{array}$

25. $\begin{array}{r} 47 \\ +\ 74 \\ \hline 111 \end{array}$

Add, regrouping as necessary. **See Examples 4 and 5.**

26. $\begin{array}{r} 97 \\ +\ 79 \\ \hline \end{array}$

27. $\begin{array}{r} 67 \\ +\ 78 \\ \hline \end{array}$

28. $\begin{array}{r} 96 \\ +\ 47 \\ \hline \end{array}$

29. $\begin{array}{r} 73 \\ +\ 29 \\ \hline \end{array}$

30. $\begin{array}{r} 68 \\ +\ 37 \\ \hline \end{array}$

31. $\begin{array}{r} 746 \\ +\ 905 \\ \hline \end{array}$

32. $\begin{array}{r} 621 \\ +\ 359 \\ \hline \end{array}$

33. $\begin{array}{r} 306 \\ +\ 848 \\ \hline \end{array}$

34. $\begin{array}{r} 798 \\ +\ 206 \\ \hline \end{array}$

35. $\begin{array}{r} 278 \\ +\ 135 \\ \hline \end{array}$

36. $\begin{array}{r} 172 \\ +\ 156 \\ \hline \end{array}$

37. $\begin{array}{r} 928 \\ +\ 843 \\ \hline \end{array}$

38. $\begin{array}{r} 686 \\ +\ 726 \\ \hline \end{array}$

39. $\begin{array}{r} 526 \\ +\ 884 \\ \hline \end{array}$

40. $\begin{array}{r} 116 \\ +\ 897 \\ \hline \end{array}$

41. $\begin{array}{r} 3574 \\ +\ 2817 \\ \hline \end{array}$ **42.** $\begin{array}{r} 6871 \\ +\ 7528 \\ \hline \end{array}$ **43.** $\begin{array}{r} 7896 \\ +\ 3728 \\ \hline \end{array}$ **44.** $\begin{array}{r} 9382 \\ +\ 7586 \\ \hline \end{array}$ **45.** $\begin{array}{r} 9625 \\ +\ 7986 \\ \hline \end{array}$

46. $\begin{array}{r} 5718 \\ 5623 \\ +\ 7436 \\ \hline \end{array}$ **47.** $\begin{array}{r} 9056 \\ 78 \\ 6089 \\ +\ 731 \\ \hline \end{array}$ **48.** $\begin{array}{r} 4022 \\ 709 \\ 8621 \\ +\ 37 \\ \hline \end{array}$ **49.** $\begin{array}{r} 18 \\ 708 \\ 9286 \\ +\ 636 \\ \hline \end{array}$ **50.** $\begin{array}{r} 1708 \\ 321 \\ 61 \\ +\ 8926 \\ \hline \end{array}$

51. $\begin{array}{r} 422 \\ 6074 \\ 435 \\ +\ 8663 \\ \hline \end{array}$ **52.** $\begin{array}{r} 6505 \\ 173 \\ 7044 \\ +\ 168 \\ \hline \end{array}$ **53.** $\begin{array}{r} 321 \\ 9603 \\ 8 \\ 21 \\ +\ 1604 \\ \hline \end{array}$ **54.** $\begin{array}{r} 7631 \\ 5983 \\ 7 \\ 36 \\ +\ 505 \\ \hline \end{array}$ **55.** $\begin{array}{r} 2109 \\ 63 \\ 16 \\ 3 \\ +\ 9887 \\ \hline \end{array}$

56. $\begin{array}{r} 322 \\ 6508 \\ 93 \\ 745 \\ 18 \\ +\ 2005 \\ \hline \end{array}$ **57.** $\begin{array}{r} 553 \\ 97 \\ 2772 \\ 437 \\ 63 \\ +\ 328 \\ \hline \end{array}$ **58.** $\begin{array}{r} 3187 \\ 810 \\ 527 \\ 76 \\ 2665 \\ +\ 317 \\ \hline \end{array}$ **59.** $\begin{array}{r} 413 \\ 85 \\ 9919 \\ 602 \\ 31 \\ +\ 1218 \\ \hline \end{array}$ **60.** $\begin{array}{r} 576 \\ 7934 \\ 60 \\ 781 \\ 5968 \\ +\ 371 \\ \hline \end{array}$

Check each addition. If an answer is incorrect, find the correct answer.
See Examples 9 and 10.

61. $\begin{array}{r} \text{2 \_ 1} \\ 832 \\ 468 \\ +\ 791 \\ \hline 2091 \end{array}$ **62.** $\begin{array}{r} \text{\_\_\_\_} \\ 326 \\ 852 \\ +\ 679 \\ \hline 1857 \end{array}$ **63.** $\begin{array}{r} \text{7 9} \\ 179 \\ 214 \\ +\ 376 \\ \hline 759 \end{array}$ **64.** $\begin{array}{r} \text{\_\_\_\_} \\ 17 \\ 296 \\ 713 \\ +\ 94 \\ \hline 1220 \end{array}$ **65.** $\begin{array}{r} \text{\_\_\_\_} \\ 4713 \\ 28 \\ 615 \\ +\ 64 \\ \hline 5420 \end{array}$

66. $\begin{array}{r} \text{\_\_\_\_\_} \\ 3\,628 \\ 72 \\ 564 \\ +\ 7\,319 \\ \hline 11{,}583 \end{array}$ **67.** $\begin{array}{r} \text{\_\_\_\_\_} \\ 678 \\ 7\,952 \\ 56 \\ 718 \\ +\ 2\,173 \\ \hline 11{,}377 \end{array}$ **68.** $\begin{array}{r} \text{\_\_\_\_\_} \\ 516 \\ 8\,760 \\ 24 \\ 189 \\ +\ 1\,723 \\ \hline 11{,}212 \end{array}$ **69.** $\begin{array}{r} \text{\_\_\_\_\_} \\ 4\,714 \\ 27 \\ 77 \\ 8\,878 \\ +\ 636 \\ \hline 14{,}332 \end{array}$ **70.** $\begin{array}{r} \text{\_\_\_\_\_} \\ 6\,715 \\ 283 \\ 9\,617 \\ 13 \\ +\ 81 \\ \hline 16{,}719 \end{array}$

71. Explain the commutative property of addition in your own words. How is this used when checking an addition problem?

72. Explain the associative property of addition. How can this be used when adding columns of numbers?

For Exercises 73–76, use the map to find the shortest route between each pair of cities.
See Examples 6 and 7.

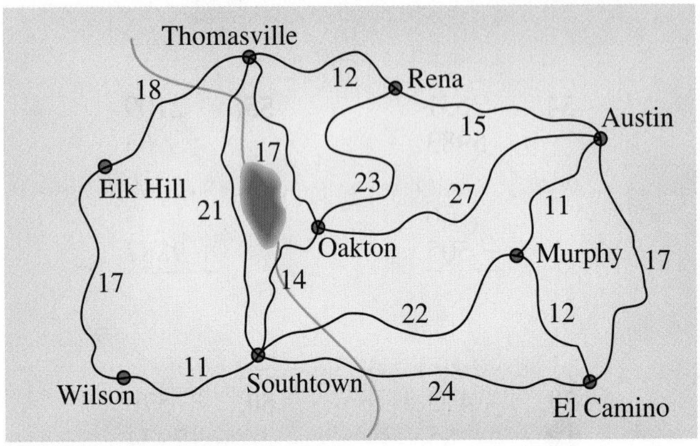

73. Southtown and Rena
GS
Southtown to Thomasville 21 miles
Thomasville to Rena + 12 miles

74. Elk Hill and Oakton
GS
Elk Hill to Thomasville 18 miles
Thomasville to Oakton + 17 miles

75. Thomasville and Murphy

76. Murphy and Thomasville

Solve each application problem.

77. The Twin Lakes Food Bank raised $3482 at a flea market and $12,860 at their annual auction. Find the total amount raised at these two events.

78. A ballpark vendor sold 185 hot dogs and 129 hamburgers. What was the total number of items sold?

79. There are 413 women and 286 men on the sales staff. How many people are on the sales staff?

80. One department in an office building has 283 employees while another department has 218 employees. How many employees are in the two departments?

81. This semester there are 13,786 students enrolled in on-campus day classes, 3497 students enrolled in night classes, and 2874 student's enrolled in on-line classes. Find the total number of students enrolled.

82. The number of tornadoes in each of the last 7 years has been 887, 223, 465, 683, 597, 214 and 1817. Find the total number of tornadoes in this 7-year period. (*Source:* Storm Prediction Center.)

Solve each problem involving perimeter. ***See Example 8.***

83. Find the total distance around a lot that has been developed as a go-cart track.

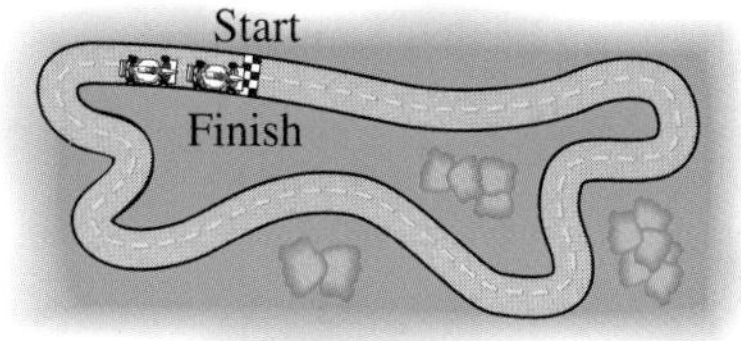

84. Because of heavy snowfall this winter, Maria needs to put new rain gutters around her entire roof. How many feet of gutters will she need?

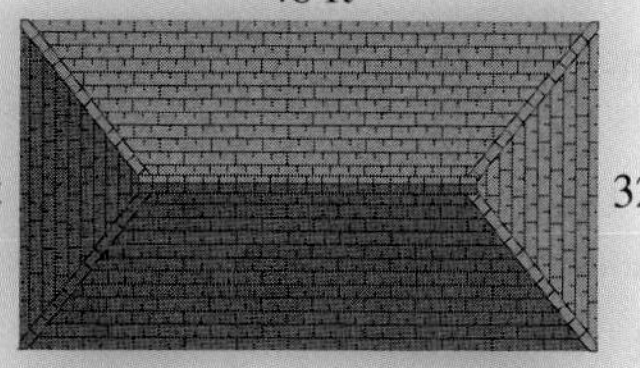

85. Martin plans to frame his back patio with redwood lumber. How many feet of lumber will he need?

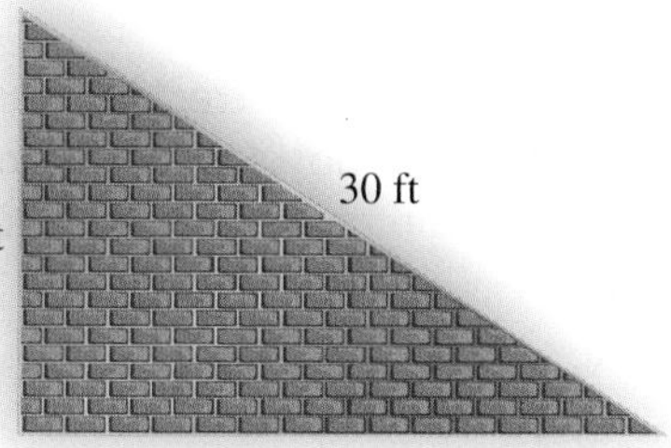

86. The university agriculture department is expanding the size of its experimental farm and wants to fence the area around the farm. How many meters of fencing will be needed?

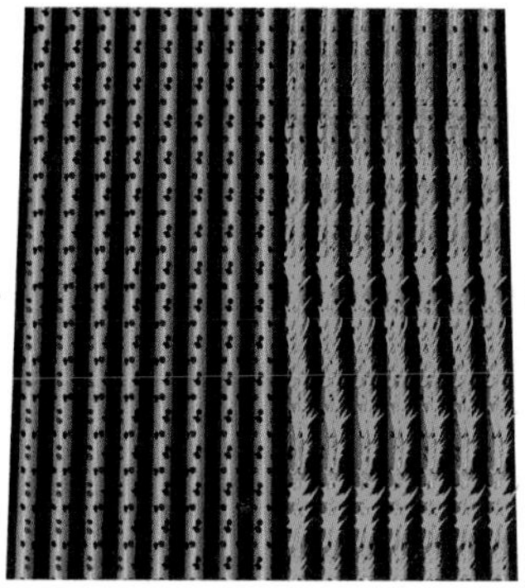

Relating Concepts (Exercises 87–94) For Individual or Group Work

Recall the place values of digits discussed in ***Section 1.1*** *and* ***work Exercises 87–94 in order.***

87. Write the largest four-digit number possible using the digits 4, 1, 9, and 2. Use each digit once.

88. Using the digits 4, 1, 9, and 2, write the smallest four-digit number possible. Use each digit once.

89. Write the largest five-digit number possible using the digits 6, 2, and 7. Use each digit at least once.

90. Using the digits 6, 2, and 7, write the smallest five-digit number possible. Use each digit at least once.

91. Write the largest seven-digit number possible using the digits 4, 3, and 9. Use each digit at least twice.

92. Using the digits 4, 3, and 9, write the smallest seven-digit number possible. Use each digit at least twice.

93. Explain your rule or procedure for writing the largest number in **Exercise 91.**

94. Explain your rule or procedure for writing the smallest number in **Exercise 92.**

1.3 Subtracting Whole Numbers

OBJECTIVES

1. Change addition problems to subtraction and subtraction problems to addition.
2. Identify the minuend, subtrahend, and difference.
3. Subtract when no regrouping (borrowing) is needed.
4. Check subtraction answers by adding.
5. Subtract with regrouping (borrowing).
6. Solve application problems with subtraction.

1 Write two subtraction problems for each addition problem.

GS **(a)** $8 + 2 = 10$

$10 - \_\_\_\_ = 8$

$10 - \_\_\_\_ = \_\_\_\_$

(b) $7 + 4 = 11$

(c) $15 + 22 = 37$

(d) $23 + 55 = 78$

Answers

1. (a) 2; $10 - 2 = 8$ or 8; 2; $10 - 8 = 2$
(b) $11 - 4 = 7$ or $11 - 7 = 4$
(c) $37 - 22 = 15$ or $37 - 15 = 22$
(d) $78 - 55 = 23$ or $78 - 23 = 55$

Suppose you have \$9, and you spend \$2 for parking. You then have \$7 left. There are two different ways of looking at these numbers.

As an addition problem:

\$2	+	\$7	=	\$9
↑ Amount spent		↑ Amount left		↑ Original amount

As a subtraction problem:

\$9	−	\$2	=	\$7
↑ Original amount	↑ Subtraction symbol	↑ Amount spent		↑ Amount left

OBJECTIVE 1 Change addition problems to subtraction and subtraction problems to addition. As shown in the box above, an addition problem can be changed to a subtraction problem and a subtraction problem can be changed to an addition problem.

EXAMPLE 1 Changing Addition Problems to Subtraction

Change each addition problem to a subtraction problem.

(a) $4 + 1 = 5$

Two subtraction problems are possible:

$$5 - 1 = 4 \quad \text{or} \quad 5 - 4 = 1$$

These figures show each subtraction problem.

(b) $8 + 7 = 15$

$$15 - 7 = 8 \quad \text{or} \quad 15 - 8 = 7$$

◀ **Work Problem 1 at the Side.**

EXAMPLE 2 Changing Subtraction Problems to Addition

Change each subtraction problem to an addition problem.

(a) $8 - 3 = 5$

$\downarrow \quad \downarrow \quad \downarrow$

$8 = 3 + 5$

It is also correct to write $8 = 5 + 3$.

Recall that changing the order of addends does not change the sum.

Continued on Next Page

(b) $18 - 13 = 5$

$18 = 13 + 5$ or $18 = 5 + 13$

(c) $29 - 13 = 16$

$29 = 13 + 16$ or $29 = 16 + 13$

Work Problem 2 at the Side. ▶

OBJECTIVE 2 Identify the minuend, subtrahend, and difference. In subtraction, as in addition, the numbers in a problem have names. For example, in the problem $8 - 5 = 3$, the number 8 is the **minuend,** 5 is the **subtrahend,** and 3 is the **difference** or answer.

$$\underset{\text{Minuend}}{8} - \underset{\text{Subtrahend}}{5} = 3 \leftarrow \text{Difference}$$

The answer in subtraction is the difference.

$$\begin{array}{r} 8 \leftarrow \text{Minuend} \\ -\ 5 \leftarrow \text{Subtrahend} \\ \hline 3 \leftarrow \text{Difference} \end{array}$$

OBJECTIVE 3 Subtract when no regrouping (borrowing) is needed. Subtract two numbers by lining up the numbers in columns so the digits in the ones place are in the same column, the tens digits are in the same column, the hundreds digits are in the same column, and so on. Next, subtract by columns, starting at the right with the ones column.

EXAMPLE 3 Subtracting Two Numbers

Subtract.

Tens digits are lined up in the same column.
Ones digits are lined up in the same column.

(a)
$$\begin{array}{r} 53 \\ -\ 21 \\ \hline 32 \end{array}$$

3 ones − 1 one = 2 ones
5 tens − 2 tens = 3 tens

Ones digits are lined up.

(b)
$$\begin{array}{r} 385 \\ -\ 165 \\ \hline 220 \end{array}$$

Subtract from right to left.

5 ones − 5 ones = 0 ones
8 tens − 6 tens = 2 tens
3 hundreds − 1 hundred = 2 hundreds

(c)
$$\begin{array}{r} 9437 \\ -\ 210 \\ \hline 9227 \end{array}$$

7 ones − 0 ones = 7 ones
3 tens − 1 ten = 2 tens
4 hundreds − 2 hundreds = 2 hundreds
9 thousands − 0 thousands = 9 thousands

Work Problem 3 at the Side. ▶

OBJECTIVE 4 Check subtraction answers by adding. Use addition to check your answer to a subtraction problem. For example, check $8 - 3 = 5$ by *adding* 3 and 5.

$$3 + 5 = 8, \quad \text{so} \quad 8 - 3 = 5 \quad \text{is correct.}$$

2 Write an addition problem for each subtraction problem.

GS **(a)** $7 - 5 = 2$

$7 = 5 +$ ____

(b) $9 - 4 = 5$

(c) $21 - 15 = 6$

(d) $58 - 42 = 16$

VOCABULARY TIP

Difference suggests comparing two things. In mathematics, this comparison is done by subtracting two numbers, the answer being the difference between them.

3 Subtract.

(a)
$$\begin{array}{r} 74 \\ -\ 43 \\ \hline \end{array}$$

(b)
$$\begin{array}{r} 68 \\ -\ 24 \\ \hline \end{array}$$

(c)
$$\begin{array}{r} 429 \\ -\ 318 \\ \hline \end{array}$$

(d)
$$\begin{array}{r} 3927 \\ -\ 2614 \\ \hline \end{array}$$

Answers

2. **(a)** 2; $7 = 5 + 2$ or $7 = 2 + 5$
(b) $9 = 4 + 5$ or $9 = 5 + 4$
(c) $21 = 15 + 6$ or $21 = 6 + 15$
(d) $58 = 42 + 16$ or $58 = 16 + 42$

3. **(a)** 31 **(b)** 44 **(c)** 111 **(d)** 1313

4 Use addition to determine whether each answer is correct. If incorrect, what should it be?

GS **(a)** $\begin{array}{r} 76 \\ -\ 45 \\ \hline 31 \end{array}$ $\qquad$ $\begin{array}{r} 45 \\ +\ \_\_ \\ \hline \_\_ \end{array}$

(b) $\begin{array}{r} 53 \\ -\ 22 \\ \hline 21 \end{array}$

(c) $\begin{array}{r} 374 \\ -\ 251 \\ \hline 113 \end{array}$

(d) $\begin{array}{r} 7531 \\ -\ 4301 \\ \hline 3230 \end{array}$

VOCABULARY TIP

Regrouping refers to both "carrying" in addition and "borrowing" in subtraction.

5 Subtract.

(a) $\begin{array}{r} 58 \\ -\ 19 \\ \hline \end{array}$

(b) $\begin{array}{r} 86 \\ -\ 38 \\ \hline \end{array}$

(c) $\begin{array}{r} 41 \\ -\ 27 \\ \hline \end{array}$

(d) $\begin{array}{r} 863 \\ -\ 47 \\ \hline \end{array}$

(e) $\begin{array}{r} 762 \\ -\ 157 \\ \hline \end{array}$

Answers

4. **(a)** 31; 76; correct **(b)** incorrect; should be 31 **(c)** incorrect; should be 123 **(d)** correct

5. **(a)** 39 **(b)** 48 **(c)** 14 **(d)** 816 **(e)** 605

EXAMPLE 4 **Checking Subtraction by Using Addition**

Use addition to check each answer. If the answer is incorrect, find the correct answer.

(a) $\begin{array}{r} 89 \\ -\ 47 \\ \hline 42 \end{array}$

Rewrite as an addition problem, as shown in **Example 2**.

Subtraction problem $\begin{array}{r} 89 \\ -\ 47 \\ \hline 42 \\ \hline 89 \end{array}$ Addition problem $\qquad$ $\begin{array}{r} 47 \\ +\ 42 \\ \hline 89 \end{array}$

Because 47 + 42 = 89, the subtraction was done correctly.

Avoid errors by checking answers.

(b) 72 − 41 = 21

Rewrite as an addition problem.

$$72 = 41 + 21$$

But, 41 + 21 = 62, **not** 72, so the subtraction was done **incorrectly.** Rework the original subtraction to get the correct answer, 31. Then, 41 + 31 = 72.

(c) $\begin{array}{r} 374 \\ -\ 141 \\ \hline 233 \end{array}$ ← Match → 141 + 233 = 374

The answer checks.

◀ Work Problem **4** at the Side.

OBJECTIVE **5** **Subtract with regrouping (borrowing).** When a digit in the minuend is less than the one directly below it, **regrouping** is necessary (also called **borrowing**).

EXAMPLE 5 **Subtracting with Regrouping**

Subtract 19 from 57.

Write the problem vertically.

$$\begin{array}{r} 57 \\ -\ 19 \\ \hline \end{array}$$

In the ones column, 7 is **less** than 9, so in order to subtract, we must regroup 1 ten as 10 ones.

5 tens − 1 ten = 4 tens ⟶ 4 17 ⟵ 1 ten = 10 ones, and 10 ones + 7 ones = 17 ones

$$\begin{array}{r} \overset{4}{\cancel{5}}\,\overset{17}{\cancel{7}} \\ -\ 1\ 9 \\ \hline \end{array}$$

Now subtract 9 ones from 17 ones in the o nes column. Then subtract 1 ten from 4 tens in the tens column.

$$\begin{array}{r} \overset{4}{\cancel{5}}\,\overset{17}{\cancel{7}} \\ -\ 1\ 9 \\ \hline 3\ 8 \end{array} \quad \text{Difference}$$

Finally, 57 − 19 = 38. Check by adding 19 and 38; you should get 57.

◀ Work Problem **5** at the Side.

EXAMPLE 6 Subtracting with Regrouping

Subtract by regrouping when necessary.

(a) $\begin{array}{r} 7856 \\ -\ 137 \\ \hline \end{array}$

Regroup 1 ten as 10 ones. — 10 ones + 6 ones = 16 ones

$\begin{array}{r} {\scriptstyle 4\ 16} \\ 78\not{5}\not{6} \\ -\ 137 \\ \hline 7719 \end{array}$ Difference

(b) $\begin{array}{r} 635 \\ -\ 546 \\ \hline \end{array}$

Regroup 1 ten as 10 ones. — 10 ones + 5 ones = 15 ones

$\begin{array}{r} {\scriptstyle 2\ 15} \\ 6\not{3}\not{5} \\ -\ 546 \\ \hline 9 \end{array}$ Need to regroup further because 2 is less than 4 in the tens column.

Regroup 1 hundred as 10 tens — 10 tens + 2 tens = 12 tens

$\begin{array}{r} {\scriptstyle 5\ 12\ 15} \\ \not{6}\not{3}\not{5} \\ -\ 546 \\ \hline 89 \end{array}$ Difference

(c) $\begin{array}{r} 647 \\ -\ 489 \\ \hline \end{array}$

$\begin{array}{r} {\scriptstyle 3\ 17} \\ 6\not{4}\not{7} \\ -\ 489 \\ \hline 8 \end{array}$ Need to regroup further because 3 is less than 8 in the tens column.

$\begin{array}{r} {\scriptstyle 5\ 13\ 17} \\ \not{6}\not{4}\not{7} \\ -\ 489 \\ \hline 158 \end{array}$ Difference

Work Problem 6 at the Side. ▶

Sometimes a minuend has zeros in some of the positions. In such cases, regrouping may be a little more complicated than what we have shown so far.

EXAMPLE 7 Regrouping with Zeros

Subtract.

$\begin{array}{r} 4607 \\ -\ 3168 \\ \hline \end{array}$

There are no tens that can be regrouped into ones. So you must first regroup 1 hundred as 10 tens.

Regroup 1 hundred as 10 tens. — Write 10 tens.

$\begin{array}{r} {\scriptstyle 5\ 10} \\ 4\not{6}\not{0}7 \\ -\ 3168 \\ \hline \end{array}$

Now we may regroup from the tens position.

$\begin{array}{r} {\scriptstyle 9} \\ {\scriptstyle 5\ 10\ 17} \\ 4\not{6}\not{0}\not{7} \\ -\ 3168 \\ \hline 9 \end{array}$

Regroup 1 ten as 10 ones. 10 tens − 1 ten = 9 tens.

10 ones + 7 ones = 17 ones

Continued on Next Page

6 Subtract.

GS **(a)** $\begin{array}{r} {\scriptstyle 8\ 12} \\ \not{9}\not{2}7 \\ -\ 43 \\ \hline \_\ \_\ 4 \end{array}$

(b) $\begin{array}{r} 675 \\ -\ 86 \\ \hline \end{array}$

(c) $\begin{array}{r} 477 \\ -\ 389 \\ \hline \end{array}$

(d) $\begin{array}{r} 1417 \\ -\ 988 \\ \hline \end{array}$

(e) $\begin{array}{r} 8739 \\ -\ 3892 \\ \hline \end{array}$

Answers

6. **(a)** 8; 8; 884 **(b)** 589 **(c)** 88 **(d)** 429 **(e)** 4847

7 Subtract.

(a) 206 − 177

(b) 703 − 415

(c) 7024 − 2632

8 Subtract.

GS (a) 308 − 159 = __ __ 9 (regrouped: 2, 10 → 9, 18)

(b) 570 − 368

(c) 1570 − 983

(d) 7001 − 5193

(e) 4000 − 1782

Answers

7. (a) 29 (b) 288 (c) 4392
8. (a) 1; 4; 149 (b) 202 (c) 587 (d) 1808 (e) 2218

Complete the problem.

$$\begin{array}{r} \overset{5}{4}\,\overset{\overset{9}{10}}{\not 6}\,\overset{17}{\not 0}\,\not 7 \\ -\ 3\ 1\ 6\ 8 \\ \hline 1\ 4\ 3\ 9 \end{array}$$ Difference

Check by adding 1439 and 3168; you should get 4607.

◀ Work Problem **7** at the Side.

EXAMPLE 8 Regrouping with Zeros

Subtract.

(a) 708 − 149

Write 10 tens.

Regroup 1 ten as 10 ones.

Regroup 1 hundred as 10 tens.

10 ones + 8 ones = 18 ones

$$\begin{array}{r} \overset{6}{\not 7}\,\overset{\overset{9}{10}}{\not 0}\,\overset{18}{\not 8} \\ -\ 1\ 4\ 9 \\ \hline 5\ 5\ 9 \end{array}$$

Remember to work from right to left.

(b) 380 − 276

Regroup 1 ten as 10 ones.

Write 10 ones.

$$\begin{array}{r} 3\,\overset{7}{\not 8}\,\overset{10}{\not 0} \\ -\ 2\ 7\ 6 \\ \hline 1\ 0\ 4 \end{array}$$

(c) 9000 − 6999

$$\begin{array}{r} \overset{8}{\not 9}\,\overset{\overset{9}{10}}{\not 0}\,\overset{\overset{9}{10}}{\not 0}\,\overset{10}{\not 0} \\ -\ 6\ 9\ 9\ 9 \\ \hline 2\ 0\ 0\ 1 \end{array}$$

Be extra careful when zeros are involved.

◀ Work Problem **8** at the Side.

As we have seen, an answer to a subtraction problem can be checked by adding.

EXAMPLE 9 Checking Subtraction by Using Addition

Use addition to check each answer.

(a)

$$\begin{array}{r} 613 \\ -\ 275 \\ \hline 338 \end{array} \quad \text{Match} \quad \begin{array}{r} \textbf{CHECK} \\ 275 \\ +\ 338 \\ \hline 613 \end{array}$$ ✓ Correct

Continued on Next Page

(b)

$$\begin{array}{r} 1915 \\ -\ 1635 \\ \hline 280 \end{array} \quad \text{Match} \quad \begin{array}{r} \text{CHECK} \\ 1635 \\ +\ 280 \\ \hline 1915 \end{array} \ \checkmark \quad \text{Correct}$$

(c)

$$\begin{array}{r} 15{,}803 \\ -\ 7\,325 \\ \hline 8\,578 \end{array} \quad \text{No Match} \quad \begin{array}{r} \text{CHECK} \\ 7\,325 \\ +\ 8\,578 \\ \hline 15{,}903 \end{array} \quad \text{Error}$$

It's always a good idea to check your work.

Rework the original problem to get the correct answer, 8478. Then, 7325 + 8478 **does** give 15,803.

Work Problem 9 at the Side. ▶

OBJECTIVE 6 Solve application problems with subtraction. As shown in the next example, subtraction can be used to solve an application problem.

EXAMPLE 10 Applying Subtraction Skills

Use the table to find how much more, on average, a person with an Associate of Arts degree earns each year than a high school graduate.

EDUCATION PAYS

The more education adults get, the higher their annual earnings.

Education Level	Average Earnings
Not a high school graduate	$33,435
High school graduate	$43,165
Some college, no degree	$50,359
Associate of Arts degree	$54,861
Bachelor's degree	$82,197
Master's degree	$99,516
Doctoral degree	$129,773
Professional degree	$166,065

Note: Average annual earnings for workers between ages 25 and 64.

Source: U.S. Census Bureau and Pearson Education, Inc.

Approach The average earnings for a person with an Associate of Arts degree is $54,861 each year. The average for a high school graduate is $43,165. Find how much more a college graduate earns by subtracting $43,165 from $54,861.

Solution

$$\begin{array}{rl} \$54{,}861 & \longleftarrow \text{Associate of Arts degree} \\ -\ \$43{,}165 & \longleftarrow \text{High school graduate} \\ \hline \$11{,}696 & \longleftarrow \text{More earnings} \end{array}$$

Education pays.

On average, a person with an Associate of Arts degree earns $11,696 more each year than a high school graduate.

Work Problem 10 at the Side. ▶

---

9 Use addition to check each answer. If the answer is incorrect, find the correct answer.

GS (a)

$$\begin{array}{r} 357 \\ -\ 168 \\ \hline 189 \end{array} \qquad \begin{array}{r} \text{CHECK} \\ 1\,6\,8 \\ +\_\ \_\ \_ \\ \hline 3\,5\,7 \end{array}$$

(b)

$$\begin{array}{r} 570 \\ -\ 328 \\ \hline 252 \end{array}$$

(c)

$$\begin{array}{r} 14{,}726 \\ -\ 8\,839 \\ \hline 5\,887 \end{array}$$

10 Use the table from **Example 10** to find, on average,

(a) how much more a person with an Associate of Arts degree earns each year than a person who is not a high school graduate.

(b) how much more a person with a Bachelor's degree earns each year than a person with an Associate of Arts degree.

Answers

9. **(a)** 1; 8; 9; correct
(b) incorrect; should be 242
(c) correct

10. **(a)** $21,426 **(b)** $27,336

1.3 Exercises

FOR EXTRA HELP

Download the MyDashBoard App

MyMathLab®

CONCEPT CHECK *Write in the number needed to complete the check.*

1. $\begin{array}{r} 48 \\ -\ 32 \\ \hline 16 \end{array}$ $\begin{array}{r} 16 \\ +\ \_\_ \\ \hline 48 \end{array}$

2. $\begin{array}{r} 17 \\ -\ 13 \\ \hline 4 \end{array}$ $\begin{array}{r} \_\_ \\ +\ 13 \\ \hline 17 \end{array}$

3. $\begin{array}{r} 86 \\ -\ 53 \\ \hline 33 \end{array}$ $\begin{array}{r} 53 \\ +\ 33 \\ \hline \_\_ \end{array}$

4. $\begin{array}{r} 78 \\ -\ 35 \\ \hline 43 \end{array}$ $\begin{array}{r} \_\_ \\ +\ 43 \\ \hline 78 \end{array}$

5. $\begin{array}{r} 77 \\ -\ 60 \\ \hline 17 \end{array}$ $\begin{array}{r} 60 \\ +\ \_\_ \\ \hline 77 \end{array}$

Work each subtraction problem. Use addition to check each answer. ***See Examples 3 and 4.***

6. $\begin{array}{r} 87 \\ -\ 63 \\ \hline \end{array}$

7. $\begin{array}{r} 335 \\ -\ 122 \\ \hline \end{array}$

8. $\begin{array}{r} 602 \\ -\ 301 \\ \hline \end{array}$

9. $\begin{array}{r} 552 \\ -\ 451 \\ \hline \end{array}$

10. $\begin{array}{r} 888 \\ -\ 215 \\ \hline \end{array}$

11. $\begin{array}{r} 7352 \\ -\ 241 \\ \hline \end{array}$

12. $\begin{array}{r} 4420 \\ -\ 310 \\ \hline \end{array}$

13. $\begin{array}{r} 5546 \\ -\ 2134 \\ \hline \end{array}$

14. $\begin{array}{r} 1875 \\ -\ 1362 \\ \hline \end{array}$

15. $\begin{array}{r} 6259 \\ -\ 4148 \\ \hline \end{array}$

16. $\begin{array}{r} 9654 \\ -\ 4323 \\ \hline \end{array}$

17. $\begin{array}{r} 24{,}392 \\ -\ 11{,}232 \\ \hline \end{array}$

18. $\begin{array}{r} 57{,}921 \\ -\ 34{,}801 \\ \hline \end{array}$

19. $\begin{array}{r} 46{,}253 \\ -\ 5\ 143 \\ \hline \end{array}$

20. $\begin{array}{r} 75{,}904 \\ -\ 3\ 702 \\ \hline \end{array}$

Use addition to check each subtraction problem. If an answer is not correct, find the correct answer. ***See Example 4.***

21. $\begin{array}{r} 54 \\ -\ 42 \\ \hline 12 \end{array}$ $\begin{array}{r} 42 \\ +\ \_\_ \\ \hline 54 \end{array}$

22. $\begin{array}{r} 87 \\ -\ 43 \\ \hline 44 \end{array}$ $\begin{array}{r} 43 \\ +\ \_\_ \\ \hline 87 \end{array}$

23. $\begin{array}{r} 89 \\ -\ 27 \\ \hline 63 \end{array}$

24. $\begin{array}{r} 47 \\ -\ 35 \\ \hline 13 \end{array}$

25. $\begin{array}{r} 382 \\ -\ 261 \\ \hline 131 \end{array}$

26. $\begin{array}{r} 754 \\ -\ 342 \\ \hline 412 \end{array}$

27. $\begin{array}{r} 4683 \\ -\ 3542 \\ \hline 1141 \end{array}$

28. $\begin{array}{r} 5217 \\ -\ 4105 \\ \hline 1132 \end{array}$

29. $\begin{array}{r} 8643 \\ -\ 1421 \\ \hline 7212 \end{array}$

30. $\begin{array}{r} 9428 \\ -\ 3124 \\ \hline 6324 \end{array}$

31. **CONCEPT CHECK** *Underline the correct answer.*

In subtraction, regrouping is necessary when the digit in the (*minuend/subtrahend*) is less value than the digit in the subtrahend that is directly (*above/below*) it.

32. **CONCEPT CHECK** *Which of the subtraction problems will require regrouping?*

A. $\begin{array}{r} 64 \\ -\ 51 \\ \hline \end{array}$

B. $\begin{array}{r} 763 \\ -\ 473 \\ \hline \end{array}$

C. $\begin{array}{r} 43{,}708 \\ -\ 22{,}607 \\ \hline \end{array}$

D. $\begin{array}{r} 6208 \\ -\ 5126 \\ \hline \end{array}$

Subtract, regrouping when necessary. ***See Examples 5–8.***

33. $\begin{array}{r} 75 \\ -\ 37 \\ \hline \end{array}$

34. $\begin{array}{r} 86 \\ -\ 28 \\ \hline \end{array}$

35. $\begin{array}{r} 94 \\ -\ 49 \\ \hline \end{array}$

36. $\begin{array}{r} 68 \\ -\ 39 \\ \hline \end{array}$

37. $\begin{array}{r} 57 \\ -\ 38 \\ \hline \end{array}$

38. $\begin{array}{r} 47 \\ -\ 29 \\ \hline \end{array}$

39. $\begin{array}{r} 828 \\ -\ 547 \\ \hline \end{array}$

40. $\begin{array}{r} 916 \\ -\ 618 \\ \hline \end{array}$

41. $\begin{array}{r} 771 \\ -\ 252 \\ \hline \end{array}$

42. $\begin{array}{r} 973 \\ -\ 788 \\ \hline \end{array}$

43. $\begin{array}{r} 7538 \\ -\ 479 \\ \hline \end{array}$ **44.** $\begin{array}{r} 5863 \\ -\ 1295 \\ \hline \end{array}$ **45.** $\begin{array}{r} 9988 \\ -\ 2399 \\ \hline \end{array}$ **46.** $\begin{array}{r} 3576 \\ -\ 1658 \\ \hline \end{array}$ **47.** $\begin{array}{r} 80 \\ -\ 73 \\ \hline \end{array}$

48. $\begin{array}{r} 60 \\ -\ 37 \\ \hline \end{array}$ **49.** $\begin{array}{r} 308 \\ -\ 289 \\ \hline \end{array}$ **50.** $\begin{array}{r} 600 \\ -\ 599 \\ \hline \end{array}$ **51.** $\begin{array}{r} 4041 \\ -\ 1208 \\ \hline \end{array}$ **52.** $\begin{array}{r} 4602 \\ -\ 2063 \\ \hline \end{array}$

53. $\begin{array}{r} 9305 \\ -\ 1530 \\ \hline \end{array}$ **54.** $\begin{array}{r} 7120 \\ -\ 6033 \\ \hline \end{array}$ **55.** $\begin{array}{r} 1580 \\ -\ 1077 \\ \hline \end{array}$ **56.** $\begin{array}{r} 3068 \\ -\ 2105 \\ \hline \end{array}$ **57.** $\begin{array}{r} 2006 \\ -\ 1850 \\ \hline \end{array}$

58. $\begin{array}{r} 8203 \\ -\ 5365 \\ \hline \end{array}$ **59.** $\begin{array}{r} 8240 \\ -\ 6056 \\ \hline \end{array}$ **60.** $\begin{array}{r} 7050 \\ -\ 6045 \\ \hline \end{array}$ **61.** $\begin{array}{r} 8503 \\ -\ 2816 \\ \hline \end{array}$ **62.** $\begin{array}{r} 16{,}004 \\ -\ \ \ 5\ 087 \\ \hline \end{array}$

63. $\begin{array}{r} 80{,}705 \\ -\ 61{,}667 \\ \hline \end{array}$ **64.** $\begin{array}{r} 81{,}000 \\ -\ 55{,}456 \\ \hline \end{array}$ **65.** $\begin{array}{r} 66{,}000 \\ -\ 34{,}444 \\ \hline \end{array}$ **66.** $\begin{array}{r} 77{,}000 \\ -\ 65{,}308 \\ \hline \end{array}$ **67.** $\begin{array}{r} 20{,}080 \\ -\ 13{,}496 \\ \hline \end{array}$

CONCEPT CHECK *Fill in each blank with the correct response.*

68. To avoid errors when solving math problems, it's a good idea to __________ your work.

69. When checking the accuracy of an answer to an addition problem, you can use __________.

70. An answer to a subtraction problem may be checked using __________.

Use addition to check each subtraction problem. If an answer is incorrect, find the correct answer. ***See Example 9.***

71. $\begin{array}{r} 9428 \\ -\ 4509 \\ \hline 4919 \end{array}$ **72.** $\begin{array}{r} 1671 \\ -\ 1325 \\ \hline 1346 \end{array}$ **73.** $\begin{array}{r} 2548 \\ -\ 2278 \\ \hline 270 \end{array}$ **74.** $\begin{array}{r} 5274 \\ -\ 1130 \\ \hline 4144 \end{array}$

75. $\begin{array}{r} 93{,}758 \\ -\ 52{,}869 \\ \hline 40{,}889 \end{array}$ **76.** $\begin{array}{r} 82{,}357 \\ -\ 14{,}396 \\ \hline 68{,}961 \end{array}$ **77.** $\begin{array}{r} 36{,}778 \\ -\ 17{,}405 \\ \hline 19{,}373 \end{array}$ **78.** $\begin{array}{r} 34{,}821 \\ -\ 17{,}735 \\ \hline 17{,}735 \end{array}$

79. An addition problem can be changed to a subtraction problem and a subtraction problem can be changed to an addition problem. Give two examples of each to demonstrate this.

80. Can you use the commutative and the associative properties in subtraction? Explain.

Solve each application problem. **See Example 10.**

81. A man burns 187 calories during 60 minutes of sitting at a computer while a woman burns 140 calories at the same activity. How many fewer calories does a woman burn than a man in 60 minutes? (*Source:* www.cookinglight.com)

82. A woman burns 302 calories during an hour of walking, while a man burns 403 calories doing the same activity. How many more calories does a man burn than a woman during an hour of walking? (*Source:* www.cookinglight.com)

83. In April 2011, there were 612 tornadoes, shattering the old record of 543. How many more tornadoes were there than the old record number? (*Source:* National Oceanic and Atmospheres Administration.)

84. With an estimated 327 deaths, the tornado outbreak in April 2011 was the third deadliest on record, behind 747 deaths in April 1925 and 332 deaths in April 1932. How many more deaths were there in the deadliest tornado outbreak than in the tornado outbreak of April 2011? (*Source:* Accu Weather.)

85. The top of each main tower of the Golden Gate Bridge is 746 feet above the water and 500 feet above the roadway. How far above the water is the roadway? (*Source:* gocalifornia.about.com)

746 feet above water
-500 feet above roadway

86. In a recent three-month period there were 81,465 Ford Explorers and 70,449 Jeep Grand Cherokees sold. Which vehicle had greater sales? By how much? (*Source:* J. D. Power and Associates.)

87. Six years ago there were 6970 bridge and lock-tender jobs across the United States. Today there are 3700 that remain. How many of these jobs have been eliminated? (*Source:* Bureau of Labor Statistics.)

88. In 1964, its first year on the market, the Ford Mustang sold for \$2500. In 2013, the Ford Mustang sold for \$28,065. Find the increase in price. (*Source:* eBay.)

89. Patriot Flag Company manufactured 14,608 U.S. flags and sold 5069. How many flags remain unsold?

90. Eye exams have been given to 14,679 children in the school district. If there are 23,156 students in the school district, how many have not received eye exams?

91. The Jordanos now pay rent of \$650 per month. If they buy a house, their housing expense will be \$913 per month. How much more will they pay per month if they buy a house?

92. A retired couple who used to receive a Social Security payment of \$1479 per month now receives \$1568 per month. Find the amount of the monthly increase.

93. The distance from New York City to Buenos Aires, Argentina is 5299 miles, while the distance from Los Angeles to Dublin, Ireland is 5158 miles. How much further is one trip than the other? (*Source:* Map Crow Travel Calculator, mapcrow info)

94. In the year 2020 it is predicted that we will need 2,820,000 nurses in the United States, while only 1,810,000 nurses will be available. Find the shortage in the number of nurses. (*Source:* American Hospital Association.)

Solve each application problem. Add or subtract as necessary.

95. This year there were 264,311 hip replacement procedures in the United States. If 125,423 of the patients were 65 years of age or older, how many patients were under age 65? (*Source:* Federal Agency for Healthcare Research and Quality.)

96. This year there were 555,800 knee surgeries performed in the United States. The number of knee surgeries performed six years ago was 328,900. How many more of these surgeries were performed this year than six years ago? (*Source:* Agency for Healthcare Research and Quality.)

SUBWAY promotes healthy food choices by offering eight sandwiches that are low in fat. The nutritional information, printed on every SUBWAY napkin, appears below and includes information to answer Exercises 97–100. (*Source:* SUBWAY.)

SUBWAY

OUR 6" SANDWICHES:	CALORIES	FAT(g)
VEGGIE DELITE®	230	3
BLACK FOREST HAM	290	5
TURKEY BREAST	280	4
ROAST BEEF	320	5
SUBWAY CLUB	310	5
TURKEY BREAST & BLACK FOREST HAM	280	4
OVEN ROASTED CHICKEN	320	5
SWEET ONION CHICKEN TERIYAKI	380	5

SUBWAY® regular 6" subs include italian or wheat bread, veggies and meat. Addition of condiments or cheese alters nutrition content.

MUSTARD (2 tsp.)	5	0
CHEESE TRIANGLES (2)	40	4
OLIVE OIL (1 tsp.)	45	5
VERSUS:		
BIG MAC®	540	29
WHOPPER®	670	40

97. How many fewer calories and grams of fat are in a 6-inch Veggie Delite sandwich than a Big Mac?

98. How many fewer calories and grams of fat are in a 6-inch Turkey Breast and Black Forest Ham sandwich than a Whopper?

99. Find the total number of calories and grams of fat in an Oven Roasted Chicken sandwich with mustard and olive oil.

100. A customer ate two sandwiches, one with the least calories and one with the most calories. Find the total number of calories and grams of fat in the two sandwiches.

1.4 Multiplying Whole Numbers

OBJECTIVES

1. Identify the parts of a multiplication problem.
2. Do chain multiplication.
3. Multiply by single-digit numbers.
4. Use multiplication shortcuts for numbers ending in zeros.
5. Multiply by numbers having more than one digit.
6. Solve application problems with multiplication.

1 Identify the factors and the product in each multiplication problem.

(a) $8 \times 5 = 40$

(b) $6(4) = 24$

(c) $7 \cdot 6 = 42$

(d) $(3)(9) = 27$

Suppose we want to know the total number of exercise bicycles available at the gym. The bicycles are arranged in four columns with three stations in each column. Adding the number 3 a total of 4 times gives 12.

$$3 + 3 + 3 + 3 = 12$$

This result can also be shown with a figure.

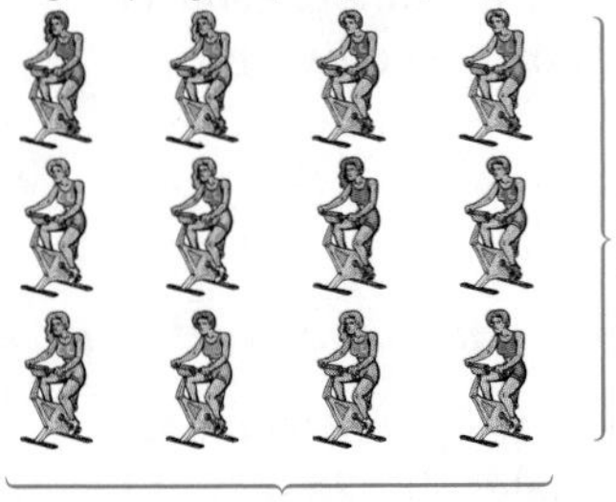

OBJECTIVE 1 Identify the parts of a multiplication problem. Multiplication is a shortcut for repeated addition. In the exercise bicycle example, instead of *adding* 3 + 3 + 3 + 3 to get 12, we can *multiply* 3 by 4 to get 12. The numbers being multiplied are called **factors.** The answer is called the **product.** For example, the product of 3 and 4 can be written with the symbol ×, a raised dot, or parentheses, as follows.

$$\begin{array}{r} 3 \\ \times\ 4 \\ \hline 12 \end{array}$$

3 ⟵ Factor (also called *multiplicand*)
4 ⟵ Factor (also called *multiplier*)
12 ⟵ Product (answer)

$3 \times 4 = 12$ *or* $3 \cdot 4 = 12$ *or* $(3)(4) = 12$ *or* $3(4) = 12$

◀ **Work Problem 1 at the Side.**

Commutative Property of Multiplication

By the **commutative property of multiplication,** the product (answer) remains the same when the order of the factors is changed. For example,

$$3 \times 5 = 15 \quad \text{and} \quad 5 \times 3 = 15.$$

Multiply numbers in any order.

EXAMPLE 1 Multiplying Two Numbers

Multiply.

(a) $3 \times 4 = 12$

(b) $6 \cdot 0 = 0$ — Multiply any number by zero and the answer is always zero.

(c) $4(8) = 32$

Continued on Next Page

Answers

1. (a) factors: 8, 5; product: 40
(b) factors: 6, 4; product: 24
(c) factors: 7, 6; product: 42
(d) factors: 3, 9; product: 27

Learning the multiplication table will help you in later chapters.

Multiplication Table

×	1	2	3	4	5	6	7	8	9
1	1	2	3	4	5	6	7	8	9
2	2	4	6	8	10	12	14	16	18
3	3	6	9	12	15	18	21	24	27
4	4	8	12	16	20	24	28	32	36
5	5	10	15	20	25	30	35	40	45
6	6	12	18	24	30	36	42	48	54
7	7	14	21	28	35	42	49	56	63
8	8	16	24	32	40	48	56	64	72
9	9	18	27	36	45	54	63	72	81

Recall that any number multiplied by 1 is always the number itself.

Work Problem 2 at the Side. ▶

OBJECTIVE 2 Do chain multiplication. Some multiplications involve more than two factors.

Associative Property of Multiplication

By the **associative property of multiplication,** grouping the factors differently does not change the product.

EXAMPLE 2 Multiplying Three Numbers

Multiply $2 \times 3 \times 5$.

$$(2 \times 3) \times 5 \qquad \text{Parentheses show what to do first.}$$

$$6 \times 5 = 30$$

Also,

$$2 \times (3 \times 5)$$

$$2 \times 15 = 30$$

Either grouping results in the same product.

Work Problem 3 at the Side. ▶

Calculator Tip

The calculator approach to **Example 2** uses chain calculations.

2 (×) 3 (×) 5 (=) **30**

A problem with more than two factors, such as the one in **Example 2**, is called a **chain multiplication** problem.

VOCABULARY TIP

Chain multiplication Just like a chain is a connected series of objects—a "chain of events," for example—chain multiplication is a series of numbers connected by the operation of multiplication.

2 Multiply.

(a) 7×4

(b) 0×9

(c) $8(5)$

(d) $6 \cdot 5$

(e) $(1)(8)$

3 Multiply.

(a) $3 \times 2 \times 5$

(b) $4 \cdot 7 \cdot 1$

(c) $(8)(3)(0)$

Answers

2. **(a)** 28 **(b)** 0 **(c)** 40 **(d)** 30 **(e)** 8
3. **(a)** 30 **(b)** 28 **(c)** 0

4 Multiply.

GS (a)
$$\begin{array}{r} \overset{1}{5}\,3 \\ \times\ \ 5 \\ \hline \_\,\_\,5 \end{array}$$

(b)
$$\begin{array}{r} 79 \\ \times\ \ 0 \\ \hline \end{array}$$

(c)
$$\begin{array}{r} 758 \\ \times\ \ 8 \\ \hline \end{array}$$

(d)
$$\begin{array}{r} 2831 \\ \times\ \ \ 7 \\ \hline \end{array}$$

(e)
$$\begin{array}{r} 4714 \\ \times\ \ \ 8 \\ \hline \end{array}$$

Answers

4. **(a)** 2; 6; 265 **(b)** 0 **(c)** 6064 **(d)** 19,817 **(e)** 37,712

OBJECTIVE 3 **Multiply by single-digit numbers.** Regrouping may be needed in multiplication problems with larger factors.

EXAMPLE 3 Multiplying with Regrouping

Multiply.

(a)
$$\begin{array}{r} 53 \\ \times\ 4 \\ \hline \end{array}$$

Start by multiplying in the ones column.

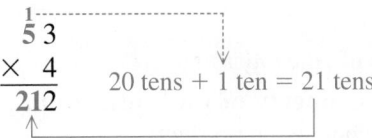

$$\begin{array}{r} \overset{1}{5}3 \\ \times\ 4 \\ \hline 2 \end{array}$$

Write 1 ten in the tens column.
$4 \times 3 =$ **12** ones
Write 2 ones in the ones column.

Next, multiply 4 times 5 tens.

$$\begin{array}{r} \overset{1}{5}\,3 \\ \times\ \ 4 \\ \hline 2 \end{array}$$

4×5 tens $=$ **20** tens

Add the 1 ten that was written at the top of the tens column.

$$\begin{array}{r} \overset{1}{5}\,3 \\ \times\ \ 4 \\ \hline 212 \end{array}$$

20 tens $+$ 1 ten $=$ 21 tens

(b)
$$\begin{array}{r} 724 \\ \times\ \ 5 \\ \hline \end{array}$$

Work as shown.

$$\begin{array}{r} \overset{1\,2}{724} \\ \times\ \ 5 \\ \hline 3620 \end{array}$$

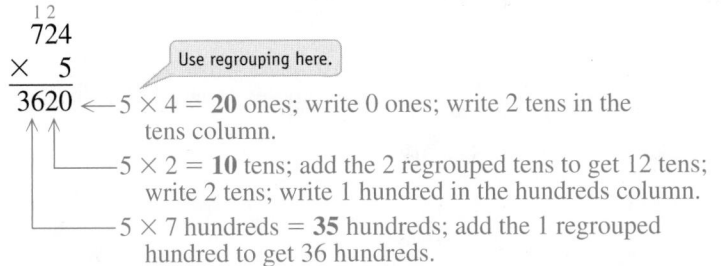

Use regrouping here.

$5 \times 4 =$ **20** ones; write 0 ones; write 2 tens in the tens column.

$5 \times 2 =$ **10** tens; add the 2 regrouped tens to get 12 tens; write 2 tens; write 1 hundred in the hundreds column.

5×7 hundreds $=$ **35** hundreds; add the 1 regrouped hundred to get 36 hundreds.

◀ **Work Problem 4 at the Side.**

OBJECTIVE 4 **Use multiplication shortcuts for numbers ending in zeros.** The product of two whole number factors is also called a **multiple** of either factor. For example, since $4 \cdot 2 = 8$, the whole number 8 is a multiple of both 4 and 2. *Multiples of 10* are very useful when multiplying. A **multiple of 10** is a whole number that ends in 0, such as 10, 20, or 30; 100, 200, or 300; 1000, 2000, or 3000; and so on. There is a short way to multiply by these multiples of 10. Look at the following examples.

$$26 \times 1 = 26$$

$$26 \times 10 = 260$$

$$26 \times 100 = 2600$$

$$26 \times 1000 = 26{,}000$$

Do you see a pattern? These examples suggest the rule that follows.

Multiplying by Multiples of 10

To multiply a whole number by 10, 100, or 1000, attach one, two, or three zeros, respectively, to the right of the whole number.

EXAMPLE 4 Using Multiples of 10 to Multiply

Multiply.

(a) $59 \times 10 = 590$ ← Attach 0.

(b) $74 \times 100 = 7400$ ← Attach 00.

(c) $803 \times 1000 = 803{,}000$ ← Attach 000.

Work Problem 5 at the Side. ▶

You can also find the product of other multiples of 10 by attaching zeros.

EXAMPLE 5 Using Multiples of 10 to Multiply

Multiply.

(a) 75×3000
Multiply 75 by 3, and then attach three zeros.

$$75 \times 3000 = 225{,}000$$

$$\begin{array}{r} 75 \\ \times\ 3 \\ \hline 225 \end{array}$$

Attach 000.

Use useful shortcuts.

(b) 150×70
Multiply 15 by 7, and then attach two zeros.

$$150 \times 70 = 10{,}500 \leftarrow \text{Attach 00.}$$

$$\begin{array}{r} 15 \\ \times\ 7 \\ \hline 105 \end{array}$$

Work Problem 6 at the Side. ▶

OBJECTIVE 5 Multiply by numbers having more than one digit. The next example shows multiplication when both factors have more than one digit.

EXAMPLE 6 Multiplying with More Than One Digit

Multiply 46 and 23.

First multiply 46 by 3.

$$\begin{array}{r} \overset{1}{4}6 \\ \times\ 3 \\ \hline 138 \end{array} \leftarrow 46 \times 3 = 138$$

Regrouping is needed here.

Continued on Next Page

5 Multiply.

GS **(a)** $63 \times 10 = 63\_\_\_$

(b) 305×100

(c) 714×1000

6 Multiply.

GS **(a)** 17×50

$$\begin{array}{r} 17 \\ \times\ 5 \\ \hline 85\_ \end{array}$$ Attach one zero.

(b) 73×400

(c) $$\begin{array}{r} 180 \\ \times\ 30 \\ \hline \end{array}$$

(d) $$\begin{array}{r} 4200 \\ \times\ 80 \\ \hline \end{array}$$

(e) $$\begin{array}{r} 800 \\ \times\ 600 \\ \hline \end{array}$$

Answers

5. **(a)** 0; 630 **(b)** 30,500 **(c)** 714,000
6. **(a)** 0; 850 **(b)** 29,200 **(c)** 5400 **(d)** 336,000 **(e)** 480,000

7 Complete each multiplication.

GS **(a)**

$$\begin{array}{r} {}^{2}_{2} \\ 35 \\ \times\ 54 \\ \hline 140 \\ 175 \\ \hline \_\_90 \end{array}$$

GS **(b)**

$$\begin{array}{r} {}^{2}_{5} \\ 76 \\ \times\ 49 \\ \hline 684 \\ 304 \\ \hline \_\_\_\_ \end{array}$$

8 Multiply.

(a) $\begin{array}{r} 52 \\ \times\ 16 \\ \hline \end{array}$

(b) $\begin{array}{r} 81 \\ \times\ 49 \\ \hline \end{array}$

(c) $\begin{array}{r} 234 \\ \times\ 73 \\ \hline \end{array}$

(d) $\begin{array}{r} 835 \\ \times\ 189 \\ \hline \end{array}$

Answers

7. **(a)** 1; 8; 1890 **(b)** 3724
8. **(a)** 832 **(b)** 3969
(c) 17,082 **(d)** 157,815

Now multiply 46 by 20.

$$\begin{array}{r} {}^{1} \\ 46 \\ \times\ 20 \\ \hline 920 \end{array} \leftarrow 46 \times 20 = 920$$

Add the results.

$$\begin{array}{rl} 46 & \\ \times\ 23 & \\ \hline 138 & \leftarrow 46 \times 3 \\ +\ 920 & \leftarrow 46 \times 20 \\ \hline 1058 & \end{array}$$

↑ Add.

Both 138 and 920 are called **partial products.** As a common practice and to save time, the 0 in 920 is usually not written.

$$\begin{array}{r} 46 \\ \times\ 23 \\ \hline 138 \\ 92 \\ \hline 1058 \end{array}$$

← 0 not written. Be very careful to place the 2 in the tens column.

◀ Work Problem **7** at the Side.

EXAMPLE 7 **Using Partial Products**

Multiply.

(a)

$$\begin{array}{rl} 233 & \\ \times\ 132 & \\ \hline \mathbf{466} & \\ 6\mathbf{99} & \text{(Tens lined up)} \\ 23\mathbf{3} & \text{(Hundreds lined up)} \\ \hline 30{,}756 & \leftarrow \text{Product} \end{array}$$

Be certain to align numbers in columns.

(b) $\begin{array}{r} 538 \\ \times\ 46 \\ \hline \end{array}$

First multiply by 6.

$$\begin{array}{r} {}^{2\,4} \\ 538 \\ \times\ 46 \\ \hline 3228 \end{array}$$

← Regrouping is needed here.

Now multiply by 4, being careful to line up the tens.

$$\begin{array}{r} {}^{1\,3} \\ {}^{2\,4} \\ 538 \\ \times\ 46 \\ \hline 3228 \\ 2152 \\ \hline 24{,}748 \end{array}$$

Finally, add the partial products.

◀ Work Problem **8** at the Side.

When 0 appears in the multiplier, be sure to move the partial products to the left to account for the position held by the 0.

EXAMPLE 8 Multiplying with Zeros

Multiply.

(a)

$$
\begin{array}{r}
137 \\
\times\,306 \\
\hline
822 \\
000 \\
411 \\
\hline
41{,}922
\end{array}
$$

(Tens lined up)
(Hundreds lined up)

(b)

$$
\begin{array}{r}
1406 \\
\times\,2001 \\
\hline
1406 \\
\mathbf{0000} \\
\mathbf{0000} \\
2812 \\
\hline
2{,}813{,}406
\end{array}
$$

← (Zeros to line up tens)
← (Zeros to line up hundreds)

Use extra caution when working with 0s.

$$
\begin{array}{r}
1406 \\
\times\,2001 \\
\hline
1406 \\
2812\mathbf{00} \\
\hline
2{,}813{,}406
\end{array}
$$

← Zeros are written so this partial product starts in the thousands column.

Note

In **Example 8(b)** in the alternative method on the right, zeros were inserted so that thousands were placed in the thousands column. This is a commonly used shortcut.

Work Problem 9 at the Side. ▶

OBJECTIVE 6 Solve application problems with multiplication. The next example shows how multiplication can be used to solve an application problem.

EXAMPLE 9 Applying Multiplication Skills

Find the total cost of 75 video games priced at $38 each.

Approach To find the cost of all the video games multiply the number of games (75) by the cost of one video game ($38).

Solution Multiply 75 by 38.

$$
\begin{array}{r}
75 \\
\times\,38 \\
\hline
600 \\
225 \\
\hline
\$2850
\end{array}
$$

The total cost of the video games is $2850.

Calculator Tip

If you are using a calculator for **Example 9**, you will do this calculation.

75 (×) 38 (=) 2850

Work Problem 10 at the Side. ▶

9 Multiply.

GS **(a)**

$$
\begin{array}{r}
28 \\
\times\,60 \\
\hline
\_\,\_ \\
168 \\
\hline
\_\,\_\,\_\,\_
\end{array}
$$

(b)
$$
\begin{array}{r}
728 \\
\times\,50 \\
\hline
\end{array}
$$

(c)
$$
\begin{array}{r}
562 \\
\times\,109 \\
\hline
\end{array}
$$

(d)
$$
\begin{array}{r}
3526 \\
\times\,6002 \\
\hline
\end{array}
$$

10 Find the total cost of the following items.

GS **(a)** 314 garden sprayers at $14 per sprayer

(b) 64 tires priced at $139 each

(c) 12 delivery vans at $28,300 per van

Answers

9. **(a)** 0; 0; 1680 **(b)** 36,400 **(c)** 61,258 **(d)** 21,163,052

10. **(a)** $4396 **(b)** $8896 **(c)** $339,600

1.4 Exercises

FOR EXTRA HELP

MyMathLab®

CONCEPT CHECK *Fill in each blank with the correct response.*

1. In a chain multiplication, if you multiply by the largest number first, rather than the smallest number first, the product will always be _____ .

2. The property described in **Exercise 1** is the _____ property of multiplication.

3. When you multiply any number by zero, the answer is always _____ .

4. When you multiply a whole number by 10, by 100 or by 1000, you can get the answer by attaching one, two, or three _____ to the _____ of the whole number.

Work each chain multiplication. See Example 2.

5. $2 \times 6 \times 2$

6. $8 \times 6 \times 1$

7. $7 \cdot 8 \cdot 0$

8. $9 \cdot 0 \cdot 5$

9. $4 \cdot 1 \cdot 6$

10. $1 \cdot 5 \cdot 7$

11. $(4)(5)(2)$

12. $(4)(1)(9)$

13. Explain in your own words the commutative property of multiplication. How do the commutative properties of addition and multiplication compare to each other?

14. Explain in your own words the associative property of multiplication. How do the associative properties of addition and multiplication compare to each other?

Multiply. See Example 3.

15. $\begin{array}{r} 35 \\ \times\ 6 \\ \hline \end{array}$

16. $\begin{array}{r} 53 \\ \times\ 7 \\ \hline \end{array}$

17. $\begin{array}{r} 34 \\ \times\ 7 \\ \hline \end{array}$

18. $\begin{array}{r} 76 \\ \times\ 5 \\ \hline \end{array}$

19. $\begin{array}{r} 642 \\ \times\ \ 5 \\ \hline \end{array}$

20. $\begin{array}{r} 472 \\ \times\ \ 4 \\ \hline \end{array}$

21. $\begin{array}{r} 624 \\ \times\ \ 3 \\ \hline \end{array}$

22. $\begin{array}{r} 852 \\ \times\ \ 7 \\ \hline \end{array}$

23. GS $\begin{array}{r} {}^{2\ 1} \\ 2\ 1\ 5\ 3 \\ \times\ \ \ \ \ 4 \\ \hline _\ _1\ 2 \end{array}$

24. GS $\begin{array}{r} {}^{1\ 2} \\ 1\ 1\ 3\ 7 \\ \times\ \ \ \ \ 3 \\ \hline _\ _1\ 1 \end{array}$

25. $\begin{array}{r} 2521 \\ \times\ \ \ 4 \\ \hline \end{array}$

26. $\begin{array}{r} 2544 \\ \times\ \ \ 3 \\ \hline \end{array}$

27. $\begin{array}{r} 2561 \\ \times\ \ \ 8 \\ \hline \end{array}$

28. $\begin{array}{r} 7326 \\ \times\ \ \ 5 \\ \hline \end{array}$

29. $\begin{array}{r} 36{,}921 \\ \times\ \ \ \ \ 7 \\ \hline \end{array}$

30. $\begin{array}{r} 28{,}116 \\ \times\ \ \ \ \ 4 \\ \hline \end{array}$

CONCEPT CHECK *Fill in each blank with the correct response.*

31. You can use multiples of 10 to multiply 86×200. First, multiply _____ $\times$ 2 to get _____ . Then, attach _____ zeros to the right of this number for a final answer of _____ .

32. You can use multiples of 10 to multiply 7800×450. First, multiply 78 $\times$ _____ to get _____ . Then, attach _____ zeros to the right of this number for a final answer of _____ .

Multiply. See Examples 4 and 5.

33. $\begin{array}{r} 80 \\ \times\ 6 \\ \hline \end{array}$

34. $\begin{array}{r} 70 \\ \times\ 5 \\ \hline \end{array}$

35. $\begin{array}{r} 740 \\ \times\ \ 3 \\ \hline \end{array}$

36. $\begin{array}{r} 400 \\ \times\ \ 8 \\ \hline \end{array}$

37. $\begin{array}{r} 600 \\ \times \quad 6 \\ \hline \end{array}$

38. $\begin{array}{r} 860 \\ \times \quad 7 \\ \hline \end{array}$

39. $\begin{array}{r} 125 \\ \times \ 30 \\ \hline \end{array}$

40. $\begin{array}{r} 246 \\ \times \ 50 \\ \hline \end{array}$

41. $\begin{array}{r} 1635 \\ \times \quad 40 \\ \hline \end{array}$

42. $\begin{array}{r} 7311 \\ \times \quad 50 \\ \hline \end{array}$

43. $\begin{array}{r} 900 \\ \times 300 \\ \hline \end{array}$

44. $\begin{array}{r} 400 \\ \times 700 \\ \hline \end{array}$

45. GS $\begin{array}{r} 43{,}000 \\ \times \ 2\,000 \\ \hline \end{array}$

$\begin{array}{r} 43 \\ \times \ 2 \\ \hline 86 \end{array}$ Attach 000,000.

46. GS $\begin{array}{r} 11{,}000 \\ \times \ 9\,000 \\ \hline \end{array}$

$\begin{array}{r} 11 \\ \times \ 9 \\ \hline 99 \end{array}$ Attach 000,000.

47. 970 • 50

48. 730 • 40

49. 800 • 900

50. 850 • 700

51. 9700 • 200

52. 10,050 • 300

Multiply. ***See Examples 6–8.***

53. $\begin{array}{r} 28 \\ \times 17 \\ \hline \end{array}$

54. $\begin{array}{r} 16 \\ \times 34 \\ \hline \end{array}$

55. $\begin{array}{r} 75 \\ \times 32 \\ \hline \end{array}$

56. $\begin{array}{r} 82 \\ \times 32 \\ \hline \end{array}$

57. $\begin{array}{r} 83 \\ \times 45 \\ \hline \end{array}$

58. (75)(21)

59. (58)(41)

60. (82)(67)

61. (67)(92)

62. (26)(33)

63. (28)(564)

64. (58)(312)

65. (619)(35)

66. (681)(47)

67. (55)(286)

68. $\begin{array}{r} 286 \\ \times 574 \\ \hline \end{array}$

69. $\begin{array}{r} 735 \\ \times \ 112 \\ \hline \end{array}$

70. $\begin{array}{r} 621 \\ \times \ 415 \\ \hline \end{array}$

71. $\begin{array}{r} 538 \\ \times \ 342 \\ \hline \end{array}$

72. $\begin{array}{r} 3228 \\ \times \ 751 \\ \hline \end{array}$

73. $\begin{array}{r} 9352 \\ \times \ 264 \\ \hline \end{array}$

74. $\begin{array}{r} 528 \\ \times 106 \\ \hline \end{array}$

75. $\begin{array}{r} 215 \\ \times 307 \\ \hline \end{array}$

76. $\begin{array}{r} 218 \\ \times \ 106 \\ \hline \end{array}$

77. $\begin{array}{r} 428 \\ \times 201 \\ \hline \end{array}$

78. $\begin{array}{r} 3706 \\ \times \ 208 \\ \hline \end{array}$

79. $\begin{array}{r} 6310 \\ \times \ 3078 \\ \hline \end{array}$

80. $\begin{array}{r} 3533 \\ \times 5001 \\ \hline \end{array}$

81. $\begin{array}{r} 2195 \\ \times 1038 \\ \hline \end{array}$

82. $\begin{array}{r} 1502 \\ \times 2009 \\ \hline \end{array}$

83. A classmate of yours is not clear on how to use a shortcut to multiply a whole number by 10, by 100, or by 1000. Write a short note explaining how this can be done.

84. Show two ways to multiply when a 0 is in the multiplier. Use the problem 291 × 307 to show this.

Solve each application problem. ***See Example 9.***

85. Carepanian Company, a health care supplier, purchased 300 cartons of Thera Bond Gym Balls. If there are 10 balls in each carton, find the total number of balls purchased.

86. A medical supply house has 30 bottles of vitamin C tablets, with each bottle containing 500 tablets. Find the total number of vitamin C tablets in the supply house.

87. The most expensive U.S. City for a hotel room is New York City, with an average cost of $194 a night. Find the cost of a 12-night stay. (*Source:* hotels.com)

88. Judge Judith Sheindlin, known as Judge Judy on court television, recently signed a four-year contract paying her $45 million each year. Find the total amount of her earnings on this contract. (*Source:* KSTE Radio News.)

89. GS The average amount of water used per person each day in the United States is 66 gallons. How much water does the average person use in one year? (1 year = 365 days). (*Source:* Oxfam International.)

365 × 66 = __ __,090

90. GS Squid are being hauled out of the Santa Barbara Channel by the ton. They are then processed, renamed calamari, and exported. Last night 27 fishing boats each hauled out 40 tons of squid. What was the total catch for the night? (*Source: Santa Barbara News Press.*)

27 × 40
27 × 4 = 108 Attach _______.

Find the total cost of the following items. ***See Examples 7–9.***

91. 75 first-aid kits at $8 per kit

92. 27 days of child care at $82 per day

93. 65 rebuilt alternators at $24 per alternator

94. 62 wheelchair cushions at $44 per cushion

95. 206 laptop computers at $548 per computer

96. 520 printers at $219 per printer

Multiply.

97. 21 • 43 • 56

98. (600)(8)(75)(40)

Use addition, subtraction, or multiplication to solve each application problem.

99. In a forest-planting project, 450 trees are planted on each acre. Find the number of trees needed to plant 85 acres.

100. The largest living land mammal is the African elephant, and the largest mammal of all time is the blue whale. An African elephant weighs 15,225 pounds and a blue whale weighs 28 times that amount. Find the weight of the blue whale.

101. New York City has a population of 8,391,881, the largest in the country. Boston, in twenty-second place, has a population of 645,169. How many more people live in New York City than in Boston? (*Source:* U.S. Census Bureau.)

102. Los Angeles, the second largest city in the country, has a population of 3,849,378. Dallas, at ninth largest, has a population of 1,232,940. Find the difference in the population of these two cities. (*Source:* U.S. Census Bureau.)

103. A medical center purchased 12 laptop computers at \$970 each and 8 printers at \$315 each. Find the total cost of this equipment.

104. In the first 17 years of food drives, the Postal Workers have collected 954 million pounds of food. This year they collected 77 million pounds of food. Find the total food collection in 18 years. (*Source:* Stamp Out Hunger, U.S. Postal Service.)

Relating Concepts (Exercises 105–114) For Individual or Group Work

Work Exercises **105–114** *in order.*

105. Add.

(a) $189 + 263$

(b) $263 + 189$

106. Your answers to **Exercise 105(a) and (b)** should be the same. This shows that the order of numbers in an addition problem does not change the sum. This is known as the __________ property of addition.

107. Add. Recall that parentheses show you what to do first.

(a) $(65 + 81) + 135$

(b) $65 + (81 + 135)$

108. Since the answers to **Exercise 107(a) and (b)** are the same, we see that grouping the numbers differently when adding does not change the sum. This is known as the __________ property of addition.

109. Multiply.

(a) 220×72

(b) 72×220

110. Since the answers to **Exercise 109(a) and (b)** are the same, we see that the product remains the same when the order of the factors is changed. This is known as the __________ property of multiplication.

111. Multiply. Recall that parentheses tell you what to do first.

(a) $(26 \times 18) \times 14$

(b) $26(18 \times 14)$

112. Since the answers to **Exercise 111(a) and (b)** are the same, we see that grouping the numbers differently when multiplying does not change the product. This is known as the __________ property of multiplication.

113. Do the commutative and associative properties apply to subtraction? Explain your answer using several examples.

114. Do you think that the commutative and associative properties will apply to division? Explain your answer using several examples.

1.5 Dividing Whole Numbers

OBJECTIVES

1. Write division problems in three ways.
2. Identify the parts of a division problem.
3. Divide 0 by a number.
4. Recognize that a number cannot be divided by 0.
5. Divide a number by itself.
6. Divide a number by 1.
7. Use short division.
8. Use multiplication to check the answer to a division problem.
9. Use tests for divisibility.

Suppose the total cost of lunch at a SUBWAY is \$18 and is to be divided equally by three friends. Each person would pay \$6, as shown here.

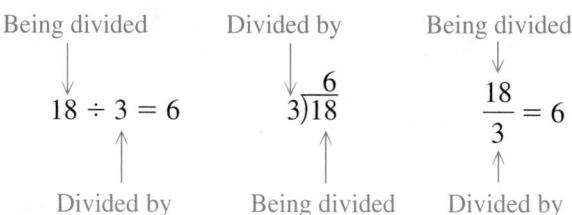

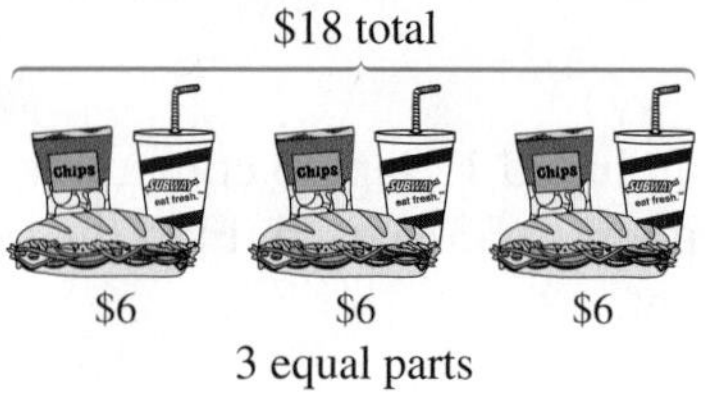

OBJECTIVE 1 **Write division problems in three ways.** Just as $3 \cdot 6$, 3×6, and $(3)(6)$ are different ways of indicating the multiplication of 3 and 6, there are several ways to write 18 divided by 3.

Being divided / Divided by	Divided by / Being divided	Being divided / Divided by
$18 \div 3 = 6$	$\begin{array}{r}6\\3\overline{)18}\end{array}$	$\dfrac{18}{3} = 6$

We will use all three division symbols, $\div$, $\overline{)\ }$, and —. In courses such as algebra, a slash symbol, /, or a fraction bar, —, is most often used.

EXAMPLE 1 **Using Division Symbols**

Write each division problem using two other symbols.

(a) $18 \div 6 = 3$

This division can also be written as shown below.

$$\begin{array}{r}3\\6\overline{)18}\end{array} \quad \text{or} \quad \frac{18}{6} = 3$$

Remember the three division symbols.

(b) $\dfrac{15}{5} = 3$ $\qquad 15 \div 5 = 3$ or $\begin{array}{r}3\\5\overline{)15}\end{array}$

(c) $\begin{array}{r}4\\5\overline{)20}\end{array}$ $\qquad 20 \div 5 = 4$ or $\dfrac{20}{5} = 4$

◀ Work Problem **1** at the Side.

OBJECTIVE 2 **Identify the parts of a division problem.** In division, the number being divided is the **dividend,** the number divided by is the **divisor,** and the answer is the **quotient.**

$$\textbf{dividend} \div \textbf{divisor} = \textbf{quotient}$$

$$\begin{array}{r}\textbf{quotient}\\\textbf{divisor}\overline{)\textbf{dividend}}\end{array} \qquad \frac{\textbf{dividend}}{\textbf{divisor}} = \textbf{quotient}$$

---

1 Write each division problem using two other symbols.

GS **(a)** $24 \div 6 = 4$

$\begin{array}{r}4\\6\overline{)24}\end{array}$ $\qquad \dfrac{24}{\_\_\_} = 4$

(b) $\begin{array}{r}4\\9\overline{)36}\end{array}$

(c) $48 \div 6 = 8$

(d) $\dfrac{42}{6} = 7$

Answers

1. (a) $\begin{array}{r}4\\6\overline{)24}\end{array}$ and $\dfrac{24}{6} = 4$

   (b) $36 \div 9 = 4$ and $\dfrac{36}{9} = 4$

   (c) $\begin{array}{r}8\\6\overline{)48}\end{array}$ and $\dfrac{48}{6} = 8$

   (d) $\begin{array}{r}7\\6\overline{)42}\end{array}$ and $42 \div 6 = 7$

EXAMPLE 2 **Identifying the Parts of a Division Problem**

Identify the dividend, divisor, and quotient.

(a) $35 \div 7 = 5$

$$\underset{\text{Dividend}}{\mathbf{35}} \div \underset{\text{Divisor}}{\mathbf{7}} = 5 \leftarrow \text{Quotient}$$

(b) $\frac{100}{20} = 5$

$$\frac{\overset{\text{Dividend}}{\mathbf{100}}}{\underset{\text{Divisor}}{\mathbf{20}}} = \mathbf{5} \leftarrow \text{Quotient}$$

The answer in a division problem is the quotient.

(c) $8\overline{)72}$ with quotient 9

9 ← Quotient
8)72 ← Dividend
↑
Divisor

Work Problem **2** at the Side. ▶

OBJECTIVE 3 **Divide 0 by a number.** If no money, or \$0, is divided equally among five people, each person gets \$0. The general rule for dividing 0 follows.

Dividing 0 by a Number

The number **0** divided by any nonzero number is **0**.

EXAMPLE 3 **Dividing 0 by a Number**

Divide.

(a) $0 \div 12 = 0$

(b) $0 \div 1728 = 0$

(c) $\frac{0}{375} = 0$

Zero divided by any nonzero number is zero.

(d) $129\overline{)0}$ with quotient 0

Work Problem **3** at the Side. ▶

Just as a subtraction such as $8 - 3 = 5$ can be written as the addition $8 = 3 + 5$, any division can be written as a multiplication. For example, $12 \div 3 = 4$ can be written as

$$3 \times 4 = 12 \quad \text{or} \quad 4 \times 3 = 12.$$

2 Identify the dividend, divisor, and quotient.

(a) $15 \div 3 = 5$

(b) $18 \div 6 = 3$

(c) $\frac{28}{7} = 4$

(d) $9\overline{)27}$ with quotient 3

3 Divide.

(a) $0 \div 5$

(b) $\frac{0}{9}$

(c) $\frac{0}{24}$

(d) $37\overline{)0}$

Answers

2. **(a)** dividend: 15; divisor: 3; quotient: 5
(b) dividend: 18; divisor: 6; quotient: 3
(c) dividend: 28; divisor: 7; quotient: 4
(d) dividend: 27; divisor: 9; quotient: 3

3. all 0

4 Write each division problem as a multiplication problem.

GS **(a)** $5\overline{)15}$ with quotient 3

$5 \cdot 3 =$ ____

____ $\cdot 5 =$ ____

(b) $\frac{32}{4} = 8$

(c) $45 \div 9 = 5$

Answers

4. (a) 15; $5 \cdot 3 = 15$ or 3; 15; $3 \cdot 5 = 15$
(b) $4 \cdot 8 = 32$ or $8 \cdot 4 = 32$
(c) $9 \cdot 5 = 45$ or $5 \cdot 9 = 45$

EXAMPLE 4 Changing Division Problems to Multiplication

Change each division problem to a multiplication problem.

(a) $\frac{20}{4} = 5$ becomes $4 \cdot 5 = 20$ or $5 \cdot 4 = 20$.

(b) $8\overline{)48}$ with quotient 6 becomes $8 \cdot 6 = 48$ or $6 \cdot 8 = 48$.

(c) $72 \div 9 = 8$ becomes $9 \cdot 8 = 72$ or $8 \cdot 9 = 72$.

◀ Work Problem **4** at the Side.

OBJECTIVE 4 **Recognize that a number cannot be divided by 0.** Division of any number by 0 cannot be done. To see why, try to find

$$9 \div 0 = ?$$

As we have just seen, any division problem can be converted to a multiplication problem so that

divisor • quotient = dividend.

If you convert the preceding problem to its multiplication counterpart, it reads as follows.

$$0 \cdot ? = 9$$

You already know that 0 times any number must always be 0. Try any number you like to replace the "?" and you'll aways get 0 instead of 9. Therefore, the division problem $9 \div 0$ cannot be done. Mathematicians say it is ***undefined*** and have agreed never to divide by 0. However, $0 \div 9$ *can* be done. Check by rewriting it as a multiplication problem.

$$0 \div 9 = 0 \quad \text{because} \quad 9 \cdot 0 = 0 \text{ is true.}$$

Dividing a Number by 0

Since dividing any number by 0 cannot be done, we say that division by **0** is ***undefined.*** It is impossible to compute an answer.

EXAMPLE 5 Dividing Numbers by 0

All the following divisions are undefined.

(a) $\frac{6}{0}$ is undefined.

(b) $0\overline{)8}$ is undefined.

(c) $18 \div 0$ is undefined. (You **cannot** divide a number by zero.)

(d) $\frac{3}{0}$ is undefined.

Division Involving 0

$$0 \div \text{nonzero number} = 0 \quad \text{and} \quad \frac{0}{\text{nonzero number}} = 0$$

but

$$\text{nonzero number} \div 0 \quad \text{and} \quad \frac{\text{nonzero number}}{0} \quad \text{are } \textbf{undefined.}$$

CAUTION

When 0 is the divisor in a problem, you write "undefined" as the answer. Never divide by 0.

Work Problem 5 at the Side. ▶

Calculator Tip

Try these two problems on your calculator. Jot down your answers.

9 (÷) 0 (=) ______ 0 (÷) 9 (=) ______

When you try to divide by 0, the calculator cannot do it, so it shows the word "Error" or the letter "E" (for error) in the display. But, when you divide 0 by 9 the calculator displays 0, which is the correct answer.

OBJECTIVE 5 **Divide a number by itself.** What happens when a number is divided by itself? For example, what is $4 \div 4$ or $97 \div 97$?

Dividing a Number by Itself

Any ***nonzero*** number divided by itself is **1.**

EXAMPLE 6 **Dividing a Nonzero Number by Itself**

Divide.

(a) $16 \div 16 = 1$

(b) $32\overline{)32}$ with quotient 1

A nonzero number divided by itself is 1.

(c) $\frac{57}{57} = 1$

Work Problem 6 at the Side. ▶

OBJECTIVE 6 **Divide a number by 1.** What happens when a number is divided by 1? For example, what is $5 \div 1$ or $86 \div 1$?

Dividing a Number by 1

Any number divided by 1 is itself.

---

5 Divide. If the division is not possible, write "undefined."

(a) $\frac{4}{0}$

(b) $\frac{0}{4}$

(c) $0\overline{)36}$

(d) $36\overline{)0}$

(e) $100 \div 0$

(f) $0 \div 100$

6 Divide.

(a) $8 \div 8$

(b) $15\overline{)15}$

(c) $\frac{37}{37}$

Answers

5. **(a)** undefined **(b)** 0 **(c)** undefined **(d)** 0 **(e)** undefined **(f)** 0

6. all 1

7 Divide.

(a) $9 \div 1$

(b) $1\overline{)18}$

(c) $\dfrac{43}{1}$

EXAMPLE 7 Dividing Numbers by 1

Divide.

(a) $5 \div 1 = 5$

(b) $\begin{array}{r} 26 \\ 1\overline{)26} \end{array}$

A number divided by 1 is itself.

(c) $\dfrac{41}{1} = 41$

◀ Work Problem **7** at the Side.

8 Divide using short division.

(a) $2\overline{)24}$

(b) $3\overline{)93}$

(c) $4\overline{)88}$

(d) $2\overline{)624}$

OBJECTIVE 7 **Use short division. Short division** is a method of dividing a number by a one-digit divisor.

EXAMPLE 8 Using Short Division

Divide using short division. $3\overline{)96}$

First, divide 9 by 3.

$$\begin{array}{r} 3 \\ 3\overline{)96} \end{array} \leftarrow \frac{9}{3} = 3$$

Next, divide 6 by 3.

$$\begin{array}{r} 32 \\ 3\overline{)96} \end{array} \leftarrow \frac{6}{3} = 2$$

◀ Work Problem **8** at the Side.

VOCABULARY TIP

Remainder In division, when the answer (quotient) is not a whole number, the portion left over (remains) is called the *remainder.*

When two numbers do not divide exactly, the leftover portion is called the **remainder.** The remainder must always be less than the divisor.

9 Divide using short division.

(a) $2\overline{)125}$

(b) $3\overline{)215}$

(c) $4\overline{)538}$

(d) $\dfrac{819}{5}$

EXAMPLE 9 Using Short Division with a Remainder

Divide 147 by 4 using short division.
Rewrite the problem.

$$4\overline{)147}$$

Because 1 cannot be divided by 4, divide 14 by 4. Notice that the 3 is placed over the 4 in 14.

Since 14 is being divided by 4, the answer (3) is placed over the 4.

$$\begin{array}{r} 3 \\ 4\overline{)14\,^{2}7} \end{array} \qquad \frac{14}{4} = 3 \text{ with 2 left over}$$

Next, divide 27 by 4. The final number left over is the remainder. Use **R** to indicate the remainder, and write the remainder to the side.

$$\begin{array}{r} 3\ \ 6\ \mathbf{R}3 \\ 4\overline{)14\,^{2}7} \end{array} \qquad \frac{27}{4} = 6 \text{ with 3 left over}$$

◀ Work Problem **9** at the Side.

Answers

7. (a) 9 (b) 18 (c) 43
8. (a) 12 (b) 31 (c) 22 (d) 312
9. (a) 62 **R**1 (b) 71 **R**2 (c) 134 **R**2 (d) 163 **R**4

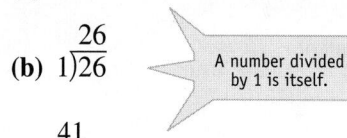

EXAMPLE 10 Dividing with a Remainder

Divide 1809 by 7.

Divide 18 by 7.

$$7\overline{)18^{4}09}\ \text{quotient } 2 \qquad \frac{18}{7} = 2 \text{ with 4 left over}$$

Divide 40 by 7.

$$7\overline{)18^{4}0^{5}9}\ \text{quotient } 2\,5 \qquad \frac{40}{7} = 5 \text{ with 5 left over}$$

Divide 59 by 7.

$$7\overline{)18^{4}0^{5}9}\ \text{quotient } 2\,5\,8\ \mathbf{R}3 \qquad \frac{59}{7} = 8 \text{ with 3 left over}$$

The remainder must be less than the divisor.

Work Problem 10 at the Side. ▶

Note

Short division takes practice but is useful when the divisor is a one-digit number.

OBJECTIVE 8 **Use multiplication to check the answer to a division problem.** **Check** the answer to a division problem as follows.

Checking Division

(divisor × quotient) + remainder = dividend

Parentheses tell you what to do first: Multiply the divisor by the quotient, then add the remainder.

EXAMPLE 11 Checking Division by Using Multiplication

Check each answer.

(a) $5\overline{)458}$ quotient 91 **R**3

(divisor × quotient) + remainder = dividend

$$(5 \times 91) + 3$$

$$455 + 3 = 458$$

Be careful! Always add the remainder when checking division.

Matches original dividend, so the division was done correctly.

Continued on Next Page

10 Divide.

(a) $4\overline{)5^{1}3^{1}0}$ quotient 1 3 _ **R** _

(b) $\frac{515}{7}$

(c) $3\overline{)1885}$

(d) $6\overline{)1415}$

Answers

10. (a) 2; 2; 132 **R**2 **(b)** 73 **R**4 **(c)** 628 **R**1 **(d)** 235 **R**5

11 Use multiplication to check each division. If an answer is incorrect, give the correct answer.

GS **(a)** $2\overline{)65}$ with quotient 32 **R**1

$2 \times 32 = 64$

____ + ____ = 65

(b) $7\overline{)586}$ with quotient 83 **R**4

(c) $3\overline{)1223}$ with quotient 407 **R**2

(d) $5\overline{)2383}$ with quotient 476 **R**3

Answers

11. **(a)** 64; 1; correct **(b)** incorrect; should be 83 **R**5 **(c)** correct **(d)** correct

(b) $6\overline{)1437}$ with quotient 239 **R4**

(divisor × quotient) + remainder = dividend

$(6 \times 239) + 4$

$1434 + 4 = \mathbf{1438}$

Does not match original dividend.

The answer does **not** check. Rework the original problem to get the correct answer, 239 **R**3. Then, $(6 \times 239) + 3$ **does** give 1437.

CAUTION

A common error when checking division is to forget to add the remainder. Be sure to add any remainder when checking a division problem.

◀ **Work Problem 11 at the Side.**

OBJECTIVE 9 **Use tests for divisibility.** It is often important to know whether a number is *divisible* by another number. You will find this useful in **Chapter 2** when writing fractions in lowest terms.

Divisibility

One whole number is **divisible** by another if the remainder is 0.

Use the following tests to decide whether one number is divisible by another number.

Tests for Divisibility

A number is divisible by

2 if it ends in 0, 2, 4, 6, or 8. These are the even numbers.

3 if the sum of its digits is divisible by 3.

4 if the last two digits make a number that is divisible by 4.

5 if it ends in 0 or 5.

6 if it is divisible by both 2 and 3.

7 has no simple test.

8 if the last three digits make a number that is divisible by 8.

9 if the sum of its digits is divisible by 9.

10 if it ends in 0.

The most commonly used tests are those for 2, 3, 5, and 10.

Divisibility by 2

A number is divisible by **2** if the number ends in 0, 2, 4, 6, or 8. All even numbers are divisible by 2.

EXAMPLE 12 Testing for Divisibility by 2

Are the following numbers divisible by 2?

(a) 986

↑ Ends in 6

All even numbers are divisible by 2.

Because the number ends in 6, which is an *even number,* the number 986 is divisible by 2.

(b) 3255 is not divisible by 2.

↑ Ends in 5, and not in 0, 2, 4, 6, or 8

Work Problem 12 at the Side. ▶

Divisibility by 3

A number is divisible by **3** if the sum of its digits is divisible by **3.**

EXAMPLE 13 Testing for Divisibility by 3

Are the following numbers divisible by 3?

(a) 4251
Add the digits.

4 2 5 1

$$4 + 2 + 5 + 1 = 12$$

If the sum of the digits is divisible by 3, the original number is divisible by 3.

Because 12 is divisible by 3, the number 4251 is also divisible by 3.

(b) 29,806
Add the digits.

2 9 8 0 6

$$2 + 9 + 8 + 0 + 6 = 25$$

Because 25 is *not* divisible by 3, the number 29,806 is *not* divisible by 3.

CAUTION

Be careful when testing for divisibility by adding the digits. This method works only for the numbers 3 and 9.

Work Problem 13 at the Side. ▶

12 Which numbers are divisible by 2?

(a) 258

(b) 307

(c) 4216

(d) 73,000

13 Which numbers are divisible by 3?

GS **(a)** 743

$7 + 4 + 3 = 14$

14 is *not* divisible by 3.

So, 743 is *not* divisible by 3.

GS **(b)** 5325

$5 + 3 + 2 + 5 =$ ____

____ is divisible by 3.

So, 5325 is divisible by ____.

(c) 374,214

(d) 205,633

Answers

12. all but b

13. 15; 15; 3; b and c

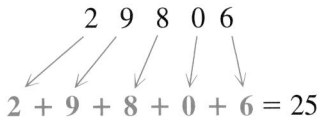

14 Which numbers are divisible by 5?

(a) 180

(b) 635

(c) 8364

(d) 206,105

15 Which numbers are divisible by 10?

(a) 270

(b) 495

(c) 5030

(d) 14,380

Answers

14. all but (c)

15. all but (b)

Divisibility by 5 and by 10

A number is divisible by **5** if it ends in 0 or 5.
A number is divisible by **10** if it ends in 0.

EXAMPLE 14 Testing for Divisibility by 5

Are the following numbers divisible by 5?

(a) 12,900 ends in 0 and is divisible by 5.

(b) 4325 ends in 5 and is divisible by 5.

If the number ends in 0 or 5, it's divisible by 5.

(c) 392 ends in 2 and is *not* divisible by 5.

◀ **Work Problem 14 at the Side.**

EXAMPLE 15 Testing for Divisibility by 10

Are the following numbers divisible by 10?

(a) 700 and 9140 both end in 0 and are divisible by 10.

If the number ends in 0, it's divisible by 10.

(b) 355 and 18,743 do not end in 0 and are *not* divisible by 10.

◀ **Work Problem 15 at the Side.**

1.5 Exercises

FOR EXTRA HELP

Download the MyDashBoard App

MyMathLab®

1. CONCEPT CHECK Write the three common symbols used to show multiplication.

2. CONCEPT CHECK Write the three common symbols used to show division.

Write each division problem using two other symbols. ***See Example 1.***

3. $24 \div 4 = 6$

4. $36 \div 3 = 12$

5. $\frac{45}{9} = 5$

6. $\frac{56}{8} = 7$

7. $2\overline{)16}$ with quotient 8

8. $8\overline{)48}$ with quotient 6

9. CONCEPT CHECK When a number is divided by 1, the answer is always the _______ itself.

10. CONCEPT CHECK When zero is divided by a number, the answer is always _______.

Divide. If the division is not possible, write "undefined." ***See Examples 3–7.***

11. $9 \div 9$

12. $36 \div 9$

13. $\frac{14}{2}$

14. $\frac{10}{0}$

15. $22 \div 0$
GS When 0 is the divisor, write ________ as the answer.

16. $6 \div 6$
GS When a number is divided by itself, write ________ as the answer.

17. $\frac{24}{1}$

18. $\frac{12}{1}$

19. $15\overline{)0}$
GS When dividing 0 by a nonzero number, the answer is ________.

20. $\frac{0}{12}$

21. $0\overline{)43}$

22. $\frac{8}{0}$

CONCEPT CHECK *Use the tests for divisibility by 2, 3, 5, and 10. Circle the numbers that will divide evenly into the given number.*

23. 8670
2 3 5 10

24. 13,785
2 3 5 10

25. 9,221,784
2 3 5 10

26. 5,409,720
2 3 5 10

Divide by using short division. Use multiplication to check each answer.
See Examples 8–10.

27. $3\overline{)75}$

28. $5\overline{)85}$

29. $7\overline{)126}$

30. $6\overline{)168}$

31. $4\overline{)1216}$

32. $5\overline{)2305}$

33. $4\overline{)2509}$

34. $8\overline{)1335}$

35. $6\overline{)9137}$

36. $9\overline{)8371}$

37. $6\overline{)1854}$

38. $8\overline{)856}$

39. $12{,}020 \div 4$

40. $8012 \div 4$

41. $30{,}036 \div 6$

42. $32{,}008 \div 8$

43. $2434 \div 3$

44. $5993 \div 7$

45. $12{,}947 \div 5$

46. $33{,}285 \div 9$

47. $\frac{21{,}040}{8}$

48. $\frac{8199}{9}$

49. $\frac{74{,}751}{6}$

50. $\frac{72{,}543}{5}$

51. $\frac{71{,}776}{7}$

52. $\frac{77{,}621}{3}$

53. $\frac{128{,}645}{7}$

54. $\frac{172{,}255}{4}$

Use multiplication to check each answer. If an answer is incorrect, find the correct answer.
See Example 11.

55. $5\overline{)1877}$ = 375 **R**2

GS **CHECK**

5 • 375 + 2 = ____

correct

56. $3\overline{)1282}$ = 427 **R**1

GS **CHECK**

3 • 427 + 1 = ____

correct

57. $3\overline{)5725}$ = 1908 **R**2

58. $5\overline{)2158}$ = 432 **R**3

59. $7\overline{)4692}$ = 650 **R**2

60. $9\overline{)5974}$ = 663 **R**5

61. $6\overline{)21{,}409}$ = 3 568 **R**2

62. $6\overline{)3192}$ = 532

63. $8\overline{)16{,}019}$ = 2 002 **R**3

64. $8\overline{)33{,}664}$ = 4 208

65. $6\overline{)69{,}140}$ = 11,523 **R**2

66. $3\overline{)82{,}598}$ = 27,532 **R**1

67. $9\overline{)86{,}655}$ = 9 628 **R**7

68. $7\overline{)50{,}809}$ = 7 258 **R**4

69. $8\overline{)222{,}576}$ = 27,822

70. $4\overline{)311{,}216}$ = 77,804

71. Explain in your own words how to check a division problem using multiplication. Be sure to tell what must be done if the quotient includes a remainder.

72. Describe the three divisibility rules that you think will be most useful and tell why.

Solve each application problem.

73. The Carnival Cruise Line has 2624 linen napkins. If it takes eight napkins to set each table, find the number of tables that can be set. (*Source: USA Today.*)

74. A school district will distribute 1620 new science books equally among 12 schools. How many books will each school receive?

75. In one 8-hour day Dreyer's Edy's can produce 76,800 ice cream drumsticks. How many are produced each hour? (*Source:* History Channel, *Modern Marvels: Snack Food Tech.*)

76. Tootsie Roll Industries produces 415,000,000 Tootsie Rolls in a 5-day week. Find the number produced each day. (*Source:* History Channel, *Modern Marvels: Snack Food Tech.*)

77. Lottery winnings of $436,500 are divided equally among nine Starbucks employees. Find the amount received by each employee.

78. How many 5-pound bags of organic whole wheat flour can be filled from a 17,175-pound bin of flour?

79. McDonald's Restaurants is hiring 660 new employees, from crew workers to managers, in one area. If 4 new employees are hired at each location, find the number of locations in this area. (*Source: Sacramento Bee.*)

80. The Wii remains the top-selling video game system. Nintendo sold 85 million Wiis in the first 5 years. Find the average number sold each year. (*Source: USA Today.*)

81. A class-action lawsuit settlement of $6,825,000 is divided evenly among six injured people. Find the amount received by each person.

82. A 12,000-square foot condominium at the edge of Central Park in Manhattan sold for a record $45,000,000. The buyer paid for the condominium in eight equal payments. Find the amount of each payment. (*Source: USA Today.*)

83. The record for picking blueberries in one day was set in the state of Maine and was 6900 pounds. Since there are 1000 blueberries in a pound, this amounted to 6,900,000 blueberries. If these berries were picked in 8 hours, find the average number of berries picked each hour. (*Source:* Discovery Channel, *Dirty Jobs.*)

84. A professional basketball player signed a 4-year contract for $21,937,500. How much is this each year?

Put a ✓ mark in the blank if the number at the left is divisible by the number at the top.
Put an X in the blank if the number is not divisible by the number at the top.
See Examples 12–15.

	2	3	5	10
85. 60	_____	_____	_____	_____
86. 35	_____	_____	_____	_____
87. 92	_____	_____	_____	_____
88. 96	_____	_____	_____	_____
89. 445	_____	_____	_____	_____
90. 897	_____	_____	_____	_____
91. 903	_____	_____	_____	_____
92. 500	_____	_____	_____	_____
93. 5166	_____	_____	_____	_____
94. 8302	_____	_____	_____	_____
95. 21,763	_____	_____	_____	_____
96. 32,472	_____	_____	_____	_____

1.6 Long Division

OBJECTIVES

1. Do long division.
2. Divide numbers ending in 0 by numbers ending in 0.
3. Use multiplication to check division answers.

If the total cost of 42 Sony iPod Docking Systems is \$3066, we can find the cost of each docking system using **long division.** Long division is used to divide by a number with more than one digit.

OBJECTIVE 1 **Do long division.** In long division, estimate the various numbers by using a **trial divisor** to get a **trial quotient.**

VOCABULARY TIP

Trial divisor A method of determining how many times the true divisor goes into the dividend by estimating, using the leading digit of the divisor.

VOCABULARY TIP

Trial quotient The quotient that results from dividing the dividend by the trial divisor.

EXAMPLE 1 Using a Trial Divisor and a Trial Quotient

Divide. $42\overline{)3066}$

Because 42 is closer to 40 than to 50, use the first digit of the divisor as a trial divisor.

42 (Using a trial divisor is a helpful tool.)

Trial divisor → 4 (first digit of 42)

Try to divide the first digit of the dividend by 4. Since 3 cannot be divided by 4, use the first *two* digits, 30.

$$\frac{30}{4} = 7 \text{ with remainder } 2$$

$$\begin{array}{r} \mathbf{7} \\ 42\overline{)3066} \end{array} \leftarrow \text{Trial quotient}$$

7 goes over the 6, because $\frac{306}{42}$ is about 7.

Multiply 7 and 42 to get 294; next, subtract 294 from 306.

$$\begin{array}{r} 7 \\ 42\overline{)3066} \\ \underline{294} \\ 12 \end{array} \quad \begin{array}{l} \\ \\ \leftarrow 7 \times 42 \\ \leftarrow 306 - 294 \end{array}$$

This number (12) must be smaller than 42, the divisor.

Bring down the 6 at the right.

$$\begin{array}{r} 7 \\ 42\overline{)3066} \\ \underline{294}\downarrow \\ 126 \end{array} \leftarrow \text{6 brought down}$$

Use the trial divisor, 4.

First two digits of 126 → $\frac{12}{4} = 3$

$$\begin{array}{r} 73 \\ 42\overline{)3066} \\ \underline{294} \\ 126 \\ \underline{126} \\ 0 \end{array} \quad \leftarrow 3 \times 42 = 126$$

The cost of each docking system is \$73.

Check the answer by multiplying 42 and 73. The product should be 3066.

1 Divide.

GS **(a)** $28\overline{)2296}$

$$\begin{array}{r} 8\_ \\ 28\overline{)2296} \\ \underline{224}\downarrow \\ 56 \\ \underline{56} \\ 0 \end{array}$$

(b) $16\overline{)1024}$

(c) $61\overline{)8784}$

(d) $\frac{2697}{93}$

2 Divide.

(a) $24\overline{)1344}$

(b) $72\overline{)4472}$

(c) $65\overline{)5416}$

(d) $89\overline{)6649}$

Answers

1. **(a)** 2; 82 **(b)** 64 **(c)** 144 **(d)** 29

2. **(a)** 56 **(b)** 62 **R**8 **(c)** 83 **R**21 **(d)** 74 **R**63

CAUTION

The ***first digit*** of the quotient in long division must be placed in the proper position over the dividend.

◀ Work Problem **1** at the Side.

EXAMPLE 2 **Dividing to Find a Trial Quotient**

Divide. $58\overline{)2730}$

Use 6 as a trial divisor, since 58 is closer to 60 than to 50.

First two digits of dividend → $\frac{27}{6} = 4$ with 3 left over

$$\begin{array}{rl} 4 & \leftarrow \text{Trial quotient} \\ 58\overline{)2730} & \\ \underline{232} & \leftarrow 4 \times 58 = 232 \\ 41 & \leftarrow 273 - 232 = 41 \text{ (smaller than 58, the divisor)} \end{array}$$

Bring down the 0.

$$\begin{array}{rl} 4 & \\ 58\overline{)2730} & \\ \underline{232}\downarrow & \\ 410 & \leftarrow \text{0 brought down} \end{array}$$

First two digits of 410 → $\frac{41}{6} = 6$ with 5 left over

$$\begin{array}{rl} 46 & \leftarrow \text{Trial quotient} \\ 58\overline{)2730} & \\ \underline{232} & \\ 410 & \\ \underline{348} & \leftarrow 6 \times 58 = 348 \\ 62 & \leftarrow \text{Greater than 58} \end{array}$$

Do not leave a remainder that is **greater** than the divisor.

The remainder, 62, is greater than the divisor, 58, so 7 should be used instead of 6.

$$\begin{array}{rl} 47\ \mathbf{R}4 & \\ 58\overline{)2730} & \\ \underline{232} & \\ 410 & \\ \underline{406} & \leftarrow 7 \times 58 = 406 \\ 4 & \leftarrow 410 - 406 \end{array}$$

Now the remainder, 4, is *less* than the divisor, 58.

◀ Work Problem **2** at the Side.

Sometimes it is necessary to write a 0 in the quotient.

EXAMPLE 3 Writing Zeros in the Quotient

Divide: $34\overline{)7068}$

Start as in **Examples 1 and 2**.

```
      2
34)7068
   68    ← 2 × 34 = 68
    2    ← 70 − 68 = 2
```

Bring down the 6.

```
      2
34)7068
   68↓
    26   ← 6 brought down
```

Since 26 cannot be divided by 34, write a 0 in the quotient as a placeholder.

```
     20  ← 0 in quotient
34)7068      Use a zero to hold a place in the quotient.
   68
    26
```

Bring down the final digit, the 8.

```
     20
34)7068
   68 ↓
    268  ← 8 brought down
```

Complete the problem.

```
    207 R30
34)7068
   68
    268
    238
     30
```

The quotient is 207 **R**30.

CAUTION

There ***must be a digit*** in the quotient (answer) above every digit in the dividend once the answer has begun. Notice in **Example 3** that a **0** was used to ensure a digit in the quotient above every digit in the dividend.

Work Problem **3** at the Side. ▶

OBJECTIVE 2 Divide numbers ending in 0 by numbers ending in 0. When the divisor and dividend both contain zeros at the far right, recall that these numbers are multiples of 10. As with multiplication, there is a short way to divide these multiples of 10. Look at the following examples.

$$26{,}000 \div 1 = 26{,}000$$
$$26{,}000 \div 10 = 2600$$
$$26{,}000 \div 100 = 260$$
$$26{,}000 \div 1000 = 26$$

Do you see a pattern? These examples suggest the following rule.

3 Divide.

GS **(a)** $17\overline{)1823}$

```
      1_7 R_
17)1823
   17↓
    12
     0↓
    123
    119
      _
```

(b) $23\overline{)4791}$

(c) $39\overline{)15{,}933}$

(d) $78\overline{)23{,}462}$

Answers

3. **(a)** 0; 4; 4; 107 **R**4 **(b)** 208 **R**7 **(c)** 408 **R**21 **(d)** 300 **R**62

4 Divide.

(a) $70 \div 10$

(b) $2600 \div 100$

(c) $505{,}000 \div 1000$

5 Divide using the shortcut of dropping zeros.

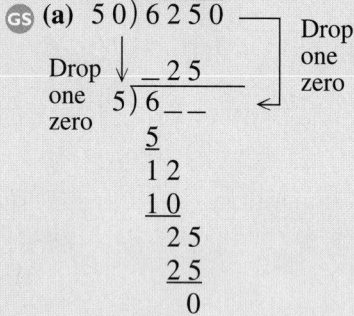

GS **(a)** $50\overline{)6250}$

Drop one zero — Drop one zero

$$\begin{array}{r} \_\,2\,5 \\ 5\overline{)6\,\_\,\_} \\ \underline{5} \\ 1\,2 \\ \underline{1\,0} \\ 2\,5 \\ \underline{2\,5} \\ 0 \end{array}$$

(b) $130\overline{)131{,}040}$

(c) $3400\overline{)190{,}400}$

Answers

4. **(a)** 7 **(b)** 26 **(c)** 505

5. **(a)** 1; 2; 5; 125 **(b)** 1008 **(c)** 56

Dividing a Whole Number by 10, by 100, or by 1000

To divide a whole number by 10, by 100, or by 1000, drop the appropriate number of zeros from the whole number.

EXAMPLE 4 Dividing by Multiples of 10

Divide.

(a) $60 \div 10 = 6$ (One 0 in divisor; 0 dropped)

(b) $3500 \div 100 = 35$ (Two zeros in divisor; 00 dropped)

The same number of zeros must be dropped from the divisor and quotient.

(c) $915{,}000 \div 1000 = 915$ (Three zeros in divisor; 000 dropped)

◀ Work Problem **4** at the Side.

Now we'll find the quotients for other multiples of 10 by dropping zeros.

EXAMPLE 5 Dividing by Multiples of 10

Divide:

(a) $40\overline{)11{,}000}$ Drop one zero from the divisor and the dividend.

$$\begin{array}{r} 275 \\ 4\overline{)1100} \\ \underline{8} \\ 30 \\ \underline{28} \\ 20 \\ \underline{20} \\ 0 \end{array}$$

Since $1100 \div 4$ is 275, then $11{,}000 \div 40$ is also 275.

(b) $3500\overline{)31{,}500}$ Drop two zeros from the divisor and the dividend.

$$\begin{array}{r} 9 \\ 35\overline{)315} \\ \underline{315} \\ 0 \end{array}$$

Since $315 \div 35$ is 9, then $31{,}500 \div 3500$ is also 9.

Note

Dropping zeros when dividing by multiples of 10 ***does not*** change the quotient (answer).

◀ Work Problem **5** at the Side.

OBJECTIVE 3 Use multiplication to check division answers. Answers in long division can be checked just as answers in short division were checked.

EXAMPLE 6 Checking Division by Using Multiplication

Check each answer.

(a)

$$48\overline{)5324} \quad \text{quotient: } 114\ \mathbf{R}43$$

```
   114
 × 48     Multiply the quotient and the divisor.
   912
  456
  5472
+   43    ← Add the remainder.
  5515    ← Result does not match dividend.
```

Remember to add the remainder when checking division.

The answer does ***not*** check. Rework the original problem to get 110 **R**44. Then (110 × 48) + 44 **does** give 5324.

(b)

```
        37                716
716)26,492              ×  37
    21 48                5 012
     5 012              21 48
     5 012              26,492   Correct
         0
```

Matches

If the answer in the check matches the dividend, the quotient is correct.

Calculator Tip

To check the answer to **Example 6(a)**, don't forget to add the remainder.

48 (×) 110 (+) 44 (=) **5324**

Add the remainder.

CAUTION

When checking a division problem, first multiply the quotient and the divisor. Then be sure to ***add any remainder*** before checking it against the original dividend.

Work Problem 6 at the Side. ▶

6 Decide whether each answer is correct. If the answer is incorrect, find the correct answer.

(a)

```
       38
  16)608
     48
     128
     128
       0
```

(b)

```
          42 R178
  426)19,170
      17 040
       1 130
         952
         178
```

(c)

```
          57 R18
  514)29,316
      25 700
       3 616
       3 598
          18
```

Answers

6. **(a)** correct **(b)** incorrect; should be 45 **(c)** correct

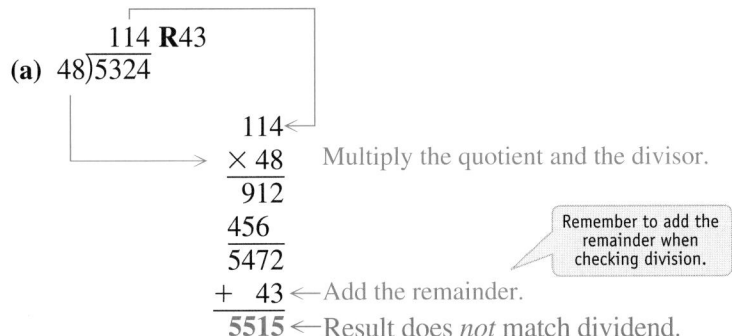

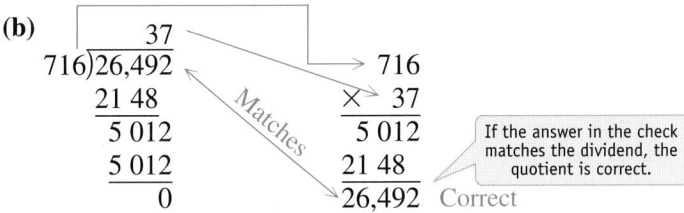

1.6 Exercises

FOR EXTRA HELP

Download the MyDashBoard App

MyMathLab®

CONCEPT CHECK *Decide where the first digit in the quotient would be located. Then, without finishing the division, you can tell which of the three choices is the correct answer. Circle your choice.*

1. $50\overline{)2650}$

5 53 530

2. $14\overline{)476}$

3 34 304

3. $18\overline{)4500}$

2 25 250

4. $35\overline{)5600}$

16 160 1600

5. $86\overline{)10,327}$

12 120 **R**7 1200

6. $46\overline{)24,026}$

5 52 522 **R**14

7. $26\overline{)28,735}$

11 110 1105 **R**5

8. $12\overline{)116,953}$

974 **R**2 9746 **R**1 97,460

9. $21\overline{)149,826}$

71 713 7134 **R**12

10. $64\overline{)208,138}$

325 **R**2 3252 **R**10 32,521

11. $523\overline{)470,800}$

9 **R**100 90 **R**100 900 **R**100

12. $230\overline{)253,230}$

11 110 1101

Divide by using long division. Use multiplication to check each answer.
See Examples 1–3, 5, and 6.

13. $18\overline{)1319}$ →

$$\begin{array}{r} 73\ \mathbf{R}\_\ \\ 18\overline{)1319} \\ \underline{126}\ \ \\ 59 \\ \underline{54} \\ 5 \end{array}$$

GS **CHECK**

$73 \cdot 18 = 1314$

$1314 + \_\_\_\_ = 1319$

14. $58\overline{)3654}$ →

$$\begin{array}{r} 6\_ \\ 58\overline{)3654} \\ \underline{348}\ \\ 174 \\ \underline{174} \\ 0 \end{array}$$

GS **CHECK**

$\_\_\_\_ \cdot 58 = 3654$

15. $23\overline{)10,963}$

16. $83\overline{)39,692}$

17. $26\overline{)62,583}$

18. $28\overline{)84,249}$

19. $74\overline{)84,819}$

20. $238\overline{)186,948}$

21. $153\overline{)509,725}$

22. $308\overline{)26,796}$

23. $420\overline{)357,000}$

24. $900\overline{)153,000}$

Use multiplication to check each answer. If an answer is incorrect, find the correct answer. ***See Example 6.***

25. $\begin{array}{r}101\,\textbf{R}4\\35\overline{)3549}\end{array}$

26. $\begin{array}{r}42\,\textbf{R}26\\64\overline{)2712}\end{array}$

27. $\begin{array}{r}658\,\textbf{R}9\\28\overline{)18{,}424}\end{array}$

28. $\begin{array}{r}239\,\textbf{R}121\\145\overline{)34{,}776}\end{array}$

29. $\begin{array}{r}62\,\textbf{R}3\\614\overline{)38{,}068}\end{array}$

30. $\begin{array}{r}174\,\textbf{R}368\\557\overline{)97{,}286}\end{array}$

Solve each application problem by using addition, subtraction, multiplication, or division as needed. ***See Examples 3–5.***

31. The first female co-host of a game show, Vanna White, is "Television's Most Frequent Clapper." On the *Wheel of Fortune* she claps 720 times per episode, or 84,240 claps each season. Find the number of episodes filmed each season. (*Source: Guinness Book of World Records* and *Spirit Magazine*.)

32. The two main towers of the Golden Gate Bridge each used 600,000 rivets in their construction. To meet today's earthquake standards, each of these rivets is being replaced with high-strength bolts. If the bolts are shipped in containers holding 160 bolts, how many containers are used to complete the work on both towers? (*Source:* Consolidated Engineering Laboratories.)

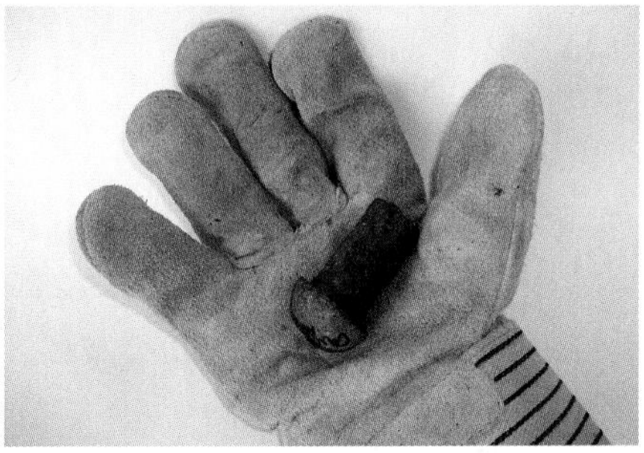

33. Don Gracey, the Mountain Timesmith, has serviced and repaired 636 clocks this year. He has worked on 272 wall clocks and 308 table clocks. The rest were standing floor clocks. Find the number of floor clocks he worked on this year.

total clocks − wall clocks − table clocks = floor clocks

636 − 272 − 308 = ________

34. There are 24,000,000 business enterprises in the United States. If 7000 of these are larger businesses (over 500 employees), find the number of businesses that are small to mid-size. (*Source:* U.S. Census Bureau.)

total business enterprises − larger businesses = small to mid-size businesses

24,000,000 − 7000 = ________

35. To complete her college education, Judy Martinez received education loans of $34,080 including interest. Find her monthly payment if the loan is to be paid off in 96 months (8 years).

36. A consultant charged $19,800 for evaluating a school's compliance with the Americans with Disabilities Act. If the consultant worked 225 hours, find the rate charged per hour.

37. Each minute there is one diamond ring sold on eBay's U.S. site. Find the number of diamond rings sold in 30 days. (*Source: Time Style and Design.*)

60 minutes • 24 hours • 30 days = ________

38. A retired milkman in Indianapolis has eaten a Twinkie every day for the last 60 years. How many Twinkies has he eaten over this time period?
Hint: 1 year = 365 days. (*Source:* History Channel, *Modern Marvels: Snack Food Tech.*)

60 years • 365 days = ___________

39. Don Gorske of Fond du Lac, Wisconsin has eaten 25,272 Big Macs over the last 39 years. (He is slim and claims a low cholesterol level.) Find

(a) the average number of Big Macs he has eaten each year.

(b) whether Gorske has eaten more or less than 2 Big Macs each day.

Hint: 1 year = 365 days
(*Source: Guinness Book of World Records.*)

40. Major League baseball teams use more than 220,020 baseballs each season. If each of the 30 major league teams uses the same number of balls in a season, how many are used by each team? (*Source: Parade, The Sunday Newspaper Magazine.*)

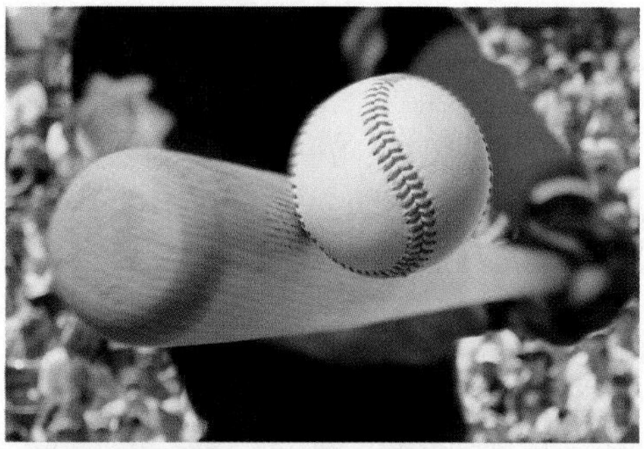

Relating Concepts (Exercises 41–48) For Individual or Group Work

Knowing and using the rules of divisibility is necessary in problem solving.
Work Exercises 43–50 in order.

41. If you have $0 and you divide this amount among three people, how much will each receive?

42. When 0 is divided by any nonzero number, the result is _____.

43. Divide.
$8 \div 0$

44. We say that division by 0 is *undefined* because it is (*possible/impossible*) to compute the answer. Give an example involving cookies that will support your answer.

45. Divide.
(a) $14 \div 1$ **(b)** $1\overline{)17}$
(c) $\frac{38}{1}$

46. Any number divided by 1 is the number itself. Is this also true when multiplying by 1? Give three examples that support your answer.

47. Divide.
(a) $32{,}000 \div 10$
(b) $32{,}000 \div 100$
(c) $32{,}000 \div 1000$

48. Write a rule that explains the shortcut for doing divisions like the ones in **Exercise 47.**

Math in the Media

THE TOLL OF WEDDING BELLS

The Royal Wedding of Prince William and Kate Middleton in 2011 was the wedding of the century. The wedding ceremony had 1900 guests and cost $34,000,000. The cake alone was $134 per slice. In the United States, the average wedding includes 150 guests. In 2008, the cost of the average wedding was $20,451 and the cost continues to grow. The graph gives most of the costs involved in a wedding in 2012. Use this information to answer the questions that follow.

Numbers in the News

'Til Debt Do You Part

With over 150 guests, the cost of an average wedding has continued to grow. Most of the money is spent on the following:

Source: Cost of a Wedding and *dudewalker.org*

1. What is the total of the costs shown in the graph?

2. How much more expensive was a wedding in 2012 compared with 2008?

3. Traditionally, the groom pays for the photography and video, the flowers, the wedding jewelry, the clergy ($500), and the groom's formal wear ($95). What is the total amount spent by the groom?

4. If you budgeted $65 per person for the wedding reception and you invited 175 guests to a wedding in 2008, how much would you have spent compared to the 2012 wedding reception costs?

5. If you budget $6000 for the wedding reception and the cost per person is $37, how many guests can you invite and how much of your budgeted amount will be left over?

6. If you budget $1000 for the wedding reception and the cost per person is $15, how many guests can you invite and how much of your budgeted amount will be left over?

7. What kind of an arithmetic problem did you work to get the answers to Problems 5 and 6? What is the mathematical term for the "left over" budget?

1.7 Rounding Whole Numbers

OBJECTIVES

1. Locate the place to which a number is to be rounded.
2. Round numbers.
3. Round numbers to estimate an answer.
4. Use front end rounding to estimate an answer.

One way to get a quick check on an answer is to *round* the numbers in the problem. **Rounding** a number means finding a number that is close to the original number, but easier to work with.

For example, the county planning commissioner might be discussing the need for more affordable housing. To demonstrate this, she probably would not need to say that the county is in need of 8235 more affordable housing units—she probably could say that the county needs 8200 or even 8000 housing units.

OBJECTIVE 1 Locate the place to which a number is to be rounded. The first step in rounding a number is to locate the *place to be rounded.*

EXAMPLE 1 Finding the Place to Which a Number Is to Be Rounded

Locate and draw a line under the place to which each number is to be rounded.

(a) Round 83 to the nearest ten. Is 83 closer to $\underline{8}0$ or to $\underline{9}0$?

$\underline{8}3$ is closer to 80. (Tens place)

83 is closer to 80 than to 90.

(b) Round 54,702 to the nearest thousand. Is it closer to $5\underline{4},000$ or to $5\underline{5},000$?

$5\underline{4},702$ is closer to 55,000. (Thousands place)

(c) Round 2,806,124 to the nearest hundred-thousand. Is it closer to $2,\underline{8}00,000$ or to $2,\underline{9}00,000$?

$2,\underline{8}06,124$ is closer to 2,800,000. (Hundred-thousands place)

◀ Work Problem 1 at the Side.

OBJECTIVE 2 Round numbers. Use the rules for rounding whole numbers.

Rounding Whole Numbers

Step 1 Locate the *place* to which the number is to be rounded. Draw a line under that place.

Step 2(a) Look only at the next digit to the right of the one you underlined. If it is *5 or more, increase* the underlined digit by 1.

Step 2(b) If the next digit to the right is *4 or less, do not change* the digit in the underlined place.

Step 3 *Change* all digits to the right of the underlined place to zeros.

EXAMPLE 2 Using Rounding Rules for 4 or Less

Round 349 to the nearest hundred.

Step 1 Locate the place to which the number is being rounded. Draw a line under that place.

$\underline{3}49$ (Hundreds place)

Continued on Next Page

VOCABULARY TIP

Rounding Rounding is a useful tool for estimating the answer to a problem.

1 Locate and draw a line under the place to which each number is to be rounded. Then answer the question.

GS **(a)** 3$\underline{7}$3 (nearest ten)
Tens place
Is 373 closer to 370 or to 380?

____

(b) 1482 (nearest thousand)

Is 1482 closer to 1000 or to 2000?

____

(c) 89,512 (nearest hundred)

Is it closer to 89,500 or to 89,600?

____

(d) 546,325 (nearest ten-thousand)

Is it closer to 540,000 or to 550,000?

____

Answers

1. (a) $3\underline{7}3$ is closer to $3\underline{7}0$.
   (b) $\underline{1}482$ is closer to $\underline{1}000$.
   (c) $89,\underline{5}12$ is closer to $89,\underline{5}00$.
   (d) $5\underline{4}6,325$ is closer to $5\underline{5}0,000$.

Step 2 Because the next digit to the right of the underlined place is 4, which is 4 or less, do *not* change the digit in the underlined place.

Next digit is 4 or less.

349

3 remains 3.

4 or less, do not change underlined digit.

Step 3 Change all digits to the right of the underlined place to zeros.

349 rounded to the nearest hundred is 300.

In other words, 349 is closer to 300 than to 400.

Work Problem 2 at the Side. ▶

EXAMPLE 3 Using Rounding Rules for 5 or More

Round 36,833 to the nearest thousand.

Step 1 Find the place to which the number is to be rounded and draw a line under that place.

36,833

Thousands

Step 2 Because the next digit to the right of the underlined place is 8, which is 5 or more, add 1 to the underlined place.

Next digit is 5 or more.

36,833

Change 6 to 7.

5 or more, add 1 to underlined digit.

Step 3 Change all digits to the right of the underlined place to zeros.

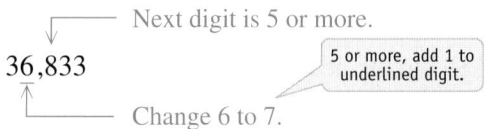

Change to 0.

36,833 rounded to the nearest thousand is 37,000.

Change 6 to 7.

In other words, 36,833 is closer to 37,000 than to 36,000.

Work Problem 3 at the Side. ▶

EXAMPLE 4 Using Rounding Rules

(a) Round 2382 to the nearest ten.

Step 1 2382

Tens place

Step 2 The next digit to the right is 2, which is 4 or less.

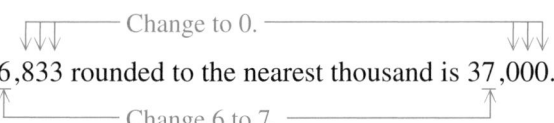

Next digit is 4 or less.

2382

Leave 8 as 8.

Step 3 2382 Change to 0.

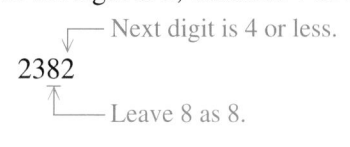

2382 rounded to the nearest ten is 2380.

In other words, 2382 is closer to 2380 than to 2390.

Continued on Next Page

2 Round to the nearest ten.

GS **(a)** 62

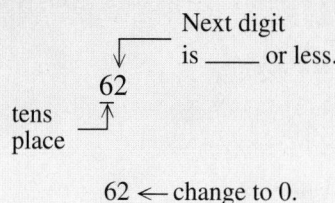

Next digit is ____ or less.

62

tens place

62 ← change to 0.

leave 6 as ____.

62 rounds to ____.

(b) 94

(c) 134

(d) 7543

3 Round to the nearest thousand.

(a) 3683

(b) 6502

(c) 84,621

(d) 55,960

Answers

2. **(a)** 4; 6; 60 **(b)** 90 **(c)** 130 **(d)** 7540

3. **(a)** 4000 **(b)** 7000 **(c)** 85,000 **(d)** 56,000

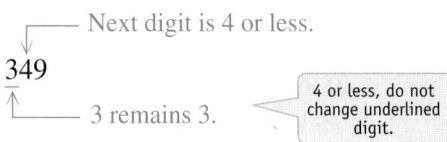

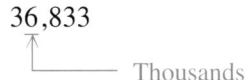

4 Round each number as indicated.

(a) 3458 to the nearest ten

(b) 6448 to the nearest hundred

GS (c) 73,077 to the nearest hundred

73,077 — Next digit is ___ or more.

Hundreds place rounds to 73, 1__ __.

(d) 85,972 to the nearest hundred

5 Round each number as indicated.

(a) 14,598 to the nearest ten-thousand

(b) 724,518,715 to the nearest million

Answers

4. (a) 3460 (b) 6400 (c) 5; 0; 0; 73,100 (d) 86,000

5. (a) 10,000 (b) 725,000,000

(b) Round 13,961 to the nearest hundred.

Step 1 13,961 — Hundreds place

Step 2 The next digit to the right is 6.

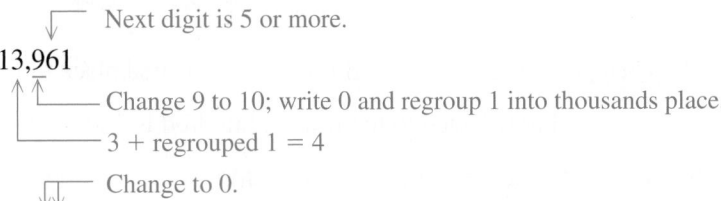

13,961
- Next digit is 5 or more.
- Change 9 to 10; write 0 and regroup 1 into thousands place.
- 3 + regrouped 1 = 4
- Change to 0.

Step 3 14,061

13,961 rounded to the nearest hundred is 14,000.
In other words, 13,961 is closer to 14,000 than to 13,900.

Note

In ***Step 2*** of **Example 4(b),** notice that the first three digits increased from 139 to 140 when we added 1 to the hundreds place.

(13,9)61 rounded to (14,0)00

◀ **Work Problem 4 at the Side.**

EXAMPLE 5 **Rounding Large Numbers**

(a) Round 37,892 to the nearest ten-thousand.

Step 1 37,892 — Ten-thousands place

Remember to *underline* the place to which you are rounding.

Step 2 The next digit to the right is 7.

37892
- Next digit is 5 or more.
- Change 3 to 4.
- Change to 0.

Step 3 47,892

37,892 rounded to the nearest ten-thousand is 40,000.

(b) Round 528,498,675 to the nearest million.

Step 1 528,498,675 — Millions place

Step 2 528,498,675
- Next digit is 4 or less.
- Leave 8 as 8.
- Change to 0.

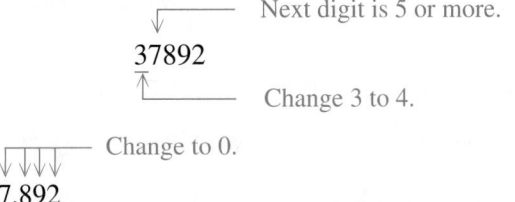

Remember to change *everything* to the right of the place you have rounded to 0.

Step 3 528,498,675

528,498,675 rounded to the nearest million is 528,000,000.

◀ **Work Problem 5 at the Side.**

Sometimes a number must be rounded to different places.

EXAMPLE 6 Rounding to Different Places

Round 648 **(a)** to the nearest ten and **(b)** to the nearest hundred.

(a) to the nearest ten

648 — Next digit is 5 or more. Tens place (4 + 1 = 5)

648 rounded to the nearest ten is 650.

(b) to the nearest hundred

648 — Next digit is 4 or less. Hundreds place stays the same.

Always start over with the original number when rounding to different places.

648 rounded to the nearest hundred is 600.

Notice that if 648 is rounded to the nearest ten (650), and then 650 is rounded to the nearest hundred, the result is 700. If, however, 648 is rounded directly to the nearest hundred, the result is 600 (not 700).

Work Problem 6 at the Side. ▶

CAUTION

Before rounding to a different place, always go back to the *original,* unrounded number.

EXAMPLE 7 Applying Rounding Rules

Round each number to the nearest ten, nearest hundred, and nearest thousand.

(a) 4358

First round 4358 to the nearest ten.

4358 — Next digit is 5 or more. Tens place (5 + 1 = 6)

4358 rounded to the nearest ten is 4360.

Now go back to 4358, the *original* number, before rounding to the nearest hundred.

4358 — Next digit is 5 or more. Hundreds place (3 + 1 = 4)

Go back to the original number.

4358 rounded to the nearest hundred is 4400.

Again, go back to the *original* number before rounding to the nearest thousand.

4358 — Next digit is 4 or less. Thousands place stays the same.

4358 rounded to the nearest thousand is 4000.

Continued on Next Page

6 Round each number to the nearest ten and to the nearest hundred.

GS **(a)** 549

nearest ten: 5 __0

nearest hundred: __ __0

(b) 458

(c) 9308

Answers

6. **(a)** 5; 550; 5; 0; 500 **(b)** 460; 500
(c) 9310; 9300

7 Round each number to the nearest ten, nearest hundred, and nearest thousand.

(a) 4078

(b) 46,364

(c) 268,328

VOCABULARY TIP

Approximately equal to The symbol for "approximately equal to" is a wavy equal sign, ≈. It almost looks like an equal sign.

8 Estimate the answers by rounding each number to the nearest ten.

(a)
$$\begin{array}{r} 16 \\ 74 \\ 58 \\ +\ 31 \\ \hline \end{array}$$

GS **(b)**
$$\begin{array}{rcr} 53 & \longrightarrow & 50 \\ -\ 19 & \longrightarrow & -\ \_0 \\ \hline & & 30 \end{array}$$

GS **(c)**
$$\begin{array}{rcr} 46 & \longrightarrow & \_\ 0 \\ \times\ 74 & \longrightarrow & \times\ 7\ \_ \\ \hline & & 35\ \_\ \_ \end{array}$$

Answers

7. **(a)** 4080; 4100; 4000
(b) 46,360; 46,400; 46,000
(c) 268,330; 268,300; 268,000

8. **(a)** 20 + 70 + 60 + 30 = 180
(b) 2; 50 − 20 = 30
(c) 5; 0; 0; 0; 70 × 50 = 3500

(b) 680,914
First, round to the nearest ten.

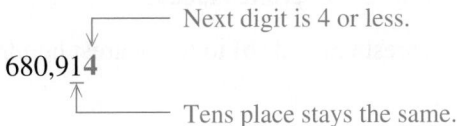

680,91**4** — Next digit is 4 or less. Tens place stays the same.

680,914 rounded to the nearest ten is 680,910.
Go back to 680,914, the *original* number, to round to the nearest hundred.

680,9**1**4 — Next digit is 4 or less. Hundreds place stays the same. (Remember to go back to the original number.)

680,914 rounded to the nearest hundred is 680,900.
Go back to the *original* number to round to the nearest thousand.

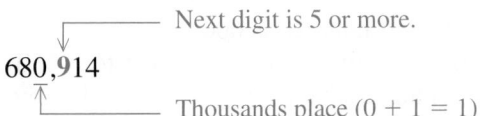

680,**9**14 — Next digit is 5 or more. Thousands place (0 + 1 = 1)

680,914 rounded to the nearest thousand is 681,000.

◀ **Work Problem 7 at the Side.**

OBJECTIVE 3 **Round numbers to estimate an answer.** Numbers may be rounded to **estimate** an answer. An estimated answer is one that is close to the exact answer and may be used as a check when the exact answer is found. The "≈" sign is often used to show that an answer has been rounded or estimated and is almost equal to the exact answer; ≈ means "approximately equal to."

EXAMPLE 8 **Using Rounding to Estimate an Answer**

Estimate each answer by rounding to the nearest ten.

(a)
$$\begin{array}{rcr} 76 & \longrightarrow & 80 \\ 53 & \longrightarrow & 50 \\ 38 & \longrightarrow & 40 \\ +\ 91 & \longrightarrow & +\ 90 \\ \hline & & 260 \end{array}$$
Rounded to the nearest ten; 260 Estimated answer

(b)
$$\begin{array}{rr} 27 & 30 \\ -\ 14 & -10 \\ \hline & 20 \end{array}$$
Rounded to the nearest ten; 20 Estimated answer

(c)
$$\begin{array}{rr} 16 & 20 \\ \times\ 21 & \times\ 20 \\ \hline & 400 \end{array}$$
Rounded to the nearest ten; 400 Estimated answer

◀ **Work Problem 8 at the Side.**

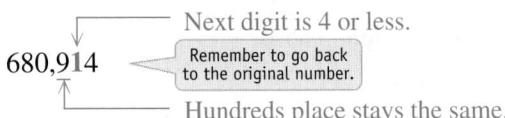

EXAMPLE 9 **Using Rounding to Estimate an Answer**

Estimate each answer by rounding to the nearest hundred.

(a)

252 →	300	Rounded to the nearest hundred
749 →	700	
576 →	600	
+ 819 →	+ 800	
	2400	Estimated answer

The hundreds position is 3 places to the left.

(b)

780	800	Rounded to the nearest hundred
− 536	− 500	
	300	Estimated answer

(c)

664	700	Rounded to the nearest hundred
× 834	× 800	
	560,000	Estimated answer

Work Problem 9 at the Side. ▶

OBJECTIVE 4 **Use front end rounding to estimate an answer.** A convenient way to estimate an answer is to use *front end rounding.* With **front end rounding,** we round to the highest possible place so that all the digits become 0 except the first one. For example, suppose you want to buy a big flat-screen television for \$2449, a home theater system for \$1759, and a reclining chair for \$525. Using front end rounding, you can estimate the total cost of these purchases.

Television	\$2449 →	2000
Home theater system	\$1759 →	2000
Reclining chair	\$525 →	+ 500
		\$4500 ← Estimated total cost

Do not round the estimated answer.

EXAMPLE 10 **Using Front End Rounding to Estimate an Answer**

Estimate each answer using front end rounding.

(a)

3825	4000	All digits changed to 0 except first digit, which is rounded
72	70	
565	600	
+ 2389	+ 2000	
	6670	Estimated answer

(b)

6712	7000	First digit rounded and all others changed to 0
− 825	− 800	
	6200	Estimated answer

Notice: Front end rounding leaves *only* one nonzero digit.

(c)

725	700	
× 86	× 90	
	63,000	Estimated answer

Work Problem 10 at the Side. ▶

---

9 Estimate the answers by rounding each number to the nearest hundred.

(a) 358
743
822
+ 978

(b) 842
− 475

(c) 723
× 478

VOCABULARY TIP

Front end rounding An estimation strategy that uses only the left-most digit of the number being rounded. This makes the number(s) the least accurate, but the easiest to work with.

10 Use front end rounding to estimate each answer.

GS (a)

36 →	4 0
3852 →	4 0 0 0
749 →	7 _ _
+ 5474 →	+ _ 0 0 0

(b) 2583
− 765

(c) 648
× 67

Answers

9. (a) 400 + 700 + 800 + 1000 = 2900
(b) 800 − 500 = 300
(c) 500 × 700 = 350,000

10. (a) 0; 0; 5; 40 + 4000 + 700 + 5000 = 9740
(b) 3000 − 800 = 2200
(c) 70 × 600 = 42,000

1.7 Exercises

FOR EXTRA HELP

Download the MyDashBoard App

MyMathLab®

CONCEPT CHECK *The following numbers have been rounded. Decide whether the number has been rounded to the nearest ten, nearest hundred, nearest thousand, or nearest ten-thousand, and fill in the blank.*

1. 624 to the nearest __________ is 620.

2. 509 to the nearest __________ is 510.

3. 86,813 to the nearest __________ is 86,800.

4. 17,211 to the nearest __________ is 17,200.

5. 78,499 to the nearest __________ is 78,000.

6. 14,314 to the nearest __________ is 14,000.

7. 12,987 to the nearest __________ is 10,000.

8. 6599 to the nearest __________ is 10,000.

Round each number as indicated. ***See Examples 1–5.***

9. 855 to the nearest ten

10. 946 to the nearest ten

11. 6771 to the nearest hundred

12. 5847 to the nearest hundred

13. 28,472 to the nearest hundred

GS 28,472
5 or more
hundreds place

14. 18,249 to the nearest hundred

GS 18,249
4 or less
hundreds place

15. 5996 to the nearest hundred

16. 4452 to the nearest hundred

17. 15,758 to the nearest thousand

18. 28,465 to the nearest thousand

19. 7,760,058,721 to the nearest billion

20. 4,468,523,628 to the nearest billion

21. 595,008 to the nearest ten-thousand

22. 725,182 to the nearest ten-thousand

23. 4,860,220 to the nearest million

24. 13,713,409 to the nearest million

Round each number to the nearest ten, nearest hundred, and nearest thousand.
See Examples 6 and 7.

	Ten	Hundred	Thousand
25. 4476	______	______	______
27. 3374	______	______	______
29. 6048	______	______	______

	Ten	Hundred	Thousand
26. 6483	______	______	______
28. 7632	______	______	______
30. 7065	______	______	______

	Ten	Hundred	Thousand
31. 5343	_____	_____	_____
33. 19,539	_____	_____	_____
35. 26,292	_____	_____	_____
37. 93,706	_____	_____	_____

	Ten	Hundred	Thousand
32. 7456	_____	_____	_____
34. 59,806	_____	_____	_____
36. 78,519	_____	_____	_____
38. 84,639	_____	_____	_____

39. Write in your own words the three steps that you would use to round a number when the digit to the right of the place to which you are rounding is *5 or more.*

40. Write in your own words the three steps that you would use to round a number when the digit to the right of the place to which you are rounding is *4 or less.*

Estimate the answer by rounding each number to the nearest ten. Then find the exact answer. ***See Example 8.***

41. *Estimate:* *Exact:*

Estimate	Rounds to	Exact
30	←	25
60	←	63
50	←	47
+ 80	←	+ 84

42. *Estimate:* *Exact:*

Estimate	Rounds to	Exact
60	←	56
20	←	24
90	←	85
+ 70	←	+ 71

43. *Estimate:* *Exact:*

Estimate	Exact
	78
− ___	− 43

44. *Estimate:* *Exact:*

Estimate	Exact
	57
− ___	− 24

45. *Estimate:* *Exact:*

Estimate	Exact
	67
× ___	× 34

46. *Estimate:* *Exact:*

Estimate	Exact
	53
× ___	× 75

Estimate the answer by rounding each number to the nearest hundred. Then find the exact answer. ***See Example 9.***

47. *Estimate:* *Exact:*

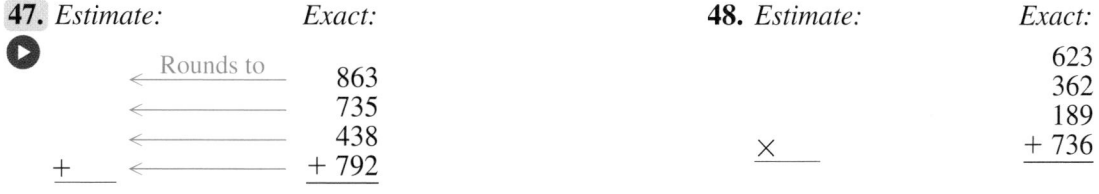

Estimate	Rounds to	Exact
	←	863
	←	735
	←	438
+ ___	←	+ 792

48. *Estimate:* *Exact:*

Estimate	Exact
	623
	362
	189
× ___	+ 736

49. *Estimate:* *Exact:*

$$\begin{array}{r} \\ -\underline{\quad\quad} \end{array} \qquad \begin{array}{r} 883 \\ -\ 448 \\ \hline \end{array}$$

50. *Estimate:* *Exact:*

$$\begin{array}{r} \\ -\underline{\quad\quad} \end{array} \qquad \begin{array}{r} 614 \\ -\ 276 \\ \hline \end{array}$$

51. *Estimate:* *Exact:*

$$\begin{array}{r} \\ \times\underline{\quad\quad} \end{array} \qquad \begin{array}{r} 752 \\ \times\ 375 \\ \hline \end{array}$$

52. *Estimate:* *Exact:*

$$\begin{array}{r} \\ \times\underline{\quad\quad} \end{array} \qquad \begin{array}{r} 845 \\ \times\ 396 \\ \hline \end{array}$$

Estimate each answer using front end rounding. Then find the exact answer. ***See Example 10.***

53. *Estimate:* *Exact:*

GS

$$\begin{array}{rcr} 8000 & \xleftarrow{\text{Rounds to}} & 8215 \\ 60 & \longleftarrow & 56 \\ 700 & \longleftarrow & 729 \\ +\ 4000 & \longleftarrow & +\ 3605 \\ \hline \end{array}$$

54. *Estimate:* *Exact:*

GS

$$\begin{array}{rcr} 3000 & \xleftarrow{\text{Rounds to}} & 2685 \\ 70 & \longleftarrow & 73 \\ 600 & \longleftarrow & 592 \\ +\ 7000 & \longleftarrow & +\ 7183 \\ \hline \end{array}$$

55. *Estimate:* *Exact:*

$$\begin{array}{r} \\ -\underline{\quad\quad} \end{array} \qquad \begin{array}{r} 687 \\ -\ 529 \\ \hline \end{array}$$

56. *Estimate:* *Exact:*

$$\begin{array}{r} \\ -\underline{\quad\quad} \end{array} \qquad \begin{array}{r} 543 \\ -\ 174 \\ \hline \end{array}$$

57. *Estimate:* *Exact:*

$$\begin{array}{r} \\ \times\underline{\quad\quad} \end{array} \qquad \begin{array}{r} 939 \\ \times\ \ 29 \\ \hline \end{array}$$

58. *Estimate:* *Exact:*

$$\begin{array}{r} \\ \times\underline{\quad\quad} \end{array} \qquad \begin{array}{r} 864 \\ \times\ \ 74 \\ \hline \end{array}$$

59. The number 3492 rounded to the nearest hundred is 3500, and 3500 rounded to the nearest thousand is 4000. But when 3492 is rounded directly to the nearest thousand it becomes 3000. Why is this true? Explain.

60. The use of rounding is helpful when estimating the answer to a problem. Why is this true? Give an example using either addition, subtraction, multiplication, or division to show how this works.

61. In 1900, the population of the United States was 76 million. Today it's 311 million. Round each of these numbers to the nearest ten-million. (*Source:* Reiman Publications and U.S. Census Bureau.)

62. Americans will eat more than 22,362,180 hot dogs in Major League ballparks this season. Round the number to the nearest ten-thousand and the nearest hundred-thousand. (*Source:* National Hot Dog and Sausage Council.)

63. There are 348,900 streets named Elm Street in the United States. Round this number to the nearest thousand and nearest ten-thousand. (*Source:* Expo Design Center.)

64. Of all the streets in the United States, the two most common candy-flavored street names are Peppermint and Chocolate. Ninety-five streets are named Peppermint and 27 are named Chocolate. Round each of these numbers to the nearest ten. (*Source:* Tel Atlas digital map database.)

65. There were 39,836,000 speeding tickets given in the United States last year. Round this number to the nearest ten-thousand, nearest hundred-thousand, and nearest million. (*Source:* Clark Howard Radio Show.)

66. Americans spend $57,463,625,000 on lottery tickets each year, or nine times as much as they spend on movie tickets. Round this number to the nearest ten-million, nearest hundred-million, and nearest billion. (*Source:* TLC-W, A Discovery Company.)

67. Ping-Pong is the fifth fastest growing sport in America. Today there are 19,265,780 players. Round this number to the nearest thousand, nearest ten-thousand, and the nearst hundred-thousand. (*Source: Parade, The Sunday Newspaper Magazine.*)

68. The highest annual production of Mitsubishi automobiles in Normal, Illinois was 221,543. Round this number to the nearest ten, nearest hundred, and nearest thousand. (*Source: Newsweek.*)

Relating Concepts (Exercises 69–75) For Individual or Group Work

To see how both rounding and front end rounding are used in solving problems, ***work Exercises 69–75 in order.***

69. A number rounded to the nearest thousand is 72,000. What is the *smallest* whole number this could have been before rounding?

70. A number rounded to the nearest thousand is 72,000. What is the *largest* whole number this could have been before rounding?

71. When front end rounding is used, a whole number rounds to 8000. What is the *smallest* possible original number?

72. When front end rounding is used, a whole number rounds to 8000. What is the *largest* possible original number?

The graph below shows the number of personal injuries in the United States each year for people participating in common activities.

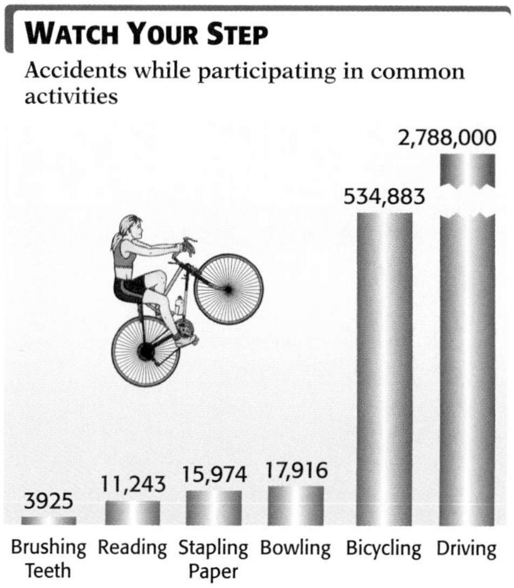

Source: AARP Magazine.

73. Round the number of accidents occurring in each activity to the nearest ten.

74. Use front end rounding to round the number of accidents in each activity.

75. **(a)** What is one advantage of using front end rounding instead of rounding to the nearest ten?

(b) What is one disadvantage?

1.8 Exponents, Roots, and Order of Operations

OBJECTIVES

1. Identify an exponent and a base.
2. Find the square root of a number.
3. Use the order of operations.

OBJECTIVE 1 Identify an exponent and a base. The product 3 • 3 can be written as 3^2 (read as "3 squared"). The small raised number 2, called an **exponent,** says to use 2 factors of 3. The number 3 is called the **base.** Writing 3^2 as 9 is called *simplifying the expression.*

EXAMPLE 1 Simplifying Expressions

Identify the exponent and the base, and then simplify each expression.

(a) 4^3 Base $\longrightarrow 4^3 \longleftarrow$ Exponent $4^3 = 4 \times 4 \times 4 = 64$

The small raised number is the exponent.

(b. $2^5 = 2 \times 2 \times 2 \times 2 \times 2 = 32$

The base is 2 and the exponent is 5.

◀ Work Problem 1 at the Side.

OBJECTIVE 2 Find the square root of a number. Because $3^2 = 9$, the number 3 is called the **square root** of 9. The square root of a number is one of two identical factors of that number. Square roots of numbers are written with the symbol $\sqrt{\ }$.

Square Root

$$\sqrt{\text{number} \cdot \text{number}} = \sqrt{\text{number}^2} = \text{number}$$

For example: $\sqrt{36} = \sqrt{6 \cdot 6} = \sqrt{6^2} = 6$ The square root of 36 is **6**.

To find the square root of 64 ask, "What number can be multiplied by itself (that is, *squared*) to give 64?" The answer is 8, so

$$\sqrt{64} = \sqrt{8 \cdot 8} = \sqrt{8^2} = 8.$$

A **perfect square** is a number that is the square of a ***whole number.*** The first few perfect squares are listed here.

Perfect Squares Table

$0 = 0^2$	$16 = 4^2$	$64 = 8^2$	$144 = 12^2$
$1 = 1^2$	$25 = 5^2$	$81 = 9^2$	$169 = 13^2$
$4 = 2^2$	$36 = 6^2$	$100 = 10^2$	$196 = 14^2$
$9 = 3^2$	$49 = 7^2$	$121 = 11^2$	$225 = 15^2$

EXAMPLE 2 Using Perfect Squares

Find each square root.

(a) $\sqrt{16}$ Because $4^2 = 16$, $\sqrt{16} = 4$.

Note that $\sqrt{16}$ is 4, not 4^2.

(b) $\sqrt{49} = 7$

(c) $\sqrt{0} = 0$

(d) $\sqrt{169} = 13$

◀ Work Problem 2 at the Side.

OBJECTIVE 3 Use the order of operations. Frequently problems may have parentheses, exponents, and square roots, and may involve more than one operation. Work these problems by following the **order of operations.**

---

1 Identify the exponent and the base, and then simplify each expression.

(a) 4^2

(b) 5^3

(c) 3^4

(d) 2^6

VOCABULARY TIP

Perfect square It is helpful to have the first 12 or so perfect squares memorized so that they are easily recognizable.

2 Find each square root.

(a) $\sqrt{4}$

(b) $\sqrt{25}$

(c) $\sqrt{36}$

(d) $\sqrt{225}$

(e) $\sqrt{1}$

Answers

1. (a) 2; 4; 16 (b) 3; 5; 125 (c) 4; 3; 81 (d) 6; 2; 64
2. (a) 2 (b) 5 (c) 6 (d) 15 (e) 1

Order of Operations

1. Do all operations inside ***parentheses*** or ***other grouping symbols.***
2. Simplify any expressions with ***exponents*** and find any ***square roots.***
3. ***Multiply*** or ***divide,*** proceeding from left to right.
4. ***Add*** or ***subtract,*** proceeding from left to right.

VOCABULARY TIP

Order of operations The first letter of the words "**P**lease **E**xcuse **M**y **D**ear **A**unt **S**ally" can help you remember the order of operations; Parentheses, Exponents, Multiplication and Division, Addition and Subtraction.

EXAMPLE 3 **Understanding the Order of Operations**

Use the order of operations to simplify each expression.

(a) $8^2 + 5 + 2$

$8^2 + 5 + 2$

$8 \cdot 8 + 5 + 2$ — Evaluate exponent first; 8^2 is $8 \cdot 8$.

$64 + 5 + 2$ — Add from left to right.

$69 + 2 = 71$

(b) $35 \div 5 \cdot 6$ — Divide first (start at left).

$7 \cdot 6 = 42$ — Multiply.

(c) $9 + (20 - 4) \cdot 3$ — Work inside parentheses first.

$9 + 16 \cdot 3$ — Multiply.

$9 + 48 = 57$ — Add last.

(d) $12 \cdot \sqrt{16} - 8(4)$ — Find the square root first.

$12 \cdot 4 - 8(4)$ — Multiply from left to right.

$48 - 32 = 16$ — Subtract last.

Work Problem 3 at the Side. ▶

EXAMPLE 4 **Using the Order of Operations**

Use the order of operations to simplify each expression.

(a) $15 - 4 + 2$ — Subtract first (start at left).

$11 + 2 = 13$ — Add.

(b) $8 + (7 - 3) \div 2$ — Work inside parentheses first.

$8 + 4 \div 2$ — Divide.

$8 + 2 = 10$ — Add last.

Add or subtract last.

(c) $4^2 \cdot 2^2 + (7 + 3) \cdot 2$ — Work inside parentheses first.

$4^2 \cdot 2^2 + 10 \cdot 2$ — Evaluate exponents.

$16 \cdot 4 + 10 \cdot 2$ — Multiply from left to right.

$64 + 20 = 84$ — Add last.

(d) $4 \cdot \sqrt{25} - 7 \cdot 2 + \frac{0}{5}$ — Find the square root first.

Zero divided by any nonzero number is zero.

$4 \cdot 5 - 7 \cdot 2 + \frac{0}{5}$ — Multiply or divide from left to right.

$20 - 14 + 0 = 6$ — Add or subtract last.

Work Problem 4 at the Side. ▶

3 Simplify each expression.

GS **(a)** $4 + 5 + 2^2$ — Evaluate exponents.

$4 + 5 + 2 \cdot 2$ — Add left to right.

$4 + 5 + \_\_\_\_$

$9 + \_\_\_\_ = \_\_\_\_$

(b) $3^2 + 2^3$

(c) $60 \div \sqrt{36} \div 2$

(d) $8 + 6(14 \div 2)$

4 Simplify each expression.

GS **(a)** $12 - 6 + 4^2$ — Evaluate exponents.

$12 - 6 + \_\_\_\_ \cdot \_\_\_\_$

$12 - 6 + 16$ — Add/subtract left to right.

$6 + 16 = \_\_\_\_$

(b) $2^3 + 3^2 - (5 \cdot 3)$

(c) $20 \div 2 + (7 - 5)$

(d) $15 \cdot \sqrt{9} - 8 \cdot \sqrt{4}$

Answers

3. **(a)** 4; 4; 13 **(b)** 17 **(c)** 5 **(d)** 50
4. **(a)** 4; 4; 22 **(b)** 2 **(c)** 12 **(d)** 29

1.8 Exercises

FOR EXTRA HELP

Download the MyDashBoard App

MyMathLab®

CONCEPT CHECK *Identify the exponent and the base.*

1. 3^2
exponent ____
base ____

2. 2^3
exponent ____
base ____

3. 5^2
exponent ____
base ____

4. 4^2
exponent ____
base ____

Identify the exponent and the base and then simplify each expression. **See Example 1.**

5. 8^2

6. 10^3

7. 15^2

8. 11^3

Use the Perfect Squares Table on the first page of this section to find each square root. **See Example 2.**

9. $\sqrt{16}$

10 $\sqrt{25}$

11. $\sqrt{64}$

12. $\sqrt{36}$

13. $\sqrt{100}$

14. $\sqrt{49}$

15. $\sqrt{144}$

16. $\sqrt{225}$

CONCEPT CHECK *Decide whether each statement is* true *or* false. *If it is* false, *explain why.*

17. The expression 5^2 means that 2 is used as a factor 5 times.

18. $4^2 = 8$

19. $1^3 = 3$

20. $6^1 = 1$

Fill in each blank. **See Example 2.**

21. $6^2 =$ ____, so $\sqrt{\quad} = 6$.
GS $6 \cdot 6 =$ ____, so $\sqrt{\_\_\_\_} = 6$.

22. $9^2 =$ ____, so $\sqrt{\quad} = 9$.
GS $9 \cdot 9 =$ ____, so $\sqrt{\_\_\_\_} = 9$.

23. $25^2 =$ ____, so $\sqrt{\quad} = 25$.

24. $50^2 =$ ____, so $\sqrt{\quad} = 50$.

25. $100^2 =$ ____, so $\sqrt{\quad} = 100$.

26. $60^2 =$ ____, so $\sqrt{\quad} = 60$.

27. Describe in your own words a perfect square. Of the two numbers 25 and 50, identify which is a perfect square and explain why.

28. Use the following list of words and phrases to write the four steps in the order of operations.

add	square root
exponents	subtract
multiply	divide

parentheses or other grouping symbols

CONCEPT CHECK *Decide whether each statement is* true *or* false. *If it is* false, *explain why.*

29. $3^2 + 8 - 5 = 12$

30. $5^2 + 5 - 6 = 24$

31. $6 + 8 \div 2 = 7$

32. $4 + 5(6 - 4) = 18$

Simplify each expression by using the order of operations. ***See Examples 3 and 4.***

33.
$$3^2 + 8 - 5$$
$$\underbrace{3^2} + 8 - 5$$
$$\underbrace{3 \cdot 3} + \underbrace{8 - 5}$$
$$9 \quad + \quad 3 =$$

34.
$$5^2 + 5 - 6$$
$$\underbrace{5^2} + 5 - 6$$
$$\underbrace{5 \cdot 5} + 5 - 6$$
$$25 \quad + 5 - 6 =$$

35. $25 \div 5(8 - 4)$

36. $36 \div 18(7 - 3)$

37. $5 \cdot 3^2 + \frac{0}{8}$

38. $8 \cdot 3^2 - \frac{10}{2}$

39. $4 \cdot 1 + 8(9 - 2) + 3$

40. $3 \cdot 2 + 7(3 + 1) + 5$

41. $2^2 \cdot 3^3 + (20 - 15) \cdot 2$

42. $4^2 \cdot 5^2 + (20 - 9) \cdot 3$

43. $5\sqrt{36} - 2(4)$

44. $2 \cdot \sqrt{100} - 3(4)$

45. $8(2) + 3 \cdot 7 - 7 =$

46. $10(3) + 6 \cdot 5 - 20$

47. $2^3 \cdot 3^2 + 3(14 - 4)$

48. $3^2 \cdot 4^2 + 2(15 - 6)$

49. $7 + 8 \div 4 + \frac{0}{7}$

50. $6 + 8 \div 2 + \frac{0}{8}$

51. $3^2 + 6^2 + (30 - 21) \cdot 2$

52. $4^2 + 5^2 + (25 - 9) \cdot 3$

53. $7 \cdot \sqrt{81} - 5 \cdot 6$

GS

$7 \cdot \underbrace{\sqrt{81}} - \underbrace{5 \cdot 6}$

$\underbrace{7 \cdot 9} - 30$

$63 - 30 =$

54. $6 \cdot \sqrt{64} - 6 \cdot 5$

GS

$6 \cdot \underbrace{\sqrt{64}} - \underbrace{6 \cdot 5}$

$\underbrace{6 \cdot 8} - 30$

$48 - 30 =$

55. $8 \cdot 2 + 5(3 \cdot 4) - 6$

56. $5 \cdot 2 + 3(5 + 3) - 6$

57. $4 \cdot \sqrt{49} - 7(5 - 2)$

58. $3 \cdot \sqrt{25} - 6(3 - 1)$

59. $7(4 - 2) + \sqrt{9}$

60. $5(4 - 3) + \sqrt{9}$

61. $7^2 + 3^2 - 8 + 5$

62. $3^2 - 2^2 + 3 - 2$

63. $5^2 \cdot 2^2 + (8 - 4) \cdot 2$

64. $5^2 \cdot 3^2 + (30 - 20) \cdot 2$

65. $5 + 9 \div 3 + 6 \cdot 3$

66. $8 + 3 \div 3 + 6 \cdot 3$

67. $8 \cdot \sqrt{49} - 6(9 - 4)$

68. $8 \cdot \sqrt{49} - 6(5 + 3)$

69. $5^2 - 4^2 + 3 \cdot 6$

70. $3^2 + 6^2 - 5 \cdot 8$

71. $8 + 8 \div 8 + 6 + \frac{5}{5}$

GS

$8 + \underbrace{8 \div 8} + 6 + \frac{5}{5}$

$8 + \quad 1 \quad + 6 + 1 =$

72. $3 + 14 \div 2 + 7 + \frac{8}{8}$

GS

$3 + \underbrace{14 \div 2} + 7 + \frac{8}{8}$

$3 + \quad 7 \quad + 7 + 1 =$

73. $6 \cdot \sqrt{25} - 7(2)$

74. $8 \cdot \sqrt{36} - 4(6)$

75. $9 \cdot \sqrt{16} - 3 \cdot \sqrt{25}$

76. $6 \cdot \sqrt{81} - 3 \cdot \sqrt{49}$

77. $7 \div 1 \cdot 8 \cdot 2 \div (21 - 5)$

78. $12 \div 4 \cdot 5 \cdot 4 \div (15 - 13)$

79. $15 \div 3 \cdot 2 \cdot 6 \div (14 - 11)$

80. $9 \div 1 \cdot 4 \cdot 2 \div (11 - 5)$

81. $6 \cdot \sqrt{25} - 4 \cdot \sqrt{16}$

82. $10 \cdot \sqrt{49} - 4 \cdot \sqrt{64}$

83. $5 \div 1 \cdot 10 \cdot 4 \div (17 - 9)$

84. $15 \div 3 \cdot 8 \cdot 9 \div (12 - 8)$

85. $8 \cdot 9 \div \sqrt{36} - 4 \div 2 + (14 - 8)$

86. $3 - 2 + 5 \cdot 4 \cdot \sqrt{144} \div \sqrt{36}$

87. $2 + 1 - 2 \cdot \sqrt{1} + 4 \cdot \sqrt{81} - 7 \cdot 2$

88. $6 - 4 + 2 \cdot 9 - 3 \cdot \sqrt{225} \div \sqrt{25}$

89. $5 \cdot \sqrt{36} \cdot \sqrt{100} \div 4 \cdot \sqrt{9} + 8$

90. $9 \cdot \sqrt{36} \cdot \sqrt{81} \div 2 + 6 - 3 - 5$

Study Skills

TAKING LECTURE NOTES

OBJECTIVES

1. **Apply note taking strategies, such as writing problems as well as explanations.**
2. **Use appropriate abbreviations in notes.**

Study the set of sample math notes in this section, and read the comments about them. Then try to incorporate the techniques into your own math note taking in class.

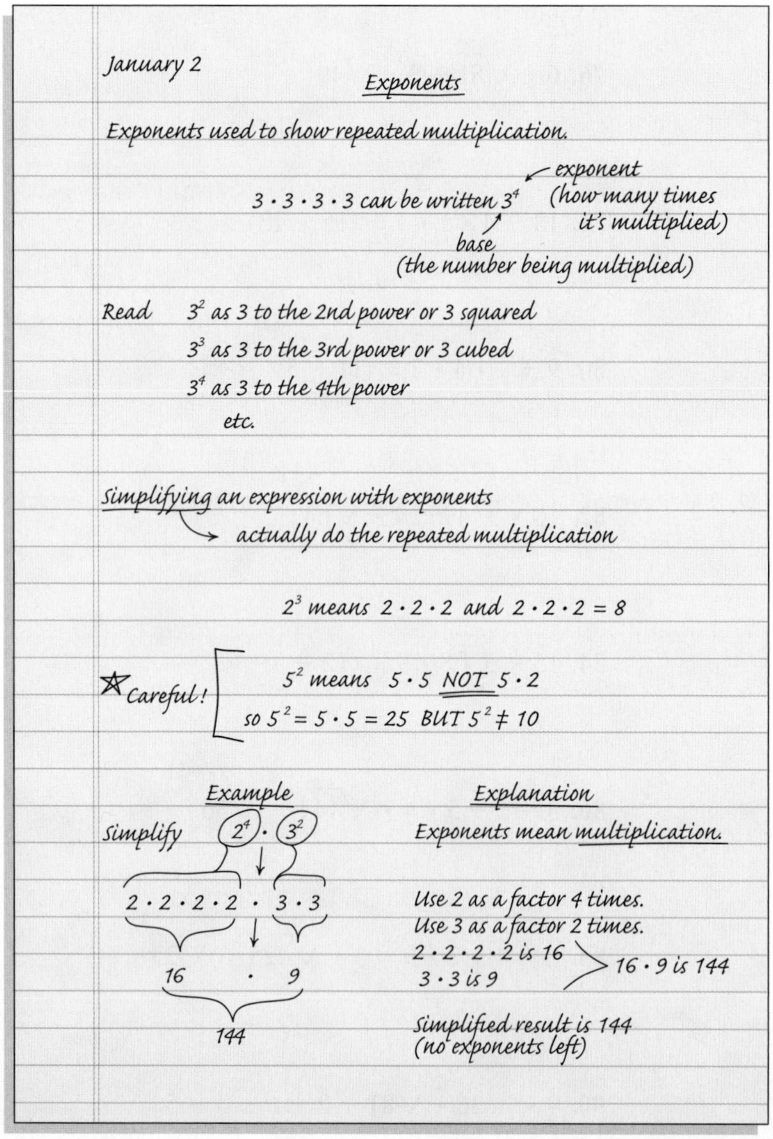

January 2

Exponents

Exponents used to show repeated multiplication.

$3 \cdot 3 \cdot 3 \cdot 3$ can be written 3^4 — *exponent (how many times it's multiplied)*; *base (the number being multiplied)*

Read 3^2 as 3 to the 2nd power or 3 squared
3^3 as 3 to the 3rd power or 3 cubed
3^4 as 3 to the 4th power
etc.

Simplifying an expression with exponents → *actually do the repeated multiplication*

2^3 means $2 \cdot 2 \cdot 2$ and $2 \cdot 2 \cdot 2 = 8$

☆ Careful! 5^2 means $5 \cdot 5$ NOT $5 \cdot 2$
so $5^2 = 5 \cdot 5 = 25$ BUT $5^2 \neq 10$

Example	*Explanation*
Simplify $2^4 \cdot 3^2$	*Exponents mean multiplication.*
$2 \cdot 2 \cdot 2 \cdot 2 \cdot 3 \cdot 3$	*Use 2 as a factor 4 times. Use 3 as a factor 2 times.*
$16 \cdot 9$	*$2 \cdot 2 \cdot 2 \cdot 2$ is 16; $3 \cdot 3$ is 9* → *$16 \cdot 9$ is 144*
144	*Simplified result is 144 (no exponents left)*

- The **date and title** of the day's lecture topic are always at the top of every page. **Always begin a new day with a new page.**
- Note the **definitions** of base and exponent are written in parentheses—don't trust your memory!
- **Skipping lines** makes the notes easier to read.
- See how the **direction word** (*simplify*) is emphasized and explained.
- A **star marks an important concept.** This is a warning to avoid future mistakes. **Note the underlining,** too, which highlights the importance.
- Notice the two columns, which allow for the example and its explanation to be close together. **Whenever you know you'll be given a series of steps to follow, try the two-column method.**
- Note the **brackets and arrows,** which clearly show how the problem is set up to be simplified.

Now Try This

Find one or two people in your math class to work with. Compare each other's lecture notes over a period of a week or so. Ask yourself the following questions as you examine the notes.

1. What are you doing in your notes to show the **main points** or larger concepts? (Such as underlining, boxing, using stars, capital letters, etc.)
2. In what ways do you **set off the explanations** for worked problems, examples, or smaller ideas (subpoints)? (Such as indenting, using arrows, circling or boxing)
3. What does **your instructor do** to show that he or she is moving from one idea to the next? (Such as saying "Next" or "Any questions," "Now," or erasing the board, etc.)
4. **How do you mark** that in your notes? (Such as skipping lines, using dashes or numbers, etc.)
5. What **explanations (in words) do you give yourself** in your notes, so when those new dendrites that you grew during lecture are fading, you can read your notes and still remember the new concepts later when you try to do your homework?
6. What **did you learn** by examining your classmates' notes?
   - ______________________________
   - ______________________________
   - ______________________________
7. What **will you try** in your own note taking? List **four** techniques that you will use next time you take notes in math class.
   - ______________________________
   - ______________________________
   - ______________________________
   - ______________________________

Why Are These Notes Brain Friendly?

The notes are **easy to look at,** and you know that the brain responds to things that are visually pleasing. Other techniques that are visually memorable are the use of spacing (the two columns), stars, underlining, and circling. All of these methods **allow your brain to take note of important concepts and steps.**

The notes are also **systematic,** which means that they use certain techniques regularly. This way, your brain easily recognizes the topic of the day, the signals that show an important point, and the steps to follow for procedures. When you develop a system that you always use in your notes, your notes are easy to understand later when you are reviewing for a test.

1.9 Reading Pictographs, Bar Graphs, and Line Graphs

OBJECTIVES

1. Read and understand a pictograph.
2. Read and understand a bar graph.
3. Read and understand a line graph.

VOCABULARY TIP

Pictograph Every pictograph has a key. The key in the graph tells the number that each picture or symbol represents.

1 Use the pictograph to answer each question.

(a) Which of the U.S. stamps had the second greatest number of sales?

Each stamp symbol represents 20 million stamps. The greatest number of symbols (6) is for the Elvis stamp. The second greatest number of symbols is for the _____ stamp.

(b) Approximately how many more Rock and Roll/Rhythm and Blues stamps were sold than Art of Disney Romance stamps?

VOCABULARY TIP

Bar graphs are typically used to display data that fit into different categories.

Answers

1. **(a)** 4; Rock and Roll/Rhythm & Blues
 (b) about 20 million (or 20,000,000) more stamps

We have all heard the saying "A picture is worth a thousand words," and there may be some truth in this. Today, so much information and data are being presented in the form of pictographs, circle graphs, bar graphs, and line graphs that it is important to be able to read and understand these tools.

OBJECTIVE 1 Read and understand a pictograph. A **pictograph** is a graph that uses pictures or symbols. It displays information that can be compared easily. However, since a symbol is used to represent a certain quantity, it can be difficult to determine the amount represented by a fraction of a symbol.

The pictograph below compares the number of U.S. postage stamps sold in the five top releases. In this pictograph, it is difficult to determine what fractional amount is represented by the partial stamps.

ELVIS IS STILL KING

Five of the most popular U.S. postage stamps of all time.

Elvis
DC Comics Super Heroes
Legends of Baseball
Art of Disney Romance
Rock & Roll/ Rhythm & Blues

= 20 million stamps

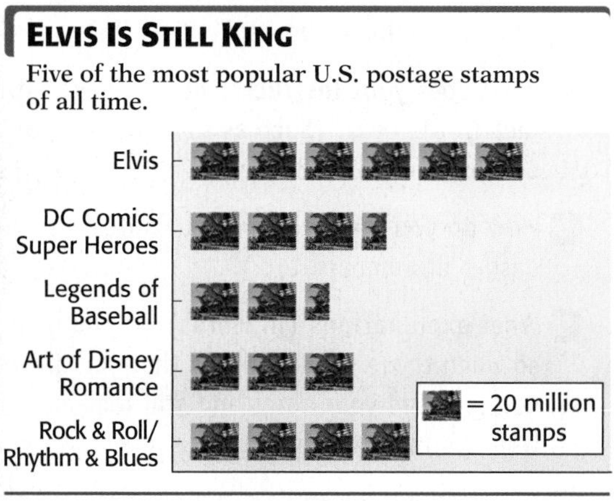

Source: United States Postal Service.

EXAMPLE 1 Using a Pictograph

Use the pictograph to answer each question. [Each stamp symbol represents 20 million stamps.]

(a) Which of the U.S. postage stamps shown has the lowest number sold?

The row representing Legends of Baseball has the fewest symbols. This means that the lowest number of U.S. postage stamps sold was Legends of Baseball.

(b) Approximately how many more Elvis stamps were sold than the Art of Disney Romance stamps?

The row representing the Elvis stamps has three more symbols than that for the Art of Disney stamps. This means that 3 • 20 million or 60,000,000 more Elvis stamps were sold than the Art of Disney Romance stamps.

◀ Work Problem **1** at the Side.

OBJECTIVE 2 Read and understand a bar graph. Bar graphs are useful for showing comparisons. For example, the following bar graph shows how many adults out of every 100 surveyed chose each money secret that they kept from their partner.

Source: American Express.

Use a ruler to help read a bar graph.

Realize that the value of each bar is an estimation.

EXAMPLE 2 Using a Bar Graph

Use the bar graph to find the number of adults who picked "Bought item partner didn't agree with" as the secret they kept from their partner.

Use a ruler or straightedge to line up the top of the bar labeled "Bought item partner didn't agree with" with the numbers on the left edge of the graph, labeled "Number of Adults." We see that 57 out of 100 adults picked "Bought item partner didn't agree with."

Work Problem **2** at the Side. ▶

OBJECTIVE 3 Read and understand a line graph. A **line graph** is often used for showing a trend. The following line graph shows the U.S. Census Bureau predictions for U.S. population growth to the year 2100.

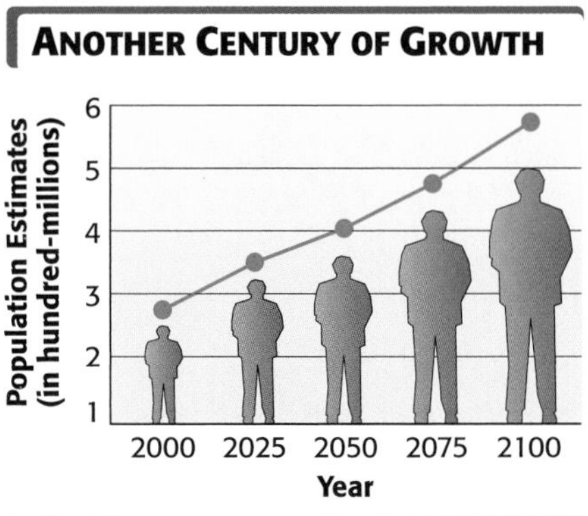

Source: U.S. Census Bureau.

Line graphs can show what is happening over time.

EXAMPLE 3 Using a Line Graph

Use the line graph to answer each question.

(a) What trend or pattern is shown in the graph?
The population will continue to increase.

(b) What is the estimated population for 2025?
Use a ruler or straightedge to line up the dot above the year labeled 2025 on the horizontal line with the numbers along the left edge of the graph. Notice that the label on the left side says "in hundred-millions." Since the 2025 dot is halfway between 3 and 4, the population in 2025 is halfway between 3 • 100,000,000 and 4 • 100,000,000 or 300,000,000 and 400,000,000. That means that the predicted population in 2025 is about 350,000,000 people.

Work Problem **3** at the Side. ▶

2 Use the bar graph to find the approximate number of adults who picked each money secret they kept from their partner.

(a) Secret account or card

(b) Misrepresented purchase price

(c) Hid purchases

(d) Hid debt

(e) Did not tell credit score

VOCABULARY TIP

Line graphs can be used to show how something changes over time. These graphs often have peaks and valleys to indicate upward and downward trends.

3 Use the line graph to find the predicted population of the United States for each year.

GS **(a)** 2050

The dot above 2050 lines up with ____. This means that the predicted population will be ___ ___ ___, 000,000.

(b) 2075

(c) 2100

Answers

2. **(a)** 27 out of 100 **(b)** 39 out of 100 **(c)** 42 out of 100 **(d)** 26 out of 100 **(e)** 12 out of 100

3. **(a)** 4; 4; 0; 0; 400,000,000 **(b)** 475,000,000 **(c)** 575,000,000

1.9 Exercises

FOR EXTRA HELP

Download the MyDashBoard App

MyMathLab®

The following pictograph shows the number of retail stores for the seven companies with the greatest number of outlets. Use the pictograph to answer Exercises 1–8. **See Example 1.**

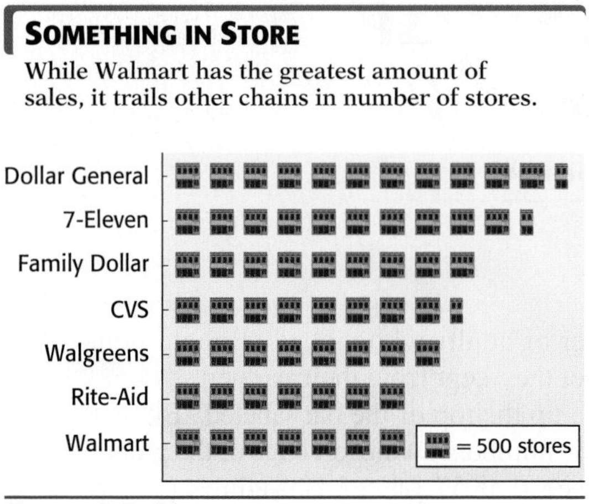

Source: T. D. Linx.

1. CONCEPT CHECK *Fill in the blanks.*
The number of pictures or symbols for the Walgreens retail stores is ______. Since each symbol represents ______ stores, Walgreens has
______ • ______ = ______ stores.

2. CONCEPT CHECK *Fill in the blanks.*
The number of pictures or symbols for the CVS retail stores is ______. Since each symbol represents ______ stores, CVS has
______ • ______ = ______ stores.

3. Find the number of Family Dollar retail stores.
Family Dollar has 9 symbols.
Each symbol is 500 stores
$9 \cdot 500 =$ __________

4. Approximately how many retail stores does 7-Eleven have?
7-Eleven has $10\frac{1}{2}$ symbols.
Each symbol is 500 stores.
$(10 \cdot 500) + \frac{1}{2} \cdot 500 =$ __________

5. Which company has the greatest number of retail stores? How many is that?

6. Which companies have the least number of retail stores? How many does each one have?

7. How many fewer stores does Family Dollar have than Dollar General?

8. How many more retail stores does Walgreens have than Walmart?

The following bar graph shows the results of a survey that was taken of 100 working adults to determine how they chose their careers. Use the bar graph to answer Exercises 11–16.
See Example 2.

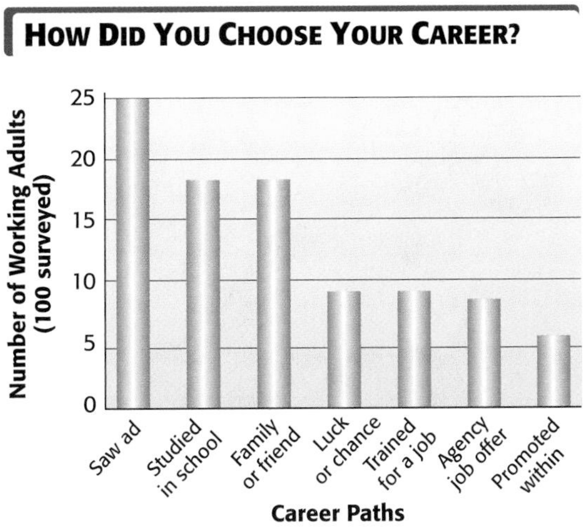

Source: Market Facts/TeleNation for Career Education Corporation.

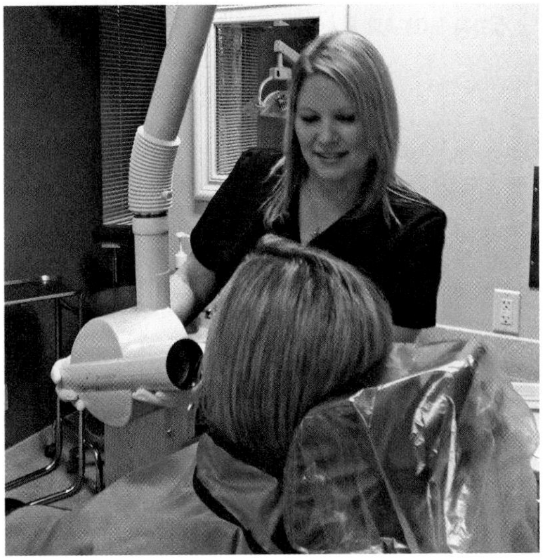

9. CONCEPT CHECK What is the number of working adults who were surveyed?

10. CONCEPT CHECK How many career paths were included in the survey?

11. How many people found their careers as a result of training for a job?

12. How many people found their careers because they studied for the career in school?

13. **(a)** Which career path was taken by the greatest number of people?

The bar graph shows that the greatest number of people ____________.

(b) How many people used this path?

From the bar graph ____________ people used this path.

14. **(a)** Which career path was taken by the least number of people?

The bar graph shows that the least number of people were ____________.

(b) How many people used this path?

From the bar graph, ____________ people used this path.

15. How many more people found their careers as a result of "Studied in school" than "Luck or chance"?

16. Find the total number of people who found their careers as a result of either "Studied in school" or "Trained for a job."

Sun Solar Products collected installation data and prepared the following line graph. Remembering that the solar installation data are shown in thousands of installations, use the line graph to answer Exercises 17–22. ***See Example 3.***

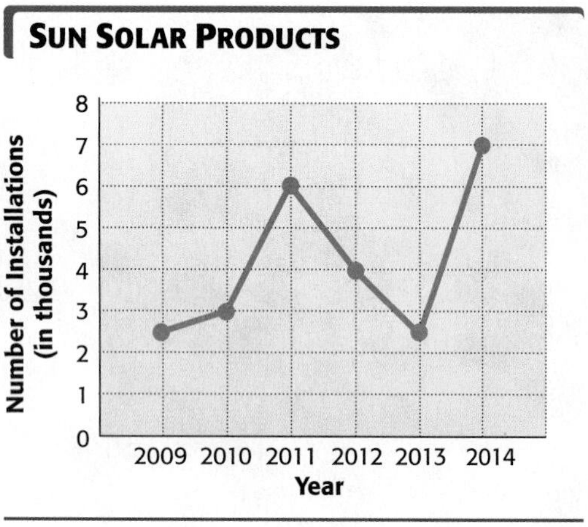

17. Which year had the greatest number of installations? How many installations were there?

The line graph shows that __________ had the greatest number.

There were 7 • 1000 = __________.

18. Which two years had the least number of installations? How many installations were there in each of those years?

The line graph shows that __________ and __________ had the least number.

The dots for both years are halfway between 2000 and 3000, or __________ installations.

19. Find the increase in the number of installations from 2013 to 2014.

20. Find the decrease in the number of installations from 2012 to 2013.

21. Give three possible explanations for the decrease in solar installations from 2011 to 2013.

22. Give three possible explanations for the increase in solar installations from 2013 to 2014.

Relating Concepts (Exercises 23–27) For Individual or Group Work

Getting a correct answer in mathematics always depends on following the order of operations. Insert grouping symbols (parentheses) so that each expression will result in the given number when simplified. ***Work Exercises 23–27 in order.***

23. $7 - 2 \cdot 3 - 6$; simplifies to 9

24. $4 + 2 \cdot 5 + 1$; simplifies to 36

25. $36 \div 3 \cdot 3 \cdot 4$; simplifies to 16

26. $56 \div 2 \cdot 2 \cdot 2 + \frac{0}{6}$; simplifies to 7

27. The Good Shepherd Ranch owns one section of land (one mile by one mile) shown below as Parcel 1. Parcels 2 and 3 are leased from neighbors.

(a) Use the order of operations to write an expression for the distance around the combined parcels.

(b) How many feet of barbed wire are needed for a three-strand barbed wire fence around the parcels?

(c) How many miles of barbed wire is this? (*Hint:* 1 mile = 5280 feet.)

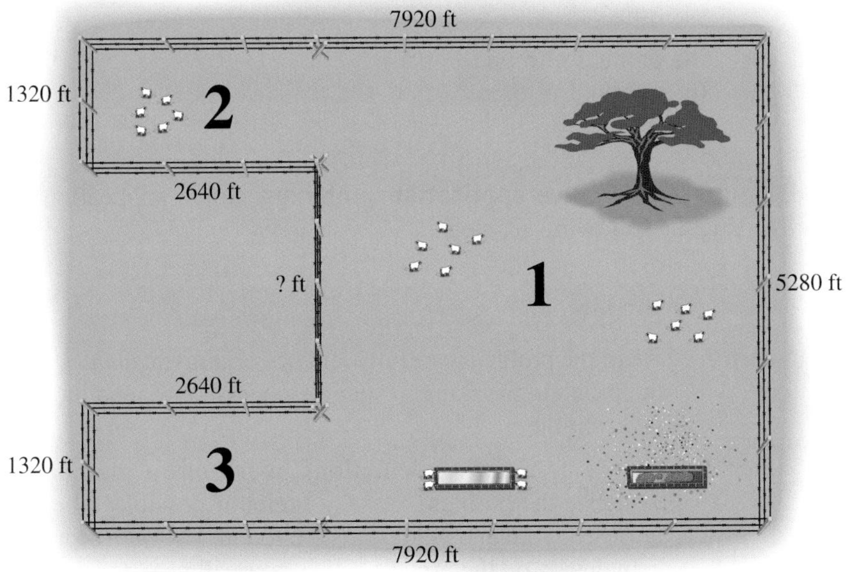

1.10 Solving Application Problems

OBJECTIVES

1. Find indicator words in application problems.
2. Solve application problems.
3. Estimate an answer.

VOCABULARY TIP

Indicator words These key words are useful in translating **word** problems from English into **math,** a necessary skill in becoming a good problem solver.

Most problems involving applications of mathematics are written in sentence form. You need to read the problem carefully to decide how to solve it.

OBJECTIVE 1 **Find indicator words in application problems.** As you read an application problem, look for **indicator words** that help you determine whether to use addition, subtraction, multiplication, or division. Some of these indicator words are shown here.

Addition	Subtraction	Multiplication	Division	Equals
plus	less	product	divided by	is
more	subtract	double	divided into	the same as
more than	subtracted from	triple	quotient	equals
added to	difference	times	goes into	equal to
increased by	less than	of	divide	yields
sum	fewer	twice	divided equally	results in
total	decreased by	twice as much	per	are
sum of	loss of			
increase of	minus			
gain of	take away			

CAUTION

The word *and* does not always indicate addition, so it does not appear as an indicator word in the table above. Notice how the "and" shows the location of different operation signs in the examples below.

The sum of 6 *and* 2 is $6 + 2$.
The difference of 6 *and* 2 is $6 - 2$.
The product of 6 *and* 2 is $6 \cdot 2$.
The quotient of 6 *and* 2 is $6 \div 2$.

OBJECTIVE 2 **Solve application problems.** Solve application problems by using the following six steps.

Solving an Application Problem

Step 1 **Read** the problem carefully and be certain you *understand* what the problem is asking. It may be necessary to read the problem several times.

Step 2 Before doing any calculations, **work out a plan** and try to visualize the problem. Draw a sketch if possible. Know which facts are given and which must be found. Use *indicator words* to help decide on the *plan* (whether you will need to add, subtract, multiply, or divide).

Always **estimate** the final answer.

Step 3 **Estimate** a *reasonable answer* by using rounding.

Step 4 **Solve** the problem by using the facts given and your plan.

Step 5 **State the answer.**

Step 6 **Check** your work. If the answer does not seem reasonable, begin again by reading the problem.

Check your answer to see if it is *reasonable*.

CAUTION

Do **NOT** make the mistake of trying the solve the problem before you know what is being asked.

OBJECTIVE 3 **Estimate an answer.** The six problem-solving steps give a systematic approach for solving word problems. Each of the steps is important, but special emphasis should be placed on Step 3, estimating a *reasonable answer*. Many times an "answer" just does not fit the problem.

What is a reasonable answer? Read the problem and try to determine the approximate size of the answer. Should the answer be part of a dollar, a few dollars, hundreds, thousands, or even millions of dollars? For example, if a problem asks for the cost of a man's shirt, would an answer of \$20 be reasonable? \$2000? \$2? \$200?

CAUTION

Always estimate the answer, then look at your final result to be sure it fits your estimate and is reasonable. This step will give greater success in problem solving.

Work Problem 1 at the Side. ▶

EXAMPLE 1 **Applying Division**

A community group has raised \$8260 for charity. Equal amounts are given to the Food Bank, Children's Center, Boy Scouts of America, and the Women's Shelter. How much did each group receive?

Step 1 **Read.** A reading of the problem shows that the four charities divided \$8260 equally.

Read and **understand** a problem before you begin.

Step 2 **Work out a plan.** The indicator words, ***divided equally,*** show that the amount each received can be found by dividing \$8260 by 4.

Step 3 **Estimate.** Round \$8260 to \$8000. Then $\$8000 \div 4 = \$2000$, so a reasonable answer would be a little greater than \$2000 each.

Step 4 **Solve.** Find the actual answer by dividing \$8260 by 4.

$$4\overline{)8260}\quad \text{quotient } 2065$$

Step 5 **State the answer.** Each charity received \$2065.

Step 6 **Check.** The exact answer of \$2065 is reasonable, as \$2065 is close to the estimated answer of \$2000. Is the exact answer of \$2065 correct? Check by multiplying.

Remember: Check your work.

$$\begin{array}{rl} \$2065 & \leftarrow \text{Amount received by each charity} \\ \times \quad 4 & \leftarrow \text{Number of charities} \\ \hline \$8260 & \leftarrow \text{Total raised; matches number given in problem} \end{array}$$

Work Problem 2 at the Side. ▶

1 Pick the most reasonable answer for each problem.

GS **(a)** A grocery clerk's hourly wage: \$1.40; \$14; \$140

____ is eliminated (low)

____ is eliminated (high)

____ is most reasonable

(b) The total length of five sport-utility vehicles: 8 ft; 18 ft; 80 ft; 800 ft

(c) The cost of heart bypass surgery: \$1000; \$100,000; \$10,000,000

2 Solve each problem.

(a) On a recent geology field trip, 84 fossils were collected. If the fossils are divided equally among John, Sean, Jenn, and Kara, how many fossils will each receive?

(b) This week there are 408 children attending a winter sports camp. If 12 children are assigned to each camp counselor, how many counselors are needed?

Answers

1. **(a)** \$1.40; \$140; \$14 **(b)** 80 ft **(c)** \$100,000

2. **(a)** 21 fossils **(b)** 34 counselors

3 Solve each problem.

(a) During the semester, Cindy Fong received the following points on examinations and quizzes: 92, 81, 83, 98, 15, 14, 15, and 12. Find her total points for the semester.

(b) Stephanie Dixon works at the telephone order desk of a catalog sales company. One week she had the following number of customer contacts: Monday, 78; Tuesday, 64; Wednesday, 118; Thursday, 102; and Friday, 196. How many customer contacts did she have that week?

Answers

3. **(a)** 410 points **(b)** 558 customer contacts

EXAMPLE 2 **Applying Addition**

One week, Andrea Abriani, operations manager, decided to total the stroller production at Safe T First Strollers. The daily figures were 7642 strollers on Monday, 8150 strollers on Tuesday, 7916 strollers on Wednesday, 8419 strollers on Thursday, and 7704 strollers on Friday. Find the total number of strollers for the week.

Step 1 **Read.** In this problem, the number of strollers for each day is given and the total strollers for the week must be found.

Step 2 **Work out a plan.** Add the daily stroller figures to arrive at the weekly total.

Step 3 **Estimate.** Because there were about 8000 strollers per day for a week of five days, a reasonable estimate would be $5 \cdot 8000 = 40{,}000$ strollers.

Step 4 **Solve.** Find the exact answer by adding the stroller numbers for the 5 days.

$$\begin{array}{r} \underline{39{,}831} \leftarrow \text{Check by adding up.} \\ 7\,642 \\ 8\,150 \\ 7\,916 \\ 8\,419 \\ +\ 7\,704 \\ \hline 39{,}831 \leftarrow \text{Number of strollers for the week} \end{array}$$

Add from bottom to top to check addition.

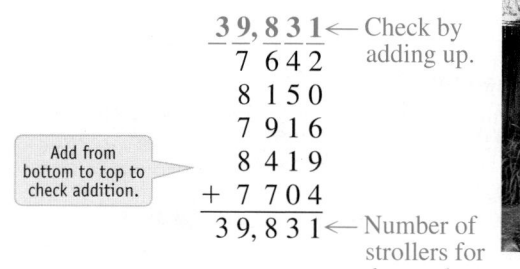

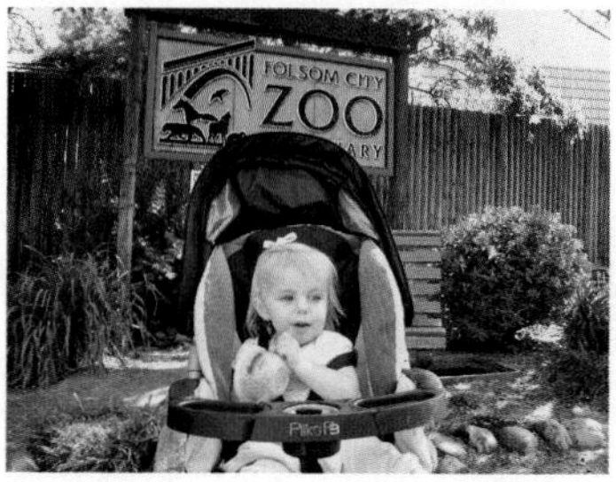

Step 5 **State the answer.** Abriani's total figure for the week was 39,831 strollers.

Step 6 **Check.** The exact answer of 39,831 strollers is close to the estimate of 40,000 strollers, so it is reasonable. Add up the columns to check the exact answer.

Calculator Tip

The calculator solution to **Example 2** uses chain calculations.

7642 (+) 8150 (+) 7916 (+) 8419 (+) 7704 (=) **39,831**

◀ **Work Problem 3 at the Side.**

EXAMPLE 3 **Determining Whether Subtraction Is Necessary**

The number of miles driven this year is 3028 fewer than the number driven last year. The miles driven last year was 16,735. Find the number of miles driven this year.

Step 1 **Read.** In this problem, the miles driven decreased from last year to this year. The miles driven last year and the decrease in miles driven are given. This year's miles driven must be found.

Step 2 **Work out a plan.** The indicator word, ***fewer,*** shows that subtraction must be used to find the number of miles driven this year.

Step 3 **Estimate.** Because the driving last year was about 17,000 miles, and the decrease in driving is about 3000 miles, a reasonable estimate would be $17{,}000 - 3000 = 14{,}000$ miles.

Continued on Next Page

Step 4 **Solve.** Find the exact answer by subtracting 3028 from 16,735.

$$\begin{array}{r} 16{,}735 \\ -\ \ 3\,028 \\ \hline 13{,}707 \end{array}$$

Step 5 **State the answer.** The driving this year is 13,707 miles.

Step 6 **Check.** The exact answer of 13,707 is reasonable, as it is close to the estimate of 14,000. Check by adding.

$$\begin{array}{rl} 13{,}707 & \leftarrow \text{miles driven this year} \\ +\ \ 3\,028 & \leftarrow \text{decrease in miles driven} \\ \hline 16{,}735 & \leftarrow \text{miles driven last year; matches number given in problem} \end{array}$$

Remember: Check your work.

Work Problem 4 at the Side. ▶

EXAMPLE 4 Solving a Two-Step Problem

In May, a landlord received $720 from each of eight tenants. After paying $2180 in expenses, how much rent money did the landlord have left?

Step 1 **Read.** The problem asks for the amount of rent remaining after expenses have been paid.

Step 2 **Work out a plan.** The wording *from each of eight tenants* indicates that the eight rents must be totaled. Since the rents are all the same, use multiplication to find the total rent received. Then, subtract expenses.

Step 3 **Estimate.** The amount of rent is about $700, making the total rent received about $700 • 8 = $5600. The expenses are about $2000. A reasonable estimate of the amount remaining is $5600 − $2000 = $3600.

Step 4 **Solve.** Find the exact amount by first multiplying $720 by 8 (the number of tenants).

$$\begin{array}{rl} \$\ 720 & \\ \times \quad\ \ 8 & \\ \hline \$5760 & \leftarrow \text{Total rent} \end{array}$$

Then subtract the $2180 in expenses from $5760.

$$\begin{array}{r} \$5760 \\ -\ \$2180 \\ \hline \$3580 \end{array}$$

The exact answer is close to the estimate and reasonable.

Step 5 **State the answer.** The amount remaining is $3580.

Step 6 **Check.** The exact answer of $3580 is reasonable, since it is close to the estimated answer of $3600. Check the answer by adding the expenses to the amount remaining and then dividing by 8.

$$\$3580 + \$2180 = \$5760$$

$$\begin{array}{r} \$720 \\ 8\overline{)5760} \end{array}$$

← Matches the rent amount given in the problem

Always check your work.

Work Problem 5 at the Side. ▶

---

4 Solve each problem.

(a) Alaska is our largest state, with an area of 663,267 square miles. The second largest state is Texas, with an area of 268,580 square miles. How much larger is Alaska than Texas? (*Source:* U.S. Census Bureau.)

(b) The Antique Military Vehicle Collectors (AMVC) had $14,863 in their club treasury bank account. After writing a check for $1180 to rent a display hall, find the amount remaining in the club account.

5 Solve each problem.

GS **(a)** An automobile insurance company purchased 318 Samsung digital cameras at a cost of $129 each. After a rebate of $2470, find the final cost of the cameras.

318 • $____ each = $41,022

$______ − $2470 = $______

(b) An Internet book company had sales of 12,628 books with a profit of $6 for each book sold. If 863 books were returned, how much profit remains?

Answers

4. **(a)** 394,687 square miles **(b)** $13,683

5. **(a)** $129; $41,022; $38,552 **(b)** $70,590

1.10 Exercises

FOR EXTRA HELP

1. CONCEPT CHECK A problem asks for the number of hours a person worked each day. Which would *not* be a reasonable answer?

A. 4 hours **B.** 8 hours **C.** 25 hours

2. CONCEPT CHECK A problem asks for the hourly earnings of a part-time student employee. Which would *not* be a reasonable answer?

A. \$9/hour **B.** \$100/hour **C.** \$12/hour

3. CONCEPT CHECK A problem asks for the cost of lunch at a fast food restaurant. Which answer is reasonable?

A. \$5 **B.** \$50 **C.** \$500

4. CONCEPT CHECK You calculate the gas mileage for your car. Which answer is reasonable? (mpg = miles per gallon)

A. 5 mpg **B.** 25 mpg **C.** 125 mpg

Solve each application problem. First use front end rounding to estimate the answer. Then find the exact answer. ***See Examples 1–4.***

5. Last week, SUBWAY sold 602 Veggie Delite sandwiches, 935 ham sandwiches, 1328 turkey breast sandwiches, 757 roast beef sandwiches, and 1586 SUBWAY club sandwiches. Find the total number of sandwiches sold.

Estimate: 600 + 900 + _____ + 800 + _____
= __________

Exact:

6. During a recent week, Radio Flyer, Inc. manufactured 32,815 Model #18 wagons, 4875 steel miniwagons, 1975 wood 40-inch wagons, 15,308 scooters, and 9815 new-design plastic wagons. Find the total number of items manufactured.

Estimate: _____ + 5000 + 2000 + 20,000
+ _____ = __________

Exact:

7. Paying ahead for a rental car can save 35% at Budget.com. They pay-at-the-counter base rate for a seven-day, full-size car rental is \$296. The rate for the same car when paying ahead using the Pay Now rate at Budget.com is \$192. How much is saved? (*Source:* Budget Car Rental.)

Estimate: \$300 − \$200 = __________

Exact:

8. The U.S. population will rise from 311 million today to 478 million by 2100. Find the expected increase in population. (*Source:* United Nations population division.)

Estimate:

Exact:

9. A packing machine can package 236 first-aid kits each hour. At this rate, find the number of first-aid kits packaged in 24 hours.

Estimate:

Exact:

10. If 450 admission tickets to a classic car show are sold each day, how many tickets are sold in a 12-day period?

Estimate: 500 × 10

Exact:

11. Clarence Hanks, coordinator of Toys for Tots, has collected 2628 toys. If his group can give the same number of toys to each of 657 children, how many toys will each child receive?

Estimate:

Exact:

12. If profits of $680,000 are divided evenly among a firm's 1000 employees, how much money will each employee receive?

Estimate:

Exact:

13. Turn down the thermostat in the winter and you can save money and energy. In the upper Midwest, setting back the thermostat from 68° to 55° at night can save $34 per month on fuel. Find the amount of money saved in five months.

Estimate:

Exact:

14. The cost of tuition and fees at a community college is $785 per quarter. If Gale Klein has five quarters remaining, find the total amount that she will need for tuition and fees.

Estimate:

Exact:

The sesquicentennial anniversary (150 years) of the beginning of the Civil War occurred in 2011. The loss of life at the conclusion of the war still stands as the highest war casualty count in American history. The table below shows the number of deaths and cause of death for both the Southern and the Northern American States. Use these data to answer Exercises 15–20.

Union (Northern) Deaths	Confederate (Southern) Deaths
Battle 110,070	Battle 94,120
Disease 250,152	Disease 164,300

Source: *The Civil War, Strange and Fascinating Facts* by Burke Davis.

15. GS Find the total number of Union deaths in the Civil War.

Estimate: 100,000 + 300,000 = ____________

Exact:

16. Find the total number of Confederate deaths in the Civil War.

Estimate:

Exact:

17. How many more Union deaths resulted from disease than battle?

Estimate:

Exact:

18. How many more Confederate deaths were caused by disease than battle?

Estimate:

Exact:

19. Find the total loss of life (North and South) in the Civil War.

Estimate:

Exact:

20. How many more Union deaths than Confederate deaths were there in the Civil War?

Estimate:

Exact:

21. Ronda Biondi decides to establish a monthly budget. She will spend $695 for rent, $340 for food, $435 for child care, $240 for transportation, $180 for other expenses, and she will put the remainder in savings. If her monthly take-home pay is $2240, find her monthly savings.

Estimate:

Exact:

22. Robert Heisner had $2874 in his bank account. He paid $308 for auto repairs, $580 for a dishwasher, and $778 for an insurance payment. Find the amount remaining in his account.

Estimate:

Exact:

23. There are 43,560 square feet in one acre. How many square feet are there in 138 acres?

Estimate:

Exact:

24. The number of gallons of water polluted each day in an industrial area is 209,670. How many gallons of water are polluted each year? (Use a 365-day year.)

Estimate:

Exact:

The Internet was used to find the following minivan optional features and the price of each feature. Use this information to answer Exercises 25–28.

Safety and Exterior Options		Interior Options	
Option	**Cost**	**Option**	**Cost**
VIP Plus Security System	$299	Carpet floor mats	$321
Roof rack crossbars	$185	Cargo nets	$51
Mudguards	$99	Interface kit for iPod	$299
Alloy wheel locks	$67	Dual screen entertainment system	$1799
Paint protection	$395	Wireless headphones	$82
Lower body moulding	$209	XM Satellite Radio	$449

Source: www.edmunds.com

25. Find the total cost of all Safety and Exterior Options listed.

Estimate:

Exact:

26. Find the total cost of all Interior Options listed.

Estimate:

Exact:

27. A new-car dealer offers an option value package that includes VIP Plus Security System, alloy wheel locks, paint protection, and lower body moulding for $785. If Jill buys the value package instead of paying for each option separately, how much will she save?

Estimate:

Exact:

28. A new-car dealer offers an option package that includes VIP Plus Security System, roof rack crossbars, paint protection, carpet floor mats, cargo nets, and XM Satellite Radio for a total of $1495. How much will Samuel save if he buys the option package instead of paying for each option separately?

Estimate:

Exact:

29. The Enabling Supply House purchased 6 wheelchairs at $1256 each and 15 speech compression recorder-players at $895 each. Find the total cost.

Estimate:

Exact:

30. A college bookstore buys 17 laptop computers at $506 each and 13 printers at $482 each. Find the total cost.

Estimate:

Exact:

31. Being able to identify indicator words is helpful in determining how to solve an application problem. Write three indicator words for each of these operations: add, subtract, multiply, and divide. Write two indicator words that mean equals.

32. Identify and explain the six steps used to solve an application problem. You may refer to the text if you need help, but use your own words.

Solve each application problem. ***See Examples 1–4.***

33. Steve Edwards, manager, decided to total his sales at SUBWAY. The daily sales figures were \$2358 on Monday, \$3056 on Tuesday, \$2515 on Wednesday, \$1875 on Thursday, \$3978 on Friday, \$3219 on Saturday, and \$3008 on Sunday. Find his total sales for the week.

34. The numbers of visitors at a war veterans' memorial during one week are 5318; 2865; 4786; 1998; 3899; 2343; and 7221. Find the total attendance for the week.

35. A car weighs 2425 pounds. If its 582-pound engine is removed and replaced with a 634-pound engine, what will the car weigh?

36. Estelle Alan has \$2324 in her preschool operating account. She spends \$734 from this account, and then the class parents raise \$568 in a rummage sale. Find the balance in the account after she deposits the money from the rummage sale.

37. In a recent survey of Reno/Lake Tahoe hotels, the cost per night at Harrah's in Reno was \$45, while the cost at Harrah's in Lake Tahoe was \$99 per night. Find the amount saved on a 7-night stay at Harrah's in Reno instead of staying at Harrah's in Lake Tahoe. (*Source:* Harrah's Casinos and Hotels.)

38. The most expensive hotel room in a recent study was the Ritz-Carlton at \$645 per night, while the least expensive was Motel 6 at \$74 per night. Find the amount saved in a 4-night stay at Motel 6 instead of staying at the Ritz-Carlton. (*Source:* Ritz-Carlton/Motel 6.)

39. A youth soccer association raised \$7588 through fund-raising projects. After expenses of \$838 were paid, the balance of the money was divided evenly among the 18 teams. How much did each team receive?

40. Feather Farms Egg Ranch collected 3545 eggs in the morning and 2575 eggs in the afternoon. If the eggs are packed in flats containing 30 eggs each, find the number of flats needed for packing.

41. A theater owner wants to provide enough seating for 1250 people. The main floor has 30 rows of 25 seats in each row. If the balcony has 25 rows, how many seats must be in each balcony row to satisfy the owner's seating requirements?

42. Jennie makes 24 grapevine wreaths per week to sell to gift shops. She works 40 weeks a year and packages six wreaths per box. If she ships equal quantities to each of five shops, find the number of boxes each store will receive.

Chapter 1 Summary

Key Terms

1.1

whole numbers The whole numbers are 0, 1, 2, 3, 4, 5, 6, 7, 8, and so on.

place value The place value of each digit in a whole number is determined by its position in the whole number.

table A table is a display of facts in rows and columns.

1.2

addition The process of finding the total is addition.

addends The numbers being added in an addition problem are addends.

sum (total) The answer in an addition problem is called the sum.

commutative property of addition The commutative property of addition states that the order of numbers in an addition problem can be changed without changing the sum.

associative property of addition The associative property of addition states that grouping the addition of numbers differently does not change the sum.

regrouping The process of regrouping is used in an addition problem when the sum of the digits in a column is greater than 9.

perimeter The perimeter is the distance around the outside edges of a flat figure.

1.3

minuend The number from which another number is being subtracted is the minuend.

subtrahend The subtrahend is the number being subtracted in a subtraction problem.

difference The answer in a subtraction problem is called the difference.

regrouping The process of regrouping is used in subtraction if a digit is less than the one directly below it.

1.4

factors The numbers being multiplied are called factors. For example, in $3 \times 4 = 12$, both 3 and 4 are factors.

product The answer in a multiplication problem is called the product.

commutative property of multiplication The commutative property of multiplication states that changing the order of the factors in a multiplication problem does not change the product.

associative property of multiplication The associative property of multiplication states that grouping the factors differently does not change the product.

chain multiplication problem A multiplication problem having more than two factors is a chain multiplication problem.

multiple The product of two whole number factors is a multiple of those numbers.

multiple of 10 A whole number that ends in 0, such as 10, 20, or 30; 100, 200, or 300.

partial products The products found when multiplying numbers having two or more digits.

1.5

dividend The number being divided by another number in a division problem is the dividend.

divisor The divisor is the number by which you are dividing in a division problem.

quotient The answer in a division problem is called the quotient.

undefined Dividing any number by 0 is *undefined*. Dividing by zero cannot be done.

short division A method of dividing a number by a one-digit divisor is short division.

remainder The remainder is the number left over when two numbers do not divide evenly.

1.6

long division The process of long division is used to divide by a number with more than one digit.

trial divisors A method of determining how many times the true divisor goes into the dividend by estimating.

trial quotient The quotient that results from dividing the dividend by the trial divisor.

1.7

rounding Rounding is used to find a number that is close to the original number, but easier to work with. Use the $\approx$ sign, which means "approximately equal to."

estimate An estimated answer is one that is close to the exact answer.

front end rounding Rounding to the highest possible place so that all the digits become zeros except the first one is front end rounding.

exponent The exponent is the small raised number (2) in the expression 3^2.

base The base is the number 3 in the expression 3^2.

1.8

square root The square root of a whole number is the number that can be multiplied by itself to produce the given number.

perfect square A number that is the square of a whole number is a perfect square.

order of operations For problems or expressions with more than one operation, the order of operations tells what to do first, second, and so on to get the correct answer.

1.9

pictograph A graph that uses pictures or symbols to show data is a pictograph.

bar graph A graph that uses bars of various heights to show quantity is a bar graph.

line graph A graph that uses dots connected by lines to show trends is a line graph.

1.10

indicator words Words in a problem that indicate the necessary operations—addition, subtraction, multiplication, or division—are indicator words.

New Symbols

$\approx$ This sign is used to show that an answer has been estimated. It means "is approximately equal to."

$\sqrt{}$ The symbol for square root.

5^2 The small raised 2 is an exponent; it tells how many times to use 5 (the base) as a factor in multiplication.

Test Your Word Power

See how well you have learned the vocabulary in this chapter.

1. When using **addends** you are performing
   A. division
   B. subtraction
   C. addition.

2. The subtrahend is the
   A. number being multiplied
   B. number being subtracted
   C. number being added.

3. A **factor** is
   A. one of two or more numbers being added
   B. one of two or more numbers being multiplied
   C. one of two or more numbers being divided.

4. The **divisor** is
   A. the number being multiplied
   B. always the largest number
   C. the number doing the dividing.

5. We use **rounding** to
   A. avoid solving a problem
   B. help estimate a reasonable answer
   C. find the remainder.

6. A **perfect square** is
   A. the square of a whole number
   B. the same as a square root
   C. similar to a perfect triangle.

Answers to Test Your Word Power

1. C; *Example:* In $2 + 3 = 5$, the 2 and the 3 are addends.

2. B; *Example:* In $5 - 4 = 1$, the 4 is the subtrahend.

3. B; *Example:* In $3 \times 5 = 15$, the numbers 3 and 5 are factors.

4. C; *Example:* In $8 \div 4 = 2$, $\frac{8}{4} = 2$, and $4\overline{)8}$ with quotient 2, the 4 is the divisor.

5. B; *Example:* We can use rounding to estimate our answer and then determine whether the exact answer is reasonable.

6. A; *Example:* 25 is a perfect square because $5^2 = 25$ and 5 is a whole number.

Quick Review

Concepts	*Examples*
1.1 Reading and Writing Whole Numbers	
Do not use the word *and* when writing a whole number. Commas help divide the periods or groups for ones, thousands, millions, and billions. A comma is not needed when a number has four digits or fewer.	795 is written *seven hundred ninety-five.* 9,768,002 is written *nine million, seven hundred sixty-eight thousand, two*

Concepts	Examples
1.2 Adding Whole Numbers Add from top to bottom, starting with the ones column and working left. To check, add from bottom to top.	$\begin{array}{r} 1140 \\ 687 \\ 26 \\ 9 \\ +\ 418 \\ \hline 1140 \end{array}$ (Add up to check.) 687, 26, 9, 418: Addends; 1140: Sum
1.2 Commutative Property of Addition Changing the order of the addends in an addition problem does not change the sum.	$2 + 4 = 6$ $4 + 2 = 6$ By the commutative property, the sum is the same.
1.2 Associative Property of Addition Grouping the addends differently when adding does not change the sum.	$(2 + 3) + 4 = 5 + 4 = 9$ $2 + (3 + 4) = 2 + 7 = 9$ By the associative property, the sum is the same.
1.3 Subtracting Whole Numbers Subtract the subtrahend from the minuend to get the difference, using regrouping when necessary. To check, add the difference to the subtrahend to get the minuend.	*Problem* $\begin{array}{r} {\scriptstyle 6\ 12\ 18} \\ 4\not{7}\not{3}\not{8} \\ -\ 649 \\ \hline 4089 \end{array}$ 4738: Minuend; 649: Subtrahend; 4089: Difference *Check* $\begin{array}{r} 4089 \\ +\ 649 \\ \hline 4738 \end{array}$
1.4 Multiplying Whole Numbers Use $\times$, $\bullet$ (a raised dot), or parentheses to indicate multiplication. The numbers being multiplied are called *factors.* The multiplicand is being multiplied by the multiplier, giving the product. When the multiplier has more than one digit, partial products must be used and added to find the product.	3×4 or $3 \bullet 4$ or $(3)(4)$ or $3(4)$ 78 Multiplicand } Factors $\times$ 24 Multiplier } Factors 312 Partial product 156 Partial product (move one position left) 1872 Product
1.4 Commutative Property of Multiplication The product in a multiplication problem remains the same when the order of the factors is changed.	$3 \times 4 = 12$ $4 \times 3 = 12$ By the commutative property, the product is the same.
1.4 Associative Property of Multiplication Grouping the factors differently when multiplying does not change the product.	$(2 \times 3) \times 4 = 6 \times 4 = 24$ $2 \times (3 \times 4) = 2 \times 12 = 24$ By the associative property, the product is the same.
1.5 Dividing Whole Numbers $\div$ and $\overline{)\ }$ mean divide. Also a —, as in $\frac{25}{5}$, means to divide the top number (dividend) by the bottom number (divisor).	22 ← Quotient Divisor → $4\overline{)88}$ ← Dividend 88 0 Dividend: $\frac{88}{4} = 22$ ← Quotient (4: Divisor) $88 \div 4 = 22$ (88: Dividend; 4: Divisor; 22: Quotient)

Concepts	Examples
1.7 Rounding Whole Numbers	
Rules for Rounding *Step 1* Locate the place to be rounded, and draw a line under it. *Step 2* If the next digit to the right is 5 or more, increase the underlined digit by 1. If the next digit is 4 or less, do not change the underlined digit. *Step 3* Change all digits to the right of the underlined place to zeros.	Round 726 to the nearest ten. Next digit is 5 or more. 7$\underline{2}$6 Tens place increases from 2 to 3. 7$\underline{2}$6 rounds to 7$\underline{3}$0. Round 1,498,586 to the nearest million. Next digit is 4 or less. $\underline{1}$,498,586 Millions place does not change. 1,498,586 rounds to 1,000,000.
1.7 Front End Rounding	
Front end rounding is rounding to the highest possible place so that all the digits become 0 except the first digit.	Round each number using front end rounding. 76 rounds to 80. 348 rounds to 300. 6512 rounds to 7000. 23,751 rounds to 20,000. 652,179 rounds to 700,000.
1.8 Order of Operations	
Problems may have several operations. Work these problems using the order of operations. **1.** Do all operations inside parentheses or other grouping symbols. **2.** Simplify any expressions with exponents and find any square roots ($\sqrt{\ }$). **3.** Multiply or divide proceeding from left to right. **4.** Add or subtract proceeding from left to right.	Simplify, using the order of operations. $7 \cdot \underbrace{\sqrt{9}} - 4 \cdot 5$ Find the square root. $\underbrace{7 \cdot 3} - \underbrace{4 \cdot 5}$ Multiply from left to right. $21 - 20 = 1$ Subtract.
1.9 Reading Pictographs, Bar Graphs, and Line Graphs	
A *pictograph* uses pictures or symbols to show data.	When reading a pictograph, be certain that you determine the quantity represented by each picture or symbol.
A *bar graph* uses bars of various heights to show quantity.	When reading a bar graph, use a straightedge to line up the top of the bar with the numbers along the left edge of the graph.
A *line graph* uses dots connected by lines to show trends.	When reading a line graph, use a straightedge to line up the dot with the numbers along the left edge of the graph.

Concepts	Examples
1.10 Application Problems	
Steps for Solving an Application Problem	Manuel earns $118 on Sunday, $87 on Monday, and $63 on Tuesday. Find his total earnings for the 3 days.
Step 1 **Read** the problem carefully, perhaps several times.	***Step 1*** The earnings for each day are given, and the total for the 3 days must be found.
Step 2 **Work out a plan** before starting. Draw a sketch if possible.	***Step 2*** Add the daily earnings to find the total.
Step 3 **Estimate** a reasonable answer.	***Step 3*** Since the earnings were about $100 + $90 + $60 = $250, a reasonable estimate would be approximately $250.
Step 4 **Solve** the problem.	***Step 4*** $\underline{\$268}$ Check by adding up $118 87 + 63 $268 Total earnings
Step 5 **State the answer.**	***Step 5*** Manuel's total earnings are $268.
Step 6 **Check** your work. If the answer is not reasonable, start over.	***Step 6*** The exact answer is reasonable, because it is close to the estimate of $250.

Chapter 1 Review Exercises

If you need help with any of these Review Exercises, look in the section indicated inside the dark blue rectangles.

1.1 *Write the digits for the given period or group in each number.*

1. 6573
thousands
ones

2. 36,215
thousands
ones

3. 105,724
thousands
ones

4. 1,768,710,618
billions
millions
thousands
ones

Rewrite each number in words.

5. 728

6. 15,310

7. 319,215

8. 62,500,005

Rewrite each number in digits.

9. ten thousand, eight

10. two hundred million, four hundred fifty-five

1.2 *Add.*

11. $\begin{array}{r} 72 \\ +\ 38 \\ \hline \end{array}$

12. $\begin{array}{r} 54 \\ +\ 67 \\ \hline \end{array}$

13. $\begin{array}{r} 807 \\ 4606 \\ +\ \ 51 \\ \hline \end{array}$

14. $\begin{array}{r} 8215 \\ 9 \\ +\ 7433 \\ \hline \end{array}$

15. $\begin{array}{r} 2130 \\ 453 \\ 8107 \\ +\ \ 296 \\ \hline \end{array}$

16. $\begin{array}{r} 5684 \\ 218 \\ 2960 \\ +\ \ 983 \\ \hline \end{array}$

17. $\begin{array}{r} 5\,732 \\ 11{,}069 \\ 37 \\ 1\,595 \\ +\ 22{,}169 \\ \hline \end{array}$

18. $\begin{array}{r} 3\,451 \\ 12{,}286 \\ 43 \\ 1\,291 \\ +\ 32{,}784 \\ \hline \end{array}$

1.3 *Subtract.*

19. $\begin{array}{r} 64 \\ -\ 28 \\ \hline \end{array}$

20. $\begin{array}{r} 46 \\ -\ 19 \\ \hline \end{array}$

21. $\begin{array}{r} 375 \\ -\ 186 \\ \hline \end{array}$

22. $\begin{array}{r} 573 \\ -\ 389 \\ \hline \end{array}$

23. $\begin{array}{r} 7416 \\ -\ 567 \\ \hline \end{array}$

24. $\begin{array}{r} 5210 \\ -\ 883 \\ \hline \end{array}$

25. $\begin{array}{r} 2210 \\ -\ 1986 \\ \hline \end{array}$

26. $\begin{array}{r} 99{,}704 \\ -\ 73{,}838 \\ \hline \end{array}$

1.4 *Multiply.*

27. $\begin{array}{r} 7 \\ \times\ 7 \\ \hline \end{array}$

28. $\begin{array}{r} 8 \\ \times\ 0 \\ \hline \end{array}$

29. 8(4)

30. 8(8)

31. (5)(9)

32. (6)(7)

33. 7 • 8

34. 9 • 9

Work each chain multiplication.

35. $5 \times 4 \times 2$

36. $9 \times 1 \times 5$

37. $4 \times 4 \times 3$

38. $2 \times 2 \times 2$

39. (6)(0)(8)

40. (7)(1)(6)

41. 6 • 1 • 8

42. 7 • 7 • 0

Multiply.

43. $\begin{array}{r} 28 \\ \times\ 3 \\ \hline \end{array}$

44. $\begin{array}{r} 46 \\ \times\ 8 \\ \hline \end{array}$

45. $\begin{array}{r} 58 \\ \times\ 9 \\ \hline \end{array}$

46. $\begin{array}{r} 98 \\ \times\ 1 \\ \hline \end{array}$

47. $\begin{array}{r} 625 \\ \times\ 8 \\ \hline \end{array}$

48. $\begin{array}{r} 374 \\ \times\ 8 \\ \hline \end{array}$

49. $\begin{array}{r} 1349 \\ \times\ 4 \\ \hline \end{array}$

50. $\begin{array}{r} 9163 \\ \times\ 5 \\ \hline \end{array}$

51. $\begin{array}{r} 7456 \\ \times\ 2 \\ \hline \end{array}$

52. $\begin{array}{r} 2880 \\ \times\ 7 \\ \hline \end{array}$

53. $\begin{array}{r} 93{,}105 \\ \times\ 5 \\ \hline \end{array}$

54. $\begin{array}{r} 21{,}873 \\ \times\ 8 \\ \hline \end{array}$

55. $\begin{array}{r} 35 \\ \times\ 25 \\ \hline \end{array}$

56. $\begin{array}{r} 74 \\ \times\ 32 \\ \hline \end{array}$

57. $\begin{array}{r} 98 \\ \times\ 12 \\ \hline \end{array}$

58. $\begin{array}{r} 68 \\ \times\ 75 \\ \hline \end{array}$

59. $\begin{array}{r} 472 \\ \times\ 33 \\ \hline \end{array}$

60. $\begin{array}{r} 392 \\ \times\ 77 \\ \hline \end{array}$

61. $\begin{array}{r} 4051 \\ \times\ 219 \\ \hline \end{array}$

62. $\begin{array}{r} 1527 \\ \times\ 328 \\ \hline \end{array}$

Find each total cost.

63. 30 scientific calculators at $12 per calculator

64. 76 subscribers at $14 per subscription

65. 318 drill bit sets at $64 per set

66. 114 earplugs at $6 per earplug

Multiply by using the shortcut for multiples of 10.

67. $\begin{array}{r} 280 \\ \times\ 50 \\ \hline \end{array}$

68. $\begin{array}{r} 340 \\ \times\ 70 \\ \hline \end{array}$

69. $\begin{array}{r} 517 \\ \times\ 400 \\ \hline \end{array}$

70. $\begin{array}{r} 637 \\ \times\ 500 \\ \hline \end{array}$

71. $\begin{array}{r} 16{,}000 \\ \times\ 8\,000 \\ \hline \end{array}$

72. $\begin{array}{r} 43{,}000 \\ \times\ 2\,100 \\ \hline \end{array}$

1.5 *Divide. If the division is not possible, write "undefined."*

73. $20 \div 4$

74. $35 \div 5$

75. $42 \div 7$

76. $18 \div 9$

77. $\frac{54}{9}$

78. $\frac{36}{9}$

79. $\frac{49}{7}$

80. $\frac{0}{6}$

81. $\frac{148}{0}$

82. $\frac{0}{23}$

83. $\frac{64}{8}$

84. $\frac{81}{9}$

1.5-1.6 *Divide.*

85. $4\overline{)328}$

86. $3\overline{)294}$

87. $6\overline{)26{,}532}$

88. $76\overline{)26{,}752}$

89. $2704 \div 18$

90. $15{,}525 \div 125$

1.7 *Round as indicated.*

91. 817 to the nearest ten

92. 15,208 to the nearest hundred

93. 20,643 to the nearest thousand

94. 67,485 to the nearest ten-thousand

Round each number to the nearest ten, nearest hundred, and nearest thousand. Remember to round from the original number.

	Ten	Hundred	Thousand
95. 3487	______	______	______
96. 20,065	______	______	______
97. 98,201	______	______	______
98. 352,118	______	______	______

1.8 *Find each square root by using the Perfect Squares Table from this chapter.*

99. $\sqrt{16}$

100. $\sqrt{49}$

101. $\sqrt{144}$

102. $\sqrt{196}$

Identify the exponent and the base, and then simplify each expression.

103. 7^3

104. 3^6

105. 5^3

106. 4^5

Simplify each expression by using the order of operations.

107. $7^2 - 15$

108. $6^2 - 10$

109. $2 \cdot 3^2 \div 2$

110. $9 \div 1 \cdot 2 \cdot 2 \div (11 - 2)$

111. $\sqrt{9} + 2(3)$

112. $6 \cdot \sqrt{16} - 6 \cdot \sqrt{9}$

1.9 *The bar graph shows the number of parents out of 100 surveyed who nag their children about performing certain household chores.*

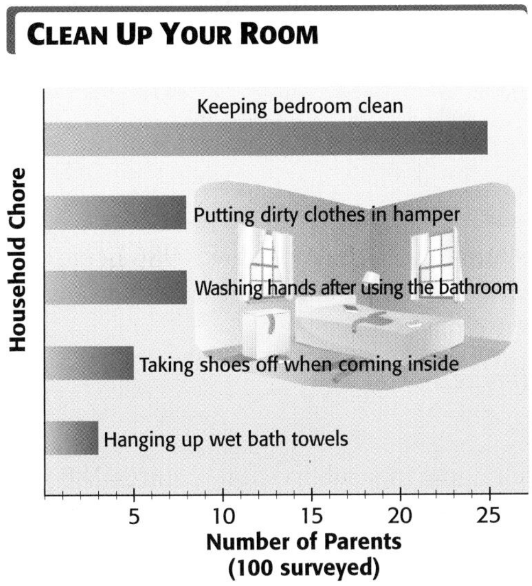

Source: Opinion Research Corporation for the Soap and Detergent Association.

113. How many parents nagged their children about washing hands after using the bathroom?

114. Find the number of parents who nagged their children about taking shoes off when coming inside.

115. Which household chore was nagged about by the greatest number of parents? How many were there?

116. Which household chore was nagged about by the least number of parents? How many did this?

1.10 *Solve each application problem. First use front end rounding to estimate the answer. Then find the exact answer.*

117. Bank of America processes 40 million checks each day. Find the number of checks processed by the bank in a year. Use a 365-day year. (*Source:* Bank of America.)

Estimate:

Exact:

118. A pulley on an evaporative cooler turns 1400 revolutions per minute. How many revolutions will the pulley turn in 60 minutes?

Estimate:

Exact:

119. The two most populated states are California, with 37,341,989 people, and Texas, with 25,268,418. Find the difference in population. (*Source:* U.S. Census Bureau.)

Estimate:

Exact:

120. The two least populated states are Alaska, with a population of 721,523, and Wyoming, with a population of 568,300. How many more people does Alaska have than Wyoming? (*Source:* U.S. Census Bureau.)

Estimate:

Exact:

121. The mechanic tells Kara Jantzi that the transmission on her Ford Escape has blown up. The cost of a new transaxle is \$2633, labor is 8 hours at \$90 per hour, and the sales tax is \$230. Find the total cost to replace her transmission. (*Source:* Undisclosed auto repair shop.)

Estimate:

Exact:

122. Moving to new home, Scott Samon rented a U-haul truck for \$55 plus \$2 per mile. After completing the move, the odometer shows that he has driven the truck 89 miles. Find the cost of the truck rental.

Estimate:

Exact:

123. Find the total cost if SUBWAY buys 32 baking ovens at \$1538 each and 28 warming ovens at \$887 each.

Estimate:

Exact:

124. A newspaper carrier has 62 customers who take the paper daily and 21 customers who take the paper on weekends only. A daily customer pays \$16 per month and a weekend-only customer pays \$7 per month. Find the total monthly collections.

Estimate:

Exact:

125. This holiday season, the average amount consumers plan to spend on holiday shopping for others is \$620. If they plan to spend \$107 on holiday shopping for themselves, how much more do they plan to spend on others than themselves? (*Source:* National Retail Federation.)

Estimate:

Exact:

126. Rachel Leach pays \$520 for rent and \$385 for her car payment. If she started with \$1924 in her bank account, how much remains in her account?

Estimate:

Exact:

127. A food canner uses 1 pound of pork for every 175 cans of pork and beans. How many pounds of pork are needed for 8750 cans?

Estimate:

Exact:

128. A stamping machine produces 986 license plates each hour. How long will it take to produce 32,538 license plates?

Estimate:

Exact:

129. Nitrogen sulfate is used in farming to enrich nitrogen-poor soil. If 625 pounds of nitrogen sulfate are spread per acre, how many acres can be spread with 32,500 pounds of nitrogen sulfate?

Estimate:

Exact:

130. Each home in a subdivision requires 180 feet of fencing. Find the number of homes that can be fenced with 5760 feet of fencing material.

Estimate:

Exact:

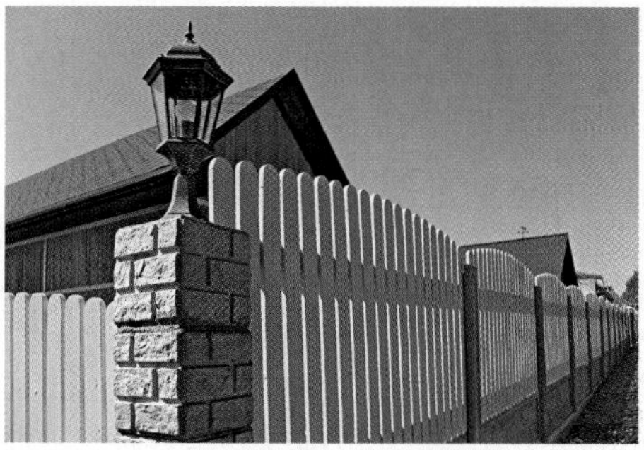

Mixed Review Exercises*

Perform the indicated operations.

131. 4(83)

132. 7(64)

133. $\begin{array}{r} 309 \\ -\ 56 \\ \hline \end{array}$

134. $\begin{array}{r} 835 \\ -\ 247 \\ \hline \end{array}$

135. $\begin{array}{r} 662 \\ +\ 379 \\ \hline \end{array}$

136. $\begin{array}{r} 789 \\ +\ 872 \\ \hline \end{array}$

137. $\begin{array}{r} 38{,}140 \\ -\ 6\,078 \\ \hline \end{array}$

138. $\begin{array}{r} 29{,}156 \\ +\ 4\,209 \\ \hline \end{array}$

139. 21 ÷ 7

140. $\frac{42}{6}$

141. $\begin{array}{r} 7\,218 \\ 3 \\ 18 \\ 1\,791 \\ 82{,}623 \\ +\ 1\,982 \\ \hline \end{array}$

142. $\begin{array}{r} 3\,812 \\ 5 \\ 22 \\ 1\,836 \\ 75{,}134 \\ +\ 2\,369 \\ \hline \end{array}$

143. $\frac{9}{0}$

144. $\frac{7}{1}$

145. 27,600 ÷ 4

*The order of exercises in this final group does not correspond to the order in which topics occur in the chapter. This random ordering should help you prepare for the chapter test in yet another way.

146. $18{,}480 \div 8$

147. $\begin{array}{r} 8430 \\ \times\ 128 \\ \hline \end{array}$

148. $\begin{array}{r} 21{,}702 \\ \times\ \quad 6 \\ \hline \end{array}$

149. $34\overline{)3672}$

150. $68\overline{)14{,}076}$

151. Rewrite 376,853 in words.

152. Rewrite 408,610 in words.

153. Round 8749 to the nearest hundred.

154. Round 400,503 to the nearest thousand.

Find each square root.

155. $\sqrt{64}$

156. $\sqrt{81}$

Find each total cost.

157. 308 pairs of knee guards at $18 per pair

158. 84 dishwashers at $370 per dishwasher

159. 208 baseball hats at $11 per hat

160. 607 boxes of avocados at $26 per box

Solve each application problem.

161. There are 52 playing cards in a deck. How many cards are there in nine decks?

162. A group of neighbors in Salem, Oregon founded an organization known as Salem Harvest. Picking fruits and vegetables to be distributed to those in need, 1700 volunteers picked 31 pounds of produce each. How many pounds did the volunteers pick? (*Source: AARP Magazine*.)

163. Push-type gasoline-powered lawn mowers cost $100 less than self-propelled mowers that you walk behind. If a self-propelled mower costs $380, find the cost of a push-type mower.

164. The Country Day School wants to raise $218,450 to construct and equip a computer lab. If $103,815 has already been raised, how much more is needed?

American River Raft Rentals lists the following daily raft rental fees. Notice that there is an additional $2 launch fee payable to the park system for each raft rented. Use this information to solve Exercises 165 and 166.

AMERICAN RIVER RAFT RENTALS

Size	Rental Fee	Launch Fee
4-person	$28	$2
6-person	$38	$2
10-person	$70	$2
12-person	$75	$2
16-person	$85	$2

Source: American River Raft Rentals.

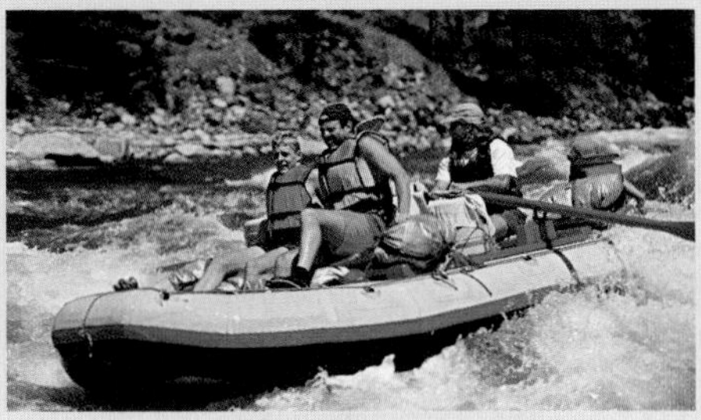

165. On a recent Tuesday the following rafts were rented: 6 4-person; 15 6-person; 10 10-person; 3 12-person; and 2 16-person. Find the total receipts, including the $2 per-raft launch fee.

166. On the 4th of July the following rafts were rented: 38 4-person; 73 6-person; 58 10-person; 34 12-person; and 18 16-person. Find the total receipts, including the $2 per-raft launch fee.

1 World Trade Center will be the tallest building in the United States and one of the world's giants. The world's tallest building is the Burj Dubai in the United Arab Emirates. The pictogram below shows the height, in feet, of some of the tallest buildings in the world. Use the information to solve Exercises 167–170.

167. How much taller is the Burj Dubai than the Empire State Building?

168. How much taller will 1 WTC in New York be than the Petronas Tower in Malaysia?

169. **(a)** What is the combined height of all the buildings shown?

(b) Is the combined height of these six buildings greater or less than two miles? How much greater or less than two miles? (*Hint:* 1 mile = 5280 feet.)

170. The height of the Willis Tower in Chicago is equivalent to the length of how many football fields? (*Hint:* A football field is 100 yards long and 1 yard = 3 feet.)

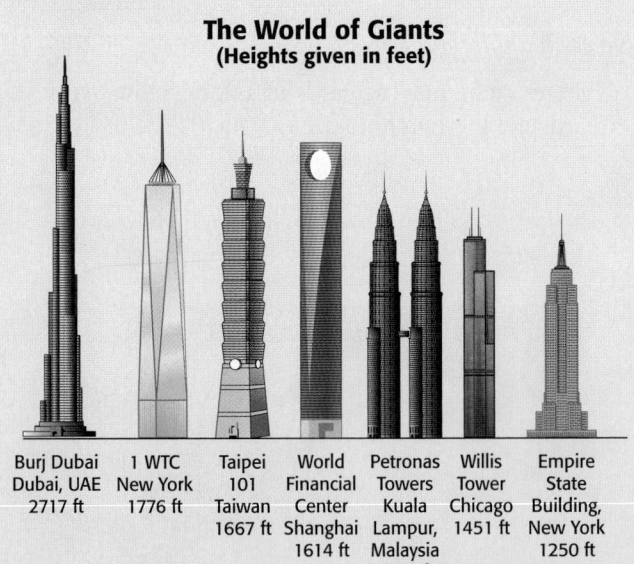

The World of Giants
(Heights given in feet)

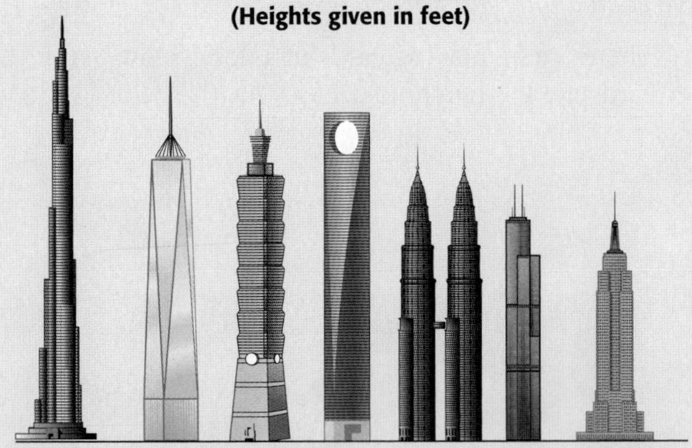

Source: Council on Tall Buildings and Urban Habitat; Silverstein Properties; The Port Authority of New York and New Jersey Associated Press.

Chapter 1 Test

CHAPTER Test Prep VIDEO

The Chapter Test Prep Videos with test solutions are available on DVD, in MyMathLab, and on YouTube—search "LialDevMath" and click on "Channels."

Write each number in words.

1. 9205

2. 25,065

3. Use digits to write four hundred twenty-six thousand, five.

Add.

4.
$$\begin{array}{r} 853 \\ 66 \\ 4022 \\ +\ 3589 \\ \hline \end{array}$$

5.
$$\begin{array}{r} 17,063 \\ 7 \\ 12 \\ 1\ 505 \\ 93,710 \\ +\quad 333 \\ \hline \end{array}$$

Subtract.

6.
$$\begin{array}{r} 9009 \\ -\ 7964 \\ \hline \end{array}$$

7.
$$\begin{array}{r} 9075 \\ -\ 2869 \\ \hline \end{array}$$

Multiply

8. $7 \times 6 \times 4$

9. $57 \cdot 3000$

10. 85(19)

11.
$$\begin{array}{r} 7381 \\ \times\ 603 \\ \hline \end{array}$$

Divide. If the division is not possible, write "undefined."

12. $16\overline{)112,752}$

13. $\dfrac{835}{0}$

14. $19,241 \div 42$

15. $280\overline{)44,800}$

Round as indicated.

16. 6347 to the nearest ten

17. 76,489 to the nearest thousand

Simplify each expression.

18. $5^2 + 8(2)$

19. $7 \cdot \sqrt{64} - 14 \cdot 2$

Solve each application problem. First use front end rounding to estimate the answer. Then find the exact answer.

20. Judy Martinez collects the following monthly rents from the tenants in her fourplex: $485, $500, $515, and $425. After she pays expenses of $785, how much does she have left?

Estimate:

Exact:

21. The major producer of ethanol made from corn is the United States. If 374 gallons of ethanol can be produced from the corn grown on one acre of land, how many acres are needed to produce 86,394 gallons. (*Source:* Earth Policy Institute.)

Estimate:

Exact:

22. Sadie Simms paid $528 for tires, $195 for brakes, and $235 for a timing belt. If this money was withdrawn from her bank account, which had a beginning balance of $1906, find her new balance.

Estimate:

Exact:

23. The technicians at a chicken ranch identify the sex of 48 baby chicks each minute for 4 hours in the morning and 36 baby chicks each minute for 3 hours in the afternoon. Find the total number of baby chicks identified. (*Source:* Discovery Channel, *Dirty Jobs.*)

Estimate:

Exact:

24. Explain in your own words the rules for rounding numbers. Give an example of rounding a number to the nearest ten-thousand.

25. List the six steps for solving application problems.

2 Multiplying and Dividing Fractions

The recipe shown below uses Jelly Belly jelly beans and will make $2\frac{1}{2}$ dozen (30) cookies. But suppose you wanted to make 3 dozen, 10 dozen, or even $3\frac{1}{2}$ dozen cookies? In this chapter we discuss multiplication and division of fractions, which you need to know when cooking or baking.

NEST COOKIES

$1\frac{1}{2}$ cups all purpose flour	$\frac{1}{2}$ cup sugar	2 cups shredded coconut
1 tsp. baking powder	1 egg white	5 oz. Jelly Belly jelly beans assorted flavors
$\frac{1}{2}$ tsp. salt	1 tsp. vanilla	
$\frac{1}{2}$ cup shortening	2 Tbs. milk	

Heat oven to 375 F. Sift together flour, baking powder, and salt and set aside. In large bowl beat shortening, sugar, egg white, and vanilla until well blended. Add flour mixture and milk until blended. Stir in coconut.

Roll dough into a ball, divide in half. Roll 15 one-inch balls from each half and place on ungreased baking sheet. Make thumb print depression in center of each ball to form nest. Bake 6 minutes. Remove from oven and place 4 Jelly Belly beans in center of each cookie. Return to oven and bake 5 more minutes. Transfer cookies to wire rack to cool. Makes 30 cookies.

2.1 Basics of Fractions

OBJECTIVES

1. Use a fraction to show how many parts of a whole are shaded.
2. Identify the numerator and denominator.
3. Identify proper and improper fractions.

In **Chapter 1** we discussed whole numbers. Many times, however, we find that parts of whole numbers are considered. One way to write parts of a whole is with **fractions.** Another way is with decimals, which is discussed in **Chapter 4.**

OBJECTIVE 1 Use a fraction to show how many parts of a whole are shaded. The number $\frac{1}{8}$ is a fraction that represents 1 of 8 equal parts. Read $\frac{1}{8}$ as "one eighth."

A fraction may be used to represent *part* of a whole.

EXAMPLE 1 Identifying Fractions

Use fractions to represent the shaded portions and the unshaded portions of each figure.

(a) The figure on the left has 6 equal parts. The 1 shaded part is represented by the fraction $\frac{1}{6}$. The *un*shaded part is $\frac{5}{6}$.

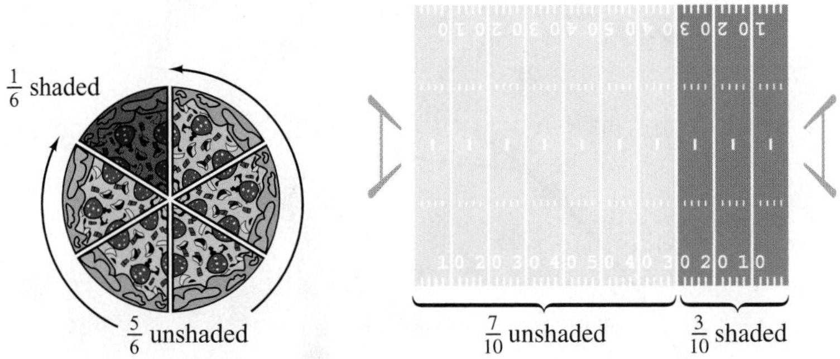

(b) The 3 shaded parts of the 10-part figure on the right are represented by the fraction $\frac{3}{10}$. The *un*shaded part is $\frac{7}{10}$.

◀ **Work Problem 1 at the Side.**

A fraction can also be used to represent more than one whole object.

EXAMPLE 2 Representing Fractions Greater Than 1

Use a fraction to represent the shaded part of each figure.

(a)

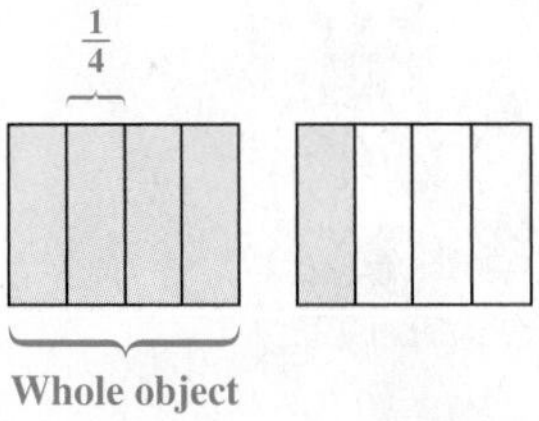

An area equal to 5 of the $\frac{1}{4}$ parts is shaded, so $\frac{5}{4}$ is shaded.

(b)

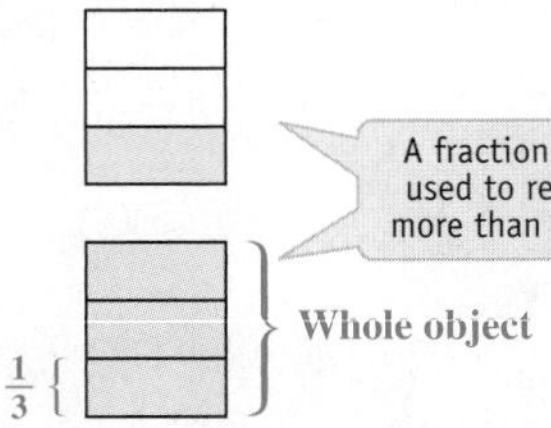

A fraction may be used to represent more than 1 whole.

An area equal to 4 of the $\frac{1}{3}$ parts is shaded, so $\frac{4}{3}$ is shaded.

◀ **Work Problem 2 at the Side.**

OBJECTIVE 2 Identify the numerator and denominator. In the fraction $\frac{2}{3}$, the number 2 is the **numerator** and 3 is the **denominator.** The line (bar) between the numerator and the denominator is the *fraction bar.*

$$\text{Fraction bar} \rightarrow \frac{2 \leftarrow \text{Numerator}}{3 \leftarrow \text{Denominator}}$$

---

1 Write fractions for the shaded portions and the unshaded portions of each figure.

(a)

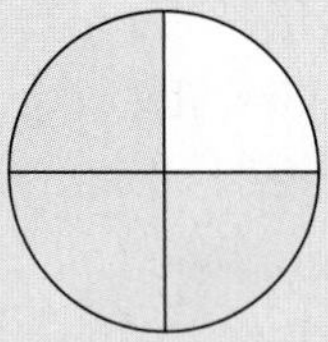

(b)

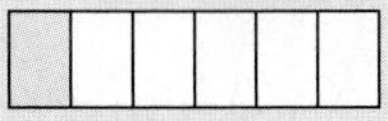

VOCABULARY TIP

Fraction The word *fraction* actually comes from the Latin *fractio*, which means "to break." Mathematically speaking, the bottom number of a fraction tells us how many parts a whole is broken into.

2 Write fractions for the shaded portions of each figure.

(a)

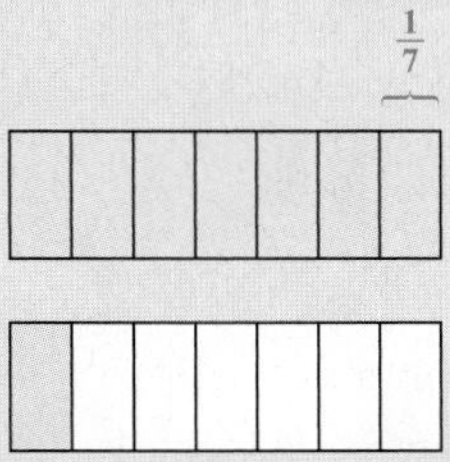

(b)

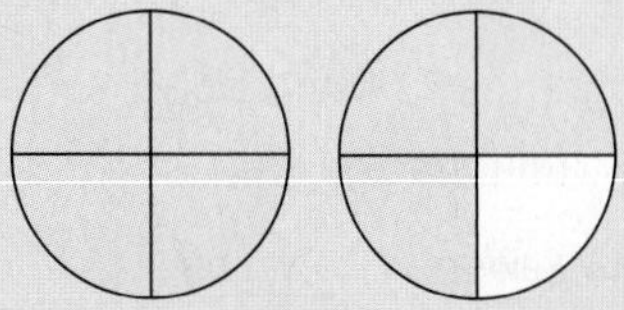

Answers

1. (a) $\frac{3}{4}; \frac{1}{4}$ (b) $\frac{1}{6}; \frac{5}{6}$
2. (a) $\frac{8}{7}$ (b) $\frac{7}{4}$

Numerators and Denominators

The **denominator** of a fraction shows the number of equivalent parts in the whole, and the **numerator** shows how many parts are being considered.

Note

A fraction bar, —, is one of the division symbols. Because division by 0 is undefined, a fraction with a denominator of 0 is also undefined.

EXAMPLE 3 Identifying Numerators and Denominators

Identify the numerator and denominator in each fraction.

(a) $\frac{3}{4}$ **(b)** $\frac{8}{5}$

$\frac{3}{4}$ 3 ← Numerator; 4 ← Denominator

$\frac{8}{5}$ 8 ← Numerator; 5 ← Denominator

Work Problem 3 at the Side. ▶

OBJECTIVE 3 Identify proper and improper fractions. Fractions can be identified as *proper* or *improper* fractions.

Proper and Improper Fractions

If the numerator of a fraction is *less* than the denominator, the fraction is a **proper fraction.** A proper fraction is less than 1 whole.

If the numerator is *greater than or equal to* the denominator, the fraction is an **improper fraction.** An improper fraction is greater than or equal to 1 whole.

Proper Fractions

$\frac{5}{8}$ $\frac{3}{5}$ $\frac{23}{24}$

Improper Fractions

$\frac{6}{5}$ $\frac{10}{10}$ $\frac{115}{112}$

EXAMPLE 4 Classifying Types of Fractions

(a) Identify all proper fractions in this list.

$$\frac{3}{4} \quad \frac{5}{9} \quad \frac{17}{5} \quad \frac{9}{7} \quad \frac{3}{3} \quad \frac{12}{25} \quad \frac{1}{9} \quad \frac{5}{3}$$

Proper fractions have a numerator that is less than the denominator. The proper fractions are shown below.

$\frac{3}{4}$ ← 3 is less than 4. $\quad \frac{5}{9} \quad \frac{12}{25} \quad \frac{1}{9}$ — A proper fraction is less than 1.

(b) Identify all improper fractions in the list in part (a).

Improper fractions have a numerator that is equal to or greater than the denominator. The improper fractions are shown below.

$\frac{17}{5}$ ← 17 is greater than 5. $\quad \frac{9}{7} \quad \frac{3}{3} \quad \frac{5}{3}$ — An improper fraction is equal to or greater than 1.

Work Problem 4 at the Side. ▶

VOCABULARY TIP

Denominator The first letter "D" of the word denominator can help you remember that it is the number "down under" the fraction bar.

VOCABULARY TIP

Numerator The numerator lies on top of the fraction bar and is the *number* of parts being considered.

3 Identify the numerator and the denominator. Draw a picture with shaded parts to show each fraction. Drawings may vary, but should have the correct number of parts.

(a) $\frac{2}{3}$ **(b)** $\frac{1}{4}$

(c) $\frac{8}{5}$ **(d)** $\frac{5}{2}$

VOCABULARY TIP

Proper fraction The everyday definition of *proper* means the correct way of doing something. If the numerator is less than the denominator, the fraction is considered in proper form, or proper for short.

4 From the following group of fractions:

$$\frac{2}{3} \quad \frac{4}{3} \quad \frac{3}{4} \quad \frac{8}{8} \quad \frac{3}{1} \quad \frac{1}{3}$$

GS **(a)** list all proper fractions;

$\frac{\_\_}{3}$ $\frac{\_\_}{4}$ $\frac{1}{\_\_}$

GS **(b)** list all improper fractions.

$\frac{4}{\_\_}$ $\frac{\_\_}{8}$ $\frac{3}{\_\_}$

Answers

3. **(a)** N: 2; D: 3 **(b)** N: 1; D: 4

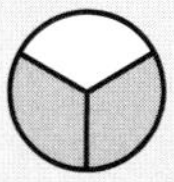

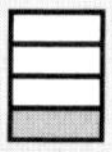

(c) N: 8; D: 5

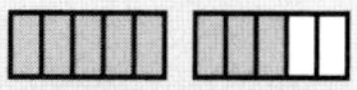

(d) N: 5; D: 2

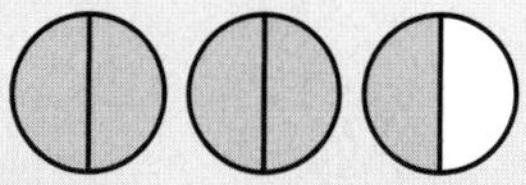

4. **(a)** $\frac{2}{3}; \frac{3}{4}; \frac{1}{3}$ **(b)** $\frac{4}{3}; \frac{8}{8}; \frac{3}{1}$

2.1 Exercises

FOR EXTRA HELP

Download the MyDashBoard App

MyMathLab®

CONCEPT CHECK *Identify the numerator and denominator.*

	Numerator	Denominator
1. $\frac{4}{5}$	________	________
2. $\frac{5}{6}$	________	________
3. $\frac{9}{8}$	________	________
4. $\frac{7}{5}$	________	________

CONCEPT CHECK *Fill in the blanks to complete each sentence.*

5. The fraction $\frac{3}{8}$ represents ______ of the ______ equal parts into which a whole is divided.

6. The fraction $\frac{7}{16}$ represents ______ of the ______ equal parts into which a whole is divided.

7. The fraction $\frac{5}{24}$ represents ______ of the ______ equal parts into which a whole is divided.

8. The fraction $\frac{24}{32}$ represents ______ of the ______ equal parts into which a whole is divided.

Write fractions to represent the shaded and unshaded portions of each figure. See ***Examples 1 and 2.***

9.

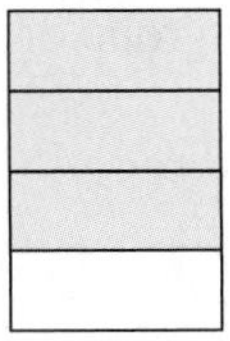

10.

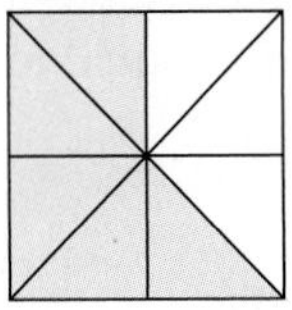

11.

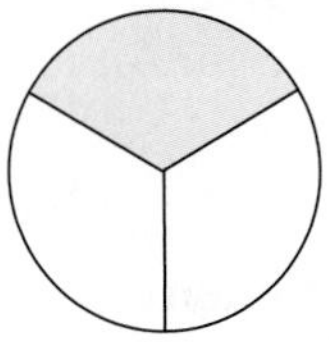

12.

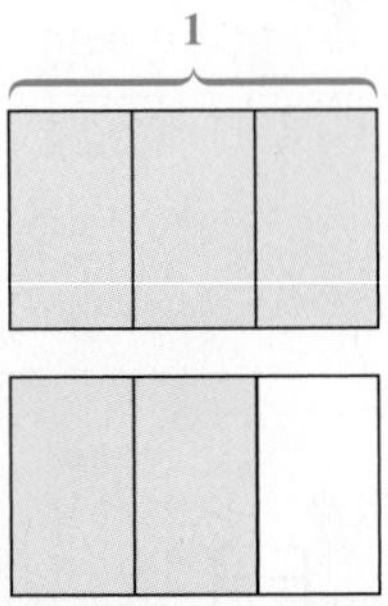

13.

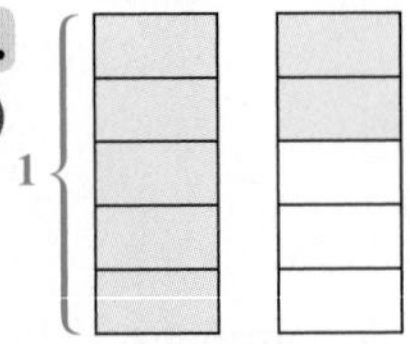

14.

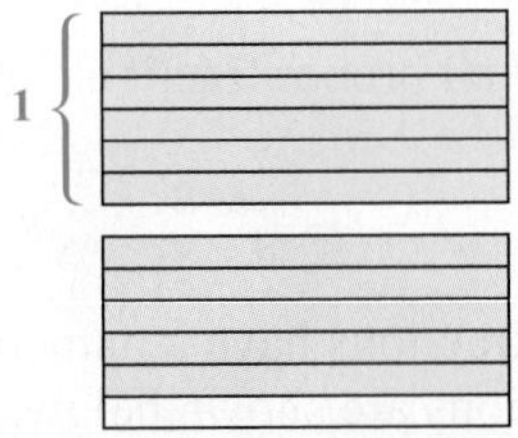

15. What fraction of these 6 bills has a life span of 2 years or greater? What fraction has a life of 4 years or less? What fraction has a life of 9 years?

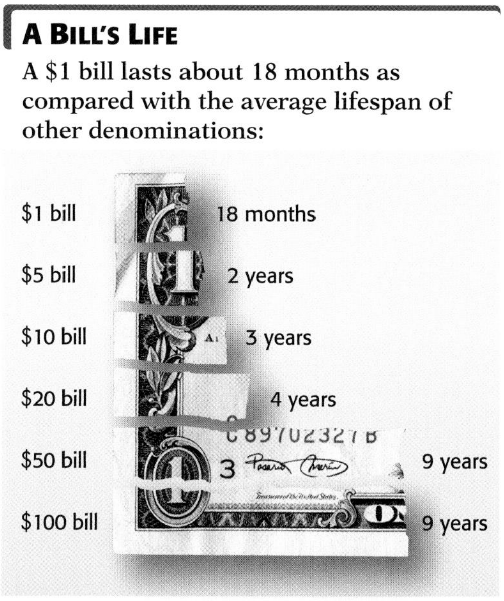

A BILL'S LIFE
A $1 bill lasts about 18 months as compared with the average lifespan of other denominations:
$1 bill
18 months
$5 bill
2 years
$10 bill
3 years
$20 bill
4 years
$50 bill
9 years
$100 bill
9 years

Source: Federal Reserve System; Bureau of Engraving and Printing.

16. What fraction of the 9 coins shown are pennies? What fraction are nickels? What fraction of the coins are dimes?

17. In an American Sign Language (ASL) class of 25 students, 8 are hearing impaired. What fraction of the students are hearing impaired?

$\frac{8}{\_\_\_\_}$ ← hearing impaired students (numerator)
← total students (denominator)

18. A supermarket has 215 shopping carts. If 76 of the shopping carts are in the parking lot and the rest are in the store, what fraction are in the store?

215 total shopping carts
− 76 in parking lot
139 shopping carts in store

Fraction of carts in store: $\frac{139}{\_\_\_\_}$

19. There are 520 rooms in a hotel. If 217 of the rooms are reserved for nonsmokers, what fraction of the rooms are for smokers?

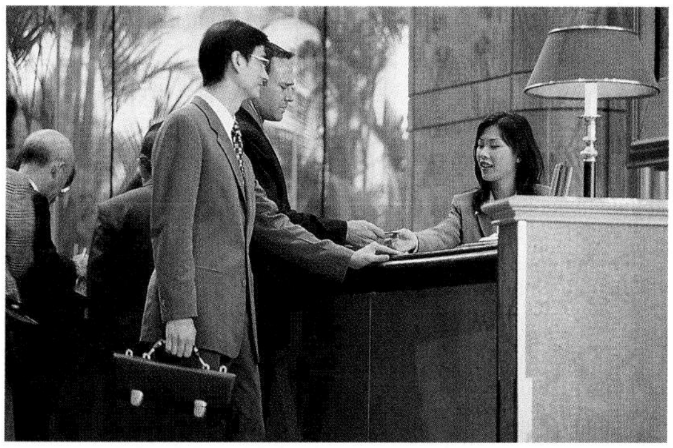

20. Of the 46 employees at the college bookstore, 15 are full-time while the rest are part-time student help. What fraction of the bookstore employees work part-time?

List the proper and improper fractions in each group. **See Example 4.**

	Proper	Improper
21. $\frac{8}{5}$ $\frac{1}{3}$ $\frac{5}{8}$ $\frac{6}{6}$ $\frac{12}{2}$ $\frac{7}{16}$	________	________
22. $\frac{1}{3}$ $\frac{3}{8}$ $\frac{16}{12}$ $\frac{10}{8}$ $\frac{6}{6}$ $\frac{3}{4}$	________	________
23. $\frac{3}{4}$ $\frac{3}{2}$ $\frac{5}{5}$ $\frac{9}{11}$ $\frac{7}{15}$ $\frac{19}{18}$	________	________
24. $\frac{12}{12}$ $\frac{15}{11}$ $\frac{13}{12}$ $\frac{11}{8}$ $\frac{17}{17}$ $\frac{19}{12}$	________	________

25. Write a fraction of your own choice. Label the parts of the fraction and write a sentence describing what each part represents. Draw a picture with shaded parts showing your fraction.

26. Give one example of a proper fraction and one example of an improper fraction. What determines whether a fraction is proper or improper? Draw pictures with shaded parts showing these fractions.

Study Skills

HOMEWORK: HOW, WHY, AND WHEN

OBJECTIVES

1. **Select an appropriate strategy for homework.**
2. **Use textbook features effectively.**

It is best for your brain if you keep up with the reading and homework in your math class. Remember that the more times you work with the information, the more dendrites you grow! So, give yourself every opportunity to read, work problems, and review your mathematics.

You have two choices for reading your math textbook. Read the short descriptions below and decide which will be best for you.

Preview before Class; Read Carefully after Class

Maddy learns best by listening to her instructor explain things. She "gets it" when she sees the instructor work problems on the board. She likes to ask questions in class and put the information in her notes. She has learned that it helps if she has *previewed* the section before the lecture, so she knows generally what to expect in class. *But after the class instruction*, when Maddy gets home, she finds that she can understand the math textbook easily. She remembers what her instructor said, and she can double-check her notes if she gets confused. So, Maddy does her **careful** reading of the section in her text **after** hearing the classroom lecture on the topic.

Read Carefully before Class

De'Lore, on the other hand, feels he learns well by reading on his own. He prefers to read the section and try working the example problems before coming to class. That way, he already knows what the instructor is going to talk about. Then, he can follow the instructor's examples more easily. It is also easier for him to take notes in class. De'Lore likes to have his questions answered right away, which he can do if he has already read the chapter section. So, De'Lore **carefully** reads the section in his text **before** he hears the classroom lecture on the topic.

Notice that there is **no one right way** to work with your textbook. You always must figure out what works best for you. Note also that both Maddy and De'Lore work with one section at a time. **The key is that you read the textbook regularly!** The rest of this activity will give you some ideas of how to make the most of your reading.

Try the following steps as you **read** your math textbook.

- Read slowly. Read only one section—or even part of a section—at a time.
- Do the sample problems in the margins **as you go.** Check them right away. The answers are at the bottom of the page.
- If your mind wanders, work problems on separate paper and write explanations in your own words.
- Make study cards as you read each section. Pay special attention to the yellow and blue boxes in the book. Make cards for new vocabulary, rules, procedures, formulas, and sample problems.
- **NOW,** you are ready to do your homework assignment!

Why Are These Reading Techniques Brain Friendly?

The steps at the left encourage you to be **actively working with the material** in your text. Your brain grows dendrites when it is doing something.

These methods require you to **try several different techniques,** not just the same thing over and over. Your brain loves variety!

Also, the techniques allow you to **take small breaks** in your learning. Those rest periods are crucial for good dendrite growth.

Study Skills

Continued from page 117

Why Are These Homework Suggestions Brain Friendly?

Your brain will grow dendrites as you study the worked examples in the text and **try doing them yourself** on separate paper. So, when you see similar problems in the homework, you will already have dendrites to work from.

Giving yourself a practice test by trying to remember the steps (without looking at your card) is an excellent way to reinforce what you are learning.

Correcting errors right away is how you learn and reinforce the correct procedures. It is hard to unlearn a mistake, so always check to see that you are on the right track.

Now Try This

Which steps for reading this book will be most helpful for you?

1. ______________________________
2. ______________________________
3. ______________________________

Homework

Instructors assign homework so you can grow your own dendrites (learn the material) and then coat the dendrites with myelin through practice (remember the material). Really! In learning, you get good at what you practice. So, completing homework every day will strengthen your neural network and prepare you for exams.

If you have read each section in your textbook according to the steps above, you will probably encounter few difficulties with the exercises in the homework. Here are some additional suggestions that will help you succeed with the homework.

- If you **have trouble with a problem,** find a similar worked example in the section. Pay attention to *every line* of the worked example to see how to get from step to step. Work it yourself too, on separate paper; don't just look at it.
- If it is **hard to remember the steps** to follow for certain procedures, write the steps on a separate card. Then write a short explanation of each step. Keep the card nearby while you do the exercises, but try *not* to look at it.
- If you **aren't sure you are working the assigned exercises correctly,** choose two or three odd-numbered problems that are a similar type and work them. Then check the answers in the Answers section of your book and see if you are doing them correctly. If you aren't, go back to the section in the text and review the examples and find out how to correct your errors. Finally, when you are sure you understand, try the assigned problems again.
- **Make sure you do some homework every day,** even if the math class does not meet each day!

Now Try This

What are your biggest homework concerns?
List your two main concerns and a **brain friendly solution** for each one.

1. **Concern:** ______________________________
   Solution: ______________________________
2. **Concern:** ______________________________
   Solution: ______________________________

2.2 Mixed Numbers

Suppose you had three whole trays of muffins and half of another tray. You would state this as a whole number and a fraction.

OBJECTIVE 1 **Identify mixed numbers.** When a whole number and a fraction are written together, the result is a **mixed number.** For example, the mixed number

$$3\frac{1}{2} \quad \text{represents} \quad 3 + \frac{1}{2}$$

or 3 wholes and $\frac{1}{2}$ of a whole. Read $3\frac{1}{2}$ as "three and one-half." As this figure shows, the mixed number $3\frac{1}{2}$ is equal to the improper fraction $\frac{7}{2}$.

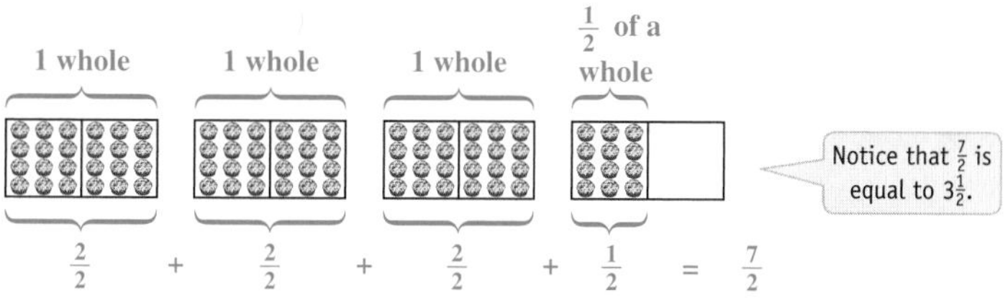

Work Problem 1 at the Side. ▶

OBJECTIVE 2 **Write mixed numbers as improper fractions.** Use the following steps to write $3\frac{1}{2}$ as an improper fraction without drawing a figure.

Step 1 Multiply 3 and 2.

$$3\frac{1}{2} \qquad 3 \cdot 2 = 6$$

Step 2 Add 1 to the product.

$$3\frac{1}{2} \qquad 6 + 1 = 7$$

Step 3 Use 7, from Step 2, as the numerator of the improper fraction and 2 as the denominator.

$$3\frac{1}{2} = \frac{7}{2}$$

Same denominator

In summary, use the following steps to *write a mixed number as an improper fraction.*

Writing a Mixed Number as an Improper Fraction

Step 1 *Multiply* the denominator of the fraction and the whole number.

Step 2 *Add* to this product the numerator of the fraction.

Step 3 Write the result of Step 2 as the *numerator* of the improper fraction and the original denominator as the *denominator.*

OBJECTIVES

1. Identify mixed numbers.
2. Write mixed numbers as improper fractions.
3. Write improper fractions as mixed numbers.

1 **(a)** Use these diagrams to write $1\frac{2}{3}$ as an improper fraction.

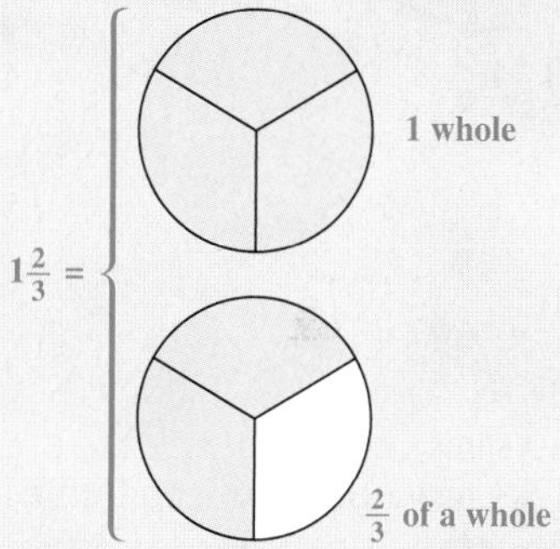

GS **(b)** Use these diagrams to write $2\frac{1}{4}$ as an improper fraction.

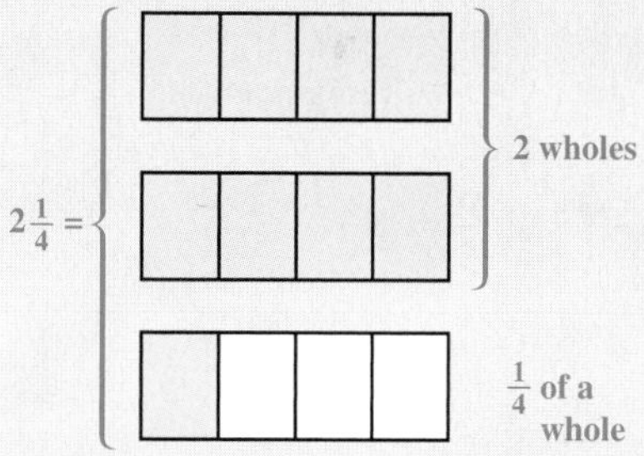

Since each of these diagrams is divided into ____ pieces, the denominator will be ____. The number of pieces shaded is ____.

VOCABULARY TIP

Mixed number A memory tip is to think "whole number times the bottom plus the top."

Answers

1. (a) $\frac{5}{3}$ **(b)** 4; 4; 9; $\frac{9}{4}$

2 Write as improper fractions.

GS (a) $6\frac{1}{2}$

$6 \cdot \_\_\_\_ = \_\_\_\_$

$12 + 1 = 13$

$6\frac{1}{2} = \frac{\_\_\_\_}{\_\_\_\_}$

(b) $7\frac{3}{4}$

(c) $4\frac{7}{8}$

(d) $8\frac{5}{6}$

3 Write as whole or mixed numbers.

GS (a) $\frac{6}{5}$

Divide ____ by ____

$$\begin{array}{r} 1 \leftarrow \text{Whole number part} \\ 5\overline{)6} \\ \underline{5} \\ 1 \leftarrow \text{Remainder} \end{array}$$

So $\frac{6}{5} = 1\frac{1}{\_\_\_\_}$

(b) $\frac{9}{4}$

(c) $\frac{35}{5}$

(d) $\frac{78}{7}$

Answers

2. (a) 2; 12; $\frac{13}{2}$ (b) $\frac{31}{4}$ (c) $\frac{39}{8}$ (d) $\frac{53}{6}$

3. (a) 6; 5; $1\frac{1}{5}$ (b) $2\frac{1}{4}$ (c) 7 (d) $11\frac{1}{7}$

EXAMPLE 1 **Writing a Mixed Number as an Improper Fraction**

Write $7\frac{2}{3}$ as an improper fraction (numerator greater than denominator).

Step 1 $7\frac{2}{3}$ $\quad 7 \cdot 3 = 21$ Multiply 7 and 3.

Step 2 $7\frac{2}{3}$ $\quad 21 + 2 = 23$ Add 2. The numerator of the improper fractions is 23.

Step 3 $7\frac{2}{3} = \frac{23}{3}$ Use the same denominator.

Always use the same denominator.

◀ **Work Problem 2 at the Side.**

OBJECTIVE 3 **Write improper fractions as mixed numbers.** Write an improper fraction as a mixed number as follows.

Writing an Improper Fraction as a Mixed Number

Write an **improper fraction** as a mixed number by dividing the numerator by the denominator. The quotient is the whole number (of the mixed number), the remainder is the numerator of the fraction part, and the denominator stays the same.

EXAMPLE 2 **Writing Improper Fractions as Mixed Numbers**

Write each improper fraction as a mixed number.

Divide numerator by denominator.

(a) $\frac{17}{5}$ Divide 17 by 5. ⟶

$$\begin{array}{r} 3 \leftarrow \text{Whole number part} \\ 5\overline{)17} \\ \underline{15} \\ 2 \leftarrow \text{Remainder} \end{array}$$

The quotient **3** is the whole number part of the mixed number. The remainder **2** is the numerator of the fraction, and the denominator stays as **5.**

$\frac{17}{5} = 3\frac{2}{5}$ ← Remainder

The denominator stays the same.

We can check this by using a diagram in which $\frac{17}{5}$ is shaded.

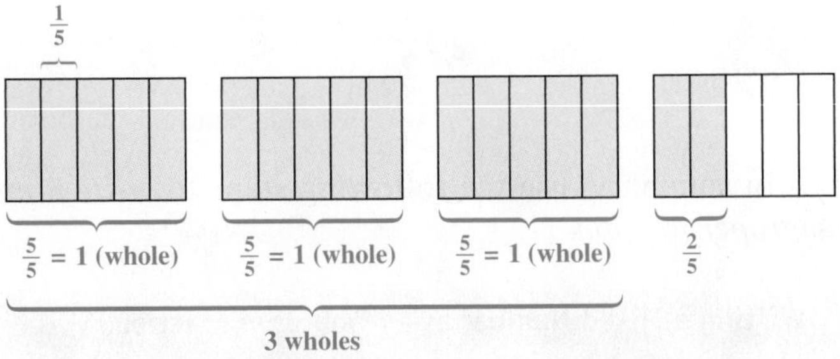

(b) $\frac{24}{4}$ Divide 24 by 4. ⟶

$$\begin{array}{r} 6 \\ 4\overline{)24} \\ \underline{24} \\ 0 \leftarrow \text{No remainder} \end{array}$$

so $\frac{24}{4} = 6$

No remainder, so no fraction part; just a whole number.

◀ **Work Problem 3 at the Side.**

2.2 Exercises

FOR EXTRA HELP

Download the MyDashBoard App

MyMathLab®

CONCEPT CHECK *Decide whether each statement is* true *or* false. *If it is* false, *explain why.*

1. The fraction $\frac{12}{12}$ is an improper fraction.

2. The fraction $\frac{2}{3}$ is a proper fraction.

3. The mixed number $7\frac{2}{5}$ can be changed to the improper fraction $\frac{14}{5}$.

4. Some mixed numbers cannot be changed to an improper fraction.

5. The mixed number $6\frac{1}{2}$ written as an improper fraction is $\frac{12}{2}$.

6. The mixed number $5\frac{5}{6}$ written as an improper fraction is $\frac{35}{6}$.

Write each mixed number as an improper fraction. ***See Example 1.***

7. $1\frac{1}{4}$

8. $2\frac{1}{2}$

9. $4\frac{3}{5}$

10. $8\frac{1}{4}$

11. $8\frac{1}{2}$

12. $1\frac{7}{11}$

13. $10\frac{1}{8}$

GS

Find the new numerator:

$10 \cdot 8 = 80$ Multiply.

$80 + 1 = 81$ Add.

$10\frac{1}{8} =$

14. $12\frac{2}{3}$

GS

Find the new numerator:

$12 \cdot 3 = 36$ Multiply.

$36 + 2 = 38$ Add.

$12\frac{2}{3} =$

15. $10\frac{3}{4}$

16. $3\frac{3}{8}$

17. $5\frac{4}{5}$

18. $2\frac{8}{9}$

19. $8\frac{3}{5}$

20. $3\frac{4}{7}$

21. $4\frac{10}{11}$

22. $11\frac{5}{8}$

23. $32\frac{3}{4}$

24. $15\frac{3}{10}$

25. $18\frac{5}{12}$

26. $19\frac{8}{11}$

27. $17\frac{14}{15}$

28. $9\frac{5}{16}$

29. $7\frac{19}{24}$

30. $9\frac{7}{12}$

CONCEPT CHECK *Decide whether each statement is* true *or* false. *If it is* false, *explain why.*

31. The improper fraction $\frac{4}{3}$ written as a mixed number is $1\frac{1}{4}$.

32. An improper fraction cannot always be written as a whole number or mixed number.

33. Some improper fractions can be written as a whole number with no fraction part.

34. The improper fraction $\frac{48}{6}$ can be written as the whole number 8.

Write each improper fraction as a whole or mixed number. **See Example 2.**

35. $\frac{4}{3}$

$$\begin{array}{r} 1 \leftarrow \text{Whole number part} \\ 3\overline{)4} \\ \underline{3} \\ 1 \leftarrow \text{Remainder} \end{array}$$

36. $\frac{11}{9}$

$$\begin{array}{r} 1 \leftarrow \text{Whole number part} \\ 9\overline{)11} \\ \underline{9} \\ 2 \leftarrow \text{Remainder} \end{array}$$

37. $\frac{9}{4}$

38. $\frac{7}{2}$

39. $\frac{54}{6}$

40. $\frac{63}{9}$

41. $\frac{38}{5}$

42. $\frac{33}{7}$

43. $\frac{63}{4}$

44. $\frac{19}{5}$

45. $\frac{47}{9}$

46. $\frac{65}{9}$

47. $\frac{65}{8}$

48. $\frac{37}{6}$

49. $\frac{84}{5}$

50. $\frac{92}{3}$

51. $\frac{112}{4}$

52. $\frac{117}{9}$

53. $\frac{183}{7}$

54. $\frac{212}{11}$

55. Your classmate asks you how to change a mixed number to an improper fraction. Write a couple of sentences and give an example to show her how this is done.

56. Explain in a sentence or two how to change an improper fraction to a mixed number. Give an example to show how this is done.

Write each mixed number as an improper fraction.

57. $250\frac{1}{2}$

58. $185\frac{3}{4}$

59. $333\frac{1}{3}$

60. $138\frac{4}{5}$

61. $522\frac{3}{8}$

62. $622\frac{1}{4}$

Write each improper fraction as a whole or mixed number.

63. $\frac{617}{4}$

64. $\frac{760}{8}$

65. $\frac{2565}{15}$

66. $\frac{2915}{16}$

67. $\frac{3917}{32}$

68. $\frac{5632}{64}$

Relating Concepts (Exercises 69–74) For Individual or Group Work

Knowing the basics of fractions is necessary in problem solving.
Work Exercises 69–74 in order.

69. Which of these fractions are proper fractions?

$\frac{2}{3}\quad \frac{4}{5}\quad \frac{8}{5}\quad \frac{3}{4}\quad \frac{6}{6}\quad \frac{7}{10}$

70. (a) The proper fractions in **Exercise 69** are the ones where the __________ is less than the __________.

(b) Draw a picture with shaded parts to show each proper fraction in **Exercise 69.**

(c) The proper fractions in **Exercise 69** are all (*less/greater*) than 1.

71. Which of these fractions are improper fractions?

$\frac{5}{5}\quad \frac{3}{4}\quad \frac{10}{3}\quad \frac{2}{3}\quad \frac{5}{6}\quad \frac{6}{5}$

72. (a) The improper fractions in **Exercise 71** are the ones where the __________ is equal to or greater than the __________.

(b) Draw a picture with shaded parts to show each improper fraction in **Exercise 71.**

(c) The improper fractions in **Exercise 71** are all equal to or (*less/greater*) than 1.

73. Identify which of these fractions can be written as whole or mixed numbers, and then write them as whole or mixed numbers.

$\frac{5}{3}\quad \frac{7}{8}\quad \frac{7}{7}\quad \frac{11}{6}\quad \frac{4}{5}\quad \frac{15}{16}$

74. (a) The fractions that can be written as whole or mixed numbers in **Exercise 73** are (*proper/improper*) fractions, and their value is always (*less than/greater than or equal to*) 1.

(b) Draw a picture with shaded parts to show each whole or mixed number in **Exercise 73.**

2.3 Factors

OBJECTIVES

1. Find factors of a number.
2. Identify prime numbers and composite numbers.
3. Find prime factorizations.

VOCABULARY TIP

Factors of a number A *factor* is a whole number that divides *exactly* into a whole number, leaving *no* remainder.

1 Find all the whole number factors of each number.

GS **(a)** 18

1, 2, ____ , 6, ____ , 18

(b) 16

(c) 36

(d) 80

VOCABULARY TIP

Prime numbers Because a prime number can be divided evenly only by itself and 1, it is the basic building block of all numbers.

2 Which of the following are prime?

4, 7, 9, 13, 17, 19, 29, 33

Answers

1. **(a)** 3; 9; 1, 2, 3, 6, 9, 18 **(b)** 1, 2, 4, 8, 16 **(c)** 1, 2, 3, 4, 6, 9, 12, 18, 36 **(d)** 1, 2, 4, 5, 8, 10, 16, 20, 40, 80

2. 7, 13, 17, 19, 29

OBJECTIVE 1 **Find factors of a number.** You will recall that numbers multiplied to give a product are called **factors.** Because $2 \cdot 5 = 10$, both 2 and 5 are factors of 10. The numbers 1 and 10 are also factors of 10, because $1 \cdot 10 = 10$. The various tests for divisibility show that 1, 2, 5, and 10 are the only whole number factors of 10. The products $2 \cdot 5$ and $1 \cdot 10$ are called **factorizations** of 10.

Note

The tests to decide whether one number is divisible by another number were shown in **Chapter 1.** You might want to review these. The tests that you will use most often are those for 2, 3, 5, and 10.

EXAMPLE 1 Using Factors

Find all possible two-number factorizations of each number.

(a) 12

$1 \cdot 12 = 12 \qquad 2 \cdot 6 = 12 \qquad 3 \cdot 4 = 12$

The factors of 12 are 1, 2, 3, 4, 6, and 12.

(b) 60

$1 \cdot 60 = 60 \qquad 2 \cdot 30 = 60$
$3 \cdot 20 = 60 \qquad 4 \cdot 15 = 60$
$5 \cdot 12 = 60 \qquad 6 \cdot 10 = 60$

The factors of a number all divide evenly into that number.

The factors of 60 are 1, 2, 3, 4, 5, 6, 10, 12, 15, 20, 30, and 60.

◀ Work Problem **1** at the Side.

OBJECTIVE 2 **Identify prime numbers and composite numbers.** Whole numbers that have only two factors are called **prime numbers.** They can only be divided evenly by themselves and 1.

Prime Numbers

A **prime number** is a whole number that has exactly ***two different*** factors, ***itself*** and ***1.***

The number 3 is a prime number, since it can be divided evenly only by itself and 1. The number 8 is **not** a prime number (it is composite), since 8 can be divided evenly by 2 and 4, as well as by itself and 1.

CAUTION

A prime number has ***only two*** different factors, itself and 1. The number 1 is not a prime number because it does not have *two different* factors; the only factor of 1 is 1.

EXAMPLE 2 Finding Prime Numbers

Which of the following numbers are prime?

2 5 11 15 27

A prime number can be divided evenly only by the number itself and by 1.

The number 15 can be divided by 3 and 5, so it is **not** prime. Also, because 27 can be divided by 3 and 9, then 27 is **not** prime. The other numbers in the list, 2, 5, and 11, are divisible only by themselves and 1, so they are prime.

◀ Work Problem **2** at the Side.

Composite Numbers

A number that is not prime is called a **composite number.** The numbers 0 and 1 are neither prime nor composite.

EXAMPLE 3 Identifying Composite Numbers

Which of the following numbers are composite?

(a) 6

Because 6 has factors of 2 and 3, as well as 6 and 1, the number 6 is composite.

(b) 11

The number 11 has only two factors, 11 and 1. It is *not* composite. (It is a prime number.)

(c) 25

Because 25 has a factor of 5, as well as 25 and 1, 25 is composite.

Work Problem 3 at the Side. ▶

OBJECTIVE 3 Find prime factorizations. For reference, here are the prime numbers less than 50.

2	3	5	7	11
13	17	19	23	29
31	37	41	43	47

These are the prime numbers less than 50.

The **prime factorization** of a number can be especially useful when we are adding or subtracting fractions and need to find a common denominator or write a fraction in lowest terms.

Prime Factorization

A **prime factorization** of a number is a factorization in which every factor is a *prime number.*

EXAMPLE 4 Determining the Prime Factorization

Find the prime factorization of 12.

Try to divide 12 by the first prime, 2.

$$12 \div 2 = 6,$$

First prime (2)

so

$$12 = 2 \cdot 6.$$

Try to divide 6 by the prime, 2.

$$6 \div 2 = 3,$$

so

$$12 = 2 \cdot \underbrace{2 \cdot 3}.$$

Factorization of 6

Because all factors are prime, the prime factorization of 12 is

$$2 \cdot 2 \cdot 3.$$

All these factors are prime numbers.

Work Problem 4 at the Side. ▶

VOCABULARY TIP

Composite number A number that is not a prime number is a composite number.

3 Which of these numbers are composite?

2, 4, 5, 6, 8, 10, 11, 13, 19, 21, 27, 28, 33, 36, 42

VOCABULARY TIP

Prime factorization Every whole number has its own unique prime factorization. The order of the factors may differ, but the factors themselves will be unique to that number.

4 Find the prime factorization of each number.

GS **(a)** 8

2 • ____ • ____

GS **(b)** 28

2 • ____ • ____

(c) 18

(d) 40

Answers

3. 4, 6, 8, 10, 21, 27, 28, 33, 36, 42

4. **(a)** 2; 2; 2 • 2 • 2 **(b)** 2; 7; 2 • 2 • 7 **(c)** 2 • 3 • 3 **(d)** 2 • 2 • 2 • 5

5 Find the prime factorization of each number. Write the factorization with exponents.

GS **(a)** 36

$2 \cdot 2 \cdot 3 \cdot$ ____

$2^2 \cdot$ ____

(b) 54

(c) 60

(d) 81

6 Write the prime factorization of each number using exponents.

(a) 48

(b) 44

(c) 90

(d) 120

(e) 180

Answers

5. **(a)** 3; 3^2; $2^2 \cdot 3^2$ **(b)** $2 \cdot 3^3$ **(c)** $2^2 \cdot 3 \cdot 5$ **(d)** 3^4

6. **(a)** $2^4 \cdot 3$ **(b)** $2^2 \cdot 11$ **(c)** $2 \cdot 3^2 \cdot 5$ **(d)** $2^3 \cdot 3 \cdot 5$ **(e)** $2^2 \cdot 3^2 \cdot 5$

CAUTION

All prime numbers are odd numbers except the number 2. Be careful, though, because *all odd numbers are not prime numbers.* For example, 9, 15, and 21 are odd numbers, but are *not* prime numbers.

EXAMPLE 5 Factoring by Using the Division Method

Find the prime factorization of 48.

1	Continue to divide until the quotient is 1.
$3\overline{)3}$	Divide 3 by 3 (second prime).
$2\overline{)6}$	Divide 6 by 2.
$2\overline{)12}$	Divide 12 by 2.
$2\overline{)24}$	Divide 24 by 2.
$2\overline{)48}$	Divide 48 by 2. (first prime)

The divisors are all prime factors.

Because all factors (divisors) are prime, the prime factorization of 48 is

$$2 \cdot 2 \cdot 2 \cdot 2 \cdot 3.$$

In **Chapter 1,** we wrote $2 \cdot 2 \cdot 2 \cdot 2$ as 2^4, so the prime factorization of 48 can be written, using exponents, as

$$48 = 2 \cdot 2 \cdot 2 \cdot 2 \cdot 3 = 2^4 \cdot 3.$$

◀ **Work Problem 5 at the Side.**

Note

When using the division method of factoring, the last quotient found is 1. The "1" is never used as a prime factor because 1 is neither prime nor composite. Besides, 1 times any number is the number itself.

EXAMPLE 6 Using Exponents with Prime Factorization

Find the prime factorization of 225.

1	Continue to divide until the quotient is 1.
$5\overline{)5}$	Divide 5 by 5.
$5\overline{)25}$	25 is not divible by 3; use 5.
$3\overline{)75}$	Divide 75 by 3.
$3\overline{)225}$	225 is not divisible by 2; use 3.

1 is not a prime factor.

All the divisors are prime factors.

Write the prime factorization.

$$225 = 3 \cdot 3 \cdot 5 \cdot 5$$

Or, using exponents,

$$225 = 3^2 \cdot 5^2$$

◀ **Work Problem 6 at the Side.**

Another method of factoring is a *factor tree*.

EXAMPLE 7 Factoring by Using a Factor Tree

Find the prime factorization of each number using a factor tree.

(a) 30

Try to divide by the first prime, 2. Write the factors under the 30. Circle the 2, since it is a prime.

Since 15 cannot be divided evenly by 2, try the next prime, 3.

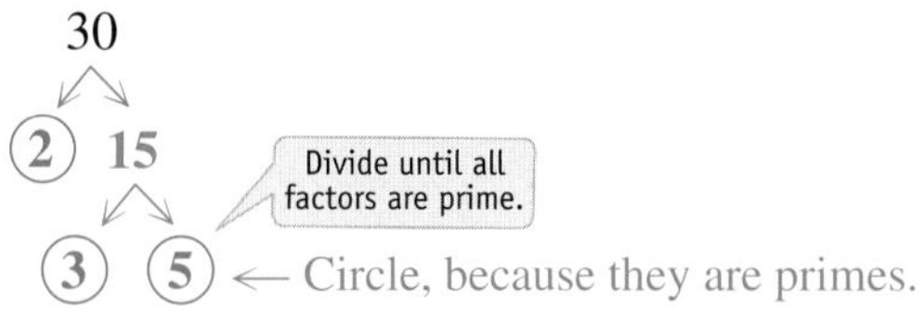

No uncircled factors remain, so the prime factorization (the circled factors) has been found.

$$30 = \mathbf{2} \cdot \mathbf{3} \cdot \mathbf{5}$$

(b) 24

Divide by 2.

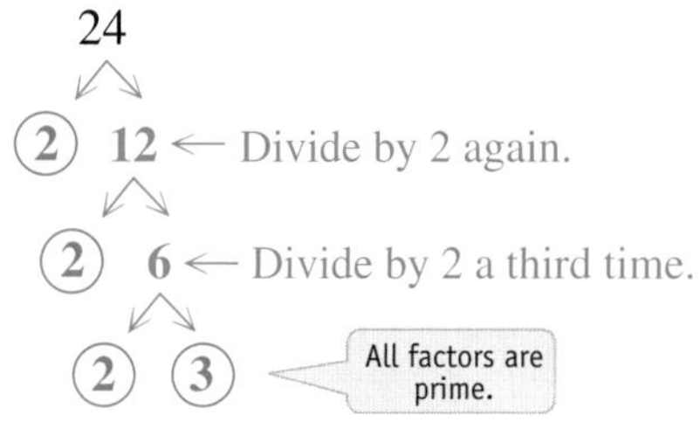

$24 = \mathbf{2} \cdot \mathbf{2} \cdot \mathbf{2} \cdot \mathbf{3}$ or, using exponents, $24 = \mathbf{2^3} \cdot \mathbf{3}$

(c) 45

Because 45 cannot be divided by 2, try 3.

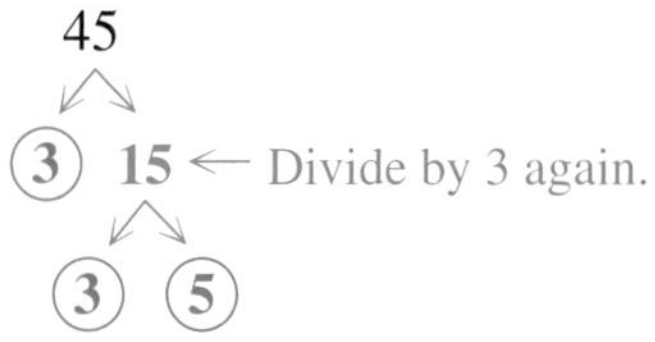

$45 = \mathbf{3} \cdot \mathbf{3} \cdot \mathbf{5}$ or, using exponents, $45 = \mathbf{3^2} \cdot \mathbf{5}$

Note

The diagrams used in **Example 7** look like tree branches, and that is why this method is referred to as using a factor tree.

Work Problem 7 at the Side. ▶

7 Complete each factor tree and give the prime factorization.

(a)

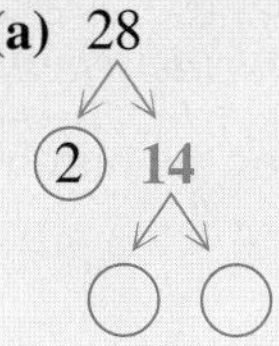

(b)

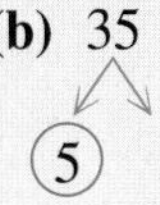

(c)

Answers

7. (a)

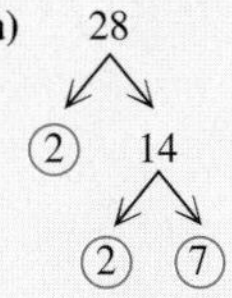

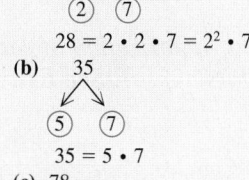

$28 = 2 \cdot 2 \cdot 7 = 2^2 \cdot 7$

(b)

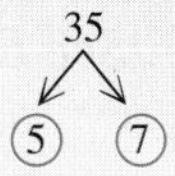

$35 = 5 \cdot 7$

(c)

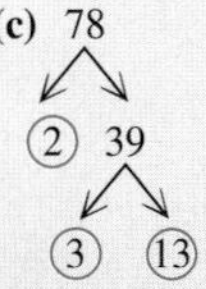

$78 = 2 \cdot 3 \cdot 13$

2.3 Exercises

FOR EXTRA HELP

Download the MyDashBoard App

MyMathLab®

CONCEPT CHECK *Decide whether the the following are* true *or* false.

1. All the factors for the number 8 are 2, 4, 6, and 8.

2. All the factors for the number 12 are 1, 2, 3, 4, 6, and 12.

3. All the factors for the number 15 are 1, 3, 5, and 15.

4. All the factors for the number 28 are 2, 4, 7, and 14.

Find all the factors of each number. **See Example 1.**

5. 48

6. 30

7. 56 Factors are: 1, 2, 4, 7, 8, 14, 28, ____

8. 72 Factors are: 1, 2, 3, 4, 6, 8, 9, 12, 18, 24, 36, ____

9. 36

10. 20

11. 40

12. 60

13. 64

14. 84

15. 82

16. 39

CONCEPT CHECK *Decide whether each number is* prime *or* composite.

17. 6

18. 9

19. 5

20. 16

21. 10

22. 13

23. 19

24. 17

25. 25

26. 48

27. 47

28. 45

CONCEPT CHECK *Choose the correct prime factorization of each number.*

29. 40
- **A.** $2 \cdot 4 \cdot 6$
- **B.** $2^3 \cdot 5$
- **C.** $2 \cdot 4 \cdot 5$

30. 36
- **A.** $3 \cdot 12$
- **B.** $2^2 \cdot 6$
- **C.** $2^2 \cdot 3^2$

31. 100
- **A.** $2^2 \cdot 5^2$
- **B.** $5^2 \cdot 4$
- **C.** $2 \cdot 5 \cdot 10$

Find the prime factorization of each number. Write answers with exponents when repeated factors appear. **See Examples 4–7.**

32. 8

33. 6

34. 20

35. 25

$5 \cdot \_\_ = 25$

36. 56

$\_\_ \cdot \_\_ \cdot 2 \cdot 7 = 56$

37. 68

38. 70

39. 72

40. 64

41. 44

42. 104

43. 100

44. 112

45. 125

46. 135

47. 180

48. 300

49. 320

50. 480

51. 360

52. 400

53. Give a definition in your own words of both a prime number and a composite number. Give three examples of each. Which whole numbers are neither prime nor composite?

54. With the exception of the number 2, all prime numbers are odd numbers. Nevertheless, all odd numbers are not prime numbers. Explain why these statements are true.

55. Explain the difference between finding all possible factors of 24 and finding the prime factorization of 24.

56. Use the division method to find the prime factorization of 36. Can you divide by 3s before you divide by 2s? Does the order of division change the answers?

Find the prime factorization of each number. Write answers using exponents.

57. 350

58. 640

59. 960

60. 1000

61. 1560

62. 2000

63. 1260

64. 2200

Relating Concepts (Exercises 65–70) For Individual or Group Work

An understanding of factors and factorization will be needed to solve fraction problems.
Work Exercises 65–70 in order.

65. A prime number is a whole number that has exactly two different factors, itself and 1. List all prime numbers less than 50.

66. Explain what it is about the numbers in **Exercise 65** that makes them prime.

67. The number 2 is an even number and a prime number. Can any other even numbers be prime numbers? Explain.

68. Can a multiple of a prime number be prime (for example, 6, 9, 12, and 15 are multiples of 3)? Explain.

69. Find the prime factorization of 2100. Do not use exponents in your answer.

70. Write the answer to **Exercise 69** using exponents for repeated factors.

2.4 Writing a Fraction in Lowest Terms

OBJECTIVES

1. Tell whether a fraction is written in lowest terms.
2. Write a fraction in lowest terms using common factors.
3. Write a fraction in lowest terms using prime factors.
4. Determine whether two fractions are equivalent.

When working problems involving fractions, we must often compare two fractions to determine whether they represent the same portion of a whole. Look at the two cases of soda.

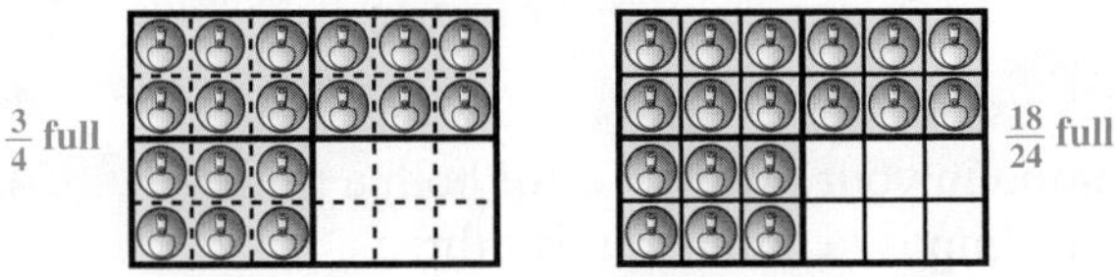

The cases of soda show areas that are $\frac{3}{4}$ full and $\frac{18}{24}$ full. Because the full areas are equivalent (the same portion), the fractions $\frac{3}{4}$ and $\frac{18}{24}$ are **equivalent fractions.** They each represent the same portion of the whole.

$$\frac{3}{4} = \frac{18}{24}$$

Because the numbers 18 and 24 both have 6 as a factor, 6 is called a **common factor** of the numbers. Other common factors of 18 and 24 are 1, 2, and 3.

◄ Work Problem 1 at the Side.

VOCABULARY TIP

Equivalent fractions The word *equivalent* means being of equal value. So, equivalent fractions have the same value, even though they may look different.

VOCABULARY TIP

Common factor A number also called the common divisor.

1 Decide whether the number in blue is a common factor of the other two numbers.

(a) 6, 12; 2 (b) 32, 64; 8

(c) 32, 56; 16 (d) 75, 81; 1

OBJECTIVE 1 Tell whether a fraction is written in lowest terms. The fraction $\frac{3}{4}$ is written in *lowest terms* because the numerator and denominator have no common factor other than 1. However, the fraction $\frac{18}{24}$ is ***not*** in lowest terms because its numerator and denominator have common factors of 6, 3, 2, and 1.

Writing a Fraction in Lowest Terms

A fraction is written in **lowest terms** when the numerator and denominator have no common factor other than 1.

VOCABULARY TIP

Lowest terms The process of reducing a fraction to lowest terms is also called *simplifying* the fraction.

2 Are the following fractions in lowest terms?

(a) $\frac{4}{5}$ (b) $\frac{6}{18}$

(c) $\frac{9}{15}$ (d) $\frac{17}{46}$

EXAMPLE 1 Understanding Lowest Terms

Are the following fractions in lowest terms?

(a) $\frac{3}{8}$ ← No common factors other than 1.

The numerator and denominator have no common factor other than 1, so the fraction is in lowest terms.

(b) $\frac{21}{36}$ ← 3 is a common factor of 21 and 36.

The numerator and denominator have a common factor of 3, so the fraction is **not in lowest terms.**

◄ Work Problem 2 at the Side.

OBJECTIVE 2 Write a fraction in lowest terms using common factors. There are two common methods for writing a fraction in lowest terms. These methods are shown in the next examples. The first method works best when the numerator and denominator are small numbers.

Answers

1. (a) yes (b) yes (c) no (d) yes
2. (a) yes (b) no (c) no (d) yes

EXAMPLE 2 Writing Fractions in Lowest Terms

Write each fraction in lowest terms.

(a) $\frac{18}{24}$

The greatest common factor of 18 and 24 is 6. Divide both numerator and denominator by 6.

$$\frac{18}{24} = \frac{18 \div 6}{24 \div 6} = \frac{3}{4}$$

Divide numerator and denominator by the greatest common factor, 6.

(b) $\frac{30}{50} = \frac{30 \div 10}{50 \div 10} = \frac{3}{5}$ Divide both numerator and denominator by 10.

(c) $\frac{24}{42} = \frac{24 \div 6}{42 \div 6} = \frac{4}{7}$ Divide both numerator and denominator by 6.

(d) $\frac{60}{72}$

Suppose we thought that 4 was the greatest common factor of 60 and 72. Dividing by 4 would give

$$\frac{60}{72} = \frac{60 \div 4}{72 \div 4} = \frac{15}{18}. \leftarrow \text{Not in lowest terms}$$

But $\frac{15}{18}$ is **not** in lowest terms, because 15 and 18 have a common factor of 3. So we divide by 3.

$$\frac{15}{18} = \frac{15 \div 3}{18 \div 3} = \frac{5}{6}$$

Continue dividing until there is no common factor other than 1.

The fraction $\frac{60}{72}$ could have been written in lowest terms in one step by dividing by 12, the greatest common factor of 60 and 72.

$$\frac{60}{72} = \frac{60 \div 12}{72 \div 12} = \frac{5}{6} \leftarrow \text{Same answer as above}$$

Continue dividing until the fraction is in lowest terms.

Note

Dividing the numerator and denominator by the same number results in an equivalent fraction.

In **Example 2,** we wrote fractions in lowest terms by dividing by a common factor. This method is summarized in the following steps.

The Method of Dividing by a Common Factor

Step 1 Find a number that will divide evenly into both the numerator and denominator. This number is a ***common factor.***

Step 2 ***Divide*** both numerator and denominator by the common factor.

Step 3 ***Check*** to see whether the new fraction has any common factors (besides 1). If it does, repeat *Steps 2 and 3.* If the only common factor is 1, the fraction is in lowest terms.

Work Problem 3 at the Side. ▶

3 Write in lowest terms.

GS **(a)** $\frac{8}{16}$

$$\frac{8}{16} = \frac{8 \div 8}{16 \div \_\_\_} = \frac{1}{\_\_\_}$$

GS **(b)** $\frac{9}{12}$

$$\frac{9}{12} = \frac{9 \div \_\_\_}{12 \div 3} = \frac{\_\_\_}{4}$$

(c) $\frac{28}{42}$

(d) $\frac{30}{80}$

(e) $\frac{16}{40}$

Answers

3. **(a)** 8; 2; $\frac{1}{2}$ **(b)** 3; 3; $\frac{3}{4}$ **(c)** $\frac{2}{3}$ **(d)** $\frac{3}{8}$ **(e)** $\frac{2}{5}$

OBJECTIVE 3 Write a fraction in lowest terms using prime factors. The method of writing a fraction in lowest terms by division works well for fractions with small numerators and denominators. For larger numbers, when common factors are not obvious, use the method of *prime factors,* which is shown in the next example.

EXAMPLE 3 Using Prime Factors

Write each fraction in lowest terms.

(a) $\dfrac{24}{42}$

Write the prime factorization of both numerator and denominator. See the previous section for help.

$$\frac{24}{42} = \frac{2 \cdot 2 \cdot 2 \cdot 3}{2 \cdot 3 \cdot 7}$$

Just as with the method used in **Example 2,** divide both numerator and denominator by any common factors. Write a 1 by each factor that has been divided.

2 ÷ 2 is 1 3 ÷ 3 is 1

$$\frac{24}{42} = \frac{\overset{1}{\cancel{2}} \cdot 2 \cdot 2 \cdot \overset{1}{\cancel{3}}}{\underset{1}{\cancel{2}} \cdot \underset{1}{\cancel{3}} \cdot 7}$$

Multiply the remaining factors in both numerator and denominator.

$$\frac{24}{42} = \frac{1 \cdot 2 \cdot 2 \cdot 1}{1 \cdot 1 \cdot 7} = \frac{4}{7}$$

$\frac{4}{7}$ is equivalent to $\frac{24}{42}$ but is written in lowest terms.

Finally, $\frac{24}{42}$ written in lowest terms is $\frac{4}{7}$.

(b) $\dfrac{162}{54}$

Write the prime factorization of both numerator and denominator.

$$\frac{162}{54} = \frac{2 \cdot 3 \cdot 3 \cdot 3 \cdot 3}{2 \cdot 3 \cdot 3 \cdot 3}$$

Now divide by the common factors. ***Do not forget to write the 1s.***

$$\frac{162}{54} = \frac{\overset{1}{\cancel{2}} \cdot \overset{1}{\cancel{3}} \cdot \overset{1}{\cancel{3}} \cdot \overset{1}{\cancel{3}} \cdot 3}{\underset{1}{\cancel{2}} \cdot \underset{1}{\cancel{3}} \cdot \underset{1}{\cancel{3}} \cdot \underset{1}{\cancel{3}}}$$

Remember to write in the 1s when dividing by a common factor.

$$= \frac{1 \cdot 1 \cdot 1 \cdot 1 \cdot 3}{1 \cdot 1 \cdot 1 \cdot 1} = \frac{3}{1} = 3$$

(c) $\dfrac{18}{90}$

$$\frac{18}{90} = \frac{\overset{1}{\cancel{2}} \cdot \overset{1}{\cancel{3}} \cdot \overset{1}{\cancel{3}}}{\underset{1}{\cancel{2}} \cdot \underset{1}{\cancel{3}} \cdot \underset{1}{\cancel{3}} \cdot 5} = \frac{1 \cdot 1 \cdot 1}{1 \cdot 1 \cdot 1 \cdot 5} = \frac{1}{5}$$

All factors of the numerator were divided, and $1 \cdot 1 \cdot 1 = 1$.

Continued on Next Page

In **Example 3,** we wrote fractions in lowest terms using prime factors. This method is summarized as follows.

The Method of Prime Factors

Step 1 Write the ***prime factorization*** of both numerator and denominator.

Step 2 Use slashes to show you are ***dividing*** both numerator and denominator by common factors.

Step 3 ***Multiply*** the remaining factors in the numerator and denominator.

Work Problem 4 at the Side. ▶

OBJECTIVE 4 Determine whether two fractions are equivalent. The next example shows how to decide whether two fractions are equivalent.

EXAMPLE 4 Determining Whether Two Fractions Are Equivalent

Determine whether each pair of fractions is equivalent. In other words, do both fractions represent the same part of a whole?

(a) $\frac{16}{48}$ and $\frac{24}{72}$

Use the method of prime factors to write each fraction in lowest terms.

$$\frac{16}{48} = \frac{\overset{1}{\cancel{2}} \cdot \overset{1}{\cancel{2}} \cdot \overset{1}{\cancel{2}} \cdot \overset{1}{\cancel{2}}}{\underset{1}{\cancel{2}} \cdot \underset{1}{\cancel{2}} \cdot \underset{1}{\cancel{2}} \cdot \underset{1}{\cancel{2}} \cdot 3} = \frac{1 \cdot 1 \cdot 1 \cdot 1}{1 \cdot 1 \cdot 1 \cdot 1 \cdot 3} = \frac{1}{3}$$

Equivalent $\left(\frac{1}{3} = \frac{1}{3}\right)$

$$\frac{24}{72} = \frac{\overset{1}{\cancel{2}} \cdot \overset{1}{\cancel{2}} \cdot \overset{1}{\cancel{2}} \cdot \overset{1}{\cancel{3}}}{\underset{1}{\cancel{2}} \cdot \underset{1}{\cancel{2}} \cdot \underset{1}{\cancel{2}} \cdot \underset{1}{\cancel{3}} \cdot 3} = \frac{1 \cdot 1 \cdot 1 \cdot 1}{1 \cdot 1 \cdot 1 \cdot 1 \cdot 3} = \frac{1}{3}$$

(b) $\frac{32}{52}$ and $\frac{64}{112}$

$$\frac{32}{52} = \frac{\overset{1}{\cancel{2}} \cdot \overset{1}{\cancel{2}} \cdot 2 \cdot 2 \cdot 2}{\underset{1}{\cancel{2}} \cdot \underset{1}{\cancel{2}} \cdot 13} = \frac{2 \cdot 2 \cdot 2}{1 \cdot 1 \cdot 13} = \frac{8}{13}$$

Not equivalent $\left(\frac{8}{13} \neq \frac{4}{7}\right)$

$$\frac{64}{112} = \frac{\overset{1}{\cancel{2}} \cdot \overset{1}{\cancel{2}} \cdot \overset{1}{\cancel{2}} \cdot \overset{1}{\cancel{2}} \cdot 2 \cdot 2}{\underset{1}{\cancel{2}} \cdot \underset{1}{\cancel{2}} \cdot \underset{1}{\cancel{2}} \cdot \underset{1}{\cancel{2}} \cdot 7} = \frac{1 \cdot 1 \cdot 1 \cdot 1 \cdot 2 \cdot 2}{1 \cdot 1 \cdot 1 \cdot 1 \cdot 7} = \frac{4}{7}$$

(c) $\frac{75}{15}$ and $\frac{60}{12}$

$$\frac{75}{15} = \frac{\overset{1}{\cancel{3}} \cdot \overset{1}{\cancel{5}} \cdot 5}{\underset{1}{\cancel{3}} \cdot \underset{1}{\cancel{5}}} = \frac{1 \cdot 1 \cdot 5}{1 \cdot 1} = 5$$

Equivalent (5 = 5)

$$\frac{60}{12} = \frac{\overset{1}{\cancel{2}} \cdot \overset{1}{\cancel{2}} \cdot \overset{1}{\cancel{3}} \cdot 5}{\underset{1}{\cancel{2}} \cdot \underset{1}{\cancel{2}} \cdot \underset{1}{\cancel{3}}} = \frac{1 \cdot 1 \cdot 1 \cdot 5}{1 \cdot 1 \cdot 1} = 5$$

Work Problem 5 at the Side. ▶

4 Use the method of prime factors to write each fraction in lowest terms.

GS **(a)** $\frac{12}{36}$

$$\frac{12}{36} = \frac{2 \cdot 2 \cdot 3}{2 \cdot \_\_\_ \cdot \_\_\_ \cdot 3}$$

$$\frac{\overset{1}{\cancel{2}} \cdot \overset{1}{\cancel{2}} \cdot \overset{1}{\cancel{3}}}{\underset{1}{\cancel{2}} \cdot \underset{1}{\cancel{2}} \cdot \underset{1}{\cancel{3}} \cdot 3} = \frac{\_\_\_}{3}$$

GS **(b)** $\frac{32}{56} = \frac{2 \cdot 2 \cdot 2 \cdot 2 \cdot \_\_\_}{2 \cdot 2 \cdot \_\_\_ \cdot 7}$

$$\frac{\overset{1}{\cancel{2}} \cdot \overset{1}{\cancel{2}} \cdot \overset{1}{\cancel{2}} \cdot 2 \cdot 2}{\underset{1}{\cancel{2}} \cdot \underset{1}{\cancel{2}} \cdot \underset{1}{\cancel{2}} \cdot 7} = \frac{4}{\_\_\_}$$

(c) $\frac{74}{111}$ **(d)** $\frac{124}{340}$

5 Is each pair of fractions equivalent?

(a) $\frac{24}{48}$ and $\frac{36}{72}$

(b) $\frac{45}{60}$ and $\frac{50}{75}$

(c) $\frac{20}{4}$ and $\frac{110}{22}$

(d) $\frac{120}{220}$ and $\frac{180}{320}$

Answers

4. (a) 2; 3; $\frac{1}{3}$ **(b)** 2; 2; 7; $\frac{4}{7}$
(c) $\frac{2}{3}$ **(d)** $\frac{31}{85}$

5. (a) equivalent **(b)** not equivalent
(c) equivalent **(d)** not equivalent

2.4 Exercises

FOR EXTRA HELP

Download the MyDashBoard App

MyMathLab®

CONCEPT CHECK *Underline the correct answer or fill in the blank in each of the following.*

1. A number can be divided by 2 if the number is an (*odd/even*) number.

2. A number can be divided by 5 if the number ends in _______ or _______.

3. Any number can be divided by 10 if the number ends in _______.

4. If the sum of a number's digits is divisible by _______, the number is divisible by 3.

Put a ✓ mark in the blank if the number at the left is divisible by the number at the top. Put an ✗ in the blank if the number is not divisible by the number at the top.

	2	3	5	10
5. 60	_____	_____	_____	_____
7. 48	_____	_____	_____	_____
9. 160	_____	_____	_____	_____
11. 138	_____	_____	_____	_____

	2	3	5	10
6. 90	_____	_____	_____	_____
8. 36	_____	_____	_____	_____
10. 175	_____	_____	_____	_____
12. 150	_____	_____	_____	_____

CONCEPT CHECK *Decide* true *or* false *whether each pair of fractions is equivalent.*

13. $\frac{6}{8} = \frac{3}{4}$

14. $\frac{7}{12} = \frac{1}{2}$

15. $\frac{3}{8} = \frac{5}{16}$

16. $\frac{4}{12} = \frac{1}{3}$

Write each fraction in lowest terms. **See Example 2.**

17. $\frac{15}{25}$ GS

$\frac{15}{25} = \frac{15 \div 5}{25 \div 5} =$

18. $\frac{32}{48}$ GS

$\frac{32}{48} = \frac{32 \div 16}{48 \div 16} =$

19. $\frac{36}{42}$

20. $\frac{22}{33}$

21. $\frac{56}{64}$

22. $\frac{21}{35}$

23. $\frac{180}{210}$

24. $\frac{72}{80}$

25. $\frac{72}{126}$

26. $\frac{73}{146}$

27. $\frac{12}{600}$

28. $\frac{8}{400}$

29. $\frac{96}{132}$

30. $\frac{165}{180}$

31. $\frac{60}{108}$

32. $\frac{112}{128}$

Write the numerator and denominator of each fraction as a product of prime factors and divide by the common factors. Then write the fraction in lowest terms. ***See Example 3.***

33. $\frac{18}{24}$ **34.** $\frac{16}{64}$ **35.** $\frac{35}{40}$

36. $\frac{20}{32}$ **37.** $\frac{90}{180}$ **38.** $\frac{36}{48}$

39. $\frac{36}{12}$ **40.** $\frac{192}{48}$

41. $\frac{72}{225}$ **42.** $\frac{65}{234}$

Write each fraction in lowest terms. Then state whether the fractions are equivalent or not equivalent. ***See Example 4.***

43. $\frac{3}{6}$ and $\frac{18}{36}$ $\quad \frac{3 \div 3}{6 \div 3} = \frac{\_\_\_}{\_\_\_}$ and $\frac{18 \div \_\_\_}{36 \div \_\_\_} = \frac{\_\_\_}{\_\_\_}$

44. $\frac{3}{8}$ and $\frac{27}{72}$ $\quad \frac{3 \div \_\_\_}{8 \div \_\_\_} = \frac{\_\_\_}{\_\_\_}$ and $\frac{27 \div 9}{72 \div 9} = \frac{\_\_\_}{\_\_\_}$

45. $\frac{10}{24}$ and $\frac{12}{30}$

46. $\frac{15}{35}$ and $\frac{18}{40}$ **47.** $\frac{15}{24}$ and $\frac{35}{52}$ **48.** $\frac{21}{33}$ and $\frac{9}{12}$

49. $\frac{14}{16}$ and $\frac{35}{40}$ **50.** $\frac{27}{90}$ and $\frac{24}{80}$ **51.** $\frac{48}{6}$ and $\frac{72}{8}$

52. $\frac{33}{11}$ and $\frac{72}{24}$ **53.** $\frac{25}{30}$ and $\frac{65}{78}$ **54.** $\frac{24}{72}$ and $\frac{30}{90}$

55. What does it mean when a fraction is written in lowest terms? Give three examples.

56. Explain what equivalent fractions are, and give an example of a pair of equivalent fractions. Show that they are equivalent.

Write each fraction in lowest terms.

57. $\frac{160}{256}$ **58.** $\frac{363}{528}$ **59.** $\frac{238}{119}$ **60.** $\frac{570}{95}$

Study Skills
USING STUDY CARDS

OBJECTIVES

1. **Create study cards for all new terms.**
2. **Create study cards for new procedures.**

You may have used "flash cards" in other classes before. In math, study cards can be helpful, too. However, they are different because the main things to remember in math are *not* necessarily terms and definitions; they are *sets of steps to follow* to solve problems (and how to know which set of steps to follow) and *concepts about how math works* (principles). So, the cards will look different but will be just as useful.

In this two-part activity, you will find four types of study cards to use in math. Look carefully at what kinds of information to put on them and where to put it. Then use them the way you would any flash card:

- to quickly review when you have a few minutes,
- to do daily reviews,
- to review before a test.

Remember, the most helpful thing about study cards is making them. While you are making them, you have to do the kind of thinking that is most brain friendly which improves your neural network of dendrites. After each card description you will find an assignment to try. It is marked **NOW TRY THIS.**

Why Are Study Cards Brain Friendly?

- Making cards is **active.**
- Cards are **visually** appealing.
- **Repetition** is good for your brain.

For details see Using Study Cards Revisited following **Section 2.7.**

New Vocabulary Cards

For **new vocabulary cards,** put the word (spelled correctly) and the page number where it is found on the front of the card. On the back, write:

- the definition (in your own words if possible),
- an example, an exception (if there are any),
- any related words, and
- a sample problem (if appropriate).

Front of Card

Prime Numbers *p. 126*

Back of Card

Definition: Whole numbers that can only be divided by themselves and 1.

★Must divide evenly, NO remainders!

Ex: 2, 3, 5, 7, 11, 13, 17 are the first few primes.
→ NOT 0 or 1
→ Used in factoring
→ Related word: composite number

Procedure ("Steps") Cards

For **procedure cards,** write the name of the procedure at the top on the front of the card. Then write each step *in words*. If you need to know abbreviations for some words, include them along with the whole words written out. On the back, put an example of the procedure, showing each step you need to take. You can review by looking at the front and practicing a new worked example, or by looking at the back and remembering what the procedure is called and what the steps are.

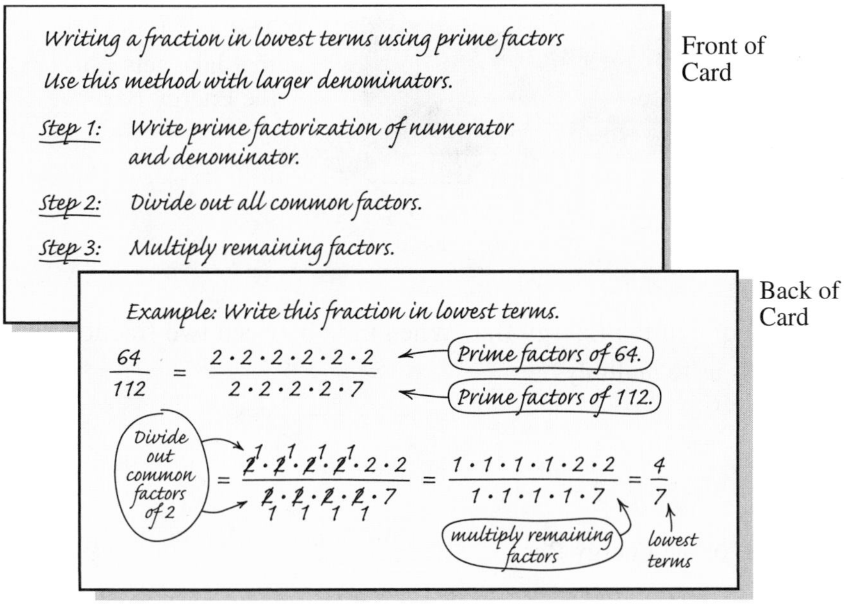

Front of Card

> *Writing a fraction in lowest terms using prime factors*
>
> *Use this method with larger denominators.*
>
> *Step 1:* *Write prime factorization of numerator and denominator.*
>
> *Step 2:* *Divide out all common factors.*
>
> *Step 3:* *Multiply remaining factors.*

Back of Card

> *Example: Write this fraction in lowest terms.*
>
> $$\frac{64}{112} = \frac{2 \cdot 2 \cdot 2 \cdot 2 \cdot 2 \cdot 2}{2 \cdot 2 \cdot 2 \cdot 2 \cdot 7}$$
>
> (Prime factors of 64.)
> (Prime factors of 112.)
>
> Divide out common factors of 2
>
> $$= \frac{\cancel{2}^1 \cdot \cancel{2}^1 \cdot \cancel{2}^1 \cdot \cancel{2}^1 \cdot 2 \cdot 2}{\cancel{2}_1 \cdot \cancel{2}_1 \cdot \cancel{2}_1 \cdot \cancel{2}_1 \cdot 7} = \frac{1 \cdot 1 \cdot 1 \cdot 1 \cdot 2 \cdot 2}{1 \cdot 1 \cdot 1 \cdot 1 \cdot 7} = \frac{4}{7}$$
>
> (multiply remaining factors) lowest terms

2.5 Multiplying Fractions

OBJECTIVES

1. Multiply fractions.
2. Use a multiplication shortcut.
3. Multiply a fraction and a whole number.
4. Find the area of a rectangle.

1 Use these figures to find $\frac{1}{4}$ of $\frac{1}{2}$.

GS

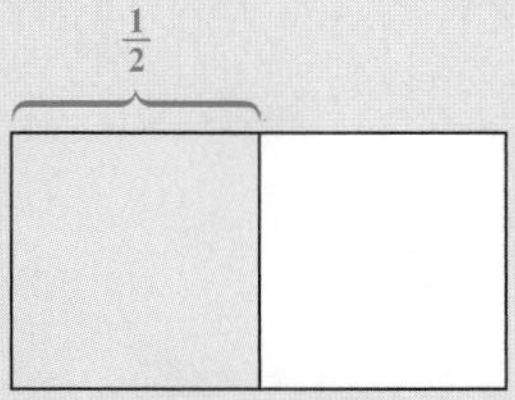

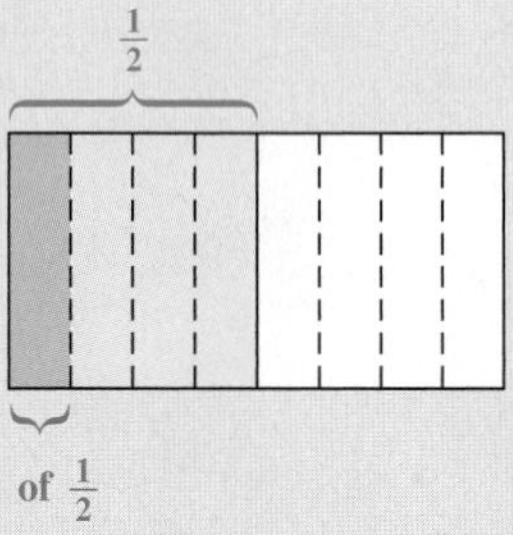

$\frac{1}{4}$ of $\frac{1}{2}$ is $\frac{\_\_\_}{\_\_\_}$.

OBJECTIVE 1 Multiply fractions. Suppose that you give $\frac{1}{2}$ of your Energy Bar to your kickboxing partner Ally. Then Ally gives $\frac{1}{2}$ of her share to Jake. How much of the Energy Bar does Jake get to eat?

Start with a sketch showing the Energy Bar cut in half (2 equal pieces).

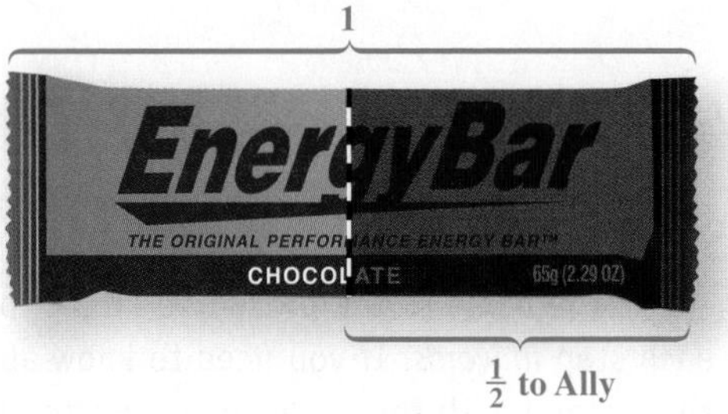

Next, take $\frac{1}{2}$ of the shaded area. (Here we are dividing $\frac{1}{2}$ into 2 equal parts and shading one darker than the other.)

The sketch shows that Jake gets $\frac{1}{4}$ of the Energy Bar.

Jake gets $\frac{1}{2}$ of $\frac{1}{2}$ of the Energy Bar. When used between two fractions, the word **of** tells us to multiply.

$$\frac{1}{2} \text{ of } \frac{1}{2} \quad \text{means} \quad \frac{1}{2} \cdot \frac{1}{2}$$

Jake's share of the Energy Bar is

$$\frac{1}{2} \cdot \frac{1}{2} = \frac{1}{4}.$$

◀ Work Problem **1** at the Side.

The rule for multiplying fractions follows.

Multiplying Fractions

Multiply two fractions by multiplying the numerators and multiplying the denominators.

Answer

1. $\frac{1}{8}$

Use this rule to find the product of $\frac{2}{3}$ and $\frac{1}{3}$, that is, to multiply $\frac{2}{3}$ by $\frac{1}{3}$.

$$\frac{2}{3} \cdot \frac{1}{3} = \frac{2 \cdot 1}{3 \cdot 3} \begin{array}{l}\leftarrow \text{Multiply numerators.} \\ \leftarrow \text{Multiply denominators.}\end{array}$$

Finish multiplying.

$$\frac{2}{3} \cdot \frac{1}{3} = \frac{2 \cdot 1}{3 \cdot 3} = \frac{2}{9} \begin{array}{l}\leftarrow 2 \cdot 1 = 2 \\ \leftarrow 3 \cdot 3 = 9\end{array}$$

Multiply numerators.

Multiply denominators.

Check that the final result is in lowest terms. $\frac{2}{9}$ is in lowest terms because 2 and 9 have no common factor other than 1.

EXAMPLE 1 Multiplying Fractions

Multiply. Write answers in lowest terms.

(a) $\frac{5}{8} \cdot \frac{3}{4}$

Multiply the numerators and multiply the denominators.

$$\frac{5}{8} \cdot \frac{3}{4} = \frac{5 \cdot 3}{8 \cdot 4} = \frac{15}{32}$$

Already in lowest terms.

Notice that 15 and 32 have no common factors other than 1, so the answer is in lowest terms.

(b) $\frac{4}{7} \cdot \frac{2}{5}$

$$\frac{4}{7} \cdot \frac{2}{5} = \frac{4 \cdot 2}{7 \cdot 5} = \frac{8}{35} \leftarrow \text{Lowest terms}$$

(c) $\frac{5}{8} \cdot \frac{3}{4} \cdot \frac{1}{2}$

$$\frac{5}{8} \cdot \frac{3}{4} \cdot \frac{1}{2} = \frac{5 \cdot 3 \cdot 1}{8 \cdot 4 \cdot 2} = \frac{15}{64} \leftarrow \text{Lowest terms}$$

Work Problem 2 at the Side. ▶

OBJECTIVE 2 Use a multiplication shortcut. A **multiplication shortcut** that can be used with fractions is shown in **Example 2.**

EXAMPLE 2 Using the Multiplication Shortcut

Multiply $\frac{5}{6}$ and $\frac{9}{10}$. Write the answer in lowest terms.

$$\frac{5}{6} \cdot \frac{9}{10} = \frac{5 \cdot 9}{6 \cdot 10} = \frac{45}{60} \leftarrow \text{Not in lowest terms}$$

The numerator and denominator have a common factor other than 1, so write the prime factorization of each number.

$$\frac{5}{6} \cdot \frac{9}{10} = \frac{5 \cdot 9}{6 \cdot 10} = \frac{5 \cdot 3 \cdot 3}{2 \cdot 3 \cdot 2 \cdot 5}$$

Write the prime factorization of each number.

Continued on Next Page

2 Multiply. Write answers in lowest terms.

GS **(a)** $\frac{1}{2} \cdot \frac{3}{4}$

$$\frac{1}{2} \cdot \frac{3}{4} = \frac{3}{\_\_\_\_} \begin{array}{l}\leftarrow 1 \cdot 3 \\ \leftarrow 2 \cdot 4\end{array}$$

(b) $\frac{3}{5} \cdot \frac{1}{3}$

(c) $\frac{5}{6} \cdot \frac{1}{2} \cdot \frac{1}{8}$

(d) $\frac{1}{2} \cdot \frac{3}{4} \cdot \frac{3}{8}$

Answers

2. (a) 8; $\frac{3}{8}$ **(b)** $\frac{1}{5}$ **(c)** $\frac{5}{96}$ **(d)** $\frac{9}{64}$

Next, divide by the common factors of 5 and 3.

$$\frac{5}{6} \cdot \frac{9}{10} = \frac{5 \cdot 9}{6 \cdot 10} = \frac{\overset{1}{\cancel{5}} \cdot \overset{1}{\cancel{3}} \cdot 3}{2 \cdot \underset{1}{\cancel{3}} \cdot 2 \cdot \underset{1}{\cancel{5}}}$$

Finally, multiply the remaining factors in the numerator and in the denominator.

$$\frac{5}{6} \cdot \frac{9}{10} = \frac{1 \cdot 1 \cdot 3}{2 \cdot 1 \cdot 2 \cdot 1} = \frac{3}{4} \leftarrow \text{Lowest terms}$$

As a shortcut, instead of writing the prime factorization of each number, find the product of $\frac{5}{6}$ and $\frac{9}{10}$ as follows.

First, divide by 5, a common factor of both 5 and 10.

$$\frac{\overset{1}{\cancel{5}}}{6} \cdot \frac{9}{\underset{2}{\cancel{10}}}$$

Divide 5 by 5 to get 1.

Divide 10 by 5 to get 2.

Next, divide by 3, a common factor of both 6 and 9.

$$\frac{\overset{1}{\cancel{5}}}{\underset{2}{\cancel{6}}} \cdot \frac{\overset{3}{\cancel{9}}}{\underset{2}{\cancel{10}}}$$

Divide 9 by 3 to get 3.

Divide 6 by 3 to get 2.

Finally, multiply numerators and multiply denominators.

$$\frac{1 \cdot 3}{2 \cdot 2} = \frac{3}{4}$$

CAUTION

When using the multiplication shortcut, you are dividing a numerator and a denominator by a common factor. Be certain that you divide a numerator and a denominator ***by the same number.*** If you do all possible divisions, your answer will be in lowest terms.

EXAMPLE 3 **Using the Multiplication Shortcut**

Use the multiplication shortcut to find each product. Write the answers in lowest terms and as mixed numbers where possible.

(a) $\frac{6}{11} \cdot \frac{7}{8}$

Divide both 6 and 8 by their common factor of 2. Notice that 7 and 11 have no common factor. Then multiply.

$$\frac{\overset{3}{\cancel{6}}}{11} \cdot \frac{7}{\underset{4}{\cancel{8}}} = \frac{3 \cdot 7}{11 \cdot 4} = \frac{21}{44} \leftarrow \text{Lowest terms}$$

(b) $\frac{7}{10} \cdot \frac{20}{21}$

Divide a numerator and a denominator by the same number.

Divide 7 and 21 by 7, then divide 10 and 20 by 10.

$$\frac{\overset{1}{\cancel{7}}}{\underset{1}{\cancel{10}}} \cdot \frac{\overset{2}{\cancel{20}}}{\underset{3}{\cancel{21}}} = \frac{1 \cdot 2}{1 \cdot 3} = \frac{2}{3} \leftarrow \text{Lowest terms}$$

Continued on Next Page

(c) $\frac{35}{12} \cdot \frac{32}{25}$

$$\frac{\overset{7}{\cancel{35}}}{\underset{3}{\cancel{12}}} \cdot \frac{\overset{8}{\cancel{32}}}{\underset{5}{\cancel{25}}} = \frac{7 \cdot 8}{3 \cdot 5} = \frac{56}{15} \quad \text{or} \quad 3\frac{11}{15} \leftarrow \text{Mixed number}$$

(d) $\frac{2}{3} \cdot \frac{8}{15} \cdot \frac{3}{4}$

$$\frac{\overset{1}{\cancel{2}}}{\underset{1}{\cancel{3}}} \cdot \frac{\overset{4}{\cancel{8}}}{15} \cdot \frac{\overset{1}{\cancel{3}}}{\underset{\underset{1}{\cancel{2}}}{\cancel{4}}} = \frac{1 \cdot 4 \cdot 1}{1 \cdot 15 \cdot 1} = \frac{4}{15} \leftarrow \text{Lowest terms}$$

This shortcut is especially helpful when the fractions involve large numbers.

Note

There is no specific order that must be used when dividing numerators and denominators, as long as both a numerator and a denominator are divided by the *same* number each time.

Work Problem 3 at the Side. ▶

OBJECTIVE 3 Multiply a fraction and a whole number. The rule for multiplying a fraction and a whole number follows.

Multiplying a Whole Number and a Fraction

Multiply a whole number and a fraction by writing the whole number as a fraction with a denominator of 1.

For example, write the whole numbers 8, 10, and 25 as follows.

$$8 = \frac{8}{1} \qquad 10 = \frac{10}{1} \qquad 25 = \frac{25}{1}$$

Write the whole number over 1.

EXAMPLE 4 **Multiplying by Whole Numbers**

Multiply. Write answers in lowest terms and as whole numbers where possible.

(a) $8 \cdot \frac{3}{4}$

Write 8 as $\frac{8}{1}$ and multiply.

$$8 \cdot \frac{3}{4} = \frac{\overset{2}{\cancel{8}}}{1} \cdot \frac{3}{\underset{1}{\cancel{4}}} = \frac{2 \cdot 3}{1 \cdot 1} = \frac{6}{1} = 6$$

$\frac{6}{1}$ is the same as $6 \div 1$, which equals 6.

Continued on Next Page

3 Use the multiplication shortcut to find each product.

GS **(a)** $\frac{3}{4} \cdot \frac{2}{5}$

$$\frac{3}{\cancel{4}} \cdot \frac{\cancel{2}}{5} =$$

(b) $\frac{6}{11} \cdot \frac{33}{21}$

(c) $\frac{20}{4} \cdot \frac{3}{40} \cdot \frac{1}{3}$

(d) $\frac{18}{17} \cdot \frac{1}{36} \cdot \frac{2}{3}$

Answers

3. (a) 2; 1; $\frac{3}{\underset{2}{\cancel{4}}} \cdot \frac{\overset{1}{\cancel{2}}}{5} = \frac{3}{10}$

(b) $\frac{\overset{2}{\cancel{6}}}{\underset{1}{\cancel{11}}} \cdot \frac{\overset{3}{\cancel{33}}}{\underset{7}{\cancel{21}}} = \frac{6}{7}$

(c) $\frac{\overset{1}{\cancel{20}}}{4} \cdot \frac{\overset{1}{\cancel{3}}}{\underset{2}{\cancel{40}}} \cdot \frac{1}{\underset{1}{\cancel{3}}} = \frac{1}{8}$

(d) $\frac{\overset{1}{\cancel{18}}}{17} \cdot \frac{1}{\underset{\underset{1}{\cancel{2}}}{\cancel{36}}} \cdot \frac{\overset{1}{\cancel{2}}}{3} = \frac{1}{51}$

4 Multiply. Write answers in lowest terms and as whole numbers or mixed numbers where possible.

GS **(a)** $8 \cdot \frac{1}{8}$

$$\frac{\overset{}{\cancel{8}}}{\_\_\_} \times \frac{1}{\underset{1}{\cancel{8}}} = \frac{1}{1} =$$

VOCABULARY TIP

Rectangle The formula for finding the area of a rectangle works for any figure with four straight sides and four right angles.

(b) $\frac{3}{4} \cdot 5 \cdot \frac{5}{3}$

(c) $\frac{3}{5} \cdot 40$

(d) $\frac{3}{25} \cdot \frac{5}{11} \cdot 99$

Answers

4. **(a)** 1; 1; 1 **(b)** $6\frac{1}{4}$ **(c)** 24 **(d)** $5\frac{2}{5}$

(b) $15 \cdot \frac{5}{6}$

$$15 \cdot \frac{5}{6} = \frac{\overset{5}{\cancel{15}}}{1} \cdot \frac{5}{\underset{2}{\cancel{6}}} = \frac{5 \cdot 5}{1 \cdot 2} = \frac{25}{2} = 12\frac{1}{2}$$

◀ **Work Problem 4 at the Side.**

OBJECTIVE 4 **Find the area of a rectangle.** To find the area of a rectangle (the amount of surface inside the rectangle), use this formula.

Area of a Rectangle

The area of a rectangle is equal to the length multiplied by the width.

Area = length • width

For example, the rectangle shown here has an area of 12 square feet (12 ft^2).

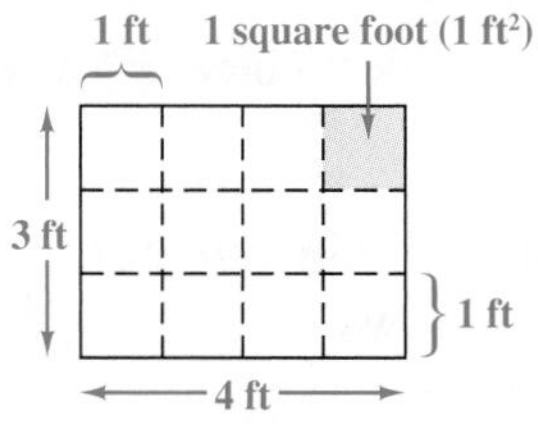

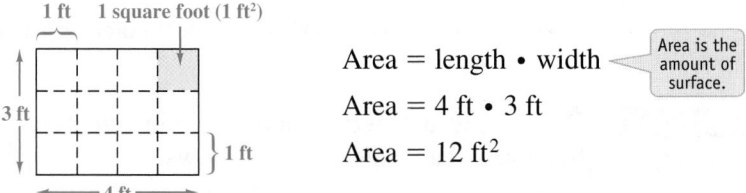

Area = length • width ← Area is the amount of surface.

Area = 4 ft • 3 ft

Area = 12 ft^2

Other units for measuring area are square inches (in.^{2}), square yards (yd^2), and square miles (mi^2). (See **Chapter 7** for more information on area.)

EXAMPLE 5 **Applying Fraction Skills**

To find the area of each rectangle, multiply its length by its width.

(a) Find the area of each rectangular shower tile.

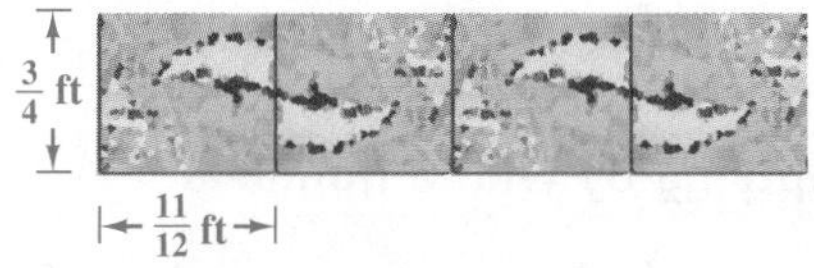

Area = length • width

$$\text{Area} = \frac{11}{12} \cdot \frac{3}{4}$$

$$= \frac{11}{\underset{4}{\cancel{12}}} \cdot \frac{\overset{1}{\cancel{3}}}{4}$$ Divide numerator and denominator by 3, so 3 ÷ 3 is 1, and 12 ÷ 3 is 4.

$$= \frac{11}{16} \text{ square foot } (ft^2)$$

Continued on Next Page

(b) Find the area of this rectangular SUV running board.

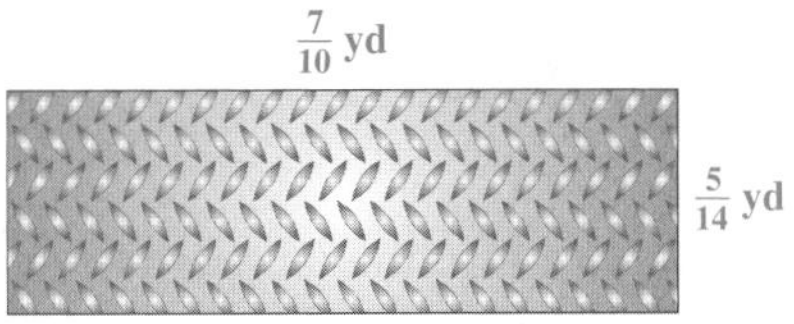

Multiply the length by the width.

$$\text{Area} = \frac{7}{10} \cdot \frac{5}{14}$$

$$= \frac{\overset{1}{\cancel{7}}}{\underset{2}{\cancel{10}}} \cdot \frac{\overset{1}{\cancel{5}}}{\underset{2}{\cancel{14}}} \quad \begin{array}{l}\text{Divide 7 and 14 by 7.}\\ \text{Divide 10 and 5 by 5.}\end{array}$$

$$= \frac{1}{4} \text{ square yard (yd}^2\text{)}$$

Work Problem 5 at the Side. ▶

5 Find the area of each rectangle.

(a)

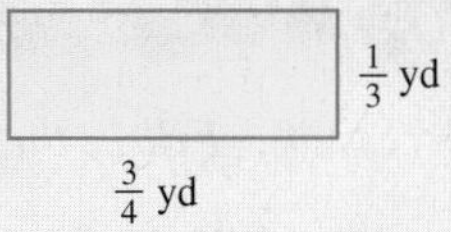

(b) a community college campus that is $\frac{3}{8}$ mile by $\frac{1}{3}$ mile

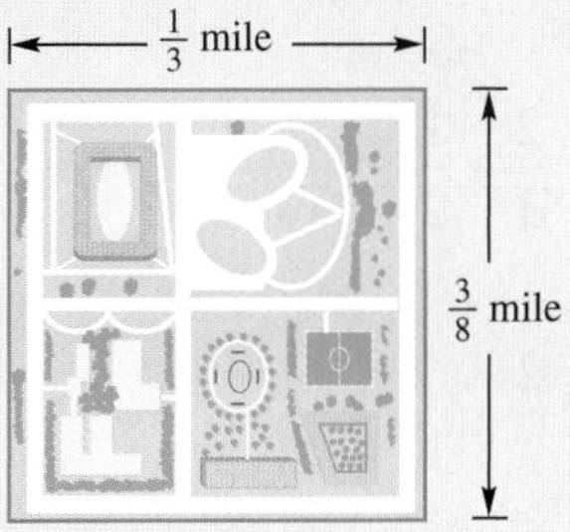

(c) a parcel of land that is $\frac{9}{7}$ mile by $\frac{7}{12}$ mile

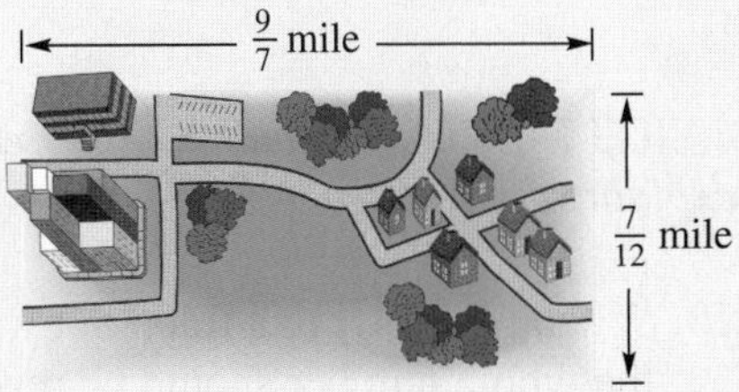

Answers

5. (a) $\frac{1}{4}$ yd^2 **(b)** $\frac{1}{8}$ mi^2 **(c)** $\frac{3}{4}$ mi^2

2.5 Exercises

FOR EXTRA HELP

Download the MyDashBoard App

MyMathLab®

CONCEPT CHECK *Fill in the blanks with the correct response.*

1. To multiply two or more fractions, you __________ the numerators and you multiply the __________.

2. To write a fraction answer in lowest terms, you must divide both the __________ and __________ by a common factor.

3. A shortcut when multiplying fractions is to __________ both a numerator and a __________ by the same number.

4. Using the shortcut when multiplying fractions should result in an answer that is in __________.

Multiply. Write answers in lowest terms. ***See Examples 1–3.***

5. $\frac{1}{3} \cdot \frac{3}{4}$

6. $\frac{2}{5} \cdot \frac{3}{4}$

7. $\frac{2}{7} \cdot \frac{1}{5}$

8. $\frac{2}{3} \cdot \frac{1}{2}$

9. $\frac{8}{5} \cdot \frac{15}{32}$

10. $\frac{5}{9} \cdot \frac{4}{3}$

11. $\frac{2}{3} \cdot \frac{7}{12} \cdot \frac{9}{14}$

$\frac{\overset{1}{\cancel{2}}}{\underset{1}{\cancel{3}}} \cdot \frac{\overset{1}{\cancel{7}}}{\underset{4}{\cancel{12}}} \cdot \frac{\overset{\overset{1}{\cancel{3}}}{\cancel{9}}}{\underset{\underset{1}{\cancel{2}}}{\cancel{14}}} =$

12. $\frac{7}{8} \cdot \frac{16}{21} \cdot \frac{1}{2}$

$\frac{\overset{1}{\cancel{7}}}{\underset{1}{\cancel{8}}} \cdot \frac{\overset{\overset{1}{\cancel{2}}}{\cancel{16}}}{\underset{3}{\cancel{21}}} \cdot \frac{1}{\underset{1}{\cancel{2}}} =$

13. $\frac{3}{4} \cdot \frac{5}{6} \cdot \frac{2}{3}$

14. $\frac{5}{8} \cdot \frac{16}{25}$

15. $\frac{6}{11} \cdot \frac{22}{15}$

16. $\frac{14}{25} \cdot \frac{65}{48} \cdot \frac{15}{28}$

17. $\frac{35}{64} \cdot \frac{32}{15} \cdot \frac{27}{72}$

18. $\frac{16}{25} \cdot \frac{35}{32} \cdot \frac{15}{64}$

19. $\frac{39}{42} \cdot \frac{7}{13} \cdot \frac{7}{24}$

CONCEPT CHECK *Decide whether each of the following is* true *or* false. *If* false, *explain why.*

20. When multiplying a fraction by a whole number, the whole number should be rewritten as the number over 1.

21. $\frac{4}{5} \cdot 8 = \frac{\overset{1}{\cancel{4}}}{5} \cdot \frac{1}{\underset{2}{\cancel{8}}} = \frac{1}{10}$

Multiply. Write answers in lowest terms and as whole or mixed numbers where possible. ***See Example 4.***

22. $5 \cdot \frac{4}{5}$

23. $20 \cdot \frac{3}{4}$

$\frac{\overset{5}{\cancel{20}}}{1} \cdot \frac{3}{\underset{1}{\cancel{4}}} = \frac{15}{1} =$

24. $36 \cdot \frac{2}{3}$

$\frac{\overset{12}{\cancel{36}}}{1} \cdot \frac{2}{\underset{1}{\cancel{3}}} = \frac{24}{1} =$

25. $36 \cdot \frac{5}{8} \cdot \frac{9}{15}$

26. $30 \cdot \frac{3}{10}$

27. $100 \cdot \frac{21}{50} \cdot \frac{3}{4}$

28. $400 \cdot \frac{7}{8}$

29. $\frac{2}{5} \cdot 200$

30. $\frac{6}{7} \cdot 245$

31. $142 \cdot \frac{2}{3}$

32. $\frac{12}{25} \cdot 430$

33. $\frac{28}{21} \cdot 640 \cdot \frac{15}{32}$

34. $\frac{21}{13} \cdot 520 \cdot \frac{7}{20}$

35. $\frac{54}{38} \cdot 684 \cdot \frac{5}{6}$

36. $\frac{76}{43} \cdot 473 \cdot \frac{5}{19}$

Find the area of each rectangle. ***See Example 5.***

37.

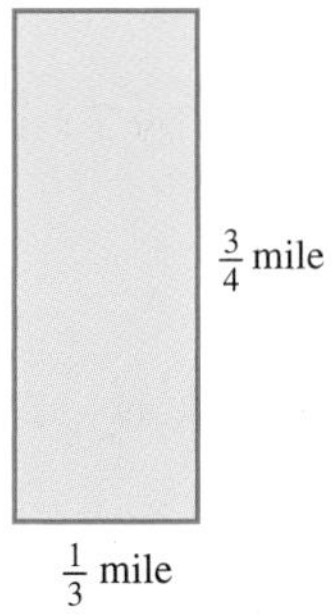

38.

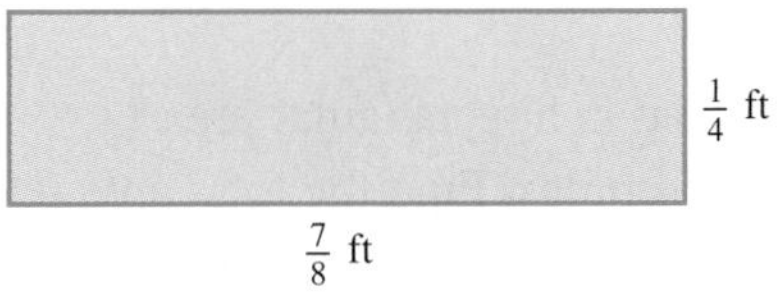

39.

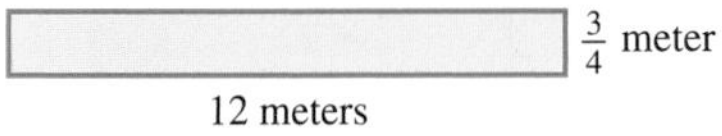

40.

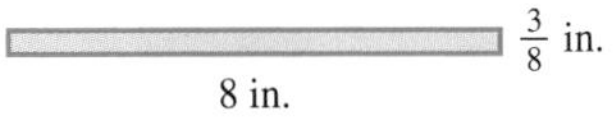

41.

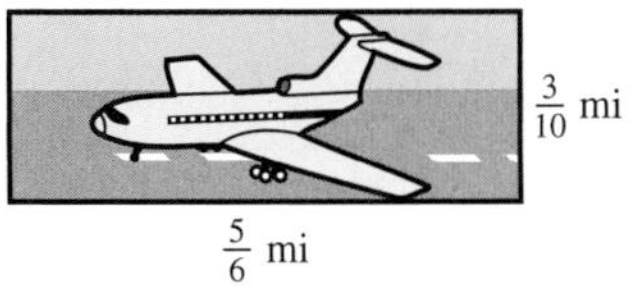

42.

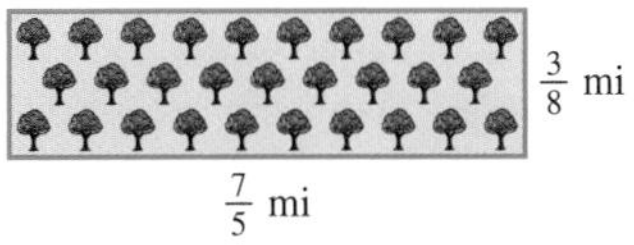

43. Write in your own words the rule for multiplying fractions. Make up an example problem to show how this works.

44. A useful shortcut when multiplying fractions is to divide a numerator and a denominator by the same number. Describe how this works and give an example.

Find the area of each rectangle in these application problems. Write answers in lowest terms and as whole or mixed numbers where possible. ***See Example 5.***

45. Find the area of a heating-duct grill having a length of 2 yd and a width of $\frac{3}{4}$ yd.
Hint: Area = length • width.

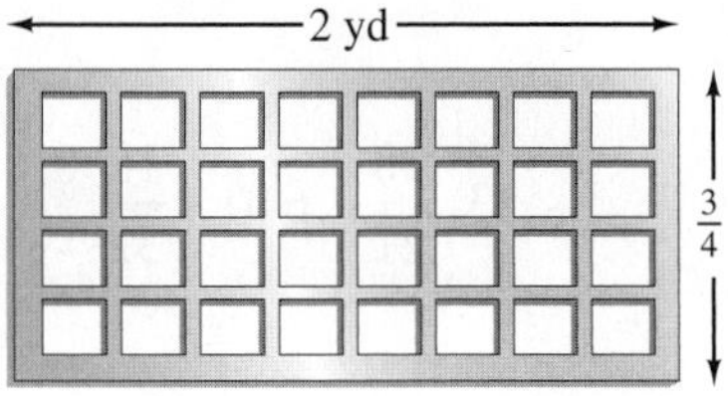

46. Find the area of a HD television having a width of 2 yd and a height of $\frac{15}{16}$ yd.

47. A wildfire is contained in a rectangular area measuring $\frac{7}{8}$ mile by 4 miles. Find the total area of the containment.

48. A rectangular flood plain is $\frac{3}{4}$ mile wide by 7 miles long. Find the area of the flood plain.

49. The Sunny Side Soccer Park is $\frac{1}{4}$ mile long and $\frac{3}{16}$ mile wide, while the Creek Side Soccer Park is $\frac{3}{8}$ mile long and $\frac{1}{8}$ mile wide. Which park has the larger area?

50. The Rocking Horse Ranch is $\frac{3}{4}$ mile long and $\frac{2}{3}$ mile wide. The Silver Spur Ranch is $\frac{5}{8}$ mile long and $\frac{4}{5}$ mile wide. Which ranch has the larger area?

Relating Concepts (Exercises 51–56) For Individual or Group Work

Front end rounding can be used to estimate an answer when multiplying fractions. ***Work Exercises 51–56 in order.*** *Round exact answers to the nearest whole number.*

The bar graph shows the number of supermarkets in several states. The greatest number are in California, while the least number are in Wyoming.

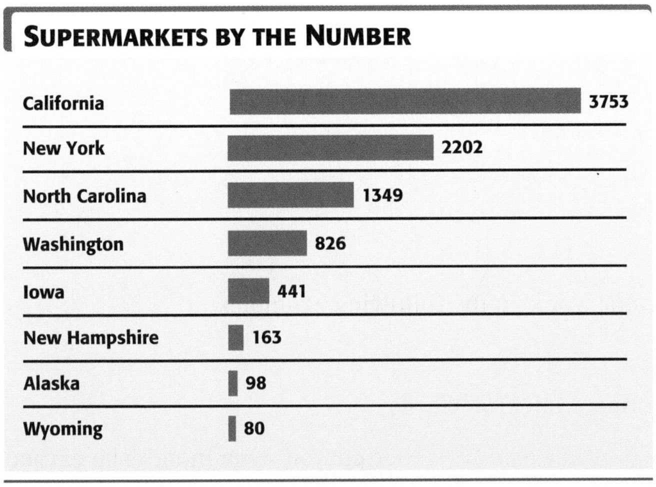

Source: The Nielsen Company.

51. Use front end rounding to estimate the total number of supermarkets in these states.

$4000 + 2000 + 1000 + 800 + 400 + 200 + 100 + 80 =$

____________.

52. Find the exact total number of supermarkets in these states.

$3753 + 2202 + 1349 + 826 + 441 + 163 + 98 + 80 =$

____________.

53. If $\frac{4}{5}$ of the supermarkets in New York are in medium to large population areas, use front end rounding to estimate, and then find the exact number of stores in these population areas.

Estimate:

Exact:

54. If $\frac{3}{8}$ of the supermarkets in New Hampshire are in shopping centers, use front end rounding to estimate, and then find the exact number of stores in these locations.

Estimate:

Exact:

Compare the estimated answers and the exact answers in ***Exercises 53 and 54.*** *How can you get an estimated answer that is closer to the exact answer? Try rounding the number of stores to some multiple of the denominator in the fraction that has two nonzero digits.*

55. Refer to **Exercise 53.** Round the number of supermarkets in New York to some multiple of the denominator in $\frac{4}{5}$ that has *two* nonzero digits. Now estimate the answer, showing your work.

56. Refer to **Exercise 54.** Round the number of supermarkets in New Hampshire to some multiple of the denominator in $\frac{3}{8}$ that has *two* nonzero digits. Now estimate the answer, showing your work.

2.6 Applications of Multiplication

OBJECTIVE

1 Solve fraction application problems using multiplication.

OBJECTIVE 1 Solve fraction application problems using multiplication. Many application problems are solved by multiplying fractions. Use the following indicator words for multiplication.

product
double
triple
times
of (when "of" follows a fraction)
twice
twice as much

Always look for indicator words.

Look for these indicator words in the following examples.

EXAMPLE 1 Applying Indicator Words

Lois Stevens gives $\frac{1}{10}$ of her income to her church. One month she earned \$2980. How much did she give to the church that month?

Step 1 **Read** the problem. The problem asks us to find the amount of money given to the church.

Step 2 **Work out a plan.** Stevens gave $\frac{1}{10}$ ***of*** her income. The indicator word is ***of***. When it follows a fraction, the word ***of*** indicates multiplication, so find the amount given to the church by multiplying $\frac{1}{10}$ and \$2980.

Step 3 **Estimate** a reasonable answer. Round the income of \$2980 to \$3000. Then divide \$3000 by 10 to find $\frac{1}{10}$ of the income (one of 10 equal parts). Our estimate is \$3000 ÷ 10 = \$300. (Recall the shortcut for dividing by 10; drop one 0 from the dividend.)

Step 4 **Solve** the problem.

$$\text{amount} = \frac{1}{\cancel{10}_{1}} \cdot \frac{\overset{298}{\cancel{2980}}}{1} = \frac{298}{1} = 298$$

Divide by the common factor of 10.

Step 5 **State the answer.** Stevens gave \$298 to her church that month.

Step 6 **Check.** The exact answer, \$298, is close to our estimate of \$300.

◀ Work Problem 1 at the Side.

EXAMPLE 2 Solving a Fraction Application Problem

Of the 39 students in Sharon Martin's high school economics class, $\frac{2}{3}$ plan to go to college. How many plan to go to college?

Step 1 **Read** the problem. The problem asks us to find the number of students who plan to go to college.

Continued on Next Page

1 Solve each problem. Use the six problem solving steps.

GS **(a)** Erich and Sabrina Means are saving $\frac{3}{8}$ of their income for the down payment on their first home. If they have a combined annual income of \$81,576, how much can they save in a year?

Estimate:

Round 81,576 to 82,000.

$\frac{1}{2} \cdot \frac{82{,}000}{1} =$ ______

Solve: to find the exact answer.

$\frac{3}{\cancel{8}_{1}} \cdot \frac{\overset{10{,}197}{\cancel{81{,}576}}}{1} =$ ______

State the answer: \$30,591 can be saved in a year.

Check: Exact answer is close to estimate.

(b) A retiring firefighter will receive $\frac{5}{8}$ of her highest annual salary as retirement income. If her highest annual salary is \$62,504, how much will she receive as retirement income?

Answers

1. (a) *Estimate:* \$41,000; *Exact:* \$30,591
   (b) \$39,065

Step 2 **Work out a plan.** Reword the problem to read

$\frac{2}{3}$ **of** the students plan to go to college.

↑ Indicator word for multiplication when it follows a fraction

Step 3 **Estimate** a reasonable answer. Round the number of students in the class from 39 to 40. Then, $\frac{1}{2}$ of 40 is 20. Since $\frac{2}{3}$ is more than $\frac{1}{2}$, our estimate is that "more than 20 students" plan to go to college.

Step 4 **Solve** the problem. Find the number who plan to go to college by multiplying $\frac{2}{3}$ and 39.

$$\text{number who plan to go} = \frac{2}{3} \cdot 39$$

$$= \frac{2}{\cancel{3}_1} \cdot \frac{\cancel{39}^{13}}{1} = \frac{26}{1} = 26$$

Step 5 **State the answer.** 26 students plan to go to college.

Check that the answer is reasonable.

Step 6 **Check.** The exact answer, 26, fits our estimate of "more than 20."

Work Problem **2** at the Side. ▶

2 At Sid's Pharmacy, $\frac{5}{16}$ of the prescriptions are paid by a third party (insurance company). If 3696 prescriptions are filled, find the number paid by a third party. Use the six problem-solving steps.

EXAMPLE 3 Finding a Fractional Part of a Fraction

In her will, a woman divides her estate into 6 equal parts. Five of the 6 parts are given to relatives. Of the sixth part, $\frac{1}{3}$ goes to the Salvation Army. What fraction of her total estate goes to the Salvation Army?

Step 1 **Read** the problem. The problem asks for the fraction of an estate that goes to the Salvation Army.

Step 2 **Work out a plan.** Reword the problem to read the Salvation Army gets $\frac{1}{3}$ **of** $\frac{1}{6}$.

↑ Indicator word for multiplication when it follows a fraction

Step 3 Estimate a reasonable answer. If the estate is divided into 6 equal parts and each of these parts was divided into 3 equal parts, we would have $6 \cdot 3 = 18$ equal parts. Our estimate is $\frac{1}{18}$.

Step 4 **Solve** the problem. The Salvation Army gets $\frac{1}{3}$ **of** $\frac{1}{6}$. Indicator word

To find the fraction that the Salvation Army is to receive, multiply $\frac{1}{3}$ and $\frac{1}{6}$.

$$\text{fraction to Salvation Army} = \frac{1}{3} \cdot \frac{1}{6}$$

$$= \frac{1}{18}$$

Step 5 **State the answer.** The Salvation Army gets $\frac{1}{18}$ of the total estate.

Step 6 **Check.** The exact answer, $\frac{1}{18}$, matches our estimate.

Remember to check your work.

Work Problem **3** at the Side. ▶

3 GS At our college, $\frac{1}{3}$ of the students speak a foreign language. Of those speaking a foreign language, $\frac{3}{4}$ speak Spanish. What fraction of the students speak Spanish? Use the six problem-solving steps.

$$\frac{1}{3} \cdot \frac{3}{4}$$

$$\frac{1}{\cancel{3}} \cdot \frac{\cancel{3}^1}{\_\_} = \_\_\_\_$$

Answers

2. 1155 prescriptions

3. 1; 4; $\frac{1}{4}$ speak Spanish

4 Solve each problem using the six problem-solving steps. Use the circle graph in **Example 4.**

(a) What fraction of the children buy food from vending machines?

GS **(b)** What number of children buy food from vending machines?

Estimate: $\frac{1}{5}$ of 2500 is 500

Solve: $\frac{1}{\cancel{5}_1} \cdot \frac{\cancel{2500}^{500}}{1} =$ ____

State the answer: 500 children

Check: Answer is the same as the estimate.

(c) What fraction of the children buy food from a convenience store or street vendor?

(d) What number of children buy food from a convenience store or street vendor?

Answers

4. (a) $\frac{1}{5}$ **(b)** 5; 500 children

(c) $\frac{1}{10}$ **(d)** 250 children

EXAMPLE 4 Using Fractions with a Circle Graph

The circle graph, or pie chart, shows where children 8 to 17 years of age are most likely to make food purchases when away from home. If 2500 children were in the survey, find the number of children who buy food in the school cafeteria.

FINDING FOOD

Children ages 8 to 17 are most likely to make food purchases at the following locations:

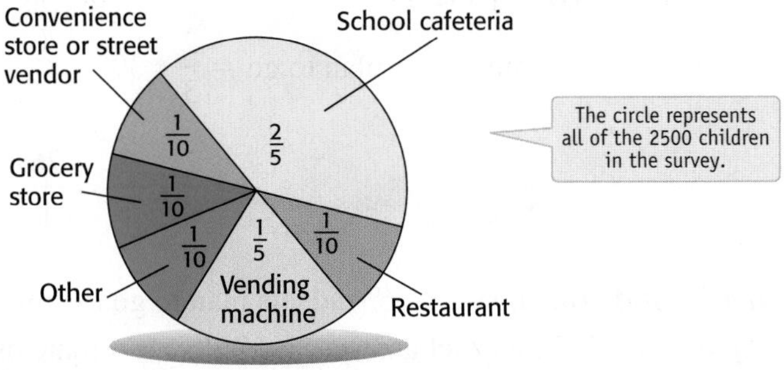

Source: Pursuant Inc. for American Dietetic Association Foundation.

Step 1 **Read** the problem. The problem asks for the number of children who buy food in the school cafeteria.

Step 2 **Work out a plan.** Reword the problem to read

$\frac{2}{5}$ **of** 2500 children buy food in the school cafeteria.

↑ Indicator word for multiplication when it follows a fraction

Step 3 **Estimate** a reasonable answer. $\frac{1}{2}$ of 2500 people is 1250 people. $\frac{2}{5}$ is less than $\frac{1}{2}$, so our estimate is "less than 1250 people."

Step 4 **Solve** the problem. Find the number who buy food in the school cafeteria by multiplying $\frac{2}{5}$ and 2500.

$$\text{number in school cafeteria} = \frac{2}{5} \cdot 2500$$

$$= \frac{2}{5} \cdot \frac{2500}{1}$$

$$= \frac{2}{\cancel{5}_1} \cdot \frac{\cancel{2500}^{500}}{1}$$ Divide both numerator and denominator by 5.

$$= \frac{1000}{1} = 1000$$

Step 5 **State the answer.** 1000 children buy food in the school cafeteria.

Step 6 **Check.** The exact answer, 1000 children, fits our estimate of "less than 1250 children."

◀ Work Problem **4** at the Side.

2.6 Exercises

FOR EXTRA HELP

Download the MyDashBoard App

MyMathLab®

1. CONCEPT CHECK Circle the words that are indicator words for multiplication.

more than	of	sum
times	twice	difference
triple	greater than	product
less than	all together	quotient
divided evenly	after a decrease	twice as much

2. CONCEPT CHECK Underline the correct answer. The final step when solving an application problem is to (*solve the problem/check your work*).

3. CONCEPT CHECK Fill in the blank. When you multiply length by width you are finding the ______ of a rectangular surface.

4. CONCEPT CHECK Fill in the blank. When calculating area, the length and the width must be in the same units of measurement. If the measurements are both in miles, the answer will be in ______ miles and shown as mi^2.

Solve each application problem. Look for indicator words. ***See Examples 1–4.***

5. A digital photo frame measures $\frac{3}{4}$ ft by $\frac{2}{3}$ ft. Find its area.

Hint: Area = length • width

6. The rectangular floor of Darby's dog house measures $\frac{14}{15}$ yd by $\frac{3}{4}$ yd. Find its area.

7. A cookie sheet is $\frac{4}{3}$ ft by $\frac{2}{3}$ ft. Find its area.

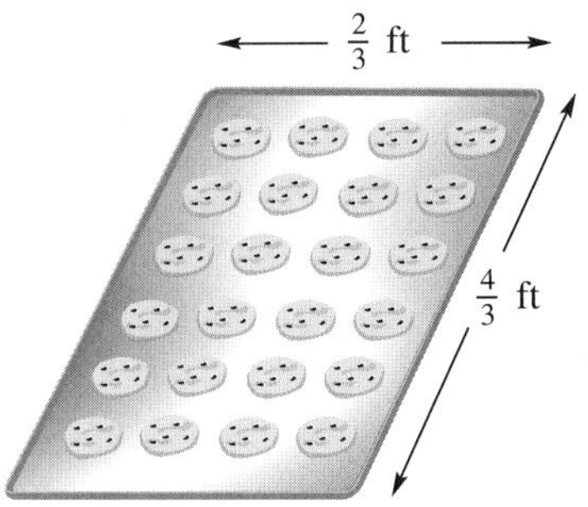

8. Each day there are 16 million people who shop at flea markets. If $\frac{2}{5}$ of these people purchase produce at the flea market, how many purchase produce? (*Source:* National Flea Market Association.)

9. Pete is helping Collin make a rectangular mahogany lamp table for Carolyn's birthday. Find the area of the top of the table if it is $\frac{4}{5}$ yd long by $\frac{3}{8}$ yd wide.

10. The average person consumes 160 bowls of cereal each year. If $\frac{3}{10}$ of the cereal is eaten in the summer months, how many bowls of cereal are eaten in the summer months? (*Source:* Target Stores.)

11. Dan Crump had expenses of \$6848 during one semester of college. His part-time job provided $\frac{3}{8}$ of the amount he needed. How much did he earn on his job?

12. The average household does 400 loads of wash each year. If $\frac{3}{8}$ of the wash loads are done in the winter months, how many wash loads are done in the winter months? (*Source:* Target Stores.)

13. The city with the most expensive daily parking fee is New York City (Midtown) at \$40. The daily parking fee in Boston (third most expensive) is $\frac{7}{8}$ as much as New York City. Find the daily parking fee in Boston. (*Source:* Colliers International.)

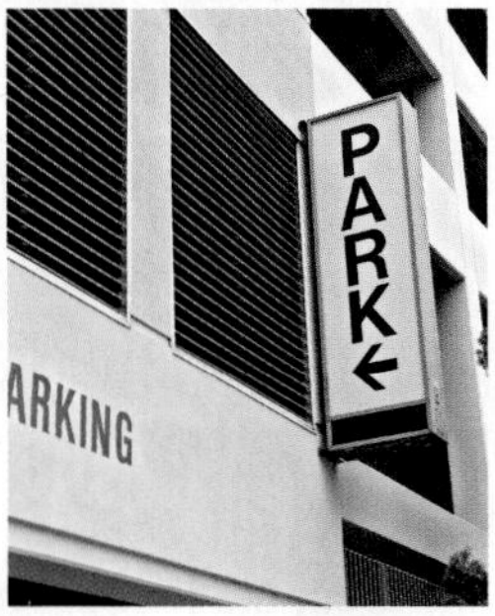

14. The daily parking fee in Boston (third most expensive) is \$35. In San Francisco (fourth most expensive), the daily parking fee is $\frac{4}{5}$ the cost of Boston. How much is the daily parking fee in San Francisco? (*Source:* Colliers International.)

15. At the Garlic Festival Fun Run, $\frac{7}{12}$ of the runners are women. If there are 1560 runners, how many of the runners are

(a) women?

(b) men?

16. A hotel has 408 rooms. Of these rooms, $\frac{9}{17}$ are nonsmoking rooms. How many rooms are

(a) nonsmoking?

(b) smoking?

Most electric cars can go 100 miles before they need to be recharged, which could take up to eight hours. The circle graph below shows how long people would be willing to wait to recharge an electric car. Use this information to work Exercises 17–20.

Time to Recharge

How long would you wait to recharge your electric car?

1020 people surveyed

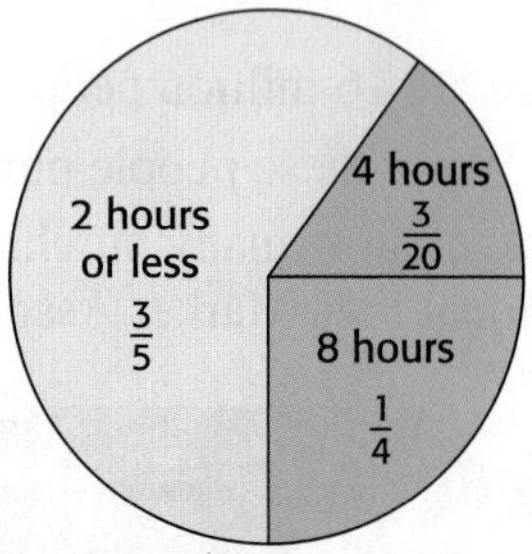

Source: Deloitte Touch Tohmatsu Limited [DTTL].

17. Which response was given by the least number of people? How many people gave this response?

18. Which response was given by the greatest number of people? How many gave this response?

19. Find the fraction and the total number of people who would be willing to wait 4 hours or less to recharge the car.

20. Find the fraction and the total number of people who would be willing to wait 4 hours or more to recharge the car.

21. Without actually adding the fractions given for all the groups in the "Time to Recharge" circle graph, explain why their sum has to be 1.

22. Refer to **Exercise 21.** Suppose you added all the fractions for the groups and did not get 1 as an answer. List some possible explanations.

The table shows the earnings for the Owens family last year, and the circle graph shows how they spent their earnings. Use this information to answer Exercises 23–28.

Month	Earnings	Month	Earnings
January	\$6075	July	\$7040
February	\$5812	August	\$5232
March	\$6488	September	\$5670
April	\$6030	October	\$7012
May	\$5820	November	\$6465
June	\$6398	December	\$7958

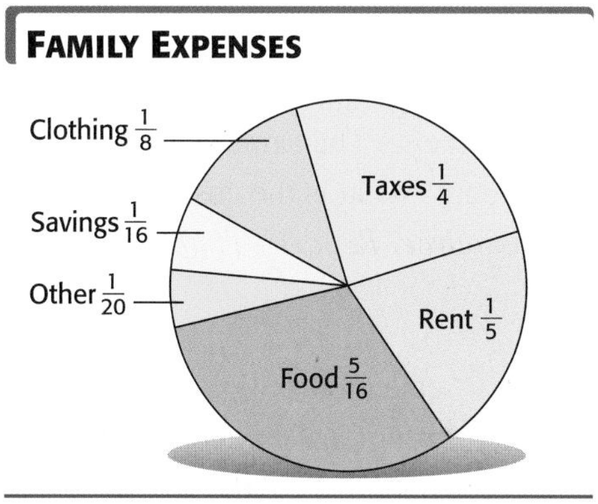

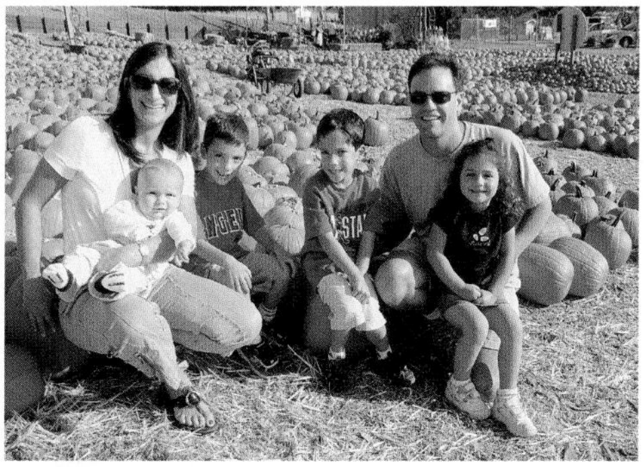

23. Find the Owens family's total income for the year.

24. How much of their annual earnings went to taxes?

25. Find the amount of their rent for the year.

26. How much did they spend for food during the year?

27. Find their annual savings.

28. How much of their annual income was spent on clothing?

29. Here is how one student solved a multiplication problem. Find the error and solve the problem correctly.

$$\frac{9}{10}\times\frac{20}{21}=\frac{\overset{3}{\cancel{9}}}{\underset{1}{\cancel{10}}}\times\frac{\overset{2}{\cancel{20}}}{\underset{3}{\cancel{21}}}=\frac{6}{3}=2$$

30. When two whole numbers are multiplied, the product is always larger than the numbers being multiplied. When two proper fractions are multiplied, the product is always smaller than the numbers being multiplied. Are these statements true? Why or why not?

Solve each application problem.

31. The cost of laser eye surgery in the United States is \$2000 for each eye. The same surgery in Thailand is $\frac{3}{8}$ of this amount. What is the cost of this procedure in Thailand? (*Source: Reader's Digest.*)

32. A knee replacement in the United States costs \$36,300, while in Mexico the same procedure costs $\frac{3}{50}$ of this amount. Find the cost of a knee replacement in Mexico. (*Source: Reader's Digest.*)

33. A collector of scale model World War II ships wants to know the length of a $\frac{1}{128}$ scale model of a ship that was 256 feet in length. Find the length of the scale model. (*Source:* Lilliput Motor Company, LTD.)

34. United Parcel Service (UPS) has trucks built to their specifications and they weigh 10,000 pounds. The company is testing a new truck that weighs $\frac{1}{10}$ less than the currently used trucks in order to save on fuel. How much does the test truck weigh?

35. LaDonna Washington is running for city council. She needs to get $\frac{2}{3}$ of her votes from senior citizens and 27,000 votes in all to win. How many votes does she need from voters other than the senior citizens?

36. The average debt of a graduate with a bachelor's degree from a four-year college is \$24,400. If $\frac{5}{16}$ of the amount was borrowed in the last year of college, find the amount borrowed in the first years. (*Source:* U.S. Department of Education.)

37. A will states that $\frac{7}{8}$ of an estate is to be divided among relatives. Of the remaining estate, $\frac{1}{4}$ goes to the American Cancer Society. What fraction of the estate goes to the American Cancer Society?

38. A couple has $\frac{2}{5}$ of their total investments in real estate. Of the remaining investments, $\frac{1}{3}$ is invested in bonds. What fraction of the total investments is in bonds?

2.7 Dividing Fractions

OBJECTIVE 1 **Find the reciprocal of a fraction.** To divide fractions, we need to know how to find the **reciprocal** of a fraction.

Reciprocal of a Fraction

Two numbers are reciprocals of each other if their product is 1. To find the reciprocal of a fraction, switch the numerator and denominator.

For example, the reciprocal of $\frac{3}{4}$ is $\frac{4}{3}$.

Fraction $\frac{3}{4} \times \frac{4}{3}$ Reciprocal

Note

Notice that you invert, or "flip," a fraction to find its reciprocal.

EXAMPLE 1 **Finding Reciprocals**

Find the reciprocal of each fraction.

Flip a fraction to find the reciprocal.

(a) The reciprocal of $\frac{1}{4}$ is $\frac{4}{1}$ because $\frac{1}{4} \cdot \frac{4}{1} = \frac{4}{4} = 1$

(b) The reciprocal of $\frac{2}{3}$ is $\frac{3}{2}$ because $\frac{2}{3} \cdot \frac{3}{2} = \frac{6}{6} = 1$

(c) The reciprocal of $\frac{3}{5}$ is $\frac{5}{3}$ because $\frac{3}{5} \cdot \frac{5}{3} = \frac{15}{15} = 1$

(d) The reciprocal of 8 is $\frac{1}{8}$ because $\frac{8}{1} \cdot \frac{1}{8} = \frac{8}{8} = 1$ Think of 8 as $\frac{8}{1}$.

Work Problem 1 at the Side. ▶

Note

Every number has a reciprocal except 0. The number 0 has no reciprocal because there is no number that can be multiplied by 0 to get 1.

$$0 \cdot (\text{reciprocal}) = 1$$

There is no number to use here that will give an answer of 1. When you multiply by 0, you always get 0.

OBJECTIVES

1. Find the reciprocal of a fraction.
2. Divide fractions.
3. Solve application problems in which fractions are divided.

VOCABULARY TIP

Reciprocal Taking the reciprocal of a number involves alternating, or switching the numerator and denominator.

1 Find the reciprocal of each fraction.

GS **(a)** $\frac{4}{5} \times \frac{5}{\_\_\_}$

GS **(b)** $\frac{3}{8} \times \frac{\_\_\_}{3}$

(c) $\frac{9}{4}$

(d) 16

Answers

1. (a) 4; $\frac{5}{4}$ **(b)** 8; $\frac{8}{3}$ **(c)** $\frac{4}{9}$ **(d)** $\frac{1}{16}$

In **Chapter 1,** we saw that the division problem $12 \div 3$ asks how many 3s are in 12. In the same way, the division problem $\frac{2}{3} \div \frac{1}{6}$ asks how many $\frac{1}{6}$s are in $\frac{2}{3}$. The figure illustrates $\frac{2}{3} \div \frac{1}{6}$.

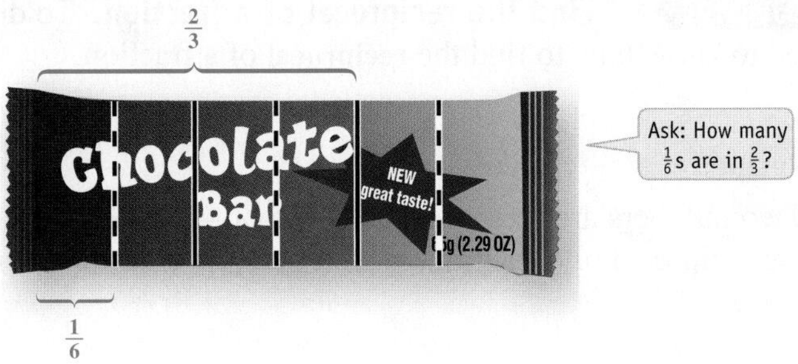

The figure shows that there are 4 of the $\frac{1}{6}$ pieces in $\frac{2}{3}$, or

$$\frac{2}{3} \div \frac{1}{6} = 4.$$

OBJECTIVE 2 Divide fractions. We will use reciprocals to divide fractions.

Dividing Fractions

To divide two fractions, multiply the first fraction by the reciprocal of the divisor (the second fraction).

EXAMPLE 2 Dividing One Fraction by Another

Divide. Write answers in lowest terms and as mixed numbers where possible.

(a) $\frac{7}{8} \div \frac{15}{16}$

The reciprocal of $\frac{15}{16}$ is $\frac{16}{15}$. ← Be sure to find the reciprocal of the divisor (the second fraction).

Reciprocals

$$\frac{7}{8} \div \frac{15}{16} = \frac{7}{8} \cdot \frac{16}{15}$$ ← Use $\frac{16}{15}$, the reciprocal of $\frac{15}{16}$, and multiply.

Change division to multiplication.

$$= \frac{7}{\overset{}{\underset{1}{\cancel{8}}}} \cdot \frac{\overset{2}{\cancel{16}}}{15}$$ Divide the numerator and denominator by the common factor of 8.

$$= \frac{7 \cdot 2}{1 \cdot 15}$$ Multiply.

$$= \frac{14}{15}$$ ← Lowest terms

Continued on Next Page

(b) $\dfrac{\frac{4}{5}}{\frac{3}{10}}$

$$\dfrac{\frac{4}{5}}{\frac{3}{10}} = \frac{4}{5} \div \frac{3}{10}$$ Rewrite by using the ÷ symbol for division.

$$= \frac{4}{\cancel{5}_1} \cdot \frac{\cancel{10}^2}{3}$$ The reciprocal of $\frac{3}{10}$ is $\frac{10}{3}$. Change ÷ to "•". Divide the numerator and denominator by the common factor of 5.

$$= \frac{4 \cdot 2}{1 \cdot 3}$$ Multiply.

$$= \frac{8}{3} = 2\frac{2}{3}$$ Rewrite the answer as a mixed number.

CAUTION

Be certain that the divisor fraction is changed to its reciprocal *before* you divide numerators and denominators by common factors.

Work Problem 2 at the Side. ▶

EXAMPLE 3 Dividing with a Whole Number

Divide. Write all answers in lowest terms and as whole or mixed numbers where possible.

(a) $5 \div \frac{1}{4}$ (*Hint:* How many quarters are in \$5?)

Write 5 as $\frac{5}{1}$. Next, use the reciprocal of $\frac{1}{4}$, which is $\frac{4}{1}$.

$$5 \div \frac{1}{4} = \frac{5}{1} \cdot \frac{4}{1}$$ Reciprocal of $\frac{1}{4}$ is $\frac{4}{1}$.

Reciprocals

$$= \frac{5 \cdot 4}{1 \cdot 1}$$ Multiply.

$$= \frac{20}{1} = 20$$ Rewrite the answer as a whole number.

Continued on Next Page

2 Divide. Write answers in lowest terms and as mixed numbers where possible.

GS **(a)** $\frac{1}{4} \div \frac{2}{3}$

$$\frac{1}{4} \cdot \frac{3}{\_\_\_} = \frac{\_\_\_}{8}$$

(b) $\frac{3}{8} \div \frac{5}{8}$

GS **(c)** $\dfrac{\frac{2}{3}}{\frac{4}{5}} = \frac{2}{3} \div \frac{4}{5}$

$$= \frac{\cancel{2}^1}{3} \cdot \frac{\_\_\_}{\cancel{4}_2}$$

$$= \frac{1 \cdot \_\_\_}{3 \cdot 2} =$$

(d) $\dfrac{\frac{5}{6}}{\frac{7}{12}}$

Answers

2. **(a)** 2; 3; $\frac{3}{8}$ **(b)** $\frac{3}{5}$ **(c)** 5; 5; $\frac{5}{6}$ **(d)** $1\frac{3}{7}$

3 Divide. Write answers in lowest terms and as whole or mixed numbers where possible.

GS (a) $10 \div \frac{1}{2}$

$= \frac{\_\_\_}{1} \cdot \frac{2}{1}$

$= \frac{10 \cdot 2}{1 \cdot 1} =$

(b) $6 \div \frac{6}{7}$

(c) $\frac{4}{5} \div 6$

(d) $\frac{3}{8} \div 4$

4 Solve each problem using the six problem-solving steps.

(a) How many $\frac{5}{6}$-ounce dispensers can be filled with 40 ounces of eye drops?

(b) Find the number of $\frac{4}{5}$-quart bottles that can be filled from a 120-quart cask.

Answers

3. (a) 10; 20 (b) 7 (c) $\frac{2}{15}$ (d) $\frac{3}{32}$

4. (a) 48 dispensers (b) 150 bottles

(b) $\frac{2}{3} \div 6$

Write 6 as $\frac{6}{1}$. The reciprocal of $\frac{6}{1}$ is $\frac{1}{6}$.

$$\frac{2}{3} \div \frac{6}{1} = \frac{2}{3} \cdot \frac{1}{6}$$

Reciprocals

Careful! Divide out common factors **after** changing to the reciprocal and to multiplication.

$$= \frac{\overset{1}{\cancel{2}}}{3} \cdot \frac{1}{\underset{3}{\cancel{6}}}$$ Divide the numerator and denominator by 2, then multiply.

$$= \frac{1 \cdot 1}{3 \cdot 3} = \frac{1}{9}$$ Lowest terms

◀ **Work Problem 3 at the Side.**

OBJECTIVE 3 Solve application problems in which fractions are divided. Many application problems require division of fractions. Recall from **Chapter 1** that typical indicator words for division are ***goes into, per, divide, divided by, divided equally,*** and ***divided into.***

EXAMPLE 4 Applying Fraction Skills

Goldie, the manager of the Burnside Deli, must fill a 10-gallon kosher dill pickle crock with salt brine. She has only a $\frac{2}{3}$-gallon container to use. How many times must she fill the $\frac{2}{3}$-gallon container and empty it into the 10-gallon crock?

Step 1 **Read** the problem. We need to find the number of times Goldie needs to use a $\frac{2}{3}$-gallon container in order to fill a 10-gallon crock.

Step 2 **Work out a plan.** We can solve the problem by finding the number of times $\frac{2}{3}$ goes into 10.

Step 3 **Estimate** a reasonable answer. Round $\frac{2}{3}$ gallon to 1 gallon. In order to fill the 10-gallon container, she would have to use the 1-gallon container 10 times, so our estimate is 10.

Step 4 **Solve** the problem.

Reciprocals

$$10 \div \frac{2}{3} = \frac{10}{1} \cdot \frac{3}{2}$$ The reciprocal of $\frac{2}{3}$ is $\frac{3}{2}$. Change " ÷ " to " • ."

$$= \frac{\overset{5}{\cancel{10}}}{1} \cdot \frac{3}{\underset{1}{\cancel{2}}}$$ Divide the numerator and denominator by 2, and then multiply.

$$= \frac{15}{1} = 15$$

Step 5 **State the answer.** Goldie must fill the container 15 times.

Step 6 **Check.** The exact answer, 15 times, is reasonably close to our estimate of 10 times.

◀ **Work Problem 4 at the Side.**

EXAMPLE 5 Applying Fraction Skills

At the Happi-Time Day Care Center, $\frac{6}{7}$ of the total budget goes to classroom operation. If there are 18 classrooms and each one receives the same amount, what fraction of the operating amount does each classroom receive?

Step 1 **Read** the problem. Since $\frac{6}{7}$ of the total budget must be split into 18 parts, we must find the fraction of the classroom operating amount received by each classroom.

Step 2 **Work out a plan.** We must divide the fraction of the total budget going to classroom operation $\left(\frac{6}{7}\right)$ by the number of classrooms (18).

Step 3 **Estimate** a reasonable answer. Round $\frac{6}{7}$ to 1. If all of the operating expenses (1 whole) were divided between 18 classrooms, each classroom would receive $\frac{1}{18}$ of the operating expenses, our estimate.

Step 4 **Solve** the problem. We solve by dividing $\frac{6}{7}$ by 18.

$$\frac{6}{7} \div 18 = \frac{6}{7} \div \frac{18}{1}$$ Write 18 as $\frac{18}{1}$.

$$= \frac{\overset{1}{\cancel{6}}}{7} \cdot \frac{1}{\underset{3}{\cancel{18}}}$$ The reciprocal of $\frac{18}{1}$ is $\frac{1}{18}$. Change "÷" to "•". Divide the numerator and denominator by 6.

$$= \frac{1}{21}$$ Multiply.

Step 5 **State the answer.** Each classroom receives $\frac{1}{21}$ of the total budget.

Step 6 **Check.** The exact answer, $\frac{1}{21}$, is close to our estimate of $\frac{1}{18}$.

Work Problem **5** at the Side. ▶

5 Solve each problem using the six problem-solving steps.

GS **(a)** The top 12 employees at Mayfield Manufacturing will divide $\frac{3}{4}$ of the annual bonus money. What fraction of the bonus money will each employee receive?

Estimate: $1 \div 12$ is $\frac{1}{12}$

Solve: $\frac{3}{4} \div 12 = \frac{3}{4} \div \frac{12}{1}$

$$= \frac{\cancel{3}}{4} \cdot \frac{1}{\cancel{12}} = \underline{\quad}$$

State the answer: = ____ of the bonus money

Check: Answer is close to estimate.

(b) A winning lottery ticket was purchased by 8 employees of United States Marketing and Promotions (USMP). They will donate $\frac{1}{5}$ of the total winnings to pay the medical expenses of a fellow employee, and then divide the remaining winnings evenly. What fraction of the prize money will each receive?

Answers

5. (a) 1; 4; $\frac{1}{16}$ of the bonus money

(b) $\frac{1}{10}$ of the lottery prize money

2.7 Exercises

FOR EXTRA HELP

Download the MyDashBoard App

MyMathLab®

CONCEPT CHECK *Fill in the blank.*

1. When you invert or flip a fraction, you have the __________ of the fraction.

2. To find the reciprocal of a whole number, you must first write the whole number over __________, and then invert it.

3. To divide by a fraction, you must first __________ the divisor and then change division to __________.

4. After completing a fraction division problem, it is best to write the answer in __________ terms.

CONCEPT CHECK *Find the reciprocal of each number.*

5. $\frac{3}{8}$

6. $\frac{2}{5}$

7. $\frac{5}{6}$

8. $\frac{12}{7}$

9. $\frac{8}{5}$

10. $\frac{13}{20}$

11. 4

12. 10

Divide. Write answers in lowest terms and as whole or mixed numbers where possible. See ***Examples 2 and 3.***

13. $\frac{1}{2} \div \frac{3}{4}$

14. $\frac{5}{8} \div \frac{7}{8}$

15. $\frac{7}{8} \div \frac{1}{3}$

16. $\frac{7}{8} \div \frac{3}{4}$

17. $\frac{3}{4} \div \frac{5}{3}$

18. $\frac{4}{5} \div \frac{9}{4}$

19. $\frac{7}{9} \div \frac{7}{36}$

20. $\frac{5}{8} \div \frac{5}{16}$

21. $\frac{15}{32} \div \frac{5}{64}$

22. $\frac{7}{12} \div \frac{14}{15}$

23. $\dfrac{\frac{13}{20}}{\frac{4}{5}}$

24. $\dfrac{\frac{9}{10}}{\frac{3}{5}}$

25. $\dfrac{\frac{5}{6}}{\frac{25}{24}}$

$\frac{5}{6} \div \frac{25}{24}$

$\frac{\overset{1}{\cancel{5}}}{\underset{1}{\cancel{6}}} \cdot \frac{\overset{4}{\cancel{24}}}{\underset{5}{\cancel{25}}} =$

26. $\dfrac{\frac{28}{15}}{\frac{21}{5}}$

$\frac{28}{15} \div \frac{21}{5}$

$\frac{\overset{4}{\cancel{28}}}{\underset{3}{\cancel{15}}} \cdot \frac{\overset{1}{\cancel{5}}}{\underset{3}{\cancel{21}}} =$

27. $12 \div \frac{2}{3}$

28. $7 \div \frac{1}{4}$

29. $\dfrac{18}{\frac{3}{4}}$

30. $\dfrac{12}{\frac{3}{4}}$

31. $\dfrac{\frac{4}{7}}{8}$

32. $\dfrac{\frac{7}{10}}{3}$

Solve each application problem by using division. ***See Examples 4 and 5.***

33. Veterinarian Jasmine Cato has $\frac{8}{9}$ quart of medication. She prescribes this medication for 4 horses in her care. If she divides the medication evenly, how much will each horse receive?

$$\frac{8}{9} \div 4 = \frac{\overset{2}{\cancel{8}}}{9} \cdot \frac{1}{\underset{1}{\cancel{4}}} =$$

34. Harold Pishke, barber, has 15 quarts of conditioning shampoo. If he wants to put this shampoo into $\frac{3}{8}$-quart containers, how many containers can be filled?

$$\frac{15}{1} \div \frac{3}{8} = \frac{\overset{5}{\cancel{15}}}{1} \cdot \frac{8}{\underset{1}{\cancel{3}}} =$$

35. Some college roommates want to make pancakes for their neighbors. They need 5 cups of flour, but have only a $\frac{1}{3}$-cup measuring cup. How many times will they need to fill their measuring cup?

36. How many $\frac{3}{8}$ pound bags of Jelly Belly jelly beans can be filled with 408 pounds of jelly beans? (*Source:* Jelly Belly Candy Company.)

37. How many $\frac{1}{8}$-ounce eye drop dispensers can be filled with 11 ounces of eye drops?

38. It is estimated that each guest at a party will eat $\frac{5}{16}$ pound of peanuts. How many guests can be served with 10 pounds of peanuts?

39. Metal fasteners used in furniture assembly weigh $\frac{5}{32}$ pound and are packaged in 25-pound cartons. Find the number of fasteners in each carton.

40. The new residential subdivision features large $\frac{3}{4}$-acre lots. How many lots are there in the 210- acre-subdivision? (Do not consider streets, sidewalks, or other improvements.)

41. Your classmate is confused on how to divide by a fraction. Write a short note telling him how this should be done.

42. If you multiply positive proper fractions, the product is less than the fractions multiplied. When you divide by a proper fraction, is the quotient less than the fractions in the problem? Prove your answer with examples.

Solve each application problem using multiplication or division.

43. The recipe for a Jelly Belly Express loafcake calls for $\frac{3}{4}$ pound of Jelly Belly jelly beans in assorted colors. If you want to make 16 cakes, how many pounds of Jelly Belly jelly beans will you need?

$$\frac{3}{4} \cdot 16 = \frac{3}{\underset{1}{\cancel{4}}} \cdot \frac{\overset{4}{\cancel{16}}}{1} =$$

44. In a recent study, it was found that one month after leaving the hospital only $\frac{7}{8}$ of the 1520 heart attack patients were still taking the life-saving drugs prescribed for them. How many of these patients were still taking their drugs? (*Source:* Dr. Michael Ho, Denver Veterans Medical Center.)

45. Broadly Plumbing finds that $\frac{3}{4}$ can of pipe joint compound is needed for each new home. How many homes can be plumbed with 156 cans of compound?

46. The Auto Technology Center at American River Community College purchases differential fluid in drums. If each car serviced needs $\frac{2}{3}$ gallon of lubricant, how many cars can be serviced with a 50-gallon drum?

47. In recordings of 186 patient visits, doctors failed to mention a new drug's side effects or how long to take the drug in $\frac{2}{3}$ of the visits.

(a) How many visits did doctors fail to discuss these issues with patients?

(b) How many times did they discuss them? (*Source:* Dr. Deijieng Tam, UCLA.)

48. Hans Bueff has completed $\frac{4}{5}$ of his kayak excursion down the Colorado River. If the trip is a total of 285 miles,

(a) how many miles has he gone?

(b) how many miles remain?

49. A dish towel manufacturer requires $\frac{3}{8}$ yard of cotton fabric for each towel. Find the number of dish towels that can be made from 912 yards of fabric.

50. A local McDonald's expects 300 applicants but only has jobs for $\frac{1}{60}$ of those who apply. How many job openings are these?

Relating Concepts (Exercises 51–56) For Individual or Group Work

Many application problems are solved using multiplication and division of fractions.
Work Exercises 51–54 in order.

51. Perhaps the most common indicator word for multiplication is the word *of* (when it follows a fraction). Circle the words in the list below that are also indicator words for multiplication.

more than	per
double	twice
times	product
less than	difference
equals	twice as much

52. Circle the words in the list below that are indicator words for division.

fewer	sum of
goes into	divide
per	quotient
equals	double
loss of	divided by

53. To divide two fractions, multiply the first fraction by the _________ of the second fraction.

54. Find the reciprocals for each number.

$\frac{3}{4}$ $\quad \frac{7}{8}$ $\quad 5$ $\quad \frac{12}{19}$

The size of an antique U.S.A. postage stamp is shown here. Use this to answer Exercises 55 and 56.

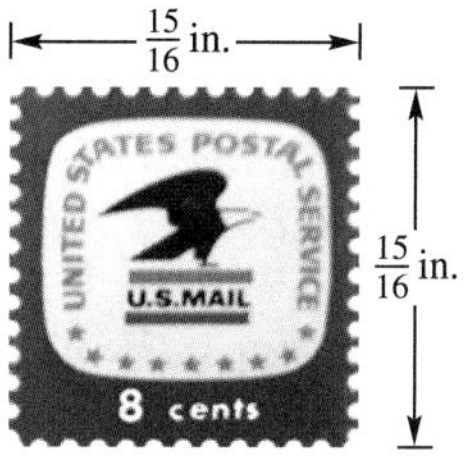

55. **(a)** Explain how to find the perimeter (distance around the edges) of any flat, equal-sided, 3-, 4-, 5-, or 6-sided figure using multiplication.

(b) Find the perimeter of the stamp using multiplication.

56. Find the area of the postage stamp. Explain how to find the area of any rectangle.

Study Skills

USING STUDY CARDS REVISITED

OBJECTIVES

1. **Create study cards for difficult problems.**
2. **Create study cards of quiz problems.**

This is the second part of the Study Cards activity. As you get further into a chapter, you can choose particular problems that will serve as a good test review. Here are two more types of study cards that will help you.

Tough Problems Card

When you are doing your homework and find yourself saying, "This is really hard," or "I'm having trouble with this," make a **tough problem** study card! On the front, write out the procedure to work the type of problem *in words*. If there are special notes (like what *not* to do), include them. On the back, work at least one example; make sure you label what you are doing.

Front of Card

Warning: *Division is NOT commutative. The order in which you write the numbers DOES matter.*

Example: $1\frac{1}{3} \div 4 = \frac{\cancel{4}^{1}}{3} \cdot \frac{1}{\cancel{4}_{1}} = \frac{1}{3}$

But $4 \div 1\frac{1}{3} = \frac{\cancel{4}^{1}}{1} \cdot \frac{3}{\cancel{4}_{1}} = \frac{3}{1} = 3$

Very different answers!

In an application problem, do NOT assume the numbers are given in the correct order. Use estimation to check that the answer is reasonable!

Back of Card

Maite painted 4 windows using $1\frac{1}{3}$ cans of paint. How much paint did she use on each window?

Try $4 \div 1\frac{1}{3}$ *(round)*

$4 \div 1 = 4$ *cans on each window* ← *Estimate*

Not reasonable — she only used $1\frac{1}{3}$ cans in all!

Need to find: paint on each window

$1\frac{1}{3}$ *cans* $\div$ $4 = \frac{\cancel{4}^{1}}{3} \cdot \frac{1}{\cancel{4}_{1}} = \frac{1}{3}$ *can*

Reasonable!

Now Try This

Choose three types of difficult problems, and work them out on *study cards*. Be sure to put the words for solving the problem on one side and the worked problem on the other side.

Practice Quiz Cards

Make up a few **quiz cards** for each type of problem you learn, and use them to prepare for a test. Choose two or three problems from the different sections of the chapter. Be sure you don't just choose the easiest problems! Put the problem **with the direction words** (like *solve, simplify, estimate*) on the front, and work the problem on the back. If you like, put the page number from the text there, too. When you review, you work the problem on a separate paper, and check it by looking at the back.

Solve this application problem.

Tiffany's monthly income is $1275. She spends $\frac{2}{5}$ of her income on rent and utilities. How much does she pay for rent and utilities?

Front of Card

Proper fraction followed by "of" indicates multiplication.

She spends $\frac{2}{5}$ of her income

$$\frac{2}{5} \cdot 1275 = \frac{2}{\underset{1}{\cancel{5}}} \cdot \frac{\overset{255}{\cancel{1275}}}{1} = \frac{510}{1} = 510$$

Tiffany spends $510 on rent and utilities.

Back of Card

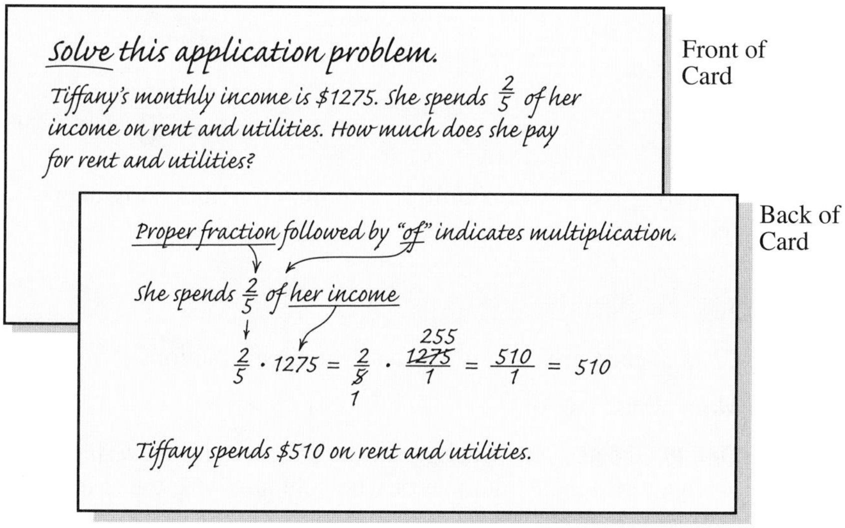

Why Are Study Cards Brain Friendly?

First, making the study cards is an **active technique** that really gets your dendrites growing. You have to make decisions about what is most important and how to put it on the card. This kind of thinking is more in depth than just memorizing, and as a result, you will understand the concepts better and remember them longer.

Second, the cards are **visually appealing** (if you write neatly and try some color). Your brain responds to pleasant visual images, and again, you will remember longer and may even be able to "picture in your mind" how your cards look. This will help you during tests.

Third, because study cards are small and portable, you can review them easily whenever you have a few minutes. Even while you're waiting for a bus or have a few minutes between classes you can take out your cards and read them to yourself. Your **brain really benefits from repetition**; each time you review your cards your dendrites are growing thicker and stronger. After a while, the information will become automatic and you will remember it for a long time.

2.8 Multiplying and Dividing Mixed Numbers

OBJECTIVES

1. Estimate the answer and multiply mixed numbers.
2. Estimate the answer and divide mixed numbers.
3. Solve application problems with mixed numbers.

Earlier in this chapter we worked with mixed numbers—a whole number and a fraction written together. Many of the fraction problems you encounter in everyday life involve mixed numbers.

OBJECTIVE 1 Estimate the answer and multiply mixed numbers. When multiplying mixed numbers, it is a good idea to estimate the answer first.

To estimate the answer, round each mixed number to the nearest whole number. If the numerator is *half* of the denominator or *more,* round up the whole number part. If the numerator is *less* than half the denominator, leave the whole number as it is.

$1\frac{5}{8}$ ← 5 is more than 4. / ← Half of 8 is 4. $\quad 1\frac{5}{8}$ rounds up to 2.

$3\frac{2}{5}$ ← 2 is less than $2\frac{1}{2}$. / ← Half of 5 is $2\frac{1}{2}$. $\quad 3\frac{2}{5}$ rounds to 3.

Round mixed numbers to the nearest whole number when estimating.

◀ Work Problem 1 at the Side.

After estimating the answer, multiply the mixed numbers by using the following steps.

Multiplying Mixed Numbers

Step 1 *Change* each mixed number to an improper fraction.

Step 2 *Multiply* as fractions.

Step 3 *Simplify* the answer, which means to write it in *lowest terms,* and change it to a mixed number or whole number where possible.

EXAMPLE 1 Multiplying Mixed Numbers

First estimate the answer. Then multiply to get an exact answer. Simplify your answers.

(a) $2\frac{1}{2} \cdot 3\frac{1}{5}$

Estimate the answer by rounding the mixed numbers.

$2\frac{1}{2}$ rounds to 3 and $3\frac{1}{5}$ rounds to 3

$3 \cdot 3 = 9$ Estimated answer

To find the exact answer, change each mixed number to an improper fraction.

Step 1

$2\frac{1}{2} = \frac{5}{2}$ and $3\frac{1}{5} = \frac{16}{5}$

Continued on Next Page

1 Round each mixed number to the nearest whole number.

GS **(a)** $4\frac{2}{3}$ ← Half of 3 is $1\frac{1}{2}$

2 is (*less than/more than*) half of 3.

$4\frac{2}{3}$ rounds to ____.

GS **(b)** $3\frac{2}{5}$ ← Half of 5 is $2\frac{1}{2}$

2 is (*less than/more than*) half of 5.

$3\frac{2}{5}$ rounds to ____.

(c) $5\frac{3}{4}$ **(d)** $4\frac{7}{12}$

(e) $1\frac{1}{2}$ **(f)** $8\frac{4}{9}$

Answers

1. **(a)** more than; 5 **(b)** less than; 3 **(c)** 6 **(d)** 5 **(e)** 2 **(f)** 8

Next, multiply.

Step 1 / *Step 2* / *Step 3*

$$2\frac{1}{2} \cdot 3\frac{1}{5} = \frac{5}{2} \cdot \frac{16}{5} = \frac{\overset{1}{\cancel{5}}}{\underset{1}{\cancel{2}}} \cdot \frac{\overset{8}{\cancel{16}}}{\underset{1}{\cancel{5}}} = \frac{1 \cdot 8}{1 \cdot 1} = \frac{8}{1} = 8$$

Remember: Change to improper fractions, then divide out common factors, and finally multiply.

The estimated answer is 9 and the exact answer is 8. The exact answer is reasonable.

(b) $3\frac{5}{8} \cdot 4\frac{4}{5}$

$3\frac{5}{8}$ rounds to 4 and $4\frac{4}{5}$ rounds to 5

$4 \cdot 5 = 20$ Estimated answer

Now find the exact answer.

Step 1 / *Step 2*

$$3\frac{5}{8} \cdot 4\frac{4}{5} = \frac{29}{8} \cdot \frac{24}{5} = \frac{29}{\underset{1}{\cancel{8}}} \cdot \frac{\overset{3}{\cancel{24}}}{5} = \frac{29 \cdot 3}{1 \cdot 5} = \frac{87}{5}$$

Step 3

$$\frac{87}{5} = 17\frac{2}{5}$$

Simplify this answer by writing it as a mixed number.

The estimate was 20, so the exact answer of $17\frac{2}{5}$ is reasonable.

(c) $1\frac{3}{5} \cdot 3\frac{1}{3}$

$1\frac{3}{5}$ rounds to 2 and $3\frac{1}{3}$ rounds to 3

$2 \cdot 3 = 6$ Estimated answer

The exact answer is shown below.

$$1\frac{3}{5} \cdot 3\frac{1}{3} = \frac{8}{\underset{1}{\cancel{5}}} \cdot \frac{\overset{2}{\cancel{10}}}{3} = \frac{8 \cdot 2}{1 \cdot 3} = \frac{16}{3} = 5\frac{1}{3}$$

The estimate was 6, so the exact answer of $5\frac{1}{3}$ is reasonable.

Work Problem 2 at the Side. ▶

2 First estimate the answer. Then multiply to find the exact answer. Simplify your answers.

GS **(a)** $3\frac{1}{4} \cdot 6\frac{2}{3}$

↓ ↓

____ • ____

= ____ *estimate*

Exact: $3\frac{1}{4} \cdot 6\frac{2}{3}$

$$\frac{13}{\underset{1}{\cancel{4}}} \cdot \frac{\overset{5}{\cancel{20}}}{3} = \frac{65}{3} = \_\_\_\_$$

(b) $4\frac{2}{3} \cdot 2\frac{3}{4}$

↓ ↓

____ • ____

= ____ *estimate*

(c) $3\frac{3}{5} \cdot 4\frac{4}{9}$

↓ ↓

____ • ____

= ____ *estimate*

(d) $5\frac{1}{4} \cdot 3\frac{3}{5}$

↓ ↓

____ • ____

= ____ *estimate*

Answers

2. **(a)** *Estimate:* 3; 7; 21; *Exact:* $21\frac{2}{3}$

(b) *Estimate:* $5 \cdot 3 = 15$; *Exact:* $12\frac{5}{6}$

(c) *Estimate:* $4 \cdot 4 = 16$; *Exact:* 16

(d) *Estimate:* $5 \cdot 4 = 20$; *Exact:* $18\frac{9}{10}$

OBJECTIVE 2 Estimate the answer and divide mixed numbers. Just as you did when multiplying mixed numbers, it is also a good idea to estimate the answer when dividing mixed numbers. To divide mixed numbers, use the following steps.

Dividing Mixed Numbers

Step 1 *Change* each mixed number to an improper fraction.

Step 2 Use the *reciprocal* of the second fraction (divisor).

Step 3 *Change* division to multiplication.

Step 4 *Simplify* the answer, which means to write it in *lowest terms,* and change it to a mixed number or whole number where possible.

Note

Recall that the reciprocal of a fraction is found by interchanging the numerator and the denominator.

EXAMPLE 2 Dividing Mixed Numbers

First estimate the answer. Then divide to find the exact answer. Simplify your exact answers.

(a) $2\frac{2}{5} \div 1\frac{1}{2}$

First estimate the answer by rounding each mixed number to the nearest whole number.

$$2\frac{2}{5} \div 1\frac{1}{2}$$

Rounded

$$2 \div 2 = 1 \quad \text{Estimated answer}$$

To find the exact answer, first change each mixed number to an improper fraction.

Step 1

$$2\frac{2}{5} \div 1\frac{1}{2} = \frac{12}{5} \div \frac{3}{2}$$

Change mixed numbers to improper fractions.

Next, use the reciprocal of the second fraction and change division to multiplication.

Step 2 Step 3 Step 4

$$\frac{12}{5} \div \frac{3}{2} = \frac{\overset{4}{\cancel{12}}}{5} \cdot \frac{2}{\underset{1}{\cancel{3}}} = \frac{4 \cdot 2}{5 \cdot 1} = \frac{8}{5} = 1\frac{3}{5} \quad \text{Exact answer simplified}$$

Reciprocals

Remember: Use the reciprocal of the *divisor* (the *second* fraction).

The estimate was 1, so the exact answer of $1\frac{3}{5}$ is reasonable.

Continued on Next Page

(b) $8 \div 3\frac{3}{5}$

$$8 \div 3\frac{3}{5}$$

Rounded

$$8 \div 4 = 2 \quad \text{Estimate}$$

Now find the exact answer.

Reciprocals

$$8 \div 3\frac{3}{5} = \frac{8}{1} \div \frac{18}{5} = \frac{\overset{4}{\cancel{8}}}{1} \cdot \frac{5}{\underset{9}{\cancel{18}}} = \frac{20}{9} = 2\frac{2}{9}$$

Write 8 as $\frac{8}{1}$.

Divide out common factors only *after* you have changed to the reciprocal and are multiplying.

The estimate was 2, so the exact answer of $2\frac{2}{9}$ is reasonable.

(c) $4\frac{3}{8} \div 5$

$$4\frac{3}{8} \div 5$$

Rounded

Reciprocals

$$4 \div 5 = \frac{4}{1} \div \frac{5}{1} = \frac{4}{1} \cdot \frac{1}{5} = \frac{4}{5} \quad \text{Estimate}$$

The exact answer is shown below.

Reciprocals

$$4\frac{3}{8} \div 5 = \frac{35}{8} \div \frac{5}{1} = \frac{\overset{7}{\cancel{35}}}{8} \cdot \frac{1}{\underset{1}{\cancel{5}}} = \frac{7}{8}$$

Write 5 as $\frac{5}{1}$.

The estimate was $\frac{4}{5}$, so the exact answer of $\frac{7}{8}$ is reasonable.

Work Problem 3 at the Side. ▶

OBJECTIVE 3 Solve application problems with mixed numbers. The next two examples show how to solve application problems involving mixed numbers.

EXAMPLE 3 Applying Multiplication Skills

The local Habitat for Humanity chapter is looking for 11 contractors who will each donate $3\frac{1}{4}$ days of labor to a community building project. How many days of labor will be donated in all?

Step 1 **Read** the problem. The problem asks for the total days of labor donated by the 11 contractors.

Step 2 **Work out a plan.** Multiply the number of contractors (11) and the amount of labor that each donates ($3\frac{1}{4}$ days).

Continued on Next Page

3 First estimate the answer. Then divide to find the exact answer. Simplify all answers.

GS **(a)** $3\frac{1}{8} \div 6\frac{1}{4}$

↓ ↓

$\underline{3} \div \underline{6}$

= ____ *estimate*

Exact: $3\frac{1}{8} \div 6\frac{1}{4} =$

$$\frac{25}{8} \div \frac{25}{4} = \frac{\overset{1}{\cancel{25}}}{\underset{2}{\cancel{8}}} \cdot \frac{\overset{1}{\cancel{4}}}{\underset{1}{\cancel{25}}} =$$

(b) $10\frac{1}{3} \div 2\frac{1}{2}$

↓ ↓

____ ÷ ____

= ____ *estimate*

(c) $8 \div 5\frac{1}{3}$

↓ ↓

____ ÷ ____

= ____ *estimate*

(d) $13\frac{1}{2} \div 18$

↓ ↓

____ ÷ ____

= ____ *estimate*

Answers

3. (a) *Estimate:* $3 \div 6 = \frac{1}{2}$; *Exact:* $\frac{1}{2}$

(b) *Estimate:* $10 \div 3 = 3\frac{1}{3}$; *Exact:* $4\frac{2}{15}$

(c) *Estimate:* $8 \div 5 = 1\frac{3}{5}$; *Exact:* $1\frac{1}{2}$

(d) *Estimate:* $14 \div 18 = \frac{7}{9}$; *Exact:* $\frac{3}{4}$

4 Use the six problem-solving steps. Simplify the answer.
GS If one car requires $2\frac{5}{8}$ quarts of paint, find the number of quarts needed to paint 16 cars.

Estimate: ____ • 16 = 48

Solve: $2\frac{5}{8} \cdot 16 = \frac{21}{8} \cdot \frac{16}{1}$

$= \frac{21}{\cancel{8}_1} \cdot \frac{\cancel{16}^2}{1}$

State the answer: ____ quarts of paint are needed.

Check: Answer is close to estimate.

5 Use the six problem-solving steps. Simplify all answers.

(a) The manufacture of one outboard engine propeller requires $4\frac{3}{4}$ pounds of brass. How many propellers can be manufactured from 57 pounds of brass?

(b) Jack Armstrong Trucking uses $21\frac{3}{4}$ quarts of motor oil for each oil change on his diesel engine truck. Find the number of oil changes that can be made with 609 quarts of oil.

Answers

4. *Estimate:* 3 • 16 = 48; *Exact:* 42 quarts
5. (a) *Estimate:* 57 ÷ 5 ≈ 11; *Exact:* 12 propellers
   (b) *Estimate:* 600 ÷ 22 ≈ 27; *Exact:* 28 oil changes

Step 3 **Estimate** a reasonable answer. Round $3\frac{1}{4}$ days to 3 days. Multiply 3 days by 11 contractors (3 • 11) to get an estimate of 33 days.

Step 4 **Solve** the problem. Find the exact answer.

$$11 \cdot 3\frac{1}{4} = 11 \cdot \frac{13}{4}$$
$$= \frac{11}{1} \cdot \frac{13}{4} = \frac{143}{4} = 35\frac{3}{4}$$

Step 5 **State the answer.** The community building project will receive $35\frac{3}{4}$ days of donated labor.

Always check to see if the answer is close to the estimate.

Step 6 **Check.** The exact answer, $35\frac{3}{4}$ days, is close to our estimate of 33 days.

◀ Work Problem **4** at the Side.

EXAMPLE 4 Applying Division Skills

A dome tent for backpacking requires $7\frac{1}{4}$ yards of nylon material. How many tents can be made from $21\frac{3}{4}$ yards of material?

Step 1 **Read** the problem. The problem asks how many tents can be made from $21\frac{3}{4}$ yards of material.

Step 2 **Work out a plan.** Divide the number of yards of cloth ($21\frac{3}{4}$ yd) by the number of yards needed for one tent ($7\frac{1}{4}$ yd).

Step 3 **Estimate** a reasonable answer.

$$21\frac{3}{4} \div 7\frac{1}{4}$$

↓ Rounded ↓

$$22 \div 7 \approx 3 \text{ tents} \quad \text{Estimate}$$

Recall ≈ means "approximately equal to."

Step 4 **Solve** the problem.

$$21\frac{3}{4} \div 7\frac{1}{4} = \frac{87}{4} \div \frac{29}{4}$$
$$= \frac{\cancel{87}^3}{\cancel{4}_1} \cdot \frac{\cancel{4}^1}{\cancel{29}_1} = \frac{3}{1} = 3 \quad \text{Matches estimate}$$

Step 5 **State the answer.** 3 tents can be made from $21\frac{3}{4}$ yards of cloth.

Step 6 **Check.** The exact answer, 3, matches our estimate.

◀ Work Problem **5** at the Side.

Note

When rounding mixed numbers to estimate the answer to a problem, the estimated answer usually varies somewhat from the exact answer. However, the importance of the estimated answer is that it will show you whether your exact answer is reasonable or not.

2.8 EXERCISES

FOR EXTRA HELP

CONCEPT CHECK *Decide whether each statement is* true *or* false. *If it is* false, *say why.*

1. When multiplying two mixed numbers, the reciprocal of the second mixed number must be used.

2. If you were dividing a mixed number by the whole number 10, the reciprocal of 10 would be $\frac{10}{1}$.

3. To round mixed numbers before estimating the answer, decide whether the numerator of the fraction part is less than or more than half of the denominator.

4. When rounding mixed numbers to estimate the answer to a problem, the estimated answer can vary quite a bit from the exact answer. However, it can still show whether the exact answer is reasonable.

First estimate the answer. Then multiply to find the exact answer. Simplify all answers.
See Example 1.

5. $4\frac{1}{2} \cdot 1\frac{3}{4}$

Estimate:

___ • ___ = ___

Exact:

6. $2\frac{1}{2} \cdot 2\frac{1}{4}$

Estimate:

___ • ___ = ___

Exact:

7. $1\frac{2}{3} \cdot 2\frac{7}{10}$

Estimate:

___ • ___ = ___

Exact:

8. $4\frac{1}{2} \cdot 2\frac{1}{4}$

Estimate:

___ • ___ = ___

Exact:

9. $3\frac{1}{9} \cdot 1\frac{2}{7}$

Estimate:

___ • ___ = ___

Exact:

10. $6\frac{1}{4} \cdot 3\frac{1}{5}$

Estimate:

___ • ___ = ___

Exact:

11. $8 \cdot 6\frac{1}{4}$

Estimate:

___ • ___ = ___

Exact:

12. $6 \cdot 2\frac{1}{3}$

Estimate:

___ • ___ = ___

Exact:

13. $4\frac{1}{2} \cdot 2\frac{1}{5} \cdot 5$

Estimate:

___ • ___ • ___ = ___

Exact:

14. $5\frac{1}{2} \cdot 1\frac{1}{3} \cdot 2\frac{1}{4}$

Estimate:

___ • ___ • ___ = ___

Exact:

15. $3 \cdot 1\frac{1}{2} \cdot 2\frac{2}{3}$

Estimate:

___ • ___ • ___ = ___

Exact:

16. $\frac{2}{3} \cdot 3\frac{2}{3} \cdot \frac{6}{11}$

Estimate:

___ • ___ • ___ = ___

Exact:

CONCEPT CHECK *Choose the best estimated answer to each problem.*

17. $3\frac{1}{4} \cdot 7\frac{5}{8}$

A. 21 **B.** 32 **C.** 28 **D.** 24

18. $5\frac{3}{4} \cdot 2\frac{2}{5}$

A. 12 **B.** 18 **C.** 10 **D.** 15

19. $2\frac{1}{8} \div 1\frac{3}{4}$

A. 2 **B.** 1 **C.** $1\frac{1}{2}$ **D.** 3

20. $8\frac{3}{5} \div 2\frac{3}{8}$

A. 4 **B.** $3\frac{1}{2}$ **C.** $4\frac{1}{2}$ **D.** 3

First estimate the answer. Then divide to find the exact answer. Simplify all answers.
See Example 2.

21. $1\frac{1}{4} \div 3\frac{3}{4}$

Estimate:

___ ÷ ___ = ___

Exact:

22. $1\frac{1}{8} \div 2\frac{1}{4}$

Estimate:

___ ÷ ___ = ___

Exact:

23. $2\frac{1}{2} \div 3$

GS

Estimate:

___ ÷ ___ = ___

Exact: $\frac{5}{2} \div \frac{3}{1}$

$\frac{5}{2} \cdot \frac{1}{3} =$

24. $2\frac{3}{4} \div 2$

GS

Estimate:

___ ÷ ___ = ___

Exact: $2\frac{3}{4} \div \frac{2}{1}$

$\frac{11}{4} \cdot \frac{1}{2} = \frac{11}{8} =$

25. $9 \div 2\frac{1}{2}$

Estimate:

___ ÷ ___ = ___

Exact:

26. $5 \div 1\frac{7}{8}$

Estimate:

___ ÷ ___ = ___

Exact:

27. $\frac{5}{8} \div 1\frac{1}{2}$

Estimate:

___ ÷ ___ = ___

Exact:

28. $\frac{3}{4} \div 2\frac{1}{2}$

Estimate:

___ ÷ ___ = ___

Exact:

29. $1\frac{7}{8} \div 6\frac{1}{4}$

Estimate:

___ ÷ ___ = ___

Exact:

30. $8\frac{2}{5} \div 3\frac{1}{2}$

Estimate:

___ ÷ ___ = ___

Exact:

31. $5\frac{2}{3} \div 6$

Estimate:

___ ÷ ___ = ___

Exact:

32. $5\frac{3}{4} \div 2$

Estimate:

___ ÷ ___ = ___

Exact:

For Exercises 33–50, first estimate the answer. Then find each exact answer and simplify it if possible. ***See Examples 3 and 4.*** *Use the recipe for Carrot Cake Cupcakes to work Exercises 33–36.*

CARROT CAKE CUPCAKES

12 paper bake cups
$1\frac{3}{4}$ cups flour
1 cup packed brown sugar
1 tsp. baking powder
1 tsp. baking soda
1 tsp. ground cinnamon
$\frac{1}{2}$ tsp. salt
1 cup shredded carrots
$\frac{3}{4}$ cup applesauce
$\frac{1}{3}$ cup vegetable oil
1 large egg
$\frac{1}{2}$ tsp. vanilla extract
1 container (16 oz.) ready-to-spread cream cheese frosting
Shredded coconut, tinted green
3 oz. Jelly Belly jelly beans, Orange Sherbet flavor

Preheat oven to 350° F. Place 12 bake cups in a muffin pan; set aside. In a large bowl, using a wire whisk, stir together flour, brown sugar, baking powder, baking soda, cinnamon, and salt. In a medium bowl, combine carrots, applesauce, oil, egg, and vanilla until blended. Add carrot mixture to flour mixture, stir well. Spoon batter into bake cups, filling $\frac{2}{3}$ full. Bake until toothpick inserted in center comes out clean, 20–25 minutes. Cool cupcakes in pan 10 minutes; remove to wire rack and cool completely. Frost with cream cheese frosting. To make carrot design on cupcakes, place gourmet Jelly Belly jelly beans in a carrot shape on each cupcake, top with green coconut for carrot top. Makes 12 cupcakes.

33. If 30 cupcakes are baked ($2\frac{1}{2}$ times the recipe), find the amount of each ingredient.

(a) Applesauce

Estimate:

Exact:

(b) Salt

Estimate:

Exact:

(c) Flour

Estimate:

Exact:

34. If 18 cupcakes are baked ($1\frac{1}{2}$ times the recipe), find the amount of each ingredient.

(a) Flour

Estimate:

Exact:

(b) Applesauce

Estimate:

Exact:

(c) Vegetable oil

Estimate:

Exact:

35. How much of each ingredient is needed if you bake one-half of the recipe?

(a) Vanilla extract

Estimate:

Exact:

(b) Applesauce

Estimate:

Exact:

(c) Flour

Estimate:

Exact:

36. How much of each ingredient is needed if you bake one-third of the recipe?

(a) Flour

Estimate:

Exact:

(b) Salt

Estimate:

Exact:

(c) Applesauce

Estimate:

Exact:

37. A new condominium conversion project requires $11\frac{3}{4}$ gallons of paint for each unit. How many units can be painted with 329 gallons of paint?

Estimate:

Exact:

38. According to an old English system of time units, a moment is $1\frac{1}{2}$ minutes. How many moments are there in an 8-hour work day? (8 hours = 480 minutes). (*Source:* hightechscience.org/funfacts.htm)

Estimate:

Exact:

39. A manufacturer of floor jacks is ordering steel tubing to make the handles for the jack shown below. How much steel tubing is needed to make 45 of these jacks? The symbol for inches is ". For example, 5" means 5 inches. (*Source:* Harbor Freight Tools.)

Estimate:

Exact:

40. A wheelbarrow manufacturer uses handles made of hardwood. Find the amount of wood that is necessary to make 182 handles. The longest dimension shown in the advertisement below is the handle length. (*Source:* Harbor Freight Tools.)

Estimate:

Exact:

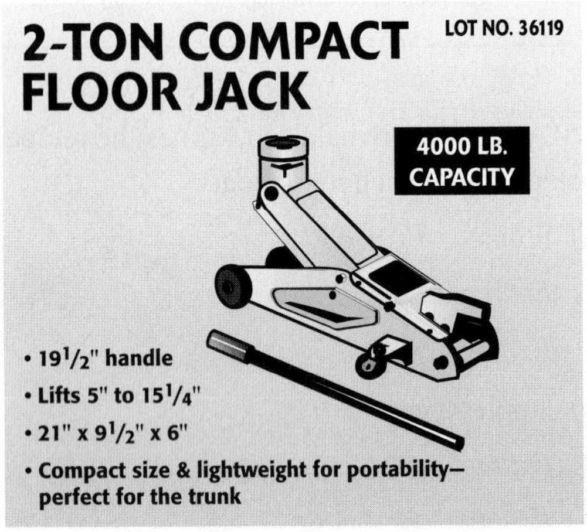

2-TON COMPACT FLOOR JACK
LOT NO. 36119
4000 LB. CAPACITY
• 19 1/2" handle
• Lifts 5" to 15 1/4"
• 21" x 9 1/2" x 6"
• Compact size & lightweight for portability—perfect for the trunk

6.0 CUBIC FT.
WHEEL BARROW
LOT NO. 46852
• Steel construction with hardwood handles
• 14" tubeless pneumatic tire
• Fully rolled edge for added tray strength
• Overall dimensions: 61 1/2" L x 27" W x 24.9" H

41. Write the three steps for multiplying mixed numbers. Use your own words.

42. Refer to **Exercise 41.** In your own words, write the additional step that must be added to the rule for multiplying mixed numbers to make it the rule for dividing mixed numbers.

43. The average cell phone contains $\frac{1}{1000}$ ounce of gold worth about $\$1\frac{2}{5}$. If you could extract the gold from all of the 130 million cell phones junked each year, how much money would you have from the sale of the gold? (*Source:* Discovery Channel, *Gold Rush Alaska.*)

Estimate:

Exact:

44. The manager of the flooring department at The Home Depot determines that each apartment unit requires $62\frac{1}{2}$ square yards of carpet. Find the number of apartment units that can be carpeted with 6750 square yards of carpet.

Estimate:

Exact:

45. Foxworthy Forest Products owns a truck that holds $1\frac{1}{4}$ cords of firewood. How many trips will he need to make to deliver 140 cords of firewood?

Estimate:

Exact:

46. BAE Roofing has completed an agreement to re-roof homes in a 20-year old residential development. Each home needs $31\frac{1}{2}$ squares of roofing material and the company has 1827 squares of material in inventory. How many homes will they be able to re-roof with this material?

Estimate:

Exact:

Use the information on bottle jacks in the advertisement to answer Exercises 47–48. The " symbol is for inches.

47. A mechanic needs a hydraulic lift that will raise a car 4 times as high as the standard jack shown.

(a) How high must it lift?

(b) Will a mechanic 6 feet tall be able to fit under the vehicle without bending down? *Hint:* 1 ft = 12 in.

(a) *Estimate:*

Exact:

(b)

48. A race car driver needs a jack that will raise a vehicle only $\frac{1}{3}$ as high as the low-profile jack pictured.

(a) How high must it lift?

(b) Will a 6-inch part fit under the car?

(a) *Estimate:*

Exact:

(b)

49. A shark can swim $5\frac{1}{2}$ times faster than a person. If a person can swim $6\frac{1}{8}$ miles per hour, how fast can a shark swim? (*Source:* Discovery Channel, *Animal Planet.*)

Estimate:

Exact:

50. A flooring contractor needs $24\frac{2}{7}$ boxes of tile to cover a kitchen floor. If there are 24 homes in a subdivision, how many boxes of tile are needed to cover all of the floors?

Estimate:

Exact:

Study Skills
REVIEWING A CHAPTER

OBJECTIVES

1. **Use the Chapter Quick Review to see examples of every type of problem.**
2. **Create study cards for vocabulary.**
3. **Practice by doing review and mixed review exercises.**
4. **Take the Chapter Test as a practice test.**

This activity is really about **preparing for tests.** Some of the suggestions are ideas that you will learn to use a little later in the term, but get started trying them out now. Often, the first chapters in your math textbook will be review, so it is good to practice some of the study techniques on material that is not too challenging.

Chapter Reviewing Techniques

Use these **chapter reviewing techniques.**

▶ **Make a study card for each vocabulary word and concept.** Include a definition, an example, a sketch, and a page reference. Include the symbol or formula if there is one. See the *Using Study Cards* activity for a quick look at some sample study cards.

▶ **Go back to the section** to find more explanations or information about any new vocabulary, formulas, or symbols.

▶ **Use the Chapter Summary** to see examples of each type of problem. Do not expect it to substitute for reading and working through the whole chapter! First, take the Test Your Word Power quiz to check your understanding of new vocabulary. The answers follow the quiz. Then read the Quick Review. **Pay special attention to the dark blue headings.** Check the explanations for the solutions to problems given. Try to think about how all the topics in the **whole chapter** are related.

▶ **Study your lecture notes** to see what your instructor has emphasized in class. Then review that material in your text.

▶ **Do the Review Exercises** to practice every type of problem.

✓ Check your answers **after** you're done with each **section of exercises.**
✓ If you get stuck on a problem, **first** check the Chapter Quick Review. If that doesn't clear up your confusion, then check the section and your lecture notes.
✓ Pay attention to **direction words** for the problems, such as *simplify, round, solve*, and *estimate*.
✓ Make **study cards for especially difficult problems.**

▶ **Do the Mixed Review exercises.** This is a good check to see if you can still do the problems when they are in mixed-up order. **Check your answers carefully** in the Answers section in the back of your book. Are your answers **exact** and **complete?** Make sure you are **labeling** answers correctly, using the right **units.** For example, does your answer need to include *$*, cm^2, *ft*, and so on?

- ▶ **Take the Chapter Test as if it is a real test.** If your instructor has skipped sections in the chapter, ask which problems to skip on the test before you start.
  - ✓ **Time yourself** just as you would for a real test.
  - ✓ **Use a calculator or notes** just as you would be permitted to (or not) on a real test.
  - ✓ **Take the test in one sitting,** just like a real test is given in one sitting.
  - ✓ **Show all your work.** Practice showing your work just the way your instructor has asked you to show it.
  - ✓ **Practice neatness.** Can someone else follow your steps?
  - ✓ **Check your answers** in the back of the book.

Notice that reviewing a chapter will take some time. Remember that it takes time for dendrites to grow! You cannot grow a good network of dendrites by rushing through a review in one night. But if you use the suggestions over a few days or evenings, you will notice that you understand the material more thoroughly and remember it longer.

Now Try This

Follow the reviewing techniques listed above for your next test. For each technique, write a comment about how it worked for you in the spaces below.

1. **Make a study card for each vocabulary word and concept.**

2. **Go back to the section** to find more explanations or information.

3. **Take the Test Your Word Power quiz and use the Quick Review** to review each concept in the chapter.

4. **Study your lecture notes** to see what your instructor has emphasized in class.

5. **Do the Review Exercises,** following the specific suggestions on the previous page.

6. **Do the Mixed Review exercises.**

7. **Take the Chapter Test** as if it is a real test.

Why Are These Review Activities Brain Friendly?

You have already become familiar with the features of your textbook. This activity requires you to make good use of them. Your **brain needs repetition** to strengthen dendrites and the connections between them. By following the steps outlined here, you will be reinforcing the concepts, procedures, and skills you need to use for tests (and for the next chapters).

The combination of techniques provides repetition in different ways. That **promotes good branching of dendrites** instead of just relying on one branch, or route, to connect to the other dendrites. A thorough review of each chapter will **solidify your dendrite connections.** It will help you be sure that you understand the concepts **completely and accurately.** Also, taking the Chapter Test will **simulate the testing situation,** which gives you practice in test taking conditions.

Study Skills

TIPS FOR TAKING MATH TESTS

OBJECTIVES

1. **Apply suggestions to tests and quizzes.**
2. **Develop a set of "best practices" to apply while testing.**

Improving Your Test Score

To Improve Your Test Score	Comments
Come prepared with a pencil, eraser, and calculator, if allowed. If you are easily distracted, sit in the corner farthest from the door.	***Working in pencil lets you erase,*** keeping your work neat and readable.
Scan the entire test, note the point value of different problems, and plan your time accordingly. Allow at least five minutes to check your work at the end of the testing time.	If you have 50 minutes to do 20 problems, $50 \div 20 = 2.5$ minutes per problem. ***Spend less time on easy ones,*** more time on problems with higher point values.
Read directions carefully, and circle any significant words. When you finish a problem, read the directions again to make sure you did what was asked.	***Pay attention to announcements*** written on the board or made by your instructor. Ask if you don't understand. You don't want to get problems wrong because you misread the directions!
Show your work. Most math teachers give partial credit if some of the steps in your work are correct, even if the final answer is wrong. ***Write neatly.*** If you like to scribble when first working or checking a problem, do it on scratch paper.	***If your instructor can't read your writing, you won't get credit for it.*** If you need more space to work, ask if you can use extra pieces of paper that you hand in with your test paper.
Check that the ***answer to an application problem is reasonable*** and makes sense. Read the problem again to make sure you've answered the question.	***Use common sense.*** Can the father really be seven years old? Would a month's rent be $32,140? Label your answer: $, years, inches, etc.
To check for careless errors, you need to ***rework the problem again, without looking at your previous work.*** Cover up your work with a piece of scratch paper, and pretend you are doing the problem for the first time. Then compare the two answers.	If you just "look over" your work, your mind can make the same mistake again without noticing it. Reworking the problem from the beginning ***forces you to rethink it.*** If possible, use a different method to solve the problem the second time.
Do not try to review up until the last minute before the test. Instead, go for a walk, do some deep breathing, and arrive just in time for the test. Ignore other students.	Listening to anxious classmates before the test ***may cause you to panic.*** Moderate exercise and deep breathing will calm your mind.

Reducing Anxiety

To Reduce Anxiety	Comments
Do a "knowledge dump" as soon as you get the test. Write important notes to yourself in a corner of the test paper: formulas, or common errors you want to watch out for.	Writing down tips and things that you've memorized ***lets you relax;*** you won't have to worry about forgetting those things and can refer to them as needed.
Do the easy problems first in order to build confidence. If you feel your anxiety starting to build, *immediately* stop for a minute, close your eyes, and take several slow, deep breaths.	Greater confidence helps you ***get the easier problems correct.*** Anxiety causes shallow breathing, which leads to confusion and reduced concentration. Deep breathing calms you.
As you work on more difficult problems, ***notice your "inner voice."*** You may have negative thoughts such as, "I can't do it," or "who cares about this test anyway." In your mind, yell, "STOP" and take several deep, slow breaths. Or, replace the negative thought with a positive one.	Here are ***examples of positive statements.*** Try writing one of them on the top of your test paper. • I know I can do it. • I can do this one step at a time. • I've studied hard, and I'll do the best I can.
If you still can't solve a difficult problem when you come back to it the second time, ***make a guess and do not change it.*** In this situation, your first guess is your best bet. Do not change an answer just because you're a little unsure. ***Change it only if you find an obvious mistake.***	If you are thinking about changing an answer, be sure you have a good reason for changing it. If you cannot find a specific error, leave your first answer alone. ***When the tests are returned, check to see if changing answers helped or hurt you.***
Read the harder problems twice. Write down *anything* that might help solve the problem: a formula, a picture, etc. If you still can't get it, circle the problem and ***come back to it later***. Do *not* erase any of the things you wrote down.	If you know even a *little* bit about the problem, write it down. The ***answer may come to you*** as you work on it, or you may get partial credit. Don't spend too long on any one problem. Your subconscious mind will work on the tough problem while you go on with the test.
Ignore students who finish early. Use the entire test time. *You do not get extra credit for finishing early.* Use the extra time to rework problems and correct careless errors.	Students who leave early are often the ones who didn't study or who are too anxious to continue working. If they bother you, ***sit as far from the door as possible.***

Why Are These Suggestions Brain Friendly?

Several suggestions address anxiety. **Reducing anxiety allows your brain to make the connections between dendrites;** in other words, you can think clearly.

Remember that **your brain continues to work on a difficult problem** even if you skip it and go on to the next one. Your subconscious mind will come through for you if you are open to the idea!

Some of the suggestions ask you to **use your common sense.** Follow the directions, show your work, write neatly, and pay attention to whether your answers really make sense.

Chapter 2 Summary

Key Terms

2.1

numerator The number above the fraction bar in a fraction is called the numerator. It shows how many of the equivalent parts are being considered.

denominator The number below the fraction bar in a fraction is called the denominator. It shows the number of equal parts in a whole.

proper fraction In a proper fraction, the numerator is smaller than the denominator. The fraction is less than 1.

improper fraction In an improper fraction, the numerator is greater than or equal to the denominator. The fraction is equal to or greater than 1.

2.2

mixed number A mixed number includes a fraction and a whole number written together.

2.3

factors Numbers that are multiplied to give a product are factors.

factorizations The numbers that can be multiplied to give a specific number (product) are factorizations of that number.

prime number A prime number is a whole number other than 0 and 1 that has exactly two factors, itself and 1.

composite number A composite number has at least one factor other than itself and 1.

prime factorization In a prime factorization, every factor is a prime number.

2.4

equivalent fractions Two fractions are equivalent when they represent the same portion of a whole.

common factor A common factor is a number that can be divided evenly into two or more whole numbers.

lowest terms A fraction is written in lowest terms when its numerator and denominator have no common factor other than 1.

2.5

multiplication shortcut When multiplying or dividing fractions, the process of dividing a numerator and denominator by a common factor can be used as a shortcut.

2.7

reciprocal Two numbers are reciprocals of each other if their product is 1. To find the reciprocal of a fraction, interchange the numerator and the denominator. Zero does not have a reciprocal.

New Formula

Area of a rectangle: Area = length • width

Test Your Word Power

See how well you have learned the vocabulary in this chapter.

1. A **numerator** is
   - **A.** a number greater than 5
   - **B.** the number above the fraction bar in a fraction
   - **C.** the number below the fraction bar in a fraction.

2. A **proper fraction**
   - **A.** has a value less than 1
   - **B.** has a whole number and a fraction
   - **C.** has a value greater than 1.

3. A **mixed number** is
   - **A.** less than 1
   - **B.** a whole number and a fraction written together
   - **C.** a number multiplied by another number.

4. A **factor** is
   - **A.** one of two or more numbers that are added to get another number
   - **B.** the answer in division
   - **C.** one of two or more numbers that are multiplied to get another number.

5. A whole number greater than 1 is **prime** if
   - **A.** it cannot be factored
   - **B.** it has just one factor
   - **C.** it has only itself and 1 as factors.

6. A **common factor** can
   - **A.** only be divided by itself and 1
   - **B.** be divided evenly into two or more whole numbers
   - **C.** never be divided by 2.

7. A fraction is in **lowest terms** when
   - **A.** it cannot be divided
   - **B.** its numerator and denominator have no common factor other than 1
   - **C.** it has a value less than 1.

8. To find the **reciprocal** of a fraction,
   - **A.** multiply it by itself
   - **B.** interchange the numerator and the denominator
   - **C.** change it to an improper fraction.

Answers to Test Your Word Power

1. B; *Example:* In $\frac{3}{8}$, the numerator is 3.

2. A; *Example:* $\frac{1}{2}$, $\frac{3}{4}$, and $\frac{7}{8}$ are all proper fractions with a value less than 1.

3. B; *Example:* $2\frac{3}{8}$ and $5\frac{3}{4}$ are mixed numbers.

4. C; *Example:* Since $3 \cdot 5 = 15$, the numbers 3 and 5 are factors of 15.

5. C; *Example:* 3, 5, and 11 are prime numbers; 4, 8, and 12 are composite numbers.

6. B; *Example:* 3 is a common factor of both 6 and 9 because it can be evenly divided into each of them.

7. B; *Example:* $\frac{3}{8}$, $\frac{4}{5}$, and $\frac{5}{6}$ are in lowest terms but $\frac{6}{8}$, $\frac{3}{6}$, and $\frac{2}{4}$ are not.

8. B; *Example:* The reciprocal of $\frac{3}{8}$ is $\frac{8}{3}$, the reciprocal of $\frac{25}{4}$ is $\frac{4}{25}$, and the reciprocal of 6 or $\frac{6}{1}$ is $\frac{1}{6}$.

Quick Review

Concepts	*Examples*
2.1 Types of Fractions	
Proper Numerator smaller than denominator; a value less than 1	$\frac{2}{3}$ $\frac{3}{4}$ $\frac{15}{16}$ $\frac{1}{8}$ Proper fractions
Improper Numerator equal to or greater than denominator; a value equal to or greater than 1	$\frac{17}{8}$ $\frac{19}{12}$ $\frac{11}{2}$ $\frac{5}{3}$ $\frac{7}{7}$ Improper fractions

Concepts	*Examples*

2.2 Converting Fractions

Mixed to Improper Multiply denominator by whole number, add numerator, and place over denominator.

$$7\frac{2}{3} = \frac{23}{3} \leftarrow (3 \cdot 7) + 2$$

Same denominator

Improper to Mixed Divide numerator by denominator and place remainder over denominator.

$$\frac{17}{5} = 3\frac{2}{5} \leftarrow \text{Remainder}$$

Same denominator

$$\begin{array}{r} 3 \\ 5\overline{)17} \\ \underline{15} \\ 2 \end{array}$$

← Divide numerator by denominator.

← Remainder

2.3 Prime Numbers

Determine whether a whole number is evenly divisible only by itself and 1. (By definition, 0 and 1 are not prime.)

The prime numbers less than 50 are 2, 3, 5, 7, 11, 13, 17, 19, 23, 29, 31, 37, 41, 43, 47.

2.3 Finding the Prime Factorization of a Number

Divide each factor by a prime number using a diagram that forms the shape of tree branches.

Find the prime factorization of 30. Use a factor tree.

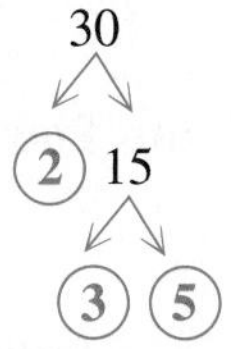

Prime factors are circled.

$30 = 2 \cdot 3 \cdot 5$

2.4 Writing Fractions in Lowest Terms

Divide the numerator and denominator by the greatest common factor.

Write $\frac{30}{42}$ in lowest terms.

$$\frac{30}{42} = \frac{30 \div 6}{42 \div 6} = \frac{5}{7}$$

2.5 Multiplying Fractions

1. Multiply the numerators and multiply the denominators.
2. Write answers in lowest terms if the multiplication shortcut was not used.

Multiply.

$$\frac{6}{11} \cdot \frac{7}{8} = \frac{\overset{3}{\cancel{6}}}{11} \cdot \frac{7}{\underset{4}{\cancel{8}}} = \frac{3 \cdot 7}{11 \cdot 4} = \frac{21}{44}$$

2.7 Finding the Reciprocal

To find the reciprocal of a fraction, interchange the numerator and denominator.

Find the reciprocal of each fraction.

$\frac{3}{4}$ The reciprocal of $\frac{3}{4}$ is $\frac{4}{3}$.

$\frac{8}{5}$ The reciprocal of $\frac{8}{5}$ is $\frac{5}{8}$.

9 The reciprocal of 9 is $\frac{1}{9}$.

Concepts	*Examples*
2.7 Dividing Fractions Use the reciprocal of the divisor (second fraction) and change division to multiplication.	Divide. $\frac{25}{36} \div \frac{15}{18} = \frac{\overset{5}{\cancel{25}}}{\underset{2}{\cancel{36}}} \cdot \frac{\overset{1}{\cancel{18}}}{\underset{3}{\cancel{15}}} = \frac{5 \cdot 1}{2 \cdot 3} = \frac{5}{6}$ Reciprocals
2.8 Multiplying Mixed Numbers First estimate the answer. Then follow these steps. ***Step 1*** *Change* each mixed number to an improper fraction. ***Step 2*** *Multiply.* ***Step 3*** *Simplify* the answer, which means to write it in *lowest terms,* and change it to a mixed number or whole number where possible.	First estimate the answer. Then multiply to get the exact answer. *Estimate:* $1\frac{3}{5} \cdot 3\frac{1}{3}$ Rounded $2 \cdot 3 = 6$ *Exact:* $1\frac{3}{5} \cdot 3\frac{1}{3} = \frac{8}{\underset{1}{\cancel{5}}} \cdot \frac{\overset{2}{\cancel{10}}}{3}$ $= \frac{8 \cdot 2}{1 \cdot 3}$ $= \frac{16}{3} = 5\frac{1}{3}$ Close to estimate
2.8 Dividing Mixed Numbers First estimate the answer. Then follow these steps. ***Step 1*** *Change* each mixed number to an improper fraction. ***Step 2*** Use the *reciprocal* of the second fraction (divisor). ***Step 3*** *Change* division to multiplication. ***Step 4*** *Simplify* the answer, which means to write it in *lowest terms,* and change it to a mixed number or whole number where possible.	First estimate the answer. Then divide to get the exact answer. *Estimate:* $3\frac{5}{9} \div 2\frac{2}{5}$ Rounded $4 \div 2 = 2$ *Exact:* $3\frac{5}{9} \div 2\frac{2}{5} = \frac{32}{9} \div \frac{12}{5}$ $= \frac{\overset{8}{\cancel{32}}}{9} \cdot \frac{5}{\underset{3}{\cancel{12}}} = \frac{40}{27}$ $= 1\frac{13}{27}$ Close to estimate

Chapter 2 Review Exercises

2.1 *Write the fraction that represents each shaded portion.*

1. 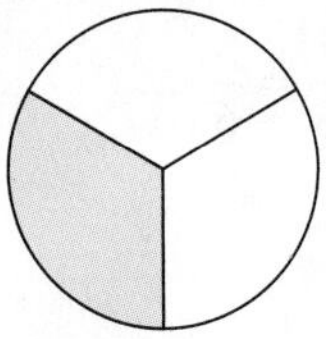

2. 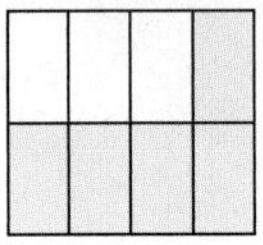

3. 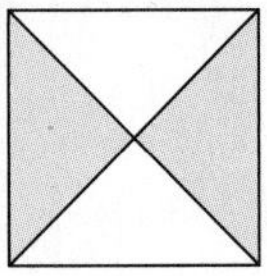

List the proper and improper fractions in each group.

	Proper	Improper
4. $\frac{1}{8}$ $\frac{4}{3}$ $\frac{5}{5}$ $\frac{3}{4}$ $\frac{2}{3}$	________	________
5. $\frac{6}{5}$ $\frac{15}{16}$ $\frac{16}{13}$ $\frac{1}{8}$ $\frac{5}{3}$	________	________

2.2 *Write each mixed number as an improper fraction. Write each improper fraction as a mixed number.*

6. $4\frac{3}{4}$ **7.** $9\frac{5}{6}$ **8.** $\frac{27}{8}$ **9.** $\frac{63}{5}$

2.3 *Find all factors of each number.*

10. 6 **11.** 24 **12.** 55 **13.** 90

Write the prime factorization of each number by using exponents.

14. 27 **15.** 150 **16.** 420

Simplify each expression.

17. 5^2 **18.** $6^2 \cdot 2^3$ **19.** $8^2 \cdot 3^3$ **20.** $4^3 \cdot 2^5$

2.4 *The purity of gold is regulated by law and is the same in all parts of the world. Exercises 21–24 show the purity of 24-kt (karat), 18-kt, 14-kt, and 10-kt gold by comparing the parts of gold to the parts of alloy (metals other than gold). Write a fraction in lowest terms to show the portion that is gold. (Source:* Costco Wholesale.)

21. 24 kt. = (24 parts gold, 0 parts alloy)

22. 18 kt. = (18 parts gold, 6 parts alloy)

23. 14 kt. = (14 parts gold, 10 parts alloy)

24. 10 kt. = (10 parts gold, 14 parts alloy)

Write the numerator and denominator of each fraction as a product of prime factors. Then, write the fraction in lowest terms.

25. $\frac{25}{60}$

26. $\frac{384}{96}$

Decide whether each pair of fractions is equivalent or not equivalent, using the method of prime factors.

27. $\frac{3}{4}$ and $\frac{48}{64}$

28. $\frac{5}{8}$ and $\frac{70}{120}$

29. $\frac{2}{3}$ and $\frac{360}{540}$

2.5–2.8 *Multiply. Write answers in lowest terms, and as mixed numbers or whole numbers where possible.*

30. $\frac{4}{5} \cdot \frac{3}{4}$

31. $\frac{3}{10} \cdot \frac{5}{8}$

32. $\frac{70}{175} \cdot \frac{5}{14}$

33. $\frac{44}{63} \cdot \frac{3}{11}$

34. $\frac{5}{16} \cdot 48$

35. $\frac{5}{8} \cdot 1000$

Divide. Write answers in lowest terms, and as mixed numbers or whole numbers where possible.

36. $\frac{2}{3} \div \frac{1}{2}$

37. $\frac{5}{6} \div \frac{1}{2}$

38. $\dfrac{\frac{15}{18}}{\frac{10}{30}}$

39. $\dfrac{\frac{3}{4}}{\frac{3}{8}}$

40. $7 \div \frac{7}{8}$

41. $18 \div \frac{3}{4}$

42. $\frac{5}{8} \div 3$

43. $\frac{2}{3} \div 5$

44. $\dfrac{\frac{12}{13}}{3}$

Find the area of each rectangle.

45.

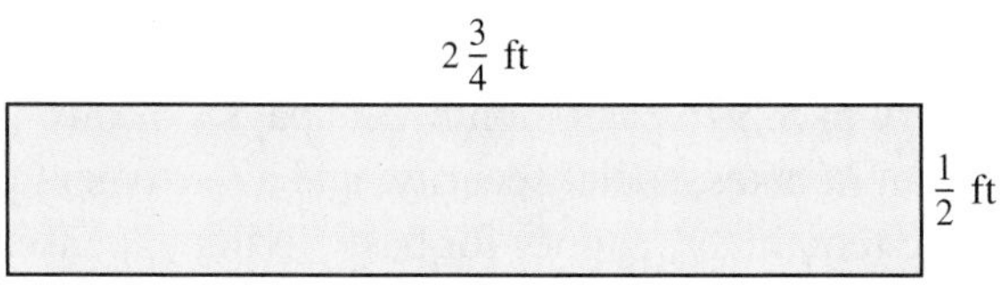

46.

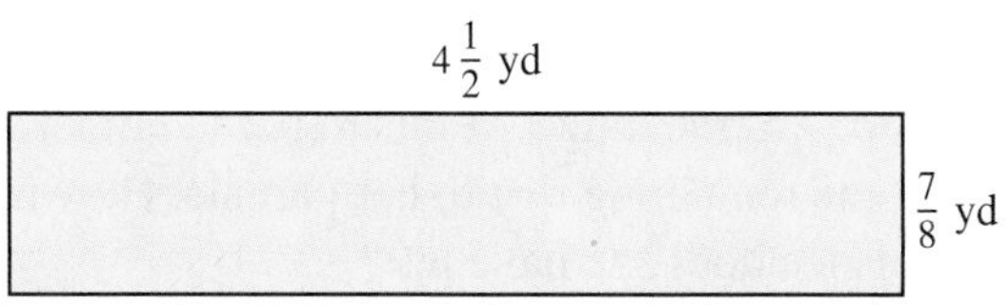

47. Ceramic tile is being installed on the floor of a meeting hall that is 108 ft long and $72\frac{3}{4}$ ft wide. Find the area.

48. Find the area of a display shelf that is a rectangle measuring 6 ft long and $\frac{11}{12}$ ft wide.

First estimate the answer. Then multiply or divide to find the exact answer. Simplify all answers.

49. $5\frac{1}{2} \cdot 1\frac{1}{4}$

Estimate:

___ • ___ = ___

Exact:

50. $2\frac{1}{4} \cdot 7\frac{1}{8} \cdot 1\frac{1}{3}$

Estimate:

___ • ___ • ___ = ___

Exact:

51. $15\frac{1}{2} \div 3$

Estimate:

___ ÷ ___ = ___

Exact:

52. $4\frac{3}{4} \div 6\frac{1}{3}$

Estimate:

___ ÷ ___ = ___

Exact:

Solve each application problem by using the six problem-solving steps.

53. Blue Diamond Almonds has 320 tons of almonds. How many $\frac{5}{8}$ ton bins will be needed to store the almonds?

54. The founding partner of Professional Networking owns $\frac{2}{5}$ of the business, while 4 other equal partners own the rest. What fraction of the total business is owned by each of the other partners?

55. How many window-blind pull cords can be made from $157\frac{1}{2}$ yards of cord if $4\frac{3}{8}$ yards of cord are needed for each blind? First estimate, and then find the exact answer.

Estimate:

Exact:

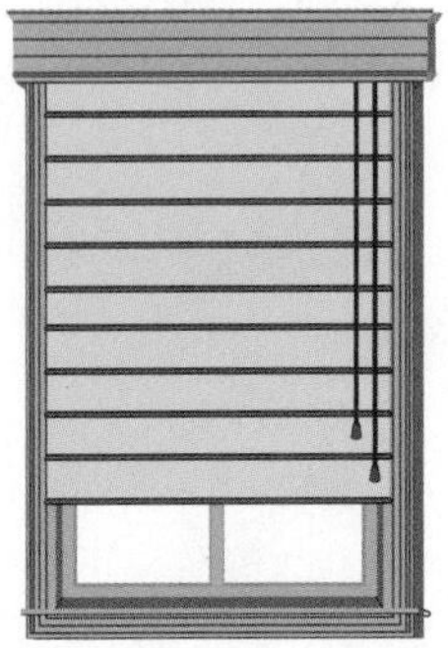

56. A gallon of water weighs $8\frac{1}{3}$ pounds. Find the weight of the water in two 50-gallon aquariums. First estimate, and then find the exact answer.

Estimate:

Exact:

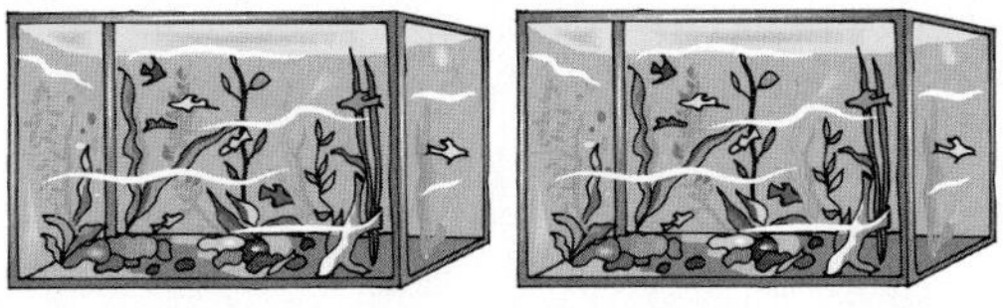

57. Ebony Wilson purchased 100 pounds of rice at the food co-op. After selling $\frac{1}{4}$ of this to her neighbor, she gave $\frac{2}{3}$ of the remaining rice to her parents. How many pounds of rice does she have left?

58. Sheila Spinney, a recent college graduate, receives a salary of \$2976 each month. She pays $\frac{3}{8}$ of this amount in taxes, social security, and a retirement plan. Of the remainder, $\frac{9}{10}$ goes for basic living expenses. How much money remains?

59. The results of a back-to-school fundraiser to purchase school supplies showed that 6 schools will divide $\frac{7}{8}$ of the total amount raised. What fraction of the total amount raised will each school receive?

60. In a morning of deep-sea fishing, 5 fishermen catch $\frac{4}{5}$ ton of salmon. If they divide the fish evenly, how much will each receive?

Mixed Review Exercises

Multiply or divide as indicated. Simplify all answers.

61. $\frac{1}{2} \cdot \frac{3}{4}$

62. $\frac{2}{3} \cdot \frac{3}{5}$

63. $12\frac{1}{2} \cdot 2\frac{1}{2}$

64. $8\frac{1}{3} \cdot 3\frac{2}{5}$

65. $\frac{\frac{4}{5}}{8}$

66. $\frac{\frac{5}{8}}{4}$

67. $\frac{15}{31} \cdot 62$

68. $3\frac{1}{4} \div 1\frac{1}{4}$

Write each mixed number as an improper fraction. Write each improper fraction as a mixed number.

69. $\frac{8}{5}$

70. $\frac{153}{4}$

71. $5\frac{2}{3}$

72. $38\frac{3}{8}$

Write the numerator and denominator of each fraction as a product of prime factors; then write the fraction in lowest terms.

73. $\frac{8}{12}$

74. $\frac{108}{210}$

Write each fraction in lowest terms.

75. $\frac{75}{90}$

76. $\frac{48}{72}$

77. $\frac{44}{110}$

78. $\frac{87}{261}$

Solve each application problem.

79. The label of the Roundup Weed and Grass Killer says to mix $2\frac{1}{2}$ ounces of the product with 1 gallon of water. How many ounces are needed for 50 gallons of water? First estimate, then find the exact answer. (*Source:* Roundup Weed and Grass Killer.)

Estimate:

Exact:

80. Valley Farms purchased some diesel fuel additive. The instructions say to use $7\frac{1}{4}$ quarts of additive for each tank of fuel. How many quarts are needed for $9\frac{1}{3}$ tanks? First estimate, then find the exact answer.

Estimate:

Exact:

81. The antique U.S.A. stamp is $1\frac{3}{4}$ in. by $\frac{7}{8}$ in. Find its area.

82. A patio table top is $\frac{7}{8}$ yard by $2\frac{1}{4}$ yards. What is the area?

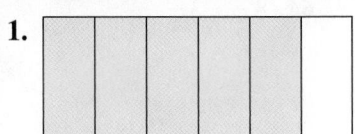

Chapter 2 Test

CHAPTER Test Prep VIDEO

The Chapter Test Prep Videos with test solutions are available on DVD, in MyMathLab*, and on* YouTube*—search "LialDevMath" and click on "Channels."*

Write a fraction to represent each shaded portion.

1.

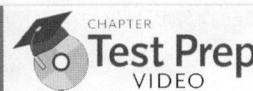

2.

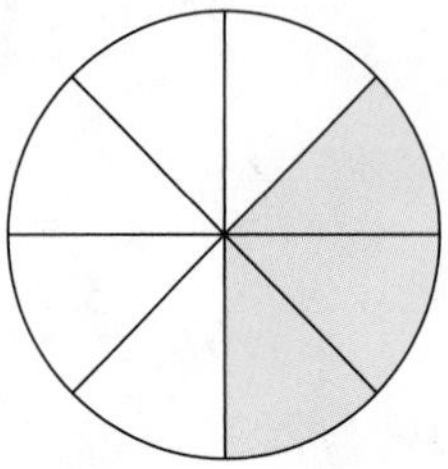

3. Identify all the proper fractions in this list:

$\frac{2}{3}$ $\frac{4}{4}$ $\frac{6}{7}$ $\frac{5}{2}$ $\frac{1}{4}$ $\frac{5}{8}$ $\frac{30}{18}$

4. Write $3\frac{3}{8}$ as an improper fraction.

5. Write $\frac{123}{4}$ as a mixed number.

6. Find all factors of 18.

Find the prime factorization of each number. Write the answers using exponents.

7. 45

8. 144

9. 500

Write each fraction in lowest terms.

10. $\frac{36}{48}$

11. $\frac{60}{72}$

12. The method of prime factors is used to write a fraction in lowest terms. Briefly explain how this is done. Use the fraction $\frac{56}{84}$ to show how this works.

13. Explain how to multiply fractions. What additional steps must be taken when dividing fractions?

Multiply or divide. Write answers in lowest terms, and as mixed numbers or whole numbers where possible.

14. $\frac{3}{4} \cdot \frac{4}{9}$

15. $54 \cdot \frac{2}{3}$

16. A rectangular barbecue grill is $\frac{15}{16}$ yard by $\frac{4}{9}$ yard. Find the area of the grill.

17. The Sierra College Conservation Club planted 8760 Douglas Fir seedlings. If $\frac{3}{8}$ of these seedlings are not expected to survive, find the number of seedlings that do survive.

18. $\frac{3}{4} \div \frac{5}{6}$

19. $\frac{7}{\frac{4}{9}}$

20. To complete a custom-designed cabinet, oak trim pieces must be cut exactly $2\frac{1}{4}$ inches long so that they can be used as dividers in a spice rack. Find the number of pieces that can be cut from a piece of oak that is 54 inches in length.

First estimate the answer. Then find the exact answer. Simplify all answers.

21. $4\frac{1}{8} \cdot 3\frac{1}{2}$

Estimate:

Exact:

22. $1\frac{5}{6} \cdot 4\frac{1}{3}$

Estimate:

Exact:

23. $9\frac{3}{5} \div 2\frac{1}{4}$

Estimate:

Exact:

24. $\frac{8\frac{1}{2}}{1\frac{3}{4}}$

Estimate:

Exact:

25. A new vaccine is synthesized at the rate of $2\frac{1}{2}$ grams per day. How many grams can be synthesized in $12\frac{1}{4}$ days?

Estimate:

Exact:

Math in the Media

RECIPES

Rachael Ray is a Food Network television host, bestselling cookbook author, and the editor of her own lifestyle magazine. Rachael Ray's recipes can be found on her Web site, www.rachaelray.com, on her television show, or in one of her many cookbooks. The recipe at the side is from her book *Rachael Ray: 30-Minute Meals.*

Alaska Burgers

1 pound 93% lean ground beef
$\frac{1}{2}$ medium Spanish onion, minced or processed
4 shakes Worcestershire sauce
$\frac{1}{4}$ teaspoon allspice (1 good pinch)
$\frac{1}{2}$ teaspoon ground cumin (two good pinches)
Cracked black pepper
$\frac{1}{3}$ pound brick of smoked cheddar cheese, cut into $\frac{1}{2}$ inch slices
4 fresh, crusty onion rolls
Thick-sliced tomato and lettuce to top

Mix beef, onion, Worcestershire, allspice, cumin, and black pepper in a bowl. Separate a quarter of the mixture. Take a slice of the smoked cheese and place it in the middle of the mixture. Form the pattie shape around the cheese filling. Patties should be no more than $\frac{3}{4}$ inch thick. Repeat with rest of mixture to have a total of 4 patties.

Heat a nonstick griddle or frying pan to medium hot. Cook burgers 5 to 6 minutes on each side. Meat should be cooked through and cheese melted. Check each burger with an instant-read thermometer for an internal temp of 170°F for well done if undercooking concerns you. Or cut into one and check the color of the meat.

Salt burgers after preparation to your taste. (Salting beef before cooking draws out juices and flavor.) Top with tomato slices and lettuce. Serves 4.

1. Following the recipe, **(a)** what is the weight of one $\frac{1}{2}$-inch slice of cheese, and **(b)** what is the thickness of a $\frac{1}{3}$-pound brick of smoked cheddar cheese?

2. According to the recipe, **(a)** how many teaspoons of allspice are in 8 good pinches, and **(b)** how many teaspoons of ground cumin are in 7 good pinches?

3. Suppose you are preparing Alaska Burgers for 18 guests. By what factor will you change the ingredient amounts?

4. You know that of the 15 guests at your next party, 5 large eaters will eat $1\frac{1}{2}$ burgers each, 5 children will eat $\frac{1}{2}$ burger each, and the rest of the guests will each eat 1 burger. **(a)** How many burgers will you need, and **(b)** by what factor will you change the ingredient amounts?

5. Fill in the blanks with the ingredient amounts needed to make 9 servings of Alaska Burgers

Lean ground beef ______________
Spanish onions ______________
Worcestershire sauce ______________
Allspice ______________
Ground cumin ______________
Cheddar cheese ______________

6. If you have $5\frac{3}{4}$ pounds of beef, how many servings of Alaska Burgers can be prepared?

3 Adding and Subtracting Fractions

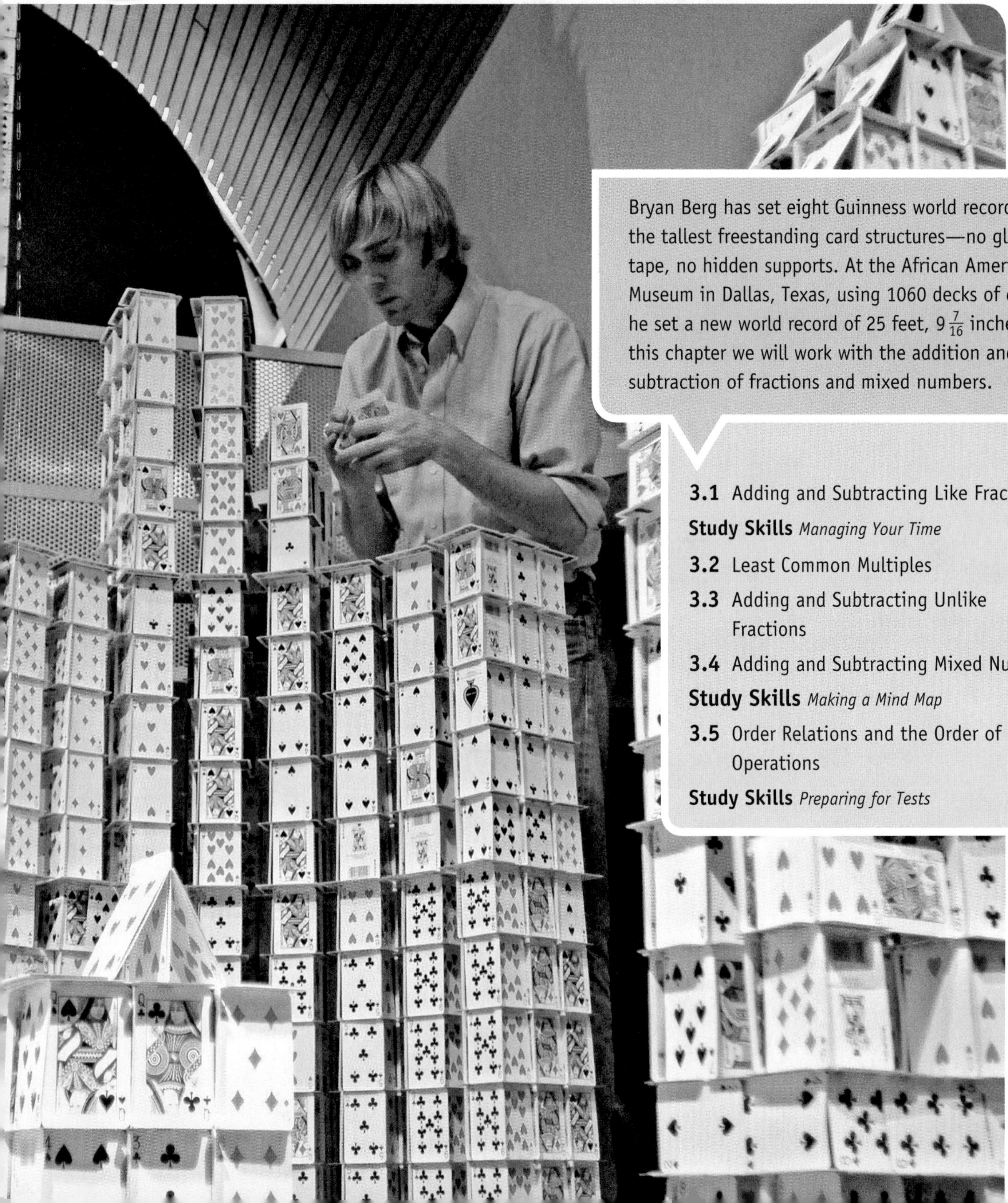

Bryan Berg has set eight Guinness world records for the tallest freestanding card structures—no glue, no tape, no hidden supports. At the African American Museum in Dallas, Texas, using 1060 decks of cards, he set a new world record of 25 feet, $9\frac{7}{16}$ inches. In this chapter we will work with the addition and subtraction of fractions and mixed numbers.

3.1 Adding and Subtracting Like Fractions

OBJECTIVES

1. Define like and unlike fractions.
2. Add like fractions.
3. Subtract like fractions.

In **Chapter 2** we looked at the basics of fractions and then practiced with multiplication and division of fractions and mixed numbers. In this chapter we will work with addition and subtraction of fractions and mixed numbers.

OBJECTIVE 1 Define like and unlike fractions. Fractions with the same denominators are **like fractions.** Fractions with different denominators are **unlike fractions.**

EXAMPLE 1 Identifying Like and Unlike Fractions

(a) $\frac{3}{4}, \frac{1}{4}, \frac{5}{4}, \frac{6}{4},$ and $\frac{4}{4}$ are **like** fractions.

All denominators are the same.

(b) $\frac{7}{12}$ and $\frac{12}{7}$ are **unlike** fractions.

Unlike fractions have different denominators.

Denominators are different.

Note

Like fractions have the *same* denominator.

◀ **Work Problem 1 at the Side.**

OBJECTIVE 2 Add like fractions. The figures below show you how to add the fractions $\frac{2}{7}$ and $\frac{4}{7}$.

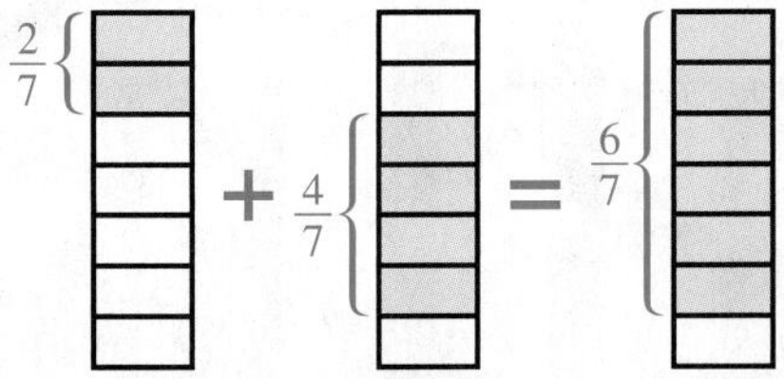

As the figures show,

$$\frac{2}{7} + \frac{4}{7} = \frac{6}{7}.$$

Add like fractions as follows.

Adding Like Fractions

Step 1 Add the numerators to find the numerator of the sum.

Step 2 Write the denominator of the like fractions as the denominator of the sum.

Step 3 Write the sum in lowest terms.

1 Next to each pair of fractions write *like* or *unlike*.

(a) $\frac{2}{5}$ $\frac{3}{5}$ ______

(b) $\frac{2}{3}$ $\frac{3}{4}$ ______

(c) $\frac{7}{12}$ $\frac{11}{12}$ ______

(d) $\frac{3}{8}$ $\frac{3}{16}$ ______

Answers

1. **(a)** like **(b)** unlike **(c)** like **(d)** unlike

EXAMPLE 2 Adding Like Fractions

Add and write the sum in lowest terms.

(a) $\frac{1}{5} + \frac{2}{5}$

Add numerators.

$\frac{3}{5}$ is already in lowest terms.

$$\frac{1}{5} + \frac{2}{5} = \frac{1+2}{5} = \frac{3}{5} \begin{matrix} \leftarrow \text{Sum of numerators} \\ \leftarrow \text{Same denominator} \end{matrix}$$

(b) $\frac{1}{12} + \frac{7}{12} + \frac{1}{12}$ Fractions are ready to add if they are *like* fractions.

Add numerators.

Step 1 $\frac{1+7+1}{12}$

Step 2 $= \frac{9}{12} \begin{matrix} \leftarrow \text{Sum of numerators} \\ \leftarrow \text{Same denominator} \end{matrix}$

Step 3 $= \frac{9 \div 3}{12 \div 3} = \frac{3}{4} \leftarrow \text{Lowest terms}$

CAUTION

Fractions may be added ***only*** if they have like denominators.

Work Problem 2 at the Side. ▶

OBJECTIVE 3 Subtract like fractions. The figures below show $\frac{7}{8}$ broken into $\frac{4}{8}$ and $\frac{3}{8}$.

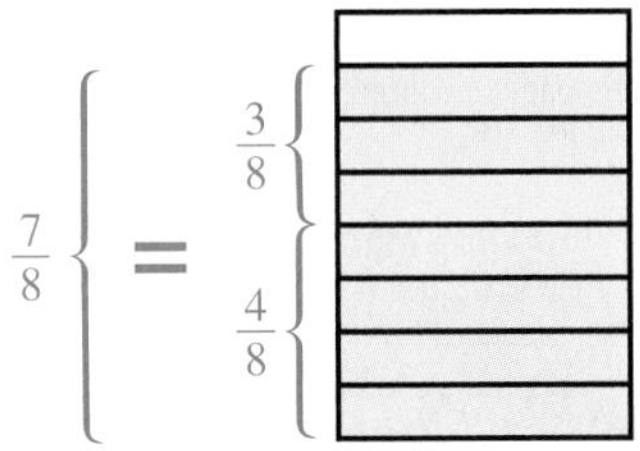

Subtracting $\frac{3}{8}$ from $\frac{7}{8}$ gives the answer $\frac{4}{8}$, or

$$\frac{7}{8} - \frac{3}{8} = \frac{4}{8}.$$

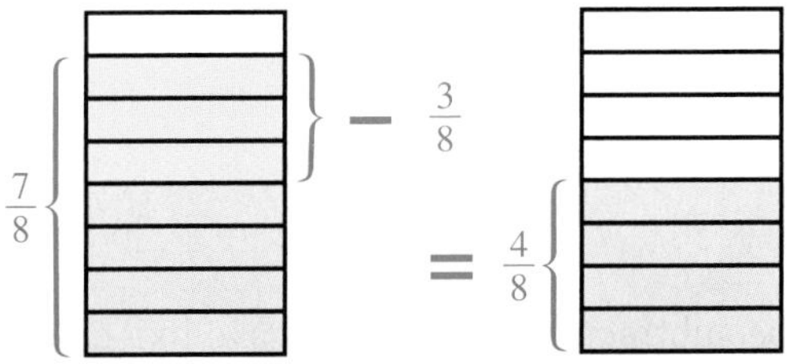

2 Add and write the sums in lowest terms.

GS **(a)** $\frac{3}{8} + \frac{1}{8}$

$$\frac{3+1}{8} = \frac{\quad}{8} = \frac{\quad}{2}$$

(b) $$\begin{array}{r} \frac{2}{9} \\ +\ \frac{5}{9} \\ \hline \end{array}$$

(c) $\frac{3}{16} + \frac{1}{16}$

(d) $\frac{3}{10} + \frac{1}{10} + \frac{4}{10}$

Answers

2. **(a)** 4; $\frac{1}{2}$ **(b)** $\frac{7}{9}$ **(c)** $\frac{1}{4}$ **(d)** $\frac{4}{5}$

3 Find the difference and simplify.

GS **(a)** $\frac{5}{6} - \frac{1}{6}$

$\frac{5-1}{6} = \frac{\quad}{6} = \frac{\quad}{3}$

(b) $\frac{16}{10} - \frac{7}{10}$

(c) $\frac{15}{3} - \frac{5}{3}$

(d) $\frac{25}{32} - \frac{6}{32}$

Write $\frac{4}{8}$ in lowest terms.

$$\frac{7}{8} - \frac{3}{8} = \frac{4 \div 4}{8 \div 4} = \frac{1}{2}$$

The steps for subtracting like fractions are very similar to those for adding like fractions.

Subtracting Like Fractions

Step 1 Subtract the numerators to find the numerator of the difference.

Step 2 Write the denominator of the like fractions as the denominator of the difference.

Step 3 Write the answer in lowest terms.

EXAMPLE 3 Subtracting Like Fractions

Find the difference and simplify the answer.

(a) $\frac{15}{16} - \frac{3}{16}$

Step 1 $\frac{15}{16} - \frac{3}{16} = \overbrace{\frac{15-3}{16}}^{\text{Subtract numerators.}}$

Step 2 $= \frac{12}{16}$ ← Difference of numerators ← Same denominator

Step 3 $= \frac{12 \div 4}{16 \div 4} = \frac{3}{4}$ ← Lowest terms

(b) $\frac{13}{4} - \frac{6}{4}$ Fractions are ready to subtract if they are *like* fractions.

$$\frac{13}{4} - \frac{6}{4} = \overbrace{\frac{13-6}{4}}^{\text{Subtract numerators.}}$$ ← Difference of numerators ← Same denominator

$$= \frac{7}{4}$$

To simplify the answer, write $\frac{7}{4}$ as a mixed number.

$$\frac{7}{4} = 1\frac{3}{4}$$ Always simplify the answer.

CAUTION

Fractions may be subtracted ***only*** if they have like denominators.

◀ Work Problem **3** at the Side.

Answers

3. **(a)** $4; \frac{2}{3}$ **(b)** $\frac{9}{10}$ **(c)** $3\frac{1}{3}$ **(d)** $\frac{19}{32}$

3.1 Exercises

FOR EXTRA HELP

Download the MyDashBoard App

MyMathLab®

CONCEPT CHECK *Identify the fractions in each pair as* like *or* unlike.

1. $\frac{3}{8}\ \ \frac{5}{8}$

2. $\frac{5}{16}\ \ \frac{1}{4}$

3. $\frac{3}{5}\ \ \frac{3}{4}$

4. $\frac{5}{12}\ \ \frac{7}{12}$

CONCEPT CHECK *Fill in the blank.*

5. In order to add or subtract fractions, they must be ___________ fractions.

6. After adding or subtracting like fractions, the answer should be written in ___________.

Find the sum and simplify the answer. ***See Example 2.***

7. $\frac{3}{8}+\frac{2}{8}$

8. $\frac{1}{5}+\frac{3}{5}$

9. $\frac{1}{4}+\frac{1}{4}$

10. $\frac{9}{10} + \frac{3}{10}$

11. $\frac{13}{12} + \frac{5}{12}$

12. $\frac{2}{9} + \frac{1}{9}$

13. $\frac{7}{12}+\frac{3}{12}=\frac{7+3}{12}=\frac{10}{12}=$

14. $\frac{4}{15}+\frac{2}{15}+\frac{5}{15}=\frac{4+2+5}{15}=$

15. $\frac{3}{8}+\frac{7}{8}+\frac{2}{8}$

16. $\frac{4}{9}+\frac{1}{9}+\frac{7}{9}$

17. $\frac{2}{54}+\frac{8}{54}+\frac{12}{54}$

18. $\frac{7}{64}+\frac{15}{64}+\frac{20}{64}$

Find the difference and simplify the answer. ***See Example 3.***

19. $\frac{7}{8}-\frac{4}{8}$

20. $\frac{2}{3}-\frac{1}{3}$

21. $\frac{10}{11}-\frac{4}{11}$

22. $\frac{4}{5}-\frac{3}{5}$

23. $\frac{9}{10}-\frac{3}{10}=\frac{9-3}{10}=\frac{6}{10}=$

24. $\frac{7}{14}-\frac{3}{14}=\frac{7-3}{14}=\frac{4}{14}=$

25. $\frac{31}{21} - \frac{7}{21}$

26. $\frac{43}{24} - \frac{13}{24}$

27. $\frac{27}{40} - \frac{19}{40}$

28. $\frac{38}{55} - \frac{16}{55}$

29. $\frac{47}{36}-\frac{5}{36}$

30. $\frac{76}{45}-\frac{21}{45}$

31. $\frac{73}{60}-\frac{7}{60}$

32. $\frac{181}{100}-\frac{31}{100}$

33. In your own words, write an explanation of how to add like fractions. Use three steps in your explanation.

34. Describe in your own words the difference between *like* fractions and *unlike* fractions. Give three examples of each type.

Solve each application problem. Write answers in lowest terms.

35. The Fair Oaks Save the Bluffs Committee raised $\frac{2}{9}$ of their target goal last year and another $\frac{5}{9}$ of the goal this year. What fraction of their goal has been raised?

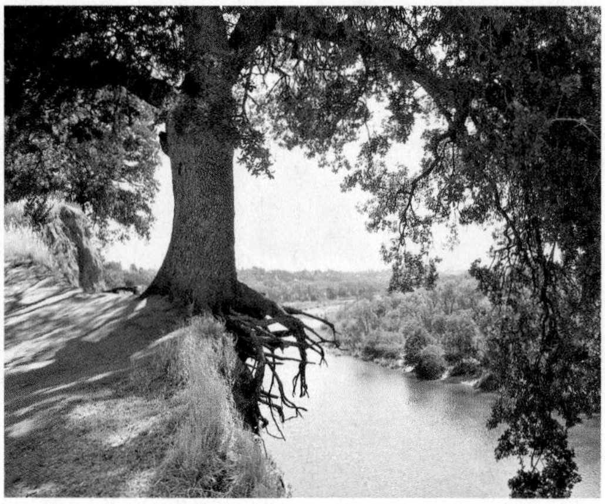

36. After an initial payment to a winner at an arcade fundraiser, the organization still owed the winner $\frac{7}{10}$ of her total winnings. If the organization pays the winner another $\frac{3}{10}$ of the winnings, what fraction is still owed?

37. GS Julie Circle, a landscaper, is working on a commercial job and completed only $\frac{1}{8}$ of the irrigation system in the first week. If she completed $\frac{5}{8}$ of the system in the second week, what fraction of the irrigation system has she completed?

$$\frac{1}{8}+\frac{5}{8}=\frac{6}{8}=$$

38. GS On a wild shopping spree, Angie Gragg spent $\frac{3}{24}$ of the day shopping in the morning and another $\frac{5}{24}$ of the day shopping after lunch. What fraction of the day did she spend shopping?

$$\frac{3}{24}+\frac{5}{24}=\frac{8}{24}=$$

39. An organic farmer purchased $\frac{9}{10}$ acre of land one year and $\frac{3}{10}$ acre the next year. She then planted carrots on $\frac{7}{10}$ acre of the land and squash on the remainder. How much land is planted with squash?

40. A forester planted $\frac{5}{12}$ acre in seedlings in the morning and $\frac{11}{12}$ acre in the afternoon. That night, $\frac{7}{12}$ acre of seedlings were destroyed by frost. How many acres of seedlings remain?

Study Skills
MANAGING YOUR TIME

OBJECTIVES

1. Create a semester schedule.
2. Create a "to do" list.

Many college students find themselves juggling a difficult schedule and multiple responsibilities. Perhaps you are going to school, working part time, and managing family demands. Here are some tips to help you develop good time management skills and habits.

- **Read the syllabus for each class.** Check on class policies, such as attendance, late homework, and make-up tests. Find out how you are graded. Keep the syllabus in your notebook.
- **Make a semester or quarter calendar.** Put test dates and major due dates for *all* your classes on the same calendar. That way you will see which weeks are the really busy ones. Try using a different color pen for each class. Your brain responds well to the use of color. A semester calendar is on the next page.
- **Make a weekly schedule.** After you fill in your classes and other regular responsibilities (such as work, picking up kids from school, etc.), block off some study periods during the day that you can guarantee you will use for studying. Aim for 2 hours of study for each 1 hour you are in class.
- **Make "To Do" lists.** Then use them by crossing off the tasks as you complete them. You might even number them in the order they need to be done (most important ones first).
- **Break big assignments into smaller chunks.** They won't seem so big that way. Make deadlines for each small part so you stay on schedule.
- **Give yourself small breaks in your studying.** Do not try to study for hours at a time! Your brain needs rest between periods of learning. Try to give yourself a 10 minute break each hour or so. You will learn more and remember it longer.
- **If you get off schedule, just try to get back on schedule tomorrow.** We all slip from time to time. All is not lost! Make a new "to do" list and start doing the most important things first.
- **Get help when you need it.** Talk with your instructor during office hours. Also, most colleges have a Learning Center, tutoring center, or counseling office. If you feel lost and overwhelmed, ask for help. Someone can help you decide what to do first and what to spend your time on right away.

Which two or three of the suggestions above will you try this week? How do you think they will help you?

1. ______________________________

2. ______________________________

3. ______________________________

Why Are These Techniques Brain Friendly?

Your brain appreciates some order. It enjoys a little routine, for example, choosing the same study time and place each day. You will find that you quickly settle in to your reading or homework.

Also, your brain **functions better when you are calm.** Too much rushing around at the last minute to get your homework and studying done sends hostile chemicals to your brain and makes it more difficult for you to learn and remember. So, a little planning can really pay off.

Building rest into your schedule is good for your brain. Remember, it takes time for dendrites to grow.

We've suggested using color on your calendars. This too, is brain friendly. Remember, your brain **likes pleasant colors and visual material** that are nice to look at. Messy and hard to read calendars will not be helpful, and you probably won't look at them often.

Study Skills

Continued from page 197

SEMESTER CALENDAR

WEEK	MON	TUES	WED	THUR	FRI	SAT	SUN
1							
2							
3							
4							
5							
6							
7							
8							
9							
10							
11							
12							
13							
14							
15							
16							

3.2 Least Common Multiples

Only *like* fractions can be added or subtracted. So, we must rewrite *unlike* fractions as *like* fractions before we can add or subtract them.

OBJECTIVE 1 Find the least common multiple (LCM). We can rewrite unlike fractions as like fractions by finding the *least common multiple* of the denominators.

Least Common Multiple (LCM)

The **least common multiple (LCM)** of two whole numbers is the smallest whole number divisible by both of those numbers.

EXAMPLE 1 Finding the Least Common Multiple (LCM)

Find the least common multiple of 6 and 9.

First, find the multiples of 6.

$$\underbrace{6 \cdot 1}_{6,} \quad \underbrace{6 \cdot 2}_{12,} \quad \underbrace{6 \cdot 3}_{18,} \quad \underbrace{6 \cdot 4}_{24,} \quad \underbrace{6 \cdot 5}_{30,} \quad \underbrace{6 \cdot 6}_{36,} \quad \underbrace{6 \cdot 7}_{42,} \quad \underbrace{6 \cdot 8}_{48, \ldots}$$

(The three dots at the end of the list show that the list continues in the same pattern without stopping.) Now, find the multiples of 9.

$$\underbrace{9 \cdot 1}_{9,} \quad \underbrace{9 \cdot 2}_{18,} \quad \underbrace{9 \cdot 3}_{27,} \quad \underbrace{9 \cdot 4}_{36,} \quad \underbrace{9 \cdot 5}_{45,} \quad \underbrace{9 \cdot 6}_{54,} \quad \underbrace{9 \cdot 7}_{63,} \quad \underbrace{9 \cdot 8}_{72, \ldots}$$

The smallest number found in *both* lists is 18, so 18 is the **least common multiple** of 6 and 9; the number 18 is the smallest whole number divisible by both 6 and 9.

Multiples of 6: 6, 12, **18**, 24, 30, 36, 42, 48, . . .

Multiples of 9: 9, **18**, 27, 36, 45, 54, 63, 72, . . .

The *smallest* number in *both* lists is the LCM.

18 is the smallest number found in both lists. **18** is the least common multiple (LCM) of 6 and 9.

Work Problem 1 at the Side. ▶

OBJECTIVE 2 Find the least common multiple using multiples of the largest number. There are several ways to find the least common multiple. If the numbers are small, the least common multiple can often be found by inspection. Can you think of a number that can be divided evenly by both 3 and 4? The number 12 will work; it is the least common multiple of 3 and 4. You can also find the least common multiple by writing multiples of the larger number.

In this case, 4 is larger than 3, so write the multiples of 4.

4, 8, 12, 16, 20, . . .

Now, check each multiple of 4 to see if it is divisible by 3.

4 is *not* divisible by 3.
8 is *not* divisible by 3.
12 *is* divisible by 3.

The first multiple of 4 that is divisible by 3 is 12, so 12 is the least common multiple of 3 and 4.

OBJECTIVES

1. Find the least common multiple (LCM).
2. Find the least common multiple using multiples of the largest number.
3. Find the least common multiple using prime factorization.
4. Find the least common multiple using an alternative method.
5. Write a fraction with an indicated denominator.

VOCABULARY TIP

Least common multiple (LCM) A common multiple can be found by simply multiplying together the numbers under consideration. To make simplifying fractions easier, it's helpful to find the **least common multiple (LCM).** This is the *smallest* number into which the original numbers divide.

1 **(a)** List the multiples of 5.

GS 5, _10_, _15_, ____, ____, ____, _35_, ____, . . .

(b) List the multiples of 8.

8, ____, ____, ____, ____, ____, ____, . . .

(c) Find the least common multiple of 5 and 8.

Answers

1. (a) 5, 10, 15, 20, 25, 30, 35, 40, . . .
(b) 8, 16, 24, 32, 40, 48, 56, . . .
(c) 40

2 Use multiples of the larger number to find the least common multiple in each set of numbers.

GS **(a)** 2 and 5

Multiples of 5 are:

5, 10, 15, 20, . . .

What is the first multiple of 5 that is divisible by 2? ____

So the LCM of 2 and 5 is ____.

(b) 3 and 9

(c) 6 and 8

(d) 4 and 7

3 Use prime factorization to find the LCM for each pair of numbers.

GS **(a)** 15 and 18

$15 = 3 \cdot 5$

$18 = \_\_\_ \cdot \_\_\_ \cdot \_\_\_$

$\text{LCM} = \_\_\_ \cdot \_\_\_ \cdot \_\_\_ \cdot \_\_\_ = \_\_\_$

(b) 12 and 20

Answers

2. **(a)** 10, 10 **(b)** 9 **(c)** 24 **(d)** 28

3. **(a)**
15 = 3 • ⑤
18 = ②•③•③
LCM = 2 • 3 • 3 • 5 = 90 (15 → 3 • 5; 18 → 2 • 3 • 3)

(b)
12 = ②•②•③
20 = 2 • 2 •⑤
LCM = 2 • 2 • 3 • 5 = 60 (12 → 2 • 2 • 3; 20 → 2 • 2 • 5)

EXAMPLE 2 Finding the Least Common Multiple (LCM)

Use multiples of the larger number to find the least common multiple of 6 and 9.

Start by writing the first few multiples of 9.

Multiples of 9

9, 18, 27, 36, 45, 54, . . .

Now check each multiple of 9 to see if it is divisible by 6. The first multiple of 9 that is divisible by 6 is 18.

9, **18**, 27, 36, 45, 54, . . .

First multiple divisible by 6, because $18 \div 6 = 3$

The least common multiple (LCM) of 6 and 9 is 18.

◄ **Work Problem 2 at the Side.**

OBJECTIVE 3 Find the least common multiple using prime factorization. **Example 2** shows how to find the least common multiple of two numbers by making a list of the multiples of the *larger* number. Although this method works well if both numbers are fairly small, it is usually easier to find the least common multiple for larger numbers by using *prime factorization,* as shown in the next example. (See **Chapter 2** for further review.)

EXAMPLE 3 Using Prime Factorization to Find the LCM

Use prime factorization to find the least common multiple of 9 and 12.

Start by finding the prime factorization of each number.

$9 = 3 \cdot 3$

$12 = 2 \cdot 2 \cdot 3$

Prime factorizations of 9 and 12.

Circle the factors where they appear the greatest number of times in either factorization.

9 = ③•③ — 3 appears most often in this factorization.

12 = ②•②• 3 — 2 appears most often in this factorization.

The LCM is the product of the circled factors.

Factors of 9

$\text{LCM} = 3 \cdot 3 \cdot 2 \cdot 2 = 36$ — The product of the circled factors is the LCM.

Factors of 12

The product of the prime factors, 36, is the least common multiple. Check to see that 36 is divisible by 9 (yes) and by 12 (yes). The smallest whole number divisible by both 9 and 12 is 36.

CAUTION

Notice that we did ***not*** repeat the factors that 9 and 12 have in common. In this case, the **3** in 2 • 2 • **3** = 12 was ***not*** used because 3 is already included in 3 • 3 = 9.

◄ **Work Problem 3 at the Side.**

EXAMPLE 4 Using Prime Factorization

Find the least common multiple of 12, 18, and 20.

Find the prime factorization of each number. Then use the prime factors to build the LCM.

$12 = ⓶ \cdot ⓶ \cdot 3$ — 2 appears most often in this factorization; circle the 2s.

$18 = 2 \cdot ⓷ \cdot ⓷$ — 3 appears most often in this factorization; circle the 3s.

$20 = 2 \cdot 2 \cdot ⓹$ — 5 appears most often in this factorization; circle the 5.

12

$\text{LCM} = 2 \cdot 2 \cdot 3 \cdot 3 \cdot 5 = 180$

18

20

Check to see that 180 is divisible by 12 (yes) and by 18 (yes) and by 20 (yes). This smallest whole number divisible by 12, 18, and 20 is 180. The LCM is 180.

Work Problem 4 at the Side. ▶

EXAMPLE 5 Finding the Least Common Multiple

Find the least common multiple for each set of numbers.

(a) 5, 6, 35

Find the prime factorization for each number.

$5 = ⓹$ — Circle either 5, but not both.

$6 = ⓶ \cdot ⓷$

$35 = 5 \cdot ⓻$

5

$\text{LCM} = 2 \cdot 3 \cdot 5 \cdot 7 = 210$

6 35

The least common multiple of 5, 6, and 35 is 210.

(b) 10, 20, 24

Find the prime factorization for each number.

$10 = 2 \cdot 5$

$20 = 2 \cdot 2 \cdot ⓹$

$24 = ⓶ \cdot ⓶ \cdot ⓶ \cdot ⓷$

10
20

$\text{LCM} = 2 \cdot 2 \cdot 2 \cdot 3 \cdot 5 = 120$

24

The least common multiple of 10, 20, and 24 is 120.

Work Problem 5 at the Side. ▶

OBJECTIVE 4 Find the least common multiple using an alternative method. Some people like the following ***alternative method*** for finding the least common multiple for larger numbers. Try both methods, and ***use the one you prefer.*** As a review, a list of the first few prime numbers follows.

First few prime numbers ⟶ 2, 3, 5, 7, 11, 13, 17

4 Find the least common multiple of the denominators in each set of fractions.

(a) $\frac{3}{8}$ and $\frac{6}{5}$

(b) $\frac{5}{6}$ and $\frac{1}{14}$

(c) $\frac{4}{9}$, $\frac{5}{18}$, and $\frac{7}{24}$

5 Find the least common multiple for each set of numbers.

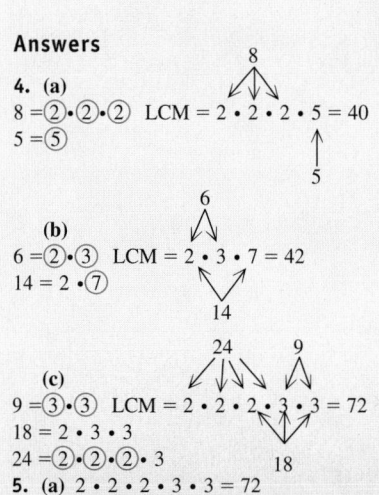

GS **(a)** 4, 8, 9

$4 = 2 \cdot 2$

$8 = 2 \cdot 2 \cdot 2$

$9 = 3 \cdot 3$

LCM = ____ • ____ • ____ • ____ • ____ = ____

(b) 3, 6, 8

(c) 15, 20, 30, 40

Answers

4. (a) $8 = ⓶ \cdot ⓶ \cdot ⓶$ $\quad \text{LCM} = 2 \cdot 2 \cdot 2 \cdot 5 = 40$ (8; 5)
$5 = ⓹$

(b) $6 = ⓶ \cdot ⓷$ $\quad \text{LCM} = 2 \cdot 3 \cdot 7 = 42$ (6; 14)
$14 = 2 \cdot ⓻$

(c) $9 = ⓷ \cdot ⓷$ $\quad \text{LCM} = 2 \cdot 2 \cdot 2 \cdot 3 \cdot 3 = 72$ (24; 9; 18)
$18 = 2 \cdot 3 \cdot 3$
$24 = ⓶ \cdot ⓶ \cdot ⓶ \cdot 3$

5. (a) $2 \cdot 2 \cdot 2 \cdot 3 \cdot 3 = 72$
(b) 24 **(c)** 120

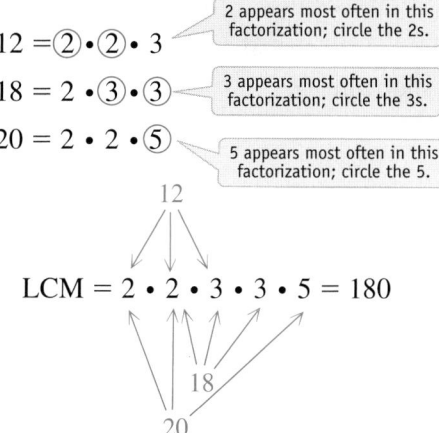

6 In the following problems, the divisions have already been worked out. Multiply the prime numbers on the left to find the least common multiple.

GS **(a)**

$$\begin{array}{r|rr} 2 & 6 & \cancel{15} \\ \hline 3 & 3 & 15 \\ \hline 5 & \cancel{1} & 5 \\ \hline & 1 & 1 \end{array}$$

$2 \cdot \_\_\_\_ \cdot \_\_\_\_ = \_\_\_\_$

(b)

$$\begin{array}{r|rr} 2 & 20 & 36 \\ \hline 2 & 10 & 18 \\ \hline 3 & \cancel{5} & 9 \\ \hline 3 & \cancel{5} & 3 \\ \hline 5 & 5 & \cancel{1} \\ \hline & 1 & 1 \end{array}$$

Answers

6. (a) 3; 5; 30 **(b)** 180

EXAMPLE 6 **Alternative Method for Finding the Least Common Multiple**

Find the least common multiple for each set of numbers.

(a) 14 and 21

Start by trying to divide 14 and 21 by the first prime number, which is 2. Use the following shortcut.

$$\begin{array}{r|rr} 2 & 14 & \cancel{21} \\ \hline & 7 & 21 \end{array}$$

14 divided by 2 is 7.

21 cannot be divided evenly by 2, so bring it down.

Because 21 cannot be divided evenly by 2, cross out 21 and bring it down. Divide by 3, the second prime.

$$\begin{array}{r|rr} 2 & 14 & \cancel{21} \\ \hline 3 & \cancel{7} & 21 \\ \hline & 7 & 7 \end{array}$$

7 cannot be divided evenly by 3, so bring it down.

21 divided by 3 is 7.

Since 7 cannot be divided evenly by the third prime, 5, skip 5. Divide by 7, the fourth prime.

$$\begin{array}{r|rr} 2 & 14 & \cancel{21} \\ \hline 3 & \cancel{7} & 21 \\ \hline 7 & 7 & 7 \\ \hline & \mathbf{1} & \mathbf{1} \end{array}$$

All quotients are 1.

Multiply these prime numbers to get the LCM.

When all quotients are 1, multiply the prime numbers on the left side.

least common multiple = $\mathbf{2 \cdot 3 \cdot 7 = 42}$

The least common multiple of 14 and 21 is 42.

(b) 6, 15, 18

Divide by 2.

$$\begin{array}{r|rrr} 2 & 6 & \cancel{15} & 18 \\ \hline & 3 & 15 & 9 \end{array}$$

Cross out 15 and bring it down.

Divide by 3.

$$\begin{array}{r|rrr} 2 & 6 & \cancel{15} & 18 \\ \hline 3 & 3 & 15 & 9 \\ \hline & 1 & 5 & 3 \end{array}$$

Divide by 3 again, since the remaining 3 can be divided.

$$\begin{array}{r|rrr} 2 & 6 & \cancel{15} & 18 \\ \hline 3 & 3 & 15 & 9 \\ \hline 3 & \cancel{1} & \cancel{5} & 3 \\ \hline & 1 & 5 & 1 \end{array}$$

Finally, divide by 5.

$$\begin{array}{r|rrr} 2 & 6 & \cancel{15} & 18 \\ \hline 3 & 3 & 15 & 9 \\ \hline 3 & \cancel{1} & \cancel{5} & 3 \\ \hline 5 & \cancel{1} & 5 & \cancel{1} \\ \hline & 1 & 1 & 1 \end{array}$$

All quotients are 1.

Multiply the prime numbers on the left side.

$2 \cdot 3 \cdot 3 \cdot 5 = 90 \leftarrow$ Least common multiple

◄ **Work Problem 6 at the Side.**

EXAMPLE 7 Find the Least Common Multiple Using Either Method

Find the least common multiple of 12, 21, and 24. For **(a)** use the prime factorization method. Then, for **(b)** use the alternative method.

(a) Find the prime factorization for each number.

$$12 = 2 \cdot 2 \cdot \textcircled{3}$$
$$21 = 3 \cdot \textcircled{7}$$
$$24 = \textcircled{2} \cdot \textcircled{2} \cdot \textcircled{2} \cdot 3$$

The product of the circled factors is the LCM.

$$\text{LCM} = 2 \cdot 2 \cdot 2 \cdot 3 \cdot 7 = 168$$

(b) Use the alternative method to find the LCM.

$$\begin{array}{r|rrr} 2 & 12 & \not{21} & 24 \\ 2 & 6 & \not{21} & 12 \\ 2 & \not{3} & \not{21} & 6 \\ 3 & 3 & 21 & 3 \\ 7 & \not{1} & 7 & \not{1} \\ \hline & 1 & 1 & 1 \end{array}$$

All quotients are 1.

The product of the prime numbers on the left is the LCM.

$$\text{LCM} = 2 \cdot 2 \cdot 2 \cdot 3 \cdot 7 = 168$$

Work Problem 7 at the Side. ▶

OBJECTIVE 5 Write a fraction with an indicated denominator. Before adding or subtracting *unlike* fractions, you must find the least common multiple, which is then used as the denominator of the fractions.

EXAMPLE 8 Writing a Fraction with an Indicated Denominator

Write the fraction $\frac{2}{3}$ with a denominator of 15.

Find a numerator, so that these fractions are equivalent.

$$\frac{2}{3} = \frac{?}{15}$$

To find the new numerator, first divide **15** by **3.**

$$\frac{2}{3} = \frac{?}{15} \qquad 15 \div 3 = 5$$

Multiply both numerator and denominator of the fraction $\frac{2}{3}$ by 5.

$$\frac{2}{3} = \frac{2 \cdot 5}{3 \cdot 5} = \frac{10}{15}$$

Multiplying by $\frac{5}{5}$ is the same as multiplying by 1, because $\frac{5}{5} = 1$.

This process is just the opposite of writing a fraction in lowest terms. Check the answer by writing $\frac{10}{15}$ in lowest terms; you should get $\frac{2}{3}$ again.

---

7 Find the least common multiple of each set of numbers. Use whichever method you prefer.

GS **(a)** 3, 6, 10

$$\begin{array}{r|rrr} 2 & \not{3} & 6 & 10 \\ 3 & 3 & 3 & \not{5} \\ 5 & \_\_ & \not{1} & \_\_ \\ \hline & 1 & 1 & 1 \end{array}$$

$2 \cdot \_\_\_\_ \cdot 5 = \_\_\_\_$

(b) 15, 40

(c) 9, 24

(d) 8, 21, 24

Answers

7. **(a)** 1; 5; 3; 30 **(b)** 120 **(c)** 72 **(d)** 168

8 Rewrite each fraction with the indicated denominator.

(a) $\frac{1}{4} = \frac{?}{16}$

(b) $\frac{5}{3} = \frac{?}{15}$

(c) $\frac{7}{16} = \frac{?}{32}$

(d) $\frac{6}{11} = \frac{?}{33}$

Answers

8. (a) $\frac{4}{16}$ **(b)** $\frac{25}{15}$ **(c)** $\frac{14}{32}$ **(d)** $\frac{18}{33}$

EXAMPLE 9 Writing Fractions with a New Denominator

Rewrite each fraction with the indicated denominator.

(a) $\frac{3}{8} = \frac{?}{48}$

Divide 48 by 8, to get 6. Now multiply both the numerator and the denominator of $\frac{3}{8}$ by 6.

$$\frac{3}{8} = \frac{3 \cdot 6}{8 \cdot 6} = \frac{18}{48}$$ Multiply numerator and denominator by 6.

Multiplying a number by 1 does *not* change the number, and $\frac{6}{6} = 1$.

That is, $\frac{3}{8} = \frac{18}{48}$. As a check, write $\frac{18}{48}$ in lowest terms. You should get $\frac{3}{8}$ again.

(b) $\frac{7}{6} = \frac{?}{42}$

Divide 42 by 6, to get 7. Next, multiply both the numerator and the denominator of $\frac{7}{6}$ by 7.

$$\frac{7}{6} = \frac{7 \cdot 7}{6 \cdot 7} = \frac{49}{42}$$ Multiply numerator and denominator by 7.

Multiplying by $\frac{7}{7}$ is the same as multiplying by 1.

This shows that $\frac{7}{6} = \frac{49}{42}$. As a check, write $\frac{49}{42}$ in lowest terms. Did you get $\frac{7}{6}$ again?

Note

In **Example 8,** on the previous page, the fraction $\frac{2}{3}$ was multiplied by $\frac{5}{5}$. In **Example 9,** the fraction $\frac{3}{8}$ was multiplied by $\frac{6}{6}$ and the fraction $\frac{7}{6}$ was multiplied by $\frac{7}{7}$. The fractions, $\frac{5}{5}$, $\frac{6}{6}$, and $\frac{7}{7}$ are all equal to 1.

$$\frac{5}{5} = 1 \qquad \frac{6}{6} = 1 \qquad \frac{7}{7} = 1$$

Recall that any number multiplied by 1 is the number itself.

◀ Work Problem **8** at the Side.

3.2 Exercises

FOR EXTRA HELP

Download the MyDashBoard App

MyMathLab®

CONCEPT CHECK *Write either* true *or* false *for each statement. If* false, *explain why.*

1. The least common multiple (LCM) of 4 and 8 is 8.

2. The least common multiple (LCM) of 6 and 5 is 25.

3. The least common multiple (LCM) of 9 and 4 is 36.

4. The least common multiple (LCM) of 3 and 7 is 28.

Use multiples of the larger number to find the least common multiple in each set of numbers. ***See Examples 1 and 2.***

5. 3 and 6

6. 2 and 4

7. 3 and 5

8. 3 and 7

9. 4 and 9

10. 6 and 8

GS

Multiples of 8

8 16 24 32 40

What is the first multiple divisible by 6? ____

11. 12 and 16

GS

Multiples of 16

16 32 48 64 80

What is the first multiple divisible by 16? ____

12. 25 and 75

13. 20 and 50

Find the least common multiple of each set of numbers. Use any method.
See Examples 3–7.

14. 4, 10

GS

Prime factorizations

$4 = \textcircled{2} \cdot \textcircled{2}$

$10 = 2 \cdot \textcircled{5}$

LCM = ____ • ____ • ____ = ____

15. 8, 10

GS

Prime factorizations

$8 = \textcircled{2} \cdot \textcircled{2} \cdot \textcircled{2}$

$10 = 2 \cdot \textcircled{5}$

LCM = ____ • ____ • ____ • ____ = ____

16. 12, 20

17. 9 and 15

18. 6, 9, 12

19. 20, 24, 30

20. 8, 9, 12, 18

21. 4, 6, 8, 10

22. 12, 15, 18, 20

23. 6, 8, 9, 27, 36

24. 8, 10, 12, 16, 36

25. 5, 6, 8, 25, 30

Rewrite each fraction with a denominator of 24. **See Examples 8 and 9.**

26. $\frac{2}{3} =$

27. $\frac{3}{8} =$

28. $\frac{3}{4} =$

29. $\frac{5}{12} =$

30. $\frac{5}{6} =$

31. $\frac{7}{8} =$

32. **CONCEPT CHECK** Circle the fraction that is equivalent to $\frac{2}{3}$.

$\frac{7}{8}$ $\frac{3}{4}$ $\frac{12}{16}$ $\frac{8}{12}$

33. **CONCEPT CHECK** Circle the fraction that is equivalent to $\frac{3}{4}$.

$\frac{4}{5}$ $\frac{18}{22}$ $\frac{21}{28}$ $\frac{9}{15}$

34. **CONCEPT CHECK** Circle the fraction that is equivalent to $\frac{9}{4}$.

$\frac{36}{16}$ $\frac{35}{20}$ $\frac{21}{8}$ $\frac{15}{6}$

35. **CONCEPT CHECK** Circle the fraction that is equivalent to $\frac{7}{8}$.

$\frac{14}{15}$ $\frac{21}{24}$ $\frac{8}{7}$ $\frac{35}{48}$

Rewrite each fraction with the indicated denominator.

36. $\frac{1}{2} = \frac{}{6}$

37. $\frac{2}{3} = \frac{}{9}$

38. $\frac{3}{4} = \frac{}{16}$

39. $\frac{7}{8} = \frac{}{32}$

40. $\frac{8}{5} = \frac{}{20}$

41. $\frac{3}{16} = \frac{}{64}$

42. $\frac{5}{8} = \frac{}{40}$

43. $\frac{9}{7} = \frac{}{56}$

44. $\frac{3}{2} = \frac{}{64}$

45. GS $\dfrac{7}{4} = \dfrac{}{48}$

$\dfrac{7}{4} = \dfrac{?}{48}$

Divide 48 by 4 to get ______.

Now multiply 7 by ______ to get the new numerator, which is ______.

46. GS $\dfrac{5}{6} = \dfrac{}{120}$

$\dfrac{5}{6} = \dfrac{?}{120}$

Divide 120 by 6 to get ______.

Now multiply 5 by ______ to get the new numerator, which is ______.

47. $\dfrac{8}{11} = \dfrac{}{132}$

48. $\dfrac{4}{15} = \dfrac{}{165}$

49. $\dfrac{3}{16} = \dfrac{}{144}$

50. $\dfrac{7}{16} = \dfrac{}{112}$

51. There are several methods for finding the least common multiple (LCM). Do you prefer the method using multiples of the largest number, the method using prime factorizations, or the alternative method for finding the least common multiple? Why? Would you ever use the other methods?

52. Explain in your own words how to write a fraction with an indicated denominator. As part of your explanation, show how to change $\frac{3}{4}$ to a fraction having 12 as a denominator. Also explain how you could check your answer.

Find the least common multiple of the denominators of each pair of fractions.

53. $\dfrac{25}{400}, \dfrac{38}{1800}$

54. $\dfrac{53}{600}, \dfrac{115}{4000}$

55. $\dfrac{109}{1512}, \dfrac{23}{392}$

56. $\dfrac{61}{810}, \dfrac{37}{1170}$

Relating Concepts (Exercises 57–64) For Individual or Group Work

Most people think that addition and subtraction of fractions are more difficult than multiplication and division of fractions. This is probably because a common denominator must be used. ***Work Exercises 57–64 in order.***

57. Fractions with the same denominators are ____________ fractions and fractions with different denominators are ____________ fractions.

58. To subtract like fractions, first subtract the ____________ to find the numerator of the difference. Write the denominator of the like fractions as the ____________ of the difference. Finally, write the answer in ____________ terms.

59. The ____________ common multiple (LCM) of two numbers is the (*smallest/largest*)whole number divisible by both those numbers.

60. The following shows the common multiples for both 8 and 10. What is the least common multiple for these two numbers?

Multiples of 8: 8, 16, 24, 32, 40, 48, 56, 64, 72, 80, 88, . . .

Multiples of 10: 10, 20, 30, 40, 50, 60, 70, 80, 90, . . .

Find the least common multiple for each set of numbers. Use whichever method you like best.

61. 5, 7, 14, 10

62. 25, 18, 30, 5

63. Explain why the least common multiple for 8, 3, 5, 4, and 10 is not 240. Find the least common multiple.

64. Explain why the least common multiple of 55 and 1760 is 1760.

3.3 Adding and Subtracting Unlike Fractions

OBJECTIVES

1. Add unlike fractions.
2. Add unlike fractions vertically.
3. Subtract unlike fractions.
4. Subtract unlike fractions vertically.

OBJECTIVE 1 Add unlike fractions. In this section, we add and subtract unlike fractions. To add unlike fractions, we must first change them to like fractions (fractions with the same denominator). For example, the figures below show $\frac{3}{8}$ and $\frac{1}{4}$.

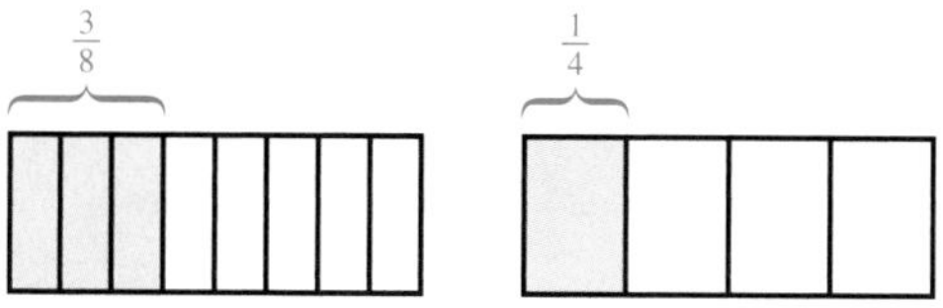

These fractions can be added by changing them to like fractions. Make like fractions by changing $\frac{1}{4}$ to the equivalent fraction $\frac{2}{8}$.

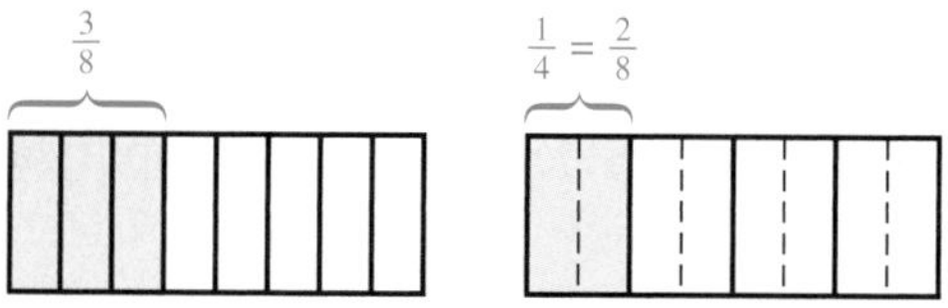

Now you can add the fractions.

Becomes ($\frac{1}{4} \to \frac{2}{8}$)

$$\frac{3}{8} + \frac{1}{4} = \frac{3}{8} + \frac{2}{8} = \frac{5}{8}$$

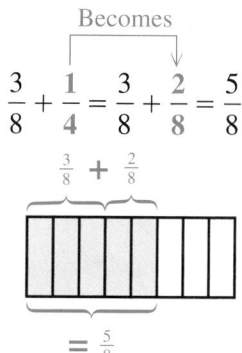

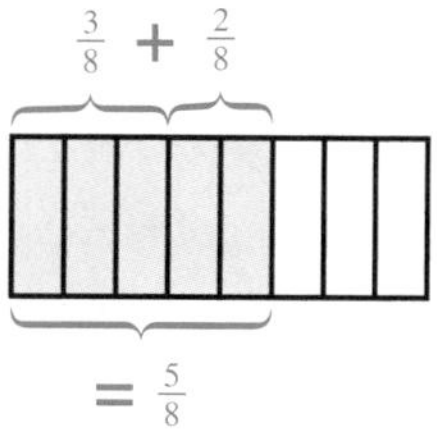

Use the following steps to add or subtract unlike fractions.

Adding or Subtracting Unlike Fractions

Step 1 Rewrite the *unlike fractions* as *like fractions* with the least common multiple as their new denominator. This new denominator is called the **least common denominator (LCD).**

Step 2 Add or subtract as with like fractions.

Step 3 Simplify the answer by writing it in lowest terms and as a whole or mixed number where possible.

VOCABULARY TIP

LCM and LCD The least common multiple **(LCM)** and the least common denominator **(LCD)** are similar. When the LCM of two or more numbers is found and then used as a new denominator, it is called the least common denominator (LCD). You must find the LCD in order to add or subtract unlike fractions.

EXAMPLE 1 Adding Unlike Fractions

Add $\frac{2}{3}$ and $\frac{1}{9}$.

The least common multiple of 3 and 9 is 9, so first rewrite the fractions as like fractions with a denominator of 9. This is the ***least common denominator (LCD)*** of 3 and 9.

Continued on Next Page

1 Add.

GS (a) $\frac{1}{2}+\frac{3}{8}$

$\frac{\_\_}{8}+\frac{3}{8}=\frac{\_\_}{8}$

(b) $\frac{3}{4}+\frac{1}{8}$

(c) $\frac{3}{5}+\frac{3}{10}$

(d) $\frac{1}{12}+\frac{5}{6}$

2 Add. Simplify all answers.

GS (a) $\frac{3}{10}+\frac{1}{5}$

$\frac{3}{10}+\frac{\_\_}{10}=\frac{\_\_}{\_\_}=\frac{\_\_}{\_\_}$

(b) $\frac{5}{8}+\frac{1}{3}$

(c) $\frac{1}{10}+\frac{1}{3}+\frac{1}{6}$

Answers

1. (a) 4; $\frac{7}{8}$ (b) $\frac{7}{8}$ (c) $\frac{9}{10}$ (d) $\frac{11}{12}$

2. (a) 2; $\frac{5}{10}$; $\frac{1}{2}$ (b) $\frac{23}{24}$ (c) $\frac{3}{5}$

Step 1 $\frac{2}{3}=\frac{?}{9}$

Divide 9 by 3, getting 3. Next, multiply numerator and denominator by 3.

$$\frac{2}{3}=\frac{2\cdot 3}{3\cdot 3}=\frac{6}{9}$$ ← $\frac{6}{9}$ is equivalent to $\frac{2}{3}$.

Now, add the like fractions $\frac{6}{9}$ and $\frac{1}{9}$.

Becomes — Both fractions must have the *same* denominator **before** you add them.

Step 2 $\frac{2}{3}+\frac{1}{9}=\frac{6}{9}+\frac{1}{9}=\frac{6+1}{9}=\frac{7}{9}$

Step 3 Step 3 is not needed because $\frac{7}{9}$ is already in lowest terms.

◀ **Work Problem 1 at the Side.**

EXAMPLE 2 Adding Fractions

Add each pair of fractions using the three steps. Simplify all answers.

(a) $\frac{1}{3}+\frac{1}{6}$

The least common multiple of 3 and 6 is 6. Rewrite both fractions as fractions with a least common denominator of 6.

Rewritten as like fractions

Step 1 $\frac{1}{3}+\frac{1}{6}=\frac{2}{6}+\frac{1}{6}$ ← 6 is the LCD (least common denominator).

Add numerators.

Step 2 $\frac{2}{6}+\frac{1}{6}=\frac{2+1}{6}=\frac{3}{6}$ ← Sum of numerators; ← Least common denominator

Step 3 $\frac{3}{6}=\frac{1}{2}$ ← Lowest terms

(b) $\frac{6}{15}+\frac{3}{10}$

The least common multiple of 15 and 10 is 30, so rewrite both fractions with a least common denominator of 30.

Rewritten as like fractions

Step 1 $\frac{6}{15}+\frac{3}{10}=\frac{12}{30}+\frac{9}{30}$ ← The least common multiple of the denominators, 30, is also the LCD.

Add numerators.

Step 2 $\frac{12}{30}+\frac{9}{30}=\frac{12+9}{30}=\frac{21}{30}$

Step 3 $\frac{21}{30}=\frac{7}{10}$ ← Lowest terms

◀ **Work Problem 2 at the Side.**

OBJECTIVE 2 Add unlike fractions vertically. Fractions can also be added vertically (one fraction written below the other).

EXAMPLE 3 Vertical Addition of Fractions

Add the following fractions vertically.

(a)
$$\begin{array}{r} \dfrac{3}{8} = \dfrac{3 \cdot 3}{8 \cdot 3} = \dfrac{9}{24} \\ + \dfrac{7}{12} = \dfrac{7 \cdot 2}{12 \cdot 2} = \dfrac{14}{24} \\ \hline \dfrac{23}{24} \end{array}$$

24 is the LCD.
Rewritten as like fractions
23 ← Add the numerators.
24 ← Denominator is 24, the LCD.

$\frac{23}{24}$ is in simplest form.

(b)
$$\begin{array}{r} \dfrac{2}{9} = \dfrac{2 \cdot 4}{9 \cdot 4} = \dfrac{8}{36} \\ + \dfrac{1}{4} = \dfrac{1 \cdot 9}{4 \cdot 9} = \dfrac{9}{36} \\ \hline \dfrac{17}{36} \end{array}$$

36 is the LCD.
Rewritten as like fractions
17 ← Add the numerators.
36 ← Denominator is 36, the LCD.

$\frac{17}{36}$ is in simplest form.

Work Problem 3 at the Side. ▶

OBJECTIVE 3 Subtract unlike fractions. The next example shows subtraction of unlike fractions.

EXAMPLE 4 Subtracting Unlike Fractions

Subtract. Simplify all answers.

As with addition, rewrite unlike fractions with a least common denominator.

(a) $\frac{3}{4} - \frac{3}{8}$

Rewritten as like fractions

Step 1 $\frac{3}{4} - \frac{3}{8} = \frac{6}{8} - \frac{3}{8}$ The LCD is 8.

Subtract numerators.

Step 2 $\frac{6}{8} - \frac{3}{8} = \frac{6-3}{8} = \frac{3}{8}$ ← Difference of numerators ← Least common denominator

Step 3 Not needed because $\frac{3}{8}$ is in lowest terms.

Continued on Next Page

3 Add the following fractions vertically.

GS (a)
$$\begin{array}{r} \dfrac{5}{8} = \dfrac{5 \cdot 3}{8 \cdot \_\_} = \dfrac{15}{\_\_} \\ + \dfrac{1}{12} = \dfrac{1 \cdot 2}{12 \cdot 2} = \dfrac{2}{24} \\ \hline \dfrac{\_\_}{\_\_} \end{array}$$

(b)
$$\begin{array}{r} \dfrac{7}{16} \\ + \dfrac{1}{4} \\ \hline \end{array}$$

(c)
$$\begin{array}{r} \dfrac{1}{8} \\ \dfrac{5}{24} \\ + \dfrac{7}{16} \\ \hline \end{array}$$

Answers

3. (a) 3; 24; $\frac{17}{24}$ (b) $\frac{11}{16}$ (c) $\frac{37}{48}$

4 Subtract. Simplify all answers.

GS **(a)** $\frac{5}{8} - \frac{1}{4}$

$\frac{5}{8} - \frac{\_\_}{8} =$

(b) $\frac{4}{5} - \frac{3}{4}$

5 Subtract vertically. Simplify all answers.

GS **(a)** $\frac{7}{8} = \frac{7 \cdot \_\_}{8 \cdot 3} = \frac{21}{\_\_}$

$- \frac{2}{3} = \frac{2 \cdot 8}{3 \cdot 8} = \frac{16}{24}$

(b) $\frac{5}{6}$

$- \frac{1}{12}$

Answers

4. **(a)** 2; $\frac{3}{8}$ **(b)** $\frac{1}{20}$

5. **(a)** 3; 24; $\frac{5}{24}$ **(b)** $\frac{3}{4}$

(b) $\frac{3}{4} - \frac{7}{12}$

Rewritten as like fractions

Step 1 $\frac{3}{4} - \frac{7}{12} = \frac{9}{12} - \frac{7}{12}$

Subtract numerators.

Step 2 $\frac{9}{12} - \frac{7}{12} = \frac{9-7}{12} = \frac{2}{12}$ ← Subtract the numerators. ← Denominator is 12, the LCD.

Step 3 $\frac{2}{12} = \frac{1}{6}$ ← Lowest terms

Always simplify the final answer.

◄ **Work Problem 4 at the Side.**

OBJECTIVE 4 Subtract unlike fractions vertically.

EXAMPLE 5 Vertical Subtraction of Fractions

Subtract the following fractions vertically. Simplify all answers.

(a)

$$\frac{4}{5} = \frac{4 \cdot 8}{5 \cdot 8} = \frac{32}{40}$$

$$-\frac{3}{8} = \frac{3 \cdot 5}{8 \cdot 5} = \frac{15}{40}$$

Rewritten as like fractions

$\frac{17}{40}$ ← Subtract numerators. ← Denominator is 40, the LCD.

$\frac{17}{40}$ is in simplest form.

(b)

$$\frac{3}{7} = \frac{3 \cdot 12}{7 \cdot 12} = \frac{36}{84}$$

$$-\frac{5}{12} = \frac{5 \cdot 7}{12 \cdot 7} = \frac{35}{84}$$

Rewritten as like fractions

$\frac{1}{84}$ ← Subtract numerators. ← Denominator is 84, the LCD.

$\frac{1}{84}$ is in simplest form.

◄ **Work Problem 5 at the Side.**

3.3 Exercises

FOR EXTRA HELP

Download the MyDashBoard App

MyMathLab®

CONCEPT CHECK *Fill in the blank with the correct response.*

1. To add or subtract unlike fractions, the first step is to rewrite the fractions as ______________ fractions.

2. To rewrite unlike fractions as like fractions, you must find the ______________ (LCD).

Add the following fractions. Simplify all answers. ***See Examples 1–3.***

3. $\frac{3}{4}+\frac{1}{8}$

4. $\frac{1}{6}+\frac{2}{3}$

5. $\frac{2}{3}+\frac{2}{9}$

6. $\frac{3}{7}+\frac{1}{14}$ LCD is 14.

$\frac{6}{14}+\frac{1}{14}$

$=\frac{7}{14}=$ ______ simplest form

7. $\frac{9}{20}+\frac{3}{10}$ LCD is 20.

$\frac{9}{20}+\frac{6}{20}$

$=\frac{15}{20}=$ ______ simplest form

8. $\frac{5}{8}+\frac{1}{4}$

9. $\frac{3}{5}+\frac{3}{8}$

10. $\frac{5}{7}+\frac{3}{14}$

11. $\frac{2}{9}+\frac{5}{12}$

12. $\frac{1}{4}+\frac{2}{9}+\frac{1}{3}$ LCD is 36.

$\frac{9}{36}+\frac{8}{36}+\frac{12}{36}=\frac{\quad}{36}$ Already in simplest form

13. $\frac{3}{7}+\frac{2}{5}+\frac{1}{10}$ LCD is 70.

$\frac{30}{70}+\frac{28}{70}+\frac{7}{70}$

$=\frac{65}{70}=$ ______ simplest form

14. $\frac{3}{10}+\frac{2}{5}+\frac{3}{20}$

15. $\frac{1}{3}+\frac{3}{8}+\frac{1}{4}$

16. $\frac{4}{15}+\frac{1}{6}+\frac{1}{3}$

17. $\frac{5}{12}+\frac{2}{9}+\frac{1}{6}$

18. $\frac{2}{3}$ $+\frac{1}{6}$

19. $\frac{1}{4}$ $+\frac{1}{8}$

20. $\frac{7}{12}$ $+\frac{1}{8}$

21. $\frac{5}{12}$ $+\frac{1}{16}$

22. $\frac{3}{7}$ $+\frac{1}{3}$

Subtract the following fractions. Simplify all answers. **See Example 4.**

23. $\frac{5}{6} - \frac{1}{3}$ LCD is 6.
GS
$\frac{5}{6} - \frac{2}{6} = \frac{3}{6} =$ simplest form

24. $\frac{3}{4} - \frac{5}{8}$ LCD is 8.
GS
$\frac{6}{8} - \frac{5}{8} =$ simplest form

25. $\frac{2}{3} - \frac{1}{6}$

26. $\frac{5}{8} - \frac{1}{4}$

27. $\frac{2}{3} - \frac{1}{5}$

28. $\frac{5}{6} - \frac{7}{9}$

29. $\frac{5}{12} - \frac{1}{4}$

30. $\frac{5}{7} - \frac{1}{3}$

31. $\frac{8}{9} - \frac{7}{15}$

32. $\begin{array}{r} \frac{4}{5} \\ -\ \frac{1}{3} \\ \hline \end{array}$

33. $\begin{array}{r} \frac{7}{8} \\ -\ \frac{4}{5} \\ \hline \end{array}$

34. $\begin{array}{r} \frac{5}{8} \\ -\ \frac{1}{3} \\ \hline \end{array}$

35. $\begin{array}{r} \frac{5}{12} \\ -\ \frac{1}{16} \\ \hline \end{array}$

36. $\begin{array}{r} \frac{7}{12} \\ -\ \frac{1}{3} \\ \hline \end{array}$

Solve each application problem.

Use the newspaper advertisement for this 4-piece chisel set to answer Exercises 37–38. (Source: Harbor Freight Tools.)

37. Find the difference in the cutting-edge width of the two chisels with the widest blades. The symbol " is for inches.

38. Find the difference in the cutting-edge width of the two chisels with the narrowest blades. The " symbol is for inches.

39. A sports and entertainment center has $\frac{4}{5}$ of its total area devoted to seating of fans and guests. If $\frac{3}{8}$ of the seating area is used for general admission seating and the rest for reserved seating, find the fraction of the total area used for reserved seating.

40. A dairy farmer must vaccinate $\frac{5}{8}$ of her cows this week. If she vaccinates $\frac{3}{16}$ of the herd on Monday, and $\frac{1}{4}$ of the herd on Wednesday, what fraction of the herd remains to be vaccinated?

41. When installing cabinets for The Home Depot, Sarah Bryn must be certain that the proper type and size of mounting screw is used. Find the total length of the screw shown.

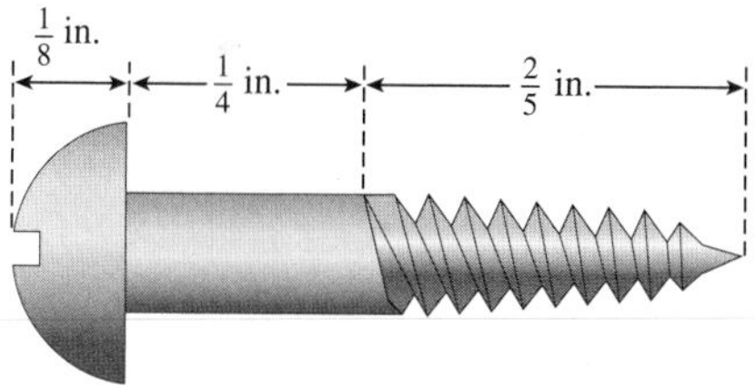

42. When installing a computer chassis, Bonnie Bottorff must be certain that the proper type and size of bolt is used. Find the total length of the bolt shown.

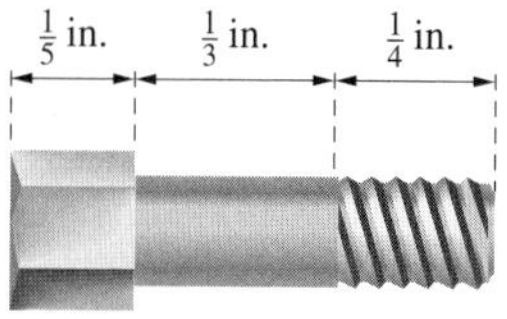

43. Bill Newton is a general contractor. He began a job with $\frac{3}{4}$ of a tank of fuel in his backhoe. He used $\frac{1}{3}$ of the tank in the morning and $\frac{3}{8}$ of the tank in the afternoon. What fraction of the tank of fuel remains?

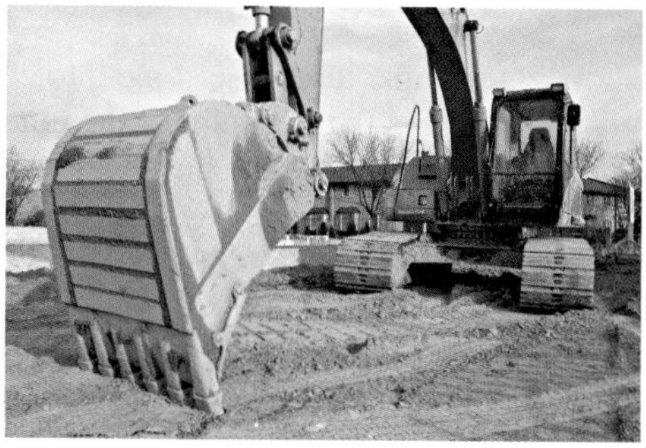

44. Cliff Dinsmore is coordinating the refurbishing of the community swimming pool. The pool was $\frac{7}{8}$ full when workers began draining it. By noon, another $\frac{3}{16}$ of the pool had been drained. An additional $\frac{1}{3}$ of the pool was drained in the afternoon. Find the fraction of the pool water remaining.

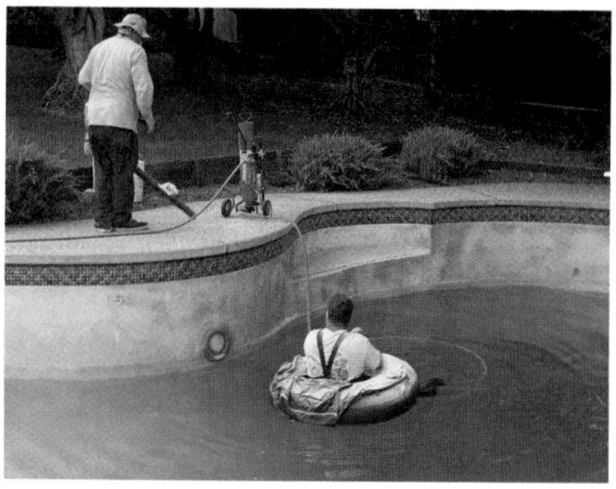

45. Step 1 in adding or subtracting unlike fractions is to rewrite the fractions so they have the least common multiple as a denominator. Explain in your own words why this is necessary.

46. Briefly list the three steps used for addition and subtraction of unlike fractions.

A survey of 1200 users of social networking sites showed that honesty was not always practiced. Refer to the circle graph to answer Exercises 47–50.

HOW HONEST ARE YOU ON SOCIAL NETWORKING SITES?

Fib a Little $\frac{1}{4}$

Totally Honest $\frac{1}{3}$

Flat-out Lie $\frac{1}{5}$

Total Fabrication $\frac{13}{60}$

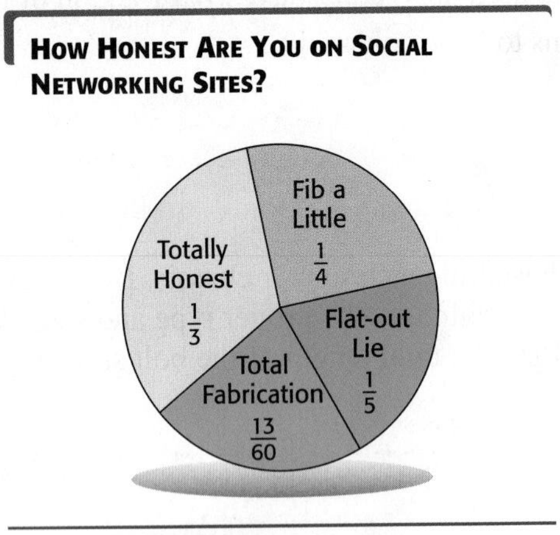

Source: USA Today.

47. What fraction of those surveyed are totally honest?

48. What fraction of those surveyed fib a little?

49. Which response was given most often? How many people gave this response? What fraction of the users gave this response and the "fib a little" response?

50. Which response was given least often? How many people gave this response? What fraction of the users gave this response and the "total fabrication" response?

51. Find the diameter of the hole in the mounting bracket shown. (The diameter is the distance across the center of the hole.)

Diameter

$\frac{3}{8}$ in. $\frac{3}{8}$ in.

$\frac{15}{16}$ in.

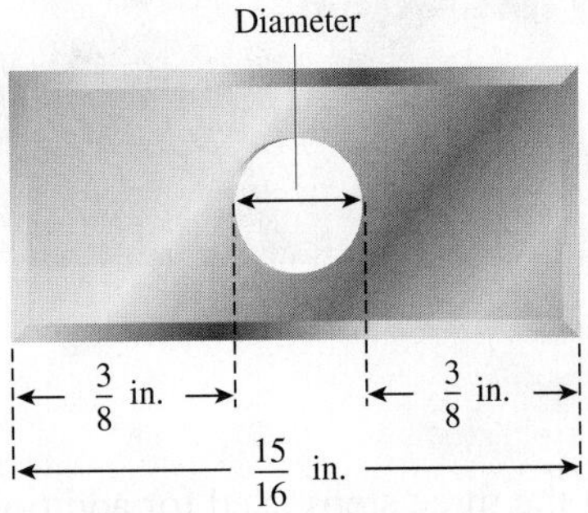

52. Chakotay is fitting a turquoise stone into a bear claw pendant. Find the diameter of the hole in the pendant. (The diameter is the distance across the center of the hole.)

$\frac{3}{16}$ in. $\frac{3}{16}$ in.

$\frac{7}{8}$ in.

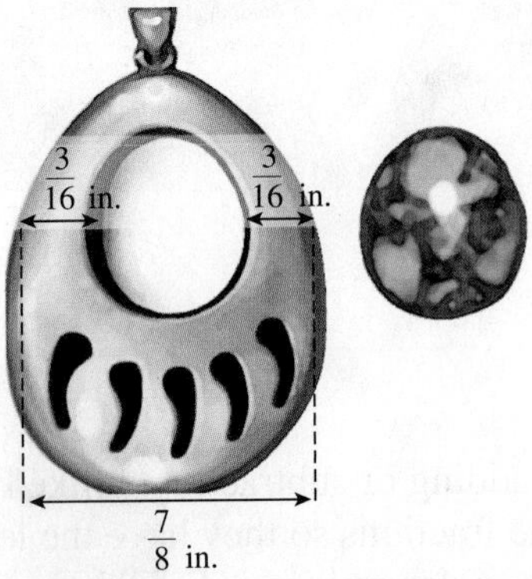

3.4 Adding and Subtracting Mixed Numbers

OBJECTIVES

1. Estimate an answer, then add or subtract mixed numbers.
2. Estimate an answer, then subtract mixed numbers by regrouping.
3. Add or subtract mixed numbers using an alternative method.

Recall that a mixed number is the sum of a whole number and a fraction. For example,

$$3\frac{2}{5} \quad \text{means} \quad 3 + \frac{2}{5}.$$

OBJECTIVE 1 Estimate an answer, then add or subtract mixed numbers. Add or subtract mixed numbers by adding or subtracting the fraction parts and then the whole number parts. It is a good idea to estimate the answer first, as we did when multiplying and dividing mixed numbers in **Chapter 2.**

Work Problem 1 at the Side. ▶

EXAMPLE 1 Adding and Subtracting Mixed Numbers

First estimate the answer. Then add or subtract to find the exact answer.

(a) $16\frac{1}{8} + 5\frac{5}{8}$

Estimate: | *Exact:*

$16 \leftarrow$ Rounds to $\; 16\frac{1}{8}$

$+\ 6 \leftarrow$ Rounds to $\; +\ 5\frac{5}{8}$

First, estimate the answer. $\;22 \qquad 21\frac{6}{8} = 21\frac{3}{4} \leftarrow$ Lowest terms

The exact answer is close to the estimate.

Sum of whole numbers; Sum of fractions

In lowest terms $\frac{6}{8}$ is $\frac{3}{4}$, so the exact answer of $21\frac{3}{4}$ is in lowest terms. The exact answer is *reasonable* because it is close to the estimate of 22.

(b) $8\frac{5}{8} - 3\frac{1}{12}$

Estimate: | *Exact:*

$9 \leftarrow$ Rounds to $\; 8\frac{5}{8} = 8\frac{15}{24}$

$-\ 3 \leftarrow$ Rounds to $\; -3\frac{1}{12} = -\ 3\frac{2}{24}$

24 is the least common denominator.

$6 \qquad 5\frac{13}{24} \leftarrow$ Lowest terms

Subtract whole numbers. Subtract fractions.

The exact answer of $5\frac{13}{24}$ is *reasonable* because it is close to the estimated answer of 6. Check by adding $5\frac{13}{24}$ and $3\frac{1}{12}$. The sum should be $8\frac{5}{8}$.

Continued on Next Page

1 As a review of mixed numbers, write each mixed number as an improper fraction and each improper fraction as a mixed number.

(a) $\frac{9}{2}$

(b) $\frac{8}{3}$

(c) $4\frac{3}{4}$

(d) $3\frac{7}{8}$

Answers

1. (a) $4\frac{1}{2}$ (b) $2\frac{2}{3}$ (c) $\frac{19}{4}$ (d) $\frac{31}{8}$

2 First estimate, and then add or subtract to find the exact answer.

GS **(a)** *Estimate:* *Exact:*

$7 \xleftarrow{\text{Rounds to}} 6\frac{7}{8} = 6\frac{7}{8}$

$+\,2 \xleftarrow{\text{Rounds to}} +\,2\frac{1}{4} = 2\frac{2}{8}$

(b) *Estimate:* *Exact:*

$\_\_ \xleftarrow{\text{Rounds to}} 4\frac{7}{9}$

$-\,\_\_ \xleftarrow{\text{Rounds to}} -2\frac{2}{3}$

3 First estimate, and then add to find the exact answer.

GS **(a)** *Estimate:* *Exact:*

$10 \xleftarrow{\text{Rounds to}} 9\frac{3}{4}$

$+\;8 \xleftarrow{\text{Rounds to}} +\,7\frac{1}{2}$

(b) *Estimate:* *Exact:*

$\_\_ \xleftarrow{\text{Rounds to}} 15\frac{4}{5}$

$+\,\_\_ \xleftarrow{\text{Rounds to}} +\,12\frac{2}{3}$

Answers

2. **(a)** $7 + 2 = 9;\ 9\frac{1}{8}$

(b) $5 - 3 = 2;\ 2\frac{1}{9}$

3. **(a)** $10 + 8 = 18;\ 17\frac{1}{4}$

(b) $16 + 13 = 29;\ 28\frac{7}{15}$

Note

When estimating, if the numerator is *half* of the denominator or *more*, round up the whole number part. If the numerator is *less* than *half* the denominator, leave the whole number part as it is.

◀ Work Problem **2** at the Side.

When you add the fraction parts of mixed numbers, the sum may be greater than 1. If this happens, simplify the fraction and regroup in the whole number column.

EXAMPLE 2 Simplify and Regroup When Adding Mixed Numbers

First estimate, and then add $9\frac{5}{8} + 13\frac{7}{8}$.

Estimate: *Exact:*

$10 \xleftarrow{\text{Rounds to}} 9\frac{5}{8}$

$+\,14 \xleftarrow{\text{Rounds to}} +\,13\frac{7}{8}$

$24 \qquad 22\frac{12}{8}$

Sum of whole numbers (22); Sum of fractions ($\frac{12}{8}$)

First, add the fractions, then the whole numbers.

The improper fraction $\frac{12}{8}$ can be written in lowest terms as $\frac{3}{2}$. Then $\frac{3}{2} = 1\frac{1}{2}$, so the simplified sum is

Becomes Becomes

$$22\frac{12}{8} = 22 + \frac{12}{8} = 22 + \frac{3}{2} = 22 + 1\frac{1}{2} = 23\frac{1}{2}.$$

The estimate was 24, so the exact answer of $23\frac{1}{2}$ is reasonable.

Note

When adding mixed numbers, first add the fraction parts, then add the whole number parts. Finally, combine the two answers and simplify.

◀ Work Problem **3** at the Side.

OBJECTIVE 2 Estimate an answer, then subtract mixed numbers by regrouping. When subtracting mixed numbers, **regrouping** is necessary when the fraction part of the first number is less than the fraction part of the second number.

EXAMPLE 3 Regroup When Subtracting Mixed Numbers

First estimate, and then subtract to find the exact answer.

(a) $7 - 2\frac{5}{6}$

Continued on Next Page

Estimate: *Exact:*

$$\begin{array}{r} 7 \\ -\,3 \\ \hline 4 \end{array} \quad \begin{array}{c} \leftarrow \text{Rounds to} \\ \leftarrow \text{Rounds to} \end{array} \quad \begin{array}{r} 7 \\ -\,2\frac{5}{6} \\ \hline \end{array}$$

There is no fraction here from which to subtract $\frac{5}{6}$.

It is **not** possible to subtract $\frac{5}{6}$ without regrouping the whole number **7** first.

Regroup 7 as 6 + 1.

$$7 = 6 + 1$$

$1 = \frac{6}{6}$

$$= 6 + \frac{6}{6}$$

$$= 6\frac{6}{6}$$

Now you can subtract.

$$\begin{array}{rcr} 7 & = & 6\frac{6}{6} \\ -\,2\frac{5}{6} & = & -\,2\frac{5}{6} \\ \hline & & 4\frac{1}{6} \end{array}$$

7 was rewritten as $6\frac{6}{6}$.

Exact answer is close to the estimate.

The estimate was 4, so the exact answer of $4\frac{1}{6}$ is reasonable.

(b) $8\frac{1}{3} - 4\frac{3}{5}$

Estimate: *Exact:*

$$\begin{array}{r} 8 \\ -\,5 \\ \hline 3 \end{array} \quad \begin{array}{c} \leftarrow \text{Rounds to} \\ \leftarrow \text{Rounds to} \end{array} \quad \begin{array}{rcr} 8\frac{1}{3} & = & 8\frac{5}{15} \\ -\,4\frac{3}{5} & = & -\,4\frac{9}{15} \\ \hline \end{array}$$

15 is the least common denominator.

It is **not** possible to subtract $\frac{9}{15}$ from $\frac{5}{15}$, so regroup the whole number **8.**

Regroup 8 as 7 + 1.

$$8\frac{5}{15} = 8 + \frac{5}{15} = 7 + 1 + \frac{5}{15}$$

$1 = \frac{15}{15}$

$$= 7 + \frac{15}{15} + \frac{5}{15}$$

$$= 7 + \frac{20}{15} \leftarrow \frac{15}{15} + \frac{5}{15}$$

$$= 7\frac{20}{15}$$

Continued on Next Page

4 First estimate and then subtract to find the exact answer.

GS (a) *Estimate:* *Exact:*

$7 \xleftarrow{\text{Rounds to}} \left\{ 7\frac{1}{3} = 7\frac{2}{6} \right.$

$-5 \xleftarrow{\text{Rounds to}} \left\{ -4\frac{5}{6} = 4\frac{5}{6} \right.$

(b) *Estimate:* *Exact:*

$\xleftarrow{\text{Rounds to}} \left\{ 4\frac{5}{8} \right.$

$- \quad \xleftarrow{\text{Rounds to}} \left\{ -2\frac{15}{16} \right.$

(c) *Estimate:* *Exact:*

$\xleftarrow{\text{Rounds to}} \left\{ 15 \right.$

$- \quad \xleftarrow{\text{Rounds to}} \left\{ -6\frac{4}{9} \right.$

5 Add or subtract by changing mixed numbers to improper fractions. Simplify answers.

GS (a) $3\frac{3}{8} = \frac{27}{8} = \frac{27}{8}$

$+2\frac{1}{2} = \frac{5}{2} = \frac{\quad}{8}$

8 is LCD.

(b) $6\frac{3}{4}$

$-4\frac{2}{3}$

Answers

4. (a) $7 - 5 = 2; 2\frac{1}{2}$

(b) $5 - 3 = 2; 1\frac{11}{16}$

(c) $15 - 6 = 9; 8\frac{5}{9}$

5. (a) $\frac{27}{8} + \frac{20}{8} = \frac{47}{8} = 5\frac{7}{8}$

(b) $\frac{25}{12} = 2\frac{1}{12}$

Now you can subtract.

$$8\frac{1}{3} = 8\frac{5}{15} = 7\frac{20}{15}$$

$$-4\frac{3}{5} = 4\frac{9}{15} = 4\frac{9}{15}$$

$$3\frac{11}{15}$$

The exact answer is $3\frac{11}{15}$ (lowest terms), which is reasonable because it is close to the estimate of 3.

◀ Work Problem **4** at the Side.

OBJECTIVE ▶ 3 Add or subtract mixed numbers using an alternative method. An alternative method for adding or subtracting mixed numbers is to first change the mixed numbers to improper fractions. Then rewrite the unlike fractions as like fractions. Finally, add or subtract the numerators and write the answer in lowest terms.

EXAMPLE 4 Adding or Subtracting Mixed Numbers

Add or subtract.

(a) $2\frac{3}{8} = \frac{19}{8} = \frac{19}{8}$

$+3\frac{3}{4} = \frac{15}{4} = \frac{30}{8}$

8 is the least common denominator.

Rewrite $2\frac{3}{8}$ as $\frac{19}{8}$ and $3\frac{3}{4}$ as $\frac{15}{4}$.

$\frac{49}{8} = 6\frac{1}{8}$ ← Answer as mixed number

(b) $4\frac{2}{3} = \frac{14}{3} = \frac{70}{15}$

$-2\frac{1}{5} = \frac{11}{5} = \frac{33}{15}$

15 is the least common denominator.

Improper fractions

$\frac{37}{15} = 2\frac{7}{15}$

Simplify the answer by writing it as a mixed number.

◀ Work Problem **5** at the Side.

Note

The advantage of this alternative method of adding or subtracting mixed numbers is that it eliminates the need to regroup. However, if the mixed numbers are large, then the numerators of the improper fractions may become so large that they are difficult to work with. In such cases, you may want to keep the numbers as mixed numbers.

3.4 Exercises

FOR EXTRA HELP

Download the MyDashBoard App

MyMathLab®

CONCEPT CHECK *Round each of the following mixed numbers to the nearest whole number.*

1. $5\frac{1}{3}$

2. $6\frac{3}{10}$

3. $8\frac{4}{5}$

4. $12\frac{1}{2}$

5. $15\frac{7}{15}$

6. $20\frac{5}{8}$

7. $16\frac{2}{3}$

8. $3\frac{5}{12}$

First estimate the answer. Then add to find the exact answer. Write answers as mixed numbers. ***See Examples 1 and 2.***

9. *Estimate:* GS 6 ← Rounds to; + 3 ← Rounds to; = 9 *Exact:* $5\frac{1}{2} = 5\frac{3}{6}$; $+\,3\frac{1}{3} = 3\frac{2}{6}$

10. *Estimate:* GS 7; + 7; = 14 *Exact:* $6\frac{3}{5} = 6\frac{6}{10}$; $+\,7\frac{1}{10} = 7\frac{1}{10}$

11. *Estimate:* + *Exact:* $7\frac{1}{3}$; $+\,4\frac{1}{6}$

12. *Estimate:* + *Exact:* $10\frac{1}{4}$; $+\,5\frac{5}{8}$

13. *Estimate:* + *Exact:* $\frac{5}{8}$; $+\,3\frac{7}{12}$

14. *Estimate:* + *Exact:* $12\frac{4}{5}$; $+\,\frac{7}{10}$

15. *Estimate:* + *Exact:* $24\frac{5}{6}$; $+\,18\frac{5}{6}$

16. *Estimate:* + *Exact:* $14\frac{6}{7}$; $+\,15\frac{1}{2}$

17. *Estimate:* + *Exact:* $33\frac{3}{5}$; $+\,18\frac{1}{2}$

18. *Estimate:* + *Exact:* $18\frac{5}{8}$; $+\,6\frac{2}{3}$

19. *Estimate:* + *Exact:* $22\frac{3}{4}$; $+\,15\frac{3}{7}$

20. *Estimate:* + *Exact:* $7\frac{1}{4}$; $+\,25\frac{7}{8}$

21. *Estimate:* *Exact:*

$$\begin{array}{r} 12\frac{8}{15} \\ 18\frac{3}{5} \\ +\ 14\frac{7}{10} \\ \hline \end{array}$$

22. *Estimate:* *Exact:*

$$\begin{array}{r} 14\frac{9}{10} \\ 8\frac{1}{4} \\ +\ 13\frac{3}{5} \\ \hline \end{array}$$

First estimate the answer. Then subtract to find the exact answer. Simplify all answers.
See Examples 1 and 3.

23. *Estimate:* *Exact:*

$$\begin{array}{rcl} 15 \leftarrow & 14\frac{7}{8} = & 14\frac{7}{8} \\ -\ 12 \leftarrow & -\ 12\frac{1}{4} = & 12\frac{2}{8} \\ \hline 3 & & \end{array}$$

24. *Estimate:* *Exact:*

$$\begin{array}{rcl} 15 \leftarrow & 14\frac{3}{4} = & 14\frac{6}{8} \\ -\ 11 \leftarrow & -\ 11\frac{3}{8} = & 11\frac{3}{8} \\ \hline 4 & & \end{array}$$

25. *Estimate:* *Exact:*

$$\begin{array}{r} 12\frac{2}{3} \\ -\ 1\frac{1}{5} \\ \hline \end{array}$$

26. *Estimate:* *Exact:*

$$\begin{array}{r} 11\frac{9}{20} \\ -\ 4\frac{3}{5} \\ \hline \end{array}$$

27. *Estimate:* *Exact:*

$$\begin{array}{r} 28\frac{3}{10} \\ -\ 6\frac{1}{15} \\ \hline \end{array}$$

28. *Estimate:* *Exact:*

$$\begin{array}{r} 15\frac{7}{20} \\ -\ 6\frac{1}{8} \\ \hline \end{array}$$

29. *Estimate:* *Exact:*

$$\begin{array}{rcl} 17 & = & 16\frac{8}{8} \\ -\ 6\frac{5}{8} & = & 6\frac{5}{8} \\ \hline \end{array}$$

30. *Estimate:* *Exact:*

$$\begin{array}{rcl} 22 & = & 21\frac{6}{6} \\ -\ 4\frac{5}{6} & = & 4\frac{5}{6} \\ \hline \end{array}$$

31. *Estimate:* *Exact:*

$$\begin{array}{r} 18\frac{3}{4} \\ -\ 5\frac{4}{5} \\ \hline \end{array}$$

32. *Estimate:* $-$ ___ *Exact:* $14\frac{5}{8} - 3\frac{2}{3}$ ___

33. *Estimate:* $-$ ___ *Exact:* $19\frac{2}{3} - 11\frac{3}{4}$ ___

34. *Estimate:* $-$ ___ *Exact:* $20\frac{3}{5} - 12\frac{7}{15}$ ___

CONCEPT CHECK *Write each mixed number as an improper fraction and each improper fraction as a mixed number.*

35. $3\frac{3}{4}$

36. $7\frac{7}{8}$

37. $\frac{12}{5}$

38. $\frac{24}{7}$

39. $5\frac{3}{8}$

40. $6\frac{2}{15}$

41. $\frac{56}{3}$

42. $\frac{29}{8}$

Add or subtract by changing mixed numbers to improper fractions. Write answers as mixed numbers when possible. ***See Example 4.***

43. GS
$$7\frac{5}{8} = \frac{61}{8} = \frac{61}{8}$$
$$+\,1\frac{1}{2} = \frac{3}{2} = \frac{12}{8}$$
$$\frac{73}{8} =$$

44. GS
$$8\frac{3}{4} = \frac{35}{4} = \frac{70}{8}$$
$$+\,1\frac{5}{8} = \frac{13}{8} = \frac{13}{8}$$
$$\frac{83}{8} =$$

45. $4\frac{2}{3} + 6\frac{5}{6}$

46. $3\frac{3}{5} + 6\frac{1}{2}$

47. $2\frac{2}{3} + 1\frac{1}{6}$

48. $4\frac{1}{2} + 2\frac{3}{4}$

49. $3\frac{1}{4} + 3\frac{2}{3}$

50. $2\frac{4}{5} + 5\frac{1}{3}$

51. $1\frac{3}{8} + 6\frac{3}{4}$

52. $1\frac{5}{12} + 1\frac{7}{8}$

53. $3\frac{1}{2} - 2\frac{2}{3}$

54. $4\frac{1}{4} - 3\frac{7}{12}$

55. $\begin{array}{r} 8\frac{3}{4} \\ -\ 5\frac{7}{8} \\ \hline \end{array}$

56. $\begin{array}{r} 5 \\ -\ 4\frac{7}{8} \\ \hline \end{array}$

57. $\begin{array}{r} 7\frac{1}{4} \\ -\ 4\frac{2}{3} \\ \hline \end{array}$

58. $\begin{array}{r} 4\frac{1}{10} \\ -\ 3\frac{7}{8} \\ \hline \end{array}$

59. $\begin{array}{r} 9\frac{1}{5} \\ -\ 3\frac{3}{4} \\ \hline \end{array}$

60. $\begin{array}{r} 9 \\ -\ 7\frac{5}{6} \\ \hline \end{array}$

61. $\begin{array}{r} 6\frac{3}{7} \\ -\ 2\frac{2}{3} \\ \hline \end{array}$

62. $\begin{array}{r} 8\frac{2}{15} \\ -\ 6\frac{1}{2} \\ \hline \end{array}$

63. In your own words, explain the steps you would take to add two large mixed numbers.

64. When subtracting mixed numbers, explain when you need to regroup. Explain how to regroup using your own example.

First estimate the answer. Then solve each application problem.

65. At the beginning of this chapter you read about Bryan Berg, who builds houses of cards. While in high school, Bryan built a house of cards $14\frac{1}{2}$ ft tall, setting his first world record. Today his current world record is $25\frac{3}{4}$ ft tall. How much taller is his current world record than his first world record? (*Source: Guinness World Records.*)

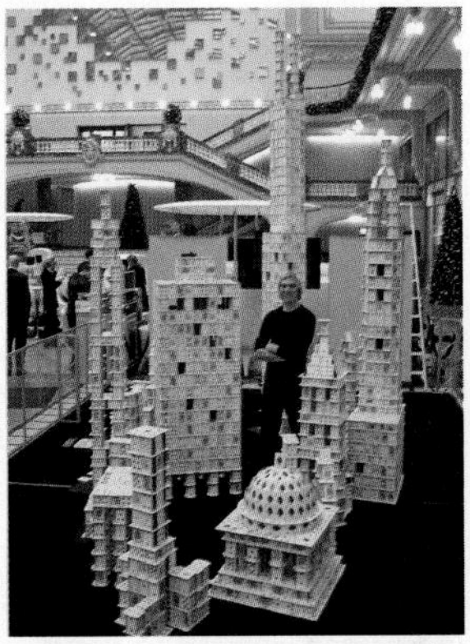

Estimate:

Exact:

66. The average age of a fast-food worker has increased over the past 12 years from $21\frac{3}{4}$ years of age to $29\frac{1}{2}$. How much older is the average fast-food worker today than 12 years ago? (*Source:* U.S. Census Bureau.)

Estimate:

Exact:

Use the newspaper advertisement for these industrial quality professional pliers to answer Exercises 67–70. The " symbol is for inches and 2-15/16" means $2\frac{15}{16}$ inches. (*Source:* Harbor Freight Tools.)

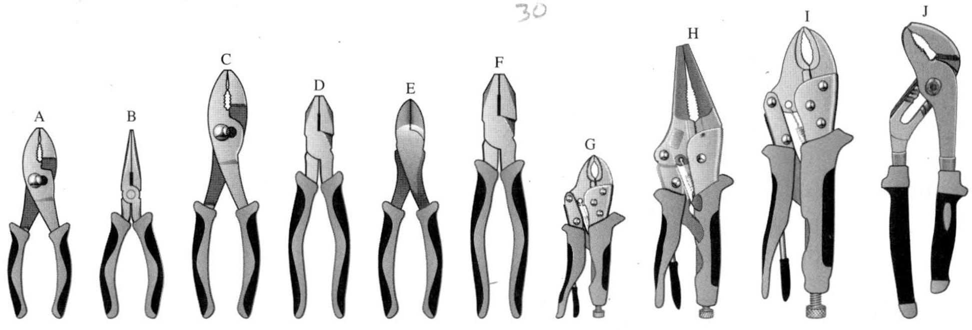

	DESCRIPTION	MAX. JAW CAP.	LOT NO.	PRICE		DESCRIPTION	MAX. JAW CAP.	LOT NO.	PRICE
A	6" SLIP JOINT	2-15/16"	94380	$2.99	F	8" LINEMAN'S	1-7/16"	94385	$6.29
B	~~6"~~ LONG NOSE	1-1/2"	94378	$4.29	G	5" CURVED JAW LOCKING	1-3/8"	94289	$3.29
C	8" SLIP JOINT	3-3/8"	94381	$4.29	H	12" LONG NOSE LOCKING	2-1/4"	94285	$6.29
D	7" LINEMAN'S	1-1/2"	94382	$5.29	I	10" CURVED JAW LOCKING	2-1/4"	94286	$7.29
E	7" DIAGONAL	7/8"	94383	$5.29	J	12" GROOVE JOINT	2"	94288	$8.29

67. How much larger is the maximum jaw capacity of the 8″ slip joint pliers than of the 6″ slip joint pliers?

Estimate:

Exact:

68. How much larger is the maximum jaw capacity of the 8″ lineman's pliers than of the 5″ curver jaw locking pliers?

Estimate:

Exact:

69. What is the difference in the maximum jaw capacity of the smallest pliers and that of the largest pliers?

Estimate:

Exact:

70. What is the difference in the maximum jaw capacity of the smallest pliers and that of the third smallest pliers?

Estimate:

Exact:

LOT NO. 1420
Storehouse
34 PC. GEAR HOSE CLAMP
ASSORTMENT
Includes: two 2-3/4", two 2-1/2", four 2-1/4", four 2", six 1-3/4", six 1-1/2", six 1-1/4", four 9/16" through 1-1/16"
SALE!
$5.97
SAVE 40%
REGULAR PRICE $9.99

A mechanic buys the 34-piece hose clamp assortment shown at the left. Use the advertisement to answer Exercises 71–72. The " symbol is for inches and 2-3/4" means $2\frac{3}{4}$ inches. (*Source:* Harbor Freight Tools.)

71. Find the difference in size between the largest hose clamp and the smallest hose clamp.

Estimate:

Exact:

72. Find the difference in size between the second to largest hose clamp and the smallest hose clamp.

Estimate:

Exact:

73. The four sides of Pam Prentiss's vegetable garden are $15\frac{1}{2}$ feet, $18\frac{3}{4}$ feet, $24\frac{1}{4}$ feet, and $30\frac{1}{2}$ feet. How many feet of fencing are needed to go around the garden?

Estimate:

Exact:

74. On a recent vacation to Canada, Janeen Cartmill drove for $7\frac{3}{4}$ hours on the first day, $5\frac{1}{4}$ hours on the second day, $6\frac{1}{2}$ hours on the third day, and 9 hours on the fourth day. How many hours did she drive altogether?

Estimate:

Exact:

75. A craftsperson must attach a lead strip around all four sides of a stained glass window before it is installed. Find the length of lead stripping needed.

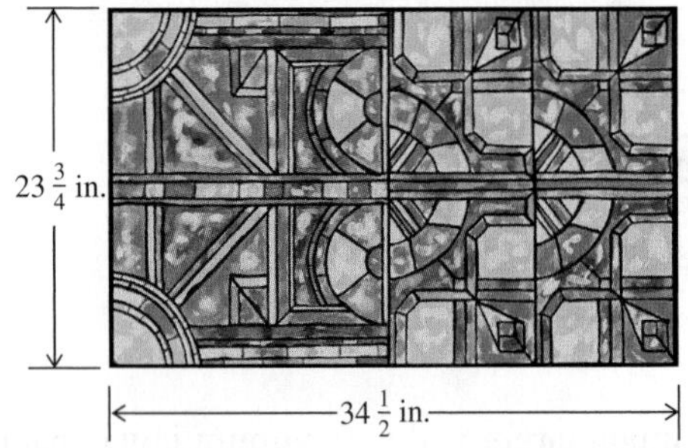

Estimate:

Exact:

76. To complete a custom order, Zak Morten of Home Depot must find the number of inches of brass trim needed to go around the four sides of the lamp base plate shown. Find the length of brass trim needed.

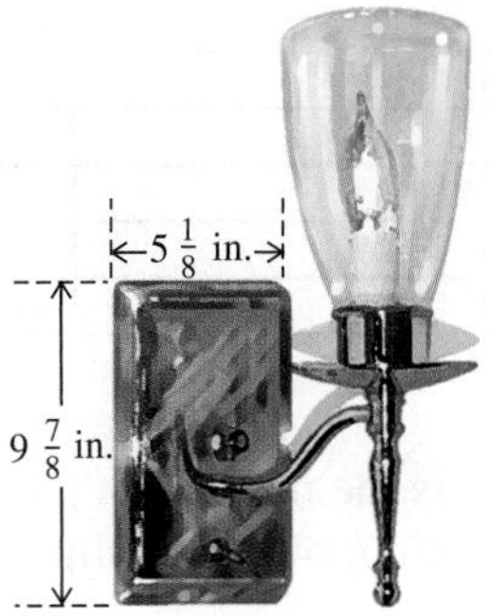

Estimate:

Exact:

77. A museum humidifier contains 100 gallons of water. The system uses $10\frac{1}{4}$ gallons of water on Monday, $13\frac{1}{2}$ gallons on Tuesday, $8\frac{7}{8}$ gallons on Wednesday, $18\frac{3}{4}$ gallons on Thursday, $12\frac{3}{8}$ gallons on Friday, $9\frac{1}{2}$ gallons on Saturday, and $14\frac{1}{8}$ gallons on Sunday. Find the total number of gallons of water remaining.

Estimate:

Exact:

78. Scott Alamo had 16 cubic yards of mulch delivered to the home he recently purchased. Using his wheelbarrow, he spread $3\frac{3}{4}$ yards of the mulch in his front yard, $4\frac{7}{8}$ yards in his back yard, $2\frac{1}{2}$ yards in the side yard, and used the remainder in the children's outdoor play area. Find the number of cubic yards of mulch remaining for the play area.

Estimate:

Exact:

79. The exercise yard at the correction center has four sides and is surrounded by $527\frac{1}{24}$ ft of security fencing. Three sides of the yard measure $107\frac{2}{3}$ ft, $150\frac{3}{4}$ ft, and $138\frac{5}{8}$ ft. Find the length of the fourth side.

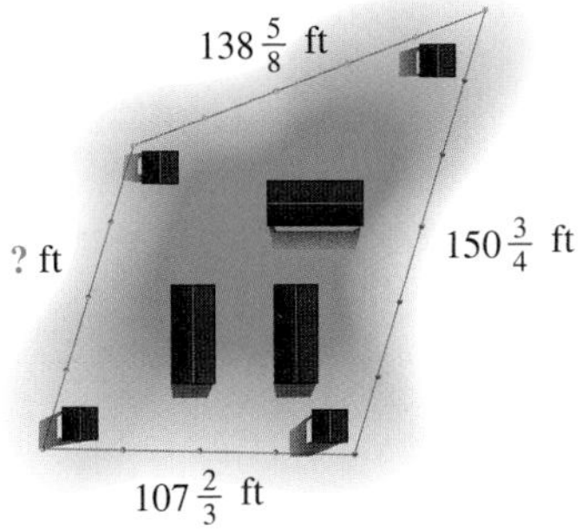

Estimate:

Exact:

80. Three sides of a parking lot are $108\frac{1}{4}$ ft, $162\frac{3}{8}$ ft, and $143\frac{1}{2}$ ft. The total distance around the lot is $518\frac{3}{4}$ ft. What is the length of the fourth side?

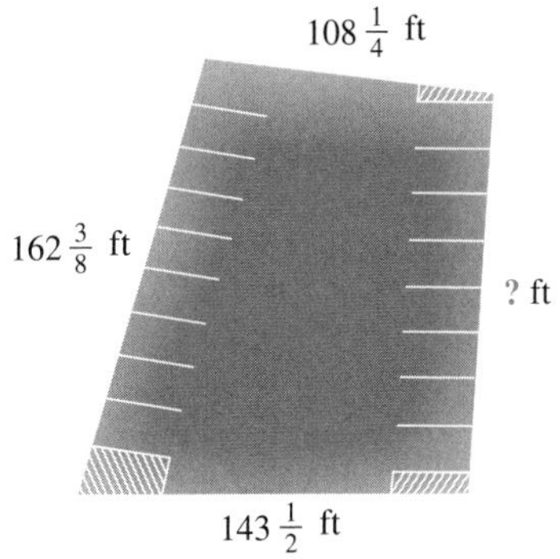

Estimate:

Exact:

81. A freight car is loaded with Morton Salt products consisting of $58\frac{1}{2}$ tons of coarse rock salt, $23\frac{5}{8}$ tons of medium rock salt, $16\frac{5}{6}$ tons of table salt, and $29\frac{1}{4}$ tons of animal salt lick blocks. The weight of the unloaded freight car is $58\frac{1}{3}$ tons. Find the weight of the loaded freight car.

Estimate:

Exact:

82. Bryan Berg, from the opening page of this chapter, built houses of cards reaching heights of $14\frac{1}{2}$ ft, $19\frac{3}{8}$ ft, $23\frac{5}{12}$ ft, and $25\frac{3}{4}$ ft (his current world record). Find the total height of these four houses of cards. (*Source: Reader's Digest.*)

Estimate:

Exact:

Find the unknown length, labeled with a question mark, in each figure.

83.

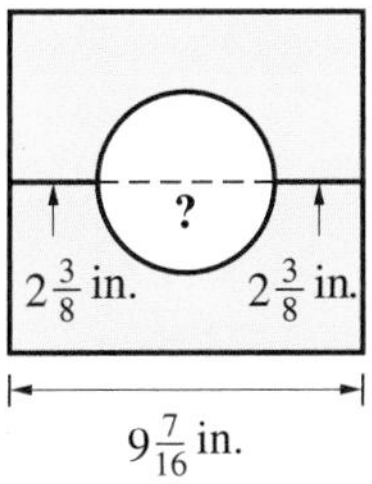

84.

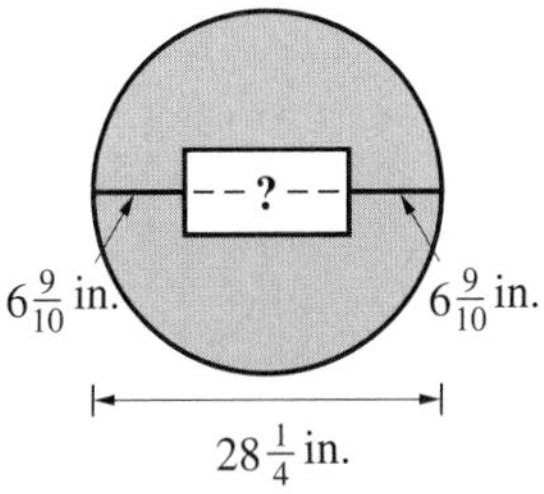

85.

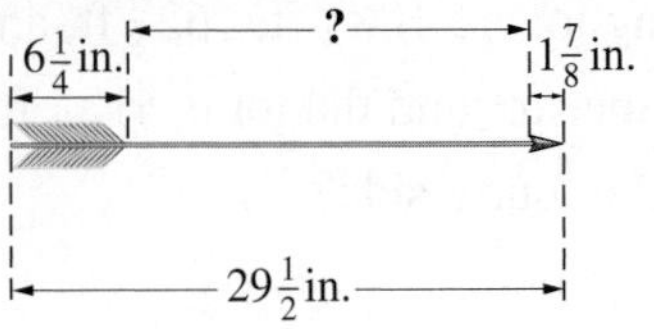

86.

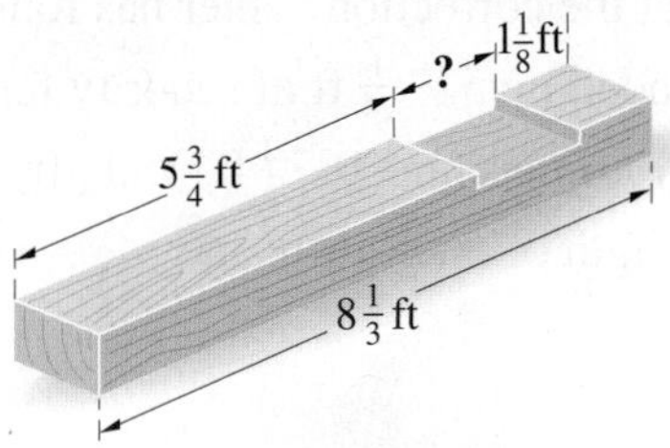

Relating Concepts (Exercises 87–92) For Individual or Group Work

Most fraction problems include fractions with different denominators.
Work Exercises 87–92 in order.

87. To add or subtract fractions, we must first rewrite them as like fractions. Rewrite each fraction with the indicated denominator.

(a) $\frac{5}{9} = \frac{}{54}$ **(b)** $\frac{7}{12} = \frac{}{48}$

(c) $\frac{5}{8} = \frac{}{40}$ **(d)** $\frac{11}{5} = \frac{}{120}$

88. When rewriting unlike fractions as like fractions with the least common multiple as a denominator, the new denominator is called the ________ ________ ________, or LCD.

89. Add or subtract as indicated. Write answers in lowest terms.

(a) $\frac{5}{8} + \frac{1}{3}$ **(b)** $\frac{19}{20} - \frac{5}{12}$

(c) $\frac{7}{12} + \frac{3}{16} + \frac{3}{24}$ **(d)** $\frac{6}{7} - \frac{2}{3}$

90. A common method for adding or subtracting mixed numbers is to add or subtract the ________ ________ and then add or subtract the whole number parts.

91. Another method for adding or subtracting mixed numbers is to first change the mixed numbers to ________ fractions. After adding or subtracting, write the answer in lowest terms and as a mixed number when possible. This method is difficult to use if the mixed numbers are (*large/small*).

92. Add or subtract these fractions as indicated. First use the method where you add or subtract fraction parts and then whole number parts. Then use the method where you change each mixed number to an improper fraction before adding or subtracting. Do you get the same answer using both methods? Which method do you prefer?

(a) $4\frac{5}{8} + 3\frac{3}{4}$ **(b)** $12\frac{2}{5} - 8\frac{7}{8}$

Study Skills
MAKING A MIND MAP

OBJECTIVES

1. **Create mind maps for appropriate concepts.**
2. **Visually show how concepts relate to each other using arrows or lines.**

Mind mapping is a visual way to show information that you have learned. It is an excellent way to review. Mapping is flexible and can be personalized, which is helpful for your memory. Your brain likes to see things that are **pleasing** to look at, **colorful,** and that **show connections** between ideas. Take advantage of that by creating maps that

- are easy to read,
- use color in a systematic way, and
- clearly show you how different concepts are related (using arrows or dotted lines, for example).

Why Is Mapping Brain Friendly?

Remember that your brain grows dendrites when you are **actively thinking** about and working with information. Making a map requires you to think hard about ***how to place the information***, ***how to show connections*** between parts of the map, and ***how color will be useful***. It also takes a lot of thinking to fill in all related details and **show how those details connect to the larger concept.** All that thinking will let your brain grow a complex, many-branched neural network of interconnected dendrites. It is time well spent.

Directions for Making a Mind Map

Here are some general directions for making a map. After you read them, work on completing the map that has been started for you on the next page. It is from **Chapters 2 and 3: Fractions.**

- To begin a mind map, write the concept in the center of a piece of paper and either circle it or draw a box around it.
- Make a line out from the center concept, and draw a box large enough to write the definition of the concept.
- Think of the other aspects (subpoints) of the concept that you have learned, such as procedures to follow or formulas. Make a separate line and box connecting each subpoint to the center.
- From each of the new boxes, add the information you've learned. You can continue making new lines and boxes or circles, or you can list items below the new information.
- Use color to highlight the major points. For example, everything related to one subpoint might be the same color. That way you can easily see related ideas.
- You may also use arrows, underlining, or small drawings to help yourself remember.

 Continued from page 229

Try This Fractions Mind Map

On a separate paper, make a map that summarizes Computations with Fractions. Follow the directions and use the starter map below.

- The longest rectangles are *instructions* for all four operations. (The first one starts "Rewrite all numbers as fractions. . ." and the second one is at the bottom of the map.)
- Notice the wavy dividing lines that separate the map into two sides.
- Your job is to complete the map by writing the steps used in multiplying and dividing fractions and the steps used in adding and subtracting fractions.

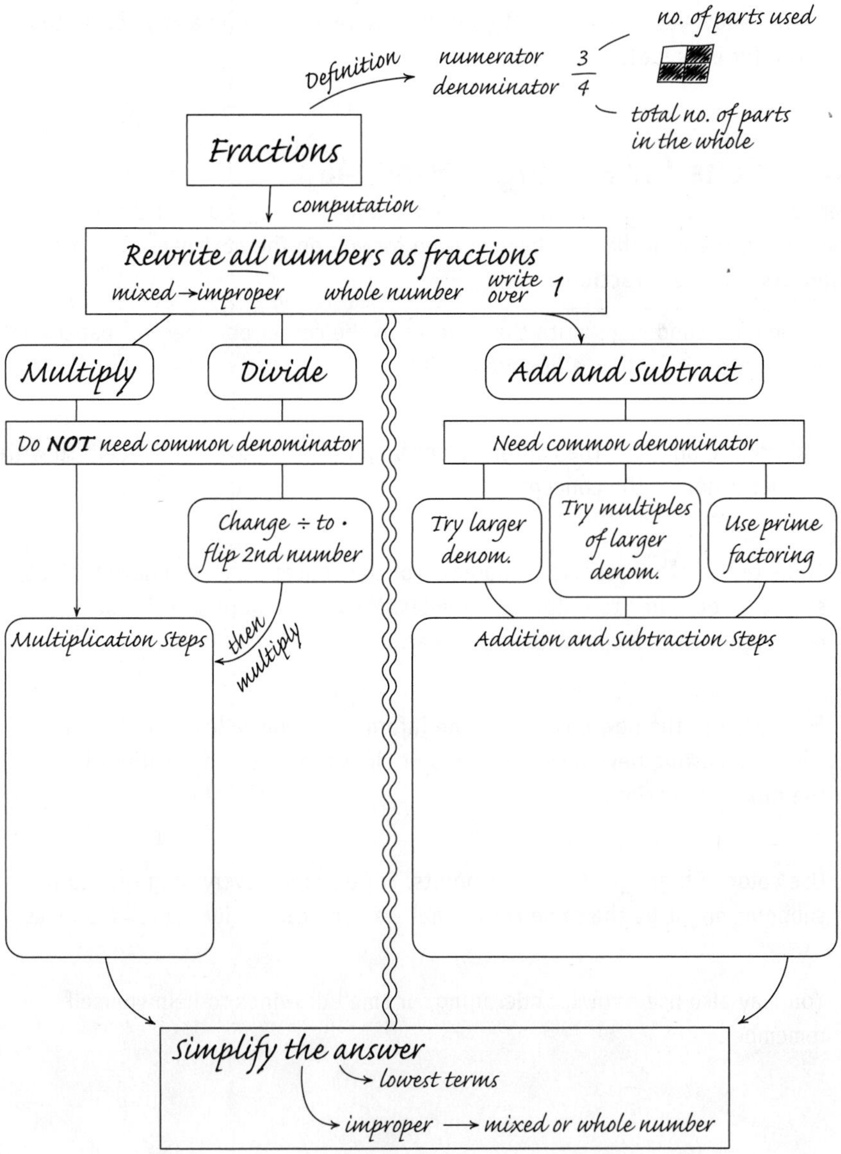

3.5 Order Relations and the Order of Operations

OBJECTIVES

1. Identify the greater or lesser of two fractions.
2. Use exponents with fractions.
3. Use the order of operations with fractions.

There are times when we want to compare the size of two numbers. For example, we might want to know which is the greater amount, the larger size, or the longer distance.

Fractions, like whole numbers, can be graphed on a number line. Fractions divide the space between whole numbers into equal parts.

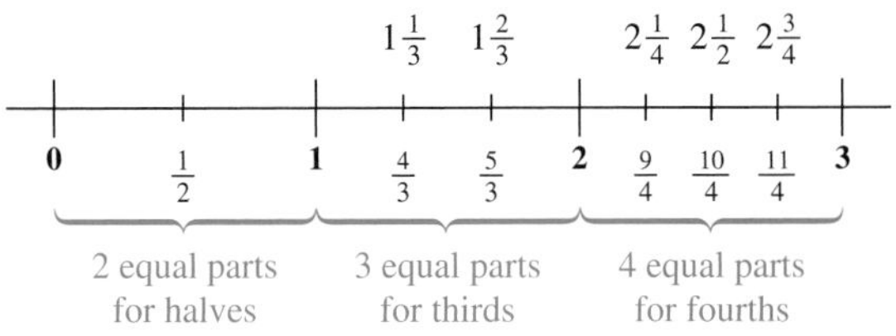

OBJECTIVE 1 Identify the greater or lesser of two fractions. To compare the size of two numbers, place the two numbers on a number line and use the following rule.

Comparing the Size of Two Numbers

The number farther to the *left* on the number line is always *less*, and the number farther to the *right* on the number line is always *greater.*

For example, on the number line above, $\frac{1}{2}$ is to the *left* of $\frac{4}{3}$, so $\frac{1}{2}$ is *less than* $\frac{4}{3}$.

Work Problem 1 at the Side. ▶

Write *order relations* using the symbols shown below.

Symbols Used to Show Order Relations

$<$ is less than $\quad$ $>$ is greater than

EXAMPLE 1 Using Less-Than and Greater-Than Symbols

Rewrite the following using $<$ and $>$ symbols.

(a) $\frac{1}{2}$ is less than $\frac{4}{3}$.

$\frac{1}{2}$ is less than $\frac{4}{3}$ is written as $\frac{1}{2} < \frac{4}{3}$.

$\frac{1}{2}$ is farther to the *left* on the number line, so it is *less* than $\frac{4}{3}$.

(b) $\frac{9}{4}$ is greater than 1.

$\frac{9}{4}$ is greater than 1 is written as $\frac{9}{4} > 1$.

$\frac{9}{4}$ is farther to the *right* on the number line, so it is *greater* than 1.

(c) $\frac{5}{3}$ is less than $\frac{11}{4}$.

$\frac{5}{3}$ is less than $\frac{11}{4}$ is written as $\frac{5}{3} < \frac{11}{4}$.

Continued on Next Page

1 Locate each fraction on the number line.

GS **(a)** $\frac{2}{3}$ Put a dot on the number line between 0 and 1.

(b) $1\frac{1}{2}$

(c) $2\frac{3}{4}$

Answer

1.

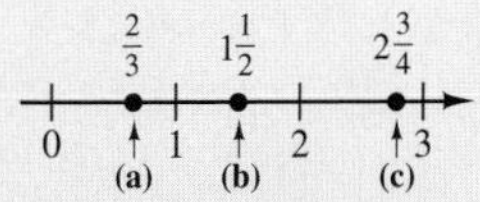

2 Use the number line on the previous page to help you write < or > in each blank to make a true statement.

GS (a) $1 \_\_\_ \frac{5}{4}$ The point on the symbol points to the lesser value.

(b) $\frac{8}{3} \_\_\_ \frac{3}{2}$

(c) $0 \_\_\_ 1$

(d) $\frac{17}{8} \_\_\_ \frac{8}{4}$

3 Write < or > in each blank to make a true statement.

GS (a) $\frac{7}{8} \_\_\_ \frac{3}{4}$ The point on the symbol points to the lesser value.

(b) $\frac{13}{8} \_\_\_ \frac{15}{9}$

(c) $\frac{9}{4} \_\_\_ \frac{7}{3}$

(d) $\frac{9}{10} \_\_\_ \frac{14}{15}$

Answers

2. (a) < (b) > (c) < (d) >
3. (a) > (b) < (c) < (d) <

Note

A number line is a very useful tool when working with order relations.

◀ Work Problem **2** at the Side.

The fraction $\frac{7}{8}$ represents 7 of 8 equal parts, while $\frac{3}{8}$ means 3 of 8 equal parts. Because $\frac{7}{8}$ represents more of the equal parts, $\frac{7}{8}$ is greater than $\frac{3}{8}$.

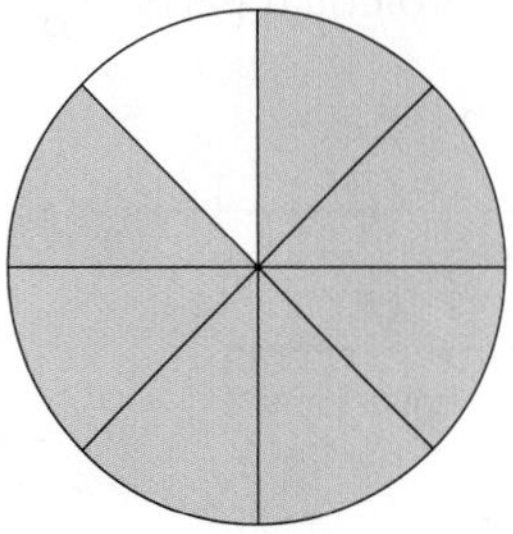

$$\frac{7}{8} > \frac{3}{8}$$

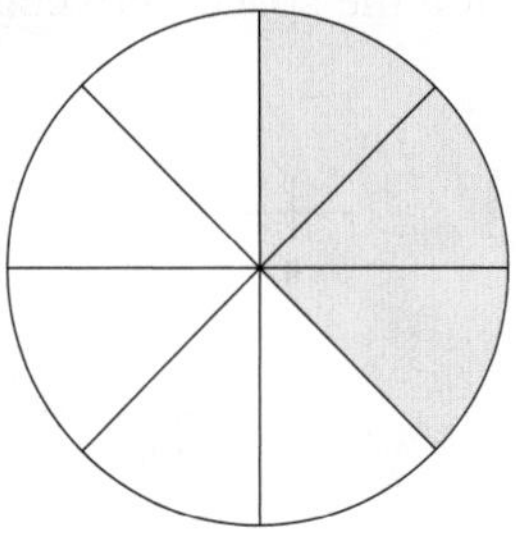

To identify the greater fraction, use the following steps.

Identifying the Greater Fraction

Step 1 Write the fractions as like fractions (same denominators).

Step 2 Compare the numerators. The fraction with the greater numerator is the greater fraction.

EXAMPLE 2 Identifying the Greater Fraction

Determine which fraction in each pair is greater.

(a) $\frac{7}{8}, \frac{9}{10}$

First, write the fractions as like fractions. The least common multiple for 8 and 10 is 40.

$$\frac{7}{8} = \frac{7 \cdot 5}{8 \cdot 5} = \frac{35}{40} \quad \text{and} \quad \frac{9}{10} = \frac{9 \cdot 4}{10 \cdot 4} = \frac{36}{40}$$

Rewrite both fractions with 40 as the denominator.

Look at the numerators. Because 36 is greater than 35, $\frac{36}{40}$ is greater than $\frac{35}{40}$. Then, because $\frac{36}{40}$ is equivalent to $\frac{9}{10}$,

$$\frac{9}{10} > \frac{7}{8} \quad \text{or} \quad \frac{7}{8} < \frac{9}{10}.$$

The greater fraction is $\frac{9}{10}$.

(b) $\frac{8}{5}, \frac{23}{15}$

The least common multiple of 5 and 15 is 15.

$$\frac{8}{5} = \frac{8 \cdot 3}{5 \cdot 3} = \frac{24}{15} \quad \text{and} \quad \frac{23}{15} = \frac{23}{15}$$

This shows that $\frac{8}{5}$ is greater than $\frac{23}{15}$, or

$$\frac{8}{5} > \frac{23}{15}.$$

◀ Work Problem **3** at the Side.

OBJECTIVE 2 Use exponents with fractions. Exponents were used in **Chapter 1** to write repeated multiplication.

$$\overset{\text{Exponent}}{3^2} = \underbrace{3 \cdot 3}_{\text{Two factors of 3}} = 9 \quad \text{and} \quad \overset{\text{Exponent}}{5^3} = \underbrace{5 \cdot 5 \cdot 5}_{\text{Three factors of 5}} = 125$$

The next example shows exponents used with fractions.

EXAMPLE 3 Using Exponents with Fractions

Simplify.

(a) $\left(\frac{1}{2}\right)^3$

$$\left(\frac{1}{2}\right)^3 = \overbrace{\frac{1}{2} \cdot \frac{1}{2} \cdot \frac{1}{2}}^{\text{Three factors of } \frac{1}{2}} = \frac{1}{8}$$

$\frac{1}{2}$ is multiplied by itself three times.

(b) $\left(\frac{5}{8}\right)^2$

$$\left(\frac{5}{8}\right)^2 = \overbrace{\frac{5}{8} \cdot \frac{5}{8}}^{\text{Two factors of } \frac{5}{8}} = \frac{25}{64}$$

(c) $\left(\frac{3}{4}\right)^2 \cdot \left(\frac{2}{3}\right)^3$

$$\left(\frac{3}{4}\right)^2 \cdot \left(\frac{2}{3}\right)^3 = \overbrace{\left(\frac{3}{4} \cdot \frac{3}{4}\right)}^{\text{Two factors of } \frac{3}{4}} \cdot \overbrace{\left(\frac{2}{3} \cdot \frac{2}{3} \cdot \frac{2}{3}\right)}^{\text{Three factors of } \frac{2}{3}}$$

$$= \frac{\overset{1}{\cancel{3}} \cdot \overset{1}{\cancel{3}} \cdot \overset{1}{\cancel{2}} \cdot \overset{1}{\cancel{2}} \cdot \overset{1}{\cancel{2}}}{\underset{2}{\cancel{4}} \cdot \underset{\underset{1}{\cancel{2}}}{\cancel{4}} \cdot \underset{1}{\cancel{3}} \cdot \underset{1}{\cancel{3}} \cdot 3}$$

Divide out all the common factors.

$$= \frac{1}{6}$$

The fraction is in lowest terms after all common factors are divided out.

Work Problem 4 at the Side. ▶

OBJECTIVE 3 Use the order of operations with fractions. Recall the *order of operations* from **Chapter 1.**

Order of Operations

1. Do all operations inside *parentheses or other grouping symbols.*
2. Simplify any expressions with *exponents* and find any *square roots.*
3. *Multiply* or *divide,* proceeding from left to right.
4. *Add* or *subtract,* proceeding from left to right.

4 Simplify.

GS **(a)** $\left(\frac{1}{2}\right)^4$

$$\overbrace{\frac{1}{2} \cdot \frac{1}{2} \cdot \frac{1}{2} \cdot \frac{\_\_}{\_\_}}^{\text{Four factors of } \frac{1}{2}} = \frac{\_\_}{\_\_}$$

(b) $\left(\frac{3}{4}\right)^2$

(c) $\left(\frac{1}{2}\right)^3 \cdot \left(\frac{2}{3}\right)^2$

(d) $\left(\frac{1}{5}\right)^2 \cdot \left(\frac{5}{3}\right)^2$

Answers

4. **(a)** 1; 2; $\frac{1}{16}$ **(b)** $\frac{9}{16}$ **(c)** $\frac{1}{18}$ **(d)** $\frac{1}{9}$

5 Simplify by using the order of operations.

GS **(a)** $\frac{5}{9} - \frac{3}{4}\left(\frac{2}{3}\right)$

$\frac{5}{9} - \frac{\overset{1}{\cancel{3}}}{\underset{2}{\cancel{4}}}\left(\frac{\overset{1}{\cancel{2}}}{\underset{1}{\cancel{3}}}\right)$ Multiply first.

$\frac{5}{9} - \frac{1}{2} = \frac{10}{18} - \frac{9}{\_\_\_} =$

(b) $\frac{3}{4}\left(\frac{2}{3} \cdot \frac{3}{5}\right)$

(c) $\frac{7}{8}\left(\frac{2}{3}\right) - \left(\frac{1}{2}\right)^2$

(d) $\dfrac{\left(\frac{5}{6}\right)^2}{\frac{4}{3}}$

Answers

5. (a) 18; $\frac{1}{18}$ (b) $\frac{3}{10}$ (c) $\frac{1}{3}$ (d) $\frac{25}{48}$

The next example shows how to apply the order of operations with fractions.

EXAMPLE 4 Using the Order of Operations with Fractions

Simplify by using the order of operations.

(a) $\frac{1}{3} + \frac{1}{2}\left(\frac{4}{5}\right)$

Multiply $\frac{1}{2}\left(\frac{4}{5}\right)$ first because multiplication and division are done *before* adding.

Do **not** add $\frac{1}{3} + \frac{1}{2}$ as the first step.

$$\frac{1}{3} + \frac{1}{\underset{1}{\cancel{2}}}\left(\frac{\overset{2}{\cancel{4}}}{5}\right) = \frac{1}{3} + \frac{2}{5}$$

Now add. The least common denominator of 3 and 5 is 15.

$$\frac{1}{3} + \frac{2}{5} = \frac{5}{15} + \frac{6}{15} = \frac{11}{15} \leftarrow \text{Lowest terms}$$

(b) $\frac{3}{8}\left(\frac{1}{2} + \frac{1}{3}\right)$

$$\frac{3}{8}\left(\frac{1}{2} + \frac{1}{3}\right) = \frac{3}{8}\left(\frac{3}{6} + \frac{2}{6}\right)$$ Work inside parentheses first.

$$= \frac{3}{8}\left(\frac{5}{6}\right)$$

$$= \frac{\overset{1}{\cancel{3}}}{8}\left(\frac{5}{\underset{2}{\cancel{6}}}\right)$$ Divide numerator and denominator by 3. Then multiply.

$$= \frac{5}{16} \leftarrow \text{Lowest terms}$$

(c) $\left(\frac{2}{3}\right)^2 - \frac{4}{5}\left(\frac{1}{2}\right)$

$$\left(\frac{2}{3}\right)^2 - \frac{4}{5}\left(\frac{1}{2}\right) = \frac{4}{9} - \frac{4}{5}\left(\frac{1}{2}\right)$$ Simplify the expression with the exponent. $\frac{2}{3} \cdot \frac{2}{3}$ is $\frac{4}{9}$.

No work inside the parentheses, so applying the exponent is next.

$$= \frac{4}{9} - \frac{\overset{2}{\cancel{4}}}{5}\left(\frac{1}{\underset{1}{\cancel{2}}}\right)$$ Multiply next.

$$= \frac{4}{9} - \frac{2}{5}$$

$$= \frac{20}{45} - \frac{18}{45}$$ Subtract last. (Least common denominator is 45.)

$$= \frac{2}{45} \leftarrow \text{Lowest terms}$$

◀ **Work Problem 5 at the Side.**

3.5 Exercises

FOR EXTRA HELP

Download the MyDashBoard App

MyMathLab®

CONCEPT CHECK *Locate each fraction in Exercises 1–12 on the following number line.*

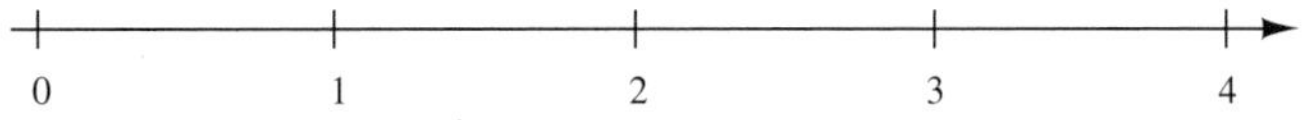

1. $\frac{1}{2}$ **2.** $\frac{1}{4}$ **3.** $\frac{3}{2}$ **4.** $\frac{5}{4}$ **5.** $\frac{7}{3}$ **6.** $\frac{11}{4}$

7. $2\frac{1}{6}$ **8.** $3\frac{4}{5}$ **9.** $\frac{7}{2}$ **10.** $\frac{7}{8}$ **11.** $3\frac{1}{4}$ **12.** $1\frac{7}{8}$

Write < or > to make a true statement. ***See Examples 1 and 2.***

13. $\frac{1}{2}$ ____ $\frac{3}{8}$ **14.** $\frac{5}{8}$ ____ $\frac{3}{4}$ **15.** $\frac{5}{6}$ ____ $\frac{11}{12}$ **16.** $\frac{13}{18}$ ____ $\frac{5}{6}$

17. $\frac{5}{12}$ ____ $\frac{3}{8}$ **18.** $\frac{17}{24}$ ____ $\frac{5}{6}$ **19.** $\frac{11}{18}$ ____ $\frac{5}{9}$ **20.** $\frac{7}{12}$ ____ $\frac{11}{20}$

CONCEPT CHECK *Write* true *or* false *for each statement, Also show how to complete each solution.*

21. GS $\left(\frac{1}{2}\right)^2 = \frac{1}{4}$

$\left(\frac{1}{2}\right)^2 = \frac{1}{2} \cdot \frac{1}{\underline{\quad}} = \frac{1}{\underline{\quad}}$

22. GS $\left(\frac{3}{8}\right)^2 = \frac{6}{16}$

$\left(\frac{3}{8}\right)^2 = \frac{3}{8} \cdot \frac{\underline{\quad}}{8} = \frac{\underline{\quad}}{64}$

23. $\left(\frac{2}{5}\right)^3 = \frac{6}{15}$

24. $\left(\frac{5}{6}\right)^3 = \frac{125}{216}$

Simplify. ***See Example 3.***

25. $\left(\frac{1}{3}\right)^2$ **26.** $\left(\frac{2}{3}\right)^2$ **27.** $\left(\frac{5}{8}\right)^2$

28. $\left(\frac{7}{8}\right)^2$ **29.** $\left(\frac{3}{4}\right)^2$ **30.** $\left(\frac{3}{5}\right)^3$

31. $\left(\frac{4}{5}\right)^3$ **32.** $\left(\frac{4}{7}\right)^3$ **33.** GS $\left(\frac{3}{2}\right)^4 = \frac{3}{2} \cdot \frac{3}{2} \cdot \frac{3}{2} \cdot \frac{3}{2} =$

34. GS $\left(\frac{4}{3}\right)^4 = \frac{4}{3} \cdot \frac{4}{3} \cdot \frac{4}{3} \cdot \frac{4}{3} =$ **35.** $\left(\frac{3}{4}\right)^4$ **36.** $\left(\frac{2}{3}\right)^5$

37. Describe in your own words what a number line is, and draw a picture of one. Be sure to include how it works and how it can be used.

38. You have used the order of operations with whole numbers and again with fractions. List from memory the steps in the order of operations.

CONCEPT CHECK *Use the order of operations to simplify each expression. Circle the correct answer.*

39. $2^4 - 4(3)$

30 4 12 2

40. $3^2 + 4(1)$

9 13 11 10

41. $3 \cdot 2^2 - \frac{6}{3}$

9 4 2 10

42. $5 \cdot 2^3 - \frac{6}{2}$

37 12 27 22

Use the order of operations to simplify each expression. ***See Example 4.***

43. $\left(\frac{1}{2}\right)^2 \cdot 4$

44. $\left(\frac{1}{4}\right)^2 \cdot 4$

45. $\left(\frac{3}{4}\right)^2 \cdot \left(\frac{1}{3}\right)$

46. $\left(\frac{2}{3}\right)^3 \cdot \left(\frac{1}{2}\right)$

47. $\left(\frac{4}{5}\right)^2 \cdot \left(\frac{5}{6}\right)^2$

48. $\left(\frac{5}{8}\right)^2 \cdot \left(\frac{4}{25}\right)^2$

49. $6\left(\frac{2}{3}\right)^2\left(\frac{1}{2}\right)^3$

50. $9\left(\frac{1}{3}\right)^3\left(\frac{4}{3}\right)^2$

51. $\frac{3}{5}\left(\frac{1}{3}\right) + \frac{2}{5}\left(\frac{3}{4}\right)$

GS

$\frac{3}{5} \cdot \frac{1}{3} + \frac{2}{5} \cdot \frac{3}{4}$

$\frac{3}{15} + \frac{6}{20}$

$\frac{12}{60} + \frac{18}{60} = \frac{30}{60} =$

52. $\frac{1}{4}\left(\frac{3}{4}\right) + \frac{3}{8}\left(\frac{4}{3}\right)$

GS

$\frac{1}{4} \cdot \frac{3}{4} + \frac{3}{8} \cdot \frac{4}{3}$

$\frac{3}{16} + \frac{12}{24}$

$\frac{9}{48} + \frac{24}{48} = \frac{33}{48} =$

53. $\frac{1}{2} + \left(\frac{1}{2}\right)^2 - \frac{3}{8}$

54. $\frac{2}{3} + \left(\frac{1}{3}\right)^2 - \frac{5}{9}$

55. $\left(\frac{1}{3}+\frac{1}{6}\right)\bullet\frac{1}{2}$

56. $\left(\frac{3}{5}-\frac{3}{20}\right)\bullet\frac{4}{3}$

57. $\frac{9}{8}\div\left(\frac{2}{3}+\frac{1}{12}\right)$

58. $\frac{6}{5}\div\left(\frac{3}{5}-\frac{3}{10}\right)$

59. GS $\left(\frac{7}{8}-\frac{3}{4}\right)\div\frac{3}{2}$

$\left(\frac{7}{8}-\frac{6}{8}\right)\div\frac{3}{2}$

$\frac{1}{8}\div\frac{3}{2}$

$\frac{1}{8}\bullet\frac{2}{3}=\frac{2}{24}=$

60. GS $\left(\frac{4}{5}-\frac{3}{10}\right)\div\frac{4}{5}$

$\left(\frac{8}{10}-\frac{3}{10}\right)\div\frac{4}{5}$

$\frac{5}{10}\div\frac{4}{5}$

$\frac{5}{10}\bullet\frac{5}{4}=\frac{25}{40}=$

61. $\frac{3}{8}\left(\frac{1}{4}+\frac{1}{2}\right)\bullet\frac{32}{3}$

62. $\frac{1}{3}\left(\frac{4}{5}-\frac{3}{10}\right)\bullet\frac{4}{2}$

63. $\left(\frac{3}{4}\right)^2-\left(\frac{1}{2}-\frac{1}{6}\right)\div\frac{4}{3}$

64. $\left(\frac{2}{3}\right)^2-\left(\frac{5}{8}-\frac{1}{2}\right)\div\frac{3}{2}$

65. $\left(\frac{7}{8}-\frac{1}{4}\right)-\frac{2}{3}\left(\frac{3}{4}\right)^2$

66. $\left(\frac{5}{6}-\frac{7}{12}\right)-\frac{3}{4}\left(\frac{1}{3}\right)^2$

67. $\left(\frac{3}{4}\right)^2\left(\frac{2}{3}-\frac{5}{9}\right)-\frac{1}{4}\left(\frac{1}{8}\right)$

68. $\left(\frac{2}{3}\right)^2\left(\frac{1}{2}-\frac{1}{8}\right)-\frac{2}{3}\left(\frac{1}{8}\right)$

Solve each application problem.

69. The population of Las Vegas, Nevada has had an increase of $\frac{11}{50}$ since the turn of the century. During this same period, the population in Atlanta, Georgia has had an increase of $\frac{5}{30}$. Which city has had a higher rate of population growth? (*Source:* Census Bureau estimates.)

70. Jacob Evan weighs $22\frac{11}{16}$ pounds and his stroller weighs $22\frac{5}{8}$ pounds. Which weighs more, Jacob or the stroller?

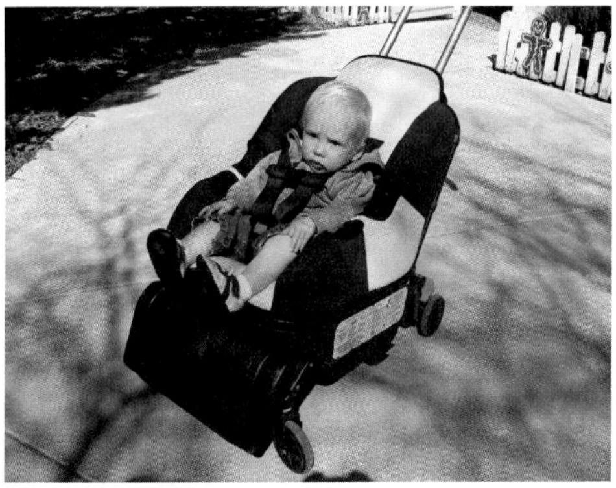

Relating Concepts (Exercises 71–80) For Individual or Group Work

You often need to use order relations and the order of operations when solving problems. ***Work Exercises 71–80 in order.***

71. When comparing the size of two numbers, the symbol __________ means **is less than** and the symbol __________ means **is greater than.**

72. (a) To identify the greater of two fractions, we must first write the fractions as __________ fractions and then compare the __________. The fraction with the greater __________ is the greater fraction.

(b) Write four pairs of fractions, all with different denominators. Write the symbol for **less than** or for **greater than** between each pair.

Use the order of operations to simplify each expression.

73. $\left(\frac{2}{3}\right)^2 - \left(\frac{4}{5} - \frac{3}{10}\right) \div \frac{5}{4}$

74. $\left(\frac{1}{2}\right)^2\left(\frac{3}{8} - \frac{1}{4}\right) + \frac{2}{3} \div \left(\frac{1}{3}\right)$

Simplify, then place the results on the number line.

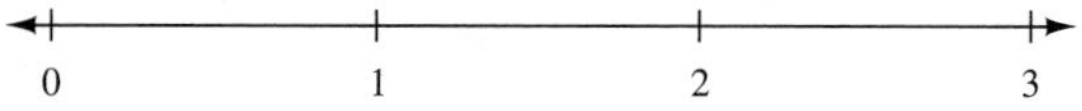

75. $\left(\frac{2}{3}\right)^2$

76. $\left(\frac{3}{2}\right)^2$

77. $\left(\frac{1}{2}\right)^3$

78. $\left(\frac{5}{3}\right)^2$

79. $4 + 2 - 2^2$

80. $\left(\frac{3}{4}\right)^2 + \left(\frac{5}{8} - \frac{1}{4}\right) \div \frac{2}{3}$

Study Skills
PREPARING FOR TESTS

OBJECTIVES

1. **Restate the importance of sleep and good nutrition as it affects learning.**
2. **Explain the effect of anxiety and stress on learning.**

Many things besides studying can improve your test scores. You may not realize that eating the right foods, and getting enough exercise and sleep, can also improve your scores. Your brain (and therefore your ability to think) is affected by the condition of your whole body. So, part of your preparation for tests includes keeping yourself in good physical shape as well as spending time on the actual course material. Try these suggestions and see the difference.

Performance Health Tips

To Improve Your Test Score	Explanation
Get ***seven to eight hours of sleep*** the night before the exam. (It's helpful to get that much sleep *every* night.)	***Fatigue and exhaustion*** reduce efficiency. They also cause poor memory and recall. If you didn't sleep much the night before a test, 20 minutes of relaxation or meditation can help.
Eat a ***small, high-energy meal*** about two hours before the test. Start the meal with a small amount of protein such as fish, chicken, or nonfat yogurt. Include carbohydrates if you like, but no high-fat foods.	Just 3 to 4 ounces of protein increases the amount of a chemical in the brain called tyrosine, which ***improves your alertness, accuracy, and motivation.*** High-fat foods dull your mind and slow down your brain.
Drink plenty of water. Don't wait until you feel thirsty; your body is already dehydrated by the time you feel it.	Research suggests that staying well hydrated improves the electrochemical communications in your brain.
Give your brain the time it needs to grow dendrites!	***Cramming doesn't work;*** your brain cannot grow dendrites that quickly. ***Studying every day*** is the way to give your brain the time it needs.

Anxiety Prevention Tips

To Prevent Anxiety	Explanation
Practice slow, deep breathing for five minutes each day. Then do a minute or two of deep breathing right before the test. Also, if you feel your anxiety building during the test, stop for a minute, close your eyes, and do some deep breathing.	When ***test anxiety*** hits, you breathe more quickly and shallowly, which causes hyperventilation. Symptoms may be confusion, inability to concentrate, shaking, dizziness, and more. Slow, deep breathing will ***calm you and prevent panic.***

Study Skills

Continued from page 239

Anxiety Prevention Tips *(Continued)*

To Prevent Anxiety	Explanation
Do 15 to 20 minutes of ***moderate exercise*** (like walking) shortly before the test. Daily exercise is even better!	***Exercise reduces stress*** and will help prevent "blanking out" on a test. Exercise also increases your alertness, clear thinking, and energy.
To help you sleep the night before the test, or any time you need to calm down, ***eat high carbohydrate foods*** such as popcorn, bread, rice, crackers, muffins, bagels, pasta, corn, baked potatoes (not fries or chips), and cereals.	Carbohydrates increase the level of a chemical in the brain called serotonin, which has a *calming effect on the mind*. It reduces feelings of tension and stress and improves your ability to concentrate. You only need to eat a small amount, like half a bagel, to get this effect.
Before the test, ***go easy on caffeinated beverages*** such as coffee, tea, and soft drinks. Do not eat candy bars or other sugary snacks.	Extra caffeine can ***make you jittery,*** "hyper," and shaky for the test. It can increase the tendency to panic. Too much sugar causes negative emotional reactions in some people.

Now Try This

What will you do to improve your next test score?
List the three or four tips you think will help you the most.

1. ______________________________
2. ______________________________
3. ______________________________
4. ______________________________

What changes will you have to make in order to try the tips you chose?

____

____

____

____

See *Tips for Taking Math Tests* and *Preparing for Your Final Exam* for more ideas about managing anxiety. (Check the Table of Contents to find their locations.)

Chapter 3 Summary

Key Terms

3.1

like fractions Fractions with the same denominator are called *like fractions.*

unlike fractions Fractions with different denominators are called *unlike fractions.*

3.2

least common multiple Given two or more whole numbers, the least common multiple is the smallest whole number that is divisible by all the numbers.

LCM The abbreviation for *least common multiple* is LCM.

3.3

least common denominator When unlike fractions are rewritten as like fractions with the least common multiple as their denominator, the new denominator is the least common denominator.

LCD The abbreviation for *least common denominator* is LCD.

3.4

regrouping when adding fractions Regrouping is used in the addition of mixed numbers when the sum of the fractions is greater than 1.

regrouping when subtracting fractions Regrouping is used in the subtraction of mixed numbers when the fraction part of the first number is less than the fraction part of the second number.

New Symbols

$<$ is less than $(2 < 5)$

$>$ is greater than $(4 > 2)$

Test Your Word Power

See how well you have learned the vocabulary in this chapter.

1. **Like fractions** are
   - **A.** fractions that are equivalent
   - **B.** fractions that have the same numerator
   - **C.** fractions that have the same denominator.

2. Two or more fractions are **unlike fractions** if
   - **A.** they are not equivalent
   - **B.** they have different numerators
   - **C.** they have different denominators.

3. The abbreviation **LCM** stands for
   - **A.** the largest common multiple
   - **B.** the longest common multiple
   - **C.** the least common multiple.

4. The **least common multiple** is
   - **A.** the smallest whole number that is divisible by each of two or more numbers
   - **B.** the smallest numerator
   - **C.** the smallest denominator.

5. The abbreviation **LCD** stands for
   - **A.** the largest common denominator
   - **B.** the least common denominator
   - **C.** the least common divisor.

6. The **least common denominator** is
   - **A.** needed when multiplying fractions
   - **B.** needed when dividing fractions
   - **C.** the least common multiple of the denominators in a fraction problem.

Answers to Test Your Word Power

1. C; *Example:* Because the fractions $\frac{3}{8}$ and $\frac{10}{8}$ both have 8 as a denominator, they are like fractions.

2. C; *Example:* The fractions $\frac{2}{3}$ and $\frac{3}{4}$ are unlike fractions because they have different denominators.

3. C; *Example:* LCM is the abbreviation for least common multiple.

4. A; *Example:* The least common multiple of 4 and 5 is 20 because 20 is the smallest number into which both 4 and 5 will divide evenly.

5. B; *Example:* LCD is the abbreviation for least common denominator.

6. C; *Example:* The least common denominator of the fractions $\frac{2}{3}$ and $\frac{1}{2}$ is 6 because 6 is the least common multiple of 3 and 2. When written using the least common denominator, $\frac{2}{3}$ and $\frac{1}{2}$ become $\frac{4}{6}$ and $\frac{3}{6}$, respectively.

Quick Review

Concepts	Examples
3.1 Adding Like Fractions Add numerators and keep the same denominator. Simplify the answer.	$\frac{3}{4}+\frac{1}{4}+\frac{5}{4}=\frac{3+1+5}{4}=\frac{9}{4}=2\frac{1}{4}$
3.1 Subtracting Like Fractions Subtract numerators and keep the same denominator. Simplify the answer.	$\frac{7}{8}-\frac{5}{8}=\frac{7-5}{8}=\frac{2}{8}=\frac{2\div 2}{8\div 2}=\frac{1}{4}$
3.2 Finding the Least Common Multiple (LCM) Method of using multiples of the larger number: List the first few multiples of the larger number. Then find the first multiple that is divisible by the smaller number.	$\frac{1}{3}+\frac{1}{4}$ 4, 8, 12, 16, . . . ← Multiples of 4 12: First multiple divisible by 3 ($12 \div 3 = 4$) The least common multiple (LCM) of 3 and 4 is 12.
3.2 Finding the Least Common Multiple (LCM) Method of prime numbers: First find the prime factorization of each number. Then use the prime factors to build the least common multiple.	$9 = (3) \cdot (3)$ $15 = 3 \cdot (5)$ $\text{LCM} = 3 \cdot 3 \cdot 5 = 45$ (Factors of 9: $3 \cdot 3$; Factors of 15: $3 \cdot 5$) The least common multiple (LCM) of 9 and 15 is 45.
3.2 Finding the Least Common Multiple (LCM) Alternative method: Start by trying to divide the numbers by the first prime number. Continue dividing by prime numbers until all quotients are 1. The product of the prime numbers is the least common multiple.	4 and 6 2 | 4 6 2 | 2 3 3 | 1 3 1 1 The least common multiple (LCM) of 4 and 6 $= 2 \cdot 2 \cdot 3 = 12$.
3.3 Adding Unlike Fractions *Step 1* Find the least common multiple (LCM). *Step 2* Rewrite the fractions with the least common multiple as the denominator. *Step 3* Add the numerators, placing the sum over the common denominator, and simplify the answer.	$\frac{1}{3}+\frac{1}{4}+\frac{1}{10}$ LCM of 3, 4, and 10 = 60 $\frac{1}{3}=\frac{20}{60}$ $\frac{1}{4}=\frac{15}{60}$ $\frac{1}{10}=\frac{6}{60}$ $\frac{20}{60}+\frac{15}{60}+\frac{6}{60}=\frac{41}{60}$ Lowest terms

Concepts	Examples
3.3 Subtracting Unlike Fractions *Step 1* Find the least common multiple (LCM). *Step 2* Rewrite the fractions with the least common multiple as the denominator. *Step 3* Subtract the numerators, place the difference over the common denominator, and simplify the answer.	$\frac{5}{8} - \frac{1}{3}$ LCM of 8 and 3 = 24 $\frac{5}{8} = \frac{15}{24}$ $\frac{1}{3} = \frac{8}{24}$ $\frac{15}{24} - \frac{8}{24} = \frac{7}{24}$ Lowest terms
3.4 Adding Mixed Numbers Round the numbers and estimate the answer. Then find the exact answer using these steps. *Step 1* Add the fractions, using a common denominator. *Step 2* Add the whole numbers. *Step 3* Combine the sums of the whole numbers and the fractions. Simplify the fraction part when necessary. Compare the exact answer to the estimate to see if it is reasonable.	*Estimate:* *Exact:* 10 ← Rounds to $9\frac{2}{3} = 9\frac{8}{12}$ + 7 ← Rounds to $+6\frac{3}{4} = 6\frac{9}{12}$ 17 $15\frac{17}{12} = 16\frac{5}{12}$ The exact answer of $16\frac{5}{12}$ is reasonable because it is close to the estimate of 17.
3.4 Subtracting Mixed Numbers Round the numbers and estimate the answer. Then find the exact answer using these steps. *Step 1* Subtract the fractions, regrouping if necessary. *Step 2* Subtract the whole numbers. *Step 3* Combine the differences of the whole numbers and the fractions. Simplify the fraction part when necessary. Compare the exact answer to the estimate to see if it is reasonable.	*Estimate:* *Exact:* 9 ← Rounds to $8\frac{5}{8} = 8\frac{15}{24} = 7\frac{39}{24}$ − 4 ← Rounds to $-3\frac{11}{12} = 3\frac{22}{24} = 3\frac{22}{24}$ 5 $4\frac{17}{24}$ The exact answer of $4\frac{17}{24}$ is reasonable because it is close to the estimate of 5.

Concepts	Examples
3.4 Adding or Subtracting Mixed Numbers Using an Alternative Method *Step 1* Change the mixed numbers to improper fractions. *Step 2* Rewrite the unlike fractions as like fractions. *Step 3* Add or subtract the numerators and simplify the answer.	Add. $2\frac{2}{3} = \frac{8}{3} = \frac{64}{24}$ $+1\frac{3}{8} = \frac{11}{8} = +\frac{33}{24}$ 24 is the least common denominator. $\frac{97}{24} = 4\frac{1}{24}$ Answer as mixed number Improper fractions Subtract. $8\frac{2}{3} = \frac{26}{3} = \frac{104}{12}$ $-5\frac{3}{4} = \frac{23}{4} = -\frac{69}{12}$ 12 is the least common denominator. $\frac{35}{12} = 2\frac{11}{12}$ Answer as mixed number Improper fractions
3.5 Identifying the Greater of Two Fractions With unlike fractions, change to like fractions first. The fraction with the greater numerator is the greater fraction. Use these symbols: < is less than > is greater than	Identify the greater fraction. $\frac{7}{8} \quad \frac{9}{10}$ $\frac{7}{8} = \frac{7 \cdot 5}{8 \cdot 5} = \frac{35}{40}$ $\frac{9}{10} = \frac{9 \cdot 4}{10 \cdot 4} = \frac{36}{40}$ $\frac{36}{40}$ is greater than $\frac{35}{40}$, so $\frac{9}{10} > \frac{7}{8}$. $\frac{9}{10}$ is greater.
3.5 Using the Order of Operations with Fractions Follow the order of operations. **1.** Do all operations inside parentheses or other grouping symbols. **2.** Simplify any expressions with exponents and find any square roots. **3.** Multiply or divide, proceeding from left to right. **4.** Add or subtract, proceeding from left to right.	Simplify by using the order of operations. $\frac{1}{2}\left(\frac{2}{3}\right) - \left(\frac{1}{4}\right)^2$ Apply the exponent first. $= \frac{1}{2}\left(\frac{2}{3}\right) - \frac{1}{16}$ Multiply next. $= \frac{1}{3} - \frac{1}{16}$ Subtract last. $= \frac{16}{48} - \frac{3}{48}$ Rewrite fractions so they have a common denominator. Subtract numerators. $= \frac{13}{48}$ ← Simplest form

Chapter 3 Review Exercises

3.1 *Add or subtract. Write answers in lowest terms.*

1. $\frac{5}{7} + \frac{1}{7}$

2. $\frac{4}{9} + \frac{3}{9}$

3. $\frac{1}{8} + \frac{3}{8} + \frac{2}{8}$

4. $\frac{5}{16} - \frac{3}{16}$

5. $\frac{5}{10} + \frac{3}{10}$

6. $\frac{5}{12} - \frac{3}{12}$

7. $\frac{36}{62} - \frac{10}{62}$

8. $\frac{68}{75} - \frac{43}{75}$

Solve each application problem. Write answers in lowest terms.

9. Jaime Villagranna earns $\frac{7}{12}$ of his income installing kitchen cabinets for Home Depot and $\frac{4}{12}$ of his income by operating his own cabinet business. What fraction of his total income comes from the two jobs?

10. At the annual Cub Scout Pinewood Derby competition, $\frac{3}{8}$ of the events were completed before lunch and $\frac{5}{8}$ after lunch. How much more was completed after lunch than before lunch?

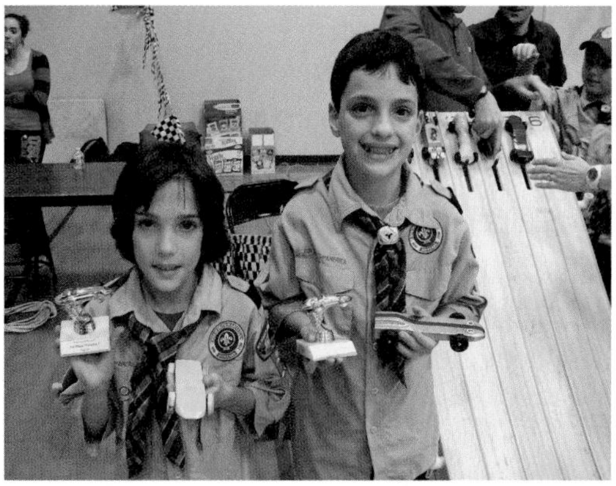

3.2 *Find the least common multiple of each set of numbers.*

11. 5, 2

12. 3, 4

13. 10, 12, 20

14. 3, 8, 4

15. 6, 8, 5, 15

16. 15, 9, 20

Rewrite each fraction using the indicated denominator.

17. $\frac{2}{3} = \frac{}{12}$

18. $\frac{3}{8} = \frac{}{56}$

19. $\frac{2}{5} = \frac{}{25}$

20. $\frac{5}{9} = \frac{}{81}$

21. $\frac{4}{5} = \frac{}{40}$

22. $\frac{5}{16} = \frac{}{64}$

3.1–3.3 *Add or subtract. Write answers in lowest terms.*

23. $\frac{1}{2} + \frac{1}{3}$

24. $\frac{1}{5} + \frac{3}{10} + \frac{3}{8}$

25. $\frac{5}{12} + \frac{5}{24}$

26. $\frac{2}{3} - \frac{1}{4}$

27. $\frac{7}{8} - \frac{1}{3}$

28. $\frac{11}{12} - \frac{4}{9}$

Solve each application problem.

29. The San Juan School District operates an after school program for students. This year $\frac{2}{5}$ of the students played after school sports, $\frac{1}{6}$ participated in arts and crafts, and $\frac{1}{3}$ spent their time in tutoring and study hall. What fraction of the total students participated in these activities?

30. When budgeting for her wedding, Kara Salzie plans to spend $\frac{1}{3}$ of her total budget on the wedding site, $\frac{3}{8}$ on food and beverages, $\frac{3}{16}$ on photos and video, and $\frac{1}{16}$ on entertainment. What portion of her budget will be spent on these four categories?

3.4 *First estimate the answer. Then add or subtract to find the exact answer. Simplify all exact answers.*

31. *Estimate:* *Exact:*

Rounds to $\quad 18\frac{5}{8}$

$+$ ___ Rounds to $\quad + 13\frac{3}{4}$

32. *Estimate:* *Exact:*

$22\frac{2}{3}$

$+$ ___ $\quad + 15\frac{4}{9}$

33. *Estimate:* *Exact:*

$12\frac{3}{5}$

$8\frac{5}{8}$

$+$ ___ $\quad + 10\frac{5}{16}$

34. *Estimate:* *Exact:*

$31\frac{3}{4}$

$-$ ___ $\quad - 14\frac{2}{3}$

35. *Estimate:* *Exact:*

34

$-$ ___ $\quad - 15\frac{2}{3}$

36. *Estimate:* *Exact:*

$215\frac{7}{16}$

$-$ ___ $\quad - 136$

Add or subtract by changing mixed numbers to improper fractions. Simplify all answers.

37. $5\frac{2}{5} + 3\frac{7}{10}$

38. $4\frac{3}{4} + 5\frac{2}{3}$

39. $5 - 1\frac{3}{4}$

40. $6\frac{1}{2} - 4\frac{5}{6}$

41. $8\frac{1}{3} - 2\frac{5}{6}$

42. $5\frac{5}{12} - 2\frac{5}{8}$

First estimate the answer and then find the exact answer for each application problem. Simplify all exact answers.

43. Two long-distance runners began an $18\frac{3}{4}$ mile run. They ran uphill $5\frac{5}{8}$ miles, downhill $7\frac{1}{3}$ miles, and the rest of the course was level. Find the distance of the level portion of the course.

Estimate:

Exact:

44. The Boys Scouts had a paper drive. They collected $28\frac{2}{3}$ tons of newspapers on Saturday and $24\frac{3}{4}$ tons on Sunday. Find the total weight of the newspapers collected.

Estimate:

Exact:

45. On a recent fishing trip to Kemmerer, Wyoming, Roy Abriani caught four fish. One fish was a German brown trout weighing $7\frac{1}{2}$ pounds, while the other three were rainbow trout weighing $2\frac{3}{4}$ pounds, $4\frac{7}{8}$ pounds, and $3\frac{3}{8}$ pounds. Find the total weight of the four fish.

Estimate:

Exact:

46. The Safeway World Championship Pumpkin Weigh-off awarded first place to the grower of a pumpkin weighing $1535\frac{3}{8}$ pounds. The second place pumpkin weighed $1475\frac{11}{16}$ pounds. How much more did the first place pumpkin weigh? (*Source:* Sacramento Bee.)

Estimate:

Exact:

3.5 *Locate each fraction in Exercises 47–50 on the number line.*

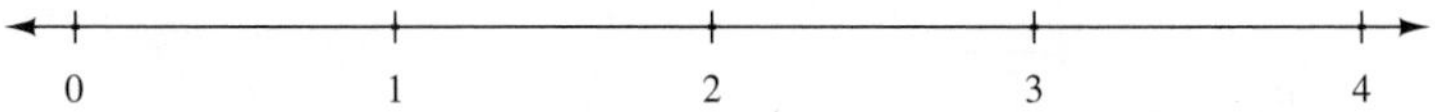

47. $\frac{3}{8}$ **48.** $\frac{7}{4}$ **49.** $\frac{8}{3}$ **50.** $3\frac{1}{5}$

Write < or > in each blank to make a true statement.

51. $\frac{2}{3}$ ___ $\frac{3}{4}$ **52.** $\frac{3}{4}$ ___ $\frac{7}{8}$ **53.** $\frac{1}{2}$ ___ $\frac{7}{15}$ **54.** $\frac{7}{10}$ ___ $\frac{8}{15}$

55. $\frac{9}{16}$ ___ $\frac{5}{8}$ **56.** $\frac{7}{20}$ ___ $\frac{8}{25}$ **57.** $\frac{19}{36}$ ___ $\frac{29}{54}$ **58.** $\frac{19}{132}$ ___ $\frac{7}{55}$

Simplify each expression.

59. $\left(\frac{1}{2}\right)^2$

60. $\left(\frac{2}{3}\right)^2$

61. $\left(\frac{3}{10}\right)^3$

62. $\left(\frac{3}{8}\right)^4$

Simplify by using the order of operations.

63. $8\left(\frac{1}{4}\right)^2$

64. $12\left(\frac{3}{4}\right)^2$

65. $\left(\frac{2}{3}\right)^2 \cdot \left(\frac{3}{8}\right)^2$

66. $\frac{7}{8} \div \left(\frac{1}{8} + \frac{3}{4}\right)$

67. $\left(\frac{1}{2}\right)^2 \cdot \left(\frac{1}{4} + \frac{1}{2}\right)$

68. $\left(\frac{1}{4}\right)^3 + \left(\frac{5}{8} + \frac{3}{4}\right)$

Mixed Review Exercises

Simplify. Use the order of operations as necessary.

69. $\frac{7}{8} - \frac{1}{8}$

70. $\frac{7}{10} - \frac{3}{10}$

71. $\frac{29}{32} - \frac{5}{16}$

72. $\frac{1}{4} + \frac{1}{8} + \frac{5}{16}$

73. $\begin{array}{r} 6\frac{2}{3} \\ -\ 4\frac{1}{2} \\ \hline \end{array}$

74. $\begin{array}{r} 9\frac{1}{2} \\ +\ 16\frac{3}{4} \\ \hline \end{array}$

75. $\begin{array}{r} 7 \\ -\ 1\frac{5}{8} \\ \hline \end{array}$

76. $\begin{array}{r} 2\frac{3}{5} \\ 8\frac{5}{8} \\ +\ \frac{5}{16} \\ \hline \end{array}$

77. $\begin{array}{r} 32\frac{5}{12} \\ -\ 17 \\ \hline \end{array}$

78. $\frac{7}{22} + \frac{3}{22} + \frac{3}{11}$

79. $\left(\frac{1}{4}\right)^2 \cdot \left(\frac{2}{5}\right)^3$

80. $\frac{3}{8} \div \left(\frac{1}{2} + \frac{1}{4}\right)$

81. $\left(\frac{2}{3}\right)^2 \cdot \left(\frac{1}{3} + \frac{1}{6}\right)$

82. $\left(\frac{2}{3}\right)^3 + \left(\frac{2}{3} - \frac{5}{9}\right)$

Write < or > in each blank to make a true statement.

83. $\frac{2}{3}$ ___ $\frac{7}{12}$

84. $\frac{8}{9}$ ___ $\frac{15}{8}$

85. $\frac{17}{30}$ ___ $\frac{36}{60}$

86. $\frac{5}{8}$ ___ $\frac{17}{30}$

Find the least common multiple of each set of numbers.

87. 12, 18

88. 6, 8, 10, 12

89. 9, 14, 21

Rewrite each fraction using the indicated denominator.

90. $\frac{2}{3} = \frac{}{27}$

91. $\frac{9}{12} = \frac{}{144}$

92. $\frac{4}{5} = \frac{}{75}$

First estimate the answer and then find the exact answer for each application problem. Simplify exact answers.

93. A cement contractor needs $13\frac{1}{2}$ ft of wire mesh for a concrete walkway and $22\frac{3}{8}$ ft of wire mesh for a driveway. If the contractor starts with a roll of wire that is $92\frac{3}{4}$ ft long, find the number of feet remaining after the two jobs have been completed.

Estimate:

Exact:

94. A baker had four 50-pound bags of sugar. She used $68\frac{1}{2}$ pounds of sugar to bake cakes, $76\frac{5}{8}$ pounds for baking pies, and $33\frac{1}{4}$ pounds for baking cookies. How many pounds of sugar remain?

Estimate:

Exact:

Chapter 3 Test

CHAPTER Test Prep VIDEO *The Chapter Test Prep Videos with test solutions are available on DVD, in MyMathLab, and on YouTube—search "LialDevMath" and click on "Channels."*

Add or subtract. Write answers in lowest terms.

1. $\frac{5}{8}+\frac{1}{8}$

2. $\frac{1}{16}+\frac{7}{16}$

3. $\frac{7}{10}-\frac{3}{10}$

4. $\frac{7}{12}-\frac{5}{12}$

Find the least common multiple of each set of numbers.

5. 2, 3, 4

6. 6, 3, 5, 15

7. 6, 9, 27, 36

Add or subtract. Write answers in lowest terms.

8. $\frac{3}{8}+\frac{1}{4}$

9. $\frac{2}{9}+\frac{5}{12}$

10. $\frac{7}{8}-\frac{2}{3}$

11. $\frac{2}{5}-\frac{3}{8}$

First estimate the answer. Then add or subtract to find the exact answer. Simplify exact answers.

12. $7\frac{2}{3}+4\frac{5}{6}$

Estimate:

Exact:

13. $16\frac{2}{5}-11\frac{2}{3}$

Estimate:

Exact:

14. $18\frac{3}{4}+9\frac{2}{5}+12\frac{1}{3}$

Estimate:

Exact:

15. $24-18\frac{3}{8}$

Estimate:

Exact:

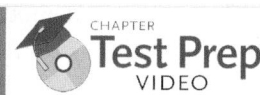

16. Most students say that "addition and subtraction of fractions are more difficult than multiplication and division of fractions." Why do you think they say this? Do you agree with these students?

17. Explain a method of estimating an answer to addition and subtraction problems involving mixed numbers. Could your estimated answer vary from the exact answer? If it did, what would the estimation accomplish?

First estimate the answer and then find the exact answer for each application problem. Simplify exact answers.

18. In one week, a kennel owner used $10\frac{3}{8}$ pounds of puppy chow, $84\frac{1}{2}$ pounds of dry kibble, $36\frac{5}{6}$ pounds of high-protein mature dog mix, and $8\frac{1}{3}$ pounds of fresh ground meat products. Find the total number of pounds used.

Estimate:

Exact:

19. A painting contractor arrived at a 6-unit apartment complex with $147\frac{1}{2}$ gallons of exterior paint. If his crew sprayed $68\frac{1}{2}$ gallons on the wood siding and rolled $37\frac{3}{8}$ gallons on the masonry exterior, find the number of gallons of paint remaining.

Estimate:

Exact:

Write < or > between each pair of fractions to make a true statement.

20. $\frac{3}{4}$ ___ $\frac{17}{24}$

21. $\frac{19}{24}$ ___ $\frac{17}{36}$

Simplify. Use the order of operations as needed.

22. $\left(\frac{1}{3}\right)^3 \cdot 54$

23. $\left(\frac{3}{4}\right)^2 - \left(\frac{7}{8} \cdot \frac{1}{3}\right)$

24. $4\left(\frac{7}{8} - \frac{7}{16}\right)$

25. $\frac{5}{6} + \frac{4}{3}\left(\frac{3}{8}\right)$

4 Decimals

Over 42 million Americans go fishing at least once a year, spending more than $2 billion on gear and tackle. This grandfather and grandson will use decimal numbers when paying for new fishing equipment. But will decimals help them catch their limit?

4.1 Reading and Writing Decimal Numbers

OBJECTIVES

1. Write parts of a whole using decimals.
2. Identify the place value of a digit.
3. Read and write decimals in words.
4. Write decimals as fractions or mixed numbers.

Fractions are used to represent parts of a whole. In this chapter, **decimals** are used as another way to show parts of a whole. For example, our money system is based on decimals. One dollar is divided into 100 equal parts. One cent ($0.01) is one of the parts, and a dime ($0.10) is 10 of the parts. Metric measurement (see **Appendix A**) is also based on decimals.

OBJECTIVE 1 Write parts of a whole using decimals. Decimals are used when a whole is divided into 10 equal parts, or into 100 or 1000 or 10,000 equal parts. In other words, decimals are fractions with denominators that are a power of 10. For example, the square below is cut into 10 equal parts. Written as a fraction, each part is $\frac{1}{10}$ of the whole. Written as a decimal, each part is 0.1. Both $\frac{1}{10}$ and 0.1 are read as "*one tenth.*"

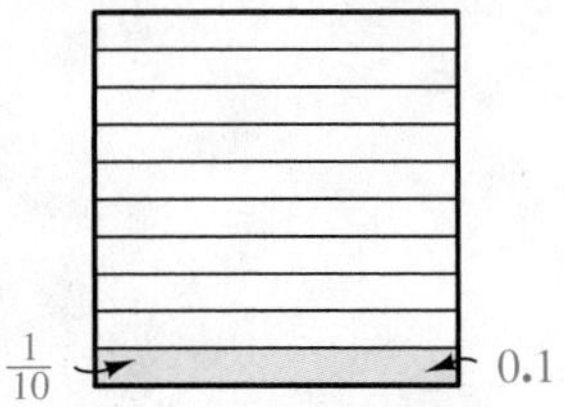

One-tenth of the square is shaded.

The dot in 0.1 is called the **decimal point.**

0.1
Decimal point ↗

The square at the right has 7 of its 10 parts shaded.

Written as a *fraction,* $\frac{7}{10}$ of the square is shaded.

Written as a *decimal,* 0.7 of the square is shaded.

Both $\frac{7}{10}$ and 0.7 are read as "*seven tenths.*"

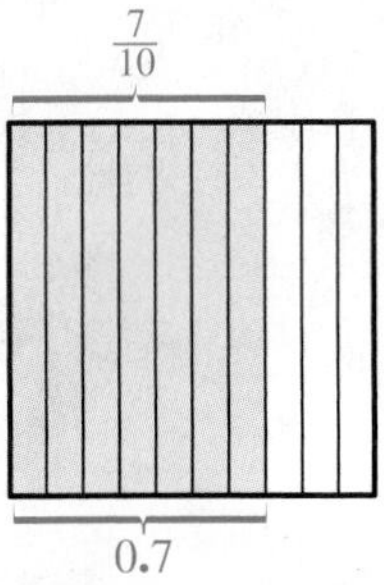

Seven-tenths of the square is shaded.

◀ **Work Problem 1 at the Side.**

The square below is cut into 100 equal parts.

Written as a *fraction,* each part is $\mathbf{\frac{1}{100}}$ of the whole.

Written as a decimal, each part is **0.01** of the whole.

Both $\frac{1}{100}$ and 0.01 are read as "*one hundredth.*"

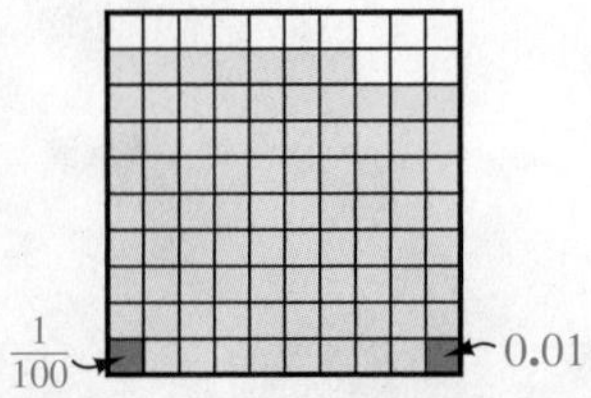

This square has 87 parts shaded.

Written as a fraction, $\mathbf{\frac{87}{100}}$ of the total area is shaded.

Written as a decimal, **0.87** of the total area is shaded.

Both $\frac{87}{100}$ and 0.87 are read as "*eighty-seven hundredths.*"

VOCABULARY TIP

Deci The prefix **deci** means **tenth**. For example, a **decimeter** is one **tenth** of a meter.

1 There are 10 dimes in one dollar. Each dime is $\frac{1}{10}$ of a dollar. Label the yellow shaded portion of each dollar as a fraction, as a decimal, and in words.

(a)

(b)

(c)

Answers

1. **(a)** $\frac{1}{10}$; 0.1; one tenth
   (b) $\frac{3}{10}$; 0.3; three tenths
   (c) $\frac{9}{10}$; 0.9; nine tenths

Work Problem **2** at the Side. ▶

Example 1 below shows several numbers written as fractions, as decimals, and in words.

EXAMPLE 1 **Using the Decimal Forms of Fractions**

Fraction	**Decimal**	**Read As**
(a) $\frac{4}{10}$	0.4	four tenths
(b) $\frac{9}{100}$	0.09	nine hundredths
(c) $\frac{71}{100}$	0.71	seventy-one hundredths
(d) $\frac{8}{1000}$	0.008	eight thousandths
(e) $\frac{45}{1000}$	0.045	forty-five thousandths
(f) $\frac{832}{1000}$	0.832	eight hundred thirty-two thousandths

Work Problem **3** at the Side. ▶

OBJECTIVE 2 **Identify the place value of a digit.** The decimal point separates the *whole number part* from the *fractional part* in a decimal number. In the chart below, you see that the **place value** names for fractional parts are similar to those on the whole number side, but end in "*ths*."

Decimal Place Value Chart

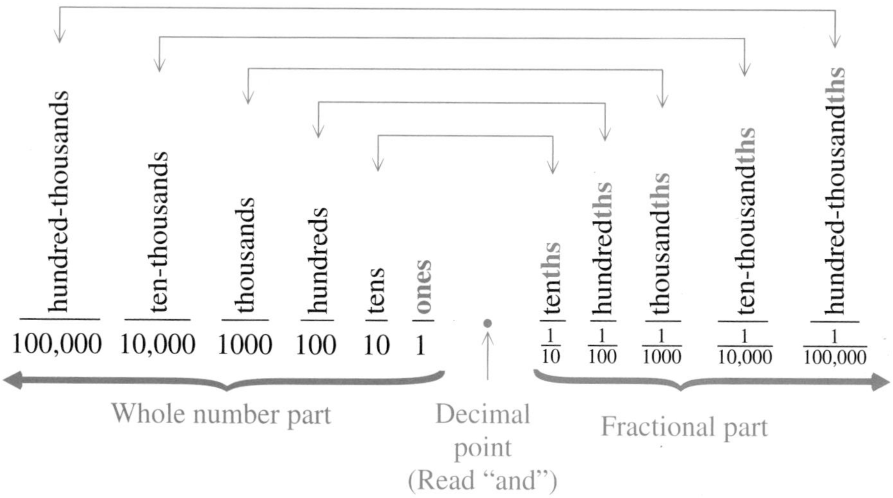

Note

Notice that the **ones** place is at the center of the place value chart. There is no "oneths" place.

Also notice that each place is 10 times the value of the place to its right.

Finally, be sure to write a **hyphen** (dash) in ten–thousand**ths** and hundred–thousand**ths**.

2 Write the portion of each square that is shaded as a fraction, as a decimal, and in words.

GS **(a)**

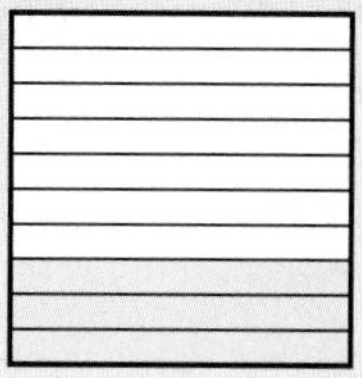

$\frac{3}{10}$ = 0.____ =

_______ tenths

(b)

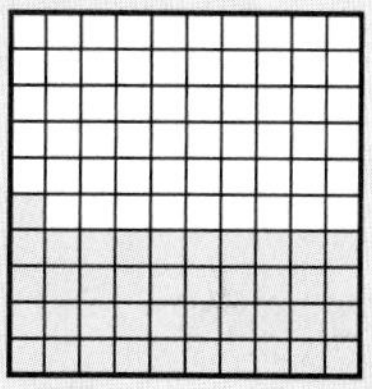

3 Write each decimal as a fraction.

GS **(a)** 0.7 = $\frac{7}{\_\_\_}$

(b) 0.2

(c) 0.03

(d) 0.69

(e) 0.047

(f) 0.351

Answers

2. **(a)** 0.3; three tenths

(b) $\frac{41}{100}$; 0.41; forty-one hundredths

3. **(a)** 10; $\frac{7}{10}$ **(b)** $\frac{2}{10}$ **(c)** $\frac{3}{100}$

(d) $\frac{69}{100}$ **(e)** $\frac{47}{1000}$ **(f)** $\frac{351}{1000}$

4 Identify the place value of each digit.

(a) 971.54

(b) 0.4

(c) 5.60

(d) 0.0835

5 Tell how to read each decimal in words.

GS **(a)** 0.6 six ________

GS **(b)** 0.46 forty-six ________

(c) 0.05 **(d)** 0.409

(e) 0.0003 **(f)** 0.2703

Answers

4. (a)

hundreds	tens	ones		tenths	hundredths
9	7	1	.	5	4

(b)

ones		tenths
0	.	4

(c)

ones		tenths	hundredths
5	.	6	0

(d)

ones		tenths	hundredths	thousandths	ten-thousandths
0	.	0	8	3	5

5. (a) six tenths
(b) forty-six hundredths
(c) five hundredths
(d) four hundred nine thousandths
(e) three ten-thousandths
(f) two thousand seven hundred three ten-thousandths

CAUTION

If a number does *not* have a decimal point, it is a *whole number.* A whole number has no fractional part. If you want to show the decimal point in a whole number, it is just to the ***right*** of the digit in the ones place. Here are two examples.

8 = 8. ↑ Decimal point

306 = 306. ↑ Decimal point

EXAMPLE 2 Identifying the Place Value of a Digit

Identify the place value of each digit.

(a) 178.36

hundreds	tens	ones		tenths	hundredths
1	7	8	.	3	6

(b) 0.00935

ones		tenths	hundredths	thousandths	ten-thousandths	hundred-thousandths
0	.	0	0	9	3	5

Notice the hyphen (dash) in ten-thousandths and in hundred-thousandths.

Notice in **Example 2(b)** that we do *not* use commas on the right side of the decimal point.

◀ **Work Problem 4 at the Side.**

OBJECTIVE 3 Read and write decimals in words. A decimal is read according to its form as a fraction.

ones		tenths
0	.	9

We read 0.9 as "nine tenths" because 0.9 is the same as $\frac{9}{10}$. Notice that 0.9 ends in the tenths place.

ones		tenths	hundredths
0	.	0	2

We read 0.02 as "two hundredths" because 0.02 is the same as $\frac{2}{100}$. Notice that 0.02 ends in the hundredths place.

EXAMPLE 3 Reading Decimal Numbers

Tell how to read each decimal in words.

(a) 0.3

Because $0.3 = \frac{3}{10}$, read the decimal as three tenths.

(b) 0.49 Read it as: forty-nine hundredths.

(c) 0.08 Read it as: eight hundredths.

Think: $0.08 = \frac{8}{100}$ so write *hundredths.*

(d) 0.918 Read it as: nine hundred eighteen thousandths.

(e) 0.0106 Read it as: one hundred six ten-thousandths.

Think: $0.0106 = \frac{106}{10{,}000}$

◀ **Work Problem 5 at the Side.**

Reading Decimal Numbers

Step 1 Read any whole number part to the ***left*** of the decimal point as you normally would.

Step 2 Read the decimal point as "***and.***"

Step 3 Read the part of the number to the ***right*** of the decimal point as if it were an ordinary whole number.

Step 4 Finish with the place value name of the rightmost digit; these names all end in "***ths.***"

Note

If there is *no whole number part,* you will use only Steps 3 and 4.

EXAMPLE 4 Reading Decimals

Read each decimal.

(a) 9 is in tenths place.

16.9

sixteen **and** nine **tenths**

Remember to say or write "and" *only* when you see a decimal point.

16.9 is read "sixteen **and** nine tenths."

(b) 5 is in hundredths place.

482.35

four hundred eighty-two **and** thirty-five **hundredths**

482.35 is read "four hundred eighty-two and thirty-five hundredths."

3 is in thousandths place.

(c) 0.063 is "sixty-three **thousandths**." (no whole number part)

(d) 11.1085 is "eleven **and** one thousand eighty-five **ten-thousandths**."

CAUTION

Use "and" when reading a decimal point. A common mistake is to read the whole number 405 as "four hundred *and* five." But there is ***no decimal point*** shown in 405, so it is read "four hundred five."

Work Problem 6 at the Side. ▶

OBJECTIVE 4 Write decimals as fractions or mixed numbers. Knowing how to read decimals will help you when writing decimals as fractions.

Writing Decimals as Fractions or Mixed Numbers

Step 1 The digits to the right of the decimal point are the numerator of the fraction.

Step 2 The denominator is 10 for tenths, 100 for hundredths, 1000 for thousandths, 10,000 for ten-thousandths, and so on.

Step 3 If the decimal has a whole number part, it will be written as a mixed number with the same whole number part.

6 Tell how to read each decimal in words.

GS **(a)** 3.8 is read

three and eight ________.

(b) 15.001

(c) 0.0073

(d) 64.309

Answers

6. **(a)** three and eight tenths
(b) fifteen and one thousandth
(c) seventy-three ten-thousandths
(d) sixty-four and three hundred nine thousandths

7 Write each decimal as a fraction or mixed number.

(a) 0.7

(b) 12.21

(c) 0.101

(d) 0.007

(e) 1.3717

8 Write each decimal as a fraction or mixed number in lowest terms.

GS (a) 0.5 Write in lowest terms.

$$\frac{5}{10} = \frac{1}{\_\_\_\_}$$

GS (b) 12.6 Write in lowest terms.

$$12.6 = 12\frac{6}{10} = 12\frac{\_\_\_\_}{\_\_\_\_}$$

(c) 0.85

(d) 3.05

(e) 0.225

(f) 420.0802

Answers

7. (a) $\frac{7}{10}$ (b) $12\frac{21}{100}$ (c) $\frac{101}{1000}$ (d) $\frac{7}{1000}$ (e) $1\frac{3717}{10{,}000}$

8. (a) $\frac{1}{2}$ (b) $12\frac{3}{5}$ (c) $\frac{17}{20}$ (d) $3\frac{1}{20}$ (e) $\frac{9}{40}$ (f) $420\frac{401}{5000}$

EXAMPLE 5 Writing Decimals as Fractions or Mixed Numbers

Write each decimal as a fraction or mixed number.

(a) 0.19

The digits to the right of the decimal point, 19, are the numerator of the fraction. The denominator is 100 for hundredths because the rightmost digit is in the hundredths place.

$$0.1\underline{9} = \frac{19}{100} \leftarrow \text{100 for hundredths}$$

Hundredths place (points to 9)

(b) 0.863

$$0.86\underline{3} = \frac{863}{1000} \leftarrow \text{1000 for thousandths}$$

Thousandths place (points to 3)

(c) 4.0099

The whole number part stays the same.

$$4.009\underline{9} = 4\frac{99}{10{,}000} \leftarrow \text{10,000 for ten-thousandths}$$

Ten-thousandths place (points to 9)

◀ Work Problem **7** at the Side.

CAUTION

After you write a decimal as a fraction or a mixed number, make sure the fraction is in lowest terms.

EXAMPLE 6 Writing Decimals as Fractions or Mixed Numbers in Lowest Terms

Write each decimal as a fraction or mixed number in lowest terms.

(a) $0.4 = \frac{4}{10} \leftarrow$ 10 for tenths

Write $\frac{4}{10}$ in lowest terms.

$$\frac{4}{10} = \frac{4 \div 2}{10 \div 2} = \frac{2}{5} \leftarrow \text{Lowest terms}$$

(b) $$0.75 = \frac{75}{100} = \frac{75 \div 25}{100 \div 25} = \frac{3}{4} \leftarrow \text{Lowest terms}$$

The whole number part stays the same.

(c) $$18.105 = 18\frac{105}{1000} = 18\frac{105 \div 5}{1000 \div 5} = 18\frac{21}{200} \leftarrow \text{Lowest terms}$$

(d) $$42.8085 = 42\frac{8085}{10{,}000} = 42\frac{8085 \div 5}{10{,}000 \div 5} = 42\frac{1617}{2000} \leftarrow \text{Lowest terms}$$

◀ Work Problem **8** at the Side.

Calculator Tip

In this book we will write a 0 in the ones place for decimal fractions. We write 0.45 instead of just .45, to emphasize that there is no whole number. Many calculators show these zeros also. Try entering [.] [4] [5]; the display probably shows 0.45 even though you did not press 0.

4.1 Exercises

FOR EXTRA HELP

Download the MyDashBoard App

MyMathLab®

CONCEPT CHECK *Circle the correct answer in Exercises 1–4.*

1. The number 11.084 has how many decimal places?
2 3 5

2. The number 0.7185 has how many decimal places?
4 5 1

3. In 22.85, the 8 means what?
8 tenths
8 hundredths
8 ones

4. In 57.213, the 3 means what?
3 thousandths
3 tenths
3 hundredths

Identify the digit that has the given place value. ***See Example 2.***

5. 70.489
tens
ones
tenths

6. 135.296
ones
tenths
tens

7. 0.83472
thousandths
ten-thousandths
tenths

8. 0.51968
tenths
ten-thousandths
hundredths

9. 149.0832
hundreds
hundredths
ones

10. 3458.712
hundreds
hundredths
tenths

11. 6285.7125
thousands
thousandths
hundredths

12. 5417.6832
thousands
thousandths
ones

Write the decimal number that has the specified place values. ***See Example 2.***

13. Fill in the blanks for this decimal number:
GS 0 ones, 5 hundredths, 1 ten, 4 hundreds, 2 tenths
410. ___ ___

14. Fill in the blanks for this decimal number:
GS 7 tens, 9 tenths, 3 ones, 6 hundredths, 8 hundreds
___ ___ ___.96

15. 3 thousandths, 4 hundredths, 6 ones, 2 ten-thousandths, 5 tenths

16. 8 ten-thousandths, 4 hundredths, 0 ones, 2 tenths, 6 thousandths

17. 4 hundredths, 4 hundreds, 0 tens, 0 tenths, 5 thousandths, 5 thousands, 6 ones

18. 7 tens, 7 tenths, 6 thousands, 6 thousandths, 3 hundreds, 3 hundredths, 2 ones

Write each decimal as a fraction or mixed number in lowest terms.
See Examples 5 and 6.

19. 0.7

20. 0.1

21. 13.4

22. 9.8

23. 0.25

24. 0.55

25. 0.66

26. 0.33

27. 10.17

28. 31.99

29. 0.06

30. 0.08

31. 0.205

32. 0.805

33. 5.002

34. 4.008

35. 0.686

36. 0.492

Tell how to read each decimal in words. **See Examples 3 and 4.**

37. 0.5

38. 0.2

39. 0.78

40. 0.55

41. 0.105

42. 0.609

43. 12.04

44. 86.09

45. 1.075

46. 4.025

For each exercise, rewrite the words as a decimal number. **See Examples 3 and 4.**

47. Fill in the blanks to write six and seven tenths as a decimal number.

____ . ____

48. Fill in the blanks to write eight and twelve hundredths as a decimal number.

____ . ____ ____

49. thirty-two hundredths

50. one hundred eleven thousandths

51. four hundred twenty and eight thousandths

52. two hundred and twenty-four thousandths

53. seven hundred three ten-thousandths

54. eight hundred and six hundredths

800.06

55. seventy-five and thirty thousandths

56. sixty and fifty hundredths

57. CONCEPT CHECK Anne read the number 4302 as "four thousand three hundred and two." Explain what is wrong with the way Anne read the number.

58. CONCEPT CHECK Jerry read the number 9.0106 as "nine and one hundred and six ten-thousandths." Explain the error he made.

The grandfather on the first page of this chapter needs to select the correct fishing line for his grandson's reel. Fishing line is sold according to how many pounds of "pull" the line can withstand before breaking. Use the table to answer Exercises 59–62. Write all fractions in lowest terms. (Note: The diameter of the fishing line is its thickness.)

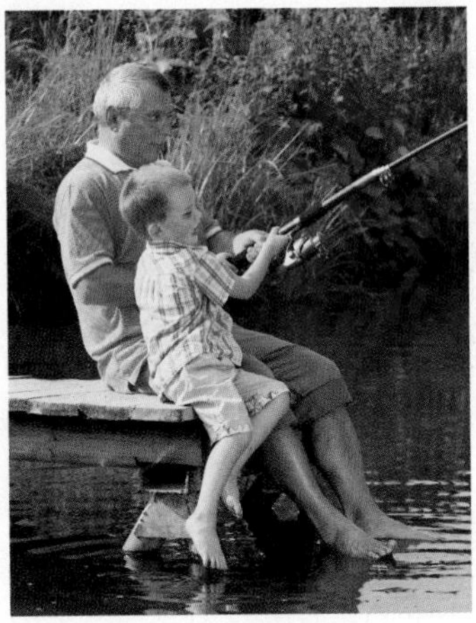

RELATING FISHING LINE DIAMETER TO TEST STRENGTH

Test Strength (pounds)	Average Diameter (inches)
4	0.008
8	0.010
12	0.013
14	0.014
17	0.015
20	0.016

Source: Berkley Outdoor Technologies Group.

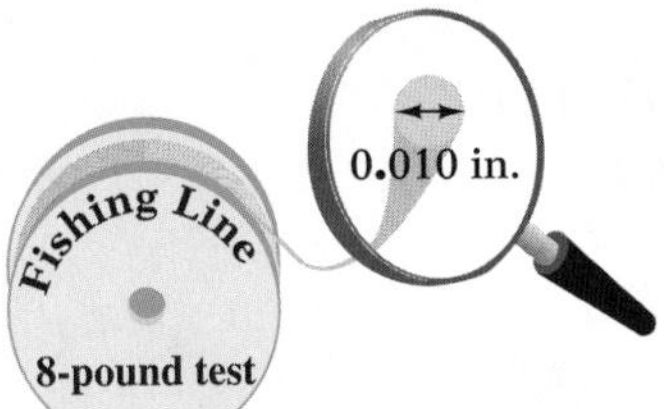

The diameter is the distance across the end of the line (or its thickness).

59. Write the diameter of 8-pound test line in words and as a fraction.

60. Write the diameter of 17-pound test line in words and as a fraction.

61. What is the test strength of the line with a diameter of $\frac{13}{1000}$ inch?

62. What is the test strength of the line with a diameter of sixteen thousandths inch?

CONCEPT CHECK *Suppose your job is to take phone orders for precision parts. Use the table, and in Exercises 63–66, write the correct part number that matches what you hear the customer say over the phone. In Exercises 67–68, write the words you would say to the customer.*

Part Number	Size in Centimeters
3-A	0.06
3-B	0.26
3-C	0.6
3-D	0.86
4-A	1.006
4-B	1.026
4-C	1.06
4-D	1.6
4-E	1.602

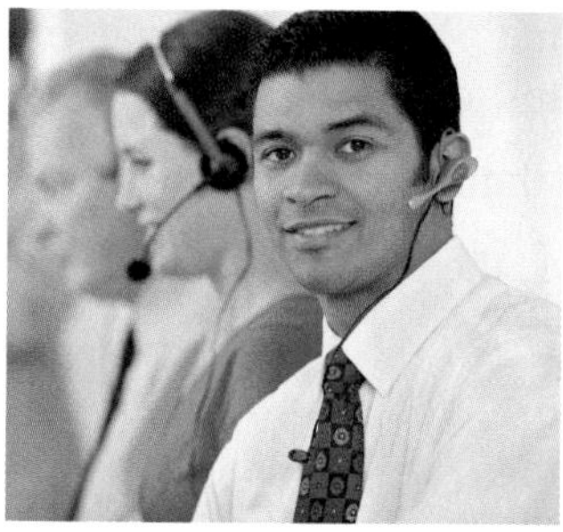

63. "Please send the six-tenths centimeter bolt."
Part number ______.

64. "The part missing from our order was the one and six hundredths size."
Part number ______.

65. "The size we need is one and six thousandths centimeters."
Part number ______.

66. "Do you still stock the twenty-six hundredths centimeter bolt?"
Part number ______.

67. "What size is part number 4-E?" Write your answer in words.

68. "What size is part number 4-B?" Write your answer in words.

Relating Concepts (Exercises 69–76) For Individual or Group Work

Use your knowledge of place value to ***work Exercises 69–76 in order.***

69. Look back at the decimal place value chart early in this section. What do you think would be the names of the next four places to the *right* of hundred-thousandths? What information did you use to come up with these names?

70. A common mistake is to think that the first place to the right of the decimal point is "oneths" and the second place is "tenths." Why might someone make that mistake? How would you explain why there is no "oneths" place?

71. Use your answer to **Exercise 69** to write 0.72436955 in words.

72. Use your answer to **Exercise 69** to write 0.000678554 in words.

73. Write 8006.500001 in words.

74. Write 20,060.000505 in words.

75. Write this decimal in numbers.
three hundred two thousand forty ten-millionths

76. Write this decimal in numbers.
nine billion, eight hundred seventy-six million, five hundred forty-three thousand, two hundred ten and one hundred million two hundred thousand three hundred billionths

4.2 Rounding Decimal Numbers

OBJECTIVES

1. Learn the rules for rounding decimals.
2. Round decimals to any given place.
3. Round money amounts to the nearest cent or nearest dollar.

In **Chapter 1,** you learned how to round whole numbers. For example, 89 rounded to the nearest ten is 90, and 8512 rounded to the nearest hundred is 8500.

OBJECTIVE 1 **Learn the rules for rounding decimals.** It is also important to be able to **round** decimals. For example, a store is selling 2 candy mints for \$0.75 but you want only one mint. The price of each mint is \$0.75 ÷ 2, which is \$0.375, but you cannot pay part of a cent. Is \$0.375 closer to \$0.37 or to \$0.38? Actually, it's exactly halfway between. When this happens in everyday situations, the rule is to round *up.* The store will charge you \$0.38 for the mint.

Rounding Decimals

Step 1 Find the place to which the rounding is being done. Draw a "cut-off" line ***after*** that place to show that you are cutting off and dropping the rest of the digits.

Step 2 Look ***only*** at the ***first*** digit you are cutting off.

Step 3(a) If this digit is ***4 or less,*** the part of the number you are keeping ***stays the same.***

Step 3(b) If this digit is ***5 or more,*** you must ***round up*** the part of the number you are keeping.

Step 4 You can use the $\approx$ symbol or the $\doteq$ symbol to indicate that the rounded number is now an approximation (close, but not exact). Both symbols mean "is approximately equal to." In this book we will use the $\approx$ symbol.

CAUTION

Do ***not*** move the decimal point when rounding.

OBJECTIVE 2 **Round decimals to any given place.** The following examples show you how to round decimals.

EXAMPLE 1 Rounding a Decimal Number

Round 14.39652 to the nearest thousandth.

Step 1 Draw a "cut-off" line after the thousandths place.

1 4 . 3 9 6 | 5 2 — You are cutting off the 5 and 2. They will be dropped.

Thousandths (points to 6)

Step 2 Look *only* at the *first* digit you are cutting off. Ignore the other digits you are cutting off.

1 4 . 3 9 6 | 5 2 — Look *only* at the 5. Ignore the 2.

Continued on Next Page

Step 3 If the first digit you are cutting off is *5 or more,* round up the part of the number you are keeping.

```
  14.396 | 5 2     First digit cut is 5 or more, so round up by adding
+  0.001 |         1 thousandth to the part you are keeping.
  14.397
```

Think: Rounding to *thousandths* means *three* decimal places.

So, 14.39652 rounded to the nearest thousandth is 14.397.
We can write 14.39652 ≈ 14.397.

CAUTION

When rounding whole numbers in **Chapter 1**, you kept all the digits but changed some to zeros. With decimals, you cut off and *drop the extra digits.* In **Example 1** above, 14.39652 rounds to 14.397, ***not*** 14.39700.

Work Problem 1 at the Side. ▶

In **Example 1**, the rounded number 14.397 had *three decimal places.* **Decimal places** are the number of digits to the *right* of the decimal point. The first decimal place is tenths, the second is hundredths, the third is thousandths, and so on.

EXAMPLE 2 Rounding Decimals to Different Places

Round to the place indicated.

(a) 5.3496 to the nearest tenth

Tenths is one decimal place.

Step 1 Draw a cut-off line after the tenths place.

```
5.3 | 4 9 6     You are cutting off the 4, 9, and 6.
  ↑             They will be dropped.
Tenths
```

Step 2

```
5.3 | 4 9 6     Look only at the 4.
        Ignore these digits.
```

Step 3

```
5.3 | 4 9 6     First digit cut is 4 or less, so the part you are keeping
5.3 ← Stays the same      stays the same.
```

5.3496 rounded to the nearest tenth is 5.3 (*one* decimal place for *tenths*).
We can write 5.3496 ≈ 5.3.
Notice: 5.3496 does ***not*** round to 5.3000, which would be ten-thousandths.

(b) 0.69738 to the nearest hundredth

Step 1

```
0.69 | 7 3 8    Draw a cut-off line after the hundredths place.
   ↑
Hundredths
```

Step 2

```
0.69 | 7 3 8    Look only at the 7.
```

Continued on Next Page

1 Round to the nearest thousandth.

(a) 0.33492

```
0.334 | 92      First digit cut is 5 or more.
    ↑
Thousandths
```

(b) 8.00851

(c) 265.42038

(d) 10.70180

Answers

1. (a) 0.335 **(b)** 8.009 **(c)** 265.420 **(d)** 10.702

2 Round to the place indicated.

GS **(a)** 0.8988 to the nearest hundredth

First digit cut is *5 or more*, so round up.

0 . 8 9 | 88

0 . 8 9 ← Keep this part.
+0 . 0 1 ← To round up, add 1 hundredth.
0 . _ _

(b) 5.8903 to the nearest hundredth

(c) 11.0299 to the nearest thousandth

(d) 0.545 to the nearest tenth

Answers

2. **(a)** 0.90 **(b)** 5.89 **(c)** 11.030 **(d)** 0.5

Step 3 0 . 6 9 | 7 3 8 — First digit cut is *5 or more*, so round up by adding 1 hundredth to the part you are keeping.

0 . 6 9 ← Keep this part.
+0 . 0 1 ← To round up, add 1 hundredth.
0 . 7 0 ← 9 + 1 is 10; write 0 and regroup 1 to the tenths place.

0.69738 rounded to the nearest hundredth is 0.70. Hundredths is *two* decimal places so you *must* write the 0 in the hundredths place. We can write 0.69738 ≈ 0.70.

Think: Rounding to *hundredths* means *two* decimal places.

CAUTION

If a *rounded* number has a 0 in the rightmost place, you *must* keep the 0. As shown above, 0.69738 rounded to the nearest hundredth is 0.70. Do ***not*** write 0.7, which is rounded to tenths instead of hundredths.

(c) 0.01806 to the nearest thousandth

0 . 0 1 8 | 0 6 — First digit cut is *4 or less*, so the part you are keeping stays the same.

0 . 0 1 8 ← Stays the same

Rounding to *thousandths* means *three* decimal places.

0.01806 rounded to the nearest thousandth is 0.018. We can write 0.01806 ≈ 0.018.

(d) 57.976 to the nearest tenth

57.9 | 76 — First digit cut is *5 or more*, so round up by adding 1 tenth to the part you are keeping.

57.9 ← Keep this part.
+ 0.1 ← To round up, add 1 tenth.
58.0 ← 9 + 1 is 10; write the 0 and regroup the 1 to the ones place.

Be sure to write the 0 in the tenths place.

57.976 rounded to the nearest tenth is 58.0. We can write 57.976 ≈ 58.0
You *must* write the 0 in the tenths place to show that the number was rounded to the nearest tenth.

CAUTION

Check that your rounded answer shows ***exactly*** the number of decimal places asked for in the problem. Be sure your answer shows *one* decimal place if you rounded to *tenths, two* decimal places for *hundredths, three* decimal places for *thousandths,* and so on.

◀ **Work Problem 2 at the Side.**

OBJECTIVE 3 Round money amounts to the nearest cent or nearest dollar. In many everyday situations, such as shopping in a store, money amounts are rounded to the nearest cent. There are 100 cents in a dollar.

$$\text{Each cent is } \frac{1}{100} \text{ of a dollar.}$$

Another way to write $\frac{1}{100}$ is 0.01. So rounding to the *nearest cent* is the same as rounding to the *nearest hundredth of a dollar.*

EXAMPLE 3 **Rounding to the Nearest Cent**

How much will you pay in each shopping situation? Round each money amount to the nearest cent.

(a) \$2.4238 (Is it closer to \$2.42 or to \$2.43?)

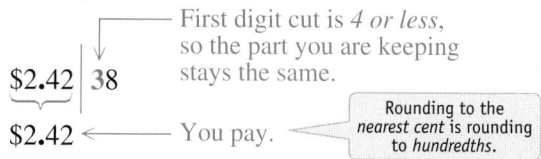

\$2.42 | 38 ← First digit cut is *4 or less*, so the part you are keeping stays the same.

\$2.42 ← You pay.

Rounding to the *nearest cent* is rounding to *hundredths*.

You pay \$2.42 because \$2.4238 is closer to \$2.42 than to \$2.43.

(b) \$0.695 (Is it closer to \$0.69 or to \$0.70?)

\$0.69 | 5 ← *5 or more;* round up.

\$0.69
\+ \$0.01 ← To round up, add 1 hundredth (1 cent).
\$0.70 ← You pay.

Work Problem 3 at the Side. ▶

Note

Some stores round *all* money amounts up to the next higher cent, even if the next digit is *4 or less*. In **Example 3(a)** above, some stores would round \$2.4238 *up* to \$2.43, even though it is closer to \$2.42.

It is also common to round money amounts to the nearest dollar. For example, you can do that on your federal and state income tax returns to make the calculations easier. Rounding to the nearest dollar is rounding to the ones place, that is, to the nearest whole number.

EXAMPLE 4 **Rounding to the Nearest Dollar**

Round to the nearest dollar.

(a) \$48.69 (Is it closer to \$48 or to \$49?)

Draw a cut-off line after the ones place. First digit cut is *5 or more*, so round up by adding \$1.

\$48 | .69

\$48
\+ 1
\$49

\$48.69 is closer to \$49 than to \$48.
So \$48.69 rounded to the nearest dollar is \$49.

Write \$49 *not* \$49.00

CAUTION

\$48.69 rounded to the nearest dollar is \$49. Write the answer as **\$49** to show that the rounding is to the *nearest dollar*. Writing \$49.00 would show rounding to the *nearest cent*.

Continued on Next Page

3 Round each money amount to the nearest cent.

GS **(a)** \$14.595

\$14.59 | 5 ← First digit cut is *5 or more*, so round up.

\$14.59
\+ 0.01 ← To round up, add 1 hundredth. (1 cent).
\$14.__ ← You pay.

(b) \$578.0663

(c) \$0.849

(d) \$0.0548

Answers

3. **(a)** \$14.60 **(b)** \$578.07 **(c)** \$0.85 **(d)** \$0.05

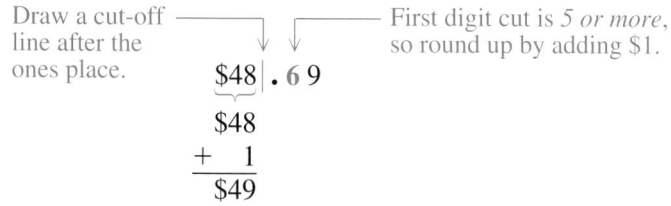

4 Round to the nearest dollar.

GS **(a)** \$29.10

First digit cut is *4 or less,* so the part you keep stays the same.

\$29 | .10

(b) \$136.49

(c) \$990.91

(d) \$5949.88

(e) \$49.60

(f) \$0.55

(g) \$1.08

Answers

4. (a) \$29 **(b)** \$136 **(c)** \$991 **(d)** \$5950 **(e)** \$50 **(f)** \$1 **(g)** \$1

(b) \$594.36 (Is it closer to \$594 or \$595?)

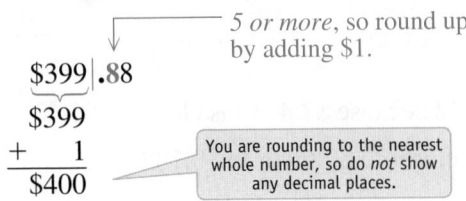

Draw a cut-off line after the ones place.

First digit cut is *4 or less*, so the part you keep stays the same.

\$594 | .36

\$594

Careful! Write \$594 (*not* \$594.00)

\$594.36 rounded to the nearest dollar is \$594.

(c) \$399.88 (Is it closer to \$399 or to \$400?)

5 or more, so round up by adding \$1.

\$399 | .88

\$399
\+ 1
\$400

You are rounding to the nearest whole number, so do *not* show any decimal places.

\$399.88 rounded to the nearest dollar is \$400.

(d) \$2689.50 (Is it closer to \$2689 or \$2690?)

5 or more, so round up by adding \$1.

\$2689 | .50

\$2689
\+ 1
\$2690

Careful! Write \$2690 (*not* \$2690.00)

\$2689.50 rounded to the nearest dollar is \$2690.

Note

When rounding \$2689.50 to the nearest dollar above, notice that it is exactly halfway between \$2689 and \$2690. When this happens in everyday situations, the rule is to round *up.* (Scientists working with technical data may use a more complicated rule when rounding numbers that are exactly in the middle.)

(e) \$0.61 (Is it closer to \$0 or to \$1?)

5 or more, so round up.

\$0 | .61

Write the rounded amount as \$1, *not* \$1.00

\$0.61 rounded to the nearest dollar is \$1.

Calculator Tip

Accountants and other people who work with money amounts often set their calculators to automatically round to two decimal places (nearest cent) or to round to zero decimal places (nearest dollar). Your calculator may have this feature.

◀ **Work Problem 4 at the Side.**

4.2 Exercises

FOR EXTRA HELP

Download the MyDashBoard App

MyMathLab®

1. CONCEPT CHECK Which digit would you look at when deciding how to round 4.8073 to the nearest tenth?
____

2. CONCEPT CHECK Which digit would you look at when deciding how to round 875.639 to the nearest hundredth?
____

3. CONCEPT CHECK Explain how to round 5.70961 to the nearest thousandth.

4. CONCEPT CHECK Explain how to round 10.028 to the nearest tenth.

Round each number to the place indicated. ***See Examples 1 and 2.***

5. 16.8 | 974 to the nearest tenth (First digit cut is *5 or more.*)

6. 193.84 | 5 to the nearest hundredth (First digit cut is *5 or more.*)

7. 0.956 | 47 to the nearest thousandth (First digit cut is *4 or less.*)

8. 96.8158 | 4 to the nearest ten-thousandth (First digit cut is *4 or less.*)

9. 0.799 to the nearest hundredth

10. 0.952 to the nearest tenth

11. 3.66062 to the nearest thousandth

12. 1.5074 to the nearest hundredth

13. 793.988 to the nearest tenth

14. 476.1196 to the nearest thousandth

15. 0.09804 to the nearest ten-thousandth

16. 176.004 to the nearest tenth

17. 9.0906 to the nearest hundredth

18. 30.1290 to the nearest thousandth

19. 82.000151 to the nearest ten-thousandth

20. 0.400594 to the nearest ten-thousandth

Nardos is grocery shopping. The store will round the amount she pays for each item to the nearest cent. Write the rounded amounts. ***See Example 3.***

21. Soup is three cans for \$2.45, so one can is \$0.81666. Nardos pays _______.

22. Orange juice is two cartons for \$3.89, so one carton is \$1.945. Nardos pays _______.

23. Facial tissue is four boxes for \$4.89, so one box is \$1.2225. Nardos pays _______.

24. Muffin mix is three packages for \$1.99, so one package is \$0.66333. Nardos pays _______.

25. Candy bars are six for \$4.19, so one bar is \$0.6983. Nardos pays _______.

26. Spaghetti is four boxes for \$4.39, so one box is \$1.0975. Nardos pays _______.

As she gets ready to do her income tax return, Ms. Chen rounds each amount to the nearest dollar. Write the rounded amounts. ***See Example 4.***

27. Income from job, \$48,649.60

28. Income from interest on bank account, \$69.58

29. Donations to charity, \$840.08

30. Federal withholding, \$6064.49

Round each money amount as indicated. ***See Examples 3 and 4.***

31. \$499.98 to the nearest dollar

32. \$9899.59 to the nearest dollar

33. \$0.996 to the nearest cent

34. \$0.09929 to the nearest cent

35. \$999.73 to the nearest dollar

36. \$9999.80 to the nearest dollar

The table lists speed records for various types of transportation. Use the table to answer Exercises 37–40.

Record	Speed (miles per hour)
Land speed record (specially built car)	763.04
Motorcycle speed record (conventional motorcycle)	252.662
Fastest roller coaster	134.8
Fastest military jet	2193.167
Boeing 757-300 airplane (regular passenger service)	509
Indianapolis 500 auto race (fastest average winning speed)	185.981
Daytona 500 auto race (fastest average winning speed)	177.602

Source: *Guinness World Records* and *World Almanac and Book of Facts.*

37. Round these speed records to the nearest whole number.

(a) Motorcycle

(b) Roller coaster

38. Round these speed records to the nearest hundredth.

(a) Daytona 500 average winning speed

(b) Indianapolis 500 average winning speed

39. Round these speed records to the nearest tenth.

(a) Indianapolis 500 average winning speed

(b) Land speed record

40. Round these speed records to the nearest hundred.

(a) military jet

(b) Boeing 757-300 airplane

Relating Concepts (Exercises 41–44) For Individual or Group Work

Use your knowledge about rounding money amounts to ***work Exercises 41–44 in order.***

41. Explain what happens when you round \$0.499 to the nearest dollar. Why does this happen?

42. Look again at **Exercise 41.** How else could you round \$0.499 that would be more helpful? What kind of guideline does this suggest about rounding to the nearest dollar?

43. Explain what happens when you round \$0.0015 to the nearest cent. Why does this happen?

44. Suppose you want to know which of these amounts is less, so you round them both to the nearest cent.

\$0.5968 \$0.6014

Explain what happens. Describe what you could do instead of rounding to the nearest cent.

4.3 Adding and Subtracting Decimal Numbers

OBJECTIVES

1. Add decimals.
2. Subtract decimals.
3. Estimate the answer when adding or subtracting decimals.

OBJECTIVE 1 Add decimals. When adding or subtracting *whole* numbers in **Chapter 1,** you lined up the numbers in columns so that you were adding ones to ones, tens to tens, and so on. A similar idea applies to adding or subtracting *decimal* numbers. With decimals, you line up the decimal points to make sure you are adding tenths to tenths, hundredths to hundredths, and so on.

Adding and Subtracting Decimals

Step 1 Write the numbers in columns with the decimal points lined up.

Step 2 If necessary, write in zeros so both numbers have the same number of decimal places. Then add or subtract as if they were whole numbers.

Step 3 Line up the decimal point in the answer directly below the decimal points in the problem.

EXAMPLE 1 Adding Decimal Numbers

Find each sum.

(a) 16.92 and 48.34

Step 1 Write the numbers in columns with the decimal points lined up.

tens	ones	.	tenths	hundredths
1	6	.	9	2
+ 4	8	.	3	4

Decimal points are lined up.

Step 2 Add as if these were whole numbers.

$$\begin{array}{r} \scriptstyle 1\,1 \\ 16.92 \\ +\,48.34 \\ \hline 65.26 \end{array}$$

Step 3 Decimal point in answer is lined up under decimal points in problem.

(b) 5.897 + 4.632 + 12.174

Write the numbers vertically with decimal points lined up. Then add.

$$\begin{array}{r} \scriptstyle 11\;21 \\ 5.897 \\ 4.632 \\ +\,12.174 \\ \hline 22.703 \end{array}$$

When you rewrite the numbers in columns, and then add, be careful to line up the decimal points.

Work Problem 1 at the Side. ▶

In **Example 1(a)** above, both numbers had *two decimal places* (two digits to the right of the decimal point). In **Example 1(b),** all the numbers had *three decimal places* (three digits to the right of the decimal point). That made it easy to add tenths to tenths, hundredths to hundredths, and so on.

---

1 Find each sum.

GS **(a)** 2.86 + 7.09

$$\begin{array}{r} 2.86 \\ +\,7.09 \\ \hline . \end{array}$$

Decimal points are lined up.

(b) 13.761 + 8.325

(c) 0.319 + 56.007 + 8.252

(d) 39.4 + 0.4 + 177.2

Answers

1. (a) 9.95 (b) 22.086 (c) 64.578 (d) 217.0

2 Find each sum.

GS **(a)** 6.54 + 9.8

$$\begin{array}{r} 6.54 \\ +\ 9.80 \\ \hline \_\_.\_\_ \end{array}$$

(b) 0.831 + 222.2 + 10

(c) 8.64 + 39.115 + 3.0076

(d) 5 + 429.823 + 0.76

If the number of decimal places does *not* match, you can write in zeros as placeholders to make them match. This is shown in **Example 2**.

EXAMPLE 2 Writing Zeros as Placeholders before Adding

Find each sum.

(a) 7.3 + 0.85

There are two decimal places in 0.85 (tenths and hundredths), so write zero in the hundredths place in 7.3 so that it has two decimal places also.

$$\begin{array}{r} 7.30 \\ +\ 0.85 \\ \hline 8.15 \end{array}$$ ← One 0 is written in.

7.30 is equivalent to 7.3 because $7\frac{30}{100}$ in lowest terms is $7\frac{3}{10}$

(b) 6.42 + 9 + 2.576

Write in zeros so that all the addends have three decimal places. Notice how the whole number 9 is written with the decimal point at the *far right* side. (If you put the decimal point on the *left* side of the 9, you would turn it into the decimal fraction 0.9.)

$$\begin{array}{r} 6.420 \\ 9.000 \\ +\ 2.576 \\ \hline 17.996 \end{array}$$

- 6.420 ← One 0 is written in.
- 9.000 ← 9 is a whole number; decimal point and three zeros are written in.
- 2.576 ← No zeros are needed.
- Decimal points are lined up.

Write the decimal point in 9 on the *right* side.

Note

Writing zeros to the right of a *decimal* number does *not* change the value of the number, as shown in **Example 2(a)** above.

◀ **Work Problem 2 at the Side.**

OBJECTIVE 2 Subtract decimals. Subtraction of decimals is done in much the same way as addition of decimals. You can check the answers to subtraction problems using addition, as you did with whole numbers.

EXAMPLE 3 Subtracting Decimal Numbers

Find each difference. Check your answers using addition.

(a) 15.82 from 28.93

Watch the order of the numbers when you see "from." Subtraction is **not** commutative like addition. So the number you are subtracting "from" goes first.

Step 1

$$\begin{array}{r} 28.93 \\ -\ 15.82 \\ \hline \end{array}$$

Line up decimal points. Then you will be subtracting hundredths from hundredths and tenths from tenths.

Continued on Next Page

Answers

2. (a) 16.34 **(b)** 233.031 **(c)** 50.7626 **(d)** 435.583

Step 2

$$\begin{array}{r} 28.93 \\ -\ 15.82 \\ \hline 13.11 \end{array}$$

Both numbers have two decimal places: no need to write in zeros.

Subtract as if they were whole numbers.

Decimal point in answer is lined up.

Step 3

Check the answer by adding 13.11 and 15.82. If the subtraction is done correctly, the sum will be 28.93.

(b) 146.35 minus 58.98

Regrouping is needed here.

$$\begin{array}{r} \overset{0}{\not 1}\,\overset{13}{\not 4}\,\overset{15}{\not 6}\,.\,\overset{12}{\not 3}\,\overset{15}{\not 5} \\ -\ \ 5\ 8\ .\ 9\ 8 \\ \hline 8\ 7\ .\ 3\ 7 \end{array}$$

Line up decimal points.

Check the answer by adding 87.37 and 58.98. If you did the subtraction correctly, the sum will be 146.35. (If it *isn't,* rework the problem.)

Work Problem 3 at the Side. ▶

EXAMPLE 4 Writing Zeros as Placeholders before Subtracting

Find each difference.

(a) 16.5 from 28.362

Use the same steps as in **Example 3** above. Remember to write in zeros so both numbers have three decimal places.

Line up decimal points.

$$\begin{array}{r} 28.362 \\ -\ 16.500 \\ \hline 11.862 \end{array}$$

16.500 is equivalent to 16.5

Write two zeros.

Subtract as usual.

Check the answer by adding.

$$\begin{array}{r} 16.500 \\ +\ 11.862 \\ \hline 28.362 \end{array}$$

Matches minuend in original problem.

(b) 59.7 − 38.914

$$\begin{array}{r} 59.700 \\ -\ 38.914 \\ \hline 20.786 \end{array}$$

Write two zeros.

Subtract as usual.

(c) 12 less 5.83

$$\begin{array}{r} 12.00 \\ -\ 5.83 \\ \hline 6.17 \end{array}$$

12.00 is equivalent to 12

Write a decimal point and two zeros.

Subtract as usual.

Work Problem 4 at the Side. ▶

OBJECTIVE 3 Estimate the answer when adding or subtracting decimals. A common error in working decimal problems by hand is to misplace the decimal point in the answer. Or, when using a calculator, you may accidentally press the wrong key. **Estimating** the answer will help you avoid these mistakes. Start by using *front end rounding* on each number (as you did with whole numbers). Here are several examples. Notice that in the rounded numbers, only the leftmost digit is something other than 0.

3.25	rounds to	3	0.812	rounds to	1
532.6	rounds to	500	26.397	rounds to	30
7094.2	rounds to	7000	351.24	rounds to	400

---

3 Find each difference. Check your answers using addition.

GS **(a)** 22.7 from 72.9

$$\begin{array}{r} 7\ 2\ .\ 9 \\ -\ 2\ 2\ .\ 7 \\ \hline \_\ \_\ .\ \_ \end{array}$$

(b) 6.425 from 11.813

(c) 20.15 − 19.67

4 Find each difference. Check your answers using addition.

GS **(a)** 18.651 from 25.3

$$\begin{array}{r} 25.300 \\ -\ 18.651 \\ \hline \end{array}$$

(b) 5.816 − 4.98

(c) 40 less 3.66

(d) 1 − 0.325

Answers

3. (a) 50.2; 50.2 + 22.7 = 72.9
(b) 5.388; 5.388 + 6.425 = 11.813
(c) 0.48; 0.48 + 19.67 = 20.15

4. (a) 6.649; 6.649 + 18.651 = 25.3
(b) 0.836; 0.836 + 4.98 = 5.816
(c) 36.34; 36.34 + 3.66 = 40
(d) 0.675; 0.675 + 0.325 = 1

5 First, use front end rounding and estimate each answer. Then add or subtract to find the exact answer.

GS **(a)** 2.83 + 5.009 + 76.1

Estimate:		*Exact:*
3	⟵	2.830
5	⟵	5.009
+ 80	⟵	76.100

(b) 11.365 from 58

Estimate: *Exact:*

(c) 398.81 + 47.658 + 4158.7

Estimate: *Exact:*

(d) Find the difference between 12.837 meters and 46.091 meters.

Estimate: *Exact:*

(e) $19.28 plus $1.53

Estimate: *Exact:*

Answers

5. **(a)** *Estimate:* 3 + 5 + 80 = 88; *Exact:* 83.939
(b) *Estimate:* 60 − 10 = 50; *Exact:* 46.635
(c) *Estimate:* 400 + 50 + 4000 = 4450; *Exact:* 4605.168
(d) *Estimate:* 50 − 10 = 40; *Exact:* 33.254 meters
(e) *Estimate:* $20 + $2 = $22; *Exact:* $20.81

EXAMPLE 5 Estimating Decimal Answers

Use front end rounding to round each number. Then add or subtract the rounded numbers to get an estimated answer. Finally, find the exact answer.

(a) Find the sum of 194.2 and 6.825.

Estimate:		*Exact:*
200	⟵ Rounds to	194.200
+ 7	⟵ Rounds to	+ 6.825
207		201.025

The estimate goes out to the hundreds place (three places to the *left* of the decimal point), and so does the exact answer. Therefore, the decimal point is probably in the correct place in the exact answer.

(b) $69.42 + $13.78

Estimate:		*Exact:*	
$70	⟵ Rounds to	$69.42	
+ 10	⟵ Rounds to	+ 13.78	
$80		$83.20	⟵ Exact answer is close to estimate, so it is reasonable.

(c) Find the difference between 0.92 ft and 8 ft.

Use subtraction to find the difference between two numbers. The larger number, 8, is written on top.

Estimate:		*Exact:*	
8	⟵ Rounds to	8.00	⟵ Write a decimal point and two zeros.
− 1	⟵ Rounds to	− 0.92	
7		7.08 ft	

0.92 has a digit in the ones place, so round to the ones place (nearest whole number).

(d) Subtract 1.8614 from 7.3.

Estimate:		*Exact:*	
7	⟵ Rounds to	7.3000	⟵ Write three zeros.
− 2	⟵ Rounds to	− 1.8614	
5		5.4386	⟵ Exact answer is close to estimate.

◀ **Work Problem 5 at the Side.**

Calculator Tip

If you are *adding* numbers, you can enter them in any order on your calculator. Try these; jot down the answers.

9.82 (+) 1.86 (=) ________ 1.86 (+) 9.82 (=) ________

The answers are the same because addition is *commutative.* (See **Chapter 1.**) But subtraction is *not* commutative. It *does* matter which number you enter first. Try these:

9.82 (−) 1.86 (=) ________ 1.86 (−) 9.82 (=) ________

The second answer has a negative sign (−) next to it. A negative number is *less* than 0. If it was shown on your bank statement, you'd be "in the hole" by $7.96. (See **Section 9.3** for more about negative numbers.)

4.3 Exercises

FOR EXTRA HELP

Download the MyDashBoard App

MyMathLab®

CONCEPT CHECK *Rewrite each addition or subtraction in columns. Write in any decimal points or zeros, as needed. Do not complete the calculation.*

1. 6.42 + 10.163

2. 7 + 9.204

3. 20 − 9.1263

4. 137.06 − 12

Find each sum or difference. ***See Examples 1–4.***

5. 5.69 + 11.79

6. 0.7759 + 9.8883

7. 8.263 − 0.5

8. 47.658 − 20.9

9. 76.5 + 0.506

10. 1.87 + 9.749

11. 21 − 0.896

12. 9 − 1.183

13. Subtract 0.291 from 0.4.

14. Subtract 0.088 from 0.35.

15. 39.76005 + 182 + 4.799 + 98.31 + 5.9999

16. 489.76 + 0.9993 + 38 + 8.55087 + 80.697

This drawing of a human skeleton shows the average length of the longest bones, in inches. Use the drawing to answer Exercises 17–20. (Source: The Human Body.)

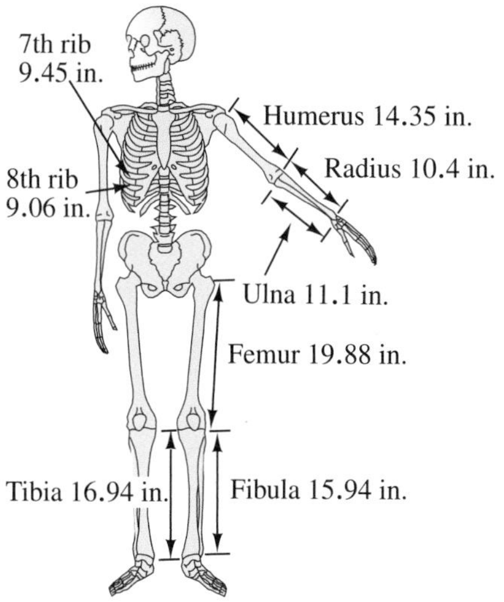

17. (a) What is the combined length of the humerus and radius bones?

(b) What is the difference in the lengths of these two bones?

18. (a) What is the total length of the femur and tibia bones?

(b) How much longer is the femur than the tibia?

19. (a) Find the sum of the lengths of the humerus, ulna, femur, and tibia.

(b) How much shorter is the 8th rib than the 7th rib?

20. (a) What is the difference in the lengths of the two bones in the lower arm?

(b) What is the difference in the lengths of the two bones in the lower leg?

21. CONCEPT CHECK Explain and correct the error that a student made when he added 0.72 + 6 + 39.5 this way.

$$\begin{array}{r} 0.72 \\ 6 \\ +\ 39.50 \\ \hline 40.28 \end{array}$$

22. CONCEPT CHECK Explain the difference between saying "subtract 2.9 from 8" and saying "2.9 minus 8."

Use front end rounding to round each number. Then add or subtract the rounded numbers to get an estimated answer. Finally, find the exact answer. ***See Example 5.***

23. The *Estimate* is already done. Now find the *Exact* answer.

Estimate: $\begin{array}{r} \$20 \\ -\ 7 \\ \hline \$13 \end{array}$ ← *Exact:* $\begin{array}{r} \$19.74 \\ -\ 6.58 \\ \hline \end{array}$

24. The *Estimate* is already done. Now find the *Exact* answer.

Estimate: $\begin{array}{r} \$30 \\ -\ 8 \\ \hline \$22 \end{array}$ ← *Exact:* $\begin{array}{r} \$27.96 \\ -\ 8.39 \\ \hline \end{array}$

25. *Estimate:* $\begin{array}{r} \\ \\ + \underline{\quad\quad} \end{array}$ *Exact:* $\begin{array}{r} 392.7 \\ 0.865 \\ +\ 21.08 \\ \hline \end{array}$

26. *Estimate:* $\begin{array}{r} \\ \\ + \underline{\quad\quad} \end{array}$ *Exact:* $\begin{array}{r} 38.55 \\ 7.716 \\ +\ 0.6 \\ \hline \end{array}$

27. What is 8.6 less 3.751?

Estimate: *Exact:*

28. What is 31.7 less 4.271?

Estimate: *Exact:*

29. *Estimate:* $\begin{array}{r} \\ \\ + \underline{\quad\quad} \end{array}$ *Exact:* $\begin{array}{r} 62.8173 \\ 539.99 \\ +\ 5.629 \\ \hline \end{array}$

30. *Estimate:* $\begin{array}{r} \\ \\ + \underline{\quad\quad} \end{array}$ *Exact:* $\begin{array}{r} 332.607 \\ 12.5 \\ +\ 823.3949 \\ \hline \end{array}$

Use your estimation skills to pick the most reasonable answer for each example. Do ***not*** *solve the problems. Circle your choice.*

31. 12 − 11.725

2.75 0.275 27.5

32. 20 − 1.37

0.1863 1.863 18. 63

33. 6.5 + 0.007

6.507 0.6507 65.07

34. 9.67 + 0.09

0.976 9.76 0.00976

35. 456.71 − 454.9

18.1 181 1.81

36. 803.25 − 0.6

802.65 0.80265 8.0265

37. 6004.003 + 52.7172

60.567202 605.67202 6056.7202

38. 128.35 + 97.0093

2253.593 225.3593 0.2253593

First use front end rounding to round each number and estimate the answer. Then find the exact answer. Use the information in the table below for Exercises 39–42.

INTERNET USERS IN SELECTED COUNTRIES

Country	Number of Users
China	420 million
United States	239.9 million
Japan	99.14 million
India	81 million
Nigeria	43.98 million
Mexico	30.6 million
Canada	26.2 million
World total	1966.5 million

***Source*:** www.internetworldstats.com

39. How many fewer Internet users are there in Mexico than in India?

Estimate:

Exact:

40. How many more users are there in China than in Nigeria?

Estimate:

Exact:

41. How many Internet users are there in all the countries listed in the table?

Estimate:

Exact:

42. Using the exact answer from **Exercise 41,** calculate the number of worldwide Internet users in countries other than the ones in the table.

Estimate:

Exact:

43. The tallest known land mammal, a prehistoric ancestor of the rhino, was 6.4 meters tall. Compare the rhino's height to the combined heights of these three NBA basketball players: Dirk Nowitzki at 2.13 meters, Kobe Bryant at 1.98 meters, and Kevin Love at 2.08 meters. Is their combined height greater or less than the rhino's height? By how much?
(*Source:* www.NBA.com/players)

Estimate:

Exact:

44. At a bakery, Sue Chee bought \$7.42 worth of muffins and \$10.09 worth of croissants for a staff party and a \$0.69 cookie for herself. How much change did she receive from two \$10 bills?

Estimate:

Exact:

45. Namiko is comparing two boxes of chicken nuggets. One box weighs 9.85 ounces and the other weighs 10.5 ounces. What is the difference in the weight of the two boxes?

Estimate:

Exact:

46. Sammy works in a veterinarian's office. He weighed two young kittens. One was 3.9 ounces and the other was 4.05 ounces. What was the difference in the weight of the two kittens?

Estimate:

Exact:

Find the perimeter of (distance around) each figure by adding the lengths of the sides.

47.

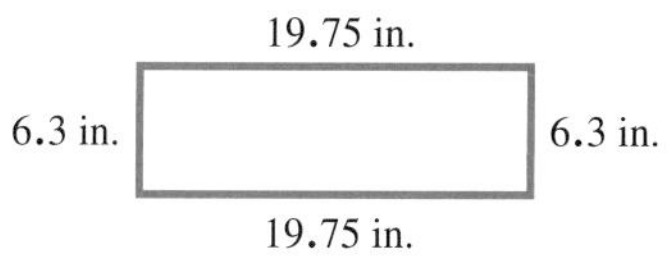

Estimate:

Exact:

48.

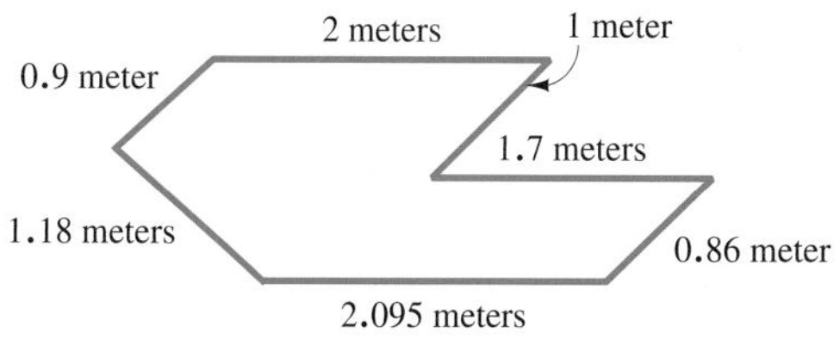

Estimate:

Exact:

The grandfather and grandson on the first page of this chapter are buying fishing equipment. They brought along the store's sale insert from the Sunday paper. Use the information below on sale prices to answer Exercises 49–52. When estimating, round to the nearest whole number.

Source: Walmart.

49. What is the difference in price between the fluorescent and regular fishing line?

Estimate:

Exact:

50. How much more does the least expensive spinning rod cost than the least expensive spinning reel?

Estimate:

Exact:

51. Find the total cost of the second highest priced spinning reel, two packages of tin split shot, and a three-tray tackle box. Sales tax for all the items was $2.31.

Estimate:

Exact:

52. The grandfather bought three bobbers on sale. He also bought some SPF15 sunscreen for $7.53 and a flotation vest for $44.96. Sales tax was $3.74. How much did he spend in all?

Estimate:

Exact:

Olivia Sanchez kept track of her expenses for one month. Use her list to answer Exercises 53–58.

MONTHLY EXPENSES

Rent	$994
Car payment	$290.78
Car repairs, gas	$205
Cable TV	$49.95
Internet access	$29.95
Electricity	$40.80
Cell phone	$57.32
Groceries	$186.81
Entertainment	$97.75
Clothing, laundry	$107

53. What were Olivia's total expenses for the month?

54. How much did Olivia pay for cell phone, cable TV, and Internet access?

55. What was the difference in the amounts spent for groceries and for the car payment?

56. Compare the amount Olivia spent on entertainment to the amount spent on car repairs and gas. What is the difference?

57. How much more did Olivia spend on rent than on all her car expenses?

58. How much less did Olivia spend on clothing and laundry than on all her car expenses?

Find the length of the dashed line in each rectangle or circle.

59.

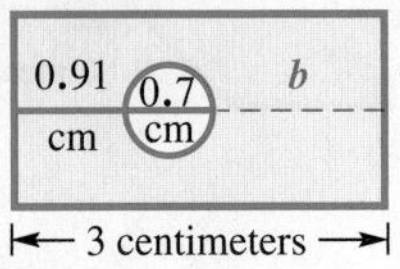

60.

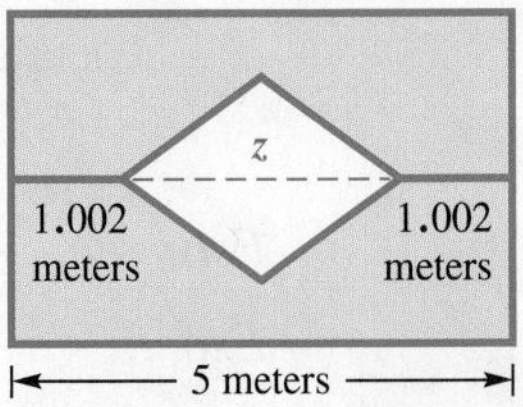

61.

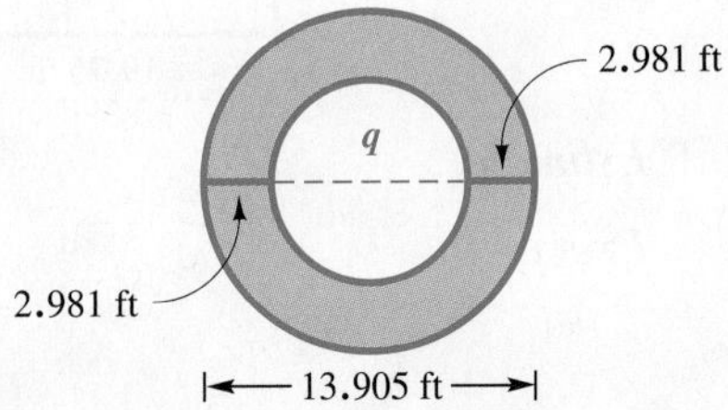

4.4 Multiplying Decimal Numbers

OBJECTIVE 1 Multiply decimals. The decimals 0.3 and 0.07 can be multiplied by writing them as fractions.

$$\underset{\text{1 decimal place}}{0.\underline{3}} \times \underset{\text{2 decimal places}}{0.\underline{0}\,\underline{7}} = \frac{3}{10} \times \frac{7}{100} = \frac{3 \times 7}{10 \times 100} = \frac{21}{1000} = \underset{\text{3 decimal places}}{0.\underline{0}\,\underline{2}\,\underline{1}}$$

1 decimal place + 2 decimal places → 3 decimal places

Can you see a way to multiply decimals without writing them as fractions? Use these steps. Remember that each number in a multiplication problem is called a *factor*, and the answer is called the *product*.

Multiplying Decimal Numbers

Step 1 Multiply the numbers (the factors) as if they were whole numbers.

Step 2 Find the *total* number of decimal places in *both* factors.

Step 3 Write the decimal point in the product (the answer) so it has the same number of decimal places as the total from Step 2. You may need to write in extra zeros on the left side of the product to get the correct number of decimal places.

Note

When multiplying decimals, you do ***not*** need to line up decimal points. (You ***do*** need to line up decimal points when adding or subtracting.)

EXAMPLE 1 Multiplying Decimal Numbers

Find the product of 8.34 and 4.2.

Step 1 Multiply the numbers as if they were whole numbers.

```
   8.3 4
×    4.2
 1 6 6 8
3 3 3 6
3 5 0 2 8
```

You do *not* have to line up decimal points when multiplying.

Step 2 Count the total number of decimal places in both factors.

```
   8.3 4   ← 2 decimal places
×    4.2   ← 1 decimal place
             3 total decimal places
 1 6 6 8
3 3 3 6
3 5 0 2 8
```

Step 3 Count over 3 places in the product and write the decimal point.

```
   8.3 4   ← 2 decimal places
×    4.2   ← 1 decimal place
             3 total decimal places
 1 6 6 8
3 3 3 6
3 5.0 2 8  ← 3 decimal places in product
```

Count over 3 places; count from *right* to *left*.

Work Problem 1 at the Side. ▶

OBJECTIVES

1. **Multiply decimals.**
2. **Estimate the answer when multiplying decimals.**

1 Find each product.

GS **(a)**

```
  2.6  ← 1 decimal place
× 0.4  ← 1 decimal place
       ← 2 decimal places in the product
```

(b)

```
  45.2
× 0.25
```

GS **(c)**

```
  0.104  ← 3 decimal places
×     7  ← 0 decimal places
         ← 3 decimal places in the product
```

(d)

```
  3.18
× 2.23
```

(e)

```
  611
× 3.7
```

Answers

1. **(a)** 1.04 **(b)** 11.300 **(c)** 0.728 **(d)** 7.0914 **(e)** 2260.7

2 Find each product.

(a) 0.04×0.09

(b) $(0.2)(0.008)$

(c) $(0.003)^2$ *Hint:* Recall that the 2 is an exponent, so multiply $(0.003)(0.003)$.

(d) $(0.0081)(0.003)$

(e) $(0.11)(0.0005)$

3 First use front end rounding and estimate the answer. Then find the exact answer.

(a) $(11.62)(4.01)$

(b) $(5.986)(33)$

(c) $8.31(4.2)$

(d) 58.6×17.4

Answers

2. **(a)** 0.0036 **(b)** 0.0016 **(c)** 0.000009 **(d)** 0.0000243 **(e)** 0.000055

3. **(a)** $(10)(4) = 40$; 46.5962 **(b)** $(6)(30) = 180$; 197.538 **(c)** $8(4) = 32$; 34.902 **(d)** $60 \times 20 = 1200$; 1019.64

EXAMPLE 2 Writing Zeros as Placeholders in the Product

Find the product: $(0.042)(0.03)$.

Start by multiplying, then count decimal places.

$$\begin{array}{rl} 0.042 & \leftarrow \text{3 decimal places} \\ \times\ 0.03 & \leftarrow \text{2 decimal places} \\ \hline 126 & \leftarrow \text{5 decimal places needed in product} \end{array}$$

After multiplying, the answer has only three decimal places, but five are needed, so write two zeros on the *left* side of the answer.

$$\begin{array}{r} 0.042 \\ \times\ 0.03 \\ \hline 00126 \end{array} \qquad \begin{array}{rl} 0.042 & \leftarrow \text{3 decimal places} \\ \times\ 0.03 & \leftarrow \text{2 decimal places} \\ \hline .00126 & \leftarrow \text{5 decimal places} \end{array}$$

Write two zeros on *left* side of answer.

Now count over 5 places and write in the decimal point.

The final product is 0.00126, which has five decimal places.

◀ Work Problem **2** at the Side.

OBJECTIVE 2 Estimate the answer when multiplying decimals. If you are doing multiplication problems by hand, estimating the answer helps you check that the decimal point is in the right place. When you are using a calculator, estimating helps you catch an error like pressing the ÷ key instead of the × key.

EXAMPLE 3 Estimating before Multiplying

First estimate the answer to $(76.34)(12.5)$ using front end rounding. Then find the exact answer.

Estimate:

$$\begin{array}{rl} 80 & \leftarrow \text{Rounds to } 76.34 \\ \times\ 10 & \leftarrow \text{Rounds to } 12.5 \\ \hline 800 & \end{array}$$

Exact:

$$\begin{array}{rl} 76.34 & \leftarrow \text{2 decimal places} \\ \times\ 12.5 & \leftarrow \text{1 decimal place} \\ \hline 38170 & \text{3 decimal places needed in product} \\ 15268 & \\ 7634 & \\ \hline 954.250 & \end{array}$$

Both the estimate and the exact answer go out to the hundreds place, so the decimal point in 954.250 is probably in the correct place.

◀ Work Problem **3** at the Side.

Calculator Tip

When working with money amounts, you may need to write a 0 in your answer. For example, try multiplying $\$3.54 \times 5$ on your calculator. Write down the result.

3.54 × 5 = ________

Notice that the result is 17.7, which is *not* the way to write a money amount. You have to write the 0 in the hundredths place: \$17.70 is correct. The calculator does not show the "extra" 0.

$$17.70 \text{ or } 17\frac{70}{100} \quad \text{simplifies to} \quad 17\frac{7}{10} \text{ or } 17.7$$

So keep an eye on your calculator—it doesn't know when you're working with money amounts.

4.4 Exercises

FOR EXTRA HELP

CONCEPT CHECK *Fill in the blanks for Exercises 1–4.*

1. In 4.2×3.46, how many decimal places will the product have? ______

2. In 8.071×2.79, how many decimal places will the product have? ______

3. In $(0.12)(0.03)$, how many zeros do you have to write in the product as placeholders? ______

4. In $(0.0006)(0.07)$, how many zeros do you have to write in the product as placeholders? ______

Find each product. ***See Example 1.***

5. 0.042×3.2

6. 0.571×2.9

7. 21.5×7.4

8. 85.4×3.5

9. $(0.666)(23.4)$

10. $(0.799)(0.896)$

11. $\$51.88 \times 665$

12. $\$736.75 \times 118$

CONCEPT CHECK *Use the fact that $72 \times 6 = 432$ to solve Exercises 13–18 by simply counting decimal places and writing the decimal point in the correct location.*

13. $72 \times 0.6 =$ 4 3 2

14. $7.2 \times 6 =$ 4 3 2

15. $(7.2)(0.06) =$ 4 3 2

16. $(0.72)(0.6) =$ 4 3 2

17. $0.72(0.06) =$ 4 3 2

18. $72(0.0006) =$ 4 3 2

Find each product. ***See Example 2.***

19. $(0.006)(0.0052)$

20. $(0.0052)(0.009)$

21. $(0.005)^2$

22. $(0.03)^2$

First use front end rounding to round each number and estimate the answer. Then find the exact answer. ***See Example 3.***

23. *Estimate:* $40 \times 5 = 200$ (39.6 Rounds to 40; 4.8 Rounds to 5) *Exact:* 39.6×4.8

24. *Estimate:* $20 \times 2 = 40$ (18.7 Rounds to 20; 2.3 Rounds to 2) *Exact:* 18.7×2.3

25. *Estimate:* ___ × ___ *Exact:* 37.1×42

26. *Estimate:* ___ × ___ *Exact:* 5.08×71

27. *Estimate:* ___ × ___ *Exact:* 6.53×4.6

28. *Estimate:* ___ × ___ *Exact:* 7.51×8.2

29. *Estimate:* (___)(___) = ___ *Exact:* $(2.809)(6.85) =$

30. *Estimate:* (___)(___) = ___ *Exact:* $(73.52)(22.34) =$

Even with most of the problem missing, you can tell whether or not these answers are reasonable. Circle reasonable *or* unreasonable. *If the answer is unreasonable, move the decimal point, or insert a decimal point, to make the answer reasonable.*

31. How much was his car payment? $28.90
reasonable
unreasonable, should be ___________

32. How many hours did she work today? 25 hours
reasonable
unreasonable, should be ___________

33. How tall is her son? 60.5 in.
reasonable
unreasonable, should be ___________

34. How much does he pay for rent now? $6.92
reasonable
unreasonable, should be ___________

35. What is the price of one gallon of milk? $419
reasonable
unreasonable, should be ___________

36. How long is the living room? 16.8 feet
reasonable
unreasonable, should be ___________

37. How much did Mrs. Brown's baby weigh? 0.095 pound
reasonable
unreasonable, should be ___________

38. What was the sale price of the jacket? $1.49
reasonable
unreasonable, should be ___________

Solve each application problem. Round money answers to the nearest cent when necessary.

39. LaTasha worked 50.5 hours over the last two weeks. She earns $18.73 per hour. How much did she make?

40. Michael's time card shows 42.2 hours at $10.03 per hour. What are his earnings?

41. Sid needs 0.6 meter of canvas material to make a carry-all bag that fits on his wheelchair. If canvas is $4.09 per meter, how much will Sid spend? (*Note:* $4.09 *per* meter means $4.09 for *one* meter.)

42. How much will Mrs. Nguyen pay for 3.5 yards of lace trim that costs $0.87 per yard?

43. Michelle filled the tank of her pickup truck with regular unleaded gas. Use the information shown on the pump to find how much she paid for gas.

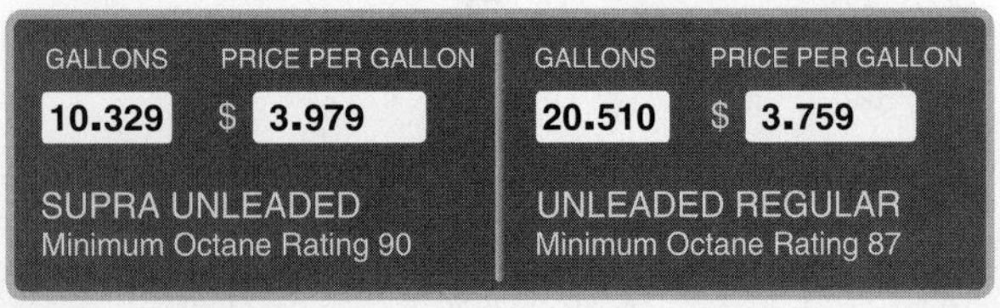

Source: Holiday.

44. Ground beef and chicken legs are on sale. Juma bought 1.7 pounds of legs. Use the information in the ad to find the amount she paid.

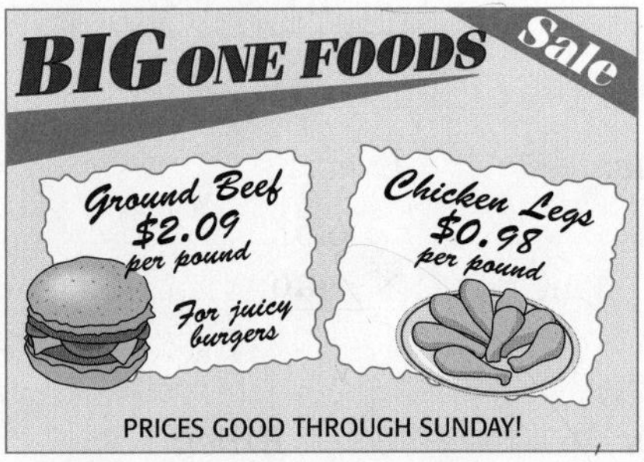

45. Ms. Rolack is a real estate broker who helps people sell their homes. Her fee is 0.07 times the price of the home. What was her fee for selling a $289,500 home?

46. Josh Hamilton of the Texas Rangers baseball team had a batting average of 0.359 in the 2010 season. He went to bat 518 times. How many hits did he make? (*Hint:* Multiply the number of times at bat by his batting average.) Round to the nearest whole number. (*Source: World Almanac and Book of Facts.*)

Paper money in the United States has not always been the same size. Shown below are the measurements of bills printed before 1929 and the measurements from 1929 on. Use this information to answer Exercises 47–50. (*Source:* www.moneyfactory.com)

Before 1929

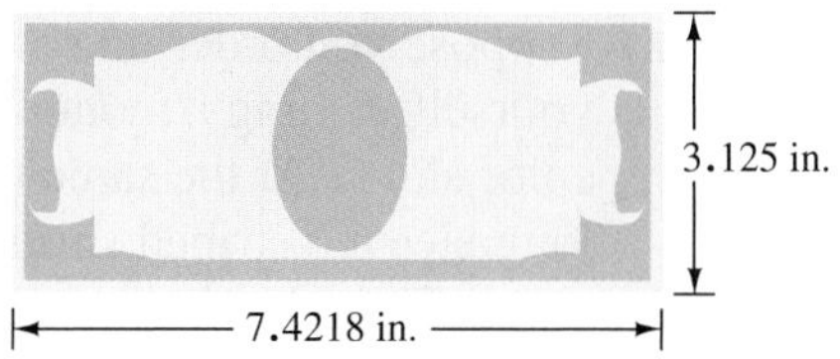

From 1929 on

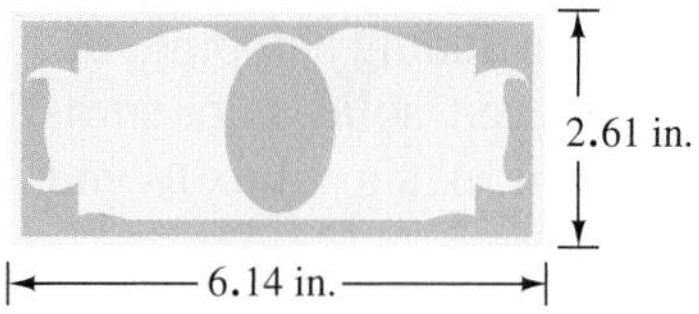

47. **(a)** Find the area of each bill, rounded to the nearest tenth. (*Hint:* Multiply to find area.)

(b) What is the difference in the rounded areas?

48. **(a)** Find the perimeter of each bill, to the nearest hundredth. (*Hint:* Add to find perimeter.)

(b) How much less is the perimeter of today's bills than the bills printed before 1929?

49. The thickness of one piece of today's paper money is 0.0043 inch.

(a) If you had a pile of 100 bills, how high would the pile be?

(b) How high would a pile of 1000 bills be?

50. **(a)** Use your answers from **Exercise 49** to find the number of bills in a pile that is 43 inches high.

(b) How much money would you have if the pile is all $20 bills?

51. Judy Lewis pays $48.96 per month for basic cable TV. The one-time installation fee was $89. How much will she pay for cable over two years? How much would she pay in two years for the deluxe cable package that costs $109.78 per month?

52. Chuck's car payment is $420.27 per month for four years. He also made a down payment of $5000 at the time he bought the car. How much will he pay altogether?

53. Barry bought 16.5 meters of rope at $0.47 per meter and three meters of wire at $1.05 per meter. How much change did he get from three $5 bills?

54. Susan bought a 46-inch LED Smart HDTV that cost $1249.97. She paid $66.59 per month for 24 months. How much could she have saved by paying for the HDTV when she bought it?

Use the information from the Look Smart online catalog to answer Exercises 55–56. Disregard sales tax.

Knit Shirt Ordering Information		
43-2A	Short-sleeved, solid colors	\$14.75 each
43-2B	Short-sleeved, stripes	\$16.75 each
43-3A	Long-sleeved, solid colors	\$18.95 each
43-3B	Long-sleeved, stripes	\$21.95 each
XXL size, add \$2 per shirt.		
Monogram, \$4.95 each. Gift box, \$5 each.		

Total Price of All Items (excluding monograms and gift boxes)	Shipping, Packing, and Handling
\$0–25.00	\$5.50
\$25.01–75.00	\$7.95
\$75.01–125.00	\$9.95
\$125.01 +	\$11.95
Shipping to each additional address, add \$4.25.	

55. **(a)** What is the total cost, including shipping, of sending three short-sleeved, solid-color shirts, size M, with monograms, in a gift box to your aunt for her birthday?

(b) How much did the monograms, gift box, and shipping add to the cost of your gift?

56. **(a)** Suppose you order one of each type of shirt for yourself, adding a monogram to each of the solid-color shirts. At the same time, you order three long-sleeved striped shirts, in the XXL size, shipped to your dad in a gift box. Find the total cost of your order.

(b) What is the difference in total cost (excluding shipping) between the shirts for yourself and the gift for your dad?

Relating Concepts (Exercises 57–58) For Individual or Group Work

Look for patterns in the multiplications as you ***work Exercises 57 and 58 in order.***

57. Do these multiplications:

$(5.96)(10) =$ ______ $(3.2)(10) =$ ______

$(0.476)(10) =$ ______ $(80.35)(10) =$ ______

$(722.6)(10) =$ ______ $(0.9)(10) =$ ______

What pattern do you see? Write a "rule" for multiplying by 10. What do you think the rule is for multiplying by 100? by 1000? Write the rules and try them out on the numbers above.

58. Do these multiplications:

$(59.6)(0.1) =$ ______ $(3.2)(0.1) =$ ______

$(0.476)(0.1) =$ ______ $(80.35)(0.1) =$ ______

$(65)(0.1) =$ ______ $(523)(0.1) =$ ______

What pattern do you see? Write a "rule" for multiplying by 0.1. What do you think the rule is for multiplying by 0.01? by 0.001? Write the rules and try them out on the numbers above.

4.5 Dividing Decimal Numbers

There are two kinds of decimal division problems: those in which a decimal is divided by a whole number, and those in which a number is divided by a decimal. First recall the parts of a division problem from **Chapter 1.**

$$\begin{array}{r} 8 \leftarrow \text{Quotient} \\ \text{Divisor} \rightarrow 4\overline{)33} \leftarrow \text{Dividend} \\ \underline{32} \\ 1 \leftarrow \text{Remainder} \end{array}$$

OBJECTIVE 1 Divide a decimal by a whole number. When the divisor is a whole number, use these steps.

Dividing Decimals by Whole Numbers

Step 1 Write the decimal point in the quotient (answer) directly above the decimal point in the dividend.

Step 2 Divide as if both numbers were whole numbers.

EXAMPLE 1 Dividing Decimals by Whole Numbers

Find each quotient. Check the quotients by multiplying.

(a) 21.93 by 3

Dividend: 21.93; Divisor: 3

Rewrite the division problem. $3\overline{)21.93}$

Step 1 Write the decimal point in the quotient directly above the decimal point in the dividend. $3\overline{)21.93}$ (Decimal points lined up)

Step 2 Divide as if the numbers were whole numbers.

$$\begin{array}{r} 7.31 \\ 3\overline{)21.93} \end{array}$$

Check by multiplying the quotient times the divisor.

CHECK

$$\begin{array}{r} 7.31 \\ \times\ \ 3 \\ \hline 21.93 \end{array}$$

Matches, so 7.31 is correct.

The quotient (answer) is 7.31.

(b) $9\overline{)470.7}$

Divisor: 9; Dividend: 470.7

Write the decimal point in the quotient above the decimal point in the dividend. Then divide as if the numbers were whole numbers.

Decimal points lined up

$$\begin{array}{r} 52.3 \\ 9\overline{)470.7} \\ \underline{45} \\ 20 \\ \underline{18} \\ 2\,7 \\ \underline{2\,7} \\ 0 \end{array}$$

Matches

CHECK

$$\begin{array}{r} 52.3 \\ \times\ \ 9 \\ \hline 470.7 \end{array}$$

Multiply the quotient by the divisor. The result should match the dividend.

The quotient is 52.3.

Work Problem 1 at the Side. ▶

OBJECTIVES

1. Divide a decimal by a whole number.
2. Divide a number by a decimal.
3. Estimate the answer when dividing decimals.
4. Use the order of operations with decimals.

1 Find each quotient. Check the quotients by multiplying.

GS **(a)**

$$\begin{array}{r} 2\,3\,.\,\_ \\ 4\overline{)9\,3\,.\,6} \\ \underline{8} \\ 1\,3 \end{array}$$

Finish the division. Then multiply the quotient by 4. You should get 93.6 if the quotient is correct.

(b) $6\overline{)6.804}$

(c) $11\overline{)278.3}$

(d) $0.51835 \div 5$

(e) $213.45 \div 15$

Answers

1. (a) 23.4; (23.4)(4) = 93.6
(b) 1.134; (1.134)(6) = 6.804
(c) 25.3; (25.3)(11) = 278.3
(d) 0.10367; (0.10367)(5) = 0.51835
(e) 14.23; (14.23)(15) = 213.45

2 Divide. Check each quotient by multiplying.

GS **(a)** $5\overline{)6.4} \rightarrow 5\overline{)6.40}$

Finish the division.

(b) 30.87 ÷ 14

(c) $\frac{259.5}{30}$

(d) 0.3 ÷ 8

Answers

2. **(a)** 1.28; (1.28)(5) = 6.40 or 6.4
(b) 2.205; (2.205)(14) = 30.870 or 30.87
(c) 8.65; (8.65)(30) = 259.50 or 259.5
(d) 0.0375; (0.0375)(8) = 0.3000 or 0.3

EXAMPLE 2 Writing Extra Zeros to Complete a Division

Divide 1.5 by 8. Check the quotient by multiplying.

Keep dividing until the remainder is 0, or until the digits in the quotient begin to repeat in a pattern. In **Example 1(b),** you ended up with a remainder of 0. But sometimes you run out of digits in the dividend before that happens. If so, write extra zeros on the right side of the dividend so you can continue dividing.

$$\begin{array}{r} 0.1 \\ 8\overline{)1.5} \\ \underline{8} \\ 7 \end{array}$$

1.5 ← All digits have been used.

7 ← Remainder is not yet 0.

Write a 0 after the 5 in the dividend so you can continue dividing. Keep writing more zeros in the dividend, if needed. Recall that writing zeros to the *right* of a decimal number does ***not*** change its value.

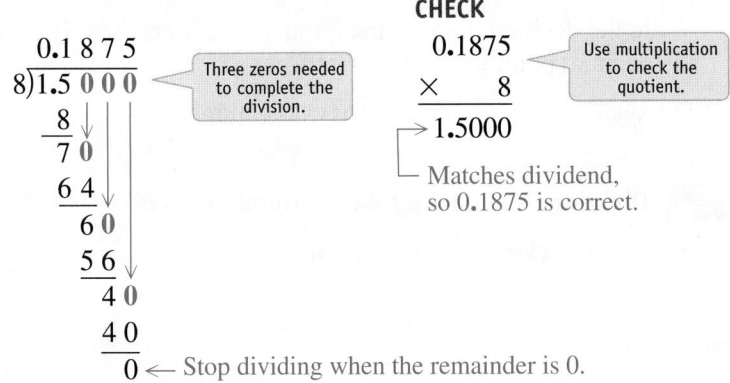

$$\begin{array}{r} 0.1875 \\ 8\overline{)1.5000} \\ \underline{8} \\ 70 \\ \underline{64} \\ 60 \\ \underline{56} \\ 40 \\ \underline{40} \\ 0 \end{array}$$

Three zeros needed to complete the division.

0 ← Stop dividing when the remainder is 0.

CHECK

$$\begin{array}{r} 0.1875 \\ \times8 \\ \hline 1.5000 \end{array}$$

Use multiplication to check the quotient.

Matches dividend, so 0.1875 is correct.

CAUTION

When dividing decimals, notice that the dividend might *not* be the greater number. In **Example 2** above, the dividend is 1.5, which is *less* than the divisor 8.

◀ **Work Problem 2 at the Side.**

Calculator Tip

In **Chapter 1,** you learned that when *multiplying* numbers, you can enter them in any order because multiplication is commutative. But division is *not* commutative. It *does* matter which number you enter first. Try **Example 2** both ways and jot down your answers.

1.5 (÷) 8 (=) ________   8 (÷) 1.5 (=) ________

Notice that the first answer, 0.1875, matches the result from **Example 2.** But the second answer is much different: 5.333333333. Be careful to enter the dividend first.

The next example shows a quotient (answer) that must be rounded because you will never get a remainder of 0.

EXAMPLE 3 **Rounding a Decimal Quotient**

Divide 4.7 by 3. Round the quotient to the nearest thousandth. Write extra zeros in the dividend so you can continue dividing.

$$\begin{array}{r l} 1.5\,6\,6\,6 & \\ 3\overline{)4.7\,0\,0\,0} & \leftarrow \text{Three zeros added so far} \\ \underline{3} & \\ 1\,7 & \\ \underline{1\,5} & \\ 2\,0 & \\ \underline{1\,8} & \\ 2\,0 & \\ \underline{1\,8} & \\ 2\,0 & \\ \underline{1\,8} & \\ 2 & \leftarrow \text{Remainder is still not 0.} \end{array}$$

Notice that the digit 6 in the answer is repeating. It will continue to do so. The remainder will *never be 0*. There are two ways to show that the answer is a **repeating decimal** that goes on forever. You can write three dots after the answer, or you can write a bar above the digits that repeat (in this case, the 6).

$$\underbrace{1.5666\ldots}_{\text{Three dots}} \quad \text{or} \quad 1.5\overline{6} \leftarrow \text{Bar above repeating digit}$$

When repeating decimals occur, round the quotient according to the directions in the problem. In this example, to round to thousandths, divide out one *more* place, to ten-thousandths.

$$4.7 \div 3 = 1.5666\ldots \quad \text{rounds to} \quad 1.567.$$

Nearest thousandth is *three* decimal places.

Check the answer by multiplying 1.567 by 3. Because 1.567 is a rounded answer, the check will not give exactly 4.7, but it should be very close.

$$(1.567)(3) = 4.701 \leftarrow \text{Does not equal exactly 4.7 because 1.567 was rounded.}$$

CAUTION

When checking quotients that you've rounded, the check will *not* match the dividend exactly, but it should be very close.

Work Problem 3 at the Side. ▶

OBJECTIVE 2 **Divide a number by a decimal.** To divide by a *decimal* divisor, first change the divisor to a whole number. Then divide as before. To see how this is done, write the problem in fraction form. Here is an example.

$$1.2\overline{)6.36} \quad \text{can be written} \quad \frac{6.36}{1.2}.$$

In **Chapter 3** you learned that multiplying the numerator and denominator by the same number gives an equivalent fraction. We want the divisor (1.2) to be a whole number. Multiplying by 10 will accomplish that.

$$\text{Decimal divisor} \rightarrow \frac{6.36}{1.2} = \frac{(6.36)(10)}{(1.2)(10)} = \frac{63.6}{12} \leftarrow \text{Whole number divisor}$$

VOCABULARY TIP

Repeating decimals To show a repeating decimal, use three dots ***or*** write a bar above the digits that repeat. Don't use both! The bar and three dots mean the same thing.

3 Divide. Round quotients to the nearest thousandth. If it is a repeating decimal, also write the answer using a bar. Check your quotients by multiplying.

GS **(a)** $13\overline{)2\,6\,7\,.\,0\,1\,0}$

Use your calculator.

$267.01 \div 13 \approx$ __________

Is the quotient a repeating decimal? ______

So round your answer to the nearest thousandth.

____

(b) $6\overline{)20.5}$

(c) $\dfrac{10.22}{9}$

(d) $16.15 \div 3$

(e) $116.3 \div 11$

Answers

3. **(a)** 20.53923; no repeating digits visible on calculator: 20.539 (rounded); $(20.539)(13) = 267.007$

(b) 3.417 (rounded); $3.41\overline{6}$; $(3.417)(6) = 20.502$

(c) 1.136 (rounded); $1.13\overline{5}$; $(1.136)(9) = 10.224$

(d) 5.383 (rounded); $5.38\overline{3}$; $(5.383)(3) = 16.149$

(e) 10.573 (rounded); $10.5\overline{72}$; $(10.573)(11) = 116.303$

4 Divide. If the quotient does not come out even, round to the nearest hundredth.

GS (a) $0.2\overline{)1.04}$

(b) $0.06\overline{)1.8072}$

(c) $0.005\overline{)32}$

(d) $8.1 \div 0.025$

(e) $\dfrac{7}{1.3}$

(f) $5.3091 \div 6.2$

The short way to multiply by 10 is to move the decimal point *one place* to the *right* in both the divisor and the dividend.

$$1.2\overline{)6.36} \quad \text{is equivalent to} \quad 12\overline{)63.6}.$$

Note

Moving the decimal points the ***same*** number of places in ***both*** the divisor and dividend will ***not*** change the answer.

Dividing by a Decimal Number

Step 1 Count the number of decimal places in the divisor and move the decimal point that many places to the *right*. (This changes the divisor to a whole number.)

Step 2 Move the decimal point in the dividend the *same* number of places to the *right*. (Write in extra zeros if needed.)

Step 3 Write the decimal point in the quotient directly above the decimal point in the dividend. Then divide as usual.

EXAMPLE 4 Dividing by Decimal Numbers

(a) $0.003\overline{)27.69}$

Move the decimal point in the divisor *three* places to the *right* so 0.003 becomes the whole number 3. To move the decimal point in the dividend the same number of places, write in an extra 0.

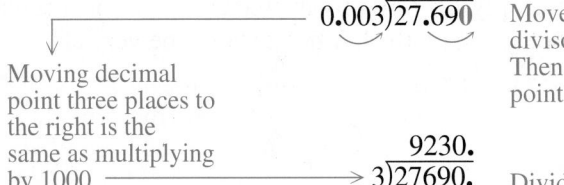

$0.003\overline{)27.690}$ Move decimal points in divisor and dividend. Then line up the decimal point in the quotient.

Moving decimal point three places to the right is the same as multiplying by 1000.

$$\begin{array}{r} 9230. \\ 3\overline{)27690.} \end{array}$$

Divide as usual.

(b) Divide 5 by 4.2. Round to the nearest hundredth.

Move the decimal point in the divisor one place to the right so 4.2 becomes the whole number 42. The decimal point in the dividend starts on the right side of 5 and is also moved one place to the right.

$$\begin{array}{r} 1.190 \\ 4.2\overline{)5.0000} \\ \underline{4\,2} \\ 8\,0 \\ \underline{4\,2} \\ 3\,8\,0 \\ \underline{3\,7\,8} \\ 2\,0 \end{array}$$

← In order to round to hundredths, divide out one *more* place, to thousandths.

Move the decimal points the *same* number of places.

Round the quotient. It is 1.19 (rounded to the nearest hundredth).

◄ Work Problem **4** at the Side.

Answers

4. (a) 5.2 (b) 30.12 (c) 6400 (d) 324 (e) 5.38 (rounded) (f) 0.86 (rounded)

OBJECTIVE 3 Estimate the answer when dividing decimals. Estimating answers helps you catch errors. Compare the estimate to your exact answer. If they are very different, work the problem again.

EXAMPLE 5 Estimating before Dividing

First use front end rounding to round each number and estimate the answer. Then divide to find the exact answer.

$$580.44 \div 2.8$$

Here is how one student solved this problem. She rounded 580.44 to 600 and rounded 2.8 to 3 to estimate the answer.

Estimate:

$$3\overline{)600} = 200$$

Exact:

```
      2 7.3
2.8)5 8 0.4 4
    5 6
      2 0 4
      1 9 6
          8 4
          8 4
            0
```

The estimate is too different. There is an error in the exact answer.

Very different; need to rework the problem

Notice that the estimate, which is in the hundreds, is very different from the exact answer, which is only in the tens. This tells the student that she needs to rework the problem. Can you find the error?
(The exact answer should be 207.3, which fits with the estimate of 200.)

Work Problem 5 at the Side. ▶

OBJECTIVE 4 Use the order of operations with decimals. Use the order of operations when a decimal problem involves more than one operation, as you did with whole numbers in **Chapter 1.**

Order of Operations

1. Do all operations inside ***parentheses*** or ***other grouping symbols.***
2. Simplify any expressions with ***exponents*** and find any ***square roots.***
3. ***Multiply*** or ***divide,*** proceeding from left to right.
4. ***Add*** or ***subtract,*** proceeding from left to right.

EXAMPLE 6 Using the Order of Operations

Use the order of operations to simplify each expression.

(a) $2.5 + 6.3^2 + 9.62$ Apply the exponent: (6.3)(6.3) is 39.69

$2.5 + 39.69 + 9.62$ Add from left to right.

$42.19 + 9.62$

51.81

(b) $1.82 + (6.7 - 5.2)(5.8)$ Work inside parentheses.

$1.82 + (1.5)(5.8)$ Multiply next.

$1.82 + 8.7$ Add last.

10.52

Continued on Next Page

5 Decide whether each answer is reasonable by using front end rounding to estimate the answer. If the exact answer is *not* reasonable, find and correct the error.

GS **(a)** $42.75 \div 3.8 = 1.125$

Estimate: $40 \div 4 = 10$

Answer of 1.125 is **not** reasonable. Rework.

$3.8\overline{)42.750}$

(b) $807.1 \div 1.76 = 458.580$ to nearest thousandth

Estimate:

(c) $48.63 \div 52 = 93.519$ to nearest thousandth

Estimate:

(d) $9.0584 \div 2.68 = 0.338$

Estimate:

Answers

5. **(a)** Exact answer should be 11.25.
(b) Estimate is $800 \div 2 = 400$; answer is reasonable.
(c) Estimate is $50 \div 50 = 1$; answer is not reasonable, should be 0.935 (rounded).
(d) Estimate is $9 \div 3 = 3$; answer is not reasonable, should be 3.38.

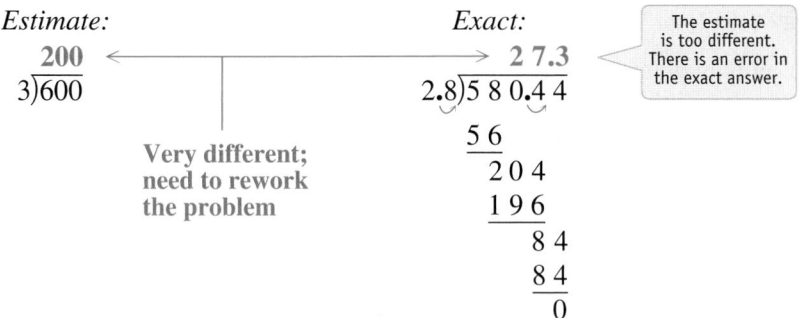

6 Use the order of operations to simplify each expression. The black brace shows you where to start.

(a) $4.6 - 0.79 + \underbrace{1.5^2}$

(b) $\underbrace{3.64 \div 1.3} \times 3.6$

(c) $0.08 + 0.6(\underbrace{3 - 2.99})$

(d) $10.85 - \underbrace{2.3(5.2)} \div 3.2$

Answers

6. **(a)** 6.06 **(b)** 10.08 **(c)** 0.086 **(d)** 7.1125

(c) $\underbrace{3.7^2} - 1.8 \div 5(1.5)$ Apply the exponent.

CAREFUL! Do **not** subtract yet.

$13.69 - \underbrace{1.8 \div 5}(1.5)$ Multiply and divide from left to right, so first divide 1.8 by 5 to get 0.36

$13.69 - \underbrace{0.36(1.5)}$ Then multiply 0.36 by 1.5

$\underbrace{13.69 - 0.54}$ Subtract last.

13.15

◀ **Work Problem 6 at the Side.**

Calculator Tip

You may want to use a calculator to check your work. Most scientific calculators that have parentheses keys ( ) can handle calculations like those in **Example 6** if you just enter the numbers in the order given. For example, the keystrokes for **Example 6(b)** on the previous page are:

Parentheses

1.82 (+) (() 6.7 (−) 5.2 ()) (×) 5.8 (=) Answer is 10.52.

Standard, four-function calculators generally do not have parentheses keys and will *not* give the correct answer if you simply enter the numbers in the order given.

Check the instruction manual that came with your calculator for information on "order of calculations" to see if your model has the rules for order of operations built into it. For a quick check, try entering this problem.

2 (+) 2 (×) 2 (=)

If the result is 6, the calculator follows the order of operations. If the result is 8, it does *not* have the rules built into it. To see why this test works, do the calculations by hand.

Follow the order of operations.	Work from left to right.
$2 + \underbrace{2 \times 2}$ Multiply before adding.	$\underbrace{2 + 2} \times 2$
$\underbrace{2 + 4}$	$\underbrace{4 \times 2}$
6 ← Correct	8 ← Incorrect

4.5 Exercises

FOR EXTRA HELP

Download the MyDashBoard App

MyMathLab®

1. CONCEPT CHECK Which problem below has a whole number for the divisor? Find it and then rewrite the problem using the $\overline{)\quad}$ symbol.

A. Divide 0.25 by 0.05. **B.** $25 \div 0.5$ **C.** $25.5 \div 5$

2. CONCEPT CHECK Which problem below has a whole number for the divisor? Find it and then rewrite the problem using the $\overline{)\quad}$ symbol.

A. $0.423 \div 0.07$ **B.** $42.3 \div 7$ **C.** Divide 423 by 0.7.

3. CONCEPT CHECK In $2.2\overline{)8.24}$, how do you make 2.2 a whole number, and how does that change 8.24? Use arrows to show how to move the decimal points.

4. CONCEPT CHECK In $5.1\overline{)10.5}$, how do you make 5.1 a whole number, and how does that change 10.5? Use arrows to show how to move the decimal points.

Find each quotient. ***See Examples 1 and 4.***

5. $7\overline{)27.3}$

6. $8\overline{)50.4}$

7. $\dfrac{4.23}{9}$

8. $\dfrac{1.62}{6}$

9. $0.05\overline{)20.01}$

10. $0.08\overline{)16.04}$

11. $1.5\overline{)54.0}$

12. $2.4\overline{)132.0}$

CONCEPT CHECK *Use the fact that* $108 \div 18 = 6$ *to work Exercises 13–16 simply by moving decimal points.*

13. $0.108 \div 1.8$

14. $10.8 \div 18$

15. $0.018\overline{)108}$

16. $0.18\overline{)1.08}$

Divide. Round quotients to the nearest hundredth if necessary. ***See Examples 3 and 4.***

17. $4.6\overline{)116.38}$

18. $2.6\overline{)4.992}$

19. $\dfrac{3.1}{0.006}$

20. $\dfrac{1.7}{0.09}$

Divide. Round quotients to the nearest thousandth.

21. $240.8 \div 9$

22. $76.43 \div 7$

23. $0.034\overline{)342.81}$

24. $0.043\overline{)1748.4}$

Decide whether each answer is reasonable *or* unreasonable *by rounding the numbers and estimating the answer. If the exact answer is not reasonable, find the correct answer.* ***See Example 5.***

25. $37.8 \div 8 = 47.25$

Estimate:

26. $345.6 \div 3 = 11.52$

Estimate:

27. $54.6 \div 48.1 = 1.135$

Estimate:

28. $2428.8 \div 4.8 = 56$

Estimate:

29. $307.02 \div 5.1 = 6.2$

Estimate:

30. $395.415 \div 5.05 = 78.3$

Estimate:

Solve each application problem. Round money answers to the nearest cent, if necessary.

31. Rob discovered that his daughter's favorite brand of tights is on sale. He decided to buy one pair as a surprise for her. How much did he pay?

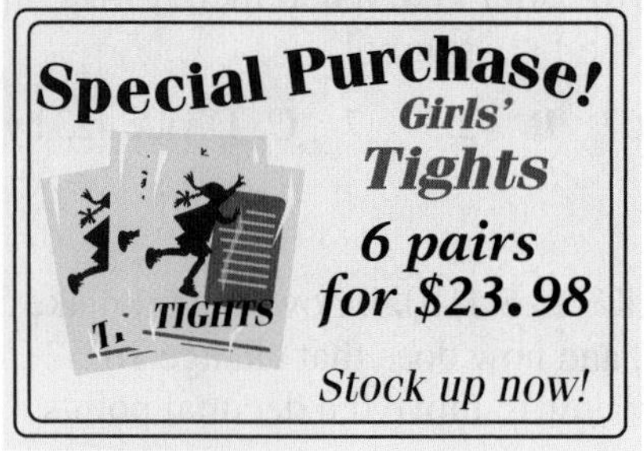

32. The bookstore has a special price on notepads. How much did Randall pay for one notepad?

33. It will take 21 equal monthly payments for Aimee to pay off her credit card balance of $1408.68. How much is she paying each month?

34. Marcella Anderson bought 2.6 meters of microfiber woven suede fabric for $33.77. How much did she pay per meter? (*Hint:* Cost *per* meter means the cost for *one* meter.)

35. Darren Jackson earned $476.80 for 40 hours of work. Find his earnings per hour.

36. Adrian Webb bought 108 patio blocks to build a backyard patio. He paid $237.60. Find the cost per block. (*Hint:* Cost *per* block means the cost for *one* block.)

37. It took 12.3 gallons of gas to fill the gas tank of Kim's car. She had driven 344.1 miles since her last fill-up. How many miles per gallon did her car get? Round to the nearest tenth.

38. Mr. Rodriquez pays $53.19 each month to Household Finance. How many months will it take him to pay off $1436.13?

39. Soup is on sale at six cans for $3.25, or you may purchase individual cans for $0.57. How much will you save per can if you buy six cans? Round to the nearest cent.

40. Nadia's diet allows her to eat 3.5 ounces of chicken nuggets. The package weighs 10.5 ounces and contains 15 nuggets. How many nuggets can Nadia eat?

Use the table of world records for the women's long jump (through the year 2010) to answer Exercises 41–46. To find an average, add up the values you are interested in and then divide the sum by the number of values. Round your answers to the nearest hundredth.

Athlete	Country	Year	Length (meters)
Galina Christyakova	USSR	1988	7.52
Jackie Joyner-Kersee	U.S.	1994	7.49
Heike Drechsler	Germany	1992	7.48
Jackie Joyner-Kersee	U.S.	1987	7.45
Jackie Joyner-Kersee	U.S.	1988	7.40
Jackie Joyner-Kersee	U.S.	1991	7.32
Jackie Joyner-Kersee	U.S.	1996	7.20
Chioma Ajunwa	Nigeria	1996	7.12
Fiona May	Italy	2000	7.09
Tatyana Lebedeva	Russia	2004	7.07

Source: CNNSI.com

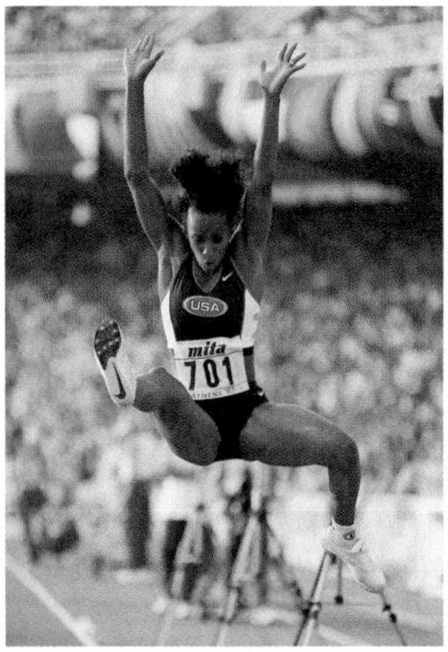

41. Find the average length of the long jumps made by Jackie Joyner-Kersee.

42. Find the average length of all the long jumps listed in the table.

43. How much longer was the fifth-longest jump than the sixth-longest jump?

44. If the first athlete in the table made five jumps of the same length, what would be the total distance jumped?

45. What was the total length jumped by the top three athletes in the table?

46. How much less was the shortest jump than the next-to-shortest jump?

47. For $2.2^2 + (9.5 - 3.1)$, list what you would do, in the correct order, to simplify the expression.
(a) First

(b) Second

(c) Third

48. For $60.41 - (0.4 + 5.07)(3)$, list what you would do, in the correct order, to simplify the expression.
(a) First

(b) Second

(c) Third

Use the order of operations to simplify each expression. ***See Example 6.***

49. $7.2 - 5.2 + 3.5^2$

50. $6.2 + 4.3^2 - 9.72$

51. $38.6 + 11.6(13.4 - 10.4)$

52. $2.25 - 1.06(4.85 - 3.95)$

53. $8.68 - 4.6(10.4) \div 6.4$

54. $25.1 + 11.4 \div 7.5(3.75)$

55. $33 - 3.2(0.68 + 9) - 1.3^2$

56. $0.6 + (1.89 + 0.11) \div 0.004(0.5)$

57. In 2010, the U.S. Treasury printed about 26,000,000 pieces of paper money each day. The printing presses run 24 hours a day. How many pieces of money are printed, to the nearest whole number:

(a) each hour?

(b) each minute?

(c) each second?

(*Source:* www.moneyfactory.com)

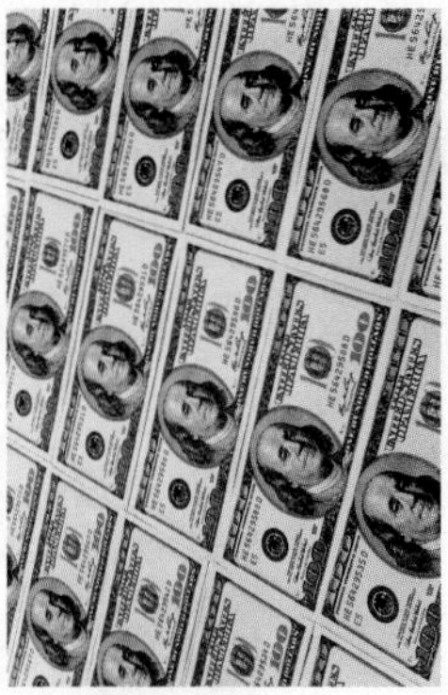

26,000,000 pieces of paper money are printed each day.

58. Mach 1 is the speed of sound. Dividing a vehicle's speed by the speed of sound gives its speed on the Mach scale. In 1997, a specially built car with two 110,000-horsepower engines broke the world land speed record by traveling 763.035 miles per hour. The speed of sound that day was 748.11 miles per hour. What was the car's Mach speed, to the nearest hundredth? (*Source:* Associated Press.)

General Mills will give a school 10¢ for each box top logo from its cereals and other products. A school can earn up to \$10,000 per year. Use this information to answer Exercises 59–62. Round your answers to the nearest whole number when necessary. (*Source:* General Mills.)

59. How many box tops would a school need to collect in one year to earn the maximum amount?

60. (Complete **Exercise 59** first.) If a school has 550 children, how many box tops would each child need to collect in one year to reach the maximum?

61. How many box tops would need to be collected during each of the 38 weeks in the school year to reach the maximum amount?

62. How many box tops would each of the 550 children need to collect during each of the 38 weeks of school to reach the maximum amount?

Relating Concepts (Exercises 63–64) For Individual or Group Work

Look for patterns as you ***work Exercises 63 and 64 in order.***

63. Do these division problems:

$3.77 \div 10 =$ ______ $9.1 \div 10 =$ ______

$0.886 \div 10 =$ ______ $30.19 \div 10 =$ ______

$406.5 \div 10 =$ ______ $6625.7 \div 10 =$ ______

(a) What pattern do you see? Write a "rule" for dividing by 10. What do you think the rule is for dividing by 100? by 1000? Write the rules and try them out on the numbers above.

(b) Compare your rules to the ones you wrote in the previous section of this chapter for **Exercise 57**. How are they different?

64. Do these division problems:

$40.2 \div 0.1 =$ ______ $7.1 \div 0.1 =$ ______

$0.339 \div 0.1 =$ ______ $15.77 \div 0.1 =$ ______

$46 \div 0.1 =$ ______ $873 \div 0.1 =$ ______

(a) What pattern do you see? Write a "rule" for dividing by 0.1. What do you think the rule is for dividing by 0.01? by 0.001? Write the rules and try them out on the numbers above.

(b) Compare your rules to the ones you wrote in the previous section of this chapter for **Exercise 58**. How are they different?

4.6 Fractions and Decimals

Writing fractions as equivalent decimals can help you do calculations more easily or compare the size of two numbers.

OBJECTIVE 1 Write fractions as equivalent decimals. Recall from **Chapter 1** that a fraction is one way to show division. For example, $\frac{3}{4}$ means $3 \div 4$. If you are doing the division by hand, write it as $4\overline{)3}$. When you do the division, the result is 0.75, the decimal equivalent of $\frac{3}{4}$.

Writing a Fraction as an Equivalent Decimal

Step 1 Divide the numerator of the fraction by the denominator.

Step 2 If necessary, round the answer to the place indicated.

Work Problem 1 at the Side. ▶

EXAMPLE 1 Writing Fractions or Mixed Numbers as Decimals

(a) Write $\frac{1}{8}$ as a decimal.

$\frac{1}{8}$ means $1 \div 8$. Write it as $8\overline{)1}$. The decimal point in the dividend is on the *right* side of the 1. Write extra zeros in the dividend so you can continue dividing until the remainder is 0.

$$\frac{1}{8} \rightarrow 1 \div 8 \rightarrow 8\overline{)1} \rightarrow \begin{array}{r} 0.125 \\ 8\overline{)1.000} \\ \underline{8} \\ 20 \\ \underline{16} \\ 40 \\ \underline{40} \\ 0 \end{array}$$

Decimal points lined up

Three extra zeros needed

Remainder is 0.

Therefore, $\frac{1}{8} = 0.125$.

To check this, write 0.125 as a fraction, then change it to lowest terms.

$$0.125 = \frac{125}{1000} \quad \text{In lowest terms} \quad \frac{125 \div 125}{1000 \div 125} = \frac{1}{8} \leftarrow \text{Original fraction}$$

Calculator Tip

When using your calculator to write fractions as decimals, enter the numbers from the top down. Remember that the order in which you enter the numbers *does* matter in division. **Example 1(a)** above works like this.

$\frac{1}{8}$ Top down Enter 1 ÷ 8 = Answer is 0.125

What happens if you enter 8 ÷ 1 = ? Do you see why that cannot possibly be correct? (Answer: $8 \div 1 = 8$. A proper fraction like $\frac{1}{8}$ *cannot* be equivalent to a whole number.)

Continued on Next Page

OBJECTIVES

1. Write fractions as equivalent decimals.
2. Compare the size of fractions and decimals.

1 Rewrite each fraction so you could do the division by hand. Do *not* complete the division.

GS **(a)** $\frac{1}{9}$ is written $9\overline{)}$

(b) $\frac{2}{3}$ is written $\overline{)}$

(c) $\frac{5}{4}$ is written $\overline{)}$

(d) $\frac{3}{10}$ is written $\overline{)}$

(e) $\frac{21}{16}$ is written $\overline{)}$

(f) $\frac{1}{50}$ is written $\overline{)}$

Answers

1. (a) $9\overline{)1}$ **(b)** $3\overline{)2}$ **(c)** $4\overline{)5}$ **(d)** $10\overline{)3}$ **(e)** $16\overline{)21}$ **(f)** $50\overline{)1}$

2 Write each fraction or mixed number as a decimal.

GS **(a)** $\frac{1}{4} \rightarrow 4\overline{)1.0\ 0}$ (quotient: 0. _ _)

GS **(b)** $2\frac{1}{2} = \frac{5}{2} \rightarrow 2\overline{)5.0}$

(c) $\frac{5}{8}$

(d) $4\frac{3}{5}$

(e) $\frac{7}{8}$

Answers

2. **(a)** 0.25 **(b)** 2.5 **(c)** 0.625 **(d)** 4.6 **(e)** 0.875

(b) Write $2\frac{3}{4}$ as a decimal.

One method is to divide 3 by 4 to get 0.75 for the fraction part. Then add the whole number part to 0.75.

$$\frac{3}{4} \rightarrow \begin{array}{r} 0.75 \\ 4\overline{)3.00} \\ \underline{2\ 8} \\ 20 \\ \underline{20} \\ 0 \end{array}$$

Fraction part ⟶ 0.75

$$\begin{array}{r} 2.00 \\ +\ 0.75 \\ \hline 2.75 \end{array}$$

2.00 ⟵ Whole number part

So, $2\frac{3}{4} = 2.75$ (Whole number parts match.)

CHECK $2.75 = 2\frac{75}{100} = 2\frac{3}{4}$ ⟵ Lowest terms

A second method is to first write $2\frac{3}{4}$ as an improper fraction and then divide numerator by denominator.

$$2\frac{3}{4} = \frac{11}{4}$$

$$\frac{11}{4} \rightarrow 11 \div 4 \rightarrow 4\overline{)11} \rightarrow \begin{array}{r} 2.75 \\ 4\overline{)11.00} \\ \underline{8} \\ 3\ 0 \\ \underline{2\ 8} \\ 20 \\ \underline{20} \\ 0 \end{array}$$

⟵ Two extra zeros needed

Whole number parts match.

So, $2\frac{3}{4} = 2.75$

$\frac{3}{4}$ **is equivalent to** $\frac{75}{100}$ **or 0.75**

◀ **Work Problem 2 at the Side.**

EXAMPLE 2 Writing a Fraction as a Decimal with Rounding

Write $\frac{2}{3}$ as a decimal and round to the nearest thousandth.

$\frac{2}{3}$ means $2 \div 3$. To round to thousandths, divide out one *more* place, to ten-thousandths.

$$\frac{2}{3} \rightarrow 2 \div 3 \rightarrow 3\overline{)2} \rightarrow \begin{array}{r} 0.6666 \\ 3\overline{)2.0000} \\ \underline{1\ 8} \\ 20 \\ \underline{18} \\ 20 \\ \underline{18} \\ 20 \\ \underline{18} \\ 2 \end{array}$$

⟵ Four zeros needed for ten-thousandths

Be careful to divide in the correct order!

Written as a repeating decimal, $\frac{2}{3} = 0.\overline{6}$. ⟵ Bar above repeating digit

Rounded to the nearest thousandth, $\frac{2}{3} \approx 0.667$.

Continued on Next Page

Calculator Tip

Try **Example 2** on your calculator. Enter 2 ÷ 3. Which answer do you get?

0.666666667 or ***0.6666666***

Many scientific calculators will show a 7 as the last digit. Because the sixes keep on repeating forever, the calculator automatically rounds in the last decimal place it has room to show. If you have a 10-digit display space, the calculator is rounding as shown below.

0.66666666666 (11 digits) rounds to 0.666666667

Next digit is 5 or more, so 6 rounds to 7.

Other calculators, especially standard, four-function ones, may *not* round. They just cut off, or *truncate,* the extra digits. Such a calculator would show 0.6666666 in the display.

Would this difference in calculators show up when changing $\frac{1}{3}$ to a decimal? Why not? (Answer: The repeating digit is 3, which is *4 or less,* so it stays as 3 whether it's rounded or not.)

Work Problem 3 at the Side. ▶

3 Write as decimals. Round to the nearest thousandth.

(a) $\frac{1}{3}$

(b) $2\frac{7}{9} = \frac{25}{9} = 25 \div 9$

Now use your calculator to do the division and round the answer to the nearest thousandth.

(c) $\frac{10}{11}$

(d) $\frac{3}{7}$

(e) $3\frac{5}{6}$

OBJECTIVE 2 Compare the size of fractions and decimals. You can use a number line to compare fractions and decimals. For example, the number line below shows the space between 0 and 1. The locations of some commonly used fractions are marked, along with their decimal equivalents.

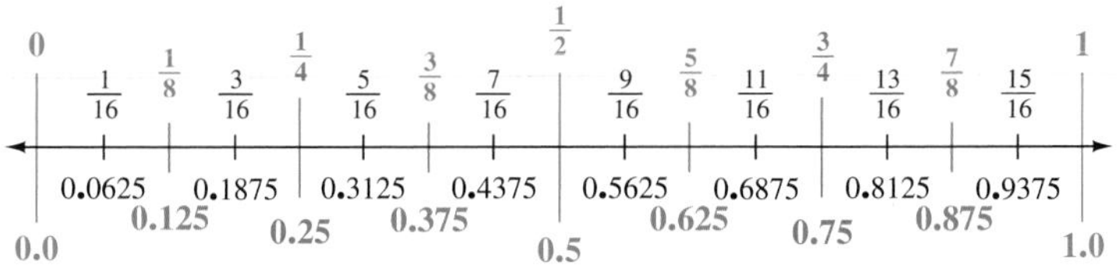

The next number line shows the locations of some commonly used fractions between 0 and 1 that are equivalent to *repeating* decimals. The decimal equivalents use a bar above repeating digits.

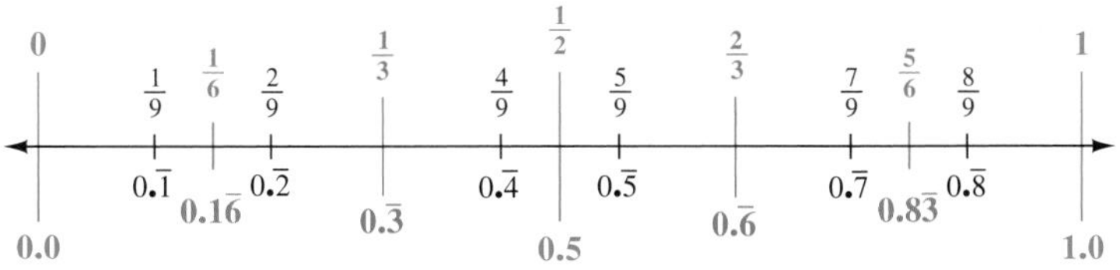

EXAMPLE 3 Using a Number Line to Compare Numbers

Use the number lines above to decide whether to write >, <, or = in the blank between each pair of numbers.

(a) 0.6875 ________ 0.625

You learned in **Chapter 3** that the number farther to the right on the number line is the greater number. Look at the first number line above. Because 0.6875 is to the *right* of 0.625, use the > symbol.

0.6875 is greater than 0.625 can be written as 0.6875 > 0.625

The larger end of the > symbol faces the greater number.

Continued on Next Page

Answers

3. All answers are rounded.
(a) 0.333 **(b)** 2.778 **(c)** 0.909
(d) 0.429 **(e)** 3.833

4 Use the number lines on the previous page to help you decide whether to write <, >, or = in each blank.

(a) 0.4375 ____ 0.5

(b) 0.75 ____ 0.6875

(c) 0.625 ____ 0.0625

(d) $\frac{2}{8}$ ____ 0.375

(e) $0.8\overline{3}$ ____ $\frac{5}{6}$

(f) $\frac{1}{2}$ ____ $0.\overline{5}$

(g) $0.\overline{1}$ ____ $0.1\overline{6}$

(h) $\frac{8}{9}$ ____ $0.\overline{8}$

(i) $0.\overline{7}$ ____ $\frac{4}{6}$

(j) $\frac{1}{4}$ ____ 0.25

5 Arrange each group in order from least to greatest.

(a) 0.7, 0.703, 0.7029

(b) 6.39, 6.309, 6.401, 6.4

(c) 1.085, $1\frac{3}{4}$, 0.9

(d) $\frac{1}{4}, \frac{2}{5}, \frac{3}{7}$, 0.428

Answers

4. (a) < (b) > (c) > (d) < (e) = (f) < (g) < (h) = (i) > (j) =

5. (a) 0.7, 0.7029, 0.703
(b) 6.309, 6.39, 6.4, 6.401
(c) 0.9, 1.085, $1\frac{3}{4}$ (d) $\frac{1}{4}, \frac{2}{5}$, 0.428, $\frac{3}{7}$

(b) $\frac{3}{4}$ ________ 0.75

On the first number line, $\frac{3}{4}$ and 0.75 are at the same point on the number line. They are equivalent.

$$\frac{3}{4} = 0.75$$

(c) 0.5 ________ $0.\overline{5}$

On the second number line, 0.5 is to the *left* of $0.\overline{5}$ (which is actually 0.555 . . .) so use the < symbol.

0.5 **is less than** $0.\overline{5}$ can be written as $0.5 < 0.\overline{5}$.

The smaller end of the < symbol points to the lesser number.

(d) $\frac{2}{6}$ ________ $0.\overline{3}$

Write $\frac{2}{6}$ in lowest terms as $\frac{1}{3}$.

On the second number line you can see that $\frac{1}{3} = 0.\overline{3}$.

◀ **Work Problem 4 at the Side.**

Fractions can also be compared by first writing each one as a decimal. The decimals can then be compared by writing each one with the same number of decimal places.

EXAMPLE 4 Arranging Numbers in Order

Write each group of numbers in order, from least to greatest.

(a) 0.49 0.487 0.4903

It is easier to compare decimals if they are all tenths, or all hundredths, and so on. Because 0.4903 has four decimal places (ten-thousandths), write zeros to the right of 0.49 and 0.487 so they also have four decimal places. Recall that writing zeros to the right of a decimal number does *not* change its value (see the section on adding decimals). Then find the least and greatest number of ten-thousandths.

0.49 = 0.4900 = **4900** ten-thousandths ⟵ 4900 is in the middle.

0.487 = 0.4870 = **4870** ten-thousandths ⟵ 4870 is the least.

0.4903 = **4903** ten-thousandths ⟵ 4903 is the greatest.

From least to greatest, the correct order is: 0.487 0.49 0.4903

(b) $2\frac{5}{8}$ 2.63 2.6

Write $2\frac{5}{8}$ as $\frac{21}{8}$ and divide $8\overline{)21}$ to get the decimal form, 2.625. Then, because 2.625 has three decimal places, write zeros so all the numbers have three decimal places.

$2\frac{5}{8}$ = 2.625 = 2 and **625** thousandths ⟵ 625 is in the middle.

2.63 = 2.630 = 2 and **630** thousandths ⟵ 630 is the greatest.

2.6 = 2.600 = 2 and **600** thousandths ⟵ 600 is the least.

From least to greatest, the correct order is: 2.6 $2\frac{5}{8}$ 2.63

◀ **Work Problem 5 at the Side.**

4.6 Exercises

FOR EXTRA HELP

CONCEPT CHECK *In Exercises 1 and 2, circle all the divisions that are correctly set up to convert the given fraction to a decimal. If a division is* **not** *set up correctly, fix it.*

1. (a) $\frac{2}{5} \rightarrow 2\overline{)5}$ **(b)** $\frac{1}{7} \rightarrow 7\overline{)1}$ **(c)** $\frac{3}{8} \rightarrow 8\overline{)3}$

2. (a) $\frac{4}{9} \rightarrow 9\overline{)4}$ **(b)** $\frac{1}{3} \rightarrow 1\overline{)3}$ **(c)** $\frac{5}{6} \rightarrow 6\overline{)5}$

CONCEPT CHECK *In Exercises 3 and 4, show the correct division set up for converting the given fraction to a decimal. Place the decimal points in the dividend and quotient. Then write the first digit in the quotient. Finally, explain what you will do next in order to continue dividing.*

3. $\frac{3}{4} \rightarrow$

4. $\frac{1}{8} \rightarrow$

CONCEPT CHECK *In Exercises 5 and 6, write* greater than *or* less than *between each pair of numbers.*

5. (a) $\frac{2}{5}$ is ____________ 0.5.

(b) $\frac{3}{4}$ is ____________ 0.6.

(c) 0.2 is ____________ $\frac{5}{8}$.

6. (a) 0.1 is ____________ $\frac{1}{5}$.

(b) $\frac{2}{3}$ is ____________ 0.5.

(c) 0.7 is ____________ $\frac{1}{2}$.

Write each fraction or mixed number as a decimal. Round to the nearest thousandth if necessary. ***See Examples 1 and 2.***

7. Finish the division. $\frac{1}{2} \rightarrow 2\overline{)1.0}$ (quotient: 0.)

8. Finish the division. $\frac{1}{4} \rightarrow 4\overline{)1.00}$ (quotient: 0.)

9. $\frac{3}{4}$

10. $\frac{1}{10}$

11. $\frac{3}{10}$

12. $\frac{7}{10}$

13. $\frac{9}{10}$

14. $\frac{4}{5}$

15. $\frac{3}{5}$

16. $\frac{2}{5}$

17. $\frac{7}{8}$

18. $\frac{3}{8}$

19. $2\frac{1}{4}$

20. $1\frac{1}{2}$

21. $14\frac{7}{10}$

22. $23\frac{3}{5}$

23. $3\frac{5}{8}$

24. $2\frac{7}{8}$

25. $6\frac{1}{3}$

26. $5\frac{2}{3}$

27. $\frac{5}{6}$

28. $\frac{1}{6}$

29. $1\frac{8}{9}$

30. $5\frac{4}{7}$

Find each decimal or fraction equivalent. Write fractions in lowest terms.

Fraction	Decimal	Fraction	Decimal
31. ____	0.4	**32.** ____	0.75
33. ____	0.625	**34.** ____	0.111
35. ____	0.35	**36.** ____	0.9
37. $\frac{7}{20}$	____	**38.** $\frac{1}{40}$	____
39. ____	0.04	**40.** ____	0.52
41. $\frac{1}{5}$	____	**42.** $\frac{1}{8}$	____
43. ____	0.09	**44.** ____	0.02

Solve each application problem.

45. The average length of a newborn baby is 20.8 inches. Charlene's baby is 20.08 inches long. Is her baby longer or shorter than the average? By how much?

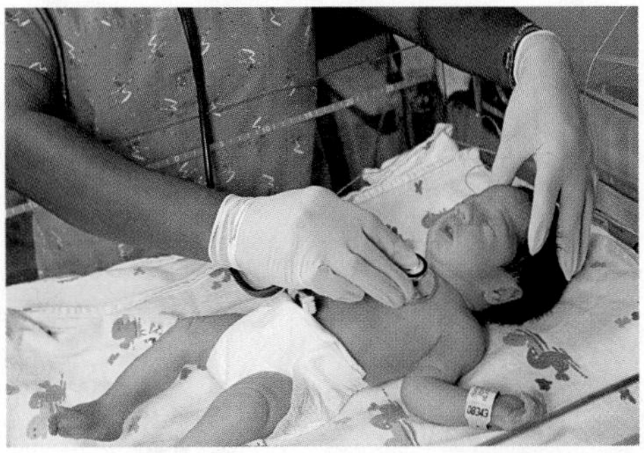

46. The patient in room 830 is supposed to get 8.3 milligrams of medicine. She was actually given 8.03 milligrams. Did she get too much or too little medicine? What was the difference?

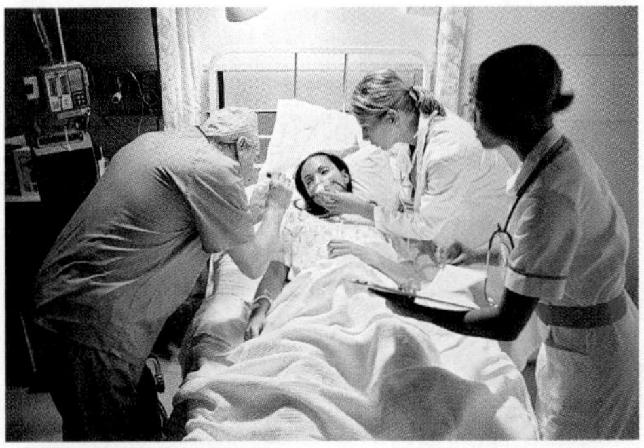

47. Ginny Brown hoped her crops would get $3\frac{3}{4}$ inches of rain this month. The newspaper said the area received 3.8 inches of rain. Was that more or less than Ginny had hoped for? By how much?

48. The rats in a medical experiment gained $\frac{3}{8}$ ounce. They were expected to gain 0.3 ounce. Was their actual gain more or less than expected? By how much?

49. The label on the bottle of vitamins says that each capsule contains 0.5 gram of calcium. When checked, each capsule had 0.505 gram of calcium. Was there too much or too little calcium? What was the difference?

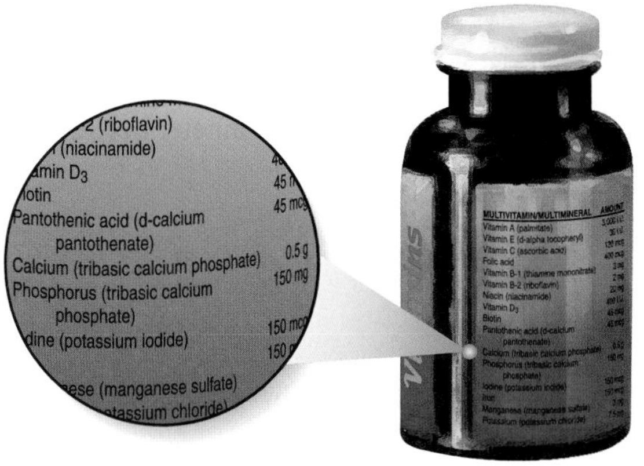

50. The glass mirror of the Hubble telescope had to be repaired in space because it would not focus properly. The problem was that the mirror's outer edge had a thickness of 0.6248 centimeter when it was supposed to be 0.625 centimeter. Was the edge too thick or too thin? By how much? (*Source:* NASA.)

51. CONCEPT CHECK Precision Medical Parts makes an artificial heart valve that must measure between 0.998 centimeter and 1.002 centimeters. Circle the lengths that are acceptable.

1.01 cm   0.9991 cm   1.0007 cm   0.99 cm

52. CONCEPT CHECK The white rats in a medical experiment must start out weighing between 2.95 ounces and 3.05 ounces. Circle the weights that can be used.

3.0 ounces   2.995 ounces   3.055 ounces   3.005 ounces

Arrange each group of numbers in order, from least to greatest. **See Example 4.**

53. 0.54, 0.5455, 0.5399

54. 0.76, 0.7, 0.7006

55. 5.8, 5.79, 5.0079, 5.804

56. 12.99, 12.5, 13.0001, 12.77

57. 0.628, 0.62812, 0.609, 0.6009

58. 0.27, 0.281, 0.296, 0.3

59. 5.8751, 4.876, 2.8902, 3.88

60. 0.98, 0.89, 0.904, 0.9

61. 0.043, 0.051, 0.006, $\frac{1}{20}$

62. 0.629, $\frac{5}{8}$, 0.65, $\frac{7}{10}$

63. $\frac{3}{8}$, $\frac{2}{5}$, 0.37, 0.4001

64. 0.1501, 0.25, $\frac{1}{10}$, $\frac{1}{5}$

Four boxes of fishing line are in the sale bin. The thicker the line, the stronger it is. The diameter of the fishing line is its thickness. Use the information on the boxes to answer Exercises 65–68.

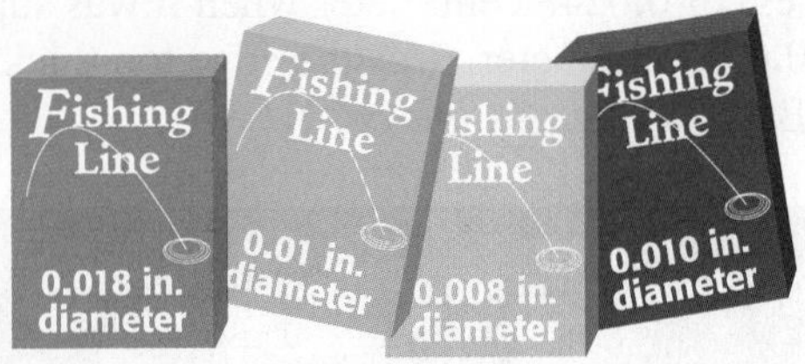

65. Which color box has the strongest line?

66. Which color box has the line with the least strength?

67. Which color box has the line that is $\frac{1}{125}$ inch in diameter?

68. What is the difference in line diameter between the blue and purple boxes?

Some rulers for technical occupations show each inch divided into tenths. Use this scale drawing for Exercises 69–74. Change the measurements on the drawing to decimals and round them to the nearest tenth of an inch.

69. Length (a) is ____________.

70. Length (b) is ____________.

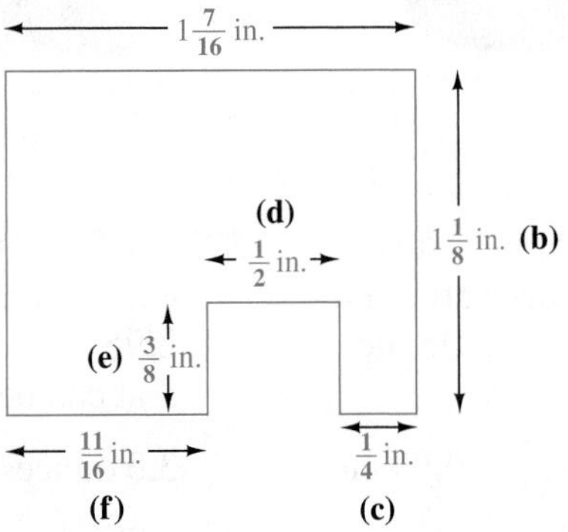

71. Length (c) is ____________.

72. Length (d) is ____________.

73. Length (e) is ____________.

74. Length (f) is ____________.

Relating Concepts (Exercises 75–78) For Individual or Group Work

Use your knowledge of fractions and decimals to **work Exercises 75–78 in order.**

75. (a) Explain how you can tell that Keith made an error *just by looking at his final answer.* Here is his work.

$$\frac{5}{9} = 9\overline{)5.0}\ \text{gives}\ 5\overline{)9.0} = 1.8 \quad \text{so} \quad \frac{5}{9} = 1.8$$

(b) Show the correct way to change $\frac{5}{9}$ to a decimal. Explain why your answer makes sense.

76. (a) How can you prove to Sandra that $2\frac{7}{20}$ is *not* equivalent to 2.035? Here is her work.

$$2\frac{7}{20} = 20\overline{)7.00} = 0.35 \quad \text{so} \quad 2\frac{7}{20} = 2.035$$

(b) What is the correct answer? Show how to prove that it is correct.

77. Ving knows that $\frac{3}{8} = 0.375$. How can he write $1\frac{3}{8}$ as a decimal *without* having to do a division? How can he write $3\frac{3}{8}$ as a decimal? $295\frac{3}{8}$? Explain your answer.

78. Iris has found a shortcut for writing mixed numbers as decimals.

$$2\frac{7}{10} = 2.7 \qquad 1\frac{13}{100} = 1.13$$

Does her shortcut work for all mixed numbers? Explain when it works and why it works.

Study Skills
ANALYZING YOUR TEST RESULTS

OBJECTIVES

1. **Determine the reason for errors.**
2. **Develop a plan to avoid test-taking errors.**
3. **Review material to correct misunderstandings.**

After taking a test, many students heave a big sigh of relief and try to forget it ever happened. Don't fall into this trap! An exam is a learning opportunity. It gives you clues about *what your instructor thinks is important*, what *concepts and skills are valued* in mathematics, and *if you are on the right track.*

Immediately After the Test

Jot down problems that caused you trouble. Find out how to solve them by checking your textbook, looking at your notes, or asking your instructor or tutor (if available). You might see those same problems again on a final exam.

After the Test Is Returned

Find out what you got wrong and why you had points deducted. Write down the problem so you can learn how to do it correctly. Sometimes you only have a short time in class to review your test. *If you need more time,* ask your instructor if you can look at the test in his or her office.

Find Out Why You Made the Errors You Made

Here is a list of typical reasons for making errors on math tests.

1. You read the directions wrong.
2. You read the question wrong or skipped over something.
3. You made a computation error (maybe even an easy one).
4. Your answer is not accurate.
5. Your answer is not complete.
6. You labeled your answer wrong. For example, you labeled it "feet" and it should have been "feet^2."
7. You didn't show your work.
8. *You didn't understand the concept.
9. *You were unable to go from words (in a word problem) to setting up the problem.
10. *You were unable to apply a procedure to a new situation.
11. You were so anxious that you made errors even when you knew the material.

The first seven errors are **test-taking errors.** They are easy to correct if you decide to carefully read test questions and directions, proofread or rework your problems, show all your work, and double check units and labels every time.

The three starred errors (*) are **test preparation errors.** Remember that to grow a complex neural network, you need to practice the kinds of problems that you will see on the tests. So, for example, if application problems are difficult for you, you must *do more application problems!* If you have practiced the study skills techniques, however, you are less likely to make these kinds of errors on tests because you will have a deeper understanding of course concepts and you will be able to remember them better.

The last error isn't really an error. **Anxiety** can play a big part in your test results. Go back to the *Preparing for Tests* activity and read the suggestions about exercise and deep breathing. Recall from the *Your Brain* **Can** *Learn Mathematics* activity that when you are anxious, your body produces adrenaline. The presence of *adrenaline in the brain blocks connections* between dendrites. If you can

Study Skills

Continued from page 303

reduce the adrenaline in your system, you will be able to *think more clearly* during your test. Just five minutes of brisk walking right before your test can help do that. Also, *practicing a relaxation technique while you do your homework* will make it more likely that you can benefit from using the technique during a test. *Deep breathing* is helpful because it gets *oxygen into your brain*. When you are anxious you tend to breathe more shallowly, which can make you feel confused and easily distracted.

Make a Plan for the Next Test

Make a plan for your next test based on your results from this test. You might review the Chapter Summary and work the problems in the Chapter Review Exercises or the Chapter Test. Ask your instructor or a tutor (if available) for more help if you are confused about any of the problems.

Now Try This

Below is a record sheet to track your progress in test-taking. Use it to find out if you make particular kinds of errors. Then you can work specifically on correcting them. Just place a check in the box when you made one of the errors. If you take more than four tests, make your own grid on separate paper.

Test-taking Errors

Test #	Read directions wrong	Read question wrong	Computation error	Not exact or accurate	Not complete	Labeled wrong	Didn't show work
1							
2							
3							
4							

Test Preparation Errors

Test #	Didn't understand concept	Didn't set up problem correctly	Couldn't apply concept to new situation
1			
2			
3			
4			

Anxiety

Test #	Felt anxious *before* the exam	Felt anxious *during* the exam	Blanked out on questions	Got questions wrong that I knew how to do
1				
2				
3				
4				

What will you do to avoid test-taking errors?

____

____

____

What will you do to avoid test preparation errors?

____

____

____

What will you do to reduce anxiety?

____

____

____

Chapter 4 Summary

Key Terms

4.1

decimals Decimals, like fractions, are used to show parts of a whole.

decimal point A decimal point is the dot that is used to separate the whole number part from the fractional part of a decimal number.

place value A place value is assigned to each place to the right or left of the decimal point. Whole numbers, such as ones and tens, are to the *left* of the decimal point. Fractional parts, such as tenths and hundredths, are to the *right* of the decimal point.

4.2

rounding Rounding is "cutting off" a number after a certain place, such as rounding to the nearest hundredth. The rounded number is less accurate than the original number. You can use the symbol "≈" to mean "is approximately equal to."

decimal places Decimal places are the number of digits to the *right* of the decimal point. For example, 6.37 has two decimal places, and 4.706 has three decimal places.

4.3

estimating Estimating is the process of rounding the numbers in a problem and getting an approximate answer. This helps you check that the decimal point is in the correct place in the exact answer.

4.5

repeating decimal A repeating decimal is a decimal number with one or more digits that repeat forever; it never ends. For example, in 0.1666 . . . , the digit 6 continues to repeat. Use three dots to indicate that it is a repeating decimal. Or, write the number with a bar above the repeating digits, as in $0.1\overline{6}$. (Use the dots or the bar, but not both.)

New Symbols

$3.8\overline{6}$ ← Bar above repeating digit(s) in a decimal number

3.866 . . . Three dots indicate a repeating decimal.

Test Your Word Power

See how well you have learned the vocabulary in this chapter.

1. **Decimal numbers** are like fractions in that they both
   A. need common denominators
   B. have decimal points
   C. represent parts of a whole.

2. **Decimal places** refer to
   A. the digits from 0 to 9
   B. digits to the left of the decimal point
   C. digits to the right of the decimal point.

3. When a decimal number is **rounded,** it
   A. always ends in 0
   B. is less accurate than the original number
   C. is less than one whole.

4. The **decimal point**
   A. separates the whole number part from the fractional part
   B. separates tenths from hundredths
   C. is at the far left side of a whole number.

5. The number $\mathbf{0.\overline{3}}$ is an example of
   A. an estimate
   B. a repeating decimal
   C. a rounded number.

6. The **place value** names on the right side of the decimal point are
   A. ones, tens, hundreds, and so on
   B. ones, tenths, hundredths, and so on
   C. tenths, hundredths, thousandths, and so on.

Answers to Test Your Word Power

1. C; *Example:* For 0.7, the whole is cut into ten equal parts, and you are interested in 7 of the parts.

2. C; *Examples:* The number 6.87 has two decimal places; 0.309 has three decimal places.

3. B; *Example:* When 0.815 is rounded to 0.8 it is accurate only to the nearest tenth, while the original number was accurate to the nearest thousandth.

4. A; *Example:* In 5.42, the whole number part is 5 ones, and the decimal part is 42 hundredths.

5. B; *Example:* The bar above the 3 in $0.\overline{3}$ indicates that the 3 repeats forever.

6. C; *Example:* In 6.219, the 2 is in the tenths place, the 1 is in the hundredths place, and the 9 is in the thousandths place.

Quick Review

Concepts / Examples

4.1 Reading and Writing Decimal Numbers

thousands	hundreds	tens	ones	decimal point	tenths	hundredths	thousandths	ten-thousandths
5	8	4	6	.	0	7	3	2

Whole number part (5 8 4 6); Fractional part (0 7 3 2)

Write each decimal in words.

15.38 — 8 is in hundredths place.

fifteen **and** thirty-eight **hundredths**

0.0103 — 3 is in ten-thousandths place.

one hundred three **ten-thousandths**

4.1 Writing Decimal Numbers as Fractions

The digits to the right of the decimal point are the numerator. The place value of the rightmost digit determines the denominator.

Always write the fractions in lowest terms.

Write 0.45 as a fraction in lowest terms.

The numerator is 45. The rightmost digit, 5, is in the hundredths place, so the denominator is 100. Then write the fraction in lowest terms.

$$\frac{45}{100} = \frac{45 \div 5}{100 \div 5} = \frac{9}{20} \quad \leftarrow \text{Lowest terms}$$

4.2 Rounding Decimal Numbers

Find the place to which you are rounding. Draw a cut-off line to the right of that place; the rest of the digits will be dropped. Look *only* at the first digit being cut. If it is *4 or less,* the part you are keeping stays the same. If it is *5 or more,* the part you are keeping rounds up. Do not move the decimal point when rounding. Write "≈" to mean "is approximately equal to."

Round 0.17952 to the nearest thousandth.

0.179|52 — First digit cut is *5 or more,* so round up.

$$\begin{array}{rl} 0.179 & \leftarrow \text{Keep this part.} \\ +\ 0.001 & \leftarrow \text{To round up, add 1 thousandth.} \\ \hline 0.180 & \end{array}$$

0.17952 rounds to 0.180. Write $0.17952 \approx 0.180$.

4.3 Adding and Subtracting Decimal Numbers

Use front end rounding to round each number and estimate the answer.

To find the exact answer, line up the decimal points. If needed, write in zeros as placeholders. Add or subtract as if they were whole numbers. Line up the decimal point in the answer directly below the decimal points in the problem.

Add 5.68 + 785.3 + 12 + 2.007.

Estimate:	*Exact:*
6	5.680
800	785.300
10	12.000
+ 2	+ 2.007
818	804.987

Use zeros as placeholders so that all numbers have three decimal places.

Line up decimal points.

The estimate and exact answer are both in the hundreds, so the decimal point is probably in the correct place.

4.4 Multiplying Decimal Numbers

Step 1 Multiply as you would for whole numbers.

Step 2 Count the total number of decimal places in both factors.

Step 3 Write the decimal point in the product so it has the same number of decimal places as the total from Step 2. You may need to write extra zeros on the left side of the product to get enough decimal places.

Multiply 0.169 × 0.21.

$$\begin{array}{rl} 0.169 & \leftarrow \text{3 decimal places} \\ \times\ 0.21 & \leftarrow \text{2 decimal places} \\ \hline 169 & \text{5 total decimal places} \\ 338 & \\ \hline .03549 & \leftarrow \text{5 decimal places in product} \end{array}$$

Write a 0 in the product so you can count over 5 decimal places. The final answer is 0.03549.

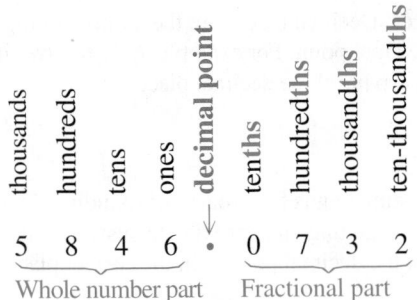

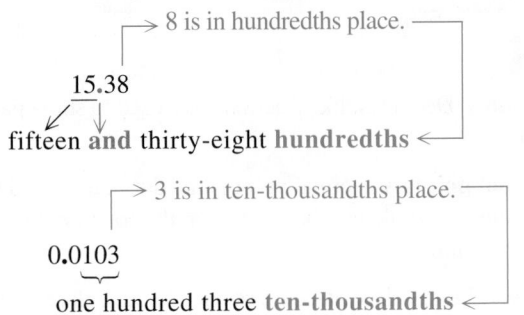

Concepts	Examples

4.5 Dividing by Decimal Numbers

Step 1 Change the divisor to a whole number by moving the decimal point to the right.

Step 2 Move the decimal point in the dividend the same number of places to the right.

Step 3 Write the decimal point in the quotient directly above the decimal point in the dividend.

Step 4 Divide as with whole numbers.

Divide 52.8 by 0.75.

```
         7 0.4
0.7 5)5 2.8 0 0
      5 2 5
          3 0 0
          3 0 0
              0
```

Move the decimal point two places to the right in the divisor and dividend. Write zeros in the dividend so you can move the decimal point and continue dividing until the remainder is 0.

To check your answer, multiply 70.4 times 0.75. If the result matches the dividend (52.8), you solved the problem correctly.

4.6 Writing Fractions as Decimal Numbers

Divide the numerator by the denominator. If necessary, round to the place indicated.

Write $\frac{1}{8}$ as a decimal.

$\frac{1}{8}$ means $1 \div 8$. Write it as $8\overline{)1}$.

The decimal point is on the right side of 1.

```
   0.125
8)1.000  <— Write a decimal point
  8          and three zeros so you
  20         can continue dividing.
  16
   40
   40
    0
```

Therefore, $\frac{1}{8}$ is equivalent to 0.125.

4.6 Comparing the Size of Fractions and Decimal Numbers

Step 1 Write any fractions as decimals.

Step 2 Write zeros so that all the numbers being compared have the same number of decimal places.

Step 3 Use < to mean "is less than," > to mean "is greater than," or list the numbers from least to greatest.

Arrange in order from least to greatest.

$$0.505 \qquad \frac{1}{2} \qquad 0.55$$

$0.505 = 505$ thousandths ← 505 is in the middle.

$\frac{1}{2} = 0.5 = 0.500 = 500$ thousandths ← 500 is least.

$0.55 = 0.550 = 550$ thousandths ← 550 is greatest.

(least) $\frac{1}{2}$ 0.505 0.55 (greatest)

Chapter 4 Review Exercises

4.1 *Name the digit that has the given place value.*

1. 243.059
tenths
hundredths

2. 0.6817
ones
tenths

3. \$5824.39
hundreds
hundredths

4. 896.503
tenths
tens

5. 20.73861
tenths
ten-thousandths

Write each decimal as a fraction or mixed number in lowest terms.

6. 0.5

7. 0.75

8. 4.05

9. 0.875

10. 0.027

11. 27.8

Write each decimal in words.

12. 0.8

13. 400.29

14. 12.007

15. 0.0306

Write each decimal in numbers.

16. eight and three tenths

17. two hundred five thousandths

18. seventy and sixty-six ten-thousandths

19. thirty hundredths

4.2 *Round each decimal to the place indicated.*

20. 275.635 to the nearest tenth

21. 72.789 to the nearest hundredth

22. 0.1604 to the nearest thousandth

23. 0.0905 to the nearest thousandth

24. 0.98 to the nearest tenth

Round each money amount to the nearest cent.

25. \$15.8333

26. \$0.698

27. \$17,625.7906

Round each income or expense item to the nearest dollar.

28. Income from pancake breakfast was \$350.48.

29. Members paid \$129.50 in dues.

30. Refreshments cost \$99.61.

31. Bank charges were \$29.37.

4.3 *First use front end rounding to round each number and estimate the answer. Then find the exact answer.*

32. *Estimate:* + ____ *Exact:*

$$\begin{array}{r} 5.81 \\ 423.96 \\ +\ 15.09 \\ \hline \end{array}$$

33. *Estimate:* + ____ *Exact:*

$$\begin{array}{r} 75.6 \\ 1.29 \\ 122.045 \\ 0.88 \\ +\ 33.7 \\ \hline \end{array}$$

34. *Estimate:* − ____ *Exact:*

$$\begin{array}{r} 308.5 \\ -\ 17.8 \\ \hline \end{array}$$

35. *Estimate:* − ____ *Exact:*

$$\begin{array}{r} 9.2 \\ -\ 7.9316 \\ \hline \end{array}$$

36. Americans' favorite household pet is a cat. There are about 93.6 million pet cats, 77.5 million pet dogs, and 16.6 million pet birds. How many more pet cats are there than pet birds? (*Source:* American Pet Products Association.)

Estimate:

Exact:

37. Jasmin started with $406 in her bank account. She paid $315.53 to the day care center and $74.67 at the grocery store. What is the new amount in her account?

Estimate:

Exact:

38. Joey spent $1.59 for toothpaste, $5.33 for vitamins, and $18.94 for a toaster. He gave the clerk three $10 bills. How much change did he get?

Estimate:

Exact:

39. Roseanne is training for a wheelchair race. She raced 2.3 kilometers on Monday, 4 kilometers on Wednesday, and 5.25 kilometers on Friday. How far did she race altogether?

Estimate:

Exact:

4.4 *First use front end rounding to round each number and estimate the answer. Then find the exact answer.*

40. *Estimate:* × ____ *Exact:*

$$\begin{array}{r} 6.138 \\ \times\ \ 3.7 \\ \hline \end{array}$$

41. *Estimate:* × ____ *Exact:*

$$\begin{array}{r} 42.9 \\ \times\ \ 3.3 \\ \hline \end{array}$$

Find each product.

42. $(5.6)(0.002)$

43. $0.071(0.005)$

4.5 *Decide whether each answer is reasonable by rounding the numbers and estimating the answer. If the exact answer is not reasonable, find and correct the error.*

44. $706.2 \div 12 = 58.85$

Estimate:

45. $26.6 \div 2.8 = 0.95$

Estimate:

Divide. Round each quotient to the nearest thousandth if necessary.

46. $3\overline{)43.4}$

47. $\dfrac{72}{0.06}$

48. $0.00048 \div 0.0012$

4.4–4.5 *Solve each application problem.*

49. Adrienne worked 46.5 hours this week. Her hourly wage is \$14.24 for the first 40 hours and 1.5 times that rate over 40 hours. Find her total earnings to the nearest dollar.

50. A book of 12 tickets costs \$35.89 at the State Fair midway. What is the cost per ticket, to the nearest cent?

51. Stock in MathTronic sells for \$3.75 per share. Kenneth is thinking of investing \$500. How many whole shares could he buy?

52. Grapes are on sale at \$0.99 per pound. How much will Ms. Lee pay for 3.5 pounds of grapes, to the nearest cent?

Simplify each expression.

53. $3.5^2 + 8.7(1.95)$

54. $11 - 3.06 \div (3.95 - 0.35)$

4.6 *Write each fraction or mixed number as a decimal. Round to the nearest thousandth when necessary.*

55. $3\frac{4}{5}$

56. $\dfrac{16}{25}$

57. $1\frac{7}{8}$

58. $\dfrac{1}{9}$

Arrange each group of numbers in order from least to greatest.

59. 3.68, 3.806, 3.6008

60. 0.215, 0.22, 0.209, 0.2102

61. $0.17, \frac{3}{20}, \frac{1}{8}, 0.159$

Mixed Review Exercises

Add, subtract, multiply, or divide as indicated.

62. $89.19 + 0.075 + 310.6 + 5$

63. 72.8×3.5

64. $1648.3 \div 0.46$ Round to the nearest thousandth.

65. $30 - 0.9102$

66. $4.38(0.007)$

67. $0.005\overline{)0.047}$

68. $72.105 + 8.2 + 95.37$

69. $81.36 \div 9$

70. $(5.6 - 1.22) + 4.8(3.15)$

71. 0.455×18

72. $(1.6)(0.58)$

73. $0.218\overline{)7.63}$

74. $21.059 - 20.8$

75. $18.3 - 3^2 \div 0.5$

Use the information in the ad to answer Exercises 76–80. Round money answers to the nearest cent. (Disregard any sales tax.)

Grand Opening Sale!
Save on Clothing for the Entire Family
Jeans for Teens
only $19.95 each
women's sizes $24.99
Athletic Shoes
regularly priced
$89.99 to $149.50
NOW just $71 to $119.60
Men's socks NOW 3 pairs for $8.99
Children's socks 6 pairs for $5
Hurry in — TWO DAYS ONLY

76. How much would one pair of men's socks cost?

77. How much more would one pair of men's socks cost than one pair of children's socks?

78. How much would Fernando pay for a dozen pair of men's socks?

79. How much would Akiko pay for five pairs of teen jeans and four pairs of women's jeans?

80. What is the difference between the cheapest sale price for athletic shoes and the highest regular price?

To decrease your risk of clogged arteries, it is recommended that you get at least 2 milligrams of vitamin B-6 each day. Use the information in the table to answer Exercises 81–82.

SOURCES OF VITAMIN B-6
(AMOUNTS IN MILLIGRAMS)

$\frac{1}{2}$ cup green peas	0.11
1 banana	0.68
1 baked potato with skin	0.7
$\frac{1}{2}$ cup strawberries	0.45
3 ounces skinless chicken	0.5
3 ounces water-packed tuna	0.2
$\frac{1}{2}$ cup chickpeas	0.57

Source: National Institutes of Health.

81. **(a)** Which food item has the highest amount of vitamin B-6?

(b) Which food item has the lowest amount?

(c) What is the difference in the amount of vitamin B-6 between the food items with the highest and lowest amounts?

82. **(a)** Suppose you ate a banana, 3 ounces of skinless chicken, and 1 cup of strawberries. How many milligrams of vitamin B-6 would you get?

(b) Did you get more or less than the recommended daily amount? By how much?

Chapter 4 Test

CHAPTER Test Prep VIDEO *The Chapter Test Prep Videos with test solutions are available on DVD, in* **MyMathLab**, *and on* **YouTube**—*search "LialDevMath" and click on "Channels."*

Write each decimal as a fraction or mixed number in lowest terms.

1. 18.4

2. 0.075

Write each decimal in words.

3. 60.007

4. 0.0208

Round each decimal to the place indicated.

5. 725.6089 to the nearest tenth

6. 0.62951 to the nearest thousandth

7. \$1.4945 to the nearest cent

8. \$7859.51 to the nearest dollar

First use front end rounding to round each number and estimate the answer. Then find the exact answer.

9. $7.6 + 82.0128 + 39.59$

Estimate:

Exact:

10. $79.1 - 3.602$

Estimate:

Exact:

11. $5.79(1.2)$

Estimate:

Exact:

12. $20.04 \div 4.8$

Estimate:

Exact:

Find the exact answer.

13. $53.1 + 4.631 + 782 + 0.031$

14. $670 - 0.996$

15. $(0.0069)(0.007)$

16. $0.15\overline{)72}$

17. Write $2\frac{5}{8}$ as a decimal. Round to the nearest thousandth, if necessary.

18. Arrange in order from least to greatest.

$0.44, 0.451, \frac{9}{20}, 0.4506$

19. Simplify this expression.

$6.3^2 - 5.9 + 3.4(0.5)$

Solve each application problem.

20. Jennifer bought a 51-inch HDTV with 3D technology. The TV had an original price of $1299.99, and she bought two pairs of 3D glasses at $49.99 each. The store gave her a $200 discount, but she had to pay $94.50 in sales tax. How much did Jennifer pay?

21. Three types of ducks that are hunted in the United States are gadwalls, wigeons, and pintails. The estimated populations of these ducks are 2.5 million gadwalls, 2.551 million wigeons, and 2.56 million pintails. List the ducks in order from the greatest number to the least. (*Source:* U.S. Fish and Wildlife Service.)

22. Mr. Yamamoto bought 1.85 pounds of cheese at $2.89 per pound. What was the total amount he paid for the cheese, to the nearest cent?

23. Loren's baby had a temperature of 102.7 degrees. Later in the day it was 99.9 degrees. How much had the baby's temperature dropped?

24. Pat bought 6.5 ft of decorative gold chain to hang a light over her dining table. She paid $24.64. What was the cost per foot, to the nearest cent?

25. Write your own application problem using decimals. Make it different from **Problems 20–24.** Then show how to solve your problem.

Math in the Media

LAWN FERTILIZER

Gotta Be Green

A lot's being written about personal responsibility these days, and the idea seems to be ending up on the front lawn—literally! Each spring, homeowners across the country gear up to green up their lawns, and the increased use of fertilizer has a lot of environmentalists concerned about the potential effects of chemical runoff into nearby rivers and streams.

Every year, according to a study conducted by the University of Minnesota's Department of Agriculture, each household in the Minneapolis/St. Paul metro area uses an average of 36 pounds of lawn fertilizer. That adds up to 25,529,295 pounds, or 12,765 tons. Add to that another 193,000 pounds of weed killer and you're looking at the total picture for keeping it green in the Twin Cities.

Source: *Minneapolis Star Tribune.*

1. Refer to the article.

(a) How many pounds of lawn fertilizer are used each year in the *entire metro area?*

(b) Do a division on your calculator to find the number of *households* in the metro area.

(c) Why does it make sense to round your answer to part (b)? How would you round it?

2. There are 2000 pounds in one ton.

(a) Find the number of tons equivalent to 25,529,295 pounds of fertilizer.

(b) Does your answer match the figure given in the article? If not, what did the author of the article do to get 12,765 tons?

(c) Is the author's figure accurate? Why or why not?

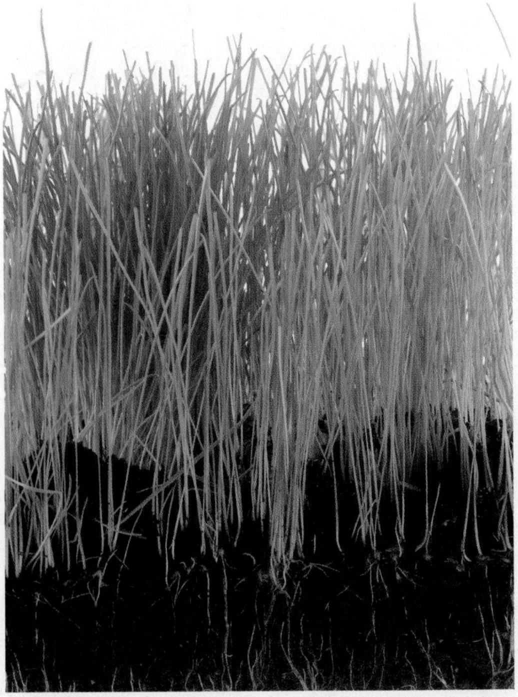

3. (a) When the average amount of lawn fertilizer per household was calculated, the answer was probably not exactly 36 pounds. List six different values that are less than 36 that would round to 36. List two values with one decimal place; two values with two decimal places; and two values with three decimal places.

(b) List six different values that are greater than 36 that would round to 36. List two values each with one, two, and three decimal places.

5 Ratio and Proportion

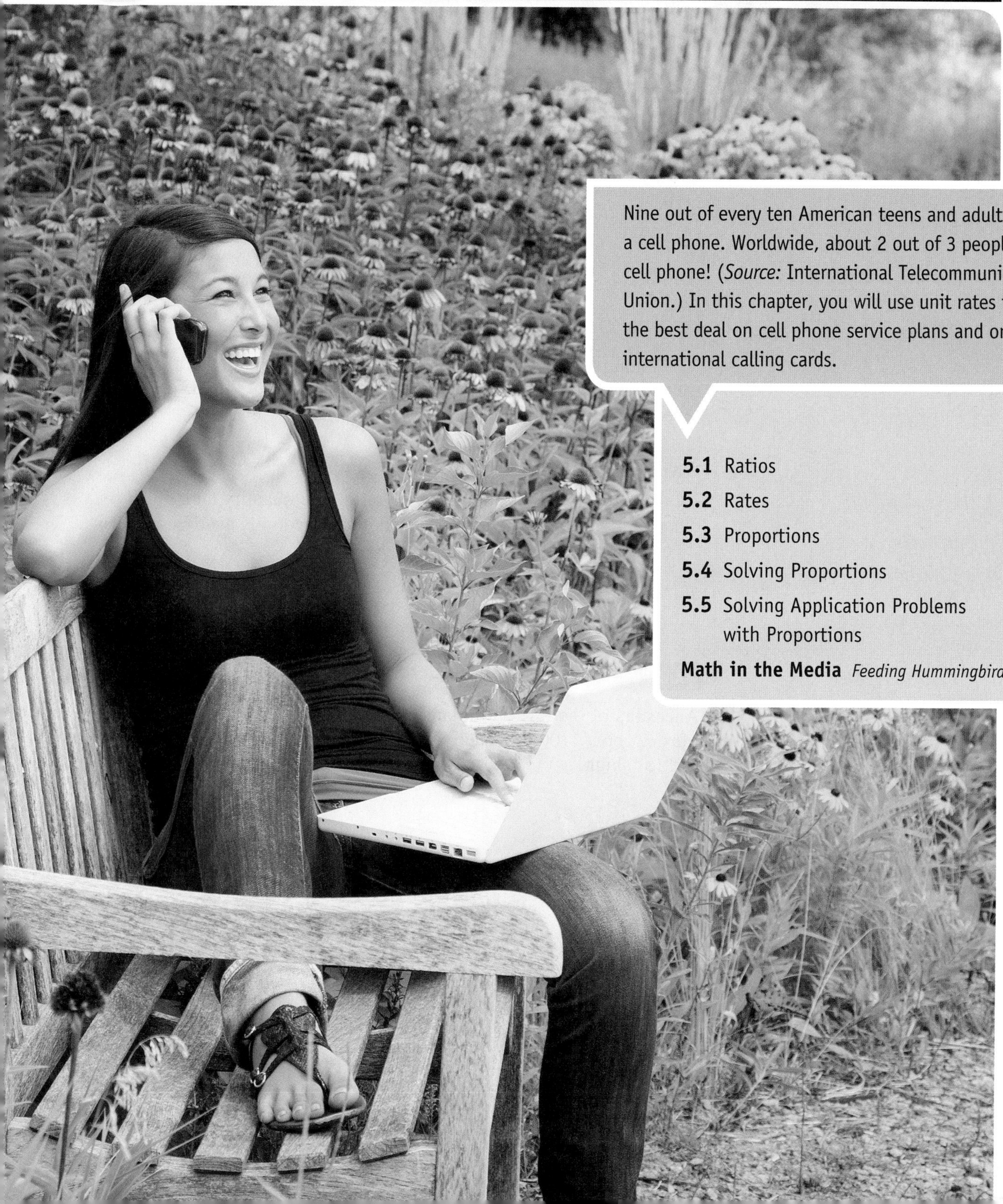

Nine out of every ten American teens and adults have a cell phone. Worldwide, about 2 out of 3 people use a cell phone! (*Source:* International Telecommunication Union.) In this chapter, you will use unit rates to find the best deal on cell phone service plans and on international calling cards.

5.1 Ratios

OBJECTIVES

1. Write ratios as fractions.
2. Solve ratio problems involving decimals or mixed numbers.
3. Solve ratio problems after converting units.

A **ratio** compares two quantities. You can compare two numbers, such as 8 and 4, or two measurements that have the *same* type of units, such as 3 *days* and 12 *days*. (*Rates* compare measurements with different types of units and are covered in the next section.)

Ratios can help you see important relationships. For example, if the ratio of your monthly expenses to your monthly income is 10 to 9, then you are spending $10 for every $9 you earn and going deeper into debt.

OBJECTIVE 1 Write ratios as fractions. A ratio can be written in three ways.

Writing a Ratio

The ratio of $7 **to** $3 can be written:

7 **to** 3 or 7:3 or $\frac{7}{3}$ ← Fraction bar indicates "**to**."

":" indicates "**to**"

Writing a ratio as a fraction is the most common method, and the one we will use here. All three ways are read, "the ratio of 7 **to** 3." The word **to** separates the quantities being compared.

Writing a Ratio as a Fraction

Order is important when writing a ratio. The quantity mentioned **first** is the **numerator**. The quantity mentioned **second** is the **denominator**. For example,

The ratio of **5** to **12** is written $\frac{5}{12}$.

EXAMPLE 1 Writing Ratios

Ancestors of the Pueblo Indians built multistory apartment towns in New Mexico about 1100 years ago. A room might measure 14 ft long, 11 ft wide, and 15 ft high.

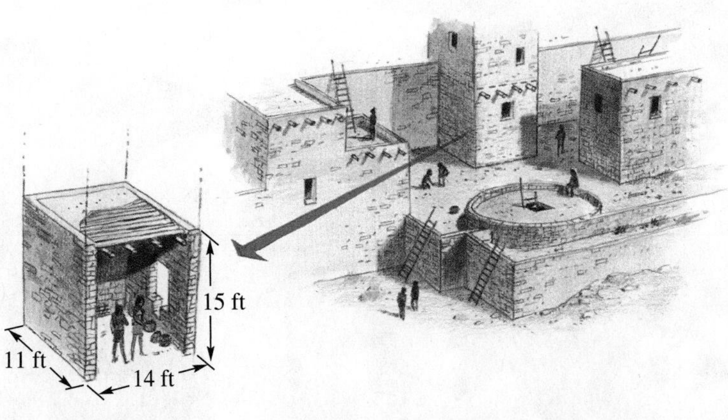

Continued on Next Page

Write each ratio as a fraction, using the room measurements.

(a) Ratio of length to width

The ratio of length to width is $\frac{14 \text{ ft}}{11 \text{ ft}} = \frac{14}{11}$.

Do *not* rewrite the ratio as $1\frac{3}{11}$.

Numerator (mentioned first) — Denominator (mentioned second)

You can divide out common *units* just like you divided out common *factors* when writing fractions in lowest terms. (See **Chapter 2.**) However, do *not* rewrite the fraction as a mixed number. Keep it as the ratio of 14 to 11.

(b) Ratio of width to height

Divide out the common units (ft).

The ratio of width to height is $\frac{11 \text{ ft}}{15 \text{ ft}} = \frac{11}{15}$.

CAUTION

Remember, the order of the numbers is important in a ratio. Look for the words "ratio of a to b." Write the ratio as $\frac{a}{b}$, ***not*** $\frac{b}{a}$. The quantity mentioned first is the numerator.

Work Problem 1 at the Side. ▶

Any ratio can be written as a fraction. Therefore, you can write a ratio in *lowest terms,* just as you do with any fraction.

EXAMPLE 2 Writing Ratios in Lowest Terms

Write each ratio in lowest terms.

(a) 60 days of sun to 20 days of rain

The ratio is $\frac{60 \text{ days}}{20 \text{ days}}$. Divide out the common units. Then write this ratio in lowest terms by dividing the numerator and denominator by 20.

$$\frac{60 \text{ days}}{20 \text{ days}} = \frac{60}{20} = \frac{60 \div 20}{20 \div 20} = \frac{3}{1} \leftarrow \text{Ratio in lowest terms}$$

So, the ratio of 60 days to 20 days is 3 to 1 or, written as a fraction, $\frac{3}{1}$. In other words, for every 3 days of sun, there is 1 day of rain.

CAUTION

In the fractions chapters you would have rewritten $\frac{3}{1}$ as 3. But a *ratio* compares *two* quantities, so you need to keep both parts of the ratio and write it as $\frac{3}{1}$.

(b) 50 ounces of medicine to 120 ounces of water

The ratio is $\frac{50 \text{ ounces}}{120 \text{ ounces}}$. Divide out the common units. Then divide the numerator and denominator by 10.

$$\frac{50 \text{ ounces}}{120 \text{ ounces}} = \frac{50}{120} = \frac{50 \div 10}{120 \div 10} = \frac{5}{12} \leftarrow \text{Ratio in lowest terms}$$

So, the ratio of 50 ounces to 120 ounces is $\frac{5}{12}$. In other words, for every 5 ounces of medicine, there are 12 ounces of water.

Continued on Next Page

VOCABULARY TIP

Ratio A **ratio** compares two quantities that have the **same** units. For example, if you are comparing two lengths, both measurements must be in feet, or both in yards, and so on. In a ratio, the common units divide out.

1 Shane spent \$14 on meat, \$5 on milk, and \$7 on fresh fruit. Write each ratio as a fraction.

GS **(a)** The ratio of amount spent on fruit to amount spent on milk.

Spent on fruit → $\frac{\quad}{5}$ ← Spent on milk

(b) The ratio of amount spent on milk to amount spent on meat.

(c) The ratio of amount spent on meat to amount spent on milk.

Answers

1. (a) 7; $\frac{7}{5}$ **(b)** $\frac{5}{14}$ **(c)** $\frac{14}{5}$

2 Write each ratio as a fraction in lowest terms.

GS **(a)** 9 hours to 12 hours

$$\frac{9 \text{ hours}}{12 \text{ hours}} = \frac{9}{12} = \frac{9 \div 3}{12 \div 3} = \frac{\_\_\_}{\_\_\_}$$

(b) 100 meters to 50 meters

(c) Write the ratio of width to length for this rectangle.

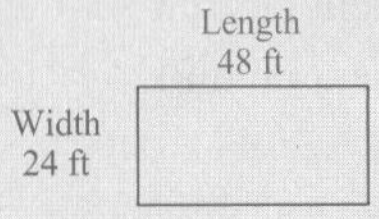

3 Write each ratio as a ratio of whole numbers in lowest terms.

(a) The price of Tamar's favorite brand of lipstick increased from \$5.50 to \$7.00. Find the ratio of the increase in price to the original price.

(b) Last week, Lance worked 4.5 hours each day. This week he cut back to 3 hours each day. Find the ratio of the decrease in hours to the original number of hours.

Answers

2. (a) $\frac{3}{4}$ (b) $\frac{2}{1}$ (c) $\frac{1}{2}$

3. (a) $\frac{1.50 \times 100}{5.50 \times 100} = \frac{150 \div 50}{550 \div 50} = \frac{3}{11}$

(b) $\frac{1.5 \times 10}{4.5 \times 10} = \frac{15 \div 15}{45 \div 15} = \frac{1}{3}$

(c) 15 people in a large van to 6 people in a small van

The ratio is $\frac{15 \text{ people}}{6 \text{ people}} = \frac{15}{6} = \frac{15 \div 3}{6 \div 3} = \frac{5}{2}$. ← Ratio in lowest terms

Note

Although $\frac{5}{2} = 2\frac{1}{2}$, ratios are *not* written as mixed numbers. Nevertheless, in **Example 2(c)** above, the ratio $\frac{5}{2}$ does mean the large van holds $2\frac{1}{2}$ times as many people as the small van.

◀ **Work Problem 2 at the Side.**

OBJECTIVE 2 Solve ratio problems involving decimals or mixed numbers. Sometimes a ratio compares two decimal numbers or two fractions. It is easier to understand if we rewrite the ratio as a ratio of two whole numbers.

EXAMPLE 3 Using Decimal Numbers in a Ratio

The price of a Sunday newspaper increased from \$1.50 to \$1.75. Find the ratio of the increase in price to the original price.

The words increase in price are mentioned first, so the increase will be the numerator. How much did the price go up? Use subtraction.

new price − original price = increase

\$1.75 − \$1.50 = \$0.25 ← Subtract first to find how much the price went up.

The words the original price are mentioned second, so the original price of \$1.50 is the denominator.

The ratio of increase in price to original price is shown below.

$$\frac{0.25}{1.50} \begin{matrix} \leftarrow \text{increase in price} \\ \leftarrow \text{original price} \end{matrix}$$

Now rewrite the ratio as a ratio of whole numbers. Recall that if you multiply both the numerator and denominator of a fraction by the same number, you get an equivalent fraction. The decimals in this example are hundredths, so multiply by 100 to get whole numbers. (If the decimals are tenths, multiply by 10. If thousandths, multiply by 1000.) Then write the ratio in lowest terms.

$$\frac{0.25}{1.50} = \frac{0.25 \times 100}{1.50 \times 100} = \frac{25}{150} = \frac{25 \div 25}{150 \div 25} = \frac{1}{6}$$ ← Ratio in lowest terms

($\frac{25}{150}$: Ratio as two whole numbers)

◀ **Work Problem 3 at the Side.**

EXAMPLE 4 Using Mixed Numbers in Ratios

Write each ratio as a comparison of whole numbers in lowest terms.

(a) 2 days to $2\frac{1}{4}$ days

Write the ratio as follows. Divide out the common units.

$$\frac{2 \text{ days}}{2\frac{1}{4} \text{ days}} = \frac{2}{2\frac{1}{4}}$$

Continued on Next Page

Next, write 2 as $\frac{2}{1}$ and $2\frac{1}{4}$ as the improper fraction $\frac{9}{4}$.

$$\frac{2}{2\frac{1}{4}} = \frac{\frac{2}{1}}{\frac{9}{4}}$$

Think: 2 • 4 = 8 and 8 + 1 = 9 so $2\frac{1}{4} = \frac{9}{4}$.

Rewrite the problem in horizontal format, using the "÷" symbol for division. Finally, multiply by the reciprocal of the divisor, as you did in **Chapter 2.**

$$\frac{\frac{2}{1}}{\frac{9}{4}} = \frac{2}{1} \div \frac{9}{4} = \frac{2}{1} \cdot \frac{4}{9} = \frac{8}{9}$$

Reciprocals

The ratio, in lowest terms, is $\frac{8}{9}$.

(b) $3\frac{1}{4}$ to $1\frac{1}{2}$

Write the ratio as $\frac{3\frac{1}{4}}{1\frac{1}{2}}$. Then write $3\frac{1}{4}$ and $1\frac{1}{2}$ as improper fractions.

Think: 3 • 4 = 12 and 12 + 1 = 13 so $3\frac{1}{4} = \frac{13}{4}$.

$$3\frac{1}{4} = \frac{13}{4} \quad \text{and} \quad 1\frac{1}{2} = \frac{3}{2}$$

Think: 1 • 2 = 2 and 2 + 1 = 3 so $1\frac{1}{2} = \frac{3}{2}$.

The ratio is shown here.

$$\frac{3\frac{1}{4}}{1\frac{1}{2}} = \frac{\frac{13}{4}}{\frac{3}{2}}$$

Rewrite as a division problem in horizontal format, using the "÷" symbol. Then multiply by the reciprocal of the divisor.

$$\frac{13}{4} \div \frac{3}{2} = \frac{13}{\underset{2}{\cancel{4}}} \cdot \frac{\overset{1}{\cancel{2}}}{3} = \frac{13}{6}$$

Ratio in lowest terms

Work Problem 4 at the Side. ▶

OBJECTIVE 3 Solve ratio problems after converting units. When a ratio compares measurements, both measurements must be in the *same* units. For example, *feet* must be compared to *feet, hours* to *hours, pints* to *pints*, and *inches* to *inches*.

EXAMPLE 5 Ratio Applications Using Measurement

(a) Write the ratio of the length of the shorter board on the left to the length of the longer board on the right. Compare in inches.

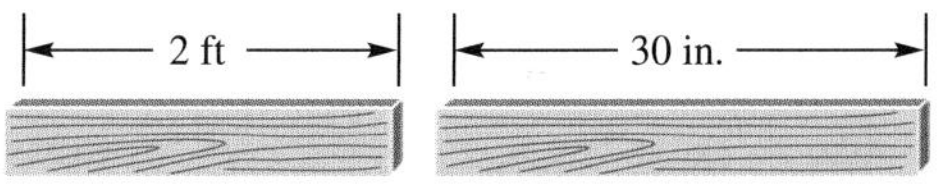

First, express 2 ft in inches. Because 1 ft has 12 in., 2 ft is

$$2 \cdot 12 \text{ in.} = 24 \text{ in.}$$

Continued on Next Page

4 Write each ratio as a ratio of whole numbers in lowest terms.

GS **(a)** $3\frac{1}{2}$ to 4

$$\frac{3\frac{1}{2}}{4} = \frac{\frac{7}{2}}{\frac{4}{1}}$$

$$= \frac{7}{2} \div \frac{4}{1} = \frac{7}{2} \cdot \frac{1}{4}$$

Reciprocals

$$= \frac{\_\_\_}{\_\_\_}$$

(b) $5\frac{5}{8}$ pounds to $3\frac{3}{4}$ pounds

(c) $3\frac{1}{2}$ in. to $\frac{7}{8}$ in.

Answers

4. (a) $\frac{7}{8}$ **(b)** $\frac{3}{2}$ **(c)** $\frac{4}{1}$

5 Write each ratio as a fraction in lowest terms. (*Hint:* Recall that it is usually easier to write the ratio using the smaller measurement unit.)

GS **(a)** 9 in. to 6 ft

Change 6 ft to inches.

$6 \cdot 12 \text{ in.} = 72 \text{ in.}$

$$\frac{9 \text{ in.}}{6 \text{ ft}} = \frac{9 \text{ in.}}{\_\_\_ \text{ in.}}$$

$$= \frac{9 \div \_\_\_}{\_\_\_ \div \_\_\_} =$$

(b) 2 days to 8 hours

(c) 7 yd to 14 ft

(d) 3 quarts to 3 gallons

(e) 25 minutes to 2 hours

(f) 4 pounds to 12 ounces

Answers

5. (a) 72 in.; $\frac{9 \div 9}{72 \div 9} = \frac{1}{8}$ (b) $\frac{6}{1}$ (c) $\frac{3}{2}$
(d) $\frac{1}{4}$ (e) $\frac{5}{24}$ (f) $\frac{16}{3}$

On the previous page, the length of the board on the left is 24 in., so the ratio of the lengths is

$$\frac{2 \text{ ft}}{30 \text{ in.}} = \frac{24 \text{ in.}}{30 \text{ in.}} = \frac{24}{30}.$$

Once the units match, you can divide them out.

Write the ratio in lowest terms.

$$\frac{24}{30} = \frac{24 \div 6}{30 \div 6} = \frac{4}{5} \leftarrow \text{Ratio in lowest terms}$$

The shorter board is $\frac{4}{5}$ the length of the longer board.

Note

Notice in the example above that we wrote the ratio using the smaller unit (inches are smaller than feet). Using the smaller unit will help you avoid working with fractions. If we wrote the ratio using feet, then, since

$$30 \text{ in.} = 2\frac{1}{2} \text{ ft},$$

the ratio in feet is

$$\frac{2 \text{ ft}}{2\frac{1}{2} \text{ ft}} = \frac{2}{1} \div \frac{5}{2} = \frac{2}{1} \cdot \frac{2}{5} = \frac{4}{5}. \leftarrow \text{Same result}$$

The ratio is the same, but it takes more steps to get the answer. Using the smaller unit is usually easier.

(b) Write the ratio of 28 days to 3 weeks.

Since it is easier to write the ratio using the smaller measurement unit, compare in *days* because days are shorter than weeks.

First express 3 weeks in days. Because 1 week has 7 days, 3 weeks is

$$3 \cdot 7 \text{ days} = 21 \text{ days}.$$

The ratio in days is shown below.

$$\frac{28 \text{ days}}{3 \text{ weeks}} = \frac{28 \text{ days}}{21 \text{ days}} = \frac{28}{21} = \frac{28 \div 7}{21 \div 7} = \frac{4}{3} \leftarrow \text{Ratio in lowest terms}$$

The following table will help you set up ratios that compare measurements. You will work with these measurements again in **Appendix A**.

Measurement Comparisons

Length	Capacity (Volume)
12 inches = 1 foot	2 cups = 1 pint
3 feet = 1 yard	2 pints = 1 quart
5280 feet = 1 mile	4 quarts = 1 gallon
Weight	**Time**
16 ounces = 1 pound	60 seconds = 1 minute
2000 pounds = 1 ton	60 minutes = 1 hour
	24 hours = 1 day
	7 days = 1 week

◀ **Work Problem 5 at the Side.**

5.1 Exercises

1. **CONCEPT CHECK** What is a ratio? Write an explanation. Then make up an example of a ratio that is different from the ones in the textbook.

2. **CONCEPT CHECK** Both quantities in a ratio have the same units. That means you can do what to the units?

3. **CONCEPT CHECK** To rewrite the ratio $\frac{80}{20}$ in lowest terms, divide both the numerator and the denominator by _____. The ratio in lowest terms is _____.

4. **CONCEPT CHECK** To rewrite the ratio $\frac{20}{75}$ in lowest terms, divide both the numerator and the denominator by _____. The ratio in lowest terms is _____.

Write each ratio as a fraction in lowest terms. ***See Examples 1 and 2.***

5. 8 days to 9 days

6. \$11 to \$15

7. \$100 to \$50

8. 35¢ to 7¢

9. 30 minutes to 90 minutes

10. 9 pounds to 36 pounds

11. 80 miles to 50 miles

12. 300 people to 450 people

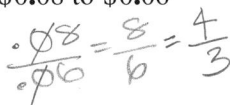

13. 6 hours to 16 hours

14. 45 books to 35 books

Write each ratio as a ratio of whole numbers in lowest terms. ***See Examples 3 and 4.***

15. \$4.50 to \$3.50

16. \$0.08 to \$0.06

$$\frac{.08}{.06} = \frac{8}{6} = \frac{4}{3}$$

17. 15 to $2\frac{1}{2} = \dfrac{15}{2\frac{1}{2}} = \dfrac{\frac{15}{1}}{\frac{5}{2}}$

$$= \frac{15}{1} \div \frac{5}{2} = \frac{\overset{3}{15}}{1} \cdot \frac{2}{\underset{1}{5}} = \underline{\quad}$$

18. 5 to $1\frac{1}{4} = \dfrac{5}{1\frac{1}{4}} = \dfrac{\frac{5}{1}}{\frac{5}{4}}$

$$= \frac{5}{1} \div \frac{5}{4} = \frac{\overset{1}{5}}{1} \cdot \frac{4}{\underset{1}{5}} = \underline{\quad}$$

19. $1\frac{1}{4}$ to $1\frac{1}{2}$

20. $2\frac{1}{3}$ to $2\frac{2}{3}$

21. **CONCEPT CHECK** When writing the ratio of 20 ounces to 3 pounds as a fraction in lowest terms, would it be easier to use ounces or pounds? _____ Explain why.

22. **CONCEPT CHECK** When writing the ratio of 4 days to 36 hours as a fraction in lowest terms, would it be easier to use days or hours? _____ Explain why.

Write each ratio as a fraction in lowest terms. For help, use the table of measurement relationships in this section. ***See Example 5.***

23. 4 ft to 30 in.
Hint: First, convert 4 ft to inches.

24. 8 ft to 4 yd
Hint: First, convert 4 yd to feet.

25. 5 minutes to 1 hour

26. 8 quarts to 5 pints

27. 15 hours to 2 days

28. 3 pounds to 6 ounces

29. 5 gallons to 5 quarts

30. 3 cups to 3 pints

The table shows the number of greeting cards sold in the United States for various occasions. Use the information to answer Exercises 31–36. Write each ratio as a fraction in lowest terms.

Holiday/Event	Cards Sold
Valentine's Day	150 million
Mother's Day	130 million
Father's Day	95 million
Graduation	60 million
Halloween	20 million
Thanksgiving	10 million

Source: Hallmark Cards.

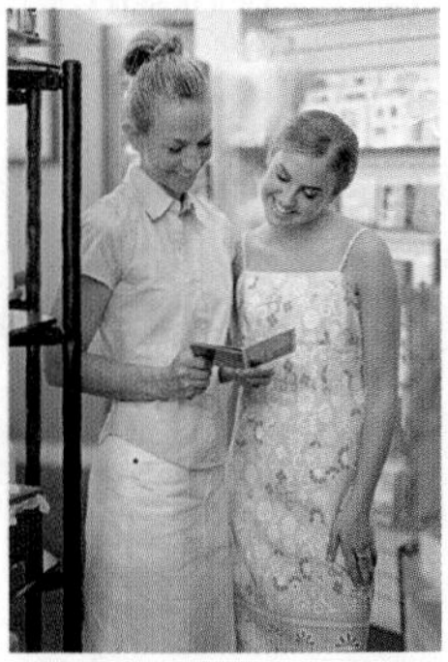

31. Find the ratio of Thanksgiving cards to graduation cards.

32. Find the ratio of Halloween cards to Mother's Day cards.

33. Find the ratio of Valentine's Day cards to Thanksgiving cards.

34. Find the ratio of Mother's Day cards to Father's Day cards.

35. Explain how you might use the information in the table if you owned a shop selling gifts and greeting cards.

36. Why is the ratio of Valentine's Day cards to graduation cards $\frac{5}{2}$? Give two possible reasons.

The bar graph shows worldwide sales of the most popular songs of all time. Use the graph to complete Exercises 37–40. Write each ratio as a fraction in lowest terms.

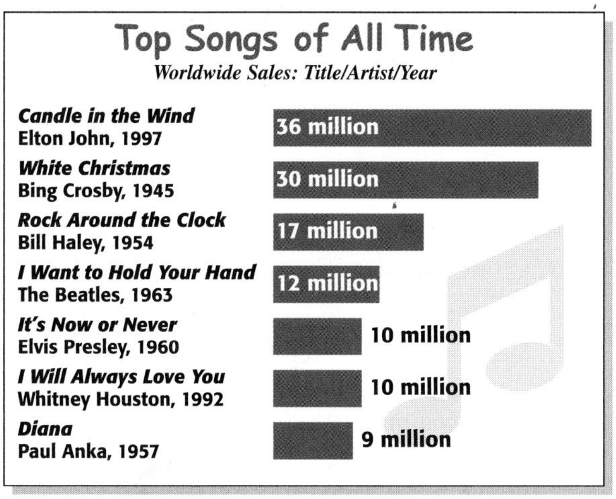

Source: The Music Information Database.

37. Write a ratio that compares the top-selling song to the second best seller, and a ratio that compares the top-selling song to the third best seller.

38. Write two ratios that compare sales of Elvis Presley's song with sales of the songs just ahead and just behind it in the graph.

39. Sales of which two songs give a ratio of $\frac{3}{1}$? There may be more than one correct answer.

40. Sales of which two songs give a ratio of $\frac{5}{6}$? There may be more than one correct answer.

For each figure, find the ratio of the length of the longest side to the length of the shortest side. Write each ratio as a fraction in lowest terms.

41.

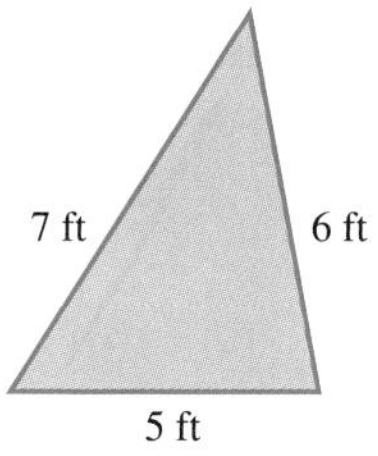

42.

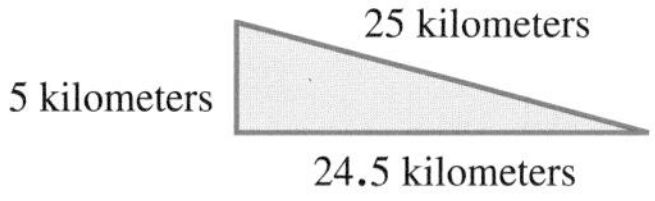

43.

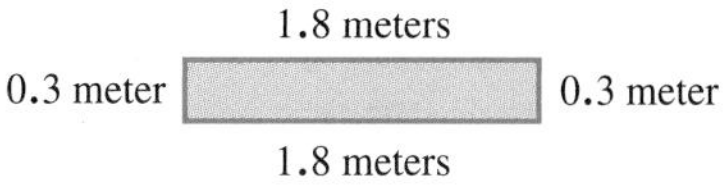

44.

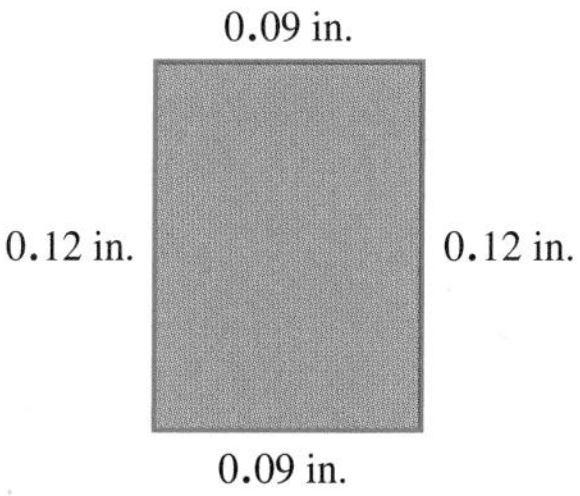

45.

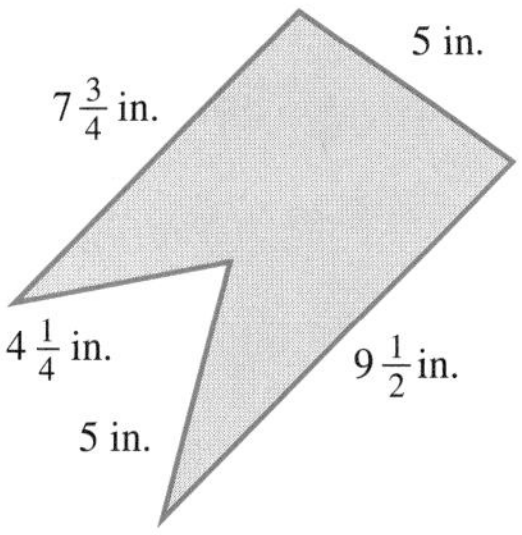

46.

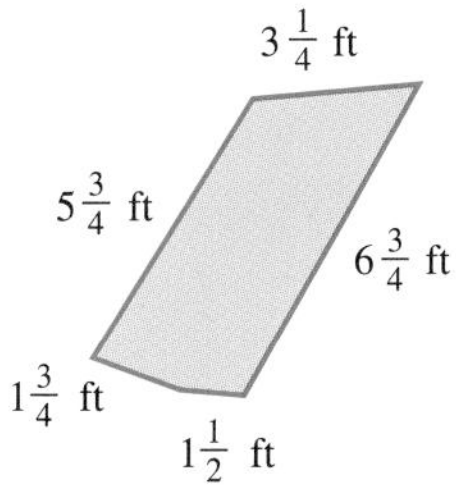

Write each ratio as a fraction in lowest terms.

47. The price of automobile engine oil has gone from \$10 to \$12.50 for the 5 quarts needed for an oil change. Find the ratio of the increase in price to the original price.

48. The price that a pharmacy pays for an antibiotic decreased from \$8.80 to \$5.60 for 10 tablets. Find the ratio of the decrease in price to the original price.

49. The first time a movie was made in Minnesota, the cast and crew spent $59\frac{1}{2}$ days filming winter scenes. The next year, another movie was filmed in $8\frac{3}{4}$ weeks. Find the ratio of the first movie's filming time to the second movie's time. Compare in weeks.

50. The percheron, a large draft horse, measures about $5\frac{3}{4}$ ft at the shoulder. The prehistoric ancestor of the horse measured only $15\frac{3}{4}$ in. at the shoulder. Find the ratio of the percheron's height to its prehistoric ancestor's height. Compare in inches. (*Source: Eyewitness Books: Horse.*)

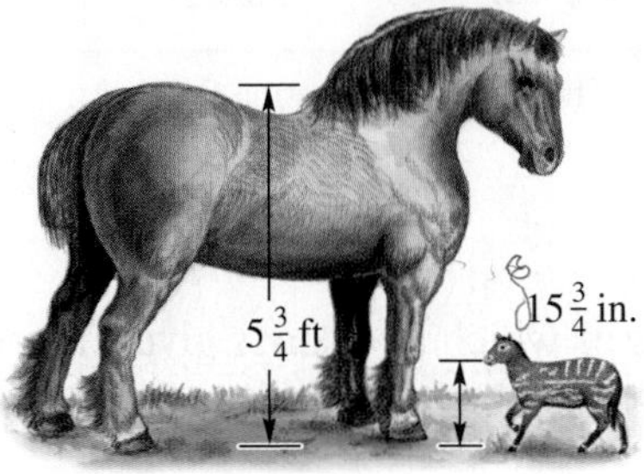

Relating Concepts (Exercises 51–54) For Individual or Group Work

Use your knowledge of ratios to ***work Exercises 51–54 in order.***

51. In this painting, what is the ratio of the length of the longest side to the length of the shortest side? What other measurements could the painting have and still maintain the same ratio?

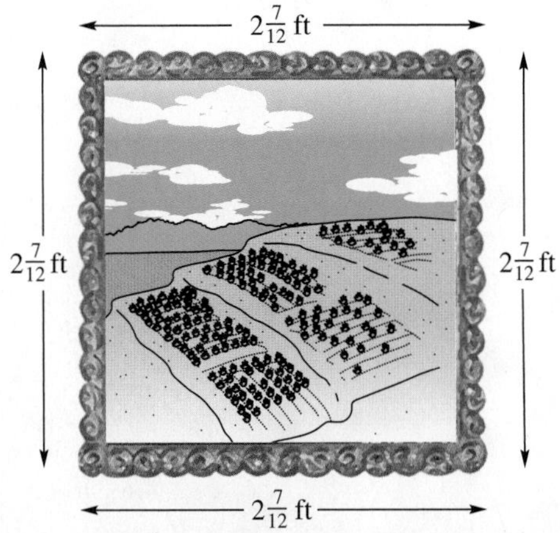

52. The ratio of my son's age to my daughter's age is 4 to 5. One possibility is that my son is 4 years old and my daughter is 5 years old. Find six other possibilities that fit the 4 to 5 ratio.

53. Amelia said that the ratio of her age to her mother's age is 5 to 3. Is this possible? Explain your answer.

54. Would you prefer that the ratio of your income to your friend's income be 1 to 3 or 3 to 1? Explain your answers.

5.2 Rates

A *ratio* compares two measurements with the **same** units, such as 9 feet to 12 feet (both length measurements). But many of the comparisons we make use measurements with **different** units.

160 dollars for 8 hours (money to time)

450 miles on 15 gallons (distance to capacity)

This type of comparison is called a **rate.**

OBJECTIVES

1. Write rates as fractions.
2. Find unit rates.
3. Find the best buy based on cost per unit.

OBJECTIVE 1 Write rates as fractions. Suppose you hiked 14 miles in 4 hours. The *rate* at which you hiked can be written as a fraction in lowest terms.

$$\frac{14 \text{ miles}}{4 \text{ hours}} = \frac{14 \text{ miles} \div 2}{4 \text{ hours} \div 2} = \frac{7 \text{ miles}}{2 \text{ hours}}$$ ← Rate in lowest terms

In a rate, you often find these words separating the quantities you are comparing.

in for on per from

CAUTION

When writing a rate, always include the units, such as miles, hours, dollars, and so on. Because the units in a rate are different, the units do *not* divide out.

VOCABULARY TIP

Ratio vs. Rate **Ratios** compare two quantities that have the **same** units, so the units divide out. However, **rates** compare two quantities with **different** units, so you must write the units in a rate.

EXAMPLE 1 Writing Rates in Lowest Terms

Write each rate as a fraction in lowest terms.

(a) 5 gallons of chemical for \$60

$$\frac{5 \text{ gallons} \div 5}{60 \text{ dollars} \div 5} = \frac{1 \text{ gallon}}{12 \text{ dollars}}$$ ← Write the units: gallons and dollars.

(b) \$1500 wages in 10 weeks

$$\frac{1500 \text{ dollars} \div 10}{10 \text{ weeks} \div 10} = \frac{150 \text{ dollars}}{1 \text{ week}}$$

(c) 2225 miles on 75 gallons of gas

$$\frac{2225 \text{ miles} \div 25}{75 \text{ gallons} \div 25} = \frac{89 \text{ miles}}{3 \text{ gallons}}$$

Be sure to write the units in a *rate*: dollars, miles, gallons, and so on.

Work Problem 1 at the Side. ▶

1 Write each rate as a fraction in lowest terms.

GS **(a)** \$6 for 30 packets

$$\frac{\$6}{30 \text{ packets}} = \frac{\$\_\_\_}{\_\_\_ \text{ packets}}$$

To write the rate in lowest terms, divide both numerator and denominator by ____.

(b) 500 miles in 10 hours

(c) 4 teachers for 90 students

OBJECTIVE 2 Find unit rates. When the *denominator* of a rate is 1, it is called a **unit rate.** We use unit rates frequently. For example, you earn \$12.75 for *1 hour* of work. This unit rate is written

\$12.75 per hour or \$12.75/hour.

Use per or a slash mark (/) when writing unit rates.

Answers

1. (a) $\frac{\$1}{5 \text{ packets}}$; divide by 6.
(b) $\frac{50 \text{ miles}}{1 \text{ hour}}$ **(c)** $\frac{2 \text{ teachers}}{45 \text{ students}}$

VOCABULARY TIP

Per means "for each." For example, if your car gets 30 miles **per** gallon, it gets 30 miles *for each gallon* of gas. Use **per** or a slash mark, like this: 30 miles/gallon.

2 Find each unit rate.

GS **(a)** \$4.35 for 3 pounds of apples

$$\frac{\$4.35}{3 \text{ pounds}}$$ ← Fraction bar indicates division.

$3\overline{)4.35}$ with quotient 1. — Finish the division.

(b) 304 miles on 9.5 gallons of gas

(c) \$850 in 5 days

(d) 24-pound turkey for 15 people

Answers

2. **(a)** \$1.45/pound **(b)** 32 miles/gallon **(c)** \$170/day **(d)** 1.6 pounds/person

EXAMPLE 2 Finding Unit Rates

Find each unit rate.

(a) 337.5 miles on 13.5 gallons of gas
Write the rate as a fraction.

$$\frac{337.5 \text{ miles}}{13.5 \text{ gallons}}$$ ← The fraction bar indicates division.

Divide 337.5 by 13.5 to find the unit rate.

$$13.5\overline{)337.5} \quad \text{quotient } 25.$$

$$\frac{337.5 \text{ miles} \div 13.5}{13.5 \text{ gallons} \div 13.5} = \frac{25 \text{ miles}}{1 \text{ gallon}}$$

Be sure to write **miles** and **gallon** in your answer.

The unit rate is 25 miles per gallon, or 25 miles/gallon.

(b) 549 miles in 18 hours

$$\frac{549 \text{ miles}}{18 \text{ hours}} \qquad \text{Divide: } 18\overline{)549.0} \quad \text{quotient } 30.5$$

The unit rate is 30.5 miles/hour.

(c) \$810 in 6 days

$$\frac{810 \text{ dollars}}{6 \text{ days}} \qquad \text{Divide: } 6\overline{)810} \quad \text{quotient } 135$$

Use *per* or a slash mark to write unit rates.

The unit rate is \$135/day.

◄ **Work Problem 2 at the Side.**

OBJECTIVE 3 Find the best buy based on cost per unit. When shopping for groceries, household supplies, and health and beauty items, you will find many different brands and package sizes. You can save money by finding the lowest *cost per unit*.

Cost per Unit

Cost per unit is a rate that tells how much you pay for *one* item or *one* unit. Examples are \$3.75 per gallon, \$47 per shirt, and \$2.98 per pound.

EXAMPLE 3 Determining the Best Buy

The local store charges the following prices for pancake syrup. Find the best buy.

Continued on Next Page

The best buy is the container with the *lowest* cost per unit. All the containers are measured in *ounces* (oz), so you first need to find the *cost per ounce* for each one. Divide the price of the container by the number of ounces in it. Round to the nearest thousandth, if necessary.

Let the *order* of the *words* help you set up the rate.

cost → **$1.28**
per (means divide) → ——
ounce → **12 ounces**

Size	Cost Per Unit (Rounded)
12 ounces	$\frac{\$1.28}{12 \text{ ounces}} \approx \$0.107$ per ounce (highest)
24 ounces	$\frac{\$1.81}{24 \text{ ounces}} \approx \$0.075$ per ounce (lowest)
36 ounces	$\frac{\$2.73}{36 \text{ ounces}} \approx \$0.076$ per ounce

The lowest cost per ounce is $0.075, so the 24-ounce container is the best buy.

Note

Earlier we rounded money amounts to the nearest hundredth (nearest cent). But when comparing unit costs, rounding to the nearest thousandth will help you see the difference between very similar unit costs. Notice that the 24-ounce and 36-ounce syrup containers above would both have rounded to $0.08 per ounce if we had rounded to hundredths.

Work Problem 3 at the Side. ▶

Calculator Tip

When using a calculator to find unit prices, remember that division is *not* commutative. In **Example 3** you wanted to find cost per ounce. Let the *order* of the *words* help you enter the numbers in the correct order.

cost	per	ounce		
↓	↓	↓		
Enter the cost.	*Per* means divide.	Enter number of ounces.		
↓	↓	↓		
$2.73	÷	**36**	=	**0.076 (rounded)**

If you entered 36 ÷ 2.73 =, you'd get the number of *ounces* per *dollar*. How could you use that information to find the best buy? (*Answer:* The best buy would be to get the greatest number of ounces per dollar.)

Finding the best buy is sometimes a complicated process. Things that affect the cost per unit can include "cents off" coupons and differences in how much use you'll get out of each unit.

3 Find the best buy (lowest cost per unit) for each purchase.

GS **(a)** 2 quarts for $3.25
3 quarts for $4.95
4 quarts for $6.48

Divide to find each unit cost.

cost → $3.25
per → ——
quart → 2 quarts

Unit cost is $ ____ per quart.

cost → $
per → ——
quart → 3 quarts

Unit cost is $ ____ per ____.

cost →
per → ——
quart →

Unit cost is $ ____ per ____.

Now compare the three unit costs. The *lowest* one is the best buy. Circle it.

(b) 6 cans of cola for $1.99
12 cans of cola for $3.49
24 cans of cola for $7

Answers

3. (a) Unit cost is $1.625 per quart
$4.95; Unit cost is $1.65 per quart
$\frac{\$6.48}{4 \text{ quarts}}$; Unit cost is $1.62 per quart
lowest unit cost is 4 quarts, at $1.62 per quart

(b) 12 cans, at $0.291 per can (rounded)

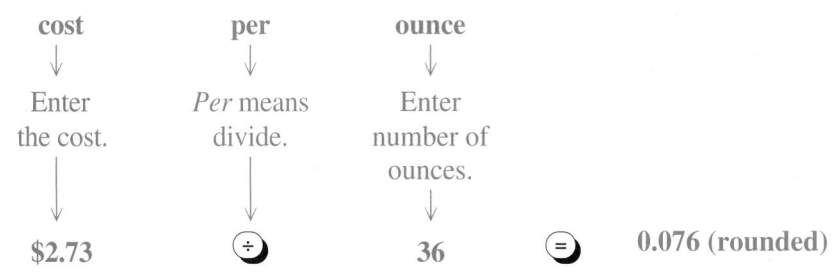

EXAMPLE 4 Solving Best Buy Applications

Solve each application problem.

(a) There are many brands of liquid laundry detergent. If you feel they all do a good job of cleaning your clothes, you can base your purchase on cost per unit. But some brands are "concentrated" so you can use less detergent for each load of clothes. The labels from two bottles are shown below. Which of the choices is the best buy?

To find Sudzy's unit cost, divide $3.99 by 64 ounces, not 50 ounces. Read the label. You're getting as many clothes washed as if you bought 64 ounces. Similarly, to find White-O's unit cost, divide $9.89 by 256 ounces (twice 128 ounces, or 2 • 128 ounces = 256 ounces).

$$\text{Sudzy} \quad \frac{\$3.99}{64 \text{ ounces}} \approx \$0.062 \text{ per ounce}$$

The best buy is the *lower* cost per ounce.

$$\text{White-O} \quad \frac{\$9.89}{256 \text{ ounces}} \approx \$0.039 \text{ per ounce}$$

White-O has the lower cost per ounce and is the better buy. (However, if you try it and it really doesn't get out all the stains, Sudzy may be worth the extra cost.)

(b) "Cents-off" coupons also affect the best buy. Suppose you are looking at these choices for "extra-strength" pain reliever. Both brands have the same amount of pain reliever in each tablet.

Brand X is $2.29 for 50 tablets.

Brand Y is $10.75 for 200 tablets.

You have a 40¢ coupon for Brand X and a 75¢ coupon for Brand Y. Which choice is the best buy?

To find the best buy, first subtract the coupon amounts, then divide to find the lower cost per ounce.

Brand X costs $2.29 − $0.40 = $1.89

$$\frac{\$1.89}{50 \text{ tablets}} \approx \$0.038 \text{ per tablet}$$

Look for the *lower* cost per tablet.

Brand Y costs $10.75 − $0.75 = $10.00

$$\frac{\$10.00}{200 \text{ tablets}} = \$0.05 \text{ per tablet}$$

Brand X has the lower cost per tablet and is the better buy.

◄ **Work Problem 4 at the Side.**

4 Solve each problem.

(a) Some batteries claim to last longer than others. If you believe these claims, which brand is the best buy?

Four-pack of AA-size batteries for $2.79

One AA-size battery for $1.19; lasts twice as long

(b) Which tube of toothpaste is the best buy? You have a coupon for 85¢ off Brand C and a coupon for 50¢ off Brand D.

Brand C is $3.89 for 6 ounces.

Brand D is $1.89 for 2.5 ounces.

Answers

4. (a) One battery that lasts twice as long (like getting two) is the better buy. The cost per unit is $0.595 per battery. The four-pack is $0.698 per battery (rounded).
(b) Brand C with the 85¢ coupon is the better buy at $0.507 per ounce (rounded). Brand D with the 50¢ coupon is $0.556 per ounce.

5.2 Exercises

FOR EXTRA HELP

Download the MyDashBoard App

MyMathLab®

Write each rate as a fraction in lowest terms. ***See Example 1.***

1. 10 cups for 6 people

2. $12 for 30 pens

3. 15 feet in 35 seconds

4. 100 miles in 30 hours

5. 72 miles on 4 gallons

6. 132 miles on 8 gallons

Find each unit rate. ***See Example 2.***

7. CONCEPT CHECK To find a unit rate, do you add, subtract, multiply, or divide? _________ Show the correct set-up to find this unit rate: $5.85 for 3 boxes.

8. CONCEPT CHECK Circle the correct answer. When finding this unit rate, 15 hospital rooms for 3 nurses, the final answer will be:

5 to 1 5 rooms for 1 nurse 5 nurses for 1 room

9. $60 in 5 hours

10. $2500 in 20 days

11. 7.5 pounds for 6 people

12. 44 bushels from 8 trees

Earl kept the following record of the gas he bought for his car. For each entry, find the number of miles he traveled and the unit rate. Round your answers to the nearest tenth.

	Date	Odometer at Start	Odometer at End	Miles Traveled	Gallons Purchased	Miles per Gallon
13.	2/4	27,432.3	27,758.2		15.5	
14.	2/9	27,758.2	28,058.1		13.4	
15.	2/16	28,058.1	28,396.7		16.2	
16.	2/20	28,396.7	28,704.5		13.3	

Source: Author's car records.

Find the best buy (based on the cost per unit) for each item. Round to the nearest thousandth, if necessary. ***See Example 3.*** (*Source:* Cub Foods, Target, Rainbow Foods.)

17. Black pepper

GS

Complete the divisions:

$$\begin{matrix}\text{cost} \rightarrow \\ \text{per} \rightarrow \\ \text{ounce} \rightarrow\end{matrix} \frac{\$2.25}{2\text{ oz}} =$$

$$\begin{matrix}\text{cost} \rightarrow \\ \text{per} \rightarrow \\ \text{ounce} \rightarrow\end{matrix} \frac{\$\quad}{4\text{ oz}} =$$

Lower unit cost (better buy) is ________, so the best buy is ________.

18. Shampoo

GS

Complete the divisions:

$$\begin{matrix}\text{cost} \rightarrow \\ \text{per} \rightarrow \\ \text{ounce} \rightarrow\end{matrix} \frac{\$3.59}{8\text{ oz}} =$$

$$\begin{matrix}\text{cost} \rightarrow \\ \text{per} \rightarrow \\ \text{ounce} \rightarrow\end{matrix} \frac{\$\quad}{12\text{ oz}} =$$

Lower unit cost (better buy) is ________, so the best buy is ________.

19. Cereal

12 ounces for $2.49
14 ounces for $2.89
18 ounces for $3.96

20. Soup (same size cans)

2 cans for $2.18
3 cans for $3.57
5 cans for $5.29

21. Chunky peanut butter

12 ounces for $1.29
18 ounces for $1.79
28 ounces for $3.39
40 ounces for $4.39

22. Baked beans

8 ounces for $0.59
16 ounces for $0.99
21 ounces for $1.29
28 ounces for $1.89

23. Suppose you are choosing between two brands of chicken noodle soup. Brand A is $0.88 per can and Brand B is $0.98 per can. The cans are the same size but Brand B has more chunks of chicken in it. Which soup is the better buy? Explain your choice.

24. A small bag of potatoes costs $0.19 per pound. A large bag costs $0.15 per pound. But there are only two people in your family, so half the large bag would probably rot before you used it up. Which bag is the better buy? Explain.

Solve each application problem. ***See Examples 2–4.***

25. Makesha lost 10.5 pounds in six weeks. What was her rate of loss in pounds per week?

26. Enrique's taco recipe uses three pounds of meat to feed 10 people. Give the rate in pounds per person.

27. Russ works 7 hours to earn $85.82. What is his pay rate per hour?

28. Find the cost of 1 gallon of Hawaiian Punch beverage if 18 gallons for a graduation party cost $55.62.

The table lists information about three long-distance calling cards. The connection fee is charged each time you make a call, no matter how long the call lasts. Use the table to answer Exercises 29–32. Round answers to the nearest thousandth when necessary.

LONG-DISTANCE CALLING CARDS (U.S.)

Card Name	Cost per Minute	Connection Fee
Penny Saver	\$0.005	\$0.39
Most Minutes	0.0025	0.49
USA Card	0.01	0.25

Source: www.noblecom.com

29. **(a)** Find the *actual* total cost, including the connection charge, for a five-minute call using each card.

(b) Find the cost per minute for this call using each card and select the best buy.

30. **(a)** Find the *actual* total cost, including the connection charge, for a 90-minute call using each card.

(b) Find the cost per minute for this call using each card and select the best buy.

31. Find the *actual* total cost and the per minute cost for a 30-minute call using each card. What do you see when you compare the unit rates for this call?

32. All the cards round calls up to the next full minute.

(a) Suppose you call the wrong number. How much would you pay for this 40-*second* call on each card?

(b) How much would you save on this call by using the USA Card instead of Most Minutes?

33. If you believe the claims that some batteries last longer, which is the better buy?

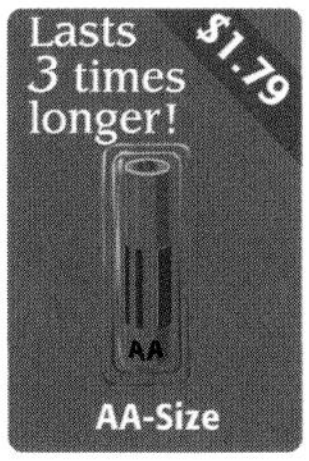

34. Which is the better buy, assuming these laundry detergents both clean equally well?

35. Three brands of cornflakes are available. Brand G is priced at \$2.39 for 10 ounces. Brand K is \$3.99 for 20.3 ounces and Brand P is \$3.39 for 16.5 ounces. You have a coupon for 50¢ off Brand P and a coupon for 60¢ off Brand G. Which cereal is the best buy based on cost per unit?

36. Two brands of facial tissue are available. Brand K is priced at \$5 for three boxes of 175 tissues each. Brand S is priced at \$1.29 per box of 125 tissues. You have a coupon for 20¢ off one box of Brand S and a coupon for 45¢ off one box of Brand K. How can you get the best buy on one box of tissue?

Relating Concepts (Exercises 37–41) For Individual or Group Work

On the first page of this chapter, we said that unit rates can help you get the best deal on cell phone service. Use the information in the table to ***work Exercises 37–41 in order.*** *Round all money answers to the nearest cent.*

CELL PHONE PLANS

Plan	Anytime Minutes	Monthly Charge*	Cost per Minute (to nearest cent)	Overage†
A	450	$39.95		45¢
B	900	$59.95		40¢

*Does not include taxes or special fees.
†Cost per extra minute if you use more minutes than your plan covers.
Source: www.Verizon.com

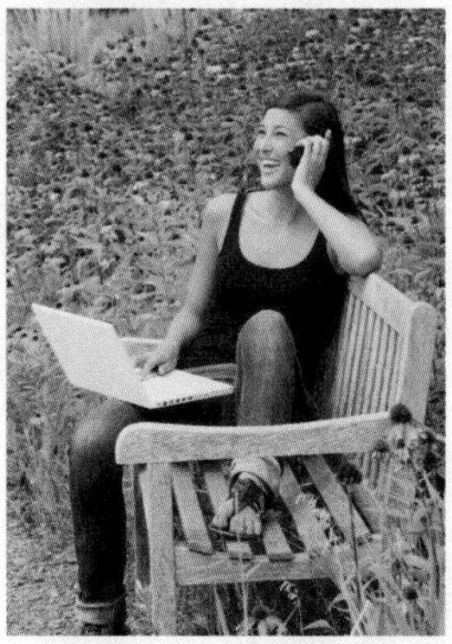

37. Find the cost per minute for each plan. Round your answers to the nearest cent and write them in the table above. Which plan is the better buy based on cost per minute?

38. In a 30-day month, what is the average number of minutes you can use your phone each day on:
(a) Plan A
(b) Plan B

39. Suppose you are on Plan A. Last month you used 2 more minutes each day than the average number of minutes you calculated in **Exercise 38.** Find the new cost per minute for last month. (Assume 30 days in last month.)

40. Suppose you are on Plan B. Last month you used 3 more minutes each day than the average number of minutes you calculated in **Exercise 38.** Find the new cost per minute for last month. (Assume 30 days in last month.)

41. The company also offers Plan C, with unlimited minutes for $69.95 per month. If you had Plan C, find the cost per minute for the situations described in **Exercises 39 and 40.**

5.3 Proportions

OBJECTIVE 1 Write proportions. A **proportion** states that two ratios (or rates) are equal. For example,

$$\frac{\$20}{4 \text{ hours}} = \frac{\$40}{8 \text{ hours}}$$

is a proportion that says the rate $\frac{\$20}{4 \text{ hours}}$ is equal to the rate $\frac{\$40}{8 \text{ hours}}$. As the amount of money doubles, the number of hours also doubles. This proportion is read:

20 dollars is to 4 hours as 40 dollars is to 8 hours.

EXAMPLE 1 Writing Proportions

Write each proportion.

(a) 6 ft is to 11 ft as 18 ft is to 33 ft.

$$\frac{6 \text{ ft}}{11 \text{ ft}} = \frac{18 \text{ ft}}{33 \text{ ft}} \quad \text{so} \quad \frac{6}{11} = \frac{18}{33}$$

The common units (ft) divide out and are not written.

(b) \$9 is to 6 liters as \$3 is to 2 liters.

$$\frac{\$9}{6 \text{ liters}} = \frac{\$3}{2 \text{ liters}}$$

The units do *not* match, so you must write them in the proportion.

Work Problem 1 at the Side. ▶

OBJECTIVE 2 Determine whether proportions are true or false. There are two ways to see whether a proportion is true. One way is to *write both of the ratios in lowest terms.*

EXAMPLE 2 Writing Both Ratios in Lowest Terms

Determine whether each proportion is true or false by writing both ratios in lowest terms.

(a) $\frac{5}{9} = \frac{18}{27}$

Write each ratio in lowest terms.

$\frac{5}{9}$ ← Already in lowest terms $\quad \frac{18 \div 9}{27 \div 9} = \frac{2}{3}$ ← Lowest terms

Because $\frac{5}{9}$ is *not* equivalent to $\frac{2}{3}$, the proportion is *false.*

(b) $\frac{16}{12} = \frac{28}{21}$

Write each ratio in lowest terms.

$$\frac{16 \div 4}{12 \div 4} = \frac{4}{3} \quad \text{and} \quad \frac{28 \div 7}{21 \div 7} = \frac{4}{3}$$

Both ratios in lowest terms.

Both ratios are equivalent to $\frac{4}{3}$, so the proportion is *true.*

Work Problem 2 at the Side. ▶

OBJECTIVES

1. Write proportions.
2. Determine whether proportions are true or false.
3. Find cross products.

1 Write each proportion.

GS **(a)** \$7 is to 3 cans as \$28 is to 12 cans.

$\frac{\$7}{3 \text{ cans}} = \frac{\_\_\_}{\_\_\_}$ Complete the proportion.

(b) 9 meters is to 16 meters as 18 meters is to 32 meters.

(c) 5 is to 7 as 35 is to 49.

(d) 10 is to 30 as 60 is to 180.

2 Determine whether each proportion is true or false by writing both ratios in lowest terms.

(a) $\frac{6}{12} = \frac{15}{30}$

(b) $\frac{20}{24} = \frac{3}{4}$

(c) $\frac{25}{40} = \frac{30}{48}$

(d) $\frac{35}{45} = \frac{12}{18}$

Answers

1. (a) $\frac{\$7}{3 \text{ cans}} = \frac{\$28}{12 \text{ cans}}$ **(b)** $\frac{9}{16} = \frac{18}{32}$
(c) $\frac{5}{7} = \frac{35}{49}$ **(d)** $\frac{10}{30} = \frac{60}{180}$

2. (a) $\frac{1}{2} = \frac{1}{2}$; true **(b)** $\frac{5}{6} \neq \frac{3}{4}$; false
(c) $\frac{5}{8} = \frac{5}{8}$; true **(d)** $\frac{7}{9} \neq \frac{2}{3}$; false

OBJECTIVE 3 **Find cross products.** Another way to test whether the ratios in a proportion are equivalent is to compare *cross products.*

Using Cross Products to Determine Whether a Proportion Is True

To see whether a proportion is true, first multiply along one diagonal, then multiply along the other diagonal, as shown here.

$$\frac{2}{5} = \frac{4}{10} \qquad \begin{array}{l} 5 \cdot 4 = 20 \\ 2 \cdot 10 = 20 \end{array} \quad \text{Cross products are equal.}$$

In this case the **cross products** are both 20. When cross products are *equal,* the proportion is *true.* If the cross products are *unequal,* the proportion is *false.*

Note

The cross products test is based on rewriting both fractions with a common denominator of 5 • 10, or 50.

$$\frac{2 \cdot 10}{5 \cdot 10} = \frac{20}{50} \quad \text{and} \quad \frac{4 \cdot 5}{10 \cdot 5} = \frac{20}{50}$$

We see that $\frac{2}{5}$ and $\frac{4}{10}$ are equivalent because both can be rewritten as $\frac{20}{50}$. The cross products test takes a shortcut by comparing only the two numerators (20 = 20).

EXAMPLE 3 Using Cross Products

Use cross products to see whether each proportion is true or false.

(a) $\frac{3}{5} = \frac{12}{20}$

Multiply along one diagonal, then multiply along the other diagonal.

$$\frac{3}{5} = \frac{12}{20} \qquad \begin{array}{l} 5 \cdot 12 = 60 \\ 3 \cdot 20 = 60 \end{array} \quad \text{Equal cross products; proportion is } \textit{true}.$$

The cross products are *equal,* so the proportion is *true.*

CAUTION

Use cross products *only* when working with *proportions.* Do ***not*** use cross products when multiplying fractions, adding fractions, or writing fractions in lowest terms.

Continued on Next Page

(b) $\dfrac{2\frac{1}{3}}{3\frac{1}{3}} = \dfrac{9}{16}$

Find the cross products.

Write $3\frac{1}{3}$ as $\frac{10}{3}$ and 9 as $\frac{9}{1}$.

$$\frac{2\frac{1}{3}}{3\frac{1}{3}} = \frac{9}{16}$$

$$3\frac{1}{3} \bullet 9 = \frac{10}{\cancel{3}_1} \bullet \frac{\cancel{9}^{3}}{1} = \frac{30}{1} = 30$$

$$2\frac{1}{3} \bullet 16 = \frac{7}{3} \bullet \frac{16}{1} = \frac{112}{3} = 37\frac{1}{3}$$

Write $2\frac{1}{3}$ as $\frac{7}{3}$ and 16 as $\frac{16}{1}$.

Unequal cross products; proportion is *false*.

The cross products are *unequal,* so the proportion is *false.*

Note

The numbers in a proportion do *not* have to be whole numbers. They can be fractions, mixed numbers, decimal numbers, and so on.

Work Problem 3 at the Side. ▶

3 Find the cross products to see whether each proportion is true or false.

GS **(a)** $\dfrac{5}{9} = \dfrac{10}{18}$ $\quad (9)(10) = \_\_\_\_$ $\quad (5)(18) = \_\_\_\_$

(b) $\dfrac{32}{15} = \dfrac{16}{8}$

(c) $\dfrac{10}{17} = \dfrac{20}{34}$

GS **(d)** $\dfrac{2.4}{6} = \dfrac{5}{12}$ $\quad (6)(5) = \_\_\_\_$ $\quad (2.4)(12) = \_\_\_\_$

(e) $\dfrac{3}{4.25} = \dfrac{24}{34}$

(f) $\dfrac{1\frac{1}{6}}{2\frac{1}{3}} = \dfrac{4}{8}$

Answers

3. **(a)** $(9)(10) = 90$; $(5)(18) = 90$; true
(b) $240 \neq 256$; false
(c) $340 = 340$; true
(d) $(6)(5) = 30$; $(2.4)(12) = 28.8$; false
(e) $102 = 102$; true
(f) $9\frac{1}{3} = 9\frac{1}{3}$; true

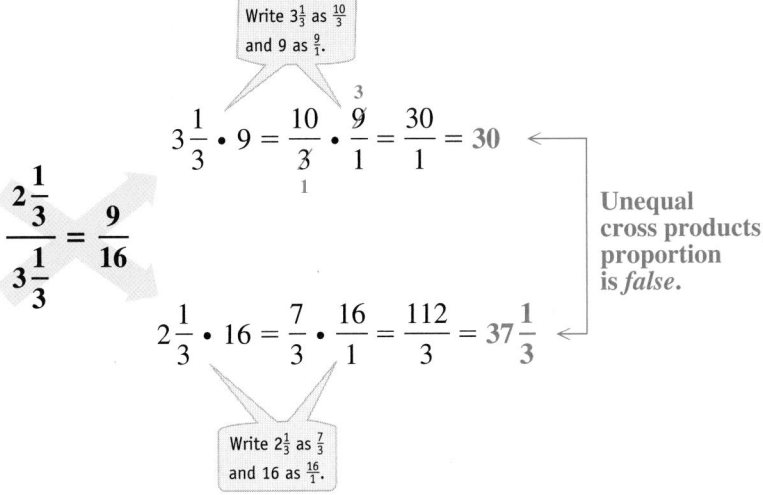

5.3 Exercises

FOR EXTRA HELP

Write each proportion. ***See Example 1.***

1. \$9 is to 12 cans as \$18 is to 24 cans.

2. 28 people is to 7 cars as 16 people is to 4 cars.

3. 200 adults is to 450 children as 4 adults is to 9 children.

4. 150 trees is to 1 acre as 1500 trees is to 10 acres.

5. 120 ft is to 150 ft as 8 ft is to 10 ft.

6. \$6 is to \$9 as \$10 is to \$15.

Determine whether each proportion is true or false by writing the ratios in lowest terms. Show the simplified ratios and then write true *or* false. ***See Example 2.***

7. $\frac{6}{10} = \frac{3}{5}$

8. $\frac{1}{4} = \frac{9}{36}$

9. $\frac{5}{8} = \frac{25}{40}$

10. $\frac{2}{3} = \frac{20}{27}$

11. $\frac{150}{200} = \frac{200}{300}$

12. $\frac{100}{120} = \frac{75}{100}$

13. $\frac{42}{15} = \frac{28}{10}$

14. $\frac{18}{16} = \frac{36}{32}$

15. $\frac{32}{18} = \frac{48}{27}$

16. $\frac{15}{48} = \frac{10}{24}$

17. $\frac{7}{6} = \frac{54}{48}$

18. $\frac{28}{21} = \frac{44}{33}$

19. **CONCEPT CHECK** What is a proportion? Explain. Then draw arrows on the proportion below and show what numbers to multiply to find the cross products.

$$\frac{30}{6} = \frac{45}{9}$$

20. **CONCEPT CHECK** Explain how to tell if a proportion is true or false after you have found the cross products.

Use cross products to determine whether each proportion is true or false. Show the cross products and then circle true *or* false. ***See Example 3.***

21. $\frac{2}{9} = \frac{6}{27}$

True False

22. $\frac{20}{25} = \frac{4}{5}$

True False

23. $\frac{20}{28} = \frac{12}{16}$

True False

24. $\frac{16}{40} = \frac{22}{55}$

True False

25. $\frac{110}{18} = \frac{160}{27}$

True False

26. $\frac{600}{420} = \frac{20}{14}$

True False

27. $\frac{3.5}{4} = \frac{7}{8}$

True False

28. $\frac{36}{23} = \frac{9}{5.75}$

True False

29. $\frac{18}{16} = \frac{2.8}{2.5}$

True False

30. $\frac{0.26}{0.39} = \frac{1.3}{1.9}$

True False

31. $\frac{6}{3\frac{2}{3}} = \frac{18}{11}$

True False

32. $\frac{16}{13} = \frac{2}{1\frac{5}{8}}$

True False

33. $\frac{2\frac{5}{8}}{3\frac{1}{4}} = \frac{21}{26}$

True False

34. $\frac{28}{17} = \frac{9\frac{1}{3}}{5\frac{2}{3}}$

True False

35. $\frac{\frac{2}{3}}{2} = \frac{2.7}{8}$ *Hint:* 2.7 is equivalent to $2\frac{7}{10}$.

True False

36. $\frac{3.75}{1\frac{1}{4}} = \frac{7.5}{2\frac{1}{2}}$

Hint: $1\frac{1}{4}$ is equivalent to 1.25 and $2\frac{1}{2}$ is equivalent to 2.5.

True False

37. $\frac{2\frac{3}{10}}{8.05} = \frac{\frac{1}{4}}{0.9}$

True False

38. $\frac{\frac{3}{5}}{6} = \frac{1.5}{\frac{7}{12}}$

True False

39. CONCEPT CHECK Suppose Joe Mauer of the Minnesota Twins had 68 hits in 200 times at bat and Joey Votto of the Cincinnati Reds was at bat 450 times and got 153 hits. Paula is trying to convince Jenny that the two men hit equally well. Show how you could use a proportion and cross products to see whether Paula is correct.

40. CONCEPT CHECK Jay worked 3.5 hours and packed 91 cartons. Craig packed 126 cartons in 5.25 hours. To see whether the men worked equally fast, Barry set up this proportion:

$$\frac{3.5}{91} = \frac{126}{5.25}.$$

Explain what is wrong with Barry's proportion and write a correct one. Is the correct proportion true or false?

5.4 Solving Proportions

OBJECTIVES

1. Find the unknown number in a proportion.
2. Find the unknown number in a proportion with mixed numbers or decimals.

OBJECTIVE 1 Find the unknown number in a proportion. Four numbers are used in a proportion. If any three of these numbers are known, the fourth can be found. For example, find the unknown number that will make this proportion true.

$$\frac{3}{5} = \frac{x}{40}$$

The x represents the unknown number. Start by finding the cross products.

$$\frac{3}{5} = \frac{x}{40} \qquad \begin{matrix} 5 \cdot x \\ 3 \cdot 40 \end{matrix} \quad \text{Cross products}$$

To make the proportion true, the cross products must be equal.

$$5 \cdot x = \underbrace{3 \cdot 40}$$

$$5 \cdot x = 120$$

The equal sign says that $5 \cdot x$ and 120 are equal. If $5 \cdot x$ and 120 are *both* divided by 5, the results will still be equal.

$$\frac{5 \cdot x}{5} = \frac{120}{5} \quad \leftarrow \text{Divide both sides by 5.}$$

On the left side, divide out the common factor of 5. Slashes indicate the divisions.

$$\frac{\overset{1}{\cancel{5}} \cdot x}{\underset{1}{\cancel{5}}} = 24$$

On the right side, divide 120 by 5 to get 24.

Multiplying by 1 does *not* change a number, so in the numerator on the left side, $1 \cdot x$ is the same as x.

$$\frac{x}{1} = 24$$

Dividing by 1 does *not* change a number, so on the left side, $\frac{x}{1}$ is the same as x.

$$x = 24 \quad \leftarrow \text{The solution is 24.}$$

The unknown number in the proportion is 24. The complete proportion is shown below.

$$\frac{3}{5} = \frac{24}{40} \quad \leftarrow x \text{ is 24.}$$

Check by finding the cross products. If they are equal, you solved the problem correctly. If they are unequal, rework the problem.

$$\frac{3}{5} = \frac{24}{40} \qquad \begin{matrix} 5 \cdot 24 = \mathbf{120} \\ 3 \cdot 40 = \mathbf{120} \end{matrix} \quad \text{Equal; proportion is true.}$$

The cross products are equal, so the solution, $x = 24$, is correct.

CAUTION

The solution is 24, which is the unknown number in the proportion. 120 is ***not*** the solution. It is the cross product you get when *checking* the solution.

Solve a proportion for an unknown number by using the following steps.

Finding an Unknown Number in a Proportion

Step 1 Find the cross products.

Step 2 Show that the cross products are equal.

Step 3 Divide both products by the number multiplied by x (the number next to x).

Step 4 Check by writing the solution in the *original* proportion and finding the cross products.

EXAMPLE 1 Solving Proportions for Unknown Numbers

Find the unknown number in each proportion. Round answers to the nearest hundredth when necessary.

(a) $\frac{16}{x} = \frac{32}{20}$

Recall that ratios can be rewritten in lowest terms. If desired, you can do that *before* finding the cross products. In this example, write $\frac{32}{20}$ in lowest terms as $\frac{8}{5}$, which gives the proportion $\frac{16}{x} = \frac{8}{5}$.

Step 1

$$\frac{\mathbf{16}}{\boldsymbol{x}} = \frac{\mathbf{8}}{\mathbf{5}} \qquad \begin{matrix} x \cdot 8 \\ 16 \cdot 5 \end{matrix} \quad \text{Find the cross products.}$$

Step 2

$x \cdot 8 = 16 \cdot 5$ ← Show that cross products are equal.

$x \cdot 8 = 80$

Step 3

$\frac{x \cdot \cancel{8}^{1}}{\cancel{8}_{1}} = \frac{80}{8}$ ← Divide both sides by 8.

The solution is 10 → $x = 10$ ← Find x. (No rounding necessary)

The unknown number in the proportion is 10.

Step 4 Write the solution in the *original* proportion and check by finding cross products.

$10 \cdot 32 = 320$

x is 10. → $\frac{\mathbf{16}}{\mathbf{10}} = \frac{\mathbf{32}}{\mathbf{20}}$ Equal; proportion is true.

$16 \cdot 20 = 320$

The solution is 10, **not** 320.

The cross products are equal, so **10 is the correct solution.**

Note

It is not necessary to write the ratios in lowest terms before solving. However, if you do, you will work with smaller numbers.

Continued on Next Page

1 Find the unknown numbers. Round to hundredths when necessary. Check your solutions by finding the cross products.

GS **(a)**

$$\frac{1}{2} = \frac{x}{12} \qquad 2 \cdot x \qquad 1 \cdot 12 = 12$$ Cross products

$2 \cdot x = 12$ Now finish the solution. Divide both sides by 2.

(b) $\frac{6}{10} = \frac{15}{x}$

(c) $\frac{28}{x} = \frac{21}{9}$

(d) $\frac{x}{8} = \frac{3}{5}$

(e) $\frac{14}{11} = \frac{x}{3}$

Answers

1. **(a)** $\frac{\overset{1}{\cancel{2}} \cdot x}{\underset{1}{\cancel{2}}} = \frac{12}{2}; x = 6$ **(b)** $x = 25$
(c) $x = 12$ **(d)** $x = 4.8$
(e) $x \approx 3.82$ (rounded to nearest hundredth)

(b) $\frac{7}{12} = \frac{15}{x}$

Step 1 $\frac{7}{12} = \frac{15}{x}$ $12 \cdot 15 = 180$; $7 \cdot x$ ← Find the cross products.

Step 2 $7 \cdot x = 180$ ← Show that cross products are equal.

Step 3 $\frac{\overset{1}{\cancel{7}} \cdot x}{\underset{1}{\cancel{7}}} = \frac{180}{7}$ ← Divide both sides by 7.

The rounded solution is 25.71 → $x \approx 25.71$ ← Rounded to nearest hundredth

When the division does not come out even, check for directions on how to round your answer. Divide out one more place, then round.

$$7\overline{)180.000} = 25.714$$ ← Divide out to thousandths so you can round to hundredths.

The unknown number in the proportion is 25.71 (rounded).

Step 4 Write the solution in the original proportion and check by finding the cross products.

$\frac{7}{12} = \frac{15}{25.71}$ $12 \cdot 15 = \mathbf{180}$; $7 \cdot 25.71 = \mathbf{179.97}$ ← Very close, but *not* equal due to rounding the solution

The cross products are slightly different because you rounded the value of x. However, they are close enough to see that the problem was done correctly and that **25.71 is the approximate solution.**

◀ **Work Problem 1 at the Side.**

OBJECTIVE 2 Find the unknown number in a proportion with mixed numbers or decimals. The next example shows how to work with mixed numbers or decimals in a proportion.

EXAMPLE 2 Solving Proportions with Mixed Numbers and Decimals

Find the unknown number in each proportion.

(a) $\frac{2\frac{1}{5}}{6} = \frac{x}{10}$ $6 \cdot x$; $2\frac{1}{5} \cdot 10$ ← Find the cross products.

Find $2\frac{1}{5} \cdot 10$.

$$2\frac{1}{5} \cdot 10 = \frac{11}{5} \cdot \frac{10}{1} = \frac{11}{\underset{1}{\cancel{5}}} \cdot \frac{\overset{2}{\cancel{10}}}{1} = \frac{22}{1} = 22$$

Changed to improper fraction

Continued on Next Page

Show that the cross products are equal.

$$6 \cdot x = 22$$

Divide both sides by 6. $\frac{\overset{1}{\cancel{6}} \cdot x}{\underset{1}{\cancel{6}}} = \frac{22}{6}$

Write the solution as a mixed number in lowest terms.

$$x = \frac{22 \div 2}{6 \div 2} = \frac{11}{3} = 3\frac{2}{3}$$

The solution is $3\frac{2}{3}$.

The unknown number is $3\frac{2}{3}$.

Write the solution in the proportion and check by finding the cross products.

$$6 \cdot 3\frac{2}{3} = \frac{\overset{2}{\cancel{6}}}{1} \cdot \frac{11}{\underset{1}{\cancel{3}}} = \frac{22}{1} = 22$$

$$\frac{2\frac{1}{5}}{6} = \frac{3\frac{2}{3}}{10}$$

Equal

$$2\frac{1}{5} \cdot 10 = \frac{11}{\underset{1}{\cancel{5}}} \cdot \frac{\overset{2}{\cancel{10}}}{1} = \frac{22}{1} = 22$$

The solution is $3\frac{2}{3}$, **not** 22.

The cross products are equal, so $\mathbf{3\frac{2}{3}}$ is the correct solution.

(b) $\frac{1.5}{0.6} = \frac{2}{x}$

Show that cross products are equal.

$$(1.5)(x) = \underbrace{(0.6)(2)}$$

$$(1.5)(x) = 1.2$$

Divide both sides by 1.5 $\frac{(\overset{1}{\cancel{1.5}})(x)}{\underset{1}{\cancel{1.5}}} = \frac{1.2}{1.5}$

$$x = \frac{1.2}{1.5}$$

Complete the division.

The solution is 0.8.

$$x = 0.8 \qquad 1.5\overline{)1.20} \text{ quotient } .8$$

So the unknown number is 0.8. Write the solution in the original proportion and check it by finding the cross products.

$$(0.6)(2) = 1.2$$

$$\frac{1.5}{0.6} = \frac{2}{0.8}$$

Equal

$$(1.5)(0.8) = 1.2$$

The solution is 0.8, **not** 1.2.

The cross products are equal, so **0.8** is the correct solution.

Work Problem 2 at the Side. ▶

2 Find the unknown numbers. Round to hundredths on the decimal problems, if necessary. Check your solutions by finding the cross products.

GS **(a)** $\frac{3\frac{1}{4}}{2} = \frac{x}{8}$ $\quad 2 \cdot x$

$$3\frac{1}{4} \cdot 8 = \frac{13}{\underset{1}{\cancel{4}}} \cdot \frac{\overset{2}{\cancel{8}}}{1} = \frac{26}{1}$$

$2 \cdot x = 26$ Finish the solution.

(b) $\frac{x}{3} = \frac{1\frac{2}{3}}{5}$

(c) $\frac{0.06}{x} = \frac{0.3}{0.4}$

(d) $\frac{2.2}{5} = \frac{13}{x}$

(e) $\frac{x}{6} = \frac{0.5}{1.2}$

(f) $\frac{0}{2} = \frac{x}{7.092}$

Answers

2. (a) $x = 13$ **(b)** $x = 1$ **(c)** $x = 0.08$ **(d)** $x \approx 29.55$ (rounded to nearest hundredth) **(e)** $x = 2.5$ **(f)** $x = 0$

5.4 Exercises

FOR EXTRA HELP

Download the MyDashBoard App

MyMathLab®

1. **CONCEPT CHECK** In $\frac{7}{10} = \frac{5}{x}$, what is the first step when finding the unknown number? Show how to do this step.

2. **CONCEPT CHECK** In $\frac{7}{10} = \frac{5}{x}$, the cross products are $7 \cdot x$ and 50. What is the next step when finding the unknown number? Show how to do this step.

Find the unknown number in each proportion. Round your answers to hundredths, if necessary. Check your answers by finding the cross products. **See Examples 1 and 2.**

3. $\frac{1}{3} = \frac{x}{12}$ (GS)

$3 \cdot x$ and $1 \cdot 12$

$3 \cdot x = \underbrace{1 \cdot 12}$

$3 \cdot x =$ ____

$\frac{\overset{1}{\cancel{3}} \cdot x}{\underset{1}{\cancel{3}}} = \frac{\_\_\_\_}{3}$

$x =$ ____

4. $\frac{x}{6} = \frac{15}{18}$ (GS)

$6 \cdot 15$ and $x \cdot 18$

$x \cdot 18 = \underbrace{6 \cdot 15}$

$x \cdot 18 =$ ____

$\frac{x \cdot \overset{1}{\cancel{18}}}{\underset{1}{\cancel{18}}} = \frac{\_\_\_\_}{18}$

$x =$ ____

5. $\frac{15}{10} = \frac{3}{x}$

6. $\frac{5}{x} = \frac{20}{8}$

7. $\frac{x}{11} = \frac{32}{4}$

8. $\frac{12}{9} = \frac{8}{x}$

9. $\frac{42}{x} = \frac{18}{39}$

10. $\frac{49}{x} = \frac{14}{18}$

11. $\frac{x}{25} = \frac{4}{20}$

12. $\frac{6}{x} = \frac{4}{8}$

13. $\frac{8}{x} = \frac{24}{30}$

14. $\frac{32}{5} = \frac{x}{10}$

15. $\frac{99}{55} = \frac{44}{x}$

16. $\frac{x}{12} = \frac{101}{147}$

17. $\frac{0.7}{9.8} = \frac{3.6}{x}$

18. $\frac{x}{3.6} = \frac{4.5}{6}$

19. $\frac{250}{24.8} = \frac{x}{1.75}$

20. $\frac{4.75}{17} = \frac{43}{x}$

Find the unknown number in each proportion. Write your answers as whole or mixed numbers when possible. ***See Example 2.***

21. $\frac{15}{1\frac{2}{3}} = \frac{9}{x}$

22. $\frac{x}{\frac{3}{10}} = \frac{2\frac{2}{9}}{1}$

23. $\frac{2\frac{1}{3}}{1\frac{1}{2}} = \frac{x}{2\frac{1}{4}}$

24. $\frac{1\frac{5}{6}}{x} = \frac{\frac{3}{14}}{\frac{6}{7}}$

Solve each proportion two different ways. First change all the numbers to decimal form and solve. Then change all the numbers to fraction form and solve; write your answers in lowest terms.

25. $\frac{\frac{1}{2}}{x} = \frac{2}{0.8}$

26. $\frac{\frac{3}{20}}{0.1} = \frac{0.03}{x}$

27. $\frac{x}{\frac{3}{50}} = \frac{0.15}{1\frac{4}{5}}$

28. $\frac{8\frac{4}{5}}{1\frac{1}{10}} = \frac{x}{0.4}$

Relating Concepts (Exercises 29–30) For Individual or Group Work

Work Exercises 29–30 in order. *First prove that the proportions are* **not** *true. Then create four true proportions for each exercise by changing one number at a time.*

29. $\frac{10}{4} = \frac{5}{3}$

30. $\frac{6}{8} = \frac{24}{30}$

5.5 Solving Application Problems with Proportions

OBJECTIVE

1 Use proportions to solve application problems.

OBJECTIVE 1 Use proportions to solve application problems. Proportions can be used to solve a wide variety of problems. Watch for problems in which you are given a ratio or rate and then are asked to find part of a corresponding ratio or rate. Remember that a ratio or rate compares two quantities and often includes one of the following indicator words.

in for on per from to

Use the six problem-solving steps you learned in **Chapter 1.**

Step 1 **Read** the problem.

Step 2 **Work out a plan.**

Step 3 **Estimate** a reasonable answer.

Step 4 **Solve** the problem.

Step 5 **State the answer.**

Step 6 **Check** your work.

EXAMPLE 1 Solving a Proportion Application

Mike's car can travel 163 miles **on** 6.4 gallons of gas. How far can it travel on a full tank of 14 gallons of gas? Round to the nearest whole mile.

Step 1 **Read** the problem. The problem asks for the number of miles the car can travel on 14 gallons of gas.

Step 2 **Work out a plan.** Decide what is being compared. This example compares miles to gallons. Write a proportion using the two rates. Be sure that *both* rates compare miles to gallons in the same order. In other words, miles is in both numerators and gallons is in both denominators. Use a letter to represent the unknown number.

Matching units

This rate compares miles to gallons. → $\frac{163 \text{ miles}}{6.4 \text{ gallons}} = \frac{x \text{ miles}}{14 \text{ gallons}}$ ← This rate compares miles to gallons.

Matching units

Step 3 **Estimate** a reasonable answer. To estimate the answer, notice that 14 gallons is a little more than *twice as much* as 6.4 gallons, so the car should travel a little more than *twice as far.* So use 2 • 163 miles = 326 miles as the estimate.

Step 4 **Solve** the problem. Ignore the units while solving for x.

$$\frac{163 \text{ miles}}{6.4 \text{ gallons}} = \frac{x \text{ miles}}{14 \text{ gallons}}$$

$$(6.4)(x) = (163)(14) \quad \text{Show that cross products are equal.}$$

$$(6.4)(x) = 2282$$

$$\frac{\overset{1}{\cancel{(6.4)}}(x)}{\underset{1}{\cancel{6.4}}} = \frac{2282}{6.4} \quad \text{Divide both sides by 6.4.}$$

$$x = 356.5625 \quad \text{Round to 357.}$$

Check the problem for rounding directions; this one asks for nearest whole number.

Continued on Next Page

Step 5 **State the answer.** Rounded to the nearest mile, the car can travel about **357 miles** on a full tank of gas.

Be sure to write **miles** in your answer.

Step 6 **Check** your work. The answer, 357 miles, is a little more than the estimate of 326 miles, so it is reasonable.

CAUTION

When setting up a proportion, do ***not*** mix up the units in the rates.

$$\text{compares miles to gallons}\left\{\frac{163 \text{ miles}}{6.4 \text{ gallons}} \neq \frac{14 \text{ gallons}}{x \text{ miles}}\right\}\text{compares gallons to miles}$$

These rates do ***not*** compare things in the same order and ***cannot*** be set up as a proportion.

Work Problem **1** at the Side. ▶

EXAMPLE 2 Solving a Proportion Application

A newspaper report says that 7 out of 10 people surveyed watch the news on TV. At that rate, how many of the 3200 people in town would you expect to watch the news?

Step 1 **Read** the problem. The problem asks how many of the 3200 people in town would be expected to watch TV news.

Step 2 **Work out a plan.** You are comparing people who watch the news to people surveyed. Set up a proportion using the two rates described in the example. Be sure that both rates make the same comparison. "People who watch the news" is mentioned first, so it should be in the numerator of *both* rates.

$$\frac{7 \leftarrow \text{People who watch news}}{10 \leftarrow \text{Total group (people surveyed)}} = \frac{x \leftarrow \text{People who watch news}}{3200 \leftarrow \text{Total group (people in town)}}$$

Step 3 **Estimate** a reasonable answer. To estimate the answer, notice that 7 out of 10 people is more than half the people, but less than all the people. Half of 3200 people is $3200 \div 2 = 1600$, so our estimate is between 1600 and 3200 people.

Step 4 **Solve** the problem. Solve for the unknown number in the proportion.

$$\frac{7}{10} = \frac{x}{3200}$$

$$10 \cdot x = 7 \cdot 3200 \quad \text{Show that cross products are equal.}$$

$$10 \cdot x = 22{,}400$$

$$\frac{\overset{1}{\cancel{10}} \cdot x}{\underset{1}{\cancel{10}}} = \frac{22{,}400}{10} \quad \text{Divide both sides by 10.}$$

$$x = 2240$$

No rounding is needed here.

Continued on Next Page

1 Set up and solve a proportion for each problem.

GS **(a)** If 2 pounds of fertilizer will cover 50 square feet of garden, how many pounds are needed for 225 square feet?

$$\frac{2 \text{ pounds}}{50 \text{ square feet}} = \frac{\_\_\_\_ \text{ pounds}}{\_\_\_\_ \text{ square feet}}$$

(b) A U.S. map has a scale of 1 inch to 75 miles. Lake Superior is 4.75 inches long on the map. What is the lake's actual length to the nearest whole mile?

(c) A cough syrup is given at the rate of 30 milliliters for each 100 pounds of body weight. How much should be given to a 34-pound child? Round to the nearest whole milliliter.

Answers

1. (a) $\frac{2 \text{ pounds}}{50 \text{ square feet}} = \frac{x \text{ pounds}}{225 \text{ square feet}}$
$x = 9$ pounds

(b) $\frac{1 \text{ inch}}{75 \text{ miles}} = \frac{4.75 \text{ inches}}{x \text{ miles}}$
$x = 356.25$ miles, rounds to 356 miles

(c) $\frac{30 \text{ milliliters}}{100 \text{ pounds}} = \frac{x \text{ milliliters}}{34 \text{ pounds}}$
$x \approx 10$ milliliters (rounded)

2 Solve each problem to find a reasonable answer. Then flip one side of your proportion to see what answer you get with an **incorrect** set-up. Explain why the second answer is **unreasonable.**

GS **(a)** A survey showed that 2 out of 3 people would like to lose weight. At this rate, how many people in a group of 150 want to lose weight?

$$\begin{matrix}\text{Lose weight} \rightarrow \\ \text{People surveyed} \rightarrow\end{matrix} \frac{2}{3} = \frac{x}{\quad} \begin{matrix}\leftarrow \text{Lose weight} \\ \leftarrow \text{People in group}\end{matrix}$$

(b) In one state, 3 out of 5 college students receive financial aid. At this rate, how many of the 4500 students at Central Community College receive financial aid?

Answers

2. (a) $\frac{2}{3} = \frac{x}{150}$; 100 people (reasonable); incorrect set-up gives 225 people (only 150 people in the group).
(b) 2700 students (reasonable); incorrect set-up gives 7500 students (only 4500 students at the college).

Step 5 **State the answer.** You would expect **2240 people** in town to watch the news on TV.

Be sure to write **people** in your answer.

Step 6 **Check** your work. The answer, 2240 people, is between 1600 and 3200, as called for in the estimate.

CAUTION

Always check that your answer is reasonable. If it is not, look at the way your proportion is set up. Be sure you have matching units in the numerators and matching units in the denominators.

For example, suppose you had set up the last proportion ***incorrectly,*** as shown here.

$$\frac{7}{10} = \frac{3200}{x} \leftarrow \textbf{Incorrect set-up}$$

$$7 \cdot x = 10 \cdot 3200$$

$$\frac{\overset{1}{\cancel{7}} \cdot x}{\underset{1}{\cancel{7}}} = \frac{32{,}000}{7}$$

$$x \approx 4571 \text{ people} \leftarrow \textbf{Unreasonable answer}$$

This answer is ***unreasonable*** because there are only 3200 people in the town; it is ***not*** possible for 4571 people to watch the news.

◀ **Work Problem 2 at the Side.**

5.5 Exercises

FOR EXTRA HELP

Download the MyDashBoard App

MyMathLab®

CONCEPT CHECK *Complete the proportion for each situation in Exercises 1 and 2. Do not solve the proportion.*

1. When Linda makes potato salad, she uses 6 potatoes and 4 hard-boiled eggs. How many eggs will she need if she uses 12 potatoes? Write the rest of the units in the proportion below.

$$\frac{6 \text{ potatoes}}{4 \text{ \_\_\_\_\_\_}} = \frac{12 \text{ \_\_\_\_\_\_}}{x \text{ \_\_\_\_\_\_}}$$

2. A potato chip bag says that 1 serving is 15 chips. Joe ate 70 chips. How many servings did he eat? Write the rest of the units in the proportion below.

$$\frac{1 \text{ \_\_\_\_\_\_}}{15 \text{ chips}} = \frac{x \text{ \_\_\_\_\_\_}}{70 \text{ \_\_\_\_\_\_}}$$

Set up and solve a proportion for each application problem. ***See Example 1.***

3. Caroline can sketch four cartoon strips in five hours. How long will it take her to sketch 18 strips?

Compares cartoon strips to hours $\left\{ \frac{4 \text{ strips}}{5 \text{ hours}} = \frac{\text{\_\_\_\_ strips}}{\text{\_\_\_\_ hours}} \right.$

4. The Cosmic Toads recorded eight songs on their first album in 26 hours. At this same rate, how long will it take them to record 14 songs for their second album?

Compares songs to hours $\left\{ \frac{8 \text{ songs}}{26 \text{ hours}} = \frac{\text{\_\_\_\_ songs}}{\text{\_\_\_\_ hours}} \right.$

5. Sixty newspapers cost \$27. Find the cost of 16 newspapers.

6. Twenty-two guitar lessons cost \$528. Find the cost of 12 lessons.

7. If three pounds of fescue grass seed cover about 350 square feet of ground, how many pounds are needed for 4900 square feet?

8. Anna earns \$1242.08 in 14 days. How much does she earn in 260 days?

9. Tom makes \$672.80 in 5 days. How much does he make in 3 days?

10. If 5 ounces of a medicine must be mixed with 8 ounces of water, how many ounces of medicine would be mixed with 20 ounces of water?

11. The bag of rice noodles below makes 7 servings. At that rate, how many ounces of noodles do you need for 12 servings, to the nearest ounce?

12. This can of sweet potatoes is enough for 4 servings. How many ounces are needed for 9 servings, to the nearest ounce?

13. Three quarts of a latex enamel paint will cover about 270 square feet of wall surface. How many quarts will you need to cover 350 square feet of wall surface in your kitchen and 100 square feet of wall surface in your bathroom?

14. One gallon of clear gloss wood finish covers about 550 square feet of surface. If you need to apply three coats of finish to 400 square feet of surface, how many gallons do you need, to the nearest tenth?

Use the floor plan shown to complete Exercises 15–18. On the plan, one inch represents four feet.

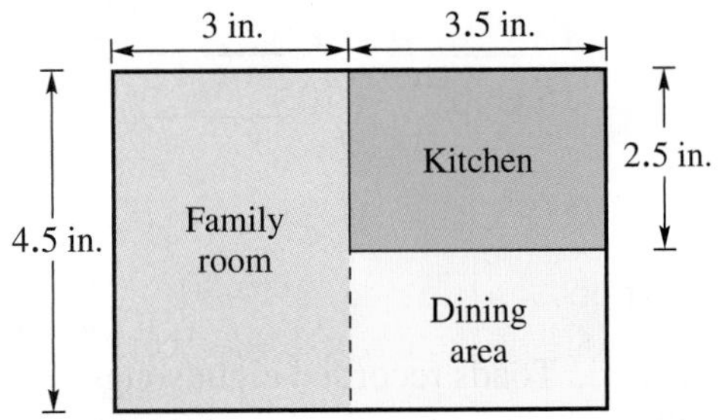

15. What is the actual length and width of the kitchen?

16. What is the actual length and width of the family room?

17. What is the actual length and width of the dining area?

18. What is the actual length and width of the entire floor plan?

The table below lists recommended amounts of food to order for 25 party guests. Use the table to answer Exercises 19 and 20. (Source: Cub Foods.)

FOOD FOR 25 GUESTS

Item	Amount
Fried chicken	40 pieces
Lasagna	14 pounds
Deli meats	4.5 pounds
Sliced cheese	$2\frac{1}{3}$ pounds
Bakery buns	3 dozen
Potato salad	6 pounds

19. How much of each food item should Nathan and Amanda order for a graduation party with 60 guests?

20. Taisha is having 20 neighbors over for a Fourth of July picnic. How much food should she buy?

21. CONCEPT CHECK Eight out of 10 students voted to take a fifteen-minute break in the middle of a two-hour class. There are 35 students in the class. How many students voted for the break? Carl set up the proportion below. Explain how to tell if Carl's answer is reasonable or not.

$$\frac{8}{10} = \frac{35}{x} \qquad 8 \cdot x = 350$$

$$\frac{\overset{1}{\cancel{8}} \cdot x}{\underset{1}{\cancel{8}}} = \frac{350}{8}$$

$$x \approx 44 \text{ students (rounded)}$$

22. CONCEPT CHECK Many experts recommend 30 minutes of exercise each day. If Vera follows that advice for 60 days, how many minutes of exercise will she get? Vera set up the proportion below. Explain how to tell if Vera's answer is reasonable or not.

$$\frac{1}{30} = \frac{x}{60} \qquad 30 \cdot x = 60$$

$$\frac{\overset{1}{\cancel{30}} \cdot x}{\underset{1}{\cancel{30}}} = \frac{60}{30}$$

$$x = 2 \text{ minutes}$$

Set up a proportion to solve each problem. Check to see whether your answer is reasonable. Then flip one side of your proportion to see what answer you get with an incorrect set-up. Explain why the second answer is unreasonable. ***See Example 2.***

23. About 7 out of 10 people entering a community college need to take a refresher math course. If there are 2950 entering students, how many will probably need refresher math? (*Source:* Minneapolis Community and Technical College.)

24. In a survey, only 3 out of 100 people like their eggs poached. At that rate, how many of the 60 customers who ordered eggs at Soon-Won's restaurant this morning asked to have them poached? Round to the nearest whole person.

25. About 1 out of 3 people choose vanilla as their favorite ice cream flavor. If 250 people attend an ice cream social, how many would you expect to choose vanilla? Round to the nearest whole person.

26. In a test of 200 sewing machines, only one had a defect. At that rate, how many of the 5600 machines shipped from the factory have defects?

Set up and solve a proportion for each problem.

27. The stock market report says that 5 stocks went up for every 6 stocks that went down. If 750 stocks went down yesterday, how many went up?

28. The human body contains 90 pounds of water for every 100 pounds of body weight. How many pounds of water are in a child who weighs 80 pounds?

29. The ratio of the length of an airplane wing to its width is 8 to 1. If the length of a wing is 32.5 meters, how wide must it be? Round to the nearest hundredth.

30. The Rosebud School District wants a student-to-teacher ratio of 19 to 1. How many teachers are needed for 1850 students? Round to the nearest whole number.

31. The number of calories you burn is proportional to your weight. A 150-pound person burns 222 calories during 30 minutes of tennis. How many calories would a 210-pound person burn, to the nearest whole number? (*Source: Wellness Encyclopedia.*)

32. Refer to **Exercise 31.** A 150-pound person burns 189 calories during 45 minutes of grocery shopping. How many calories would a 115-pound person burn, to the nearest whole number? (*Source: Wellness Encyclopedia.*)

33. At 3 P.M., Coretta's shadow is 1.05 meters long. Her height is 1.68 meters. At the same time, a tree's shadow is 6.58 meters long. How tall is the tree? Round to the nearest hundredth.

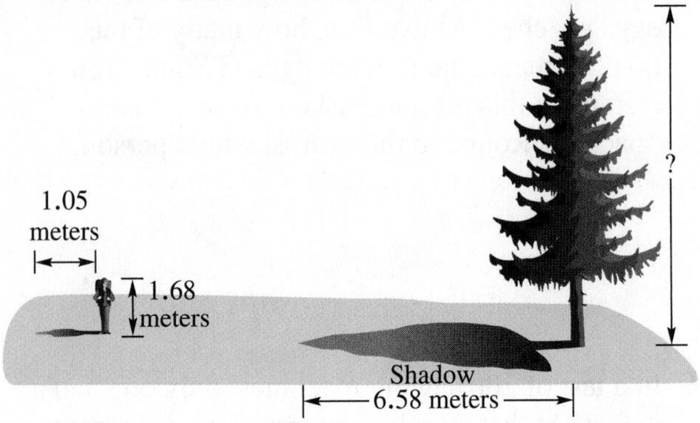

34. Refer to **Exercise 33.** Later in the day, Coretta's shadow was 2.95 meters long. How long a shadow did the tree have at that time? Round to the nearest hundredth.

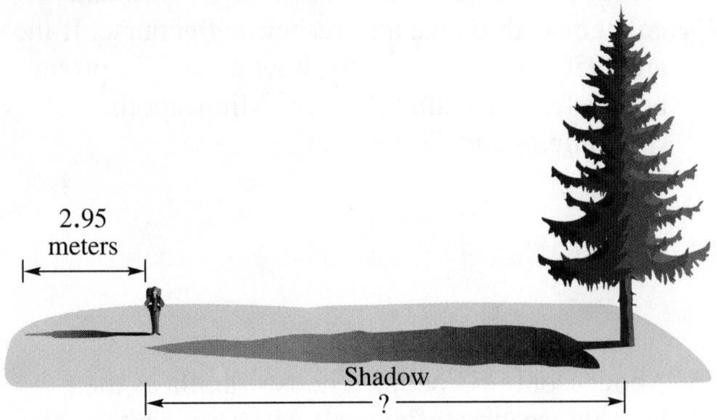

35. Can you set up a proportion to solve this problem? Explain why or why not. Jim is 25 years old and weighs 180 pounds. How much will he weigh when he is 50 years old?

36. Write your own application problem that can be solved by setting up a proportion. Also show the proportion and the steps needed to solve your problem.

37. A survey of college students shows that 4 out of 5 drink coffee. Of the students who drink coffee, 1 out of 8 adds cream to it. How many of the 56,100 students at Ohio State University would be expected to use cream in their coffee?

38. About 9 out of 10 adults think it is a good idea to exercise regularly. But of the ones who think it is a good idea, only 1 in 6 actually exercises at least three times a week. At this rate, how many of the 300 employees in our company exercise regularly?

39. The nutrition information on a bran cereal box says that a $\frac{1}{3}$-cup serving provides 80 calories and 8 grams of dietary fiber. At that rate, how many calories and grams of fiber are in a $\frac{1}{2}$-cup serving? (*Source:* Kraft Foods, Inc.)

40. A $\frac{3}{4}$-cup serving of whole grain penne pasta has 210 calories and 6 grams of dietary fiber. How many calories and grams of fiber would be in a 1-cup serving? (*Source:* Barilla Pasta.)

Relating Concepts (Exercises 41–42) For Individual or Group Work

A box of instant mashed potatoes has the list of ingredients shown in the table. Use this information to ***work Exercises 41–42 in order.***

Ingredient	For 12 Servings
Water	$3\frac{1}{2}$ cups
Margarine	6 Tbsp
Milk	$1\frac{1}{2}$ cups
Potato flakes	4 cups

Source: General Mills.

41. Find the amount of each ingredient needed for 6 servings. Show *two* different methods for finding the amounts. One method should use proportions.

42. Find the amount of each ingredient needed for 18 servings. Show *two* different methods for finding the amounts, one using proportions and one using your answers from **Exercise 41.**

Chapter 5 Summary

Key Terms

5.1

ratio A ratio compares two quantities having the same units. For example, the ratio of 6 apples to 11 apples is written in fraction form as $\frac{6}{11}$. The common units (apples) divide out.

5.2

rate A rate compares two measurements with different units. Examples are 96 dollars for 8 hours, or 450 miles on 18 gallons.

unit rate A unit rate has 1 in the denominator.

cost per unit Cost per unit is a rate that tells how much you pay for one item or one unit. The lowest cost per unit is the best buy.

5.3

proportion A proportion states that two ratios or rates are equal.

cross products Multiply along one diagonal and then multiply along the other diagonal to find the cross products of a proportion. If the cross products are equal, the proportion is true.

Test Your Word Power

See how well you have learned the vocabulary in this chapter.

1. A **ratio**
   - **A.** can be written only as a fraction
   - **B.** compares two quantities that have the same units
   - **C.** compares two quantities that have different units.

2. A **rate**
   - **A.** can be written only as a decimal
   - **B.** compares two quantities that have the same units
   - **C.** compares two quantities that have different units.

3. A **unit rate**
   - **A.** has a numerator of 1
   - **B.** has a denominator of 1
   - **C.** is found by cross multiplying.

4. **Cost per unit** is
   - **A.** the best buy
   - **B.** a ratio written in lowest terms
   - **C.** the price of one item or one unit.

5. A **proportion**
   - **A.** shows that two ratios or rates are equal
   - **B.** contains only whole numbers or decimals
   - **C.** always has one unknown number.

6. **Cross products** are
   - **A.** equal when a proportion is false
   - **B.** used to find the best buy
   - **C.** equal when a proportion is true.

Answers to Test Your Word Power

1. B; *Example:* The ratio of 3 miles to 4 miles is $\frac{3}{4}$; the common units (miles) divide out.

2. C; *Example:* \$4.50 for 3 pounds is a rate comparing dollars to pounds.

3. B; *Example:* $\frac{\$1.79}{1 \text{ pound}}$ is a unit rate. We write it as \$1.79 per pound or \$1.79/pound.

4. C; *Example:* \$3.95 per gallon tells the price of one gallon (one unit).

5. A; *Example:* The proportion $\frac{5}{6} = \frac{25}{30}$ says that $\frac{5}{6}$ is equal to $\frac{25}{30}$.

6. C; *Example:* The cross products for $\frac{5}{6} = \frac{25}{30}$ are $6 \cdot 25 = 150$ and $5 \cdot 30 = 150$.

Quick Review

Concepts	Examples
5.1 Writing a Ratio A ratio compares two quantities that have the same units. A ratio is usually written as a fraction with the number that is mentioned first in the numerator. The common units divide out and are not written in the answer. Check that the fraction is in lowest terms.	Write this ratio as a fraction in lowest terms. 60 ounces of medicine to 160 ounces of medicine $\frac{60\text{ ounces}}{160\text{ ounces}} = \frac{60 \div 20}{160 \div 20} = \frac{3}{8}$ ← Ratio in lowest terms Divide out common units.
5.1 Using Mixed Numbers in a Ratio If a ratio has mixed numbers, change the mixed numbers to improper fractions. Rewrite the problem in horizontal format using the "÷" symbol for division. Finally, multiply by the reciprocal of the divisor.	Write as a ratio of whole numbers in lowest terms. $2\frac{1}{2}$ to $3\frac{3}{4}$ $\frac{2\frac{1}{2}}{3\frac{3}{4}} = \frac{\frac{5}{2}}{\frac{15}{4}}$ Reciprocal $= \frac{5}{2} \div \frac{15}{4} = \frac{5}{2} \bullet \frac{4}{15}$ $= \frac{\overset{1}{\cancel{5}}}{\underset{1}{\cancel{2}}} \bullet \frac{\overset{2}{\cancel{4}}}{\underset{3}{\cancel{15}}} = \frac{2}{3}$ ← Ratio in lowest terms
5.1 Using Measurements in Ratios When a ratio compares measurements, both measurements must be in the *same* units. It is usually easier to compare the measurements using the smaller unit, for example, inches instead of feet.	Write as a ratio in lowest terms. 8 in. to 6 ft Compare using the smaller unit, inches. Because 1 ft has 12 in., 6 ft is $6 \bullet 12\text{ in.} = 72\text{ in.}$ The ratio is shown below. $\frac{8\text{ in.}}{72\text{ in.}} = \frac{8 \div 8}{72 \div 8} = \frac{1}{9}$ Divide out common units.
5.2 Writing Rates A rate compares two measurements with different units. The units do *not* divide out, so you must write them as part of the rate.	Write the rate as a fraction in lowest terms. 475 miles in 10 hours $\frac{475\text{ miles} \div 5}{10\text{ hours} \div 5} = \frac{95\text{ miles}}{2\text{ hours}}$ ← Must write units: miles and hours
5.2 Finding a Unit Rate A unit rate has 1 in the denominator. To find the unit rate, divide the numerator by the denominator. Write unit rates using the word **per** or a / mark.	Write as a unit rate: \$1278 in 9 days. $\frac{\$1278}{9\text{ days}}$ ← The fraction bar indicates division. $9\overline{)1278}$ = 142 so $\frac{\$1278 \div 9}{9\text{ days} \div 9} = \frac{\$142}{1\text{ day}}$ Write the answer as \$142 **per** day or \$142/day.

Concepts	Examples

5.2 Finding the Best Buy

The best buy is the item with the lowest cost per unit. Divide the price by the number of units. Round to thousandths, if necessary. Then compare to find the lowest cost per unit.

Find the best buy on grapes. You have a coupon for 50¢ off on 2 pounds or 75¢ off on 3 pounds.

2 pounds for \$2.75

3 pounds for \$4.15

Find the cost per unit (cost per pound) after subtracting the coupon.

2 pounds cost \$2.75 − \$0.50 = \$2.25.

$$\frac{\$2.25}{2} = \$1.125 \text{ per pound}$$

3 pounds cost \$4.15 − \$0.75 = \$3.40.

$$\frac{\$3.40}{3} \approx \$1.133 \text{ per pound}$$

The lower cost per pound is \$1.125, so 2 pounds of grapes is the best buy.

5.3 Writing Proportions

A proportion states that two ratios or rates are equal. The proportion "5 is to 6 as 25 is to 30" is written as shown below.

$$\frac{5}{6} = \frac{25}{30}$$

To see whether a proportion is true or false, multiply along one diagonal, then multiply along the other diagonal. If the two cross products are equal, the proportion is true. If the two cross products are unequal, the proportion is false.

Write as a proportion: 8 is to 40 as 32 is to 160.

$$\frac{8}{40} = \frac{32}{160}$$

Is this proportion true or false?

$$\frac{6}{8\frac{1}{2}} = \frac{24}{34}$$

Find the cross products.

$$\frac{6}{8\frac{1}{2}} = \frac{24}{34}$$

$$8\frac{1}{2} \cdot 24 = \frac{17}{\underset{1}{\cancel{2}}} \cdot \frac{\overset{12}{\cancel{24}}}{1} = 204$$

$$6 \cdot 34 = 204$$

Equal

The cross products are equal, so the proportion is true.

5.4 Solving Proportions

Solve for an unknown number in a proportion by using the steps shown on the next page.

Find the unknown number.

$$\frac{12}{x} = \frac{6}{8}$$

$$\frac{12}{x} = \frac{3}{4}$$

Write $\frac{6}{8}$ in lowest terms as $\frac{3}{4}$.

(Continued)

Concepts

5.4 Solving Proportions (*Continued*)

Step 1 Find the cross products. (If desired, you can rewrite the ratios in lowest terms before finding the cross products.)

Step 2 Show that the cross products are equal.

Step 3 Divide both products by the number multiplied by x (the number next to x).

Step 4 Check by writing the solution in the *original* proportion and finding the cross products.

Examples

Step 1 $\frac{\mathbf{12}}{\boldsymbol{x}} = \frac{\mathbf{3}}{\mathbf{4}}$ $x \cdot 3$ and $12 \cdot 4$ Find cross products.

Step 2 $x \cdot 3 = \underbrace{12 \cdot 4}$ Show that cross products are equal.

$x \cdot 3 = 48$

Step 3 $\frac{x \cdot \overset{1}{\cancel{3}}}{\underset{1}{\cancel{3}}} = \frac{48}{3}$ Divide both sides by 3.

$x = 16$ The solution is 16.

Step 4

x is 16. → $\frac{\mathbf{12}}{\mathbf{16}} = \frac{\mathbf{6}}{\mathbf{8}}$ $16 \cdot 6 = \mathbf{96}$ and $12 \cdot 8 = \mathbf{96}$ Equal

The cross products are equal, so **16 is the correct solution** (**not** 96).

5.5 Solving Application Problems with Proportions

Decide what is being compared. Set up and solve a proportion using the two rates described in the problem. Be sure that *both* rates compare things in the *same order*. Use a letter, like x, to represent the unknown number.

Use the six problem-solving steps.

Step 1 **Read** the problem carefully.

Step 2 **Work out a plan.**

Step 3 **Estimate** a reasonable answer.

Step 4 **Solve** the problem.

If 3 pounds of grass seed cover 450 square feet of lawn, how much seed is needed for 1500 square feet of lawn?

Step 1 The problem asks for the pounds of grass seed needed for 1500 square feet of lawn.

Step 2 Pounds of seed is compared to square feet of lawn. Set up and solve a proportion using the two given rates. Be sure that pounds of seed is in both numerators and square feet of lawn is in both denominators.

Step 3 Because 1500 square feet is about three times as much lawn as 450 square feet, about three times as much seed is needed. So, 3 • 3 pounds = 9 pounds as our estimate.

Step 4 With the proportion set up correctly, solve for the unknown number.

Matching units

$$\frac{3 \text{ pounds}}{450 \text{ square feet}} = \frac{x \text{ pounds}}{1500 \text{ square feet}}$$

Matching units

Concepts	Examples
5.5 Solving Application Problems with Proportions (*Continued*)	Both sides compare pounds to square feet. Ignore the units while finding the cross products and solving for *x*. $450 \cdot x = \underbrace{3 \cdot 1500}$ Show that cross products are equal. $450 \cdot x = 4500$ $\frac{\overset{1}{\cancel{450}} \cdot x}{\underset{1}{\cancel{450}}} = \frac{4500}{450}$ Divide both sides by 450. $x = 10$
Step 5 **State the answer.**	***Step 5*** **10 pounds** of grass seed are needed.
Step 6 **Check** your work.	***Step 6*** The exact answer, 10 pounds of seed, is close to our estimate of 9 pounds, so it is reasonable.

Chapter 5 Review Exercises

5.1 *Write each ratio as a fraction in lowest terms. Change to the same units when necessary, using the table of measurement comparisons in the first section of this chapter. Use the information in the graph to answer Exercises 1–3.*

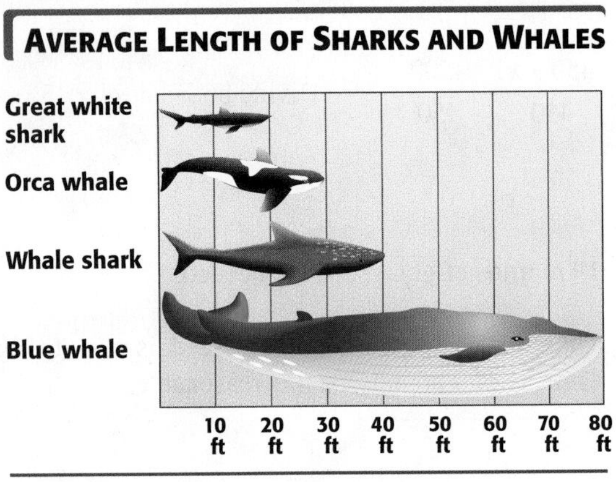

Source: Grolier Multimedia Encyclopedia.

1. Ratio of orca whale's length to whale shark's length

2. Ratio of blue whale's length to great white shark's length

3. Which two animals' lengths give a ratio of $\frac{1}{2}$? There are several answers.

4. \$2.50 to \$1.25

5. \$0.30 to \$0.45

6. $1\frac{2}{3}$ cups to $\frac{2}{3}$ cup

7. $2\frac{3}{4}$ miles to $16\frac{1}{2}$ miles

8. 5 hours to 100 minutes

9. 9 in. to 2 ft

10. 1 ton to 1500 pounds

11. 8 hours to 3 days

12. Jake sold \$350 worth of his kachina figures. Ramona sold \$500 worth of her pottery. What is the ratio of Ramona's sales to Jake's sales?

13. Ms. Wei's new car gets 35 miles per gallon. Her old car got 25 miles per gallon. Find the ratio of the new car's mileage to the old car's mileage.

14. This fall, 6000 students are taking math courses and 7200 students are taking English courses. Find the ratio of math students to English students.

5.2 *Write each rate as a fraction in lowest terms.*

15. \$88 for 8 dozen

16. 96 children in 40 families

17. When entering data into his computer, Patrick can type four pages in 20 minutes. Give his rate in pages per minute and minutes per page.

18. Elena made \$60 in three hours. Give her earnings in dollars per hour and hours per dollar.

Find the best buy.

19. Minced onion
8 ounces for \$4.98
3 ounces for \$2.49
2 ounces for \$1.89

20. Dog food; you have a coupon for \$1 off on 8 pounds or more.
35.2 pounds for \$36.96
17.6 pounds for \$18.69
3.5 pounds for \$4.25

5.3 *Use either the method of writing in lowest terms or the method of finding cross products to decide whether each proportion is true or false. Show your work and then write* true *or* false.

21. $\frac{6}{10} = \frac{9}{15}$

22. $\frac{6}{48} = \frac{9}{36}$

23. $\frac{47}{10} = \frac{98}{20}$

24. $\frac{64}{36} = \frac{96}{54}$

25. $\frac{1.5}{2.4} = \frac{2}{3.2}$

26. $\frac{3\frac{1}{2}}{2\frac{1}{3}} = \frac{6}{4}$

5.4 *Find the unknown number in each proportion. Round answers to the nearest hundredth, if necessary.*

27. $\frac{4}{42} = \frac{150}{x}$

28. $\frac{16}{x} = \frac{12}{15}$

29. $\frac{100}{14} = \frac{x}{56}$

30. $\frac{5}{8} = \frac{x}{20}$

31. $\frac{x}{24} = \frac{11}{18}$

32. $\frac{7}{x} = \frac{18}{21}$

33. $\frac{x}{3.6} = \frac{9.8}{0.7}$

34. $\frac{13.5}{1.7} = \frac{4.5}{x}$

35. $\frac{0.82}{1.89} = \frac{x}{5.7}$

5.5 *Set up and solve a proportion for each application problem.*

36. The ratio of cats to dogs at the animal shelter is 3 to 5. If there are 45 dogs, how many cats are there?

37. Danielle had 8 hits in 28 times at bat during last week's games. If she continues to hit at the same rate, how many hits will she gets in 161 times at bat?

38. If 3.5 pounds of ground beef cost $9.77, what will 5.6 pounds cost? Round to the nearest cent.

39. About 4 out of 10 students are expected to vote in campus elections. There are 8247 students. How many are expected to vote? Round to the nearest whole number.

40. The scale on Brian's model railroad is 1 in. to 16 ft. One of the scale model boxcars is 4.25 in. long. What is the length of a real boxcar in feet?

41. Marvette makes necklaces to sell at a local gift shop. She made 2 dozen necklaces in $16\frac{1}{2}$ hours. How long will it take her to make 40 necklaces?

42. A 180-pound person burns 284 calories playing basketball for 25 minutes. At this rate, how many calories would the person burn in 45 minutes, to the nearest whole number? (*Source: Wellness Encyclopedia.*)

43. In the hospital pharmacy, Michiko sees that a medicine is to be given at the rate of 3.5 milligrams for every 50 pounds of body weight. How much medicine should be given to a patient who weighs 210 pounds?

Mixed Review Exercises

Find the unknown number in each proportion. Round answers to the nearest hundredth, if necessary.

44. $\frac{x}{45} = \frac{70}{30}$

45. $\frac{x}{52} = \frac{0}{20}$

46. $\frac{64}{10} = \frac{x}{20}$

47. $\frac{15}{x} = \frac{65}{100}$

48. $\frac{7.8}{3.9} = \frac{13}{x}$

49. $\frac{34.1}{x} = \frac{0.77}{2.65}$

Find cross products to decide whether each proportion is true or false. Show the cross products and then circle true *or* false.

50. $\frac{55}{18} = \frac{80}{27}$

True False

51. $\frac{5.6}{0.6} = \frac{18}{1.94}$

True False

52. $\frac{\frac{1}{5}}{2} = \frac{1\frac{1}{6}}{11\frac{2}{3}}$

True False

Write each ratio as a fraction in lowest terms. Change to the same units when necessary.

53. 4 dollars to 10 quarters

54. $4\frac{1}{8}$ in. to 10 in.

55. 10 yd to 8 ft

56. \$3.60 to \$0.90

57. 12 eggs to 15 eggs

58. 37 meters to 7 meters

59. 3 pints to 4 quarts

60. 15 minutes to 3 hours

61. $4\frac{1}{2}$ miles to $1\frac{3}{10}$ miles

62. Nearly 7 out of 8 fans buy something to drink at rock concerts. How many of the 28,500 fans at today's concert would be expected to buy a beverage? Round to the nearest hundred fans.

63. Emily spent \$150 on car repairs and \$400 on car insurance. What is the ratio of the amount spent on insurance to the amount spent on repairs?

64. Antonio is choosing among three packages of plastic wrap. Is the best buy 25 ft for \$0.78; 75 ft for \$1.99; or 100 ft for \$2.59? He has a coupon for 50¢ off either of the larger two packages.

65. On this scale drawing of a backyard patio, 0.5 in. represents 6 ft. If the patio measures 1.75 in. long and 1.25 in. wide on the drawing, what will be the actual length and width of the patio when it is built?

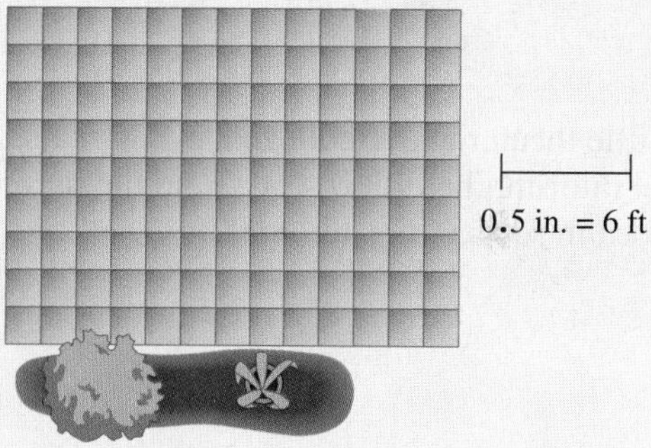

66. A lawn mower uses 0.8 gallon of gas every 3 hours. How long can the mower run on 2 gallons of gas?

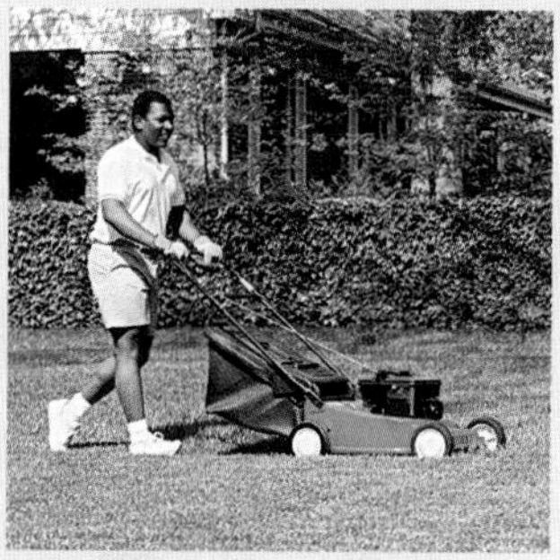

67. An antibiotic is given at the rate of $1\frac{1}{2}$ teaspoons for every 24 pounds of body weight. How much should be given to an infant who weighs 8 pounds?

68. Charles made 251 points during 169 minutes of playing time last year. At that same rate, how many points would you expect him to make if he plays 14 minutes in tonight's game? Round to the nearest whole number.

69. Refer to **Exercise 67.** Explain each step you took in solving the problem. Be sure to tell how you decided which way to set up the proportion and how you checked your answer.

70. A vitamin supplement for cats is given at the rate of 1000 milligrams for a 5-pound cat. (*Source:* St. Jon Pet Care Products.)

(a) How much should be given to a 7-pound cat?

(b) How much should be given to an 8-ounce kitten?

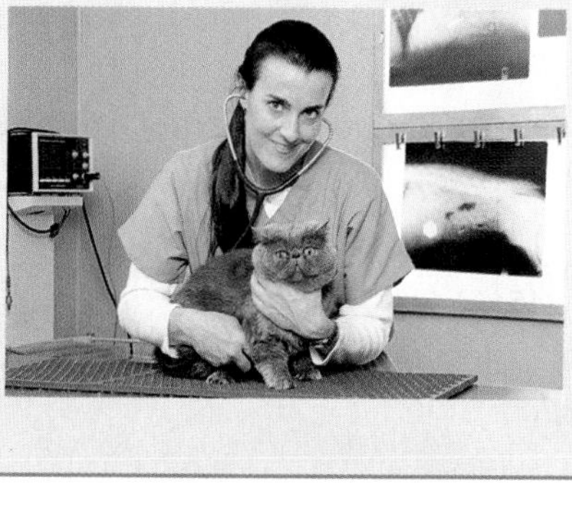

Chapter 5 Test

CHAPTER Test Prep VIDEO

The Chapter Test Prep Videos with test solutions are available on DVD, in MyMathLab, and on YouTube—search "LialDevMath" and click on "Channels."

Write each rate or ratio as a fraction in lowest terms. Change to the same units when necessary.

1. 16 fish to 20 fish

2. 300 miles on 15 gallons

3. \$15 for 75 minutes

4. 3 hours to 40 minutes

5. The little theater at our college has 320 seats. The auditorium has 1200 seats. Find the ratio of auditorium seats to theater seats.

6. Use the information in the table about Quizno's honey mustard chicken sub sandwich to find the best buy.

Size	Length of Sub	Price
Small	5 inches	\$ 5.99
Regular	8 inches	\$ 6.89
Large	11 inches	\$10

Source: Quizno's.

7. Find the best buy on spaghetti sauce. You have a coupon for 75¢ off Brand X and a coupon for 50¢ off Brand Y.

26 ounces of Brand X for \$3.89

16 ounces of Brand Y for \$1.89

14 ounces of Brand Z for \$1.29

8. Suppose the ratio of your income last year to your income this year is 3 to 2. Explain what this means. Give an example of the dollars earned last year and this year that fits the 3 to 2 ratio.

Determine whether each proportion is true or false. Show your work and then write true *or* false.

9. $\frac{6}{14} = \frac{18}{45}$

10. $\frac{8.4}{2.8} = \frac{2.1}{0.7}$

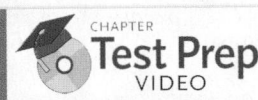

Find the unknown number in each proportion. In Problems 11–13, round the answers to the nearest hundredth, if necessary.

11. $\frac{5}{9} = \frac{x}{45}$

12. $\frac{3}{1} = \frac{8}{x}$

13. $\frac{x}{20} = \frac{6.5}{0.4}$

14. $\frac{2\frac{1}{3}}{x} = \frac{\frac{8}{9}}{4}$

Set up and solve a proportion for each application problem.

15. Pedro entered 18 orders into his computer in thirty minutes at his job. At that rate, how many orders could he enter in forty minutes?

16. Just 0.8 ounce of wildflower seeds is enough for 50 square feet of ground. What weight of seeds is needed for a garden with 225 square feet? (*Source:* White Swan Ltd.)

17. About 2 out of every 15 people are left-handed. How many of the 650 students in our school would you expect to be left-handed? Round to the nearest whole number.

18. A student set up the proportion for **Problem 17** this way and arrived at an answer of 4875.

$\frac{2}{15} = \frac{650}{x}$

CHECK $\frac{2}{15} = \frac{650}{4875}$ $15 \cdot 650 = 9750$ $2 \cdot 4875 = 9750$

Because the cross products are equal, the student said the answer is correct. Is the student right? Explain why or why not.

19. A medication is given at the rate of 8.2 grams for every 50 pounds of body weight. How much should be given to a 145-pound person? Round to the nearest tenth of a gram.

20. On a scale model, 1 in. represents 8 ft. If a building in the model is 7.5 in. tall, what is the actual height of the building in feet?

Math in the Media

FEEDING HUMMINGBIRDS

After getting a hummingbird feeder, the next step is to fill it! You have two choices at this point: you can either buy one of the commercial mixtures or you can make your own solution. See the recipe at the right.

The concentration of the sugar is important. The 1-to-4 ratio of sugar to water is recommended because it approximates the ratio of sugar to water found in the nectar of many hummingbird flowers.

Boiling the solution helps slow down fermentation. Sugar-and-water solutions are subject to rapid spoiling, especially in hot weather.

Source: *The Hummingbird Book.*

Recipe for Homemade Mixture:
1 part sugar (not honey)
4 parts water
Boil for 1 to 2 minutes. Cool.
Store extra in refrigerator.

A recipe can be used to make as much of a mixture as you need as long as the ingredients are kept proportional. Use the recipe for a homemade mixture of sugar water for hummingbird feeders to answer these problems.

1. What is the ratio of sugar to water in the recipe? What is the ratio of water to sugar?

2. Complete each table.

Sugar	Water
1 cup	4 cups
	5 cups
	6 cups
	7 cups
2 cups	8 cups

Sugar	Water
1 cup	4 cups
	3 cups
	2 cups
	1 cup

3. How much water would you need if you used

(a) 3 cups of sugar?

(b) 4 cups of sugar?

(c) $\frac{1}{3}$ cup of sugar?

4. As you change the amounts of water and sugar, should you change the length of time that you boil the mixture? Explain your answer.

6 Percent

In this chapter, you will look at the many ways percent is used in your daily life. For example, sales tax, commission rates, interest rates on savings and investments, automobile loans, home loans, and other installment loans are almost always given as percents.

6.1 Basics of Percent

OBJECTIVES

1. Learn the meaning of percent.
2. Write percents as decimals.
3. Write decimals as percents.
4. Understand 100%, 200%, and 300%.
5. Use 50%, 10%, and 1%.

VOCABULARY TIP

Percent means "per 100" or "out of 100," so any **percent** can be converted to a decimal by dividing the percent by 100. For example, $30\% = \frac{30}{100} = 0.30$.

1 Write as percents.

(a) In a group of 100 adults, 74 keep fit by walking. What percent are walking?

(b) The sales tax is $6 per $100. What percent is this?

(c) Out of 100 Americans, 32 picked football as their favorite sport. What percent picked football?

Answers
1. **(a)** 74% **(b)** 6% **(c)** 32%

Notice that the figure below has one hundred squares of equal size. Eleven of the squares are shaded. The shaded portion is $\frac{11}{100}$, or 0.11, of the total figure.

Shaded portion is 11 out of 100 parts, or $\frac{11}{100}$, or 0.11 or 11%.

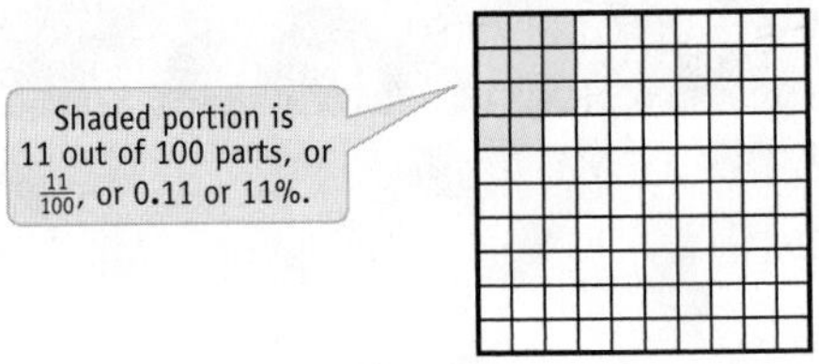

The shaded portion is also 11% of the total, or "eleven parts out of 100 parts." Read **11%** as "eleven percent."

OBJECTIVE 1 **Learn the meaning of percent.** As we just saw, a percent is a ratio with a denominator of 100.

The Meaning of Percent

Percent means *per one hundred*. The "%" symbol is used to show the number of parts out of one hundred parts.

EXAMPLE 1 Understanding Percent

(a) If *43 out of 100* students are men, then *43 per 100* or $\frac{43}{100}$ or **43%** of the students are men.

(b) If a person pays a tax of $7 on every $100 of purchases, then the tax rate is $7 per $100. The ratio is $\frac{7}{100}$ and the percent of tax is **7%**.

◀ **Work Problem 1** at the Side.

OBJECTIVE 2 **Write percents as decimals.** If 8% means 8 parts out of 100 parts or $\frac{8}{100}$, then p% means p parts out of 100 parts or $\frac{p}{100}$. Because $\frac{p}{100}$ is another way to write the division $p \div 100$, we have

$$p\% = \frac{p}{100} = p \div 100.$$

Writing a Percent as a Decimal

$$p\% = \frac{p}{100} \qquad \text{or} \qquad p\% = p \div 100$$

As a fraction As a decimal

EXAMPLE 2 Writing Percents as Decimals

Write each percent as a decimal.

(a) 47%

$$p\% = p \div 100$$

0.47 is $\frac{47}{100}$, which is equivalent to 47%.

$$47\% = 47 \div 100 = 0.47 \leftarrow \text{Decimal form}$$

Continued on Next Page

(b) 76% $76\% = 76 \div 100 = 0.76$ ← Decimal form

(c) 28.2% $28.2\% = 28.2 \div 100 = 0.282$ ← Decimal form

(d) 100% $100\% = 100 \div 100 = 1.00$ ← Decimal form

CAUTION

In **Example 2(d)** above, notice that 100% is 1.00, or 1, which is a whole number. Whenever you have a percent that is *100% or greater*, the equivalent decimal number will be *1 or greater than 1*.

Work Problem 2 at the Side. ▶

The answers in **Example 2** above suggest these steps for writing a percent as a decimal.

Writing a Percent as a Decimal

Step 1 Drop the percent symbol.

Step 2 Divide by 100.

Note

Recall from **Chapter 4** that a quick way to divide a number by 100 is to move the decimal point **two places to the left.**

EXAMPLE 3 **Writing Percents as Decimals by Moving the Decimal Point**

Write each percent as a decimal by moving the decimal point two places to the left.

(a) 17%

$17\% = 17.\%$ Decimal point starts at far right side.

0.17 ← Percent symbol is dropped. *(Step 1)*

Decimal point is moved two places to the left. *(Step 2)*

$17\% = 0.17$

(b) 160%

1.60 is equivalent to 1.6 because $1\frac{60}{100}$ simplifies to $1\frac{6}{10}$.

$160\% = 160.\% = 1.60$ or 1.6 Decimal point starts at far right side.

(c) 4.9%

0 is attached so the decimal point can be moved two places to the left.

.049

$4.9\% = 0.049$

Continued on Next Page

2 Write each percent as a decimal.

GS **(a)** 68%

$68\% = 68 \div$ ____ = ____

(b) 34%

(c) 58.5%

(d) 175%

(e) 200%

Answers

2. **(a)** 100; 0.68 **(b)** 0.34 **(c)** 0.585 **(d)** 1.75 **(e)** 2.00 or 2

3 Write each percent as a decimal.

GS **(a)** 96%

Decimal point starts here.

96% = 96.%

= .96% Move decimal point two places to the left.

= ____

(b) 6%

(c) 24.8%

GS **(d)** 0.9%

0.9% = 0.___ ___9

Answers

3. **(a)** 0.96 **(b)** 0.06 **(c)** 0.248
(d) 0; 0; 0.009

(d) 0.6%

0.6% = 0.006 — Two zeros are attached so the decimal point can be moved two places to the left.

CAUTION

Look at **Example 3(d)** above, where 0.6% is less than 1%. Because 0.6% is $\frac{6}{10}$ of 1%, it is *less than 1%*. Any fraction of a percent is *less than 1%*.

◀ **Work Problem 3 at the Side.**

OBJECTIVE 3 Write decimals as percents. You can write any decimal as a percent. For example, the decimal 0.78 is the same as the fraction

$$\frac{78}{100}.$$

This fraction means 78 out of 100 parts, or 78%. The following steps give the same result.

Writing a Decimal as a Percent

Step 1 Multiply by 100.

Step 2 Attach a percent symbol.

Note

A quick way to divide or multiply a number by 100 is to move the decimal point two places to the left or two places to the right, respectively.

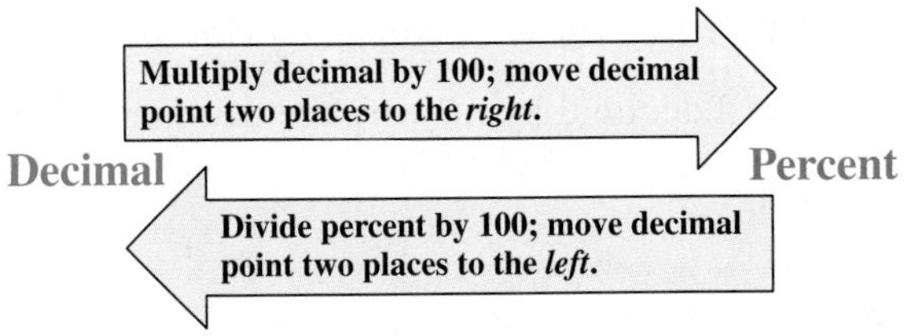

Multiply decimal by 100; move decimal point two places to the right.
Decimal
Percent
Divide percent by 100; move decimal point two places to the left.

EXAMPLE 4 Writing Decimals as Percents by Moving the Decimal Point

Write each decimal as a percent by moving the decimal point two places to the right.

(a) 0.21

0.21

Decimal point is moved two places to the right. *(Step 1)*

0.21 = 21% ← Percent symbol is attached. *(Step 2)* — Remember to attach the percent (%) symbol.

Decimal point is not written with whole number percents.

Continued on Next Page

(b) 0.529 = 52.9%

(c) 1.92 = 192%

Move the decimal point two places to the right. Then attach a % symbol.

(d) 2.5

2.50 — 0 is attached so the decimal point can be moved two places to the right.

2.5 = 250% ← Attach % symbol.

When necessary, attach zeros so you can move the decimal point.

(e) 3

3. = 3.00 — Two zeros are attached so the decimal point can be moved two places to the right.

so 3 = 300% ← Attach % symbol.

CAUTION

Look at **Examples 4(c), 4(d), and 4(e)** above, where 1.92, 2.5, and 3 are greater than 1. Because the number 1 is equivalent to 100%, all numbers greater than 1 will be *greater than 100%.*

Work Problem 4 at the Side. ▶

OBJECTIVE 4 Understand 100%, 200%, and 300%. When working with percents, it is helpful to have several reference points. 100%, 200%, and 300% are three such helpful reference points.

100% means 100 parts out of 100 parts. That's ***all*** of the parts. If 100% of the 18 people attending last week's meeting attended this week's meeting, then 18 people (***all*** of them) attended this week.

If attendance at the meeting this week is 200% of last week's attendance of 18 people, then this week's attendance is 36 people, or *two* times as many people (2 • 18 = 36). Likewise, if attendance is 300% of last week's attendance, then *three* times as many people, or 54 people, attended (3 • 18 = 54).

EXAMPLE 5 Finding 100%, 200%, and 300% of a Number

Answer the following.

Notice that 100% of something is all of it (the whole thing).

(a) What is 100% of 82 people?

100% is ***all*** of the people. So, 100% of 82 people is 82 people.

(b) What is 200% of $63?

200% is twice (2 times) as much money.

So, 200% of $63 is 2 • $63 = $126.

(c) What is 300% of 32 employees?

300% is 3 times as many employees.

So, 300% of 32 employees is 3 • 32 = 96 employees.

Work Problem 5 at the Side. ▶

---

4 Write each decimal as a percent.

(a) 0.74 (b) 0.15

(c) 0.09

GS (d) 0.617

0.__ __ __

So, 0.617 = ____ %.

(e) 0.834 (f) 5.34

(g) 2.8 (h) 4

5 Answer the following.

(a) What is 100% of $7.80?

(b) What is 100% of 1850 workers?

GS (c) What is 200% of 24 photographs?

200% is twice (____ times).

So, 2 • 24 = ______.

(d) What is 300% of 8 miles?

Answers

4. (a) 74% (b) 15% (c) 9%
(d) 6; 1; 7; 61.7% (e) 83.4% (f) 534%
(g) 280% (h) 400%

5. (a) $7.80 (b) 1850 workers
(c) 2; 48 photographs (d) 24 miles

6 Answer the following.

GS **(a)** What is 50% of 200 patients?

50% is half of the patients.

$\frac{1}{2}$ of 200 patients is ______.

(b) What is 50% of 64 tweets?

GS **(c)** What is 10% of 3850 elm trees?

10% is $\frac{1}{10}$ of the trees. Move the decimal point ____ place to the ____.

10% of 3850. is ______.

(d) What is 10% of 7 pounds?

GS **(e)** What is 1% of 240 ft?

1% is $\frac{1}{100}$ of the length.

Move the decimal point ____ places to the ____.

1% of 2 40. is ______.

(f) What is 1% of $3000?

OBJECTIVE 5 **Use 50%, 10%, and 1%.** 50% means 50 parts out of 100 parts, which is ***half*** of the parts ($\frac{50}{100} = \frac{1}{2}$). So, 50% of $18 is $9 (***half*** of the money).

When using 10%, we have 10 parts out of 100 parts, which is $\frac{1}{10}$ of the parts ($\frac{10}{100} = \frac{1}{10}$). To find 10% or $\frac{1}{10}$ of a number, we move the decimal point **one** place to the left. 10% of $285 is $28.50 (because $28.5. = $28.50).

To find 1% of a number ($\frac{1}{100}$), we move the decimal point **two** places to the left. 1% of $198 is $1.98 (because $1.98. = $1.98).

EXAMPLE 6 Finding 50%, 10%, and 1% of a Number

Answer the following.

(a) What is 50% of 24 hours?

Think: 50% of something is $\frac{50}{100}$ or $\frac{1}{2}$ of it.

50% is half of the hours.

So, 50% of 24 hours is **12 hours**.

(b) What is 10% of 280 pages?

10% is $\frac{1}{10}$ of the pages. Move the decimal point *one* place to the left.

So, 10% of 280. pages is **28 pages**.

(c) What is 1% of $540?

1% is $\frac{1}{100}$ of the money. Move the decimal point *two* places to the left.

So, 1% of $540. is **$5.40**.

◄ **Work Problem 6 at the Side.**

Note

Two other frequently used percents are 25% and 75%.

25% means 25 parts out of 100 parts, which is $\frac{1}{4}$ of the parts $\left(\frac{25}{100} = \frac{1}{4}\right)$. For example, 25% of $32 is $\frac{1}{4}$ of the money, or $8.

75% means 75 parts out of 100 parts, which is $\frac{3}{4}$ of the parts $\left(\frac{75}{100} = \frac{3}{4}\right)$. For example, 75% of $32 is $\frac{3}{4}$ of the money, or $24.

Answers

6. **(a)** 100 patients **(b)** 32 tweets **(c)** one; left; 385 elm trees **(d)** 0.7 pound **(e)** two; left; 2.4 ft **(f)** $30

6.1 Exercises

FOR EXTRA HELP

Download the MyDashBoard App

MyMathLab®

CONCEPT CHECK *Fill in each blank with the correct response.*

1. To write a percent as a decimal, drop the ________ symbol and then divide by ________.

2. A quick way to divide a number by 100 is to move the ________ two places to the ________.

Write each percent as a decimal. ***See Examples 2 and 3.***

3. 12%

GS 12.% = __.12

4. 57%

GS 57.% = 0.__ __

5. 70%

6. 40%

7. 25%

8. 35%

9. 140%

10. 250%

11. 5.5%

12. 6.7%

13. 100%

14. 600%

15. 0.5%

16. 0.25%

17. 0.35%

18. 0.75%

CONCEPT CHECK *Fill in each blank with the correct response.*

19. To write a decimal as a percent, multiply by ______ and then attach a ______ symbol.

20. A quick way to multiply a number by 100 is to move the ________ two places to the ________.

Write each decimal as a percent. ***See Example 4.***

21. 0.6

GS 0.6 = 0.60

So, 0.6 = 60%.

22. 0.9

GS 0.9 = 0.__ __

So, 0.9 = __ __%.

23. 0.01

24. 0.07

25. 0.375

26. 0.625

27. 2

28. 5

29. 3.7

30. 2.2

31. 0.0312

32. 0.0625

33. 4.162

34. 8.715

35. 0.0028

36. 0.0064

37. Fractions, decimals, and percents are all used to describe a part of something. The use of percents is much more common than that of fractions and decimals. Why do you suppose this is true?

38. List five uses of percent that are or will be part of your life. Consider the activities of working, shopping, saving, and planning for the future.

Over the next 10 years, the projected growth in the workforce for several occupations is shown. Write each percent as a decimal and each decimal as a percent. ***See Examples 2–4.*** *(Source: Bloomberg BusinessWeek.)*

39. Truck drivers 13%

40. Registered nurses 22.2%

41. Physicians and surgeons 0.218

42. Waiters and waitresses 0.064

43. Postsecondary teachers 15.1%

44. Security guards 14.2%

45. Customer service representatives 0.177

46. Carpenters 0.129

47. Cooks 14.6%

48. Management analysts 23.9%

Write each percent as a decimal and each decimal as a percent. ***See Examples 2–4.***

49. Only 0.08 of the total population has the professional training.

50. The church building fund has 0.8 of the money needed.

51. The patient's blood pressure was 30% above normal.

52. Success with the diet was 170% greater than anticipated.

CONCEPT CHECK *Fill in the blanks. Remember that 100% is all of something, 200% is two times as many, and 300% is three times as many.* ***See Example 5.***

53. The Saturday morning tae kwon do class has 12 children enrolled. If 100% of the children are present, how many children are there? __________

54. When 500 adults were asked, "Do you think your taxes are too high," 100% said yes. How many said yes? __________

55. Last year we had 210 employees. This year we have 200% of that number. How many employees do we have this year? ______________

56. Last season Active Sports sold 380 fishing licenses. This season they sold 200% of that number. How many fishing licenses did they sell this season?

____

57. Last week 90 chairs were used for the meeting. This week we need 300% of that number of chairs. We'll need ______________.

57. One month ago she had 28 friends on Facebook. Now she has 300% of that number. The new number of friends is ______________.

Fill in the blanks. Remember that 50% is half of something, 10% is found by moving the decimal point one place to the left, and 1% is found by moving the decimal point two places to the left. ***See Example 6.***

59. Jacob owes $755 for tuition. Financial aid will pay 50% of the cost. Financial aid will pay ______________.

60. Linda Redding needs $2320 for school and living expenses this semester. She applied for a student loan to cover 50% of this amount. The amount of the loan is ______________.

61. Only 10% of 8200 commuters are carpooling to work. How many commuters carpool? ______________

Move the decimal point one place to the left. 8200.

62. Sarah Bryn expects that 10% of the 240 dozen plants in her greenhouse will not be sold. The expected number of unsold plants is ______________.

63. The naturalist said that 1% of the 2600 plants in the park are poisonous. How many plants are poisonous?

____

64. Of the 4800 accidents, only 1% were caused by mechanical failure. How many accidents were caused by mechanical failure? ______________

Move the decimal point two places to the left. 4800.

65. **(a)** Describe a shortcut method of finding 100% of a number.

(b) Show an example using your shortcut.

66. **(a)** Describe a shortcut method of finding 50% of a number.

(b) Show an example using your shortcut.

67. **(a)** Describe a shortcut method of finding 200% of a number.

(b) Show an example using your shortcut.

68. **(a)** Describe a shortcut method of finding 300% of a number.

(b) Show an example using your shortcut.

69. **(a)** Describe a shortcut method of finding 10% of a number.

(b) Show an example using your shortcut.

70. **(a)** Describe a shortcut method of finding 1% of a number.

(b) Show an example using your shortcut.

More than 7.4 million households dress up their pets (dogs and cats) in Halloween costumes. The bar graph shows the ranking of the top pet costumes and the percent of pet owners selecting each costume. Use this graph to answer Exercises 71–74. Write each answer as a percent and as a decimal. (Source: BIGresearch survey of 8877 adult pet owners.*)*

71. What portion of the pet owners selected the devil costume for their pet?

72. What portion of the pet owners selected the pirate costume for their pet?

73. **(a)** What was the third-most-popular costume?

(b) Write the portion of the pet owners who selected this costume.

74. **(a)** What was the second-most-popular costume?

(b) Write the portion of the pet owners who selected this costume.

PETS GET DRESSED UP

This year, 7.4 million households plan to put their furry friends into a Halloween costume. Top outfits:

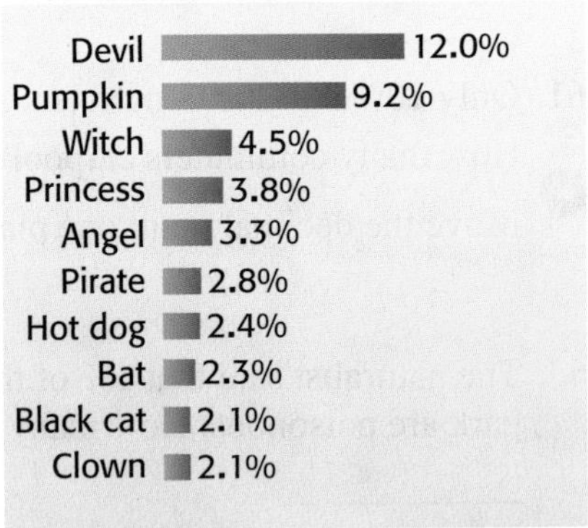

Source: BIGresearch survey of 8877 adult pet owners.

Motorists were asked which traffic violation they felt should be given the biggest fine. The circle graph shows the percent of motorists who chose each traffic violation. Use this graph to answer Exercises 75–78. Write each answer as a percent and a decimal.

75. What portion of the motorists chose "Tailgating" for the biggest fine?

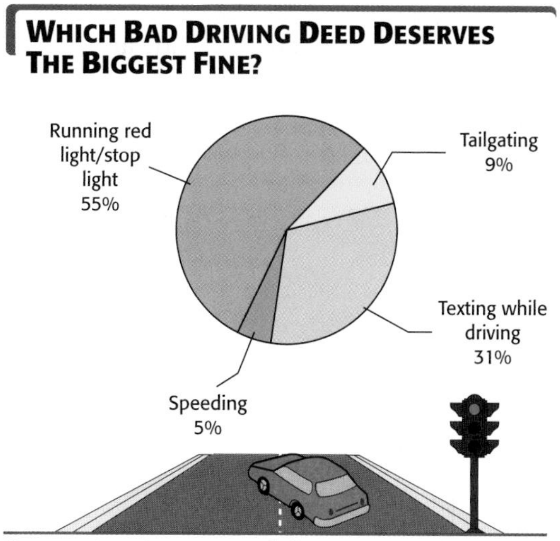

Source: American Automobile Association (AAA).

76. What portion of the motorists chose "Texting while driving" for the biggest fine?

77. (a) Which violation was chosen least often for the biggest fine?

(b) What portion was this?

78. (a) Which violation was chosen most often for the biggest fine?

(b) What portion was this?

Food researchers suspect that childhood obesity starts early. The pictograph shows the percent of children 19 to 24 months old who eat each type of food at least once a day. Use this graph to answer Exercises 79–82. Write each answer as a percent and as a decimal.

79. What portion of the children eat french fries?

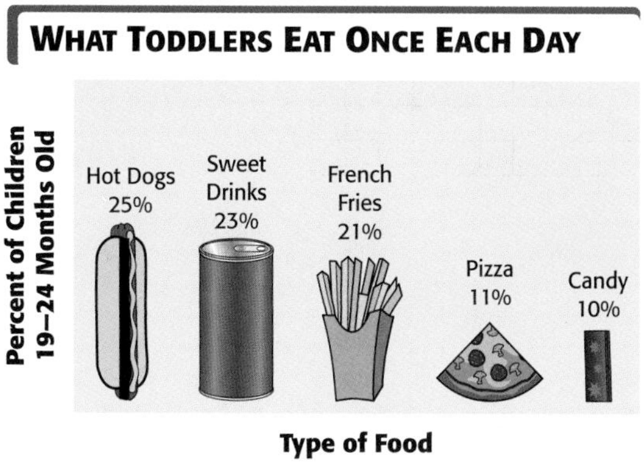

Source: Mathematica Policy Research for Gerber Products Company.

80. What portion of the children eat pizza?

81. (a) Which food is eaten by the lowest portion of children?

(b) What portion is this?

82. (a) Which food is eaten by the highest portion of children?

(b) What portion is this?

Write a percent for both the shaded and unshaded portions of each figure.

83. 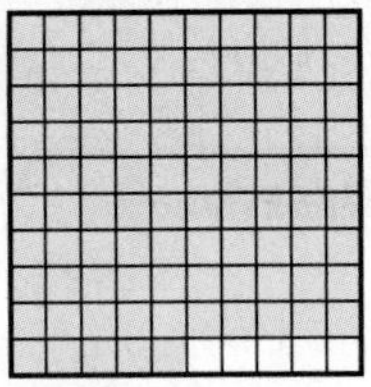

95 out of 100 parts are shaded = ____ %

5 out of 100 parts are not shaded = ____ %

84. 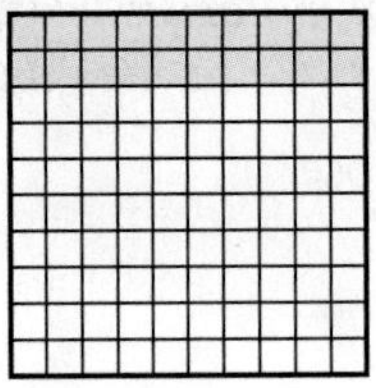

20 out of 100 parts are shaded = ____ %

80 out of 100 parts are not shaded = ____ %

85. 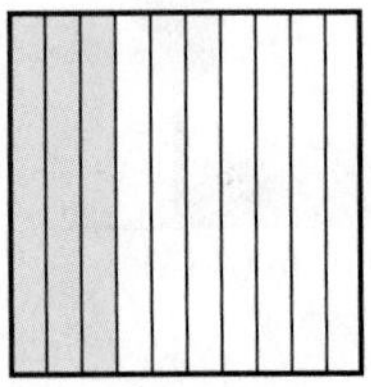

86. 

87. 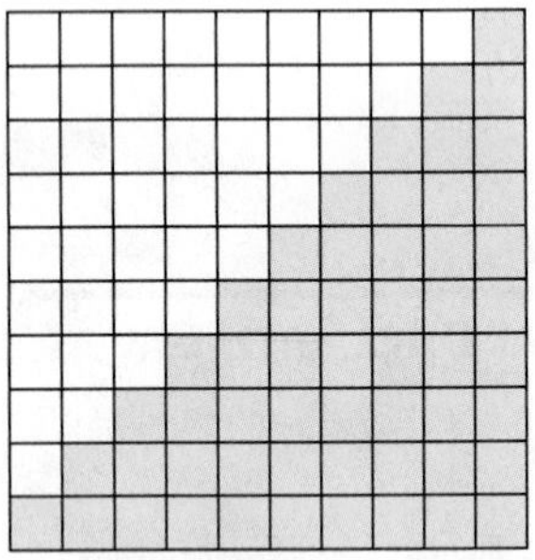

88. 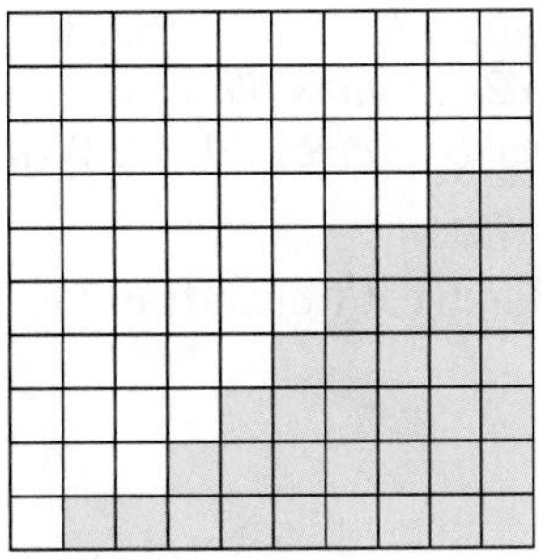

89. 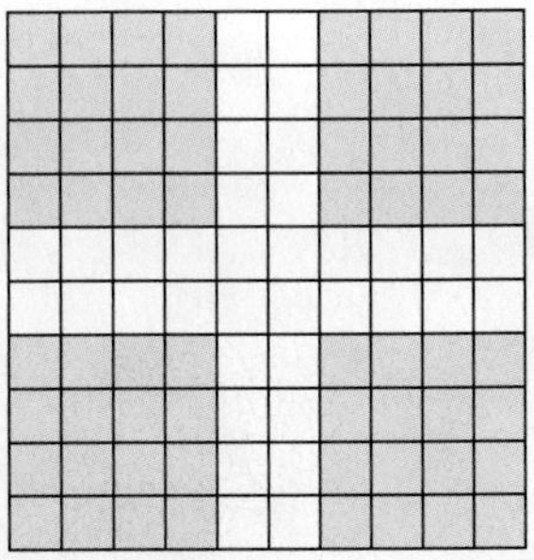

90. 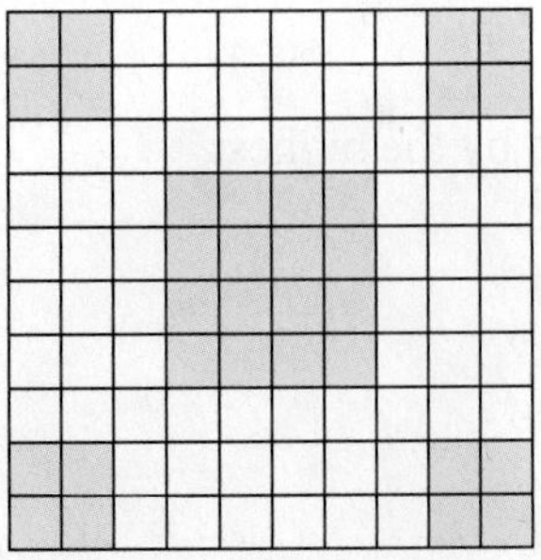

6.2 Percents and Fractions

OBJECTIVES

① Write percents as fractions.

② Write fractions as percents.

③ Use the table of percent equivalents.

OBJECTIVE ① **Write percents as fractions.** Percents can be written as fractions by using what we learned in the previous section.

Writing a Percent as a Fraction

$$p\% = \frac{p}{100}, \quad \text{as a fraction}$$

EXAMPLE 1 Writing Percents as Fractions

Write each percent as a fraction or mixed number in lowest terms.

(a) 25%

As we saw in the last section, 25% can be written as a decimal.

$$25\% = 25 \div 100 = 0.25 \quad \text{\small Percent symbol dropped}$$

Because 0.25 means 25 hundredths,

$$0.25 = \frac{25}{100} = \frac{25 \div 25}{100 \div 25} = \frac{1}{4}. \leftarrow \text{Lowest terms}$$

It is not necessary, however, to write 25% as a decimal first.

$$25\% = \frac{25}{100} \quad \text{\small 25 per 100}$$

$$= \frac{1}{4} \leftarrow \text{Lowest terms}$$

(b) 76%

The percent becomes the numerator.

Write 76% as $\frac{76}{100}$.

The *denominator* is always 100 because percent means *parts per 100.*

Write $\frac{76}{100}$ in lowest terms.

To write a fraction in lowest terms, divide numerator and denominator by the *same number.*

$$\frac{76 \div 4}{100 \div 4} = \frac{19}{25} \leftarrow \text{Lowest terms}$$

(c) 150%

$$150\% = \frac{150}{100} = \frac{150 \div 50}{100 \div 50} = \frac{3}{2} = 1\frac{1}{2} \leftarrow \text{Mixed number}$$

Lowest terms

Note

Remember that percent means *per 100.*

Work Problem ① at the Side. ▶

① Write each percent as a fraction or mixed number in lowest terms.

(a) 50%

$$50\% = \frac{50}{100} = \frac{50 \div 50}{\underline{} \div 50}$$

$$= \underline{}$$

(b) 75%

(c) 48%

(d) 23%

(e) 125%

$$125\% = \frac{}{100}$$

$$= \frac{ \div 25}{100 \div 25}$$

$$= \frac{}{4} = \underline{}$$

(f) 250%

Answers

1. (a) $100; \frac{1}{2}$ **(b)** $\frac{3}{4}$ **(c)** $\frac{12}{25}$ **(d)** $\frac{23}{100}$

 (e) $125; 125; 5; 1\frac{1}{4}$ **(f)** $\frac{5}{2} = 2\frac{1}{2}$

2 Write each percent as a fraction in lowest terms.

GS **(a)** 37.5%

$$37.5\% = \frac{37.5}{100} = \frac{37.5\ (10)}{100\ (\_\_\_)} = \frac{375\ \div\ \_\_\_}{1000\ \div\ 125} = \_\_\_$$

(b) 62.5%

(c) 4.5%

(d) $66\frac{2}{3}\%$

(e) $10\frac{1}{3}\%$

(f) $87\frac{1}{2}\%$

Answers

2. (a) 10; 125; $\frac{3}{8}$ (b) $\frac{5}{8}$ (c) $\frac{9}{200}$ (d) $\frac{2}{3}$ (e) $\frac{31}{300}$ (f) $\frac{7}{8}$

The next example shows how to write decimal and fraction percents as fractions.

EXAMPLE 2 Writing Decimal or Fraction Percents as Fractions

Write each percent as a fraction in lowest terms.

(a) 15.5%

Write 15.5 over 100.

$$15.5\% = \frac{15.5}{100}$$

The denominator is 100 because percent means **parts per 100.**

To get a whole number in the numerator, multiply the numerator and denominator by 10. (Recall that multiplying by $\frac{10}{10}$ is the same as multiplying by 1.)

$$\frac{15.5}{100} = \frac{15.5\,(10)}{100\,(10)} = \frac{155}{1000}$$

Move the decimal point 1 place to the *right* to multiply by 10.

Now write the fraction in lowest terms.

$$\frac{155\ \div 5}{1000\ \div 5} = \frac{31}{200}$$

(b) $33\frac{1}{3}\%$

Write $33\frac{1}{3}$ over 100.

$$33\frac{1}{3}\% = \frac{33\frac{1}{3}}{100}$$

When there is a mixed number in the numerator, rewrite the mixed number as an improper fraction.

$$\frac{33\frac{1}{3}}{100} = \frac{\frac{100}{3}}{100}$$

Write $33\frac{1}{3}$ as $\frac{100}{3}$.

Next, rewrite the division problem in a horizontal form. Finally, multiply by the reciprocal of the divisor.

Reciprocals

$$\frac{\frac{100}{3}}{100} = \frac{100}{3} \div 100 = \frac{100}{3} \div \frac{100}{1} = \frac{\overset{1}{\cancel{100}}}{3} \cdot \frac{1}{\underset{1}{\cancel{100}}} = \frac{1}{3} \leftarrow \text{Lowest terms}$$

Note

In **Example 2(a)** at the top of the page, we could have changed 15.5% to $15\frac{1}{2}\%$ and then written it as the improper fraction $\frac{31}{2}$ over 100. But it is usually easier *not* to change decimals to fractions and to work with decimal percents as they are.

◀ **Work Problem 2 at the Side.**

OBJECTIVE 2 Write fractions as percents. We will use the formula from the beginning of this section to write fractions as percents.

$$p\% = \frac{p}{100}$$

EXAMPLE 3 Writing Fractions as Percents

Write each fraction as a percent. Round to the nearest tenth if necessary.

(a) $\frac{3}{5}$

Write $\frac{3}{5}$ as a percent by solving for p in the proportion below.

$$\frac{3}{5} = \frac{p}{100}$$

Find cross products and show that they are equal.

$$5 \cdot p = 3 \cdot 100$$

$$5 \cdot p = 300$$

$$\frac{\overset{1}{\cancel{5}} \cdot p}{\underset{1}{\cancel{5}}} = \frac{300}{5} \quad \text{Divide both sides by 5.}$$

$$p = 60$$

This result means that $\frac{3}{5} = \frac{60}{100}$ or 60%.

Note

Solving proportions can be reviewed in **Chapter 5.**

(b) $\frac{7}{8}$

Write a proportion.

$$\frac{7}{8} = \frac{p}{100}$$

$$8 \cdot p = 7 \cdot 100 \quad \text{Show that cross products are equal.}$$

$$8 \cdot p = 700$$

$$\frac{\overset{1}{\cancel{8}} \cdot p}{\underset{1}{\cancel{8}}} = \frac{700}{8} \quad \text{Divide both sides by 8.}$$

$$p = 87.5$$

So, $\frac{7}{8} = 87.5\%$.

Note

If you think of $\frac{700}{8}$ as an improper fraction, changing it to a mixed number gives an answer of $87\frac{1}{2}$. So $\frac{7}{8} = 87.5\%$ or $87\frac{1}{2}\%$.

Continued on Next Page

3 Write as percents. Round to the nearest tenth if necessary.

GS **(a)** $\frac{1}{4}$

$\frac{1}{4} = \frac{p}{100}$

$4 \cdot p = 1 \cdot$ ____

$\frac{\overset{1}{\cancel{4}} \cdot p}{\underset{1}{\cancel{4}}} = \frac{100}{4}$

$p =$ ____

(b) $\frac{3}{10}$

(c) $\frac{6}{25}$

(d) $\frac{5}{8}$

(e) $\frac{1}{6}$

(f) $\frac{2}{9}$

Answers

3. **(a)** 100; 25% **(b)** 30% **(c)** 24% **(d)** 62.5% **(e)** 16.7% (rounded); $16\frac{2}{3}\%$ (exact) **(f)** 22.2% (rounded); $22\frac{2}{9}\%$ (exact)

(c) $\frac{5}{6}$

Start with a proportion.

$$\frac{5}{6} = \frac{p}{100}$$

$6 \cdot p = 5 \cdot 100$ Show that cross products are equal.

$6 \cdot p = 500$

$\frac{\overset{1}{\cancel{6}} \cdot p}{\underset{1}{\cancel{6}}} = \frac{500}{6}$ Divide both sides by 6.

$p = 83.\overline{3}$ — A bar over the 3 indicates that the decimal keeps repeating 83.3333 forever.

$p \approx 83.3$ Round to the nearest tenth.

So, $\frac{5}{6} = 83.\overline{3}\% \approx 83.3\%$ (rounded) — The "≈" symbol shows that $83.\overline{3}\%$ is rounded.

Note

You can change $\frac{500}{6}$ to a mixed number to get an exact answer of $83\frac{1}{3}\%$.

◀ **Work Problem 3 at the Side.**

OBJECTIVE 3 **Use the table of percent equivalents.** Knowing how to find the equivalents of fractions, decimals, and percents is important. However, the table on the next page shows common fractions and mixed numbers and their decimal and percent equivalents.

EXAMPLE 4 **Using the Table of Percent Equivalents**

Find the following in the table.

(a) $\frac{1}{6}$ as a percent

Find $\frac{1}{6}$ in the "fraction" column. The equivalent percent is 16.7% (rounded) or $16\frac{2}{3}\%$ (exact).

(b) 0.375 as a fraction

Look in the "decimal" column for 0.375. The equivalent fraction is $\frac{3}{8}$.

(c) $\frac{7}{8}$ as a percent

Find $\frac{7}{8}$ in the "fraction" column. The equivalent percent is 87.5% or $87\frac{1}{2}\%$.

Calculator Tip

Example 4 (c) above can be solved on a calculator as shown below.

7 (÷) 8 (=) 0.875 (×) 100 (=) 87.5

Multiplying by 100 changes the decimal to a percent.

Or, if your calculator has a percent key, follow these steps.

7 (÷) 8 (%) 87.5

Press % key.

On scientific calculators, you may need to press the (2nd) key to access the % function, and some models require pressing (=) to get the answer.

Note

When a fraction like $\frac{1}{3}$ is changed to a decimal, it is a *repeating decimal* that goes on forever, 0.333333…. In the table below these decimals are rounded to the nearest thousandth. When the decimal is then changed to a percent, it will be to the nearest tenth of a percent. Decimals that do not repeat are usually not rounded.

Work Problem 4 at the Side. ▶

PERCENT, DECIMAL, AND FRACTION EQUIVALENTS

Percent (*rounded to tenths when necessary*)	Decimal	Fraction
1%	0.01	$\frac{1}{100}$
5%	0.05	$\frac{1}{20}$
10%	0.1	$\frac{1}{10}$
12.5% or $12\frac{1}{2}$%	0.125	$\frac{1}{8}$
16.7% (rounded) or $16\frac{2}{3}$% (exact)	$0.1\overline{6}$ rounds to 0.167.	$\frac{1}{6}$
20%	0.2	$\frac{1}{5}$
25%	0.25	$\frac{1}{4}$
33.3% (rounded) or $33\frac{1}{3}$% (exact)	$0.\overline{3}$ rounds to 0.333.	$\frac{1}{3}$
37.5% or $37\frac{1}{2}$%	0.375	$\frac{3}{8}$
40%	0.4	$\frac{2}{5}$
50%	0.5	$\frac{1}{2}$
60%	0.6	$\frac{3}{5}$
62.5% or $62\frac{1}{2}$%	0.625	$\frac{5}{8}$
66.7% (rounded) or $66\frac{2}{3}$% (exact)	$0.\overline{6}$ rounds to 0.667.	$\frac{2}{3}$
75%	0.75	$\frac{3}{4}$
80%	0.8	$\frac{4}{5}$
87.5% or $87\frac{1}{2}$%	0.875	$\frac{7}{8}$
100%	1.0	1
150%	1.5	$1\frac{1}{2}$
200%	2.0	2

4 Find the following fractions, mixed numbers, decimals, and percents in the table on this page. If you already know the answer or can solve for the answer quickly, don't use the table.

(a) $\frac{3}{4}$ as a percent

(b) 10% as a fraction

(c) $0.\overline{6}$ as a fraction

(d) $37\frac{1}{2}$% as a fraction

(e) $\frac{7}{8}$ as a percent

(f) $\frac{1}{2}$ as a percent

(g) $33\frac{1}{3}$% as a fraction

(h) $1\frac{1}{2}$ as a percent

Answers

4. **(a)** 75% **(b)** $\frac{1}{10}$ **(c)** $\frac{2}{3}$ **(d)** $\frac{3}{8}$ **(e)** 87.5% **(f)** 50% **(g)** $\frac{1}{3}$ **(h)** 150%

6.2 Exercises

FOR EXTRA HELP

Download the MyDashBoard App

MyMathLab®

CONCEPT CHECK *Is each percent equivalent to the given fraction? Write* true *or* false.

1. $25\% = \frac{1}{2}$

2. $30\% = \frac{3}{10}$

3. $75\% = \frac{3}{4}$

4. $80\% = \frac{5}{6}$

Write each percent as a fraction in lowest terms and as a mixed number when possible.
See Examples 1 and 2.

5. 85%
GS
$0.85 = \frac{85}{100} = \frac{17}{\_\_\_}$

6. 45%
GS
$0.45 = \frac{45}{100} =$

7. 62.5%

8. 87.5%

9. 6.25%

10. 43.75%

11. $16\frac{2}{3}\%$

12. $66\frac{2}{3}\%$

13. $6\frac{2}{3}\%$

14. $46\frac{2}{3}\%$

15. 0.5%

16. 0.8%

17. 180%

18. 140%

19. 375%

20. 225%

CONCEPT CHECK *Is each fraction equivalent to the given percent? Write* true *or* false.

21. $\frac{1}{2} = 50\%$

22. $\frac{1}{100} = 10\%$

23. $\frac{4}{5} = 75\%$

24. $\frac{3}{10} = 30\%$

Write each fraction as a percent. Round percents to the nearest tenth if necessary.
See Example 3.

25. $\frac{7}{10} = \frac{p}{100}$
GS
Solve for p.
$p =$ ____%

26. $\frac{3}{4} = \frac{p}{100}$
GS
Solve for p.
$p =$ ____%

27. $\frac{37}{100}$

28. $\frac{63}{100}$

29. $\frac{5}{8}$

30. $\frac{1}{8}$

31. $\frac{7}{8}$

32. $\frac{3}{8}$

33. $\frac{12}{25}$

34. $\frac{15}{25}$

35. $\frac{23}{50}$

36. $\frac{18}{50}$

37. $\frac{7}{20}$

38. $\frac{9}{20}$

39. $\frac{5}{6}$

40. $\frac{1}{6}$

41. $\frac{5}{9}$

42. $\frac{7}{9}$

43. $\frac{1}{7}$

44. $\frac{5}{7}$

45. CONCEPT CHECK Which of the following is equal to $\frac{3}{4}$?

A. 65% **B.** 0.75 **C.** 0.5 **D.** 25%

46. CONCEPT CHECK Which of the following is equal to 80%?

A. 0.85 **B.** $\frac{3}{5}$ **C.** $\frac{7}{10}$ **D.** $\frac{4}{5}$

Complete the chart. Round decimals to the nearest thousandth and percents to the nearest tenth if necessary. ***See Examples 3 and 4.***

	Fraction	Decimal	Percent
47.	1/2	0.5	50%
48.	4/5	0.8	80%
49.	7/8	0.875	87.5%
50.		0.6	60%
51.	$\frac{1}{6}$	0.167	16.7%
52.	$\frac{1}{3}$	0.333	33.3%
53.		0.7	

	Fraction	Decimal	Percent
54.	______	______	37.5%
55.	______	______	12.5%
56.	______	0.625	______
57.	$\frac{2}{3}$	______	______
58.	$\frac{5}{6}$	______	______
59.	$\frac{3}{50}$	______	______
60.	$\frac{3}{10}$	______	______
61.	$\frac{8}{100}$	______	______
62.	______	______	100%
63.	$\frac{1}{200}$	______	______
64.	$\frac{1}{400}$	______	______

	Fraction	Decimal	Percent
65.	________	2.5	________
66.	________	1.7	________
67.	$3\frac{1}{4}$	________	________
68.	$2\frac{4}{5}$	________	________

69. Select a decimal percent and write it as a fraction. Select a different fraction and write it as a percent. Write an explanation of each step of your work.

70. Prepare a table showing fraction, decimal, and percent equivalents for five fractions and mixed numbers of your choice.

In the following application problems, write the answer as a fraction in lowest terms, as a decimal, and as a percent.

71. Many pet owners say they have used the Internet to find pet information. Of 500 people who used the Internet for this purpose, 90 said they used it when buying a pet. What portion used the Internet when buying a pet? (*Source:* American Animal Hospital Association.)

72. About $\frac{1}{3}$ of all books purchased last year were for children. Of these children's books, 27 of every 100 purchased included a coloring activity. What portion of the children's books included a coloring activity? (*Source:* Consumer Research Study on Book Publishing, the American Booksellers, and the Book Industry Study Group.)

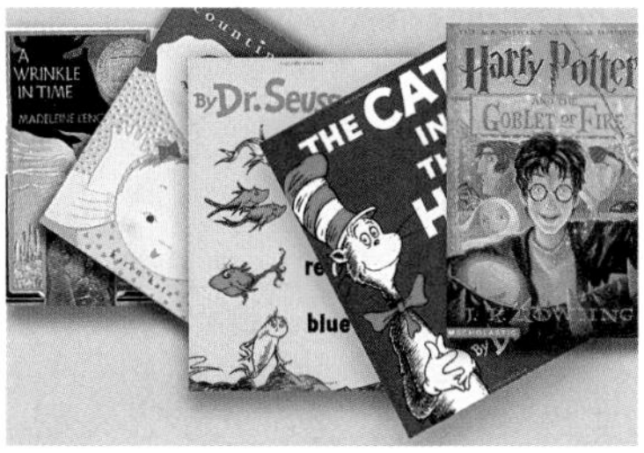

73. Only 13 out of every 100 adults ages 19–50 consume the recommended 1000 milligrams of calcium daily. What portion consumes the recommended daily amount? (*Source:* Market Facts for Milk Mustache.)

74. In a survey on how people learn to parent, 360 parents out of 800 said they were most influenced by relatives, friends, and spouses. What portion learned to parent this way? (*Source:* Bama Research.)

75. In a recent survey, 750 workers were asked if they would hire their own boss if they were in charge. A total of 150 workers said no, they would not. What portion of the workers said no? (*Source:* Marlin's 13th annual Attitudes in the Workplace Survey.)

76. When 1500 adults were asked what was the most important factor to consider when relocating after retirement, 675 said that climate was most important. What portion consider climate to be most important? (*Source:* Longevity Alliance Retirement and Relocation Survey.)

77. An insurance office has 80 employees. If 64 of the employees have iPhones, find

(a) the portion of the employees who have iPhones.

(b) the portion of the employees who do not have iPhones.

78. For every 640 tons of food purchased, 160 tons are discarded by Americans. (*Source:* U.S. Department of Agriculture.)

(a) What portion of the food is discarded?

(b) What portion of the food is not discarded?

79. An antibiotic is used to treat 380 people. If 342 people do not have side effects from the antibiotic, find the portion that do have side effects.

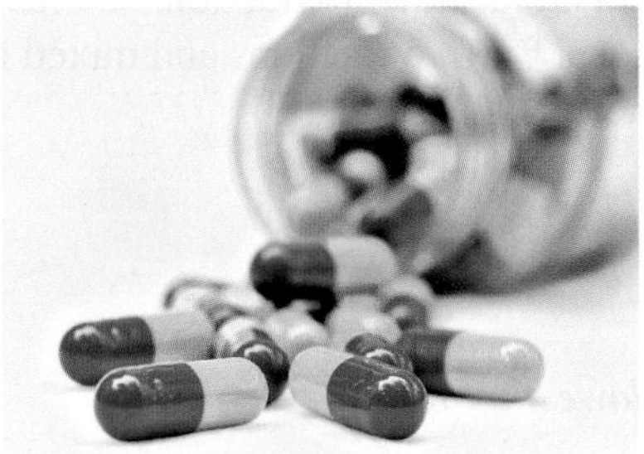

80. A survey of 340 doctors found that 119 of the doctors used a basic system of electronic recordkeeping.

(a) What portion of the doctors used electronic recordkeeping?

(b) What portion did not use it?

The circle graph shows the number of people in a survey of 5400 adults who picked each season as their favorite. Use this graph to answer Exercises 81–84, giving each answer as a fraction, as a decimal, and as a percent.

WEATHER OR NOT

Americans Pick Their Favorite Season (5400 adults surveyed)

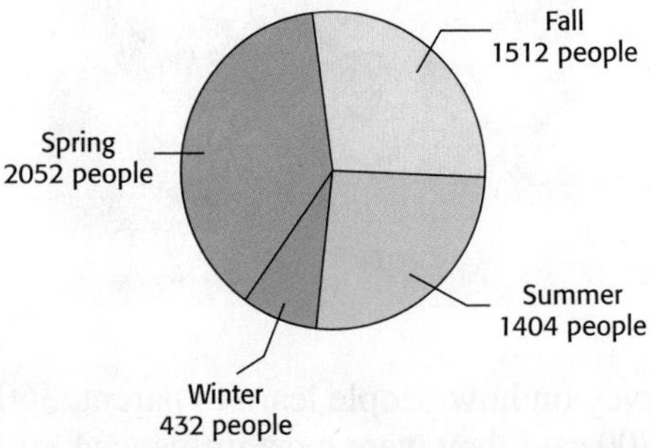

Source: Strategy One Survey for Back to Nature Foods.

81. What portion of the adults picked Winter as their favorite season?

82. What portion of the adults picked Spring as their favorite season?

83. Find the portion of the adults who say Summer is their favorite season.

84. Find the portion of the adults who say Fall is their favorite season.

Relating Concepts (Exercises 85–94) For Individual or Group Work

To review the basics of percent, ***work Exercises 85–94 in order.***

85. 100% of a number means all of the parts or __________ parts out of __________ parts. 200% means two times as many parts and 300% means three times as many parts.

86. Fill in the blanks.

(a) 100% of 765 workers is __________.

(b) 200% of 48 letters is __________.

(c) 300% of 7 DVDs is __________.

87. 50% of a number is __________ parts out of __________ parts, which is __________ of the parts.

88. 10% of a number is __________ parts out of __________ parts and can be found quickly by moving the decimal point __________ place to the __________.

89. 1% of a number is __________ part out of __________ parts and can be found quickly by moving the decimal point __________ places to the __________.

90. Fill in the blanks.

(a) 50% of 1050 homes is __________.

(b) 10% of 370 printers is __________.

(c) 1% of $8 is __________.

In Exercises 91–94, use the shortcut methods for finding 1%, 10%, 50%, 100%, 200%, and 300%.

91. Describe a shortcut method for finding 15% of a number. Use your method to find 15% of $160.

92. Describe a shortcut method for finding 150% of a number. Use your method to find 150% of $160.

93. Explain a shortcut method for finding 90% of a number. Show how to find 90% of $450 using your method.

94. Explain a shortcut method for finding 210% of a number. Show how to find 210% of $800 using your method.

6.3 Using the Percent Proportion and Identifying the Components in a Percent Problem

OBJECTIVES

1. Learn the percent proportion.
2. Solve for an unknown value in a percent proportion.
3. Identify the percent.
4. Identify the whole.
5. Identify the part.

There are two ways to solve percent problems. One method uses proportions and is discussed in this and the next section. The other method uses the percent equation and is explained later in this chapter.

OBJECTIVE 1 **Learn the percent proportion.** We have seen that a statement of two equivalent ratios is called a proportion.

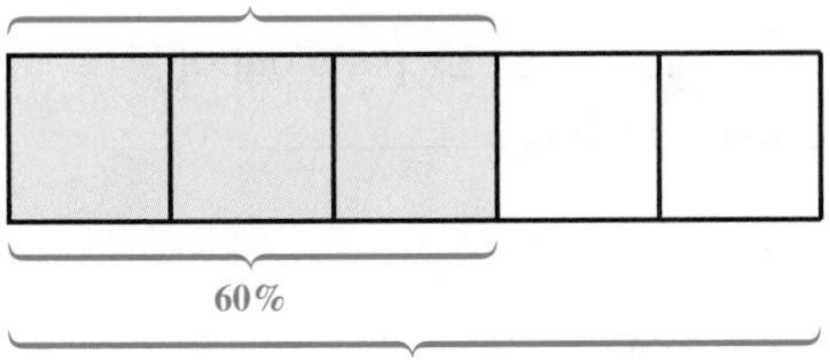

For example, the fraction $\frac{3}{5}$ is the same as the ratio 3 to 5, and 60% is the same as the ratio 60 to 100. As the figure above shows, these two ratios are equivalent and make a proportion.

◀ **Work Problem 1 at the Side.**

The percent proportion can be used to solve percent problems.

Percent Proportion

Part is to *whole* as *percent* is to *100*.

$$\frac{\text{part}}{\text{whole}} = \frac{\text{percent}}{100} \leftarrow \text{Always 100 because percent means } per\ 100$$

In the figure at the top of the page, the **whole** is 5 (the entire quantity), the **part** is 3 (the part of the whole), and the **percent** is 60. Write the percent proportion as follows.

$$\frac{\text{part} \rightarrow 3}{\text{whole} \rightarrow 5} = \frac{60 \leftarrow \text{percent}}{100 \leftarrow 100}$$

Remember: Percent means per 100.

OBJECTIVE 2 **Solve for an unknown value in a percent proportion.** As shown in **Chapter 5**, if any three of the four values in a proportion are known, the fourth can be found by solving the proportion.

EXAMPLE 1 **Using the Percent Proportion**

Use the percent proportion to solve for x, representing the unknown value.

(a) part = 12, percent = 25; find the whole.

$$\frac{\text{part}}{\text{whole}} = \frac{\text{percent}}{100}$$ Percent proportion

Percent

$$\text{Part} \rightarrow \frac{12}{x} = \frac{25}{100} \quad \text{or} \quad \frac{12}{x} = \frac{1}{4}$$

Whole (unknown) → x

$\frac{25}{100}$ in lowest terms is $\frac{1}{4}$.

Continued on Next Page

1 As a review of proportions, use the method of comparing cross products to decide whether each proportion is *true* or *false*. Show the cross products.

GS **(a)** $\frac{1}{2} = \frac{25}{50}$

$\frac{1}{2} = \frac{25}{50}$ $\quad 2 \cdot 25 =$ ____ $\quad 1 \cdot 50 =$ ____

The proportion is (true/false).

(b) $\frac{3}{4} = \frac{150}{200}$

(c) $\frac{7}{8} = \frac{180}{200}$

(d) $\frac{112}{41} = \frac{332}{123}$

Answers

1. (a) 50; 50; true (b) 600 = 600; true
(c) 1440 ≠ 1400; false
(d) 13,612 ≠ 13,776; false

First find the cross products.

The unknown in this proportion is the *whole*.

$$\frac{12}{x} = \frac{1}{4} \qquad x \cdot 1 \qquad 12 \cdot 4$$

Show that the cross products are equal.

$$x \cdot 1 = 12 \cdot 4$$
$$x = 48$$

The whole is 48.

CAUTION

In a **proportion,** you **cannot** divide out a common factor from the numerator of one ratio and the denominator of the other ratio. Dividing out common factors is done ***only*** when you are multiplying fractions.

(b) part = 30, whole = 50; find the percent.
Use the percent proportion.

Percent (unknown) — The unknown in this proportion is the percent.

Part → 30, Whole → 50

$$\frac{30}{50} = \frac{x}{100} \quad \text{Percent proportion}$$

$$\frac{3}{5} = \frac{x}{100} \quad \text{Write } \tfrac{30}{50} \text{ as } \tfrac{3}{5} \text{ in lowest terms.}$$

$$5 \cdot x = 3 \cdot 100 \quad \text{Find the cross products.}$$

$$5 \cdot x = 300$$

$$\frac{\overset{1}{\cancel{5}} \cdot x}{\underset{1}{\cancel{5}}} = \frac{300}{5} \quad \text{Divide both sides by 5.}$$

$$x = 60$$

The percent is 60, written as 60%.

Remember to write the % symbol when solving for an unknown percent.

(c) whole = 150, percent = 18; find the part.

Percent ↓; Part (unknown) → x; Whole → 150

$$\frac{x}{150} = \frac{18}{100} \quad \text{or} \quad \frac{x}{150} = \frac{9}{50} \quad \text{Write } \tfrac{18}{100} \text{ as } \tfrac{9}{50} \text{ in lowest terms.}$$

$$x \cdot 50 = 150 \cdot 9 \quad \text{Find the cross products.}$$

$$x \cdot 50 = 1350$$

$$\frac{x \cdot \overset{1}{\cancel{50}}}{\underset{1}{\cancel{50}}} = \frac{1350}{50} \quad \text{Divide both sides by 50.}$$

$$x = 27$$

The part is 27.

Work Problem 2 at the Side. ▶

2 Use the percent proportion $\left(\frac{\text{part}}{\text{whole}} = \frac{\text{percent}}{100}\right)$ and solve for the unknown value.

GS **(a)** part = 12, percent = 16

$$\frac{12}{x} = \frac{16}{100} \quad \text{or} \quad \frac{12}{x} = \frac{4}{25}$$

$$x \cdot 4 = 12 \cdot \underline{\quad}$$

$$x \cdot 4 = \underline{\quad}$$

$$\frac{x \cdot \overset{1}{\cancel{4}}}{\underset{1}{\cancel{4}}} = \frac{300}{4}$$

$$x = \underline{\quad}$$

(b) part = 30, whole = 120

(c) whole = 210, percent = 20

(d) whole = 4000, percent = 32

(e) part = 74, whole = 185

Answers

2. (a) 25; 300; 75; whole = 75
(b) percent = 25 (so, the percent is 25%)
(c) part = 42 **(d)** part = 1280
(e) percent = 40 (so, the percent is 40%)

3 Identify the percent.

GS **(a)** Of the 900 blood glucose tests, 25% will be completed by Adrian.

The number preceding the ____ symbol is the percent.

So, ____ is the percent.

(b) Of the 620 preschool students, 65% will be served breakfast and lunch.

(c) Find the amount of sales tax by multiplying $590 and $6\frac{1}{2}$ percent.

(d) 8500 tons of recyclables is 42% of what number of tons?

(e) What percent of the 1450 parents use child care?

Answers

3. **(a)** % or percent; 25 **(b)** 65 **(c)** $6\frac{1}{2}$ **(d)** 42 **(e)** The percent is unknown.

As a help in solving percent problems, keep in mind this basic idea.

Percent Problems

All percent problems involve a comparison between a part of something and the whole.

Solving these problems requires identifying the three components of a percent proportion: part, whole, and percent.

OBJECTIVE 3 Identify the percent. Look for the percent first. It is the easiest to identify.

Percent

The **percent** is the ratio of a part to a whole, with 100 as the denominator. In a problem, the percent appears with the word **percent** or with the symbol **"%"** after it.

EXAMPLE 2 Finding the Percent in Percent Problems

Find the percent in the following.

(a) 32% of the 900 men were too large for the imported car.

32% → Percent — Look for the percent symbol (%) or the word *percent*.

The percent is 32. The number 32 appears with the percent symbol (%).

(b) $150 is 25 percent of what number?

25 percent → Percent

The percent is 25 because 25 appears with the word *percent.*

(c) What percent of 7000 pounds is 3500 pounds?

What percent → Percent (unknown)

The word *percent* has no number with it, so the percent is the unknown part of the problem.

◀ Work Problem **3** at the Side.

OBJECTIVE 4 Identify the whole. Next, look for the whole.

Whole

The **whole** is the entire quantity. In a percent problem, the whole often appears after the word **of**.

EXAMPLE 3 **Finding the Whole in Percent Problems**

Identify the whole in the following.

(a) 32% of the 900 men were too large for the imported car.

the 900 men → Whole

The whole is 900. The number 900 appears after the word *of.*

(b) $150 is 25 percent of what number?

what number → Whole — The whole is the unknown part of the problem.

(c) What percent of 7000 pounds is 3500 pounds?

7000 → Whole

Work Problem 4 at the Side. ▶

OBJECTIVE 5 **Identify the part.** Finally, look for the part.

Part

The **part** is the portion being compared with the whole.

Note

If you have trouble identifying the part, find the percent and whole first. The remaining number is the part.

EXAMPLE 4 **Finding the Part in Percent Problems**

Identify the part. Then set up the percent proportion. (Do **not** solve the proportions.)

(a) 54% of 700 students is 378 students.
First find the percent and the whole.

54% of 700 students is 378 students.

54% → Percent; with % symbol; 700 → Whole; follows "of"

Find the percent and then the whole. The remaining number, 378, is the part.

The remaining number, 378, is the part.

54% of 700 students is 378 students.

54% → Percent; 700 → Whole; 378 → Part

$$\frac{\text{Part} \rightarrow 378}{\text{Whole} \rightarrow 700} = \frac{54 \leftarrow \text{Percent}}{100 \leftarrow \text{Always } 100}$$

(b) $150 is 25% of what number?

25% → Percent; what number → Whole (unknown)

$$\frac{\text{Part} \rightarrow 150}{\text{Whole} \rightarrow \text{unknown}} = \frac{25 \leftarrow \text{Percent}}{100 \leftarrow \text{Always } 100}$$

$150 is the remaining number, so the part is $150.

(c) 85% of 7000 is what number?

85% → Percent; 7000 → Whole; what number → Part (unknown)

$$\frac{\text{Part} \rightarrow \text{unknown}}{\text{Whole} \rightarrow 7000} = \frac{85 \leftarrow \text{Percent}}{100 \leftarrow \text{Always } 100}$$

Work Problem 5 at the Side. ▶

4 Identify the whole.

(a) Of the 900 blood glucose tests, 25% will be completed by Adrian.

(b) Of the 620 preschool students, 65% will be served breakfast and lunch.

(c) 8500 tons of recyclables is 42% of what number of tons?

5 Identify the part, then set up the percent proportion.

(a) Of the 900 blood glucose tests, 25% or 225 will be completed by Adrian.

(b) Of the 620 preschool students, 65% or 403 will be served breakfast and lunch.

(c) 8500 tons of recyclables is 42% of what number of tons?

Answers

4. **(a)** 900 **(b)** 620
(c) what number (an unknown)

5. **(a)** 225; $\frac{\text{Part} \rightarrow 225}{\text{Whole} \rightarrow 900} = \frac{25 \leftarrow \text{Percent}}{100 \leftarrow \text{Always } 100}$

(b) 403; $\frac{\text{Part} \rightarrow 403}{\text{Whole} \rightarrow 620} = \frac{65 \leftarrow \text{Percent}}{100 \leftarrow \text{Always } 100}$

(c) 8500; $\frac{\text{Part} \rightarrow 8500}{\text{Whole} \rightarrow \text{unknown}} = \frac{42 \leftarrow \text{Percent}}{100 \leftarrow \text{Always } 100}$

6.3 Exercises

FOR EXTRA HELP

Download the MyDashBoard App

MyMathLab®

CONCEPT CHECK *Identify the component that is unknown. Circle the correct answer. Then, set up the percent proportion. Do **not** solve the proportion.*

1. part = 5, percent = 10
part whole percent

2. part 1.5, whole = 4.5
part whole percent

3. part = 36, whole = 24
part whole percent

4. whole = 72, percent = 30
part whole percent

5. whole = 160, percent = 35
part whole percent

6. part = 20, percent = 25
part whole percent

Find the unknown value in the percent proportion $\frac{part}{whole} = \frac{percent}{100}$. *Round to the nearest tenth if necessary. If the answer is a percent, be sure to include a percent symbol (%).* ***See Example 1.***

7. part = 30, percent = 20

8. part = 25, percent = 25

9. part = 28, percent = 40

10. part = 11, percent = 5

11. part = 15, whole = 60

12. part = 105, whole = 35

13. part = 9.25, whole = 27.75

14. part = 12.8, whole = 9.6

15. whole = 52, percent = 50

16. whole = 115, percent = 38

17. whole = 94.4, part = 25

18. whole = 89.6, part = 50

Solve each problem. If the answer is a percent, be sure to include a percent sign (%). ***See Examples 2–4.***

19. Find the whole if the part is 46 and the percent is 40.

$$\frac{\text{part}}{\text{whole}} = \frac{\text{percent}}{100}$$

$$\frac{46}{x} = \frac{40}{100} \qquad 40 \cdot x \qquad 4600$$

$40 \cdot x = 4600$

$x =$ ______

20. The percent is 45 and the whole is 160. Find the part.

$$\frac{\text{part}}{\text{whole}} = \frac{\text{percent}}{100}$$

$$\frac{x}{160} = \frac{45}{100} \qquad 7200 \qquad 100 \cdot x$$

$100 \cdot x = 7200$

$x =$ ______

21. The whole is 5000 and the part is 20. Find the percent.

22. Suppose the part is 15 and the whole is 2500. Find the percent.

23. Find the percent if the whole is 4300 and the part is $107\frac{1}{2}$.

24. What is the part, if the percent is $12\frac{3}{4}$ and the whole is 5600?

25. The whole is 6480 and the part is 19.44. Find the percent.

26. Suppose the part is 281.25 and the percent is $1\frac{1}{4}$. Find the whole.

CONCEPT CHECK *Fill in each blank with the correct response.*

27. In a percent problem, the percent can be identified because it appears with the word ____________ or with the ____________ symbol after it.

28. In a percent problem, the whole is the ____________ quantity and often appears after the word ____________.

29. In a percent problem, the part is the portion being compared with the ____________.

30. To solve a percent proportion, you must cross ____________.

In Exercises 31–42, set up the percent proportion, and write "unknown" for any value that is not given. Recall that the percent proportion is $\frac{\textit{part}}{\textit{whole}} = \frac{\textit{percent}}{100}$*. Do* ***not*** *try to solve for the unknowns.* ***See Examples 2–4.***

31. 10% of how many bicycles is 60 bicycles?

Part → 60 / Whole → unknown = ___ ← Percent / ← Always 100

$$\frac{60}{\text{unknown}} = \frac{\quad}{100}$$

32. 58% of how many preschoolers is 203 preschoolers?

Part → ___ / Whole → ___ = 58 ← Percent / 100 ← Always 100

$$\frac{\quad}{\quad} = \frac{58}{100}$$

33. What % of \$800 is \$600?

34. What % of \$1500 is \$1395?

35. What is 25% of \$970?

36. What is 61% of 830 homes?

37. 12 injections is 20% of what number of injections?

38. 410 pallets is $33\frac{1}{3}$% of how many pallets.

39. 54.34 is 3.25% of what number?

40. 16.74 is 11.9% of what number?

41. 0.68% of \$487 is what amount?

42. What amount is 6.21% of \$704.35?

43. Identify the three components in a percent problem. In your own words, write a sentence telling how you will identify each of these three components.

44. Write one short sentence using numbers and words. The sentence should include a percent, a whole, and a part. Identify each of these three components.

*Set up the percent proportion for each application problem. Do **not** try to solve for any unknowns.*

45. (GS) Fry's Electronics sold 1262 computers in a recent promotion. If 730 of these computers were laptop computers, what percent of the computers were laptops?

Part → $\frac{730}{}$ = $\frac{\_\_\_\_}{100}$ ← Percent

Whole → ← Always 100

46. (GS) Ivory Soap is $99\frac{44}{100}$% pure. If a bar of Ivory Soap weighs 4 ounces, how many ounces are pure? (*Source:* Procter & Gamble.)

Part → $\frac{\text{unknown}}{}$ = $\frac{\_\_\_\_}{100}$ ← Percent

Whole → ← Always 100

47. Of the 142 people attending a movie theater, 86 bought buttered popcorn. What percent bought buttered popcorn?

48. On her first check from the Pizza Hut Restaurant, 15% was withheld from Maria's total earnings of $225. What amount was withheld?

49. Of the customers buying a salad at McDonald's, 23% prefer Newman's Own Light Salad Dressing. If the total number of salad customers is 610, find the number who prefer Newman's Own Light Salad Dressing.

50. At a community college campus, it was found that 2322 of the students work full-time. If this was 27% of the students, find the total number of students on campus. (*Source*: American Association of Community Colleges.)

51. A survey of 8600 full-time community college students found that 4300 of them work part-time. What percent of the full-time students work part-time? (*Source*: American Association of Community Colleges.)

52. There have been 36 cups of coffee served from a banquet-sized coffee pot. If this is 30% of the capacity of the pot, find the capacity of the pot.

53. In a recent survey of 480 adults, 55% said that they would prefer to have their wedding at a religious site. How many said they would prefer the religious site? (*Source:* National Family Opinion Research.)

54. Sue Ann needs 64 credits to graduate. If she has completed 48 of the credits needed, what percent of the credits has she already completed?

55. In a poll of 822 people, 49.5% said that they get their news from television. Find the number of people who said they get their news from television. (*Source: Brill's Content.*)

56. The sales tax on a new car is $1575. If the sales tax rate is 7%, find the price of the car before the sales tax is added.

57. When asked "What co-worker behaviors annoy you the most," 12% of those surveyed said "Being perpetually late." Find the number of people in the survey if 168 people gave this response. (*Source*: Accountemps.)

58. The state troopers tested 924 cars for safety. There were 231 cars that failed the safety test for one or more reasons. Find the percent of cars that failed the test.

59. Jerry Azzaro has listed 680 antique toys on eBay. If 45% of these were antique toy trains, find the number that were toy trains.

60. In a survey of 27,400 American sports fans, ages 12 and up, 3562 picked Major League baseball as their favorite spectator sport. What percent picked Major League baseball as their favorite sport? (*Source*: ESPN Sports Poll.)

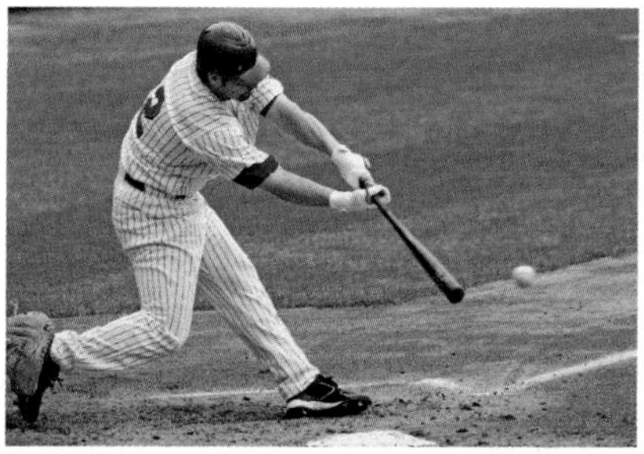

6.4 Using Proportions to Solve Percent Problems

OBJECTIVES

1. Use the percent proportion to find the part.
2. Find the whole using the percent proportion.
3. Find the percent using the percent proportion.

This is the percent proportion that you learned about in the previous section.

$$\frac{\text{part}}{\text{whole}} = \frac{\text{percent}}{100}$$

Recall that if one of the values is unknown, you can find it by solving the percent proportion.

OBJECTIVE 1 Use the percent proportion to find the part. The first example shows how to use the percent proportion to find the part.

EXAMPLE 1 Finding the Part with the Percent Proportion

Find 15% of \$160.

Here the percent is 15 and the whole is 160. (Recall that the whole often comes after the word ***of***.) Now find the part. Let x represent the unknown part.

$$\frac{\text{part}}{\text{whole}} = \frac{\text{percent}}{100} \quad \text{so} \quad \frac{x}{160} = \frac{15}{100} \quad \text{or} \quad \frac{x}{160} = \frac{3}{20}$$

Write $\frac{15}{100}$ as $\frac{3}{20}$ in lowest terms.

Find the cross products in the proportion and show that they are equal.

$x \cdot 20 = 160 \cdot 3$ Cross products

$x \cdot 20 = 480$

$\frac{x \cdot \overset{1}{\cancel{20}}}{\underset{1}{\cancel{20}}} = \frac{480}{20}$ Divide both sides by 20.

$x = 24$ The unknown part is 24.

15% of \$160 is **\$24**.

◀ **Work Problem 1 at the Side.**

1 Use the percent proportion to find the part.

GS (a) 8% of 400 patients

$$\frac{x}{400} = \frac{8}{100} \quad \text{or} \quad \frac{x}{400} = \frac{2}{25}$$

$x \cdot 25 = 400 \cdot$ ____

$x \cdot 25 =$ ____

$\frac{x \cdot \overset{1}{\cancel{25}}}{\underset{1}{\cancel{25}}} = \frac{800}{25}$

$x =$ ______

(b) 15% of \$3220

(c) 7% of 2700 miles

(d) 48% of 1580 kilowatts

Just as with some of the fraction application problems in **Chapter 2**, the word ***of*** may be an indicator word meaning ***multiply.*** Here is an example.

15% of 160

↓

15% • 160

In this type of example, there is another way to find the part.

Finding the Part Using Multiplication

To find the part:

Step 1 Identify the percent. Write the percent as a decimal.

Step 2 Multiply this decimal by the whole.

Answers

1. (a) 2; 800; 32 patients (b) \$483 (c) 189 miles (d) 758.4 kilowatts

EXAMPLE 2 **Finding the Part Using Multiplication**

Use multiplication to find the part.

(a) Find 42% of 830 yards.

Step 1 Here, the percent is 42. Write 42% as the decimal 0.42.

Step 2 Multiply 0.42 and the whole, which is 830.

$$\text{part} = (0.42)(830)$$
$$= 348.6 \text{ yd}$$

When the percent and whole are given, the part must be found.

It is a good idea to estimate the answer, to make sure no mistakes were made with decimal points. Round 42% to 40% or 0.4, and round 830 to 800. Next, 40% of 800 is

$$(0.4)(800) = 320 \leftarrow \text{Estimate}$$

so the exact answer of 348.6 is reasonable.

(b) Find 25% of 1680 cars.
Identify the percent as 25. Write 25% in decimal form as 0.25. Now, multiply 0.25 and 1680.

$$\text{part} = (0.25)(1680) = 420 \text{ cars} \quad \text{Multiply.}$$

You can also use a shortcut to find the answer. Since 25% means 25 parts out of 100 parts, this is the same as $\frac{1}{4}$ of the whole ($\frac{25}{100} = \frac{1}{4}$). Do you see a shortcut here? You can find $\frac{1}{4}$ of a number by dividing the number by 4. So, this shortcut gives us the exact answer, $1680 \div 4 = 420$.

(c) Find 140% of 60 miles.
In this problem, the percent is 140. Write 140% as the decimal 1.40. Next, multiply 1.40 and 60.

$$\text{part} = (1.40)(60) = 84 \text{ miles} \quad \text{Multiply.}$$

You can estimate the answer by realizing that 140% is close to 150% (which is $1\frac{1}{2}$) and $1\frac{1}{2}$ times 60 is 90. So, 84 miles is a reasonable answer.

(d) Find 0.4% of 50 kilometers.

$$\text{part} = (0.004)(50) = 0.2 \text{ kilometer} \quad \text{Multiply.}$$

↑ Write 0.4% as a decimal.

Estimate the answer by realizing that 0.4% is less than 1%.

$$1\% \text{ of } 50 \text{ kilometers} = 50. = 0.5 \text{ kilometer}$$

So our exact answer should be *less than* 0.5 kilometer, and 0.2 kilometer fits this requirement.

Work Problem 2 at the Side. ▶

EXAMPLE 3 **Solving for the Part in an Application Problem**

Raley's Markets has 850 employees. Of these employees, 28% are students. How many of the employees are students? Use the six problem-solving steps.

Continued on Next Page

2 Use multiplication to find the part.

GS **(a)** 55% of 10,000 x-rays

Write 55% in decimal form as 0.55.

part = (0.55) (________)

= ________

(b) 16% of 120 miles

(c) 135% of 60 dosages

(d) 0.5% of $238

Answers

2. **(a)** 10,000; 5500 x-rays **(b)** 19.2 miles **(c)** 81 dosages **(d)** $1.19

3 Use the six problem-solving steps to solve each problem.

GS **(a)** One day on Jacob's mail route there were 2920 pieces of mail. If 45% of those were advertising pieces, find the number of advertising pieces.

Notice the blue word of as an indicator for multiplication. Write 45% in decimal form as _____.

part = (____) • (2920)

= ________

(b) There are 9750 students at the college. If 12% of them wear glasses or contact lenses, how many students wear glasses or contact lenses?

Answers

3. **(a)** 0.45; 0.45; 1314 advertising pieces
(b) 1170 wear glasses or contact lenses

Step 1 **Read** the problem. The problem asks us to find the number of employees who are students.

Step 2 **Work out a plan.** Look for the word *of* as an indicator word for multiplication.

28% of the employees are students.
↑ Indicator word

The total number of employees is 850, so the whole is 850. The percent is 28. To find the number of students, find the part.

Step 3 **Estimate** a reasonable answer. You can estimate the answer by rounding 28% to 25% and 850 to 900. Remember that 25% is 25 parts out of 100, which is equivalent to $\frac{1}{4}$. So divide 900 by 4.

$900 \div 4 = 225$ students ← Estimate

Step 4 **Solve** the problem.

part = $(0.28)(850) = 238$ Multiply.
↑ Write 28% as a decimal.

Notice that the decimal point was moved two places to the left.

Step 5 **State the answer.** Raley's Markets has 238 student employees.

Step 6 **Check.** The exact answer, 238 students, is close to our estimate of 225 students.

◀ Work Problem **3** at the Side.

Calculator Tip

If you are using a calculator, you could solve **Example 3** above like this.

0.28 (×) 850 (=) 238

Or, if your calculator has a % key, check the instructions for an alternate method using the % key.

OBJECTIVE 2 Find the whole using the percent proportion. The next example shows how to use the percent proportion to find the whole.

Note

Remember, the *whole* is the entire quantity.

EXAMPLE 4 Finding the Whole with the Percent Proportion

(a) 8 iPods is 4% of what number of iPods?

Here the percent is 4, the whole is unknown, and the part is 8. Use the percent proportion to find the whole. Let x represent the unknown whole.

$$\frac{8}{x} = \frac{4}{100} \quad \text{or} \quad \frac{8}{x} = \frac{1}{25}$$

Write $\frac{4}{100}$ as $\frac{1}{25}$ in lowest terms.

$x \cdot 1 = 8 \cdot 25$ Cross products

$x = 200$

8 iPods is 4% of **200 iPods**.

Continued on Next Page

(b) 135 tourists is 15% of what number of tourists?
The percent is 15 and the part is 135.

$$\frac{135}{x} = \frac{15}{100} \quad \begin{matrix}\text{Part} \rightarrow 135,\ 15 \leftarrow \text{Percent} \\ \text{Whole (unknown)} \rightarrow x,\ 100 \leftarrow \text{Always 100}\end{matrix}$$

If the part and percent are given, the *whole* must be found.

$$\frac{135}{x} = \frac{3}{20} \quad \text{Write } \tfrac{15}{100} \text{ as } \tfrac{3}{20} \text{ in lowest terms.}$$

$$x \bullet 3 = 135 \bullet 20 \quad \text{Cross products}$$

$$x \bullet 3 = 2700$$

$$\frac{x \bullet \overset{1}{\cancel{3}}}{\underset{1}{\cancel{3}}} = \frac{2700}{3} \quad \text{Divide both sides by 3.}$$

$$x = 900$$

135 tourists is 15% of **900 tourists.**

Work Problem 4 at the Side. ▶

EXAMPLE 5 Applying the Percent Proportion

At Newark Salt Works, 78 employees are absent because of illness. If this is 5% of the total number of employees, how many employees does the company have? Use the six problem-solving steps.

Step 1 **Read** the problem. The problem asks for the total number of employees.

Step 2 **Work out a plan.** From the information in the problem, the percent is 5 and the part of the total number of employees is 78. The total number of employees or entire quantity, which is the whole, is the unknown.

Step 3 **Estimate** a reasonable answer. Round the number of employees from 78 to 80. Then, 5% is equivalent to the fraction $\frac{1}{20}$, and 80 is $\frac{1}{20}$ of the total number of employees.

$$80 \bullet 20 = 1600 \text{ employees} \leftarrow \text{Estimate}$$

Step 4 **Solve** the problem. Use the percent proportion to find the whole (the total number of employees).

$$\frac{78}{x} = \frac{5}{100} \quad \begin{matrix}\text{Part} \rightarrow 78,\ 5 \leftarrow \text{Percent} \\ \text{Whole (unknown)} \rightarrow x,\ 100 \leftarrow \text{Always 100}\end{matrix}$$

$$\frac{78}{x} = \frac{1}{20} \quad \text{Write } \tfrac{5}{100} \text{ as } \tfrac{1}{20} \text{ in lowest terms.}$$

$$x \bullet 1 = 78 \bullet 20 \quad \text{Cross products}$$

$$x = 1560$$

Step 5 **State the answer.** The company has **1560 employees.**

Step 6 **Check.** The exact answer, 1560 employees, is close to our estimate of 1600 employees.

Continued on Next Page

4 Use the percent proportion to find the unknown whole.

GS **(a)** 75 *American Idol* contestants are only 5% of what number who auditioned?

$$\frac{75}{x} = \frac{5}{100} \quad \text{Write } \tfrac{5}{100} \text{ in lowest terms.}$$

$$\frac{75}{x} = \frac{1}{\_\_\_\_}$$

$$x \bullet 1 = 75 \bullet 20$$

$$x = \_\_\_\_\_\_$$

(b) 28 antiques is 35% of what number of antiques?

(c) 387 customers is 36% of what number of customers?

(d) 292.5 miles is 37.5% of what number of miles?

Answers

4. **(a)** 20; 1500 auditioned **(b)** 80 antiques **(c)** 1075 customers **(d)** 780 miles

5 Use the six problem-solving steps and the percent proportion to solve each problem.

GS **(a)** A freeze resulted in a loss of 52% of an avocado crop. If the loss was 182 tons, find the total number of tons in the crop.

$$\frac{182}{x} = \frac{52}{100} \quad \text{Write } \tfrac{52}{100} \text{ in lowest terms.}$$

$$\frac{182}{x} = \frac{13}{\_\_\_\_}$$

$$\frac{x \cdot \overset{1}{\cancel{13}}}{\underset{1}{\cancel{13}}} = \frac{4550}{13}$$

$x =$ ________

(b) A factory batch of cake mix contains 900 pounds of sugar, which is 18%, by weight, of the entire batch. What is the total weight of the batch?

Answers

5. **(a)** 25; 350 tons **(b)** 5000 pounds

Note

To estimate the answer to **Example 5** on the previous page, 5% was changed to its fraction equivalent, $\frac{1}{20}$. Because 80 (rounded) is $\frac{1}{20}$ of the total employees, 80 was multiplied by 20 to get 1600, the estimated answer.

◀ **Work Problem 5 at the Side.**

OBJECTIVE 3 Find the percent using the percent proportion. If the part and the whole are known, the percent proportion can be used to find the percent.

EXAMPLE 6 Using the Percent Proportion to Find the Percent

(a) 13 coupons is what percent of 52 coupons?
The whole is 52 (follows *of*) and the part is 13. Next, find the percent.

The whole often follows the word "of".

$$\frac{\text{part}}{\text{whole}} = \frac{\text{percent}}{100}$$

$$\begin{array}{r} \text{Part} \rightarrow \\ \text{Whole} \rightarrow \end{array} \frac{13}{52} = \frac{x}{100} \begin{array}{l} \leftarrow \text{Percent (unknown)} \\ \leftarrow \text{Always 100} \end{array}$$

Write $\frac{13}{52}$ as $\frac{1}{4}$ in lowest terms.

$$\frac{1}{4} = \frac{x}{100}$$

Find the cross products.

$$4 \cdot x = 1 \cdot 100 \quad \text{Cross products}$$

$$\frac{\overset{1}{\cancel{4}} \cdot x}{\underset{1}{\cancel{4}}} = \frac{100}{4} \quad \text{Divide both sides by 4.}$$

$$x = 25$$

13 coupons is **25%** of 52 coupons.

(b) What percent of \$500 is \$100?
The whole is 500 (follows *of*) and the part is 100.

$$\frac{100}{500} = \frac{x}{100} \leftarrow \text{Percent (unknown)}$$

Write $\frac{100}{500}$ as $\frac{1}{5}$ in lowest terms.

$$\frac{1}{5} = \frac{x}{100}$$

$$5 \cdot x = 1 \cdot 100 \quad \text{Cross products}$$

$$5 \cdot x = 100$$

$$\frac{\overset{1}{\cancel{5}} \cdot x}{\underset{1}{\cancel{5}}} = \frac{100}{5} \quad \text{Divide both sides by 5.}$$

$$x = 20$$

Remember to write the % symbol in the answer.

20% of \$500 is \$100.

Continued on Next Page

CAUTION

When finding the percent, be sure to label your answer with the percent symbol (%).

Work Problem 6 at the Side. ▶

EXAMPLE 7 **Applying the Percent Proportion**

A roof is expected to last 20 years before needing replacement. If the roof is now 15 years old, what percent of the roof's life has been used?

Step 1 **Read** the problem. The problem asks for the percent of the roof's life that is already used.

Step 2 **Work out a plan.** The expected life of the roof is the entire quantity or *whole,* which is 20. The *part* of the roof's life that is already used is 15. Use the percent proportion to find the percent of the roof's life used.

Step 3 **Estimate** a reasonable answer. Since the roof is 15 years old, it is $\frac{15}{20}$ or $\frac{3}{4}$ used. Remember that $\frac{3}{4}$ is equivalent to 75%, so our estimate is 75%.

Step 4 **Solve** the problem. Let x represent the unknown percent.

$$\frac{15}{20} = \frac{x}{100} \quad \text{or} \quad \frac{3}{4} = \frac{x}{100}$$

Part → 15; Whole → 20. Write $\frac{15}{20}$ as $\frac{3}{4}$ in lowest terms.

$$4 \cdot x = 3 \cdot 100 \quad \text{Cross products}$$

$$4 \cdot x = 300$$

$$\frac{\overset{1}{\cancel{4}} \cdot x}{\underset{1}{\cancel{4}}} = \frac{300}{4} \quad \text{Divide both sides by 4.}$$

$$x = 75$$

Step 5 **State the answer.** 75% of the roof's life has been used.

Step 6 **Check.** The exact answer, 75%, matches our estimate of 75%.

Work Problem 7 at the Side. ▶

EXAMPLE 8 **Applying the Percent Proportion**

Rainfall this year was 33 inches, while normal rainfall is only 30 inches. What percent of normal rainfall is this year's rainfall?

Step 1 **Read** the problem. The problem asks us to find what percent this year's rainfall is of normal rainfall.

Step 2 **Work out a plan.** The normal rainfall is the *whole,* which is 30. This year's rainfall is *all of normal rainfall and more,* or 33 (part = 33). You need to find the percent that this year's rainfall is of normal rainfall.

Continued on Next Page

6 Use the percent proportion to solve each problem.

GS **(a)** \$21 is what percent of \$105?

Write $\frac{21}{105}$ in lowest terms.

$$\frac{21}{105} = \frac{x}{100}$$

$$\frac{1}{\_\_\_} = \frac{x}{100}$$

$$5 \cdot x = 100$$

$$\frac{\overset{1}{\cancel{5}} \cdot x}{\underset{1}{\cancel{5}}} = \frac{100}{5}$$

$$x = \_\_\_\%$$

(b) What percent of 320 Internet companies is 48 Internet companies?

(c) What percent of 2280 court trials is 1026 trials?

7 Solve each problem.

(a) The bid price on an auction item is \$289 while the minimum acceptable price is \$425. The bid price is what percent of the minimum?

(b) A laboratory technician completes 80 tests in one day. If 52 of these tests were completed in the morning, what percent of the tests were completed in the morning?

Answers

6. (a) 5; 20% **(b)** 15% **(c)** 45%

7. (a) 68% **(b)** 65%

8 Solve each problem.

GS **(a)** A new Toyota Prius Hybrid gets 32 miles per gallon on the highway and 48 miles per gallon around town. What percent of the highway mileage does the car get around town?

Write $\frac{48}{32}$ in lowest terms.

$$\frac{48}{32} = \frac{x}{100}$$

$$\frac{3}{2} = \frac{x}{100}$$

$$2 \cdot x = 3 \cdot 100$$

$$\frac{\overset{1}{\cancel{2}} \cdot x}{\underset{1}{\cancel{2}}} = \frac{\_\_\_\_}{2}$$

$x =$ ____%

(b) The service department set a goal of 360 service calls this week. If they made 432 service calls, find the percent of their goal that they achieved.

Step 3 **Estimate** a reasonable answer. The increase in rainfall is 3 inches and the whole is 30 inches. The increase is $\frac{3}{30}$ or $\frac{1}{10}$ which is 10%. The whole is 100%, so 100% + 10% = 110%, our estimate.

Step 4 **Solve** the problem. Let x represent the unknown percent.

$$\frac{33}{30} = \frac{x}{100} \quad \text{or} \quad \frac{11}{10} = \frac{x}{100}$$ Write $\frac{33}{30}$ as $\frac{11}{10}$ in lowest terms.

$$10 \cdot x = 11 \cdot 100$$ Cross products

$$10 \cdot x = 1100$$

$$\frac{\overset{1}{\cancel{10}} \cdot x}{\underset{1}{\cancel{10}}} = \frac{1100}{10}$$ Divide both sides by 10.

$$x = 110$$

If the part is *greater than* the whole, the percent is greater than 100.

Step 5 **State the answer.** This year's rainfall is **110%** of normal rainfall.

Step 6 **Check.** The exact answer, 110%, matches our estimate of 110%.

◀ **Work Problem 8 at the Side.**

Note

In **Example 8**, the part (33 inches) is *greater* than the whole (30 inches). This can occur when there is an increase or a gain and we are comparing this *greater* amount (part) to the original amount (whole).

Answers

8. **(a)** 300; 150% **(b)** 120%

6.4 Exercises

FOR EXTRA HELP

Download the MyDashBoard App

MyMathLab®

1. **CONCEPT CHECK** To find the part using the multiplication shortcut, use the formula

part = ________ • ________.

2. **CONCEPT CHECK** Solve for the part when the percent is 38 and the whole is 200. Circle the correct answer.

7.6 76 760 7600

Find the part using the multiplication shortcut. ***See Example 2.***

3. 35% of 120 test tubes
part = (0.35)(120)
part = ____________

4. 20% of 1800 rentals
part = (0.20)(1800)
part = ____________

5. 45% of 4080 military personnel

6. 12% of 3650 websites

7. 4% of 120 ft

8. 9% of $150

9. 150% of 210 files

10. 130% of 60 trees

11. 52.5% of 1560 trucks

12. 38.2% of 4250 loads

13. 2% of $164

14. 6% of $434

15. 225% of 680 tables

16. 110% of 150 apartments

17. 17.5% of 1040 cell phones

18. 46.1% of 843 kilograms

19. 0.9% of $2400

20. 0.3% of $1400

CONCEPT CHECK *Set up a percent proportion for each application problem. Do* ***not*** *try to solve for any unknowns.*

21. 80 e-mails is 25% of what number of e-mails?

22. 32 medical exams is 5% of what number of medical exams?

Find the whole using the percent proportion. ***See Example 4.***

23. 30% of what number of hay bales is 48 hay bales?
part is 48; percent is 30; whole is unknown

$$\frac{48}{x} = \frac{30}{100} \text{ or } \frac{48}{x} = \frac{3}{10}$$

$3 \cdot x = 480$ Find cross products.

$$\frac{\overset{1}{\cancel{3}} \cdot x}{\underset{1}{\cancel{3}}} = \frac{480}{3}$$ Divide both sides by 3.

$x =$ ____

24. 55% of what number of experiments is 209 experiments?

$$\frac{209}{x} = \frac{55}{100} \text{ or } \frac{209}{x} = \frac{11}{20}$$

$11 \cdot x = 4180$ Find cross products.

$$\frac{\overset{1}{\cancel{11}} \cdot x}{\underset{1}{\cancel{11}}} = \frac{4180}{11}$$ Divide both sides by 11.

$x =$ ____

25. 495 successful students is 90% of what number of students?

26. 168 text messages is 28% of what number of text messages?

27. 462 mountain bikes is 140% of what number of mountain bikes?

28. 1496 graduates is 110% of what number of graduates?

29. $12\frac{1}{2}\%$ of what number is 350?
($Hint$: Write $12\frac{1}{2}\%$ as 12.5%.)

30. $5\frac{1}{2}\%$ of what number is 176?
($Hint$: Write $5\frac{1}{2}\%$ as 5.5%.)

CONCEPT CHECK *Set up a percent proportion for each application problem. Do* **not** *try to solve for any unknowns.*

31. 18 bean burritos is what percent of 36 bean burritos?

32. 62 hospital rooms is what percent of 248 hospital rooms?

Find the percent using the percent proportion. Round answers to the nearest tenth if necessary. ***See Example 6.***

33. 780 hybrid cars is what percent of 1500 hybrid cars?

34. 14 tweets is what percent of 700 tweets?

35. 27 downloaded songs is what percent of 1800 downloaded songs?
GS

part = 27; whole = 1800

$$\frac{27}{1800} = \frac{x}{100} \quad \text{or} \quad \frac{3}{200} = \frac{x}{100}$$

$$200 \cdot x = 300$$

$$\frac{\overset{1}{\cancel{200}} \cdot x}{\underset{1}{\cancel{200}}} = \frac{300}{200}$$

$x =$ ____

The percent is ____.

36. 60 cartons is what percent of 2400 cartons?
GS

$$\frac{60}{2400} = \frac{x}{100} \quad \text{or} \quad \frac{1}{40} = \frac{x}{100}$$

$$40 \cdot x = 100$$

$$\frac{\overset{1}{\cancel{40}} \cdot x}{\underset{1}{\cancel{40}}} = \frac{100}{40}$$

$x =$ ____

The percent is ____.

37. What percent of \$344 is \$64?

38. What percent of \$398 is \$14?

39. What percent of 250 tires is 23 tires?

40. What percent of 105 employees is 54 employees?

41. A student turned in the following answers on a test. You can tell that two of the answers are incorrect without even working the problems. Find the incorrect answers and explain how you identified them (without actually solving the problems).

50% of \$84 is _\$42_.

150% of \$30 is _\$20_.

25% of \$16 is _\$32_.

100% of \$217 is _\$217_.

42. Write a percent problem on any topic you choose. Be sure to include only two of the three components so that you can solve for the third component. Identify each component of the problem and then solve it.

Solve each application problem. Round percent answers to the nearest tenth if necessary. ***See Examples 3, 5, 7, and 8.***

43. Bonnie Boehme, who works part-time, earns \$240 per week and has 22% of this amount withheld for taxes, Social Security, and Medicare. Find

(a) the amount withheld.

(b) the amount remaining after the withholdings.

44. An estimated 29.5% of automobile crashes are caused by driver distractions such as using a cell phone or texting. (*Source:* National Conference of State Legislatures.) If there are 16,450 automobile crashes in a study, find

(a) the number of crashes caused by driver distractions.

(b) the number caused by other factors. Round to the nearest whole number.

The graph below shows when people check their e-mail. This data was gathered from 3020 adults and those surveyed could give more than one response. Use this graph to answer Exercises 45–48. Round to the nearest number.

45. Find the number of people who check their e-mail while working.

46. How many people check their e-mail while in bed?

47. Find the number of people who check their e-mail as soon as they wake up.

48. How many people check their e-mail during meals?

49. Each day in Los Angeles there are 48 million commuter trips. (*Source:* CNN.) If only 2% of these trips use public transportation, find

(a) the number of trips using public transportation.

(b) the number of trips not using public transportation.

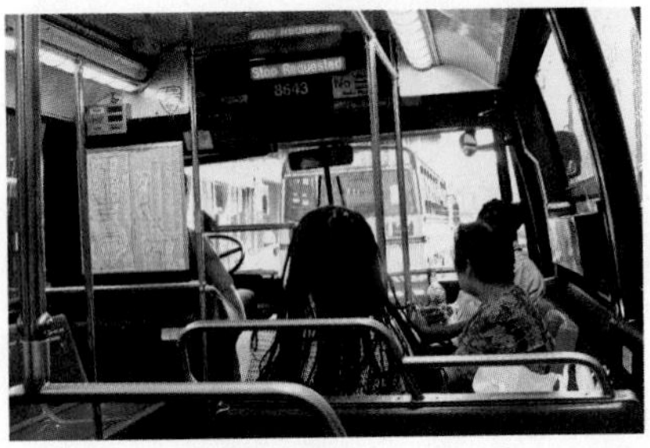

50. At work, 68% of those surveyed said that they recycle bottles and cans while 56% recycle paper products and newspapers. (*Source*: Microsystems by Harris Interactive.) If 1750 workers were surveyed,

(a) how many recycle bottles and cans?

(b) how many recycle paper products and newspapers?

The bar graph below shows the percent of children 6–11 years of age who are neglecting dental hygiene. Use the information to answer Exercises 51–54.

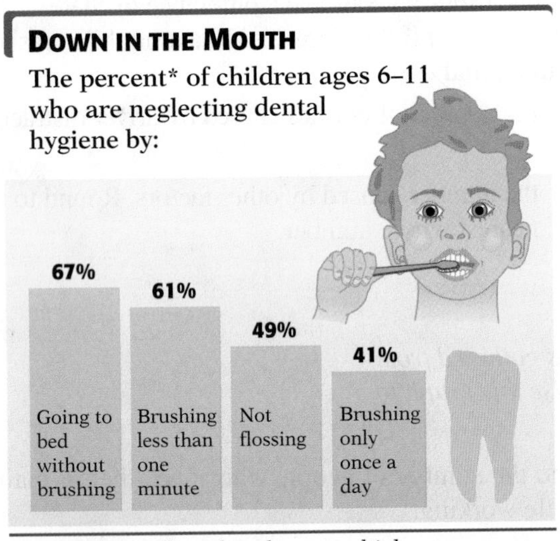

*Respondents allowed to choose multiple answers.
Source: Services for Crest.

51. What percent of the children brush their teeth before going to bed?

52. What percent of the children floss their teeth?

53. If 3400 children answered the questions for this survey, how many of the children brush less than one minute?

54. How many of the 3400 children in the survey brush only once a day?

55. A recent study examined 48,000 military jobs, such as Army attack helicopter pilot or Navy gunner's mate. It was found that only 960 of these jobs are filled by women. What percent of these jobs are filled by women? (*Source:* Rand's National Defense Research Institute.)

56. There are more than 55,000 words in *Webster's Dictionary,* but most educated people can identify only 20,000 of these words. What percent of the words in the dictionary can these people identify?

57. Ebony Durrant has 7.5% of her earnings deposited into her retirement plan. If $240 per month is deposited in the plan, find her monthly and yearly earnings.

58. The number of federal income tax returns that were filed electronically 5 years ago was 68.3 million, or 51% of all returns. Find the total number of income tax returns that were filed 5 years ago. Round to the nearest tenth of a million. (*Source:* Internal Revenue Service.)

The circle graph shows the percent of various ice cream brands purchased by the 1582 Americans in a recent survey. Use this information to answer Exercises 59–62.

Source: Information Resources Inc.; NPD Group.

59. Of the specific brands purchased (not "Other" or "Private Label"), which brand was purchased most often?

60. What percent of ice cream purchases were "Private Label" or "Other" brands?

61. Find the number of people in the survey who said they purchase Häagen-Dazs. Round to the nearest whole number.

62. How many more people said they purchase Blue Bell brand than Ben & Jerry's brand? Round to the nearest whole number.

The bar graph below shows the percent of drivers in each age group who were stopped by police last year. Use the information to answer Exercises 63–66.

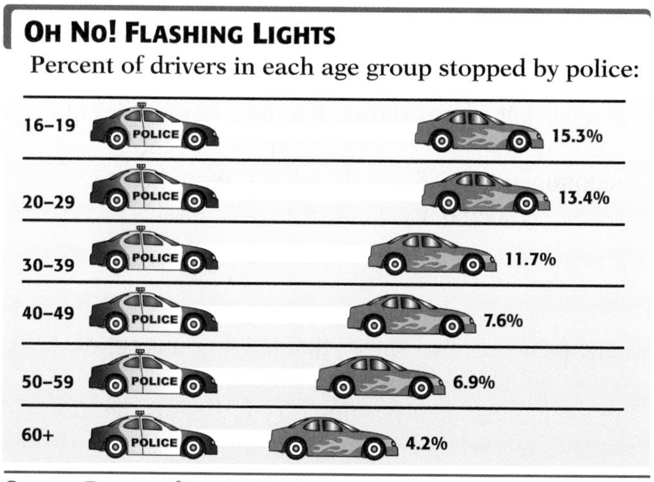

Source: Bureau of Justice Statistics.

63. What percent of drivers in the 16–19 age group were not pulled over by police?

64. What percent of drivers in the 60+ age group were not pulled over by police?

65. If 8000 drivers in the study were in the 20–29 age group, how many were pulled over by police?

66. How many of the 6000 drivers in the 30–39 age group were pulled over by police?

67. Marketing Intelligence Service says that there were 15,401 new products introduced last year. If 86% of the products introduced last year failed to reach their business objectives, find the number of products that were successful. (Round to the nearest whole number.)

68. The income earned on an investment is 8.5% of the amount invested. If the income is $12,750, find the amount of the investment.

69. Kathy West owns Banjo Shirt Company and sells handmade clothing online. If 85% of her 1540 customers paid for their orders using Pay Pal, how many of her customers used some other method of payment?

70. A family of four with a monthly income of $2900 spends 90% of its earnings and saves the rest. Find
(a) the monthly savings.
(b) the annual savings of this family.

Relating Concepts (Exercises 71–78) For Individual or Group Work

Knowing and using the percent proportion is useful when solving percent problems.
Work Exercises 71–78 in order.

71. In the percent proportion, part is to ______ as percent is to ______.

72. All percent problems involve a comparison between a part of something and the ______.

Use this Ramen Noodles label of nutrition facts to answer Exercises 73–78. Read the label very carefully. Round to the nearest tenth of a percent.

Ramen Noodles

Nutrition Facts
Serving Size 1/2 Pkg. (15 oz/42.5 g)
Servings Per Package 2
Calories 190
Calories from Fat 70

Amount/serving	%DV*	Amount/serving	%DV*
Total Fat 8g	12%	Total Carbohydrates 27g	9%
Saturated Fat 4g	20%	Dietary Fiber Less Than 1g	3%
Cholesterol 0mg	0%	Sugars Less Than 1g	
Sodium 670mg	28%	Protein 4g	

Vitamin A 0% • Vitamin C 0% • Calcium 2% • Iron 4%

*Percent Daily Values (DV) are based on a 2,000 calorie diet.
Calories Per Gram
Fat 9 • Carbohydrates 4 • Protein 4

Source: Ramen Noodles package.

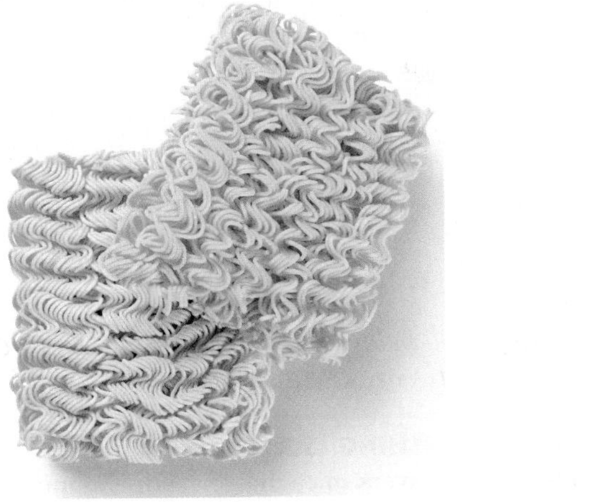

73. How many calories per serving are from total carbohydrates?

74. The package label shows that the 27 grams (g) of carbohydrates in one serving are 9% of the recommended Daily Value (DV). What is the recommended daily value of carbohydrates?

75. Find the number of grams of total fat that are needed to meet the recommended Daily Value (DV) of fat. Round to the nearest gram.

76. The package label shows that the 4 grams (g) of saturated fat in one serving is 20% of the recommended Daily Value (DV). What is the recommended daily value of saturated fat?

77. Will a person eating two packages of Ramen Noodles in one day exceed their recommended daily value of sodium? Explain your answer.

78. How many packages of Ramen Noodles must be eaten in a day to meet the recommended daily value of fiber? Would this be possible? Would this result in good nutrition? (Round to the nearest whole package.)

6.5 Using the Percent Equation

OBJECTIVES

1. **Use the percent equation to find the part.**
2. **Find the whole using the percent equation.**
3. **Find the percent using the percent equation.**

In the last section you used a proportion to solve percent problems. In this section we show another way to solve these problems by using the **percent equation.** The percent equation is a rearrangement of the percent proportion.

Percent Equation

part = percent • whole

Be sure to write the percent as a decimal before using the equation.

When using the percent proportion, we did *not* have to write the percent as a decimal because 100 was used in the denominator of the proportion. However, because there is no 100 in the percent *equation, we must* first write the percent as a decimal by dividing by 100.

Some of the examples solved earlier will be reworked using the percent equation. For comparison, you can look back in the previous section to see how these same problems were solved using proportions.

OBJECTIVE 1 Use the percent equation to find the part. The first example shows how to find the part.

EXAMPLE 1 Finding the Part

(a) Find 15% of \$160.

Write 15% as the decimal 0.15. The whole, which comes after the word *of,* is 160. Next, use the percent equation. Let x represent the unknown part.

part = percent • whole

$$x = (0.15)(160)$$

Multiply 0.15 and 160.

15% *must* be written as the decimal 0.15.

$$x = 24$$

15% of \$160 is \$24.

(b) Find 110% of 80 cases.

Write 110% as the decimal 1.10. The whole is 80. Let x represent the unknown part.

part = percent • whole

$$x = (1.10)(80)$$

$$x = 88$$

Write 110% as a decimal. 110% = 1.10

110% of 80 cases is 88 cases.

(c) Find 0.4% of 250 patients.

Write 0.4% as the decimal 0.004. The whole is 250. Let x represent the unknown part.

part = percent • whole

$$x = (0.004)(250)$$

$$x = 1$$

Write 0.4% as a decimal. 0.4% = 0.004

0.4% of 250 patients is 1 patient.

Continued on Next Page

1 Use the percent equation to find the part.

GS **(a)** 15% of 880 policyholders

part = percent • whole

part = (____) (____)

↑

Write 15% in decimal form.

part = ____

(b) 23% of 840 gallons

(c) 120% of \$220

GS **(d)** 135% of \$1080

part = percent • whole

part = (____) (____)

(e) 0.5% of 1200 fruit cups

(f) 0.25% of 1600 lab tests

Answers

1. **(a)** (0.15)(880); 132 policyholders **(b)** 193.2 gallons **(c)** \$264 **(d)** (1.35)(1080); \$1458 **(e)** 6 fruit cups **(f)** 4 lab tests

To estimate the answer, think of 0.4% as approximately 0.5% or $\frac{1}{2}$ of 1%. Because 1% is $\frac{1}{100}$, you can use the shortcut from earlier in this chapter: Move the decimal point two places to the left.

$$1\% \text{ of } 250. = 2.5$$

Since 1% of 250 is 2.5, then 0.5% of 250 is 1.25 (because 2.5 ÷ 2 = 1.25). So, the exact answer of 1 patient is reasonable.

◀ **Work Problem 1 at the Side.**

CAUTION

When using the percent equation, the percent must always be *changed to a decimal* before multiplying.

OBJECTIVE 2 Find the whole using the percent equation. The next example shows how to use the percent equation to find the whole.

Note

When the word *of* follows a percent, it is an indicator word for *multiply.*

EXAMPLE 2 Solving for the Whole

(a) 8 tables is 4% of what number of tables?

The part is 8 and the percent is 4% or the decimal 0.04. The whole is unknown.

8 is **4% of** what number?

The word "of" can be used to identity the whole.

Now, use the percent equation.

part = percent • whole

$8 = (0.04)(x)$ Let x represent the unknown whole. Write 4% in decimal form as 0.04.

$\frac{8}{0.04} = \frac{\overset{1}{\cancel{(0.04)}}(x)}{\underset{1}{\cancel{0.04}}}$ Divide both sides by 0.04.

$200 = x$ ← Whole

8 tables is 4% of **200 tables.**

(b) 135 tourists is 15% of what number of tourists?

Write 15% as 0.15. The part is 135. Use the percent equation to find the whole.

part = percent • whole

$135 = (0.15)(x)$ Let x represent the unknown whole. Write 15% in decimal form as 0.15.

$\frac{135}{0.15} = \frac{\overset{1}{\cancel{(0.15)}}(x)}{\underset{1}{\cancel{0.15}}}$ Divide both sides by 0.15.

$900 = x$ ← Whole

135 tourists is 15% of **900 tourists.**

Continued on Next Page

(c) $8\frac{1}{2}\%$ of what number is 102?

Write $8\frac{1}{2}\%$ as 8.5%, or the decimal 0.085. The part is 102. Use the percent equation.

Write 8.5% as a decimal. 8.5% = 0.085%

$$\text{part} = \text{percent} \cdot \text{whole}$$

$$102 = (0.085)(x) \quad \text{Let } x \text{ represent the unknown whole.}$$

$$\frac{102}{0.085} = \frac{\overset{1}{\cancel{(0.085)}}(x)}{\underset{1}{\cancel{0.085}}} \quad \text{Divide both sides by 0.085.}$$

$$1200 = x \leftarrow \text{Whole}$$

102 is $8\frac{1}{2}\%$ of **1200**.

Estimate the answer. Notice that $8\frac{1}{2}\%$ is close to 10%. If 102 is 10% of a number, then the number is 10 times 102, or 1020. So the exact answer, 1200, is reasonable.

CAUTION

In **Example 2(c)** above, $8\frac{1}{2}\%$ was first written as 8.5%, which is the decimal form of $8\frac{1}{2}\%$ ($8\frac{1}{2} = 8.5$). **The percent sign still remained** in 8.5%. Then 8.5% was changed to the decimal 0.085 so the equation could be solved.

Work Problem 2 at the Side. ▶

OBJECTIVE 3 Find the percent using the percent equation. The final example shows how to use the percent equation to find the percent.

EXAMPLE 3 Finding the Percent

(a) 13 auto mechanics is what percent of 52 auto mechanics?

Because 52 follows *of,* the whole is 52. The part is 13, and the percent is unknown. Use the percent equation.

$$\text{part} = \text{percent} \cdot \text{whole}$$

$$13 = x \cdot 52 \quad \text{Let } x \text{ represent the unknown percent.}$$

$$\frac{13}{52} = \frac{x \cdot \overset{1}{\cancel{52}}}{\underset{1}{\cancel{52}}} \quad \text{Divide both sides by 52.}$$

$$0.25 = x$$

0.25 is 25%. — You *must* write the decimal answer as a percent. 0.25 = 25%

13 auto mechanics is **25%** of 52 auto mechanics.

The equation can also be set up using *of* as an indicator word for multiplication and *is* as an indicator word for "is equal to."

$$13 \text{ is what percent of } 52?$$
$$13 = \quad x \quad \cdot 52$$

$$13 = x \cdot 52 \quad \text{Same equation as above}$$

Continued on Next Page

2 Find the whole using the percent equation.

GS **(a)** 18 dancers is 45% of what number of dancers?

part = percent • whole

$18 = (\_\_\_\_)(x)$

$$\frac{18}{0.45} = \frac{\overset{1}{\cancel{0.45}}(x)}{\underset{1}{\cancel{0.45}}}$$

$\_\_\_\_ = x$

(b) 67.5 containers is 27% of what number of containers?

(c) 666 inoculations is 45% of what number of inoculations?

(d) $5\frac{1}{2}\%$ of what number of policies is 66 policies?

Answers

2. **(a)** 0.45; 40; 40 dancers **(b)** 250 containers **(c)** 1480 inoculations **(d)** 1200 policies

3 Find the percent using the percent equation.

GS **(a)** What percent of 35 Facebook friends is 7 Facebook friends?

part = percent • whole

$7 = x \cdot$ ____

$\frac{7}{35} = \frac{x \cdot \overset{1}{\cancel{35}}}{\underset{1}{\cancel{35}}}$

$0.20 = x$

0.20 is ____%

(b) 34 post office boxes is what percent of 85 post office boxes?

(c) What percent of 920 invitations is 1288 invitations?

(d) 9 world-class runners is what percent of 1125 runners?

Answers

3. **(a)** 35; 20% **(b)** 40% **(c)** 140% **(d)** 0.8%

(b) What percent of \$500 is \$100?
The whole is 500 and the part is 100. Let x represent the unknown percent.

part = percent • whole

$100 = x \cdot 500$ Let x represent the unknown percent.

$\frac{100}{500} = \frac{x \cdot \overset{1}{\cancel{500}}}{\underset{1}{\cancel{500}}}$ Divide both sides by 500.

$0.20 = x$

0.20 is 20%. (Write the decimal as a percent. 0.20 = 20%)

20% of \$500 is \$100.

(c) What percent of \$300 is \$390?
The whole is 300 and the part is 390. Let x represent the unknown percent.

part = percent • whole

$390 = x \cdot 300$ Let x represent the unknown percent.

$\frac{390}{300} = \frac{x \cdot \overset{1}{\cancel{300}}}{\underset{1}{\cancel{300}}}$ Divide both sides by 300.

$1.3 = x$

1.3 is 130%. (Write the decimal as a percent. 1.3 = 1.30 = 130%)

130% of \$300 is \$390.

(d) 6 ladders is what percent of 1200 ladders?
Since 1200 follows *of,* the whole is 1200. The part is 6.

part = percent • whole

$6 = x \cdot 1200$ Let x represent the unknown percent.

$\frac{6}{1200} = \frac{x \cdot \overset{1}{\cancel{1200}}}{\underset{1}{\cancel{1200}}}$ Divide both sides by 1200.

$0.005 = x$

0.005 is 0.5%. (Write the decimal as a percent. 0.005 = 0.5%)

6 ladders is **0.5%** of 1200 ladders.

You can estimate the answer because 1% of 1200 ladders is found by moving the decimal point two places to the left in 1200, resulting in 12. Since 6 ladders is half of 12 ladders, our answer should be $\frac{1}{2}$ of 1% or 0.5%. Our exact answer matches the estimate.

CAUTION

When you use the percent equation to solve for an unknown percent, the answer will always be in decimal form. Notice that in **Example 3(a), (b), (c), and (d)** above, **the decimal answer had to be changed to a percent** by multiplying by 100 and attaching the percent symbol. The answers became: (a) 0.25 = 25%; (b) 0.20 = 20%; (c) 1.3 = 130%; and (d) 0.005 = 0.5%.

◀ **Work Problem 3 at the Side.**

6.5 Exercises

FOR EXTRA HELP

Download the MyDashBoard App

MyMathLab®

CONCEPT CHECK *When solving for the part in a percent problem, the percent and the whole must be identified. Write either* percent *or* whole *next to the values taken from each problem.*

1. 46% of 780 text messages

46 ______ 780 ______

2. 14% of 1800 forums

1800 ______ 14 ______

Find the part using the percent equation. **See Example 1.**

3. 25% of 1080 blood donors

part = percent • whole

$x = (0.25)(1080)$

$x =$ ____

4. 19% of 700 MP3 players

part = percent • whole

$x = (0.19)(700)$

$x =$ ____

5. 45% of 3000 bath towels

6. 75% of 360 dosages

7. 32% of 260 quarts

8. 44% of 430 liters

9. 140% of 2500 air bags

10. 145% of 580 hamburgers

11. 12.4% of 8300 meters

12. 26.4% of 4700 miles

13. 0.8% of $520

14. 0.3% of $480

CONCEPT CHECK *When solving for the whole in a percent problem, the part and the percent must be identified. Write either* part *or* percent *next to the values taken from each problem.*

15. 70% of what number of backpackers is 476 backpackers?

70 ______ 476 ______

16. 270 lab tests is 45% of what number of lab tests?

270 ______ 45 ______

Find the whole using the percent equation. **See Example 2.**

17. 24 patients is 15% of what number of patients?

part = percent • whole

$24 = (0.15)(x)$

$$\frac{24}{0.15} = \frac{\overset{1}{\cancel{(0.15)}}(x)}{\underset{1}{\cancel{(0.15)}}}$$

$160 = x$

18. 32 classrooms is 20% of what number of classrooms?

part = percent • whole

$32 = (0.2)(x)$

$$\frac{32}{0.2} = \frac{\overset{1}{\cancel{(0.2)}}(x)}{\underset{1}{\cancel{(0.2)}}}$$

$160 = x$

19. 40% of what number of salads is 130 salads?

20. 75% of what number of wrenches is 675 wrenches?

21. $12\frac{1}{2}\%$ of what number of people is 135 people?

22. $18\frac{1}{2}\%$ of what number of batteries is 370 batteries?

23. $1\frac{1}{4}\%$ of what number of gallons is 3.75 gallons?

24. $2\frac{1}{4}\%$ of what number of files is 9 files?

CONCEPT CHECK *Fill in each blank with the correct response.*

25. When using the percent equation to solve for percent, you must always change the decimal answer to ______ by moving the decimal point ______ places to the ______ and attaching a ______ symbol.

26. When using the percent equation to solve for part or whole, you must always change the percent to a decimal by moving the ______ point ______ places to the ______ and dropping the ______ symbol.

Find the percent using the percent equation. ***See Example 3.***

27. 114 tuxedos is what percent of 150 tuxedos?

GS

part = percent • whole

$114 = x \cdot 150$

$\frac{114}{150} = \frac{x \cdot \overset{1}{\cancel{150}}}{\underset{1}{\cancel{150}}}$

$0.76 = x$

change 0.76 to a percent = ______

28. 75 iPods is what percent of 125 iPods?

GS

part = percent • whole

$75 = x \cdot 125$

$\frac{75}{125} = \frac{x \cdot \overset{1}{\cancel{125}}}{\underset{1}{\cancel{125}}}$

$0.6 = x$

0.60 is ______

29. What percent of 160 liters is 2.4 liters?

30. What percent of 600 meters is 7.5 meters?

31. 170 cartons is what percent of 68 cartons?

32. 612 orders is what percent of 425 orders?

33. When using the percent equation, the percent must always be changed to a decimal before doing any calculations. Show and explain how to change a fraction percent to a decimal. Use $2\frac{1}{2}\%$ in your explanation.

34. Suppose a problem on your homework assignment was, "Find $\frac{1}{2}\%$ of \$1300." Your classmates got answers of \$0.65, \$6.50, \$65, and \$650. Which answer is correct? How and why are they getting all of these answers? Explain.

Solve each application problem.

35. A study of office workers found that 27% would like more storage space. If there are 14 million office workers, how many want more storage space? (*Source:* Steelcase Workplace Index.)

36. Most shampoos contain 75% to 90% water. If a 16-ounce bottle of shampoo contains 78% water, find the number of ounces of water in the bottle. Round to the nearest tenth of an ounce.

37. The results of a recent poll are shown below. If 12,500 people were included in the poll, find

(a) the number who would rather admit their age.

(b) the number who would rather admit their weight.

(c) the number who would rather admit their salary.

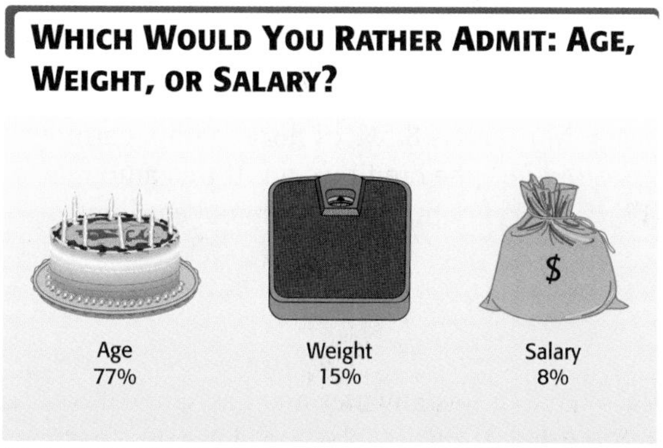

Source: Parade magazine.

38. Three Spam Mobiles travel throughout the country promoting Spam and Spam Lite. By using these kitchens on wheels, the annual goal is to give 1.5 million taste samples to the public. If 58.6% of the goal has been met, find the number of samples that have been given. (*Source:* Hormel Foods Corporation.)

39. In the United States, 98% of all households have a refrigerator. (*Source:* American Housing Authority.) Out of 18,000 households,

(a) how many are expected to have a refrigerator?

(b) how many are expected to not have a refrigerator?

40. In the United States, 56% of the households have a dishwasher. (*Source:* American Housing Authority.) Out of 214,500 households,

(a) how many are expected to have a dishwasher?

(b) how many are expected to not have a dishwasher?

41. In the United States, 2 of the 50 states (Alaska and Louisiana) do not have any drive-in movies. The remaining states do have drive-in movies. (*Source: USA Today.*)

(a) What percent of the states do not have drive-in movies?

(b) What percent have drive-in movies?

42. In a recent survey of 3450 attorneys, 897 said that they take work home with them during the week and 1311 said that they take work home with them on the weekend. Find

(a) the percent of the attorneys who take work home during the week.

(b) the percent of attorneys who take work home on the weekend.

43. In a survey of 1250 Americans, 461 rated their health as excellent. What percent of these Americans rate their health as excellent? Round to the nearest tenth of a percent. (*Source:* National Health Interview Survey.)

44. General Nutrition Center now has 3200 stores and plans to add 450 more stores. Find the percent of additional stores that they have planned. Round to the nearest tenth of a percent.

The graph shows the type of day care used by families for their preschool children. Assume that 7800 families were surveyed to gather the data. Use this graph to answer Exercises 45–48.

WHERE THE KIDS GO

Day care for preschoolers:

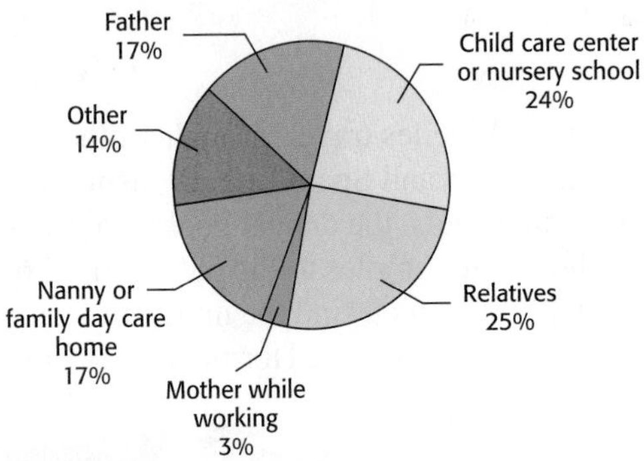

Source: U.S. Census Bureau.

45. (a) Which type of day care was used most often?

(b) Find the number of families who used this type of day care.

46. (a) Which type of day care was used least often?

(b) What number of families used this type of day care?

47. How many families used a nanny or family day care home?

48. Find the number of families who left their preschoolers with a child care center or nursery school.

49. Last year, college seniors who graduated carried an average of $22,650 in student loan debt. This year the average amount of loan debt has increased 6%. Find the average amount of student loan debt this year. (*Source:* U.S. Department of Education.)

50. Cyndy Mason has 8.5% of her monthly earnings deposited into the credit union. If this amounts to $131.75 per month, find her annual earnings.

51. The Chevy Camaro was introduced in 1967. Sales that year were 220,917 Camaros, which was 46.2% of the number of Ford Mustangs sold in the same year. Find the number of Mustangs sold in 1967. Round to the nearest whole number.

52. Chris Goodwin is a waiter and has sales of $822.25 on Saturday. If this is 28.6% of his sales for the week, find his weekly sales.

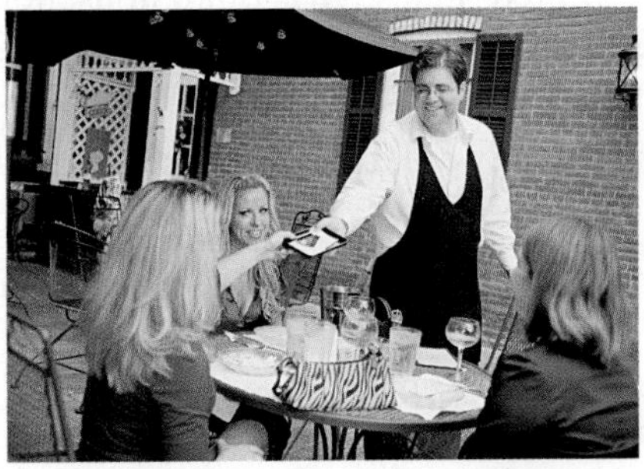

53. J & K Mustang has increased the sale of auto parts by $32\frac{1}{2}\%$ over last year. If the sale of parts last year amounted to $385,200, find the amount of sales this year.

54. An ad for steel-belted radial tires promises 15% better mileage. If mileage had been 25.6 miles per gallon in the past, what mileage could be expected after these tires are installed? Round to the nearest tenth of a mile.

55. A Polaris Vac-Sweep is priced at $524 with an allowed trade-in of $125 on an old unit. If sales tax of $7\frac{3}{4}\%$ is charged on the price of the new Polaris unit before the trade-in, find the total cost to the customer after receiving the trade-in. (*Hint:* Trade-in is subtracted last.)

56. General Motors car sales in China were 36.7% greater than last year's sales of 629,778 cars. Find this year's sales. Round to the nearest whole number. (*Source:* General Motors Corporation.)

6.6 Solving Application Problems with Percent

Percent has many applications in our daily lives. This section discusses percent as it applies to sales tax, commissions, discounts, and the percent of change (increase and decrease).

OBJECTIVES

1. Find sales tax.
2. Find commissions.
3. Find the discount and sale price.
4. Find the percent of change.

OBJECTIVE 1 **Find sales tax.** States, countries, and cities often collect taxes on sales to customers. The **sales tax** is a percent of the cost of an item. The following formula for finding sales tax is based on the percent equation.

Sales Tax Formula

$$\underbrace{\text{amount of sales tax}}_{\text{part}} = \underbrace{\text{rate of tax}}_{\text{percent}} \cdot \underbrace{\text{cost of item}}_{\text{whole}}$$

VOCABULARY TIP

Sales tax The sales tax is always part of the cost of an item. The cost of the item is the **whole,** the sales tax rate is the **percent,** and the amount of the sales tax is the **part.**

EXAMPLE 1 Solving for Sales Tax

Office Max sells a laptop computer for $499. If the sales tax rate is 5%, how much tax is paid? What is the total cost of the laptop computer? Use the six problem-solving steps.

Step 1 **Read** the problem. The problem asks for the total cost of the laptop computer, including the sales tax.

Step 2 **Work out a plan.** Use the sales tax formula to find the amount of sales tax. Write the tax rate (5%) as a decimal (0.05). The cost of the item is $499. Use the letter a to represent the unknown *amount* of tax. Add the sales tax to the cost of the item.

Step 3 **Estimate** a reasonable answer. Round $499 to $500. Recall that 5% is equivalent to $\frac{1}{20}$, so divide $500 by 20 to estimate the tax.

$$\$500 \div 20 = \$25 \text{ tax}$$

The total estimated cost is $500 + $25 = $525. ← Estimate

Step 4 **Solve** the problem.

$$\underbrace{\text{amount of sales tax}}_{\text{part}} = \underbrace{\text{rate of tax}}_{\text{percent}} \cdot \underbrace{\text{cost of item}}_{\text{whole}}$$

$$a = (5\%)(\$499)$$

Write 5% as a decimal 0.05. → $a = (0.05)(\$499)$

$$a = \$24.95 \quad \text{Sales tax}$$

The tax paid on the laptop computer is $24.95. The customer would pay a total cost of $499 + $24.95 = $523.95.

Step 5 **State the answer.** The total cost of the laptop computer is $523.95.

Step 6 **Check.** The exact answer, $523.95, is close to our estimate of $525.

Work Problem 1 at the Side. ▶

1 Suppose the sales tax rate in your state is 6%. Find the amount of the tax and the total you would pay for each item.

GS **(a)** $29 Little League bat

Sales tax = (0.06)($____)
= $____

Total cost = $29 + $____
= $____

(b) $349 home theater system

(c) $1287 leather chair and ottoman

(d) $24,500 pickup truck

Answers

1. **(a)** $29; $1.74; $1.74; $30.74
(b) $20.94; $369.94 **(c)** $77.22; $1364.22
(d) $1470; $25,970

2 Find the rate of sales tax.

GS **(a)** The tax on a $320 patio set is $25.60.

$\$25.60 = r \cdot$ ____

$$\frac{25.6}{320} = \frac{r \cdot \overset{1}{\cancel{320}}}{\underset{1}{\cancel{320}}}$$

$r = 0.08 =$ ____

(b) The tax on a $96 park bench is $6.24.

(c) The tax on a $22,995 Dodge Charger is $919.80.

Answers

2. (a) $320; 8% **(b)** 6.5% or $6\frac{1}{2}\%$ **(c)** 4%

EXAMPLE 2 **Finding the Sales Tax Rate**

The sales tax on a $24,200 Ford Edge is $1573. Find the rate of the sales tax.

Step 1 **Read** the problem. This problem asks us to find the sales tax rate.

Step 2 **Work out a plan.** Use the sales tax formula.

sales tax = rate of tax • cost of item

Solve for the rate of tax, which is the percent. The cost of the Ford Edge (the whole) is $24,200, and the amount of sales tax (the part) is $1573. Use the letter *r* to represent the unknown *rate* of tax (the percent).

Step 3 **Estimate** a reasonable answer. Round $24,200 to $24,000 and round $1573 to $1600. The sales tax is $\frac{1600}{24,000}$ or $\frac{1}{15}$ of the cost of the car. So divide 1 by 15 to estimate the percent (rate) of sales tax.

$$\frac{1}{15} = 0.06\overline{6} \approx 7\% \leftarrow \text{Rounded estimate}$$

Step 4 **Solve** the problem.

sales tax = rate of tax • cost of item

$$\$1573 = (r) \quad (\$24,200)$$

$$\frac{1573}{24,200} = \frac{(r)(\overset{1}{\cancel{24,200}})}{\underset{1}{\cancel{24,200}}} \quad \text{Divide both sides by 24,200.}$$

$$0.065 = r$$

0.065 is 6.5%.

Write the decimal as a percent. 0.065 = 6.5%.

Step 5 **State the answer.** The sales tax rate is 6.5% or $6\frac{1}{2}\%$.

Step 6 **Check.** The exact answer, $6\frac{1}{2}\%$, is close to our estimate of 7%.

◀ **Work Problem 2 at the Side.**

Note

You can use the sales tax formula to find the amount of sales tax, the cost of an item, or the rate of sales tax (the percent).

OBJECTIVE 2 **Find commissions.** Many salespeople are paid by *commission* rather than an hourly wage. If you are paid by **commission,** you are paid a certain percent of your total sales dollars. The formula below for finding the commission is based on the percent equation.

Commission Formula

part	=	percent	•	whole
↓		↓		↓
amount of commission	=	rate of commission	•	amount of sales

EXAMPLE 3 **Determining the Amount of Commission**

Lynn Cochran had automotive tool sales of \$42,500 last month. If Cochran's commission rate is 9%, find the amount of his commission.

Step 1 **Read** the problem. The problem asks for the amount of commission that Cochran earned.

Step 2 **Work out a plan.** Use the commission formula. Write the rate of commission (9%) as a decimal (0.09). The amount of Cochran's sales (\$42,500) is the whole. Use the letter c to represent the unknown *amount* of commission.

Step 3 **Estimate** a reasonable answer. Round the commission rate of 9% to 10%. Round the amount of sales from \$42,500 to \$40,000. Since 10% is equivalent to $\frac{1}{10}$, divide \$40,000 by 10 to estimate the amount of commission: \$40,000 ÷ 10 gives \$4000 as our estimate.

Step 4 **Solve** the problem.

amount of commission = rate of commission • amount of sales

$$c = (9\%)(\$42{,}500)$$

$$c = (0.09)(\$42{,}500)$$

$$c = \$3825 \quad \text{Amount of commission}$$

Step 5 **State the answer.** Cochran earned a commission of \$3825.

Step 6 **Check.** The exact answer, \$3825, is close to our estimate of \$4000.

Work Problem 3 at the Side. ▶

EXAMPLE 4 **Finding the Rate of Commission**

Carol Merrigan earned a commission of \$510 for selling \$17,000 worth of shipping supplies. Find the rate of commission.

Step 1 **Read** the problem. We must find the rate (percent) of commission.

Step 2 **Work out a plan.** You could use the commission formula. Another approach is to use the percent proportion. The *whole* is \$17,000, the *part* is \$510, and the *percent* (rate of commission) is the unknown.

Step 3 **Estimate** a reasonable answer. Round the commission, \$510, to \$500, and round \$17,000 to \$20,000. The commission in fraction form is $\frac{\$500}{\$20{,}000}$, which simplifies to $\frac{1}{40}$. Changing $\frac{1}{40}$ to a percent gives $2\frac{1}{2}\%$ (rounded), as our estimate.

Step 4 **Solve** the problem.

$$\frac{\text{part}}{\text{whole}} = \frac{x}{100} \leftarrow \text{Percent (unknown)}$$

Think: The rate of the commission is the percent.

$$\frac{510}{17{,}000} = \frac{x}{100}$$

$$17{,}000 \cdot x = 510 \cdot 100 \quad \text{Cross products}$$

$$\frac{\overset{1}{\cancel{17{,}000}} \cdot x}{\underset{1}{\cancel{17{,}000}}} = \frac{51{,}000}{17{,}000} \quad \text{Divide both sides by 17,000.}$$

$$x = 3$$

Continued on Next Page

3 Find the amount of commission.

GS **(a)** Jill Owens sells dental equipment at a commission rate of 11% and has sales for the month of \$38,700.

Use the letter c to represent the amount of commission.

$c = (11\%)(\$\_\_\_\_\_\_\_)$

$c = (0.11)(38{,}700)$

$c = \$\_\_\_\_\_\_\_$

(b) Last month Alyssa Paige sold a home for \$189,500 and earned a commission of 6%.

Answers

3. **(a)** \$38,700; \$4257 **(b)** \$11,370

4 Find the rate of commission.

GS **(a)** A commission of $450 is earned on the sale of computer products worth $22,500.

$$\frac{450}{22{,}500} = \frac{x}{100}$$

$$22{,}500 \cdot x = 450 \cdot 100$$

$$\frac{\overset{1}{\cancel{22{,}500}} \cdot x}{\underset{1}{\cancel{22{,}500}}} = \frac{45{,}000}{22{,}500}$$

$$x = \_\_\_\_$$

(b) Jamal Story earned $2898 for selling office furniture worth $32,200.

5 Find the amount of the discount and the sale price.

GS **(a)** An Easy-Boy leather recliner originally priced at $950 is offered at a 42% discount.

Discount = (0.42) ($950)

= $399

Sale price = $950 − $____

= $____

(b) Walmart has women's sweater sets on sale at 35% off. One sweater set was originally priced at $30.

Answers

4. (a) 2% (b) 9%

5. (a) $399; $551 (b) $10.50; $19.50

Step 5 **State the answer.** The rate of commission is 3%.

Step 6 **Check.** The exact answer of 3% is close to our estimate of $2\frac{1}{2}$%.

◀ **Work Problem 4 at the Side.**

OBJECTIVE 3 Find the discount and sale price. Most of us prefer buying things when they are on sale. A store will reduce prices, or **discount,** to attract additional customers. Use the following formula to find the discount and the sale price.

Discount Formula and Sale Price Formula

amount of discount = rate (or percent) of discount • original price

sale price = original price − amount of discount

EXAMPLE 5 Finding a Sale Price

Whitings Oak Furniture Store has a home theater cabinet with an original price of $840 on sale at 15% off. Find the sale price of the cabinet.

Step 1 **Read** the problem. This problem asks for the price of a home theater cabinet after a discount of 15%.

Step 2 **Work out a plan.** The problem is solved in two steps. First, find the amount of the discount, that is, the amount that will be "taken off" (subtracted), by multiplying the original price ($840) by the rate of the discount (15%). The second step is to subtract the amount of discount from the original price. This gives you the sale price, which is what you will actually pay for the home theater cabinet.

Step 3 **Estimate** a reasonable answer. Round the original price from $840 to $800, and the rate of discount from 15% to 20%. Since 20% is equivalent to $\frac{1}{5}$, the estimated discount is $800 ÷ 5 = $160, so the estimated sale price is $800 − $160 = $640.

Step 4 **Solve** the problem. First find the exact amount of the discount.

amount of discount = rate of discount • original price

a = (0.15) ($840)   Write 15% as a decimal.

a = $126   Amount of discount

Now find the sale price of the home theater cabinet by subtracting the amount of the discount ($126) from the original price.

sale price = original price − amount of discount

= $840 − $126   (Remember to subtract the discount from the original price to get the sale price.)

= $714   Sale price

Step 5 **State the answer.** The sale price of the home theater cabinet is $714.

Step 6 **Check.** The exact answer, $714, is close to our estimate of $640.

◀ **Work Problem 5 at the Side.**

Calculator Tip

In **Example 5** on the previous page, you can use a scientific calculator to find the amount of discount and subtract the discount from the original price, all in one step.

840 (−) .15 (×) 840 (=) 714

Original price ↑ 840; Amount of discount: .15 × 840; Sale price ↑ 714

A scientific calculator observes the order of operations, so it will automatically do the multiplication before the subtraction.

VOCABULARY TIP

Percent of increase or decrease The change from the original value is the amount of increase or decrease. The amount of increase or decrease is the **part** and the original value is **always the whole.**

OBJECTIVE 4 Find the percent of change. We are often interested in looking at increases or decreases in sales, production, population, and many other items. This type of problem involves finding the *percent of change.* Use the following steps to find the **percent of increase.**

Finding the Percent of Increase

Step 1 Use subtraction to find the amount of increase.

Step 2 Use the percent proportion to find the percent of increase.

$$\frac{\textbf{amount of increase (part)}}{\textbf{original value (whole)}} = \frac{\textbf{percent}}{\textbf{100}}$$

EXAMPLE 6 Finding the Percent of Increase

Attendance at county parks climbed from 18,300 last month to 56,730 this month. Find the percent of increase.

Step 1 **Read** the problem. The problem asks for the percent of increase.

Step 2 **Work out a plan.** Subtract the attendance last month (18,300) from the attendance this month (56,730) to find the amount of increase in attendance. Next, use the percent proportion. The whole is 18,300 (last month's original attendance), the part is 38,430 (amount of increase in attendance), and the percent is unknown.

Step 3 **Estimate** a reasonable answer. Round 18,300 to 20,000 and 56,730 to 60,000. The amount of increase is 60,000 − 20,000 = 40,000. Since 40,000 (the increase) is *twice* as large as the original amount, the estimated percent of increase is 200%.

Step 4 **Solve** the problem.

$$56{,}730 - 18{,}300 = 38{,}430$$

Subtract to find the **amount of increase** in attendance.

Amount of increase →

$$\frac{38{,}430}{18{,}300} = \frac{x}{100}$$

Percent proportion

Use the *original* value of 18,300 (**not** 56,730).

Solve this proportion to find that $x = 210$.

Step 5 **State the answer.** The percent of increase is 210%.

Step 6 **Check.** The exact answer, 210%, is close to our estimate of 200%.

Work Problem 6 at the Side. ▶

6 Find the percent of increase.

(a) A manufacturer of snowboards increased production from 14,100 units last year to 19,035 this year.

19,035 − 14,100 = ______ increase

Increase →

$$\frac{\_\_\_\_}{14{,}100} = \frac{x}{100}$$

(Original value → 14,100)

$$\frac{\overset{1}{\cancel{14{,}100}} \cdot x}{\underset{1}{\cancel{14{,}100}}} = \frac{493{,}500}{14{,}100}$$

$$x = 35$$

Percent of increase is ____.

(b) The number of flu cases rose from 496 cases last week to 620 this week.

Answers

6. (a) 4935; 4935; 35% (b) 25%

7 Find the percent of decrease.

GS **(a)** The number of service calls fell from 380 last month to 285 this month.

$380 - 285 = \_\_\_\_$ decrease

Decrease → $\dfrac{\_\_\_\_}{380} = \dfrac{x}{100}$ ← Original value

$$\frac{\overset{1}{\cancel{380}} \cdot x}{\underset{1}{\cancel{380}}} = \frac{9500}{380}$$

$$x = 25$$

Percent of decrease is ____.

(b) The number of workers applying for unemployment fell from 4850 last month to 3977 this month.

Use the following steps to find the **percent of decrease.**

Finding the Percent of Decrease

Step 1 Use subtraction to find the amount of decrease.

Step 2 Use the percent proportion to find the percent of decrease.

$$\frac{\textbf{amount of decrease (part)}}{\textbf{original value (whole)}} = \frac{\textbf{percent}}{\textbf{100}}$$

EXAMPLE 7 **Finding the Percent of Decrease**

The number of production employees this week fell to 1406 people from 1480 people last week. Find the percent of decrease.

Step 1 **Read** the problem. The problem asks for the percent of decrease.

Step 2 **Work out a plan.** Subtract the number of employees this week (1406) from the number of employees last week (1480) to find the amount of decrease. Then, use the percent proportion. The whole is 1480 (last week's *original* number of employees), the part is 74 (amount of decrease in employees), and the percent is unknown.

Step 3 **Estimate** a reasonable answer. Estimate the answer by rounding 1406 to 1400 and 1480 to 1500. The decrease is $1500 - 1400 = 100$. Since 100 is $\frac{1}{15}$ of 1500, our estimate is $1 \div 15 \approx 0.07$ or 7%.

Step 4 **Solve** the problem.

$$1480 - 1406 = 74$$

Subtract to find the **amount of decrease** in number of employees.

Amount of decrease → $\dfrac{74}{1480} = \dfrac{x}{100}$ Percent proportion

Use the original value of 1480 (**not** 1406).

Solve this proportion to find that $x = 5$.

Step 5 **State the answer.** The percent of decrease is 5%.

Step 6 **Check.** The exact answer, 5%, is close to our estimate of 7%.

CAUTION

When solving for percent of increase or decrease, the ***whole is always the original value*** or ***value before the change occurred.*** The part is the change in values, that is, how much something went up or went down.

◀ **Work Problem 7 at the Side.**

Answers

7. **(a)** 95; 95; 25% **(b)** 18%

6.6 Exercises

FOR EXTRA HELP — Download the MyDashBoard App — MyMathLab®

CONCEPT CHECK *Underline the correct answer.*

1. When solving a sales tax problem, the cost of an item is the (*tax rate/whole*), the sales tax rate is the (*percent/whole*), and the amount of sales tax is the (*part/whole*).

CONCEPT CHECK *Fill in each blank with the correct response.*

2. To find the amount of sales tax, multiply the sales tax ________ by the ________. The total cost is the cost of the item plus the ________ tax.

Find the amount of sales tax or the tax rate and the total cost (amount of sale + amount of tax = total cost). Round money answers to the nearest cent if necessary. ***See Examples 1 and 2.***

	Cost of Item	Tax Rate	Amount of Tax	Total Cost
3.	\$6	4%	________	________
4.	\$45	5%	________	________
5.	\$425	________	\$12.75	________
6.	\$84	________	\$5.88	________
7.	\$12,229	$5\frac{1}{2}\%$	________	________
8.	\$11,789	$7\frac{1}{2}\%$	________	________

CONCEPT CHECK *Underline the correct answer.*

9. When solving commission problems, the whole is the (*commission/sales amount*), the percent is the (*commission rate/sales amount*), and the part is the (*commission/sales amount*).

CONCEPT CHECK *Fill in each blank with the correct response.*

10. To find the commission, multiply the rate of ________ by the ________.

Find the commission earned or the rate of commission. Round money answers to the nearest cent if necessary. ***See Examples 3 and 4.***

	Sales	Rate of Commission	Commission
11.	\$280	8%	________
12.	\$660	10%	________

GS Commission = rate of commission • sales

$= (10\%)(\$\_\_\_\_) = (0.\_\_\_)(\$660)$

	Sales	Rate of Commission	Commission
13.	\$3000	________	\$600

GS $\frac{600}{3000} = \frac{x}{100}$ or $\frac{1}{5} = \frac{x}{100}$; $\frac{5x}{5} = \frac{100}{5}$; $x = 20$

	Sales	Rate of Commission	Commission
14.	\$7800	________	\$1170
15.	\$6183.50	3%	________
16.	\$4416.70	7%	________

CONCEPT CHECK *Underline the correct answer.*

17. When solving retail sales discount problems, the *original* price is always the (*part/whole*), the rate of discount is the (*percent/whole*), and the amount of discount is the (*percent/part*).

CONCEPT CHECK *Fill in each blank with the correct response.*

18. To find the sale price in a retail discount problem, subtract the amount of ________ from the ________ price.

Find the amount or rate of discount and the sale price after the discount. Round money answers to the nearest cent if necessary. ***See Example 5.***

	Original Price	Rate of Discount	Amount of Discount	Sale Price
19.	$199.99	10%	________	________
20.	$29.95	15%	________	________
21.	$180	________	$54	________
22.	$38	________	$9.50	________
23.	$58.40	15%	________	________
24.	$99.80	30%	________	________

25. You are trying to decide between Company A paying a 10% commission and Company B paying an 8% commission. For which company would you prefer to work? What considerations other than commission rate would be important to you?

26. Give four examples of where you might use the percent of increase or the percent of decrease in your own personal activities. Think in terms of work, school, home, hobbies, and sports.

Solve each application problem. Round money answers to the nearest cent and rates to the nearest tenth of a percent if necessary. ***See Examples 1–7.***

Country Store has a unique selection of merchandise that it sells by mail and over the Internet. Use the shipping and insurance delivery chart at the right and a sales tax rate of 5% to solve Exercises 27–30. There is no sales tax on shipping and insurance. (*Source:* Country Store Catalog.)

SHIPPING AND INSURANCE DELIVERY CHART

Up to $15.00	add $4.99
$15.01 to $25.00	add $6.99
$25.01 to $35.00	add $7.99
$35.01 to $50.00	add $8.99
$50.01 to $70.00	add $10.99
$70.01 to $99.99	add $12.99
$100.00 or more	add $14.99

27. Find the total cost of six Small Fry Handi-Pan electric skillets priced at $29.99 each.

28. A customer ordered five sets of flour-sack towels priced at $12.99 per set. What is the total cost?

29. Find the total cost of three pop-up hampers at $9.99 each and four nonstick mini doughnut pans at $10.99 each.

30. What is the total cost of five coach lamp bird feeders at $19.99 each and six garden weather centers at $14.99 each?

31. An Anderson wood-frame French door is priced at $1980 with a sales tax of $99. Find the rate of sales tax.

32. Six gallons of Dutch Boy Dirt Fighter exterior paint cost $155.94 plus sales tax of $7.80. Find the sales tax rate. (*Source*: Orchard Supply Hardware.)

33. Today there are 635,000 women motorcyclists in the United States, up from 467,400 just eight years ago. Find the percent of increase in the number of women motorcyclists. (*Source:* Motorcycle Industry Council.)

Increase $= 635{,}000 - 467{,}400 = 167{,}600$

$$\begin{matrix}\text{increase} \rightarrow \\ \text{original} \rightarrow\end{matrix} \frac{167{,}600}{467{,}400} = \frac{x}{100}$$

34. This year Jo O'Neill has earned \$46,750, up from \$32,500 last year. Find the percent of increase in her annual earnings.

Increase $= \$46{,}750 - \$32{,}500 = \$14{,}250$

$$\begin{matrix}\text{increase} \rightarrow \\ \text{original} \rightarrow\end{matrix} \frac{14{,}250}{32{,}500} = \frac{x}{100}$$

35. Misty Downes sold her Toyota Prius, which got 48 miles per gallon on the highway, and purchased a Chevrolet Cruze, which gets 36 miles per gallon. What is the percent of decrease in mileage? (*Source:* edmonds.com)

36. Rich Williams is considering selling his Corvette, which gets 26 miles per gallon on the highway, and purchasing a new Corvette ZR1, which gets 20 miles per gallon. If he buys the new Corvette, find the percent of decrease in mileage. (*Source:* edmonds.com)

37. A preholiday sale at Macy's offers 60% off on all women's wear. What is the sale price of a wool coat normally priced at \$335?

38. What is the sale price of a \$1098 Kenmore washer/dryer set with a discount of 25%?

A weekly sales report for the top four sales people at Active Sports is shown below. Use this information to answer Exercises 39–42.

Employee	Sales	Rate of Commission	Commission
Strong, A.	\$18,960	3%	______
Ferns, K.	\$21,460	3%	______
Keyes, B.	\$17,680	______	\$707.20
Vargas, K.	\$23,104	______	\$1152.20

39. Find the commission for Strong.

40. Find the commission for Ferns.

41. What is the rate of commission for Keyes?

42. What is the rate of commission for Vargas?

43. A Sony Micro Hi Fi Component System was priced at \$390 and is on sale at 22% off. Find the discount and the sale price.

44. A Honda Pilot is offered at 12% off the manufacturer's suggested retail price. Find the discount and the sale price of this SUV, originally priced at \$32,500.

45. The price of a video game was marked down from \$45.50 to \$35.49. Find the percent of discount.

46. In the past five years, the cost of generating electricity from the sun has been brought down from 24 cents per kilowatt hour to 8 cents (less than the newest nuclear power plants). Find the percent of decrease.

47. College students are offered a 6% discount on a dictionary that sells for \$18.50. If the sales tax is 6%, find the cost of the dictionary, including the sales tax.

48. A fax machine priced at \$398 is marked down 7% to promote the new model. If the sales tax is also 7%, find the cost of the fax machine, including sales tax.

49. A plumbing supply sales representative has sales of \$380,450 and is paid a commission of 2% of sales. If 20% of his commission is deducted for office space and other expenses, how much does the sales representative get?

50. The local real estate agents' association collects a fee of 2% on all money received by its members. The members charge 6% of the selling price of a property as their fee. How much does the association get, if its members sell property worth a total of \$8,680,000?

51. What is the total price of a ski boat with an original price of \$15,321, if it is sold at a 15% discount? The sales tax rate is $7\frac{3}{4}\%$.

52. A commercial security alarm system originally priced at \$10,800 is discounted 22%. Find the total price of the system if the sales tax rate is $7\frac{1}{4}\%$.

Relating Concepts (Exercises 53–58) For Individual or Group Work

Knowing how to use the percent equation is important when solving application problems involving sales tax. ***Work Exercises 53–58 in order.***

53. The percent equation is

part = ______ • ______.

54. The formula used to find sales tax is an application of the percent equation. The sales tax formula is

sales tax = ____________ • ____________.

In the United States there are certain items on which an excise tax is charged in addition to a sales tax. A table of federal excise taxes is shown here. Use this table to answer Exercises 55–58. (Excise tax is calculated on the amount of the sale ***before*** *sales tax is added.) Round answers to the nearest cent.*

FEDERAL EXCISE TAXES*

Product or Service	Rate	Product or Service	Rate
Local telephone service	3%	Bows and arrows	11%
Cigarettes	36¢ per pack	Gasoline	18.4¢/gal
Tires	\$0.0945 for each 10 pounds of the maximum rated load capacity over 3500 pounds	Diesel fuel and kerosene	24.4¢/gal
		Aviation fuel	19.4¢/gal
		Truck and trailer, chassis and bodies	12%
Air transportation	7.5%		
International air travel	\$15.40 per person	Inland waterways fuel	20¢/gal
Air freight	6.25%	Ship passenger tax	\$3/passenger
Fishing rods	10%	Vaccines	78¢/dose

*In addition to the federal excise taxes shown here, there are a number of additional excise taxes that apply to alcoholic beverages, tobacco products, and firearms.
Source: Publication 510, I.R.S., Excise Taxes.

55. Julia Lauren purchased a fly fishing rod for \$59 and a spinning rod for \$36. Use the federal excise tax table and a sales tax rate of $6\frac{1}{2}\%$ to find the cost of the equipment, including both taxes. Sales tax is not charged on the \$9.50 federal excise tax. (Round to the nearest cent.)

56. Refer to **Exercise 55.** Calculate the two taxes separately and then add them together. Now, add the two tax rates together and then find the tax. Are your answers the same? Why or why not? (*Hint:* Recall the commutative and associative properties of multiplication.)

57. The price of an international airline ticket is \$1248. Use the federal excise tax table and a sales tax rate of $7\frac{3}{4}\%$ to find the total cost of one ticket. Sales tax is not charged on the \$15.40 federal excise tax. (Round to the nearest cent.)

58. Refer to **Exercise 57.** Can the federal excise tax be added to the sales tax rate to find the total tax? Why or why not?

6.7 Simple Interest

OBJECTIVES

1. Find the simple interest on a loan.
2. Find the total amount due on a loan.

When we open a savings account, we are actually lending money to the bank or credit union. The bank or credit union will in turn lend this money to individuals and businesses. These people then become borrowers. The bank or credit union pays a fee to the savings account holders and charges a higher fee to its borrowers. These fees are called *interest.*

Interest is a fee paid or a charge made for lending or borrowing money. The amount of money borrowed is called the **principal.** The charge for interest is often given as a percent, called the interest rate or **rate of interest.** The rate of interest is assumed to be *per year,* unless stated otherwise. Time is always expressed in years or fractions of a year.

VOCABULARY TIP

Interest is what the borrower pays for the use of the lender's money.

OBJECTIVE 1 Find the simple interest on a loan. In most cases, interest on a loan is computed on the *original principal* and is called **simple interest.** We use the following **interest formula** to find simple interest.

Formula for Simple Interest

Interest = principal • rate • time

The formula is usually written using letters.

$$I = p \cdot r \cdot t$$

Note

Simple interest is used for most short-term business loans, most real estate loans, and many automobile and consumer loans.

EXAMPLE 1 Finding Simple Interest for a Year

Find the interest on \$5000 at 3% for 1 year.

The amount borrowed, or principal (p), is \$5000. The interest rate (r) is 3%, which is 0.03 as a decimal, and the time of the loan (t) is 1 year. Use the formula.

$$I = p \cdot r \cdot t$$

$$I = (5000)(0.03)(1)$$ Notice that "1" is used as the time for 1 year.

$$I = \$150$$ The interest rate is always written in decimal form.

The interest is \$150.

Work Problem 1 at the Side. ▶

EXAMPLE 2 Finding Simple Interest for More Than a Year

Find the interest on \$4200 at 4% for three and a half years.

The principal (p) is \$4200. The rate (r) is 4%, or 0.04 as a decimal, and the time (t) is $3\frac{1}{2}$ or 3.5 years. Use the formula.

$$I = p \cdot r \cdot t$$

$$I = (4200)(0.04)(3.5)$$ 3.5 years is equivalent to $3\frac{1}{2}$ years because 3.5 is $3\frac{5}{10}$ which simplifies to $3\frac{1}{2}$.

$$I = 588$$ 4% is changed to the decimal 0.04.

The interest is \$588.

Work Problem 2 at the Side. ▶

1 Find the interest.

GS (a) \$1000 at 3% for 1 year

$= p \cdot r \cdot t$

$= (1000)(\_\_\_\_)(1)$

$= \_\_\_\_$

(b) \$3650 at 2% for 1 year

VOCABULARY TIP

Rate of interest The number with the % symbol or **"percent"** after it is the rate of interest.

2 Find the interest.

GS (a) \$820 at 1% for $3\frac{1}{2}$ years

$I = p \cdot r \cdot t$

$= (\_\_\_\_)(\_\_\_\_)(3.5)$

$= \_\_\_\_$

(b) \$4850 at 4% for $2\frac{1}{2}$ years

(c) \$16,800 at 3% for $2\frac{3}{4}$ years

Answers

1. (a) 0.03; \$30 (b) \$73
2. (a) \$820; 0.01; \$28.70 (b) \$485 (c) \$1386

3 Find the interest.

GS **(a)** \$1800 at 3% for 4 months

$I = p \cdot r \cdot t$

$= (1800)(0.03)(\_\_\_\_)$

$= \_\_\_\_$

(b) \$28,000 at $7\frac{1}{2}\%$ for 3 months

4 Find the total amount due on each loan.

GS **(a)** \$3800 at $6\frac{1}{2}\%$ for 6 months

$I = p \cdot r \cdot t$

$= (3800)(0.065)(\_\_\_\_)$

$= \_\_\_\_$

Total amount due = principle + interest

= \$3800 + \$123.50

= ____

(b) \$12,400 at 5% for 5 years

(c) \$2400 at $4\frac{1}{2}\%$ for $2\frac{3}{4}$ years

Answers

3. (a) $\frac{4}{12}$ or $\frac{1}{3}$; \$18 **(b)** \$525

4. (a) $\frac{6}{12}$ or $\frac{1}{2}$; \$123.50; \$3923.50 **(b)** \$15,500 **(c)** \$2697

CAUTION

It is best to rewrite fractions of percents or fractions of years in their decimal form. In **Example 2**, $3\frac{1}{2}$ years is rewritten as 3.5 years.

Interest rates are given ***per year.*** For loan periods of less than one year, be careful to express time as a fraction of a year.

If time is given in months, for example, use a denominator of 12, because there are 12 months in a year. A loan for 9 months would be for $\frac{9}{12}$ of a year.

EXAMPLE 3 Finding Simple Interest for Less Than 1 Year

Find the interest on \$840 at $4\frac{1}{2}\%$ for 9 months.

The principal is \$840. The rate is $4\frac{1}{2}\%$ or 0.045 as a decimal, and the time is $\frac{9}{12}$ of a year. Use the formula $I = p \cdot r \cdot t$.

$$I = (840)(0.045)\left(\frac{9}{12}\right)$$

9 months $= \frac{9}{12}$ of a year

Rewrite $4\frac{1}{2}\%$ as 0.045

$$= (37.8)\left(\frac{3}{4}\right)$$

Write $\frac{9}{12}$ in lowest terms as $\frac{3}{4}$.

$$= \frac{(37.8)(3)}{4}$$

$$= \frac{113.4}{4} = 28.35$$

The interest is \$28.35.

Calculator Tip

The calculator solution to **Example 3** above uses chain calculations.

840 × .045 × 9 ÷ 12 = 28.35

◀ **Work Problem 3 at the Side.**

OBJECTIVE 2 Find the total amount due on a loan. When a loan is repaid, the interest is added to the original principal to find the total amount due.

Formula for Total Amount Due

Total amount due = principal + interest

EXAMPLE 4 Calculating the Total Amount Due

A loan of \$3240 was made at 6% for 3 months. Find the total amount due.

First find the interest. Then add the principal and the interest to find the total amount due.

$$I = (3240)(0.06)\left(\frac{3}{12}\right) \quad \text{3 months} = \tfrac{3}{12} \text{ of a year.}$$

$$I = \$48.60$$

The interest is \$48.60.

Remember that the *total amount due* is the amount of the loan plus the interest.

Total amount due = principal + interest

= \$3240 + \$48.60 = \$3288.60

The total amount due is \$3288.60.

◀ **Work Problem 4 at the Side.**

6.7 Exercises

FOR EXTRA HELP

Download the
MyDashBoard App

MyMathLab®

CONCEPT CHECK *Rewrite all interest rates and times in decimal form.*

1. $3\frac{1}{2}\%$ **2.** $7\frac{1}{2}\%$ **3.** $2\frac{1}{4}$ years **4.** $5\frac{3}{4}$ years

Find the interest. **See Examples 1 and 2.**

Principal	Rate	Time in Years	Interest
5. $100	6%	1	_____

$I = p \cdot r \cdot t$
$= (100)(0.06)(1)$

6. $200	3%	1	_____

$I = p \cdot r \cdot t$
$= (200)(0.03)(1)$

7. $700	5%	3	_____
8. $900	2%	4	_____
9. $2300	$4\frac{1}{2}\%$	$2\frac{1}{2}$	_____

$I = p \cdot r \cdot t$
$= (2300)(0.045)(2.5)$

10. $4700	$5\frac{1}{2}\%$	$1\frac{1}{2}$	_____

$I = p \cdot r \cdot t$
$= (4700)(0.055)(1.5)$

11. $10,800	$7\frac{1}{2}\%$	$2\frac{3}{4}$	_____
12. $12,400	$1\frac{1}{2}\%$	$3\frac{3}{4}$	_____

CONCEPT CHECK *Change all times in months to fractions of a year and to decimal form.*

13. 3 months **14.** 6 months **15.** 9 months **16.** 15 months

Find the interest. Round to the nearest cent if necessary. **See Example 3.**

Principal	Rate	Time in Months	Interest
17. $400	3%	6	_____

$I = p \cdot r \cdot t$

$= (400)(0.03)\left(\frac{1}{2}\right)$ 6 months $= \frac{1}{2}$ year

Principal	Rate	Time in Months	Interest
18. \$780	5%	24	________
19. \$940	3%	18	________
20. \$178	4%	12	________
21. \$1225	$5\frac{1}{2}\%$	3	________
22. \$2660	$7\frac{1}{2}\%$	3	________
23. \$15,300	$7\frac{1}{4}\%$	7	________
24. \$13,700	$3\frac{3}{4}\%$	11	________

Exercise 18 (GS):

$I = p \cdot r \cdot t$

$= (780)(0.05)(2)$ 24 months = 2 years

Find the total amount due on the following loans. Round to the nearest cent if necessary. ***See Example 4.***

Principal	Rate	Time	Total Amount Due
25. \$200	5%	1 year	________
26. \$400	2%	6 months	________
27. \$740	6%	9 months	________
28. \$1180	3%	2 years	________

Exercise 25 (GS):

$I = p \cdot r \cdot t$

$= (200)(0.05)(1)$

Total amount due = principal + interest

= \$200 + interest

Exercise 26 (GS):

$I = p \cdot r \cdot t$

$= (400)(0.02)\left(\frac{1}{2}\right)$ 6 months $= \frac{1}{2}$ year

Total amount due = principal + interest

= \$400 + interest

	Principal	Rate	Time	Total Amount Due
29.	$1800	9%	18 months	______________
30.	$9000	6%	7 months	______________
31.	$3250	$3\frac{1}{2}\%$	6 months	______________
32.	$7600	$4\frac{1}{2}\%$	1 year	______________
33.	$16,850	$7\frac{1}{2}\%$	9 months	______________
34.	$19,450	$5\frac{1}{2}\%$	6 months	______________

35. The amount of interest paid on savings accounts and charged on loans can vary from one institution to another. However, when the amount of interest is calculated, three factors are used in the calculation. Name these three factors and describe them in your own words.

36. Interest rates are usually given as a rate per year (annual rate). Explain what must be done when time is given in months. Write your own problem where time is given in months and then show how to solve it.

Solve each application problem. Round to the nearest cent if necessary.

37. Paul Plescia has a savings account of $8642 at his credit union. If the account pays interest of 2%, how much interest will he earn in 2 years?

38. Esther Albert, a professional dancer, deposits $68,000 of her earnings at 4% for 5 years. How much interest will she earn?

39. To increase distribution of her office supply products, Stella Glitter borrowed $150,000 at 7% for 30 months. Find the amount of interest on this loan.

40. The Jidobu family invests $18,000 at 3% for 9 months. What amount of interest will the family earn?

41. To complete her fourth semester of college, Yohko Kitagawa borrowed \$4650 at 5%. If she repaid the loan in 5 months, find the total amount due.

42. Sarah Brynski borrows \$5500 from her dad for a used car. The loan will be paid back with 4% interest at the end of 9 months. Find the total amount due.

43. Nicholas Thomas deposits \$14,800 in his school credit union account for 10 months. If the credit union pays $2\frac{1}{4}\%$ interest, find the amount of interest he will earn.

44. Sid and Shirley Kordell, owners of the Nut House, borrow \$54,000 to update their store. If the loan is for 42 months at $7\frac{1}{4}\%$, find the amount of interest they will owe.

45. After the sale of her home, the buyer still owes Tiina Luig a balance of \$8800. If the interest rate on this amount is $7\frac{1}{4}\%$, find the amount of interest she will earn every 3 months.

46. Pat Carper owes \$1900 in taxes. She is charged a penalty of $9\frac{1}{4}\%$ annual interest and pays the taxes and penalty after 6 months. Find the amount of the penalty.

47. Business has been good at Rocky's Grill, and the owner deposits \$5400 into an account paying $2\frac{3}{4}\%$ interest. Find

(a) the amount of interest.

(b) the total amount in the account at the end of $1\frac{1}{2}$ years.

48. Moises Guardado, the owner of MG Painting, borrowed \$29,400 to purchase another truck and other equipment for his business. The loan is for $3\frac{1}{2}$ years at $5\frac{1}{4}\%$ interest. Find

(a) the amount of interest.

(b) the total amount he must repay.

49. Jake's Exotic Pets bought four new aviary environments (bird cages) at a cost of \$980 each. The owner borrowed 70% of the cost of the cages at an interest rate of $7\frac{1}{2}\%$. Find the total amount owed at the end of $2\frac{1}{2}$ years.

50. The owners of Baily and Daughters Excavating purchased four earth movers at a cost of \$485,000 each. If they borrowed 80% of the total purchase price for $2\frac{1}{2}$ years at $10\frac{1}{2}\%$ interest, find the total amount due.

6.8 Compound Interest

The interest we studied in the last section was *simple interest* (interest only on the original principal). A common type of interest used with savings accounts and most investments is **compound interest** or interest paid on past interest as well as on the principal.

OBJECTIVES

1. Understand compound interest.
2. Understand compound amount.
3. Find the compound amount.
4. Use a compound interest table.
5. Find the compound amount and the amount of interest.

OBJECTIVE 1 **Understand compound interest.** Suppose that you make a single deposit of \$1000 in a savings account that earns 5% per year. What will happen to your savings over 3 years? At the end of the first year, 1 year's interest on the original deposit is calculated. Use the simple interest formula.

Interest = principal • rate • time

Year 1 (\$1000)(0.05)(1) = \$50

Add the interest to the \$1000 to find the amount in your account at the end of the first year. \$1000 + \$50 = \$1050 (In year 1, interest is calculated on principal.)

The interest for the second year is found on \$1050, that is, the interest is **compounded.**

Year 2 (\$1050)(0.05)(1) = \$52.50

Add this interest to the \$1050 to find the amount in your account at the end of the second year. \$1050 + \$52.50 = \$1102.50.

The interest for the third year is found on \$1102.50.

Year 3 (\$1102.50)(0.05)(1) ≈ \$55.13 (In year 2 and thereafter, interest is calculated on the principal and all past interest.)

Add this interest to the \$1102.50. So, \$1102.50 + \$55.13 = \$1157.63.

At the end of 3 years, you will have **\$1157.63** in your savings account. The \$1157.63 that you have in your account is called the **compound amount.**

If you had earned only *simple* interest for 3 years, your interest would be as follows.

I = (\$1000)(0.05)(3)
= \$150 ← Simple interest

At the end of 3 years, you would have \$1000 + \$150 = \$1150 in your account. Compounding the interest increased your earnings by \$7.63 because \$1157.63 − \$1150 = \$7.63.

With *compound* interest, the interest earned during the second year is greater than that earned during the first year, and the interest earned during the third year is greater than that earned during the second year. This happens because the interest earned each year is *added* to the principal, and the new total is used to calculate the amount of interest in the next year.

Compound Interest

Interest paid on principal plus past interest is called **compound interest.**

VOCABULARY TIP

Simple interest With simple interest, there is usually only one interest calculation for a specific period of time at a specific rate of interest.

VOCABULARY TIP

Compound interest/Compounded *Compound* means to mix, combine, or put together. When interest has been compounded, it means that it has been combined with the original principal and all previous interest on the principal.

VOCABULARY TIP

Compound amount The compound amount at the end of any compound interest period is the total of the original principal and all the interest combined.

1 Find the compound amount given the following deposits. Round to the nearest cent if necessary.

GS **(a)** \$400 at 4% for 2 years

yr 1: (\$400)(0.04)(1) = \$16

\$400 + \$16 = \$416

yr 2: (\$416)(0.04)(1) = ______

\$416 + \$16.64 = ______

(b) \$2000 at 2% for 3 years

2 Find the compound amount by multiplying the original deposit by 100% plus the compound interest rate. Round to the nearest cent if necessary.

GS **(a)** \$1800 at 2% for 3 years

(1800)(1.02)(1.02)(1.02) =

______

(b) \$900 at 3% for 2 years

(c) \$2500 at 5% for 4 years

Answers

1. (a) \$16.64; \$432.64 (b) \$2122.42 (rounded)
2. (a) \$1910.17 (rounded) (b) \$954.81
(c) \$3038.77 (rounded)

OBJECTIVE 2 **Understand compound amount.** Find the compound amount as shown below.

EXAMPLE 1 Finding the Compound Amount

Mavis Chamski deposits \$3400 in an account that pays 3% interest compounded annually for 4 years. Find the compound amount. Round to the nearest cent when necessary.

Year	Interest	Compound Amount
1	(\$3400)(0.03)(1) = **\$102** \$3400 + **\$102** =	**\$3502**
2	(\$3502)(0.03)(1) = **\$105.06** \$3502 + **\$105.06** =	**\$3607.06**
3	(\$3607.06)(0.03)(1) ≈ **\$108.21** \$3607.06 + **\$108.21** =	**\$3715.27**
4	(\$3715.27)(0.03)(1) ≈ **\$111.46** \$3715.27 + **\$111.46** =	**\$3826.73**

The compound amount is \$3826.73.

◀ Work Problem **1** at the Side.

OBJECTIVE 3 **Find the compound amount.** A more efficient way of finding the compound amount is to add the interest rate to 100% and then multiply by the original deposit. Notice that in **Example 1** above, at the end of the first year, you will have \$3400 (100% of the original deposit) plus 3% (of the original deposit) or 103% (because 100% + 3% = 103%).

EXAMPLE 2 Finding the Compound Amount

Find the compound amount in **Example 1** using multiplication.

Year 1 Year 2 Year 3 Year 4

(\$3400)(1.03)(1.03)(1.03)(1.03) ≈ \$3826.73

↑ Original deposit; 100% + 3% = 103% = 1.03; ↑ Compound amount

This method works well with a calculator.

Our answer, \$3826.73, is the same as in **Example 1** above.

◀ Work Problem **2** at the Side.

Note

By adding the compound interest rate to 100%, we can then multiply by the original deposit. This will give us the compound amount at the end of each compound interest period.

Calculator Tip

If you use a calculator for **Example 2** above, you can use the y^x key (exponent key). (On some calculators, you use the ^ key instead.)

3400 × 1.03 y^x 4 = 3826.73 (rounded)

The 4 following the y^x key represents the number of compound interest periods.

OBJECTIVE 4 Use a compound interest table. The calculation of compound interest can be quite tedious. For this reason, compound interest tables have been developed.

Suppose you deposit \$1 in a savings account today that earns 4% compounded annually and you allow the deposit to remain for 3 years. The diagram below shows the compound amount at the end of each of the 3 years.

Compound Amount

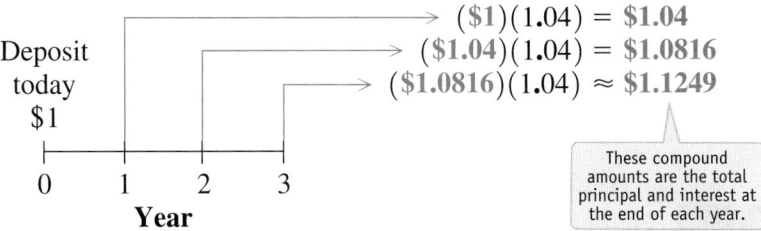

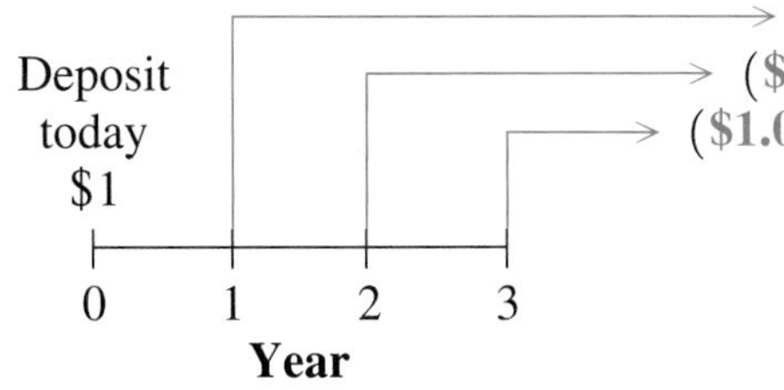

$(\$1)(1.04) = \$1.04$

$(\$1.04)(1.04) = \$1.0816$

$(\$1.0816)(1.04) \approx \$1.1249$

These compound amounts are the total principal and interest at the end of each year.

Using the compound amounts for \$1, a table can be formed. Look at the table below and find the column headed 4%. The first three numbers for years 1, 2, and 3 are the same as those we have calculated for \$1 at 4% for 3 years. This table, giving the compound amounts on a \$1 deposit for given lengths of time and interest rates, can be used for finding the compound amount on any amount of deposit.

COMPOUND INTEREST TABLE

Time Periods	2.00%	2.50%	3.00%	3.50%	4.00%	4.50%	5.00%	5.50%	6.00%	Time Periods
1	1.0200	1.0250	1.0300	1.0350	1.0400	1.0450	1.0500	1.0550	1.0600	1
2	1.0404	1.0506	1.0609	1.0712	1.0816	1.0920	1.1025	1.1130	1.1236	2
3	1.0612	1.0769	1.0927	1.1087	1.1249	1.1412	1.1576	1.1742	1.1910	3
4	1.0824	1.1038	1.1255	1.1475	1.1699	1.1925	1.2155	1.2388	1.2625	4
5	1.1040	1.1314	1.1593	1.1877	1.2167	1.2462	1.2763	1.3070	1.3382	5
6	1.1261	1.1597	1.1941	**1.2293**	1.2653	1.3023	1.3401	1.3788	1.4185	6
7	1.1486	1.1887	1.2299	1.2723	1.3159	1.3609	1.4071	1.4547	1.5036	7
8	1.1717	1.2184	1.2668	1.3168	1.3686	1.4221	1.4775	1.5347	1.5938	8
9	1.1951	1.2489	1.3048	1.3629	1.4233	1.4861	1.5513	1.6191	1.6895	9
10	1.2190	1.2801	1.3439	1.4106	1.4802	1.5530	1.6289	1.7081	1.7908	10
11	1.2434	1.3121	1.3842	1.4600	1.5395	1.6229	1.7103	1.8021	1.8983	11
12	1.2682	1.3449	1.4258	1.5111	1.6010	1.6959	1.7959	1.9012	2.0122	12

EXAMPLE 3 Using a Compound Interest Table

Find the compound amount using the compound interest table.

A ruler or straightedge helps align the column and row.

(a) \$1 is deposited at a 5% interest rate for 10 years.

Look down the column headed 5%, and across to row 10 (because 10 years = 10 time periods). At the intersection of the column and row, read the compound amount, 1.6289.

(b) \$1 is deposited at $3\frac{1}{2}\%$ for 6 years.

The intersection of the $3\frac{1}{2}\%$ (3.50%) column and row 6 shows **1.2293** as the compound amount.

Work Problem 3 at the Side. ▶

3 Find the compound amount using the compound interest table.

GS **(a)** \$1 at 3% for 6 years

$(1)(1.1941) =$ ________

↑

Look down the 3% column in the table to row 6.

(b) \$1 at 2% for 8 years

(c) \$1 at $4\frac{1}{2}\%$ for 12 years

Answers

3. (a) \$1.19 (rounded) **(b)** \$1.17 (rounded) **(c)** \$1.70 (rounded)

4 Use the compound interest table to find the compound amount and the interest.

GS **(a)** \$4000 at 3% for 10 years

Table shows 1.3439.

(4000)(1.3439) = ________

________ − 4000

= ________ interest

(b) \$12,600 at $3\frac{1}{2}$% for 8 years

(c) \$32,700 at $4\frac{1}{2}$% for 12 years

OBJECTIVE 5 Find the compound amount and the amount of interest.

Find the compound amount and interest as follows.

Finding the Compound Amount and the Interest

Compound Amount

Multiply the principal by the compound amount for \$1 (from the table on the previous page).

Interest

Find the interest earned on a deposit by subtracting the original deposit from the compound amount.

EXAMPLE 4 Finding Compound Amount and Interest

Use the compound interest table to find the compound amount and the interest.

(a) \$1000 at $5\frac{1}{2}$% interest for 12 years

Look in the table on the previous page for $5\frac{1}{2}$% (5.50%) and 12 periods to find the number **1.9012** but do *not* round it. Multiply this number and the principal of \$1000.

Never round the numbers found in the table.

(\$1000) (**1.9012**) = **\$1901.20**

The account will contain \$1901.20 after 12 years.

Find the interest by subtracting the original deposit from the compound amount.

Compound amount → **\$1901.20** − Original deposit → \$1000 = Interest → **\$901.20**

(b) \$6400 at 2% for 7 years

Look in the table for 2% and 7 periods to find **1.1486**. Multiply.

(\$6400) (**1.1486**) = **\$7351.04** Compound amount

Subtract the original deposit from the compound amount.

\$7351.04 − \$6400 = **\$951.04** Interest

Remember: Compound amount − Principal = Interest.

A total of \$951.04 in interest was earned.

◀ **Work Problem 4 at the Side.**

Answers

4. **(a)** \$5375.60; \$5375.60; \$1375.60
(b) \$16,591.68; \$3991.68
(c) \$55,455.93; \$22,755.93

6.8 Exercises

FOR EXTRA HELP

Download the MyDashBoard App

MyMathLab®

1. CONCEPT CHECK *Write* true *or* false.

For short periods of time, five years or less, the amount of interest earned will be the same whether you use simple or compound interest.

2. CONCEPT CHECK *Underline the correct answer.*

When using compound interest, the amount of (*interest/principal*) earned at the end of a compound interest period will be added to the amount of (*interest/principal*) at the beginning of that period.

Find the compound amount given the following deposits. Calculate the interest each year, then add it to the previous year's amount. ***See Example 1.***

3. \$500 at 4% for 2 years

$I = p \cdot r \cdot t$

Year 1 $\begin{cases} (\$500)(0.04)(1) = \$20 \\ \$500 + \$20 = \$520 \end{cases}$

Year 2 $\begin{cases} (\$520)(0.04)(1) = \$20.80 \\ \$520 + \$20.80 = \underline{\qquad\qquad} \end{cases}$

4. \$1500 at 5% for 3 years

$I = p \cdot r \cdot t$

Year 1 $\begin{cases} (\$1500)(0.05)(1) = \$75 \\ \$1500 + \$75 = \$1575 \end{cases}$

Year 2 $\begin{cases} (\$1575)(0.05)(1) = \$78.75 \\ \$1575 + \$78.75 = \$1653.75 \end{cases}$

Year 3 $\begin{cases} (\$1653.75)(0.05)(1) = \$82.69 \text{ (rounded)} \\ \$1653.75 + \$82.69 = \underline{\qquad\qquad} \end{cases}$

5. \$1800 at 3% for 3 years

6. \$2000 at 2% for 3 years

7. \$3500 at 7% for 4 years

8. \$5500 at 6% for 4 years

9. CONCEPT CHECK *Fill in the blanks.*

To find the compound amount for a deposit of \$2000 at 2% for 3 years, you can multiply

(2000)_____ _____ _____ = \$2122.42 (rounded)

10. CONCEPT CHECK *Fill in the blanks.*

To find the compound amount for a deposit of \$3000 at 5% for 2 years, you can multiply

(\$3000)_____ _____ = \$3307.50

Find each compound amount by multiplying the original deposit by 100% plus the compound rate given in the following. ***See Example 2.*** *Round answers to the nearest cent if necessary.*

11. \$1000 at 5% for 2 years

Year 1 Year 2

$(\$1000)(1.05)(1.05) = \underline{\qquad\qquad}$

12. \$500 at 4% for 3 years

Year 1 Year 2 Year 3

$(\$500)(1.04)(1.04)(1.04) = \underline{\qquad\qquad}$

13. \$1400 at 6% for 5 years

14. \$2500 at 3% for 4 years

15. \$1180 at 7% for 8 years

16. \$12,800 at 6% for 7 years

17. \$10,940 at 4% for 6 years

18. \$15,710 at 2% for 8 years

CONCEPT CHECK *In the compound interest table, locate the compound amount for \$1 for each of the following situations.*

19. 3% for 6 years

20. 4.50% for 10 years

21. 2% for 8 years

22. 5.50% for 3 years

Use the table to find the compound amount and the interest. Interest is compounded annually. Round answers to the nearest cent if necessary. ***See Examples 3 and 4.***

23. \$1000 at 4% for 5 years

GS Table shows 1.2167.

(\$1000)(1.2167) = \$1216.70 compound amount

\$1216.70 − \$1000 = ______

(Compound amount − Original deposit = Interest)

24. \$10,000 at 3% for 4 years

GS Table shows 1.1255.

(\$10,000)(1.1255) = \$11,255 compound amount

\$11,255 − \$10,000 = ______

(Compound amount − Original deposit = Interest)

25. \$8000 at 2% for 10 years

26. \$7800 at 5% for 8 years

27. \$8428.17 at $4\frac{1}{2}$% for 6 years

28. \$10,472.88 at $5\frac{1}{2}$% for 12 years

29. Write a definition for compound interest. Describe in your own words what compound interest means to you.

30. What is the difference between the compound amount and the interest?

Use the compound interest table in this section to solve each application problem. Round answers to the nearest cent if necessary. ***See Examples 3 and 4.***

31. After winning $50,000 in the Veterans of Foreign Wars (VFW) lottery, Jerry Murray opened a money market account. He deposited all of his winnings and earned 4% interest compounded annually. Find the amount he will have (compound amount) in the account at the end of 5 years.

32. Yen Lee borrowed $32,800 from her family to open a small restaurant named Shanghai Winds. Her plan is to repay the loan at the end of 4 years at $3\frac{1}{2}\%$ interest compounded annually. Find the total amount that she must repay.

33. Al Granard lends $76,000 to the owner of Rick's Limousine Service. He will be repaid at the end of 9 years at 6% interest compounded annually. Find

(a) the total amount that he should be repaid.

(b) the amount of interest earned.

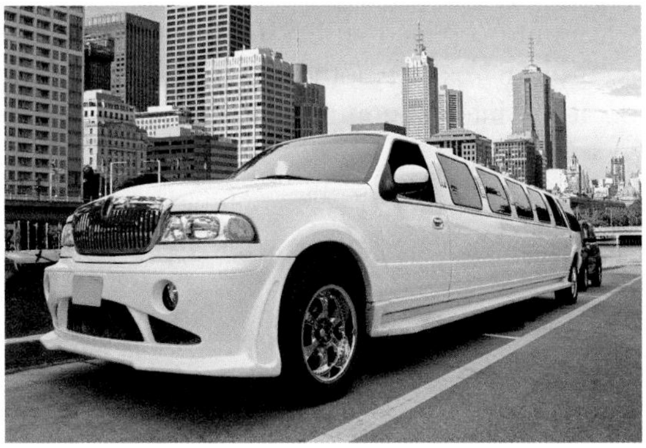

34. Recent sales of their music gave Bob and Cindy Kilpatrick $28,500 in additional income. If they invest all of it at 4% interest compounded annually for 10 years, find

(a) the total amount they will have at the end of 10 years.

(b) the amount of interest earned.

35. Jennifer Barrister deposits $30,000 at 6% interest compounded annually. Two years after she makes the first deposit, she deposits another $40,000, also at 6% compounded annually.

(a) What total amount will she have five years after her first deposit?

(b) What amount of interest will she have earned?

36. Christine Campbell invested $25,000 at 4% interest compounded annually. Three years after she made the first deposit, she deposited another $25,000, also at 4% compounded annually.

(a) What total amount will she have five years after her first deposit?

(b) What amount of interest will she earn?

Relating Concepts (Exercises 37–42) For Individual or Group Work

Knowing how to solve interest problems is important to businesspeople and consumers alike. ***Work Exercises 37–40 in order.***

37. Simple interest calculation is used for most short-term business loans, most real estate loans, and many automobile and consumer loans. The formula for simple interest is

Interest = ______ • ______ • ______

or $I =$ ______ • ______ • ______.

38. When a loan is repaid, the interest is added to the original principal. The formula for the total amount due is

Amount due = ______ + ______.

39. Compound interest is paid on most savings accounts and many other types of investments. Compound interest is interest calculated on ______ plus past ______.

40. The compound amount is the total amount in an account at the end of a period of time. Compound amount is the original ______ + compound ______.

Use the compound interest table to answer Exercises 41 and 42.

41. Julie Maxey has two choices. She can invest $4350 at 6% simple interest for 6 years, or she can invest the same amount at 6% interest compounded annually for 6 years.

(a) Find the difference in the amount of interest earned in these two accounts.

(b) If the length of time is doubled from 6 to 12 years, will the difference in the interest earned also double?

(c) Use your own example to determine that what you found in part (b) is true with a different interest rate and length of time.

42. One account is opened with $10,000 at 5% simple interest for 10 years. Another account is opened with $9500 at 5% interest compounded annually for 10 years.

(a) At the end of the 10 years, which account has the higher balance and by how much?

(b) What does this tell you about compound interest? Why?

Study Skills

PREPARING FOR YOUR FINAL EXAM

OBJECTIVES

1. **Create a final exam week plan.**
2. **Break studying into chunks and study over several days.**
3. **Practice all types of problems.**

Your math final exam is likely to be a **comprehensive exam.** This means that it will cover material from the **entire term.** The end of the term will be less stressful if you **make a plan** for how you will prepare for each of your exams.

First, figure out the **score you need to earn on the final exam** to get the course grade you want. Check your course syllabus for grading policies, or ask your instructor if you are not sure of them. This allows you to set a goal for yourself.

How many points do you need to earn on your mathematics final exam to get the grade you want? ______________________________

____

Create a Plan

Second, create a **final exam week plan for your work and personal life.** If you need to make an adjustment in your work schedule, do it in advance, so you aren't scrambling at the last minute. If you have family members to care for, you might want to enlist some help from others so you can spend extra time studying. Try to plan in advance so you don't create additional stress for yourself. You will have to set some priorities, and studying has to be at the top of the list! Although life doesn't stop for finals, some things can be ignored for a short time. You don't want to "burn out" during final exam week; **get enough sleep and healthy food so you can perform your best.**

What adjustments in your personal life do you need to make for final exam week? ______________________________

____

Study and Review

Third, use the following suggestions to guide your studying and reviewing.

- **Know exactly which chapters and sections will be on the final exam.**
- **Divide up the chapters,** and decide how much you will review each day.
- Begin your reviewing **several days** before the exam.
- **Use returned quizzes and tests** to review earlier material (if you have them).
- **Practice all types of problems,** but emphasize the types that are most difficult for you. Use the **Cumulative Reviews** that are at the end of each chapter in your textbook.
- **Rewrite your notes or make mind maps** to create summaries.
- **Make study cards for all types of problems.** Be sure to use the same **direction words** (such as *simplify, solve, estimate*) that your exam will use. Carry the cards with you and review them whenever you have a few spare minutes.

Study Skills

Continued from page 439

Managing Stress

Of course, a week of final exams produces stress. **Students who develop skills for reducing and managing stress do better on their final exams and are less likely to "bomb" an exam.** You already know the damaging effect of adrenaline on your ability to think clearly. But several days (or weeks) of elevated stress is also harmful to your brain and your body. You will feel better if you make a conscious effort to reduce your stress level. Even if it takes you away from studying for a little while each day, the time will be well spent.

Reducing Physical Stress

Examples of ways to reduce **physical stress** are listed below. Can you add any of your own ideas to the list?

Which techniques will you try?

___

___

___

- *Laugh until your eyes water.* Watching your favorite funny movie, exchanging a joke with a friend, or viewing a comedy bit on the Internet are all ways to generate a healthy laugh. Laughing raises the level of *calming* chemicals (endorphins) in your brain.
- *Exercise for 20 to 30 minutes.* If you normally exercise regularly, do NOT stop during final exam week! Exercising helps relax muscles, diffuses adrenaline, and raises the level of endorphins in your body. If you don't exercise much, get some gentle exercise, such as a daily walk, to help you relax.
- *Practice deep breathing.* Several minutes of deep, smooth breathing will calm you. Close your eyes too.
- *Visualize a relaxing scene.* Choose something that you find peaceful and picture it. Imagine what it feels like and sounds like. Try to put yourself in the picture.
- If you feel stress in your muscles, such as in your shoulders or back, *slowly squeeze the muscles as much as you can, and then release them.* Sometimes we don't realize we are clenching our teeth or holding tension in our shoulders until we consciously work with them. Try to notice what it feels like when they are relaxed and loose. Squeezing and then releasing muscles is also something you can do during an exam if you feel yourself tightening up.

Reducing Mental Stress

Mental stress reduction is also a powerful tool both before and during an exam. In addition to these suggestions, do you have any of your own techniques?

Which techniques will you try?

___

___

___

- *Talk positively to yourself.* Tell yourself you will get through it.
- *Reward yourself.* Give yourself small breaks, a little treat—something that makes you happy—every day of final exam week.
- *Make a list of things to do* and feel the sense of accomplishment when you cross each item off.
- When you take time to relax or exercise, *make sure you are relaxing your mind too.* Use your mind for something *completely* different from the kind of thinking you do when you study. Plan your garden, play your favorite music, walk your dog, read a good book.
- *Visualize.* Picture yourself completing exams and projects successfully. Picture yourself taking the test calmly and confidently.

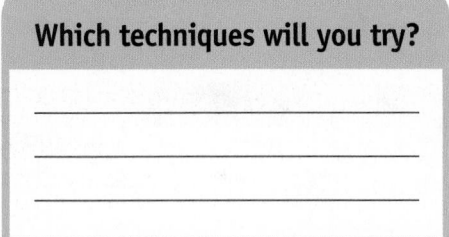

Chapter 6 Summary

Key Terms

6.1

percent Percent means "per one hundred." A percent is a ratio with a denominator of 100.

6.3

percent proportion The proportion $\frac{\text{part}}{\text{whole}} = \frac{\text{percent}}{100}$ is used to solve percent problems.

whole The whole in a percent problem is the entire quantity, the total, or the base.

part The part in a percent problem is the portion being compared with the whole.

6.5

percent equation The percent equation is: part = percent • whole. It is another way to solve percent problems.

6.6

sales tax Sales tax is a percent of the cost of an item charged as a tax.

commission Commission is a percent of the dollar value of total sales paid to a salesperson.

discount Discount is often expressed as a percent of the original price; it is then deducted from the original price, resulting in the sale price.

percent of increase or decrease Percent of increase or decrease is the amount of change (increase or decrease) expressed as a percent of the original amount.

6.7

interest Interest is a fee paid or a charge made for lending or borrowing money.

principal Principal is the amount of money on which interest is earned.

rate of interest Often referred to as "rate," it is the charge for interest and is given as a percent.

simple interest Interest that is computed only on the original principal is simple interest.

interest formula The interest formula is used to calculate interest. It is: interest = principal • rate • time, or $I = p \cdot r \cdot t$.

6.8

compound interest Compound interest is interest paid both on the past interest and on the principal.

compounding Interest that is compounded once each year is compounded annually.

compound amount Compound amount is the total amount in an account, including compound interest and the original principal.

New Symbols

% percent (per one hundred)

New Formulas

To write percents as decimals: $p\% = p \div 100$ **To write percents as fractions:** $p\% = \frac{p}{100}$

Percent proportion: $\frac{\text{part}}{\text{whole}} = \frac{\text{percent}}{100}$ **Percent equation:** part = percent • whole

Amount of sales tax: amount of sales tax = rate of tax • cost of item

Amount of commission: amount of commission = rate of commission • amount of sales

Amount of discount: amount of discount = rate of discount • original price

Sale price: sale price = original price − amount of discount

Percent of increase: $\frac{\text{amount of increase}}{\text{original value}} = \frac{\text{percent}}{100}$

Percent of decrease: $\frac{\text{amount of decrease}}{\text{original value}} = \frac{\text{percent}}{100}$

Interest: Interest = principal • rate • time or $I = p \cdot r \cdot t$

Total amount due: total amount due = principal + interest

Test Your Word Power

See how well you have learned the vocabulary in this chapter.

1. To write a **percent as a decimal,** you drop the percent symbol
   - **A.** after finding the decimal point
   - **B.** and move the decimal point two places to the right
   - **C.** and move the decimal point two places to the left.

2. To write a **decimal as a percent,** you attach the percent symbol
   - **A.** after removing the decimal point
   - **B.** after moving the decimal point two places to the right
   - **C.** after moving the decimal point two places to the left.

3. **Percent** means
   - **A.** the same as interest
   - **B.** per one thousand
   - **C.** per one hundred.

4. When you use $\frac{\textbf{part}}{\textbf{whole}} = \frac{\textbf{percent}}{\textbf{100}}$ to solve percent problems, you are using the
   - **A.** simple interest formula
   - **B.** percent proportion
   - **C.** percent equation.

5. The **percent equation** is
   - **A.** part = percent • whole
   - **B.** $I = p \bullet r \bullet t$
   - **C.** $p\% = \frac{p}{100}$.

6. In a **percent of increase** problem, the increase is a percent of
   - **A.** the largest amount
   - **B.** the original amount
   - **C.** the new or most recent amount.

7. In the formula $\boldsymbol{I = p \bullet r \bullet t}$, the $\boldsymbol{p}$ stands for
   - **A.** proportion
   - **B.** principal
   - **C.** percent.

8. The term **rate** in an interest problem represents the
   - **A.** whole
   - **B.** percent
   - **C.** part.

Answers to Test Your Word Power

1. C; *Example:* 50% written as a decimal is 0.50. or 0.5.
2. B; *Example:* 0.25 written as a percent is 0.25 or 25%.
3. C; *Example:* 8% means 8 per 100.
4. B; *Example:* Part = 4, and whole = 25. To find the percent, $\frac{4}{25} = \frac{x}{100}$; $x = 16$ or 16%.
5. A; *Example:* Percent = 25, and whole = 300. To find the part, $(0.25)(300) = 75$.
6. B; *Example:* Original value = \$200, and amount of increase = \$40. To find the percent of increase, $\frac{40}{200} = \frac{x}{100}$; $x = 0.20$ or 20% increase.
7. B; *Example:* Principal (p) = \$800, rate (r) = 5%, and time (t) = 1 year. Then, $I = p \bullet r \bullet t$ so $I = (\$800)(0.05)(1) = \$40$.
8. B; *Example:* Principal (p) = \$1650, rate (r) = 4%, and time $(t) = \frac{1}{2}$ year. To find the interest, $I = (\$1650)(0.04)\left(\frac{1}{2}\right) = \$33$.

Quick Review

Concepts	*Examples*
6.1 Basics of Percent **Writing a Percent as a Decimal** To write a percent as a decimal, drop the percent symbol and move the decimal point two places to the left. **Writing a Decimal as a Percent** To write a decimal as a percent, move the decimal point two places to the right and attach a % symbol.	50% (.50%) = 0.50 or just 0.5 3%(.03%) = 0.03 12.5% (12.5%) = 0.125 0.75 (0.75) = 75% 0.875(0.875) = 87.5% 3.6 (3.60) = 360%
6.2 Writing a Fraction as a Percent Use a proportion and solve for p to change a fraction to percent.	$\frac{2}{5} = \frac{p}{100}$ Proportion $5 \bullet p = 2 \bullet 100$ Cross products $5 \bullet p = 200$ $\frac{\overset{1}{\cancel{5}} \bullet p}{\underset{1}{\cancel{5}}} = \frac{200}{5}$ Divide both sides by 5. $p = 40$ $\frac{2}{5} = 40\%$ Attach % symbol.

Concepts	Examples

6.3 Learning the Percent Proportion

Part is to whole as percent is to 100.

$$\frac{\text{part}}{\text{whole}} = \frac{\text{percent}}{100} \leftarrow \text{Always 100 because percent means "per 100."}$$

Use the percent proportion to solve for the unknown value.
part = 30, whole = 50; find the percent.

Percent (unknown)

$$\text{Part} \rightarrow \frac{30}{50} = \frac{x}{100} \leftarrow \text{Always 100}$$
Whole →

$$\frac{3}{5} = \frac{x}{100}$$ Write $\frac{30}{50}$ as $\frac{3}{5}$ in lowest terms.

$$5 \cdot x = 3 \cdot 100$$ Cross products

$$5 \cdot x = 300$$

$$\frac{\overset{1}{\cancel{5}} \cdot x}{\underset{1}{\cancel{5}}} = \frac{300}{5}$$ Divide both sides by 5.

$$x = 60$$

The percent is 60, which is written as 60%.

6.3 Identifying Percent, Whole, and Part in a Percent Problem

The percent appears with the word **percent** or with the symbol %.

The whole often appears after the word **of**. The whole is the entire quantity or total.

The part is the portion of the total. If the percent and the whole are found first, the remaining number is the part.

Find the percent, whole, and part in the following.

10% **of** the 500 pies is how many pies?
Percent: 10%; Whole: 500; Part (unknown): how many pies

20 cats is 5% **of** what number of cats?
Part: 20; Percent: 5%; Whole (unknown): what number of cats

What percent **of** $220 is $33?
Percent (unknown): What percent; Whole: $220; Part: $33

6.4 Applying the Percent Proportion

Read the problem and identify the percent, whole, and part. Use the percent proportion to solve for the unknown quantity.

A liquid mixture in a tank contains 35% distilled water. If 28 gallons of distilled water are in the tank when it is full, find the capacity of the tank.

percent = 35 and part = 28

Use the percent proportion to find the whole.

$$\frac{\text{part}}{x} = \frac{\text{percent}}{100}$$ ← Whole (unknown) → x

$$\frac{28}{x} = \frac{35}{100}$$

$$\frac{28}{x} = \frac{7}{20}$$ Write $\frac{35}{100}$ as $\frac{7}{20}$ in lowest terms.

$$x \cdot 7 = 560$$ Cross products

$$\frac{x \cdot \overset{1}{\cancel{7}}}{\underset{1}{\cancel{7}}} = \frac{560}{7}$$ Divide both sides by 7.

$$x = 80$$

The capacity of the tank is 80 gallons.

Concepts

Examples

6.5 Using the Percent Equation

The percent equation is part = percent • whole. Identify the percent, whole, and part and solve for the unknown quantity. Always write the percent as a decimal before using the equation.

Solve each problem.

(a) Find 20% of 220 applicants.

$$\text{part (unknown)} = \text{percent} \cdot \text{whole}$$
$$x = (0.2)(220)$$
$$x = 44$$

20% of 220 applicants is 44 applicants.

(b) 8 balls is 4% of what number of balls?

$$\text{part} = \text{percent} \cdot \text{whole (unknown)}$$
$$8 = (0.04)(x)$$
$$\frac{8}{0.04} = \frac{\overset{1}{\cancel{(0.04)}}(x)}{\underset{1}{\cancel{0.04}}}$$
$$x = 200$$

8 balls is 4% of 200 balls.

(c) \$13 is what percent of \$52?

$$\text{part} = \text{percent (unknown)} \cdot \text{whole}$$
$$13 = x \cdot 52$$
$$\frac{13}{52} = \frac{x \cdot \overset{1}{\cancel{52}}}{\underset{1}{\cancel{52}}}$$
$$x = 0.25 = 25\%$$

\$13 is 25% of \$52.

6.6 Solving Application Problems with Proportions

To solve for **sales tax,** use this formula.

amount of sales tax = rate of tax • cost of item

The price of a 46-inch plasma HD television is \$699, and the sales tax is 5%. Find the sales tax.

$$\text{amount of sales tax} = (5\%)(\$699)$$
$$= (0.05)(\$699) = \$34.95$$

To find **commissions,** use this formula.

amount of commission = rate of commission • amount of sales

The sales are \$92,000 with a commission rate of 3%. Find the commission.

$$\text{amount of commission} = (3\%)(\$92{,}000)$$
$$= (0.03)(\$92{,}000)$$
$$= \$2760$$

To find the **discount** and the **sale price,** use these formulas.

amount of discount = rate of discount • original price

sale price = original price − amount of discount

A gas oven originally priced at \$480 is offered at a 25% discount. Find the amount of the discount and the sale price.

$$\text{discount} = (0.25)(\$480) = \$120$$
$$\text{sale price} = \$480 - \$120 = \$360$$

Concepts	Examples

6.6 Solving Application Problems with Proportions *(Continued)*

To find the **percent of change,** subtract to find the amount of change (increase or decrease), which is the part. The whole is the *original* value or value *before* the change.

The number of parking violations rose from 1980 violations to 2277. Find the percent of increase.

$$2277 - 1980 = 297 \quad \text{Increase}$$

$$\begin{array}{r} \text{Increase} \rightarrow \\ \text{Original value} \rightarrow \end{array} \frac{297}{1980} = \frac{\textbf{percent}}{100}$$

Solve the proportion to find that the percent = 15, so the percent of increase is 15%.

6.7 Finding Simple Interest

Use the formula $I = p \cdot r \cdot t$

Interest = principal • rate • time

Time (t) is in years. When the time is given in months, use a fraction with 12 in the denominator because there are 12 months in a year.

\$2800 is deposited at 2% for 3 months. Find the amount of interest.

$$\begin{aligned} I &= p \cdot r \cdot t \\ &= \underbrace{(2800)(0.02)}\left(\frac{3}{12}\right) \\ &= \quad (56) \quad \left(\frac{1}{4}\right) = \frac{(56)(1)}{4} = \$14 \end{aligned}$$

6.8 Finding Compound Amount and Compound Interest

There are three methods for finding the compound amount.

1. Calculate the interest for each compound interest period, then add it back to the principal.

2. Multiply the original deposit by 100% plus the compound interest rate.

3. Use the compound interest table to find the interest on \$1. Then, multiply the table value by the principal.

The interest is found with this formula.

Interest = compound amount − original deposit

Find the compound amount and interest if \$1500 is deposited at 5% interest for 3 years.

1.

	Interest	Compound Amount
Year 1	(\$1500)(0.05)(1) = **\$75**	\$1500 + **\$75** = **\$1575**
Year 2	(\$1575)(0.05)(1) = **\$78.75**	\$1575 + **\$78.75** = **\$1653.75**
Year 3	(\$1653.75)(0.05) ≈ **\$82.69**	\$1653.75 + **\$82.69** = **\$1736.44**

2. (\$1500)(1.05)(1.05)(1.05) ≈ \$1736.44

↑ Original deposit; (1.05)(1.05)(1.05): 100% + 5% = 105% = 1.05; ↑ Compound amount

3. Locate 5% across the top of the table and 3 periods at the left. The table value is **1.1576**.

compound amount = (\$1500)(**1.1576**) = **\$1736.40***

interest = \$1736.40 − \$1500 = \$236.40

* The difference in the compound amount results from rounding in the table.

Chapter 6 Review Exercises

6.1 *Write each percent as a decimal and each decimal as a percent.*

1. 35%
2. 150%
3. 99.44%
4. 0.085%
5. 3.15
6. 0.02
7. 0.875
8. 0.002

6.2 *Write each percent as a fraction or mixed number in lowest terms and each fraction as a percent.*

9. 15%
10. 37.5%
11. 175%
12. 0.25%
13. $\frac{3}{4}$
14. $\frac{5}{8}$
15. $3\frac{1}{4}$
16. $\frac{1}{200}$

Complete this chart.

Fraction	Decimal	Percent
$\frac{1}{8}$	**17.** ____	**18.** ____
19. ____	0.25	**20.** ____
21. ____	**22.** ____	180%

6.3 *Find the unknown value in the percent proportion* $\frac{\textit{part}}{\textit{whole}} = \frac{\textit{percent}}{100}$.

23. part = 25, percent = 10
24. whole = 480, percent = 5

Identify each component and then set up each problem using the percent proportion $\frac{\textit{part}}{\textit{whole}} = \frac{\textit{percent}}{100}$. *Do **not** try to solve for the unknown value.*

25. 35% of 820 mailboxes is 287 mailboxes.
26. 73 DVDs is what percent of 90 DVDs?
27. Find 14% of 160 mountain bikes.
28. 418 curtains is 16% of what number of curtains?
29. A golfer lost three of his eight golf balls. What percent were lost?
30. Only 88% of the door keys cut will operate properly. If there are 1280 keys cut, find the number of keys that will operate properly.

6.4 *Find the part using the percent proportion or the multiplication shortcut.*

31. 18% of 950 programs

32. 60% of 1450 reference books

33. 0.6% of 5200 acres

34. 0.2% of 1400 kilograms

Find the whole using the percent proportion.

35. 105 crates is 14% of what number of crates?

36. 348 test tubes is 15% of what number of test tubes?

37. 677.6 miles is 140% of what number of miles?

38. 2.5% of what number of cases is 425 cases?

Find the percent using the percent proportion. Round percent answers to the nearest tenth if necessary.

39. 649 tulip bulbs is what percent of 1180 tulip bulbs?

40. What percent of 1620 dinner rolls is 85 dinner rolls?

41. What percent of 380 pairs of socks is 36 pairs?

42. What percent of 650 soup cans is 200 soup cans?

6.1–6.4 *Solve each application problem. Round percent answers to the nearest tenth if necessary.*

43. Last year there was a total of 63 shark attacks on humans in the world (6 were fatal). This year there was an increase of 25.4% in the number of these attacks. Find the number of shark attacks on humans this year. (Round to the nearest whole number.) (*Source*: MSNBC.com)

44. Each week 3200 people shop at the three Crescent City Farmers Markets in New Orleans, Louisiana. If 896 of these shoppers are seniors, what percent of the shoppers are seniors? (*Source: USA Today.*)

6.5 *Use the percent equation to answer each question.*

45. 32% of $454 is what amount?

46. 155% of 120 trucks is how many trucks?

47. 0.128 ounce is what percent of 32 ounces?

48. 304.5 meters is what percent of 174 meters?

49. 33.6 miles is 28% of what number of miles?

50. $92 is 16% of what amount?

6.6 *Find the amount of sales tax or the tax rate and the total cost. Round to the nearest cent if necessary.*

Cost of Item	Tax Rate	Amount of Tax	Total Cost
51. $630	5%	________	________
52. $780	________	$58.50	________

Find the commission earned or the rate of commission.

Sales	Rate of Commission	Commission
53. $3450	8%	________
54. $65,300	________	$3265

Find the amount or rate of discount and the sale price. Round to the nearest cent if necessary.

Original Price	Rate of Discount	Amount of Discount	Sale Price
55. $112.50	30%	________	________
56. $252	________	$63	________

6.7 *Find the simple interest due on each loan.*

	Principal	Rate	Time in Years	Interest
57.	$200	4%	1	________
58.	$1080	5%	$1\frac{1}{4}$	________

Find the simple interest paid on each investment.

	Principal	Rate	Time in Months	Interest
59.	$400	$3\frac{1}{2}\%$	3	________
60.	$1560	$6\frac{1}{2}\%$	18	________

Find the total amount due on each simple interest loan.

	Principal	Rate	Time	Total Amount Due
61.	$750	$5\frac{1}{2}\%$	2 years	________
62.	$1560	3%	9 months	________

6.8 *Find the compound amount and the interest. Interest is compounded annually. You may use the compound interest table. Round answers to the nearest cent if necessary.*

	Principal	Rate	Time in Years	Compound Amount	Interest
63.	$4000	3%	10	________	________
64.	$1870	4%	4	________	________
65.	$3600	$4\frac{1}{2}\%$	3	________	________
66.	$12,500	$5\frac{1}{2}\%$	5	________	________

Mixed Review Exercises

Find the unknown value in the percent proportion $\frac{part}{whole} = \frac{percent}{100}$.

67. whole = 80, percent = 15

68. part = 738, percent = 45

Use the percent proportion or percent equation to answer each question.

69. 12% of 194 meters is how many meters?

70. 327 cars is what percent of 218 cars?

71. 0.6% of $85 is what amount?

72. 99 employees is 5% of what number of employees?

73. 76 chickens is what percent of 190 chickens?

74. 214.484 liters is 43% of what number of liters?

Write each percent as a decimal and each decimal as a percent.

75. 55%

76. 300%

77. 5

78. 4.71

79. 8.6%

80. 0.621

81. 0.375%

82. 0.0006

Write each percent as a fraction in lowest terms and each fraction or mixed number as a percent.

83. $\frac{3}{4}$

84. 42%

85. 87.5%

86. $\frac{3}{8}$

87. $32\frac{1}{2}\%$

88. $\frac{3}{5}$

89. 0.25%

90. $3\frac{3}{4}$

Solve each application problem. Round percent answers to the nearest tenth and money answers to the nearest cent if necessary.

91. Eva Jacob deposits $20,500 in her credit union savings account. If she earns $6\frac{1}{2}\%$ simple interest for 30 months, how much interest will be earned?

92. Hap Pishke, owner of Mardi Gras Barbers, borrows $14,750 to remodel his shop. He agrees to an 8% simple interest rate and will repay the loan in 18 months. Find the total amount due.

93. A study of 1005 adults found that 965 of them have a smoke alarm in their home and 461 have a carbon monoxide detector. (*Source*: Liberty Mutual International Association of Firefighters.)

(a) Find the percent of adults who have a smoke alarm in their home. Round to the nearest tenth of a percent.

(b) Find the percent of adults who have a carbon monoxide detector in their home. Round to the nearest tenth of a percent.

94. Alan Zagorin borrows \$148,000 to expand his business, located at the end of historic Route 66. The loan has an interest rate of 5% compounded annually and will be repaid in 4 years.

(a) Find the compound amount of this loan at the end of 4 years. Do not use the table.

(b) Find the amount of interest that he owes.

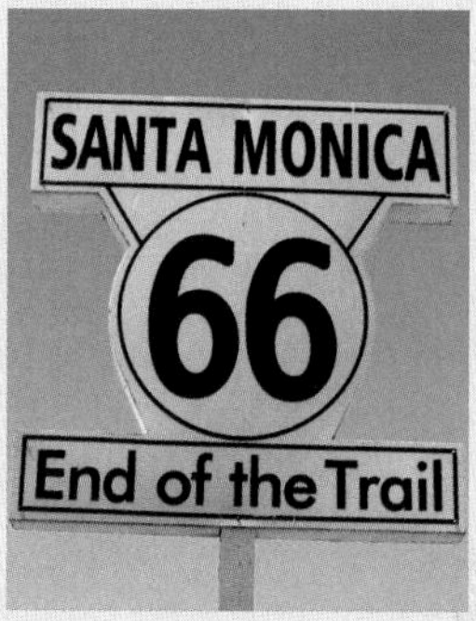

95. Tom Dugally, a real estate agent, sold two properties, one for \$125,000 and the other for \$290,000. He receives a commission of $1\frac{1}{2}\%$ of total sales. Find the commission that he earned.

96. Vending machines on campus must include healthy food choices such as fruits, fruit juices, and healthy snacks. Sales of healthy foods in the vending machines increased from 4320 items last month to 5107 items this month. Find the percent of increase.

97. A Sears Kenmore washer/dryer set priced at \$958 is marked down 18%. If the sales tax is 8%, find the cost of the washer/dryer set, including the sales tax.

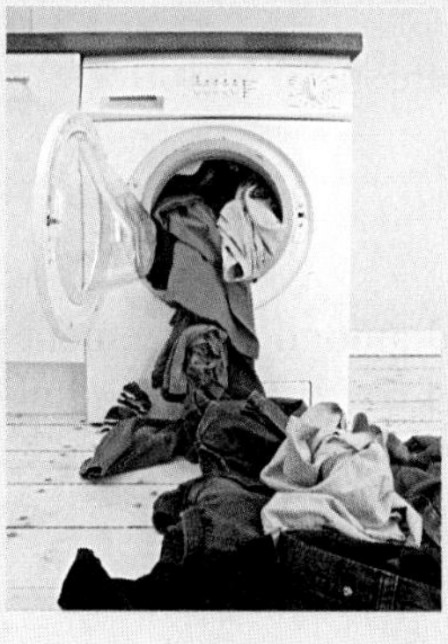

98. In a recent insurance company study of boaters who had lost items overboard, 88 boaters or 8% said that they lost their cell phones. Find the total number of boaters in the survey. (*Source:* Progressive Groups of Insurance Companies.)

99. Jack and Jill Ahearn begin to budget 25% for rent, 20% for food, 7% for education, 5% for clothing, 10% for transportation, 12% for travel and recreation, 4% for miscellaneous, and the remainder for savings. Jack takes home \$2850 per month, and Jill takes home \$42,300 per year. How much money will the couple save in a year?

100. The mileage on a hybrid car dropped from 42.8 miles per gallon in the city to 28.5 miles per gallon on the highway. Find the percent of decrease.

Chapter 6 Test

CHAPTER Test Prep VIDEO

The Chapter Test Prep Videos with test solutions are available on DVD, in MyMathLab, and on YouTube —search "LialDevMath" and click on "Channels."

Write each percent as a decimal and each decimal as a percent.

1. 65%

2. 0.8

3. 1.75

4. 0.875

5. 300%

6. 2%

Write each percent as a fraction in lowest terms.

7. 12.5%

8. 0.25%

Write each fraction or mixed number as a percent.

9. $\frac{3}{5}$

10. $\frac{5}{8}$

11. $2\frac{1}{2}$

Solve each problem.

12. 32 sacks is 4% of what number of sacks?

13. $680 is what percent of $3400?

14. There are still 100,000 households in the United States that do not have electricity. If this is 0.08% of the homes, find the total number of households. (*Source: Time* magazine.)

15. The price of a diamond engagement ring is $3240 plus sales tax of $6\frac{1}{2}$%. Find the total cost of the engagement ring including sales tax.

16. An insurance company pays its salespeople on commission. If a commission of $628 is earned on insurance sales of $7850, find the rate of commission.

17. Attendance at the homecoming game decreased from 5760 fans last year to 4320 fans this year. Find the percent of decrease.

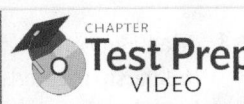

18. A problem includes last year's salary, this year's salary, and asks for the percent of increase. Explain how you would identify the part, the whole, and the percent in the problem. Show the percent proportion that you would use.

19. Write the formula used to find interest. Explain the difference in what to do if the time is expressed in months or in years. Write a problem that involves finding interest for 9 months and another problem that involves finding interest for $2\frac{1}{2}$ years. Use your own numbers for the principal and the rate. Show how to solve your problems.

Find the amount of discount and the sale price.

	Original Price	Rate of Discount	Amount of Discount	Sale Price
20.	$96	12%	________	________
21.	$280	32.5%	________	________

Find the simple interest on each loan.

	Principal	Rate	Time	Simple Interest
22.	$4200	6%	$1\frac{1}{2}$ years	________
23.	$6400	9%	4 months	________

24. Sita Lalchandani takes out a short-term loan of $19,200 to pay for her daughter's medical school expenses. The interest rate on the loan is 7%, and the loan is for 15 months. Find the total amount needed to repay the loan.

25. The River City School PTA Emergency Fund deposited $4000 at 6% interest compounded annually. Two years after the first deposit, they deposit another $5000, also at 6% interest compounded annually. Use the compound interest table.

(a) What total amount will they have 4 years after their first deposit? Round to the nearest dollar.

(b) What amount of interest will they have earned?

Math in the Media

EDUCATIONAL TAX INCENTIVES

The government sponsors tax incentive programs to make education more affordable. To qualify for the programs, you must have an adjusted gross income below a certain level (most recently $90,000). You can find specific information at the Internal Revenue Service website (irs.gov).

- The Hope Scholarship offers 100% of the first $2000 spent for certain expenses, such as tuition and books, during the first year of college, plus 25% of the next $2000. The scholarship money is payable as a tax refund. The student cannot have completed the first two years of post-secondary education and must meet certain educational goals and workload criteria.

Suppose you are paying your own educational costs, and your adjusted gross income meets the guidelines to qualify for the Hope Scholarship. Your goals are to earn an Associate of Arts degree from a community college and then transfer to a state university to complete a Bachelor's degree. Tuition costs for resident students at American River Community College in California are used as an example of educational expenses.

Residents of the college district pay an enrollment fee of $46 per semester hour plus a parking permit fee of $30 each semester. Assume that you must study a total of 15 semester hours in developmental work in mathematics, reading, and writing, and an additional 60 semester hours to complete an Associate of Arts degree. You decide to limit your course load to 15 credit hours each semester. Assume that one course is 3 semester hours, and you will have to purchase books at an approximate cost of $95 per course.

1. How many semesters and how many courses will it take you to finish the requirements for an Associate of Arts degree?

2. What is the total cost to complete the Associate of Arts degree for (**a**) books and (**b**) tuition and fees?

3. (**a**) Calculate the total cost for enrollment fees, parking fees, and books to complete the Associate of Arts degree (5 semesters). (**b**) What is the maximum tax incentive payable to you under the Hope Scholarship?

7 Geometry

In this chapter, you'll see how geometry is used in many different ways, both in our personal lives and on the job. For example, predicting hurricanes uses the geometry of circles.

7.1 Lines and Angles

OBJECTIVES

1. Identify and name lines, line segments, and rays.
2. Identify parallel and intersecting lines.
3. Identify and name angles.
4. Classify angles as right, acute, straight, or obtuse.
5. Identify perpendicular lines.
6. Identify complementary angles and supplementary angles and find the measure of a complement or supplement of a given angle.
7. Identify congruent angles and vertical angles and use this knowledge to find the measures of angles.
8. Identify corresponding angles and alternate interior angles and use this knowledge to find the measures of angles.

Geometry starts with the idea of a point. A **point** can be described as a location in space. It has no length or width. A point is represented by a dot and is named by writing a capital letter next to the dot.

P

Point P

OBJECTIVE 1 Identify and name lines, line segments, and rays. A **line** is a straight row of points that goes on forever in both directions. A line is drawn by using arrowheads to show that it never ends. The line is named by using the letters of any two points on the line.

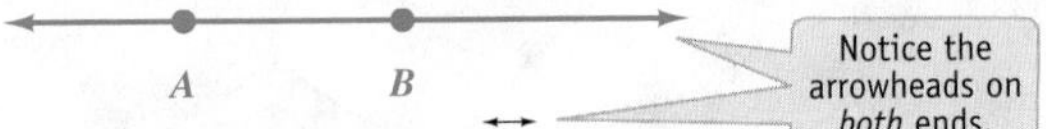
A
B
Notice the arrowheads on *both* ends.

Line AB, written $\overleftrightarrow{AB}$

A piece of a line that has two endpoints is called a **line segment.** A line segment is named for its endpoints. The segment with endpoints P and Q is shown below. It can be named $\overline{PQ}$ or $\overline{QP}$.

P
Q
There are *no* arrowheads.

Line segment PQ, written $\overline{PQ}$

A **ray** is a part of a line that has only one endpoint and goes on forever in one direction. A ray is named by using the endpoint and some other point on the ray. The endpoint is always mentioned first.

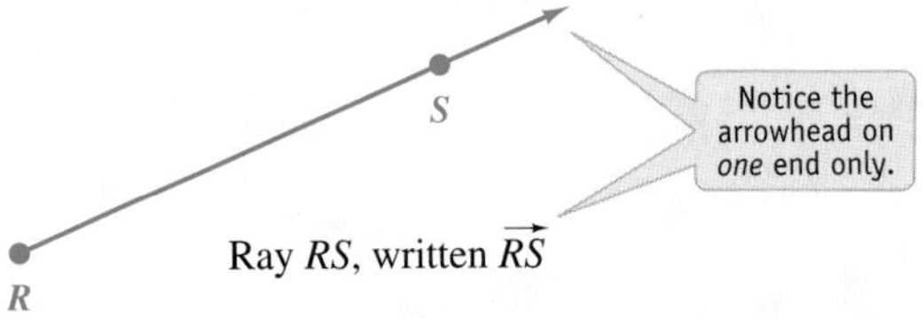
S
R
Notice the arrowhead on *one* end only.

Ray RS, written $\overrightarrow{RS}$

EXAMPLE 1 Identifying and Naming Lines, Rays, and Line Segments

Identify each figure below as a line, line segment, or ray and name it using the appropriate symbol.

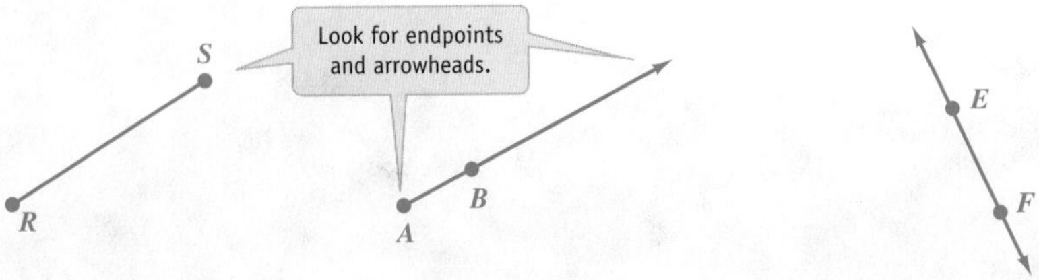
Look for endpoints and arrowheads.
S
R
A
B
E
F

(a) **(b)** **(c)**

Figure **(a)** has two endpoints, so it is a *line segment* named $\overline{RS}$ or $\overline{SR}$.
Figure **(b)** starts at point A and goes on forever in one direction, so it is a *ray* named $\overrightarrow{AB}$.
Figure **(c)** goes on forever in both directions, so it is a *line* named $\overleftrightarrow{EF}$ or $\overleftrightarrow{FE}$.

◀ **Work Problem 1 at the Side.**

1 Identify each figure as a line, line segment, or ray, and name it.

(a) *Hint:* There are *two* endpoints.

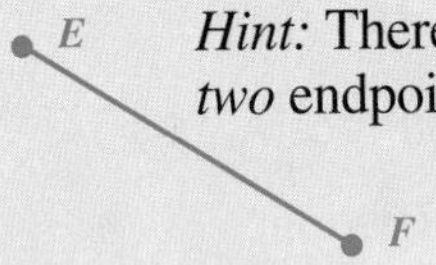
E
F

(b)

R
S

(c)

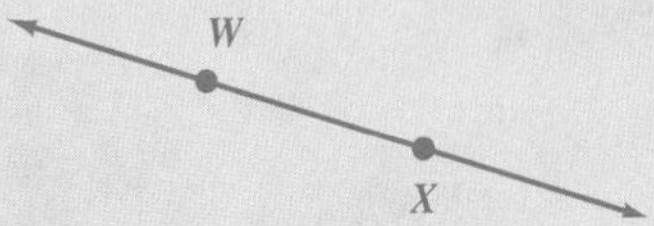
W
X

(d)

C
D

Answers

1. **(a)** line segment named $\overline{EF}$ or $\overline{FE}$
   (b) ray named $\overrightarrow{SR}$
   (c) line named $\overleftrightarrow{WX}$ or $\overleftrightarrow{XW}$
   (d) line segment named $\overline{CD}$ or $\overline{DC}$

OBJECTIVE 2 Identify parallel and intersecting lines. A *plane* is an infinitely large, flat surface. A floor or a wall is part of a plane. Lines that are in the *same plane,* but that never intersect (never cross), are called **parallel lines,** while lines that cross are called **intersecting lines.** (Think of an intersection, where two streets cross each other.)

EXAMPLE 2 Identifying Parallel and Intersecting Lines

Label each pair of lines as appearing to be parallel or as intersecting.

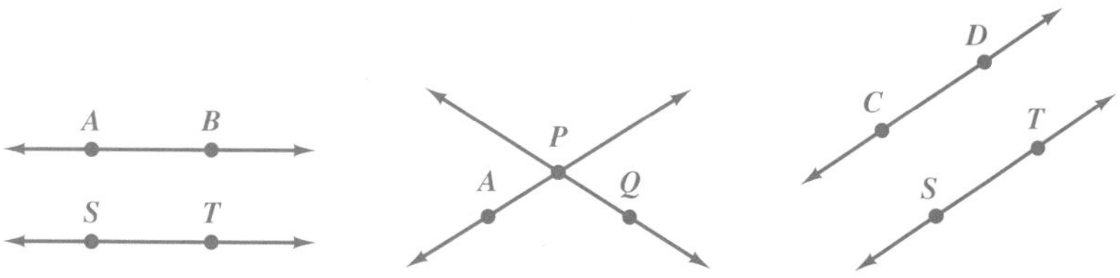

(a) **(b)** **(c)**

The lines in Figures **(a)** and **(c)** do not intersect; they appear to be *parallel lines.*

The lines in Figure **(b)** cross at P, so they are *intersecting lines.*

CAUTION

Appearances may be deceiving! Do not assume that lines are parallel unless it is stated that they are parallel.

Work Problem 2 at the Side. ▶

OBJECTIVE 3 Identify and name angles. An **angle** is made up of two rays that start at a common endpoint. This common endpoint is called the *vertex.*

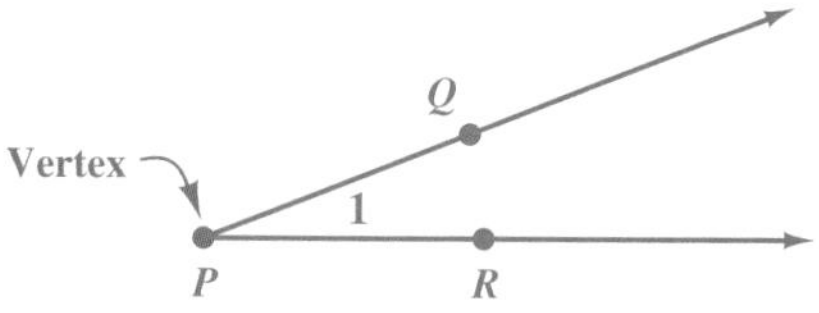

$\overrightarrow{PQ}$ and $\overrightarrow{PR}$ are called the *sides* of the angle. The angle can be named in four different ways, as shown below.

$\angle 1$ $\quad$ $\angle P$ $\quad$ $\angle QPR$ $\quad$ $\angle RPQ$

$\angle P$: Vertex alone

$\angle QPR$, $\angle RPQ$: Vertex in the middle

Naming an Angle

To name an angle, write the vertex alone or write the vertex in the middle of two other points, one from each side. If two or more angles have the *same vertex,* as in **Example 3** on the next page, do *not* use the vertex alone to name an angle.

VOCABULARY TIPS

Parallel *Hint:* The double "**ll**" in **parallel** can remind you that parallel lines never cross.

Intersecting The root word **sect** means "to cut." Intersecting lines cut across each other.

2 Label each pair of lines as appearing to be parallel or as intersecting.

GS **(a)**

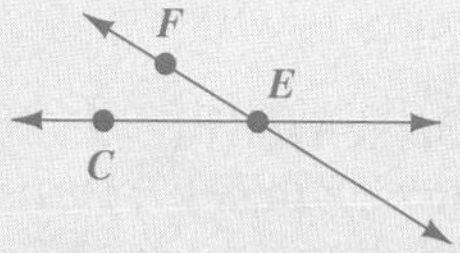

These lines cross, so they are ______ lines.

(b)

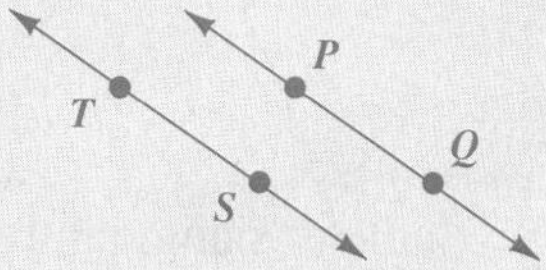

(c)

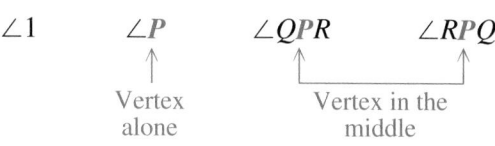

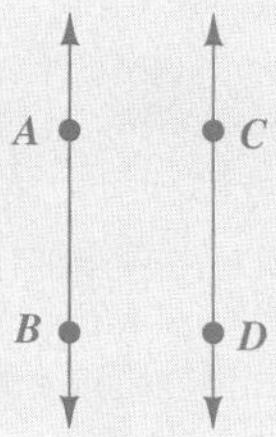

Answers

2. **(a)** intersecting **(b)** appear to be parallel **(c)** appear to be parallel

3 (a) Name the highlighted angle in three different ways.

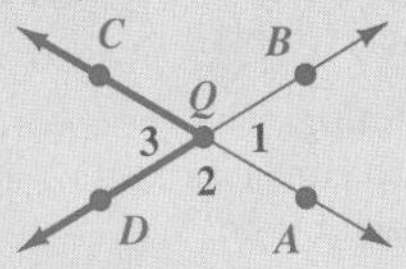

(b) Darken the rays that make up $\angle ZTW$.

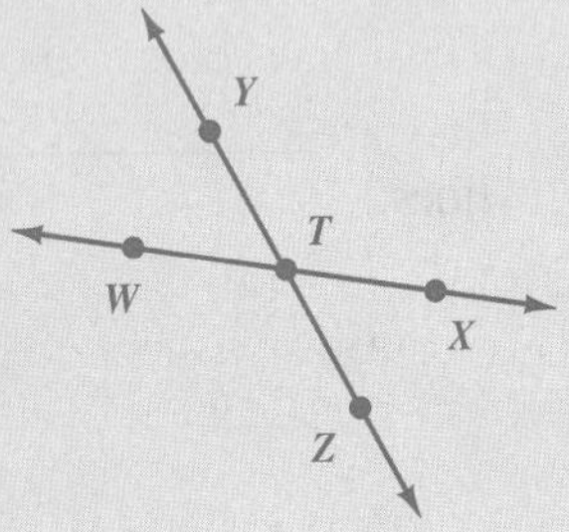

(c) Name this angle in four different ways.

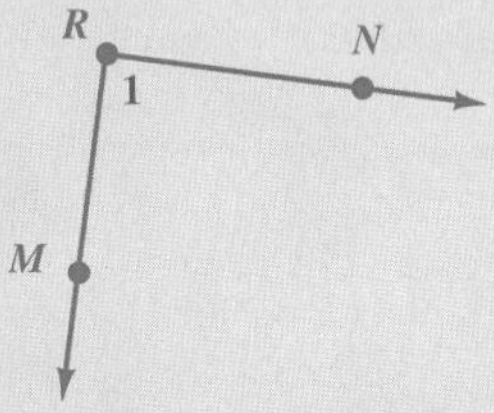

Answers

3. **(a)** $\angle 3$, $\angle CQD$, $\angle DQC$

(b)

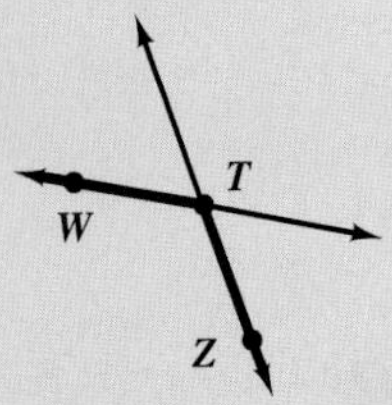

(c) $\angle 1$, $\angle R$, $\angle MRN$, $\angle NRM$

EXAMPLE 3 Identifying and Naming an Angle

Name the highlighted angle in three different ways.

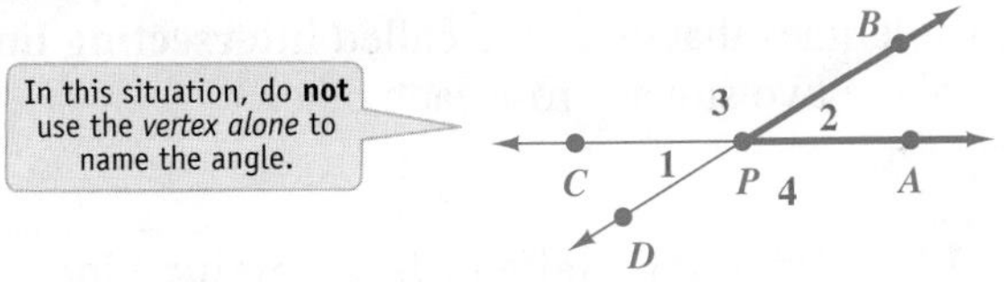

The angle can be named $\angle BPA$, $\angle APB$, or $\angle 2$. It *cannot* be named $\angle P$, using the vertex alone, because four different angles have P as their vertex.

◀ Work Problem 3 at the Side.

OBJECTIVE 4 Classify angles as right, acute, straight, or obtuse. Angles can be measured in **degrees.** The symbol for degrees is a small, raised circle °. Think of the minute hand on a clock as a ray of an angle. Suppose it is at 12:00. During one hour of time, the minute hand moves around in a complete circle. It moves 360 ***degrees***, or 360°. In half an hour, at 12:30, the minute hand has moved halfway around the circle, or 180°. An angle of 180° is called a **straight angle.** When two rays go in opposite directions and form a straight line, then the rays form a straight angle.

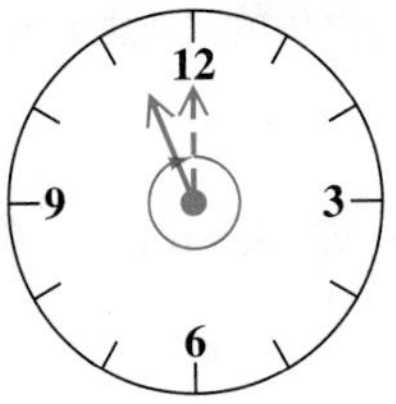

Complete circle
360°

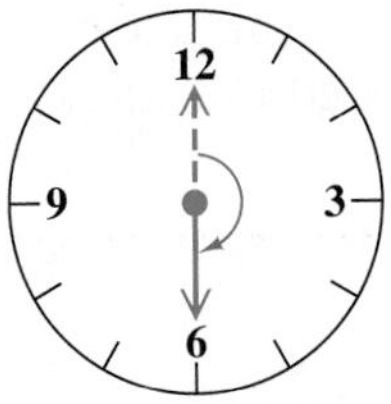

Straight angle
(half a circle)
180°

In a quarter of an hour, at 12:15, the minute hand has moved $\frac{1}{4}$ of the way around the circle, or 90°. An angle of 90° is called a **right angle.** The rays of a right angle form one corner of a square. So, to show that an angle is a **right angle**, we draw a **small square** at the vertex.

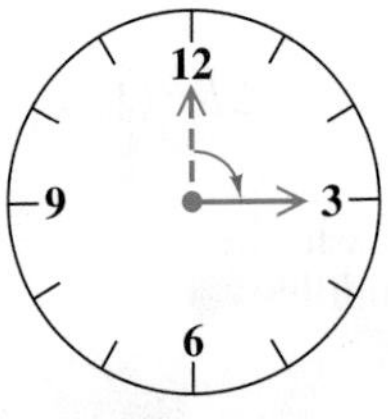

Right angle
($\frac{1}{4}$ of a circle)
90°

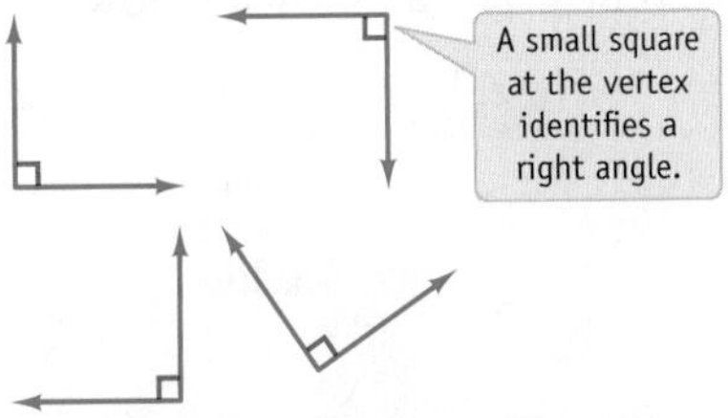

An angle that measures 1° is shown below. You can see that an angle of 1° is very small.

 1° angle

Some other terms used to describe angles are shown below.

Acute angles measure less than 90°.

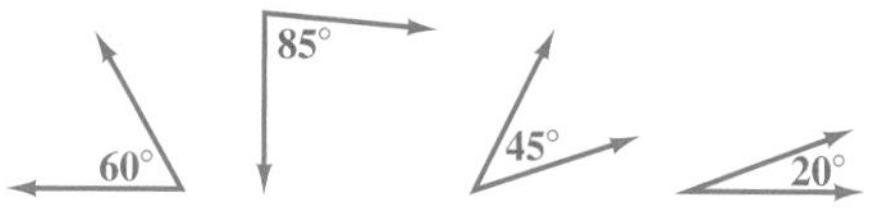

Examples of acute angles

Obtuse angles measure more than 90° but less than 180°.

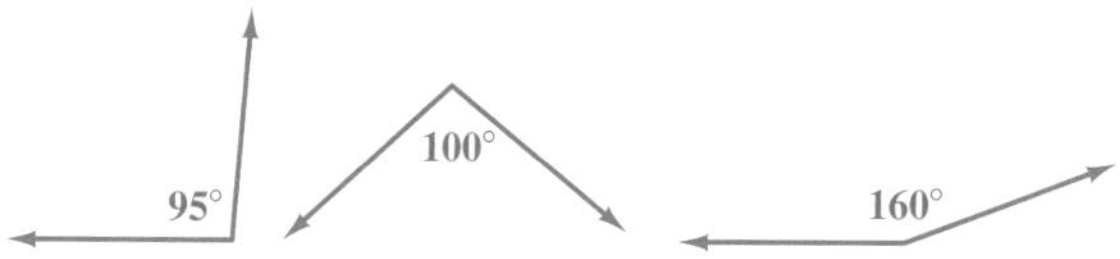

Examples of obtuse angles

Section 8.1 shows you how to use a tool called a *protractor* to measure the number of degrees in an angle.

Classifying Angles: Four Types of Angles

Acute angles measure less than 90°.

Right angles measure *exactly* 90°.

Obtuse angles measure more than 90° but less than 180°.

Straight angles measure *exactly* 180°.

Note

Angles can also be measured in radians, which you will learn about in a later math course.

EXAMPLE 4 Classifying an Angle

Label each angle as acute, right, obtuse, or straight.

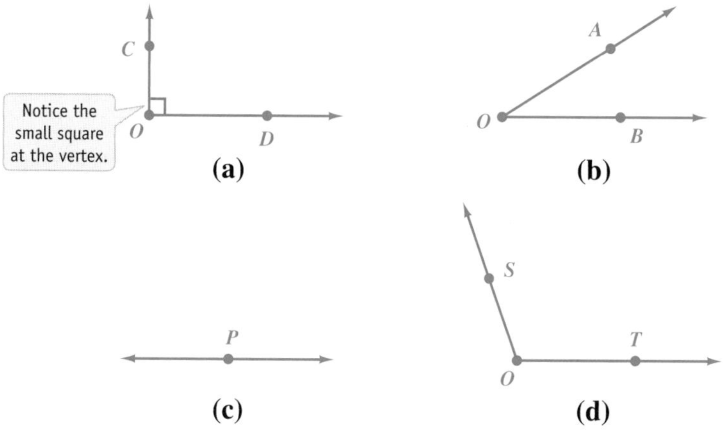

Figure **(a)** shows a *right angle* (exactly 90° and identified by a small square at the vertex).

Figure **(b)** shows an *acute angle* (less than 90°).

Figure **(c)** shows a *straight angle* (exactly 180°).

Figure **(d)** shows an *obtuse angle* (more than 90° but less than 180°).

Work Problem 4 at the Side. ▶

4 Label each angle as acute, right, obtuse, or straight. State the number of degrees in the right angle and in the straight angle.

(a)

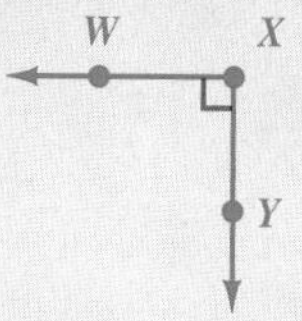

Hint: Notice the small red square at the vertex.

(b)

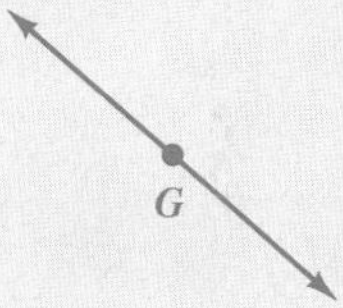

(c)

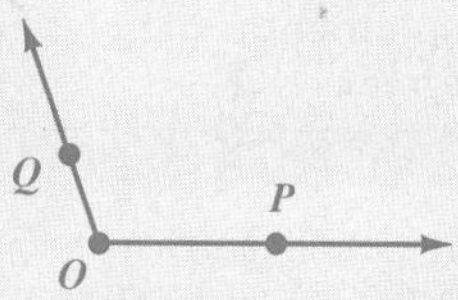

(d)

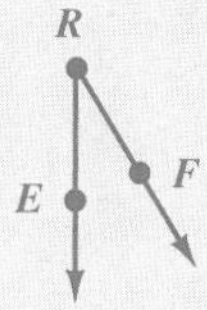

Answers

4. (a) right; 90° (b) straight; 180°
(c) obtuse (d) acute

❺ Which pair of lines is perpendicular? (*Hint:* Look for a right angle.) How can you describe the other pair of lines?

(a)

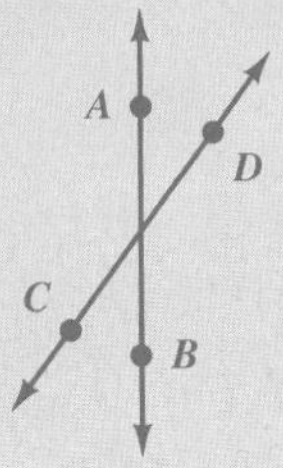

(b)

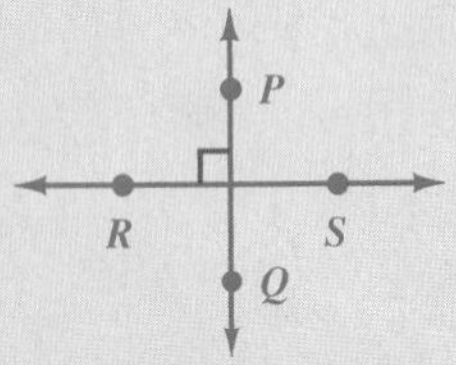

VOCABULARY TIP

Complementary angles *Hint:* The letter "**C**" in complementary can remind you of "corner," like a 90° angle. **Complementary angles** add up to 90 degrees.

❻ Identify each pair of complementary angles.

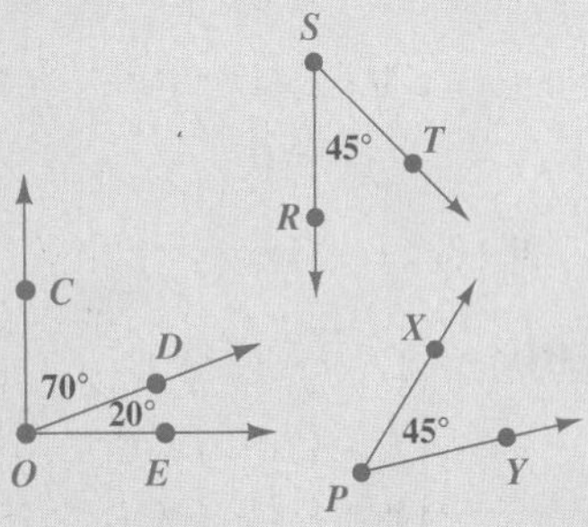

Answers

5. Figure **(b)** shows perpendicular lines; Figure **(a)** shows intersecting lines.
6. ∠*COD* and ∠*DOE*; ∠*RST* and ∠*XPY*

OBJECTIVE ❺ Identify perpendicular lines. Two lines are called **perpendicular lines** if they intersect to form a right angle.

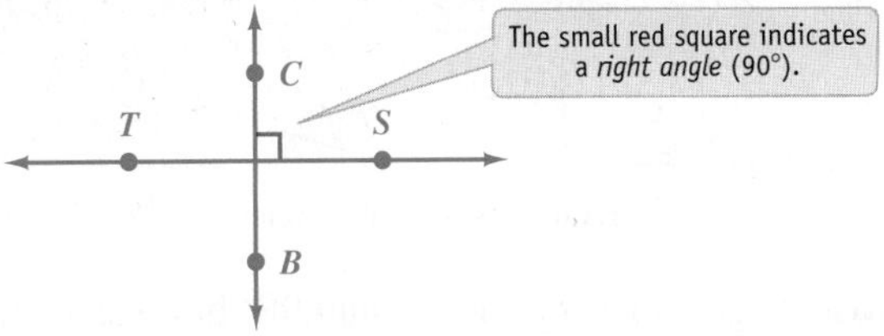

$\overleftrightarrow{CB}$ and $\overleftrightarrow{ST}$ are **perpendicular** lines because they intersect at right angles, as indicated by the small red square in the figure.

Perpendicular lines can be written in the following way: $\overleftrightarrow{CB} \perp \overleftrightarrow{ST}$.

EXAMPLE 5 **Identifying Perpendicular Lines**

Which pairs of lines are perpendicular?

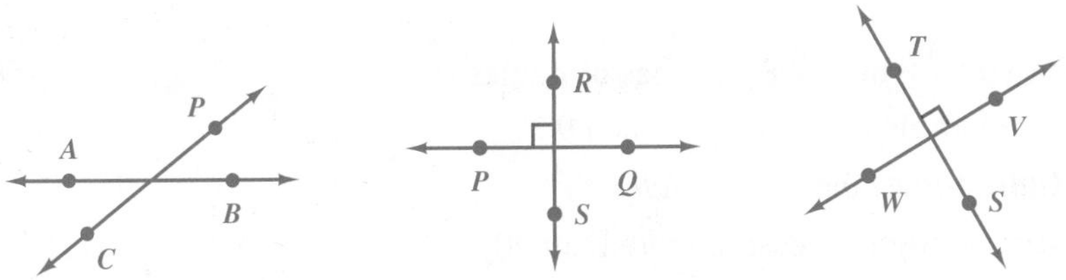

(a) **(b)** **(c)**

The lines in Figures **(b)** and **(c)** are *perpendicular* to each other, because they intersect at right angles.

The lines in Figure **(a)** are *intersecting lines,* but they are *not* perpendicular because they do *not* form a right angle.

◀ **Work Problem ❺ at the Side.**

OBJECTIVE ❻ Identify complementary angles and supplementary angles and find the measure of a complement or supplement of a given angle. Two angles are called **complementary angles** if the sum of their measures is 90°. If two angles are complementary, each angle is the *complement* of the other.

EXAMPLE 6 **Identifying Complementary Angles**

Identify each pair of complementary angles.

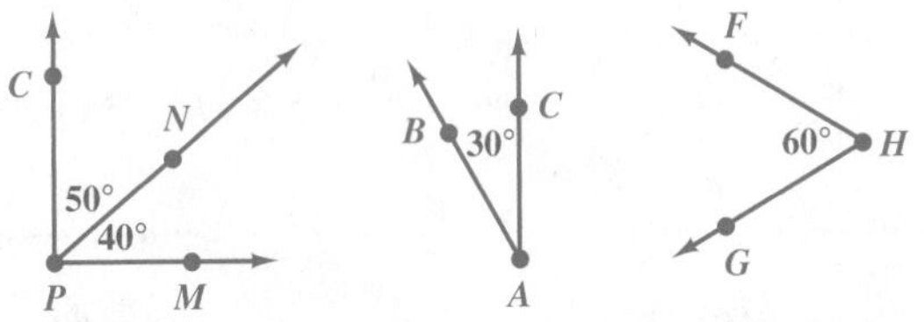

∠*MPN* (40°) and ∠*NPC* (50°) are complementary angles because

$$40° + 50° = \mathbf{90°}.$$

∠*CAB* (30°) and ∠*FHG* (60°) are complementary angles because

$$30° + 60° = \mathbf{90°}.$$

◀ **Work Problem ❻ at the Side.**

EXAMPLE 7 **Finding the Complement of Angles**

Find the complement of each angle.

Subtract from 90° to find the complement.

(a) 30°

Find the complement of 30° by subtracting. $90° - 30° = \mathbf{60°}$ ← Complement

(b) 75°

Find the complement of 75° by subtracting. $90° - 75° = \mathbf{15°}$ ← Complement

Work Problem 7 at the Side. ▶

Two angles are called **supplementary angles** if the sum of their measures is 180°. If two angles are supplementary, each angle is the *supplement* of the other.

EXAMPLE 8 **Identifying Supplementary Angles**

Identify each pair of supplementary angles.

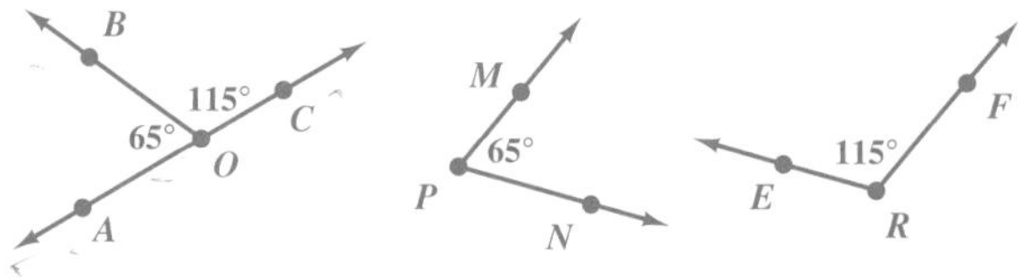

$\angle BOA$ and $\angle BOC$, because $65° + 115° = 180°$

$\angle BOA$ and $\angle ERF$, because $65° + 115° = 180°$

$\angle BOC$ and $\angle MPN$, because $115° + 65° = 180°$

$\angle MPN$ and $\angle ERF$, because $65° + 115° = 180°$

Work Problem 8 at the Side. ▶

EXAMPLE 9 **Finding the Supplement of Angles**

Find the supplement of each angle.

Subtract from 180° to find the supplement.

(a) 70°

Find the supplement of 70° by subtracting. $180° - 70° = \mathbf{110°}$ ← Supplement

(b) 140°

Find the supplement of 140° by subtracting. $180° - 140° = \mathbf{40°}$ ← Supplement

Work Problem 9 at the Side. ▶

OBJECTIVE 7 **Identify congruent angles and vertical angles and use this knowledge to find the measures of angles.** Two angles are called **congruent angles** if they measure the same number of degrees. If two angles are congruent, this is written as $\angle A \cong \angle B$ and read as, "angle *A* is congruent to angle *B*." Here is an example.

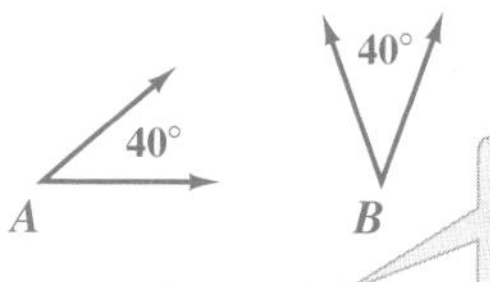

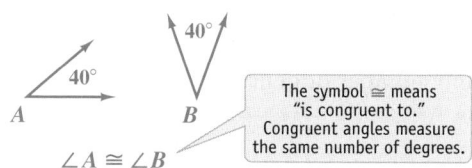

The symbol ≅ means "is congruent to." Congruent angles measure the same number of degrees.

$\angle A \cong \angle B$

Example of congruent angles

---

7 Find the complement of each angle.

GS **(a)** 35° $90° - 35° =$ ____

(b) 80°

VOCABULARY TIP

Supplementary angles *Hint:* The letter "**S**" in supplementary can remind you of "straight," like a 180° angle. **Supplementary angles** add up to 180 degrees.

8 Identify each pair of supplementary angles. (*Hint:* There are four pairs.)

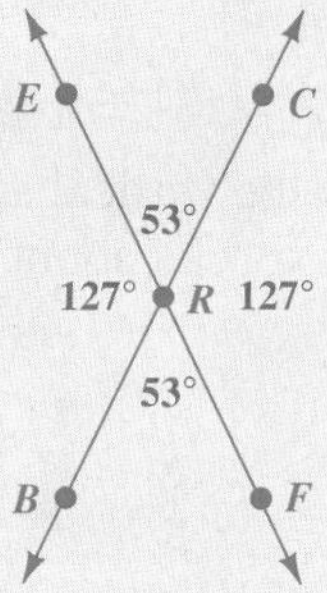

9 Find the supplement of each angle.

GS **(a)** 175° $180° - 175° =$ ____

(b) 30°

Answers

7. **(a)** 55° **(b)** 10°

8. $\angle CRF$ and $\angle BRF$; $\angle CRE$ and $\angle ERB$; $\angle BRF$ and $\angle BRE$; $\angle CRE$ and $\angle CRF$

9. **(a)** 5° **(b)** 150°

10 Identify the angles that are congruent.

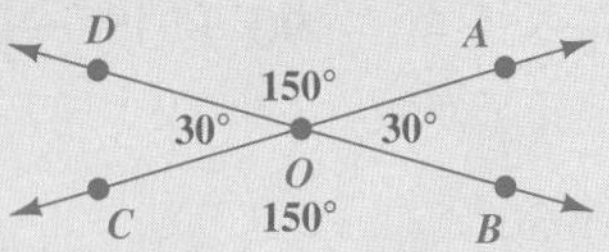

Hint: Congruent angles measure the same number of degrees.

11 Identify the vertical angles. What is special about vertical angles?

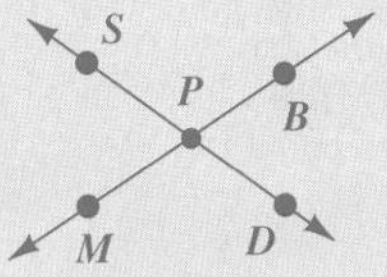

Answers

10. $\angle BOC \cong \angle AOD$; $\angle AOB \cong \angle DOC$

11. $\angle SPB$ and $\angle MPD$; $\angle BPD$ and $\angle SPM$; vertical angles are congruent (they measure the same number of degrees).

EXAMPLE 10 Identifying Congruent Angles

Identify the angles that are congruent.

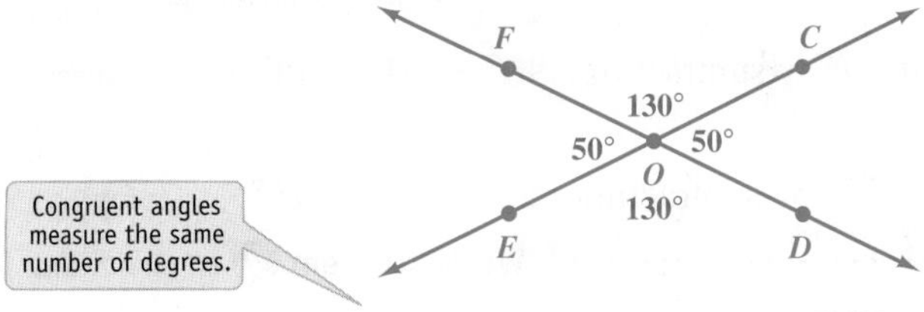

$\angle FOC \cong \angle EOD$ and $\angle COD \cong \angle EOF$

◀ **Work Problem 10 at the Side.**

Angles that share a common side and a common vertex are called *adjacent* angles, such as $\angle FOC$ and $\angle COD$ in **Example 10** above. Angles that do *not* share a common side are called *nonadjacent* angles. Two nonadjacent angles formed by two intersecting lines are called **vertical angles.**

EXAMPLE 11 Identifying Vertical Angles

Identify the vertical angles in this figure.

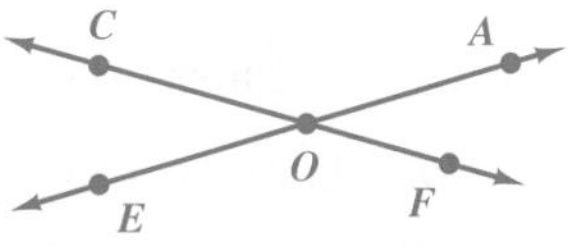

$\angle AOF$ and $\angle COE$ are vertical angles because they do *not* share a common side and they are formed by two intersecting lines ($\overleftrightarrow{CF}$ and $\overleftrightarrow{EA}$).

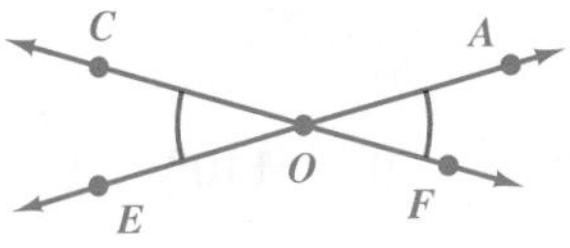

$\angle COA$ and $\angle EOF$ are also vertical angles.

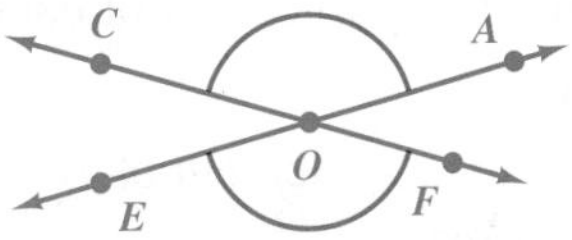

◀ **Work Problem 11 at the Side.**

Look back at **Example 10** at the top of the page. Notice that the two *congruent* angles that measure 130° are also *vertical* angles. Also, the two congruent angles that measure 50° are vertical angles. This illustrates the following property.

Vertical Angles Are Congruent

If two angles are *vertical* angles, they are *congruent;* that is, they measure the same number of degrees.

EXAMPLE 12 Finding the Measures of Vertical Angles

In the figure below, find the measure of each unlabeled angle.

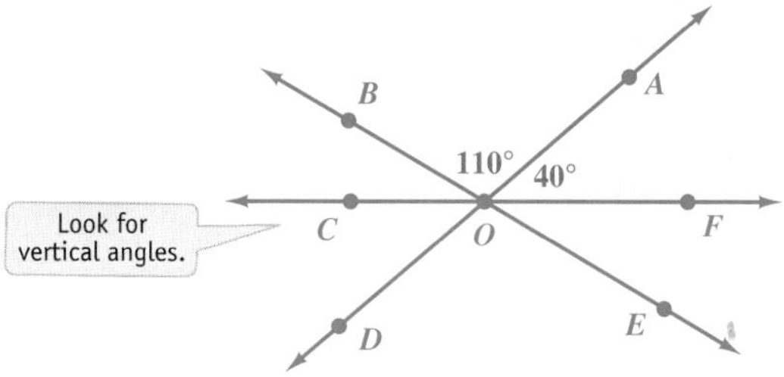

(a) $\angle COD$

$\angle COD$ and $\angle AOF$ are vertical angles, so they are congruent. This means they measure the same number of degrees.

The measure of $\angle AOF$ is 40° so the measure of $\angle COD$ is **40°** also.

(b) $\angle DOE$

$\angle DOE$ and $\angle BOA$ are vertical angles, so they are congruent.

The measure of $\angle BOA$ is 110° so the measure of $\angle DOE$ is **110°** also.

(c) $\angle COB$

Look at $\angle COB$, $\angle BOA$, and $\angle AOF$. Notice that $\overrightarrow{OC}$ and $\overrightarrow{OF}$ go in opposite directions. Therefore, $\angle COF$ is a straight angle and measures 180°. To find the measure of $\angle COB$, subtract the sum of the other two angles from 180°.

$$180° - (110° + 40°) = 180° - (150°) = 30°$$

The measure of $\angle COB$ is **30°.**

(d) $\angle EOF$

$\angle EOF$ and $\angle COB$ are vertical angles, so they are congruent. We know from part (c) above that the measure of $\angle COB$ is 30° so the measure of $\angle EOF$ is **30°** also.

Work Problem 12 at the Side. ▶

OBJECTIVE 8 Identify corresponding angles and alternate interior angles and use this knowledge to find the measures of angles. We can also find congruent angles (angles with the same measure) when two *parallel lines* are crossed by a third line, called a *transversal.* When a transversal crosses two *parallel* lines, eight angles are formed, as shown below. There are special names for certain pairs of angles.

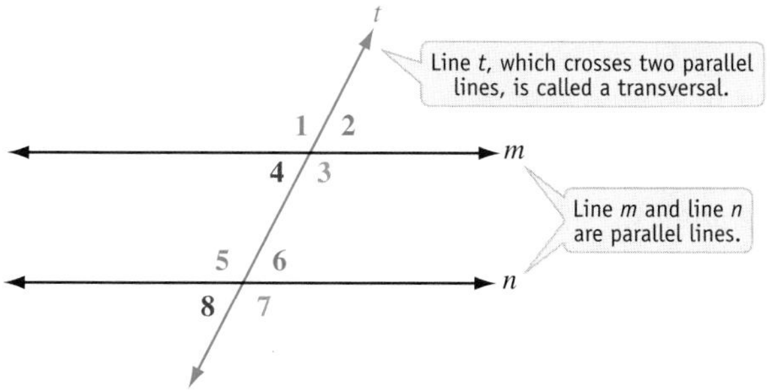

$\angle 1$ and $\angle 5$ are called **corresponding angles.** Notice that they are both on the same side of the transversal (line t) and in the same relative position.

12 In the figure below, find the measure of each unlabeled angle. Write the angle measures on the figure.

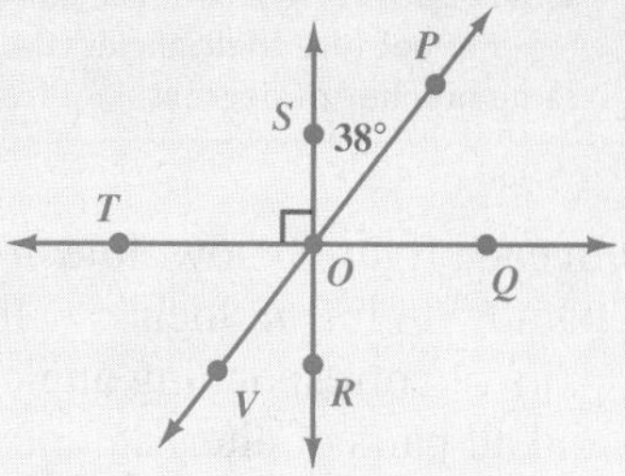

(a) $\angle TOS$

$\angle TOS$ has a small red square at the vertex, so it is a right angle and measures _____.

(b) $\angle QOR$

$\angle QOR$ and $\angle TOS$ are vertical angles, so they are congruent. That means $\angle QOR$ also measures _____.

(c) $\angle VOR$

(d) $\angle POQ$

(e) $\angle TOV$

Answers

12. **(a)** 90° **(b)** 90° **(c)** 38° **(d)** 52° **(e)** 52°

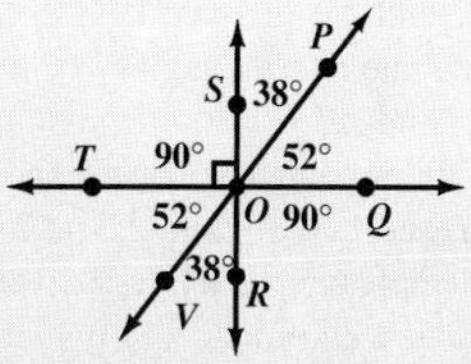

Corresponding angles are congruent, so $\angle 1$ and $\angle 5$ measure the same number of degrees. There are four pairs of corresponding angles.

$\angle 1$ and $\angle 5$ are corresponding angles, so $\angle 1 \cong \angle 5$.

$\angle 2$ and $\angle 6$ are corresponding angles, so $\angle 2 \cong \angle 6$.

$\angle 3$ and $\angle 7$ are corresponding angles, so $\angle 3 \cong \angle 7$.

$\angle 4$ and $\angle 8$ are corresponding angles, so $\angle 4 \cong \angle 8$.

When a transversal crosses two parallel lines, angles 3, 4, 5, and 6 are called *interior angles*. You can see that they are "inside" the *parallel* lines.

$\angle 3$ and $\angle 5$ are alternate interior angles.

$\angle 4$ and $\angle 6$ are alternate interior angles.

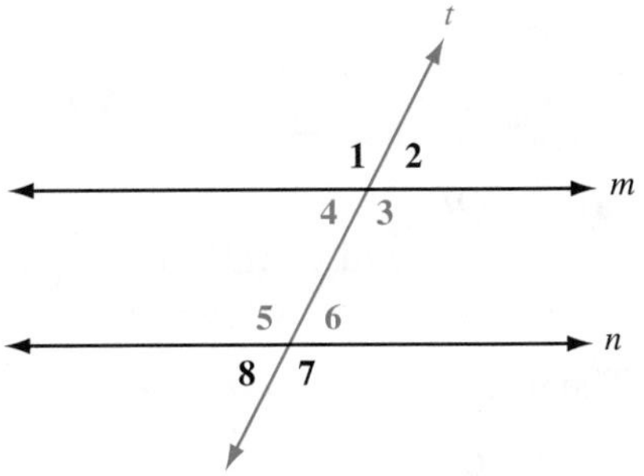

When two lines are *parallel*, then **alternate interior angles** *are congruent* (they have the same measure). Notice that alternate interior angles are on opposite (alternate) sides of the transversal.

$\angle 3 \cong \angle 5$ and $\angle 4 \cong \angle 6$

Angles Formed by Parallel Lines and a Transversal

When two parallel lines are crossed by a transversal:
1. Corresponding angles are congruent, and
2. Alternate interior angles are congruent.

EXAMPLE 13 **Identifying Corresponding Angles and Alternate Interior Angles**

In each figure, line *m* is parallel to line *n*. Identify all pairs of corresponding angles and all pairs of alternate interior angles.

(a)

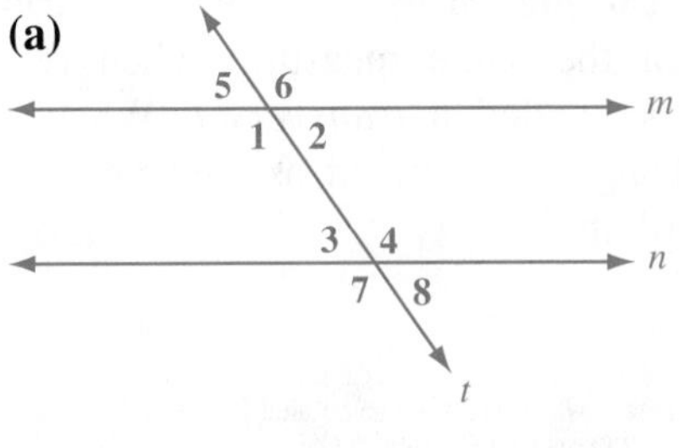

There are four pairs of corresponding angles:

$\angle 5$ and $\angle 3$ $\quad$ $\angle 6$ and $\angle 4$
$\angle 1$ and $\angle 7$ $\quad$ $\angle 2$ and $\angle 8$

Alternate interior angles:
$\angle 1$ and $\angle 4$ $\quad$ $\angle 2$ and $\angle 3$

(b)

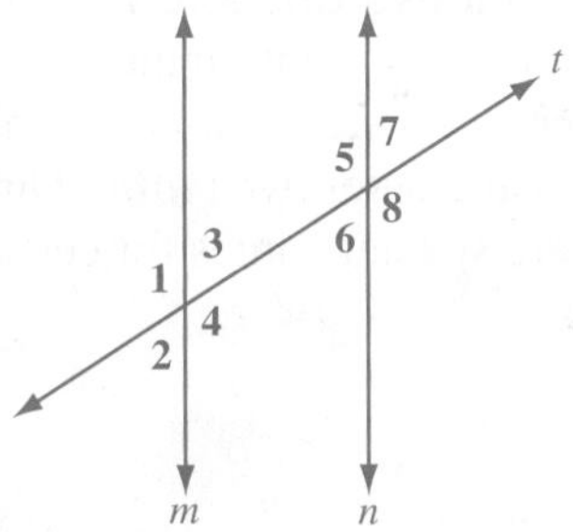

Corresponding angles:

$\angle 1$ and $\angle 5$ $\quad$ $\angle 3$ and $\angle 7$
$\angle 2$ and $\angle 6$ $\quad$ $\angle 4$ and $\angle 8$

Alternate interior angles:
$\angle 3$ and $\angle 6$ $\quad$ $\angle 4$ and $\angle 5$

◀ Work Problem 13 at the Side.

VOCABULARY TIPS

Trans- means "across." A **transversal** cuts *across* two parallel lines.

Co-* or *cor- means "together" or "with." **Corresponding** angles are paired *together* because their angles have the same number of degrees.

13 In each figure below, line *m* is parallel to line *n*. Identify all pairs of corresponding angles and all pairs of alternate interior angles.

GS **(a)**

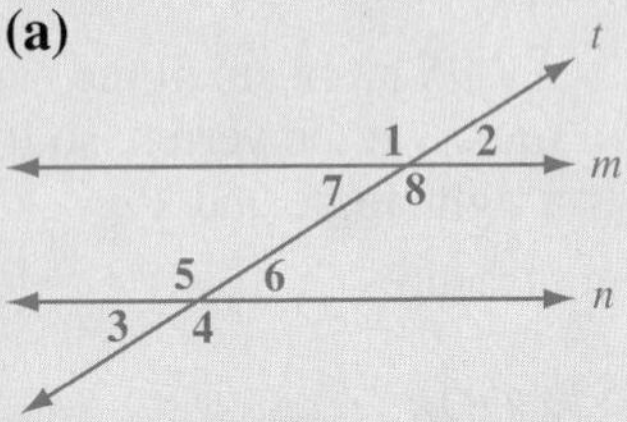

$\angle 1$ and $\angle 5$ are *corresponding* angles. (Find three more pairs.)

$\angle 7$ and $\angle 6$ are *alternate interior* angles. (Find one more pair.)

(b)

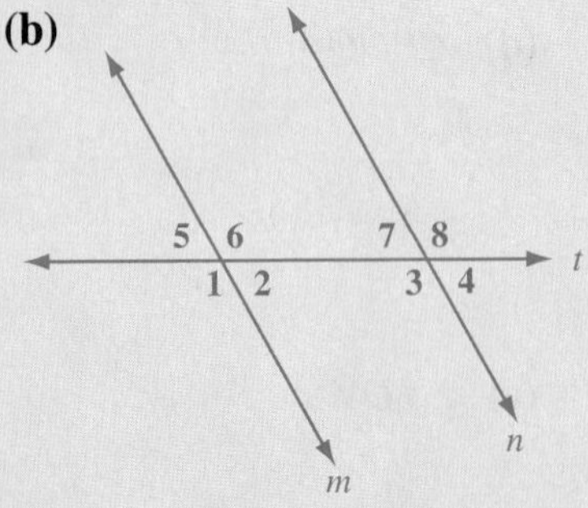

Answers

13. (a) corresponding angles: $\angle 1$ and $\angle 5$; $\angle 2$ and $\angle 6$; $\angle 7$ and $\angle 3$; $\angle 8$ and $\angle 4$
alternate interior angles: $\angle 7$ and $\angle 6$; $\angle 8$ and $\angle 5$

(b) corresponding angles: $\angle 5$ and $\angle 7$; $\angle 6$ and $\angle 8$; $\angle 1$ and $\angle 3$; $\angle 2$ and $\angle 4$
alternate interior angles: $\angle 6$ and $\angle 3$; $\angle 2$ and $\angle 7$

Recall that two angles are supplementary angles if the sum of their measures is 180°. Also remember that two rays that form a 180° angle form a straight line. Now you can combine your knowledge about supplementary angles with the information on parallel lines.

EXAMPLE 14 Working with Parallel Lines

In the figure at the right, line m is parallel to line n and the measure of $\angle 4$ is 70°. Find the measures of the other angles.

Look for corresponding angles and for alternate interior angles.

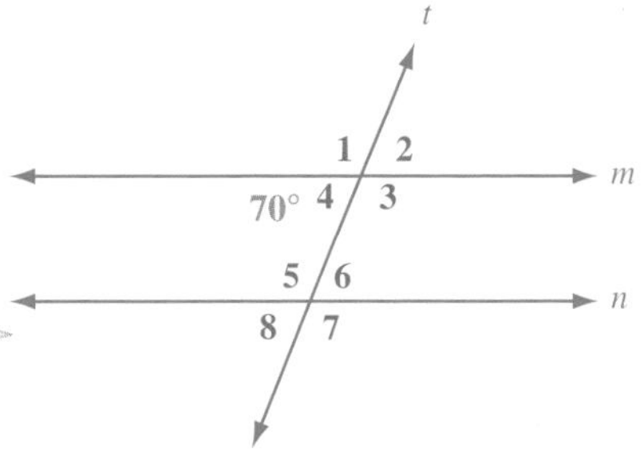

As you find the measure of each angle, write it on the figure.

$\angle 4 \cong \angle 8$ (corresponding angles), so the measure of $\angle 8$ is also 70°.

$\angle 4 \cong \angle 6$ (alternate interior angles), so the measure of $\angle 6$ is also 70°.

$\angle 6 \cong \angle 2$ (corresponding angles), so the measure of $\angle 2$ is also 70°.

Notice that the exterior sides of $\angle 4$ and $\angle 3$ form a straight line, that is, a straight angle of 180°. Therefore, $\angle 4$ and $\angle 3$ are supplementary angles and the sum of their measures is 180°. If $\angle 4$ is 70° then $\angle 3$ must be 110° because $180° - 70° = 110°$. So the measure of $\angle 3$ is 110°.

$\angle 3 \cong \angle 7$ (corresponding angles), so the measure of $\angle 7$ is also 110°.

$\angle 3 \cong \angle 5$ (alternate interior angles), so the measure of $\angle 5$ is also 110°.

$\angle 5 \cong \angle 1$ (corresponding angles), so the measure of $\angle 1$ is also 110°.

With the measures of all the angles labeled, you can double-check that each pair of angles that forms a straight angle also adds up to 180°.

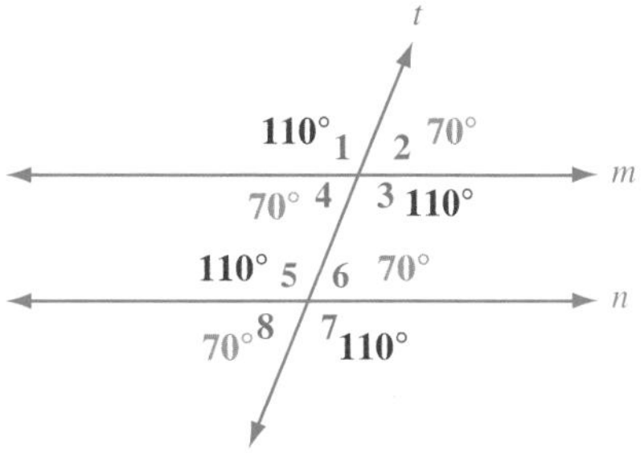

Work Problem 14 at the Side. ▶

14 In each figure below, line m is parallel to line n.

GS **(a)** The measure of $\angle 6$ is 150°. Find the measures of the other angles.

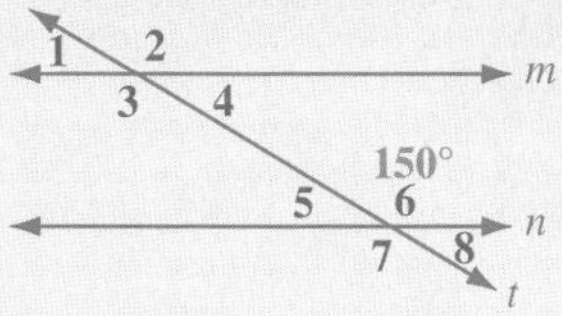

$\angle 6 \cong \angle 2$ (corresponding angles) so $\angle 2$ is also ____

(b) The measure of $\angle 1$ is 45°. Find the measures of the other angles.

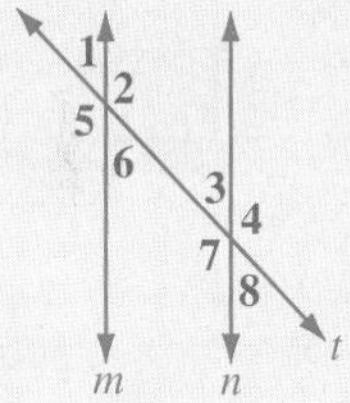

Answers

14. (a) $\angle 2$ is also 150°.

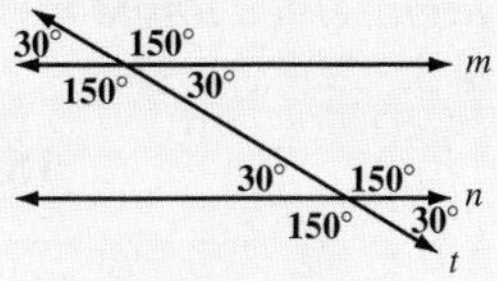

(b)

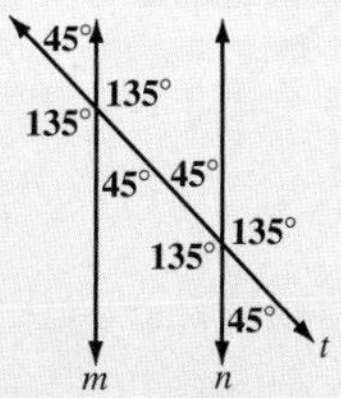

7.1 Exercises

FOR EXTRA HELP

Download the MyDashBoard App

MyMathLab®

1. **CONCEPT CHECK** Explain the difference between a line, a line segment, and a ray. Draw a picture of each one.

2. **CONCEPT CHECK** Explain the difference between acute, obtuse, straight, and right angles. Draw a picture of each type of angle.

Identify each figure as a line, line segment, *or* ray *and name it using the appropriate symbol.* ***See Example 1.***

3.
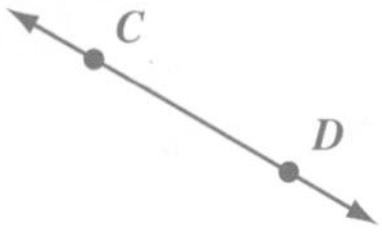

4.

5.
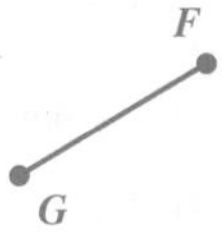

6.

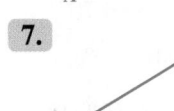

7.
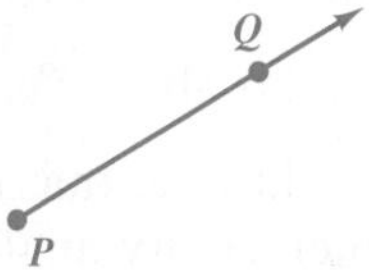

8.
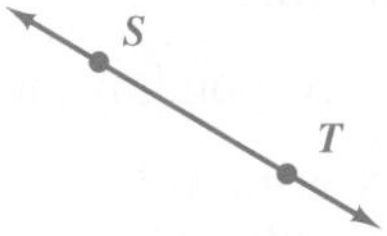

Label each pair of lines as appearing to be parallel, *as* perpendicular, *or as* intersecting. ***See Examples 2 and 5.***

9.
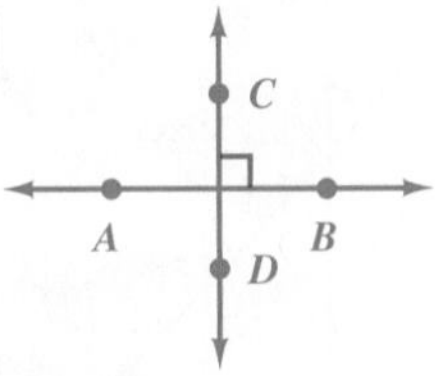

10.
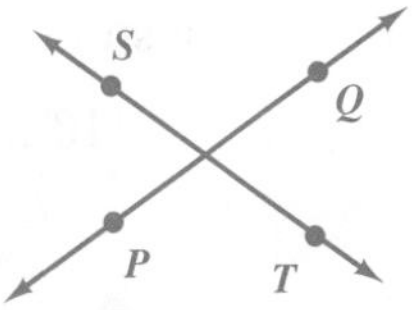

11.
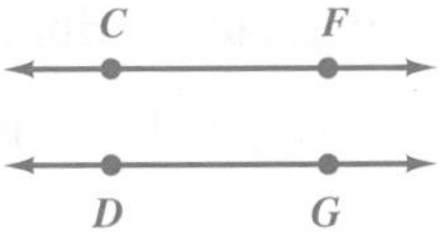

12.
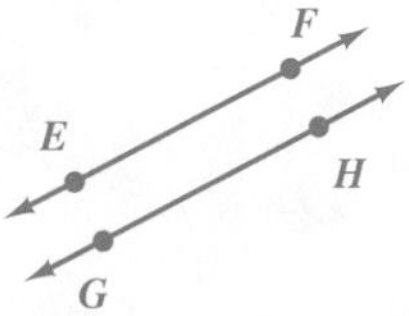

13.
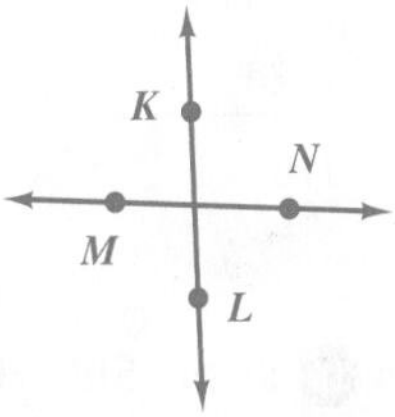

14.
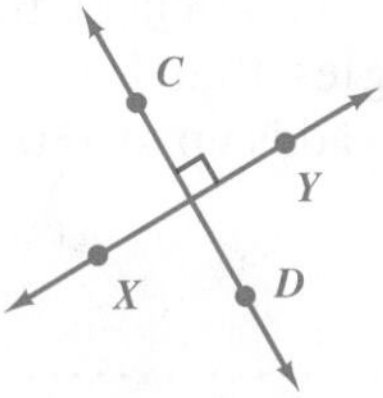

Name each highlighted angle by using the three-letter form of identification. ***See Example 3.***

15.
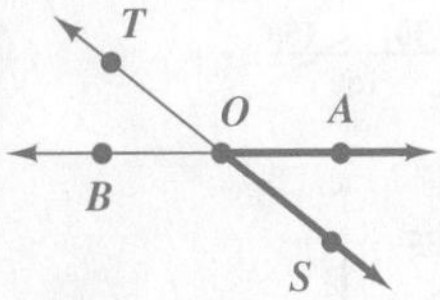

16.
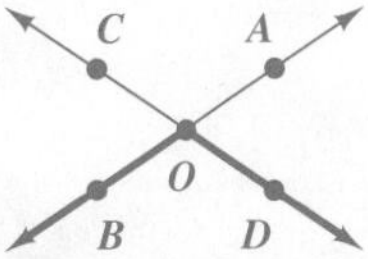

17.
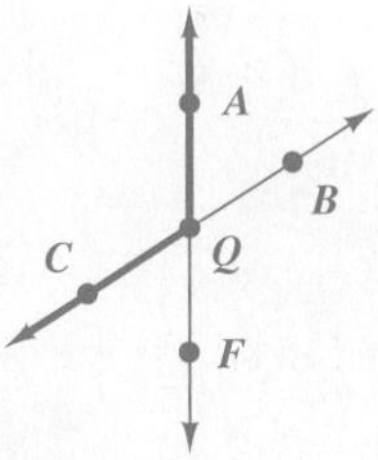

18.
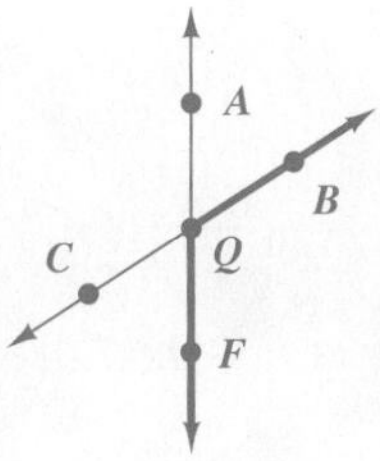

Label each angle as acute, right, obtuse, *or* straight. *For right angles and straight angles, indicate the number of degrees in the angle.* ***See Example 4.***

19.

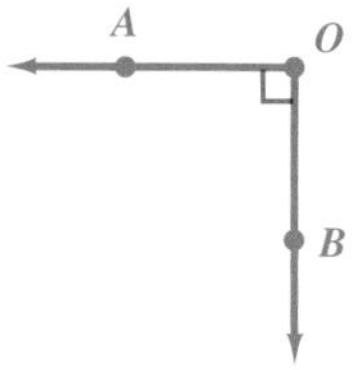

20.

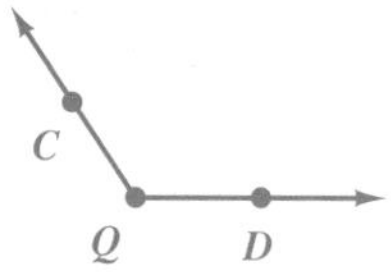

21.

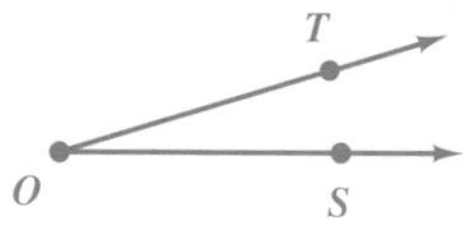

22.

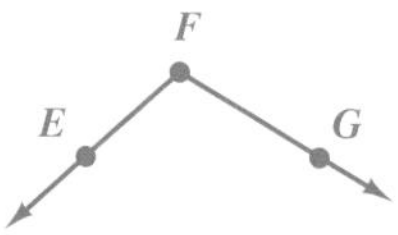

23.

24.

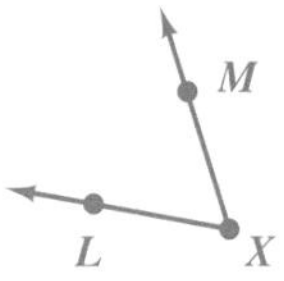

Identify each pair of complementary angles. ***See Example 6.***

25.

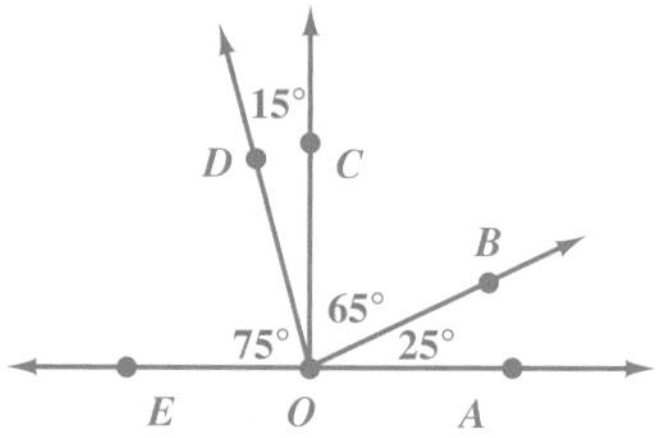

26.

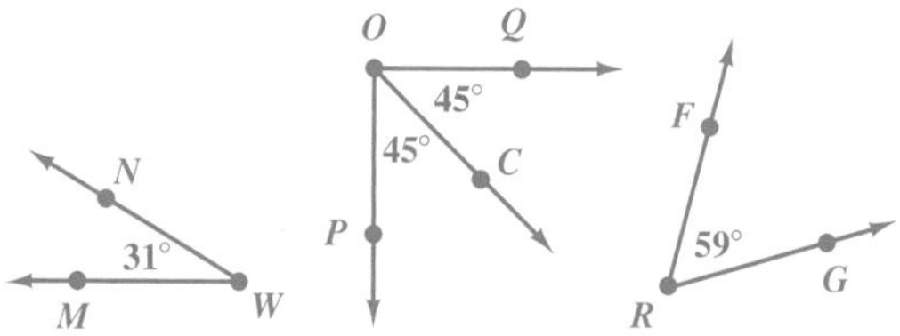

Identify each pair of supplementary angles. ***See Example 8.***

27.

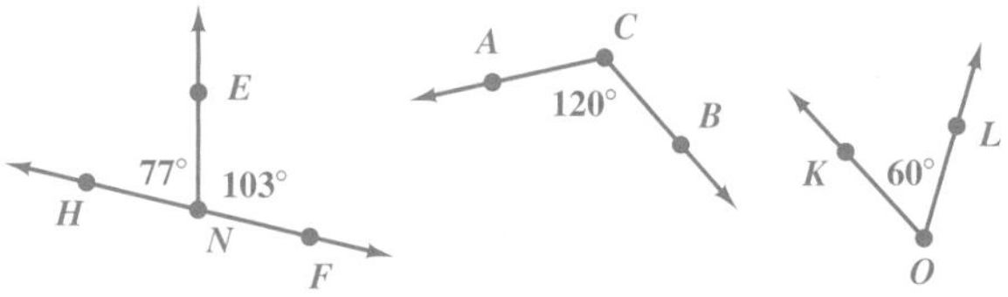

28.

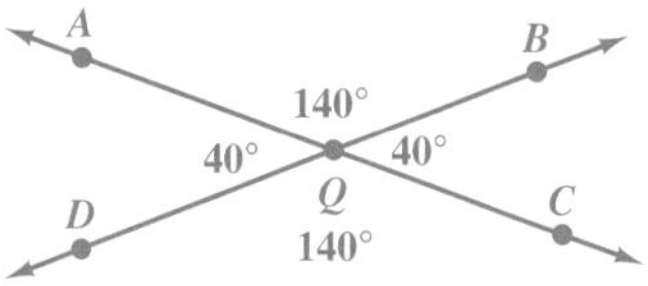

Find the complement of each angle. ***See Example 7.***

29. 40°
$90° - 40° = \_\_\_\_$

30. 35°
$90° - 35° = \_\_\_\_$

31. 86°

32. 59°

Find the supplement of each angle. ***See Example 9.***

33. 130°
$180° - 130° = \_\_\_\_$

34. 75°
$180° - 75° = \_\_\_\_$

35. 90°

36. 5°

In Exercises 37 and 38, identify the angles that are congruent. ***See Examples 10 and 11.***

37.

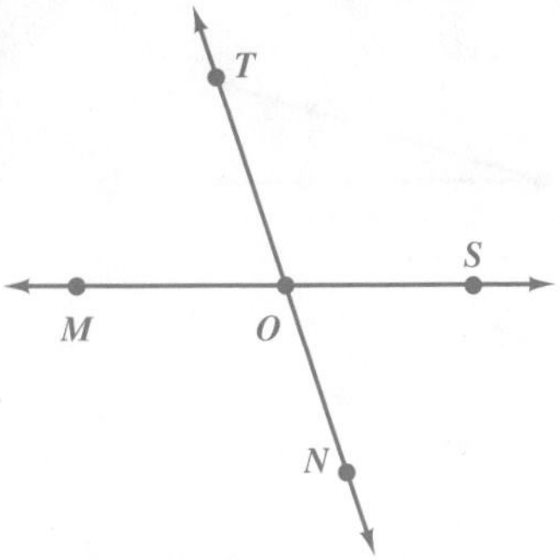

38.

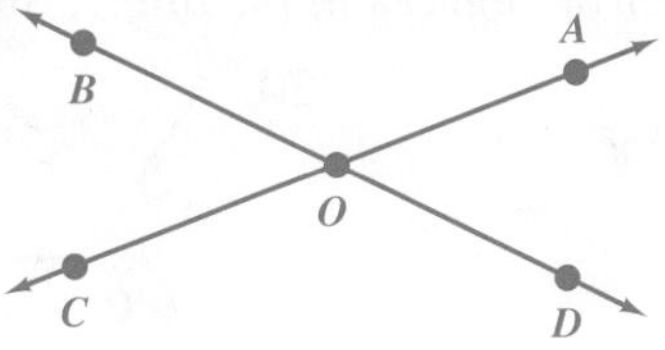

In Exercises 39 and 40, find the measure of each of the angles. ***See Example 12.***

39. In the figure below, ∠*AOH* measures 37° and ∠*COE* measures 63°.

Hint: ∠*AOC* completes a straight line (180° angle). Subtract 180° − (63° + 37°) to find ∠ *AOC*.

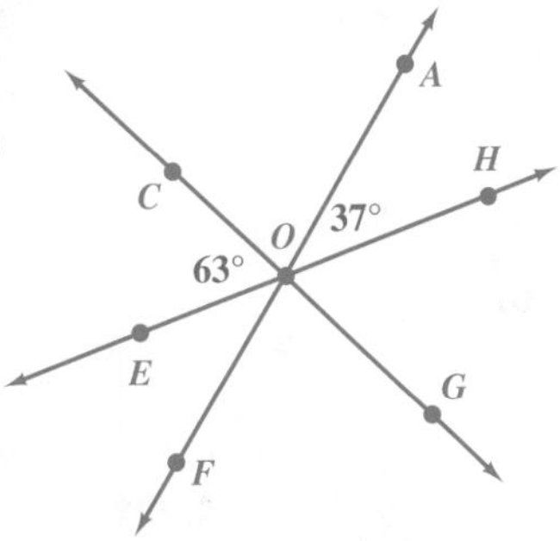

40. In the figure below, ∠*POU* measures 105° and ∠*UOT* measures 40°.

Hint: ∠*TOS* completes a straight line (180° angle). Subtract 180° − (105° + 40°) to find ∠*TOS*.

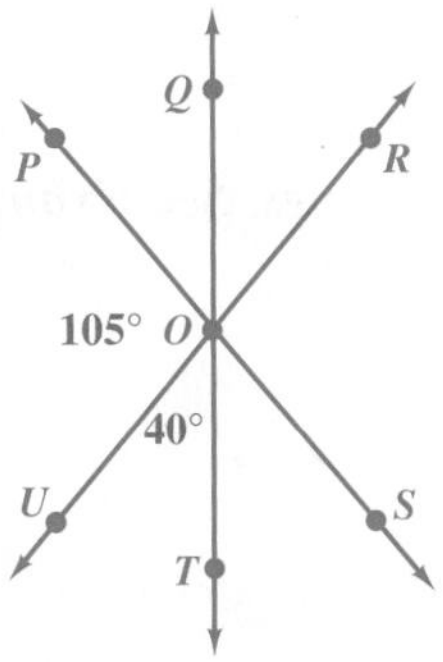

CONCEPT CHECK *Use the figure to work Exercises 41–46. Decide whether each statement is* true *or* false. *If it is true, explain why. If it is false, rewrite it to make a true statement.*

41. ∠*UST* is 90°.

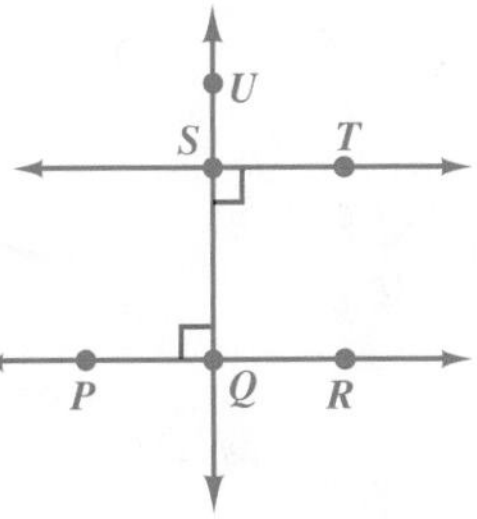

42. $\overleftrightarrow{SQ}$ and $\overleftrightarrow{PQ}$ are perpendicular.

43. The measure of ∠*USQ* is less than the measure of ∠*PQR*.

44. $\overleftrightarrow{ST}$ and $\overleftrightarrow{PR}$ are intersecting.

45. $\overleftrightarrow{QU}$ and $\overleftrightarrow{TS}$ are parallel.

46. ∠*UST* and ∠*UQR* measure the same number of degrees.

In each figure, line m is parallel to line n. Identify all pairs of corresponding angles and all pairs of alternate interior angles. ***See Example 13.***

47.

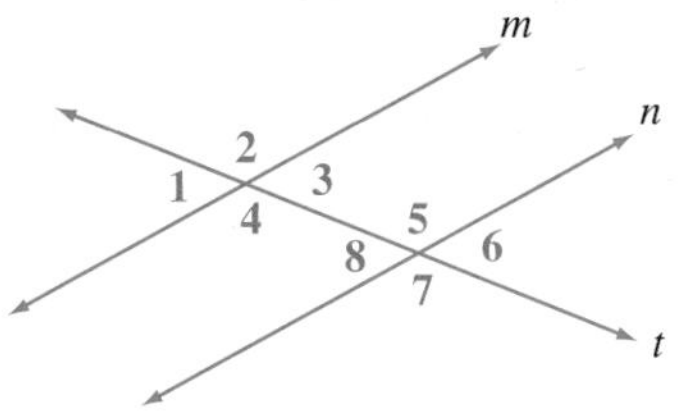

48.

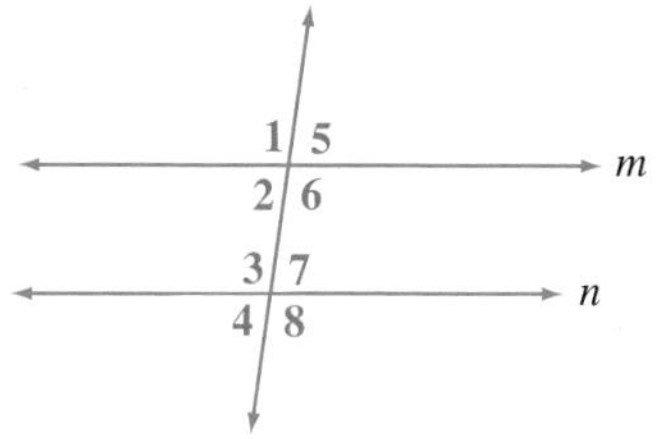

In each figure, line m is parallel to line n. Find the measure of each angle.
See Example 14.

49. ∠8 measures 130°.

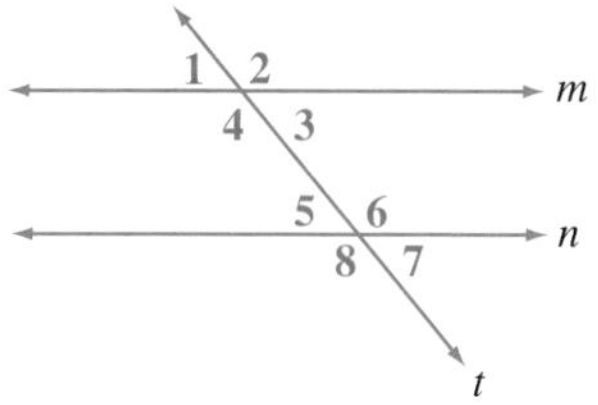

50. ∠2 measures 80°.

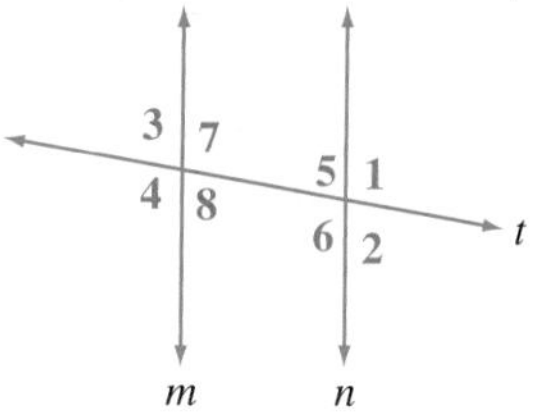

51. ∠6 measures 47°.

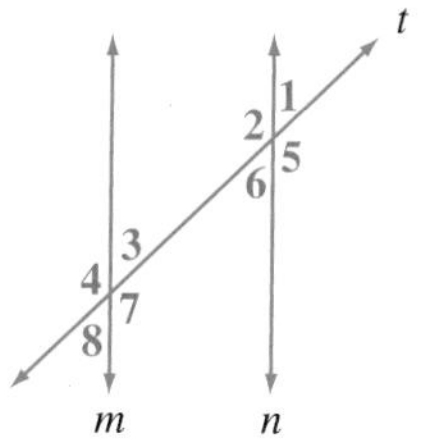

52. ∠2 measures 108°.

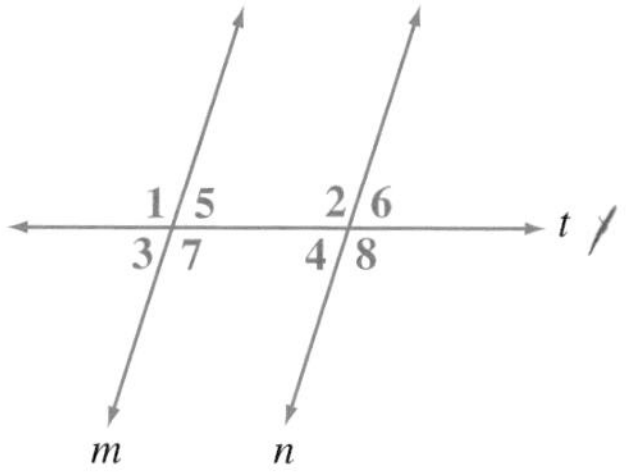

53. ∠6 measures 114°.

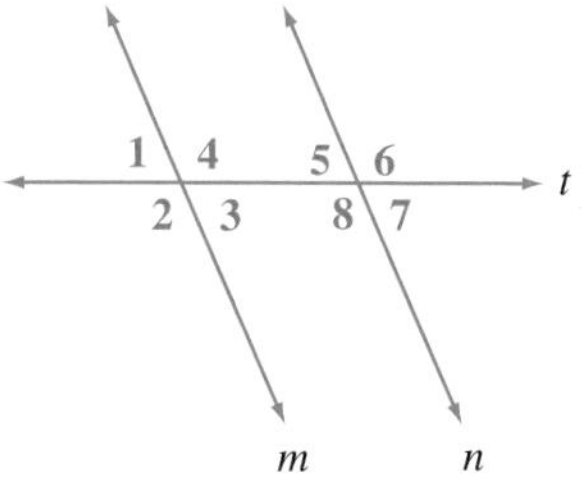

54. ∠3 measures 59°.

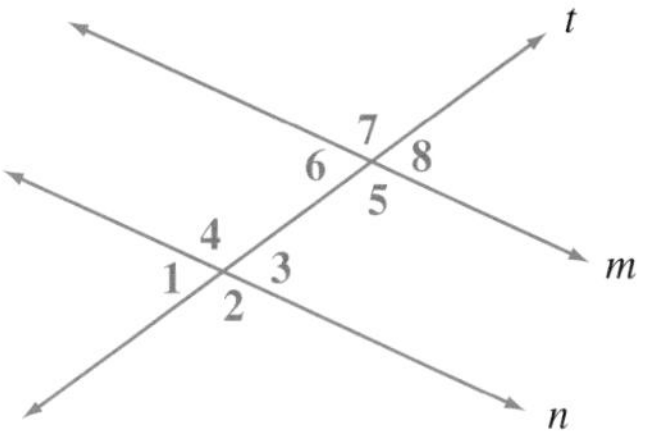

In each figure, $\overrightarrow{BA}$ *is parallel to* $\overrightarrow{CD}$. *Find the measure of each numbered angle.*

55.

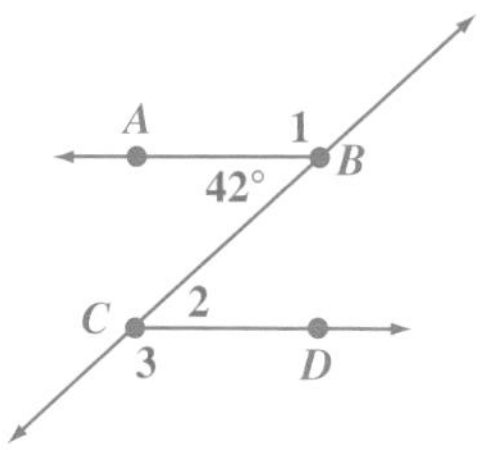

56.

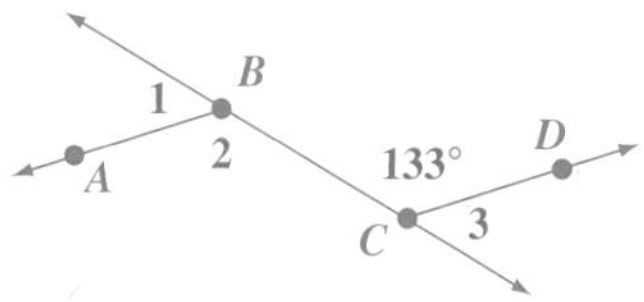

7.2 Rectangles and Squares

OBJECTIVES

1. Find the perimeter and area of a rectangle.
2. Find the perimeter and area of a square.
3. Find the perimeter and area of a composite figure.

A **rectangle** is a figure with four sides that meet to form 90° angles. Each set of opposite sides is *parallel* and *congruent* (has the same length).

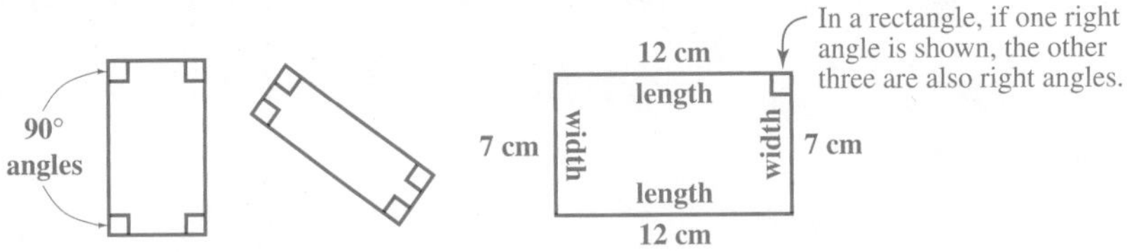

Each longer side of a rectangle is called the length (l) and each shorter side is called the width (w).

◀ Work Problem 1 at the Side.

OBJECTIVE 1 Find the perimeter and area of a rectangle. The distance around the outside edges of a flat figure is the **perimeter** of the figure. Think of how much fence you would need to put around the sides of a garden plot, or how far you would walk if you walked around the outside edges of your living room. In either case you would add up the lengths of the sides. Look at the rectangle above that has the lengths of the sides labeled. To find its perimeter, you add the lengths of the sides.

$$\text{Perimeter} = 12\text{ cm} + 12\text{ cm} + 7\text{ cm} + 7\text{ cm} = 38\text{ cm}$$

Because the two long sides are both 12 cm, and the two short sides are both 7 cm, you can also use this formula.

Finding the Perimeter of a Rectangle

$$\text{Perimeter of a rectangle} = \text{length} + \text{length} + \text{width} + \text{width}$$

$$P = (2 \cdot \text{length}) + (2 \cdot \text{width})$$

$$P = 2 \cdot l + 2 \cdot w$$

EXAMPLE 1 Finding the Perimeter of a Rectangle

Find the perimeter of each rectangle.

(a)

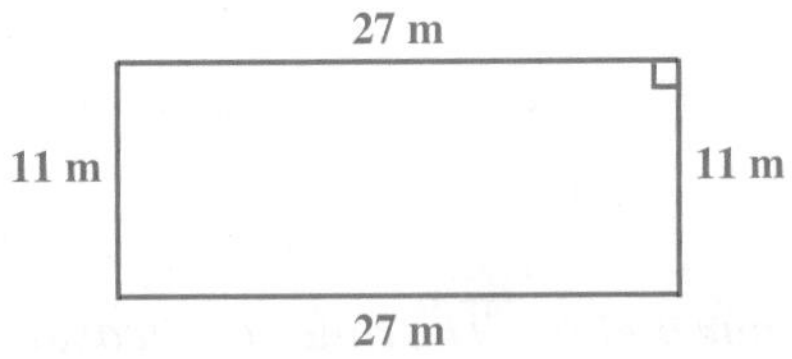

The length of this rectangle is 27 m and the width is 11 m.

Use the formula $P = 2 \cdot l + 2 \cdot w$

$P = 2 \cdot \quad l \quad + 2 \cdot \quad w$ — Replace l with 27 m and w with 11 m.

$P = \underbrace{2 \cdot 27\text{ m}} + \underbrace{2 \cdot 11\text{ m}}$ — Do the multiplications first.

$P = \quad 54\text{ m} \quad + \quad 22\text{ m}$ — Add last.

$P = 76\text{ m}$

The perimeter of the rectangle (the distance you would walk around the outside edges of the rectangle) is 76 m.

Continued on Next Page

VOCABULARY TIP

Rect means "right." It reminds you that all the angles in a **rect**angle are "right" angles.

1 Identify all the rectangles.

(a)

(b)

(c)

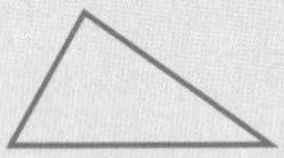

(d)

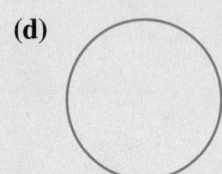

(e)

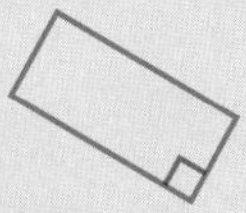

(f)

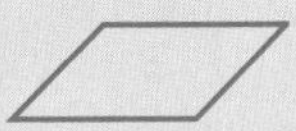

(g)

Answers

1. (a), (b), and (e) are rectangles; (c), (d), (f), and (g) are not.

As a check, you can add up the lengths of the four sides.

$$P = 27 \text{ m} + 27 \text{ m} + 11 \text{ m} + 11 \text{ m}$$

$$P = 76 \text{ m} \leftarrow \text{Same result as using the formula}$$

(b) A rectangle 8.9 ft by 12.3 ft

You can use the formula, as shown below.

$$P = 2 \cdot l + 2 \cdot w$$

$$P = \underbrace{2 \cdot 12.3 \text{ ft}} + \underbrace{2 \cdot 8.9 \text{ ft}}$$

$$P = 24.6 \text{ ft} + 17.8 \text{ ft}$$

$$P = 42.4 \text{ ft}$$

Be sure to write **ft** in the answer.

Or, you can add up the lengths of the four sides.

$$P = 12.3 \text{ ft} + 12.3 \text{ ft} + 8.9 \text{ ft} + 8.9 \text{ ft}$$

$$P = 42.4 \text{ ft} \leftarrow \text{Same result as using the formula}$$

Either method will give you the correct result.

Work Problem 2 at the Side. ▶

The *perimeter* of a rectangle is the distance around the *outside edges.* The **area** of a rectangle is the amount of surface *inside* the rectangle. We measure area by seeing how many squares of a certain size are needed to cover the surface inside the rectangle. Think of covering the floor of a rectangular living room with carpet. Carpet is measured in square yards, that is, square pieces that measure 1 yard along each side. Here is a drawing of a living room floor.

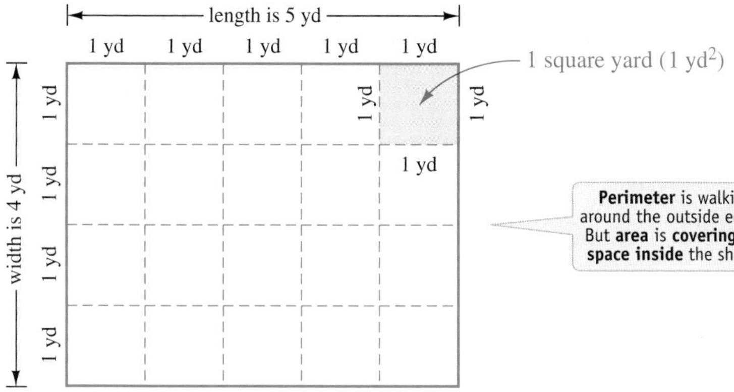

Perimeter is walking around the outside edges. But **area** is **covering the space inside** the shape.

You can see from the drawing that it takes 20 squares to cover the floor. We say that the area of the floor is 20 *square yards.* A shorter way to write square yards is yd^2.

20 square yards can be written as 20 yd^2.

To find the number of squares, you can count them, or you can multiply the number of squares in the length (5) times the number of squares in the width (4) to get 20. The formula is given below.

Finding the Area of a Rectangle

Area of a rectangle = length • width

$$A = l \cdot w$$

Remember to use ***square units*** when measuring area.

VOCABULARY TIP

Perimeter Remember, in **perimeter, peri** means "around," and **meter** means "measure." To find the perimeter, you measure around the outside edges of a flat figure. Write the perimeter using **linear units,** such as feet or yards, for example, 10 **ft** or 15 **yd.**

2 Find the perimeter of each rectangle by using the formula or by adding the lengths of the sides.

GS **(a)**

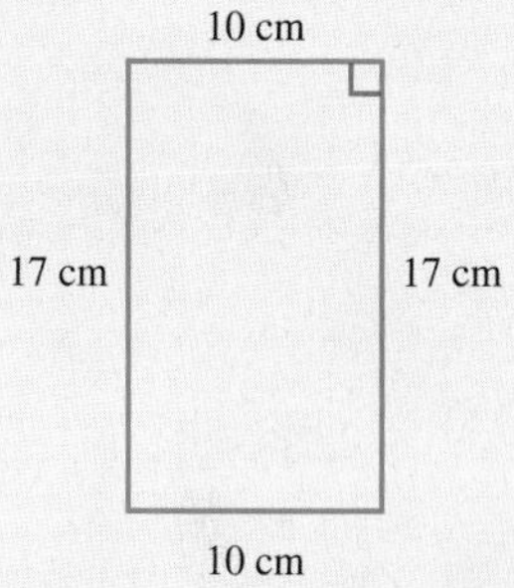

Try adding the lengths of the sides:

$P = 17 \text{ cm} + 17 \text{ cm} +$ ______

\+ ______

$P =$ ______

GS **(b)**

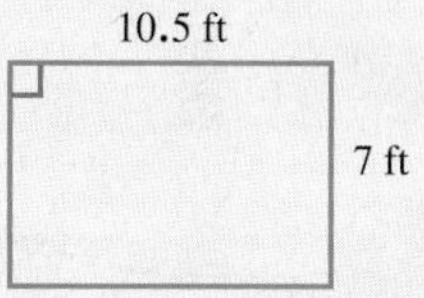

Try using the formula:

$P = 2 \cdot$ ______ $+ 2 \cdot$ ______

$P =$ ______ + ______

$P =$ ______

(c) 6 m wide and 11 m long

(d) 0.9 km by 2.8 km

Answers

2. **(a)** $P = 17 \text{ cm} + 17 \text{ cm} + 10 \text{ cm} + 10 \text{ cm}$; $P = 54 \text{ cm}$
(b) $P = 2 \cdot 10.5 \text{ ft} + 2 \cdot 7 \text{ ft}$; $P = 21 \text{ ft} + 14 \text{ ft}$; $P = 35 \text{ ft}$
(c) $P = 34 \text{ m}$ **(d)** $P = 7.4 \text{ km}$

Squares of many sizes can be used to measure area. For smaller areas, you might use the ones shown below.

Area is measured in **square units.**

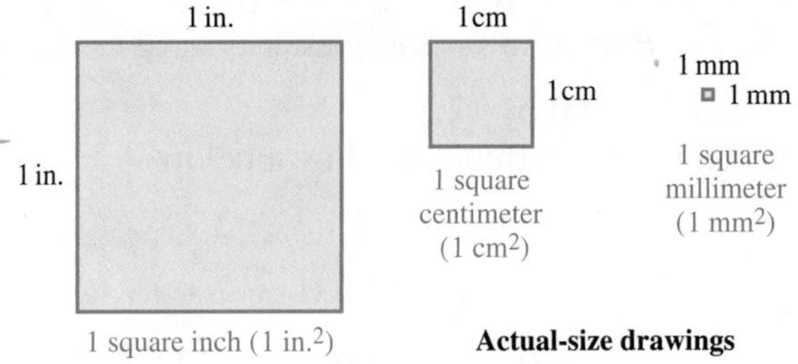

Other sizes of squares that are often used to measure area are listed here, but they are too large to draw on this page.

1 square meter (1 m^2)	1 square foot (1 ft^2)
1 square kilometer (1 km^2)	1 square yard (1 yd^2)
	1 square mile (1 mi^2)

CAUTION

The raised 2 in 4^2 means that you multiply 4 • 4 to get 16. The raised 2 in cm^2 or yd^2 is a short way to write the word *square*. When you see 5 cm^2, say "five square centimeters." Do *not* multiply 5 • 5. The exponent applies to cm, *not* to the number.

EXAMPLE 2 Finding the Area of a Rectangle

Find the area of each rectangle.

(a)

For **area**, find how many squares are needed to **cover the space inside** the rectangle.

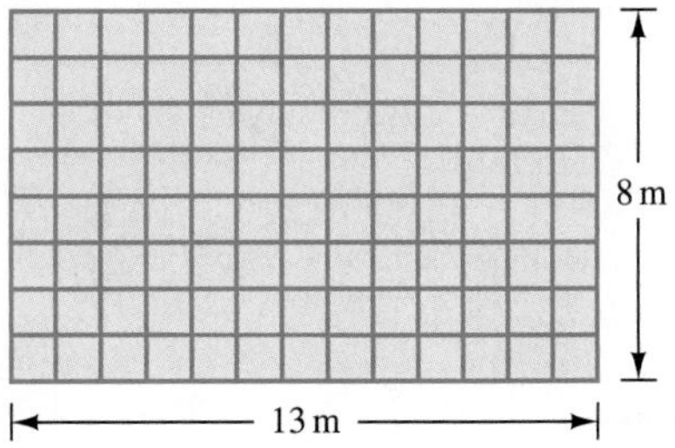

The length of this rectangle is 13 m and the width is 8 m. Use the formula $A = l \cdot w$.

$A = l \cdot w$ Replace l with 13 m and w with 8 m.

$A = \mathbf{13\ m \cdot 8\ m}$ Multiply.

$A = 104$ square meters

Write $\mathbf{m^2}$ in the answer.

"Square meters" can be written as m^2, so the area is 104 m^2.

(b) A rectangle measuring 7 cm by 21 cm

First make a sketch of the rectangle. The length is 21 cm (the longer measurement) and the width is 7 cm. Then use the formula for the area of a rectangle, $A = l \cdot w$.

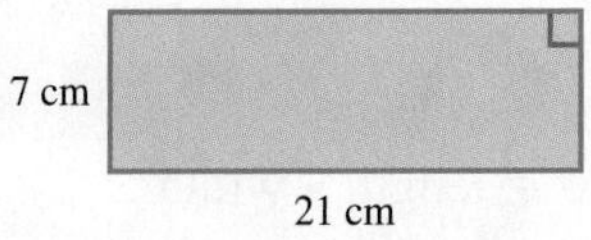

$A = 21\text{ cm} \cdot 7\text{ cm}$

$A = 147\ \mathbf{cm^2}$

Square units for area; write $\mathbf{cm^2}$ in the answer.

The area of the rectangle is 147 cm^2.

Continued on Next Page

CAUTION

The units for **area** will always be **square units** (cm^2, m^2, yd^2, mi^2, and so on). The units for **perimeter** will always be **linear units** (cm, m, yd, mi, and so on) *not* square units.

Work Problem 3 at the Side. ▶

OBJECTIVE 2 Find the perimeter and area of a square. A **square** is a rectangle with all sides the same length. Two squares are shown below. Notice the 90° angles.

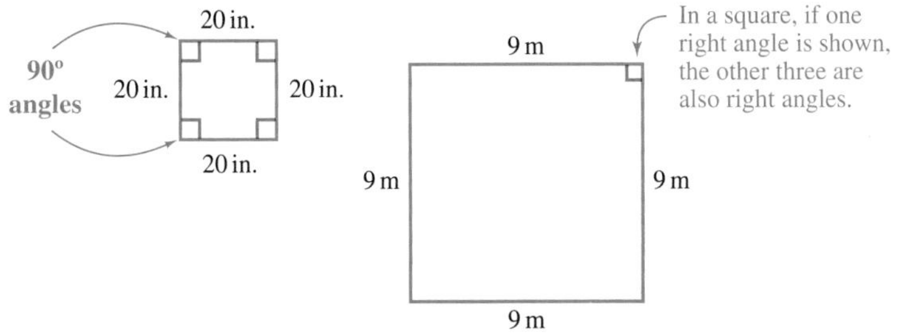

To find the *perimeter* of (distance around) the square on the right, you could add 9 m + 9 m + 9 m + 9 m to get 36 m. A shorter way is to multiply the length of one side times 4, because all four sides are the same length.

Finding the Perimeter of a Square

Perimeter of a square = side + side + side + side

or, $P = 4 \cdot \text{side}$

$P = 4 \cdot s$

As with a rectangle, you can multiply length times width to find the *area* of (surface inside) a square. Because the length and the width are the same in a square, the formula is written as shown below.

Finding the Area of a Square

Area of a square = side • side

$A = s \cdot s$

$A = s^2$

Remember to use ***square units*** when measuring area.

EXAMPLE 3 Finding the Perimeter and Area of a Square

(a) Find the perimeter of the square shown above where each side measures 9 m.

Use the formula.

$P = 4 \cdot s$

$P = 4 \cdot 9 \text{ m}$

$P = 36 \text{ m}$

Or add up the four sides.

$P = 9 \text{ m} + 9 \text{ m} + 9 \text{ m} + 9 \text{ m}$

$P = 36 \text{ m}$ — This is **perimeter**, so write **m** in the answer.

Same answer

Continued on Next Page

VOCABULARY TIP

Area The **area** is the amount of surface **inside** a flat shape. Write the area using **square units,** such as square inches or square feet—for example, 10 $\textbf{in.}^2$ or 12 $\textbf{ft}^2$.

3 Find the area of each rectangle.

GS **(a)**

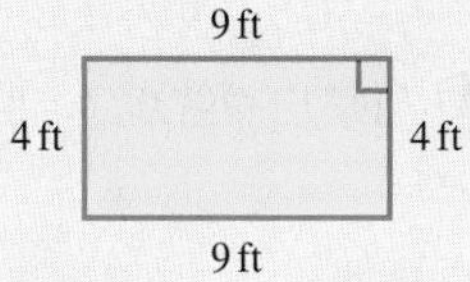

$A = l \cdot w$

$A = 9 \text{ ft} \cdot 4 \text{ ft}$

$A =$ ________

Write $\textbf{ft}^2$ as part of your answer.

GS **(b)** A rectangle is 6 m long and 0.5 m wide. (First make a sketch of the rectangle and label the lengths of the sides.)

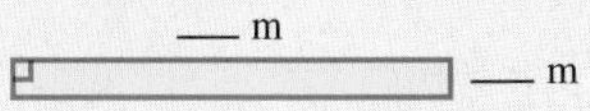

(c) A rectangular patio measures 3.5 yd by 2.5 yd. (First make a sketch of the patio and label the lengths of the sides.)

Answers

3. **(a)** $A = 36 \text{ ft}^2$
(b) $A = 3 \text{ m}^2$

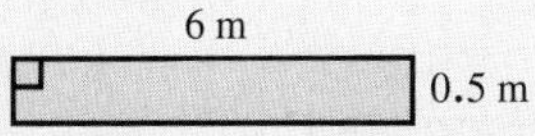

(c) $A = 8.75 \text{ yd}^2$

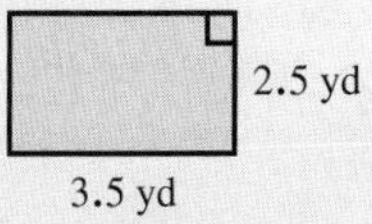

(b) Find the area of the same square where each side measures 9 m.

$$A = s^2$$
$$A = s \cdot s$$
$$A = 9\text{ m} \cdot 9\text{ m}$$
$$A = 81\text{ m}^2$$

This is **area,** so write $\mathbf{m^2}$ in the answer.

CAUTION

Be careful! s^2 means $s \cdot s$. It does ***not*** mean $s \cdot 2$. In **Example 3(b)** above, s is 9 m, so s^2 is 9 m • 9 m = 81 m². It is ***not*** 9m • 2 = 18 m.

◀ **Work Problem 4 at the Side.**

OBJECTIVE 3 Find the perimeter and area of a composite figure. As with any other shape, you can find the perimeter of (distance around) an irregular shape by adding up the lengths of the sides. To find the area (surface inside the shape), try to break it up into pieces that are squares or rectangles. Find the area of each piece and then add them together.

CAUTION

Perimeter is the ***distance around the outside edges*** of a flat shape. It is always measured in *linear units* such as cm, m, yd, and so on.

Area is the amount of ***surface inside*** a flat shape. It is always measured in *square units* such as cm², m², yd², and so on.

EXAMPLE 4 Finding the Perimeter and Area of a Composite Figure

The floor of a room has the shape shown below.

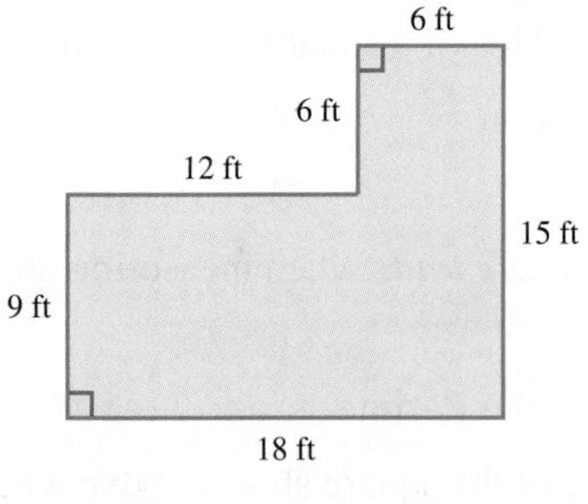

(a) Suppose you want to put a new wallpaper border along the top of all the walls. How much material do you need?

Find the **perimeter** of the room by adding up the lengths of the sides.

$$P = 9\text{ ft} + 12\text{ ft} + 6\text{ ft} + 6\text{ ft} + 15\text{ ft} + 18\text{ ft}$$
$$P = 66\text{ ft}$$

You need 66 ft of wallpaper border.

Continued on Next Page

4 Find the perimeter and area of each square.

(a)

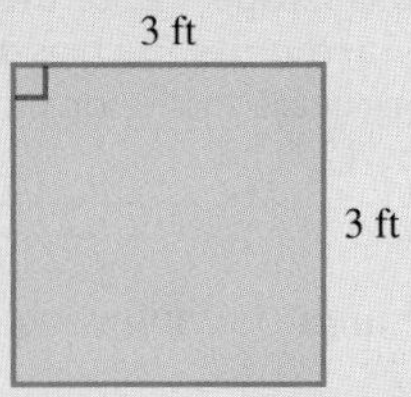

For perimeter of a square,

$P = 4 \cdot s$

$P = 4 \cdot$ ______

$P =$ ______

For area of a square,

$A = s^2$ or $A = s \cdot s$

$A =$ ______ $\cdot$ ______

$A =$ ______

(b) 10.5 cm on each side. (Make a sketch of the square.)

(c) 2.1 mi on a side. (Make a sketch of the square.)

Answers

4. **(a)** $P = 4 \cdot 3$ ft; $P = 12$ ft; $A = 3\text{ ft} \cdot 3\text{ ft}$; $A = 9\text{ ft}^2$

(b) $P = 42$ cm; $A = 110.25\text{ cm}^2$

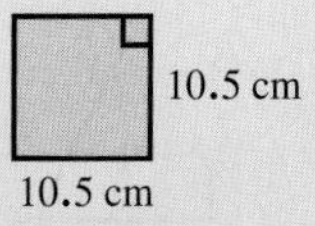

(c) $P = 8.4$ mi; $A = 4.41\text{ mi}^2$

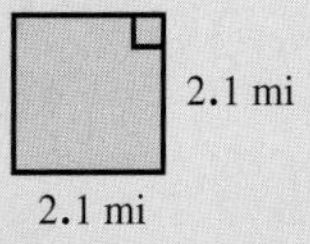

(b) The carpet you like costs \$20.50 per square yard. How much will it cost to carpet the room?

First change the measurements from feet to yards, because the carpet is sold in square yards. There are 3 ft in 1 yd, so multiply by the unit fraction that allows you to divide out feet. Let's start with 9 ft.

$$\frac{\overset{3}{\cancel{9}}\ \cancel{\text{ft}}}{1} \cdot \frac{1\text{ yd}}{\underset{1}{\cancel{3}}\ \cancel{\text{ft}}} = 3\text{ yd}$$

Divide out ft.
Divide 9 and 3 by 3.

Use the same unit fraction to change the other measurements to yards.

$$\frac{\overset{4}{\cancel{12}}\ \cancel{\text{ft}}}{1} \cdot \frac{1\text{ yd}}{\underset{1}{\cancel{3}}\ \cancel{\text{ft}}} = 4\text{ yd} \qquad \frac{\overset{2}{\cancel{6}}\ \cancel{\text{ft}}}{1} \cdot \frac{1\text{ yd}}{\underset{1}{\cancel{3}}\ \cancel{\text{ft}}} = 2\text{ yd}$$

$$\frac{\overset{5}{\cancel{15}}\ \cancel{\text{ft}}}{1} \cdot \frac{1\text{ yd}}{\underset{1}{\cancel{3}}\ \cancel{\text{ft}}} = 5\text{ yd} \qquad \frac{\overset{6}{\cancel{18}}\ \cancel{\text{ft}}}{1} \cdot \frac{1\text{ yd}}{\underset{1}{\cancel{3}}\ \cancel{\text{ft}}} = 6\text{ yd}$$

Next, break up the room into two pieces. Use just the measurements for the length and width of each piece.

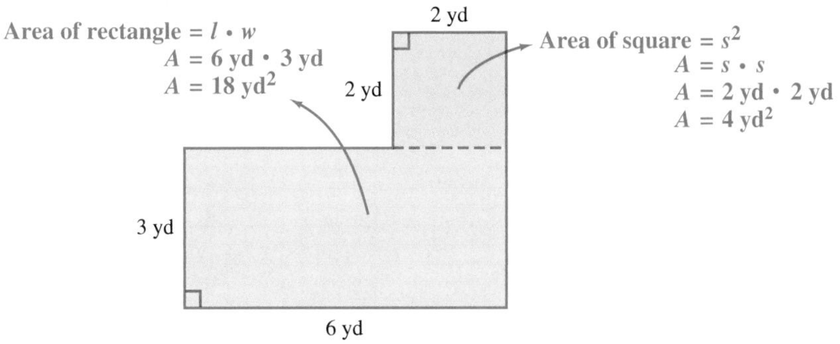

Total area = 18 yd² + 4 yd² = 22 yd² ← This is **area**, so it is measured in **yd²**.

Multiply to find the cost of the carpet.

$$\text{Cost} = \frac{22\ \cancel{\text{yd}^2}}{1} \cdot \frac{\$20.50}{1\ \cancel{\text{yd}^2}} = \$451.00$$

It will cost \$451.00 to carpet the room.

You could have cut the room into two rectangles as shown below. The total area is the same.

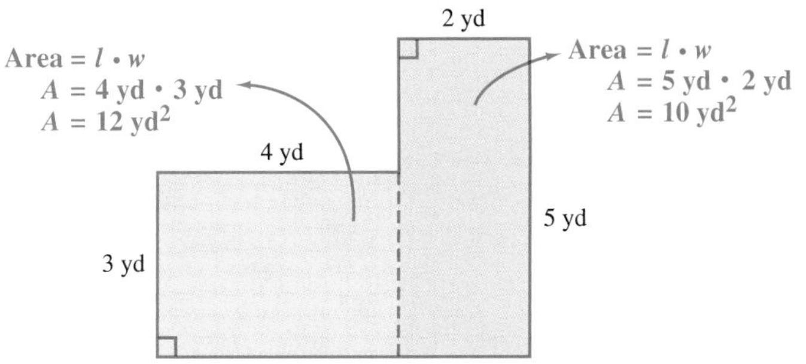

Total area = 12 yd² + 10 yd² = 22 yd² ← Same answer as above

Work Problem 5 at the Side. ▶

---

5 Carpet costs \$19.95 per square yard. Find the cost of carpeting each room. Round your answers to the nearest cent if necessary.

GS **(a)**

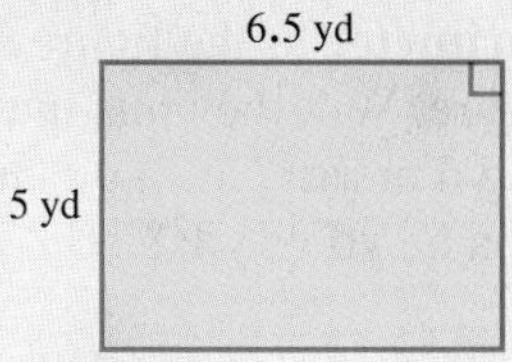

$A = l \cdot w$

$A = 6.5\text{ yd} \cdot 5\text{ yd}$

$A = \_\_\_\_\text{ yd}^2$

$$\text{Cost} = \frac{\_\_\_\_\ \cancel{\text{yd}^2}}{1} \cdot \frac{\$19.95}{1\ \cancel{\text{yd}^2}}$$

Cost = ________

(b)

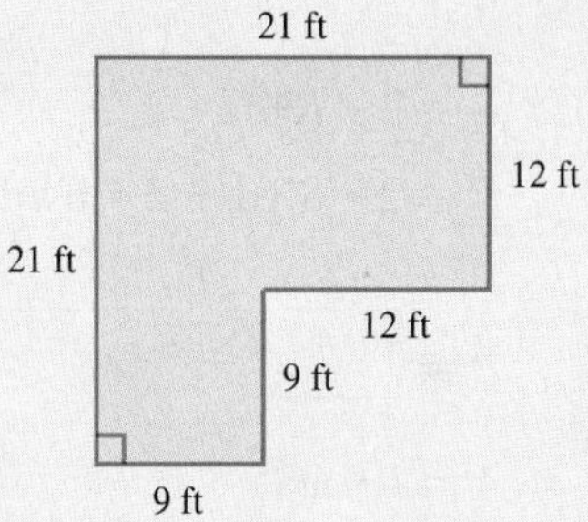

(c) A rectangular classroom is 24 ft long and 18 ft wide. (Make a sketch of the classroom.)

Answers

5. (a) $A = 32.5\text{ yd}^2$; Cost = \$648.38 (rounded)
(b) 37 yd²; Cost = \$738.15
(c)

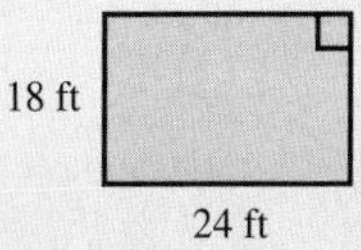

48 yd²; Cost = \$957.60

7.2 Exercises

FOR EXTRA HELP — Download the MyDashBoard App — MyMathLab®

1. **CONCEPT CHECK** Look at the figure in **Exercise 3** below. Explain the difference between finding the **perimeter** of the figure and finding the **area** of the figure. Will the units in the answer for the perimeter be **yd** or **yd²**? Will the units in the answer for the area be **yd** or **yd²**?

2. **CONCEPT CHECK** Explain what is special about a rectangle. Explain what is special about a square. Make a drawing of each figure and label all the sides with measurements that you choose.

Find the perimeter and area of each rectangle or square. ***See Examples 1–3.***

3.

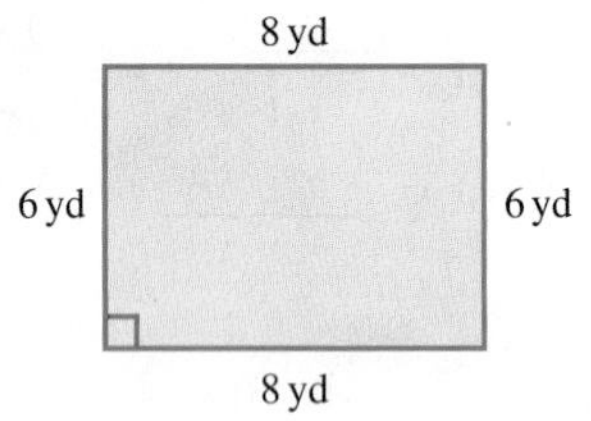

$P = 2 \cdot l + 2 \cdot w$

$P = 2 \cdot 8 \text{ yd} + 2 \cdot 6 \text{ yd}$

$P =$ ______ + ______

$P =$ ______

$A = l \cdot w$

$A = 8 \text{ yd} \cdot 6 \text{ yd} =$ ______

4.

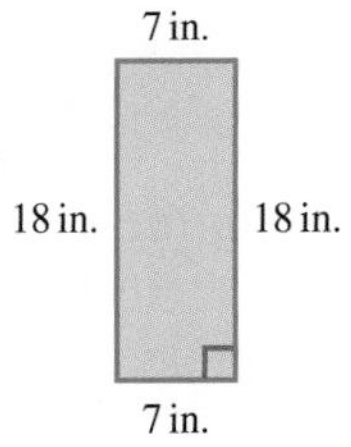

$P = 2 \cdot l + 2 \cdot w$

$P = 2 \cdot 18 \text{ in.} + 2 \cdot 7 \text{ in.}$

$P =$ ______ + ______

$P =$ ______

$A = l \cdot w$

$A = 18 \text{ in.} \cdot 7 \text{ in.} =$ ______

5.

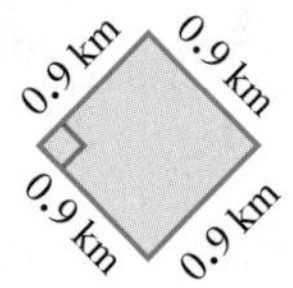

6.

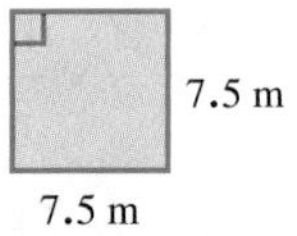

Draw a sketch of each square or rectangle and label the lengths of the sides. Then find the perimeter and the area. (Sketches may vary; show your sketches to your instructor.)

7. 10 ft by 10 ft

8. 8 cm by 17 cm

9. A storage building that is 76.1 ft by 22 ft

10. A science lab measuring 12 m by 12 m

11. A square nature preserve 3 mi wide

12. A square of cardboard 20.3 cm on a side

Find the perimeter and area of each figure. ***See Example 4.***

13.

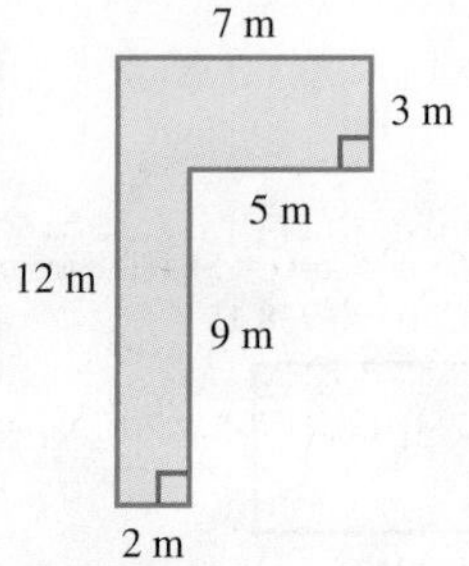

14.

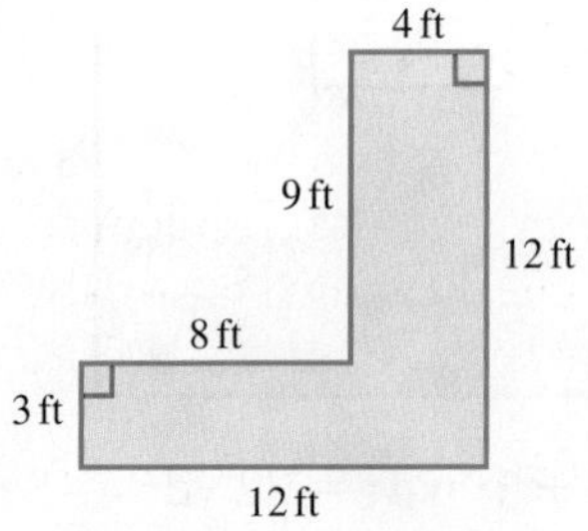

15.

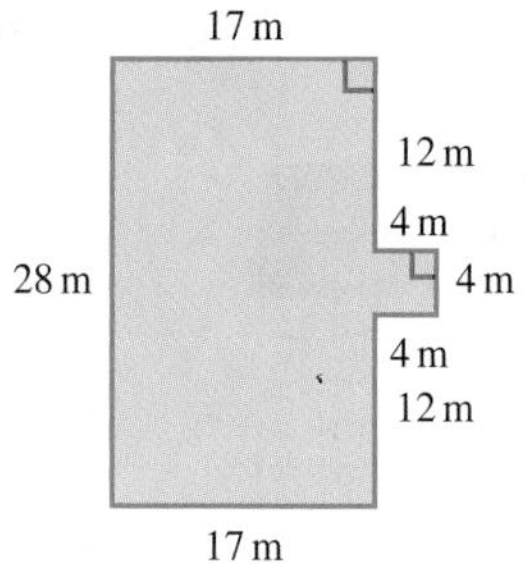

16.

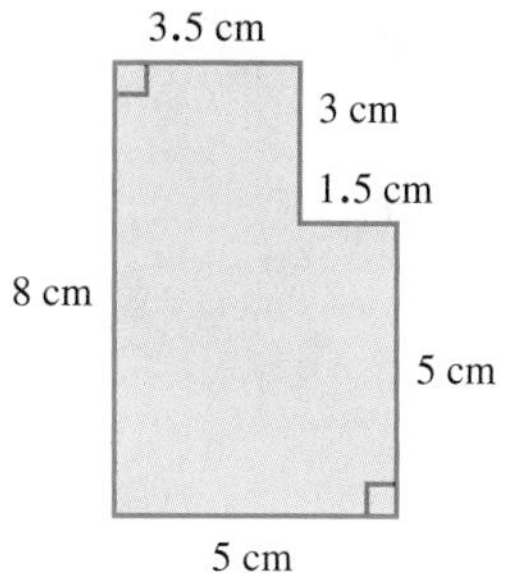

First find the length of the unlabeled side in each figure. Then find the perimeter and area of each figure.

17.

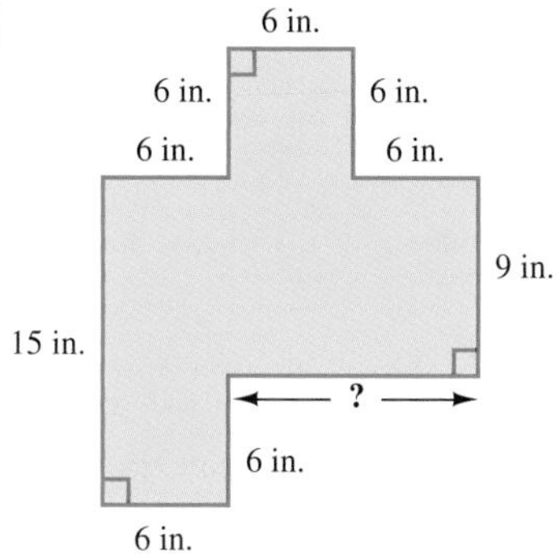

18.

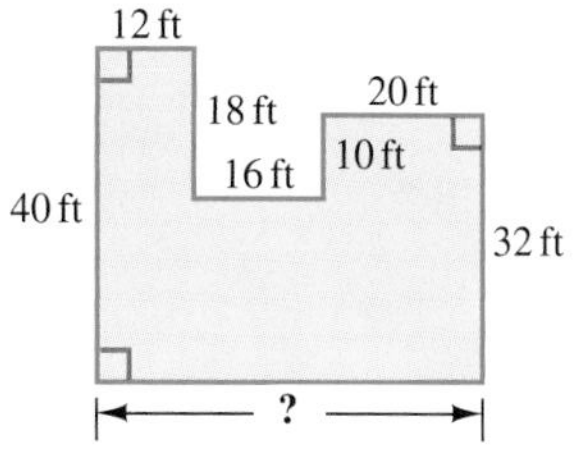

Solve each application problem. In Exercises 19–24, draw a sketch for each problem and label it with the appropriate measurements. (Sketches may vary; show your sketches to your instructor.)

19. Gymnastic floor exercises are performed on a square mat that is 12 meters on a side. Find the perimeter and area of a mat. (*Source:* nist.gov)

20. A regulation volleyball court is 18 meters by 9 meters. Find the perimeter and area of a regulation court. (*Source:* nist.gov)

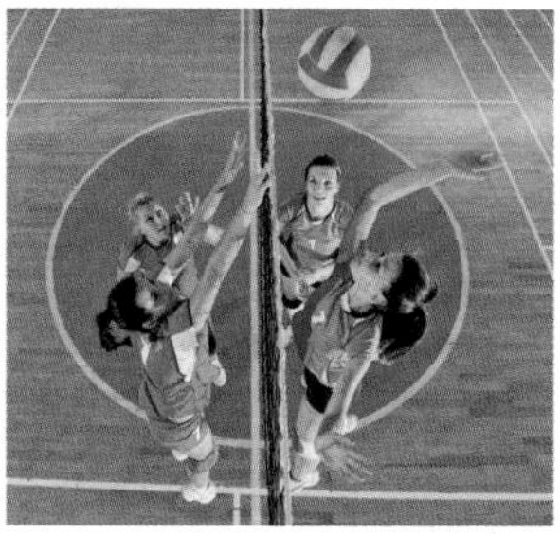

21. SooJin's laptop computer has a rectangular screen that is 12.3 in. wide and 7.8 in. high. The screen on her notebook computer measures 10.3 in. by 7.5 in. Find the area of each screen to the nearest whole number. Then find the difference in the areas.

22. A page in this book measures 27.5 cm from top to bottom and 21 cm from side to side. Find the perimeter and the area of the page.

23. Tyra's kitchen is 4.4 m wide and 5.1 m long. She is pasting a decorative border strip that costs $4.99 per meter around the top edge of all the walls. How much will she spend?

24. Mr. and Mrs. Gomez are buying carpet for their square-shaped bedroom that is 5 yd wide. The carpet is $23 per square yard and padding and installation is another $6 per square yard. How much will they spend in all?

25. Regulation soccer fields for teens and adults can measure from 50 to 80 yards wide and 100 to 120 yards long, depending on the age and skill level of the players. Find the area of the smallest soccer field and the area of the largest soccer field. What is the difference in the playing room between the smallest and largest fields?

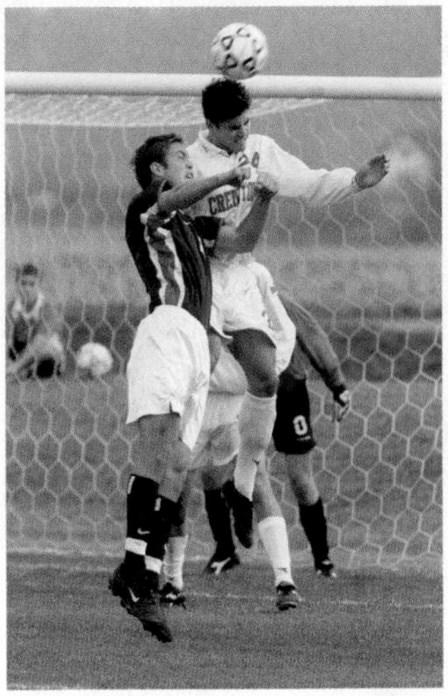

26. The table below shows information on two tents for camping.

Tents	Coleman Family Dome	Eddie Bauer Dome Tent
Dimensions	13 ft × 13 ft	12 ft × 12 ft
Sleeps	8 campers	6 campers
Sale price	$127	$99

Source: Target.com

(a) For the Coleman tent, find the perimeter, area, and number of square feet of floor space for each camper. Round to the nearest whole number.

(b) Find the same information for the Eddie Bauer tent.

27. A regulation football field is rectangular, 100 yd long (excluding end zones), and has an area of 5300 yd^2. Find the width of the field. (*Source:* National Football League.)

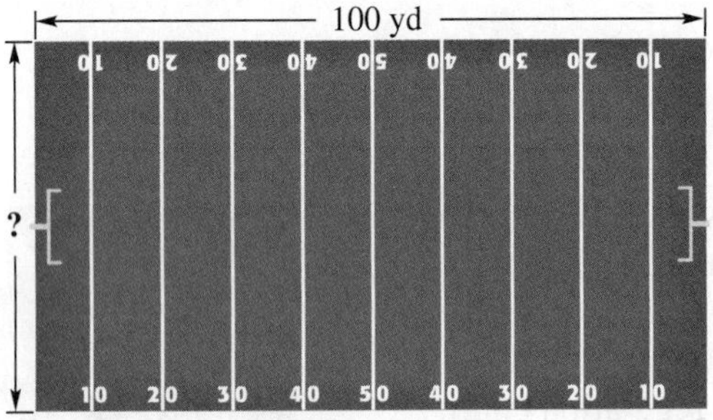

28. There are 15,130 ft^2 of ice on the surface of the rectangular playing area for a major league hockey game (excluding the area behind the goal lines). If the playing area is 85 ft wide, how long is it? (*Source:* National Hockey League.)

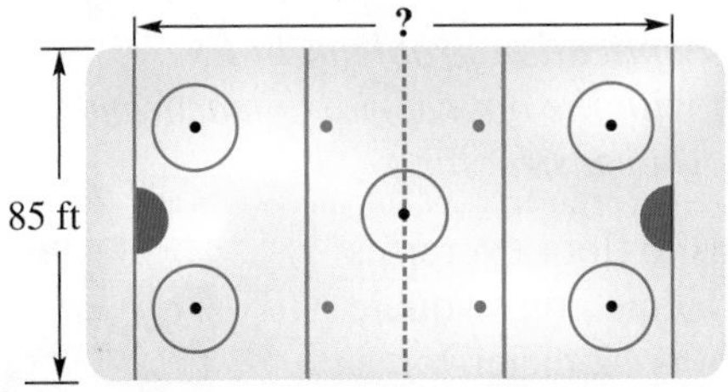

29. A rectangular lot is 124 ft by 172 ft. County rules require that nothing be built on land within 12 ft of any edge of the lot. First, add labels to the sketch of the lot, showing the land that cannot be built on. Then find the area of the land that cannot be built on.

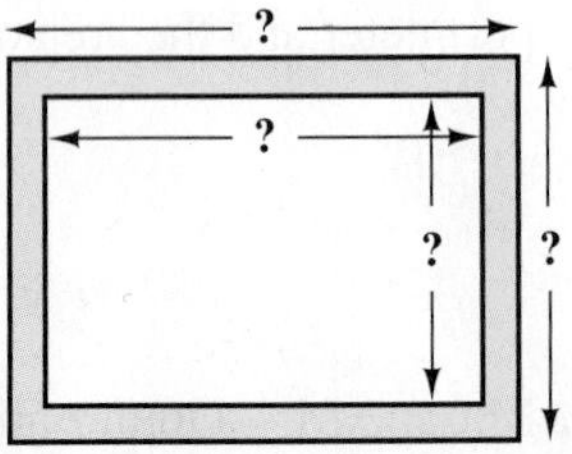

30. Find the cost of fencing needed for this rectangular field. Fencing along the country roads costs $4.25 per foot. Fencing for the other two sides costs $2.75 per foot.

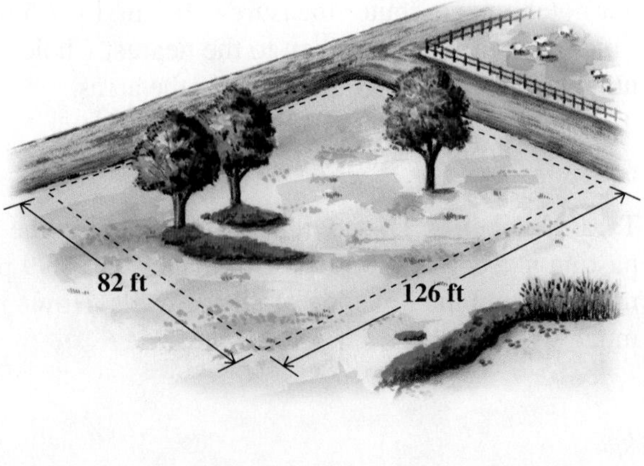

Relating Concepts (Exercises 31–36) For Individual or Group Work

Use your knowledge of perimeter and area to ***work Exercises 31–36 in order.***

31. Suppose you have 12 ft of fencing to make a square or rectangular garden plot. Draw sketches of *all* the possible plots that use exactly 12 ft of fencing and label the lengths of the sides. Use only *whole number* lengths. (*Hint:* There are three possibilities.)

32. (a) Find the area of each plot in **Exercise 31.**

(b) Which plot has the greatest area?

33. Repeat **Exercise 31** using 16 ft of fencing. Be sure to draw *all* possible plots that have whole number lengths for the sides.

34. (a) Find the area of each plot in **Exercise 33.**

(b) Compare your results to those from **Exercise 32.** What do you notice about the plots with the greatest area?

35. (a) Draw a sketch of a rectangular plot 3 ft by 2 ft. Find the perimeter and area.

(b) Suppose you *double* the length of the plot and *double* the width. Draw a sketch of the enlarged plot and find the perimeter and area.

(c) The *perimeter* of the enlarged plot is how many times greater than the perimeter of the original plot? The *area* of the enlarged plot is how many times greater than the original area?

36. (a) Refer to **Exercise 35(a).** Suppose you *triple* the length and width of the original plot. Draw a sketch of the enlarged plot and find the perimeter and area.

(b) How many times greater is the *perimeter* of the enlarged plot? How many times greater is the *area* of the enlarged plot?

(c) Suppose you make the length and width *four times greater* in the enlarged plot. What would you predict will happen to the perimeter and area, compared to the original plot?

7.3 Parallelograms and Trapezoids

OBJECTIVES

1. Find the perimeter and area of a parallelogram.
2. Find the perimeter and area of a trapezoid.

A **parallelogram** is a four-sided figure with opposite sides parallel, such as the ones shown below. Notice that opposite sides have the same length.

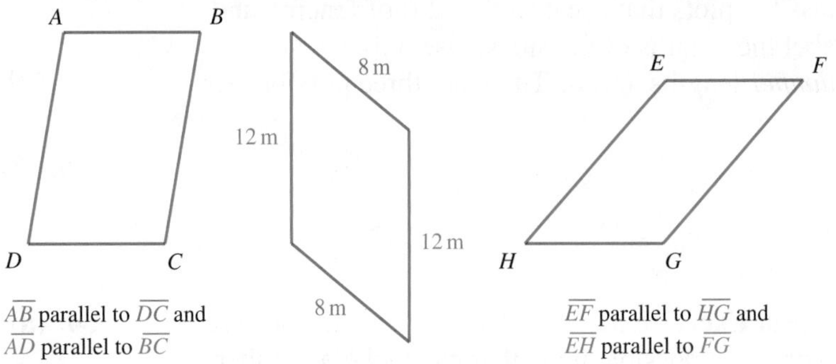

$\overline{AB}$ parallel to $\overline{DC}$ and $\overline{AD}$ parallel to $\overline{BC}$

$\overline{EF}$ parallel to $\overline{HG}$ and $\overline{EH}$ parallel to $\overline{FG}$

OBJECTIVE 1 Find the perimeter and area of a parallelogram. Perimeter is the distance around a flat shape, so the easiest way to find the perimeter of a parallelogram is to add the lengths of the four sides.

EXAMPLE 1 Finding the Perimeter of a Parallelogram

Find the perimeter of the middle parallelogram above.

$$P = 12 \text{ m} + 12 \text{ m} + 8 \text{ m} + 8 \text{ m} = 40 \text{ m}$$

This is **perimeter**, so write **m** in the answer, **not** m^2.

◀ **Work Problem 1 at the Side.**

1 Find the perimeter of each parallelogram.

GS **(a)**

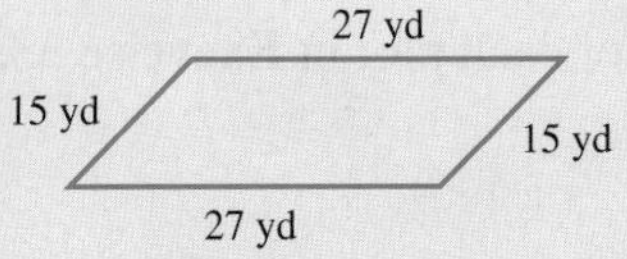

$P = 27 \text{ yd} + 27 \text{ yd} + 15 \text{ yd}$
$+ \_\_\_ \text{ yd}$

$P = \_\_\_\_\_\_$

(b)

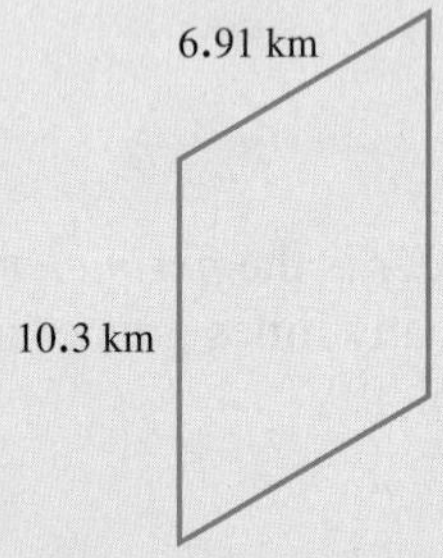

To find the area of a parallelogram, first draw a dashed line inside the figure as shown here.

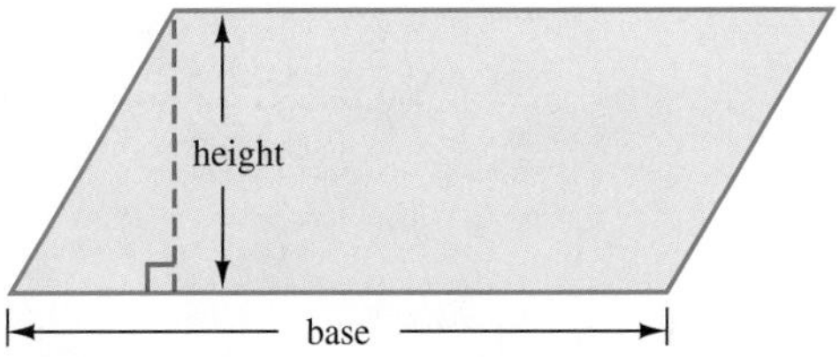

Try this yourself by tracing this parallelogram onto a piece of paper.

The length of the dashed line is the *height* of the parallelogram. It forms a *right angle* with the base. The height is the shortest distance between the base and the opposite side.

Now cut off the triangle created on the left side of the parallelogram above and move it to the right side, as shown below.

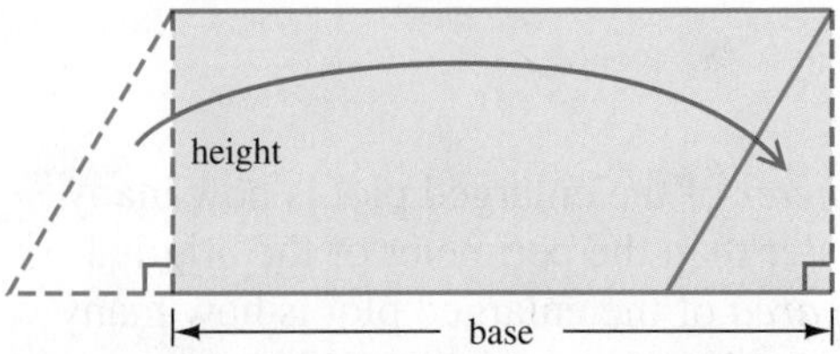

The parallelogram has been made into a rectangle. You can see that the area of the parallelogram and the rectangle are the same.

Equal areas:
Area of the rectangle = length • width
↓ ↓ ↓
Area of the parallelogram = base • height

Answers

1. (a) 15 yd; $P = 84$ yd
(b) $P = 34.42$ km

Finding the Area of a Parallelogram

Area of a parallelogram = base • height

$$A = b \bullet h$$

Remember to use ***square units*** when measuring area.

EXAMPLE 2 Finding the Area of Parallelograms

Find the area of each parallelogram.

(a)

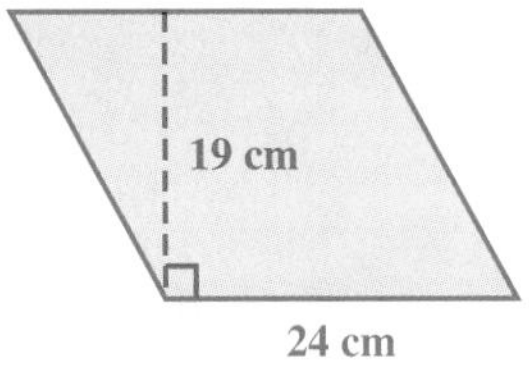

The base is 24 cm and the height is 19 cm. Use the formula $A = b \bullet h$.

$A = b \bullet h$

$A = 24 \text{ cm} \bullet 19 \text{ cm}$

$A = 456 \text{ cm}^2$ — This is **area**; write **cm²** in the answer.

(b)

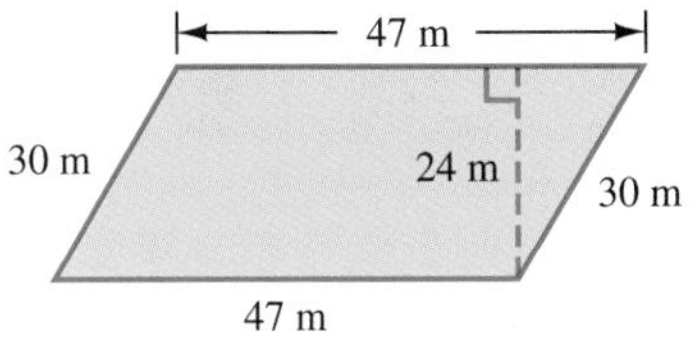

$A = 47 \text{ m} \bullet 24 \text{ m}$

$A = 1128 \text{ m}^2$ — Write **m²** in the answer.

Notice that you do *not* use the 30 m sides when finding the area. But you would use them when finding the *perimeter* of the parallelogram.

Work Problem 2 at the Side. ▶

OBJECTIVE 2 Find the perimeter and area of a trapezoid. A **trapezoid** is a four-sided figure with exactly one pair of parallel sides, such as the figures shown below. Unlike parallelograms, opposite sides of a trapezoid might *not* have the same length.

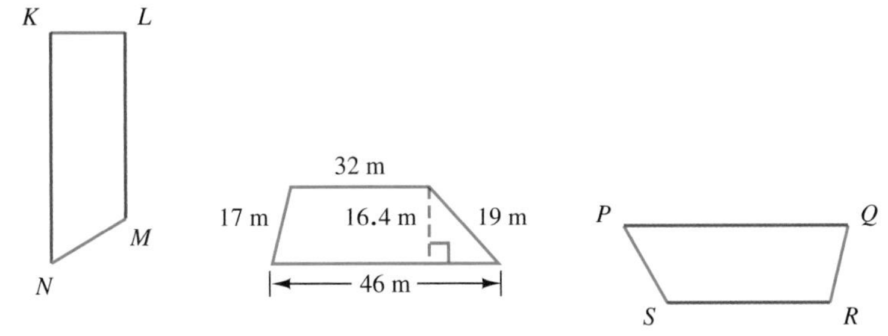

$\overline{KN}$ is parallel to $\overline{LM}$.

The 32 m and 46 m sides are parallel.

$\overline{PQ}$ is parallel to $\overline{SR}$.

EXAMPLE 3 Finding the Perimeter of a Trapezoid

Find the perimeter of the middle trapezoid above.

You can find the perimeter of any flat shape by adding the lengths of the sides.

$$P = 17 \text{ m} + 32 \text{ m} + 19 \text{ m} + 46 \text{ m}$$

$$P = 114 \text{ m}$$

The height of 16.4 m is **not** one of the sides in the perimeter.

Notice that the height (16.4 m) is *not* part of the perimeter, because the height is *not* one of the *outside edges* of the shape.

Work Problem 3 at the Side. ▶

2 Find the area of each parallelogram.

GS **(a)**

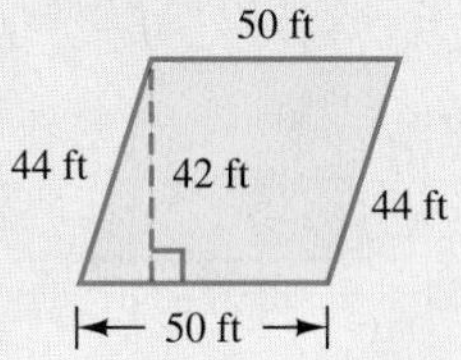

$A = b \bullet h$

$A = 50 \text{ ft} \bullet 42 \text{ ft}$

$A =$ ________

(b)

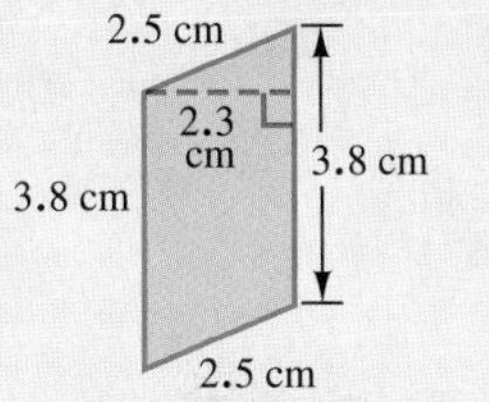

3 Find the perimeter of each trapezoid.

(a)

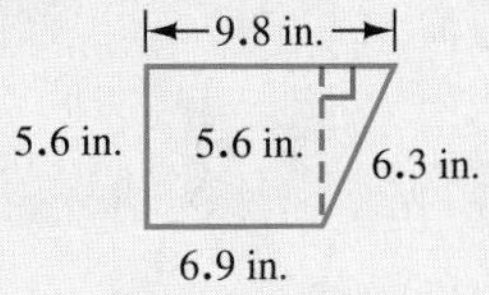

(b)

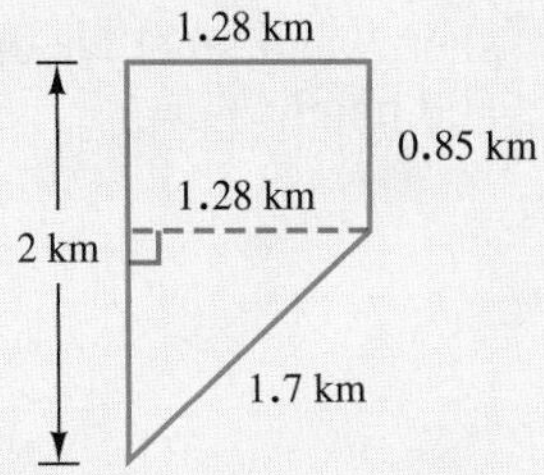

Answers

2. **(a)** $A = 2100 \text{ ft}^2$ **(b)** $A = 8.74 \text{ cm}^2$
3. **(a)** $P = 28.6$ in. **(b)** $P = 5.83$ km

4 Find the area of each trapezoid.

GS **(a)**

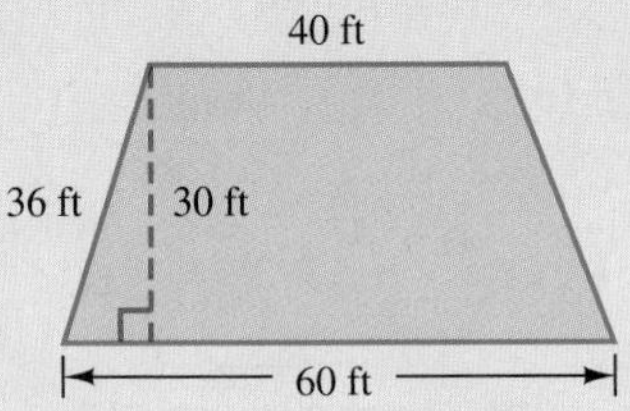

$A = \frac{1}{2} \cdot \quad h \quad \cdot (b \quad + \quad B)$

$A = \frac{1}{2} \cdot 30 \text{ ft} \cdot (40 \text{ ft} + 60 \text{ ft})$

$A = \frac{1}{\cancel{2}_{1}} \cdot \overset{15}{\cancel{30}} \text{ ft} \cdot (100 \text{ ft})$

$A =$ ________

(b)

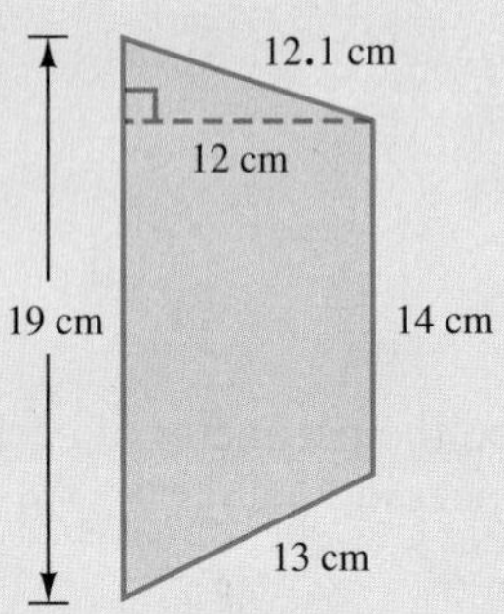

(c) A trapezoid with height 4.7 m, short base 9 m, and long base 10.5 m (First draw a sketch and label the bases and height.)

Use this formula to find the *area* of a trapezoid.

Finding the Area of a Trapezoid

$$\text{Area} = \frac{1}{2} \cdot \text{height} \cdot (\text{short base} + \text{long base})$$

$$A = \frac{1}{2} \cdot h \cdot (b + B)$$

$$\text{or} \quad A = 0.5 \cdot h \cdot (b + B)$$

Remember to use ***square units*** when measuring area.

EXAMPLE 4 Finding the Area of a Trapezoid

Find the area of this trapezoid. The short base and long base are the *parallel* sides.

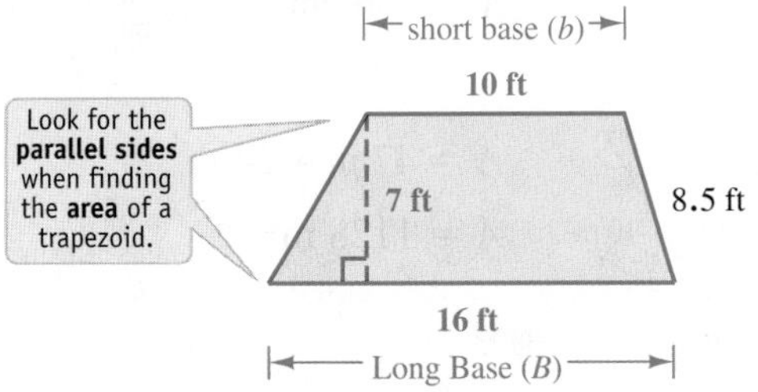

The height (h) is **7 ft**, the short base (b) is **10 ft**, and the long base (B) is **16 ft**. You do *not* need the lengths of the other two sides to find the area.

$A = \frac{1}{2} \cdot \quad h \quad \cdot \quad (b \quad + \quad B)$

$A = \frac{1}{2} \cdot 7 \text{ ft} \cdot (10 \text{ ft} + 16 \text{ ft})$ Work inside parentheses first.

$A = \frac{1}{\cancel{2}_{1}} \cdot 7 \text{ ft} \cdot (\overset{13}{\cancel{26}} \text{ ft})$

$A = 91 \text{ ft}^2$ This is **area**, so write $\textbf{ft}^2$ in the answer.

You can also use 0.5, the decimal equivalent for $\frac{1}{2}$, in the formula.

$A = 0.5 \cdot h \cdot (b \quad + B)$

$A = 0.5 \cdot 7 \cdot (\underbrace{10 + 16})$

$A = 0.5 \cdot 7 \cdot \quad 26$

$A = 91 \text{ ft}^2$ Same answer as above

Calculator Tip

Use the parentheses keys on your scientific calculator to work **Example 4** above.

0.5 (×) 7 (×) (( ) 10 (+) 16 ( )) (=) 91

What happens if you do *not* use the parentheses keys? What order of operations will the calculator follow then? (Answer: The calculator will multiply 0.5 times 7 times 10, and then add 16, giving an *incorrect* answer of 51.)

◀ **Work Problem 4 at the Side.**

Answers

4. **(a)** $A = 1500 \text{ ft}^2$ **(b)** $A = 198 \text{ cm}^2$
(c) $A = 45.825 \text{ m}^2$

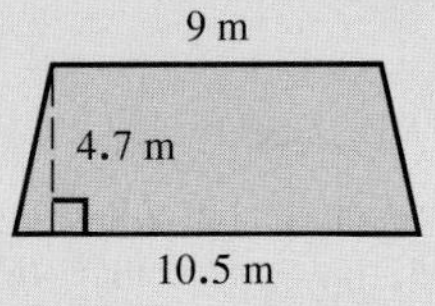

EXAMPLE 5 Finding the Area of a Composite Figure

Find the area of this figure.

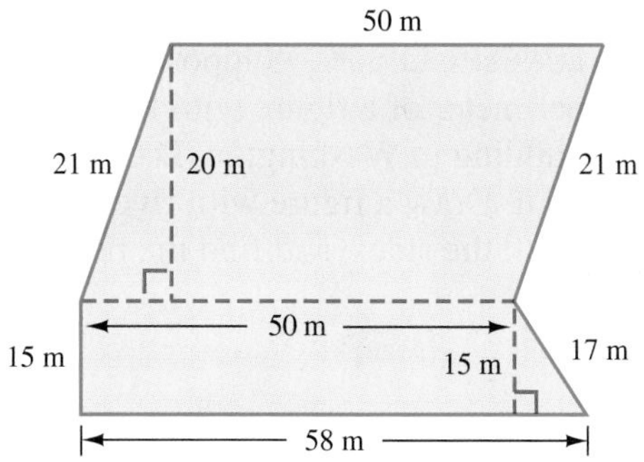

Break the figure into two pieces, a parallelogram (top part) and a trapezoid (bottom part). Find the area of each piece, and then add the areas.

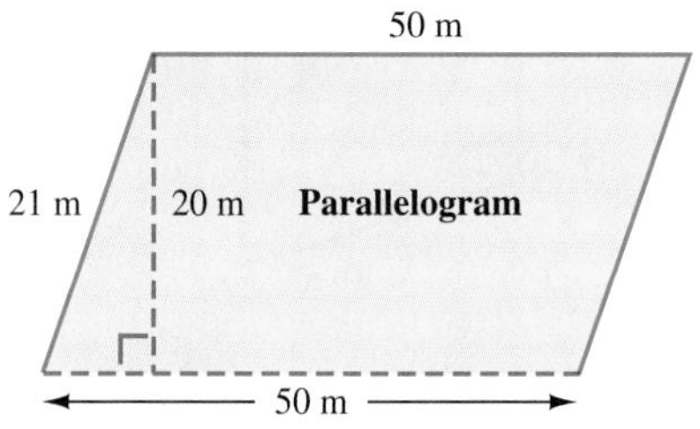

Area of parallelogram

$A = b \cdot h$

$A = 50 \text{ cm} \cdot 20 \text{ cm}$

$A = \mathbf{1000\ m^2}$

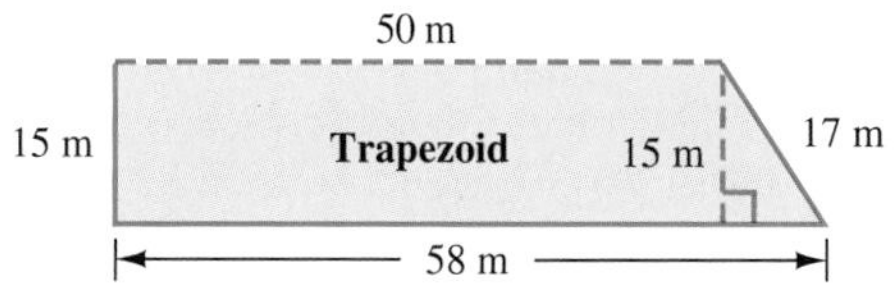

Area of trapezoid

$A = \frac{1}{2} \cdot h \cdot (b + B)$

$A = 0.5 \cdot 15 \text{ m} \cdot (50 \text{ m} + 58 \text{ m})$

$A = \mathbf{810\ m^2}$

Total area $= 1000 \text{ m}^2 + 810 \text{ m}^2 = \mathbf{1810\ m^2}$ ← This is **area**, so write **m²** in the answer.

The area of the figure is 1810 m².

Work Problem **5** at the Side. ▶

EXAMPLE 6 Applying Knowledge of Area

Suppose the figure in **Example 5** above represents the floor plan of a hotel lobby. What is the cost of labor to install tile on the floor if the labor charge is \$35.11 per square meter?

From **Example 5,** the floor area is 1810 m². To find the labor cost, multiply the number of square meters times the cost of labor per square meter.

$$\text{Cost} = \frac{1810 \cancel{\text{m}^2}}{1} \cdot \frac{\$35.11}{1 \cancel{\text{m}^2}}$$

$$\text{Cost} = \$63{,}549.10$$

The cost of the labor is \$63,549.10.

Work Problem **6** at the Side. ▶

5 Find the area of each floor.

(a)

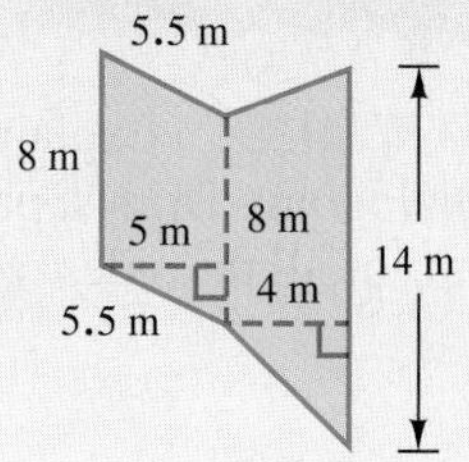

Hint: The left side of the floor is a parallelogram; the right side is a trapezoid.

(b)

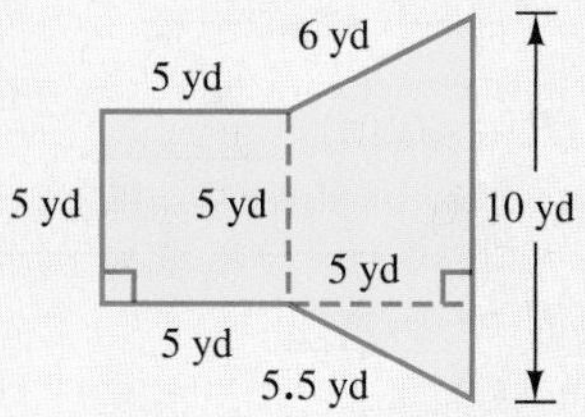

6 Find the cost of carpeting the floors in **Margin Problem 5** above. The cost of carpet is as follows:

(a) Floor (a) is \$18.50 per square meter.

(b) Floor (b) is \$28 per square yard.

Answers

5. (a) $A = 40 \text{ m}^2 + 44 \text{ m}^2 = 84 \text{ m}^2$
   (b) $A = 25 \text{ yd}^2 + 37.5 \text{ yd}^2 = 62.5 \text{ yd}^2$
6. (a) \$1554 (b) \$1750

7.3 Exercises

FOR EXTRA HELP

Download the MyDashBoard App

MyMathLab®

1. **CONCEPT CHECK** Explain what is special about a parallelogram. Draw a parallelogram and label the lengths of all the sides.

2. **CONCEPT CHECK** Suppose you want to find the perimeter of a figure with five sides, like the Pentagon building in Washington, D.C. Explain how you would do it. Draw a figure with five sides, label the lengths of all the sides, and find the perimeter.

Find the perimeter of each figure. ***See Examples 1 and 3.***

3. GS

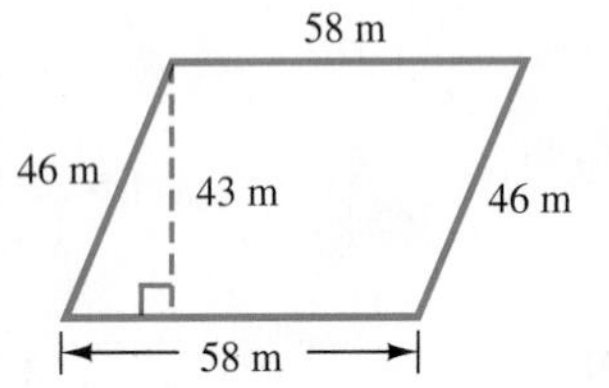

$P = 46\text{ m} + \_\_\_\_\_\_ + \_\_\_\_\_\_ + \_\_\_\_\_\_$

$P = \_\_\_\_\_\_$

4. GS

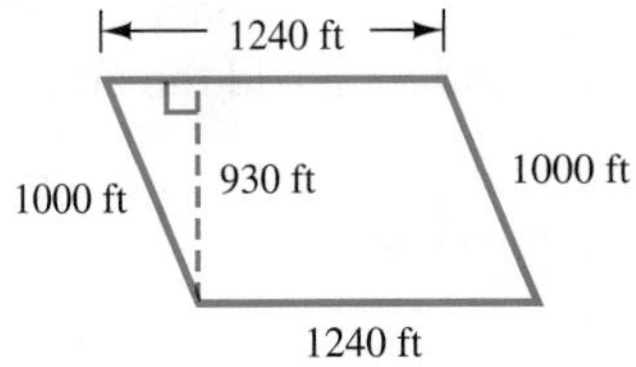

$P = 1000\text{ ft} + \_\_\_\_\_\_ + \_\_\_\_\_\_ + \_\_\_\_\_\_$

$P = \_\_\_\_\_\_$

5.

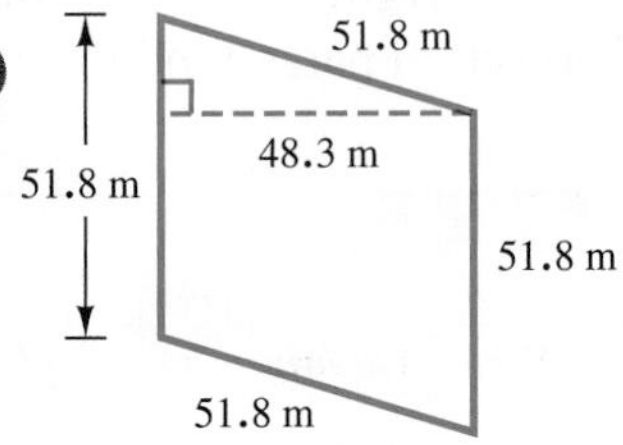

6.

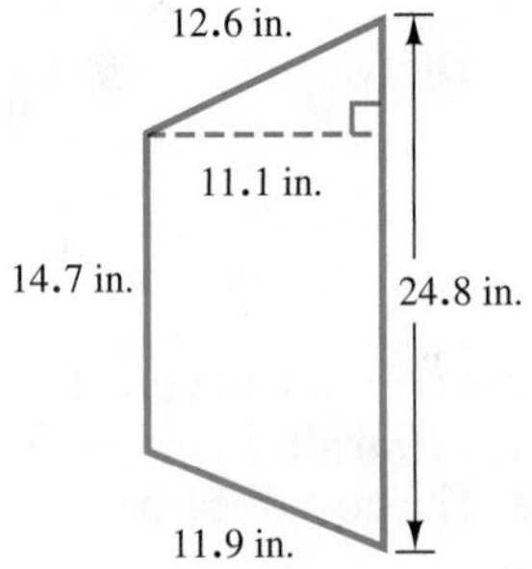

7.

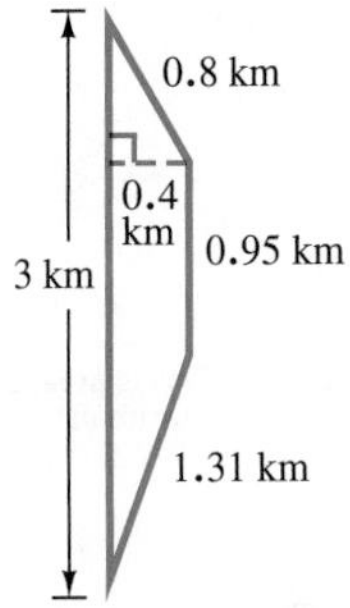

8.

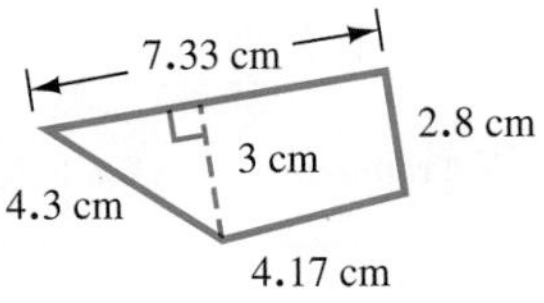

Find the area of each figure. ***See Examples 2 and 4.***

9.

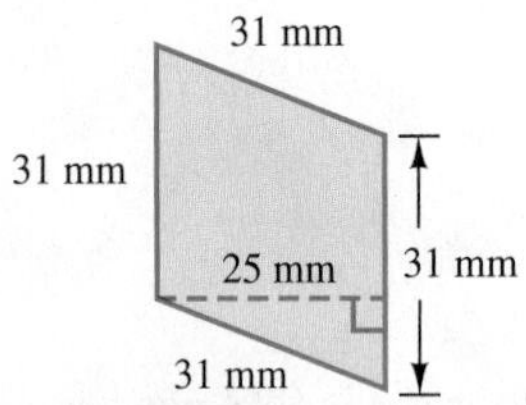

$A = b \cdot h$

10.

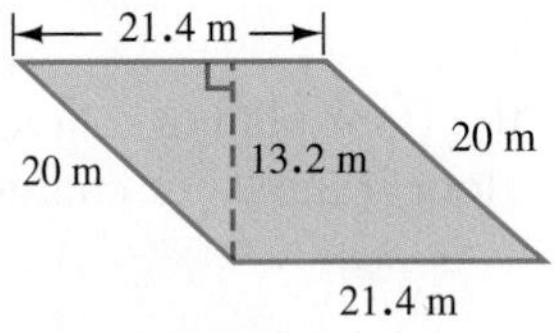

$A = b \cdot h$

11.

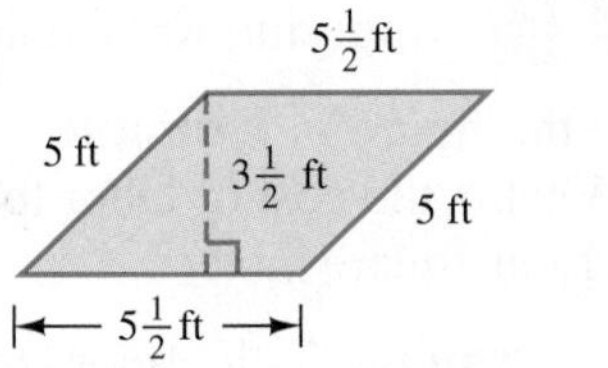

12.

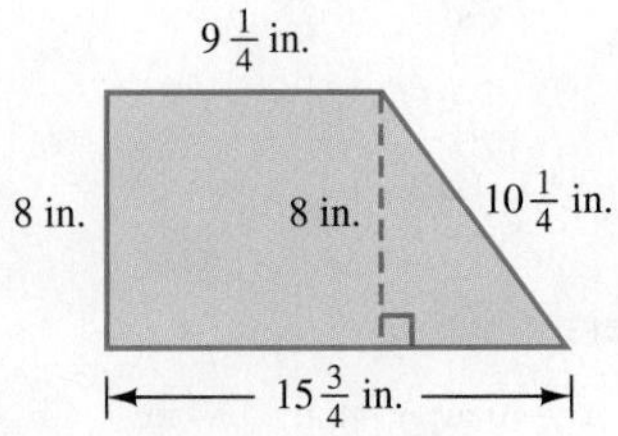

13.

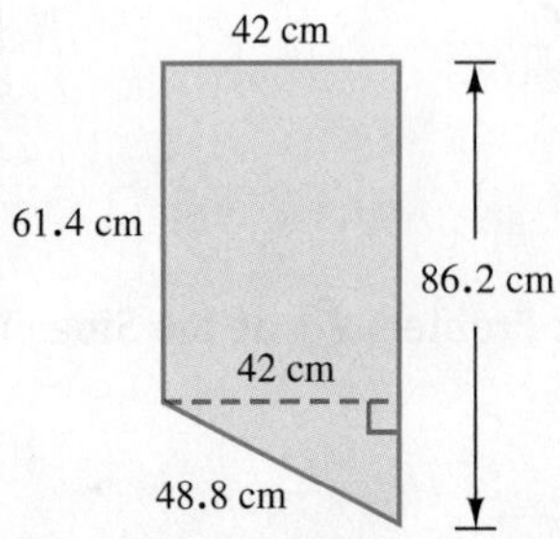

14.

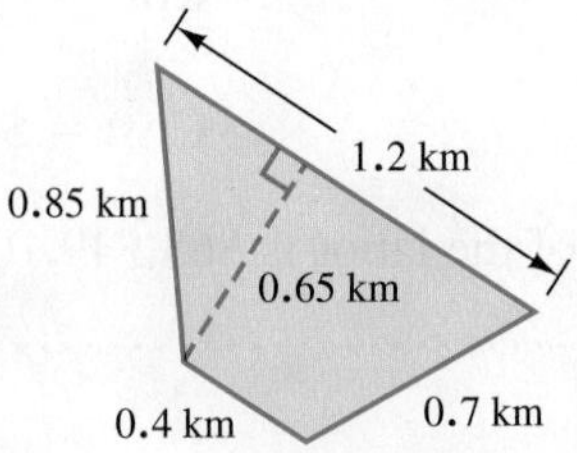

CONCEPT CHECK *Find* ***two*** *errors in each student's solution below. Write a sentence explaining each error. Then show how to work the problem correctly.*

15.

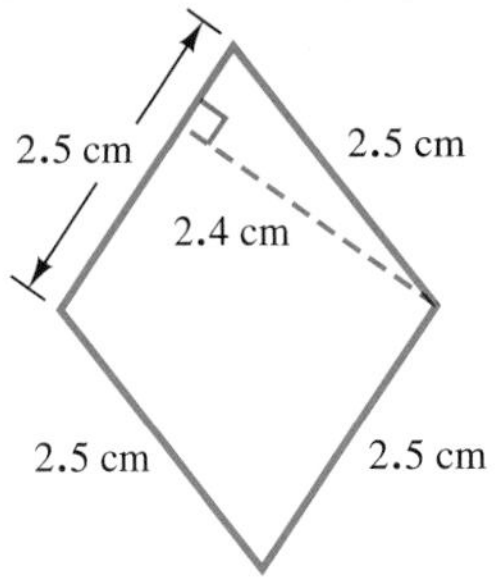

$P = 2.5\text{ cm} + 2.4\text{ cm} + 2.5\text{ cm} + 2.5\text{ cm} + 2.5\text{ cm}$

$P = 12.4\text{ cm}^2$

16.

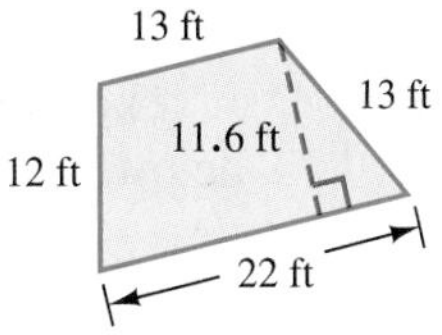

$A = (0.5)(11.6\text{ ft}) \cdot (12\text{ ft} + 13\text{ ft})$

$A = 145\text{ ft}$

Solve each application problem. First label the bases and heights on the sketches. ***See Example 6.***

17. The backyard of a new home is shaped like a trapezoid with a height of 45 ft and bases of 80 ft and 110 ft. What is the cost of putting sod on the yard if the landscaper charges $0.33 per square foot for sod?

18. The Dockland office building in Hamburg, Germany, is seven stories tall. Each of the seven floors is in the shape of a parallelogram with a height of 22 m and a base of 88 m. Find the total area of all seven floors.

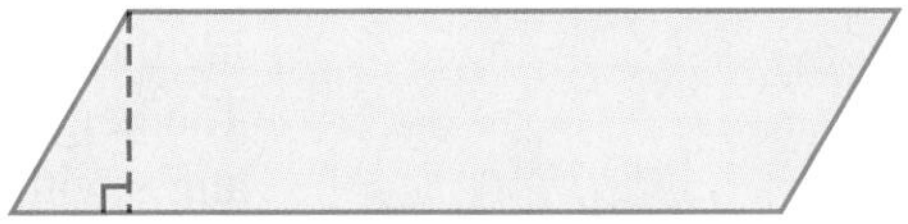

19. A piece of fabric for a quilt design is in the shape of a parallelogram. The base is 5 in. and the height is 3.5 in. What is the total area of the 25 parallelogram pieces needed for the quilt?

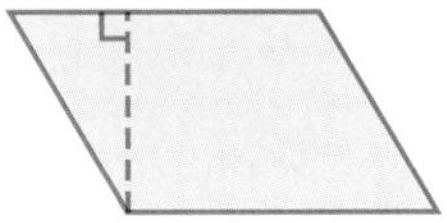

20. An accountant is paying $832 per month to rent an office in an old building. Her office is shaped like a trapezoid, with bases of 32 ft and 20 ft and a height of 20 ft. How much rent is she paying per square foot?

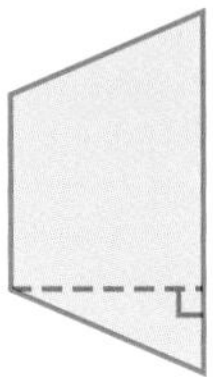

Find the area of each figure. ***See Example 5.***

21.

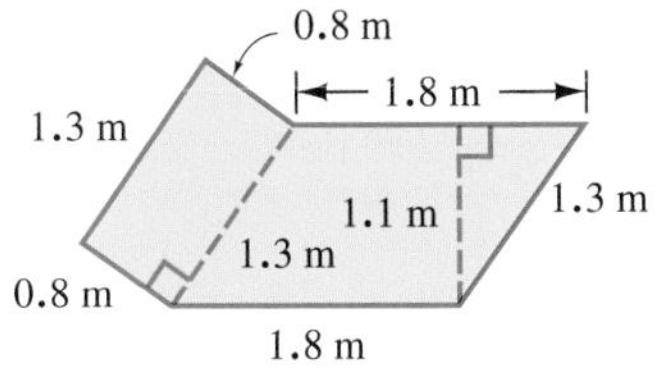

22.

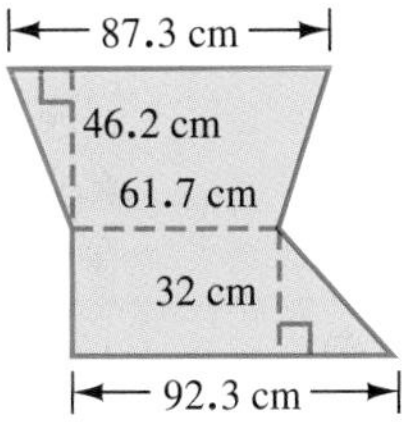

23.

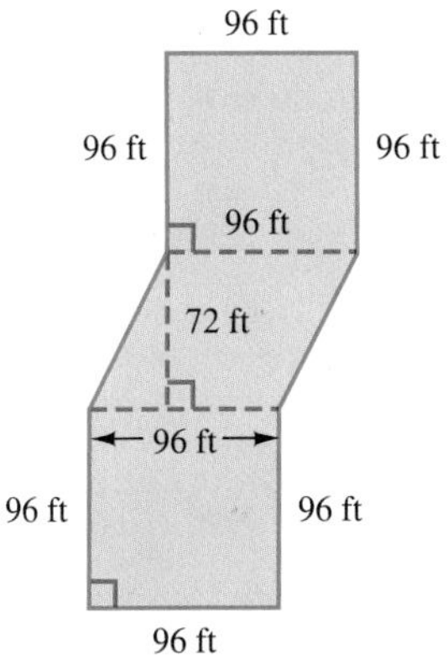

7.4 Triangles

OBJECTIVES

1. Find the perimeter of a triangle.
2. Find the area of a triangle.
3. Given the measures of two angles in a triangle, find the measure of the third angle.

A **triangle** is a figure with exactly three sides. Some examples are shown below.

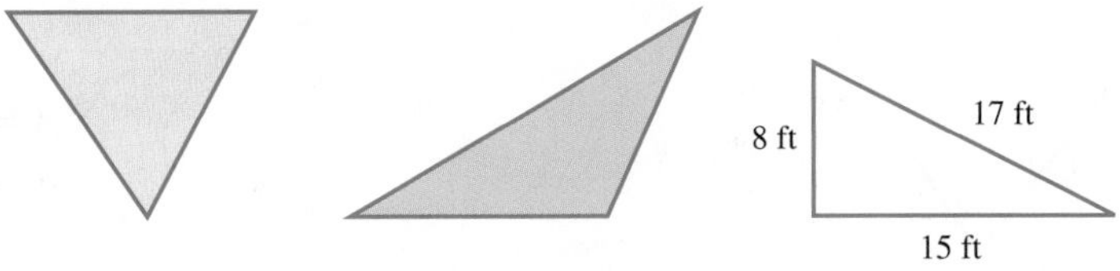

VOCABULARY TIP

Tri- means three. Just as a **tri**cycle has three wheels, a **tri**angle has three sides.

OBJECTIVE 1 Find the perimeter of a triangle. To find the perimeter of a triangle (the distance around the edges), add the lengths of the three sides.

EXAMPLE 1 Finding the Perimeter of a Triangle

Find the perimeter of the triangle above on the right.

$$P = 8\text{ ft} + 15\text{ ft} + 17\text{ ft}$$

$$P = 40\text{ ft}$$

This is **perimeter**, so write **ft** in the answer.

◀ **Work Problem 1 at the Side.**

1 Find the perimeter of each triangle.

GS **(a)**

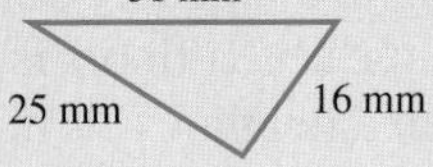

$P = 31\text{ mm} + 25\text{ mm}$

$+$ ______

$P =$ ______

(b)

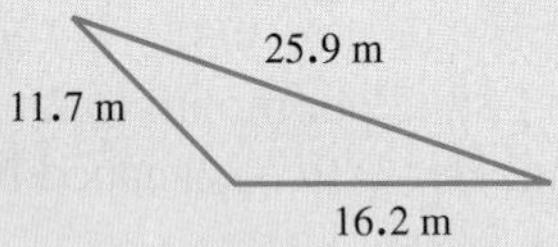

(c) A triangle with sides of $6\frac{1}{2}$ yd, $9\frac{3}{4}$ yd, and $11\frac{1}{4}$ yd

As with parallelograms, you can find the *height* of a triangle by measuring the distance from one vertex of the triangle to the opposite side (the base). The height line must be *perpendicular* to the base; that is, it must form a right angle with the base. Sometimes you have to extend the base in order to draw the height perpendicular to it, as shown below in the figure on the right.

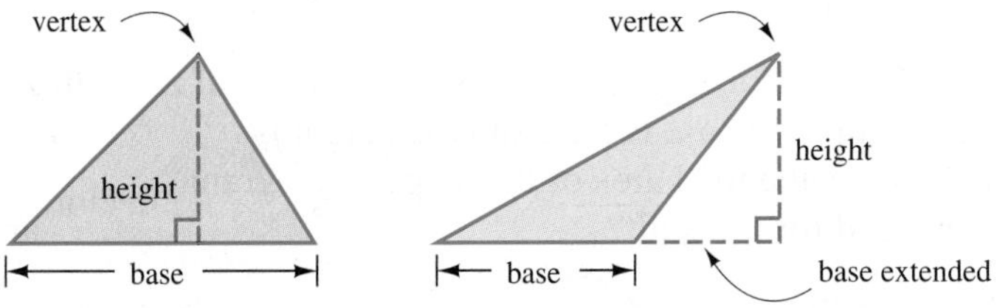

If you cut out two identical triangles and turn one upside down, you can fit them together to form a parallelogram, as shown below.

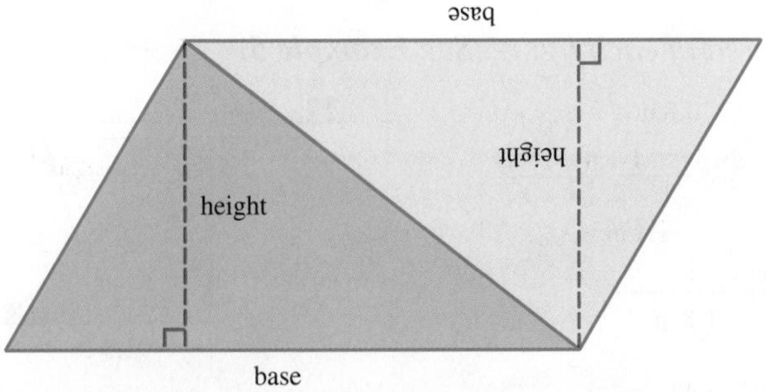

Recall from the last section that the area of the parallelogram is *base* times *height*. Because each triangle is *half* of the parallelogram, the area of one triangle is

$$\frac{1}{2}\text{ of base times height.}$$

Answers

1. **(a)** 16 mm; $P = 72$ mm
(b) $P = 53.8$ m **(c)** $P = 27\frac{1}{2}$ yd or 27.5 yd

OBJECTIVE 2 **Find the area of a triangle.** Use the following formula to find the *area* of a triangle.

Finding the Area of a Triangle

$$\text{Area of a triangle} = \frac{1}{2} \cdot \text{base} \cdot \text{height}$$

$$A = \frac{1}{2} \cdot b \cdot h$$

$$\text{or} \quad A = 0.5 \cdot b \cdot h$$

Remember to use ***square units*** when measuring area.

EXAMPLE 2 Finding the Area of Triangles

Find the area of each triangle.

(a)

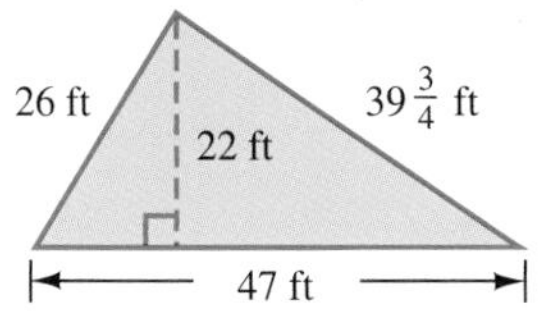

The base is 47 ft and the height is 22 ft. You do *not* need the 26 ft or $39\frac{3}{4}$ ft sides to find the area.

$$A = \frac{1}{2} \cdot b \cdot h$$

$$A = \frac{1}{\cancel{2}_1} \cdot 47\text{ ft} \cdot \cancel{22}^{11}\text{ ft}$$ Divide out common factor of 2.

$$A = 517\text{ ft}^2$$ Square units for area

This is *area*, so write **ft²** in the answer.

(b)

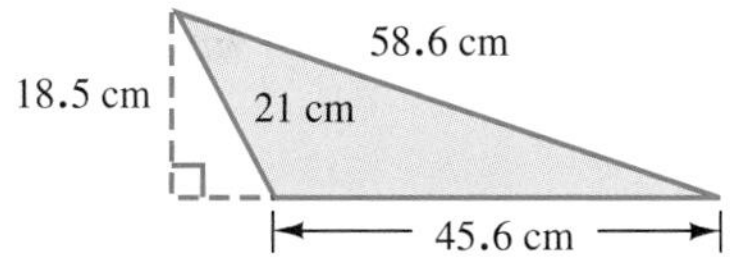

$A = 0.5 \cdot 45.6\text{ cm} \cdot 18.5\text{ cm}$

$A = 421.8\text{ cm}^2$

The base must be extended to draw the height. However, still use 45.6 cm for b in the formula. Because the measurements are decimal numbers, it is easier to use 0.5 (the decimal equivalent of $\frac{1}{2}$) in the formula.

(c)

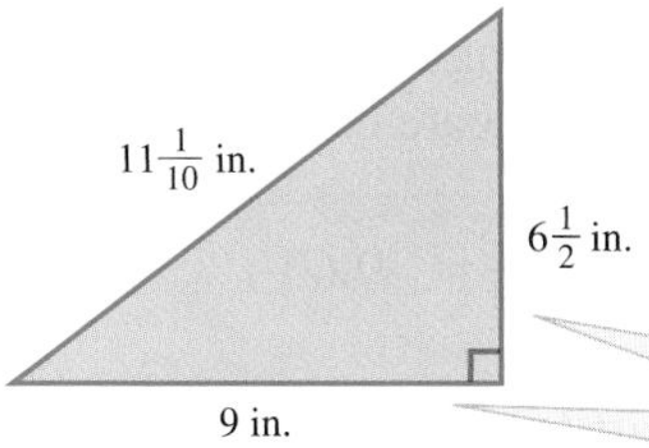

The measurements for this triangle are mixed numbers. Change them to improper fractions or use the equivalent decimal numbers.

Two sides of the triangle are perpendicular to each other, so use those sides as the base and the height.

Using fractions $A = \frac{1}{2} \cdot 9\text{ in.} \cdot 6\frac{1}{2}\text{ in.} = \frac{1}{2} \cdot \frac{9\text{ in.}}{1} \cdot \frac{13\text{ in.}}{2} = \frac{117}{4}\text{ in.}^2 = 29\frac{1}{4}\text{ in.}^2$

Using decimals $A = 0.5 \cdot 9\text{ in.} \cdot 6.5\text{ in.} = 29.25\text{ in.}^2$ ← Equivalent

Work Problem 2 at the Side. ▶

---

2 Find the area of each triangle.

GS **(a)**

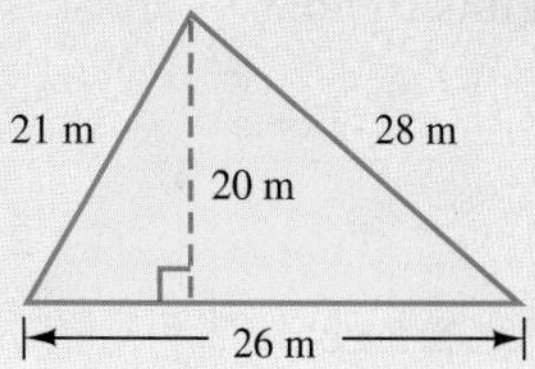

$A = \frac{1}{2} \cdot b \cdot h$

$A = \frac{1}{\cancel{2}_1} \cdot \cancel{26}^{13}\text{ m} \cdot 20\text{ m}$

$A =$ ______

(b)

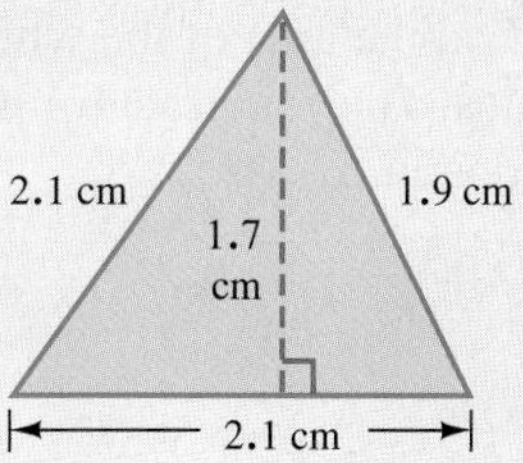

(c)

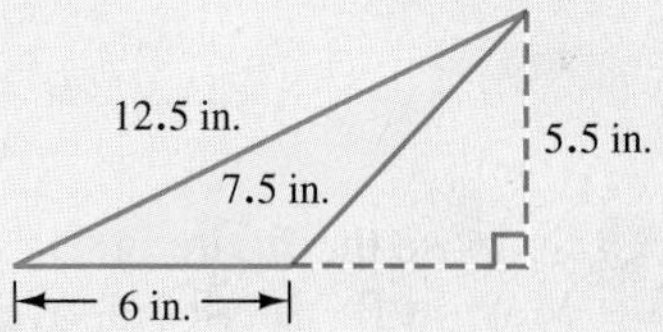

GS **(d)**

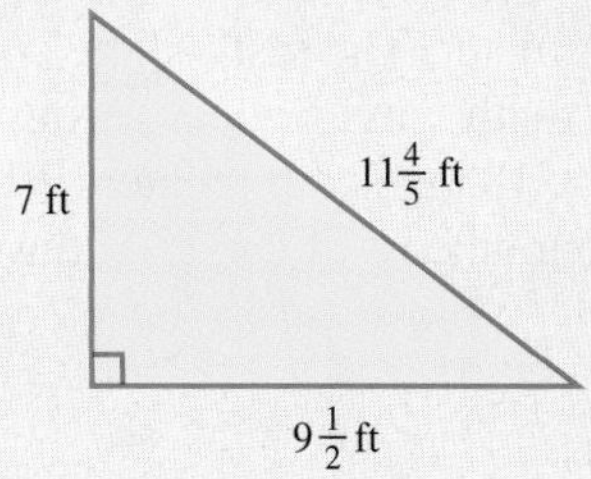

Hint: The base and height are perpendicular. So the base is $9\frac{1}{2}$ ft, or use 9.5 ft. The height is 7 ft.

Answers

2. **(a)** $A = 260\text{ m}^2$ **(b)** $A = 1.785\text{ cm}^2$
(c) $A = 16.5\text{ in.}^2$
(d) $A = 33.25\text{ ft}^2$ or $33\frac{1}{4}\text{ ft}^2$

3 Find the area of the shaded part in this figure.

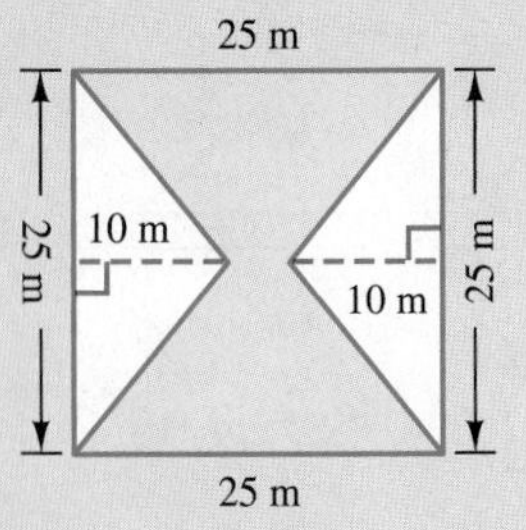

The entire figure is a square. Find the area of the square.

$A = 25 \text{ m} \cdot 25 \text{ m}$

$A =$ _______ for entire figure.

Now find the area of the white triangles. Then subtract to find the shaded area.

4 Suppose the figure in **Margin Problem 3** above is an auditorium floor plan. The shaded part will be covered with carpet costing \$27 per square meter. The rest will be covered with vinyl floor covering costing \$18 per square meter. What is the total cost of covering the floor?

Answers

3. $A = 625 \text{ m}^2$ for entire figure.
For shaded portion,
$A = 625 \text{ m}^2 - 125 \text{ m}^2 - 125 \text{ m}^2 = 375 \text{ m}^2$.
4. \$10,125 + \$2250 + \$2250 = \$14,625

EXAMPLE 3 Using the Concept of Area

Find the area of the shaded part in this figure.

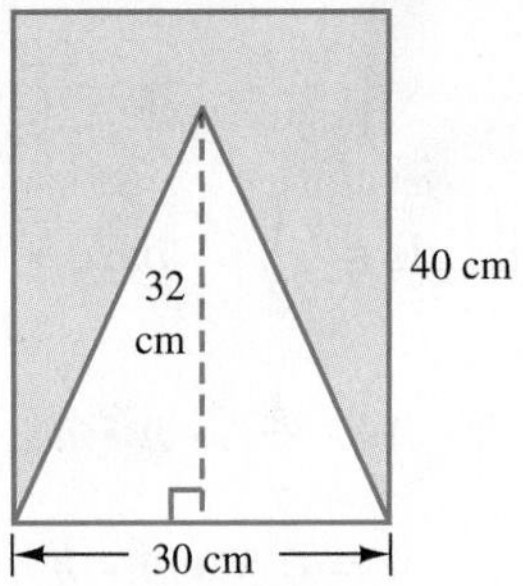

The *entire* figure is a rectangle. Find the area of the rectangle.

$$A = l \cdot w$$

$$A = 30 \text{ cm} \cdot 40 \text{ cm}$$

$$A = 1200 \text{ cm}^2$$

The *un*shaded part is a triangle. Find the area of the triangle.

$$A = \frac{1}{\cancel{2}_1} \cdot \cancel{30}^{15} \text{ cm} \cdot 32 \text{ cm}$$

$$A = 480 \text{ cm}^2$$

Subtract to find the area of the shaded part.

$$A = \underbrace{1200 \text{ cm}^2}_{\text{Entire area}} - \underbrace{480 \text{ cm}^2}_{\text{Unshaded part}} = \underbrace{720 \text{ cm}^2}_{\text{Shaded part}}$$

The area of the shaded part of the figure is 720 cm^2. ◀ This is *area*, so write **cm^2** in the answer.

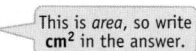

◀ **Work Problem 3 at the Side.**

EXAMPLE 4 Applying the Concept of Area

The Department of Transportation cuts triangular signs out of rectangular pieces of metal using the measurements shown above in **Example 3.** If the metal costs \$0.02 per square centimeter, how much does the metal cost for the sign? What is the cost of the metal that is *not* used?

From **Example 3** above, the area of the triangle (the sign) is 480 cm^2. Multiply the area of the sign times the cost per square centimeter.

$$\text{cost of sign} = \frac{480\ \cancel{\text{cm}^2}}{1} \cdot \frac{\$0.02}{1\ \cancel{\text{cm}^2}} = \$9.60$$

The metal that is *not* used is the *shaded* part from **Example 3.** The unused area is 720 cm^2.

$$\text{cost of unused metal} = \frac{720\ \cancel{\text{cm}^2}}{1} \cdot \frac{\$0.02}{1\ \cancel{\text{cm}^2}} = \$14.40$$

The cost of the metal for the triangular sign is \$9.60. The unused metal costs \$14.40.

◀ **Work Problem 4 at the Side.**

OBJECTIVE 3 Given the measures of two angles in a triangle, find the measure of the third angle. The *tri* in *tri*angle means *three.* So the name tells you that a triangle has three angles. The sum of the measures of the three angles in any triangle is *always* 180°. You can see it by drawing a triangle, cutting off the three angles, and rearranging them to make a straight angle (180°).

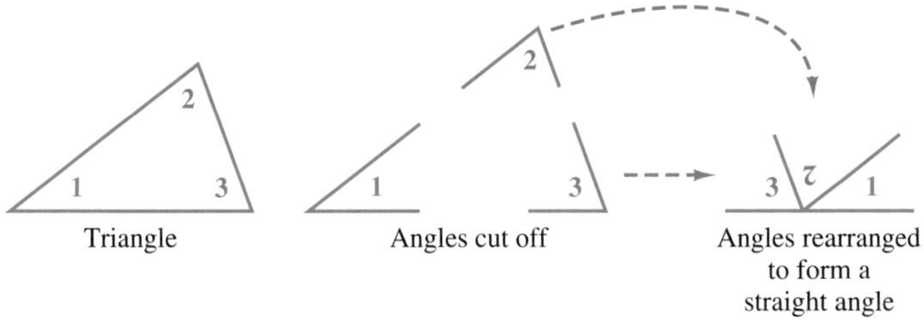

Finding the Unknown Angle Measurement in a Triangle

Step 1 Add the number of degrees in the measures of the two given angles.

Step 2 Subtract the sum from 180°.

EXAMPLE 5 Finding an Angle Measurement in Triangles

Find the number of degrees in the indicated angle.

(a) Angle R

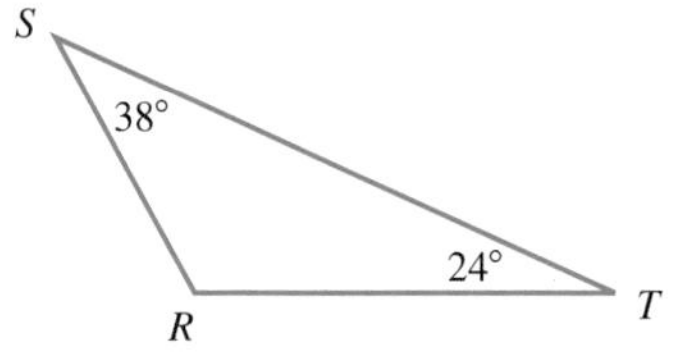

Step 1 Add the two angle measurements you are given.

$38° + 24° = 62°$

Step 2 Subtract the sum from 180°.

$180° - 62° = 118°$

$\angle R$ measures 118°.

(b) Angle F

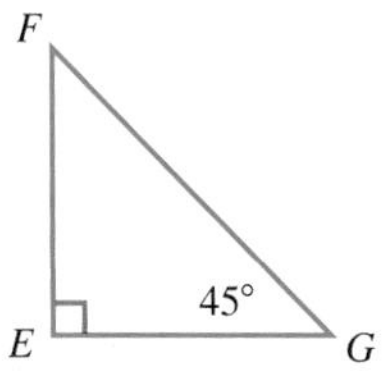

$\angle E$ is a right angle, so it measures 90°.

Step 1 $90° + 45° = 135°$

Step 2 $180° - 135° = 45°$

$\angle F$ measures 45°.

Write a small, raised circle for "degrees."

Work Problem 5 at the Side. ▶

5 Find the number of degrees in the third angle of each triangle.

GS **(a)**

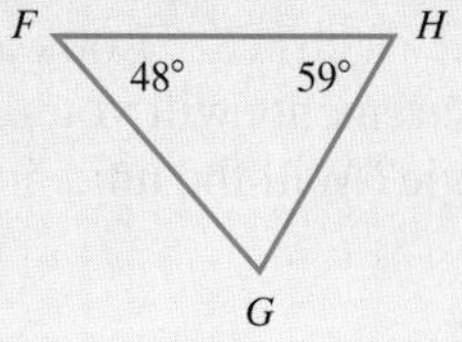

Step 1 48° + 59° = ____

Step 2 180° − ____ = ____

$\angle G$ measures ____

(b)

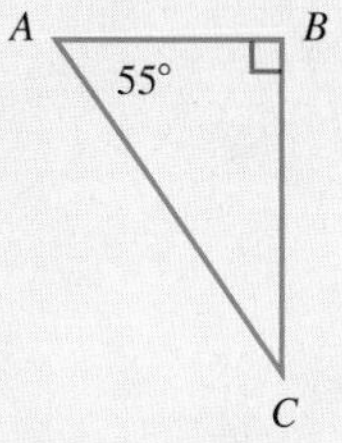

(c)

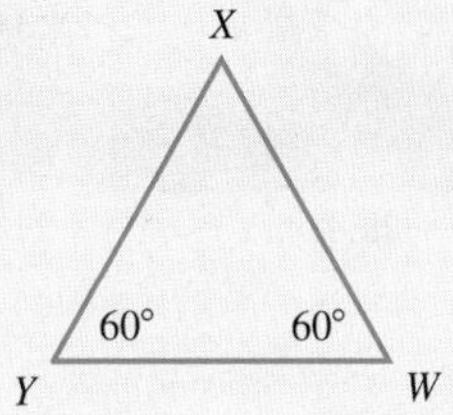

Answers

5. **(a)** 107°; 180° − 107° = 73°; $\angle G$ measures 73°
(b) $\angle C$ measures 35°
(c) $\angle X$ measures 60°

7.4 Exercises

FOR EXTRA HELP

Download the MyDashBoard App

MyMathLab®

1. CONCEPT CHECK Look at **Exercise 3** below. Which measurements will you use to find the area of the triangle? Will the units in your answer be m or m^2?

2. CONCEPT CHECK Look at **Exercise 3** below. Which measurements will you use to find the perimeter of the triangle? Will the units in your answer be m or m^2?

Find the perimeter and area of each triangle. ***See Examples 1 and 2.***

3.

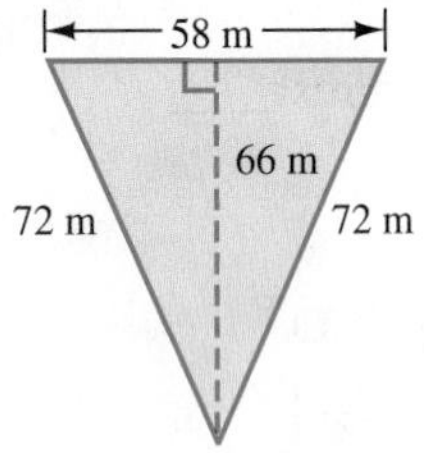

4.

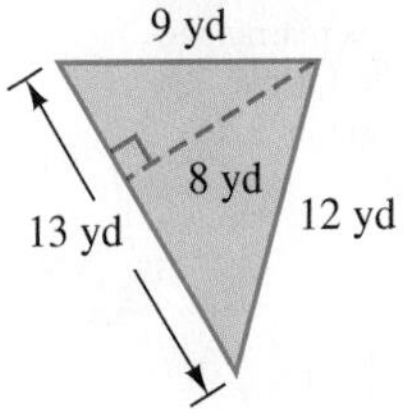

5.

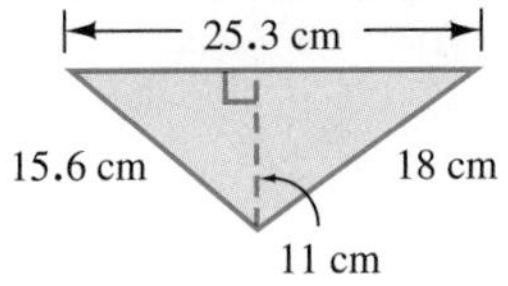

6.

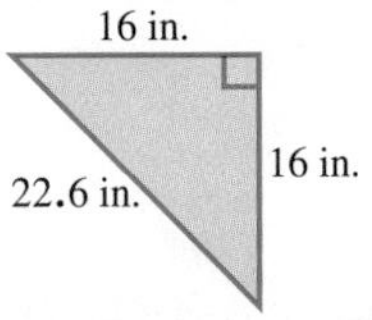

7.

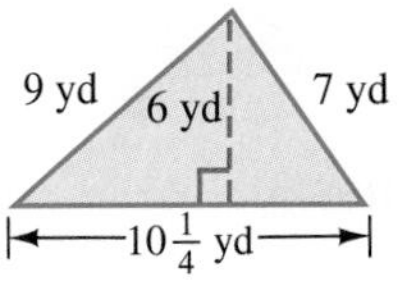

8.

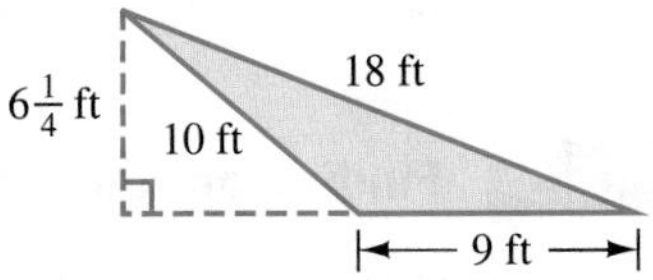

9.

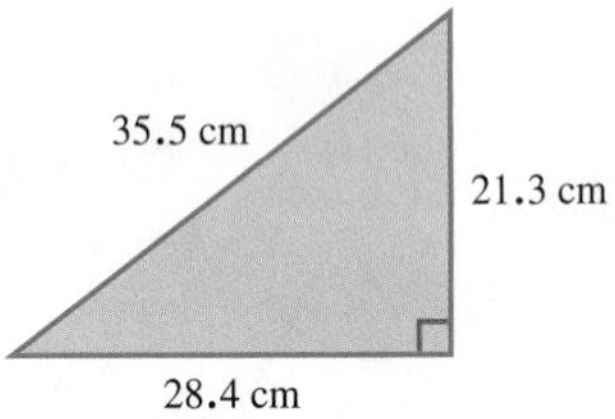

10.

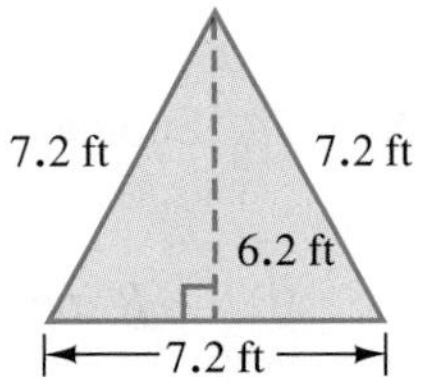

Find the shaded area in each figure. ***See Example 3.***

11. GS

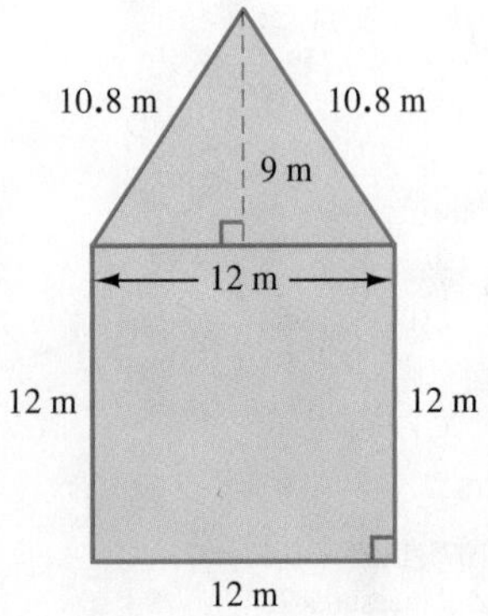

Find the area of the triangle.

$A = 0.5 \cdot 12\text{ m} \cdot$ ______

$A =$ ______

Find the area of the square.

$A = 12\text{ m} \cdot$ ______

$A =$ ______

Now add to find the area of the entire figure.

12. GS

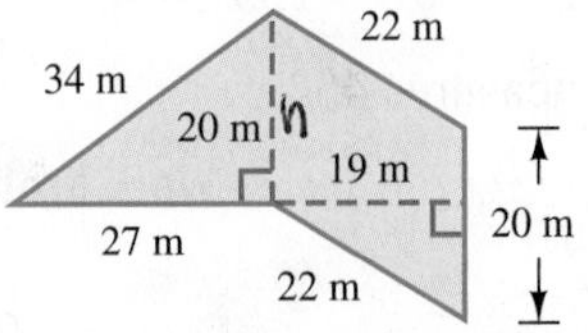

Find the area of the triangle on the left.

$A = 0.5 \cdot 27\text{ m} \cdot$ ______

$A =$ ______

Find the area of the parallelogram on the right.

$A = 20\text{ m} \cdot$ ______

$A =$ ______

Now add to find the area of the entire figure.

13.

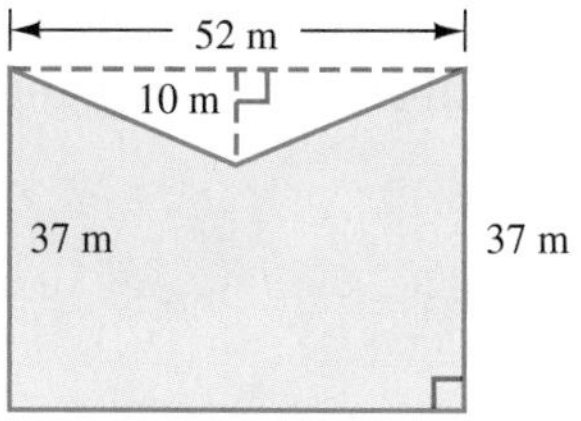

14.

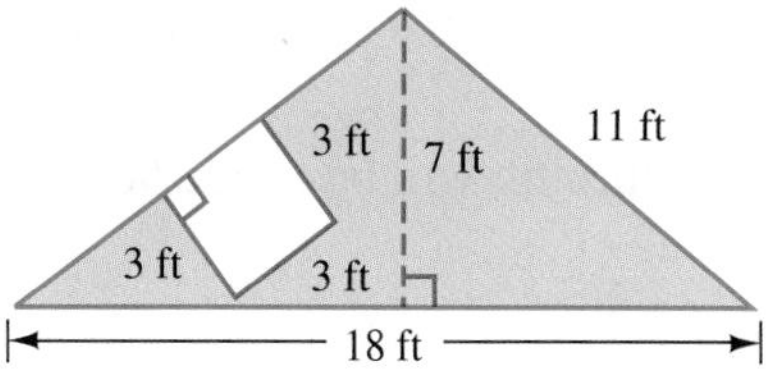

Find the number of degrees in the third angle of each triangle. ***See Example 5.***

15.

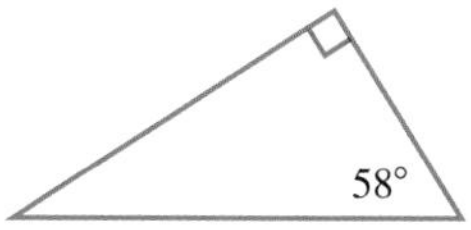

16.

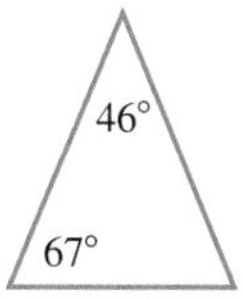

17.

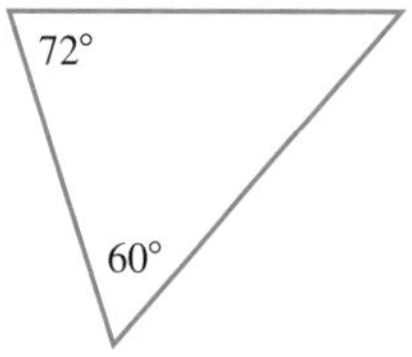

18.

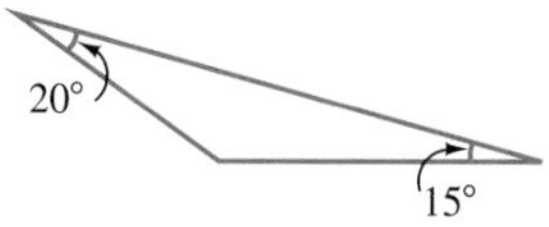

19. CONCEPT CHECK Can a triangle have two right angles? Explain your answer.

20. CONCEPT CHECK In your own words, explain where the $\frac{1}{2}$ comes from in the formula for area of a triangle. Draw a sketch to illustrate your explanation.

Solve each application problem. ***See Example 4.***

21. A triangular tent flap measures $3\frac{1}{2}$ ft along the base and has a height of $4\frac{1}{2}$ ft. How much canvas is needed to make the flap?

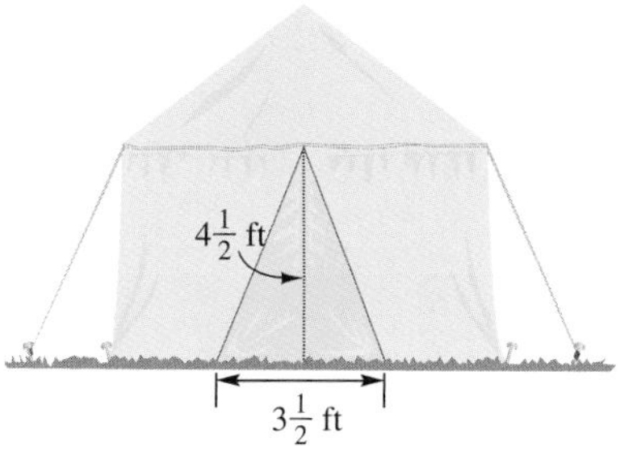

22. A wooden sign in the shape of a right triangle has perpendicular sides measuring 1.5 m and 1.2 m. How much surface area does the sign have?

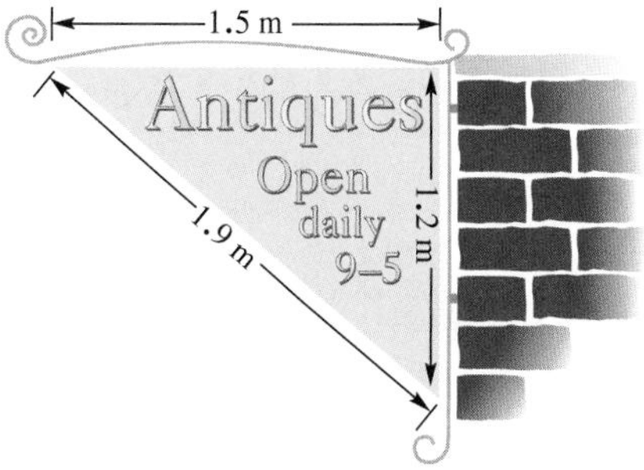

23. A triangular space between three streets has the measurements shown below.

(a) How much new curbing will be needed to go around the space?

(b) How much sod will be needed to cover the space?

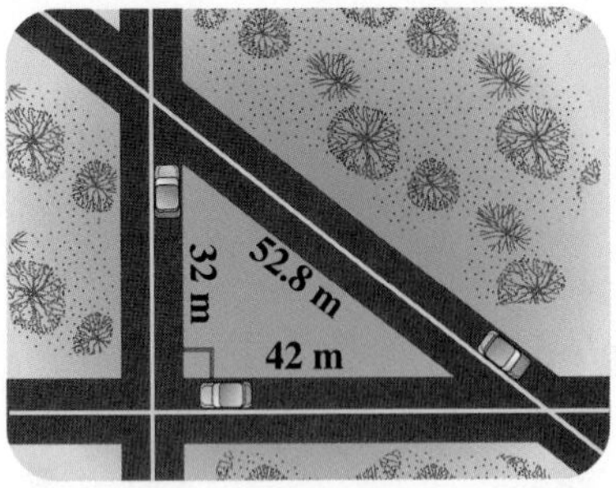

24. Each gable end of a new house has a span of 36 ft and a rise of 9.5 ft. What is the total area of both gable ends of the house?

25. All sides of the house below are congruent and all roof sections are congruent.

(a) Find the area of one side of the house.

(b) Find the area of one roof section.

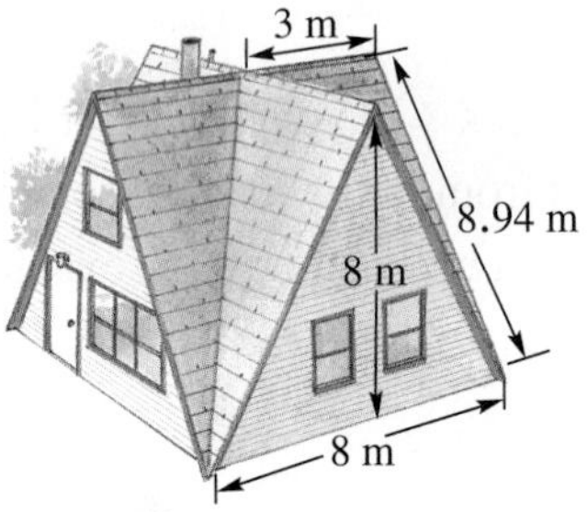

26. The sketch shows the plan for an office building. The shaded area will be a parking lot. What is the cost of building the parking lot if the contractor charges $35 per square yard for materials and labor?

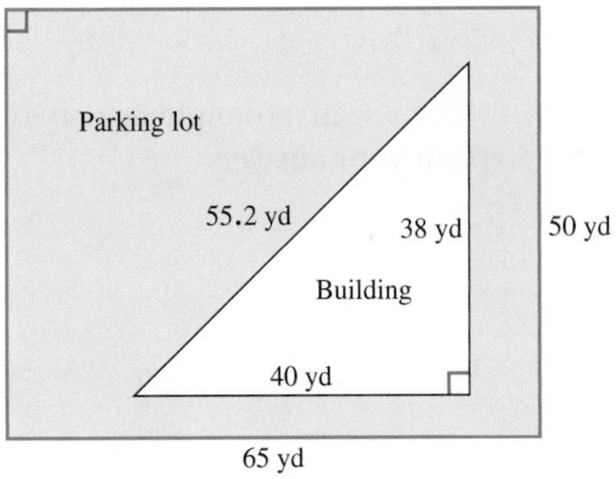

27. A city lot with an unusual shape is shown below.

(a) How much frontage (distance along streets) does the lot have?

(b) What is the area of the lot?

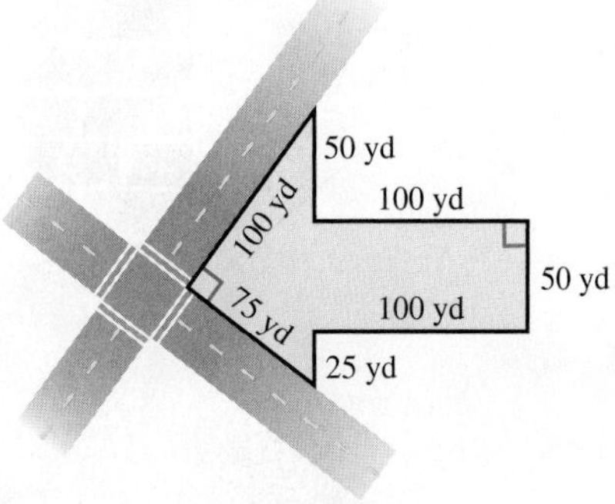

28. A car dealership wants three nylon pennants to hang in its front window. The height of the two smaller pennants is 3.5 ft, and the larger pennant has a height of 6.5 ft. How much nylon fabric is needed for all the pennants?

7.5 Circles

OBJECTIVES

1. Find the radius and diameter of a circle.
2. Find the circumference of a circle.
3. Find the area of a circle.
4. Become familiar with Latin and Greek prefixes used in math terminology.

OBJECTIVE 1 Find the radius and diameter of a circle. Suppose you start with one dot on a piece of paper. Then you draw many dots that are each 2 cm away from the first dot. If you draw enough dots (points) you'll end up with a *circle*. Each point on the circle is exactly 2 cm away from the *center* of the circle. The 2 cm distance is called the *radius*, *r*, of the circle. The distance across the circle (passing through the center) is called the *diameter*, *d*, of the circle.

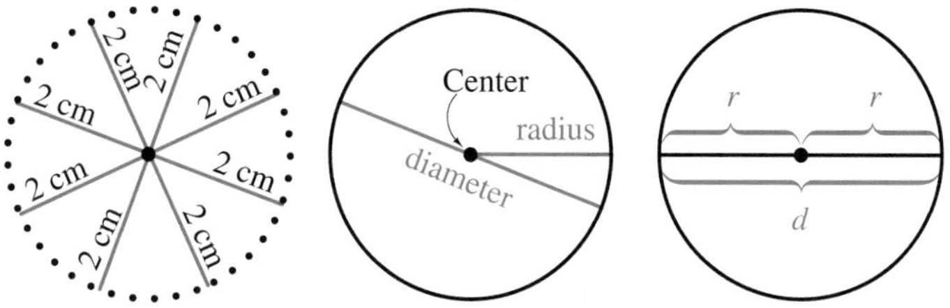

VOCABULARY TIP

Diameter **Dia** means "through," and **meter** means "measure." So, the **diameter** of a circle is a line through the circle, passing through the center. Recall that **radius** comes from the root word meaning "spoke," like the spoke of a wheel. So the **radius** is a line from the center of a circle out to the edge of the circle.

Circle, Radius, and Diameter

A **circle** is a two-dimensional (flat) figure with all points the same distance from a fixed center point.

The **radius** (r) is the distance from the center of the circle to any point on the circle.

The **diameter** (d) is the distance across the circle passing through the center.

Using the circle above on the right as a model, you can see some relationships between the radius and diameter.

Finding the Diameter and Radius of a Circle

$$\text{diameter} = 2 \cdot \text{radius}$$

$$d = 2 \cdot r$$

$$\text{and} \quad \text{radius} = \frac{\text{diameter}}{2} \quad \text{or} \quad r = \frac{d}{2}$$

EXAMPLE 1 **Finding the Diameter and Radius of a Circle**

Find the unknown length of the diameter or radius in each circle.

(a)

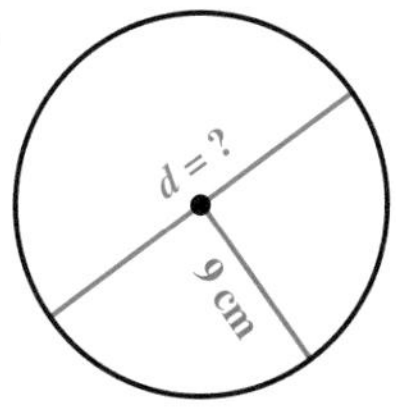

Because the radius is 9 cm, the diameter is twice as long.

$d = 2 \cdot r$

$d = 2 \cdot 9 \text{ cm}$

$d = 18 \text{ cm}$ ← Write **cm** in your answer, **not** cm^2.

Continued on Next Page

1 Find the unknown length of the diameter or radius in each circle.

GS **(a)**

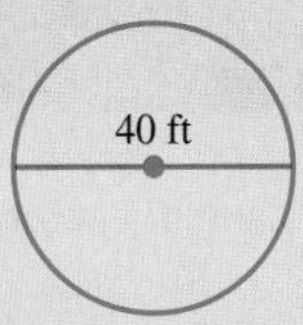

radius (r) is unknown

$$r = \frac{d}{2}$$

$$r = \frac{40 \text{ ft}}{2} = \underline{\qquad}$$

(b)

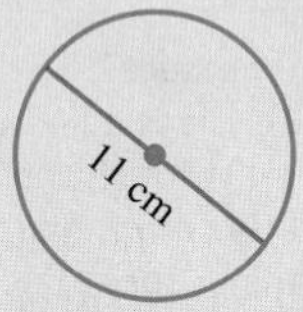

(c)

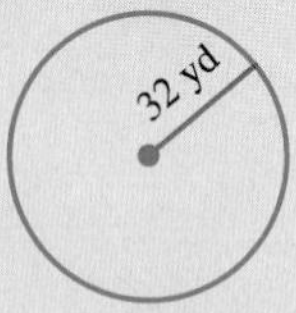

(d)

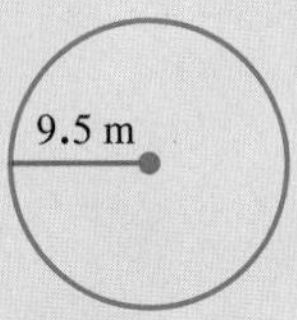

Answers

1. **(a)** $r = 20$ ft **(b)** $r = 5.5$ cm **(c)** $d = 64$ yd **(d)** $d = 19$ m

(b)

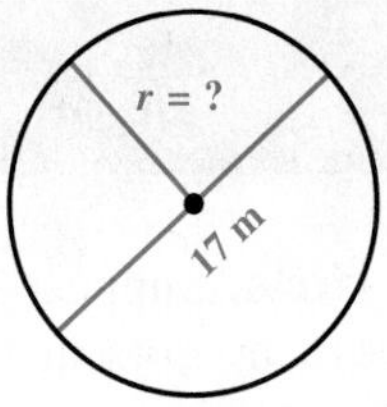

The radius is half the diameter.

$$r = \frac{d}{2}$$

$$r = \frac{\textbf{17 m}}{2}$$

$$r = 8.5 \text{ m} \quad \text{or} \quad 8\frac{1}{2} \text{ m}$$

$8.5 = 8\frac{1}{2}$ because $8\frac{5}{10}$ simplifies to $8\frac{1}{2}$.

◀ **Work Problem 1 at the Side.**

OBJECTIVE 2 **Find the circumference of a circle.** The perimeter of a circle is called its **circumference.** Circumference is the distance around the edge of a circle.

The diameter of the can in the drawing is about 10.6 cm, and the circumference of the can is about 33.3 cm. Dividing the circumference of the circle by the diameter gives an interesting result.

$$\frac{\text{circumference}}{\text{diameter}} = \frac{33.3}{10.6} \approx 3.14 \qquad \text{Rounded to the nearest hundredth}$$

Dividing the circumference of *any* circle by its diameter *always* gives an answer close to 3.14. This means that going around the edge of any circle is a little more than 3 times as far as going straight across the circle.

This ratio of circumference to diameter is called π, (the Greek letter **pi**, pronounced PIE). There is no decimal that is exactly equal to π, but here is the *approximate* value.

$$\pi \approx 3.14159265359$$

Rounding the Value of *Pi* (π)

We usually round π to 3.14. Therefore, calculations involving π will give approximate answers and should be written using the $\approx$ symbol.

Use the following formulas to find the *circumference* of a circle.

Finding the Circumference (Distance around a Circle)

$$\text{Circumference} = \pi \cdot \text{diameter}$$

$$C = \pi \cdot d$$

or, because $d = 2 \cdot r$, then $C = \pi \cdot 2 \cdot r$ usually written $C = 2 \cdot \pi \cdot r$

Remember to use linear units such as ft, yd, m, and cm when measuring circumference (**not** square units).

EXAMPLE 2 Finding the Circumference of Circles

Find the circumference of each circle. Use 3.14 as the approximate value for π. Round answers to the nearest tenth.

(a)

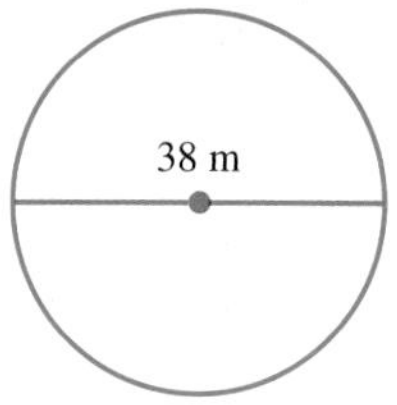

The *diameter* is 38 m, so use the formula with d in it.

$C = \pi \cdot d$

$C \approx 3.14 \cdot 38$ m — Write **m** in the answer, **not** m².

$C \approx 119.3$ m — Rounded

(b)

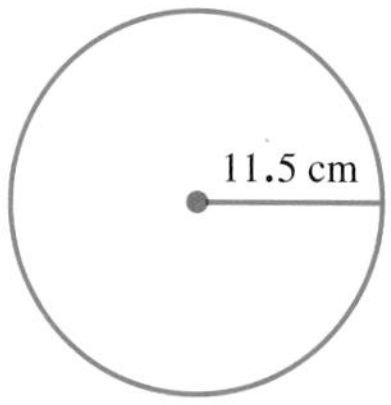

In this example, the *radius* is labeled, so it is easier to use the formula with r in it.

$C = 2 \cdot \pi \cdot r$

$C \approx 2 \cdot 3.14 \cdot 11.5$ cm — Write **cm** in the answer, **not** cm².

$C \approx 72.2$ cm — Rounded

Calculator Tip

Many *scientific* calculators have a (π) key. Try pressing it. With a 10-digit display, you'll see the value of π to the nearest billionth.

(π) (=) 3.141592654

But this is still an approximate value, although it is more precise than rounding π to 3.14 Try finding the circumference in **Example 2(a)** above using the (π) key.

(π) (×) 38 (=) Answer is 119.3805208; round to 119.4

When you used 3.14 as the approximate value of π, the result rounded to 119.3 so the answers are slightly different. In this book, we will use 3.14 instead of the (π) key. Our measurements of radius and diameter are given as whole numbers or with tenths, so it is acceptable to round π to hundredths. And, some students may be using standard calculators without a (π) key or doing the calculations by hand.

Work Problem 2 at the Side. ▶

OBJECTIVE 3 Find the area of a circle. To find the formula for the area of a circle, start by cutting two circles into many pie-shaped pieces.

Circumference (distance around) is $2 \cdot \pi \cdot r$.

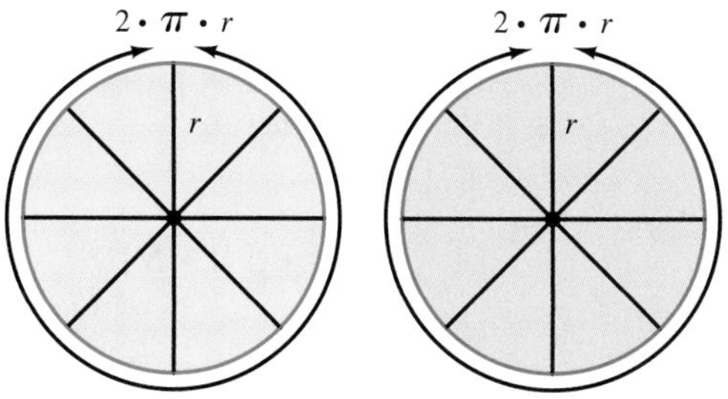

Unfold the circles, much as you might "unfold" a peeled orange, and put them together as shown here.

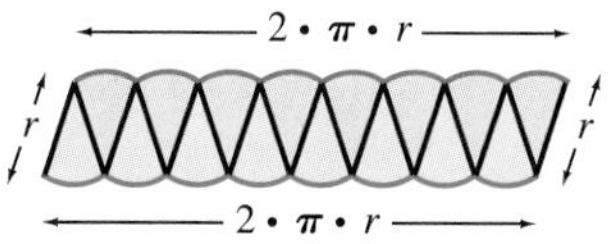

VOCABULARY TIP

Circumference **Circum** means "around, as in a circle." So, the **circumference** is the distance around the outside edge of a circle. It is like the perimeter of the circle.

2 Find the circumference of each circle. Use 3.14 as the approximate value for π. Round answers to the nearest tenth.

GS **(a)**

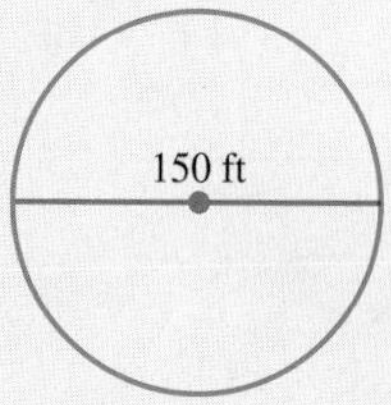

diameter is known; use

$C = \pi \cdot d$

$C \approx 3.14 \cdot 150$ ft

$C \approx$ ________

(b)

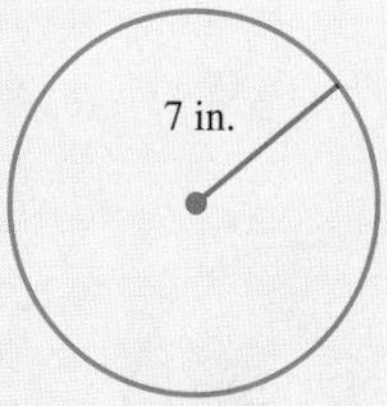

radius is known, use

$C = 2 \cdot \pi \cdot r$

(c) diameter 0.9 km

(d) radius 4.6 m

Answers

2. **(a)** $C \approx 471$ ft **(b)** $C \approx 44.0$ in. (rounded)
(c) $C \approx 2.8$ km (rounded)
(d) $C \approx 28.9$ m (rounded)

3 Find the area of each circle. Use 3.14 for π. Round your answers to the nearest tenth.

GS **(a)**

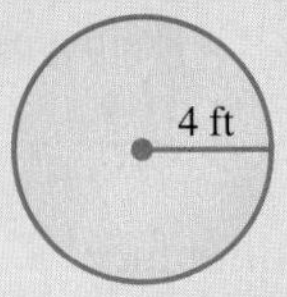

radius is 4 ft

$A = \pi \cdot r^2$

$A = \pi \cdot r \cdot r$

$A \approx 3.14 \cdot 4 \text{ ft} \cdot 4 \text{ ft}$

$A \approx$ ________

(b)

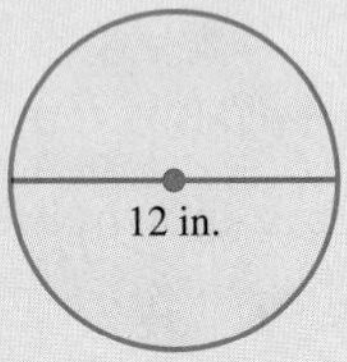

Hint: The diameter is 12 in. so $r =$ ____ in.

(c)

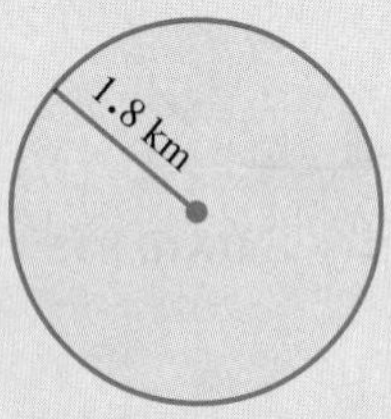

(d)

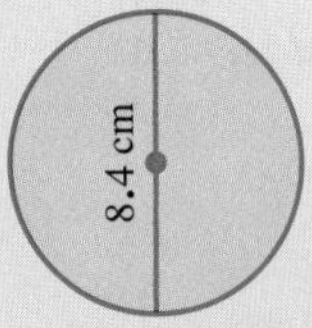

Answers

3. All answers are rounded.
(a) $A \approx 50.2 \text{ ft}^2$
(b) $r = 6$ in.; $A \approx 113.0 \text{ in.}^2$
(c) $A \approx 10.2 \text{ km}^2$ **(d)** $A \approx 55.4 \text{ cm}^2$

The figure is approximately a parallelogram with height r (the radius of the original circle) and base $2 \cdot \pi \cdot r$ (the circumference of the original circle). The area of the "parallelogram" is base times height.

$$\text{Area} = b \cdot h$$
$$\text{Area} = \underbrace{2 \cdot \pi \cdot r}_{b} \cdot r$$
$$\text{Area} = 2 \cdot \pi \cdot r^2 \leftarrow \text{Recall that } r \cdot r \text{ is } r^2.$$

Because the "parallelogram" was formed from *two* circles, the area of *one* circle is half as much.

$$\frac{1}{2} \cdot 2 \cdot \pi \cdot r^2 = 1 \cdot \pi \cdot r^2 \quad \text{or simply} \quad \pi \cdot r^2$$

Finding the Area of a Circle

Area of a circle $= \pi \cdot$ radius $\cdot$ radius

$$A = \pi \cdot r^2$$

Remember to use ***square units*** when measuring area.

EXAMPLE 3 Finding the Area of Circles

Find the area of each circle. Use 3.14 for π. Round your answers to the nearest tenth.

(a) A circle with a radius of 8.2 cm
Use the formula $A = \pi \cdot r^2$, which means $\pi \cdot r \cdot r$.

$$A = \pi \cdot r \cdot r$$
$$A \approx 3.14 \cdot 8.2 \text{ cm} \cdot 8.2 \text{ cm}$$
$$A \approx 211.1 \text{ cm}^2 \quad \text{Rounded}$$

This is **area**, so write **cm²** in the answer.

(b)

To use the area formula, you need to know the *radius* (r). In this circle, the *diameter* is 10 ft. First find the radius.

You *cannot* use the *diameter* measurement in the area formula, so find the *radius*.

$$r = \frac{d}{2}$$
$$r = \frac{10 \text{ ft}}{2} = 5 \text{ ft}$$

Now find the area.

$$A \approx 3.14 \cdot 5 \text{ ft} \cdot 5 \text{ ft}$$
$$A \approx 78.5 \text{ ft}^2$$

Write square units (**ft²**) in your answer.

CAUTION

When finding *circumference,* you can start with either the radius or the diameter. When finding *area,* you must use the *radius.* If you are given the diameter, divide it by 2 to find the radius. Then find the area.

◀ **Work Problem 3 at the Side.**

Calculator Tip

You can use your calculator to find the area of the circle in **Example 3(a)** on the previous page. The first method works on both scientific and standard calculators:

3.14 (×) 8.2 (×) 8.2 (=) Answer is 211.1336.

You round the answer to 211.1 (nearest tenth).

On a *scientific* calculator you can also use the (x^2) key, which automatically squares the number you enter (that is, multiplies the number times itself):

3.14 (×) 8.2 (x^2) 67.24 (=) Answer is 211.1336.

Appears automatically on some calculators; 8.2 × 8.2 is 67.24.

Next, we find the area of a *semicircle,* which is half the area of a circle.

EXAMPLE 4 Finding the Area of a Semicircle

Find the area of the semicircle. Use 3.14 for π. Round your answer to the nearest tenth.

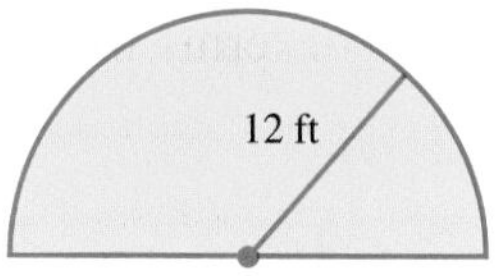

First, find the area of a whole circle with a radius of 12 ft.

$$A = \pi \cdot r \cdot r$$

$$A \approx 3.14 \cdot 12 \text{ ft} \cdot 12 \text{ ft}$$

$$A \approx 452.16 \text{ ft}^2 \leftarrow \text{Do not round yet.}$$

Divide the area of the whole circle by 2 to find the area of the semicircle.

$$\frac{452.16 \text{ ft}^2}{2} = 226.08 \text{ ft}^2$$

The *last* step is rounding 226.08 to the nearest tenth.

Area of semicircle $\approx 226.1 \text{ ft}^2$ Rounded

This is area, so write **ft²** in the answer.

Work Problem 4 at the Side. ▶

EXAMPLE 5 Applying the Concept of Circumference

A circular rug is 8 ft in diameter. The cost of fringe for the edge is \$2.25 per foot. What will it cost to add fringe to the rug? Use 3.14 for π.

$$\text{Circumference} = \pi \cdot d$$

$$C \approx 3.14 \cdot 8 \text{ ft}$$

$$C \approx 25.12 \text{ ft}$$

cost = cost per foot • circumference

$$\text{cost} = \frac{\$2.25}{1 \cancel{\text{ft}}} \cdot \frac{25.12 \cancel{\text{ft}}}{1}$$

The common units divide out.

cost = \$56.52

The cost of adding fringe to the rug is \$56.52.

Work Problem 5 at the Side. ▶

VOCABULARY TIP

Semicircle **Semi** means "half."

4 Find the area of each semicircle. Use 3.14 for π. Round your answers to the nearest tenth.

(a)

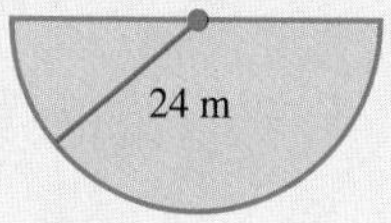

(b)

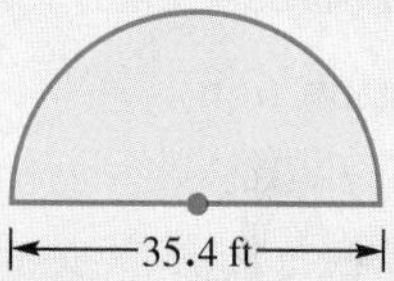

(c)

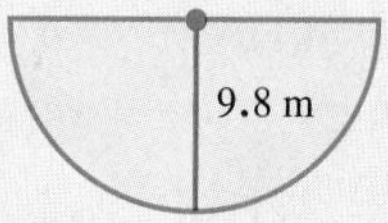

5 Find the cost of binding around the edge of a circular rug that is 3 m in diameter. The binder charges \$4.50 per meter. Use 3.14 for π.

Answers

4. All answer are rounded.
(a) $A \approx 904.3 \text{ m}^2$ **(b)** $A \approx 491.9 \text{ ft}^2$
(c) $A \approx 150.8 \text{ m}^2$

5. \$42.39

6 Find the cost of covering the underside of the rug in **Margin Problem 5** with a nonslip rubber backing. The rubber backing costs \$3.89 per square meter.

7 **(a)** Here are some prefixes you have seen in this textbook. List at least one math term and one nonmathematical word that use each prefix.

dia- (through):

fract- (break):

par- (beside):

per- (divide):

peri- (around);

rad- (ray):

rect- (right):

sub- (below):

(b) How could you use your knowledge of prefixes to remember the difference between perimeter and area?

Answers

6. \$27.48 (rounded)

7. **(a)** Some possibilities are:
*dia*meter; *dia*gonal
*fract*ion; *fract*ure
*par*allel; *par*amedic
*per*cent; *per* capita
*peri*meter; *peri*scope
*rad*ius; *rad*iate
*rect*angle; *rect*ify
*sub*tract; *sub*marine

(b) *Peri-* in perimeter means "around," so perimeter is the distance *around* the edges of a shape.

EXAMPLE 6 **Applying the Concept of Area**

Find the cost of covering the rug in **Example 5** (on the previous page) with a plastic cover. The material for the cover costs \$1.50 per square foot. Use 3.14 for π.

First find the radius. $r = \frac{d}{2} = \frac{8 \text{ ft}}{2} = 4 \text{ ft}$

Then find the area. $A = \pi \cdot r^2$

$A \approx 3.14 \cdot 4 \text{ ft} \cdot 4 \text{ ft}$

$A \approx 50.24 \text{ ft}^2$

$$\text{cost} = \frac{\$1.50}{1 \cancel{\text{ft}^2}} \cdot \frac{50.24 \cancel{\text{ft}^2}}{1} = \$75.36$$

The cost of the plastic cover is \$75.36.

◀ Work Problem **6** at the Side.

OBJECTIVE 4 **Become familiar with Latin and Greek prefixes used in math terminology.** Many English words are built from Latin or Greek root words and prefixes. Knowing the meaning of the more common ones can help you figure out the meaning of terms in many subject areas, including mathematics.

EXAMPLE 7 **Using Prefixes to Understand Math Terms**

(a) Listed below are some Latin and Greek root words and prefixes with their meanings in parentheses. You've already seen math terms in this textbook that use these prefixes. List at least one math term and one nonmathematical word that use each prefix or root word.

cent- (100): ***cent*imeter; *cent*ury**

circum- (around): ***circum*ference; *circum*vent**

de- (down): ***de*nominator; *de*cline**

dec- (10): ***dec*imal; *Dec*ember** (originally the 10th month in the old calendar)

There are many answers. These are some of the possibilities.

(b) Suppose you have trouble remembering which part of a fraction is the denominator. How could your knowledge of prefixes help in this situation?

The *de-* prefix in ***de*nominator** means "down" so the denominator is the number *down* below the fraction bar.

◀ Work Problem **7** at the Side.

Note

Here are some additional prefixes and root words and their meanings that you will see in the rest of this chapter, in later chapters of this book, and in other math classes. Examples of a math term and a nonmathematical word are shown for each one.

equ- (equal): ***equ*ation; *equ*inox**

lateral (side): **quadri*lateral*; bi*lateral***

hemi- (half): ***hemi*sphere; *hemi*trope**

re- (back or again): ***re*ciprocal; *re*duce**

7.5 Exercises

FOR EXTRA HELP

Download the MyDashBoard App

MyMathLab®

1. CONCEPT CHECK Fill in the blanks. Choose from height, perimeter, area, base, ft, and ft^2.

Finding the circumference of a circle is like finding the ________ of a triangle. If the circle's radius and diameter are measured in feet, then the units for the circumference will be ________.

2. CONCEPT CHECK Fill in the blanks. Choose from radius, diameter, circumference, divide by 2, and multiply by 2.

To use the formula for finding the area of a circle, you need to know the length of the ________. If you are given the diameter of a circle, what should you do to it to find the radius? ________

Find the unknown length in each circle. ***See Example 1.***

3. GS

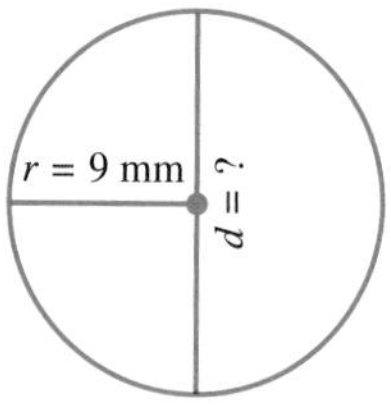

$d = 2 \cdot r$

$d = 2 \cdot$ ______

$d =$ ______

4. GS

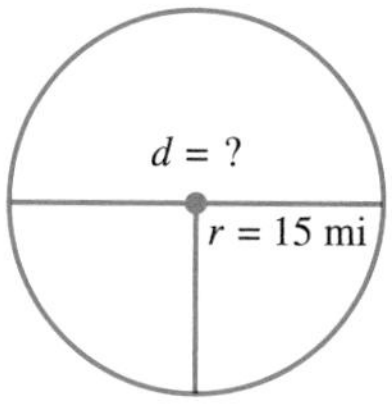

$d = 2 \cdot r$

$d = 2 \cdot$ ______

$d =$ ______

5.

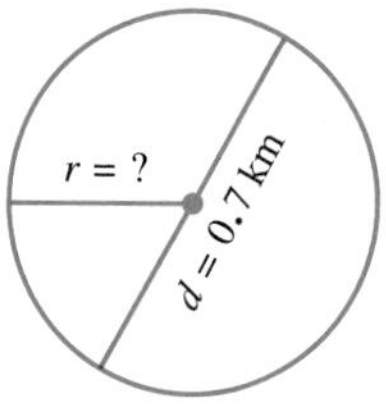

6.

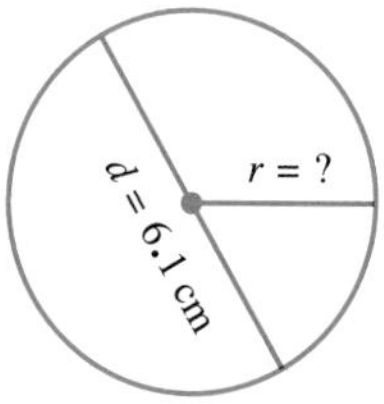

Find the circumference and area of each circle. Use 3.14 as the approximate value for π. Round your answers to the nearest tenth. ***See Examples 2 and 3.***

7.

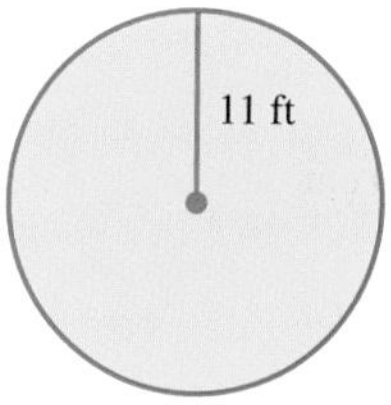

8.

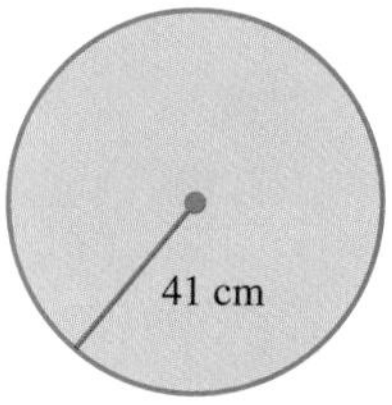

9.

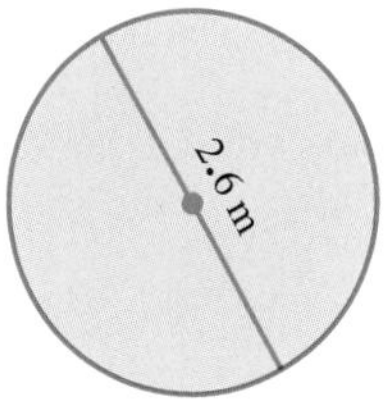

10.

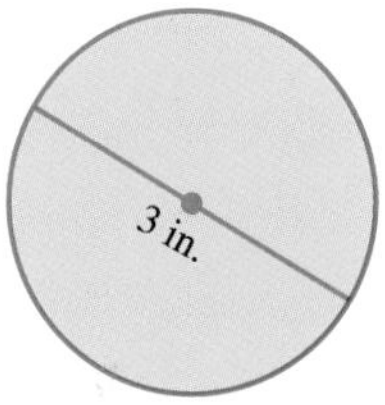

Find the circumference and area of circles having the following diameters. Use 3.14 as the approximate value of π. Round your answers to the nearest tenth. ***See Examples 2 and 3.***

11. GS $d = 15$ cm

$C = \pi \cdot d$

$C \approx 3.14 \cdot 15$ cm

$C \approx$ ______

To find the **area,** first find the **radius.** Then use $A = \pi \cdot r \cdot r$

12. GS $d = 39$ ft

$C = \pi \cdot d$

$C \approx 3.14 \cdot 39$ ft

$C \approx$ ______

To find the **area,** first find the **radius.** Then use $A = \pi \cdot r \cdot r$

13. $d = 7\frac{1}{2}$ ft

14. $d = 4\frac{1}{2}$ yd

Find each shaded area. Note that Exercises 15–18 all contain semicircles. Use 3.14 as the approximate value of π. Round your answers to the nearest tenth if necessary. ***See Example 4.***

15.

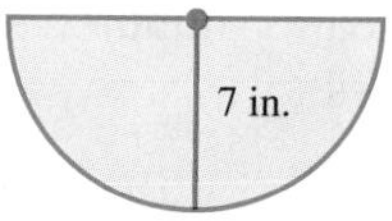

16.

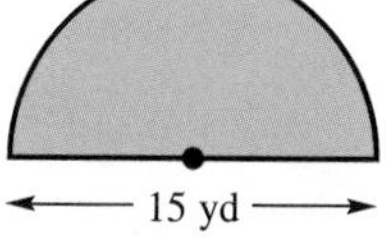

17.

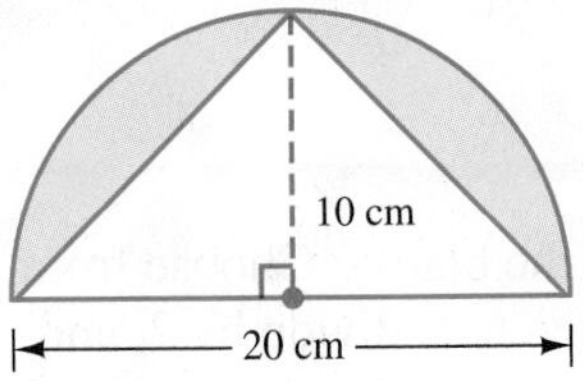

18.

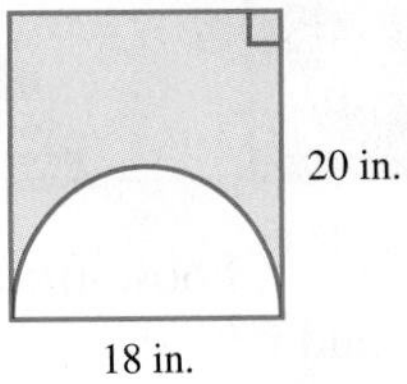

19. **CONCEPT CHECK** How would you explain π to a friend who is not in your math class? Write an explanation.

20. **CONCEPT CHECK** Make up a test question that requires the use of π. Show how to solve your problem.

Solve each application problem. Use 3.14 as the approximate value of π. Round your answers to the nearest tenth. ***See Examples 5 and 6.***

21. An irrigation system moves around a center point to water a circular area for crops. If the irrigation system is 50 yd long, how large is the watered area?

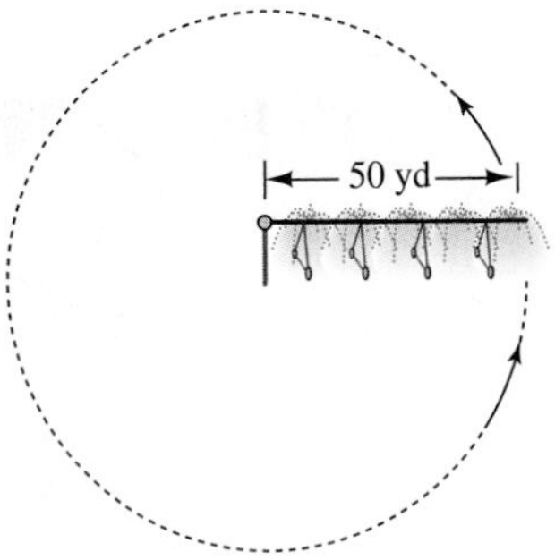

22. If you swing a ball held at the end of a string 20 cm long, how far will the ball travel on each turn?

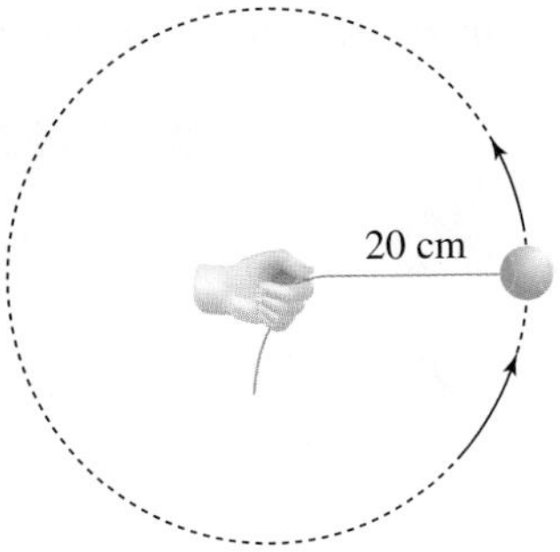

23. A Michelin Cross Terrain SUV tire has an overall diameter of 29.10 inches. How far will a point on the tire tread move in one complete turn? (*Source:* Michelin.) *Bonus question:* How many revolutions does the tire make per mile? Round to the nearest whole number.

24. In August 2011, Hurricane Irene moved along the East Coast of the United States, from North Carolina to New York City. Some 65 million people live along this coast, the greatest number of people threatened by a hurricane in American history. The huge circular storm had a diameter of 500 miles. By contrast, Hurricane Katrina, which flooded New Orleans in 2005, had a diameter of 210 miles. What was the area of each hurricane, to the nearest hundred square miles? (*Source*: Associated Press).

Circular hurricane as seen from a weather satellite.

For Exercises 25–30, first draw a circle and label the radius or diameter. Then solve the problem. Use 3.14 as the approximate value of π and round answers to the nearest tenth, money answers to the nearest cent.

25. A radio station can be heard 150 miles in all directions during evening hours. How many square miles are in the station's broadcast area? (*Hint:* Use the drawing of the broadcast area to help you.)

The radius is 150 miles.

26. An earthquake was felt by people 900 km away in all directions from the epicenter (the source of the earthquake). How much area was affected by the quake?

27. The diameter of Diana Hestwood's wristwatch is 1 in. and the radius of the clock face on her kitchen wall is 3 in. Find the circumference and the area of the watch face and the clock face.

28. The diameter of the largest known ball of twine is 12 ft 9 in. The sign posted near the ball says it has a circumference of 40 ft. Is the sign correct? *Hint:* First change 9 in. to feet and add it to 12 ft. (*Source: Guinness World Records.*)

29. Blaine Fenstad wants to buy a pair of two-way radios. Some models have a range of 2 miles under ideal conditions. More expensive models have a range of 5 miles. What is the difference in the area covered by the 2-mile and 5-mile models? (*Source:* Best Buy.)

30. The National Audubon Society holds an end-of-year bird count. Volunteers count all the birds they see in a circular area during a 24-hour period. Each circle has a diameter of 15 miles. About 1700 circular areas are counted across the United States each December. What is the total area covered by the count?

31. To find the age of a living tree, a forestry specialist measures the circumference of the tree. (See photo.)

(a) If the circumference of a tree is 144 cm, what is the diameter?

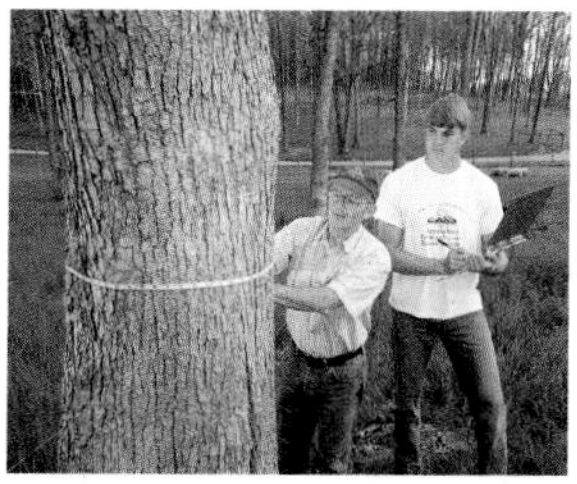

(b) Explain how you solved part (a).

32. In Atlanta, Interstate 285 circles the city and is known as the "perimeter." If the circumference of the circle made by the highway is 62.8 miles, find

(a) the diameter of the circle.

(b) the area inside the circle.

(*Source: Greater Atlanta Newcomer's Guide.*)

33. Find the cost of sod, at $0.49 per square foot, for this playing field that has a semicircle on each end.

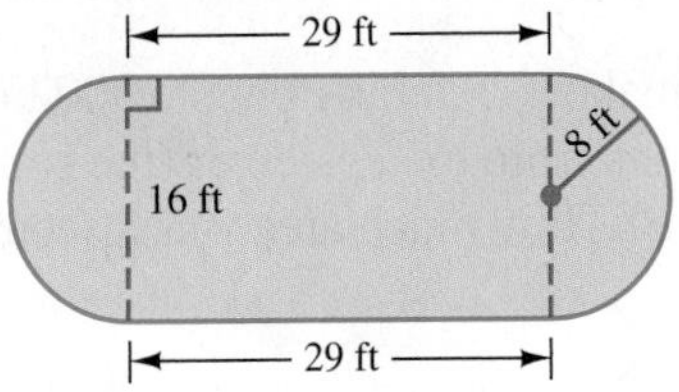

34. Find the area of this skating rink.

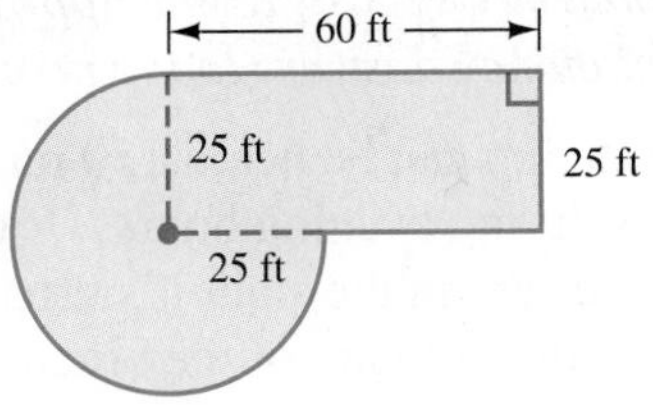

Use the information about prefixes in ***Example 7*** *to answer Exercises 35 and 36.*

35. Explain how you could use the information about prefixes to remember the difference between radius, diameter, and circumference.

36. Explain how you could use the information about prefixes to avoid confusion between parallel and perpendicular lines.

Relating Concepts (Exercises 37–42) For Individual or Group Work

Use the table below to ***work Exercises 37–42 in order.***

Find the best buy for each type of pizza. The best buy is the lowest cost per square inch of pizza. All the pizzas are circular, and the measurement given on the menu board is the diameter of the pizza in inches. Use 3.14 as the approximate value of π. Round the area to the nearest tenth. Round cost per square inch to the nearest thousandth.

Pizza Menu	Small 10"	Medium 12"	Large 14"
Cheese only	$ 7	$ 9	$11
"The Works"	$12	$14	$16
Pepperoni	$ 8	$10	$12

Source: Papa John's.

37. Find the area of a small pizza.

38. Find the area of a medium pizza.

39. Find the area of a large pizza.

40. What is the cost per square inch for each size of cheese pizza? Which size is the best buy?

41. What is the cost per square inch for each size of "The Works" pizza? Which size is the best buy?

42. You use a coupon for $2 off any small pizza. What is the cost per square inch for each size of pepperoni pizza? Which size is the best buy?

7.6 Volume and Surface Area

OBJECTIVES

1. Find the volume of a rectangular solid.
2. Find the volume of a sphere.
3. Find the volume of a cylinder.
4. Find the volume of a cone and a pyramid.
5. Find the surface area of a rectangular solid.
6. Find the surface area of a cylinder.

OBJECTIVE 1 **Find the volume of a rectangular solid.** A shoe box and a cereal box are examples of three-dimensional (or solid) figures. The three dimensions are length, width, and height. (A rectangle or square is a two-dimensional figure. The two dimensions are length and width.)

If you want to know how much a shoe box will hold, you find its *volume.* We measure volume by seeing how many cubes of a certain size will fill the space inside the box. Three sizes of *cubic units* are shown below. Notice that all the edges of a cube have the same length and all the sides meet at right angles.

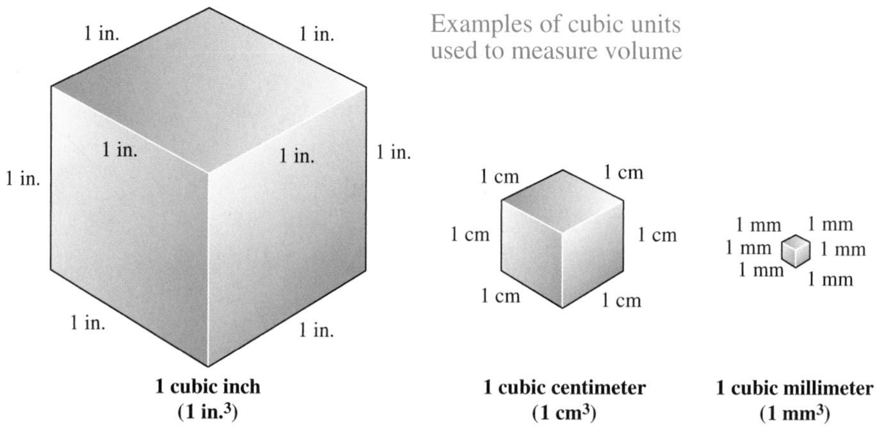

Some other sizes of cubes that are used to measure volume are 1 cubic foot (1 ft^3), 1 cubic yard (1 yd^3), and 1 cubic meter (1 m^3).

CAUTION

The raised 3 in 4^3 means that you multiply $4 \cdot 4 \cdot 4$ to get 64. The raised 3 in cm^3 or ft^3 is a short way to write the word *cubic.* When you see 5 cm^3, say "five *cubic* centimeters." Do *not* multiply $5 \cdot 5 \cdot 5$. The exponent applies to cm, *not* to the number.

Volume

Volume is a measure of the space inside a solid shape. The volume of a solid is how many cubic units it takes to fill the solid.

Use the formula below to find the *volume* of *rectangular solids* (box-like shapes).

Finding the Volume of Rectangular Solids

Volume of a rectangular solid = length • width • height

$$V = l \cdot w \cdot h$$

Remember to use ***cubic units*** when measuring volume.

1 Find the volume of each box. Round your answers to the nearest tenth if necessary.

GS **(a)**

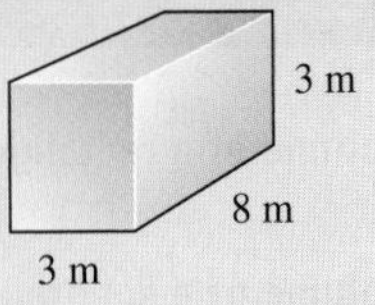

$V = l \cdot w \cdot h$

$V = 3\text{ m} \cdot 8\text{ m} \cdot$ ____ m

$V =$ ________

(b)

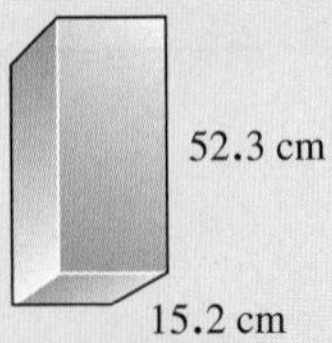

(c) Length $6\frac{1}{4}$ ft, width $3\frac{1}{2}$ ft, height 2 ft

Answers

1. **(a)** 3 m; $V = 72\text{ m}^3$
   (b) $V \approx 18{,}602.1\text{ cm}^3$ (rounded)
   (c) $V = 43\frac{3}{4}\text{ ft}^3$ (exactly) or $V \approx 43.8\text{ ft}^3$ (rounded)

EXAMPLE 1 Finding the Volume of Rectangular Solids

Find the volume of each box.

(a)

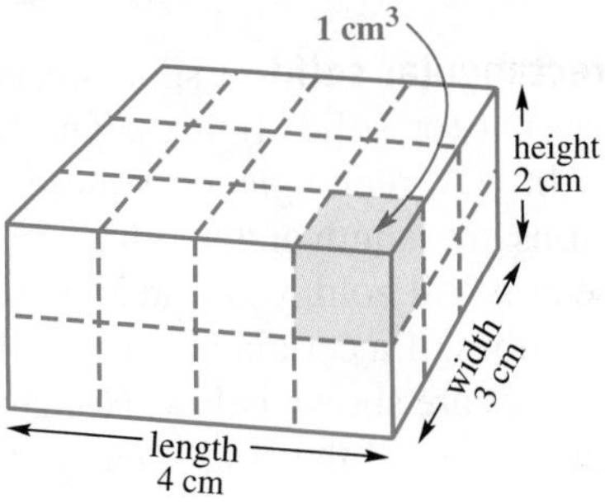

Each cube that fits in the box is 1 cubic centimeter (1 cm^3). To find the volume, you can count the number of cubes.

Bottom layer has 12 cubes. Top layer has 12 cubes. } total of 24 cubes (24 cm^3)

Or, you can use the formula for rectangular solids.

$$V = l \cdot w \cdot h$$
$$V = 4\text{ cm} \cdot 3\text{ cm} \cdot 2\text{ cm}$$
$$V = 24\text{ cm}^3 \quad \text{Cubic units for volume}$$

(b)

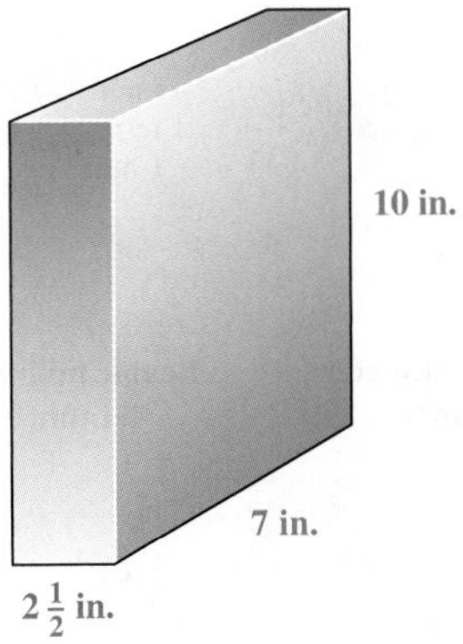

Use the formula $V = l \cdot w \cdot h$.

$$V = 7\text{ in.} \cdot 2\frac{1}{2}\text{ in.} \cdot 10\text{ in.}$$

$$V = \frac{7\text{ in.}}{1} \cdot \frac{5\text{ in.}}{\cancel{2}_1} \cdot \frac{\cancel{10}^{5}\text{ in.}}{1} = 175\text{ in.}^3$$

Write **in.³** for volume.

If you like, use 2.5 instead of $2\frac{1}{2}$; they are equivalent.

$$V = 7\text{ in.} \cdot 2.5\text{ in.} \cdot 10\text{ in.} = 175\text{ in.}^3$$

◄ **Work Problem 1 at the Side.**

OBJECTIVE 2 Find the volume of a sphere. A *sphere* is shown below. Examples of spheres include baseballs, oranges, and Earth. (They aren't perfect spheres, but they're close.)

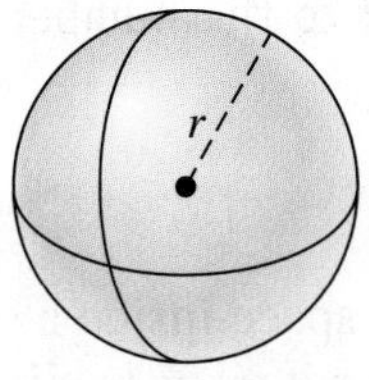

As with circles, the *radius* of a sphere is the distance from the center to the edge of the sphere. Use this formula to find the *volume* of a *sphere*.

Finding the Volume of a Sphere

$$\text{Volume of a sphere} = \frac{4}{3} \cdot \pi \cdot r \cdot r \cdot r$$

$$V = \frac{4}{3} \cdot \pi \cdot r^3 \quad \text{or} \quad \frac{4 \cdot \pi \cdot r^3}{3}$$

Remember to use ***cubic units*** when measuring volume.

EXAMPLE 2 Finding the Volume of Spheres

Find the volume of each sphere with the help of a calculator. Use 3.14 as the approximate value of π. Round your answers to the nearest tenth.

(a)

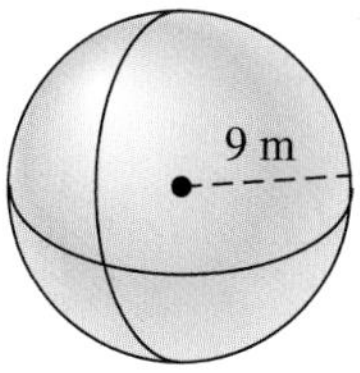

$$V = \frac{4}{3} \cdot \pi \cdot r^3$$ (r^3 means $r \cdot r \cdot r$.)

$$V \approx \frac{4 \cdot 3.14 \cdot 9 \text{ m} \cdot 9 \text{ m} \cdot 9 \text{ m}}{3}$$

$V \approx 3052.08 \text{ m}^3$ Now round to tenths.

$V \approx 3052.1 \text{ m}^3$ Cubic units for volume

(b)

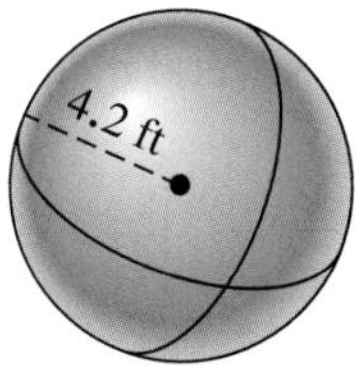

$$V \approx \frac{4 \cdot 3.14 \cdot 4.2 \text{ ft} \cdot 4.2 \text{ ft} \cdot 4.2 \text{ ft}}{3}$$

$V \approx 310.18176 \text{ ft}^3$ Now round to tenths.

$V \approx 310.2 \text{ ft}^3$ Cubic units for volume

Calculator Tip

You can find the volume of the sphere in **Example 2(b)** above on your calculator. The first method works on both scientific and standard calculators:

4 (×) 3.14 (×) 4.2 (×) 4.2 (×) 4.2 (÷) 3 (=) Answer is 310.18176.

Round the answer to 310.2 ft³.

On a *scientific* calculator you can use the (y^x) key (or the (^) key on some models) to calculate r^3 (to multiply the radius times itself three times).

4 (×) 3.14 (×) 4.2 (y^x) 3 (÷) 3 (=) Answer is 310.18176.
(4.2 (y^x) 3 is r^3)

Recall that we are using 3.14 as the approximate value for π instead of using the (π) key.

You can also use the (y^x) key with other exponents. For example:

To find 2^5, press 2 (y^x) 5 (=) Answer is 32.

To find 6^4, press 6 (y^x) 4 (=) Answer is 1296.

Work Problem 2 at the Side. ▶

Half a sphere is called a *hemisphere*. The volume of a hemisphere is *half* the volume of a sphere. Use the following formula to find the *volume* of a *hemisphere*.

Finding the Volume of a Hemisphere

$$\text{Volume of a hemisphere} = \frac{1}{\cancel{2}_1} \cdot \frac{\cancel{4}^2}{3} \cdot \pi \cdot r^3$$

$$V = \frac{2}{3} \cdot \pi \cdot r^3 \quad \text{or} \quad \frac{2 \cdot \pi \cdot r^3}{3}$$

Remember to use ***cubic units*** when measuring volume.

2 Find the volume of each sphere. Use 3.14 for π. Round your answers to the nearest tenth.

GS **(a)**

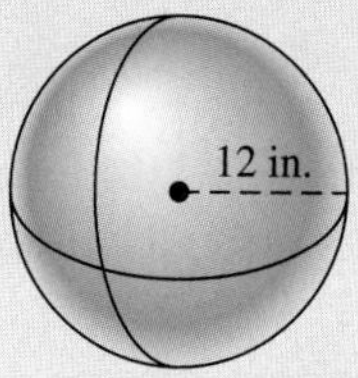

$$V = \frac{4}{3} \cdot \pi \cdot r^3$$

$$V \approx \frac{4 \cdot 3.14 \cdot 12 \text{ in.} \cdot 12 \text{ in.} \cdot 12 \text{ in.}}{3}$$

$V \approx$ ______ Round to nearest tenth.

(b)

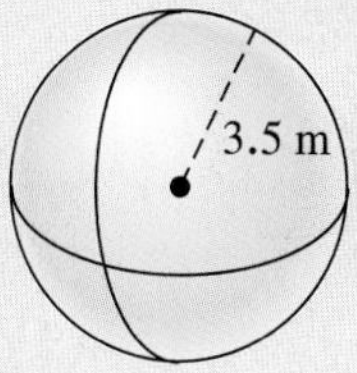

(c) Sphere with a radius of 2.7 cm

Answers

2. **(a)** $V \approx 7234.6 \text{ in.}^3$ (rounded)
(b) $V \approx 179.5 \text{ m}^3$ (rounded)
(c) $V \approx 82.4 \text{ cm}^3$ (rounded)

VOCABULARY TIP

Hemisphere **Hemi** means "half." So a **hemi**sphere is **half** of a sphere.

3 Find the volume of each hemisphere. Use 3.14 for π. Round answers to the nearest tenth.

(a)

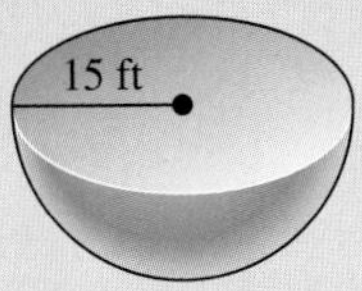

$$V = \frac{2 \cdot \pi \cdot r \cdot r \cdot r}{3}$$

(b)

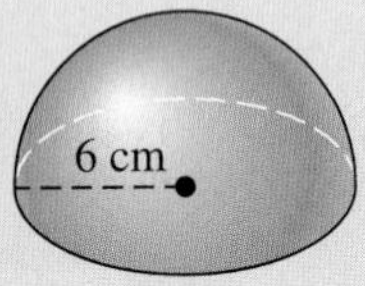

4 Find the volume of each cylinder. Use 3.14 for π. Round your answers to the nearest tenth.

GS **(a)**

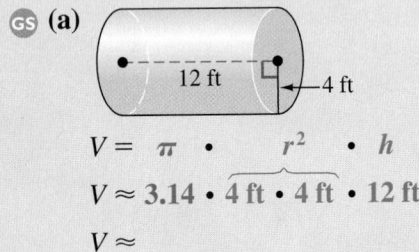

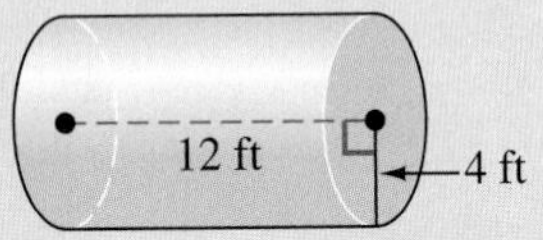

$V = \pi \cdot r^2 \cdot h$

$V \approx 3.14 \cdot \overbrace{4\text{ ft} \cdot 4\text{ ft}} \cdot 12\text{ ft}$

$V \approx$ ______

(b)

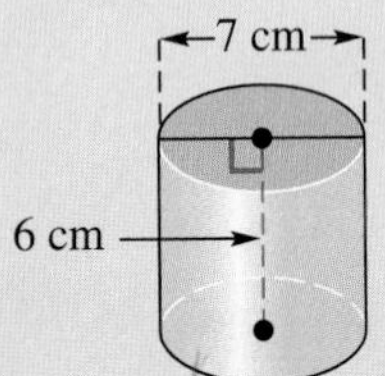

Hint: Find the **radius** first.

Answers

3. (a) $V \approx 7065\text{ ft}^3$
   (b) $V \approx 452.2\text{ cm}^3$ (rounded)
4. (a) $V \approx 602.9\text{ ft}^3$ (rounded)
   (b) $V \approx 230.8\text{ cm}^3$ (rounded)

EXAMPLE 3 Finding the Volume of a Hemisphere

Find the volume of the hemisphere with the help of a calculator. Use 3.14 for π. Round your answer to the nearest tenth.

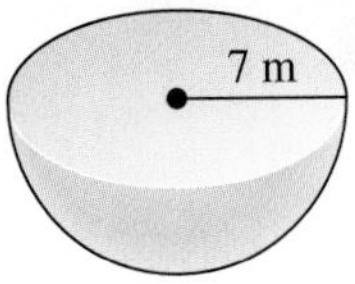

$$V = \frac{2 \cdot \pi \cdot r^3}{3}$$

$$V \approx \frac{2 \cdot 3.14 \cdot 7\text{ m} \cdot 7\text{ m} \cdot 7\text{ m}}{3}$$

Write m^3 for volume.

$V \approx 718.0\text{ m}^3$ Rounded to nearest tenth

◀ Work Problem **3** at the Side.

OBJECTIVE 3 Find the volume of a cylinder.

Here are several *cylinders*.

The height must be perpendicular to the circular top and bottom of the cylinder.

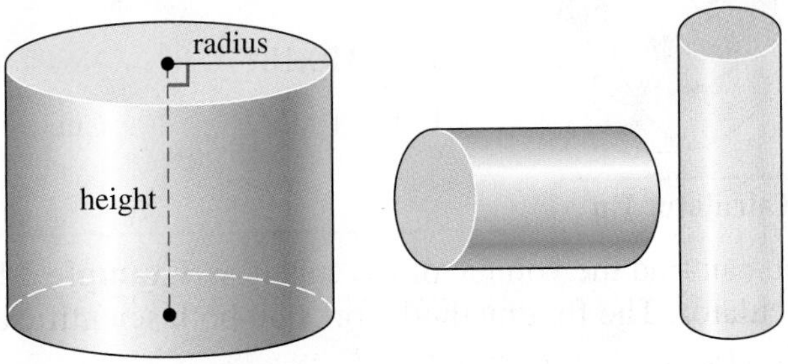

These are called *right circular cylinders* because the top and bottom are circles, and the side makes a right angle with the top and bottom. Examples of cylinders are a soup can, a home water heater, and a piece of pipe.

Use the formula below to find the *volume* of a *cylinder*. Notice that the first part of the formula, $\pi \cdot r^2$, is the area of the circular base.

Finding the Volume of a Cylinder

Volume of a cylinder $= \pi \cdot r \cdot r \cdot h$

$$V = \pi \cdot r^2 \cdot h$$

Remember to use ***cubic units*** when measuring volume.

EXAMPLE 4 Finding the Volume of Cylinders

Find the volume of each cylinder. Use 3.14 as the approximate value of π. Round your answers to the nearest tenth if necessary.

(a)

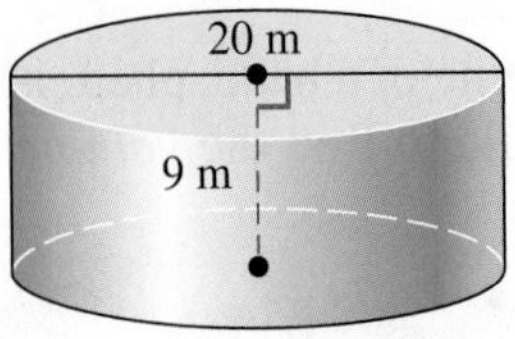

The diameter is 20 m so the radius is $\frac{20\text{ m}}{2} = 10\text{ m}$. The height is 9 m. Use the formula to find the volume.

$V = \pi \cdot r^2 \cdot h$

$V \approx 3.14 \cdot \overbrace{10\text{ m} \cdot 10\text{ m}} \cdot 9\text{ m}$

$V \approx 2826\text{ m}^3$ Cubic units for volume

(b)

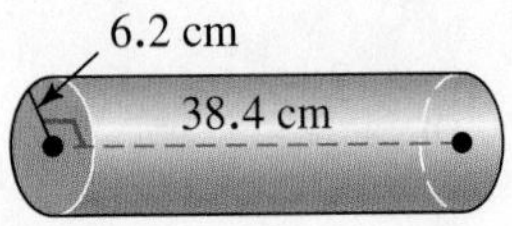

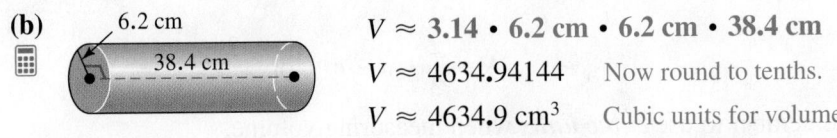

$V \approx 3.14 \cdot 6.2\text{ cm} \cdot 6.2\text{ cm} \cdot 38.4\text{ cm}$

$V \approx 4634.94144$ Now round to tenths.

$V \approx 4634.9\text{ cm}^3$ Cubic units for volume

◀ Work Problem **4** at the Side.

OBJECTIVE 4 Find the volume of a cone and a pyramid. A cone and a pyramid are shown below. Notice that the height line is perpendicular to the base in both solids.

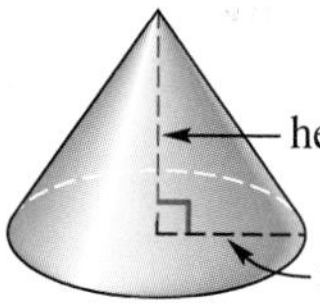

Cone

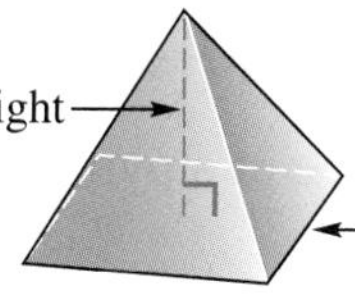

Pyramid

Use the formula below to find the *volume* of a cone.

Finding the Volume of a Cone

$$\text{Volume of a cone} = \frac{1}{3} \cdot B \cdot h$$

$$\text{or} \quad V = \frac{B \cdot h}{3}$$

where B is the area of the circular base of the cone and h is the height of the cone.

Remember to use ***cubic units*** when measuring volume.

EXAMPLE 5 Finding the Volume of a Cone

Find the volume of the cone. Use 3.14 for π. Round your answer to the nearest tenth.

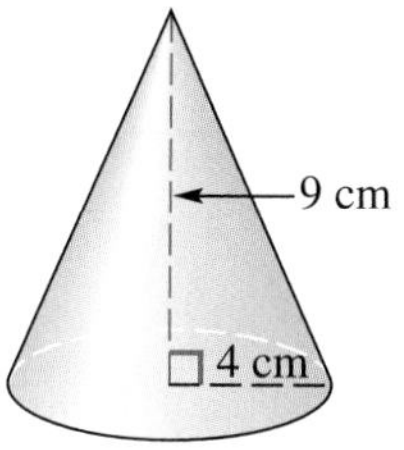

First find the value of B in the formula, which is the *area of the circular base.* Recall that the formula for the area of a circle is $\pi \cdot r^2$.

$B = \pi \cdot r \cdot r$

$B \approx 3.14 \cdot 4 \text{ cm} \cdot 4 \text{ cm}$

$B \approx \mathbf{50.24 \text{ cm}^2}$ ← Do **not** round to tenths yet.

Now find the volume of the cone. The height is 9 cm.

$$V = \frac{B \cdot h}{3}$$

$$V \approx \frac{\mathbf{50.24 \text{ cm}^2 \cdot 9 \text{ cm}}}{3}$$

$V \approx 150.72 \text{ cm}^3$ Now round to tenths.

$V \approx 150.7 \text{ cm}^3$ Cubic units for volume

This is **volume,** so write **cm³** in the answer.

Work Problem 5 at the Side. ▶

5 Find the volume of a cone with base radius 2 ft and height 11 ft. Use 3.14 for π. Round your answer to the nearest tenth.

First find the area of the circular base.

$B = \pi \cdot r \cdot r$

$B \approx 3.14 \cdot 2 \text{ ft} \cdot 2 \text{ ft}$

$B \approx$ ______ Do **not** round yet.

Now find the volume of the cone.

$$V = \frac{B \cdot h}{3}$$

$$V \approx \frac{\text{\_\_\_\_\_\_} \cdot \mathbf{11 \text{ ft}}}{3}$$

$V \approx$ ______ Round to nearest tenth.

Answer

5. $B \approx 12.56 \text{ ft}^2$

$$V \approx \frac{12.56 \text{ ft}^2 \cdot 11 \text{ ft}}{3}$$

$V \approx 46.1 \text{ ft}^3$ (rounded)

6 Find the volume of a pyramid with a square base 10 m by 10 m. The height of the pyramid is 8 m. Round your answer to the nearest tenth.

GS First find the area of the square base.

$B = 10\text{ m} \cdot 10\text{ m}$

$B = \_\_\_\_\_\_$

Now find the volume of the pyramid.

$$V = \frac{B \cdot h}{3}$$

$$V = \frac{\_\_\_\_\_\_ \cdot 8\text{ m}}{3}$$

$V \approx \_\_\_\_\_\_$ Round to nearest tenth.

Use the same formula to find the *volume* of a *pyramid* as you did to find the *volume* of a *cone.*

Finding the Volume of a Pyramid

$$\text{Volume of a pyramid} = \frac{1}{3} \cdot B \cdot h$$

$$\text{or} \quad V = \frac{B \cdot h}{3}$$

where B is the area of the square or rectangular base of the pyramid and h is the height of the pyramid.

Remember to use ***cubic units*** when measuring volume.

Note

In this book, we will work only with pyramids that have a base with four sides (square or rectangle). In later math courses you may work with pyramids that have a base with three sides (triangle), five sides (pentagon), six sides (hexagon), and so on.

EXAMPLE 6 Finding the Volume of a Pyramid

Find the volume of this pyramid with a rectangular base. Round your answer to the nearest tenth.

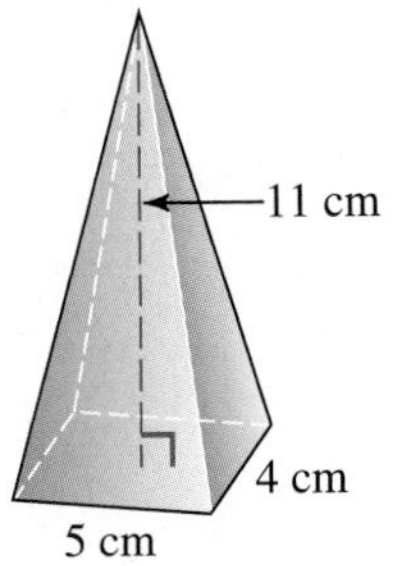

First find the value of B in the formula, which is the *area of the rectangular base.* Recall that the area of a rectangle is found by multiplying length times width.

$B = 5\text{ cm} \cdot 4\text{ cm}$

$B = 20\text{ cm}^2$ ← First find the *area* of the rectangular base.

Now find the volume of the pyramid.

$$V = \frac{B \cdot h}{3}$$

$$V = \frac{20\text{ cm}^2 \cdot 11\text{ cm}}{3}$$

$V \approx 73.3\text{ cm}^3$ Rounded to nearest tenth

This is **volume**, so write $\textbf{cm}^3$ in the answer.

◀ **Work Problem 6 at the Side.**

Answer

6. $B = 100\text{ m}^2$

$$V = \frac{100\text{ m}^2 \cdot 8\text{ m}}{3}$$

$V \approx 266.7\text{ m}^3$ (rounded)

OBJECTIVE 5 Find the surface area of a rectangular solid. Earlier in this section, you found the *volume* of a rectangular solid. For example, the volume of the cereal box shown below is $V = lwh = (7 \text{ in.})(2 \text{ in.})(10 \text{ in.}) = 140 \text{ in.}^3$ But if your company makes cereal boxes, you also need to know how much cardboard is needed for each box. You need to find the *surface area* of the box.

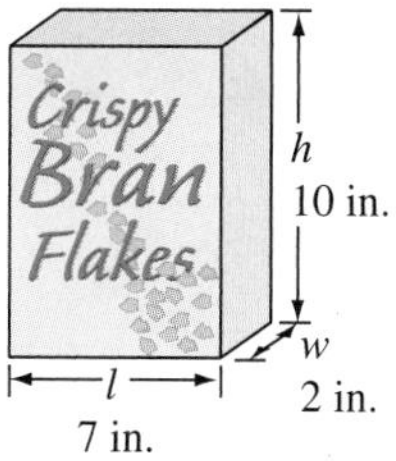

Unfold the box and flatten it out.

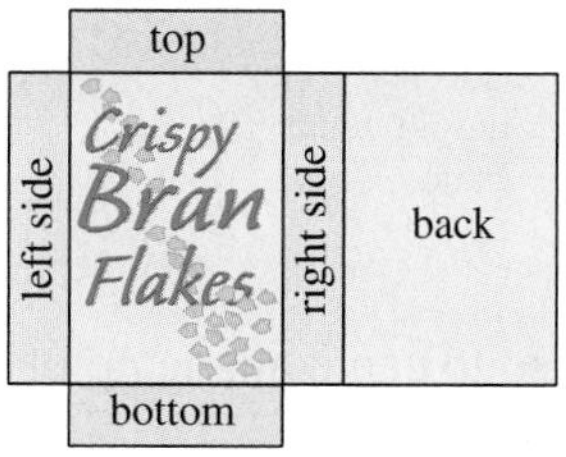

The unfolded box is made up of six rectangles: front, back, top, bottom, left side, right side.

Surface area is the area on the surface of a three-dimensional object (a solid). For a rectangular solid like the cereal box, the surface area is the sum of the areas of the six rectangular sides. Notice that the top and bottom have the same area, the front and back have the same area, and the left and right sides have the same area.

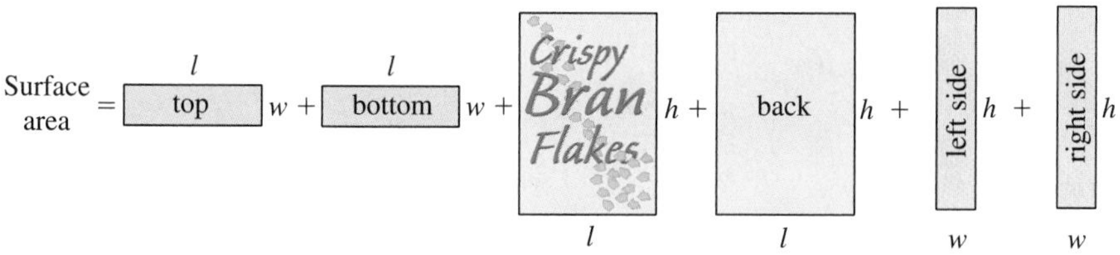

$$S = l \cdot w \quad + \quad l \cdot w \quad + \quad l \cdot h \quad + \quad l \cdot h \quad + w \cdot h + w \cdot h$$

$$S = \quad 2lw \quad + \quad 2lh \quad + \quad 2wh$$

Finding the Surface Area of a Rectangular Solid

Surface area $= (2 \cdot l \cdot w) + (2 \cdot l \cdot h) + (2 \cdot w \cdot h)$

$$S = \quad 2lw \quad + \quad 2lh \quad + \quad 2wh$$

Use ***square units*** when measuring *surface* area.

EXAMPLE 7 Finding the Volume and Surface Area of a Rectangular Solid

Find the volume and surface area of this shipping carton.

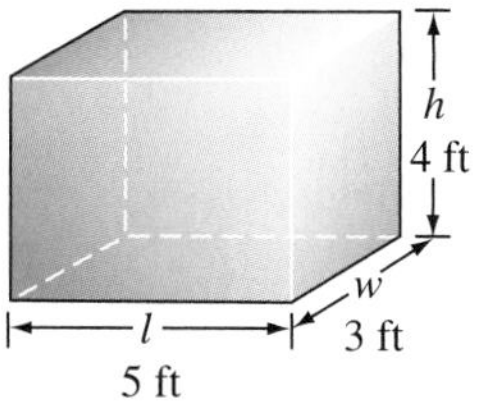

First find the volume.

$V = lwh$

$V = 5 \text{ ft} \cdot 3 \text{ ft} \cdot 4 \text{ ft}$ Write **ft³** for **volume.**

$V = 60 \text{ ft}^3$ ← **Cubic** units for **volume**

Next find the surface area.

$$S = \quad 2lw \quad + \quad 2lh \quad + \quad 2wh$$

$$S = (2 \cdot 5 \text{ ft} \cdot 3 \text{ ft}) + (2 \cdot 5 \text{ ft} \cdot 4 \text{ ft}) + (2 \cdot 3 \text{ ft} \cdot 4 \text{ ft})$$

$S = \quad 30 \text{ ft}^2 \quad + \quad 40 \text{ ft}^2 \quad + \quad 24 \text{ ft}^2$ Write **ft²** for **area.**

$S = 94 \text{ ft}^2$ ← **Square** units for **area**

Work Problem 7 at the Side. ▶

---

7 Find the volume and surface area of each rectangular solid.

GS **(a)**

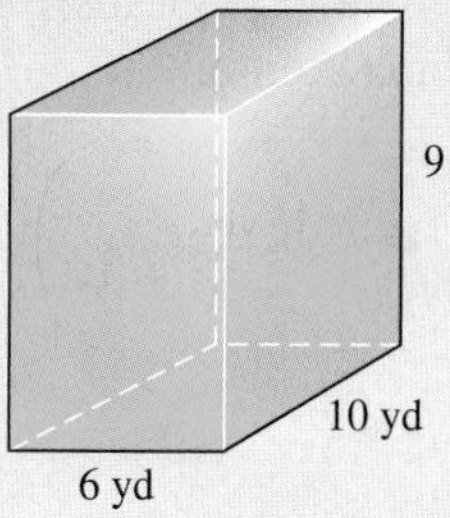

$V = lwh$

$V =$ ____ • ____ • ____

$V =$ ______

$S = 2lw + 2lh + 2wh$

(b)

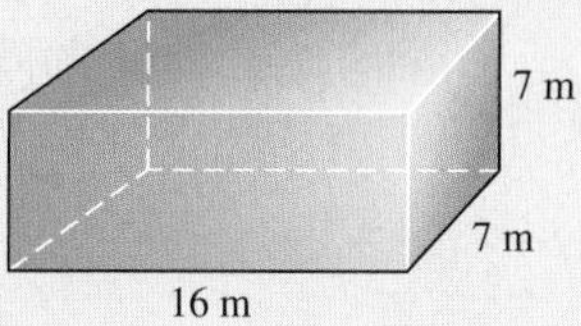

Answers

7. (a) $V = 6 \text{ yd} \cdot 10 \text{ yd} \cdot 9 \text{ yd}$;
$V = 540 \text{ yd}^3$ (*cubic* yd for *volume*)
$S = 408 \text{ yd}^2$ (*square* yd for *area*)
(b) $V = 784 \text{ m}^3$
$S = 546 \text{ m}^2$

8 Find the volume and surface area of each cylinder. Use 3.14 for π. Round your answers to the nearest tenth.

GS **(a)**

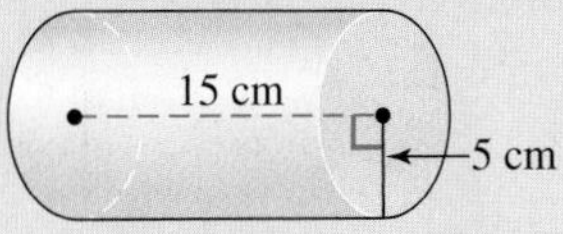

$V = \pi r^2 h$

$V \approx 3.14 \bullet$ ____ $\bullet$ ____ $\bullet$ ____

$V \approx$ ______

$S \approx 2\pi rh + 2\pi r^2$

(b)

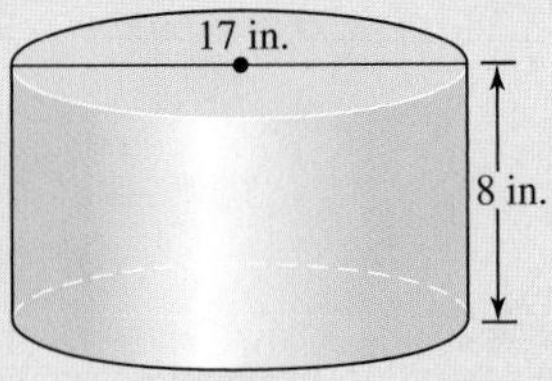

Hint: You are given the *diameter* of the cylinder. Start by finding the *radius*.

OBJECTIVE 5 **Find the surface area of a cylinder.** You can use the same idea of "unfolding" a shape to find the surface area of a cylinder, such as the soup can shown below. Finding the surface area will tell you how much aluminum you need to make the can.

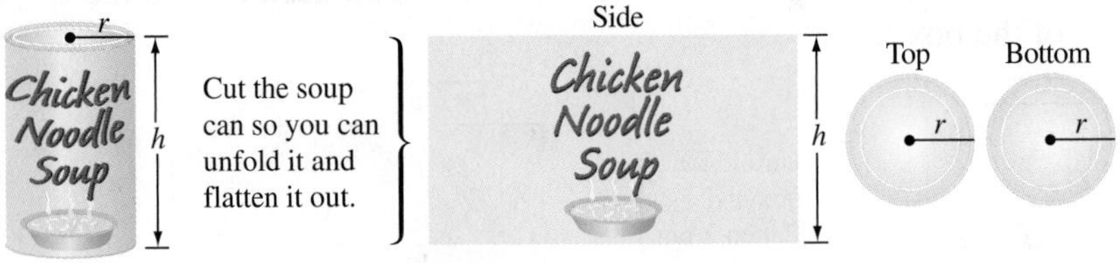

The unfolded soup can is made up of a rectangular side, a circular top, and a circular bottom.

Remember that the formula for the area of a circle is πr^2.

Surface area =

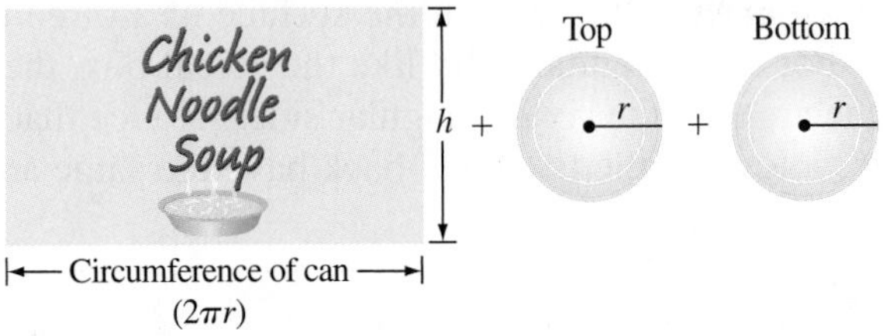

$$S = \underbrace{2\pi r \bullet h}_{2\pi rh} + \underbrace{\pi r^2 + \pi r^2}_{2\pi r^2}$$

$$S = 2\pi rh + 2\pi r^2$$

Finding the Surface Area of a Right Circular Cylinder

Surface area $= (2 \bullet \pi \bullet r \bullet h) + (2 \bullet \pi \bullet r \bullet r)$

$$S = 2\pi rh + 2\pi r^2$$

Remember that area is measured in square units, so use ***square units*** when measuring *surface* area.

EXAMPLE 8 **Finding the Volume and Surface Area of a Right Circular Cylinder**

Find the volume and surface area of this water tank. Use 3.14 as the approximate value for π. Round your answers to the nearest tenth when necessary.

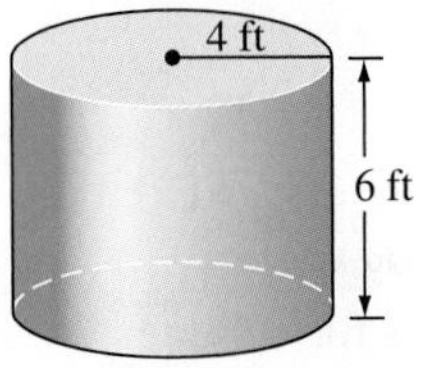

First find the volume using $V = \pi r^2 h$.

$V \approx \mathbf{3.14 \bullet 4\ ft \bullet 4\ ft \bullet 6\ ft}$

$V \approx 301.44\ ft^3$ ← Now round to tenths.

$V \approx 301.4\ ft^3$ ← **Cubic** units for **volume**

(ft³ for volume)

Now find the surface area.

$S = 2\pi rh + 2\pi r^2$

$S \approx (2 \bullet \mathbf{3.14 \bullet 4\ ft \bullet 6\ ft}) + (2 \bullet \mathbf{3.14 \bullet 4\ ft \bullet 4\ ft})$

$S \approx 150.72\ ft^2 + 100.48\ ft^2$

$S \approx 251.2\ ft^2$ ← **Square** units for **area** (ft² for area)

◀ **Work Problem 8 at the Side.**

Answers

8. **(a)** $V \approx 3.14 \bullet 5\text{ cm} \bullet 5\text{ cm} \bullet 15\text{ cm}$;
$V \approx 1177.5\text{ cm}^3$ (*cubic* units for *volume*)
$S \approx 628\text{ cm}^2$ (*square* units for *area*)
(b) $V \approx 1814.9\text{ in.}^3$
$S \approx 880.8\text{ in.}^2$

7.6 Exercises

FOR EXTRA HELP

Download the MyDashBoard App

MyMathLab®

1. CONCEPT CHECK Look at the solid shape in **Exercise 3** below. Explain why this solid is **not** a cube. When you find the volume of this solid, what will be the units in your answer?

2. CONCEPT CHECK Give two examples of everyday objects that are rectangular solids; two objects that are cylinders; and two objects that are spheres.

Name each solid and find its volume. For Exercises 3–6, also and the surface area. Use 3.14 as the approximate value of π. Round your answers to the nearest tenth if necessary. ***See Examples 1–8.***

3.

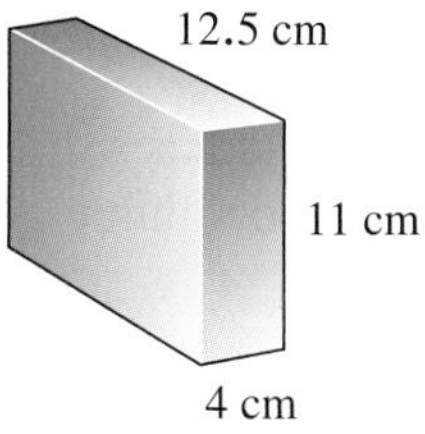

4.

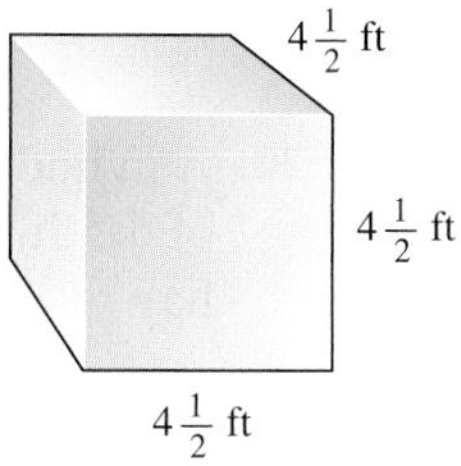

5.

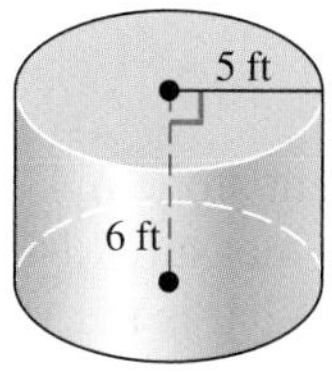

6.

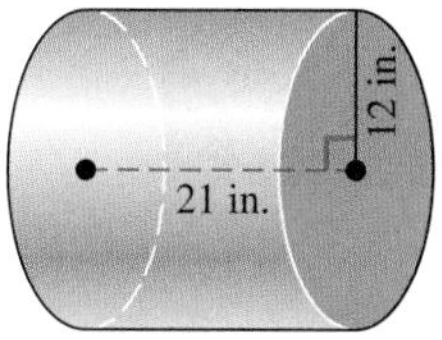

7.

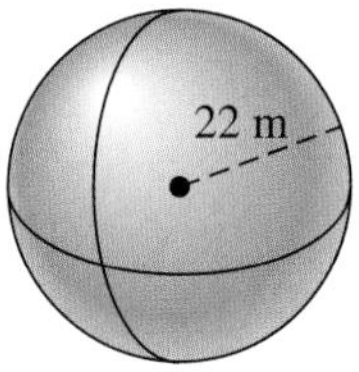

8.

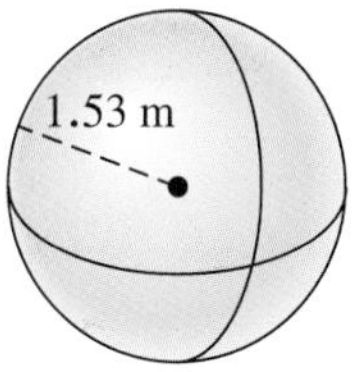

9.

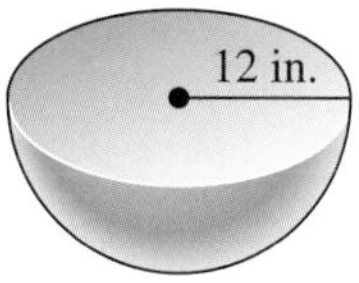

10.

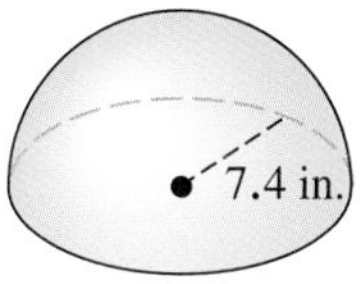

11.

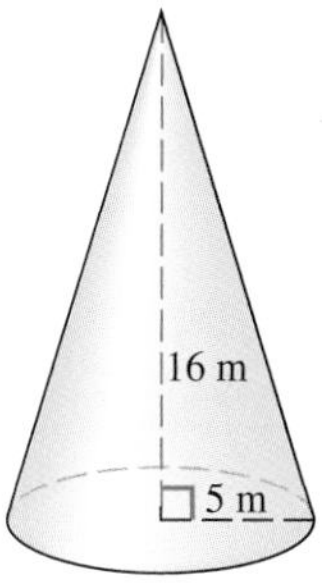

12.

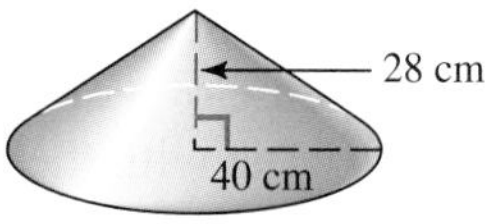

13.

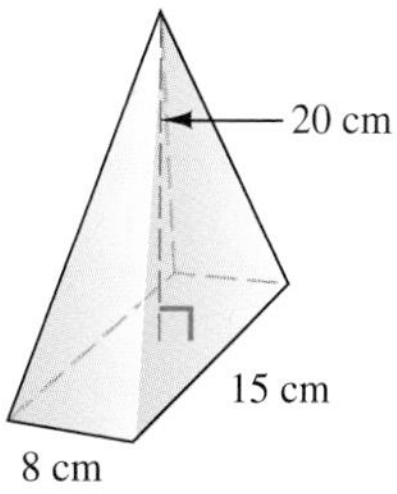

14.

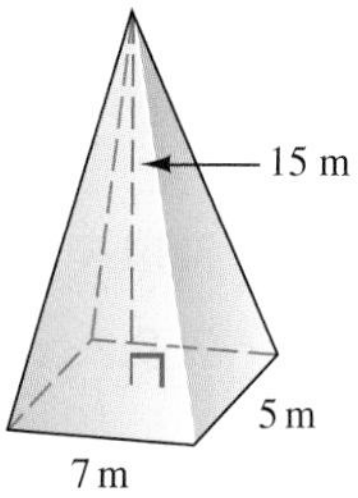

Solve each application problem. Use 3.14 as the approximate value of π. Round your final answers to the nearest tenth if necessary.

15. A pencil box measures 3 in. by 8 in. by $\frac{3}{4}$ in. high. Find the volume of the box. (*Source:* Faber Castell.)
Hint: A pencil box is a rectangular solid.

16. A train is being loaded with shipping crates. Each one is 6 m long, 3.4 m wide, and 2 m high. How much space will each crate take?
Hint: A shipping crate is a rectangular solid.

17. An oil candle globe made of hand-blown glass has a diameter of 16.8 cm. What is the volume of the globe?
Hint: A globe is another word for a sphere. First find the **radius** of the globe.

18. A metal sphere used as part of a fountain has a diameter of $6\frac{1}{2}$ ft. Find its volume.
Hint: First find the **radius** of the sphere.

19. One of the ancient stone pyramids in Egypt has a square base that measures 145 m on each side. The height is 93 m. What is the volume of the pyramid? (*Source: The Columbia Encyclopedia.*)

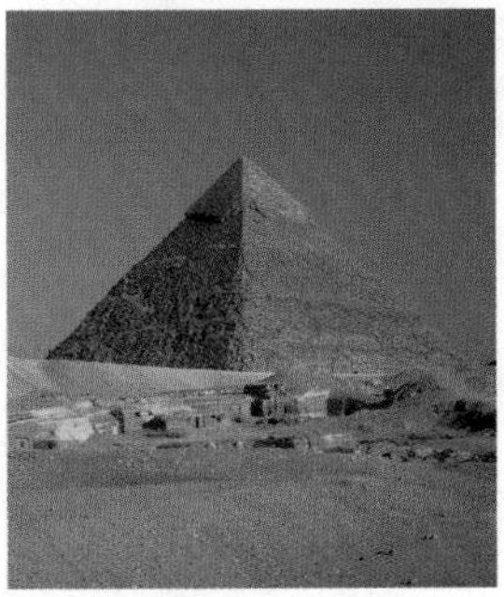

20. A cylindrical roll of hay in a farm field has a diameter of 4.5 ft and is 3.5 ft long. What is the volume of the hay roll?

21. A city sewer pipe has a diameter of 5 ft and a length of 200 ft. Find the volume of the pipe.

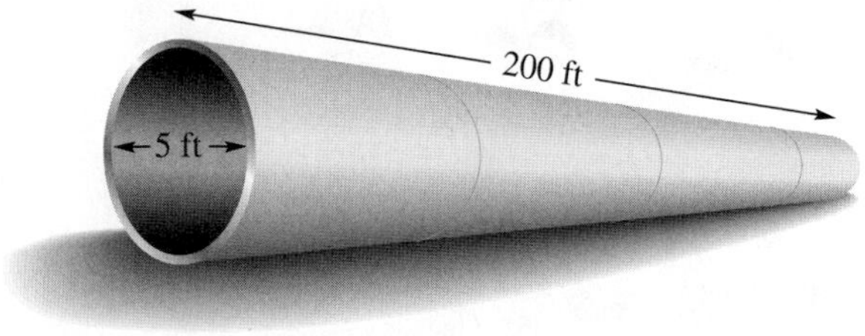

22. An ice cream cone has a diameter of 2 in. and a height of 4 in. Find its volume.

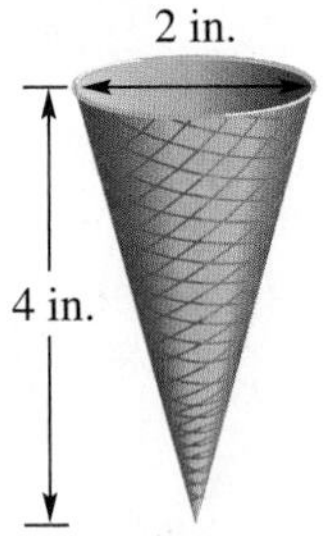

23. **CONCEPT CHECK** Explain the *two* errors made by a student in finding the volume of a cylinder with a diameter of 7 cm and a height of 5 cm. Find the correct answer.

$$V \approx 3.14 \cdot 7 \cdot 7 \cdot 5$$
$$V \approx 769.3 \text{ cm}^2$$

24. **CONCEPT CHECK** Compare the steps in finding the volume of a cylinder and a cone. How are they similar? Suppose you know the volume of a cylinder. How can you find the volume of a cone with the same radius and height by doing just a one-step calculation?

7.7 Pythagorean Theorem

OBJECTIVES

1. Find square roots using the square root key on a calculator.
2. Find the unknown length in a right triangle.
3. Solve application problems involving right triangles.

In **Section 7.2** you used this formula for the area of a square, $A = s^2$. The blue square below has an area of 25 cm^2 because 5 cm • 5 cm = 25 cm^2.

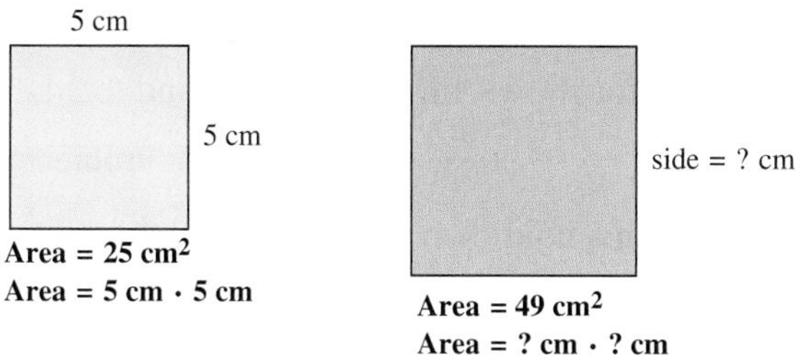

The red square above has an area of 49 cm^2. To find the length of a side, ask yourself, "What number can be multiplied by itself to give 49?" Because 7 • 7 = 49, the length of each side is 7 cm.

Also, because 7 • 7 = 49, we say that 7 is the *square root* of 49, or $\sqrt{49} = 7$. Also, $\sqrt{81} = 9$, because 9 • 9 = 81. (See **Section 1.8** for further review.)

Work Problem **1** at the Side. ▶

A number that has a whole number as its square root is called a *perfect square.* For example, 9 is a perfect square because $\sqrt{9} = 3$, and 3 is a whole number.

Some of the perfect squares are listed below.

Some of the Perfect Squares

$\sqrt{1} = 1$	$\sqrt{16} = 4$	$\sqrt{49} = 7$	$\sqrt{100} = 10$
$\sqrt{4} = 2$	$\sqrt{25} = 5$	$\sqrt{64} = 8$	$\sqrt{121} = 11$
$\sqrt{9} = 3$	$\sqrt{36} = 6$	$\sqrt{81} = 9$	$\sqrt{144} = 12$

OBJECTIVE 1 Find square roots using the square root key on a calculator. If a number is *not* a perfect square, then you can find its *approximate* square root by using a calculator with a square root key.

Calculator Tip

To find a square root, use the (√) key, or, on some calculators, the (√x) key. Try these.

To find $\sqrt{16}$ press: 16 (√x) Answer is 4.

To find $\sqrt{7}$ press: 7 (√x) Answer is 2.645751311.

On some scientific calculators, you press the √ key first, then enter the number. (Check your model's operating instructions.)

For $\sqrt{7}$, your calculator shows 2.645751311, an *approximate* answer. (Some calculators show more or fewer digits.) We will be rounding to the nearest thousandth, so $\sqrt{7} \approx 2.646$. To check, multiply 2.646 times 2.646. Do you get 7 as the result? No, you get 7.001316, which is very close to 7. The difference is due to rounding.

1 Find each square root.

GS **(a)** $\sqrt{36}$

What number, multiplied by itself, gives 36? ____

So, $\sqrt{36} =$ ____

(b) $\sqrt{25}$

(c) $\sqrt{9}$

(d) $\sqrt{100}$

(e) $\sqrt{121}$

Answers

1. (a) 6; 6 (b) 5 (c) 3 (d) 10 (e) 11

2 Use a calculator with a square root key to find each square root. Round to the nearest thousandth if necessary.

(a) $\sqrt{11}$

(b) $\sqrt{40}$

(c) $\sqrt{56}$

(d) $\sqrt{196}$

(e) $\sqrt{147}$

Answers

2. (a) $\sqrt{11} \approx 3.317$ (b) $\sqrt{40} \approx 6.325$ (c) $\sqrt{56} \approx 7.483$ (d) $\sqrt{196} = 14$ (e) $\sqrt{147} \approx 12.124$

EXAMPLE 1 Finding the Square Root of Numbers

Use a calculator to find each square root. Round answers to the nearest thousandth.

Your calculator may show more or fewer digits.

(a) $\sqrt{35}$ Calculator shows 5.916079783; round to 5.916.

(b) $\sqrt{124}$ Calculator shows 11.13552873; round to 11.136.

(c) $\sqrt{200}$ Calculator shows 14.14213562; round to 14.142.

◀ Work Problem **2** at the Side.

OBJECTIVE 2 Find the unknown length in a right triangle. One place you will use square roots is when working with the *Pythagorean Theorem.* This theorem applies only to *right* triangles (triangles with a 90° angle). The longest side of a right triangle is called the **hypotenuse.** It is opposite the right angle. The other two sides are called *legs.* The legs form the right angle.

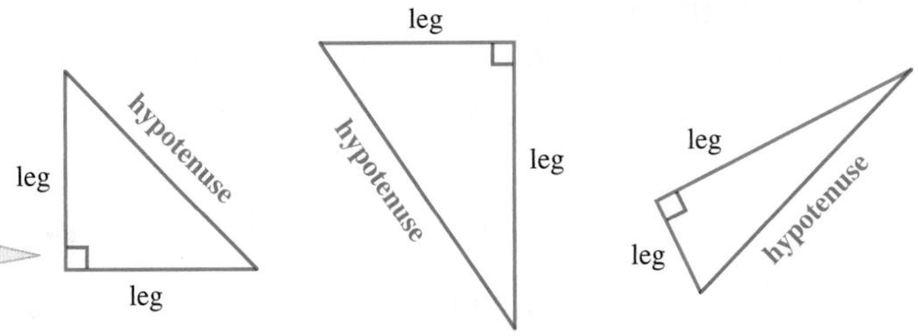

Examples of right triangles

Pythagorean Theorem

$$(\textbf{hypotenuse})^2 = (\textbf{leg})^2 + (\textbf{leg})^2$$

In other words, square the length of each side. After you have squared all the sides, the sum of the squares of the two legs will equal the square of the hypotenuse. An example is shown below.

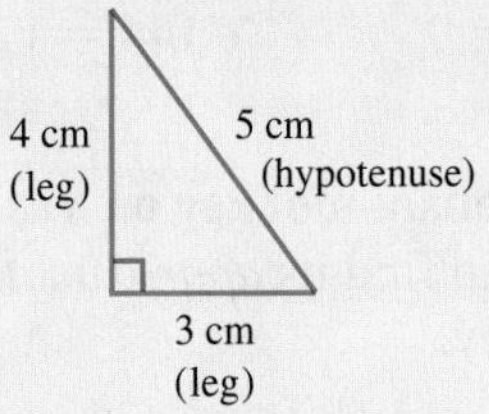

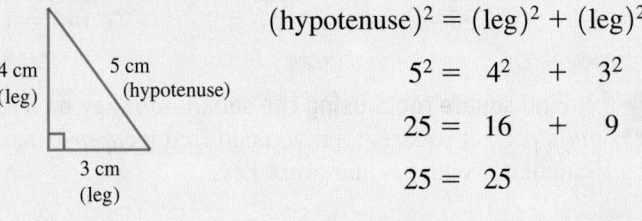

$$(\text{hypotenuse})^2 = (\text{leg})^2 + (\text{leg})^2$$
$$5^2 = 4^2 + 3^2$$
$$25 = 16 + 9$$
$$25 = 25$$

The theorem is named after Pythagoras, a Greek mathematician who lived about 2500 years ago. He and his followers may have used triangular floor tiles to prove the theorem, as shown below.

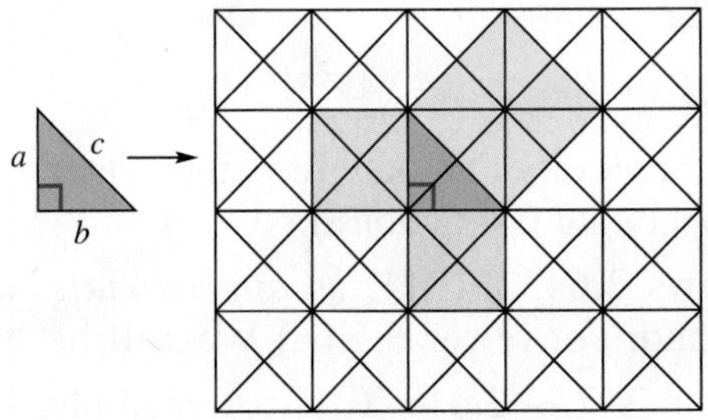

The green right triangle in the center of the floor tiles has sides *a*, *b*, and *c*. The pink square drawn on side *a* contains four triangular tiles. The pink square on side *b* contains four tiles. The blue square on side *c* contains eight tiles. The number of tiles in the square on side *c* equals the sum of the number of tiles in the squares on sides *a* and *b*, that is, 8 tiles = 4 tiles + 4 tiles. As a result, you often see the Pythagorean Theorem written as $c^2 = a^2 + b^2$.

If you know the lengths of any two sides in a right triangle, you can use the Pythagorean Theorem to find the length of the third side.

Formulas Based on the Pythagorean Theorem

To find the hypotenuse, use this formula:

$$\textbf{hypotenuse} = \sqrt{(\textbf{leg})^2 + (\textbf{leg})^2}$$

To find a leg, use this formula:

$$\textbf{leg} = \sqrt{(\textbf{hypotenuse})^2 - (\textbf{leg})^2}$$

CAUTION

Remember: A small square drawn in one angle of a triangle indicates a right angle. You can use the Pythagorean Theorem *only* on triangles that have a right angle.

EXAMPLE 2 Finding the Unknown Length in Right Triangles

Find the unknown length in each right triangle. Round answers to the nearest tenth if necessary.

(a)

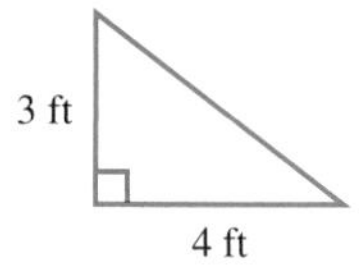

The **unknown length** is the side opposite the right angle, which is the **hypotenuse.** Use the formula for finding the hypotenuse.

$\text{hypotenuse} = \sqrt{(\text{leg})^2 + (\text{leg})^2}$ Find the hypotenuse.

$\text{hypotenuse} = \sqrt{(3)^2 + (4)^2}$ Legs are 3 and 4.

$= \sqrt{9 + 16}$ $3 \cdot 3$ is 9 and $4 \cdot 4$ is 16.

$= \sqrt{25}$

$= 5$ This is the length of a side, so write **ft** in the answer (**not** ft^2).

The hypotenuse is 5 ft long.

(b)

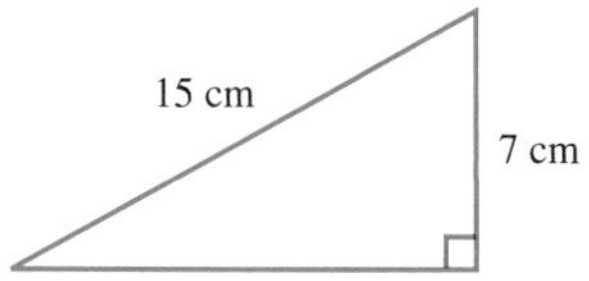

We *do* know the length of the hypotenuse (15 cm), so it is **the length of one of the legs that is unknown.** Use the formula for finding a leg.

$\text{leg} = \sqrt{(\text{hypotenuse})^2 - (\text{leg})^2}$ Find a leg.

$\text{leg} = \sqrt{(15)^2 - (7)^2}$ Hypotenuse is 15; one leg is 7.

$= \sqrt{225 - 49}$ $15 \cdot 15$ is 225 and $7 \cdot 7$ is 49.

$= \sqrt{176}$ Use calculator to find $\sqrt{176}$.

≈ 13.3 Round 13.26649916 to 13.3.

The length of the leg is approximately 13.3 cm. Write **cm** in the answer, **not** cm^2.

Work Problem 3 at the Side. ▶

3 Find the unknown length in each right triangle. Round your answers to the nearest tenth if necessary.

GS **(a)**

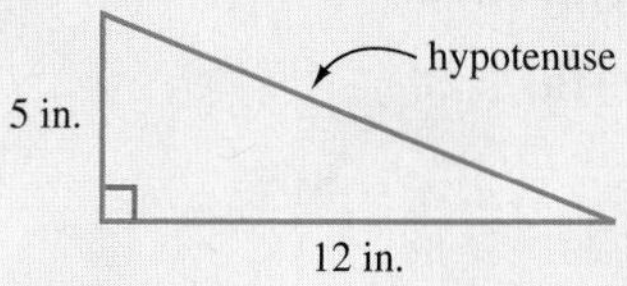

$\text{hypotenuse} = \sqrt{(5)^2 + (12)^2}$

$= \sqrt{25 + \_\_\_\_}$

$= \sqrt{\_\_\_\_} = \_\_\_\_$

GS **(b)**

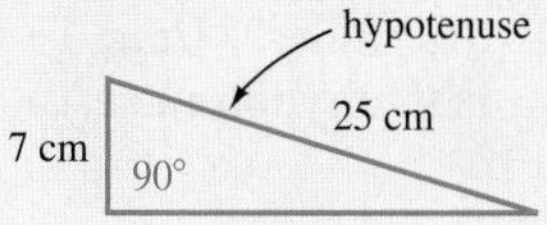

$\text{leg} = \sqrt{(25)^2 - (7)^2}$

$= \sqrt{625 - \_\_\_\_}$

$= \sqrt{\_\_\_\_} = \_\_\_\_$

(c)

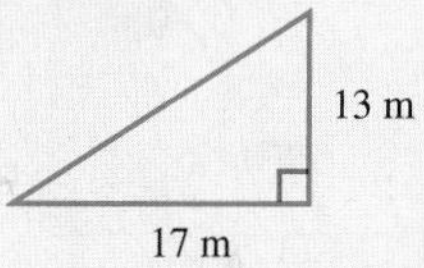

(d)

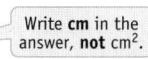

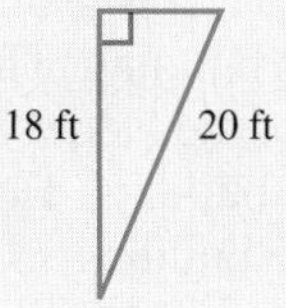

Answers

3. **(a)** $\sqrt{25 + 144} = \sqrt{169} = 13$ in.
(b) $\sqrt{625 - 49} = \sqrt{576} = 24$ cm
(c) $\sqrt{458} \approx 21.4$ m **(d)** $\sqrt{76} \approx 8.7$ ft

4 These problems show ladders leaning against buildings. Find the unknown lengths. Round to the nearest tenth of a foot if necessary.

(a)

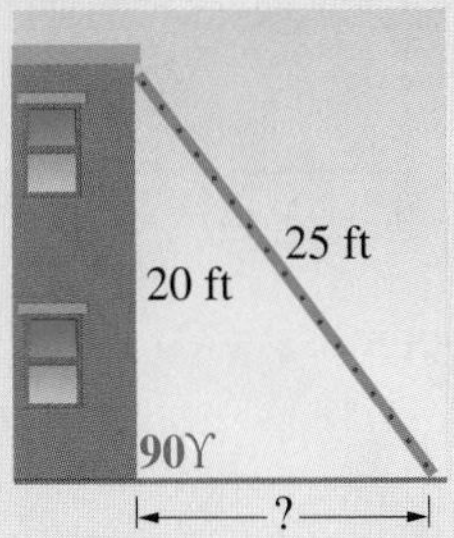

How far away from the building is the bottom of the ladder? (*Hint:* The ladder is the **hypotenuse.**)

(b)

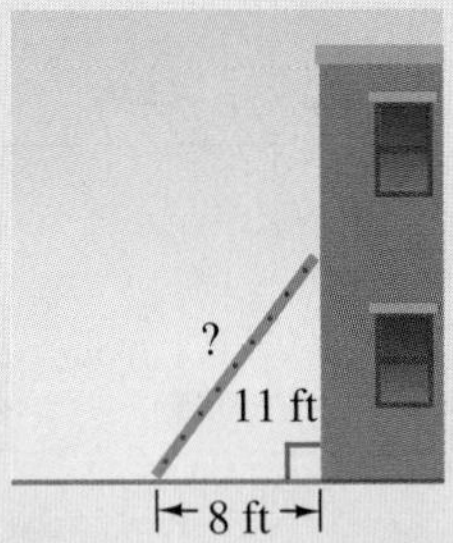

How long is the ladder?

(c) A 17 ft ladder is leaning against a building. The bottom of the ladder is 10 ft from the building. How high up on the building will the ladder reach? (*Hint:* Start by drawing the building and the ladder.)

Answers

4. **(a)** leg $= \sqrt{225} = 15$ ft
(b) hypotenuse $= \sqrt{185} \approx 13.6$ ft
(c) leg $= \sqrt{189} \approx 13.7$ ft

OBJECTIVE 3 Solve application problems involving right triangles. The next example shows an application of the Pythagorean Theorem.

EXAMPLE 3 Using the Pythagorean Theorem

A television antenna is on the roof of a house, as shown below. Find the length of the support wire. Round your answer to the nearest tenth of a meter if necessary.

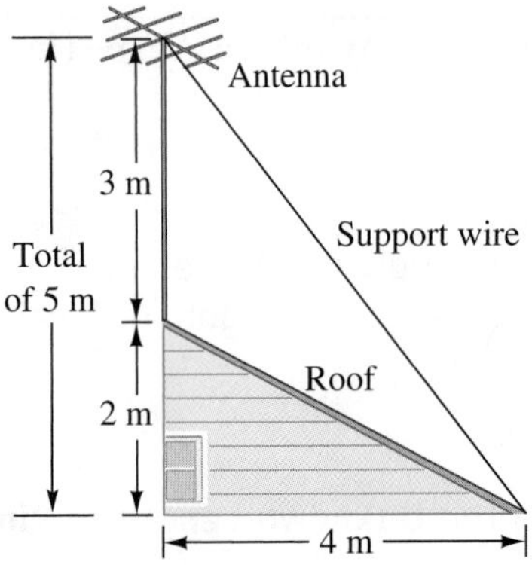

A right triangle is formed. The total length of the leg on the left is 3 m + 2 m = 5 m.

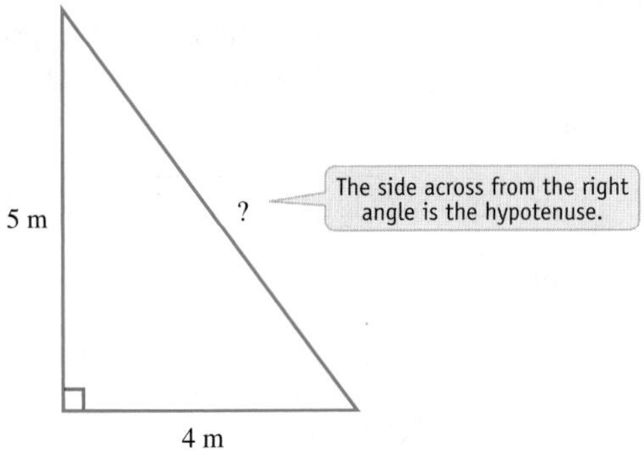

Notice that the support wire is opposite the right angle, so it is the hypotenuse of the right triangle.

$$\text{hypotenuse} = \sqrt{(\text{leg})^2 + (\text{leg})^2} \quad \text{Find the hypotenuse.}$$

$$\text{hypotenuse} = \sqrt{(5)^2 + (4)^2} \quad \text{Legs are 5 and 4.}$$

$$= \sqrt{25 + 16} \quad 5^2 \text{ is 25 and } 4^2 \text{ is 16.}$$

$$= \sqrt{41} \quad \text{Use a calculator to find } \sqrt{41}.$$

$$\approx 6.4 \quad \text{Round 6.403124237 to 6.4}$$

The length of the support wire is approximately 6.4 m. — This is **length**, so write **m** in the answer (**not** m^2).

CAUTION

You use the Pythagorean Theorem to find the ***length*** of one side, *not* the area of the triangle. Your answer will be in linear units, such as ft, yd, cm, m, and so on (*not* ft^2, yd^2, cm^2, m^2).

◀ **Work Problem 4 at the Side.**

7.7 Exercises

FOR EXTRA HELP

Download the MyDashBoard App

MyMathLab®

Find each square root. Starting with Exercise 5, use the square root key on a calculator. Round your answers to the nearest thousandth if necessary. **See Example 1.**

1. $\sqrt{16}$

2. $\sqrt{4}$

3. $\sqrt{64}$

4. $\sqrt{81}$

5. $\sqrt{11}$

6. $\sqrt{23}$

7. $\sqrt{5}$

8. $\sqrt{2}$

9. $\sqrt{73}$

10. $\sqrt{80}$

11. $\sqrt{101}$

12. $\sqrt{125}$

13. $\sqrt{190}$

14. $\sqrt{160}$

15. $\sqrt{1000}$

16. $\sqrt{2000}$

17. **CONCEPT CHECK** You know that $\sqrt{25} = 5$ and $\sqrt{36} = 6$. Using just that information (no calculator), describe how you could *estimate* $\sqrt{30}$. How would you estimate $\sqrt{26}$ or $\sqrt{35}$? Now check your estimates using a calculator.

18. **CONCEPT CHECK** Explain the relationship between *squaring* a number and finding the *square root* of a number. Include two examples to illustrate your explanation.

Find the unknown length in each right triangle. Use a calculator to find square roots. Round your answers to the nearest tenth if necessary. **See Example 2.**

19. GS

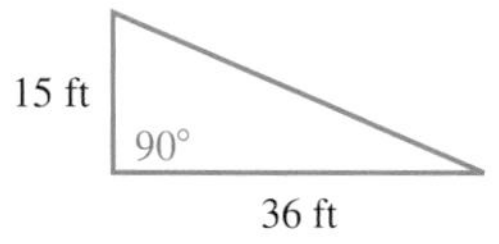

Hint: The unknown length is the **hypotenuse.**

hypotenuse $= \sqrt{(15)^2 + (36)^2}$

20. GS

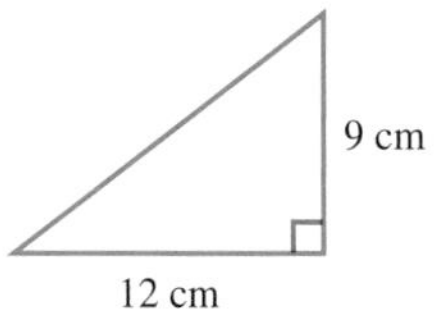

Hint: The unknown length is the **hypotenuse.**

hypotenuse $= \sqrt{(9)^2 + (12)^2}$

21.

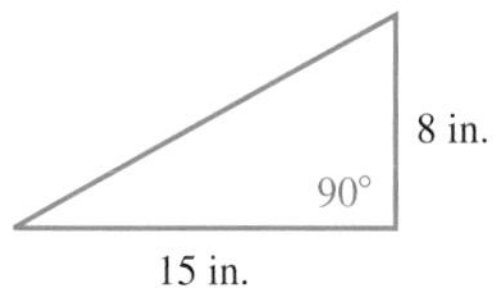

22.

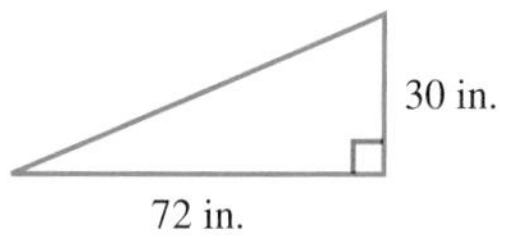

23. GS

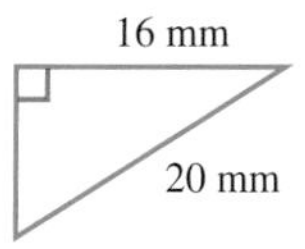

Hint: The unknown length is a **leg.**

leg $= \sqrt{(20)^2 - (16)^2}$

24. GS

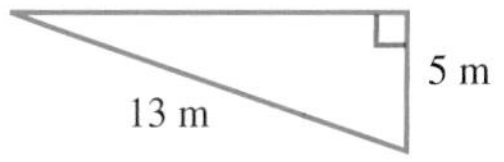

Hint: The unknown length is a **leg.**

leg $= \sqrt{(13)^2 - (5)^2}$

25.

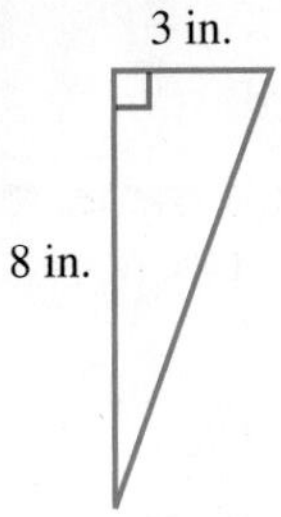

26.

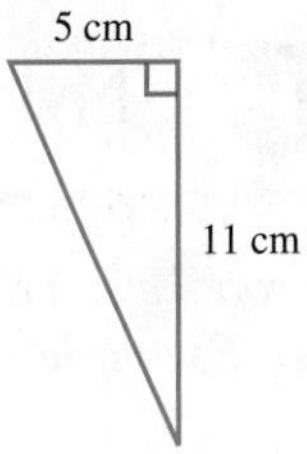

27.

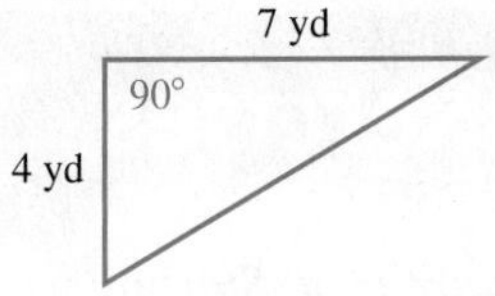

28.

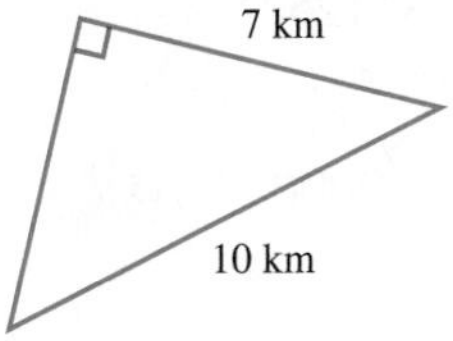

29.

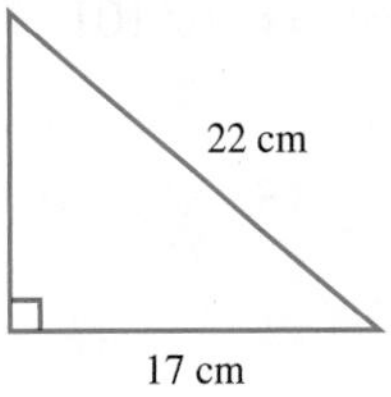

30.

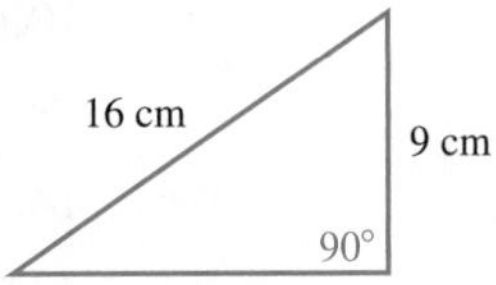

31.

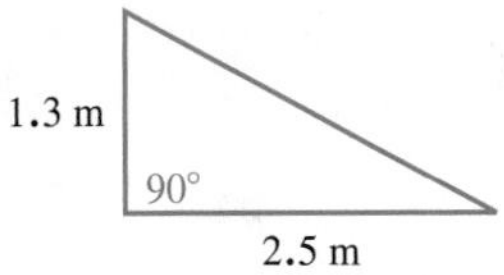

32.

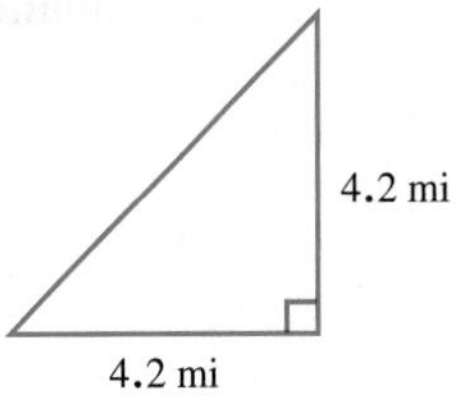

33.

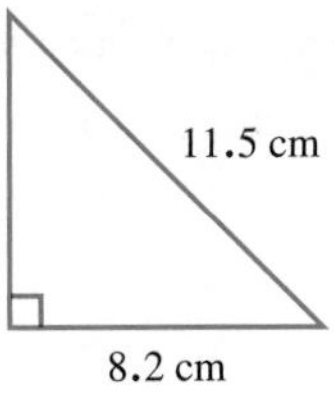

34.

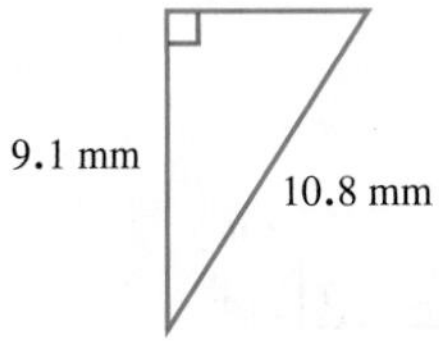

35.

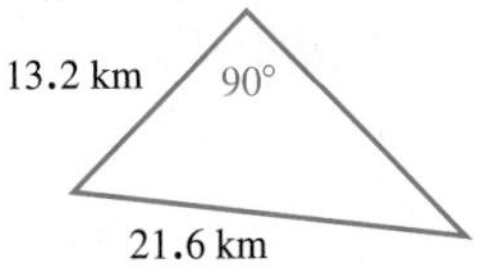

36.

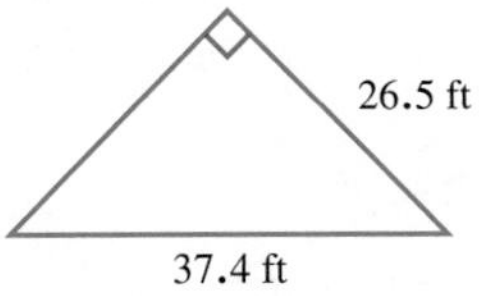

37. CONCEPT CHECK Explain the *two* errors made by a student in solving this problem. Also find the correct answer. Round to the nearest tenth.

$$? = \sqrt{(9)^2 + (7)^2}$$

$$= \sqrt{18 + 14}$$

$$= \sqrt{32} \approx 5.657 \text{ in.}$$

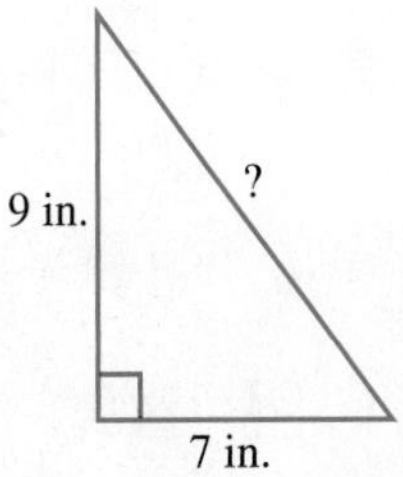

38. CONCEPT CHECK Explain the *two* errors made by a student in solving this problem. Also find the correct answer. Round to the nearest tenth.

$$? = \sqrt{(13)^2 + (20)^2}$$

$$= \sqrt{169 + 400}$$

$$= \sqrt{569} \approx 23.9 \text{ m}^2$$

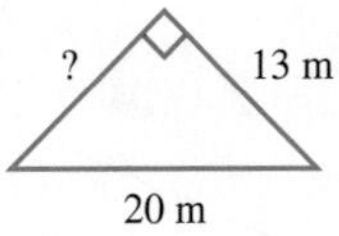

Solve each application problem. Round your answers to the nearest tenth if necessary.
See Example 3.

39. Find the length of this loading ramp.
Hint: The ramp is the **hypotenuse** in a right triangle.

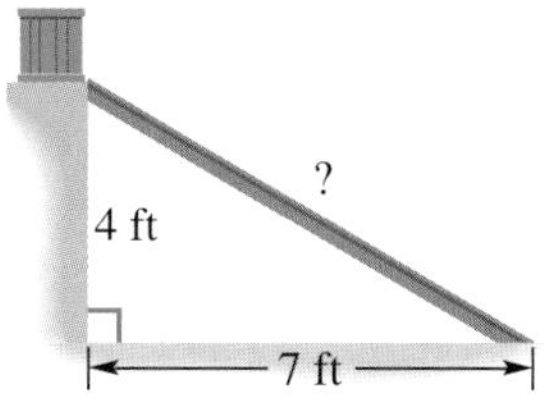

40. Find the unknown length in this window frame.
Hint: The unknown length is the **hypotenuse** in a right triangle.

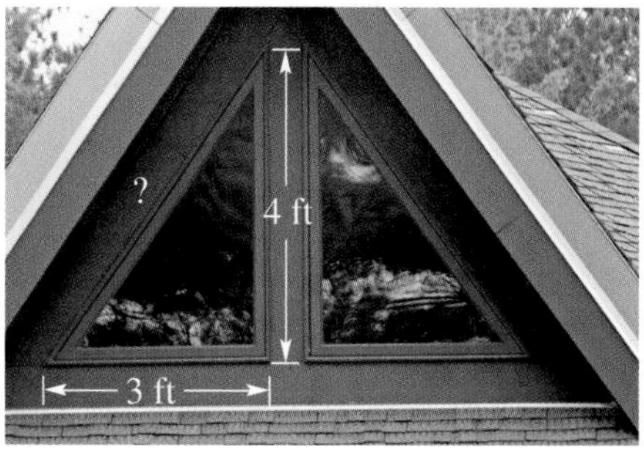

41. How high is the airplane above the ground?
Hint: The height of the plane is a **leg** in the right triangle.

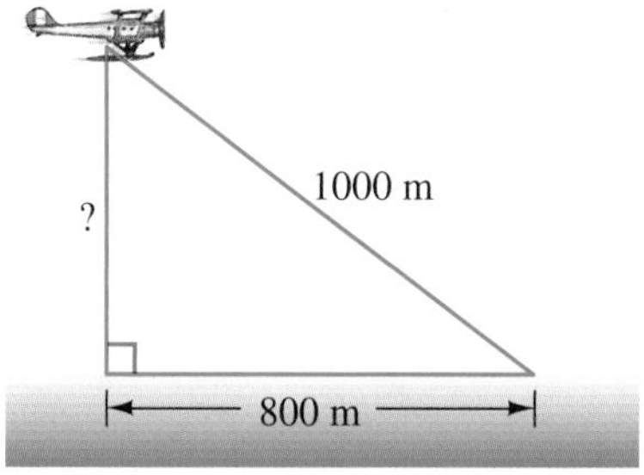

42. Find the height of this farm silo.
Hint: The height of the silo is a **leg** in the right triangle.

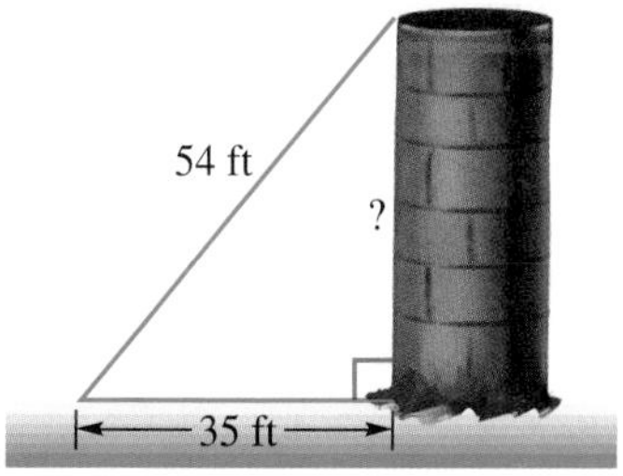

43. How long is the diagonal brace on this rectangular gate?

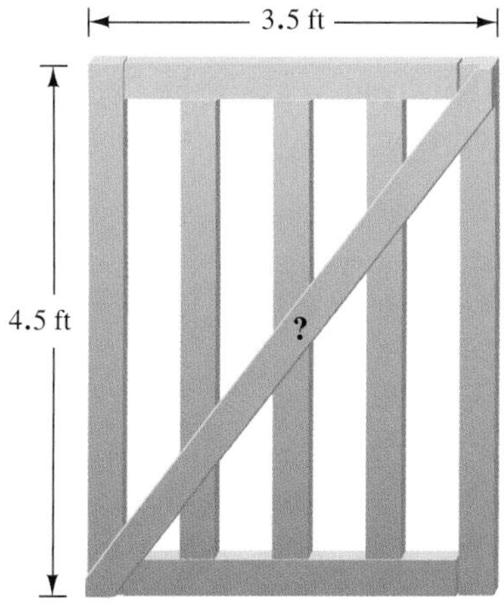

44. Find the height of this rectangular television screen.

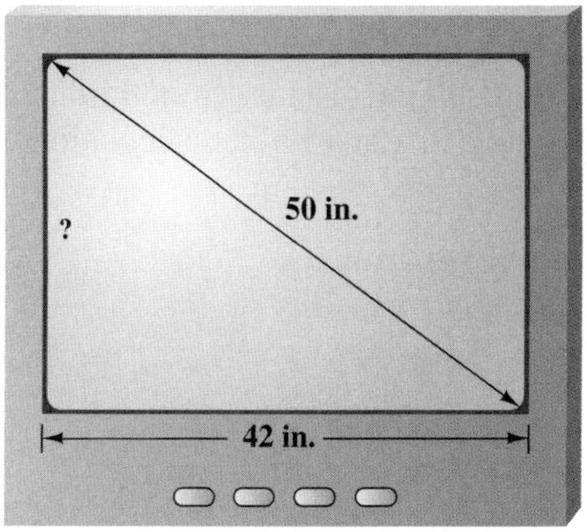

45. To reach his ladylove, a knight placed a 12 ft ladder against the castle wall. If the base of the ladder is 3 ft from the building, how high on the castle will the top of the ladder reach? Draw a sketch of the castle and ladder and solve the problem.

46. William drove his car 15 miles north, then made a 90° right turn and drove 7 miles east. How far is he, in a straight line, from his starting point? Draw a sketch to illustrate the problem and solve it.

Relating Concepts (Exercises 47–50) For Individual or Group Work

Use your knowledge of the Pythagorean Theorem to **work Exercises 47–50 in order.** *Round answers to the nearest tenth.*

47. A major league baseball diamond is a square shape measuring 90 ft on each side. If the catcher throws a ball from home plate to second base, how far is he throwing the ball? (*Source:* American League of Professional Baseball Clubs.)

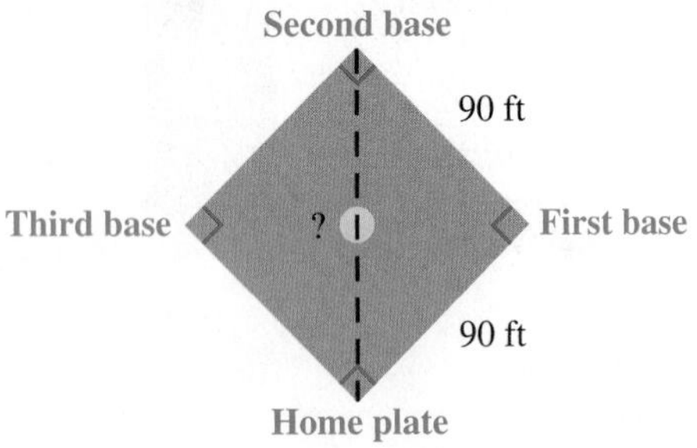

48. A softball diamond is only 60 ft on each side. (*Source:* Amateur Softball Association.)

(a) Draw a sketch of the softball diamond and label the bases and the lengths of the sides.

(b) How far is it to throw a ball from home plate to second base?

49. Look back at your answer to **Exercise 47.** Explain how you can tell the distance from third base to first base without doing any further calculations.

50. Show how you could set up a proportion to answer **Exercise 48** instead of using the Pythagorean Theorem. (You'll need your answer from **Exercise 47.**)

7.8 Congruent and Similar Triangles

OBJECTIVES

1. Identify corresponding parts of congruent triangles.
2. Prove that triangles are congruent using ASA, SSS, or SAS.
3. Identify corresponding parts of similar triangles.
4. Find the unknown lengths of sides in similar triangles.
5. Solve application problems involving similar triangles.

Two useful concepts in geometry are *congruence* and *similarity*. If two figures are *identical,* both in *shape* and in *size,* we say the figures are **congruent,** or perfect duplicates of each other. This is like getting two exact copies of a photo. If two figures have the *same shape* but are *different sizes,* we say the figures are **similar,** like getting a photo and then enlarging it. We'll explore the ideas of congruence and similarity using triangles.

OBJECTIVE 1 Identify corresponding parts of congruent triangles.

The two triangles below are *congruent* because they are the *same shape* and the *same size*.

One way to name this triangle is $\triangle ABC$. One way to name this triangle is $\triangle DEF$.

If you pick up $\triangle ABC$ and slide it over on top of $\triangle DEF$, the triangles would be a perfect match. $\angle 1$ would be on top of $\angle 4$, so they are *corresponding angles*. Similarly, $\angle 2$ and $\angle 5$ are corresponding angles, and $\angle 3$ and $\angle 6$ are corresponding angles. Corresponding angles have the same measure, as indicated below.

$$m\angle 1 = m\angle 4 \qquad m\angle 2 = m\angle 5 \qquad m\angle 3 = m\angle 6$$

The abbreviation for measure is m, so $m\angle 1$ is read, "the measure of angle 1."

When you put $\triangle ABC$ on top of $\triangle DEF$, you would also see that side AB is on top of side DE. We say that $\overline{AB}$ and $\overline{DE}$ are *corresponding sides*. Similarly, $\overline{BC}$ and $\overline{EF}$ are corresponding sides, and $\overline{AC}$ and $\overline{DF}$ are corresponding sides. You would see that corresponding sides have the same length.

$$AB = DE \qquad BC = EF \qquad AC = DF$$

Because corresponding angles have the same measure, and corresponding sides have the same length, we know that $\triangle ABC$ **is congruent to** $\triangle DEF$. We can write this as $\triangle ABC \cong \triangle DEF$.

Congruent Triangles

If two triangles are congruent, then

1. Corresponding angles have the same measure, and
2. Corresponding sides have the same length.

EXAMPLE 1 Identifying Corresponding Parts in Congruent Triangles

Each pair of triangles is congruent. List the corresponding angles and corresponding sides.

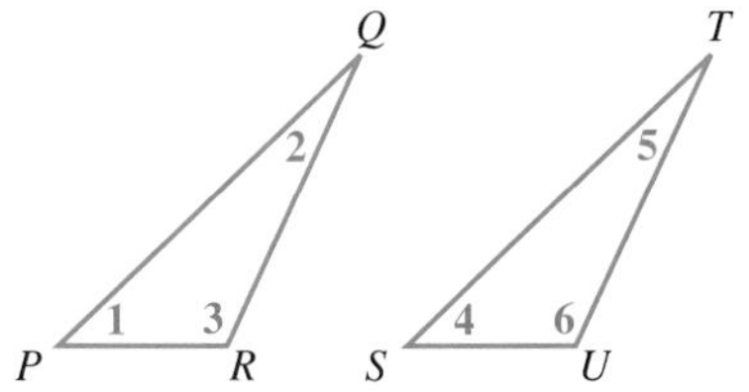

(a) If you slid $\triangle PQR$ on top of $\triangle STU$, the two triangles would match.

The corresponding parts are:

$\angle 1$ and $\angle 4$ $\quad \overline{PQ}$ and $\overline{ST}$

$\angle 2$ and $\angle 5$ $\quad \overline{PR}$ and $\overline{SU}$

$\angle 3$ and $\angle 6$ $\quad \overline{QR}$ and $\overline{TU}$

Continued on Next Page

1 Each pair of triangles is congruent. List the corresponding angles and the corresponding sides.

(a)

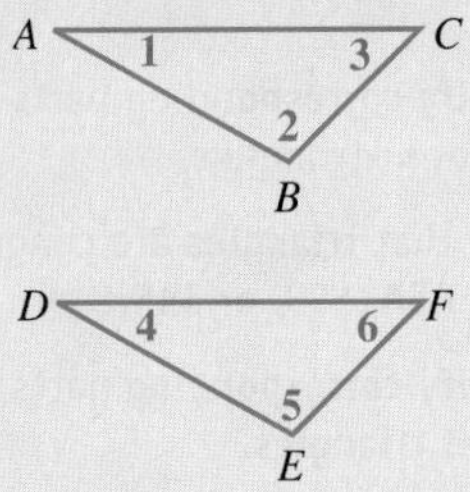

(b)

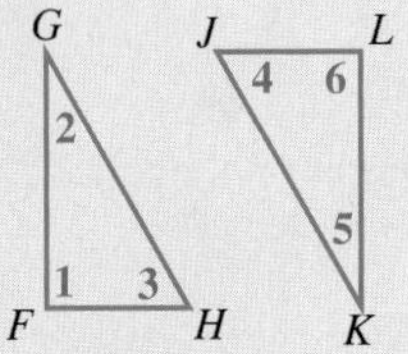

Hint: Rotate $\triangle FGH$, then slide it on top of $\triangle JLK$.

(c)

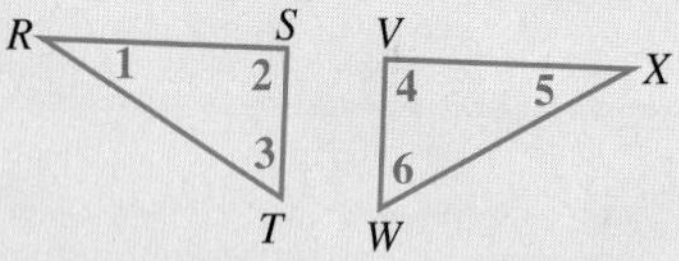

Hint: Flip $\triangle RST$ over, then slide it on top of $\triangle VWX$.

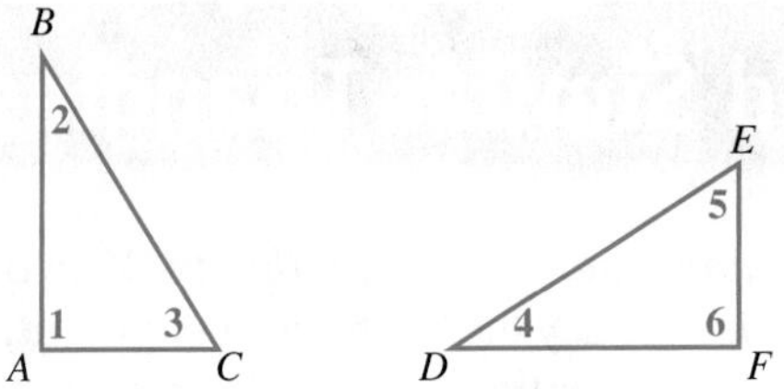

(b) If you slide $\triangle ABC$ on top of $\triangle DEF$, it wouldn't match. But if you *rotate* $\triangle ABC$ before sliding it on top of $\triangle DEF$, it *will* match.

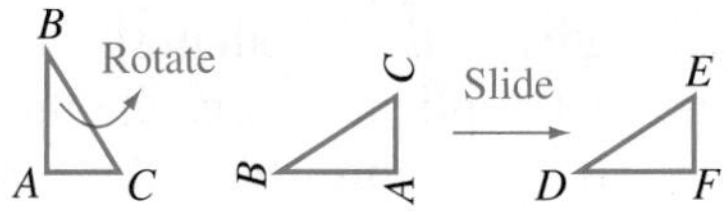

Be careful identifying corresponding parts when one triangle is rotated or flipped.

The corresponding parts are:

$\angle 1$ and $\angle 6$ $\quad \overline{BC}$ and $\overline{DE}$

$\angle 2$ and $\angle 4$ $\quad \overline{BA}$ and $\overline{DF}$

$\angle 3$ and $\angle 5$ $\quad \overline{CA}$ and $\overline{EF}$

◀ **Work Problem 1 at the Side.**

OBJECTIVE 2 Prove that triangles are congruent using ASA, SSS, or SAS. One way to prove that two triangles are congruent would be to measure all the angles and all the sides. If the measures of the corresponding angles and sides are equal, then the triangles are congruent. But here are three quicker methods to prove that two triangles are congruent.

Proving That Two Triangles Are Congruent

1. Angle–Side–Angle (ASA) Method

If two angles and the side between them on one triangle measure the same as the corresponding parts on another triangle, the triangles are congruent.

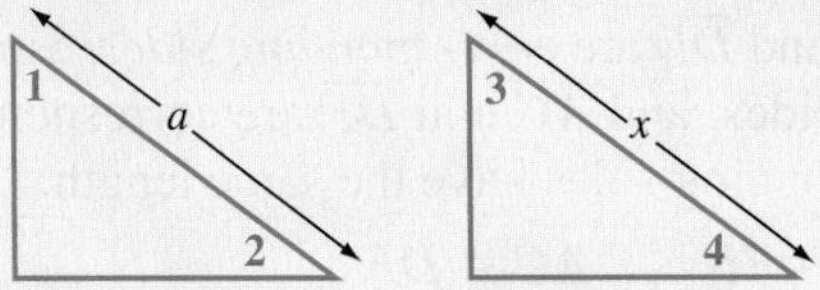

If $m\angle 1 = m\angle 3$ and $m\angle 2 = m\angle 4$ and $a = x$, then the two triangles are congruent.

2. Side–Side–Side (SSS) Method

If three sides of one triangle measure the same as the corresponding sides of another triangle, the triangles are congruent.

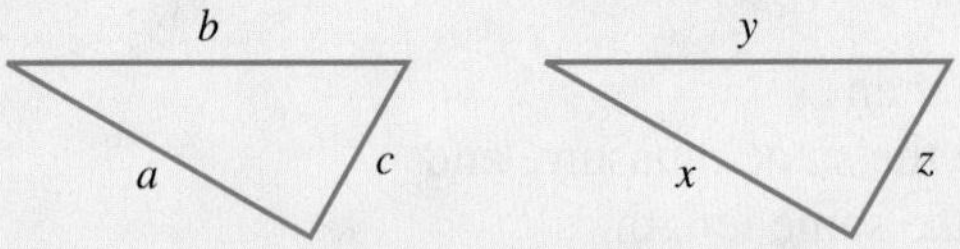

If $a = x$ and $b = y$ and $c = z$, then the two triangles are congruent.

3. Side–Angle–Side (SAS) Method

If two sides and the angle between them on one triangle measure the same as the corresponding parts on another triangle, the triangles are congruent.

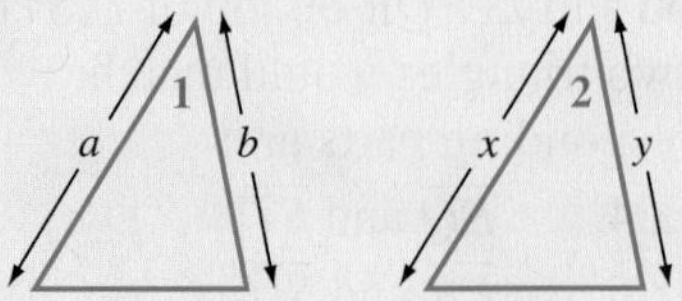

If $a = x$ and $b = y$ and $m\angle 1 = m\angle 2$, then the two triangles are congruent.

Answers

1. (a) $\angle 1$ and $\angle 4$, $\angle 2$ and $\angle 5$, $\angle 3$ and $\angle 6$; $\overline{AC}$ and $\overline{DF}$, $\overline{AB}$ and $\overline{DE}$, $\overline{BC}$ and $\overline{EF}$

(b) $\angle 1$ and $\angle 6$, $\angle 2$ and $\angle 5$, $\angle 3$ and $\angle 4$; $\overline{GF}$ and $\overline{KL}$, $\overline{FH}$ and $\overline{LJ}$, $\overline{GH}$ and $\overline{KJ}$

(c) $\angle 1$ and $\angle 5$, $\angle 2$ and $\angle 4$, $\angle 3$ and $\angle 6$; $\overline{RS}$ and $\overline{XV}$, $\overline{RT}$ and $\overline{XW}$, $\overline{ST}$ and $\overline{VW}$

EXAMPLE 2 **Proving That Two Triangles Are Congruent**

Explain which method can be used to prove that each pair of triangles is congruent. Choose from ASA, SSS, and SAS.

(a)

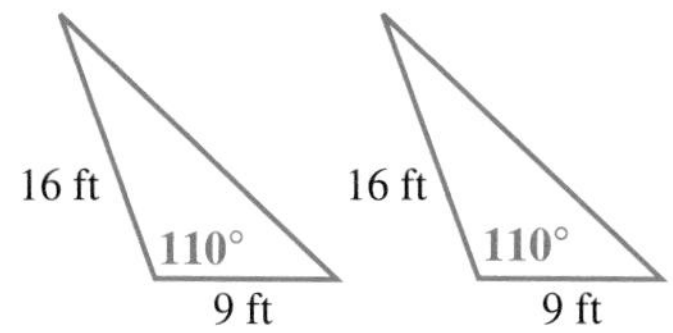

(b)

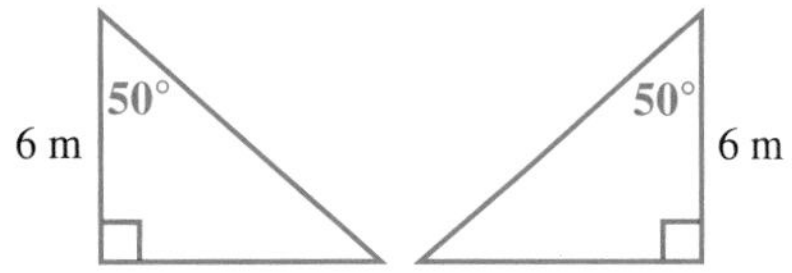

(c)

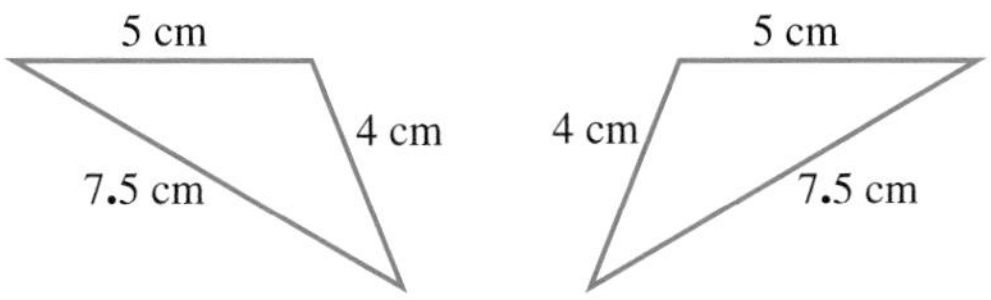

(a) On both triangles, two corresponding sides and the angle between them measure the same, so the Side–Angle–Side (SAS) method can be used to prove that the triangles are congruent.

(b) On both triangles, two corresponding angles and the side between them measure the same, so the Angle–Side–Angle (ASA) method can be used to prove that the triangles are congruent.

(c) Each pair of corresponding sides has the same length, so the Side–Side–Side (SSS) method can be used to prove that the triangles are congruent.

Work Problem 2 at the Side. ▶

OBJECTIVE 3 **Identify corresponding parts of similar triangles.** Now that you've worked with *congruent* triangles, let's look at *similar* triangles. Remember that congruent triangles match exactly, both in shape and in size. Similar triangles, on the other hand, have the same shape but are *different sizes*. Three pairs of similar triangles are shown here.

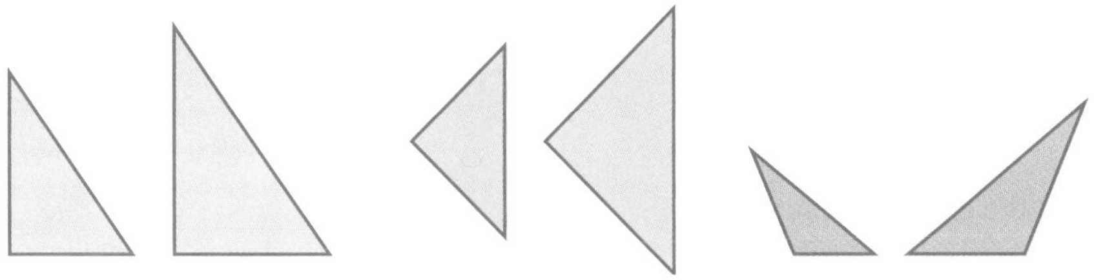

Each pair of triangles has the same shape because the corresponding angles have the same measure. But the corresponding sides are *not* the same length, so the triangles are of *different sizes*.

Two similar triangles are shown to the right. Notice that corresponding angles have the same measure, but corresponding sides have different lengths. $\overline{RQ}$ corresponds to $\overline{CB}$. Also, $\overline{RP}$ corresponds to $\overline{CA}$, and $\overline{QP}$ corresponds to $\overline{BA}$.

Notice that each side in the smaller triangle is *half* the length of the corresponding side in the larger triangle.

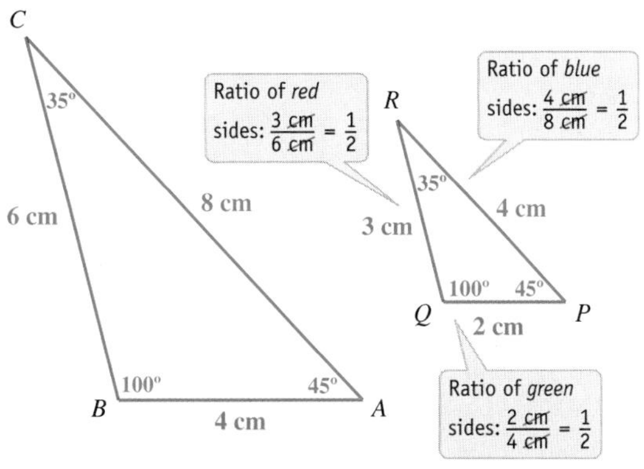

Work Problem 3 at the Side. ▶

2 State which method can be used to prove that each pair of triangles is congruent.

(a)

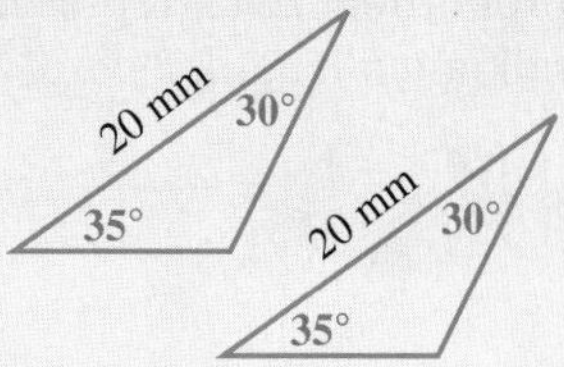

(b)

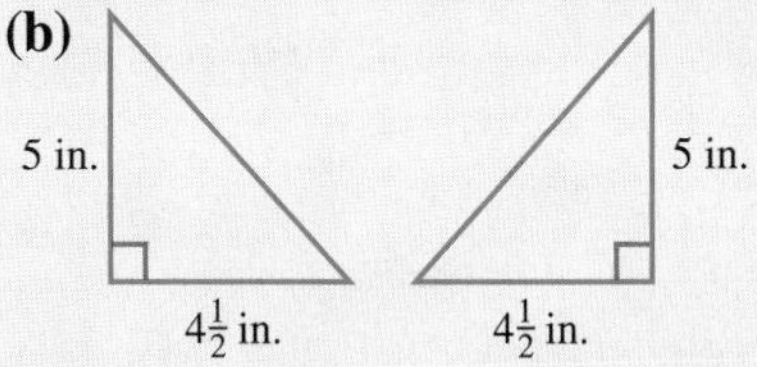

(c)

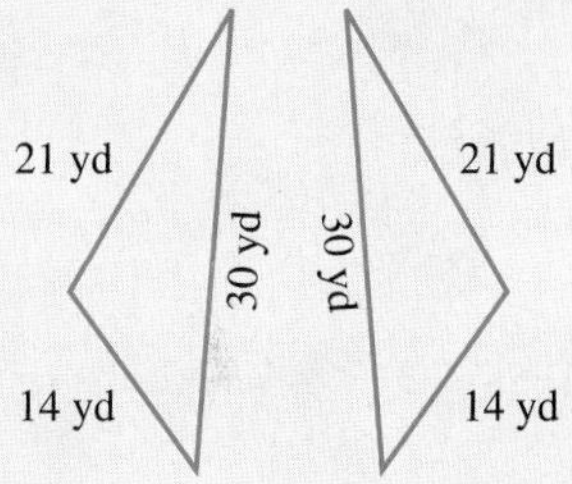

3 Identify corresponding angles and sides in these similar triangles.

(a)

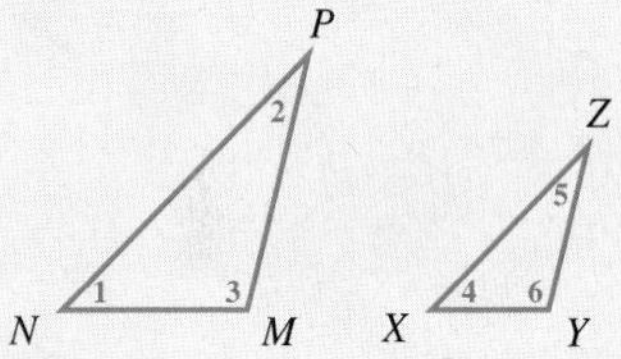

Angles:	**Sides:**
1 and _____	$\overline{PN}$ and _____
2 and _____	$\overline{PM}$ and _____
3 and _____	$\overline{NM}$ and _____

(b)

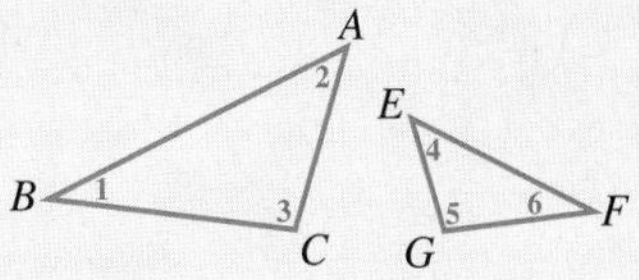

Angles:	**Sides:**
1 and _____	$\overline{AB}$ and _____
2 and _____	$\overline{BC}$ and _____
3 and _____	$\overline{AC}$ and _____

Answers

2. **(a)** ASA **(b)** SAS **(c)** SSS

3. **(a)** 4; 5; 6; $\overline{ZX}$; $\overline{ZY}$; $\overline{XY}$

(b) 6; 4; 5; $\overline{EF}$; $\overline{FG}$; $\overline{EG}$

4 Find the length of $\overline{EF}$ in **GS** **Example 3** at the right by setting up and solving a proportion. Let x represent the unknown length.

$$\begin{matrix} EF \rightarrow \\ CB \rightarrow \end{matrix} \frac{x}{33} = \frac{\quad}{\quad} \begin{matrix} \leftarrow ED \\ \leftarrow CA \end{matrix}$$

OBJECTIVE 4 Find the unknown lengths of sides in similar triangles. Similar triangles are useful because of the following definition.

Similar Triangles

If two triangles are similar, then

1. Corresponding angles have the same measure, and
2. The *ratios* of the lengths of corresponding sides are equal.

EXAMPLE 3 Finding the Unknown Lengths of Sides in Similar Triangles

Find the length of $\overline{DF}$ in the smaller triangle. Assume the triangles are similar.

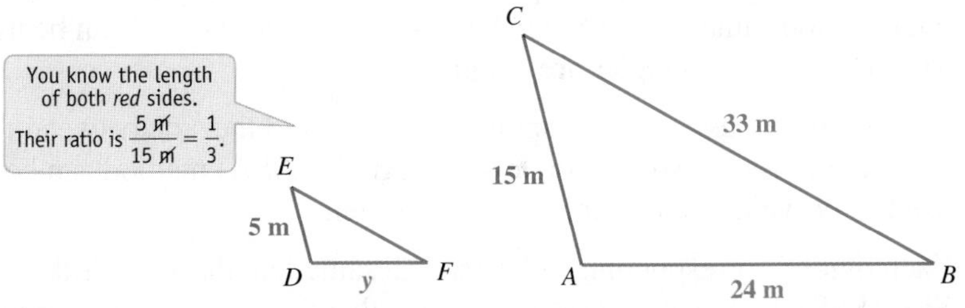

The length you want to find in the smaller triangle is $\overline{DF}$, and it corresponds to $\overline{AB}$ in the *larger* triangle. Then, notice that $\overline{ED}$ in the smaller triangle corresponds to $\overline{CA}$ in the larger triangle, and you know both of their lengths. Since the *ratios* of the lengths of corresponding sides are equal, you can set up a proportion. (Recall that a proportion states that two ratios are equal.)

$$\text{Corresponding sides} \left\{ \begin{matrix} DF \rightarrow \\ AB \rightarrow \end{matrix} \frac{y}{24} = \frac{5}{15} \begin{matrix} \leftarrow ED \\ \leftarrow CA \end{matrix} \right\} \text{Corresponding sides}$$

$$\frac{y}{24} = \frac{1}{3}$$

Write $\frac{5}{15}$ in lowest terms as $\frac{1}{3}$.

Find the cross products.

$$24 \cdot 1 = 24$$

$$\frac{y}{24} = \frac{1}{3}$$

$$y \cdot 3$$

$$y \cdot 3 = 24 \quad \text{Show that the cross products are equal.}$$

$$\frac{y \cdot \overset{1}{\cancel{3}}}{\underset{1}{\cancel{3}}} = \frac{24}{3} \quad \text{Divide both sides by 3.}$$

$$y = 8$$

Write **m** in the answer.

$\overline{DF}$ has a length of 8 m.

◀ **Work Problem 4 at the Side.**

Answer

4. $\frac{x}{33} = \frac{5}{15}$; $x = 11$ m

EXAMPLE 4 Finding an Unknown Length and the Perimeter

Find the perimeter of the smaller triangle. Assume the triangles are similar.

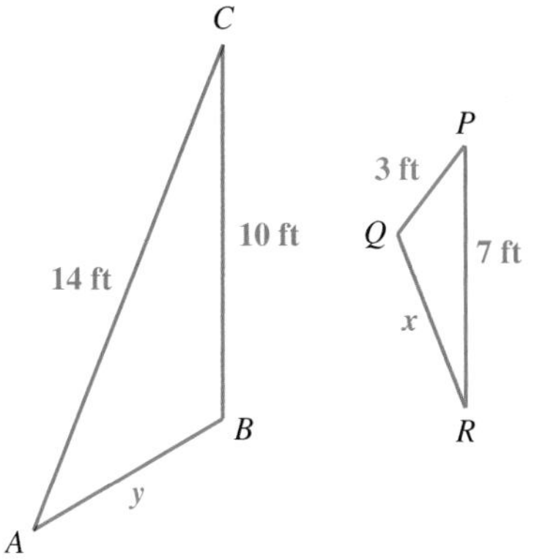

You know the length of both *red* sides. Their ratio is $\frac{7 \text{ ft}}{14 \text{ ft}} = \frac{1}{2}$.

First find x, the length of $\overline{QR}$ in the smaller triangle, then add the lengths of all three sides to find the perimeter.

The smaller triangle is turned "upside down" compared to the larger triangle, so be careful when identifying corresponding sides. $\overline{PR}$ is the longest side in the smaller triangle, and $\overline{AC}$ is the longest side in the larger triangle. So $\overline{PR}$ and $\overline{AC}$ are corresponding sides and you know both of their lengths. $\overline{QR}$, the length you want to find in the smaller triangle, corresponds to $\overline{BC}$ in the larger triangle. The ratios of the lengths of corresponding sides are equal, so you can set up a proportion.

$$\frac{QR \rightarrow x}{BC \rightarrow 10} = \frac{7 \leftarrow PR}{14 \leftarrow AC}$$

$$\frac{x}{10} = \frac{1}{2}$$ Write $\frac{7}{14}$ in lowest terms as $\frac{1}{2}$.

Find the cross products.

$$\frac{x}{10} = \frac{1}{2} \qquad 10 \bullet 1 = 10 \qquad x \bullet 2$$

$$x \bullet 2 = 10$$ Show that the cross products are equal.

$$\frac{x \bullet \overset{1}{\cancel{2}}}{\underset{1}{\cancel{2}}} = \frac{10}{2}$$ Divide both sides by 2.

$$x = 5$$ Write **ft** for the length.

$\overline{QR}$ has a length of 5 ft.

Now add the lengths of all three sides to find the perimeter of the smaller triangle.

Perimeter = 5 ft + 3 ft + 7 ft = 15 ft

Work Problem 5 at the Side. ▶

5 (a) Find the perimeter of triangle *ABC* in **Example 4** at the left.

Let y represent the unknown length of side *AB*.

$$\frac{PQ \rightarrow 3}{AB \rightarrow y} = \frac{\quad \leftarrow PR}{\quad \leftarrow AC}$$

(b) Find the perimeter of each triangle. Assume the triangles are similar.

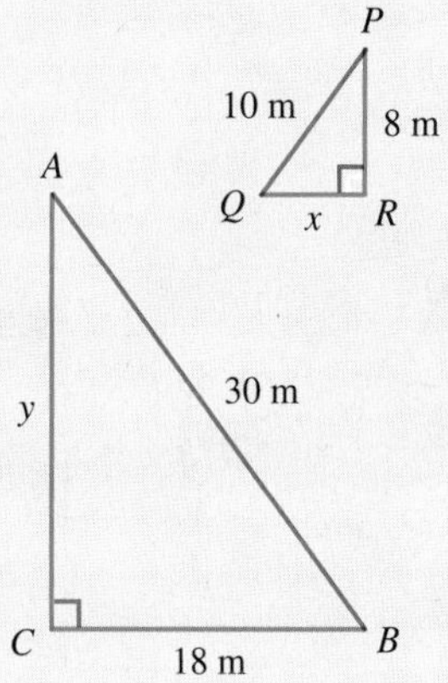

Answers

5. (a) $\frac{3}{y} = \frac{7}{14}$; $y = 6$ ft;
Perimeter = 14 ft + 10 ft + 6 ft = 30 ft
(b) $x = 6$ m, Perimeter = 24 m;
$y = 24$ m, Perimeter = 72 m

6 Find the height of each flagpole.

GS (a)

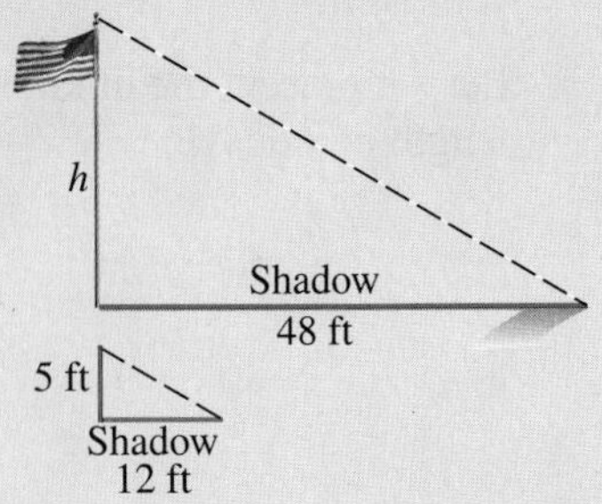

$$\left.\begin{matrix}\text{Longer}\\\text{height}\end{matrix}\right\} \to \frac{h}{5} = \underline{} \left\{\begin{matrix}\text{Longer}\\\text{shadow}\end{matrix}\right.$$

$$\left.\begin{matrix}\text{Shorter}\\\text{height}\end{matrix}\right\} \to 5 \qquad \left\{\begin{matrix}\text{Shorter}\\\text{shadow}\end{matrix}\right.$$

(b)

7.2 m

5 m

h

12.5 m

OBJECTIVE 5 **Solve application problems involving similar triangles.**
The next example shows an application of similar triangles.

| **EXAMPLE 5** | **Using Similar Triangles in an Application** |

A flagpole casts a shadow 99 m long at the same time that a pole 10 m tall casts a shadow 18 m long. Find the height of the flagpole.

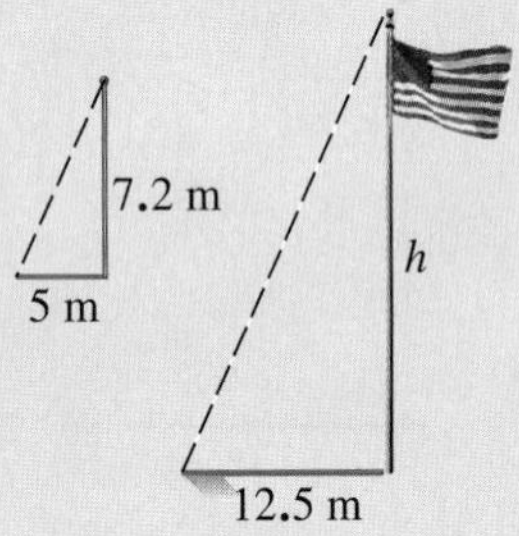

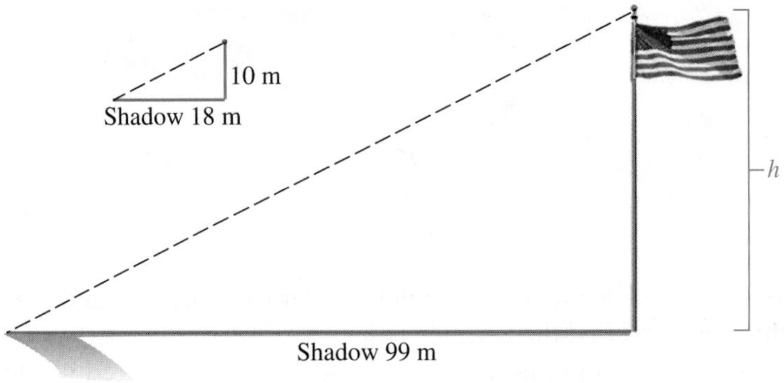

The triangles shown are similar, so write a proportion to find h.

$$\text{Height in larger triangle} \to \frac{h}{10} = \frac{99}{18} \leftarrow \text{Shadow in larger triangle}$$
$$\text{Height in smaller triangle} \to \qquad \qquad \leftarrow \text{Shadow in smaller triangle}$$

Find the cross products and show that they are equal.

$$h \cdot 18 = 10 \cdot 99$$
$$h \cdot 18 = 990$$

$$\frac{h \cdot \overset{1}{\cancel{18}}}{\underset{1}{\cancel{18}}} = \frac{990}{18} \qquad \text{Divide both sides by 18.}$$

$$h = 55$$

The flagpole is 55 m high.

> **Note**
>
> There are several other correct ways to set up the proportion in **Example 5** above. One way is to simply flip the ratios on *both* sides of the equal sign.
>
> $$\frac{10}{h} = \frac{18}{99}$$
>
> But there is another option, shown below.
>
> $$\text{Height in larger triangle} \to \frac{h}{99} = \frac{10}{18} \leftarrow \text{Height in smaller triangle}$$
> $$\text{Shadow in larger triangle} \to \qquad \qquad \leftarrow \text{Shadow in smaller triangle}$$
>
> Notice that both ratios compare *height* to *shadow* in the same order. The ratio on the left describes the larger triangle, and the ratio on the right describes the smaller triangle.

◀ **Work Problem 6** at the Side.

Answers

6. (a) $\frac{h}{5} = \frac{48}{12}$; $h = 20$ ft **(b)** $h = 18$ m

7.8 Exercises

FOR EXTRA HELP

Download the MyDashBoard App

MyMathLab®

1. **CONCEPT CHECK** Look up the word *congruent* in a dictionary. What is the nonmathematical definition of this word? Describe two examples of congruent objects at home, school, or work.

2. **CONCEPT CHECK** Look up the word *similar* in a dictionary. What is the nonmathematical definition of this word? Describe two examples of similar objects at home, school, or work.

Each pair of triangles is congruent. List the corresponding angles and the corresponding sides. **See Example 1.**

3. 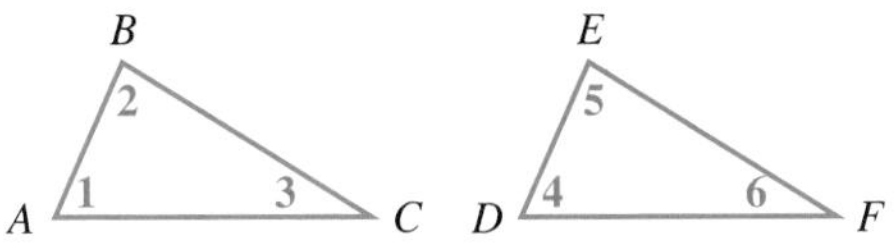

4. 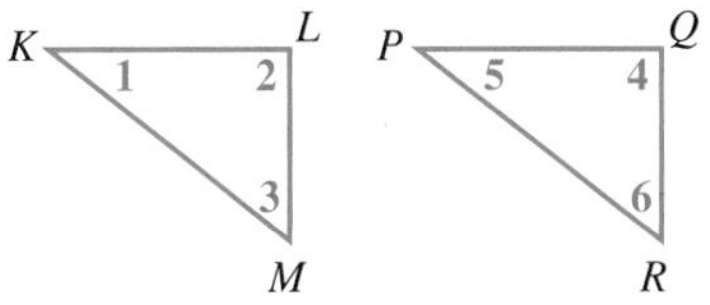

5. 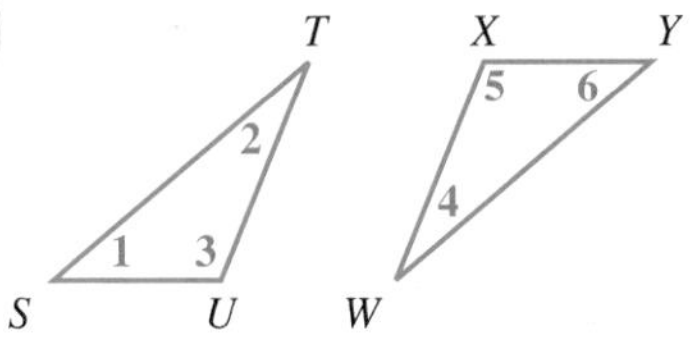

6. 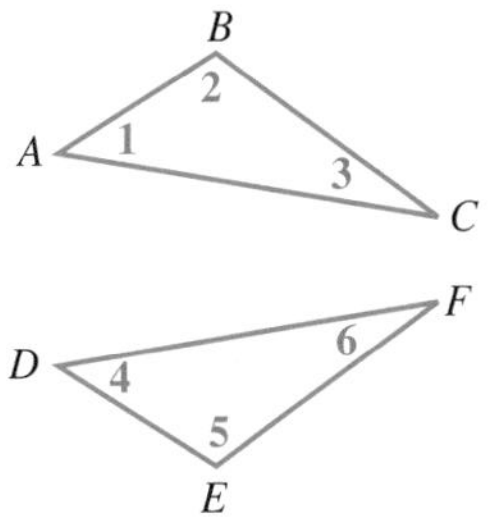

7. 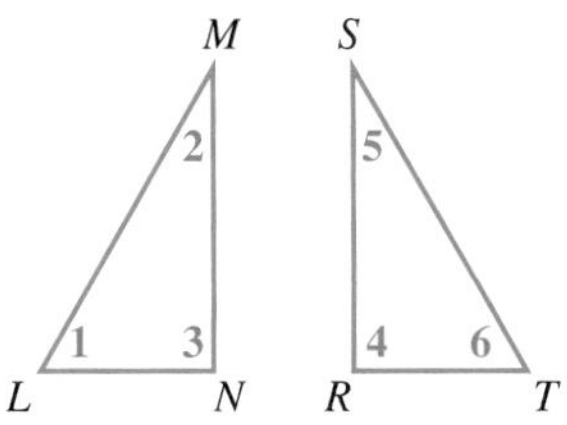

8. 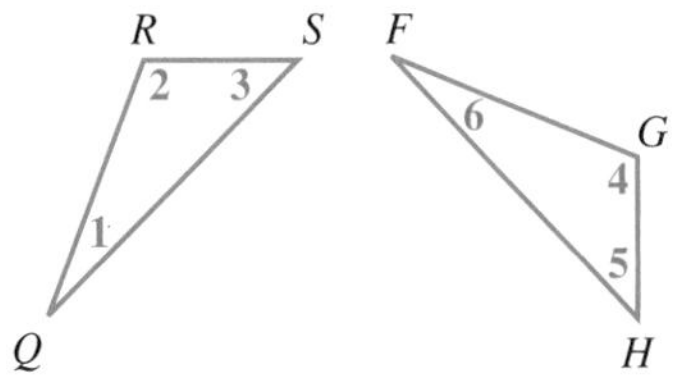

State which of these methods can be used to prove that each pair of triangles is congruent; Angle–Side–Angle (ASA), Side–Side–Side (SSS), or Side–Angle–Side (SAS).
See Example 2.

9.

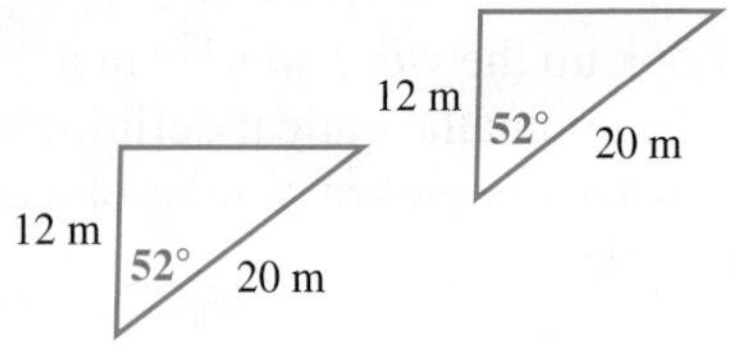

10.

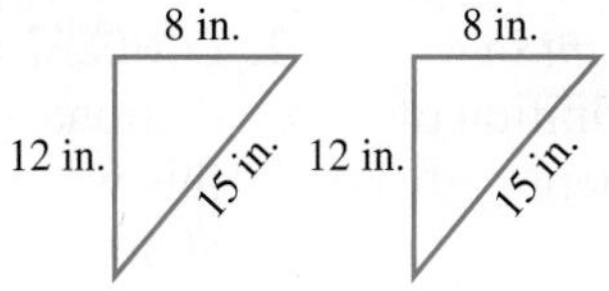

11.

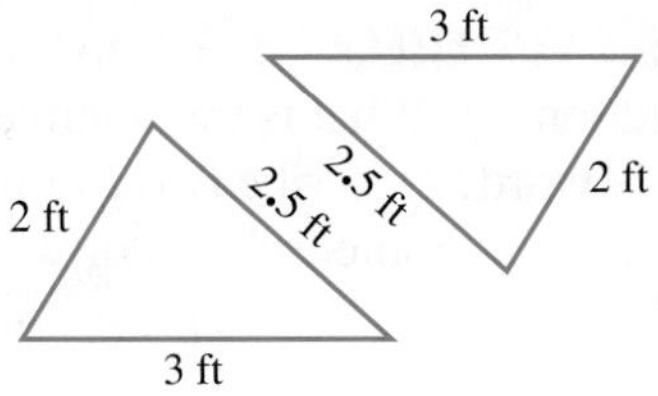

12.

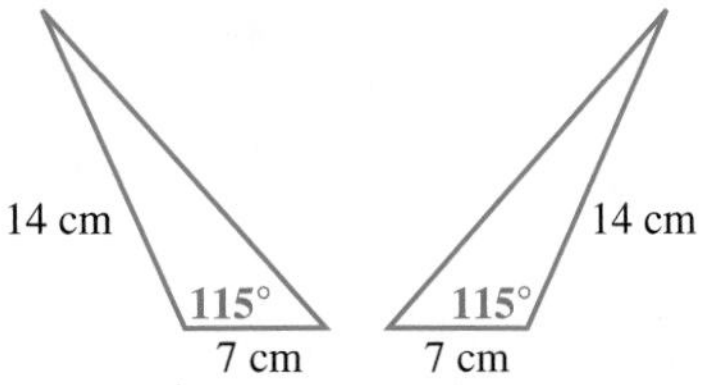

13.

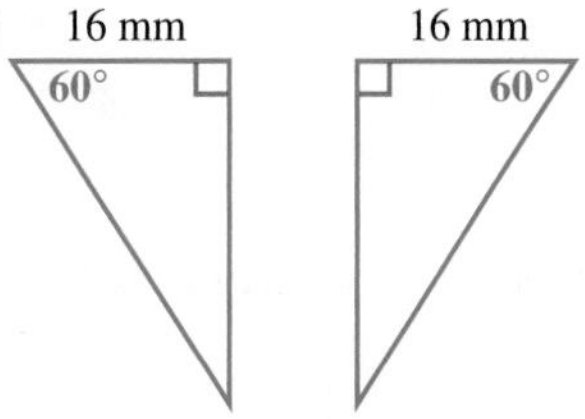

14.

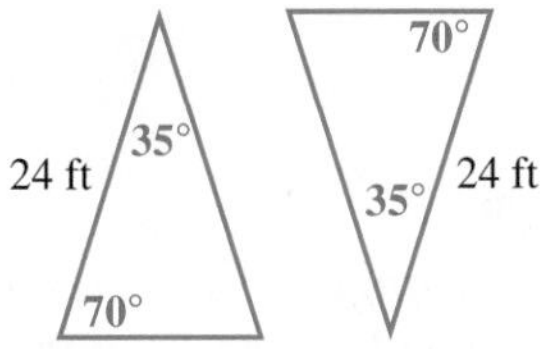

Given the information in Exercises 15–18, show how you can prove that the indicated triangles are congruent. Note: A midpoint divides a segment into two congruent parts.

15.

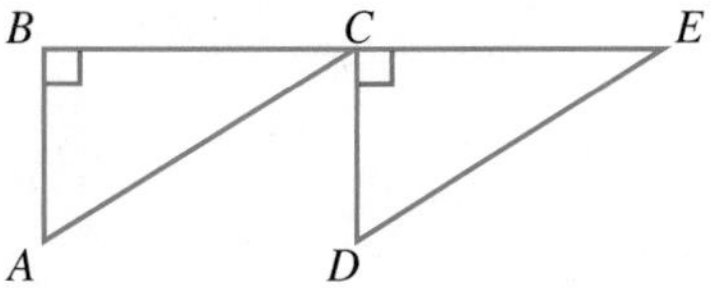

C is the midpoint of $\overline{BE}$ and $CD = BA$.
Prove that $\triangle ABC \cong \triangle DCE$.

16.

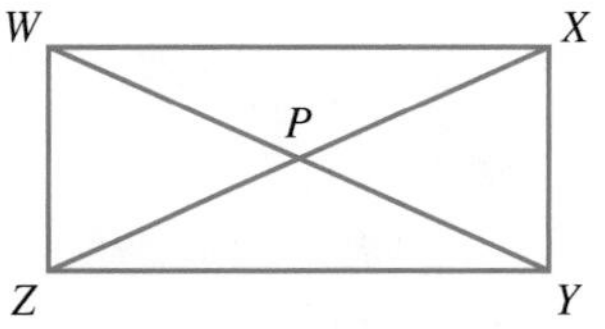

P is the midpoint of both $\overline{WY}$ and $\overline{XZ}$; $WZ = XY$.
Prove that $\triangle WPZ \cong \triangle YPX$.

17.

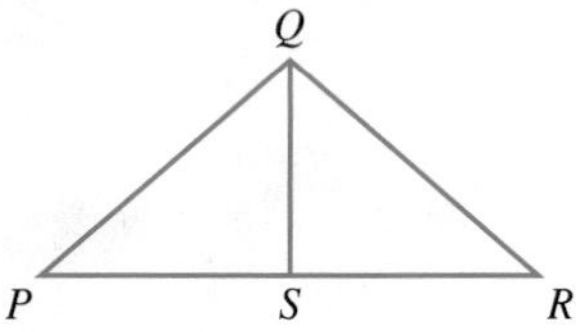

$\overline{QS} \perp \overline{PR}$ and *S* is the midpoint of $\overline{PR}$.
Prove that $\triangle PQS \cong \triangle RQS$.

18.

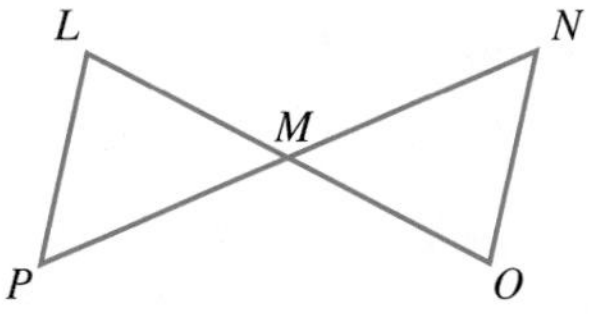

M is the midpoint of both $\overline{LO}$ and $\overline{PN}$.
Prove that $\triangle PLM \cong \triangle NOM$.

In Exercises 19–24, find the unknown lengths in each pair of similar triangles.
See Example 3.

19. GS

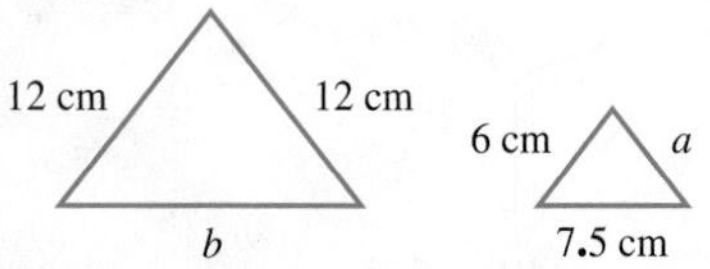

Solve two proportions.

$$\frac{a}{12} = \frac{6}{12} \qquad \frac{7.5}{b} = \frac{}{}$$

20. GS

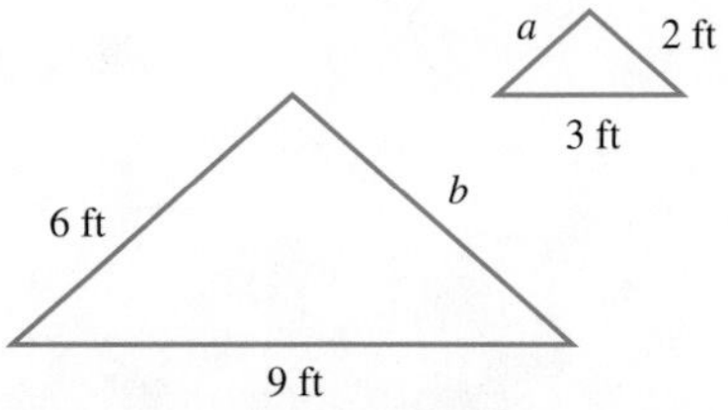

Solve two proportions.

$$\frac{a}{6} = \frac{3}{\_\_\_} \qquad \frac{2}{b} = \frac{}{}$$

21.

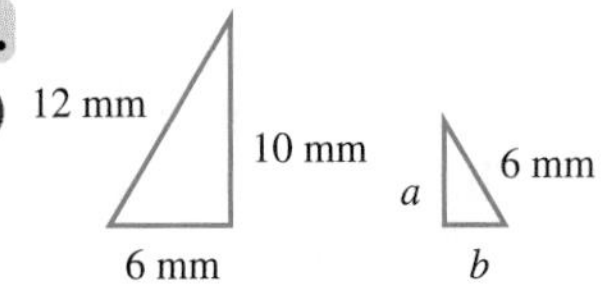

22.

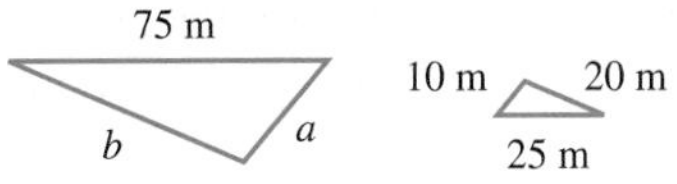

23.

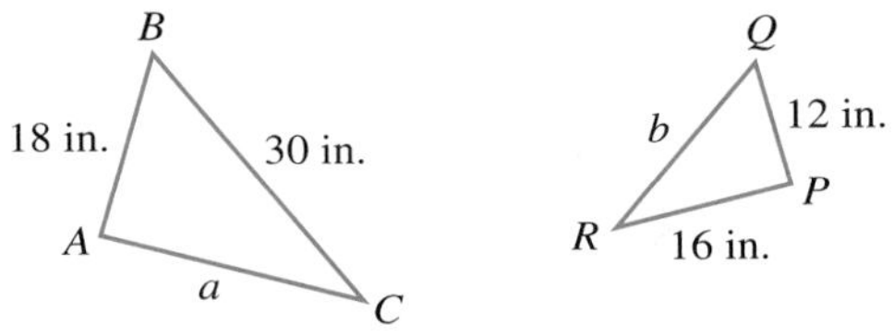

24.

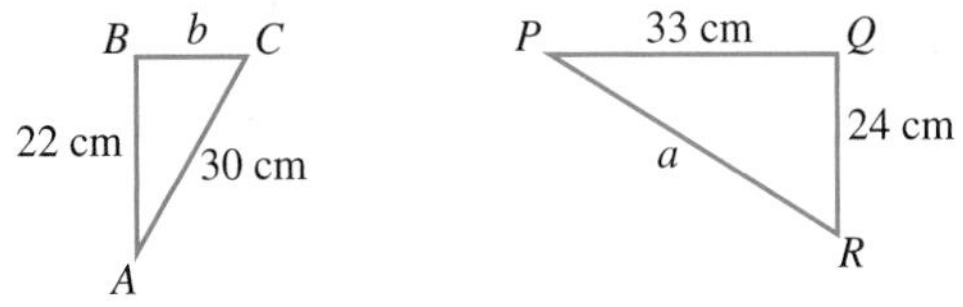

In Exercises 25 and 26, find the perimeter of each triangle. Assume the triangles are similar. ***See Example 4.***

25.

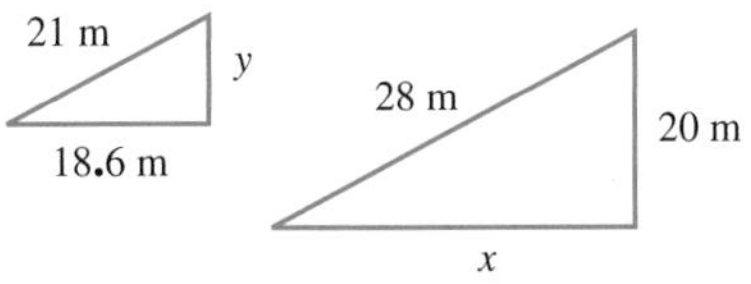

26.

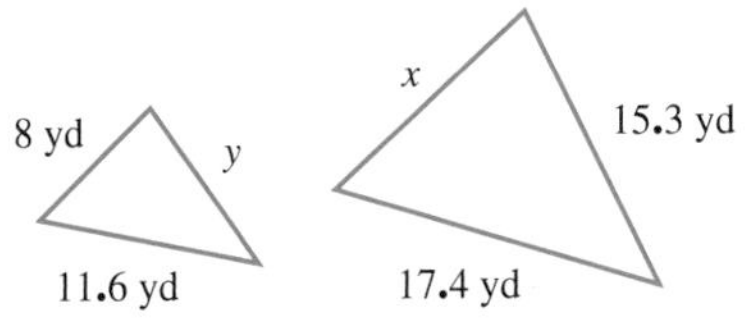

27. Triangles *CDE* and *FGH* are similar. Find the perimeter and area of triangle *FGH*. *Note:* The heights of similar triangles have the same ratio as corresponding sides. Round to the nearest tenth when necessary.

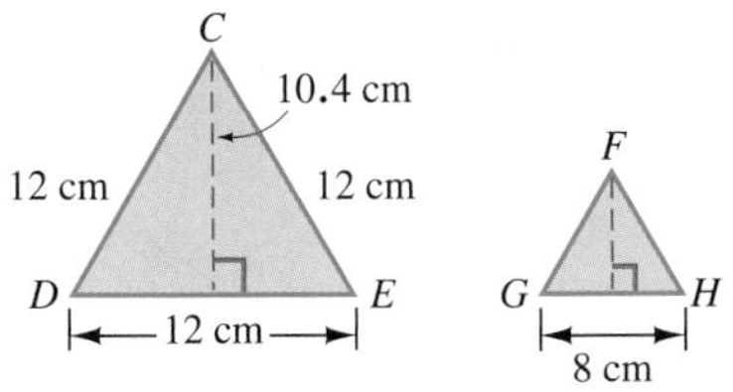

28. Triangles *JKL* and *MNO* are similar. Find the perimeter and area of triangle *MNO*.

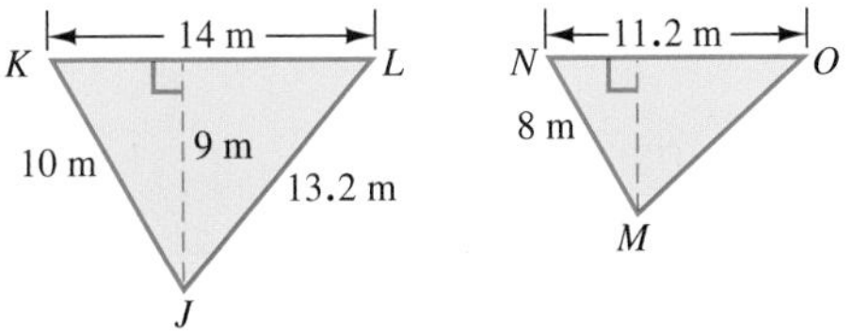

Solve each application problem. **See Example 5.**

29. The height of the house shown here can be found by comparing its shadow to the shadow cast by a 3-foot stick. Find the height of the house by writing a proportion and solving it.

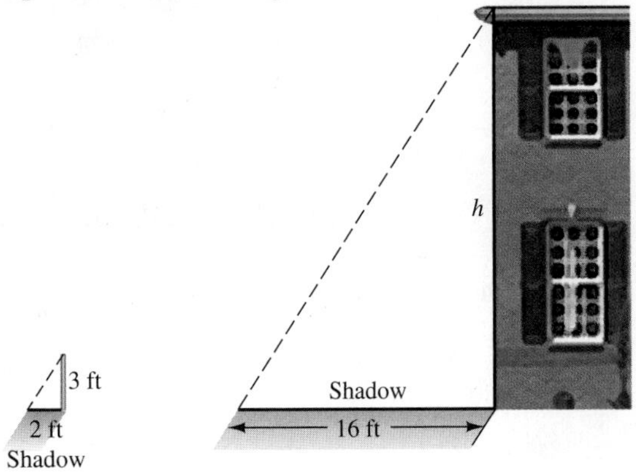

30. A fire lookout tower provides an excellent view of the surrounding countryside. The height of the tower can be found by lining up the top of the tower with the top of a 2-meter stick. Use similar triangles to find the height of the tower.

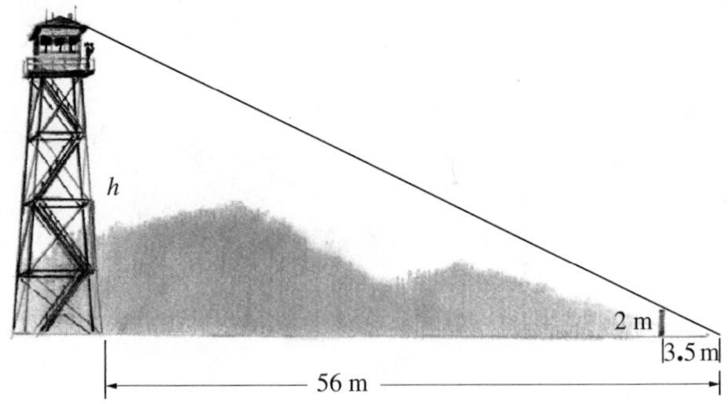

Find the unknown length in Exercises 31–34. Round your answers to the nearest tenth.
Note: When a line is drawn parallel to one side of a triangle, the smaller triangle that is formed will be similar to the original triangle. In Exercises 31–32, the red segments are parallel.

31.

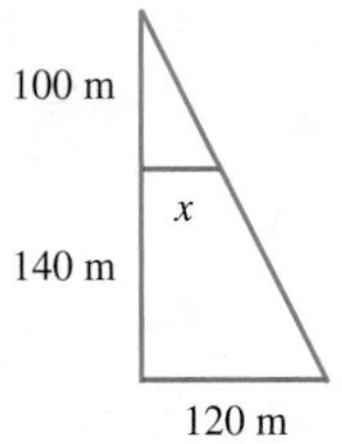

Hint: Redraw the two triangles and label the sides.

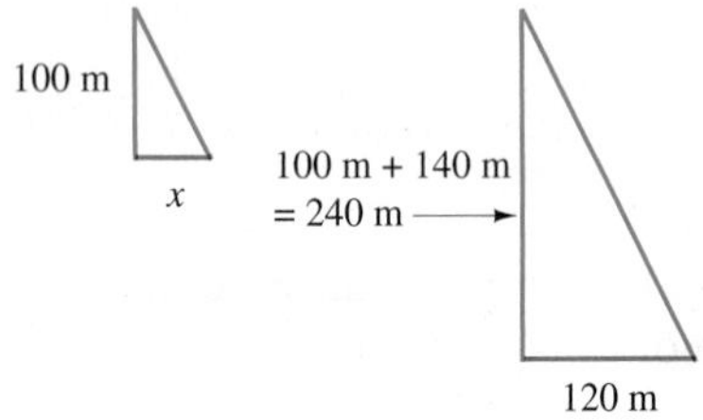

32.

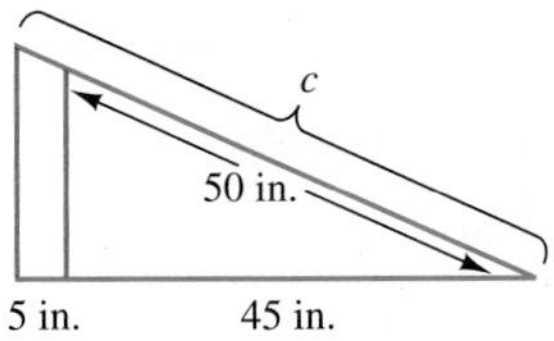

Hint: Redraw the two triangles and label the sides.

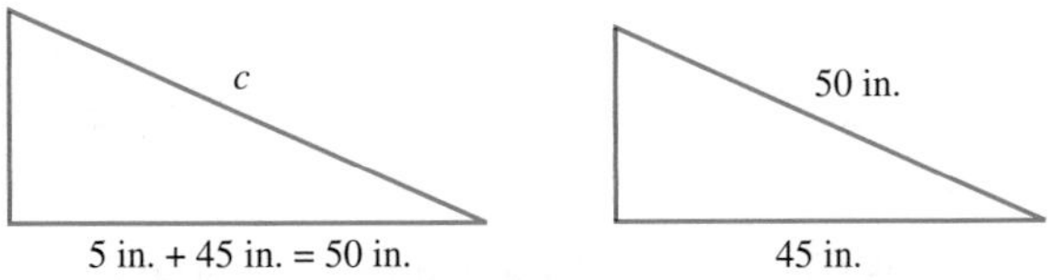

33. Use similar triangles and a proportion to find the length of the lake shown here. (*Hint:* The side 100 m long in the smaller triangle corresponds to the side of 100 m + 120 m = 220 m in the larger triangle.)

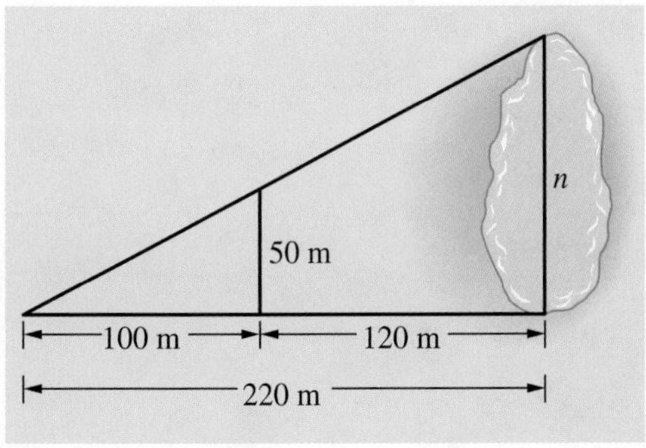

34. To find the height of the tree, find y and then add $5\frac{1}{2}$ ft for the distance from the ground to the eye level of the person.

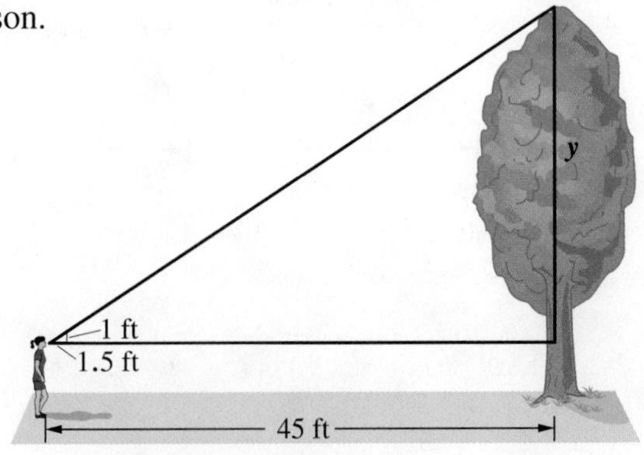

Distance from person to tree

Chapter 7 Summary

Key Terms

7.1

point A point is a location in space. *Example:* Point P at the right.

line A line is a straight row of points that goes on forever in both directions. *Example:* Line AB, written $\overleftrightarrow{AB}$, at the right.

line segment A line segment is a piece of a line with two endpoints. *Example:* Line segment PQ, written $\overline{PQ}$, at the right.

ray A ray is a part of a line that has one endpoint and extends forever in one direction. *Example:* Ray RS, written $\overrightarrow{RS}$, at the right.

parallel lines Parallel lines are two lines in the same plane that never intersect (never cross). *Example:* $\overleftrightarrow{AB}$ is parallel to $\overleftrightarrow{ST}$ at the right.

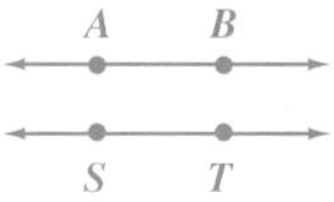

intersecting lines Intersecting lines cross. *Example:* $\overleftrightarrow{RQ}$ intersects $\overleftrightarrow{AB}$ at point P at the right.

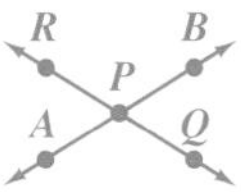

angle An angle is made up of two rays that have a common endpoint called the vertex. *Example:* Angle 1 at the right.

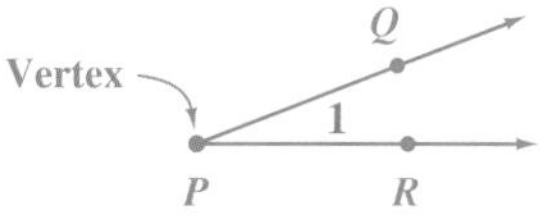

degrees Degrees are used to measure angles; a complete circle is 360 degrees, written 360°.

straight angle A straight angle is an angle that measures *exactly* 180°; its sides form a straight line. *Example:* Angle G at the right.

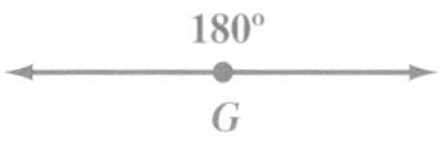

right angle A right angle is an angle that measures *exactly* 90°, identified by a small square at the vertex. *Example:* Angle AOB at the right.

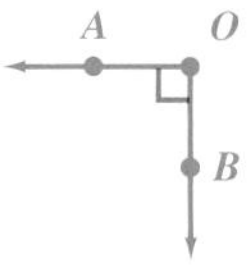

acute angle An acute angle is an angle that measures less than 90°. *Example*: Angle E at the right.

obtuse angle An obtuse angle is an angle that measures more than 90° but less than 180°. *Example:* Angle F at the right.

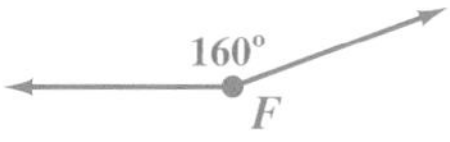

perpendicular lines Perpendicular lines are two lines that intersect to form a right angle. *Example:* $\overleftrightarrow{PQ}$ is perpendicular to $\overleftrightarrow{RS}$ at the right.

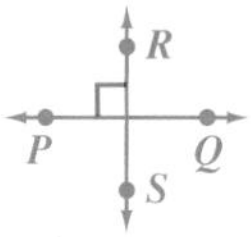

complementary angles Complementary angles are two angles whose measures add up to 90°.

supplementary angles Supplementary angles are two angles whose measures add up to 180°.

congruent angles Congruent angles are angles that measure the same number of degrees.

vertical angles Vertical angles are two nonadjacent congruent angles formed by two intersecting lines. *Example:* $\angle COA$ and $\angle EOF$ are vertical angles at the right.

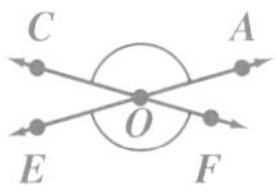

corresponding angles Corresponding angles are formed when two parallel lines are crossed by a transversal; corresponding angles are congruent and are on the same side of the transversal and in the same relative position. *Example:* In the figure at the right, line m is parallel to line n. The pairs of corresponding angles are $\angle 1$ and $\angle 5$, $\angle 2$ and $\angle 6$, $\angle 3$ and $\angle 7$, $\angle 4$ and $\angle 8$.

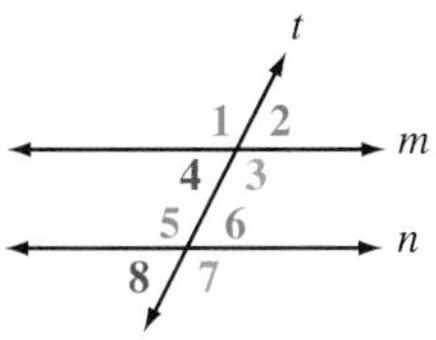

alternate interior angles When two parallel lines are crossed by a transversal, there are two pairs of alternate interior angles and each pair is congruent. They are on opposite sides of the transversal. *Example:* In the figure at the right, line m is parallel to line n. The pairs of alternate interior angles are $\angle 3$ and $\angle 5$, $\angle 4$ and $\angle 6$.

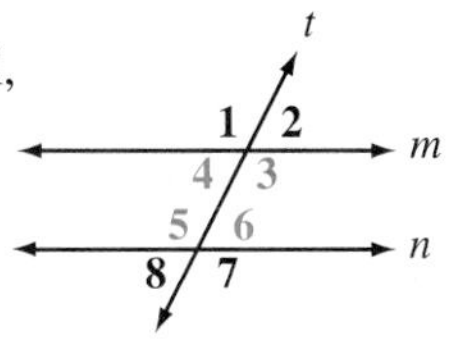

7.2–7.4

perimeter Perimeter is the distance around the outside edges of a flat shape. It is measured in linear units such as ft, yd, cm, m, km, and so on.

7.2–7.6

area Area is the surface inside a two-dimensional (flat) shape. It is measured by determining the number of squares of a certain size needed to cover the surface inside the shape. Some of the commonly used units for measuring area are square inches ($in.^2$), square feet (ft^2), square yards (yd^2), square centimeters (cm^2), and square meters (m^2).

7.2

rectangle A rectangle is a four-sided figure with all sides meeting at 90° angles. The opposite sides are the same length. *Example:* The rectangle measuring 12 cm by 7 cm at the right.

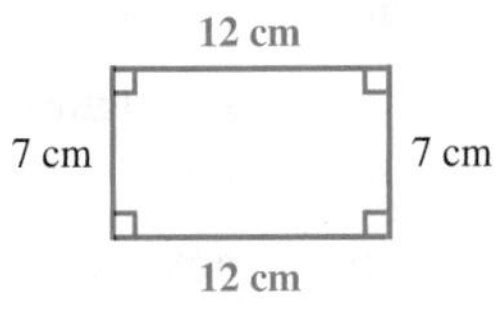

square A square is a rectangle with all four sides the same length. *Example:* The square with a side measurement of 20 inches at the right.

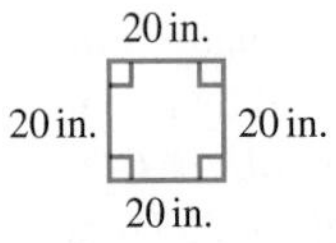

7.3

parallelogram A parallelogram is a four-sided figure with both pairs of opposite sides parallel and equal in length. *Example:* The parallelogram at the right with sides measuring 8 m and 12 m.

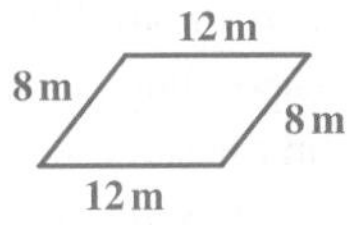

trapezoid A trapezoid is a four-sided figure with exactly one pair of parallel sides. *Example:* Trapezoid *PQRS* at the right; $\overline{PQ}$ is parallel to $\overline{SR}$.

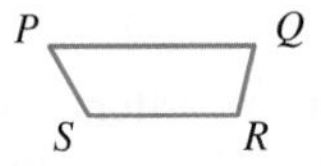

7.4

triangle A triangle is a figure with exactly three sides. *Example:* Triangle *ABC* at the right.

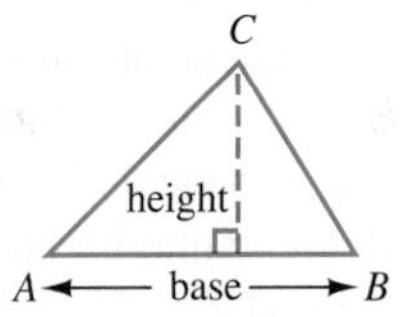

7.5

circle A circle is a figure with all points the same distance from a fixed center point. *Example:* See figure at the right.

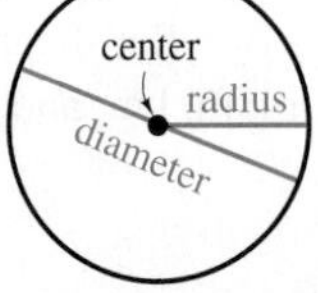

radius Radius is the distance from the center of a circle to any point on the circle. *Example:* See the red radius in the circle at the right.

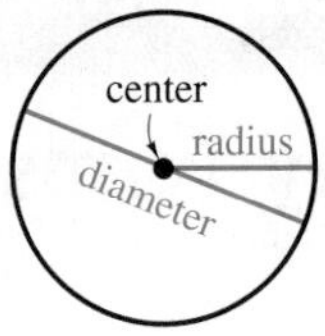

diameter Diameter is the distance across a circle, passing through the center. *Example:* See the blue diameter in the circle at the right.

circumference Circumference is the distance around a circle.

π (pi) π is the ratio of the circumference to the diameter of any circle. It is approximately equal to 3.14.

7.6

volume Volume is a measure of the space inside a three-dimensional (solid) shape. Volume is measured in cubic units such as $in.^3$, ft^3, yd^3, mm^3, cm^3, and so on.

surface area Surface area is the area on the surface of a three-dimensional object (a solid). Surface area is measured in square units.

7.7

hypotenuse The hypotenuse is the side of a right triangle opposite the 90° angle; it is the longest side. *Example:* See the red side in the triangle at the right.

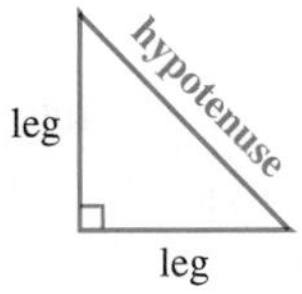

7.8

congruent figures Congruent figures are identical both in shape and in size.

similar figures Similar figures have the same shape but not necessarily the same size.

congruent triangles Congruent triangles are triangles with the same shape and the same size; corresponding angles measure the same number of degrees and corresponding sides have the same length.

similar triangles Similar triangles are triangles with the same shape but not necessarily the same size; corresponding angles measure the same number of degrees, and the *ratios* of the lengths of corresponding sides are equal.

New Symbols

$\overleftrightarrow{AB}$	line AB
$\overline{EF}$	line segment EF
$\overrightarrow{RS}$	ray RS
$\angle MRN$	angle MRN
1°	one degree

Right angle: (90° angle) — small square in corner

$\perp$	is perpendicular to
$\cong$	is congruent to
π	Greek letter pi; ratio of circumference to diameter of any circle

Square units for measuring area	in.^{2}	ft^2	yd^2	mi^2
	mm^2	cm^2	m^2	km^2
Cubic units for measuring volume	in.^{3}	ft^3	yd^3	
	mm^3	cm^3	m^3	

New Formulas

Perimeter of a rectangle: $P = 2 \cdot l + 2 \cdot w$

Area of a rectangle: $A = l \cdot w$

Perimeter of a square: $P = 4 \cdot s$

Area of a square: $A = s^2$ or $A = s \cdot s$

Area of a parallelogram: $A = b \cdot h$

Area of a trapezoid: $A = \frac{1}{2} \cdot h \cdot (b + B)$

or $A = 0.5 \cdot h \cdot (b + B)$

Area of a triangle: $A = \frac{1}{2} \cdot b \cdot h$

or $A = 0.5 \cdot b \cdot h$

Diameter of a circle: $d = 2 \cdot r$

Radius of a circle: $r = \frac{d}{2}$

Circumference of a circle: $C = \pi \cdot d$

or $C = 2 \cdot \pi \cdot r$

Area of a circle: $A = \pi \cdot r^2$

Area of semicircle: $A = \frac{\pi \cdot r^2}{2}$

Volume of rectangular solid: $V = l \cdot w \cdot h$

Volume of a sphere: $V = \frac{4}{3} \cdot \pi \cdot r^3$

or $V = \frac{4 \cdot \pi \cdot r \cdot r \cdot r}{3}$

Volume of a hemisphere: $V = \frac{2}{3} \cdot \pi \cdot r^3$

or $V = \frac{2 \cdot \pi \cdot r \cdot r \cdot r}{3}$

Volume of a cylinder: $V = \pi \cdot r^2 \cdot h$

Volume of a cone: $V = \frac{1}{3} \cdot B \cdot h$ or $V = \frac{B \cdot h}{3}$

Volume of a pyramid: $V = \frac{1}{3} \cdot B \cdot h$ or $V = \frac{B \cdot h}{3}$

Surface area of a rectangular solid: $S = 2lw + 2lh + 2wh$

Surface area of a cylinder: $S = 2\pi rh + 2\pi r^2$

Right triangle: hypotenuse $= \sqrt{(\text{leg})^2 + (\text{leg})^2}$

leg $= \sqrt{(\text{hypotenuse})^2 - (\text{leg})^2}$

Note: Use 3.14 as the approximate value of π.

Test Your Word Power

See how well you have learned the vocabulary in this chapter.

1 Two angles that are **complementary**
A. have measures that add up to 180°
B. form a straight angle
C. have measures that add up to 90°.

2 The **perimeter** of a flat shape is
A. measured in square units
B. the distance around the outside edges
C. measured in cubic units.

3 An **obtuse angle**
A. is formed by perpendicular lines
B. measures more than 90° but less than 180°
C. measures less than 90°.

4 The **hypotenuse** is
A. the height line in a parallelogram
B. the longest side in a right triangle
C. the distance across a circle, passing through the center.

5 π is the ratio of
A. the diameter to the radius of a circle
B. the circumference to the diameter of a circle
C. the diameter to the circumference of a circle.

6 **Perpendicular lines**
A. intersect to form a right angle
B. intersect to form an acute angle
C. never intersect.

7 In a pair of **similar triangles,**
A. corresponding sides have the same length
B. all the angles have the same measure
C. the ratios of the lengths of corresponding sides are equal.

8 The **area of a rectangle** is found by
A. multiplying length times width
B. adding the lengths of the sides
C. using the formula $V = l \cdot w \cdot h$.

Answers to Test Your Word Power

1. C; *Example:* If $\angle 1$ measures 35° and $\angle 2$ measures 55°, the angles are complementary because 35° + 55° = 90°.

2. B; *Example:* If a square measures 5 ft on each side, then the perimeter is 5 ft + 5 ft + 5 ft + 5 ft = 20 ft.

3. B; *Examples:* Angles that measure 91°, 120°, and 175° are all obtuse angles.

4. B; *Example:* In triangle *ABC* at the right, side *AC* is the hypotenuse; sides *AB* and *BC* are the legs.

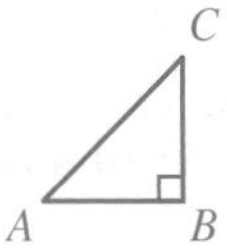

5. B; *Example:* The ratio of a circumference of 12.57 cm to a diameter of 4 cm is $\frac{12.57}{4} \approx 3.14$ (rounded).

6. A; *Example:* $\overleftrightarrow{EF}$ is perpendicular to $\overleftrightarrow{GH}$, at the right.

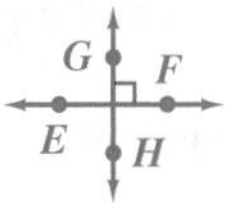

7. C; *Example:* Triangle *ABC* is similar to triangle *DEF,* so the ratios of corresponding sides are equal.

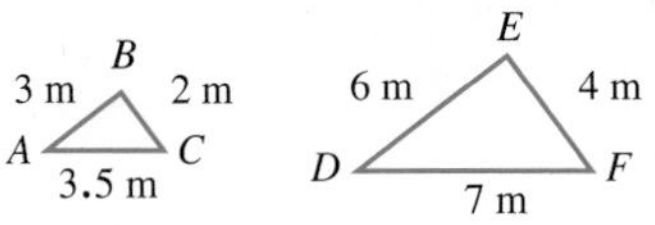

$\frac{AB}{DE} = \frac{3\ \cancel{m}}{6\ \cancel{m}} = \frac{1}{2}$ $\frac{BC}{EF} = \frac{2\ \cancel{m}}{4\ \cancel{m}} = \frac{1}{2}$ $\frac{AC}{DF} = \frac{3.5\ \cancel{m}}{7\ \cancel{m}} = \frac{1}{2}$

8. A; *Example:* In a rectangle with a length of 8 in. and a width of 5 in., Area = 8 in. • 5 in. = 40 in.^{2}

Quick Review

Concepts	Examples

7.1 Lines

A *line* is a straight row of points that goes on forever in both directions.

If a piece of a line has one endpoint, it is a *ray.*

If a piece of a line has two endpoints, it is a *line segment.*

Identify each of the following as a line, line segment, or ray and name it using the appropriate symbol.

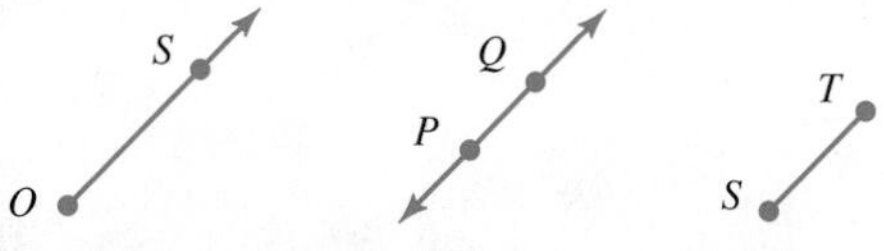

(a) **(b)** **(c)**

Figure **(a)** shows a ray named $\overrightarrow{OS}$.

Figure **(b)** shows a line named $\overleftrightarrow{PQ}$ or $\overleftrightarrow{QP}$.

Figure **(c)** shows a line segment named $\overline{ST}$ or $\overline{TS}$.

(Continued)

Concepts	Examples

7.1 Lines (*Continued*)

If two lines intersect at right angles, they are *perpendicular.*

If two lines in the same plane never intersect, they are *parallel.* Label each pair of lines as appearing to be parallel or as perpendicular.

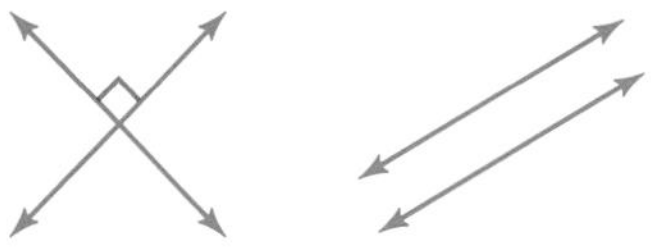

(a) **(b)**

Figure **(a)** shows perpendicular lines (they intersect at 90°).

Figure **(b)** shows lines that appear to be parallel (they never intersect).

7.1 Angles

If the sum of the measures of two angles is 90°, they are *complementary.*

If the sum of the measures of two angles is 180°, they are *supplementary.*

If two angles measure the same number of degrees, the angles are *congruent.* The symbol for congruent is $\cong$.

Two nonadjacent angles formed by two intersecting lines are called *vertical angles.* Vertical angles are congruent.

Find the complement and supplement of a 35° angle.

$$90^\circ - 35^\circ = 55^\circ \quad \text{(the complement)}$$
$$180^\circ - 35^\circ = 145^\circ \quad \text{(the supplement)}$$

Identify the vertical angles in this figure. Which angles are congruent?

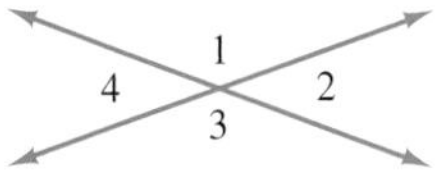

$\angle 1$ and $\angle 3$ are vertical angles.

$\angle 2$ and $\angle 4$ are vertical angles.

Vertical angles are congruent, so $\angle 1 \cong \angle 3$ and $\angle 2 \cong \angle 4$.

7.1 Parallel Lines

When two parallel lines are crossed by a transversal, corresponding angles are congruent, and alternate interior angles are congruent. Use this information to find the measures of the other angles.

Read $m\angle 1$ as "the measure of angle 1."

Line m is parallel to line n and the measure of $\angle 4$ is 125°. Find the measures of the other angles.

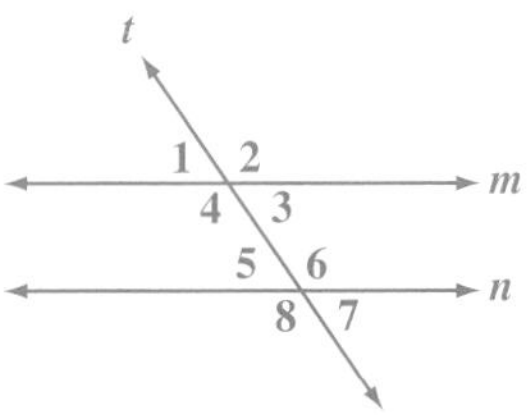

$\angle 4 \cong \angle 8$ (corresponding angles), so $m\angle 8 = 125^\circ$.

$\angle 4 \cong \angle 6$ (alternate interior angles), so $m\angle 6 = 125^\circ$.

$\angle 6 \cong \angle 2$ (corresponding angles), so $m\angle 2 = 125^\circ$.

$\angle 4$ and $\angle 3$ are supplements, so $m\angle 3 = 180^\circ - 125^\circ = 55^\circ$.

$\angle 3 \cong \angle 7$ (corresponding angles), so $m\angle 7 = 55^\circ$.

$\angle 3 \cong \angle 5$ (alternate interior angles), so $m\angle 5 = 55^\circ$.

$\angle 5 \cong \angle 1$ (corresponding angles), so $m\angle 1 = 55^\circ$.

Concepts | **Examples**

7.2 Rectangles and Squares

Use this formula to find the perimeter of a *rectangle.*

$$P = 2 \cdot l + 2 \cdot w$$

Use this formula to find the area of a rectangle.

$$A = l \cdot w$$

Area is measured in **square units.**

Use these formulas to find the perimeter and area of a *square.*

$$P = 4 \cdot s$$
$$A = s^2$$

Area is measured in **square units.**

Find the perimeter and area of this rectangle.

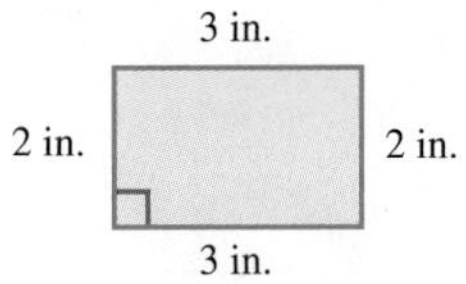

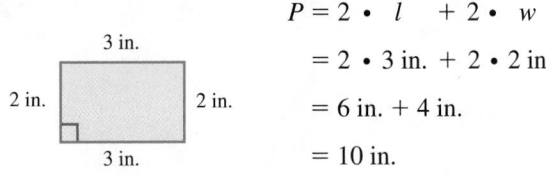

$$P = 2 \cdot l + 2 \cdot w$$
$$= 2 \cdot 3 \text{ in.} + 2 \cdot 2 \text{ in.}$$
$$= 6 \text{ in.} + 4 \text{ in.}$$
$$= 10 \text{ in.}$$

$$A = l \cdot w = 3 \text{ in.} \cdot 2 \text{ in.} = 6 \text{ in.}^2$$

Find the perimeter and area of this square.

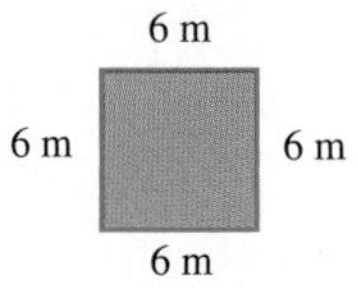

$$P = 4 \cdot s = 4 \cdot 6 \text{ m} = 24 \text{ m}$$
$$A = s^2 = s \cdot s = 6 \text{ m} \cdot 6 \text{ m} = 36 \text{ m}^2$$

7.3 Parallelograms

Use these formulas to find the perimeter and area of a *parallelogram.*

$$P = \text{sum of the lengths of the sides}$$
$$A = b \cdot h$$

Area is measured in **square units.**

Find the perimeter and area of this parallelogram.

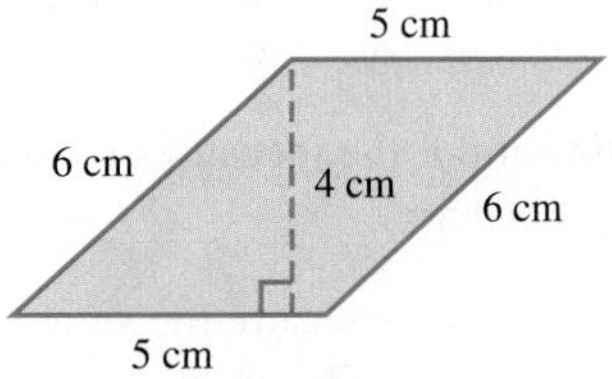

$$P = 5 \text{ cm} + 6 \text{ cm} + 5 \text{ cm} + 6 \text{ cm} = 22 \text{ cm}$$
$$A = 5 \text{ cm} \cdot 4 \text{ cm} = 20 \text{ cm}^2$$

7.3 Trapezoids

Use these formulas to find the perimeter and area of a *trapezoid.*

$$P = \text{sum of the lengths of the sides}$$
$$A = \frac{1}{2} \cdot h \cdot (b + B),$$

where b is the short base and B is the long base.

Area is measured in **square units.**

Find the perimeter and area of this trapezoid.

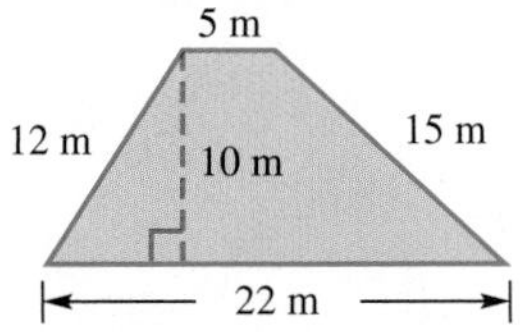

$$P = 5 \text{ m} + 15 \text{ m} + 22 \text{ m} + 12 \text{ m} = 54 \text{ m}$$
$$A = \frac{1}{\underset{1}{\cancel{2}}} \cdot \overset{5}{\cancel{10}} \text{ m} \cdot (5\text{m} + 22 \text{ m})$$
$$= 5 \text{ m} \cdot (27 \text{ m}) = 135 \text{ m}^2$$

7.4 Triangles

Use these formulas to find the perimeter and area of a *triangle.*

$$P = \text{sum of the lengths of the sides}$$
$$A = \frac{1}{2} \cdot b \cdot h$$
$$\downarrow$$
$$\text{or } A = 0.5 \cdot b \cdot h$$

Area is measured in **square units.**

Find the perimeter and area of this triangle.

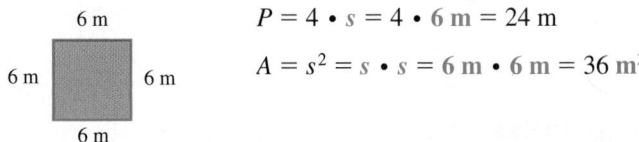

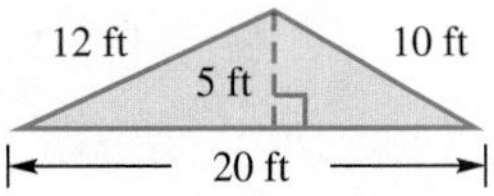

$$P = 12 \text{ ft} + 10 \text{ ft} + 20 \text{ ft} = 42 \text{ ft}$$
$$A = \frac{1}{2} \cdot b \cdot h$$
$$= \frac{1}{\underset{1}{\cancel{2}}} \cdot \overset{10}{\cancel{20}} \text{ ft} \cdot 5 \text{ ft} = 50 \text{ ft}^2$$
$$\text{or } A = 0.5 \cdot 20 \text{ ft} \cdot 5 \text{ ft} = 50 \text{ ft}^2$$

Concepts	*Examples*

7.5 Circles

Use this formula to find the *diameter* of a circle when you are given the radius.

$$d = 2 \cdot r$$

Find the diameter of a circle if the radius is 7 yd.

$$d = 2 \cdot r = 2 \cdot 7 \text{ yd} = 14 \text{ yd}$$

Use this formula to find the *radius* of a circle when you are given the diameter.

$$r = \frac{d}{2}$$

Find the radius of a circle if the diameter is 5 cm.

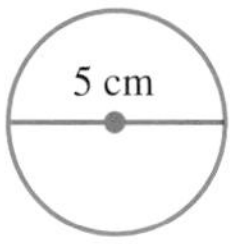

$$r = \frac{d}{2} = \frac{5 \text{ cm}}{2} = 2.5 \text{ cm}$$

Use these formulas to find the *circumference* of a circle.

When you know the radius, use $C = 2 \cdot \pi \cdot r$.

When you know the diameter, use $C = \pi \cdot d$.

Use 3.14 as the approximate value for π.

Find the circumference of a circle with a radius of 3 cm.

$C = 2 \cdot \pi \cdot r$

$C \approx 2 \cdot 3.14 \cdot 3 \text{ cm} \approx 18.8 \text{ cm}$ ← Rounded

Use this formula to find the *area* of a circle.

$$A = \pi \cdot r^2$$

Area is measured in **square units**.

Find the area of this circle.

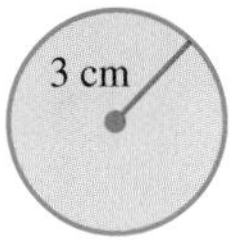

$A = \pi \cdot r^2$

$A \approx 3.14 \cdot 3 \text{ cm} \cdot 3 \text{ cm}$

$A \approx 28.3 \text{ cm}^2$ ← Rounded; square units for area

7.6 Volume of a Rectangular Solid

Use this formula to find the volume of *rectangular solids* (box-like solids).

$$V = l \cdot w \cdot h$$

Volume is measured in **cubic units**.

Find the volume of this box.

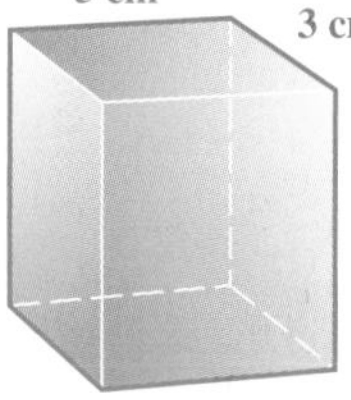

$V = l \cdot w \cdot h$

$V = 5 \text{ cm} \cdot 3 \text{ cm} \cdot 6 \text{ cm}$

$V = 90 \text{ cm}^3$ Cubic units for volume

7.6 Volume of a Sphere

Use this formula to find the volume of a *sphere* (a ball-shaped solid).

$$V = \frac{4}{3} \cdot \pi \cdot r^3$$

$$\text{or} \quad V = \frac{4 \cdot \pi \cdot r \cdot r \cdot r}{3},$$

where r is the radius of the sphere.

Volume is measured in **cubic units**.

Find the volume of a sphere with a radius of 5 m.

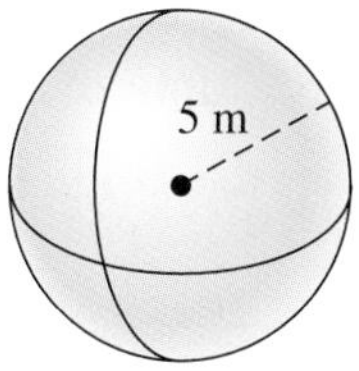

$$V = \frac{4 \cdot \pi \cdot r^3}{3}$$

$$V \approx \frac{4 \cdot 3.14 \cdot 5 \text{ cm} \cdot 5 \text{ cm} \cdot 5 \text{ cm}}{3}$$

$V \approx 523.3 \text{ m}^3$ ← Rounded; cubic units for volume

Concepts | **Examples**

7.6 Volume of a Hemisphere

Use this formula to find the volume of a *hemisphere* (half of a sphere).

$$V = \frac{2}{3} \cdot \pi \cdot r^3$$

$$\text{or} \quad V = \frac{2 \cdot \pi \cdot r \cdot r \cdot r}{3},$$

where r is the radius of the hemisphere.

Volume is measured in **cubic units**.

Find the volume of a hemisphere with a radius of 20 cm.

$$V = \frac{2 \cdot \pi \cdot r^3}{3}$$

$$V \approx \frac{\mathbf{2 \cdot 3.14 \cdot 20\ cm \cdot 20\ cm \cdot 20\ cm}}{3}$$

$V \approx$ **16,746.7 cm³** ← Rounded; cubic units for volume

7.6 Volume of a Cylinder

Use this formula to find the volume of a *cylinder*.

$$V = \pi \cdot r^2 \cdot h,$$

where r is the radius of the circular base and h is the height of the cylinder.

Volume is measured in **cubic units**.

Find the volume of this cylinder.

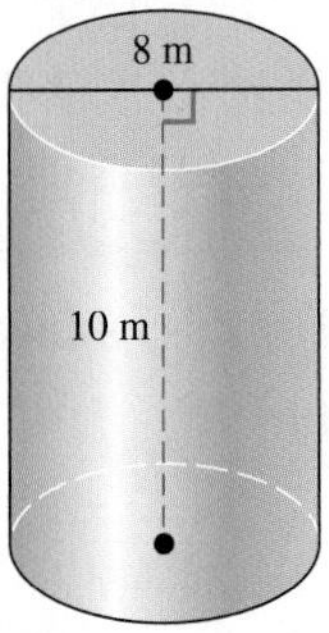

First, find the radius. $r = \frac{8\text{ m}}{2} = 4\text{ m}$

$V = \pi \cdot r^2 \cdot h$

$V \approx$ **3.14 • 4 m • 4 m • 10 m**

$V \approx$ **502.4 m³** ← Cubic units for volume

7.6 Volume of a Cone

Use this formula to find the volume of a *cone*.

$$V = \frac{1}{3} \cdot B \cdot h$$

$$\text{or} \quad V = \frac{B \cdot h}{3},$$

where B is the area of the circular base and h is the height of the cone.

Volume is measured in **cubic units**.

Find the volume of this cone.

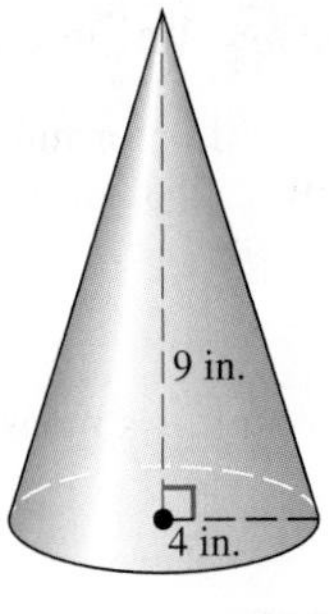

Area of circular *Base* $\approx$ 3.14 • 4 in. • 4 in.

$B \approx 50.24\text{ in.}^2$

$$V = \frac{B \cdot h}{3}$$

$$V \approx \frac{\mathbf{50.24\ in.^2 \cdot 9\ in.}}{3}$$

$V \approx$ **150.7 in.³** ← Rounded; cubic units for volume

7.6 Volume of a Pyramid

Use this formula to find the volume of a *pyramid*.

$$V = \frac{1}{3} \cdot B \cdot h$$

$$\text{or} \quad V = \frac{B \cdot h}{3},$$

where B is the area of the square or rectangular base and h is the height of the pyramid.

Volume is measured in **cubic units**.

Find the volume of this pyramid.

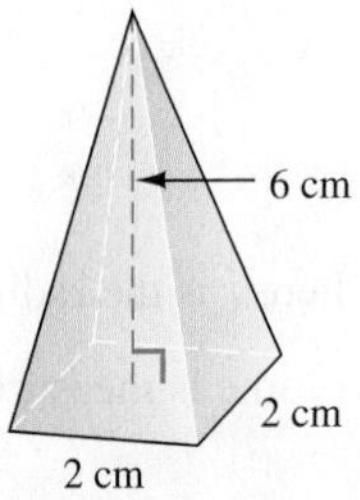

Area of square *Base* = 2 cm • 2 cm

$B = 4\text{ cm}^2$

$$V = \frac{B \cdot h}{3}$$

$$V = \frac{\mathbf{4\ cm^2 \cdot 6\ cm}}{3}$$

$V =$ **8 cm³** ← Cubic units for volume

Concepts	Examples

7.6 Surface Area of a Rectangular Solid

Use this formula to find the surface area of a rectangular solid.

Surface area = $(2 \cdot l \cdot w) + (2 \cdot l \cdot h) + (2 \cdot w \cdot h)$

or $S = 2lw + 2lh + 2wh,$

where l is the length, w is the width, and h is the height of the solid.

Surface area is measured in **square units**.

Find the surface area of this packing crate.

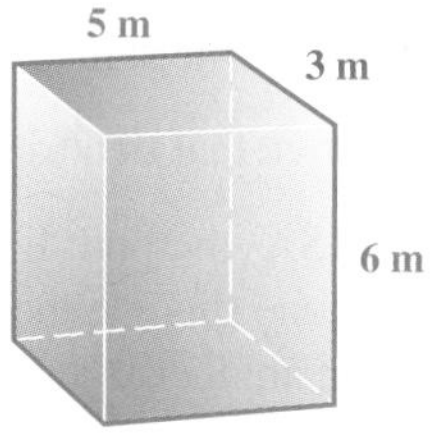

$S = 2lw + 2lh + 2wh$

$S = (2 \cdot 5\text{ m} \cdot 3\text{ m}) + (2 \cdot 5\text{ m} \cdot 6\text{ m}) + (2 \cdot 3\text{ m} \cdot 6\text{ m})$

$S = 30\text{ m}^2 + 60\text{ m}^2 + 36\text{ m}^2$

$S = 126\text{ m}^2 \leftarrow$ Square units for surface area

7.6 Surface Area of a Cylinder

Use this formula to find the surface area of a cylinder.

Surface area = $(2 \cdot \pi \cdot r \cdot h) + (2 \cdot \pi \cdot r \cdot r)$

or $S = 2\pi rh + 2\pi r^2,$

where r is the radius of the circular base and h is the height of the cylinder. Use 3.14 as the approximate value of π.

Surface area is measured in **square units**.

Find the surface area of a hot water tank with a height of 4.5 ft and a diameter of 1.8 ft. Round your answer to the nearest tenth.

First, find the radius. $r = \dfrac{1.8\text{ ft}}{2} = 0.9\text{ ft}$

$S = 2\pi rh + 2\pi r^2$

$S \approx (2 \cdot 3.14 \cdot 0.9\text{ ft} \cdot 4.5\text{ ft}) + (2 \cdot 3.14 \cdot 0.9\text{ ft} \cdot 0.9\text{ ft})$

$S \approx 25.434\text{ ft}^2 + 5.0868\text{ ft}^2$

$S \approx 30.5208\text{ ft}^2$ Now round to tenths.

$S \approx 30.5\text{ ft}^2 \leftarrow$ Square units for surface area

7.7 Finding the Square Root of a Number

Use the square root key on a calculator, (√) or (√x). Round to the nearest thousandth if necessary.

$\sqrt{64} = 8$ A perfect square

$\sqrt{43} \approx 6.557$ 6.557438524 is rounded to nearest thousandth.

7.7 Finding the Unknown Length in a Right Triangle

To find the *hypotenuse,* use this formula.

$$\text{hypotenuse} = \sqrt{(\text{leg})^2 + (\text{leg})^2}$$

The hypotenuse is the side opposite the right angle; it is the longest side in a right triangle.

Find the unknown length. Round to the nearest tenth.

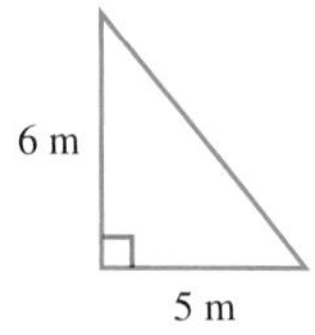

$$\text{hypotenuse} = \sqrt{(6)^2 + (5)^2} = \sqrt{36 + 25} = \sqrt{61} \approx 7.8\text{ m}$$

To find a *leg,* use this formula.

$$\text{leg} = \sqrt{(\text{hypotenuse})^2 - (\text{leg})^2}$$

The legs are the sides that form the right angle.

Find the unknown length. Round to the nearest tenth.

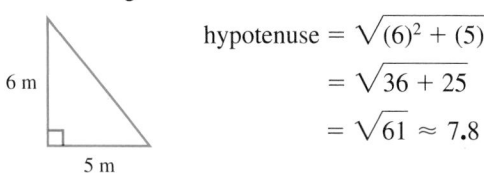

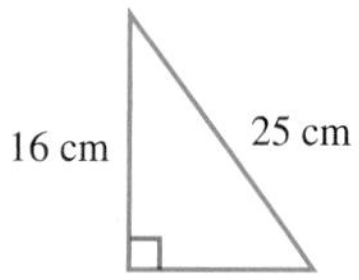

$$\text{leg} = \sqrt{(25)^2 - (16)^2} = \sqrt{625 - 256} = \sqrt{369} \approx 19.2\text{ cm}$$

Concepts

7.8 Proving That Two Triangles Are Congruent

Congruent triangles are identical both in shape and in size. This means that corresponding angles have the same measure and corresponding sides have the same length.

Here are three ways to prove that two triangles are congruent.

1. Angle–Side–Angle (ASA) method: If two angles and the side between them on one triangle measure the same as the corresponding parts on another triangle, the triangles are congruent.
2. Side–Side–Side (SSS) method: If three sides of one triangle measure the same as the corresponding sides of another triangle, the triangles are congruent.
3. Side–Angle–Side (SAS) method: If two sides and the angle between them on one triangle measure the same as the corresponding parts on another triangle, the triangles are congruent.

Examples

State which method can be used to prove that each pair of triangles is congruent.

(a)

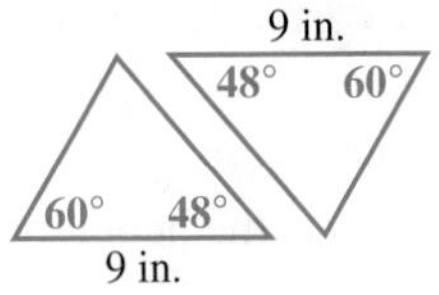

(b)

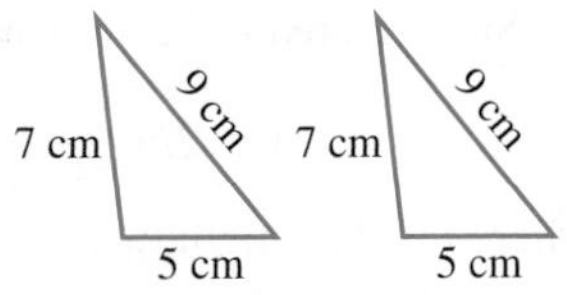

(c)

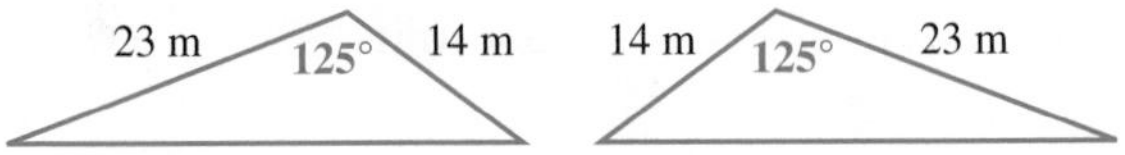

(a) On both triangles, two corresponding angles and the side between them measure the same, so use ASA.
(b) Each pair of corresponding sides has the same length, so use SSS.
(c) On both triangles, two corresponding sides and the angle between them measure the same, so use SAS.

7.8 Finding the Unknown Lengths in Similar Triangles

Use the fact that in similar triangles, the *ratios* of the lengths of corresponding sides are equal. Write a proportion. Then find the cross products and show that they are equal. Finish solving for the unknown length.

Find the unknown lengths in this pair of similar triangles.

$$\frac{x}{8} = \frac{5}{10}$$

$$x \cdot 10 = 8 \cdot 5$$

$$\frac{x \cdot \overset{1}{\cancel{10}}}{\underset{1}{\cancel{10}}} = \frac{40}{10}$$

$$x = 4 \text{ m}$$

$$\frac{y}{12} = \frac{5}{10}$$

$$y \cdot 10 = 12 \cdot 5$$

$$\frac{y \cdot \overset{1}{\cancel{10}}}{\underset{1}{\cancel{10}}} = \frac{60}{10}$$

$$y = 6 \text{ m}$$

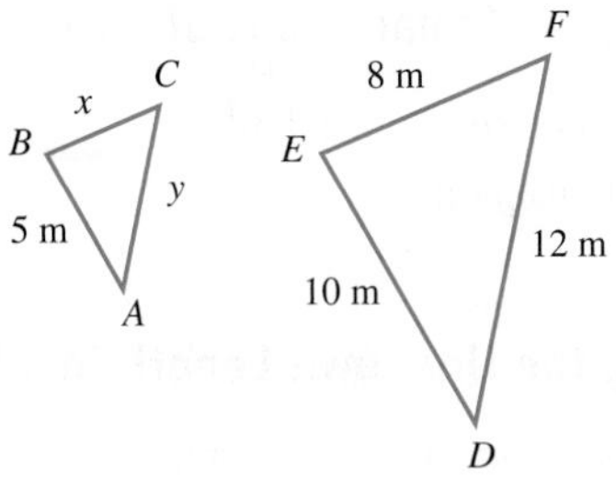

Chapter 7 Review Exercises

7.1 *Identify each figure as a line, line segment, or ray, and name it using the appropriate symbol.*

1.

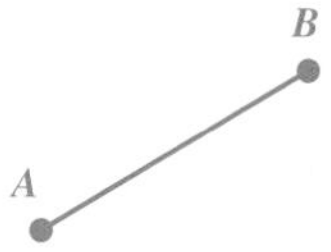

2.

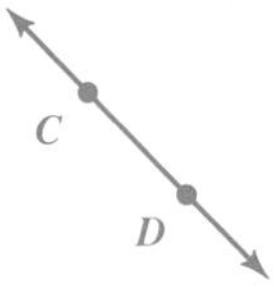

3.

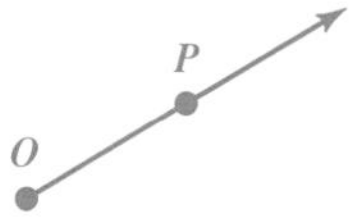

Label each pair of lines as appearing to be parallel, as perpendicular, or as intersecting.

4.

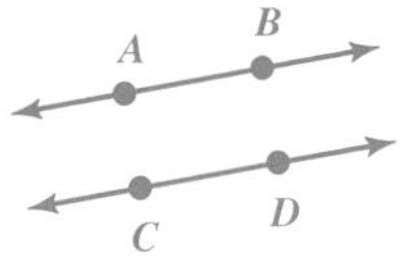

5.

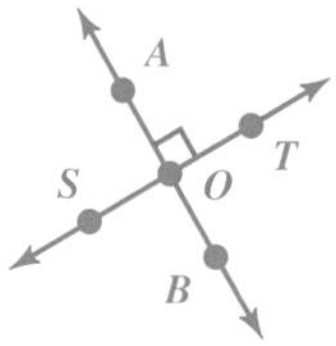

6.

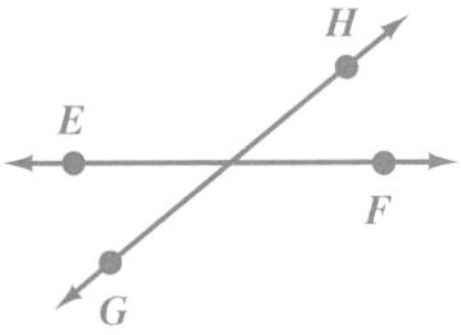

Label each angle as an acute, right, obtuse, *or* straight angle. *For right and straight angles, indicate the number of degrees in the angle.*

7.

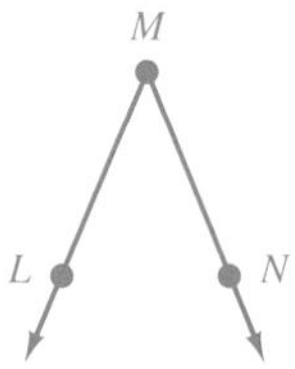

8.

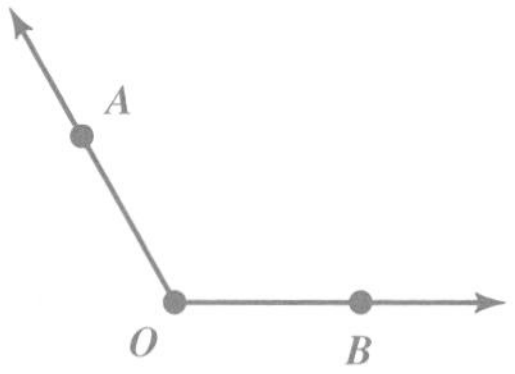

9.

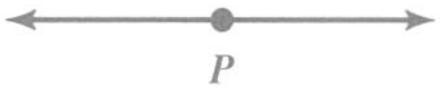

10.

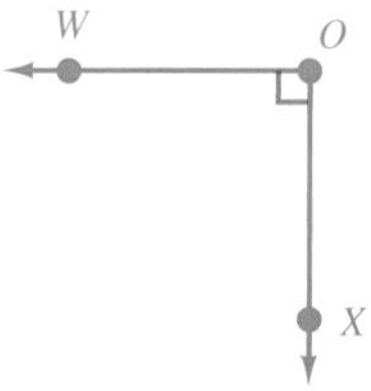

Name the pairs of supplementary angles in each figure.

11.

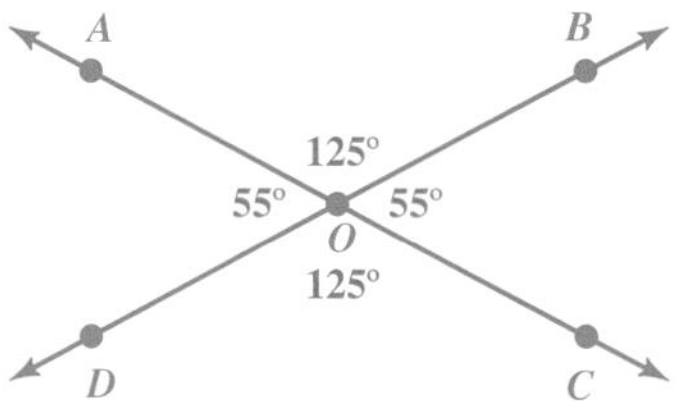

12.

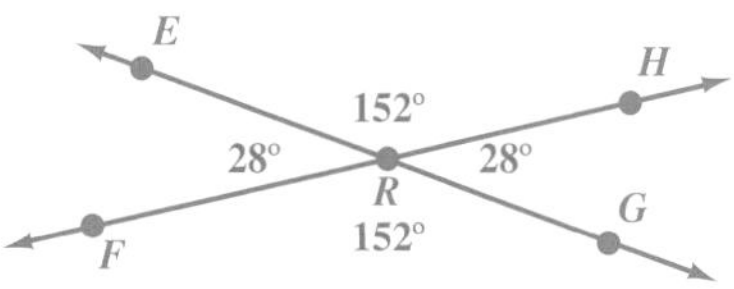

Find the complement or supplement of each angle.

13. Find each complement.

(a) 80°

(b) 45°

(c) 7°

14. Find each supplement.

(a) 155°

(b) 90°

(c) 33°

15. In the figure below, $\angle 2$ measures $60°$. Find the measure of each of the other angles.

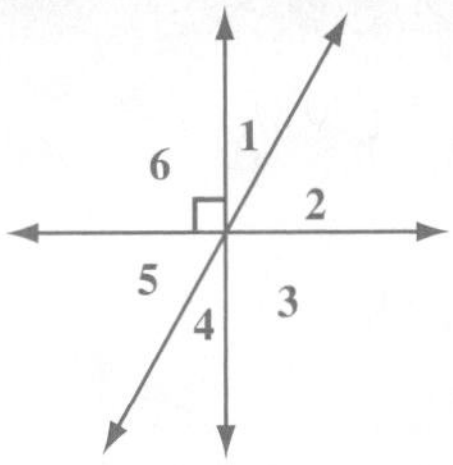

16. Line m is parallel to line n and $\angle 8$ measures $160°$. Find the measures of the other angles.

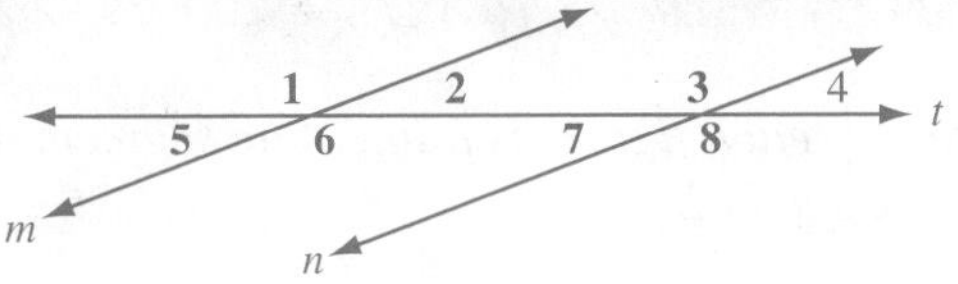

7.2 *Find the perimeter of each rectangle or square.*

17.

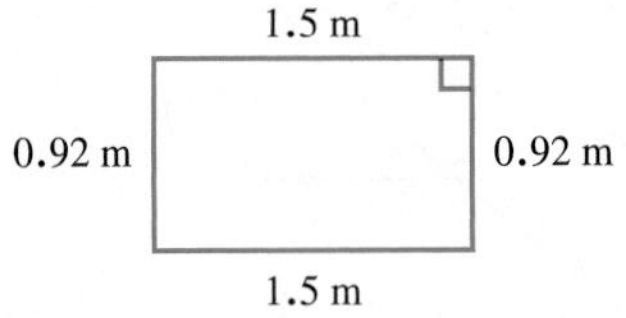

18.

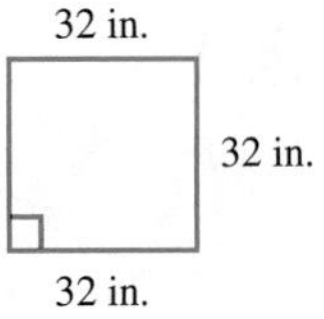

19. A square-shaped pillow measures 38 cm along each side. How much lace is needed to trim all the edges?

20. A rectangular garden plot is $8\frac{1}{2}$ ft wide and 12 ft long. How much fencing is needed to surround the garden?

Find the area of each rectangle or square. Round your answers to the nearest tenth when necessary.

21.

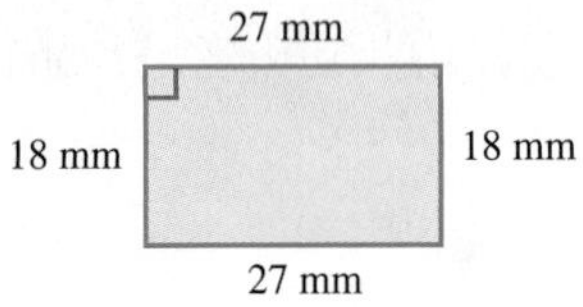

22.

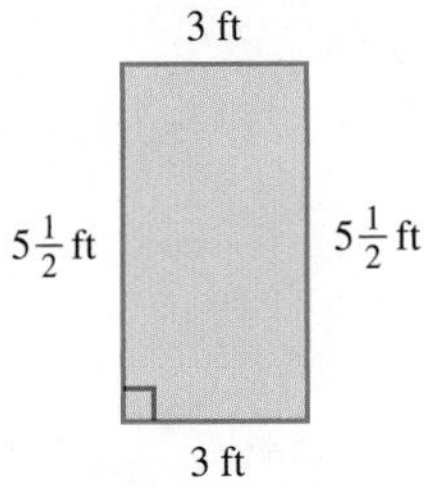

23.

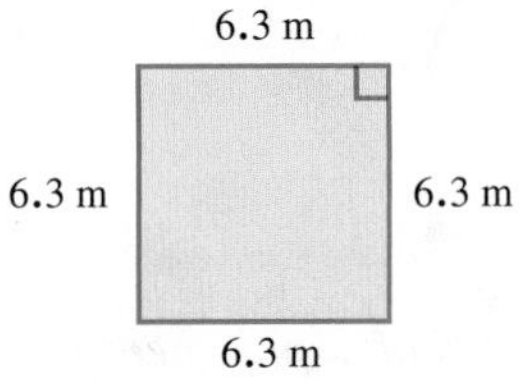

7.3 *Find the perimeter and area of each parallelogram or trapezoid. Round your answers to the nearest tenth when necessary.*

24.

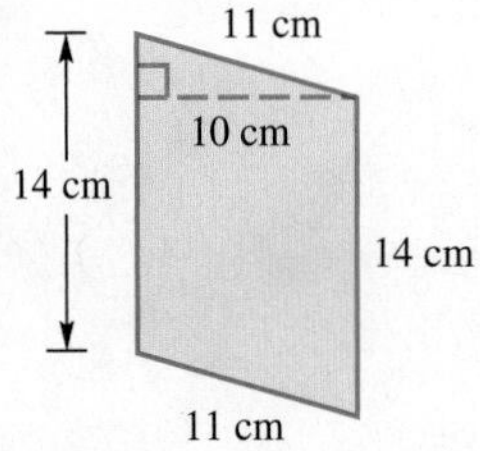

25.

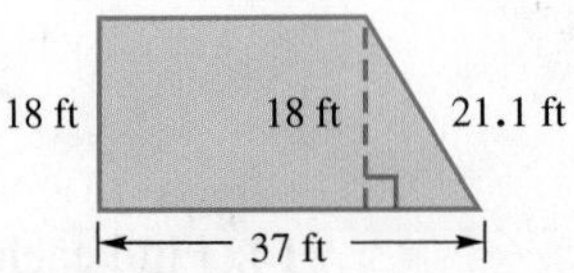

26.

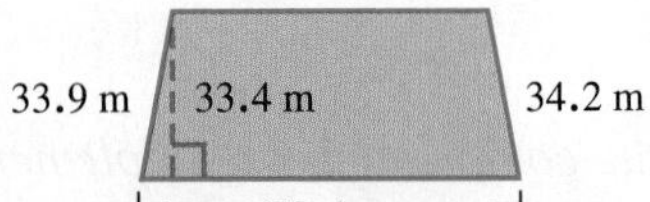

7.4 *Find the perimeter and area of each triangle.*

27.

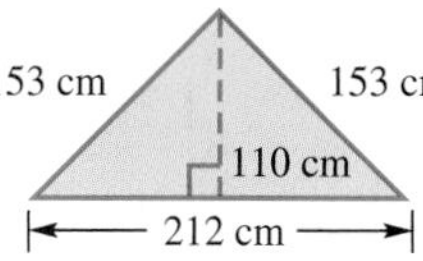

28.

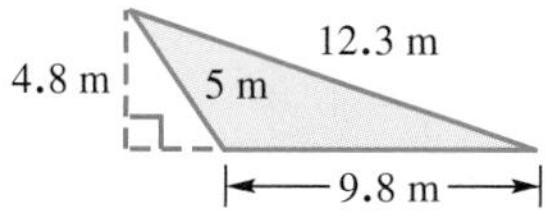

29.

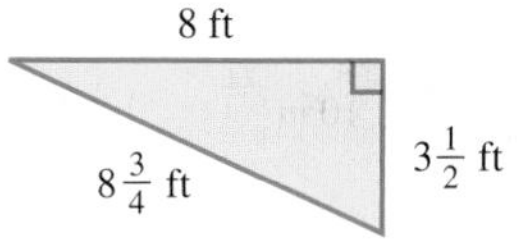

Find the number of degrees in the third angle of each triangle.

30.

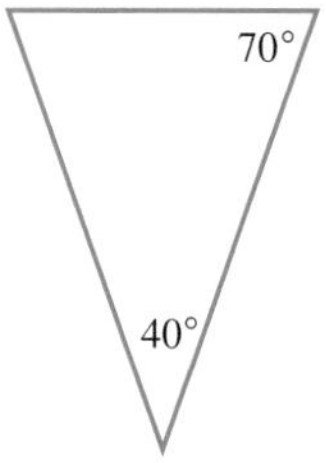

31.

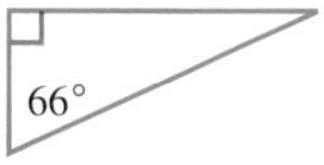

7.5 *Find the unknown length.*

32. The radius of a circular irrigation field is 68.9 m. What is the diameter of the field?

33. The diameter of a juice can is 3 in. What is the radius of the can?

Find the circumference and area of each circle. Use 3.14 as the approximate value for π. Round your answers to the nearest tenth.

34.

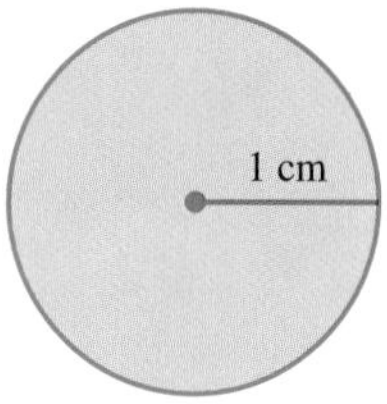

35.

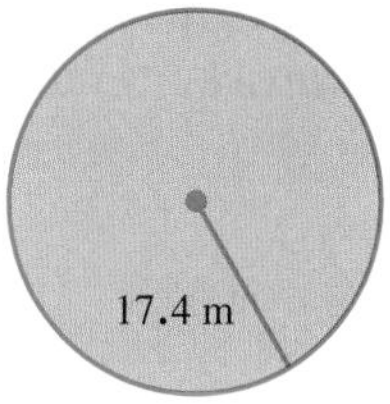

36.

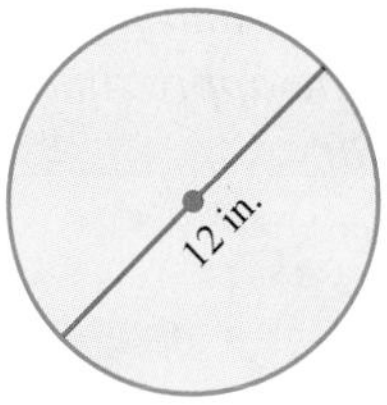

7.2–7.5 *Find each shaded area. Note that Exercises 44 and 45 contain semicircles. Use 3.14 as the approximate value for π. Round your answers to the nearest tenth when necessary.*

37.

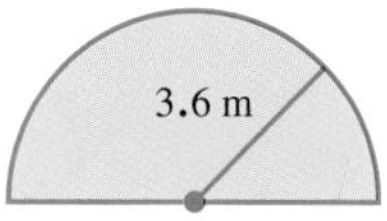

38.

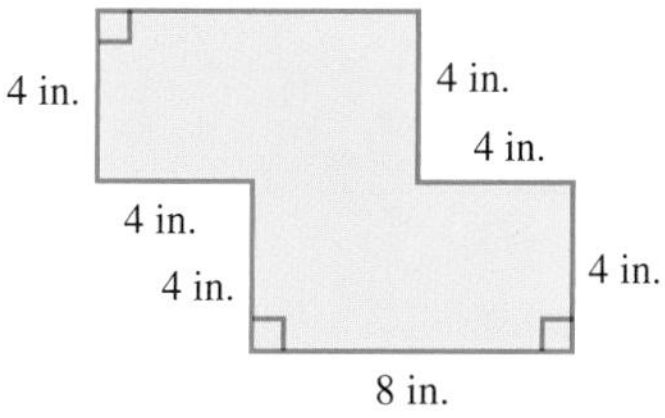

39.

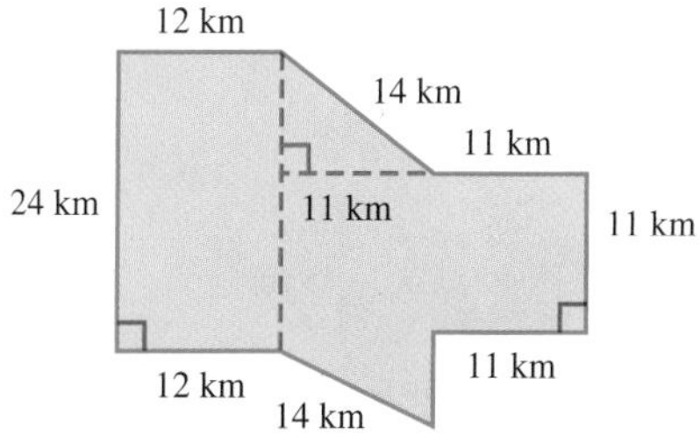

40.

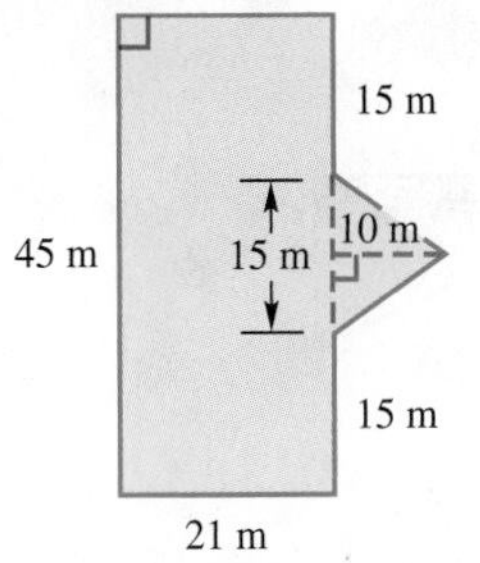

41.

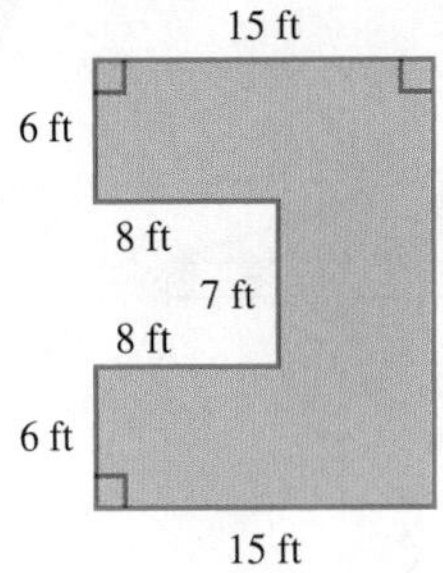

42.

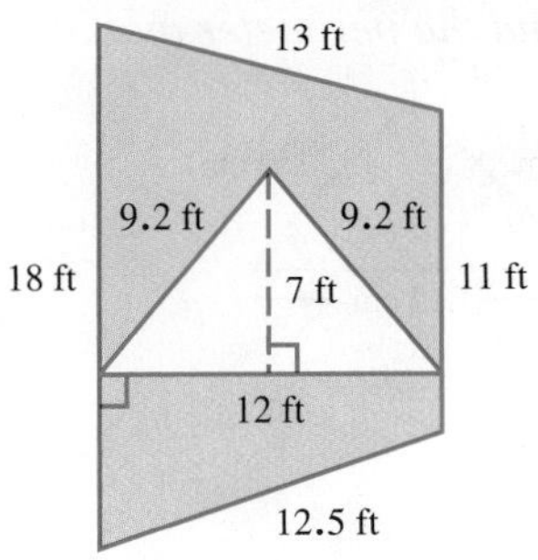

43.

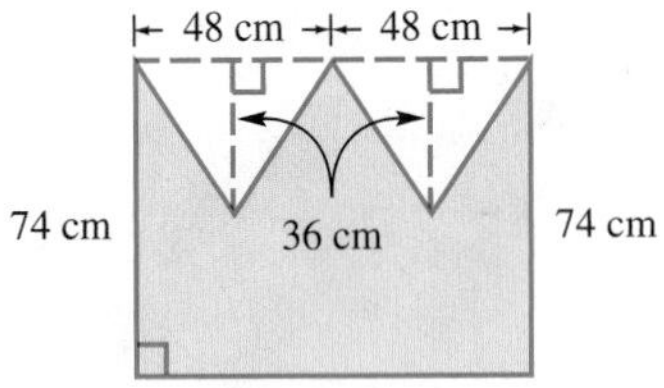

44.

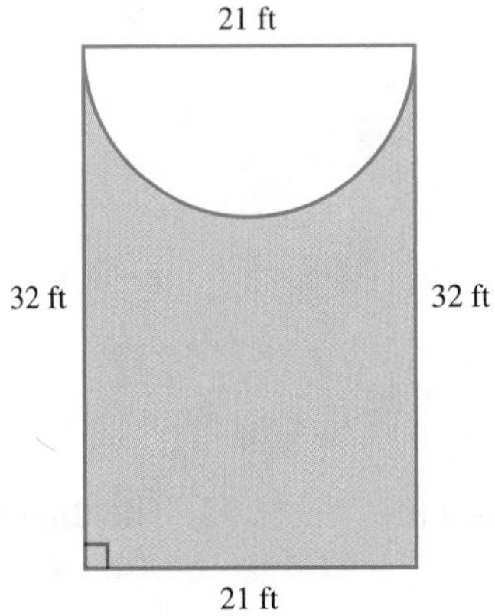

45.

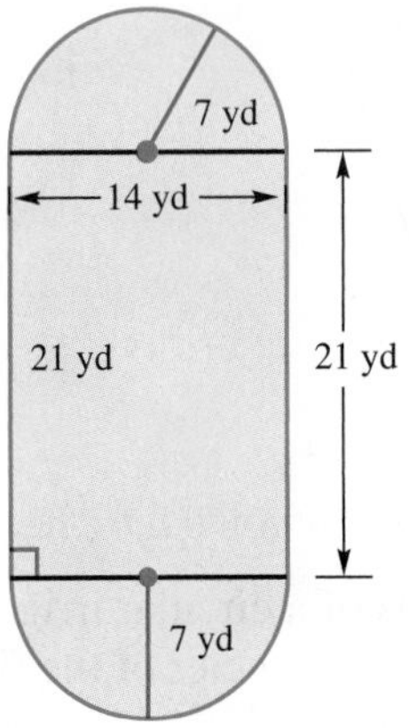

7.6 *Name each solid and find its volume. For Exercises 46–50, also find the surface area. Use 3.14 as the approximate value for π. Round your answers to the nearest tenth when necessary.*

46.

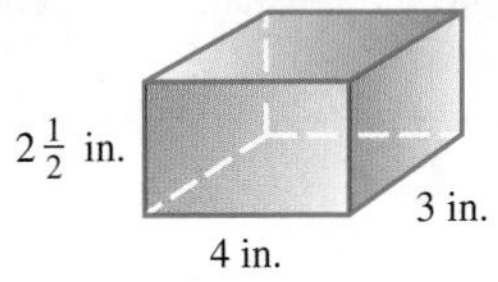

47.

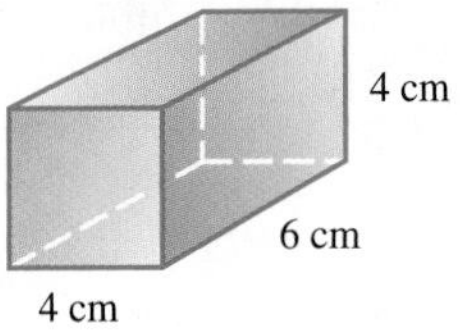

48.

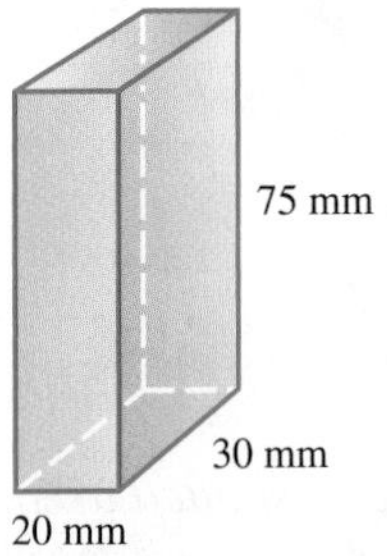

49.

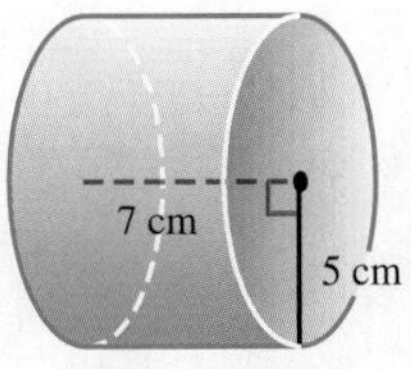

50.

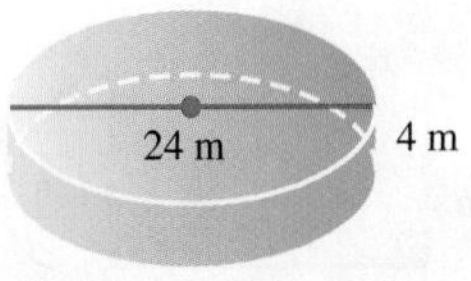

51.

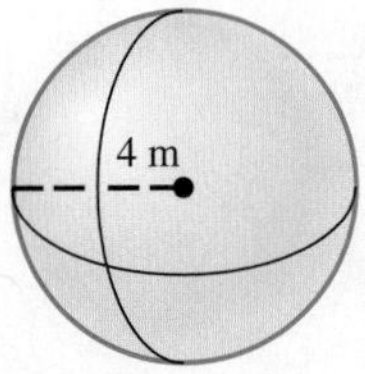

52.

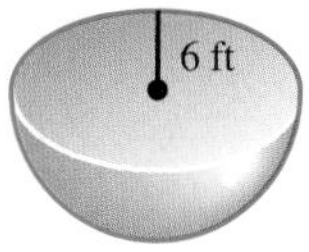

53.

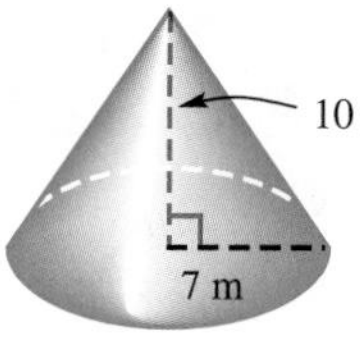

54.

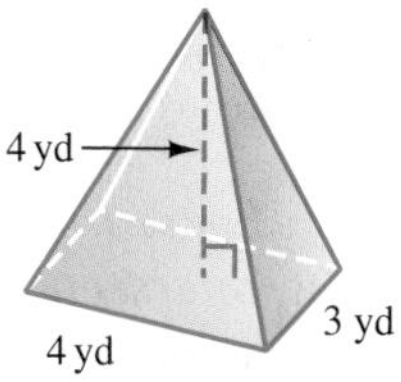

7.7 *Find each square root. Round your answers to the nearest thousandth when necessary.*

55. $\sqrt{49}$

56. $\sqrt{8}$

57. $\sqrt{3000}$

58. $\sqrt{144}$

59. $\sqrt{58}$

60. $\sqrt{625}$

61. $\sqrt{105}$

62. $\sqrt{80}$

Find the unknown length in each right triangle. Use a calculator to find square roots. Round your answers to the nearest tenth when necessary.

63.

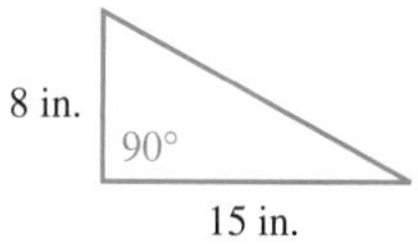

64.

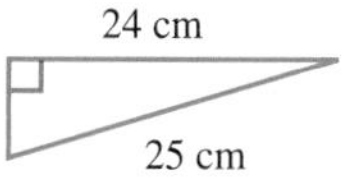

65.

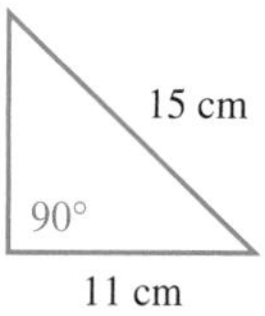

66.

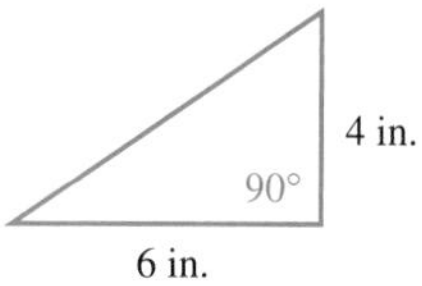

67.

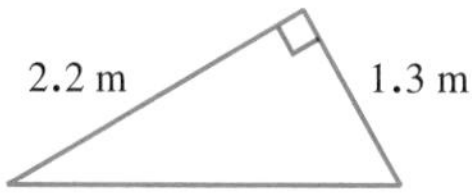

68.

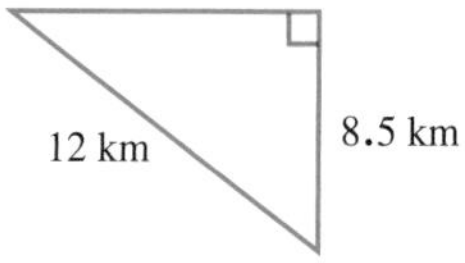

7.8 *State which method can be used to prove that each pair of triangles is congruent.*

69.

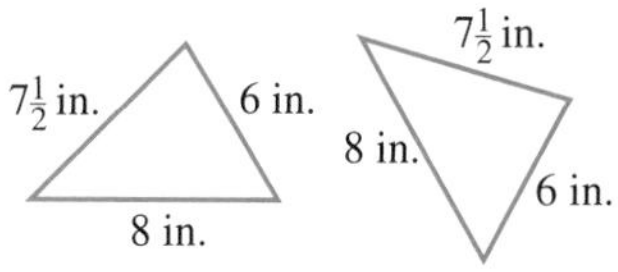

70.

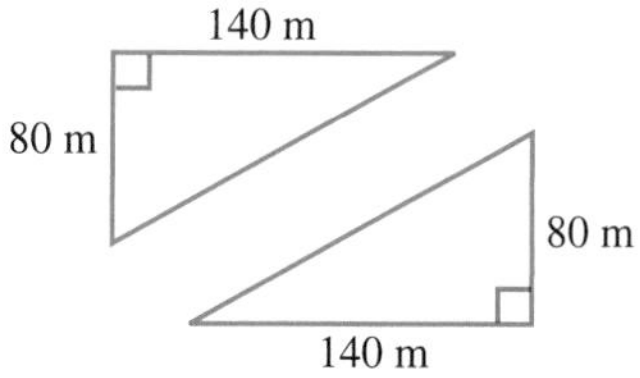

71.

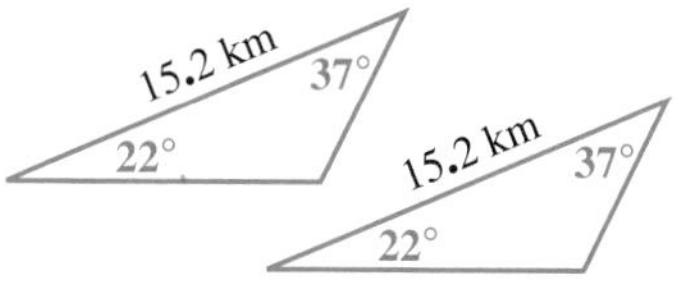

Find the unknown lengths in each pair of similar triangles. Then find the perimeter of the larger triangle in each pair.

72.

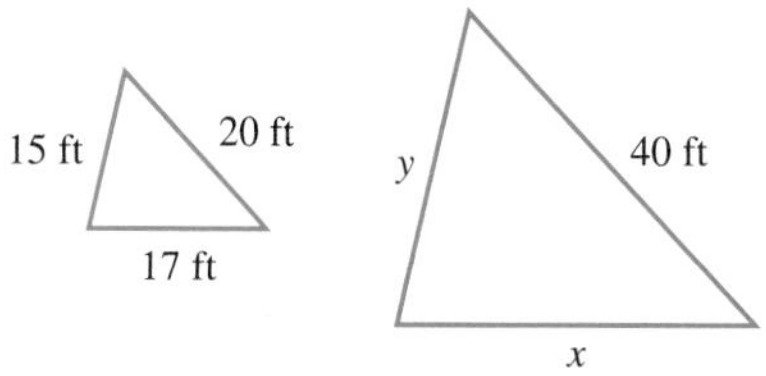

73.

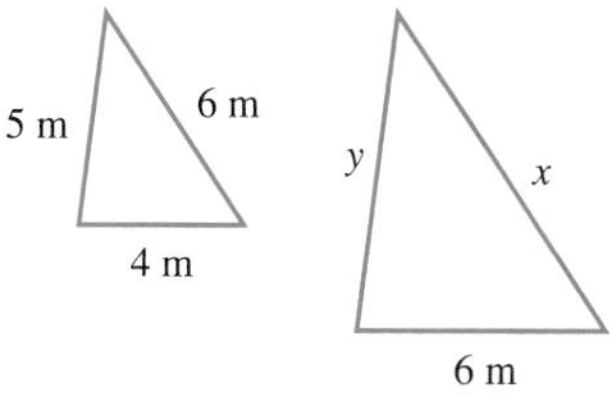

74.

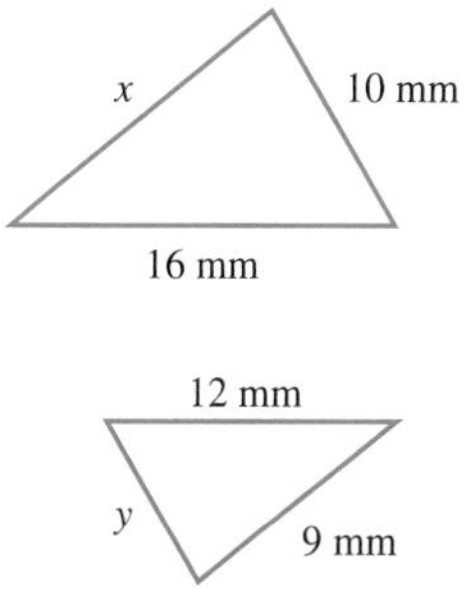

Mixed Review Exercises

Label each figure. Choose from these labels: line segment, ray, parallel lines, perpendicular lines, intersecting lines, acute angle, right angle, straight angle, obtuse angle. Indicate the number of degrees in the right angle and the straight angle.

75.

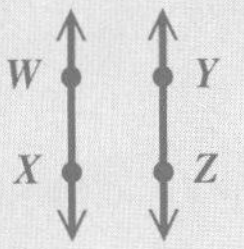

76.

77.

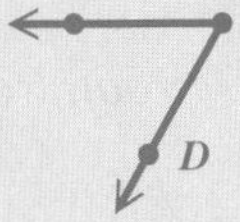

78.

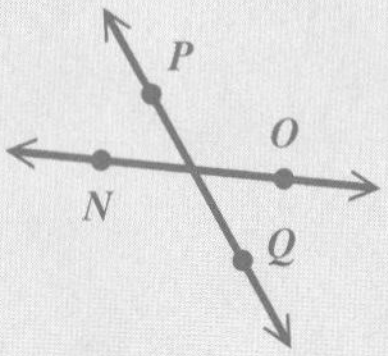

79.

80.

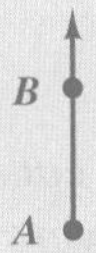

81.

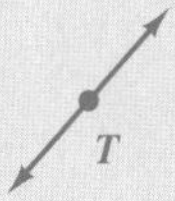

82.

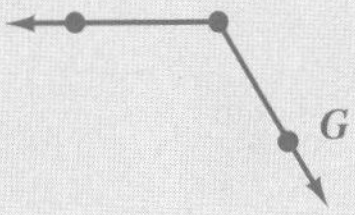

83.

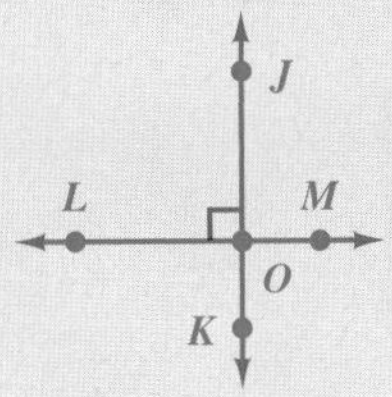

84. What is the complement of an angle measuring 9°?

85. What is the supplement of an angle measuring 42°?

Name each figure and find its perimeter (or circumference) and area. Use 3.14 as the approximate value for π. Round your answers to the nearest tenth when necessary.

86.

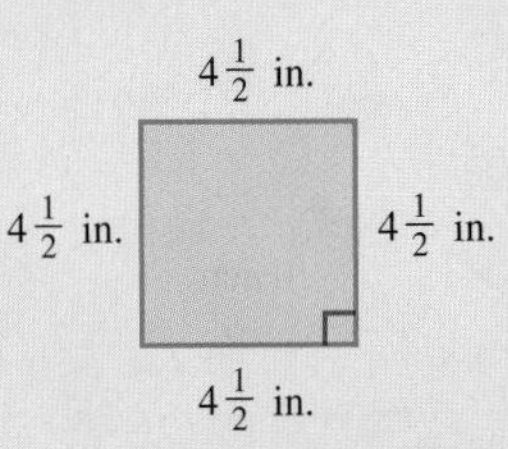

87.

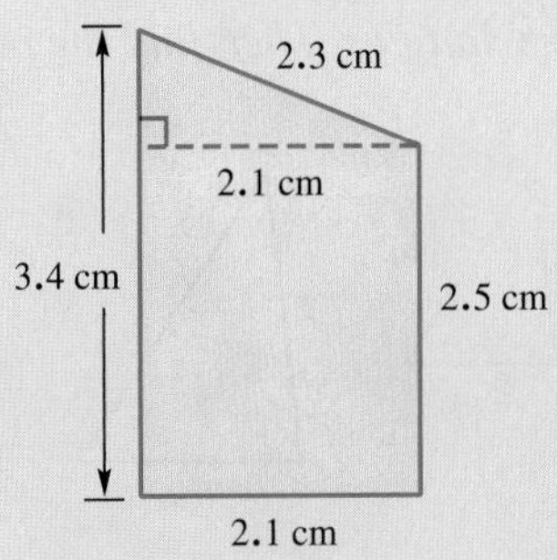

88.

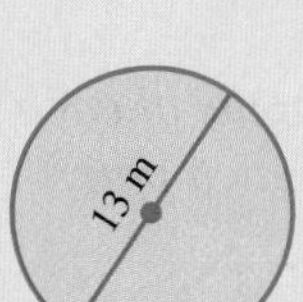

89.

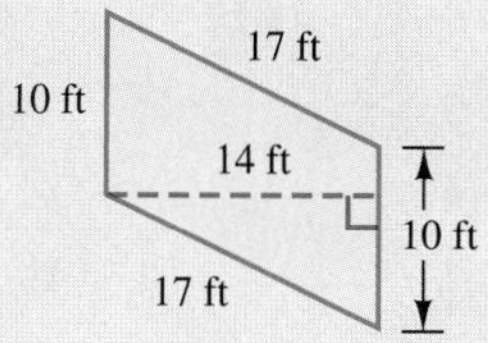

90.

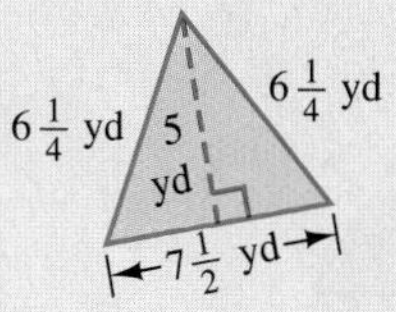

91.

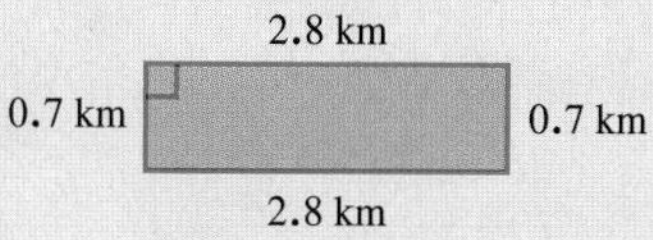

92.

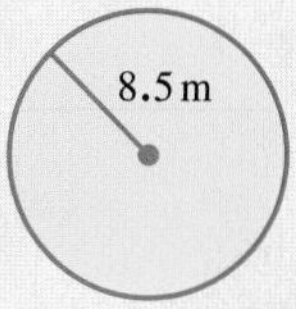

93.

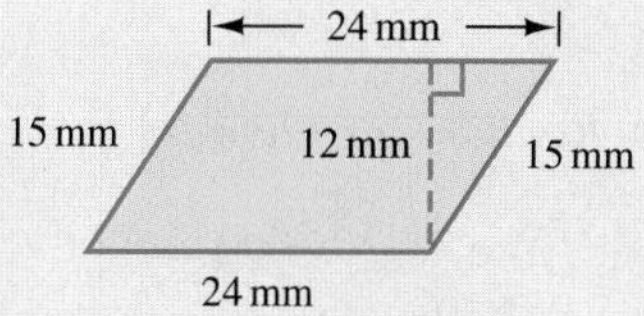

94.

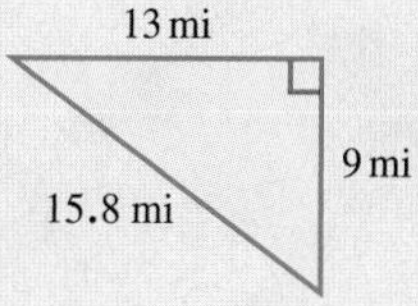

Find the perimeter and area of each figure. In Exercise 95, assume that all angles are 90°.

95.

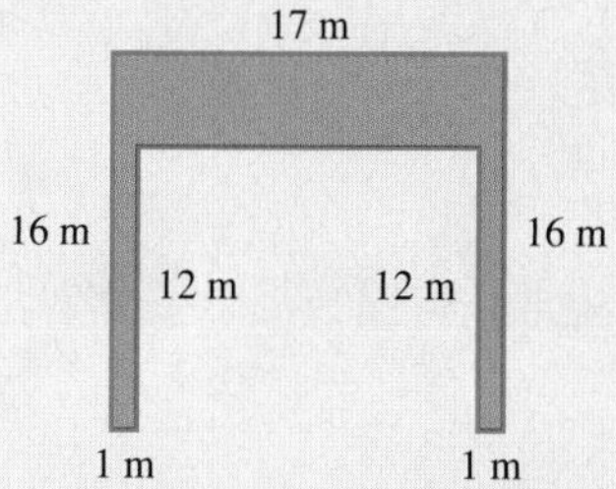

96.

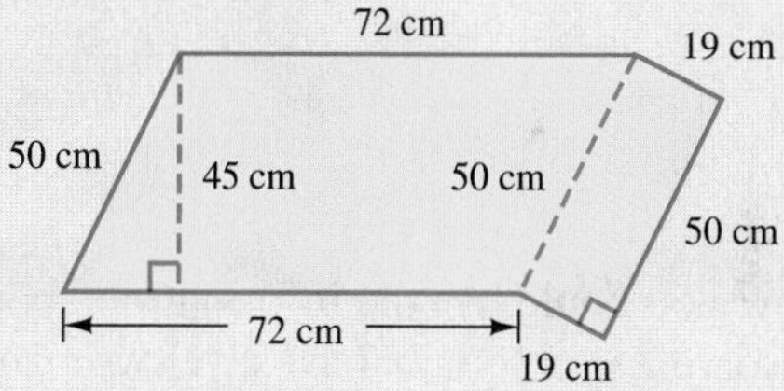

Name each solid and find its volume. Use 3.14 as the approximate value for π. Round your answers to the nearest tenth when necessary.

97.

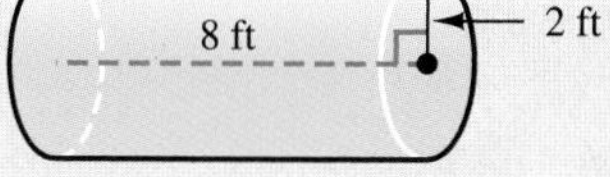

98.

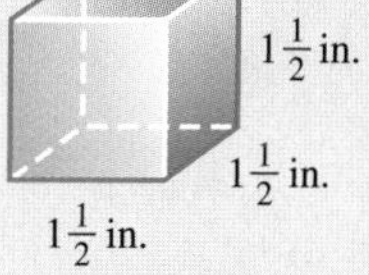

99.

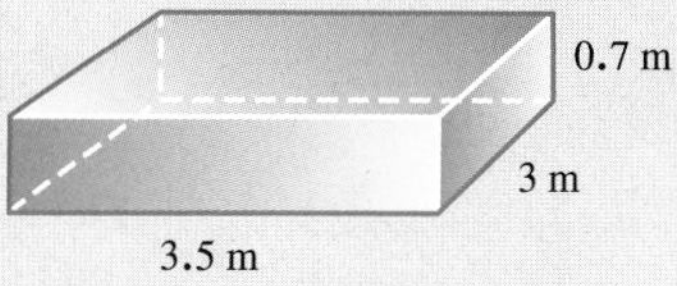

100.

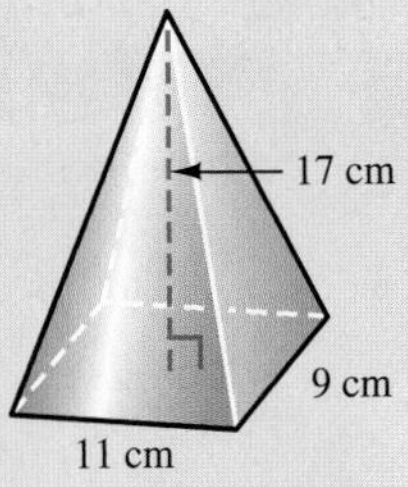

101.

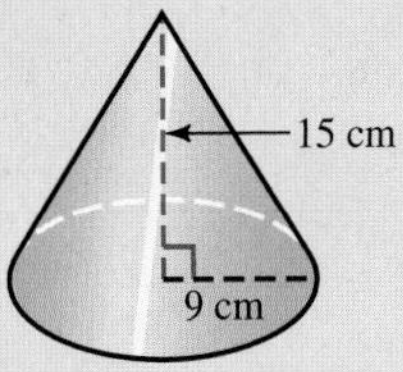

102.

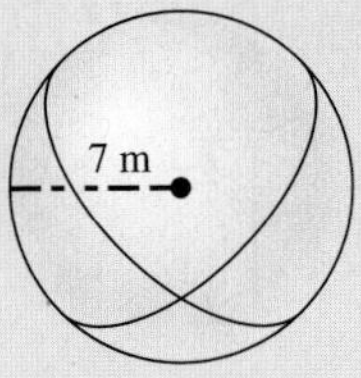

Find the unknown angle or side measurement. Round your answers to the nearest tenth when necessary.

103.

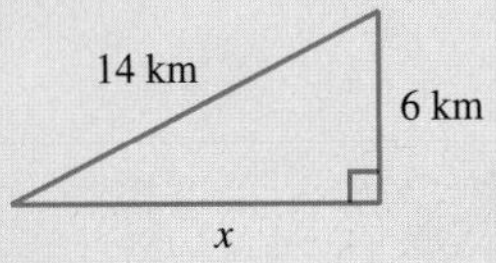

104.

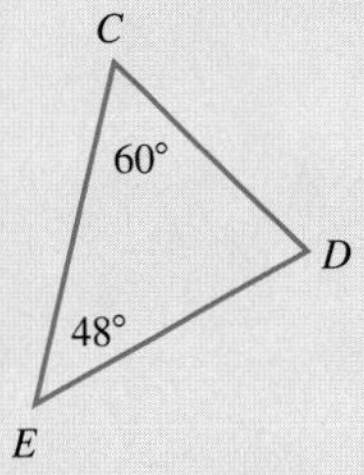

105. Similar triangles

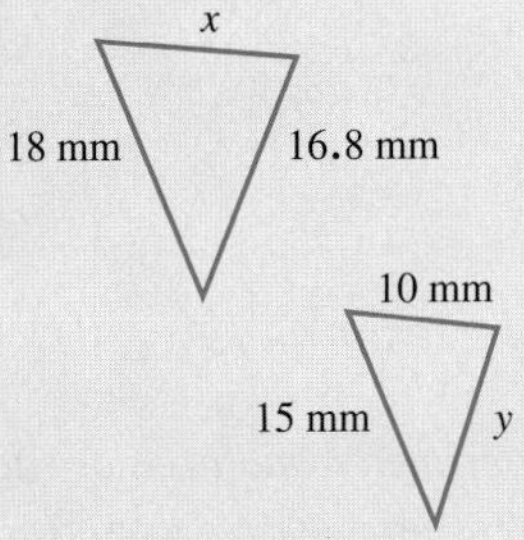

106. Explain how you could use the information about prefixes from earlier in this chapter to solve a problem that asks, "How many decades are in two centuries?"

Chapter 7 Test

Chapter Test Prep VIDEO

The Chapter Test Prep Videos with test solutions are available on DVD, in MyMathLab, and on YouTube—search "LialDevMath" and click on "Channels."

Choose the figure that matches each label. For right and straight angles, indicate the number of degrees in the angle.

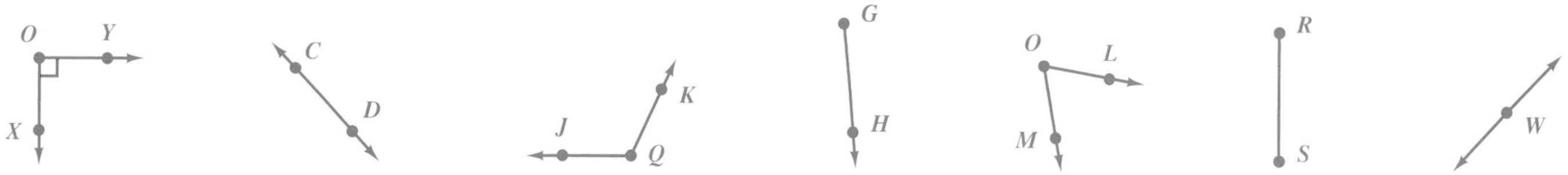

A. B. C. D. E. F. G.

1. Acute angle is figure ____.

2. Right angle is figure ____ and its measure is ____.

3. Ray is figure ____.

4. Straight angle is figure ____ and its measure is ____.

5. Write a definition of parallel lines and a definition of perpendicular lines. Make a sketch to illustrate each definition.

6. Find the complement of an 81° angle.

7. Find the supplement of a 20° angle.

8. Find the measure of each unlabeled angle in the figure at the right.

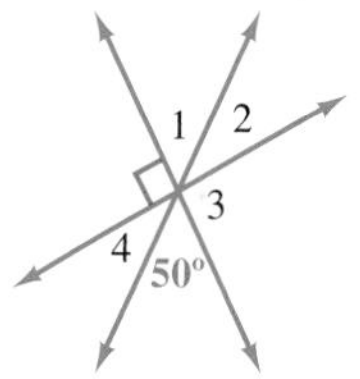

Name each figure and find its perimeter and area.

9.

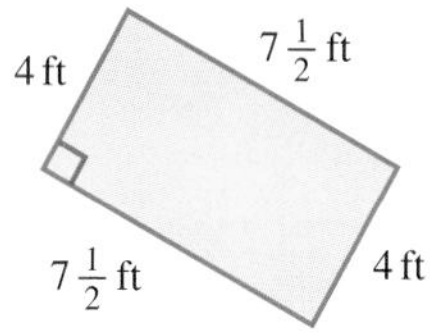

10.

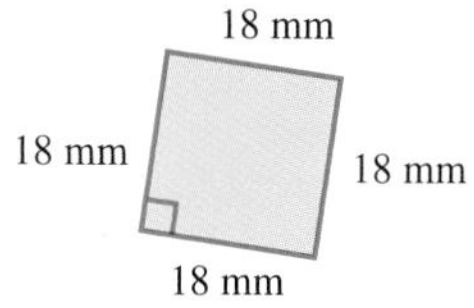

11.

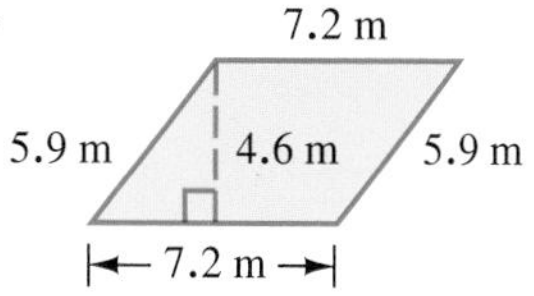

12.

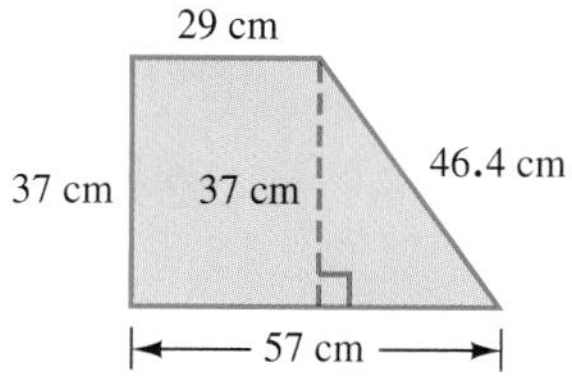

Find the perimeter and area of each triangle.

13.

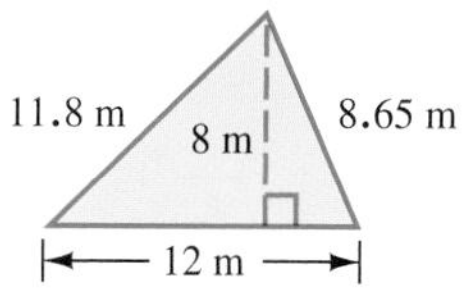

14.

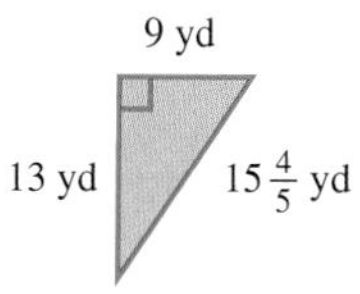

15. A triangle has angles that measure 90° and 35°. What does the third angle measure?

In Problems 16–22, use 3.14 as the approximate value for π. Round your answers to the nearest tenth when necessary.

16. Find the radius.

17. Find the circumference.

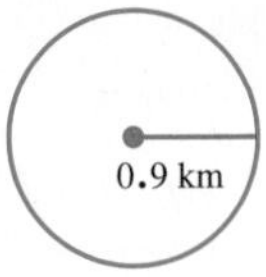

Find the area of each figure.

18.

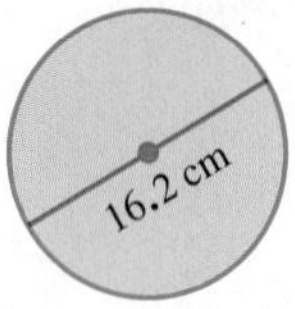

19.

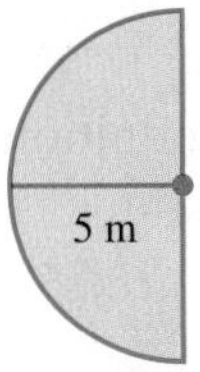

Name each solid and find its volume. In Problems 20 and 22, also find the surface area.

20.

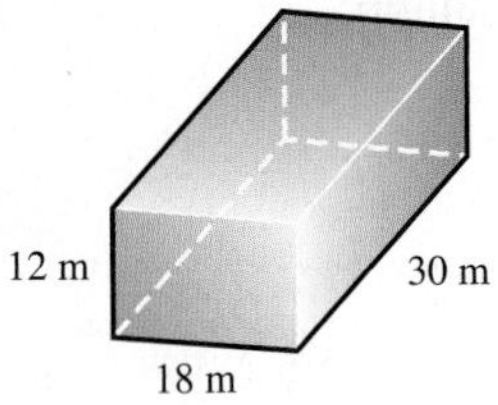

21.

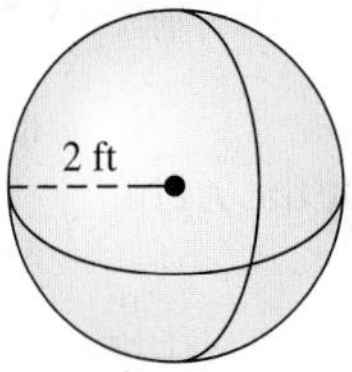

22.

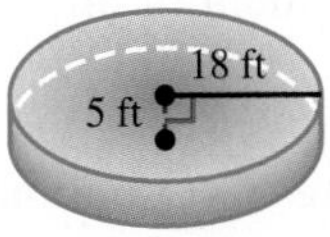

Find the unknown lengths. Round your answers to the nearest tenth when necessary.

23.

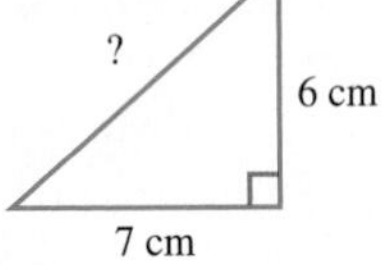

24. Similar triangles

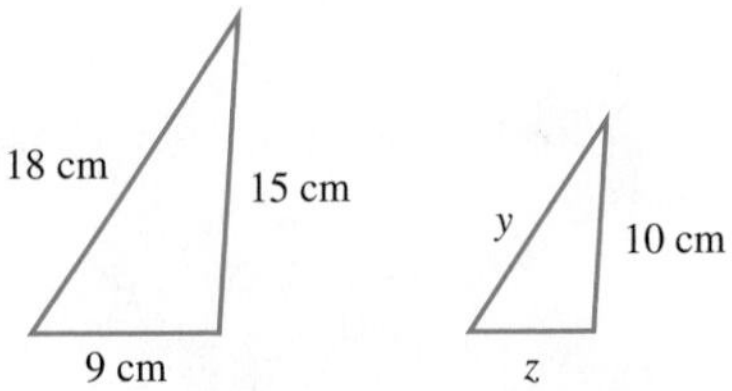

25. Explain the difference between cm, cm^2, and cm^3. In what types of geometry problems might you use each of these units?

8 Statistics

The old saying “A picture is worth a thousand words” was never more true than when applied to the understanding of data. We are constantly being bombarded with facts and numbers. In this chapter you will improve your ability to interpret and understand essential information.

8.1 Circle Graphs

OBJECTIVES

1. Read and understand a circle graph.
2. Use a circle graph.
3. Draw a circle graph.

The word *statistics* originally came from words that mean *state numbers.* State numbers refer to numerical information, or *data,* gathered by the government such as the number of births, deaths, or marriages in a population. Today, the word *statistics* has a much broader application; data from the fields of economics, social science, and business can all be organized and studied under the branch of mathematics called *statistics.*

VOCABULARY TIP

Circle graphs, also known as **pie graphs** or **pie charts**, are the best type of graph for showing the relative size of different items being compared.

OBJECTIVE 1 Read and understand a circle graph. It can be hard to understand a large collection of data. The graphs described in this section help you make sense of such data. For example, a **circle graph** shows how a total amount is divided into parts. The circle graph below shows you how 24 hours in the life of a college student are divided among different activities.

THE DAY OF A COLLEGE STUDENT

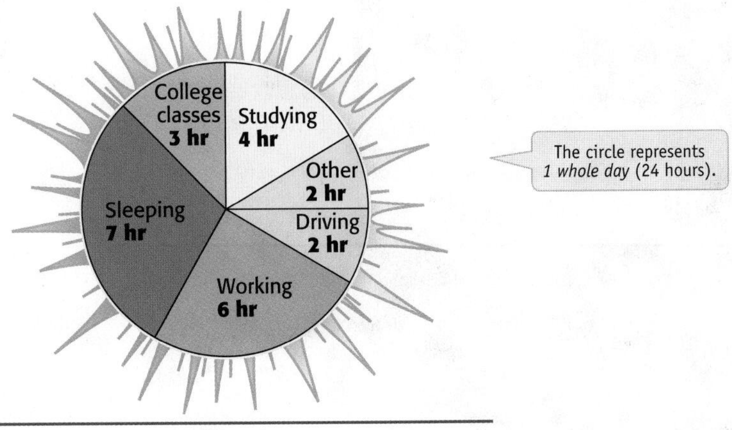

◀ Work Problem 1 at the Side.

1 Use the circle graph to answer each question.

(a) The greatest number of hours is spent on which activity?

(b) How many more hours are spent working than studying?

(c) Find the total number of hours spent studying, working, and attending classes.

OBJECTIVE 2 Use a circle graph. The above circle graph uses pie-shaped pieces called *sectors* to show the amount of time spent on each activity (the total must be one day, which is 24 hours). The circle graph can therefore be used to compare the time spent on any one activity to the total number of hours in the day.

EXAMPLE 1 Using a Circle Graph

Find the ratio of hours spent in college classes to the total number of hours in the day. Write the ratio as a fraction in lowest terms. (See **Section 5.1.**)

The circle graph shows that 3 of the 24 hours in a day are spent in class. The ratio of class time to the hours in a day is shown below.

$$\frac{3 \text{ hours (college classes)}}{24 \text{ hours (whole day)}} = \frac{3 \text{ hours}}{24 \text{ hours}} = \frac{3 \div 3}{24 \div 3} = \frac{1}{8} \leftarrow \text{Lowest terms}$$

◀ Work Problem 2 at the Side.

2 Use the circle graph to find each ratio. Write the ratios as fractions in lowest terms.

GS **(a)** Hours spent driving to whole day

$$\frac{2 \text{ hr driving}}{\_\_\_\_ \text{ hr in whole day}} = \_\_\_\_$$

(b) Hours spent sleeping to whole day

(c) Hours spent attending class and studying to whole day

(d) Hours spent driving and working to whole day

The circle graph above can also be used to find the ratio of the time spent on one activity to the time spent on any other activity.

Answers

1. **(a)** sleeping **(b)** 2 hr **(c)** 13 hr

2. **(a)** 24; $\frac{1}{12}$ **(b)** $\frac{7}{24}$ **(c)** $\frac{7}{24}$ **(d)** $\frac{1}{3}$

EXAMPLE 2 **Finding a Ratio from a Circle Graph**

Use the circle graph about a student's day to find the ratio of study time to class time. Write the ratio as a fraction in lowest terms.

The circle graph shows 4 hours spent studying and 3 hours spent in class. The ratio of study time to class time is shown below.

The common units (hours) divide out.

$$\frac{4 \text{ hours (study)}}{3 \text{ hours (class)}} = \frac{4 \cancel{\text{hours}}}{3 \cancel{\text{hours}}} = \frac{4}{3} \leftarrow \text{Lowest terms}$$

Work Problem 3 at the Side. ▶

A circle graph often shows data as percents. For example, suppose that the yearly vending machine snack food sales in the United States were \$36 billion. The circle graph below shows how sales were divided among various types of snack foods. The entire circle represents the total \$36 billion in sales. Each sector represents the sales of one snack item as a percent of the total sales (the total must be 100%).

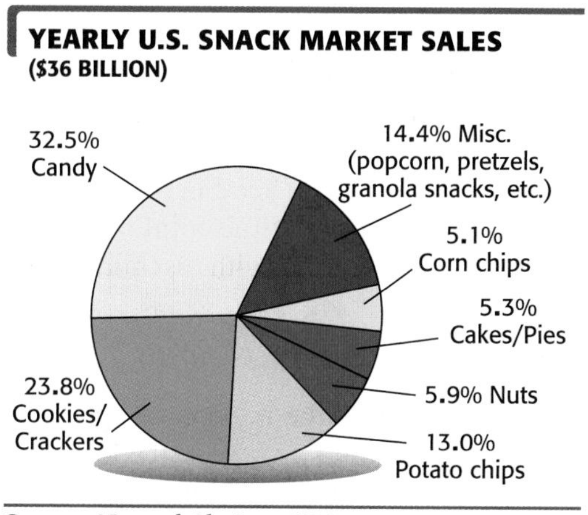

Source: Natural Choice—USA.

The circle represents 100% of the snack sales. (100% of something is all of it.)

EXAMPLE 3 **Calculating an Amount Using a Circle Graph**

Use the circle graph above on vending machine snack sales to find the amount spent on candy for the year.

Recall the percent equation.

$$\text{part} = \text{percent} \cdot \text{whole}$$

The total sales are \$36 billion, so the whole is \$36 billion. The percent is 32.5% or, as a decimal, 0.325. Find the part.

$$\text{part} = \text{percent} \cdot \text{whole}$$

32.5% = 0.325

$$x = (0.325)(36 \text{ billion})$$

$$x = 11.7 \text{ billion}$$

The amount spent on candy was \$11.7 billion or \$11,700,000,000.

Work Problem 4 at the Side. ▶

3 Use the circle graph on a student's day to find the following ratios. Write the ratios as fractions in lowest terms.

GS **(a)** Hours spent in class to hours spent studying

$$\frac{\_\_\_\text{ hr (class)}}{4 \text{ hr (study)}}$$

GS **(b)** Hours spent working to hours spent sleeping

$$\frac{6 \text{ hr}}{\_\_\_\text{ hr}}$$

(c) Hours spent driving to hours spent working

(d) Hours spent in class to hours spent for "Other"

4 Use the circle graph on vending machine snack sales to find the following.

GS **(a)** Amount spent on corn chips; 5.1% is 0.051 as a decimal

$$x = (0.051)(36 \text{ billion})$$

$$x = \_\_\_\_\_\_$$

(b) The amount spent on misc. (popcorn, pretzels, granola snacks, etc.)

(c) The amount spent on cakes/pies

(d) The amount spent on cookies/crackers

Answers

3. **(a)** $3; \frac{3}{4}$ **(b)** $7; \frac{6}{7}$ **(c)** $\frac{1}{3}$ **(d)** $\frac{3}{2}$

4. **(a)** \$1.836 billion or \$1,836,000,000
(b) \$5.184 billion or \$5,184,000,000
(c) \$1.908 billion or \$1,908,000,000
(d) \$8.568 billion or \$8,568,000,000

OBJECTIVE 3 Draw a circle graph. In a recent year, Goodwill Industries donors helped fund programs that let nearly 1 million people take their first steps toward new and better jobs and financial independence. The following table shows those who were served.

A HELPING HAND

Group Served	Percent of Total
People with disabilities	25%
Welfare recipients	15%
Working poor	10%
Ex-offenders	10%
At-risk youth	5%
Unemployed	35%
Total	**100%**

Source: Goodwill Industries.

You can show these percents visually by using a circle graph. The entire circle will represent all of the groups served (all 100%).

EXAMPLE 4 Drawing a Circle Graph

Using the data in the table, find the number of degrees in the sector that would represent the "People with disabilities" and begin constructing a circle graph.

Recall that a complete circle has 360°. (See **Section 7.1.**) Because the "People with disabilities" make up 25% of the total number of people, the number of degrees needed for the "People with disabilities" sector of the circle graph is 25% of 360°.

$$(360°)(25\%) = (360°)(0.25) = 90°$$

Use a tool called a **protractor** to make a circle graph. First, using a straightedge, draw a line from the center of a circle to the left edge. Place the hole in the protractor over the center of the circle, making sure that 0 on the protractor lines up with the line that was drawn. Find 90° and make a mark as shown in the illustration. Then remove the protractor and use the straightedge to draw a line from the center of the circle to the 90° mark at the edge of the circle. This sector is 90° and represents "People with disabilities." Label the sector with the group name and percent.

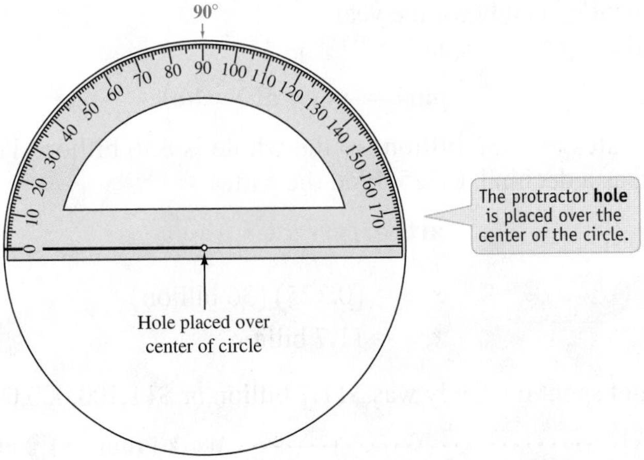

Continued on Next Page

To draw the "Welfare recipients" sector, begin by finding the number of degrees in the sector. From the table, you see that "Welfare recipients" represent 15% of the total.

$$(360°)(15\%) = (360°)(0.15) = 54°$$

Again, place the hole of the protractor over the center of the circle, but this time align 0 on the second line that was drawn. Make a mark at 54° and draw a line as before. This sector is 54° and represents those who are "Welfare recipients."

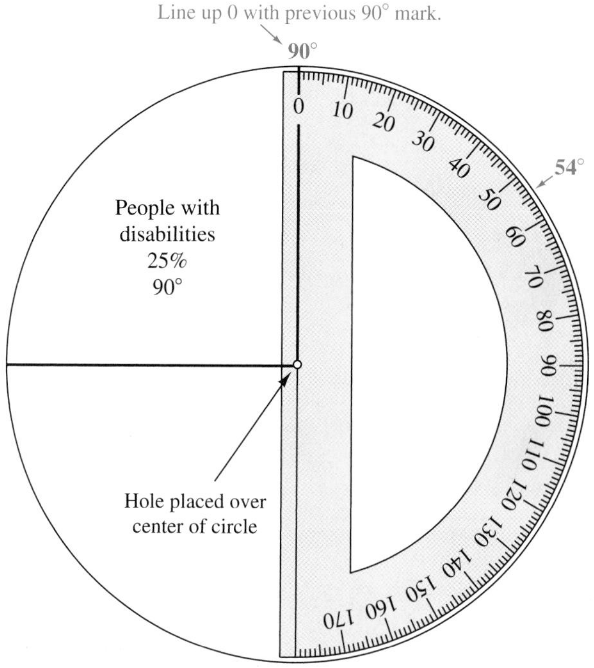

CAUTION

You must be certain that the hole in the protractor is placed over the exact center of the circle each time you measure the size of a sector.

Work Problem 5 at the Side. ▶

Use this circle for **Margin Problem 5** at the side.

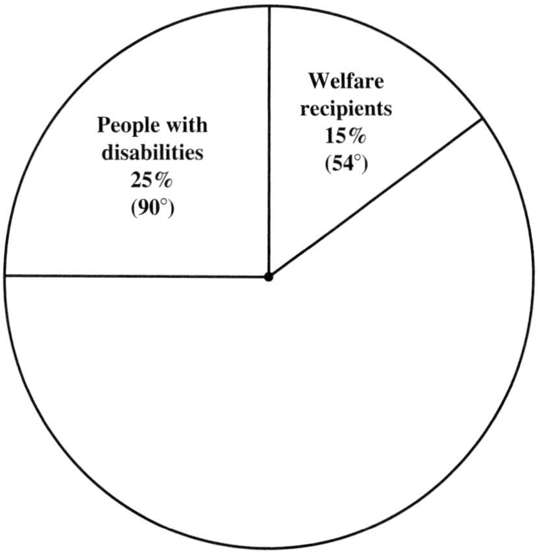

Draw the rest of the sectors using the protractor.

5 Using the information in the table on the groups served, find the number of degrees needed for each sector. Complete the circle graph at the bottom. Label each sector with the group name and percent.

GS **(a)** Working poor

$(360°)(10\%)$

$= (360°)(\_\_\_\_)$

$= \_\_\_\_$

(b) Ex-offenders

(c) At-risk youth

(d) Unemployed

Answers

5. (a) 0.10; 36° **(b)** 36° **(c)** 18° **(d)** 126°

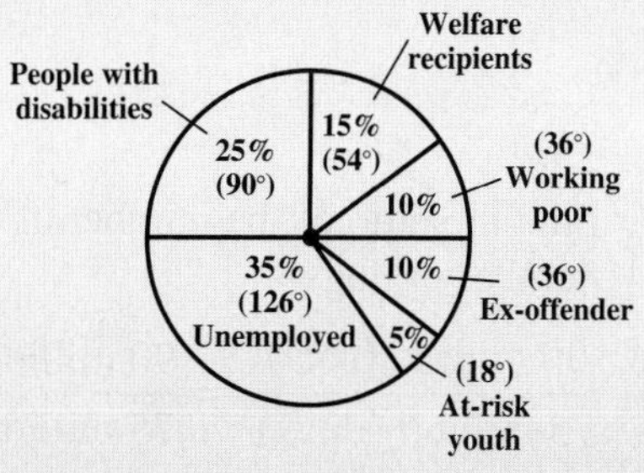

8.1 Exercises

FOR EXTRA HELP

Download the MyDashBoard App

MyMathLab®

CONCEPT CHECK. *Fill in each blank with the correct response.*

1. A circle graph shows how a total amount is divided into __________.

2. When making a circle graph, a tool called a __________ is used to measure degrees in the circle.

This circle graph shows the number of pets owned in the United States. Use the circle graph to answer Exercises 3–8. Write ratios as fractions in lowest terms.
See Examples 1 and 2.

MORE PETS THAN PEOPLE

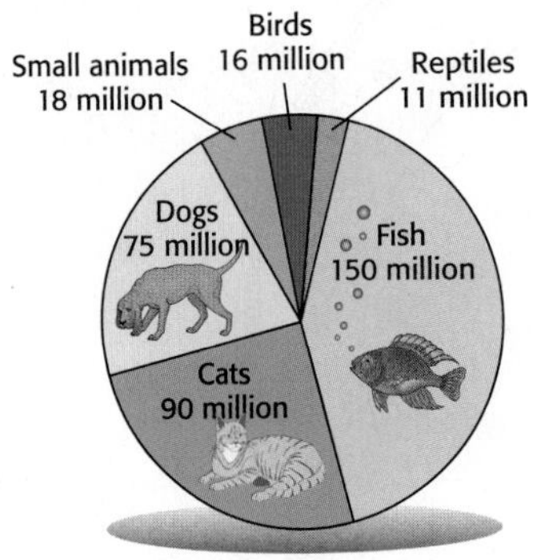

Source: American Pet Product Manufacturers Association.

3. Find the number of pets owned in the United States.

4. Which type of pet is owned by the greatest number of people? How many of these pets are owned?

5. Find the ratio of the number of cats owned to the total number of pets.

6. Find the ratio of the number of small animals owned to the total number of pets.

7. (GS) Find the ratio of the number of cats owned to the number of dogs.

$$\frac{90 \text{ million (cats)}}{75 \text{ million (dogs)}} = \frac{90 \not{\text{million}}}{75 \not{\text{million}}} = \frac{90 \div 15}{75 \div 15} =$$

8. (GS) Find the ratio of the number of fish owned to the number of cats.

$$\frac{150 \text{ million (fish)}}{90 \text{ million (cats)}} = \frac{150 \not{\text{million}}}{90 \not{\text{million}}} = \frac{150 \div 30}{90 \div 30} =$$

This circle graph shows how Americans spend their time in a typical day. Use this circle graph to answer Exercises 9–16. ***See Examples 1 and 2.*** *(Hint: There are 60 minutes in one hour and 24 hours in one day.)*

ALL IN A DAY

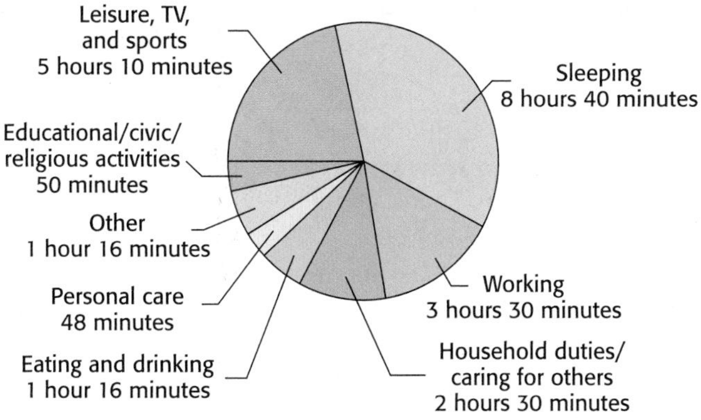

Source: Bureau of Labor Statistics.

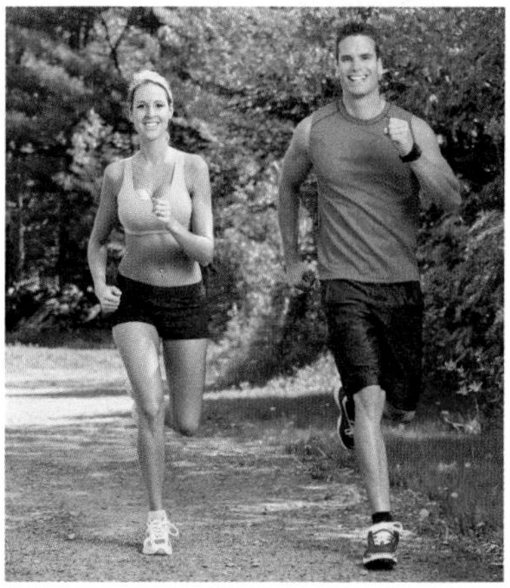

9. Americans spend most of their time doing which activity? How much time do they spend?

10. Americans spend the second most amount of time doing which activity? How much time do they spend?

Answer Exercises 11–16 by writing a ratio as a fraction in lowest terms.

11. Americans spend how much time sleeping compared to the total number of minutes in a day?
(*Hint:* First find the total minutes in a day by multiplying $(24)(60) =$ ______ minutes.)

12. Americans spend how much time on personal care compared to the total number of minutes in a day?

13. Americans spend how much time on educational/ civic/religious activities compared to eating and drinking?

14. Americans spend how much time on personal care compared to sleeping?

15. Americans spend how much time working compared to household duties/caring for others?

16. Americans spend how much time on leisure, TV, and sports compared to working?

This circle graph shows the favorite hot dog toppings in the United States. Each topping is expressed as a percent of the 3200 people in the survey. Use the graph to find the number of people in the survey who favored each of the toppings in Exercises 17–22. ***See Example 3.***

FAVORITE HOT DOG TOPPINGS

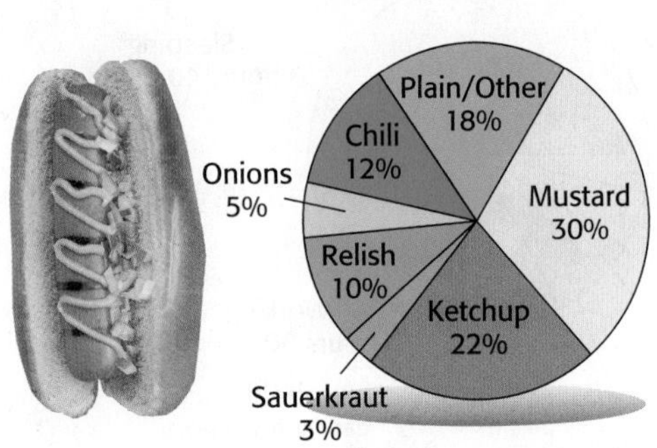

Source: National Hot Dog and Sausage Council.

17. Onions

part = percent • whole

$x = (0.05)(3200)$

$x =$ ______

18. Ketchup

19. Sauerkraut

20. Relish

21. Mustard

22. Chili

J.D. Power and Associates estimates that by 2016 there will be 64,722 electric cars sold in the United States. The primary benefit to the driver is that an electric car will cost considerably less to operate than a traditional gasoline-powered car. The circle graph shows the number of miles an electric car would need to travel between charges before a person would consider buying one. If 3021 people were surveyed for this study, find the number of people giving each response in Exercises 23–28. Round answers to the nearest whole number. ***See Example 3.***

GOING ELECTRIC

How far must it go for you to buy an electric car?

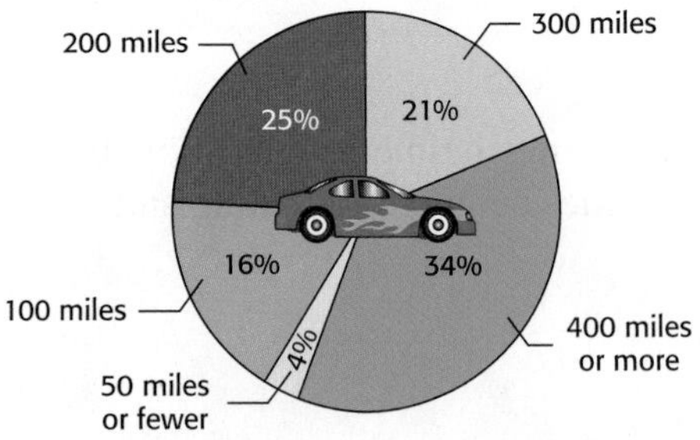

Source: J.D. Power and Associates.

23. Need the car to go 400 miles or more

24. Need the car to go 200 miles

25. Need the car to go 50 miles or fewer

26. Need the car to go 300 miles

27. Need the car to go 100 miles

28. How many people would consider buying an electric car if it will go 300 miles or more between charges?

29. Describe the procedure for determining how large each sector must be to represent each of the items in a circle graph.

30. A protractor is the tool used to draw a circle graph. Give a brief explanation of what the protractor does and how you would use it to measure and draw each sector in the circle graph.

During one month the Orangevale Parks and Recreation District spent $5460 for the activities shown in the following table. Find all numbers missing from the table.

Item	**Dollar Amount**	**Percent of Total**	**Degrees of a Circle**
31. Adult sports	$1365	25%	________
32. Children's sports	$1092	________	72°
33. Day camp	$546	________	________
34. Senior fitness	$546	10%	________
35. Annual egg hunt	$819	15%	________
36. Arts and crafts	$273	________	________
37. Mommy and baby exercise	$819	________	54°

38. Draw a circle graph by using the information from **Exercises 31–37.** Label each sector in your graph. **See Example 4.**

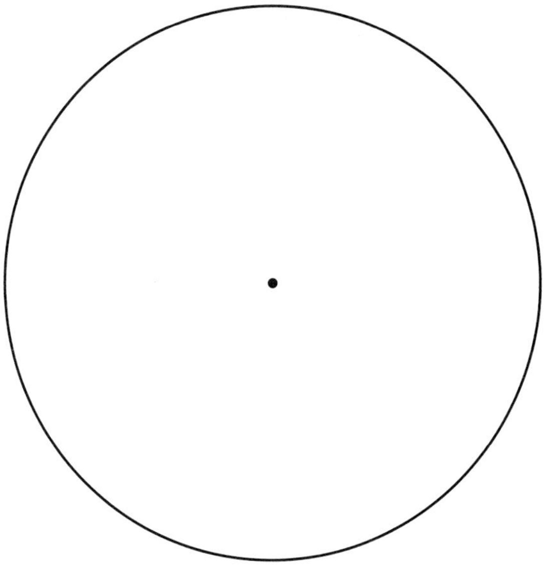

39. White Water Rafting Company divides its annual sales into five categories as follows.

Category	Annual Sales
Adventure classes	$12,500
Grocery and provision sales	$40,000
Equipment rentals	$60,000
Rafting tours	$50,000
Equipment sales	$37,500

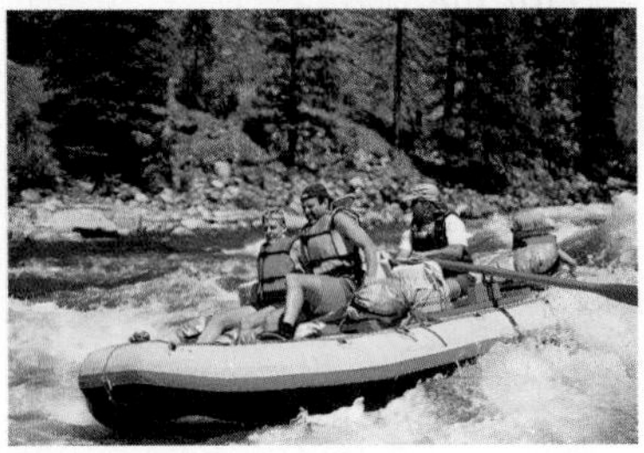

(a) Find the total sales for the year.

(b) Find the number of degrees in a circle graph for each item.

(c) Make a circle graph showing the percent for each category. Label each sector in your graph.

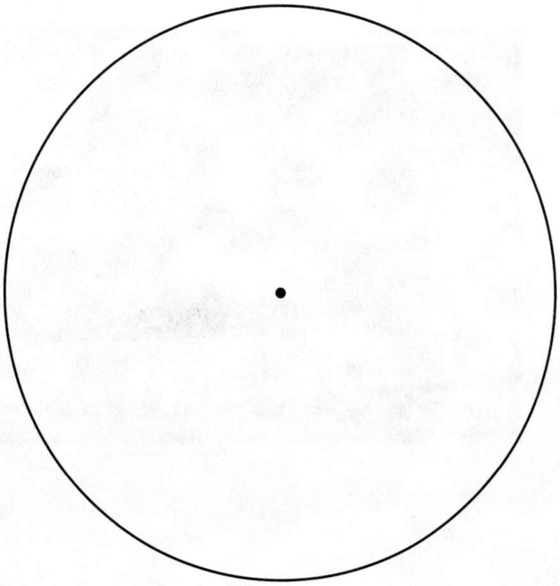

40. Online retail sales in the United States are growing rapidly. The top online retail sales categories are 50% for travel; 20% for apparel, accessories, and footwear; 15% for computer hardware and software; 10% for autos and auto parts; and 5% for home furnishings. (*Source:* Forester Research.)

(a) Find the number of degrees in a circle graph for each online retail sales category.

(b) Draw a circle graph showing the percent for each category. Label each sector in your graph.

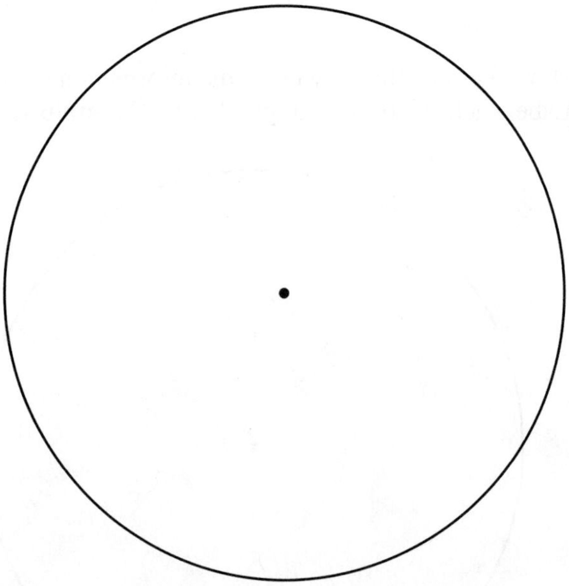

41. Match.com and Chadwick Martin Bailey surveyed adults who met and married within the last three years. Each person was asked how he or she met his or her spouse. The results are shown in the figure on the right.

(a) Use this information to complete the table and draw a circle graph. Round to the nearest whole percent and to the nearest degree. Label each sector in your graph.

FINDING YOUR MATCH

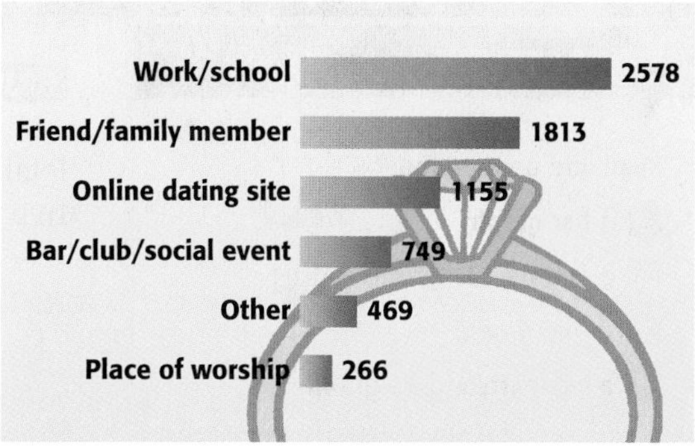

Source: Match.com and Chadwick Martin Bailey.

How They Met	Number of Adults	Percent of Total	Number of degrees
Work/school			
Place of worship			
Online dating site			
Friend/family member			
Bar/club/social event			
Other			

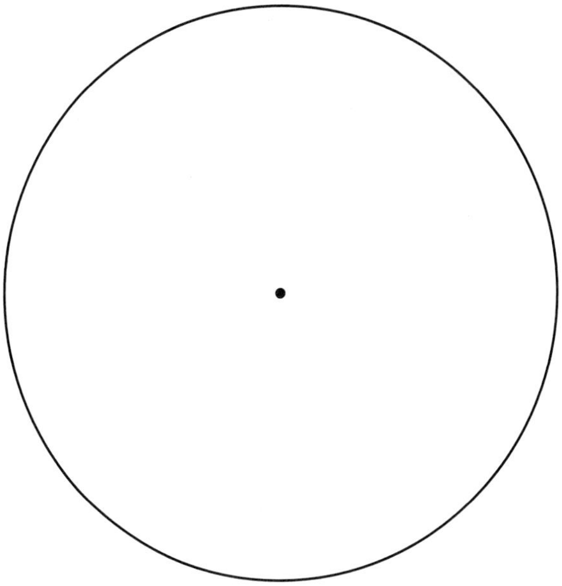

(b) Add up the percents. Is the total 100%? Explain why or why not.

(c) Add up the degrees. Is the total 360°? Explain why or why not.

8.2 Bar Graphs and Line Graphs

OBJECTIVES

Read and understand

1. a bar graph;
2. a double-bar graph;
3. a line graph;
4. a comparison line graph.

VOCABULARY TIP

Bar Graphs are typically used to display data that fit into different categories. The bars are usually separated by spaces.

1 Use the bar graph in the text to find the number of members in the Fitness Centers in each of these years.

GS **(a)** 2010

Bar rises to ____.

Multiply ____ by 1000.

There were ________ members.

(b) 2009

(c) 2013

(d) 2012

Answers

1. (a) 20; 20; 20,000 members
(b) 25,000 members **(c)** 35,000 members
(d) 30,000 members

OBJECTIVE 1 Read and understand a bar graph. **Bar graphs** are useful when showing comparisons. For example, the bar graph below compares the total number of members in all the Fitness Center locations during each of five years.

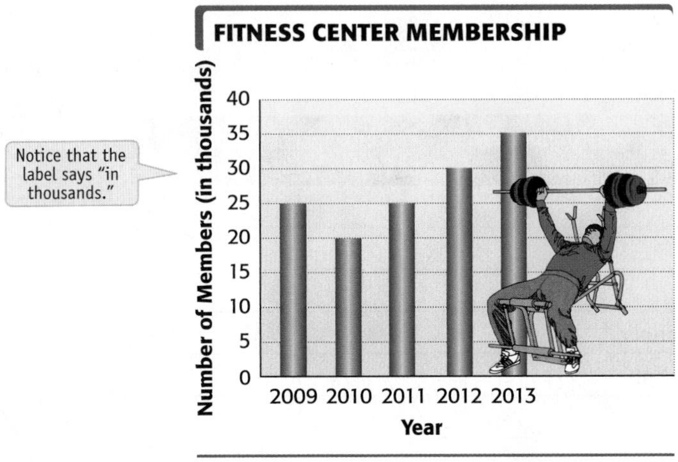

EXAMPLE 1 Using a Bar Graph

How many members did the Fitness Center have in 2011?

The bar for 2011 rises to 25. Notice the label along the left side of the graph that says "Number of Members (in thousands)." The phrase *in thousands* means you have to multiply 25 by 1000 to get 25,000. So, there were 25,000 (**not** 25) members in the Fitness Center locations in 2011.

◄ Work Problem **1** at the Side.

OBJECTIVE 2 Read and understand a double-bar graph. A **double-bar graph** can be used to compare two sets of data. The graph below shows the number of new E-flix subscribers each quarter for two different years.

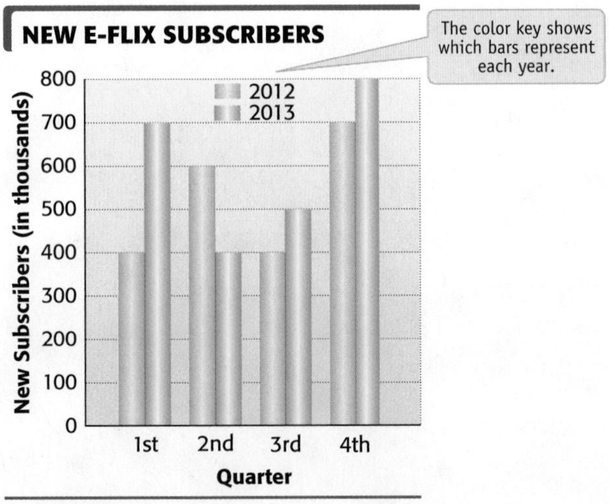

EXAMPLE 2 Reading a Double-Bar Graph

Use the double-bar graph on the previous page to find the following.

(a) The number of new E-flix subscribers in the second quarter of 2012

There are two bars for the second quarter. The color code in the upper center of the graph tells you that the **red bars** represent 2012. So the **red bar** on the *left* is for the 2nd quarter of 2012. It rises to 600. Multiply 600 by 1000 because the label on the left side of the graph says *in thousands.* So there were 600,000 new E-flix subscribers for the second quarter in 2012.

(b) The number of new E-flix subscribers in the second quarter of 2013

The **green bar** for the second quarter rises to 400, and 400 times 1000 is 400,000. So, in the second quarter of 2013, there were 400,000 new E-flix subscribers.

CAUTION

Use a ruler or straightedge to line up the top of the bar with the number on the left side of the graph.

Work Problem **2** at the Side. ▶

OBJECTIVE 3 **Read and understand a line graph.** A **line graph** is often useful for showing a trend. The line graph below shows the number of trout stocked along the Feather River over a 5-month period. Each dot indicates the number of trout stocked during the month directly below that dot.

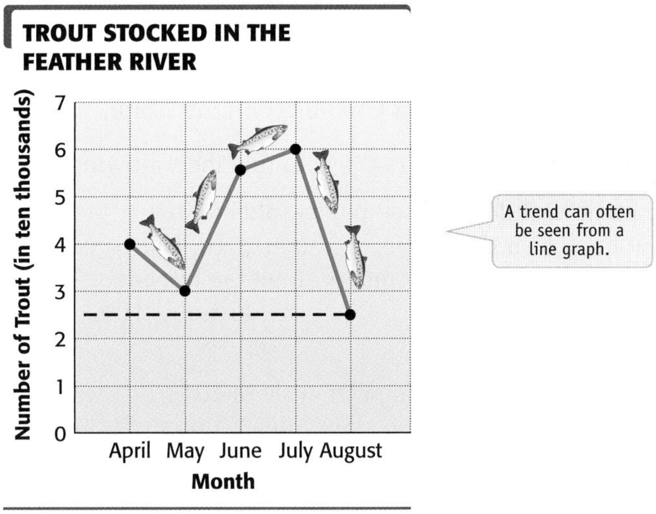

EXAMPLE 3 Understanding a Line Graph

Use the line graph to find the following.

(a) In which month were the least number of trout stocked?

The lowest point on the graph is the dot directly over August, so the least number of trout were stocked in August.

(b) How many trout were stocked in August?

Use a ruler or straightedge to line up the August dot with the numbers along the left edge of the graph. The August dot is halfway between the 2 and the 3. Notice that the label on the left side says *in ten-thousands.* So August is halfway between $(2 \cdot 10{,}000)$ and $(3 \cdot 10{,}000)$. It is halfway between 20,000 and 30,000. That means 25,000 trout were stocked in August.

Work Problem **3** at the Side. ▶

2 Use the double-bar graph to find the number of new E-flix subscribers in 2012 and 2013 for each quarter.

GS **(a)** 1st quarter

2012 is the red bar. Multiply $(400)(1000) =$ ________.

2013 is the green bar. Multiply $(700)(1000) =$ ________.

(b) 3rd quarter

(c) 4th quarter

(d) Find the greatest number of new subscribers. Identify the quarter and the year in which that occurred.

VOCABULARY TIP

Line Graphs can be used to show how something changes over time. These graphs are most often characterized by peaks and valleys to indicate upward and downward trends.

3 Use the line graph in the text to find the number of trout stocked in each month.

(a) June

(b) May

(c) April

(d) July

Answers

2. **(a)** 400,000; 700,000 subscribers
(b) 400,000; 500,000 subscribers
(c) 700,000; 800,000 subscribers
(d) 800,000 subscribers; 4th quarter 2013

3. **(a)** 55,000 trout **(b)** 30,000 trout
(c) 40,000 trout **(d)** 60,000 trout

OBJECTIVE 4 **Read and understand a comparison line graph.** Two sets of data can also be compared by drawing two line graphs together as a **comparison line graph.** For example, the line graph below compares the number of desktop computers and the number of laptop computers sold during each of 5 years by a major electronics retailer.

COMPUTE THIS!

Sales of Laptop Computers vs. Sales of Desktop Computers.

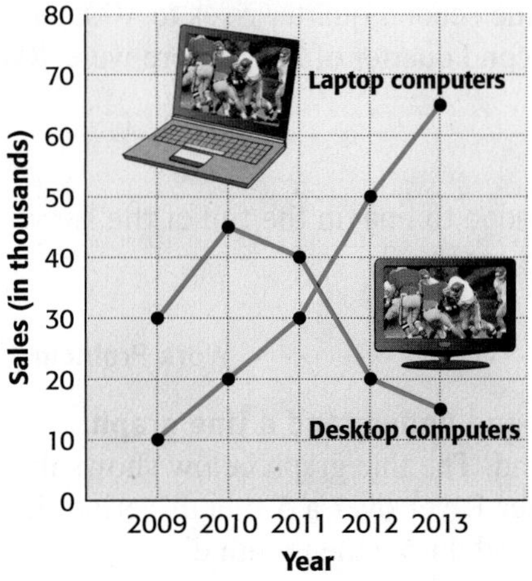

Use a ruler to line up each dot with the numbers along the left edge.

EXAMPLE 4 Interpreting a Comparison Line Graph

Use the comparison line graph above to find the following.

(a) The number of desktop computers sold in 2010

Find the dot on the **blue line** above 2010. Use a ruler or straightedge to line up the dot with the numbers along the left edge. The dot is halfway between 40 and 50, which is 45. Then, 45 times 1000 is 45,000 desktop computers sold in 2010.

(b) The number of laptop computers sold in 2013

The **red line** on the graph shows that 65 times 1000, or 65,000 laptop computers were sold in 2013.

Note

Both the double-bar graph and the comparison line graph are used to compare two sets of data.

◀ **Work Problem 4 at the Side.**

4 Use the comparison line graph in the text to find the following.

GS **(a)** The number of desktop computers sold in 2009, 2011, 2012, and 2013

2009: (30) (1000) = ______

2011: (40) (1000) = ______

2012: (20) (1000) = ______

2013: (15) (1000) = ______

(b) The number of laptop computers sold in 2009, 2010, 2011, and 2012

(c) The first full year in which the number of laptop computers sold was greater than the number of desktop computers sold

Answers

4. **(a)** 30,000; 40,000; 20,000; 15,000 desktop computers
(b) 10,000; 20,000; 30,000; 50,000 laptop computers **(c)** 2012

8.2 Exercises

FOR EXTRA HELP

Download the MyDashBoard App

MyMathLab®

CONCEPT CHECK *Label each statement as* true *or* false.

1. A double-bar graph cannot be used to compare two sets of data.

2. A line graph is often useful for showing a trend.

The American Farm Bureau Federation reports that the average adult in the United States will work 40 days (rounded to the nearest day) to earn enough to pay the annual household food bill. This was found by multiplying the average percent of household income spent on food by 365 (the number of days in a year). This bar graph shows the percent of income spent in various countries of the world. Use this graph to answer Exercises 3–8. ***See Example 1.***

TAKING A BITE OUT OF HOUSEHOLD INCOME

The average American adult will work 40 days each year to earn enough to pay the household food bill. Percent of household income spent on food in:

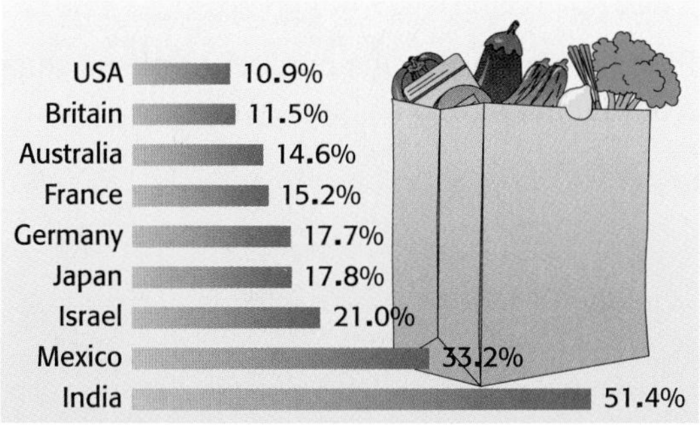

Source: American Farm Bureau Federation.

3. In which country is the highest percent of income spent on food? What percent is this?

4. In which country is the lowest percent of income spent on food? What percent is this?

5. List all countries in the graph in which less than 15% of household income is spent, on average, for food.

6. List all countries in the graph in which more than 20% of household income is spent, on average, for food.

7. How many days each year will the average adult have to work to earn enough to pay for food in Mexico? Round to the nearest day.

   33.2% of 365 days

   $x = (0.332)(365 \text{ days})$

   $x =$ ______ days (rounded)

8. How many days each year will the average adult have to work to earn enough to pay for food in Israel? Round to the nearest day.

   21.0% of 365 days

   $x = (0.21)(365)$

   $x =$ ______ days (rounded)

This double-bar graph shows the number of outdoor plants shipped by Capital Growers during the first six months of 2012 and 2013. Use this graph to answer Exercises 9–14.
See Example 2.

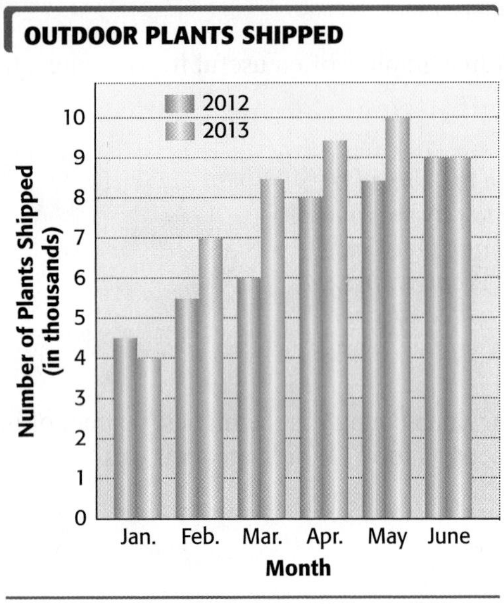

9. In which month in 2013 were the greatest number of plants shipped? What was the total number of plants shipped in that month?

10. How many plants were shipped in January of 2012?

11. How many more plants were shipped in February of 2013 than in February of 2012?

12. How many fewer plants were shipped in March of 2012 than in March of 2013?

13. Find the increase in the number of plants shipped from February 2012 to April 2013.

14. Find the increase in the number of plants shipped from January 2013 to June 2013.

This double-bar graph shows sales of super unleaded and supreme unleaded gasoline at a service station for each of 5 years. Use this graph to answer Exercises 15–20.
See Example 2.

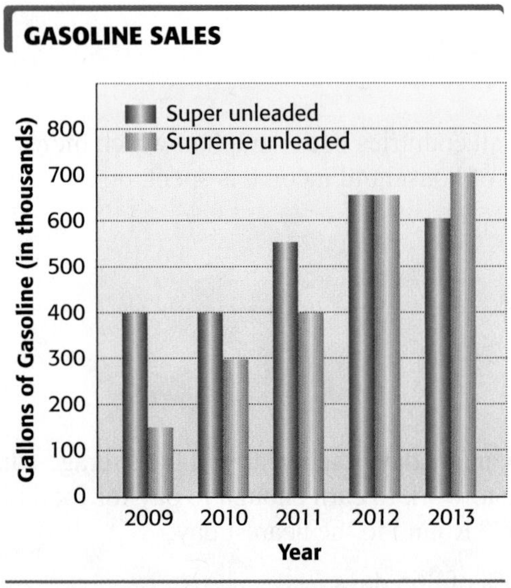

15. How many gallons of supreme unleaded gasoline were sold in 2009?

16. How many gallons of super unleaded gasoline were sold in 2012?

17. In which year did the greatest difference in sales between super unleaded and supreme unleaded gasoline occur? Find the difference.

18. In which year did the sales of supreme unleaded gasoline surpass the sales of super unleaded gasoline?

19. Find the increase in supreme unleaded gasoline sales from 2009 to 2013.

20. Find the increase in super unleaded gasoline sales from 2009 to 2013.

This line graph shows how the personal computer (PC) has evolved over three decades of its existence. What began as a technician's dream is now a common tool of business and home life. Use this line graph to answer Exercises 21–26. ***See Example 3.***

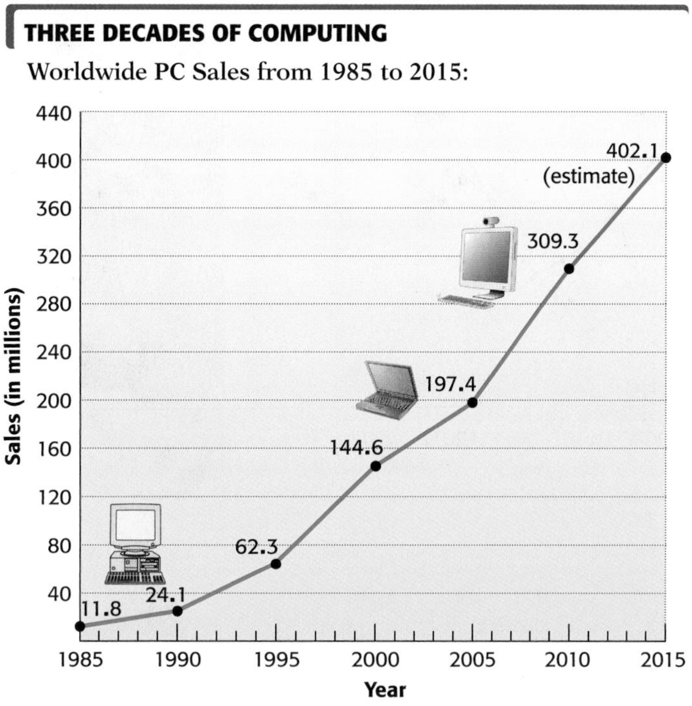

Source: Gartner Dataquest and ARS Technic.

21. Find the number of PCs sold in 1990.

In 1990: (24.1) (1,000,000) = ______________

22. What was the number of PCs sold in 1995?

In 1995: (62.3) (1,000,000) = ______________

23. Find the increase in the estimated number of PCs to be sold in 2015 from the number sold in 2005.

$\underbrace{(402.1)}_{2015}(1{,}000{,}000) - \underbrace{(197.4)}_{2005}(1{,}000{,}000) =$

24. How many more PCs were sold in 2000 than in 1990?

$\underbrace{(144.6)}_{2000}(1{,}000{,}000) - \underbrace{(24.1)}_{1990}(1{,}000{,}000) =$

25. Give two possible explanations for the increase in the number of PCs sold.

26. Give two possible conditions that could result in a decrease in PC sales in the future.

This comparison line graph shows the number of MP3 players sold by two different chain stores during each of 5 years. Use this graph to find the annual number of MP3 players sold each year in Exercises 27–34. **See Example 4.**

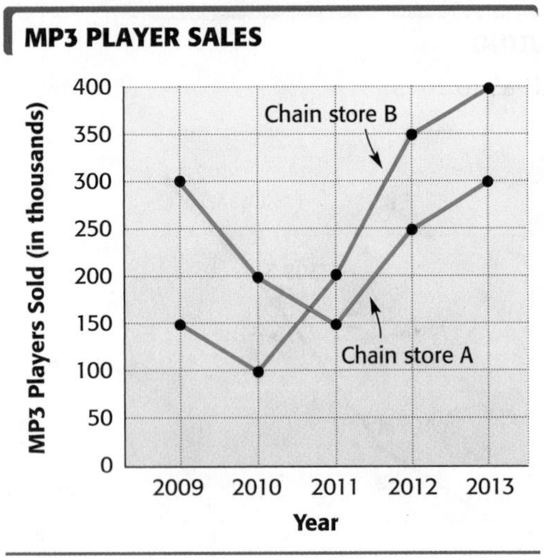

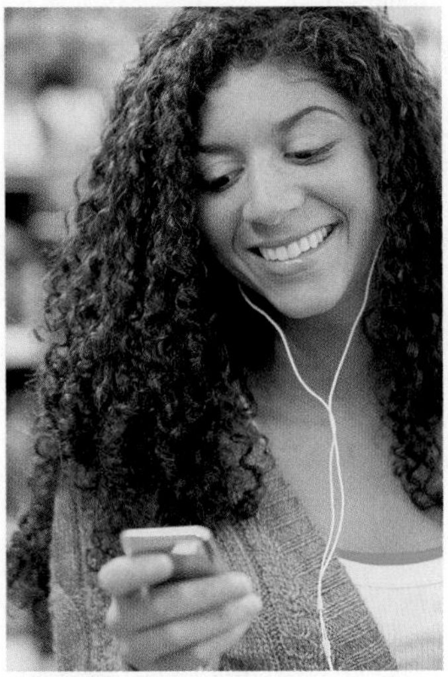

27. Store A in 2013

28. Store A in 2012

29. Store A in 2011

30. Store B in 2013

31. Store B in 2012

32. Store B in 2011

33. Looking at the comparison line graph above, which store would you like to own? Explain why. Based on the graph, what amount of MP3 player sales would you predict for your store in 2014?

34. In the comparison line graph above, Store B used to have lower sales than Store A. What might have happened to cause this change? Give two possible explanations.

35. Explain in your own words why a bar graph or a line graph (not a double-bar graph or comparison line graph) can be used to show only one set of data.

36. The double-bar graph and the comparison line graph are both useful for comparing two sets of data. Explain how this works and give your own example.

This comparison line graph shows the sales and profits of Refresh Salon and Day Spa for each of 4 years. Use the graph to answer Exercises 37–44. **See Example 4.**

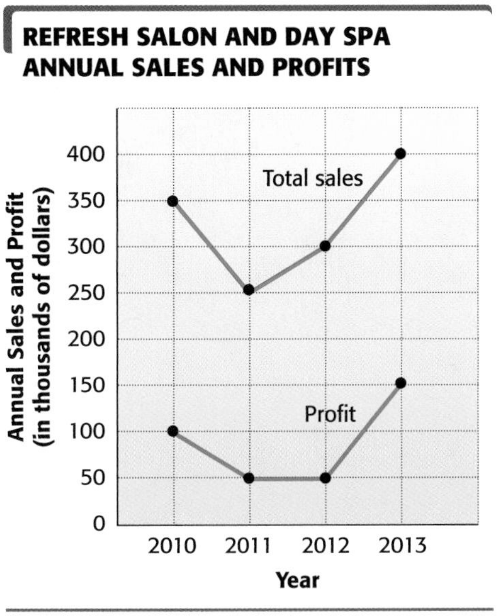

37. Total sales in 2013

38. Total sales in 2012

39. Total sales in 2011

40. Profit in 2013

41. Profit in 2012

42. Profit in 2011

43. Give two possible explanations for the decrease in sales from 2010 to 2011 and two possible explanations for the increase in sales from 2011 to 2013.

44. Based on the graph, what conclusion can you make about the relationship between sales and profits?

Relating Concepts (Exercises 45–50) For Individual or Group Work

The following information includes statistics regarding Life Savers. Use this information to ***work Exercises 45–50 in order.***

ROLL WITH IT

Sweet-toothed fans recently voted to change three of the original flavors. The new five-flavor roll includes cherry, watermelon, pineapple, raspberry, and blackberry. The orange, lemon, and lime flavors have been replaced.

Life Savers by the numbers

Year first flavor invented: **1912 (Pep-O-Mint)**
Year five-flavor roll invented: **1935**
Total number of flavors today: **25**
Number of candies per roll: **14**
Life Saver candies produced daily: **3 million**
Pounds of sugar used per day: **250,000**
Number of miniature rolls given out at Halloween: **88 million**

Source: *USA Today* and Wrigley Company.

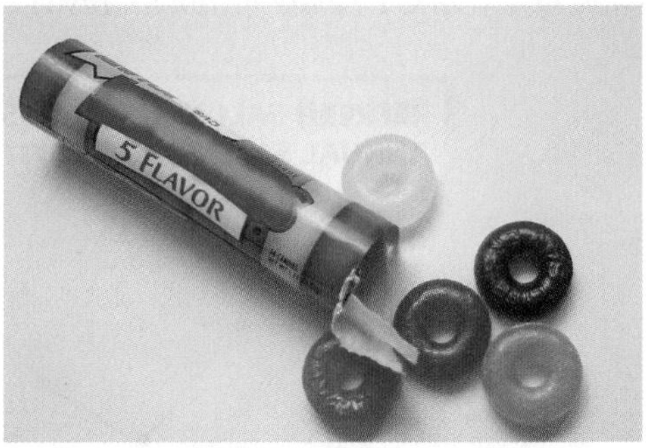

45. The first Life Savers flavor was Pep-O-Mint. How long after the Pep-O-Mint Life Saver was invented was the five-flavor roll invented?

46. In addition to the five flavors, how many more flavors of Life Savers are there?

47. Find the number of rolls of Life Savers produced daily. Round to the nearest whole number.

48. Use your answer from **Exercise 47** to find the amount of sugar in one roll of Life Savers. Round to the nearest hundredth of a pound.

49. Check your work in **Exercise 48.** Is the answer reasonable? Explain why or why not.

50. Name three possible causes of errors in statistics.

8.3 Frequency Distributions and Histograms

OBJECTIVES

1. Understand a frequency distribution.
2. Arrange data in class intervals.
3. Read and understand a histogram.

The owner of Towne Insurance Agency has kept track of her personal phone sales call activity over the past 50 weeks. The number of sales calls made for each of the weeks is given below. Read down the columns, beginning with the left column, for successive weeks of the year.

75	65	40	50	45	30	30	35	45	25
75	70	60	55	30	25	44	30	35	30
75	70	50	30	50	20	30	30	20	25
60	62	45	45	48	40	35	25	20	25
75	45	50	40	35	40	40	30	27	40

OBJECTIVE 1 Understand a frequency distribution. A long list of numbers can be confusing. You can make the data easier to read by putting them in a special type of table called a **frequency distribution.**

VOCABULARY TIP

Frequency distribution is a way of organizing data by listing a set of values and their frequency (how many times each one occurs).

EXAMPLE 1 Preparing a Frequency Distribution

Using the data above, construct a table that shows each possible number of sales calls. Then go through the original data and place a *tally* mark (|) in the tally column next to each corresponding value. Total the tally marks and place the totals in the third column. The result is a frequency distribution table.

Number of Sales Calls	Tally	Frequency	Number of Sales Calls	Tally	Frequency
20	III	3	48	I	1
25	~~IIII~~	5	50	IIII	4
27	I	1	55	I	1
30	~~IIII~~ IIII	9	60	II	2
35	IIII	4	62	I	1
40	~~IIII~~ I	6	65	I	1
44	I	1	70	II	2
45	~~IIII~~	5	75	IIII	4

Work Problem 1 at the Side. ▶

1 Use the frequency distribution table in the text to find the following.

(a) The least number of sales calls made in a week

(b) The most common number of sales calls made in a week

(c) The number of weeks in which 35 calls were made

(d) The number of weeks in which 45 calls were made

OBJECTIVE 2 Arrange data in class intervals. The frequency distribution given in **Example 1** above contains a great deal of information—perhaps too much to digest. It can be simplified by combining the number of sales calls into groups, forming the class intervals shown in the left column below.

GROUPED DATA

Class Intervals (Number of Sales Calls)	Class Frequency (Number of Weeks)
20–29	9
30–39	13
40–49	13
50–59	5
60–69	4
70–79	6

In the table above, look at how many weeks had 20 to 29 sales calls: $3 + 5 + 1 = 9$ weeks.

Answers

1. (a) 20 calls **(b)** 30 calls **(c)** 4 weeks **(d)** 5 weeks

Note

The number of class intervals in the left column of the grouped data table varies. Grouped data usually have between 5 and 15 class intervals.

EXAMPLE 2 Analyzing a Frequency Distribution

Use the grouped data for the insurance agency (on the preceding page) to answer the following questions.

(a) During how many weeks were fewer than 30 calls made?

The first interval in the grouped data table (20–29) is the number of weeks during which fewer than 30 calls were made. Therefore, the owner made fewer than 30 calls during 9 weeks out of the 50 weeks shown.

(b) During how many weeks were 40 or more calls made?

The last four intervals in the grouped data table are the number of weeks during which 40 or more calls were made.

$$13 + 5 + 4 + 6 = 28 \text{ weeks}$$

◀ **Work Problem 2 at the Side.**

OBJECTIVE 3 Read and understand a histogram. The results in the grouped data table have been used to draw the special bar graph below, which is called a **histogram.** In a histogram, the width of each bar represents a range of numbers (*class interval*). The height of each bar in a histogram gives the *class frequency,* that is, the number of occurrences in each class interval.

SALES CALL DATA FOR THE PAST 50 WEEKS
(GROUPED DATA)

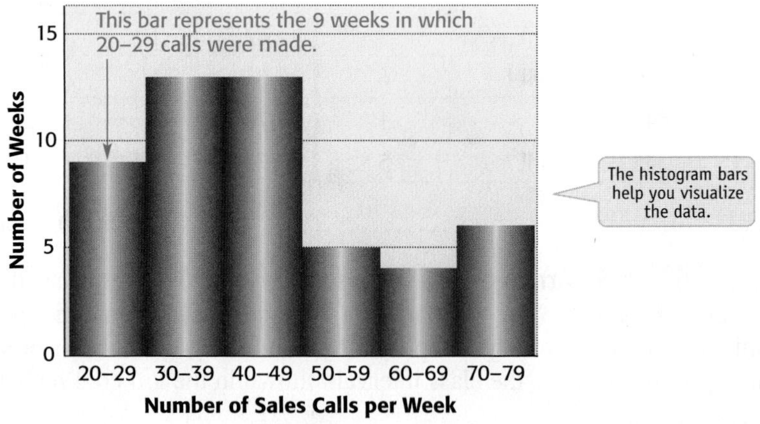

EXAMPLE 3 Using a Histogram

Use the histogram to find the number of weeks in which fewer than 40 calls were made.

The histogram shows that 20–29 calls were made during 9 of the weeks, and 30–39 calls were made during 13 of the weeks. So, the number of weeks in which fewer than 40 calls were made is 9 + 13 = 22 weeks.

◀ **Work Problem 3 at the Side.**

2 Use the grouped data for the insurance agency on the previous page to answer each question.

GS **(a)** During how many weeks were fewer than 50 calls made?

Number of Calls	Number of Weeks
20–29	9
30–39	13
40–49	13

So, 9 + 13 + ____ = ____ weeks

(b) During how many weeks were 50 or more calls made?

VOCABULARY TIP

Histograms are special bar graphs used to illustrate a frequency distribution. The bars are not separated as they generally are in bar graphs.

3 Use the histogram at the right to answer each question.

GS **(a)** During how many weeks were 60 or more calls made?

Number of Calls	Number of Weeks
60–69	4
70–79	6

4 + ____ = ____ weeks

(b) During how many weeks were fewer than 60 calls made?

Answers

2. **(a)** 13; 35 weeks **(b)** 15 weeks
3. **(a)** 6; 10 weeks **(b)** 40 weeks

8.3 Exercises

FOR EXTRA HELP

Download the MyDashBoard App

MyMathLab®

CONCEPT CHECK *Fill in each blank with the correct response.*

1. A large number of data can be easier to understand if they are in a special type of __________ called a frequency distribution.

2. A histogram is a special type of __________.

Launched in February 2004, Facebook is the most used social networking site in the world. Its mass appeal to people of all demographics has made Facebook an integral part of our lives. The ages of Facebook's users are shown below in the histogram. Use the histogram to answer Exercises 3–8. ***See Example 3.***

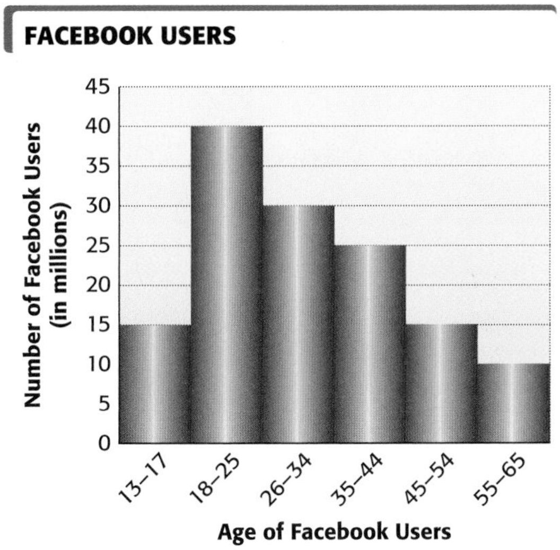

3. The greatest number of users are in which age group? How many users are in that group?

4. The least number of users are in which age group? How many users are in that group?

5. Find the number of users 35 years of age and over.

6. Find the number of users ages 13 to 17.

7. How many users are 35 to 44 years of age?

8. How many users are 26 to 34 years of age?

This histogram shows the annual earnings for the part-time employees of Rally World Amusement Park. Use this histogram to answer Exercises 9–14. ***See Example 3.***

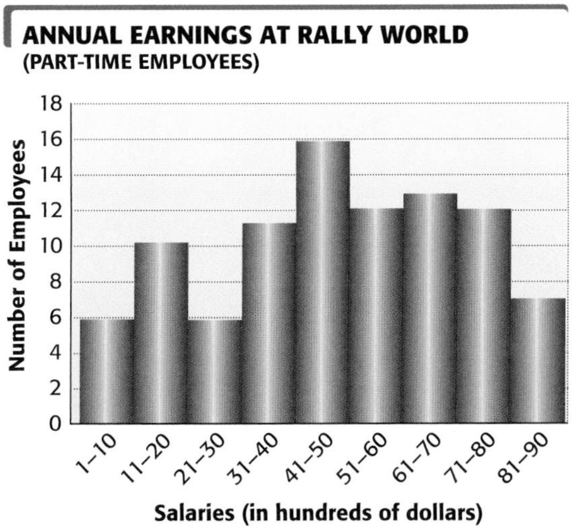

9. GS The greatest number of employees are in which earnings group? How many are in that group?

The tallest bar for employees is in the $______ to $______ group. ______ employees are in that group.

10. GS The fewest number of employees are in which earnings groups? How many are in each group?

The shortest bars for employees are in the $ ______ to $ ______ group and the $______ to $______ groups. ______ employees are in each of these groups.

11. Find the number of employees who earn $3100 to $4000.

12. Find the number of employees who earn $1100 to $2000.

13. How many employees earn $5000 or less?

GS Earning $5000 or less: 6 + 10 + 6 + 11 + 16 = ______ employees.

14. How many employees earn $6100 or more?

GS Earning $6100 or more: 13 + ______ + 7 = ______ employees.

15. Describe class interval and class frequency. How are they used when preparing a histogram?

16. What might be a problem of using two few or too many class intervals?

This list shows the number of new accounts opened annually by the employees of the Schools Credit Union. Use it to complete the table. ***See Example 1.***

186	191	144	198	147	158	174
193	142	155	174	162	151	178
145	151	199	182	147	195	146

	Class Intervals (Number of New Accounts)	Tally	Class Frequency (Number of Employees)
17.	140–149	______	______
18.	150–159	______	______
19.	160–169	______	______
20.	170–179	______	______
21.	180–189	______	______
22.	190–199	______	______

A college professor asked her 30 students how many hours they worked each week. Use her list of student responses to complete the following table. ***See Examples 1–3.***

14	8	12	28	33	14
6	34	17	20	13	20
25	33	32	4	7	14
0	6	10	8	35	31
25	4	32	18	0	24

	Class Intervals (Number of Hours Worked)	Tally	Class Frequency (Number of Students)
23.	0–5	_____	_____
24.	6–10	_____	_____
25.	11–15	_____	_____
26.	16–20	_____	_____
27.	21–25	_____	_____
28.	26–30	_____	_____
29.	31–35	_____	_____

30. Construct a histogram by using the data in **Exercises 23–29.**

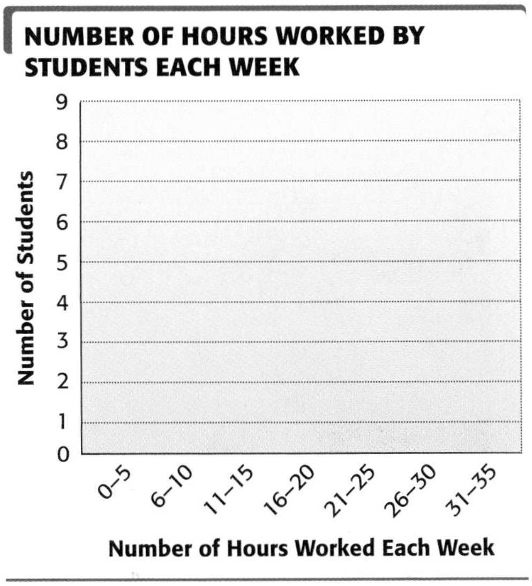

The popular Los Angeles Lakers were a leading NBA (National Basketball Association) team in the 2011 season. Listed below are the number of points scored by the Lakers during their pre-season and regular season games. Use these numbers to complete the following table. ***See Example 1.***

92	120	112	118	92	84	79	108	100	95	92	106	101	139	87
88	105	108	103	96	99	80	99	92	96	113	108	88	112	85
98	112	121	112	99	103	82	101	101	114	75	90	96	102	86
102	114	99	117	113	109	103	109	100	88	89	90	91	110	106
94	107	112	98	115	93	102	112	107	101	99	92	106	96	102
74	124	116	96	87	120	85	115	120	93	104	99	84	90	116

	Class Intervals (Points Scored)	Tally	Class Frequency (Number of Games)
31.	70–79	________	________
32.	80–89	________	________
33.	90–99	________	________
34.	100–109	________	________
35.	110–119	________	________
36.	120–129	________	________
37.	130–139	________	________

38. Make a histogram showing the results from **Exercises 31–37.**

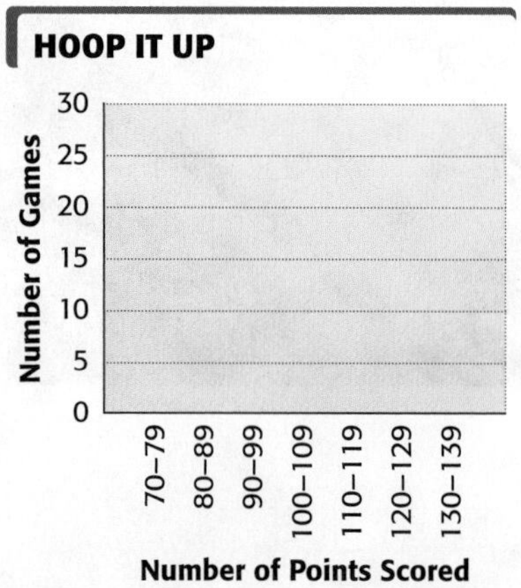

Source: NBA.com

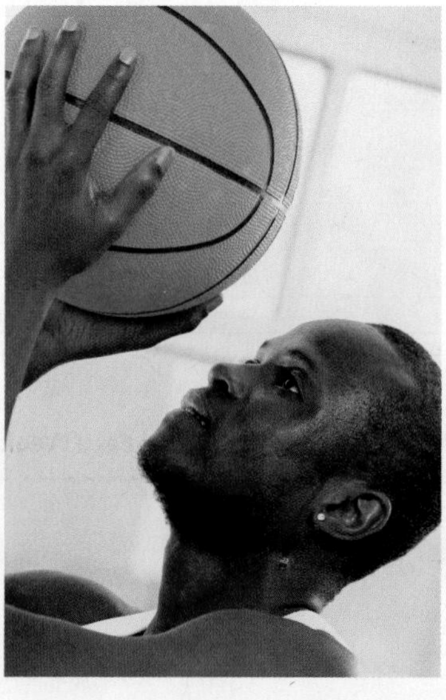

8.4 Mean, Median, and Mode

OBJECTIVES

1. Find the mean of a list of numbers.
2. Find a weighted mean.
3. Find the median.
4. Find the mode.

Businesses, governments, laboratories, colleges, and others working with lists of numbers are often faced with the problem of analyzing a great volume of raw data. The measures discussed in this section are helpful for such analyses.

OBJECTIVE 1 **Find the mean of a list of numbers.** When analyzing data, one of the first things to look for is a *measure of central tendency*—a single number that we can use to represent the entire list of numbers. One such measure is the *average*, or **mean.** The mean can be found with the following formula.

Finding the Mean (Average)

$$\text{mean} = \frac{\text{sum of all values}}{\text{number of values}}$$

EXAMPLE 1 Finding the Mean

David had test scores of 84, 90, 95, 98, and 88. Find the mean (average) of his scores.

Use the formula for finding the mean. Add up all the test scores and then divide by the number of tests.

$$\text{mean} = \frac{84 + 90 + 95 + 98 + 88}{5}$$ (Add up all the test scores.) (Divide by the number of tests.)

$$\text{mean} = \frac{455}{5}$$ Divide.

$$\text{mean} = 91$$

David has a mean (average) score of 91.

Work Problem 1 at the Side. ▶

EXAMPLE 2 Applying the Average or Mean

Milk sales at a local 7-Eleven for each of the days last week were

\$86, \$103, \$118, \$117, \$126, \$158, and \$149

Find the mean milk sales (rounded to the nearest cent) as shown below.

$$\text{mean} = \frac{\$86 + \$103 + \$118 + \$117 + \$126 + \$158 + \$149}{7}$$

$$\text{mean} = \frac{\$857}{7}$$

$$\text{mean} \approx \$122.43 \text{ (rounded)}$$

The mean daily sales amount for milk was \$122.43.

Work Problem 2 at the Side. ▶

VOCABULARY TIP

Mean The **mean** is the mathematical average.

1 Andrew has math test scores of 96, 98, 84, 88, 82, and 92. Find his mean (average) score.

$$\frac{96 + 98 + 84 + 88 + \_\_ + \_\_}{6}$$

$$= \frac{\_\_}{6} = \_\_$$

2 Find the mean for each list of numbers. Round to the nearest cent if necessary.

(a) Monthly Starbucks purchases:

\$50.28, \$85.16, \$110.50, \$78, \$120.70, \$58.64, \$73.80, \$86.24, \$67.85, \$96.56, \$138.65, \$48.90

(b) The attendance at eight major league home games this season:

48,076; 58,595; 37,874; 46,289; 29,235; 59,311; 45,675; 39,721

Answers

1. 82; 92; 540; 90

2. (a) $\frac{\$1015.28}{12} \approx \$84.61$ (rounded)

(b) $\frac{364{,}776}{8} = 45{,}597$ people

3 The numbers below show the amount that high school student Matthew Ryan spent for lunch and snacks and the number of days he spent that amount. Find the weighted mean.

Value	Frequency
$ 2	4
$ 4	6
$ 6	5
$ 8	6
$10	12
$12	5
$14	8
$16	4

OBJECTIVE 2 Find a weighted mean. Some items in a list might appear more than once. In this case, we find a **weighted mean,** in which each value is "weighted" by multiplying it by the number of times it occurs.

EXAMPLE 3 Understanding the Weighted Mean

The following table shows the family size and the number of families (frequency) who were given groceries in one morning at the Twin Lakes Food Bank. Find the weighted mean for family size.

Family Size	Frequency	
1	7	← 7 families had 1 person.
2	12	
3	8	
4	8	
5	5	
6	9	← 9 families had 6 people.
7	1	
8	6	

The same number of people were in more than one family: for example, there were 2 people in 12 of the families and there were 4 people in 8 of the families. There were 7 people in just 1 of the families. To find the mean, multiply the family size by its frequency. Then add the products. Next, add the numbers in the frequency column to find the total number of families.

Size	Frequency	Product
1	7	$(1 \cdot 7) = 7$
2	12	$(2 \cdot 12) = 24$
3	8	$(3 \cdot 8) = 24$
4	8	$(4 \cdot 8) = 32$
5	5	$(5 \cdot 5) = 25$
6	9	$(6 \cdot 9) = 54$
7	1	$(7 \cdot 1) = 7$
8	6	$(8 \cdot 6) = 48$
Totals	56	221

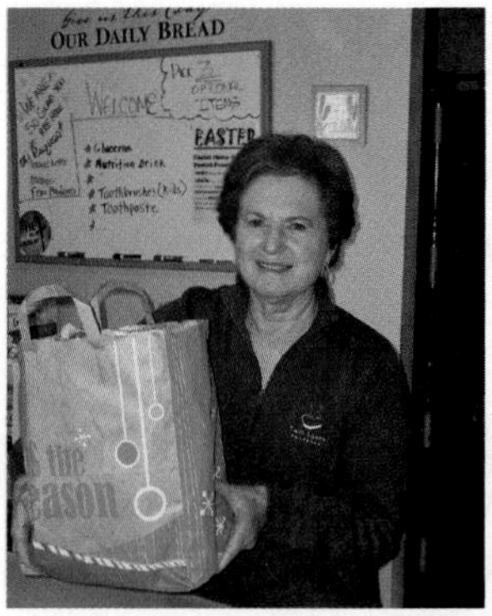

Finally, divide the totals. Round to the nearest hundredth.

$$\text{mean} = \frac{221}{56} \approx 3.95 \text{ (rounded)}$$

The mean family size of those using the food bank was 3.95 people.

◄ **Work Problem 3 at the Side.**

VOCABULARY TIP

Weighted mean is used when calculating grade point average. See **Example 4.**

Answer

3. $\text{mean} = \frac{\$466}{50} = \$9.32$

A common use of the weighted mean is to find a student's *grade point average* (GPA), as shown by the next example.

EXAMPLE 4 Applying the Weighted Mean

Find the grade point average for a student earning the following grades. Assume A = 4, B = 3, C = 2, D = 1, and F = 0. The number of credits determines how many times the grade is counted (the frequency).

Course	Credits	Grade	Credits • Grade
Mathematics	4	A (= 4)	4 • 4 = 16
Speech	3	C (= 2)	3 • 2 = 6
English	3	B (= 3)	3 • 3 = 9
Computer science	2	A (= 4)	2 • 4 = 8
Art history	2	D (= 1)	2 • 1 = 2
Totals	14		41

It is common to round grade point averages to the nearest hundredth. So the grade point average for this student is rounded to 2.93.

$$\text{GPA} = \frac{41}{14} \approx 2.93$$

Work Problem **4** at the Side. ▶

OBJECTIVE 3 Find the median. Because it can be affected by extremely high or low numbers, the mean is often a poor indicator of central tendency for a list of numbers. In cases like this, another measure of central tendency, called the *median,* can be used. The **median** divides a group of numbers in half; half the numbers lie above the median, and half lie below the median.

Find the median by listing the numbers *in order* from *least* to *greatest.* If the list contains an *odd* number of items, the median is the *middle number.*

EXAMPLE 5 Finding the Median for an Odd Number of Items

Find the median for the following list of prices for women's T-shirts.

$9, $23, $15, $8, $18, $12, $24

First arrange the numbers in numerical order from least to greatest.

Least ⟶ 8, 9, 12, 15, 18, 23, 24 ⟵ Greatest

Next, find the middle number in the list.

8, 9, 12, 15, 18, 23, 24

Three are below. Middle number Three are above.

Remember to list the numbers from least to greatest **before** finding the median.

The median price is $15.

Work Problem **5** at the Side. ▶

4 Find the grade point average (GPA) for Laticia Perez, who earned the following grades last semester. Round to the nearest hundredth.

Course	Credits	Grade
Mathematics	3	A (= 4)
P.E.	1	C (= 2)
English	3	C (= 2)
Keyboarding	2	B (= 3)
Biology	4	B (= 3)

VOCABULARY TIP

Median The word **median** actually means "middle." It may help to remember that the grassy area in the middle of two roadways is called a highway median, just as the middle number in a set of data points is called the numerical median.

5 GS Find the median for the following weights of Halloween pumpkins.

14 lb, 18 lb, 10 lb, 17 lb, 15 lb, 19 lb, 20 lb

First list the weights from least to greatest.

10, __, __, __, __, __, 20
↑ least ↑ greatest

Which weight is in the middle of the list? ______

Answers

4. GPA ≈ 2.92
5. 10, 14, 15, 17, 18, 19, 20; 17 lb (the middle number when the numbers are arranged from least to greatest.)

6 Find the median for the following list of measurements.

125 m, 87 m, 96 m, 108 m, 136 m, 74 m

VOCABULARY TIP

Mode is the most common or that which appears most often.

7 Find the mode for each list of numbers.

GS **(a)** Number of laps swum:

20, 30, 26, 24, 26

↑ ↑

Which number occurs most often? ____

(b) Wait time for Disneyland rides (in minutes):

38, 45, 52, 60, 38, 45

(c) Monthly commissions of salespeople:

\$1706, \$1289, \$1653, \$1892, \$1301, \$1782

Answers

6. $\frac{96 + 108}{2} = \frac{204}{2} = 102\text{ m}$

7. (a) 26 laps
(b) bimodal, 38 minutes and 45 minutes (this list has two modes)
(c) no mode (no number occurs more than once)

If a list contains an *even* number of items, there is no single middle number. In this case, the median is defined as the mean (average) of the ***middle two*** numbers.

EXAMPLE 6 **Finding the Median for an Even Number of Items**

Find the median for the following list of ages.

74, 7, 15, 13, 25, 28, 47, 59, 32, 68

First arrange the numbers in numerical order from least to greatest. Then find the middle two numbers.

Least ⟶ 7, 13, 15, 25, 28, 32, 47, 59, 68, 74 ⟵ Greatest

Middle two numbers (28, 32)

The median age is the mean of the two middle numbers.

$$\text{median} = \frac{28 + 32}{2} = \frac{60}{2} = \textbf{30 years}$$

◀ Work Problem **6** at the Side.

OBJECTIVE 4 **Find the mode.** Another important statistical measure is the **mode,** the number that occurs ***most often*** in a list of numbers. For example, if the test scores for 10 students were

74, 81, 39, 74, 82, 80, 100, 92, 74, and 85,

To find the mode, you do **not** have to list the numbers from least to greatest.

then the mode is 74. Three students earned a score of 74, so 74 appears more times on the list than any other score. (It is not necessary to place the numbers in numerical order when looking for the mode.)

A list can have two modes; such a list is sometimes called **bimodal.** If no number occurs more frequently than any other number, the list has *no mode.*

EXAMPLE 7 **Finding the Mode**

Find the mode for each list of numbers.

(a) 51, 32, 49, 73, 49, 90

The **mode** is the value that occurs most often.

The number 49 occurs more often than any other number; therefore, 49 is the mode.

(b) 482, 485, 483, 485, 487, 487, 489

Because both 485 and 487 occur twice, each is a mode. This list is *bimodal.*

(c) \$10,708; \$11,519; \$10,972; \$12,546; \$13,905; \$12,182

No number occurs more than once. This list has *no mode.*

◀ Work Problem **7** at the Side.

Measures of Central Tendency

The **mean** is the sum of all the values divided by the number of values. It is the mathematical average.

The **median** is the middle number (or the average of the middle two numbers) in a group of values that are listed from least to greatest. It divides a group of numbers in half.

The **mode** is the value that occurs most often in a group of values.

8.4 Exercises

FOR EXTRA HELP

Download the MyDashBoard App

MyMathLab®

CONCEPT CHECK *Label each statement as* true *or* false.

1. Another word for *mean* is *average*.

2. The mode divides a group of numbers in half; half of the numbers lie above the mode, and half below the mode.

CONCEPT CHECK *Fill in each blank with the correct response.*

3. A common use of the weighted mean is to find a student's ______________.

4. The mode is the number that occurs ______________ in a list of numbers.

Find the mean for each list of numbers. Round answers to the nearest tenth if necessary. ***See Example 1.***

5. Hours spent playing video games each week:

9, 14, 18, 11, 12, 20

6. Minutes of cell phone use each day:

53, 77, 38, 29, 46, 48, 52

7. Inches of rain per month of 3.1, 1.5, 2.8, 0.8, 4.1

GS

$$\frac{3.1 + 1.5 + 2.8 + 0.8 + 4.1}{\_\_\_\_} =$$

8. Algebra quiz scores of 32, 26, 30, 19, 51, 46, 38, 39

GS

$$\frac{32 + 26 + 30 + 19 + 51 + 46 + 38 + 39}{\_\_\_\_} =$$

9. Annual salaries of \$38,500; \$39,720; \$42,183; \$21,982; \$43,250

10. Numbers of students enrolled in community colleges: 27,500; 18,250; 17,357; 14,298; 33,110

Solve each application problem. ***See Example 2.***

11. The Sunrise Pharmacy filled prescriptions that cost the following amounts: \$18.38, \$168.75, \$28.63, \$72.85, \$39.60, \$183.74, \$15.82, \$33.18, \$87.45, \$98.72, and \$50.70. Find the average cost (mean) of the prescriptions sold.

12. In one evening, a waitress collected the following checks from her dinner customers: \$30.10, \$42.80, \$91.60, \$51.20, \$88.30, \$21.90, \$43.70, \$51.20. Find the average (mean) dinner check amount.

Find the weighted mean. Round answers to the nearest tenth. ***See Example 3.***

13.

GS

Customers Each Hour	Frequency	
8	2	$(8 \cdot 2) =$
11	12	$(11 \cdot 12) =$
15	5	$(15 \cdot 5) =$
26	1	$(26 \cdot 1) =$ ____
		Total →

$$\frac{\text{sum of products}}{\text{total customers}} =$$

14.

GS

Deliveries Each Week	Frequency	
4	1	$(4 \cdot 1) =$
8	3	$(8 \cdot 3) =$
16	5	$(16 \cdot 5) =$
20	1	$(20 \cdot 1) =$ ____
		Total →

$$\frac{\text{sum of products}}{\text{total deliveries}} =$$

15.

Fish per Boat	Frequency
12	4
13	2
15	5
19	3
22	1
23	5

16.

Patients per Clinic	Frequency
25	1
26	2
29	5
30	4
32	3
33	5

Solve each application problem. ***See Example 3.***

17. On a popular television game show, contestants won various amounts of money. The table below shows how many contestants won each amount of money over a two-month period. Find the weighted mean for the amount of money won.

Game Show Winnings	Number of Contestants
\$ 5000	5
\$ 10,000	10
\$ 25,000	7
\$ 50,000	5
\$ 100,000	12
\$ 250,000	5
\$ 500,000	5
\$1,000,000	1

18. A national health survey provided the information for this table. It shows how often each of the adults in the survey engages in a vigorous, leisure time activity. Find the weighted mean to determine the weekly hours of vigorous activity for the adults surveyed, to the nearest tenth.

Hours per week	Number of Adults
0	224
1	86
2	62
3	45
4	25
5	18
6	24
7	16

Find the median for each list of numbers. ***See Examples 5 and 6.***

19. Number of wedding guests: 125, 100, 150, 135, 114

20. Number of Google searches (in thousands):
140, 85, 122, 114, 98

21. Calories in fast-food menu items:
501, 412, 521, 515, 298, 621, 346, 528

298, 346, 412, $\underbrace{501, 515,}_{\text{middle two numbers}}$ 521, 528, 621

$$\frac{501 + 515}{2} = \_\_\_\_ \text{ calories}$$

22. Number of cars in the parking lot each day:
520, 523, 513, 1283, 338, 509, 290, 420, 320, 980

290, 320, 338, 420, $\underbrace{509, 513,}_{\text{middle two numbers}}$ 520, 523, 980, 1283

$$\frac{509 + 513}{2} = \_\_\_\_ \text{ cars}$$

Find the mode or modes for each list of numbers. ***See Example 7.***

23. Porosity of soil samples:
21%, 18%, 21%, 28%, 22%, 21%, 25%

24. Daily low temperatures (in degrees Fahrenheit):
21, 32, 46, 32, 49, 32, 49

25. Ages of residents (in years) at Leisure Village:
74, 68, 68, 68, 75, 75, 74, 74, 70

26. Number of pages read:
86, 84, 79, 75, 88, 66, 72, 85, 71

27. When is the median a better measure of central tendency than the mean to describe a set of data? Make up a list of numbers to illustrate your explanation. Calculate both the mean and the median.

28. Suppose you own a hat shop and can order a certain hat in only one size. You look at last year's sales to decide on the size to order. Should you find the mean, median, or mode for these sales? Explain your answer.

Find the grade point average for students earning the following grades. Assume A = 4, B = 3, C = 2, D = 1, and F = 0. *Round answers to the nearest hundredth.* **See Example 4.**

29. GS

Credits	Grade
4	B (= 3)
2	C (= 2)
2	A (= 4)
1	C (= 2)
3	D (= 1)

Credits • Grade

$4 \cdot 3 = 12$

$2 \cdot 2 = 4$

$2 \cdot 4 = 8$

$1 \cdot 2 = 2$

$3 \cdot 1 = 3$

29

$$\text{GPA} = \frac{29}{12} =$$

30. GS

Credits	Grade
1	C (= 2)
3	A (= 4)
4	B (= 3)
3	C (= 2)
2	A (= 4)

Credits • Grade

$1 \cdot 2 = 2$

$3 \cdot 4 = 12$

$4 \cdot 3 = 12$

$3 \cdot 2 = 6$

$2 \cdot 4 = 8$

40

$$\text{GPA} = \frac{40}{13} =$$

31.

Credits	Grade
4	B
2	A
5	C
1	F
3	B

32.

Credits	Grade
3	A
3	B
4	B
3	C
3	C

33.

Credits	Grade
2	A
3	C
4	A
1	C
4	B

34.

Credits	Grade
3	A
2	A
5	B
4	A
1	A

Relating Concepts (Exercises 35–44) For Individual or Group Work

Gluco Industries manufactures and sells glucose monitors and other diabetes-related products. The number of sales calls made over an 8-week period by two sales representatives, Scott Samuels and Rob Stricker, is shown below. Use this information to **work Exercises 35–44 in order.**

Week	Number of Calls	
	Samuels	Stricker
1	39	21
2	15	22
3	40	20
4	22	23
5	13	19
6	22	24
7	17	25
8	8	22

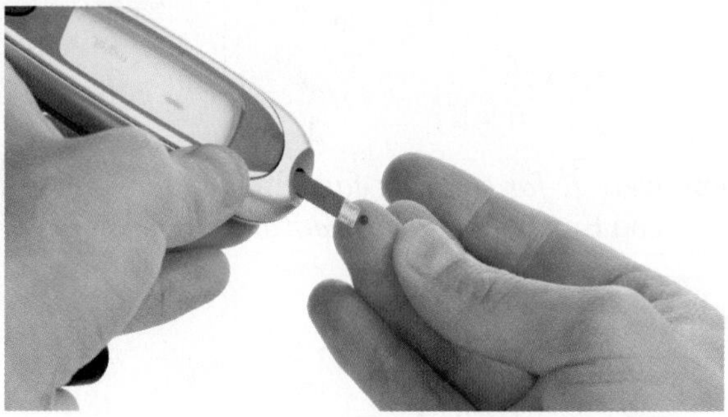

35. Find the total number of sales calls made by each of the sales representatives.

36. Find the mean number of sales calls made by Samuels and by Stricker.

37. Find the median number of sales calls made by Samuels and by Stricker.

38. Find the mode for the number of sales calls for each of the sales representatives.

39. How do the mean, median, and mode for the two sales representatives compare?

40. Describe how the pattern in the number of weekly sales calls made by Samuels differs from the pattern in Stricker's weekly sales calls.

To show the variation *or* spread *of the number of sales calls made by each of the sales representatives requires some* **measure of the dispersion,** *or spread of the numbers around the mean. A common measure of the dispersion is the* **range.** *The range is the* difference between the greatest value and the least value in the set of numbers.

41. Find the range for the number of sales calls made by Samuels.

42. Find the range for the number of sales calls made by Stricker.

43. Are the performance data for these two sales representatives sufficient to accurately determine which is the best? What else might you want to know?

44. List three possible explanations for the wide variation (range) in the number of sales calls for Samuels.

Chapter 8 Summary

Key Terms

8.1

circle graph A circle graph shows how a total amount is divided into parts or sectors. It is based on percents of 360°.

protractor A protractor is a device (usually in the shape of a half-circle) used to measure the number of degrees in angles or parts of a circle.

8.2

bar graph A bar graph uses bars of various heights or lengths to show quantity or frequency.

double-bar graph A double-bar graph compares two sets of data by showing two sets of bars.

line graph A line graph uses dots connected by lines to show trends.

comparison line graph A comparison line graph shows how two sets of data relate to each other by showing a line graph for each set of data.

8.3

frequency distribution A frequency distribution is a table that includes a column showing each possible number in the data collected. The original data are then entered in another column using a tally mark for each corresponding value. The tally marks are counted and the totals are placed in a third column.

histogram A histogram is a bar graph in which the width of each bar represents a range of numbers (class interval) and the height represents the quantity or frequency of items that fall within the interval.

8.4

mean The mean is the sum of all the values divided by the number of values. It is often called the *average.*

weighted mean The weighted mean is a mean calculated so that each value is multiplied by its frequency.

median The median is the middle number in a group of values that are listed from least to greatest. It divides a group of values in half. If there are an even number of values, the median is the mean (average) of the two middle values.

mode The mode is the value that occurs most often in a group of values.

bimodal A list of numbers is bimodal when it has two modes. The two values occur the same number of times.

dispersion The dispersion is the variation, or spread, of the numbers around the mean.

range The range is a common measure of the dispersion of numbers. It is the difference between the greatest value and the least value in the set of numbers.

New Formula

Mean or average: $\text{mean} = \dfrac{\text{sum of all values}}{\text{number of values}}$

Test Your Word Power

See how well you have learned the vocabulary in this chapter.

1 A **circle graph**
A. uses bars of various heights to show quantity or frequency
B. shows how a total amount is divided into parts or sectors
C. uses bars of various widths to represent a range of numbers.

2 A **bar graph**
A. uses bars of various heights to show quantity or frequency
B. shows how a total amount is divided into parts or sectors
C. uses dots connected by lines to show trends.

3 A **histogram** is a graph in which
A. two sets of data are compared using two sets of bars
B. dots are connected by lines to show trends
C. the width of each bar represents a range of numbers and the height represents the frequency of items within that range.

4 A **protractor** is a device used to
A. construct a histogram
B. calculate measures of central tendency
C. measure the number of degrees in angles or parts of a circle.

5 The **mean** is
A. calculated so that each value is multiplied by its frequency
B. the sum of all values divided by the number of values
C. the middle number in a group of values that are listed from least to greatest.

6 The **mode** is
A. calculated so that each value is multiplied by its frequency
B. the middle number in a group of values that are listed from least to greatest
C. the value that occurs most often in a group of values.

Answers to Test Your Word Power

1. B; *Example:*

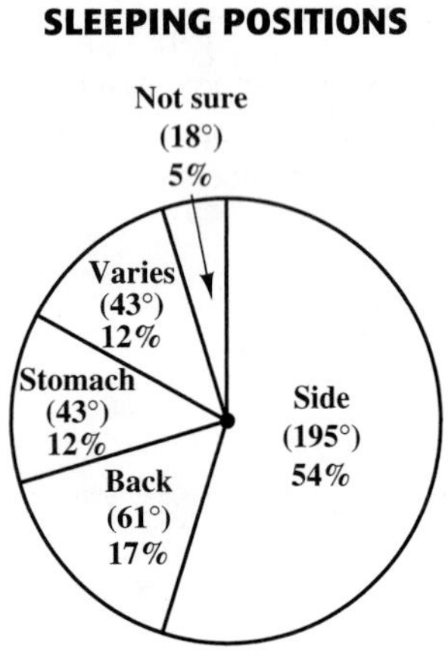

2. A; *Example:*

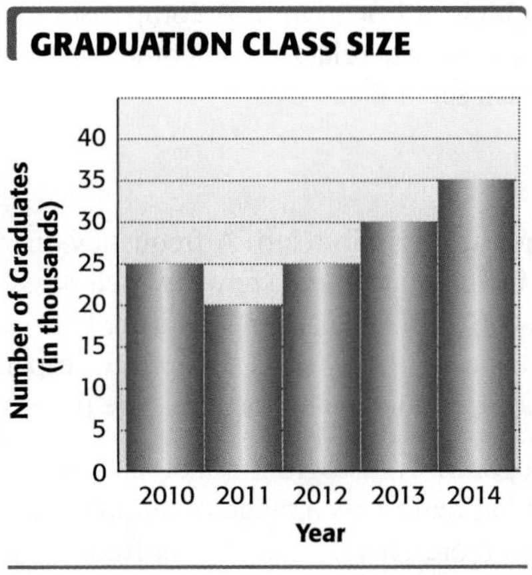

3. C; *Example:*

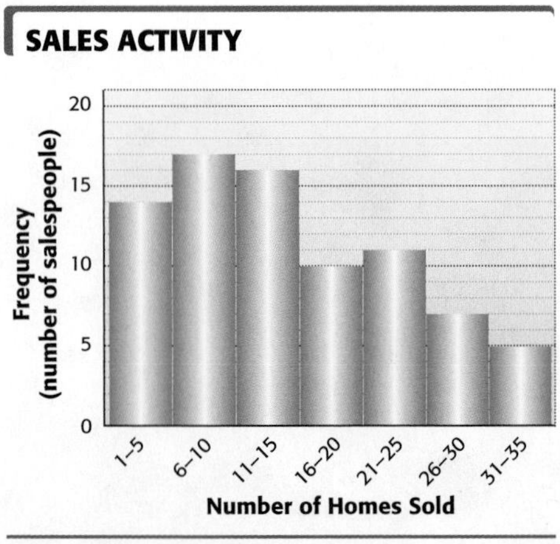

4. C; *Example:*

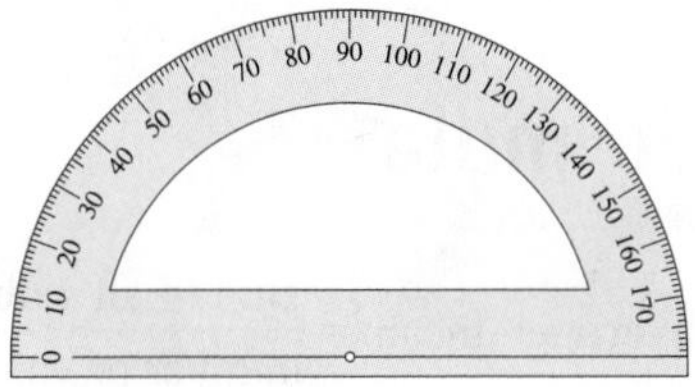

5. B; *Example:* The mean of the values \$5, \$9, \$7, \$5, \$2, and \$8, is

$$\frac{\$5 + \$9 + \$7 + \$5 + \$2 + \$8}{6} = \frac{\$36}{6} = \$6.$$

6. C; *Example:* The mode of the values \$5, \$9, \$7, \$5, \$2, and \$8 is \$5 because \$5 appears twice in the list.

Quick Review

Concepts	Examples

8.1 Constructing a Circle Graph

Step 1 Determine the percent of the total for each item.

Step 2 Find the number of degrees out of 360° that each percent represents.

Step 3 Use a protractor to measure the number of degrees for each item in the circle.

Item	Amount	Percent of Total	Sector Size
Leather interior	\$1600	$\frac{\$1600}{\$8000} = \frac{1}{5} = 20\%$ so $360° \cdot 20\% = 360 \cdot 0.20$	= 72°
Wheels/tires	\$2400	$\frac{\$2400}{\$8000} = \frac{3}{10} = 30\%$ so $360° \cdot 30\% = 360 \cdot 0.30$	= 108°
Sun roof	\$1200	$\frac{\$1200}{\$8000} = \frac{3}{20} = 15\%$ so $360° \cdot 15\% = 360 \cdot 0.15$	= 54°
Sport package	\$2800	$\frac{\$2800}{\$8000} = \frac{7}{20} = 35\%$ so $360° \cdot 35\% = 360 \cdot 0.35$	= 126°

Construct a circle graph from the following table, which lists the costs of options on a new luxury sports car.

Item	Amount
Leather interior	\$1600
Wheels/tires	\$2400
Sun roof	\$1200
Sport package	\$2800
Total	\$8000

SPORTY CAR OPTION COSTS

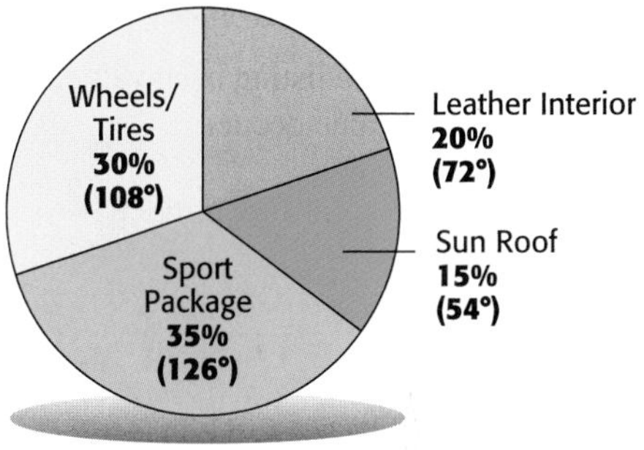

8.2 Reading a Bar Graph

The height of the bar is used to show the quantity or frequency (number) in a specific category. Use a ruler or straightedge to line up the top of each bar with the numbers on the left side of the graph.

Use the bar graph below to determine the number of students who earned each letter grade.

STUDENT GRADES

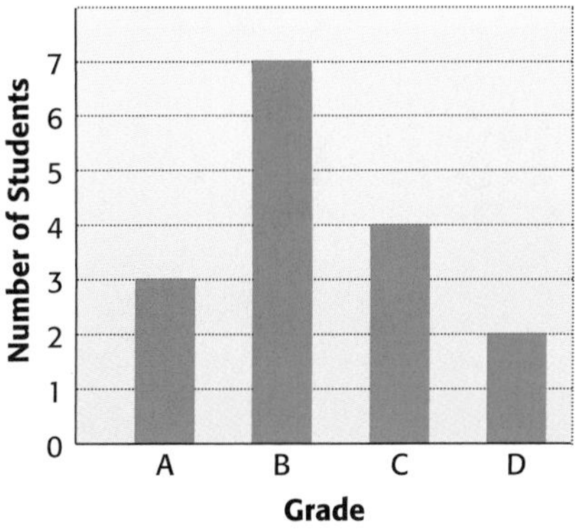

Grade of A: 3 students; B: 7 students; C: 4 students; D: 2 students.

Concepts	Examples

8.2 Reading a Line Graph

A dot is used to show the number or quantity in a specific class. The dots are connected with lines. This kind of graph is used to show a trend.

The line graph below shows the annual sales for the Fabric Supply Center for each of 4 years.

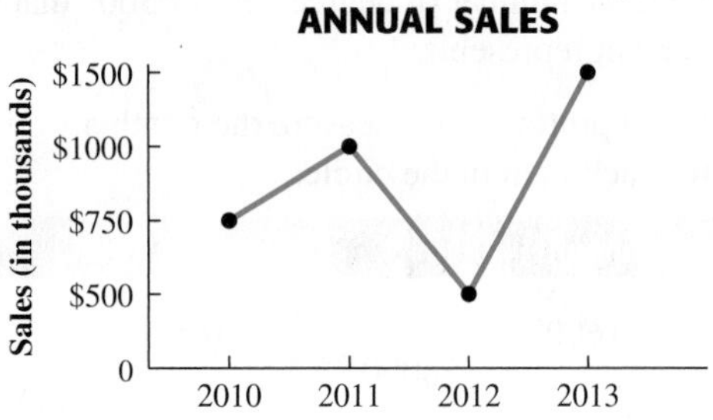

Find the sales in 2012.
The dot above 2012 lines up with $500 on the left edge. Then $500 • 1000 = $500,000 in sales.

8.3 Preparing a Frequency Distribution and a Histogram from Raw Data

Step 1 Construct a table listing each value, and the number of times this value occurs.

Draw a histogram for these student quiz scores.

12	15	15	14
13	20	10	12
11	9	10	12
17	20	16	17
14	18	19	13

Quiz Score	Tally	Frequency	
9	I	1	**1st class interval**
10	II	2	
11	I	1	
12	III	3	**2nd class interval**
13	II	2	
14	II	2	
15	II	2	**3rd class interval**
16	I	1	
17	II	2	
18	I	1	**4th class interval**
19	I	1	
20	II	2	

Concepts	Examples

8.3 Preparing a Frequency Distribution and a Histogram from Raw Data (*Continued*)

Step 2 Divide the data into groups, categories, or classes.

Class Interval (Quiz Scores)	Frequency (Number of Students)
9–11	4
12–14	7
15–17	5
18–20	4

Step 3 Draw bars representing these groups to make a histogram.

STUDENT QUIZ SCORES

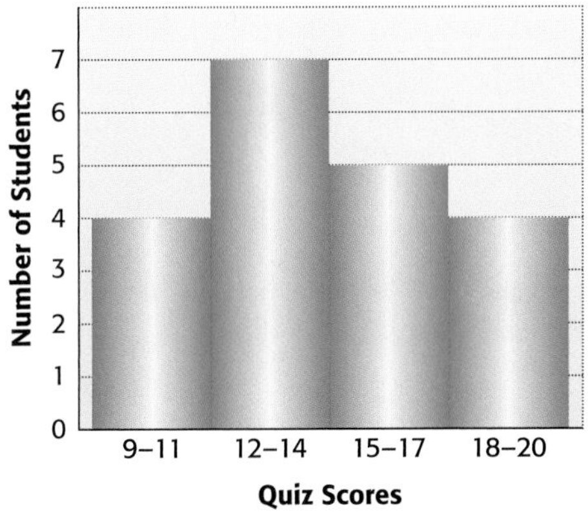

8.4 Finding the Mean (Average) of a Set of Numbers

Step 1 Add all values to obtain a total.

Step 2 Divide the total by the number of values.

Use this formula:

$$\text{mean (average)} = \frac{\text{sum of all values}}{\text{number of values}}$$

The test scores for Kathy West in her algebra course were as follows:

80	92	92	94
76	88	84	93

Find West's mean (average) test score to the nearest tenth.

$$\text{mean} = \frac{80 + 92 + 92 + 94 + 76 + 88 + 84 + 93}{8}$$

$$= \frac{699}{8} \approx 87.4 \text{ (rounded)}$$

West's mean test score is approximately 87.4.

Concepts	Examples

8.4 Finding the Weighted Mean

Step 1 Multiply each value by its frequency.

Step 2 Add all the products from Step 1.

Step 3 Divide the sum in Step 2 by the total number of pieces of data.

This table shows the distribution of the number of school-age children in a survey of 30 families.

Number of School-Age Children	Frequency (Number of Families)
0	12
1	6
2	7
3	3
4	2
Total of 30 families	

Find the mean number of school-age children per family. Round to the nearest hundredth.

Value	Frequency	Product
0	12	$(0 \cdot 12) = 0$
1	6	$(1 \cdot 6) = 6$
2	7	$(2 \cdot 7) = 14$
3	3	$(3 \cdot 3) = 9$
4	2	$(4 \cdot 2) = 8$
Totals	**30**	**37**

$$\text{mean} = \frac{37}{30} \approx 1.23 \text{ (rounded)}$$

The mean number of school-age children per family is approximately 1.23.

8.4 Finding the Median of a Set of Numbers

Step 1 Arrange the data from least to greatest.

Step 2 Select the middle value, or, if there is an even number of values, find the average of the two middle values.

Find the median for Kathy West's test scores from the previous page.

The data arranged from least to greatest is:

76 80 84 88 92 92 93 94

(76 ← Least; 88 and 92 ← Middle values; 94 ← Greatest)

The middle two values are 88 and 92. The average of these two values is

$$\frac{88 + 92}{2} = 90$$

West's median test score is 90.

8.4 Finding the Mode of a Set of Values

Find the value that appears most often in the list of values. If no value appears more than once, there is no mode. If two different values appear the same number of times, the list is bimodal.

Find the mode for West's test scores shown above.

The most frequently occurring score is 92 (it occurs twice). Therefore, the mode is 92.

Chapter 8 Review Exercises

8.1 *The number of girls participating in high school sports in the United States exceeds 3 million. The circle graph shows the number of high school girls' teams in the most popular sports.*

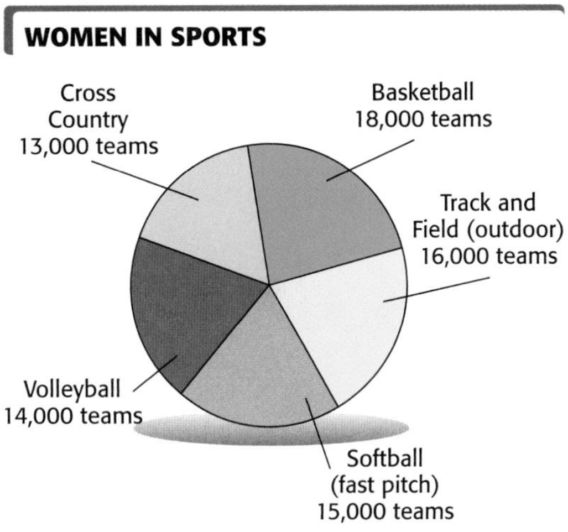

Source: National Federation of State High School Associations.

Use the circle graph to answer Exercises 1–6. Write ratios as fractions in lowest terms.

1. What women's sport has the greatest number of teams? How many are there?

2. Number of track and field teams to the total number of teams.

3. Number of softball teams to the total number of teams.

4. Number of volleyball teams to the total number of teams.

5. Number of basketball teams to the number of track and field teams.

6. Number of track and field teams to the number of volleyball teams.

8.2 *This bar graph shows the most frequently offered "work perks" and the percent of the responding companies offering them. The survey was conducted online and included* 4800 *companies ranging in size from* 2 *to* 5000 *employees. Use this graph to find the number of companies offering each work perk listed in Exercises 7–10 and to answer Exercises 11 and 12.* (*Source:* Work Perks Survey, Ceridian Employer Services.)

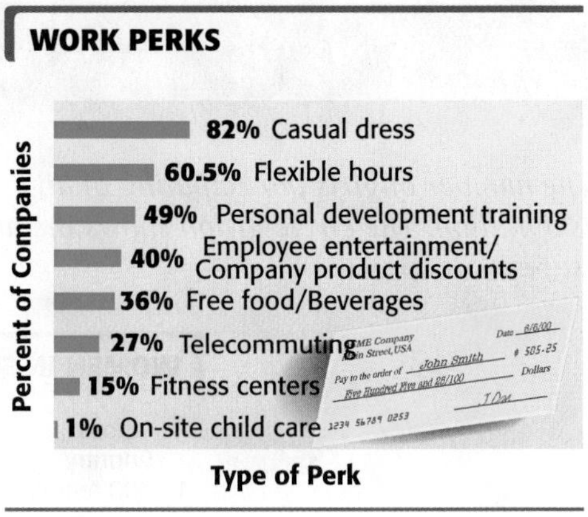

7. Casual dress

8. Free food/Beverages

9. Fitness centers

10. Flexible hours

11. Which two work perks do companies offer least often? Give one possible explanation why these work perks are not offered.

12. Which two work perks do companies offer most often? Give one possible explanation why these work perks are so popular.

This double-bar graph shows the number of acre-feet of water in Lake Natoma for each of the first six months of 2012 *and* 2013. *Use this graph to answer Exercises 13–18.*

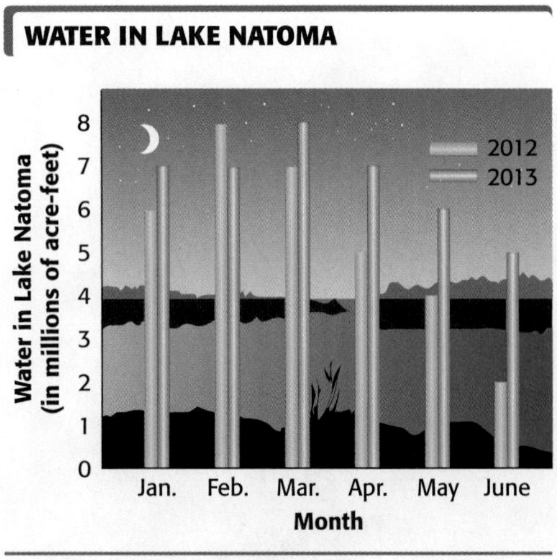

13. During which month in 2013 was the greatest amount of water in the lake? How much was there?

14. During which month in 2012 was the least amount of water in the lake? How much was there?

15. How many acre-feet of water were in the lake in June of 2013?

16. How many acre-feet of water were in the lake in May of 2012?

17. Find the decrease in the amount of water in the lake from March 2012 to June 2012.

18. Find the decrease in the amount of water in the lake from April 2013 to June 2013.

This comparison line graph shows the annual floor-covering sales of two different home improvement centers during each of 5 years. Use this graph to find the amount of annual floor-covering sales in each year shown in Exercises 19–22 and to answer Exercises 23 and 24.

ANNUAL FLOOR-COVERING SALES

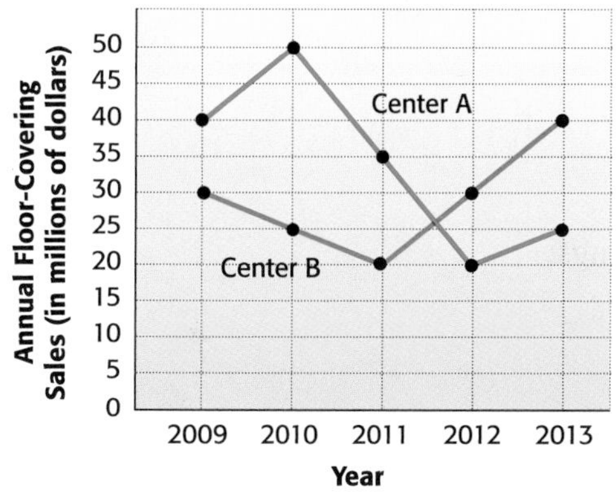

19. Center A in 2010

20. Center A in 2012

21. Center B in 2011

22. Center B in 2013

23. What trend do you see in center A's sales from 2010 to 2013? Why might this have happened?

24. What trend do you see in center B's sales starting in 2011? Why might this have happened?

8.4 *Find the mean for each list of numbers. Round answers to the nearest tenth if necessary.*

25. Digital cameras sold:

18, 12, 15, 24, 9, 42, 54, 87, 21, 3

26. Number of harassment complaints filed:

31, 9, 8, 22, 46, 51, 48, 42, 53, 42

Find the weighted mean for each list. Round to the nearest hundredth if necessary.

27.

Dollar Value	Frequency
$42	3
$47	7
$53	2
$55	3
$59	5

28.

Total Points	Frequency
243	1
247	3
251	5
255	7
263	4
271	2
279	2

Find the median for each list of numbers.

29. The number of accident forms filed:

43, 37, 13, 68, 54, 75, 28, 35, 39

30. Commissions of:

$576, $578, $542, $151, $559, $565, $525, $590

Find the mode or modes for each list of numbers.

31. Running shoes priced at:

$79, $56, $110, $79, $72, $86, $79

32. Boat launchings:

18, 25, 63, 32, 28, 37, 32, 26, 18

Mixed Review Exercises

In her senior year of college Sue Hogan had expenses of $17,920. This amount was spent as shown below. In order to make a circle graph, find all the missing numbers in Exercises 33–37.

Item	Dollar Amount	Percent of Total	Degrees of Circle
33. Books and supplies	$1792	10%	______
34. Rent	$6272	______	126°
35. Food	$3584	______	______
36. Tuition/fees	$4480	______	______
37. Miscellaneous	$1792	______	______

38. Draw a circle graph using the information in **Exercises 33–37.**

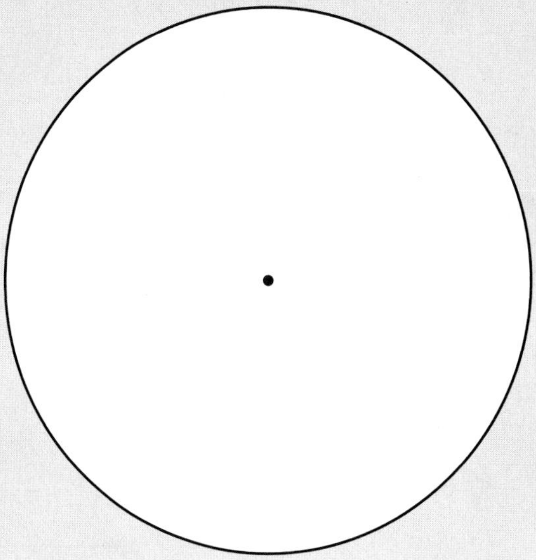

Find the mean for each list of numbers. Round answers to the nearest tenth if necessary.

39. Number of volunteers for the project:

48, 72, 52, 148, 180

40. Number of flu vaccinations in a day:

122, 135, 146, 159, 128, 147, 168, 139, 158

Find the mode or modes for each list of numbers.

41. Job applicants meeting the qualifications:

48, 43, 46, 47, 48, 48, 43

42. Number of two-bedroom apartments in each building:

26, 31, 31, 37, 43, 51, 31, 43, 43

Find the median for each list of numbers.

43. Hours worked:

4.7, 3.2, 2.9, 5.3, 7.1, 8.2, 9.4, 1.0

44. Number of e-mails each day:

35, 51, 9, 2, 17, 12, 46, 23, 3, 19, 39, 27

Here are the scores of 40 *students on a mathematics exam. Complete the table.*

78	89	36	59	78	99	92	86
73	78	85	57	99	95	82	76
63	93	53	76	92	79	72	62
74	81	77	76	59	84	76	94
58	37	76	54	80	30	45	38

	Class Intervals (Scores)	Tally	Class Frequency (Number of Students)
45.	30–39	________	________
46.	40–49	________	________
47.	50–59	________	________
48.	60–69	________	________
49.	70–79	________	________
50.	80–89	________	________
51.	90–99	________	________

52. Construct a histogram by using the data in **Exercises 45–51.**

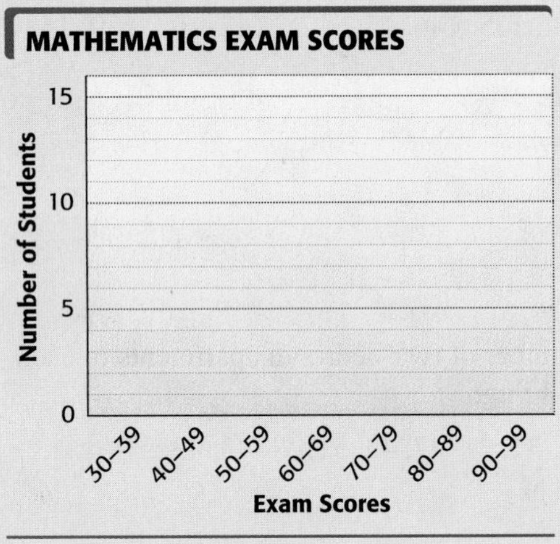

Find each weighted mean. Round answers to the nearest tenth if necessary.

53.

Test Score	Frequency
46	4
54	10
62	8
70	12
78	10

54.

Units Sold	Frequency
104	6
112	14
115	21
119	13
123	22
127	6
132	9

Chapter 8 Test

CHAPTER Test Prep VIDEO *The Chapter Test Prep Videos with test solutions are available on DVD, in MyMathLab, and on YouTube—search "LialDevMath" and click on "Channels."*

The circle graph shows the sources of electricity generated in the United States. If the total cost of all electricity generated in one year was $328 billion, find the dollar amount spent on electricity generated by each of these sources. Round to the nearest tenth of a billion.

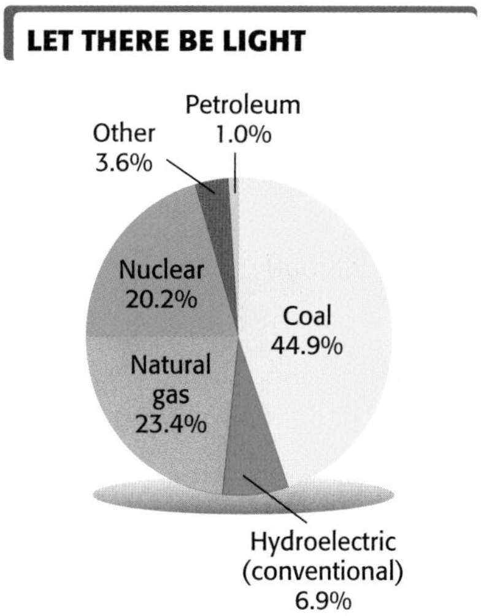

Source: U.S. Energy Information Administration.

1. Nuclear

2. Hydroelectric (conventional)

3. Petroleum

4. Natural gas

5. Coal

6. Other

During a one-year period, Big 5 Sporting Goods had the following sales in each department. Find all numbers missing from the table.

Item	Dollar Amount	Percent of Total	Degrees of a Circle
7. Team sports	\$432,000	30%	______
8. Golf	\$144,000	10%	______
9. Hunting and fishing	\$288,000	20%	______
10. Athletic shoes	\$504,000	35%	______
11. Water sports	\$72,000	______	18°

12. Draw a circle graph using the information in **Exercises 7–11** and a protractor. Label each sector of the graph.

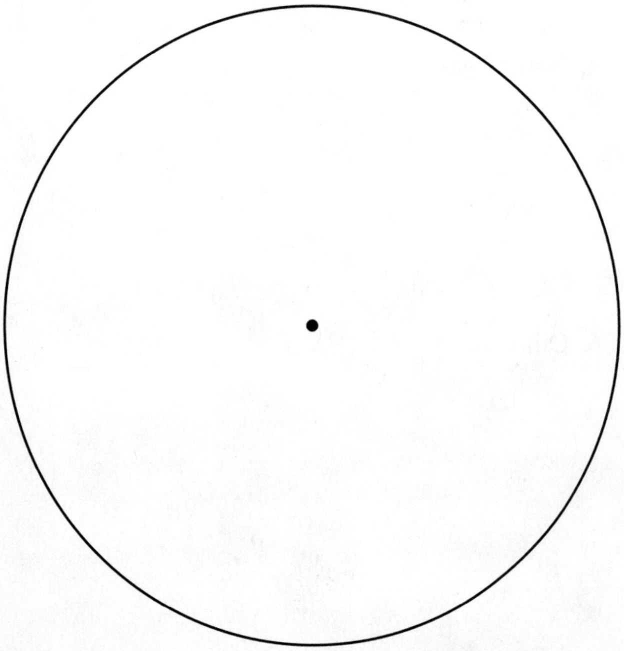

Here are the profits for each of the past 20 weeks from Alan's Snack Bar vending machines. Complete the table.

\$142	\$137	\$125	\$132	\$147	\$129	\$151	\$172	\$175	\$129
\$159	\$148	\$173	\$160	\$152	\$174	\$169	\$163	\$149	\$173

Profit	Tally	Number of Weeks
13. \$120–129	______	______
14. \$130–139	______	______
15. \$140–149	______	______
16. \$150–159	______	______
17. \$160–169	______	______
18. \$170–179	______	______

19. Use the information in **Exercises 13–18** to draw a histogram.

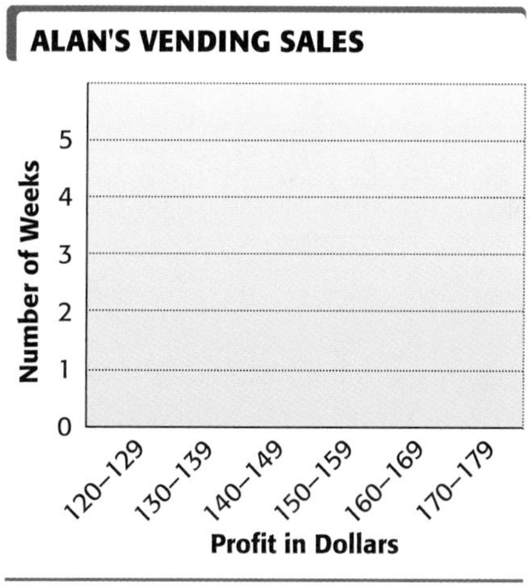

Find the mean for each list of numbers. Round answers to the nearest tenth if necessary.

20. Number of miles run each week while training:

52, 61, 68, 69, 73, 75, 79, 84, 91, 98

21. Weight in pounds for the largest bass caught in the lake:

11, 14, 12, 14, 20, 16, 17, 18

22. Airplane speeds in miles per hour:

458, 432, 496, 491, 500, 508, 512, 396, 492, 504

23. Explain why a weighted mean is used to find a student's grade point average. Calculate your own grade point average for last semester or quarter. If you are a new student, make up a grade point average problem of your own and solve it. Round to the nearest hundredth.

24. Explain in your own words the procedure for finding the median when there are an odd number of values in a list. Make up a problem with a list of five numbers and solve for the median.

Find the weighted mean for the following. Round answers to the nearest whole number if necessary.

25.

Cost	Frequency
$12	5
$20	6
$22	8
$28	4
$38	6
$48	2

26.

Points Scored	Frequency
150	15
160	17
170	21
180	28
190	19
200	7

Find the median for each list of numbers.

27. Highest daily temperatures in Alaska in degrees Fahrenheit:

32, 41, 28, 28, 37, 35, 16, 31

28. The lengths of steel beams in meters:

7.6, 11.4, 6.2, 12.5, 31.7, 22.8, 9.1, 10.0, 9.5

Find the mode or modes for each list of numbers.

29. Blood sample amounts in milliliters:

72, 46, 52, 37, 28, 18, 52, 61

30. Hot tub temperatures in degrees Fahrenheit:

96, 104, 103, 104, 103, 104, 91, 74, 103

Chapters 1–8 Cumulative Review Exercises

1. Write these numbers in words.

(a) 45.0203

(b) 30,000,650,008

2. Write these numbers using digits.

(a) One hundred sixty million, five hundred

(b) Seventy-five thousandths

Round each number as indicated.

3. 46,908 to the nearest hundred

4. 6.197 to the nearest hundredth

5. 0.66148 to the nearest thousandth

6. 9951 to the nearest hundred

First use front end rounding to round each number and estimate the answer. Then find the exact answer.

7. *Estimate:* $\underline{-\quad\quad}$ *Exact:* $\begin{array}{r} 75{,}078 \\ -\ 46{,}090 \\ \hline \end{array}$

8. *Estimate:* $\underline{-\quad}$ *Exact:* $\begin{array}{r} 7.8 \\ -\ 3.5029 \\ \hline \end{array}$

9. *Estimate:* $\underline{\times\quad\quad}$ *Exact:* $\begin{array}{r} 6538 \\ \times\ \ 708 \\ \hline \end{array}$

10. *Estimate:* $\underline{\times\quad}$ *Exact:* $\begin{array}{r} 65.3 \\ \times\ 8.7 \\ \hline \end{array}$

11. *Estimate:* $\overline{)\quad\quad}$ *Exact:* $43\overline{)38{,}786}$

12. *Estimate:* $\overline{)\ }$ *Exact:* $0.8\overline{)6.76}$

Simplify. Write answers in lowest terms and as whole or mixed numbers when possible.

13. $4\frac{3}{5} + 5\frac{2}{3}$

14. $6\frac{2}{3} - 4\frac{3}{4}$

15. $\left(9\frac{3}{5}\right)\left(4\frac{5}{8}\right)$

16. $22\left(\frac{2}{5}\right)$

17. $3\frac{1}{3} \div 8\frac{3}{4}$

18. $\frac{2}{3}\left(\frac{7}{8} - \frac{3}{4}\right)$

19. $4 + 10 \div 2 + 7(2)$

20. $\sqrt{81} - 4(2) + 9$

21. $2^2 \cdot 3^3$

22. $\frac{2}{3}\left(\frac{7}{8} - \frac{1}{2}\right)$

23. $\frac{7}{8} \div \left(\frac{3}{4} + \frac{1}{8}\right)$

24. $\left(\frac{5}{6} - \frac{5}{12}\right) - \left(\frac{1}{2}\right)^2 \cdot \frac{2}{3}$

Write each fraction as a decimal. Round to the nearest thousandth if necessary.

25. $\frac{3}{4}$

26. $\frac{3}{8}$

27. $\frac{7}{12}$

28. $\frac{11}{20}$

Write in order, from least to greatest.

29. 0.218, 0.22, 0.199, 0.207, 0.2215

30. 0.6319, $\frac{5}{8}$, 0.608, $\frac{13}{20}$, 0.58

Simplify each expression.

31. $10 - 0.329$

32. $0.7 + 85 + 7.903$

33. $3.2(2.5)$

34. $25.2 \div 0.56$

Write each ratio in lowest terms. Be sure to make all necessary conversions.

35. $5\frac{1}{2}$ in. to 44 in.

36. 3 hr to 45 min

Find the unknown number in each proportion.

37. $\frac{1}{5} = \frac{x}{30}$

38. $\frac{15}{x} = \frac{390}{156}$

39. $\frac{200}{135} = \frac{24}{x}$

40. $\frac{x}{208} = \frac{6.5}{26}$

Solve each percent problem.

41. Find 5.4% of 6000 homes.

42. $8\frac{1}{2}$% of what number of people is 238 people?

43. What percent of \$555 is \$1443?

Write each percent as a decimal. Write each decimal as a percent.

44. 3%

45. 200%

46. 0.87

47. 3.8

Write each percent as a fraction or mixed number in lowest terms. Write each fraction or mixed number as a percent.

48. 8%

49. 62.5%

50. 175%

51. $\frac{7}{8}$

52. $4\frac{1}{5}$

Solve each application problem.

53. Diane McKinney paid \$29.90 in sales tax on a \$460 purchase. What was the tax rate?

54. While enrolled in college full-time, Theresa Goldsmith is a part-time waitress. Her tips average 14%, and on a recent shift she had total sales of \$837. Find the amount of her tips.

55. A box spring and mattress originally priced at \$456 is discounted 45%. Find the amount of discount and the sale price.

56. A loan of \$46,300 is made at 3% simple interest for 9 months. Find the total amount to be repaid.

Set up and solve a proportion for each problem.

57. Gerri Junso gives school readiness tests to 9 children in 4 hours. Find the number of children she can test in 20 hours.

58. If 12.5 ounces of Roundup weed and grass killer is needed to make 5 gallons of spray, how much Roundup is needed for 102 gallons of spray?

Name each figure and find its perimeter (or circumference) and the area. Use 3.14 as the approximate value of π. Round answers to the nearest tenth, if necessary.

59.

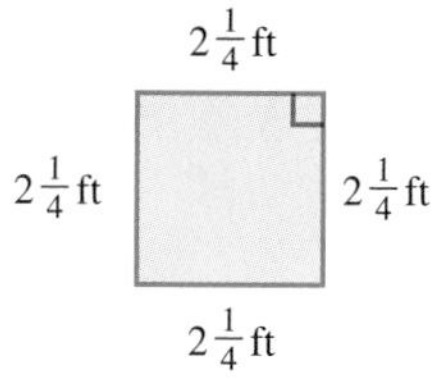

60.

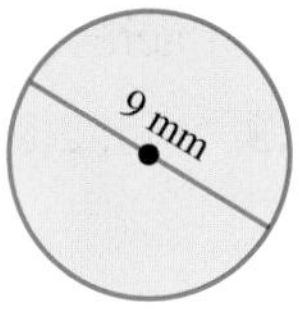

61.

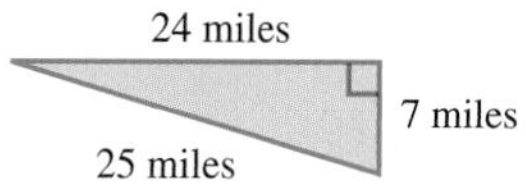

62.

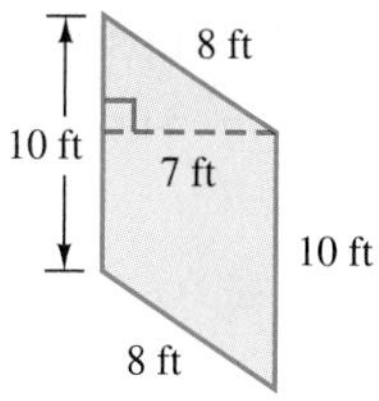

Find the unknown length, perimeter, area, or volume. Use 3.14 as the approximate value of π and round answers to the nearest tenth.

63. Find the perimeter and the area.

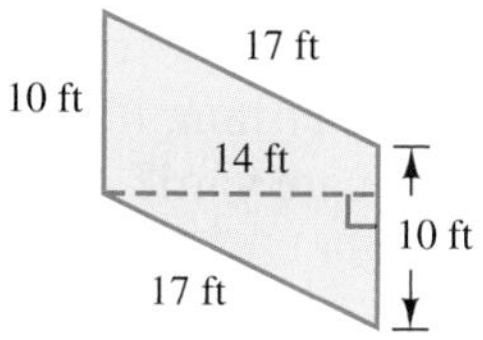

64. Find the circumference and the area.

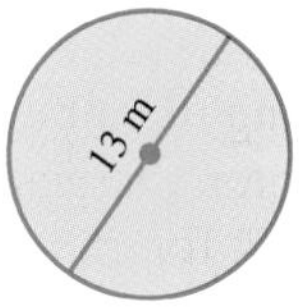

65. Find the length of the third side and the area.

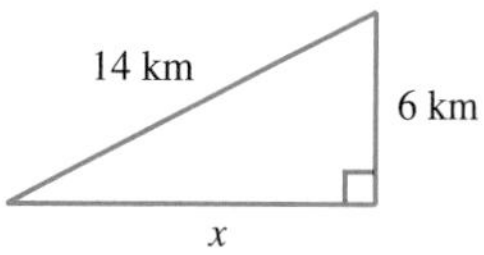

66. Find the volume and surface area.

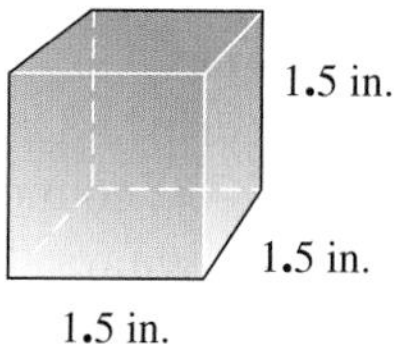

Find the unknown length, perimeter, area, or volume. When necessary, use 3.14 as the approximate value of π and round answers to the nearest tenth.

67. Find the perimeter and the area.

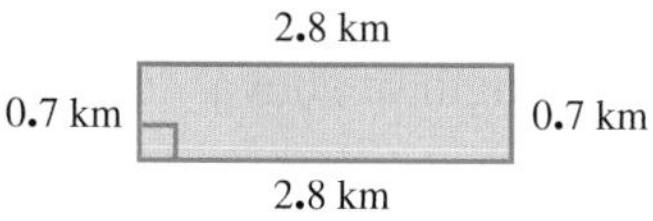

68. Find the diameter, circumference, and area.

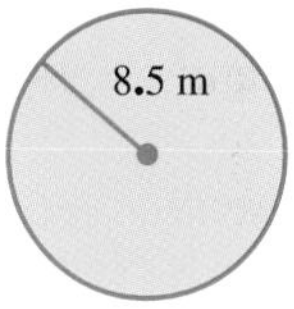

69. Find the volume and surface area.

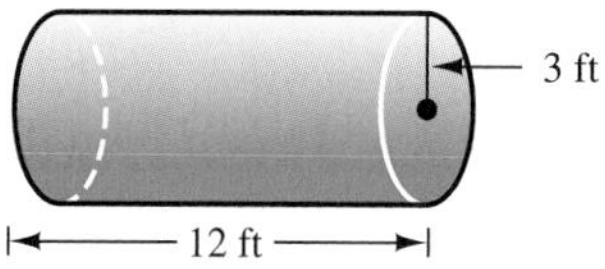

In Exercises 70 and 71, name each solid and find its volume. Use 3.14 as the approximate value for π. In Exercise 72, find the unknown length. Round answers to the nearest tenth if necessary.

70.

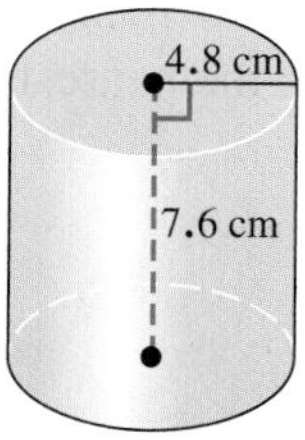

71.

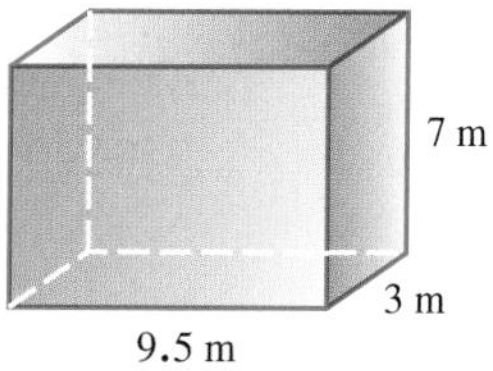

72.

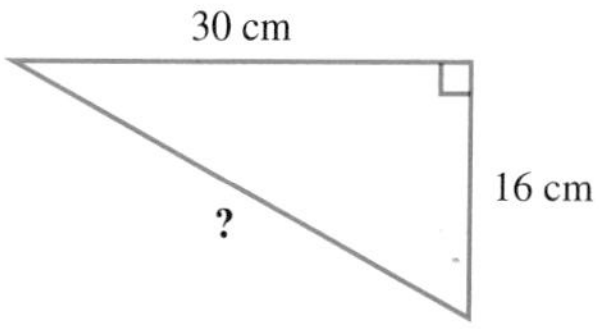

Find the mean, the median, and the mode for each list of numbers. Round to the nearest tenth if necessary.

73. Cable hookups per installer:

16, 37, 27, 31, 19, 25, 15, 38, 43, 19

74. Number of acres plowed each hour:

10.3, 4.3, 1.65, 2.85, 5.3, 5.7, 2.3, 4.35, 2.85

Solve each application problem.

75. In a study of 2082 workers it was found that only 874 of them used all of their paid time-off. What percent used all of their time-off? Round to the nearest tenth of a percent. (*Source:* Hudson Time-Off Survey.)

76. The average increase in residential winter heating bills will be 9.8%. (*Source:* Energy Information Administration.) If the increase amounts to \$87.20, find

(a) the average heating bill before the increase,

(b) the average heating bill after the increase.

77. Breathe Right® nasal strips are sold in four package sizes: 12 nasal strips \$6.50; 24 nasal strips \$7.50; 30 nasal strips \$8.95; 38 nasal strips \$9.95. You have a \$2-off coupon for the 12-strip size and a \$1-off coupon for the 30-strip size. Which choice is the best buy?

78. The sketch below shows the plans for a lobby in a large commercial complex. What is the cost of carpeting the lobby, excluding the atrium, if the contractor charges \$43.50 per square yard? Use 3.14 for π.

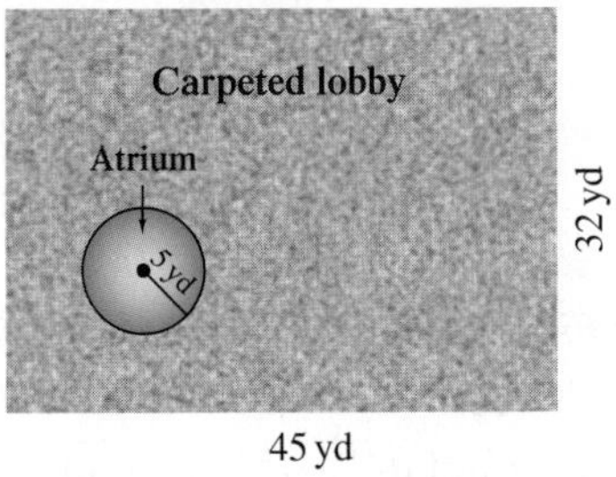

79. A Folger's coffee can has a diameter of 15.6 cm and a height of 16 cm. Find the volume of the can. Use 3.14 for π and round your answer to the nearest tenth. (*Source:* Folger's.)

80. Steven bought a $4\frac{1}{2}$ yd length of canvas material to repair the tents used by the scout troop. He used $1\frac{2}{3}$ yd on one tent and $1\frac{3}{4}$ yd on another. What length of material is left?

81. A packing crate measures 2.4 m long, 1.2 m wide, and 1.2 m high. A trucking company wants crates that hold 4 m^3. The crate's volume is how much more or less than 4 m^3?

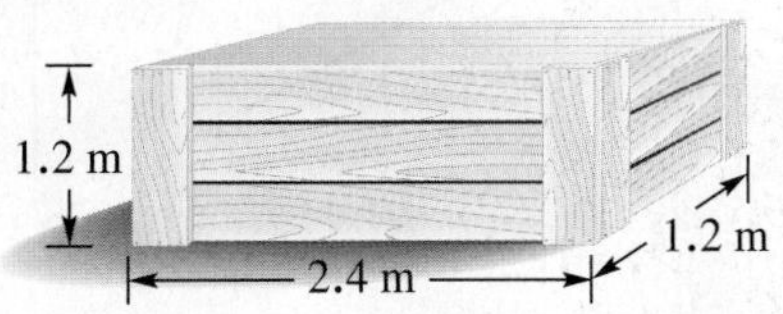

82. Jackie drove her car 364 miles on 14.5 gallons of gas. Maya used 16.3 gallons to drive 406 miles. Naomi drove 300 miles on 11.9 gallons. Which car had the highest number of miles per gallon? How many miles per gallon did that car get, rounded to the nearest tenth?

9 The Real Number System

Positive and *negative numbers*, shown here to indicate gains and losses, are examples of *real numbers,* the subject of this chapter.

9.1 Exponents, Order of Operations, and Inequality

OBJECTIVES

1. Use exponents.
2. Use the rules for order of operations.
3. Use more than one grouping symbol.
4. Know the meanings of $\neq$, $<$, $>$, $\leq$, and $\geq$.
5. Translate word statements to symbols.
6. Write statements that change the direction of inequality symbols.

OBJECTIVE 1 Use exponents. In **Section 2.3,** we factored a number as the product of its prime factors. For example,

81 can be written as $3 \cdot 3 \cdot 3 \cdot 3$, $\cdot$ indicates multiplication.

where the factor 3 appears four times. Repeated factors are written in an abbreviated form by using an *exponent.*

$$\underbrace{\mathbf{3 \cdot 3 \cdot 3 \cdot 3}}_{\text{4 factors of 3}} = \mathbf{3}^{4}$$

(Exponent: 4; Base: 3)

The number 4 is the **exponent,** or **power,** and 3 is the **base** in the **exponential expression** 3^4. The exponent tells how many times the base is used as a factor. We read 3^4 as **"3 to the fourth power,"** or simply **"3 to the fourth."** ***A number raised to the first power is simply that number.*** For example,

$6^1 = 6$ and $(2.5)^1 = 2.5$. In general, $a^1 = a$.

EXAMPLE 1 Evaluating Exponential Expressions

Find the value of each exponential expression.

(a) 5^2 means $5 \cdot 5$, which equals 25.

5 is used as a factor 2 times.

Read 5^2 as "5 to the second power" or, more commonly, "5 squared."

(b) 6^3 means $6 \cdot 6 \cdot 6$, which equals 216.

6 is used as a factor 3 times.

Read 6^3 as "6 to the third power" or, more commonly, "6 cubed."

(c) 2^5 means $2 \cdot 2 \cdot 2 \cdot 2 \cdot 2$, which equals 32. 2 is used as a factor 5 times.

Read 2^5 as "2 to the fifth power."

(d) $\left(\frac{2}{3}\right)^3$ means $\frac{2}{3} \cdot \frac{2}{3} \cdot \frac{2}{3}$, which equals $\frac{8}{27}$. $\frac{2}{3}$ is used as a factor 3 times.

(e) $(0.3)^2$ means $0.3(0.3)$, which equals 0.09. 0.3 is used as a factor 2 times.

◀ Work Problem 1 at the Side.

CAUTION

Squaring, or raising a number to the second power, is not the same as doubling the number. **For example,**

$$3^2 \text{ means } 3 \cdot 3, \textit{ not } 2 \cdot 3.$$

Thus $3^2 = 9$, not 6. Similarly, cubing, or raising a number to the third power, does *not* mean tripling the number.

---

1 Find the value of each exponential expression.

GS **(a)** 6^2 means ____ $\cdot$ ____, which equals ____.

(b) 3^5

GS **(c)** $\left(\frac{3}{4}\right)^2$ means ____ $\cdot$ ____, which equals ____.

(d) $\left(\frac{1}{2}\right)^4$

GS **(e)** $(0.4)^3$ means ________, which equals ____.

Answers

1. (a) 6; 6; 36 (b) 243 (c) $\frac{3}{4}; \frac{3}{4}; \frac{9}{16}$ (d) $\frac{1}{16}$ (e) 0.4(0.4)(0.4); 0.064

OBJECTIVE 2 **Use the rules for order of operations.** When a problem involves more than one operation, we often use **grouping symbols.** If no grouping symbols are used, we apply the rules for order of operations.

Consider the expression $5 + 2 \cdot 3$. To show that the multiplication should be performed before the addition, we use parentheses to group $2 \cdot 3$.

$$5 + (2 \cdot 3) \quad \text{equals} \quad 5 + 6, \quad \text{or} \quad 11.$$

If addition is to be performed first, the parentheses should group $5 + 2$.

$$(5 + 2) \cdot 3 \quad \text{equals} \quad 7 \cdot 3, \quad \text{or} \quad 21.$$

Other grouping symbols are brackets [ ], braces { }, and fraction bars. (For example, in $\frac{8-2}{3}$, the expression $8 - 2$ is considered to be grouped in the numerator.)

To work problems with more than one operation, use the following rules for **order of operations.** This order is used by most calculators and computers.

Order of Operations

If grouping symbols are present, simplify within them, innermost first (and above and below fraction bars separately), in the following order.

Step 1 Apply all **exponents.**

Step 2 Do any **multiplications** or **divisions** in the order in which they occur, working from left to right.

Step 3 Do any **additions** or **subtractions** in the order in which they occur, working from left to right.

If no grouping symbols are present, start with Step 1.

Another way to show multiplication besides with a raised dot is with parentheses.

$$3(7) \text{ means } \quad 3 \cdot 7, \quad \text{or} \quad 21.$$

$$3(4 + 5) \quad \text{means} \quad \text{“3 times the sum of 4 and 5.”}$$

When simplifying $3(4 + 5)$, the sum in parentheses must be found first, then the product.

EXAMPLE 2 **Using the Rules for Order of Operations**

Find the value of each expression.

(a) $24 - 12 \div 3$ Be careful. Divide first.

$= 24 - 4$ Divide.

$= 20$ Subtract.

(b) $9(6 + 11)$

$= 9(17)$ Work inside parentheses.

$= 153$ Multiply.

(c) $6 \cdot 8 + 5 \cdot 2$

$= 48 + 10$ Multiply, working from left to right.

$= 58$ Add.

Continued on Next Page

2 Label the order in which each **GS** expression should be evaluated. Then find the value of each expression.

(a) $7 + 3 \cdot 8$

② ①

$= 7 + \_\_\_\_$

$= \_\_\_\_$

(b) $2 \cdot 9 + 7 \cdot 3$

○ ○ ○

$= \_\_\_\_ + \_\_\_\_$

$= \_\_\_\_$

(c) $7 \cdot 6 - 3(8 + 1)$

○ ○○ ○

(d) $2 + 3^2 - 5$

○○○

3 Find the value of each expression.

(a) $9[(4 + 8) - 3]$

(b) $\dfrac{2(7 + 8) + 2}{3 \cdot 5 + 1}$

Answers

2. **(a)** 24; 31 **(b)** ①, ③, ②; 18; 21; 39
(c) ②, ④, ③, ①; 15
(d) ②, ①, ③; 6
3. **(a)** 81 **(b)** 2

(d) $2(5 + 6) + 7 \cdot 3$

Start here.

$= 2(11) + 7 \cdot 3$ Work inside parentheses.

$= 22 + 21$ Multiply.

$= 43$ Add.

(e) $9 + 2^3 - 5$

$2^3 = 2 \cdot 2 \cdot 2$, not $2 \cdot 3$

$= 9 + 8 - 5$ Apply the exponent.

$= 12$ Add, and then subtract.

◀ **Work Problem 2 at the Side.**

OBJECTIVE 3 Use more than one grouping symbol. In an expression such as $2(8 + 3(6 + 5))$, we often use brackets, [ ], in place of the outer pair of parentheses.

EXAMPLE 3 Using Brackets and Fraction Bars as Grouping Symbols

Find the value of each expression.

Start here.

(a) $2[8 + 3(6 + 5)]$

$= 2[8 + 3(11)]$ Add inside parentheses.

$= 2[8 + 33]$ Multiply inside brackets.

$= 2[41]$ Add inside brackets.

$= 82$ Multiply.

(b) $\dfrac{4(5 + 3) + 3}{2(3) - 1}$ Simplify the numerator and denominator separately.

$= \dfrac{4(8) + 3}{2(3) - 1}$ Work inside parentheses.

$= \dfrac{32 + 3}{6 - 1}$ Multiply.

$= \dfrac{35}{5}$ Add and subtract.

$= 7$ Divide.

◀ **Work Problem 3 at the Side.**

Note

The expression $\frac{4(5 + 3) + 3}{2(3) - 1}$ in **Example 3(b)** can be written as a quotient.

$$[4(5 + 3) + 3] \div [2(3) - 1]$$

The fraction bar "groups" the numerator and denominator separately.

Calculator Tip

Calculators follow the order of operations. Try some of the examples to see that your calculator gives the same answers. Use the parentheses keys to insert parentheses where they are needed, such as around the numerator and the denominator in **Example 3(b).**

OBJECTIVE 4 Know the meanings of $\neq$, $<$, $>$, $\leq$, and $\geq$. So far, we have used only the symbols of arithmetic, such as $+$, $-$, $\cdot$, and $\div$ and the equality symbol $=$. The equality symbol with a slash through it means "is *not* equal to."

$7 \neq 8$ 7 is not equal to 8.

If two numbers are not equal, then one of the numbers must be less than the other. The symbol $<$ represents "is less than."

$7 < 8$ 7 is less than 8.

The symbol $>$ means "is greater than."

$8 > 2$ 8 is greater than 2.

To keep the meanings of the symbols $<$ and $>$ clear, remember that the symbol always points to the lesser number.

Lesser number $\rightarrow$ **8** < 15

$15 >$ **8** $\leftarrow$ Lesser number

Work Problem 4 at the Side. ▶

The symbol $\leq$ means "is less than or equal to."

$5 \leq 9$ 5 is less than or equal to 9.

If either the $<$ part or the $=$ part is true, then the inequality $\leq$ is true. The statement $5 \leq 9$ is true. Also, $8 \leq 8$ is true because $8 = 8$ is true.

The symbol $\geq$ means "is greater than or equal to."

$9 \geq 5$ 9 is greater than or equal to 5.

EXAMPLE 4 Using the Symbols $\leq$ and $\geq$

Determine whether each statement is *true* or *false.*

(a) $15 \leq 20$ The statement $15 \leq 20$ is true because $15 < 20$.

(b) $12 \geq 12$ Since $12 = 12$, this statement is true.

(c) $15 \leq 20 \cdot 2$ The statement $15 \leq 20 \cdot 2$ is true, because $15 < 40$.

(d) $\frac{6}{15} \geq \frac{2}{3}$

$\frac{6}{15} \geq \frac{10}{15}$ Find a common denominator.

The statements $\frac{6}{15} > \frac{10}{15}$ and $\frac{6}{15} = \frac{10}{15}$ are false. Because at least one of them is false, $\frac{6}{15} \geq \frac{2}{3}$ is also false.

Work Problem 5 at the Side. ▶

4 Write each statement in words. Then decide whether it is *true* or *false*.

(a) $7 < 5$

(b) $12 > 6$

(c) $4 \neq 10$

5 Determine whether each statement is *true* or *false*.

(a) $30 \leq 40$

(b) $25 \geq 10$

(c) $40 \leq 10$

(d) $21 \leq 21$

(e) $9 \cdot 3 \geq 28$

(f) $\frac{4}{7} \leq \frac{5}{8}$

Answers

4. **(a)** Seven is less than five. False
(b) Twelve is greater than six. True
(c) Four is not equal to ten. True

5. **(a)** true **(b)** true **(c)** false
(d) true **(e)** false **(f)** true

6 Write each word statement in symbols.

(a) Nine is equal to eleven minus two.

(b) Seventeen is less than thirty.

(c) Eight is not equal to ten.

(d) Fourteen is greater than twelve.

(e) Thirty is less than or equal to fifty.

(f) Two is greater than or equal to two.

7 Write each statement as another true statement with the inequality symbol reversed.

(a) $8 < 10$

(b) $3 > 1$

(c) $9 \leq 15$

(d) $6 \geq 2$

Answers

6. (a) $9 = 11 - 2$ (b) $17 < 30$ (c) $8 \neq 10$ (d) $14 > 12$ (e) $30 \leq 50$ (f) $2 \geq 2$

7. (a) $10 > 8$ (b) $1 < 3$ (c) $15 \geq 9$ (d) $2 \leq 6$

OBJECTIVE 5 Translate word statements to symbols.

EXAMPLE 5 Translating from Words to Symbols

Write each word statement in symbols.

(a) Twelve is equal to ten plus two. $12 = 10 + 2$

(b) Nine is less than ten. $9 < 10$
Compare this with "9 less than 10," which is written $10 - 9$.

(c) Fifteen is not equal to eighteen. $15 \neq 18$

(d) Seven is greater than four. $7 > 4$

(e) Thirteen is less than or equal to forty. $13 \leq 40$

(f) Six is greater than or equal to six. $6 \geq 6$

◀ **Work Problem 6 at the Side.**

OBJECTIVE 6 Write statements that change the direction of inequality symbols. Any statement with $<$ can be converted to one with $>$, and any statement with $>$ can be converted to one with $<$. ***We do this by reversing both the order of the numbers and the direction of the symbol.***

Interchange numbers.

$6 < 10$ becomes $10 > 6$.

Reverse symbol.

EXAMPLE 6 Converting between Inequality Symbols

Parts (a)–(c) show the same statements written in two equally correct ways. In each inequality, the symbol points toward the lesser number.

(a) $5 > 2,\ 2 < 5$ **(b)** $3 \leq 8,\ 8 \geq 3$ **(c)** $12 \geq 5,\ 5 \leq 12$

◀ **Work Problem 7 at the Side.**

Symbol	Meaning	Example
$=$	Is equal to	$0.5 = \frac{1}{2}$ means 0.5 is equal to $\frac{1}{2}$.
$\neq$	Is not equal to	$3 \neq 7$ means 3 is not equal to 7.
$<$	Is less than	$6 < 10$ means 6 is less than 10.
$>$	Is greater than	$15 > 14$ means 15 is greater than 14.
$\leq$	Is less than or equal to	$4 \leq 8$ means 4 is less than or equal to 8.
$\geq$	Is greater than or equal to	$1 \geq 0$ means 1 is greater than or equal to 0.

CAUTION

Equality and inequality symbols are used to write mathematical ***sentences,*** while operation symbols ($+$, $-$, $\cdot$, and $\div$) are used to write mathematical ***expressions*** that represent a number. Compare the following.

Sentence: $4 < 10$ ← Gives the relationship between 4 and 10

Expression: $4 + 10$ ← Tells how to operate on 4 and 10 to get 14

9.1 Exercises

FOR EXTRA HELP

Download the MyDashBoard App

MyMathLab®

CONCEPT CHECK *Decide whether each statement is* true *or* false. *If it is* false, *explain why.*

1. An exponent tells how many times its base is used as a factor.

2. Some grouping symbols are $+$, $-$, $\cdot$, and $\div$.

3. When evaluated, $4 + 3(8 - 2)$ is equal to 42.

4. $3^3 = 9$

5. The statement "4 is 12 less than 16" is interpreted $4 = 12 - 16$.

6. The statement "6 is 4 less than 10" is interpreted $6 < 10 - 4$.

7. **CONCEPT CHECK** When evaluating $(4^2 + 3^3)^4$, what is the *last* exponent that would be applied? Why?

8. **CONCEPT CHECK** Which are not grouping symbols—parentheses, brackets, fraction bars, exponents?

Find the value of each exponential expression. **See Example 1.**

9. 7^2

10. 4^2

11. 12^2

12. 14^2

13. 4^3

14. 5^3

15. 10^3

16. 11^3

17. 3^4

18. 6^4

19. 4^5

20. 3^5

21. $\left(\frac{2}{3}\right)^4$

22. $\left(\frac{3}{4}\right)^3$

23. $(0.04)^3$

24. $(0.05)^4$

Find the value of each expression. **See Examples 2 and 3.**

25. $13 + 9 \cdot 5$
GS ○ ○
$= 13 + \_\_\_\_$
$= \_\_\_\_$

26. $11 + 7 \cdot 6$
GS ○ ○
$= 11 + \_\_\_\_$
$= \_\_\_\_$

27. $20 - 4 \cdot 3 + 5$
GS ○ ○ ○
$= 20 - \_\_\_\_ + 5$
$= \_\_\_\_ + 5$
$= \_\_\_\_$

28. $18 - 7 \cdot 2 + 6$
GS ○ ○ ○
$= 18 - \_\_\_\_ + 6$
$= \_\_\_\_ + 6$
$= \_\_\_\_$

29. $9 \cdot 5 - 13$
GS ○ ○

30. $7 \cdot 6 - 11$
GS ○ ○

31. $18 - 2 + 3$
GS ○ ○

32. $22 - 8 + 9$
GS ○ ○

33. $\frac{1}{4} \cdot \frac{2}{3} + \frac{2}{5} \cdot \frac{11}{3}$

34. $\frac{9}{4} \cdot \frac{2}{3} + \frac{4}{5} \cdot \frac{5}{3}$

35. $9 \cdot 4 - 8 \cdot 3$

36. $11 \cdot 4 + 10 \cdot 3$

37. $2.5(1.9) + 4.3(7.3)$

38. $4.3(1.2) + 2.1(8.5)$

39. $10 + 40 \div 5 \cdot 2$

40. $12 + 35 \div 7 \cdot 3$

41. $18 - 2(3 + 4)$

42. $30 - 3(4 + 2)$

43. $5[3 + 4(2^2)]$

44. $4^2[(13 + 4) - 8]$

45. $\left(\frac{3}{2}\right)^2\left[\left(11 + \frac{1}{3}\right) - 6\right]$

46. $6\left[\frac{3}{4} + 8\left(\frac{1}{2}\right)^3\right]$

47. $\frac{8 + 6(3^2 - 1)}{3 \cdot 2 - 2}$

48. $\frac{8 + 2(8^2 - 4)}{4 \cdot 3 - 10}$

49. $\frac{4(7 + 2) + 8(8 - 3)}{6(4 - 2) - 2^2}$

50. $\frac{6(5 + 1) - 9(1 + 1)}{5(8 - 4) - 2^3}$

CONCEPT CHECK *Insert one pair of parentheses so that the left side of each equation is equal to the right side.*

51. $3 \cdot 6 + 4 \cdot 2 = 60$

52. $2 \cdot 8 - 1 \cdot 3 = 42$

53. $10 - 7 - 3 = 6$

54. $15 - 10 - 2 = 7$

55. $8 + 2^2 = 100$

56. $4 + 2^2 = 36$

Tell whether each statement is true *or* false. *In Exercises 59–68, first simplify each expression involving an operation.* **See Example 4.**

57. $8 \geq 17$

58. $10 \geq 41$

59. $17 \leq 18 - 1$

60. $12 \geq 10 + 2$

61. $6 \cdot 8 + 6 \cdot 6 \geq 0$

62. $4 \cdot 20 - 16 \cdot 5 \geq 0$

63. $6[5 + 3(4 + 2)] \leq 70$

64. $6[2 + 3(2 + 5)] \leq 135$

65. $\frac{9(7 - 1) - 8 \cdot 2}{4(6 - 1)} > 3$

66. $\frac{2(5 + 3) + 2 \cdot 2}{2(4 - 1)} > 1$

67. $8 \leq 4^2 - 2^2$

68. $10^2 - 8^2 > 6^2$

Write each word statement in symbols. **See Example 5.**

69. Fifteen is equal to five plus ten.

70. Twelve is equal to twenty minus eight.

71. Nine is greater than five minus four.

72. Ten is greater than six plus one.

73. Sixteen is not equal to nineteen.

74. Three is not equal to four.

75. Two is less than or equal to three.

76. Five is less than or equal to nine.

Write each statement in words and decide whether it is true *or* false. *(Hint: To compare fractions, write them with the same denominator.)*

77. $7 < 19$

78. $9 < 10$

79. $\frac{1}{3} \neq \frac{3}{10}$

80. $\frac{10}{7} \neq \frac{3}{2}$

81. $8 \geq 11$

82. $4 \leq 2$

Write each statement as another true statement with the inequality symbol reversed. ***See Example 6.***

83. $5 < 30$

84. $8 > 4$

85. $12 \geq 3$

86. $25 \leq 41$

87. $2.5 \geq 1.3$

88. $4.1 \leq 5.3$

89. $\frac{4}{5} > \frac{3}{4}$

90. $\frac{8}{3} < \frac{11}{4}$

One way to measure a person's cardiofitness is to calculate how many METs, or metabolic units, he or she can reach at peak exertion. One MET is the amount of energy used when sitting quietly. To calculate ideal METs, we can use the following expressions.

$$14.7 - \text{age} \cdot 0.13 \quad \text{For women}$$

$$14.7 - \text{age} \cdot 0.11 \quad \text{For men}$$

(Source: New England Journal of Medicine.)

91. A 40-yr-old woman wishes to calculate her ideal MET.

(a) Write the expression, using her age.

(b) Calculate her ideal MET. (*Hint:* Use the rules for order of operations.)

(c) Researchers recommend that a person reach approximately 85% of his or her MET when exercising. Calculate 85% of the ideal MET from part (b). Then refer to the following table. What activity can the woman do that is approximately this value?

Activity	METs	Activity	METs
Golf (with cart)	2.5	Skiing (water or downhill)	6.8
Walking (3 mph)	3.3	Swimming	7.0
Mowing lawn (power)	4.5	Walking (5 mph)	8.0
Ballroom or square dancing	5.5	Jogging	10.2
Cycling	5.7	Skipping rope	12.0

Source: Harvard School of Public Health.

(d) Repeat parts (a)−(c) for a 55-yr-old man.

92. Repeat parts (a)−(c) of **Exercise 91** for your age and gender. For yourself 5 yr from now.

The table shows the number of pupils per teacher in U.S. public schools in selected states during the 2009–2010 school year. Use this table to answer the questions in Exercises 93–96.

State	Pupils per Teacher
Alaska	15.3
Texas	14.7
California	22.2
Wyoming	12.7
Maine	11.8
Idaho	18.4
Missouri	14.1

Source: National Center for Education Statistics.

93. Which states had a number greater than 14.1?

94. Which states had a number that was at most 14.7?

95. Which states had a number not less than 14.1?

96. Which states had a number greater than 22.2?

9.2 Variables, Expressions, and Equations

OBJECTIVES

1. Evaluate algebraic expressions, given values for the variables.
2. Translate word phrases to algebraic expressions.
3. Identify solutions of equations.
4. Translate sentences to equations.
5. Distinguish between *equations* and *expressions*.

A **variable** is a symbol, usually a letter, used to represent an unknown number. A **constant** is a fixed, unchanging number.

x, y, or z Variables | 5, $\frac{3}{4}$, $8\frac{1}{2}$, 10.8 Constants

An **algebraic expression** is a collection of constants, variables, operation symbols, and grouping symbols, such as parentheses, square brackets, or fraction bars.

$x + 5$, $2m - 9$, $8p^2 + 6(p - 2)$ Algebraic expressions

$2m$ means $2 \cdot m$, the product of 2 and m.

$6(p - 2)$ means the product of 6 and $p - 2$.

OBJECTIVE 1 Evaluate algebraic expressions, given values for the variables. An algebraic expression can have different numerical values for different values of the variables.

EXAMPLE 1 Evaluating Expressions

Find the value of each algebraic expression for $m = 5$ and then for $m = 9$.

(a) $8m$

$= 8 \cdot 5$ Let $m = 5$.

$= 40$ Multiply.

$8m$

$= 8 \cdot 9$ Let $m = 9$.

$= 72$ Multiply.

(b) $3m^2$

$= 3 \cdot 5^2$ Let $m = 5$. ($5^2 = 5 \cdot 5$)

$= 3 \cdot 25$ Square 5.

$= 75$ Multiply.

$3m^2$

$= 3 \cdot 9^2$ Let $m = 9$. ($9^2 = 9 \cdot 9$)

$= 3 \cdot 81$ Square 9.

$= 243$ Multiply.

◀ Work Problem 1 at the Side.

CAUTION

$3m^2$ means $3 \cdot m^2$, **not** $3m \cdot 3m$. See **Example 1(b).**

Unless parentheses are used, the exponent refers only to the variable or number just before it. We use parentheses to write $3m \cdot 3m$ with exponents as $(3m)^2$.

EXAMPLE 2 Evaluating Expressions

Find the value of each expression for $x = 5$ and $y = 3$.

(a) $2x + 5y$ (We could write this with parentheses as $2(5) + 5(3)$.)

$= 2 \cdot 5 + 5 \cdot 3$ Replace x with 5 and y with 3.

$= 10 + 15$ Multiply. (Follow the rules for order of operations.)

$= 25$ Add.

Continued on Next Page

1 Find the value of each expression for $p = 3$.

GS **(a)** $6p$

$= 6 \cdot$ ____

$=$ ____

GS **(b)** $p + 12$

$=$ ____ $+ 12$

$=$ ____

GS **(c)** $5p^2$

$= 5 \cdot$ ____^{2}

$= 5 \cdot$ ____

$=$ ____

(d) $16p$

(e) $2p^3$

Answers

1. **(a)** 3; 18 **(b)** 3; 15 **(c)** 3; 9; 45 **(d)** 48 **(e)** 54

(b) $\dfrac{9x - 8y}{2x - y}$

$= \dfrac{9 \cdot 5 - 8 \cdot 3}{2 \cdot 5 - 3}$ Replace x with 5 and y with 3.

$= \dfrac{45 - 24}{10 - 3}$ Multiply.

$= \dfrac{21}{7}$ Subtract.

$= 3$ Divide.

(c) $x^2 - 2y^2$

$= 5^2 - 2 \cdot 3^2$ Replace x with 5 and y with 3. ($3^2 = 3 \cdot 3$; $5^2 = 5 \cdot 5$)

$= 25 - 2 \cdot 9$ Apply the exponents.

$= 25 - 18$ Multiply.

$= 7$ Subtract.

Work Problem **2** at the Side. ▶

2 Find the value of each expression for $x = 6$ and $y = 9$.

GS **(a)** $4x + 7y$

$= 4 \cdot \_\_\_\_ + 7 \cdot \_\_\_\_$

$= \_\_\_\_ + \_\_\_\_$

$= \_\_\_\_$

(b) $\dfrac{4x - 2y}{x + 1}$

(c) $2x^2 + y^2$

OBJECTIVE 2 Translate word phrases to algebraic expressions.

Problem-Solving Hint

Sometimes variables must be used to change word phrases into algebraic expressions. This process will be important in later chapters when we solve applied problems.

EXAMPLE 3 Using Variables to Write Word Phrases as Algebraic Expressions

Write each word phrase as an algebraic expression, using x as the variable.

(a) The sum of a number and 9

$x + 9$, or $9 + x$ "Sum" is the answer to an addition problem.

(b) 7 minus a number

$7 - x$ "Minus" indicates subtraction.

$x - 7$ is incorrect. We cannot subtract in either order and get the same result.

(c) A number subtracted from 12

$12 - x$ Be careful with order.

Compare this result with "12 subtracted from a number," which is $x - 12$.

(d) The product of 11 and a number

$11 \cdot x$, or $11x$

Continued on Next Page

Answers

2. **(a)** 6; 9; 24; 63; 87 **(b)** $\frac{6}{7}$ **(c)** 153

(e) 5 **divided by** a number

$$5 \div x, \quad \text{or} \quad \frac{5}{x}$$

$\frac{x}{5}$ is *not* correct here.

(f) The **product of** 2 and the **difference** between a number and 8

We are multiplying 2 times "something." This "something" is the difference between a number and 8, written $x - 8$. We use parentheses around this difference.

$$2 \cdot (x - 8), \quad \text{or} \quad 2(x - 8)$$

$8 - x$, which means the difference between 8 and a number, is not correct.

◀ **Work Problem 3 at the Side.**

OBJECTIVE 3 Identify solutions of equations. An **equation** is a statement that two expressions are equal. ***An equation always includes the equality symbol, =.***

$$\left.\begin{matrix} x + 4 = 11, & 2y = 16, & 4p + 1 = 25 - p, \\ \frac{3}{4}x + \frac{1}{2} = 0, & z^2 = 4, & 4(m - 0.5) = 2m \end{matrix}\right\} \text{Equations}$$

To **solve** an equation, we must find all values of the variable that make the equation true. Such values of the variable are the **solutions** of the equation.

EXAMPLE 4 Deciding Whether a Number Is a Solution of an Equation

Decide whether the given number is a solution of the equation.

(a) $5p + 1 = 36$; 7

$$5p + 1 = 36$$
$$5 \cdot 7 + 1 \stackrel{?}{=} 36 \quad \text{Let } p = 7.$$
$$35 + 1 \stackrel{?}{=} 36 \quad \text{Multiply.}$$
$$36 = 36 \checkmark \quad \text{True—the left side of the equation equals the right side.}$$

Be careful. Multiply first.

The number 7 is a solution of the equation.

(b) $9m - 6 = 32$; $\frac{14}{3}$

$$9m - 6 = 32$$
$$9 \cdot \frac{14}{3} - 6 \stackrel{?}{=} 32 \quad \text{Let } m = \tfrac{14}{3}.$$
$$42 - 6 \stackrel{?}{=} 32 \quad \text{Multiply.}$$
$$36 = 32 \quad \text{False—the left side does } \textit{not} \text{ equal the right side.}$$

The number $\frac{14}{3}$ is not a solution of the equation.

◀ **Work Problem 4 at the Side.**

3 Write each word phrase as an algebraic expression. Use x as the variable.

(a) The sum of 5 and a number

(b) A number minus 4

(c) A number subtracted from 48

(d) The product of 6 and a number

(e) 9 multiplied by the sum of a number and 5

4 Decide whether the given number is a solution of the equation.

(a) $p - 1 = 3$; 2

(b) $2k + 3 = 15$; 7

(c) $7p - 11 = 5$; $\frac{16}{7}$

Answers

3. (a) $5 + x$ (b) $x - 4$ (c) $48 - x$ (d) $6x$ (e) $9(x + 5)$
4. (a) no (b) no (c) yes

OBJECTIVE 4 Translate sentences to equations. Sentences given in words are translated as equations.

EXAMPLE 5 Translating Sentences to Equations

Write each word sentence as an equation. Use x as the variable.

(a) Twice the sum of a number and four is six.

"Twice" means two times. The word *is* suggests equals. With x representing the number, translate as follows.

Twice	the sum of a number and four	is	six.
↓	↓	↓	↓
$2 \cdot$	$(x + 4)$	$=$	6

The word *is* translates as =.

Use x for the unknown number.

$$2(x + 4) = 6$$

(b) Nine more than five times a number is 49.

Start with $5x$, and then add 9 to it.	The word *is* translates as =.		
↓	↓		
$5x + 9$	$=$	49	$5 \cdot x = 5x$

$$5x + 9 = 49$$

(c) Seven less than three times a number is equal to eleven.

Three times a number	less	seven	is equal to	eleven.
↓	↓	↓	↓	↓
$3x$	$-$	7	$=$	11

$$3x - 7 = 11$$

Work Problem 5 at the Side. ▶

OBJECTIVE 5 Distinguish between equations and expressions. Students often have trouble distinguishing between equations and expressions. ***An equation is a sentence—it has something on the left side, an = symbol, and something on the right side. An expression is a phrase that represents a number.***

$$\underbrace{4x + 5}_{\text{Left side}} = \underbrace{9}_{\text{Right side}}$$

Equation

$$4x + 5$$

Expression

EXAMPLE 6 Distinguishing between Equations and Expressions

Decide whether each is an *equation* or an *expression.*

(a) $2x - 3$ Ask, "Is there an equality symbol?" The answer is no, so this is an expression.

(b) $2x - 3 = 8$ Because there is an equality symbol with something on either side of it, this is an equation.

(c) $5x^2 + 2y^2$ There is no equality symbol. This is an expression.

Work Problem 6 at the Side. ▶

---

5 Write each word sentence as an equation. Use x as the variable.

GS **(a)** Three times the sum of a number and 13 is 19.

___ (___ + ___) = ___

(b) Five times a number subtracted from 21 is 15.

(c) Five less than six times a number is equal to nineteen.

6 Decide whether each is an *equation* or an *expression.*

(a) $2x + 5y - 7$

(b) $\dfrac{3x - 1}{5}$

(c) $2x + 5 = 7$

(d) $\dfrac{x}{x - 3} = 4x$

Answers

5. **(a)** $3; x; 13; 19$ **(b)** $21 - 5x = 15$ **(c)** $6x - 5 = 19$

6. **(a)** expression **(b)** expression **(c)** equation **(d)** equation

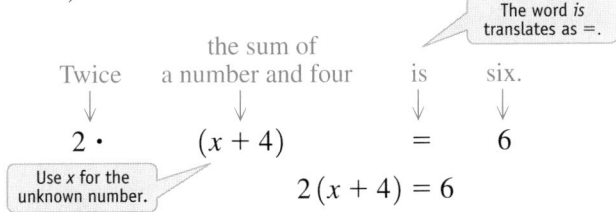

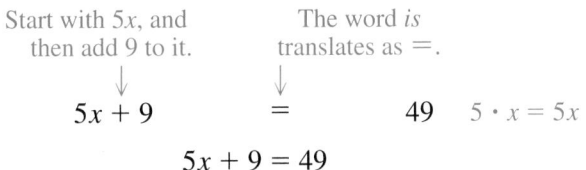

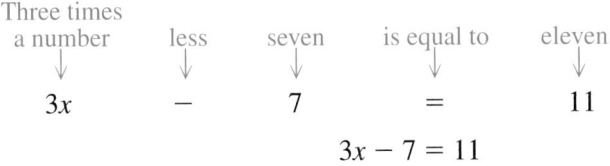

9.2 Exercises

FOR EXTRA HELP

CONCEPT CHECK *Choose the letter(s) of the correct response.*

1. The expression $8x^2$ means ______.

A. $8 \cdot x \cdot 2$ **B.** $8 \cdot x \cdot x$ **C.** $8 + x^2$ **D.** $8x^2 \cdot 8x^2$

2. If $x = 2$ and $y = 1$, then the value of xy is ______.

A. $\frac{1}{2}$ **B.** 1 **C.** 2 **D.** 3

3. The sum of 15 and a number x is represented by ______.

A. $15 + x$ **B.** $15 - x$ **C.** $x - 15$ **D.** $15x$

4. Which of the following are expressions?

A. $6x = 7$ **B.** $6x + 7$ **C.** $6x - 7$ **D.** $6x - 7 = 0$

CONCEPT CHECK *Complete each statement.*

5. For $x = 3$, the value of $x + 8$ is ______.

6. For $x = 1$ and $y = 2$, the value of $5xy$ is ______.

7. "The sum of 13 and x" is represented by the expression ______. For $x = 3$, the value of this expression is ______.

8. $2x + 6$ is an (*equation / expression*), while $2x + 6 = 8$ is an (*equation / expression*).

In Exercises 9–12, give a short explanation.

9. If the words *more than* in **Example 5(b)** were changed to *less than,* how would the equation be changed?

10. Why is $2x^3$ not the same as $2x \cdot 2x \cdot 2x$? Explain, using an exponent to write $2x \cdot 2x \cdot 2x$.

11. There are many pairs of values of x and y for which $2x + y$ can be evaluated. Name two such pairs and then evaluate the expression.

12. When evaluating the expression $4x^2$ for $x = 3$, explain why 3 must be squared *before* multiplying by 4.

Find the value of each expression for **(a)** $x = 4$ *and* **(b)** $x = 6$. ***See Example 1.***

13. $4x^2$
(a) **(b)**

14. $5x^2$
(a) **(b)**

15. $\dfrac{3x - 5}{2x}$
(a) **(b)**

16. $\dfrac{4x - 1}{3x}$
(a) **(b)**

17. $\dfrac{6.459x}{2.7}$ (to the nearest thousandth)
(a) **(b)**

18. $\dfrac{0.74x^2}{0.85}$ (to the nearest thousandth)
(a) **(b)**

19. $3x^2 + x$
(a) **(b)**

20. $2x + x^2$
(a) **(b)**

Find the value of each expression for **(a)** $x = 2$ *and* $y = 1$ *and* **(b)** $x = 1$ *and* $y = 5$.
See Example 2.

21. $3(x + 2y)$
(a) **(b)**

22. $2(2x + y)$
(a) **(b)**

23. $x + \dfrac{4}{y}$
(a) **(b)**

24. $y + \dfrac{8}{x}$
(a) **(b)**

25. $\dfrac{x}{2} + \dfrac{y}{3}$
(a) **(b)**

26. $\dfrac{x}{5} + \dfrac{y}{4}$
(a) **(b)**

27. $\dfrac{2x + 4y - 6}{5y + 2}$
(a) **(b)**

28. $\dfrac{4x + 3y - 1}{2x + y}$
(a) **(b)**

29. $2y^2 + 5x$
(a) **(b)**

30. $6x^2 + 4y$
(a) **(b)**

31. $\dfrac{3x + y^2}{2x + 3y}$
(a) **(b)**

32. $\dfrac{x^2 + 1}{4x + 5y}$
(a) **(b)**

33. $0.841x^2 + 0.32y^2$
(a) **(b)**

34. $0.941x^2 + 0.2y^2$
(a) **(b)**

Write each word phrase as an algebraic expression, using x as the variable. ***See Example 3.***

35. Twelve times a number

36. Thirteen added to a number

37. Two subtracted from a number

38. Eight subtracted from a number

39. One-third of a number, subtracted from seven

40. One-fifth of a number, subtracted from fourteen

41. The difference between twice a number and 6

42. The difference between 6 and half a number

43. 12 divided by the sum of a number and 3

44. The difference between a number and 5, divided by 12

45. The product of 6 and four less than a number

46. The product of 9 and five more than a number

47. Suppose that the directions on a quiz read, "Evaluate each equation for $x = 3$." How can you justify politely correcting the person who wrote these directions?

48. Suppose that the directions on a test read, "Solve the following expressions." How can you justify politely correcting the person who wrote these directions?

Decide whether the given number is a solution of the equation. ***See Example 4.***

49. Is 7 a solution of $x - 5 = 12$?

50. Is 10 a solution of $x + 6 = 15$?

51. Is 1 a solution of $5x + 2 = 7$?

52. Is 1 a solution of $3x + 5 = 8$?

53. Is $\frac{1}{5}$ a solution of $6x + 4x + 9 = 11$?

54. Is $\frac{12}{5}$ a solution of $2x + 3x + 8 = 20$?

55. Is 3 a solution of $2y + 3(y - 2) = 14$?

56. Is 2 a solution of $6x + 2(x + 3) = 14$?

57. Is $\frac{1}{3}$ a solution of $\frac{z+4}{2-z} = \frac{13}{5}$?

58. Is $\frac{13}{4}$ a solution of $\frac{x+6}{x-2} = \frac{37}{5}$?

59. Is 4.3 a solution of $3r^2 - 2 = 53.47$?

60. Is 3.7 a solution of $2x^2 + 1 = 28.38$?

Write each word sentence as an equation. Use x as the variable. ***See Example 5.***

61. The sum of a number and 8 is 18.

62. A number minus three equals 1.

63. Five more than twice a number is 5.

64. The product of 2 and the sum of a number and 5 is 14.

65. Sixteen minus three-fourths of a number is 13.

66. The sum of six-fifths of a number and 2 is 14.

67. Three times a number is equal to 8 more than twice the number.

68. Twelve divided by a number equals $\frac{1}{3}$ times that number.

Identify each as an expression *or an* equation. ***See Example 6.***

69. $3x + 2(x - 4)$

70. $5y - (3y + 6)$

71. $7t + 2(t + 1) = 4$

72. $9r + 3(r - 4) = 2$

73. $x + y = 9$

74. $x + y - 9$

75. $\frac{3x - 8}{2}$

76. $\frac{4x + 3}{2} = 11$

Relating Concepts (Exercises 77–80) For Individual or Group Work

A ***mathematical model*** *is an equation that describes the relationship between two quantities. For example, the life expectancy of Americans at birth can be approximated by the equation*

$$y = 0.174x - 270,$$

where x is a year between 1950 and 2008 and y is age in years. (Source: Centers for Disease Control and Prevention.)

Use this model to approximate life expectancy (to the nearest year) in each of the following years.

77. 1950

78. 1975

79. 1995

80. 2008

9.3 Real Numbers and the Number Line

OBJECTIVES

1. Classify numbers and graph them on number lines.
2. Tell which of two real numbers is less than the other.
3. Find the additive inverse of a real number.
4. Find the absolute value of a real number.

A **set** is a collection of objects. In mathematics, these objects are usually numbers. The objects that belong to the set are its **elements.** They are written between braces.

$\{1, 2, 3, 4, 5\}$ The set of numbers 1, 2, 3, 4, and 5

OBJECTIVE 1 Classify numbers and graph them on number lines. The set of numbers used for counting is the *natural numbers*. The set of *whole numbers* includes 0 with the natural numbers.

Natural Numbers and Whole Numbers

$\{1, 2, 3, 4, 5, \ldots\}$ is the set of **natural numbers** (or **counting numbers**).

$\{0, 1, 2, 3, 4, 5, \ldots\}$ is the set of **whole numbers.**

We can represent numbers on a **number line** like the one in **Figure 1.**

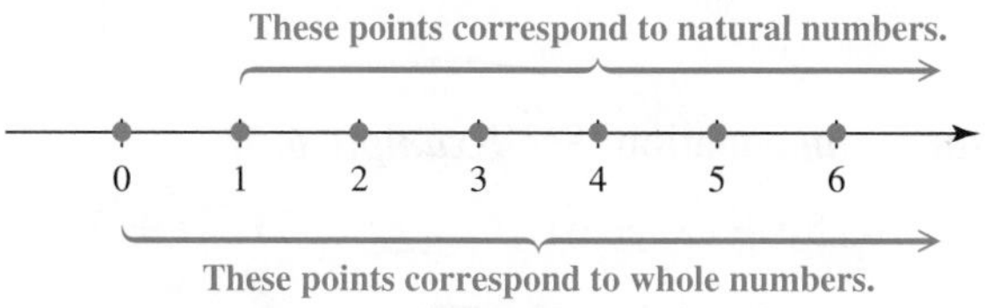

Figure 1

To draw a number line, choose any point on the line and label it 0. Then choose any point to the right of 0 and label it 1. Use the distance between 0 and 1 as the scale to locate, and then label, other points.

The natural numbers are located to the right of 0 on the number line. For each natural number, we can place a corresponding number to the left of 0, labeling the points -1, -2, -3, and so on, as shown in **Figure 2.** Each is the **opposite,** or **negative,** of a natural number. The natural numbers, their opposites, and 0 form the set of *integers*.

Integers

$\{\ldots, -3, -2, -1, 0, 1, 2, 3, \ldots\}$ is the set of **integers.**

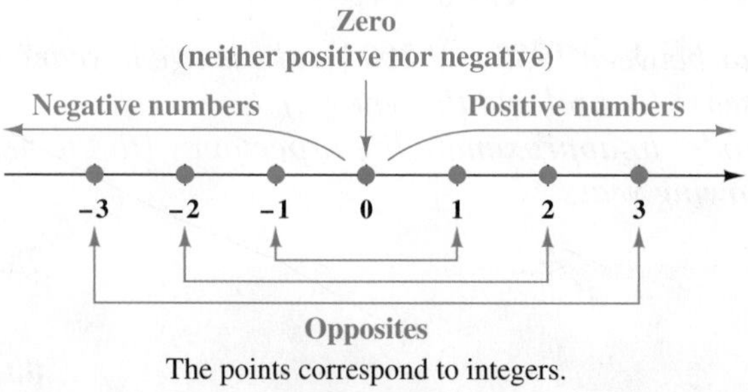

Figure 2

Positive numbers and *negative numbers* are **signed numbers.**

EXAMPLE 1 **Using Negative Numbers in Applications**

Use an integer to express the boldface italic number in each application.

(a) The lowest Fahrenheit temperature ever recorded in meteorological records was ***129°*** below zero at Vostok, Antarctica, on July 21, 1983. (*Source: World Almanac and Book of Facts.*)
Use -129 because "below zero" indicates a negative number.

(b) The shore surrounding the Dead Sea is ***1348*** ft below sea level. (*Source: World Almanac and Book of Facts.*)
Again, "below sea level" indicates a negative number, -1348.

Work Problem 1 at the Side. ▶

Fractions, discussed in **Chapters 2 and 3,** are *rational numbers.*

Rational Numbers

$\{x \mid x$ is a quotient of two integers, with denominator not $0\}$ is the set of **rational numbers.**

(Read the part in the braces as "the set of all numbers x such that x is a quotient of two integers, with denominator not 0.")

Note

The set symbolism used in the definition of rational numbers,

$$\{x \mid x \text{ has a certain property}\},$$

is called **set-builder notation.** This notation is convenient to use when it is not possible to list all the elements of a set.

Since any number that can be written as the quotient of two integers (that is, as a fraction) is a rational number, ***all integers, mixed numbers, terminating (or ending) decimals, and repeating decimals are rational.*** The table gives examples.

Rational Number	Equivalent Quotient of Two Integers
-5	$\frac{-5}{1}$ (means $-5 \div 1$)
$1\frac{3}{4}$	$\frac{7}{4}$ (means $7 \div 4$)
0.23 (terminating decimal)	$\frac{23}{100}$ (means $23 \div 100$)
0.3333…, or $0.\overline{3}$ (repeating decimal)	$\frac{1}{3}$ (means $1 \div 3$)
4.7	$\frac{47}{10}$ (means $47 \div 10$)

To **graph** a number, we place a dot on the number line at the point that corresponds to the number. The number is the **coordinate** of the point. Think of the graph of a set of numbers as a picture of the set.

1 Use an integer to express the boldface italic number(s) in each application.

(a) Erin discovers that she has spent $***53*** more than she has in her checking account.

(b) The record high Fahrenheit temperature in the United States was ***134°*** in Death Valley, California, on July 10, 1913. (*Source: World Almanac and Book of Facts.*)

(c) A football team gained ***5*** yd, then lost ***10*** yd on the next play.

Answers

1. (a) -53 **(b)** 134 **(c)** $5, -10$

2 Graph each number on the number line.

$-3, \ \frac{17}{8}, \ -2.75, \ 1\frac{1}{2}, \ -\frac{3}{4}$

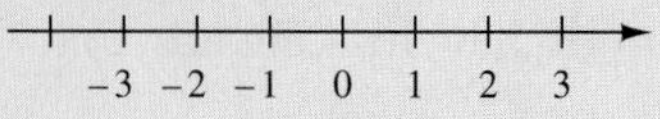

EXAMPLE 2 Graphing Numbers

Graph each number on a number line.

$$-\frac{3}{2}, \quad -\frac{2}{3}, \quad \frac{1}{2}, \quad 1\frac{1}{3}, \quad \frac{23}{8}, \quad 3\frac{1}{4}$$

To locate the improper fractions on the number line, write them as mixed numbers or decimals. The graph is shown in **Figure 3.**

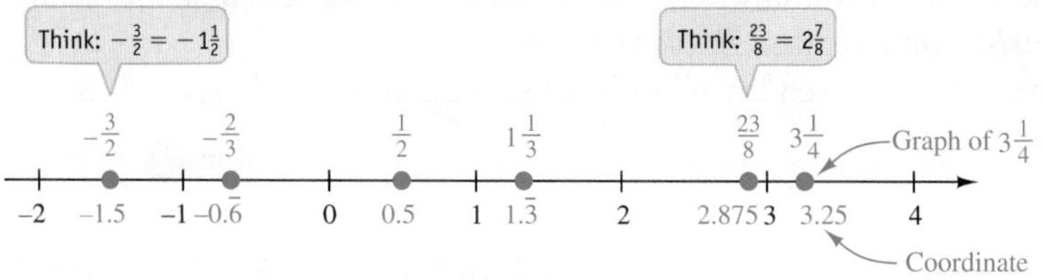

Figure 3

◀ **Work Problem 2 at the Side.**

The square root of 2, written $\sqrt{2}$, cannot be written as a quotient of two integers. Because of this, $\sqrt{2}$ is an *irrational number.* (See **Figure 4.**)

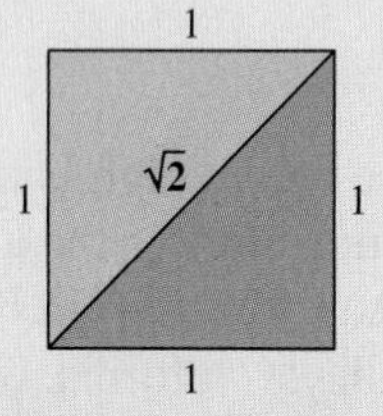

This square has diagonal of length $\sqrt{2}$. The number $\sqrt{2}$ is an irrational number.

Figure 4

Irrational Numbers

$\{x \mid x$ is a nonrational number represented by a point on the number line$\}$ is the set of **irrational numbers.**

The decimal form of an irrational number neither terminates nor repeats.

Both rational and irrational numbers can be represented by points on the number line. Together, they form the **real numbers.** See **Figure 5.**

Real Numbers

$\{x \mid x$ is a rational or an irrational number$\}$ is the set of **real numbers.**

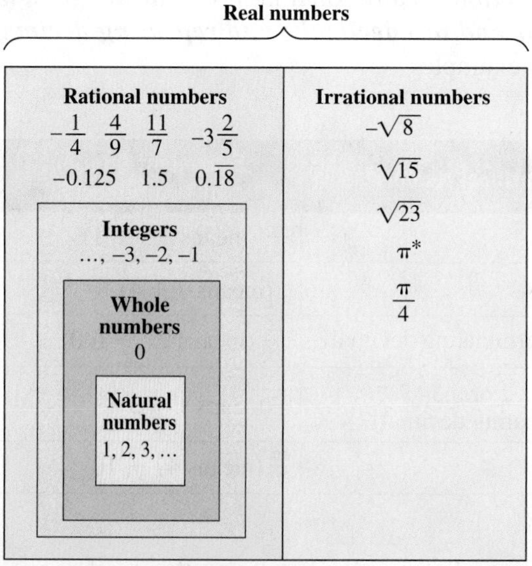

Figure 5

*The value of π (pi) is approximately 3.141592654. The decimal digits continue forever with no repeated pattern.

Answer

2.

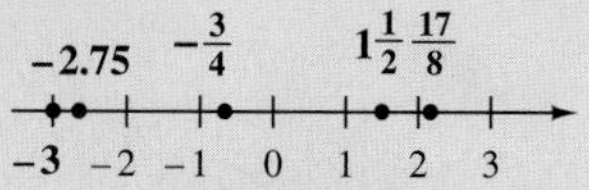

EXAMPLE 3 **Determining Whether a Number Belongs to a Set**

List the numbers in the following set that belong to each set of numbers.

$$\left\{-5,\ -\frac{2}{3},\ 0,\ 0.\overline{6},\ \sqrt{2},\ 3\frac{1}{4},\ 5,\ 5.8\right\}$$

(a) Natural numbers: 5

(b) Whole numbers: 0 and 5
The whole numbers consist of the natural (counting) numbers and 0.

(c) Integers: -5, 0, and 5

(d) Rational numbers: $-5, -\frac{2}{3}, 0, 0.\overline{6}\left(\text{or } \frac{2}{3}\right), 3\frac{1}{4}\left(\text{or } \frac{13}{4}\right), 5, \text{ and } 5.8\left(\text{or } \frac{58}{10}\right)$
Each of these numbers can be written as the quotient of two integers.

(e) Irrational numbers: $\sqrt{2}$

(f) Real numbers: All the numbers in the set are real numbers.

Work Problem **3** at the Side. ▶

OBJECTIVE 2 **Tell which of two real numbers is less than the other.** Given any two positive integers, you probably can tell which number is less than the other. Positive numbers decrease as the corresponding points on the number line go to the left. For example, $8 < 12$ because 8 is to the left of 12 on the number line. This ordering is extended to all real numbers by definition.

Ordering of Real Numbers

For any two real numbers *a* and *b*, ***a* is less than *b*** if *a* is to the left of *b* on a number line.

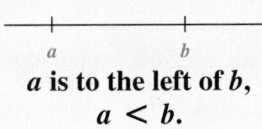

***a* is to the left of *b*,**
$\boldsymbol{a < b}$.

We can also say that, for any real numbers *a* and *b*, ***a* is *greater than* b** if *a* is to the right of *b* on the number line. Any negative number is less than 0, and any negative number is less than any positive number. Also, 0 is less than any positive number.

EXAMPLE 4 **Determining the Order of Real Numbers**

Is the statement $-3 < -1$ *true* or *false*?

Locate -3 and -1 on a number line, as shown in **Figure 6.** Because -3 is to the left of -1 on the number line, -3 is less than -1. The statement $-3 < -1$ is true.

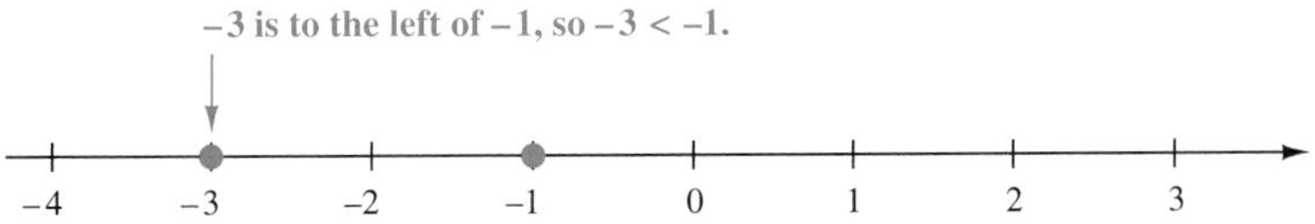

Figure 6

Work Problem **4** at the Side. ▶

3 List the numbers in the following set that belong to each set of numbers.

$$\left\{-7,\ -\frac{4}{5},\ 0,\ \sqrt{3},\ 2.7,\ 13\right\}$$

(a) Whole numbers

(b) Integers

(c) Rational numbers

(d) Irrational numbers

4 Tell whether each statement is *true* or *false*.

GS **(a)** $-2 < 4$
-2 is to the (*left* / *right*) of 4 on a number line, so the statement is (*true* / *false*).

GS **(b)** $6 > -3$
6 is to the (*left* / *right*) of -3 on a number line, so the statement is (*true* / *false*).

(c) $-9 < -12$

(d) $-4 \geq -1$

(e) $-6 \leq 0$

Answers

3. **(a)** 0, 13 **(b)** -7, 0, 13
(c) $-7, -\frac{4}{5}, 0, 2.7, 13$ **(d)** $\sqrt{3}$

4. **(a)** left; true **(b)** right; true **(c)** false
(d) false **(e)** true

5 Find the additive inverse of each number.

(a) 15

(b) −9

(c) −12

(d) $\frac{1}{2}$

(e) 0

Answers

5. (a) −15 (b) 9 (c) 12
(d) $-\frac{1}{2}$ (e) 0

OBJECTIVE 3 Find the additive inverse of a real number. By a property of the real numbers, for any real number x (except 0), there is exactly one number on a number line the same distance from 0 as x, but on the *opposite* side of 0. See **Figure 7.** Such pairs of numbers are *additive inverses,* or *opposites,* of each other.

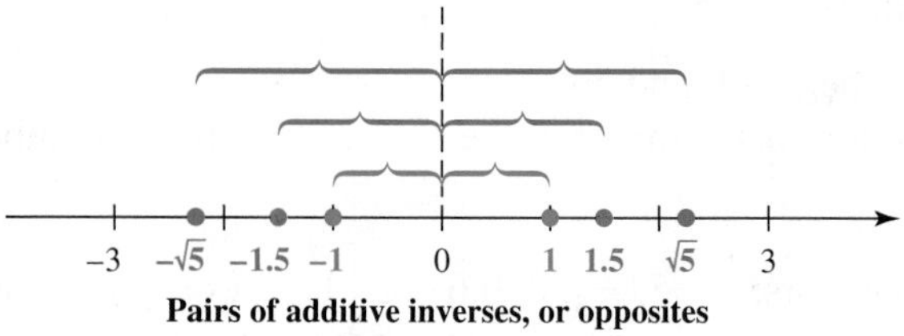

Figure 7

Additive Inverse

The **additive inverse** of a number x is the number that is the same distance from 0 on a number line as x, but on the *opposite* side of 0. (The number 0 is its own additive inverse.)

We indicate the additive inverse of a number by writing the symbol − in front of the number. For example, the additive inverse of 7 is −7 (read "negative 7"). We could write the additive inverse of −4 as −(−4), but we know that 4 is the opposite of −4. Since a number can have only one additive inverse, −(−4) and 4 must represent the same number, so

$$-(-4) = 4.$$

This idea can be generalized.

Double Negative Rule

For any real number a, $\quad \boldsymbol{-(-a) = a.}$

The following table shows several numbers and their additive inverses.

Number	Additive Inverse
−4	−(−4), or 4
0	0
5	−5
$-\frac{2}{3}$	$\frac{2}{3}$
0.52	−0.52

The table suggests the following rule.

The additive inverse of a nonzero number is found by changing the sign of the number.

◀ Work Problem 5 at the Side.

OBJECTIVE 4 Find the absolute value of a real number. Because additive inverses are the same distance from 0 on a number line, a number and its additive inverse have the same *absolute value*. The **absolute value** of a real number x, written $|x|$ and read **"the absolute value of x,"** can be defined as the distance between 0 and the number on a number line.

$|2| = 2$ The distance between 2 and 0 on a number line is 2 units.

$|-2| = 2$ The distance between -2 and 0 on a number line is also 2 units.

Distance is a physical measurement, which is never negative. ***Therefore, the absolute value of a number is never negative.***

Absolute Value

For any real number x,

$$|x| = \begin{cases} x & \text{if } x \geq 0 \\ -x & \text{if } x < 0. \end{cases}$$

By this definition, if x is a positive number or 0, then its absolute value is x itself. For example, since 8 is a positive number,

$$|8| = 8 \quad \text{and} \quad |0| = 0.$$

If x is a negative number, then its absolute value is the additive inverse of x.

$|-8| = -(-8) = 8$ The additive inverse of -8 is 8.

EXAMPLE 5 Finding the Absolute Value

Simplify by finding the absolute value.

(a) $|0| = 0$ **(b)** $|5| = 5$ **(c)** $|-5| = -(-5) = 5$

(d) $-|5| = -(5) = -5$ **(e)** $-|-5| = -(5) = -5$

(f) $|8 - 2| = |6| = 6$ **(g)** $-|8 - 2| = -|6| = -6$

Parts (f) and (g) show that absolute value bars are grouping symbols. We perform any operations inside absolute value symbols *before* finding the absolute value.

Work Problem 6 at the Side. ▶

6 Simplify by finding the absolute value.

(a) $|-6|$

(b) $|9|$

(c) $-|15|$

GS **(d)** $-|-9|$

$= -\_\_\_\_$

$= \_\_\_\_$

(e) $|9 - 4|$

GS **(f)** $-|32 - 2|$

$= -|\_\_\_\_|$

$= \_\_\_\_$

Answers

6. **(a)** 6 **(b)** 9 **(c)** -15 **(d)** 9; -9 **(e)** 5 **(f)** 30; -30

9.3 Exercises

FOR EXTRA HELP

Download the MyDashBoard App

MyMathLab®

CONCEPT CHECK *Complete each statement.*

1. The number ___________ is a whole number, but not a natural number.

2. The natural numbers, their additive inverses, and 0 form the set of ___________.

3. The additive inverse of every negative number is a (*negative* / *positive*) number.

4. If x and y are real numbers with $x > y$, then x lies to the (*left* / *right*) of y on a number line.

5. A rational number is the ___________ of two integers, with the ___________ not equal to 0.

6. Decimals that neither terminate nor repeat are ___________ numbers.

Use an integer to express each boldface italic number representing a change in the following applications. **See Example 1.**

7. Between July 1, 2008, and July 1, 2009, the population of the United States increased by approximately ***2,632,000.*** (*Source:* U.S. Census Bureau.)

8. From 2010 to 2011, the mean SAT critical reading score for Illinois students increased by ***14,*** while the mathematics score increased by ***17.*** (*Source:* The College Board.)

9. From 2005 to 2009, the paid circulation of daily newspapers in the United States went from 53,345 thousand to 46,278 thousand, representing a decrease of ***7067*** thousand. (*Source: Editor and Publisher International Yearbook,* 2010.)

10. In 1935, there were 15,295 banks in the United States. By 2010, the number was 7821, representing a decrease of ***7474*** banks. (*Source:* Federal Deposit Insurance Corporation.)

Graph each group of numbers on a number line. **See Example 2.**

11. $0, 3, -5, -6$

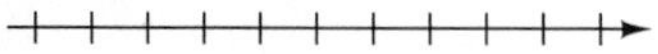

12. $2, 6, -2, -1$

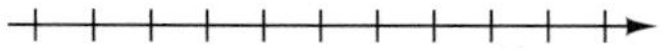

13. $-2, -6, -4, 3, 4$

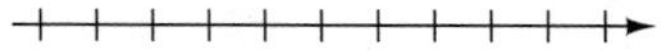

14. $-5, -3, -2, 0, 4$

15. $\frac{1}{4}, 2\frac{1}{2}, -3\frac{4}{5}, -4, -\frac{13}{8}$

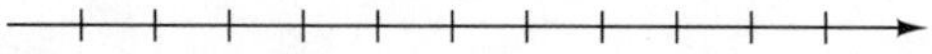

16. $5\frac{1}{4}, \frac{41}{9}, -2\frac{1}{3}, 0, -3\frac{2}{5}$

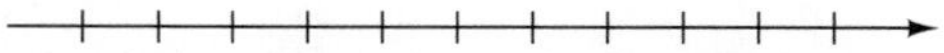

List all numbers from each set that are **(a)** *natural numbers,* **(b)** *whole numbers,* **(c)** *integers,* **(d)** *rational numbers,* **(e)** *irrational numbers,* **(f)** *real numbers.* **See Example 3.**

17. $\left\{-9, -\sqrt{7}, -1\frac{1}{4}, -\frac{3}{5}, 0, \sqrt{5}, 3, 5.9, 7\right\}$

18. $\left\{-5.3, -5, -\sqrt{3}, -1, -\frac{1}{9}, 0, 1.2, 4, \sqrt{12}\right\}$

19. $\left\{\frac{7}{9}, -2.\overline{3}, \sqrt{3}, 0, -8\frac{3}{4}, 11, -6, \pi\right\}$

20. $\left\{1\frac{5}{8}, -0.\overline{4}, \sqrt{6}, 9, -12, 0, \sqrt{10}, 0.026\right\}$

CONCEPT CHECK *In Exercises 21–26, give a number that satisfies the given condition.*

21. An integer between 3.6 and 4.6

22. A rational number between 2.8 and 2.9

23. A whole number that is not positive and is less than 1

24. A whole number greater than 3.5

25. An irrational number that is between $\sqrt{12}$ and $\sqrt{14}$

26. A real number that is neither negative nor positive

CONCEPT CHECK *In Exercises 27–32, decide whether each statement is* true *or* false.

27. Every natural number is positive.

28. Every whole number is positive.

29. Every integer is a rational number.

30. Every rational number is a real number.

31. Some numbers are both rational and irrational.

32. Every terminating decimal is a rational number.

CONCEPT CHECK *Give three numbers between* -6 *and* 6 *that satisfy each given condition.*

33. Positive real numbers but not integers

34. Real numbers but not positive numbers

35. Real numbers but not whole numbers

36. Rational numbers but not integers

37. Real numbers but not rational numbers

38. Rational numbers but not negative numbers

Select the lesser number in each pair. ***See Example 4.***

39. $-11, -4$

40. $-9, -16$

41. $-21, 1$

42. $-57, 3$

43. $0, -100$

44. $-215, 0$

45. $-\frac{2}{3}, -\frac{1}{4}$

46. $-\frac{3}{8}, -\frac{9}{16}$

Decide whether each statement is true *or* false. ***See Example 4.***

47. $8 < -16$

48. $12 < -24$

49. $-3 < -2$

50. $-10 < -9$

For each number, (a) find its additive inverse and (b) find its absolute value.

51. -2 (a) (b)

52. -8 (a) (b)

53. 6 (a) (b)

54. 11 (a) (b)

55. $-\frac{3}{4}$ (a) (b)

56. $-\frac{1}{3}$ (a) (b)

57. 4.95 (a) (b)

58. 0.007 (a) (b)

59. CONCEPT CHECK Match each expression in Column I with its value in Column II. Choices in Column II may be used once, more than once, or not at all.

I	II
(a) $\lvert -9 \rvert$	A. 9
(b) $-(-9)$	B. -9
(c) $-\lvert -9 \rvert$	C. Neither A nor B
(d) $-\lvert -(-9) \rvert$	D. Both A and B

60. CONCEPT CHECK Fill in the blanks with the correct values: The opposite of -5 is ____, while the absolute value of -5 is ____. The additive inverse of -5 is ____, while the additive inverse of the absolute value of -5 is ____.

Simplify by finding the absolute value. ***See Example 5.***

61. $\lvert -7 \rvert$

62. $\lvert -3 \rvert$

63. $-\lvert 12 \rvert$

64. $-\lvert 23 \rvert$

65. $-\left\lvert -\frac{2}{3} \right\rvert$

66. $-\left\lvert -\frac{4}{5} \right\rvert$

67. $\lvert 13 - 4 \rvert$

68. $\lvert 8 - 7 \rvert$

Decide whether each statement is true *or* false. ***See Examples 4 and 5.***

69. $\lvert -8 \rvert < 7$

70. $\lvert -6 \rvert \geq -\lvert 6 \rvert$

71. $4 \leq \lvert 4 \rvert$

72. $-\lvert -3 \rvert > 2$

The table shows the change in the Consumer Price Index (CPI) for selected categories of goods and services from 2008 to 2009 and from 2009 to 2010. Use the table to answer Exercises 73–76.

Category	Change from 2008 to 2009	Change from 2009 to 2010
Apparel	1.2	−0.6
Food	3.9	1.6
Energy	−43.6	18.3
Medical care	11.5	12.8
Transportation	−16.2	14.1

Source: U.S. Bureau of Labor Statistics.

73. Which category for which period represents the greatest decrease?

74. Which category for which period represents the least change?

75. Which has lesser absolute value, the change for Transportation from 2008 to 2009 or from 2009 to 2010?

76. Which has greater absolute value, the change for Energy from 2008 to 2009 or from 2009 to 2010?

9.4 Adding Real Numbers

OBJECTIVES

1. **Add two numbers with the same sign.**
2. **Add numbers with different signs.**
3. **Add mentally.**
4. **Use the rules for order of operations with real numbers.**
5. **Translate words and phrases that indicate addition.**

OBJECTIVE 1 Add two numbers with the same sign. Recall that the answer to an addition problem is the **sum.** A number line can be used to add real numbers.

EXAMPLE 1 Adding Two Positive Numbers on a Number Line

Use a number line to find the sum $2 + 3$.

Step 1 Start at 0 and draw an arrow 2 units to the *right*. See **Figure 8.**

Step 2 From the right end of that arrow, draw another arrow 3 units to the right.

The number below the end of this second arrow is 5, so $2 + 3 = 5$.

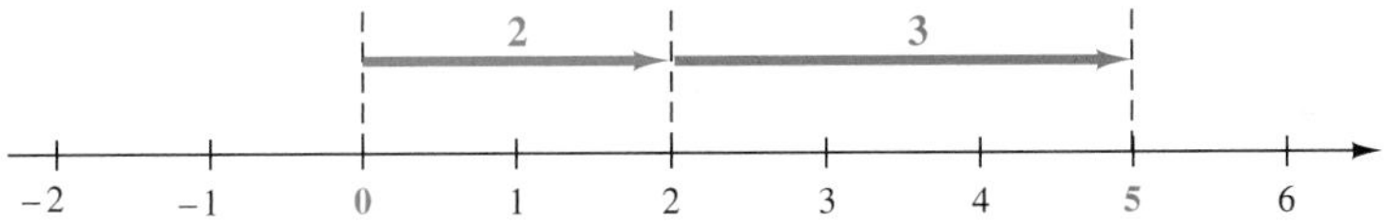

Figure 8

Work Problem **1** at the Side. ▶

1 Use a number line to find the sum $1 + 4$.

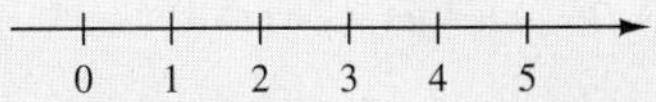

EXAMPLE 2 Adding Two Negative Numbers on a Number Line

Use a number line to find the sum $-2 + (-4)$. (We put parentheses around the -4 due to the $+$ and $-$ next to each other.)

Step 1 Start at 0 and draw an arrow 2 units to the *left*. See **Figure 9**.

Step 2 From the left end of the first arrow, draw a second arrow 4 units to the *left* to represent the addition of a *negative* number.

The number below the end of this second arrow is -6, so $-2 + (-4) = -6$.

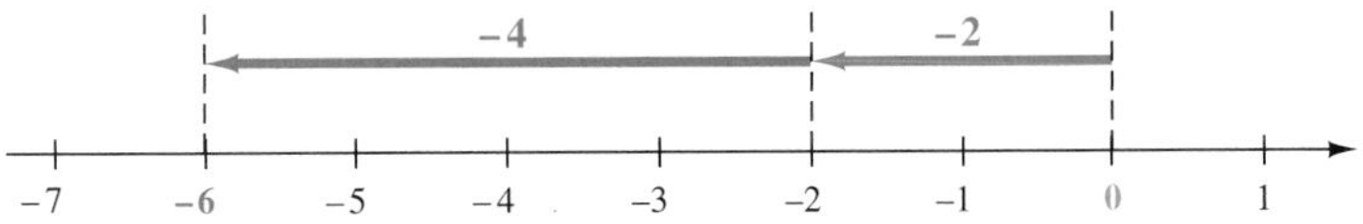

Figure 9

Work Problem **2** at the Side. ▶

2 Use a number line to find the sum $-2 + (-5)$.

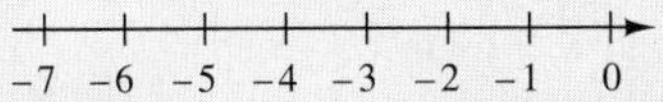

In **Example 2,** we found that the sum of the two negative numbers -2 and -4 is a negative number whose distance from 0 is the sum of the distance of -2 from 0 and the distance of -4 from 0. ***That is, the sum of two negative numbers is the opposite of the sum of their absolute values.***

$$\begin{aligned} &-2 + (-4) \\ &= -(|-2| + |-4|) \\ &= -(2 + 4) \\ &= -6 \end{aligned}$$

Answers

1. $1 + 4 = 5$

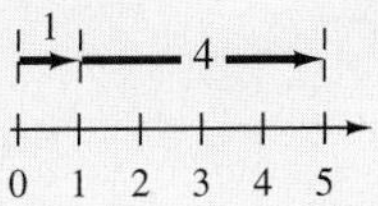

2. $-2 + (-5) = -7$

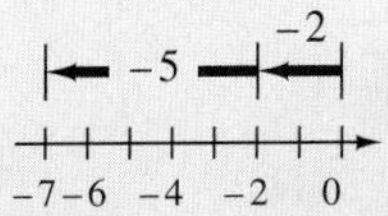

3 Find each sum.

(a) $-7 + (-3)$

(b) $-12 + (-18)$

(c) $-15 + (-4)$

4 Use a number line to find each sum.

(a) $6 + (-3)$

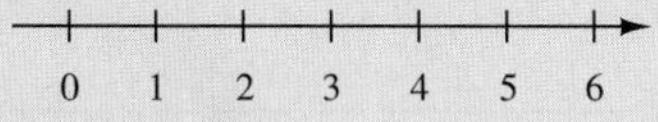

(b) $-5 + 1$

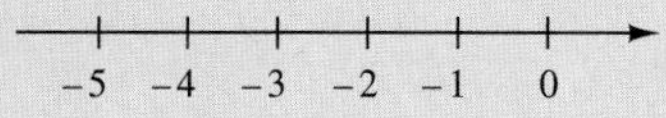

5 Find each sum.

(a) $-24 + 11$

(b) $30 + (-8)$

Adding Real Numbers with the Same Sign

To add two numbers with the same sign, add the absolute values of the numbers. Give the result the same sign as the numbers being added.

Example: $-4 + (-3) = -7$

EXAMPLE 3 Adding Two Negative Numbers

Find each sum.

(a) $-2 + (-9)$
$= -11$

(b) $-8 + (-12)$
$= -20$

(c) $-15 + (-3)$
$= -18$

The sum of two negative numbers is negative.

◀ Work Problem **3** at the Side.

OBJECTIVE 2 Add numbers with different signs.

EXAMPLE 4 Adding Numbers with Different Signs

Use the number line to find the sum $-2 + 5$.

Step 1 Start at 0 and draw an arrow 2 units to the left. See **Figure 10.**

Step 2 From the left end of this arrow, draw a second arrow 5 units to the right.

The number below the end of the second arrow is 3, so $-2 + 5 = 3$.

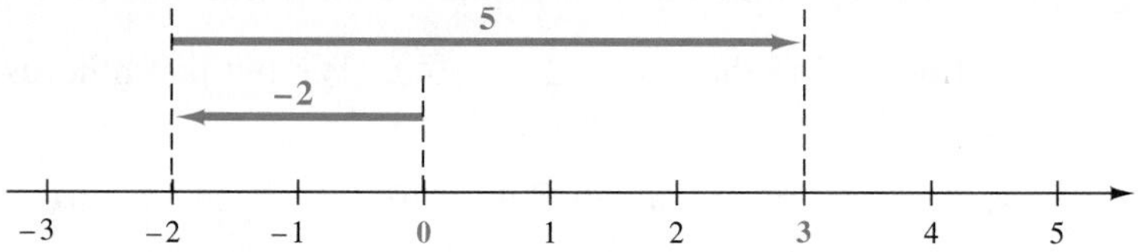

Figure 10

◀ Work Problem **4** at the Side.

Adding Real Numbers with Different Signs

To add two numbers with different signs, find the absolute values of the numbers. Subtract the lesser absolute value from the greater. Give the answer the same sign as the number with the greater absolute value.

Example: $-12 + 6 = -6$

EXAMPLE 5 Adding Numbers with Different Signs

Find the sum $-12 + 5$.

Find the absolute value of each number.

$$|-12| = 12 \quad \text{and} \quad |5| = 5$$

Then find the difference between these absolute values: $12 - 5 = 7$. The sum will be negative, since $|-12| > |5|$.

$$-12 + 5 = -7$$

◀ Work Problem **5** at the Side.

Answers

3. **(a)** -10 **(b)** -30 **(c)** -19

4. **(a)** $6 + (-3) = 3$

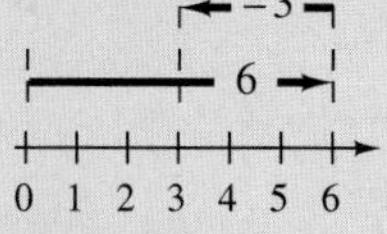

(b) $-5 + 1 = -4$

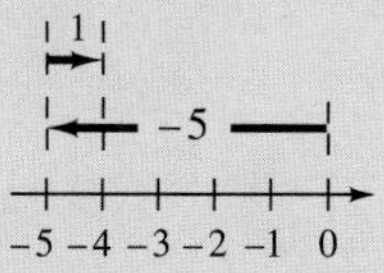

5. **(a)** -13 **(b)** 22

Calculator Tip

The (−) or (+/−) key is used to input a negative number in some scientific calculators.

OBJECTIVE 3 Add mentally.

EXAMPLE 6 Adding a Positive Number and a Negative Number

Check each answer by adding mentally. If necessary, use a number line.

(a) $7 + (-4)$
$= 3$

(b) $-8 + 12$
$= 4$

(c) $-\frac{1}{2} + \frac{1}{8}$
$= -\frac{4}{8} + \frac{1}{8}$
$= -\frac{3}{8}$

(d) $\frac{5}{6} + \left(-1\frac{1}{3}\right)$
$= \frac{5}{6} + \left(-\frac{4}{3}\right)$
$= \frac{5}{6} + \left(-\frac{8}{6}\right)$
$= -\frac{3}{6}, \text{ or } -\frac{1}{2}$

(e) $-4.6 + 8.1$
$= 3.5$

(f) $-16 + 16$
$= 0$

(g) $42 + (-42)$
$= 0$

Work Problem 6 at the Side. ▶

Adding Signed Numbers

Same sign Add the absolute values of the numbers. Give the sum the same sign as the numbers being added.

Different signs Find the absolute values of the numbers, and subtract the lesser absolute value from the greater. Give the answer the sign of the number having the greater absolute value.

OBJECTIVE 4 Use the rules for order of operations with real numbers. When a problem involves square brackets, [ ], we do the calculations inside the brackets until a single number is obtained.

EXAMPLE 7 Adding with Brackets

Find each sum.

(a) $-3 + [4 + (-8)]$ (Start here.)
$= -3 + (-4)$
$= -7$

(b) $8 + [(-2 + 6) + (-3)]$
$= 8 + [4 + (-3)]$
$= 8 + 1$
$= 9$

Follow the order of operations.

Work Problem 7 at the Side. ▶

6 Check each answer by adding mentally. If necessary, use a number line.

(a) $-8 + 2 = -6$

(b) $-15 + 4 = -11$

(c) $17 + (-10) = 7$

(d) $\frac{3}{4} + \left(-1\frac{3}{8}\right) = -\frac{5}{8}$

(e) $-9.5 + 3.8 = -5.7$

(f) $37 + (-37) = 0$

7 GS Label the order in which each expression should be evaluated. Then find each sum.

(a) $2 + [7 + (-3)]$
② ①
$= 2 +$ ____
$=$ ____

(b) $6 + [(-2 + 5) + 7]$
◯ ◯ ◯
$= 6 + [$____ $+ 7]$
$= 6 +$ ____
$=$ ____

(c) $-9 + [-4 + (-8 + 6)]$
◯ ◯ ◯

Answers

6. All are correct.

7. (a) 4; 6 **(b)** ③, ①, ②; 3; 10; 16
(c) ③, ②, ①; −15

8 Write a numerical expression for each phrase, and simplify the expression.

(a) 4 more than -12

(b) The sum of 6 and -7

(c) -12 added to -31

(d) 7 increased by the sum of 8 and -3

9 Solve the problem.

A football team lost 8 yd on first down, lost 5 yd on second down, and then gained 7 yd on third down. How many yards did the team gain or lose altogether on these plays?

Answers

8. **(a)** $-12 + 4$; -8 **(b)** $6 + (-7)$; -1
(c) $-31 + (-12)$; -43
(d) $7 + [8 + (-3)]$; 12

9. The team lost 6 yd.

OBJECTIVE 5 Translate words and phrases that indicate addition. Problem solving requires translating words and phrases into symbols. We began this process in **Section 9.1.** The table lists key words and phrases that indicate addition.

Word or Phrase Indicating Addition	Example	Numerical Expression and Simplification
Sum of	The *sum of* -3 and 4	$-3 + 4$, which equals 1
Added to	5 *added to* -8	$-8 + 5$, which equals -3
More than	12 *more than* -5	$(-5) + 12$, which equals 7
Increased by	-6 *increased by* 13	$-6 + 13$, which equals 7
Plus	3 *plus* 14	$3 + 14$, which equals 17

EXAMPLE 8 Translating Words and Phrases (Addition)

Write a numerical expression for each phrase, and simplify the expression.

(a) The **sum of** -8 and 4 and 6

$-8 + 4 + 6$ simplifies to $-4 + 6$, which equals 2.

Add in order from left to right.

(b) 3 **more than** -5, **increased by** 12

$(-5 + 3) + 12$ simplifies to $-2 + 12$, which equals 10.

Here we *simplified* each expression by performing the operations.

◀ **Work Problem 8 at the Side.**

In applications, gains may be interpreted as positive numbers and losses as negative numbers.

EXAMPLE 9 Interpreting Gains and Losses

A football team gained 3 yd on first down, lost 12 yd on second down, and then gained 13 yd on third down. How many yards did the team gain or lose altogether on these plays?

$3 + (-12) + 13$ Gains plus losses

$= [3 + (-12)] + 13$ Add from left to right.

$= (-9) + 13$

$= 4$ Add.

The team gained 4 yd altogether on these plays.

◀ **Work Problem 9 at the Side.**

9.4 Exercises

FOR EXTRA HELP

Download the MyDashBoard App

MyMathLab®

CONCEPT CHECK *Complete each of the following.*

1. The sum of two negative numbers will always be a (*positive / negative*) number. Give a number-line illustration using the sum $-2 + (-3) =$ ____.

2. The sum of a number and its opposite will always be ________.

3. When adding a positive number and a negative number, where the negative number has the greater absolute value, the sum will be a (*positive / negative*) number. Give a number-line illustration using the sum $-4 + 2 =$ ____.

4. To simplify the expression $8 + [-2 + (-3 + 5)]$, one should begin by adding ____ and ____, according to the rules for order of operations.

CONCEPT CHECK *By the rules for order of operations, what is the first step you would use to simplify each expression?*

5. $4[3(-2 + 5) - 1]$

6. $[-4 + 7(-6 + 2)]$

7. $9 + ([-1 + (-3)] + 5)$

8. $[(-8 + 4) + (-6)] + 5$

Find each sum. **See Examples 1–7.**

9. $6 + (-4)$

10. $8 + (-5)$

11. $12 + (-15)$

12. $4 + (-8)$

13. $-7 + (-3)$

14. $-11 + (-4)$

15. $-10 + (-3)$

16. $-16 + (-7)$

17. $-12.4 + (-3.5)$

18. $-21.3 + (-2.5)$

19. $10 + [-3 + (-2)]$
$= 10 + (\_\_\_\_)$
$= \_\_\_\_$

20. $13 + [-4 + (-5)]$
$= 13 + (\_\_\_\_)$
$= \_\_\_\_$

21. $5 + [14 + (-6)]$

22. $7 + [3 + (-14)]$

23. $-3 + [5 + (-2)]$

24. $-7 + [10 + (-3)]$

25. $-8 + [3 + (-1) + (-2)]$

26. $-7 + [5 + (-8) + 3]$

27. $\frac{9}{10}+\left(-\frac{3}{5}\right)$

28. $\frac{5}{8}+\left(-\frac{17}{12}\right)$

29. $-\frac{1}{6}+\frac{2}{3}$

30. $-\frac{6}{25}+\frac{19}{20}$

31. $2\frac{1}{2}+\left(-3\frac{1}{4}\right)$

32. $6\frac{1}{2}+\left(-4\frac{3}{8}\right)$

33. $7.8+(-9.4)$

34. $14.7+(-10.1)$

35. $-7.1+[3.3+(-4.9)]$

36. $-9.5+[-6.8+(-1.3)]$

37. $[-8+(-3)]+[-7+(-7)]$
$=$ ____ $+$ (____)
$=$ ____

38. $[-5+(-4)]+[9+(-2)]$
$=$ ____ $+$ ____
$=$ ____

39. $\left(-\frac{1}{2}+0.25\right)+\left(-\frac{3}{4}+0.75\right)$

40. $\left(-\frac{3}{2}+0.75\right)+\left(-\frac{1}{2}+2.25\right)$

Perform each operation, and then determine whether the statement is true *or* false. *Try to do all work mentally.* ***See Examples 6 and 7.***

41. $-11+13=13+(-11)$

42. $16+(-9)=-9+16$

43. $-10+6+7=-3$

44. $-12+8+5=-1$

45. $\frac{7}{3}+\left(-\frac{1}{3}\right)+\left(-\frac{6}{3}\right)=0$

46. $-\frac{3}{2}+1+\frac{1}{2}=0$

47. $|-8+10|=-8+(-10)$

48. $|-4+6|=-4+(-6)$

49. $2\frac{1}{5}+\left(-\frac{6}{11}\right)=-\frac{6}{11}+2\frac{1}{5}$

50. $-1\frac{1}{2}+\frac{5}{8}=\frac{5}{8}+\left(-1\frac{1}{2}\right)$

51. $-7+[-5+(-3)]=[(-7)+(-5)]+3$

52. $6+[-2+(-5)]=[(-4)+(-2)]+5$

Write a numerical expression for each phrase, and simplify the expression. ***See Example 8.***

53. The sum of -5 and 12 and 6

54. The sum of -3 and 5 and -12

55. 14 added to the sum of -19 and -4

56. -2 added to the sum of -18 and 11

57. The sum of -4 and -10, increased by 12

58. The sum of -7 and -13, increased by 14

59. $\frac{2}{7}$ more than the sum of $\frac{5}{7}$ and $-\frac{9}{7}$

60. 0.85 more than the sum of -1.25 and -4.75

Solve each problem. ***See Example 9.***

61. Nathaniel owed his older sister Jenna \$24 for his share of the bill when they took their mother out to dinner for her birthday. He later borrowed \$38 from his younger sister Ilana to buy two DVDs. What positive or negative number represents Nathaniel's financial situation with his siblings?

62. Bonika's checking account balance is \$54.00. She then takes a gamble by writing a check for \$89.00. What is her new balance? (Write the balance as a signed number.)

63. The surface, or rim, of a canyon is at altitude 0. On a hike down into the canyon, a party of hikers stops for a rest at 130 m below the surface. They then descend another 54 m. What is their new altitude? (Write the altitude as a signed number.)

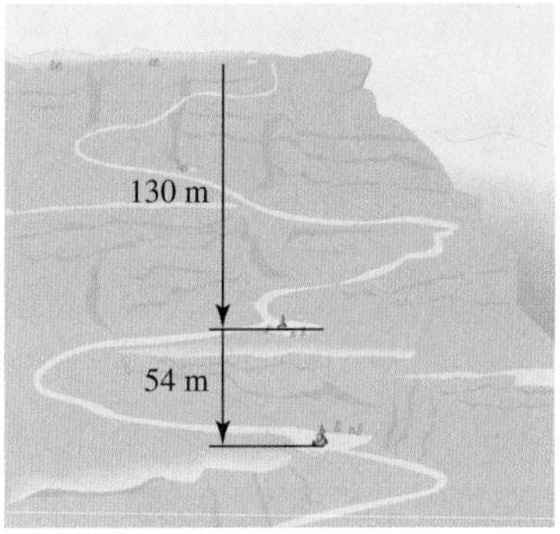

64. A pilot announces to the passengers that the current altitude of their plane is 34,000 ft. Because of some unexpected turbulence, the pilot is forced to descend 2100 ft. What is the new altitude of the plane? (Write the altitude as a signed number.)

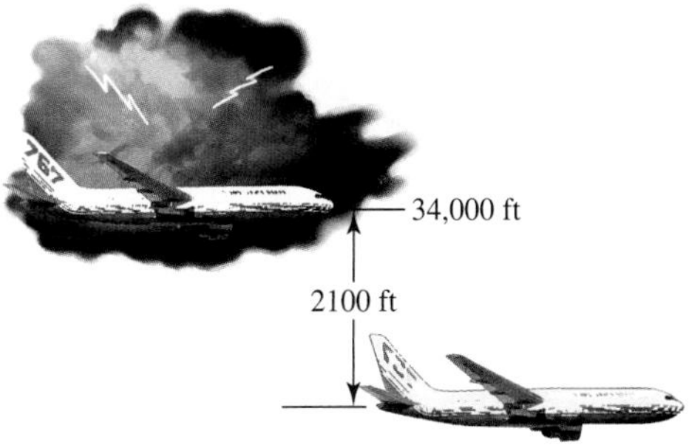

65. J. D. Patin enjoys playing Triominoes every Wednesday night. Last Wednesday, on four successive turns, his scores were -19, 28, -5, and 13. What was his final score for the four turns?

66. Gay Aguillard also enjoys playing Triominoes. On five successive turns, her scores were -13, 15, -12, 24, and 14. What was her total score for the five turns?

67. On three consecutive passes, a quarterback passed for a gain of 6 yd, was sacked for a loss of 12 yd, and passed for a gain of 43 yd. What positive or negative number represents the total net yardage for the plays?

68. On a series of three consecutive running plays, a running back gained 4 yd, lost 3 yd, and lost 2 yd. What positive or negative number represents his total net yardage for the series of plays?

69. The lowest temperature ever recorded in Arkansas was −29°F. The highest temperature ever recorded there was 149°F more than the lowest. What was this highest temperature? (*Source: National Climatic Data Center.*)

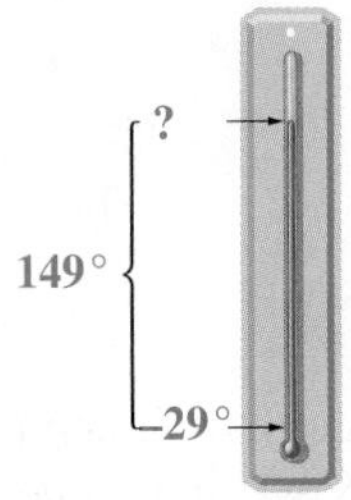

70. On January 23, 1943, the temperature rose 49°F in two minutes in Spearfish, South Dakota. If the starting temperature was −4°F, what was the temperature two minutes later?

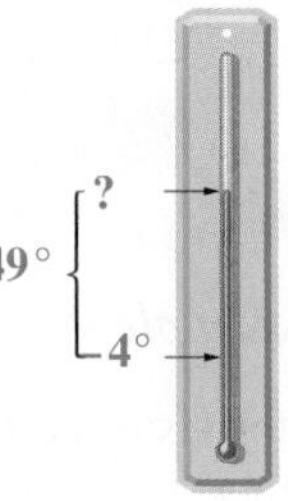

71. Dana Weightman owes $153 to a credit card company. She makes a $14 purchase with the card, and then pays $60 on the account. What is her current balance as a signed number?

72. A female polar bear weighed 660 lb when she entered her winter den. She lost 45 lb during each of the first two months of hibernation, and another 205 lb before leaving the den with her two cubs in March. How much did she weigh when she left the den?

73. Based on the 2010 Census, New York lost 2 seats in the U.S. House of Representatives. Illinois, Iowa, Missouri, and Massachusetts were among the states that each lost 1 seat. Write a signed number that represents the total change in number of seats for these five states. (*Source:* U.S. Census Bureau.)

74. Based on the 2010 Census, Ohio and New York lost the most seats in the U.S. House of Representatives, each losing 2 seats. The states gaining the most seats were Texas with 4 and Florida with 2. Write a signed number that represents the algebraic sum of these changes. (*Source:* U.S. Census Bureau.)

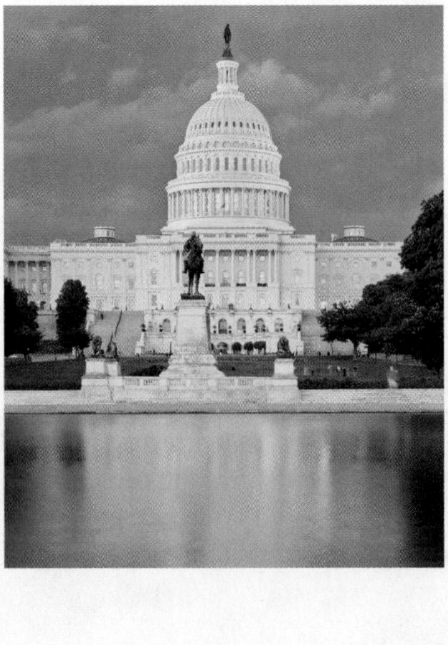

9.5 Subtracting Real Numbers

OBJECTIVES

1. Subtract two numbers on a number line.
2. Use the definition of subtraction.
3. Work subtraction problems that involve brackets.
4. Translate words and phrases that indicate subtraction.

OBJECTIVE 1 Subtract two numbers on a number line. Recall that the answer to a subtraction problem is a **difference.** In the subtraction $x - y$, x is the **minuend** and y is the **subtrahend.**

$$\underset{\text{Minuend}}{x} - \underset{\text{Subtrahend}}{y} = \underset{\text{Difference}}{z}$$

EXAMPLE 1 Subtracting Numbers on a Number Line

Use a number line to find the difference $7 - 4$.

Step 1 Start at 0 and draw an arrow 7 units to the *right*. See **Figure 11.**

Step 2 From the right end of the first arrow, draw a second arrow 4 units to the *left* to represent the subtraction of a positive number.

The number below the end of this second arrow is 3, so $7 - 4 = 3$.

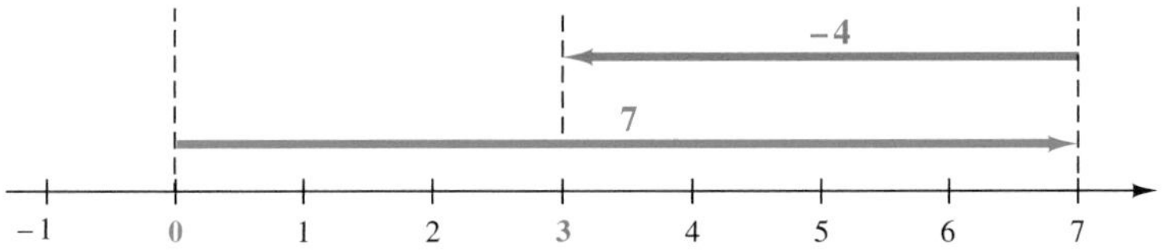

Figure 11

Work Problem 1 at the Side. ▶

1 Use a number line to find each difference.

(a) $5 - 1$

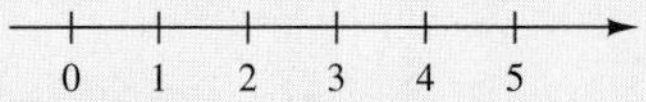

(b) $6 - 2$

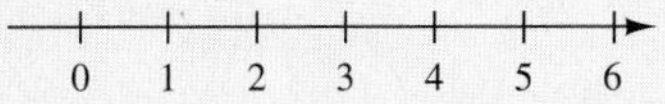

OBJECTIVE 2 Use the definition of subtraction. The procedure used in **Example 1** to find $7 - 4$ is the same procedure for finding $7 + (-4)$.

$$7 - 4 \quad \text{is equal to} \quad 7 + (-4).$$

This shows that *subtracting* a positive number from a larger positive number is the same as *adding* the opposite of the smaller number to the larger.

Definition of Subtraction

For any real numbers a and b,

$$a - b \quad \text{is defined as} \quad a + (-b).$$

To subtract b from a, add the additive inverse (or opposite) of b to a. In words, change the subtrahend to its opposite and add.

Example: $4 - 9$

$= 4 + (-9)$

$= -5$

Subtracting Signed Numbers

Step 1 Change the subtraction symbol to an addition symbol, and change the sign of the subtrahend.

Step 2 Add, as in **Section 9.4.**

Answers

1. (a) $5 - 1 = 4$

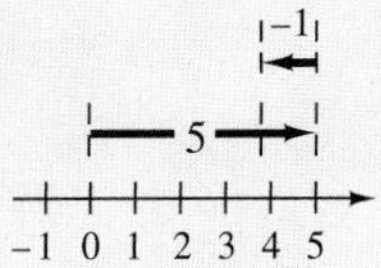

(b) $6 - 2 = 4$

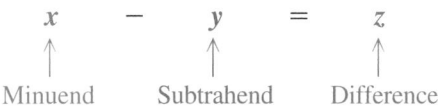

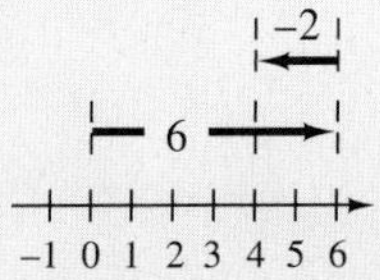

2 Subtract.

GS **(a)** $6 - 10$

$= 6 + (\_\_\_)$

$= \_\_\_$

(b) $-2 - 4$

GS **(c)** $3 - (-5)$

$= 3 + \_\_\_$

$= \_\_\_$

(d) $-8 - (-12)$

(e) $\frac{5}{4} - \left(-\frac{3}{7}\right)$

Answers

2. **(a)** -10; -4 **(b)** -6 **(c)** 5; 8 **(d)** 4 **(e)** $\frac{47}{28}$

EXAMPLE 2 **Using the Definition of Subtraction**

Subtract.

No change; Change − to +.; Opposite of 3

(a) $12 - 3 = 12 + (-3) = 9$

(b) $5 - 7 = 5 + (-7) = -2$ (-7 has the greater absolute value, so the sum is negative.)

(c) $-8 - 15 = -8 + (-15) = -23$

No change; Change − to +.; Opposite of −5

(d) $-3 - (-5) = -3 + (5) = 2$

(e) $\frac{3}{8} - \left(-\frac{4}{5}\right)$

$= \frac{15}{40} - \left(-\frac{32}{40}\right)$ Find a common denominator.

$= \frac{15}{40} + \frac{32}{40}$ Use the definition of subtraction.

$= \frac{47}{40}$ Add the fractions.

◀ **Work Problem 2 at the Side.**

Uses of the Symbol −

We use the symbol − for three purposes.

1. It can represent ***subtraction,*** as in $9 - 5 = 4$.
2. It can represent ***negative numbers,*** such as -10, -2, and -3.
3. It can represent ***the opposite (or additive inverse) of a number,*** as in "the opposite (or additive inverse) of 8 is -8."

We may see more than one use in the same problem, such as $-6 - (-9)$, where -9 is subtracted from -6. The meaning of the symbol depends on its position in the algebraic expression.

OBJECTIVE 3 **Work subtraction problems that involve brackets.** As before, first perform any operations inside the parentheses and brackets.

EXAMPLE 3 **Subtracting with Grouping Symbols**

Perform each operation.

(a) $-6 - [2 - (8 + 3)]$ (Work from the inside out.)

$= -6 - [2 - 11]$ Add inside the parentheses.

$= -6 - [2 + (-11)]$ Definition of subtraction

$= -6 - (-9)$ Add.

$= -6 + 9$ Definition of subtraction

$= 3$ Add.

Continued on Next Page

(b) $5-\left[\left(-\frac{1}{3}-\frac{1}{2}\right)-(4-1)\right]$ Work from the inside out.

$=5-\left[\left(-\frac{1}{3}+\left(-\frac{1}{2}\right)\right)-3\right]$ Start within each set of parentheses inside the brackets.

$=5-\left[\left(-\frac{5}{6}\right)-3\right]$ Use 6 as the common denominator; $-\frac{1}{3}+\left(-\frac{1}{2}\right)=-\frac{2}{6}+\left(-\frac{3}{6}\right)=-\frac{5}{6}$

$=5-\left[\left(-\frac{5}{6}\right)+(-3)\right]$ Definition of subtraction

$=5-\left[\left(-\frac{5}{6}\right)+\left(-\frac{18}{6}\right)\right]$ Use 6 as the common denominator; $-\frac{3}{1}\cdot\frac{6}{6}=-\frac{18}{6}$

$=5-\left(-\frac{23}{6}\right)$ Add inside the brackets.

$=5+\frac{23}{6}$ Definition of subtraction

$=\frac{30}{6}+\frac{23}{6}$ Write 5 as an improper fraction.

$=\frac{53}{6}$ Add fractions.

Work Problem 3 at the Side. ▶

OBJECTIVE 4 Translate words and phrases that indicate subtraction.
The table lists words and phrases that indicate subtraction in problem solving.

Word, Phrase, or Sentence Indicating Subtraction	Example	Numerical Expression and Simplification
Difference between	The *difference between* −3 and −8	$-3-(-8)$ simplifies to $-3+8$, which equals 5
Subtracted from*	12 *subtracted from* 18	$18-12$, which equals 6
From . . . , subtract	*From* 12, *subtract* 8.	$12-8$ simplifies to $12+(-8)$, which equals 4
Less	6 *less* 5	$6-5$, which equals 1
Less than*	6 *less than* 5	$5-6$ simplifies to $5+(-6)$, which equals -1
Decreased by	9 *decreased by* −4	$9-(-4)$ simplifies to $9+4$, which equals 13
Minus	8 *minus* 5	$8-5$, which equals 3

*Be careful with order when translating.

CAUTION

When subtracting two numbers, be careful to write them in the correct order, because, in general,

$$x-y\neq y-x. \quad \text{For example, } 5-3\neq 3-5.$$

Think carefully before interpreting an expression involving subtraction.

3 Label the order in which each expression should be evaluated. Then perform each operation.

(a) $2-[(-3)-(4+6)]$
○ ○ ○

(b) $[(5-7)+3]-8$
○ ○ ○

(c) $6-[(-1-4)-2]$
○ ○ ○

Answers

3. (a) ③, ②, ①; 15
(b) ①, ②, ③; −7
(c) ③, ①, ②; 13

4 Write a numerical expression for each phrase, and simplify the expression.

(a) The difference between -5 and -12

(b) -2 subtracted from the sum of 4 and -4

(c) 7 less than -2

(d) 9, decreased by 10 less than 7

5 Solve the problem.

The highest elevation in Argentina is Mt. Aconcagua, which is 6960 m above sea level. The lowest point in Argentina is the Valdes Peninsula, 40 m below sea level. Find the difference between the highest and lowest elevations.

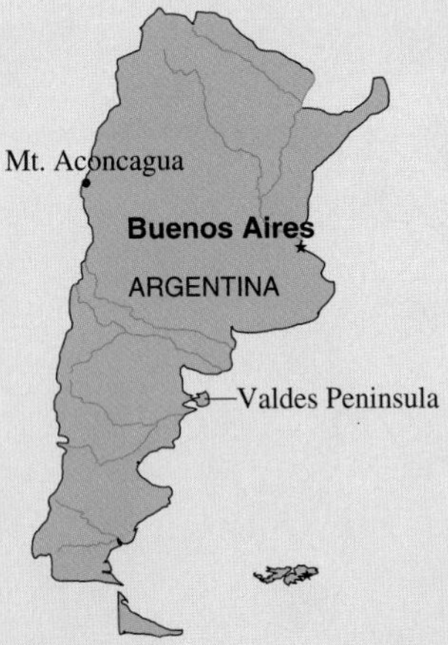

Answers

4. (a) $-5-(-12)$; 7
(b) $[4+(-4)]-(-2)$; 2
(c) $-2-7$; -9
(d) $9-(7-10)$; 12
5. 7000 m

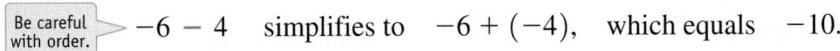

EXAMPLE 4 Translating Words and Phrases (Subtraction)

Write a numerical expression for each phrase, and simplify the expression.

(a) The **difference between** -8 and 5

When "difference between" is used, write the numbers in the order they are given.

$$-8-5 \quad \text{simplifies to} \quad -8+(-5), \quad \text{which equals} \quad -13.$$

(b) 4 **subtracted from** the sum of 8 and -3

Here the operation of addition is also used, as indicated by the word *sum*. First, add 8 and -3. Next, subtract 4 from this sum. The expression is

$$[8+(-3)]-4 \quad \text{simplifies to} \quad 5-4, \quad \text{which equals} \quad 1.$$

(c) 4 **less than** -6

Here 4 must be taken *from* -6, so write -6 first.

Be careful with order.

$$-6-4 \quad \text{simplifies to} \quad -6+(-4), \quad \text{which equals} \quad -10.$$

Notice that "4 less than -6" differs from "4 *is less than* -6." The statement "4 is less than -6" is symbolized as $4<-6$ (which is a false statement).

(d) 8, **decreased by** 5 **less than** 12

First, write "5 less than 12" as $12-5$. Next, subtract $12-5$ from 8.

$$8-(12-5) \quad \text{simplifies to} \quad 8-7, \quad \text{which equals} \quad 1.$$

◀ **Work Problem 4 at the Side.**

EXAMPLE 5 Solving an Applied Problem Involving Subtraction

The record high temperature of 134°F in the United States was recorded at Death Valley, California, in 1913. The record low was -80°F, at Prospect Creek, Alaska, in 1971. See **Figure 12.** What is the difference between these highest and lowest temperatures? (*Source:* National Climatic Data Center.)

We must subtract the lowest temperature from the highest temperature.

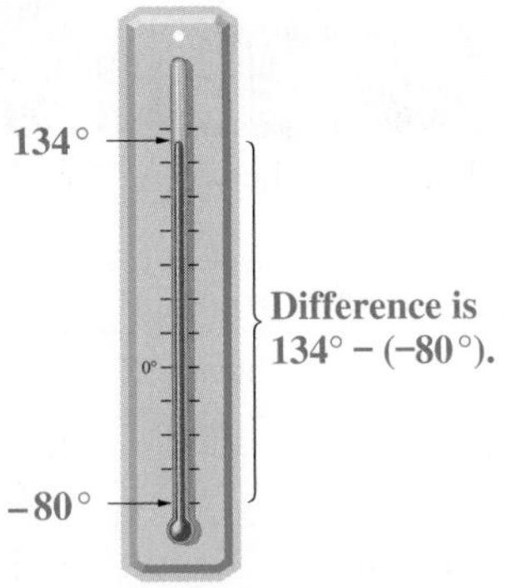

Figure 12

Order of numbers matters in subtraction.

$$134-(-80)$$
$$=134+80 \quad \text{Use the definition of subtraction.}$$
$$=214 \quad \text{Add.}$$

The difference between the two temperatures is 214°F.

◀ **Work Problem 5 at the Side.**

9.5 Exercises

FOR EXTRA HELP

Download the MyDashBoard App

MyMathLab®

CONCEPT CHECK *Fill in each blank with the correct response.*

1. By the definition of subtraction, in order to perform the subtraction $-6-(-8)$, we must add the opposite of _______ to _______ to get _______.

2. By the rules for order of operations, to simplify $8-[3-(-4-5)]$, the first step is to subtract _______ from _______.

3. "The difference between 7 and 12" translates as _______, while "the difference between 12 and 7" translates as _______.

4. To subtract b from a, add the _______ _______, or _______, of b to a.

5. $-8-4$ can be rewritten as $-8+($_______$)$.

6. $-19-22$ can be rewritten as $-19+($_______$)$.

Perform the indicated operations. ***See Examples 1–3.***

7. $-7-3$
$=-7+($____$)$
$=$ ____

8. $-12-5$
$=-12+($____$)$
$=$ ____

9. $-10-6$

10. $-13-16$

11. $7-(-4)$
$=7+$ ____
$=$ ____

12. $9-(-6)$
$=9+$ ____
$=$ ____

13. $6-(-13)$

14. $13-(-3)$

15. $-7-(-3)$

16. $-8-(-6)$

17. $3-(4-6)$

18. $6-(7-14)$

19. $-3-(6-9)$

20. $-4-(5-12)$

21. $\frac{1}{2}-\left(-\frac{1}{4}\right)$

22. $\frac{1}{3}-\left(-\frac{4}{3}\right)$

23. $-\frac{3}{4}-\frac{5}{8}$

24. $-\frac{5}{6}-\frac{1}{2}$

25. $\frac{5}{8}-\left(-\frac{1}{2}-\frac{3}{4}\right)$

26. $\frac{9}{10}-\left(-\frac{1}{8}-\frac{3}{10}\right)$

27. $4.4-(-9.2)$

28. $6.7-(-12.6)$

29. $-7.4-4.5$

30. $-5.4-9.6$

KEVIN

31. $8-(-3)-9+6$

32. $12-(-8)-25+9$

33. $-5.2-(8.4-10.8)$

34. $-9.6-(3.5-12.6)$

35. $[(-3.1)-4.5]-(0.8-2.1)$

36. $[(-7.8)-9.3]-(0.6-3.5)$

37. $-12-[(9-2)-(-6-3)]$

38. $-4-[(6-9)-(-7-4)]$

39. $\left(-\frac{3}{8}-\frac{2}{3}\right)-\left(-\frac{9}{8}-3\right)$

40. $\left(-\frac{3}{4}-\frac{5}{2}\right)-\left(-\frac{1}{8}-1\right)$

41. $[-12.25-(8.34+3.57)]-17.88$

42. $[-34.99+(6.59-12.25)]-8.33$

Write a numerical expression for each phrase and simplify. ***See Example 4.***

43. The difference between 4 and -8

44. The difference between 7 and -14

45. 8 less than -2

46. 9 less than -13

47. The sum of 9 and -4, decreased by 7

48. The sum of 12 and -7, decreased by 14

49. 12 less than the difference between 8 and -5

50. 19 less than the difference between 9 and -2

Solve each problem. ***See Example 5.***

51. The lowest temperature ever recorded in Illinois was -36°F on January 5, 1999. The lowest temperature ever recorded in Utah was on February 1, 1985, and was 33°F lower than Illinois's record low. What is the record low temperature for Utah? (*Source:* National Climatic Data Center.)

52. The lowest temperature ever recorded in South Carolina was -19°F on January 21, 1985. The lowest temperature ever recorded in Wisconsin was 36° lower than South Carolina's record low. What is the record low temperature in Wisconsin? (*Source:* National Climatic Data Center.)

53. The top of Mount Whitney, visible from Death Valley, has an altitude of 14,494 ft above sea level. The bottom of Death Valley is 282 ft below sea level. Using 0 as sea level, find the difference between these two elevations. (*Source: World Almanac and Book of Facts.*)

54. The height of Mount Pumasillo in Peru is 20,492 ft. The depth of the Peru-Chile Trench in the Pacific Ocean is 26,457 ft below sea level. Find the difference between these two elevations. (*Source: World Almanac and Book of Facts.*)

55. Samir owed his brother \$10. He later borrowed \$70. What positive or negative number represents his present financial status?

56. Francesca has \$15 in her purse, and Emilio has a debt of \$12. Find the difference between these amounts.

57. A chemist is running an experiment under precise conditions. At first, she runs it at $-174.6°$F. She then lowers the temperature by 2.3°F. What is the new temperature for the experiment?

58. At 2:00 A.M., a plant worker found that a dial reading was 7.904. At 3:00 A.M., she found the reading to be -3.291. Find the difference between these two readings.

59. In August, Kari Heen began with a checking account balance of \$904.89. Her checks and deposits for August are given below.

Checks	Deposits
\$35.84	\$85.00
\$26.14	\$120.76
\$3.12	

Assuming no other transactions, what was her account balance at the end of August?

60. In September, Derek Bowen began with a checking account balance of \$904.89. His checks and deposits for September are given below.

Checks	Deposits
\$41.29	\$80.59
\$13.66	\$276.13
\$84.40	

Assuming no other transactions, what was his account balance at the end of September?

61. A certain Greek mathematician was born in 426 B.C. His father was born 43 years earlier. In what year was his father born?

62. A certain Roman philosopher was born in 325 B.C. Her mother was born 35 years earlier. In what year was her mother born?

63. Kim Falgout owes $870.00 on her MasterCard account. She returns two items costing $35.90 and $150.00 and receives credits for these on the account. Next, she makes a purchase of $82.50, and then two more purchases of $10.00 each. She makes a payment of $500.00. She then incurs a finance charge of $37.23. How much does she still owe?

64. Charles Vosburg owes $679.00 on his Visa account. He returns three items costing $36.89, $29.40, and $113.55 and receives credits for these on the account. Next, he makes purchases of $135.78 and $412.88, and two purchases of $20.00 each. He makes a payment of $400. He then incurs a finance charge of $24.57. How much does he still owe?

65. José Martinez enjoys diving in Lake Okoboji. He dives to 34 ft below the surface of the lake. His partner, Sean O'Malley, dives to 40 ft below the surface, but then ascends 20 ft. What is the vertical distance between José and Sean?

66. Rhonda Alessi also enjoys diving. She dives to 12 ft below the surface of False River. Her sister, Sandy, dives to 20 ft below the surface, but then ascends 10 ft. What is the vertical distance between Rhonda and Sandy?

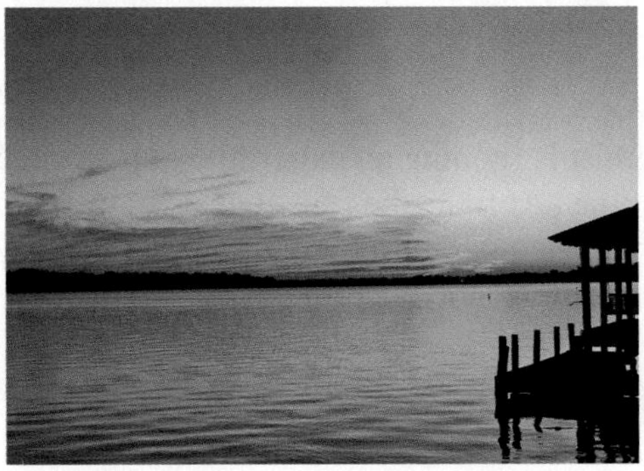

67. The height of Mt. Foraker is 17,400 ft, while the depth of the Java Trench is 23,376 ft. What is the vertical distance between the top of Mt. Foraker and the bottom of the Java Trench? (*Source: World Almanac and Book of Facts.*)

68. The height of Mt. Wilson in Colorado is 14,246 ft, while the depth of the Cayman Trench is 24,721 ft. What is the vertical distance between the top of Mt. Wilson and the bottom of the Cayman Trench? (*Source: World Almanac and Book of Facts.*)

69. In 2005, Americans saved -0.5% of their after-tax incomes, the first negative personal savings rate since 1933. In the fourth quarter of 2011, they saved 3.9%. Find the difference between the two amounts for 2011 and 2005.

70. Refer to **Exercise 69.** How is it possible that Americans had a negative personal savings rate in 2005?

71. In 2000, the federal budget had a surplus of \$236 billion. In 2010, the federal budget had a deficit of \$1294 billion. Find the difference between the amounts for 2000 and 2010. (*Source:* U.S. Office of Management and Budget.)

72. In 1998, undergraduate college students had an average (mean) credit card balance of \$1879. The average balance increased \$869 by 2000, then dropped \$579 by 2004, and then increased \$1004 by 2008. What was the average credit card balance of undergraduate college students in 2008? (*Source:* Sallie Mae.)

Median sales prices for existing single-family homes in the United States for the years 2005 through 2009 are shown in the table. Complete the table, determining the change from one year to the next by subtraction.

	Year	Median Sales Price	Change from Previous Year
	2005	\$219,900	—
73.	2006	\$221,900	
74.	2007	\$219,000	
75.	2008	\$198,100	
76.	2009	\$175,500	

Source: National Association of Realtors.

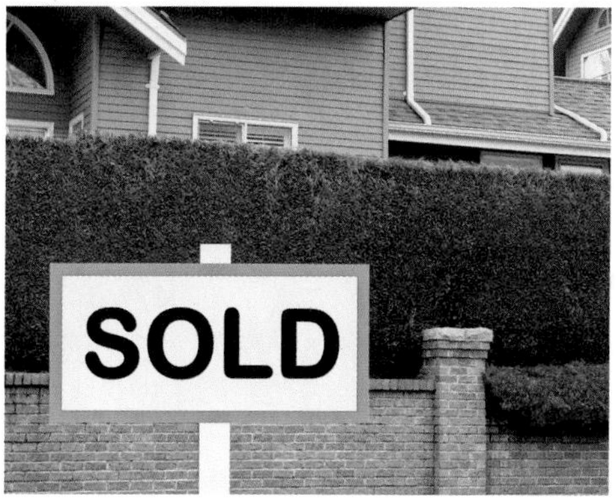

The table shows the heights of some selected mountains. Use the information given to answer Exercises 77 and 78.

77. How much higher is Mt. Wilson than Pikes Peak?

78. If Mt. Wilson and Pikes Peak were stacked one on top of the other, how much higher would they be than Mt. Foraker?

Mountain	Height (in feet)
Foraker	17,400
Wilson	14,246
Pikes Peak	14,110

Source: *World Almanac and Book of Facts.*

CONCEPT CHECK *In Exercises 79–84, suppose that x represents a positive number and y represents a negative number. Determine whether the given expression must represent a positive number or a negative number.*

79. $y - x$

80. $x - y$

81. $x + |y|$

82. $y - |x|$

83. $|x| + |y|$

84. $-(|x| + |y|)$

9.6 Multiplying and Dividing Real Numbers

OBJECTIVES

1. Find the product of a positive number and a negative number.
2. Find the product of two negative numbers.
3. Use the reciprocal of a number to apply the definition of division.
4. Use the rules for order of operations when multiplying and dividing signed numbers.
5. Evaluate expressions involving variables.
6. Translate words and phrases involving multiplication and division.
7. Translate simple sentences into equations.

The result of multiplication is the **product.** We know that the product of two positive numbers is positive. We also know that the product of 0 and any positive number is 0, so we extend that property to all real numbers.

Multiplication Property of 0

For any real number a, the following hold.

$$a \cdot 0 = 0 \quad \text{and} \quad 0 \cdot a = 0$$

OBJECTIVE 1 Find the product of a positive number and a negative number. Observe the following pattern.

$$3 \cdot 5 = 15$$
$$3 \cdot 4 = 12$$
$$3 \cdot 3 = 9$$
$$3 \cdot 2 = 6$$
$$3 \cdot 1 = 3$$
$$3 \cdot 0 = 0$$
$$3 \cdot (-1) = ?$$

The products decrease by 3.

What should $3(-1)$ equal? Since multiplication can also be considered repeated addition, the product $3(-1)$ represents the sum

$$-1 + (-1) + (-1), \quad \text{which equals} \quad -3,$$

so the product should be -3, which fits the pattern. Also, $3(-2)$ represents the sum

$$-2 + (-2) + (-2), \quad \text{which equals} \quad -6.$$

◀ Work Problem 1 at the Side.

These results suggest the following rule.

Multiplying Numbers with Different Signs

The product of a positive number and a negative number is negative.

Examples: $6(-3) = -18$ and $-3(6) = -18$

EXAMPLE 1 Multiplying a Positive Number and a Negative Number

Find each product using the multiplication rule given in the box.

(a) $8(-5)$
$= -(8 \cdot 5)$
$= -40$

(b) $-9\left(\frac{1}{3}\right)$
$= -3$

(c) $-6.2(4.1)$
$= -25.42$

◀ Work Problem 2 at the Side.

1 Find each product by finding the sum of three numbers.

(a) $3(-3)$

(b) $3(-4)$

(c) $3(-5)$

2 Find each product.

(a) $2(-6)$

(b) $7(-8)$

(c) $-9(2)$

(d) $-16\left(\frac{5}{32}\right)$

(e) $4.56(-10)$

Answers

1. (a) -9 (b) -12 (c) -15
2. (a) -12 (b) -56 (c) -18 (d) $-\frac{5}{2}$ (e) -45.6

OBJECTIVE 2 Find the product of two negative numbers. The product of two positive numbers is positive, and the product of a positive number and a negative number is negative. What about the product of two negative numbers? Look at another pattern.

$$-5(4) = -20$$
$$-5(3) = -15$$
$$-5(2) = -10$$
$$-5(1) = -5$$
$$-5(0) = 0$$
$$-5(-1) = ?$$

The products increase by 5.

The numbers in color on the left side of the equality symbols decrease by 1 for each step down the list. The products on the right increase by 5 for each step down the list. To maintain this pattern, $-5(-1)$ should be 5 more than $-5(0)$, or 5 more than 0, so

$$-5(-1) = 5.$$

The pattern continues with

$$-5(-2) = 10$$
$$-5(-3) = 15$$
$$-5(-4) = 20$$
$$-5(-5) = 25, \quad \text{and so on.}$$

This pattern suggests the next rule.

Multiplying Two Negative Numbers

The product of two negative numbers is positive.

Example: $-5(-4) = 20$

EXAMPLE 2 **Multiplying Two Negative Numbers**

Find each product using the multiplication rule given in the box.

(a) $-9(-2)$
$= 18$

(b) $-6(-12)$
$= 72$

(c) $-2(4)(-1)$
$= -8(-1)$
$= 8$

(d) $3(-5)(-2)$
$= -15(-2)$
$= 30$

Work Problem 3 at the Side. ▶

Multiplying Signed Numbers

The product of two numbers having the *same* sign is *positive.*

The product of two numbers having *different* signs is *negative.*

3 Find each product.

(a) $-5(-6)$

(b) $-7(-3)$

(c) $-8(-5)$

(d) $-11(-2)$

(e) $-17(3)(-7)$

(f) $-41(2)(-13)$

Answers

3. (a) 30 (b) 21 (c) 40 (d) 22
(e) 357 (f) 1066

4 Complete the table.

	Number	Reciprocal
(a)	6	____
(b)	-2	____
(c)	$\frac{2}{3}$	____
(d)	$-\frac{1}{4}$	____
(e)	0.75	____
(f)	0	____

5 Find each quotient.

(a) $\dfrac{42}{7}$

(b) $\dfrac{-36}{(-2)(-3)}$

(c) $\dfrac{-12.56}{-0.4}$

(d) $\dfrac{10}{7} \div \left(-\dfrac{24}{5}\right)$

Answers

4. (a) $\frac{1}{6}$ **(b)** $\frac{1}{-2}$, or $-\frac{1}{2}$ **(c)** $\frac{3}{2}$
(d) -4 **(e)** $\frac{4}{3}$ **(f)** none

5. (a) 6 **(b)** -6 **(c)** 31.4
(d) $-\frac{25}{84}$

OBJECTIVE 3 **Use the reciprocal of a number to apply the definition of division.** Recall that the result of division is the **quotient.** The *quotient* of two numbers involves multiplying by the *reciprocal* of the second number, which is the *divisor*.

Reciprocals

Pairs of numbers whose product is 1 are **reciprocals, or multiplicative inverses,** of each other.

The following table shows several numbers and their reciprocals.

Number	Reciprocal
4	$\frac{1}{4}$
-5	$\frac{1}{-5}$, or $-\frac{1}{5}$
0.3, or $\frac{3}{10}$	$\frac{10}{3}$
$-\frac{5}{8}$	$-\frac{8}{5}$

A number and its reciprocal have a product of 1. For example,

$$4 \cdot \tfrac{1}{4} = \tfrac{4}{4}, \text{ or } 1.$$

0 has no reciprocal because the product of 0 and any number is 0, and cannot be 1.

◀ **Work Problem 4 at the Side.**

The quotient of a and b is the product of a and the reciprocal of b.

Definition of Division

The quotient $\frac{a}{b}$ of real numbers a and b, with $b \neq 0$, is

$$\frac{a}{b} = a \cdot \frac{1}{b}.$$

Example: $\dfrac{8}{-4}$ means $8\left(-\dfrac{1}{4}\right)$, which equals -2.

Since division is defined in terms of multiplication, the rules for multiplying signed numbers also apply to dividing them.

EXAMPLE 3 **Using the Definition of Division**

Find each quotient using the definition of division.

(a) $\dfrac{12}{3}$

$= 12 \cdot \dfrac{1}{3}$ $\frac{a}{b} = a \cdot \frac{1}{b}$

$= 4$

(b) $\dfrac{5(-2)}{2}$

$= -10 \cdot \dfrac{1}{2}$

$= -5$

(c) $\dfrac{-1.47}{-7}$

$= -1.47 \cdot \left(-\dfrac{1}{7}\right)$

$= 0.21$

(d) $-\dfrac{2}{3} \div \left(-\dfrac{5}{4}\right)$

$= -\dfrac{2}{3} \cdot \left(-\dfrac{4}{5}\right)$

$= \dfrac{8}{15}$

◀ **Work Problem 5 at the Side.**

In **Example 3(a),** $\frac{12}{3} = 4,$ since $4 \cdot 3 = 12.$ Multiply to check a division problem.

This relationship between multiplication and division allows us to investigate division involving 0. Consider the quotient $\frac{0}{3}$.

$$\frac{0}{3} = 0, \quad \text{since} \quad 0 \cdot 3 = 0.$$

Now consider $\frac{3}{0}$.

$$\frac{3}{0} = ?$$

We need to find a number that when multiplied by 0 will equal 3, that is, $? \cdot 0 = 3$. *No* real number satisfies this equation, since the product of any real number and 0 must be 0. ***Thus, division by 0 is undefined.***

Division Involving 0

For any real number x, with $x \neq 0$,

$$\frac{0}{x} = 0 \quad \text{and} \quad \frac{x}{0} \text{ is undefined.}$$

Examples: $\frac{0}{-10} = 0$ and $\frac{-10}{0}$ is undefined.

Work Problem 6 at the Side. ▶

When dividing fractions, multiplying by the reciprocal of the divisor works well. However it is often easier to divide in the usual way, and then determine the sign of the answer.

Dividing Signed Numbers

The quotient of two numbers having the *same* sign is *positive.*

The quotient of two numbers having *different* signs is *negative.*

Examples: $\frac{-15}{-5} = 3,$ $\frac{15}{-5} = -3,$ and $\frac{-15}{5} = -3$

EXAMPLE 4 **Dividing Signed Numbers**

Find each quotient.

(a) $\frac{8}{-2} = -4$ **(b)** $\frac{-10}{2} = -5$ **(c)** $\frac{-4.5}{-0.09} = 50$

(d) $-\frac{1}{8} \div \left(-\frac{3}{4}\right)$

$= -\frac{1}{8} \cdot \left(-\frac{4}{3}\right)$ Multiply by the reciprocal of the divisor.

$= \frac{1}{6}$ Write in lowest terms.

Work Problem 7 at the Side. ▶

6 Decide whether each quotient is undefined or equal to 0.

(a) $\frac{-3}{0}$

(b) $\frac{0}{-53}$

7 Find each quotient.

(a) $\frac{-8}{-2}$

(b) $\frac{-16.4}{2.05}$

(c) $\frac{1}{4} \div \left(-\frac{2}{3}\right)$

(d) $\frac{12}{-4}$

Answers

6. (a) undefined **(b)** 0

7. (a) 4 **(b)** −8 **(c)** $-\frac{3}{8}$ **(d)** −3

8 Label the order in which each expression should be evaluated. Then perform the indicated operations.

GS **(a)** $-3(4) - 2(6)$

○ ○○

= ___ − ___

= ___

GS **(b)** $-8[-1 - (-4)(-5)]$

○ ○ ○

$= -8[-1 - \_\_\_]$

$= -8[\_\_\_]$

= ___

(c) $\dfrac{6(-4) - 2(5)}{3(2 - 7)}$

(d) $\dfrac{-6(-8) + 3(9)}{-2[4 - (-3)]}$

Answers

8. (a) ①, ③, ②; −12; 12; −24

(b) ③, ②, ①; 20; −21; 168

(c) $\frac{34}{15}$ **(d)** $-\frac{75}{14}$

From the definitions of multiplication and division of real numbers,

$$\frac{-40}{8} = -5 \quad \text{and} \quad \frac{40}{-8} = -5, \quad \text{so} \quad \frac{-40}{8} = \frac{40}{-8}.$$

Based on this example, the quotient of a positive number and a negative number can be written in any of the following three forms.

Equivalent Forms

For any positive real numbers a and b, the following equivalences hold.

$$\frac{-a}{b} = \frac{a}{-b} = -\frac{a}{b}$$

Similarly, the quotient of two negative numbers can be expressed as the quotient of two positive numbers.

Equivalent Forms

For any positive real numbers a and b, the following equivalence holds.

$$\frac{-a}{-b} = \frac{a}{b}$$

OBJECTIVE 4 Use the rules for order of operations when multiplying and dividing signed numbers.

EXAMPLE 5 Using the Rules for Order of Operations

Simplify.

(a) $-9(2) - (-3)(2)$

$= -18 - (-6)$ Multiply.

$= -18 + 6$ Definition of subtraction

$= -12$ Add.

(b) $-6(-2) - 3(-4)$

$= 12 - (-12)$ Multiply.

$= 12 + 12$ Definition of subtraction

$= 24$ Add.

(c) $\dfrac{5(-2) - 3(4)}{2(1 - 6)}$

$= \dfrac{-10 - 12}{2(-5)}$ Simplify the numerator and denominator separately.

$= \dfrac{-22}{-10}$ Subtract in the numerator. Multiply in the denominator.

$= \dfrac{11}{5}$ Write in lowest terms.

◀ Work Problem 8 at the Side.

OBJECTIVE 5 Evaluate expressions involving variables. To *evaluate* an expression means to find its *value*.

EXAMPLE 6 Evaluating Expressions for Numerical Values

Evaluate each expression for $x = -1$, $y = -2$, and $m = -3$.

(a) $(3x + 4y)(-2m)$

Use parentheses around substituted negative values to avoid errors.

$= [3(-1) + 4(-2)][-2(-3)]$ Substitute the given values for the variables.

$= [-3 + (-8)][6]$ Multiply.

$= [-11]6$ Add inside the brackets.

$= -66$ Multiply.

(b) $2x^2 - 3y^2$

Think: $(-2)^2 = -2(-2)$

$= 2(-1)^2 - 3(-2)^2$ Substitute.

Think: $(-1)^2 = -1(-1)$

$= 2(1) - 3(4)$ Apply the exponents.

$= 2 - 12$ Multiply.

$= -10$ Subtract.

(c) $\dfrac{4y^2 + x}{m}$

$= \dfrac{4(-2)^2 + (-1)}{-3}$ Substitute.

$= \dfrac{4(4) + (-1)}{-3}$ Apply the exponent.

$= \dfrac{16 + (-1)}{-3}$ Multiply.

$= \dfrac{15}{-3}$, or -5 Add, and then divide.

(d) $\left(\frac{3}{4}x + \frac{5}{8}y\right)\left(-\frac{1}{2}m\right)$

$= \left[\frac{3}{4}(-1) + \frac{5}{8}(-2)\right] \cdot \left[-\frac{1}{2}(-3)\right]$ Substitute.

$= \left[-\frac{3}{4} + \left(-\frac{5}{4}\right)\right] \cdot \left[\frac{3}{2}\right]$ Multiply inside the brackets.

$= \left[-\frac{8}{4}\right] \cdot \left[\frac{3}{2}\right]$ Add inside the brackets.

$= (-2)\left(\frac{3}{2}\right)$ Divide.

$= -3$ Multiply.

Work Problem 9 at the Side. ▶

9 Evaluate each expression.

GS **(a)** $2x - 7(y + 1)$, for $x = -4$ and $y = 3$

$2x - 7(y + 1)$

$= 2(\_\_\_\_) - 7(\_\_\_\_ + 1)$

$= \_\_\_\_ - 7(\_\_\_\_)$

$= \_\_\_\_ - \_\_\_\_$

$= \_\_\_\_$

(b) $2x^2 - 4y^2$, for $x = -2$ and $y = -3$

(c) $\dfrac{4x - 2y}{-3x}$, for $x = 2$ and $y = -1$

(d) $\left(\frac{2}{5}x - \frac{5}{6}y\right)\left(-\frac{1}{3}z\right)$, for $x = 10$, $y = 6$, and $z = -9$

Answers

9. (a) -4; 3; -8; 4; -8; 28; -36
(b) -28 **(c)** $-\frac{5}{3}$ **(d)** -3

10 Write a numerical expression for each phrase, and simplify the expression.

(a) The product of 6 and the sum of -5 and -4

(b) Three times the difference between 4 and -6

(c) Three-fifths of the sum of 2 and -7

(d) 20% of the sum of 9 and -4

(e) Triple the product of 5 and 6

Answers

10. **(a)** $6[(-5)+(-4)]$; -54
(b) $3[4-(-6)]$; 30
(c) $\frac{3}{5}[2+(-7)]$; -3
(d) $0.20[9+(-4)]$; 1
(e) $3(5 \cdot 6)$; 90

OBJECTIVE 6 Translate words and phrases involving multiplication and division. The table gives words and phrases that indicate multiplication.

Word or Phrase Indicating Multiplication	Example	Numerical Expression and Simplification
Product of	The *product of* -5 and -2	$-5(-2)$, which equals 10
Times	13 *times* -4	$13(-4)$, which equals -52
Twice (meaning "2 times")	*Twice* 6	$2(6)$, which equals 12
Of (used with fractions)	$\frac{1}{2}$ *of* 10	$\frac{1}{2}(10)$, which equals 5
Percent of	12% *of* -16	$0.12(-16)$, which equals -1.92
As much as	$\frac{2}{3}$ *as much as* 30	$\frac{2}{3}(30)$, which equals 20

EXAMPLE 7 Translating Words and Phrases (Multiplication)

Write a numerical expression for each phrase, and simplify the expression.

(a) The **product of** 12 and the sum of 3 and -6

$12[3+(-6)]$ simplifies to $12[-3]$, which equals -36.

(b) **Twice** the difference between 8 and -4

$2[8-(-4)]$ simplifies to $2[12]$, which equals 24.

(c) Two-thirds **of** the sum of -5 and -3

$\frac{2}{3}[-5+(-3)]$ simplifies to $\frac{2}{3}[-8]$, which equals $-\frac{16}{3}$.

(d) 15**% of** the difference between 14 and -2

$0.15[14-(-2)]$ simplifies to $0.15[16]$, which equals 2.4.

Remember that 15% = 0.15.

(e) **Double** the product of 3 and 4

$2 \cdot (3 \cdot 4)$ simplifies to $2(12)$, which equals 24.

◀ **Work Problem 10 at the Side.**

The word *quotient* refers to the answer in a division problem. In algebra, a quotient is usually represented with a fraction bar. The table gives some phrases associated with division.

Phrase Indicating Division	Example	Numerical Expression and Simplification
Quotient of	The *quotient of* -24 and 3	$\frac{-24}{3}$, which equals -8
Divided by	-16 *divided by* -4	$\frac{-16}{-4}$, which equals 4
Ratio of	The *ratio of* 2 to 3	$\frac{2}{3}$

When translating a phrase involving division, we write the first number named as the numerator and the second as the denominator.

EXAMPLE 8 **Translating Words and Phrases (Division)**

Write a numerical expression for each phrase, and simplify the expression.

(a) The **quotient of** 14 and the sum of −9 and 2

$\frac{14}{-9+2}$ simplifies to $\frac{14}{-7}$, which equals −2.

"Quotient" indicates division.

(b) The product of 5 and −6, **divided by** the difference between −7 and 8

$\frac{5(-6)}{-7-8}$ simplifies to $\frac{-30}{-15}$, which equals 2.

Work Problem 11 at the Side. ▶

OBJECTIVE 7 **Translate simple sentences into equations.** We can use words and phrases to translate sentences into equations.

EXAMPLE 9 **Translating Sentences into Equations**

Write each sentence in symbols, using x to represent the number.

(a) Three **times** a number **is** −18.

The word *times* indicates multiplication. The word *is* translates as =.

$3 \cdot x = -18$, or $3x = -18$ $\quad 3 \cdot x = 3x$

(b) The **sum** of a number and 9 **is** 12.

$$x + 9 = 12$$

(c) The **difference between** a number and 5 **is** 0.

$$x - 5 = 0$$

(d) The **quotient of** 24 and a number **is** −2.

$$\frac{24}{x} = -2$$

Work Problem 12 at the Side. ▶

CAUTION

In **Examples 7 and 8,** the *phrases* translate as *expressions*, while in **Example 9,** the *sentences* translate as *equations*. ***An expression is a phrase. An equation is a sentence with something on the left side, an = symbol, and something on the right side.***

$\frac{5(-6)}{-7-8}$ ← Expression $\qquad 3x = -18$ ← Equation

11 Write a numerical expression for each phrase, and simplify the expression.

(a) The quotient of 20 and the sum of 8 and −3

(b) The product of −9 and 2, divided by the difference between 5 and −1

12 Write each sentence in symbols, using x to represent the number.

(a) Twice a number is −6.

(b) The difference between −8 and a number is −11.

(c) The sum of 5 and a number is 8.

(d) The quotient of a number and −2 is 6.

Answers

11. **(a)** $\frac{20}{8+(-3)}$; 4 **(b)** $\frac{-9(2)}{5-(-1)}$; −3

12. **(a)** $2x = -6$ **(b)** $-8 - x = -11$ **(c)** $5 + x = 8$ **(d)** $\frac{x}{-2} = 6$

9.6 Exercises

FOR EXTRA HELP

Download the MyDashBoard App

MyMathLab®

CONCEPT CHECK *Fill in each blank with one of the following:*

greater than 0, less than 0, *or* equal to 0.

1. The product or the quotient of two numbers with the same sign is ___________.

2. The product or the quotient of two numbers with different signs is ___________.

3. If three negative numbers are multiplied together, the product is ___________.

4. If two negative numbers are multiplied and then their product is divided by a negative number, the result is ___________.

5. If a negative number is squared and the result is added to a positive number, the final answer is ___________.

6. The reciprocal of a negative number is ___________.

Find each product. **See Examples 1 and 2.**

7. $-7(4)$

8. $-8(5)$

9. $-5(-6)$

10. $-4(-20)$

11. $-8(0)$

12. $0(-12)$

13. $-\frac{3}{8}\left(-\frac{20}{9}\right)$

14. $-\frac{5}{4}\left(-\frac{6}{25}\right)$

15. $-6.8(0.35)$

16. $-4.6(0.24)$

17. $-6\left(-\frac{1}{4}\right)$

18. $-8\left(-\frac{1}{2}\right)$

Find each quotient. **See Examples 3 and 4** *and the discussion of division involving 0.*

19. $\frac{-15}{5}$

20. $\frac{-18}{6}$

21. $\frac{20}{-10}$

22. $\frac{28}{-4}$

23. $\frac{-160}{-10}$

24. $\frac{-260}{-20}$

25. $\frac{0}{-3}$

26. $\frac{0}{-5}$

27. $\frac{-10.252}{0}$

28. $\frac{-29.584}{0}$

29. $\left(-\frac{3}{4}\right) \div \left(-\frac{1}{2}\right)$

30. $\left(-\frac{3}{16}\right) \div \left(-\frac{5}{8}\right)$

31. CONCEPT CHECK Which expression is undefined?

A. $\frac{5-5}{5+5}$ **B.** $\frac{5+5}{5+5}$ **C.** $\frac{5-5}{5-5}$ **D.** $\frac{5-5}{5}$

32. CONCEPT CHECK Which expression is undefined?

A. $13 \div 0$ **B.** $13 \div 13$ **C.** $0 \div 13$ **D.** $13 \cdot 0$

Perform each indicated operation. ***See Example 5.***

33. $\frac{-5(-6)}{9-(-1)}$

$= \frac{\_\_\_\_}{\_\_\_\_}$

$= \_\_\_\_$

34. $\frac{-12(-5)}{7-(-5)}$

$= \frac{\_\_\_\_}{\_\_\_\_}$

$= \_\_\_\_$

35. $\frac{-21(3)}{-3-6}$

36. $\frac{-40(3)}{-2-3}$

37. $\frac{-10(2)+6(2)}{-3-(-1)}$

38. $\frac{8(-1)+6(-2)}{-6-(-1)}$

39. $\frac{-27(-2)-(-12)(-2)}{-2(3)-2(2)}$

40. $\frac{-13(-4)-(-8)(-2)}{(-10)(2)-4(-2)}$

41. $\frac{3^2-4^2}{7(-8+9)}$

42. $\frac{5^2-7^2}{2(3+3)}$

43. $\frac{4(2^3-5)-5(-3^3+21)}{3[6-(-2)]}$

44. $\frac{-3(-2^4+10)+4(2^5-12)}{-2[8-(-7)]}$

Evaluate each expression for $x=6$, $y=-4$, *and* $a=3$. ***See Example 6.***

45. $6x-5y+4a$

46. $5x-2y+3a$

47. $(5x-2y)(-2a)$

48. $(2x+y)(3a)$

49. $\left(\frac{5}{6}x+\frac{3}{2}y\right)\left(-\frac{1}{3}a\right)$

50. $\left(\frac{1}{3}x-\frac{4}{5}y\right)\left(-\frac{1}{5}a\right)$

51. $(6-x)(5+y)(3+a)$

52. $(-5+x)(-3+y)(3-a)$

53. $5x-4a^2$

54. $-2y^2+3a$

55. $\frac{xy+9a}{x+y-2}$

56. $\frac{2y^2-x}{a-3}$

Write a numerical expression for each phrase and simplify. ***See Examples 7 and 8.***

57. The product of 4 and -7, added to -12

58. The product of -9 and 2, added to 9

59. Twice the product of -8 and 2, subtracted from -1

60. Twice the product of -1 and 6, subtracted from -4

61. The product of -3 and the difference between 3 and -7

62. The product of 12 and the difference between 9 and -8

63. Three-tenths of the sum of -2 and -28

64. Four-fifths of the sum of -8 and -2

65. The quotient of -20 and the sum of -8 and -2

66. The quotient of -12 and the sum of -5 and -1

67. The sum of -18 and -6, divided by the product of 2 and -4

68. The sum of 15 and -3, divided by the product of 4 and -3

69. The product of $-\frac{2}{3}$ and $-\frac{1}{5}$, divided by $\frac{1}{7}$

70. The product of $-\frac{1}{2}$ and $\frac{3}{4}$, divided by $-\frac{2}{3}$

Write each sentence in symbols, using x to represent the number. ***See Example 9.***

71. Nine times a number is -36.

72. Seven times a number is -42.

73. The quotient of a number and 4 is -1.

74. The quotient of a number and 3 is -3.

75. $\frac{9}{11}$ less than a number is 5.

76. $\frac{1}{2}$ less than a number is 2.

77. When 6 is divided by a number, the result is -3.

78. When 15 is divided by a number, the result is -5.

In 2011, the following question and expression appeared on boxes of Swiss Miss Chocolate: On average, how many mini-marshmallows are in one serving?

$$3 + 2 \times 4 \div 2 - 3 \times 7 - 4 + 47$$

79. The box gave 92 as the answer. What is the *correct* answer? (*Source*: Swiss Miss Chocolate box.)

80. Explain the error that somebody at the company made in calculating the answer.

Relating Concepts (Exercises 81–86) For Individual or Group Work

To find the ***average*** *of a group of numbers, we add the numbers and then divide this sum by the number of terms added.* ***Work Exercises 81–84 in order,*** *to find the average of* 23, 18, 13, -4, *and* -8. *Then find the averages in* ***Exercises 85 and 86.***

81. Find the sum of the given group of numbers.

82. How many numbers are in the group?

83. Divide your answer for **Exercise 81** by your answer for **Exercise 82.** Give the quotient as a mixed number.

84. What is the average of the given group of numbers?

85. What is the average of all integers between -10 and 14, including both -10 and 14?

86. What is the average of all integers between -15 and -10, including both -15 and -10?

9.7 Properties of Real Numbers

OBJECTIVES

1. Use the commutative properties.
2. Use the associative properties.
3. Use the identity properties.
4. Use the inverse properties.
5. Use the distributive property.

In the basic properties covered in this section, a, b, and c represent real numbers.

OBJECTIVE 1 Use the commutative properties. The word *commute* means to go back and forth. Many people commute to work or to school. If you travel from home to work and follow the same route from work to home, you travel the same distance each time.

The **commutative properties** say that if two numbers are added or multiplied in either order, the result is the same.

Commutative Properties

$a + b = b + a$	Addition
$ab = ba$	Multiplication

EXAMPLE 1 Using the Commutative Properties

Use a commutative property to complete each statement.

(a) $-8 + 5 = 5 + \underline{\ ?\ }$

Notice that the "order" changed.

$-8 + 5 = 5 + (-8)$ Commutative property of addition

(b) $(-2)7 = \underline{\ ?\ }(-2)$

$-2(7) = 7(-2)$ Commutative property of multiplication

◀ Work Problem 1 at the Side.

1 Complete each statement. Use a commutative property.

(a) $x + 9 = 9 + \_\_\_\_$

(b) $-12(4) = \_\_\_\_(-12)$

(c) $5x = x \cdot \_\_\_\_$

OBJECTIVE 2 Use the associative properties. When we *associate* one object with another, we think of those objects as being grouped together.

The **associative properties** say that when we add or multiply three numbers, we can group the first two together or the last two together and get the same answer.

Associative Properties

$(a + b) + c = a + (b + c)$	Addition
$(ab)c = a(bc)$	Multiplication

EXAMPLE 2 Using the Associative Properties

Use an associative property to complete each statement.

(a) $-8 + (1 + 4) = (-8 + \underline{\ ?\ }) + 4$

The "order" is the same. The "grouping" changed.

$-8 + (1 + 4) = (-8 + 1) + 4$ Associative property of addition

(b) $[2 \cdot (-7)] \cdot 6 = 2 \cdot \underline{\ ?\ }$

$[2 \cdot (-7)] \cdot 6 = 2 \cdot [(-7) \cdot 6]$ Associate property of multiplication

◀ Work Problem 2 at the Side.

2 Complete each statement. Use an associative property.

(a) $(9 + 10) + (-3)$
$= 9 + [\_\_\_\_ + (-3)]$

(b) $-5 + (2 + 8)$
$= (\_\_\_\_\_\_\_\_) + 8$

(c) $10 \cdot [-8 \cdot (-3)]$
$= \_\_\_\_\_\_\_\_$

Answers

1. (a) x (b) 4 (c) 5
2. (a) 10 (b) $-5 + 2$
(c) $[10 \cdot (-8)] \cdot (-3)$

By the associative property, the sum (or product) of three numbers will be the same no matter how the numbers are "associated" in groups. Parentheses can be left out if a problem contains only addition (or multiplication). For example,

$$(-1+2)+3 \quad \text{and} \quad -1+(2+3) \quad \text{can be written as} \quad -1+2+3.$$

EXAMPLE 3 **Distinguishing between Properties**

Decide whether each statement is an example of a commutative property, an associative property, or both.

(a) $(2+4)+5=2+(4+5)$

The order of the three numbers is the same on both sides of the equality symbol. The only change is in the *grouping*, or association, of the numbers. This is an example of the associative property.

(b) $6 \cdot (3 \cdot 10) = 6 \cdot (10 \cdot 3)$

The same numbers, 3 and 10, are grouped on each side. On the left, the 3 appears first, but on the right, the 10 appears first. Since the only change involves the *order* of the numbers, this is an example of the commutative property.

(c) $(8+1)+7=8+(7+1)$

Both the order and the grouping are changed. On the left, the order of the three numbers is 8, 1, and 7. On the right, it is 8, 7, and 1. On the left, the 8 and 1 are grouped. On the right, the 7 and 1 are grouped. Therefore, *both* properties are used.

Work Problem 3 at the Side. ▶

3 Decide whether each statement is an example of a commutative property, an associative property, or both.

(a) $2 \cdot (4 \cdot 6) = (2 \cdot 4) \cdot 6$

(b) $(2 \cdot 4) \cdot 6 = (4 \cdot 2) \cdot 6$

(c) $(2+4)+6=4+(2+6)$

Example 4 illustrates that we can sometimes use the commutative and associative properties to rearrange and regroup numbers to simplify calculations.

EXAMPLE 4 **Using the Commutative and Associative Properties**

Find each sum or product.

(a) $23+41+2+9+25$

$= (41+9)+(23+2)+25$

$= 50+25+25$ Use the commutative and associative properties.

$= 100$

(b) $25(69)(4)$

$= 25(4)(69)$

$= 100(69)$

$= 6900$

Work Problem 4 at the Side. ▶

4 Find each sum or product.

(a) $5+18+29+31+12$

(b) $5(37)(20)$

OBJECTIVE 3 **Use the identity properties.** If a child wears a costume on Halloween, the child's appearance is changed, but his or her *identity* is unchanged. The identity of a real number is left unchanged when identity properties are applied.

The **identity properties** say that the sum of 0 and any number equals that number, and the product of 1 and any number equals that number.

Identity Properties

$a+0=a$	and	$0+a=a$	Addition
$a \cdot 1 = a$	and	$1 \cdot a = a$	Multiplication

Answers

3. (a) associative (b) commutative (c) both
4. (a) 95 (b) 3700

5 Use an identity property to complete each statement.

(a) $9 + 0 = \_\_\_\_$

(b) $\_\_\_\_ + (-7) = -7$

(c) $\_\_\_\_ \cdot 1 = 5$

6 In part (a), write in lowest terms. In part (b), perform the operation.

(a) $\dfrac{85}{105}$

(b) $\dfrac{9}{10} - \dfrac{53}{50}$

Answers

5. (a) 9 (b) 0 (c) 5

6. (a) $\dfrac{17}{21}$ (b) $-\dfrac{4}{25}$

The number 0 leaves the identity, or value, of any real number unchanged by addition, so 0 is the **identity element for addition**, or the **additive identity.** Since multiplication by 1 leaves any real number unchanged, 1 is the **identity element for multiplication**, or the **multiplicative identity.**

EXAMPLE 5 Using the Identity Properties

Use an identity property to complete each statement.

(a) $-3 + \underline{\ ?\ } = -3$

$-3 + 0 = -3$

Identity property for addition

(b) $\underline{\ ?\ } \cdot \dfrac{1}{2} = \dfrac{1}{2}$

$1 \cdot \dfrac{1}{2} = \dfrac{1}{2}$

Identity property for multiplication

◀ **Work Problem 5 at the Side.**

We use the identity property for multiplication to write fractions in lowest terms and to find common denominators.

EXAMPLE 6 Using the Identity Property for Multiplication

In part (a), write in lowest terms. In part (b), perform the operation.

(a) $\dfrac{49}{35}$

$= \dfrac{7 \cdot 7}{5 \cdot 7}$ Factor.

$= \dfrac{7}{5} \cdot \dfrac{7}{7}$ Write as a product.

$= \dfrac{7}{5} \cdot 1$ Property of 1

$= \dfrac{7}{5}$ Identity property

(b) $\dfrac{3}{4} + \dfrac{5}{24}$

$= \dfrac{3}{4} \cdot 1 + \dfrac{5}{24}$ Identity property

$= \dfrac{3}{4} \cdot \dfrac{6}{6} + \dfrac{5}{24}$ Use $1 = \frac{6}{6}$ to get a common denominator.

$= \dfrac{18}{24} + \dfrac{5}{24}$ Multiply.

$= \dfrac{23}{24}$ Add.

◀ **Work Problem 6 at the Side.**

OBJECTIVE 4 Use the inverse properties. Each day before you go to work or school, you probably put on your shoes. Before you go to sleep at night, you probably take them off. These operations from everyday life are examples of *inverse* operations.

The **inverse properties** of addition and multiplication lead to the additive and multiplicative identities, respectively. Recall that $-a$ is the **additive inverse,** or **opposite,** of a and $\frac{1}{a}$ is the **multiplicative inverse,** or **reciprocal,** of the nonzero number a. The sum of the numbers a and $-a$ is 0, and the product of the nonzero numbers a and $\frac{1}{a}$ is 1.

Inverse Properties

$a + (-a) = 0$ and $-a + a = 0$ Addition

$a \cdot \dfrac{1}{a} = 1$ and $\dfrac{1}{a} \cdot a = 1 \quad (a \neq 0)$ Multiplication

EXAMPLE 7 Using the Inverse Properties

Use an inverse property to complete each statement.

(a) $\underline{\ ?\ } + \frac{1}{2} = 0$ **(b)** $4 + \underline{\ ?\ } = 0$ **(c)** $-0.75 + \frac{3}{4} = \underline{\ ?\ }$

$-\frac{1}{2} + \frac{1}{2} = 0$ $4 + (-4) = 0$ $-0.75 + \frac{3}{4} = 0$

The inverse property for addition is used in parts (a)–(c).

(d) $\underline{\ ?\ } \cdot \frac{5}{2} = 1$ **(e)** $-5(\underline{\ ?\ }) = 1$ **(f)** $4(0.25) = \underline{\ ?\ }$

$\frac{2}{5} \cdot \frac{5}{2} = 1$ $-5\left(-\frac{1}{5}\right) = 1$ $4(0.25) = 1$

The inverse property for multiplication is used in parts (d)–(f).

Work Problem 7 at the Side. ▶

7 Complete each statement so that it is an example of either an identity property or an inverse property. Tell which property is used.

(a) $-6 + \_\_\_\_ = 0$

(b) $\frac{4}{3} \cdot \_\_\_\_ = 1$

(c) $-\frac{1}{9} \cdot (\_\_\_\_) = 1$

(d) $275 + \_\_\_\_ = 275$

(e) $-0.75 + \frac{3}{4} = \_\_\_\_$

(f) $0.2(5) = \_\_\_\_$

OBJECTIVE 5 **Use the distributive property.** The word *distribute* means "to give out from one to several." Consider the following expressions.

$$2(5+8) \text{ equals } 2(13), \text{ or } 26.$$

$$2(5) + 2(8) \text{ equals } 10 + 16, \text{ or } 26.$$

Since both expressions equal 26,

$$2(5+8) = 2(5) + 2(8).$$

This result is an example of the *distributive property of multiplication with respect to addition,* the only property involving *both* addition and multiplication. With this property, a product can be changed to a sum. This idea is illustrated by the divided rectangle in **Figure 13.**

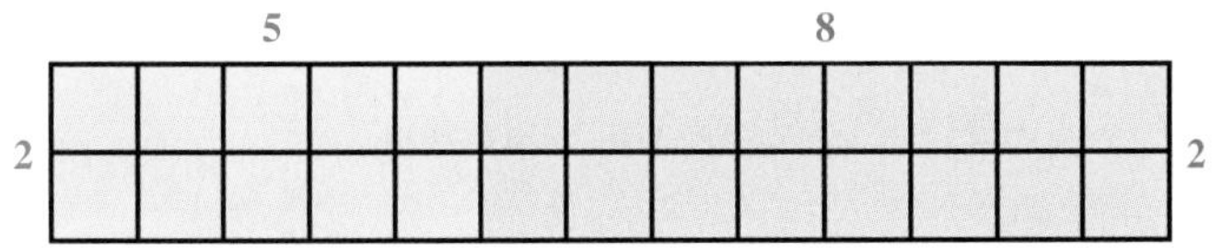

The area of the left part is 2(5) = 10.
The area of the right part is 2(8) = 16.
The total area is 2(5 + 8) = 26 or the total area is
2(5) + 2(8) = 10 + 16 = 26.
Thus, 2(5 + 8) = 2(5) + 2(8).

Figure 13

The **distributive property** says that multiplying a number a by a sum of numbers $b + c$ gives the same result as multiplying a by b and a by c and then adding the two products.

Distributive Property

$$a(b+c) = ab + ac \quad \text{and} \quad (b+c)a = ba + ca$$

As the arrows show, the a outside the parentheses is "distributed" over the b and c inside.

Answers

7. (a) 6; inverse (b) $\frac{3}{4}$; inverse
(c) -9; inverse (d) 0; identity
(e) 0; inverse (f) 1; inverse

8 Use the distributive property to rewrite each expression.

GS **(a)** $2(p+5)$

$= 2 \cdot \_\_\_ + 2 \cdot \_\_\_$

$= \_\_\_ + \_\_\_$

(b) $-4(y+7)$

(c) $5(m-4)$

(d) $7(2y+7k-9m)$

GS **(e)** $9 \cdot k + 9 \cdot 5$

$= \_\_\_(\_\_\_ + \_\_\_)$

(f) $3a - 3b$

Answers

8. **(a)** p; 5; $2p$; 10 **(b)** $-4y-28$ **(c)** $5m-20$ **(d)** $14y+49k-63m$ **(e)** 9; k; 5 **(f)** $3(a-b)$

The distributive property is also valid for subtraction.

$$a(b-c) = ab - ac \quad \text{and} \quad (b-c)a = ba - ca$$

The distributive property also can be extended to more than two numbers.

$$a(b+c+d) = ab + ac + ad$$

The distributive property can also be written "in reverse."

$$ab + ac = a(b+c)$$

We will use this form in the next section to combine like terms.

EXAMPLE 8 Using the Distributive Property

Use the distributive property to rewrite each expression.

(a) $5(9+6)$

$= 5 \cdot 9 + 5 \cdot 6$ Distributive property

$= 45 + 30$ Multiply.

$= 75$ Add.

Multiply first.

(b) $4(x+5+y)$

$= 4x + 4 \cdot 5 + 4y$ Distributive property

$= 4x + 20 + 4y$ Multiply.

(c) $-2(x+3)$

$= -2x + (-2)(3)$ Distributive property

$= -2x + (-6)$ Multiply.

$= -2x - 6$ Definition of subtraction

(d) $3(k-9)$

Be careful here.

$= 3[k + (-9)]$ Definition of subtraction

$= 3k + 3(-9)$ Distributive property

$= 3k - 27$ Multiply.

(e) $8(3r + 11t + 5z)$

$= 8(3r) + 8(11t) + 8(5z)$ Distributive property

$= (8 \cdot 3)r + (8 \cdot 11)t + (8 \cdot 5)z$ Associative property

$= 24r + 88t + 40z$ Multiply.

(f) $6 \cdot 8 + 6 \cdot 2$

$= 6(8+2)$ Distributive property in reverse

$= 6(10)$ Add.

$= 60$ Multiply.

(g) $4x - 4m$

$= 4(x-m)$ Distributive property in reverse

◀ **Work Problem 8 at the Side.**

The symbol $-a$ may be interpreted as $-1 \cdot a$. Using this result and the distributive property, we can remove (or clear) parentheses from some expressions.

EXAMPLE 9 **Using the Distributive Property to Remove (Clear) Parentheses**

Write each expression without parentheses.

(a) $-(2y + 3)$

$= -1 \cdot (2y + 3)$ — $-a = -1 \cdot a$

$= -1 \cdot 2y + (-1) \cdot 3$ — Distributive property

$= -2y - 3$ — Multiply.

The − symbol indicates a factor of −1.

(b) $-(-9w - 2)$

$= -1(-9w - 2)$

$= -1(-9w) - 1(-2)$

$= 9w + 2$

We can also interpret the negative sign in front of the parentheses to mean the *opposite* of each of the terms within the parentheses.

(c) $-(-x - 3y + 6z)$

$= -1(-1x - 3y + 6z)$ — Be careful with signs.

$= -1(-1x) - 1(-3y) - 1(6z)$ — Distributive property

$= x + 3y - 6z$ — $-1(-1x) = 1x = x$

Work Problem 9 at the Side. ▶

Here is a summary of the basic properties of real numbers.

Properties of Addition and Multiplication

For any real numbers a, b, and c, the following properties hold.

Commutative properties $\quad a + b = b + a \qquad ab = ba$

Associative properties $\quad (a + b) + c = a + (b + c)$

$$(ab)c = a(bc)$$

Identity properties There is a real number 0 such that

$$a + 0 = a \quad \text{and} \quad 0 + a = a.$$

There is a real number 1 such that

$$a \cdot 1 = a \quad \text{and} \quad 1 \cdot a = a.$$

Inverse properties For each real number a, there is a single real number $-a$ such that

$$a + (-a) = 0 \quad \text{and} \quad (-a) + a = 0.$$

For each nonzero real number a, there is a single real number $\frac{1}{a}$ such that

$$a \cdot \frac{1}{a} = 1 \quad \text{and} \quad \frac{1}{a} \cdot a = 1.$$

Distributive property $\quad a(b + c) = ab + ac$

$$(b + c)a = ba + ca$$

9 Write each expression without parentheses.

GS **(a)** $-(3k - 5)$

$= \_\_\_\_ \cdot (3k - 5)$

$= \_\_\_\_ \cdot 3k + (-1)(\_\_\_\_)$

$= \_\_\_\_ + \_\_\_\_$

(b) $-(2 - r)$

(c) $-(-5y + 8)$

(d) $-(-z + 4)$

(e) $-(-t - 4u + 5v)$

Answers

9. (a) -1; -1; -5; $-3k$; 5 **(b)** $-2 + r$ **(c)** $5y - 8$ **(d)** $z - 4$ **(e)** $t + 4u - 5v$

9.7 Exercises

FOR EXTRA HELP

Download the MyDashBoard App

MyMathLab®

1. **CONCEPT CHECK** Match each item in Column I with the correct choice(s) from Column II. Choices may be used once, more than once, or not at all.

I	II
(a) Identity element for addition	**A.** $(5 \cdot 4) \cdot 3 = 5 \cdot (4 \cdot 3)$
(b) Identity element for multiplication	**B.** 0
(c) Additive inverse of a	**C.** $-a$
(d) Multiplicative inverse, or reciprocal, of the nonzero number a	**D.** -1
(e) The number that is its own additive inverse	**E.** $5 \cdot 4 \cdot 3 = 60$
(f) The two numbers that are their own multiplicative inverses	**F.** 1
(g) The only number that has no multiplicative inverse	**G.** $(5 \cdot 4) \cdot 3 = 3 \cdot (5 \cdot 4)$
(h) An example of the associative property	**H.** $5(4 + 3) = 5 \cdot 4 + 5 \cdot 3$
(i) An example of the commutative property	**I.** $\frac{1}{a}$
(j) An example of the distributive property	

2. **CONCEPT CHECK** Fill in the blanks: The commutative property allows us to change the _____ of the terms in a sum or the factors in a product. The associative property allows us to change the _______ of the terms in a sum or the factors in a product.

CONCEPT CHECK *Tell whether or not the following everyday activities are commutative.*

3. Washing your face and brushing your teeth

4. Putting on your left sock and putting on your right sock

5. Preparing a meal and eating a meal

6. Starting a car and driving away in a car

7. Putting on your socks and putting on your shoes

8. Getting undressed and taking a shower

9. **CONCEPT CHECK** Use parentheses to show how the associative property can be used to give two different meanings to the phrase "foreign sales clerk."

10. **CONCEPT CHECK** Use parentheses to show how the associative property can be used to give two different meanings to the phrase "defective merchandise counter."

Use the commutative or the associative property to complete each statement. State which property is used. ***See Examples 1 and 2.***

11. $-15 + 9 = 9 +$ _____

12. $6 + (-2) = -2 +$ _____

13. $-8 \cdot 3 =$ _____ $\cdot (-8)$

14. $-12 \cdot 4 = 4 \cdot$ _____

15. $(3 + 6) + 7 = 3 + ($_____ $+ 7)$

16. $(-2 + 3) + 6 = -2 + ($_____ $+ 6)$

17. $7 \cdot (2 \cdot 5) = ($_____ $\cdot 2) \cdot 5$

18. $8 \cdot (6 \cdot 4) = (8 \cdot$ _____$) \cdot 4$

19. CONCEPT CHECK Evaluate $25 - (6 - 2)$ and evaluate $(25 - 6) - 2$. Do you think subtraction is associative?

20. CONCEPT CHECK Evaluate $180 \div (15 \div 3)$ and evaluate $(180 \div 15) \div 3$. Do you think division is associative?

21. CONCEPT CHECK Complete the table and the statements beside it.

Number	Additive Inverse	Multiplicative Inverse
5		
-10		
$-\frac{1}{2}$		
$\frac{3}{8}$		
x		
$-y$		

In general, a number and its additive inverse have (*the same / opposite*) signs.

A number and its multiplicative inverse have (*the same / opposite*) signs.

22. CONCEPT CHECK The following conversation actually took place between one of the authors of this book and his son, Jack, when Jack was 4 years old.

DADDY: "Jack, what is 3 + 0?"
JACK: "3."
DADDY: "Jack, what is 4 + 0?"
JACK: "4. And Daddy, *string* plus zero equals *string*!"

What property of addition did Jack recognize?

Decide whether each statement is an example of a commutative, associative, identity, or inverse property, or of the distributive property. ***See Examples 1, 2, 3, and 5–8.***

23. $\frac{2}{3}(-4) = -4\left(\frac{2}{3}\right)$

24. $6\left(-\frac{5}{6}\right) = \left(-\frac{5}{6}\right)6$

25. $-6 + (12 + 7) = (-6 + 12) + 7$

26. $(-8 + 13) + 2 = -8 + (13 + 2)$

27. $-6 + 6 = 0$

28. $12 + (-12) = 0$

29. $\left(\frac{2}{3}\right)\left(\frac{3}{2}\right) = 1$

30. $\left(\frac{5}{8}\right)\left(\frac{8}{5}\right) = 1$

31. $2.34 \cdot 1 = 2.34$

32. $-8.456 \cdot 1 = -8.456$

33. $(4 + 17) + 3 = 3 + (4 + 17)$

34. $(-8 + 4) + (-12) = -12 + (-8 + 4)$

35. $6(x + y) = 6x + 6y$

36. $14(t + s) = 14t + 14s$

37. $-\frac{5}{9} = -\frac{5}{9} \cdot \frac{3}{3} = -\frac{15}{27}$

38. $\frac{13}{12} = \frac{13}{12} \cdot \frac{7}{7} = \frac{91}{84}$

39. $5(2x) + 5(3y) = 5(2x + 3y)$

40. $3(5t) - 3(7r) = 3(5t - 7r)$

41. **CONCEPT CHECK** Suppose that a student simplifies the expression $-3(4 - 6)$ as shown.

$$-3(4 - 6)$$
$$= -3(4) - 3(6)$$
$$= -12 - 18$$
$$= -30$$

What Went Wrong? Work the problem correctly.

42. Explain how the procedure of changing $\frac{3}{4}$ to $\frac{9}{12}$ requires the use of the multiplicative identity element, 1.

Write a new expression that is equal to the given expression, using the given property. Then simplify the new expression if possible. ***See Examples 1, 2, 5, 7, and 8.***

43. $r + 7$; commutative

44. $t + 9$; commutative

45. $s + 0$; identity

46. $w + 0$; identity

47. $-6(x + 7)$; distributive

48. $-5(y + 2)$; distributive

49. $(w + 5) + (-3)$; associative

50. $(b + 8) + (-10)$; associative

Use the properties of this section to perform the operations. ***See Example 4.***

51. $50(67)2$

52. $26 + 8 - 26 + 12$

53. $-\frac{3}{8} + \frac{2}{5} + \frac{8}{5} + \frac{3}{8}$

54. $-\frac{5}{12} - \frac{3}{7} + \frac{5}{12} + \frac{10}{7}$

55. $\left(-\frac{8}{5}\right)(0.77)\left(-\frac{5}{8}\right)$

56. $\frac{9}{7}(-0.38)\left(\frac{7}{9}\right)$

Use the distributive property to rewrite each expression. Simplify if possible.
See Example 8.

57. $4(t+3)$

58. $5(w+4)$

59. $-8(r+3)$

60. $-11(x+4)$

61. $-5(y-4)$
$= -5(\_\_\_) - 5(\_\_\_)$
$= \_\_\_\_\_$

62. $-9(g-4)$
$= -9(\_\_\_) - 9(\_\_\_)$
$= \_\_\_\_\_$

63. $-\frac{4}{3}(12y+15z)$

64. $-\frac{2}{5}(10b+20a)$

65. $8 \cdot z + 8 \cdot w$

66. $4 \cdot s + 4 \cdot r$

67. $5 \cdot 3 + 5 \cdot 17$

68. $15 \cdot 6 + 5 \cdot 6$

69. $7(2v) + 7(5r)$

70. $13(5w) + 13(4p)$

71. $8(3r + 4s - 5y)$

72. $2(5u - 3v + 7w)$

73. $-3(8x + 3y + 4z)$

74. $-5(2x - 5y + 6z)$

Use the distributive property to write each expression without parentheses.
See Example 9.

75. $-(4t + 5m)$

76. $-(9x + 12y)$

77. $-(-5c - 4d)$

78. $-(-13x - 15y)$

79. $-(-3q + 5r - 8s)$

80. $-(-4z + 5w - 9y)$

9.8 Simplifying Expressions

OBJECTIVES

1. Simplify expressions.
2. Identify terms and numerical coefficients.
3. Identify like terms.
4. Combine like terms.
5. Simplify expressions from word phrases.

1 Simplify each expression.

(a) $9k + 12 - 5$

GS **(b)** $7(3p + 2q)$

$= 7(\_\_\_\_) + 7(\_\_\_\_)$

$= \_\_\_\_ + \_\_\_\_$

(c) $2 + 5(3z - 1)$

(d) $-3 - (2 + 5y)$

Answers

1. **(a)** $9k + 7$ **(b)** $3p$; $2q$; $21p$; $14q$ **(c)** $15z - 3$ **(d)** $-5 - 5y$

OBJECTIVE 1 Simplify expressions. We now simplify expressions using the properties of addition and multiplication introduced in **Section 9.7.**

EXAMPLE 1 Simplifying Expressions

Simplify each expression.

(a) $4x + 8 + 9$ simplifies to $4x + 17$.

(b) $4(3m - 2n)$ (To simplify, we clear the parentheses.)

$= 4(3m) - 4(2n)$ Distributive property

$= 12m - 8n$ Associative property

(c) $6 + 3(4k + 5)$ (Don't start by adding.)

$= 6 + 3(4k) + 3(5)$ Distributive property

$= 6 + 12k + 15$ Multiply.

$= 21 + 12k$ Add.

(d) $5 - (2y - 8)$

$= 5 - 1(2y - 8)$ $-a = -1 \cdot a$ (Be careful with signs.)

$= 5 - 2y + 8$ Distributive property

$= 13 - 2y$ Add.

◀ Work Problem 1 at the Side.

Note

In **Examples 1(c) and 1(d),** we mentally used the commutative and associative properties to add in the last step. In practice, these steps are often not written out.

OBJECTIVE 2 Identify terms and numerical coefficients. A **term** is a number (constant), a variable, or a product or quotient of numbers and variables raised to powers.

$$9x, \quad 15y^2, \quad -3, \quad -8m^2n, \quad \frac{2}{p}, \quad \text{and} \quad k \qquad \text{Terms}$$

In the term $9x$, the **numerical coefficient,** or simply **coefficient,** of the variable x is 9. Additional examples are shown in the table on the next page.

CAUTION

It is important to be able to distinguish between ***terms*** and ***factors.*** Consider the following expressions.

$8x^3 + 12x^2$ This expression has **two terms,** $8x^3$ and $12x^2$. **Terms** are separated by a + or – symbol.

$(8x^3)(12x^2)$ This is a **one-term** expression. The **factors** $8x^3$ and $12x^2$ are multiplied.

Term	Numerical Coefficient
$-7y$	-7
$34r^3$	34
$-26x^5yz^4$	-26
$-k$, or $-1 \cdot k$	-1
r, or $1r$	1
$\frac{3x}{8} = \frac{3}{8}x$	$\frac{3}{8}$

Work Problem **2** at the Side. ▶

OBJECTIVE 3 Identify like terms. Terms with exactly the same variables that have the same exponents are **like terms.** Here are some examples.

Like Terms	Unlike Terms	
$9t$ and $4t$	$4y$ and $7t$	Different variables
$6x^2$ and $-5x^2$	$17x$ and $-8x^2$	Different exponents
$-2pq$ and $11pq$	$4xy^2$ and $4xy$	Different exponents
$3x^2y$ and $5x^2y$	$-7wz^3$ and $2xz^3$	Different variables

Work Problem **3** at the Side. ▶

OBJECTIVE 4 Combine like terms. Recall the distributive property.

$x(y+z) = xy + xz$ can also be written "in reverse" as $xy + xz = x(y+z)$.

This last form, which may be used to find the sum or difference of like terms, provides justification for **combining like terms.**

EXAMPLE 2 Combining Like Terms

Combine like terms in each expression.

(a) $-9m + 5m$
$= (-9+5)m$
$= -4m$

(b) $6r + 3r + 2r$
$= (6+3+2)r$
$= 11r$

(c) $4x + x$
$= 4x + 1x$ $\quad x = 1x$
$= (4+1)x$
$= 5x$

(d) $16y^2 - 9y^2$
$= (16-9)y^2$
$= 7y^2$

(e) $32y + 10y^2$ These unlike terms cannot be combined.

Work Problem **4** at the Side. ▶

Simplifying an Expression

An expression has been simplified when the following conditions have been met.

- All grouping symbols have been removed.
- All like terms have been combined.
- Operations have been performed, when possible.

2 Give the numerical coefficient of each term.

(a) $15q$

(b) $-2m^3$

(c) $-18m^7q^4$

(d) $-r$

(e) $\dfrac{5x}{4}$

3 Identify each pair of terms as *like* or *unlike*.

(a) $9x, 4x$

(b) $-8y^3, 12y^2$

(c) $5x^2y^4, 5x^4y^2$

(d) $7x^2y^4, -7x^2y^4$

(e) $13kt, 4tk$

4 Combine like terms.

(a) $4k + 7k$

(b) $4r - r$

(c) $5z + 9z - 4z$

(d) $8p + 8p^2$

Answers

2. **(a)** 15 **(b)** -2 **(c)** -18 **(d)** -1 **(e)** $\frac{5}{4}$

3. **(a)** like **(b)** unlike **(c)** unlike **(d)** like **(e)** like

4. **(a)** $11k$ **(b)** $3r$ **(c)** $10z$ **(d)** cannot be combined

5 Simplify.

GS **(a)** $10p + 3(5 + 2p)$

$= 10p + \_\_\_(5) + 3(\_\_\_)$

$= 10p + \_\_\_ + \_\_\_$

$= \_\_\_\_$

(b) $7z - 2 - (1 + z)$

(c) $-(3k^2 + 5k) + 7(k^2 - 4k)$

(d) $-\frac{3}{4}(x - 8) - \frac{1}{3}x$

6 Translate each phrase into a mathematical expression and simplify.

(a) Three times a number, subtracted from the sum of the number and 8

(b) Twice a number added to the sum of 6 and the number

Answers

5. **(a)** 3; $2p$; 15; $6p$; $16p + 15$ **(b)** $6z - 3$ **(c)** $4k^2 - 33k$ **(d)** $-\frac{13}{12}x + 6$

6. **(a)** $(x + 8) - 3x$; $-2x + 8$ **(b)** $(6 + x) + 2x$; $3x + 6$

CAUTION

Remember that only like terms may be combined.

EXAMPLE 3 Simplifying Expressions Involving Like Terms

Simplify each expression.

(a) $14y + 2(6 + 3y)$

$= 14y + 2(6) + 2(3y)$ Distributive property

$= 14y + 12 + 6y$ Multiply.

$= 20y + 12$ Combine like terms.

(b) $9k - 6 - 3(2 - 5k)$ Be careful with signs.

$= 9k - 6 - 3(2) - 3(-5k)$ Distributive property

$= 9k - 6 - 6 + 15k$ Multiply.

$= 24k - 12$ Combine like terms.

(c) $-(2 - r) + 10r$

$= -1(2 - r) + 10r$ $\quad -(2 - r) = -1(2 - r)$

$= -1(2) - 1(-r) + 10r$ Distributive property

Be careful with signs.

$= -2 + r + 10r$ Multiply.

$= -2 + 11r$ Combine like terms.

(d) $-\frac{2}{3}(x - 6) - \frac{1}{6}x$

$= -\frac{2}{3}x - \frac{2}{3}(-6) - \frac{1}{6}x$ Distributive property

$= -\frac{2}{3}x + 4 - \frac{1}{6}x$ Multiply.

$= -\frac{4}{6}x + 4 - \frac{1}{6}x$ Get a common denominator.

$= -\frac{5}{6}x + 4$ Combine like terms,

◀ Work Problem **5** at the Side.

OBJECTIVE 5 Simplify expressions from word phrases.

EXAMPLE 4 Translating Words into a Mathematical Expression

Translate the phrase into a mathematical expression and simplify.

The sum of 9, five times a number, four times the number, and six times the number

The word "sum" indicates that the terms should be added. Use x for the number.

$9 + 5x + 4x + 6x$ simplifies to $9 + 15x$. Combine like terms.

This is an expression to be simplified, *not* an equation to be solved.

◀ Work Problem **6** at the Side.

9.8 Exercises

FOR EXTRA HELP

CONCEPT CHECK *In Exercises 1–4, choose the letter of the correct response.*

1. Which expression is a simplified form of $-(6x - 3)$?

A. $-6x - 3$ **B.** $-6x + 3$ **C.** $6x - 3$ **D.** $6x + 3$

2. Which is an example of a pair of like terms?

A. $6t, 6w$ **B.** $-8x^2y, 9xy^2$ **C.** $5ry, 6yr$ **D.** $-5x^2, 2x^3$

3. Which is an example of a term with numerical coefficient 5?

A. $5x^3y^7$ **B.** x^5 **C.** $\frac{x}{5}$ **D.** 5^2xy^3

4. Which is a correct translation for "six times a number, subtracted from the product of eleven and the number" (if x represents the number)?

A. $6x - 11x$ **B.** $11x - 6x$ **C.** $(11 + x) - 6x$ **D.** $6x - (11 + x)$

Simplify each expression. ***See Examples 1 and 2.***

5. $3x + 12x$

6. $4y + 9y$

7. $12b + b$

8. $19x + x$

9. $4r + 19 - 8$

10. $7t + 18 - 4$

11. $5 + 2(x - 3y)$

12. $8 + 3(s - 6t)$

13. $-2 - (5 - 3p)$

14. $-10 - (7 - 14r)$

15. $-\frac{5}{6}(y + 12) - \frac{1}{2}y$

16. $-\frac{2}{3}(w + 15) - \frac{1}{4}w$

Give the numerical coefficient. ***See Objective 2.***

17. $-12k$

18. $-23y$

19. $5m^2$

20. $-3n^6$

21. xw

22. pq

23. $-x$

24. $-t$

25. 74

26. 98

27. $-\frac{3}{8}x$

28. $-\frac{5}{4}z$

29. $\frac{x}{2}$

30. $\frac{x}{6}$

31. $\frac{2x}{5}$

32. $\frac{8x}{9}$

33. $-1.28r^2$

34. $-2.985t^3$

Identify each group of terms as like *or* unlike. ***See Objective 3.***

35. $8r, -13r$

36. $-7a, 12a$

37. $5z^4, 9z^3$

38. $8x^5, -10x^3$

39. $4, 9, -24$

40. $7, 17, -83$

41. x, y

42. t, s

Simplify each expression. **See Examples 2 and 3.**

43. $-5 - 2(x - 3)$
$= -5 - 2(\_\_\_) - 2(\_\_\_)$
$= -5 - \_\_\_ + \_\_\_$
$= \_\_\_\_$

44. $-8 - 3(2x + 4)$
$= -8 - 3(\_\_\_) - 3(\_\_\_)$
$= -8 - \_\_\_ - \_\_\_$
$= \_\_\_\_$

45. $-\frac{4}{3} + 2t + \frac{1}{3}t - 8 - \frac{8}{3}t$

46. $-\frac{5}{6} + 8x + \frac{1}{6}x - 7 - \frac{7}{6}$

47. $-5.3r + 4.9 - (2r + 0.7) + 3.2r$

48. $2.7b + 5.8 - (3b + 0.5) - 4.4b$

49. $2y^2 - 7y^3 - 4y^2 + 10y^3$

50. $9x^4 - 7x^6 + 12x^4 + 14x^6$

51. $13p + 4(4 - 8p)$

52. $5x + 3(7 - 2x)$

53. $\frac{1}{2}(2x + 4) - \frac{1}{3}(9x - 6)$

54. $\frac{1}{4}(8x + 16) - \frac{1}{5}(20x - 15)$

55. $-\frac{2}{3}(5x + 7) - \frac{1}{3}(4x + 8)$

56. $-\frac{3}{4}(7x + 9) - \frac{1}{4}(5x + 7)$

57. $-\frac{4}{3}(y - 12) - \frac{1}{6}y$

58. $-\frac{7}{5}(t - 15) - \frac{1}{2}t$

59. $-5(5y - 9) + 3(3y + 6)$

60. $-3(2t + 4) + 8(2t - 4)$

61. $-3(2r - 3) + 2(5r + 3)$

62. $-4(5y - 7) + 3(2y - 5)$

63. $8(2k - 1) - (4k - 3)$

64. $6(3p - 2) - (5p + 1)$

65. $-2(-3k + 2) - (5k - 6) - 3k - 5$

66. $-2(3r - 4) - (6 - r) + 2r - 5$

67. $-4(-3x + 3) - (6x - 4) - 2x + 1$

68. $-5(8x + 2) - (5x - 3) - 3x + 17$

69. $-7.5(2y + 4) - 2.9(3y - 6)$

70. $8.4(6t - 6) + 2.4(9 - 3t)$

Write each phrase as a mathematical expression. Use x to represent the number. Combine like terms when possible. ***See Example 4.***

71. Five times a number, added to the sum of the number and three

72. Six times a number, added to the sum of the number and six

73. A number multiplied by -7, subtracted from the sum of 13 and six times the number

74. A number multiplied by 5, subtracted from the sum of 14 and eight times the number

75. Six times a number added to -4, subtracted from twice the sum of three times the number and 4

76. Nine times a number added to 6, subtracted from triple the sum of 12 and 8 times the number

77. Write the expression $9x - (x + 2)$ using words, as in **Exercises 71–76.**

78. Write the expression $2(3x + 5) - 2(x + 4)$ using words, as in **Exercises 71–76.**

79. CONCEPT CHECK A student simplified the expression $7x - 2(3 - 2x)$ as shown.

$$\begin{aligned} &7x - 2(3 - 2x) \\ &= 7x - 2(3) - 2(2x) \\ &= 7x - 6 - 4x \\ &= 3x - 6 \end{aligned}$$

What Went Wrong? Find the correct simplified answer.

80. CONCEPT CHECK A student simplified the expression $3 + 2(4x - 5)$ as shown.

$$\begin{aligned} &3 + 2(4x - 5) \\ &= 5(4x - 5) \\ &= 5(4x) + 5(-5) \\ &= 20x - 25 \end{aligned}$$

What Went Wrong? Find the correct simplified answer.

Relating Concepts (Exercises 81–84) For Individual or Group Work

A manufacturer has fixed costs of \$1000 *to produce widgets. Each widget costs* \$5 *to make. The fixed cost to produce gadgets is* \$750, *and each gadget costs* \$3 *to make.*
Work Exercises 81–84 in order.

81. Write an expression for the cost to make x widgets. (*Hint:* The cost will be the sum of the fixed cost and the cost per item times the number of items.)

82. Write an expression for the cost to make y gadgets.

83. Write an expression for the total cost to make x widgets and y gadgets.

84. Simplify the expression you wrote in **Exercise 83.**

Chapter 9 Summary

Key Terms

9.1

exponent An exponent, or **power,** is a number that indicates how many times a factor is repeated.

3^4 ← Exponent; ↑ Base } Exponential expression

base The base is the number that is a repeated factor when written with an exponent.

exponential expression A number written with an exponent is an exponential expression.

9.2

variable A variable is a symbol, usually a letter, used to represent an unknown number.

constant A constant is a fixed, unchanging number.

algebraic expression An algebraic expression is a collection of numbers (constants), variables, operation symbols, and grouping symbols.

equation An equation is a statement that says two expressions are equal.

solution A solution of an equation is any value of the variable that makes the equation true.

9.3

natural numbers The set of natural numbers is $\{1, 2, 3, 4, \ldots\}$.

whole numbers The set of whole numbers is $\{0, 1, 2, 3, 4, \ldots\}$.

number line A number line shows the ordering of real numbers on a line.

additive inverse The additive inverse of a number *a* is the number that is the same distance from 0 on the number line as *a*, but on the opposite side of 0. This number is also called the opposite of *a*.

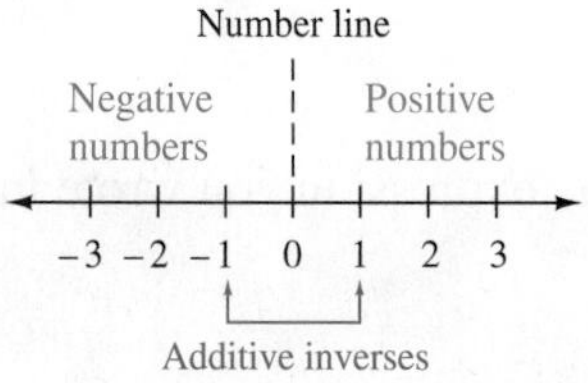

integers The set of integers is $\{\ldots, -3, -2, -1, 0, 1, 2, 3, \ldots\}$.

negative number A negative number is located to the *left* of 0 on the number line.

positive number A positive number is located to the *right* of 0 on the number line.

signed numbers Positive numbers and negative numbers are signed numbers.

rational numbers A rational number is a number that can be written as the quotient of two integers, with denominator not 0.

set-builder notation Set-builder notation uses a variable and a description to describe a set. It is often used to describe sets whose elements cannot easily be listed.

coordinate The number that corresponds to a point on the number line is the coordinate of that point.

irrational numbers An irrational number is a real number that is not a rational number.

real numbers Real numbers are numbers that can be represented by points on the number line (that is, all rational and irrational numbers).

absolute value The absolute value of a number is the distance between 0 and the number on a number line.

9.4

sum The answer to an addition problem is the sum.

9.5

minuend In the operation $a - b$, a is the minuend.

subtrahend In the operation $a - b$, b is the subtrahend.

difference The answer to a subtraction problem is the difference.

$15 - 7 = 8$ ← Difference; 15 ↑ Minuend; 7 ↑ Subtrahend

9.6

product The answer to a multiplication problem is the product.

quotient The answer to a division problem is the quotient.

reciprocal Pairs of numbers whose product is 1 are reciprocals, or **multiplicative inverses,** of each other.

9.7

identity element for addition When the identity element for addition, which is 0, is added to a number, the number is unchanged.

identity element for multiplication When a number is multiplied by the identity element for multiplication, which is 1, the number is unchanged.

9.8

term A term is a number (constant), a variable, or a product or quotient of a number and one or more variables raised to powers.

$$\underbrace{\overset{\text{Numerical coefficient}}{\overset{\downarrow}{-7}}x^2}_{\text{Term}}$$

numerical coefficient The numerical factor of a term is its numerical coefficient, or **coefficient.**

like terms Terms with exactly the same variables (including the same exponents) are like terms.

New Symbols

a^n	n factors of a	$a(b)$, $(a)b$, $(a)(b)$, $a \cdot b$, or ab	a times b
$=$	is equal to	$\frac{a}{b}$, a/b, or $a \div b$	a divided by b
$\neq$	is not equal to	$\{\ \}$	set braces
		$\{x \mid x$ **has a certain property**$\}$	set-builder notation
$<$	is less than	$[\ ]$	square brackets
$\leq$	is less than or equal to	$\lvert x \rvert$	absolute value of x
$>$	is greater than	$-x$	additive inverse, or opposite, of x
$\geq$	is greater than or equal to	$\frac{1}{x}$	multiplicative inverse, or reciprocal, of x (where $x \neq 0$)

Test Your Word Power

See how well you have learned the vocabulary in this chapter.

1. A **product** is
   - **A.** the answer in an addition problem
   - **B.** the answer in a multiplication problem
   - **C.** one of two or more numbers that are added to get another number
   - **D.** one of two or more numbers that are multiplied to get another number.

2. An **exponent** is
   - **A.** a symbol that tells how many numbers are being multiplied
   - **B.** a number raised to a power
   - **C.** a number that tells how many times a factor is repeated
   - **D.** one of two or more numbers that are multiplied.

3. A **variable** is
   - **A.** a symbol used to represent an unknown number
   - **B.** a value that makes an equation true
   - **C.** a solution of an equation
   - **D.** the answer in a division problem.

4. An **integer** is
   - **A.** a positive or negative number
   - **B.** a natural number, its opposite, or zero
   - **C.** any number that can be graphed on a number line
   - **D.** the quotient of two numbers.

5. A **coordinate** is
   - **A.** the number that corresponds to a point on a number line
   - **B.** the graph of a number
   - **C.** any point on a number line
   - **D.** the distance from 0 on a number line.

6. The **absolute value** of a number is
   - **A.** the graph of the number
   - **B.** the reciprocal of the number
   - **C.** the opposite of the number
   - **D.** the distance between 0 and the number on a number line.

7. A **term** is
   - **A.** a numerical factor
   - **B.** a number, a variable, or a product or quotient of numbers and variables raised to powers
   - **C.** one of several variables with the same exponents
   - **D.** a sum of numbers and variables raised to powers.

8. The **subtrahend** in $a - b = c$ is
   - **A.** a
   - **B.** b
   - **C.** c
   - **D.** $a - b$.

Answers to Test Your Word Power

1. B; *Example:* The product of 2 and 5, or 2 times 5, is 10.
2. C; *Example:* In 2^3, the number 3 is the exponent (or power), so 2 is a factor three times; $2^3 = 2 \cdot 2 \cdot 2 = 8$.
3. A; *Examples:* x, y, z
4. B; *Examples:* $-9, 0, 6$
5. A; *Example:* The point graphed three units to the right of 0 on a number line has coordinate 3.
6. D; *Examples:* $|2| = 2$ and $|-2| = 2$
7. B; *Examples:* $6, \frac{x}{2}, -4ab^2$
8. B; *Example:* In $5 - 3 = 2$, 5 is the minuend, 3 is the subtrahend, and 2 is the difference.

Quick Review

Concepts | **Examples**

9.1 Exponents, Order of Operations, and Inequality

Order of Operations
Simplify within any parentheses or brackets and above and below fraction bars, using the following steps.

Step 1 Apply all exponents.

Step 2 Multiply or divide from left to right.

Step 3 Add or subtract from left to right.

$$\frac{9(2+6)}{2} - 2(2^3 + 3)$$

$= \frac{9(8)}{2} - 2(8+3)$ Add inside parentheses. Apply the exponent.

$= 36 - 2(8+3)$ Multiply and divide.

$= 36 - 2(11)$ Add inside parentheses.

$= 36 - 22$ Multiply.

$= 14$ Subtract.

9.2 Variables, Expressions, and Equations

To evaluate an expression means to find its value. Evaluate an expression with a variable by substituting a given number for the variable.

Evaluate $2x + y^2$ for $x = 3$ and $y = -4$.

$2x + y^2$

$= 2(3) + (-4)^2$ Substitute.

$= 6 + 16$ Multiply and apply the exponent.

$= 22$ Add.

Values of a variable that make an equation true are solutions of the equation.

Is 2 a solution of $5x + 3 = 18$?

$5(2) + 3 \stackrel{?}{=} 18$ Let $x = 2$.

$13 = 18$ False

2 is not a solution.

9.3 Real Numbers and the Number Line

Ordering Real Numbers
a is less than b if a is to the left of b on a number line.

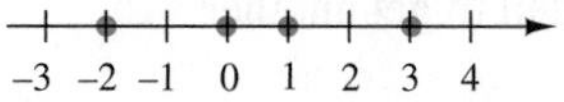

$-2 < 3$ $\quad 3 > 0$ $\quad 1 < 3$

The additive inverse, or opposite, of a is $-a$.

$-(5) = -5$ $\quad -(-7) = 7$ $\quad -0 = 0$

The absolute value of a, written $|a|$, is the distance between a and 0 on a number line.

$|13| = 13$ $\quad |0| = 0$ $\quad |-5| = 5$

9.4 Adding Real Numbers

To add two numbers with the *same sign*, add their absolute values. The sum has that same sign.

$9 + 4 = 13$

$-8 + (-5) = -13$

To add two numbers with *different signs*, subtract the lesser absolute value from the greater. The sum has the sign of the number with greater absolute value.

$7 + (-12) = -5$

$-5 + 13 = 8$

Concepts	*Examples*
9.5 Subtracting Real Numbers	
For any real numbers a and b, $a - b$ is defined as $a + (-b)$.	$5-(-2) = 5+2 = 7$ $-3-4 = -3+(-4) = -7$ $-2-(-6) = -2+6 = 4$
9.6 Multiplying and Dividing Real Numbers	
The product (or quotient) of two numbers having the *same sign* is *positive*.	$6 \cdot 5 = 30$ $-7(-8) = 56$ $\frac{-24}{-6} = 4$
The product (or quotient) of two numbers having *different signs* is *negative*.	$-6(5) = -30$ $6(-5) = -30$
To divide a by b, multiply a by the reciprocal of b.	$-18 \div 9 = \frac{-18}{9} = -2$ $49 \div (-7) = \frac{49}{-7} = -7$ $10 \div \frac{2}{3} = 10 \cdot \frac{3}{2} = 15$
0 divided by a nonzero number is 0. Division by 0 is undefined.	$\frac{0}{5} = 0$ $\frac{5}{0}$ is undefined.
9.7 Properties of Real Numbers	
Commutative Properties $a+b=b+a$, $ab=ba$	$7+(-1) = -1+7$ $5(-3) = (-3)5$
Associative Properties $(a+b)+c = a+(b+c)$, $(ab)c = a(bc)$	$(3+4)+8 = 3+(4+8)$ $[-2(6)]4 = -2[6(4)]$
Identity Properties $a+0=a$ $0+a=a$; $a \cdot 1 = a$ $1 \cdot a = a$	$-7+0=-7$ $0+(-7)=-7$ $9 \cdot 1 = 9$ $1 \cdot 9 = 9$
Inverse Properties $a+(-a)=0$ $-a+a=0$; $a \cdot \frac{1}{a} = 1$ $\frac{1}{a} \cdot a = 1$ $(a \neq 0)$	$7+(-7)=0$ $-7+7=0$ $-2\left(-\frac{1}{2}\right)=1$ $-\frac{1}{2}(-2)=1$
Distributive Properties $a(b+c)=ab+ac$; $(b+c)a=ba+ca$; $a(b-c)=ab-ac$	$5(4+2)=5(4)+5(2)$ $(4+2)5=4(5)+2(5)$ $9(5-4)=9(5)-9(4)$
9.8 Simplifying Expressions	
Only like terms may be combined. We use the distributive property.	$4(3+2x)-6(5-x)$ $= 12+8x-30+6x$ $= 14x-18$

Chapter 9 Review Exercises

9.1 *Find the value of each exponential expression.*

1. 5^4

2. $(0.03)^4$

3. 0.21^3

4. $\left(\frac{5}{2}\right)^3$

Find the value of each expression.

5. $8 \cdot 5 - 13$

6. $5[4^2 + 3(2^3)]$

7. $\frac{7(3^2 - 5)}{16 - 2 \cdot 6}$

8. $\frac{5(6^2 - 2^4)}{3 \cdot 5 + 10}$

Write each word sentence in symbols.

9. Thirteen is less than seventeen.

10. Five plus two is not equal to ten.

11. Write $6 < 15$ in words.

12. Construct a false statement that involves addition on the left side, the symbol $\geq$, and division on the right side.

9.2 *Evaluate each expression for $x = 6$ and $y = 3$.*

13. $2x + 6y$

14. $4(3x - y)$

15. $\frac{x}{3} + 4y$

16. $\frac{x^2 + 3}{3y - x}$

Change each word phrase to an algebraic expression. Use x to represent the number.

17. Six added to a number

18. A number subtracted from eight

19. Nine subtracted from six times a number

20. Three-fifths of a number added to 12

Decide whether the given number is a solution of the equation.

21. $5x + 3(x + 2) = 22;\ \ 2$

22. $\frac{x + 5}{3x} = 1;\ \ 6$

Change each word sentence to an equation. Use x to represent the number.

23. Six less than twice a number is 10.

24. The product of a number and 4 is 8.

Identify each of the following as either an equation *or an* expression.

25. $5r - 8(r + 7) = 2$

26. $2y + (5y - 9) + 2$

9.3 *Graph each group of numbers on a number line.*

27. $-4, -\frac{1}{2}, 0, 2.5, 5$

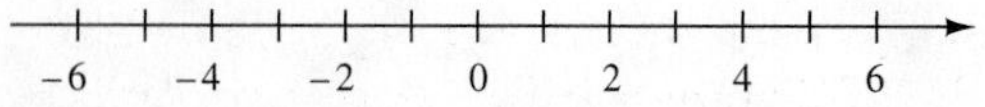

28. $-3\frac{1}{4}, \frac{14}{5}, -1\frac{1}{8}, \frac{5}{6}$

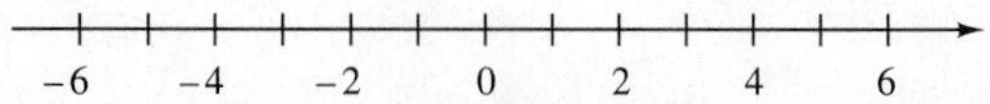

Select the lesser number in each pair.

29. $-10, 5$

30. $-8, -9$

31. $-\frac{2}{3}, -\frac{3}{4}$

32. $0, -|23|$

Decide whether each statement is true *or* false.

33. $12 > -13$

34. $0 > -5$

35. $-9 < -7$

36. $-13 > -13$

Simplify by finding the absolute value.

37. $-|3|$

38. $-|-19|$

39. $-|9 - 2|$

40. $|15 - 6|$

9.4 *Find each sum.*

41. $-10 + 4$

42. $14 + (-18)$

43. $-8 + (-9)$

44. $\frac{4}{9} + \left(-\frac{5}{4}\right)$

45. $[-6 + (-8) + 8] + [9 + (-13)]$

46. $(-4 + 7) + (-11 + 3) + (-15 + 1)$

Write a numerical expression for each phrase, and simplify the expression.

47. 19 added to the sum of -31 and 12

48. 13 more than the sum of -4 and -8

Solve each problem.

49. Like many people, Otis Taylor neglects to keep up his checkbook balance. When he finally balanced his account, he found that the balance was $-\$23.75$, so he deposited \$50.00. What is his new balance?

50. The low temperature in Yellowknife, in the Canadian Northwest Territories, one January day was $-26°$F. It rose $16°$ to its high that day. What was the high temperature?

9.5 *Find each difference.*

51. $-7 - 4$

52. $-12 - (-11)$

53. $5 - (-2)$

54. $-\frac{3}{7} - \frac{4}{5}$

55. $2.56 - (-7.75)$

56. $(-10 - 4) - (-2)$

57. $(-3 + 4) - (-1)$

58. $|5 - 9| - |-3 + 6|$

Write a numerical expression for each phrase, and simplify the expression.

59. The difference between -4 and -6

60. Five less than the sum of 4 and -8

61. The difference between 18 and -23, decreased by 15

62. Nineteen, decreased by 12 less than -7

Solve each problem.

63. The quaterback for the Chicago Bears passed for a gain of 8 yd, was sacked for a loss of 12 yd, and then threw a 42 yd touchdown pass. What positive or negative number represents the total net yardage for the plays?

64. On Tuesday, January 10, 2012, the Dow Jones Industrial Average closed at 12,462.47, up 69.78 from the previous day. What was the closing price on Monday, January 9, 2012? (*Source: The New York Times.*)

The table shows the number of people naturalized in the United States (that is, made citizens of the United States) for the years 2003 through 2010. In Exercises 65–68, use a signed number to represent the change in the number of people naturalized for each time period.

Year	Number of People (in thousands)
2003	463
2004	337
2005	604
2006	702
2007	1383
2008	526
2009	570
2010	710

Source: U.S. Department of Homeland Security.

65. 2003–2004

66. 2004–2005

67. 2006–2007

68. 2007–2008

9.6 *Perform the indicated operations.*

69. $(-12)(-3)$

70. $15(-7)$

71. $-\frac{4}{3}\left(-\frac{3}{8}\right)$

72. $-4.8(-2.1)$

73. $5(8 - 12)$

74. $(5 - 7)(8 - 3)$

75. $2(-6) - (-4)(-3)$

76. $3(-10) - 5$

77. $\frac{-36}{-9}$

78. $\frac{220}{-11}$

79. $-\frac{1}{2} \div \frac{2}{3}$

80. $-33.9 \div (-3)$

81. $\frac{-5(3) - 1}{8 - 4(-2)}$

82. $\frac{5(-2) - 3(4)}{-2[3 - (-2)] + 10}$

83. $\frac{10^2 - 5^2}{8^2 + 3^2 - (-2)}$

84. $\frac{4^2 - 8 \cdot 2}{(-1.2)^2 - (-0.56)}$

Evaluate each expression for $x = -5$, $y = 4$, and $z = -3$.

85. $6x - 4z$

86. $5x + y - z$

87. $5x^2$

88. $z^2(3x - 8y)$

Write a numerical expression for each phrase, and simplify the expression.

89. Nine less than the product of -4 and 5

90. Five-sixths of the sum of 12 and -6

91. The quotient of 12 and the sum of 8 and -4

92. The product of -20 and 12, divided by the difference between 15 and -15

Translate each sentence to an equation, using x to represent the number.

93. The quotient of a number and the sum of the number and 5 is -2.

94. 3 less than 8 times a number is -7.

9.7 *Decide whether each statement is an example of a commutative, associative, identity, or inverse property, or of the distributive property.*

95. $6 + 0 = 6$

96. $5 \cdot 1 = 5$

97. $-\frac{2}{3}\left(-\frac{3}{2}\right) = 1$

98. $17 + (-17) = 0$

99. $3x + 3y = 3(x + y)$

100. $w(xy) = (wx)y$

101. $5 + (-9 + 2) = [5 + (-9)] + 2$

102. $(1 + 2) + 3 = 3 + (1 + 2)$

Use the distributive property to rewrite each expression. Simplify if possible.

103. $7y + y$

104. $-12(4 - t)$

105. $3(2s) + 3(4y)$

106. $-(-4r + 5s)$

9.8 *Use the distributive property as necessary and combine like terms.*

107. $16p^2 - 8p^2 + 9p^2$

108. $4r^2 - 3r + 10r + 12r^2$

109. $-8(5k - 6) + 3(7k + 2)$

110. $2s - (-3s + 6)$

111. $-7(2t - 4) - 4(3t + 8) - 19(t + 1)$

112. $3.6t^2 + 9t - 8.1(6t^2 + 4t)$

Mixed Review Exercises

Perform the indicated operations.

113. $\frac{15}{2} \cdot \left(-\frac{4}{5}\right)$

114. $\left(-\frac{5}{6}\right)^2$

115. $-|(-7)(-4)| - (-2)$

116. $\frac{6(-4) + 2(-12)}{5(-3) + (-3)}$

117. $\frac{3}{8} - \frac{5}{12}$

118. $\frac{12^2 + 2^2 - 8}{10^2 - (-4)(-15)}$

119. $\frac{8^2 + 6^2}{7^2 + 1^2}$

120. $-16(-3.5) - 7.2(-3)$

121. $2\frac{5}{6} - 4\frac{1}{3}$

122. $-8 + [(-4 + 17) - (-3 - 3)]$

123. $-\frac{12}{5} \div \frac{9}{7}$

124. $(-8 - 3) - 5(2 - 9)$

125. $[-7 + (-2) - (-3)] + [8 + (-13)]$

126. $[(-2) + 7 - (-5)] + [-4 - (-10)]$

Solve each problem.

127. The highest temperature ever recorded in Iowa was 118°F at Keokuk on July 20, 1934. The lowest temperature ever recorded in the state was at Elkader on February 3, 1996, and was 165° lower than the highest temperature. What is the record low temperature for Iowa? (*Source:* National Climatic Data Center.)

128. Humpback whales love to heave their 45-ton bodies out of the water. This is called *breaching*. Chantelle, a researcher based on the island of Maui, noticed that one of her favorite whales, "Pineapple," breached 15 ft above the surface of the ocean while her mate cruised 12 ft below the surface. What is the difference between these two heights?

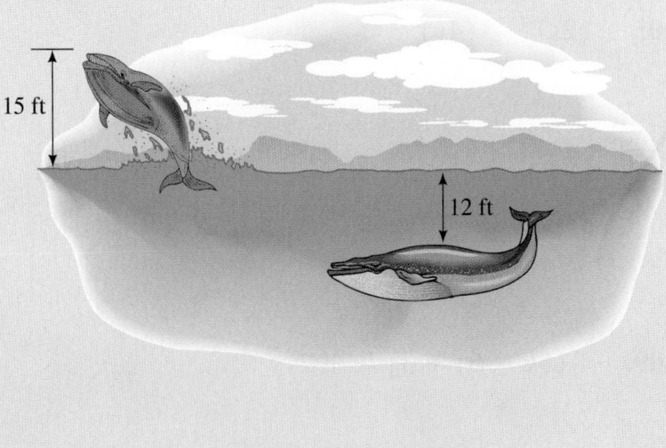

Chapter 9 Test

Decide whether each statement is true *or* false.

1. $4[-20+7(-2)] \leq -135$

2. $\left(\frac{1}{2}\right)^2 + \left(\frac{2}{3}\right)^2 = \left(\frac{1}{2}+\frac{2}{3}\right)^2$

3. Graph the numbers -1, -3, $|-4|$, and $|-1|$ on the number line.

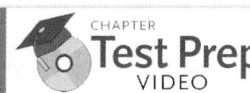

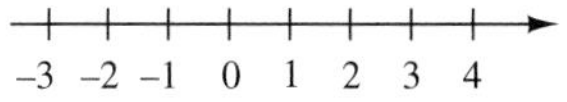

Select the lesser number from each pair.

4. $6, -|-8|$

5. $-0.742, -1.277$

6. Write in symbols: The quotient of -6 and the sum of 2 and -8. Simplify the expression.

7. If a and b are both negative, is $\frac{a+b}{a \cdot b}$ positive or negative?

Perform the indicated operations whenever possible.

8. $-2-(5-17)+(-6)$

9. $-5\frac{1}{2}+2\frac{2}{3}$

10. $4^2+(-8)-(2^3-6)$

11. $-6.2-[-7.1+(2.0-3.1)]$

12. $(-5)(-12)+4(-4)+(-8)^2$

13. $\dfrac{-7-|-6+2|}{-5-(-4)}$

14. $\dfrac{30(-1-2)}{-9[3-(-2)]-12(-2)}$

In Exercises 15 and 16, evaluate each expression for $x=-2$ and $y=4$.

15. $3x-4y^2$

16. $\dfrac{5x+7y}{3(x+y)}$

Solve each problem.

17. The highest Fahrenheit temperature ever recorded in Idaho was 118°F, while the lowest was −60°F. What is the difference between these highest and lowest temperatures? (*Source: World Almanac and Book of Facts.*)

18. For a certain system of rating relief pitchers, 3 points are awarded for a save, 3 points are awarded for a win, 2 points are subtracted for a loss, and 2 points are subtracted for a blown save. If a pitcher has 4 saves, 3 wins, 2 losses, and 1 blown save, how many points does he have?

19. For 2009, the U.S. federal government collected $2.10 trillion in revenues, but spent $3.52 trillion. Write the federal budget deficit as a signed number. (*Source: The Gazette.*)

Match each statement in Column I with the property it illustrates in Column II.

I	II
20. $3x + 0 = 3x$	**A.** Commutative property
21. $(5 + 2) + 8 = 8 + (5 + 2)$	**B.** Associative property
22. $-3(x + y) = -3x + (-3y)$	**C.** Inverse property
23. $-5 + (3 + 2) = (-5 + 3) + 2$	**D.** Identity property
24. $-\frac{5}{3}\left(-\frac{3}{5}\right) = 1$	**E.** Distributive property

Simplify.

25. $8x + 4x - 6x + x + 14x$

26. $-2(3x^2 + 4) - 3(x^2 + 2x)$

27. Which properties are used to show that $-(3x + 1) = -3x - 1$?

28. Consider the expression $-6[5 + (-2)]$.

(a) Evaluate it by first working within the brackets.

(b) Evaluate it by using the distributive property.

(c) Why must the answers in parts (a) and (b) be the same?

10 Equations, Inequalities, and Applications

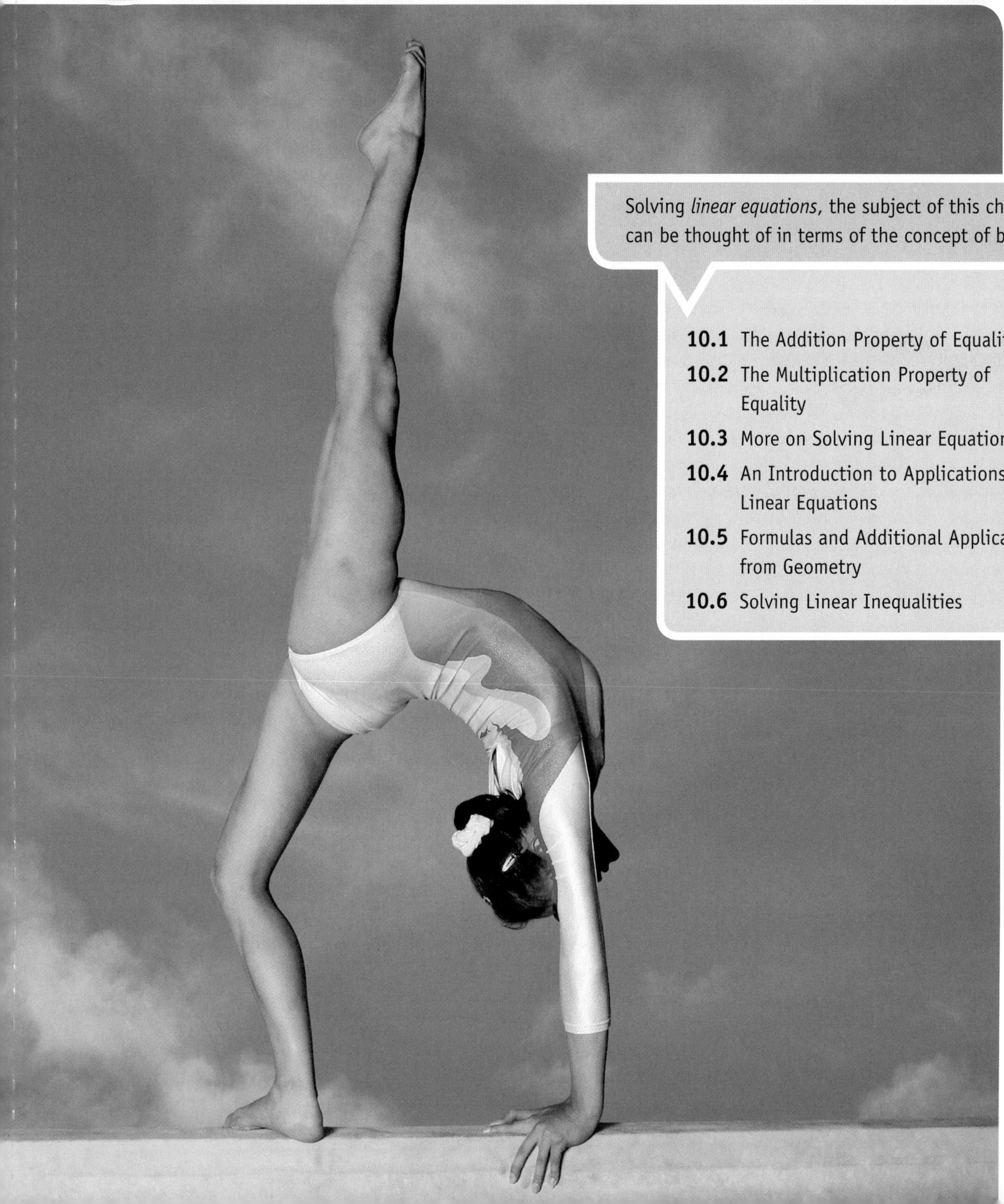

Solving *linear equations*, the subject of this chapter, can be thought of in terms of the concept of balance.

10.1 The Addition Property of Equality

OBJECTIVES

1. Identify linear equations.
2. Use the addition property of equality.
3. Simplify, and then use the addition property of equality.

An *equation* is a statement asserting that two algebraic expressions are equal. In this chapter, we *solve* equations.

CAUTION

Remember that an equation always includes an equality symbol.

Equation	Expression
Left side → $x - 5 = 2$ ← Right side	$x - 5$
An equation can be solved.	An expression **cannot** be solved. (It can often be *evaluated* or *simplified*.)

OBJECTIVE 1 Identify linear equations.

Linear Equation in One Variable

A **linear equation in one variable** can be written in the form

$$Ax + B = C,$$

where A, B, and C are real numbers, with $A \neq 0$.

$4x + 9 = 0$, $2x - 3 = 5$, and $x = 7$ Linear equations

$x^2 + 2x = 5$, $x^3 - x = 6$, and $|2x + 6| = 0$ *Non*linear equations

Recall that a **solution** of an equation is a number that makes the equation true when it replaces the variable. An equation is solved by finding its **solution set,** the set of all solutions. Equations that have exactly the same solution sets are **equivalent equations.** A linear equation in x is solved by using a series of steps to produce a simpler equivalent equation of the form

$$x = \text{a number} \quad \text{or} \quad \text{a number} = x.$$

OBJECTIVE 2 Use the addition property of equality. In the equation $x - 5 = 2$, both $x - 5$ and 2 represent the same number because that is the meaning of the equality symbol. To solve the equation, we change the left side from $x - 5$ to just x, as follows.

$x - 5 = 2$	Given equation
$x - 5 + 5 = 2 + 5$	Add 5 to *each* side to keep them equal.
$x + 0 = 7$	Additive inverse property
$x = 7$	Additive identity property

Add 5. It is the opposite (additive inverse) of -5, and $-5 + 5 = 0$.

The solution is 7. We check by replacing x with 7 in the original equation.

CHECK

$x - 5 = 2$	Original equation
$7 - 5 \stackrel{?}{=} 2$	Let $x = 7$.
$2 = 2$ ✓	True

The left side equals the right side.

Since the final equation is true, 7 checks as the solution and $\{7\}$ is the solution set. We write a solution set using set braces $\{\ \}$.

To solve the equation $x - 5 = 2$, we added the same number, 5, to each side. The **addition property of equality** justifies this step.

Addition Property of Equality

If A, B, and C represent real numbers, then the equations

$$A = B \quad \text{and} \quad A + C = B + C$$

are equivalent equations.

In words, we can add the same number to each side of an equation without changing the solution set.

In this property, any quantity that represents a real number C can be added to each side of an equation to obtain an equivalent equation.

Note

Equations can be thought of in terms of a balance. Thus, adding the *same* quantity to each side does not affect the balance. See **Figure 1.**

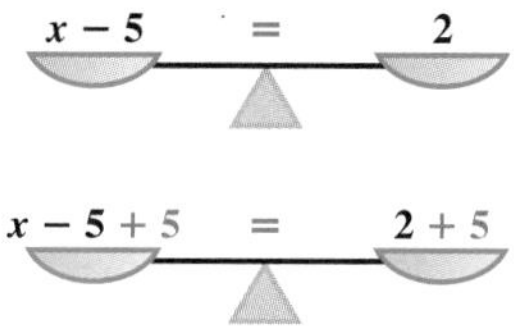

Figure 1

EXAMPLE 1 Using the Addition Property of Equality

Solve $x - 16 = 7$.

Our goal is to get an equivalent equation of the form x = **a number.**

$$x - 16 = 7$$

$$x - 16 + 16 = 7 + 16 \quad \text{Add 16 to each side.}$$

$$x = 23 \quad \text{Combine like terms.}$$

CHECK

$$x - 16 = 7 \quad \text{Original equation}$$

$$23 - 16 \stackrel{?}{=} 7 \quad \text{Substitute 23 for } x.$$

$$7 = 7 \checkmark \quad \text{True}$$

7 is *not* the solution.

Since a true statement results, 23 is the solution and $\{23\}$ is the solution set.

Work Problem 1 at the Side. ▶

EXAMPLE 2 Using the Addition Property of Equality

Solve $x - 2.9 = -6.4$.

Our goal is to isolate x.

$$x - 2.9 = -6.4$$

$$x - 2.9 + 2.9 = -6.4 + 2.9 \quad \text{Add 2.9 to each side.}$$

$$x = -3.5$$

CHECK

$$x - 2.9 = -6.4 \quad \text{Original equation}$$

$$-3.5 - 2.9 \stackrel{?}{=} -6.4 \quad \text{Let } x = -3.5.$$

$$-6.4 = -6.4 \checkmark \quad \text{True}$$

Since a true statement results, the solution set is $\{-3.5\}$.

Work Problem 2 at the Side. ▶

1 Solve.

GS **(a)** Fill in the blanks.

$$x - 12 = 9$$

$$x - 12 + \_\_\_\_ = 9 + \_\_\_\_$$

$$x = \_\_\_\_$$

CHECK $x - 12 = 9$

$$\_\_\_\_ - 12 \stackrel{?}{=} 9$$

$$\_\_\_\_ = 9 \quad (\textit{True / False})$$

The solution set is ____.

(b) $x - 25 = -18$

(c) $x - 8 = -13$

2 Solve.

(a) $x - 3.7 = -8.1$

(b) $a - 4.1 = 6.3$

Answers

1. (a) 12; 12; 21; 21; 9; True; $\{21\}$ (b) $\{7\}$ (c) $\{-5\}$
2. (a) $\{-4.4\}$ (b) $\{10.4\}$

In **Section 9.5,** subtraction was defined as addition of the opposite. Thus, we can also use the following rule when solving an equation.

The same number may be *subtracted* from each side of an equation without changing the solution set.

For example, to solve the equation $x + 4 = 10$, we *subtract* 4 from each side, which is the same as adding -4. The result is $x = 6$.

EXAMPLE 3 Using the Addition Property of Equality

Solve $-7 = x + 22$.

Here the variable x is on the right side of the equation.

$-7 = x + 22$ — The variable can be isolated on *either* side.

$-7 - 22 = x + 22 - 22$ — Subtract 22 from each side.

$-29 = x$, or $x = -29$ — Rewrite; a number $= x$, or $x =$ a number.

CHECK $-7 = x + 22$ — Original equation

$-7 \stackrel{?}{=} -29 + 22$ — Let $x = -29$.

$-7 = -7$ ✓ — True

The check confirms that the solution set is $\{-29\}$.

◀ **Work Problem 3 at the Side.**

Note

In **Example 3,** what happens if we subtract $-7 - 22$ incorrectly, obtaining $x = -15$ (instead of $x = -29$) as the last line of the solution? A check should indicate an error.

CHECK $-7 = x + 22$ — Original equation from **Example 3**

$-7 \stackrel{?}{=} -15 + 22$ — Let $x = -15$.

$-7 = 7$ — False (The left side does *not* equal the right side.)

The false statement indicates that -15 is *not* a solution of the equation. If this happens, rework the problem.

EXAMPLE 4 Subtracting a Variable Term

Solve $\frac{3}{5}k + 15 = \frac{8}{5}k$.

$$\frac{3}{5}k + 15 = \frac{8}{5}k$$ — We must get the terms with k on the same side of the = symbol.

$$\frac{3}{5}k + 15 - \frac{3}{5}k = \frac{8}{5}k - \frac{3}{5}k$$ — Subtract $\frac{3}{5}k$ from each side.

$15 = 1k$ — $\frac{3}{5}k - \frac{3}{5}k = 0$; $\frac{8}{5}k - \frac{3}{5}k = \frac{5}{5}k = 1k$ (From now on we will skip this step.)

$15 = k$ — Multiplicative identity property

Check by substituting 15 in the original equation. The solution set is $\{15\}$.

---

3 Solve.

GS **(a)** $-3 = a + 2$

Our goal is to isolate ____.

On each side of the equation, (*add / subtract*) 2.

Now solve.

(b) $22 = -16 + r$

Answers

3. (a) a; subtract; $\{-5\}$ **(b)** $\{38\}$

What happens if we solve the equation in **Example 4** by first subtracting $\frac{8}{5}k$ from each side?

$$\frac{3}{5}k + 15 = \frac{8}{5}k \quad \text{Original equation from Example 4}$$

$$\frac{3}{5}k + 15 - \frac{8}{5}k = \frac{8}{5}k - \frac{8}{5}k \quad \text{Subtract } \tfrac{8}{5}k \text{ from each side.}$$

$$15 - k = 0 \quad \tfrac{3}{5}k - \tfrac{8}{5}k = -\tfrac{5}{5}k = -1k = -k;\ \tfrac{8}{5}k - \tfrac{8}{5}k = 0$$

$$15 - k - 15 = 0 - 15 \quad \text{Subtract 15 from each side.}$$

$$-k = -15 \quad \text{Combine like terms; additive inverse}$$

This result gives the value of $-k$, but not of k itself. However, it does say that the additive inverse of k is -15, which means that k must be 15.

$$k = 15 \quad \text{Same result as in Example 4}$$

(This result can also be justified using the multiplication property of equality, covered in **Section 10.2.**) We can make the following generalization.

If a is a number and $-x = a$, then $x = -a$.

Work Problem 4 at the Side. ▶

EXAMPLE 5 Using the Addition Property of Equality Twice

Solve $8 - 6p = -7p + 5$.

We must get all terms with variables on the same side of the equation and all terms without variables on the other side of the equation.

$$8 - 6p = -7p + 5$$

$$8 - 6p + 7p = -7p + 5 + 7p \quad \text{Add } 7p \text{ to each side.}$$

$$8 + p = 5 \quad \text{Combine like terms.}$$

$$8 + p - 8 = 5 - 8 \quad \text{Subtract 8 from each side.}$$

$$p = -3 \quad \text{Combine like terms.}$$

CHECK Substitute -3 for p in the original equation.

$$8 - 6p = -7p + 5 \quad \text{Original equation}$$

$$8 - 6(-3) \stackrel{?}{=} -7(-3) + 5 \quad \text{Let } p = -3.$$

Use parentheses when substituting to avoid errors.

$$8 + 18 \stackrel{?}{=} 21 + 5 \quad \text{Multiply.}$$

$$26 = 26 \ \checkmark \quad \text{True}$$

The check results in a true statement, so the solution set is $\{-3\}$.

Work Problem 5 at the Side. ▶

Note

There are often several equally correct ways to solve an equation. In **Example 5,** we could begin by adding $6p$, instead of $7p$, to each side. Combining like terms and subtracting 5 from each side gives $3 = -p$. (Try this.) If $3 = -p$, then $-3 = p$, and the variable has been isolated on the right side of equation. The same solution results.

---

4 **(a)** Solve $5m + 4 = 6m$.

(b) Solve $\frac{7}{2}m + 1 = \frac{9}{2}m$.

(c) What is the solution set of $-x = 6$?

(d) What is the solution set of $-x = -12$?

5 Solve.

GS **(a)** $10 - a = -2a + 9$

$10 - a +$ ____ $= -2a + 9 + 2a$

$10 + a = 9$

$10 + a -$ ____ $= 9 -$ ____

$a =$ ____

The solution set is ____.

(b) $6x - 8 = 12 + 5x$

(c) $8t + 6 = 6 + 7t$

Answers

4. **(a)** $\{4\}$ **(b)** $\{1\}$ **(c)** $\{-6\}$ **(d)** $\{12\}$

5. **(a)** $2a$; 10; 10; -1; $\{-1\}$ **(b)** $\{20\}$ **(c)** $\{0\}$

6 Solve.

(a) $4x + 6 + 2x - 3 = 9 + 5x - 4$

(b) $9r + 4r + 6 - 2 = 9r + 4 + 3r$

7 Solve.

(a) $4(r + 1) - (3r + 5) = 1$

GS (b) $2(5 + 2m) - 3(m - 4) = 29$

$2(5) + 2(2m) - \_\_\_\_ m - \_\_\_\_(\_\_\_\_) = 29$

$10 + 4m - \_\_\_\_ + \_\_\_\_ = 29$

Now complete the solution.

Answers

6. (a) {2} (b) {0}
7. (a) {2} (b) 3; 3; −4; 3*m*; 12; {7}

OBJECTIVE 3 Simplify, and then use the addition property of equality.

EXAMPLE 6 Combining Like Terms When Solving

Solve $3t - 12 + t + 2 = 5 + 3t + 2$.

$3t - 12 + t + 2 = 5 + 3t + 2$	Combine like terms on each side.
$4t - 10 = 7 + 3t$	
$4t - 10 - 3t = 7 + 3t - 3t$	Subtract $3t$ from each side.
$t - 10 = 7$	Combine like terms.
$t - 10 + 10 = 7 + 10$	Add 10 to each side.
$t = 17$	Combine like terms.

CHECK

$3t - 12 + t + 2 = 5 + 3t + 2$	Original equation
$3(17) - 12 + 17 + 2 \stackrel{?}{=} 5 + 3(17) + 2$	Let $t = 17$.
$51 - 12 + 17 + 2 \stackrel{?}{=} 5 + 51 + 2$	Multiply.
$58 = 58$ ✓	True

The check results in a true statement, so the solution set is $\{17\}$.

◀ Work Problem 6 at the Side.

CAUTION

The final line of the check does not give the solution to the equation. It gives a confirmation that the solution found is correct.

EXAMPLE 7 Using the Distributive Property When Solving

Solve $3(2 + 5x) - (1 + 14x) = 6$.

$3(2 + 5x) - (1 + 14x) = 6$

Be sure to distribute to *all* terms within the parentheses.

$3(2 + 5x) - 1(1 + 14x) = 6$	$-(1 + 14x) = -1(1 + 14x)$
$3(2) + 3(5x) - 1(1) - 1(14x) = 6$	Distributive property
$6 + 15x - 1 - 14x = 6$ (Be careful here.)	Multiply.
$x + 5 = 6$	Combine like terms.
$x + 5 - 5 = 6 - 5$	Subtract 5 from each side.
$x = 1$	Combine like terms.

Check by substituting 1 for x in the original equation. The solution set is $\{1\}$.

◀ Work Problem 7 at the Side.

CAUTION

Be careful to apply the distributive property correctly in a problem like that in **Example 7**, or a sign error may result.

10.1 Exercises

FOR EXTRA HELP

Download the MyDashBoard App

MyMathLab®

CONCEPT CHECK *In Exercises 1–4, fill in each blank with the correct response.*

1. An equation includes a(n) ___________ symbol, while a(n) ___________ does not.

2. A(n) ___________ equation in one variable can be written in the form $Ax + B$ ____ C.

3. Equations that have exactly the same solution set are ___________ equations.

4. The addition property of equality states that the same expression may be ___________ or ___________ each side of an equation without changing the solution set.

5. CONCEPT CHECK Substitute to determine which number is a solution of the equation $2x - 5 = 3x$.

   A. 0 B. 5 C. −5 D. −1

6. CONCEPT CHECK Which of the following are *not* linear equations in one variable?

   A. $x^2 - 5x + 6 = 0$ B. $x^3 = x$

   C. $3x - 4 = 0$ D. $7x - 6x = 3 + 9x$

7. CONCEPT CHECK Decide whether each is an *expression* or an *equation.* If it is an expression, simplify it. If it is an equation, solve it.

   (a) $5x + 8 - 4x + 7$ (b) $-6m + 12 + 7m - 5$

   (c) $5x + 8 - 4x = 7$ (d) $-6m + 12 + 7m = -5$

8. Explain how to check a solution of an equation.

Solve each equation, and check your solution. **See Examples 1–5.**

9. $x - 4 = 8$

10. $x - 8 = 9$

11. $x - 5 = -8$

12. $x - 7 = -9$

13. $r + 9 = 13$

14. $t + 6 = 10$

15. $x + 26 = 17$

16. $x + 45 = 24$

17. $x + \frac{1}{4} = -\frac{1}{2}$

18. $x + \frac{2}{3} = -\frac{1}{6}$

19. $x - 8.4 = -2.1$

20. $z - 15.5 = -5.1$

21. $t + 12.3 = -4.6$

22. $x + 21.5 = -13.4$

23. $7 + r = -3$

24. $8 + k = -4$

25. $2 = p + 15$

26. $3 = z + 17$

27. $-\frac{1}{3} = x - \frac{3}{5}$

28. $-\frac{1}{4} = x - \frac{2}{3}$

29. $3x = 2x + 7$

30. $5x = 4x + 9$

31. $8x - 3 = 9x$

32. $4x - 1 = 5x$

33. $10x + 4 = 9x$

34. $8t + 5 = 7t$

35. $\frac{9}{7}r - 3 = \frac{2}{7}r$

36. $\frac{8}{5}w - 6 = \frac{3}{5}w$

37. $5.6x + 2 = 4.6x$ **38.** $9.1x + 5 = 8.1x$ **39.** $3p + 6 = 10 + 2p$ **40.** $8x + 4 = -6 + 7x$

41. $5 - x = -2x - 11$ **42.** $3 - 8x = -9x - 1$ **43.** $-4z + 7 = -5z + 9$ **44.** $-6q + 3 = -7q + 10$

Solve each equation, and check your solution. ***See Examples 6 and 7.***

45. $3x + 6 - 10 = 2x - 2$

46. $8k - 4 + 6 = 7k + 1$

47. $6x + 5 + 7x + 3 = 12x + 4$

48. $4x - 3 - 8x + 1 = -5x + 9$

49. $10x + 5x + 7 - 8 = 12x + 3 + 2x$

50. $7p + 4p + 13 - 7 = 7p + 9 + 3p$

51. $5.2q - 4.6 - 7.1q = -0.9q - 4.6$

52. $-4.0x + 2.7 - 1.6x = -4.6x + 2.7$

53. $\frac{5}{7}x + \frac{1}{3} = \frac{2}{5} - \frac{2}{7}x + \frac{2}{5}$

54. $\frac{6}{7}s - \frac{3}{4} = \frac{4}{5} - \frac{1}{7}s + \frac{1}{6}$

55. $(5x + 6) - (3 + 4x) = 10$

56. $(8r - 3) - (7r + 1) = -6$

57. $2(p + 5) - (9 + p) = -3$

58. $4(k - 6) - (3k + 2) = -5$

59. $-6(2x + 1) + (13x - 7) = 0$

60. $-5(3w - 3) + (16w + 1) = 0$

61. $10(-2x + 1) = -19(x + 1)$

62. $2(-3r + 2) = -5(r - 3)$

63. $-2(8p + 2) - 3(2 - 7p) = 2(4 + 2p)$

64. $4(3 - z) - 5(1 - 2z) = 7(3 + z)$

65. **CONCEPT CHECK** Write an equation that requires the use of the addition property of equality, in which 6 must be added to each side to solve the equation and the solution is a negative number.

66. **CONCEPT CHECK** Write an equation that requires the use of the addition property of equality, in which $\frac{1}{2}$ must be subtracted from each side and the solution is a positive number.

10.2 The Multiplication Property of Equality

OBJECTIVES

1. Use the multiplication property of equality.
2. Simplify, and then use the multiplication property of equality.

OBJECTIVE 1 Use the multiplication property of equality. The addition property of equality is not enough to solve some equations.

$$3x + 2 = 17$$

$$3x + 2 - 2 = 17 - 2 \quad \text{Subtract 2 from each side.}$$

$$3x = 15 \quad \text{Combine like terms.}$$

The coefficient of x here is 3, not 1 as desired. The **multiplication property of equality** is needed to change $3x = 15$ to an equation of the form

$$x = \text{a number.}$$

Since $3x = 15$, both $3x$ and 15 must represent the same number. Multiplying both $3x$ and 15 by the same number will result in an equivalent equation.

Multiplication Property of Equality

If A, B, and C (where $C \neq 0$) represent real numbers, then the equations

$$A = B \quad \text{and} \quad AC = BC$$

are equivalent equations.

In words, we can multiply each side of an equation by the same nonzero number without changing the solution set.

In $3x = 15$, we must change $3x$ to $1x$, or x. To do this, we multiply each side of the equation by $\frac{1}{3}$, the reciprocal of 3, because $\frac{1}{3} \cdot 3 = \frac{3}{3} = 1$.

$$3x = 15$$

$$\frac{1}{3}(3x) = \frac{1}{3}(15) \quad \text{Multiply each side by } \tfrac{1}{3}.$$

$$\left(\frac{1}{3} \cdot 3\right)x = \frac{1}{3}(15) \quad \text{Associative property}$$

$$1x = 5 \quad \text{Multiplicative inverse property}$$

$$x = 5 \quad \text{Multiplicative identity property}$$

The product of a number and its reciprocal is 1.

The solution is 5. We can check this result in the original equation.

Work Problem 1 at the Side. ▶

Just as the addition property of equality permits *subtracting* the same number from each side of an equation, the multiplication property of equality permits *dividing* each side of an equation by the same nonzero number.

$$3x = 15$$

$$\frac{3x}{3} = \frac{15}{3} \quad \text{Divide each side by 3.}$$

$$x = 5 \quad \text{Same result as above}$$

We can divide each side of an equation by the same nonzero number without changing the solution. ***Do not, however, divide each side by a variable, since the variable might be equal to 0.***

1 Check that 5 is the solution of $3x = 15$. (GS)

$$3x = 15$$

$$3(\_\_\_\_) \stackrel{?}{=} 15$$

$$\_\_\_\_ = 15 \quad (\textit{True / False})$$

The solution set is ____.

Answer

1. 5; 15; True; $\{5\}$

2 Solve.

(a) Fill in the blanks.

$$7x = 91$$

$$\frac{7x}{\_\_\_\_} = \frac{91}{\_\_\_\_}$$

$$x = \_\_\_\_$$

CHECK $7x = 91$

$$7(\_\_\_\_) \stackrel{?}{=} 91$$

$$\_\_\_\_ = 91 \quad (True / False)$$

The solution set is ____.

(b) $3r = -12$

(c) $15x = 75$

3 Solve.

(a) $2m = 15$

(b) $-6x = 14$

(c) $10z = -45$

Answers

2. (a) 7; 7; 13; 13; 91; True; $\{13\}$ (b) $\{-4\}$ (c) $\{5\}$

3. (a) $\left\{\frac{15}{2}\right\}$ (b) $\left\{-\frac{7}{3}\right\}$ (c) $\left\{-\frac{9}{2}\right\}$

Note

In practice, it is usually easier to multiply on each side if the coefficient of the variable is a fraction, and divide on each side if the coefficient is an integer or a decimal. For example, to solve

$$\frac{3}{4}x = 12, \quad \text{it is easier to multiply by } \frac{4}{3} \text{ than to divide by } \frac{3}{4}.$$

On the other hand, to solve

$$5x = 20, \quad \text{it is easier to divide by 5 than to multiply by } \frac{1}{5}.$$

EXAMPLE 1 Dividing Each Side of an Equation by a Nonzero Number

Solve $5x = 60$.

$$5x = 60$$ Our goal is to isolate x.

$$\frac{5x}{5} = \frac{60}{5}$$ Divide each side by 5, the coefficient of x.

Dividing by 5 is the same as multiplying by $\frac{1}{5}$.

$$x = 12 \qquad \frac{5x}{5} = \frac{5}{5}x = 1x = x$$

CHECK Substitute 12 for x in the original equation.

$$5x = 60 \qquad \text{Original equation}$$

$$5(12) \stackrel{?}{=} 60 \qquad \text{Let } x = 12.$$

$$60 = 60 \checkmark \qquad \text{True}$$

60 is *not* the solution.

Since a true statement results, the solution set is $\{12\}$.

◄ Work Problem **2** at the Side.

EXAMPLE 2 Using the Multiplication Property of Equality

Solve $-25p = 30$.

$$-25p = 30$$

$$\frac{-25p}{-25} = \frac{30}{-25} \qquad \text{Divide each side by } -25, \text{ the coefficient of } p.$$

$$p = -\frac{30}{25} \qquad \frac{a}{-b} = -\frac{a}{b}$$

$$p = -\frac{6}{5} \qquad \text{Write in lowest terms.}$$

CHECK

$$-25p = 30 \qquad \text{Original equation}$$

$$\frac{-25}{1}\left(-\frac{6}{5}\right) \stackrel{?}{=} 30 \qquad \text{Let } p = -\frac{6}{5}.$$

$$30 = 30 \checkmark \qquad \text{True}$$

The check confirms that the solution set is $\left\{-\frac{6}{5}\right\}$.

◄ Work Problem **3** at the Side.

EXAMPLE 3 **Using the Multiplication Property of Equality**

Solve $6.09 = 2.1x$.

$$6.09 = 2.1x$$ Isolate x on the right.

$$\frac{6.09}{2.1} = \frac{2.1x}{2.1}$$ Divide each side by 2.1.

$$2.9 = x, \quad \text{or} \quad x = 2.9$$ Use a calculator if necessary.

Check to confirm that the solution set is $\{2.9\}$.

Work Problem 4 at the Side. ▶

EXAMPLE 4 **Using the Multiplication Property of Equality**

Solve $\frac{x}{4} = 3$.

$$\frac{x}{4} = 3$$

$$\frac{1}{4}x = 3 \qquad \frac{x}{4} = \frac{1x}{4} = \frac{1}{4}x$$

$$4 \cdot \frac{1}{4}x = 4 \cdot 3$$ Multiply each side by 4, the reciprocal of $\frac{1}{4}$.

$$x = 12$$ Multiplicative inverse property; multiplicative identity property ($4 \cdot \frac{1}{4}x = 1x = x$)

CHECK

$$\frac{x}{4} = 3$$ Original equation

$$\frac{12}{4} \stackrel{?}{=} 3$$ Let $x = 12$.

$$3 = 3 \checkmark$$ True

Since a true statement results, the solution set is $\{12\}$.

Work Problem 5 at the Side. ▶

EXAMPLE 5 **Using the Multiplication Property of Equality**

Solve $\frac{3}{4}h = 6$.

$$\frac{3}{4}h = 6$$

$$\frac{4}{3} \cdot \frac{3}{4}h = \frac{4}{3} \cdot 6$$ Multiply each side by $\frac{4}{3}$, the reciprocal of $\frac{3}{4}$.

$$1 \cdot h = \frac{4}{3} \cdot \frac{6}{1}$$ Multiplicative inverse property

$$h = 8$$ Multiplicative identity property; multiply fractions.

Check to confirm that the solution set is $\{8\}$.

Work Problem 6 at the Side. ▶

---

4 Solve.

(a) $-0.7m = -5.04$

(b) $-63.75 = 12.5k$

5 Solve.

GS **(a)** $\frac{x}{5} = 5$

What is the coefficient of x here? ____ By what number should we multiply each side? ____ Now solve.

(b) $\frac{p}{4} = -6$

6 Solve.

GS **(a)** $-\frac{5}{6}t = -15$

By what number should we multiply each side? ____ Now solve.

(b) $\frac{3}{5}k = -21$

Answers

4. **(a)** $\{7.2\}$ **(b)** $\{-5.1\}$

5. **(a)** $\frac{1}{5}$; 5; $\{25\}$ **(b)** $\{-24\}$

6. **(a)** $-\frac{6}{5}$; $\{18\}$ **(b)** $\{-35\}$

7 Solve.

(a) $-m = 2$

(b) $-p = -7$

In **Section 10.1,** we obtained the equation $-k = -15$ in our alternative solution to **Example 4.** We reasoned that since this equation says that the additive inverse (or opposite) of k is -15, then k must equal 15. We can also use the multiplication property of equality to obtain the same result.

EXAMPLE 6 **Using the Multiplication Property of Equality When the Coefficient of the Variable Is −1**

Solve $-k = -15$.

$$-k = -15$$

$$-1 \cdot k = -15 \qquad -k = -1 \cdot k$$

$$-1(-1 \cdot k) = -1(-15) \qquad \text{Multiply each side by } -1.$$

$$[-1(-1)] \cdot k = 15 \qquad \text{Associative property; multiply.}$$

$$1 \cdot k = 15 \qquad \text{Multiplicative inverse property}$$

$$k = 15 \qquad \text{Multiplicative identity property}$$

CHECK

$$-k = -15 \qquad \text{Original equation}$$

$$-(15) \stackrel{?}{=} -15 \qquad \text{Let } k = 15.$$

$$-15 = -15 \checkmark \qquad \text{True}$$

The solution, 15, checks, so the solution set is $\{15\}$.

◀ Work Problem **7** at the Side.

8 Solve.

(a) $7m - 5m = -12$

(b) $4r - 9r = 20$

OBJECTIVE 2 **Simplify, and then use the multiplication property of equality.**

EXAMPLE 7 **Combining Like Terms When Solving**

Solve $5m + 6m = 33$.

$$5m + 6m = 33$$

$$11m = 33 \qquad \text{Combine like terms.}$$

$$\frac{11m}{11} = \frac{33}{11} \qquad \text{Divide each side by 11.}$$

$$m = 3 \qquad \text{Multiplicative identity property}$$

CHECK

$$5m + 6m = 33 \qquad \text{Original equation}$$

$$5(3) + 6(3) \stackrel{?}{=} 33 \qquad \text{Let } m = 3.$$

$$15 + 18 \stackrel{?}{=} 33 \qquad \text{Multiply.}$$

$$33 = 33 \checkmark \qquad \text{True}$$

Since a true statement results, the solution set is $\{3\}$.

◀ Work Problem **8** at the Side.

Answers

7. **(a)** $\{-2\}$ **(b)** $\{7\}$

8. **(a)** $\{-6\}$ **(b)** $\{-4\}$

10.2 Exercises

FOR EXTRA HELP

Download the MyDashBoard App

MyMathLab®

1. **CONCEPT CHECK** Tell whether you would use the addition or multiplication property of equality to solve each equation. *Do not actually solve.*

   (a) $3x = 12$ (b) $3 + x = 12$

   (c) $-x = 4$ (d) $-12 = 6 + x$

2. **CONCEPT CHECK** Which equation does *not* require the use of the multiplication property of equality?

   A. $3x - 5x = 6$ B. $-\frac{1}{4}x = 12$

   C. $5x - 4x = 7$ D. $\frac{x}{3} = -2$

CONCEPT CHECK *By what number is it necessary to multiply each side of each equation in order to isolate x on the left side? Do not actually solve.*

3. $\frac{2}{3}x = 8$
4. $\frac{4}{5}x = 6$
5. $\frac{x}{10} = 3$
6. $\frac{x}{100} = 8$
7. $-\frac{9}{2}x = -4$
8. $-\frac{8}{3}x = -11$
9. $-x = 0.36$
10. $-x = 0.29$

CONCEPT CHECK *By what number is it necessary to divide each side of each equation in order to isolate x on the left side? Do not actually solve.*

11. $6x = 5$
12. $7x = 10$
13. $-4x = 13$
14. $-13x = 6$
15. $0.12x = 48$
16. $0.21x = 63$
17. $-x = 23$
18. $-x = 49$

19. **CONCEPT CHECK** In the solution of a linear equation, the next-to-the-last step reads "$-x = -\frac{3}{4}$." Which of the following would be the solution of this equation?

   A. $-\frac{3}{4}$ B. $\frac{3}{4}$ C. -1 D. $\frac{4}{3}$

20. **CONCEPT CHECK** Which of the following is the solution of the equation

   $$-x = -24?$$

   A. 24 B. -24 C. 1 D. -1

Solve each equation, and check your solution. ***See Examples 1–7.***

21. $5x = 30$
22. $7x = 56$
23. $2m = 15$
24. $3m = 10$
25. $3a = -15$
26. $5k = -70$
27. $10t = -36$
28. $4s = -34$
29. $-6x = -72$
30. $-8x = -64$
31. $2r = 0$

   $\frac{2r}{\_\_\_} = \frac{0}{\_\_\_}$

   $r = \_\_\_$

   Solution set: ____

32. $5x = 0$

   $\frac{5x}{\_\_\_} = \frac{0}{\_\_\_}$

   $x = \_\_\_$

   Solution set: ____

33. $-x = 12$

34. $-t = 14$

35. $0.2t = 8$

36. $0.9x = 18$

37. $-2.1m = 25.62$

38. $-3.9a = 31.2$

39. $\frac{1}{4}x = -12$

40. $\frac{1}{5}p = -3$

41. $\frac{z}{6} = 12$

42. $\frac{x}{5} = 15$

43. $\frac{x}{7} = -5$

44. $\frac{k}{8} = -3$

45. $\frac{2}{7}p = 4$

46. $\frac{3}{8}x = 9$

47. $-\frac{7}{9}c = \frac{3}{5}$

48. $-\frac{5}{6}d = \frac{4}{9}$

49. $4x + 3x = 21$

50. $9x + 2x = 121$

51. $3r - 5r = 10$

52. $9p - 13p = 24$

53. $\frac{2}{5}x - \frac{3}{10}x = 2$

54. $\frac{2}{3}x - \frac{5}{9}x = 4$

55. $5m + 6m - 2m = 63$

56. $11r - 5r + 6r = 168$

57. $x + x - 3x = 12$
(*Hint:* What is the coefficient of each x? ____)

58. $z - 3z + z = -16$
(*Hint:* What is the coefficient of each z? ____)

59. $-6x + 4x - 7x = 0$

60. $-5x + 4x - 8x = 0$

61. $0.9w - 0.5w + 0.1w = -3$

62. $0.5x - 0.6x + 0.3x = -1$

63. CONCEPT CHECK Write an equation that requires the use of the multiplication property of equality, where each side must be multiplied by $\frac{2}{3}$ and the solution is a negative number.

64. CONCEPT CHECK Write an equation that requires the use of the multiplication property of equality, where each side must be divided by 100 and the solution is not an integer.

Write an equation using the information given in the problem. Use x as the variable. Then solve the equation.

65. When a number is multiplied by -4, the result is 10. Find the number.

66. When a number is multiplied by 4, the result is 6. Find the number.

67. When a number is divided by -5, the result is 2. Find the number.

68. If twice a number is divided by 5, the result is 4. Find the number.

10.3 More on Solving Linear Equations

OBJECTIVES

1. Learn and use the four steps for solving a linear equation.
2. Solve equations that have no solution or infinitely many solutions.
3. Solve equations with fractions or decimals as coefficients.
4. Write expressions for two related unknown quantities.

In this section, we solve linear equations using *both* properties of equality introduced in **Sections 10.1 and 10.2.**

Work Problem **1** at the Side. ▶

OBJECTIVE 1 Learn and use the four steps for solving a linear equation. We use the following method.

Solving a Linear Equation

Step 1 **Simplify each side separately.** Clear (eliminate) parentheses, fractions, and decimals, using the distributive property as needed, and combine like terms.

Step 2 **Isolate the variable term on one side.** Use the addition property so that the variable term is on one side of the equation and a number is on the other.

Step 3 **Isolate the variable.** Use the multiplication property to get the equation in the form x = a number, or a number = x. (Other letters may be used for the variable.)

Step 4 **Check.** Substitute the proposed solution into the original equation to see if a true statement results. If not, rework the problem.

Remember that when we solve an equation, our primary goal is to isolate the variable on one side of the equation.

EXAMPLE 1 **Using Both Properties of Equality to Solve a Linear Equation**

Solve $-6x + 5 = 17$.

Step 1 There are no parentheses, fractions, or decimals in this equation, so this step is not necessary.

Our goal is to isolate x.

$$-6x + 5 = 17$$

Step 2

$$-6x + 5 - 5 = 17 - 5 \quad \text{Subtract 5 from each side.}$$

$$-6x = 12 \quad \text{Combine like terms.}$$

Step 3

$$\frac{-6x}{-6} = \frac{12}{-6} \quad \text{Divide each side by } -6.$$

$$x = -2$$

Step 4 Check by substituting -2 for x in the original equation.

CHECK

$$-6x + 5 = 17 \quad \text{Original equation}$$

$$-6(-2) + 5 \stackrel{?}{=} 17 \quad \text{Let } x = -2.$$

$$12 + 5 \stackrel{?}{=} 17 \quad \text{Multiply.}$$

$$17 = 17 \ \checkmark \quad \text{True}$$

17 is *not* the solution.

The solution, -2, checks, so the solution set is $\{-2\}$.

Work Problem **2** at the Side. ▶

1 As a review, tell whether you would use the addition or multiplication property of equality to solve each equation. *Do not actually solve.*

(a) $7 + x = -9$

(b) $-13x = 26$

(c) $-x = \frac{3}{4}$

(d) $-12 = x - 4$

2 Solve.

(a) $-5p + 4 = 19$

(b) $7 + 2m = -3$

Answers

1. **(a)** and **(d):** addition property of equality
(b) and **(c):** multiplication property of equality
2. **(a)** $\{-3\}$ **(b)** $\{-5\}$

3 Solve.

GS **(a)** $5 - 8k = 2k - 5$

Begin with Step 2.

$$5 - 8k - 2k = 2k - 5 - \_\_\_\_$$

$$5 - \_\_\_\_ = -5$$

$$5 - 10k - \_\_\_\_ = -5 - \_\_\_\_$$

$$-10k = -10$$

$$\frac{-10k}{\_\_\_\_} = \frac{-10}{\_\_\_\_}$$

$$k = \_\_\_\_$$

Check by substituting ____ for k in the original equation.

The solution set is ____.

(b) $2q + 3 = 4q - 9$

(c) $6x + 7 = -8 + 3x$

Answers

3. **(a)** $2k$; $10k$; 5; 5; -10; -10; 1; 1; $\{1\}$
(b) $\{6\}$ **(c)** $\{-5\}$

EXAMPLE 2 **Using Both Properties of Equality to Solve a Linear Equation**

Solve $3x + 2 = 5x - 8$.

Step 1 There are no parentheses, fractions, or decimals in the equation, so we begin with Step 2.

$$3x + 2 = 5x - 8$$ Our goal is to isolate x.

Step 2	$3x + 2 - 5x = 5x - 8 - 5x$	Subtract $5x$ from each side.
	$-2x + 2 = -8$	Combine like terms.
	$-2x + 2 - 2 = -8 - 2$	Subtract 2 from each side.
	$-2x = -10$	Combine like terms.
Step 3	$\frac{-2x}{-2} = \frac{-10}{-2}$	Divide each side by -2.
	$x = 5$	

Step 4 Check by substituting 5 for x in the original equation.

CHECK	$3x + 2 = 5x - 8$	Original equation
	$3(5) + 2 \stackrel{?}{=} 5(5) - 8$	Let $x = 5$.
	$15 + 2 \stackrel{?}{=} 25 - 8$	Multiply.
	$17 = 17$ ✓	True

The solution, 5, checks, so the solution set is $\{5\}$.

Note

Remember that a variable can be isolated on either side of an equation. In **Example 2,** x will be isolated on the right if we begin by subtracting $3x$, instead of $5x$, from each side of the equation.

$3x + 2 = 5x - 8$	Equation from **Example 2**
$3x + 2 - 3x = 5x - 8 - 3x$	Subtract $3x$ from each side.
$2 = 2x - 8$	Combine like terms.
$2 + 8 = 2x - 8 + 8$	Add 8 to each side.
$10 = 2x$	Combine like terms.
$\frac{10}{2} = \frac{2x}{2}$	Divide each side by 2.
$5 = x$	The same solution results.

There are often several equally correct ways to solve an equation.

◀ **Work Problem 3 at the Side.**

EXAMPLE 3 Solving a Linear Equation

Solve $4(k-3)-k=k-6$.

Step 1 Clear the parentheses using the distributive property.

$$4(k-3)-k=k-6$$

$$4(k)+4(-3)-k=k-6 \quad \text{Distributive property}$$

$$4k-12-k=k-6 \quad \text{Multiply.}$$

$$3k-12=k-6 \quad \text{Combine like terms.}$$

Step 2

$$3k-12-k=k-6-k \quad \text{Subtract } k.$$

$$2k-12=-6 \quad \text{Combine like terms.}$$

$$2k-12+12=-6+12 \quad \text{Add 12.}$$

$$2k=6 \quad \text{Combine like terms.}$$

Step 3

$$\frac{2k}{2}=\frac{6}{2} \quad \text{Divide by 2.}$$

$$k=3$$

Step 4 Check by substituting 3 for k in the original equation.

CHECK

$$4(k-3)-k=k-6 \quad \text{Original equation}$$

$$4(3-3)-3\stackrel{?}{=}3-6 \quad \text{Let } k=3.$$

$$4(0)-3\stackrel{?}{=}3-6 \quad \text{Work inside the parentheses.}$$

$$-3=-3 \checkmark \quad \text{True}$$

The solution, 3, checks, so the solution set is $\{3\}$.

Work Problem 4 at the Side. ▶

EXAMPLE 4 Solving a Linear Equation

Solve $8a-(3+2a)=3a+1$.

Step 1

$$8a-(3+2a)=3a+1$$

$$8a-1(3+2a)=3a+1 \quad \text{Multiplicative identity property}$$

$$8a-3-2a=3a+1 \quad \text{Distributive property}$$

Be careful with signs.

$$6a-3=3a+1 \quad \text{Combine like terms.}$$

Step 2

$$6a-3-3a=3a+1-3a \quad \text{Subtract } 3a.$$

$$3a-3=1 \quad \text{Combine like terms.}$$

$$3a-3+3=1+3 \quad \text{Add 3.}$$

$$3a=4 \quad \text{Combine like terms.}$$

Step 3

$$\frac{3a}{3}=\frac{4}{3} \quad \text{Divide by 3.}$$

$$a=\frac{4}{3}$$

Step 4 Check that the solution set is $\{\frac{4}{3}\}$.

---

4 Solve.

GS **(a)** $7(p-2)+p=2p+4$

Step 1 ____ the parentheses.

$$\_\_\_\_(p)+7(\_\_\_\_)+p=2p+4$$

$$\_\_\_\_-\_\_\_\_+p=2p+4$$

$$\_\_\_\_-14=2p+4$$

Now complete the solution. Give the solution set.

(b) $11+3(x+1)=5x+16$

Answers

4. (a) Clear; 7; −2; $7p$; 14; $8p$; The solution set is $\{3\}$.
(b) $\{-1\}$

5 Solve.

(a) $7m - (2m - 9) = 39$

(b) $5x - (x + 9) = x - 4$

6 Solve.

(a) $2(4 + 3r) = 3(r + 1) + 11$

(b) $2 - 3(2 + 6z) = 4(z + 1) - 8$

Answers

5. (a) $\{6\}$ (b) $\left\{\frac{5}{3}\right\}$

6. (a) $\{2\}$ (b) $\{0\}$

CAUTION

In an expression such as $8a - (3 + 2a)$ in **Example 4,** the $-$ sign acts like a factor of -1 and affects the sign of *every* term within the parentheses.

$$8a - (3 + 2a) \leftarrow \text{Left side of the equation in Example 4}$$
$$= 8a - 1(3 + 2a)$$
$$= 8a + (-1)(3 + 2a)$$
$$= 8a - 3 - 2a$$

Change to $-$ in *both* terms.

◀ **Work Problem 5 at the Side.**

EXAMPLE 5 Solving a Linear Equation

Solve $4(4 - 3x) = 32 - 8(x + 2)$.

Do *not* subtract 8 from 32 here. Be careful with signs.

Step 1	$4(4 - 3x) = 32 - 8(x + 2)$	
	$16 - 12x = 32 - 8x - 16$	Distributive property
	$16 - 12x = 16 - 8x$	Combine like terms.
Step 2	$16 - 12x + 8x = 16 - 8x + 8x$	Add $8x$.
	$16 - 4x = 16$	Combine like terms.
	$16 - 4x - 16 = 16 - 16$	Subtract 16.
	$-4x = 0$	Combine like terms.
Step 3	$\frac{-4x}{-4} = \frac{0}{-4}$	Divide by -4.
	$x = 0$	
Step 4 CHECK	$4(4 - 3x) = 32 - 8(x + 2)$	Original equation
	$4[4 - 3(0)] \stackrel{?}{=} 32 - 8(0 + 2)$	Let $x = 0$.
	$4(4 - 0) \stackrel{?}{=} 32 - 8(2)$	Work inside the brackets and parentheses.
	$4(4) \stackrel{?}{=} 32 - 16$	Subtract and multiply.
	$16 = 16$ ✓	True

Since the solution 0 checks, the solution set is $\{0\}$.

◀ **Work Problem 6 at the Side.**

OBJECTIVE 2 Solve equations that have no solution or infinitely many solutions. Each equation so far has had exactly one solution. An equation with exactly one solution is a **conditional equation** because it is only true under certain conditions. Some equations have no solution or infinitely many solutions.

EXAMPLE 6 **Solving an Equation That Has Infinitely Many Solutions**

Solve $5x - 15 = 5(x - 3)$.

$$5x - 15 = 5(x - 3)$$

$$5x - 15 = 5x - 15 \quad \text{Distributive property}$$

$$5x - 15 - 5x = 5x - 15 - 5x \quad \text{Subtract } 5x.$$

Notice that the variable "disappeared."

$$-15 = -15 \quad \text{Combine like terms.}$$

$$-15 + 15 = -15 + 15 \quad \text{Add 15.}$$

$$0 = 0 \quad \text{True}$$

Solution set: {all real numbers}

Since the last statement $(0 = 0)$ is true, *any* real number is a solution. We could have predicted this from the second line in the solution,

$5x - 15 = 5x - 15$ ← This is true for *any* value of x.

Try several values for x in the original equation to see that they all satisfy it.

An equation with both sides exactly the same, like $0 = 0$, is an **identity.** An identity is true for all replacements of the variables. As shown above, we write the solution set as **{all real numbers}.**

CAUTION

In **Example 6,** do not write $\{0\}$ as the solution set of the equation. While 0 is a solution, there are infinitely many other solutions. ***For*** $\{0\}$ ***to be the solution set, the last line must include a variable, such as x, and read x = 0 (as in Example 5), not 0 = 0.***

EXAMPLE 7 **Solving an Equation That Has No Solution**

Solve $2x + 3(x + 1) = 5x + 4$.

$$2x + 3(x + 1) = 5x + 4$$

$$2x + 3x + 3 = 5x + 4 \quad \text{Distributive property}$$

$$5x + 3 = 5x + 4 \quad \text{Combine like terms.}$$

$$5x + 3 - 5x = 5x + 4 - 5x \quad \text{Subtract } 5x.$$

Again, the variable "disappeared."

$$3 = 4 \quad \text{False}$$

There is no solution. Solution set: $\emptyset$

A false statement $(3 = 4)$ results. A **contradiction** is an equation that has no solution. Its solution set is the **empty set,** or **null set,** symbolized $\emptyset$.

Work Problem 7 at the Side. ▶

CAUTION

Do not write $\{\emptyset\}$ to represent the empty set.

OBJECTIVE 3 **Solve equations with fractions or decimals as coefficients.** To avoid messy computations, we clear an equation of fractions by multiplying each side by the least common denominator (LCD) of all the fractions in the equation. Doing this will give an equation with only *integer* coefficients.

7 Solve.

(a) $2(x - 6) = 2x - 12$

(b) $3x + 6(x + 1) = 9x - 4$

(c) $-4x + 12 = 3 - 4(x - 3)$

Answers

7. (a) {all real numbers} **(b)** $\emptyset$ **(c)** $\emptyset$

8 Solve.

(a) $\frac{1}{4}x - 4 = \frac{3}{2}x + \frac{3}{4}x$

(b) $\frac{1}{2}x + \frac{5}{8}x = \frac{3}{4}x - 6$

Answers

8. (a) $\{-2\}$ **(b)** $\{-16\}$

EXAMPLE 8 Solving an Equation with Fractions as Coefficients

Solve $\frac{2}{3}x - \frac{1}{2}x = -\frac{1}{6}x - 2$.

Step 1 $\frac{2}{3}x - \frac{1}{2}x = -\frac{1}{6}x - 2$ (Pay particular attention here.)

$6\left(\frac{2}{3}x - \frac{1}{2}x\right) = 6\left(-\frac{1}{6}x - 2\right)$ Multiply each side by 6, the LCD.

$6\left(\frac{2}{3}x\right) + 6\left(-\frac{1}{2}x\right) = 6\left(-\frac{1}{6}x\right) + 6(-2)$ Distributive property; multiply *each* term inside the parentheses by 6.

$4x - 3x = -x - 12$ Multiply. (The fractions have been cleared.)

$x = -x - 12$ Combine like terms.

Step 2 $x + x = -x - 12 + x$ Add x.

$2x = -12$ Combine like terms.

Step 3 $\frac{2x}{2} = \frac{-12}{2}$ Divide by 2.

$x = -6$

Step 4 **CHECK** $\frac{2}{3}x - \frac{1}{2}x = -\frac{1}{6}x - 2$ Original equation

$\frac{2}{3}(-6) - \frac{1}{2}(-6) \stackrel{?}{=} -\frac{1}{6}(-6) - 2$ Let $x = -6$.

$-4 + 3 \stackrel{?}{=} 1 - 2$ Multiply.

$-1 = -1$ ✓ True

The solution, -6, checks. The solution set is $\{-6\}$.

◀ Work Problem **8** at the Side.

CAUTION

When clearing an equation of fractions, be sure to multiply every term on each side of the equation by the LCD.

EXAMPLE 9 Solving an Equation with Fractions as Coefficients

Solve $\frac{1}{3}(x + 5) - \frac{3}{5}(x + 2) = 1$.

Step 1 $\frac{1}{3}(x + 5) - \frac{3}{5}(x + 2) = 1$

$15\left[\frac{1}{3}(x + 5) - \frac{3}{5}(x + 2)\right] = 15(1)$ To clear the fractions, multiply by 15, the LCD.

$15\left[\frac{1}{3}(x + 5)\right] + 15\left[-\frac{3}{5}(x + 2)\right] = 15(1)$ Distributive property

$5(x + 5) - 9(x + 2) = 15$ Multiply.

$5x + 25 - 9x - 18 = 15$ Distributive property

($15\left[\frac{1}{3}(x + 5)\right] = 15 \cdot \frac{1}{3} \cdot (x + 5) = 5(x + 5)$)

$-4x + 7 = 15$ Combine like terms.

Continued on Next Page

Step 2 $-4x + 7 - 7 = 15 - 7$ Subtract 7.

$-4x = 8$ Combine like terms.

Step 3 $\frac{-4x}{-4} = \frac{8}{-4}$ Divide by -4.

$x = -2$

Step 4 Check to confirm that $\{-2\}$ is the solution set.

Work Problem 9 at the Side. ▶

CAUTION

Be sure you understand how to multiply by the LCD to clear an equation of fractions. ***Study Step 1 in Examples 8 and 9 carefully.***

EXAMPLE 10 Solving an Equation with Decimals as Coefficients

Solve $0.1t + 0.05(20 - t) = 0.09(20)$.

Step 1 The decimals here are expressed as tenths (0.1) and hundredths $(0.05$ and $0.09)$. We choose the least exponent on 10 needed to eliminate the decimal points. In this equation, we use $10^2 = 100$.

$$0.1t + 0.05(20 - t) = 0.09(20)$$

$$0.10t + 0.05(20 - t) = 0.09(20) \quad 0.1 = 0.10$$

$$100[0.10t + 0.05(20 - t)] = 100[0.09(20)] \quad \text{Multiply by 100.}$$

$$100(0.10t) + 100[0.05(20 - t)] = 100[0.09(20)] \quad \text{Distributive property}$$

$$10t + 5(20 - t) = 9(20) \quad \text{Multiply.}$$

$$10t + 5(20) + 5(-t) = 180 \quad \text{Distributive property}$$

$$10t + 100 - 5t = 180 \quad \text{Multiply.}$$

$$5t + 100 = 180 \quad \text{Combine like terms.}$$

Step 2 $5t + 100 - 100 = 180 - 100$ Subtract 100.

$5t = 80$ Combine like terms.

Step 3 $\frac{5t}{5} = \frac{80}{5}$ Divide by 5.

$t = 16$

Step 4 Check to confirm that $\{16\}$ is the solution set.

Work Problem 10 at the Side. ▶

Note

In **Example 10,** multiplying by 100 is accomplished by moving the decimal point two places to the right.

$$0.10t + 0.05(20 - t) = 0.09(20)$$

$$10t + 5(20 - t) = 9(20) \quad \text{Multiply by 100.}$$

9 Solve.

$$\frac{1}{4}(x + 3) - \frac{2}{3}(x + 1) = -2$$

10 Solve.

$$0.06(100 - x) + 0.04x = 0.05(92)$$

Answers

9. $\{5\}$
10. $\{70\}$

11 Perform each translation.

(a) Two numbers have a sum of 36. One of the numbers is represented by r. Write an expression for the other number.

(b) Two numbers have a product of 18. One of the numbers is represented by x. Write an expression for the other number.

Answers

11. (a) $36 - r$ **(b)** $\frac{18}{x}$

The table summarizes the solution sets of the equations in this section.

Type of Equation	Final Equation in Solution	Number of Solutions	Solution Set
Conditional **(See Examples 1–5, 8–10.)**	$x =$ a number	One	{a number}
Identity **(See Example 6.)**	A true statement with no variable, such as $0 = 0$	Infinite	{all real numbers}
Contradiction **(See Example 7.)**	A false statement with no variable, such as $3 = 4$	None	$\emptyset$

OBJECTIVE 4 Write expressions for two related unknown quantities.

Problem-Solving Hint

When we solve applied problems, we must often write *expressions* to relate unknown quantities. We then use these expressions to write the *equation* needed to solve the application. The next example provides preparation for doing this in **Section 10.4.**

EXAMPLE 11 Translating Phrases into Algebraic Expressions

Perform each translation.

(a) Two numbers have a sum of 23. If one of the numbers is represented by x, find an expression for the other number.

First, suppose that the sum of two numbers is 23, and one of the numbers is 10. To find the other number, we would subtract 10 from 23.

$$23 - 10 \leftarrow \text{This gives 13 as the other number.}$$

Instead of using 10 as one of the numbers, we use x. The other number would be obtained in the same way—by subtracting x from 23.

$$23 - x$$

$x - 23$ is *not* correct. Subtraction is *not* commutative.

To check, we find the sum of the two numbers.

$$x + (23 - x) = 23, \quad \text{as required.}$$

(b) Two numbers have a product of 24. If one of the numbers is represented by x, find an expression for the other number.

Suppose that one of the numbers is 4. To find the other number, we would divide 24 by 4.

$$\frac{24}{4} \leftarrow \text{This gives 6 as the other number. The product } 6 \cdot 4 \text{ is 24.}$$

In the same way, if x is one of the numbers, then we divide 24 by x to find the other number.

$$\frac{24}{x} \leftarrow \text{The other number}$$

◀ **Work Problem 11 at the Side.**

10.3 Exercises

FOR EXTRA HELP

Download the MyDashBoard App

MyMathLab®

CONCEPT CHECK *Based on the methods of this section, fill in each blank to indicate what we should do first to solve the given equation. Do not actually solve.*

1. $7x + 8 = 1$

Use the ___________ property of equality to ___________ 8 from each side.

2. $7x - 5x + 15 = 8 + x$

On the ___________ side, combine ___________ terms.

3. $3(2t - 4) = 20 - 2t$

Use the ___________ property to clear ___________ on the left side of the equation.

4. $\frac{3}{4}z = -15$

Use the ___________ property of equality to multiply each side by ___________ to obtain z on the left.

5. $\frac{2}{3}x - \frac{1}{6} = \frac{3}{2}x + 1$

Clear ___________ by multiplying by ___________, the LCD.

6. $0.9x + 0.3(x + 12) = 6$

Clear ___________ by multiplying each side by ___________ to obtain $9x$ as the first term on the left.

7. **CONCEPT CHECK** Suppose that when solving three linear equations, we obtain the final results shown in parts (a)–(c). Fill in the blanks in parts (a)–(c), and then match each result with the solution set in choices A–C for the *original* equation.

(a) $6 = 6$ (The original equation is a(n) ___________.)

(b) $x = 0$ (The original equation is a(n) ___________ equation.)

(c) $-5 = 0$ (The original equation is a(n) ___________.)

A. $\{0\}$

B. $\{\text{all real numbers}\}$

C. $\emptyset$

CONCEPT CHECK *In Exercises 8–10, give the letter of the correct choice.*

8. Which linear equation does *not* have all real numbers as solutions?

A. $5x = 4x + x$ **B.** $2(x + 6) = 2x + 12$ **C.** $\frac{1}{2}x = 0.5x$ **D.** $3x = 2x$

9. The expression $12\left[\frac{1}{6}(x + 2) - \frac{2}{3}(x + 1)\right]$ is equivalent to which of the following?

A. $-6x - 4$ **B.** $2x - 8$ **C.** $-6x + 1$ **D.** $-6x + 4$

10. The expression $100[0.03(x - 10)]$ is equivalent to which of the following?

A. $0.03x - 0.3$ **B.** $3x - 3$ **C.** $3x - 10$ **D.** $3x - 30$

Solve each equation, and check your solution. **See Examples 1–7.**

11. $3x + 2 = 14$ **12.** $4x + 3 = 27$ **13.** $-5z - 4 = 21$ **14.** $-7w - 4 = 10$

15. $4p - 5 = 2p$

16. $6q - 2 = 3q$

17. $5m + 8 = 7 + 3m$

18. $4r + 2 = r - 6$

19. $10p + 6 = 12p - 4$

20. $-5x + 8 = -3x + 10$

21. $7r - 5r + 2 = 5r - r$

22. $9p - 4p + 6 = 7p - p$

23. $x + 3 = -(2x + 2)$

24. $2x + 1 = -(x + 3)$

25. $3(4x - 2) + 5x = 30 - x$

26. $5(2m + 3) - 4m = 8m + 27$

27. $-2p + 7 = 3 - (5p + 1)$

28. $4x + 9 = 3 - (x - 2)$

29. $11x - 5(x + 2) = 6x + 5$

30. $6x - 4(x + 1) = 2x + 4$

31. $-(8x - 2) + 5x - 6 = -4$

32. $-(7x - 5) + 4x - 7 = -2$

33. $4(2x - 1) = -6(x + 3)$

34. $6(3w + 5) = 2(10w + 10)$

35. $6(4x - 1) = 12(2x + 3)$

36. $6(2x + 8) = 4(3x - 6)$

37. $3(2x - 4) = 6(x - 2)$

38. $3(6 - 4x) = 2(-6x + 9)$

39. $24 - 4(7 - 2t) = 4(t - 1)$

40. $8 - 2(2 - x) = 4(x + 1)$

Solve each equation, and check your solution. **See Examples 8–10.**

41. $-\frac{2}{7}r + 2r = \frac{1}{2}r + \frac{17}{2}$

(*Hint:* Clear the fractions by multiplying each side of the equation by the LCD, ____.)

42. $\frac{3}{5}t - \frac{1}{10}t = t - \frac{5}{2}$

(*Hint:* Clear the fractions by multiplying each side of the equation by the LCD, ____.)

43. $\frac{3}{4}x - \frac{1}{3}x + 5 = \frac{5}{6}x$

44. $\frac{1}{5}x - \frac{2}{3}x - 2 = -\frac{2}{5}x$

45. $\frac{1}{7}(3x + 2) - \frac{1}{5}(x + 4) = 2$

46. $\frac{1}{4}(3x - 1) + \frac{1}{6}(x + 3) = 3$

47. $\frac{1}{9}(x + 18) + \frac{1}{3}(2x + 3) = x + 3$

48. $-\frac{1}{4}(x - 12) + \frac{1}{2}(x + 2) = x + 4$

49. $-\frac{5}{6}q - \left(q - \frac{1}{2}\right) = \frac{1}{4}(q + 1)$

50. $\frac{2}{3}k - \left(k + \frac{1}{4}\right) = \frac{1}{12}(k + 4)$

51. $0.3(30) + 0.15x = 0.2(30 + x)$
(*Hint:* Clear the decimals by multiplying each side of the equation by ____.)

52. $0.2(60) + 0.05x = 0.1(60 + x)$
(*Hint:* Clear the decimals by multiplying each side of the equation by ____.)

53. $0.92x + 0.98(12 - x) = 0.96(12)$

54. $1.00x + 0.05(12 - x) = 0.10(63)$

55. $0.02(5000) + 0.03x = 0.025(5000 + x)$

56. $0.06(10{,}000) + 0.08x = 0.072(10{,}000 + x)$

Solve each equation, and check your solution. ***See Examples 1–10.***

57. $-3(5z + 24) + 2 = 2(3 - 2z) - 4$

58. $-2(2s - 4) - 8 = -3(4s + 4) - 1$

59. $-(6k - 5) - (-5k + 8) = -3$

60. $-(4x + 2) - (-3x - 5) = 3$

61. $8(t - 3) + 4t = 6(2t + 1) - 10$

62. $9(v + 1) - 3v = 2(3v + 1) - 8$

63. $4(x + 3) = 2(2x + 8) - 4$

64. $4(x + 8) = 2(2x + 6) + 20$

65. $\frac{1}{3}(x + 3) + \frac{1}{6}(x - 6) = x + 3$

66. $\frac{1}{2}(x + 2) + \frac{3}{4}(x + 4) = x + 5$

67. $0.3(x + 15) + 0.4(x + 25) = 25$

68. $0.1(x + 80) + 0.2x = 14$

Write the answer to each problem in terms of the variable. ***See Example 11.***

69. Two numbers have a sum of 12. One of the numbers is q. What expression represents the other number?

70. Two numbers have a sum of 26. One of the numbers is r. What expression represents the other number?

71. The product of two numbers is 9. One of the numbers is z. What expression represents the other number?

72. The product of two numbers is 13. One of the numbers is k. What expression represents the other number?

73. A football player gained x yards rushing. On the next down, he gained 29 yd. What expression represents the number of yards he gained altogether?

74. A football player gained y yards on a punt return. On the next return, he gained 25 yd. What expression represents the number of yards he gained altogether?

75. Monica is a years old. What expression represents her age 12 yr from now? 2 yr ago?

76. Chandler is b years old. What expression represents his age 3 yr ago? 5 yr from now?

77. Tom has r quarters. Express the value of the quarters in cents.

78. Jean has y dimes. Express the value of the dimes in cents.

79. A bank teller has t dollars, all in \$5 bills. What expression represents the number of \$5 bills the teller has?

80. A clerk has v dollars, all in \$10 bills. What expression represents the number of \$10 bills the clerk has?

81. A plane ticket costs x dollars for an adult and y dollars for a child. Find an expression that represents the total cost for 3 adults and 2 children.

82. A concert ticket costs p dollars for an adult and q dollars for a child. Find an expression that represents the total cost for 4 adults and 6 children.

10.4 An Introduction to Applications of Linear Equations

OBJECTIVES

1. Learn the six steps for solving applied problems.
2. Solve problems involving unknown numbers.
3. Solve problems involving sums of quantities.
4. Solve problems involving consecutive integers.
5. Solve problems involving complementary and supplementary angles.

OBJECTIVE 1 Learn the six steps for solving applied problems. While there is not one specific method, we suggest the following.

Solving an Applied Problem

Step 1 **Read** the problem, several times if necessary. What information is given? What is to be found?

Step 2 **Assign a variable** to represent the unknown value. Use a sketch, diagram, or table, as needed. Express any other unknown values in terms of the variable.

Step 3 **Write an equation** using the variable expression(s).

Step 4 **Solve** the equation.

Step 5 **State the answer.** Label it appropriately. Does the answer seem reasonable?

Step 6 **Check** the answer in the words of the *original* problem.

OBJECTIVE 2 Solve problems involving unknown numbers.

EXAMPLE 1 Finding the Value of an Unknown Number

The product of 4, and a number decreased by 7, is 100. What is the number?

Step 1 **Read** the problem carefully. We are asked to find a number.

Step 2 **Assign a variable** to represent the unknown quantity.

Let $x =$ the number.

Step 3 **Write an equation.**

Writing a "word equation" is often helpful.

The product of 4,	and	a number	decreased by	7,	is	100.
↓		↓	↓	↓	↓	↓
$4 \cdot$		$(x$	$-$	$7)$	$=$	100

Note the careful use of parentheses.

Is, are, was, and *were* translate as =.

Because of the commas in the given problem, writing the equation as $4x - 7 = 100$ is *incorrect*. The equation $4x - 7 = 100$ corresponds to the statement "The product of 4 and a number, decreased by 7, is 100."

Step 4 **Solve** the equation.

$$4(x-7) = 100 \quad \text{Equation from Step 3}$$
$$4x - 28 = 100 \quad \text{Distributive property}$$
$$4x - 28 + 28 = 100 + 28 \quad \text{Add 28.}$$
$$4x = 128 \quad \text{Combine like terms.}$$
$$\frac{4x}{4} = \frac{128}{4} \quad \text{Divide by 4.}$$
$$x = 32$$

Continued on Next Page

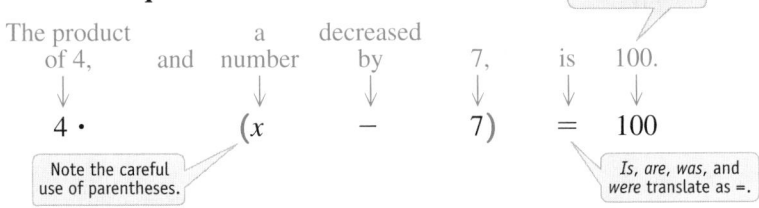

1 Solve each problem.

GS (a) If 5 is added to a number, the result is 7 less than 3 times the number. Find the number.

Step 1
We must find a ________.

Step 2
Let $x =$ the ________.

Step 3

If 5 is added to a number,	the result is	7 less than 3 times the number.
↓	↓	↓
________	=	________

Complete Steps 4–6 to solve the problem. Give the answer.

(b) If 5 is added to the product of 9 and a number, the result is 19 less than the number. Find the number.

Answers

1. (a) number; number; $x + 5$ (or $5 + x$); $3x - 7$; The number is 6.
(b) -3

Step 5 **State the answer.** The number is 32.

Step 6 **Check.** The number 32 decreased by 7 is 25. The product of 4 and 25 is 100, as required. The answer, 32, is correct.

◀ Work Problem 1 at the Side.

OBJECTIVE 3 Solve problems involving sums of quantities.

Problem-Solving Hint

In general, to solve problems involving sums of quantities, choose a variable to represent one of the unknowns. ***Then represent the other quantity in terms of the same variable.*** (See **Example 11** in **Section 10.3.**)

EXAMPLE 2 Finding Numbers of Olympic Medals

At the 2010 Winter Olympics in Vancouver, Canada, the United States won 14 more medals than Norway. The two countries won a total of 60 medals. How many medals did each country win? (*Source: World Almanac and Book of Facts.*)

Step 1 **Read** the problem. We are given the total number of medals and asked to find the number each country won.

Step 2 **Assign a variable.**

Let $x =$ the number of medals Norway won.

Then $x + 14 =$ the number of medals the United States won.

Step 3 **Write an equation.**

The total	is	the number of medals Norway won	plus	the number of medals the U.S. won.
↓	↓	↓	↓	↓
60	=	x	+	$(x + 14)$

Step 4 **Solve** the equation.

$$60 = 2x + 14 \quad \text{Combine like terms.}$$

$$60 - 14 = 2x + 14 - 14 \quad \text{Subtract 14.}$$

$$46 = 2x \quad \text{Combine like terms.}$$

$$\frac{46}{2} = \frac{2x}{2} \quad \text{Divide by 2.}$$

$$23 = x, \quad \text{or} \quad x = 23$$

Step 5 **State the answer.** The variable x represents the number of medals Norway won, so Norway won 23 medals. Then the number of medals the United States won is

$$x + 14 = 23 + 14, \quad \text{or} \quad 37.$$

Step 6 **Check.** Since the United States won 37 medals and Norway won 23, the total number of medals was $37 + 23 = 60$. Because $37 - 23 = 14$, the United States won 14 more medals than Norway. This information agrees with what is given in the problem, so the answer checks.

Problem-Solving Hint

The problem in **Example 2** could also be solved by letting x represent the number of medals the United States won. Then $x - 14$ would represent the number of medals Norway won. The equation would be different.

$$60 = x + (x - 14) \quad \text{Alternative equation for } \textbf{Example 2}$$

The solution of this equation is 37, which is the number of U.S. medals. The number of Norwegian medals would be $37 - 14 = 23$. ***The answers are the same,*** whichever approach is used, even though the equation and its solution are different.

Work Problem 2 at the Side. ▶

EXAMPLE 3 **Analyzing a Gasoline/Oil Mixture**

A lawn trimmer uses a mixture of gasoline and oil. The mixture contains 16 oz of gasoline for each 1 ounce of oil. If the tank holds 68 oz of the mixture, how many ounces of oil and how many ounces of gasoline does it require when it is full?

Step 1 **Read** the problem. We must find how many ounces of oil and gasoline are needed to fill the tank.

Step 2 **Assign a variable.**

Let x = the number of ounces of oil required.

Then $16x$ = the number of ounces of gasoline required.

A diagram like the following is sometimes helpful.

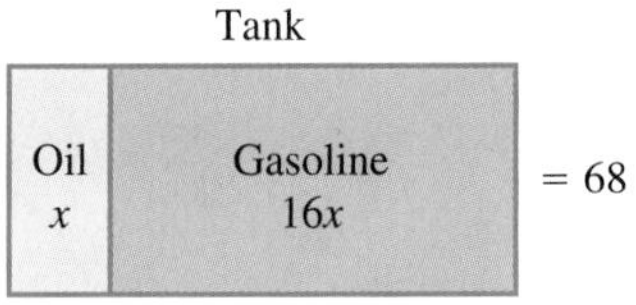

Step 3 **Write an equation.**

Amount of gasoline	plus	amount of oil	is	total amount in tank.
↓	↓	↓	↓	↓
$16x$	$+$	x	$=$	68

Step 4 **Solve.**

$$17x = 68 \quad \text{Combine like terms.}$$

$$\frac{17x}{17} = \frac{68}{17} \quad \text{Divide by 17.}$$

$$x = 4$$

Step 5 **State the answer.** When full, the lawn trimmer requires 4 oz of oil and $16x = 16(4)$ oz, or 64 oz of gasoline.

Step 6 **Check.** Since $4 + 64 = 68$, and $64 = 16(4)$, the answer checks.

Work Problem 3 at the Side. ▶

2 Solve each problem.

GS **(a)** The 150-member Iowa legislature includes 18 fewer Democrats than Republicans. (No other parties are represented.) How many Democrats and Republicans are there in the legislature? (*Source*: www.legis.iowa.gov)

Step 1
We must find the number of Democrats and ________.

Step 2
Let x = the number of Republicans.

Then ________ = the number of ________.

Complete Steps 3–6 to solve the problem. Give the equation and the answer.

(b) At the 2010 Winter Olympics, Germany won 19 more medals than China. The two countries won a total of 41 medals. How many medals did each country win? (*Source: World Almanac and Book of Facts.*)

3 Solve the problem.

At a club meeting, each member brought two nonmembers. If a total of 27 people attended, how many were members and how many were nonmembers?

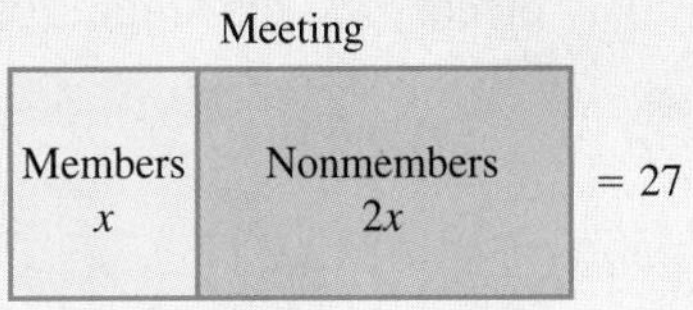

Answers

2. **(a)** Republicans; $x - 18$; Democrats; $x + (x - 18) = 150$; Republicans: 84; Democrats: 66
(b) Germany: 30; China: 11

3. members: 9; nonmembers: 18

4 Solve each problem.

(a) Over a 6-hr period, a basketball player spent twice as much time lifting weights as practicing free throws and 2 hr longer watching game films than practicing free throws. How many hours did he spend on each task?

Steps 1 and 2
The unknown found in both pairs of comparisons is the time spent _______.

Let $x =$ the time spent practicing free throws.

Then ____ = the time spent lifting weights, and

____ = the time spent watching game films.

Step 3
Write an equation.

________________ = 6

Complete Steps 4–6 to solve the problem. Give the answer.

(b) A piece of pipe is 50 in. long. It is cut into three pieces. The longest piece is 10 in. longer than the middle-sized piece, and the shortest piece measures 5 in. less than the middle-sized piece. Find the lengths of the three pieces.

Answers

4. (a) practicing free throws; $2x$; $x + 2$; $x + 2x + (x + 2)$; practicing free throws: 1 hr; lifting weights: 2 hr; watching game films: 3 hr
(b) longest: 25 in.; middle: 15 in.; shortest: 10 in.

Problem-Solving Hint

Sometimes we must find three unknown quantities. ***When the three unknowns are compared in pairs, let the variable represent the unknown found in both pairs.***

EXAMPLE 4 **Dividing a Board into Pieces**

A project calls for three pieces of wood. The longest piece must be twice the length of the middle-sized piece. The shortest piece must be 10 in. shorter than the middle-sized piece. If a board 70 in. long is to be used, how long must each piece be?

Step 1 **Read** the problem. Three lengths must be found.

Step 2 **Assign a variable.** Since the middle-sized piece appears in both pairs of comparisons, let x represent the length, in inches, of the middle-sized piece.

Let $x =$ the length of the middle-sized piece.

Then $2x =$ the length of the longest piece,

and $x - 10 =$ the length of the shortest piece.

A sketch is helpful here. See **Figure 2.**

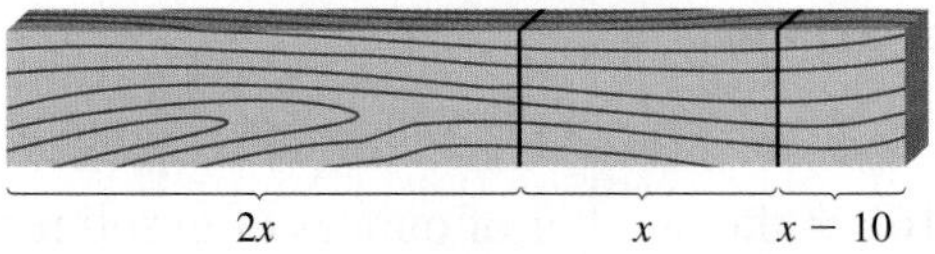

Figure 2

Step 3 **Write an equation.**

Longest	plus	middle-sized	plus	shortest	is	total length.
↓	↓	↓	↓	↓	↓	↓
$2x$	$+$	x	$+$	$(x - 10)$	$=$	70

Step 4 **Solve.**

$$4x - 10 = 70 \quad \text{Combine like terms.}$$

$$4x - 10 + \mathbf{10} = 70 + \mathbf{10} \quad \text{Add 10.}$$

$$4x = 80 \quad \text{Combine like terms.}$$

$$\frac{4x}{4} = \frac{80}{4} \quad \text{Divide by 4.}$$

$$x = 20$$

Step 5 **State the answer.** The middle-sized piece is 20 in. long, the longest piece is $2(20) = 40$ in. long, and the shortest piece is $20 - 10 = 10$ in. long.

Step 6 **Check.** The sum of the lengths is 70 in. All conditions of the problem are satisfied.

◀ Work Problem **4** at the Side.

OBJECTIVE 4 Solve problems involving consecutive integers. Two integers that differ by 1 are **consecutive integers.** For example, 3 and 4, 6 and 7, and -2 and -1 are pairs of consecutive integers. See **Figure 3.**

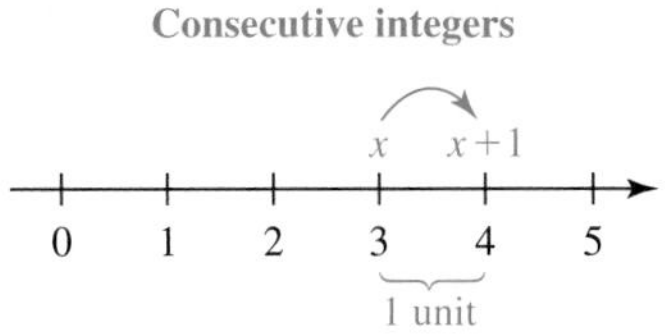

Figure 3

In general, if x represents an integer, then $x + 1$ represents the next greater consecutive integer.

EXAMPLE 5 Finding Consecutive Integers

Two pages that face each other in this book have 305 as the sum of their page numbers. What are the page numbers?

Step 1 **Read** the problem. Because the two pages face each other, they must have page numbers that are consecutive integers.

Step 2 **Assign a variable.**

Let $x =$ the lesser page number.

Then $x + 1 =$ the greater page number.

Step 3 **Write an equation.** The sum of the page numbers is 305.

$$x + (x + 1) = 305$$

Step 4 **Solve.**

$$2x + 1 = 305 \quad \text{Combine like terms.}$$
$$2x = 304 \quad \text{Subtract 1.}$$
$$x = 152 \quad \text{Divide by 2.}$$

Step 5 **State the answer.** The lesser page number is 152, and the greater is $152 + 1 = 153$. (Your book is opened to these two pages.)

Step 6 **Check.** The sum of 152 and 153 is 305. The answer is correct.

Work Problem 5 at the Side. ▶

Consecutive *even* integers, such as 2 and 4, and 8 and 10, differ by 2. Similarly, **consecutive *odd* integers,** such as 1 and 3, and 9 and 11, also differ by 2. See **Figure 4.**

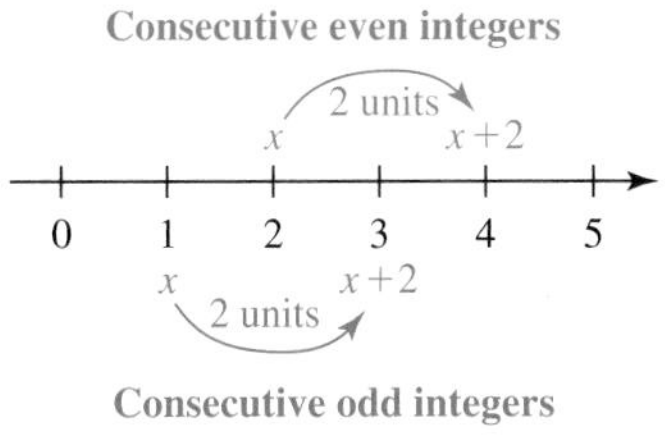

Figure 4

In general, if x represents an even (or odd) integer, then $x + 2$ represents the next greater consecutive even (or odd) integer, respectively.

5 Solve the problem.

Two pages that face each other in this book have a sum of 569. What are the page numbers?

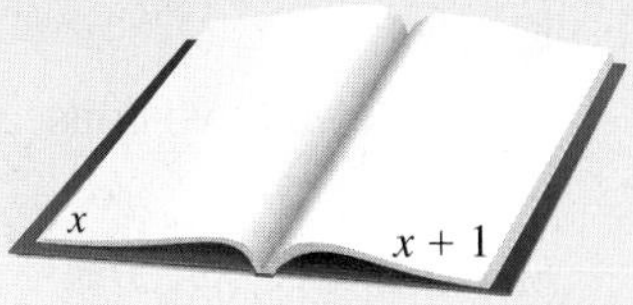

Answer

5. 284, 285

In this book, we list consecutive integers in increasing order.

Problem-Solving Hint

When solving consecutive integer problems, if $x =$ the lesser integer, then the following apply.

For two consecutive integers, use	$\mathbf{x, \quad x + 1.}$
For two consecutive *even* integers, use	$\mathbf{x, \quad x + 2.}$
For two consecutive *odd* integers, use	$\mathbf{x, \quad x + 2.}$

EXAMPLE 6 Finding Consecutive Odd Integers

If the lesser of two consecutive odd integers is doubled, the result is 7 more than the greater of the two integers. Find the two integers.

Step 1 **Read** the problem. We must find two consecutive odd integers.

Step 2 **Assign a variable.**

Let $x =$ the lesser consecutive odd integer.

Then $x + 2 =$ the greater consecutive odd integer.

Step 3 **Write an equation.**

If the lesser is doubled,	the result is	7	more than	the greater.
↓	↓	↓	↓	↓
$2x$	$=$	7	$+$	$x + 2$

Step 4 **Solve.**

$2x = 9 + x$ Combine like terms.

$x = 9$ Subtract x.

Step 5 **State the answer.** The lesser integer is 9. The greater is $9 + 2 = 11$.

Step 6 **Check.** When 9 is doubled, we get 18, which is 7 more than the greater odd integer, 11. The answers are correct.

◀ **Work Problem 6 at the Side.**

OBJECTIVE 5 Solve problems involving complementary and supplementary angles. An angle can be measured by a unit called the degree (°), which is $\frac{1}{360}$ of a complete rotation. Two angles whose sum is 90° are *complements* of each other, or **complementary.** An angle that measures 90° is a **right angle.** Two angles whose sum is 180° are *supplements* of each other, or **supplementary.** One angle *supplements* the other to form a **straight angle** of 180°. See **Figure 5.**

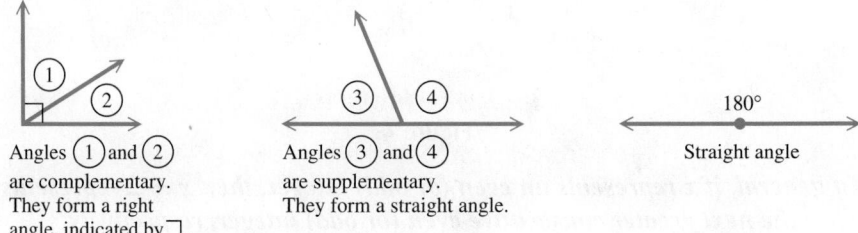

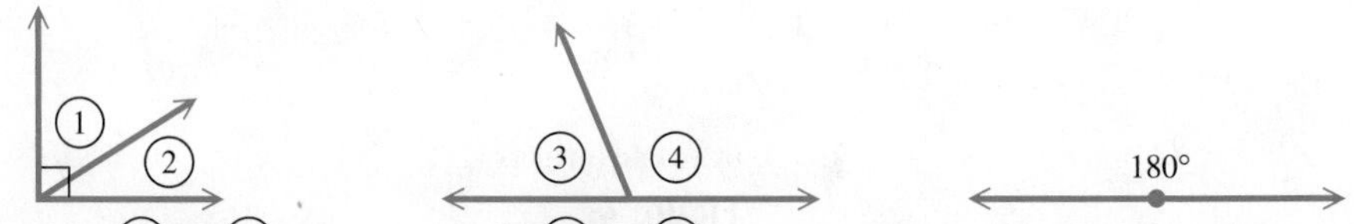

Angles (1) and (2) are complementary. They form a right angle, indicated by ⌉.

Angles (3) and (4) are supplementary. They form a straight angle.

Straight angle

Figure 5

6 Solve each problem.

GS **(a)** The sum of two consecutive even integers is 254. Find the integers.

Step 1
We must find two consecutive ______.

Step 2
Let $x =$ the lesser of the two ______ even integers.

Then ______ = the greater of the two consecutive even integers.

Step 3

The lesser consecutive even integer	+	the greater consecutive even integer	is	the total.
↓	↓	↓	↓	↓
______	+	______	=	______

Complete Steps 4–6 to solve the problem. Give the answer.

(b) Find two consecutive odd integers such that the sum of twice the lesser and three times the greater is 191.

Answers

6. (a) even integers; consecutive; $x + 2$; x; $x + 2$; 254; 126 and 128
(b) 37, 39

Problem-Solving Hint

If x represents the degree measure of an angle, then

$\mathbf{90 - x}$ represents the degree measure of its complement,

and $\mathbf{180 - x}$ represents the degree measure of its supplement.

Work Problem 7 at the Side. ▶

EXAMPLE 7 Finding the Measure of an Angle

Find the measure of an angle whose complement is five times its measure.

Step 1 **Read** the problem. We must find the measure of an angle, given information about the measure of its complement.

Step 2 **Assign a variable.**

Let $x =$ the degree measure of the angle.

Then $90 - x =$ the degree measure of its complement.

Step 3 **Write an equation.**

Measure of the complement	is	5 times the measure of the angle.
↓	↓	↓
$90 - x$	$=$	$5x$

Step 4 **Solve.**

$$90 - x + x = 5x + x \quad \text{Add } x.$$

$$90 = 6x \quad \text{Combine like terms.}$$

$$\frac{90}{6} = \frac{6x}{6} \quad \text{Divide by 6.}$$

$$15 = x, \quad \text{or} \quad x = 15$$

Step 5 **State the answer.** The measure of the angle is 15°.

Step 6 **Check.** If the angle measures 15°, then its complement measures $90° - 15° = 75°$, which is equal to five times 15°, as required.

Work Problem 8 at the Side. ▶

EXAMPLE 8 Finding the Measure of an Angle

Find the measure of an angle whose supplement is 10° more than twice its complement.

Step 1 **Read** the problem. We must find the measure of an angle, given information about its complement and its supplement.

Step 2 **Assign a variable.**

Let $x =$ the degree measure of the angle.

Then $90 - x =$ the degree measure of its complement,

and $180 - x =$ the degree measure of its supplement.

Continued on Next Page

7 Fill in the blank below each figure. Then solve.

(a) Find the complement of an angle that measures 26°.

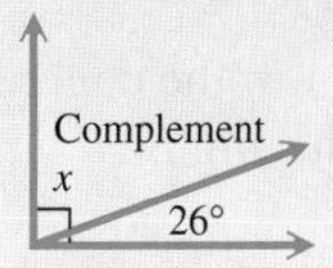

$x + 26 =$ ______

(b) Find the supplement of an angle that measures 92°.

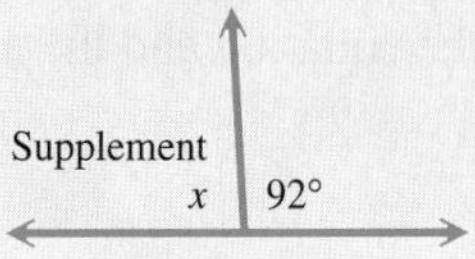

$x + 92 =$ ______

8 Solve each problem. Give the equation and the answer. (Let $x =$ the degree measure of the angle.)

(a) Find the measure of an angle whose complement is eight times its measure.

(b) Find the measure of an angle whose supplement is twice its measure.

Answers

7. **(a)** 90; 64° **(b)** 180; 88°

8. **(a)** $90 - x = 8x$; 10°

(b) $180 - x = 2x$; 60°

9 Solve each problem.

GS **(a)** Find the measure of an angle such that twice its complement is 30° less than its supplement.

Let $x =$ the degree measure of the angle.

Then ________ = the degree measure of its complement,

and ________ = the degree measure of its supplement.

Complete the solution. Give the equation and the answer.

(b) Find the measure of an angle whose supplement is 46° less than three times its complement.

Answers

9. (a) $90 - x$; $180 - x$; $2(90 - x) = (180 - x) - 30$; 30°
(b) 22°

We can visualize this information using a sketch. See **Figure 6.**

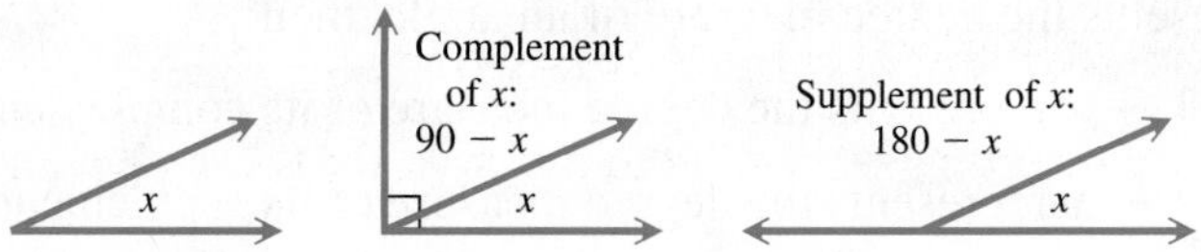

Figure 6

Step 3 **Write an equation.**

Supplement	is	10	more than	twice	its complement.
↓	↓	↓	↓	↓	↓
$180 - x$	$=$	10	$+$	2 ·	$(90 - x)$

Step 4 **Solve.**

Be sure to use parentheses here.

$$180 - x = 10 + 2(90 - x)$$

$$180 - x = 10 + 180 - 2x \quad \text{Distributive property}$$

$$180 - x = 190 - 2x \quad \text{Combine like terms.}$$

$$180 - x + \mathbf{2x} = 190 - 2x + \mathbf{2x} \quad \text{Add } 2x.$$

$$180 + x = 190 \quad \text{Combine like terms.}$$

$$180 + x - \mathbf{180} = 190 - \mathbf{180} \quad \text{Subtract 180.}$$

$$x = 10$$

Step 5 **State the answer.** The measure of the angle is 10°.

Step 6 **Check.** The complement of 10° is 80° and the supplement of 10° is 170°. Also, 170° is equal to 10° more than twice 80° (that is, $170 = 10 + 2(80)$ is true). Therefore, the answer is correct.

◀ **Work Problem 9 at the Side.**

10.4 Exercises

FOR EXTRA HELP

Download the MyDashBoard App

MyMathLab®

1. Give the six steps introduced in this section for solving application problems.

Step 1: ______________________

Step 2: ______________________

Step 3: ______________________

Step 4: ______________________

Step 5: ______________________

Step 6: ______________________

2. CONCEPT CHECK List some of the words that translate as "=" when writing an equation to solve an applied problem.

CONCEPT CHECK *In Exercises 3–6, which choice would* **not** *be a reasonable answer? Justify your response.*

3. A problem requires finding the number of cars on a dealer's lot.

A. 0 **B.** 45 **C.** 1 **D.** $6\frac{1}{2}$

4. A problem requires finding the number of hours a light bulb is on during a day.

A. 0 **B.** 4.5 **C.** 13 **D.** 25

5. A problem requires finding the distance traveled in miles.

A. -10 **B.** 1.8 **C.** $10\frac{1}{2}$ **D.** 50

6. A problem requires finding the time in minutes.

A. 0 **B.** 10.5 **C.** -5 **D.** 90

CONCEPT CHECK *Fill in each blank with the correct response.*

7. Consecutive integers differ by ______, such as 15 and ______, and -8 and ______.

8. Consecutive odd integers are ______ integers that differ by ______, such as ______ and 13. Consecutive even integers are ______ integers that differ by ______, such as 12 and ______.

9. Two angles whose measures sum to 90° are ______ angles. Two angles whose measures sum to 180° are ______ angles.

10. A right angle has measure ______. A straight angle has measure ______.

CONCEPT CHECK *Answer each question.*

11. Is there an angle that is equal to its supplement? Is there an angle that is equal to its complement? If the answer is yes to either question, give the measure of the angle.

12. If x represents an integer, how can you express the next *smaller* consecutive integer in terms of x? The next *smaller* even integer?

Solve each problem. In each case, give the equation using x as the variable, and give the answer. **See Example 1.**

13. The product of 8, and a number increased by 6, is 104. What is the number?

14. The product of 5, and 3 more than twice a number, is 85. What is the number?

15. If 2 is added to five times a number, the result is equal to 5 more than four times the number. Find the number.

16. If four times a number is added to 8, the result is three times the number, added to 5. Find the number.

17. Two less than three times a number is equal to 14 more than five times the number. What is the number?

18. Nine more than five times a number is equal to 3 less than seven times the number. What is the number?

19. If 2 is subtracted from a number and this difference is tripled, the result is 6 more than the number. Find the number.

20. If 3 is added to a number and this sum is doubled, the result is 2 more than the number. Find the number.

21. The sum of three times a number and 7 more than the number is the same as the difference between -11 and twice the number. What is the number?

22. If 4 is added to twice a number and this sum is multiplied by 2, the result is the same as if the number is multiplied by 3 and 4 is added to the product. What is the number?

GS *In Exercises 23 and 24, complete the six problem-solving steps to solve each problem.* **See Example 2.**

23. Pennsylvania and New York were among the states with the most remaining drive-in movie screens in 2012. Pennsylvania had 3 more screens than New York, and there were 59 screens total in the two states. How many drive-in movie screens remained in each state? (*Source:* www.Drive-Ins.com)

Step 1 **Read.** What are we asked to find?

____

____

Step 2 **Assign a variable.** Let $x =$ the number of screens in New York.

Then $x + 3 =$ ________________

____

Step 3 **Write an equation.**

____ + ____ = 59

Step 4 **Solve** the equation.

$x =$ ____

Step 5 **State the answer.** New York had ____ screens. Pennsylvania had ____ + 3 = ____ screens.

Step 6 **Check.** The number of screens in Pennsylvania was ____ more than the number of ____________. The total number of screens was 28 + ____ = ____.

24. Two of the most watched episodes in television were the final episodes of *M*A*S*H* and *Cheers*. The total number of viewers for these two episodes was about 92 million, with 8 million more people watching the *M*A*S*H* episode than the *Cheers* episode. How many people watched each episode? (*Source:* Nielsen Media Research.)

Step 1 **Read.** What are we asked to find?

____

____

Step 2 **Assign a variable.** Let $x =$ the number of people who watched the *Cheers* episode.

Then $x + 8 =$ ________________

____

Step 3 **Write an equation.**

____ + ____ = ____

Step 4 **Solve** the equation.

$x =$ ____

Step 5 **State the answer.** ____ million people watched the *Cheers* episode. ____ + 8 = ____ million watched the *M*A*S*H** episode.

Step 6 **Check.** The number of people who watched the *M*A*S*H** episode was ____ million more than the number who watched the *Cheers* episode. The total number of viewers in millions was 42 + ____ = ____.

Solve each problem. ***See Example 2.***

25. The total number of Democrats and Republicans in the U.S. House of Representatives during the 112th session (2011–2012) was 435. There were 49 more Republicans than Democrats. How many members of each party were there? (*Source: World Almanac and Book of Facts.*)

26. During the 112th session, the U.S. Senate had a total of 98 Democrats and Republicans. There were 4 more Democrats than Republicans. How many Democrats and Republicans were there in the Senate? (*Source: World Almanac and Book of Facts.*)

27. Bon Jovi and Roger Waters had the two top-grossing North American concert tours in 2010, together generating $197.7 million in ticket sales. If Roger Waters took in $18.7 million less than Bon Jovi, how much did each tour generate? (*Source:* Pollstar.)

28. In the United States in 2010, Honda Accord sales were 30 thousand less than Toyota Camry sales, and 596 thousand of these two cars were sold. How many of each make of car were sold? (*Source:* www.wikipedia.org)

29. In the 2010–2011 NBA regular season, the Los Angeles Lakers won 7 more than twice as many games as they lost. The Lakers played 82 games. How many wins and losses did the team have? (*Source:* www.nba.com)

30. In the 2011 regular MLB season, the Detroit Tigers won 39 fewer than twice as many games as they lost. The Tigers played 162 regular season games. How many wins and losses did the team have? (*Source:* www.mlb.com)

31. A one-cup serving of orange juice contains 3 mg less than four times the amount of vitamin C as a one-cup serving of pineapple juice. Servings of the two juices contain a total of 122 mg of vitamin C. How many milligrams of vitamin C are in a serving of each type of juice? (*Source:* U.S. Agriculture Department.)

32. A one-cup serving of pineapple juice has 9 more than three times as many calories as a one-cup serving of tomato juice. Servings of the two juices contain a total of 173 calories. How many calories are in a serving of each type of juice? (*Source:* U.S. Agriculture Department.)

Solve each problem. ***See Example 3.***

33. U.S. five-cent coins are made from a combination of nickel and copper. For every 1 lb of nickel, 3 lb of copper are used. How many pounds of copper would be needed to make 560 lb of five-cent coins? (*Source:* The United States Mint.)

34. The recipe for a special whole-grain bread calls for 1 oz of rye flour for every 4 oz of whole-wheat flour. How many ounces of each kind of flour should be used to make a loaf of bread weighing 32 oz?

35. A medication contains 9 mg of active ingredients for every 1 mg of inert ingredients. How much of each kind of ingredient would be contained in a single 250-mg caplet?

36. A recipe for salad dressing uses 2 oz of olive oil for each 1 oz of red wine vinegar. If 42 oz of salad dressing are needed, how many ounces of each ingredient should be used?

37. The value of a "Mint State-63" (uncirculated) 1950 Jefferson nickel minted at Denver is $\frac{4}{3}$ the value of a similar condition 1944 nickel minted at Philadelphia. Together, the value of the two coins is \$28.00. What is the value of each coin? (*Source:* Yeoman, R., *A Guide Book of United States Coins.*)

38. In one day, a store sold $\frac{8}{5}$ as many DVDs as Blu-ray discs. The total number of DVDs and Blu-ray discs sold that day was 273. How many DVDs were sold?

39. The world's largest taco contained approximately 1 kg of onion for every 6.6 kg of grilled steak. The total weight of these two ingredients was 617.6 kg. To the nearest tenth of a kilogram, how many kilograms of each ingredient were used? (*Source: Guinness World Records.*)

40. As of mid-2011, the combined population of China and India was estimated at 2.5 billion. If there were about 88% as many people living in India as China, what was the population of each country, to the nearest tenth of a billion? (*Source:* U.S. Census Bureau.)

Solve each problem. ***See Example 4.***

41. An office manager booked 55 airline tickets. He booked 7 more tickets on American Airlines than United Airlines. On Southwest Airlines, he booked 4 more than twice as many tickets as on United. How many tickets did he book on each airline?

42. A mathematics textbook editor spent 7.5 hr making telephone calls, writing e-mails, and attending meetings. She spent twice as much time attending meetings as making telephone calls and 0.5 hr longer writing e-mails than making telephone calls. How many hours did she spend on each task?

43. Nagaraj Nanjappa has a party-length submarine sandwich 59 in. long. He wants to cut it into three pieces so that the middle piece is 5 in. longer than the shortest piece and the shortest piece is 9 in. shorter than the longest piece. How long should the three pieces be?

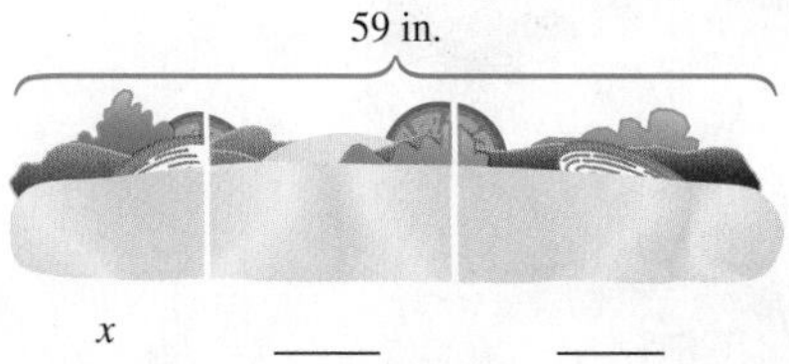

44. A three-foot-long deli sandwich must be split into three pieces so that the middle piece is twice as long as the shortest piece and the shortest piece is 8 in. shorter than the longest piece. How long should the three pieces be? (*Hint:* How many inches are in 3 ft?)

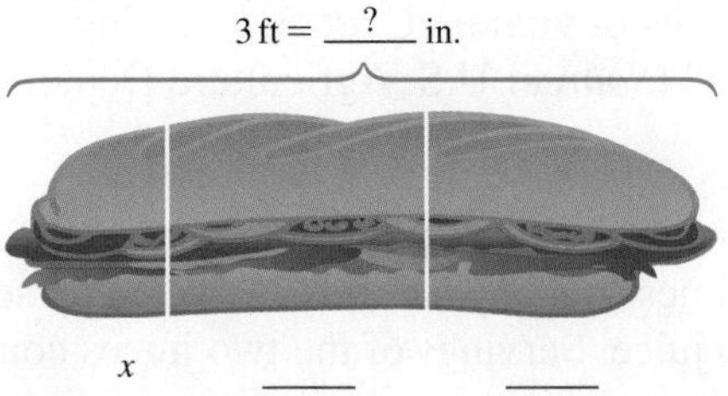

45. The United States earned 104 medals at the 2012 Summer Olympics. The number of silver medals earned was the same as the number of bronze medals. The number of gold medals was 17 more than the number of silver medals. How many of each kind of medal did the United States earn? (*Source:* www.london2012.com)

46. China earned 88 medals at the 2012 Summer Olympics. The number of gold medals earned was 15 more than the number of bronze medals. The number of silver medals earned was 4 more than the number of bronze medals. How many of each kind of medal did China earn? (*Source:* www.london2012.com)

47. Venus is 31.2 million mi farther from the sun than Mercury, while Earth is 57 million mi farther from the sun than Mercury. If the total of the distances from these three planets to the sun is 196.2 million mi, how far away from the sun is Mercury? (All distances given here are mean (*average*) distances.) (*Source: The New York Times Almanac.*)

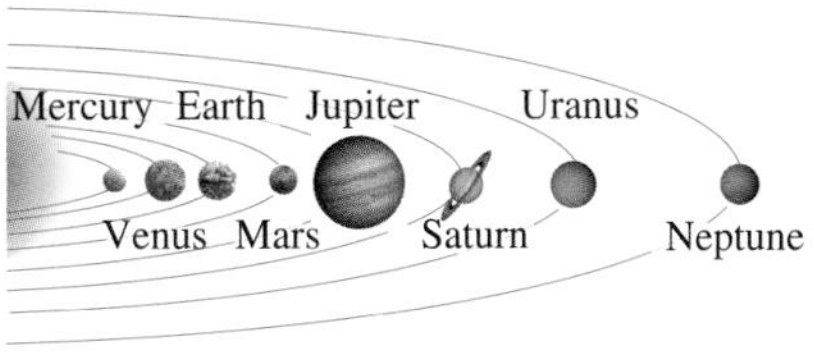

48. Saturn, Jupiter, and Uranus have a total of 153 known satellites (moons). Jupiter has 2 more satellites than Saturn, and Uranus has 35 fewer satellites than Saturn. How many known satellites does Uranus have? (*Source:* http://solarsystem.nasa.gov)

49. The sum of the measures of the angles of any triangle is 180°. In triangle *ABC*, angles *A* and *B* have the same measure, while the measure of angle *C* is 60° greater than each of *A* and *B*. What are the measures of the three angles?

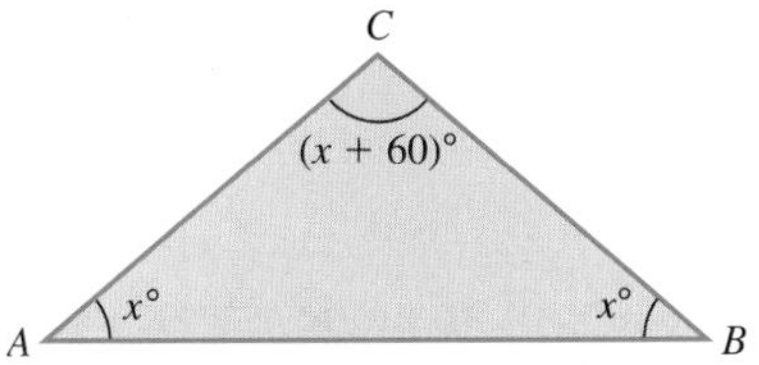

50. In triangle *ABC*, the measure of angle *A* is 141° more than the measure of angle *B*. The measure of angle *B* is the same as the measure of angle *C*. Find the measure of each angle. (*Hint:* See **Exercise 49.**)

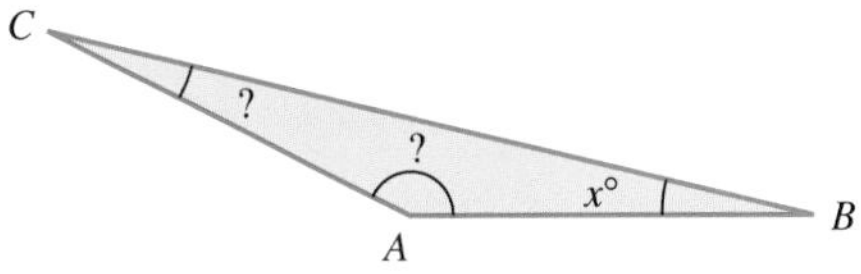

Solve each problem. ***See Examples 5 and 6.***

51. The numbers on two consecutively numbered gym lockers have a sum of 137. What are the locker numbers?

52. The sum of two consecutive check numbers is 357. Find the numbers.

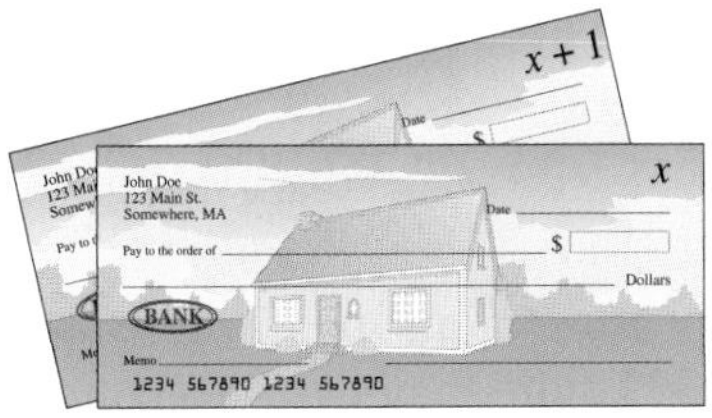

53. Two pages that are back-to-back in this book have 293 as the sum of their page numbers. What are the page numbers?

54. Two hotel rooms have room numbers that are consecutive integers. The sum of the numbers is 515. What are the two room numbers?

55. Find two consecutive even integers such that the lesser added to three times the greater gives a sum of 46.

56. Find two consecutive even integers such that six times the lesser added to the greater gives a sum of 86.

57. Find two consecutive odd integers such that 59 more than the lesser is four times the greater.

58. Find two consecutive odd integers such that twice the greater is 17 more than the lesser.

59. When the lesser of two consecutive integers is added to three times the greater, the result is 43. Find the integers.

60. If five times the lesser of two consecutive integers is added to three times the greater, the result is 59. Find the integers.

61. If the sum of three consecutive even integers is 60, what is the first of the three even integers? (*Hint:* If x and $x + 2$ represent the first two consecutive even integers, how would we represent the third consecutive even integer?)

62. If the sum of three consecutive odd integers is 69, what is the third of the three odd integers? (*Hint:* If x and $x + 2$ represent the first two consecutive odd integers, how would we represent the third consecutive odd integer?)

Solve each problem. ***See Examples 7 and 8.***

63. Find the measure of an angle whose complement is four times its measure.

64. Find the measure of an angle whose complement is five times its measure.

65. Find the measure of an angle whose supplement is eight times its measure.

66. Find the measure of an angle whose supplement is three times its measure.

67. Find the measure of an angle whose supplement measures 39° more than twice its complement.

68. Find the measure of an angle whose supplement measures 38° less than three times its complement.

69. Find the measure of an angle such that the difference between the measures of its supplement and three times its complement is 10°.

70. Find the measure of an angle such that the sum of the measures of its complement and its supplement is 160°.

10.5 Formulas and Additional Applications from Geometry

OBJECTIVES

1. Solve a formula for one variable, given the values of the other variables.
2. Use a formula to solve an applied problem.
3. Solve problems involving vertical angles and straight angles.
4. Solve a formula for a specified variable.

A **formula** is an equation in which variables are used to describe a relationship. For example, formulas exist for finding perimeters and areas of geometric figures, for calculating money earned on bank savings, and for converting among measurements.

$$P = 4s, \quad A = \pi r^2, \quad I = prt, \quad F = \frac{9}{5}C + 32 \qquad \text{Formulas}$$

Many of the formulas used in this book are given on the last two pages of the book.

OBJECTIVE 1 Solve a formula for one variable, given the values of the other variables. In **Example 1,** we use the idea of *area.* The **area** of a plane (two-dimensional) geometric figure is a measure of the surface covered by the figure. Area is measured in square units.

EXAMPLE 1 Using Formulas to Evaluate Variables

Find the value of the remaining variable in each formula.

(a) $A = LW$; $A = 64, L = 10$

This formula gives the area of a rectangle. See **Figure 7.** Substitute the given values for A and L into the formula.

$A = LW$ — Solve for W.

$64 = 10W$ — Let $A = 64$ and $L = 10$.

$\frac{64}{10} = \frac{10W}{10}$ — Divide by 10.

$6.4 = W$

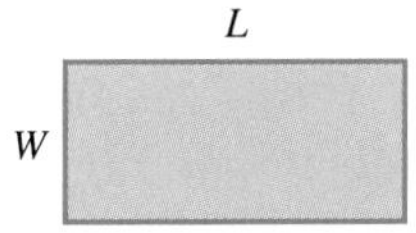

Rectangle
$A = LW$

Figure 7

The width is 6.4. Since $10(6.4) = 64$, the given area, the answer checks.

(b) $A = \frac{1}{2}h(b + B)$; $A = 210, B = 27, h = 10$

This formula gives the area of a trapezoid. See **Figure 8.**

$A = \frac{1}{2}h(b + B)$ — Solve for b.

$210 = \frac{1}{2}(10)(b + 27)$ — Let $A = 210, h = 10, B = 27$.

$210 = 5(b + 27)$ — Multiply $\frac{1}{2}(10)$.

$210 = 5b + 135$ — Distributive property

$210 - 135 = 5b + 135 - 135$ — Subtract 135.

$75 = 5b$ — Combine like terms.

$\frac{75}{5} = \frac{5b}{5}$ — Divide by 5.

$15 = b$

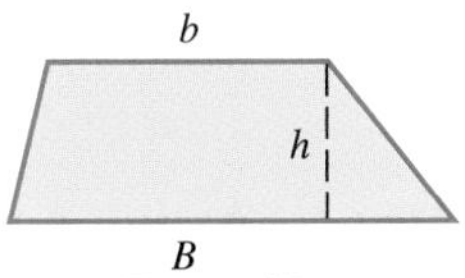

Trapezoid
$A = \frac{1}{2}h(b + B)$

Figure 8

The length of the shorter parallel side, b, is 15. Since $\frac{1}{2}(10)(15 + 27) = 210$, the given area, the answer checks.

Work Problem 1 at the Side. ▶

1 Find the value of the remaining variable in each formula.

(a) $A = bh$
(area of a parallelogram);
$A = 96, h = 8$

Given values for A and h, solve for the value of ____.

GS **(b)** $I = prt$ (simple interest);
$I = 246, r = 0.06$ (that is, 6%), $t = 2$

$I = prt$

$\_\_\_ = p(\_\_\_)(\_\_\_)$

$246 = \_\_\_ p$

$\frac{246}{\_\_\_} = \frac{0.12p}{0.12}$

$\_\_\_ = p$

(c) $P = 2L + 2W$
(perimeter of a rectangle);
$P = 126, W = 25$

Answers

1. (a) b; $b = 12$
(b) 246; 0.06; 2; 0.12; 0.12; 2050
(c) $L = 38$

OBJECTIVE 2 Use a formula to solve an applied problem. Examples 2 and 3 use the idea of *perimeter*. The **perimeter** of a plane (two-dimensional) geometric figure is the distance around the figure. For a polygon (e.g., a rectangle, square, or triangle), it is the sum of the lengths of its sides.

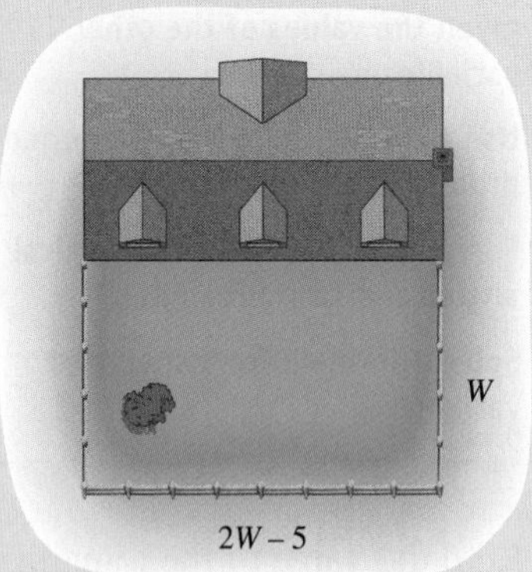

Figure 9

EXAMPLE 2 Finding the Dimensions of a Rectangular Yard

Kari Heen's backyard is in the shape of a rectangle. The length is 5 m less than twice the width, and the perimeter is 80 m. Find the dimensions of the yard.

Step 1 **Read** the problem. We must find the dimensions of the yard.

Step 2 **Assign a variable.** Let $W =$ the width of the lot, in meters. Since the length is 5 m less than twice the width, the length is given by $L = 2W - 5$. See **Figure 9.**

Step 3 **Write an equation.** Use the formula for the perimeter of a rectangle.

$$P = 2L + 2W$$

$$\begin{array}{ccccc} \text{Perimeter} & = & 2 \cdot \text{Length} & + & 2 \cdot \text{Width} \\ \downarrow & & \downarrow & & \downarrow \\ 80 & = & 2(2W-5) & + & 2W \end{array}$$ Substitute $2W - 5$ for length L.

Step 4 **Solve.**

$$\begin{aligned} 80 &= 4W - 10 + 2W && \text{Distributive property} \\ 80 &= 6W - 10 && \text{Combine like terms.} \\ 80 + 10 &= 6W - 10 + 10 && \text{Add 10.} \\ 90 &= 6W && \text{Combine like terms.} \\ \frac{90}{6} &= \frac{6W}{6} && \text{Divide by 6.} \\ 15 &= W \end{aligned}$$

Step 5 **State the answer.** The width is 15 m. The length is $2(15) - 5 = 25$ m.

Step 6 **Check.** If the width of the yard is 15 m and the length is 25 m, the perimeter is $2(25) + 2(15) = 50 + 30 = 80$ m, as required.

◀ **Work Problem 2 at the Side.**

2 Solve the problem.

A farmer has 800 m of fencing material to enclose a rectangular field. The width of the field is 175 m. Find the length of the field.

EXAMPLE 3 Finding the Dimensions of a Triangle

The longest side of a triangle is 3 ft longer than the shortest side. The medium side is 1 ft longer than the shortest side. If the perimeter of the triangle is 16 ft, what are the lengths of the three sides?

Step 1 **Read** the problem. We are given the perimeter of a triangle and must find the lengths of the three sides.

Step 2 **Assign a variable.** The shortest side is mentioned in each pair of comparisons.

Let $s =$ the length of the shortest side, in feet.

Then $s + 1 =$ the length of the medium side, in feet,

and $s + 3 =$ the length of the longest side, in feet.

It is a good idea to draw a sketch. See **Figure 10.**

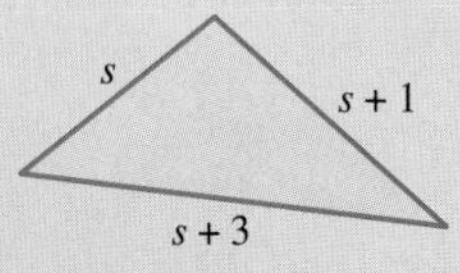

Figure 10

Continued on Next Page

Answer

2. 225 m

Step 3 **Write an equation.** Use the formula for the perimeter of a triangle.

$$P = a + b + c \quad \text{Perimeter of a triangle}$$

$$16 = s + (s + 1) + (s + 3) \quad \text{Substitute.}$$

Step 4 **Solve.** $16 = 3s + 4$ Combine like terms.

$$12 = 3s \quad \text{Subtract 4.}$$

$$4 = s \quad \text{Divide by 3.}$$

Step 5 **State the answer.** Since s represents the length of the shortest side, its measure is 4 ft.

$$s + 1 = 4 + 1 = 5 \text{ ft} \quad \text{Length of the medium side}$$

$$s + 3 = 4 + 3 = 7 \text{ ft} \quad \text{Length of the longest side}$$

Step 6 **Check.** The medium side, 5 ft, is 1 ft longer than the shortest side, and the longest side, 7 ft, is 3 ft longer than the shortest side. Furthermore, the perimeter is

$$4 + 5 + 7 = 16 \text{ ft}, \quad \text{as required.}$$

Work Problem 3 at the Side. ▶

3 Solve the problem.

The longest side of a triangle is 1 in. longer than the medium side. The medium side is 5 in. longer than the shortest side. If the perimeter is 32 in., what are the lengths of the three sides?

EXAMPLE 4 Finding the Height of a Triangular Sail

The area of a triangular sail of a sailboat is 126 ft^2. (Recall that ft^2 means "square feet.") The base of the sail is 12 ft. Find the height of the sail.

Step 1 **Read** the problem. We must find the height of the triangular sail.

Step 2 **Assign a variable.** Let h = the height of the sail, in feet. See **Figure 11.**

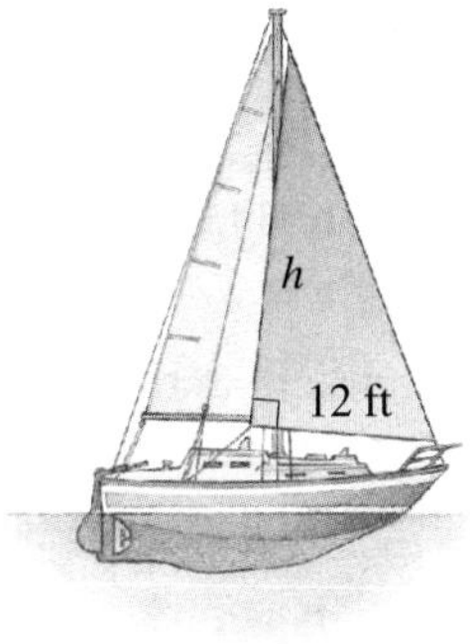

Figure 11

Step 3 **Write an equation.** Use the formula for the area of a triangle.

$$A = \frac{1}{2}bh \quad \text{A is the area, b is the base, and h is the height.}$$

$$126 = \frac{1}{2}(12)h \quad \text{Substitute } A = 126,\ b = 12.$$

Step 4 **Solve.** $126 = 6h$ Multiply.

$$21 = h \quad \text{Divide by 6.}$$

Step 5 **State the answer.** The height of the sail is 21 ft.

Step 6 **Check** to see that the values $A = 126$, $b = 12$, and $h = 21$ satisfy the formula for the area of a triangle.

Work Problem 4 at the Side. ▶

4 Solve the problem.

The area of a triangle is 120 m^2. The height is 24 m. Find the length of the base of the triangle.

Answers

3. 7 in.; 12 in.; 13 in.
4. 10 m

5 Find the measure of each marked angle.

(a)

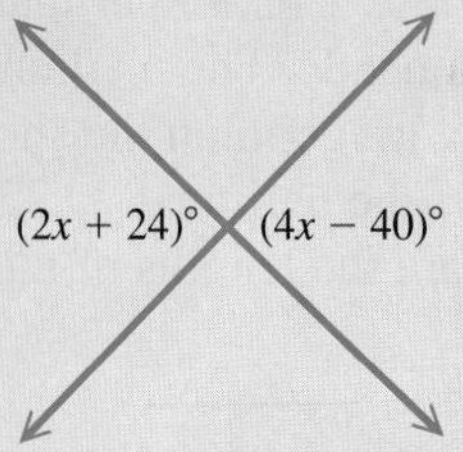

(b)

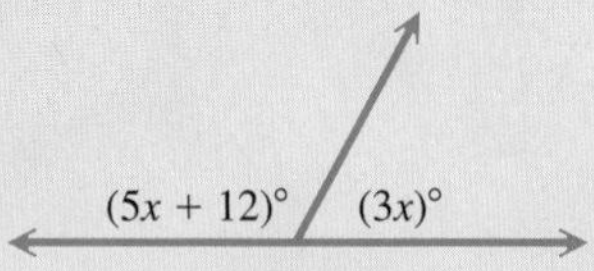

GS **(c)**

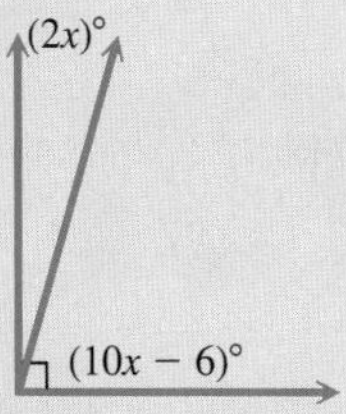

Because of the ⌝ symbol, the marked angles are ________ angles and have a sum of ____. The equation to use is

________ = 90.

Solve this equation to find that $x =$ ____. Then substitute to find the measure of each angle.

$2x = 2(\_\_\_\_)$

$= \_\_\_\_$

$10x - 6 = 10(\_\_\_\_) - 6$

$= \_\_\_\_$

The two angles measure ________.

Answers

5. **(a)** Both measure 88°. **(b)** 117° and 63° **(c)** complementary; 90°; $2x + (10x - 6)$; 8; 8; 16; 8; 74; 16° and 74°

OBJECTIVE 3 Solve problems involving vertical angles and straight angles. **Figure 12** shows two intersecting lines forming angles that are numbered ①, ②, ③, and ④. Angles ① and ③ lie "opposite" each other. They are **vertical angles.** Another pair of vertical angles is ② and ④. ***Vertical angles have equal measures.***

Now look at angles ① and ②. When their measures are added, we get 180°, the measure of a **straight angle.** There are three other angle pairs that form straight angles: ② and ③, ③ and ④, and ① and ④.

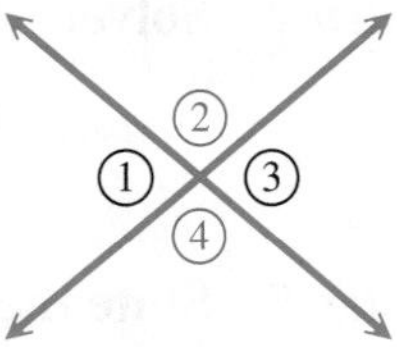

Figure 12

EXAMPLE 5 Finding Angle Measures

Refer to the appropriate figure in each part.

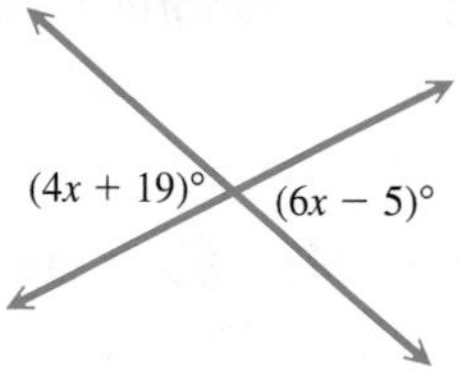

Figure 13

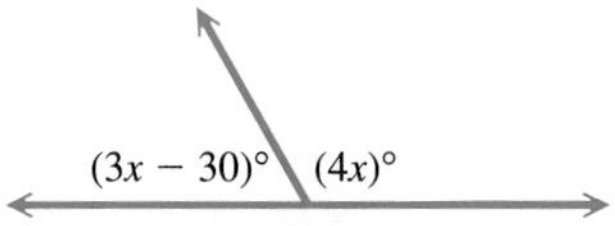

Figure 14

(a) Find the measure of each marked angle in **Figure 13.**

Since the marked angles are vertical angles, they have equal measures.

$$4x + 19 = 6x - 5 \quad \text{Set } 4x + 19 \text{ equal to } 6x - 5.$$
$$19 = 2x - 5 \quad \text{Subtract } 4x.$$
$$24 = 2x \quad \text{Add 5.}$$
$$12 = x \quad \text{Divide by 2.}$$

This is *not* the answer.

Since $x = 12$, one angle has measure $4(12) + 19 = 67$ degrees. The other has the same measure, since $6(12) - 5 = 67$ as well. Each angle measures 67°.

(b) Find the measure of each marked angle in **Figure 14.**

The measures of the marked angles must add to 180° because together they form a straight angle. (They are also *supplements* of each other.)

$$(3x - 30) + 4x = 180 \quad \text{Supplementary angles sum to 180°.}$$
$$7x - 30 = 180 \quad \text{Combine like terms.}$$
$$7x = 210 \quad \text{Add 30.}$$
$$x = 30 \quad \text{Divide by 7.}$$

Don't stop here!

Replace x with 30 in the measure of each marked angle.

$$3x - 30 = 3(30) - 30 = 90 - 30 = 60$$
$$4x = 4(30) = 120$$

The measures of the angles add to 180°, as required.

The two angle measures are 60° and 120°.

CAUTION

In **Example 5,** the answer is *not* the value of x. ***Remember to substitute the value of the variable into the expression given for each angle.***

◀ **Work Problem 5 at the Side.**

OBJECTIVE 4 Solve a formula for a specified variable. Sometimes we want to rewrite a formula in terms of a *different* variable in the formula. For example, consider $A = LW$, the formula for the area of a rectangle.

How can we rewrite $A = LW$ in terms of W?

The process whereby we do this is called **solving for a specified variable,** or **solving a literal equation.**

To solve a formula for a specified variable, we use the *same* steps that we used to solve an equation with just one variable. Consider the parallel reasoning to solve each of the following for x.

$3x + 4 = 13$		$ax + b = c$	
$3x + 4 - 4 = 13 - 4$	Subtract 4.	$ax + b - b = c - b$	Subtract b.
$3x = 9$		$ax = c - b$	
$\frac{3x}{3} = \frac{9}{3}$	Divide by 3.	$\frac{ax}{a} = \frac{c-b}{a}$	Divide by a.
$x = 3$		$x = \frac{c-b}{a}$	

When we solve a formula for a specified variable, we treat the specified variable as if it were the ONLY variable in the equation, and treat the other variables as if they were numbers.

EXAMPLE 6 Solving for a Specified Variable

Solve $A = LW$ for W.

Think of undoing what has been done to W. Since W is multiplied by L, undo the multiplication by dividing each side of $A = LW$ by L.

$$A = LW$$ Our goal is to isolate W.

$$\frac{A}{L} = \frac{LW}{L}$$ Divide by L.

$$\frac{A}{L} = W, \quad \text{or} \quad W = \frac{A}{L}$$ $\frac{LW}{L} = \frac{L}{L} \cdot W = 1 \cdot W = W$

Work Problem 6 at the Side. ▶

EXAMPLE 7 Solving for a Specified Variable

Solve $P = 2L + 2W$ for L.

$$P = 2L + 2W$$ Our goal is to isolate L.

$$P - 2W = 2L + 2W - 2W$$ Subtract $2W$.

$$P - 2W = 2L$$ Combine like terms.

$$\frac{P - 2W}{2} = \frac{2L}{2}$$ Divide by 2.

$$\frac{P - 2W}{2} = L, \quad \text{or} \quad L = \frac{P - 2W}{2}$$ $\frac{2L}{2} = \frac{2}{2} \cdot L = 1 \cdot L = L$

Work Problem 7 at the Side. ▶

6 Solve each formula for the specified variable.

(a) $W = Fd$ for F

GS **(b)** $I = prt$ for t

Our goal is to isolate ____.

$$I = prt$$

$$\frac{I}{\_\_\_\_} = \frac{prt}{\_\_\_\_}$$

$$\_\_\_\_ = t$$

7 Solve for the specified variable.

GS **(a)** $Ax + By = C$ for A

Our goal is to isolate ____. To do this, subtract ____ from each side. Then ____ each side by ____.

Show these steps and write the formula solved for A.

(b) $Ax + By = C$ for B

Answers

6. (a) $F = \frac{W}{d}$ **(b)** t; pr; pr; $\frac{I}{pr}$

7. (a) A; By; divide; x; $A = \frac{C - By}{x}$

(b) $B = \frac{C - Ax}{y}$

8 Solve each formula for the specified variable.

(a) $A = p + prt$ for t

(b) $x = u + zs$ for z

9 Solve each equation for y.

(a) $5x + y = 3$

(b) $x - 2y = 8$

Answers

8. **(a)** $t = \frac{A - p}{pr}$ **(b)** $z = \frac{x - u}{s}$

9. **(a)** $y = -5x + 3$ **(b)** $y = \frac{1}{2}x - 4$

EXAMPLE 8 Solving for a Specified Variable

Solve $F = \frac{9}{5}C + 32$ for C.

Our goal is to isolate C.

$$F = \frac{9}{5}C + 32$$ This is the formula for converting temperatures from Celsius to Fahrenheit.

$$F - 32 = \frac{9}{5}C + 32 - 32$$ Subtract 32.

$$F - 32 = \frac{9}{5}C$$

Be sure to use parentheses.

$$\frac{5}{9}(F - 32) = \frac{5}{9} \cdot \frac{9}{5}C$$ Multiply by $\frac{5}{9}$.

$$\frac{5}{9}(F - 32) = C$$ $\frac{5}{9} \cdot \frac{9}{5}C = 1C = C$

$$C = \frac{5}{9}(F - 32)$$ This is the formula for converting temperatures from Fahrenheit to Celsius.

◀ **Work Problem 8 at the Side.**

EXAMPLE 9 Solving for a Specified Variable

Solve each equation for y.

(a) $2x - y = 7$ Our goal is to isolate y.

$$2x - y - 2x = 7 - 2x$$ Subtract $2x$.

$$-y = 7 - 2x$$ Combine like terms.

$$-1(-y) = -1(7 - 2x)$$ Multiply by -1.

$$y = -7 + 2x$$ Multiply; distributive property

$$y = 2x - 7$$ $-a + b = b - a$

We could have added y and subtracted 7 from each side of the equation to isolate y on the right, giving $2x - 7 = y$, a different form of the same result. There is often more than one way to solve for a specified variable.

(b) $-3x + 2y = 6$

$$-3x + 2y + 3x = 6 + 3x$$ Add $3x$.

$$2y = 3x + 6$$ Combine like terms; commutative property

$$\frac{2y}{2} = \frac{3x + 6}{2}$$ Divide by 2.

Be careful here.

$$y = \frac{3x}{2} + \frac{6}{2}$$ $\frac{a + b}{c} = \frac{a}{c} + \frac{b}{c}$

$\frac{3x}{2} = \frac{3}{2} \cdot \frac{x}{1} = \frac{3}{2}x$

$$y = \frac{3}{2}x + 3$$ Simplify.

Although we could have given our answer as $y = \frac{3x + 6}{2}$, in preparation for our work in **Chapter 11,** we simplified further as shown in the last two lines of the solution.

◀ **Work Problem 9 at the Side.**

10.5 Exercises

FOR EXTRA HELP

CONCEPT CHECK *Give a one-sentence definition of each term.*

1. Perimeter of a plane geometric figure

2. Area of a plane geometric figure

CONCEPT CHECK *Decide whether perimeter or area would be used to solve a problem concerning the measure of the quantity.*

3. Sod for a lawn

4. Carpeting for a bedroom

5. Baseboards for a living room

6. Fencing for a yard

7. Fertilizer for a garden

8. Tile for a bathroom

9. Determining the cost of planting rye grass in a lawn for the winter

10. Determining the cost of replacing a linoleum floor with a wood floor

In the following exercises a formula is given, along with the values of all but one of the variables in the formula. Find the value of the variable that is not given. In Exercises 21–24, use 3.14 as an approximation for π (pi). ***See Example 1.***

11. $P = 2L + 2W$ (perimeter of a rectangle); $L = 8$, $W = 5$

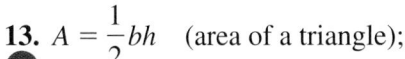

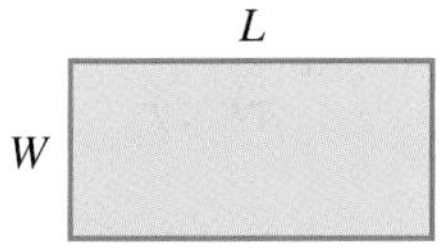

12. $P = 2L + 2W$; $L = 6$, $W = 4$

13. $A = \frac{1}{2}bh$ (area of a triangle); $b = 8$, $h = 16$

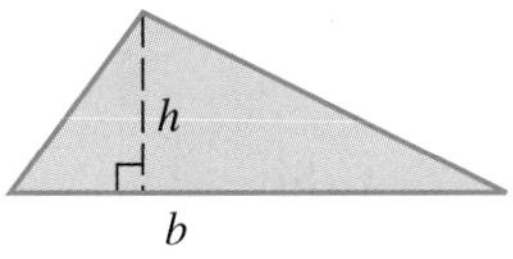

14. $A = \frac{1}{2}bh$; $b = 10$, $h = 14$

15. $P = a + b + c$ (perimeter of a triangle); $P = 12$, $a = 3$, $c = 5$

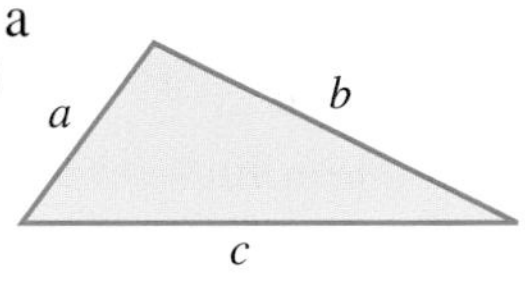

16. $P = a + b + c$; $P = 15$, $a = 3$, $b = 7$

17. $d = rt$ (distance formula); $d = 252$, $r = 45$

18. $d = rt$; $d = 100$, $t = 2.5$

19. $A = \frac{1}{2}h(b + B)$ (area of a trapezoid); $A = 91$, $b = 12$, $B = 14$

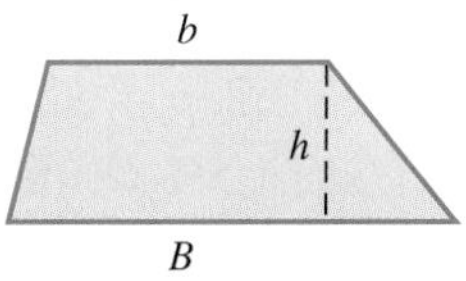

20. $A = \frac{1}{2}h(b + B)$; $A = 75$, $b = 19$, $B = 31$

21. $C = 2\pi r$ (circumference of a circle); $C = 16.328$

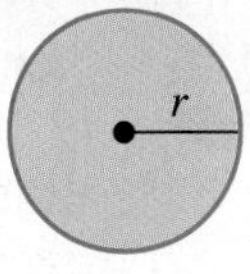

22. $C = 2\pi r$; $C = 8.164$

23. $A = \pi r^2$ (area of a circle); $r = 4$

24. $A = \pi r^2$; $r = 12$

The ***volume*** *of a three-dimensional object is a measure of the space occupied by the object. For example, we would need to know the volume of a gasoline tank in order to know how many gallons of gasoline it would take to completely fill the tank. Volume is measured in cubic units.*

In Exercises 25–30, a formula for the volume (V) of a three-dimensional object is given, along with values for the other variables. Evaluate V. (Use 3.14 as an approximation for π.) ***See Example 1.***

25. $V = LWH$ (volume of a rectangular box); $L = 10$, $W = 5$, $H = 3$

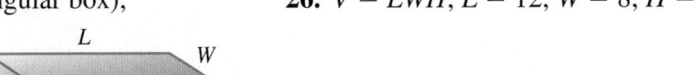

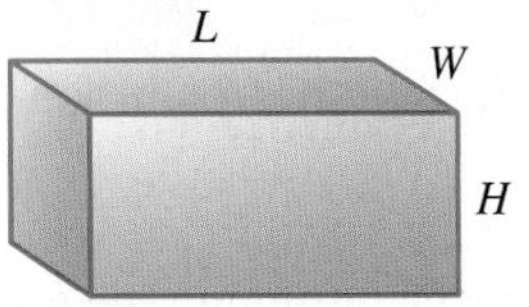

26. $V = LWH$; $L = 12$, $W = 8$, $H = 4$

27. $V = \frac{1}{3}Bh$ (volume of a pyramid); $B = 12$, $h = 13$

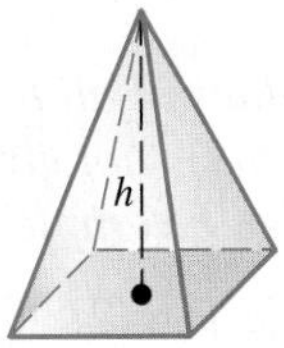

28. $V = \frac{1}{3}Bh$; $B = 36$, $h = 4$

29. $V = \frac{4}{3}\pi r^3$ (volume of a sphere); $r = 12$

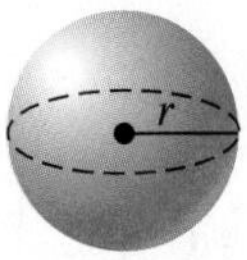

30. $V = \frac{4}{3}\pi r^3$; $r = 6$

Simple interest I in dollars is calculated using the following formula.

$$I = prt \quad \text{Simple interest formula}$$

Here, p represents the principal, or amount, in dollars that is invested or borrowed, r represents the annual interest rate, expressed as a percent, and t represents time, in years.

In Exercises 31–36, find the value of the remaining variable in the simple interest formula. ***See Example 1.*** *(Hint: Write percents as decimals. See* ***Section 6.1*** *to review how to do this if necessary.)*

31. $p = \$7500$, $r = 4\%$, $t = 2$ yr

32. $p = \$3600$, $r = 3\%$, $t = 4$ yr

33. $I = \$33$, $r = 2\%$, $t = 3$ yr

34. $I = \$270$, $r = 5\%$, $t = 6$ yr

35. $I = \$180$, $p = \$4800$, $r = 2.5\%$
(*Hint:* 2.5% written as a decimal is _____.)

36. $I = \$162$, $p = \$2400$, $r = 1.5\%$
(*Hint:* 1.5% written as a decimal is _____.)

Solve each perimeter problem. ***See Examples 2 and 3.***

37. The length of a rectangle is 9 in. more than the width. The perimeter is 54 in. Find the length and the width of the rectangle.

38. The width of a rectangle is 3 ft less than the length. The perimeter is 62 ft. Find the length and the width of the rectangle.

39. The perimeter of a rectangle is 36 m. The length is 2 m more than three times the width. Find the length and the width of the rectangle.

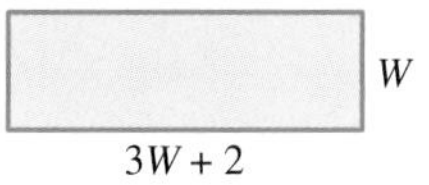

40. The perimeter of a rectangle is 36 yd. The width is 18 yd less than twice the length. Find the length and the width of the rectangle.

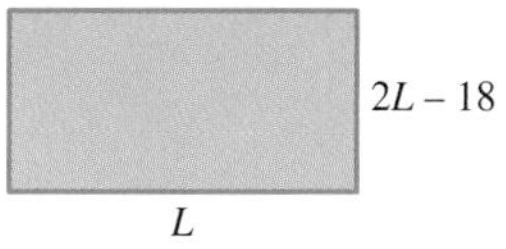

41. The longest side of a triangle is 3 in. longer than the shortest side. The medium side is 2 in. longer than the shortest side. If the perimeter of the triangle is 20 in., what are the lengths of the three sides?

42. The perimeter of a triangle is 28 ft. The medium side is 4 ft longer than the shortest side, while the longest side is twice as long as the shortest side. What are the lengths of the three sides?

43. Two sides of a triangle have the same length. The third side measures 4 m less than twice that length. The perimeter of the triangle is 24 m. Find the lengths of the three sides.

44. A triangle is such that its medium side is twice as long as its shortest side and its longest side is 7 yd less than three times its shortest side. The perimeter of the triangle is 47 yd. What are the lengths of the three sides?

Use a formula to write an equation for each application, and then solve. (Use 3.14 *as an approximation for* π.) ***Formulas are found inside the back cover of this book. See Examples 2–4.***

45. One of the largest fashion catalogues in the world was published in Hamburg, Germany. Each of the 212 pages in the catalogue measured 1.2 m by 1.5 m. What was the perimeter of a page? What was the area? (*Source: Guinness World Records.*)

46. One of the world's largest mandalas (sand paintings) measures 12.24 m by 12.24 m. What is the perimeter of the sand painting? To the nearest hundredth of a square meter, what is the area? (*Source: Guinness World Records.*)

47. The area of a triangular road sign is 70 ft^2. If the base of the sign measures 14 ft, what is the height of the sign?

48. The area of a triangular advertising banner is 96 ft^2. If the height of the banner measures 12 ft, find the measure of the base.

49. A prehistoric ceremonial site dating to about 3000 B.C. was discovered at Stanton Drew in southwestern England. The site, which is larger than Stonehenge, is a nearly perfect circle, consisting of nine concentric rings that probably held upright wooden posts. Around this timber temple is a wide, encircling ditch enclosing an area with a diameter of 443 ft. Find this enclosed area to the nearest thousand square feet. (*Source: Archaeology.*)

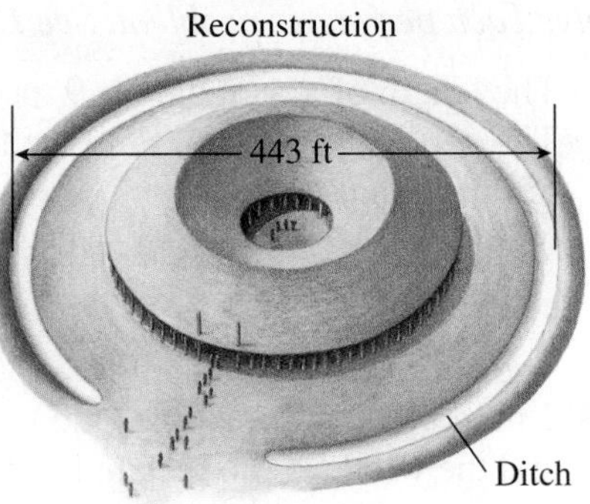

50. The Rogers Centre in Toronto, Canada, is the first stadium with a hard-shell, retractable roof. The steel dome is 630 ft in diameter. To the nearest foot, what is the circumference of this dome? (*Source:* www.ballparks.com)

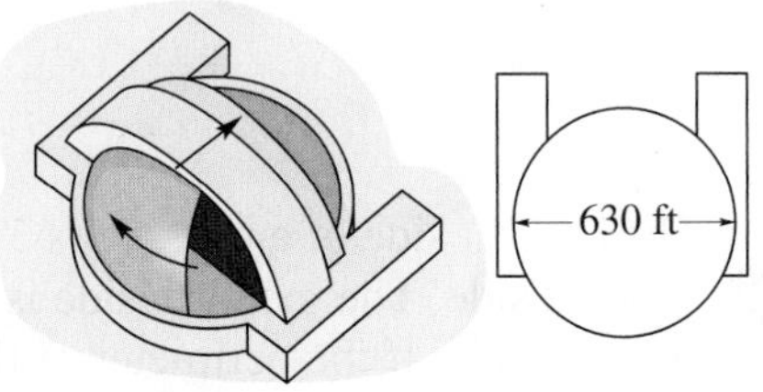

51. One of the largest drums ever constructed was made from Japanese cedar and cowhide, with radius 7.87 ft. What was the area of the circular face of the drum? What was the circumference of the drum? Round your answers to the nearest hundredth. (*Source: Guinness World Records.*)

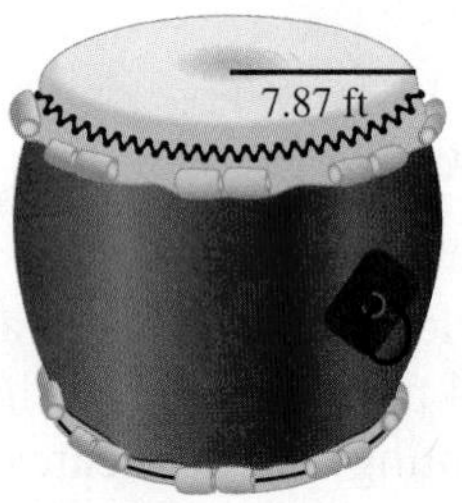

52. A drum played at the Royal Festival Hall in London had radius 6.5 ft. What was the area of the circular face of the drum? What was the circumference of the drum? (*Source: Guinness World Records.*)

53. The survey plat depicted here shows two lots that form a trapezoid. The measures of the parallel sides are 115.80 ft and 171.00 ft. The height of the trapezoid is 165.97 ft. Find the combined area of the two lots. Round your answer to the nearest hundredth of a square foot.

54. Lot A in the figure is in the shape of a trapezoid. The parallel sides measure 26.84 ft and 82.05 ft. The height of the trapezoid is 165.97 ft. Find the area of Lot A. Round your answer to the nearest hundredth of a square foot.

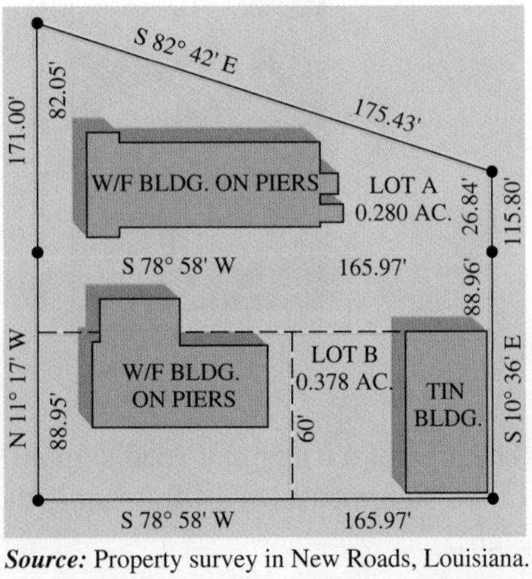

Source: Property survey in New Roads, Louisiana.

55. The U.S. Postal Service requires that any box sent by Priority Mail® have length plus girth (distance around) totaling no more than 108 in. The maximum volume that meets this condition is contained by a box with a square end 18 in. on each side. What is the length of the box? What is the maximum volume? (*Source:* United States Postal Service.)

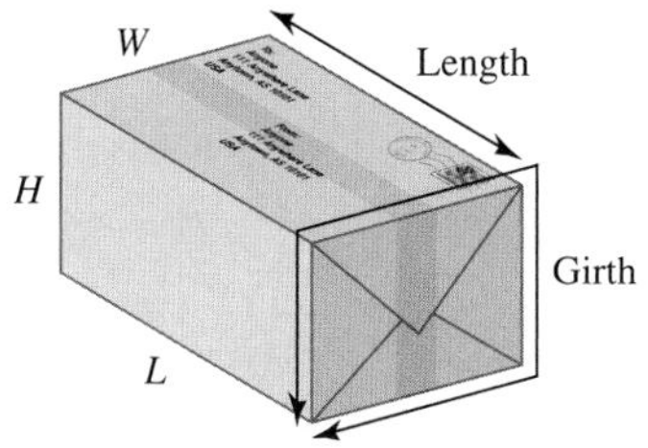

56. One of the world's largest sandwiches, made by Wild Woody's Chill and Grill in Roseville, Michigan, was 12 ft long, 12 ft wide, and $17\frac{1}{2}$ in. $(1\frac{11}{24}\text{ ft})$ thick. What was the volume of the sandwich? (*Source: Guinness World Records.*)

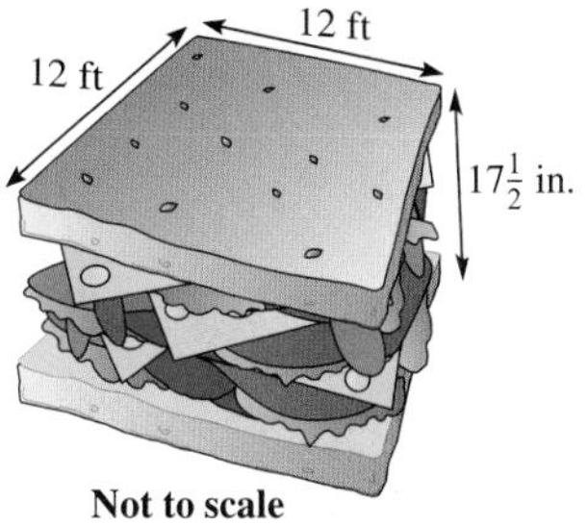

Find the measure of each marked angle. ***See Example 5.***

57.

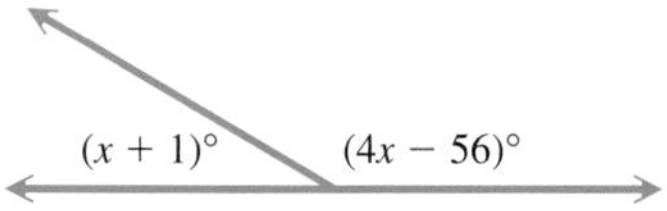

58.

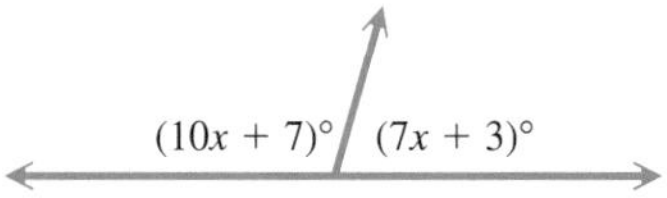

59.

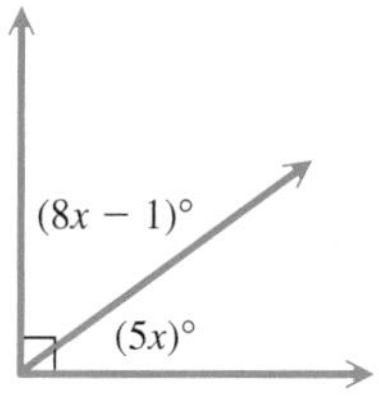

60.

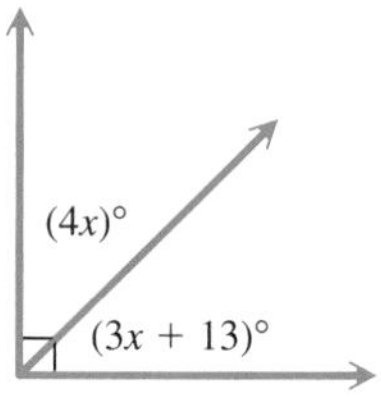

61.

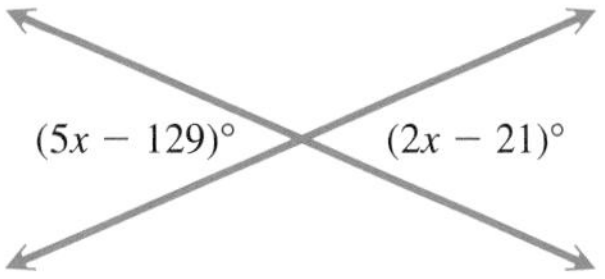

62.

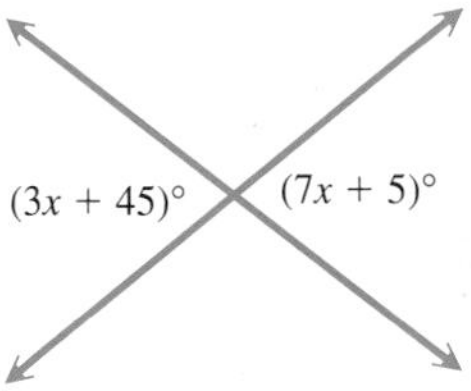

63.

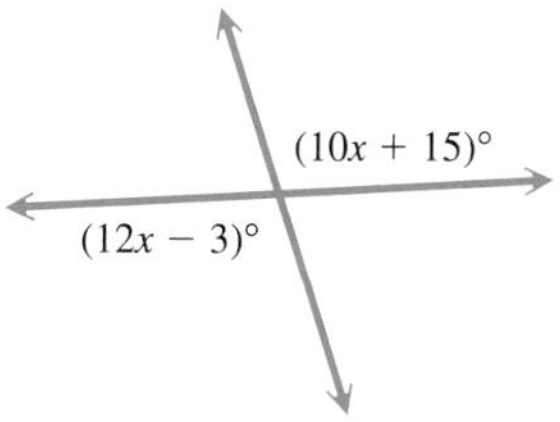

64.

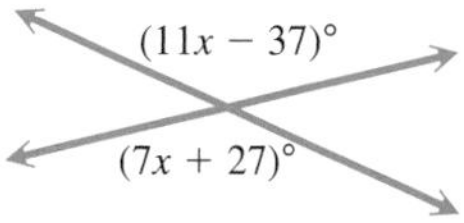

Solve each formula for the specified variable. ***See Examples 6–8.***

65. $d = rt$ for t

66. $d = rt$ for r

67. $V = LWH$ for H

68. $V = LWH$ for L

69. $P = a + b + c$ for b

70. $P = a + b + c$ for c

71. $C = 2\pi r$ for r

72. $C = \pi d$ *for d*

73. $I = prt$ for r

74. $I = prt$ for p

75. $A = \frac{1}{2}bh$ for h

76. $A = \frac{1}{2}bh$ for b

77. $V = \frac{1}{3}\pi r^2 h$ for h

78. $V = \pi r^2 h$ for h

79. $P = 2L + 2W$ for W

80. $A = p + prt$ for r

81. $y = mx + b$ for m

82. $y = mx + b$ for x

83. $Ax + By = C$ for y

84. $Ax + By = C$ for x

85. $M = C(1 + r)$ for r

86. $C = \frac{5}{9}(F - 32)$ for F

87. $P = 2(a + b)$ for a

88. $P = 2(a + b)$ for b

Solve each equation for y. ***See Example 9.***

89. $6x + y = 4$

90. $3x + y = 6$

91. $5x - y = 2$

92. $4x - y = 1$

93. $-3x + 5y = -15$

94. $-2x + 3y = -9$

95. $x - 3y = 12$

96. $x - 5y = 10$

10.6 Solving Linear Inequalities

OBJECTIVES

1. Graph intervals on a number line.
2. Use the addition property of inequality.
3. Use the multiplication property of inequality.
4. Solve linear inequalities by using both properties of inequality.
5. Use inequalities to solve applied problems.
6. Solve linear inequalities with three parts.

An **inequality** relates algebraic expressions using the symbols

$<$	"is less than,"	$\leq$	"is less than or equal to,"
$>$	"is greater than,"	$\geq$	"is greater than or equal to."

In each case, the interpretation is based on reading the symbol from left to right.

Linear Inequality in One Variable

A **linear inequality in one variable** can be written in the form

$$Ax + B < C, \quad Ax + B \leq C, \quad Ax + B > C, \quad \text{or} \quad Ax + B \geq C,$$

where A, B, and C represent real numbers, and $A \neq 0$.

Some examples of linear inequalities in one variable follow.

$$x + 5 < 2, \quad z - \frac{3}{4} \geq 5, \quad \text{and} \quad 2k + 5 \leq 10 \qquad \text{Linear inequalities}$$

We solve a linear inequality by finding all of its real number solutions. For example, the set

$$\{x \mid x \leq 2\}$$ Set-builder notation **(Section 9.3)**

The set of | all x | such that | x is less than or equal to 2

includes *all real numbers* that are less than or equal to 2, not just the *integers* less than or equal to 2.

OBJECTIVE 1 Graph intervals on a number line. Graphing is a good way to show the solution set of an inequality. To graph all real numbers belonging to the set $\{x \mid x \leq 2\}$, we place a square bracket at 2 on a number line and draw an arrow extending from the bracket to the left (since all numbers *less than* 2 are also part of the graph). See **Figure 15.**

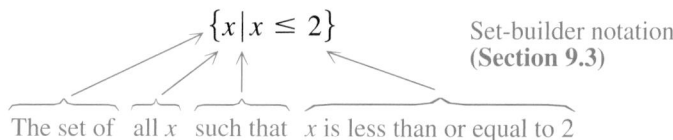

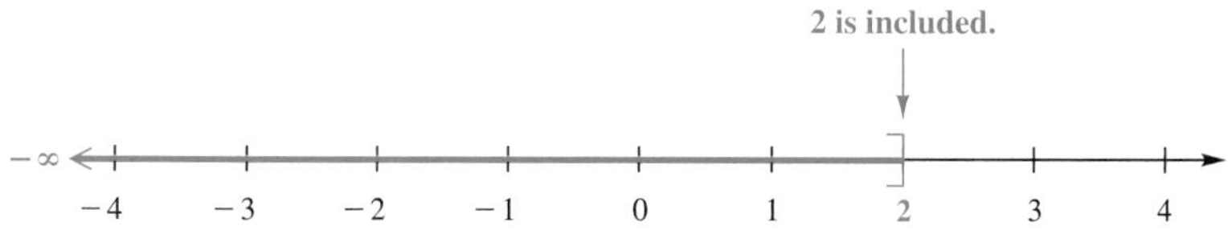

Figure 15 Graph of the interval $(-\infty, 2]$

The set of numbers less than or equal to 2 is an example of an **interval** on the number line. We can write this interval using **interval notation,** which uses the **infinity symbols** ∞ or $-\infty$.

$$(-\infty, 2] \qquad \text{Interval notation}$$

The negative infinity symbol $-\infty$ here does not indicate a number, but shows that the interval includes *all* real numbers less than 2. Again, the square bracket indicates that 2 is part of the solution.

1 Write each inequality in interval notation, and graph the interval.

(a) $x \leq 3$

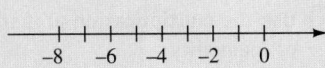

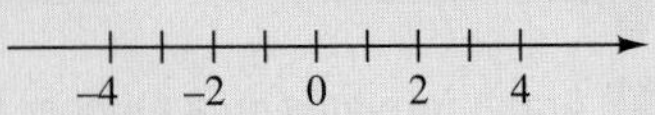

(b) $x > -4$

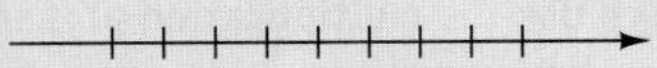

(c) $x \leq -\frac{3}{4}$

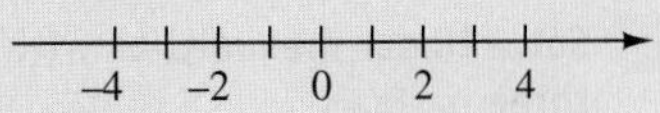

2 Write each inequality in interval notation and graph the interval.

(a) $-4 \geq x$

(b) $0 < x$

Answers

1. (a) $(-\infty, 3]$

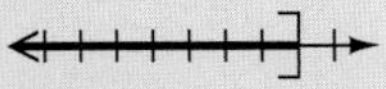

(b) $(-4, \infty)$

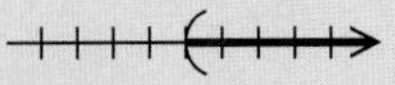

(c) $\left(-\infty, -\frac{3}{4}\right]$

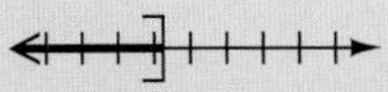

2. (a) $(-\infty, -4]$

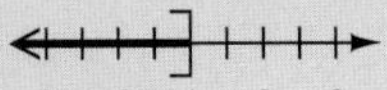

(b) $(0, \infty)$

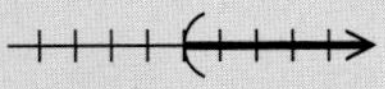

EXAMPLE 1 Graphing an Interval on a Number Line

Graph $x > -5$.

The statement $x > -5$ says that x can represent any value greater than -5 but cannot equal -5, written $(-5, \infty)$. We graph this interval by placing a parenthesis at -5 and drawing an arrow to the right, as in **Figure 16.** The parenthesis at -5 indicates that -5 is *not* part of the graph.

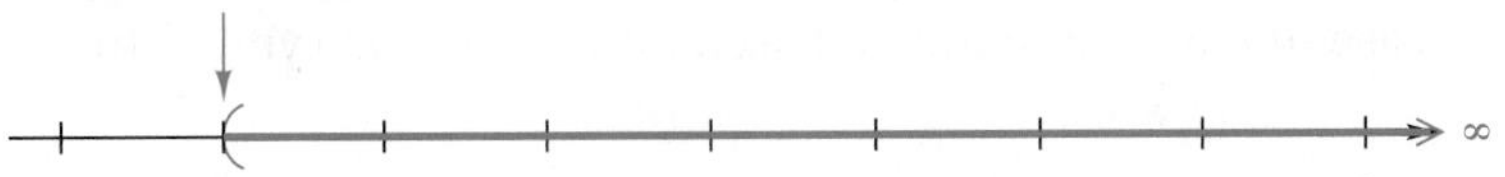

Figure 16 Graph of the interval $(-5, \infty)$

◀ **Work Problem 1 at the Side.**

Keep the following important concepts regarding interval notation in mind.

1. A parenthesis indicates that an endpoint is *not included* in a solution set.
2. A bracket indicates that an endpoint is *included* in a solution set.
3. A parenthesis is *always* used next to an infinity symbol, $-\infty$ or ∞.
4. The set of all real numbers is written in interval notation as $(-\infty, \infty)$.

EXAMPLE 2 Graphing an Interval on a Number Line

Graph $3 > x$.

The statement $3 > x$ means the same as $x < 3$. ***The inequality symbol continues to point to the lesser value.*** The graph of $x < 3$, written in interval notation as $(-\infty, 3)$, is shown in **Figure 17.**

Figure 17 Graph of the interval $(-\infty, 3)$

◀ **Work Problem 2 at the Side.**

The table summarizes solution sets of linear inequalities.

Set-Builder Notation	Interval Notation	Graph
$\{x \mid x < a\}$	$(-\infty, a)$	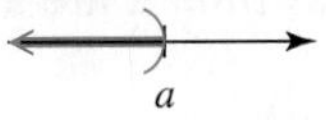
$\{x \mid x \leq a\}$	$(-\infty, a]$	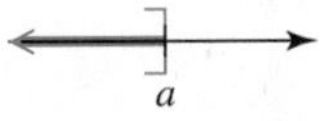
$\{x \mid x > a\}$	(a, ∞)	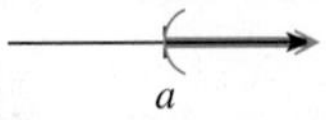
$\{x \mid x \geq a\}$	$[a, \infty)$	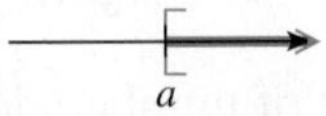
$\{x \mid x \text{ is a real number}\}$	$(-\infty, \infty)$	

OBJECTIVE 2 Use the addition property of inequality. Consider the true inequality $2 < 5$. If 4 is added to each side, the result is

$$2 + 4 < 5 + 4 \quad \text{Add 4.}$$

$$6 < 9, \quad \text{True}$$

also a true sentence. This suggests the **addition property of inequality.**

Addition Property of Inequality

For any real numbers A, B, and C, the inequalities

$$A < B \quad \text{and} \quad A + C < B + C$$

have exactly the same solutions.

In words, the same number may be added to each side of an inequality without changing the solutions.

As with the addition property of equality, the same number may be subtracted from each side of an inequality.

EXAMPLE 3 **Using the Addition Property of Inequality**

Solve $7 + 3k \geq 2k - 5$, and graph the solution set.

$$7 + 3k \geq 2k - 5$$

$$7 + 3k - 2k \geq 2k - 5 - 2k \qquad \text{Subtract } 2k.$$

$$7 + k \geq -5 \qquad \text{Combine like terms.}$$

$$7 + k - 7 \geq -5 - 7 \qquad \text{Subtract 7.}$$

$$k \geq -12 \qquad \text{Combine like terms.}$$

The solution set is $[-12, \infty)$. Its graph is shown in **Figure 18.**

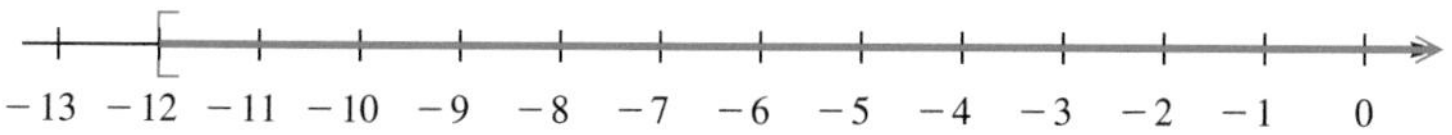

Figure 18

Work Problem 3 at the Side. ▶

Note

Because an inequality has many solutions, we cannot check all of them by substitution as we did with the single solution of an equation. To check the solutions in the **Example 3** solution set $[-12, \infty)$, we use a multistep process. First, we substitute -12 for k in the related *equation*.

CHECK

$$7 + 3k = 2k - 5 \qquad \text{Related equation}$$

$$7 + 3(-12) \stackrel{?}{=} 2(-12) - 5 \qquad \text{Let } k = -12.$$

$$7 - 36 \stackrel{?}{=} -24 - 5 \qquad \text{Multiply.}$$

$$-29 = -29 \ \checkmark \qquad \text{True}$$

A true statement results, so -12 is indeed the "boundary" point. Now we test a number other than -12 from the interval $[-12, \infty)$. We choose 0 since it is easy to substitute.

CHECK

$$7 + 3k \geq 2k - 5 \qquad \text{Original inequality}$$

$$7 + 3(0) \stackrel{?}{\geq} 2(0) - 5 \qquad \text{Let } k = 0.$$

$$7 \geq -5 \ \checkmark \qquad \text{True}$$

0 is easy to substitute.

Again, a true statement results, so the checks confirm that solutions to the inequality are in the interval $[-12, \infty)$. Any number "outside" the interval $[-12, \infty)$, that is, any number in $(-\infty, -12)$, will give a false statement when tested. (Try this.)

3 Solve each inequality, and graph the solution set.

(a) $-1 + 8r < 7r + 2$

(b) $5 + 5x \geq 4x + 3$

Answers

3. (a) $(-\infty, 3)$

(b) $[-2, \infty)$

4 Work each of the following.

(a) Multiply each side of

$$-3 < 7$$

by 2 and then by -5. Reverse the direction of the inequality symbol if necessary to make a true statement.

(b) Multiply each side of

$$3 > -7$$

by 2 and then by -5. Reverse the direction of the inequality symbol if necessary to make a true statement.

(c) Multiply each side of

$$-7 < -3$$

by 2 and then by -5. Reverse the direction of the inequality symbol if necessary to make a true statement.

Answers

4. (a) $-6 < 14$; $15 > -35$
(b) $6 > -14$; $-15 < 35$
(c) $-14 < -6$; $35 > 15$

OBJECTIVE 3 Use the multiplication property of inequality. Consider the true inequality $3 < 7$. Multiply each side by the positive number 2.

$$3 < 7$$

$$2(3) < 2(7) \quad \text{Multiply by 2.}$$

$$6 < 14 \quad \text{True}$$

Now multiply each side of $3 < 7$ by the negative number -5.

$$3 < 7$$

$$-5(3) < -5(7) \quad \text{Multiply by } -5.$$

$$-15 < -35 \quad \text{False}$$

To get a true statement when multiplying each side by -5, ***we must reverse the direction of the inequality symbol.***

$$3 < 7$$

$$-5(3) > -5(7) \quad \text{Multiply by } -5. \text{ Reverse the direction of the symbol.}$$

$$-15 > -35 \quad \text{True}$$

◀ **Work Problem 4 at the Side.**

These examples suggest the **multiplication property of inequality.**

Multiplication Property of Inequality

Let A, B, and C (where $C \neq 0$) be real numbers.

1. If C is ***positive,*** then the inequalities

$$A < B \quad \text{and} \quad AC < BC$$

have exactly the same solutions.

2. If C is ***negative,*** then the inequalities

$$A < B \quad \text{and} \quad AC > BC$$

have exactly the same solutions.

In words, each side of an inequality may be multiplied by the same positive number without changing the solutions. ***If the multiplier is negative, we must reverse the direction of the inequality symbol.***

As with the multiplication property of equality, the same nonzero number may be divided into each side.

Note the following differences for positive and negative numbers.

1. When each side of an inequality is multiplied or divided by a ***positive number,*** the direction of the inequality symbol *does not change.*
2. When each side of an inequality is multiplied or divided by a ***negative number, reverse the direction of the inequality symbol.***

EXAMPLE 4 Using the Multiplication Property of Inequality

Solve $3r < -18$, and graph the solution set.

We divide each side by 3, a positive number, so the direction of the inequality symbol *does not* change. ***(It does not matter that the number on the right side of the inequality is negative.)***

$$3r < -18$$

$$\frac{3r}{3} < \frac{-18}{3} \quad \text{Divide by 3.}$$

3 is *positive*. Do NOT reverse the direction of the symbol.

$$r < -6$$

The graph of the solution set, $(-\infty, -6)$, is shown in **Figure 19.**

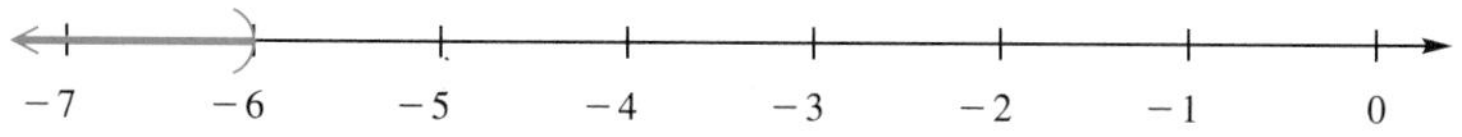

Figure 19

EXAMPLE 5 Using the Multiplication Property of Inequality

Solve $-4t \geq 8$, and graph the solution set.

Here each side of the inequality must be divided by -4, a negative number, which *does* require changing the direction of the inequality symbol.

$$-4t \geq 8$$

$$\frac{-4t}{-4} \leq \frac{8}{-4} \quad \text{Divide by } -4. \text{ Reverse the symbol.}$$

-4 is *negative*. Change $\geq$ to $\leq$.

$$t \leq -2$$

The solution set, $(-\infty, -2]$, is graphed in **Figure 20.**

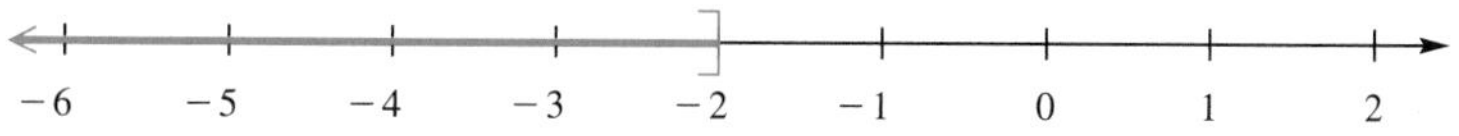

Figure 20

Work Problem 5 at the Side. ▶

OBJECTIVE 4 Solve linear inequalities by using both properties of inequality. The steps to solve a linear inequality are summarized below.

Solving a Linear Inequality

Step 1 **Simplify each side separately.** Use the distributive property to clear parentheses and combine like terms on each side as needed.

Step 2 **Isolate the variable term on one side.** Use the addition property of inequality to get all terms with variables on one side of the inequality and all numbers on the other side.

Step 3 **Isolate the variable.** Use the multiplication property of inequality to write the inequality as $x <$ a number or $x >$ a number.

Remember: Reverse the direction of the inequality symbol only when multiplying or dividing each side of an inequality by a negative number.

5 Solve each inequality. Graph the solution set.

GS **(a)** $9x < -18$

$$\frac{9x}{9} \;(<\,/\,>)\; \frac{-18}{\_\_\_\_}$$

$x \;(<\,/\,>)$ ____

The solution set is ____.

GS **(b)** $-2r > -12$

$$\frac{-2r}{\_\_\_\_} \;(<\,/\,>)\; \frac{-12}{\_\_\_\_}$$

$r \;(<\,/\,>)$ ____

The solution set is ____.

(c) $-5p \leq 0$

Answers

5. **(a)** $<$; 9; $<$; -2; $(-\infty, -2)$

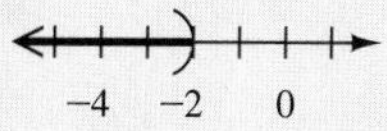

(b) -2; $<$; -2; $<$; 6; $(-\infty, 6)$

(c) $[0, \infty)$

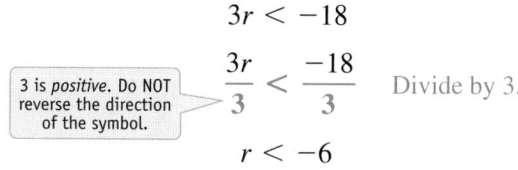

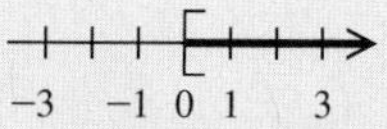

6 Solve.

$7x - 6 + 1 \geq 5x - x + 2$

Graph the solution set.

EXAMPLE 6 Solving a Linear Inequality

Solve $3x + 2 - 5 > -x + 7 + 2x$. Graph the solution set.

Step 1 Combine like terms and simplify.

$$3x + 2 - 5 > -x + 7 + 2x$$
$$3x - 3 > x + 7$$

Step 2 Use the addition property of inequality.

$$3x - 3 - x > x + 7 - x \quad \text{Subtract } x.$$
$$2x - 3 > 7 \quad \text{Combine like terms.}$$
$$2x - 3 + 3 > 7 + 3 \quad \text{Add 3.}$$
$$2x > 10 \quad \text{Combine like terms.}$$

Step 3 Use the multiplication property of inequality.

$$\frac{2x}{2} > \frac{10}{2} \quad \text{Divide by 2.}$$

Because 2 is positive, keep the symbol >.

$$x > 5$$

The solution set is $(5, \infty)$. Its graph is shown in **Figure 21.**

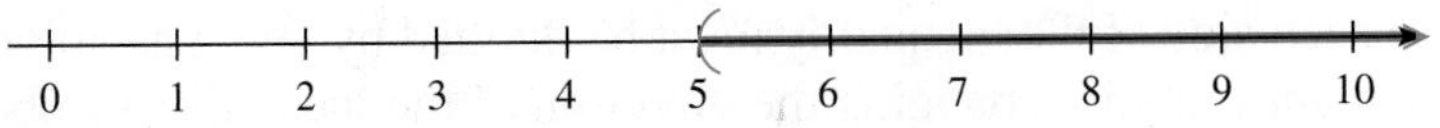

Figure 21

◀ **Work Problem 6 at the Side.**

7 Solve.

$-15 - (2x + 1) \geq 4(x - 1) - 3x$

Graph the solution set.

EXAMPLE 7 Solving a Linear Inequality

Solve $5(k - 3) - 7k \geq 4(k - 3) + 9$. Graph the solution set.

Step 1 $5(k - 3) - 7k \geq 4(k - 3) + 9$ — Start by clearing parentheses.

$$5k - 15 - 7k \geq 4k - 12 + 9 \quad \text{Distributive property}$$
$$-2k - 15 \geq 4k - 3 \quad \text{Combine like terms.}$$

Step 2

$$-2k - 15 - 4k \geq 4k - 3 - 4k \quad \text{Subtract } 4k.$$
$$-6k - 15 \geq -3 \quad \text{Combine like terms.}$$
$$-6k - 15 + 15 \geq -3 + 15 \quad \text{Add 15.}$$
$$-6k \geq 12 \quad \text{Combine like terms.}$$

Step 3

$$\frac{-6k}{-6} \leq \frac{12}{-6} \quad \text{Divide by } -6. \text{ Reverse the symbol.}$$

Because −6 is negative, change ≥ to ≤.

$$k \leq -2$$

The solution set is $(-\infty, -2]$. Its graph is shown in **Figure 22.**

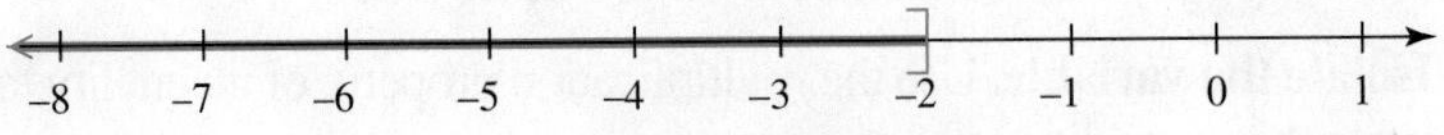

Figure 22

◀ **Work Problem 7 at the Side.**

Answers

6. $\left[\frac{7}{3}, \infty\right)$

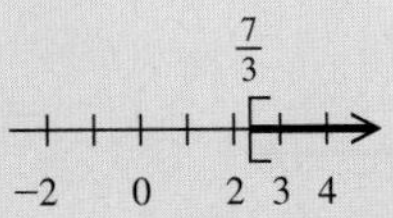

7. $(-\infty, -4]$

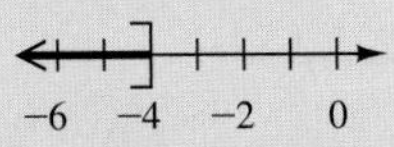

OBJECTIVE 5 Use inequalities to solve applied problems. The table gives some common phrases that suggest inequality.

Phrase/Word	Example	Inequality
Is greater than	A number *is greater than* 4	$x > 4$
Is less than	A number *is less than* -12	$x < -12$
Exceeds	A number *exceeds* 3.5	$x > 3.5$
Is at least	A number *is at least* 6	$x \geq 6$
Is at most	A number *is at most* 8	$x \leq 8$

Work Problem 8 at the Side. ▶

The next example uses the idea of finding the average of a number of scores. ***In general, to find the average of n numbers, add the numbers and divide by n.*** We use the six problem-solving steps from **Section 10.4,** changing Step 3 to "Write an inequality."

EXAMPLE 8 Finding an Average Test Score

Brent has scores of 86, 88, and 78 (out of a possible 100) on each of his first three tests in geometry. If he wants an average of at least 80 after his fourth test, what are the possible scores he can make on that test?

Step 1 **Read** the problem again.

Step 2 **Assign a variable.**

Let x = Brent's score on his fourth test.

Step 3 **Write an inequality.**

Average ↓ is at least ↓ 80. ↓

$$\frac{86 + 88 + 78 + x}{4} \geq 80$$

To find his average after four tests, add the test scores and divide by 4.

Step 4 **Solve.**

$$\frac{252 + x}{4} \geq 80 \quad \text{Add the known scores.}$$

$$4\left(\frac{252 + x}{4}\right) \geq 4(80) \quad \text{Multiply by 4.}$$

$$252 + x \geq 320$$

$$252 + x - 252 \geq 320 - 252 \quad \text{Subtract 252.}$$

$$x \geq 68 \quad \text{Combine like terms.}$$

Step 5 **State the answer.** He must score 68 or more on the fourth test to have an average of *at least* 80.

Step 6 **Check.**

$$\frac{86 + 88 + 78 + 68}{4} = \frac{320}{4} = 80$$

To complete the check, also show that any number greater than 68 (but less than or equal to 100) makes the average greater than 80.

Work Problem 9 at the Side. ▶

8 Translate each statement into an inequality, using x as the variable.

(a) The total cost is less than \$10.

(b) Chicago received at most 5 in. of snow.

(c) The car's speed exceeded 60 mph.

(d) You must be at least 18 yr old to vote.

9 Solve each problem.

(a) Kristine has grades of 98 and 85 on her first two tests in algebra. If she wants an average of at least 92 after her third test, what score must she make on that test?

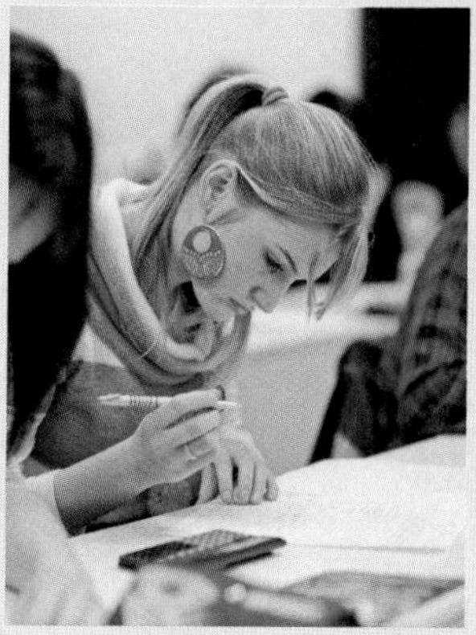

(b) Maggie has scores of 98, 86, and 88 on her first three tests in algebra. If she wants an average of at least 90 after her fourth test, what score must she make on her fourth test?

Answers

8. **(a)** $x < 10$ **(b)** $x \leq 5$
(c) $x > 60$ **(d)** $x \geq 18$
9. **(a)** 93 or more **(b)** 88 or more

OBJECTIVE 6 Solve linear inequalities with three parts. An inequality that says that one number is *between* two other numbers is a **three-part inequality.** For example,

$-3 < 5 < 7$ says that 5 is between −3 and 7.

EXAMPLE 9 Graphing a Three-Part Inequality

Write the inequality $-3 \le x < 2$ in interval notation, and graph the interval.

The statement is read "−3 is less than or equal to *x and x* is less than 2." We want the set of numbers that are *between* −3 and 2, with −3 included and 2 excluded. In interval notation, we write $[-3, 2)$, using a square bracket at −3 because −3 is part of the graph and a parenthesis at 2 because 2 is not part of the graph. See **Figure 23.**

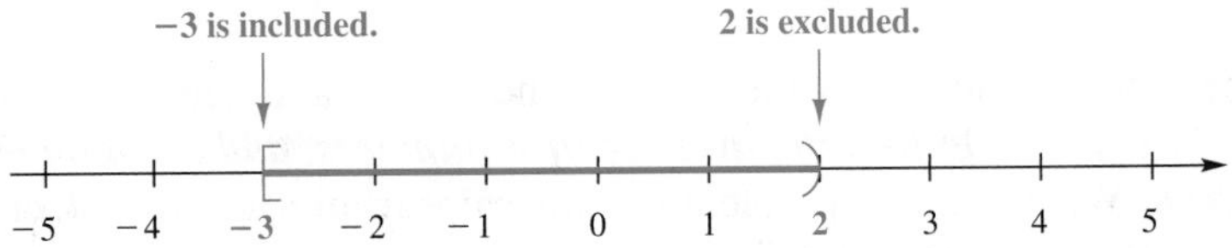

Figure 23 Graph of the interval $[-3, 2)$

◀ **Work Problem 10 at the Side.**

The three-part inequality $3 < x + 2 < 8$ says that $x + 2$ is between 3 and 8. We solve this inequality as follows.

$$3 - 2 < x + 2 - 2 < 8 - 2 \quad \text{Subtract 2 from } each \text{ part.}$$
$$1 < x < 6$$

The idea is to get the inequality in the form

a number < *x* < another number.

CAUTION

Three-part inequalities are written so that the symbols point in the same direction and both point toward the lesser number. It would be *wrong* to write an inequality as $8 < x + 2 < 3$, since this would imply that $8 < 3$, a false statement.

EXAMPLE 10 Solving a Three-Part Inequality

Solve each inequality, and graph the solution set.

(a)
$$4 \le 3x - 5 < 10$$
$$4 + 5 \le 3x - 5 + 5 < 10 + 5 \quad \text{Add 5 to each part.}$$
$$9 \le 3x < 15$$
$$\frac{9}{3} \le \frac{3x}{3} < \frac{15}{3} \quad \text{Divide each part by 3.}$$
$$3 \le x < 5$$

Remember to divide all *three* parts by 3.

The solution set is $[3, 5)$. Its graph is shown in **Figure 24.**

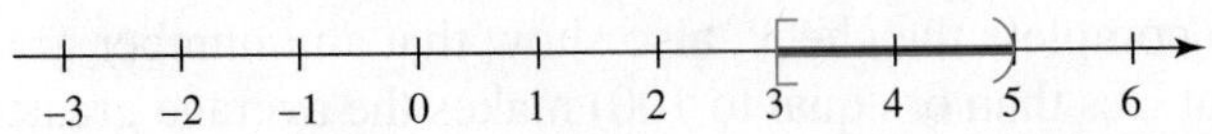

Figure 24

Continued on Next Page

10 Write each inequality in interval notation and graph the interval.

(a) $-7 < x < -2$

(b) $-6 < x \le -4$

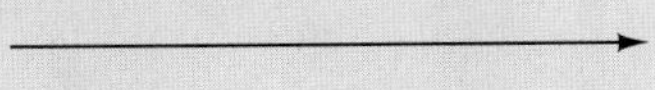

Answers

10. (a) $(-7, -2)$

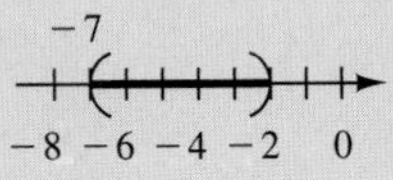

(b) $(-6, -4]$

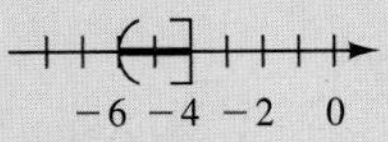

(b)

$$-4 \leq \frac{2}{3}m - 1 < 8$$ Work with all three parts at the same time.

$$3(-4) \leq 3\left(\frac{2}{3}m - 1\right) < 3(8)$$ Multiply each part by 3 to clear the fraction.

$$-12 \leq 2m - 3 < 24$$ Multiply. Use the distributive property.

$$-12 + 3 \leq 2m - 3 + 3 < 24 + 3$$ Add 3 to each part.

$$-9 \leq 2m < 27$$

$$\frac{-9}{2} \leq \frac{2m}{2} < \frac{27}{2}$$ Divide each part by 2.

$$-\frac{9}{2} \leq m < \frac{27}{2}$$

The solution set is $\left[-\frac{9}{2}, \frac{27}{2}\right)$. Its graph is shown in **Figure 25.**

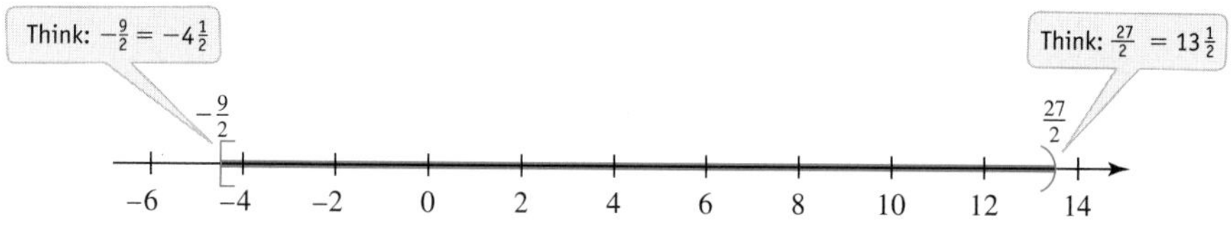

Figure 25

Work Problem 11 at the Side. ▶

11 Solve each inequality, and graph the solution set.

(a) $2 \leq 3x - 1 \leq 8$

(b) $-4 \leq \frac{3}{2}x - 1 \leq 0$

Note

The inequality in **Example 10(b),** $-4 \leq \frac{2}{3}m - 1 < 8$, can also be solved by first adding 1 to each part and then multiplying each part by $\frac{3}{2}$. Try this.

The table summarizes solution sets of three-part inequalities.

Set-Builder Notation	Interval Notation	Graph
$\{x \mid a < x < b\}$	(a, b)	
$\{x \mid a < x \leq b\}$	$(a, b]$	
$\{x \mid a \leq x < b\}$	$[a, b)$	
$\{x \mid a \leq x \leq b\}$	$[a, b]$	

Answers

11. (a) $[1, 3]$

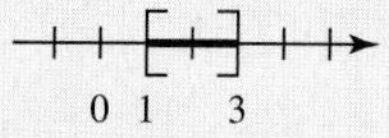

(b) $\left[-2, \frac{2}{3}\right]$

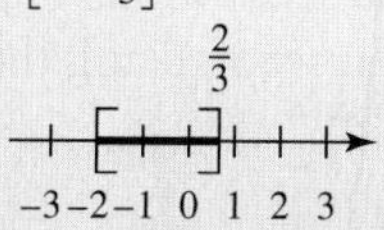

10.6 Exercises

FOR EXTRA HELP

Download the MyDashBoard App

MyMathLab®

CONCEPT CHECK *Work each problem.*

1. When graphing an inequality, use a parenthesis if the inequality symbol is ________ or ________. Use a square bracket if the inequality symbol is ________ or ________.

2. *True* or *false*? In interval notation, a square bracket is sometimes used next to an infinity symbol.

3. In interval notation, the set $\{x \mid x > 0\}$ is written ______.

4. In interval notation, the set of all real numbers is written ______.

CONCEPT CHECK *Write an inequality using the variable x that corresponds to each graph of solutions on a number line.*

5. 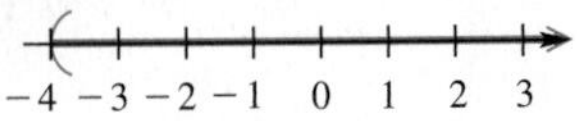

6. 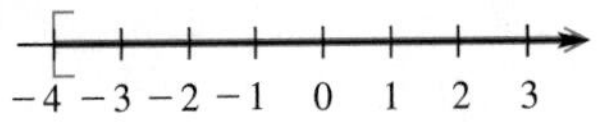

7. 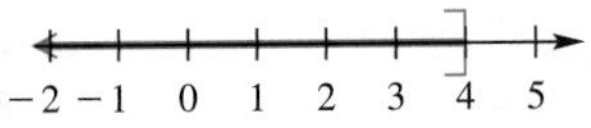

8. 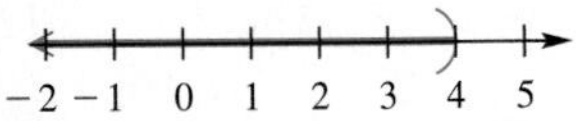

9. 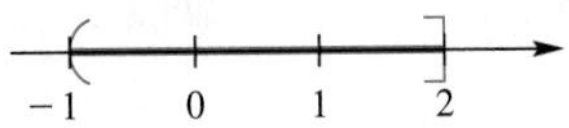

10. 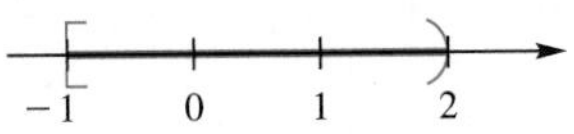

Write each inequality in interval notation, and graph the interval. **See Examples 1, 2, and 9.**

11. $k \leq 4$

12. $r \leq -10$

13. $x > -3$

14. $x > 3$

15. $8 \leq x \leq 10$

16. $3 \leq x \leq 5$

17. $0 < x \leq 10$

18. $-3 \leq x < 5$

Solve each inequality. Write the solution set in interval notation, and graph it. **See Example 3.**

19. $z - 8 \geq -7$

20. $p - 3 \geq -11$

21. $2k + 3 \geq k + 8$

22. $3x + 7 \geq 2x + 11$

23. $3n + 5 < 2n - 1$

24. $5x - 2 < 4x - 5$

25. Under what conditions must the inequality symbol be reversed when using the multiplication property of inequality?

26. Explain the steps you would use to solve the inequality $-5x > 20$.

Solve each inequality. Write the solution set in interval notation, and graph it.
See Examples 4 and 5.

27. $3x < 18$

28. $5x < 35$

29. $2x \geq -20$

30. $6m \geq -24$

31. $-8t > 24$

32. $-7x > 49$

33. $-x \geq 0$

34. $-k < 0$

35. $-\frac{3}{4}r < -15$

36. $-\frac{7}{8}t < -14$

37. $-0.02x \leq 0.06$

38. $-0.03v \geq -0.12$

Solve each inequality. Write the inequality in interval notation, and graph it.
See Examples 3–7.

39. $8x + 9 \leq -15$

40. $6x + 7 \leq -17$

41. $-4x - 3 < 1$

42. $-5x - 4 < 6$

43. $5r + 1 \geq 3r - 9$

44. $6t + 3 < 3t + 12$

45. $6x + 3 + x < 2 + 4x + 4$

46. $-4w + 12 + 9w \geq w + 9 + w$

47. $-x + 4 + 7x \leq -2 + 3x + 6$

48. $14x - 6 + 7x > 4 + 10x - 10$

49. $5(t - 1) > 3(t - 2)$

50. $7(m - 2) < 4(m - 4)$

51. $5(x + 3) - 6x \leq 3(2x + 1) - 4x$

52. $2(x - 5) + 3x < 4(x - 6) + 1$

53. $\frac{2}{3}(p + 3) > \frac{5}{6}(p - 4)$

54. $\frac{7}{9}(n - 4) \leq \frac{4}{3}(n + 5)$

55. $4x - (6x + 1) \leq 8x + 2(x - 3)$

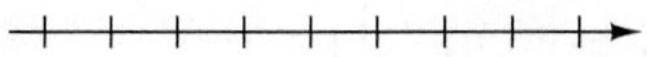

56. $2x - (4x + 3) < 6x + 3(x + 4)$

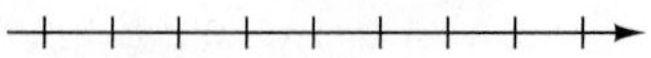

57. $5(2k + 3) - 2(k - 8) > 3(2k + 4) + k - 2$

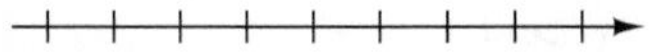

58. $2(3z - 5) + 4(z + 6) \geq 2(3z + 2) + 3z - 15$

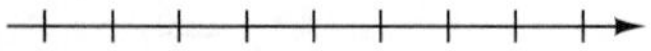

CONCEPT CHECK *Translate each statement into an inequality. Use x as the variable.*

59. You must be at least 16 yr old to drive.

60. Less than 1 in. of rain fell.

61. Denver received more than 8 in. of snow.

62. A full-time student must take at least 12 credits.

63. Tracy could spend at most $20 on a gift.

64. The wind speed exceeded 40 mph.

Solve each problem. ***See Example 8.***

65. John Douglas has grades of 84 and 98 on his first two history tests. What must he score on his third test so that his average is at least 90?

66. Elizabeth Gainey has scores of 74 and 82 on her first two algebra tests. What must she score on her third test so that her average is at least 80?

67. A student has scores of 87, 84, 95, and 79 on four quizzes. What must she score on the fifth quiz to have an average of at least 85?

68. Another student has scores of 82, 93, 94, and 86 on four quizzes. What must he score on the fifth quiz to have an average of at least 90?

69. When 2 is added to the difference between six times a number and 5, the result is greater than 13 added to 5 times the number. Find all such numbers.

70. When 8 is subtracted from the sum of three times a number and 6, the result is less than 4 more than the number. Find all such numbers.

71. The formula for converting Celsius temperature to Fahrenheit is $F = \frac{9}{5}C + 32$. The Fahrenheit temperature of Providence, Rhode Island, has never exceeded 104°. How would you describe this using Celsius temperature?

72. The formula for converting Fahrenheit temperature to Celsius is $C = \frac{5}{9}(F - 32)$. If the Celsius temperature on a certain day in San Diego, California, is never more than 25°, how would you describe the corresponding Fahrenheit temperature?

73. For what values of x would the rectangle have perimeter of at least 400?

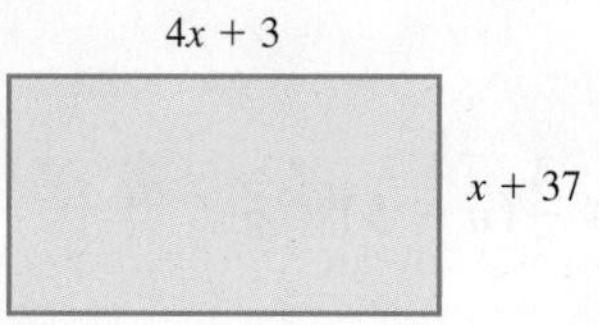

74. For what values of x would the triangle have perimeter of at least 72?

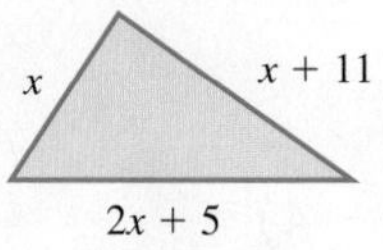

75. An international phone call costs \$2.00, plus \$0.30 per minute or fractional part of a minute. If x represents the number of minutes of the length of the call, then $2 + 0.30x$ represents the cost of the call. If Jorge has \$5.60 to spend on a call, what is the maximum total time he can use the phone?

76. At the Speedy Gas 'n Go, a car wash costs \$4.50, and gasoline is selling for \$3.40 per gal. Terri Hoelker has \$43.60 to spend, and her car is so dirty that she must have it washed. What is the maximum number of gallons of gasoline that she can purchase?

Solve each inequality. Write the solution set in interval notation, and graph it. ***See Example 10.***

77. $-5 \le 2x - 3 \le 9$

78. $-7 \le 3x - 4 \le 8$

79. $10 < 7p + 3 < 24$

80. $-8 \le 3r - 1 \le -1$

81. $-12 < -1 + 6m \le -5$

82. $-14 \le 1 + 5q < 3$

83. $-12 \le \frac{1}{2}z + 1 \le 4$

84. $-6 \le 3 + \frac{1}{3}x \le 5$

85. $1 \le \frac{2}{3}p + 3 \le 7$

86. $2 < \frac{3}{4}x + 6 < 12$

Relating Concepts (Exercises 87–90) For Individual or Group Work

Work Exercises 87–90 in order, *to see the connection between the solution of an equation and the solutions of the corresponding inequalities.*

In Exercises 87 and 88, solve and graph the solutions.

87. $3x + 2 = 14$

88. **(a)** $3x + 2 < 14$

(b) $3x + 2 > 14$

89. Now graph all the solutions together on the following number line.

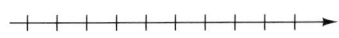

Describe the graph.

90. Based on the results from **Exercises 87–89,** if we were to graph the solutions of

$$-4x + 3 = -1, \quad -4x + 3 > -1, \quad \text{and} \quad -4x + 3 < -1$$

on the same number line, describe the graph. Give this solution set using interval notation.

Chapter 10 Summary

Key Terms

10.1

linear equation A linear equation in one variable is an equation that can be written in the form $Ax + B = C$, where A, B, and C are real numbers, with $A \neq 0$.

solution set The set of all solutions of an equation is its solution set.

equivalent equations Equations that have exactly the same solution sets are equivalent equations.

10.3

conditional equation A conditional equation is an equation that is true for some values of the variable and false for others.

identity An identity is an equation that is true for all values of the variable.

contradiction A contradiction is an equation with no solution.

10.4

consecutive integers Two integers that differ by 1 are consecutive integers.

consecutive even (or odd) integers Two even (or odd) integers that differ by 2 are consecutive even (or odd) integers.

complementary angles Two angles whose measures have a sum of 90° are complementary angles.

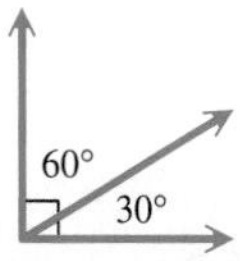

right angle A right angle measures 90°.

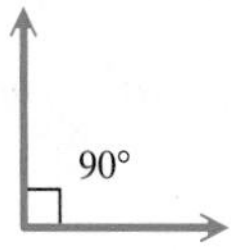

supplementary angles Two angles whose measures have a sum of 180° are supplementary angles.

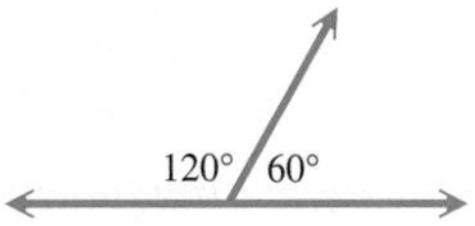

straight angle A straight angle measures 180°.

10.5

formula A formula is an equation in which variables are used to describe a relationship.

area The area of a plane geometric figure is a measure of the surface covered by the figure.

perimeter The perimeter of a plane geometric figure is the distance around the figure.

vertical angles Vertical angles are angles formed by intersecting lines. They have the same measure. (In the figure, ① and ③ are vertical angles, as are ② and ④.)

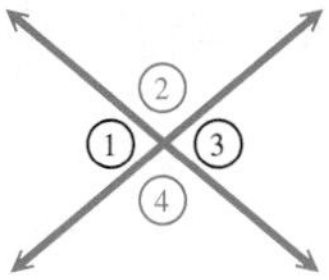

10.6

inequality Inequalities are statements relating algebraic expressions using $<$, $\leq$, $>$, or $\geq$.

linear inequality A linear inequality in one variable can be written in the form $Ax + B < C$, $Ax + B \leq C$, $Ax + B > C$, or $Ax + B \geq C$, where A, B, and C are real numbers, with $A \neq 0$.

interval An interval is a portion of a number line.

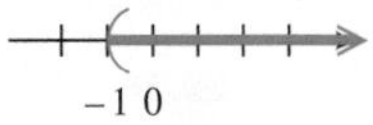

The interval $(-1, \infty)$

interval notation Interval notation is a special notation that uses parentheses ( ) and/or brackets [ ] to describe an interval on a number line.

three-part inequality An inequality that says that one number is between two other numbers is a three-part inequality.

New Symbols

$\emptyset$	empty (null) set	(a, b)	interval notation for $a < x < b$	∞	infinity
1°	one degree	$[a, b]$	interval notation for $a \leq x \leq b$	$-\infty$	negative infinity
⅂	right angle			$(-\infty, \infty)$	set of all real numbers

Test Your Word Power

See how well you have learned the vocabulary in this chapter.

1. A **solution** of an equation is a number that
   - **A.** makes an expression undefined
   - **B.** makes the equation false
   - **C.** makes the equation true
   - **D.** makes an expression equal to 0.

2. **Complementary angles** are angles
   - **A.** formed by two parallel lines
   - **B.** whose sum is 90°
   - **C.** whose sum is 180°
   - **D.** formed by perpendicular lines.

3. **Supplementary angles** are angles
   - **A.** formed by two parallel lines
   - **B.** whose sum is 90°
   - **C.** whose sum is 180°
   - **D.** formed by perpendicular lines.

Answers to Test Your Word Power

1. C; *Example:* 8 is the solution of $2x + 5 = 21$.

2. B; *Example:* Angles with measures 35° and 55° are complementary angles.

3. C; *Example:* Angles with measures 112° and 68° are supplementary angles.

Quick Review

Concepts	*Examples*
10.1 The Addition Property of Equality The same number may be added to (or subtracted from) each side of an equation without changing the solution set.	Solve. $x - 6 = 12$ $x - 6 + 6 = 12 + 6$ Add 6. $x = 18$ Combine like terms. Solution set: $\{18\}$
10.2 The Multiplication Property of Equality Each side of an equation may be multiplied (or divided) by the same nonzero number without changing the solution set.	Solve. $\frac{3}{4}x = -9$ $\frac{4}{3} \cdot \frac{3}{4}x = \frac{4}{3}(-9)$ Multiply by $\frac{4}{3}$. $x = -12$ Solution set: $\{-12\}$

Concepts	Examples
10.3 More on Solving Linear Equations	
Step 1 **Simplify each side separately.**	Solve. $2x + 2(x + 1) = 14 + x$ $2x + 2x + 2 = 14 + x$ Distributive property $4x + 2 = 14 + x$ Combine like terms.
Step 2 **Isolate the variable term on one side.**	$4x + 2 - x - 2 = 14 + x - x - 2$ Subtract x. Subtract 2. $3x = 12$ Combine like terms.
Step 3 **Isolate the variable.**	$\frac{3x}{3} = \frac{12}{3}$ Divide by 3. $x = 4$
Step 4 **Check.**	CHECK $2(4) + 2(4 + 1) \stackrel{?}{=} 14 + 4$ Let $x = 4$. $18 = 18$ ✓ True Solution set: $\{4\}$
10.4 An Introduction to Applications of Linear Equations	
Step 1 **Read.**	One number is 5 more than another. Their sum is 21. What are the numbers? We are looking for two numbers.
Step 2 **Assign a variable.**	Let $x =$ the lesser number. Then $x + 5 =$ the greater number.
Step 3 **Write an equation.**	$x + (x + 5) = 21$
Step 4 **Solve the equation.**	$2x + 5 = 21$ Combine like terms. $2x = 16$ Subtract 5. $x = 8$ Divide by 2.
Step 5 **State the answer.**	The numbers are 8 and 13.
Step 6 **Check.**	13 is 5 more than 8, and $8 + 13 = 21$. The answer checks.
10.5 Formulas and Additional Applications from Geometry	
To find the value of one of the variables in a formula, given values for the others, substitute the known values into the formula.	Find L if $A = LW$, given that $A = 24$ and $W = 3$. $24 = L \cdot 3$ $A = 24, W = 3$ $\frac{24}{3} = \frac{L \cdot 3}{3}$ Divide by 3. $8 = L$
To solve a formula for one of the variables, isolate that variable by treating the other variables as numbers and using the steps for solving equations.	Solve $P = 2a + 2b$ for b. $P - 2a = 2a + 2b - 2a$ Subtract $2a$. $P - 2a = 2b$ Combine like terms. $\frac{P - 2a}{2} = \frac{2b}{2}$ Divide by 2. $\frac{P - 2a}{2} = b$, or $b = \frac{P - 2a}{2}$

Concepts

10.6 Solving Linear Inequalities

Step 1 **Simplify each side separately.**

Step 2 **Isolate the variable term on one side.**

Step 3 **Isolate the variable.**

Be sure to reverse the direction of the inequality symbol when multiplying or dividing by a negative number.

To solve a three-part inequality such as

$$4 < 2x + 6 < 8,$$

work with all three parts at the same time.

Examples

Solve and graph the solution set.

$$\begin{aligned} 3(1-x)+5-2x &> 9-6 & & \\ 3-3x+5-2x &> 9-6 & & \text{Distributive property} \\ 8-5x &> 3 & & \text{Combine like terms.} \\ 8-5x-8 &> 3-8 & & \text{Subtract 8.} \\ -5x &> -5 & & \text{Combine like terms.} \\ \frac{-5x}{-5} &< \frac{-5}{-5} & & \text{Divide by } -5. \text{ Change } > \text{ to } <. \\ x &< 1 & & \end{aligned}$$

Solution set: $(-\infty, 1)$

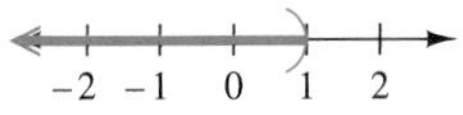

Solve. $4 < 2x + 6 < 8$

$$\begin{aligned} 4-6 < 2x+6-6 &< 8-6 & & \text{Subtract 6.} \\ -2 < 2x &< 2 & & \\ \frac{-2}{2} < \frac{2x}{2} &< \frac{2}{2} & & \text{Divide by 2.} \\ -1 < x &< 1 & & \end{aligned}$$

Solution set: $(-1, 1)$

Chapter 10 Review Exercises

10.1–10.3 *Solve each equation. Check the solution.*

1. $x - 7 = 2$

2. $4r - 6 = 10$

3. $5x + 8 = 4x + 2$

4. $8t = 7t + \frac{3}{2}$

5. $(4r - 8) - (3r + 12) = 0$

6. $7(2x + 1) = 6(2x - 9)$

7. $-\frac{6}{5}y = -18$

8. $\frac{1}{2}r - \frac{1}{6}r + 3 = 2 + \frac{1}{6}r + 1$

9. $3x - (-2x + 6) = 4(x - 4) + x$

10. $0.10(x + 80) + 0.20x = 8 + 0.30x$

10.4 *Solve each problem.*

11. If 7 is added to five times a number, the result is equal to three times the number. Find the number.

12. If 4 is subtracted from twice a number, the result is 36. Find the number.

13. The land area of Hawaii is 5213 mi^2 greater than that of Rhode Island. Together, the areas total 7637 mi^2. What is the area of each state?

14. The height of Seven Falls in Colorado is $\frac{5}{2}$ the height (in feet) of Twin Falls in Idaho. The sum of the heights is 420 ft. Find the height of each.

15. The supplement of an angle measures 10 times the measure of its complement. What is the measure of the angle (in degrees)?

16. Find two consecutive odd integers such that when the lesser is added to twice the greater, the result is 24 more than the greater integer.

10.5 *A formula is given in each exercise, along with the values for all but one of the variables. Find the value of the variable that is not given. (For Exercises 19 and 20, use 3.14 as an approximation for π.)*

17. $A = \frac{1}{2}bh$; $A = 44$, $b = 8$

18. $A = \frac{1}{2}h(b + B)$; $b = 3$, $B = 4$, $h = 8$

19. $C = 2\pi r$; $C = 29.83$

20. $V = \frac{4}{3}\pi r^3$; $r = 9$

Solve each formula for the specified variable.

21. $A = bh$ for h

22. $A = \frac{1}{2}h(b + B)$ for h

Solve each equation for y.

23. $x + y = 11$

24. $3x - 2y = 12$

Find the measure of each marked angle.

25.

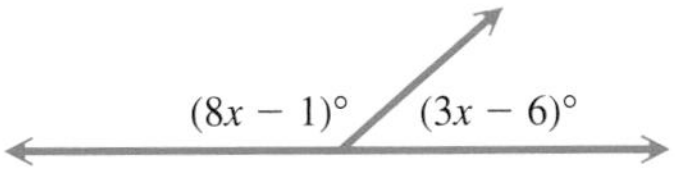

26.

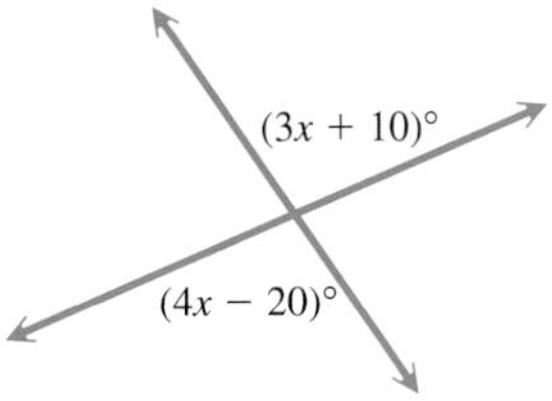

Solve each application of geometry.

27. A cinema screen in Indonesia has length 92.75 ft and width 70.5 ft. What is the perimeter? What is the area? (*Source: Guinness World Records.*)

28. A Montezuma cypress in Mexico is 137 ft tall and has a circumference of about 146.9 ft. What is the diameter of the tree? What is its radius? Use 3.14 as an approximation for π. Round answers to the nearest hundredth. (*Source: Guinness World Records.*)

10.6 *Write each inequality in interval notation, and graph it.*

29. $p \geq -4$

30. $x < 7$

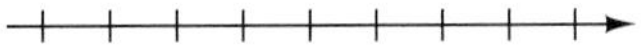

31. $-5 \leq k < 6$

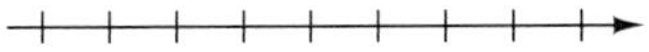

32. $r \geq \frac{1}{2}$

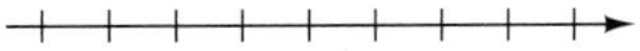

Solve each inequality. Write the solution set in interval notation, and graph it.

33. $x + 6 \geq 3$

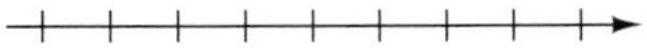

34. $5t < 4t + 2$

35. $-6x \leq -18$

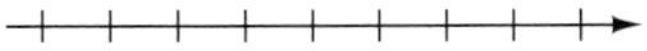

36. $8(k - 5) - (2 + 7k) \geq 4$

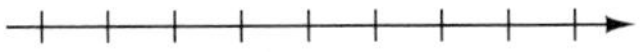

37. $4x - 3x > 10 - 4x + 7x$

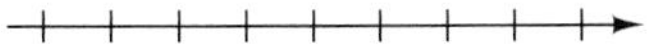

38. $3(2w + 5) + 4(8 + 3w) < 5(3w + 2) + 2w$

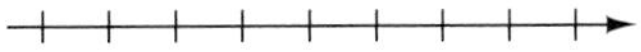

39. $-3 \leq 2x + 1 < 4$

40. $8 < 3x + 5 \leq 20$

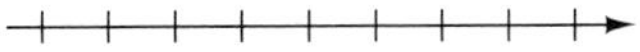

41. Justin Sudak has grades of 94 and 88 on his first two calculus tests. What possible scores on a third test will give him an average of at least 90?

42. If nine times a number is added to 6, the result is at most 3. Find all such numbers.

Mixed Review Exercises

Solve.

43. $\frac{x}{7} = \frac{x-5}{2}$

44. $d = 2r$ for r

45. $-2x > -4$

46. $2k - 5 = 4k + 13$

47. $0.05x + 0.02x = 4.9$

48. $2 - 3(t - 5) = 4 + t$

49. $9x - (7x + 2) = 3x + (2 - x)$

50. $\frac{1}{3}s + \frac{1}{2}s + 7 = \frac{5}{6}s + 5 + 2$

51. In 2010, the top-selling frozen pizza brands, DiGiorno and Red Baron, together had sales of \$937.5 million. Red Baron's sales were \$399.9 million less than DiGiorno's. What were sales in millions for each brand? (*Source:* www.aibonline.org)

52. Find the measure of each marked angle.

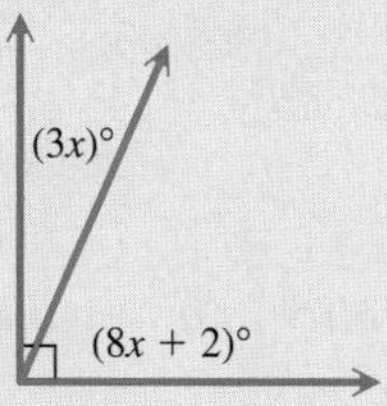

53. Of the 26 medals earned by Canada during the 2010 Winter Olympic games, there were two times as many gold as silver medals and 9 more gold than bronze medals. How many of each medal did Canada earn? (*Source: World Almanac and Book of Facts.*)

54. In triangle *DEF*, the measure of angle *E* is twice the measure of angle *D*. Angle *F* has measure 18° less than six times the measure of angle *D*. Find the measure of each angle.

55. The perimeter of a triangle is 96 m. One side is twice as long as another, and the third side is 30 m long. What is the length of the longest side?

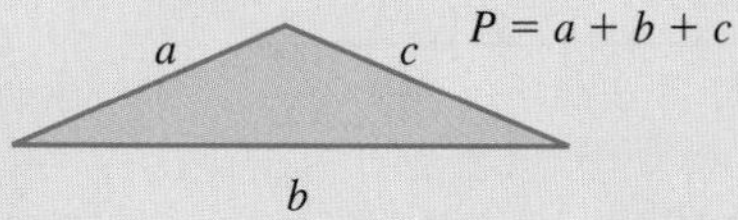

56. The perimeter of a rectangle is 288 ft. The length is 4 ft longer than the width. Find the width.

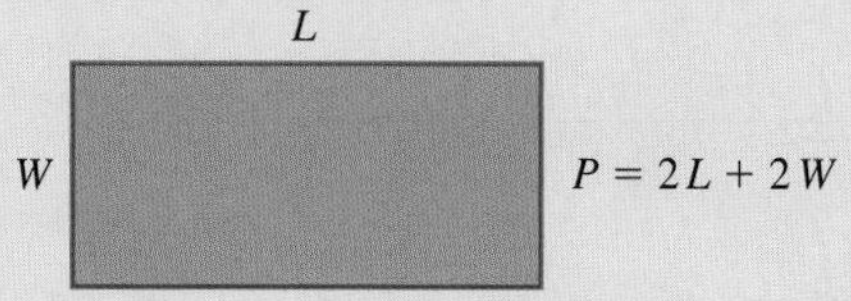

Chapter 10 Test

CHAPTER **Test Prep** VIDEO *The Chapter Test Prep Videos with test solutions are available on DVD, in MyMathLab, and on YouTube—search "LialDevMath" and click on "Channels."*

Solve each equation, and check your solution.

1. $3x - 7 = 11$

2. $5x + 9 = 7x + 21$

3. $2 - 3(x - 5) = 3 + (x + 1)$

4. $2.3x + 13.7 = 1.3x + 2.9$

5. $-\frac{4}{7}x = -12$

6. $-8(2x + 4) = -4(4x + 8)$

7. $0.06(x + 20) + 0.08(x - 10) = 4.6$

8. $7 - (m - 4) = -3m + 2(m + 1)$

Solve each problem.

9. The Chicago Bulls finished with the best record for the 2010–2011 NBA regular season. They won 2 more than three times as many games as they lost. They played 82 games. How many games did they win and lose? (*Source:* www.nba.com)

10. Three islands in the Hawaiian island chain are Hawaii (the Big Island), Maui, and Kauai. Together, their areas total 5300 mi^2. The island of Hawaii is 3293 mi^2 larger than the island of Maui, and Maui is 177 mi^2 larger than Kauai. What is the area of each island?

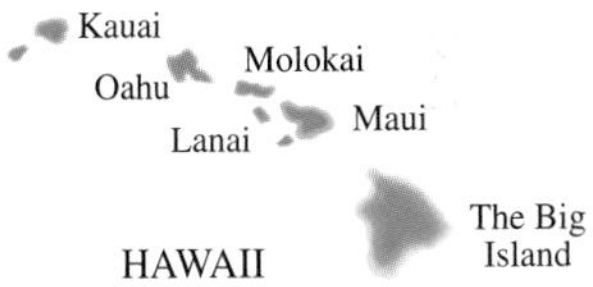

11. If the lesser of two consecutive even integers is tripled, the result is 20 more than twice the greater integer. Find the two integers.

12. Find the measure of an angle if its supplement measures 10° more than three times its complement.

13. The formula for the perimeter of a rectangle is $P = 2L + 2W$.

(a) Solve for W.

(b) If $P = 116$ and $L = 40$, find the value of W.

14. Solve the following equation for y.

$$5x - 4y = 8$$

Find the measure of each marked angle.

15.

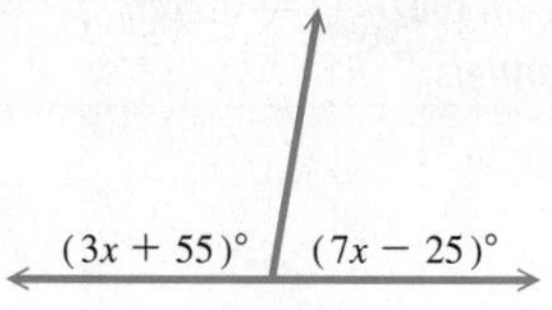

16.

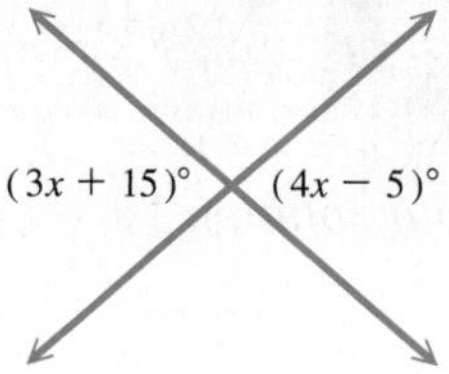

17. Write an inequality involving x that describes the numbers graphed.

(a)

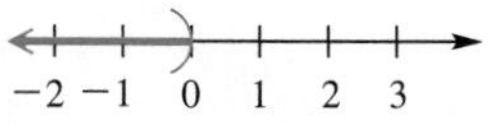

(b)

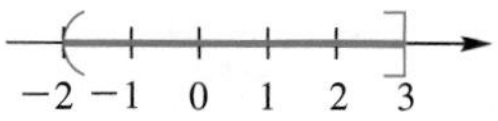

Solve each inequality. Write the solution set in interval notation, and graph it.

18. $-3x > -33$

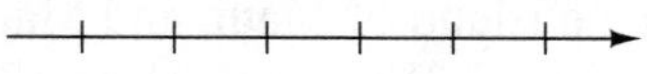

19. $-0.04x \leq 0.12$

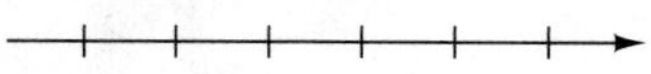

20. $-4x + 2(x - 3) \geq 4x - (3 + 5x) - 7$

21. $-10 < 3x - 4 \leq 14$

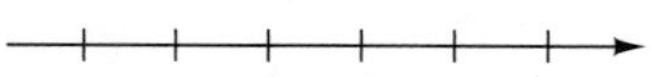

22. Shania Johnson has scores of 76 and 81 on her first two algebra tests. If she wants an average of at least 80 after her third test, what score must she make on her third test?

11 Graphs of Linear Equations and Inequalities in Two Variables

We determine location on a map using *coordinates,* a concept that is based on a *rectangular coordinate system,* one of the topics of this chapter.

11.1 Linear Equations in Two Variables; The Rectangular Coordinate System

OBJECTIVES

1. Interpret graphs.
2. Write a solution as an ordered pair.
3. Decide whether a given ordered pair is a solution of a given equation.
4. Complete ordered pairs for a given equation.
5. Complete a table of values.
6. Plot ordered pairs.

1 Refer to the bar graph in **Figure 1.**

(a) Which years had per-capita health care spending less than $7000?

(b) Estimate per-capita health care spending in 2006 and 2007.

(c) Describe the change in per-capita health care spending from 2006 to 2007.

Answers

1. (a) 2004, 2005, 2006
(b) 2006: about $6750; 2007: about $7000
(c) Per-capita health care spending increased by about $250 from 2006 to 2007.

As we saw in **Section 8.1,** circle graphs (pie charts) provide a convenient way to organize and communicate information. Along with *bar graphs* and *line graphs*, they can be used to analyze data, make predictions, or simply to entertain us.

OBJECTIVE 1 **Interpret graphs.** A **bar graph** is used to show comparisons. It consists of a series of bars (or simulations of bars) arranged either vertically or horizontally. In a bar graph, values from two categories are paired with each other.

EXAMPLE 1 Interpreting a Bar Graph

The bar graph in **Figure 1** shows annual per-capita spending on health care in the United States for the years 2004 through 2009.

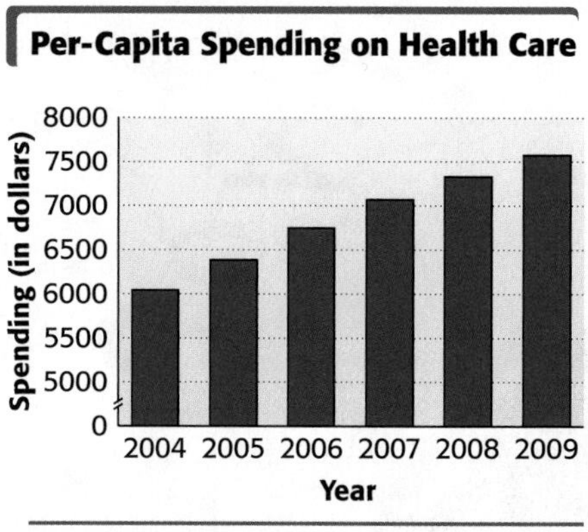

Source: U.S. Centers for Medicare and Medicaid Services.

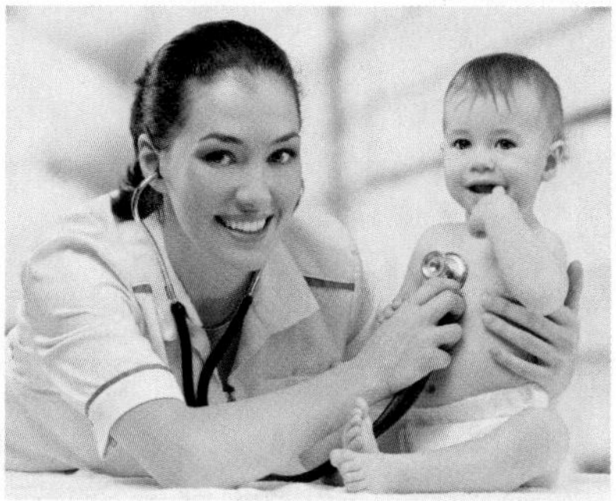

Figure 1

(a) In what years was per-capita health care spending greater than $7000?

Locate 7000 on the vertical axis and follow the line across to the right. Three years—2007, 2008, and 2009—have bars that extend above the line for 7000, so per-capita health care spending was greater than $7000 in those years.

(b) Estimate per-capita health care spending in 2004 and 2008.

Locate the top of the bar for 2004 and move horizontally across to the vertical scale to see that it is about 6000. Per-capita health care spending for 2004 was about $6000.

Similarly, follow the top of the bar for 2008 across to the vertical scale to see that it lies a little more than halfway between 7000 and 7500, so per-capita health care spending in 2008 was about $7300.

(c) Describe the change in per-capita spending as the years progressed.

As the years progressed, per-capita spending on health care increased steadily, from about $6000 in 2004 to about $7500 in 2009.

◀ **Work Problem 1 at the Side.**

A **line graph** is used to show changes or trends in data over time. To form a line graph, we connect a series of points representing data with line segments.

EXAMPLE 2 Interpreting a Line Graph

The line graph in **Figure 2** shows average prices of a gallon of regular unleaded gasoline in the United States for the years 2004 through 2011.

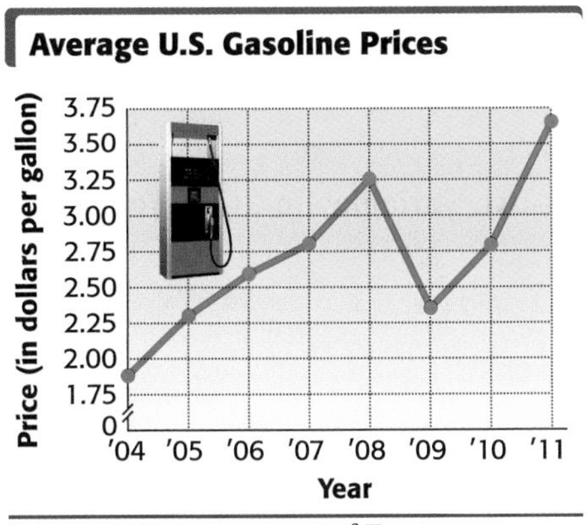

Source: U.S. Department of Energy.

Figure 2

(a) Between which years did the average price of a gallon of gasoline decrease?

The line between 2008 and 2009 falls, so the average price of a gallon of gasoline decreased from 2008 to 2009.

(b) What was the general trend in the average price of a gallon of gasoline from 2004 through 2008?

The line graph rises from 2004 to 2008, so the average price of a gallon of gasoline increased over those years.

(c) Estimate the average price of a gallon of gasoline in 2004 and 2008. About how much did the price increase between 2004 and 2008?

Move up from 2004 on the horizontal scale to the point plotted for 2004. This point is about halfway between the lines on the vertical scale for \$1.75 and \$2.00. Halfway between \$1.75 and \$2.00 would be about \$1.88. So, it cost about \$1.88 for a gallon of gasoline in 2004.

Similarly, locate the point plotted for 2008. Moving across to the vertical scale, the graph indicates that the price for a gallon of gasoline in 2008 was about \$3.25.

Between 2004 and 2008, the average price of a gallon of gasoline increased by about

$$\$3.25 - \$1.88 = \$1.37.$$

Work Problem 2 at the Side. ▶

The line graph in **Figure 2** relates years to average prices for a gallon of gasoline. We can also represent these two related quantities using a table of data, as shown in the margin. In table form, we can see more precise data rather than estimating it. Trends in the data are easier to see from the graph, which gives a "picture" of the data.

2 Refer to the line graph in **Figure 2.**

(a) Estimate the average price of a gallon of regular unleaded gasoline in 2005.

(b) About how much did the average price of a gallon of gasoline increase from 2005 to 2008?

(c) About how much did the average price of a gallon of gasoline decrease from 2008 to 2009?

Year	Average Price (in dollars per gallon)
2004	1.88
2005	2.30
2006	2.59
2007	2.80
2008	3.25
2009	2.35
2010	2.79
2011	3.65

Source: U.S. Department of Energy.

Answers

2. **(a)** about \$2.30 **(b)** about \$0.95 **(c)** about \$0.90

3 Write each solution as an ordered pair.

GS **(a)** $x = 5$ and $y = 7$

x-value y-value
↓ ↓
(____, ____)

GS **(b)** $y = 6$ and $x = -1$

x-value y-value
↓ ↓
(____, ____)

(c) $y = 4$ and $x = -3$

(d) $x = \frac{2}{3}$ and $y = -12$

(e) $y = 1.5$ and $x = -2.4$

(f) $x = 0$ and $y = 0$

Answers

3. (a) 5; 7 (b) −1; 6 (c) $(-3, 4)$
(d) $\left(\frac{2}{3}, -12\right)$ (e) $(-2.4, 1.5)$
(f) $(0, 0)$

We can extend these ideas to the subject of this chapter, *linear equations in **two** variables*. A linear equation in two variables, one for each of the quantities being related, can be used to represent the data in a table or graph. ***The graph of a linear equation in two variables is a line.***

Linear Equation in Two Variables

A **linear equation in two variables** is an equation that can be written in the form

$$Ax + By = C,$$

where A, B, and C are real numbers and A and B are not both 0.

Some examples of linear equations in two variables in this form, called *standard form*, follow.

$$3x + 4y = 9, \quad x - y = 0, \quad \text{and} \quad x + 2y = -8$$ Linear equations in two variables

Note

Other linear equations in two variables, such as

$$y = 4x + 5 \quad \text{and} \quad 3x = 7 - 2y,$$

are not written in standard form, but could be algebraically rewritten in this form. We discuss the forms of linear equations in **Section 11.4.**

OBJECTIVE 2 **Write a solution as an ordered pair.** Recall from **Section 9.2** that a *solution* of an equation is a number that makes the equation true when it replaces the variable. For example, the linear equation in *one* variable $x - 2 = 5$ has solution 7, since replacing x with 7 gives a true statement.

A solution of a linear equation in two variables requires two numbers, one for each variable. For example, a true statement results when we replace x with 2 and y with 13 in the equation $y = 4x + 5$ since

$$13 = 4(2) + 5. \quad \text{Let } x = 2 \text{ and } y = 13.$$

The pair of numbers $x = 2$ and $y = 13$ gives one solution of the equation $y = 4x + 5$. The phrase "$x = 2$ and $y = 13$" is abbreviated

x-value → (2, 13) ← y-value
Ordered pair

with the x-value, 2, and the y-value, 13, given as a pair of numbers written inside parentheses. ***The x-value is always given first.*** A pair of numbers such as $(2, 13)$ is called an **ordered pair.**

CAUTION

The ordered pairs $(2, 13)$ and $(13, 2)$ are *not* the same. In the first pair, $x = 2$ and $y = 13$. In the second pair, $x = 13$ and $y = 2$. ***The order in which the numbers are written in an ordered pair is important.*** For two ordered pairs to be equal, their x-values must be equal and their y-values must be equal.

◀ **Work Problem 3 at the Side.**

OBJECTIVE 3 Decide whether a given ordered pair is a solution of a given equation. We substitute the x- and y-values of an ordered pair into a linear equation in two variables to see whether the ordered pair is a solution.

EXAMPLE 3 Deciding Whether Ordered Pairs Are Solutions

Decide whether each ordered pair is a solution of the equation $2x + 3y = 12$.

(a) $(3, 2)$

Substitute 3 for x and 2 for y in the given equation $2x + 3y = 12$.

$$2x + 3y = 12$$
$$2(3) + 3(2) \stackrel{?}{=} 12 \quad \text{Let } x = 3 \text{ and } y = 2.$$
$$6 + 6 \stackrel{?}{=} 12 \quad \text{Multiply.}$$
$$12 = 12 \checkmark \quad \text{True}$$

This result is true, so $(3, 2)$ is a solution of $2x + 3y = 12$.

(b) $(-2, -7)$

$$2x + 3y = 12$$
$$2(-2) + 3(-7) \stackrel{?}{=} 12 \quad \text{Let } x = -2 \text{ and } y = -7.$$
$$-4 + (-21) \stackrel{?}{=} 12 \quad \text{Multiply.}$$
$$-25 = 12 \quad \text{False}$$

Use parentheses to avoid errors.

This result is false, so $(-2, -7)$ is *not* a solution of $2x + 3y = 12$.

Work Problem 4 at the Side. ▶

OBJECTIVE 4 Complete ordered pairs for a given equation. We substitute a number for one variable to find the value of the other variable.

EXAMPLE 4 Completing Ordered Pairs

Complete each ordered pair for the equation $y = 4x + 5$.

(a) $(7, \_\_\_)$

The x-value always comes first.

In this ordered pair, $x = 7$. Replace x with 7 in the given equation.

$$y = 4x + 5$$
$$y = 4(7) + 5 \quad \text{Let } x = 7.$$
$$y = 28 + 5 \quad \text{Multiply.}$$
$$y = 33 \quad \text{Add.}$$

Solve for the value of y.

The ordered pair is $(7, 33)$.

(b) $(\_\_\_, -3)$

In this ordered pair, $y = -3$. Replace y with -3 in the given equation.

$$y = 4x + 5$$
$$-3 = 4x + 5 \quad \text{Let } y = -3.$$
$$-8 = 4x \quad \text{Subtract 5 from each side.}$$
$$-2 = x \quad \text{Divide each side by 4.}$$

Solve for the value of x.

The ordered pair is $(-2, -3)$.

Work Problem 5 at the Side. ▶

---

4 Decide whether each ordered pair is a solution of the equation $5x + 2y = 20$.

GS **(a)** $(0, 10)$

In this ordered pair, $x = \_\_\_$ and $y = \_\_\_$.

$$5x + 2y = 20$$
$$5(\_\_\_) + 2(\_\_\_) \stackrel{?}{=} 20$$
$$\_\_\_ + 20 \stackrel{?}{=} 20$$
$$\_\_\_ = 20$$

Is $(0, 10)$ a solution? (yes / no)

(b) $(2, -5)$

(c) $(3, 2)$

(d) $(-4, 20)$

5 Complete each ordered pair for the equation $y = 2x - 9$.

GS **(a)** $(5, \_\_\_)$

In this ordered pair, $x = \_\_\_$.

We must find the corresponding value of ___.

$$y = 2x - 9$$
$$y = 2(\_\_\_) - 9$$
$$y = \_\_\_ - 9$$
$$y = \_\_\_$$

The ordered pair is ___.

(b) $(2, \_\_\_)$

(c) $(\_\_\_, 7)$

(d) $(\_\_\_, -13)$

Answers

4. **(a)** 0; 10; 0; 10; 0; 20; yes
(b) no **(c)** no **(d)** yes

5. **(a)** 5; y; 5; 10; 1; $(5, 1)$
(b) $(2, -5)$ **(c)** $(8, 7)$ **(d)** $(-2, -13)$

6 Complete the table of values for each equation. Write the results as ordered pairs.

GS **(a)** $2x - 3y = 12$

x	y
0	
	0
3	
	-3

From the first row of the table, let $x =$ ____.

$$2x - 3y = 12$$
$$2(\_\_\_\_) - 3y = 12$$
$$\_\_\_\_ - 3y = 12$$
$$-3y = 12$$
$$\frac{-3y}{\_\_\_\_} = \frac{12}{\_\_\_\_}$$
$$y = \_\_\_\_$$

Write ____ for y in the first row of the table. Repeat this process to complete the rest of the table.

(b) $x = -1$

x	y
	-4
	0
-1	2

(c) $y = 4$

x	y
-3	4
2	
5	

Answers

6. (a) 0; 0; 0; -3; -3; -4; -4

x	y
0	-4
6	0
3	-2
$\frac{3}{2}$	-3

$(0, -4)$, $(6, 0)$, $(3, -2)$, $(\frac{3}{2}, -3)$

(b)

x	y
-1	-4
-1	0
-1	2

$(-1, -4)$, $(-1, 0)$, $(-1, 2)$

(c)

x	y
-3	4
2	4
5	4

$(-3, 4)$, $(2, 4)$, $(5, 4)$

OBJECTIVE 5 Complete a table of values. Ordered pairs are often displayed in a **table of values.** Although we usually write tables of values vertically, they may be written horizontally.

EXAMPLE 5 Completing Tables of Values

Complete the table of values for each equation. Then write the results as ordered pairs.

(a) $x - 2y = 8$

x	y		Ordered Pairs
2		⟶	(2, ____)
10		⟶	(10, ____)
	0	⟶	(____, 0)
	-2	⟶	(____, -2)

From the first row of the table, let $x = 2$ in the equation. From the second row of the table, let $x = 10$.

If	$x = 2,$	If	$x = 10,$
then	$x - 2y = 8$	then	$x - 2y = 8$
becomes	$2 - 2y = 8$	becomes	$10 - 2y = 8$
	$-2y = 6$		$-2y = -2$
	$y = -3.$		$y = 1.$

The first two ordered pairs are $(2, -3)$ and $(10, 1)$. From the third and fourth rows of the table, let $y = 0$ and $y = -2$, respectively.

If	$y = 0,$	If	$y = -2,$
then	$x - 2y = 8$	then	$x - 2y = 8$
becomes	$x - 2(0) = 8$	becomes	$x - 2(-2) = 8$
	$x - 0 = 8$		$x + 4 = 8$
	$x = 8.$		$x = 4.$

The last two ordered pairs are $(8, 0)$ and $(4, -2)$. The completed table of values and corresponding ordered pairs follow.

Write y-values in the second column.

Write x-values in the first column.

x	y		Ordered Pairs
2	-3	⟶	$(2, -3)$
10	1	⟶	$(10, 1)$
8	0	⟶	$(8, 0)$
4	-2	⟶	$(4, -2)$

Each ordered pair is a solution of the given equation $x - 2y = 8$.

(b) $x = 5$

x	y
	-2
	6
	3

The given equation is $x = 5$. No matter which value of y is chosen, the value of x is *always* 5.

x	y		Ordered Pairs
5	-2	⟶	$(5, -2)$
5	6	⟶	$(5, 6)$
5	3	⟶	$(5, 3)$

◀ **Work Problem 6 at the Side.**

Note

We can think of $x = 5$ in **Example 5(b)** as an equation in two variables by rewriting $x = 5$ as

$$x + 0y = 5.$$

This form shows that, for any value of y, the value of x is 5. Similarly, $y = 4$ in **Margin Problem 6(c)** on the preceding page is the same as

$$0x + y = 4.$$

OBJECTIVE 6 Plot ordered pairs. In **Section 10.3,** we saw that linear equations in *one* variable have either one, zero, or an infinite number of real number solutions. These solutions can be graphed on *one* number line. For example, the linear equation in one variable $x - 2 = 5$ has solution 7, which is graphed on the number line in **Figure 3.**

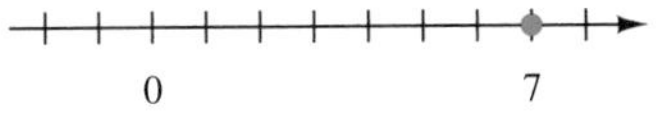

Figure 3

Every linear equation in *two* variables has an infinite number of ordered pairs as solutions. To graph these solutions, represented as the ordered pairs (x, y), we need *two* number lines, one for each variable. These two number lines are drawn at right angles as in **Figure 4.** The horizontal number line is the **x-axis,** and the vertical line is the **y-axis.** Together, the x-axis and y-axis form a **rectangular coordinate system.** The point at which the x-axis and y-axis intersect is the **origin.**

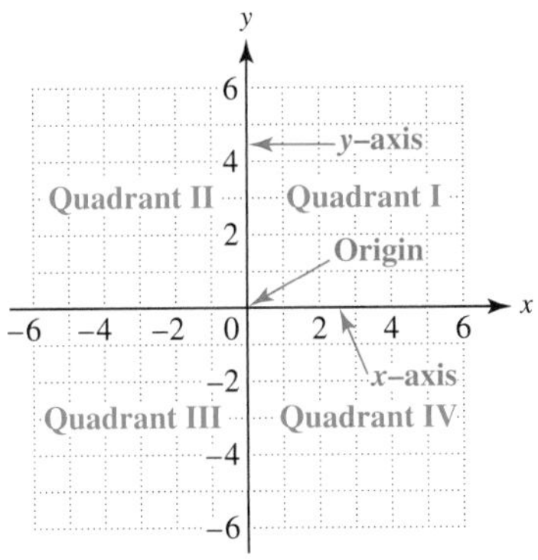

Rectangular coordinate system

Figure 4

The rectangular coordinate system is divided into four regions, or **quadrants.** These quadrants are numbered counterclockwise. See **Figure 4.**

Work Problem 7 at the Side. ▶

The x-axis and y-axis determine a **plane**—a flat surface (as illustrated by a sheet of paper). By referring to the two axes, every point in the plane can be associated with an ordered pair. The numbers in the ordered pair are the **coordinates** of the point.

Note

In a plane, *both* numbers in the ordered pair are needed to locate a point. The ordered pair is a name for the point.

7 Name the quadrant in which each point in the figure is located.

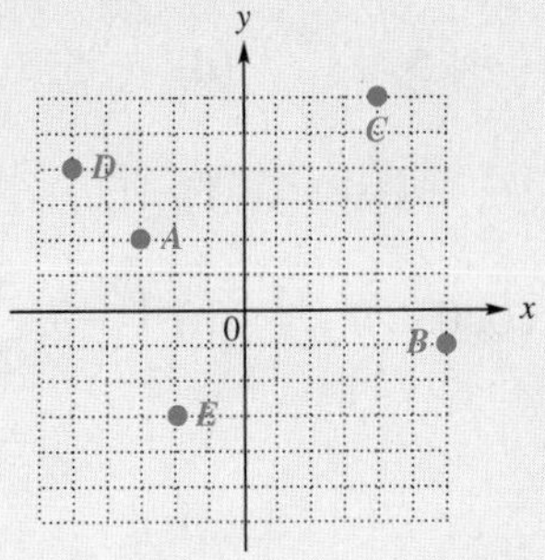

René Descartes (1596–1650)

The rectangular coordinate system is also called the **Cartesian coordinate system,** in honor of René Descartes, the French mathematician credited with its invention.

Answer

7. *A*: II; *B*: IV; *C*: I; *D*: II; *E*: III

8 Plot the given points in a coordinate system.

(a) $(3, 5)$ **(b)** $(-2, 6)$

(c) $(-4.5, 0)$ **(d)** $(-5, -2)$

(e) $(6, -2)$ **(f)** $(0, -6)$

(g) $(0, 0)$ **(h)** $\left(-3, \frac{5}{2}\right)$

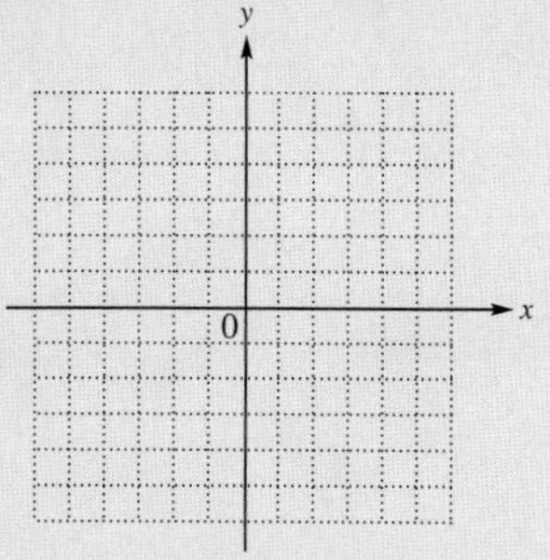

Indicate the points that lie in each quadrant.

quadrant I: ________

quadrant II: ________

quadrant III: ________

quadrant IV: ________

no quadrant: ________

EXAMPLE 6 Plotting Ordered Pairs

Plot the given points in a coordinate system.

(a) $(2, 3)$ **(b)** $(-1, -4)$ **(c)** $(-2, 3)$ **(d)** $(3, -2)$ **(e)** $\left(\frac{3}{2}, 2\right)$

(f) $(4, -3.75)$ **(g)** $(5, 0)$ **(h)** $(0, -3)$ **(i)** $(0, 0)$

The point $(2, 3)$ from part (a) is **plotted** (graphed) in **Figure 5.** The other points are plotted in **Figure 6.** In each case, we begin at the origin.

Step 1 Move right or left the number of units that corresponds to the x-coordinate in the ordered pair—***right if the x-coordinate is positive or left if it is negative.***

Step 2 Then turn and move up or down the number of units that corresponds to the y-coordinate—***up if the y-coordinate is positive or down if it is negative.***

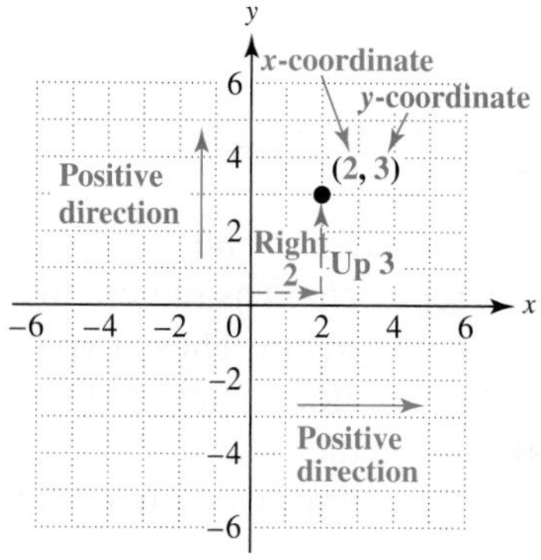

Figure 5

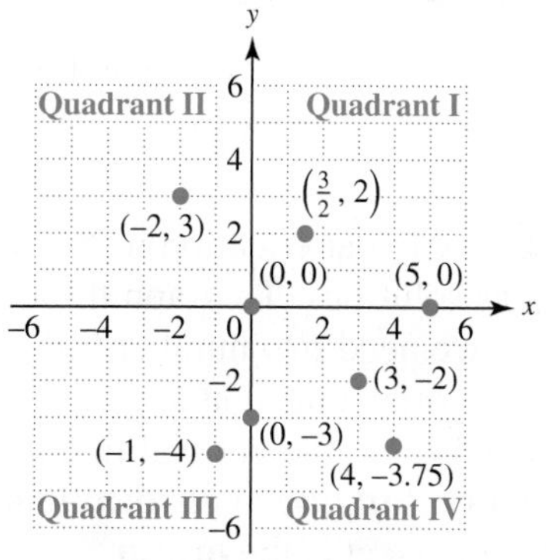

Figure 6

Notice in **Figure 6** that the point $(-2, 3)$ is in quadrant II, whereas the point $(3, -2)$ is in quadrant IV. ***The order of the coordinates is important. The x-coordinate is always given first in an ordered pair.***

To plot the point $\left(\frac{3}{2}, 2\right)$, think of the improper fraction $\frac{3}{2}$ as the mixed number $1\frac{1}{2}$ and move $\frac{3}{2}\left(\text{or } 1\frac{1}{2}\right)$ units to the right along the x-axis. Then turn and go 2 units up, parallel to the y-axis. The point $(4, -3.75)$ is plotted similarly, by approximating the location of the decimal y-coordinate.

The point $(5, 0)$ lies on the x-axis since the y-coordinate is 0. The point $(0, -3)$ lies on the y-axis since the x-coordinate is 0. The point $(0, 0)$ is at the origin. ***Points on the axes themselves are not in any quadrant.***

◀ **Work Problem 8 at the Side.**

We can use a linear equation in two variables to mathematically describe, or *model*, certain real-life situations, as shown in the next example.

EXAMPLE 7 Using a Linear Model

The annual number of twin births in the United States from 2001 through 2007 can be approximated by the linear equation

Number of twin births → y; Year → x

$$y = 2.929x - 5738,$$

which relates x, the year, and y, the number of twin births in thousands. (*Source:* National Center for Health Statistics.)

Continued on Next Page

Answers

8.

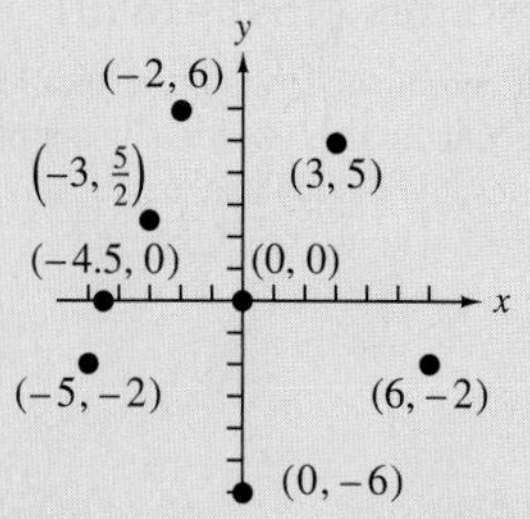

quadrant I: $(3, 5)$

quadrant II: $(-2, 6), \left(-3, \frac{5}{2}\right)$

quadrant III: $(-5, -2)$

quadrant IV: $(6, -2)$

no quadrant: $(-4.5, 0), (0, 0), (0, -6)$

(a) Complete the table of values for the given linear equation.

x (Year)	y (Number of Twin Births, in thousands)
2001	
2004	
2007	

To find y for $x = 2001$, substitute into the equation.

$$y = 2.929x - 5738$$

$$y = 2.929(2001) - 5738 \quad \text{Let } x = 2001.$$

$$y \approx 123 \quad \text{Use a calculator.}$$

≈ means "is approximately equal to."

In 2001, there were about 123 thousand (or 123,000) twin births.

Work Problem 9 at the Side. ▶

Including the results from **Margin Problem 9** gives the completed table that follows.

x (Year)	y (Number of Twin Births, in thousands)		Ordered Pairs (x, y)
2001	123	⟶	(2001, 123)
2004	132	⟶	(2004, 132)
2007	141	⟶	(2007, 141)

Here each year x is paired with a number of twin births y (in thousands).

(b) Graph the ordered pairs found in part (a).
See **Figure 7.** A graph of ordered pairs of data is a **scatter diagram.**

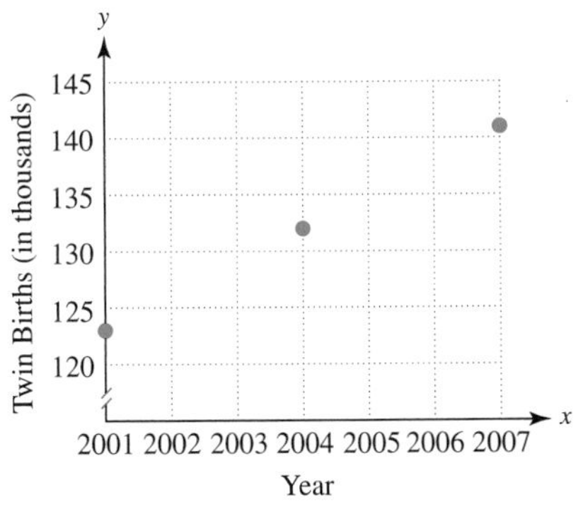

Figure 7

Notice the axis labels and scales. Each square represents 1 unit in the horizontal direction and 5 units in the vertical direction. We show a break in the y-axis, to indicate the jump from 0 to 120.

A scatter diagram can indicate whether two quantities are related. In **Figure 7,** the plotted points could be connected to approximate a straight ***line***, so the variables x (year) and y (number of twin births) have a ***line***ar relationship. The increase in the number of twin births is also reflected.

CAUTION

The equation in **Example 7** is valid only for the years 2001 through 2007. ***Do not assume that it would provide reliable data for other years.***

9 Refer to the linear equation in **Example 7.** Round to the nearest whole number.

GS **(a)** Find the y-value for $x = 2004$.

$y = 2.929x - 5738$

$y = 2.929(\_\_\_\_) - \_\_\_\_$

$y \approx \_\_\_\_$

(b) Find the y-value for $x = 2007$. Interpret your result.

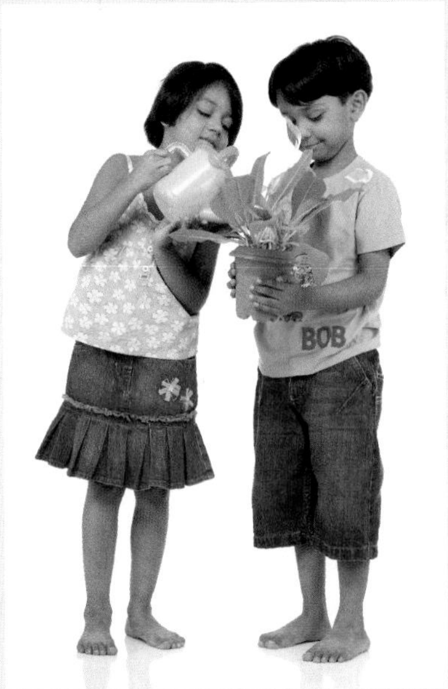

Answers

9. (a) 2004; 5738; 132
(b) 141; In 2007, there were about 141 thousand (or 141,000) twin births in the United States.

11.1 Exercises

FOR EXTRA HELP — Download the MyDashBoard App — MyMathLab®

CONCEPT CHECK *Complete each statement.*

1. The symbol (x, y) (*does / does not*) represent an ordered pair, while the symbols $[x, y]$ and $\{x, y\}$ (*do / do not*) represent ordered pairs.

2. The point whose graph has coordinates $(-4, 2)$ is in quadrant ____.

3. The point whose graph has coordinates $(0, 5)$ lies on the ____-axis.

4. The ordered pair $(4,$ ____$)$ is a solution of the equation $y = 3$.

5. The ordered pair $($____$, -2)$ is a solution of the equation $x = 6$.

6. The ordered pair $(3, 2)$ is a solution of the equation $2x - 5y =$ ____.

7. Do $(4, -1)$ and $(-1, 4)$ represent the same ordered pair? Explain.

8. Explain why it would be easier to find the corresponding y-value for $x = \frac{1}{3}$ in the equation $y = 6x + 2$ than it would be for $x = \frac{1}{7}$.

CONCEPT CHECK *Fill in each blank with the word* positive *or the word* negative.

The point with coordinates (x, y) is in

9. quadrant III if x is __________ and y is __________.

10. quadrant II if x is __________ and y is __________.

11. quadrant IV if x is __________ and y is __________.

12. quadrant I if x is __________ and y is __________.

The bar graph shows total U.S. milk production in billions of pounds for the years 2004 through 2010. Use the bar graph to work Exercises 13–16. **See Example 1.**

13. In what years was U.S. milk production greater than 185 billion pounds?

14. In what years was U.S. milk production about the same?

15. Estimate U.S. milk production in 2004 and 2010.

16. Describe the change in U.S. milk production from 2004 to 2010.

U.S. Milk Production

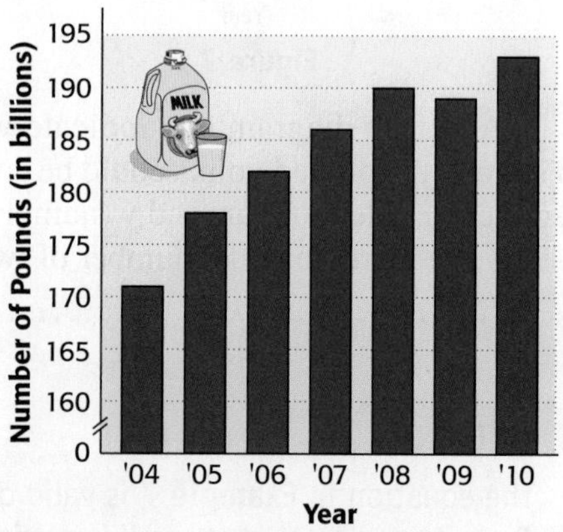

Source: U.S. Department of Agriculture.

The line graph shows the number of new and used passenger cars (in millions) imported into the United States over the years 2005 through 2010. Use the line graph to work Exercises 17–20. **See Example 2.**

17. Over which two consecutive years did the number of imported cars increase the most? About how much was this increase?

18. Estimate the number of cars imported during 2007, 2008, and 2009.

19. Describe the trend in car imports from 2006 to 2009.

20. During which year(s) were fewer than 6 millions cars imported into the United States?

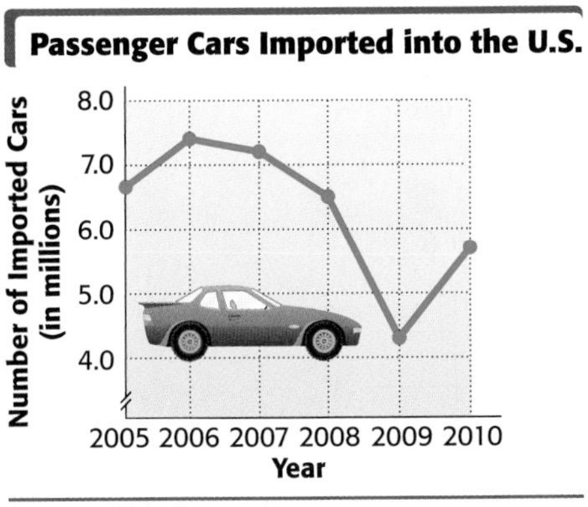

Source: U.S. Census Bureau.

Decide whether each ordered pair is a solution of the given equation. **See Example 3.**

21. $x + y = 9$; $(0, 9)$

22. $x + y = 8$; $(0, 8)$

23. $2x - y = 6$; $(4, 2)$

24. $2x + y = 5$; $(3, -1)$

25. $4x - 3y = 6$; $(2, 1)$

26. $5x - 3y = 15$; $(5, 2)$

27. $y = \frac{2}{3}x$; $(-6, -4)$

28. $y = -\frac{1}{4}x$; $(-8, 2)$

29. $x = -6$; $(5, -6)$

30. $x = -9$; $(8, -9)$

31. $y = 2$; $(2, 4)$

32. $y = 7$; $(7, -2)$

Complete each ordered pair for the equation $y = 2x + 7$. **See Example 4.**

33. (2, ____)

34. (0, ____)

35. (____, 0)

36. (____, −3)

Complete each ordered pair for the equation $y = -4x - 4$. **See Example 4.**

37. (0, ____)

38. (____, 0)

39. (____, 16)

40. (____, 24)

Complete each table of values. Write the results as ordered pairs. **See Example 5.**

41. $2x + 3y = 12$

x	y
0	
	0
	8

42. $4x + 3y = 24$

x	y
0	
	0
	4

43. $3x - 5y = -15$

x	y
0	
	0
	−6

44. $4x - 9y = -36$

x	y
	0
0	
	8

45. $x = -9$

x	y
	6
	2
	−3

46. $x = 12$

x	y
	3
	8
	0

47. $y = -6$

x	y
8	
4	
−2	

48. $y = -10$

x	y
4	
0	
−4	

49. $x - 8 = 0$

x	y
	8
	3
	0

50. $x + 4 = 0$

x	y
	4
	0
	−4

51. $y + 2 = 0$

x	y
9	
2	
0	

52. $y + 1 = 0$

x	y
1	
0	
−1	

Give the ordered pairs for the points labeled A–F in the figure. (All coordinates are integers.) Tell the quadrant in which each point is located. **See Example 6.**

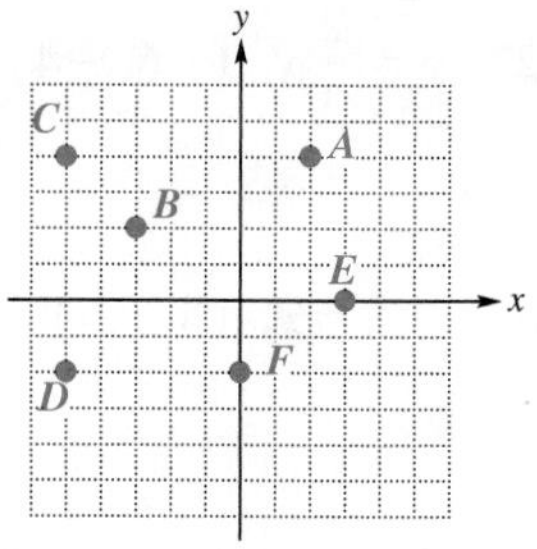

53. *A*

54. *B*

55. *C*

56. *D*

57. *E*

58. *F*

59. A point (x, y) has the property that $xy < 0$. In which quadrant(s) must the point lie? Explain.

60. A point (x, y) has the property that $xy > 0$. In which quadrant(s) must the point lie? Explain.

Plot each ordered pair on the rectangular coordinate system provided. **See Example 6.**

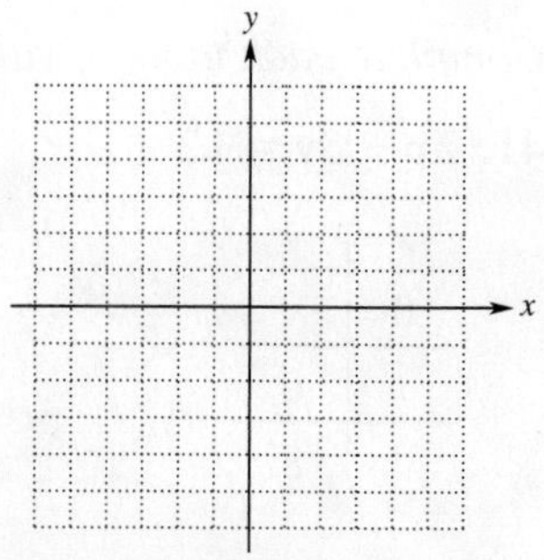

61. $(6, 2)$

62. $(5, 3)$

63. $(-4, 2)$

64. $(-3, 5)$

65. $\left(-\frac{4}{5}, -1\right)$

66. $\left(-\frac{3}{2}, -4\right)$

67. $(3, -1.75)$

68. $(5, -4.25)$

69. $(0, 4)$

70. $(0, -3)$

71. $(4, 0)$

72. $(-3, 0)$

Complete each table of values, and then plot the ordered pairs. ***See Examples 5 and 6.***

73. $x - 2y = 6$

x	y
0	
	0
2	
	−1

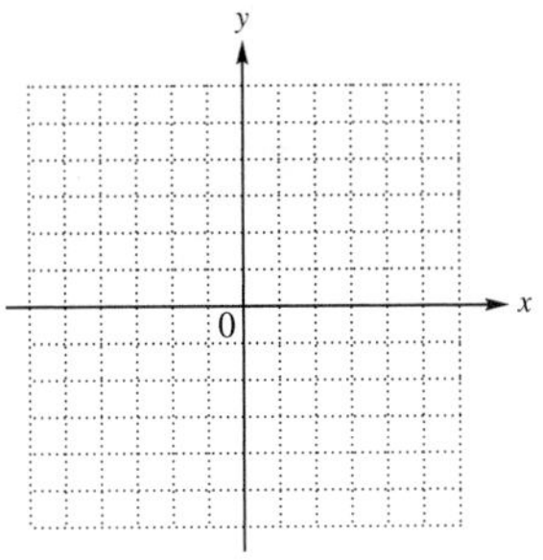

74. $2x - y = 4$

x	y
0	
	0
1	
	−6

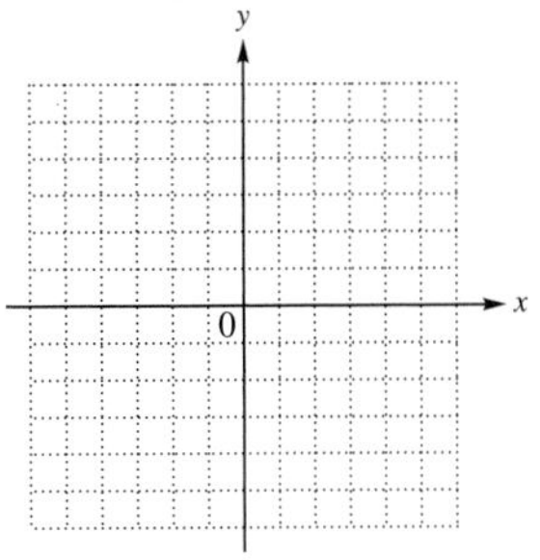

75. $3x - 4y = 12$

x	y
0	
	0
−4	
	−4

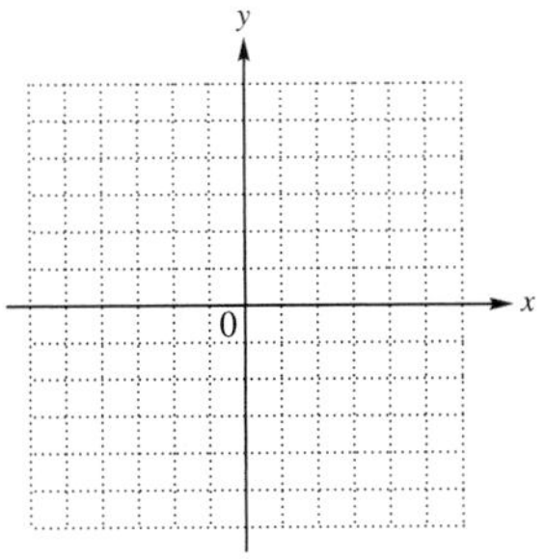

76. $2x - 5y = 10$

x	y
0	
	0
−5	
	−3

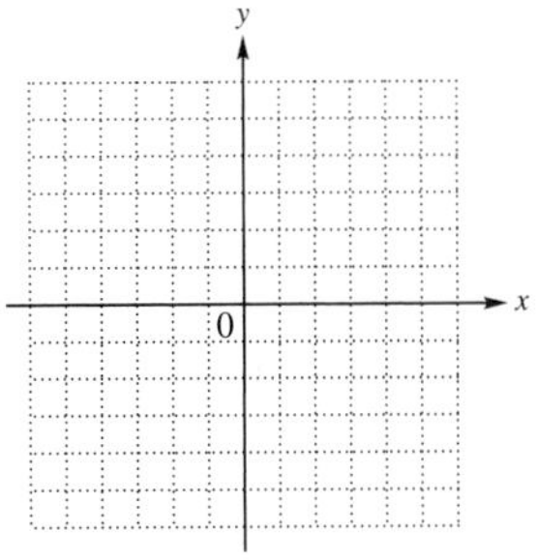

77. $y + 4 = 0$

x	y
0	
5	
−2	
−3	

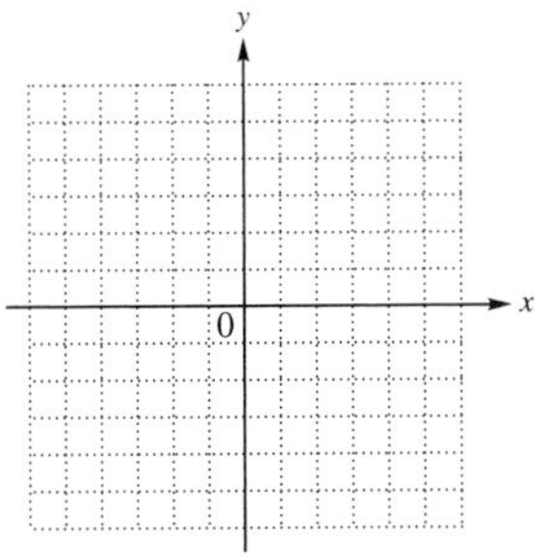

78. $x - 5 = 0$

x	y
	1
	0
	6
	−4

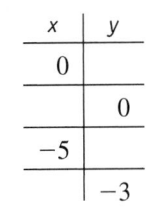

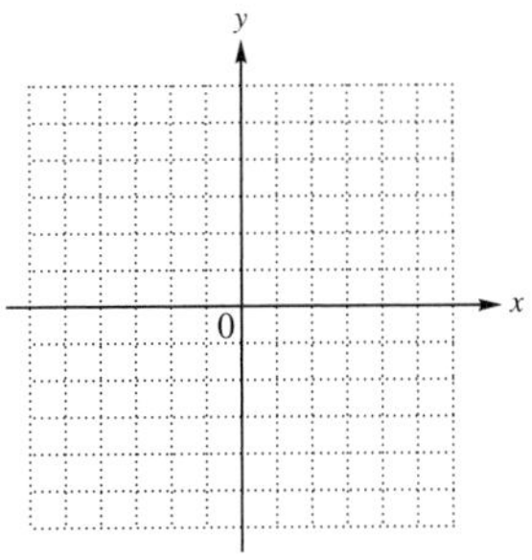

79. Describe the pattern indicated by the plotted points in **Exercises 73–78.**

80. Answer each question.

(a) A line through the plotted points in **Exercise 77** would be horizontal. What do you notice about the y-coordinates of the ordered pairs?

(b) A line through the plotted points in **Exercise 78** would be vertical. What do you notice about the x-coordinates of the ordered pairs?

Work each problem. ***See Example 7.***

81. Suppose that it costs a flat fee of \$20 plus \$5 per day to rent a pressure washer. Therefore, the cost y in dollars to rent the pressure washer for x days is given by

$$y = 5x + 20.$$

Express each of the following as an ordered pair.

(a) When the washer is rented for 5 days, the cost is \$45. (*Hint:* What does x represent? What does y represent?)

(b) We paid \$50 when we returned the washer, so we must have rented it for 6 days.

82. Suppose that it costs \$5000 to start up a business selling snow cones. Furthermore, it costs \$0.50 per cone in labor, ice, syrup, and overhead. Then the cost y in dollars to make x snow cones is given by

$$y = 0.50x + 5000.$$

Express each of the following as an ordered pair.

(a) When 100 snow cones are made, the cost is \$5050. (*Hint:* What does x represent? What does y represent?)

(b) When the cost is \$6000, the number of snow cones made is 2000.

83. The table shows the rate (in percent) at which 2-year college students (public) complete a degree within 3 years.

Year	Percent
2007	27.1
2008	29.3
2009	28.3
2010	28.0
2011	26.9

Source: ACT.

(a) Write the data from the table as ordered pairs (x, y), where x represents the year and y represents the percent.

(b) What would the ordered pair (2000, 32.4) mean in the context of this problem?

(c) Make a scatter diagram of the data using the ordered pairs from part (a).

2-YEAR COLLEGE STUDENTS COMPLETING A DEGREE WITHIN 3 YEARS

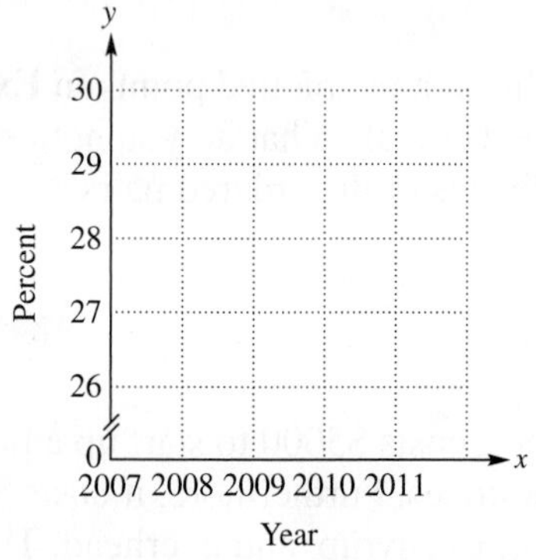

(d) Describe the pattern indicated by the points on the scatter diagram. What is happening to the rates at which 2-year college students complete a degree within 3 years?

84. The table shows the number of U.S. students studying abroad (in thousands) for recent academic years.

Academic Year	Number of Students (in thousands)
2005	224
2006	242
2007	262
2008	260
2009	271

Source: Institute of International Education.

(a) Write the data from the table as ordered pairs (x, y), where x represents the year and y represents the number of U.S. students studying abroad, in thousands.

(b) What does the ordered pair (2000, 154) mean in the context of this problem?

(c) Make a scatter diagram of the data using the ordered pairs from part (a).

U.S. STUDENTS STUDYING ABROAD

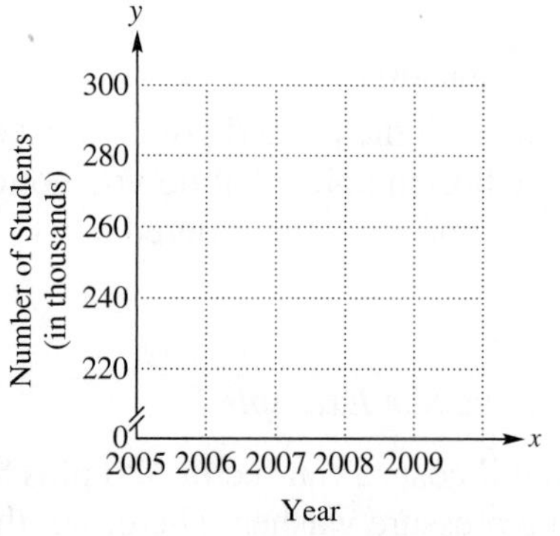

(d) Describe the pattern indicated by the points on the scatter diagram. What is the trend in the number of U.S. students studying abroad?

85. The maximum benefit for the heart from exercising occurs if the heart rate is in the target heart rate zone. The lower limit of this target zone can be approximated by the linear equation

$$y = -0.5x + 108,$$

where x represents age and y represents heartbeats per minute. (*Source:* www.fitresource.com)

(a) Complete the table of values for this linear equation.

Age	Heartbeats (per minute)
20	
40	
60	
80	

(b) Write the data from the table of values as ordered pairs.

(c) Make a scatter diagram of the data. Do the points lie in a linear pattern?

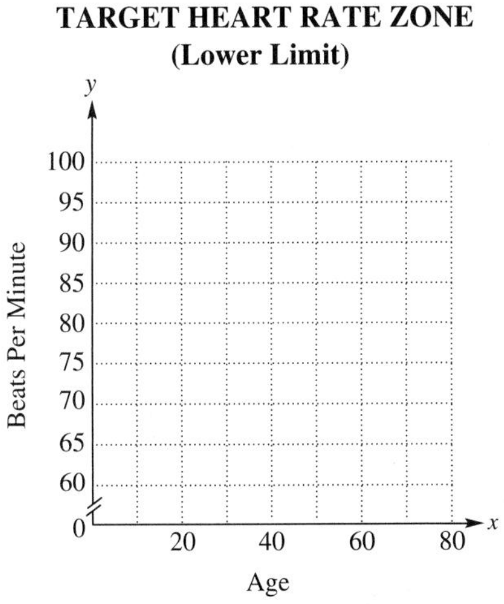

86. (See **Exercise 85.**) The upper limit of the target heart rate zone can be approximated by the linear equation

$$y = -0.8x + 173,$$

where x represents age and y represents heartbeats per minute. (*Source:* www.fitresource.com)

(a) Complete the table of values for this linear equation.

Age	Heartbeats (per minute)
20	
40	
60	
80	

(b) Write the data from the table of values as ordered pairs.

(c) Make a scatter diagram of the data. Describe the pattern indicated by the data.

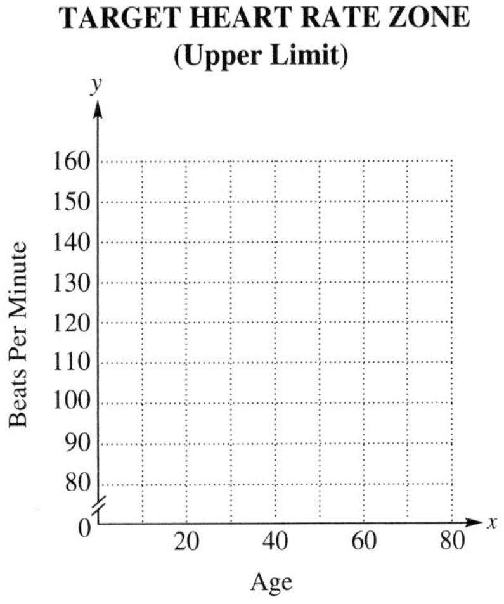

87. See **Exercises 85 and 86.** What is the target heart rate zone for age 20? Age 40?

88. See **Exercises 85 and 86.** What is the target heart rate zone for age 60? Age 80?

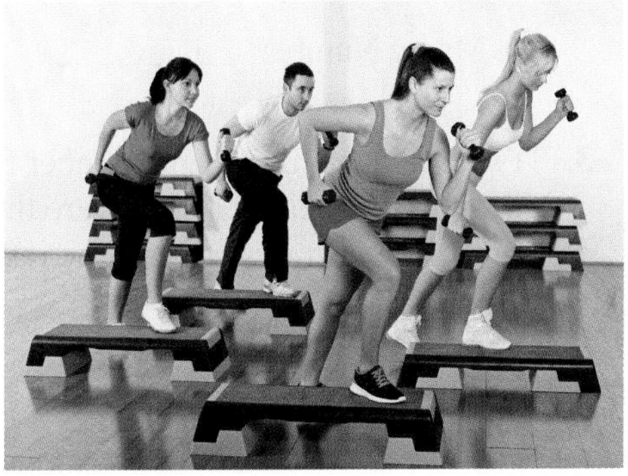

Math in the Media

THE MAGIC NUMBER IN SPORTS

The climax of any sports season is the playoffs. Baseball fans eagerly debate predictions of which team will win the pennant for their division. The *magic number* for each first-place team is often reported in media outlets. The **magic number** (sometimes called the **elimination number**) is the combined number of wins by the first-place team and losses by the second-place team that would clinch the title for the first-place team.

American League

East	W	L	PCT	GB
New York	87	57	.604	—
Boston	85	60	.586	2.5
Tampa Bay	80	64	.556	7.0
Toronto	73	73	.500	15.0
Baltimore	58	86	.403	29.0

Central	W	L	PCT	GB
Detroit	83	62	.572	—
Chicago	73	71	.507	9.5
Cleveland	71	72	.497	11.0
Kansas City	61	86	.415	23.0
Minnesota	59	86	.407	24.0

West	W	L	PCT	GB
Texas	82	64	.562	—
Los Angeles	80	65	.552	1.5
Oakland	66	79	.455	15.5
Seattle	61	84	.421	20.5

Source: mlb.com

To calculate the magic number, consider the following conditions.

The number of wins for the first-place team (W_1) *plus the magic number* (M) *is one more than the sum of the number of wins to date* (W_2) *and the number of games remaining in the season* (N_2) *for the second-place team.*

1. First, use the variable definitions to write an equation involving the magic number. Second, solve the equation for the magic number and write a formula for it.

2. The American League standings on September 10, 2011, are shown above. There were 162 regulation games in the 2011 season. Find the magic number for each team. The number of games remaining in the season for the second-place team is calculated as

$$N_2 = 162 - (W_2 + L_2),$$

where L_2 represents the number of losses for the second-place team.

(a) AL East: New York vs Boston

Magic Number ___________

(b) AL Central: Detroit vs Chicago

Magic Number ___________

(c) AL West: Texas vs Los Angeles

Magic Number ___________

3. Try to calculate the magic number for Seattle vs Texas. (Treat Seattle as if it were the second-place team.) How can we interpret the result?

11.2 Graphing Linear Equations in Two Variables

OBJECTIVES

1. Graph linear equations by plotting ordered pairs.
2. Find intercepts.
3. Graph linear equations of the form $Ax + By = 0$.
4. Graph linear equations of the form $y = k$ or $x = k$.
5. Use a linear equation to model data.

OBJECTIVE 1 Graph linear equations by plotting ordered pairs. There are infinitely many ordered pairs that satisfy an equation in two variables. We find these ordered-pair solutions by choosing as many values of x (or y) as we wish and then completing each ordered pair.

For example, consider the equation $x + 2y = 7$. If we choose $x = 1$, then we can substitute to find the corresponding value of y.

$$x + 2y = 7 \quad \text{Given equation}$$
$$1 + 2y = 7 \quad \text{Let } x = 1.$$
$$2y = 6 \quad \text{Subtract 1.}$$
$$y = 3 \quad \text{Divide by 2.}$$

If $x = 1$, then $y = 3$, and the ordered pair $(1, 3)$ is a solution of the equation.

$$1 + 2(3) = 7 \checkmark \quad (1, 3) \text{ is a solution.}$$

Work Problem 1 at the Side. ▶

Figure 8 shows a graph of all the ordered-pair solutions found above and in **Margin Problem 1** for $x + 2y = 7$.

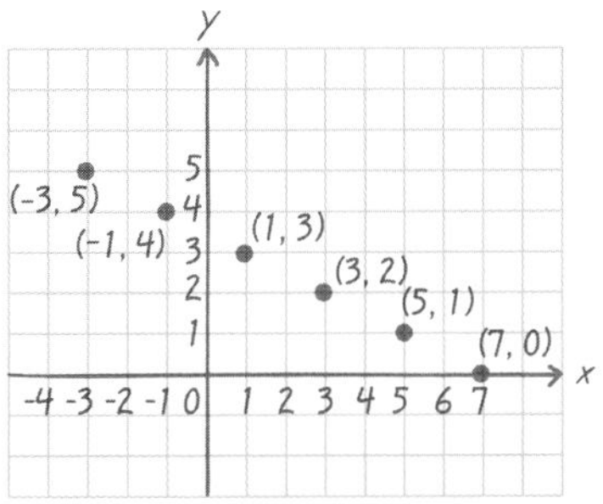

Figure 8

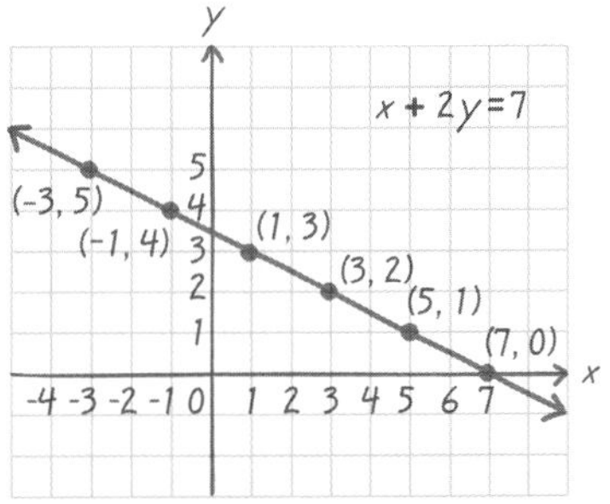

Figure 9

Notice that the points plotted in **Figure 8** all appear to lie on a straight line, as shown in **Figure 9.** In fact, the following is true.

Every point on the line represents a solution of the equation $x + 2y = 7$, and every solution of the equation corresponds to a point on the line.

This line gives a "picture" of all the solutions of the equation $x + 2y = 7$. The line extends indefinitely in both directions, as suggested by the arrowhead on each end, and is the **graph** of the equation $x + 2y = 7$. The process of plotting the ordered pairs and drawing the line through the corresponding points is called **graphing.**

The preceding discussion can be generalized.

Graph of a Linear Equation

The graph of any linear equation in two variables is a straight line.

Notice that the word ***line*** appears in the term "***line***ar equation."

1 Complete each ordered pair to find additional solutions of the equation $x + 2y = 7$.

GS **(a)** $(-3, \_\_\_)$

Substitute ____ for x in the equation and solve for ____.

(b) $(-1, \_\_\_)$

(c) $(3, \_\_\_)$

(d) $(5, \_\_\_)$

(e) $(7, \_\_\_)$

Answers

1. (a) -3; y; $(-3, 5)$ **(b)** $(-1, 4)$ **(c)** $(3, 2)$ **(d)** $(5, 1)$ **(e)** $(7, 0)$

2 Graph the linear equation.

GS $x + y = 6$

x	y
0	
	0
2	

From the first row of the table, let $x =$ ____.

$$x + y = 6$$
$$\_\_\_\_ + y = 6$$
$$y = \_\_\_\_$$

Write ____ for the y-value in the first row of the table.

The first ordered pair is ____.

Repeat this process to complete the table. Then plot the corresponding points, and draw the ____ through them.

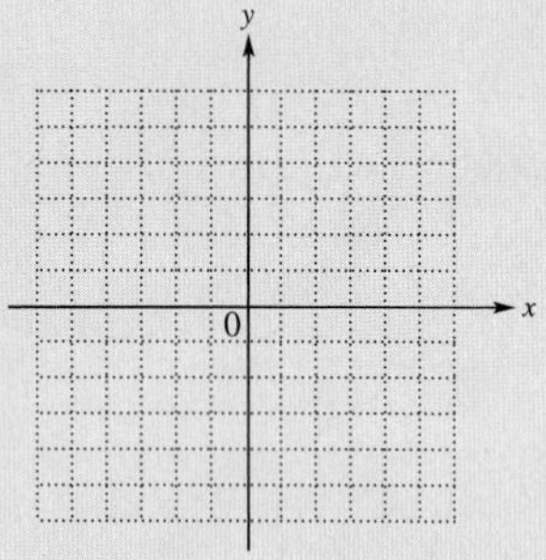

Answer

2. 0; 0; 6; 6; (0, 6); line

x	y
0	6
6	0
2	4

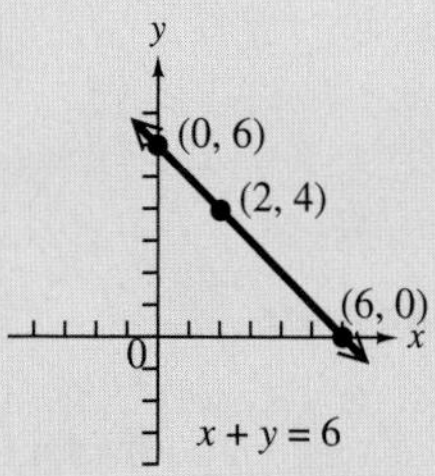

EXAMPLE 1 Graphing a Linear Equation

Graph $4x - 5y = 20$.

At least two different points are needed to draw the graph. First let $x = 0$ and then let $y = 0$ to determine two ordered pairs.

$4x - 5y = 20$	0 is easy to substitute.	$4x - 5y = 20$	0 is easy to substitute.
$4(0) - 5y = 20$		$4x - 5(0) = 20$	
$0 - 5y = 20$	Multiply.	$4x - 0 = 20$	Multiply.
$-5y = 20$	Subtract.	$4x = 20$	Subtract.
$y = -4$	Divide by -5.	$x = 5$	Divide by 4.

The ordered pairs are $(0, -4)$ and $(5, 0)$. Find a third ordered pair (as a check) by choosing a number *other than 0* for x or y. We choose $y = 2$.

$4x - 5y = 20$	We arbitrarily let $y = 2$. Other numbers could be used for y, or for x, instead.
$4x - 5(2) = 20$	
$4x - 10 = 20$	Multiply.
$4x = 30$	Add 10.
$x = \frac{30}{4}$, or $\frac{15}{2}$	Divide by 4. Write in lowest terms.

This gives the ordered pair $(\frac{15}{2}, 2)$, or $(7\frac{1}{2}, 2)$. We plot the three ordered pairs $(0, -4)$, $(5, 0)$, and $(7\frac{1}{2}, 2)$, and draw a line through them. See **Figure 10.**

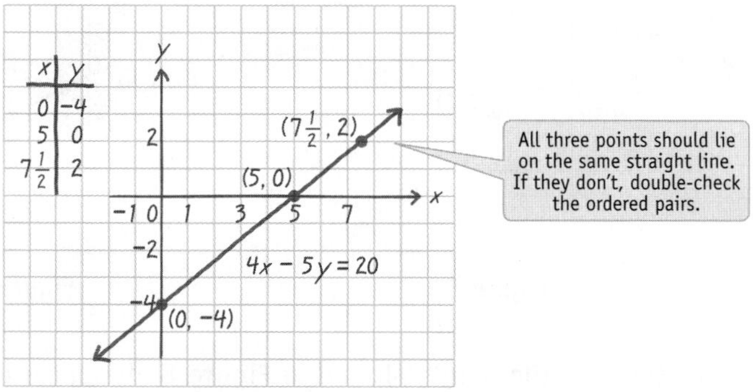

Figure 10

◀ Work Problem **2** at the Side.

OBJECTIVE 2 Find intercepts. In **Figure 10,** the graph crosses, or intersects, the y-axis at $(0, -4)$ and the x-axis at $(5, 0)$. For this reason, $(0, -4)$ is called the ***y*-intercept,** and $(5, 0)$ is called the ***x*-intercept** of the graph. The intercepts are particularly useful for graphing linear equations.

Finding Intercepts

To find the x-intercept, let $y = 0$ in the given equation and solve for x. Then $(x, 0)$ is the x-intercept.

To find the y-intercept, let $x = 0$ in the given equation and solve for y. Then $(0, y)$ is the y-intercept.

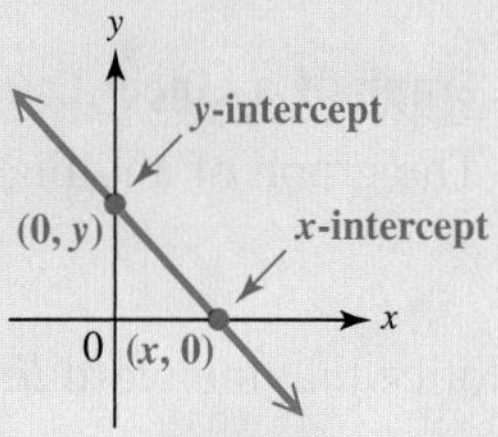

EXAMPLE 2 Graphing a Linear Equation Using Intercepts

Graph $2x + y = 4$ using intercepts.

To find the y-intercept, let $x = 0$.

$2x + y = 4$

$2(\mathbf{0}) + y = 4$ Let $x = 0$.

$0 + y = 4$ Multiply.

$\mathbf{y = 4}$ y-intercept is $(0, 4)$.

To find the x-intercept, let $y = 0$.

$2x + y = 4$

$2x + \mathbf{0} = 4$ Let $y = 0$.

$2x = 4$ Add.

$x = 2$ x-intercept is $(2, 0)$.

The intercepts are $(0, 4)$ and $(2, 0)$. To find a third point (as a check), we let $x = 1$.

$2x + y = 4$

$2(\mathbf{1}) + y = 4$ Let $x = 1$.

$2 + y = 4$ Multiply.

$y = 2$ Subtract 2.

This gives the ordered pair $(1, 2)$. The graph, with the two intercepts in red, is shown in **Figure 11.**

	x	y
y-intercept →	0	4
x-intercept →	2	0
	1	2

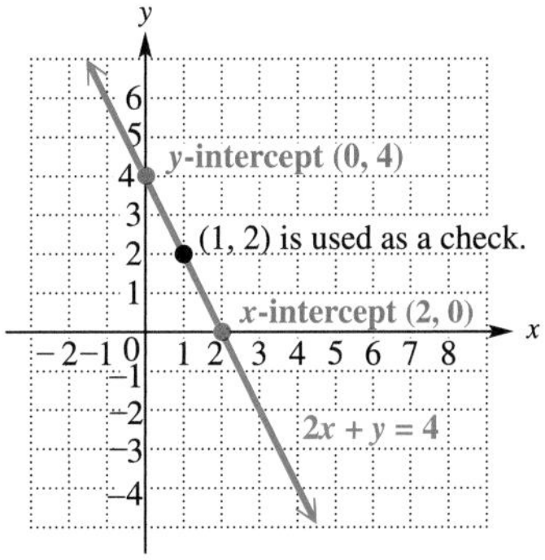

Figure 11

Work Problem 3 at the Side. ▶

EXAMPLE 3 Graphing a Linear Equation Using Intercepts

Graph $y = -\frac{3}{2}x + 3$.

Although this linear equation is not in standard form $Ax + By = C$, it *could* be written in that form. The intercepts can be found by first letting $x = 0$ and then letting $y = 0$.

$y = -\frac{3}{2}x + 3$

$y = -\frac{3}{2}(\mathbf{0}) + 3$ Let $x = 0$.

$y = 0 + 3$ Multiply.

$\mathbf{y = 3}$ Add.

$y = -\frac{3}{2}x + 3$

$\mathbf{0} = -\frac{3}{2}x + 3$ Let $y = 0$.

$\frac{3}{2}x = 3$ Add $\frac{3}{2}x$.

$\mathbf{x = 2}$ Multiply by $\frac{2}{3}$.

This gives the intercepts $(0, 3)$ and $(2, 0)$.

Continued on Next Page

3 Graph the linear equation using intercepts. (Be sure to get a third point as a check.)

$$5x + 2y = 10$$

Find the y-intercept by letting $x =$ ____ in the equation.

The y-intercept is ____. Find the x-intercept by letting $y =$ ____ in the equation.

The x-intercept is ____.

Find a third point with $x = 4$ as a check, and graph the equation.

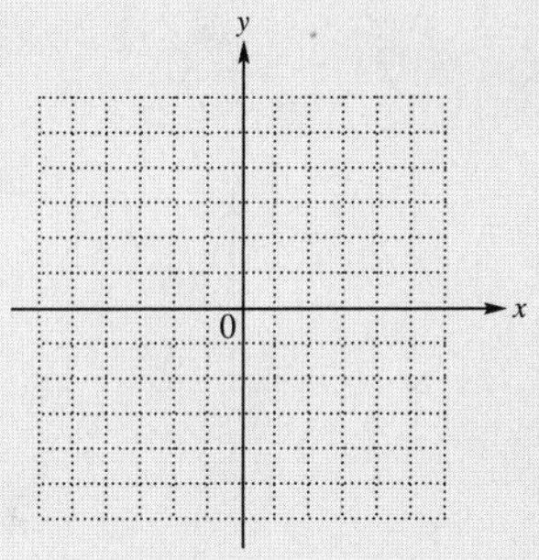

Answer

3. 0; (0, 5); 0; (2, 0)

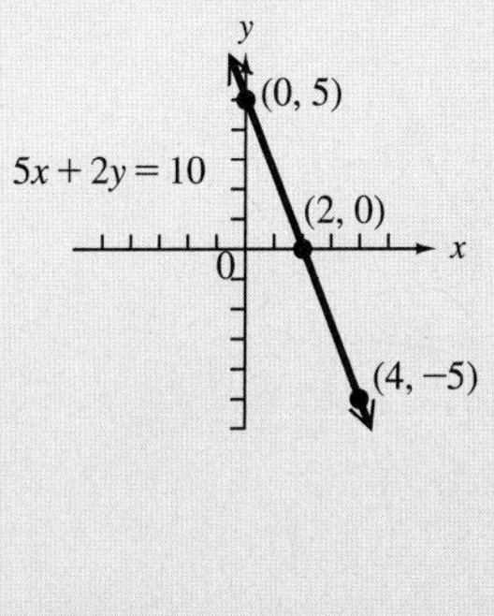

4 Graph the linear equation using intercepts.

$$y = \frac{2}{3}x - 2$$

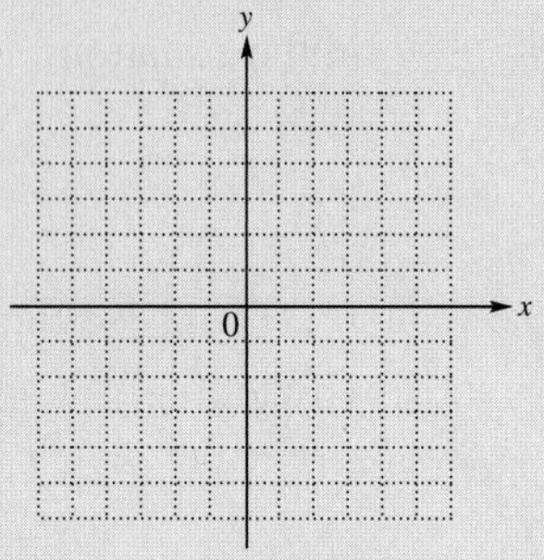

5 Graph $2x - y = 0$.

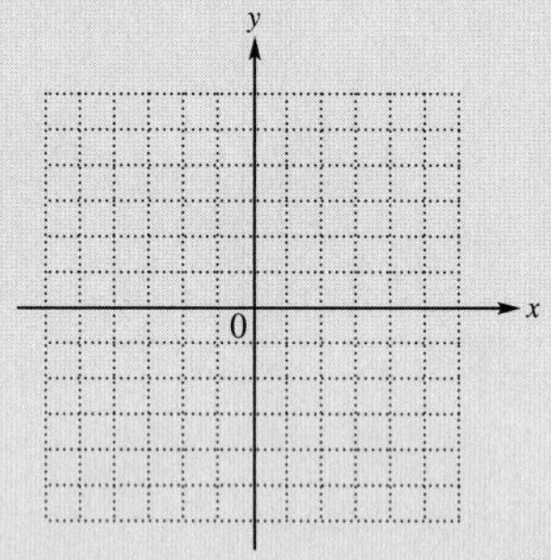

To find a third point, we let $x = -2$.

$$y = -\frac{3}{2}x + 3$$ Choosing a multiple of 2 makes multiplying by $-\frac{3}{2}$ easier.

$$y = -\frac{3}{2}(-2) + 3 \quad \text{Let } x = -2.$$

$$y = 3 + 3 \quad \text{Multiply.}$$

$$y = 6 \quad \text{Add.}$$

This gives the ordered pair $(-2, 6)$. We plot this ordered pair and both intercepts and draw a line through them, as shown in **Figure 12.**

x	y
0	3
2	0
−2	6

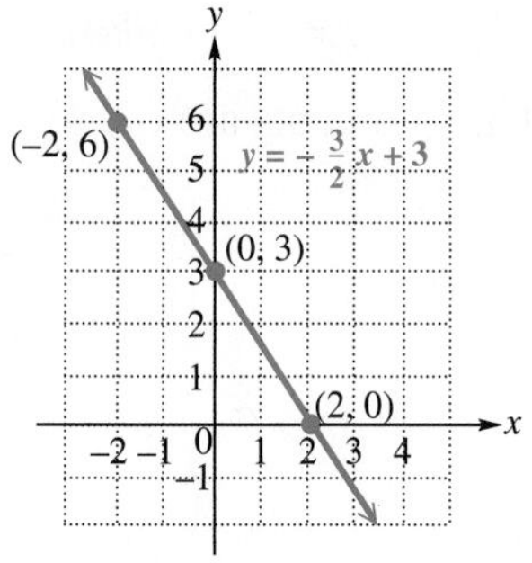

Figure 12

◀ **Work Problem 4 at the Side.**

OBJECTIVE 3 Graph linear equations of the form $Ax + By = 0$.

EXAMPLE 4 Graphing an Equation with x- and y-Intercepts (0, 0)

Graph $x - 3y = 0$.

To find the y-intercept, let $x = 0$.

$$x - 3y = 0$$

$$0 - 3y = 0 \quad \text{Let } x = 0.$$

$$-3y = 0 \quad \text{Subtract.}$$

$$y = 0 \quad y\text{-intercept is } (0, 0).$$

To find the x-intercept, let $y = 0$.

$$x - 3y = 0$$

$$x - 3(0) = 0 \quad \text{Let } y = 0.$$

$$x - 0 = 0 \quad \text{Multiply.}$$

$$x = 0 \quad x\text{-intercept is } (0, 0).$$

The x- and y-intercepts are the *same* point, (0, 0). We must select *two other values* for x or y to find two other points. We choose $x = 6$ and $x = -6$.

$$x - 3y = 0$$

$$6 - 3y = 0 \quad \text{Let } x = 6.$$

$$-3y = -6 \quad \text{Subtract 6.}$$

$$y = 2 \quad \text{Gives } (6, 2)$$

$$x - 3y = 0$$

$$-6 - 3y = 0 \quad \text{Let } x = -6.$$

$$-3y = 6 \quad \text{Add 6.}$$

$$y = -2 \quad \text{Gives } (-6, -2)$$

We use $(-6, -2)$, $(0, 0)$, and $(6, 2)$ to draw the graph in **Figure 13.**

x	y
0	0
6	2
−6	−2

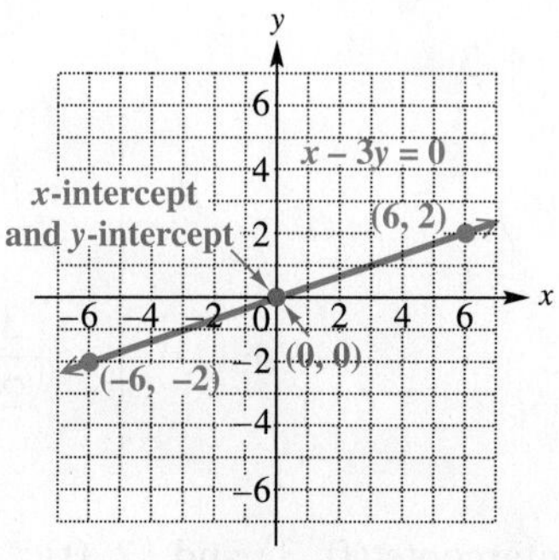

Figure 13

◀ **Work Problem 5 at the Side.**

Answers

4.

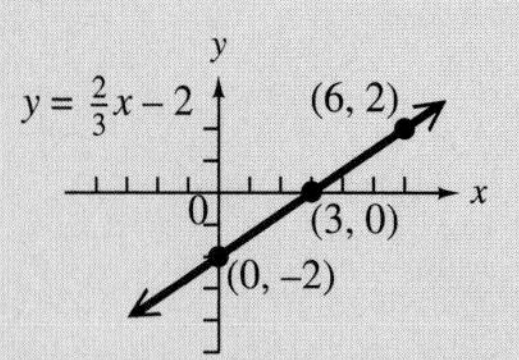

5.

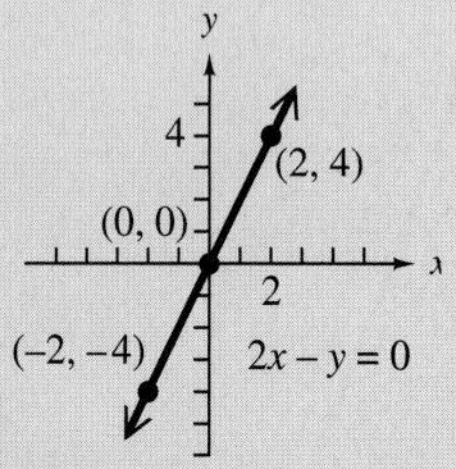

Line through the Origin

The graph of a linear equation of the form

$$Ax + By = 0,$$

where A and B are nonzero real numbers, passes through the origin $(0, 0)$.

OBJECTIVE 4 Graph linear equations of the form $y = k$ or $x = k$.

Consider the following linear equations.

$y = -4$ can be written as $0x + y = -4$.

$x = 3$ can be written as $x + 0y = 3$.

When the coefficient of x or y is 0, the graph is a horizontal or vertical line.

EXAMPLE 5 Graphing a Horizontal Line ($y = k$)

Graph $y = -4$.

For any value of x, y is always equal to -4. Three ordered pairs that satisfy the equation are shown in the table of values. Drawing a line through these points gives the **horizontal line** shown in **Figure 14.** The y-intercept is $(0, -4)$. There is no x-intercept.

x can be *any* real number.

x	y
-2	-4
0	-4
3	-4

y must be -4.

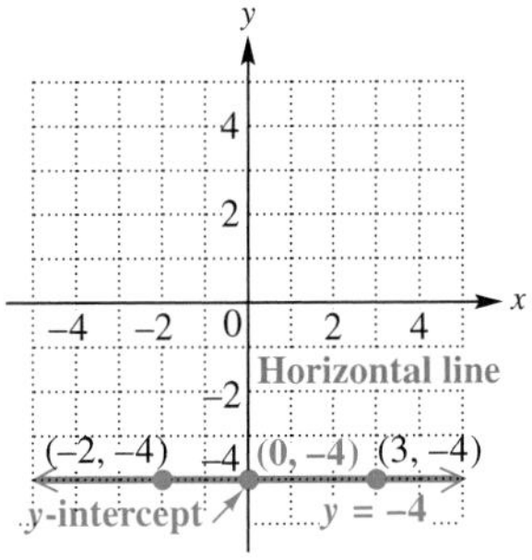

Figure 14

Work Problem 6 at the Side. ▶

EXAMPLE 6 Graphing a Vertical Line ($x = k$)

Graph $x - 3 = 0$.

First we add 3 to each side of the equation $x - 3 = 0$ to get the equivalent equation $x = 3$. All ordered-pair solutions of this equation have x-coordinate 3. Any number can be used for y. We show three ordered pairs that satisfy the equation in the table of values. The graph is the **vertical line** shown in **Figure 15.** The x-intercept is $(3, 0)$. There is no y-intercept.

y can be *any* real number.

x	y
3	3
3	0
3	-2

x must be 3.

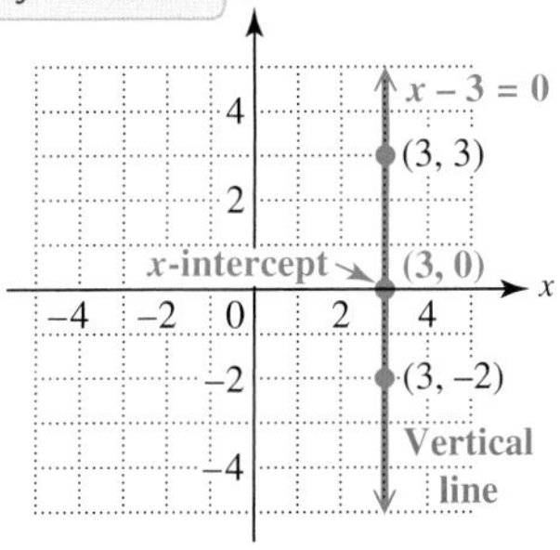

Figure 15

Work Problem 7 at the Side. ▶

6 Graph $y = -5$.

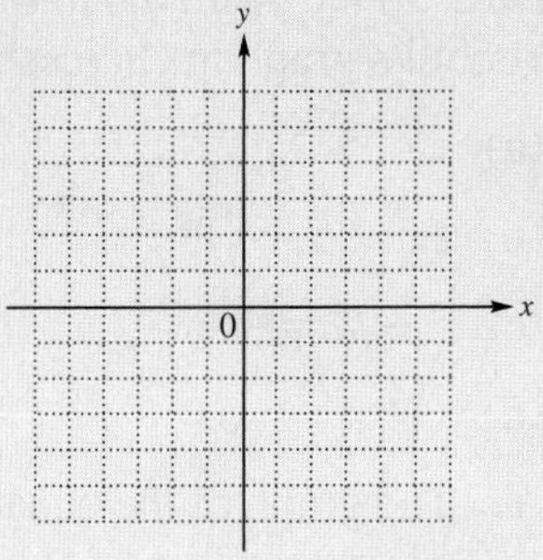

This is the graph of a ________ line. There is no x-intercept.

7 Graph $x + 4 = 6$.

GS First subtract ____ from each side to obtain the equivalent equation $x =$ ____. Now graph this equation.

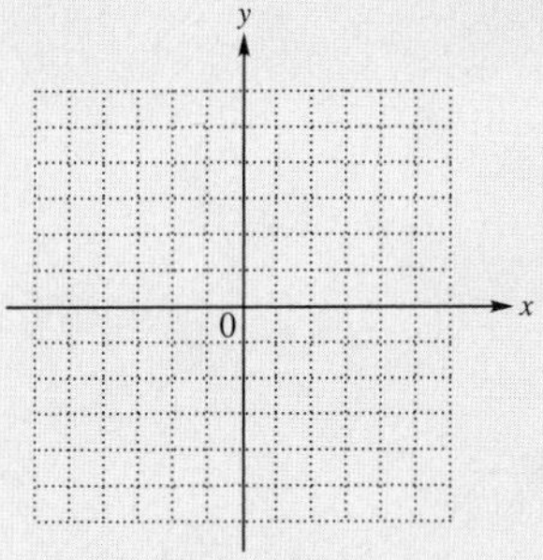

This is the graph of a ________ line. There is no y-intercept.

Answers

6.

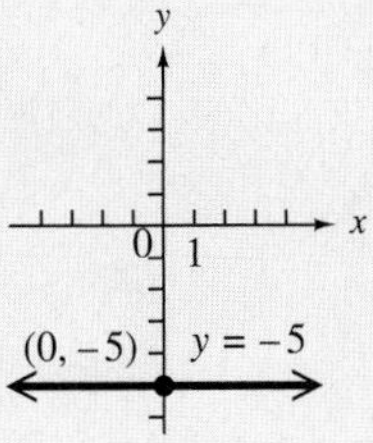

horizontal

7. 4; 2

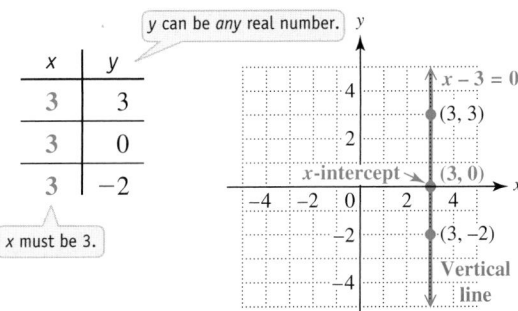

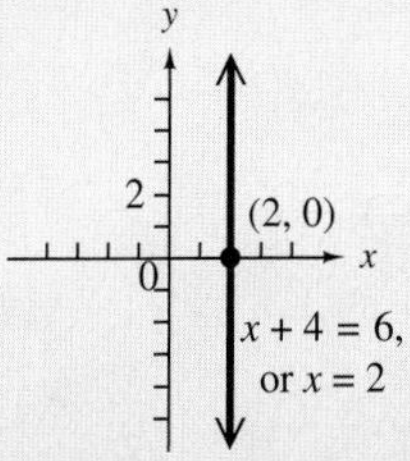

vertical

8 Match each linear equation in parts (a)–(d) with the information about its graph in choices A–D.

(a) $x = 5$

(b) $2x - 5y = 8$

(c) $y - 2 = 3$

(d) $x + 4y = 0$

A. The graph of the equation is a horizontal line.

B. The graph of the equation passes through the origin.

C. The graph of the equation is a vertical line.

D. The graph of the equation passes through $(9, 2)$.

Answers

8. **(a)** C **(b)** D **(c)** A **(d)** B

From **Examples 5 and 6,** we make the following observations.

Horizontal and Vertical Lines

The graph of the linear equation $\mathbf{y = k}$**,** where k is a real number, is the **horizontal line** with y-intercept $(0, k)$. There is no x-intercept (unless the horizontal line is the x-axis itself).

The graph of the linear equation $\mathbf{x = k}$**,** where k is a real number, is the **vertical line** with x-intercept $(k, 0)$. There is no y-intercept (unless the vertical line is the y-axis itself).

The x-axis is the horizontal line given by the equation y = 0, and the y-axis is the vertical line given by the equation x = 0.

CAUTION

The equations of horizontal and vertical lines are often confused with each other. The graph of $y = k$ is parallel to the x-axis and the graph of $x = k$ is parallel to the y-axis (for $k \neq 0$).

A summary of the forms of linear equations from this section follows.

Graphing a Linear Equation

Equation	*To Graph*	*Example*
$y = k$	Draw a horizontal line, through $(0, k)$.	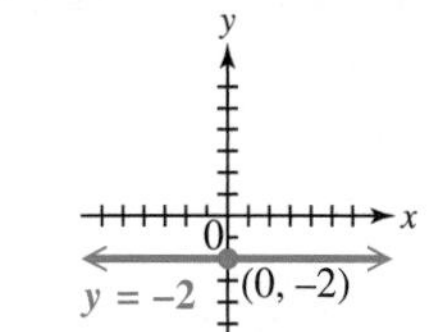
$x = k$	Draw a vertical line, through $(k, 0)$.	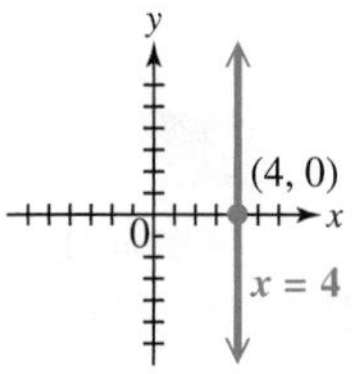
$Ax + By = 0$	The graph passes through $(0, 0)$. To find additional points that lie on the graph, choose any value for x or y, except 0.	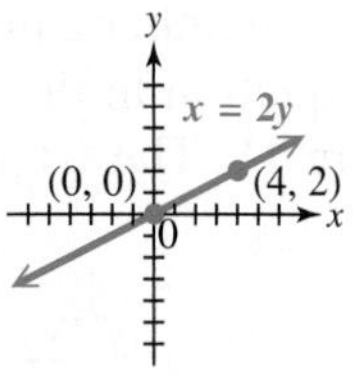
$Ax + By = C$ (where $A, B, C \neq 0$)	Find any two points on the line. A good choice is to find the intercepts. Let $x = 0$, and find the corresponding value of y. Then let $y = 0$, and find x. As a check, get a third point by choosing a value for x or y that has not yet been used.	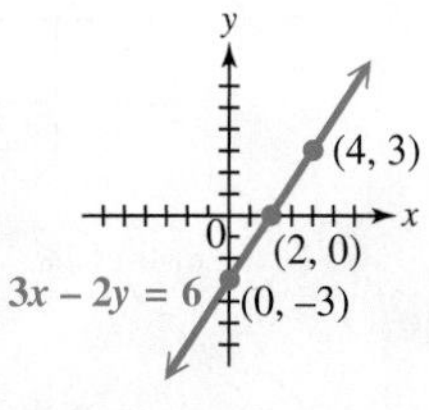

◀ **Work Problem 8 at the Side.**

OBJECTIVE 5 **Use a linear equation to model data.**

EXAMPLE 7 **Using a Linear Equation to Model Credit Card Debt**

The amount of credit card debt y in billions of dollars in the United States from 2000 through 2008 can be modeled by the linear equation

$$y = 32.0x + 684,$$

where $x = 0$ represents the year 2000, $x = 1$ represents 2001, and so on. (*Source:* The Nilson Report.)

(a) Use the equation to approximate credit card debt in the years 2000, 2004, and 2008.

Substitute the appropriate value for each year x to find credit card debt in that year.

For 2000: $y = 32.0(0) + 684$ — Replace x with 0.
$y = 684$ billion dollars — Multiply, and then add.

For 2004: $y = 32.0(4) + 684$ — $2004 - 2000 = 4$
$y = 812$ billion dollars — Replace x with 4.

For 2008: $y = 32.0(8) + 684$ — $2008 - 2000 = 8$
$y = 940$ billion dollars — Replace x with 8.

(b) Write the information from part (a) as three ordered pairs, and use them to graph the given linear equation.

Since x represents the year and y represents the debt, the ordered pairs are (0, 684), (4, 812), and (8, 940). See **Figure 16.** (Arrowheads are not included with the graphed line, since the data are for the years 2000 to 2008 only—that is, from $x = 0$ to $x = 8$.)

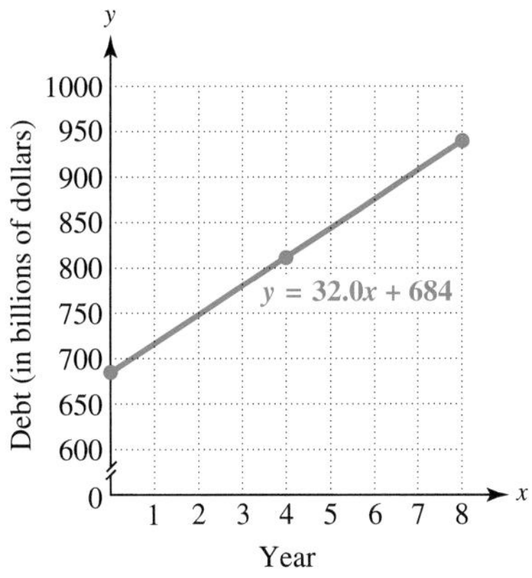

Figure 16

(c) Use the graph and the equation to approximate credit card debt in 2002.

For 2002, $x = 2$. On the graph, find 2 on the horizontal axis, move up to the graphed line and then across to the vertical axis. It appears that credit card debt in 2002 was about 750 billion dollars.

To use the equation, substitute 2 for x.

$y = 32.0x + 684$ — Given linear equation
$y = 32.0(2) + 684$ — Let $x = 2$.
$y = 748$ billion dollars — Multiply, and then add.

This result for 2002 is close to our estimate of 750 billion dollars from the graph.

Work Problem 9 at the Side. ▶

9 Use **(a)** the graph and **(b)** the equation in **Example 7** to approximate credit card debt in 2006.

Answers

9. (a) about 875 billion dollars
(b) 876 billion dollars

11.2 Exercises

FOR EXTRA HELP

Download the MyDashBoard App

MyMathLab®

CONCEPT CHECK *Fill in each blank with the correct response.*

1. A linear equation in two variables can be written in the form $Ax +$ ______ $=$ ______, where A, B, and C are real numbers and A and B are not both ______.

2. The graph of any linear equation in two variables is a straight ______. Every point on the line represents a ______ of the equation.

3. CONCEPT CHECK Match the information about each graph in Column I with the correct linear equation in Column II.

I	II
(a) The graph of the equation has y-intercept $(0, -4)$.	**A.** $3x + y = -4$
(b) The graph of the equation has $(0, 0)$ as x-intercept and y-intercept.	**B.** $x - 4 = 0$
(c) The graph of the equation does not have an x-intercept.	**C.** $y = 4x$
(d) The graph of the equation has x-intercept $(4, 0)$.	**D.** $y = 4$

4. CONCEPT CHECK Which of these equations have a graph with only one intercept?

A. $x + 8 = 0$ **B.** $x - y = 3$ **C.** $x + y = 0$ **D.** $y = 4$

CONCEPT CHECK *Identify the intercepts of each graph. (All coordinates are integers.)*

5.

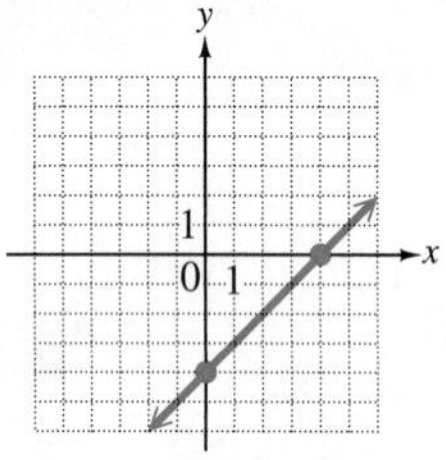

6.

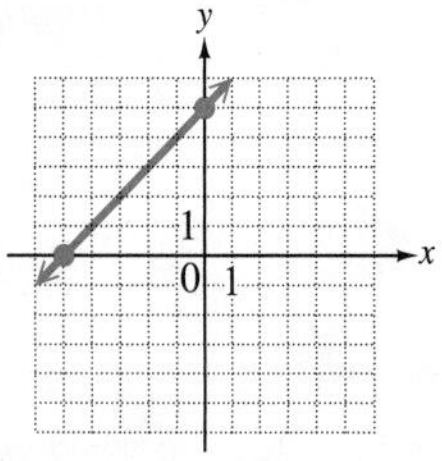

7.

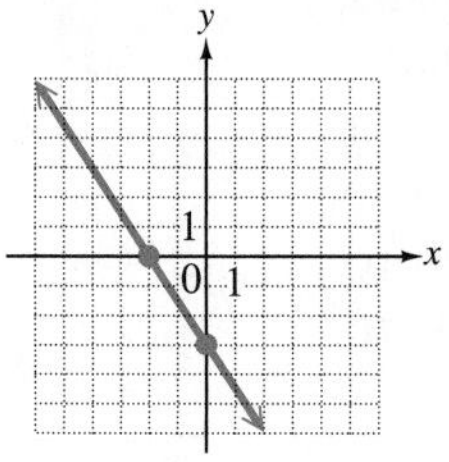

8.

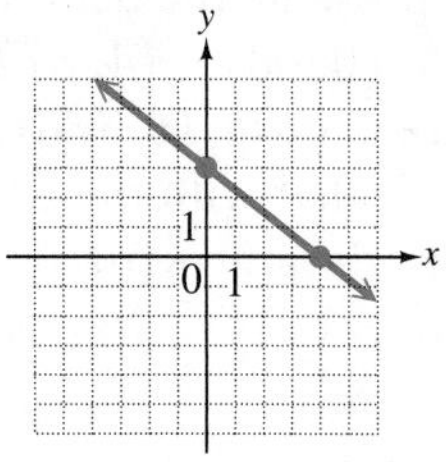

Complete the given ordered pairs for each equation. Then graph each equation by plotting the points and drawing the line through them. ***See Examples 1 and 2.***

9. $x + y = 5$

(0, ____), (____, 0), (2, ____)

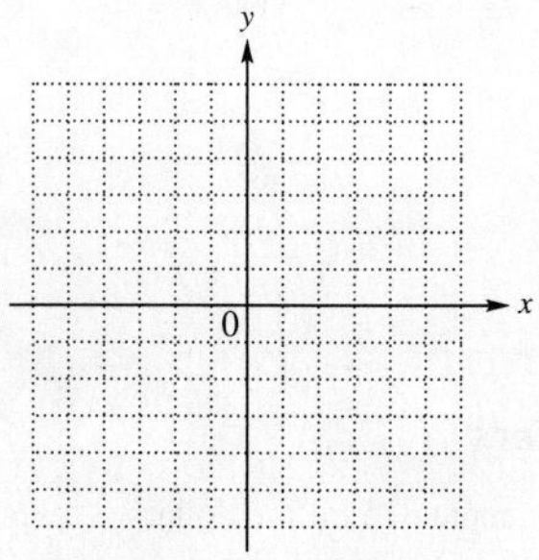

10. $x - y = 2$

(0, ____), (____, 0), (5, ____)

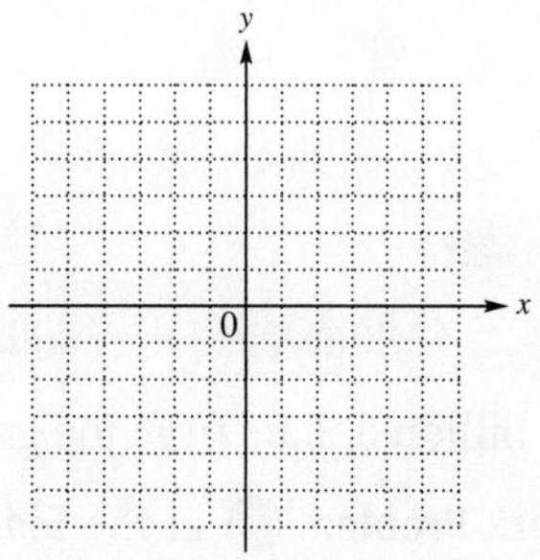

11. $y = \frac{2}{3}x + 1$

(0, ____), (3, ____), (−3, ____)

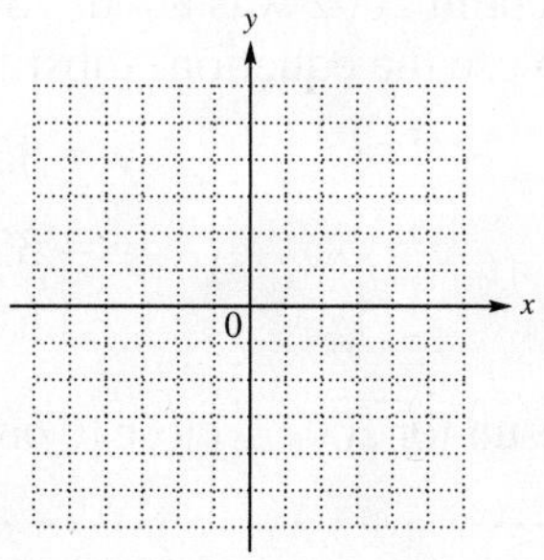

12. $y = -\frac{3}{4}x + 2$

$(0, \_\_\_)$, $(4, \_\_\_)$, $(-4, \_\_\_)$

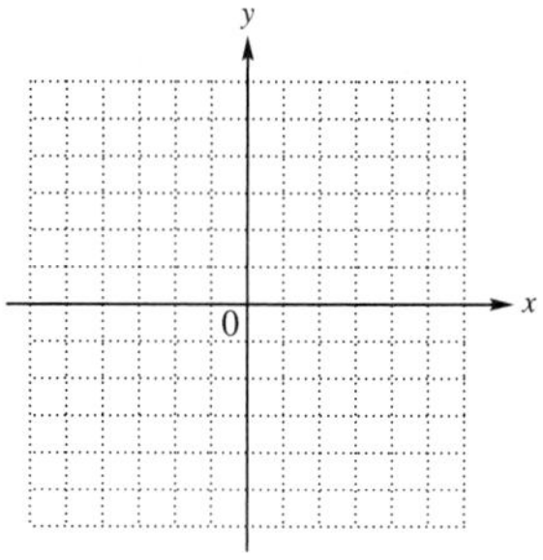

13. $3x = -y - 6$

$(0, \_\_\_)$, $(\_\_\_, 0)$, $\left(-\frac{1}{3}, \_\_\_\right)$

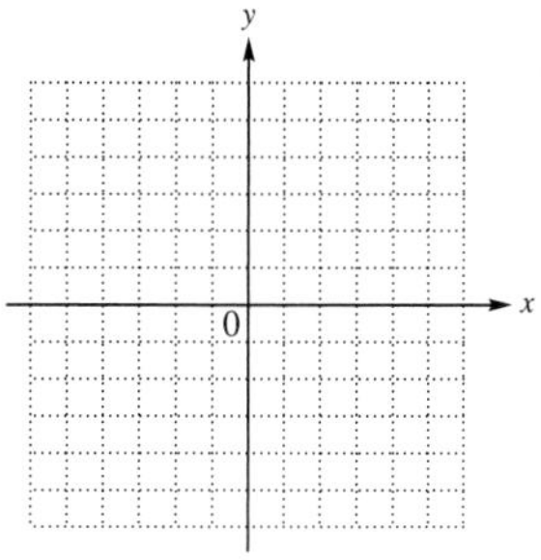

14. $x = 2y + 3$

$(\_\_\_, 0)$, $(0, \_\_\_)$, $\left(\_\_\_, \frac{1}{2}\right)$

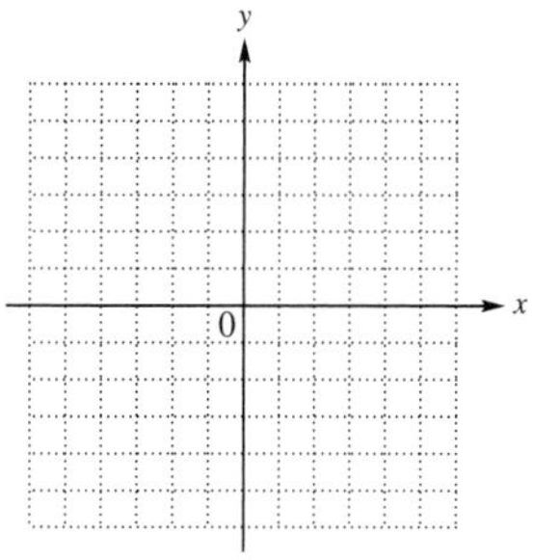

Find the intercepts for the graph of each equation. ***See Example 2.***

15. $x - y = 8$

GS

To find the y-intercept,
let $x =$ _______.
The y-intercept is _______.
To find the x-intercept,
let $y =$ _______.
The x-intercept is _______.

16. $x - y = 10$

GS

To find the y-intercept,
let $x =$ _______.
The y-intercept is _______.
To find the x-intercept,
let $y =$ _______.
The x-intercept is _______.

17. $2x - 3y = 24$

y-intercept: _______
x-intercept: _______

18. $-3x + 8y = 48$

y-intercept: _______
x-intercept: _______

19. $x + 6y = 0$

y-intercept: _______
x-intercept: _______

20. $3x - y = 0$

y-intercept: _______
x-intercept: _______

Graph each linear equation using intercepts. ***See Examples 1–6.***

21. $y = x - 2$

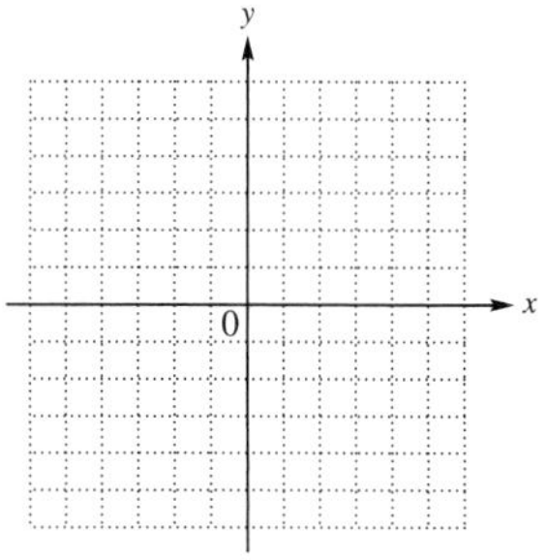

22. $y = -x + 6$

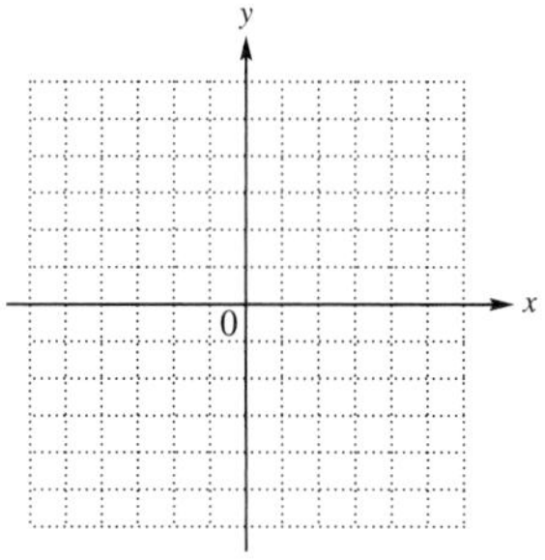

23. $x - y = 4$

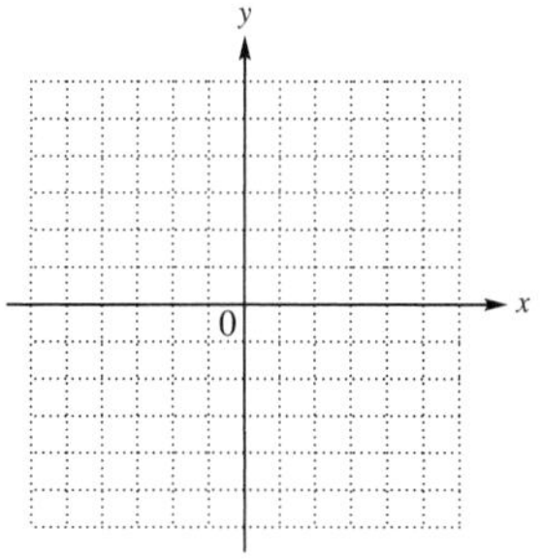

24. $x - y = 5$

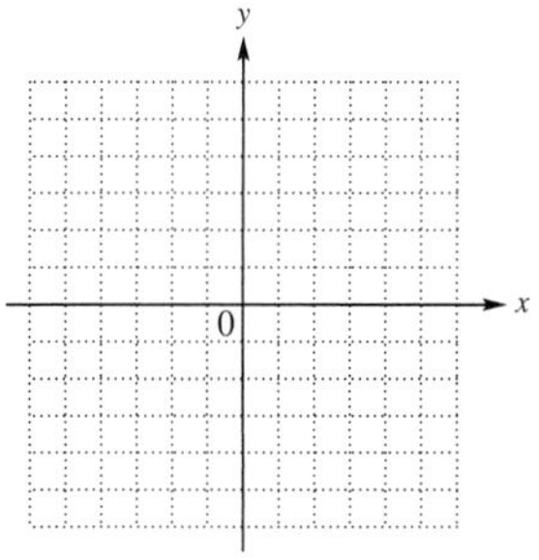

25. $2x + y = 6$

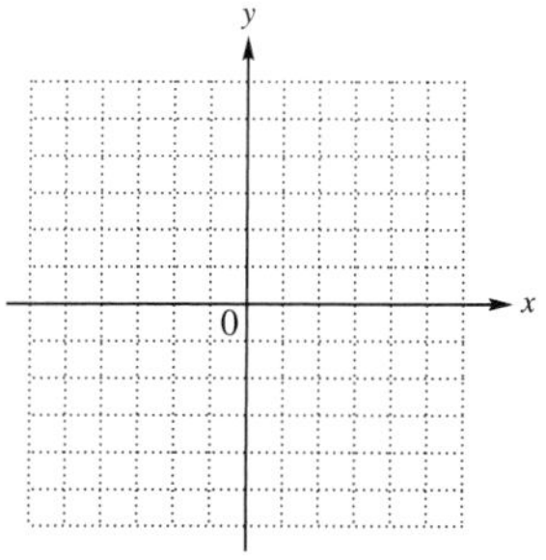

26. $-3x + y = -6$

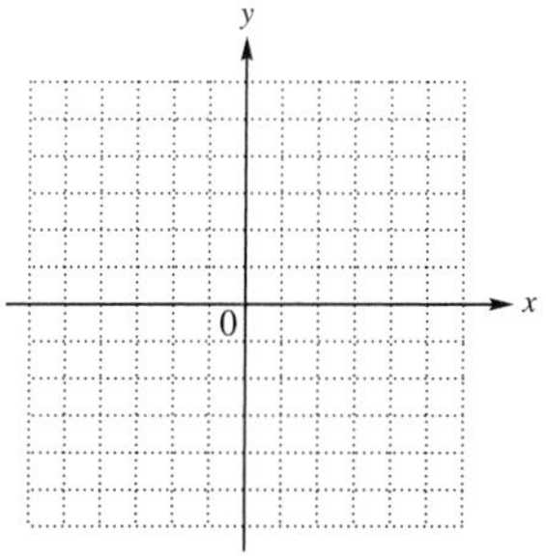

27. $y = 2x - 5$

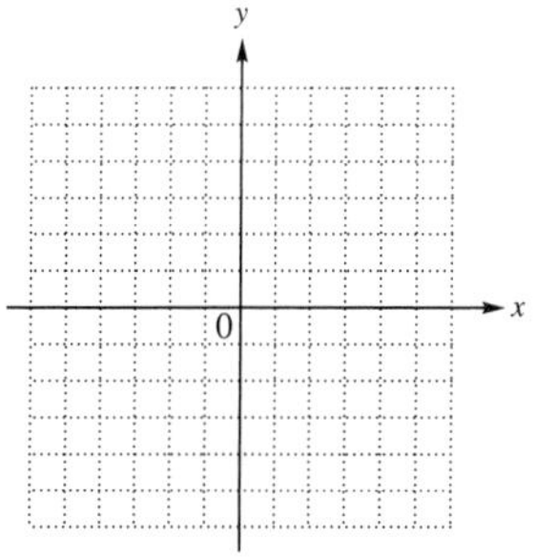

28. $y = 4x + 3$

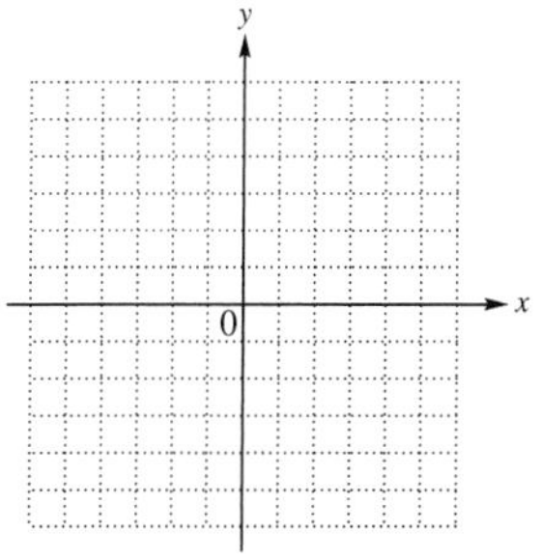

29. $3x + 7y = 14$

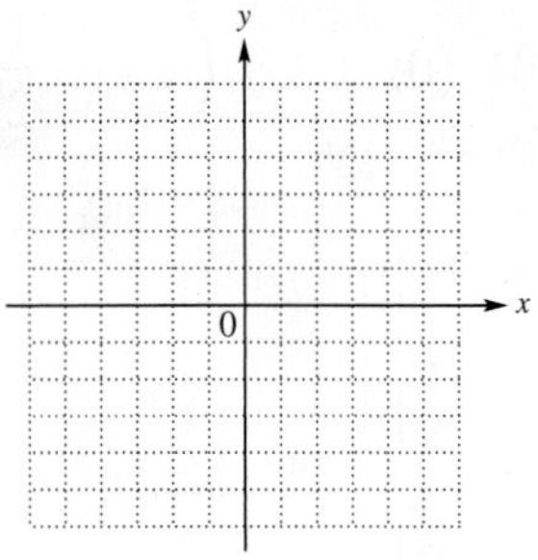

30. $6x - 5y = 18$

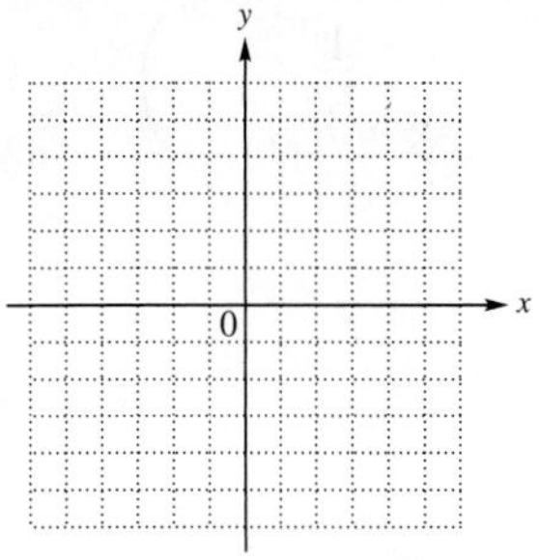

31. $y - 2x = 0$

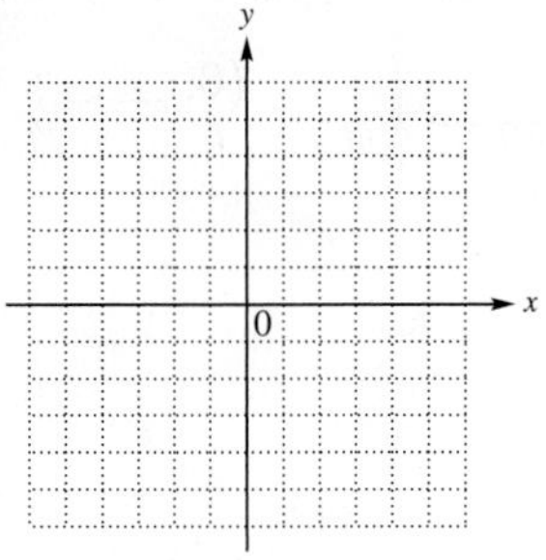

32. $y + 3x = 0$

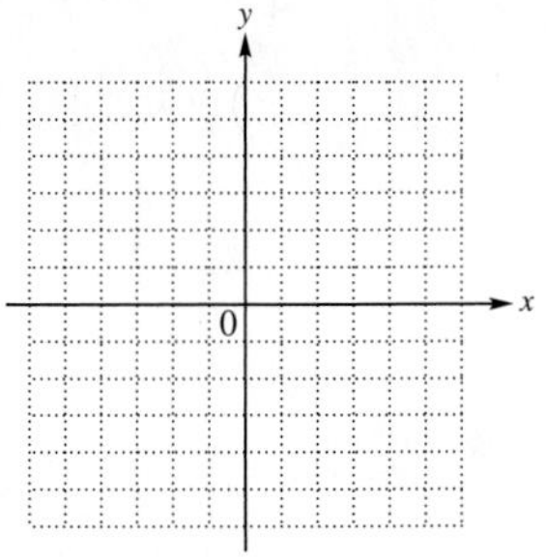

33. $y = -6x$

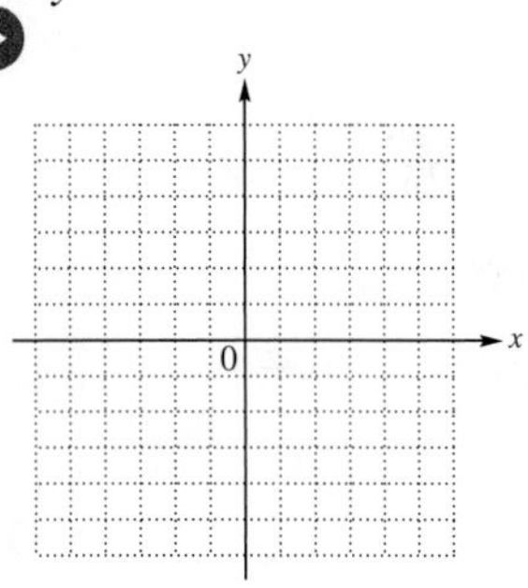

34. $y = 4x$

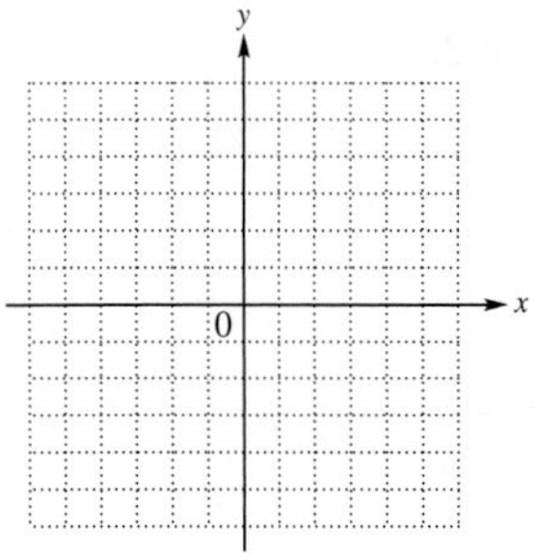

35. $x = -2$

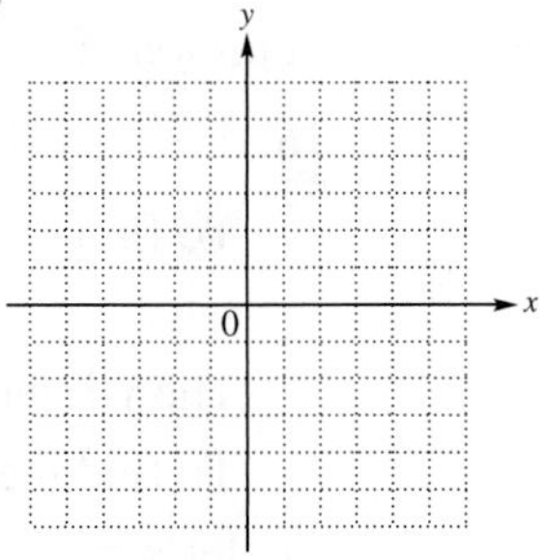

36. $x = 4$

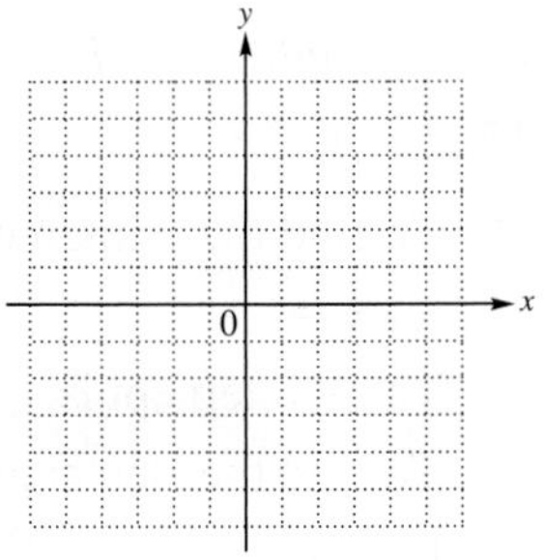

37. $y - 3 = 0$

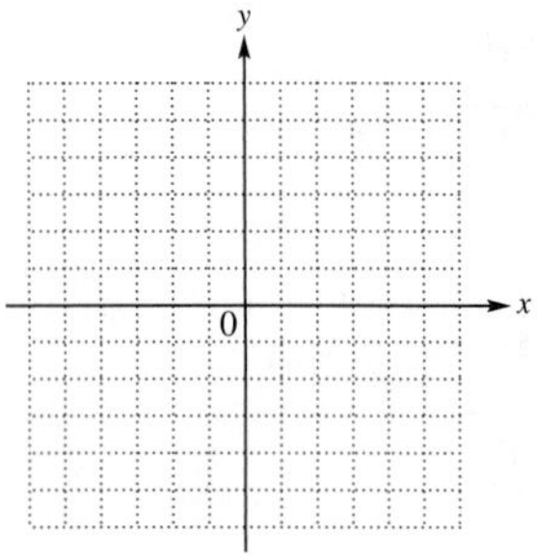

38. $y + 1 = 0$

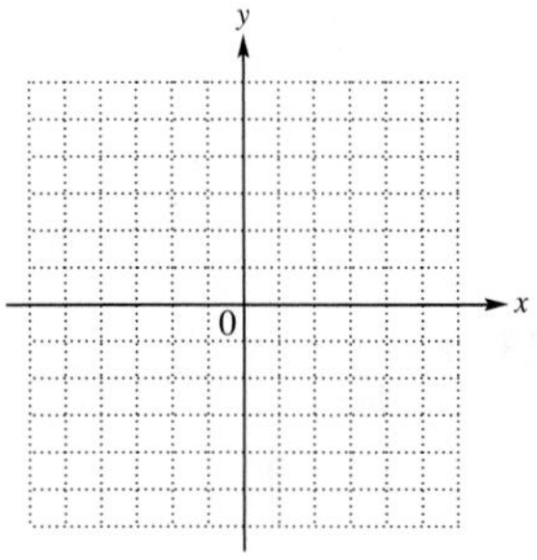

39. $-3y = 15$

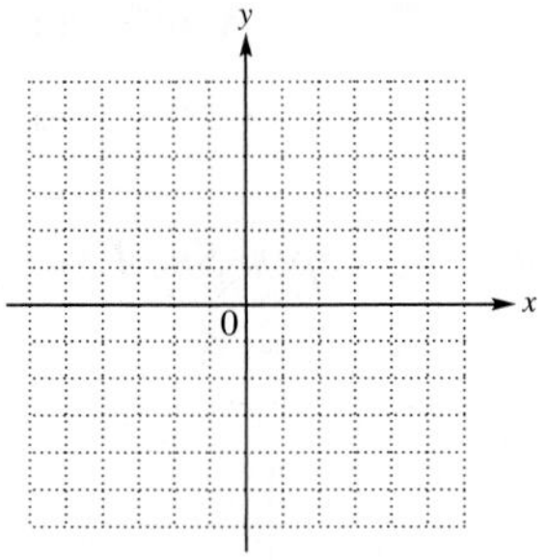

40. $-2y = 12$

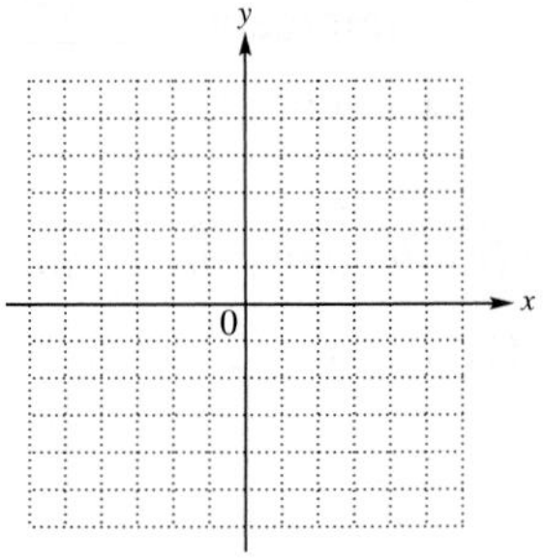

41. CONCEPT CHECK Match each equation in parts (a)–(d) with its graph in choices A–D.

(a) $x = -2$ **(b)** $y = -2$ **(c)** $x = 2$ **(d)** $y = 2$

A.

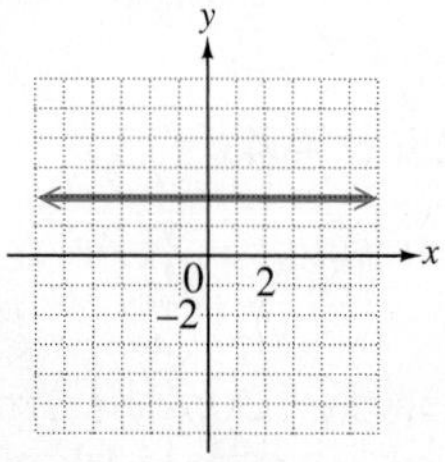

B.

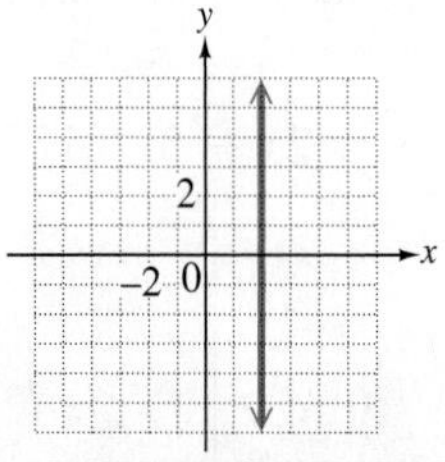

C.

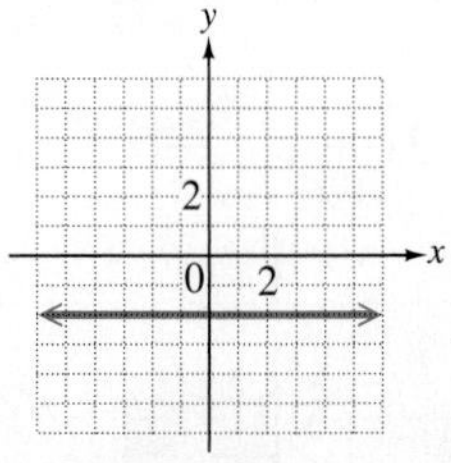

D.

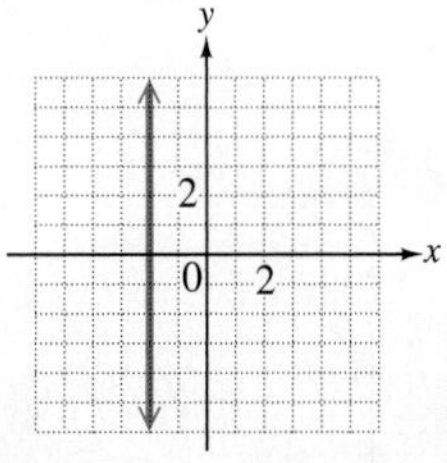

42. CONCEPT CHECK What is the equation of the x-axis? What is the equation of the y-axis?

In Exercises 43–46, describe what the graph of each linear equation will look like on the coordinate plane. (Hint: Rewrite the equation if necessary so that it is in a more recognizable form.)

43. $3x = y - 9$

44. $x - 10 = 1$

45. $3y = -6$

46. $2x = 4y$

Solve each problem. ***See Example 7.***

47. The height y of a woman is related to the length of her radius bone x and is approximated by the linear equation

$$y = 3.9x + 73.5,$$

where both x and y are in centimeters.

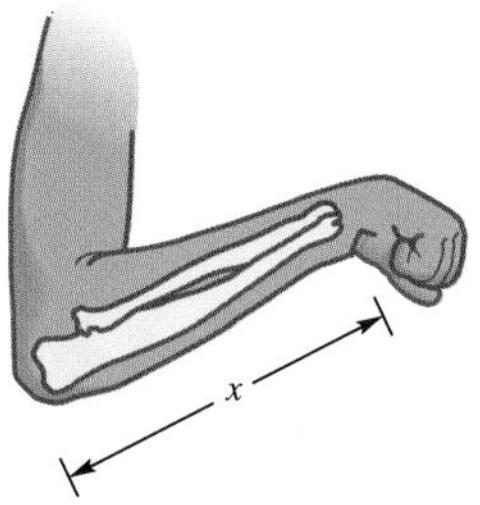

(a) Use the equation to find the approximate heights of women with radius bones of lengths 20 cm, 22 cm, and 26 cm.

(b) Write the information from part (a) as three ordered pairs.

(c) Graph the equation using the data from part (b).

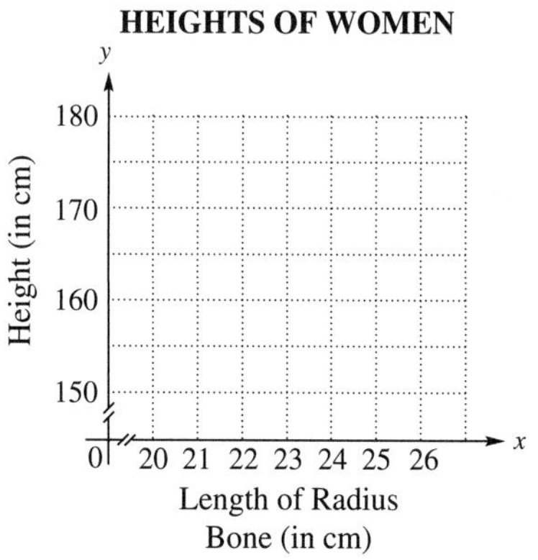

(d) Use the graph to estimate the length of the radius bone in a woman who is 167 cm tall. Then use the equation to find the length of this radius bone to the nearest centimeter. (*Hint:* Substitute for y in the equation.)

48. The weight y (in pounds) of a man taller than 60 in. can be roughly approximated by the linear equation

$$y = 5.5x - 220,$$

where x is the height of the man in inches.

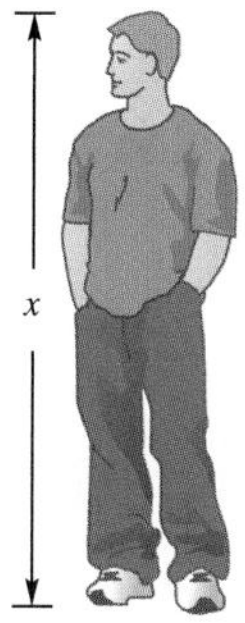

(a) Use the equation to approximate the weights of men whose heights are 62 in., 66 in., and 72 in.

(b) Write the information from part (a) as three ordered pairs.

(c) Graph the equation using the data from part (b).

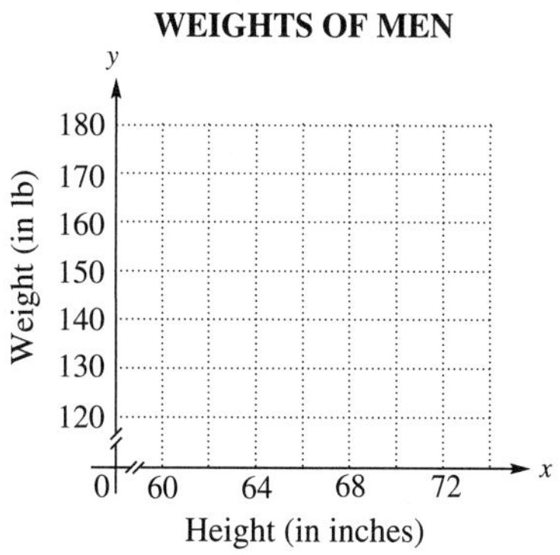

(d) Use the graph to estimate the height of a man who weighs 155 lb. Then use the equation to find the height of this man to the nearest inch. (*Hint:* Substitute for y in the equation.)

49. As a fundraiser, a school club is selling posters. The printer charges a \$25 set-up fee, plus \$0.75 for each poster. The cost y in dollars to print x posters is given by the linear equation

$$y = 0.75x + 25.$$

(a) What is the cost y in dollars to print 50 posters? To print 100 posters?

(b) Find the number of posters x if the printer billed the club for costs of \$175.

(c) Write the information from parts (a) and (b) as three ordered pairs.

(d) Use the data from part (c) to graph the equation.

POSTER COSTS

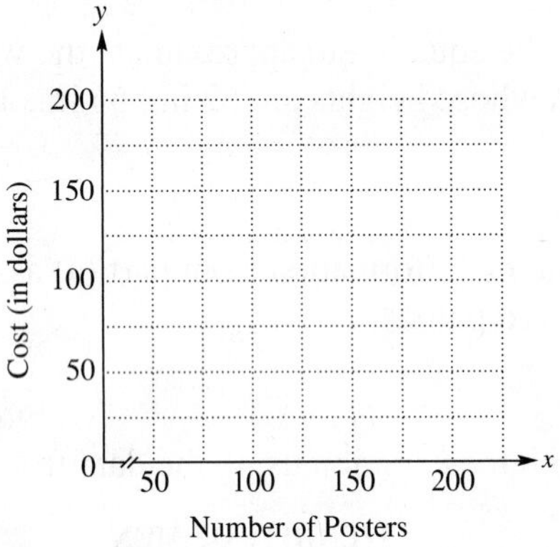

50. A gas station is selling gasoline for \$4.50 per gallon and charges \$7 for a car wash. The cost y in dollars for x gallons of gasoline and a car wash is given by the linear equation

$$y = 4.50x + 7.$$

(a) What is the cost y in dollars for 9 gallons of gasoline and a car wash? For 4 gallons of gasoline and a car wash?

(b) Find the number of gallons of gasoline x if the cost for the gasoline and a car wash is \$43.

(c) Write the information from parts (a) and (b) as three ordered pairs.

(d) Use the data from part (c) to graph the equation.

GASOLINE AND CAR WASH COSTS

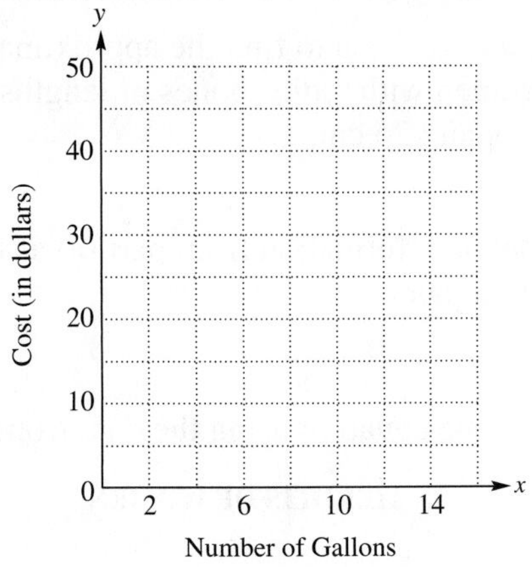

51. The graph shows the value of a certain sport-utility vehicle over the first 5 yr of ownership. Use the graph to do the following.

(a) Determine the initial value of the SUV.

(b) Find the **depreciation** (loss in value) from the original value after the first 3 yr.

(c) What is the annual or yearly depreciation in each of the first 5 yr?

(d) What does the ordered pair (5, 5000) mean in the context of this problem?

SUV VALUE

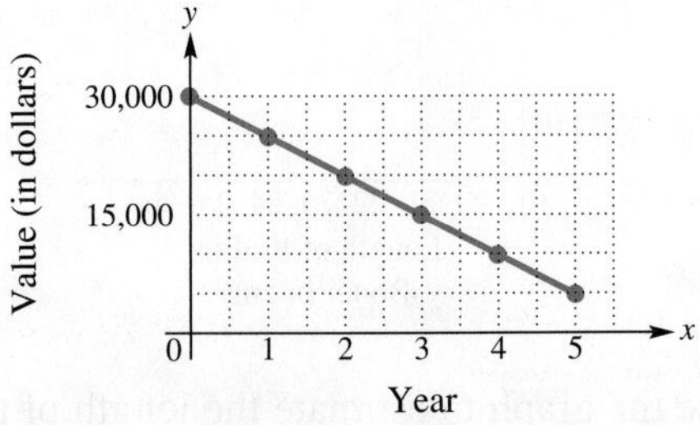

52. Demand for an item is often closely related to its price. As price increases, demand decreases, and as price decreases, demand increases. Suppose demand for a video game is 2000 units when the price is \$40, and demand is 2500 units when the price is \$30.

(a) Let x be the price and y be the demand for the game. Graph the two given pairs of prices and demands.

VIDEO GAME PRICE/DEMAND

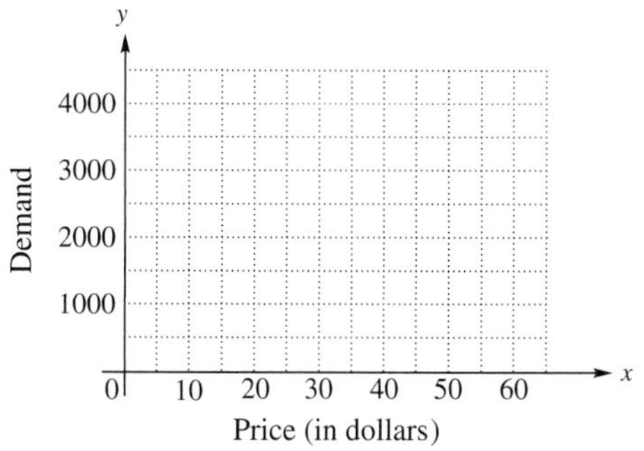

(b) Assume the relationship is linear. Draw a line through the two points from part (a). From your graph, estimate the demand if the price drops to \$20.

(c) Use the graph to estimate the price if the demand is 3500 units.

(d) Write the prices and demands from parts (b) and (c) as ordered pairs.

53. U.S. per capita consumption of cheese increased for the years 1980 through 2009 as shown in the graph. If $x = 0$ represents 1980, $x = 5$ represents 1985, and so on, per capita consumption y in pounds can be modeled by the linear equation

$$y = 0.5249x + 18.54.$$

Cheese Consumption

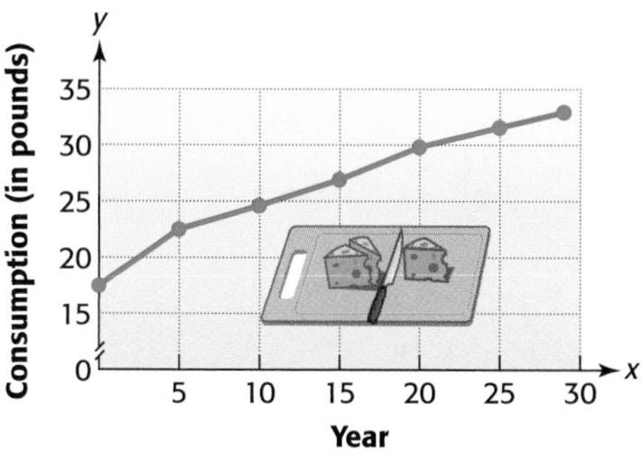

Source: U.S. Department of Agriculture.

(a) Use the equation to approximate cheese consumption in 1990, 2000, and 2009 to the nearest tenth.

(b) Use the graph to estimate cheese consumption for the same years.

(c) How do the approximations using the equation compare to the estimates from the graph?

54. The number of U.S. marathon finishers y (in thousands) from 1990 through 2010 are shown in the graph and modeled by the linear equation

$$y = 13.36x + 220.8,$$

where $x = 0$ corresponds to 1990, $x = 5$ corresponds to 1995, and so on.

U.S. Marathon Finishers

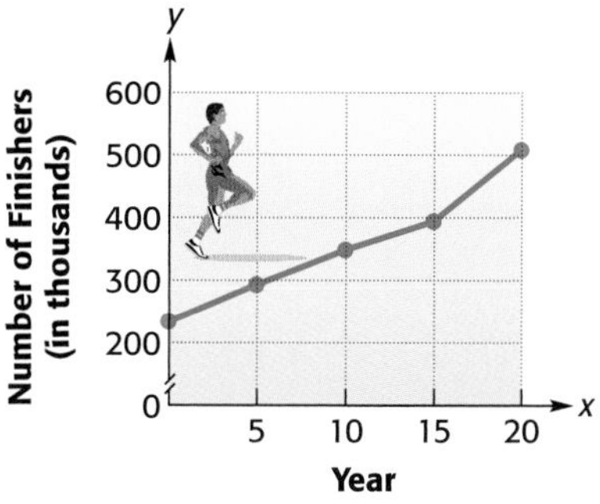

Source: Running U.S.A.

(a) Use the equation to approximate the number of U.S. marathon finishers in 1990, 2000, and 2010 to the nearest thousand.

(b) Use the graph to estimate the number of U.S. marathon finishers for the same years.

(c) How do the approximations using the equation compare to the estimates from the graph?

11.3 The Slope of a Line

OBJECTIVES

1. Find the slope of a line, given two points.
2. Find the slope from the equation of a line.
3. Use slope to determine whether two lines are parallel, perpendicular, or neither.

An important characteristic of the lines we graphed in the previous section is their slant or "steepness," as viewed from *left to right*. See **Figure 17**.

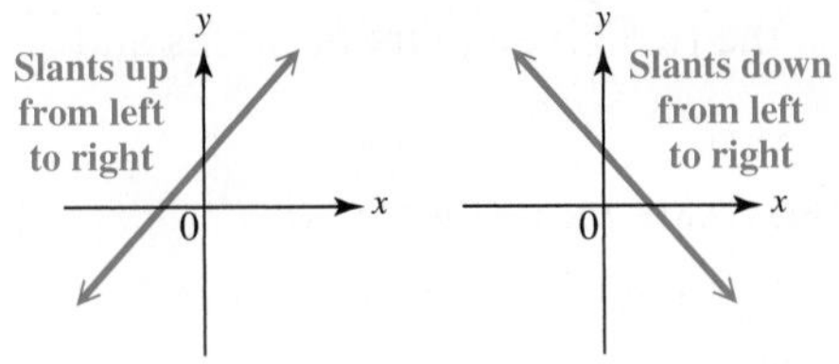

Figure 17

One way to measure the steepness of a line is to compare the vertical change in the line to the horizontal change while moving along the line from one fixed point to another. This measure of steepness is the *slope* of the line.

OBJECTIVE 1 Find the slope of a line, given two points. To find the steepness, or slope, of the line in **Figure 18,** we begin at point Q and move to point P. The vertical change, or **rise,** is the change in the y-values, which is the difference

$$6 - 1 = 5 \text{ units.}$$

The horizontal change, or **run,** from Q to P is the change in the x-values, which is the difference

$$5 - 2 = 3 \text{ units.}$$

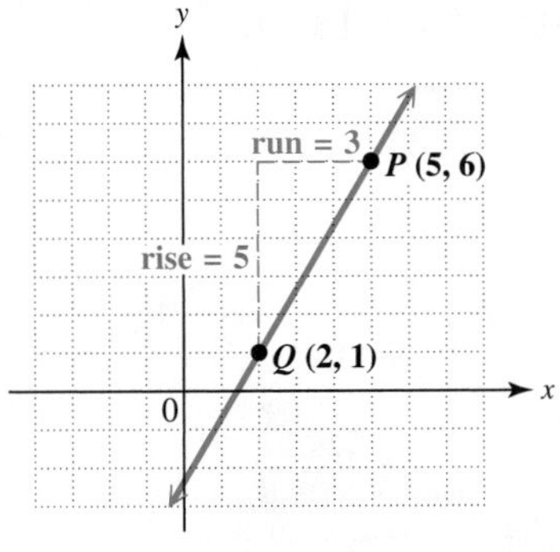

Figure 18

Remember from **Section 5.1** that one way to compare two numbers is by using a ratio. **Slope** is the ratio of the vertical change in y to the horizontal change in x. The line in **Figure 18** has

$$\text{slope} = \frac{\textbf{vertical change in } y \textbf{ (rise)}}{\textbf{horizontal change in } x \textbf{ (run)}} = \frac{5}{3}.$$

To confirm this ratio, we can count grid squares. We start at point Q in **Figure 18** and count *up* 5 grid squares to find the vertical change (rise). To find the horizontal change (run) and arrive at point P, we count to the *right* 3 grid squares. The slope is $\frac{5}{3}$, as found above.

Slope is a single number that allows us to determine the direction in which a line is slanting from left to right, as well as how much slant there is to the line.

EXAMPLE 1 Finding the Slope of a Line

Find the slope of the line in **Figure 19.**

We use the two points shown on the line. The vertical change is the difference in the y-values, or $-1 - 3 = -4$, and the horizontal change is the difference in the x-values, or $6 - 2 = 4$. Thus, the line has

$$\text{slope} = \frac{\text{change in } y \text{ (rise)}}{\text{change in } x \text{ (run)}} = \frac{-4}{4}, \text{ or } -1.$$

Counting grid squares, we begin at point P and count *down* 4 grid squares. Because we counted down, we write the vertical change as a negative number, -4 here. Then we count to the *right* 4 grid squares to reach point Q. The slope is $\frac{-4}{4}$, or -1.

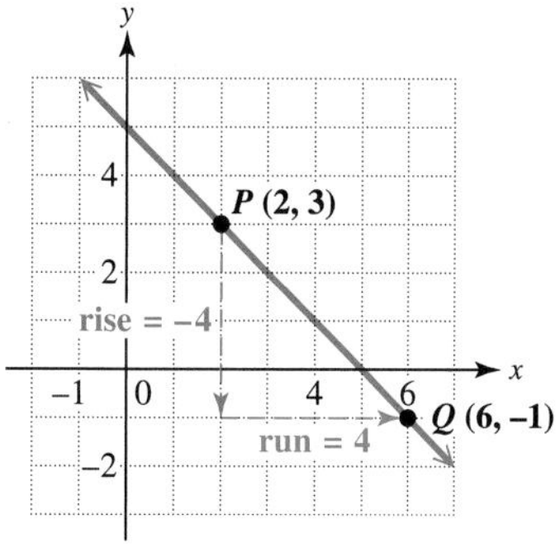

Figure 19

Work Problem 1 at the Side. ▶

Note

The slope of a line is the same for any two points on the line. To see this, refer to **Figure 19.** Find two points, say $(3, 2)$ and $(5, 0)$, which also lie on the line. If we start at $(3, 2)$ and count *down* 2 units and then to the *right* 2 units, we arrive at $(5, 0)$. The slope is $\frac{-2}{2}$, or -1, the same slope we found in **Example 1.**

The concept of slope is used in many everyday situations. See **Figure 20.** A highway with a 10%, or $\frac{1}{10}$, grade (or slope) rises 1 m for every 10 m horizontally. Architects specify the pitch of a roof by using slope. A $\frac{5}{12}$ roof means that the roof rises 5 ft vertically for every 12 ft that it runs horizontally. The slope of a stairwell indicates the ratio of the vertical rise to the horizontal run. The slope of the stairwell is $\frac{8}{12}$, or $\frac{2}{3}$.

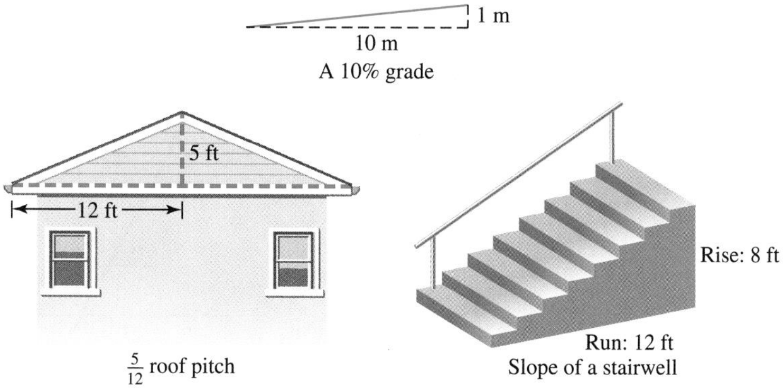

Figure 20

We can generalize the preceding discussion and find the slope of a line through two nonspecific points (x_1, y_1) and (x_2, y_2). This notation is called **subscript notation.** Read x_1 as **"x-sub-one"** and x_2 as **"x-sub-two."** See **Figure 21** on the next page.

1 Find the slope of each line.

GS **(a)**

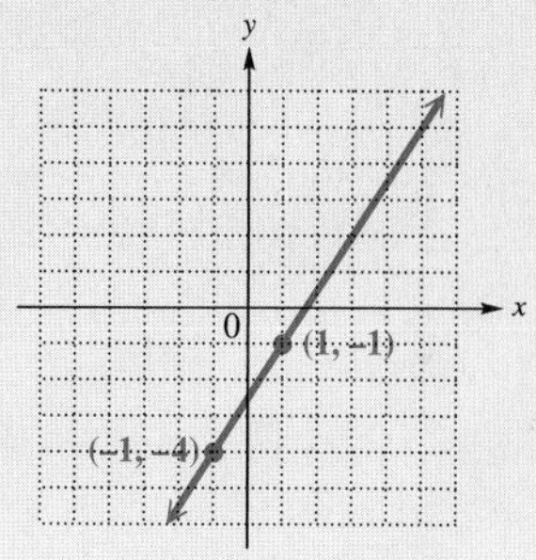

Begin at $(-1, -4)$.

Count (*up* / *down*) ____ units.

Then count to the (*right* / *left*) ____ units.

$$\text{slope} = \frac{\text{vertical change in } y}{\text{horizontal change in } x} = \frac{\_\_\_}{\_\_\_}$$

The slope is ____.

(b)

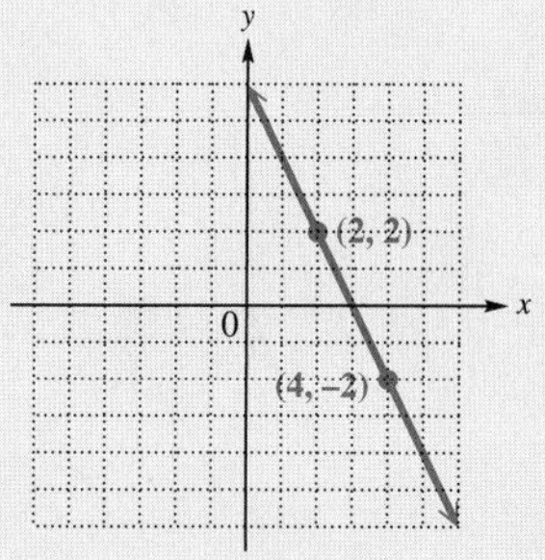

Answers

1. **(a)** up; 3; right; 2; 3; 2; $\frac{3}{2}$
   (b) -2

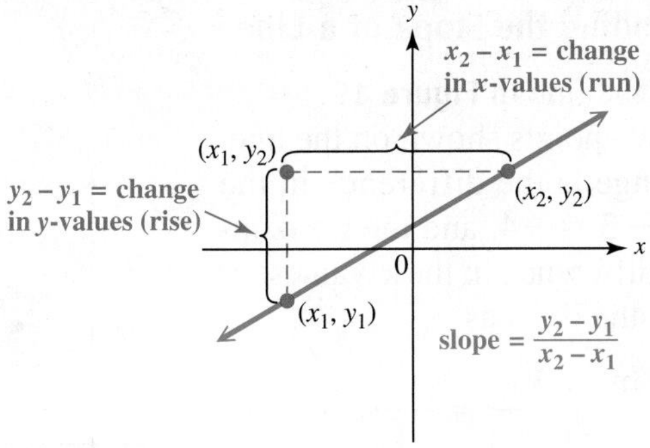

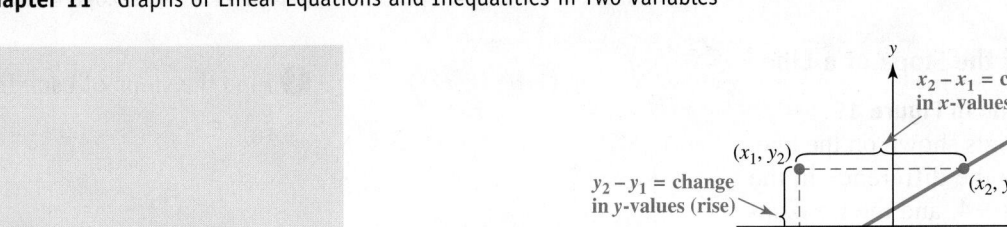

Figure 21

Moving along the line from the point (x_1, y_1) to the point (x_2, y_2), we see that y changes by $y_2 - y_1$ units. This is the vertical change (rise). Similarly, x changes by $x_2 - x_1$ units, which is the horizontal change (run). The slope of the line is the ratio of $y_2 - y_1$ to $x_2 - x_1$.

Note

Subscript notation is used to identify a point. It does *not* indicate any operation. Note the difference between x_2, a nonspecific value, and x^2, which means $x \cdot x$. Read x_2 as "x-sub-two," *not* "x squared."

Slope Formula

The **slope** m of the line passing through the points (x_1, y_1) and (x_2, y_2) is defined as follows. (Traditionally, the letter m represents slope.)

$$m = \frac{\textbf{change in } y}{\textbf{change in } x} = \frac{y_2 - y_1}{x_2 - x_1} \quad (\text{where } x_1 \neq x_2)$$

The slope gives the change in y for each unit of change in x.

EXAMPLE 2 **Finding Slopes of Lines Given Two Points**

Find the slope of each line.

(a) The line passing through $(-4, 7)$ and $(1, -2)$

Label the given points, and then apply the slope formula.

$$\begin{matrix} (x_1, y_1) & & (x_2, y_2) \\ \downarrow\ \downarrow & & \downarrow\ \downarrow \\ (-4, 7) & \text{and} & (1, -2) \end{matrix}$$

$$\text{slope } m = \frac{y_2 - y_1}{x_2 - x_1} = \frac{-2 - 7}{1 - (-4)}$$

Substitute carefully.

$$= \frac{-9}{5}, \text{ or } -\frac{9}{5}$$

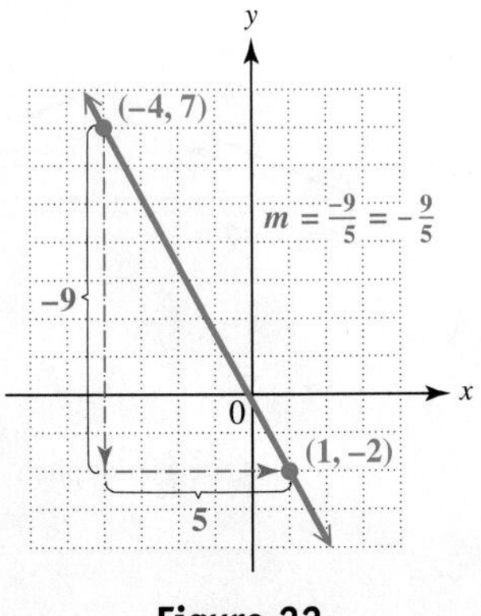

Figure 22

Count grid squares in **Figure 22** to confirm that the slope is $-\frac{9}{5}$.

Continued on Next Page

(b) The line passing through $(-9, -2)$ and $(12, 5)$

Label the points, and then apply the slope formula.

$$\begin{matrix} (x_1, y_1) & & (x_2, y_2) \\ \downarrow\ \downarrow & & \downarrow\ \downarrow \\ (-9, -2) & \text{and} & (12, 5) \end{matrix}$$

$$\text{slope } m = \frac{y_2 - y_1}{x_2 - x_1} = \frac{5-(-2)}{12-(-9)}$$

$$= \frac{7}{21}, \text{ or } \frac{1}{3} \quad \text{Write in lowest terms.}$$

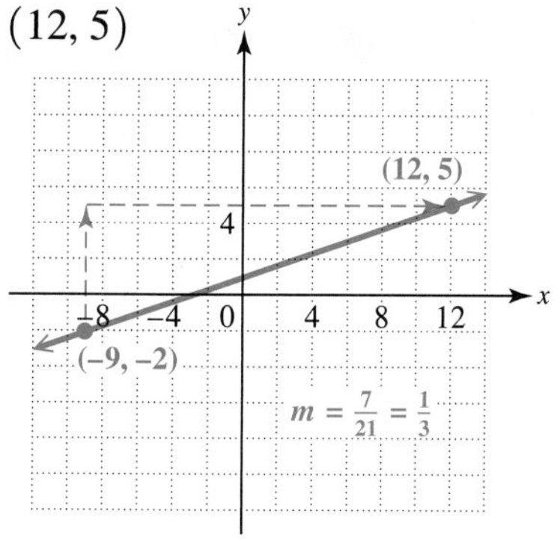

Figure 23

Confirm this calculation using **Figure 23.**

The same slope is obtained if we label the points in reverse order. ***It makes no difference which point is identified as (x_1, y_1) or (x_2, y_2).***

$$\begin{matrix} (x_2, y_2) & & (x_1, y_1) \\ \downarrow\ \downarrow & & \downarrow\ \downarrow \\ (-9, -2) & \text{and} & (12, 5) \end{matrix}$$

$$\text{slope } m = \frac{y_2 - y_1}{x_2 - x_1} = \frac{-2-5}{-9-12} \quad \text{Substitute.}$$

$$= \frac{-7}{-21}, \text{ or } \frac{1}{3} \quad \text{The same slope results.}$$

Work Problem 2 at the Side. ▶

The lines in **Figures 22 and 23** suggest the following generalization.

Orientation of Lines with Positive and Negative Slopes

A line with positive slope rises (slants up) from left to right.

A line with negative slope falls (slants down) from left to right.

EXAMPLE 3 Finding the Slope of a Horizontal Line

Find the slope of the line passing through $(-5, 4)$ and $(2, 4)$.

$$\begin{matrix} (x_1, y_1) & & (x_2, y_2) \\ \downarrow\ \downarrow & & \downarrow\ \downarrow \\ (-5, 4) & \text{and} & (2, 4) \end{matrix} \quad \text{Label the points.}$$

$$m = \frac{y_2 - y_1}{x_2 - x_1} = \frac{4-4}{2-(-5)} \quad \text{Substitute in the slope formula.}$$

$$= \frac{0}{7}, \text{ or } 0 \quad \textbf{Slope 0}$$

As shown in **Figure 24,** the line through these two points is horizontal, with equation $y = 4$. ***All horizontal lines have slope 0,*** since the difference in their y-values is always 0.

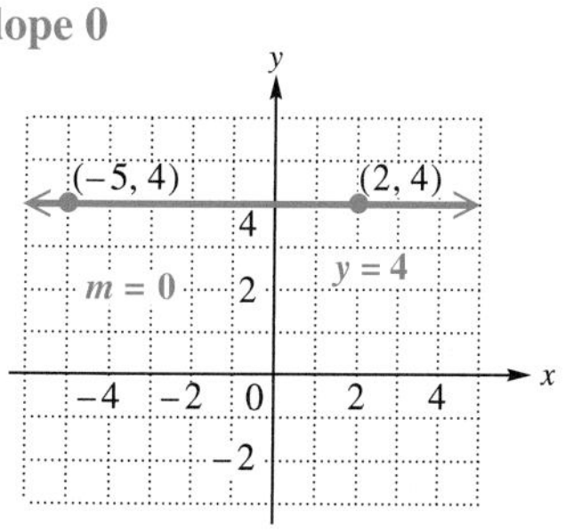

Figure 24

2 Find the slope of each line.

GS **(a)** The line passing through $(6, -2)$ and $(5, 4)$

Label the points.

$$\begin{matrix} (x_1, y_1) & & (\_\_, \_\_) \\ \downarrow\ \downarrow & & \downarrow\ \downarrow \\ (6, -2) & \text{and} & (5,\ 4) \end{matrix}$$

$$\text{slope } m = \frac{y_2 - y_1}{x_2 - x_1}$$

$$= \frac{\_\_ - (\_\_)}{\_\_ - 6}$$

$$= \frac{6}{\_\_}, \text{ or } \_\_$$

(b) The line passing through $(-3, 5)$ and $(-4, -7)$

(c) The line passing through $(6, -8)$ and $(-2, 4)$

(Find this slope in two different ways as in **Example 2(b).**)

Answers

2. (a) x_2; y_2; 4; -2; 5; -1; -6

(b) 12 **(c)** $-\frac{3}{2}$; $-\frac{3}{2}$

3 Find the slope of each line.

(a) The line passing through $(2, 5)$ and $(-1, 5)$

(b) The line passing through $(3, 1)$ and $(3, -4)$

4 Find the slope of each line.

GS **(a)** The line with equation $y = -1$

This is the equation of a (*horizontal* / *vertical*) line. It has (0 / *undefined*) slope.

(b) The line with equation $x - 4 = 0$

Rewrite this equation in an equivalent form.

$$x - 4 = 0$$

$$x = \_\_\_\_$$

This is the equation of a (*horizontal* / *vertical*) line. It has (0 / *undefined*) slope.

Answers

3. **(a)** 0 **(b)** undefined

4. **(a)** horizontal; 0
(b) 4; vertical; undefined

EXAMPLE 4 **Finding the Slope of a Vertical Line**

Find the slope of the line passing through $(6, 2)$ and $(6, -4)$.

$$\begin{matrix} (x_1, y_1) & & (x_2, y_2) \\ \downarrow\ \downarrow & & \downarrow\ \downarrow \\ (6, 2) & \text{and} & (6, -4) \end{matrix} \quad \text{Label the points.}$$

$$m = \frac{y_2 - y_1}{x_2 - x_1} = \frac{-4 - 2}{6 - 6} \quad \text{Substitute in the slope formula.}$$

$$= \frac{-6}{0} \quad \textbf{Undefined slope}$$

Because division by 0 is undefined, this line has undefined slope. (This is why the slope formula at the beginning of this section had the restriction $x_1 \neq x_2$.) The graph in **Figure 25** shows that this line is vertical, with equation $x = 6$. All points on a vertical line have the same x-value, so ***the slope of any vertical line is undefined.***

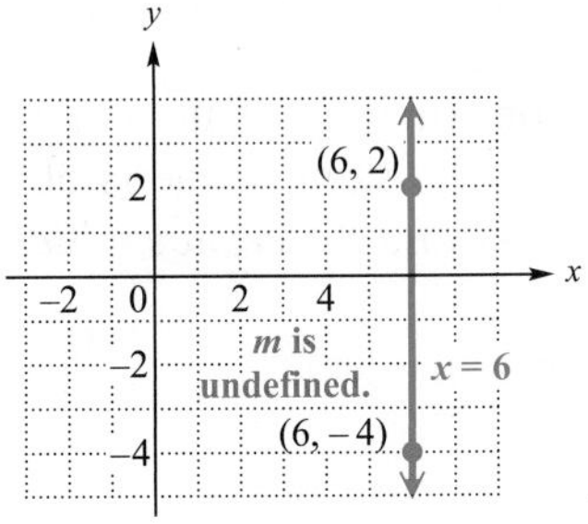

Figure 25

◀ **Work Problem 3 at the Side.**

Slopes of Horizontal and Vertical Lines

Horizontal lines, which have equations of the form $y = k$ (where k is a constant (number)), have **slope 0.**

Vertical lines, which have equations of the form $x = k$ (where k is a constant (number)), have **undefined slope.**

◀ **Work Problem 4 at the Side.**

Figure 26 summarizes the four cases for slopes of lines.

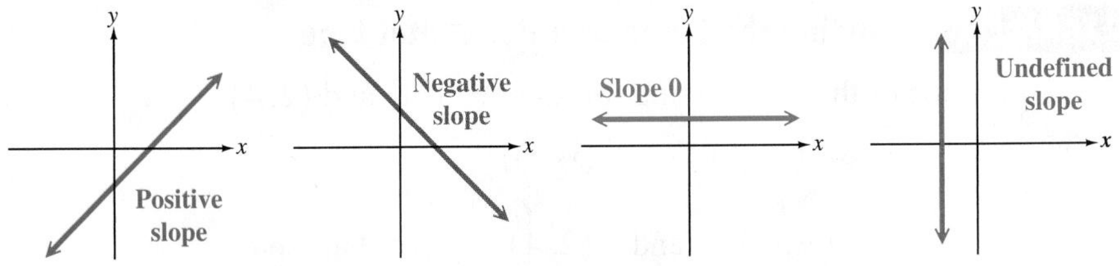

Slopes of lines

Figure 26

OBJECTIVE 2 **Find the slope from the equation of a line.** Consider the linear equation $y = -3x + 5$. We can find the slope of the line using any two points on the line. We get these two points by first choosing two different values of x and then finding the corresponding values of y. We arbitrarily choose $x = -2$ and $x = 4$.

$y = -3x + 5$		$y = -3x + 5$	
$y = -3(-2) + 5$	Let $x = -2$.	$y = -3(4) + 5$	Let $x = 4$.
$y = 6 + 5$	Multiply.	$y = -12 + 5$	Multiply.
$y = 11$	Add.	$y = -7$	Add.

The ordered pairs are $(-2, 11)$ and $(4, -7)$.

Now we apply the slope formula using the two points $(-2, 11)$ and $(4, -7)$.

$$m = \frac{-7 - 11}{4 - (-2)} = \frac{-18}{6} = -3$$

The slope, $m = -3$, is the same number as the coefficient of x in the equation $y = -3x + 5$. It can be shown that this always happens, *as long as the equation is solved for y*. This fact is used to find the slope of a line from its equation.

Finding the Slope of a Line from Its Equation

Step 1 Solve the equation for y. (See **Section 10.5.**)

Step 2 The slope is given by the coefficient of x.

EXAMPLE 5 Finding Slopes from Equations

Find the slope of each line.

(a) $2x - 5y = 4$

Step 1 Solve the equation for y.

$$2x - 5y = 4 \quad \text{Isolate } y \text{ on one side.}$$

$$-5y = 4 - 2x \quad \text{Subtract } 2x.$$

$$-5y = -2x + 4 \quad \text{Commutative property}$$

$$y = \frac{2}{5}x - \frac{4}{5} \quad \text{Divide } each \text{ term by } -5.$$

$\frac{-2x}{-5} = \frac{-2}{-5}x = \frac{2}{5}x$

↑ Slope

Step 2 The slope is given by the coefficient of x, so the slope is $\frac{2}{5}$.

(b)

$$8x + 4y = 1 \quad \text{Solve for } y.$$

$$4y = 1 - 8x \quad \text{Subtract } 8x.$$

$$4y = -8x + 1 \quad \text{Commutative property}$$

$$y = -2x + \frac{1}{4} \quad \text{Divide } each \text{ term by 4.}$$

The slope of this line is given by the coefficient of x, which is -2.

(c)

$$3y + x = -3$$

We leave out the step showing the commutative property.

$$3y = -x - 3 \quad \text{Subtract } x.$$

$$y = \frac{-x}{3} - 1 \quad \text{Divide } each \text{ term by 3.}$$

$$y = -\frac{1}{3}x - 1 \qquad \frac{-x}{3} = \frac{-1x}{3} = -\frac{1}{3}x$$

The slope is $-\frac{1}{3}$, *not* $\frac{-x}{3}$ or $-\frac{x}{3}$.

The coefficient of x is $-\frac{1}{3}$, so the slope of this line is $-\frac{1}{3}$.

Work Problem 5 at the Side. ▶

5 Find the slope of each line.

GS **(a)** $y = -\frac{7}{2}x + 1$

Since this equation is already solved for y, Step 1 is not needed. The slope, which is given by the ________ of ____, can be read directly from the equation. The slope is ____.

GS **(b)** $4y = 4x - 3$

Solve for y here by dividing each term by ____ to get the equation

$y =$ ________.

The coefficient of x is ____, so the slope is ____.

(c) $3x + 2y = 9$

(d) $5y - x = 10$

Answers

5. (a) coefficient; x; $-\frac{7}{2}$ **(b)** 4; $x - \frac{3}{4}$; 1; 1 **(c)** $-\frac{3}{2}$ **(d)** $\frac{1}{5}$

OBJECTIVE 3 Use slope to determine whether two lines are parallel, perpendicular, or neither. Two lines in a plane that never intersect are **parallel.** We use slopes to tell whether two lines are parallel.

Figure 27 shows the graphs of $x + 2y = 4$ and $x + 2y = -6$. These lines appear to be parallel. We solve each equation for y to find the slope.

$x + 2y = 4$	$x + 2y = -6$
$2y = -x + 4$ Subtract x.	$2y = -x - 6$ Subtract x.
$y = \frac{-x}{2} + 2$ Divide by 2.	$y = \frac{-x}{2} - 3$ Divide by 2.
$y = -\frac{1}{2}x + 2$ ($\frac{-x}{2} = \frac{-1x}{2} = -\frac{1}{2}x$) ↑ Slope	$y = -\frac{1}{2}x - 3$ (The slope is $-\frac{1}{2}$, *not* $-\frac{x}{2}$.) ↑ Slope

Each line has slope $-\frac{1}{2}$. ***Nonvertical parallel lines always have equal slopes.***

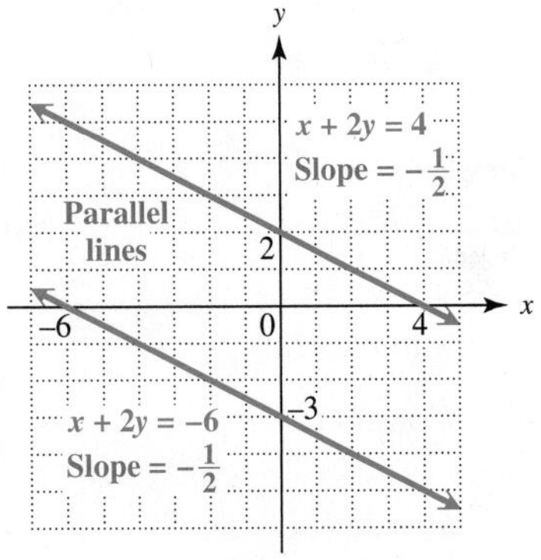

Figure 27

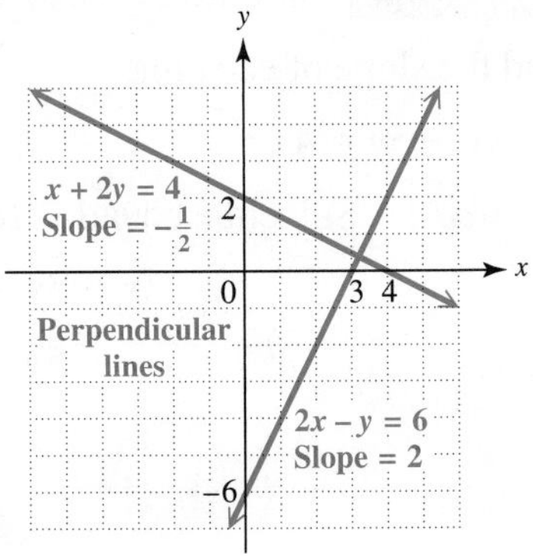

Figure 28

Figure 28 shows the graphs of $x + 2y = 4$ and $2x - y = 6$. These lines appear to be **perpendicular** (that is, they intersect at a 90° angle). As shown on the left above, solving $x + 2y = 4$ for y gives $y = -\frac{1}{2}x + 2$, with slope $-\frac{1}{2}$. We must solve $2x - y = 6$ for y.

$$2x - y = 6$$

$$-y = -2x + 6 \quad \text{Subtract } 2x.$$

$$y = 2x - 6 \quad \text{Multiply by } -1.$$

↑ Slope

The product of the two slopes $-\frac{1}{2}$ and 2 is

$$-\frac{1}{2}(2) = -1.$$

This condition is true in general. ***The product of the slopes of two perpendicular lines, neither of which is vertical, is always* −1.** This means that the slopes of perpendicular lines are negative (or opposite) reciprocals—if one slope is the nonzero number a, then the other is $-\frac{1}{a}$. The table in the margin shows several examples.

Number	Negative Reciprocal
$\frac{3}{4}$	$-\frac{4}{3}$
$\frac{1}{2}$	$-\frac{2}{1}$, or -2
-6, or $-\frac{6}{1}$	$\frac{1}{6}$
-0.4, or $-\frac{4}{10}$	$\frac{10}{4}$, or 2.5

***The product of each number and its negative reciprocal is* −1.**

Slopes of Parallel and Perpendicular Lines

Two lines with the same slope are parallel.

Two lines whose slopes have a product of -1 are perpendicular.

EXAMPLE 6 Deciding Whether Two Lines Are Parallel or Perpendicular

Decide whether each pair of lines is *parallel, perpendicular,* or *neither.*

(a) $x + 2y = 7$ and $-2x + y = 3$

Find the slope of each line by first solving each equation for y.

$x + 2y = 7$

$2y = -x + 7$ Subtract x.

$y = -\frac{1}{2}x + \frac{7}{2}$ Divide by 2.

The slope is $-\frac{1}{2}$.

$-2x + y = 3$

$y = 2x + 3$ Add $2x$.

The slope is 2.

Because the slopes are not equal, the lines are not parallel. Check the product of the slopes: $-\frac{1}{2}(2) = -1$. The two lines are perpendicular because the product of their slopes is -1. See **Figure 29.**

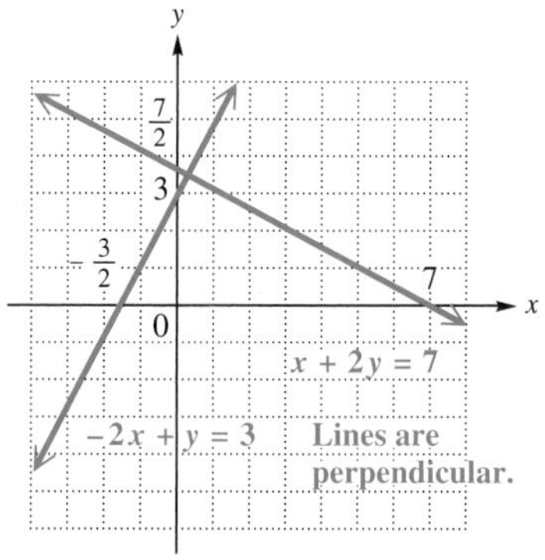

Figure 29

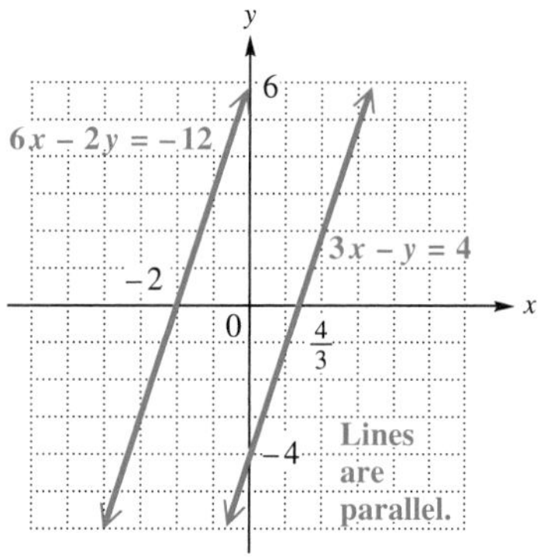

Figure 30

(b) $3x - y = 4$ and $6x - 2y = -12$ — Solve each equation for y. → $y = 3x - 4$ and $y = 3x + 6$

Both lines have slope 3, so the lines are parallel. See **Figure 30.**

(c) $4x + 3y = 6$ and $2x - y = 5$ — Solve each equation for y. → $y = -\frac{4}{3}x + 2$ and $y = 2x - 5$

The slopes are $-\frac{4}{3}$ and 2. Because $-\frac{4}{3} \neq 2$ and $-\frac{4}{3}(2) \neq -1$, these lines are neither parallel nor perpendicular.

(d) $5x - y = 1$ and $x - 5y = -10$ — Solve each equation for y. → $y = 5x - 1$ and $y = \frac{1}{5}x + 2$

$5(\frac{1}{5}) = 1$, *not* -1.

Here the slopes are 5 and $\frac{1}{5}$. The lines are not parallel, nor are they perpendicular.

Work Problem 6 at the Side. ▶

6 Decide whether each pair of lines is *parallel, perpendicular,* or *neither.*

(a) $x + y = 6$
$x + y = 1$

(b) $3x - y = 4$
$x + 3y = 9$

(c) $2x - y = 5$
$2x + y = 3$

(d) $3x - 7y = 35$
$7x - 3y = -6$

Answers

6. **(a)** parallel **(b)** perpendicular **(c)** neither **(d)** neither

11.3 Exercises

FOR EXTRA HELP

Download the MyDashBoard App

MyMathLab®

1. **CONCEPT CHECK** Slope is used to measure the ________ of a line. Slope is the (*horizontal* / *vertical*) change compared to the (*horizontal* / *vertical*) change while moving along the line from one point to another.

2. **CONCEPT CHECK** Slope is the ________ of the vertical change in ________, called the (*rise* / *run*), to the horizontal change in ________, called the (*rise* / *run*).

3. Look at the graph at the right, and answer the following.

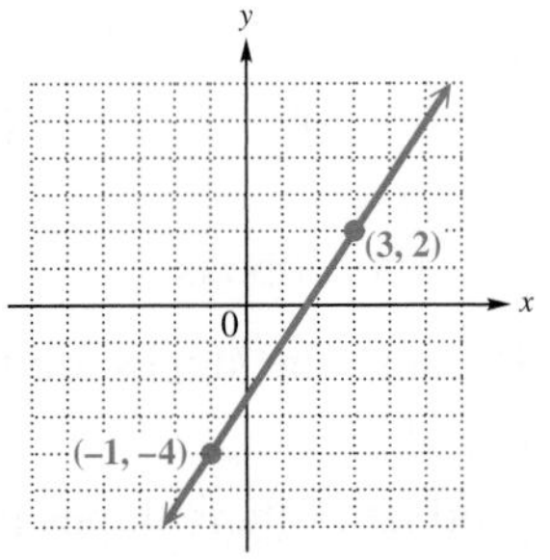

(a) Start at the point $(-1, -4)$ and count vertically up to the horizontal line that goes through the other plotted point. What is this vertical change? (Remember: "up" means positive, "down" means negative.) ______

(b) From this new position, count horizontally to the other plotted point. What is this horizontal change? (Remember: "right" means positive, "left" means negative.) ______

(c) What is the quotient of the numbers found in parts (a) and (b)? ______
What do we call this number? ____________

(d) If we were to *start* at the point $(3, 2)$ and *end* at the point $(-1, -4)$ would the answer to part (c) be the same? Explain.

4. **CONCEPT CHECK** Match the graph of each line in parts (a)–(d) with its slope in choices A–D.

(a)

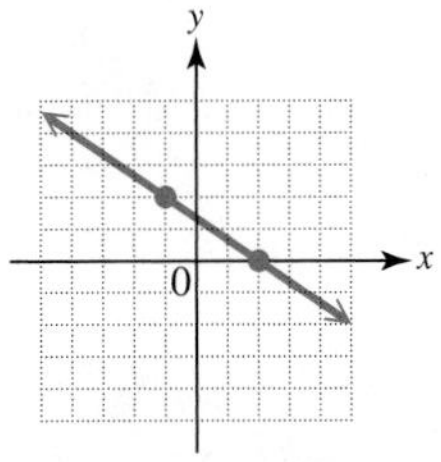

(b)

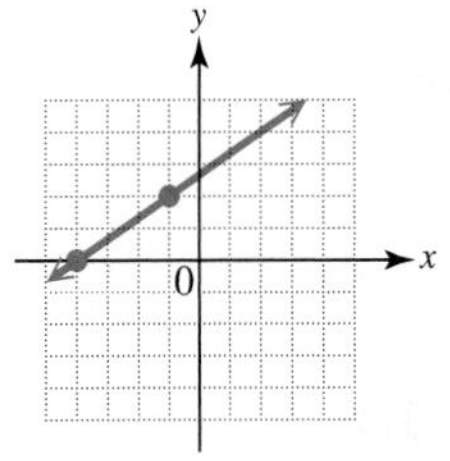

(c)

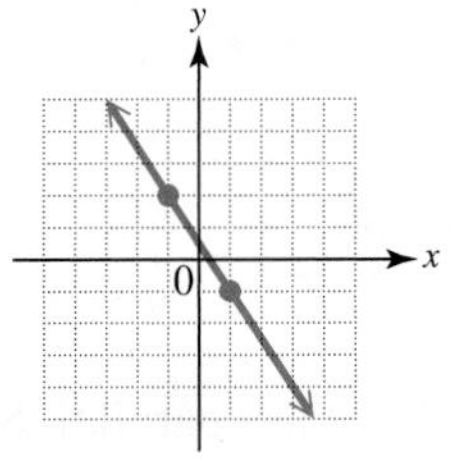

(d)

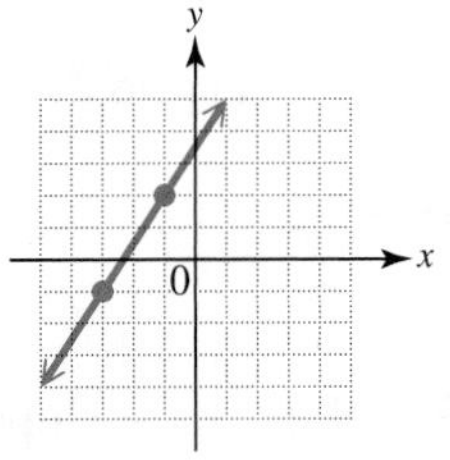

A. $\frac{2}{3}$ **B.** $\frac{3}{2}$ **C.** $-\frac{2}{3}$ **D.** $-\frac{3}{2}$

Use the coordinates of the indicated points to find the slope of each line. ***See Example 1.***

5.

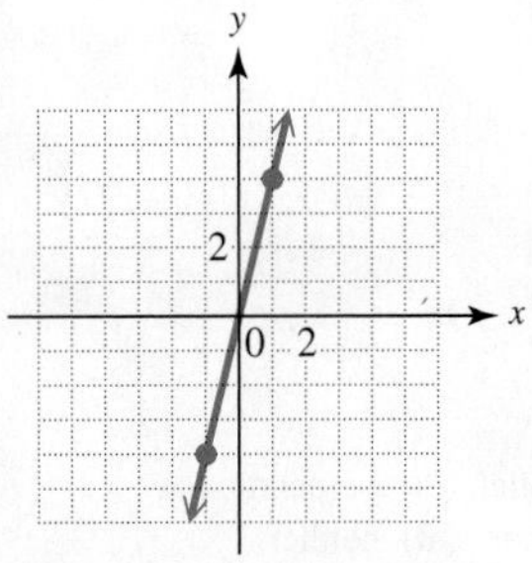

6.

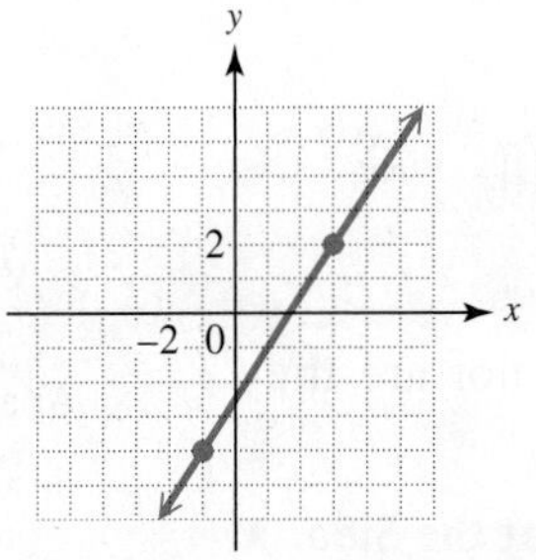

7.

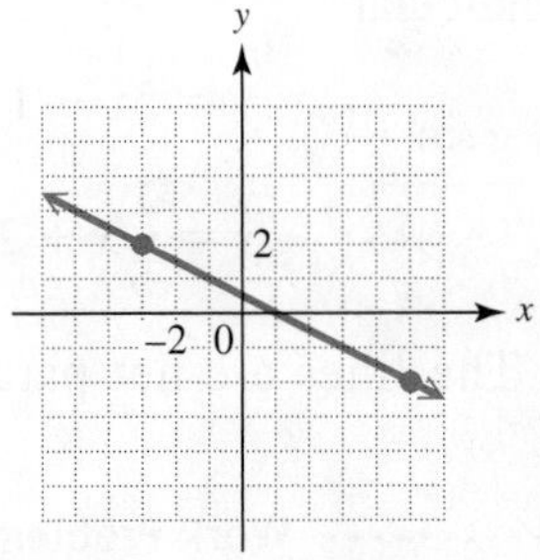

8.

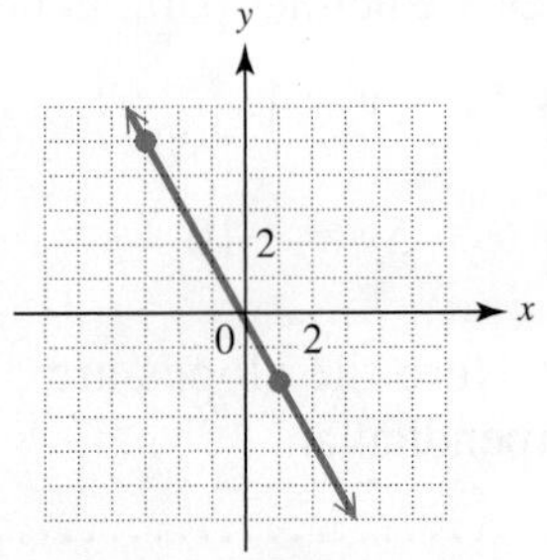

9. **CONCEPT CHECK** On the given axes, sketch the graph of a straight line with the indicated slope.

(a) Negative **(b)** Positive **(c)** Undefined **(d)** Zero

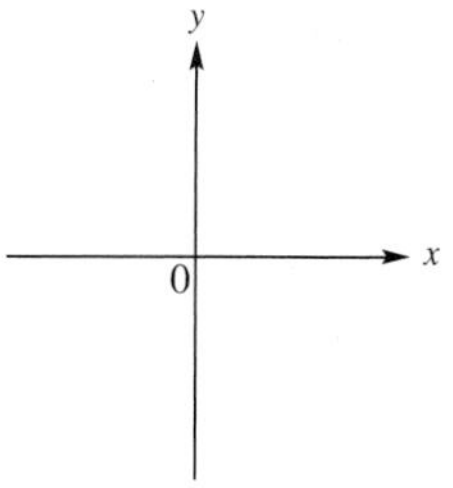

10. **CONCEPT CHECK** Decide whether the line with the given slope m *rises from left to right, falls from left to right,* is *horizontal*, or is *vertical.*

(a) $m = -4$ **(b)** $m = 0$ **(c)** m is undefined. **(d)** $m = \frac{3}{7}$

CONCEPT CHECK *The figure at the right shows a line that has a positive slope (because it rises from left to right) and a positive y-value for the y-intercept (because it intersects the y-axis above the origin).*

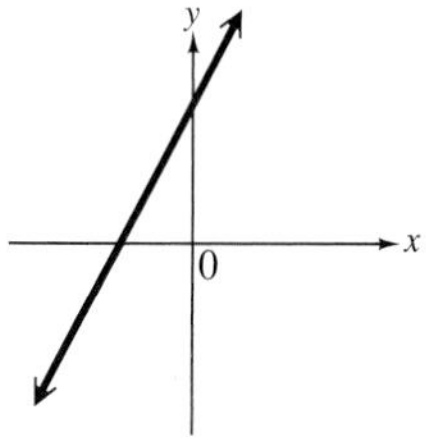

For each figure in Exercises 11–16, decide whether **(a)** *the slope is* positive, negative, *or* 0 *and whether* **(b)** *the y-value of the y-intercept is* positive, negative, *or* 0.

11. **(a)** __________
(b) __________

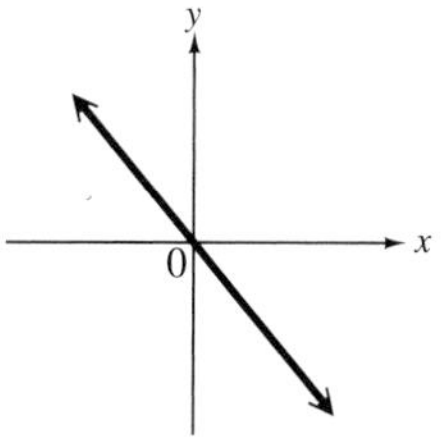

12. **(a)** __________
(b) __________

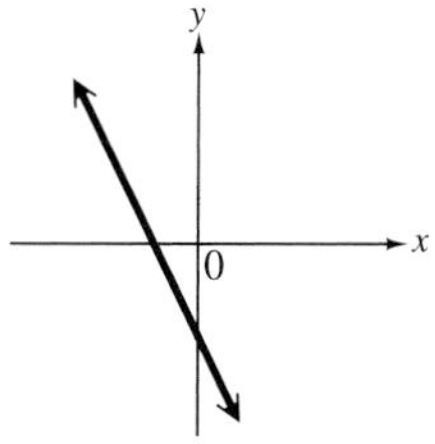

13. **(a)** __________
(b) __________

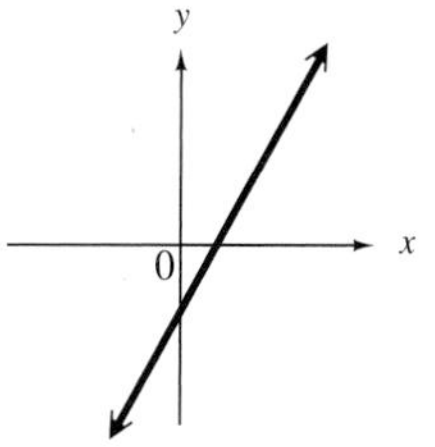

14. **(a)** __________
(b) __________

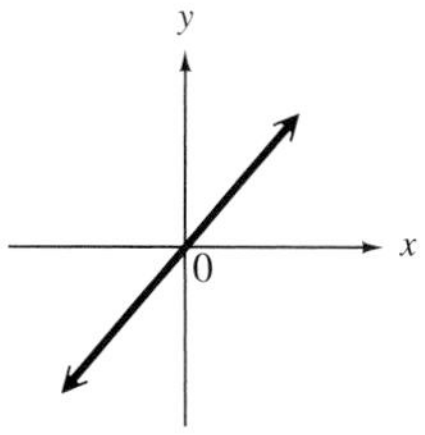

15. **(a)** __________
(b) __________

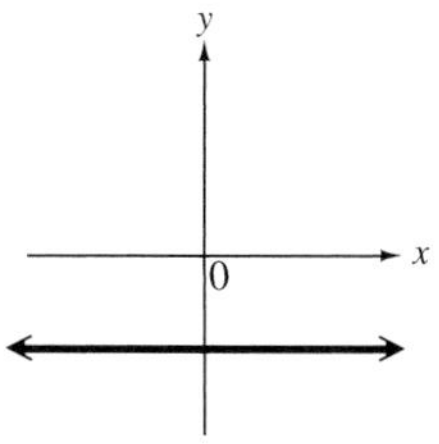

16. **(a)** __________
(b) __________

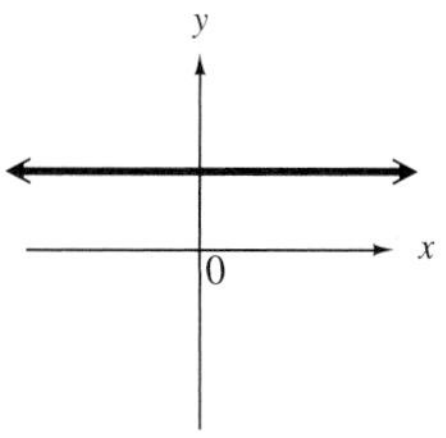

CONCEPT CHECK *Answer each question.*

17. What is the slope (or grade) of this hill?

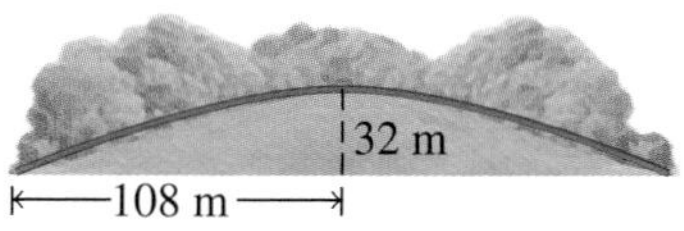

18. What is the slope (or pitch) of this roof?

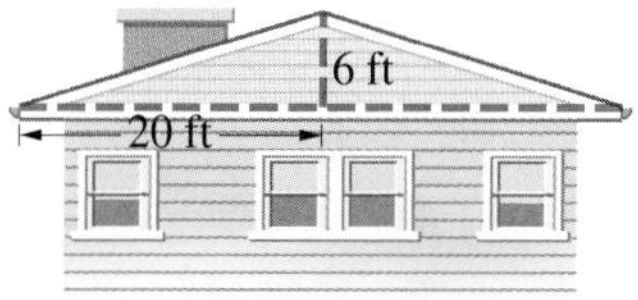

19. What is the slope of the slide? (*Hint:* The slide *drops* 8 ft vertically as it extends 12 ft horizontally.)

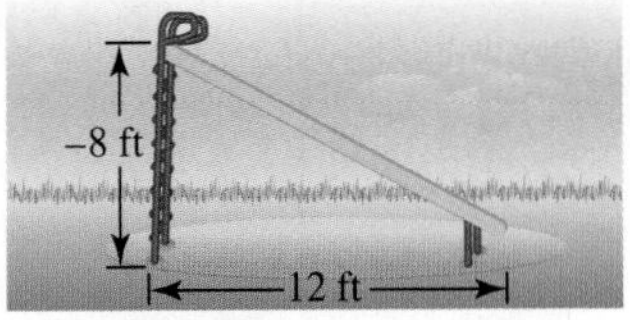

20. What is the slope (or grade) of this ski slope? (*Hint:* The ski slope *drops* 25 ft vertically as it extends 100 ft horizontally.)

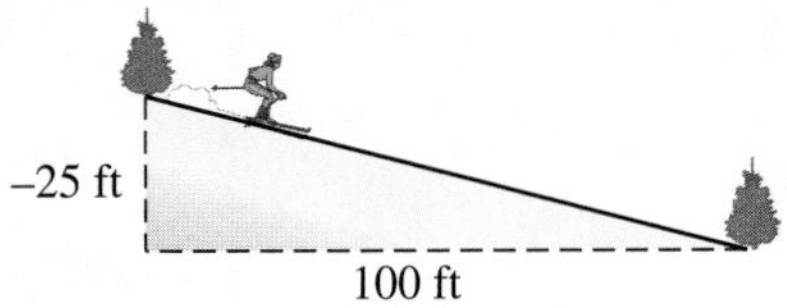

21. CONCEPT CHECK A student found the slope of the line through the points $(2, 5)$ and $(-1, 3)$ as follows.

$$\frac{3-5}{2-(-1)} = \frac{-2}{3}, \quad \text{or} \quad -\frac{2}{3}$$

What Went Wrong? Give the correct slope.

22. CONCEPT CHECK A student found the slope of the line through the points $(-2, 4)$ and $(6, -1)$ as follows.

$$\frac{6-(-2)}{-1-4} = \frac{8}{-5}, \quad \text{or} \quad -\frac{8}{5}$$

What Went Wrong? Give the correct slope.

Find the slope of the line passing through each pair of points. ***See Examples 2–4.***

23. $(1, -2)$ and $(-3, -7)$

24. $(4, -1)$ and $(-2, -8)$

25. $(0, 3)$ and $(-2, 0)$

26. $(-8, 0)$ and $(0, -5)$

27. $(-2, 4)$ and $(-3, 7)$

28. $(-4, -5)$ and $(-5, -8)$

29. $(4, 3)$ and $(-6, 3)$

30. $(6, -5)$ and $(-12, -5)$

31. $(-12, 3)$ and $(-12, -7)$

32. $(-8, 6)$ and $(-8, -1)$

33. $(4.8, 2.5)$ and $(3.6, 2.2)$

34. $(3.1, 2.6)$ and $(1.6, 2.1)$

35. $\left(-\frac{7}{5}, \frac{3}{10}\right)$ and $\left(\frac{1}{5}, -\frac{1}{2}\right)$

36. $\left(-\frac{4}{3}, \frac{1}{2}\right)$ and $\left(\frac{1}{3}, -\frac{5}{6}\right)$

Find the slope of each line. ***See Example 5.***

37. $y = 5x + 12$

38. $y = 2x - 3$

39. $4y = x + 1$

40. $2y = -x + 4$

41. $3x - 2y = 3$

42. $-6x + 4y = 4$

43. $-2y - 3x = 5$

44. $-4y - 3x = 2$

45. $y = 6$

46. $y = 4$

47. $x = -2$

48. $x = 5$

49. $x - y = 0$

50. $x + y = 0$

For each pair of equations, give the slope of each line, and then determine whether the two lines are parallel, perpendicular, *or* neither. ***See Example 6.***

51. $-4x + 3y = 4$
$-8x + 6y = 0$

52. $2x + 5y = 4$
$4x + 10y = 1$

53. $5x - 3y = -2$
$3x - 5y = -8$

54. $8x - 9y = 6$
$8x + 6y = -5$

55. $3x - 5y = -1$
$5x + 3y = 2$

56. $3x - 2y = 6$
$2x + 3y = 3$

Relating Concepts (Exercises 57–62) For Individual or Group Work

Figure A *gives the percent of first-time full-time freshmen at 4-year colleges and universities who planned to major in the Biological Sciences.* **Figure B** *shows the percent of the same group of students who planned to major in Business.*

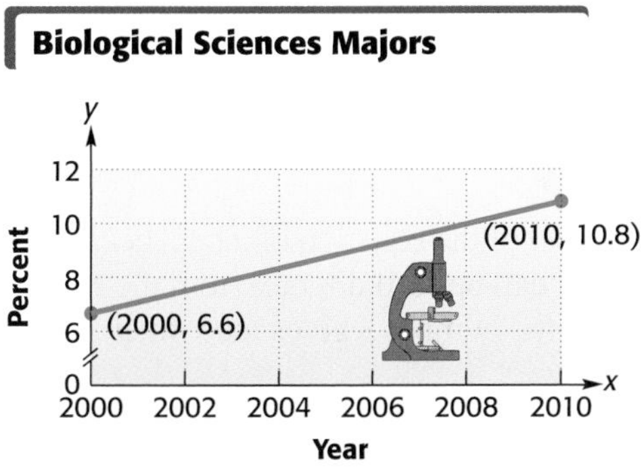

Source: Higher Education Research Institute, UCLA.

Figure A

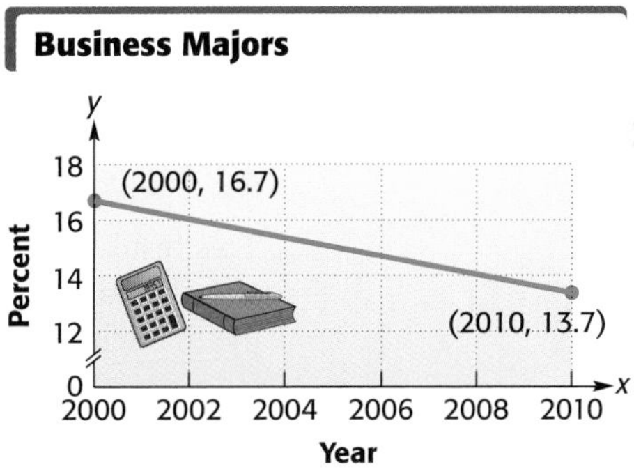

Source: Higher Education Research Institute, UCLA.

Figure B

Work Exercises 57–62 in order.

57. Use the given ordered pairs to find the slope of the line in **Figure A.**

58. The slope of the line in **Figure A** is (*positive / negative*). This means that during the period represented, the percent of freshmen planning to major in the Biological Sciences (*increased / decreased*).

59. The slope of a line represents the *rate of change.* Based on **Figure A,** what was the increase in the percent of freshmen *per year* who planned to major in the Biological Sciences during the period shown?

60. Use the given ordered pairs to find the slope of the line in **Figure B.**

61. The slope of the line in **Figure B** is (*positive / negative*). This means that during the period represented, the percent of freshmen planning to major in Business (*increased / decreased*).

62. Based on **Figure B,** what was the decrease in the percent of freshmen *per year* who planned to major in Business?

11.4 Writing and Graphing Equations of Lines

OBJECTIVES

1. Use the slope-intercept form of the equation of a line.
2. Graph a line by using its slope and a point on the line.
3. Write an equation of a line by using its slope and any point on the line.
4. Write an equation of a line by using two points on the line.
5. Write an equation of a line that fits a data set.

1 Identify the slope and y-intercept of the line with each equation.

(a) $y = 2x - 6$

(b) $y = -\frac{3}{5}x - 9$

(c) $y = -\frac{x}{3} + \frac{7}{3}$

(d) $y = -x$

Answers

1. (a) slope: 2; y-intercept: $(0, -6)$
(b) slope: $-\frac{3}{5}$; y-intercept: $(0, -9)$
(c) slope: $-\frac{1}{3}$; y-intercept: $\left(0, \frac{7}{3}\right)$
(d) slope: -1; y-intercept: $(0, 0)$

OBJECTIVE 1 Use the slope-intercept form of the equation of a line. In **Section 11.3,** we found the slope (steepness) of a line by solving the equation of the line for y. In that form, the slope is the coefficient of x. For example, the graph of the equation

$$y = 2x + 3 \quad \text{has slope} \quad 2.$$

What does the number 3 represent? To find out, suppose a line has slope m and y-intercept $(0, b)$. We can find an equation of this line by choosing another point (x, y) on the line, as shown in **Figure 31.** Then we use the slope formula.

$$m = \frac{y - b}{x - 0} \quad \begin{matrix}\leftarrow \text{Change in } y\text{-values} \\ \leftarrow \text{Change in } x\text{-values}\end{matrix}$$

$$m = \frac{y - b}{x} \quad \text{Subtract in the denominator.}$$

$$mx = y - b \quad \text{Multiply by } x.$$

$$mx + b = y \quad \text{Add } b.$$

$$y = mx + b \quad \text{Rewrite.}$$

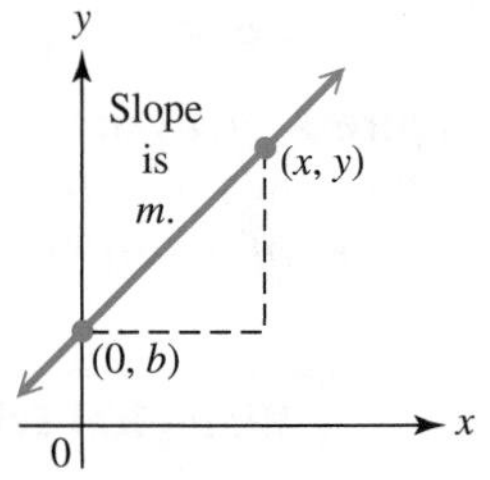

Figure 31

This result is the *slope-intercept form* of the equation of a line, because both the slope and the y-intercept of the line can be read directly from the equation. For the line with equation $y = 2x + 3$, the number 3 gives the y-intercept $(0, 3)$.

Slope-Intercept Form

The **slope-intercept form** of the equation of a line with slope m and y-intercept $(0, b)$ is given as follows.

$$y = mx + b$$

Slope ↑ (at m); ↑ $(0, b)$ is the y-intercept.

The intercept given is the y-intercept.

EXAMPLE 1 Identifying Slopes and y-Intercepts

Identify the slope and y-intercept of the line with each equation.

(a) $y = -4x + 1$
Slope ↑ (-4); ↑ y-intercept $(0, 1)$

(b) $y = x - 8$ can be written as $y = 1x + (-8)$.
Slope ↑ (1); ↑ y-intercept $(0, -8)$

(c) $y = 6x$ can be written as $y = 6x + 0$.
Slope ↑ (6); ↑ y-intercept $(0, 0)$

(d) $y = \frac{x}{4} - \frac{3}{4}$ can be written as $y = \frac{1}{4}x + \left(-\frac{3}{4}\right)$.
Slope ↑ ($\frac{1}{4}$); ↑ y-intercept $\left(0, -\frac{3}{4}\right)$

◀ Work Problem **1** at the Side.

EXAMPLE 2 **Writing an Equation of a Line**

Write an equation of the line with slope $\frac{2}{3}$ and y-intercept $(0, -1)$.

Here, $m = \frac{2}{3}$ and $b = -1$, so we can write the following equation.

Slope → m; y-intercept $(0, b)$ → b

$y = mx + b$ Slope-intercept form

$y = \frac{2}{3}x + (-1)$, or $y = \frac{2}{3}x - 1$ Substitute for m and b.

Work Problem 2 at the Side. ▶

OBJECTIVE ▶ 2 Graph a line by using its slope and a point on the line. We can use the slope and y-intercept to graph a line.

Graphing a Line by Using Its Slope and *y*-Intercept

Step 1 Write the equation in slope-intercept form $y = mx + b$, if necessary, by solving for y.

Step 2 Identify the y-intercept. Graph the point $(0, b)$.

Step 3 Identify slope m of the line. Use the geometric interpretation of slope ("rise over run") to find another point on the graph by counting from the y-intercept.

Step 4 Join the two points with a line to obtain the graph.

EXAMPLE 3 **Graphing Lines by Using Slopes and *y*-Intercepts**

Graph each equation by using the slope and y-intercept.

(a) $y = \frac{2}{3}x - 1$

Step 1 The equation is in slope-intercept form.

$$y = \frac{2}{3}x - 1$$

Slope ($\frac{2}{3}$); Value of b in y-intercept $(0, b)$ (-1)

Step 2 The y-intercept is $(0, -1)$. Graph this point. See **Figure 32.**

Step 3 The slope is $\frac{2}{3}$. By definition,

$$\text{slope } m = \frac{\textbf{change in } y \textbf{ (rise)}}{\textbf{change in } x \textbf{ (run)}} = \frac{2}{3}.$$

From the y-intercept, count up 2 units and to the right 3 units to obtain the point $(3, 1)$.

Step 4 Draw the line through the points $(0, -1)$ and $(3, 1)$ to obtain the graph in **Figure 32.**

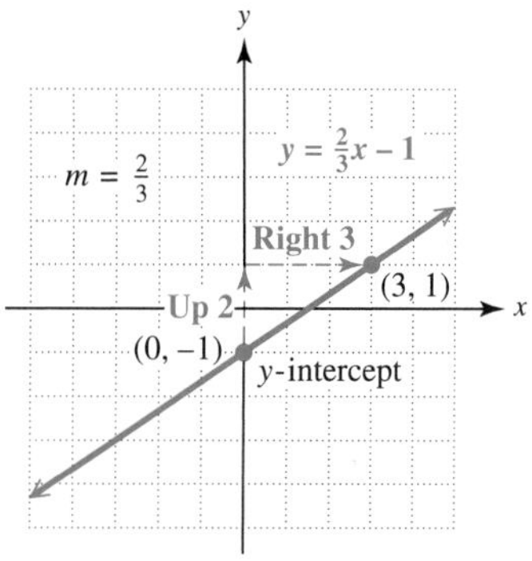

Figure 32

Continued on Next Page

2 Write an equation of the line with the given slope and y-intercept.

(a) slope $\frac{1}{2}$; y-intercept $(0, -4)$

(b) slope -1; y-intercept $(0, 8)$

(c) slope 3; y-intercept $(0, 0)$

(d) slope 0; y-intercept $(0, 2)$

(e) slope 1; y-intercept $(0, 0.75)$

Answers

2. (a) $y = \frac{1}{2}x - 4$ **(b)** $y = -x + 8$ **(c)** $y = 3x$ **(d)** $y = 2$ **(e)** $y = x + 0.75$

3 Graph $3x - 4y = 8$ by using the slope and y-intercept.

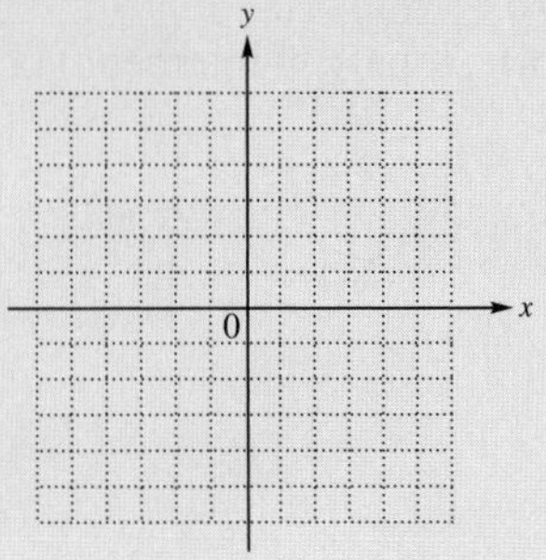

4 Graph the line passing through the point $(2, -3)$, with slope $-\frac{1}{3}$.

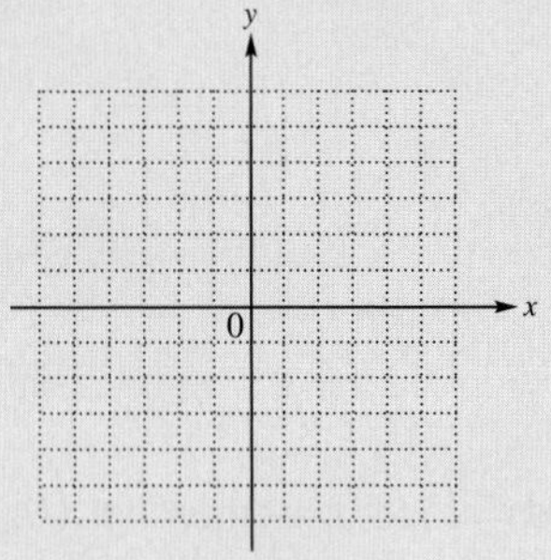

(b) $3x + 4y = 8$

Step 1 Solve for y to write the equation in slope-intercept form.

$$3x + 4y = 8$$

Isolate y on one side.

$$4y = -3x + 8 \qquad \text{Subtract } 3x.$$

Slope-intercept form $\longrightarrow$ $y = -\frac{3}{4}x + 2$ Divide *each* term by 4.

Step 2 The y-intercept is $(0, 2)$. Graph this point. See **Figure 33.**

Step 3 The slope is $-\frac{3}{4}$, which can be written as either $\frac{-3}{4}$ or $\frac{3}{-4}$. We use $\frac{-3}{4}$ here.

$$m = \frac{\text{change in } y \text{ (rise)}}{\text{change in } x \text{ (run)}} = \frac{-3}{4}$$

From the y-intercept, count *down* 3 units (because of the negative sign) and to the right 4 units, to obtain the point $(4, -1)$.

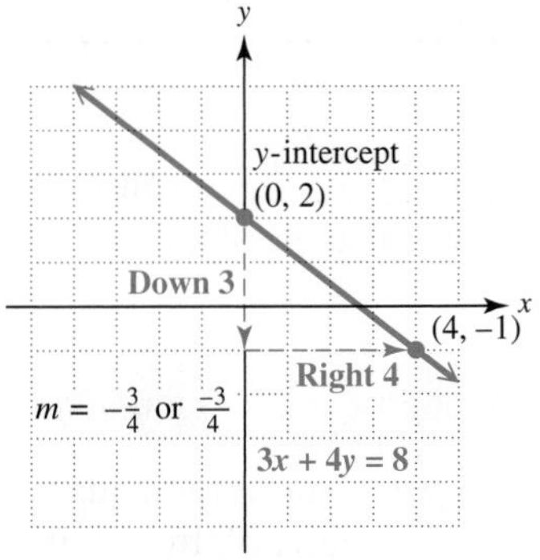

Figure 33

Step 4 Draw the line through the two points $(0, 2)$ and $(4, -1)$ to obtain the graph in **Figure 33.**

◀ **Work Problem 3 at the Side.**

Note

In Step 3 of **Example 3(b),** we could use $\frac{3}{-4}$ for the slope. From the y-intercept, count up 3 units and to the *left* 4 units (because of the negative sign) to obtain the point $(-4, 5)$. Verify that this produces the same line.

EXAMPLE 4 Graphing a Line by Using Its Slope and a Point

Graph the line passing through the point $(-2, 3)$, with slope -4.

First, locate the point $(-2, 3)$. See **Figure 34.** Then write the slope -4 as

$$\text{slope } m = \frac{\text{change in } y}{\text{change in } x} = -4 = \frac{-4}{1}.$$

Locate another point on the line by counting *down* 4 units from $(-2, 3)$ and then to the right 1 unit. Finally, draw the line through this new point P and the given point $(-2, 3)$. See **Figure 34.**

We could have written the slope as $\frac{4}{-1}$ instead. In this case, we would move up 4 units from $(-2, 3)$ and then to the *left* 1 unit. Verify that this produces the same line.

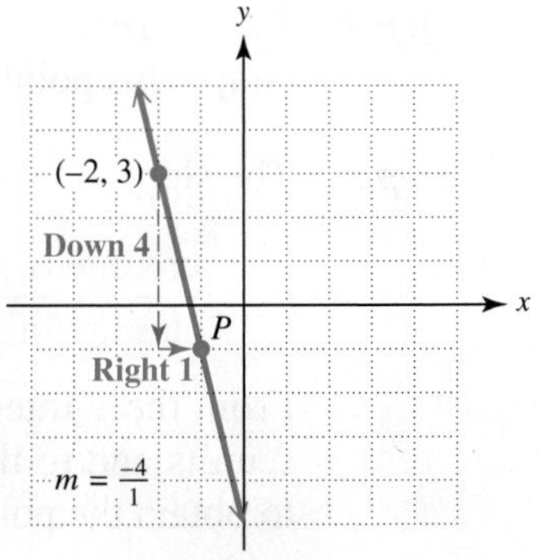

Figure 34

◀ **Work Problem 4 at the Side.**

Answers

3.

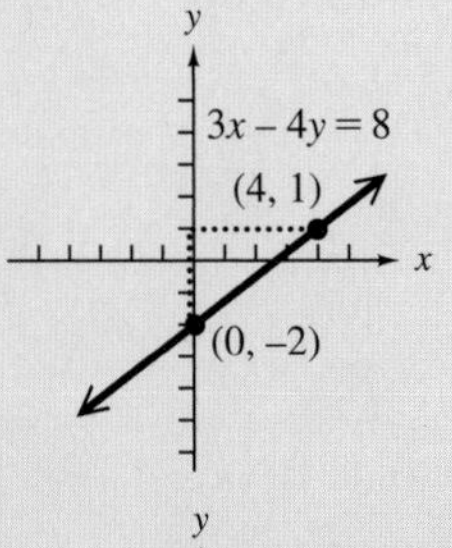

4.

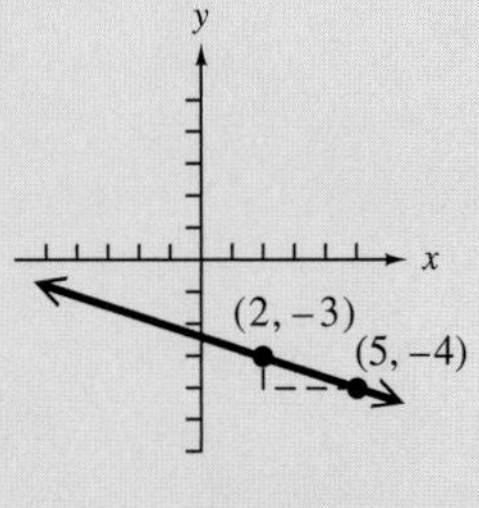

OBJECTIVE 3 Write an equation of a line by using its slope and any point on the line. We can use the slope-intercept form to do this.

EXAMPLE 5 Using the Slope-Intercept Form to Write an Equation of a Line

Write an equation, in slope-intercept form, of the line having slope 4 passing through the point $(2, 5)$.

Since the line passes through the point $(2, 5)$, we can substitute $x = 2$, $y = 5$, and the given slope $m = 4$ into $y = mx + b$ and solve for b.

$y = mx + b$	Slope-intercept form
$5 = 4(2) + b$	Let $x = 2$, $y = 5$, and $m = 4$.
$5 = 8 + b$	Multiply.
$-3 = b$	Subtract 8.

$(0, b)$ is the y-intercept. Don't stop here.

Now substitute the values of m and b into slope-intercept form.

$y = mx + b$	Slope-intercept form
$y = 4x - 3$	Let $m = 4$ and $b = -3$.

Work Problem 5 at the Side. ▶

Note

Slope-intercept form is an especially useful form for a linear equation because of the information we can determine from it. It is the form used by graphing calculators and the one that describes *a linear function.*

There is another form that can be used to write the equation of a line. To develop this form, let m represent the slope of a line and let (x_1, y_1) represent a given point on the line. Let (x, y) represent any other point on the line. See **Figure 35.**

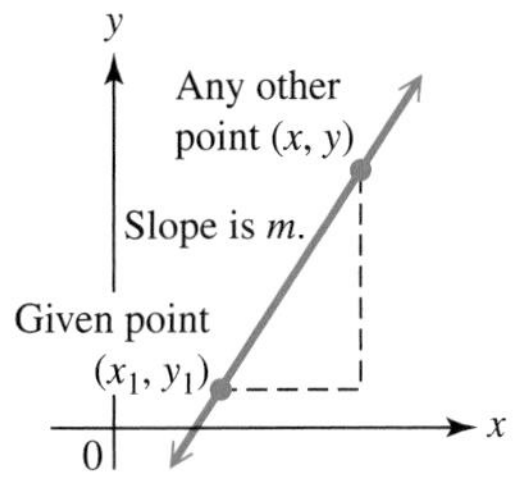

Figure 35

$m = \dfrac{y - y_1}{x - x_1}$	Definition of slope
$m(x - x_1) = y - y_1$	Multiply each side by $x - x_1$.
$y - y_1 = m(x - x_1)$	Rewrite.

This result is the **point-slope form** of the equation of a line.

Point-Slope Form

The point-slope form of the equation of a line with slope m passing through the point (x_1, y_1) is given as follows.

$$\mathbf{y - y_1 = m(x - x_1)}$$

(Slope: m; Given point: (x_1, y_1))

5 Write an equation, in slope-intercept form, of each line.

GS **(a)** The line having slope -2 passing through the point $(-1, 4)$

Substitute values and solve for b.

$y = mx + b$

$\_\_\_ = \_\_\_(\_\_\_) + b$

$4 = \_\_\_ + b$

$\_\_\_ = b$

An equation of the line is $y =$ ________.

(b) The line having slope 3 passing through the point $(-2, 1)$

Answers

5. (a) $4; -2; -1; 2; 2; -2x + 2$
(b) $y = 3x + 7$

6 Write an equation of each line. Give the final answer in slope-intercept form.

GS **(a)** The line passing through $(-1, 3)$, with slope -2

$$y - y_1 = m(x - x_1)$$

$$y - \_\_\_ = \_\_\_[x - (\_\_\_)]$$

$$y - 3 = -2(x + \_\_\_)$$

$$y - 3 = -2x - \_\_\_$$

$$y = \_\_\_\_$$

(b) The line passing through $(5, 2)$, with slope $-\frac{1}{3}$

EXAMPLE 6 Using Point-Slope Form to Write Equations

Write an equation of each line. Give the final answer in slope-intercept form.

(a) The line passing through $(-2, 4)$, with slope -3

The given point is $(-2, 4)$ so $x_1 = -2$ and $y_1 = 4$. Also, $m = -3$. Substitute these values into the point-slope form.

Only y_1, m, and x_1 are replaced with numbers.

$y - y_1 = m(x - x_1)$	Point-slope form
$y - 4 = -3[x - (-2)]$	Let $x_1 = -2$, $y_1 = 4$, $m = -3$.
$y - 4 = -3(x + 2)$	Definition of subtraction
$y - 4 = -3x - 6$	Distributive property
$y = -3x - 2$	Add 4.

(b) The line passing through $(4, 2)$, with slope $\frac{3}{5}$

$y - y_1 = m(x - x_1)$	Point-slope form
$y - 2 = \frac{3}{5}(x - 4)$	Let $x_1 = 4$, $y_1 = 2$, $m = \frac{3}{5}$.
$y - 2 = \frac{3}{5}x - \frac{12}{5}$	Distributive property
$y = \frac{3}{5}x - \frac{12}{5} + \frac{10}{5}$	Add $2 = \frac{10}{5}$.
$y = \frac{3}{5}x - \frac{2}{5}$	Combine like terms.

We did not clear fractions after the substitution step because we want the equation in slope-intercept form—that is, solved for y.

◀ **Work Problem 6 at the Side.**

OBJECTIVE 4 Write an equation of a line by using two points on the line. Many of the linear equations in **Sections 11.1–11.3** were given in the form

$$Ax + By = C,$$

called **standard form,** where A, B, and C are real numbers and A and B are not both 0. In most cases, A, B, and C are rational numbers. For consistency in this book, we give answers so that A, B, and C are integers with greatest common factor 1 and $A \geq 0$. (If $A = 0$, then we give $B > 0$.)

Note

The definition of standard form is not the same in all texts. A linear equation can be written in many different, yet equally correct, ways. For example,

$$3x + 4y = 12, \quad 6x + 8y = 24, \quad \text{and} \quad -9x - 12y = -36$$

all represent the same set of ordered pairs. When giving answers, $3x + 4y = 12$ is preferable to the other forms because the greatest common factor of 3, 4, and 12 is 1 and $A \geq 0$.

Answers

6. (a) 3; -2; -1; 1; 2; $-2x + 1$

(b) $y = -\frac{1}{3}x + \frac{11}{3}$

EXAMPLE 7 **Writing an Equation of a Line by Using Two Points**

Write an equation of the line passing through the points $(3, 4)$ and $(-2, 5)$. Give the final answer in slope-intercept form and then in standard form.

First, find the slope of the line.

$$\overset{(x_1,\, y_1)}{(3, 4)} \quad \text{and} \quad \overset{(x_2,\, y_2)}{(-2, 5)} \qquad \text{Label the points.}$$

$$\text{slope } m = \frac{y_2 - y_1}{x_2 - x_1} = \frac{5 - 4}{-2 - 3} \qquad \text{Apply the slope formula.}$$

$$= \frac{1}{-5}, \text{ or } -\frac{1}{5} \qquad \text{Simplify the fraction.}$$

Now use (x_1, y_1), here $(3, 4)$, and either slope-intercept or point-slope form.

$$y - y_1 = m(x - x_1) \qquad \text{We choose point-slope form.}$$

$$y - 4 = -\frac{1}{5}(x - 3) \qquad \text{Let } x_1 = 3, y_1 = 4, m = -\tfrac{1}{5}.$$

$$y - 4 = -\frac{1}{5}x + \frac{3}{5} \qquad \text{Distributive property}$$

$$y = -\frac{1}{5}x + \frac{3}{5} + \frac{20}{5} \qquad \text{Add } 4 = \tfrac{20}{5} \text{ to each side.}$$

$$\text{Slope-intercept form} \longrightarrow y = -\frac{1}{5}x + \frac{23}{5} \qquad \text{Combine like terms.}$$

$$5y = -x + 23 \qquad \text{Multiply by 5 to clear fractions.}$$

$$\text{Standard form} \longrightarrow x + 5y = 23 \qquad \text{Add } x.$$

The same result would be found using $(-2, 5)$ for (x_1, y_1).

Work Problem **7** at the Side. ▶

7 Write an equation of the line passing through each pair of points. Give the final answer in slope-intercept form and then in standard form.

(a) $(2, 5)$ and $(-1, 6)$

(b) $(-3, 1)$ and $(2, 4)$

Summary of Forms of Linear Equations

Equation	*Description*	*Example*
$x = k$	**Vertical line** Slope is undefined. x-intercept is $(k, 0)$.	$x = 3$
$y = k$	**Horizontal line** Slope is 0. y-intercept is $(0, k)$.	$y = 3$
$y = mx + b$	**Slope-intercept form** Slope is m. y-intercept is $(0, b)$.	$y = \frac{3}{2}x - 6$
$y - y_1 = m(x - x_1)$	**Point-slope form** Slope is m. Line passes through (x_1, y_1).	$y + 3 = \frac{3}{2}(x - 2)$
$Ax + By = C$	**Standard form** Slope is $-\frac{A}{B}$. x-intercept is $\left(\frac{C}{A}, 0\right)$. y-intercept is $\left(0, \frac{C}{B}\right)$.	$3x - 2y = 12$

Answers

7. (a) $y = -\frac{1}{3}x + \frac{17}{3}$; $x + 3y = 17$

(b) $y = \frac{3}{5}x + \frac{14}{5}$; $3x - 5y = -14$

OBJECTIVE 5 Write an equation of a line that fits a data set. If a given set of data fits a linear pattern—that is, its graph consists of points lying close to a straight line—we can write a linear equation that models the data.

Year	Cost (in dollars)
1	3766
3	4645
5	5491
7	6185
9	7050
11	8244

Source: The College Board.

EXAMPLE 8 Writing an Equation of a Line That Describes Data

The table in the margin lists the average annual cost (in dollars) of tuition and fees for instate students at public 4-year colleges and universities for selected years. Year 1 represents 2001, year 3 represents 2003, and so on. Plot the data and find an equation that approximates it.

Letting y represent the cost in year x, we plot the data as shown in **Figure 36.**

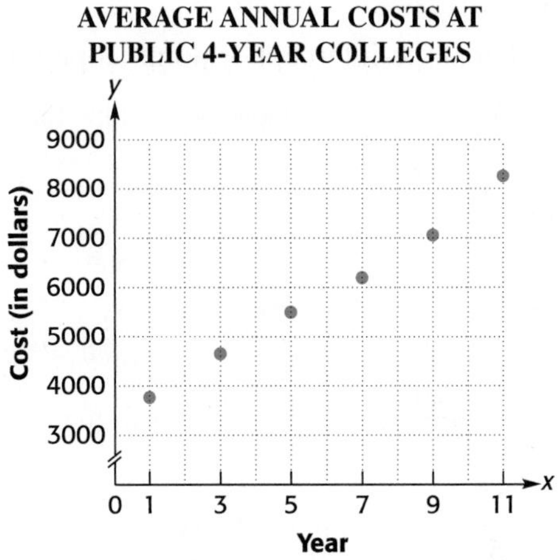

Figure 36

The points appear to lie approximately in a straight line. To find an equation of the line, we choose the ordered pairs $(5, 5491)$ and $(7, 6185)$ from the table and determine the slope of the line through these points.

$$m = \frac{y_2 - y_1}{x_2 - x_1} = \frac{6185 - 5491}{7 - 5} = 347 \qquad \text{Let } (7, 6185) = (x_2, y_2) \text{ and } (5, 5491) = (x_1, y_1).$$

The slope, 347, is positive, indicating that tuition and fees *increased* \$347 each year. Use the slope and the point $(5, 5491)$ in the slope-intercept form.

$y = mx + b$ — Slope-intercept form

Solve for b, the y-value of the y-intercept.

$5491 = 347(5) + b$ — Substitute for x, y, and m.

$5491 = 1735 + b$ — Multiply.

$3756 = b$ — Subtract 1735.

Thus, $m = 347$ and $b = 3756$, so an equation of the line is

$$y = 347x + 3756.$$

To see how well this equation approximates the ordered pairs in the data table, let $x = 3$ (for 2003) and find y.

$y = 347x + 3756$ — Equation of the line

$y = 347(3) + 3756$ — Substitute 3 for x.

$y = 4797$ — Multiply, and then add.

The corresponding value in the table for $x = 3$ is 4645, so the equation approximates the data reasonably well.

◀ **Work Problem 8 at the Side.**

8 Use the points $(1, 3766)$ and $(9, 7050)$ to find an equation in slope-intercept form that approximates the data of **Example 8.** How well does this equation approximate the cost in 2007?

Answer

8. $y = 410.5x + 3355.5$;
The equation gives $y = 6229$ when $x = 7$, which is a little high but a reasonable approximation.

11.4 Exercises

FOR EXTRA HELP

Download the MyDashBoard App

MyMathLab®

CONCEPT CHECK *Work each problem.*

1. Match each description of a line in Column I with the correct equation in Column II.

I	II
(a) Slope -2, passes through the point $(4, 1)$	**A.** $y = 4x$
(b) Slope -2, y-intercept $(0, 1)$	**B.** $y = \frac{1}{4}x$
(c) Passes through the points $(0, 0)$ and $(4, 1)$	**C.** $y = -2x + 1$
(d) Passes through the points $(0, 0)$ and $(1, 4)$	**D.** $y - 1 = -2(x - 4)$

2. Which equations are equivalent to $2x - 3y = 6$?

A. $y = \frac{2}{3}x - 2$ **B.** $-2x + 3y = -6$ **C.** $y = -\frac{3}{2}x + 3$ **D.** $y - 2 = \frac{2}{3}(x - 6)$

3. Match each equation with the graph that would most closely resemble its graph.

(a) $y = x + 3$ **(b)** $y = -x + 3$ **(c)** $y = x - 3$ **(d)** $y = -x - 3$

A.

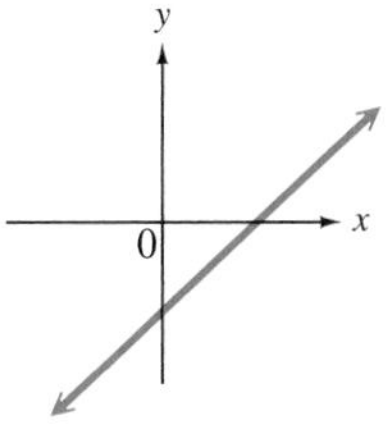

B.

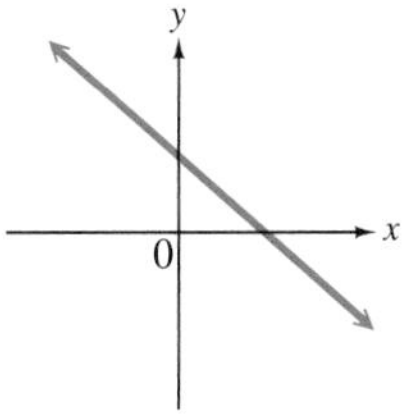

C.

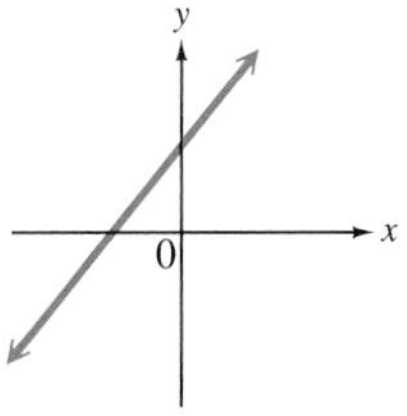

D.

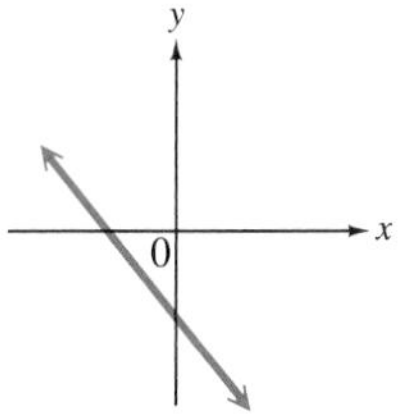

4. (a) What is the common name given to the vertical line whose x-intercept is the origin?

(b) What is the common name given to the line with slope 0 whose y-intercept is the origin?

Identify the slope and y-intercept of the line with each equation. ***See Example 1.***

5. $y = \frac{5}{2}x - 4$

6. $y = \frac{7}{3}x - 6$

7. $y = -x + 9$

8. $y = x + 1$

9. $y = \frac{x}{5} - \frac{3}{10}$

10. $y = \frac{x}{7} - \frac{5}{14}$

Write the equation of the line with the given slope and y-intercept. ***See Example 2.***

11. slope 4, y-intercept $(0, -3)$

12. slope -5, y-intercept $(0, 6)$

13. $m = -1$, y-intercept $(0, -7)$

14. $m = 1$, y-intercept $(0, -9)$

15. slope 0, *y*-intercept $(0, 3)$

16. slope 0, *y*-intercept $(0, -4)$

17. undefined slope, *y*-intercept $(0, -2)$

18. undefined slope, *y*-intercept $(0, 5)$

CONCEPT CHECK *Use the geometric interpretation of slope (rise divided by run, from* ***Section 11.3****) to find the slope of each line. Then, by identifying the y-intercept from the graph, write the slope-intercept form of the equation of the line.*

19.

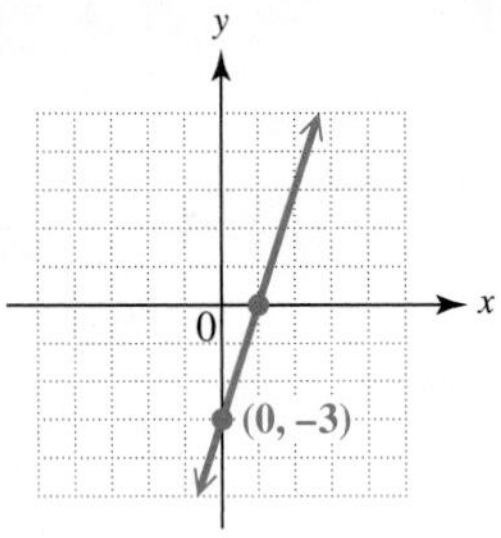

20.

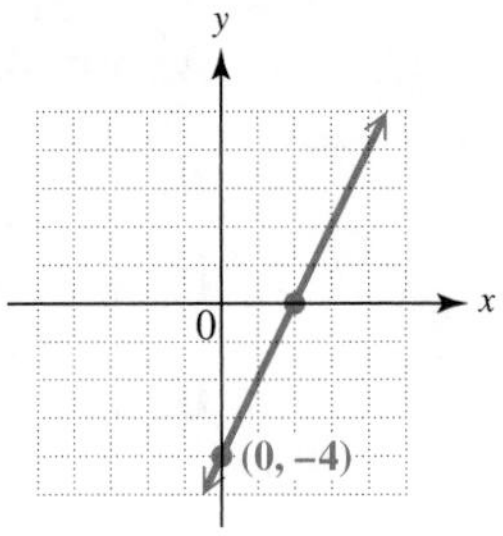

21.

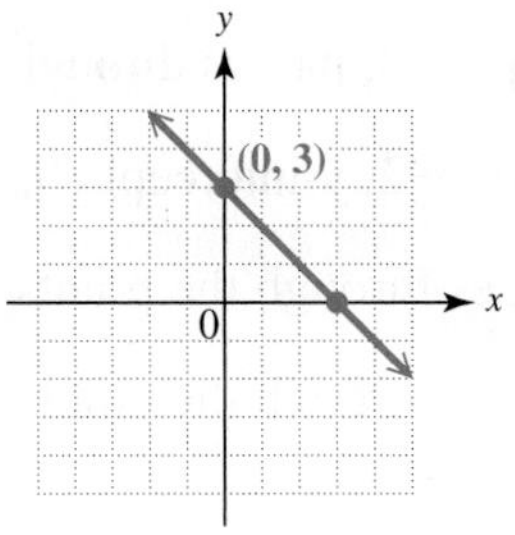

22.

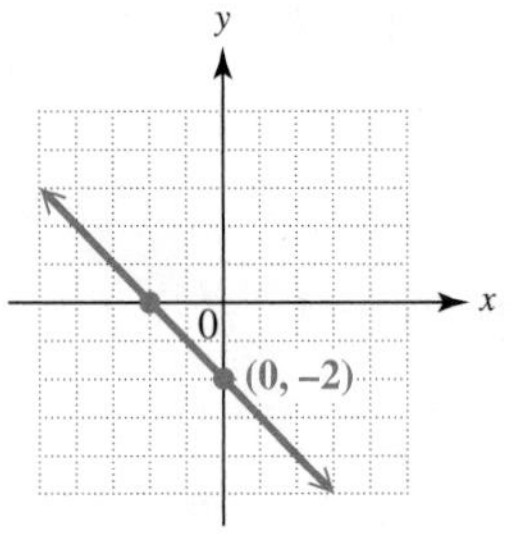

23.

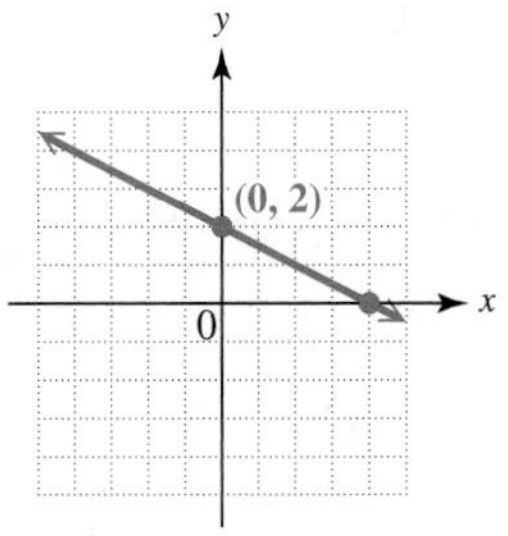

24.

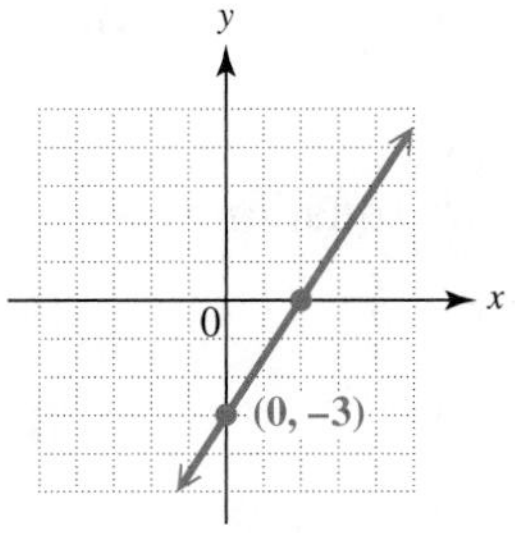

Graph each equation by identifying the slope and y-intercept, and using their definitions to find two points on the line. ***See Example 3.***

25. $y = 3x + 2$

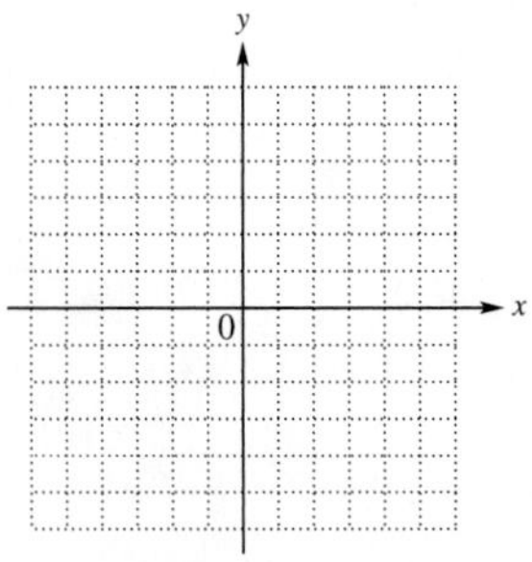

26. $y = 4x - 4$

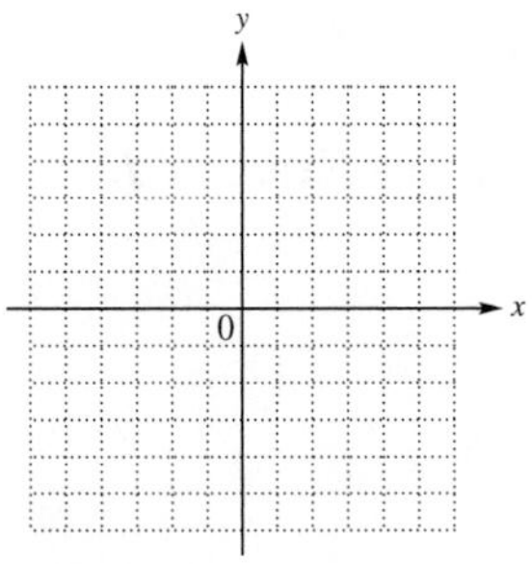

27. $y = \frac{3}{4}x - 1$

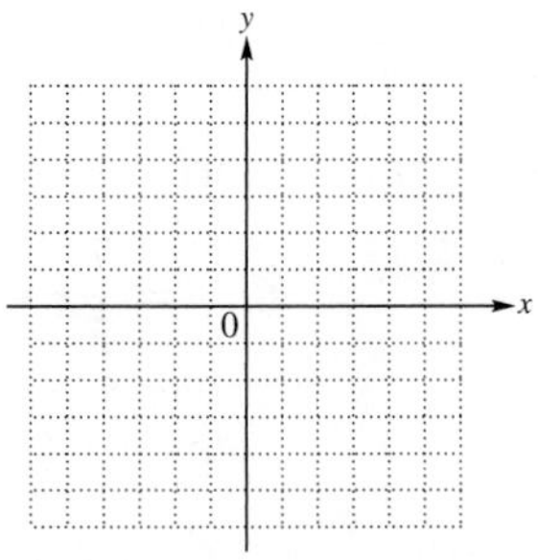

28. $y = \frac{3}{2}x + 2$

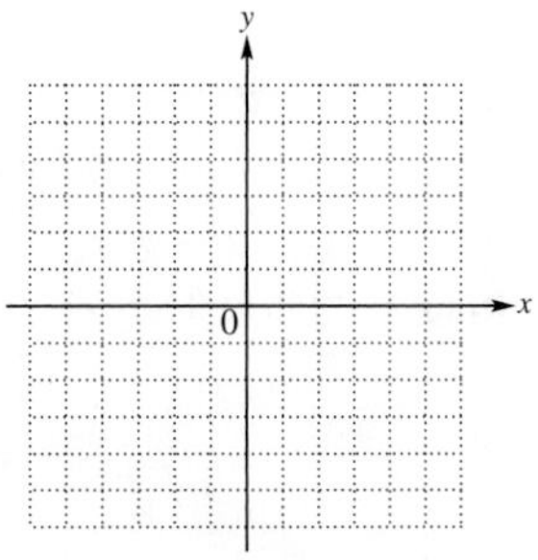

29. $2x + y = -5$

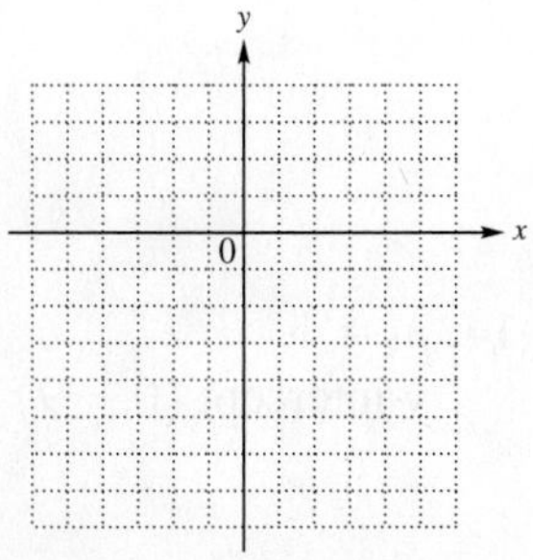

30. $3x + y = -2$

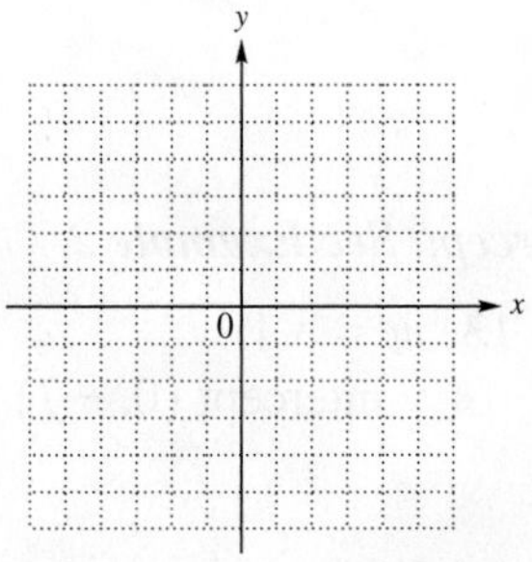

31. $x + 2y = 4$

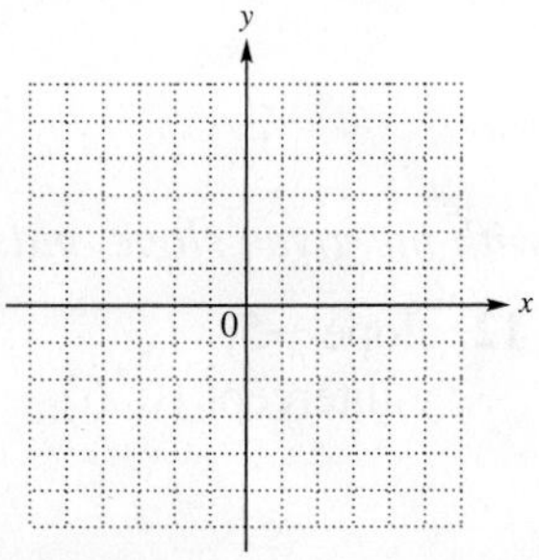

32. $x + 3y = 12$

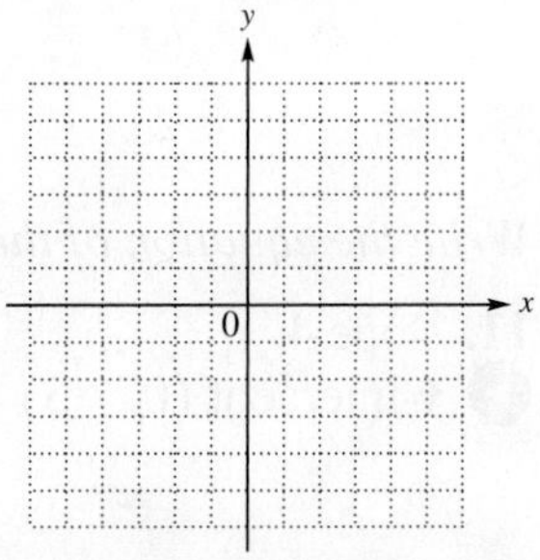

Graph each line passing through the given point and having the given slope. (In Exercises 37–40, recall the types of lines having slope 0 and undefined slope.) Give the slope-intercept form of the equation of the line if possible. **See Examples 4 and 5.**

33. $(-2, 3)$, $m = \frac{1}{2}$

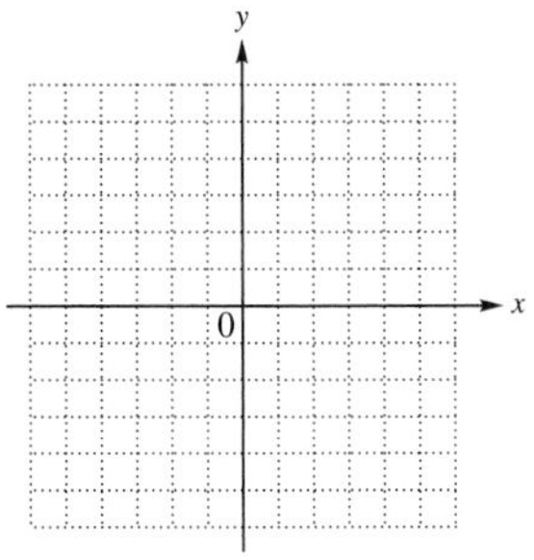

34. $(-4, -1)$, $m = \frac{3}{4}$

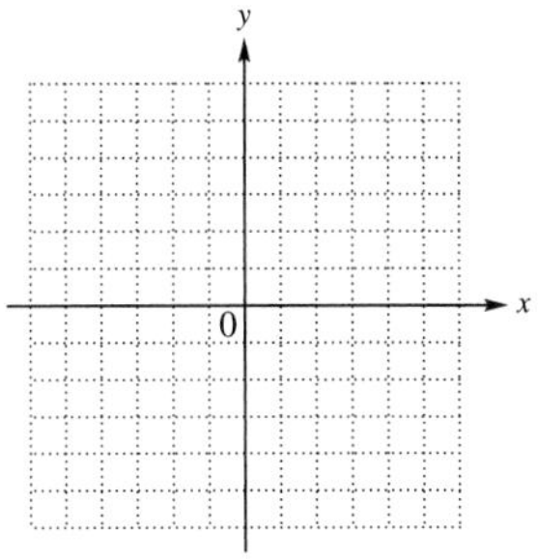

35. $(1, -5)$, $m = -\frac{2}{5}$

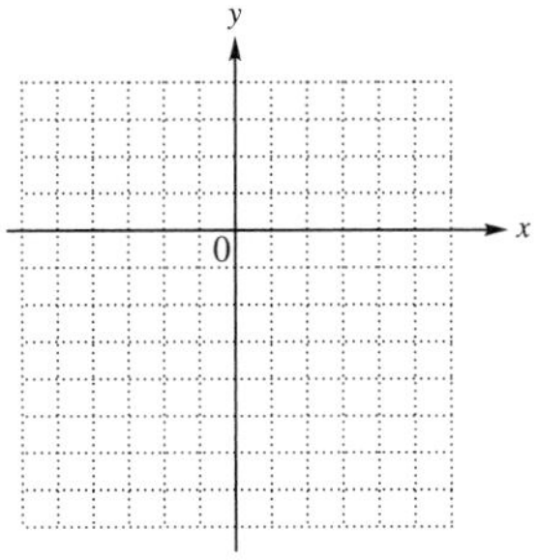

36. $(2, -1)$, $m = -\frac{1}{3}$

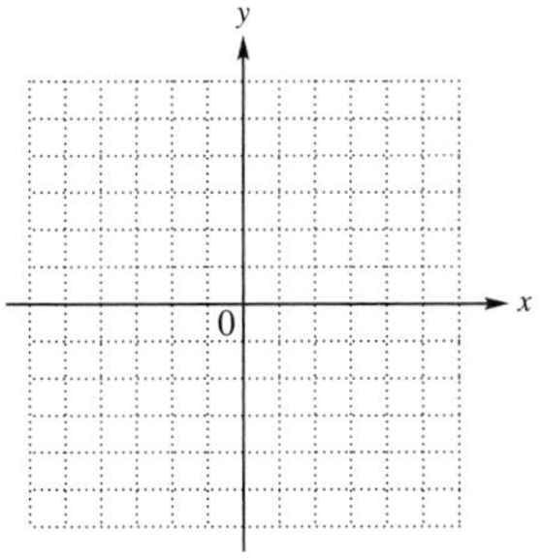

37. $(3, 2)$, $m = 0$

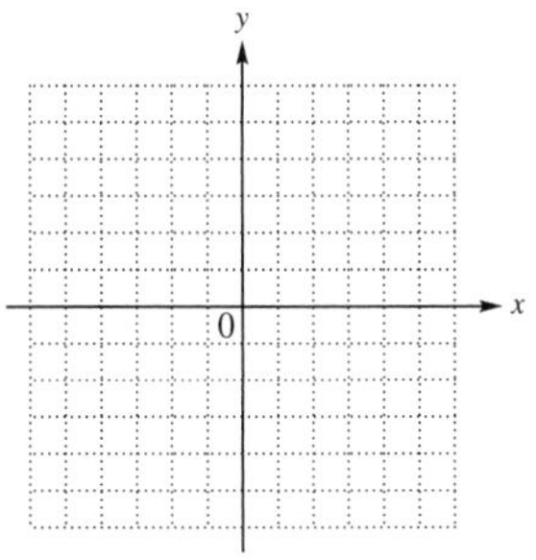

38. $(-2, 3)$, $m = 0$

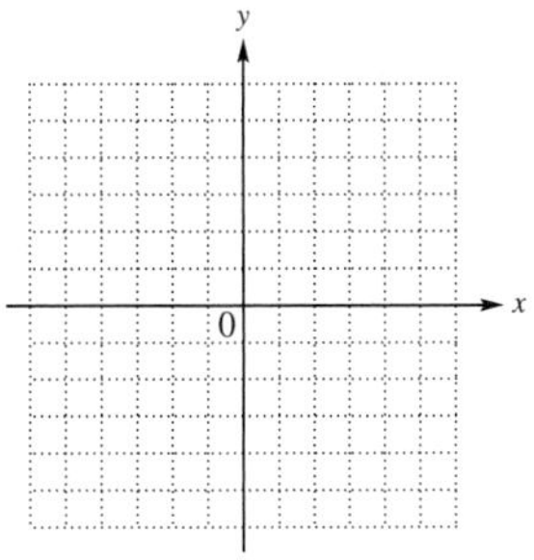

39. $(3, -2)$, undefined slope

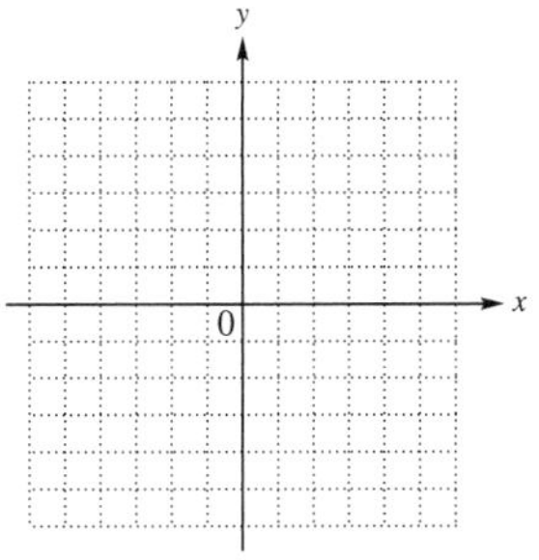

40. $(2, 4)$, undefined slope

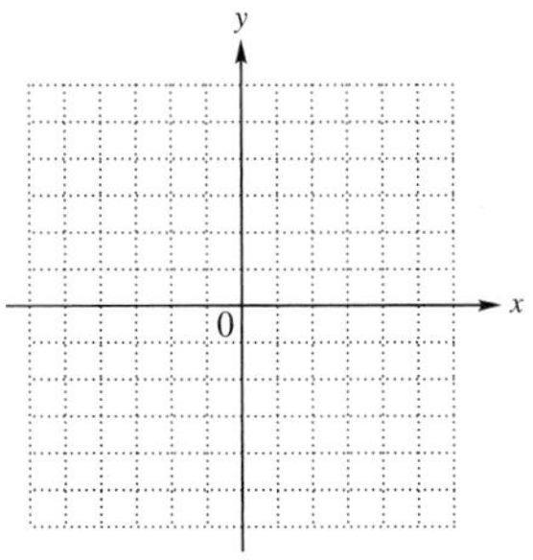

41. $(0, 0)$, $m = \frac{2}{3}$

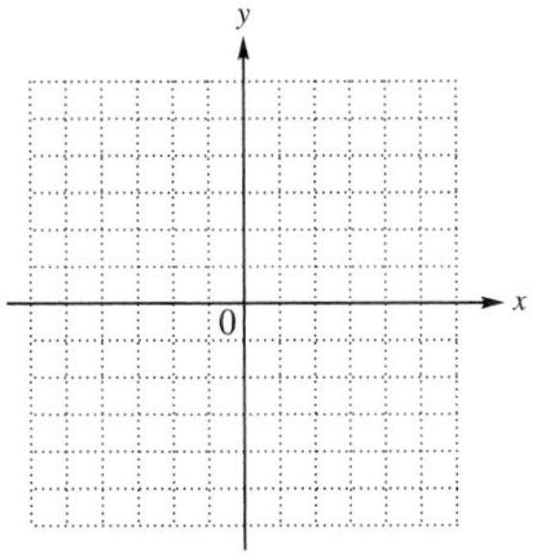

42. $(0, 0)$, $m = \frac{5}{2}$

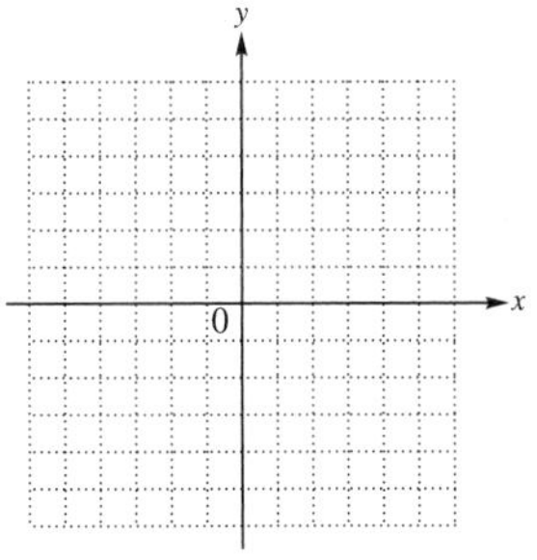

Write an equation of the line passing through the given point and having the given slope. Give the final answer in slope-intercept form. **See Examples 5 and 6.**

43. $(4, 1)$, $m = 2$

44. $(2, 7)$, $m = 3$

45. $(3, -10)$, $m = -2$

46. $(2, -5)$, $m = -4$

47. $(-2, 5)$, $m = \frac{2}{3}$

48. $(-4, 1)$, $m = \frac{3}{4}$

*Write an equation of the line passing through each pair of points. Give the final answer in **(a)** slope-intercept form and **(b)** standard form. See **Example 7**.*

49. $(8, 5)$ and $(9, 6)$

50. $(4, 10)$ and $(6, 12)$

51. $(-1, -7)$ and $(-8, -2)$

52. $(-2, -1)$ and $(3, -4)$

53. $(0, -2)$ and $(-3, 0)$

54. $(-4, 0)$ and $(0, 2)$

55. $\left(\frac{1}{2}, \frac{3}{2}\right)$ and $\left(-\frac{1}{4}, \frac{5}{4}\right)$

56. $\left(-\frac{2}{3}, \frac{8}{3}\right)$ and $\left(\frac{1}{3}, \frac{7}{3}\right)$

*Write an equation of the line satisfying the given conditions. Give the final answer in slope-intercept form. (Hint: Recall the relationships among slopes of parallel and perpendicular lines. See **Section 11.3**.)*

57. Perpendicular to $x - 2y = 7$, y-intercept $(0, -3)$

58. Parallel to $5x = 2y + 10$, y-intercept $(0, 4)$

59. Passing through $(2, 3)$, parallel to $4x - y = -2$

60. Passing through $(4, 2)$, perpendicular to $x - 3y = 7$

61. Passing through $(2, -3)$, parallel to $3x = 4y + 5$

62. Passing through $(-1, 4)$, perpendicular to $2x + 3y = 8$

*The cost y to produce x items is, in some cases, expressed as $y = mx + b$. The number b gives the **fixed cost** (the cost that is the same no matter how many items are produced), and the number m gives the **variable cost** (the cost to produce an additional item). Use this information to work Exercises 63 and 64.*

63. It costs \$400 to start up a business selling campaign buttons. Each button costs \$0.25 to produce.

(a) What is the fixed cost?

(b) What is the variable cost?

(c) Write the cost equation.

(d) What will be the cost to produce 100 campaign buttons, based on the cost equation?

(e) How many campaign buttons will be produced if total cost is \$775?

64. It costs \$2000 to purchase a copier, and each copy costs \$0.02 to make.

(a) What is the fixed cost?

(b) What is the variable cost?

(c) Write the cost equation.

(d) What will be the cost to produce 10,000 copies, based on the cost equation?

(e) How many copies will be produced if total cost is \$2600?

Solve each problem. ***See Example 8.***

65. The table lists the average annual cost *y* (in dollars) of tuition and fees at 2-year public colleges for selected years *x*, where $x = 1$ represents 2007, $x = 2$ represents 2008, and so on.

Year *x*	Cost *y* (in dollars)
1	2294
2	2372
3	2558
4	2727
5	2963

Source: The College Board.

(a) Write five ordered pairs for the data.

(b) Plot the ordered pairs. Do the points lie approximately in a straight line?

AVERAGE ANNUAL COSTS AT 2-YEAR COLLEGES

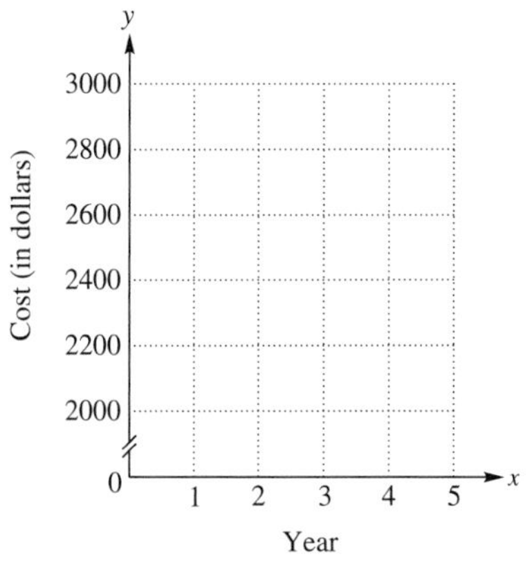

(c) Use the ordered pairs $(2, 2372)$ and $(5, 2963)$ to find the equation of a line that approximates the data. Write the equation in slope-intercept form.

(d) Use the equation from part (c) to estimate the average annual cost at 2-year colleges in 2012 to the nearest dollar. (*Hint:* What is the value of *x* for 2012?)

66. The table gives heavy-metal nuclear waste (in thousands of metric tons) from spent reactor fuel awaiting permanent storage. (*Source: Scientific American.*)

Year *x*	Waste *y* (in thousands of tons)
1995	32
2000	42
2010*	61
2020*	76

*Estimates by the U.S. Department of Energy.

Let $x = 0$ represent 1995, $x = 5$ represent 2000 (since $2000 - 1995 = 5$), and so on.

(a) For 1995, the ordered pair is $(0, 32)$. Write ordered pairs for the data for the other years given in the table.

(b) Plot the ordered pairs (x, y). Do the points lie approximately in a straight line?

HEAVY-METAL NUCLEAR WASTE AWAITING STORAGE

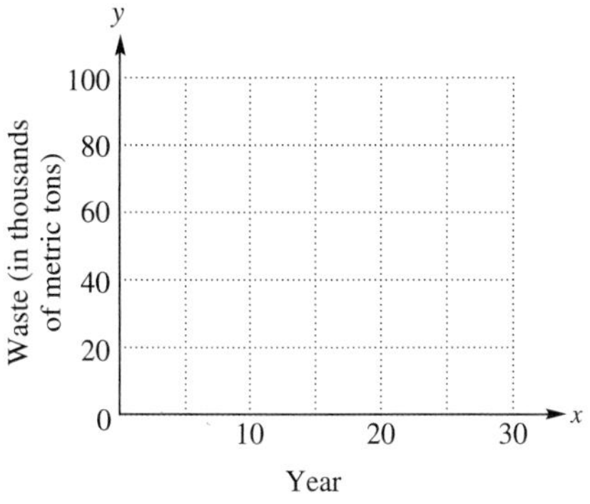

(c) Use the ordered pairs $(0, 32)$ and $(25, 76)$ to find the equation of a line that approximates the data. Write the equation in slope-intercept form.

(d) Use the equation from part (c) to estimate the amount of nuclear waste in 2015. (*Hint:* What is the value of *x* for 2015?)

The points on the graph show the number of colleges that teamed up with banks to issue student ID cards which doubled as debit cards from 2002 through 2007. The graph of a linear equation that models the data is also shown.

67. Use the ordered pairs shown on the graph to write an equation of the line that models the data. Give the equation in slope-intercept form.

68. Use the equation from **Exercise 67** to estimate the number of colleges that teamed up with banks to offer debit IDs in 2004, the year with unavailable data.

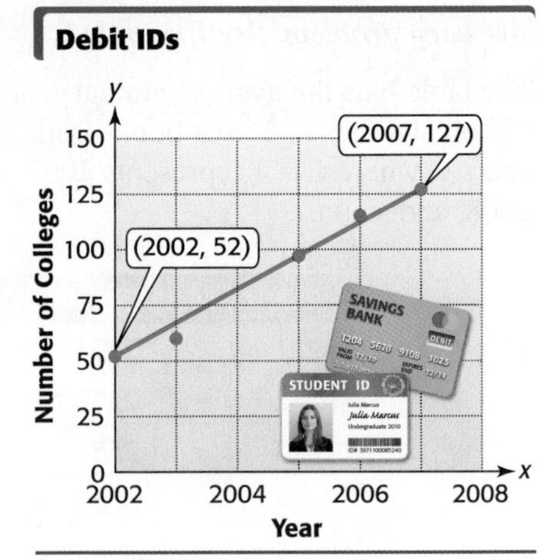

Source: CR80News.
* Data for 2004 unavailable.

Relating Concepts (Exercises 69–76) For Individual or Group Work

If we think of ordered pairs of the form (C, F), then the two most common methods of measuring temperature, Celsius and Fahrenheit, can be related as follows:

When $C = 0, F = 32,$ *and when* $C = 100, F = 212.$

Work Exercises 69–76 in order.

69. Write two ordered pairs relating these two temperature scales.

70. Find the slope of the line through the two points.

71. Use the point-slope form to find an equation of the line. (Your variables should be C and F rather than x and y.)

72. Write an equation for F in terms of C.

73. Use the equation from **Exercise 72** to write an equation for C in terms of F.

74. Use the equation from **Exercise 72** to find the Fahrenheit temperature when $C = 30$.

75. Use the equation from **Exercise 73** to find the Celsius temperature when $F = 50$.

76. For what temperature is $F = C$? (Use the photo shown at the right to confirm your answer.)

11.5 Graphing Linear Inequalities in Two Variables

OBJECTIVES

1. Graph linear inequalities in two variables.
2. Graph an inequality with a boundary line through the origin.

In **Section 11.2** we graphed linear equations, such as

$$2x + 3y = 6.$$

Now we extend this work to include *linear inequalities in two variables,* such as

$$2x + 3y \leq 6.$$

(Recall that $\leq$ is read "is less than or equal to.")

Linear Inequality in Two Variables

An inequality that can be written in the form

$$Ax + By < C, \quad Ax + By > C, \quad Ax + By \leq C, \quad \text{or} \quad Ax + By \geq C,$$

where A, B, and C are real numbers and A and B are not both 0, is a **linear inequality in two variables.**

OBJECTIVE 1 **Graph linear inequalities in two variables.** Consider the graph in **Figure 37.**

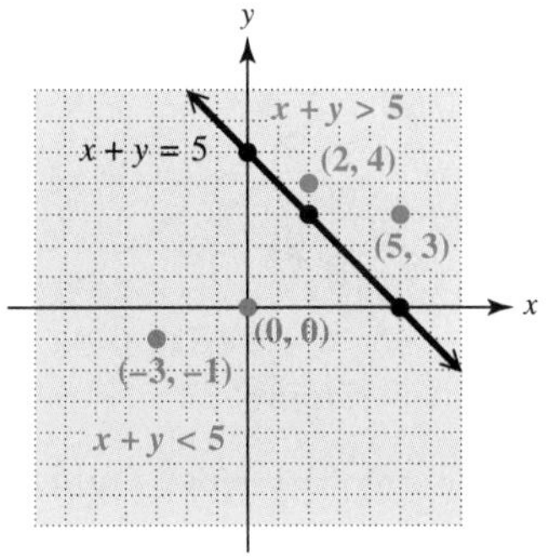

Figure 37

The graph of the line $x + y = 5$ in **Figure 37** divides the points in the rectangular coordinate system into three sets.

1. Those points that lie on the line itself and satisfy the equation $x + y = 5$ (like $(0, 5)$, $(2, 3)$, and $(5, 0)$)
2. Those points that lie in the region above the line and satisfy the inequality $x + y > 5$ (like $(5, 3)$ and $(2, 4)$)
3. Those points that lie in the region below the line and satisfy the inequality $x + y < 5$ (like $(0, 0)$ and $(-3, -1)$)

The graph of the line $x + y = 5$ is the **boundary line** for the inequalities

$$x + y > 5 \quad \text{and} \quad x + y < 5.$$

Graphs of linear inequalities in two variables are regions in the real number plane that may or may not include boundary lines.

1 Use $(0, 0)$ as a test point, and **GS** shade the appropriate region for the linear inequality.

$$3x + 4y \leq 12$$

$$3(\_\_\_\_) + 4(\_\_\_\_) \overset{?}{\leq} 12$$

$$\_\_\_\_ \overset{?}{\leq} 12 \text{ (True / False)}$$

Shade the region of the graph that (*includes / does not include*) the test point $(0, 0)$.

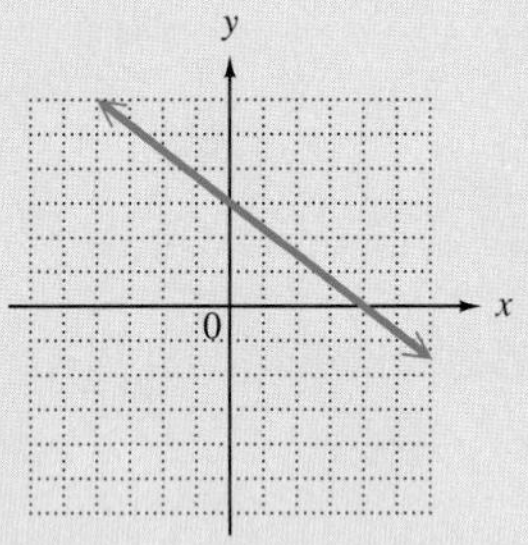

2 Graph $4x - 5y \leq 20$.

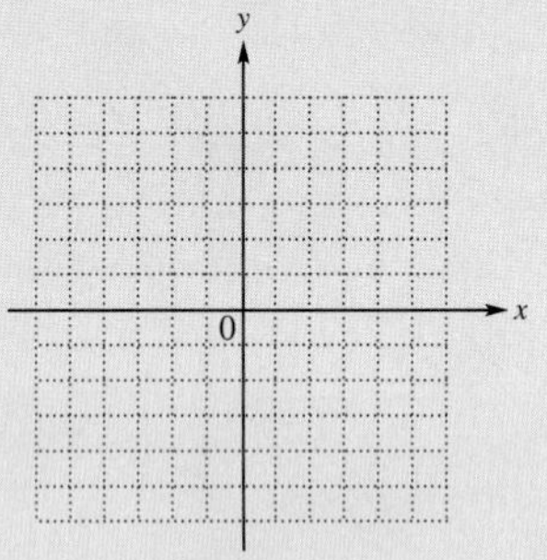

Answers

1. 0; 0; 0; True; includes

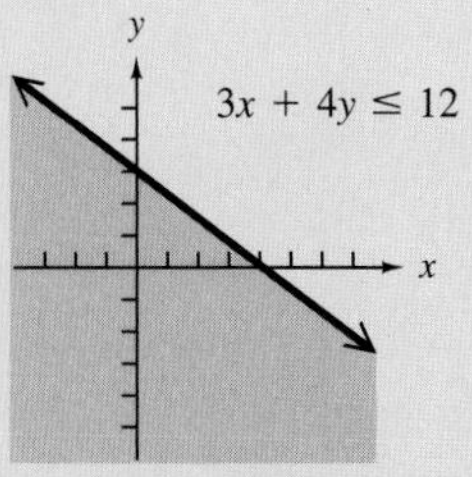

2.

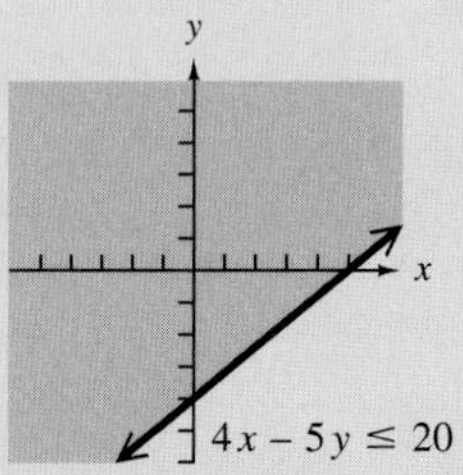

EXAMPLE 1 Graphing a Linear Inequality

Graph $2x + 3y \leq 6$.

The inequality $2x + 3y \leq 6$ means that

$$2x + 3y < 6 \quad \text{or} \quad 2x + 3y = 6.$$

We begin by graphing the equation $2x + 3y = 6$, a line with intercepts $(0, 2)$ and $(3, 0)$ as shown in **Figure 38.** This boundary line divides the plane into two regions, one of which satisfies the inequality. To find the correct region, we choose a test point *not* on the boundary line and substitute it into the given inequality to see whether the resulting statement is true or false. The point $(0, 0)$ is a convenient choice.

$2x + 3y \leq 6$	Original inequality
$2(0) + 3(0) \overset{?}{\leq} 6$	Let $x = 0$ and $y = 0$.
$0 + 0 \overset{?}{\leq} 6$	Multiply.
$0 \leq 6$	True

Use (0, 0) as a test point.

Since the last statement is true, we shade the region that includes the test point $(0, 0)$. See **Figure 38.** The shaded region, along with the boundary line, is the desired graph.

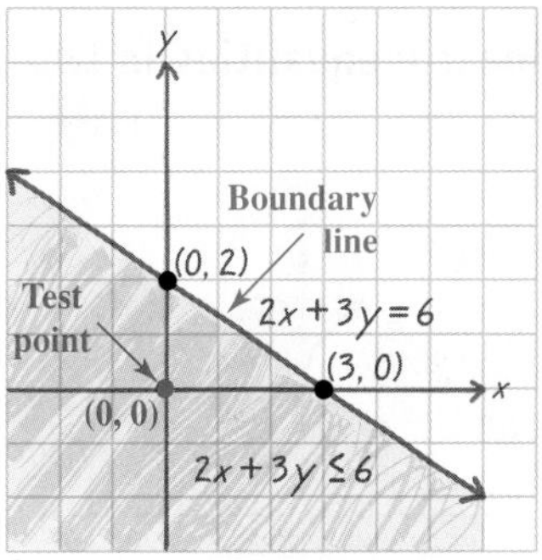

Figure 38

◀ **Work Problems 1 and 2 at the Side.**

Note

Alternatively in **Example 1,** we can find the required region by solving the given inequality for y.

$2x + 3y \leq 6$	Inequality from **Example 1**
$3y \leq -2x + 6$	Subtract $2x$.
$y \leq -\frac{2}{3}x + 2$	Divide each term by 3.

Ordered pairs in which y is equal to $-\frac{2}{3}x + 2$ are on the boundary line, so pairs in which y *is less than* $-\frac{2}{3}x + 2$ will be *below* that line. (As we move *down* vertically, the y-values *decrease*.) This gives the same region that we shaded in **Figure 38.** (Ordered pairs in which y *is greater than* $-\frac{2}{3}x + 2$ will be *above* the boundary line.)

EXAMPLE 2 **Graphing a Linear Inequality**

Graph $x - y > 5$.

This inequality does *not* involve equality. Therefore, the points on the line $x - y = 5$ do *not* belong to the graph. However, the line still serves as a boundary for two regions, one of which satisfies the inequality.

To graph the inequality, first graph the equation $x - y = 5$. Use a *dashed line* to show that the points on the line are *not* solutions of the inequality $x - y > 5$. See **Figure 39.** Then choose a test point to see which region satisfies the inequality.

$x - y > 5$ Original inequality

(0, 0) is a convenient test point.

$0 - 0 \overset{?}{>} 5$ Let $x = 0$ and $y = 0$.

$0 > 5$ False

Since $0 > 5$ is false, the graph of the inequality is the region that *does not* contain $(0, 0)$. Shade the *other* region, as shown in **Figure 39.**

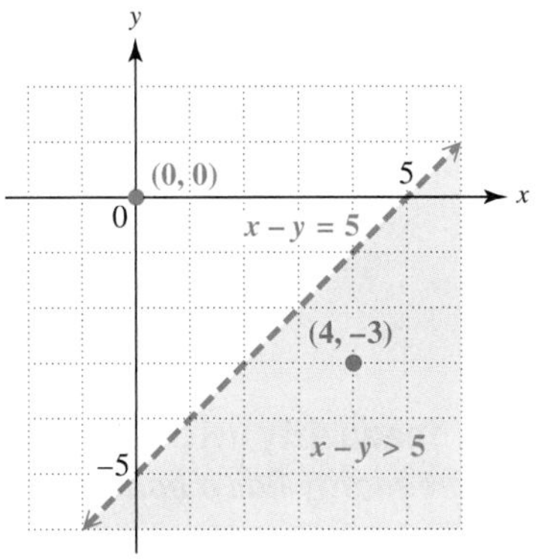

Figure 39

CHECK To check that the proper region is shaded, we test a point in the shaded region. For example, we use $(4, -3)$.

$x - y > 5$

$4 - (-3) \overset{?}{>} 5$ Let $x = 4$ and $y = -3$.

Use parentheses to avoid errors.

$7 > 5$ ✓ True

This verifies that the correct region is shaded in **Figure 39.**

Work Problems 3 and 4 at the Side. ▶

A summary of the steps used to graph a linear inequality in two variables follows.

Graphing a Linear Inequality in Two Variables

Step 1 **Graph the boundary.** Graph the line that is the boundary of the region. Use the methods of **Section 11.2.** Draw a solid line if the inequality involves $\leq$ or $\geq$. Draw a dashed line if the inequality involves $<$ or $>$.

Step 2 **Shade the appropriate side.** Use any point not on the line as a test point. Substitute for x and y in the *inequality*. If a true statement results, shade the region containing the test point. If a false statement results, shade the other region.

3 Use $(0, 0)$ as a test point and shade the appropriate region for the linear inequality.

$$3x + 5y > 15$$

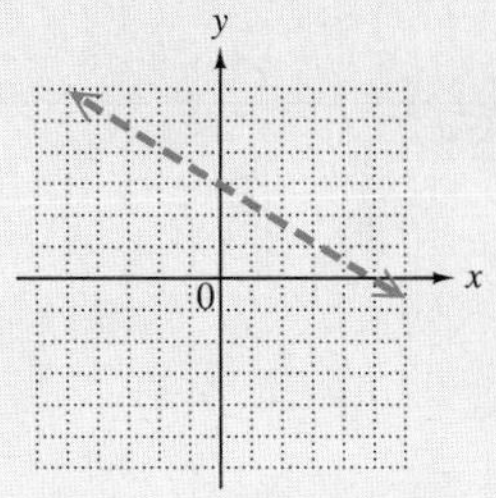

4 Graph $x + 2y > 6$.

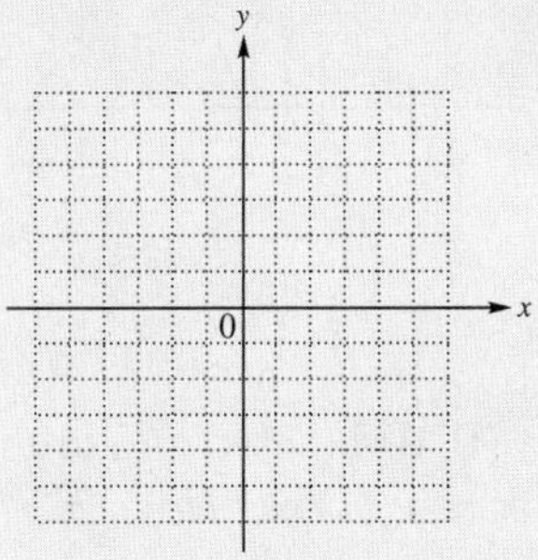

Answers

3.

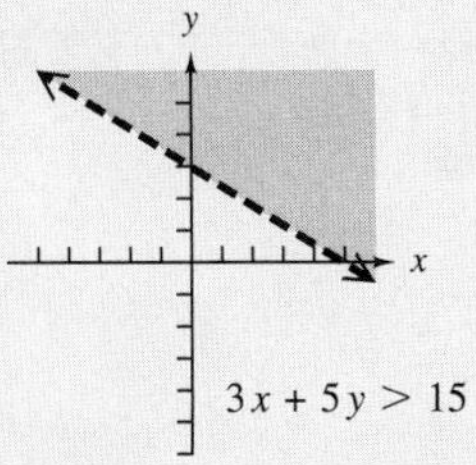

4.

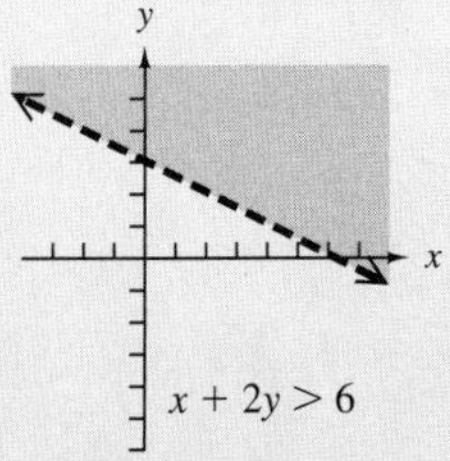

5 Graph $y < 4$.

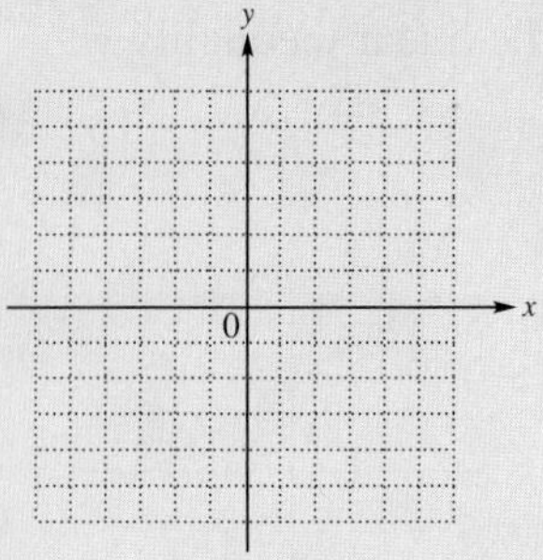

6 Graph $x \geq -3y$.

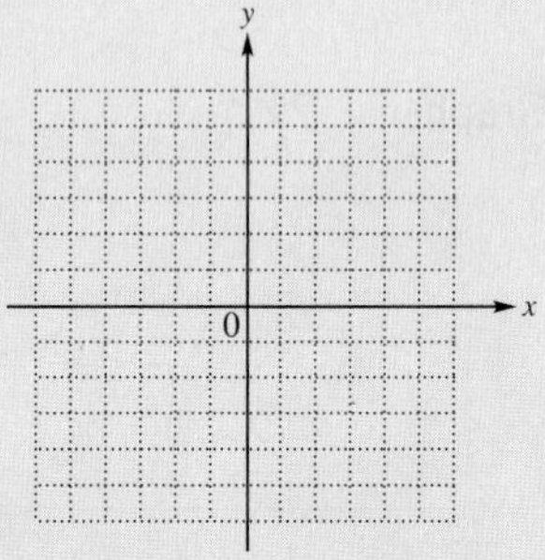

Can $(0, 0)$ be used as a test point here? (*Yes* / *No*)

EXAMPLE 3 **Graphing a Linear Inequality with a Vertical Boundary Line**

Graph $x < 3$.

First graph $x = 3$, a vertical line through the point $(3, 0)$. Use a dashed line, and choose $(0, 0)$ as a test point.

$x < 3$ Original inequality

$0 \overset{?}{<} 3$ Let $x = 0$.

$0 < 3$ True

Because $0 < 3$ is true, we shade the region containing $(0, 0)$, as in **Figure 40.**

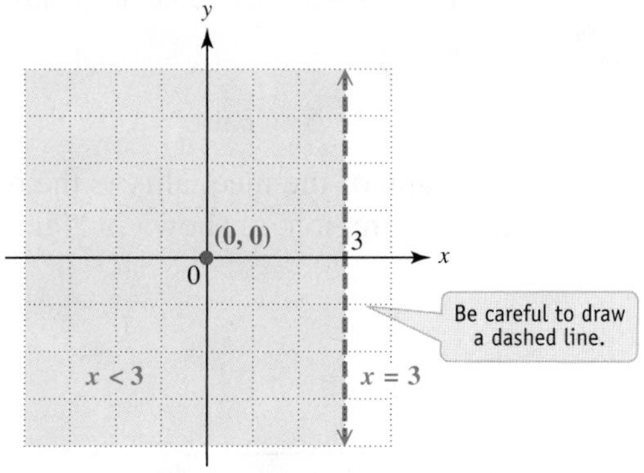

Figure 40

◀ **Work Problem 5 at the Side.**

OBJECTIVE 2 Graph an inequality with a boundary line through the origin. ***If the graph of an inequality has a boundary line through the origin, (0, 0) cannot be used as a test point.***

EXAMPLE 4 **Graphing a Linear Inequality with a Boundary Line through the Origin**

Graph $x \leq 2y$.

We graph $x = 2y$ using a solid line through $(0, 0)$, $(6, 3)$, and $(4, 2)$. Because $(0, 0)$ is *on* the line $x = 2y$, it cannot be used as a test point. Instead, we choose a test point *off* the line, say $(1, 3)$.

$x \leq 2y$ Original inequality

$1 \overset{?}{\leq} 2(3)$ Let $x = 1$ and $y = 3$.

$1 \leq 6$ True

Because $1 \leq 6$ is true, we shade the side of the graph containing the test point $(1, 3)$. See **Figure 41.**

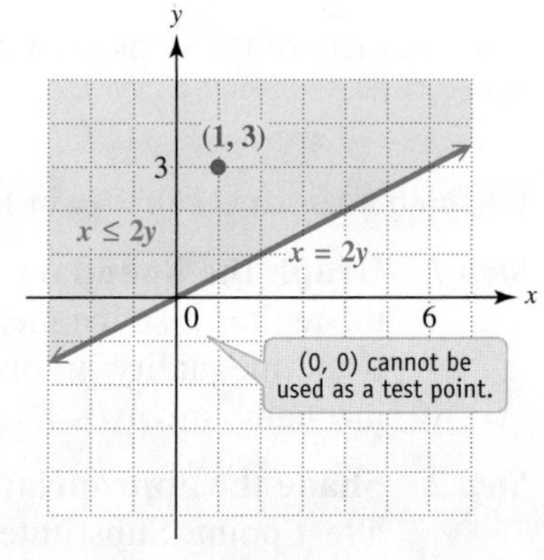

Figure 41

◀ **Work Problem 6 at the Side.**

Answers

5.

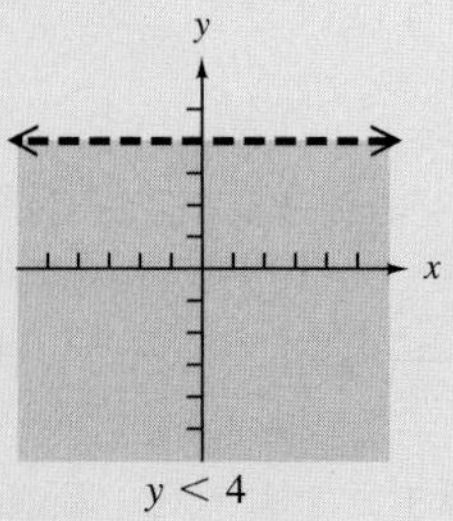

6. No

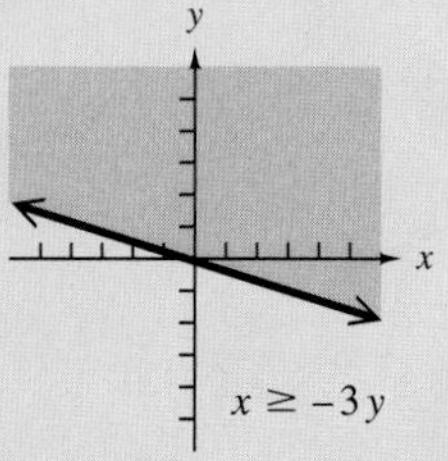

11.5 Exercises

FOR EXTRA HELP

Download the MyDashBoard App

MyMathLab®

CONCEPT CHECK *The statements in Exercises 1–4 were taken from various web sites. Each includes a phrase that can be symbolized with one of the inequality symbols* $<$, $\leq$, $>$, *or* $\geq$. *Give the inequality symbol for the bold faced words.*

1. In 2009, carbon dioxide emissions from the consumption of fossil fuels were **more than** 5400 million metric tons in the United States and **more than** 7700 million metric tons in China. (*Source:* www.eia.gov)

2. The number of motor vehicle deaths in the United States in 2009 was 3615 **less than** the number in 2008. (*Source:* www.nhtsa.gov)

3. As of January 2012, American Airlines passengers were allowed one carry-on bag, with dimensions (length + width + height) of **at most** 45 in. (*Source:* www.aa.com)

4. As of January 2012, the usage charge on many of AT&T's calling plans was **at least** \$9.99 per month. (*Source:* www.consumer.att.com)

CONCEPT CHECK *Decide whether each statement is* true *or* false. *If false, explain why.*

5. The point $(4, 0)$ lies on the graph of $3x - 4y < 12$.

6. The point $(4, 0)$ lies on the graph of $3x - 4y \leq 12$.

7. The point $(0, 0)$ can be used as a test point to determine which region to shade when graphing the linear inequality $x + 4y > 0$.

8. When graphing the linear inequality $3x + 2y \geq 12$, use a dashed line for the boundary line.

9. Both points $(4, 1)$ and $(0, 0)$ lie on the graph of $3x - 2y \geq 0$.

10. The graph of $y > x$ does not contain points in quadrant IV.

In Exercises 11–16, the straight-line boundary has been drawn. Complete each graph by shading the correct region. ***See Examples 1–4.***

11. $x + y \geq 4$

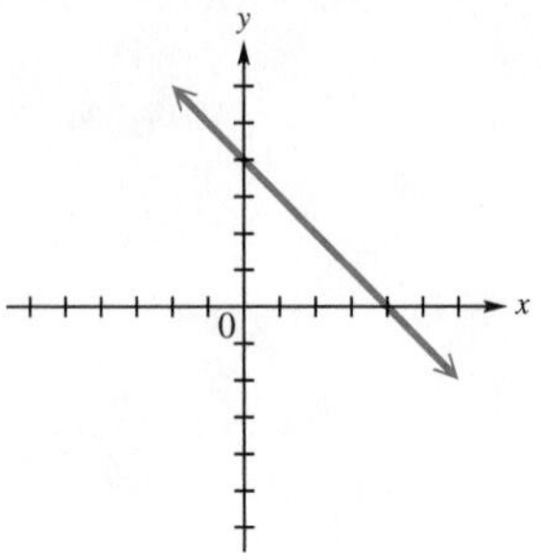

12. $x + y \leq 2$

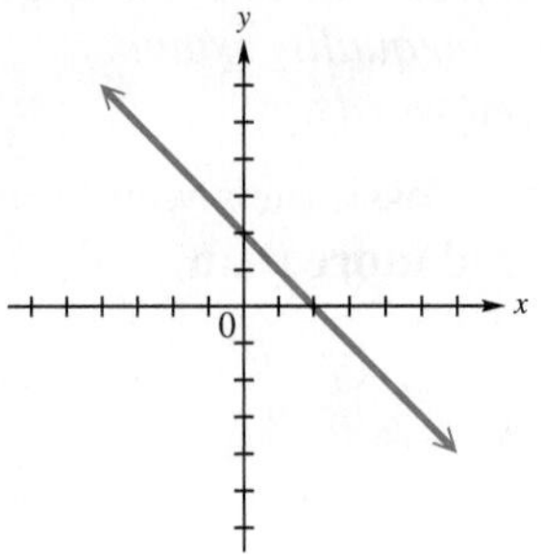

13. $x + 2y \geq 7$

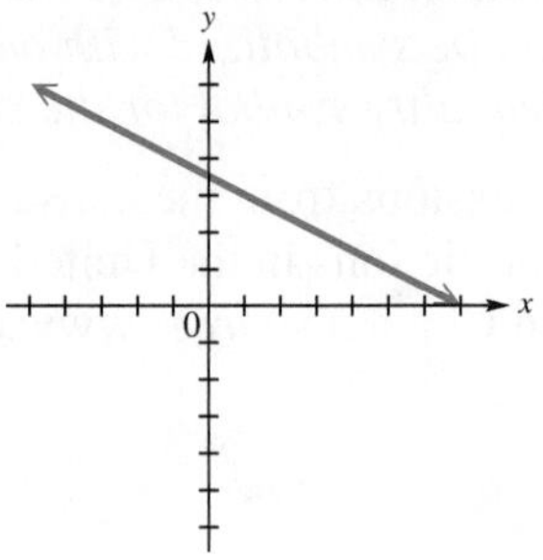

14. $2x + y \geq 5$

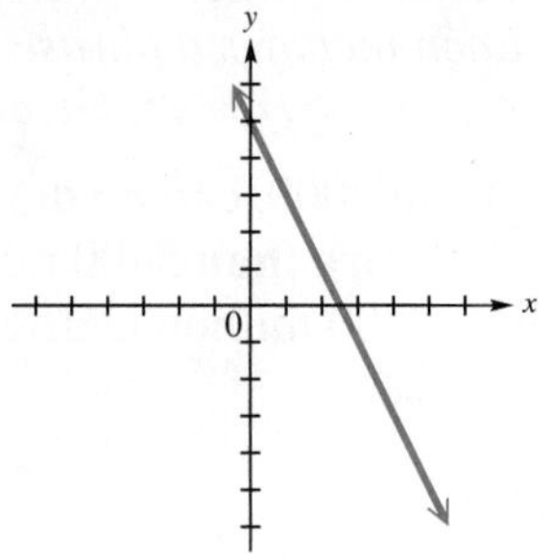

15. $-3x + 4y > 12$

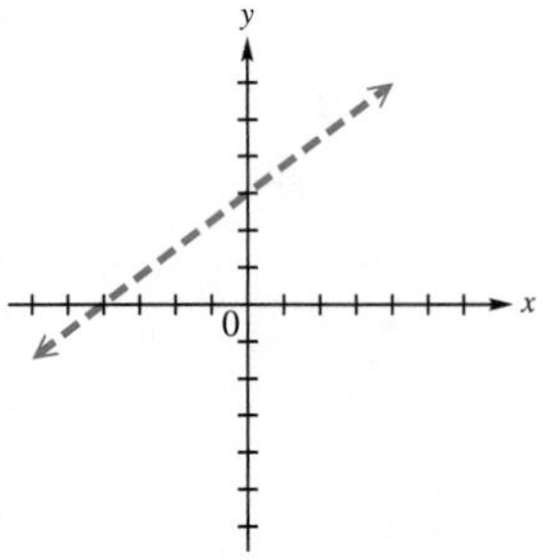

16. $4x - 5y < 20$

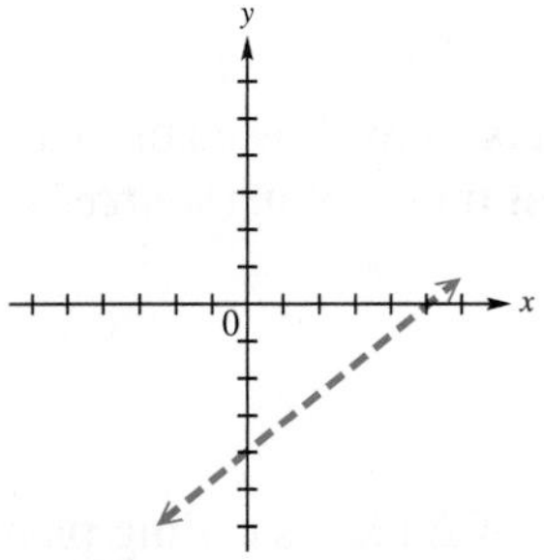

17. $x > 4$

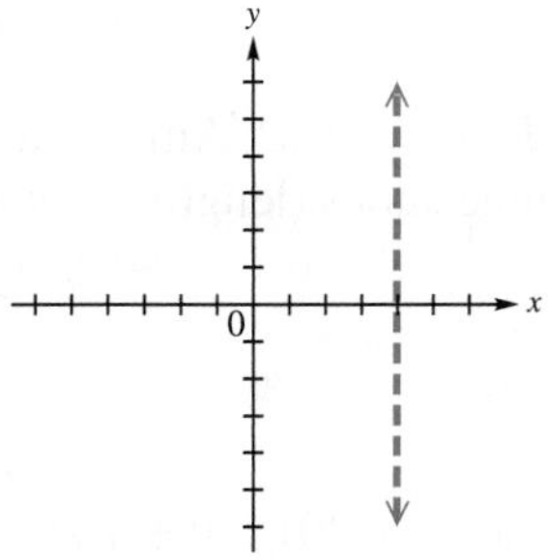

18. $y < -1$

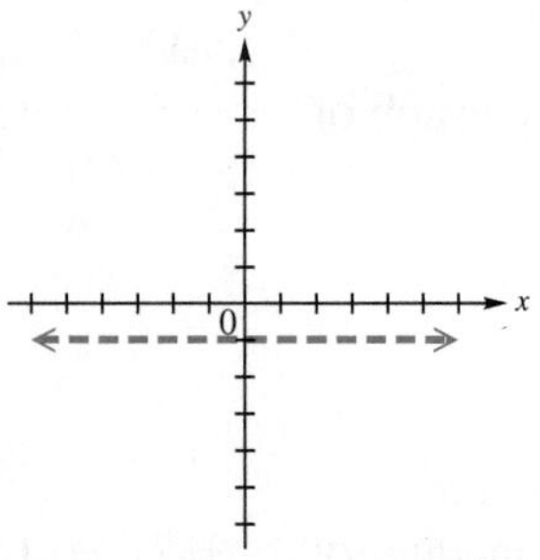

19. $x \geq -y$

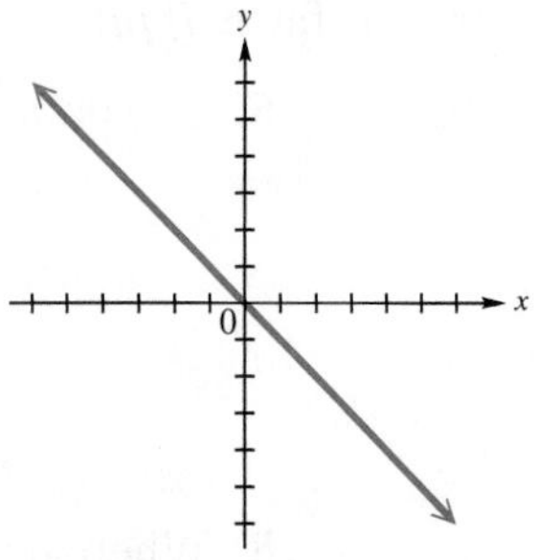

20. $x > 3y$

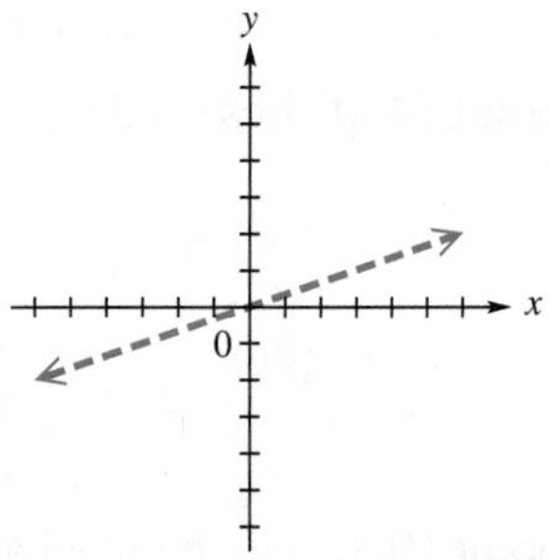

21. Explain how to determine whether to use a dashed line or a solid line when graphing a linear inequality in two variables.

22. Explain why the point $(0, 0)$ is not an appropriate choice for a test point when graphing an inequality whose boundary goes through the origin.

Graph each linear inequality. ***See Examples 1–4.***

23. $x + y \leq 5$

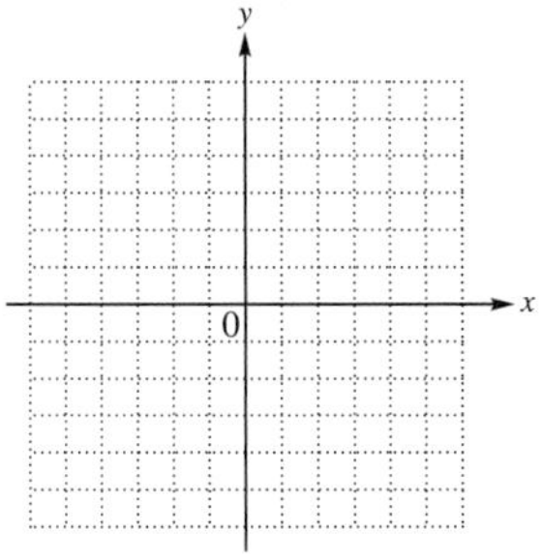

24. $x + y \geq 3$

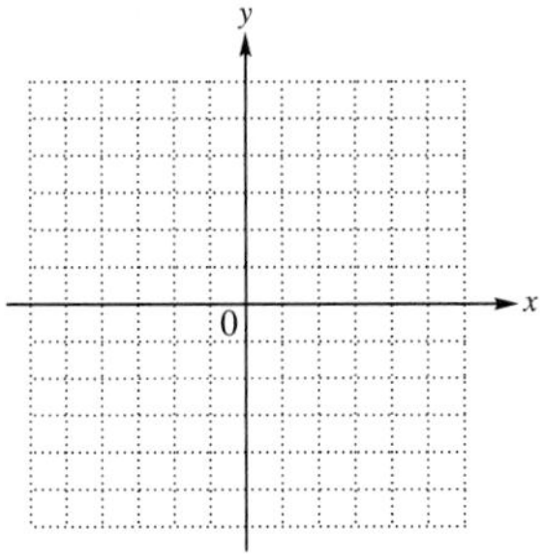

25. $x + 2y < 4$

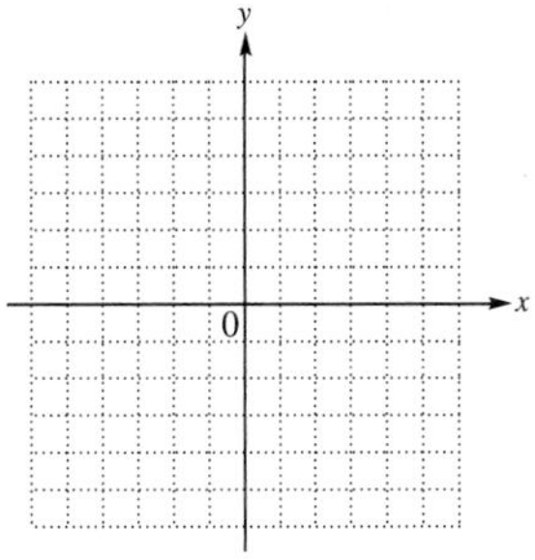

26. $x + 3y > 6$

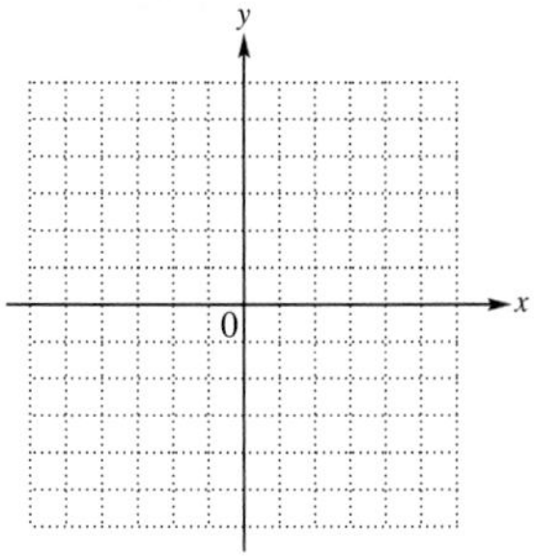

27. $2x + 6 > -3y$

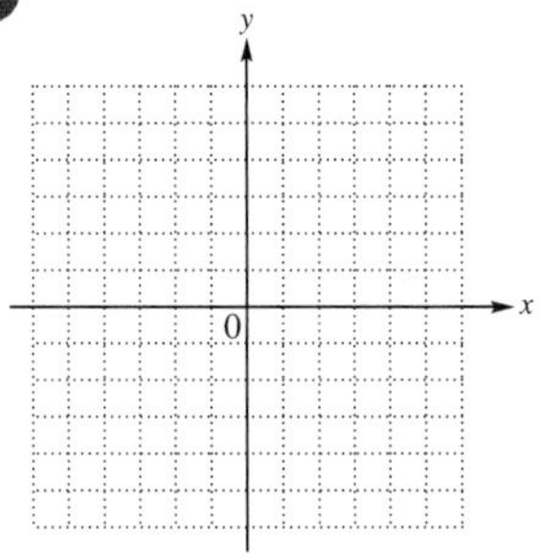

28. $-4y > 3x - 12$

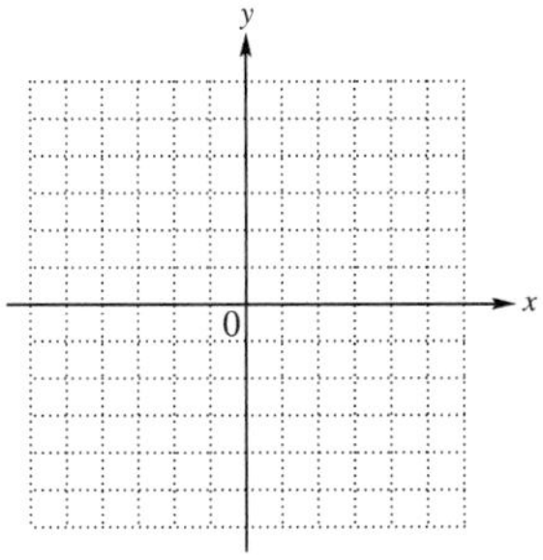

29. $y \geq 2x + 1$

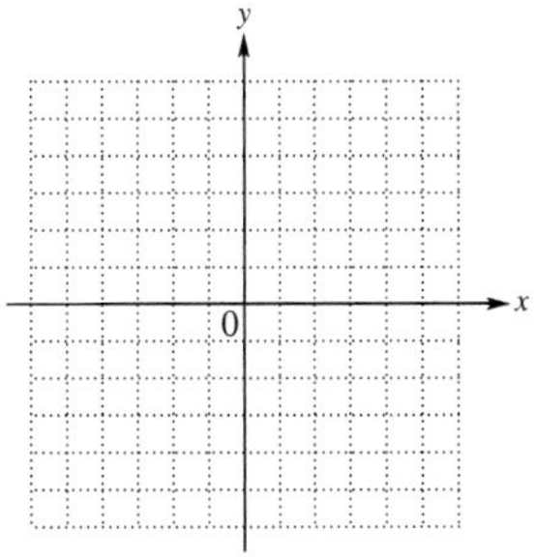

30. $y < -3x + 1$

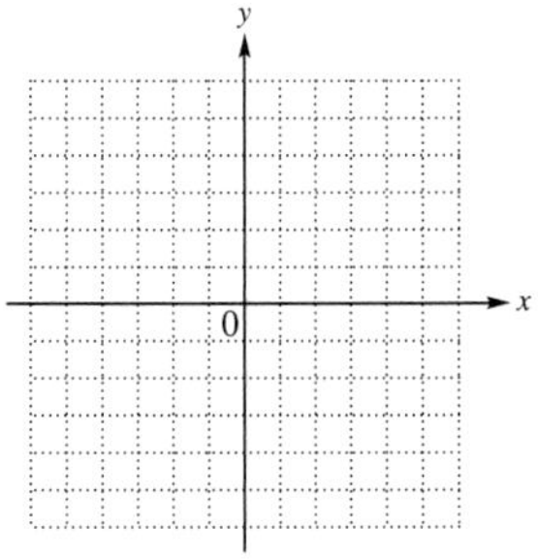

31. $x \leq -2$

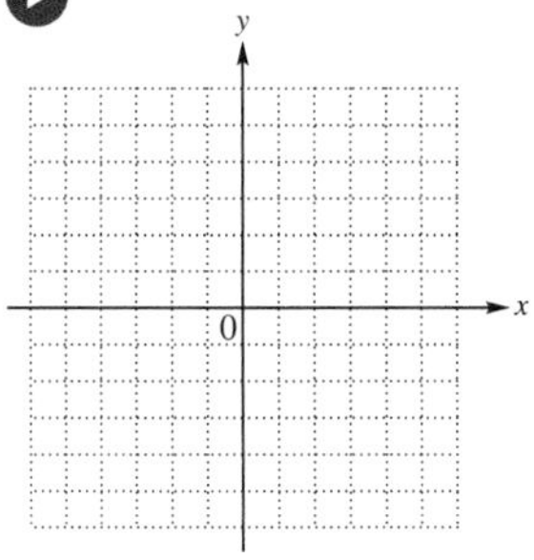

32. $x \geq 1$

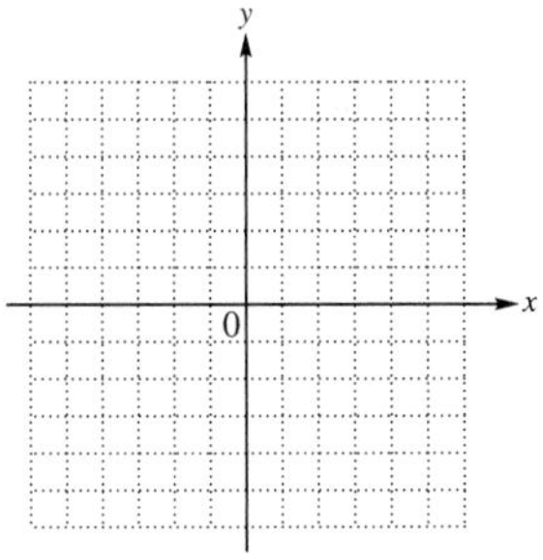

33. $y < 5$

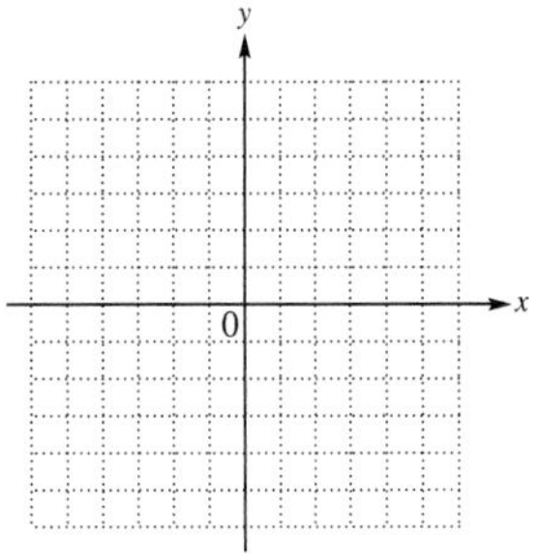

34. $y < -3$

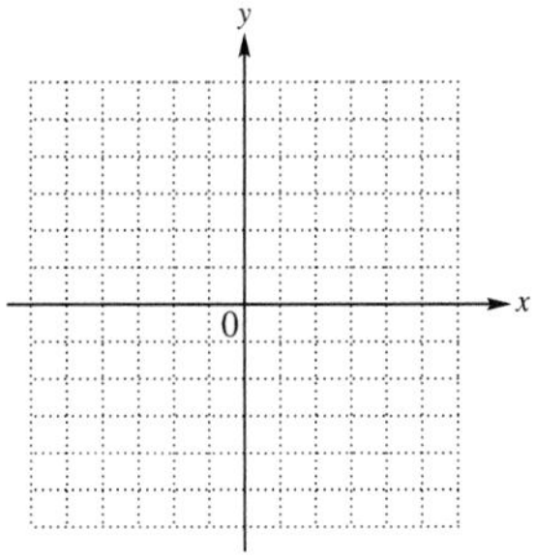

35. $y \geq 4x$

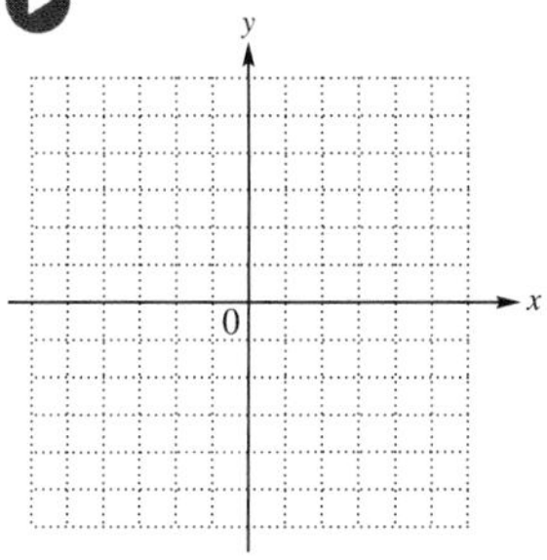

36. $y \leq 2x$

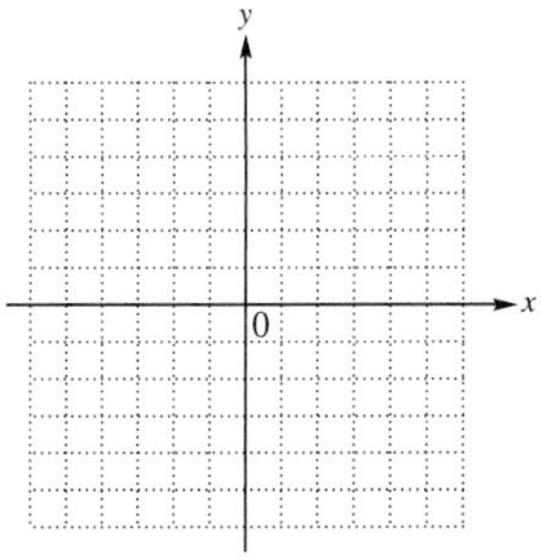

37. $x < -2y$

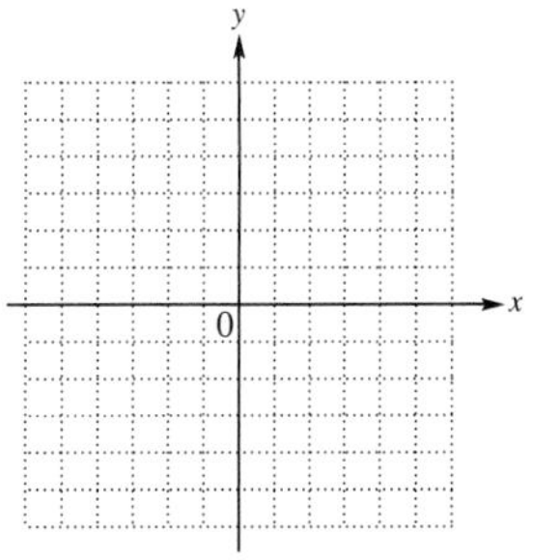

38. $x > -5y$

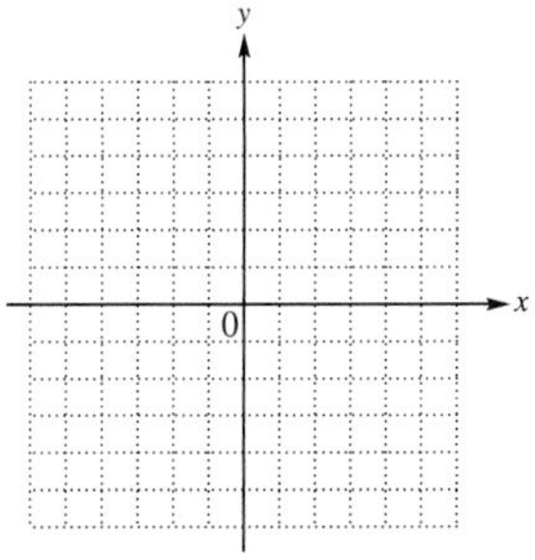

Chapter 11 Summary

Key Terms

11.1

bar graph A bar graph is a series of bars used to show comparisons between two categories of data.

line graph A line graph consists of a series of points that are connected with line segments and is used to show changes or trends in data.

linear equation in two variables An equation that can be written in the form $Ax + By = C$ is a linear equation in two variables. (A and B are real numbers that cannot both be 0.)

ordered pair A pair of numbers written between parentheses in which order is important is an ordered pair.

table of values A table showing selected ordered pairs of numbers that satisfy an equation is a table of values.

x	y
0	4
2	0
1	2

Table of values for $2x + y = 4$

***x*-axis** The horizontal axis in a coordinate system is the x-axis.

***y*-axis** The vertical axis in a coordinate system is the y-axis.

rectangular (Cartesian) coordinate system An x-axis and y-axis at right angles form a coordinate system.

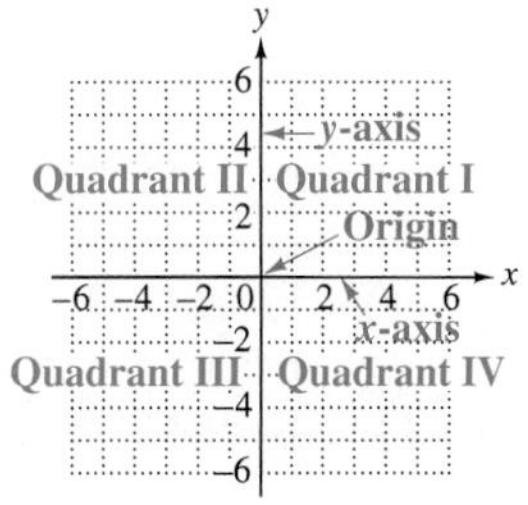

origin The point at which the x-axis and y-axis intersect is the origin.

quadrants A coordinate system divides the plane into four regions or quadrants.

plane A flat surface determined by two intersecting lines is a plane.

coordinates The numbers in an ordered pair are the coordinates of the corresponding point.

plot To plot an ordered pair is to find the corresponding point on a coordinate system.

scatter diagram A graph of ordered pairs of data is a scatter diagram.

11.2

graph The graph of an equation is the set of all points that correspond to the ordered pairs that satisfy the equation.

graphing The process of plotting the ordered pairs that satisfy a linear equation and drawing a line through them is called graphing.

***y*-intercept** If a graph intersects the y-axis at k, then the y-intercept is $(0, k)$.

***x*-intercept** If a graph intersects the x-axis at k, then the x-intercept is $(k, 0)$.

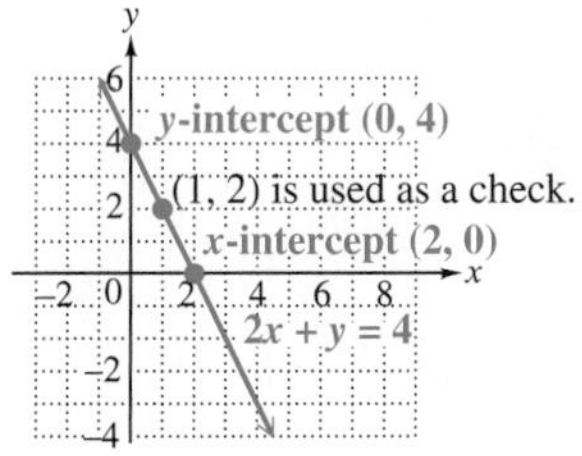

Graph of $2x + y = 4$

11.3

rise Rise is the vertical change between two different points on a line.

run Run is the horizontal change between two different points on a line.

slope The slope of a line is the ratio of the change in y compared to the change in x when moving along the line from one point to another.

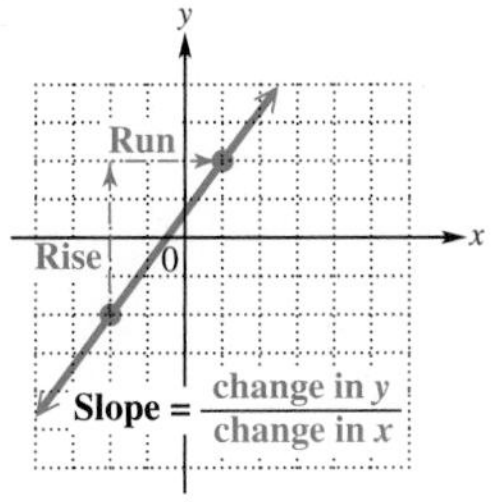

parallel lines Two lines in a plane that never intersect are parallel.

perpendicular lines Perpendicular lines intersect at a 90° angle.

11.5

linear inequality in two variables An inequality that can be written in the form $Ax + By < C$, $Ax + By > C$, $Ax + By \le C$, or $Ax + By \ge C$ is a linear inequality in two variables.

boundary line In the graph of a linear inequality, the boundary line separates the region that satisfies the inequality from the region that does not satisfy the inequality.

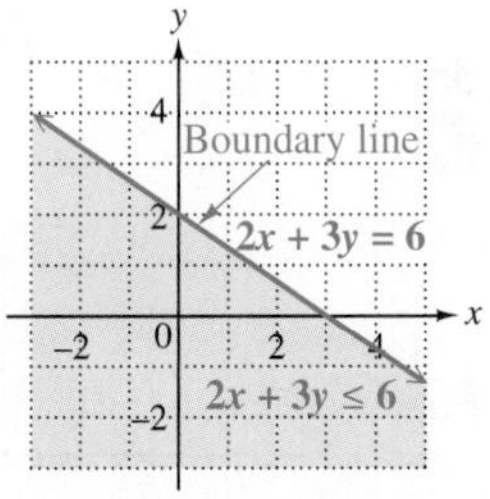

New Symbols

(x, y) ordered pair

(x_1, y_1) subscript notation; x-sub-one, y-sub-one

m slope

Test Your Word Power

See how well you have learned the vocabulary in this chapter.

1. An **ordered pair** is a pair of numbers written
   - **A.** in numerical order between brackets
   - **B.** between parentheses or brackets
   - **C.** between parentheses in which order is important
   - **D.** between parentheses in which order does not matter.

2. The **coordinates** of a point are
   - **A.** the numbers in the corresponding ordered pair
   - **B.** the solution of an equation
   - **C.** the values of the x- and y-intercepts
   - **D.** the graph of the point.

3. An **intercept** is
   - **A.** the point where the x-axis and y-axis intersect
   - **B.** a pair of numbers written in parentheses in which order is important
   - **C.** one of the four regions determined by a rectangular coordinate system
   - **D.** the point where a graph intersects the x-axis or the y-axis.

4. The **slope** of a line is
   - **A.** the measure of the run over the rise of the line
   - **B.** the distance between two points on the line
   - **C.** the ratio of the change in y to the change in x along the line
   - **D.** the horizontal change compared to the vertical change of two points on the line.

5. Two lines in a plane are **parallel** if
   - **A.** they represent the same line
   - **B.** they never intersect
   - **C.** they intersect at a 90° angle
   - **D.** one has a positive slope and one has a negative slope.

6. Two lines in a plane are **perpendicular** if
   - **A.** they represent the same line
   - **B.** they never intersect
   - **C.** they intersect at a 90° angle
   - **D.** one has a positive slope and one has a negative slope.

Answers to Test Your Word Power

1. C; *Examples:* $(0, 3)$, $(3, 8)$, $(4, 0)$
2. A; *Example:* The point associated with the ordered pair $(1, 2)$ has x-coordinate 1 and y-coordinate 2.
3. D; *Example:* The graph of the equation $4x - 3y = 12$ has x-intercept $(3, 0)$ and y-intercept $(0, -4)$.
4. C; *Example:* The line through $(3, 6)$ and $(5, 4)$ has slope $\frac{4-6}{5-3} = \frac{-2}{2} = -1$.
5. B; *Example:* See **Figure 27** in **Section 11.3.**
6. C; *Example:* See **Figure 28** in **Section 11.3.**

Quick Review

Concepts	*Examples*

11.1 Linear Equations in Two Variables; The Rectangular Coordinate System

Concepts: An ordered pair is a solution of an equation if it makes the equation a true statement.

Examples: Is $(2, -5)$ or $(0, -6)$ a solution of $4x - 3y = 18$?

$$4(2) - 3(-5) \stackrel{?}{=} 18$$
$$8 + 15 \stackrel{?}{=} 18$$
$$23 = 18 \quad \text{False}$$

$(2, -5)$ is not a solution.

$$4(0) - 3(-6) \stackrel{?}{=} 18$$
$$0 + 18 \stackrel{?}{=} 18$$
$$18 = 18 \checkmark \quad \text{True}$$

$(0, -6)$ is a solution.

Concepts: If a value of either variable in an equation is given, the value of the other variable can be found by substitution.

Examples: Complete the ordered pair $(0, \_\_\_\_)$ for $3x = y + 4$.

$$3(0) = y + 4 \quad \text{Let } x = 0.$$
$$0 = y + 4 \quad \text{Multiply.}$$
$$-4 = y \quad \text{Subtract 4.}$$

The ordered pair is $(0, -4)$.

(continued)

Concepts	*Examples*

11.1 Linear Equations in Two Variables; The Rectangular Coordinate System *(continued)*

To plot the ordered pair $(-3, 4)$, start at the origin, move 3 units to the left, and from there move 4 units up.

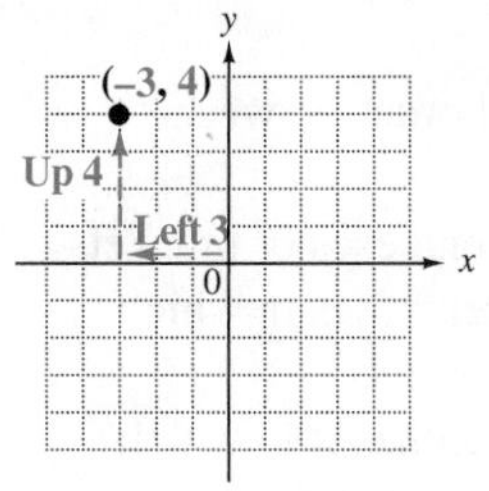

11.2 Graphing Linear Equations in Two Variables

To graph a linear equation, follow these steps.

Step 1 Find at least two ordered pairs that are solutions of the equation. (The intercepts are good choices.)

Step 2 Plot the corresponding points.

Step 3 Draw a straight line through the points.

Graph $x - 2y = 4$.

x	y
0	−2
4	0
−2	−3

Intercepts (rows $(0, -2)$ and $(4, 0)$)

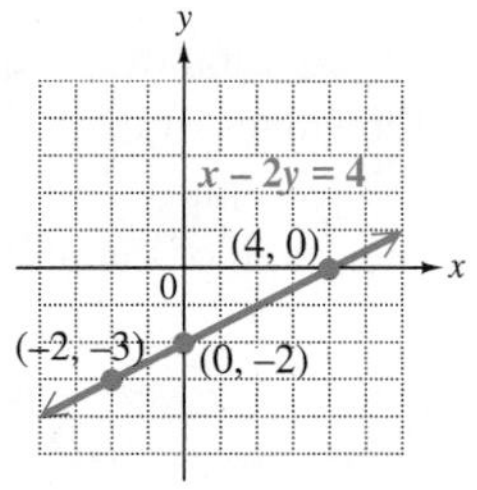

The graph of $\mathbf{Ax + By = 0}$ passes through the origin. In this case, find and plot at least one other point that satisfies the equation. Then draw the line through these points.

The graph of $\mathbf{y = k}$ is a horizontal line through $(0, k)$.

The graph of $\mathbf{x = k}$ is a vertical line through $(k, 0)$.

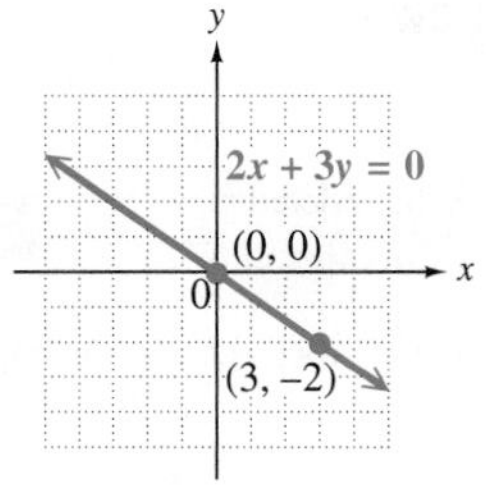

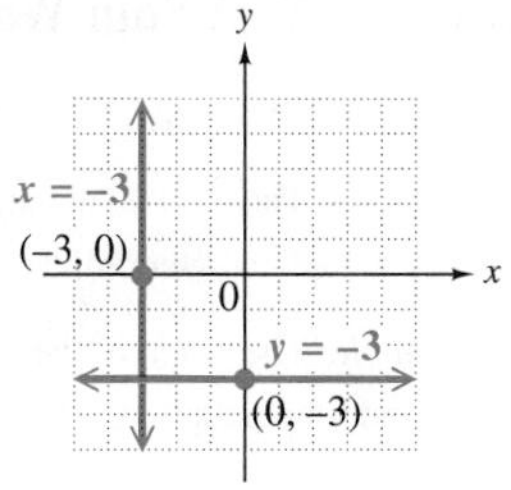

11.3 The Slope of a Line

The slope m of the line passing through the points (x_1, y_1) and (x_2, y_2) is defined as follows.

$$m = \frac{\textbf{change in } y}{\textbf{change in } x} = \frac{y_2 - y_1}{x_2 - x_1} \quad (\text{where } x_1 \neq x_2)$$

The line through $(-2, 3)$ and $(4, -5)$ has slope

$$m = \frac{-5 - 3}{4 - (-2)} = \frac{-8}{6} = -\frac{4}{3}.$$

Horizontal lines have slope 0.

Vertical lines have undefined slope.

The line $y = \mathbf{-2}$ has slope **0**.

The line $x = \mathbf{4}$ has **undefined slope.**

To find the slope of a line from its equation, solve for y. The slope is the coefficient of x.

Find the slope of the graph of $3x - 4y = 12$.

$$-4y = -3x + 12 \qquad \text{Add } -3x.$$

$$y = \frac{3}{4}x - 3 \qquad \text{Divide by } -4.$$

↑ Slope

Parallel lines have the same slope.

The lines $y = 3x - 1$ and $y = 3x + 4$ are parallel because both have slope **3**.

The slopes of perpendicular lines are negative reciprocals (that is, their product is -1).

The lines $y = -3x - 1$ and $y = \frac{1}{3}x + 4$ are perpendicular because their slopes are -3 and $\frac{1}{3}$, and $-3\left(\frac{1}{3}\right) = \mathbf{-1}$.

Concepts	Examples

11.4 Writing and Graphing Equations of Lines

Slope-Intercept Form
$y = mx + b$
m is the slope.
$(0, b)$ is the y-intercept.

Write an equation in slope-intercept form of the line with slope 2 and y-intercept $(0, -5)$.

$$y = 2x - 5$$

Point-Slope Form
$y - y_1 = m(x - x_1)$
m is the slope.
(x_1, y_1) is a point on the line.

Write an equation of the line with slope $-\frac{1}{2}$ through $(-4, 5)$.

$$y - y_1 = m(x - x_1) \quad \text{Point-slope form}$$
$$y - 5 = -\frac{1}{2}[x - (-4)] \quad \text{Substitute.}$$
$$y - 5 = -\frac{1}{2}(x + 4) \quad \text{Definition of subtraction}$$
$$y - 5 = -\frac{1}{2}x - 2 \quad \text{Distributive property}$$
$$y = -\frac{1}{2}x + 3 \quad \text{Add 5.}$$

Standard Form
$Ax + By = C$
A, B, and C are real numbers and A and B are not both 0.

The equation $y = -\frac{1}{2}x + 3$ is written in standard form as

$$x + 2y = 6. \quad A = 1, B = 2, \text{ and } C = 6.$$

11.5 Graphing Linear Inequalities in Two Variables

Step 1 Graph the line that is the boundary of the region. Draw a solid line if the inequality is $\leq$ or $\geq$. Draw a dashed line if the inequality is $<$ or $>$.

Step 2 Use any point not on the line as a test point. Substitute for x and y in the inequality. If the result is true, shade the region containing the test point. If the result is false, shade the other region.

Graph $2x + y \leq 5$.

Graph the line $2x + y = 5$. Make it solid because the symbol $\leq$ includes equality.

Use $(0, 0)$ as a test point.

$$2(0) + 0 \overset{?}{\leq} 5$$
$$0 \leq 5 \quad \text{True}$$

Shade the side of the line containing $(0, 0)$.

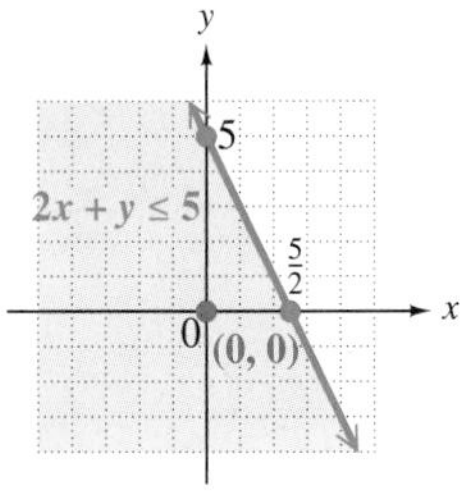

Chapter 11 Review Exercises

11.1 *The percents of first-year college students at two-year public institutions who returned for a second year for the years 2006 through 2011 are shown in the graph.*

1. Write two ordered pairs of the form (year, percent) for the data shown in the graph for the years 2006 and 2011.

2. What does the ordered pair (2000, 52.0) mean in the context of these problems?

3. Over which two consecutive years did the percent stay the same? Estimate this percent.

4. Over which pairs of consecutive years did the percent increase? Estimate the increases.

Percents of Students Who Return for Second Year (2-Year Public Institutions)

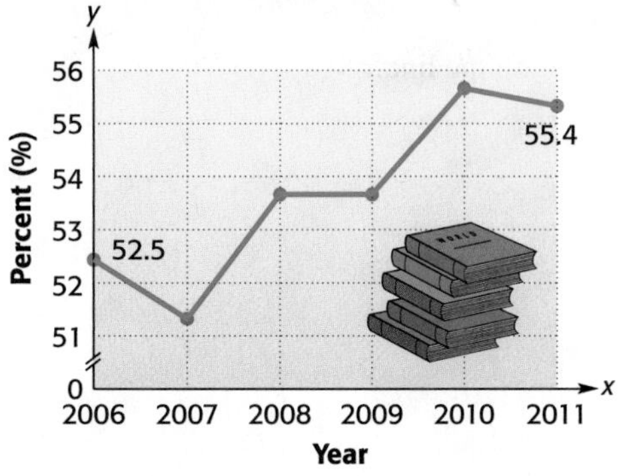

Source: ACT.

Complete the given ordered pairs for each equation.

5. $y = 3x + 2$; $(-1, \_\_\_)$, $(0, \_\_\_)$, $(\_\_\_, 5)$

6. $4x + 3y = 6$; $(0, \_\_\_)$, $(\_\_\_, 0)$, $(-2, \_\_\_)$

7. $x = 3y$; $(0, \_\_\_)$, $(8, \_\_\_)$, $(\_\_\_, -3)$

8. $x - 7 = 0$; $(\_\_\_, -3)$, $(\_\_\_, 0)$, $(\_\_\_, 5)$

Decide whether each ordered pair is a solution of the given equation.

9. $x + y = 7$; $(2, 5)$

10. $2x + y = 5$; $(-1, 3)$

11. $3x - y = 4$; $\left(\frac{1}{3}, -3\right)$

12. $x = -1$; $(0, -1)$

Name the quadrant in which each point lies. Then plot each ordered pair on the given coordinate system.

13. $(2, 3)$

14. $(-4, 2)$

15. $(3, 0)$

16. $(0, -6)$

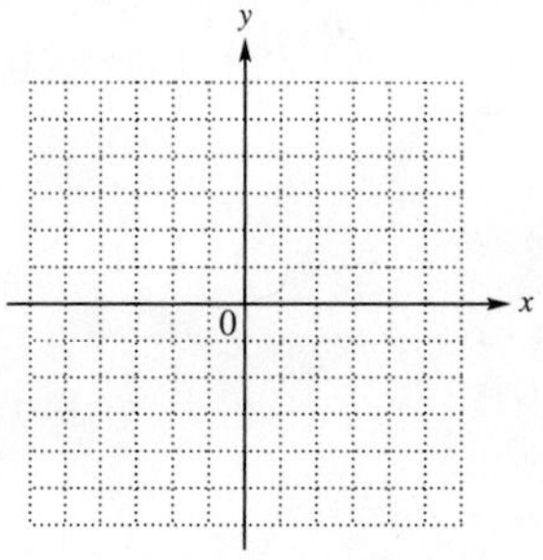

17. If $x > 0$ and $y < 0$, in what quadrant(s) must (x, y) lie? Explain.

18. On what axis does the point $(k, 0)$ lie for any real value of k? The point $(0, k)$? Explain.

11.2 *Graph each linear equation using intercepts.*

19. $2x - y = 3$

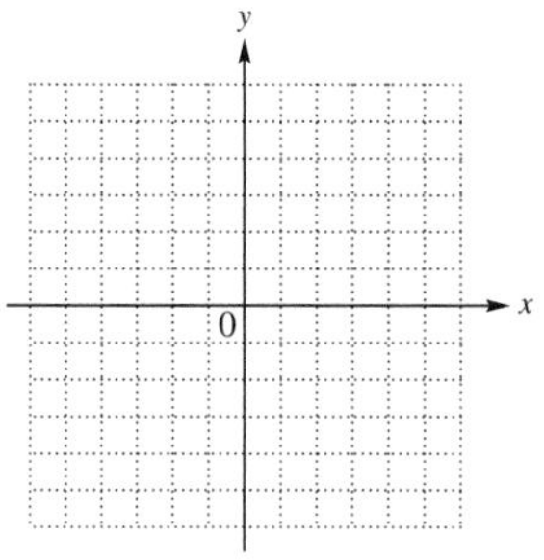

20. $x + 2y = -4$

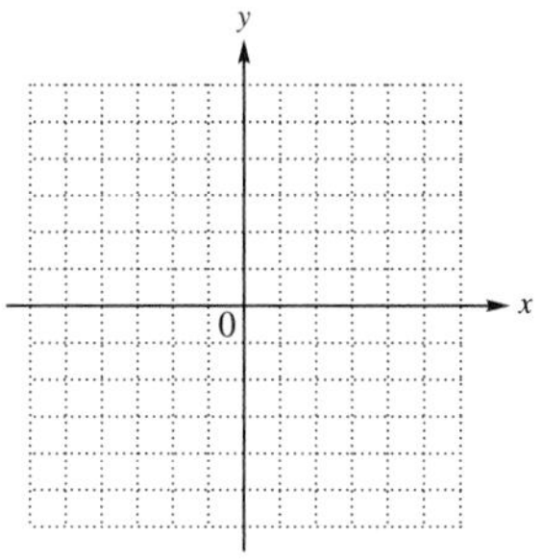

21. $x + y = 0$

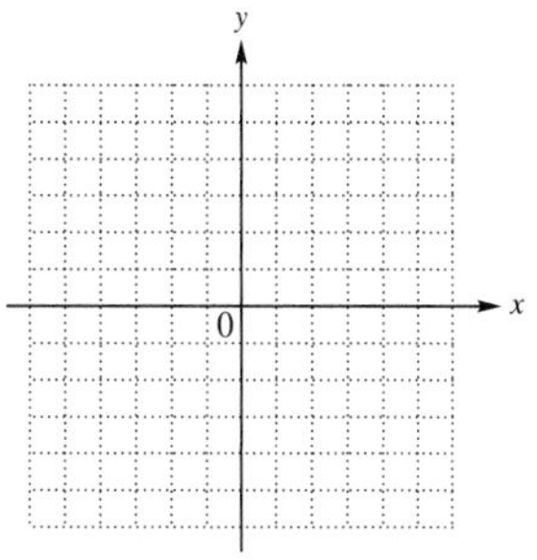

22. $y = 3x + 4$

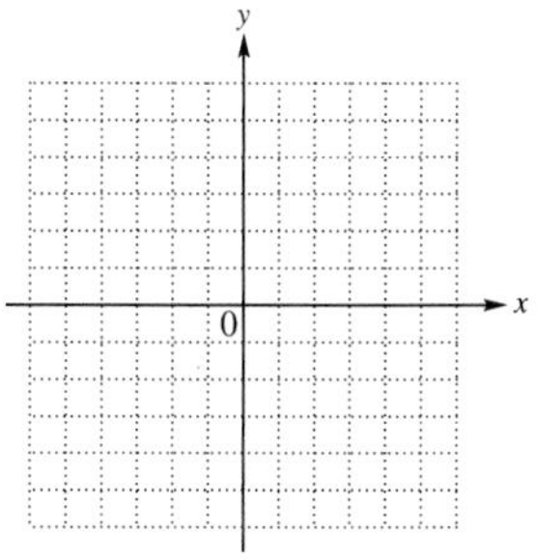

11.3 *Find the slope of each line.*

23. The line passing through $(2, 3)$ and $(-4, 6)$

24. The line passing through $(0, 0)$ and $(-3, 2)$

25. The line passing through $(0, 6)$ and $(1, 6)$

26. The line passing through $(2, 5)$ and $(2, 8)$

27. $y = 3x - 4$

28. $y = \frac{2}{3}x + 1$

29.

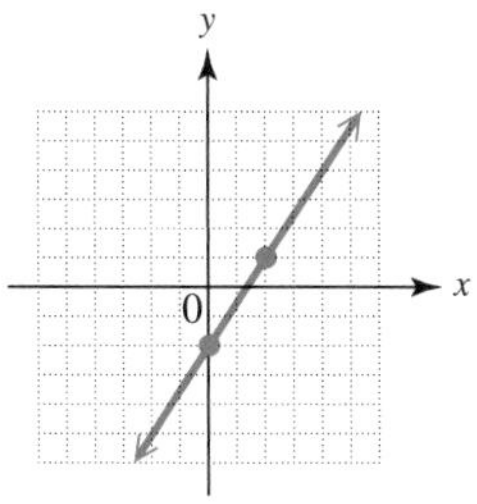

30.

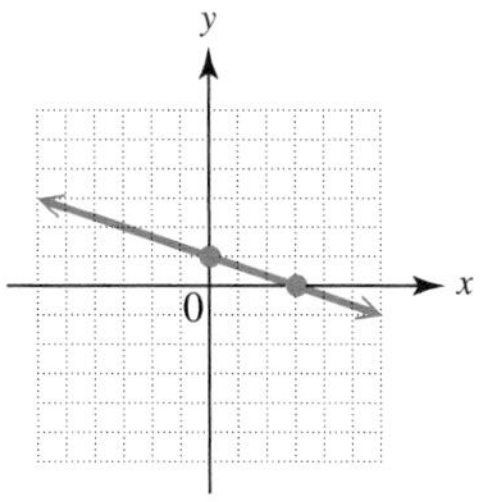

31. The line having these points

x	y
0	1
2	4
6	10

32. $x = 0$

33. $y = 4$

34. **(a)** A line parallel to the graph of $y = 2x + 3$

(b) A line perpendicular to the graph of $y = -3x + 3$

Decide whether each pair of lines is parallel, perpendicular, *or* neither.

35. $3x + 2y = 6$
$6x + 4y = 8$

36. $x - 3y = 1$
$3x + y = 4$

37. $x - 2y = 8$
$x + 2y = 8$

38. CONCEPT CHECK What is the slope of a line perpendicular to a line with undefined slope?

11.4 *Write an equation of each line. Give the final answer in slope-intercept form (if possible).*

39. $m = -1, b = \frac{2}{3}$

40. The line in **Exercise 30**

41. The line passing through $(4, -3), m = 1$

42. The line passing through $(-1, 4), m = \frac{2}{3}$

43. The line passing through $(1, -1), m = -\frac{3}{4}$

44. The line passing through $(2, 1)$ and $(-2, 2)$

45. The line passing through $(-4, 1)$, slope 0

46. The line passing through $\left(\frac{1}{3}, -\frac{3}{4}\right)$, undefined slope

47. Consider the linear equation $x + 3y = 15$.

(a) Write it in the form $y = mx + b$.

(b) What is the slope? What is the y-intercept?

(c) Use the slope and y-intercept to graph the line. Indicate two points on the graph.

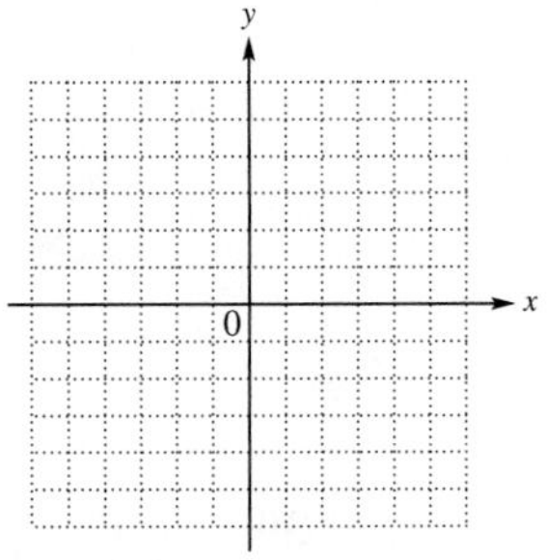

11.5 *Graph each linear inequality.*

48. $3x + 5y > 9$

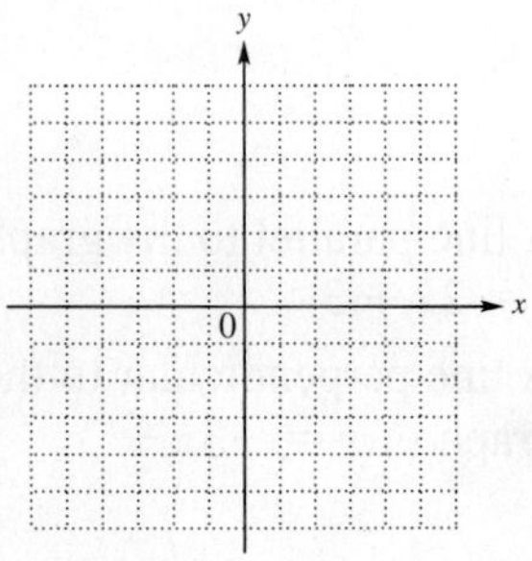

49. $2x - 3y > -6$

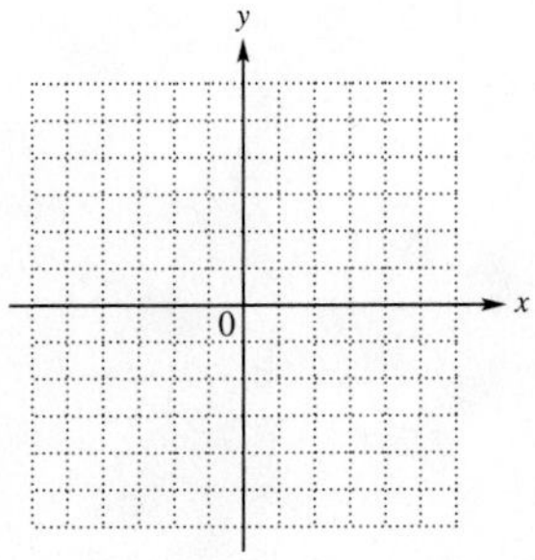

50. $x \geq -4$

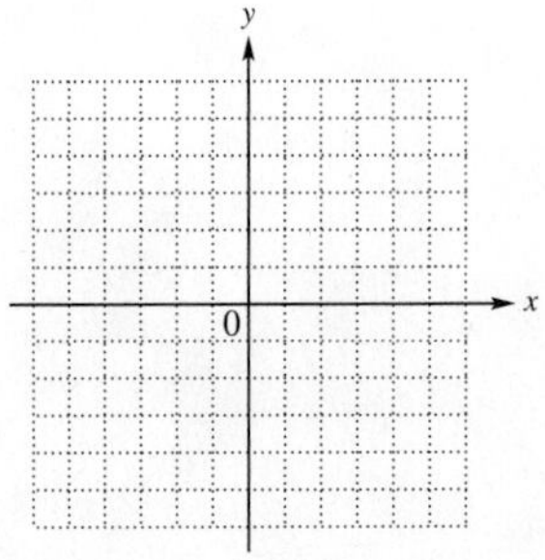

Mixed Review Exercises

CONCEPT CHECK *In Exercises 51–56, match each statement to the appropriate graph or graphs in choices A–D. Graphs may be used more than once.*

A.

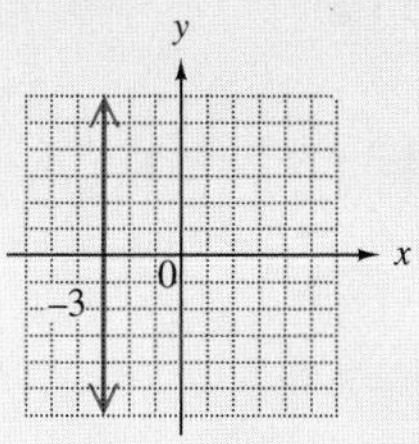

B.

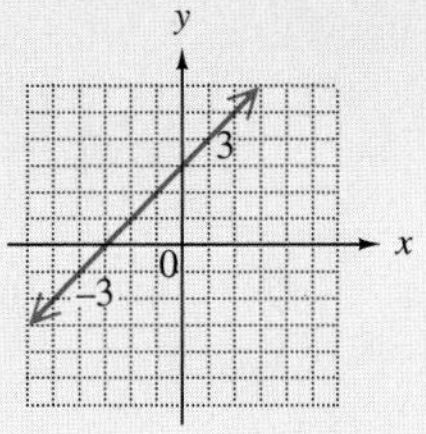

C.

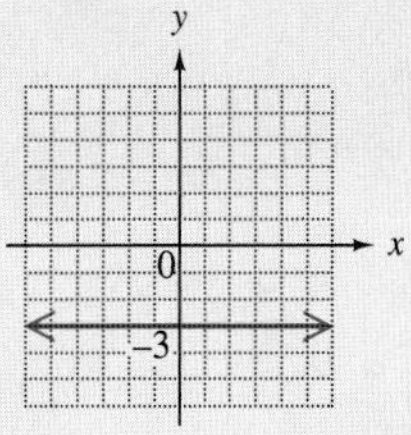

D.

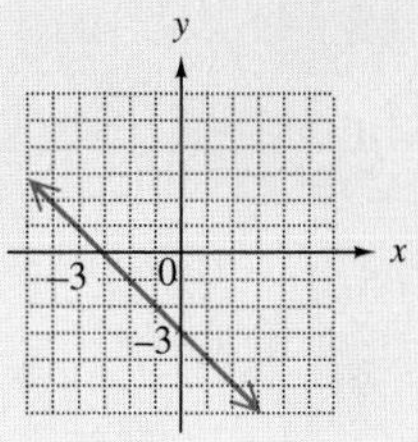

51. The line shown in the graph has undefined slope.

52. The graph of the equation has y-intercept $(0, -3)$.

53. The graph of the equation has x-intercept $(-3, 0)$.

54. The line shown in the graph has negative slope.

55. The graph is that of the equation $y = -3$.

56. The line shown in the graph has slope 1.

Find the intercepts and the slope of each line. Then use them to graph the line.

57. $y = -2x - 5$

y-intercept: ______

x-intercept: ______

slope: ______

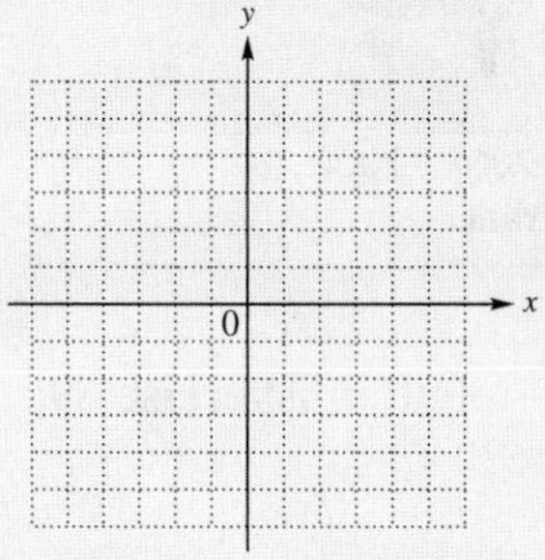

58. $x + 3y = 0$

y-intercept: ______

x-intercept: ______

slope: ______

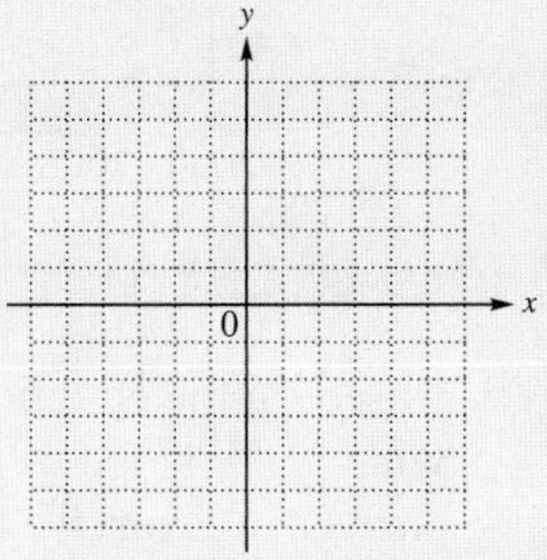

59. $y - 5 = 0$

y-intercept: ______

x-intercept: ______

slope: ______

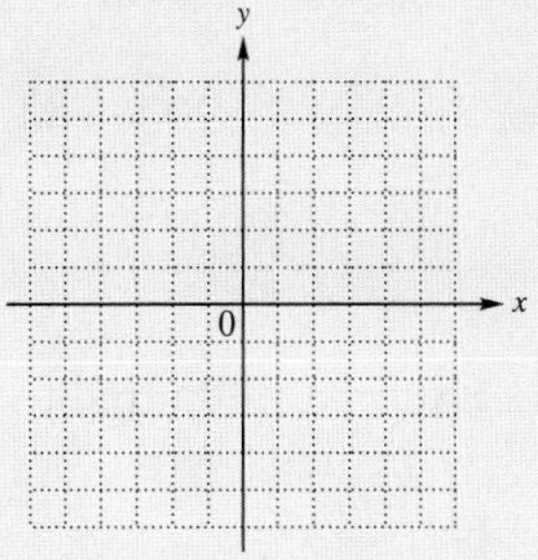

60. $x + 4 = 3$

y-intercept: ______

x-intercept: ______

slope: ______

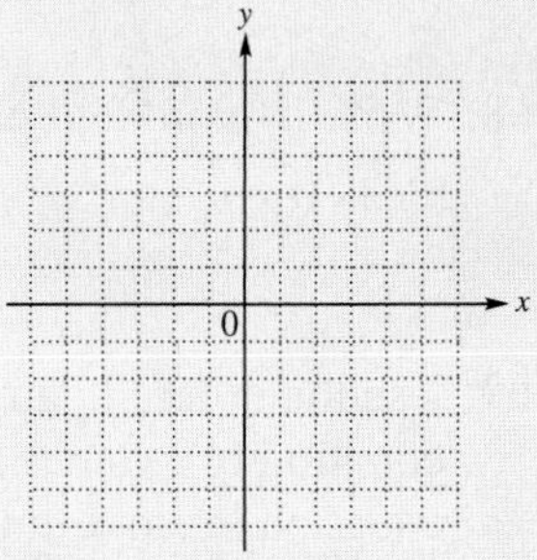

Write an equation of each line. Give the final answer in ***(a)*** *slope-intercept form and* ***(b)*** *standard form.*

61. $m = -\frac{1}{4}, b = -\frac{5}{4}$

62. The line passing through $(8, 6)$, $m = -3$

63. The line passing through $(3, -5)$ and $(-4, -1)$

Graph each inequality.

64. $y < -4x$

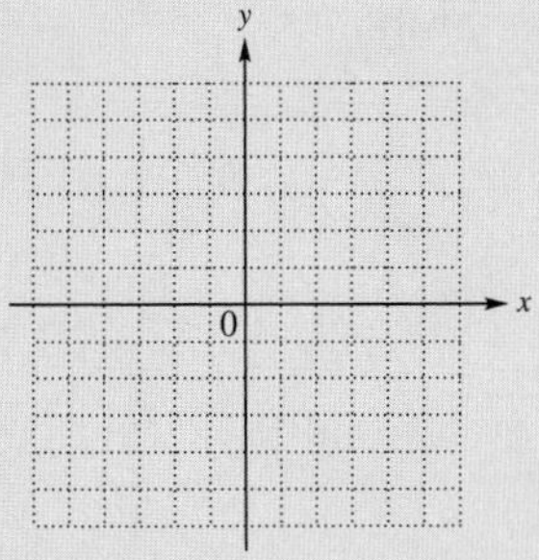

65. $x - 2y \leq 6$

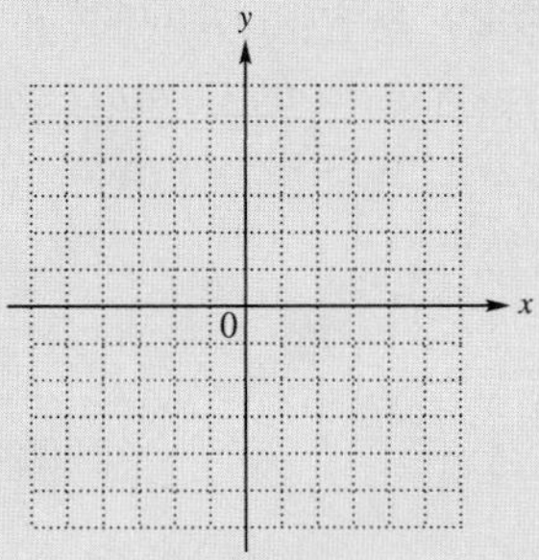

Relating Concepts (Exercises 66–72) *For Individual or Group Work*

The percents of four-year college students in public schools who earned a degree within five years of entry between 2004 and 2009 are shown in the graph. Use the graph to ***work Exercises 66–72 in order.***

Percents of Students Graduating Within 5 Years (Public Institutions)

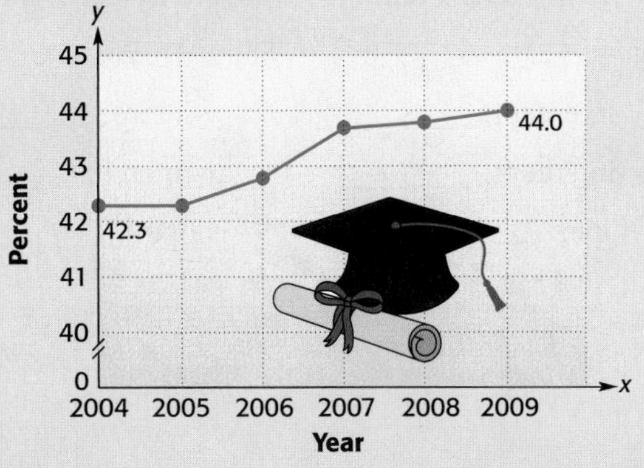

Source: ACT.

66. What was the increase in percent from 2004 to 2009?

67. Since the points of the graph lie approximately in a linear pattern, a straight line can be used to model the data. Will this line have positive or negative slope? Explain.

68. Write two ordered pairs for the data for 2004 and 2009.

69. Use the ordered pairs from **Exercise 68** to find the equation of a line that models the data. Write the equation in slope-intercept form.

70. Based on the equation you found in **Exercise 69,** what is the slope of the line? Does it agree with your answer in **Exercise 67?**

71. Use the equation from **Exercise 69** to approximate the percents for 2005 through 2008 to the nearest tenth, and complete the table.

Year	Percent
2005	
2006	
2007	
2008	

72. Use the equation from **Exercise 69** to estimate the percent for 2010. Can we be sure that this estimate is accurate?

Chapter 11 Test

CHAPTER Test Prep VIDEO *The Chapter Test Prep Videos with test solutions are available on DVD, in MyMathLab, and on YouTube—search "LialDevMath" and click on "Channels."*

The line graph shows the overall unemployment rate in the U.S. civilian labor force for the years 2003 through 2010. Use the graph to work Exercises 1–3.

1. Between which pairs of consecutive years did the unemployment rate decrease?

2. What was the general trend in the unemployment rate between 2007 and 2010?

3. Estimate the overall unemployment rate in 2008 and 2009. About how much did the unemployment rate increase between 2008 and 2009?

Unemployment Rate

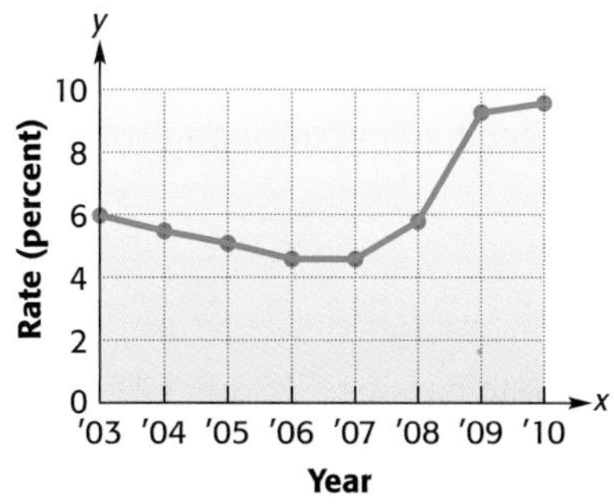

Source: Bureau of Labor Statistics.

Graph each linear equation. Give the intercepts.

4. $3x + y = 6$

y-intercept: ______

x-intercept: ______

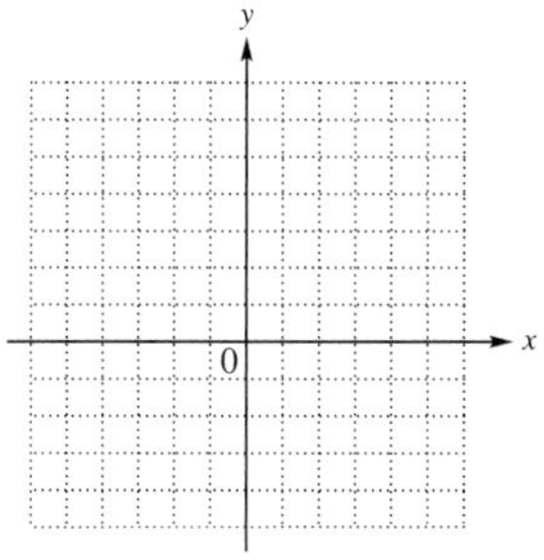

5. $y - 2x = 0$

y-intercept: ______

x-intercept: ______

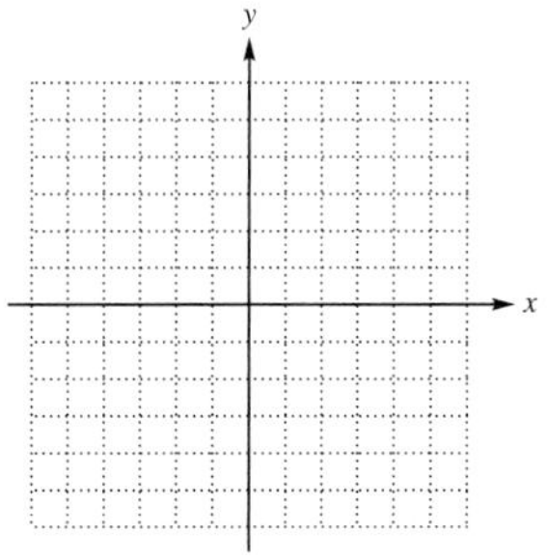

6. $x + 3 = 0$

y-intercept: ______

x-intercept: ______

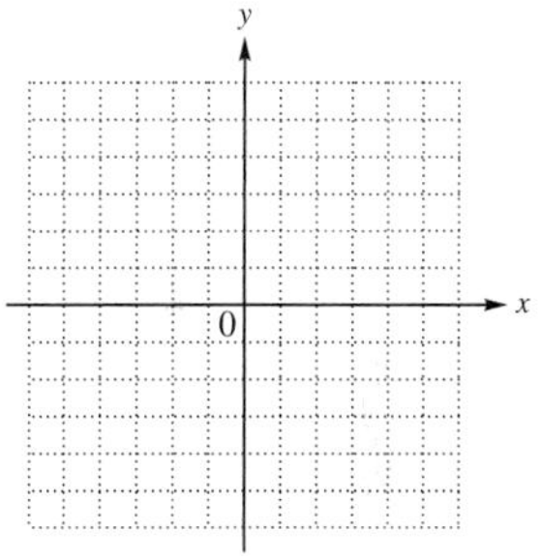

7. $y = 1$

y-intercept: ______

x-intercept: ______

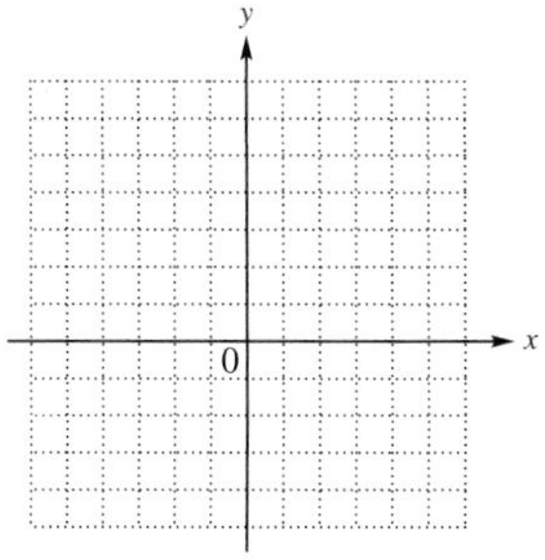

8. $x - y = 4$

y-intercept: ______

x-intercept: ______

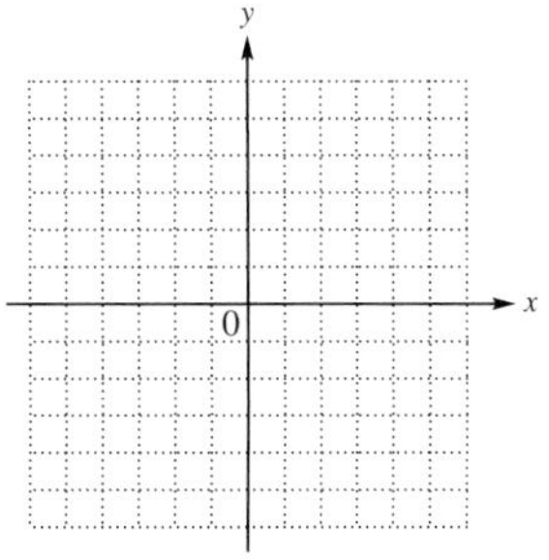

Find the slope of each line.

9. The line passing through $(-4, 6)$ and $(-1, -2)$

10. $2x + y = 10$

11. $x + 12 = 0$

12.

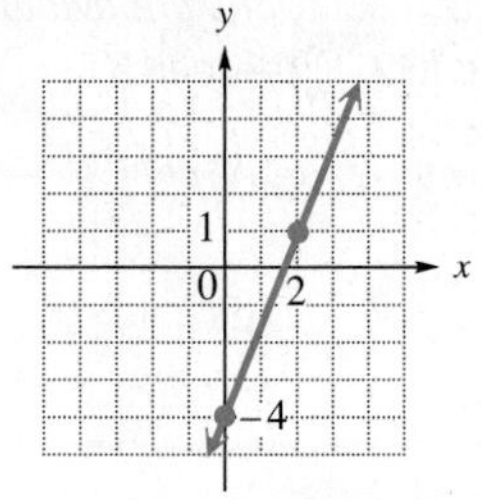

13. A line parallel to the graph of $y - 4 = 6$

Write an equation for each line. Give the final answer in slope-intercept form.

14. The line passing through $(-1, 4)$, $m = 2$

15. The line in **Exercise 12**

16. The line passing through $(2, -6)$ and $(1, 3)$

17. x-intercept: $(3, 0)$; y-intercept: $\left(0, \frac{9}{2}\right)$

Graph each linear inequality.

18. $x + y \leq 3$

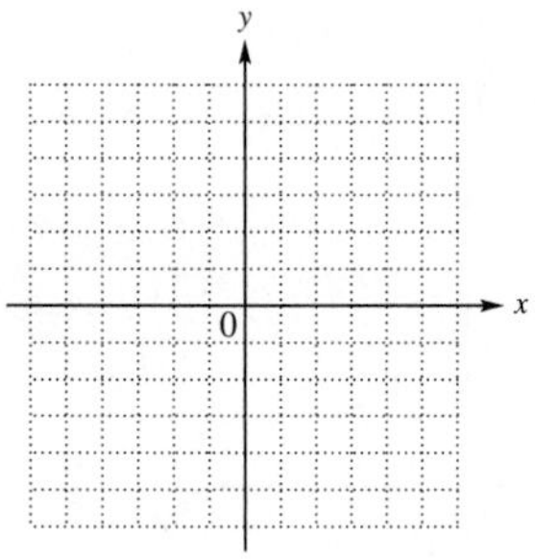

19. $3x - y > 0$

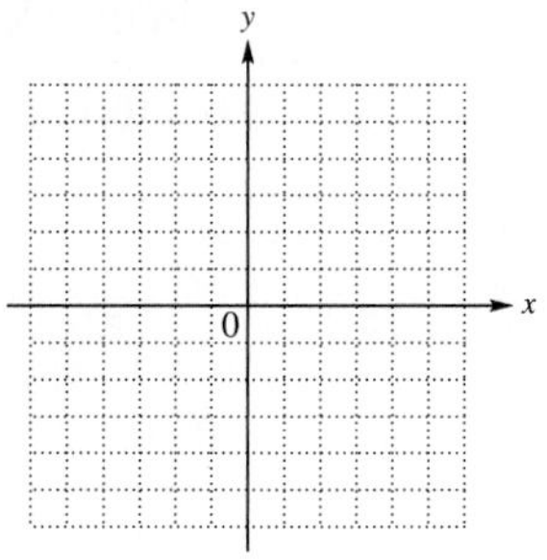

The graph shows worldwide snowmobile sales from 2000 through 2011, where 2000 corresponds to $x = 0$. Use the graph to work Exercises 20–24.

20. Is the slope of the line in the graph positive or negative? Explain.

Source: www.snowmobile.org

21. Write two ordered pairs for the data points shown in the graph. Use them to find the slope of the line to the nearest tenth.

22. Use the ordered pairs and slope from **Exercise 21** to find an equation of a line that models the data. Write the equation in slope-intercept form.

23. Use the equation from **Exercise 22** to approximate worldwide snowmobile sales for 2007. How does your answer compare to the actual sales of 160.3 thousand?

24. What does the ordered pair (11, 123) mean in the context of this problem?

12 Exponents and Polynomials

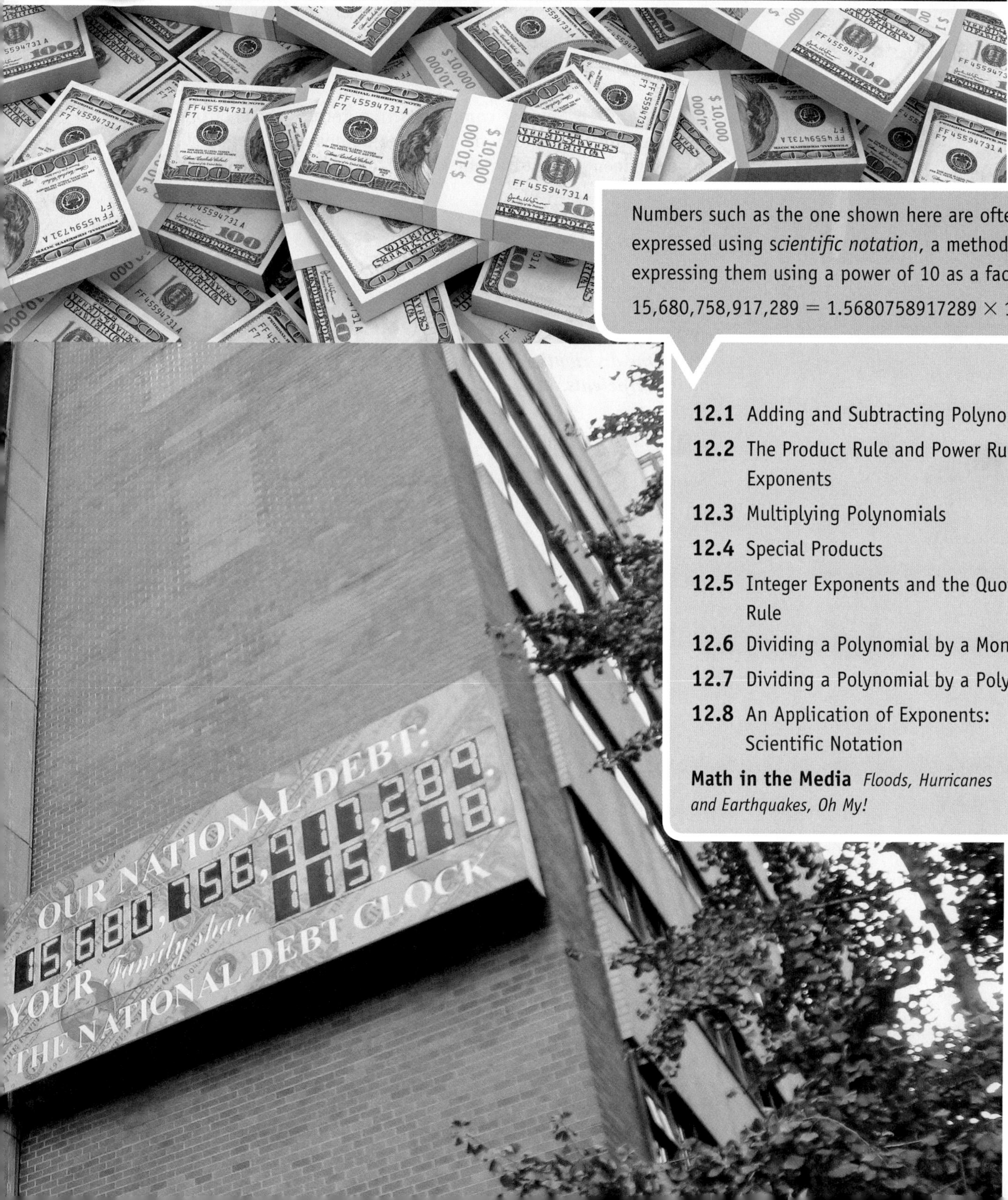

Numbers such as the one shown here are often expressed using *scientific notation*, a method of expressing them using a power of 10 as a factor: $15{,}680{,}758{,}917{,}289 = 1.5680758917289 \times 10^{13}$.

12.1 Adding and Subtracting Polynomials

OBJECTIVES

1. Review combining like terms.
2. Know the vocabulary for polynomials.
3. Evaluate polynomials.
4. Add polynomials.
5. Subtract polynomials.
6. Add and subtract polynomials with more than one variable.

Recall from **Section 9.8** that in an algebraic expression such as

$$4x^3 + 6x^2 + 5x + 8,$$

the quantities that are added, $4x^3$, $6x^2$, $5x$, and 8, are **terms.** In the term $4x^3$, the number 4 is the **numerical coefficient,** or simply the **coefficient,** of x^3. In the same way, 6 is the coefficient of x^2 in the term $6x^2$, and 5 is the coefficient of x in the term $5x$. The ***constant*** term 8 can be thought of as $8 \cdot 1 = 8x^0$, so 8 is the coefficient in the term 8. Other examples are given in the table at the side.

Term	Numerical Coefficient
$-7y$	-7
$34r^3$	34
$-26x^5yz^4$	-26
$-k = -1k$	-1
$r = 1r$	1
$\frac{3x}{8} = \frac{3}{8}x$	$\frac{3}{8}$
$\frac{x}{3} = \frac{1x}{3} = \frac{1}{3}x$	$\frac{1}{3}$

OBJECTIVE 1 Review combining like terms. Recall from **Section 9.8** that **like terms** have exactly the same combination of variables, with the same exponents on the variables. ***Only the coefficients may differ.***

Examples of like terms:

$19m^5$ and $14m^5$

$-37y^9$ and y^9

$3pq$ and $-2pq$

$2xy^2$ and $-xy^2$

Examples of unlike terms:

$7x$ and $7y$

z^4 and z

$2pq$ and $2p$

$-4xy^2$ and $5x^2y$

Using the distributive property, we combine, or add, like terms by adding their coefficients.

EXAMPLE 1 Combining Like Terms

Simplify each expression by combining like terms.

(a) $-4x^3 + 6x^3$
$= (-4 + 6)x^3$
$= 2x^3$

(b) $8rs - 13rs + 9rs$
$= (8 - 13 + 9)rs$
$= 4rs$

(c) $9x^6 - 14x^6 + x^6$ ($x^6 = 1x^6$)
$= (9 - 14 + 1)x^6$
$= -4x^6$

(d) $y + \frac{2}{3}y$
$= 1y + \frac{2}{3}y$ ($y = 1y$)
$= \left(\frac{3}{3} + \frac{2}{3}\right)y$
$= \frac{5}{3}y$

(e) $12m^2 + 5m + 4m^2$
$= (12 + 4)m^2 + 5m$
$= 16m^2 + 5m$

(f) $5u + 11v$
These are unlike terms. They cannot be combined.

◀ Work Problem 1 at the Side.

CAUTION

In **Example 1(e)**, we cannot combine $16m^2$ and $5m$. These two terms are unlike because the exponents on the variables are different. ***Unlike terms have different variables or different exponents on the same variables.***

1 Combine like terms.

(a) $5x^4 + 7x^4$

(b) $9pq + 3pq - 2pq$

(c) $r^2 + 3r + 5r^2$

(d) $x + \frac{1}{2}x$

(e) $8t + 6w$

(f) $3x^4 - 3x^2$

Answers

1. **(a)** $12x^4$ **(b)** $10pq$ **(c)** $6r^2 + 3r$ **(d)** $\frac{3}{2}x$
(e) These are unlike terms. They cannot be combined.
(f) These are unlike terms. They cannot be combined.

OBJECTIVE 2 Know the vocabulary for polynomials. A **polynomial in x** is a term or the sum of a finite number of terms of the form ax^n, for any real number a and any whole number n. For example,

$$16x^8 - 7x^6 + 5x^4 - 3x^2 + 4$$ Polynomial in x (The 4 can be written as $4x^0$.)

is a polynomial in x. It is written in **descending powers,** because the exponents on x decrease from left to right. By contrast, the expression

$$2x^3 - x^2 + \frac{4}{x}$$ Not a polynomial

is not a polynomial in x. A variable appears in a denominator.

Note

We can define *polynomial* using any variable and not just x, as in **Example 1(d).** Polynomials may have terms with more than one variable, as in **Example 1(b).**

Work Problem 2 at the Side. ▶

The **degree of a term** is the sum of the exponents on the variables. The **degree of a polynomial** is the greatest degree of any nonzero term of the polynomial. The table gives several examples.

Term	Degree	Polynomial	Degree
$3x^4$	4	$3x^4 - 5x^2 + 6$	4
$5x$, or $5x^1$	1	$5x + 7$	1
-7, or $-7x^0$	0	$x^2y + xy - 5y^2$	3
$2x^2y$, or $2x^2y^1$	$2 + 1 = 3$	$x^5 + 3x^6$	6

A polynomial with only one term is a **monomial.** (*Mono-* means "one," as in *mono*rail.) A polynomial with exactly two terms is a **binomial.** (*Bi-* means "two," as in *bi*cycle.) A polynomial with exactly three terms is a **trinomial.** (*Tri-* means "three," as in *tri*angle.)

$9m$, $-6y^5$, a^2, and 6 Monomials

$-9x^4 + 9x^3$, $8m^2 + 6m$, and $3m^5 - 9m^2$ Binomials

$9m^3 - 4m^2 + 6$, $\frac{19}{3}y^2 + \frac{8}{3}y + 5$, and $-3m^5 - 9m^2 + 2$ Trinomials

EXAMPLE 2 Classifying Polynomials

Simplify each polynomial if possible. Then give the degree and tell whether the polynomial is a *monomial*, a *binomial*, a *trinomial*, or *none of these*.

(a) $2x^3 + 5$ We cannot simplify further. This is a binomial of degree 3.

(b) $4x - 5x + 2x$

$= x$ Combine like terms to simplify.

The degree is 1 (since $x = x^1$). The simplified polynomial is a monomial.

Work Problem 3 at the Side. ▶

2 Choose all descriptions that apply for each of the expressions in parts (a)–(d).

A. Polynomial

B. Polynomial written in descending powers

C. Not a polynomial

(a) $3m^5 + 5m^2 - 2m + 1$

(b) $2p^4 + p^6$

(c) $\frac{1}{x} + 2x^2 + 3$

(d) $x - 3$

3 Simplify each polynomial if possible. Then give the degree and tell whether the polynomial is a *monomial,* a *binomial,* a *trinomial*, or *none of these.*

(a) $3x^2 + 2x - 4$

(b) $x^3 + 4x^3$

(c) $x^8 - x^7 + 2x^8$

Answers

2. (a) A and B **(b)** A **(c)** C **(d)** A and B

3. (a) degree 2; trinomial

(b) degree 3; monomial (simplify to $5x^3$)

(c) degree 8; binomial (simplify to $3x^8 - x^7$)

4 Find the value of $2x^3 + 8x - 6$ in each case.

GS **(a)** For $x = -1$

$$\begin{aligned} &2x^3 + 8x - 6 \\ &= 2(\_\_)^3 + 8(\_\_) - 6 \\ &= 2(\_\_) + 8(-1) - 6 \\ &= \_\_\_ - \_\_\_ - 6 \\ &= \_\_\_ \end{aligned}$$

(b) For $x = 4$

(c) For $x = -2$

5 Add each pair of polynomials.

(a) $4x^3 - 3x^2 + 2x$ and $6x^3 + 2x^2 - 3x$

(b) $x^2 - 2x + 5$ and $4x^2 - 2$

Answers

4. (a) -1; -1; -1; -2; 8; -16
(b) 154 (c) -38

5. (a) $10x^3 - x^2 - x$
(b) $5x^2 - 2x + 3$

OBJECTIVE 3 Evaluate polynomials. When we evaluate an expression, we find its value. A polynomial usually represents different numbers for different values of the variable.

EXAMPLE 3 Evaluating a Polynomial

Find the value of $3x^4 + 5x^3 - 4x - 4$ for **(a)** $x = -2$ and for **(b)** $x = 3$.

(a)

$3x^4 + 5x^3 - 4x - 4$	
$= 3(-2)^4 + 5(-2)^3 - 4(-2) - 4$	Substitute -2 for x.
$= 3(16) + 5(-8) - 4(-2) - 4$	Apply the exponents.
$= 48 - 40 + 8 - 4$	Multiply.
$= 12$	Add and subtract.

Use parentheses to avoid errors.

(b)

$3x^4 + 5x^3 - 4x - 4$	
$= 3(3)^4 + 5(3)^3 - 4(3) - 4$	Let $x = 3$.
$= 3(81) + 5(27) - 4(3) - 4$	Apply the exponents.
$= 243 + 135 - 12 - 4$	Multiply.
$= 362$	Add and subtract.

◀ Work Problem **4** at the Side.

OBJECTIVE 4 Add polynomials.

Adding Polynomials

To add two polynomials, add like terms.

EXAMPLE 4 Adding Polynomials Vertically

(a) Add $6x^3 - 4x^2 + 3$ and $-2x^3 + 7x^2 - 5$.

$$\begin{array}{r} 6x^3 - 4x^2 + 3 \\ \underline{-2x^3 + 7x^2 - 5} \end{array}$$

Write like terms in columns.

Now add, column by column.

$$\begin{array}{ccc} 6x^3 & -4x^2 & 3 \\ \underline{-2x^3} & \underline{7x^2} & \underline{-5} \\ 4x^3 & 3x^2 & -2 \end{array}$$

Combine the coefficients only. Do *not* add the exponents.

Add the three sums together to obtain the answer.

$$4x^3 + 3x^2 + (-2) = 4x^3 + 3x^2 - 2$$

(b) Add $2x^2 - 4x + 3$ and $x^3 + 5x$.
Write like terms in columns and add column by column.

$$\begin{array}{rrrr} & 2x^2 & - 4x & + 3 \\ \underline{x^3} & & \underline{+ 5x} & \\ x^3 & + 2x^2 & + x & + 3 \end{array}$$

Leave spaces for missing terms.

◀ Work Problem **5** at the Side.

The polynomials in **Example 4** also could be added horizontally.

EXAMPLE 5 **Adding Polynomials Horizontally**

Find each sum.

(a) Add $6x^3 - 4x^2 + 3$ and $-2x^3 + 7x^2 - 5$.

$(6x^3 - 4x^2 + 3) + (-2x^3 + 7x^2 - 5) = 4x^3 + 3x^2 - 2$ Combine like terms. Same answer found in **Example 4(a)**

(b) $(2x^2 - 4x + 3) + (x^3 + 5x)$

$= x^3 + 2x^2 - 4x + 5x + 3$ Commutative property

$= x^3 + 2x^2 + x + 3$ Combine like terms.

Work Problem 6 at the Side. ▶

OBJECTIVE ▶ 5 **Subtract polynomials.** In **Section 9.5,** the difference $x - y$ was defined as $x + (-y)$. (We find the difference $x - y$ by adding x and the opposite of y.)

$7 - 2$ is equivalent to $7 + (-2)$, which equals 5.

$-8 - (-2)$ is equivalent to $-8 + 2$, which equals -6.

A similar method is used to subtract polynomials.

Subtracting Polynomials

To subtract two polynomials, change all the signs of the subtrahend (second polynomial) and add the result to the minuend (first polynomial).

EXAMPLE 6 **Subtracting Polynomials Horizontally**

Perform each subtraction.

(a) $(5x - 2) - (3x - 8)$

$= (5x - 2) + [-(3x - 8)]$ Definition of subtraction

$= (5x - 2) + [-1(3x - 8)]$ $-a = -1a$

$= (5x - 2) + (-3x + 8)$ Distributive property

$= 2x + 6$ Combine like terms.

(b) Subtract $6x^3 - 4x^2 + 2$ from $11x^3 + 2x^2 - 8$.

$(11x^3 + 2x^2 - 8) - (6x^3 - 4x^2 + 2)$ Be careful to write the problem in the correct order.

$= (11x^3 + 2x^2 - 8) + (-6x^3 + 4x^2 - 2)$

$= 5x^3 + 6x^2 - 10$ Answer

CHECK To check a subtraction problem, use the following fact.

If $a - b = c$, then $a = b + c$.

Here, add $6x^3 - 4x^2 + 2$ and $5x^3 + 6x^2 - 10$.

$(6x^3 - 4x^2 + 2) + (5x^3 + 6x^2 - 10)$

$= 11x^3 + 2x^2 - 8$ ✓

Work Problem 7 at the Side. ▶

6 Find each sum.

(a) $(2x^4 - 6x^2 + 7)$
$+ (-3x^4 + 5x^2 + 2)$

(b) $(3x^2 + 4x + 2)$
$+ (6x^3 - 5x - 7)$

7 Subtract, and check your answers by addition.

GS **(a)** $(14y^3 - 6y^2 + 2y - 5)$
$- (2y^3 - 7y^2 - 4y + 6)$
$= (\_\_\_\_\_\_\_\_)$
$+ (\_\_\_\_\_\_\_\_)$
$= \_\_\_\_\_\_\_\_$

(b) Subtract

$\left(-\frac{3}{2}y^2 + \frac{4}{3}y + 6\right)$

from $\left(\frac{7}{2}y^2 - \frac{11}{3}y + 8\right)$.

Answers

6. **(a)** $-x^4 - x^2 + 9$
(b) $6x^3 + 3x^2 - x - 5$

7. **(a)** $14y^3 - 6y^2 + 2y - 5$; $-2y^3 + 7y^2 + 4y - 6$; $12y^3 + y^2 + 6y - 11$
(b) $5y^2 - 5y + 2$

Subtraction also can be done in columns. We use vertical subtraction in **Section 12.7** when we study polynomial division.

EXAMPLE 7 Subtracting Polynomials Vertically

Subtract by columns: $(14y^3 - 6y^2 + 2y - 5) - (2y^3 - 7y^2 - 4y + 6)$.

$$\begin{array}{r} 14y^3 - 6y^2 + 2y - 5 \\ \underline{2y^3 - 7y^2 - 4y + 6} \end{array}$$ Arrange like terms in columns.

Change all signs in the second row, and then add.

$$\begin{array}{rl} 14y^3 - 6y^2 + 2y - \ \ 5 & \\ \underline{-2y^3 + 7y^2 + 4y - \ \ 6} & \text{Change signs.} \\ 12y^3 + \ \ y^2 + 6y - 11 & \text{Add.} \end{array}$$

◀ Work Problem 8 at the Side.

EXAMPLE 8 Adding and Subtracting More Than Two Polynomials

Perform the indicated operations to simplify the expression

$$(4 - x + 3x^2) - (2 - 3x + 5x^2) + (8 + 2x - 4x^2).$$

Rewrite, changing the subtraction to adding the opposite.

$$\begin{aligned} &(4 - x + 3x^2) - (2 - 3x + 5x^2) + (8 + 2x - 4x^2) \\ &= (4 - x + 3x^2) + (-2 + 3x - 5x^2) + (8 + 2x - 4x^2) \\ &= (2 + 2x - 2x^2) + (8 + 2x - 4x^2) && \text{Combine like terms.} \\ &= 10 + 4x - 6x^2 && \text{Combine like terms.} \end{aligned}$$

◀ Work Problem 9 at the Side.

OBJECTIVE 6 Add and subtract polynomials with more than one variable. Polynomials in more than one variable are added and subtracted by combining like terms, just as with single-variable polynomials.

EXAMPLE 9 Adding and Subtracting Multivariable Polynomials

Add or subtract as indicated.

(a) $(4a + 2ab - b) + (3a - ab + b)$

$= 4a + 2ab - b + 3a - ab + b$

$= 7a + ab$ Combine like terms.

(b) $(2x^2y + 3xy + y^2) - (3x^2y - xy - 2y^2)$

$= 2x^2y + 3xy + y^2 - 3x^2y + xy + 2y^2$ Be careful with signs.

$= -x^2y + 4xy + 3y^2$

◀ Work Problem 10 at the Side.

8 Subtract by columns.

$(4y^3 - 16y^2 + 2y)$
$\quad - (12y^3 - 9y^2 + 16)$

9 Perform the indicated operations.

$(6p^4 - 8p^3 + 2p - 1)$
$\quad - (-7p^4 + 6p^2 - 12)$
$\quad + (p^4 - 3p + 8)$

10 Add or subtract.

(a) $(3mn + 2m - 4n)$
$\quad + (-mn + 4m + n)$

(b) $(5p^2q^2 - 4p^2 + 2q)$
$\quad - (2p^2q^2 - p^2 - 3q)$

Answers

8. $-8y^3 - 7y^2 + 2y - 16$
9. $14p^4 - 8p^3 - 6p^2 - p + 19$
10. **(a)** $2mn + 6m - 3n$
(b) $3p^2q^2 - 3p^2 + 5q$

12.1 Exercises

FOR EXTRA HELP

Download the MyDashBoard App

MyMathLab®

CONCEPT CHECK *Complete each statement.*

1. In the term $7x^5$, the coefficient is ______ and the exponent is ______.

2. The expression $5x^3 - 4x^2$ has (*one / two / three*) term(s).

3. The degree of the term $-4x^8$ is ______.

4. The polynomial $4x^2 - y^2$ (*is / is not*) an example of a trinomial.

5. When $x^2 + 10$ is evaluated for $x = 4$, the result is ______.

6. ______ is an example of a monomial with coefficient 5, in the variable x, having degree 9.

For each polynomial, determine the number of terms, and give the coefficient of each term. ***See Objective 2.***

7. $6x^4$

8. $-9y^5$

9. t^4

10. s^7

11. $\frac{x}{5}$

12. $\frac{z}{8}$

13. $-19r^2 - r$

14. $2y^3 - y$

15. $x - 8x^2 + \frac{2}{3}x^3$

16. $v - 2v^3 + \frac{3}{4}v^2$

In each polynomial, combine like terms whenever possible. Write the result with descending powers. ***See Example 1.***

17. $-3m^5 + 5m^5$

18. $-4y^3 + 3y^3$

19. $2r^5 + (-3r^5)$

20. $-19y^2 + 9y^2$

21. $\frac{1}{2}x^4 + \frac{1}{6}x^4$

22. $\frac{3}{10}x^6 + \frac{1}{5}x^6$

23. $-0.5m^2 + 0.2m^5$

24. $-0.9y + 0.9y^2$

25. $-3x^5 + 2x^5 - 4x^5$

26. $6x^3 - 8x^3 + 9x^3$

27. $-4p^7 + 8p^7 + 5p^9$

28. $-3a^8 + 4a^8 - 3a^2$

29. $-4y^2 + 3y^2 - 2y^2 + y^2$

30. $3r^5 - 8r^5 + r^5 + 2r^5$

For each polynomial, first simplify, if possible, and write it with descending powers. Then give the degree of the resulting polynomial, and tell whether it is a monomial, *a* binomial, *a* trinomial, *or* none of these. ***See Example 2.***

31. $6x^4 - 9x$

32. $7t^3 - 3t$

33. $5m^4 - 3m^2 + 6m^5 - 7m^3$

34. $6p^5 + 4p^3 - 8p^4 + 10p^2$

35. $\frac{5}{3}x^4 - \frac{2}{3}x^4 + \frac{1}{3}x^2 - 4$

36. $\frac{4}{5}r^6 + \frac{1}{5}r^6 - r^4 + \frac{2}{5}r$

37. $0.8x^4 - 0.3x^4 - 0.5x^4 + 7$

38. $1.2t^3 - 0.9t^3 - 0.3t^3 + 9$

39. $2.5x^2 + 0.5x + x^2 - x - 2x^2$

40. $8.3y - 9.2y^3 - 2.6y + 4.8y^3 - 6.7y^2$

Find the value of each polynomial for **(a)** $x = 2$ *and for* **(b)** $x = -1$. **See Example 3.**

41. $-2x + 3$

42. $5x - 4$

43. $2x^2 + 5x + 1$

44. $-3x^2 + 14x - 2$

45. $2x^5 - 4x^4 + 5x^3 - x^2$

46. $x^4 - 6x^3 + x^2 + 1$

47. $-4x^5 + x^2$

48. $2x^6 - 4x$

Add or subtract as indicated. **See Examples 4 and 7.**

49. Add.

$$\begin{array}{r} 3m^2 + 5m \\ \underline{2m^2 - 2m} \end{array}$$

50. Add.

$$\begin{array}{r} 4a^3 - 4a^2 \\ \underline{6a^3 + 5a^2} \end{array}$$

51. Subtract.

$$\begin{array}{r} 12x^4 - x^2 \\ \underline{8x^4 + 3x^2} \end{array}$$

52. Subtract.

$$\begin{array}{r} 13y^5 - y^3 \\ \underline{7y^5 + 5y^3} \end{array}$$

53. Add.

$$\begin{array}{r} \frac{2}{3}x^2 + \frac{1}{5}x + \frac{1}{6} \\ \underline{\frac{1}{2}x^2 - \frac{1}{3}x + \frac{2}{3}} \end{array}$$

54. Add.

$$\begin{array}{r} \frac{4}{7}y^2 - \frac{1}{5}y + \frac{7}{9} \\ \underline{\frac{1}{3}y^2 - \frac{1}{3}y + \frac{2}{5}} \end{array}$$

55. Subtract.

$$\begin{array}{r} 12m^3 - 8m^2 + 6m + 7 \\ \underline{5m^2 \qquad\quad - 4} \end{array}$$

56. Subtract.

$$\begin{array}{r} 5a^4 - 3a^3 + 2a^2 - a + 6 \\ \underline{-6a^4 \qquad\quad - a^2 + a - 1} \end{array}$$

57. Subtract.

$$\begin{array}{r} 4.3x^3 - 6.1x^2 - 3.0x - 5 \\ \underline{1.4x^3 - 2.6x^2 - 1.5x + 4} \end{array}$$

58. Subtract.

$$\begin{array}{r} 8.7z^3 + 4.2z^2 - 9.0z - 7 \\ \underline{5.9z^3 - 6.6z^2 + 3.5z - 4} \end{array}$$

Perform the indicated operations. ***See Examples 5, 6, and 8.***

59. $(2r^2 + 3r - 12) + (6r^2 + 2r)$

60. $(3r^2 + 5r - 6) + (2r - 5r^2)$

61. $(8m^2 - 7m) - (3m^2 + 7m - 6)$

62. $(x^2 + x) - (3x^2 + 2x - 1)$

63. $(16x^3 - x^2 + 3x) + (-12x^3 + 3x^2 + 2x)$

64. $(-2b^6 + 3b^4 - b^2) + (b^6 + 2b^4 + 2b^2)$

65. $(7y^4 + 3y^2 + 2y) - (18y^5 - 5y^3 + y)$

66. $(8t^5 + 3t^3 + 5t) - (19t^4 - 6t^2 + t)$

67. $[(8m^2 + 4m - 7) - (2m^3 - 5m + 2)] - (m^2 + m)$

68. $[(9b^3 - 4b^2 + 3b + 2) - (-2b^3 + b)] - (8b^3 + 6b + 4)$

69. Subtract $9x^2 - 3x + 7$ from $-2x^2 - 6x + 4$.

70. Subtract $-5w^3 + 5w^2 - 7$ from $6w^3 + 8w + 5$.

Find a polynomial that represents the perimeter of each square, rectangle, or triangle.

71.

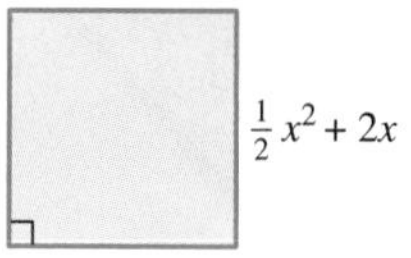

72.

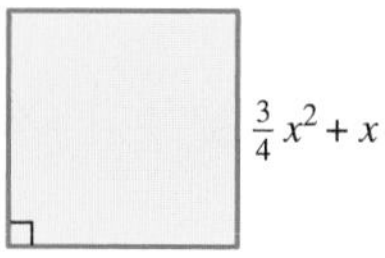

73.

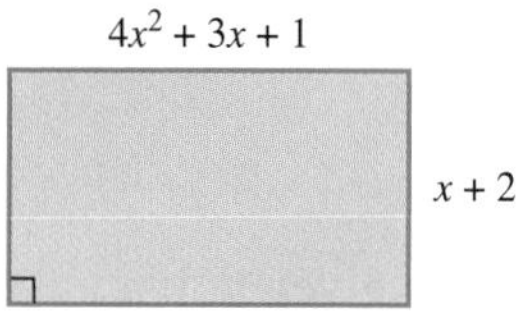

74.

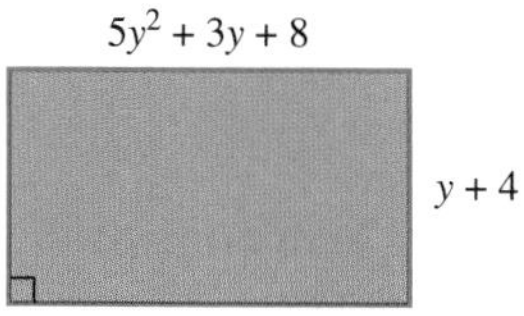

75.

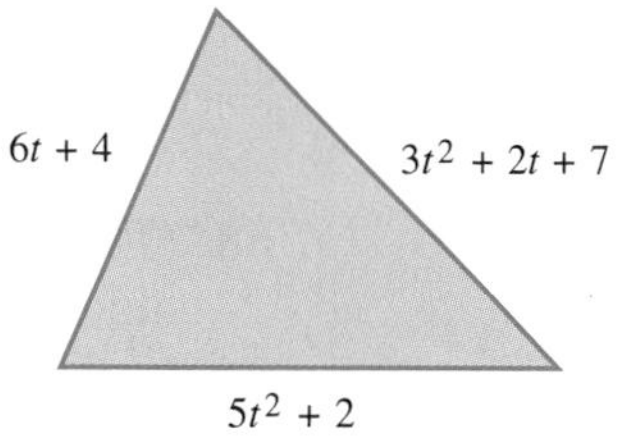

76.

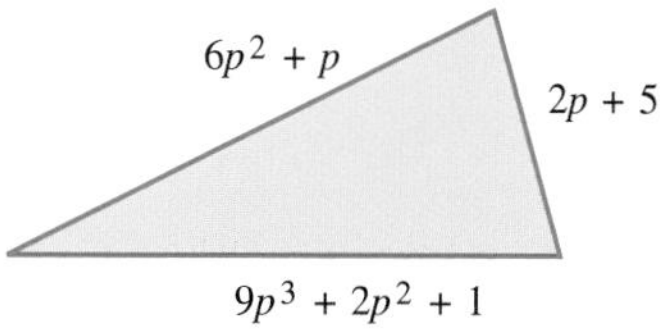

Add or subtract as indicated. ***See Example 9.***

77. $(9a^2b - 3a^2 + 2b) + (4a^2b - 4a^2 - 3b)$

78. $(4xy^3 - 3x + y) + (5xy^3 + 13x - 4y)$

79. $(2c^4d + 3c^2d^2 - 4d^2) - (c^4d + 8c^2d^2 - 5d^2)$

80. $(3k^2h^3 + 5kh + 6k^3h^2) - (2k^2h^3 - 9kh + k^3h^2)$

81. Subtract.

$$\begin{array}{r} 9m^3n - 5m^2n^2 + 4mn^2 \\ \underline{-3m^3n + 6m^2n^2 + 8mn^2} \end{array}$$

82. Subtract.

$$\begin{array}{r} 12r^5t + 11r^4t^2 - 7r^3t^3 \\ \underline{-8r^5t + 10r^4t^2 + 3r^3t^3} \end{array}$$

Find **(a)** *a polynomial that represents the perimeter of each triangle and* **(b)** *the measures of the angles of the triangle. (Hint: In part (b), the sum of the measures of the angles of any triangle is* 180°.*)*

83.

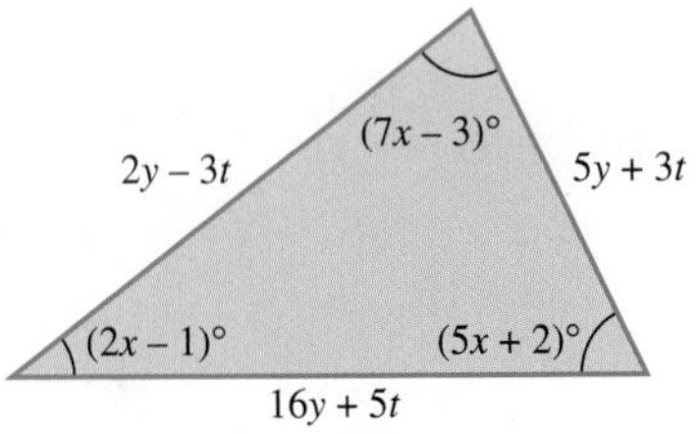

84.

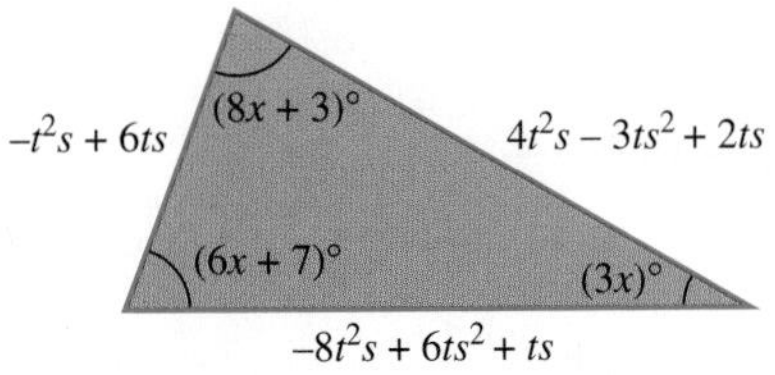

Relating Concepts (Exercises 85–88) For Individual or Group Work

A polynomial can model the distance in feet that a car going approximately 68 *mph will skid in t seconds. If we let D represent this distance, then*

$$D = 100t - 13t^2.$$

Each time we evaluate this polynomial for a value of t, we get one and only one output value D. This idea is basic to the concept of a ***function,*** *an important concept in mathematics. Exercises 85–88 illustrate this idea with this polynomial and two others.* ***Work them in order.***

85. Evaluate the given polynomial for $t = 5$. Use the result to fill in the blanks:

In _______ seconds, the car will skid _______ feet.

86. Use the polynomial equation $D = 100t - 13t^2$ to find the distance the car will skid in 1 sec. Write an ordered pair of the form (t, D).

87. (This is a business application of functions.) If it costs \$15 plus \$2 per day to rent a chain saw, the binomial $2x + 15$ gives the cost in dollars to rent the chain saw for x days. Evaluate this polynomial for $x = 6$. Use the result to fill in the blanks:

If the saw is rented for _______ days, the cost is _______.

88. (This is a physics application of functions.) If an object is projected upward under certain conditions, its height in feet is given by the trinomial $-16t^2 + 60t + 80$, where t is in seconds. Evaluate this trinomial for $t = 2.5$, and then use the result to fill in the blanks:

If _______ seconds have elapsed, the height of the object is _______ feet.

12.2 The Product Rule and Power Rules for Exponents

OBJECTIVES

1. Use exponents.
2. Use the product rule for exponents.
3. Use the rule $(a^m)^n = a^{mn}$.
4. Use the rule $(ab)^m = a^m b^m$.
5. Use the rule $\left(\frac{a}{b}\right)^m = \frac{a^m}{b^m}$.
6. Use combinations of the rules for exponents.
7. Use the rules for exponents in a geometry problem.

OBJECTIVE 1 Use exponents. In **Section 9.1,** we used exponents to write repeated products. In the expression 5^2, the number 5 is the **base** and 2 is the **exponent,** or **power.** The expression 5^2 is an **exponential expression.** Although we do not usually write the exponent when it is 1, in general,

$$a^1 = a, \quad \text{for any quantity } a.$$

EXAMPLE 1 Using Exponents

Write $3 \cdot 3 \cdot 3 \cdot 3 \cdot 3$ in exponential form and evaluate.

Since 3 occurs as a factor five times, the base is 3 and the exponent is 5. The exponential expression is 3^5, read "3 to the fifth power," or simply "3 to the fifth."

$$\underbrace{3 \cdot 3 \cdot 3 \cdot 3 \cdot 3}_{\text{5 factors of 3}} \quad \text{means} \quad 3^5, \quad \text{which equals} \quad 243.$$

Work Problem **1** at the Side. ▶

EXAMPLE 2 Evaluating Exponential Expressions

Name the base and the exponent of each expression. Then evaluate.

Expression	Base	Exponent	Value
(a) 5^4	5	4	$5 \cdot 5 \cdot 5 \cdot 5$, which equals 625
(b) -5^4	5	4	$-1 \cdot (5 \cdot 5 \cdot 5 \cdot 5)$, which equals -625
(c) $(-5)^4$	-5	4	$(-5)(-5)(-5)(-5)$, which equals 625

CAUTION

Compare **Examples 2(b) and 2(c).** In -5^4, the absence of parentheses shows that the exponent 4 applies only to the base 5, not -5. In $(-5)^4$, the parentheses show that the exponent 4 applies to the base -5. In summary, $-a^n$ and $(-a)^n$ are not necessarily the same.

Expression	Base	Exponent	Example
$-a^n$	a	n	$-3^2 = -(3 \cdot 3) = -9$
$(-a)^n$	$-a$	n	$(-3)^2 = (-3)(-3) = 9$

Work Problem **2** at the Side. ▶

OBJECTIVE 2 Use the product rule for exponents. To develop the product rule, we use the definition of an exponent.

$$\begin{aligned} &2^4 \cdot 2^3 \\ &= \underbrace{(2 \cdot 2 \cdot 2 \cdot 2)}_{\text{4 factors}}\underbrace{(2 \cdot 2 \cdot 2)}_{\text{3 factors}} \\ &= \underbrace{2 \cdot 2 \cdot 2 \cdot 2 \cdot 2 \cdot 2 \cdot 2}_{4 + 3 = 7 \text{ factors}} \\ &= 2^7 \end{aligned}$$

1 Write each product in exponential form and evaluate.

(a) $2 \cdot 2 \cdot 2 \cdot 2$

(b) $(-3)(-3)(-3)$

2 Name the base and the exponent of each expression. Then evaluate.

(a) 2^5 **(b)** -2^5

(c) $(-2)^5$ **(d)** 4^2

(e) -4^2 **(f)** $(-4)^2$

Answers

1. **(a)** 2^4; 16 **(b)** $(-3)^3$; -27
2. **(a)** 2; 5; 32 **(b)** 2; 5; -32 **(c)** -2; 5; -32 **(d)** 4; 2; 16 **(e)** 4; 2; -16 **(f)** -4; 2; 16

3 Simplify by using the product rule, if possible.

GS **(a)** $8^2 \cdot 8^5$

$= 8^{\_\_+\_\_}$

$= \_\_\_\_$

(b) $(-7)^5(-7)^3$

GS **(c)** $y^3 \cdot y$

$= y^3 \cdot y^{\_\_}$

$= y^{\_\_+\_\_}$

$= \_\_\_\_$

(d) $z^2z^5z^6$

(e) $4^2 \cdot 3^5$

(f) $6^4 + 6^2$

Answers

3. **(a)** 2; 5; 8^7 **(b)** $(-7)^8$
(c) 1; 3; 1; y^4 **(d)** z^{13}
(e) The product rule does not apply. (The product is 3888.)
(f) The product rule does not apply. (The sum is 1332.)

Also,

$$6^2 \cdot 6^3$$
$$= (6 \cdot 6)(6 \cdot 6 \cdot 6)$$
$$= 6 \cdot 6 \cdot 6 \cdot 6 \cdot 6$$
$$= 6^5.$$

Generalizing from these examples, we have the following.

$2^4 \cdot 2^3$ is equal to 2^{4+3}, which equals 2^7.

$6^2 \cdot 6^3$ is equal to 6^{2+3}, which equals 6^5.

In each case, adding the exponents gives the exponent of the product, suggesting the **product rule for exponents.**

Product Rule for Exponents

For any positive integers m and n, $\boldsymbol{a^m \cdot a^n = a^{m+n}}$.
(Keep the same base and add the exponents.)

Example: $6^2 \cdot 6^5 = 6^{2+5} = 6^7$

CAUTION

Do not multiply the bases when using the product rule. ***Keep the same base and add the exponents.*** For example,

$6^2 \cdot 6^5$ is equal to 6^7, ***not*** 36^7.

EXAMPLE 3 Using the Product Rule

Use the product rule for exponents to simplify, if possible.

(a) $6^3 \cdot 6^5$

$= 6^{3+5}$ Product rule

$= 6^8$ Add the exponents.

(b) $(-4)^7(-4)^2$

$= (-4)^{7+2}$ Product rule

$= (-4)^9$ Add the exponents.

(c) $x^2 \cdot x$

$= x^2 \cdot x^1$ $a = a^1$, for all a.

$= x^{2+1}$ Product rule

$= x^3$ Add the exponents.

(d) $m^4m^3m^5$

$= m^{4+3+5}$ Product rule

$= m^{12}$ Add the exponents.

(e) $2^3 \cdot 3^2$ (Think: 3^2 means $3 \cdot 3$.) (Think: 2^3 means $2 \cdot 2 \cdot 2$.)

$= 8 \cdot 9$ Evaluate 2^3 and 3^2.

$= 72$ Multiply.

The product rule does not apply because the bases are different.

(f) $2^3 + 2^4$

$= 8 + 16$ Evaluate 2^3 and 2^4.

$= 24$ Add.

The product rule does not apply. This is a *sum,* not a *product.*

◀ **Work Problem 3 at the Side.**

EXAMPLE 4 **Using the Product Rule**

Multiply $2x^3$ and $3x^7$.

$2x^3 \cdot 3x^7$ ($2x^3 = 2 \cdot x^3$; $3x^7 = 3 \cdot x^7$)

$= (2 \cdot 3) \cdot (x^3 \cdot x^7)$ Commutative and associative properties

$= 6x^{3+7}$ Multiply; product rule

$= 6x^{10}$ Add the exponents.

Work Problem 4 at the Side. ▶

CAUTION

Be sure that you understand the difference between *adding* and *multiplying* exponential expressions. For example,

$8x^3 + 5x^3$ means $(8+5)x^3$, which equals $13x^3$,

but $(8x^3)(5x^3)$ means $(8 \cdot 5)x^{3+3}$, which equals $40x^6$.

OBJECTIVE 3 **Use the rule $(a^m)^n = a^{mn}$.** We can simplify an expression such as $(5^2)^4$ with the product rule for exponents, as follows.

$(5^2)^4$

$= 5^2 \cdot 5^2 \cdot 5^2 \cdot 5^2$ Definition of exponent

$= 5^{2+2+2+2}$ Product rule

$= 5^8$ Add the exponents.

Observe that $2 \cdot 4 = 8$. This example suggests **power rule (a) for exponents.**

Power Rule (a) for Exponents

For any positive integers m and n, $\quad (a^m)^n = a^{mn}$.
(Raise a power to a power by multiplying exponents.)

Example: $(3^2)^4 = 3^{2 \cdot 4} = 3^8$

EXAMPLE 5 **Using Power Rule (a)**

Use power rule (a) for exponents to simplify.

(a) $(2^5)^3$
$= 2^{5 \cdot 3}$
$= 2^{15}$

(b) $(5^7)^2$
$= 5^{7 \cdot 2}$
$= 5^{14}$

(c) $(x^2)^5$
$= x^{2 \cdot 5}$ Power rule (a)
$= x^{10}$ Multiply.

Work Problem 4 at the Side. ▶

OBJECTIVE 4 **Use the rule $(ab)^m = a^m b^m$.** We can rewrite the expression $(4x)^3$ as shown below.

$(4x^3)$

$= (4x)(4x)(4x)$ Definition of exponent

$= 4 \cdot 4 \cdot 4 \cdot x \cdot x \cdot x$ Commutative and associative properties

$= 4^3x^3$ Definition of exponent

---

4 Multiply.

GS **(a)** $5m^2 \cdot 2m^6$
$= (5 \cdot \_\_\_\_) \cdot (m^{\_\_} \cdot m^{\_\_})$
$= \_\_\_\_ \, m^{\_\_+\_\_}$
$= \_\_\_\_$

(b) $3p^5 \cdot 9p^4$

(c) $-7p^5 \cdot (3p^8)$

5 Simplify.

GS **(a)** $(5^3)^4$
$= 5^{\_\_ \cdot \_\_}$
$= \_\_\_\_$

(b) $(6^2)^5$

(c) $(3^2)^4$

(d) $(a^6)^5$

Answers

4. **(a)** 2; 2; 6; 10; 2; 6; $10m^8$
(b) $27p^9$ **(c)** $-21p^{13}$

5. **(a)** 3; 4; 5^{12} **(b)** 6^{10} **(c)** 3^8 **(d)** a^{30}

6 Simplify.

GS **(a)** $(2ab)^4$

$= 2\_\_ a\_\_ b\_\_$

$= \_\_\_\_$

(b) $5(mn)^3$

(c) $(3a^2b^4)^5$

(d) $(-5m^2)^3$

The example $(4x)^3 = 4^3x^3$ suggests **power rule (b) for exponents.**

Power Rule (b) for Exponents

For any positive integer m, $(ab)^m = a^mb^m$.
(Raise a product to a power by raising each factor to the power.)

Example: $(2p)^5 = 2^5p^5$

EXAMPLE 6 **Using Power Rule (b)**

Use power rule (b) for exponents to simplify.

(a) $(3xy)^2$

$= 3^2x^2y^2$ Power rule (b)

$= 9x^2y^2$ $3^2 = 3 \cdot 3 = 9$

(b) $9(pq)^2$

$= 9(p^2q^2)$ Power rule (b)

$= 9p^2q^2$ Multiply.

(c) $5(2m^2p^3)^4$

$= 5[2^4(m^2)^4(p^3)^4]$ Power rule (b)

$= 5(2^4m^8p^{12})$ Power rule (a)

$= 80m^8p^{12}$ $5 \cdot 2^4 = 5 \cdot 16 = 80$

(d) $(-5^6)^3$

$= (-1 \cdot 5^6)^3$ $-a = -1 \cdot a$

$= (-1)^3(5^6)^3$ Power rule (b)

$= -1 \cdot 5^{18}$ Power rule (a)

$= -5^{18}$ Multiply.

Raise −1 to the designated power.

◀ Work Problem **6** at the Side.

CAUTION

Power rule (b) does not apply to a sum.

$(4x)^2 = 4^2x^2$, *but* $(4 + x)^2 \neq 4^2 + x^2$.

OBJECTIVE 5 Use the rule $\left(\frac{a}{b}\right)^m = \frac{a^m}{b^m}$. Since the quotient $\frac{a}{b}$ can be written as $a \cdot \frac{1}{b}$, we can use power rule (b), together with some of the properties of real numbers, to state **power rule (c) for exponents.**

Power Rule (c) for Exponents

For any positive integer m, $\left(\frac{a}{b}\right)^m = \frac{a^m}{b^m}$ **(where $b \neq 0$).**

(Raise a quotient to a power by raising both the numerator and the denominator to the power.)

Example: $\left(\frac{5}{3}\right)^2 = \frac{5^2}{3^2}$

Answers

6. **(a)** 4; 4; 4; $16a^4b^4$ **(b)** $5m^3n^3$ **(c)** $243a^{10}b^{20}$ **(d)** $-125m^6$

EXAMPLE 7 **Using Power Rule (c)**

Use power rule (c) for exponents to simplify.

(a) $\left(\frac{2}{3}\right)^5$

$= \frac{2^5}{3^5}$ Power rule (c)

$= \frac{32}{243}$ Simplify.

(b) $\left(\frac{m}{n}\right)^4$

$= \frac{m^4}{n^4}$ (where $n \neq 0$)

(c) $\left(\frac{1}{5}\right)^4$

$= \frac{1^4}{5^4}$ Power rule (c)

$= \frac{1}{625}$ Simplify.

Note

In **Example 7(c),** we used the fact that $1^4 = 1$ since $1 \cdot 1 \cdot 1 \cdot 1 = 1$.

In general, $\mathbf{1^n = 1}$, ***for any integer n.***

Work Problem 7 at the Side. ▶

The rules for exponents discussed in this section should be *memorized.*

Rules for Exponents

For positive integers m and n, the following are true.

		Examples
Product rule	$a^m \cdot a^n = a^{m+n}$	$6^2 \cdot 6^5 = 6^{2+5} = 6^7$
Power rules (a)	$(a^m)^n = a^{mn}$	$(3^2)^4 = 3^{2 \cdot 4} = 3^8$
(b)	$(ab)^m = a^m b^m$	$(2p)^5 = 2^5 p^5$
(c)	$\left(\frac{a}{b}\right)^m = \frac{a^m}{b^m}$ $(b \neq 0)$	$\left(\frac{5}{3}\right)^2 = \frac{5^2}{3^2}$

OBJECTIVE 6 **Use combinations of the rules for exponents.**

EXAMPLE 8 **Using Combinations of Rules**

Simplify each expression.

(a) $\left(\frac{2}{3}\right)^2 \cdot 2^3$

$= \frac{2^2}{3^2} \cdot \frac{2^3}{1}$ Power rule (c)

$= \frac{2^2 \cdot 2^3}{3^2 \cdot 1}$ Multiply fractions.

$= \frac{2^{2+3}}{3^2}$ Product rule

$= \frac{2^5}{3^2}$, or $\frac{32}{9}$

(b) $(5x)^3(5x)^4$

$= (5x)^7$ Product rule

$= 5^7x^7$ Power rule (b)

Continued on Next Page

7 Simplify. Assume that all variables represent nonzero real numbers.

GS **(a)** $\left(\frac{5}{2}\right)^4$

$= \frac{5^{—}}{2^{—}}$

$=$ ____

(b) $\left(\frac{p}{q}\right)^2$

(c) $\left(\frac{r}{t}\right)^3$

(d) $\left(\frac{1}{3}\right)^5$

(e) $\left(\frac{1}{x}\right)^{10}$

Answers

7. (a) 4; 4; $\frac{625}{16}$ **(b)** $\frac{p^2}{q^2}$ **(c)** $\frac{r^3}{t^3}$

(d) $\frac{1}{243}$ **(e)** $\frac{1}{x^{10}}$

8 Simplify.

GS **(a)** $(2m)^5(2m)^3$

$= (2m)^{\_\_}$

$= \_\_\_\_$

(b) $\left(\frac{5k^3}{3}\right)^2$

(c) $\left(\frac{1}{5}\right)^4(2x)^2$

(d) $(-3xy^2)^3(x^2y)^4$

9 Find a polynomial that represents the area of each figure.

(a)

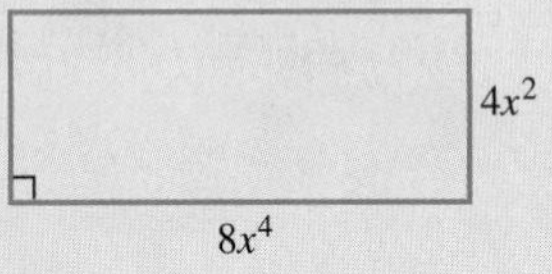

(b)

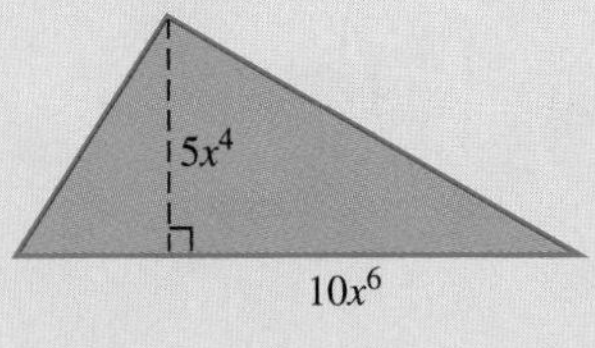

Answers

8. **(a)** 8; 2^8m^8, or $256m^8$ **(b)** $\frac{5^2k^6}{3^2}$, or $\frac{25k^6}{9}$ **(c)** $\frac{2^2x^2}{5^4}$, or $\frac{4x^2}{625}$ **(d)** $-3^3x^{11}y^{10}$, or $-27x^{11}y^{10}$

9. **(a)** $32x^6$ **(b)** $25x^{10}$

(c) $(2x^2y^3)^4(3xy^2)^3$

$= 2^4(x^2)^4(y^3)^4 \cdot 3^3x^3(y^2)^3$	Power rule (b)
$= 2^4x^8y^{12} \cdot 3^3x^3y^6$	Power rule (a)
$= 2^4 \cdot 3^3x^8x^3y^{12}y^6$	Commutative and associative properties
$= 16 \cdot 27x^{11}y^{18}$	Product rule
$= 432x^{11}y^{18}$	Multiply.

Notice that $(2x^2y^3)^4$ means $2^4x^{2\cdot4}y^{3\cdot4}$, ***not*** $(2 \cdot 4)\,x^{2\cdot4}y^{3\cdot4}$.

Do *not* multiply the coefficient 2 and the exponent 4.

(d) $(-x^3y)^2(-x^5y^4)^3$

Think of the negative sign in each factor as −1.

$= (-1x^3y)^2(-1x^5y^4)^3$	$-a = -1 \cdot a$
$= (-1)^2(x^3)^2(y^2) \cdot (-1)^3(x^5)^3(y^4)^3$	Power rule (b)
$= (-1)^2(x^6)(y^2) \cdot (-1)^3(x^{15})(y^{12})$	Power rule (a)
$= (-1)^5(x^{6+15})(y^{2+12})$	Product rule
$= -1x^{21}y^{14}$	Add the exponents.
$= -x^{21}y^{14}$	$-1 \cdot a = -a$

◀ **Work Problem 8 at the Side.**

OBJECTIVE 7 Use the rules for exponents in a geometry problem.

EXAMPLE 9 Using Area Formulas

Find a polynomial that represents the area in **(a) Figure 1** and **(b) Figure 2**.

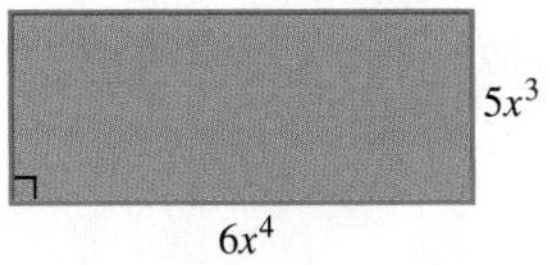

Figure 1

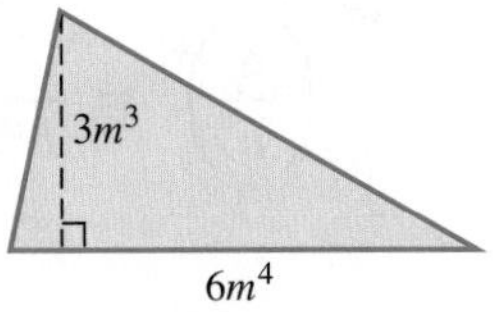

Figure 2

(a) For **Figure 1,** use the formula for the area of a rectangle, $A = LW$.

$A = (6x^4)(5x^3)$	Substitute.
$A = 6 \cdot 5 \cdot x^{4+3}$	Commutative property and product rule
$A = 30x^7$	Multiply. Add the exponents.

(b) **Figure 2** is a triangle with base $6m^4$ and height $3m^3$.

$A = \frac{1}{2}bh$	Area formula
$A = \frac{1}{2}(6m^4)(3m^3)$	Substitute.
$A = \frac{1}{2}(6 \cdot 3 \cdot m^{4+3})$	Commutative property and product rule
$A = 9m^7$	Multiply. Add the exponents.

◀ **Work Problem 9 at the Side.**

12.2 Exercises

FOR EXTRA HELP

Download the MyDashBoard App

MyMathLab®

CONCEPT CHECK *Decide whether each statement is* true *or* false.

1. $3^3 = 9$
2. $(-2)^4 = 2^4$
3. $(a^2)^3 = a^5$
4. $\left(\frac{1}{4}\right)^2 = \frac{1}{4^2}$

5. CONCEPT CHECK What exponent is understood on the base x in the expression xy^2?

6. CONCEPT CHECK How are the expressions 3^2, 5^3, and 7^4 read?

Write each expression using exponents. ***See Example 1.***

7. $t \cdot t \cdot t \cdot t \cdot t \cdot t \cdot t$
8. $w \cdot w \cdot w \cdot w \cdot w \cdot w$
9. $\left(\frac{1}{2}\right)\left(\frac{1}{2}\right)\left(\frac{1}{2}\right)\left(\frac{1}{2}\right)\left(\frac{1}{2}\right)$
10. $\left(-\frac{1}{4}\right)\left(-\frac{1}{4}\right)\left(-\frac{1}{4}\right)\left(-\frac{1}{4}\right)$
11. $(-8p)(-8p)$
12. $(-7x)(-7x)(-7x)$

Identify the base and the exponent for each exponential expression. In Exercises 13–16, also evaluate the expression. ***See Example 2.***

13. 3^5
14. 2^7
15. $(-3)^5$
16. $(-2)^7$
17. $(-6x)^4$
18. $(-8x)^4$
19. $-6x^4$
20. $-8x^4$

Use the product rule for exponents to simplify each expression, if possible. If it does not apply, say so. Write each answer in exponential form. ***See Examples 3 and 4.***

21. $5^2 \cdot 5^6$
22. $3^6 \cdot 3^7$
23. $4^2 \cdot 4^7 \cdot 4^3$
24. $5^3 \cdot 5^8 \cdot 5^2$
25. $(-7)^3(-7)^6$
26. $(-9)^8(-9)^5$
27. $t^3t^8t^{13}$
28. $n^5n^6n^9$
29. $(-8r^4)(7r^3)$
30. $(10a^7)(-4a^3)$
31. $(-6p^5)(-7p^5)$
32. $(-5w^8)(-9w^8)$
33. $3^8 + 3^9$
34. $4^{12} + 4^5$
35. $5^8 \cdot 3^8$
36. $6^3 \cdot 8^3$

Use the power rules for exponents to simplify each expression. ***See Examples 5–7.***

37. $(4^3)^2$
38. $(8^3)^6$
39. $(t^4)^5$
40. $(y^6)^5$
41. $(7r)^3$
42. $(11x)^4$
43. $(-5^2)^6$
44. $(-9^4)^8$
45. $(-8^3)^5$
46. $(-7^5)^7$
47. $(5xy)^5$
48. $(9pq)^6$
49. $8(qr)^3$
50. $4(vw)^5$
51. $\left(\frac{1}{2}\right)^3$
52. $\left(\frac{1}{3}\right)^5$

53. $\left(\frac{a}{b}\right)^3 \quad (b \neq 0)$ **54.** $\left(\frac{r}{t}\right)^4 \quad (t \neq 0)$ **55.** $\left(\frac{9}{5}\right)^8$ **56.** $\left(\frac{12}{7}\right)^6$

57. $(-2x^2y)^3$ **58.** $(-5m^4p^2)^3$ **59.** $(3a^3b^2)^2$ **60.** $(4x^3y^5)^4$

Simplify each expression. ***See Example 8.***

61. $\left(\frac{5}{2}\right)^3 \cdot \left(\frac{5}{2}\right)^2$ **62.** $\left(\frac{3}{4}\right)^5 \cdot \left(\frac{3}{4}\right)^6$ **63.** $\left(\frac{9}{8}\right)^3 \cdot 9^2$ **64.** $\left(\frac{8}{5}\right)^4 \cdot 8^3$

65. $(2x)^9(2x)^3$ **66.** $(6y)^5(6y)^8$ **67.** $(-6p)^4(-6p)$ **68.** $(-13q)^3(-13q)$

69. $(6x^2y^3)^5$ **70.** $(5r^5t^6)^7$ **71.** $(x^2)^3(x^3)^5$ **72.** $(y^4)^5(y^3)^5$

73. $(2w^2x^3y)^2(x^4y)^5$ **74.** $(3x^4y^2z)^3(yz^4)^5$ **75.** $(-r^4s)^2(-r^2s^3)^5$ **76.** $(-ts^6)^4(-t^3s^5)^3$

77. $\left(\frac{5a^2b^5}{c^6}\right)^3 \quad (c \neq 0)$ **78.** $\left(\frac{6x^3y^9}{z^5}\right)^4 \quad (z \neq 0)$ **79.** $(-5m^3p^4q)^2(p^2q)^3$

80. $(-a^4b^5)(-6a^3b^3)^2$ **81.** $(2x^2y^3z)^4(xy^2z^3)^2$ **82.** $(4q^2r^3s^5)^3(qr^2s^3)^4$

83. **CONCEPT CHECK** A student simplified $(10^2)^3$ as 1000^6. **What Went Wrong?**

84. **CONCEPT CHECK** A student simplified $(3x^2y^3)^4$ as $12x^8y^{12}$. **What Went Wrong?**

Find a polynomial that represents the area of each figure. In Exercise 88, leave π *in your answer.* ***See Example 9.*** *(If necessary, refer to the formulas inside the back cover. The ⌉ in the figures indicates* 90° *(right) angles.)*

85.
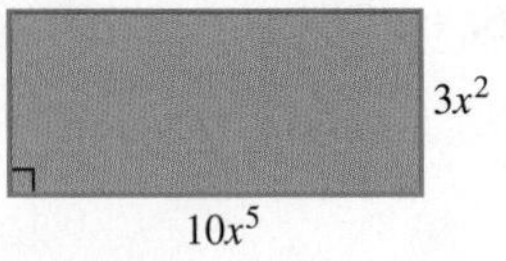

86.
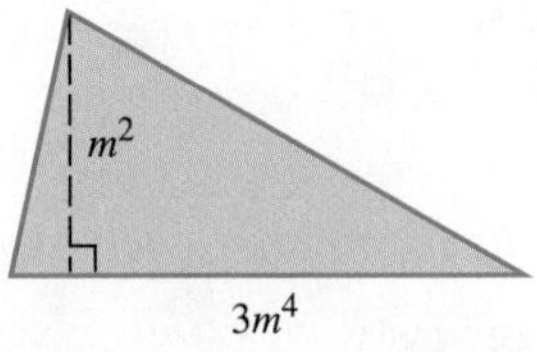

87.
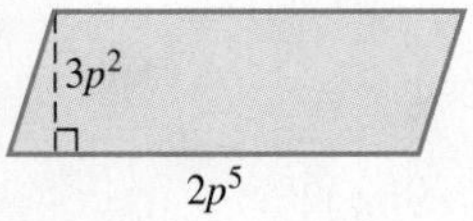

88.
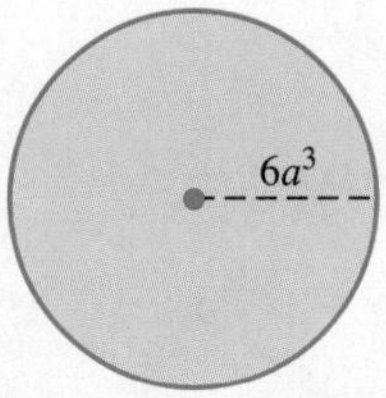

12.3 Multiplying Polynomials

OBJECTIVES

1. **Multiply a monomial and a polynomial.**
2. **Multiply two polynomials.**
3. **Multiply binomials by the FOIL method.**

OBJECTIVE 1 Multiply a monomial and a polynomial. In **Section 12.2,** we found the product of two monomials as follows.

$(-8m^6)(-9n^6)$

$= (-8)(-9)(m^6)(n^6)$ Commutative and associative properties

$= 72m^6n^6$ Multiply.

CAUTION

Do not confuse addition of terms with multiplication of terms.

$7q^5 + 2q^5$	$(7q^5)(2q^5)$
$= (7 + 2)q^5$	$= 7 \cdot 2q^{5+5}$
$= 9q^5$	$= 14q^{10}$

To find the product of a monomial and a polynomial with more than one term, we use the distributive property and multiplication of monomials.

EXAMPLE 1 **Multiplying Monomials and Polynomials**

Find each product.

(a) $4x^2(3x + 5)$

$= 4x^2(3x) + 4x^2(5)$ Distributive property

$= 12x^3 + 20x^2$ Multiply monomials.

(b) $-8m^3(4m^3 + 3m^2 + 2m - 1)$

$= -8m^3(4m^3) + (-8m^3)(3m^2)$

$+ (-8m^3)(2m) + (-8m^3)(-1)$ Distributive property

$= -32m^6 - 24m^5 - 16m^4 + 8m^3$ Multiply monomials.

Work Problem 1 at the Side. ▶

1 Find each product.

GS **(a)** $5m^3(2m + 7)$

$= 5m^3(\_\_\_) + 5m^3(\_\_\_)$

$= \_\_\_\_\_$

(b) $2x^4(3x^2 + 2x - 5)$

(c) $-4y^2(3y^3 + 2y^2 - 4y + 8)$

OBJECTIVE 2 Multiply two polynomials. To find the product of the polynomials $x^2 + 3x + 5$ and $x - 4$, think of $x - 4$ as a single quantity and use the distributive property as follows.

$(x^2 + 3x + 5)(x - 4)$

$= x^2(x - 4) + 3x(x - 4) + 5(x - 4)$ Distributive property

$= x^2(x) + x^2(-4) + 3x(x) + 3x(-4) + 5(x) + 5(-4)$

Distributive property again

$= x^3 - 4x^2 + 3x^2 - 12x + 5x - 20$ Multiply monomials.

$= x^3 - x^2 - 7x - 20$ Combine like terms.

Multiplying Polynomials

To multiply two polynomials, multiply each term of the second polynomial by each term of the first polynomial and add the products.

Answers

1. (a) $2m$; 7; $10m^4 + 35m^3$
   (b) $6x^6 + 4x^5 - 10x^4$
   (c) $-12y^5 - 8y^4 + 16y^3 - 32y^2$

EXAMPLE 2 Multiplying Two Polynomials

Multiply $(m^2 + 5)(4m^3 - 2m^2 + 4m)$.

$(m^2 + 5)(4m^3 - 2m^2 + 4m)$ — Multiply each term of the second ploynomial by each term of the first.

$= m^2(4m^3) + m^2(-2m^2) + m^2(4m) + 5(4m^3) + 5(-2m^2) + 5(4m)$ — Distributive property

$= 4m^5 - 2m^4 + 4m^3 + 20m^3 - 10m^2 + 20m$ — Distributive property again

$= 4m^5 - 2m^4 + 24m^3 - 10m^2 + 20m$ — Combine like terms.

◀ **Work Problem 2 at the Side.**

EXAMPLE 3 Multiplying Polynomials Vertically

Multiply $(x^3 + 2x^2 + 4x + 1)(3x + 5)$ vertically.

$$\begin{array}{r} x^3 + 2x^2 + 4x + 1 \\ 3x + 5 \\ \hline \end{array}$$

Write the polynomials vertically.

Begin by multiplying each of the terms in the top row by 5.

$$\begin{array}{rl} x^3 + 2x^2 + 4x + 1 & \\ 3x + 5 & \\ \hline 5x^3 + 10x^2 + 20x + 5 & 5(x^3 + 2x^2 + 4x + 1) \end{array}$$

Now multiply each term in the top row by $3x$. Then add like terms.

$$\begin{array}{rl} x^3 + 2x^2 + 4x + 1 & \\ 3x + 5 & \\ \hline 5x^3 + 10x^2 + 20x + 5 & \\ 3x^4 + 6x^3 + 12x^2 + 3x \phantom{{}+5} & 3x(x^3 + 2x^2 + 4x + 1) \\ \hline 3x^4 + 11x^3 + 22x^2 + 23x + 5 & \text{Add in columns.} \end{array}$$

This process is similar to multiplication of whole numbers.

Place *like* terms in columns so they can be added.

The product is $3x^4 + 11x^3 + 22x^2 + 23x + 5$.

◀ **Work Problem 3 at the Side.**

We can use a rectangle to model polynomial multiplication. For example, to find the product

$$(2x + 1)(3x + 2),$$

we label a rectangle with each term as shown below on the left. Then we write the product of each pair of monomials in the appropriate box as shown on the right.

	$3x$	2
$2x$		
1		

	$3x$	2
$2x$	$6x^2$	$4x$
1	$3x$	2

The product of the binomials is the sum of these four monomial products.

$$\begin{aligned} &(2x + 1)(3x + 2) \\ &= 6x^2 + 4x + 3x + 2 \\ &= 6x^2 + 7x + 2 \quad \text{Combine like terms.} \end{aligned}$$

◀ **Work Problem 4 at the Side.**

---

2 Multiply.

(a) $(m + 3)(m^2 - 2m + 1)$

(b) $(6p^2 + 2p - 4)(3p^2 - 5)$

3 Multiply vertically.

$$\begin{array}{r} 3x^2 + 4x - 5 \\ x + 4 \\ \hline \end{array}$$

4 Use the rectangle method to find each product.

GS **(a)** $(4x + 3)(x + 2)$

	x	2
$4x$	$4x^2$	
3		

(b) $(x + 5)(x^2 + 3x + 1)$

Answers

2. **(a)** $m^3 + m^2 - 5m + 3$
(b) $18p^4 + 6p^3 - 42p^2 - 10p + 20$

3. $3x^3 + 16x^2 + 11x - 20$

4. **(a)**

	x	2
$4x$	$4x^2$	$8x$
3	$3x$	6

; $4x^2 + 11x + 6$

(b) $x^3 + 8x^2 + 16x + 5$

OBJECTIVE 3 Multiply binomials by the FOIL method. When multiplying binomials, the **FOIL method** reduces the rectangle method to a systematic approach without the rectangle. Consider this example.

$$(x+3)(x+5)$$

$= (x+3)x + (x+3)5$	Distributive property
$= x(x) + 3(x) + x(5) + 3(5)$	Distributive property again
$= x^2 + 3x + 5x + 15$	Multiply.
$= x^2 + 8x + 15$	Combine like terms.

The letters of the word FOIL refer to the positions of the terms.

$(x+3)(x+5)$ Multiply the **First terms:** $x(x)$. **F**

$(x+3)(x+5)$ Multiply the **Outer terms:** $x(5)$. This is the **outer product.** **O**

$(x+3)(x+5)$ Multiply the **Inner terms:** $3(x)$. This is the **inner product.** **I**

$(x+3)(x+5)$ Multiply the **Last terms:** $3(5)$. **L**

We add the outer product, $5x$, and the inner product, $3x$, mentally so that the three terms of the answer can be written without extra steps.

$$(x+3)(x+5)$$
$$= x^2 + 8x + 15$$

Multiplying Binomials by the FOIL Method

Step 1 Multiply the two **F**irst terms of the binomials to get the first term of the answer.

Step 2 Find the **O**uter product and the **I**nner product and combine them (when possible) to get the middle term of the answer.

Step 3 Multiply the two **L**ast terms of the binomials to get the last term of the answer.

F $= x^2$ **L** $= 15$

$(x+3)(x+5)$ $\quad$ $(x+3)(x+5)$
$\quad = x^2 + 8x + 15$

I $= 3x$
O $= 5x$
$8x$ Add.

Add the terms found in Steps 1–3.

Work Problem 5 at the Side.

5 For the product

$$(2p-5)(3p+7),$$

find and simplify the following.

(a) Product of first terms

___(___)
= ___

(b) Outer product

___(___)
= ___

(c) Inner product

___(___)
= ___

(d) Product of last terms

___(___)
= ___

(e) Complete product in simplified form

____

Answers

5. **(a)** $2p$; $3p$; $6p^2$ **(b)** $2p$; 7; $14p$
(c) -5; $3p$; $-15p$ **(d)** -5; 7; -35
(e) $6p^2 - p - 35$

EXAMPLE 4 Using the FOIL Method

Use the FOIL method to find the product $(x+8)(x-6)$.

Step 1 **F** Multiply the **first** terms: $x(x) = x^2$.

Step 2 **O** Find the **outer** product: $x(-6) = -6x$.

I Find the **inner** product: $8(x) = 8x$.

Add the outer and inner products mentally: $-6x + 8x = \mathbf{2x}$.

Step 3 **L** Multiply the **last** terms: $8(-6) = \mathbf{-48}$.

The product $(x+8)(x-6)$ is $\mathbf{x^2 + 2x - 48}$. Add the terms found in Steps 1–3.

Shortcut: $(x+8)(x-6)$ — First, Last; Inner, Outer

$(x+8)(x-6)$ — x^2, -48; $8x$, $-6x$; $2x$ Add.

$(x+8)(x-6)$
$= x^2 + 2x - 48$

◄ **Work Problem 6 at the Side.**

EXAMPLE 5 Using the FOIL Method

Multiply $(9x-2)(3y+1)$.

First	$(9x-2)(3y+1)$	$27xy$
Outer	$(9x-2)(3y+1)$	$9x$
Inner	$(9x-2)(3y+1)$	$-6y$
Last	$(9x-2)(3y+1)$	-2

These unlike terms *cannot* be added.

F O I L

The product $(9x-2)(3y+1)$ is $27xy + 9x - 6y - 2$.

◄ **Work Problem 7 at the Side.**

EXAMPLE 6 Using the FOIL Method

Find each product.

(a) $(2k+5y)(k+3y)$

$= 2k(k) + 2k(3y) + 5y(k) + 5y(3y)$ FOIL method

$= 2k^2 + 6ky + 5ky + 15y^2$ Multiply.

$= 2k^2 + 11ky + 15y^2$ Combine like terms.

(b) $(7p+2q)(3p-q)$

$= 21p^2 - 7pq + 6pq - 2q^2$

$= 21p^2 - pq - 2q^2$

(c) $2x^2(x-3)(3x+4)$

$= 2x^2(3x^2 - 5x - 12)$

$= 6x^4 - 10x^3 - 24x^2$

◄ **Work Problem 8 at the Side.**

Note

Alternatively, **Example 6(c)** can be simplified as follows.

$2x^2(x-3)(3x+4)$ Multiply $2x^2$ and $x-3$ first.

$= (2x^3 - 6x^2)(3x+4)$ Multiply that product and $3x+4$.

$= 6x^4 - 10x^3 - 24x^2$ Same answer

---

6 Use the FOIL method to find each product.

GS **(a)** $(m+4)(m-3)$
$= m(\_\_\_\_) + m(\_\_\_\_)$
$+ 4(\_\_\_\_) + 4(\_\_\_\_)$
$= \_\_\_\_\_\_\_\_$

(b) $(y+7)(y+2)$

(c) $(r-8)(r-5)$

7 Find the product.

$(4x-3)(2y+5)$

8 Find each product.

GS **(a)** $(6m+5)(m-4)$
$= 6m(\_\_\_\_) + 6m(\_\_\_\_)$
$+ 5(\_\_\_\_) + 5(\_\_\_\_)$
$= \_\_\_\_\_\_\_\_$

(b) $(3r+2t)(3r+4t)$

GS **(c)** $y^2(8y+3)(2y+1)$
$= y^2(\_\_\_\_\_\_\_\_)$
$= \_\_\_\_\_\_\_\_$

Answers

6. **(a)** m; -3; m; -3; $m^2 + m - 12$
(b) $y^2 + 9y + 14$
(c) $r^2 - 13r + 40$

7. $8xy + 20x - 6y - 15$

8. **(a)** m; -4; m; -4; $6m^2 - 19m - 20$
(b) $9r^2 + 18rt + 8t^2$
(c) $16y^2 + 14y + 3$; $16y^4 + 14y^3 + 3y^2$

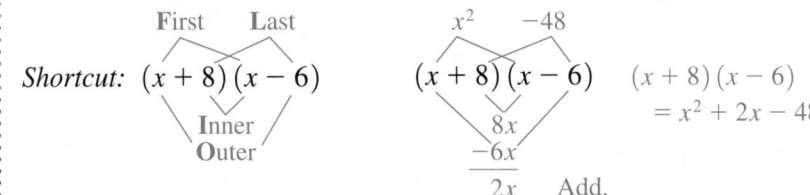

12.3 Exercises

FOR EXTRA HELP

Download the MyDashBoard App

MyMathLab®

CONCEPT CHECK *In Exercises 1 and 2, match each product in Column I with the correct polynomial in Column II.*

I	II
1. **(a)** $5x^3(6x^7)$	**A.** $125x^{21}$
(b) $-5x^7(6x^3)$	**B.** $30x^{10}$
(c) $(5x^7)^3$	**C.** $-216x^9$
(d) $(-6x^3)^3$	**D.** $-30x^{10}$

I	II
2. **(a)** $(x-5)(x+4)$	**A.** $x^2+9x+20$
(b) $(x+5)(x+4)$	**B.** $x^2-9x+20$
(c) $(x-5)(x-4)$	**C.** x^2-x-20
(d) $(x+5)(x-4)$	**D.** x^2+x-20

CONCEPT CHECK *Fill in each blank with the correct response.*

3. In multiplying a monomial by a polynomial, such as in $4x(3x^2+7x^3)=4x(3x^2)+4x(7x^3)$, the first property that is used is the __________ property.

4. The FOIL method can only be used to multiply two polynomials when both polynomials are __________.

Find each product. ***See Section 5.2.***

5. $5p(3q^2)$

6. $4a^3(3b^2)$

7. $(-6m^3)(3n^2)$

8. $(9r^3)(-2s^2)$

9. $y^5 \cdot 9y \cdot y^4$

10. $x^2 \cdot 3x^3 \cdot 2x$

11. $(4x^3)(2x^2)(-x^5)$

12. $(7t^5)(3t^4)(-t^8)$

Find each product. ***See Example 1.***

13. $-2m(3m+2)$

14. $-5p(6+3p)$

15. $\frac{3}{4}p(8-6p+12p^3)$

16. $\frac{4}{3}x(3+2x+5x^3)$

17. $2y^5(5y^4+2y+3)$

18. $2m^4(3m^2+5m+6)$

19. $2y^3(3y^3+2y+1)$

20. $2m^4(3m^2+5m+1)$

21. $-4r^3(-7r^2+8r-9)$

22. $-9a^5(-3a^6-2a^4+8a^2)$

23. $3a^2(2a^2-4ab+5b^2)$

24. $4z^3(8z^2+5zy-3y^2)$

25. $7m^3n^2(3m^2+2mn-n^3)$

GS $= 7m^3n^2(\_\_\_\_) + 7m^3n^2(\_\_\_\_) - 7m^3n^2(\_\_\_\_)$

$= \_\_\_\_\_\_\_\_\_\_\_\_$

26. $2p^2q(3p^2q^2-5p+2q^2)$

GS $= \_\_\_\_(3p^2q^2) + \_\_\_\_(-5p) + \_\_\_\_(2q^2)$

$= \_\_\_\_\_\_\_\_\_\_\_\_$

Find each product. **See Examples 2 and 3.**

27. $(6x + 1)(2x^2 + 4x + 1)$

28. $(9a + 2)(9a^2 + a + 1)$

29. $(2r - 1)(3r^2 + 4r - 4)$

30. $(9y - 2)(8y^2 - 6y + 1)$

31. $(4m + 3)(5m^3 - 4m^2 + m - 5)$

32. $(y + 4)(3y^3 - 2y^2 + y + 3)$

33. $(5x^2 + 2x + 1)(x^2 - 3x + 5)$

34. $(2m^2 + m - 3)(m^2 - 4m + 5)$

35. $(6x^4 - 4x^2 + 8x)\left(\frac{1}{2}x + 3\right)$

36. $(8y^6 + 4y^4 - 12y^2)\left(\frac{3}{4}y^2 + 2\right)$

Find each product using the rectangle method shown in the text.

37. $(x + 3)(x + 4)$

	x	4
x		
3		

Product: ____________________

38. $(x + 5)(x + 2)$

	___	___
x		
5		

Product: ____________________

39. $(2x + 1)(x^2 + 3x + 2)$

	x^2	$3x$	2
$2x$			
1			

Product: ____________________

40. $(x + 4)(3x^2 + 2x + 1)$

	___	$2x$	___
___			
___			

Product: ____________________

Find each product. Use the FOIL method. In Exercises 61–66, use the FOIL method and the distributive property. **See Examples 4–6.**

41. $(m + 7)(m + 5)$

42. $(x + 4)(x + 7)$

43. $(n - 2)(n + 3)$

44. $(r - 6)(r + 8)$

45. $(4r + 1)(2r - 3)$

46. $(5x + 2)(2x - 7)$

47. $(3x + 2)(3x - 2)$

48. $(7x + 3)(7x - 3)$

49. $(3q + 1)(3q + 1)$

50. $(4w + 7)(4w + 7)$

51. $(5x + 7)(3y - 8)$

52. $(4x + 3)(2y - 1)$

53. $(3t + 4s)(2t + 5s)$

54. $(8v + 5w)(2v + 3w)$

55. $(-0.3t + 0.4)(t + 0.6)$

56. $(-0.5x + 0.9)(x - 0.2)$

57. $\left(x - \frac{2}{3}\right)\left(x + \frac{1}{4}\right)$

58. $\left(y + \frac{3}{5}\right)\left(y - \frac{1}{2}\right)$

59. $\left(-\frac{5}{4} + 2r\right)\left(-\frac{3}{4} - r\right)$

60. $\left(-\frac{8}{3} + 3k\right)\left(-\frac{2}{3} - k\right)$

61. $x(2x - 5)(x + 3)$

62. $m(4m - 1)(2m + 3)$

63. $3y^3(2y + 3)(y - 5)$

64. $5t^4(t + 3)(3t - 1)$

65. $-8r^3(5r^2 + 2)(5r^2 - 2)$

66. $-5t^4(2t^4 + 1)(2t^4 - 1)$

*Find polynomials that represent **(a)** the area and **(b)** the perimeter of each square or rectangle. (If necessary, refer to the formulas inside the back cover.)*

67.

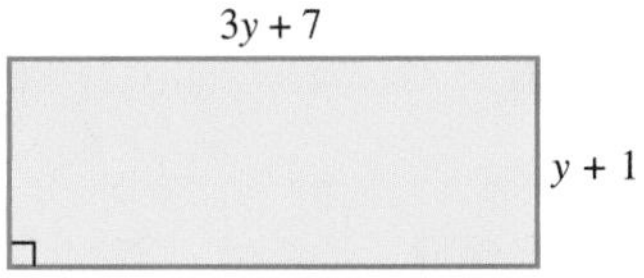

68.

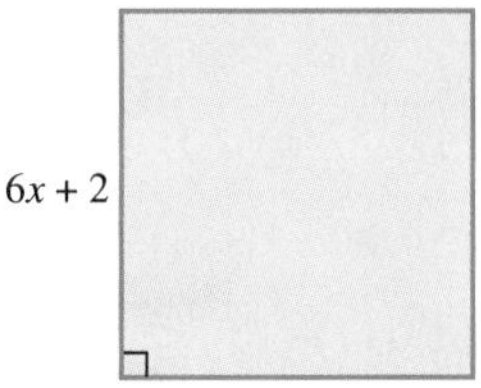

Relating Concepts (Exercises 69–74) For Individual or Group Work

Work Exercises 69–74 in order. *(All units are in yards.)*

69. Find a polynomial that represents the area of the rectangle.

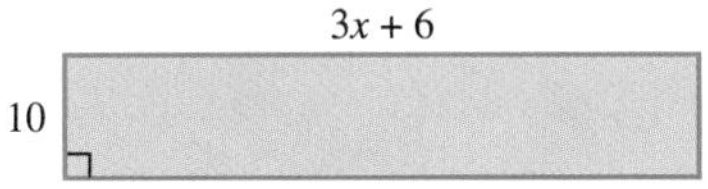

70. Suppose you know that the area of the rectangle is 600 yd^2. Use this information and the polynomial from **Exercise 69** to write an equation in x, and solve it.

71. Refer to **Exercise 70.** What are the dimensions of the rectangle?

72. Use the result of **Exercise 71** to find the perimeter of the rectangle.

73. Suppose the rectangle represents a lawn and it costs \$3.50 per square yard to lay sod on the lawn. How much will it cost to sod the entire lawn?

74. Again, suppose the rectangle represents a lawn and it costs \$9.00 per yard to fence the lawn. How much will it cost to fence the lawn?

12.4 Special Products

OBJECTIVES

1. Square binomials.
2. Find the product of the sum and difference of two terms.
3. Find greater powers of binomials.

1 Consider the binomial $x + 4$.

GS **(a)** What is the first term of the binomial? ____

Square it. ____

(b) What is the last term of the binomial? ____

Square it. ____

(c) Find twice the product of the two terms of the binomial.

$2(\_\_\_)(\_\_\_) = \_\_\_$

(d) Use the results of parts (a)–(c) to find $(x + 4)^2$.

____

Answers

1. **(a)** x; x^2 **(b)** 4; 16 **(c)** x; 4; $8x$ **(d)** $x^2 + 8x + 16$

OBJECTIVE 1 **Square binomials.** The square of a binomial can be found quickly by using the method shown in **Example 1.**

EXAMPLE 1 Squaring a Binomial

Find $(m + 3)^2$.

$(m + 3)^2$ means $(m + 3)(m + 3)$.

$$(m + 3)(m + 3)$$
$$= m^2 + 3m + 3m + 9 \quad \text{FOIL method}$$
$$= m^2 + 6m + 9 \quad \text{Combine like terms.}$$

This result has the squares of the first and the last terms of the binomial.

$$m^2 = m^2 \quad \text{and} \quad 3^2 = 9$$

The middle term, 6m, is twice the product of the two terms of the binomial, since the outer and inner products are $m(3)$ and $3(m)$.

$$m(3) + 3(m) = 2(m)(3) = 6m$$

◀ Work Problem 1 at the Side.

Example 1 suggests the following rules.

Square of a Binomial

The square of a binomial is a trinomial consisting of

the square of the first term + twice the product of the two terms + the square of the last term.

For a and b, the following are true.

$$(a + b)^2 = a^2 + 2ab + b^2$$
$$(a - b)^2 = a^2 - 2ab + b^2$$

EXAMPLE 2 Squaring Binomials

Square each binomial.

$$(a - b)^2 = a^2 - 2 \cdot a \cdot b + b^2$$

(a) $(5z - 1)^2 = (5z)^2 - 2(5z)(1) + (1)^2$

$$= 5^2z^2 - 10z + 1 \quad (5z)^2 = 5^2z^2 = 25z^2 \text{ by power rule (b).}$$
$$= 25z^2 - 10z + 1$$

(b) $(3b + 5r)^2$

$$= (3b)^2 + 2(3b)(5r) + (5r)^2 \quad (a + b)^2 = a^2 + 2ab + b^2$$
$$= 9b^2 + 30br + 25r^2 \quad (3b)^2 = 3^2b^2 = 9b^2 \text{ and } (5r)^2 = 5^2r^2 = 25r^2$$

Continued on Next Page

(c) $(2a - 9x)^2$

$= (2a)^2 - 2(2a)(9x) + (9x)^2$ $\quad (a - b)^2 = a^2 - 2ab + b^2$

$= 4a^2 - 36ax + 81x^2$ $\quad (2a)^2 = 2^2a^2 = 4a^2$ and $(9x)^2 = 9^2x^2 = 81x^2$

(d) $\left(4m + \frac{1}{2}\right)^2$

$= (4m)^2 + 2(4m)\left(\frac{1}{2}\right) + \left(\frac{1}{2}\right)^2$ $\quad (a + b)^2 = a^2 + 2ab + b^2$

$= 16m^2 + 4m + \frac{1}{4}$ $\quad (4m)^2 = 4^2m^2 = 16m^2$ and $\left(\frac{1}{2}\right)^2 = \frac{1}{4}$

(e) $x(4x - 3)^2$ (Remember the middle term.)

$= x(16x^2 - 24x + 9)$ $\quad$ Square the binomial.

$= 16x^3 - 24x^2 + 9x$ $\quad$ Distributive property

In the square of a sum, all of the terms are positive, as in Examples 2(b) and (d). In the square of a difference, the middle term is negative, as in Examples 2(a) and (c).

CAUTION

A common error when squaring a binomial is to forget the middle term of the product. In general, remember the following.

$$(a + b)^2 = a^2 + 2ab + b^2, \quad \textbf{not} \quad a^2 + b^2,$$

and

$$(a - b)^2 = a^2 - 2ab + b^2, \quad \textbf{not} \quad a^2 - b^2.$$

Work Problem 2 at the Side. ▶

OBJECTIVE 2 Find the product of the sum and difference of two terms. In binomial products of the form $(a + b)(a - b)$, one binomial is the sum of two terms, and the other is the difference of the *same* two terms. Consider the following.

$(x + 2)(x - 2)$

$= x^2 - 2x + 2x - 4$ $\quad$ FOIL method

$= x^2 - 4$ $\quad$ Combine like terms.

Thus, the product of $a + b$ and $a - b$ is the **difference of two squares.**

Product of the Sum and Difference of Two Terms

$$(a + b)(a - b) = a^2 - b^2$$

Note

The expressions $a + b$ and $a - b$, the sum and difference of the *same* two terms, are **conjugates.** In the example above, $x + 2$ and $x - 2$ are conjugates.

2 Square each binomial.

GS **(a)** $(t - 6)^2$

$= (\_\_\_)^2 - 2(\_\_\_)(\_\_\_) + (\_\_\_)^2$

$= \_\_\_\_\_\_\_\_$

(b) $(2m - p)^2$

GS **(c)** $(4p + 3q)^2$

$= (\_\_\_)^2 + \_\_\_(\_\_\_)3q + (\_\_\_)^2$

$= \_\_\_\_\_\_\_\_$

(d) $(5r - 6s)^2$

(e) $\left(3k - \frac{1}{2}\right)^2$

(f) $x(2x + 7)^2$

Answers

2. **(a)** t; t; 6; 6; $t^2 - 12t + 36$
(b) $4m^2 - 4mp + p^2$
(c) $4p$; 2; $4p$; $3q$; $16p^2 + 24pq + 9q^2$
(d) $25r^2 - 60rs + 36s^2$
(e) $9k^2 - 3k + \frac{1}{4}$
(f) $4x^3 + 28x^2 + 49x$

3 Find each product.

GS **(a)** $(y+3)(y-3)$

$= (\_\_\_)^2 - (\_\_\_)^2$

$= \_\_\_\_$

(b) $(10m+7)(10m-7)$

GS **(c)** $(7p+2q)(7p-2q)$

$= (\_\_\_)^2 - (\_\_\_)^2$

$= \_\_\_\_$

(d) $\left(3r-\frac{1}{2}\right)\left(3r+\frac{1}{2}\right)$

(e) $3x(x^3-4)(x^3+4)$

Answers

3. **(a)** y; 3; y^2-9 **(b)** $100m^2-49$ **(c)** $7p$; $2q$; $49p^2-4q^2$ **(d)** $9r^2-\frac{1}{4}$ **(e)** $3x^7-48x$

EXAMPLE 3 **Finding the Product of the Sum and Difference of Two Terms**

Find each product.

(a) $(x+4)(x-4)$

Use the rule for the product of the sum and difference of two terms.

$(x+4)(x-4)$

$= x^2-4^2$ $\quad (a+b)(a-b)=a^2-b^2$

$= x^2-16$ $\quad$ Square 4.

(b) $\left(\frac{2}{3}-w\right)\left(\frac{2}{3}+w\right)$

$= \left(\frac{2}{3}+w\right)\left(\frac{2}{3}-w\right)$ $\quad$ Commutative property

$= \left(\frac{2}{3}\right)^2 - w^2$ $\quad (a+b)(a-b)=a^2-b^2$

$= \frac{4}{9} - w^2$ $\quad$ Square $\frac{2}{3}$.

(c) $x(x+2)(x-2)$

$= x(x^2-4)$ $\quad$ Find the product of the sum and difference of two terms.

$= x^3-4x$ $\quad$ Distributive property

EXAMPLE 4 **Finding the Product of the Sum and Difference of Two Terms**

Find each product.

$$\begin{array}{cccc} (a & +\, b) & (a & -\, b) \\ \downarrow & \downarrow & \downarrow & \downarrow \end{array}$$

(a) $(5m+3)(5m-3)$

Use the rule for the product of the sum and difference of two terms.

$(5m+3)(5m-3)$

$= (5m)^2-3^2$ $\quad (a+b)(a-b)=a^2-b^2$

$= 25m^2-9$ $\quad$ Apply the exponents.

Be careful to square $5m$ correctly.

(b) $(4x+y)(4x-y)$

$= (4x)^2-y^2$

$= 16x^2-y^2$

$(4x)^2 = 4^2x^2 = 16x^2$

(c) $\left(z-\frac{1}{4}\right)\left(z+\frac{1}{4}\right)$

$= z^2 - \left(\frac{1}{4}\right)^2$

$= z^2 - \frac{1}{16}$

(d) $2p(p^2+3)(p^2-3)$

$= 2p(p^4-9)$ $\quad$ Multiply the conjugates.

$= 2p^5-18p$ $\quad$ Distributive property

◀ **Work Problem 3 at the Side.**

The product rules of this section will be important in **Chapters 13 and 14** and should be *memorized.*

OBJECTIVE 3 Find greater powers of binomials. The methods used in the previous section and this section can be combined to find greater powers of binomials.

EXAMPLE 5 Finding Greater Powers of Binomials

Find each product.

(a) $(x + 5)^3$

$= (x + 5)^2(x + 5)$ — $a^3 = a^2 \cdot a$

$= (x^2 + 10x + 25)(x + 5)$ — Square the binomial.

$= x^3 + 10x^2 + 25x + 5x^2 + 50x + 125$ — Multiply polynomials.

$= x^3 + 15x^2 + 75x + 125$ — Combine like terms.

(b) $(2y - 3)^4$

$= (2y - 3)^2(2y - 3)^2$ — $a^4 = a^2 \cdot a^2$

$= (4y^2 - 12y + 9)(4y^2 - 12y + 9)$ — Square each binomial.

$= 16y^4 - 48y^3 + 36y^2 - 48y^3 + 144y^2 - 108y + 36y^2 - 108y + 81$ — Multiply polynomials.

$= 16y^4 - 96y^3 + 216y^2 - 216y + 81$ — Combine like terms.

(c) $-2r(r + 2)^3$

$= -2r(r + 2)(r + 2)^2$ — $a^3 = a \cdot a^2$

$= -2r(r + 2)(r^2 + 4r + 4)$ — Square the binomial.

$= -2r(r^3 + 4r^2 + 4r + 2r^2 + 8r + 8)$ — Multiply polynomials.

$= -2r(r^3 + 6r^2 + 12r + 8)$ — Combine like terms.

$= -2r^4 - 12r^3 - 24r^2 - 16r$ — Distributive property

Work Problem 4 at the Side. ▶

4 Find each product.

GS **(a)** $(m + 1)^3$

$= (\_\_\_\_\_)^2(\_\_\_\_\_)$

$= (\_\_\_\_\_\_\_)(m + 1)$

$= \_\_\_\_\_\_\_\_\_$

(b) $(3k - 2)^4$

(c) $-3x(x - 4)^3$

Answers

4. **(a)** $m + 1$; $m + 1$; $m^2 + 2m + 1$; $m^3 + 3m^2 + 3m + 1$
(b) $81k^4 - 216k^3 + 216k^2 - 96k + 16$
(c) $-3x^4 + 36x^3 - 144x^2 + 192x$

12.4 Exercises

FOR EXTRA HELP

CONCEPT CHECK *Fill in each blank with the correct response.*

1. The square of a binomial is a trinomial consisting of the _______ of the first term + _______ the _______ of the two terms + the _______ of the last term.

2. In the square of a sum, all of the terms are _______, while in the square of a difference, the middle term is _______.

3. The product of the sum and difference of two terms is the _______ of the _______ of the two terms.

4. The word ________ describes two expressions that represent the sum and the difference of the same two terms.

5. Consider the binomial square $(2x + 3)^2$.

(a) What is the simplest form of the square of the first term, $(2x)^2$?

(b) What is the simplest form of twice the product of the two terms, $2(2x)(3)$?

(c) What is the simplest form of the square of the last term, 3^2?

(d) Write the final product, which is a trinomial, using your results from parts (a)–(c).

6. Repeat **Exercise 5** for the binomial square $(3x - 2)^2$.

Square each binomial. **See Examples 1 and 2.**

7. $(p + 2)^2$ **8.** $(r + 5)^2$ **9.** $(z - 5)^2$ **10.** $(x - 3)^2$

11. $\left(x - \frac{3}{4}\right)^2$ **12.** $\left(y + \frac{5}{8}\right)^2$ **13.** $(v + 0.4)^2$ **14.** $(w - 0.9)^2$

15. $(4x - 3)^2$ **16.** $(9y - 4)^2$ **17.** $(10z + 6)^2$ **18.** $(5y + 2)^2$

19. $(2p + 5q)^2$ **20.** $(8a - 3b)^2$ **21.** $(0.8t + 0.7s)^2$ **22.** $(0.7z - 0.3w)^2$

23. $\left(5x + \frac{2}{5}y\right)^2$ **24.** $\left(6m - \frac{4}{5}n\right)^2$ **25.** $t(3t - 1)^2$ **26.** $x(2x + 5)^2$

27. $3t(4t + 1)^2$ **28.** $2x(7x - 2)^2$ **29.** $-(4r - 2)^2$ **30.** $-(3y - 8)^2$

31. Consider the product $(7x + 3y)(7x - 3y)$.

(a) What is the simplest form of the product of the first terms, $7x(7x)$?

(b) Multiply the outer terms, $7x(-3y)$. Then multiply the inner terms, $3y(7x)$. Add the results. What is this sum?

(c) What is the simplest form of the product of the last terms, $3y(-3y)$?

(d) Write the final product using your answers in parts (a) and (c). Why is the sum found in part (b) omitted here?

32. Repeat **Exercise 31** for the product $(5x + 7y)(5x - 7y)$.

Find each product. ***See Examples 3 and 4.***

33. $(q + 2)(q - 2)$

34. $(x + 8)(x - 8)$

35. $\left(r - \frac{3}{4}\right)\left(r + \frac{3}{4}\right)$

36. $\left(q - \frac{7}{8}\right)\left(q + \frac{7}{8}\right)$

37. $(s + 2.5)(s - 2.5)$

38. $(t + 1.4)(t - 1.4)$

39. $(2w + 5)(2w - 5)$

40. $(3z + 8)(3z - 8)$

41. $(10x + 3y)(10x - 3y)$

42. $(13r + 2z)(13r - 2z)$

43. $(2x^2 - 5)(2x^2 + 5)$

44. $(9y^2 - 2)(9y^2 + 2)$

45. $\left(7x + \frac{3}{7}\right)\left(7x - \frac{3}{7}\right)$

46. $\left(9y + \frac{2}{3}\right)\left(9y - \frac{2}{3}\right)$

47. $p(3p + 7)(3p - 7)$

48. $q(5q - 1)(5q + 1)$

Find each product. ***See Example 5.***

49. $(m - 5)^3$

50. $(p + 3)^3$

51. $(y + 2)^3$

52. $(x - 7)^3$

53. $(2a + 1)^3$

54. $(3m - 1)^3$

55. $(3r - 2t)^4$

56. $(2z + 5y)^4$

57. $3x^2(x - 3)^3$

58. $4p^3(p + 4)^3$

59. $-8x^2y(x + y)^4$

60. $-5uv^2(u - v)^4$

Determine a polynomial that represents the area of each figure. In Exercise 65, leave π in your answer. (If necessary, refer to the formulas inside the back cover.)

61.

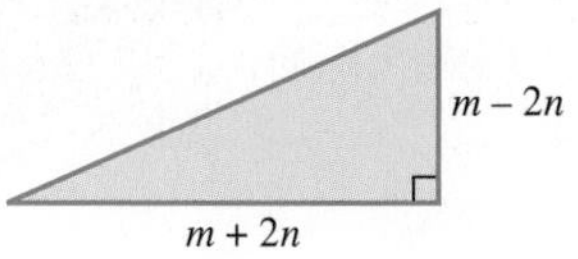

62.

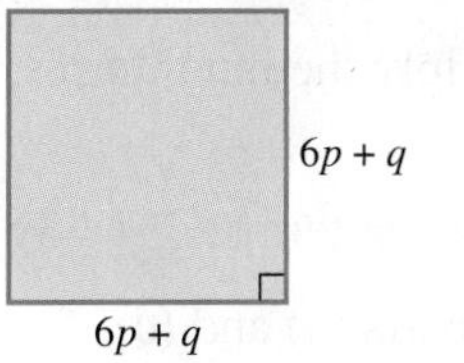

63.

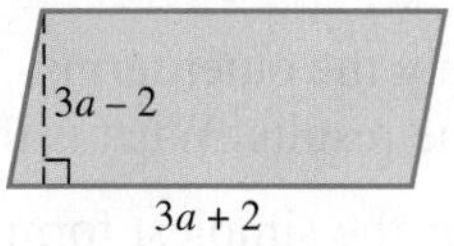

64.

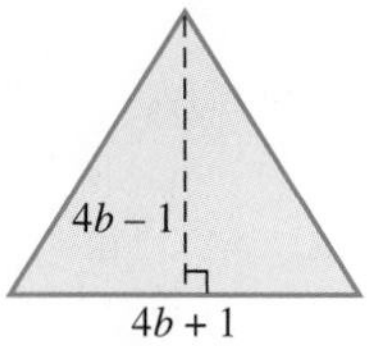

65.

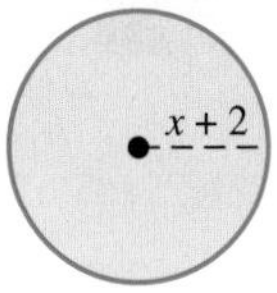

66.

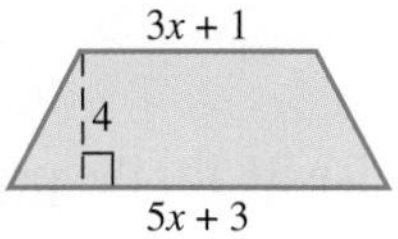

In Exercises 67 and 68, refer to the figure shown here.

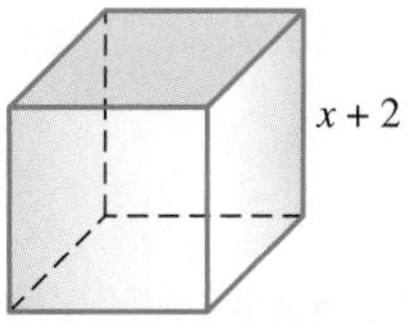

67. Find a polynomial that represents the volume of the cube (in cubic units).

68. If the value of x is 6, what is the volume of the cube (in cubic units)?

Relating Concepts (Exercises 69–78) For Individual or Group Work

Special products can be illustrated by using areas of rectangles. Use the figure and ***work Exercises 69–74 in order,*** *to justify the special product* $(a + b)^2 = a^2 + 2ab + b^2$.

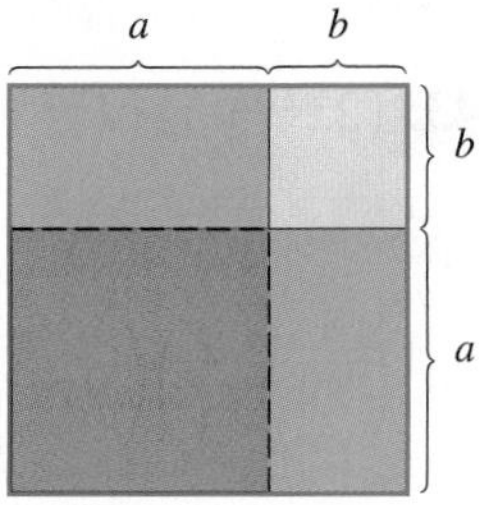

69. Express the area of the large square as the square of a binomial.

70. Give the monomial that represents the area of the red square.

71. Give the monomial that represents the sum of the areas of the blue rectangles.

72. Give the monomial that represents the area of the yellow square.

73. What is the sum of the monomials you obtained in **Exercises 70–72?**

74. Explain why the binomial square found in **Exercise 69** must equal the polynomial found in **Exercise 73.**

To understand how the special product $(a + b)^2 = a^2 + 2ab + b^2$ *can be applied to a purely numerical problem,* ***work Exercises 75–78 in order.***

75. Evaluate 35^2 using either traditional paper-and-pencil methods or a calculator.

76. The number 35 can be written as $30 + 5$. Therefore, $35^2 = (30 + 5)^2$. Use the special product for squaring a binomial with $a = 30$ and $b = 5$ to write an expression for $(30 + 5)^2$. Do not simplify at this time.

77. Use the order of operations to simplify the expression found in **Exercise 76.**

78. How do the answers in **Exercises 75 and 77** compare?

12.5 Integer Exponents and the Quotient Rule

OBJECTIVES

1. Use 0 as an exponent.
2. Use negative numbers as exponents.
3. Use the quotient rule for exponents.
4. Use combinations of rules.

Consider the following list.

$$2^4 = 16$$
$$2^3 = 8$$
$$2^2 = 4$$

Each time we decrease the exponent by 1, the value is divided by 2 (the base). Using this pattern, we can continue the list to lesser and lesser integer exponents.

$$2^1 = 2$$
$$2^0 = 1$$
$$2^{-1} = \frac{1}{2}$$

Work Problem 1 at the Side. ▶

From the preceding list and the answers to **Margin Problem 1,** it appears that we should define 2^0 as 1 and negative exponents as reciprocals.

OBJECTIVE 1 Use 0 as an exponent. We want the definitions of 0 and negative exponents to satisfy the rules for exponents from **Section 12.2.** For example, if $6^0 = 1$,

$$6^0 \cdot 6^2 = 1 \cdot 6^2 = 6^2 \quad \text{and} \quad 6^0 \cdot 6^2 = 6^{0+2} = 6^2,$$

so the product rule is satisfied. The power rules are also valid for a 0 exponent. Thus, we define a 0 exponent as follows.

Zero Exponent

For any nonzero real number a, $\mathbf{a^0 = 1.}$

Example: $17^0 = 1$

EXAMPLE 1 Using Zero Exponents

Evaluate.

(a) $60^0 = 1$

(b) $(-60)^0 = 1$

(c) $-60^0 = -(1)$, or -1

(d) $y^0 = 1 \quad (y \neq 0)$

(e) $6y^0 = 6(1)$, or $6 \quad (y \neq 0)$

(f) $(6y)^0 = 1 \quad (y \neq 0)$

Work Problem 2 at the Side. ▶

CAUTION

Look again at **Examples 1(b) and 1(c).** In $(-60)^0$, the base is -60, and since any nonzero base raised to the 0 exponent is 1, $(-60)^0 = 1$. In -60^0, which can be written $-(60)^0$, the base is 60, so $-60^0 = -1$.

1 Continue the list of exponentials using -2, -3, and -4 as exponents.

$2^{-2} =$ ____

$2^{-3} =$ ____

$2^{-4} =$ ____

2 Evaluate.

(a) 28^0

(b) $(-16)^0$

(c) -7^0

(d) $m^0 \quad (m \neq 0)$

(e) $-2p^0 \quad (p \neq 0)$

(f) $(5r)^0 \quad (r \neq 0)$

Answers

1. $2^{-2} = \frac{1}{4}$; $2^{-3} = \frac{1}{8}$; $2^{-4} = \frac{1}{16}$

2. (a) 1 (b) 1 (c) -1 (d) 1 (e) -2 (f) 1

OBJECTIVE 2 **Use negative numbers as exponents.** Review the lists at the beginning of this section and **Margin Problem 1.** Since $2^{-2} = \frac{1}{4}$ and $2^{-3} = \frac{1}{8}$, we can think that 2^{-n} should perhaps equal $\frac{1}{2^n}$. Is the product rule valid in such cases? For example, if we multiply 6^{-2} by 6^2, we get

$$6^{-2} \cdot 6^2 = 6^{-2+2} = 6^0 = 1.$$

The expression 6^{-2} behaves as if it were the reciprocal of 6^2, because their product is 1. The reciprocal of 6^2 may be written $\frac{1}{6^2}$, leading us to define 6^{-2} as $\frac{1}{6^2}$, and generalize accordingly.

Negative Exponents

For any nonzero real number a and any integer n, $\quad \boldsymbol{a^{-n} = \frac{1}{a^n}}.$

Example: $3^{-2} = \frac{1}{3^2}$

By definition, a^{-n} and a^n are reciprocals, since

$$a^n \cdot a^{-n} = a^n \cdot \frac{1}{a^n} = 1.$$

Since $1^n = 1$, the definition of a^{-n} can also be written

$$a^{-n} = \frac{1}{a^n} = \frac{1^n}{a^n} = \left(\frac{1}{a}\right)^n.$$

For example, $\quad 6^{-3} = \left(\frac{1}{6}\right)^3 \quad$ and $\quad \left(\frac{1}{3}\right)^{-2} = 3^2.$

EXAMPLE 2 Using Negative Exponents

Simplify by writing with positive exponents. Assume that all variables represent nonzero real numbers.

(a) $3^{-2} = \frac{1}{3^2}$, or $\frac{1}{9}$ $\quad a^{-n} = \frac{1}{a^n}$

(b) $5^{-3} = \frac{1}{5^3}$, or $\frac{1}{125}$ $\quad a^{-n} = \frac{1}{a^n}$

(c) $\left(\frac{1}{2}\right)^{-3} = 2^3$, or 8 $\quad \frac{1}{2}$ and 2 are reciprocals.

Notice that we can change the base to its reciprocal if we also change the sign of the exponent.

(d) $\left(\frac{2}{5}\right)^{-4}$

$= \left(\frac{5}{2}\right)^4$ $\quad \frac{2}{5}$ and $\frac{5}{2}$ are reciprocals.

$= \frac{5^4}{2^4}$ $\quad$ Power rule (c)

$= \frac{625}{16}$ $\quad$ Apply the exponents.

(e) $\left(\frac{4}{3}\right)^{-5}$

$= \left(\frac{3}{4}\right)^5$ $\quad \frac{4}{3}$ and $\frac{3}{4}$ are reciprocals.

$= \frac{3^5}{4^5}$ $\quad$ Power rule (c)

$= \frac{243}{1024}$ $\quad$ Apply the exponents.

Continued on Next Page

(f) $4^{-1} - 2^{-1}$

$= \frac{1}{4} - \frac{1}{2}$ Apply the exponents.

$= \frac{1}{4} - \frac{2}{4}$ Find a common denominator.

$= -\frac{1}{4}$ Subtract.

(g) $3p^{-2}$

$= \frac{3}{1} \cdot \frac{1}{p^2}$ $a^{-n} = \frac{1}{a^n}$

$= \frac{3}{p^2}$ Multiply.

(h) $\frac{1}{x^{-4}}$

$= \frac{1^{-4}}{x^{-4}}$ $1^n = 1$, for any integer n

$= \left(\frac{1}{x}\right)^{-4}$ Power rule (c)

$= x^4$ $\frac{1}{x}$ and x are reciprocals.

(i) x^3y^{-4}

$= \frac{x^3}{1} \cdot \frac{1}{y^4}$ $a^{-n} = \frac{1}{a^n}$

$= \frac{x^3}{y^4}$ Multiply.

In general, $\frac{1}{a^{-n}} = a^n$**.**

CAUTION

A negative exponent does not indicate a negative number. Negative exponents lead to reciprocals.

Expression	**Example**	
a^{-n}	$3^{-2} = \frac{1}{3^2} = \frac{1}{9}$	Not negative
$-a^{-n}$	$-3^{-2} = -\frac{1}{3^2} = -\frac{1}{9}$	Negative

Work Problem 3 at the Side. ▶

Consider the following.

$$\frac{2^{-3}}{3^{-4}}$$

$$= \frac{\frac{1}{2^3}}{\frac{1}{3^4}} \quad \text{Definition of negative exponent}$$

$$= \frac{1}{2^3} \div \frac{1}{3^4} \quad \frac{a}{b} \text{ means } a \div b.$$

$$= \frac{1}{2^3} \cdot \frac{3^4}{1} \quad \text{To divide, multiply by the reciprocal of the divisor.}$$

$$= \frac{3^4}{2^3} \quad \text{Multiply.}$$

Therefore, $\frac{2^{-3}}{3^{-4}} = \frac{3^4}{2^3}$.

3 Simplify by writing with positive exponents. Assume that all variables represent nonzero real numbers.

(a) 4^{-3}

(b) 6^{-2}

(c) $\left(\frac{1}{4}\right)^{-2}$

(d) $\left(\frac{2}{3}\right)^{-2}$

(e) $2^{-1} + 5^{-1}$

(f) $7m^{-5}$

(g) $\frac{1}{z^{-6}}$

(h) p^2q^{-5}

Answers

3. **(a)** $\frac{1}{4^3}$, or $\frac{1}{64}$ **(b)** $\frac{1}{6^2}$, or $\frac{1}{36}$
(c) 4^2, or 16 **(d)** $\left(\frac{3}{2}\right)^2$, or $\frac{9}{4}$
(e) $\frac{1}{2} + \frac{1}{5} = \frac{7}{10}$ **(f)** $\frac{7}{m^5}$ **(g)** z^6 **(h)** $\frac{p^2}{q^5}$

4 Simplify. Assume that all variables represent nonzero real numbers.

GS **(a)** $\dfrac{7^{-1}}{5^{-4}}$

$= \dfrac{5\_\_}{7\_\_}$

$= \_\_\_\_$

(b) $\dfrac{x^{-3}}{y^{-2}}$

(c) $\dfrac{4h^{-5}}{m^{-2}k}$

GS **(d)** $\left(\dfrac{3m}{p}\right)^{-2}$

$= \left(\dfrac{p}{3m}\right)^{\_\_}$

$= \dfrac{p\_\_}{3\_\_m\_\_}$

$= \_\_\_\_$

Answers

4. **(a)** 4; 1; $\dfrac{625}{7}$ **(b)** $\dfrac{y^2}{x^3}$ **(c)** $\dfrac{4m^2}{h^5k}$

(d) 2; 2; 2; 2; $\dfrac{p^2}{9m^2}$

Changing from Negative to Positive Exponents

For any nonzero numbers a and b, and any integers m and n,

$$\frac{a^{-m}}{b^{-n}} = \frac{b^n}{a^m} \quad \text{and} \quad \left(\frac{a}{b}\right)^{-m} = \left(\frac{b}{a}\right)^m.$$

Examples: $\dfrac{3^{-5}}{2^{-4}} = \dfrac{2^4}{3^5}$ and $\left(\dfrac{4}{5}\right)^{-3} = \left(\dfrac{5}{4}\right)^3$

EXAMPLE 3 **Changing from Negative to Positive Exponents**

Simplify. Assume that all variables represent nonzero real numbers.

(a) $\dfrac{4^{-2}}{5^{-3}} = \dfrac{5^3}{4^2}$, or $\dfrac{125}{16}$

(b) $\dfrac{m^{-5}}{p^{-1}} = \dfrac{p^1}{m^5}$, or $\dfrac{p}{m^5}$

(c) $\dfrac{a^{-2}b}{3d^{-3}} = \dfrac{bd^3}{3a^2}$ Notice that b in the numerator and the coefficient 3 in the denominator are not affected.

(d) $\left(\dfrac{x}{2y}\right)^{-4}$

$= \left(\dfrac{2y}{x}\right)^4$ Negative-to-positive rule

$= \dfrac{2^4y^4}{x^4}$ Power rules (b) and (c)

$= \dfrac{16y^4}{x^4}$ Apply the exponent.

◀ Work Problem **4** at the Side.

CAUTION

Be careful. We cannot use the rule $\dfrac{a^{-m}}{b^{-n}} = \dfrac{b^n}{a^m}$ to change negative exponents to positive exponents if the exponents occur in a *sum* or *difference* of terms. For example,

$$\frac{5^{-2} + 3^{-1}}{7 - 2^{-3}} \text{ would be written with positive exponents as } \frac{\frac{1}{5^2} + \frac{1}{3}}{7 - \frac{1}{2^3}}.$$

OBJECTIVE 3 **Use the quotient rule for exponents.** Consider a quotient of two exponential expressions with the same base.

$$\frac{6^5}{6^3} = \frac{6 \cdot 6 \cdot 6 \cdot 6 \cdot 6}{6 \cdot 6 \cdot 6} = 6^2$$

The difference between the exponents, $5 - 3 = 2$, is the exponent in the quotient. Also,

$$\frac{6^2}{6^4} = \frac{6 \cdot 6}{6 \cdot 6 \cdot 6 \cdot 6} = \frac{1}{6^2} = 6^{-2}.$$

Here, $2 - 4 = -2$. These examples suggest the **quotient rule for exponents.**

Quotient Rule for Exponents

For any nonzero real number a and any integers m and n,

$$\frac{a^m}{a^n} = a^{m-n}.$$

(Keep the same base and subtract the exponents.)

Example: $\frac{5^8}{5^4} = 5^{8-4} = 5^4$

CAUTION

A common **error** is to write $\frac{5^8}{5^4} = 1^{8-4} = 1^4$. ***This is incorrect.*** By the quotient rule, the quotient must have the *same base*, which is 5 here.

$$\frac{5^8}{5^4} = 5^{8-4} = 5^4$$

We can confirm this by writing out the factors.

$$\frac{5^8}{5^4} = \frac{5 \cdot 5 \cdot 5 \cdot 5 \cdot 5 \cdot 5 \cdot 5 \cdot 5}{5 \cdot 5 \cdot 5 \cdot 5} = 5^4$$

EXAMPLE 4 **Using the Quotient Rule**

Simplify. Assume that all variables represent nonzero real numbers.

(a) $\frac{5^8}{5^6} = 5^{8-6} = 5^2 = 25$ (Keep the same base.)

(b) $\frac{4^2}{4^9} = 4^{2-9} = 4^{-7} = \frac{1}{4^7}$

(c) $\frac{5^{-3}}{5^{-7}} = 5^{-3-(-7)} = 5^4 = 625$ (Be careful with signs.)

(d) $\frac{q^5}{q^{-3}} = q^{5-(-3)} = q^8$

(e) $\frac{3^2x^5}{3^4x^3}$

$= \frac{3^2}{3^4} \cdot \frac{x^5}{x^3}$

$= 3^{2-4} \cdot x^{5-3}$ Quotient rule

$= 3^{-2}x^2$ Subtract.

$= \frac{x^2}{3^2}$, or $\frac{x^2}{9}$

(f) $\frac{(m+n)^{-2}}{(m+n)^{-4}}$

$= (m+n)^{-2-(-4)}$

$= (m+n)^{-2+4}$

$= (m+n)^2 \quad (m \neq -n)$

The restriction $m \neq -n$ is necessary to prevent a denominator of 0 in the original expression. Division by 0 is undefined.

(g) $\frac{7x^{-3}y^2}{2^{-1}x^2y^{-5}}$

$= \frac{7 \cdot 2^1y^2y^5}{x^2x^3}$ Definition of negative exponent

$= \frac{14y^7}{x^5}$ Multiply; product rule

Work Problem 5 at the Side. ▶

5 Simplify. Assume that all variables represent nonzero real numbers.

GS (a) $\frac{5^{11}}{5^8}$

$= 5^{\_\_-\_\_}$

$= 5^{\_\_}$

$= \_\_\_\_$

(b) $\frac{4^7}{4^{10}}$

GS (c) $\frac{6^{-5}}{6^{-2}}$

$= 6^{\_\_-(\_\_)}$

$= 6^{\_\_}$

$= \frac{1}{6^{\_\_}}$

$= \_\_\_\_$

(d) $\frac{8^4m^9}{8^5m^{10}}$

(e) $\frac{3^{-1}(x+y)^{-3}}{2^{-2}(x+y)^{-4}} \quad (x \neq -y)$

Answers

5. (a) 11; 8; 3; 125 (b) $\frac{1}{64}$
(c) -5; -2; -3; 3; $\frac{1}{216}$
(d) $\frac{1}{8m}$ (e) $\frac{4}{3}(x+y)$

The definitions and rules for exponents given in this section and **Section 12.2** are summarized here.

Definitions and Rules for Exponents

For any integers m and n, the following are true.

		Examples
Product rule	$a^m \cdot a^n = a^{m+n}$	$7^4 \cdot 7^5 = 7^{4+5} = 7^9$
Zero exponent	$a^0 = 1 \quad (a \neq 0)$	$(-3)^0 = 1$
Negative exponent	$a^{-n} = \frac{1}{a^n} \quad (a \neq 0)$	$5^{-3} = \frac{1}{5^3}$
Quotient rule	$\frac{a^m}{a^n} = a^{m-n} \quad (a \neq 0)$	$\frac{2^2}{2^5} = 2^{2-5} = 2^{-3} = \frac{1}{2^3}$
Power rules (a)	$(a^m)^n = a^{mn}$	$(4^2)^3 = 4^{2 \cdot 3} = 4^6$
(b)	$(ab)^m = a^m b^m$	$(3k)^4 = 3^4 k^4$
(c)	$\left(\frac{a}{b}\right)^m = \frac{a^m}{b^m} \quad (b \neq 0)$	$\left(\frac{2}{3}\right)^2 = \frac{2^2}{3^2}$
Negative-to-positive rules	$\frac{a^{-m}}{b^{-n}} = \frac{b^n}{a^m} \quad (a, b \neq 0)$	$\frac{2^{-4}}{5^{-3}} = \frac{5^3}{2^4}$
	$\left(\frac{a}{b}\right)^{-m} = \left(\frac{b}{a}\right)^m$	$\left(\frac{4}{7}\right)^{-2} = \left(\frac{7}{4}\right)^2$

OBJECTIVE 4 **Use combinations of rules.** We sometimes need to use more than one rule to simplify an expression.

EXAMPLE 5 Using a Combination of Rules

Simplify each expression. Assume that all variables represent nonzero real numbers.

(a) $\frac{(4^2)^3}{4^5}$

$= \frac{4^6}{4^5}$ Power rule (a)

$= 4^{6-5}$ Quotient rule

$= 4^1$ Subtract the exponents.

$= 4$ $4^1 = 4$

(b) $(2x)^3(2x)^2$

$= (2x)^5$ Product rule

$= 2^5 x^5$ Power rule (b)

$= 32x^5$ $2^5 = 32$

Continued on Next Page

(c) $\left(\frac{2x^3}{5}\right)^{-4}$

$= \left(\frac{5}{2x^3}\right)^4$ Negative-to-positive rule

$= \frac{5^4}{2^4x^{12}}$ Power rules (a)–(c)

$= \frac{625}{16x^{12}}$ Apply the exponents.

(d) $\left(\frac{3x^{-2}}{4^{-1}y^3}\right)^{-3}$

$= \frac{3^{-3}x^6}{4^3y^{-9}}$ Power rules (a)–(c)

$= \frac{x^6y^9}{4^3 \cdot 3^3}$ Negative-to-positive rule

$= \frac{x^6y^9}{1728}$ $4^3 \cdot 3^3 = 64 \cdot 27 = 1728$

(e) $\frac{(4m)^{-3}}{(3m)^{-4}}$

$= \frac{4^{-3}m^{-3}}{3^{-4}m^{-4}}$ Power rule (b)

$= \frac{3^4m^4}{4^3m^3}$ Negative-to-positive rule

$= \frac{3^4m^{4-3}}{4^3}$ Quotient rule

$= \frac{3^4m}{4^3}$ Subtract.

$= \frac{81m}{64}$ Apply the exponents.

Note

Since the steps can be done in several different orders, there are many equally correct ways to simplify expressions like those in **Examples 5(c) through 5(e).**

Work Problem 6 at the Side. ▶

6 Simplify each expression. Assume that all variables represent nonzero real numbers.

(a) $\frac{(3^4)^2}{3^3}$

(b) $(4x)^2(4x)^4$

(c) $\left(\frac{5}{2z^4}\right)^{-3}$

(d) $\left(\frac{6y^{-4}}{7^{-1}z^5}\right)^{-2}$

(e) $\frac{(6x)^{-1}}{(3x^2)^{-2}}$

Answers

6. **(a)** 243 **(b)** 4^6x^6, or $4096x^6$ **(c)** $\frac{8z^{12}}{125}$ **(d)** $\frac{y^8z^{10}}{1764}$ **(e)** $\frac{3x^3}{2}$

12.5 Exercises

FOR EXTRA HELP

Download the MyDashBoard App

MyMathLab®

CONCEPT CHECK *Decide whether each expression is positive, negative, or 0.*

1. $(-2)^{-3}$ **2.** $(-3)^{-2}$ **3.** -2^4 **4.** -3^6

5. $\left(\frac{1}{4}\right)^{-2}$ **6.** $\left(\frac{1}{5}\right)^{-2}$ **7.** $1 - 5^0$ **8.** $1 - 7^0$

CONCEPT CHECK *In Exercises 9 and 10, match each expression in Column I with the equivalent expression in Column II. Choices in Column II may be used once, more than once, or not at all. (In Exercise 9, $x \neq 0$.)*

	I	II
9.	**(a)** x^0	**A.** 0
	(b) $-x^0$	**B.** 1
	(c) $7x^0$	**C.** -1
	(d) $(7x)^0$	**D.** 7
	(e) $-7x^0$	**E.** -7
	(f) $(-7x)^0$	**F.** $\frac{1}{7}$

	I	II
10.	**(a)** -2^{-4}	**A.** 8
	(b) $(-2)^{-4}$	**B.** 16
	(c) 2^{-4}	**C.** $-\frac{1}{16}$
	(d) $\frac{1}{2^{-4}}$	**D.** -8
	(e) $\frac{1}{-2^{-4}}$	**E.** -16
	(f) $\frac{1}{(-2)^{-4}}$	**F.** $\frac{1}{16}$

Decide whether each expression is equal to 0, 1, or −1. ***See Example 1.***

11. 9^0 **12.** 5^0 **13.** $(-4)^0$ **14.** $(-10)^0$ **15.** -9^0

16. -5^0 **17.** $(-2)^0 - 2^0$ **18.** $(-8)^0 - 8^0$ **19.** $\frac{0^{10}}{10^0}$ **20.** $\frac{0^5}{5^0}$

Evaluate each expression. ***See Examples 1 and 2.***

21. $7^0 + 9^0$ **22.** $8^0 + 6^0$ **23.** 4^{-3} **24.** 5^{-4} **25.** $\left(\frac{1}{2}\right)^{-4}$

26. $\left(\frac{1}{3}\right)^{-3}$ **27.** $\left(\frac{6}{7}\right)^{-2}$ **28.** $\left(\frac{2}{3}\right)^{-3}$ **29.** $(-3)^{-4}$ **30.** $(-4)^{-3}$

31. $5^{-1} + 3^{-1}$ **32.** $6^{-1} + 2^{-1}$ **33.** $-2^{-1} + 3^{-2}$ **34.** $(-3)^{-2} + (-4)^{-1}$

Simplify by writing each expression with positive exponents. Assume that all variables represent nonzero real numbers. **See Examples 2–4.**

35. $\frac{9^4}{9^5}$ **36.** $\frac{7^3}{7^4}$ **37.** $\frac{6^{-3}}{6^2}$ **38.** $\frac{4^{-2}}{4^3}$ **39.** $\frac{1}{6^{-3}}$

40. $\frac{1}{5^{-2}}$ **41.** $\frac{2}{r^{-4}}$ **42.** $\frac{3}{s^{-8}}$ **43.** $\frac{4^{-3}}{5^{-2}}$ **44.** $\frac{6^{-2}}{5^{-4}}$

45. p^5q^{-8} **46.** $x^{-8}y^4$ **47.** $\frac{r^5}{r^{-4}}$ **48.** $\frac{a^6}{a^{-4}}$ **49.** $\frac{6^4x^8}{6^5x^3}$

50. $\frac{3^8y^5}{3^{10}y^2}$ **51.** $\frac{6y^3}{2y}$ **52.** $\frac{5m^2}{m}$ **53.** $\frac{3x^5}{3x^2}$ **54.** $\frac{10p^8}{2p^4}$

55. $\frac{x^{-3}y}{4z^{-2}}$ **56.** $\frac{p^{-5}q^{-4}}{9r^{-3}}$ **57.** $\frac{(a+b)^{-3}}{(a+b)^{-4}}$ **58.** $\frac{(x+y)^{-8}}{(x+y)^{-9}}$

Simplify by writing each expression with positive exponents. Assume that all variables represent nonzero real numbers. **See Example 5.**

59. $\frac{(7^4)^3}{7^9}$ **60.** $\frac{(5^3)^2}{5^2}$ **61.** $x^{-3} \cdot x^5 \cdot x^{-4}$ **62.** $y^{-8} \cdot y^5 \cdot y^{-2}$

63. $\frac{(3x)^{-2}}{(4x)^{-3}}$ **64.** $\frac{(2y)^{-3}}{(5y)^{-4}}$ **65.** $\left(\frac{x^{-1}y}{z^2}\right)^{-2}$ **66.** $\left(\frac{p^{-4}q}{r^{-3}}\right)^{-3}$

67. $(6x)^4(6x)^{-3}$ **68.** $(10y)^9(10y)^{-8}$ **69.** $\frac{(m^7n)^{-2}}{m^{-4}n^3}$ **70.** $\frac{(m^8n^{-4})^2}{m^{-2}n^5}$

71. $\frac{5x^{-3}}{(4x)^2}$ **72.** $\frac{-3k^5}{(2k)^2}$ **73.** $\left(\frac{2p^{-1}q}{3^{-1}m^2}\right)^2$ **74.** $\left(\frac{4xy^2}{x^{-1}y}\right)^{-2}$

75. CONCEPT CHECK A student simplified $\frac{16^3}{2^2}$ as shown.

$$\frac{16^3}{2^2} = \left(\frac{16}{2}\right)^{3-2} = 8^1 = 8$$

What Went Wrong? Give the correct answer.

76. CONCEPT CHECK A student simplified 5^{-4} as shown.

$$5^{-4} = -5^4 = -625$$

What Went Wrong? Give the correct answer.

12.6 Dividing a Polynomial by a Monomial

OBJECTIVE

1 Divide a polynomial by a monomial.

OBJECTIVE 1 Divide a polynomial by a monomial. We add two fractions with a common denominator as follows.

$$\frac{a}{c}+\frac{b}{c}=\frac{a+b}{c}$$

In reverse, this statement gives a rule for dividing a polynomial by a monomial.

Dividing a Polynomial by a Monomial

To divide a polynomial by a monomial, divide each term of the polynomial by the monomial.

$$\frac{a+b}{c}=\frac{a}{c}+\frac{b}{c} \quad (\text{where } c \neq 0)$$

Examples: $\frac{2+5}{3}=\frac{2}{3}+\frac{5}{3}$ and $\frac{x+3z}{2y}=\frac{x}{2y}+\frac{3z}{2y}$ $(y \neq 0)$

The parts of a division problem are named here.

$$\text{Dividend} \rightarrow \frac{12x^2+6x}{6x} = 2x+1 \leftarrow \text{Quotient}$$
$$\text{Divisor} \rightarrow 6x$$

EXAMPLE 1 Dividing a Polynomial by a Monomial

Divide $5m^5 - 10m^3$ by $5m^2$.

$$\frac{5m^5-10m^3}{5m^2}$$

$$=\frac{5m^5}{5m^2}-\frac{10m^3}{5m^2}$$ Use the preceding rule, with + replaced by −.

$$= m^3 - 2m$$ Quotient rule

CHECK Multiply. $5m^2 \cdot (m^3 - 2m) = 5m^5 - 10m^3$ ✓

(Divisor: $5m^2$; Quotient: m^3-2m; Original polynomial (Dividend): $5m^5-10m^3$)

Because division by 0 is undefined, the quotient $\frac{5m^5-10m^3}{5m^2}$ is undefined if $m = 0$. From now on, we assume that no denominators are 0.

◀ Work Problem 1 at the Side.

EXAMPLE 2 Dividing a Polynomial by a Monomial

Divide.

$$\frac{16a^5-12a^4+8a^2}{4a^3}$$

$$=\frac{16a^5}{4a^3}-\frac{12a^4}{4a^3}+\frac{8a^2}{4a^3}$$ Divide each term by $4a^3$.

This becomes $\frac{2}{a}$, not $2a$.

$$=4a^2-3a+\frac{2}{a}$$ Quotient rule

Continued on Next Page

1 Divide.

(a) $\frac{6p^4+18p^7}{3p^2}$

$= \frac{\quad}{3p^2} + \frac{\quad}{3p^2}$

$=$ ______

(b) $\frac{12m^6+18m^5+30m^4}{6m^2}$

(c) $(18r^7 - 9r^2) \div (3r)$

Answers

1. (a) $6p^4$; $18p^7$; $2p^2 + 6p^5$
(b) $2m^4 + 3m^3 + 5m^2$ (c) $6r^6 - 3r$

The quotient $4a^2 - 3a + \frac{2}{a}$ is *not* a polynomial because of the expression $\frac{2}{a}$, which has a variable in the denominator. While the sum, difference, and product of two polynomials are always polynomials, the quotient of two polynomials may not be.

CHECK $4a^3\left(4a^2 - 3a + \dfrac{2}{a}\right)$ Divisor × Quotient should equal Dividend.

$= 4a^3(4a^2) + 4a^3(-3a) + 4a^3\left(\dfrac{2}{a}\right)$ Distributive property

$= 16a^5 - 12a^4 + 8a^2$ ✓ Dividend

Work Problem 2 at the Side. ▶

EXAMPLE 3 Dividing a Polynomial by a Monomial with a Negative Coefficient

Divide $-7x^3 + 12x^4 - 4x$ by $-4x$.

Write the polynomial in descending powers as $12x^4 - 7x^3 - 4x$ before dividing.

Write in descending powers.

$$\frac{12x^4 - 7x^3 - 4x}{-4x}$$

$$= \frac{12x^4}{-4x} - \frac{7x^3}{-4x} - \frac{4x}{-4x}$$ Divide each term by $-4x$.

$$= -3x^3 - \frac{7x^2}{-4} - (-1)$$ Quotient rule

$$= -3x^3 + \frac{7}{4}x^2 + 1$$ Be sure to include the 1 in the answer.

CHECK $-4x\left(-3x^3 + \dfrac{7}{4}x^2 + 1\right)$ Divisor × Quotient should equal Dividend.

$= -4x(-3x^3) - 4x\left(\dfrac{7}{4}x^2\right) - 4x(1)$ Distributive property

$= 12x^4 - 7x^3 - 4x$ ✓ Dividend

Work Problem 3 at the Side. ▶

EXAMPLE 4 Dividing a Polynomial by a Monomial

Divide $180x^4y^{10} - 150x^3y^8 + 120x^2y^6 - 90xy^4 + 100y$ by $-30xy^2$.

$$\frac{180x^4y^{10} - 150x^3y^8 + 120x^2y^6 - 90xy^4 + 100y}{-30xy^2}$$

$$= \frac{180x^4y^{10}}{-30xy^2} - \frac{150x^3y^8}{-30xy^2} + \frac{120x^2y^6}{-30xy^2} - \frac{90xy^4}{-30xy^2} + \frac{100y}{-30xy^2}$$

$$= -6x^3y^8 + 5x^2y^6 - 4xy^4 + 3y^2 - \frac{10}{3xy}$$

Check by multiplying the quotient by the divisor.

Work Problem 4 at the Side. ▶

2 Divide.

GS (a) $\dfrac{20x^4 - 25x^3 + 5x}{5x^2}$

$= \dfrac{\_\_}{\_\_} - \dfrac{\_\_}{\_\_}$

$+ \dfrac{\_\_}{\_\_}$

$= \_\_\_\_\_\_$

(b) $\dfrac{50m^4 - 30m^3 + 20m}{10m^3}$

3 Divide.

(a) $\dfrac{-9y^6 + 8y^7 - 11y - 4}{y^2}$

(b) $\dfrac{-8p^4 - 6p^3 - 12p^5}{-3p^3}$

4 Divide.

$\dfrac{45x^4y^3 + 30x^3y^2 - 60x^2y}{-15x^2y}$

Answers

2. (a) The complete expression is $\dfrac{20x^4}{5x^2} - \dfrac{25x^3}{5x^2} + \dfrac{5x}{5x^2}$; $4x^2 - 5x + \dfrac{1}{x}$

(b) $5m - 3 + \dfrac{2}{m^2}$

3. (a) $8y^5 - 9y^4 - \dfrac{11}{y} - \dfrac{4}{y^2}$

(b) $4p^2 + \dfrac{8p}{3} + 2$

4. $-3x^2y^2 - 2xy + 4$

12.6 Exercises

FOR EXTRA HELP

Download the MyDashBoard App

MyMathLab®

CONCEPT CHECK *In Exercises 1–4, complete each statement.*

1. In the statement $\frac{6x^2 + 8}{2} = 3x^2 + 4$, _______ is the dividend, _______ is the divisor, and _______ is the quotient.

2. The expression $\frac{3x + 12}{x}$ is undefined if $x =$ _______.

3. To check the division shown in **Exercise 1,** multiply _______ by _______ and show that the product is _______.

4. The expression $5x^2 - 3x + 6 + \frac{2}{x}$ (*is / is not*) a polynomial.

5. Explain why the division problem $\frac{16m^3 - 12m^2}{4m}$ can be performed using the method of this section, while the division problem $\frac{4m}{16m^3 - 12m^2}$ cannot.

6. **CONCEPT CHECK** Evaluate $\frac{5y + 6}{2}$ for $y = 2$. Evaluate $5y + 3$ for $y = 2$. Is $\frac{5y + 6}{2}$ equivalent to $5y + 3$?

7. **CONCEPT CHECK** A polynomial in the variable x has degree 6 and is divided by a monomial in the variable x having degree 4. What is the degree of the quotient?

8. **CONCEPT CHECK** If $-60x^5 - 30x^4 + 20x^3$ is divided by $3x^2$, what is the sum of the coefficients of the third- and second-degree terms in the quotient?

Perform each division. ***See Examples 1–4.***

9. $\frac{12m^4 - 6m^3}{6m^2}$

10. $\frac{35n^5 - 5n^2}{5n}$

11. $\frac{60x^4 - 20x^2 + 10x}{2x}$

12. $\frac{120x^6 - 60x^3 + 80x^2}{2x}$

13. $\frac{20m^5 - 10m^4 + 5m^2}{-5m^2}$

14. $\frac{12t^5 - 6t^3 + 6t^2}{-6t^2}$

15. $\frac{8t^5 - 4t^3 + 4t^2}{2t}$

16. $\frac{8r^4 - 4r^3 + 6r^2}{2r}$

17. $\frac{4a^5 - 4a^2 + 8}{4a}$

18. $\frac{5t^8 + 5t^7 + 15}{5t}$

19. $\frac{12x^5 - 4x^4 + 6x^3}{-6x^2}$

20. $\frac{24x^6 - 14x^5 + 32x^4}{-4x^2}$

21. $\frac{4x^2 + 20x^3 - 36x^4}{4x^2}$

22. $\dfrac{5x^2 - 30x^4 + 30x^5}{5x^2}$

23. $\dfrac{-3x^3 - 4x^4 + 2x}{-3x^2}$

24. $\dfrac{-8x + 6x^3 - 5x^4}{-3x^2}$

25. $\dfrac{27r^4 - 36r^3 - 6r^2 + 3r - 2}{3r}$

$= \dfrac{\_\_\_}{3r} - \dfrac{\_\_\_}{3r} - \dfrac{\_\_\_}{3r} + \dfrac{\_\_\_}{3r} - \dfrac{\_\_\_}{3r}$

$= \_\_\_\_\_\_\_\_\_\_$

26. $\dfrac{8k^4 - 12k^3 - 2k^2 - 2k - 3}{2k}$

$= \dfrac{8k^4}{\_\_\_} - \dfrac{12k^3}{\_\_\_} - \dfrac{2k^2}{\_\_\_} - \dfrac{2k}{\_\_\_} - \dfrac{3}{\_\_\_}$

$= \_\_\_\_\_\_\_\_\_\_$

27. $\dfrac{2m^5 - 6m^4 + 8m^2}{-2m^3}$

28. $\dfrac{6r^5 - 8r^4 + 10r^2}{-2r^4}$

29. $(120x^{11} - 60x^{10} + 140x^9 - 100x^8) \div (10x^{12})$

30. $(120x^{12} - 84x^9 + 60x^8 - 36x^7) \div (12x^9)$

31. $(20a^4b^3 - 15a^5b^2 + 25a^3b) \div (-5a^4b)$

32. $(16y^5z - 8y^2z^2 + 12yz^3) \div (-4y^2z^2)$

Use an area formula to answer each question. (If necessary, refer to the formulas inside the back cover.)

33. What expression represents the length of the rectangle?

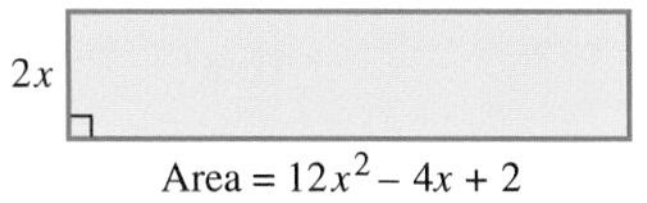

34. What expression represents the length of the base of the triangle?

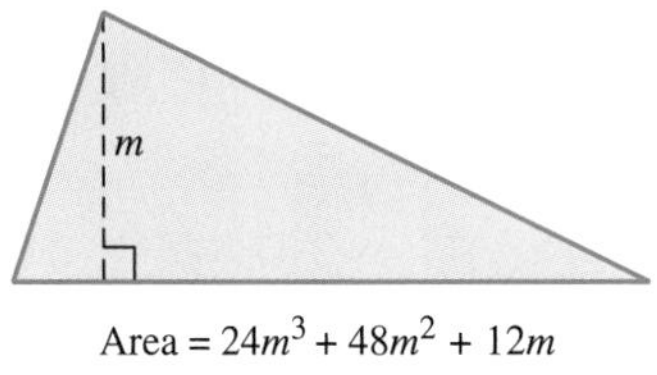

35. **CONCEPT CHECK** What polynomial, when divided by $5x^3$, yields $3x^2 - 7x + 7$ as a quotient?

36. **CONCEPT CHECK** The quotient of a certain polynomial and $-12y^3$ is $6y^3 - 5y^2 + 2y - 3 + \frac{7}{y}$. What is this polynomial?

12.7 Dividing a Polynomial by a Polynomial

OBJECTIVES

1. Divide a polynomial by a polynomial.
2. Apply division to a geometry problem.

OBJECTIVE 1 Divide a polynomial by a polynomial. We use a method of "long division" to divide a polynomial by a polynomial (other than a monomial). ***Both polynomials must be written in descending powers.***

Dividing Whole Numbers	Dividing Polynomials
Step 1	
Divide 6696 by 27. $27\overline{)6696}$	Divide $8x^3 - 4x^2 - 14x + 15$ by $2x + 3$. $2x + 3\overline{)8x^3 - 4x^2 - 14x + 15}$
Step 2	
66 divided by 27 = 2. $2 \cdot 27 = 54$ $\begin{array}{r} 2 \\ 27\overline{)6696} \\ \underline{54} \end{array}$	$8x^3$ divided by $2x = 4x^2$. $4x^2(2x + 3) = 8x^3 + 12x^2$ $\begin{array}{r} 4x^2 \\ 2x + 3\overline{)8x^3 - 4x^2 - 14x + 15} \\ \underline{8x^3 + 12x^2} \end{array}$
Step 3	
Subtract. Then bring down the next digit. $\begin{array}{r} 2 \\ 27\overline{)6696} \\ \underline{54}\downarrow \\ 129 \end{array}$	Subtract. Then bring down the next term. $\begin{array}{r} 4x^2 \\ 2x + 3\overline{)8x^3 - 4x^2 - 14x + 15} \\ \underline{8x^3 + 12x^2} \downarrow \\ -16x^2 - 14x \end{array}$ ***(To subtract two polynomials, change the signs of the second and then add.)***
Step 4	
129 divided by 27 = 4. $4 \cdot 27 = 108$ $\begin{array}{r} 24 \\ 27\overline{)6696} \\ \underline{54} \\ 129 \\ \underline{108} \end{array}$	$-16x^2$ divided by $2x = -8x$. $-8x(2x + 3) = -16x^2 - 24x$ $\begin{array}{r} 4x^2 - 8x \\ 2x + 3\overline{)8x^3 - 4x^2 - 14x + 15} \\ \underline{8x^3 + 12x^2} \\ -16x^2 - 14x \\ \underline{-16x^2 - 24x} \end{array}$
Step 5	
Subtract. Then bring down the next digit. $\begin{array}{r} 24 \\ 27\overline{)6696} \\ \underline{54} \\ 129 \\ \underline{108}\downarrow \\ 216 \end{array}$	Subtract. Then bring down the next term. $\begin{array}{r} 4x^2 - 8x \\ 2x + 3\overline{)8x^3 - 4x^2 - 14x + 15} \\ \underline{8x^3 + 12x^2} \\ -16x^2 - 14x \\ \underline{-16x^2 - 24x} \downarrow \\ 10x + 15 \end{array}$

(continued)

Step 6

216 divided by 27 = **8**.

8 · 27 = **216**

$$
\begin{array}{r}
248 \\
27\overline{)6696} \\
\underline{54} \\
129 \\
\underline{108} \\
216 \\
\underline{216} \\
\end{array}
$$

Remainder → 0

6696 divided by 27 is 248.

10x divided by 2x = **5**.

5(2x + 3) = 10x + 15

$$
\begin{array}{r}
4x^2 - 8x + 5 \\
2x + 3\overline{)8x^3 - 4x^2 - 14x + 15} \\
\underline{8x^3 + 12x^2} \\
-16x^2 - 14x \\
\underline{-16x^2 - 24x} \\
10x + 15 \\
\underline{10x + 15} \\
\end{array}
$$

Remainder → 0

$8x^3 - 4x^2 - 14x + 15$ divided by $2x + 3$ is $4x^2 - 8x + 5$.

Step 7 Multiply to check.

CHECK 27 · 248 = 6696 ✓

CHECK $(2x + 3)(4x^2 - 8x + 5)$
$= 8x^3 - 4x^2 - 14x + 15$ ✓

EXAMPLE 1 **Dividing a Polynomial by a Polynomial**

Divide $5x + 4x^3 - 8 - 4x^2$ by $2x - 1$.

The dividend, the first polynomial here, must be written in descending powers as $4x^3 - 4x^2 + 5x - 8$. Then divide by $2x - 1$.

$$
\begin{array}{r}
2x^2 - x + 2 \\
2x - 1\overline{)4x^3 - 4x^2 + 5x - 8} \\
\underline{4x^3 - 2x^2} \\
-2x^2 + 5x \\
\underline{-2x^2 + x} \\
4x - 8 \\
\underline{4x - 2} \\
-6 \leftarrow \text{Remainder}
\end{array}
$$

Write in descending powers.

To subtract, add the opposite.

Step 1 $4x^3$ divided by $2x$ is $2x^2$. $2x^2(2x - 1) = 4x^3 - 2x^2$

Step 2 Subtract. Bring down the next term.

Step 3 $-2x^2$ divided by $2x$ is $-x$. $-x(2x - 1) = -2x^2 + x$

Step 4 Subtract. Bring down the next term.

Step 5 $4x$ divided by $2x$ is 2. $2(2x - 1) = 4x - 2$

Step 6 Subtract. The remainder is -6. Write the remainder as the numerator of a fraction that has $2x - 1$ as its denominator. Because of the nonzero remainder, the answer is not a polynomial.

Dividend → $\dfrac{4x^3 - 4x^2 + 5x - 8}{2x - 1}$ = $\underbrace{2x^2 - x + 2}_{\substack{\text{Quotient} \\ \text{polynomial}}} + \underbrace{\dfrac{-6}{2x - 1}}_{\substack{\text{Fractional} \\ \text{part of} \\ \text{quotient}}}$ ← Remainder

Divisor → ← Divisor

Continued on Next Page

Step 7 Multiply to check.

CHECK $(2x-1)\left(2x^2-x+2+\dfrac{-6}{2x-1}\right)$ Multiply Divisor × (Quotient including the Remainder).

$$= (2x-1)(2x^2) + (2x-1)(-x) + (2x-1)(2) + (2x-1)\left(\frac{-6}{2x-1}\right)$$

$$= 4x^3 - 2x^2 - 2x^2 + x + 4x - 2 - 6$$

$$= 4x^3 - 4x^2 + 5x - 8 \checkmark$$

◀ Work Problem **1** at the Side.

CAUTION

Remember to include "$+\frac{\text{remainder}}{\text{divisor}}$" as part of the answer.

EXAMPLE 2 Dividing into a Polynomial with Missing Terms

Divide x^3-1 by $x-1$.

Here the dividend, x^3-1, is missing the x^2-term and the x-term. We use 0 as the coefficient for each missing term. Thus, $x^3-1=x^3+0x^2+0x-1$.

$$\begin{array}{r} x^2 + x + 1 \\ x-1\overline{)x^3+0x^2+0x-1} \\ \underline{x^3-x^2} \\ x^2+0x \\ \underline{x^2-x} \\ x-1 \\ \underline{x-1} \\ 0 \end{array}$$

Insert placeholders for the missing terms.

The remainder is 0. The quotient is x^2+x+1.

CHECK $(x-1)(x^2+x+1)$

$$= x^3 + x^2 + x - x^2 - x - 1$$

$$= x^3 - 1 \checkmark \quad \text{Divisor} \times \text{Quotient} = \text{Dividend}$$

◀ Work Problem **2** at the Side.

EXAMPLE 3 Dividing by a Polynomial with Missing Terms

Divide $x^4+2x^3+2x^2-x-1$ by x^2+1.

Since x^2+1 has a missing x-term, write it as x^2+0x+1.

$$\begin{array}{r} x^2+2x+1 \\ x^2+0x+1\overline{)x^4+2x^3+2x^2-x-1} \\ \underline{x^4+0x^3+x^2} \\ 2x^3+x^2-x \\ \underline{2x^3+0x^2+2x} \\ x^2-3x-1 \\ \underline{x^2+0x+1} \\ -3x-2 \leftarrow \text{Remainder} \end{array}$$

Insert a placeholder for the missing term.

Continued on Next Page

1 Divide.

(a) $(x^3+x^2+4x-6)\div(x-1)$

(b) $\dfrac{p^3-2p^2-5p+9}{p+2}$

2 Divide.

(a) $\dfrac{r^2-5}{r+4}$

(b) $(x^3-8)\div(x-2)$

Answers

1. (a) x^2+2x+6
(b) $p^2-4p+3+\dfrac{3}{p+2}$
2. (a) $r-4+\dfrac{11}{r+4}$
(b) x^2+2x+4

When the result of subtracting ($-3x - 2$, in this case) is a constant or a polynomial of degree less than the divisor ($x^2 + 0x + 1$), that constant or polynomial is the remainder. We write the answer as follows.

$$x^2 + 2x + 1 + \frac{-3x - 2}{x^2 + 1}$$

Remember to include "$+ \frac{\text{remainder}}{\text{divisor}}$."

CHECK Show that $(x^2 + 1)\left(x^2 + 2x + 1 + \frac{-3x - 2}{x^2 + 1}\right)$ gives the original dividend, $x^4 + 2x^3 + 2x^2 - x - 1$. ✓

Work Problem 3 at the Side. ▶

3 Divide.

(a) $(2x^4 + 3x^3 - x^2 + 6x + 5) \div (x^2 - 1)$

(b) $\dfrac{2m^5 + m^4 + 6m^3 - 3m^2 - 18}{m^2 + 3}$

EXAMPLE 4 Dividing a Polynomial When the Quotient Has Fractional Coefficients

Divide $4x^3 + 2x^2 + 3x + 2$ by $4x - 4$.

$\frac{6x^2}{4x} = \frac{3}{2}x$ $\qquad$ $\frac{9x}{4x} = \frac{9}{4}$

$$\begin{array}{r}
x^2 + \frac{3}{2}x + \frac{9}{4} \\
4x - 4\overline{)4x^3 + 2x^2 + 3x + 2} \\
\underline{4x^3 - 4x^2} \qquad\qquad \\
6x^2 + 3x \qquad \\
\underline{6x^2 - 6x} \qquad \\
9x + 2 \\
\underline{9x - 9} \\
11
\end{array}$$

The answer is $x^2 + \frac{3}{2}x + \frac{9}{4} + \frac{11}{4x - 4}$.

Work Problem 4 at the Side. ▶

4 Divide $3x^3 + 7x^2 + 7x + 10$ by $3x + 6$.

OBJECTIVE 2 Apply division to a geometry problem.

EXAMPLE 5 Using an Area Formula

The area of the rectangle in **Figure 3** is given by $(x^3 + 4x^2 + 8x + 8)$ sq. units. The width is given by $(x + 2)$ units. What is its length?

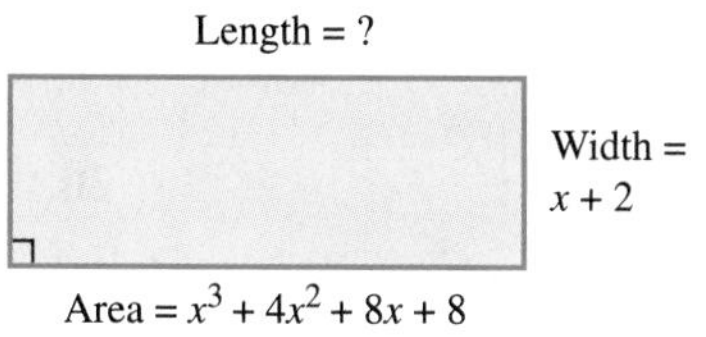

Figure 3

For a rectangle, $A = LW$. Solving for L gives $L = \frac{A}{W}$. Divide the area, $x^3 + 4x^2 + 8x + 8$, by the width, $x + 2$.

Work Problem 5 at the Side. ▶

The quotient from **Margin Problem 5,** $x^2 + 2x + 4$, represents the length of the rectangle in units.

5 Divide $x^3 + 4x^2 + 8x + 8$ by $x + 2$.

Answers

3. **(a)** $2x^2 + 3x + 1 + \frac{9x + 6}{x^2 - 1}$

(b) $2m^3 + m^2 - 6$

4. $x^2 + \frac{1}{3}x + \frac{5}{3}$

5. $x^2 + 2x + 4$

12.7 Exercises

FOR EXTRA HELP

Download the MyDashBoard App

MyMathLab®

CONCEPT CHECK *Answer each question.*

1. In the division problem $(4x^4 + 2x^3 - 14x^2 + 19x + 10) \div (2x + 5) = 2x^3 - 4x^2 + 3x + 2$, which polynomial is the divisor? Which is the quotient?

2. When dividing one polynomial by another, how do you know when to stop dividing?

3. In dividing $12m^2 - 20m + 3$ by $2m - 3$, what is the first step?

4. In the division in **Exercise 3,** what is the second step?

Perform each division. ***See Example 1.***

5. $\dfrac{x^2 - x - 6}{x - 3}$

6. $\dfrac{m^2 - 2m - 24}{m - 6}$

7. $\dfrac{2y^2 + 9y - 35}{y + 7}$

8. $\dfrac{2y^2 + 9y + 7}{y + 1}$

9. $\dfrac{p^2 + 2p + 20}{p + 6}$

10. $\dfrac{x^2 + 11x + 16}{x + 8}$

11. $(r^2 - 8r + 15) \div (r - 3)$

12. $(t^2 + 2t - 35) \div (t - 5)$

13. $\dfrac{4a^2 - 22a + 32}{2a + 3}$

14. $\dfrac{9w^2 + 6w + 10}{3w - 2}$

15. $\dfrac{8x^3 - 10x^2 - x + 3}{2x + 1}$

16. $\dfrac{12t^3 - 11t^2 + 9t + 18}{4t + 3}$

Perform each division. ***See Examples 2–4.***

17. $\dfrac{3y^3 + y^2 + 2}{y + 1}$

18. $\dfrac{2r^3 - 6r - 36}{r - 3}$

19. $\dfrac{2x^3 + x + 2}{x + 1}$

20. $\dfrac{3x^3 + x + 5}{x + 1}$

21. $\dfrac{3k^3 - 4k^2 - 6k + 10}{k^2 - 2}$

22. $\dfrac{5z^3 - z^2 + 10z + 2}{z^2 + 2}$

23. $(x^4 - x^2 - 2) \div (x^2 - 2)$

24. $(r^4 + 2r^2 - 3) \div (r^2 - 1)$

25. $\dfrac{x^4 - 1}{x^2 - 1}$

26. $\dfrac{y^3 + 1}{y + 1}$

27. $\dfrac{6p^4 - 15p^3 + 14p^2 - 5p + 10}{3p^2 + 1}$

28. $\dfrac{6r^4 - 10r^3 - r^2 + 15r - 8}{2r^2 - 3}$

29. $\dfrac{2x^5 + x^4 + 11x^3 - 8x^2 - 13x + 7}{2x^2 + x - 1}$

30. $\dfrac{4t^5 - 11t^4 - 6t^3 + 5t^2 - t + 3}{4t^2 + t - 3}$

31. $(10x^3 + 13x^2 + 4x + 1) \div (5x + 5)$

32. $(6x^3 - 19x^2 - 19x - 4) \div (2x - 8)$

Work each problem. ***See Example 5.*** *(If necessary, refer to the formulas inside the back cover.)*

33. Give the length of the rectangle.

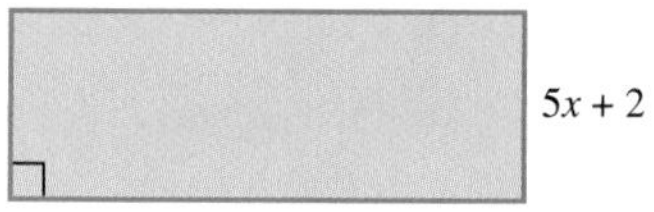

The area is $(5x^3 + 7x^2 - 13x - 6)$ sq. units.

34. Find the measure of the base of the parallelogram.

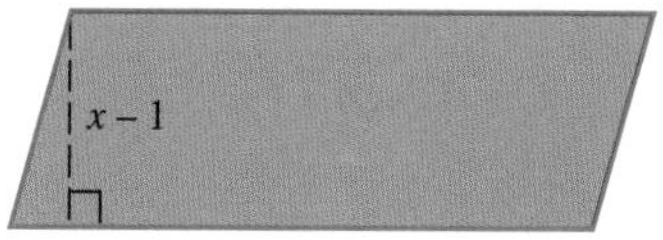

The area is $(2x^3 + 2x^2 - 3x - 1)$ sq. units.

Relating Concepts (Exercises 35–38) For Individual or Group Work

We can find the value of a polynomial in x for a given value of x by substituting that number for x. Surprisingly, we can accomplish the same thing by division. For example, to find the value of $2x^2 - 4x + 3$ *for* $x = -3$, *we would divide* $2x^2 - 4x + 3$ *by* $x - (-3)$. *The remainder will give the value of the polynomial for* $x = -3$.
Work Exercises 35–38 in order.

35. Find the value of $2x^2 - 4x + 3$ for $x = -3$ by substitution.

36. Divide $2x^2 - 4x + 3$ by $x + 3$. Give the remainder.

37. Compare your answers to **Exercises 35 and 36.** What do you notice?

38. Choose another polynomial and evaluate it both ways for some value of the variable. Do the answers agree?

12.8 An Application of Exponents: Scientific Notation

OBJECTIVES

1. Express numbers in scientific notation.
2. Convert numbers in scientific notation to numbers without exponents.
3. Use scientific notation in calculations.

OBJECTIVE 1 Express numbers in scientific notation. Numbers occurring in science are often extremely large (such as the distance from Earth to the sun, 93,000,000 mi) or extremely small (the wavelength of yellow-green light, approximately 0.0000006 m). Because of the difficulty of working with many zeros, scientists often express such numbers with exponents, using a form called *scientific notation.*

Scientific Notation

A number is written in **scientific notation** when it is expressed in the form

$$a \times 10^n,$$

where $1 \le |a| < 10$ and n is an integer.

In **scientific notation,** there is always one nonzero digit to the left of the decimal point. This is shown in the following, where the first number is in scientific notation.

$3.19 \times 10^1 = 3.19 \times 10 = 31.9$ — Decimal point moves 1 place to the right.

$3.19 \times 10^2 = 3.19 \times 100 = 319.$ — Decimal point moves 2 places to the right.

$3.19 \times 10^3 = 3.19 \times 1000 = 3190.$ — Decimal point moves 3 places to the right.

$3.19 \times 10^{-1} = 3.19 \times 0.1 = 0.319$ — Decimal point moves 1 place to the left.

$3.19 \times 10^{-2} = 3.19 \times 0.01 = 0.0319$ — Decimal point moves 2 places to the left.

$3.19 \times 10^{-3} = 3.19 \times 0.001 = 0.00319$ — Decimal point moves 3 places to the left.

Note

In scientific notation, a multiplication cross (or symbol) $\times$ is commonly used.

A number in scientific notation is always written with the decimal point after the first nonzero digit and then multiplied by the appropriate power of 10. For example, 56,200 is written 5.62×10^4, since

$$56{,}200 = 5.62 \times \mathbf{10{,}000} = 5.62 \times \mathbf{10^4}.$$

Other examples include

	42,000,000	written	4.2×10^7,
	0.000586	written	5.86×10^{-4},
and	2,000,000,000	written	2×10^9.

It is not necessary to write 2.0.

To write a positive number in scientific notation, follow the steps given on the next page. (For a negative number, follow these steps using the *absolute value* of the number. Then make the result negative.)

Writing a Positive Number in Scientific Notation

Step 1 Move the decimal point to the right of the first nonzero digit.

Step 2 Count the number of places you moved the decimal point.

Step 3 The number of places in Step 2 is the absolute value of the exponent on 10.

Step 4 The exponent on 10 is positive if the original number is greater than the number in Step 1. The exponent is negative if the original number is less than the number in Step 1. If the decimal point is not moved, the exponent is 0.

EXAMPLE 1 Using Scientific Notation

Write each number in scientific notation.

(a) 93,000,000

Move the decimal point to follow the first nonzero digit (the 9). Count the number of places the decimal point was moved.

93,000,000. ← Decimal point
7 places

The number will be written in scientific notation as 9.3×10^n. To find the value of n, first compare the original number, 93,000,000, with 9.3. Since 93,000,000 is *greater* than 9.3, we must multiply by a *positive* power of 10 so that the product 9.3×10^n will equal the larger number.

Since the decimal point was moved 7 places, and since n is positive,

$$93{,}000{,}000 = 9.3 \times 10^7.$$

(b) $63{,}200{,}000{,}000 = 6.3200000000 = 6.32 \times 10^{10}$
10 places

(c) $3.021 = 3.021 \times 10^0$

(d) 0.00462

Move the decimal point to the right of the first nonzero digit and count the number of places the decimal point was moved.

0.00462 3 places

Since 0.00462 is *less* than 4.62, the exponent must be *negative*.

$$0.00462 = 4.62 \times 10^{-3}$$

(e) $-0.0000762 = -7.62 \times 10^{-5}$
5 places
Remember the *negative* sign.

Work Problem 1 at the Side. ▶

Note

When writing a positive number in scientific notation, think as follows.

1. If the original number is "large," like 93,000,000, use a *positive* exponent on 10, since positive is greater than negative.
2. If the original number is "small," like 0.00462, use a *negative* exponent on 10, since negative is less than positive.

1 Write each number in scientific notation.

GS **(a)** 63,000

The first nonzero digit is ____.

The decimal point should be moved ____ places.

$63{,}000 = \_\_\_\_ \times 10^{—}$

(b) 5,870,000

(c) 7.0065

GS **(d)** 0.0571

The first nonzero digit is ____.

The decimal point should be moved ____ places.

$0.0571 = \_\_\_\_ \times 10^{—}$

(e) −0.00062

Answers

1. **(a)** 6; 4; 6.3; 4 **(b)** 5.87×10^6 **(c)** 7.0065×10^0 **(d)** 5; 2; 5.71; −2 **(e)** -6.2×10^{-4}

2 Write without exponents.

GS **(a)** 4.2×10^3

Move the decimal point

____ places to the ____.

$4.2 \times 10^3 =$ ________

(b) 8.7×10^5

GS **(c)** 6.42×10^{-3}

Move the decimal point

____ places to the ____.

$6.42 \times 10^{-3} =$ ________

(d) -5.27×10^{-1}

3 Perform each calculation. Write answers in scientific notation and also without exponents.

(a) $(2.6 \times 10^4)(2 \times 10^{-6})$

(b) $(3 \times 10^5)(5 \times 10^{-2})$

(c) $\dfrac{4.8 \times 10^2}{2.4 \times 10^{-3}}$

Answers

2. **(a)** 3; right; 4200 **(b)** 870,000
(c) 3; left; 0.00642 **(d)** −0.527

3. **(a)** 5.2×10^{-2}; 0.052
(b) 1.5×10^4; 15,000
(c) 2×10^5; 200,000

OBJECTIVE **2** **Convert numbers in scientific notation to numbers without exponents.** To convert a number written in scientific notation to a number without exponents, work in reverse.

Multiplying a positive number by a positive power of 10 will make the number greater. Multiplying by a negative power of 10 will make the number less.

EXAMPLE 2 Writing Numbers without Exponents

Write each number without exponents.

(a) 6.2×10^3

Since the exponent is positive, make 6.2 greater by moving the decimal point 3 places to the right. It is necessary to attach two 0s.

$$6.2 \times 10^3 = 6.200 = 6200$$

(b) $4.283 \times 10^5 = 4.28300 = 428{,}300$ Move 5 places to the right. Attach 0s as necessary.

(c) $-9.73 \times 10^{-2} = -09.73 = -0.0973$ Move 2 places to the left.

The exponent tells the number of places and the direction in which the decimal point is moved.

◀ **Work Problem 2 at the Side.**

OBJECTIVE **3** **Use scientific notation in calculations.** The next example uses scientific notation with products and quotients.

EXAMPLE 3 Multiplying and Dividing with Scientific Notation

Perform each calculation. Write answers in scientific notation and also without exponents.

(a) $(7 \times 10^3)(5 \times 10^4)$

$= (7 \times 5)(10^3 \times 10^4)$ Commutative and associative properties

$= 35 \times 10^7$ Multiply and use the product rule.

$= (3.5 \times 10^1) \times 10^7$ Write 35 in scientific notation.

$= 3.5 \times (10^1 \times 10^7)$ Associative property

$= 3.5 \times 10^8$ Product rule

$= 350{,}000{,}000$ Write without exponents.

Don't stop. This number is *not* in scientific notation, since 35 is not between 1 and 10.

(b) $\dfrac{4 \times 10^{-5}}{2 \times 10^3}$

$= \dfrac{4}{2} \times \dfrac{10^{-5}}{10^3}$

$= 2 \times 10^{-8}$ Divide and use the quotient rule.

$= 0.00000002$ Write without exponents.

◀ **Work Problem 3 at the Side.**

Calculator Tip

Calculators usually have a key labeled EE or EXP for scientific notation. See your owner's manual for more information.

Note

Multiplying or dividing numbers written in scientific notation may produce an answer in the form $a \times 10^0$. Since $10^0 = 1$, $a \times 10^0 = a$. For example,

$$(8 \times 10^{-4})(5 \times 10^4) = 40 \times 10^0 = 40. \quad 10^0 = 1$$

Also, if $a = 1$, then $a \times 10^n = 10^n$. For example, we could write 1,000,000 as 10^6 instead of 1×10^6.

EXAMPLE 4 Using Scientific Notation to Solve an Application

A *nanometer* is a very small unit of measure that is equivalent to about 0.00000003937 in. About how much would 700,000 nanometers measure in inches? (*Source: World Almanac and Book of Facts.*)

Write each number in scientific notation, and then multiply.

$700{,}000(0.00000003937)$

$= (7 \times 10^5)(3.937 \times 10^{-8})$ Write in scientific notation.

$= (7 \times 3.937)(10^5 \times 10^{-8})$ Properties of real numbers

$= 27.559 \times 10^{-3}$ Multiply; product rule

Don't stop here.

$= (2.7559 \times 10^1) \times 10^{-3}$ Write 27.559 in scientific notation.

$= 2.7559 \times 10^{-2}$ Product rule

$= 0.027559$ Write without exponents.

Thus, 700,000 nanometers would measure

$$2.7559 \times 10^{-2} \text{ in., or } 0.027559 \text{ in.}$$

Work Problem 4 at the Side. ▶

EXAMPLE 5 Using Scientific Notation to Solve an Application

In 2010, the national debt was $\$1.3562 \times 10^{13}$ (which is more than \$13 trillion). The population of the United States was approximately 309 million that year. About how much would each person have had to contribute in order to pay off the national debt? (*Source:* U.S. Department of the Treasury; U.S. Census Bureau.)

Write the population in scientific notation. Then divide to obtain the per person contribution.

$\dfrac{1.3562 \times 10^{13}}{309{,}000{,}000}$

$= \dfrac{1.3562 \times 10^{13}}{3.09 \times 10^8}$ Write 309 million in scientific notation.

$= \dfrac{1.3562}{3.09} \times 10^5$ Quotient rule

$\approx 0.4389 \times 10^5$ Divide. Round to 4 decimal places.

$= 43{,}890$ Write without exponents.

Each person would have to pay about \$43,890.

Work Problem 5 at the Side. ▶

4 The speed of light is approximately 3.0×10^5 km per sec. How far does light travel in 6.0×10^1 sec? (*Source: World Almanac and Book of Facts.*)

5 If the speed of light is approximately 3.0×10^5 km per sec, how many seconds does it take light to travel approximately 1.5×10^8 km from the Sun to Earth? (*Source: World Almanac and Book of Facts.*)

Answers

4. 1.8×10^7 km, or 18,000,000 km
5. 5×10^2 sec, or 500 sec

12.8 Exercises

FOR EXTRA HELP

Download the MyDashBoard App

MyMathLab®

CONCEPT CHECK *Match each number written in scientific notation in Column I with the correct choice from Column II. Not all choices in Column II will be used.*

I	II
1. (a) 4.6×10^{-4}	**A.** 46,000
(b) 4.6×10^{4}	**B.** 460,000
(c) 4.6×10^{5}	**C.** 0.00046
(d) 4.6×10^{-5}	**D.** 0.000046
	E. 4600

I	II
2. (a) 1×10^{9}	**A.** 1 billion
(b) 1×10^{6}	**B.** 100 million
(c) 1×10^{8}	**C.** 1 million
(d) 1×10^{10}	**D.** 10 billion
	E. 100 billion

CONCEPT CHECK *Determine whether or not the given number is written in scientific notation as defined in* ***Objective 1.*** *If it is not, write it as such.*

3. 4.56×10^{3}

4. 7.34×10^{5}

5. 5,600,000

6. 34,000

7. 0.004

8. 0.0007

9. 0.8×10^{2}

10. 0.9×10^{3}

11. Explain what it means for a number to be written in scientific notation.

12. Explain how to multiply a number by a positive power of ten. Then explain how to multiply a number by a negative power of ten.

Write each number in scientific notation. ***See Example 1.***

13. 5,876,000,000

14. 9,994,000,000

15. 82,350

16. 78,330

17. 0.000007

18. 0.0000004

19. −0.00203

20. −0.0000578

Write each number without exponents. ***See Example 2.***

21. 7.5×10^{5}

22. 8.8×10^{6}

23. 5.677×10^{12}

24. 8.766×10^{9}

25. 1×10^{12}

26. 1×10^{7}

27. -6.21×10^{0}

28. -8.56×10^{0}

29. 7.8×10^{-4}

30. 8.9×10^{-5}

31. 5.134×10^{-9}

32. 7.123×10^{-10}

Perform the indicated operations. Write the answers in scientific notation and then without exponents. ***See Example 3.***

33. $(2 \times 10^{8})(3 \times 10^{3})$

34. $(3 \times 10^{7})(3 \times 10^{3})$

35. $(5 \times 10^{4})(3 \times 10^{2})$

36. $(8 \times 10^5)(2 \times 10^3)$

37. $(4 \times 10^{-6})(2 \times 10^3)$

38. $(3 \times 10^{-7})(2 \times 10^2)$

39. $(6 \times 10^3)(4 \times 10^{-2})$

40. $(7 \times 10^5)(3 \times 10^{-4})$

41. $(9 \times 10^4)(7 \times 10^{-7})$

42. $(6 \times 10^4)(8 \times 10^{-8})$

43. $(3.15 \times 10^{-4})(2.04 \times 10^8)$

44. $(4.92 \times 10^{-3})(2.25 \times 10^7)$

45. $\dfrac{9 \times 10^{-5}}{3 \times 10^{-1}}$

46. $\dfrac{12 \times 10^{-4}}{4 \times 10^{-3}}$

47. $\dfrac{8 \times 10^{3}}{2 \times 10^{2}}$

48. $\dfrac{15 \times 10^{4}}{3 \times 10^{3}}$

49. $\dfrac{2.6 \times 10^{-3}}{2 \times 10^{2}}$

50. $\dfrac{9.5 \times 10^{-1}}{5 \times 10^{3}}$

51. $\dfrac{4 \times 10^{5}}{8 \times 10^{2}}$

52. $\dfrac{3 \times 10^{9}}{6 \times 10^{5}}$

53. $\dfrac{2.6 \times 10^{-3} \times 7.0 \times 10^{-1}}{2 \times 10^{2} \times 3.5 \times 10^{-3}}$

54. $\dfrac{9.5 \times 10^{-1} \times 2.4 \times 10^{4}}{5 \times 10^{3} \times 1.2 \times 10^{-2}}$

55. $\dfrac{(1.65 \times 10^{8})(5.24 \times 10^{-2})}{(6 \times 10^{4})(2 \times 10^{7})}$

56. $\dfrac{(3.25 \times 10^{7})(6.48 \times 10^{-4})}{(5 \times 10^{2})(4 \times 10^{6})}$

Each statement taken from a book or newspaper contains a number in boldface italic type. If the number is in scientific notation, write it without exponents. If the number is not in scientific notation, write it as such. **See Examples 1 and 2.**

57. The muon, a close relative of the electron produced by the bombardment of cosmic rays against the upper atmosphere, has a half-life of 2 millionths of a second ($\boldsymbol{2 \times 10^{-6}}$ s). (*Source:* Hewitt, Paul G., *Conceptual Physics.*)

58. There are 13 red balls and 39 black balls in a box. Mix them up and draw 13 out one at a time without returning any ball . . . the probability that the 13 drawings each will produce a red ball is . . . $\boldsymbol{1.6 \times 10^{-12}}$. (*Source:* Weaver, Warren, *Lady Luck.*)

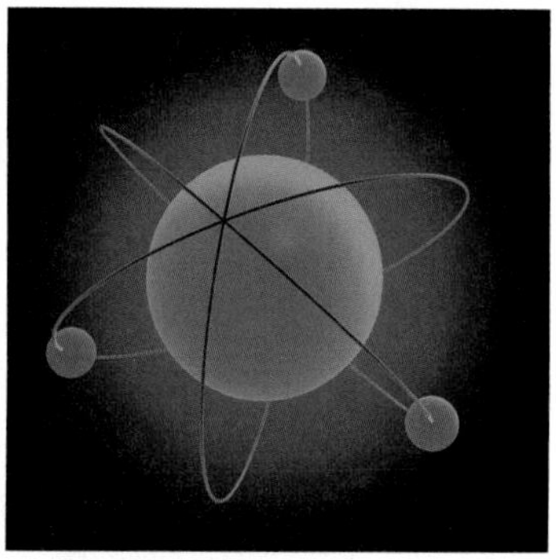

59. An electron and a positron attract each other in two ways: the electromagnetic attraction of their opposite electric charges, and the gravitational attraction of their two masses. The electromagnetic attraction is

4,200,000,000,000,000,000,000,000,000,000,000,000,000,000,000

times as strong as the gravitational. (*Source:* Asimov, I., *Isaac Asimov's Book of Facts.*)

60. A googol is

10,000,000,000,000,000,000,000,000,000,000,000,000,000,000,000,000,000,000,000,000,000,000,000,000,000,000,000,000,000,000,000,000,000,000,000.

The Web search engine Google is named after a googol. Sergey Brin, president and cofounder of Google, Inc., was a mathematics major. He chose the name Google to describe the vast reach of this search engine. (*Source: The Gazette.*)

Use scientific notation to calculate the answer to each problem. In Exercises 63–65 and 69–70, give answers without exponents. ***See Examples 4 and 5.***

61. The Double Helix Nebula, a conglomeration of dust and gas stretching across the center of the Milky Way galaxy, is 25,000 light-years from Earth. If one light-year is about 6,000,000,000,000 mi, about how many miles is the Double Helix Nebula from Earth? (*Source:* www.spitzer.caltech.edu)

62. Pollux, one of the brightest stars in the night sky, is 33.7 light-years from Earth. If one light-year is about 6,000,000,000,000 mi (that is, 6 trillion mi), about how many miles is Pollux from Earth? (*Source: World Almanac and Book of Facts.*)

63. In 2011, the population of the United States was about 311.6 million. To the nearest dollar, calculate how much each person in the United States would have had to contribute in order to make one person a trillionaire (that is, to give that person \$1,000,000,000,000). (*Source:* U.S. Census Bureau.)

64. In 2010, the U.S. government collected about \$3767 per person in individual income taxes. If the population at that time was 309,000,000, how much did the government collect in taxes for 2010? (*Source: World Almanac and Book of Facts.*)

65. In 2011, Congress raised the debt limit to $\$1.64 \times 10^{13}$. When this national debt limit is reached, about how much is it for every man, woman, and child in the country? Use 311 million as the population of the United States. (*Source:* www.tradingnrg.org)

66. In 2010, the state of Minnesota had about 8.1×10^4 farms with an average of 3.44×10^2 acres per farm. What was the total number of acres devoted to farmland in Minnesota that year? (*Source:* U.S. Department of Agriculture.)

67. Venus is 6.68×10^7 mi from the Sun. If light travels at a speed of 1.86×10^5 mi per sec, how long does it take light to travel from the Sun to Venus? (*Source: World Almanac and Book of Facts.*)

68. The distance to Earth from Pluto is 4.58×10^9 km. *Pioneer 10* transmitted radio signals from Pluto to Earth at the speed of light, 3.00×10^5 km per sec. How long (in seconds) did it take for the signals to reach Earth?

69. During the 2010–2011 season, Broadway shows grossed a total of 1.08×10^9 dollars. Total attendance for the season was 1.25×10^7. What was the average ticket price for a Broadway show? (*Source:* The Broadway League.)

70. In 2010, 10.6×10^9 dollars were spent to attend motion pictures in the United States and Canada. The total number of tickets sold that year totaled 1.34 billion. What was the average ticket price? (*Source:* Motion Picture Association of America.)

Chapter 12 Summary

Key Terms

12.1

term A term is a number, a variable, or a product or quotient of a number and one or more variables raised to powers.

like terms Terms with exactly the same variables (including the same exponents) are like terms.

polynomial A polynomial is a term or the sum of a finite number of terms with whole number exponents.

descending powers A polynomial in x is written in descending powers if the exponents on x in its terms are in decreasing order.

degree of a term The degree of a term is the sum of the exponents on the variables.

degree of a polynomial The degree of a polynomial is the greatest degree of any term of the polynomial.

monomial A monomial is a polynomial with exactly one term.

binomial A binomial is a polynomial with exactly two terms.

trinomial A trinomial is a polynomial with exactly three terms.

12.2

exponential expression A number written with an exponent is an exponential expression.

3^4 ← Exponent; 3 ← Base; 3^4: Exponential expression

12.3

FOIL method The FOIL method is used to find the product of two binomials. The letters of the word **FOIL** originate as follows: Multiply the **F**irst terms, multiply the **O**uter terms (to get the outer product), multiply the **I**nner terms (to get the inner product), and multiply the **L**ast terms.

outer product The outer product of $(a + b)(c + d)$ is ad.

inner product The inner product of $(a + b)(c + d)$ is bc.

12.4

conjugate The conjugate of $a + b$ is $a - b$.

12.8

scientific notation A number written as $a \times 10^n$, where $1 \leq |a| < 10$ and n is an integer, is in scientific notation.

New Symbols

x^{-n} x to the negative n power

Test Your Word Power

See how well you have learned the vocabulary in this chapter.

1. A **polynomial** is an algebraic expression made up of
   A. a term or a finite product of terms with positive coefficients and exponents
   B. a term or a finite sum of terms with real coefficients and whole number exponents
   C. the product of two or more terms with positive exponents
   D. the sum of two or more terms with whole number coefficients and exponents.

2. The **degree of a term** is
   A. the number of variables in the term
   B. the product of the exponents on the variables
   C. the least exponent on the variables
   D. the sum of the exponents on the variables.

3. A **trinomial** is a polynomial with
   A. only one term
   B. exactly two terms
   C. exactly three terms
   D. more than three terms.

4. A **binomial** is a polynomial with
   A. only one term
   B. exactly two terms
   C. exactly three terms
   D. more than three terms.

5. A **monomial** is a polynomial with
   A. only one term
   B. exactly two terms
   C. exactly three terms
   D. more than three terms.

6. **FOIL** is a method for
   A. adding two binomials
   B. adding two trinomials
   C. multiplying two binomials
   D. multiplying two trinomials.

Answers to Test Your Word Power

1. B; *Example:* $5x^3 + 2x^2 - 7$
2. D; *Examples:* The term 6 has degree 0, $3x$ has degree 1, $-2x^8$ has degree 8, and $5x^2y^4$ has degree 6.
3. C; *Example:* $2a^2 - 3ab + b^2$
4. B; *Example:* $3t^3 + 5t$
5. A; *Examples:* -5 and $4xy^5$
6. C; *Example:* $(m + 4)(m - 3)$

$$\begin{aligned} &\quad\ \text{F} \quad\ \ \text{O} \quad\ \ \text{I} \quad\ \ \text{L} \\ &= m(m) - 3m + 4m + 4(-3) \\ &= m^2 + m - 12 \end{aligned}$$

Quick Review

Concepts	Examples
12.1 Adding and Subtracting Polynomials	
Addition Add like terms.	Add. $\begin{array}{r} 2x^2 + 5x - 3 \\ \underline{5x^2 - 2x + 7} \\ 7x^2 + 3x + 4 \end{array}$
Subtraction Change the signs of the terms in the second polynomial and add to the first polynomial.	Subtract. $(2x^2 + 5x - 3) - (5x^2 - 2x + 7)$ $= (2x^2 + 5x - 3) + (-5x^2 + 2x - 7)$ $= -3x^2 + 7x - 10$
12.2 The Product Rule and Power Rules for Exponents	
For any integers m and n, the following hold.	Perform the operations by using the rules for exponents.
Product rule $\quad a^m \cdot a^n = a^{m+n}$	$2^4 \cdot 2^5 = 2^{4+5} = 2^9$
Power rules (a) $(a^m)^n = a^{mn}$	$(3^4)^2 = 3^{4 \cdot 2} = 3^8$
(b) $(ab)^m = a^m b^m$	$(6a)^5 = 6^5a^5$
(c) $\left(\frac{a}{b}\right)^m = \frac{a^m}{b^m}$ (where $b \neq 0$)	$\left(\frac{2}{3}\right)^4 = \frac{2^4}{3^4}$
12.3 Multiplying Polynomials	
Multiply each term of the first polynomial by each term of the second polynomial. Then add like terms.	Multiply. $\begin{array}{rrrrr} & 3x^3 & - 4x^2 & + 2x & - 7 \\ & & & 4x & + 3 \\ \hline & 9x^3 & - 12x^2 & + 6x & - 21 \\ 12x^4 & - 16x^3 & + 8x^2 & - 28x & \\ \hline 12x^4 & - 7x^3 & - 4x^2 & - 22x & - 21 \end{array}$
FOIL method *Step 1* Multiply the two **F**irst terms to get the first term of the answer. *Step 2* Find the **O**uter product and the **I**nner product and mentally add them, when possible, to get the middle term of the answer. *Step 3* Multiply the two **L**ast terms to get the last term of the answer. Add the terms found in Steps 1–3.	Multiply $(2x + 3)(5x - 4)$. $2x(5x) = 10x^2$ F $2x(-4) + 3(5x) = 7x$ O, I $3(-4) = -12$ L The product is $10x^2 + 7x - 12$.
12.4 Special Products	
Square of a Binomial $(a + b)^2 = a^2 + 2ab + b^2$ $(a - b)^2 = a^2 - 2ab + b^2$	Multiply. $(3x + 1)^2$ $= (3x)^2 + 2(3x)(1) + 1^2$ $= 9x^2 + 6x + 1$ $(2m - 5n)^2$ $= (2m)^2 - 2(2m)(5n) + (5n)^2$ $= 4m^2 - 20mn + 25n^2$

Concepts	Examples

12.4 Special Products *(continued)*

Product of the Sum and Difference of Two Terms

$$(a + b)(a - b) = a^2 - b^2$$

$$(4a + 3)(4a - 3)$$
$$= (4a)^2 - 3^2$$
$$= 16a^2 - 9$$

12.5 Integer Exponents and the Quotient Rule

If $a, b \neq 0$, for integers m and n, the following hold.

Zero exponent $\quad a^0 = 1$

Negative exponent $\quad a^{-n} = \dfrac{1}{a^n}$

Quotient rule $\quad \dfrac{a^m}{a^n} = a^{m-n}$

Negative-to-positive rules $\quad \dfrac{a^{-m}}{b^{-n}} = \dfrac{b^n}{a^m} \quad \left(\dfrac{a}{b}\right)^{-m} = \left(\dfrac{b}{a}\right)^m$

Simplify by using the rules for exponents.

$$15^0 = 1$$

$$5^{-2} = \dfrac{1}{5^2} = \dfrac{1}{25}$$

$$\dfrac{4^8}{4^3} = 4^{8-3} = 4^5$$

$$\dfrac{6^{-2}}{7^{-3}} = \dfrac{7^3}{6^2} \quad \left(\dfrac{5}{3}\right)^{-4} = \left(\dfrac{3}{5}\right)^4$$

12.6 Dividing a Polynomial by a Monomial

Divide each term of the polynomial by the monomial.

$$\dfrac{a + b}{c} = \dfrac{a}{c} + \dfrac{b}{c}$$

Divide.

$$\dfrac{4x^3 - 2x^2 + 6x - 8}{2x}$$

$$= \dfrac{4x^3}{2x} - \dfrac{2x^2}{2x} + \dfrac{6x}{2x} - \dfrac{8}{2x} \quad \text{Divide each term in the dividend by } 2x, \text{ the divisor.}$$

$$= 2x^2 - x + 3 - \dfrac{4}{x}$$

12.7 Dividing a Polynomial by a Polynomial

Use "long division."

Divide.

$$
\begin{array}{r}
2x - 5 \\
3x + 4\overline{)6x^2 - 7x - 21} \\
\underline{6x^2 + 8x} \\
-15x - 21 \\
\underline{-15x - 20} \\
-1 \leftarrow \text{Remainder}
\end{array}
$$

The answer is $2x - 5 + \dfrac{-1}{3x + 4}$.

12.8 An Application of Exponents: Scientific Notation

To write a positive number in scientific notation

$$a \times 10^n,$$

move the decimal point to follow the first nonzero digit.

1. If moving the decimal point makes the number less, n is positive.
2. If it makes the number greater, n is negative.
3. If the decimal point is not moved, n is 0.

For a negative number, follow these steps using the *absolute value* of the number. Then make the result negative.

Write in scientific notation.

$$247 = 2.47 \times 10^2$$
$$0.0051 = 5.1 \times 10^{-3}$$

Write without exponents.

$$3.25 \times 10^5 = 325,000$$
$$8.44 \times 10^{-6} = 0.00000844$$

Chapter 12 Review Exercises

12.1 *Combine terms where possible in each polynomial. Write the answer in descending powers of the variable. Give the degree of the answer. Identify the polynomial as a* monomial, *a* binomial, *a* trinomial, *or* none of these.

1. $9m^2 + 11m^2 + 2m^2$

2. $-4p + p^3 - p^2 + 8p + 2$

3. $12a^5 - 9a^4 + 8a^3 + 2a^2 - a + 3$

4. $-7y^5 - 8y^4 - y^5 + y^4 + 9y$

Add or subtract as indicated.

5. Add.
$$\begin{array}{r} -2a^3 + 5a^2 \\ \underline{-3a^3 - a^2} \end{array}$$

6. Add.
$$\begin{array}{r} 4r^3 - 8r^2 + 6r \\ \underline{-2r^3 + 5r^2 + 3r} \end{array}$$

7. Subtract.
$$\begin{array}{r} 6y^2 - 8y + 2 \\ \underline{-5y^2 + 2y - 7} \end{array}$$

8. Subtract.
$$\begin{array}{r} -12k^4 - 8k^2 + 7k - 5 \\ \underline{k^4 + 7k^2 + 11k + 1} \end{array}$$

9. $(2m^3 - 8m^2 + 4) + (8m^3 + 2m^2 - 7)$

10. $(-5y^2 + 3y + 11) + (4y^2 - 7y + 15)$

11. $(6p^2 - p - 8) - (-4p^2 + 2p + 3)$

12. $(12r^4 - 7r^3 + 2r^2) - (5r^4 - 3r^3 + 2r^2 + 1)$

12.2 *Simplify each expression.*

13. $4^3 \cdot 4^8$

14. $(-5)^6(-5)^5$

15. $(-8x^4)(9x^3)$

16. $(2x^2)(5x^3)(x^9)$

17. $(19x)^5$

18. $(-4y)^7$

19. $5(pt)^4$

20. $\left(\frac{7}{5}\right)^6$

21. $(3x^2y^3)^3$

22. $(t^4)^8(t^2)^5$

23. $(6x^2z^4)^2(x^3yz^2)^4$

24. $\left(\frac{2m^3n}{p^2}\right)^3$

25. Find a polynomial that represents the volume of the figure. (If necessary, refer to the formulas inside the back cover.)

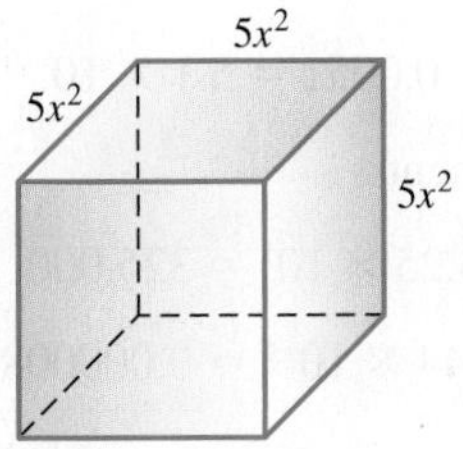

26. Explain why the product rule for exponents does not apply to the expression

$$7^2 + 7^4.$$

12.3 *Find each product.*

27. $5x(2x + 14)$

28. $-3p^3(2p^2 - 5p)$

29. $(3r - 2)(2r^2 + 4r - 3)$

30. $(2y + 3)(4y^2 - 6y + 9)$

31. $(5p^2 + 3p)(p^3 - p^2 + 5)$

32. $(x + 6)(x - 3)$

33. $(3k - 6)(2k + 1)$

34. $(6p - 3q)(2p - 7q)$

35. $(m^2 + m - 9)(2m^2 + 3m - 1)$

12.4 *Find each product.*

36. $(a + 4)^2$

37. $(3p - 2)^2$

38. $(2r + 5s)^2$

39. $(r + 2)^3$

40. $(2x - 1)^3$

41. $(2z + 7)(2z - 7)$

42. $(6m - 5)(6m + 5)$

43. $(5a + 6b)(5a - 6b)$

44. $(2x^2 + 5)(2x^2 - 5)$

45. **CONCEPT CHECK** The square of a binomial leads to a polynomial with how many terms? The product of the sum and difference of two terms leads to a polynomial with how many terms?

46. Explain why $(a + b)^2$ is not equivalent to $a^2 + b^2$.

12.5 *Evaluate each expression.*

47. $5^0 + 8^0$

48. 2^{-5}

49. $\left(\frac{6}{5}\right)^{-2}$

50. $4^{-2} - 4^{-1}$

Simplify each expression. Assume that all variables represent nonzero numbers.

51. $\frac{6^{-3}}{6^{-5}}$

52. $\frac{x^{-7}}{x^{-9}}$

53. $\frac{p^{-8}}{p^4}$

54. $\frac{r^{-2}}{r^{-6}}$

55. $(2^4)^2$

56. $(9^3)^{-2}$

57. $(5^{-2})^{-4}$

58. $(8^{-3})^4$

59. $\frac{(m^2)^3}{(m^4)^2}$

60. $\frac{y^4 \cdot y^{-2}}{y^{-5}}$

61. $\frac{r^9 \cdot r^{-5}}{r^{-2} \cdot r^{-7}}$

62. $(-5m^3)^2$

63. $(2y^{-4})^{-3}$

64. $\frac{ab^{-3}}{a^4b^2}$

65. $\frac{(6r^{-1})^2 \cdot (2r^{-4})}{r^{-5}(r^2)^{-3}}$

66. $\frac{(2m^{-5}n^2)^3(3m^2)^{-1}}{m^{-2}n^{-4}(m^{-1})^2}$

12.6 *Perform each division.*

67. $\dfrac{-15y^4}{-9y^2}$ **68.** $\dfrac{-12x^3y^2}{6xy}$ **69.** $\dfrac{6y^4 - 12y^2 + 18y}{-6y}$ **70.** $\dfrac{2p^3 - 6p^2 + 5p}{2p^2}$

71. $(5x^{13} - 10x^{12} + 20x^7 - 35x^5) \div (-5x^4)$ **72.** $(-10m^4n^2 + 5m^3n^3 + 6m^2n^4) \div (5m^2n)$

12.7 *Perform each division.*

73. $(2r^2 + 3r - 14) \div (r - 2)$ **74.** $\dfrac{12m^2 - 11m - 10}{3m - 5}$

75. $\dfrac{10a^3 + 5a^2 - 14a + 9}{5a^2 - 3}$ **76.** $\dfrac{2k^4 + 4k^3 + 9k^2 - 8}{2k^2 + 1}$

12.8 *Write each number in scientific notation.*

77. 48,000,000 **78.** 28,988,000,000 **79.** 0.000065 **80.** 0.0000000824

Write each number without exponents.

81. 2.4×10^4 **82.** 7.83×10^7 **83.** 8.97×10^{-7} **84.** 9.95×10^{-12}

Perform the indicated operations. Write the answers in scientific notation and then without exponents.

85. $(2 \times 10^{-3})(4 \times 10^5)$ **86.** $\dfrac{8 \times 10^4}{2 \times 10^{-2}}$

87. $\dfrac{12 \times 10^{-5} \times 5 \times 10^4}{4 \times 10^3 \times 6 \times 10^{-2}}$ **88.** $\dfrac{2.5 \times 10^5 \times 4.8 \times 10^{-4}}{7.5 \times 10^8 \times 1.6 \times 10^{-5}}$

89. A computer can perform 466,000,000 calculations per second. How many calculations can it perform per minute? Per hour?

90. In theory, there are 1×10^9 possible Social Security numbers. The population of the United States is about 3×10^8. How many Social Security numbers are available for each person? (*Source:* U.S. Census Bureau.)

Mixed Review Exercises

Perform each indicated operation. Assume that all variables represent nonzero real numbers.

91. $19^0 - 3^0$

92. $(3p)^4(3p^{-7})$

93. 7^{-2}

94. $(-7 + 2k)^2$

95. $\dfrac{2y^3 + 17y^2 + 37y + 7}{2y + 7}$

96. $\left(\dfrac{6r^2s}{5}\right)^4$

97. $-m^5(8m^2 + 10m + 6)$

98. $\left(\dfrac{1}{2}\right)^{-5}$

99. $(25x^2y^3 - 8xy^2 + 15x^3y) \div (5x)$

100. $(6r^{-2})^{-1}$

101. $(2x + y)^3$

102. $2^{-1} + 4^{-1}$

103. $(a + 2)(a^2 - 4a + 1)$

104. $(5y^3 - 8y^2 + 7) - (-3y^3 + y^2 + 2)$

105. $(2r + 5)(5r - 2)$

106. $(12a + 1)(12a - 1)$

107. What polynomial represents the area of this rectangle? The perimeter?

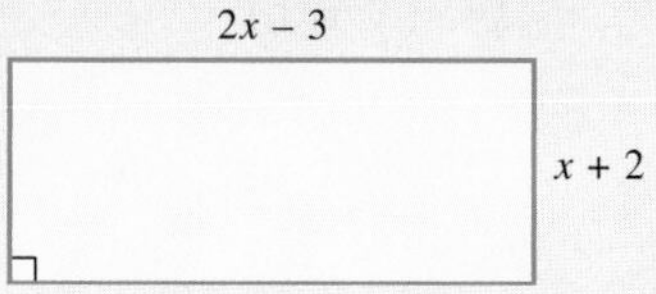

108. What polynomial represents the perimeter of this square? The area?

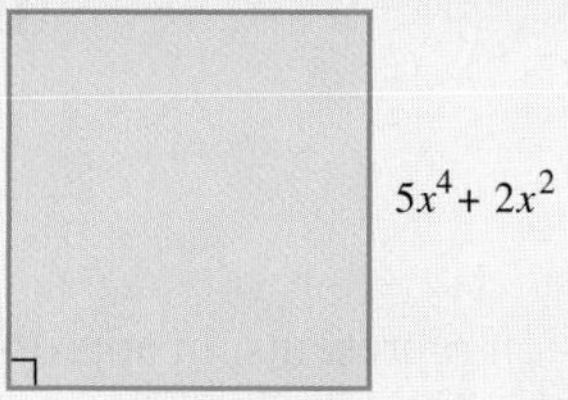

109. **CONCEPT CHECK** One of your friends in class simplified

$$\frac{6x^2 - 12x}{6} \quad \text{as} \quad x^2 - 12x.$$

What Went Wrong? Give the correct answer.

110. **CONCEPT CHECK** What polynomial, when multiplied by $6m^2n$, gives the following product?

$$12m^3n^2 + 18m^6n^3 - 24m^2n^2$$

Chapter 12 Test

CHAPTER Test Prep VIDEO

The Chapter Test Prep Videos with test solutions are available on DVD, in MyMathLab, and on YouTube—search "LialDevMath" and click on "Channels."

For each polynomial, combine like terms when possible and write the polynomial in descending powers of the variable. Give the degree of the simplified polynomial. Decide whether the simplified polynomial is a monomial, *a* binomial, *a* trinomial, *or* none of these.

1. $5x^2 + 8x - 12x^2$

2. $13n^3 - n^2 + n^4 + 3n^4 - 9n^2$

Perform the indicated operations.

3. $(5t^4 - 3t^2 + 7t + 3) - (t^4 - t^3 + 3t^2 + 8t + 3)$

4. $(2y^2 - 8y + 8) + (-3y^2 + 2y + 3) - (y^2 + 3y - 6)$

5. Subtract.

$$\begin{array}{r} 9t^3 - 4t^2 + 2t + 2 \\ \underline{9t^3 + 8t^2 - 3t - 6} \end{array}$$

6. $(-2)^3(-2)^2$

7. $\left(\frac{6}{m^2}\right)^3 \quad (m \neq 0)$

8. $3x^2(-9x^3 + 6x^2 - 2x + 1)$

9. $(2r - 3)(r^2 + 2r - 5)$

10. $(t - 8)(t + 3)$

11. $(4x + 3y)(2x - y)$

12. $(5x - 2y)^2$

13. $(10v + 3w)(10v - 3w)$

14. $(x + 1)^3$

15. What expression represents, in appropriate units, the perimeter of this square? The area?

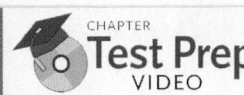

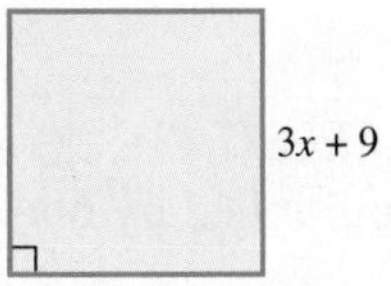

16. Determine whether each expression represents a number that is *positive, negative,* or *zero.*

(a) 3^{-4} **(b)** $(-3)^4$ **(c)** -3^4 **(d)** 3^0 **(e)** $(-3)^0 - 3^0$ **(f)** $(-3)^{-3}$

Evaluate each expression.

17. 5^{-4}

18. $(-3)^0 + 4^0$

19. $4^{-1} + 3^{-1}$

Perform the indicated operations. In Exercises 20 and 21, write each answer using only positive exponents. Assume that variables represent nonzero numbers.

20. $\dfrac{8^{-1} \cdot 8^4}{8^{-2}}$

21. $\dfrac{(x^{-3})^{-2}(x^{-1}y)^2}{(xy^{-2})^2}$

22. $\dfrac{8y^3 - 6y^2 + 4y + 10}{2y}$

23. $(-9x^2y^3 + 6x^4y^3 + 12xy^3) \div (3xy)$

24. $\dfrac{2x^2 + x - 36}{x - 4}$

25. $(3x^3 - x + 4) \div (x - 2)$

26. Write each number in scientific notation.

(a) 344,000,000,000 **(b)** 0.00000557

27. Write each number without exponents.

(a) 2.96×10^7 **(b)** 6.07×10^{-8}

28. A satellite galaxy of our own Milky Way, known as the Large Magellanic Cloud, is ***1000*** light-years across. A light-year is equal to ***5,890,000,000,000*** mi. (*Source:* "Images of Brightest Nebula Unveiled," *USA Today.*)

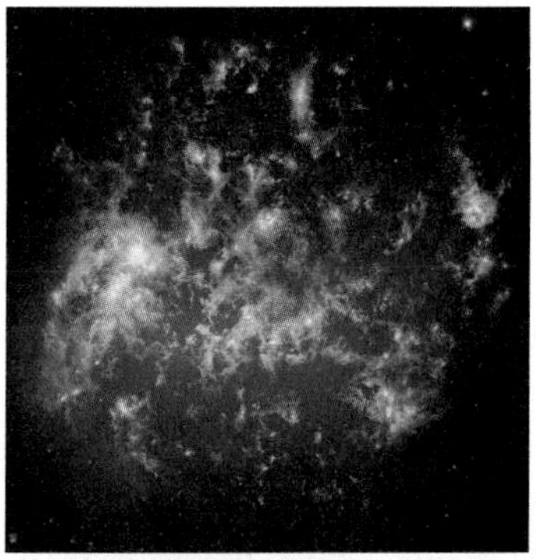

(a) Write the two boldface italic numbers in scientific notation.

(b) How many miles across is the Large Magellanic Cloud?

Math in the Media

FLOODS, HURRICANES, AND EARTHQUAKES, OH MY!

Charles F. Richter devised a scale in 1935 to compare the intensities, or relative powers, of earthquakes. The **intensity** of an earthquake is measured relative to the intensity of a standard **zero-level** earthquake of intensity I_0. The relationship is equivalent to $I = I_0 \times 10^R$, where R is the **Richter scale** measure. For example, if an earthquake has magnitude 5.0 on the Richter scale, then its intensity is calculated as $I = I_0 \times 10^{5.0} = I_0 \times 100{,}000$, which is 100,000 times as intense as a zero-level earthquake.

To compare an earthquake that measures 8.0 on the Richter scale to one that measures 5.0, find the ratio of the two intensities.

$$\frac{\text{intensity } 8.0}{\text{intensity } 5.0} = \frac{I_0 \times 10^{8.0}}{I_0 \times 10^{5.0}} = \frac{10^8}{10^5} = 10^{8-5} = 10^3 = 1000$$

Therefore, an earthquake that measures 8.0 is 1000 times as intense as one that measures 5.0.

The Gazette

Quake rattles Iowans

Chances of 'big one' happening here remote UNI professor says

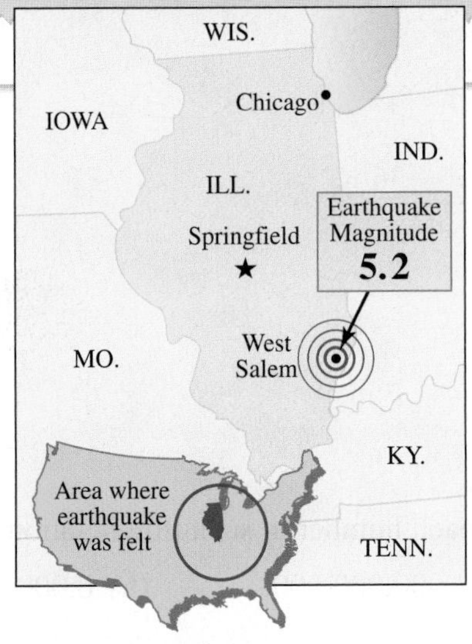

Source: WSRI; USGS.

Year	Earthquake	Richter Scale Measurement
2011	Northeast Japan	9.0
2005	Northern Sumatra, Indonesia	8.6
2004	West coast of Northern Sumatra	9.1
2003	Southeastern Iran	6.6
1998	Balleny Islands region	8.1
1906	San Francisco, CA	7.7

Source: U.S. Geological Survey.

1. Compare the intensity of the 2004 west coast of northern Sumatra earthquake to that of the 1998 Balleny Islands region earthquake.

2. Compare the intensity of the 2005 northern Sumatra, Indonesia earthquake to that of the 2003 southeastern Iran earthquake.

3. Compare the intensity of the 2011 northeast Japan earthquake to that of the 1906 San Francisco earthquake, the most powerful to strike the United States. (*Hint*: Use the exponential key of a scientific calculator to compute the required power of 10.)

4. Suppose an earthquake measures a value of x on the Richter scale. How would the intensity of a second earthquake compare if its Richter scale measure is $x + 4.0$? How would it compare if its Richter scale measure is $x - 1.0$?

13 Factoring and Applications

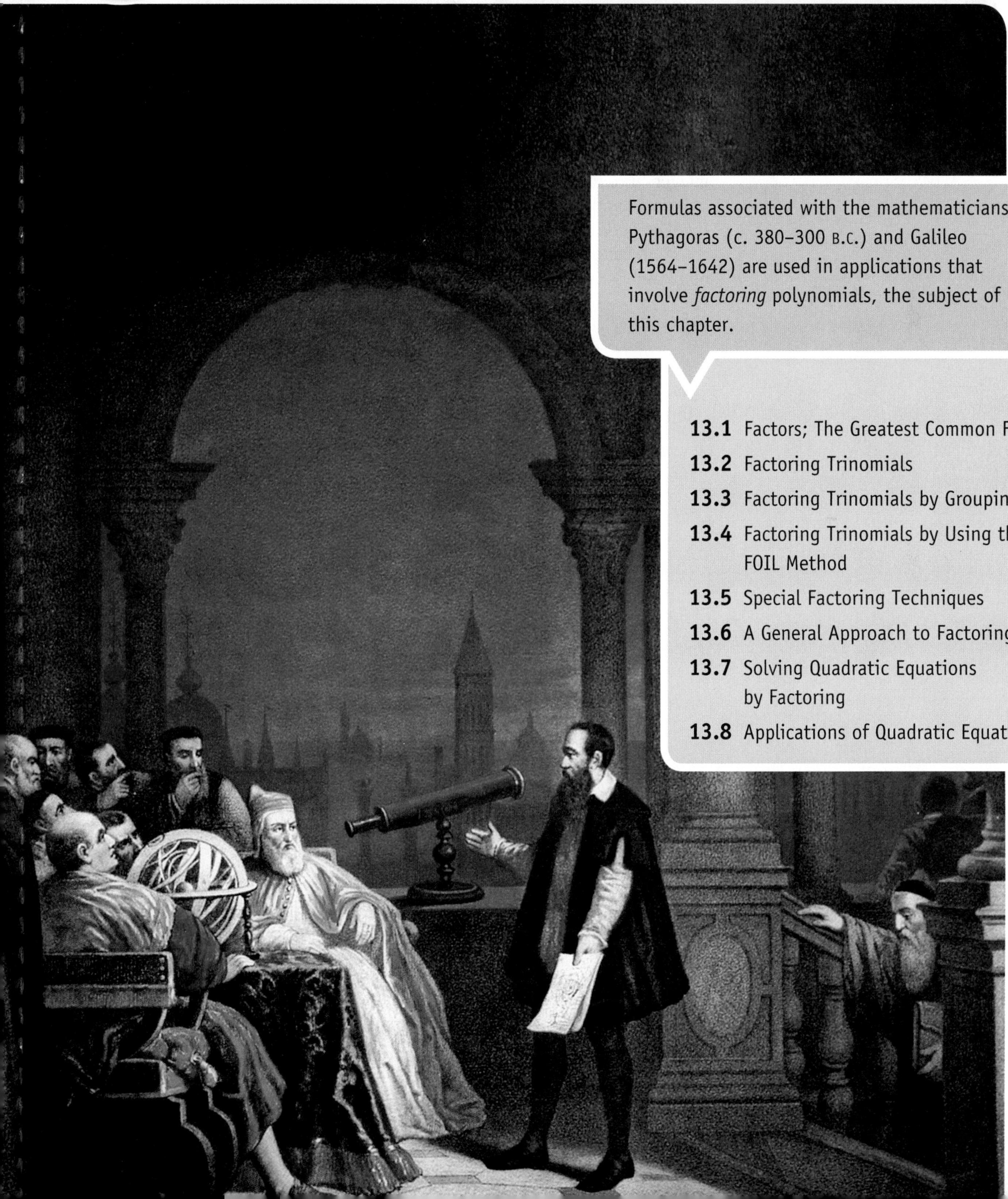

Formulas associated with the mathematicians Pythagoras (c. 380–300 B.C.) and Galileo (1564–1642) are used in applications that involve *factoring* polynomials, the subject of this chapter.

13.1 Factors; The Greatest Common Factor

OBJECTIVES

1. Find the greatest common factor of a list of numbers.
2. Find the greatest common factor of a list of variable terms.
3. Factor out the greatest common factor.
4. Factor by grouping.

To **factor** a number means to write it as the product of two or more numbers. The product is the **factored form** of the number. Consider an example.

$$12 = \underbrace{\overbrace{6 \cdot 2}^{\text{Factors}}}_{\text{Factored form}}$$

Factoring is a process that "undoes" multiplying. We multiply $6 \cdot 2$ to get 12, but we factor 12 by writing it as $6 \cdot 2$.

OBJECTIVE 1 Find the greatest common factor of a list of numbers. An integer that is a factor of two or more integers is a **common factor** of those integers. For example, 6 is a common factor of 18 and 24 because 6 is a factor of both 18 and 24. Other common factors of 18 and 24 are 1, 2, and 3.

The **greatest common factor (GCF)** of a list of integers is the largest common factor of those integers. This means 6 is the greatest common factor of 18 and 24, since it is the largest of their common factors.

Note

Factors of a number are also ***divisors*** of the number. The ***greatest common factor*** is the same as the ***greatest common divisor.*** Here are some divisibility rules for deciding what numbers divide into a given number.

A Whole Number Divisible by	Must Have the Following Property
2	Ends in 0, 2, 4, 6, or 8
3	Sum of digits divisible by 3
4	Last two digits form a number divisible by 4
5	Ends in 0 or 5
6	Divisible by both 2 and 3
8	Last three digits form a number divisible by 8
9	Sum of digits divisible by 9
10	Ends in 0

EXAMPLE 1 Finding the Greatest Common Factor for Numbers

Find the greatest common factor for each list of numbers.

(a) 30, 45

$$30 = 2 \cdot 3 \cdot 5$$
$$45 = 3 \cdot 3 \cdot 5$$

Write the prime factored form of each number.

Use each prime the **least** ***number of times it appears in*** **all** ***the factored forms.*** There is no 2 in the prime factored form of 45, so there will be no 2 in the greatest common factor. The least number of times 3 appears in all the factored forms is 1. The least number of times 5 appears is also 1.

$$\text{GCF} = 3^1 \cdot 5^1 = 15 \qquad 3^1 = 3 \text{ and } 5^1 = 5.$$

Continued on Next Page

(b) 72, 120, 432

$$72 = 2 \cdot 2 \cdot 2 \cdot 3 \cdot 3$$

$$120 = 2 \cdot 2 \cdot 2 \cdot 3 \cdot 5$$ Write the prime factored form of each number.

$$432 = 2 \cdot 2 \cdot 2 \cdot 2 \cdot 3 \cdot 3 \cdot 3$$

The least number of times 2 appears in all the factored forms is 3, and the least number of times 3 appears is 1. There is no 5 in the prime factored form of either 72 or 432.

$$\text{GCF} = 2^3 \cdot 3^1 = 24$$ $2^3 = 8$ and $3^1 = 3$.

(c) 10, 11, 14

$$10 = 2 \cdot 5$$

$$11 = 11$$ Write the prime factored form of each number.

$$14 = 2 \cdot 7$$

There are no primes common to all three numbers, so the GCF is 1.

Work Problem 1 at the Side. ▶

OBJECTIVE 2 Find the greatest common factor of a list of variable terms. The terms x^4, x^5, x^6, and x^7 have x^4 as the greatest common factor because the least exponent on the variable x in the factored forms is 4.

$$x^4 = 1 \cdot x^4, \quad x^5 = x \cdot x^4, \quad x^6 = x^2 \cdot x^4, \quad x^7 = x^3 \cdot x^4$$

$$\text{GCF} = x^4$$

Note

The exponent on a variable in the GCF is the* least *exponent that appears on that variable in* all *the terms.

EXAMPLE 2 Finding the Greatest Common Factor for Variable Terms

Find the greatest common factor for each list of terms.

(a) $21m^7, \quad 18m^6, \quad 45m^8, \quad 24m^5$

$$21m^7 = 3 \cdot 7 \cdot m^7$$

$$18m^6 = 2 \cdot 3 \cdot 3 \cdot m^6$$

$$45m^8 = 3 \cdot 3 \cdot 5 \cdot m^8$$

$$24m^5 = 2 \cdot 2 \cdot 2 \cdot 3 \cdot m^5$$

Here, 3 is the greatest common factor of the coefficients 21, 18, 45, and 24. The least exponent on m is 5.

$$\text{GCF} = 3m^5$$

(b) $x^4y^2, \quad x^7y^5, \quad x^3y^7, \quad y^{15}$

$$x^4y^2 = x^4 \cdot y^2$$

$$x^7y^5 = x^7 \cdot y^5$$

$$x^3y^7 = x^3 \cdot y^7$$

$$y^{15} = y^{15}$$

There is no x in the last term, y^{15}, so x will not appear in the greatest common factor. There is a y in each term, however, and 2 is the least exponent on y.

$$\text{GCF} = y^2$$

---

1 Find the greatest common factor for each list of numbers.

GS **(a)** 30, 20, 15

$30 = 2 \cdot 3 \cdot 5$

$20 = 2 \cdot$ ____ $\cdot$ ____

$15 = 3 \cdot$ ____

GCF = ____

(b) 42, 28, 35

(c) 12, 18, 26, 32

(d) 10, 15, 21

Answers

1. **(a)** 2; 5; 5; 5 **(b)** 7 **(c)** 2 **(d)** 1

2 Find the greatest common factor for each list of terms.

GS **(a)** $6m^4, 9m^2, 12m^5$

$6m^4 = 2 \cdot \_\_\_\_ \cdot m^4$

$9m^2 = 3 \cdot \_\_\_\_ \cdot \_\_\_\_$

$12m^5 = 2 \cdot 2 \cdot \_\_\_\_ \cdot \_\_\_\_$

GCF = ____

(b) $12p^5, 18q^4$

(c) y^4z^2, y^6z^8, z^9

(d) $12p^{11}, 17q^5$

Answers

2. **(a)** $3; 3; m^2; 3; m^5; 3m^2$
(b) 6 **(c)** z^2 **(d)** 1

Finding the Greatest Common Factor (GCF)

Step 1 **Factor.** Write each number in prime factored form.

Step 2 **List common factors.** List each prime number or each variable that is a factor of every term in the list. (If a prime does not appear in one of the prime factored forms, it *cannot* appear in the greatest common factor.)

Step 3 **Choose least exponents.** Use as exponents on the common prime factors the *least* exponents from the prime factored forms.

Step 4 **Multiply** the primes from Step 3. If there are no primes left after Step 3, the greatest common factor is 1.

◀ **Work Problem 2 at the Side.**

OBJECTIVE 3 Factor out the greatest common factor. Writing a polynomial (a sum) in factored form as a product is called **factoring** the polynomial. For example, the polynomial

$$3m + 12$$

has two terms, $3m$ and 12. The greatest common factor of these two terms is 3. We can write $3m + 12$ so that each term is a product with 3 as one factor.

$$3m + 12$$
$$= 3 \cdot m + 3 \cdot 4 \quad \text{GCF} = 3$$
$$= 3(m + 4) \quad \text{Distributive property}$$

The factored form of $3m + 12$ is $3(m + 4)$. This process is called **factoring out the greatest common factor.**

CAUTION

The polynomial $3m + 12$ is *not* in factored form when written as

$$3 \cdot m + 3 \cdot 4. \quad \text{Not in factored form}$$

The terms are factored, but the polynomial is not. The factored form of $3m + 12$ is the *product*

$$3(m + 4). \quad \text{In factored form}$$

EXAMPLE 3 Factoring Out the Greatest Common Factor

Write in factored form by factoring out the greatest common factor.

(a) $5y^2 + 10y$

$= 5y(y) + 5y(2)$ GCF = $5y$

$= 5y(y + 2)$ Distributive property

CHECK Multiply the factored form.

$5y(y + 2)$

$= 5y(y) + 5y(2)$ Distributive property

$= 5y^2 + 10y$ ✓ Original polynomial

Continued on Next Page

(b) $20m^5 + 10m^4 - 15m^3$

$= 5m^3(4m^2) + 5m^3(2m) - 5m^3(3)$ GCF $= 5m^3$

$= 5m^3(4m^2 + 2m - 3)$ Factor out $5m^3$.

CHECK $5m^3(4m^2 + 2m - 3)$

$= 5m^3(4m^2) + 5m^3(2m) + 5m^3(-3)$ Distributive property

$= 20m^5 + 10m^4 - 15m^3$ ✓ Original polynomial

(c) $x^5 + x^3$

$= x^3(x^2) + x^3(1)$ GCF $= x^3$

$= x^3(x^2 + 1)$ Don't forget the 1.

Check mentally by distributing x^3 over each term inside the parentheses.

(d) $20m^7p^2 - 36m^3p^4$

$= 4m^3p^2(5m^4) - 4m^3p^2(9p^2)$ GCF $= 4m^3p^2$

$= 4m^3p^2(5m^4 - 9p^2)$ Factor out $4m^3p^2$.

Check mentally by distributing $4m^3p^2$ over each term inside the parentheses.

CAUTION

Be sure to include the 1 in a problem like **Example 3(c)**. ***Check that the factored form can be multiplied out to give the original polynomial.***

Work Problem 3 at the Side. ▶

EXAMPLE 4 Factoring Out a Negative Common Factor

Write $-8x^4 + 16x^3 - 4x^2$ in factored form.

We can factor out either $4x^2$ or $-4x^2$ here. We usually factor out $-4x^2$ so that the coefficient of the first term in the trinomial factor will be positive.

$-8x^4 + 16x^3 - 4x^2$ Be careful with signs.

$= -4x^2(2x^2) - 4x^2(-4x) - 4x^2(1)$ $-4x^2$ is a common factor.

$= -4x^2(2x^2 - 4x + 1)$ Factor out $-4x^2$.

CHECK $-4x^2(2x^2 - 4x + 1)$

$= -4x^2(2x^2) - 4x^2(-4x) - 4x^2(1)$ Distributive property

$= -8x^4 + 16x^3 - 4x^2$ ✓ Original polynomial

Work Problem 4 at the Side. ▶

Note

Whenever we factor a polynomial in which the coefficient of the first term is negative, we will factor out the negative common factor, even if it is just -1. However, it would also be correct to factor out $4x^2$ in **Example 4** to obtain

$$4x^2(-2x^2 + 4x - 1).$$

3 Write in factored form by factoring out the greatest common factor.

GS **(a)** $4x^2 + 6x$

$= \_\_\_(\_\_\_) + 2x(\_\_\_)$

$= \_\_\_(\_\_\_ + \_\_\_)$

(b) $10y^5 - 8y^4 + 6y^2$

GS **(c)** $m^7 + m^9$

$= \_\_\_(\_\_\_) + \_\_\_(m^2)$

$= \_\_\_(\_\_\_ + \_\_\_)$

(d) $15x^3 - 10x^2 + 5x$

(e) $8p^5q^2 + 16p^6q^3 - 12p^4q^7$

4 Write

$$-14a^3b^2 - 21a^2b^3 + 7ab$$

in factored form by factoring out a negative common factor.

Answers

3. (a) $2x$; $2x$; 3; $2x$; $2x$; 3
(b) $2y^2(5y^3 - 4y^2 + 3)$
(c) m^7; 1; m^7; m^7; 1; m^2
(d) $5x(3x^2 - 2x + 1)$
(e) $4p^4q^2(2p + 4p^2q - 3q^5)$

4. $-7ab(2a^2b + 3ab^2 - 1)$

5 Write in factored form by factoring out the greatest common factor.

GS **(a)** $r(t-4)+5(t-4)$
$= (t-4)(\_\_\_\_)$

(b) $x(x+2)+7(x+2)$

(c) $y^2(y+2)-3(y+2)$

(d) $x(x-1)-5(x-1)$

EXAMPLE 5 Factoring Out a Common Binomial Factor

Write in factored form by factoring out the greatest common factor.

(a) $a(a+3)+4(a+3)$ (Same) — The binomial $a+3$ is the greatest common factor.

$= (a+3)(a+4)$ — Factor out $a+3$.

(b) $x^2(x+1)-5(x+1)$

$= (x+1)(x^2-5)$ — Factor out $x+1$.

◀ **Work Problem 5 at the Side.**

Note

In factored forms like those in **Example 5,** the order of the factors does not matter because of the commutative property of multiplication.

$(a+3)(a+4)$ can also be written $(a+4)(a+3)$.

OBJECTIVE 4 Factor by grouping. ***When a polynomial has four terms, common factors can sometimes be used to factor by grouping.***

EXAMPLE 6 Factoring by Grouping

Factor by grouping.

(a) $2x+6+ax+3a$

Group the first two terms and the last two terms, since the first two terms have a common factor of 2 and the last two terms have a common factor of *a*.

$2x+6+ax+3a$

$= (2x+6)+(ax+3a)$ — Group the terms.

$= 2(x+3)+a(x+3)$ — Factor each group.

The expression is still not in factored form because it is the *sum* of two terms. Now, however, $x+3$ is a common factor and can be factored out.

$= 2(x+3)+a(x+3)$ — $x+3$ is a common factor.

$= (x+3)(2+a)$ — Factor out $x+3$.

($(2+a)(x+3)$ is also correct.)

The final result is in factored form because it is a ***product.***

CHECK $(x+3)(2+a)$

$= x(2)+x(a)+3(2)+3(a)$ — Multiply using the FOIL method (**Section 12.3**).

$= 2x+ax+6+3a$ — Simplify.

$= 2x+6+ax+3a$ ✓ — Rearrange terms to obtain the original polynomial.

Continued on Next Page

Answers

5. **(a)** $r+5$ **(b)** $(x+2)(x+7)$
(c) $(y+2)(y^2-3)$ **(d)** $(x-1)(x-5)$

(b) $6ax + 24x + a + 4$

$= (6ax + 24x) + (a + 4)$ Group the terms.

$= 6x(a + 4) + 1(a + 4)$ Factor each group.

Remember the 1.

$= (a + 4)(6x + 1)$ Factor out $a + 4$.

CHECK $(a + 4)(6x + 1)$

$= 6ax + a + 24x + 4$ FOIL method

$= 6ax + 24x + a + 4$ ✓ Rearrange terms to obtain the original polynomial.

(c) $2x^2 - 10x + 3xy - 15y$

$= (2x^2 - 10x) + (3xy - 15y)$ Group the terms.

$= 2x(x - 5) + 3y(x - 5)$ Factor each group.

$= (x - 5)(2x + 3y)$ Factor out the common factor, $x - 5$.

CHECK $(x - 5)(2x + 3y)$

$= 2x^2 + 3xy - 10x - 15y$ FOIL method

$= 2x^2 - 10x + 3xy - 15y$ ✓ Original polynomial

(d) $t^3 + 2t^2 - 3t - 6$

Write a + sign between the groups.

$= (t^3 + 2t^2) + (-3t - 6)$ Group the terms.

$= t^2(t + 2) - 3(t + 2)$ Factor out -3 so there is a common factor, $t + 2$; $-3(t + 2) = -3t - 6$.

Be careful with signs.

$= (t + 2)(t^2 - 3)$ Factor out $t + 2$.

Check by multiplying using the FOIL method.

CAUTION

Be careful with signs when grouping in a problem like **Example 6(d).** It is wise to check the factoring in the second step, as shown in the example side comment, before continuing.

Work Problem 6 at the Side. ▶

Factoring a Polynomial with Four Terms by Grouping

Step 1 **Group terms.** Collect the terms into two groups so that each group has a common factor.

Step 2 **Factor within groups.** Factor out the greatest common factor from each group.

Step 3 **Factor the entire polynomial.** Factor a common binomial factor from the results of Step 2.

Step 4 **If necessary, rearrange terms.** If Step 2 does not result in a common binomial factor, try a different grouping.

6 Factor by grouping.

GS **(a)** $pq + 5q + 2p + 10$

$= (\_\_ + 5q) + (\_\_ + 10)$

$= \_\_(p + 5) + \_\_(p + 5)$

$= (\_\_\_\_)(\_\_\_\_)$

(b) $2xy + 3y + 2x + 3$

(c) $2a^2 - 4a + 3ab - 6b$

(d) $x^3 + 3x^2 - 5x - 15$

Answers

6. **(a)** pq; $2p$; q; 2; $p + 5$; $q + 2$
(b) $(2x + 3)(y + 1)$
(c) $(a - 2)(2a + 3b)$
(d) $(x + 3)(x^2 - 5)$

7 Factor by grouping.

(a) $6y^2 - 20w + 15y - 8yw$

(b) $9mn - 4 + 12m - 3n$

(c) $12p^2 - 28q - 16pq + 21p$

EXAMPLE 7 Rearranging Terms before Factoring by Grouping

Factor by grouping.

(a) $10x^2 - 12y + 15x - 8xy$

Factoring out the common factor 2 from the first two terms and the common factor x from the last two terms gives the following.

$$10x^2 - 12y + 15x - 8xy$$
$$= 2(5x^2 - 6y) + x(15 - 8y)$$

This does not lead to a common factor, so we try rearranging the terms. There is usually more than one way to do this. We try the following.

$$10x^2 - 12y + 15x - 8xy$$

$= 10x^2 - 8xy - 12y + 15x$	Commutative property
$= (10x^2 - 8xy) + (-12y + 15x)$	Group the terms.
$= 2x(5x - 4y) + 3(-4y + 5x)$	Factor each group.
$= 2x(5x - 4y) + 3(5x - 4y)$	Rewrite $-4y + 5x$.
$= (5x - 4y)(2x + 3)$	Factor out $5x - 4y$.

CHECK $(5x - 4y)(2x + 3)$

$= 10x^2 + 15x - 8xy - 12y$	FOIL method
$= 10x^2 - 12y + 15x - 8xy$ ✓	Original polynomial

(b) $2xy + 12 - 3y - 8x$

We must rearrange the terms to get two groups that each have a common factor. Trial and error suggests the following grouping.

$$2xy + 12 - 3y - 8x$$

Write a + sign between the two groups.

$= (2xy - 3y) + (-8x + 12)$	Group the terms.
$= y(2x - 3) - 4(2x - 3)$	Factor each group; $-4(2x - 3) = -8x + 12$
$= (2x - 3)(y - 4)$	Factor out $2x - 3$.

Be careful with signs.

Since the quantities in parentheses in the second step must be the same, we factored out -4 rather than 4.

CHECK $(2x - 3)(y - 4)$

$= 2xy - 8x - 3y + 12$	FOIL method
$= 2xy + 12 - 3y - 8x$ ✓	Original polynomial

CAUTION

Use negative signs carefully when grouping, as in **Example 7(b)**, or a sign error will occur. ***Always check by multiplying.***

◀ **Work Problem 7 at the Side.**

Answers

7. (a) $(2y + 5)(3y - 4w)$
(b) $(3m - 1)(3n + 4)$
(c) $(3p - 4q)(4p + 7)$

1.1 Exercises

FOR EXTRA HELP

Download the MyDashBoard App

MyMathLab®

CONC CHECK *Complete each statement.*

1. Tactor a number or quantity means to write it as a _____. Factoring is the opposite, or inverse, prcs of ________.

2. An integer or variable expression that is a factor of two or more terms is a ________ ________. For example, 12 (*is* / *is not*) a common factor of both 36 and 72 since it ________ evenly into both integers.

Find threatest common factor for each list of numbers. **See Example 1.**

3. 12,

4. 18, 24

5. 40, 20, 4

6. 50, 30, 5

7. 18, , 36, 48

8. 15, 30, 45, 75

9. 4, 9, 12

10. 9, 16, 24

Find threatest common factor for each list of terms. **See Example 2.**

11. $16y4$

12. $18w, 27$

13. $30x^3, 40x^6, 50x^7$

14. $60z\ 70z^8, 90z^9$

15. x^4y^3, xy^2

16. a^4b^5, a^3b

17. $42a^3, 36a, 90b, 48ab$

18. $45c^3d, 75c, 90d, 105cd$

19. $12m^2, 18m^5n^4, 36m^8n^3$

20. $25p^5r^7, 30p^7r^8, 50p^5r^3$

CONCEP CHECK *An expression is factored when it is written as a product, not a sum. Determe whether each expression is* factored *or* not factored.

21. $2k^2 5k)$

22. $2k^2(5k + 1)$

23. $2k^2 + (5k + 1)$

24. $(2k^2 + 5k) + 1$

25. COCEPT CHECK A student factored $18x^3y^2 + 9xy$ as $xy(2x^2y)$. **What Went Wrong?** Factor correctly.

26. How can we check an answer when we factor a polynomial?

GS *Comrlete each factoring by writing each polynomial as the product of two factors.* **See Exmple 3.**

27. $9n^4$
$= 3m^2(\_\_\_\_)$

28. $12p^5$
$= 6p^3(\_\_\_\_)$

29. $-8z^9$
$= -4z^5(\_\_\_\_)$

30. $-15k^{11}$
$= -5k^8(\_\_\_\_)$

31. $6m^4n^5$
$= 3m^3n(\_\_\_\_)$

32. $27a^3b^2$
$= 9a^2b(\_\_\_\_)$

33. $12y + 24$
$= 12(\_\_\_\_\_\_)$

34. $18p + 36$
$= 18(\_\_\_\_\_\_)$

35. $10a^2 - 20a$
$= 10a(\_\_\_\_\_\_)$

36. $15x^2 - 30x$
$= 15x(\_\_\_\_\_\_)$

37. $8x^2y + 12x^3y^2$
$= 4x^2y(\_\_\_\_\_\_)$

38. $18s^3t^2 + 10st$
$= 2st(\_\_\_\_\_\_)$

Write in factored form by factoring out the greatest common factor (or a negative common factor if the coefficient of the term of greatest degree is negative). ***See Examples 3–5.***

39. $x^2 - 4x$

40. $m^2 - 7m$

41. $6t^2 + 15t$

42. $8x^2 + 6x$

43. $m^3 - m^2$

44. $p^3 - p^2$

45. $-12x^3 - 6x^2$

46. $-21b^3 + 7b$

47. $65y^{10} + 35y^6$

48. $100a^5 + 16a^3$

49. $11w^3 - 100$

50. $13z^5 - 80$

51. $8mn^3 + 24m^2n^3$

52. $19p^2y + 38p^2y^3$

53. $-4x^3 + 10x^2 - 6x$

54. $-9z^3 + 6z^2 - 12z$

55. $13y^8 + 26y^4 - 39y^2$

56. $5x^5 + 25x^4 - 20x^3$

57. $45q^4p^5 + 36qp^6 + 81q^2p^3$

58. $125a^3z^5 + 60a^4z^4 - 85z^2$

59. $c(x + 2) + d(x + 2)$

60. $r(5 - x) + t(5 - x)$

61. $a^2(2a + b) - b(2a +$

62. $3x(x^2 + 5) - y(x^2 + 5)$

63. $q(p + 4) - 1(p + 4)$

64. $y^2(x - 4) + 1(x - 4)$

CONCEPT CHECK *Students often have difficulty when factoring by grouping because they are not able to tell when the polynomial is completely factored. For example,*

$$5y(2x - 3) + 8t(2x - 3) \quad \text{Not in factored form}$$

is not in factored form, because it is the **sum** *of two terms:* $5y(2x - 3)$ *and* $8t(2x - 3)$. *However, because* $2x - 3$ *is a common factor of these two terms, the expression can now be factored.*

$$(2x - 3)(5y + 8t) \quad \text{In factored form}$$

The factored form is a **product** *of the two factors* $2x - 3$ *and* $5y + 8t$.

Determine whether each expression is in factored form *or is* not in factored form. *If it is not in factored form, factor it if possible.*

65. $8(7t + 4) + x(7t + 4)$

66. $3r(5x - 1) + 7(5x - 1)$

67. $(8 + x)(7t + 4)$

68. $(3r + 7)(5x - 1)$

69. $18x^2(y + 4) + 7(y - 4)$

70. $12k^3(s - 3) + 7(s + 3)$

Factor by grouping. ***See Examples 6 and 7.***

71. $5m + mn + 20 + 4n$

72. $ts + 5t + 2s + 10$

73. $6xy - 21x + 8y - 28$

74. $2mn - 8n + 3m - 12$

75. $3xy + 9x + y + 3$

76. $6n + 4mn + 3 + 2m$

77. $7z^2 + 14z - az - 2a$

78. $2b^2 + 3b - 8ab - 12a$

79. $18r^2 + 12ry - 3xr - 2xy$

80. $5m^2 + 15mp - 2mr - 6pr$

81. $w^3 + w^2 + 9w + 9$

82. $y^3 + y^2 + 6y + 6$

83. $3a^3 + 6a^2 - 2a - 4$

84. $10x^3 + 15x^2 - 8x - 12$

85. $16m^3 - 4m^2p^2 - 4mp + p^3$

86. $10t^3 - 2t^2s^2 - 5ts + s^3$

87. $y^2 + 3x + 3y + xy$

88. $m^2 + 14p + 7m + 2mp$

89. $2z^2 + 6w - 4z - 3wz$

90. $2a^2 + 20b - 8a - 5ab$

91. $5m - 6p - 2mp + 15$

92. $7y - 9x - 3xy + 21$

93. $18r^2 - 2ty + 12ry - 3rt$

94. $12a^2 - 4bc + 16ac - 3ab$

Relating Concepts (Exercises 95–98) For Individual or Group Work

In many cases, the choice of which pairs of terms to group when factoring by grouping can be made in different ways. To see this for ***Example 7(b), work Exercises 95–98 in order.***

95. Start with the polynomial from **Example 7(b),** $2xy + 12 - 3y - 8x$, and rearrange the terms as follows: $2xy - 8x - 3y + 12$. What property from **Section 9.7** allows this?

96. Group the first two terms and the last two terms of the rearranged polynomial in **Exercise 95.** Then factor each group.

97. Is your result from **Exercise 96** in factored form? Explain your answer.

98. If your answer to **Exercise 97** is *no*, factor the polynomial. Is the result the same as the one shown for **Example 7(b)**?

13.2 Factoring Trinomials

OBJECTIVES

1. Factor trinomials with a coefficient of 1 for the second-degree term.
2. Factor such trinomials after factoring out the greatest common factor.

1 **(a)** Complete the table to find pairs of positive integers whose product is 6. Then find the sum of each pair.

GS

Factors of 6	Sums of Factors
6, ___	6 + ___ = ___
___, 2	___ + 2 = ___

(b) Which pair of factors from the table in part (a) has a sum of 5?

Answers

1. **(a)** 1; 1; 7; 3; 3; 5 **(b)** 3, 2

Using the FOIL method, we can find the product of the binomials $k - 3$ and $k + 1$.

$$(k-3)(k+1) = k^2 - 2k - 3 \quad \text{Multiplying}$$

Suppose instead that we are given the polynomial $k^2 - 2k - 3$ and want to rewrite it as the product $(k-3)(k+1)$.

$$k^2 - 2k - 3 = (k-3)(k+1) \quad \text{Factoring}$$

Recall from **Section 13.1** that this process is called *factoring* the polynomial. Factoring reverses or, "undoes," multiplying.

OBJECTIVE 1 **Factor trinomials with a coefficient of 1 for the second-degree term.** When factoring polynomials with integer coefficients, we use only integers in the factors. For example, we can factor $x^2 + 5x + 6$ by finding integers m and n such that

$$x^2 + 5x + 6 \quad \text{is written as} \quad (x+m)(x+n).$$

To find these integers m and n, we multiply the two binomials on the right.

$$\begin{aligned}&(x+m)(x+n)\\ &= x^2 + nx + mx + mn && \text{FOIL method}\\ &= x^2 + (n+m)x + mn && \text{Distributive property}\end{aligned}$$

Comparing this result with $x^2 + 5x + 6$ shows that we must find integers m and n having a sum of 5 and a product of 6.

Product of m and n is 6.

$$x^2 + 5x + 6 = x^2 + (n + m)x + mn$$

Sum of m and n is 5.

Because many pairs of integers have a sum of 5, it is best to begin by listing those pairs of integers whose product is 6. Both 5 and 6 are positive, so we consider only pairs in which both integers are positive.

◀ **Work Problem 1 at the Side.**

From **Margin Problem 1,** we see that the numbers 6 and 1 and the numbers 3 and 2 both have a product of 6, but only the pair 3 and 2 has a sum of 5. So 3 and 2 are the required integers.

$$x^2 + 5x + 6 \quad \text{is factored as} \quad (x+3)(x+2).$$

Check by multiplying the binomials using the FOIL method. ***Make sure that the sum of the outer and inner products produces the correct middle term.***

CHECK $(x+3)(x+2) = x^2 + 5x + 6$ ✓ Correct

$3x$

$\underline{2x}$

$5x$ Add.

This method of factoring can be used only for trinomials that have 1 as the coefficient of the second-degree (squared) term.

EXAMPLE 1 Factoring a Trinomial (All Positive Terms)

Factor $m^2 + 9m + 14$.

Look for two integers whose product is 14 and whose sum is 9. List the pairs of integers whose products are 14, and examine the sums. Only positive integers are needed since all signs in $m^2 + 9m + 14$ are positive.

Factors of 14	Sums of Factors	
14, 1	$14 + 1 = 15$	
7, 2	$7 + 2 = 9$	Sum is 9.

From the table, 7 and 2 are the required integers, since $7 \cdot 2 = 14$ and $7 + 2 = 9$.

$m^2 + 9m + 14$ factors as $(m + 7)(m + 2)$. ($(m + 2)(m + 7)$ is also correct.)

CHECK $(m + 7)(m + 2)$

$= m^2 + 2m + 7m + 14$ FOIL method

$= m^2 + 9m + 14$ ✓ Original polynomial

Work Problem 2 at the Side. ▶

Note

Remember that because of the commutative property of multiplication, the order of the factors does not matter. ***Always check by multiplying.***

EXAMPLE 2 Factoring a Trinomial (Negative Middle Term)

Factor $x^2 - 9x + 20$.

Find two integers whose product is 20 and whose sum is -9. Since the numbers we are looking for have a *positive product* and a *negative sum,* we consider only pairs of negative integers.

Factors of 20	Sums of Factors	
$-20, -1$	$-20 + (-1) = -21$	
$-10, -2$	$-10 + (-2) = -12$	
$-5, -4$	$-5 + (-4) = -9$	Sum is -9.

The required integers are -5 and -4.

$x^2 - 9x + 20$ factors as $(x - 5)(x - 4)$. (The order of the factors does not matter.)

CHECK $(x - 5)(x - 4)$

$= x^2 - 4x - 5x + 20$ FOIL method

$= x^2 - 9x + 20$ ✓ Original polynomial

Work Problem 3 at the Side. ▶

2 Factor each trinomial.

GS **(a)** $y^2 + 12y + 20$

Find two integers whose product is ____ and whose sum is ____. Complete the table.

Factors of 20	Sums of Factors
20, 1	$20 + 1 = 21$
10, ___	10 + ___ = ___
5, ___	5 + ___ = ___

Which pair of factors has the required sum? ______

Now factor the trinomial.

(b) $x^2 + 9x + 18$

3 Factor each trinomial.

GS **(a)** $t^2 - 12t + 32$

Find two integers whose product is ____ and whose sum is ____. Complete the table.

Factors of 32	Sums of Factors
$-32, -1$	$-32 + (-1) = -33$
-16, ___	-16 + (___) = ___
-8, ___	-8 + (___) = ___

Which pair of factors has the required sum? ______

Now factor the trinomial.

(b) $y^2 - 10y + 24$

Answers

2. **(a)** 20; 12; 2; 2; 12; 4; 4; 9; 10 and 2; $(y + 10)(y + 2)$
(b) $(x + 3)(x + 6)$

3. **(a)** 32; -12; -2; -2; -18; -4; -4; -12; -8 and -4; $(t - 8)(t - 4)$
(b) $(y - 6)(y - 4)$

4 Factor each trinomial.

(a) $z^2 + z - 30$

Factors of -30	Sums of Factors

(b) $x^2 + x - 42$

5 Factor each trinomial.

(a) $a^2 - 9a - 22$

Factors of -22	Sums of Factors

(b) $r^2 - 6r - 16$

Answers

4. **(a)** $(z+6)(z-5)$ **(b)** $(x+7)(x-6)$
5. **(a)** $(a-11)(a+2)$ **(b)** $(r-8)(r+2)$

EXAMPLE 3 Factoring a Trinomial (Negative Last (Constant) Term)

Factor $x^2 + x - 6$.

We must find two integers whose product is -6 and whose sum is 1 (since the coefficient of x, or $1x$, is 1). To get a *negative product,* the pairs of integers must have different signs.

Once we find the required pair, we can stop listing factors.

Factors of -6	Sums of Factors	
$6, -1$	$6 + (-1) = 5$	
$-6, 1$	$-6 + 1 = -5$	
$3, -2$	$3 + (-2) = 1$	Sum is 1.

The required integers are 3 and -2.

$$x^2 + x - 6 \quad \text{factors as} \quad (x+3)(x-2).$$

CHECK $(x+3)(x-2)$

$= x^2 - 2x + 3x - 6$ FOIL method

$= x^2 + x - 6$ ✓ Original polynomial

◀ **Work Problem 4 at the Side.**

EXAMPLE 4 Factoring a Trinomial (Two Negative Terms)

Factor $p^2 - 2p - 15$.

Find two integers whose product is -15 and whose sum is -2. Because the constant term, -15, is negative, we need pairs of integers with different signs.

Factors of -15	Sums of Factors	
$15, -1$	$15 + (-1) = 14$	
$-15, 1$	$-15 + 1 = -14$	
$5, -3$	$5 + (-3) = 2$	
$-5, 3$	$-5 + 3 = -2$	Sum is -2.

The required integers are -5 and 3.

$$p^2 - 2p - 15 \quad \text{factors as} \quad (p-5)(p+3).$$

CHECK Multiply $(p-5)(p+3)$ to obtain $p^2 - 2p - 15$. ✓

Note

In **Examples 1–4,** we listed factors in descending order (disregarding their signs) when we were looking for the required pair of integers. This helps avoid skipping the correct combination.

◀ **Work Problem 5 at the Side.**

Some trinomials cannot be factored using only integers. Such trinomials are **prime polynomials.**

EXAMPLE 5 **Deciding Whether Polynomials Are Prime**

Factor each trinomial.

(a) $x^2 - 5x + 12$

As in **Example 2,** both factors must be negative to give a positive product, 12, and a negative sum, −5. List pairs of negative integers whose product is 12, and examine the sums.

Factors of 12	Sums of Factors	
$-12, -1$	$-12 + (-1) = -13$	
$-6, -2$	$-6 + (-2) = -8$	
$-4, -3$	$-4 + (-3) = -7$	No sum is −5.

None of the pairs of integers has a sum of −5. Therefore, the trinomial $x^2 - 5x + 12$ *cannot be factored using only integers.* It is a *prime polynomial.*

(b) $k^2 - 8k + 11$

There is no pair of integers whose product is 11 and whose sum is −8, so $k^2 - 8k + 11$ is a prime polynomial.

Work Problem 6 at the Side. ▶

Factoring $x^2 + bx + c$

Find two integers whose product is c and whose sum is b.

1. Both integers must be positive if b and c are positive. (See **Example 1.**)
2. Both integers must be negative if c is positive and b is negative. (See **Example 2.**)
3. One integer must be positive and one must be negative if c is negative. (See **Examples 3 and 4.**)

EXAMPLE 6 **Factoring a Trinomial with Two Variables**

Factor $z^2 - 2bz - 3b^2$.

Here, the coefficient of z in the middle term is $-2b$, so we need to find two expressions whose product is $-3b^2$ and whose sum is $-2b$.

Factors of $-3b^2$	Sums of Factors	
$3b, -b$	$3b + (-b) = 2b$	
$-3b, b$	$-3b + b = -2b$	Sum is $-2b$.

$$z^2 - 2bz - 3b^2 \text{ factors as } (z - 3b)(z + b).$$

CHECK $(z - 3b)(z + b)$

$= z^2 + zb - 3bz - 3b^2$ FOIL method

$= z^2 + 1bz - 3bz - 3b^2$ Identity and commutative properties

$= z^2 - 2bz - 3b^2$ ✓ Combine like terms.

Work Problem 7 at the Side. ▶

6 Factor each trinomial, if possible.

(a) $x^2 + 5x + 8$

(b) $r^2 - 3r - 4$

(c) $m^2 - 2m + 5$

7 Factor each trinomial.

GS **(a)** $b^2 - 3ab - 4a^2$

We need two expressions whose product is ____ and whose sum is ____.

Complete the table.

Factors of $-4a^2$	Sums of Factors
$4a$, ___	$4a + (\_\_) = \_\_$
$-4a$, ___	$-4a + \_\_ = \_\_$
$2a, -2a$	$2a + (-2a) = 0$

Which pair of factors has the required sum? ________

Now factor the trinomial.

(b) $r^2 - 6rs + 8s^2$

Answers

6. (a) prime **(b)** $(r - 4)(r + 1)$ **(c)** prime

7. (a) $-4a^2$; $-3a$; $-a$; $-a$; $3a$; a; a; $-3a$; $-4a$ and a; $(b - 4a)(b + a)$ **(b)** $(r - 4s)(r - 2s)$

8 Factor each trinomial completely.

GS **(a)** $2p^3 + 6p^2 - 8p$

$= \_\_\_(\_\_\_ + 3p - \_\_\_)$

$= \_\_\_(\_\_\_\_)(\_\_\_\_)$

(b) $3y^4 - 27y^3 + 60y^2$

(c) $-3x^4 + 15x^3 - 18x^2$

Answers

8. **(a)** $2p$; p^2; 4; $2p$; $p + 4$; $p - 1$
(b) $3y^2(y - 5)(y - 4)$
(c) $-3x^2(x - 3)(x - 2)$

OBJECTIVE 2 Factor such trinomials after factoring out the greatest common factor. If the terms of a trinomial have a common factor, first factor it out. Then factor the remaining trinomial as in **Examples 1–6.**

EXAMPLE 7 Factoring a Trinomial with a Common Factor

Factor $4x^5 - 28x^4 + 40x^3$.

There is no second-degree term. Look for a common factor.

$$4x^5 - 28x^4 + 40x^3$$
$$= 4x^3(x^2 - 7x + 10) \quad \text{Factor out the greatest common factor, } 4x^3.$$

Now factor $x^2 - 7x + 10$. The integers -5 and -2 have a product of 10 and a sum of -7.

Include $4x^3$.
$$= 4x^3(x - 5)(x - 2) \quad \text{Completely factored form}$$

CHECK $4x^3(x - 5)(x - 2)$
$$= 4x^3(x^2 - 2x - 5x + 10) \quad \text{FOIL method}$$
$$= 4x^3(x^2 - 7x + 10) \quad \text{Combine like terms.}$$
$$= 4x^5 - 28x^4 + 40x^3 \checkmark \quad \text{Distributive property}$$

◀ Work Problem **8** at the Side.

CAUTION

When factoring, always look for a common factor first. Remember to include the common factor as part of the answer. Check by multiplying out the completely factored form.

13.2 Exercises

FOR EXTRA HELP

Download the MyDashBoard App

MyMathLab®

CONCEPT CHECK *Answer each question.*

1. When factoring a trinomial in x as $(x + a)(x + b)$, what must be true of a and b, if the coefficient of the constant term of the trinomial is negative?

2. In **Exercise 1,** what must be true of a and b if the coefficient of the constant term is positive?

3. Which one of the following is the correct factored form of $x^2 - 12x + 32$?

A. $(x - 8)(x + 4)$ **B.** $(x + 8)(x - 4)$

C. $(x - 8)(x - 4)$ **D.** $(x + 8)(x + 4)$

4. What would be the first step in factoring

$$2x^3 + 8x^2 - 10x?$$ (See **Example 7.**)

5. What polynomial can be factored as

$$(a + 9)(a + 4)?$$

6. What polynomial can be factored as

$$(y - 7)(y + 3)?$$

7. What is meant by a *prime polynomial?*

8. How can we check our work when factoring a trinomial? Does the check ensure that the trinomial is *completely* factored?

In Exercises 9–12, list all pairs of integers with the given product. Then find the pair whose sum is given. ***See the tables in Examples 1–4.***

9. Product: 12; Sum: 7

10. Product: 18; Sum: 9

11. Product: -24; Sum: -5

12. Product: -36; Sum: -16

GS *Complete each factoring.* ***See Examples 1–4.***

13. $p^2 + 11p + 30$
$= (p + 5)(\_\_\_\_\_)$

14. $x^2 + 10x + 21$
$= (x + 7)(\_\_\_\_\_)$

15. $x^2 + 15x + 44$
$= (x + 4)(\_\_\_\_\_)$

16. $r^2 + 15r + 56$
$= (r + 7)(\_\_\_\_\_)$

17. $x^2 - 9x + 8$
$= (x - 1)(\_\_\_\_\_)$

18. $t^2 - 14t + 24$
$= (t - 2)(\_\_\_\_\_)$

19. $y^2 - 2y - 15$
$= (y + 3)(\_\_\_\_\_)$

20. $t^2 - t - 42$
$= (t + 6)(\_\_\_\_\_)$

21. $x^2 + 9x - 22$
$= (x - 2)(\_\_\_\_\_)$

22. $x^2 + 6x - 27$
$= (x - 3)(\_\_\_\_\_)$

23. $y^2 - 7y - 18$
$= (y + 2)(\_\_\_\_\_)$

24. $y^2 - 2y - 24$
$= (y + 4)(\_\_\_\_\_)$

Factor completely. If a polynomial cannot be factored, write prime.
See Examples 1–5.

25. $y^2 + 9y + 8$

26. $a^2 + 9a + 20$

27. $b^2 + 8b + 15$

28. $x^2 + 6x + 8$

29. $m^2 + m - 20$

30. $p^2 + 4p - 5$

31. $x^2 + 3x - 40$

32. $d^2 + 4d - 45$

33. $x^2 + 4x + 5$

34. $t^2 + 11t + 12$

35. $y^2 - 8y + 15$

36. $y^2 - 6y + 8$

37. $z^2 - 15z + 56$

38. $x^2 - 13x + 36$

39. $r^2 - r - 30$

40. $q^2 - q - 42$

41. $a^2 - 8a - 48$

42. $m^2 - 10m - 24$

Factor completely. ***See Example 6.***

43. $r^2 + 3ra + 2a^2$

44. $x^2 + 5xa + 4a^2$

45. $x^2 + 4xy + 3y^2$

46. $p^2 + 9pq + 8q^2$

47. $t^2 - tz - 6z^2$

48. $a^2 - ab - 12b^2$

49. $v^2 - 11vw + 30w^2$

50. $v^2 - 11vx + 24x^2$

51. $a^2 + 2ab - 15b^2$

52. $m^2 + 4mn - 12n^2$

Factor completely. See ***Example 7.***

53. $4x^2 + 12x - 40$

54. $5y^2 - 5y - 30$

55. $2t^3 + 8t^2 + 6t$

56. $3t^3 + 27t^2 + 24t$

57. $-2x^6 - 8x^5 + 42x^4$

58. $-4y^5 - 12y^4 + 40y^3$

59. $a^5 + 3a^4b - 4a^3b^2$

60. $z^{10} - 4z^9y - 21z^8y^2$

61. $5m^5 + 25m^4 - 40m^2$

62. $12k^5 - 6k^3 + 10k^2$

63. $m^3n - 10m^2n^2 + 24mn^3$

64. $y^3z + 3y^2z^2 - 54yz^3$

13.3 Factoring Trinomials by Grouping

OBJECTIVE 1 **Factor trinomials by grouping when the coefficient of the second-degree term is not 1.** We now extend our work to factor a trinomial such as $2x^2 + 7x + 6$, where the coefficient of x^2 is *not* 1.

OBJECTIVE

1 **Factor trinomials by grouping when the coefficient of the second-degree term is not 1.**

EXAMPLE 1 **Factoring a Trinomial by Grouping (Coefficient of the Second-Degree Term Not 1)**

Factor $2x^2 + 7x + 6$.

To factor this trinomial, we look for two positive integers whose product is $2 \cdot 6 = 12$ and whose sum is 7.

$$\underset{\text{Product is } 2 \cdot 6 = 12.}{\overset{\text{Sum is 7.}}{2x^2 + 7x + 6}}$$

The required integers are 3 and 4, since $3 \cdot 4 = 12$ and $3 + 4 = 7$. We use these integers to write the middle term $7x$ as $3x + 4x$.

$$2x^2 + 7x + 6$$
$$= 2x^2 + \underset{7x}{\underbrace{3x + 4x}} + 6$$
$$= (2x^2 + 3x) + (4x + 6) \qquad \text{Group the terms.}$$
$$= x(2x + 3) + 2(2x + 3) \qquad \text{Factor each group.}$$
$$\qquad\qquad \text{Must be the same}$$
$$= (2x + 3)(x + 2) \qquad \text{Factor out } 2x + 3.$$

CHECK Multiply $(2x + 3)(x + 2)$ to obtain $2x^2 + 7x + 6.$ ✓

··· **Work Problem** 1 **at the Side.** ▶

1 **(a)** In **Example 1,** we factored
$$2x^2 + 7x + 6$$
by writing $7x$ as $3x + 4x$. This trinomial can also be factored by writing $7x$ as $4x + 3x$.

Complete the following.
$$2x^2 + 7x + 6$$
$$= 2x^2 + 4x + 3x + 6$$
$$= (2x^2 + \underline{\quad}) + (3x + \underline{\quad})$$
$$= 2x(x + \underline{\quad}) + 3(x + \underline{\quad})$$
$$= (\underline{\quad})(2x + 3)$$

(b) Is the answer in part (a) the same as in **Example 1**? (Remember that the order of the factors does not matter.)

(c) Factor $2z^2 + 5z + 3$ by grouping.

EXAMPLE 2 **Factoring Trinomials by Grouping**

Factor each trinomial.

(a) $6r^2 + r - 1$

We must find two integers with a product of $6(-1) = -6$ and a sum of 1 (since the coefficient of r, or $1r$, is 1). The integers are -2 and 3. We write the middle term r as $-2r + 3r$.

$$6r^2 + r - 1$$
$$= 6r^2 - 2r + 3r - 1 \qquad r = -2r + 3r$$
$$= (6r^2 - 2r) + (3r - 1) \qquad \text{Group the terms.}$$
$$= 2r(3r - 1) + 1(3r - 1) \qquad \text{The binomials must be the same.}$$
$$\qquad\qquad \text{Remember the 1.}$$
$$= (3r - 1)(2r + 1) \qquad \text{Factor out } 3r - 1.$$

CHECK Multiply $(3r - 1)(2r + 1)$ to obtain $6r^2 + r - 1.$ ✓

···································· **Continued on Next Page**

Answers

1. **(a)** $4x$; 6; 2; 2; $x + 2$ **(b)** yes
 (c) $(2z + 3)(z + 1)$

2 Factor each trinomial by grouping.

(a) $2m^2 + 7m + 3$

(b) $5p^2 - 2p - 3$

(c) $15k^2 - km - 2m^2$

3 Factor the trinomial completely.

GS $4x^2 - 2x - 30$

$= \_\_(2x^2 - \_\_ - 15)$

$= \_\_(2x^2 - \_\_ + \_\_ - 15)$

$= 2[(2x^2 - \_\_) + (5x - 15)]$

$= 2[\_\_(\_\_) + 5(x - 3)]$

$= \_\_\_\_\_\_$

4 Factor each trinomial completely.

(a) $18p^4 + 63p^3 + 27p^2$

(b) $6a^2 + 3ab - 18b^2$

Answers

2. (a) $(2m + 1)(m + 3)$
(b) $(5p + 3)(p - 1)$
(c) $(5k - 2m)(3k + m)$

3. 2; x; 2; $6x$; $5x$; $6x$; $2x$; $x - 3$; $2(x - 3)(2x + 5)$

4. (a) $9p^2(2p + 1)(p + 3)$
(b) $3(2a - 3b)(a + 2b)$

(b) $12z^2 - 5z - 2$

Look for two integers whose product is $12(-2) = -24$ and whose sum is -5. The required integers are 3 and -8.

$$12z^2 - 5z - 2$$

$$= 12z^2 + 3z - 8z - 2 \quad -5z = 3z - 8z$$

$$= (12z^2 + 3z) + (-8z - 2) \quad \text{Group the terms.}$$

$$= 3z(4z + 1) - 2(4z + 1) \quad \text{Factor each group.}$$

Be careful with signs.

$$= (4z + 1)(3z - 2) \quad \text{Factor out } 4z + 1.$$

CHECK Multiply $(4z + 1)(3z - 2)$ to obtain $12z^2 - 5z - 2$. ✓

(c) $10m^2 + mn - 3n^2$

Two integers whose product is $10(-3) = -30$ and whose sum is 1 are -5 and 6. Rewrite the trinomial with four terms.

$$10m^2 + mn - 3n^2$$

$$= 10m^2 - 5mn + 6mn - 3n^2 \quad mn = -5mn + 6mn$$

$$= 5m(2m - n) + 3n(2m - n) \quad \text{Group the terms. Factor each group.}$$

$$= (2m - n)(5m + 3n) \quad \text{Factor out } 2m - n.$$

CHECK Multiply $(2m - n)(5m + 3n)$ to obtain $10m^2 + mn - 3n^2$. ✓

◀ **Work Problem 2 at the Side.**

EXAMPLE 3 Factoring a Trinomial with a Common Factor by Grouping

Factor $28x^5 - 58x^4 - 30x^3$.

$$28x^5 - 58x^4 - 30x^3$$

$$= 2x^3(14x^2 - 29x - 15) \quad \text{Factor out the greatest common factor, } 2x^3.$$

To factor $14x^2 - 29x - 15$, find two integers whose product is $14(-15) = -210$ and whose sum is -29. Factoring 210 into prime factors helps find these integers.

$$210 = 2 \cdot 3 \cdot 5 \cdot 7$$

Combine the prime factors in pairs in different ways, using one positive factor and one negative factor to get -210. The factors 6 and -35 have the correct sum, -29.

$$28x^5 - 58x^4 - 30x^3$$

$$= 2x^3(14x^2 - 29x - 15) \quad \text{From above}$$

Remember the common factor.

$$= 2x^3(14x^2 + 6x - 35x - 15) \quad -29x = 6x - 35x$$

$$= 2x^3[(14x^2 + 6x) + (-35x - 15)] \quad \text{Group the terms.}$$

$$= 2x^3[2x(7x + 3) - 5(7x + 3)] \quad \text{Factor each group.}$$

$$= 2x^3[(7x + 3)(2x - 5)] \quad \text{Factor out } 7x + 3.$$

$$= 2x^3(7x + 3)(2x - 5)$$

Check by multiplying.

◀ **Work Problems 3 and 4 at the Side.**

13.3 Exercises

FOR EXTRA HELP

Download the MyDashBoard App

MyMathLab®

1. CONCEPT CHECK Which pair of integers would be used to rewrite the middle term when factoring $12y^2 + 5y - 2$ by grouping?

   A. $-8, 3$ B. $8, -3$ C. $-6, 4$ D. $6, -4$

2. CONCEPT CHECK Which pair of integers would be used to rewrite the middle term when factoring $20b^2 - 13b + 2$ by grouping?

   A. $10, 3$ B. $-10, -3$ C. $8, 5$ D. $-8, -5$

GS *The middle term of each trinomial has been rewritten. Now factor by grouping.* ***See Examples 1 and 2.***

3. $m^2 + 8m + 12$
   $= m^2 + 6m + 2m + 12$

4. $x^2 + 9x + 14$
   $= x^2 + 7x + 2x + 14$

5. $a^2 + 3a - 10$
   $= a^2 + 5a - 2a - 10$

6. $y^2 - 2y - 24$
   $= y^2 + 4y - 6y - 24$

7. $10t^2 + 9t + 2$
   $= 10t^2 + 5t + 4t + 2$

8. $6x^2 + 13x + 6$
   $= 6x^2 + 9x + 4x + 6$

9. $15z^2 - 19z + 6$
   $= 15z^2 - 10z - 9z + 6$

10. $12p^2 - 17p + 6$
    $= 12p^2 - 9p - 8p + 6$

11. $8s^2 + 2st - 3t^2$
    $= 8s^2 - 4st + 6st - 3t^2$

12. $3x^2 - xy - 14y^2$
    $= 3x^2 - 7xy + 6xy - 14y^2$

13. $15a^2 + 22ab + 8b^2$
    $= 15a^2 + 10ab + 12ab + 8b^2$

14. $25m^2 + 25mn + 6n^2$
    $= 25m^2 + 15mn + 10mn + 6n^2$

GS *Complete the steps to factor each trinomial by grouping.* ***See Examples 1 and 2.***

15. $2m^2 + 11m + 12$

    (a) Find two integers whose product is ____ • ____ = ____ and whose sum is ____.

    (b) The required integers are ____ and ____.

    (c) Write the middle term $11m$ as ____ + ____.

    (d) Rewrite the given trinomial using four terms.

    (e) Factor the polynomial in part (d) by grouping.

    (f) Check by multiplying.

16. $6y^2 - 19y + 10$

    (a) Find two integers whose product is ____ • ____ = ____ and whose sum is ____.

    (b) The required integers are ____ and ____.

    (c) Write the middle term $-19y$ as ____ + ____.

    (d) Rewrite the given trinomial using four terms.

    (e) Factor the polynomial in part (d) by grouping.

    (f) Check by multiplying.

*Factor each trinomial by grouping. **See Examples 1–3.***

17. $2x^2 + 7x + 3$

18. $3y^2 + 13y + 4$

19. $4r^2 + r - 3$

20. $4r^2 + 3r - 10$

21. $8m^2 - 10m - 3$

22. $20x^2 - 28x - 3$

23. $21m^2 + 13m + 2$

24. $38x^2 + 23x + 2$

25. $6b^2 + 7b + 2$

26. $6w^2 + 19w + 10$

27. $12y^2 - 13y + 3$

28. $15a^2 - 16a + 4$

29. $16 + 16x + 3x^2$

30. $18 + 65x + 7x^2$

31. $24x^2 - 42x + 9$

32. $48b^2 - 74b - 10$

33. $2m^3 + 2m^2 - 40m$

34. $3x^3 + 12x^2 - 36x$

35. $-32z^5 + 20z^4 + 12z^3$

36. $-18x^5 - 15x^4 + 75x^3$

37. $12p^2 + 7pq - 12q^2$

38. $6m^2 - 5mn - 6n^2$

39. $6a^2 - 7ab - 5b^2$

40. $25g^2 - 5gh - 2h^2$

41. CONCEPT CHECK On a quiz, a student factored $16x^2 - 24x + 5$ by grouping as follows.

$$\begin{aligned} &16x^2 - 24x + 5 \\ &= 16x^2 - 4x - 20x + 5 \\ &= 4x(4x - 1) - 5(4x - 1) \quad \text{His answer} \end{aligned}$$

He thought his answer was correct since it checked by multiplying. **What Went Wrong?** What is the correct factored form?

42. CONCEPT CHECK On the same quiz, another student factored $3k^3 - 12k^2 - 15k$ by first factoring out the common factor $3k$ to get $3k(k^2 - 4k - 5)$. Then she wrote the following.

$$\begin{aligned} &k^2 - 4k - 5 \\ &= k^2 - 5k + k - 5 \\ &= k(k - 5) + 1(k - 5) \\ &= (k - 5)(k + 1) \quad \text{Her answer} \end{aligned}$$

What Went Wrong? What is the correct factored form?

13.4 Factoring Trinomials by Using the FOIL Method

OBJECTIVE

1 **Factor trinomials by using the FOIL method.**

OBJECTIVE 1 Factor trinomials by using the FOIL method. This section shows an alternative method of factoring trinomials that uses trial and error.

EXAMPLE 1 Factoring a Trinomial Using FOIL (Coefficient of the Second-Degree Term Not 1)

Factor $2x^2 + 7x + 6$ (the trinomial in **Section 13.3, Example 1**).

We want to write $2x^2 + 7x + 6$ as the product of two binomials.

$$2x^2 + 7x + 6$$
$$= (\_\_\_\_)(\_\_\_\_)$$ We use the FOIL method in reverse.

The product of the two first terms of the binomials is $2x^2$. The possible factors of $2x^2$ are $2x$ and x, or $-2x$ and $-x$. Since all terms of the trinomial are positive, we consider only positive factors. Thus, we have the following.

$$2x^2 + 7x + 6$$
$$= (2x\_\_\_\_)(x\_\_\_\_)$$

The product of the two last terms of the binomials is 6. It can be factored as $1 \cdot 6$, $6 \cdot 1$, $2 \cdot 3$, or $3 \cdot 2$. We want the pair that gives the correct middle term, $7x$. Begin by trying 1 and 6 in $(2x\_\_\_\_)(x\_\_\_\_)$.

$(2x + 1)(x + 6)$ Incorrect

x

$12x$

$13x$ Add.

Now try the pair 6 and 1 in $(2x\_\_\_\_)(x\_\_\_\_)$.

$(2x + 6)(x + 1)$ Incorrect

$6x$

$2x$

$8x$ Add.

Since $2x + 6 = 2(x + 3)$, the terms of the binomial $2x + 6$ have a factor of 2, while the terms of $2x^2 + 7x + 6$ have no common factor other than 1. The product $(2x + 6)(x + 1)$ cannot be correct.

If the terms of the original polynomial have greatest common factor 1, then each of its factors will also have terms with GCF 1.

We try the pair 2 and 3 in $(2x\_\_\_\_)(x\_\_\_\_)$. Because of the common factor 2 in the terms of $2x + 2$, the product $(2x + 2)(x + 3)$ will not work. Finally, we try the pair 3 and 2 in $(2x\_\_\_\_)(x\_\_\_\_)$.

$(2x + 3)(x + 2) = 2x^2 + 7x + 6$ Correct

$3x$

$4x$

$7x$ Add.

Thus, $2x^2 + 7x + 6$ factors as $(2x + 3)(x + 2)$.

CHECK Multiply $(2x + 3)(x + 2)$ to obtain $2x^2 + 7x + 6$. ✓

Work Problem 1 at the Side. ▶

1 Factor each trinomial.

GS **(a)** $2p^2 + 9p + 9$

The possible factors of the first term $2p^2$ are ____ and p.

The possible factors of the last term 9 are 9 and ____, or ____ and 3.

Try various combinations to find the pair of factors that gives the correct middle term, ____.

Possible Pairs of Factors	Middle Term	Correct?
$(2p + 9)(p + 1)$	$11p$	(*Yes*/*No*)
$(2p + 1)(p + 9)$	____	(*Yes*/*No*)
$(2p + 3)(\_\_\_\_)$	____	(*Yes*/*No*)

$2p^2 + 9p + 9$ factors as ________.

(b) $8y^2 + 22y + 5$

Answers

1. **(a)** $2p$; 1; 3; $9p$; No; $19p$; No; $p + 3$; $9p$; Yes; $(2p + 3)(p + 3)$
(b) $(4y + 1)(2y + 5)$

2 Factor each trinomial.

(a) $6p^2 + 19p + 10$

(b) $8x^2 + 14x + 3$

EXAMPLE 2 **Factoring a Trinomial Using FOIL (All Positive Terms)**

Factor $8p^2 + 14p + 5$.

The number 8 has several possible pairs of factors, but 5 has only 1 and 5 or -1 and -5, so we begin by considering the factors of 5. We ignore the negative factors, since all coefficients in the trinomial are positive. If $8p^2 + 14p + 5$ can be factored, the factors will have this form.

$$(\_\_\_\_ + 5)(\_\_\_\_ + 1)$$

The possible pairs of factors of $8p^2$ are $8p$ and p, or $4p$ and $2p$. We try various combinations, checking to see if the middle term is $14p$.

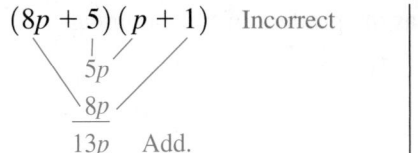

$(8p + 5)(p + 1)$ Incorrect: $5p + 8p = 13p$ Add.

$(p + 5)(8p + 1)$ Incorrect: $40p + p = 41p$ Add.

$(4p + 5)(2p + 1)$ Correct: $10p + 4p = 14p$ Add.

This last combination produces $14p$, the correct middle term.

$$8p^2 + 14p + 5 \quad \text{factors as} \quad (4p + 5)(2p + 1).$$

CHECK Multiply $(4p + 5)(2p + 1)$ to obtain $8p^2 + 14p + 5$. ✓

◀ **Work Problem 2 at the Side.**

3 Factor each trinomial.

(a) $4y^2 - 11y + 6$

(b) $9x^2 - 21x + 10$

EXAMPLE 3 **Factoring a Trinomial Using FOIL (Negative Middle Term)**

Factor $6x^2 - 11x + 3$.

Since 3 has only 1 and 3 or -1 and -3 as factors, it is better here to begin by factoring 3. The last term of the trinomial $6x^2 - 11x + 3$ is positive and the middle term has a negative coefficient, so we consider only negative factors. We need two negative factors because the *product* of two negative factors is positive and their *sum* is negative, as required.

We try -3 and -1 as factors of 3.

$$(\_\_\_\_ - 3)(\_\_\_\_ - 1)$$

The factors of $6x^2$ may be either $6x$ and x, or $2x$ and $3x$.

$(6x - 3)(x - 1)$ Incorrect: $-3x + (-6x) = -9x$ Add.

$(2x - 3)(3x - 1)$ Correct: $-9x + (-2x) = -11x$ Add.

The factors $2x$ and $3x$ produce $-11x$, the correct middle term.

$$6x^2 - 11x + 3 \quad \text{factors as} \quad (2x - 3)(3x - 1).$$

Check by multiplying.

◀ **Work Problem 3 at the Side.**

Answers

2. **(a)** $(3p + 2)(2p + 5)$
(b) $(4x + 1)(2x + 3)$

3. **(a)** $(4y - 3)(y - 2)$
(b) $(3x - 5)(3x - 2)$

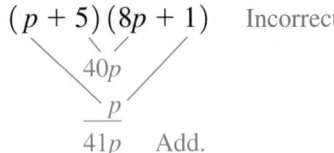

Note

Also in **Example 3,** our initial attempt to factor $6x^2 - 11x + 3$ as $(6x - 3)(x - 1)$ *cannot* be correct since the terms of $6x - 3$ have a common factor of 3, while those of the original polynomial do not.

EXAMPLE 4 **Factoring a Trinomial Using FOIL (Negative Constant Term)**

Factor $8x^2 + 6x - 9$.

The integer 8 has several possible pairs of factors, as does -9. Since the constant term is negative, one positive factor and one negative factor of -9 are needed. Since the coefficient of the middle term is relatively small, it is wise to avoid large factors, such as 8 or 9. We try $4x$ and $2x$ as factors of $8x^2$, and 3 and -3 as factors of -9, and check the middle term.

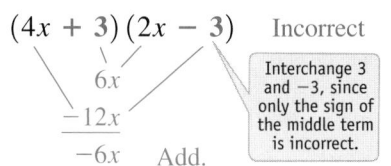

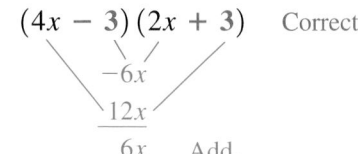

$(4x + 3)(2x - 3)$ Incorrect

$6x$
$-12x$
$-6x$ Add.

Interchange 3 and -3, since only the sign of the middle term is incorrect.

$(4x - 3)(2x + 3)$ Correct

$-6x$
$12x$
$6x$ Add.

The combination on the right produces $6x$, the correct middle term.

$$8x^2 + 6x - 9 \quad \text{factors as} \quad (4x - 3)(2x + 3).$$

Check by multiplying.

Work Problem **4** at the Side. ▶

EXAMPLE 5 **Factoring a Trinomial with Two Variables**

Factor $12a^2 - ab - 20b^2$.

There are several pairs of factors of $12a^2$, including

$$12a \text{ and } a, \quad 6a \text{ and } 2a, \quad \text{and} \quad 4a \text{ and } 3a.$$

There are also many possible pairs of factors of $-20b^2$, including

$$20b \text{ and } -b, \quad -20b \text{ and } b, \quad 10b \text{ and } -2b,$$
$$-10b \text{ and } 2b, \quad 4b \text{ and } -5b, \quad \text{and} \quad -4b \text{ and } 5b.$$

Once again, since the coefficient of the middle term is relatively small, avoid the larger factors. Try the factors $6a$ and $2a$, and $4b$ and $-5b$.

$$(6a + 4b)(2a - 5b) \quad \text{Incorrect}$$

This cannot be correct, since the terms of $6a + 4b$ have 2 as a common factor, while the terms of the given trinomial do not. Try $3a$ and $4a$ with $4b$ and $-5b$.

$$(3a + 4b)(4a - 5b)$$
$$= 12a^2 + ab - 20b^2 \quad \text{Incorrect}$$

Here the middle term is ab, rather than $-ab$. Interchange the signs of the last two terms in the factors.

$$(3a - 4b)(4a + 5b)$$
$$= 12a^2 - ab - 20b^2 \quad \text{Correct}$$

Thus, $12a^2 - ab - 20b^2$ factors as $(3a - 4b)(4a + 5b)$.

Work Problem **5** at the Side. ▶

4 Factor each trinomial, if possible.

GS **(a)** $6x^2 + 5x - 4$
$= (3x + \_\_\_)(\_\_\_ - \_\_\_)$

(b) $6m^2 - 11m - 10$

(c) $4x^2 - 3x - 7$

(d) $3y^2 + 8y - 6$

5 Factor each trinomial.

GS **(a)** $2x^2 - 5xy - 3y^2$
$= (2x + \_\_\_)(\_\_\_ - \_\_\_)$

(b) $8a^2 + 2ab - 3b^2$

Answers

4. **(a)** 4; $2x$; 1
(b) $(2m - 5)(3m + 2)$
(c) $(4x - 7)(x + 1)$
(d) prime

5. **(a)** y; x; $3y$
(b) $(4a + 3b)(2a - b)$

6 Factor each trinomial.

GS **(a)** $36z^3 - 6z^2 - 72z$

$= \_\_\_\_(\_\_\_\_ - \_\_\_\_ - 12)$

$= \_\_\_\_(\_\_\_\_\_\_)(\_\_\_\_\_\_)$

(b) $10x^3 + 45x^2 - 90x$

(c) $-24x^3 + 32x^2 + 6x$

Answers

6. **(a)** $6z$; $6z^2$; z; $6z$; $3z + 4$; $2z - 3$
(b) $5x(2x - 3)(x + 6)$
(c) $-2x(6x + 1)(2x - 3)$

EXAMPLE 6 Factoring Trinomials with Common Factors

Factor each trinomial.

(a) $15y^3 + 55y^2 + 30y$

$$15y^3 + 55y^2 + 30y$$
$$= 5y(3y^2 + 11y + 6) \quad \text{Factor out the greatest common factor, } 5y.$$

Now factor $3y^2 + 11y + 6$. Try $3y$ and y as factors of $3y^2$ and 2 and 3 as factors of 6.

$$(3y + 2)(y + 3)$$
$$= 3y^2 + 11y + 6 \quad \text{Correct}$$

This leads to the completely factored form.

$$15y^3 + 55y^2 + 30y$$
$$= 5y(3y + 2)(y + 3)$$

Remember the common factor.

CHECK $5y(3y + 2)(y + 3)$

$= 5y(3y^2 + 9y + 2y + 6)$ FOIL method

$= 5y(3y^2 + 11y + 6)$ Combine like terms.

$= 15y^3 + 55y^2 + 30y$ ✓ Distributive property

(b) $-24a^3 - 42a^2 + 45a$

The common factor could be $3a$ or $-3a$. If we factor out $-3a$, the first term of the trinomial will be positive, which makes it easier to factor the remaining trinomial.

$$-24a^3 - 42a^2 + 45a$$
$$= -3a(8a^2 + 14a - 15) \quad \text{Factor out } -3a.$$
$$= -3a(4a - 3)(2a + 5) \quad \text{Use trial and error.}$$

CHECK $-3a(4a - 3)(2a + 5)$

$= -3a(8a^2 + 14a - 15)$ FOIL method; Combine like terms.

$= -24a^3 - 42a^2 + 45a$ ✓ Distributive property

CAUTION

Remember to include the common factor in the final factored form.

◀ **Work Problem 6 at the Side.**

13.4 Exercises

FOR EXTRA HELP

Download the MyDashBoard App

MyMathLab®

CONCEPT CHECK *Decide which is the correct factored form of the given polynomial.*

1. $2x^2 - x - 1$

 A. $(2x - 1)(x + 1)$ B. $(2x + 1)(x - 1)$

2. $3a^2 - 5a - 2$

 A. $(3a + 1)(a - 2)$ B. $(3a - 1)(a + 2)$

3. $4y^2 + 17y - 15$

 A. $(y + 5)(4y - 3)$ B. $(2y - 5)(2y + 3)$

4. $12c^2 - 7c - 12$

 A. $(6c - 2)(2c + 6)$ B. $(4c + 3)(3c - 4)$

5. $4k^2 + 13mk + 3m^2$

 A. $(4k + m)(k + 3m)$ B. $(4k + 3m)(k + m)$

6. $2x^2 + 11x + 12$

 A. $(2x + 3)(x + 4)$ B. $(2x + 4)(x + 3)$

Complete each factoring. ***See Examples 1–6.***

7. $6a^2 + 7ab - 20b^2$

 $= (3a - 4b)(\_\_\_\_\_\_)$

8. $9m^2 - 3mn - 2n^2$

 $= (3m + n)(\_\_\_\_\_\_)$

9. $2x^2 + 6x - 8$

 $= 2(\_\_\_\_\_\_\_\_\_\_)$

 $= 2(\_\_\_\_\_\_)(\_\_\_\_\_\_)$

10. $3x^2 - 9x - 30$

 $= 3(\_\_\_\_\_\_\_\_\_\_)$

 $= 3(\_\_\_\_\_\_)(\_\_\_\_\_\_)$

11. $4z^3 - 10z^2 - 6z$

 $= 2z(\_\_\_\_\_\_\_\_\_\_)$

 $= 2z(\_\_\_\_\_\_)(\_\_\_\_\_\_)$

12. $15r^3 - 39r^2 - 18r$

 $= 3r(\_\_\_\_\_\_\_\_\_\_)$

 $= 3r(\_\_\_\_\_\_)(\_\_\_\_\_\_)$

13. CONCEPT CHECK A student factoring the trinomial

$$12x^2 + 7x - 12$$

wrote $(4x + 4)$ as one binomial factor. **What Went Wrong?** Factor the trinomial correctly.

14. CONCEPT CHECK Another student completely factored the trinomial

$$4x^2 + 10x - 6$$

as $(4x - 2)(x + 3)$. **What Went Wrong?** Factor the trinomial completely.

Factor each trinomial completely. ***See Examples 1–6.***

15. $3a^2 + 10a + 7$

16. $7r^2 + 8r + 1$

17. $2y^2 + 7y + 6$

18. $5z^2 + 12z + 4$

19. $15m^2 + m - 2$

20. $6x^2 + x - 1$

21. $12s^2 + 11s - 5$

22. $20x^2 + 11x - 3$

23. $10m^2 - 23m + 12$

24. $6x^2 - 17x + 12$

25. $8w^2 - 14w + 3$

26. $9p^2 - 18p + 8$

27. $20y^2 - 39y - 11$

28. $10x^2 - 11x - 6$

29. $3x^2 - 15x + 16$

30. $2t^2 + 13t - 18$

31. $20x^2 + 22x + 6$

32. $36y^2 + 81y + 45$

33. $-40m^2q - mq + 6q$

34. $-15a^2b - 22ab - 8b$

35. $15n^4 - 39n^3 + 18n^2$

36. $24a^4 + 10a^3 - 4a^2$

37. $-15x^2y^2 + 7xy^2 + 4y^2$

38. $-14a^2b^3 - 15ab^3 + 9b^3$

39. $5a^2 - 7ab - 6b^2$

40. $6x^2 - 5xy - y^2$

41. $12s^2 + 11st - 5t^2$

42. $25a^2 + 25ab + 6b^2$

43. $6m^6n + 7m^5n^2 + 2m^4n^3$

44. $12k^3q^4 - 4k^2q^5 - kq^6$

If a trinomial has a negative coefficient for the second-degree term, such as $-2x^2 + 11x - 12$, *it may be easier to factor by first factoring out* -1.

$$-2x^2 + 11x - 12$$
$$= -1(2x^2 - 11x + 12) \quad \text{Factor out } -1.$$
$$= -1(2x - 3)(x - 4) \quad \text{Factor the trinomial.}$$

Use this method to factor each trinomial in Exercises 45–50.

45. $-x^2 - 4x + 21$

46. $-x^2 + x + 72$

47. $-3x^2 - x + 4$

48. $-5x^2 + 2x + 16$

49. $-2a^2 - 5ab - 2b^2$

50. $-3p^2 + 13pq - 4q^2$

Relating Concepts (Exercises 51–56) For Individual or Group Work

Often there are several different equivalent forms of an answer that are all correct. **Work Exercises 51–56 in order,** *to see this for factoring problems.*

51. Factor the integer 35 as the product of two prime numbers.

52. Factor the integer 35 as the product of the negatives of two prime numbers.

53. Verify that $6x^2 - 11x + 4$ factors as $(3x - 4)(2x - 1)$.

54. Verify that $6x^2 - 11x + 4$ factors as $(4 - 3x)(1 - 2x)$.

55. Compare the two valid factored forms in **Exercises 53 and 54.** How do the factors in each case compare?

56. Suppose you know that the correct factored form of a particular trinomial is $(7t - 3)(2t - 5)$. From **Exercises 51–55,** what is another valid factored form?

13.5 Special Factoring Techniques

OBJECTIVES

1. Factor a difference of squares.
2. Factor a perfect square trinomial.
3. Factor a difference of cubes.
4. Factor a sum of cubes.

By reversing the rules for multiplication of binomials from **Section 12.4,** we obtain rules for factoring polynomials in certain forms.

OBJECTIVE 1 Factor a difference of squares. The formula for the product of the sum and difference of the same two terms is

$$(a+b)(a-b)=a^2-b^2.$$

Reversing this rule leads to the following special factoring rule.

Factoring a Difference of Squares

$$a^2-b^2=(a+b)(a-b)$$

For example,

$$m^2-16$$
$$=m^2-4^2$$
$$=(m+4)(m-4).$$

The following must be true for a binomial to be a difference of squares.

1. Both terms of the binomial must be squares, such as

$$x^2,\quad 9y^2=(3y)^2,\quad 25=5^2,\quad 1=1^2,\quad m^4=(m^2)^2.$$

2. The terms of the binomial must have different signs (one positive and one negative).

EXAMPLE 1 Factoring Differences of Squares

Factor each binomial, if possible.

$$a^2 - b^2 = (a+b)\,(a-b)$$
$$\downarrow\quad\downarrow\qquad\downarrow\quad\downarrow\quad\downarrow\quad\downarrow$$

(a) $x^2-49=x^2-7^2=(x+7)(x-7)$

(b) $y^2-m^2=(y+m)(y-m)$

(c) x^2-8

Because 8 is not the square of an integer, this binomial is not a difference of squares. It is a prime polynomial.

(d) p^2+16

Since p^2+16 is a *sum* of squares, it is not equal to $(p+4)(p-4)$. Also, we use the FOIL method and try the following.

$(p-4)(p-4)$	$(p+4)(p+4)$
$=p^2-8p+16$, not p^2+16.	$=p^2+8p+16$, not p^2+16.

Thus, p^2+16 is a prime polynomial.

Work Problem **1** at the Side. ▶

CAUTION

As Example 1(d) suggests, after any common factor is removed, a sum of squares cannot be factored.

1 Factor each binomial, if possible.

GS **(a)** x^2-81

$=(x+\_\_\_\_)(x-\_\_\_\_)$

(b) p^2-100

(c) t^2-s^2

(d) y^2-10

(e) x^2+36

(f) x^2+y^2

Answers

1. (a) 9; 9
(b) $(p+10)(p-10)$
(c) $(t+s)(t-s)$
(d) prime (e) prime (f) prime

2 Factor each difference of squares.

GS **(a)** $9m^2 - 49$

$= (\_\_\_)^2 - \_\_\_^2$

$= (\_\_\_\_)(\_\_\_\_)$

(b) $64a^2 - 25$

(c) $25a^2 - 64b^2$

3 Factor completely.

GS **(a)** $50r^2 - 32$

$= \_\_\_(\_\_\_\_)$

$= \_\_\_[(\_\_\_)^2 - \_\_\_^2]$

$= \_\_\_(\_\_\_\_)(\_\_\_\_)$

(b) $27y^2 - 75$

(c) $k^4 - 49$

(d) $81r^4 - 16$

Answers

2. **(a)** $3m$; 7; $3m + 7$; $3m - 7$
(b) $(8a + 5)(8a - 5)$
(c) $(5a + 8b)(5a - 8b)$

3. **(a)** 2; $25r^2 - 16$; 2; $5r$; 4; 2; $5r + 4$; $5r - 4$
(b) $3(3y + 5)(3y - 5)$
(c) $(k^2 + 7)(k^2 - 7)$
(d) $(9r^2 + 4)(3r + 2)(3r - 2)$

EXAMPLE 2 **Factoring Differences of Squares**

Factor each difference of squares.

$$a^2 \;-\; b^2 = (a + b)(a - b)$$

(a) $25m^2 - 16 = (5m)^2 - 4^2 = (5m + 4)(5m - 4)$

(b) $49z^2 - 64$

$= (7z)^2 - 8^2$ Write each term as a square.

$= (7z + 8)(7z - 8)$ Factor the difference of squares.

(c) $9x^2 - 4z^2$

$= (3x)^2 - (2z)^2$ Write each term as a square.

$= (3x + 2z)(3x - 2z)$ Factor the difference of squares.

◀ **Work Problem 2 at the Side.**

Note

Always check a factored form by multiplying.

EXAMPLE 3 **Factoring More Complex Differences of Squares**

Factor completely.

(a) $81y^2 - 36$

$= 9(9y^2 - 4)$ Factor out the GCF, 9.

$= 9[(3y)^2 - 2^2]$ Write each term as a square.

$= 9(3y + 2)(3y - 2)$ Factor the difference of squares.

(b) $p^4 - 36$

$= (p^2)^2 - 6^2$ Write each term as a square.

$= (p^2 + 6)(p^2 - 6)$ Factor the difference of squares.

Neither binomial can be factored further.

(c) $m^4 - 16$

$= (m^2)^2 - 4^2$ Write each term as a square.

$= (m^2 + 4)(m^2 - 4)$ Factor the difference of squares.

Don't stop here.

$= (m^2 + 4)(m + 2)(m - 2)$ Factor the difference of squares again.

CAUTION

Factor again when any of the factors is a difference of squares, as in **Example 3(c).** Check by multiplying.

◀ **Work Problem 3 at the Side.**

OBJECTIVE 2 **Factor a perfect square trinomial.** The expressions 144, $4x^2$, and $81m^6$ are **perfect squares** because

$$144 = 12^2, \quad 4x^2 = (2x)^2, \quad \text{and} \quad 81m^6 = (9m^3)^2.$$

A **perfect square trinomial** is a trinomial that is the square of a binomial. For example, $x^2 + 8x + 16$ is a perfect square trinomial because it is the square of the binomial $x + 4$.

$$\begin{aligned} &x^2 + 8x + 16 \\ &= (x + 4)(x + 4) \\ &= (x + 4)^2 \end{aligned}$$

On the one hand, a necessary condition for a trinomial to be a perfect square is that *two of its terms must be perfect squares*. For this reason,

$$16x^2 + 4x + 15$$

is *not* a perfect square trinomial because only the term $16x^2$ is a perfect square.

On the other hand, even if two of the terms are perfect squares, the trinomial may *not* be a perfect square trinomial. For example,

$$x^2 + 6x + 36$$

has two perfect square terms, but it is *not* a perfect square trinomial.

We can multiply to see that the square of a binomial gives one of the following perfect square trinomials.

Factoring Perfect Square Trinomials

$$a^2 + 2ab + b^2 = (a + b)^2$$
$$a^2 - 2ab + b^2 = (a - b)^2$$

The middle term of a perfect square trinomial is always twice the product of the two terms in the squared binomial. (See Section 12.4.) Use this to check any attempt to factor a trinomial that appears to be a perfect square.

EXAMPLE 4 **Factoring a Perfect Square Trinomial**

Factor $x^2 + 10x + 25$.

The x^2-term is a perfect square, and so is 25. Try to factor the trinomial

$$x^2 + 10x + 25 \quad \text{as} \quad (x + 5)^2.$$

To check, take twice the product of the two terms in the squared binomial.

$$\mathbf{2 \cdot x \cdot 5 = 10x} \leftarrow \text{Middle term of } x^2 + 10x + 25$$

Twice; First term of binomial; Last term of binomial

Since $10x$ is the middle term of the given trinomial, the trinomial is a perfect square.

$$x^2 + 10x + 25 \quad \text{factors as} \quad (x + 5)^2.$$

Work Problem 4 at the Side. ▶

4 Factor each trinomial.

GS **(a)** $p^2 + 14p + 49$

The term p^2 is a perfect square. Since $49 = \_\_\_\_^2$, 49 (*is / is not*) a perfect square. Try to factor the trinomial

$$p^2 + 14p + 49$$

as $\quad (p + \_\_\_\_)^2$.

Check by taking twice the ______ of the two terms of the squared binomial.

$$2 \cdot p \cdot \_\_\_\_ = \_\_\_\_$$

The result (*is / is not*) the middle term of the given trinomial. The trinomial is a perfect square and factors as ______.

(b) $m^2 + 8m + 16$

(c) $x^2 + 2x + 1$

Answers

4. **(a)** 7; is; 7; product; 7; $14p$; is; $(p + 7)^2$
(b) $(m + 4)^2$ **(c)** $(x + 1)^2$

5 Factor each trinomial.

(a) $p^2 - 18p + 81$

(b) $16a^2 + 56a + 49$

(c) $121p^2 + 110p + 100$

(d) $64x^2 - 48x + 9$

(e) $27y^3 + 72y^2 + 48y$

Answers

5. **(a)** $(p-9)^2$ **(b)** $(4a+7)^2$ **(c)** prime **(d)** $(8x-3)^2$ **(e)** $3y(3y+4)^2$

EXAMPLE 5 Factoring Perfect Square Trinomials

Factor each trinomial.

(a) $x^2 - 22x + 121$

The first and last terms are perfect squares ($121 = 11^2$ or $(-11)^2$). Check to see whether the middle term of $x^2 - 22x + 121$ is twice the product of the first and last terms of the binomial $x - 11$.

$$\underbrace{2}_{\text{Twice}} \cdot \underbrace{x}_{\text{First term}} \cdot \underbrace{(-11)}_{\text{Last term}} = \mathbf{-22x} \leftarrow \text{Middle term of } x^2 - 22x + 121$$

Thus, $x^2 - 22x + 121$ is a perfect square trinomial.

$$x^2 - 22x + 121 \quad \text{factors as} \quad (x - 11)^2.$$

(Same sign)

Notice that the sign of the second term in the squared binomial is the same as the sign of the middle term in the trinomial.

(b) $9m^2 - 24m + 16 = (3m)^2 + 2(3m)(-4) + (-4)^2 = (3m - 4)^2$

(Twice: 2; First term: $3m$; Last Term: -4)

(c) $25y^2 + 20y + 16$

The first and last terms are perfect squares.

$$25y^2 = (5y)^2 \quad \text{and} \quad 16 = 4^2$$

Twice the product of the first and last terms of the binomial $5y + 4$ is

$$2 \cdot 5y \cdot 4 = \mathbf{40y},$$

which is *not* the middle term of

$$25y^2 + \mathbf{20y} + 16.$$

This trinomial is not a perfect square. In fact, the trinomial cannot be factored even with the methods of the previous sections. It is a prime polynomial.

(d) $12z^3 + 60z^2 + 75z$

$= 3z(4z^2 + 20z + 25)$ — Factor out the common factor, $3z$.

$= 3z[(2z)^2 + 2(2z)(5) + 5^2]$ — $4z^2 + 20z + 25$ is a perfect square trinomial.

$= 3z(2z + 5)^2$ — Factor.

Note

1. The sign of the second term in the squared binomial is always the same as the sign of the middle term in the trinomial.
2. The first and last terms of a perfect square trinomial must be *positive*, because they are squares. For example, the polynomial $x^2 - 2x - 1$ cannot be a perfect square because the last term is negative.
3. Perfect square trinomials can also be factored using grouping or the FOIL method. Using the method of this section is often easier.

◄ Work Problem **5** at the Side.

OBJECTIVE 3 Factor a difference of cubes. We factored a difference of squares at the beginning of this section. We can also factor a **difference of cubes.**

Factoring a Difference of Cubes

$$a^3 - b^3 = (a - b)(a^2 + ab + b^2)$$

This pattern should be memorized. To see that the pattern is correct, multiply

$$(a - b)(a^2 + ab + b^2),$$

as shown in the margin.

$$\begin{array}{r} a^2 + ab + b^2 \\ a - b \\ \hline -a^2b - ab^2 - b^3 \\ a^3 + a^2b + ab^2 \quad\quad \\ \hline a^3 \quad\quad\quad\quad - b^3 \end{array}$$

Multiply vertically. **(Section 12.3)**

Add.

Notice the pattern of the terms in the factored form of $a^3 - b^3$.

- $a^3 - b^3$ factors as (a binomial factor) · (a trinomial factor).
- The binomial factor has the difference of the cube roots of the given terms.
- The terms in the trinomial factor are all positive.
- The terms in the binomial factor determine the trinomial factor.

$$a^3 - b^3 = (a - b)(\underbrace{a^2}_{\text{First term squared}} + \underbrace{ab}_{\text{Positive product of the terms}} + \underbrace{b^2}_{\text{Second term squared}})$$

EXAMPLE 6 Factoring Differences of Cubes

Factor each difference of cubes.

(a) $m^3 - 125$ Use the pattern for a difference of cubes.

$$\begin{aligned} a^3 - b^3 &= (a - b)(a^2 + ab + b^2) \\ m^3 - 125 = m^3 - 5^3 &= (m - 5)(m^2 + 5m + 5^2) \\ &= (m - 5)(m^2 + 5m + 25) \end{aligned}$$

(b) $8p^3 - 27$

$= (2p)^3 - 3^3$ $\quad$ $8p^3 = (2p)^3$ and $27 = 3^3$.

$= (2p - 3)[(2p)^2 + (2p)3 + 3^2]$ $\quad$ Let $a = 2p$ and $b = 3$.

$= (2p - 3)(4p^2 + 6p + 9)$ $\quad$ Apply the exponents. Multiply.

$(2p)^2 = 2^2p^2 = 4p^2$

(c) $4m^3 - 32n^3$

$= 4(m^3 - 8n^3)$ $\quad$ Factor out the common factor.

$= 4[m^3 - (2n)^3]$ $\quad$ $8n^3 = (2n)^3$

$= 4(m - 2n)[m^2 + m(2n) + (2n)^2]$ $\quad$ Let $a = m$ and $b = 2n$.

$= 4(m - 2n)(m^2 + 2mn + 4n^2)$ $\quad$ Apply the exponents. Multiply.

Work Problem 6 at the Side. ▶

6 Factor each difference of cubes.

GS **(a)** $t^3 - 64$

$= \_\_\_^3 - \_\_\_^3$

$= (\_\_\_ - \_\_\_) \cdot (\_\_\_^2 + \_\_\_ + \_\_\_^2)$

$= \_\_\_\_\_\_$

(b) $2x^3 - 54$

(c) $8k^3 - y^3$

Answers

6. (a) t; 4; t; 4; t; $4t$; 4; $(t - 4)(t^2 + 4t + 16)$
(b) $2(x - 3)(x^2 + 3x + 9)$
(c) $(2k - y)(4k^2 + 2ky + y^2)$

7 Factor each sum of cubes.

GS **(a)** $x^3 + 8$

$= \_\_\_^3 + \_\_\_^3$

$= (\_\_\_ + \_\_\_) \cdot$
$(\_\_\_^2 - \_\_\_ + \_\_\_^2)$

$= \_\_\_\_\_\_\_\_$

(b) $64y^3 + 1$

(c) $27m^3 + 343n^3$

CAUTION

A common error in factoring $a^3 - b^3 = (a - b)(a^2 + ab + b^2)$ is to try to factor $a^2 + ab + b^2$. This is usually not possible.

OBJECTIVE 4 Factor a sum of cubes. A sum of squares, such as $m^2 + 25$, cannot be factored using real numbers, but a **sum of cubes** can.

Factoring a Sum of Cubes

$$a^3 + b^3 = (a + b)(a^2 - ab + b^2)$$

Observe the positive and negative signs in these factoring patterns.

Positive

$a^3 - b^3 = (a - b)(a^2 + ab + b^2)$ Difference of cubes

Same sign Opposite sign

The only difference between the patterns is the positive and negative signs.

Positive

$a^3 + b^3 = (a + b)(a^2 - ab + b^2)$ Sum of cubes

Same sign Opposite sign

EXAMPLE 7 Factoring Sums of Cubes

Factor each sum of cubes.

(a) $k^3 + 27$

$= k^3 + 3^3$ $\quad 27 = 3^3$

$= (k + 3)(k^2 - 3k + 3^2)$ $\quad$ Let $a = k$ and $b = 3$.

$= (k + 3)(k^2 - 3k + 9)$ $\quad$ Apply the exponent.

(b) $8m^3 + 125p^3$

$= (2m)^3 + (5p)^3$ $\quad 8m^3 = (2m)^3$; $125p^3 = (5p)^3$

$= (2m + 5p)[(2m)^2 - (2m)(5p) + (5p)^2]$ $\quad$ Let $a = 2m$ and $b = 5p$.

$= (2m + 5p)(4m^2 - 10mp + 25p^2)$ $\quad$ Apply the exponents. Multiply.

◀ Work Problem **7** at the Side.

Answers

7. (a) x; 2; x; 2; x; $2x$; 2;
$(x + 2)(x^2 - 2x + 4)$
(b) $(4y + 1)(16y^2 - 4y + 1)$
(c) $(3m + 7n)(9m^2 - 21mn + 49n^2)$

13.5 Exercises

FOR EXTRA HELP

Download the MyDashBoard App

MyMathLab®

CONCEPT CHECK *Work each problem.*

1. To help factor a difference of squares, complete the following list of squares.

$1^2 =$ ______	$2^2 =$ ______	$3^2 =$ ______	$4^2 =$ ______	$5^2 =$ ______
$6^2 =$ ______	$7^2 =$ ______	$8^2 =$ ______	$9^2 =$ ______	$10^2 =$ ______
$11^2 =$ ______	$12^2 =$ ______	$13^2 =$ ______	$14^2 =$ ______	$15^2 =$ ______
$16^2 =$ ______	$17^2 =$ ______	$18^2 =$ ______	$19^2 =$ ______	$20^2 =$ ______

2. To use the factoring techniques described in this section, it is helpful to recognize fourth powers of integers. Complete the following list of fourth powers.

$1^4 =$ ______ $2^4 =$ ______ $3^4 =$ ______ $4^4 =$ ______ $5^4 =$ ______

3. Which of the following are differences of squares?

A. $x^2 - 4$ **B.** $y^2 + 9$ **C.** $2a^2 - 25$ **D.** $9m^2 - 1$

4. On a quiz, a student indicated *prime* when asked to factor $4x^2 + 16$, since she said that a sum of squares cannot be factored. **What Went Wrong?**

Factor each binomial completely. ***See Examples 1–3.***

5. $y^2 - 25$

6. $t^2 - 16$

7. $x^2 - 144$

8. $x^2 - 400$

9. $m^2 - 12$

10. $k^2 - 18$

11. $m^2 + 64$

12. $k^2 + 49$

13. $9r^2 - 4$

14. $4x^2 - 9$

15. $36x^2 - 16$

16. $32a^2 - 8$

17. $196p^2 - 225$

18. $361q^2 - 400$

19. $16r^2 - 25a^2$

20. $49m^2 - 100p^2$

21. $100x^2 + 49$

22. $81w^2 + 16$

23. $p^4 - 49$

24. $r^4 - 25$

25. $x^4 - 1$

26. $y^4 - 10{,}000$

27. $p^4 - 256$

28. $x^4 - 625$

29. **CONCEPT CHECK** Which of the following are perfect square trinomials?

A. $y^2 - 13y + 36$ B. $x^2 + 6x + 9$ C. $4z^2 - 4z + 1$ D. $16m^2 + 10m + 1$

30. In the polynomial $9y^2 + 14y + 25$, the first and last terms are perfect squares. Can the polynomial be factored? If it can, factor it. If it cannot, explain why it is not a perfect square trinomial.

Factor each trinomial completely. It may be necessary to factor out the greatest common factor first. **See Examples 4 and 5.**

31. $w^2 + 2w + 1$

32. $p^2 + 4p + 4$

33. $x^2 - 8x + 16$

34. $x^2 - 10x + 25$

35. $x^2 - 10x + 100$

36. $x^2 - 18x + 36$

37. $2x^2 + 24x + 72$

38. $3y^2 + 48y + 192$

39. $4x^2 + 12x + 9$

40. $25x^2 + 10x + 1$

41. $16x^2 - 40x + 25$

42. $36y^2 - 60y + 25$

43. $49x^2 - 28xy + 4y^2$

44. $4z^2 - 12zw + 9w^2$

45. $64x^2 + 48xy + 9y^2$

46. $9t^2 + 24tr + 16r^2$

47. $-50h^3 + 40h^2y - 8hy^2$

48. $-18x^3 - 48x^2y - 32xy^2$

Although we usually factor polynomials using integers, we can apply the same concepts to factoring using fractions and decimals.

$$z^2 - \frac{9}{16}$$

$$= z^2 - \left(\frac{3}{4}\right)^2 \qquad \frac{9}{16} = \left(\frac{3}{4}\right)^2$$

$$= \left(z + \frac{3}{4}\right)\left(z - \frac{3}{4}\right) \qquad \text{Factor the difference of squares.}$$

Factor each binomial or trinomial.

49. $p^2 - \frac{1}{9}$

50. $q^2 - \frac{1}{4}$

51. $4m^2 - \frac{9}{25}$

52. $100b^2 - \frac{4}{49}$

53. $x^2 - 0.64$

54. $y^2 - 0.36$

55. $t^2 + t + \frac{1}{4}$

56. $m^2 + \frac{2}{3}m + \frac{1}{9}$

57. $x^2 - 1.0x + 0.25$

58. $y^2 - 1.4y + 0.49$

CONCEPT CHECK *Work each problem.*

59. To help factor the sum or difference of cubes, complete the following list of cubes.

$1^3 =$ ____ $2^3 =$ ____ $3^3 =$ ____ $4^3 =$ ____ $5^3 =$ ____

$6^3 =$ ____ $7^3 =$ ____ $8^3 =$ ____ $9^3 =$ ____ $10^3 =$ ____

60. The following powers of x are all perfect cubes: $x^3, x^6, x^9, x^{12}, x^{15}$. Based on this observation, we may make a conjecture that if the power of a variable is divisible by ____ (with 0 remainder), then we have a perfect cube.

61. Which of the following are differences of cubes?

A. $9x^3 - 125$ **B.** $x^3 - 16$ **C.** $x^3 - 1$ **D.** $8x^3 - 27y^3$

62. Which of the following are sums of cubes?

A. $x^3 + 1$ **B.** $x^3 + 36$ **C.** $12x^3 + 27$ **D.** $64x^3 + 216y^3$

Factor. Use your answers in ***Exercises 59 and 60*** *as necessary.* ***See Examples 6 and 7.***

63. $x^3 + 1$

64. $m^3 + 8$

65. $x^3 - 1$

66. $m^3 - 8$

67. $p^3 + q^3$

68. $w^3 + z^3$

69. $y^3 - 216$

70. $x^3 - 343$

71. $k^3 + 1000$

72. $p^3 + 512$

73. $27x^3 - 1$

74. $64y^3 - 27$

75. $125x^3 + 8$

76. $216x^3 + 125$

77. $y^3 - 8x^3$

78. $w^3 - 216z^3$

79. $27x^3 - 64y^3$

80. $125m^3 - 8n^3$

81. $8p^3 + 729q^3$

82. $27x^3 + 1000y^3$

83. $16t^3 - 2$

84. $3p^3 - 81$

85. $40w^3 + 135$

86. $32z^3 + 500$

87. $x^3 + y^6$

88. $p^9 + q^3$

89. $125k^3 - 8m^9$

90. $125c^6 - 216d^3$

Relating Concepts (Exercises 91–98) For Individual or Group Work

A binomial may be both a difference of squares and a difference of cubes. One example of such a binomial is $x^6 - 1$. Using the techniques of this section, one factoring method will give the completely factored form, while the other will not. ***Work Exercises 91–98 in order,*** *to determine the method to use.*

91. Factor $x^6 - 1$ as the difference of two squares.

92. The factored form obtained in **Exercise 91** consists of a difference of cubes multiplied by a sum of cubes. Factor each binomial further.

93. Now start over and factor $x^6 - 1$ as a difference of cubes.

94. The factored form obtained in **Exercise 93** consists of a binomial that is a difference of squares and a trinomial. Factor the binomial further.

95. Compare the results in **Exercises 92 and 94.** Which one of these is the completely factored form?

96. Verify that the trinomial in the factored form in **Exercise 94** is the product of the two trinomials in the factored form in **Exercise 92.**

97. Use the results of **Exercises 91–96** to complete the following statement.

In general, if I must choose between factoring first using the method for a difference of squares or the method for a difference of cubes, I should choose the ______________ method to eventually obtain the completely factored form.

98. Find the *completely* factored form of $x^6 - 729$ using the knowledge gained in **Exercises 91–97.**

13.6 A General Approach to Factoring

OBJECTIVES

1. Factor out any common factor.
2. Factor binomials.
3. Factor trinomials.
4. Factor polynomials of more than three terms.

A polynomial is *completely factored* when it is written as a product of prime polynomials with integer coefficients.

Factoring a Polynomial

Step 1 **Factor out any common factor.**

Step 2 **If the polynomial is a binomial,** check to see if it is the difference of squares, the difference of cubes, or the sum of cubes.

If the polynomial is a trinomial, check to see if it is a perfect square trinomial. If it is not, factor as in **Sections 13.2–13.4.**

If the polynomial has more than three terms, try to factor by grouping as in **Section 13.1.**

Step 3 **If any of the factors can be factored further, do so.**

Step 4 **Check the factored form by multiplying.**

OBJECTIVE 1 **Factor out any common factor.** ***This step is always the same, regardless of the number of terms in the polynomial.***

EXAMPLE 1 Factoring Out a Common Factor

Factor each polynomial.

(a) $8m^2p^2 + 4mp$

$= 4mp(2mp + 1)$ GCF $= 4mp$

(b) $5x(a + b) - y(a + b)$

$= (a + b)(5x - y)$ Factor out $a + b$.

Work Problem 1 at the Side. ▶

OBJECTIVE 2 **Factor binomials.** Check for the following patterns.

Factoring a Binomial (Two Terms)

Difference of squares	$x^2 - y^2 = (x + y)(x - y)$
Difference of cubes	$x^3 - y^3 = (x - y)(x^2 + xy + y^2)$
Sum of cubes	$x^3 + y^3 = (x + y)(x^2 - xy + y^2)$

EXAMPLE 2 Factoring Binomials

Factor each binomial if possible.

(a) $64m^2 - 9n^2$

$= (8m)^2 - (3n)^2$ Difference of squares

$= (8m + 3n)(8m - 3n)$

(b) $8p^3 - 27$

$= (2p)^3 - 3^3$ Difference of cubes

$= (2p - 3)[(2p)^2 + (2p)(3) + 3^2]$

$= (2p - 3)(4p^2 + 6p + 9)$

(c) $1000m^3 + 1$

$= (10m)^3 + 1^3$ Sum of cubes

$= (10m + 1)[(10m)^2 - (10m)(1) + 1^2]$

$= (10m + 1)(100m^2 - 10m + 1)$

(d) $25m^2 + 121$ is prime. It is the *sum* of squares. There is no common factor.

Work Problem 2 at the Side. ▶

1 Factor each polynomial.

(a) $8x - 80$

(b) $2x^3 + 10x^2 - 2x$

(c) $12m(p - q) - 7n(p - q)$

2 Factor each binomial if possible.

(a) $36x^2 - y^2$

(b) $4t^2 + 1$

(c) $125x^3 - 27y^3$

(d) $x^3 + 343y^3$

Answers

1. **(a)** $8(x - 10)$
   (b) $2x(x^2 + 5x - 1)$
   (c) $(p - q)(12m - 7n)$
2. **(a)** $(6x + y)(6x - y)$
   (b) prime
   (c) $(5x - 3y)(25x^2 + 15xy + 9y^2)$
   (d) $(x + 7y)(x^2 - 7xy + 49y^2)$

3 Factor each trinomial.

(a) $16m^2 + 56m + 49$

(b) $r^2 + 18r + 72$

(c) $8t^2 - 13t + 5$

(d) $6x^2 - 3x - 63$

4 Factor each polynomial.

(a) $20 - 5m - 12n + 3mn$

(b) $p^3 - 2pq^2 + p^2q - 2q^3$

(c) $9x^2 + 24x + 16 - y^2$

Answers

3. (a) $(4m + 7)^2$
(b) $(r + 6)(r + 12)$
(c) $(8t - 5)(t - 1)$
(d) $3(2x - 7)(x + 3)$

4. (a) $(4 - m)(5 - 3n)$
(b) $(p + q)(p^2 - 2q^2)$
(c) $(3x + 4 + y)(3x + 4 - y)$

The binomial $25m^2 + 625$ is a sum of squares. It *can* be factored as $25(m^2 + 25)$ because its terms have greatest common factor 25.

OBJECTIVE 3 Factor trinomials. Consider the following.

Factoring a Trinomial (Three Terms)

For a **trinomial,** decide if it is a perfect square trinomial.

$$x^2 + 2xy + y^2 = (x + y)^2 \quad \text{or} \quad x^2 - 2xy + y^2 = (x - y)^2$$

If not, use the methods of **Sections 13.2–13.4.**

EXAMPLE 3 Factoring Trinomials

Factor each trinomial.

(a) $p^2 + 10p + 25$
$= (p + 5)^2$ Perfect square trinomial

(b) $49z^2 - 42z + 9$
$= (7z - 3)^2$ Perfect square trinomial

(c) $y^2 - 5y - 6$
$= (y - 6)(y + 1)$ The numbers -6 and 1 have a product of -6 and a sum of -5.

(d) $2k^2 - k - 6$
$= (2k + 3)(k - 2)$

(e) $28z^2 + 6z - 10$
$= 2(14z^2 + 3z - 5)$
$= 2(7z + 5)(2z - 1)$

◀ Work Problem **3** at the Side.

OBJECTIVE 4 Factor polynomials of more than three terms. Consider factoring by grouping as in **Section 13.1.**

EXAMPLE 4 Factoring Polynomials of More than Three Terms

Factor each polynomial.

(a) $4 - 2q - 6p + 3pq$
$= (4 - 2q) + (-6p + 3pq)$ Group the terms.
$= 2(2 - q) - 3p(2 - q)$ Factor each group.
$= (2 - q)(2 - 3p)$ Factor out $2 - q$.

(b) $20k^3 + 4k^2 - 45k - 9$
$= (20k^3 + 4k^2) + (-45k - 9)$ Group the terms. (Be careful with signs.)
$= 4k^2(5k + 1) - 9(5k + 1)$ Factor each group.
$= (5k + 1)(4k^2 - 9)$ $5k + 1$ is a common factor.
$= (5k + 1)(2k + 3)(2k - 3)$ Difference of squares

(c) $4a^2 + 4a + 1 - b^2$
$= (4a^2 + 4a + 1) - b^2$ Group the first three terms.
$= (2a + 1)^2 - b^2$ Perfect square trinomial
$= (2a + 1 + b)(2a + 1 - b)$ Difference of squares

◀ Work Problem **4** at the Side.

13.6 Exercises

FOR EXTRA HELP

Download the MyDashBoard App

MyMathLab®

CONCEPT CHECK *In Exercises 1 and 2, match each polynomial in Column I with the method or methods for factoring it in Column II. The choices in Column II may be used once, more than once, or not at all.*

I	II
1. (a) $49x^2 - 81y^2$	**A.** Factor out the GCF.
(b) $125z^3 + 1$	**B.** Factor a difference of squares.
(c) $88r^2 - 55s^2$	**C.** Factor a difference of cubes.
(d) $8a^3 - b^6$	**D.** Factor a sum of cubes.
(e) $50x^2 - 128y^4$	**E.** The polynomial is prime.

I	II
2. (a) $ab - 5a + 3b - 15$	**A.** Factor out the GCF.
(b) $z^2 - 3z + 6$	**B.** Factor a perfect square trinomial.
(c) $x^2 - 12x + 36 - 4p^2$	**C.** Factor by grouping.
(d) $r^2 - 24r + 144$	**D.** Factor into two distinct binomials.
(e) $2y^2 + 36y + 162$	**E.** The polynomial is prime.

The following exercises are of mixed variety. Factor each polynomial. ***See Examples 1–4.***

3. $32m^9 + 16m^5 + 24m^3$

4. $2m^2 - 10m - 48$

5. $14k^3 + 7k^2 - 70k$

6. $9z^2 + 64$

7. $m^2 - 3mn - 4n^2$

8. $49z^2 - 16y^2$

9. $100n^2r^2 + 30nr^3 - 50n^2r$

10. $16x^2 + 20x$

11. $20 + 5m + 12n + 3mn$

12. $x^3 + 1000$

13. $y^4 - 81$

14. $m^2 + 2m - 15$

15. $8p^3 - 125$

16. $32z^3 + 56z^2 - 16z$

17. $p^2 - 24p + 144$

18. $9m^2 - 64$

19. $16z^2 - 8z + 1$

20. $6y^2 - 6y - 12$

21. $x^2 + 2x + 16$

22. $p^2 - 17p + 66$

23. $p^3 + 64$

24. $k^2 + 100$

25. $2a^3 + a^2 - 14a - 7$

26. $a^2 - 3ab - 28b^2$

27. $16r^2 + 24rm + 9m^2$

28. $3k^2 + 4k - 4$

29. $n^2 - 12n - 35$

30. $a^4 - 625$

31. $16k^2 - 48k + 36$

32. $36y^6 - 42y^5 - 120y^4$

33. $8k^2 - 2kh - 3h^2$

34. $54m^2 - 24z^2$

35. $10z^2 - 7z - 6$

36. $125m^3 - 216n^3$

37. $9y^2 + 12y - 5$

38. $9u^2 + 66uv + 121v^2$

39. $36x^2 + 32x + 9$

40. $10m^2 + 25m - 60$

41. $4 - 2q - 6p + 3pq$

42. $k^2 - 121$

43. $1000z^3 + 512$

44. $m^3 + 4m^2 - 6m - 24$

45. $100a^2 - 81y^2$

46. $8a^2 + 23ab - 3b^2$

47. $a^2 + 8a + 16$

48. $4y^2 - 25$

49. $2x^2 + 5x + 6$

50. $-3x^3 + 12xy^2$

51. $25a^2 - 70ab + 49b^2$

52. $-4x^2 + 24xy - 36y^2$

53. $2m^2 - 15n - 5mn + 6m$

54. $y^6 + 5y^4 - 3y^2 - 15$

55. $54m^3 - 2000$

56. $12p^3 - 54p^2 - 30p$

13.7 Solving Quadratic Equations by Factoring

OBJECTIVES

1. Solve quadratic equations by factoring.
2. Solve other equations by factoring.

Galileo Galilei developed theories to explain physical phenomena. According to legend, Galileo dropped objects of different weights from the Leaning Tower of Pisa to disprove the belief that heavier objects fall faster than lighter objects. He developed a formula for freely falling objects described by

$$d = 16t^2,$$

where d is the distance in feet that an object falls (disregarding air resistance) in t seconds, regardless of weight.

The equation $d = 16t^2$ is a *quadratic equation*. A quadratic equation contains a second-degree (squared) term and no terms of greater degree.

Galileo Galilei (1564–1642)

Quadratic Equation

A **quadratic equation** is an equation that can be written in the form

$$ax^2 + bx + c = 0,$$

where a, b, and c are real numbers, with $a \neq 0$. The given form is called **standard form.**

$x^2 + 5x + 6 = 0, \quad 2t^2 - 5t = 3, \quad y^2 = 4$ Quadratic equations

In these examples, only $x^2 + 5x + 6 = 0$ is in standard form.

Work Problems 1 and 2 at the Side. ▶

We have factored many quadratic *expressions* of the form $ax^2 + bx + c$. We use factored quadratic expressions to solve quadratic *equations*.

OBJECTIVE 1 Solve quadratic equations by factoring. We do this using the **zero-factor property.**

Zero-Factor Property

If a and b are real numbers and $ab = 0$, then $a = 0$ or $b = 0$.

In words, if the product of two numbers is 0, then at least one of the numbers must be 0. One number *must* be 0, but both *may* be 0.

EXAMPLE 1 Using the Zero-Factor Property

Solve each equation.

(a) $(x + 3)(2x - 1) = 0$

The product $(x + 3)(2x - 1)$ is equal to 0. By the zero-factor property, the only way that the product of these two factors can be 0 is if at least one of the factors equals 0. Therefore, either $x + 3 = 0$ or $2x - 1 = 0$.

$x + 3 = 0$ or $2x - 1 = 0$ Zero-factor property

$x = -3$ or $2x = 1$ Solve each equation.

$x = \frac{1}{2}$ Divide each side by 2.

Continued on Next Page

1 Which of the following equations are quadratic equations?

A. $y^2 - 4y - 5 = 0$

B. $x^3 - x^2 + 16 = 0$

C. $2z^2 + 7z = -3$

D. $x + 2y = -4$

2 Write each quadratic equation in standard form.

(a) $x^2 - 3x = 4$

(b) $y^2 = 9y - 8$

Answers

1. A, C
2. (a) $x^2 - 3x - 4 = 0$
   (b) $y^2 - 9y + 8 = 0$

3 Solve each equation. Check your solutions.

GS **(a)** $(x-5)(x+2)=0$

To solve, use the zero-factor property.

$\_\_\_\_ = 0$ or $\_\_\_\_ = 0$

Solve these two equations, obtaining

$x = \_\_\_\_$ or $x = \_\_\_\_$.

CHECK Let $x = 5$.

$$(x-5)(x+2)=0$$
$$(\_\_\_\_-5)(\_\_\_\_+2) \stackrel{?}{=} 0$$
$$\_\_\_\_(7)=0$$

(*True / False*)

To check the other solution, substitute ____ for x in the original equation. A (*true / false*) statement results.

The solution set is $\{\_\_\_\_, \_\_\_\_\}$.

(b) $(3x-2)(x+6)=0$

(c) $z(2z+5)=0$

Answers

3. **(a)** $x-5$; $x+2$; 5; -2; 5; 5; 0; True; -2; true; 5; -2
(b) $\left\{-6, \frac{2}{3}\right\}$ **(c)** $\left\{-\frac{5}{2}, 0\right\}$

The original equation, $(x+3)(2x-1)=0$, has two solutions, -3 and $\frac{1}{2}$. *Check* these solutions by substituting -3 for x in this equation. ***Then start over*** and substitute $\frac{1}{2}$ for x.

CHECK Let $x = -3$.

$$(x+3)(2x-1)=0$$
$$(-3+3)[2(-3)-1] \stackrel{?}{=} 0$$
$$0(-7) \stackrel{?}{=} 0$$
$$0 = 0 \checkmark \text{ True}$$

Let $x = \frac{1}{2}$.

$$(x+3)(2x-1)=0$$
$$\left(\frac{1}{2}+3\right)\left(2\cdot\frac{1}{2}-1\right) \stackrel{?}{=} 0$$
$$\frac{7}{2}(1-1) \stackrel{?}{=} 0$$
$$0 = 0 \checkmark \text{ True}$$

Both -3 and $\frac{1}{2}$ result in true equations, so the solution set is $\{-3, \frac{1}{2}\}$.

(b)

$$y(3y-4)=0$$

$y=0$ or $3y-4=0$ Zero-factor property

Don't forget that 0 is a solution.

$3y=4$ Add 4.

$y=\frac{4}{3}$ Divide by 3.

Check these solutions by substituting each one in the original equation. The solution set is $\{0, \frac{4}{3}\}$.

◀ Work Problem **3** at the Side.

Note

The word *or* as used in **Example 1** means "one or the other or both."

If the polynomial in an equation is not already factored, first make sure that the equation is in standard form. Then factor and solve.

EXAMPLE 2 Solving Quadratic Equations

Solve each equation.

(a) $x^2-5x=-6$

First, write the equation in standard form by adding 6 to each side.

Don't factor x out at this step.

$$x^2-5x=-6$$

$x^2-5x+6=0$ Add 6.

Now factor x^2-5x+6. Find two numbers whose product is 6 and whose sum is -5. These two numbers are -2 and -3, so we factor as follows.

$(x-2)(x-3)=0$ Factor.

$x-2=0$ or $x-3=0$ Zero-factor property

$x=2$ or $x=3$ Solve each equation.

Continued on Next Page

CHECK Let $x = 2$.

$$x^2 - 5x = -6$$
$$2^2 - 5(2) \stackrel{?}{=} -6$$
$$4 - 10 \stackrel{?}{=} -6$$
$$-6 = -6 \checkmark \text{ True}$$

Let $x = 3$.

$$x^2 - 5x = -6$$
$$3^2 - 5(3) \stackrel{?}{=} -6$$
$$9 - 15 \stackrel{?}{=} -6$$
$$-6 = -6 \checkmark \text{ True}$$

Both solutions check, so the solution set is $\{2, 3\}$.

(b) $y^2 = y + 20$ — Write this equation in standard form.

Standard form ⟶ $y^2 - y - 20 = 0$	Subtract y and 20.
$(y-5)(y+4) = 0$	Factor.
$y - 5 = 0$ or $y + 4 = 0$	Zero-factor property
$y = 5$ or $y = -4$	Solve each equation.

Check by substituting in the original equation. The solution set is $\{-4, 5\}$.

Work Problem 4 at the Side. ▶

4 Solve each equation. Check your solutions.

(a) $m^2 - 3m - 10 = 0$

(b) $r^2 + 2r = 8$

Solving a Quadratic Equation by Factoring

Step 1 **Write the equation in standard form,** that is, with all terms on one side of the equality symbol in descending powers of the variable and 0 on the other side.

Step 2 **Factor** completely.

Step 3 **Use the zero-factor property** to set each factor with a variable equal to 0.

Step 4 **Solve** the resulting equations.

Step 5 **Check** each solution in the original equation.

5 Solve each equation. Check your solutions.

GS **(a)** $10a^2 - 5a - 15 = 0$

$\_\_\_\_(\_\_\_\_ - a - 3) = 0$

Divide each side by ____.

$2a^2 - a - 3 = 0$

$(2a - 3)(\_\_\_\_\_) = 0$

$2a - 3 = \_\_\_\_$ or $\_\_\_\_\_ = 0$

$a = \_\_\_\_$ or $a = \_\_\_\_$

Check by substituting the solutions in the original equation.

The solution set is _____.

(b) $4x^2 - 2x = 42$

EXAMPLE 3 Solving a Quadratic Equation (Common Factor)

Solve $4p^2 + 40 = 26p$.

$4p^2 + 40 = 26p$	
$4p^2 - 26p + 40 = 0$	Standard form
$2(2p^2 - 13p + 20) = 0$	Factor out 2.
$2p^2 - 13p + 20 = 0$	Divide each side by 2.
$(2p-5)(p-4) = 0$	Factor.
$2p - 5 = 0$ or $p - 4 = 0$	Zero-factor property
$p = \frac{5}{2}$ or $p = 4$	Solve each equation.

This 2 is *not* a solution of the equation.

Check that the solution set is $\left\{\frac{5}{2}, 4\right\}$ by substituting in the original equation.

Work Problem 5 at the Side. ▶

CAUTION

A common error is to include the common factor 2 as a solution in **Example 3.** ***Only factors containing variables lead to solutions.***

Answers

4. **(a)** $\{-2, 5\}$ **(b)** $\{-4, 2\}$

5. **(a)** 5; $2a^2$; 5; $a + 1$; 0; $a + 1$; $\frac{3}{2}$; -1; $\left\{-1, \frac{3}{2}\right\}$, or $\left\{\frac{3}{2}, -1\right\}$

(b) $\left\{-3, \frac{7}{2}\right\}$

6 Solve each equation. Check your solutions.

(a) $49m^2 - 9 = 0$

(b) $m^2 = 3m$

(c) $p(4p + 7) = 2$

EXAMPLE 4 Solving Quadratic Equations

Solve each equation.

(a)

$$16m^2 - 25 = 0 \quad \text{Factor the difference of squares. (Section 13.5)}$$

$$(4m + 5)(4m - 5) = 0$$

$$4m + 5 = 0 \quad \text{or} \quad 4m - 5 = 0 \quad \text{Zero-factor property}$$

$$4m = -5 \quad \text{or} \quad 4m = 5 \quad \text{Solve each equation.}$$

$$m = -\frac{5}{4} \quad \text{or} \quad m = \frac{5}{4} \quad \text{Divide by 4.}$$

Check the solutions $-\frac{5}{4}$ and $\frac{5}{4}$ in the original equation. The solution set is $\left\{-\frac{5}{4}, \frac{5}{4}\right\}$.

(b)

$$y^2 = 2y$$

$$y^2 - 2y = 0 \quad \text{Standard form}$$

Don't forget to set the variable factor y equal to 0.

$$y(y - 2) = 0 \quad \text{Factor.}$$

$$y = 0 \quad \text{or} \quad y - 2 = 0 \quad \text{Zero-factor property}$$

$$y = 2 \quad \text{Solve.}$$

The solution set is $\{0, 2\}$.

(c)

$$k(2k + 5) = 3$$

To be in standard form, 0 must be on one side.

$$2k^2 + 5k = 3 \quad \text{Multiply.}$$

$$\text{Standard form} \longrightarrow 2k^2 + 5k - 3 = 0 \quad \text{Subtract 3.}$$

$$(2k - 1)(k + 3) = 0 \quad \text{Factor.}$$

$$2k - 1 = 0 \quad \text{or} \quad k + 3 = 0 \quad \text{Zero-factor property}$$

$$2k = 1 \quad \text{or} \quad k = -3 \quad \text{Solve each equation.}$$

$$k = \frac{1}{2}$$

The solution set is $\left\{-3, \frac{1}{2}\right\}$.

CAUTION

In **Example 4(b)**, it is tempting to begin by dividing both sides of

$$y^2 = 2y$$

by y to get $y = 2$. We do not get the other solution, 0, if we divide by a variable. (We *may* divide each side of an equation by a *nonzero* real number, however. In **Example 3** we divided each side by 2.)

In **Example 4(c)**, we cannot use the zero-factor property to solve

$$k(2k + 5) = 3$$

in its given form because of the 3 on the right side of the equation. ***The zero-factor property applies only to a product that equals 0.***

◀ **Work Problem 6 at the Side.**

Answers

6. (a) $\left\{-\frac{3}{7}, \frac{3}{7}\right\}$ (b) $\{0, 3\}$ (c) $\left\{-2, \frac{1}{4}\right\}$

EXAMPLE 5 **Solving Quadratic Equations (Double Solutions)**

Solve each equation.

(a) $z^2 - 22z + 121 = 0$

$(z-11)^2 = 0$ Factor the perfect square trinomial.

$(z-11)(z-11) = 0$ $a^2 = a \cdot a$

$z - 11 = 0$ or $z - 11 = 0$ Zero-factor property

Because the two factors are identical, they both lead to the same solution, called a **double solution.**

$z = 11$ Add 11.

CHECK $z^2 - 22z + 121 = 0$ Original equation

$11^2 - 22(11) + 121 \stackrel{?}{=} 0$ Let $z = 11$.

$121 - 242 + 121 \stackrel{?}{=} 0$ Simplify.

$0 = 0$ ✓ True

The solution set is $\{11\}$.

(b) $9t^2 - 30t = -25$

$9t^2 - 30t + 25 = 0$ Standard form

$(3t-5)^2 = 0$ Factor the perfect square trinomial.

$3t - 5 = 0$ or $3t - 5 = 0$ Zero-factor property

$3t = 5$ Add 5.

$t = \frac{5}{3}$ $\frac{5}{3}$ is a double solution.

CHECK $9t^2 - 30t = -25$ Original equation

$9\left(\frac{5}{3}\right)^2 - 30\left(\frac{5}{3}\right) \stackrel{?}{=} -25$ Let $t = \frac{5}{3}$.

$9\left(\frac{25}{9}\right) - 30\left(\frac{5}{3}\right) \stackrel{?}{=} -25$ Apply the exponent.

$25 - 50 \stackrel{?}{=} -25$ Multiply.

$-25 = -25$ ✓ True

The solution set is $\left\{\frac{5}{3}\right\}$.

Work Problem 7 at the Side. ▶

CAUTION

Each equation in **Example 5** has only *one* distinct solution. ***We will write a double solution only once in a solution set.***

7 Solve each equation. Check your solutions.

GS **(a)** $x^2 + 16x = -64$

$x^2 + 16x +$ ____ $= 0$

(______$)^2 = 0$

______ $= 0$ or $x + 8 =$ ____

Solve to obtain $x =$ ____,

so -8 is a ______ solution.

The solution set is ____.

(b) $4x^2 - 4x + 1 = 0$

(c) $4z^2 + 20z = -25$

Answers

7. (a) 64; $x + 8$; $x + 8$; 0; -8; double; $\{-8\}$

(b) $\left\{\frac{1}{2}\right\}$ **(c)** $\left\{-\frac{5}{2}\right\}$

8 Solve each equation. Check your solutions.

(a) $r^3 - 16r = 0$

(b) $x^3 - 3x^2 - 18x = 0$

9 Solve each equation. Check your solutions.

(a) $(m + 3)(m^2 - 11m + 10) = 0$

(b) $(2x + 5)(4x^2 - 9) = 0$

Answers

8. (a) $\{-4, 0, 4\}$ (b) $\{-3, 0, 6\}$

9. (a) $\{-3, 1, 10\}$ (b) $\left\{-\frac{5}{2}, -\frac{3}{2}, \frac{3}{2}\right\}$

Note

Not all quadratic equations can be solved by factoring. A more general method for solving such equations is given in **Chapter 17.**

OBJECTIVE 2 Solve other equations by factoring. We can extend the zero-factor property to solve equations that involve more than two factors with variables. (These equations will have at least one term greater than second degree. They are *not* quadratic equations.)

EXAMPLE 6 Solving an Equation with More Than Two Variable Factors

Solve $6z^3 - 6z = 0$.

$$6z^3 - 6z = 0$$

$$6z(z^2 - 1) = 0 \quad \text{Factor out } 6z.$$

$$6z(z + 1)(z - 1) = 0 \quad \text{Factor } z^2 - 1.$$

By an extension of the zero-factor property, this product can equal 0 only if at least one of the factors equals 0. Write and solve three equations, one for each factor with a variable.

$$6z = 0 \quad \text{or} \quad z + 1 = 0 \quad \text{or} \quad z - 1 = 0 \quad \text{Zero-factor property}$$

$$z = 0 \quad \text{or} \quad z = -1 \quad \text{or} \quad z = 1 \quad \text{Solve each equation.}$$

Check by substituting, in turn, 0, −1, and 1 in the original equation. The solution set is $\{-1, 0, 1\}$.

◀ **Work Problem 8 at the Side.**

EXAMPLE 7 Solving an Equation with a Quadratic Factor

Solve $(2x - 1)(x^2 - 9x + 20) = 0$.

$$(2x - 1)(x^2 - 9x + 20) = 0$$

$$(2x - 1)(x - 5)(x - 4) = 0 \quad \text{Factor } x^2 - 9x + 20.$$

$$2x - 1 = 0 \quad \text{or} \quad x - 5 = 0 \quad \text{or} \quad x - 4 = 0 \quad \text{Zero-factor property}$$

$$x = \frac{1}{2} \quad \text{or} \quad x = 5 \quad \text{or} \quad x = 4 \quad \text{Solve each equation.}$$

Check to verify that the solution set is $\left\{\frac{1}{2}, 4, 5\right\}$.

◀ **Work Problem 9 at the Side.**

CAUTION

In **Example 7,** it would be unproductive to begin by multiplying the two factors together. The zero-factor property requires the *product* of two or more factors to equal 0. ***Always consider first whether an equation is given in the appropriate form to apply the zero-factor property.***

13.7 Exercises

FOR EXTRA HELP

Download the MyDashBoard App

MyMathLab®

CONCEPT CHECK *In Exercises 1–4, fill in each blank with the correct response.*

1. A quadratic equation in x is an equation that can be put into the form ___________ $= 0$.

2. The form $ax^2 + bx + c = 0$ is called ___________ form.

3. If a quadratic equation is in standard form, to solve the equation we should begin by attempting to ___________ the polynomial.

4. If the product of two numbers is 0, then at least one of the numbers is ___________. This is the ___________-___________ property.

CONCEPT CHECK *Work each problem.*

5. Identify each equation as *linear* or *quadratic*.

(a) $2x - 5 = 6$ **(b)** $x^2 - 5 = -4$

(c) $x^2 + 2x - 1 = 2x^2$ **(d)** $5^2x + 2 = 0$

6. The number 9 is a *double solution* of the equation

$$(x - 9)^2 = 0.$$

Why is this so?

7. Look at this "solution." **What Went Wrong?**

$$2x(3x - 4) = 0$$

$$x = 2 \quad \text{or} \quad x = 0 \quad \text{or} \quad 3x - 4 = 0$$

$$x = \frac{4}{3}$$

The solution set is $\{2, 0, \frac{4}{3}\}$.

8. Look at this "solution." **What Went Wrong?**

$$x(7x - 1) = 0$$

$$7x - 1 = 0 \quad \text{Zero-factor property}$$

$$x = \frac{1}{7}$$

The solution set is $\{\frac{1}{7}\}$.

Solve each equation, and check your solutions. ***See Example 1.***

9. $(x + 5)(x - 2) = 0$

10. $(x - 1)(x + 8) = 0$

11. $(2m - 7)(m - 3) = 0$

12. $(6k + 5)(k + 4) = 0$

13. $(2x + 1)(6x - 1) = 0$

14. $(3x + 2)(10x - 1) = 0$

15. $t(6t + 5) = 0$

16. $w(4w + 1) = 0$

17. $2x(3x - 4) = 0$

18. $6x(4x + 9) = 0$

19. $(x - 9)(x - 9) = 0$

20. $(2x + 1)(2x + 1) = 0$

Solve each equation, and check your solutions. **See Examples 2–7.**

21. $y^2 + 3y + 2 = 0$

22. $p^2 + 8p + 7 = 0$

23. $y^2 - 3y + 2 = 0$

24. $r^2 - 4r + 3 = 0$

25. $x^2 = 24 - 5x$

26. $t^2 = 2t + 15$

27. $x^2 = 3 + 2x$

28. $m^2 = 4 + 3m$

29. $z^2 + 3z = -2$

30. $p^2 - 2p = 3$

31. $m^2 + 8m + 16 = 0$

32. $b^2 - 6b + 9 = 0$

33. $3x^2 + 5x - 2 = 0$

34. $6r^2 - r - 2 = 0$

35. $12p^2 = 8 - 10p$

36. $18x^2 = 12 + 15x$

37. $9s^2 + 12s = -4$

38. $36x^2 + 60x = -25$

39. $y^2 - 9 = 0$

40. $m^2 - 100 = 0$

41. $16k^2 - 49 = 0$

42. $4w^2 - 9 = 0$

43. $n^2 = 121$

44. $x^2 = 400$

45. $x^2 = 7x$

46. $t^2 = 9t$

47. $6r^2 = 3r$

48. $10y^2 = -5y$

49. $g(g - 7) = -10$

50. $r(r - 5) = -6$

51. $z(2z + 7) = 4$

52. $b(2b + 3) = 9$

53. $2(y^2 - 66) = -13y$

54. $3(t^2 + 4) = 20t$

55. $5x^3 - 20x = 0$

56. $3x^3 - 48x = 0$

57. $9y^3 - 49y = 0$

58. $16r^3 - 9r = 0$

59. $(2r + 5)(3r^2 - 16r + 5) = 0$

60. $(3m + 4)(6m^2 + m - 2) = 0$

61. $(2x + 7)(x^2 + 2x - 3) = 0$

62. $(x + 1)(6x^2 + x - 12) = 0$

63. $x^3 + x^2 - 20x = 0$

64. $y^3 - 6y^2 + 8y = 0$

65. $r^4 = 2r^3 + 15r^2$

66. $x^3 = 3x + 2x^2$

67. $3x(x + 1) = (2x + 3)(x + 1)$

68. $2x(x + 3) = (3x + 1)(x + 3)$

69. Galileo's formula describing the motion of freely falling objects is

$$d = 16t^2.$$

The distance d in feet an object falls depends on the time t elapsed, in seconds. (This is an example of a **function,** introduced in **Section 17.5.**)

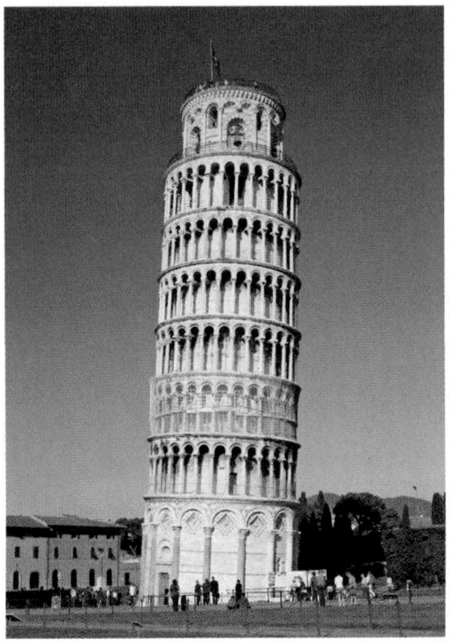

(a) Use Galileo's formula and complete the following table. (*Hint:* Substitute each given value into the formula and solve for the unknown value.)

t in seconds	0	1	2	3	___	___
d in feet	0	16	___	___	256	576

(b) When $t = 0$, $d = 0$. Explain this in the context of the problem.

70. In **Exercise 69,** when you substituted 256 for d and solved for t, you should have found two solutions: 4 and -4. Why doesn't -4 make sense as an answer?

13.8 Applications of Quadratic Equations

OBJECTIVES

1. Solve problems involving geometric figures.
2. Solve problems involving consecutive integers.
3. Solve problems by applying the Pythagorean theorem.
4. Solve problems by using given quadratic models.

We can use factoring to solve quadratic equations that arise in applications by following the same six problem-solving steps given in **Section 10.4.**

Solving an Applied Problem

Step 1 **Read** the problem carefully. What information is given? What is to be found?

Step 2 **Assign a variable** to represent the unknown value. Use a sketch, diagram, or table, as needed. Express any other unknown values in terms of the variable.

Step 3 **Write an equation** using the variable expression(s).

Step 4 **Solve** the equation.

Step 5 **State the answer.** Label it appropriately. Does it seem reasonable?

Step 6 **Check** the answer in the words of the *original* problem.

OBJECTIVE 1 **Solve problems involving geometric figures.** Refer to the formulas inside the back cover of the text, if necessary.

EXAMPLE 1 Solving an Area Problem

The Monroes want to plant a rectangular garden in their yard. The width of the garden will be 4 ft less than its length, and they want it to have an area of 96 ft^2. (ft^2 means square feet.) Find the length and width of the garden.

Step 1 **Read** the problem carefully. We need to find the dimensions of a garden with area 96 ft^2.

Step 2 **Assign a variable.**

Let $x =$ the length of the garden.

Then $x - 4 =$ the width. (The width is 4 ft less than the length.)

See **Figure 1.**

Figure 1

Step 3 **Write an equation.** The area of a rectangle is given by

$$\text{Area} = \text{Length} \times \text{Width}. \quad \text{Area formula}$$

Substitute 96 for area, x for length, and $x - 4$ for width.

$$A = LW$$

$$96 = x(x - 4) \quad \text{Let } A = 96,\ L = x,\ W = x - 4.$$

Continued on Next Page

Step 4 **Solve.** $96 = x(x-4)$ — Equation from Step 3

$96 = x^2 - 4x$ — Distributive property

$x^2 - 4x - 96 = 0$ — Standard form

$(x-12)(x+8) = 0$ — Factor.

$x - 12 = 0$ or $x + 8 = 0$ — Zero-factor property

$x = 12$ or $x = -8$ — Solve each equation.

Step 5 **State the answer.** The solutions are 12 and −8. A rectangle cannot have a side of negative length, so we discard −8. The length of the garden will be 12 ft. The width will be 12 − 4 = 8 ft.

Step 6 **Check.** The width of the garden is 4 ft less than the length, and the area is $12 \cdot 8 = 96 \text{ ft}^2$, as required.

Work Problem 1 at the Side. ▶

1 Solve each problem.

(a) The length of a rectangular room is 2 m more than the width. The area of the floor is 48 m^2. Find the length and width of the room.

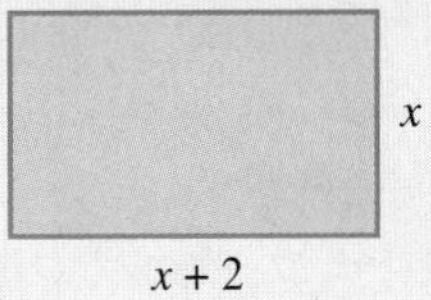

Give the equation using x as the variable, and give the answer.

(b) The length of each side of a square is increased by 4 in. The sum of the areas of the original square and the larger square is 106 in^2. What is the length of a side of the original square?

Problem-Solving Hint

When solving applied problems, always check solutions against physical facts and discard any answers that are not appropriate.

OBJECTIVE 2 Solve problems involving consecutive integers. Recall from **Section 10.4** that **consecutive integers** are integers that are next to each other on a number line, such as 1 and 2, or −11 and −10. See **Figure 2.**

Consecutive odd integers are *odd* integers that are next to each other, such as 1 and 3, or −13 and −11. **Consecutive even integers** are defined similarly, so 2 and 4 are consecutive even integers, as are −10 and −8. See **Figure 3.** (We list consecutive integers in increasing order from left to right.)

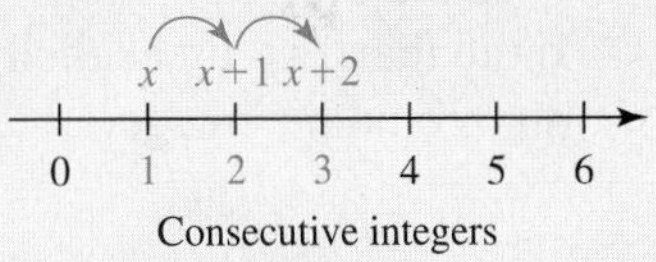

Figure 2

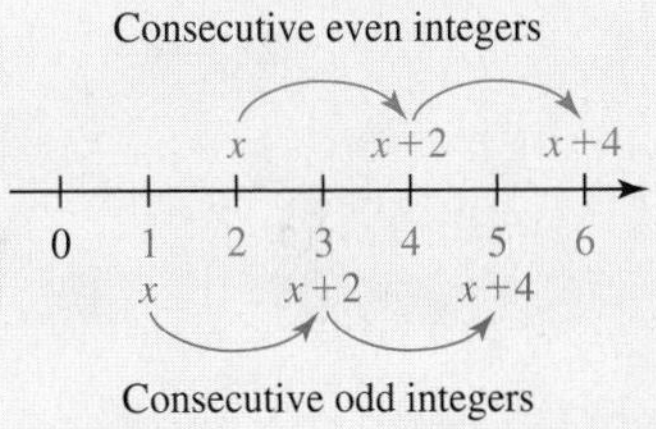

Figure 3

Problem-Solving Hint

If x represents the lesser integer, then the following apply.

For two consecutive integers, use **x, $x + 1$.**

For three consecutive integers, use **x, $x + 1$, $x + 2$.**

For two consecutive even or odd integers, use **x, $x + 2$.**

For three consecutive even or odd integers, use **x, $x + 2$, $x + 4$.**

EXAMPLE 2 Solving a Consecutive Integer Problem

The product of the numbers on two consecutive post-office boxes is 210. Find the box numbers.

Step 1 **Read** the problem. Note that the boxes are numbered consecutively.

Step 2 **Assign a variable.** See **Figure 4.**

Let x = the first box number.

Then $x + 1$ = the next consecutive box number.

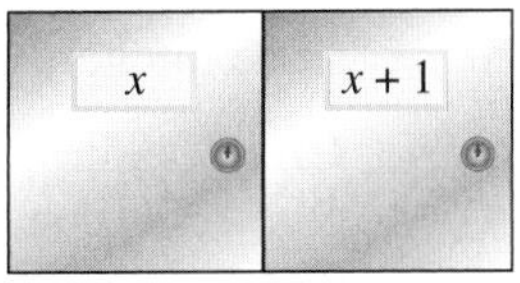

Figure 4

Continued on Next Page

Answers

1. **(a)** $48 = (x+2)x$; length: 8 m; width: 6 m
(b) 5 in.

2 Solve the problem.

The product of the numbers on two consecutive lockers at a health club is 132. Find the locker numbers.

3 Solve each problem. Give the equation using x to represent the least integer, and give the answer.

(a) The product of two consecutive even integers is 4 more than two times their sum. Find the integers.

(b) Find three consecutive odd integers such that the product of the least and greatest is 16 more than the middle integer.

Answers

2. 11 and 12

3. **(a)** $x(x+2) = 4 + 2(x+x+2)$; 4 and 6 or -2 and 0

(b) $x(x+4) = 16 + (x+2)$; 3, 5, 7

Step 3 **Write an equation.** The product of the box numbers is 210.

$$x(x+1) = 210$$

Step 4 **Solve.**

$$x^2 + x = 210 \quad \text{Distributive property}$$
$$x^2 + x - 210 = 0 \quad \text{Standard form}$$
$$(x+15)(x-14) = 0 \quad \text{Factor.}$$
$$x + 15 = 0 \quad \text{or} \quad x - 14 = 0 \quad \text{Zero-factor property}$$
$$x = -15 \quad \text{or} \quad x = 14 \quad \text{Solve each equation.}$$

Step 5 **State the answer.** The solutions are -15 and 14. We discard -15 since a box number cannot be negative. If $x = 14$, then $x + 1 = 15$, so the boxes have numbers 14 and 15.

Step 6 **Check.** The numbers 14 and 15 are consecutive and their product is $14 \cdot 15 = 210$, as required.

◀ **Work Problem 2 at the Side.**

EXAMPLE 3 Solving a Consecutive Integer Problem

The product of two consecutive odd integers is 1 less than five times their sum. Find the integers.

Step 1 **Read** carefully. This problem is a little more complicated.

Step 2 **Assign a variable.** We must find two consecutive *odd* integers.

Let x = the lesser integer.

Then $x + 2$ = the next greater odd integer.

Step 3 **Write an equation.**

The product	is	five times the sum	less 1.
↓	↓	↓	↓
$x(x+2)$	$=$	$5(x+x+2)$	$-\ 1$

Step 4 **Solve.**

$$x^2 + 2x = 5x + 5x + 10 - 1 \quad \text{Distributive property}$$
$$x^2 + 2x = 10x + 9 \quad \text{Combine like terms.}$$
$$x^2 - 8x - 9 = 0 \quad \text{Standard form}$$
$$(x-9)(x+1) = 0 \quad \text{Factor.}$$
$$x - 9 = 0 \quad \text{or} \quad x + 1 = 0 \quad \text{Zero-factor property}$$
$$x = 9 \quad \text{or} \quad x = -1 \quad \text{Solve each equation.}$$

Step 5 **State the answer.** We need to find two consecutive odd integers.

If $x = 9$ is the lesser, then $x + 2 = 9 + 2 = 11$ is the greater.

If $x = -1$ is the lesser, then $x + 2 = -1 + 2 = 1$ is the greater.

Do not discard the solution -1 here. There are two sets of answers since integers can be positive *or* negative.

Step 6 **Check.** The product of the first pair of integers is $9 \cdot 11 = 99$. One less than five times their sum is $5(9 + 11) - 1 = 99$. Thus, 9 and 11 satisfy the problem. Repeat the check with -1 and 1.

◀ **Work Problem 3 at the Side.**

CAUTION

Do *not* use x, $x + 1$, $x + 3$ to represent consecutive odd integers. To see why, let $x = 3$. Then $x + 1 = 3 + 1$ or 4, and $x + 3 = 3 + 3$ or 6. The numbers 3, 4, and 6 are not consecutive odd integers.

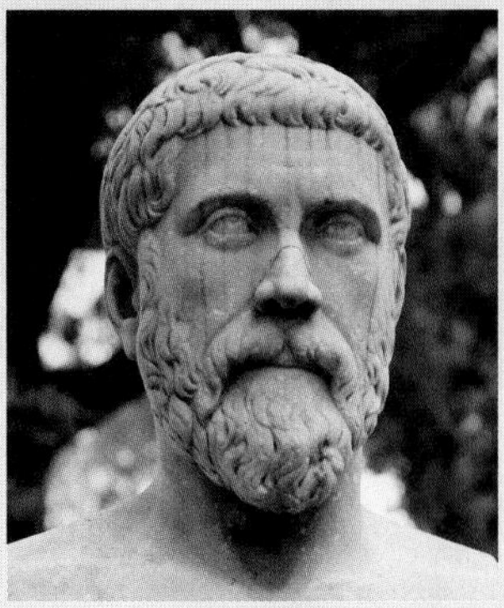

Pythagoras (c. 380–300 B.C.)

OBJECTIVE 3 **Solve problems by applying the Pythagorean theorem.**

The next example requires the Pythagorean theorem from geometry.

Pythagorean Theorem

If a right triangle (a triangle with a 90° angle) has longest side of length c and two other sides of lengths a and b, then

$$a^2 + b^2 = c^2.$$

The longest side, the **hypotenuse,** is opposite the right angle. The two shorter sides are the **legs** of the triangle.

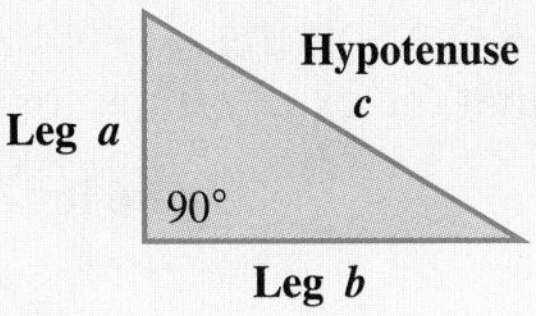

EXAMPLE 4 Applying the Pythagorean Theorem

Amy and Kevin leave their office, with Amy traveling north and Kevin traveling east. When Kevin is 1 mi farther than Amy from the office, the distance between them is 2 mi more than Amy's distance from the office. Find their distances from the office and the distance between them.

Step 1 **Read** the problem again. We must find three distances.

Step 2 **Assign a variable.**

Let x = Amy's distance from the office.

Then $x + 1$ = Kevin's distance from the office,

and $x + 2$ = the distance between them.

Place these expressions on a right triangle, as in **Figure 5.**

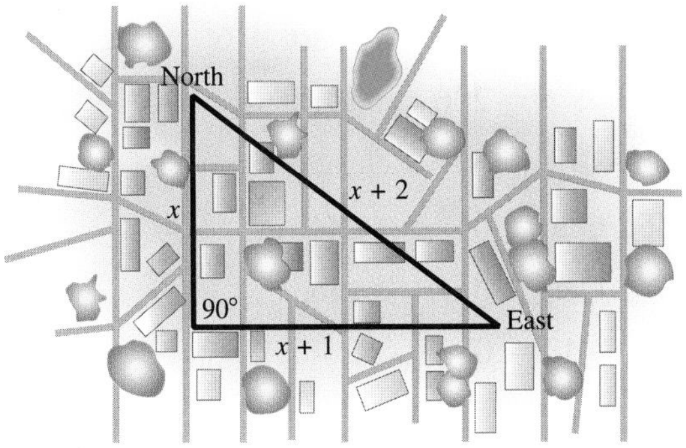

Figure 5

Step 3 **Write an equation.** Substitute into the Pythagorean theorem.

$$a^2 + b^2 = c^2$$

$$x^2 + (x + 1)^2 = (x + 2)^2$$

Be careful to substitute properly.

Continued on Next Page

4 Solve the problem.

The hypotenuse of a right triangle is 3 in. longer than the longer leg. The shorter leg is 3 in. shorter than the longer leg. Find the lengths of the sides of the triangle.

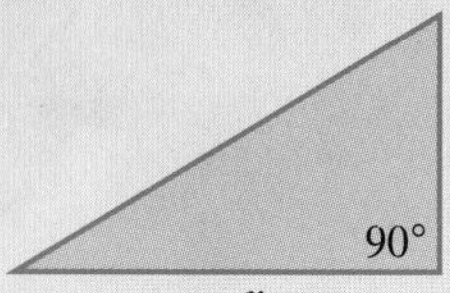

Step 4 **Solve.** $x^2 + x^2 + 2x + 1 = x^2 + 4x + 4$

$$x^2 - 2x - 3 = 0 \quad \text{Standard form}$$
$$(x-3)(x+1) = 0 \quad \text{Factor.}$$
$$x - 3 = 0 \quad \text{or} \quad x + 1 = 0 \quad \text{Zero-factor property}$$
$$x = 3 \quad \text{or} \quad x = -1 \quad \text{Solve each equation.}$$

Step 5 **State the answer.** Since -1 cannot represent a distance, 3 is the only possible answer. Amy's distance is 3 mi, Kevin's distance is $3 + 1 = 4$ mi, and the distance between them is $3 + 2 = 5$ mi.

Step 6 **Check.** Since $3^2 + 4^2 = 5^2$ is true, the answer is correct.

◀ **Work Problem 4 at the Side.**

Problem-Solving Hint

When solving a problem involving the Pythagorean theorem, be sure that the expressions for the sides of the triangle are properly placed.

$$\textbf{(one leg)}^2 + \textbf{(other leg)}^2 = \textbf{hypotenuse}^2$$

5 Solve each problem.

(a) Refer to **Example 5.** How long will it take for the ball to reach a height of 50 ft?

(b) The number y of impulses fired after a nerve has been stimulated is modeled by

$$y = -x^2 + 2x + 60,$$

where x is in milliseconds (ms) after the stimulation. When will 45 impulses occur? Do we get two solutions? Why is only one answer acceptable?

OBJECTIVE 4 **Solve problems by using given quadratic models.** In **Examples 1–4,** we wrote quadratic equations to model, or mathematically describe, various situations and then solved the equations. Now we are given the quadratic models and must use them to determine data.

EXAMPLE 5 **Finding the Height of a Ball**

A tennis player can hit a ball 180 ft per sec (123 mph). If she hits a ball directly upward, the height h of the ball in feet at time t in seconds is modeled by the quadratic equation

$$h = -16t^2 + 180t + 6.$$

How long will it take for the ball to reach a height of 206 ft?

A height of 206 ft means $h = 206$, so we substitute 206 for h in the equation and then solve for t.

$$h = -16t^2 + 180t + 6$$
$$\mathbf{206} = -16t^2 + 180t + 6 \quad \text{Let } h = 206.$$
$$-16t^2 + 180t + 6 = 206 \quad \text{Interchange sides.}$$
$$-16t^2 + 180t - 200 = 0 \quad \text{Standard form}$$
$$4t^2 - 45t + 50 = 0 \quad \text{Divide by } -4.$$
$$(4t-5)(t-10) = 0 \quad \text{Factor.}$$
$$4t - 5 = 0 \quad \text{or} \quad t - 10 = 0 \quad \text{Zero-factor property}$$
$$t = \frac{5}{4} \quad \text{or} \quad t = 10 \quad \text{Solve each equation.}$$

Since we found two acceptable answers, the ball will be 206 ft above the ground twice (once on its way up and once on its way down)—at $\frac{5}{4}$ sec and at 10 sec after it is hit. See **Figure 6.**

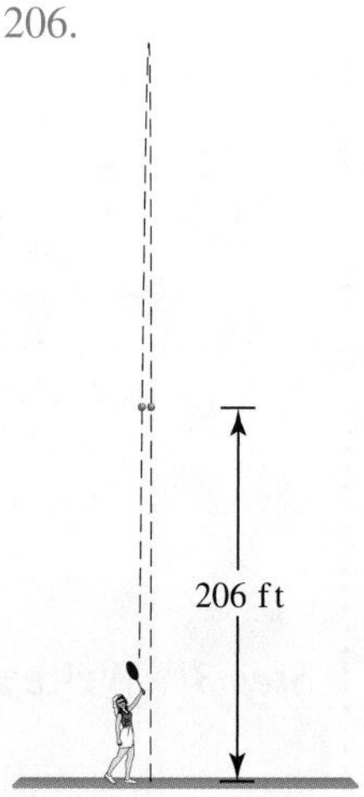

Figure 6

◀ **Work Problem 5 at the Side.**

Answers

4. 9 in., 12 in., 15 in.

5. **(a)** $\frac{1}{4}$ sec and 11 sec

(b) After 5 ms; There are two solutions, -3 and 5; Only one answer makes sense here because a negative answer is not appropriate.

EXAMPLE 6 **Modeling the Foreign-Born Population of the United States**

The foreign-born population of the United States over the years 1930–2010 can be modeled by the quadratic equation

$$y = 0.009665x^2 - 0.4942x + 15.12,$$

where $x = 0$ represents 1930, $x = 10$ represents 1940, and so on, and y is the number of people in millions. (*Source:* U.S. Census Bureau.)

(a) Use the model to find the foreign-born population in 1980 to the nearest tenth of a million.

Since $x = 0$ represents 1930, $x = 50$ represents 1980. Substitute 50 for x in the equation.

$$y = 0.009665(50)^2 - 0.4942(50) + 15.12 \quad \text{Let } x = 50.$$

$$y = 14.6 \quad \text{Round to the nearest tenth.}$$

In 1980, the foreign-born population of the United States was about 14.6 million.

(b) Repeat part (a) for 2010.

$$y = 0.009665(80)^2 - 0.4942(80) + 15.12 \quad \text{For 2010, let } x = 80.$$

$$y = 37.4 \quad \text{Round to the nearest tenth.}$$

In 2010, the foreign-born population of the United States was about 37.4 million.

(c) The model used in parts (a) and (b) was developed using the data in the table below. How do the results in parts (a) and (b) compare to the actual data from the table?

Year	Foreign-Born Population (millions)
1930	14.2
1940	11.6
1950	10.3
1960	9.7
1970	9.6
1980	14.1
1990	19.8
2000	28.4
2010	37.6

From the table, the actual value for 1980 is 14.1 million. By comparison, our answer in part (a), 14.6 million, is slightly high. For 2010, the actual value is 37.6 million, so our answer of 37.4 million in part (b) is slightly low, but a good estimate.

Work Problem **6** at the Side. ▶

6 Use the model in **Example 6** to find the foreign-born population of the United States in 1990. Give your answer to the nearest tenth of a million. How does it compare to the actual value from the table?

Answer

6. 20.3 million; The actual value is 19.8 million, so our answer using the model is somewhat high.

13.8 Exercises

1. **CONCEPT CHECK** To review the six problem-solving steps first introduced in **Section 2.4,** complete each statement.

Step 1: _______ the problem carefully.

Step 2: Assign a _______ to represent the unknown value.

Step 3: Write a(n) _______ using the variable expression(s).

Step 4: _______ the equation.

Step 5: State the _______.

Step 6: _______ the answer in the words of the _______ problem.

2. **CONCEPT CHECK** A student solves an applied problem and gets 6 or -3 for the length of the side of a square. Which of these answers is reasonable? Why?

In Exercises 3–6, a figure and a corresponding geometric formula are given. Using x as the variable, complete Steps 3–6 for each problem. (Refer to the steps in **Exercise 1** *as needed.)*

3. 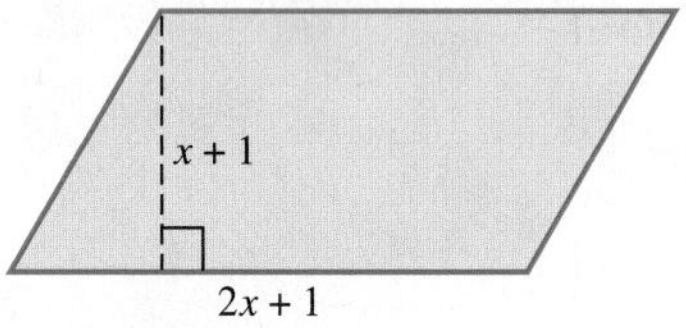

Area of a parallelogram: $A = bh$

The area of this parallelogram is 45 sq. units. Find its base and height.

Step 3: $45 =$ ____________

Step 4: $x =$ ____ or $x =$ ____

Step 5: base: ____ units; height: ____ units

Step 6: _______ $= 45$

4. 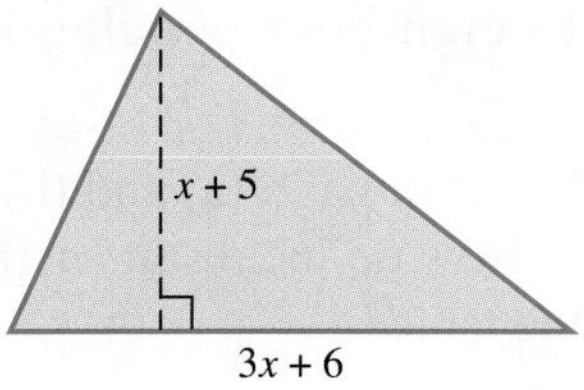

Area of a triangle: $A = \frac{1}{2}bh$

The area of this triangle is 60 sq. units. Find its base and height.

Step 3: $60 =$ ____________

Step 4: $x =$ ____ or $x =$ ____

Step 5: base: ____ units; height: ____ units

Step 6: _______ $= 60$

5. 

Volume of a rectangular jewelry box: $V = LWH$

The volume of this box is 192 cu. units. Find its length and width.

Step 3: ____ = ____$(x + 2)$

Step 4: $x =$ ____ or $x =$ ____

Step 5: length: ____ units; width: ____ units

Step 6: _______ $\cdot 4 =$ ____

6. 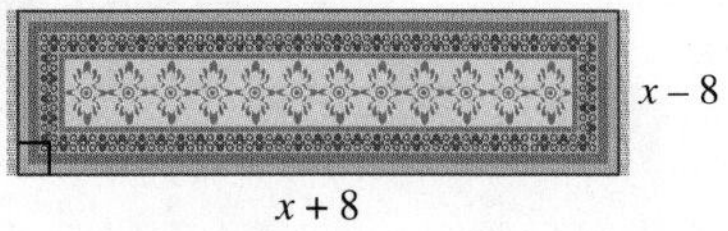

Area of a rectangular rug: $A = LW$

The area of this rug is 80 sq. units. Find its length and width.

Step 3: ____ $= (x + 8)$ _______

Step 4: $x =$ ____ or $x =$ ____

Step 5: length: ____ units; width: ____ units

Step 6: _______ $= 80$

Solve each problem. Check answers to be sure they are reasonable. Refer to the formulas inside the back cover. ***See Example 1.***

7. The length of a standard jewel case is 2 cm more than its width. The area of the rectangular top of the case is 168 cm^2. Find the length and width of the jewel case.

8. A standard DVD case is 6 cm longer than it is wide. The area of the rectangular top of the case is 247 cm^2. Find the length and width of the case.

9. The area of a triangle is 30 $in.^2$. The base of the triangle measures 2 in. more than twice the height of the triangle. Find the measures of the base and the height.

10. A certain triangle has its base equal in measure to its height. The area of the triangle is 72 m^2. Find the equal base and height measure.

11. The dimensions of a flat-panel monitor are such that its length is 3 in. more than its width. If the length were doubled and if the width were decreased by 1 in., the area would be increased by 150 $in.^2$. What are the length and width of the flat panel?

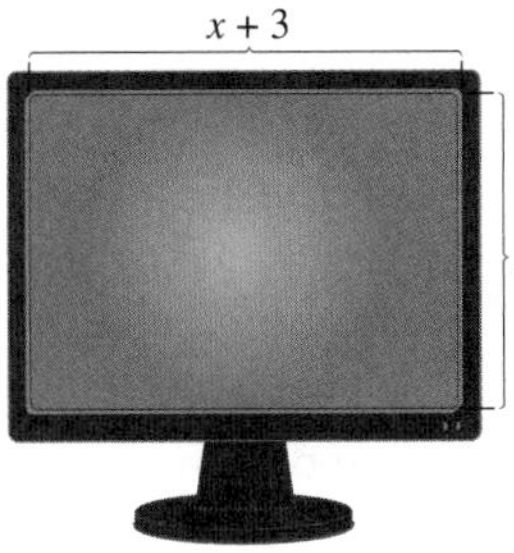

12. A computer keyboard is 11 in. longer than it is wide. If the length were doubled and if 2 in. were added to the width, the area would be increased by 198 $in.^2$. What are the length and width of the keyboard?

13. A 10-gal aquarium is 3 in. higher than it is wide. Its length is 21 in., and its volume is 2730 $in.^3$. What are the height and width of the aquarium?

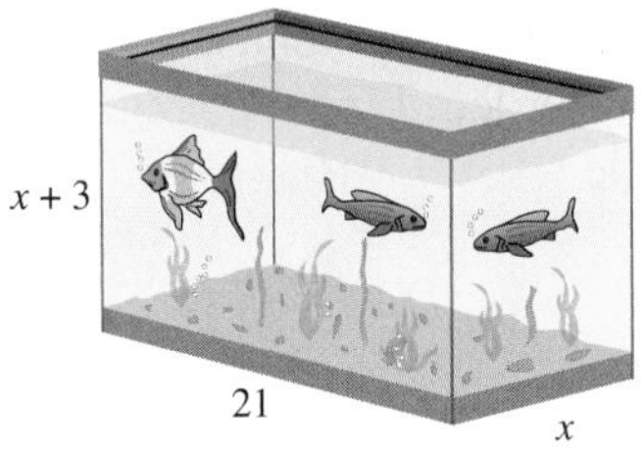

14. A toolbox is 2 ft high, and its width is 3 ft less than its length. If its volume is 80 ft^3, find the length and width of the box.

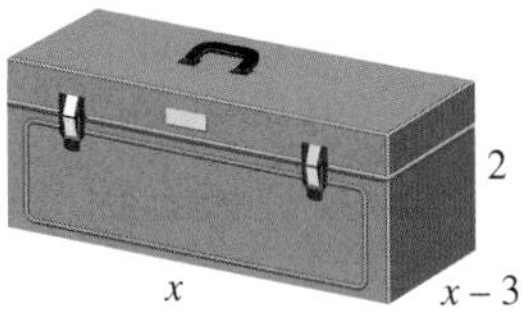

15. A square mirror has sides measuring 2 ft less than the sides of a square painting. If the difference between their areas is 32 ft^2, find the lengths of the sides of the mirror and the painting.

16. The sides of one square have length 3 m more than the sides of a second square. If the area of the larger square is subtracted from 4 times the area of the smaller square, the result is 36 m^2. What are the lengths of the sides of each square?

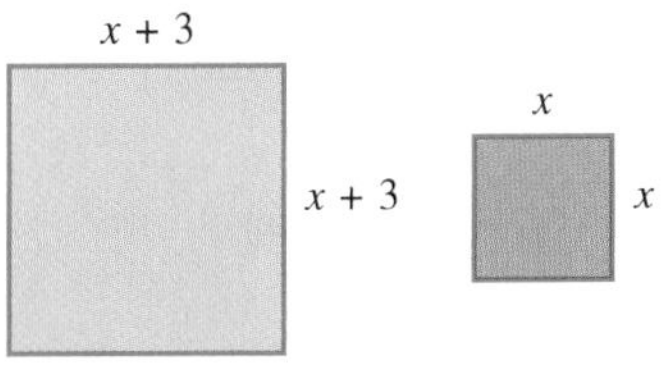

Solve each problem about consecutive integers. **See Examples 2 and 3.**

17. The product of the numbers on two consecutive volumes of research data is 420. Find the volume numbers.

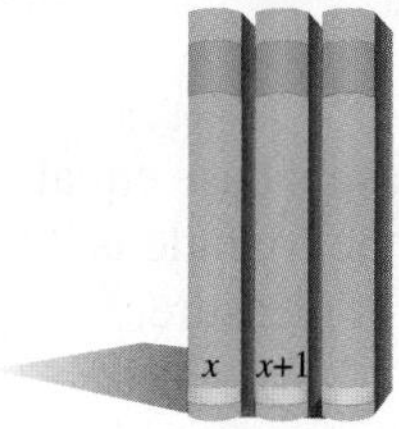

18. The product of the page numbers on two facing pages of a book is 600. Find the page numbers.

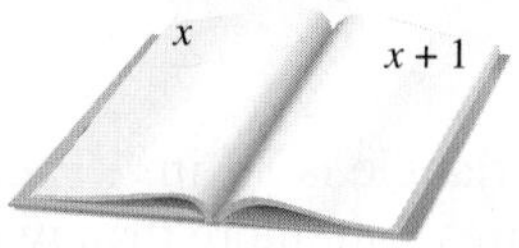

19. The product of two consecutive integers is 11 more than their sum. Find the integers.

20. The product of two consecutive integers is 4 less than four times their sum. Find the integers.

21. Find two consecutive odd integers such that their product is 15 more than three times their sum.

22. Find two consecutive odd integers such that five times their sum is 23 less than their product.

23. Find three consecutive even integers such that the sum of the squares of the lesser two is equal to the square of the greatest.

24. Find three consecutive even integers such that the square of the sum of the lesser two is equal to twice the greatest.

25. Find three consecutive odd integers such that 3 times the sum of all three is 18 more than the product of the first and second integers.

26. Find three consecutive odd integers such that the sum of all three is 42 less than the product of the second and third integers.

Use the Pythagorean theorem to solve each problem. **See Example 4.**

27. The hypotenuse of a right triangle is 1 cm longer than the longer leg. The shorter leg is 7 cm shorter than the longer leg. Find the length of the longer leg of the triangle.

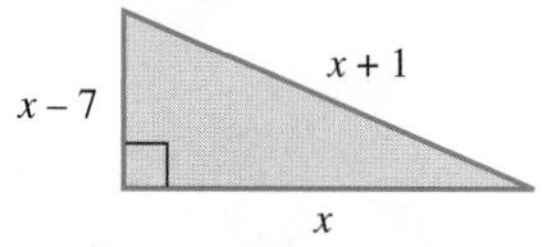

28. The longer leg of a right triangle is 1 m longer than the shorter leg. The hypotenuse is 1 m shorter than twice the shorter leg. Find the length of the shorter leg of the triangle.

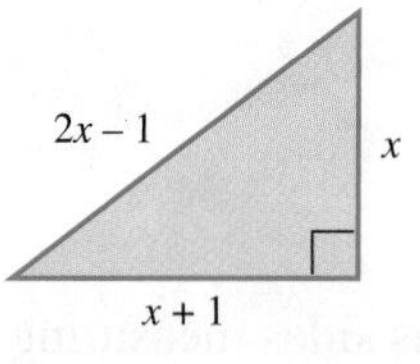

29. Terri works due north of home. Her husband Denny works due east. They leave for work at the same time. By the time Terri is 5 mi from home, the distance between them is 1 mi more than Denny's distance from home. How far from home is Denny?

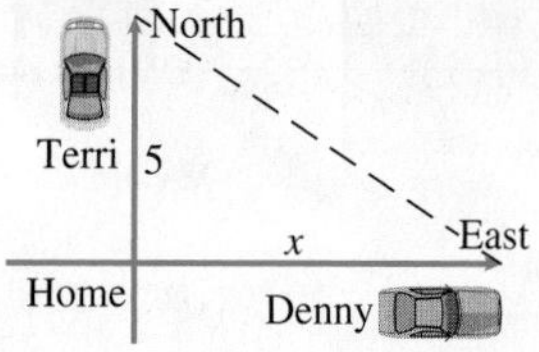

30. Two cars left an intersection at the same time. One traveled north. The other traveled 14 mi farther, but to the east. How far apart were they then, if the distance between them was 4 mi more than the distance traveled east?

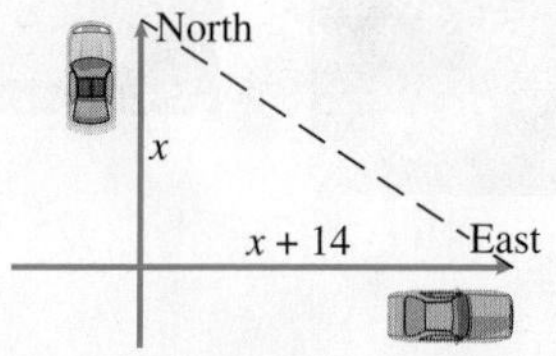

31. A ladder is leaning against a building. The distance from the bottom of the ladder to the building is 4 ft less than the length of the ladder. How high up the side of the building is the top of the ladder if that distance is 2 ft less than the length of the ladder?

32. A lot has the shape of a right triangle with one leg 2 m longer than the other. The hypotenuse is 2 m less than twice the length of the shorter leg. Find the length of the shorter leg.

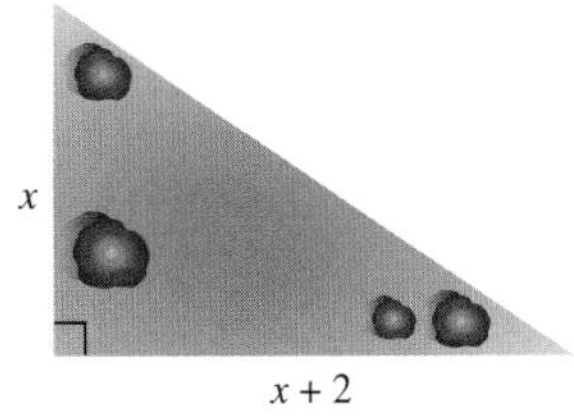

Solve each problem. ***See Example 5.***

33. An object projected from a height of 48 ft with an initial velocity of 32 ft per sec after t seconds has height

$$h = -16t^2 + 32t + 48.$$

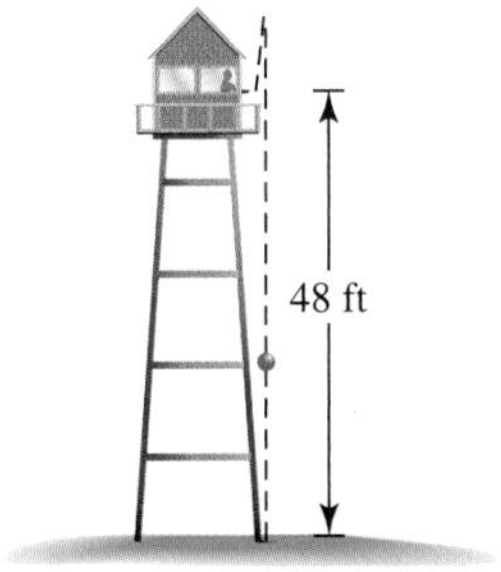

(a) After how many seconds is the height 64 ft? (*Hint:* Let $h = 64$ and solve.)

(b) After how many seconds is the height 60 ft?

(c) After how many seconds does the object hit the ground? (*Hint:* When the object hits the ground, $h = 0$.)

(d) The quadratic equation from part (c) has two solutions, yet only one of them is appropriate for answering the question. Why is this so?

34. If an object is projected upward from ground level with an initial velocity of 64 ft per sec, its height h in feet t seconds later is

$$h = -16t^2 + 64t.$$

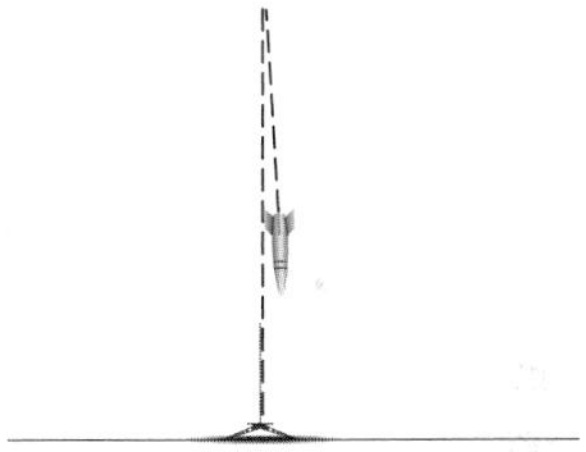

(a) After how many seconds is the height 48 ft? (*Hint:* Let $h = 48$ and solve.)

(b) The object reaches its maximum height 2 sec after it is projected. What is this maximum height?

(c) After how many seconds does the object hit the ground? (*Hint:* When the object hits the ground, $h = 0$.)

(d) The quadratic equation from part (c) has two solutions, yet only one of them is appropriate for answering the question. Why is this so?

If an object is projected upward with an initial velocity of 128 ft per sec, its height h in feet after t seconds is

$$h = 128t - 16t^2.$$

Find the height of the object after each period of time. **See Example 5.**

35. 1 sec

36. 2 sec

37. 4 sec

38. How long does it take the object just described to return to the ground?

Solve each problem. **See Example 6.**

39. The table shows the number of cellular phone subscribers (in millions) in the United States.

Year	Subscribers (in millions)
1994	24
1996	44
1998	69
2000	109
2002	141
2004	182
2006	233
2008	286
2010	303

Source: CTIA-The Wireless Association.

We used the data to develop the quadratic equation

$$y = 0.347x^2 + 13.14x + 17.98,$$

which models the number of cellular phone subscribers y (in millions) in the year x, where $x = 0$ represents 1994, $x = 2$ represents 1996, and so on.

(a) Use the model to find the number of subscribers in 2000 to the nearest million. How does the result compare to the actual data in the table?

(b) What value of x corresponds to 2008?

(c) Use the model to find the number of subscribers in 2008 to the nearest million. How does the result compare to the actual data in the table?

(d) Assuming that the trend in the data continues, use the quadratic equation to estimate the number of subscribers in 2013 to the nearest million.

40. Annual revenue in billions of dollars for eBay is shown in the table.

Year	Annual Revenue (in billions of dollars)
2001	0.749
2002	1.21
2003	2.17
2004	3.27
2005	4.55
2006	5.97
2007	7.67
2008	8.54

Source: eBay.

Using the data, we developed the quadratic equation

$$y = 0.06x^2 + 0.75x + 0.55$$

to model eBay revenues y in year x, where $x = 0$ represents 2001, $x = 1$ represents 2002, and so on.

(a) Use the model to find annual revenue for eBay in 2005 and 2008. How do the results compare with the actual data in the table?

(b) Use the model to estimate annual revenue for eBay in 2009, to the nearest tenth.

(c) Actual revenue in 2009 was \$8.73 billion. How does the result from part (b) compare with the actual revenue in 2009?

(d) Should the quadratic equation be used to estimate eBay revenue for years after 2008? Explain.

Chapter 13 Summary

Key Terms

13.1

factor For integers a and b, if $a \cdot b = c$, then a and b are factors of c.

factored form An expression is in factored form when it is written as a product.

greatest common factor (GCF) The greatest common factor is the largest quantity that is a factor of each of a group of quantities.

factoring The process of writing a polynomial as a product is called factoring.

13.2

prime polynomial A prime polynomial is a polynomial that cannot be factored using only integers.

13.5

perfect square trinomial A perfect square trinomial is a trinomial that can be factored as the square of a binomial.

13.7

quadratic equation A quadratic equation is an equation that can be written in the form $ax^2 + bx + c = 0$, with $a \neq 0$.

standard form The form $ax^2 + bx + c = 0$ is the standard form of a quadratic equation.

13.8

hypotenuse The longest side of a right triangle, opposite the right angle, is the hypotenuse.

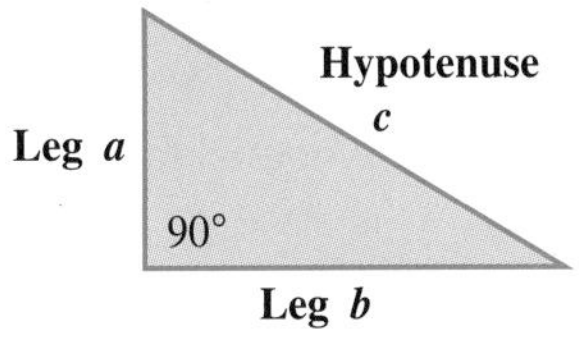

legs The two shorter sides of a right triangle are the legs.

Test Your Word Power

See how well you have learned the vocabulary in this chapter.

1. **Factoring** is
   - **A.** a method of multiplying polynomials
   - **B.** the process of writing a polynomial as a product
   - **C.** the answer in a multiplication problem
   - **D.** a way to add the terms of a polynomial.

2. A polynomial is in **factored form** when
   - **A.** it is prime
   - **B.** it is written as a sum
   - **C.** the second-degree term has a coefficient of 1
   - **D.** it is written as a product.

3. The **greatest common factor** of a polynomial is
   - **A.** the least integer that divides evenly into all its terms
   - **B.** the least expression that is a factor of all its terms
   - **C.** the greatest expression that is a factor of all its terms
   - **D.** the variable that is common to all its terms.

4. A **perfect square trinomial** is a trinomial
   - **A.** that can be factored as the square of a binomial
   - **B.** that cannot be factored
   - **C.** that is multiplied by a binomial
   - **D.** where all terms are perfect squares.

5. A **quadratic equation** is an equation that can be written in the form
   - **A.** $y = mx + b$
   - **B.** $ax^2 + bx + c = 0\ (a \neq 0)$
   - **C.** $Ax + By = C$
   - **D.** $x = k$.

6. A **hypotenuse** is
   - **A.** either of the two shorter sides of a triangle
   - **B.** the shortest side of a right triangle
   - **C.** the side opposite the right angle in a right triangle
   - **D.** the longest side in any triangle.

Answers to Test Your Word Power

1. B; *Example:* $x^2 - 5x - 14$ factors as $(x - 7)(x + 2)$.

2. D; *Example:* The factored form of $x^2 - 5x - 14$ is $(x - 7)(x + 2)$.

3. C; *Example:* The greatest common factor of $8x^2$, $22xy$, and $16x^3y^2$ is $2x$.

4. A; *Example:* $a^2 + 2a + 1$ is a perfect square trinomial. Its factored form is $(a + 1)^2$.

5. B; *Examples:* $y^2 - 3y + 2 = 0$, $x^2 - 9 = 0$, $2m^2 = 6m + 8$

6. C; *Example:* See the triangle included in the Key Terms above.

Quick Review

Concepts	Examples

13.1 Factors; The Greatest Common Factor

Finding the Greatest Common Factor (GCF)

Step 1 Write each number in prime factored form.

Step 2 List each prime number or each variable that is a factor of every term in the list.

Step 3 Use as exponents on the common prime factors the *least* exponents from the prime factored forms.

Step 4 Multiply the primes from Step 3.

Find the greatest common factor of $4x^2y$, $6x^2y^3$, and $2xy^2$.

$$4x^2y = 2 \cdot 2 \cdot x^2 \cdot y$$
$$6x^2y^3 = 2 \cdot 3 \cdot x^2 \cdot y^3$$
$$2xy^2 = 2 \cdot x \cdot y^2$$

The greatest common factor is $2xy$.

Factoring by Grouping

Step 1 Group the terms.

Step 2 Factor out the greatest common factor from each group.

Step 3 Factor out a common binomial factor from the results of Step 2.

Step 4 If necessary, rearrange terms and try a different grouping.

Factor by grouping.

$2a^2 + 2ab + a + b$

$= (2a^2 + 2ab) + (a + b)$ Group the terms.

$= 2a(a + b) + 1(a + b)$ Factor each group.

$= (a + b)(2a + 1)$ Factor out $a + b$.

13.2 Factoring Trinomials

To factor $x^2 + bx + c$, find m and n such that $mn = c$ and $m + n = b$.

$$\begin{array}{c} mn = c \\ \downarrow \\ x^2 + bx + c \\ \uparrow \\ m + n = b \end{array}$$

Then $x^2 + bx + c$ factors as $(x + m)(x + n)$.

Check by multiplying.

Factor $x^2 + 6x + 8$.

$$\begin{array}{c} mn = 8 \\ \downarrow \\ x^2 + 6x + 8 \\ \uparrow \\ m + n = 6 \end{array}$$

Here, $m = 2$ and $n = 4$.

$x^2 + 6x + 8$ factors as $(x + 2)(x + 4)$.

CHECK $(x + 2)(x + 4)$

$= x^2 + 4x + 2x + 8$ FOIL method

$= x^2 + 6x + 8$ ✓ Combine like terms.

13.3 Factoring Trinomials by Grouping

To factor $ax^2 + bx + c$, find m and n such that $mn = ac$ and $m + n = b$.

$$\begin{array}{c} m + n = b \\ \downarrow \\ ax^2 + bx + c \\ \uparrow \quad \uparrow \\ mn = ac \end{array}$$

Then factor $ax^2 + mx + nx + b$ by grouping.

Factor $3x^2 + 14x - 5$.

$3 \cdot (-5) = -15$ (product of the first and last coefficients)

Find two integers with a product of $3(-5) = -15$ and a sum of 14. The integers are -1 and 15.

$3x^2 + 14x - 5$

$= 3x^2 - x + 15x - 5$ $14x = -x + 15x$

$= (3x^2 - x) + (15x - 5)$ Group the terms.

$= x(3x - 1) + 5(3x - 1)$ Factor each group.

$= (3x - 1)(x + 5)$ Factor out $3x - 1$.

Concepts	Examples

13.4 Factoring Trinomials by Using the FOIL Method

To factor $ax^2 + bx + c$ by trial and error, use the FOIL method in reverse.

Factor $3x^2 + 14x - 5$.

The only positive factors of 3 are 3 and 1, and -5 has possible factors of 1 and -5, or -1 and 5. Using trial and error,

$$3x^2 + 14x - 5 \quad \text{factors as} \quad (3x - 1)(x + 5).$$

13.5 Special Factoring Techniques

Difference of Squares

$$a^2 - b^2 = (a + b)(a - b)$$

Perfect Square Trinomials

$$a^2 + 2ab + b^2 = (a + b)^2$$

$$a^2 - 2ab + b^2 = (a - b)^2$$

Difference of Cubes

$$x^3 - y^3 = (x - y)(x^2 + xy + y^2)$$

Sum of Cubes

$$x^3 + y^3 = (x + y)(x^2 - xy + y^2)$$

Factor.

$$4x^2 - 9 = (2x + 3)(2x - 3)$$

$$9x^2 + 6x + 1 = (3x + 1)^2 \qquad 4x^2 - 20x + 25 = (2x - 5)^2$$

$$8 - 27a^3 = (2 - 3a)(4 + 6a + 9a^2)$$

$$64z^3 + 1 = (4z + 1)(16z - 4z + 1)$$

13.7 Solving Quadratic Equations by Factoring

Zero-Factor Property

If a and b are real numbers and $ab = 0$, then $a = 0$ or $b = 0$.

If $(x - 2)(x + 3) = 0$, then $x - 2 = 0$ or $x + 3 = 0$.

Solving a Quadratic Equation by Factoring

Step 1 Write the equation in standard form.

Step 2 Factor.

Step 3 Use the zero-factor property.

Step 4 Solve the resulting equations.

Step 5 Check.

Solve $2x^2 = 7x + 15$.

$2x^2 - 7x - 15 = 0$	Standard form
$(2x + 3)(x - 5) = 0$	Factor.
$2x + 3 = 0$ or $x - 5 = 0$	Zero-factor property
$2x = -3$ or $x = 5$	Solve each equation.
$x = -\frac{3}{2}$	

The solutions $-\frac{3}{2}$ and 5 satisfy the original equation.

The solution set is $\{-\frac{3}{2}, 5\}$.

13.8 Applications of Quadratic Equations

Pythagorean Theorem
In a right triangle, the square of the hypotenuse equals the sum of the squares of the legs.

$$a^2 + b^2 = c^2$$

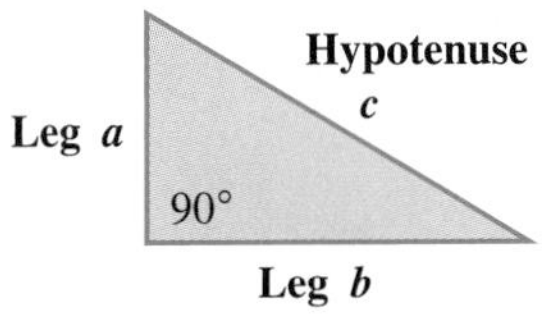

In a right triangle, one leg measures 2 ft longer than the other. The hypotenuse measures 4 ft longer than the shorter leg. Find the lengths of the three sides of the triangle.

Let x = the length of the shorter leg. Then

$$x^2 + (x + 2)^2 = (x + 4)^2.$$

Solve this equation to get $x = 6$ or $x = -2$. Discard -2 as a solution. Check that the sides measure

$$6 \text{ ft}, \quad 6 + 2 = 8 \text{ ft}, \quad \text{and} \quad 6 + 4 = 10 \text{ ft}.$$

Chapter 13 Review Exercises

13.1 *Factor out the greatest common factor or factor by grouping.*

1. $15t + 45$

2. $60z^3 + 30z$

3. $44x^3 + 55x^2$

4. $100m^2n^3 - 50m^3n^4 + 150m^2n^2$

5. $2xy - 8y + 3x - 12$

6. $6y^2 + 9y + 4xy + 6x$

13.2 *Factor completely.*

7. $x^2 + 10x + 21$

8. $y^2 - 13y + 40$

9. $q^2 + 6q - 27$

10. $r^2 - r - 56$

11. $x^2 + x + 1$

12. $3x^2 + 6x + 6$

13. $r^2 - 4rs - 96s^2$

14. $p^2 + 2pq - 120q^2$

15. $-8p^3 + 24p^2 + 80p$

16. $3x^4 + 30x^3 + 48x^2$

17. $m^2 - 3mn - 18n^2$

18. $y^2 - 8yz + 15z^2$

19. $p^7 - p^6q - 2p^5q^2$

20. $-3r^5 + 6r^4s + 45r^3s^2$

13.3–13.4

21. CONCEPT CHECK To begin factoring $6r^2 - 5r - 6$, what are the possible first terms of the two binomial factors, if we consider only positive integer coefficients?

22. CONCEPT CHECK What is the first step you would use to factor $2z^3 + 9z^2 - 5z$?

Factor completely.

23. $2k^2 - 5k + 2$

24. $3r^2 + 11r - 4$

25. $6r^2 - 5r - 6$

26. $10z^2 - 3z - 1$

27. $5t^2 - 11t + 12$

28. $24x^5 - 20x^4 + 4x^3$

29. $-6x^2 + 3x + 30$

30. $10r^3s + 17r^2s^2 + 6rs^3$

31. $-30y^3 - 5y^2 + 10y$

32. $4z^2 - 5z + 7$

33. $-3m^3n + 19m^2n + 40mn$

34. $14a^2 - 27ab - 20b^2$

13.5

35. CONCEPT CHECK Which one of the following is a difference of squares?

A. $32x^2 - 1$ **B.** $4x^2y^2 - 25z^2$

C. $x^2 + 36$ **D.** $25y^3 - 1$

36. CONCEPT CHECK Which one of the following is a perfect square trinomial?

A. $x^2 + x + 1$ **B.** $y^2 - 4y + 9$

C. $4x^2 + 10x + 25$ **D.** $x^2 - 20x + 100$

Factor completely.

37. $n^2 - 64$

38. $25b^2 - 121$

39. $49y^2 - 25w^2$

40. $144p^2 - 36q^2$

41. $x^2 + 100$

42. $z^2 + 10z + 25$

43. $9t^2 - 42t + 49$

44. $16m^2 + 40mn + 25n^2$

45. $125x^3 - 1$

46. $1000p^3 + 27$

13.7

Solve each equation, and check the solutions.

47. $(4t + 3)(t - 1) = 0$

48. $(x + 7)(x - 4)(x + 3) = 0$

49. $x(2x - 5) = 0$

50. $z^2 + 4z + 3 = 0$

51. $m^2 - 5m + 4 = 0$

52. $x^2 = -15 + 8x$

53. $3z^2 - 11z - 20 = 0$

54. $81t^2 - 64 = 0$

55. $y^2 = 8y$

56. $n(n - 5) = 6$

57. $t^2 - 14t + 49 = 0$

58. $t^2 = 12(t - 3)$

59. $(5z + 2)(z^2 + 3z + 2) = 0$

60. $x^2 = 9$

61. $64x^3 - 9x = 0$

62. $(2r + 1)(12r^2 + 5r - 3) = 0$

63. $25w^2 - 90w + 81 = 0$

64. $r(r - 7) = 30$

13.8

Solve each problem.

65. The length of a rug is 6 ft more than the width. The area is 40 ft^2. Find the length and width of the rug.

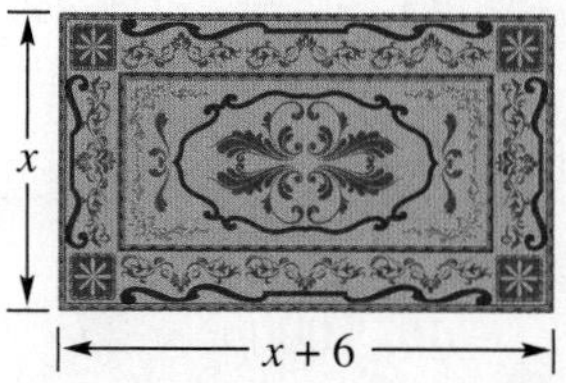

66. The surface area S of a box is given by

$$S = 2WH + 2WL + 2LH.$$

A treasure chest from a sunken galleon has dimensions (in feet) as shown in the figure. Its surface area is 650 ft^2. Find its width.

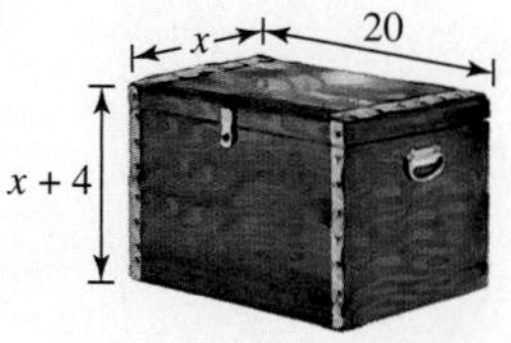

67. The product of two consecutive integers is 29 more than their sum. What are the integers?

68. The product of the lesser two of three consecutive integers is equal to 23 plus the greatest. Find the integers.

69. Two cars left an intersection at the same time. One traveled west, and the other traveled 14 mi less, but to the south. How far apart were they then, if the distance between them was 16 mi more than the distance traveled south?

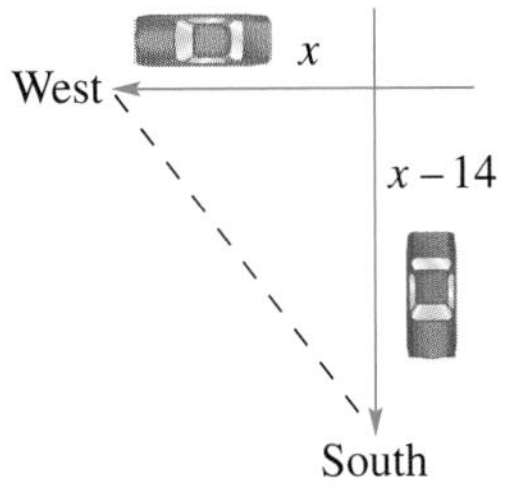

70. The triangular sail of a schooner has an area of 30 m^2. The height of the sail is 4 m more than the base. Find the length of the base of the sail.

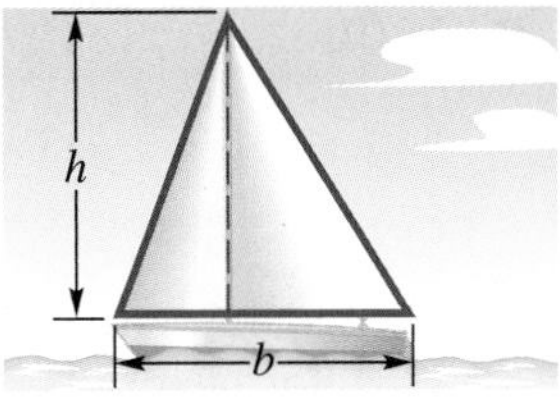

71. The floor plan for a house is a rectangle with length 7 m more than its width. The area is 170 m^2. Find the width and length of the house.

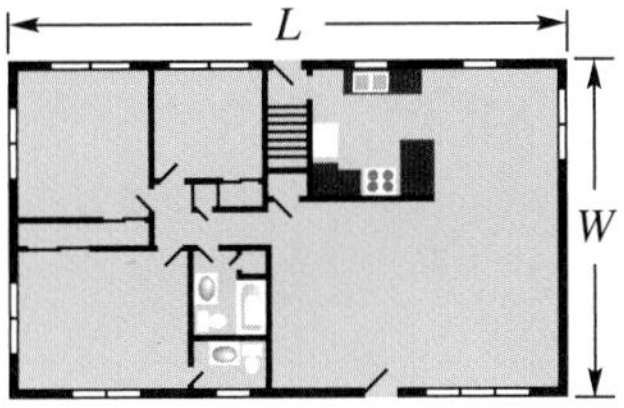

72. If an object is dropped, the distance d in feet it falls in t seconds (disregarding air resistance) is given by the quadratic equation

$$d = 16t^2.$$

Find the distance an object would fall in each of the following times.

(a) 4 sec **(b)** 8 sec

The numbers of alternative-fueled vehicles, in thousands, in use in the United States for selected years over the period 2001–2009 are given in the table.

Year	Number (in thousands)
2001	425
2003	534
2005	592
2007	696
2009	826

Source: Energy Information Administration.

To model the number of vehicles y in year x, we developed the quadratic equation

$$y = 1.57x^2 + 32.5x + 400.$$

Here we used $x = 1$ for 2001, $x = 2$ for 2002, and so on.

73. Use the model to find the number of alternative-fueled vehicles in 2005, to the nearest thousand. How does the result compare with the actual data in the table?

74. What estimate for 2010 is given by the model? Why might the estimate for 2010 be unreliable?

Mixed Review Exercises

75. CONCEPT CHECK Which of the following is *not* factored completely?

A. $3(7t)$ **B.** $3x(7t+4)$ **C.** $(3+x)(7t+4)$ **D.** $3(7t+4)+x(7t+4)$

76. CONCEPT CHECK A student factored $6x^2+16x-32$ as $(2x+8)(3x-4)$. Tell why the polynomial is not factored completely, and give the completely factored form.

Factor completely.

77. $15m^2+20mp-12m-16p$

78. $24ab^3c^2-56a^2bc^3+72a^2b^2c$

79. $z^2-11zx+10x^2$

80. $3k^2+11k+10$

81. y^4-625

82. $6m^3-21m^2-45m$

83. $25a^2+15ab+9b^2$

84. $8z^3+64y^3$

85. $-12r^2-8rq+15q^2$

86. $100a^2-9$

87. $49t^2+56t+16$

Solve each equation, and check the solutions.

88. $t(t-7)=0$

89. $x^2+3x=10$

90. $25x^2+20x+4=0$

Solve each problem.

91. A lot is in the shape of a right triangle. The hypotenuse is 3 m longer than the longer leg. The longer leg is 6 m longer than twice the length of the shorter leg. Find the lengths of the sides of the lot.

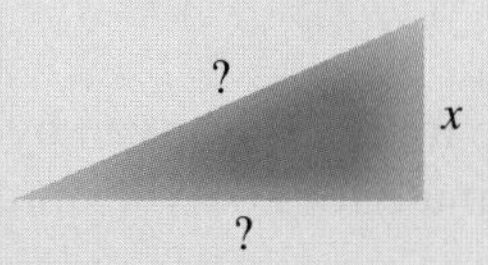

92. A pyramid has a rectangular base with a length that is 2 m more than the width. The height of the pyramid is 6 m, and its volume is 48 m^3. Find the length and width of the base.

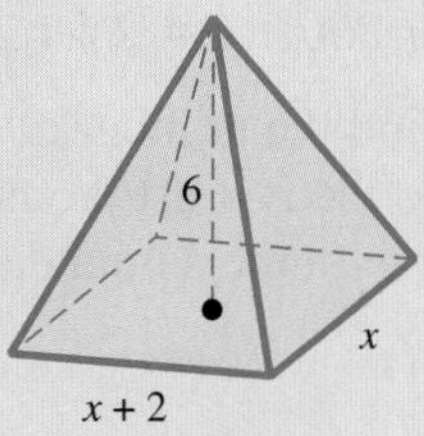

Chapter 13 Test

CHAPTER Test Prep VIDEO

The Chapter Test Prep Videos with test solutions are available on DVD, in MyMathLab*, and on* YouTube*—search "LialDevMath" and click on "Channels."*

1. Which one of the following is the correct, completely factored form of $2x^2 - 2x - 24$?

A. $(2x + 6)(x - 4)$ **B.** $(x + 3)(2x - 8)$

C. $2(x + 4)(x - 3)$ **D.** $2(x + 3)(x - 4)$

Factor each polynomial completely. If the polynomial is prime, *say so.*

2. $12x^2 - 30x$

3. $2m^3n^2 + 3m^3n - 5m^2n^2$

4. $2ax - 2bx + ay - by$

5. $x^2 - 9x + 14$

6. $2x^2 + x - 3$

7. $6x^2 - 19x - 7$

8. $3x^2 - 12x - 15$

9. $10z^2 - 17z + 3$

10. $t^2 + 6t + 10$

11. $x^2 + 36$

12. $y^2 - 49$

13. $81a^2 - 121b^2$

14. $x^2 + 16x + 64$

15. $4x^2 - 28xy + 49y^2$

16. $-2x^2 - 4x - 2$

17. $4t^3 + 32t^2 + 64t$

18. $x^4 - 81$

19. $x^3 - 512$

20. $8k^3 + 64$

Solve each equation.

21. $(x + 3)(x - 9) = 0$

22. $2r^2 - 13r + 6 = 0$

23. $25x^2 - 4 = 0$

24. $x(x - 20) = -100$

25. $t^2 = 3t$

26. $(s + 8)(6s^2 + 13s - 5) = 0$

Solve each problem.

27. The length of a rectangular flower bed is 3 ft less than twice its width. The area of the bed is 54 ft^2. Find the dimensions of the flower bed.

28. Find two consecutive integers such that the square of the sum of the two integers is 11 more than the lesser integer.

29. A carpenter needs to cut a brace to support a wall stud, as shown in the figure. The brace should be 7 ft less than three times the length of the stud. If the brace will be anchored on the floor 15 ft away from the stud, how long should the brace be?

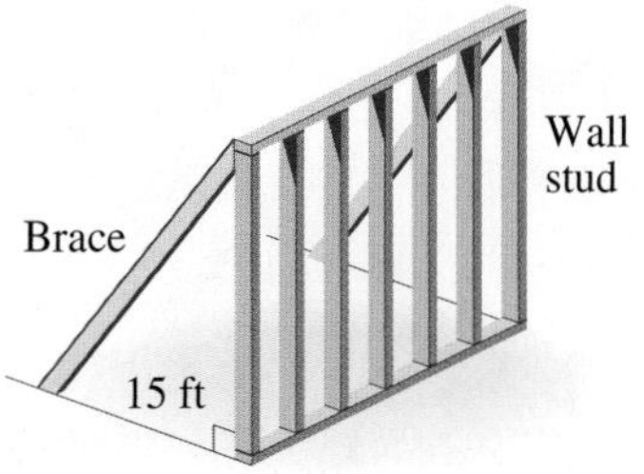

30. The number of pieces of first-class mail sent through the U.S. Postal Service (USPS) from 2001 through 2010 can be approximated by the quadratic equation

$$y = -0.3470x^2 + 1.374x + 100.7,$$

where $x = 1$ represents 2001, $x = 2$ represents 2002, and so on, and y is the number of pieces of mail in billions. Use the model to estimate the number of pieces of first-class mail sent through the USPS in 2009. Round your answer to the nearest billion. (*Source:* U.S. Postal Service.)

14 Rational Expressions and Applications

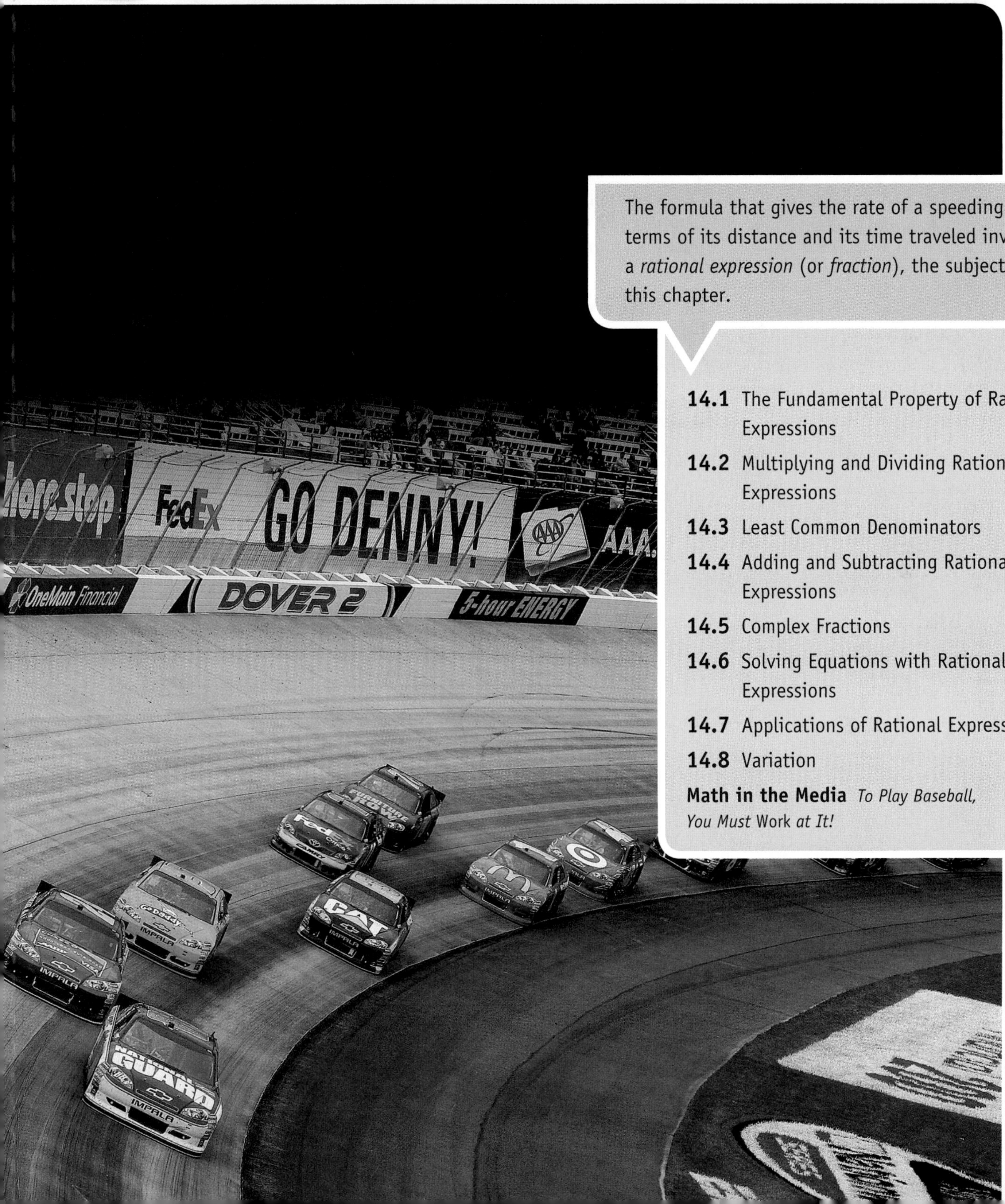

The formula that gives the rate of a speeding car in terms of its distance and its time traveled involves a *rational expression* (or *fraction*), the subject of this chapter.

14.1 The Fundamental Property of Rational Expressions

OBJECTIVES

1. Find the numerical value of a rational expression.
2. Find the values of the variable for which a rational expression is undefined.
3. Write rational expressions in lowest terms.
4. Recognize equivalent forms of rational expressions.

1 Find the numerical value of each rational expression for $x = -3$, $x = 0$, and $x = 3$.

GS **(a)** $\dfrac{x}{2x+1}$

For $x = -3$:

$$\frac{x}{2x+1} = \frac{\_\_\_}{2(\_\_\_)+1} = \frac{\_\_\_}{\_\_\_+1} = \_\_\_$$

(b) $\dfrac{2x+6}{x-3}$

Answers

1. (a) For $x = -3$: -3; -3; -3; -6; $\frac{3}{5}$; for $x = 0$, final answer 0; for $x = 3$, final answer $\frac{3}{7}$
(b) 0; -2; undefined

The quotient of two integers (with denominator not 0), such as $\frac{2}{3}$ or $-\frac{3}{4}$, is a *rational number*. In the same way, the quotient of two polynomials with denominator not equal to 0 is a *rational expression*.

Rational Expression

A **rational expression** is an expression of the form $\frac{P}{Q}$, where P and Q are polynomials, with $Q \neq 0$.

$$\frac{-6x}{x^3+8}, \quad \frac{9x}{y+3}, \quad \text{and} \quad \frac{2m^3}{8} \qquad \text{Rational expressions}$$

Our work with rational expressions will require much of what we learned in **Chapters 12 and 13** on polynomials and factoring, as well as the rules for fractions from **Chapters 2 and 3.**

OBJECTIVE 1 Find the numerical value of a rational expression. Remember that to *evaluate* an expression means to find its *value*. We use substitution to evaluate a rational expression for a given value of the variable.

EXAMPLE 1 Evaluating Rational Expressions

Find the numerical value of $\frac{3x+6}{2x-4}$ for each value of x.

(a) $x = 1$

$$\frac{3x+6}{2x-4} = \frac{3(1)+6}{2(1)-4} \quad \text{Let } x = 1.$$

$$= \frac{9}{-2}, \text{ or } -\frac{9}{2} \quad \frac{a}{-b} = -\frac{a}{b}$$

(b) $x = 0$

$$\frac{3x+6}{2x-4} = \frac{3(0)+6}{2(0)-4} \quad \text{Let } x = 0.$$

$$= \frac{6}{-4}, \text{ or } -\frac{3}{2} \quad \text{Lowest terms}$$

(c) $x = 2$

$$\frac{3x+6}{2x-4} = \frac{3(2)+6}{2(2)-4} \quad \text{Let } x = 2.$$

$$= \frac{12}{0}$$

The expression is undefined for $x = 2$.

(d) $x = -2$

$$\frac{3x+6}{2x-4} = \frac{3(-2)+6}{2(-2)-4} \quad \text{Let } x = -2.$$

$$= \frac{0}{-8}, \text{ or } 0 \quad \frac{0}{b} = 0$$

◀ Work Problem 1 at the Side.

Note

The numerator of a rational expression may be any real number. If the numerator equals 0 and the denominator does not equal 0, then the rational expression equals 0. **See Example 1(d).**

OBJECTIVE 2 Find the values of the variable for which a rational expression is undefined. In the definition of a rational expression $\frac{P}{Q}$, Q cannot equal 0. ***The denominator of a rational expression cannot equal 0 because division by 0 is undefined.***

For instance, in the rational expression

$$\frac{8x^2}{x-3}, \leftarrow \text{Denominator cannot equal 0.}$$

the variable x can take on any value except 3. When x is 3, the denominator becomes $3 - 3 = 0$, making the expression undefined. Thus, x cannot equal 3. We indicate this restriction by writing $x \neq 3$.

Determining When a Rational Expression Is Undefined

Step 1 Set the denominator of the rational expression equal to 0.

Step 2 Solve this equation.

Step 3 The solutions of the equation are the values that make the rational expression undefined. The variable *cannot* equal these values.

EXAMPLE 2 Finding Values That Make Rational Expressions Undefined

Find any values of the variable for which each rational expression is undefined.

(a) $\frac{x+5}{3x+2}$ We must find any value of x that makes the *denominator* equal to 0, since division by 0 is undefined.

Step 1 Set the denominator equal to 0.

$$3x + 2 = 0$$

Step 2 Solve.

$$3x = -2 \quad \text{Subtract 2.}$$

$$x = -\frac{2}{3} \quad \text{Divide by 3.}$$

Step 3 The given expression is undefined for $-\frac{2}{3}$, so $x \neq -\frac{2}{3}$.

(b) $\frac{9m^2}{m^2 - 5m + 6}$

$$m^2 - 5m + 6 = 0 \quad \text{Set the denominator equal to 0.}$$

$$(m-2)(m-3) = 0 \quad \text{Factor.}$$

$$m - 2 = 0 \quad \text{or} \quad m - 3 = 0 \quad \text{Zero-factor property}$$

$$m = 2 \quad \text{or} \quad m = 3 \quad \text{Solve for } m.$$

The given expression is undefined for 2 and 3, so $m \neq 2$, $m \neq 3$.

(c) $\frac{2r}{r^2+1}$ This denominator will not equal 0 for any value of r, because r^2 is always greater than or equal to 0, and adding 1 makes the sum greater than or equal to 1. There are no values for which this expression is undefined.

Work Problem 2 at the Side. ▶

2 Find any values of the variable for which each rational expression is undefined. Write answers with the symbol $\neq$.

GS **(a)** $\frac{x+2}{x-5}$

Step 1 ______ = 0

Step 2 $x =$ ____

Step 3 The given expression is undefined for ____. Thus, $x\ (= / \neq)\ 5$.

(b) $\frac{3r}{r^2 + 6r + 8}$

(c) $\frac{-5m}{m^2 + 4}$

Answers

2. (a) $x - 5$; 5; 5; $\neq$ **(b)** $r \neq -4$, $r \neq -2$
(c) never undefined

3 Write each expression in lowest terms.

(a) $\dfrac{15}{40}$

(b) $\dfrac{5x^4}{15x^2}$

(c) $\dfrac{6p^3}{2p^2}$

Answers

3. (a) $\dfrac{3}{8}$ (b) $\dfrac{x^2}{3}$ (c) $3p$

OBJECTIVE 3 Write rational expressions in lowest terms. A fraction such as $\frac{2}{3}$ is said to be in *lowest terms.*

Lowest Terms

A rational expression $\frac{P}{Q}$ $(Q \neq 0)$ is in **lowest terms** if the greatest common factor of its numerator and denominator is 1.

We use the **fundamental property of rational expressions** to write a rational expression in lowest terms.

Fundamental Property of Rational Expressions

If $\frac{P}{Q}$ (where $Q \neq 0$) is a rational expression and if K represents any polynomial (where $K \neq 0$), then the following is true.

$$\frac{PK}{QK} = \frac{P}{Q}$$

This property is based on the identity property for multiplication.

$$\frac{PK}{QK} = \frac{P}{Q} \cdot \frac{K}{K} = \frac{P}{Q} \cdot 1 = \frac{P}{Q}$$

EXAMPLE 3 Writing in Lowest Terms

Write each expression in lowest terms.

(a) $\dfrac{30}{72}$

Begin by factoring.

$$= \frac{2 \cdot 3 \cdot 5}{2 \cdot 2 \cdot 2 \cdot 3 \cdot 3}$$

(b) $\dfrac{14k^2}{2k^3}$

Write k^2 as $k \cdot k$ and k^3 as $k \cdot k \cdot k$.

$$= \frac{2 \cdot 7 \cdot k \cdot k}{2 \cdot k \cdot k \cdot k}$$

Group any factors common to the numerator and denominator.

(a) $$= \frac{5 \cdot (2 \cdot 3)}{2 \cdot 2 \cdot 3 \cdot (2 \cdot 3)}$$

(b) $$= \frac{7(2 \cdot k \cdot k)}{k(2 \cdot k \cdot k)}$$

Use the fundamental property.

(a) $$= \frac{5}{2 \cdot 2 \cdot 3}, \quad \text{or} \quad \frac{5}{12}$$

(b) $$= \frac{7}{k}$$

◀ Work Problem **3** at the Side.

Writing a Rational Expression in Lowest Terms

Step 1 **Factor** the numerator and denominator completely.

Step 2 **Use the fundamental property** to divide out any common factors.

EXAMPLE 4 **Writing in Lowest Terms**

Write each rational expression in lowest terms.

(a) $\dfrac{3x-12}{5x-20}$

$x \neq 4$, since the denominator is 0 for this value.

$= \dfrac{3(x-4)}{5(x-4)}$ Factor. (Step 1)

$= \dfrac{3}{5}$ Fundamental property (Step 2)

The given expression is equal to $\frac{3}{5}$ for all values of x, where $x \neq 4$ (since the denominator of the original rational expression is 0 when x is 4).

(b) $\dfrac{2y^2-8}{2y+4}$

$y \neq -2$, since the denominator is 0 for this value.

$= \dfrac{2(y^2-4)}{2(y+2)}$ Factor. (Step 1)

$= \dfrac{2(y+2)(y-2)}{2(y+2)}$ Factor the numerator completely.

$= y-2$ Fundamental property (Step 2)

(c) $\dfrac{m^2+2m-8}{2m^2-m-6}$

$m \neq -\frac{3}{2}$, $m \neq 2$

$= \dfrac{(m+4)(m-2)}{(2m+3)(m-2)}$ Factor. (Step 1)

$= \dfrac{m+4}{2m+3}$ Fundamental property (Step 2)

From now on, we write statements of equality of rational expressions with the understanding that they apply only to those real numbers that make neither denominator equal to 0.

CAUTION

Rational expressions cannot be written in lowest terms until after the numerator and denominator have been factored. Only common factors can be divided out, not common terms.

$\dfrac{6x+9}{4x+6} = \dfrac{3(2x+3)}{2(2x+3)} = \dfrac{3}{2}$ — Divide out the common factor.

$\dfrac{6+x}{4x}$ ← Numerator cannot be factored. Already in lowest terms

Work Problem 4 at the Side. ▶

4 Write each rational expression in lowest terms.

GS **(a)** $\dfrac{4y+2}{6y+3}$

$= \dfrac{2(\_\_\_\_)}{3(\_\_\_\_)}$

$= \_\_\_\_$

(b) $\dfrac{8p+8q}{5p+5q}$

(c) $\dfrac{x^2+4x+4}{4x+8}$

(d) $\dfrac{a^2-b^2}{a^2+2ab+b^2}$

Answers

4. (a) $2y+1$; $2y+1$; $\frac{2}{3}$ **(b)** $\frac{8}{5}$

(c) $\frac{x+2}{4}$ **(d)** $\frac{a-b}{a+b}$

5 Write $\frac{x-y}{y-x}$ from **Example 5** in lowest terms by factoring -1 from the numerator. How does the result compare to the result in **Example 5?**

EXAMPLE 5 Writing in Lowest Terms (Factors Are Opposites)

Write $\frac{x-y}{y-x}$ in lowest terms.

To get a common factor, the denominator $y-x$ can be factored as follows.

$y - x$ — We are factoring out -1, **NOT** multiplying by it.

$= \mathbf{-1}(-y+x)$ Factor out -1.

$= -1(x-y)$ Commutative property

With this result in mind, we simplify as follows.

$$\frac{x-y}{y-x}$$

$$= \frac{1(x-y)}{-1(x-y)} \quad y-x=-1(x-y) \text{ from above.}$$

$$= \frac{1}{-1}, \text{ or } -1 \quad \text{Fundamental property}$$

◀ **Work Problem 5 at the Side.**

Note

The numerator *or* the denominator could have been factored in the first step in **Example 5.** Factor -1 from the numerator, and confirm that the result is the same.

CAUTION

Although x and y appear in both the numerator and denominator in **Example 5,** we cannot use the fundamental property right away because they are *terms,* not *factors.* ***Terms are added, while factors are multiplied.***

In **Example 5,** notice that $y-x$ is the **opposite** (or **additive inverse**) of $x-y$. A general rule for this situation follows.

Quotient of Opposites

If the numerator and the denominator of a rational expression are opposites, such as in $\frac{x-y}{y-x}$, then the rational expression is equal to -1.

Based on this result, the following are true.

Numerator and denominator are opposites. $\frac{q-7}{7-q} = -1$ and $\frac{-5a+2b}{5a-2b} = -1$

However, the following expression cannot be simplified further.

$\frac{x-2}{x+2}$ ← Numerator and denominator are *not* opposites.

Answer

5. The result is -1, the same as in **Example 5.**

EXAMPLE 6 Writing in Lowest Terms (Factors Are Opposites)

Write each rational expression in lowest terms.

(a) $\dfrac{2-m}{m-2}$

Since $2-m$ and $m-2$ (or $-2+m$) are opposites, this expression equals -1.

(b) $\dfrac{4x^2-9}{6-4x}$

$= \dfrac{(2x+3)(2x-3)}{2(3-2x)}$ Factor the numerator and denominator.

$= \dfrac{(2x+3)(2x-3)}{2(-1)(2x-3)}$ Write $3-2x$ as $-1(2x-3)$.

$= \dfrac{2x+3}{2(-1)}$ Fundamental property

$= \dfrac{2x+3}{-2}$, or $-\dfrac{2x+3}{2}$ $\quad \frac{a}{-b} = -\frac{a}{b}$

(c) $\dfrac{3+r}{3-r}$ $\quad 3-r$ is *not* the opposite of $3+r$.

The rational expression is already in lowest terms.

Work Problem 6 at the Side. ▶

OBJECTIVE 4 Recognize equivalent forms of rational expressions. The common fraction $-\frac{5}{6}$ can also be written as $\frac{-5}{6}$ and as $\frac{5}{-6}$.

Consider the final rational expression from **Example 6(b).**

$$-\frac{2x+3}{2}$$

The $-$ sign representing the factor -1 is in front of the expression, aligned with the fraction bar. The factor -1 may instead be placed in the numerator or in the denominator. Some other equivalent forms of this rational expression are

Use parentheses.

$$\frac{-(2x+3)}{2} \quad \text{and} \quad \frac{2x+3}{-2}.$$

By the distributive property,

$$\frac{-(2x+3)}{2} \quad \text{can also be written} \quad \frac{-2x-3}{2}.$$

Multiply *each* term in the parentheses by -1.

CAUTION

$\frac{-2x+3}{2}$ is *not* an equivalent form of $\frac{-(2x+3)}{2}$. The sign preceding 3 in the numerator of $\frac{-2x+3}{2}$ should be $-$ rather than $+$. ***Be careful to apply the distributive property correctly.***

6 Write each rational expression in lowest terms.

(a) $\dfrac{5-y}{y-5}$

(b) $\dfrac{m-n}{n-m}$

GS **(c)** $\dfrac{25x^2-16}{12-15x}$

$= \dfrac{(5x+4)(\_\_\_\_)}{3(\_\_\_\_)}$

$= \dfrac{(5x+4)(5x-4)}{3(\_\_\_\_)(5x-4)}$

$= \_\_\_\_$

(d) $\dfrac{9-k}{9+k}$

Answers

6. **(a)** -1 **(b)** -1

(c) $5x-4$; $4-5x$; -1; $\dfrac{5x+4}{-3}$, or $-\dfrac{5x+4}{3}$

(d) already in lowest terms

7 Decide whether each rational expression is equivalent to

$$-\frac{2x-6}{x+3}.$$

(a) $\frac{-(2x-6)}{x+3}$

(b) $\frac{-2x+6}{x+3}$

(c) $\frac{-2x-6}{x+3}$

(d) $\frac{2x-6}{-(x+3)}$

(e) $\frac{2x-6}{-x-3}$

(f) $\frac{2x-6}{x-3}$

Answers

7. **(a)** equivalent **(b)** equivalent **(c)** not equivalent **(d)** equivalent **(e)** equivalent **(f)** not equivalent

EXAMPLE 7 Writing Equivalent Forms of a Rational Expression

Write four equivalent forms of the following rational expression.

$$-\frac{3x+2}{x-6}$$

If we apply the negative sign to the numerator, we obtain these equivalent forms.

$$\text{①}\rightarrow \frac{-(3x+2)}{x-6}, \quad \text{and, by the distributive property,} \quad \frac{-3x-2}{x-6} \leftarrow \text{②}$$

If we apply the negative sign to the denominator of the given rational expression, we obtain two more forms.

$$\text{③}\rightarrow \frac{3x+2}{-(x-6)} \quad \text{and, by distributing again,} \quad \frac{3x+2}{-x+6} \leftarrow \text{④}$$

CAUTION

Recall that $-\frac{5}{6} \neq \frac{-5}{-6}$. Thus, in **Example 7,** it would be incorrect to distribute the negative sign in $-\frac{3x+2}{x-6}$ to *both* the numerator *and* the denominator. (Doing this would actually lead to the *opposite* of the original expression.)

◀ **Work Problem 7 at the Side.**

14.1 Exercises

1. CONCEPT CHECK Fill in each blank with the correct response.

(a) The rational expression $\frac{x+5}{x-3}$ is undefined when $x =$ _______, and is equal to 0 when $x =$ _______.

(b) The rational expression $\frac{p-q}{q-p}$ is undefined when $p =$ _______, and in all other cases is equal to _______.

2. CONCEPT CHECK Make the correct choice to complete each statement.

(a) $\frac{4-r^2}{4+r^2}$ (*is* / *is not*) equal to -1.

(b) $\frac{5+2x}{3-x}$ and $\frac{-5-2x}{x-3}$ (*are* / *are not*) equivalent rational expressions.

3. Define *rational expression* in your own words, and give an example.

4. Why can't the denominator of a rational expression equal 0?

Find the numerical value of each rational expression for **(a)** $x = 2$ *and* **(b)** $x = -3$. ***See Example 1.***

5. $\frac{5x-2}{4x}$

6. $\frac{3x+1}{5x}$

7. $\frac{x^2-4}{2x+1}$

8. $\frac{2x^2-4x}{3x-1}$

9. $\frac{(-3x)^2}{4x+12}$

10. $\frac{(-2x)^3}{3x+9}$

11. $\frac{5x+2}{2x^2+11x+12}$

12. $\frac{7-3x}{3x^2-7x+2}$

Find any value(s) of the variable for which each rational expression is undefined. Write answers with the symbol $\neq$. ***See Example 2.***

13. $\frac{2}{5y}$

14. $\frac{7}{3z}$

15. $\frac{x+1}{x+6}$

16. $\frac{m+2}{m+5}$

17. $\frac{4x^2}{3x-5}$

18. $\frac{2x^3}{3x-4}$

19. $\frac{m+2}{m^2+m-6}$

20. $\frac{r-5}{r^2-5r+4}$

21. $\frac{x^2-3x}{4}$

22. $\frac{x^2-4x}{6}$

23. $\frac{3x}{x^2+2}$

24. $\frac{4q}{q^2+9}$

Write each rational expression in lowest terms. ***See Examples 3 and 4.***

25. $\frac{18r^3}{6r}$

26. $\frac{27p^2}{3p}$

27. $\frac{4(y-2)}{10(y-2)}$

28. $\frac{15(m-1)}{9(m-1)}$

29. $\dfrac{(x+1)(x-1)}{(x+1)^2}$

30. $\dfrac{(t+5)(t-3)}{(t-1)(t+5)}$

31. $\dfrac{7m+14}{5m+10}$

32. $\dfrac{8z-24}{4z-12}$

33. $\dfrac{m^2-n^2}{m+n}$

34. $\dfrac{a^2-b^2}{a-b}$

35. $\dfrac{12m^2-3}{8m-4}$

36. $\dfrac{20p^2-45}{6p-9}$

37. $\dfrac{3m^2-3m}{5m-5}$

38. $\dfrac{6t^2-6t}{2t-2}$

39. $\dfrac{9r^2-4s^2}{9r+6s}$

40. $\dfrac{16x^2-9y^2}{12x-9y}$

41. $\dfrac{2x^2-3x-5}{2x^2-7x+5}$

42. $\dfrac{3x^2+8x+4}{3x^2-4x-4}$

43. $\dfrac{zw+4z-3w-12}{zw+4z+5w+20}$

44. $\dfrac{km+4k+4m+16}{km+4k+5m+20}$

45. $\dfrac{ac-ad+bc-bd}{ac-ad-bc+bd}$

46. $\dfrac{rt-ru-st+su}{rt-ru+st-su}$

Write each rational expression in lowest terms. ***See Examples 5 and 6.***

47. $\dfrac{6-t}{t-6}$

48. $\dfrac{2-k}{k-2}$

49. $\dfrac{m^2-1}{1-m}$

50. $\dfrac{a^2-b^2}{b-a}$

51. $\dfrac{q^2-4q}{4q-q^2}$

52. $\dfrac{z^2-5z}{5z-z^2}$

53. $\dfrac{p+6}{p-6}$

54. $\dfrac{5-x}{5+x}$

Write four equivalent forms for each rational expression. ***See Example 7.***

55. $-\dfrac{x+4}{x-3}$

56. $-\dfrac{x+6}{x-1}$

57. $-\dfrac{2x-3}{x+3}$

58. $-\dfrac{5x-6}{x+4}$

59. $-\dfrac{3x-1}{5x-6}$

60. $-\dfrac{2x-9}{7x-1}$

61. The area of the rectangle is represented by $x^4 + 10x^2 + 21$. What is the width? (*Hint:* Use $W = \frac{A}{L}$.)

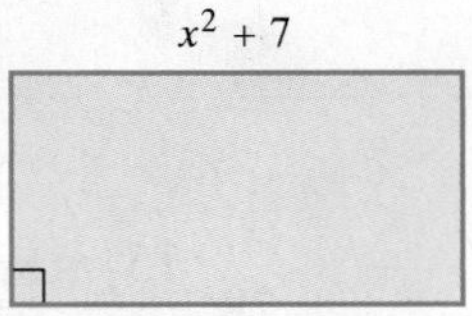

62. The volume of the box is represented by
$$(x^2 + 8x + 15)(x + 4).$$
Find the polynomial that represents the area of the bottom of the box.

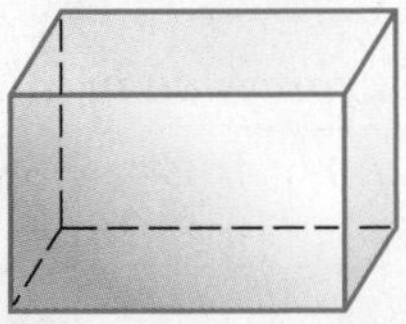

14.2 Multiplying and Dividing Rational Expressions

OBJECTIVES

1. **Multiply rational expressions.**
2. **Find reciprocals.**
3. **Divide rational expressions.**

OBJECTIVE 1 Multiply rational expressions. The product of two fractions is found by multiplying the numerators and multiplying the denominators. Rational expressions are multiplied in the same way.

Multiplying Rational Expressions

The product of the rational expressions $\frac{P}{Q}$ and $\frac{R}{S}$ is defined as follows.

$$\frac{P}{Q} \cdot \frac{R}{S} = \frac{PR}{QS}$$

In words: To multiply rational expressions, multiply the numerators and multiply the denominators.

EXAMPLE 1 Multiplying Rational Expressions

Multiply. Write each answer in lowest terms.

(a) $\frac{3}{10} \cdot \frac{5}{9}$ | (b) $\frac{6}{x} \cdot \frac{x^2}{12}$

Indicate the product of the numerators and the product of the denominators.

$= \frac{3 \cdot 5}{10 \cdot 9}$ | $= \frac{6 \cdot x^2}{x \cdot 12}$

Leave the products in factored form. Factor the numerator and denominator to further identify any common factors. Then use the fundamental property to write each product in lowest terms.

$= \frac{3 \cdot 5}{2 \cdot 5 \cdot 3 \cdot 3}$ | $= \frac{6 \cdot x \cdot x}{2 \cdot 6 \cdot x}$

$= \frac{1}{6}$ Remember to write 1 in the numerator. | $= \frac{x}{2}$

Work Problem 1 at the Side. ▶

EXAMPLE 2 Multiplying Rational Expressions

Multiply. Write the answer in lowest terms.

$$\frac{x+y}{2x} \cdot \frac{x^2}{(x+y)^2}$$

$= \frac{(x+y)x^2}{2x(x+y)^2}$ Multiply numerators. Multiply denominators. (Use parentheses here around $x+y$.)

$= \frac{(x+y)x \cdot x}{2x(x+y)(x+y)}$ Factor. Identify the common factors.

$= \frac{x}{2(x+y)}$ $\frac{(x+y)x}{x(x+y)} = 1$; Write in lowest terms.

Work Problem 2 at the Side. ▶

---

1 Multiply. Write each answer in lowest terms.

(a) $\frac{2}{7} \cdot \frac{5}{10}$

(b) $\frac{3m^2}{2} \cdot \frac{10}{m}$

(c) $\frac{8p^2q}{3} \cdot \frac{9}{q^2p}$

2 Multiply. Write each answer in lowest terms.

(a) $\frac{a+b}{5} \cdot \frac{30}{2(a+b)}$

(b) $\frac{3(p-q)}{q^2} \cdot \frac{q}{2(p-q)^2}$

Answers

1. (a) $\frac{1}{7}$ (b) $15m$ (c) $\frac{24p}{q}$
2. (a) 3 (b) $\frac{3}{2q(p-q)}$

3 Multiply. Write each answer in lowest terms.

(a) $\dfrac{x^2+7x+10}{3x+6} \cdot \dfrac{6x-6}{x^2+2x-15}$

(b) $\dfrac{m^2+4m-5}{m+5} \cdot \dfrac{m^2+8m+15}{m-1}$

4 Find the reciprocal of each rational expression.

(a) $\dfrac{5}{8}$

(b) $\dfrac{6b^5}{3r^2b}$

(c) $\dfrac{t^2-4t}{t^2+2t-3}$

Answers

3. **(a)** $\dfrac{2(x-1)}{x-3}$ **(b)** $(m+5)(m+3)$

4. **(a)** $\dfrac{8}{5}$ **(b)** $\dfrac{3r^2b}{6b^5}$ **(c)** $\dfrac{t^2+2t-3}{t^2-4t}$

EXAMPLE 3 **Multiplying Rational Expressions**

Multiply. Write the answer in lowest terms.

$$\frac{x^2+3x}{x^2-3x-4} \cdot \frac{x^2-5x+4}{x^2+2x-3}$$

$$= \frac{(x^2+3x)(x^2-5x+4)}{(x^2-3x-4)(x^2+2x-3)} \quad \text{Definition of multiplication}$$

$$= \frac{x(x+3)(x-4)(x-1)}{(x-4)(x+1)(x+3)(x-1)} \quad \text{Factor.}$$

$$= \frac{x}{x+1} \quad \text{Divide out the common factors.}$$

The quotients $\frac{x+3}{x+3}$, $\frac{x-4}{x-4}$, and $\frac{x-1}{x-1}$ are all equal to 1, justifying the final product $\frac{x}{x+1}$.

◀ **Work Problem 3 at the Side.**

OBJECTIVE 2 **Find reciprocals.** If the product of two rational expressions is 1, the rational expressions are **reciprocals** (or **multiplicative inverses**) of each other. The reciprocal of a rational expression is found by interchanging the numerator and the denominator.

$$\frac{2x-1}{x-5} \text{ has reciprocal } \frac{x-5}{2x-1}.$$

EXAMPLE 4 **Finding Reciprocals of Rational Expressions**

Find the reciprocal of each rational expression.

(a) $\dfrac{4p^3}{9q}$ has reciprocal $\dfrac{9q}{4p^3}$. Interchange the numerator and denominator.

(b) $\dfrac{k^2-9}{k^2-k-20}$ has reciprocal $\dfrac{k^2-k-20}{k^2-9}$.

◀ **Work Problem 4 at the Side.**

OBJECTIVE 3 **Divide rational expressions.** Suppose we have $\frac{7}{8}$ gal of milk and want to find how many quarts we have. Since 1 qt is $\frac{1}{4}$ gal, we ask, "How many $\frac{1}{4}$s are there in $\frac{7}{8}$?" This would be interpreted as follows.

$$\frac{7}{8} \div \frac{1}{4}, \quad \text{or} \quad \frac{\frac{7}{8}}{\frac{1}{4}} \leftarrow \text{The fraction bar means division.}$$

The fundamental property of rational expressions can be applied to rational number values of P, Q, and K.

$$\frac{P}{Q} = \frac{P \cdot K}{Q \cdot K} = \frac{\frac{7}{8} \cdot 4}{\frac{1}{4} \cdot 4} = \frac{\frac{7}{8} \cdot 4}{1} = \frac{7}{8} \cdot \frac{4}{1}$$

Let $P = \frac{7}{8}$, $Q = \frac{1}{4}$, and $K = 4$. (K is the reciprocal of Q.)

So, to divide $\frac{7}{8}$ by $\frac{1}{4}$, we multiply $\frac{7}{8}$ by the reciprocal of $\frac{1}{4}$, namely $\frac{4}{1}$, or 4. Since $\frac{7}{8}(4) = \frac{7}{2}$, there are $\frac{7}{2}$ qt, or $3\frac{1}{2}$ qt, in $\frac{7}{8}$ gal.

Dividing Rational Expressions

If $\frac{P}{Q}$ and $\frac{R}{S}$ are any two rational expressions, with $\frac{R}{S} \neq 0$, then their quotient is defined as follows.

$$\frac{P}{Q} \div \frac{R}{S} = \frac{P}{Q} \cdot \frac{S}{R} = \frac{PS}{QR}$$

In words: To divide one rational expression by another rational expression, multiply the first rational expression (dividend) by the reciprocal of the second rational expression (divisor).

EXAMPLE 5 Dividing Rational Expressions

Divide. Write each answer in lowest terms.

(a) $\frac{5}{8} \div \frac{7}{16}$

(b) $\frac{y}{y+3} \div \frac{4y}{y+5}$

Multiply the first expression by the reciprocal of the second.

(a)

$= \frac{5}{8} \cdot \frac{16}{7}$ ← Reciprocal of $\frac{7}{16}$

$= \frac{5 \cdot 16}{8 \cdot 7}$ Multiply.

$= \frac{5 \cdot 8 \cdot 2}{8 \cdot 7}$ Factor 16.

$= \frac{10}{7}$ Lowest terms

(b)

$= \frac{y}{y+3} \cdot \frac{y+5}{4y}$ ← Reciprocal of $\frac{4y}{y+5}$

$= \frac{y(y+5)}{(y+3)(4y)}$ Multiply.

$= \frac{y+5}{4(y+3)}$ Lowest terms

Work Problem **5** at the Side. ▶

EXAMPLE 6 Dividing Rational Expressions

Divide. Write the answer in lowest terms.

$$\frac{(3m)^2}{(2p)^3} \div \frac{6m^3}{16p^2}$$

$= \frac{(3m)^2}{(2p)^3} \cdot \frac{16p^2}{6m^3}$ Multiply by the reciprocal.

$= \frac{9m^2}{8p^3} \cdot \frac{16p^2}{6m^3}$ Power rule for exponents

($(3m)^2 = 3^2m^2$; $(2p)^3 = 2^3p^3$)

$= \frac{9 \cdot 16m^2p^2}{8 \cdot 6p^3m^3}$ Multiply numerators. Multiply denominators.

$= \frac{3 \cdot 3 \cdot 8 \cdot 2 \cdot m^2 \cdot p^2}{8 \cdot 3 \cdot 2 \cdot p^2 \cdot p \cdot m^2 \cdot m}$ Factor.

$= \frac{3}{pm}$, or $\frac{3}{mp}$ Lowest terms

Work Problem **6** at the Side. ▶

---

5 Divide. Write each answer in lowest terms.

(a) $\frac{3}{4} \div \frac{5}{16}$

(b) $\frac{r}{r-1} \div \frac{3r}{r+4}$

GS **(c)** $\frac{6x-4}{3} \div \frac{15x-10}{9}$

$= \frac{6x-4}{3} \cdot \frac{\_\_\_\_}{\_\_\_\_}$

$= \frac{2(\_\_\_\_)}{3} \cdot \frac{\_\_\_\_}{5(\_\_\_\_)}$

$= \frac{2(3x-2) \cdot 3 \cdot 3}{\_\_\_\_ \cdot 5(3x-2)}$

$= \_\_\_\_$

6 Divide. Write each answer in lowest terms.

(a) $\frac{5a^2b}{2} \div \frac{10ab^2}{8}$

(b) $\frac{(3t)^2}{w} \div \frac{3t^2}{5w^4}$

Answers

5. (a) $\frac{12}{5}$ **(b)** $\frac{r+4}{3(r-1)}$

(c) 9; $15x - 10$; $3x - 2$; 9; $3x - 2$; 3; $\frac{6}{5}$

6. (a) $\frac{2a}{b}$ **(b)** $15w^3$

7 Divide. Write each answer in lowest terms.

GS **(a)** $\dfrac{y^2 + 4y + 3}{y + 3} \div \dfrac{y^2 - 4y - 5}{y - 3}$

Rewrite as multiplication, and complete the problem.

__________ • __________

(b) $\dfrac{4x(x + 3)}{2x + 1} \div \dfrac{-x^2(x + 3)}{4x^2 - 1}$

8 Divide. Write each answer in lowest terms.

(a) $\dfrac{ab - a^2}{a^2 - 1} \div \dfrac{a - b}{a - 1}$

(b) $\dfrac{x^2 - 9}{2x + 6} \div \dfrac{9 - x^2}{4x - 12}$

EXAMPLE 7 Dividing Rational Expressions

Divide. Write the answer in lowest terms.

$$\frac{x^2 - 4}{(x + 3)(x - 2)} \div \frac{(x + 2)(x + 3)}{-2x}$$

$$= \frac{x^2 - 4}{(x + 3)(x - 2)} \cdot \frac{-2x}{(x + 2)(x + 3)} \quad \text{Multiply by the reciprocal.}$$

$$= \frac{-2x(x^2 - 4)}{(x + 3)(x - 2)(x + 2)(x + 3)} \quad \text{Multiply numerators. Multiply denominators.}$$

$$= \frac{-2x(x + 2)(x - 2)}{(x + 3)(x - 2)(x + 2)(x + 3)} \quad \text{Factor the numerator.}$$

$$= \frac{-2x}{(x + 3)^2}, \quad \text{or} \quad -\frac{2x}{(x + 3)^2} \quad \text{Lowest terms; } \frac{-a}{b} = -\frac{a}{b}$$

◀ Work Problem **7** at the Side.

EXAMPLE 8 Dividing Rational Expressions (Factors Are Opposites)

Divide. Write the answer in lowest terms.

$$\frac{m^2 - 4}{m^2 - 1} \div \frac{2m^2 + 4m}{1 - m}$$

$$= \frac{m^2 - 4}{m^2 - 1} \cdot \frac{1 - m}{2m^2 + 4m} \quad \text{Multiply by the reciprocal.}$$

$$= \frac{(m^2 - 4)(1 - m)}{(m^2 - 1)(2m^2 + 4m)} \quad \text{Multiply numerators. Multiply denominators.}$$

$$= \frac{(m + 2)(m - 2)(1 - m)}{(m + 1)(m - 1)(2m)(m + 2)} \quad \text{Factor. } 1 - m \text{ and } m - 1 \text{ are opposites.}$$

$$= \frac{-1(m - 2)}{2m(m + 1)} \quad \text{From **Section 14.1**, } \frac{1 - m}{m - 1} = -1.$$

$$= \frac{-m + 2}{2m(m + 1)}, \quad \text{or} \quad \frac{2 - m}{2m(m + 1)} \quad \text{Distribute } -1 \text{ in the numerator. Rewrite } -m + 2 \text{ as } 2 - m.$$

◀ Work Problem **8** at the Side.

In summary, follow these steps to multiply or divide rational expressions.

Multiplying or Dividing Rational Expressions

Step 1 **Note the operation.** If the operation is division, use the definition of division to rewrite as multiplication.

Step 2 **Multiply** numerators and multiply denominators.

Step 3 **Factor** all numerators and denominators completely.

Step 4 **Write in lowest terms** using the fundamental property.

Steps 2 and 3 may be interchanged based on personal preference.

Answers

7. (a) $\dfrac{y^2 + 4y + 3}{y + 3} \cdot \dfrac{y - 3}{y^2 - 4y - 5}$; $\dfrac{y - 3}{y - 5}$
(b) $-\dfrac{4(2x - 1)}{x}$

8. (a) $\dfrac{-a}{a + 1}$ (b) $\dfrac{-2x + 6}{3 + x}$

14.2 Exercises

FOR EXTRA HELP

Download the MyDashBoard App

MyMathLab®

1. **CONCEPT CHECK** Match each multiplication problem in Column I with the correct product in Column II.

I	II
(a) $\frac{5x^3}{10x^4} \cdot \frac{10x^7}{2x}$	A. $\frac{2}{5x^5}$
(b) $\frac{10x^4}{5x^3} \cdot \frac{10x^7}{2x}$	B. $\frac{5x^5}{2}$
(c) $\frac{5x^3}{10x^4} \cdot \frac{2x}{10x^7}$	C. $\frac{1}{10x^7}$
(d) $\frac{10x^4}{5x^3} \cdot \frac{2x}{10x^7}$	D. $10x^7$

2. **CONCEPT CHECK** Match each division problem in Column I with the correct quotient in Column II.

I	II
(a) $\frac{5x^3}{10x^4} \div \frac{10x^7}{2x}$	A. $\frac{5x^5}{2}$
(b) $\frac{10x^4}{5x^3} \div \frac{10x^7}{2x}$	B. $10x^7$
(c) $\frac{5x^3}{10x^4} \div \frac{2x}{10x^7}$	C. $\frac{2}{5x^5}$
(d) $\frac{10x^4}{5x^3} \div \frac{2x}{10x^7}$	D. $\frac{1}{10x^7}$

Multiply. Write each answer in lowest terms. ***See Examples 1 and 2.***

3. $\frac{10m^2}{7} \cdot \frac{14}{15m}$

4. $\frac{36z^3}{6} \cdot \frac{28}{z^2}$

5. $\frac{16y^4}{18y^5} \cdot \frac{15y^5}{y^2}$

6. $\frac{20x^5}{-2x^2} \cdot \frac{8x^4}{35x^3}$

7. $\frac{2(c+d)}{3} \cdot \frac{18}{6(c+d)^2}$

8. $\frac{4(y-2)}{x} \cdot \frac{3x}{6(y-2)^2}$

9. $\frac{(x-y)^2}{2} \cdot \frac{24}{3(x-y)}$

10. $\frac{(a+b)^2}{5} \cdot \frac{30}{2(a+b)}$

Find the reciprocal of each rational expression. ***See Example 4.***

11. $\frac{3p^3}{16q}$

12. $\frac{6x^4}{9y^2}$

13. $\frac{r^2+rp}{7}$

14. $\frac{16}{9a^2+36a}$

15. $\frac{z^2+7z+12}{z^2-9}$

16. $\frac{p^2-4p+3}{p^2-3p}$

Divide. Write each answer in lowest terms. ***See Examples 5 and 6.***

17. $\frac{9z^4}{3z^5} \div \frac{3z^2}{5z^3}$

18. $\frac{35q^8}{9q^5} \div \frac{25q^6}{10q^5}$

19. $\frac{4t^4}{2t^5} \div \frac{(2t)^3}{-6}$

20. $\frac{-12a^6}{3a^2} \div \frac{(2a)^3}{27a}$

21. $\frac{3}{2y-6} \div \frac{6}{y-3}$

22. $\frac{4m+16}{10} \div \frac{3m+12}{18}$

23. $\frac{(x-3)^2}{6x} \div \frac{x-3}{x^2}$

24. $\frac{2a}{a+4} \div \frac{a^2}{(a+4)^2}$

Multiply or divide. Write each answer in lowest terms. ***See Examples 3, 7, and 8.***

25. GS $\dfrac{5x-15}{3x+9} \cdot \dfrac{4x+12}{6x-18}$

$= \dfrac{5(\_\_\_\_)}{3(\_\_\_\_)} \cdot \dfrac{4(\_\_\_\_)}{6(\_\_\_\_)}$

$= \_\_\_\_$

26. GS $\dfrac{8r+16}{24r-24} \cdot \dfrac{6r-6}{3r+6}$

$= \dfrac{8(\_\_\_\_)}{24(\_\_\_\_)} \cdot \dfrac{6(\_\_\_\_)}{3(\_\_\_\_)}$

$= \_\_\_\_$

27. $\dfrac{2-t}{8} \div \dfrac{t-2}{6}$

28. $\dfrac{4}{m-2} \div \dfrac{16}{2-m}$

29. $\dfrac{5-4x}{5+4x} \cdot \dfrac{4x+5}{4x-5}$

30. $\dfrac{5-x}{5+x} \cdot \dfrac{x+5}{x-5}$

31. $\dfrac{6(m-2)^2}{5(m+4)^2} \cdot \dfrac{15(m+4)}{2(2-m)}$

32. $\dfrac{7(q-1)}{3(q+1)^2} \cdot \dfrac{6(q+1)}{3(1-q)^2}$

33. $\dfrac{p^2+4p-5}{p^2+7p+10} \div \dfrac{p-1}{p+4}$

34. $\dfrac{z^2-3z+2}{z^2+4z+3} \div \dfrac{z-1}{z+1}$

35. $\dfrac{2k^2-k-1}{2k^2+5k+3} \div \dfrac{4k^2-1}{2k^2+k-3}$

36. $\dfrac{2m^2-5m-12}{m^2+m-20} \div \dfrac{4m^2-9}{m^2+4m-5}$

37. $\dfrac{2k^2+3k-2}{6k^2-7k+2} \cdot \dfrac{4k^2-5k+1}{k^2+k-2}$

38. $\dfrac{2m^2-5m-12}{m^2-10m+24} \div \dfrac{4m^2-9}{m^2-9m+18}$

39. $\dfrac{m^2+2mp-3p^2}{m^2-3mp+2p^2} \div \dfrac{m^2+4mp+3p^2}{m^2+2mp-8p^2}$

40. $\dfrac{r^2+rs-12s^2}{r^2-rs-20s^2} \div \dfrac{r^2-2rs-3s^2}{r^2+rs-30s^2}$

41. $\left(\dfrac{x^2+10x+25}{x^2+10x} \cdot \dfrac{10x}{x^2+15x+50}\right) \div \dfrac{x+5}{x+10}$

42. $\left(\dfrac{m^2-12m+32}{8m} \cdot \dfrac{m^2-8m}{m^2-8m+16}\right) \div \dfrac{m-8}{m-4}$

Solve each problem.

43. If the rational expression $\frac{5x^2y^3}{2pq}$ represents the area of a rectangle and $\frac{2xy}{p}$ represents the length, what rational expression represents the width?

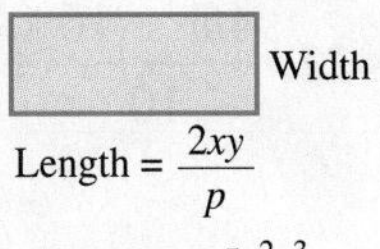

The area is $\dfrac{5x^2y^3}{2pq}$.

44. If the rational expression $\frac{12a^3b^4}{5cd}$ represents the area of a rectangle and $\frac{9ab^2}{c}$ represents the width, what rational expression represents the length?

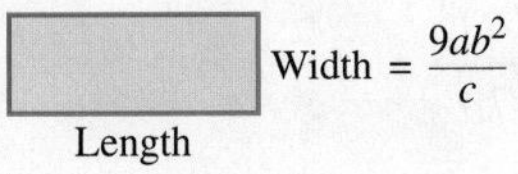

The area is $\dfrac{12a^3b^4}{5cd}$.

14.3 Least Common Denominators

OBJECTIVES

1. Find the least common denominator for a list of fractions.
2. Write equivalent rational expressions.

OBJECTIVE 1 Find the least common denominator for a list of fractions. Adding or subtracting rational expressions often requires a **least common denominator (LCD).** The LCD is the simplest expression that is divisible by all of the denominators in all of the expressions. For example, the fractions

$$\frac{2}{9} \text{ and } \frac{5}{12} \text{ have LCD 36,}$$

because 36 is the least positive number divisible by both 9 and 12.

We can often find least common denominators by inspection. In other cases, we find the LCD by a procedure similar to that used in **Section 13.1.**

Finding the Least Common Denominator (LCD)

Step 1 **Factor** each denominator into prime factors.

Step 2 **List each different denominator factor** the *greatest* number of times it appears in any of the denominators.

Step 3 **Multiply** the denominator factors from Step 2 to get the LCD.

When each denominator is factored into prime factors, every prime factor must be a factor of the least common denominator.

EXAMPLE 1 Finding Least Common Denominators

Find the LCD for each pair of fractions.

(a) $\frac{1}{24}, \frac{7}{15}$ **(b)** $\frac{1}{8x}, \frac{3}{10x}$

Step 1 Write each denominator in factored form with numerical coefficients in prime factored form.

(a) $24 = 2 \cdot 2 \cdot 2 \cdot 3 = 2^3 \cdot 3$

$15 = 3 \cdot 5$

(b) $8x = 2 \cdot 2 \cdot 2 \cdot x = 2^3 \cdot x$

$10x = 2 \cdot 5 \cdot x$

Step 2 We find the LCD by taking each different factor the *greatest* number of times it appears as a factor in any of the denominators.

(a) The factor 2 appears three times in one product and not at all in the other, so the greatest number of times 2 appears is three. The greatest number of times both 3 and 5 appear is one.

(b) Here, 2 appears three times in one product and once in the other, so the greatest number of times 2 appears is three. The greatest number of times 5 appears is one. The greatest number of times x appears is one.

Step 3

(a) $\text{LCD} = 2 \cdot 2 \cdot 2 \cdot 3 \cdot 5$

$= 2^3 \cdot 3 \cdot 5$

$= 120$

(b) $\text{LCD} = 2 \cdot 2 \cdot 2 \cdot 5 \cdot x$

$= 2^3 \cdot 5 \cdot x$

$= 40x$

Work Problem 1 at the Side. ▶

1 Find the LCD for each pair of fractions.

GS **(a)** $\frac{7}{10}, \frac{1}{25}$

Step 1 $10 = 2 \cdot$ ____

$25 =$ ____ $\cdot$ ____

Complete Step 2 and Step 3.

The LCD is ____.

(b) $\frac{7}{20p}, \frac{11}{30p}$

(c) $\frac{4}{5x}, \frac{12}{10x}$

Answers

1. (a) 5; 5; 5; 50 **(b)** $60p$ **(c)** $10x$

2 Find the LCD.

(a) $\frac{4}{16m^3n}, \frac{5}{9m^5}$

(b) $\frac{3}{25a^2}, \frac{2}{10a^3b}$

3 Find the LCD for the fractions in each list.

(a) $\frac{7}{3a}, \frac{11}{a^2 - 4a}$

(b) $\frac{2m}{m^2 - 3m + 2}, \frac{5m - 3}{m^2 + 3m - 10}, \frac{4m + 7}{m^2 + 4m - 5}$

(c) $\frac{6}{x - 4}, \frac{3x - 1}{4 - x}$

Answers

2. **(a)** $144m^5n$ **(b)** $50a^3b$

3. **(a)** $3a(a - 4)$
(b) $(m - 1)(m - 2)(m + 5)$
(c) either $x - 4$ or $4 - x$

EXAMPLE 2 Finding the LCD

Find the LCD for $\frac{5}{6r^2}$ and $\frac{3}{4r^3}$.

Step 1 Factor each denominator.

$$6r^2 = 2 \cdot 3 \cdot r^2$$
$$4r^3 = 2^2 \cdot r^3$$

Step 2 The greatest number of times 2 appears is two, the greatest number of times 3 appears is one, and the greatest number of times r appears is three.

Step 3 $$\text{LCD} = 2^2 \cdot 3 \cdot r^3 = 12r^3$$

◀ **Work Problem 2 at the Side.**

CAUTION

When finding the LCD, use each factor the *greatest* number of times it appears in any *single* denominator, not the *total* number of times it appears. For instance, the greatest number of times r appears as a factor in one denominator in **Example 2** is 3, *not* 5.

EXAMPLE 3 Finding LCDs

Find the LCD for the fractions in each list.

(a) $\frac{6}{5m}, \frac{4}{m^2 - 3m}$

$$\left.\begin{aligned} 5m &= 5 \cdot m \\ m^2 - 3m &= m(m - 3) \end{aligned}\right\} \text{Factor each denominator.}$$

Use each different factor the greatest number of times it appears.

$$\text{LCD} = 5 \cdot m \cdot (m - 3) = 5m(m - 3)$$

Be sure to include m as a factor in the LCD.

Because m is not a *factor* of $m - 3$, both factors, m and $m - 3$, must appear in the LCD.

(b) $\frac{1}{r^2 - 4r - 5}, \frac{3}{r^2 - r - 20}, \frac{1}{r^2 - 10r + 25}$

$$\left.\begin{aligned} r^2 - 4r - 5 &= (r - 5)(r + 1) \\ r^2 - r - 20 &= (r - 5)(r + 4) \\ r^2 - 10r + 25 &= (r - 5)^2 \end{aligned}\right\} \text{Factor each denominator.}$$

Use each different factor the greatest number of times it appears as a factor.

$$\text{LCD} = (r + 1)(r + 4)(r - 5)^2$$

Be sure to include the exponent 2 on the factor $r - 5$.

(c) $\frac{1}{q - 5}, \frac{3}{5 - q}$

The expressions $q - 5$ and $5 - q$ are opposites of each other because

$$-(q - 5) = -q + 5 = 5 - q.$$

Therefore, either $q - 5$ or $5 - q$ can be used as the LCD.

◀ **Work Problem 3 at the Side.**

OBJECTIVE 2 Write equivalent rational expressions. Once the LCD has been found, the next step in preparing to add or subtract two rational expressions is to use the fundamental property to write equivalent rational expressions.

Writing a Rational Expression with a Specified Denominator

Step 1 **Factor** both denominators.

Step 2 **Decide what factor(s) the denominator must be multiplied by** in order to equal the specified denominator.

Step 3 **Multiply** the rational expression by that factor divided by itself. (That is, multiply by 1.)

EXAMPLE 4 Writing Equivalent Rational Expressions

Write each rational expression as an equivalent expression with the indicated denominator.

(a) $\frac{3}{8}=\frac{?}{40}$ **(b)** $\frac{9k}{25}=\frac{?}{50k}$

Step 1 For each example, first factor the denominator on the right. Then compare the denominator on the left with the one on the right to decide what factors are missing.

(a) $\frac{3}{8}=\frac{?}{5\cdot 8}$

Step 2 A factor of 5 is missing.

Step 3 Multiply $\frac{3}{8}$ by $\frac{5}{5}$.

$$\frac{3}{8}=\frac{3}{8}\cdot\frac{5}{5}=\frac{15}{40}$$

$\frac{5}{5}=1$

(b) $\frac{9k}{25}=\frac{?}{25\cdot 2k}$

Factors of 2 and k are missing.

Multiply $\frac{9k}{25}$ by $\frac{2k}{2k}$.

$$\frac{9k}{25}=\frac{9k}{25}\cdot\frac{2k}{2k}=\frac{18k^2}{50k}$$

$\frac{2k}{2k}=1$

Work Problem 4 at the Side. ▶

EXAMPLE 5 Writing Equivalent Rational Expressions

Write each rational expression as an equivalent expression with the indicated denominator.

(a) $\frac{8}{3x+1}=\frac{?}{12x+4}$

$\frac{8}{3x+1}=\frac{?}{4(3x+1)}$ Factor the denominator on the right.

The missing factor is 4, so multiply the fraction on the left by $\frac{4}{4}$.

$\frac{8}{3x+1}\cdot\frac{4}{4}=\frac{32}{12x+4}$ Fundamental property

Continued on Next Page

4 Write each rational expression as an equivalent expression with the indicated denominator.

(a) $\frac{3}{4}=\frac{?}{36}$

GS **(b)** $\frac{7k}{5}=\frac{?}{30p}$

Step 1 Factor the denominator on the right.

$\frac{7k}{5}=\frac{?}{5\cdot \_\_\_\_}$

Step 2 Factors of ____ and ____ are missing.

Step 3

$\frac{7k}{5}\cdot\frac{\_\_\_\_}{\_\_\_\_}=\frac{\_\_\_\_}{30p}$

Answers

4. (a) $\frac{27}{36}$

(b) $6p$; 6; p; $6p$; $6p$; $42kp$

5 Write each rational expression as an equivalent expression with the indicated denominator.

(a) $\dfrac{9}{2a+5} = \dfrac{?}{6a+15}$

(b) $\dfrac{5k+1}{k^2+2k} = \dfrac{?}{k^3+k^2-2k}$

(b) $\dfrac{12p}{p^2+8p} = \dfrac{?}{p^3+4p^2-32p}$

Factor the denominator in each rational expression.

$$\frac{12p}{p(p+8)} = \frac{?}{p(p+8)(p-4)}$$

p^3+4p^2-32p
$= p(p^2+4p-32)$
$= p(p+8)(p-4)$

The factor $p-4$ is missing, so multiply $\frac{12p}{p(p+8)}$ by $\frac{p-4}{p-4}$.

$$= \frac{12p}{p(p+8)} \cdot \frac{p-4}{p-4} \quad \text{Fundamental property}$$

$$= \frac{12p(p-4)}{p(p+8)(p-4)} \quad \text{Multiply numerators. Multiply denominators.}$$

$$= \frac{12p^2-48p}{p^3+4p^2-32p} \quad \text{Multiply the factors.}$$

Note

In the next section we add and subtract rational expressions, which sometimes requires the steps illustrated in **Examples 4 and 5.** While it may be beneficial to leave the equivalent expression in factored form, we multiplied out the factors in the numerator and the denominator in **Example 5(b),**

$$\frac{12p(p-4)}{p(p+8)(p-4)},$$

in order to give the answer,

$$\frac{12p^2-48p}{p^3+4p^2-32p},$$

in the same form as the original problem.

◀ **Work Problem 5 at the Side.**

Answers

5. (a) $\dfrac{27}{6a+15}$

(b) $\dfrac{(5k+1)(k-1)}{k^3+k^2-2k}$, or $\dfrac{5k^2-4k-1}{k^3+k^2-2k}$

14.3 Exercises

FOR EXTRA HELP

Download the MyDashBoard App

MyMathLab®

CONCEPT CHECK *Choose the correct response in Exercises 1–4.*

1. Suppose that the greatest common factor of a and b is 1. Then the least common denominator for $\frac{1}{a}$ and $\frac{1}{b}$ is

A. a **B.** b **C.** ab **D.** 1.

2. If a is a factor of b, then the least common denominator for $\frac{1}{a}$ and $\frac{1}{b}$ is

A. a **B.** b **C.** ab **D.** 1.

3. The least common denominator for $\frac{11}{20}$ and $\frac{1}{2}$ is

A. 40 **B.** 2 **C.** 20 **D.** none of these.

4. To write the rational expression $\frac{1}{(x-4)^2(y-3)}$ with denominator $(x-4)^3(y-3)^2$, we must multiply both the numerator and the denominator by

A. $(x-4)(y-3)$ **B.** $(x-4)^2$

C. $x-4$ **D.** $(x-4)^2(y-3)$.

Find the LCD for the fractions in each list. **See Examples 1 and 2.**

5. $\frac{2}{15}, \frac{3}{10}, \frac{7}{30}$

6. $\frac{5}{24}, \frac{7}{12}, \frac{9}{28}$

7. $\frac{3}{x^4}, \frac{5}{x^7}$

8. $\frac{2}{y^5}, \frac{3}{y^6}$

9. $\frac{5}{36q}, \frac{17}{24q}$

10. $\frac{4}{30p}, \frac{9}{50p}$

11. $\frac{6}{21r^3}, \frac{8}{12r^5}$

12. $\frac{9}{35t^2}, \frac{5}{49t^6}$

13. $\frac{5}{12x^2y}, \frac{7}{24x^3y^2}, \frac{-11}{6xy^4}$

14. $\frac{3}{10a^4b}, \frac{7}{15ab}, \frac{-7}{30ab^7}$

15. $\frac{1}{x^9y^{14}z^{21}}, \frac{3}{x^8y^{14}z^{22}}, \frac{5}{x^{10}y^{13}z^{23}}$

16. $\frac{2}{t^{15}s^{12}u^9}, \frac{4}{t^{14}s^{12}u^8}, \frac{9}{t^{13}s^{11}u^{10}}$

Find the LCD for the fractions in each list. **See Examples 1–3.**

17. $\frac{9}{28m^2}, \frac{3}{12m-20}$

18. $\frac{15}{27a^3}, \frac{8}{9a-45}$

19. $\frac{7}{5b-10}, \frac{11}{6b-12}$

20. $\frac{3}{7x^2+21x}, \frac{1}{5x^2+15x}$

21. $\frac{5}{c-d}, \frac{8}{d-c}$

22. $\frac{4}{y-x}, \frac{7}{x-y}$

23. $\dfrac{3}{k^2+5k}, \dfrac{2}{k^2+3k-10}$

24. $\dfrac{1}{z^2-4z}, \dfrac{4}{z^2-3z-4}$

25. $\dfrac{6}{a^2+6a}, \dfrac{-5}{a^2+3a-18}$

26. $\dfrac{8}{y^2-5y}, \dfrac{-5}{y^2-2y-15}$

27. $\dfrac{5}{p^2+8p+15}, \dfrac{3}{p^2-3p-18}, \dfrac{2}{p^2-p-30}$

28. $\dfrac{10}{y^2-10y+21}, \dfrac{2}{y^2-2y-3}, \dfrac{5}{y^2-6y-7}$

Write each rational expression as an equivalent expression with the indicated denominator. ***See Examples 4 and 5.***

29. $\dfrac{4}{11}=\dfrac{?}{55}$

30. $\dfrac{6}{7}=\dfrac{?}{42}$

31. $\dfrac{-5}{k}=\dfrac{?}{9k}$

32. $\dfrac{-3}{q}=\dfrac{?}{6q}$

33. $\dfrac{13}{40y}=\dfrac{?}{80y^3}$

34. $\dfrac{5}{27p}=\dfrac{?}{108p^4}$

35. $\dfrac{5t^2}{6r}=\dfrac{?}{42r^4}$

36. $\dfrac{8y^2}{3x}=\dfrac{?}{30x^3}$

37. $\dfrac{5}{2(m+3)}=\dfrac{?}{8(m+3)}$

38. $\dfrac{7}{4(y-1)}=\dfrac{?}{16(y-1)}$

39. $\dfrac{-4t}{3t-6}=\dfrac{?}{12-6t}$

40. $\dfrac{-7k}{5k-20}=\dfrac{?}{40-10k}$

41. $\dfrac{14}{z^2-3z}=\dfrac{?}{z(z-3)(z-2)}$

42. $\dfrac{12}{x(x+4)}=\dfrac{?}{x(x+4)(x-9)}$

43. $\dfrac{2(b-1)}{b^2+b}=\dfrac{?}{b^3+3b^2+2b}$

44. $\dfrac{3(c+2)}{c(c-1)}=\dfrac{?}{c^3-5c^2+4c}$

14.4 Adding and Subtracting Rational Expressions

OBJECTIVES

1. Add rational expressions having the same denominator.
2. Add rational expressions having different denominators.
3. Subtract rational expressions.

OBJECTIVE 1 Add rational expressions having the same denominator. We find the sum of two such rational expressions with the same procedure that we used in **Section 3.1** for adding two fractions having the same denominator.

Adding Rational Expressions (Same Denominator)

The rational expressions $\frac{P}{Q}$ and $\frac{R}{Q}$ (where $Q \neq 0$) are added as follows.

$$\frac{P}{Q} + \frac{R}{Q} = \frac{P + R}{Q}$$

In words: To add rational expressions with the same denominator, add the numerators and keep the same denominator.

EXAMPLE 1 Adding Rational Expressions (Same Denominator)

Add. Write each answer in lowest terms.

(a) $\frac{4}{9} + \frac{2}{9}$

The denominators are the same, so the sum is found by adding the two numerators and keeping the same (common) denominator.

$= \frac{4 + 2}{9}$ Add.

$= \frac{6}{9}$

$= \frac{2 \cdot 3}{3 \cdot 3}$ Factor.

$= \frac{2}{3}$ Lowest terms

(b) $\frac{3x}{x + 1} + \frac{3}{x + 1}$

$= \frac{3x + 3}{x + 1}$ Add.

$= \frac{3(x + 1)}{x + 1}$ Factor.

$= 3$ Lowest terms

Work Problem 1 at the Side. ▶

1 Add. Write each answer in lowest terms.

(a) $\frac{7}{15} + \frac{3}{15}$

(b) $\frac{3}{y + 4} + \frac{2}{y + 4}$

(c) $\frac{x}{x + y} + \frac{1}{x + y}$

(d) $\frac{a}{a + b} + \frac{b}{a + b}$

(e) $\frac{x^2}{x + 1} + \frac{x}{x + 1}$

OBJECTIVE 2 Add rational expressions having different denominators. As in **Section 3.1,** we use the following steps to add two rational expressions having different denominators.

Adding Rational Expressions (Different Denominators)

Step 1 **Find the least common denominator (LCD).**

Step 2 **Write each rational expression** as an equivalent rational expression with the LCD as the denominator.

Step 3 **Add** the numerators to get the numerator of the sum. The LCD is the denominator of the sum.

Step 4 **Write in lowest terms** using the fundamental property.

Answers

1. (a) $\frac{2}{3}$ **(b)** $\frac{5}{y + 4}$ **(c)** $\frac{x + 1}{x + y}$
(d) 1 **(e)** x

2 Add. Write each answer in lowest terms.

(a) $\frac{1}{10} + \frac{1}{15}$

GS **(b)** $\frac{6}{5x} + \frac{9}{2x}$

$= \frac{6(\_\_\_)}{5x(\_\_\_)} + \frac{9(\_\_\_)}{2x(\_\_\_)}$

$= \frac{12 + 45}{\_\_\_}$

$= \_\_\_\_$

(c) $\frac{m}{3n} + \frac{2}{7n}$

Answers

2. (a) $\frac{1}{6}$ **(b)** 2; 2; 5; 5; $10x$; $\frac{57}{10x}$

(c) $\frac{7m + 6}{21n}$

EXAMPLE 2 Adding Rational Expressions (Different Denominators)

Add. Write each answer in lowest terms.

(a) $\frac{1}{12} + \frac{7}{15}$ **(b)** $\frac{2}{3y} + \frac{1}{4y}$

Step 1 First find the LCD, using the methods of the previous section.

(a)	(b)
$12 = 2 \cdot 2 \cdot 3 = 2^2 \cdot 3$	$3y = 3 \cdot y$
$15 = 3 \cdot 5$	$4y = 2 \cdot 2 \cdot y = 2^2 \cdot y$
$\text{LCD} = 2^2 \cdot 3 \cdot 5 = 60$	$\text{LCD} = 2^2 \cdot 3 \cdot y = 12y$

Step 2 Now write each rational expression as an equivalent expression with the LCD (either 60 or $12y$) as the denominator.

$$\frac{1}{12} + \frac{7}{15} \quad \text{The LCD is 60.}$$

$$= \frac{1(5)}{12(5)} + \frac{7(4)}{15(4)}$$

$$= \frac{5}{60} + \frac{28}{60}$$

$$\frac{2}{3y} + \frac{1}{4y} \quad \text{The LCD is } 12y.$$

$$= \frac{2(4)}{3y(4)} + \frac{1(3)}{4y(3)}$$

$$= \frac{8}{12y} + \frac{3}{12y}$$

Step 3 Add the numerators. The LCD is the denominator.

Step 4 Write in lowest terms if necessary.

$$= \frac{5 + 28}{60}$$

$$= \frac{33}{60}, \quad \text{or} \quad \frac{11}{20}$$

$$= \frac{8 + 3}{12y}$$

$$= \frac{11}{12y}$$

◀ **Work Problem 2 at the Side.**

EXAMPLE 3 Adding Rational Expressions

Add. Write the answer in lowest terms.

$$\frac{2x}{x^2 - 1} + \frac{-1}{x + 1}$$

Step 1 Since the denominators are different, find the LCD.

$$\left.\begin{array}{l} x^2 - 1 = (x + 1)(x - 1) \\ x + 1 \text{ is prime.} \end{array}\right\} \quad \text{The LCD is } (x + 1)(x - 1).$$

Step 2 Write each rational expression as an equivalent expression with the LCD as the denominator.

$$\frac{2x}{x^2 - 1} + \frac{-1}{x + 1} \qquad \text{The LCD is } (x + 1)(x - 1).$$

$$= \frac{2x}{(x + 1)(x - 1)} + \frac{-1(x - 1)}{(x + 1)(x - 1)} \qquad \text{Multiply the second fraction by } \tfrac{x - 1}{x - 1}.$$

$$= \frac{2x}{(x + 1)(x - 1)} + \frac{-x + 1}{(x + 1)(x - 1)} \qquad \text{Distributive property}$$

Continued on Next Page

Step 3 $= \dfrac{2x - x + 1}{(x+1)(x-1)}$ Add numerators. Keep the same denominator.

$= \dfrac{x+1}{(x+1)(x-1)}$ Combine like terms in the numerator.

Step 4 $= \dfrac{1(x+1)}{(x+1)(x-1)}$ Identity property for multiplication

Remember to write 1 in the numerator.

$= \dfrac{1}{x-1}$ Divide out the common factors.

Work Problem 3 at the Side. ▶

EXAMPLE 4 **Adding Rational Expressions**

Add. Write the answer in lowest terms.

$$\frac{2x}{x^2+5x+6} + \frac{x+1}{x^2+2x-3}$$

$= \dfrac{2x}{(x+2)(x+3)} + \dfrac{x+1}{(x+3)(x-1)}$ Factor the denominators.

$= \dfrac{2x(x-1)}{(x+2)(x+3)(x-1)} + \dfrac{(x+1)(x+2)}{(x+2)(x+3)(x-1)}$ The LCD is $(x+2)(x+3)(x-1)$.

$= \dfrac{2x(x-1) + (x+1)(x+2)}{(x+2)(x+3)(x-1)}$ Add numerators. Keep the same denominator.

$= \dfrac{2x^2 - 2x + x^2 + 3x + 2}{(x+2)(x+3)(x-1)}$ Multiply.

$= \dfrac{3x^2 + x + 2}{(x+2)(x+3)(x-1)}$ Combine like terms.

The numerator cannot be factored here, so the expression is in lowest terms.

Work Problem 4 at the Side. ▶

EXAMPLE 5 **Adding Rational Expressions (Denominators Are Opposites)**

Add. Write the answer in lowest terms.

$\dfrac{y}{y-2} + \dfrac{8}{2-y}$ The denominators are opposites.

$= \dfrac{y}{y-2} + \dfrac{8(-1)}{(2-y)(-1)}$ Multiply $\frac{8}{2-y}$ by $\frac{-1}{-1}$ to get a common denominator.

$= \dfrac{y}{y-2} + \dfrac{-8}{-2+y}$ Multiply. Apply the distributive property in the denominator.

$= \dfrac{y}{y-2} + \dfrac{-8}{y-2}$ Rewrite $-2 + y$ as $y - 2$.

$= \dfrac{y-8}{y-2}$ Add numerators. Keep the same denominator.

Work Problem 5 at the Side. ▶

3 Add. Write each answer in lowest terms.

(a) $\dfrac{2p}{3p+3} + \dfrac{5p}{2p+2}$

(b) $\dfrac{4}{y^2-1} + \dfrac{6}{y+1}$

(c) $\dfrac{-2}{p+1} + \dfrac{4p}{p^2-1}$

4 Add. Write each answer in lowest terms.

(a) $\dfrac{2k}{k^2-5k+4} + \dfrac{3}{k^2-1}$

(b) $\dfrac{4m}{m^2+3m+2} + \dfrac{2m-1}{m^2+6m+5}$

5 Add. Write the answer in lowest terms.

$$\frac{m}{2m-3n} + \frac{n}{3n-2m}$$

Answers

3. (a) $\dfrac{19p}{6(p+1)}$ **(b)** $\dfrac{2(3y-1)}{(y+1)(y-1)}$

(c) $\dfrac{2}{p-1}$

4. (a) $\dfrac{(2k-3)(k+4)}{(k-4)(k-1)(k+1)}$

(b) $\dfrac{6m^2+23m-2}{(m+2)(m+1)(m+5)}$

5. $\dfrac{m-n}{2m-3n}$, or $\dfrac{n-m}{3n-2m}$

6 Subtract. Write each answer in lowest terms.

(a) $\dfrac{3}{m^2} - \dfrac{2}{m^2}$

GS **(b)** $\dfrac{x}{2x+3} - \dfrac{3x+4}{2x+3}$

$= \dfrac{x - (\_\_\_\_)}{2x+3}$

$= \dfrac{x - \_\_\_\_ - \_\_\_\_}{2x+3}$

$= \dfrac{\_\_\_\_}{2x+3}$

If the numerator is factored, the answer can be written as ______.

(c) $\dfrac{5t}{t-1} - \dfrac{5+t}{t-1}$

Answers

6. (a) $\dfrac{1}{m^2}$

(b) $3x+4$; $3x$; 4; $-2x-4$; $\dfrac{-2(x+2)}{2x+3}$

(c) $\dfrac{4t-5}{t-1}$

Note

If we had chosen to use $2 - y$ as the common denominator, the final answer would be in the form $\frac{8-y}{2-y}$, which is equivalent to $\frac{y-8}{y-2}$.

$$\frac{y}{y-2} + \frac{8}{2-y}$$ See **Example 5.**

$$= \frac{y(-1)}{(y-2)(-1)} + \frac{8}{2-y}$$ Multiply $\frac{y}{y-2}$ by $\frac{-1}{-1}$.

$$= \frac{-y}{2-y} + \frac{8}{2-y}$$ In the first denominator, $(y-2)(-1) = -y + 2 = 2 - y$.

$$= \frac{-y+8}{2-y}, \quad \text{or} \quad \frac{8-y}{2-y}$$ Add numerators. Keep the same denominator.

OBJECTIVE 3 Subtract rational expressions. To subtract rational expressions having the same denominator, use the following rule.

Subtracting Rational Expressions (Same Denominator)

The rational expressions $\frac{P}{Q}$ and $\frac{R}{Q}$ (where $Q \neq 0$) are subtracted as follows.

$$\frac{P}{Q} - \frac{R}{Q} = \frac{P-R}{Q}$$

In words: To subtract rational expressions with the same denominator, subtract the numerators and keep the same denominator.

EXAMPLE 6 Subtracting Rational Expressions (Same Denominator)

Subtract. Write the answer in lowest terms.

$$\frac{2m}{m-1} - \frac{m+3}{m-1}$$

Use parentheses around the numerator of the subtrahend.

$$= \frac{2m - (m+3)}{m-1}$$ Subtract numerators. Keep the same denominator.

Be careful with signs.

$$= \frac{2m - m - 3}{m-1}$$ Distributive property

$$= \frac{m-3}{m-1}$$ Combine like terms.

CAUTION

Sign errors often occur in subtraction problems like the one in **Example 6.** The numerator of the fraction being subtracted must be treated as a single quantity. ***Be sure to use parentheses after the subtraction symbol.***

◀ **Work Problem 6 at the Side.**

EXAMPLE 7 **Subtracting Rational Expressions (Different Denominators)**

Subtract. Write the answer in lowest terms.

$$\frac{9}{x-2}-\frac{3}{x}$$ The LCD is $x(x-2)$.

$$=\frac{9x}{x(x-2)}-\frac{3(x-2)}{x(x-2)}$$ Write each expression with the LCD.

$$=\frac{9x-3(x-2)}{x(x-2)}$$ Subtract numerators. Keep the same denominator.

Be careful here.

$$=\frac{9x-3x+6}{x(x-2)}$$ Distributive property

$$=\frac{6x+6}{x(x-2)}, \quad \text{or} \quad \frac{6(x+1)}{x(x-2)}$$ Combine like terms. Factor the numerator.

Note

We factored the final numerator in **Example 7** to get $\frac{6(x+1)}{x(x-2)}$. The fundamental property does not apply, however, since there are no common factors to divide out. The answer is in lowest terms.

Work Problem 7 at the Side. ▶

EXAMPLE 8 **Subtracting Rational Expressions (Denominators Are Opposites)**

Subtract. Write the answer in lowest terms.

$$\frac{3x}{x-5}-\frac{2x-25}{5-x}$$ The denominators are opposites. We choose $x-5$ as the common denominator.

$$=\frac{3x}{x-5}-\frac{(2x-25)(-1)}{(5-x)(-1)}$$ Multiply $\frac{2x-25}{5-x}$ by $\frac{-1}{-1}$ to get a common denominator.

$$=\frac{3x}{x-5}-\frac{-2x+25}{x-5}$$ $(2x-25)(-1)=-2x+25=25-2x$; $(5-x)(-1)=-5+x=x-5$

Use parentheses.

$$=\frac{3x-(-2x+25)}{x-5}$$ Subtract numerators.

Be careful with signs.

$$=\frac{3x+2x-25}{x-5}$$ Distributive property

$$=\frac{5x-25}{x-5}$$ Combine like terms.

$$=\frac{5(x-5)}{x-5}$$ Factor.

$$=5$$ Divide out the common factor.

Work Problem 8 at the Side. ▶

7 Subtract. Write each answer in lowest terms.

(a) $\frac{1}{k+4}-\frac{2}{k}$

(b) $\frac{6}{a+2}-\frac{1}{a-3}$

8 Subtract. Write each answer in lowest terms.

(a) $\frac{5}{x-1}-\frac{3x}{1-x}$

(b) $\frac{2y}{y-2}-\frac{1+y}{2-y}$

Answers

7. **(a)** $\frac{-k-8}{k(k+4)}$ **(b)** $\frac{5(a-4)}{(a+2)(a-3)}$

8. **(a)** $\frac{5+3x}{x-1}$, or $\frac{-5-3x}{1-x}$

(b) $\frac{3y+1}{y-2}$, or $\frac{-3y-1}{2-y}$

9 Subtract. Write each answer in lowest terms.

(a) $\dfrac{4y}{y^2-1}-\dfrac{5}{y^2+2y+1}$

GS **(b)** $\dfrac{3r}{r-5}-\dfrac{4}{r^2-10r+25}$

The LCD is ________.

Complete the subtraction.

10 Subtract. Write each answer in lowest terms.

(a) $\dfrac{2}{p^2-5p+4}-\dfrac{3}{p^2-1}$

(b) $\dfrac{q}{2q^2+5q-3}-\dfrac{3q+4}{3q^2+10q+3}$

Answers

9. (a) $\dfrac{4y^2-y+5}{(y+1)^2(y-1)}$

(b) $(r-5)^2$; $\dfrac{3r-19}{(r-5)^2}$

10. (a) $\dfrac{14-p}{(p-4)(p-1)(p+1)}$

(b) $\dfrac{-3q^2-4q+4}{(2q-1)(q+3)(3q+1)}$

EXAMPLE 9 Subtracting Rational Expressions

Subtract. Write the answer in lowest terms.

$$\frac{6x}{x^2-2x+1}-\frac{1}{x^2-1}$$

$$=\frac{6x}{(x-1)^2}-\frac{1}{(x-1)(x+1)}$$ Factor the denominators.

From the factored denominators, we identify the LCD, $(x-1)^2(x+1)$. ***We use the factor x − 1 twice*** because it appears twice in the first denominator.

$$=\frac{6x(x+1)}{(x-1)^2(x+1)}-\frac{1(x-1)}{(x-1)(x-1)(x+1)}$$ Fundamental property

$$=\frac{6x(x+1)-1(x-1)}{(x-1)^2(x+1)}$$ Subtract numerators.

$$=\frac{6x^2+6x-x+1}{(x-1)^2(x+1)}$$ Distributive property

$$=\frac{6x^2+5x+1}{(x-1)^2(x+1)}, \quad\text{or}\quad \frac{(2x+1)(3x+1)}{(x-1)^2(x+1)}$$ Combine like terms. Factor the numerator.

◀ Work Problem **9** at the Side.

EXAMPLE 10 Subtracting Rational Expressions

Subtract. Write the answer in lowest terms.

$$\frac{q}{q^2-4q-5}-\frac{3}{2q^2-13q+15}$$

$$=\frac{q}{(q+1)(q-5)}-\frac{3}{(q-5)(2q-3)}$$ Factor the denominators. The LCD is $(q+1)(q-5)(2q-3)$.

$$=\frac{q(2q-3)}{(q+1)(q-5)(2q-3)}-\frac{3(q+1)}{(q+1)(q-5)(2q-3)}$$ Fundamental property

$$=\frac{q(2q-3)-3(q+1)}{(q+1)(q-5)(2q-3)}$$ Subtract numerators.

$$=\frac{2q^2-3q-3q-3}{(q+1)(q-5)(2q-3)}$$ Distributive property

$$=\frac{2q^2-6q-3}{(q+1)(q-5)(2q-3)}$$ Combine like terms.

The numerator cannot be factored, so the final answer is in lowest terms.

◀ Work Problem **10** at the Side.

14.4 Exercises

FOR EXTRA HELP

Download the MyDashBoard App

MyMathLab®

CONCEPT CHECK *Match each expression in Column I with the correct sum or difference in Column II.*

I

1. $\frac{x}{x+6}+\frac{6}{x+6}$ **2.** $\frac{2x}{x-6}-\frac{12}{x-6}$

3. $\frac{6}{x-6}-\frac{x}{x-6}$ **4.** $\frac{6}{x+6}-\frac{x}{x+6}$

5. $\frac{x}{x+6}-\frac{6}{x+6}$ **6.** $\frac{1}{x}+\frac{1}{6}$

7. $\frac{1}{6}-\frac{1}{x}$ **8.** $\frac{1}{6x}-\frac{1}{6x}$

II

A. 2 **B.** $\frac{x-6}{x+6}$

C. -1 **D.** $\frac{6+x}{6x}$

E. 1 **F.** 0

G. $\frac{x-6}{6x}$ **H.** $\frac{6-x}{x+6}$

Add or subtract. Write each answer in lowest terms. ***See Examples 1 and 6.***

9. $\frac{4}{m}+\frac{7}{m}$

10. $\frac{5}{p}+\frac{11}{p}$

11. $\frac{a+b}{2}-\frac{a-b}{2}$

12. $\frac{x-y}{2}-\frac{x+y}{2}$

13. $\frac{5}{y+4}-\frac{1}{y+4}$

14. $\frac{4}{y+3}-\frac{1}{y+3}$

15. $\frac{5m}{m+1}-\frac{1+4m}{m+1}$

16. $\frac{4x}{x+2}-\frac{2+3x}{x+2}$

17. $\frac{x^2}{x+5}+\frac{5x}{x+5}$

18. $\frac{t^2}{t-3}+\frac{-3t}{t-3}$

19. $\frac{y^2-3y}{y+3}+\frac{-18}{y+3}$

$= \frac{\underline{\qquad\qquad}}{y+3}$

$= \frac{(y+\underline{\quad})(y-\underline{\quad})}{y+3}$

$= \underline{\qquad}$

20. $\frac{r^2-8r}{r-5}+\frac{15}{r-5}$

$= \frac{\underline{\qquad\qquad}}{\underline{\qquad}}$

$= \frac{(r-\underline{\quad})(r-5)}{\underline{\qquad}}$

$= \underline{\qquad}$

21. Explain with an example how to add or subtract rational expressions with the same denominator.

22. Explain with an example how to add or subtract rational expressions with different denominators.

Add or subtract. Write each answer in lowest terms. ***See Examples 2, 3, 4, and 7.***

23. $\frac{z}{5}+\frac{1}{3}$

24. $\frac{p}{8}+\frac{3}{5}$

25. $\frac{5}{7}-\frac{r}{2}$

26. $\frac{10}{9}-\frac{z}{3}$

27. $-\frac{3}{4}-\frac{1}{2x}$

28. $-\frac{5}{8}-\frac{3}{2a}$

29. $\frac{3}{5x}+\frac{9}{4x}$

30. $\frac{3}{2x}+\frac{4}{7x}$

31. $\frac{x+1}{6}+\frac{3x+3}{9}$

32. $\frac{2x-6}{4}+\frac{x+5}{6}$

33. $\frac{x+3}{3x}+\frac{2x+2}{4x}$

34. $\frac{x+2}{5x}+\frac{6x+3}{3x}$

35. $\frac{2}{x+3}+\frac{1}{x}$

The LCD is _______.

Complete the addition.

36. $\frac{3}{x-4}+\frac{2}{x}$

The LCD is _______.

Complete the addition.

37. $\frac{1}{k+5}-\frac{2}{k}$

38. $\frac{3}{m+1}-\frac{4}{m}$

39. $\frac{x}{x-2}+\frac{-8}{x^2-4}$

40. $\frac{2x}{x-1}+\frac{-4}{x^2-1}$

41. $\frac{x}{x-2}+\frac{4}{x+2}$

42. $\frac{2x}{x-1}+\frac{3}{x+1}$

43. $\frac{t}{t+2}+\frac{5-t}{t}-\frac{4}{t^2+2t}$

44. $\frac{2p}{p-3}+\frac{2+p}{p}-\frac{-6}{p^2-3p}$

45. CONCEPT CHECK What are the two possible LCDs that could be used for the following sum?

$$\frac{10}{m-2}+\frac{5}{2-m}$$

46. CONCEPT CHECK If one form of the correct answer to a sum or difference of rational expressions is $\frac{4}{k-3}$, what would be an alternative form of the answer if the denominator is $3-k$?

Add or subtract. Write each answer in lowest terms. ***See Examples 5 and 8.***

47. $\frac{4}{x-5}+\frac{6}{5-x}$

48. $\frac{10}{m-2}+\frac{5}{2-m}$

49. $\frac{-1}{1-y}+\frac{3-4y}{y-1}$

50. $\frac{-4}{p-3}-\frac{p+1}{3-p}$

51. $\frac{2}{x-y^2}+\frac{7}{y^2-x}$

52. $\frac{-8}{p-q^2}+\frac{3}{q^2-p}$

53. $\frac{x}{5x-3y}-\frac{y}{3y-5x}$

54. $\frac{t}{8t-9s}-\frac{s}{9s-8t}$

55. $\frac{3}{4p-5}+\frac{9}{5-4p}$

56. $\frac{8}{3-7y}-\frac{2}{7y-3}$

In each subtraction problem, the rational expression that follows the subtraction sign has a numerator with more than one term. ***Be careful with signs*** *and find each difference.* ***See Examples 6–10.***

57. $\frac{2m}{m-n}-\frac{5m+n}{2m-2n}$

58. $\frac{5p}{p-q}-\frac{3p+1}{4p-4q}$

59. $\dfrac{5}{x^2-9}-\dfrac{x+2}{x^2+4x+3}$

60. $\dfrac{1}{a^2-1}-\dfrac{a-1}{a^2+3a-4}$

61. $\dfrac{2q+1}{3q^2+10q-8}-\dfrac{3q+5}{2q^2+5q-12}$

62. $\dfrac{4y-1}{2y^2+5y-3}-\dfrac{y+3}{6y^2+y-2}$

Perform the indicated operations. ***See Examples 1–10.***

63. $\dfrac{4}{r^2-r}+\dfrac{6}{r^2+2r}-\dfrac{1}{r^2+r-2}$

64. $\dfrac{6}{k^2+3k}-\dfrac{1}{k^2-k}+\dfrac{2}{k^2+2k-3}$

65. $\dfrac{x+3y}{x^2+2xy+y^2}+\dfrac{x-y}{x^2+4xy+3y^2}$

66. $\dfrac{m}{m^2-1}+\dfrac{m-1}{m^2+2m+1}$

67. $\dfrac{r+y}{18r^2+9ry-2y^2}+\dfrac{3r-y}{36r^2-y^2}$

68. $\dfrac{2x-z}{2x^2+xz-10z^2}-\dfrac{x+z}{x^2-4z^2}$

69. Refer to the rectangle in the figure.

(a) Find an expression that represents its perimeter. Give the simplified form.

(b) Find an expression that represents its area. Give the simplified form.

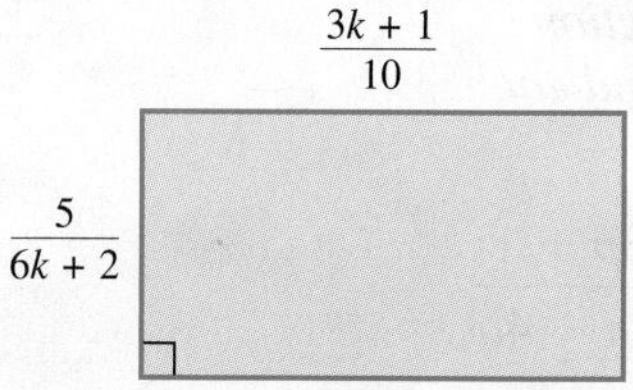

70. Refer to the triangle in the figure.

(a) Find an expression that represents its perimeter. Give the simplified form.

(b) Find an expression that represents its area. Give the simplified form.

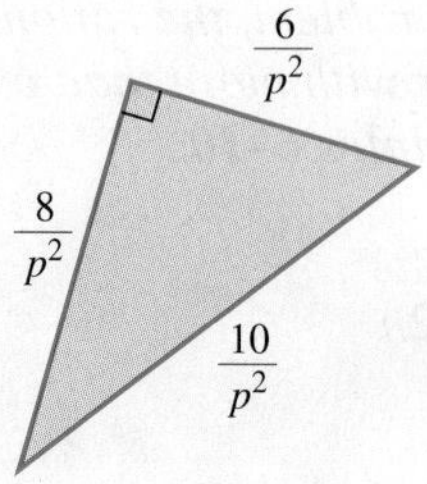

14.5 Complex Fractions

OBJECTIVES

1. Simplify a complex fraction by writing it as a division problem (Method 1).
2. Simplify a complex fraction by multiplying numerator and denominator by the least common denominator (Method 2).

The quotient of two mixed numbers in arithmetic, such as $2\frac{1}{2} \div 3\frac{1}{4}$, can be written as a fraction.

$$2\frac{1}{2} \div 3\frac{1}{4}$$

$$= \frac{2\frac{1}{2}}{3\frac{1}{4}}$$

$$= \frac{2 + \frac{1}{2}}{3 + \frac{1}{4}}$$

We do this to illustrate a *complex fraction.*

In algebra, some rational expressions also have fractions in the numerator, or denominator, or both.

Complex Fraction

A rational expression with one or more fractions in the numerator, or denominator, or both, is a **complex fraction.**

$$\frac{2 + \frac{1}{2}}{3 + \frac{1}{4}}, \quad \frac{\frac{3x^2 - 5x}{6x^2}}{2x - \frac{1}{x}}, \quad \text{and} \quad \frac{3 + x}{5 - \frac{2}{x}} \qquad \text{Complex fractions}$$

The parts of a complex fraction are named as follows.

$$\frac{\frac{2}{p} - \frac{1}{q}}{\frac{3}{p} + \frac{5}{q}}$$

← Numerator of complex fraction (top: $\frac{2}{p} - \frac{1}{q}$)
← Main fraction bar
← Denominator of complex fraction (bottom: $\frac{3}{p} + \frac{5}{q}$)

OBJECTIVE ▶ 1 **Simplify a complex fraction by writing it as a division problem (Method 1).** Since the main fraction bar represents division in a complex fraction, one method of simplifying a complex fraction involves division.

Method 1 for Simplifying a Complex Fraction

Step 1 Write both the numerator and denominator as single fractions.

Step 2 Change the complex fraction to a division problem.

Step 3 Perform the indicated division.

1 Simplify each complex fraction using Method 1.

GS **(a)** $\dfrac{\frac{2}{5}+\frac{1}{4}}{\frac{1}{2}+\frac{1}{3}}$

Step 1
Write the numerator as a single fraction. ____

Write the denominator as a single fraction. ____

Step 2
Write the equivalent fraction as a division problem.

____

Step 3
Write the division problem as a multiplication problem.

____

Multiply and write the answer in lowest terms. ____

(b) $\dfrac{6+\frac{1}{x}}{5-\frac{2}{x}}$

(c) $\dfrac{9-\frac{4}{p}}{\frac{2}{p}+1}$

Answers

1. (a) $\frac{13}{20}$; $\frac{5}{6}$; $\frac{13}{20} \div \frac{5}{6}$; $\frac{13}{20} \cdot \frac{6}{5}$; $\frac{39}{50}$
(b) $\frac{6x+1}{5x-2}$ (c) $\frac{9p-4}{2+p}$

EXAMPLE 1 Simplifying Complex Fractions (Method 1)

Simplify each complex fraction.

(a) $\dfrac{\frac{2}{3}+\frac{5}{9}}{\frac{1}{4}+\frac{1}{12}}$ **(b)** $\dfrac{6+\frac{3}{x}}{\frac{x}{4}+\frac{1}{8}}$

Step 1 First, write each numerator as a single fraction.

(a)	(b)
$\frac{2}{3}+\frac{5}{9}$	$6+\frac{3}{x}$
$=\frac{2(3)}{3(3)}+\frac{5}{9}$	$=\frac{6}{1}+\frac{3}{x}$
$=\frac{6}{9}+\frac{5}{9}$	$=\frac{6x}{x}+\frac{3}{x}$
$=\mathbf{\frac{11}{9}}$	$=\mathbf{\frac{6x+3}{x}}$

Do the same thing with each denominator.

(a)	(b)
$\frac{1}{4}+\frac{1}{12}$	$\frac{x}{4}+\frac{1}{8}$
$=\frac{1(3)}{4(3)}+\frac{1}{12}$	$=\frac{x(2)}{4(2)}+\frac{1}{8}$
$=\frac{3}{12}+\frac{1}{12}$	$=\frac{2x}{8}+\frac{1}{8}$
$=\mathbf{\frac{4}{12}}$	$=\mathbf{\frac{2x+1}{8}}$

Step 2 Write the equivalent complex fraction as a division problem.

(a)	(b)
$\dfrac{\frac{11}{9}}{\frac{4}{12}}$	$\dfrac{\frac{6x+3}{x}}{\frac{2x+1}{8}}$
$=\frac{11}{9} \div \frac{4}{12}$	$=\frac{6x+3}{x} \div \frac{2x+1}{8}$

Step 3 Now use the definition of division and multiply by the reciprocal. Then write in lowest terms using the fundamental property.

(a)	(b)
$=\frac{11}{9} \cdot \frac{12}{4}$	$=\frac{6x+3}{x} \cdot \frac{8}{2x+1}$
$=\frac{11 \cdot 3 \cdot 4}{3 \cdot 3 \cdot 4}$	$=\frac{3(2x+1)}{x} \cdot \frac{8}{2x+1}$
$=\frac{11}{3}$	$=\frac{24}{x}$

◄ Work Problem **1** at the Side.

EXAMPLE 2 **Simplifying a Complex Fraction (Method 1)**

Simplify the complex fraction.

$$\frac{\frac{xp}{q^3}}{\frac{p^2}{qx^2}}$$

The numerator and denominator are single fractions, so use the definition of division and then the fundamental property.

$$\frac{xp}{q^3} \div \frac{p^2}{qx^2}$$
$$= \frac{xp}{q^3} \cdot \frac{qx^2}{p^2}$$
$$= \frac{x^3}{q^2p}$$

Work Problem 2 at the Side. ▶

EXAMPLE 3 **Simplifying a Complex Fraction (Method 1)**

Simplify the complex fraction.

$$\frac{\frac{3}{x+2} - 4}{\frac{2}{x+2} + 1}$$

$$= \frac{\frac{3}{x+2} - \frac{4(x+2)}{x+2}}{\frac{2}{x+2} + \frac{1(x+2)}{x+2}}$$ Write both second terms with a denominator of $x + 2$.

$$= \frac{\frac{3 - 4(x+2)}{x+2}}{\frac{2 + 1(x+2)}{x+2}}$$ Subtract in the numerator. Add in the denominator.

$$= \frac{\frac{3 - 4x - 8}{x+2}}{\frac{2 + x + 2}{x+2}}$$ Be careful with signs. Distributive property

$$= \frac{\frac{-5 - 4x}{x+2}}{\frac{4 + x}{x+2}}$$ Combine like terms.

$$= \frac{-5 - 4x}{x+2} \cdot \frac{x+2}{4+x}$$ Multiply by the reciprocal.

$$= \frac{-5 - 4x}{4 + x}$$ Divide out the common factor.

Work Problem 3 at the Side. ▶

OBJECTIVE 2 **Simplify a complex fraction by multiplying numerator and denominator by the least common denominator (Method 2).** If we multiply both the numerator and the denominator of a complex fraction by the LCD of all the fractions within the complex fraction, the result will no longer be complex. This is Method 2.

2 Simplify each complex fraction using Method 1.

(a) $\dfrac{\frac{rs^2}{t}}{\frac{r^2s}{t^2}}$

(b) $\dfrac{\frac{m^2n^3}{p}}{\frac{m^4n}{p^2}}$

3 Simplify using Method 1.

$$\frac{\frac{2}{x-1} + \frac{1}{x+1}}{\frac{3}{x-1} - \frac{4}{x+1}}$$

Answers

2. **(a)** $\frac{st}{r}$ **(b)** $\frac{n^2p}{m^2}$

3. $\frac{3x+1}{-x+7}$

4 Simplify each complex fraction using Method 2.

GS **(a)** $\dfrac{\frac{2}{3}-\frac{1}{4}}{\frac{4}{9}+\frac{1}{2}}$

Step 1
The LCD for ____, ____, ____, and ____ is ____.

Refer to Step 2 in **Example 4,** and simplify the complex fraction.

(b) $\dfrac{2-\frac{6}{a}}{3+\frac{4}{a}}$

(c) $\dfrac{\frac{p}{5-p}}{\frac{4p}{2p+1}}$

Method 2 for Simplifying a Complex Fraction

Step 1 Find the LCD of all fractions within the complex fraction.

Step 2 Multiply both the numerator and the denominator of the complex fraction by this LCD using the distributive property as necessary. Write in lowest terms.

EXAMPLE 4 **Simplifying Complex Fractions (Method 2)**

Simplify each complex fraction.

(a) $\dfrac{\frac{2}{3}+\frac{5}{9}}{\frac{1}{4}+\frac{1}{12}}$ (In **Example 1,** we simplified these same fractions using Method 1.)

(b) $\dfrac{6+\frac{3}{x}}{\frac{x}{4}+\frac{1}{8}}$

Step 1 Find the LCD for all denominators in the complex fraction.

The LCD for 3, 9, 4, and 12 is 36. | The LCD for x, 4, and 8 is $8x$.

Step 2 Multiply the numerator and denominator of the complex fraction by the LCD.

(a)

$$\dfrac{\frac{2}{3}+\frac{5}{9}}{\frac{1}{4}+\frac{1}{12}}$$

$$=\dfrac{36\left(\frac{2}{3}+\frac{5}{9}\right)}{36\left(\frac{1}{4}+\frac{1}{12}\right)}$$

$$=\dfrac{36\left(\frac{2}{3}\right)+36\left(\frac{5}{9}\right)}{36\left(\frac{1}{4}\right)+36\left(\frac{1}{12}\right)}$$

Multiply *each* term by 36.

$$=\frac{24+20}{9+3}$$

$$=\frac{44}{12}$$

$$=\frac{4\cdot 11}{4\cdot 3}$$

$$=\frac{11}{3}$$

(b)

$$\dfrac{6+\frac{3}{x}}{\frac{x}{4}+\frac{1}{8}}$$

$$=\dfrac{8x\left(6+\frac{3}{x}\right)}{8x\left(\frac{x}{4}+\frac{1}{8}\right)}$$

$$=\dfrac{8x(6)+8x\left(\frac{3}{x}\right)}{8x\left(\frac{x}{4}\right)+8x\left(\frac{1}{8}\right)}$$

Multiply *each* term by $8x$.

$$=\frac{48x+24}{2x^2+x}$$

$$=\frac{24(2x+1)}{x(2x+1)}$$

$$=\frac{24}{x}$$

◀ Work Problem **4** at the Side.

Answers

4. (a) 3; 4; 9; 2; 36; $\frac{15}{34}$

(b) $\frac{2a-6}{3a+4}$ (c) $\frac{2p+1}{4(5-p)}$

EXAMPLE 5 **Simplifying a Complex Fraction (Method 2)**

Simplify the complex fraction.

$$\frac{\frac{3}{5m}-\frac{2}{m^2}}{\frac{9}{2m}+\frac{3}{4m^2}}$$ The LCD for $5m$, m^2, $2m$, and $4m^2$ is $20m^2$.

$$=\frac{20m^2\left(\frac{3}{5m}-\frac{2}{m^2}\right)}{20m^2\left(\frac{9}{2m}+\frac{3}{4m^2}\right)}$$ Multiply numerator and denominator by $20m^2$.

$$=\frac{20m^2\left(\frac{3}{5m}\right)-20m^2\left(\frac{2}{m^2}\right)}{20m^2\left(\frac{9}{2m}\right)+20m^2\left(\frac{3}{4m^2}\right)}$$ Distributive property

$$=\frac{12m-40}{90m+15}, \quad \text{or} \quad \frac{4(3m-10)}{5(18m+3)}$$ Multiply and factor.

Work Problem **5** at the Side. ▶

Some students use Method 1 for problems like **Example 2,** which is the quotient of two fractions, and Method 2 for problems like **Examples 1, 3, 4, and 5,** which have sums or differences in the numerators or denominators.

EXAMPLE 6 **Simplifying Complex Fractions**

Simplify each complex fraction. Use either method.

(a) $$\frac{\frac{1}{y}+\frac{2}{y+2}}{\frac{4}{y}-\frac{3}{y+2}}$$ There are sums and differences in the numerator and denominator. Use Method 2.

$$=\frac{\left(\frac{1}{y}+\frac{2}{y+2}\right)y(y+2)}{\left(\frac{4}{y}-\frac{3}{y+2}\right)y(y+2)}$$ Multiply numerator and denominator by the LCD, $y(y+2)$.

$$=\frac{\left(\frac{1}{y}\right)y(y+2)+\left(\frac{2}{y+2}\right)y(y+2)}{\left(\frac{4}{y}\right)y(y+2)-\left(\frac{3}{y+2}\right)y(y+2)}$$ Distributive property

$$=\frac{1(y+2)+2y}{4(y+2)-3y}$$ Multiply and simplify.

$$=\frac{y+2+2y}{4y+8-3y}$$ Distributive property

$$=\frac{3y+2}{y+8}$$ Combine like terms.

Continued on Next Page

5 Simplify using Method 2.

GS

$$\frac{\frac{2}{5x}-\frac{3}{x^2}}{\frac{7}{4x}+\frac{1}{2x^2}}$$

The LCD for ____, ____, ____, and ____ is ____.

Simplify the complex fraction.

Answer

5. $5x$; x^2; $4x$; $2x^2$; $20x^2$; $\frac{8x-60}{35x+10}$, or $\frac{4(2x-15)}{5(7x+2)}$

6 Simplify each complex fraction. Use either method.

(a) $\dfrac{\frac{1}{x}+\frac{2}{x-1}}{\frac{2}{x}-\frac{4}{x-1}}$

(b) $\dfrac{1-\frac{2}{x}-\frac{15}{x^2}}{1+\frac{5}{x}+\frac{6}{x^2}}$

(c) $\dfrac{\frac{2x+3}{x-4}}{\frac{4x^2-9}{x^2-16}}$

(b) $\dfrac{1-\frac{2}{x}-\frac{3}{x^2}}{1-\frac{5}{x}+\frac{6}{x^2}}$ There are sums and differences in the numerator and denominator. Use Method 2.

$=\dfrac{\left(1-\frac{2}{x}-\frac{3}{x^2}\right)x^2}{\left(1-\frac{5}{x}+\frac{6}{x^2}\right)x^2}$ Multiply numerator and denominator by the LCD, x^2.

$=\dfrac{x^2-2x-3}{x^2-5x+6}$ Distributive property

$=\dfrac{(x-3)(x+1)}{(x-3)(x-2)}$ Factor.

$=\dfrac{x+1}{x-2}$ Divide out the common factor.

(c) $\dfrac{\frac{x+2}{x-3}}{\frac{x^2-4}{x^2-9}}$ This is a quotient of two rational expressions. Use Method 1.

$=\dfrac{x+2}{x-3}\div\dfrac{x^2-4}{x^2-9}$ Write as a division problem.

$=\dfrac{x+2}{x-3}\cdot\dfrac{x^2-9}{x^2-4}$ Multiply by the reciprocal.

$=\dfrac{(x+2)(x+3)(x-3)}{(x-3)(x+2)(x-2)}$ Multiply and then factor.

$=\dfrac{x+3}{x-2}$ Divide out the common factors.

◀ **Work Problem 6 at the Side.**

Answers

6. (a) $\dfrac{3x-1}{-2x-2}$ **(b)** $\dfrac{x-5}{x+2}$ **(c)** $\dfrac{x+4}{2x-3}$

14.5 Exercises

FOR EXTRA HELP

CONCEPT CHECK *In Exercises 1 and 2, answer each question.*

1. In a fraction, what operation does the fraction bar represent?

2. What property of real numbers justifies Method 2 of simplifying complex fractions?

3. Consider the complex fraction $\dfrac{\frac{1}{2}-\frac{1}{3}}{\frac{5}{6}-\frac{1}{12}}$. Answer each part, outlining Method 1 for simplifying this complex fraction.

(a) To combine the terms in the numerator, we must find the LCD of $\frac{1}{2}$ and $\frac{1}{3}$. What is this LCD? _______ Determine the simplified form of the numerator of the complex fraction. _______

(b) To combine the terms in the denominator, we must find the LCD of $\frac{5}{6}$ and $\frac{1}{12}$. What is this LCD? _______ Determine the simplified form of the denominator of the complex fraction. _______

(c) Now use the results from parts (a) and (b) to write the complex fraction as a division problem using the symbol $\div$. _______

(d) Perform the operation from part (c) to obtain the final simplification. _______

4. Consider the same complex fraction given in **Exercise 1:** $\dfrac{\frac{1}{2}-\frac{1}{3}}{\frac{5}{6}-\frac{1}{12}}$. Answer each part, outlining Method 2 for simplifying this complex fraction.

(a) We must determine the LCD of all the fractions within the complex fraction. What is this LCD? _______

(b) Multiply every term in the complex fraction by the LCD found in part (a), but at this time do not combine the terms in the numerator and the denominator. _______

(c) Now combine the terms from part (b) to obtain the simplified form of the complex fraction. _______

Simplify each complex fraction. Use either method. **See Examples 1–6.**

5. $\dfrac{-\frac{4}{3}}{\frac{2}{9}}$ **6.** $\dfrac{-\frac{5}{6}}{\frac{5}{4}}$ **7.** $\dfrac{\frac{p}{q^2}}{\frac{p^2}{q}}$ **8.** $\dfrac{\frac{a}{x}}{\frac{a^2}{2x}}$

9. $\dfrac{\frac{x}{y^2}}{\frac{x^2}{y}}$ **10.** $\dfrac{\frac{p^4}{r}}{\frac{p^2}{r^2}}$ **11.** $\dfrac{\frac{4a^4b^3}{3a}}{\frac{2ab^4}{b^2}}$ **12.** $\dfrac{\frac{2r^4t^2}{3t}}{\frac{5r^2t^5}{3r}}$

13. $\dfrac{\frac{m+2}{3}}{\frac{m-4}{m}}$ **14.** $\dfrac{\frac{q-5}{q}}{\frac{q+5}{3}}$ **15.** $\dfrac{\frac{2}{x}-3}{\frac{2-3x}{2}}$ **16.** $\dfrac{6+\frac{2}{r}}{\frac{3r+1}{4}}$

17. $\dfrac{\frac{1}{x}+x}{\frac{x^2+1}{8}}$ **18.** $\dfrac{\frac{3}{m}-m}{\frac{3-m^2}{4}}$ **19.** $\dfrac{a-\frac{5}{a}}{a+\frac{1}{a}}$

20. $\dfrac{q+\frac{1}{q}}{q+\frac{4}{q}}$ **21.** $\dfrac{\frac{1}{2}+\frac{1}{p}}{\frac{2}{3}+\frac{1}{p}}$ **22.** $\dfrac{\frac{3}{4}-\frac{1}{r}}{\frac{1}{5}+\frac{1}{r}}$

23. $\dfrac{\frac{2}{p^2}-\frac{3}{5p}}{\frac{4}{p}+\frac{1}{4p}}$ **24.** $\dfrac{\frac{2}{m^2}-\frac{3}{m}}{\frac{2}{5m^2}+\frac{1}{3m}}$ **25.** $\dfrac{\frac{t}{t+2}}{\frac{4}{t^2-4}}$

26. $\dfrac{\frac{m}{m+1}}{\frac{3}{m^2-1}}$ **27.** $\dfrac{\frac{1}{k+1}-1}{\frac{1}{k+1}+1}$ **28.** $\dfrac{\frac{2}{p-1}+2}{\frac{3}{p-1}-2}$

29. $\dfrac{2+\frac{1}{x}-\frac{28}{x^2}}{3+\frac{13}{x}+\frac{4}{x^2}}$ **30.** $\dfrac{4-\frac{11}{x}-\frac{3}{x^2}}{2-\frac{1}{x}-\frac{15}{x^2}}$ **31.** $\dfrac{\frac{1}{m-1}+\frac{2}{m+2}}{\frac{2}{m+2}-\frac{1}{m-3}}$

32. $\dfrac{\frac{5}{r+3}-\frac{1}{r-1}}{\frac{2}{r+2}+\frac{3}{r+3}}$ **33.** $2-\dfrac{2}{2+\frac{2}{2+2}}$ **34.** $3-\dfrac{2}{4+\frac{2}{4-2}}$

14.6 Solving Equations with Rational Expressions

OBJECTIVES

1. Distinguish between operations with rational expressions and equations with terms that are rational expressions.
2. Solve equations with rational expressions.
3. Solve a formula for a specified variable.

OBJECTIVE 1 Distinguish between operations with rational expressions and equations with terms that are rational expressions. Before solving equations with rational expressions, we emphasize the difference between sums and differences of terms with rational coefficients, or rational *expressions*, and *equations* with terms that are rational expressions.

Sums and differences are expressions to simplify. Equations are solved.

EXAMPLE 1 Distinguishing between Expressions and Equations

Identify each of the following as an *expression* or an *equation*. Then simplify the expression or solve the equation.

(a) $\frac{3}{4}x - \frac{2}{3}x$ This is a difference of two terms. It represents an *expression* to simplify since there is no equality symbol.

$= \frac{3 \cdot 3}{3 \cdot 4}x - \frac{4 \cdot 2}{4 \cdot 3}x$ The LCD is 12. Write each coefficient with this LCD.

$= \frac{9}{12}x - \frac{8}{12}x$ Multiply.

$= \frac{1}{12}x$ Combine like terms, using the distributive property: $\frac{9}{12}x - \frac{8}{12}x = \left(\frac{9}{12} - \frac{8}{12}\right)x$.

(b) $\frac{3}{4}x - \frac{2}{3}x = \frac{1}{2}$ Because there is an equality symbol, this is an *equation* to be solved.

$12\left(\frac{3}{4}x - \frac{2}{3}x\right) = 12\left(\frac{1}{2}\right)$ Use the multiplication property of equality to clear fractions. Multiply by 12, the LCD.

$12\left(\frac{3}{4}x\right) - 12\left(\frac{2}{3}x\right) = 12\left(\frac{1}{2}\right)$ Distributive property

Multiply each term by 12.

$9x - 8x = 6$ Multiply.

$x = 6$ Combine like terms.

CHECK $\frac{3}{4}x - \frac{2}{3}x = \frac{1}{2}$ Original equation

$\frac{3}{4}(6) - \frac{2}{3}(6) \stackrel{?}{=} \frac{1}{2}$ Let $x = 6$.

$\frac{9}{2} - 4 \stackrel{?}{=} \frac{1}{2}$ Multiply.

$\frac{1}{2} = \frac{1}{2}$ ✓ True

Since a true statement results, $\{6\}$ is the solution set of the equation.

Work Problem 1 at the Side. ▶

1 Identify each as an *expression* or an *equation*. Then simplify the expression or solve the equation.

(a) $\frac{2x}{3} - \frac{4x}{9}$

(b) $\frac{2x}{3} - \frac{4x}{9} = 2$

Answers

1. (a) expression; $\frac{2x}{9}$ **(b)** equation; $\{9\}$

2 Solve each equation, and check your solutions.

(a) $\frac{x}{5} + 3 = \frac{3}{5}$

(b) $\frac{x}{2} - \frac{x}{3} = \frac{5}{6}$

Answers

2. **(a)** $\{-12\}$ **(b)** $\{5\}$

The ideas of **Example 1** can be summarized as follows.

Uses of the LCD

When adding or subtracting rational expressions, keep the LCD throughout the simplification. (See **Example 1(a)**.)

When solving an equation with terms that are rational expression, multiply each side by the LCD so that denominators are eliminated. (See **Example 1(b)**.)

OBJECTIVE 2 Solve equations with rational expressions. When an equation involves fractions as in **Example 1(b),** we use the multiplication property of equality to clear it of fractions. When we choose the LCD of all the denominators as the multiplier, the resulting equation contains no fractions.

EXAMPLE 2 Solving an Equation with Rational Expressions

Solve, and check the solution.

$$\frac{x}{3} + \frac{x}{4} = 10 + x$$

$$12\left(\frac{x}{3} + \frac{x}{4}\right) = 12(10 + x) \quad \text{Multiply by the LCD, 12, to clear fractions.}$$

$$12\left(\frac{x}{3}\right) + 12\left(\frac{x}{4}\right) = 12(10) + 12x \quad \text{Distributive property}$$

$$4x + 3x = 120 + 12x \quad \text{Multiply.}$$

$$7x = 120 + 12x \quad \text{Combine like terms.}$$

$$-5x = 120 \quad \text{Subtract } 12x.$$

$$x = -24 \quad \text{Divide by } -5.$$

CHECK

$$\frac{x}{3} + \frac{x}{4} = 10 + x \quad \text{Original equation}$$

$$\frac{-24}{3} + \frac{-24}{4} \stackrel{?}{=} 10 - 24 \quad \text{Let } x = -24.$$

$$-8 + (-6) \stackrel{?}{=} -14 \quad \text{Divide. Subtract.}$$

$$-14 = -14 \checkmark \quad \text{True}$$

The solution set is $\{-24\}$.

◄ **Work Problem 2 at the Side.**

CAUTION

In **Example 2,** we used the multiplication property of equality to multiply each side of an *equation* by the LCD. In **Section 14.5,** we used the fundamental property to multiply a *fraction* by another fraction that had the LCD as both its numerator and denominator. ***Be careful not to confuse these procedures.***

EXAMPLE 3 Solving an Equation with Rational Expressions

Solve, and check the solution.

$$\frac{p}{2} - \frac{p-1}{3} = 1$$

$$6\left(\frac{p}{2} - \frac{p-1}{3}\right) = 6 \cdot 1 \qquad \text{Multiply by the LCD, 6.}$$

$$\mathbf{6}\left(\frac{p}{2}\right) - \mathbf{6}\left(\frac{p-1}{3}\right) = 6 \qquad \text{Distributive property}$$

$$3p - 2(p-1) = 6 \qquad \boxed{\text{Use parentheses around } p-1 \text{ to avoid errors.}}$$

$$3p - 2(p) - 2(-1) = 6 \qquad \text{Distributive property}$$

$$\boxed{\text{Be careful with signs.}} \qquad 3p - 2p + 2 = 6 \qquad \text{Multiply.}$$

$$p = 4 \qquad \text{Combine like terms. Subtract 2.}$$

Check that {4} is the solution set by replacing p with 4 in the original equation.

················· **Work Problem ❸ at the Side. ▶**

Recall that division by 0 is undefined. ***Therefore, when solving an equation with rational expressions that have variables in the denominator, the solution cannot be a number that makes the denominator equal 0.***

A value of the variable that appears to be a solution after both sides of an equation with rational expressions are multiplied by a variable expression is a **proposed solution**. ***All proposed solutions must be checked in the original equation.***

EXAMPLE 4 Solving an Equation with Rational Expressions

Solve, and check the proposed solution.

$$\frac{x}{x-2} = \frac{2}{x-2} + 2 \qquad \begin{array}{l}x \text{ cannot equal 2, since}\\ \text{2 causes both denomi-}\\ \text{nators to equal 0.}\end{array}$$

$$(x-2)\left(\frac{x}{x-2}\right) = (x-2)\left(\frac{2}{x-2} + 2\right) \qquad \begin{array}{l}\text{Multiply each side}\\ \text{by the LCD, } x-2.\end{array}$$

$$(x-2)\left(\frac{x}{x-2}\right) = (x-2)\left(\frac{2}{x-2}\right) + (x-2)(2) \qquad \text{Distributive property}$$

$$x = 2 + 2x - 4 \qquad \text{Simplify.}$$

$$x = -2 + 2x \qquad \text{Combine like terms.}$$

$$-x = -2 \qquad \text{Subtract } 2x.$$

$$\begin{array}{l}\text{Proposed}\\ \text{solution}\end{array} \rightarrow x = 2 \qquad \text{Divide by } -1.$$

CHECK $\qquad \dfrac{x}{x-2} = \dfrac{2}{x-2} + 2 \qquad \text{Original equation}$

$$\frac{2}{2-2} \overset{?}{=} \frac{2}{2-2} + 2 \qquad \text{Let } x = 2.$$

$$\boxed{\begin{array}{l}\text{Division by 0}\\ \text{is undefined.}\end{array}} \qquad \frac{2}{0} \overset{?}{=} \frac{2}{0} + 2 \qquad \text{Subtract in the denominators.}$$

Thus, the proposed solution 2 must be rejected, and the solution set is $\emptyset$.

················· **Work Problem ❹ at the Side. ▶**

❸ Solve each equation, and check the solutions.

(GS) **(a)** $\dfrac{k}{6} - \dfrac{k+1}{4} = -\dfrac{1}{2}$

Multiply by the LCD, ____.

$$\underline{}\left(\frac{k}{6} - \frac{k+1}{4}\right) = \underline{}\left(-\frac{1}{2}\right)$$

Complete the solution.

(b) $\dfrac{2m-3}{5} - \dfrac{m}{3} = -\dfrac{6}{5}$

❹ Solve the equation, and check the proposed solution.

$$1 - \frac{2}{x+1} = \frac{2x}{x+1}$$

Answers

3. **(a)** 12; 12; 12; {3} **(b)** {−9}
4. $\emptyset$ (When the equation is solved, −1 is a proposed solution. However, because $x = -1$ leads to a 0 denominator in the original equation, there is no solution.)

5 Solve each equation, and check the proposed solutions.

(a) $\frac{4}{x^2 - 3x} = \frac{1}{x^2 - 9}$

GS **(b)** $\frac{2}{p^2 - 2p} = \frac{3}{p^2 - p}$

Step 1
Factor the denominators.

$p^2 - 2p = p(\_\_\_\_\_\_)$

$p^2 - p = p(\_\_\_\_\_\_)$

The LCD is ________.

The numbers ____, ____, and ____ cannot be solutions.

Complete the steps to solve the equation.

Answers

5. (a) $\{-4\}$
(b) $p - 2$; $p - 1$; $p(p-2)(p-1)$; 0; 1; 2 (Order of the numbers 0, 1, and 2 does not matter.); $\{4\}$

A proposed solution that is not an actual solution of the original equation, such as 2 in **Example 4,** is an **extraneous solution,** or **extraneous value.** Some students like to determine which numbers cannot be solutions *before* solving the equation, as we did at the beginning of **Example 4.**

Solving an Equation with Rational Expressions

Step 1 **Multiply each side of the equation by the LCD.** (This clears the equation of fractions.) Be sure to distribute to *every* term on *both* sides of the equation.

Step 2 **Solve** the resulting equation for proposed solutions.

Step 3 **Check** each proposed solution by substituting it in the original equation. Reject any that cause a denominator to equal 0.

EXAMPLE 5 Solving an Equation with Rational Expressions

Solve, and check the proposed solution.

$$\frac{2}{x^2 - x} = \frac{1}{x^2 - 1}$$

Step 1 $\frac{2}{x(x-1)} = \frac{1}{(x+1)(x-1)}$ Factor the denominators to find the LCD, $x(x+1)(x-1)$.

Notice that 0, -1, and 1 cannot be solutions of this equation. Otherwise, a denominator will equal 0.

$$x(x+1)(x-1)\frac{2}{x(x-1)} = x(x+1)(x-1)\frac{1}{(x+1)(x-1)}$$ Multiply by the LCD.

Step 2 $2(x+1) = x$ Divide out the common factors.

$2x + 2 = x$ Distributive property

$x + 2 = 0$ Subtract x.

Proposed solution $\rightarrow$ $x = -2$ Subtract 2.

Step 3 The proposed solution is -2, which does not make any denominator equal 0.

CHECK $\frac{2}{x^2 - x} = \frac{1}{x^2 - 1}$ Original equation

$\frac{2}{(-2)^2 - (-2)} \stackrel{?}{=} \frac{1}{(-2)^2 - 1}$ Let $x = -2$.

$\frac{2}{4 + 2} \stackrel{?}{=} \frac{1}{4 - 1}$ Apply the exponents; Definition of subtraction

$\frac{1}{3} = \frac{1}{3}$ ✓ True

The solution set is $\{-2\}$.

◀ Work Problem **5** at the Side.

EXAMPLE 6 Solving an Equation with Rational Expressions

Solve, and check the proposed solution.

$$\frac{2m}{(m^2-4)} + \frac{1}{m-2} = \frac{2}{m+2}$$

$$\frac{2m}{(m+2)(m-2)} + \frac{1}{m-2} = \frac{2}{m+2}$$ Factor the denominator on the left to find the LCD, $(m+2)(m-2)$.

Notice that -2 and 2 cannot be solutions of the equation.

$$(m+2)(m-2)\left(\frac{2m}{(m+2)(m-2)} + \frac{1}{m-2}\right)$$

$$= (m+2)(m-2)\frac{2}{m+2}$$ Multiply by the LCD.

$$(m+2)(m-2)\frac{2m}{(m+2)(m-2)} + (m+2)(m-2)\frac{1}{m-2}$$

$$= (m+2)(m-2)\frac{2}{m+2}$$ Distributive property

$2m + m + 2 = 2(m-2)$ Divide out the common factors.

$3m + 2 = 2m - 4$ Combine like terms; distributive property

$m = -6$ Subtract $2m$. Subtract 2.

A check will verify that $\{-6\}$ is the solution set.

Work Problem 6 at the Side. ▶

EXAMPLE 7 Solving an Equation with Rational Expressions

Solve, and check the proposed solution(s).

$$\frac{1}{x-1} + \frac{1}{2} = \frac{2}{x^2-1}$$

$x \neq 1, -1$ or a denominator is 0.

$$\frac{1}{x-1} + \frac{1}{2} = \frac{2}{(x+1)(x-1)}$$

Factor the denominator on the right. The LCD is $2(x+1)(x-1)$.

$$2(x+1)(x-1)\left(\frac{1}{x-1} + \frac{1}{2}\right) = 2(x+1)(x-1)\frac{2}{(x+1)(x-1)}$$

Multiply by the LCD.

$$2(x+1)(x-1)\frac{1}{x-1} + 2(x+1)(x-1)\frac{1}{2}$$

$$= 2(x+1)(x-1)\frac{2}{(x+1)(x-1)}$$

Distributive property

$2(x+1) + (x+1)(x-1) = 2(2)$ Divide out the common factors.

$2x + 2 + x^2 - 1 = 4$ Distributive property; multiply.

Write in standard form. $x^2 + 2x - 3 = 0$ Subtract 4. Combine like terms.

Continued on Next Page

6 Solve each equation, and check the proposed solutions.

(a) $\dfrac{2p}{p^2-1} = \dfrac{2}{p+1} - \dfrac{1}{p-1}$

(b) $\dfrac{8r}{4r^2-1} = \dfrac{3}{2r+1} + \dfrac{3}{2r-1}$

Answers

6. **(a)** $\{-3\}$ **(b)** $\{0\}$

7 Solve the equation, and check the proposed solution(s).

$$\frac{2}{3x+1} - \frac{1}{x} = \frac{-6x}{3x+1}$$

8 Solve each equation, and check the proposed solution(s).

(a) $\frac{1}{x-2} + \frac{1}{5} = \frac{2}{5(x^2-4)}$

(b) $\frac{6}{5a+10} - \frac{1}{a-5} = \frac{4}{a^2-3a-10}$

Answers

7. $\left\{\frac{1}{2}\right\}$

8. **(a)** $\{-4, -1\}$ **(b)** $\{60\}$

$$(x+3)(x-1) = 0 \quad \text{Factor } x^2+2x-3.$$

$$x+3=0 \quad \text{or} \quad x-1=0 \quad \text{Zero-factor property}$$

$$x = -3 \quad \text{or} \quad x = 1 \longleftarrow \text{Proposed solutions}$$

Since 1 makes an original denominator equal 0, the proposed solution 1 is an extraneous value. Check that -3 is a solution.

CHECK

$$\frac{1}{x-1} + \frac{1}{2} = \frac{2}{x^2-1} \quad \text{Original equation}$$

$$\frac{1}{-3-1} + \frac{1}{2} \stackrel{?}{=} \frac{2}{(-3)^2-1} \quad \text{Let } x = -3.$$

$$\frac{1}{-4} + \frac{1}{2} \stackrel{?}{=} \frac{2}{9-1} \quad \text{Subtract. Apply the exponent.}$$

$$\frac{1}{4} = \frac{1}{4} \checkmark \quad \text{True}$$

The check shows that $\{-3\}$ is the solution set.

◀ **Work Problem 7 at the Side.**

EXAMPLE 8 Solving an Equation with Rational Expressions

Solve, and check the proposed solution(s).

$$\frac{1}{k^2+4k+3} + \frac{1}{2k+2} = \frac{3}{4k+12}$$

$$\frac{1}{(k+1)(k+3)} + \frac{1}{2(k+1)} = \frac{3}{4(k+3)} \quad \text{Factor each denominator. The LCD is } 4(k+1)(k+3).$$

$k \neq -1, -3$

$$4(k+1)(k+3)\left(\frac{1}{(k+1)(k+3)} + \frac{1}{2(k+1)}\right) = 4(k+1)(k+3)\frac{3}{4(k+3)} \quad \text{Multiply by the LCD.}$$

$$4(k+1)(k+3)\frac{1}{(k+1)(k+3)} + 2 \cdot 2(k+1)(k+3)\frac{1}{2(k+1)} = 4(k+1)(k+3)\frac{3}{4(k+3)} \quad \text{Distributive property}$$

Do not add 4 + 2 here.

$$4 + 2(k+3) = 3(k+1) \quad \text{Divide out the common factors.}$$

$$4 + 2k + 6 = 3k + 3 \quad \text{Distributive property}$$

$$2k + 10 = 3k + 3 \quad \text{Combine like terms.}$$

$$10 = k + 3 \quad \text{Subtract } 2k.$$

$$7 = k \quad \text{Subtract 3.}$$

The proposed solution, 7, does not make an original denominator equal 0. A check shows that the algebra is correct, so $\{7\}$ is the solution set.

◀ **Work Problem 8 at the Side.**

OBJECTIVE 3 Solve a formula for a specified variable. When solving a formula for a specified variable, ***remember to treat the variable for which you are solving as if it were the only variable, and all others as if they were constants.***

EXAMPLE 9 Solving for Specified Variables

Solve each formula for the specified variable.

(a) $a = \frac{v - w}{t}$ for v

Our goal is to isolate v.

$$a = \frac{v - w}{t}$$

$$at = v - w \quad \text{Multiply by } t.$$

$$at + w = v, \quad \text{or} \quad v = at + w \quad \text{Add } w.$$

CHECK Substitute $at + w$ for v in the original equation.

$$a = \frac{v - w}{t} \quad \text{Original equation}$$

$$a = \frac{at + w - w}{t} \quad \text{Let } v = at + w.$$

$$a = \frac{at}{t} \quad \text{Combine like terms.}$$

$$a = a \checkmark \quad \text{True}$$

(b) $F = \frac{k}{d - D}$ for d

$$F = \frac{k}{d - D}$$

We must isolate d.

$$F(d - D) = \frac{k}{d - D}(d - D) \quad \text{Multiply by } d - D \text{ to clear the fraction.}$$

$$F(d - D) = k \quad \text{Simplify.}$$

$$Fd - FD = k \quad \text{Distributive property}$$

$$Fd = k + FD \quad \text{Add } FD.$$

$$d = \frac{k + FD}{F} \quad \text{Divide by } F.$$

We can write an equivalent form of this answer as follows.

$$d = \frac{k + FD}{F} \quad \text{Answer from above}$$

$$d = \frac{k}{F} + \frac{FD}{F} \quad \text{Definition of addition of fractions: } \frac{a+b}{c} = \frac{a}{c} + \frac{b}{c}$$

$$d = \frac{k}{F} + D \quad \text{Divide out the common factor from } \frac{FD}{F}.$$

Either form of the answer is correct.

Work Problem 9 at the Side. ▶

---

9 Solve each formula for the specified variable.

GS **(a)** $r = \frac{A - p}{pt}$ for A

The goal is to isolate ____.

Multiply by ____ to obtain the equation ________.

Add ____ to obtain the equation

________ $= A$, or ________.

(b) $p = \frac{x - y}{z}$ for y

(c) $z = \frac{x}{x + y}$ for y

Answers

9. (a) A; pt; $rpt = A - p$; p; $p + prt$; $A = p + prt$

(b) $y = x - pz$

(c) $y = \frac{x - zx}{z}$, or $y = \frac{x}{z} - x$

10 Solve $\frac{2}{x} = \frac{1}{y} + \frac{1}{z}$ for the specified variable.

(a) for z

(b) for y

EXAMPLE 10 Solving for a Specified Variable

Solve the formula $\frac{1}{a} = \frac{1}{b} + \frac{1}{c}$ for c.

$$\frac{1}{a} = \frac{1}{b} + \frac{1}{c}$$ Goal: Isolate c, the specified variable.

$$abc\left(\frac{1}{a}\right) = abc\left(\frac{1}{b} + \frac{1}{c}\right)$$ Multiply by the LCD, abc.

$$abc\left(\frac{1}{a}\right) = abc\left(\frac{1}{b}\right) + abc\left(\frac{1}{c}\right)$$ Distributive property

$$bc = ac + ab$$ Simplify.

$$bc - ac = ab$$ Subtract ac to get both terms with c on the same side.

Pay careful attention here.

$$c(b - a) = ab$$ Factor out c.

$$c = \frac{ab}{b - a}$$ Divide by $b - a$.

CAUTION

Students often have trouble in the step that involves factoring out the variable for which they are solving. In **Example 10,** we needed to get both terms with c on the same side of the equation. This allowed us to factor out c on the left, and then isolate it by dividing each side by $b - a$.

When solving an equation for a specified variable, be sure that the specified variable appears alone on only one side of the equality symbol in the final equation.

◀ Work Problem 10 at the Side.

Answers

10. **(a)** $z = \frac{xy}{2y - x}$ **(b)** $y = \frac{xz}{2z - x}$

14.6 Exercises

FOR EXTRA HELP

Download the MyDashBoard App

MyMathLab®

CONCEPT CHECK *Fill in each blank with the correct response.*

1. A value of the variable that appears to be a solution after both sides of an equation with rational expressions are multiplied by a variable expression is a(n) ________ solution. It must be checked in the ________ equation to determine whether it is an actual solution.

2. A proposed solution that is not an actual solution of an original equation is a(n) ________ solution, or ________ value.

Identify each as an expression or an equation. Then simplify the expression or solve the equation. ***See Example 1.***

3. $\frac{7}{8}x + \frac{1}{5}x$

4. $\frac{4}{7}x + \frac{3}{5}x$

5. $\frac{7}{8}x + \frac{1}{5}x = 1$

6. $\frac{4}{7}x + \frac{3}{5}x = 1$

7. $\frac{3}{5}y - \frac{7}{10}y$

8. $\frac{3}{5}y - \frac{7}{10}y = 1$

Solve each equation, and check your solutions. ***See Examples 2 and 3.***

9. $\frac{2}{3}x + \frac{1}{2}x = -7$

10. $\frac{1}{4}x - \frac{1}{3}x = 1$

11. $\frac{3x}{5} - 6 = x$

12. $\frac{5t}{4} + t = 9$

13. $\frac{4m}{7} + m = 11$

14. $a - \frac{3a}{2} = 1$

15. $\frac{z-1}{4} = \frac{z+3}{3}$

16. $\frac{r-5}{2} = \frac{r+2}{3}$

17. $\frac{3p+6}{8} = \frac{3p-3}{16}$

18. $\frac{2z+1}{5} = \frac{7z+5}{15}$

19. $\frac{2x+3}{-6} = \frac{3}{2}$

20. $\frac{4x+3}{6} = \frac{5}{2}$

21. $\frac{q+2}{3} + \frac{q-5}{5} = \frac{7}{3}$

22. $\frac{b+7}{8} - \frac{b-2}{3} = \frac{4}{3}$

23. $\frac{t}{6} + \frac{4}{3} = \frac{t-2}{3}$

24. $\frac{x}{2} = \frac{5}{4} + \frac{x-1}{4}$

25. $\frac{3m}{5} - \frac{3m-2}{4} = \frac{1}{5}$

26. $\frac{8p}{5} = \frac{3p-4}{2} + \frac{5}{2}$

When solving an equation with variables in denominators, we must determine the values that cause these denominators to equal 0, so that we can reject these values if they appear as proposed solutions. Find all values for which at least one denominator is equal to 0. Write answers using the symbol ≠. Do not solve. **See Examples 4–8.**

27. $\frac{3}{x+2} - \frac{5}{x} = 1$

28. $\frac{7}{x} + \frac{9}{x-4} = 5$

29. $\frac{-1}{(x+3)(x-4)} = \frac{1}{2x+1}$

30. $\frac{8}{(x-7)(x+3)} = \frac{7}{3x-10}$

31. $\frac{4}{x^2+8x-9} + \frac{1}{x^2-4} = 0$

32. $\frac{-3}{x^2+9x-10} - \frac{12}{x^2-49} = 0$

Solve each equation, and check your solutions. **See Examples 4–8.**

33. $\frac{5-2x}{x} = \frac{1}{4}$

34. $\frac{2x+3}{x} = \frac{3}{2}$

35. $\frac{k}{k-4} - 5 = \frac{4}{k-4}$

36. $\frac{-5}{a+5} = \frac{a}{a+5} + 2$

37. $\frac{3}{x-1} + \frac{2}{4x-4} = \frac{7}{4}$

The LCD is ______.
Complete the steps to solve the equation.

38. $\frac{2}{p+3} + \frac{3}{8} = \frac{5}{4p+12}$

The LCD is ______.
Complete the steps to solve the equation.

39. $\frac{x}{3x+3} = \frac{2x-3}{x+1} - \frac{2x}{3x+3}$

40. $\frac{2k+3}{k+1} - \frac{3k}{2k+2} = \frac{-2k}{2k+2}$

41. $\frac{2}{m} = \frac{m}{5m+12}$

42. $\frac{x}{4-x} = \frac{2}{x}$

43. $\frac{5x}{14x+3} = \frac{1}{x}$

44. $\frac{m}{8m+3} = \frac{1}{3m}$

45. $\frac{2}{z-1} - \frac{5}{4} = \frac{-1}{z+1}$

46. $\frac{5}{p-2} = 7 - \frac{10}{p+2}$

47. $\frac{4}{x^2-3x} = \frac{1}{x^2-9}$

48. $\frac{2}{t^2-4} = \frac{3}{t^2-2t}$

49. $\frac{-2}{z+5} + \frac{3}{z-5} = \frac{20}{z^2-25}$

50. $\frac{3}{r+3} - \frac{2}{r-3} = \frac{-12}{r^2-9}$

51. $\frac{1}{x+4} + \frac{x}{x-4} = \frac{-8}{x^2-16}$

52. $\frac{x}{x-3} + \frac{4}{x+3} = \frac{18}{x^2-9}$

53. $\frac{4}{3x+6} - \frac{3}{x+3} = \frac{8}{x^2+5x+6}$

54. $\frac{-13}{t^2+6t+8} + \frac{4}{t+2} = \frac{3}{2t+8}$

55. $\frac{3x}{x^2+5x+6} = \frac{5x}{x^2+2x-3} - \frac{2}{x^2+x-2}$

56. $\frac{x+4}{x^2-3x+2} - \frac{5}{x^2-4x+3} = \frac{x-4}{x^2-5x+6}$

In Exercises 57 and 58, answer each question.

57. If you are solving a formula for the letter k, and your steps lead to the equation $kr - mr = km$, what would be your next step?

58. If you are solving a formula for the letter k, and your steps lead to the equation $kr - km = mr$, what would be your next step?

Solve each formula for the specified variable. ***See Example 9.***

59. $m = \frac{kF}{a}$ for F

60. $I = \frac{kE}{R}$ for E

61. $m = \frac{kF}{a}$ for a

62. $I = \frac{kE}{R}$ for R

63. $m = \frac{y - b}{x}$ for y

64. $y = \frac{C - Ax}{B}$ for C

65. $I = \frac{E}{R + r}$ for R

66. $I = \frac{E}{R + r}$ for r

67. $h = \frac{2A}{B + b}$ for b

68. $h = \frac{2A}{B + b}$ for B

69. $d = \frac{2S}{n(a + L)}$ for a

70. $d = \frac{2S}{n(a + L)}$ for L

Solve each equation for the specified variable. ***See Example 10.***

71. $\frac{2}{r} + \frac{3}{s} + \frac{1}{t} = 1$ for t

72. $\frac{5}{p} + \frac{2}{q} + \frac{3}{r} = 1$ for r

73. $\frac{1}{a} - \frac{1}{b} - \frac{1}{c} = 2$ for c

74. $\frac{-1}{x} + \frac{1}{y} + \frac{1}{z} = 4$ for y

75. $9x + \frac{3}{z} = \frac{5}{y}$ for z

76. $-3t - \frac{4}{p} = \frac{6}{s}$ for p

14.7 Applications of Rational Expressions

OBJECTIVES

1. Solve problems about numbers.
2. Solve problems about distance, rate, and time.
3. Solve problems about work.

For applications that lead to rational equations, the six-step problem-solving method of **Section 10.4** still applies.

OBJECTIVE 1 Solve problems about numbers.

EXAMPLE 1 Solving a Problem about an Unknown Number

If the same number is added to both the numerator and the denominator of the fraction $\frac{2}{5}$, the result is equivalent to $\frac{2}{3}$. Find the number.

Step 1 **Read** the problem carefully. We are trying to find a number.

Step 2 **Assign a variable.**

Let x = the number added to the numerator and the denominator.

Step 3 **Write an equation.** The fraction

$$\frac{2+x}{5+x}$$

represents the result of adding the same number x to both the numerator and the denominator. This result is equivalent to $\frac{2}{3}$, so we write the following equation.

$$\frac{2+x}{5+x} = \frac{2}{3}$$

Step 4 **Solve.**

$$3(5+x)\frac{2+x}{5+x} = 3(5+x)\frac{2}{3} \quad \text{Multiply by the LCD, } 3(5+x).$$

$$3(2+x) = 2(5+x) \quad \text{Divide out the common factors.}$$

$$6+3x = 10+2x \quad \text{Distributive property}$$

$$x = 4 \quad \text{Subtract } 2x. \text{ Subtract } 6.$$

Step 5 **State the answer.** The number is 4.

Step 6 **Check** the solution in the words of the original problem. If 4 is added to both the numerator and the denominator of $\frac{2}{5}$, the result is $\frac{2+4}{5+4} = \frac{6}{9}$, or $\frac{2}{3}$, as required.

Work Problem 1 at the Side. ▶

OBJECTIVE 2 Solve problems about distance, rate, and time.

If an automobile travels at an average rate of 65 mph for 2 hr, then it travels

$$65 \times 2 = 130 \text{ mi.} \quad rt = d, \text{ or } d = rt \text{ (Relationship between distance, rate, and time)}$$

By solving, in turn, for r and t in the formula $d = rt$, we obtain two other equivalent forms of the formula. The three forms are given below.

Distance, Rate, and Time Relationship

$$d = rt \qquad r = \frac{d}{t} \qquad t = \frac{d}{r}$$

---

1 Solve each problem.

GS (a) A certain number is added to the numerator and subtracted from the denominator of $\frac{5}{8}$. The new fraction equals the reciprocal of $\frac{5}{8}$. Find the number.

Step 1
We are trying to find a(n) ________.

Step 2
Let x = the number ________ to the numerator and ________ from the denominator.

Step 3
The expression ________ represents the new fraction.

The reciprocal of $\frac{5}{8}$ is ____. Write an equation.

____

Complete Steps 4–6 to solve the problem. Give the answer.

(b) The denominator of a fraction is 1 more than the numerator. If 6 is added to the numerator and subtracted from the denominator, the result is $\frac{15}{4}$. Find the original fraction.

Answers

1. (a) number; added; subtracted; $\frac{5+x}{8-x}$; $\frac{8}{5}$; $\frac{5+x}{8-x} = \frac{8}{5}$; The number is 3.
(b) $\frac{9}{10}$

2 Solve each problem.

(a) A new world record in the men's 100-m dash was set by Usain Bolt of Jamaica in 2009. He ran it in 9.58 sec. What was his rate in meters per second, to the nearest hundredth? (*Source: World Almanac and Book of Facts.*)

(b) A new world record for the women's 3000-m steeplechase was set by Gulnara Samitova of Russia in 2008. Her rate was 5.568 m per sec. To the nearest second, what was her time? (*Source: World Almanac and Book of Facts.*)

(c) A small plane flew from Chicago to St. Louis averaging 145 mph. The trip took 2 hr. What is the distance between Chicago and St. Louis?

EXAMPLE 2 Finding Distance, Rate, or Time

Solve each problem using a form of the distance formula.

(a) The speed (rate) of sound is 1088 ft per sec at sea level at 32°F. Find the distance sound travels in 5 sec under these conditions.

We must find distance, given rate and time, using $d = rt$ (or $rt = d$).

$$\underset{\text{Rate}}{1088} \times \underset{\text{Time}}{5} = \underset{\text{Distance}}{5440 \text{ ft}}$$

(b) The winner of the first Indianapolis 500 race (in 1911) was Ray Harroun, driving a Marmon Wasp at an average rate of 74.60 mph. (*Source: World Almanac and Book of Facts.*) How long did it take him to complete the 500 mi?

We must find time, given rate and distance, using $t = \frac{d}{r}$ (or $\frac{d}{r} = t$).

$$\frac{500 \;\leftarrow \text{Distance}}{74.60 \;\leftarrow \text{Rate}} = 6.70 \text{ hr (rounded)} \leftarrow \text{Time}$$

To convert 0.70 hr to minutes, we multiply by 60 to get $0.70(60) = 42$. It took Harroun about 6 hr, 42 min, to complete the race.

(c) At the 2008 Olympic Games, Australian swimmer Leisel Jones set an Olympic record of 65.17 sec in the women's 100-m breaststroke swimming event. (*Source: World Almanac and Book of Facts.*) Find her rate.

We must find rate, given distance and time, using $r = \frac{d}{t}$ (or $\frac{d}{t} = r$).

$$\frac{100 \;\leftarrow \text{Distance}}{65.17 \;\leftarrow \text{Time}} = 1.53 \text{ m per sec (rounded)} \leftarrow \text{Rate}$$

◀ **Work Problem 2 at the Side.**

Problem-Solving Hint

Many applied problems use the formulas just discussed. The next two examples show how to solve typical applications of the formula $d = rt$.

A helpful strategy for solving such problems is to ***first make a sketch*** showing what is happening in the problem. ***Then make a table*** using the information given, along with the unknown quantities. The table will help organize the information, and the sketch will help set up the equation.

EXAMPLE 3 Solving a Problem about Distance, Rate, and Time

Two cars leave Iowa City, Iowa, at the same time and travel east on Interstate 80. One travels at a constant rate of 55 mph. The other travels at a constant rate of 63 mph. In how many hours will the distance between them be 24 mi?

Step 1 **Read** the problem. We must find the time it will take for the distance between the cars to be 24 mi.

Continued on Next Page

Answers

2. **(a)** 10.44 m per sec
(b) 539 sec, or 8 min, 59 sec
(c) 290 mi

Step 2 **Assign a variable.** We are looking for time.

Let $t =$ the number of hours until the distance between the cars is 24 mi.

The sketch in **Figure 1** shows what is happening in the problem.

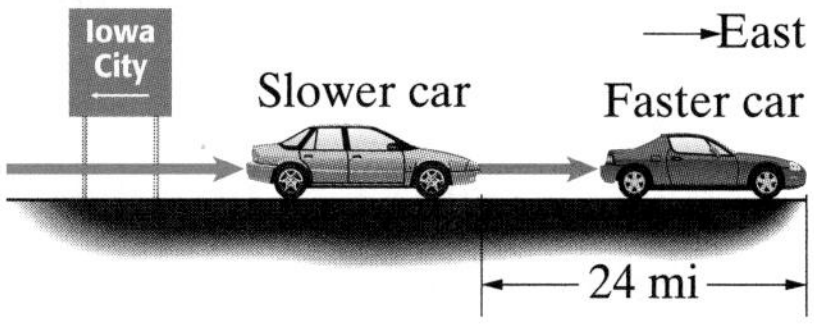

Figure 1

To construct a table, we fill in the information given in the problem, using t for the time traveled by each car. We multiply rate by time to get the expressions for distances traveled.

	Rate	Time	Distance
Faster Car	63	t	$63t$
Slower Car	55	t	$55t$

The quantities $63t$ and $55t$ represent the two distances. The *difference* between the larger distance and the smaller distance is 24 mi.

Step 3 **Write an equation.**

$$63t - 55t = 24$$

Step 4 **Solve.**

$$8t = 24 \quad \text{Combine like terms.}$$

$$t = 3 \quad \text{Divide by 8.}$$

Step 5 **State the answer.** It will take the cars 3 hr to be 24 mi apart.

Step 6 **Check.** After 3 hr, the faster car will have traveled

$$63 \times 3 = 189 \text{ mi}$$

and the slower car will have traveled

$$55 \times 3 = 165 \text{ mi.}$$

The difference is

$$189 - 165 = 24, \quad \text{as required.}$$

Problem-Solving Hint

In problems involving distance, rate, and time like the one in **Example 3,** once we have filled in two pieces of information in each row of the table, we can automatically fill in the third piece of information, using the appropriate form of the distance formula. Then we set up the equation based on our sketch and the information in the table.

Work Problem 3 at the Side. ▶

3 Solve each problem.

GS **(a)** From a point on a straight road, Lupe and Maria ride bicycles in opposite directions. Lupe rides 10 mph and Maria rides 12 mph. In how many hours will they be 55 mi apart?

Steps 1 and 2
Let $x =$ the number of ________ until the distance between Lupe and Maria is ________.

Step 3
Complete the table.

	Rate	× Time	= Distance
Maria	10	t	_____
Lupe	_____	t	_____

Because Lupe and Maria are traveling in opposite directions we must ________ the distances they travel to find the distance between them. Write an equation.

____

Complete Steps 4–6 to solve the problem. Give the answer.

(b) At a given hour, two steamboats leave a city in the same direction on a straight canal. One travels at 18 mph, and the other travels at 25 mph. In how many hours will the boats be 35 mi apart?

Answers

3. (a) hours; 55 mi

	Rate	× Time	= Distance
Maria	10	t	$10t$
Lupe	12	t	$12t$

add; $10t + 12t = 55$; $2\frac{1}{2}$ hr

(b) 5 hr

EXAMPLE 4 Solving a Problem about Distance, Rate, and Time

The Tickfaw River has a current of 3 mph. A motorboat takes as long to go 12 mi downstream as to go 8 mi upstream. What is the rate of the boat in still water?

Step 1 **Read** the problem. We want the rate (speed) of the boat in still water.

Step 2 **Assign a variable.**

Let $x =$ the rate of the boat in still water.

Because the current pushes the boat when the boat is going downstream, the rate of the boat downstream will be the *sum* of the rate of the boat and the rate of the current, $(x + 3)$ mph.

Because the current slows down the boat when the boat is going upstream, the boat's rate upstream is given by the *difference* between the rate of the boat in still water and the rate of the current, $(x - 3)$ mph. See **Figure 2.**

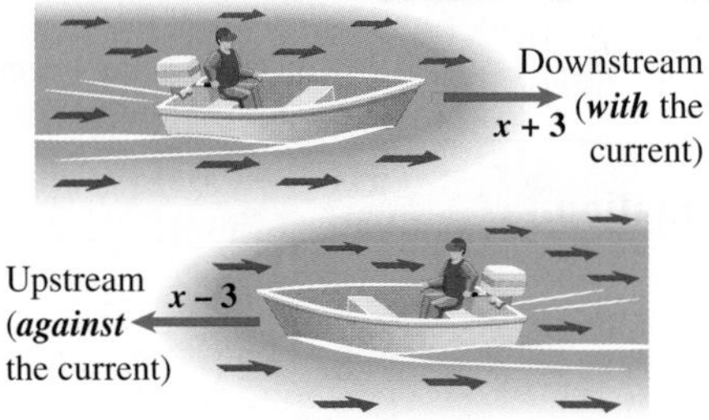

Figure 2

This information is summarized in the following table.

	d	r	t
Downstream	12	$x + 3$	
Upstream	8	$x - 3$	

Fill in the times by using the formula $t = \frac{d}{r}$.

The time downstream is the distance divided by the rate.

$$t = \frac{d}{r} = \frac{12}{x + 3} \quad \text{Time downstream}$$

The time upstream is also the distance divided by the rate.

$$t = \frac{d}{r} = \frac{8}{x - 3} \quad \text{Time upstream}$$

	d	r	t
Downstream	12	$x + 3$	$\frac{12}{x + 3}$
Upstream	8	$x - 3$	$\frac{8}{x - 3}$

Times are equal.

Step 3 **Write an equation.**

$$\frac{12}{x + 3} = \frac{8}{x - 3}$$

The time downstream equals the time upstream, so the two times from the table must be equal.

Continued on Next Page

Step 4 **Solve.** $\frac{12}{x+3} = \frac{8}{x-3}$ Equation from Step 3

$(x+3)(x-3)\frac{12}{x+3} = (x+3)(x-3)\frac{8}{x-3}$ Multiply by the LCD, $(x+3)(x-3)$.

$12(x-3) = 8(x+3)$ Divide out the common factors.

$12x - 36 = 8x + 24$ Distributive property

$4x = 60$ Subtract $8x$. Add 36.

$x = 15$ Divide by 4.

Step 5 **State the answer.** The rate of the boat in still water is 15 mph.

Step 6 **Check.** First find the rate of the boat downstream, which is $15 + 3 = 18$ mph. Divide 12 mi by 18 mph to find the time.

$$t = \frac{d}{r} = \frac{12}{18} = \frac{2}{3}\text{ hr}$$

The rate of the boat upstream is $15 - 3 = 12$ mph. Divide 8 mi by 12 mph to find the time.

$$t = \frac{d}{r} = \frac{8}{12} = \frac{2}{3}\text{ hr}$$

The time upstream equals the time downstream, as required.

Work Problem 4 at the Side. ▶

OBJECTIVE 3 Solve problems about work. Suppose that you can mow your lawn in 4 hr. Then after 1 hr, you will have mowed $\frac{1}{4}$ of the lawn. After 2 hr, you will have mowed $\frac{2}{4}$, or $\frac{1}{2}$, of the lawn, and so on. This idea is generalized as follows.

Rate of Work

If a job can be completed in t units of time, then the rate of work is

$$\frac{1}{t}\text{ job per unit of time.}$$

Problem-Solving Hint

Recall that the formula $d = rt$ says that distance traveled is equal to rate of travel multiplied by time traveled. Similarly, the fractional part of a job accomplished is equal to the rate of work multiplied by the time worked.

In the lawn mowing example, after 3 hr, the fractional part of the job done is found as follows.

$$\underbrace{\frac{1}{4}}_{\text{Rate of work}} \cdot \underbrace{3}_{\text{Time worked}} = \underbrace{\frac{3}{4}}_{\text{Fractional part of job done}}$$

After 4 hr, $\frac{1}{4}(4) = 1$ whole job has been done.

4 Solve each problem.

GS **(a)** A boat can go 10 mi against the current in the same time it can go 30 mi with the current. The current is flowing at 4 mph. Find the rate of the boat with no current.

Steps 1 and 2
Let x = the ________ of the boat with no current.

Step 3
Complete the table.

	d	r	t
Against the Current	10	____	____
With the Current	30	____	____

How are the times going with the current and against the current related?

____

Write an equation.

____

Complete Steps 4–6 to solve the problem. Give the answer.

(b) An airplane, maintaining a constant airspeed, takes as long to go 450 mi with the wind as it does to go 375 mi against the wind. If the wind is blowing at 15 mph, what is the rate of the plane?

Answers

4. (a) rate

	d	r	t
Against the Current	10	$x-4$	$\frac{10}{x-4}$
With the Current	30	$x+4$	$\frac{30}{x+4}$

They are the same; $\frac{10}{x-4} = \frac{30}{x+4}$; 8 mph

(b) 165 mph

EXAMPLE 5 Solving a Problem about Work Rates

With spraying equipment, Mateo can paint the woodwork in a small house in 8 hr. His assistant, Chet, needs 14 hr to complete the same job painting by hand. If both Mateo and Chet work together, how long will it take them to paint the woodwork?

Step 1 **Read** the problem again. We are looking for time working together.

Step 2 **Assign a variable.**

Let $x =$ the number of hours it will take for Mateo and Chet to paint the woodwork, working together.

Begin by making a table. Based on the previous discussion, Mateo's rate alone is $\frac{1}{8}$ job per hour, and Chet's rate is $\frac{1}{14}$ job per hour.

	Rate	Time Working Together	Fractional Part of the Job Done When Working Together
Mateo	$\frac{1}{8}$	x	$\frac{1}{8}x$
Chet	$\frac{1}{14}$	x	$\frac{1}{14}x$

Sum is 1 whole job.

Step 3 **Write an equation.**

$$\underbrace{\text{Fractional part done by Mateo}}_{\frac{1}{8}x} + \underbrace{\text{Fractional part done by Chet}}_{\frac{1}{14}x} = \underbrace{1 \text{ whole job}}_{1}$$

Together, Mateo and Chet complete 1 whole job. Add the fractional parts and set the sum equal to 1.

Step 4 **Solve.**

$$56\left(\frac{1}{8}x + \frac{1}{14}x\right) = 56(1) \qquad \text{Multiply by the LCD, 56.}$$

$$56\left(\frac{1}{8}x\right) + 56\left(\frac{1}{14}x\right) = 56(1) \qquad \text{Distributive property}$$

$$7x + 4x = 56 \qquad \text{Multiply.}$$

$$11x = 56 \qquad \text{Combine like terms.}$$

$$x = \frac{56}{11} \qquad \text{Divide by 11.}$$

Step 5 **State the answer.** Working together, Mateo and Chet can paint the woodwork in $\frac{56}{11}$ hr, or $5\frac{1}{11}$ hr.

Step 6 **Check.** Substitute $\frac{56}{11}$ for x in the equation from Step 3.

$$\frac{1}{8}x + \frac{1}{14}x = 1 \qquad \text{Equation from Step 3}$$

$$\frac{1}{8}\left(\frac{56}{11}\right) + \frac{1}{14}\left(\frac{56}{11}\right) \stackrel{?}{=} 1 \qquad \text{Let } x = \frac{56}{11}.$$

$$\frac{7}{11} + \frac{4}{11} = 1 \ \checkmark \ \text{True}$$

Our answer, $\frac{56}{11}$ hr, or $5\frac{1}{11}$ hr, is correct.

Problem-Solving Hint

A common error students make when solving a work problem like that in **Example 5** is to add the two times.

$$8 \text{ hr} + 14 \text{ hr} = \mathbf{22 \text{ hr}} \leftarrow \text{Incorrect answer}$$

The answer 22 hr is unreasonable, because the slower worker (Chet) can do the job *alone* in 14 hr. The correct answer *must* be less than 14 hr.

Keep in mind that the correct time for the answer must also be less than that of the *faster* worker (Mateo, at 8 hr) because he is getting help. Based on this reasoning, does our answer of $5\frac{1}{11}$ hr in **Example 5** make sense?

Work Problem 5 at the Side. ▶

An alternative approach is to consider the part of the job that can be done in 1 hr. For instance, in **Example 5** Mateo can do the entire job in 8 hr, and Chet can do it in 14 hr. Thus, their work rates, as we saw in **Example 5,** are $\frac{1}{8}$ and $\frac{1}{14}$, respectively. Since it takes them x hours to complete the job when working together, in 1 hr they can paint $\frac{1}{x}$ of the woodwork.

The amount painted by Mateo in 1 hr plus the amount painted by Chet in 1 hr must equal the amount they can do together. This leads to the following alternative equation.

Amount by Chet ↓

$$\text{Amount by Mateo} \rightarrow \frac{1}{8} + \frac{1}{14} = \frac{1}{x} \leftarrow \text{Amount together}$$

Compare this with the equation in Step 3 of **Example 5.** Multiplying each side by $56x$ leads to the same equation found in the third line of Step 4 in the example.

$$7x + 4x = 56$$

The same solution results.

5 Solve each problem.

GS **(a)** Michael can paint a room, working alone, in 8 hr. Lindsay can paint the same room, working alone, in 6 hr. How long will it take them if they work together?

Steps 1 and 2
Let $x =$ the number of ________ it will take for Michael and Lindsay to paint the room, working ________.

Step 3
Complete the table.

	Rate	Time Working Together	Fractional Part of the Job Done When Working Together
Michael	____	x	____
Lindsay	____	x	____

Together, Michael and Lindsay complete ____ whole job(s). Write an equation.

____

Complete Steps 4–6 to solve the problem. Give the answer.

(b) Roberto can detail his Camaro in 2 hr working alone. His brother Marco can do the job in 3 hr working alone. How long would it take them if they worked together?

Answers

5. (a) hours; together

	Rate	Time Working Together	Fractional Part of the Job Done When Working Together
Michael	$\frac{1}{8}$	x	$\frac{1}{8}x$
Lindsay	$\frac{1}{6}$	x	$\frac{1}{6}x$

1; $\frac{1}{8}x + \frac{1}{6}x = 1$; $\frac{24}{7}$ hr, or $3\frac{3}{7}$ hr

(b) $\frac{6}{5}$ hr, or $1\frac{1}{5}$ hr

14.7 Exercises

FOR EXTRA HELP

Download the MyDashBoard App

MyMathLab®

1. **CONCEPT CHECK** If a migrating hawk travels m mph in still air, what is its rate when it flies into a steady headwind of 5 mph? What is its rate with a tailwind of 5 mph?

2. **CONCEPT CHECK** Suppose Stephanie walks D miles at R mph in the same time that Wally walks d miles at r mph. Give an equation relating D, R, d, and r.

CONCEPT CHECK *Answer each question.*

3. If it takes Elayn 10 hr to do a job, what is her rate?

4. If it takes Clay 12 hr to do a job, how much of the job does he do in 8 hr?

GS *Use Steps 2 and 3 of the six-step problem solving method to set up an equation to use to solve each problem. (Remember that Step 1 is to read the problem carefully.) Do not actually solve the equation.* ***See Example 1.***

5. The numerator of the fraction $\frac{5}{6}$ is increased by an amount so that the value of the resulting fraction is equivalent to $\frac{13}{3}$. By what amount was the numerator increased?
   (a) Let $x =$ __________. *(Step 2)*
   (b) Write an expression for "the numerator of the fraction $\frac{5}{6}$ is increased by an amount." __________
   (c) Set up an equation to solve the problem. *(Step 3)* __________

6. If the same number is added to the numerator and subtracted from the denominator of $\frac{23}{12}$, the resulting fraction is equivalent to $\frac{3}{2}$. What is the number?
   (a) Let $x =$ __________. *(Step 2)*
   (b) Write an expression for "a number is added to the numerator of $\frac{23}{12}$." __________ Then write an expression for "the same number is subtracted from the denominator of $\frac{23}{12}$." __________
   (c) Set up an equation to solve the problem. *(Step 3)* __________

In each problem, state what x represents, write an equation, and answer the question. ***See Example 1.***

7. In a certain fraction, the denominator is 4 less than the numerator. If 3 is added to both the numerator and the denominator, the resulting fraction is equivalent to $\frac{3}{2}$. What was the original fraction?

8. In a certain fraction, the denominator is 6 more than the numerator. If 3 is added to both the numerator and the denominator, the resulting fraction is equivalent to $\frac{5}{7}$. What was the original fraction (*not* written in lowest terms)?

9. The denominator of a certain fraction is three times the numerator. If 2 is added to the numerator and subtracted from the denominator, the resulting fraction is equivalent to 1. What was the original fraction (*not* written in lowest terms)?

10. The numerator of a certain fraction is four times the denominator. If 6 is added to both the numerator and the denominator, the resulting fraction is equivalent to 2. What was the original fraction (*not* written in lowest terms)?

11. One-sixth of a number is 5 more than the same number. What is the number?

12. One-third of a number is 2 more than one-sixth of the same number. What is the number?

13. A quantity, its $\frac{3}{4}$, its $\frac{1}{2}$, and its $\frac{1}{3}$, added together, become 93. What is the quantity? (*Source: Rhind Mathematical Papyrus.*)

14. A quantity, its $\frac{2}{3}$, its $\frac{1}{2}$, and its $\frac{1}{7}$, added together, become 33. What is the quantity? (*Source: Rhind Mathematical Papyrus.*)

Solve each problem. ***See Example 2.***

15. In 2007, British explorer and endurance swimmer Lewis Gordon Pugh became the first person to swim at the North Pole. He swam 0.6 mi at 0.0319 mi per min in waters created by melted sea ice. What was his time (to three decimal places)? (*Source: The Gazette.*)

16. In the 2008 Summer Olympics, Britta Steffen of Germany won the women's 100-m freestyle swimming event. Her rate was 1.8825 m per sec. What was her time (to two decimal places)? (*Source: World Almanac and Book of Facts.*)

17. The winner of the 2012 Daytona 500 (mile) race was Matt Kenseth, who drove his Ford to victory with a rate of 140.256 mph. What was his time (to the nearest thousandth of an hour)? (*Source: World Almanac and Book of Facts.*)

18. In 2011, the late Dan Wheldon drove his Dellara-Honda Racing car to victory in the Indianapolis 500 (mile) race. His rate was 170.265 mph. What was his time (to the nearest thousandth of an hour)? (*Source: World Almanac and Book of Facts.*)

19. Tirunesh Dibaba of Ethiopia won the women's 5000-m race in the 2008 Olympics with a time of 15.911 min. What was her rate (to three decimal places)? (*Source: World Almanac and Book of Facts.*)

20. The winner of the women's 1500-m run in the 2008 Olympics was Nancy Jebet Langat of Kenya with a time of 4.004 min. What was her rate (to three decimal places)? (*Source: World Almanac and Book of Facts.*)

Complete the table and write an equation to use to solve each problem. Do not actually solve the equation. ***See Examples 3 and 4.***

21. Luvenia can row 4 mph in still water. She takes as long to row 8 mi upstream as 24 mi downstream. How fast is the current?

Let $x =$ the rate of the current.

	d	r	t
Upstream	8	$4 - x$	____
Downstream	24	$4 + x$	____

22. Julio flew his airplane 500 mi against the wind in the same time it took him to fly it 600 mi with the wind. If the rate of the wind was 10 mph, what was the average rate of his plane in still air?

Let $x =$ the rate of the plane in still air.

	d	r	t
Against the Wind	500	$x - 10$	____
With the Wind	600	$x + 10$	____

Solve each problem. ***See Examples 3 and 4.***

23. A boat can go 20 mi against a current in the same time that it can go 60 mi with the current. The rate of the current is 4 mph. Find the rate of the boat in still water.

24. A plane flies 350 mi with the wind in the same time that it can fly 310 mi against the wind. The plane has a still-air rate of 165 mph. Find the wind speed.

25. The sanderling is a small shorebird about 6.5 in. long, with a thin, dark bill and a wide, white wing stripe. If a sanderling can fly 30 mi with the wind in the same time it can fly 18 mi against the wind when the wind speed is 8 mph, what is the rate of the bird in still air? (*Source:* U.S. Geological Survey.)

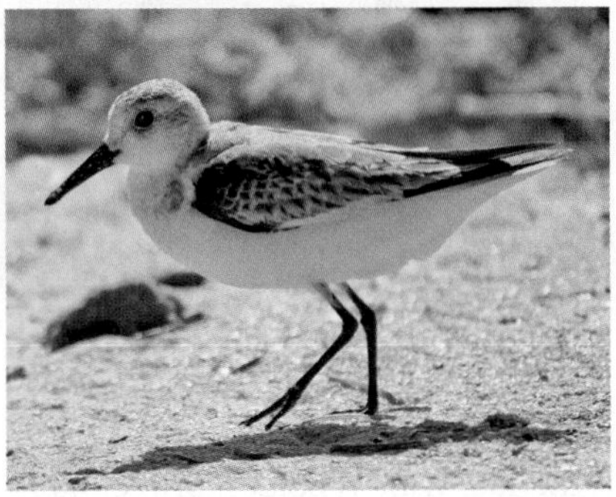

26. Airplanes usually fly faster from west to east than from east to west because the prevailing winds go from west to east. The air distance between Chicago and London is about 4000 mi, while the air distance between New York and London is about 3500 mi. If a jet can fly eastbound from Chicago to London in the same time it can fly westbound from London to New York in a 35-mph wind, what is the rate of the plane in still air? (*Source:* www.geobytes.com)

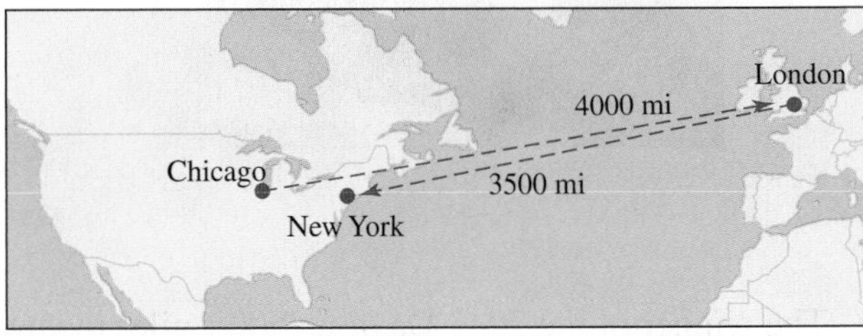

27. Perian Herring's boat goes 12 mph. Find the rate of the current of the river if she can go 6 mi upstream in the same amount of time she can go 10 mi downstream.

28. Bridget Jarvis can travel 8 mi upstream in the same time it takes her to go 12 mi downstream. Her boat goes 15 mph in still water. What is the rate of the current?

Complete the table and write an equation to use to solve each problem. Do not actually solve the equation. ***See Example 5.***

29. Eric Jensen can tune up his Chevy in 2 hr working alone. His son Oscar can do the job in 3 hr working alone. How long would it take them if they worked together? Let $x =$ the time working together.

	Rate	Time Working Together	Fractional Part of the Job Done When Working Together
Eric	____	x	____
Oscar	____	x	____

30. Working alone, Kyle can paint a room in 8 hr. Julianne can paint the same room working alone in 6 hr. How long will it take them if they work together? Let $x =$ the time working together.

	Rate	Time Working Together	Fractional Part of the Job Done When Working Together
Kyle	____	x	____
Julianne	____	x	____

Solve each problem. ***See Example 5.***

31. A high school mathematics teacher gave a geometry test. Working alone, it would take her 4 hr to grade the tests. Her student teacher would take 6 hr to grade the same tests. How long would it take them to grade these tests if they work together?

32. Geraldo and Luisa Hernandez operate a small laundry. Luisa, working alone, can clean a day's laundry in 9 hr. Geraldo can clean a day's laundry in 8 hr. How long would it take them if they work together?

33. Todd's copier can do a printing job in 7 hr. Scott's copier can do the same job in 12 hr. How long would it take to do the job using both copiers?

34. A pump can pump the water out of a flooded basement in 10 hr. A smaller pump takes 12 hr. How long would it take to pump the water from the basement using both pumps?

35. Hilda can paint a room in 6 hr. Working together with Brenda, they can paint the room in $3\frac{3}{4}$ hr. How long would it take Brenda to paint the room by herself?

36. Grant can completely mess up his room in 15 min. If his cousin Wade helps him, they can completely mess up the room in $8\frac{4}{7}$ min. How long would it take Wade to mess up the room by himself?

37. An inlet pipe can fill a swimming pool in 9 hr, and an outlet pipe can empty the pool in 12 hr. Through an error, both pipes are left open. How long will it take to fill the pool?

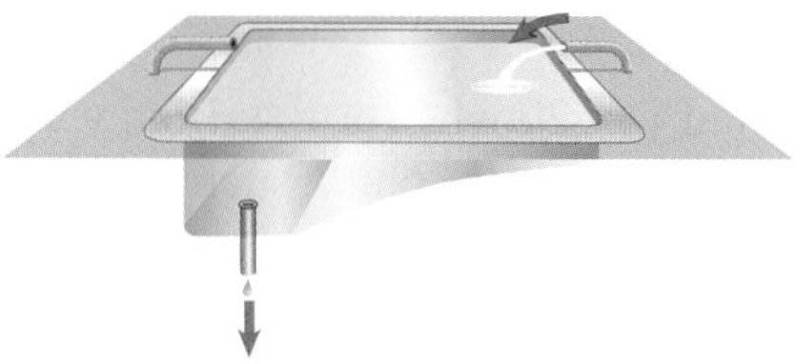

38. One pipe can fill a swimming pool in 6 hr, and another pipe can do it in 9 hr. How long will it take the two pipes working together to fill the pool $\frac{3}{4}$ full?

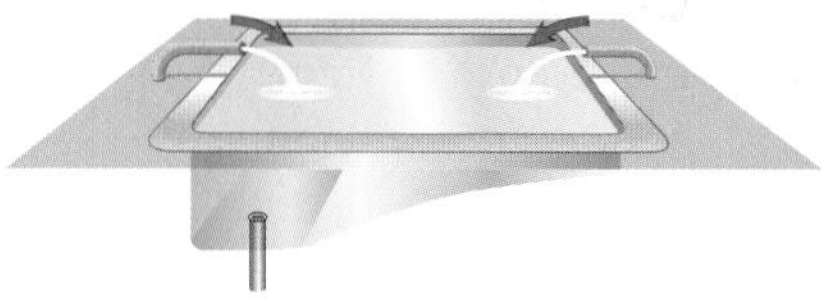

Solve each problem. (*Source:* Mary Jo Boyer, *Math for Nurses,* Wolter Kluwer.)

39. Nurses use Young's Rule to calculate the pediatric (child's) dose P of a medication, given a child's age c in years and a normal adult dose a. This rule applies to children ages 1 to 12 yr old.

$$P = \frac{c}{c + 12} \cdot a$$

The normal adult dose for milk of magnesia is 30 mL. Use Young's Rule to calculate the correct dose to give a 6-yr-old boy.

40. Nurses use Clark's Rule to calculate the pediatric (child's) dose P of a medication, given a child's weight w in pounds and a normal adult dose a. This rule applies to children ages 2 yr or older.

$$P = \frac{w}{150} \cdot a$$

The normal adult dose for ibuprofen is 200 mg. Use Clark's Rule to calculate the correct dose to give a 4-yr-old girl who weighs 30 lb.

14.8 Variation

OBJECTIVES

1. Solve direct variation problems.
2. Solve inverse variation problems.

1 Solve each problem.

(a) If z varies directly as t, and $z = 11$ when $t = 4$, find z when $t = 32$.

GS **(b)** The circumference of a circle varies directly as the radius. A circle with a radius of 7 cm has a circumference of 43.96 cm. Find the circumference if the radius is 11 cm.

Let C = the ________ and r = the ________.

$C = k \cdot$ ____

$43.96 = k \cdot$ ____

$k =$ ____

The variation equation for this problem is $C =$ ____ r. Now find the value of C when $r =$ ____.
If the radius is 11 cm, the circumference is ____.

Answers

1. (a) 88
(b) circumference; radius; r; 7; 6.28; 6.28; 11; 69.08 cm

OBJECTIVE 1 Solve direct variation problems. Suppose that gasoline costs \$4.50 per gal. Then 1 gal costs \$4.50, 2 gal cost 2(\$4.50) = \$9.00, 3 gal cost 3(\$4.50) = \$13.50, and so on. Each time, the total cost is obtained by multiplying the number of gallons by the price per gallon. In general, if k equals the price per gallon and x equals the number of gallons, then the total cost y is equal to kx.

As the *number of gallons* **increases,** the *total cost* **increases**.

The following is also true.

As the *number of gallons* **decreases,** the *total cost* **decreases.**

The preceding discussion is an example of *variation*. ***Two variables vary directly if one is a constant multiple of the other.***

Direct Variation

y* varies directly as *x if there is a constant k such that the following is true.

$$y = kx$$

Also, y is said to be *proportional to x*. The constant k in the equation for direct variation is a numerical value, such as 4.50 in the gasoline price discussion. This value is the **constant of variation.**

EXAMPLE 1 Using Direct Variation

Suppose y varies directly as x, and $y = 20$ when $x = 4$. Find y when $x = 9$.

Since y varies directly as x, there is a constant k such that $y = kx$. We let $y = 20$ and $x = 4$ in this equation and solve for k.

$y = kx$ Equation for direct variation

$20 = k \cdot 4$ Substitute the given values.

$k = 5 \longleftarrow$ Constant of variation

Since $y = kx$ and $k = 5$, we have the following.

$y = 5x$ Let $k = 5$.

Now we can find the value of y when $x = 9$.

$y = 5x = 5 \cdot 9 = 45$ Let $x = 9$.

Thus, $y = 45$ when $x = 9$.

◀ Work Problem 1 at the Side.

OBJECTIVE 2 Solve inverse variation problems. In direct variation, where $k > 0$, as x increases, y increases. Similarly, as x decreases, y decreases. Another type of variation is *inverse variation.* With inverse variation, where $k > 0$,

As one variable ***increases,*** the other variable ***decreases.***

For example, in a closed space, volume decreases as pressure increases, as illustrated by a trash compactor. See **Figure 3.** As the compactor presses down, the pressure on the trash increases. In turn, the trash occupies a smaller space.

Inverse Variation

y* varies inversely as *x if there exists a constant k such that the following is true.

$$y = \frac{k}{x}$$

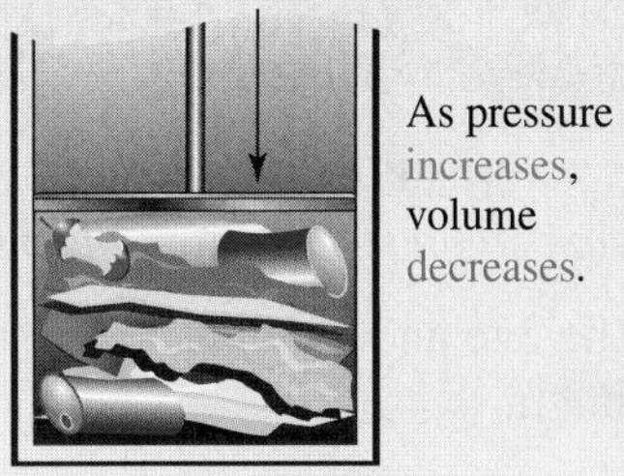
As pressure increases, volume decreases.

Figure 3

EXAMPLE 2 Using Inverse Variation

Suppose y varies inversely as x, and $y = 3$ when $x = 8$. Find y when $x = 6$.

Since y varies inversely as x, there is a constant k such that $y = \frac{k}{x}$. We know that $y = 3$ when $x = 8$, so we can find k.

$$y = \frac{k}{x} \quad \text{Equation for inverse variation}$$

$$3 = \frac{k}{8} \quad \text{Substitute the given values.}$$

$$k = 24 \longleftarrow \text{Constant of variation}$$

Since $y = \frac{24}{x}$, we let $x = 6$ and solve for y.

$$y = \frac{24}{x} = \frac{24}{6} = 4 \quad \text{Let } x = 6.$$

Therefore, when $x = 6$, $y = 4$.

Work Problem 2 at the Side. ▶

EXAMPLE 3 Using Inverse Variation

In the manufacturing of a certain medical syringe, the cost of producing the syringe varies inversely as the number produced. If 10,000 syringes are produced, the cost is \$2 per unit. Find the cost per unit to produce 25,000 syringes.

Let x = the number of syringes produced

and c = the cost per unit.

Since c varies inversely as x, there is a constant k such that $c = \frac{k}{x}$.

$$c = \frac{k}{x} \quad \text{Equation for inverse variation}$$

$$2 = \frac{k}{10{,}000} \quad \text{Substitute the given values.}$$

$$k = 20{,}000 \quad \text{Multiply by 10,000. Rewrite.}$$

Now use $c = \frac{k}{x}$ to find the value of c when $x = 25{,}000$.

$$c = \frac{20{,}000}{25{,}000} = 0.80 \quad \text{Let } k = 20{,}000 \text{ and } x = 25{,}000.$$

The cost per syringe to make 25,000 syringes is \$0.80.

Work Problem 3 at the Side. ▶

2 Solve the problem.

Suppose z varies inversely as t, and $z = 8$ when $t = 2$. Find z when $t = 32$.

3 Solve the problem.

If the cost of producing pairs of rubber gloves varies inversely as the number of pairs produced, and 5000 pairs can be produced for \$0.50 per pair, how much will it cost per pair to produce 10,000 pairs?

Answers

2. $\frac{1}{2}$

3. \$0.25

14.8 Exercises

FOR EXTRA HELP

Download the MyDashBoard App

MyMathLab®

CONCEPT CHECK *Use personal experience or intuition to determine whether the situation suggests direct or inverse variation.**

1. The rate and the distance traveled by a pickup truck in 3 hr

2. The number of different lottery tickets you buy and your probability of winning that lottery

3. The number of days from now until December 25 and the magnitude of the frenzy of Christmas shopping

4. Your age and the probability that you believe in Santa Claus

5. The amount of gasoline that you pump and the amount of empty space left in your tank

6. The surface area of a balloon and its diameter

7. The amount of pressure put on the accelerator of a car and the rate of the car

8. The number of days until the end of the baseball season and the number of home runs that Jose Bautista has hit during the season

9. **CONCEPT CHECK** Make the correct choice.

(a) If the constant of variation is positive and y varies directly as x, then as x increases, y (*increases / decreases*).

(b) If the constant of variation is positive and y varies inversely as x, then as x increases, y (*increases / decreases*).

10. Bill Veeck was the owner of several major league baseball teams in the 1950s and 1960s. He was known to often sit in the stands and enjoy games with his paying customers. Here is a quote attributed to him:

"*I have discovered in 20 years of moving around a ballpark, that the knowledge of the game is usually in inverse proportion to the price of the seats.*"

Explain in your own words the meaning of this statement. (To prove his point, Veeck once allowed the fans to vote on managerial decisions.)

CONCEPT CHECK *Determine whether each equation represents* direct *or* inverse *variation.*

11. $y = \frac{3}{x}$

12. $y = \frac{5}{x}$

13. $y = 50x$

14. $y = 200x$

*The authors thank Linda Kodama for suggesting these exercises.

Solve each problem involving direct variation. **See Example 1.**

15. If z varies directly as x, and $z = 30$ when $x = 8$, find z when $x = 4$.

16. If y varies directly as x, and $x = 27$ when $y = 6$, find x when $y = 2$.

17. If d varies directly as r, and $d = 200$ when $r = 40$, find d when $r = 60$.

18. If d varies directly as t, and $d = 150$ when $t = 3$, find d when $t = 5$.

Solve each problem involving inverse variation. **See Example 2.**

19. If z varies inversely as x, and $z = 50$ when $x = 2$, find z when $x = 25$.

20. If x varies inversely as y, and $x = 3$ when $y = 8$, find y when $x = 4$.

21. If m varies inversely as r, and $m = 12$ when $r = 8$, find m when $r = 16$.

22. If p varies inversely as q, and $p = 7$ when $q = 6$, find p when $q = 2$.

Solve each variation problem. **See Examples 1–3.**

23. For a given base, the area of a triangle varies directly as its height. Find the area of a triangle with a height of 6 in., if the area is 10 in.^{2} when the height is 4 in.

24. The interest on an investment varies directly as the rate of interest. If the interest is \$51 when the interest rate is 4%, find the interest when the rate is 3.2%.

25. Hooke's law for an elastic spring states that the distance a spring stretches varies directly with the force applied. If a force of 75 lb stretches a certain spring 16 in., how much will a force of 200 lb stretch the spring?

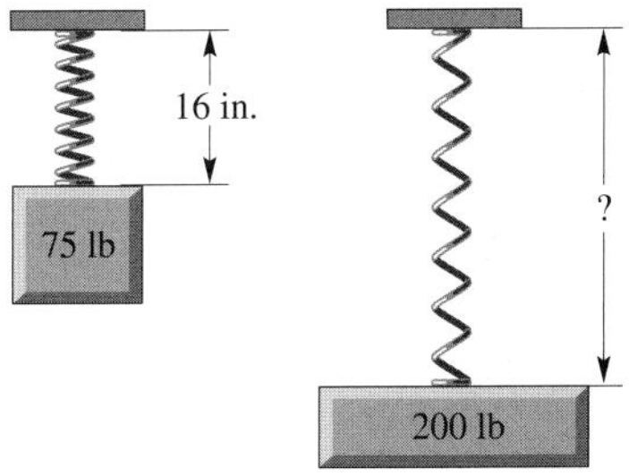

26. The pressure exerted by water at a given point varies directly with the depth of the point beneath the surface of the water. Water exerts 4.34 lb per in.^{2} for every 10 ft traveled below the water's surface. What is the pressure exerted on a scuba diver at 20 ft?

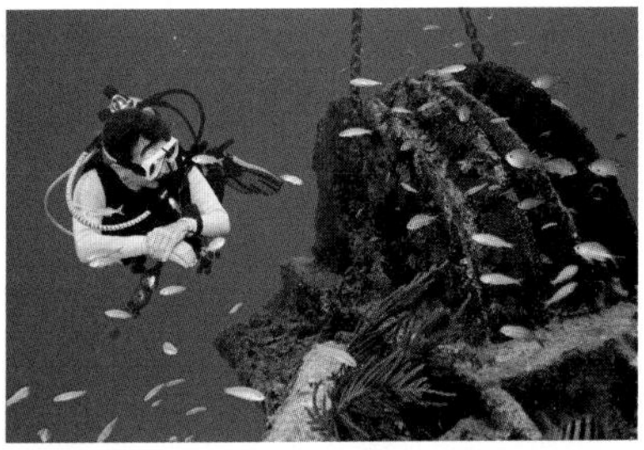

27. For a constant area, the length of a rectangle varies inversely as the width. The length of a rectangle is 27 ft when the width is 10 ft. Find the width of a rectangle with the same area if the length is 18 ft.

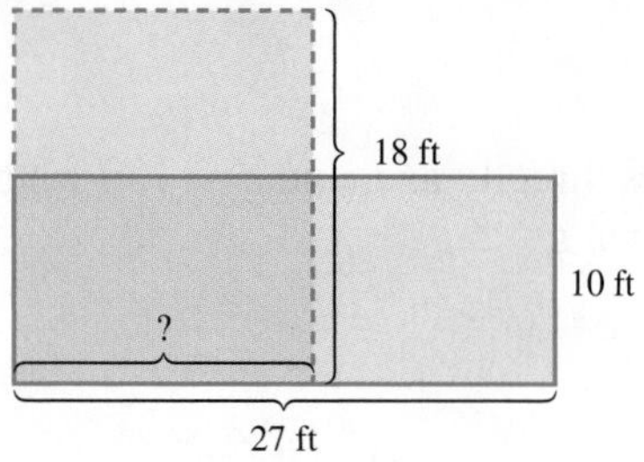

28. Over a specified distance, rate varies inversely with time. If a Dodge Viper on a test track goes a certain distance in one-half minute at 160 mph, what rate is needed to go the same distance in three-fourths minute?

29. If the temperature is constant, the pressure of a gas in a container varies inversely as the volume of the container. If the pressure is 10 lb per ft^2 in a container with volume 3 ft^3, what is the pressure in a container with volume 1.5 ft^3?

30. The current in a simple electrical circuit varies inversely as the resistance. If the current is 20 amps when the resistance is 5 ohms, find the current when the resistance is 8 ohms.

31. In the inversion of raw sugar, the rate of change of the amount of raw sugar varies directly as the amount of raw sugar remaining. The rate is 200 kg per hr when there are 800 kg left. What is the rate of change per hour when only 100 kg are left?

32. The force required to compress a spring varies directly as the change in the length of the spring. If a force of 12 lb is required to compress a certain spring 3 in., how much force is required to compress the spring 5 in.?

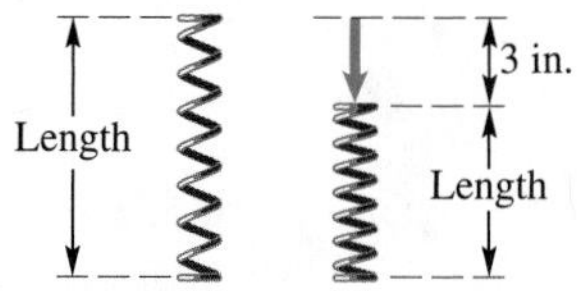

33. In rectangles of constant area, length and width vary inversely. When the length is 24, the width is 2. What is the width when the length is 12?

34. The speed of a pulley varies inversely as its diameter. One kind of pulley, with diameter 6 in., turns at 150 revolutions per minute. Find the number of revolutions per minute for a similar pulley with diameter 10 in.

Chapter 14 Summary

Key Terms

14.1

rational expression The quotient of two polynomials with denominator not 0 is a rational expression.

lowest terms A rational expression is written in lowest terms if the greatest common factor of its numerator and denominator is 1.

14.3

least common denominator (LCD) The simplest expression that is divisible by all denominators is the least common denominator.

14.5

complex fraction A rational expression with one or more fractions in the numerator, denominator, or both, is a complex fraction.

14.6

proposed solution A value of the variable that appears to be a solution after both sides of an equation with rational expressions are multiplied by a variable expression is a proposed solution.

extraneous solution (extraneous value) A proposed solution that is not an actual solution of a given equation is an extraneous solution, or extraneous value.

14.8

direct variation y varies directly as x if there is a constant k such that $y = kx$.

constant of variation In the equation $y = kx$ or $y = \frac{k}{x}$, the number k is the constant of variation.

inverse variation y varies inversely as x if there is a constant k such that $y = \frac{k}{x}$.

Test Your Word Power

See how well you have learned the vocabulary in this chapter.

1. A **rational expression** is
   A. an algebraic expression made up of a term or the sum of a finite number of terms with real coefficients and whole number exponents
   B. a polynomial equation of degree 2
   C. an expression with one or more fractions in the numerator, denominator, or both
   D. the quotient of two polynomials with denominator not 0.

2. A **complex fraction** is
   A. an algebraic expression made up of a term or the sum of a finite number of terms with real coefficients and whole number exponents
   B. a polynomial equation of degree 2
   C. a rational expression with one or more fractions in the numerator, denominator, or both
   D. the quotient of two polynomials with denominator not 0.

3. If two positive quantities x and y are in **direct variation,** and the constant of variation is positive, then
   A. as x increases, y decreases
   B. as x increases, y increases
   C. as x increases, y remains constant
   D. as x decreases, y remains constant.

4. If two positive quantities x and y are in **inverse variation,** and the constant of variation is positive, then
   A. as x increases, y decreases
   B. as x increases, y increases
   C. as x increases, y remains constant
   D. as x decreases, y remains constant.

Answers to Test Your Word Power

1. D; *Examples:* $-\frac{3}{4y}$, $\frac{5x^3}{x+2}$, $\frac{a+3}{a^2-4a-5}$

2. C; *Examples:* $\frac{\frac{2}{3}}{\frac{4}{7}}$, $\frac{x-\frac{1}{y}}{x+\frac{1}{y}}$, $\frac{\frac{2}{a+1}}{a^2-1}$

3. B; *Example:* The equation $y = 3x$ represents direct variation. When $x = 2$, $y = 6$. If x increases to 3, then y increases to $3(3) = 9$.

4. A; *Example:* The equation $y = \frac{3}{x}$ represents inverse variation. When $x = 1$, $y = 3$. If x increases to 2, then y decreases to $\frac{3}{2}$, or $1\frac{1}{2}$.

Quick Review

Concepts	Examples

14.1 The Fundamental Property of Rational Expressions

To find the value(s) for which a rational expression is undefined, set the denominator equal to 0 and solve the equation.

Find the values for which $\frac{x-4}{x^2-16}$ is undefined.

$$x^2 - 16 = 0$$

$$(x-4)(x+4) = 0 \quad \text{Factor.}$$

$$x - 4 = 0 \quad \text{or} \quad x + 4 = 0 \quad \text{Zero-factor property}$$

$$x = 4 \quad \text{or} \quad x = -4 \quad \text{Solve for } x.$$

The rational expression is undefined for 4 and -4, so $x \neq 4$ and $x \neq -4$.

Writing a Rational Expression in Lowest Terms

Step 1 Factor the numerator and denominator.

Step 2 Use the fundamental property to divide out common factors from the numerator and denominator.

Write the expression in lowest terms.

$$\frac{x^2-1}{(x-1)^2}$$

$$= \frac{(x-1)(x+1)}{(x-1)(x-1)} \quad \text{Factor.}$$

$$= \frac{x+1}{x-1} \quad \text{Lowest terms}$$

There are often several different equivalent forms of a rational expression.

Give four equivalent forms of $-\frac{x-1}{x+2}$.

$$① \rightarrow \frac{-(x-1)}{x+2}, \quad \text{or} \quad \frac{-x+1}{x+2} \leftarrow ② \quad \text{Distribute the negative sign in the numerator.}$$

$$③ \rightarrow \frac{x-1}{-(x+2)}, \quad \text{or} \quad \frac{x-1}{-x-2} \leftarrow ④ \quad \text{Distribute the negative sign in the denominator.}$$

14.2 Multiplying and Dividing Rational Expressions

Multiplying or Dividing Rational Expressions

Step 1 Note the operation. If the operation is division, use the definition of division to rewrite as multiplication.

Step 2 Multiply numerators and multiply denominators.

Step 3 Factor numerators and denominators completely.

Step 4 Write in lowest terms, using the fundamental property.

Steps 2 and 3 may be interchanged based on personal preference.

Multiply. $\frac{3x+9}{x-5} \cdot \frac{x^2-3x-10}{x^2-9}$

$$= \frac{(3x+9)(x^2-3x-10)}{(x-5)(x^2-9)} \quad \text{Multiply numerators and denominators.}$$

$$= \frac{3(x+3)(x-5)(x+2)}{(x-5)(x+3)(x-3)} \quad \text{Factor.}$$

$$= \frac{3(x+2)}{x-3} \quad \text{Lowest terms}$$

Divide. $\frac{2x+1}{x+5} \div \frac{6x^2-x-2}{x^2-25}$

$$= \frac{(2x+1)(x^2-25)}{(x+5)(6x^2-x-2)} \quad \text{Multiply by the reciprocal of the divisor.}$$

$$= \frac{(2x+1)(x+5)(x-5)}{(x+5)(2x+1)(3x-2)} \quad \text{Factor.}$$

$$= \frac{x-5}{3x-2} \quad \text{Lowest terms}$$

Concepts	Examples

14.3 Least Common Denominators

Finding the LCD

Step 1 Factor each denominator into prime factors.

Step 2 List each different factor the greatest number of times it appears.

Step 3 Multiply the factors from Step 2 to get the LCD.

Find the LCD for $\frac{3}{k^2 - 8k + 16}$ and $\frac{1}{4k^2 - 16k}$.

$$\left.\begin{aligned} k^2 - 8k + 16 &= (k-4)^2 \\ 4k^2 - 16k &= 4k(k-4) \end{aligned}\right\} \text{Factor each denominator.}$$

$$\text{LCD} = (k-4)^2 \cdot 4 \cdot k$$
$$= 4k(k-4)^2$$

Writing Equivalent Rational Expressions

Step 1 Factor both denominators.

Step 2 Decide what factors the denominator must be multiplied by to equal the specified denominator.

Step 3 Multiply the rational expression by that factor divided by itself. (That is, multiply by 1.)

Find the numerator. $\frac{5}{2z^2 - 6z} = \frac{?}{4z^3 - 12z^2}$

$$\frac{5}{2z(z-3)} = \frac{?}{4z^2(z-3)}$$

$2z(z-3)$ must be multiplied by $2z$.

$$\frac{5}{2z(z-3)} \cdot \frac{2z}{2z}$$

$$= \frac{10z}{4z^2(z-3)}, \quad \text{or} \quad \frac{10z}{4z^3 - 12z^2}$$

14.4 Adding and Subtracting Rational Expressions

Adding Rational Expressions

Step 1 Find the LCD.

Step 2 Rewrite each rational expression with the LCD as denominator.

Step 3 Add the numerators to get the numerator of the sum. The LCD is the denominator of the sum.

Step 4 Write in lowest terms.

Add. $\frac{2}{3m+6} + \frac{m}{m^2 - 4}$

$$\left.\begin{aligned} 3m + 6 &= 3(m+2) \\ m^2 - 4 &= (m+2)(m-2) \end{aligned}\right\} \text{The LCD is } 3(m+2)(m-2).$$

$$= \frac{2(m-2)}{3(m+2)(m-2)} + \frac{3m}{3(m+2)(m-2)}$$ Write with the LCD.

$$= \frac{2m - 4 + 3m}{3(m+2)(m-2)}$$ Add numerators. Keep the same denominator.

$$= \frac{5m - 4}{3(m+2)(m-2)}$$ Combine like terms.

Subtracting Rational Expressions

Follow the same steps as for addition, but subtract in Step 3.

Subtract. $\frac{6}{k+4} - \frac{2}{k}$ The LCD is $k(k+4)$.

$$= \frac{6k}{(k+4)k} - \frac{2(k+4)}{k(k+4)}$$ Write with the LCD.

$$= \frac{6k - 2(k+4)}{k(k+4)}$$ Subtract numerators. Keep the same denominator.

$$= \frac{6k - 2k - 8}{k(k+4)}$$ Distributive property

$$= \frac{4k - 8}{k(k+4)}$$ Combine like terms.

Concepts	Examples

14.5 Complex Fractions

Simplifying Complex Fractions

Method 1 Simplify the numerator and denominator separately. Then divide the simplified numerator by the simplified denominator.

Method 2 Multiply the numerator and denominator of the complex fraction by the LCD of all the fractions in the numerator and denominator of the complex fraction. Write in lowest terms.

Simplify.

Method 1

$$\frac{\frac{1}{a}-a}{1-a}=\frac{\frac{1}{a}-\frac{a^2}{a}}{1-a}=\frac{\frac{1-a^2}{a}}{1-a}=\frac{1-a^2}{a}\div(1-a)=\frac{1-a^2}{a}\cdot\frac{1}{1-a}=\frac{(1-a)(1+a)}{a(1-a)}=\frac{1+a}{a}$$

Method 2

$$\frac{\frac{1}{a}-a}{1-a}=\frac{\left(\frac{1}{a}-a\right)a}{(1-a)a}=\frac{\frac{a}{a}-a^2}{(1-a)a}=\frac{1-a^2}{(1-a)a}=\frac{(1+a)(1-a)}{(1-a)a}=\frac{1+a}{a}$$

14.6 Solving Equations with Rational Expressions

Solving Equations with Rational Expressions

Step 1 Multiply each side of the equation by the LCD. (This clears the equation of fractions.)

Solve $\frac{x}{x-3}+\frac{4}{x+3}=\frac{18}{x^2-9}$.

$$\frac{x}{x-3}+\frac{4}{x+3}=\frac{18}{(x-3)(x+3)} \qquad \text{Factor.}$$

The LCD is $(x-3)(x+3)$. Note that 3 and -3 cannot be solutions, as they cause a denominator to equal 0.

$$(x-3)(x+3)\left(\frac{x}{x-3}+\frac{4}{x+3}\right)=(x-3)(x+3)\frac{18}{(x-3)(x+3)}$$

Multiply by the LCD.

Step 2 Solve the resulting equation.

$$x(x+3)+4(x-3)=18 \qquad \text{Distributive property}$$

$$x^2+3x+4x-12=18 \qquad \text{Distributive property}$$

$$x^2+7x-30=0 \qquad \text{Standard form}$$

$$(x-3)(x+10)=0 \qquad \text{Factor.}$$

$$x-3=0 \quad \text{or} \quad x+10=0 \qquad \text{Zero-factor property}$$

Reject → $x=3$ or $x=-10$ Solve for x.

Step 3 Check each proposed solution by substituting it in the original equation. Reject any value that causes an original denominator to equal 0.

Since 3 causes denominators to equal 0, the only solution is -10. Thus, $\{-10\}$ is the solution set.

Concepts	Examples

14.7 Applications of Rational Expressions

Solving Problems about Distance, Rate, and Time

Use the formulas relating d, r, and t.

$$d = rt, \quad r = \frac{d}{t}, \quad t = \frac{d}{r}$$

Solving Problems about Work

Step 1 **Read** the problem carefully.

Step 2 **Assign a variable.** State what the variable represents. Put the information from the problem in a table. If a job is done in t units of time, then the rate is $\frac{1}{t}$.

Step 3 **Write an equation.** The sum of the fractional parts should equal 1 (whole job).

Step 4 **Solve** the equation.

Step 5 **State the answer.**

Step 6 **Check.**

It takes the regular mail carrier 6 hr to cover her route. A substitute takes 8 hr to cover the same route. How long would it take them to cover the route together?

Let $x =$ the number of hours to cover the route together.

The rate of the regular carrier is $\frac{1}{6}$ job per hour, and the rate of the substitute is $\frac{1}{8}$ job per hour. Multiply rate by time to get the fractional part of the job done.

	Rate	Time	Part of the Job Done
Regular	$\frac{1}{6}$	x	$\frac{1}{6}x$
Substitute	$\frac{1}{8}$	x	$\frac{1}{8}x$

$\frac{1}{6}x + \frac{1}{8}x = 1$ — The parts add to 1 whole job.

$24\left(\frac{1}{6}x + \frac{1}{8}x\right) = 24(1)$ — The LCD is 24.

$4x + 3x = 24$ — Distributive property

$7x = 24$ — Combine like terms.

$x = \frac{24}{7}$ — Divide by 7.

It would take them $\frac{24}{7}$ hr, or $3\frac{3}{7}$ hr, to cover the route together. The solution checks because $\frac{1}{6}\left(\frac{24}{7}\right) + \frac{1}{8}\left(\frac{24}{7}\right) = 1$.

14.8 Variation

Solving Variation Problems

Step 1 Write the variation equation. Use

$y = kx$ — Direct variation

or $y = \frac{k}{x}$. — Inverse variation

Step 2 Find k by substituting the given values of x and y into the equation.

Step 3 Write the equation with the value of k from Step 2 and the given value of x or y. Solve for the remaining variable.

If y varies inversely as x, and $y = 4$ when $x = 9$, find y when $x = 6$.

$y = \frac{k}{x}$ — Equation for inverse variation

$4 = \frac{k}{9}$ — Substitute given values.

$k = 36$ — Solve for k.

$y = \frac{36}{x}$ — Let $k = 36$.

$y = \frac{36}{6}$, or 6 — Let $x = 6$.

Chapter 14 Review Exercises

14.1 *Find the value(s) of the variable for which each rational expression is undefined. Write answers with the symbol $\neq$.*

1. $\dfrac{4}{x-3}$

2. $\dfrac{x+3}{2x}$

3. $\dfrac{m-2}{m^2-2m-3}$

4. $\dfrac{2k+1}{3k^2+17k+10}$

Find the numerical value of each rational expression for **(a)** $x=-2$ *and* **(b)** $x=4$.

5. $\dfrac{x^2}{x-5}$

6. $\dfrac{4x-3}{5x+2}$

7. $\dfrac{3x}{x^2-4}$

8. $\dfrac{x-1}{x+2}$

Write each rational expression in lowest terms.

9. $\dfrac{5a^3b^3}{15a^4b^2}$

10. $\dfrac{m-4}{4-m}$

11. $\dfrac{4x^2-9}{6-4x}$

12. $\dfrac{4p^2+8pq-5q^2}{10p^2-3pq-q^2}$

Write four equivalent expressions for each fraction.

13. $-\dfrac{4x-9}{2x+3}$

14. $-\dfrac{8-3x}{3-6x}$

14.2 *Multiply or divide. Write each answer in lowest terms.*

15. $\dfrac{8x^2}{12x^5}\cdot\dfrac{6x^4}{2x}$

16. $\dfrac{9m^2}{(3m)^4}\div\dfrac{6m^5}{36m}$

17. $\dfrac{x-3}{4}\cdot\dfrac{5}{2x-6}$

18. $\dfrac{2r+3}{r-4}\cdot\dfrac{r^2-16}{6r+9}$

19. $\dfrac{3q+3}{5-6q}\div\dfrac{4q+4}{2(5-6q)}$

20. $\dfrac{y^2-6y+8}{y^2+3y-18}\div\dfrac{y-4}{y+6}$

21. $\dfrac{2p^2+13p+20}{p^2+p-12}\cdot\dfrac{p^2+2p-15}{2p^2+7p+5}$

22. $\dfrac{3z^2+5z-2}{9z^2-1}\cdot\dfrac{9z^2+6z+1}{z^2+5z+6}$

14.3 *Find the least common denominator for the fractions in each list.*

23. $\frac{1}{8}, \frac{5}{12}, \frac{7}{32}$

24. $\frac{4}{9y}, \frac{7}{12y^2}, \frac{5}{27y^4}$

25. $\frac{1}{m^2 + 2m}, \frac{4}{m^2 + 7m + 10}$

26. $\frac{3}{x^2 + 4x + 3}, \frac{5}{x^2 + 5x + 4}, \frac{2}{x^2 + 7x + 12}$

Write each rational expression as an equivalent expression with the indicated denominator.

27. $\frac{5}{8} = \frac{?}{56}$

28. $\frac{10}{k} = \frac{?}{4k}$

29. $\frac{3}{2a^3} = \frac{?}{10a^4}$

30. $\frac{9}{x - 3} = \frac{?}{18 - 6x}$

31. $\frac{-3y}{2y - 10} = \frac{?}{50 - 10y}$

32. $\frac{4b}{b^2 + 2b - 3} = \frac{?}{(b + 3)(b - 1)(b + 2)}$

14.4 *Add or subtract. Write each answer in lowest terms.*

33. $\frac{10}{x} + \frac{5}{x}$

34. $\frac{6}{3p} - \frac{12}{3p}$

35. $\frac{9}{k} - \frac{5}{k - 5}$

36. $\frac{4}{y} + \frac{7}{7 + y}$

37. $\frac{m}{3} - \frac{2 + 5m}{6}$

38. $\frac{12}{x^2} - \frac{3}{4x}$

39. $\frac{5}{a - 2b} + \frac{2}{a + 2b}$

40. $\frac{4}{k^2 - 9} - \frac{k + 3}{3k - 9}$

41. $\frac{8}{z^2 + 6z} - \frac{3}{z^2 + 4z - 12}$

42. $\frac{11}{2p - p^2} - \frac{2}{p^2 - 5p + 6}$

14.5 *Simplify each complex fraction.*

43. $\dfrac{\frac{a^4}{b^2}}{\frac{a^3}{b}}$

44. $\dfrac{\frac{y - 3}{y}}{\frac{y + 3}{4y}}$

45. $\dfrac{\frac{3m + 2}{m}}{\frac{2m - 5}{6m}}$

46. $\dfrac{\frac{1}{p} - \frac{1}{q}}{\frac{1}{q - p}}$

47. $\dfrac{x + \frac{1}{w}}{x - \frac{1}{w}}$

48. $\dfrac{\frac{1}{r + t} - 1}{\frac{1}{r + t} + 1}$

14.6 *Solve each equation. Check the solutions.*

49. $\frac{k}{5} - \frac{2}{3} = \frac{1}{2}$

50. $\frac{4-z}{z} + \frac{3}{2} = \frac{-4}{z}$

51. $\frac{x}{2} - \frac{x-3}{7} = -1$

52. $\frac{3y-1}{y-2} = \frac{5}{y-2} + 1$

53. $\frac{3}{m-2} + \frac{1}{m-1} = \frac{7}{m^2-3m+2}$

Solve for the specified variable.

54. $m = \frac{Ry}{t}$ for t

55. $x = \frac{3y-5}{4}$ for y

56. $\frac{1}{r} - \frac{1}{s} = \frac{1}{t}$ for t

14.7 *Solve each problem.*

57. In a certain fraction, the denominator is 5 less than the numerator. If 5 is added to both the numerator and the denominator, the resulting fraction is equivalent to $\frac{5}{4}$. Find the original fraction (*not* written in lowest terms).

58. The denominator of a certain fraction is six times the numerator. If 3 is added to the numerator and subtracted from the denominator, the resulting fraction is equivalent to $\frac{2}{5}$. Find the original fraction (*not* written in lowest terms).

59. On June 25, 2011, Marco Andretti won the fifth Iowa Corn Indy 250. He drove a Dallara-Honda the 218.75 mi distance with an average rate of 118.671 mph. What was his time (to the nearest thousandth of an hour)? (*Source:* www.indycar.com)

60. In the 2010 Winter Olympics in Vancouver, Canada, Sven Kramer of the Netherlands won the men's 5000-m speed skating event in 6.243 min. What was his rate (to three decimal places)? (*Source: World Almanac and Book of Facts.*)

61. Zachary and Samuel are brothers who share a bedroom. By himself, Zachary can completely mess up their room in 20 min, while it would take Samuel only 12 min to do the same thing. How long would it take them to mess up the room together?

62. A man can plant his garden in 5 hr, working alone. His daughter can do the same job in 8 hr. How long would it take them if they worked together?

14.8 *Solve each problem.*

63. If y varies directly as x, and $x = 12$ when $y = 5$, find x when $y = 3$.

64. If a parallelogram has a fixed area, the height varies inversely as the base. A parallelogram has a height of 8 cm and a base of 12 cm. Find the height if the base is changed to 24 cm.

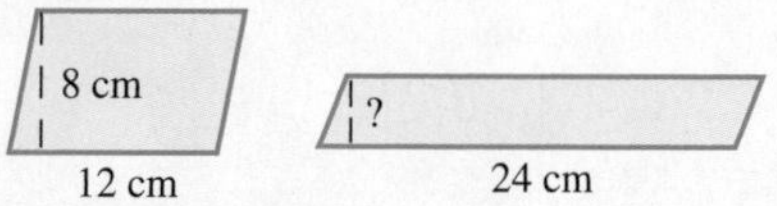

Mixed Review Exercises

Perform the indicated operations. Write each answer in lowest terms.

65. $\frac{4}{m-1} - \frac{3}{m+1}$

66. $\frac{8p^5}{5} \div \frac{2p^3}{10}$

67. $\frac{r-3}{8} \div \frac{3r-9}{4}$

68. $\frac{\frac{5}{x} - 1}{\frac{5-x}{3x}}$

69. $\frac{4}{z^2 - 2z + 1} - \frac{3}{z^2 - 1}$

70. $\frac{2x^2 + 5x - 12}{4x^2 - 9} \cdot \frac{x^2 - 3x - 28}{x^2 + 8x + 16}$

Solve each equation, and check the solutions.

71. $\frac{5t}{6} = \frac{2t-1}{3} + 1$

72. $\frac{2}{z} - \frac{z}{z+3} = \frac{1}{z+3}$

73. $\frac{2x}{x^2 - 16} - \frac{2}{x-4} = \frac{4}{x+4}$

74. Solve $a = \frac{v - w}{t}$ for w.

Solve each problem.

75. If the same number is added to both the numerator and denominator of the fraction $\frac{4}{11}$, the result is equivalent to $\frac{1}{2}$. Find the number.

76. Working together, Seema and Satish can clean their house in 3 hr. Working alone, it takes Seema 5 hr to clean the house. How long would it take Satish to clean the house by himself?

77. Anne Kelly flew her plane 400 km with the wind in the same time it took her to go 200 km against the wind. The wind speed is 50 km per hr. Find the rate of the plane in still air. Use x as the variable.

	d	r	t
With the Wind	400	______	______
Against the Wind	200	______	______

78. At a given hour, two steamboats leave a city in the same direction on a straight canal. One travels at 18 mph, and the other travels at 25 mph. In how many hours will the boats be 70 mi apart?

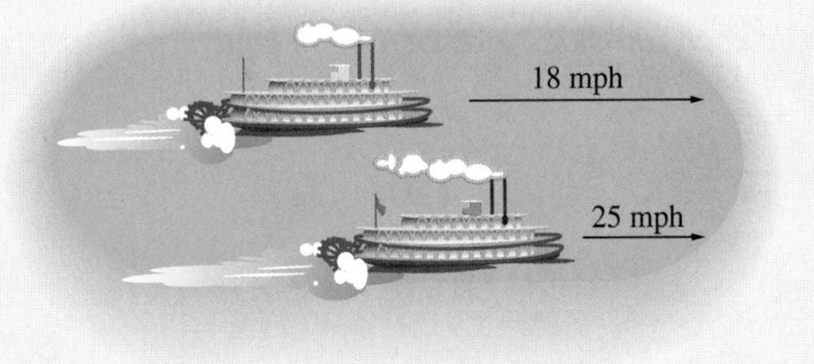

79. The current in a simple electrical circuit varies inversely as the resistance. If the current is 80 amps when the resistance is 10 ohms, find the current if the resistance is 16 ohms.

80. **CONCEPT CHECK** If a rectangle of fixed area has its length increased, then its width decreases. Is this an example of direct or inverse variation?

Chapter 14 Test

CHAPTER Test Prep VIDEO

The Chapter Test Prep Videos with test solutions are available on DVD, in MyMathLab, and on YouTube — search "LialDevMath" and click on "Channels."

1. Find any values for which $\dfrac{3x-1}{x^2-2x-8}$ is undefined. Write your answer with the symbol $\neq$.

2. Find the numerical value of $\dfrac{6r+1}{2r^2-3r-20}$ for each value of r.

(a) $r=-2$ **(b)** $r=4$

3. Write four rational expressions equivalent to $-\dfrac{6x-5}{2x+3}$.

Write each rational expression in lowest terms.

4. $\dfrac{-15x^6y^4}{5x^4y}$

5. $\dfrac{6a^2+a-2}{2a^2-3a+1}$

Multiply or divide. Write each answer in lowest terms.

6. $\dfrac{5(d-2)}{9} \div \dfrac{3(d-2)}{5}$

7. $\dfrac{6k^2-k-2}{8k^2+10k+3} \cdot \dfrac{4k^2+7k+3}{3k^2+5k+2}$

8. $\dfrac{4a^2+9a+2}{3a^2+11a+10} \div \dfrac{4a^2+17a+4}{3a^2+2a-5}$

9. $\dfrac{x^2-10x+25}{9-6x+x^2} \cdot \dfrac{x-3}{5-x}$

Find the least common denominator for each list of fractions.

10. $\dfrac{-3}{10p^2}, \dfrac{21}{25p^3}, \dfrac{-7}{30p^5}$

11. $\dfrac{r+1}{2r^2+7r+6}, \dfrac{-2r+1}{2r^2-7r-15}$

Write each rational expression as an equivalent expression with the indicated denominator.

12. $\dfrac{15}{4p} = \dfrac{?}{64p^3}$

13. $\dfrac{3}{6m-12} = \dfrac{?}{42m-84}$

Add or subtract. Write each answer in lowest terms.

14. $\dfrac{4x+2}{x+5} + \dfrac{-2x+8}{x+5}$

15. $\dfrac{-4}{y+2} + \dfrac{6}{5y+10}$

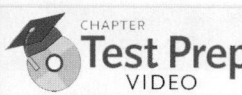

16. $\dfrac{x+1}{3-x} - \dfrac{x^2}{x-3}$

17. $\dfrac{3}{2m^2-9m-5} - \dfrac{m+1}{2m^2-m-1}$

Simplify each complex fraction.

18. $\dfrac{\frac{2p}{k^2}}{\frac{3p^2}{k^3}}$

19. $\dfrac{\frac{1}{x+3}-1}{1+\frac{1}{x+3}}$

Solve each equation.

20. $\dfrac{3x}{x+1} = \dfrac{3}{2x}$

21. $\dfrac{2}{x-1} - \dfrac{2}{3} = \dfrac{-1}{x+1}$

22. $\dfrac{2x}{x-3} + \dfrac{1}{x+3} = \dfrac{-6}{x^2-9}$

23. Solve the formula $F = \dfrac{k}{d-D}$ for D.

Solve each problem.

24. If the same number is added to the numerator and subtracted from the denominator of $\frac{5}{6}$, the resulting fraction is equivalent to $\frac{1}{10}$. What is the number?

25. A boat travels 7 mph in still water. It takes as long to go 20 mi upstream as 50 mi downstream. Find the rate of the current.

26. A man can paint a room in his house, working alone, in 5 hr. His wife can do the job in 4 hr. How long will it take them to paint the room if they work together?

27. Under certain conditions, the length of time that it takes for fruit to ripen during the growing season varies inversely as the average maximum temperature during the season. If it takes 25 days for fruit to ripen with an average maximum temperature of 80°F, find the number of days it would take at 75°F. Round your answer to the nearest whole number.

28. If x varies directly as y, and $x = 12$ when $y = 4$, find x when $y = 9$.

Math in the Media

TO PLAY BASEBALL, YOU MUST *WORK* AT IT!

In the 1994 movie *Little Big League,* the young Billy Heywood inherits the Minnesota Twins baseball team and becomes its manager. He leads the team to the Division Championship and then to the playoffs. But before the final playoff game, the biggest game of the year, he can't keep his mind on his job because a homework problem is giving him trouble.

> *If Joe can paint a house in 3 hours, and Sam can paint the same house in 5 hours, how long does it take for them to do it together?*

With the help of one of his players, he is able to solve the problem, and the team goes on to victory.

1. Use the method described in **Example 5** of **Section 14.7** to solve this problem.

2. Before the player was able to solve the problem correctly, Billy got "help" from some of the other players. The incorrect answers they gave him are included in the following.

(a) 15 hr **(b)** 8 hr **(c)** 4 hr

Explain the faulty reasoning behind each of these incorrect answers.

3. The player who gave Billy the correct answer solved the problem as follows:

> *Using the simple formula a times b over a plus b, we get our answer of one and seven-eighths.*

Show that if it takes one person a hours to complete one job and another b hours to complete the same job, then the expression stated by the player,

$$\frac{a \cdot b}{a + b}$$

actually does give the number of hours it would take them to do the job together. (*Hint:* Refer to **Example 5** and use a and b rather than 8 and 14. Then solve the resulting formula for x.)

15 Systems of Linear Equations and Inequalities

The point of intersection of two lines can be found using a *system of linear equations*, the subject of this chapter.

15.1 Solving Systems of Linear Equations by Graphing

OBJECTIVES

1. Decide whether a given ordered pair is a solution of a system.
2. Solve linear systems by graphing.
3. Solve special systems by graphing.
4. Identify special systems without graphing.

A **system of linear equations**, also called a **linear system,** consists of two or more linear equations with the same variables.

$$2x + 3y = 4 \qquad x + 3y = 1 \qquad x - y = 1$$
$$3x - y = -5 \qquad -y = 4 - 2x \qquad y = 3$$

Linear systems

In the system on the right, think of $y = 3$ as an equation in two variables by writing it as $0x + y = 3$.

OBJECTIVE 1 Decide whether a given ordered pair is a solution of a system. A **solution of a system** of linear equations is an ordered pair that makes both equations true at the same time. A solution of an equation is said to *satisfy* the equation.

EXAMPLE 1 Determining Whether an Ordered Pair Is a Solution

Determine whether the ordered pair $(4, -3)$ is a solution of each system.

(a) $x + 4y = -8$

$3x + 2y = 6$

To decide whether $(4, -3)$ is a solution of the system, substitute 4 for x and -3 for y in each equation.

$x + 4y = -8$		$3x + 2y = 6$	
$4 + 4(-3) \stackrel{?}{=} -8$	Substitute.	$3(4) + 2(-3) \stackrel{?}{=} 6$	Substitute.
$4 + (-12) \stackrel{?}{=} -8$	Multiply.	$12 + (-6) \stackrel{?}{=} 6$	Multiply.
$-8 = -8$ ✓	True	$6 = 6$ ✓	True

Because $(4, -3)$ satisfies *both* equations, it is a solution of the system.

(b) $2x + 5y = -7$

$3x + 4y = 2$

Again, substitute 4 for x and -3 for y in each equation.

$2x + 5y = -7$		$3x + 4y = 2$	
$2(4) + 5(-3) \stackrel{?}{=} -7$	Substitute.	$3(4) + 4(-3) \stackrel{?}{=} 2$	Substitute.
$8 + (-15) \stackrel{?}{=} -7$	Multiply.	$12 + (-12) \stackrel{?}{=} 2$	Multiply.
$-7 = -7$ ✓	True	$0 = 2$	False

The ordered pair $(4, -3)$ is *not* a solution of this system because it does not satisfy the second equation.

◀ **Work Problem 1 at the Side.**

OBJECTIVE 2 Solve linear systems by graphing. The set of all ordered pairs that are solutions of a system is its **solution set.** One way to find the solution set of a system of two linear equations is to graph both equations on the same axes. Any intersection point would be on both lines and would therefore be a solution of *both* equations. ***Thus, the coordinates of any point where the lines intersect give a solution of the system.***

---

1 Determine whether the given ordered pair is a solution of the system.

GS **(a)** $(2, 5)$

$3x - 2y = -4$

$5x + y = 15$

$3x - 2y = -4$

$3(\_\_\_\_) - 2(\_\_\_\_) \stackrel{?}{=} -4$

$5x + y = 15$

$5(2) + \_\_\_\_ \stackrel{?}{=} \_\_\_\_$

$(2, 5)$ (*is* / *is not*) a solution.

(b) $(1, -2)$

$x - 3y = 7$

$4x + y = 5$

$(1, -2)$ (*is* / *is not*) a solution.

Answers

1. **(a)** 2; 5; 5; 15; is **(b)** is not

The graph in **Figure 1** shows that the solution of the system in **Example 1(a)** is the intersection point $(4, -3)$. ***The solution (point of intersection) is always written as an ordered pair.*** Because *two different* straight lines can intersect at no more than one point, there can never be more than one solution for such a system.

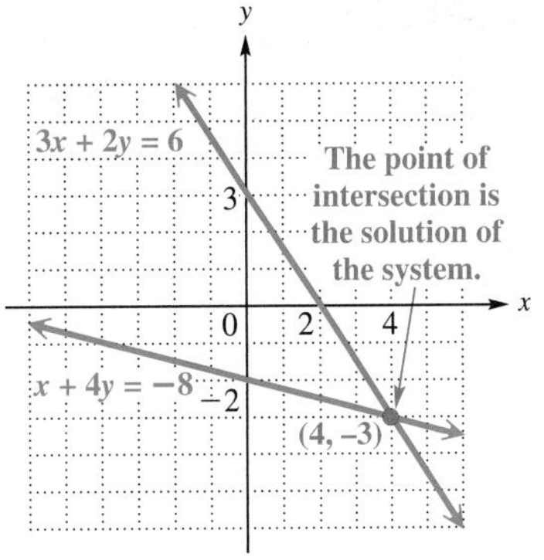

Figure 1

EXAMPLE 2 Solving a System by Graphing

Solve the system of equations by graphing both equations on the same axes.

$$2x + 3y = 4$$
$$3x - y = -5$$

We graph these two equations by plotting several points for each line. Recall from **Section 11.2** that the intercepts are often convenient choices. As review, we show finding the intercepts for $2x + 3y = 4$.

To find the y-intercept, let $x = 0$.

$$2x + 3y = 4$$
$$2(0) + 3y = 4 \quad \text{Let } x = 0.$$
$$3y = 4$$
$$y = \frac{4}{3} \quad \text{Divide by 3.}$$

To find the x-intercept, let $y = 0$.

$$2x + 3y = 4$$
$$2x + 3(0) = 4 \quad \text{Let } y = 0.$$
$$2x = 4$$
$$x = 2 \quad \text{Divide by 2.}$$

The tables show the intercepts and a check point for each graph.

$2x + 3y = 4$

x	y	
0	$\frac{4}{3}$	← y-intercept
2	0	← x-intercept
−2	$\frac{8}{3}$	Find a third ordered pair as a check.

$3x - y = -5$

x	y	
0	5	← y-intercept
$-\frac{5}{3}$	0	← x-intercept
−2	−1	

The lines in **Figure 2** suggest that the graphs intersect at the point $(-1, 2)$. We check by substituting -1 for x and 2 for y in *both* equations.

CHECK

$$2x + 3y = 4$$
$$2(-1) + 3(2) \stackrel{?}{=} 4$$
$$4 = 4 \quad ✓ \text{ True}$$
$$3x - y = -5$$
$$3(-1) - 2 \stackrel{?}{=} -5$$
$$-5 = -5 \quad ✓ \text{ True}$$

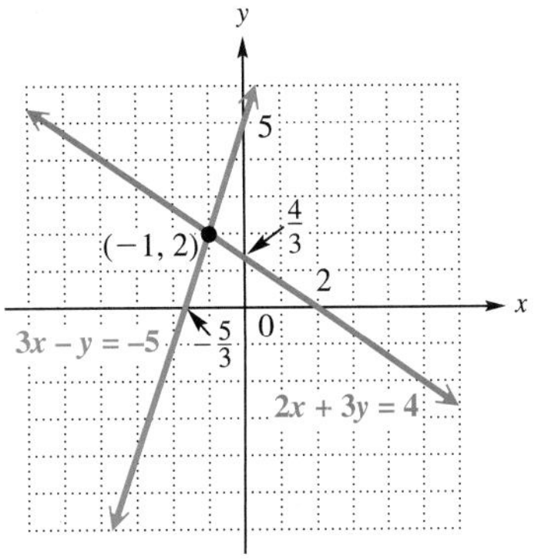

Figure 2

Because $(-1, 2)$ satisfies both equations, the solution set of this system is $\{(-1, 2)\}$.

Work Problem 2 at the Side. ▶

---

2 Solve each system of equations by graphing both equations on the same axes. Check each solution.

GS **(a)** $5x - 3y = 9$
$x + 2y = 7$

One of the lines is already graphed. Complete the table, and graph the other line.

$5x - 3y = 9$

x	y
0	−3
$\frac{9}{5}$	0
3	2

$x + 2y = 7$

x	y
0	___
___	0
−1	___

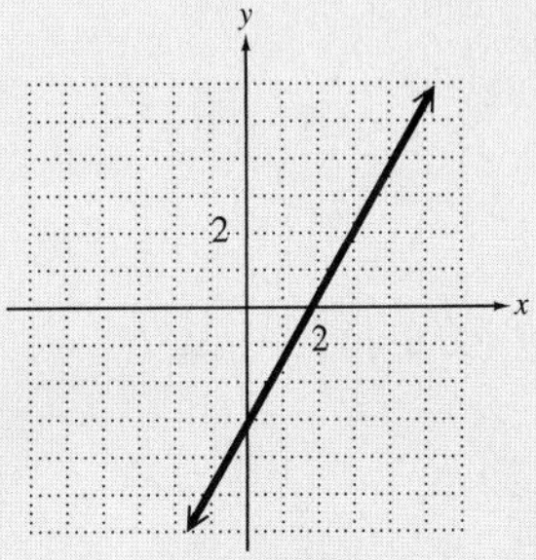

The point of intersection of the lines is the ordered pair ________, so the solution set of the system of equations is ________.

(b) $x + y = 4$
$2x - y = -1$

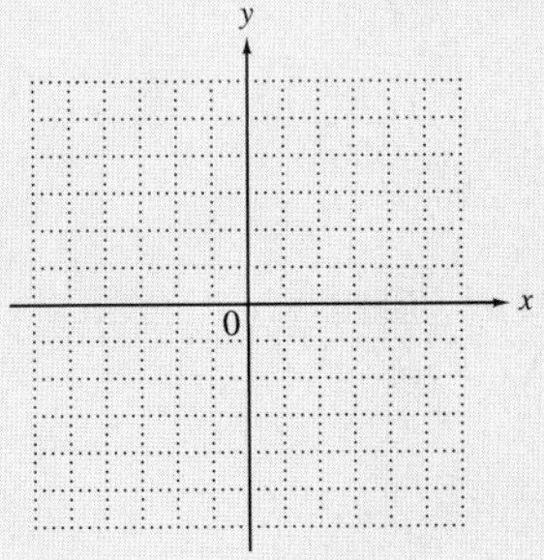

Answers

2. **(a)** $\frac{7}{2}$; 7; 4; $(3, 2)$; $\{(3, 2)\}$ **(b)** $\{(1, 3)\}$

Note

We can also write each equation in a system in slope-intercept form and use the slope and y-intercept to graph each line. For **Example 2,**

$2x + 3y = 4$ becomes $y = -\frac{2}{3}x + \frac{4}{3}$. y-intercept $\left(0, \frac{4}{3}\right)$; slope $-\frac{2}{3}$

$3x - y = -5$ becomes $y = 3x + 5$. y-intercept $(0, 5)$; slope 3, or $\frac{3}{1}$

Confirm that graphing these equations results in the same lines and the same solution shown in **Figure 2** on the preceding page.

Solving a Linear System by Graphing

Step 1 **Graph each equation** of the system on the same coordinate axes.

Step 2 **Find the coordinates of the point of intersection** of the graphs if possible, and write it as an ordered pair. This is the solution of the system.

Step 3 **Check** the ordered-pair solution in *both* of the original equations. If it satisfies *both* equations, write the solution set.

CAUTION

With the graphing method, it may not be possible to determine from the graph the exact coordinates of the point that represents the solution, particularly if these coordinates are not integers. The graphing method does, however, show geometrically how solutions are found and is useful when approximate answers will suffice.

OBJECTIVE 3 Solve special systems by graphing. The graphs of the equations in a system may not intersect at all or may be the same line.

EXAMPLE 3 Solving Special Systems

Solve each system by graphing.

(a) $2x + y = 2$

$2x + y = 8$

The graphs of these lines are shown in **Figure 3.** The two lines are parallel and have no points in common. For such a system, there is no solution. We write the solution set as $\emptyset$.

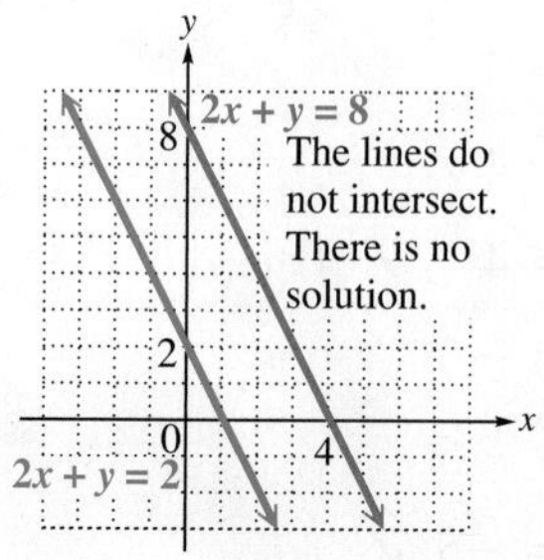

Figure 3

Continued on Next Page

(b) $2x + 5y = 1$

$6x + 15y = 3$

The graphs of these two equations are the same line. See **Figure 4.** We can obtain the second equation by multiplying each side of the first equation by 3. In this case, every point on the line is a solution of the system, and the solution set contains an infinite number of ordered pairs, each of which satisfies both equations of the system.

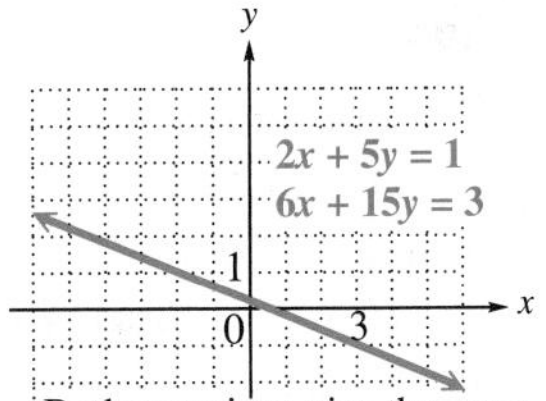

Both equations give the same graph. There is an infinite number of solutions.

Figure 4

We write the solution set as

$$\{(x, y) \mid 2x + 5y = 1\},$$

This is the first equation in the system. See the Note on the next page.

read "the set of ordered pairs (x, y) such that $2x + 5y = 1$." Recall from **Section 9.3** that this notation is called **set-builder notation.**

Work Problem 3 at the Side. ▶

The system in **Example 2** has exactly one solution. A system with at least one solution is a **consistent system.** A system of equations with no solution, such as the one in **Example 3(a),** is an **inconsistent system.**

The equations in **Example 2** are **independent equations** with different graphs. The equations of the system in **Example 3(b)** have the same graph and are equivalent. Because they are different forms of the same equation, these equations are **dependent equations.**

Three Cases for Solutions of Linear Systems with Two Variables

Case 1 The graphs intersect at exactly one point, which gives the (single) ordered-pair solution of the system. The **system is consistent** and the **equations are independent.** See **Figure 5(a).**

Case 2 The graphs are parallel lines, so there is no solution and the solution set is $\emptyset$. The **system is inconsistent** and the **equations are independent.** See **Figure 5(b).**

Case 3 The graphs are the same line. There is an infinite number of solutions, and the solution set is written in set-builder notation as

$$\{(x, y) \mid \_\_\_\_\_\_\},$$

where one of the equations follows the | symbol. The **system is consistent** and the **equations are dependent.** See **Figure 5(c).**

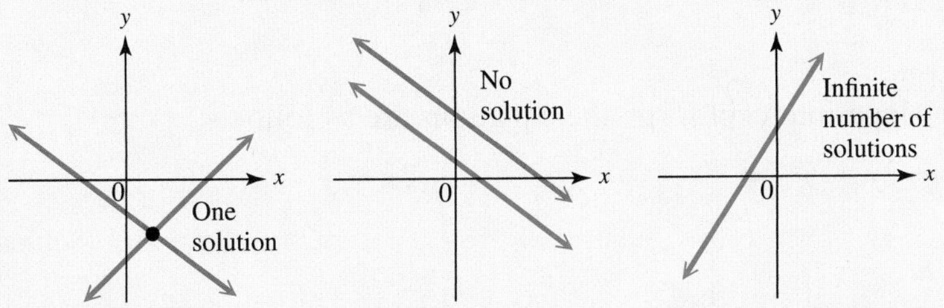

The system is consistent. The equations are independent.
(a)

The system is inconsistent. The equations are independent.
(b)

The system is consistent. The equations are dependent.
(c)

Figure 5

3 Solve each system of equations by graphing both equations on the same axes.

GS **(a)** $3x - y = 4$

$6x - 2y = 12$

One of the lines is already graphed.

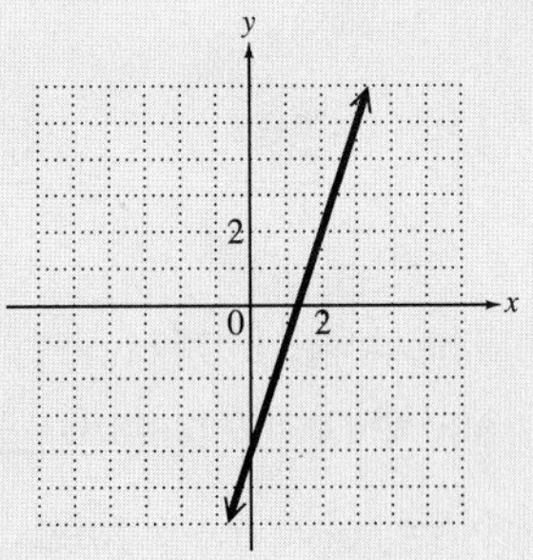

(b) $x - 3y = -2$

$2x - 6y = -4$

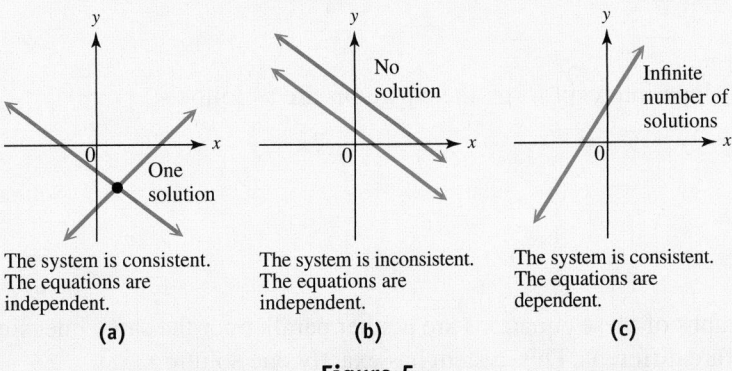

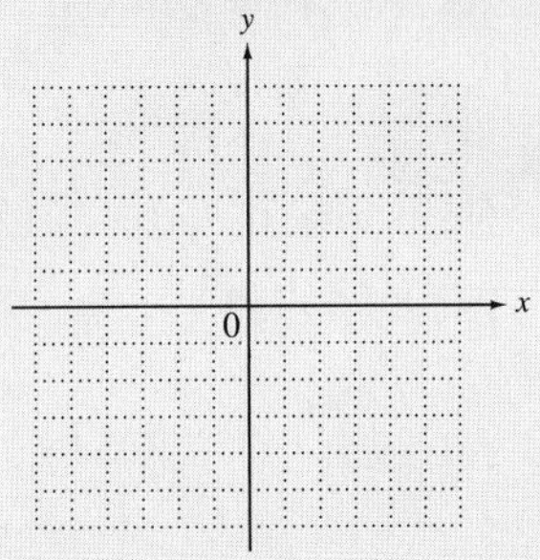

Answers

3. (a) $\emptyset$ **(b)** $\{(x, y) \mid x - 3y = -2\}$

4 Describe each system without graphing. State the number of solutions.

GS **(a)** $2x - 3y = 5$

$3y = 2x - 7$

Solve the first equation for ____.

$2x - 3y = 5$

$-3y = \_\_\_\_ + 5$

$y = \_\_\_\_ - \_\_\_\_$

The slope of this line is ____, and the y-intercept is ____.

Now solve the second equation for y, and complete the solution.

(b) $-x + 3y = 2$

$2x - 6y = -4$

(c) $6x + y = 3$

$2x - y = -11$

Answers

4. (a) y; $-2x$; $\frac{2}{3}x$; $\frac{5}{3}$; $\frac{2}{3}$; $\left(0, -\frac{5}{3}\right)$
The equations represent parallel lines. The system has no solution.
(b) The equations represent the same line. The system has an infinite number of solutions.
(c) The equations represent lines that are neither parallel nor the same line. The system has exactly one solution.

Note

When a system has an infinite number of solutions, as in Case 3, either equation of the system can be used to write the solution set. ***We prefer to use the equation in standard form with integer coefficients having greatest common factor 1.*** If neither of the given equations is in this form, use an *equivalent* equation that is in standard form with integer coefficients having greatest common factor 1.

OBJECTIVE 4 Identify special systems without graphing.

EXAMPLE 4 Identifying the Three Cases by Using Slopes

Describe each system without graphing. State the number of solutions.

(a) $3x + 2y = 6$

$-2y = 3x - 5$

Write each equation in slope-intercept form, $y = mx + b$.

$3x + 2y = 6$	Solve for y.	$-2y = 3x - 5$	Solve for y.
$2y = -3x + 6$	Subtract $3x$.	$y = -\frac{3}{2}x + \frac{5}{2}$	Divide *each* term by -2.
$y = -\frac{3}{2}x + 3$	Divide *each* term by 2.		

Both equations have slope $-\frac{3}{2}$ but they have different y-intercepts, $(0, 3)$ and $\left(0, \frac{5}{2}\right)$. Lines with the same slope are parallel (**Section 11.3**), so these equations have graphs that are parallel lines. Thus, the system has no solution.

(b) $2x - y = 4$

$x = \frac{y}{2} + 2$

Again, write each equation in slope-intercept form.

$2x - y = 4$		$x = \frac{y}{2} + 2$	
$-y = -2x + 4$	Subtract $2x$.	$\frac{y}{2} + 2 = x$	Interchange sides.
$\mathbf{y = 2x - 4}$	Multiply by -1.	$\mathbf{y = 2x - 4}$	Subtract 2. Multiply by 2.

The equations are exactly the same—their graphs are the same line. Thus, the system has an infinite number of solutions.

(c) $x - 3y = 5$

$2x + y = 8$

In slope-intercept form, the equations are as follows.

$x - 3y = 5$		$2x + y = 8$	
$-3y = -x + 5$	Subtract x.	$y = -2x + 8$	Subtract $2x$.
$y = \frac{1}{3}x - \frac{5}{3}$	Divide by -3.		

The graphs of these equations are neither parallel nor the same line, since the slopes are different. This system has exactly one solution.

◀ **Work Problem 4 at the Side.**

15.1 Exercises

FOR EXTRA HELP

Download the MyDashBoard App

MyMathLab®

CONCEPT CHECK *Complete each statement. The following terms may be used once, more than once, or not at all.*

consistent	system of linear equations	inconsistent	solution
ordered pair	independent	linear equation	dependent

1. A(n) __________ consists of two or more linear equations with the (*same / different*) variables.

2. A solution of a system of linear equations is a(n) __________ that makes all equations of the system (*true / false*) at the same time.

3. The equations of two parallel lines form a(n) __________ system that has (*one / no / infinitely many*) solution(s). The equations are __________ because their graphs are different.

4. If the graphs of a linear system intersect in one point, the point of intersection is the __________ of the system. The system is __________ and the equations are independent.

5. If two equations of a linear system have the same graph, the equations are __________. The system is __________ and has (*one / no / infinitely many*) solution(s).

6. If a linear system is inconsistent, the graphs of the two equations are (*intersecting / parallel / the same*) line(s). The system has no __________.

CONCEPT CHECK *Work each problem.*

7. A student determined that the ordered pair $(1, -2)$ is a solution of the following system. His reasoning was that the ordered pair satisfies the first equation $x + y = -1$, since $1 + (-2) = -1$. **What Went Wrong?**

$$x + y = -1$$
$$2x + y = 4$$

8. The following system has infinitely many solutions. Write its solution set, using set-builder notation.

$$6x - 4y = 8$$
$$3x - 2y = 4$$

9. Which ordered pair could be a solution of the system graphed? Why is it the only valid choice?

A. $(2, 2)$
B. $(-2, 2)$
C. $(-2, -2)$
D. $(2, -2)$

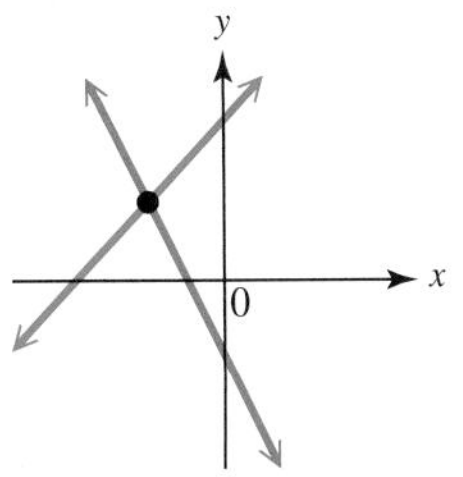

10. Which ordered pair could be a solution of the system graphed? Why is it the only valid choice?

A. $(2, 0)$
B. $(0, 2)$
C. $(-2, 0)$
D. $(0, -2)$

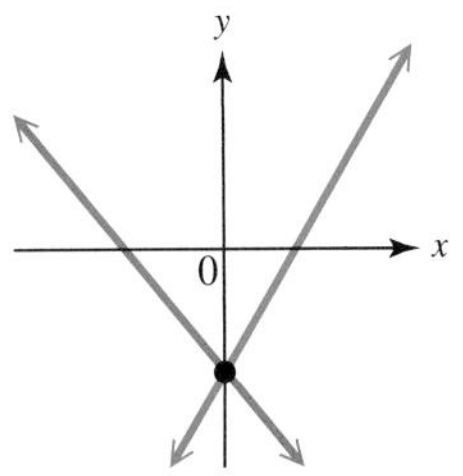

Decide whether the given ordered pair is a solution of the given system. ***See Example 1.***

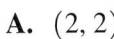

11. $(2, -3)$
$$x + y = -1$$
$$2x + 5y = 19$$

12. $(4, 3)$
$$x + 2y = 10$$
$$3x + 5y = 3$$

13. $(-1, -3)$
$$3x + 5y = -18$$
$$4x + 2y = -10$$

14. $(-9, -2)$
$$2x - 5y = -8$$
$$3x + 6y = -39$$

15. $(7, -2)$
$4x = 26 - y$
$3x = 29 + 4y$

16. $(9, 1)$
$2x = 23 - 5y$
$3x = 24 + 3y$

17. $(6, -8)$
$-2y = x + 10$
$3y = 2x + 30$

18. $(-5, 2)$
$5y = 3x + 20$
$3y = -2x - 4$

Solve each system of equations by graphing. If the system is inconsistent or the equations are dependent, say so. ***See Examples 2 and 3.***

19. $x - y = 2$
$x + y = 6$

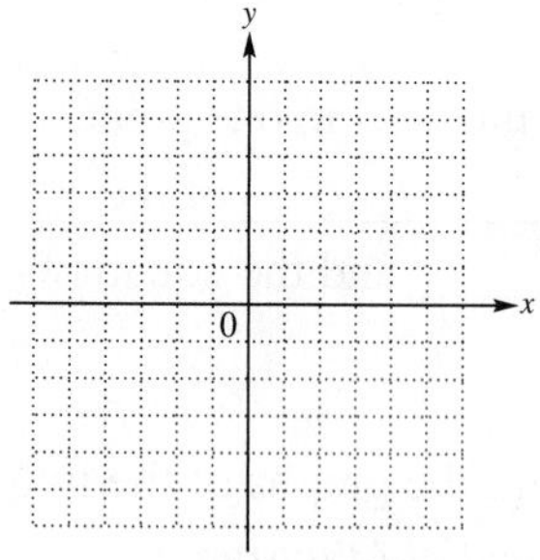

20. $x - y = 3$
$x + y = -1$

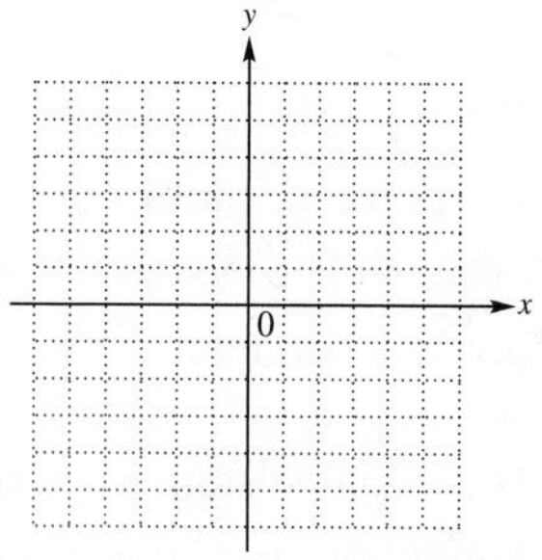

21. $x + y = 4$
$y - x = 4$

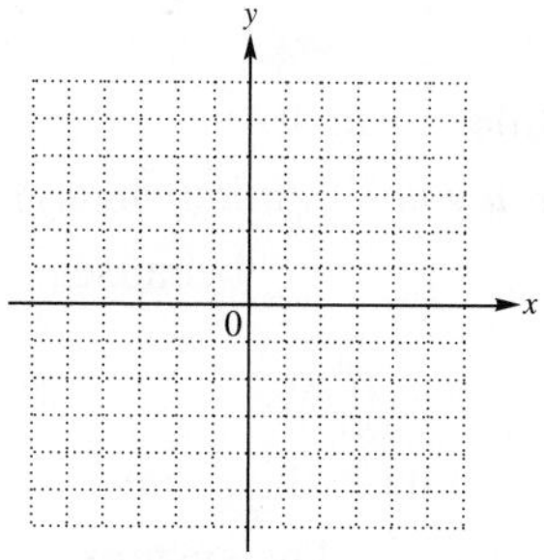

22. $x + y = -5$
$x - y = 5$

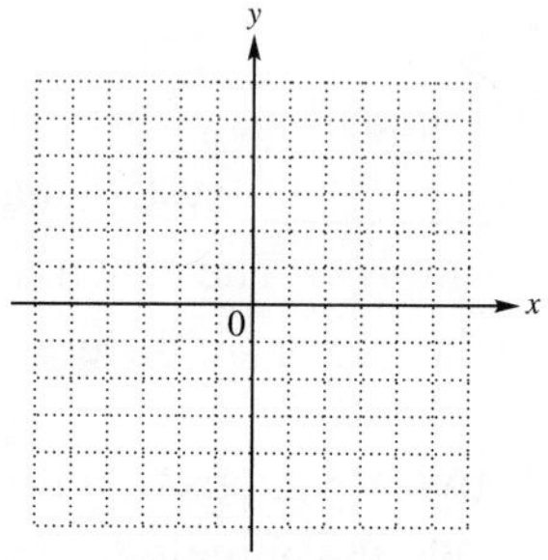

23. $x - 2y = 6$
$x + 2y = 2$

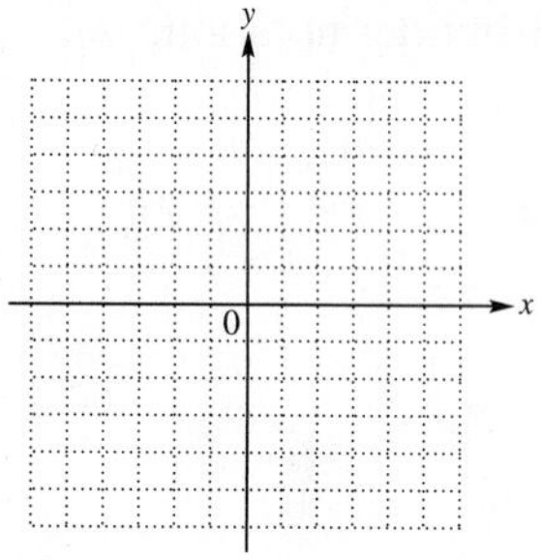

24. $2x - y = 4$
$4x + y = 2$

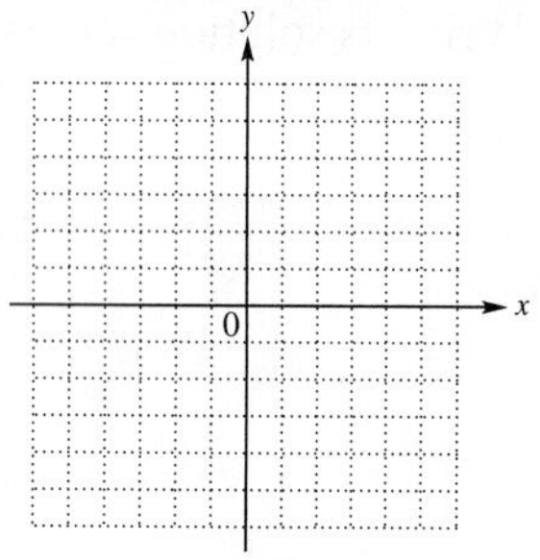

25. $3x - 2y = -3$
$-3x - y = -6$

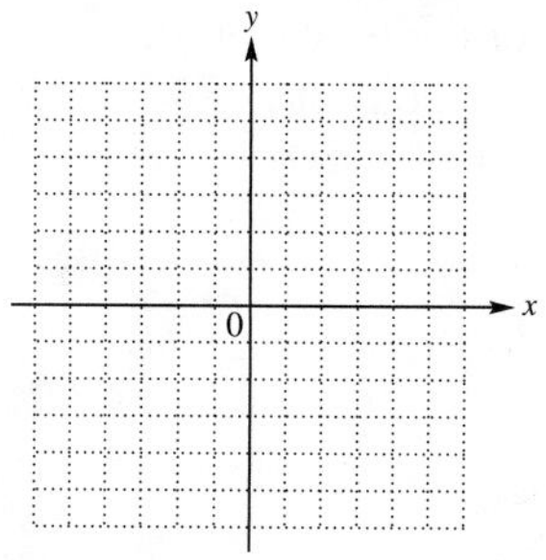

26. $2x - y = 4$
$2x + 3y = 12$

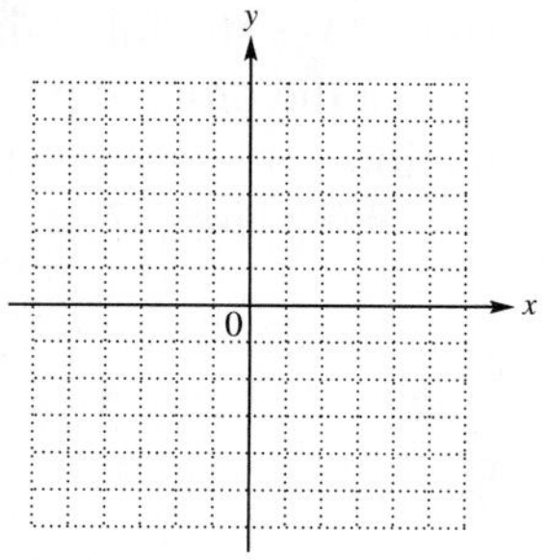

27. $2x - 3y = -6$
$y = -3x + 2$

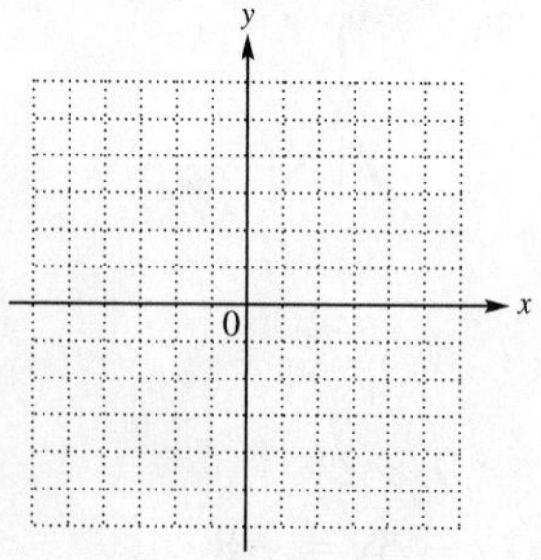

28. $-3x + y = -3$
$y = x - 3$

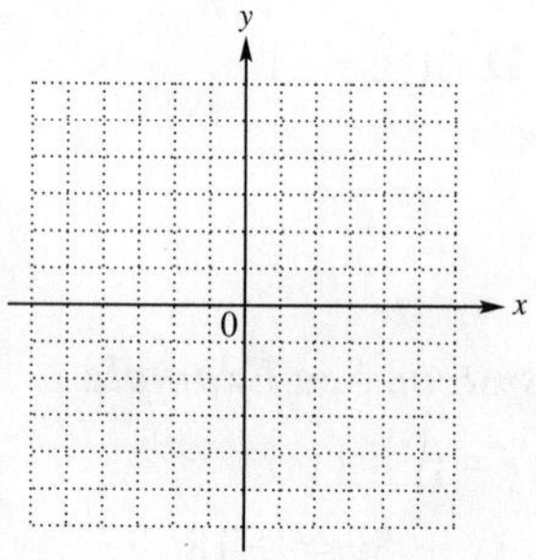

29. $x + 2y = 6$
$2x + 4y = 8$

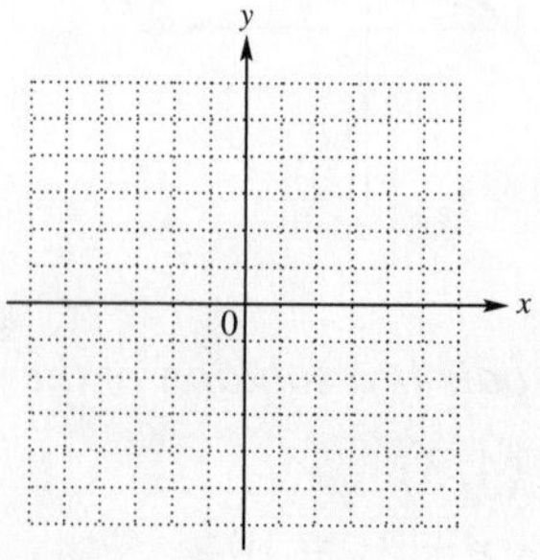

30. $2x - y = 6$
$6x - 3y = 12$

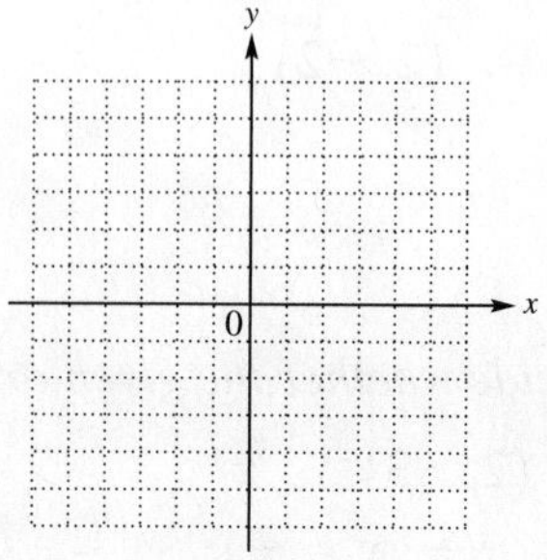

31. $5x - 3y = 2$
$10x - 6y = 4$

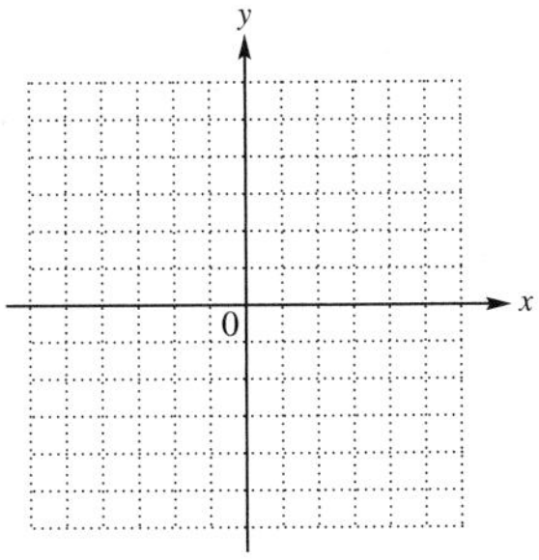

32. $2x - 5y = 8$
$4x - 10y = 16$

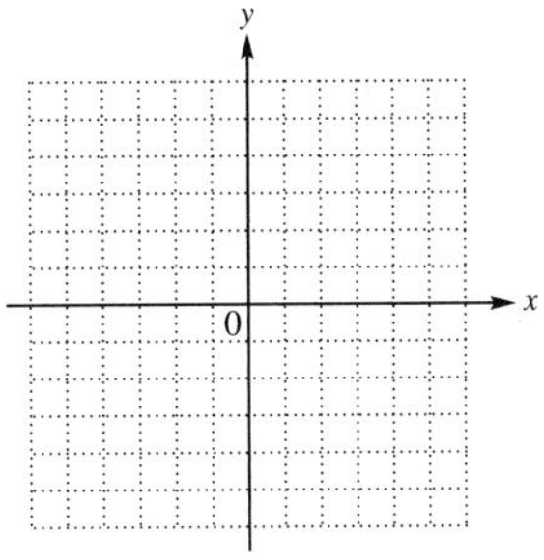

33. $3x - 4y = 24$
$y = -\frac{3}{2}x + 3$

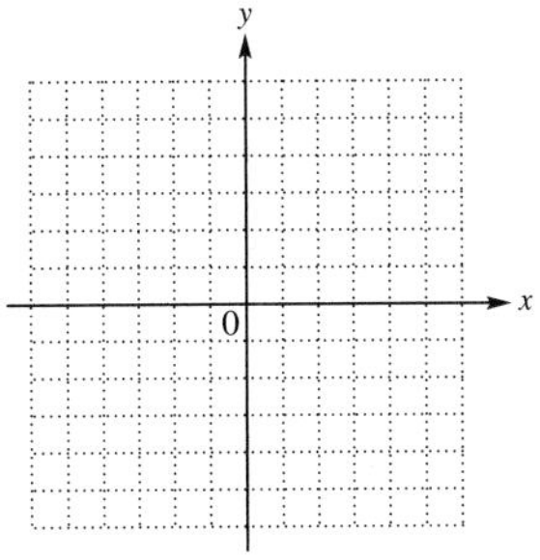

34. $3x - 2y = 12$
$y = -4x + 5$

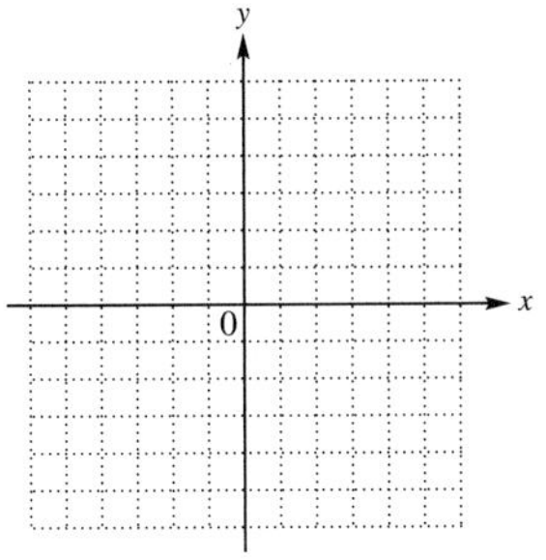

35. $4x - 2y = 8$
$2x = y + 4$

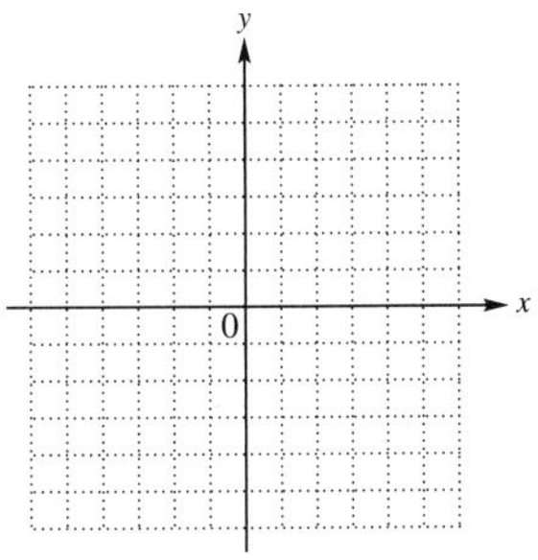

36. $3x = 5 - y$
$6x + 2y = 10$

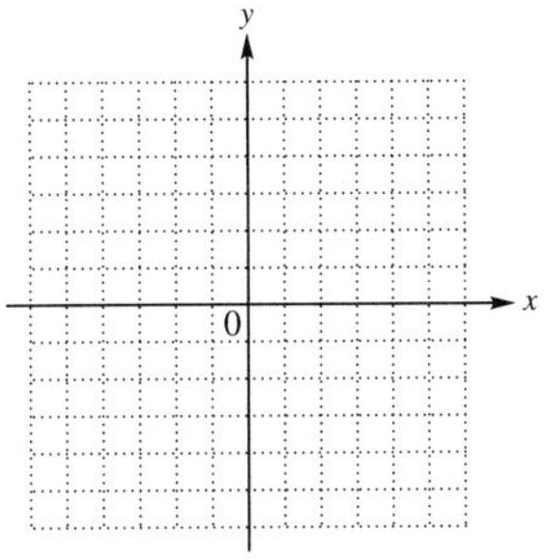

37. $3x = y + 5$
$6x - 5 = 2y$

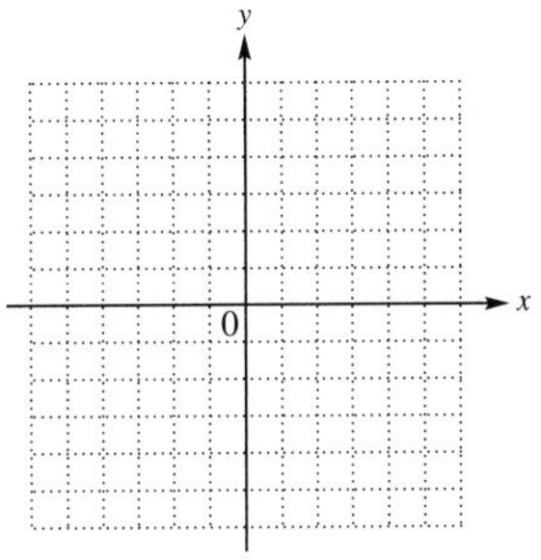

38. $2x = y - 4$
$4x - 2y = -4$

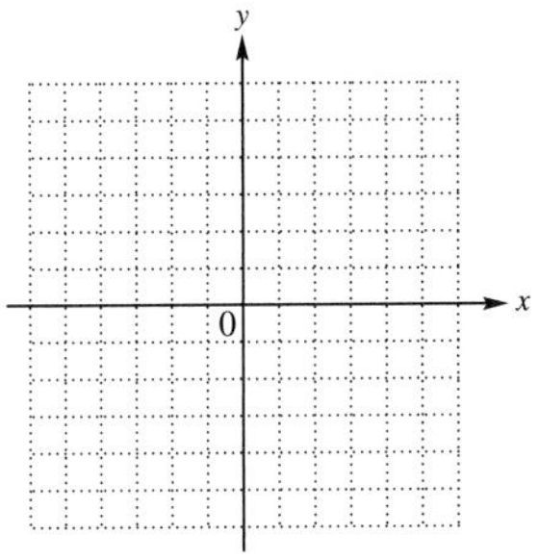

Without graphing, answer the following questions for each linear system. ***See Example 4.***

(a) Is the system inconsistent, are the equations dependent, or neither?
(b) Is the graph a pair of intersecting lines, a pair of parallel lines, or one line?
(c) Does the system have one solution, no solution, or an infinite number of solutions?

39. $y - x = -5$
$x + y = 1$

40. $2x + y = 6$
$x - 3y = -4$

41. $x + 2y = 0$
$4y = -2x$

42. $4x - 6y = 10$
$-6x + 9y = -15$

43. $5x + 4y = 7$
$10x + 8y = 4$

44. $y = 3x$
$y + 3 = 3x$

45. $x - 3y = 5$
$2x + y = 8$

46. $2x + 3y = 12$
$2x - y = 4$

Economics deals with ***supply and demand.*** *Typically, as the price of an item increases, the demand for the item decreases while the supply increases. If supply and demand can be described by straight-line equations, the point at which the lines intersect determines the* ***equilibrium supply*** *and* ***equilibrium demand.***

The price per unit, p, and the demand, x, for a particular aluminum siding are related by the linear equation $p = 60 - \frac{3}{4}x$*, while the price and the supply are related by the linear equation* $p = \frac{3}{4}x$*. See the figure. Use the graph to work Exercises 47 and 48.*

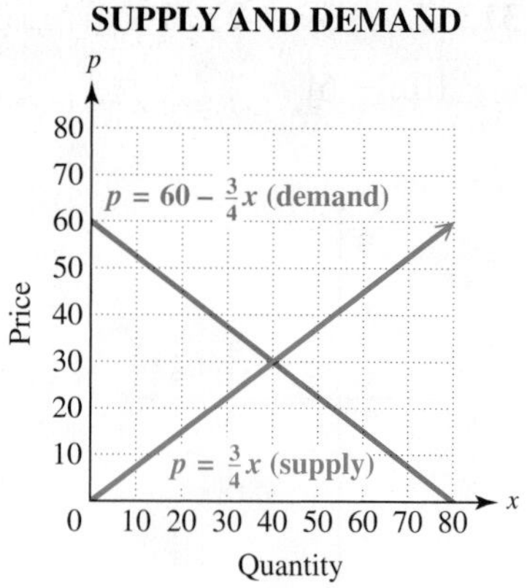

47. (a) At what value of x does supply equal demand?

(b) At what value of p does supply equal demand?

48. When $x > 40$, does demand exceed supply or does supply exceed demand?

The per-capita consumption of beef and poultry in the United States in selected years is shown in the graph. Use the graph to work Exercises 49 and 50.

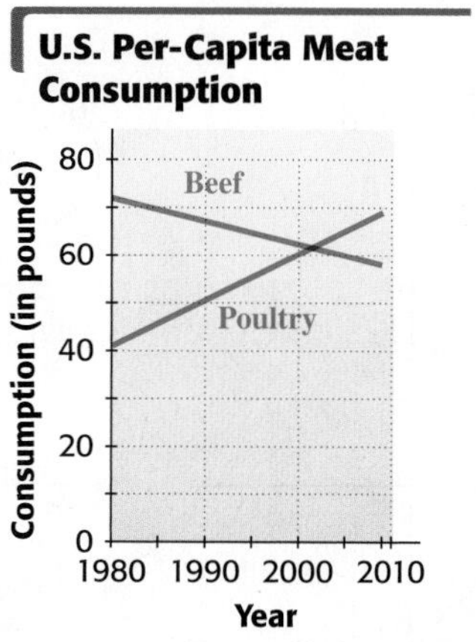

Source: U.S. Department of Agr

49. For which years did per-capita consumption of beef exceed that of poultry?

50. Estimate the year in which the per-capita consumption of beef and poultry was about the same. About how many pounds of each kind of meat were consumed by the average American in that year?

The graph shows sales of music CDs and digital downloads of single songs (in millions) in the United States over selected years. Use the graph to work Exercises 51–54.

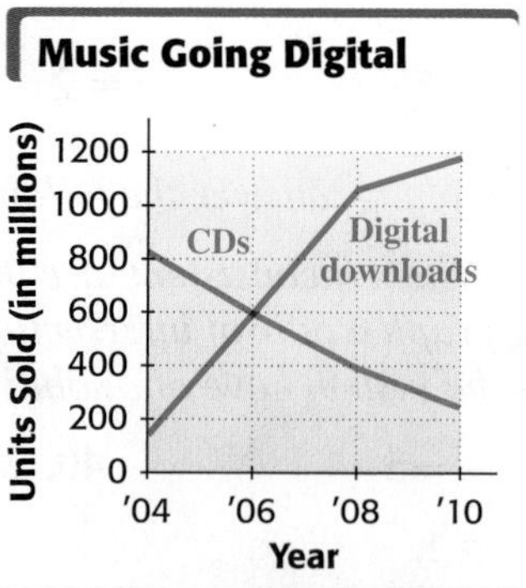

Source: Recording Industry Association of America.

51. In what year did Americans purchase about the same number of CDs as single digital downloads? How many units was this?

52. Express the point of intersection of the two graphs as an ordered pair of the form (year, units in millions).

53. Describe the trend in sales of music CDs over the years 2004 to 2010. If a straight line were used to approximate its graph, would the line have positive, negative, or zero slope? Explain.

54. If a straight line were used to approximate the graph of sales of digital downloads over the years 2004 to 2010, would the line have positive, negative, or zero slope? Explain.

15.2 Solving Systems of Linear Equations by Substitution

OBJECTIVES

1. Solve linear systems by substitution.
2. Solve special systems by substitution.
3. Solve linear systems with fractions and decimals.

OBJECTIVE 1 **Solve linear systems by substitution.**

Work Problem 1 at the Side. ▶

As we see in **Margin Problem 1,** graphing to solve a system of equations has a serious drawback: It is difficult to accurately find a solution such as $\left(\frac{11}{3}, -\frac{4}{9}\right)$ from a graph.

As a result, there are algebraic methods for solving systems of equations. The **substitution method,** which gets its name from the fact that an expression in one variable is *substituted* for the other variable, is one such method.

EXAMPLE 1 **Using the Substitution Method**

Solve the system by the substitution method.

$$3x + 5y = 26 \quad (1)$$
$$y = 2x \quad (2)$$

We number the equations for reference in our discussion.

Equation (2) is already solved for y. This equation says that $y = 2x$, so we substitute $2x$ for y in equation (1).

$3x + 5y = 26$	(1)	
$3x + 5(2x) = 26$	Let $y = 2x$.	
$3x + 10x = 26$	Multiply.	
$13x = 26$	Combine like terms.	
$x = 2$	Divide by 13.	Don't stop here.

Now we can find the value of y by substituting 2 for x in either equation. We choose equation (2) since the substitution is easier.

$y = 2x$	(2)
$y = 2(2)$	Let $x = 2$.
$y = 4$	Multiply.

We check the solution $(2, 4)$ by substituting 2 for x and 4 for y in *both* equations.

CHECK

$3x + 5y = 26$	(1)	$y = 2x$	(2)
$3(2) + 5(4) \stackrel{?}{=} 26$	Substitute.	$4 \stackrel{?}{=} 2(2)$	Substitute.
$6 + 20 \stackrel{?}{=} 26$	Multiply.	$4 = 4$ ✓	True
$26 = 26$ ✓	True		

Since $(2, 4)$ satisfies both equations, the solution set of the system is $\{(2, 4)\}$.

Work Problem 2 at the Side. ▶

CAUTION

A system is not completely solved until values for both x and y are found. Write the solution set as a set containing an ordered pair.

1 Solve the system by graphing.

$$2x + 3y = 6$$
$$x - 3y = 5$$

Can you determine the answer? Why or why not?

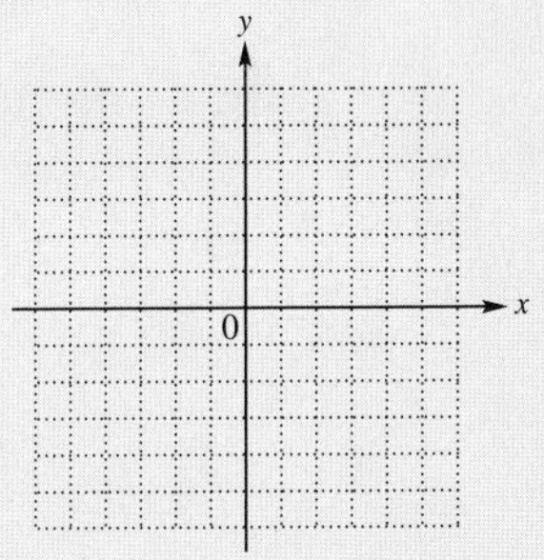

2 (GS) Fill in the blanks to solve by the substitution method. Check the solution.

$$3x + 5y = 69$$
$$y = 4x$$

$3x + 5(\_\_\_\_) = 69$

$3x + \_\_\_\_ = 69$

$\_\_\_\_ = 69$

$x = \_\_\_\_$

$y = 4(\_\_\_\_) = \_\_\_\_$

The solution set is ____.

Answers

1. The answer cannot be determined from the graph because it is too difficult to read the exact coordinates.
2. $4x$; $20x$; $23x$; 3; 3; 12; $\{(3, 12)\}$

3 Solve each system by the substitution method. Check each solution.

(a) $2x + 7y = -12$

$x = 3 - 2y$

(b) $x = y - 3$

$4x + 9y = 1$

Answers

3. **(a)** $\{(15, -6)\}$ **(b)** $\{(-2, 1)\}$

EXAMPLE 2 **Using the Substitution Method**

Solve the system by the substitution method.

$$2x + 5y = 7 \quad (1)$$
$$x = -1 - y \quad (2)$$

Equation (2) gives x in terms of y. We substitute $-1 - y$ for x in equation (1).

$$2x + 5y = 7 \quad (1)$$
$$2(-1 - y) + 5y = 7 \quad \text{Let } x = -1 - y.$$
$$-2 - 2y + 5y = 7 \quad \text{Distributive property}$$

Distribute 2 to *both* −1 and −y.

$$-2 + 3y = 7 \quad \text{Combine like terms.}$$
$$3y = 9 \quad \text{Add 2.}$$
$$\mathbf{y = 3} \quad \text{Divide by 3.}$$

To find x, substitute 3 for y in equation (2).

$$x = -1 - y \quad (2)$$
$$x = -1 - \mathbf{3} \quad \text{Let } y = 3.$$
$$x = \mathbf{-4} \quad \text{Subtract.}$$

Write the x-coordinate first.

We check the solution $(-4, 3)$ in *both* equations.

CHECK

(1)		(2)	
$2x + 5y = 7$	(1)	$x = -1 - y$	(2)
$2(-4) + 5(3) \stackrel{?}{=} 7$	Substitute.	$-4 \stackrel{?}{=} -1 - 3$	Substitute.
$-8 + 15 \stackrel{?}{=} 7$	Multiply.	$-4 = -4$ ✓	True
$7 = 7$ ✓	True		

Both results are true, so the solution set of the system is $\{(-4, 3)\}$.

◀ **Work Problem 3 at the Side.**

CAUTION

Even though we found y first in **Example 2,** ***the x-coordinate is always written first in the ordered-pair solution of a system.***

To solve a system by substitution, follow these steps.

Solving a Linear System by Substitution

Step 1 **Solve one equation for either variable.** If one of the variables has coefficient 1 or −1, choose it, since it usually makes the substitution easier.

Step 2 **Substitute** for that variable in the other equation. The result should be an equation with just one variable.

Step 3 **Solve** the equation from Step 2.

Step 4 **Substitute** the result from Step 3 into the equation from Step 1 to find the value of the other variable.

Step 5 **Check** the ordered-pair solution in both of the *original* equations. Then write the solution set.

EXAMPLE 3 Using the Substitution Method

Solve the system by the substitution method.

$$2x = 4 - y \quad (1)$$
$$5x + 3y = 10 \quad (2)$$

Step 1 We must solve one of the equations for either x or y. Because the coefficient of y in equation (1) is -1, we avoid fractions by choosing this equation and solving for y.

$$2x = 4 - y \quad (1)$$
$$y + 2x = 4 \quad \text{Add } y.$$
$$y = -2x + 4 \quad \text{Subtract } 2x.$$

Step 2 Now substitute $-2x + 4$ for y in equation (2).

$$5x + 3y = 10 \quad (2)$$
$$5x + 3(-2x + 4) = 10 \quad \text{Let } y = -2x + 4.$$

Step 3 Solve the equation from Step 2.

$$5x - 6x + 12 = 10 \quad \text{Distributive property}$$
$$-x + 12 = 10 \quad \text{Combine like terms.}$$
$$-x = -2 \quad \text{Subtract 12.}$$
$$x = 2 \quad \text{Multiply by } -1.$$

Distribute 3 to *both* $-2x$ and 4.

Step 4 Equation (1) solved for y is $y = -2x + 4$. Substitute 2 for x.

$$y = -2(2) + 4 \quad \text{Substitute in (1).}$$
$$y = 0 \quad \text{Multiply, and then add.}$$

Step 5 Check that $(2, 0)$ is the solution.

CHECK

$$2x = 4 - y \quad (1)$$
$$2(2) \stackrel{?}{=} 4 - 0 \quad \text{Substitute.}$$
$$4 = 4 \checkmark \quad \text{True}$$

$$5x + 3y = 10 \quad (2)$$
$$5(2) + 3(0) \stackrel{?}{=} 10 \quad \text{Substitute.}$$
$$10 = 10 \checkmark \quad \text{True}$$

Since both results are true, the solution set of the system is $\{(2, 0)\}$.

Work Problem 4 at the Side. ▶

OBJECTIVE 2 Solve special systems by substitution.

EXAMPLE 4 Solving an Inconsistent System by Substitution

Use substitution to solve the system.

$$x = 5 - 2y \quad (1)$$
$$2x + 4y = 6 \quad (2)$$

Because equation (1) is already solved for x, we substitute $5 - 2y$ for x in equation (2).

$$2x + 4y = 6 \quad (2)$$
$$2(5 - 2y) + 4y = 6 \quad \text{Let } x = 5 - 2y.$$
$$10 - 4y + 4y = 6 \quad \text{Distributive property}$$
$$10 = 6 \quad \text{False}$$

Continued on Next Page

4 Solve each system by the substitution method. Check each solution.

GS **(a)** Fill in the blanks to solve.

$$x + 4y = -1$$
$$2x - 5y = 11$$

Solve the first equation for x.

$$x = -1 - \_\_\_\_$$

Substitute into the second equation to find y.

$$2(\_\_\_\_) - 5y = 11$$
$$-2 - 8y - 5y = 11$$
$$-2 - \_\_\_\_ y = 11$$
$$\_\_\_\_ y = 13$$
$$y = \_\_\_\_$$

Find x.

$$x = -1 - 4y$$
$$x = -1 - 4(\_\_\_\_)$$
$$x = \_\_\_\_$$

The solution set is ____.

(b) $2x + 5y = 4$
$x + y = -1$

Answers

4. **(a)** $4y$; $-1 - 4y$; 13; -13; -1; -1; 3; $\{(3, -1)\}$
(b) $\{(-3, 2)\}$

5 Use substitution to solve the system.

$$8x - 2y = 1$$
$$y = 4x - 8$$

6 Solve each system by substitution.

(a) $7x - 6y = 10$
$-14x + 20 = -12y$

(b) $8x - y = 4$
$y = 8x + 4$

Answers

5. $\emptyset$
6. (a) $\{(x, y) \mid 7x - 6y = 10\}$ (b) $\emptyset$

A false result, here $10 = 6$, means that the equations in the system

$$x = 5 - 2y$$
$$2x + 4y = 6$$

have graphs that are parallel lines. The system is inconsistent and has no solution, so the solution set is $\emptyset$. See **Figure 6.**

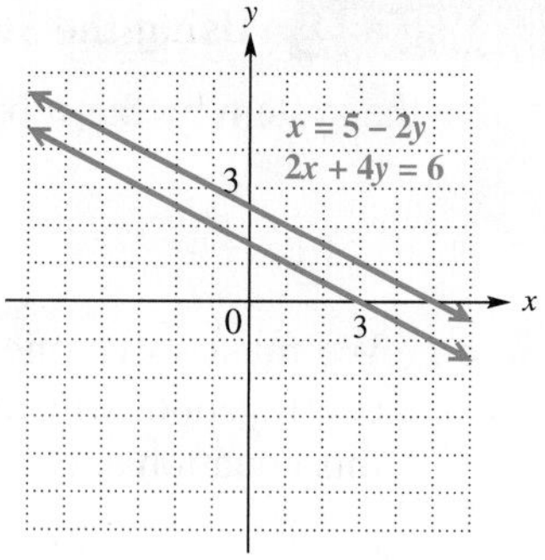

Figure 6

◀ **Work Problem 5 at the Side.**

CAUTION

It is a common error to give "false" as the answer to an inconsistent system. The correct response is $\emptyset$.

EXAMPLE 5 Solving a System with Dependent Equations by Substitution

Solve the system by the substitution method.

$$3x - y = 4 \quad (1)$$
$$-9x + 3y = -12 \quad (2)$$

Begin by solving equation (1) for y to get $y = 3x - 4$. Substitute $3x - 4$ for y in equation (2) and solve the resulting equation.

$$-9x + 3y = -12 \quad (2)$$
$$-9x + 3(3x - 4) = -12 \quad \text{Let } y = 3x - 4.$$
$$-9x + 9x - 12 = -12 \quad \text{Distributive property}$$
$$0 = 0 \quad \text{Add 12. Combine like terms.}$$

This true result means that every solution of one equation is also a solution of the other, so the system has an infinite number of solutions—all the ordered pairs corresponding to points that lie on the common graph. The solution set is

$$\{(x, y) \mid 3x - y = 4\}.$$

A graph of the equations of this system is shown in **Figure 7.**

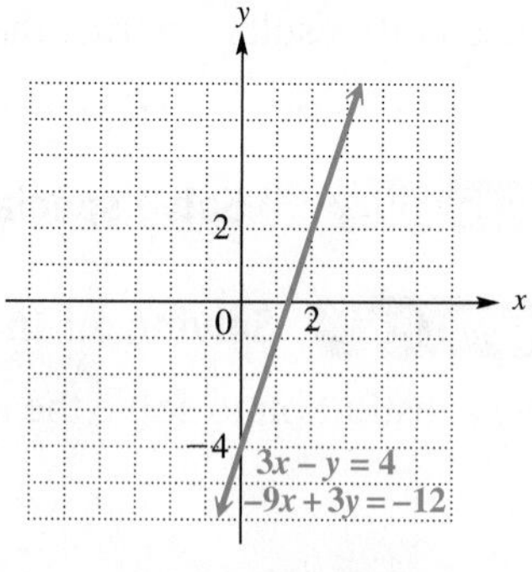

Figure 7

◀ **Work Problem 6 at the Side.**

CAUTION

It is a common error to give "true" as the solution of a system of dependent equations. Write the solution set in set-builder notation using the equation in the system (or an equivalent equation) that is in standard form, with integer coefficients having greatest common factor 1.

OBJECTIVE 3 Solve linear systems with fractions and decimals.

EXAMPLE 6 Using the Substitution Method with Fractions

Solve the system by the substitution method.

$$3x + \frac{1}{4}y = 2 \qquad (1)$$

$$\frac{1}{2}x + \frac{3}{4}y = -\frac{5}{2} \qquad (2)$$

Clear equation (1) of fractions by multiplying each side by 4.

$$4\left(3x + \frac{1}{4}y\right) = 4(2) \qquad \text{Multiply by 4, the common denominator.}$$

$$4(3x) + 4\left(\frac{1}{4}y\right) = 4(2) \qquad \text{Distributive property}$$

$$12x + y = 8 \qquad (3)$$

Now clear equation (2) of fractions by multiplying each side by 4.

$$\frac{1}{2}x + \frac{3}{4}y = -\frac{5}{2} \qquad (2)$$

$$4\left(\frac{1}{2}x + \frac{3}{4}y\right) = 4\left(-\frac{5}{2}\right) \qquad \text{Multiply by 4, the common denominator.}$$

$$4\left(\frac{1}{2}x\right) + 4\left(\frac{3}{4}y\right) = 4\left(-\frac{5}{2}\right) \qquad \text{Distributive property.}$$

$$2x + 3y = -10 \qquad (4)$$

The given system of equations has been simplified to an equivalent system.

$$12x + y = 8 \qquad (3)$$

$$2x + 3y = -10 \qquad (4)$$

To solve this system by substitution, solve equation (3) for y.

$$12x + y = 8 \qquad (3)$$

$$y = -12x + 8 \qquad \text{Subtract } 12x.$$

Now substitute this result for y in equation (4).

$$2x + 3y = -10 \qquad (4)$$

$$2x + 3(-12x + 8) = -10 \qquad \text{Let } y = -12x + 8.$$

$$2x - 36x + 24 = -10 \qquad \text{Distributive property}$$

$$-34x = -34 \qquad \text{Combine like terms. Subtract 24.}$$

$$x = 1 \qquad \text{Divide by } -34.$$

Distribute 3 to both $-12x$ and 8.

Substitute 1 for x in $y = -12x + 8$ (equation (3) solved for y).

$$y = -12(1) + 8 = -4$$

Check $(1, -4)$ in both of the original equations. The solution set of the system is $\{(1, -4)\}$.

Work Problem 7 at the Side. ▶

---

7 Solve each system by the substitution method.

GS **(a)** $\frac{2}{3}x + \frac{1}{2}y = 6 \qquad (1)$

$\frac{1}{2}x - \frac{3}{4}y = 0 \qquad (2)$

First clear all fractions. The LCD for the denominators 3 and 2 in (1) is ____.

$\_\_\_\left(\frac{2}{3}x + \frac{1}{2}y\right) = \_\_\_ \qquad (6)$

$\_\_\_\left(\frac{2}{3}x\right) + 6(\_\_\_) = 36$

$\_\_\_ + 3y = 36$

Clear fractions in (2) by multiplying both sides by ____, the LCD of denominators 2 and 4.

Complete the solution of the system. The solution set is ____.

(b) $x + \frac{1}{2}y = \frac{1}{2}$

$\frac{1}{6}x - \frac{1}{3}y = \frac{4}{3}$

Answers

7. **(a)** 6; 6; 6; 6; $\frac{1}{2}y$; $4x$; 4; $\{(6, 4)\}$

(b) $\{(2, -3)\}$

8 Solve the system by the substitution method.

$$0.2x + 0.3y = 0.5$$
$$0.3x - 0.1y = 1.3$$

EXAMPLE 7 **Using the Substitution Method with Decimals**

Solve the system by the substitution method.

$$0.5x + 2.4y = 4.2 \quad (1)$$
$$-0.1x + 1.5y = 5.1 \quad (2)$$

Clear each equation of decimals, as in **Section 10.3,** by multiplying by 10.

$$10(0.5x + 2.4y) = 10(4.2) \quad \text{Multiply equation (1) by 10.}$$
$$10(0.5x) + 10(2.4y) = 10(4.2) \quad \text{Distributive property}$$
$$5x + 24y = 42 \quad (3)$$

$$10(-0.1x + 1.5y) = 10(5.1) \quad \text{Multiply equation (2) by 10.}$$
$$10(-0.1x) + 10(1.5y) = 10(5.1) \quad \text{Distributive property}$$
$$-x + 15y = 51 \quad (4)$$

$10(-0.1x) = -1x = -x$

Now solve the equivalent system of equations by substitution.

$$5x + 24y = 42 \quad (3)$$
$$-x + 15y = 51 \quad (4)$$

Equation (4) can be solved for x.

$$x = 15y - 51 \quad \text{Equation (4) solved for } x$$

Substitute this result for x in equation (3).

$$5x + 24y = 42 \quad (3)$$
$$5(15y - 51) + 24y = 42 \quad \text{Let } x = 15y - 51.$$
$$75y - 255 + 24y = 42 \quad \text{Distributive property}$$
$$99y = 297 \quad \text{Combine like terms. Add 255.}$$
$$y = 3 \quad \text{Divide by 99.}$$

Since equation (4) solved for x is $x = 15y - 51$, substitute 3 for y.

$$x = 15(3) - 51 = -6$$

Check $(-6, 3)$ in both of the original equations. The solution set is $\{(-6, 3)\}$.

◄ **Work Problem 8 at the Side.**

Answer

8. $\{(4, -1)\}$

15.2 Exercises

FOR EXTRA HELP

Download the MyDashBoard App

MyMathLab®

CONCEPT CHECK *Work each problem.*

1. A student solves the following system and finds that $x = 3$, which is correct. The student gives $\{3\}$ as the solution set. **What Went Wrong?**

$$5x - y = 15$$
$$7x + y = 21$$

2. A student solves the following system and obtains the statement $0 = 0$. The student gives the solution set as $\{(0, 0)\}$. **What Went Wrong?**

$$x + y = 4$$
$$2x + 2y = 8$$

3. When we use the substitution method, how can we tell that a system has no solution?

4. When we use the substitution method, how can we tell that a system has an infinite number of solutions?

Solve each system by the substitution method. Check each solution. ***See Examples 1–5.***

5. $x + y = 12$
$y = 3x$

6. $x + 3y = -28$
$y = -5x$

7. $3x + 2y = 27$
$x = y + 4$

8. $4x + 3y = -5$
$x = y - 3$

9. $3x + 5y = 14$
$x - 2y = -10$

10. $5x + 2y = -1$
$2x - y = -13$

11. $3x + 4 = -y$
$2x + y = 0$

12. $2x - 5 = -y$
$x + 3y = 0$

13. $7x + 4y = 13$
$x + y = 1$

14. $3x - 2y = 19$
$x + y = 8$

15. $3x - y = 5$
$y = 3x - 5$

16. $4x - y = -3$
$y = 4x + 3$

17. $2x + 8y = 3$
$x = 8 - 4y$

18. $2x + 10y = 3$
$x = 1 - 5y$

19. $2y = 4x + 24$
$2x - y = -12$

20. $2y = 14 - 6x$
$3x + y = 7$

21. $6x - 8y = 6$
$2y = -2 + 3x$

22. $3x + 2y = 6$
$6x = 8 + 4y$

Solve each system by the substitution method. Check each solution. ***See Examples 6 and 7.***

23. $\frac{1}{5}x + \frac{2}{3}y = -\frac{8}{5}$
$3x - y = 9$

24. $\frac{1}{3}x - \frac{1}{2}y = -\frac{2}{3}$
$4x + y = 6$

25. $\frac{1}{2}x + \frac{1}{3}y = -\frac{1}{3}$
$\frac{1}{2}x + 2y = -7$

26. $\frac{1}{6}x + \frac{1}{6}y = 1$
$-\frac{1}{2}x - \frac{1}{3}y = -5$

27. $\frac{x}{5} + 2y = \frac{16}{5}$
$\frac{3x}{5} + \frac{y}{2} = -\frac{7}{5}$

28. $\frac{x}{2} + \frac{y}{3} = \frac{7}{6}$
$\frac{x}{4} - \frac{3y}{2} = \frac{9}{4}$

29. $0.1x + 0.9y = -2$
$0.5x - 0.2y = 4.1$

30. $0.2x - 1.3y = -3.2$
$-0.1x + 2.7y = 9.8$

31. $0.8x - 0.1y = 1.3$
$2.2x + 1.5y = 8.9$

32. $0.3x - 0.1y = 2.1$
$0.6x + 0.3y = -0.3$

Relating Concepts (Exercises 33–36) For Individual or Group Work

A system of linear equations can be used to model the cost and the revenue of a business. ***Work Exercises 33–36 in order.***

33. Suppose that to start a business manufacturing and selling bicycles, it costs \$5000. Each bicycle will cost \$400 to manufacture. Explain why the linear equation

$$y_1 = 400x + 5000 \quad (y_1 \text{ in dollars})$$

gives the *total* cost to manufacture x bicycles.

34. We decide to sell each bicycle for \$600. Write an equation using y_2 (in dollars) to express the revenue when we sell x bicycles.

35. Form a system from the two equations in **Exercises 33 and 34.** Then solve the system, assuming $y_1 = y_2$, that is, cost = revenue.

36. The value of x in **Exercise 35** is the number of bicycles it takes to *break even.* Fill in the blanks: When ______ bicycles are sold, the break-even point is reached. At that point, we have spent ______ dollars and taken in ______ dollars.

15.3 Solving Systems of Linear Equations by Elimination

OBJECTIVES

1. Solve linear systems by elimination.
2. Multiply when using the elimination method.
3. Use an alternative method to find the second value in a solution.
4. Solve special systems by elimination.

OBJECTIVE 1 **Solve linear systems by elimination.** Recall that adding the same quantity to each side of an equation results in equal sums.

$$\text{If} \quad A = B, \quad \text{then} \quad A + C = B + C.$$

We can take this addition a step further. Adding *equal* quantities, rather than the *same* quantity, to each side of an equation also results in equal sums.

$$\text{If} \quad A = B \quad \text{and} \quad C = D, \quad \text{then} \quad A + C = B + D.$$

Using the addition property of equality to solve systems is called the **elimination method.**

EXAMPLE 1 Using the Elimination Method

Use the elimination method to solve the system.

$$x + y = 5 \quad (1)$$
$$x - y = 3 \quad (2)$$

Each equation in this system is a statement of equality, so the sum of the left sides equals the sum of the right sides. Adding vertically in this way gives the following.

$$\begin{aligned} x + y &= 5 && (1) \\ x - y &= 3 && (2) \\ \hline 2x \quad &= 8 && \text{Add left sides and add right sides.} \\ x &= 4 && \text{Divide by 2.} \end{aligned}$$

Notice that y has been eliminated. The result, $x = 4$, gives the x-value of the ordered-pair solution of the given system. To find the y-value of the solution, substitute 4 for x in either of the two equations of the system. We choose equation (1).

$$\begin{aligned} x + y &= 5 && (1) \\ 4 + y &= 5 && \text{Let } x = 4. \\ y &= 1 && \text{Subtract 4.} \end{aligned}$$

Check the solution, (4, 1), by substituting 4 for x and 1 for y in both equations of the given system.

CHECK

$x + y = 5$ (1)		$x - y = 3$ (2)	
$4 + 1 \stackrel{?}{=} 5$	Substitute.	$4 - 1 \stackrel{?}{=} 3$	Substitute.
$5 = 5$ ✓	True	$3 = 3$ ✓	True

Since *both* results are true, the solution set of the system is $\{(4, 1)\}$.

Work Problem 1 at the Side. ▶

1 Solve each system by the elimination method. Check each solution.

GS **(a)** Fill in the blanks to solve the following system.

$$\begin{aligned} x + y &= 8 && (1) \\ x - y &= 2 && (2) \\ \hline \_\_\_\_ &= \_\_\_\_ && \text{Add.} \\ x &= \_\_\_\_ \end{aligned}$$

Find y.

$$\begin{aligned} x - y &= 2 && (2) \\ \_\_\_\_ - y &= 2 \\ -y &= \_\_\_\_ \\ y &= \_\_\_\_ \end{aligned}$$

The solution set is ____.

(b) $3x - y = 7$
$2x + y = 3$

CAUTION

A system is not completely solved until values for both x and y are found. Do not stop after finding the value of only one variable. Remember to write the solution set as a set containing an ordered pair.

Answers

1. **(a)** $2x$; 10; 5; 5; -3; 3; $\{(5, 3)\}$
(b) $\{(2, -1)\}$

With the elimination method, the idea is to *eliminate* one of the variables. ***To do this, one pair of variable terms in the two equations must have coefficients that are opposites.***

Solving a Linear System by Elimination

Step 1 **Write both equations in standard form $Ax + By = C$.**

Step 2 **Transform so that the coefficients of one pair of variable terms are opposites.** Multiply one or both equations by appropriate numbers so that the sum of the coefficients of either the x- or y-terms is 0.

Step 3 **Add** the new equations to eliminate a variable. The sum should be an equation with just one variable.

Step 4 **Solve** the equation from Step 3 for the remaining variable.

Step 5 **Substitute** the result from Step 4 into *either* of the original equations, and solve for the other variable.

Step 6 **Check** the ordered-pair solution in *both* of the *original* equations. Then write the solution set.

It does not matter which variable is eliminated first. Usually we choose the one that is more convenient to work with.

EXAMPLE 2 **Using the Elimination Method**

Solve the system.

$$y + 11 = 2x \quad (1)$$
$$5x = y + 26 \quad (2)$$

Step 1 Write both equations in standard form $Ax + By = C$.

$$-2x + y = -11 \quad \text{Subtract } 2x \text{ and } 11 \text{ in equation (1).}$$
$$5x - y = 26 \quad \text{Subtract } y \text{ in equation (2).}$$

Step 2 Because the coefficients of y are 1 and -1, adding will eliminate y. It is not necessary to multiply either equation by a number.

Step 3 Add the two equations.

$$\begin{array}{rcl} -2x + y &=& -11 \\ 5x - y &=& 26 \\ \hline 3x \quad &=& 15 \end{array} \quad \text{Add in columns.}$$

Step 4 Solve.

$$x = 5 \quad \text{Divide by 3.}$$

Step 5 Find the value of y by substituting 5 for x in either of the original equations.

$$y + 11 = 2x \quad (1)$$
$$y + 11 = 2(5) \quad \text{Let } x = 5.$$
$$y + 11 = 10 \quad \text{Multiply.}$$
$$y = -1 \quad \text{Subtract 11.}$$

Continued on Next Page

Step 6 Check the ordered-pair solution $(5, -1)$ by substituting $x = 5$ and $y = -1$ into both of the original equations.

CHECK

$y + 11 = 2x$	(1)	$5x = y + 26$	(2)
$-1 + 11 \stackrel{?}{=} 2(5)$	Substitute.	$5(5) \stackrel{?}{=} -1 + 26$	Substitute.
$10 = 10$ ✓	True	$25 = 25$ ✓	True

Since $(5, -1)$ is a solution of *both* equations, the solution set is $\{(5, -1)\}$.

Work Problem 2 at the Side. ▶

OBJECTIVE 2 Multiply when using the elimination method. Sometimes we need to multiply each side of one or both equations in a system by some number so that adding the equations will eliminate a variable.

EXAMPLE 3 Using the Elimination Method

Solve the system.

$$2x + 3y = -15 \quad (1)$$
$$5x + 2y = 1 \quad (2)$$

Adding the two equations gives $7x + 5y = -14$, which does not eliminate either variable. However, we can multiply each equation by a suitable number so that the coefficients of one of the two variables are opposites. For example, to eliminate x, multiply each side of equation (1) by 5, and each side of equation (2) by -2.

$$10x + 15y = -75 \quad \text{Multiply equation (1) by 5.}$$
$$-10x - 4y = -2 \quad \text{Multiply equation (2) by } -2.$$
$$11y = -77 \quad \text{Add.}$$
$$y = -7 \quad \text{Divide by 11.}$$

The coefficients of x are opposites.

Substituting -7 for y in either equation (1) or (2) gives $x = 3$. Check that the solution set of the system is $\{(3, -7)\}$.

Work Problem 3 at the Side. ▶

OBJECTIVE 3 Use an alternative method to find the second value in a solution. Sometimes it is easier to find the value of the second variable in a solution by using the elimination method twice.

EXAMPLE 4 Finding the Second Value Using an Alternative Method

Solve the system.

$$4x = 9 - 3y \quad (1)$$
$$5x - 2y = 8 \quad (2)$$

Write equation (1) in standard form by adding $3y$ to each side.

$$4x + 3y = 9 \quad (3)$$
$$5x - 2y = 8 \quad (4)$$

One way to proceed is to eliminate y by multiplying each side of equation (3) by 2 and each side of equation (2) by 3, and then adding.

Continued on Next Page

2 Solve each system by the elimination method. Check each solution.

(a) $2x - y = 2$

$4x + y = 10$

(b) $8x - 5y = 32$

$4x + 5y = 4$

3 **(a)** Solve the system in **Example 3** by first eliminating the variable y. Check the solution.

(b) Solve the system, and check the solution.

$6x + 7y = 4$

$5x + 8y = -1$

Answers

2. **(a)** $\{(2, 2)\}$ **(b)** $\left\{\left(3, -\frac{8}{5}\right)\right\}$

3. **(a)** $\{(3, -7)\}$ **(b)** $\{(3, -2)\}$

$$\begin{aligned} 8x + 6y &= 18 && \text{Multiply equation (3) } 4x + 3y = 9 \text{ by 2.} \\ 15x - 6y &= 24 && \text{Multiply equation (2) } 5x - 2y = 8 \text{ by 3.} \\ \hline 23x \quad &= 42 && \text{Add.} \\ x &= \frac{42}{23} && \text{Divide by 23.} \end{aligned}$$

The coefficients of y are opposites.

Substituting $\frac{42}{23}$ for x in one of the given equations would give y, but the arithmetic would be complicated. Instead, solve for y by starting again with the original equations in standard form (equations (3) and (2)) and eliminating x.

$$\begin{aligned} 20x + 15y &= 45 && \text{Multiply equation (3) } 4x + 3y = 9 \text{ by 5.} \\ -20x + 8y &= -32 && \text{Multiply equation (2) } 5x - 2y = 8 \text{ by } -4. \\ \hline 23y &= 13 && \text{Add.} \\ y &= \frac{13}{23} && \text{Divide by 23.} \end{aligned}$$

The coefficients of x are opposites.

Check that the solution set is $\left\{\left(\frac{42}{23}, \frac{13}{23}\right)\right\}$.

◀ Work Problem 4 at the Side.

4 Solve each system of equations.

(a) $5x = 7 + 2y$

$5y = 5 - 3x$

(b) $3y = 8 + 4x$

$6x = 9 - 2y$

Note

When the value of the first variable is a fraction, the method used in **Example 4** helps avoid arithmetic errors. This method could be used to solve any system.

5 Solve each system by the elimination method.

(a) $4x + 3y = 10$

$2x + \frac{3}{2}y = 12$

(b) $4x - 6y = 10$

$-10x + 15y = -25$

OBJECTIVE 4 Solve special systems by elimination.

EXAMPLE 5 Solve Special Systems Using the Elimination Method

Solve each system by the elimination method.

(a)

$$\begin{aligned} 2x + 4y &= 5 && (1) \\ 4x + 8y &= -9 && (2) \end{aligned}$$

Multiply each side of equation (1) by -2. Then add the two equations.

$$\begin{aligned} -4x - 8y &= -10 && \text{Multiply equation (1) by } -2. \\ 4x + 8y &= -9 && (2) \\ \hline \mathbf{0} &= \mathbf{-19} && \text{False} \end{aligned}$$

The false statement $0 = -19$ indicates that the system has solution set $\emptyset$.

(b)

$$\begin{aligned} 3x - y &= 4 && (1) \\ -9x + 3y &= -12 && (2) \end{aligned}$$

Multiply each side of equation (1) by 3. Then add the two equations.

$$\begin{aligned} 9x - 3y &= 12 && \text{Multiply equation (1) by 3.} \\ -9x + 3y &= -12 && (2) \\ \hline \mathbf{0} &= \mathbf{0} && \text{True} \end{aligned}$$

A true statement occurs when the equations are equivalent. This indicates that every solution of one equation is also a solution of the other. The solution set is $\{(x, y) \mid 3x - y = 4\}$.

◀ Work Problem 5 at the Side.

Answers

4. (a) $\left\{\left(\frac{45}{31}, \frac{4}{31}\right)\right\}$ (b) $\left\{\left(\frac{11}{26}, \frac{42}{13}\right)\right\}$

5. (a) $\emptyset$
(b) $\{(x, y) \mid 2x - 3y = 5\}$
(*Note:* To write the solution set, we divided each term of the equation $4x - 6y = 10$ by 2 so that the coefficients would have greatest common factor 1.)

15.3 Exercises

FOR EXTRA HELP

Download the MyDashBoard App

MyMathLab®

CONCEPT CHECK *In Exercises 1–4, answer* true *or* false *for each statement. If false, tell why.*

1. The ordered pair (0, 0) *must* be a solution of a system in the following form.

$$Ax + By = 0$$
$$Cx + Dy = 0$$

2. To eliminate the y-terms in the following system, we should multiply equation (2) by 3 and then add the result to equation (1).

$$2x + 12y = 7 \quad (1)$$
$$3x + 4y = 1 \quad (2)$$

3. The following system has solution set $\emptyset$.

$$x + y = 1$$
$$x + y = 2$$

4. The ordered pair $(4, -5)$ cannot be a solution of a system that contains the following equation.

$$5x - 4y = 0$$

Solve each system by the elimination method. Check each solution. ***See Examples 1 and 2.***

5. $x + y = 2$
$2x - y = -5$

6. $3x - y = -12$
$x + y = 4$

7. $2x + y = -5$
$x - y = 2$

8. $2x + y = -15$
$-x - y = 10$

9. $3x + 2y = 0$
$-3x - y = 3$

10. $5x - y = 5$
$-5x + 2y = 0$

11. $6x - y = -1$
$5y = 17 + 6x$

12. $y = 9 - 6x$
$-6x + 3y = 15$

Solve each system by the elimination method. Check each solution. ***See Examples 3–5.***

13. $2x - y = 12$
$3x + 2y = -3$

14. $x + y = 3$
$-3x + 2y = -19$

15. $x + 3y = 19$
$2x - y = 10$

16. $4x - 3y = -19$
$2x + y = 13$

17. $x + 4y = 16$
$3x + 5y = 20$

18. $2x + y = 8$
$5x - 2y = -16$

19. $5x - 3y = -20$
$-3x + 6y = 12$

20. $4x + 3y = -28$
$5x - 6y = -35$

21. $2x - 8y = 0$
$4x + 5y = 0$

22. $3x - 15y = 0$
$6x + 10y = 0$

23. $x + y = 7$
$x + y = -3$

24. $x - y = 4$
$x - y = -3$

25. $-x + 3y = 4$
$-2x + 6y = 8$

26. $6x - 2y = 24$
$-3x + y = -12$

27. $2x + 3y = 21$
$5x - 2y = -14$

28. $5x + 4y = 12$
$3x + 5y = 15$

29. $3x - 7 = -5y$
$5x + 4y = -10$

30. $2x + 3y = 13$
$6 + 2y = -5x$

31. $2x + 3y = 0$
$4x + 12 = 9y$

32. $4x - 3y = -2$
$5x + 3 = 2y$

33. $24x + 12y = -7$
$16x - 17 = 18y$

34. $9x + 4y = -3$
$6x + 7 = -6y$

35. $5x - 2y = 3$
$10x - 4y = 5$

36. $3x - 5y = 1$
$6x - 10y = 4$

37. $6x + 3y = 0$
$-18x - 9y = 0$

38. $3x - 5y = 0$
$9x - 15y = 0$

39. $3x = 3 + 2y$
$-\frac{4}{3}x + y = \frac{1}{3}$

40. $3x = 27 + 2y$
$x - \frac{7}{2}y = -25$

41. $\frac{1}{5}x + y = \frac{6}{5}$

$\frac{1}{10}x + \frac{1}{3}y = \frac{5}{6}$

42. $\frac{1}{3}x + \frac{1}{2}y = \frac{13}{6}$

$\frac{1}{2}x - \frac{1}{4}y = -\frac{3}{4}$

43. $2.4x + 1.7y = 7.6$

$1.2x - 0.5y = 9.2$

44. $0.5x + 3.4y = 13$

$1.5x - 2.6y = -25$

Relating Concepts (Exercises 45–48) For Individual or Group Work

The graph shows average U.S. movie theater ticket prices from 2001 through 2010. In 2001, the average ticket price was \$5.66, *as represented on the graph by the point* $P(2001, 5.66)$. *In 2010, the average ticket price was* \$7.89, *as represented on the graph by the point* $Q(2010, 7.89)$. ***Work Exercises 45–48 in order.***

Average Movie Ticket Price

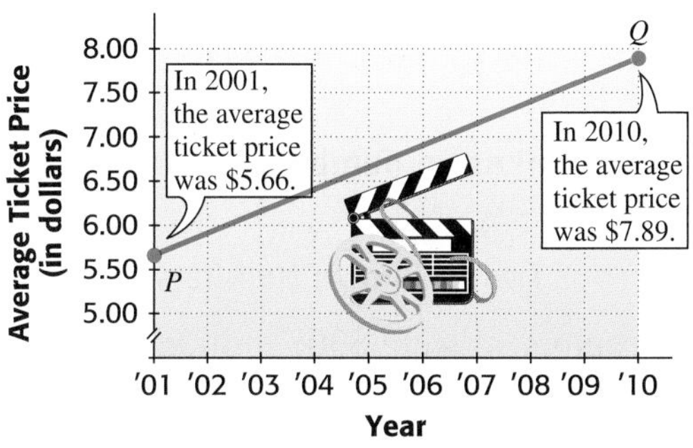

Source: Motion Picture Association of America.

45. The line segment has an equation that can be written in the form $y = ax + b$. Using the coordinates of point P with $x = 2001$ and $y = 5.66$, write an equation in the variables a and b.

46. Using the coordinates of point Q with $x = 2010$ and $y = 7.89$, write a second equation in the variables a and b.

47. Write the system of equations formed from the two equations in **Exercises 45 and 46.** Solve the system, giving the values of a and b to three decimal places. (*Hint*: Eliminate b using the elimination method.)

48. (a) What is the equation of the segment PQ?

(b) Let $x = 2008$ in the equation from part (a), and solve for y (to two decimal places). How does the result compare with the actual figure of \$7.18?

15.4 Applications of Linear Systems

OBJECTIVES

1. **Solve problems about unknown numbers.**
2. **Solve problems about quantities and their costs.**
3. **Solve problems about mixtures.**
4. **Solve problems about distance, rate (or speed), and time.**

We modify the six-step problem-solving method from **Section 10.4** slightly to allow for two variables and two equations.

Solving an Applied Problem with Two Variables

Step 1 **Read** the problem carefully. What information is given? What is to be found?

Step 2 **Assign variables** to represent the unknown values. Use a sketch, diagram, or table, as needed. Write down what each variable represents.

Step 3 **Write two equations** using both variables.

Step 4 **Solve** the system of two equations.

Step 5 **State the answer.** Label it appropriately. Does it seem reasonable?

Step 6 **Check** the answer in the words of the *original* problem.

OBJECTIVE 1 Solve problems about unknown numbers.

EXAMPLE 1 Solving a Problem about Two Unknown Numbers

In 2011, consumer sales of sports equipment were \$307 million more for snow skiing than for snowboarding. Together, total equipment sales for these two sports were \$931 million. What were equipment sales for each sport? (*Source:* National Sporting Goods Association.)

Step 1 **Read** the problem carefully. We must find equipment sales (in millions of dollars) for snow skiing and for snowboarding. We know how much more equipment sales were for snow skiing than for snowboarding. Also, we know the total sales.

Step 2 **Assign variables.**

Let x = equipment sales for skiing (in millions of dollars),

and y = equipment sales for snowboarding (in millions of dollars).

Step 3 **Write two equations.**

$x = 307 + y$ Equipment sales for skiing were \$307 million more than equipment sales for snowboarding. (1)

$x + y = 931$ Total sales were \$931 million. (2)

Step 4 **Solve** the system from Step 3. We use the substitution method since the first equation is already solved for x.

$x + y = 931$ (2)

$(307 + y) + y = 931$ Let $x = 307 + y$.

$307 + 2y = 931$ Combine like terms.

$2y = 624$ Subtract 307.

Don't stop here. $y = 312$ Divide by 2.

Continued on Next Page

To find the value of x, we substitute 312 for y in either equation (1) or (2).

$$x = 307 + \mathbf{312} \qquad \text{Let } y = 312 \text{ in equation (1).}$$

$$x = \mathbf{619} \qquad \text{Add.}$$

Step 5 **State the answer.** Equipment sales for skiing were \$**619** million, and equipment sales for snowboarding were \$**312** million.

Step 6 **Check** the answer in the original problem. Since

$$619 = 307 + 312 \quad \text{and} \quad 619 + 312 = 931,$$

the answer satisfies the information given in the problem.

Work Problem **1** at the Side. ▶

CAUTION

If an applied problem asks for *two* values as in **Example 1,** be sure to give both of them in your answer.

OBJECTIVE 2 **Solve problems about quantities and their costs.**

EXAMPLE 2 Solving a Problem about Quantities and Costs

For a production of *Jersey Boys* at the August Wilson Theatre in New York, main floor tickets cost \$127 and rear mezzanine tickets cost \$97. The members of a club spent a total of \$3150 for 30 tickets. How many tickets of each kind did they buy? (*Source*: www.broadway.com)

Step 1 **Read** the problem several times.

Step 2 **Assign variables.**

Let $x =$ the number of main floor tickets,

and $y =$ the number of rear mezzanine tickets.

Summarize the information given in the problem in a table. The entries in the first two rows of the Total Value column were found by multiplying Number of Tickets by Price per Ticket.

	Number of Tickets	Price per Ticket (in dollars)	Total Value (in dollars)
Main Floor	x	127	$127x$
Mezzanine	y	97	$97y$
Total	30	XXXXXX	3150

Step 3 **Write two equations.**

$$x + y = 30 \qquad \text{Total number of tickets was 30.} \quad (1)$$

$$127x + 97y = 3150 \qquad \text{Total value of tickets was \$3150.} \quad (2)$$

Step 4 **Solve** the system formed in Step 3 using the elimination method.

$$x + y = 30 \quad (1)$$

$$127x + 97y = 3150 \quad (2)$$

Continued on Next Page

1 Solve each problem.

GS **(a)** In 2011, consumer sales of sports equipment were \$26 million less for tennis than for archery. Together, total equipment sales for these two sports were \$876 million. What were equipment sales for each sport? (*Source*: National Sporting Goods Association.)

Step 1
We must find ________.

Step 2
Let $x =$ equipment sales for tennis (in millions of dollars), and $y =$ equipment sales for ________ (in millions of dollars).

Step 3
Write two equations.

$x =$ ________ (1)
(See the first sentence in the problem.)

________ (2)
(See the second sentence in the problem.)

Complete Steps 4–6 to solve the problem. Give the answer.

(b) Two of the most popular movies of 2011 were *Fast Five* and *Cars 2*. Together, their domestic gross was about \$401 million. *Cars 2* grossed \$18 million less than *Fast Five*. How much did each movie gross? (*Source:* www.boxofficemojo.com)

Answers

1. (a) equipment sales for each sport; archery; $y - 26$; $x + y = 876$; equipment sales for tennis: \$425 million; equipment sales for archery: \$451 million
   (b) *Fast Five:* \$209.5 million; *Cars 2:* \$191.5 million

2 For a production of *The Lion King* playing at the Minskoff Theatre in New York, orchestra tickets cost \$129 and rear mezzanine tickets cost \$87. If a group of 18 people attended the show and spent a total of \$1776 for their tickets, how many of each kind of ticket did they buy? (*Source:* www.broadway.com)

(a) Complete the table.

	Number of Tickets	Price per Ticket (dollars)	Total Value (dollars)
Orchestra	x	____	____
Mezzanine	y	____	____
Total	____	XXX	____

(b) Write a system of equations.

(c) Use the system of equations to solve the problem. Check your answer in the words of the original problem.

$$\begin{aligned} -97x - 97y &= -2910 && \text{Multiply equation (1) by } -97. \\ 127x + 97y &= 3150 && (2) \\ \hline 30x &= 240 && \text{Add.} \end{aligned}$$

Main floor tickets $\longrightarrow$ $x = 8$ Divide by 30.

Substitute 8 for x in equation (1).

$$x + y = 30 \quad (1)$$

$$8 + y = 30 \quad \text{Let } x = 8.$$

Mezzanine tickets $\longrightarrow$ $y = 22$ Subtract 8.

Step 5 **State the answer.** The club members bought 8 main floor tickets and 22 rear mezzanine tickets.

Step 6 **Check.** The sum of 8 and 22 is 30, so the total number of tickets is correct. Since 8 tickets were purchased at \$127 each and 22 at \$97 each, the total spent on all the tickets is

$$\$127(8) + \$97(22) = \$3150, \quad \text{as required.}$$

◀ **Work Problem 2 at the Side.**

OBJECTIVE 3 **Solve problems about mixtures.** Mixture problems that involve percent can be solved using a system of two equations in two variables.

EXAMPLE 3 **Solving a Mixture Problem Involving Percent**

A pharmacist needs 100 L of 50% alcohol solution. She has a 30% alcohol solution and an 80% alcohol solution, which she can mix. How many liters of each will be required to make the 100 L of a 50% alcohol solution?

Step 1 **Read** the problem. Note the percent of each solution and of the mixture.

Step 2 **Assign variables.**

Let x = the number of liters of 30% alcohol needed,

and y = the number of liters of 80% alcohol needed.

Liters of Mixture	Percent (as a decimal)	Liters of Pure Alcohol
x	0.30	$0.30x$
y	0.80	$0.80y$
100	0.50	$0.50(100)$

Summarize the information in a table. Percents are written as decimals.

Figure 8 gives an idea of what is happening in this problem.

Figure 8

Continued on Next Page

Answers

2. (a)

	Number of Tickets	Price per Ticket (dollars)	Total Value (dollars)
Orchestra	x	129	$129x$
Mezzanine	y	87	$87y$
Total	18	XXX	1776

(b) $x + y = 18$
$129x + 87y = 1776$

(c) orchestra: 5; mezzanine: 13

Step 3 **Write two equations.** The total number of liters in the final mixture will be 100, which gives one equation.

$$x + y = 100$$

To find the amount of pure alcohol in each mixture, multiply the number of liters by the concentration. (Refer to the table.) The amount of pure alcohol in the 30% solution added to the amount of pure alcohol in the 80% solution will equal the amount of pure alcohol in the final 50% solution. This gives a second equation.

$$0.30x + 0.80y = 0.50(100)$$

These two equations form a system.

Be sure to write two equations.

$$x + y = 100 \quad (1)$$
$$0.30x + 0.80y = 50 \quad (2) \quad 0.50(100) = 50$$

Step 4 **Solve** the system using the substitution method. Solving equation (1) for x gives $x = 100 - y$. Substitute $100 - y$ for x in equation (2).

$$0.30x + 0.80y = 50 \quad (2)$$
$$0.30(\mathbf{100 - y}) + 0.80y = 50 \quad \text{Let } x = 100 - y.$$
$$30 - 0.30y + 0.80y = 50 \quad \text{Distributive property}$$
$$30 + 0.50y = 50 \quad \text{Combine like terms.}$$
$$0.50y = 20 \quad \text{Subtract 30.}$$
$$\text{Liters of 80\% solution} \longrightarrow y = \mathbf{40} \quad \text{Divide by 0.50.}$$

Equation (1) solved for x is $x = 100 - y$. Substitute 40 for y to find the value of x.

$$x = 100 - \mathbf{40} \quad \text{Let } y = 40.$$
$$\text{Liters of 30\% solution} \longrightarrow x = \mathbf{60} \quad \text{Subtract.}$$

Step 5 **State the answer.** The pharmacist should use 60 L of the 30% solution and 40 L of the 80% solution.

Step 6 **Check** the answer in the original problem. Since

$$60 + 40 = 100 \quad \text{and} \quad 0.30(60) + 0.80(40) = 50,$$

this mixture will give the 100 L of 50% solution, as required.

Work Problems 3 and 4 at the Side. ▶

Note

In Example 3, we could have used the elimination method. Also, we could have cleared decimals by multiplying each side of equation (2) by 10.

OBJECTIVE ▶ 4 Solve problems about distance, rate (or speed), and time.
If an automobile travels at an average rate of 50 mph for 2 hr, then it travels

$$50 \times 2 = 100 \text{ mi.}$$

This is an example of the basic relationship between distance, rate, and time,

distance = rate × time, given by the formula $\boldsymbol{d = rt}$.

Work Problem 5 at the Side. ▶

3 GS How many liters of a 25% alcohol solution must be mixed with a 12% solution to get 13 L of a 15% solution?

(a) Complete the table.

Liters	Percent (as a decimal)	Liters of Pure Alcohol
x	0.25	$0.25x$
y	0.12	_______
13	0.15	_______

(b) Write a system of equations, and use it to solve the problem.

4 Solve the problem.

Joe needs 60 milliliters (mL) of 20% acid solution for a chemistry experiment. The lab has on hand only 10% and 25% solutions. How much of each should he mix to get the desired amount of 20% solution?

5 Solve using the distance formula $d = rt$.

A small plane traveled from Stockholm, Sweden, to Oslo, Norway, averaging 244 km per hr. The trip took 1.7 hr. To the nearest kilometer, what is the distance between the two cities?

Answers

3. (a)

Liters	Percent (as a decimal)	Liters of Pure Alcohol
x	0.25	$0.25x$
y	0.12	$0.12y$
13	0.15	$0.15(13)$

(b) $x + y = 13$
$0.25x + 0.12y = 0.15(13)$;
3 L of 25%; 10 L of 12%

4. 20 mL of 10%; 40 mL of 25%

5. 415 km

6 Two cars that were 450 mi apart traveled toward each other. They met after 5 hr. If one car traveled twice as fast as the other, what were their rates?

(a) Complete this table.

	r	t	d
Faster Car	x	5	___
Slower Car	y	5	___

(b) Write a system of equations, and use it to solve the problem.

7 Solve the problem.

From a truck stop, two trucks travel in opposite directions on a straight highway. In 3 hr they are 405 mi apart. Find the rate of each truck if one travels 5 mph faster than the other.

Answers

6. (a)

	r	t	d
Faster Car	x	5	$5x$
Slower Car	y	5	$5y$

(b) $5x + 5y = 450$
$x = 2y$;
faster car: 60 mph; slower car: 30 mph

7. faster truck: 70 mph; slower truck: 65 mph

EXAMPLE 4 Solving a Problem about Distance, Rate, and Time

Two executives in cities 400 mi apart drive to a business meeting at a location on the line between their cities. They meet after 4 hr. Find the rate (speed) of each car if one car travels 20 mph faster than the other.

Step 1 **Read** the problem carefully.

Step 2 **Assign variables.**

Let $x =$ the rate of the faster car,

and $y =$ the rate of the slower car.

Make a table using the formula $d = rt$, and draw a sketch. See **Figure 9.**

	r	t	d
Faster Car	x	4	$x \cdot 4$, or $4x$
Slower Car	y	4	$y \cdot 4$, or $4y$

Since each car travels for 4 hr, the time t for each car is 4. Find d, using $d = rt$ (or $rt = d$).

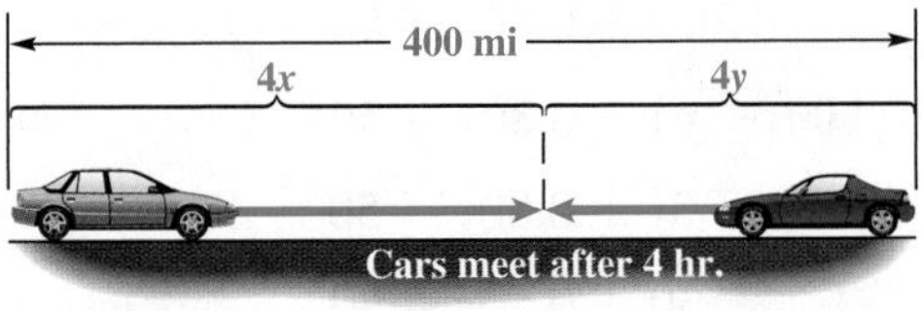

Figure 9

Step 3 **Write two equations.** The total distance traveled by both cars is 400 mi, which gives equation (1). The faster car goes 20 mph faster than the slower car, which gives equation (2).

$$4x + 4y = 400 \quad (1)$$
$$x = 20 + y \quad (2)$$

Step 4 **Solve** the system of equations by substitution. Replace x with $20 + y$ in equation (1) and then solve for y.

$$4x + 4y = 400 \quad (1)$$
$$4(20 + y) + 4y = 400 \quad \text{Let } x = 20 + y.$$
$$80 + 4y + 4y = 400 \quad \text{Distributive property}$$
$$80 + 8y = 400 \quad \text{Combine like terms.}$$
$$8y = 320 \quad \text{Subtract 80.}$$
$$\text{Slower car} \rightarrow y = 40 \quad \text{Divide by 8.}$$

To find x, substitute 40 for y in equation (2), $x = 20 + y$.

$$x = 20 + 40 \quad \text{Let } y = 40.$$
$$\text{Faster car} \rightarrow x = 60 \quad \text{Add.}$$

Step 5 **State the answer.** The rates of the cars are 40 mph and 60 mph.

Step 6 **Check** the answer. Since each car travels for 4 hr, total distance is

$$4(60) + 4(40) = 240 + 160 = 400 \text{ mi}, \quad \text{as required.}$$

◀ **Work Problems 6 and 7 at the Side.**

CAUTION

Be careful. ***When you use two variables to solve a problem, you must write two equations.***

EXAMPLE 5 Solving a Problem about Distance, Rate, and Time

A plane flies 560 mi in 1.75 hr traveling with the wind. The return trip against the same wind takes the plane 2 hr. Find the rate (speed) of the plane and the wind speed.

Step 1 **Read** the problem several times.

Step 2 **Assign variables.**

Let x = the rate of the plane,

and y = the wind speed.

When the plane is traveling *with* the wind, the wind "pushes" the plane. In this case, the rate (speed) of the plane is the *sum* of the rate of the plane and the wind speed, $(x + y)$ mph. See **Figure 10.**

When the plane is traveling *against* the wind, the wind "slows" the plane down. In this case, the rate (speed) of the plane is the *difference* between the rate of the plane and the wind speed, $(x - y)$ mph. Again, see **Figure 10.**

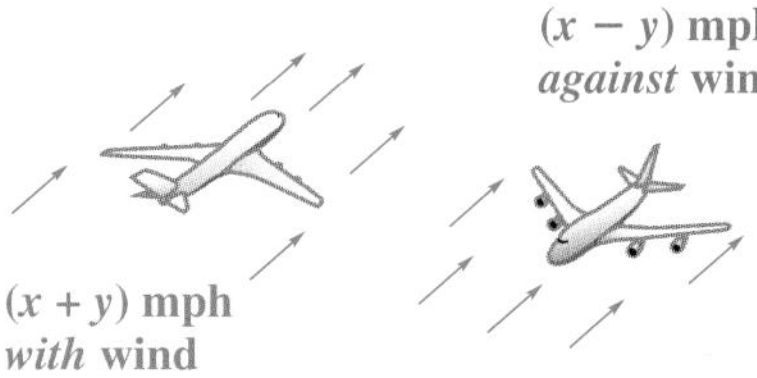

Figure 10

	r	t	d
With Wind	$x + y$	1.75	560
Against Wind	$x - y$	2	560

Summarize this information in a table. Use the formula $d = rt$ (or $rt = d$).

Step 3 **Write two equations.** Refer to the table to do this.

$$(x + y)1.75 = 560 \xrightarrow{\text{Divide by 1.75.}} x + y = 320 \quad (1)$$

$$(x - y)2 = 560 \xrightarrow{\text{Divide by 2.}} x - y = 280 \quad (2)$$

Step 4 **Solve** the system of equations (1) and (2) using elimination.

$$\begin{aligned} x + y &= 320 && (1) \\ x - y &= 280 && (2) \\ \hline 2x \quad &= 600 && \text{Add.} \\ x &= 300 && \text{Divide by 2.} \end{aligned}$$

Since $x + y = 320$ and $x = 300$, it follows that $y = 20$.

Step 5 **State the answer.** The rate of the plane is 300 mph, and the wind speed is 20 mph.

Step 6 **Check.** The answer seems reasonable, and true statements result when the values are substituted into the equations of the system.

Work Problem 8 at the Side. ▶

8 Solve each problem.

GS **(a)** In 1 hr, Gigi can row 2 mi against the current or 10 mi with the current. Find the rate of the current and Gigi's rate in still water.

Steps 1 and 2
Let x = the rate of the current,

and y = _______ in still water.

Then her rate *against* the current is (_______) mph,

and her rate with the current is (_______) mph.

Complete the table.

	r	t	d
Against Current	___	___	___
With Current	___	___	___

Step 3
Write two equations.

$(y - x) \cdot 1 =$ ____ (1)
(See the first row of the table.)

_______________ (2)
(See the second row of the table.)

Complete Steps 4–6 to solve the problem. Give the answer.

(b) In 1 hr, a boat travels 15 mph with the current and 9 mph against the current. Find the rate of the boat and the rate of the current.

Answers

8. (a) Gigi's rate; $y - x$; $y + x$

	r	t	d
Against Current	$y - x$	1	2
With Current	$y + x$	1	10

2; $(y + x) \cdot 1 = 10$;
rate of the current: 4 mph;
Gigi's rate: 6 mph

(b) rate of the boat: 12 mph;
rate of the current: 3 mph

15.4 Exercises

FOR EXTRA HELP

Download the MyDashBoard App

MyMathLab®

CONCEPT CHECK *Choose the correct response in Exercises 1–8.*

1. Which expression represents the monetary value of x 20-dollar bills?

A. $\frac{x}{20}$ dollars **B.** $\frac{20}{x}$ dollars **C.** $(20 + x)$ dollars **D.** $20x$ dollars

2. Which expression represents the cost of t pounds of candy that sells for \$1.95 per lb?

A. \$1.95t **B.** $\frac{\$1.95}{t}$ **C.** $\frac{t}{\$1.95}$ **D.** $\$1.95 + t$

3. Which expression represents the amount of interest earned on d dollars invested at an interest rate of 2% for 1 yr?

A. $2d$ dollars **B.** $0.02d$ dollars **C.** $0.2d$ dollars **D.** $200d$ dollars

4. Suppose that x liters of a 40% acid solution are mixed with y liters of a 35% solution to obtain 100 L of a 38% solution. One equation in a system for solving this problem is $x + y = 100$. Which one of the following is the other equation?

A. $0.35x + 0.40y = 0.38(100)$ **B.** $0.40x + 0.35y = 0.38(100)$

C. $35x + 40y = 38$ **D.** $40x + 35y = 0.38(100)$

5. According to *Natural History* magazine, the speed of a cheetah is 70 mph. If a cheetah runs for x hours, how many miles does the cheetah cover?

A. $(70 + x)$ miles **B.** $(70 - x)$ miles **C.** $\frac{70}{x}$ miles **D.** $70x$ miles

6. How far does a car travel in 2.5 hr if it travels at an average rate of x miles per hour?

A. $(x + 2.5)$ miles **B.** $\frac{2.5}{x}$ miles **C.** $\frac{x}{2.5}$ miles **D.** $2.5x$ miles

7. What is the rate of a plane that travels at 560 mph *with* a wind of r mph?

A. $\frac{r}{560}$ mph **B.** $(560 - r)$ mph **C.** $(560 + r)$ mph **D.** $(r - 560)$ mph

8. What is the rate of a plane that travels at 560 mph *against* a wind of r mph?

A. $(560 + r)$ mph **B.** $\frac{560}{r}$ mph **C.** $(560 - r)$ mph **D.** $(r - 560)$ mph

GS *In Exercises 9 and 10, refer to the six-step problem-solving method. Fill in the blanks for Steps 2 and 3, and then complete the solution by applying Steps 4–6.*

9. The sum of two numbers is 98 and the difference between them is 48. Find the two numbers.

Step 1 **Read** the problem carefully.

Step 2 **Assign variables.**

Let $x =$ the first number

and $y =$ ____________.

Step 3 **Write two equations.**

First equation: $x + y = 98$

Second equation: ________

10. The sum of two numbers is 201 and the difference between them is 11. Find the two numbers.

Step 1 **Read** the problem carefully.

Step 2 **Assign variables.**

Let $x =$ ____________

and $y =$ the second number.

Step 3 **Write two equations.**

First equation: ________

Second equation: $x - y = 11$

Solve each problem using a system of equations. **See Example 1.**

11. Two of the longest-running shows on Broadway are *The Phantom of the Opera* and *Cats*. As of August 21, 2011, there had been a total of 17,288 performances of the two shows, with 2318 more performances of *The Phantom of the Opera* than *Cats*. How many performances were there of each show? (*Source:* The Broadway League.)

12. During Broadway runs of *A Chorus Line* and *Beauty and the Beast,* there have been 676 fewer performances of *Beauty and the Beast* than of *A Chorus Line.* There were a total of 11,598 performances of the two shows. How many performances were there of each show? (*Source:* The Broadway League.)

13. The two domestic top-grossing movies of 2011 were *Harry Potter and the Deathly Hallows Part 2* and *Transformers: Dark of the Moon*. The Transformers movie grossed $29 million less than the Harry Potter movie, and together the two films took in $733 million. How much did each of these movies earn? (*Source:* www.boxofficemojo.com)

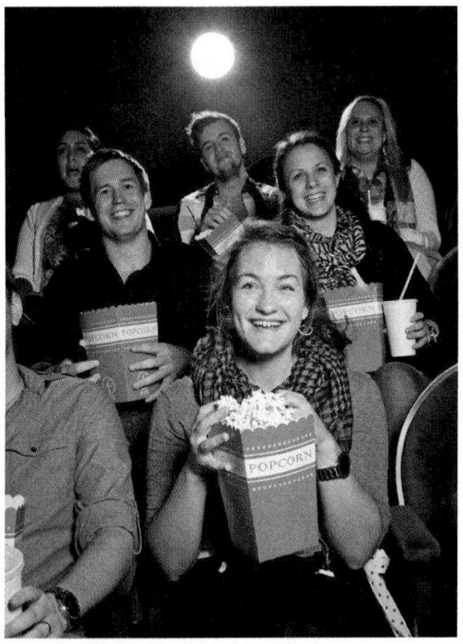

14. During their opening weekends, the movies *Harry Potter and the Deathly Hallows Part 2* and *Transformers: Dark of the Moon* grossed a total of $267 million, with the Harry Potter movie grossing $71 million more than the Transformers movie. How much did each of these movies earn during their opening weekends? (*Source:* www.boxofficemojo.com)

15. The Terminal Tower in Cleveland, Ohio, is 239 ft shorter than the Key Tower, also in Cleveland. The total of the heights of the two buildings is 1655 ft. Find the heights of the buildings. (*Source: World Almanac and Book of Facts.*)

16. The total of the heights of the Chase Tower and the One America Tower, both in Indianapolis, Indiana, is 1234 ft. The Chase Tower is 168 ft taller than the One America Tower. Find the heights of the two buildings. (*Source: World Almanac and Book of Facts.*)

17. GS An official playing field (including end zones) for the Indoor Football League has length 38 yd longer than its width. The perimeter of the rectangular field is 188 yd. Find the length and width of the field. (*Source:* Indoor Football League.)

Steps 1 and 2 **Read** carefully, and **assign** ________.
Let $x =$ the length (in yards),
and $y =$ the ________ (in yards).

Step 3 **Write two equations.**
Equation (1): See the first sentence in the problem. Express the length in terms of the width.

$x =$ ________ (1)

Equation (2): See the second sentence in the problem. Perimeter of a rectangle equals twice the ________ plus ________ the width.

$2x +$ ________ $=$ ________ (2)

Now complete the solution.

18. GS Pickleball is a combination of badminton, tennis, and ping pong. The perimeter of the rectangular-shaped court is 128 ft. The width is 24 ft shorter than the length. Find the length and width of the court. (*Source:* www.sportsknowhow.com)

Steps 1 and 2 **Read** carefully, and **assign** ________.
Let $x =$ the ________ (in feet),
and $y =$ the width (in feet).

Step 3 **Write two equations.**
Equation (1): Perimeter of a rectangle equals ________ the length plus twice the ________.

________ $+ 2y =$ ________ (1)

Equation (2): See the third sentence in the problem. Express the width in terms of the length.

$y =$ ________ (2)

Now complete the solution.

Suppose that x units of a product cost C dollars to manufacture and earn revenue of R dollars. The value of x, where the expressions for C and R are equal, is the ***break-even quantity,*** *the number of units that produce 0 profit.*

In Exercises 19 and 20, ***(a)*** *find the break-even quantity, and* ***(b)*** *decide whether the product should be produced based on if it will earn a profit. (Profit equals revenue minus cost.)*

19. $C = 85x + 900$; $R = 105x$;
No more than 38 units can be sold.

20. $C = 105x + 6000$; $R = 255x$;
No more than 400 units can be sold.

For each problem, complete any tables. Then solve the problem using a system of equations. ***See Example 2.***

21. A motel clerk counts his \$1 and \$10 bills at the end of a day. He finds that he has a total of 74 bills having a combined monetary value of \$326. Find the number of bills of each denomination that he has.

Number of Bills	Denomination of Bill (in dollars)	Total Value (in dollars)
x	1	________
y	10	________
74	XXXXXX	326

22. Carly is a bank teller. At the end of a day, she has a total of 69 \$5 and \$10 bills. The total value of the money is \$590. How many of each denomination does she have?

Number of Bills	Denomination of Bill (in dollars)	Total Value (in dollars)
x	5	$5x$
y	10	________
________	XXXXXX	________

23. Tracy Sudak bought each of her seven nephews a DVD of *Moneyball* or a Blu-ray disc of *The Concert for George*. Each DVD cost \$16.99. Each Blu-ray disc cost \$22.42. Tracy spent a total of \$129.79. How many of each did she buy? (*Source:* www.amazon.com)

24. Terry Wong bought each of his five nieces a DVD of *Dolphin Tale* or a Blu-ray disc of *Treasure Buddies*. Each DVD cost \$14.99. Each Blu-ray disc cost \$24.99. Terry spent a total of \$84.95. How many of each did he buy? (*Source:* www.amazon.com)

25. GS Maria Lopez has twice as much money invested at 5% simple annual interest as she does at 4%. If her yearly income from these two investments is \$350, how much does she have invested at each rate?

Amount Invested (in dollars)	Rate of Interest	Interest for One Year (in dollars)
x	5%, or 0.05	$0.05x$
y	______	______
XXXXX	XXXXX	350

Equation (1): $x =$ ____

Equation (2): $0.05x +$ ____ $=$ ____

26. GS Charles Miller invested in two accounts, one paying 3% simple annual interest and the other paying 2% interest. He earned a total of \$880 interest. If he invested three times as much in the 3% account as in the 2% account, how much did he invest at each rate?

Amount Invested (in dollars)	Rate of Interest	Interest for One Year (in dollars)
x	______	______
y	______	______
XXXXX	XXXX	______

Equation (1): ____ + ____ = 880

Equation (2): $x =$ ____

27. The two top-grossing North American concert tours in 2011 were U2 and Taylor Swift. Based on average ticket prices, it cost a total of \$912 to buy six tickets for U2 and five tickets for Taylor Swift. Three tickets for U2 and four tickets for Taylor Swift cost \$564. How much did an average ticket cost for each tour? (*Source:* Pollstar.)

28. Two other popular North American concert tours in 2011 were Kenny Chesney and Lady Gaga. Based on average ticket prices, it cost a total of \$875 to buy eight tickets for Kenny Chesney and three tickets for Lady Gaga. Four tickets for Kenny Chesney and five tickets for Lady Gaga cost \$777. How much did an average ticket cost for each tour? (*Source:* Pollstar.)

For each problem, complete any tables. Then solve the problem using a system of equations. ***See Example 3.***

29. A 40% dye solution is to be mixed with a 70% dye solution to get 120 L of a 50% solution. How many liters of the 40% and 70% solutions will be needed?

Liters of Solution	Percent (as a decimal)	Liters of Pure Dye
x	0.40	______
y	0.70	______
120	0.50	______

30. A 90% antifreeze solution is to be mixed with a 75% solution to make 120 L of a 78% solution. How many liters of the 90% and 75% solutions will be used?

Liters of Solution	Percent (as a decimal)	Liters of Pure Antifreeze
x	0.90	______
y	0.75	______
120	0.78	______

31. Ahmad Hashemi wishes to mix coffee worth \$6 per lb with coffee worth \$3 per lb to get 90 lb of a mixture worth \$4 per lb. How many pounds of the \$6 and the \$3 coffees will be needed?

Number of Pounds	Dollars per Pound	Cost (in dollars)
x	6	____
y	____	____
90	____	____

32. Mariana Coanda wishes to blend candy selling for \$1.20 per lb with candy selling for \$1.80 per lb to get a mixture that will be sold for \$1.40 per lb. How many pounds of the \$1.20 and the \$1.80 candies should be used to get 45 lb of the mixture?

Number of Pounds	Dollars per Pound	Cost (in dollars)
x	____	____
y	1.80	____
45	____	____

33. How many pounds of nuts selling for \$6 per lb and raisins selling for \$3 per lb should Kelli Hammer combine to obtain 60 lb of a trail mix selling for \$5 per lb?

34. Callie Daniels is preparing cheese trays. She uses some cheeses that sell for \$8 per lb and others that sell for \$12 per lb. How many pounds of each cheese should she use in order for the mixed cheeses on the trays to weigh a total of 56 lb and sell for \$10.50 per lb?

For each problem, complete any tables. Then solve the problem using a system of equations. ***See Examples 4 and 5.***

35. Two trains start from towns 495 mi apart and travel toward each other on parallel tracks. They pass each other 4.5 hr later. If one train travels 10 mph faster than the other, find the rate of each train.

	r	t	d
Train 1	x	____	____
Train 2	y	____	____

Equation (1): $4.5x +$ ____ $=$ ____

Equation (2): $x =$ ____ $+$ ____

36. Two trains that are 495 mi apart travel toward each other. They pass each other 5 hr later. If one train travels half as fast as the other, what are their rates?

	r	t	d
Train 1	x	____	____
Train 2	____	____	$5y$

Equation (1): ____ $+ 5y =$ ____

Equation (2): $x =$ ____y, or ____ $= y$

37. **RAGBRAI®**, the Des Moines **R**egister's **A**nnual **G**reat **B**icycle **R**ide **A**cross **I**owa, is the longest and oldest touring bicycle ride in the world. Suppose a cyclist began the 471-mi ride on July 22, 2012, in western Iowa at the same time that a car traveling toward it left eastern Iowa. If the bicycle and the car met after 7.5 hr and the car traveled 35.8 mph faster than the bicycle, find the average rate of each. (*Source:* www.ragbrai.com)

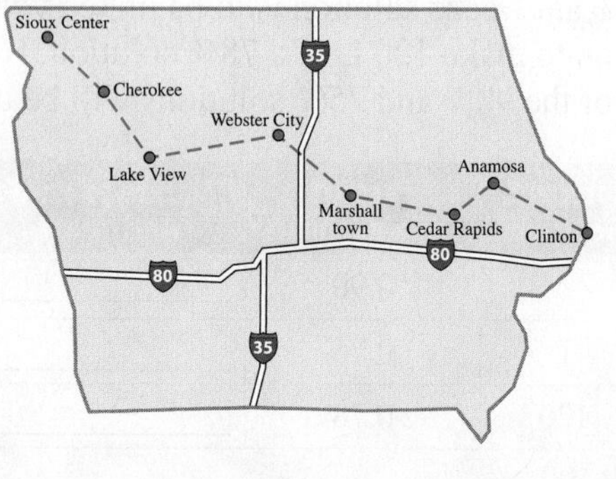

38. In 2010, Atlanta's Hartsfield Airport was the nation's busiest. Suppose two planes leave the airport at the same time, one traveling east and the other traveling west. If the planes are 2100 mi apart after 2 hr and one plane travels 50 mph faster than the other, find the rate of each plane. (*Source:* Airports Council International.)

39. Toledo and Cincinnati are 200 mi apart. A car leaves Toledo traveling toward Cincinnati, and another car leaves Cincinnati at the same time, traveling toward Toledo. The car leaving Toledo averages 15 mph faster than the other, and they meet after 1 hr, 36 min. What are the rates of the cars?

40. Kansas City and Denver are 600 mi apart. Two cars start from these cities, traveling toward each other. They meet after 6 hr. Find the rate of each car if one travels 30 mph slower than the other.

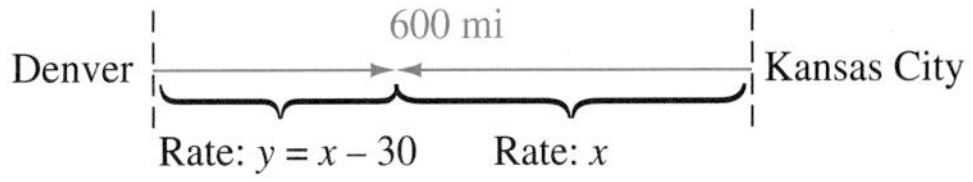

41. At the beginning of a bicycle ride for charity, Roberto and Juana are 30 mi apart. If they leave at the same time and ride in the same direction, Roberto overtakes Juana in 6 hr. If they ride toward each other, they meet in 1 hr. What are their rates?

42. Mr. Abbot left Farmersville in a plane at noon to travel to Exeter. Mr. Baker left Exeter in his automobile at 2 P.M. to travel to Farmersville. It is 400 mi from Exeter to Farmersville. If the sum of their rates was 120 mph, and if they crossed paths at 4 P.M., find the rate of each.

43. A boat takes 3 hr to go 24 mi upstream. It can go 36 mi downstream in the same time. Find the rate of the current and the rate of the boat in still water.

Let $x =$ the rate of the boat in still water and $y =$ the rate of the current.

	r	t	d
Downstream	$x + y$	____	36
Upstream	$x - y$	____	____

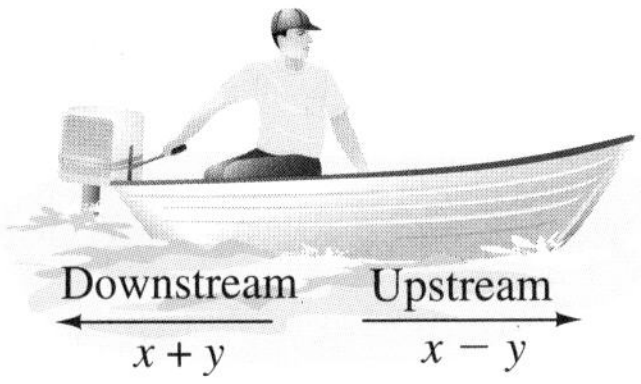

44. It takes a boat $1\frac{1}{2}$ hr to go 12 mi downstream, and 6 hr to return. Find the rate of the boat in still water and the rate of the current.

Let $x =$ the rate of the boat in still water and $y =$ the rate of the current.

	r	t	d
Downstream	$x + y$	$\frac{3}{2}$	____
Upstream	____	6	____

45. If a plane can travel 440 mph against the wind and 500 mph with the wind, find the wind speed and the rate of the plane in still air.

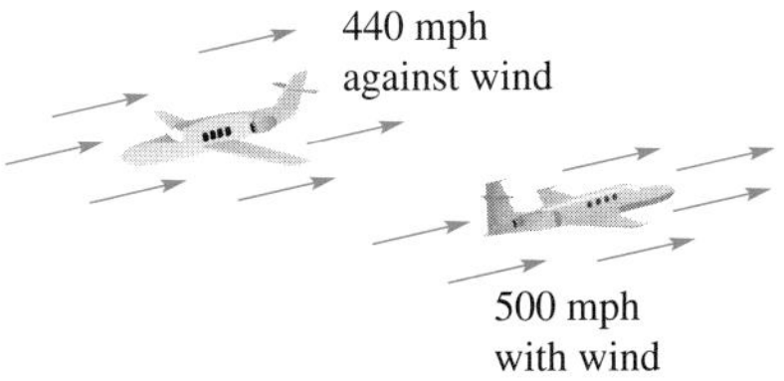

46. A small plane travels 200 mph with the wind and 120 mph against it. Find the wind speed and the rate of the plane in still air.

15.5 Solving Systems of Linear Inequalities

OBJECTIVE

1 Solve systems of linear inequalities by graphing.

We graphed the solutions of a linear inequality in **Section 11.5.** For example, to graph the solutions of $x + 3y > 12$, we first graph the boundary line $x + 3y = 12$ by finding and plotting a few ordered pairs that satisfy the equation. (The x- and y-intercepts are good choices.) Because the $>$ symbol does not include equality, the points on the line do *not* satisfy the inequality, and we graph it using a dashed line. To decide which region includes the points that are solutions, we choose a test point not on the line.

$$x + 3y > 12 \quad \text{Original inequality}$$

(0, 0) is a convenient test point.

$$0 + 3(0) \overset{?}{>} 12 \quad \text{Let } x = 0 \text{ and } y = 0.$$

$$0 > 12 \quad \text{False}$$

This false result indicates that the solutions are those points on the side of the line that does *not* include (0, 0), as shown in **Figure 11.**

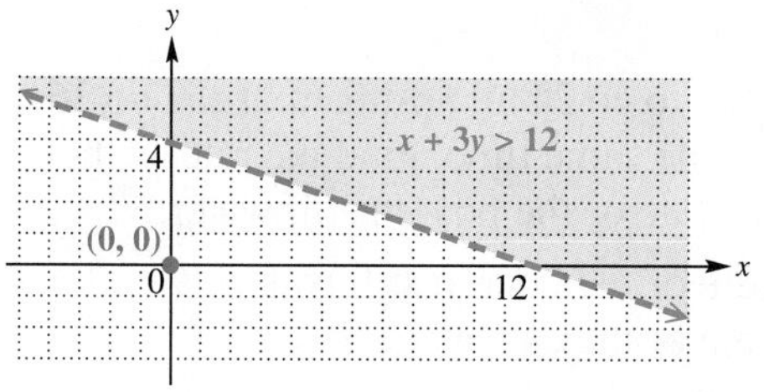

Figure 11

OBJECTIVE 1 **Solve systems of linear inequalities by graphing.** A **system of linear inequalities** consists of two or more linear inequalities. The **solution set of a system of linear inequalities** includes all points that make all inequalities of the system true at the same time.

Solving a System of Linear Inequalities

Step 1 **Graph each linear inequality.** Use the method of **Section 11.5.**

Step 2 **Choose the intersection.** Indicate the solution set of the system by shading the intersection of the graphs (the region where the graphs overlap).

EXAMPLE 1 Solving a System of Two Linear Inequalities

Graph the solution set of the system.

$$3x + 2y \leq 6$$
$$2x - 5y \geq 10$$

Step 1 To graph $3x + 2y \leq 6$, graph the solid boundary line $3x + 2y = 6$ using the intercepts $(0, 3)$ and $(2, 0)$. Determine the region to shade.

$$3x + 2y \leq 6$$

$$3(0) + 2(0) \overset{?}{\leq} 6 \quad \text{Use (0, 0) as a test point.}$$

$$0 \leq 6 \quad \text{True}$$

Shade the region containing $(0, 0)$. See **Figure 12(a)** on the next page.

Continued on Next Page

Now graph $2x - 5y \geq 10$ with solid boundary line $2x - 5y = 10$ using the intercepts $(0, -2)$ and $(5, 0)$. Determine the region to shade.

$$2x - 5y \geq 10$$

$$2(0) - 5(0) \stackrel{?}{\geq} 10 \quad \text{Use } (0, 0) \text{ as a test point.}$$

$$0 \geq 10 \quad \text{False}$$

Shade the region that does *not* contain $(0, 0)$. See **Figure 12(b).**

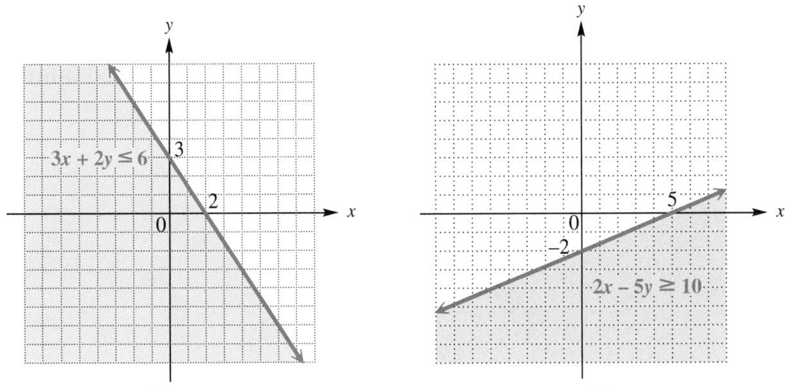

(a) **(b)**

Figure 12

Step 2 The solution set of this system includes all points in the intersection (overlap) of the graphs of the two inequalities. As shown in **Figure 13,** this intersection is the gray shaded region and portions of the two boundary lines that surround it.

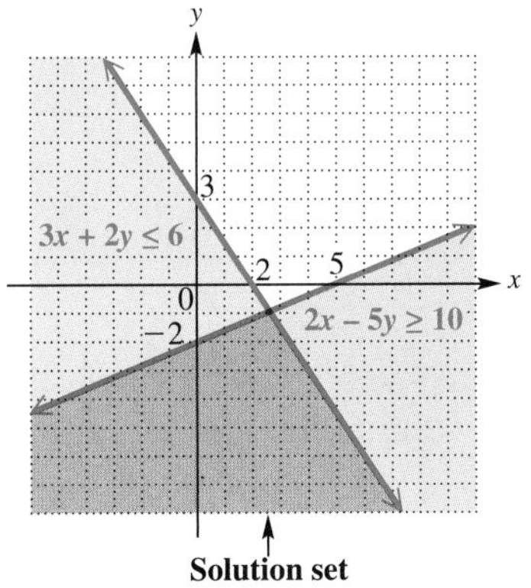

Figure 13

CHECK To confirm that the correct region of intersection is shaded, select a test point in the gray shaded region, say $(0, -4)$, and substitute it into *both* inequalities of the given system to make sure that true statements result. Try this. (Using an ordered pair that has one coordinate 0 makes the substitution easier.) ✓

Work Problem 1 at the Side. ▶

Note

We usually do all the work on one set of axes. In the remaining examples, only one graph is shown. Be sure that the region of the final solution set is clearly indicated.

1 Graph the solution set of each system.

GS **(a)** $x - 2y \leq 8$

$3x + y \geq 6$

To get you started, the graphs of $x - 2y = 8$ and $3x + y = 6$ are shown.

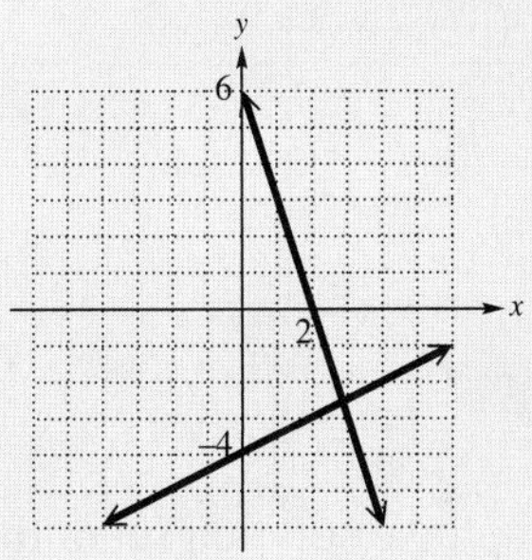

(b) $4x - 2y \leq 8$

$x + 3y \geq 3$

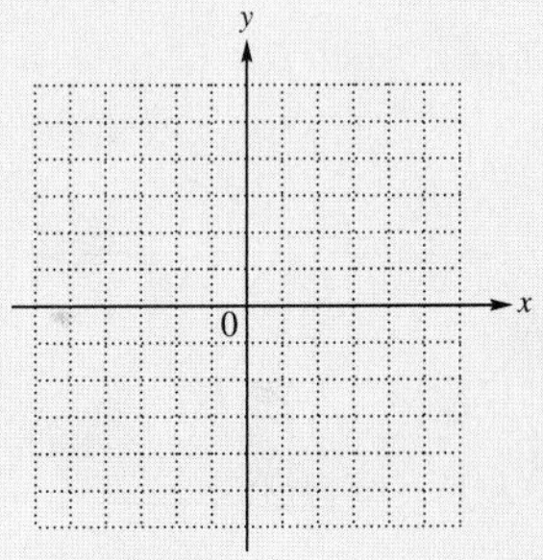

Answers

1. (a)

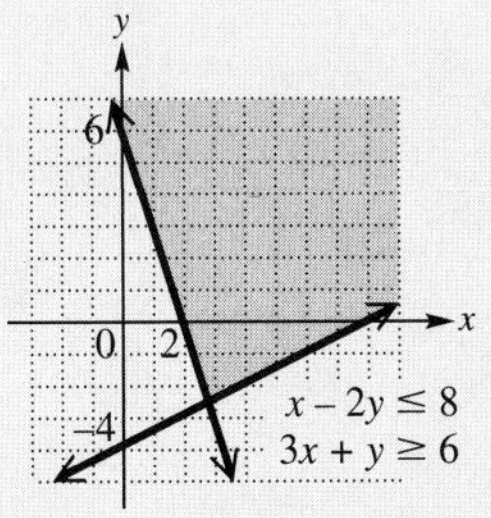

(b)

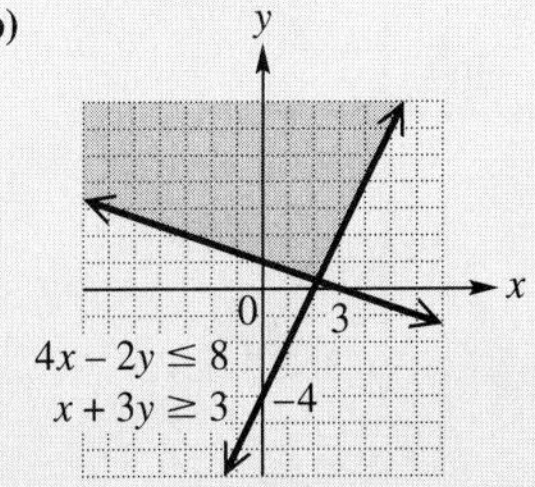

2 Graph the solution set of the system.

$$x + 2y < 0$$
$$3x - 4y < 12$$

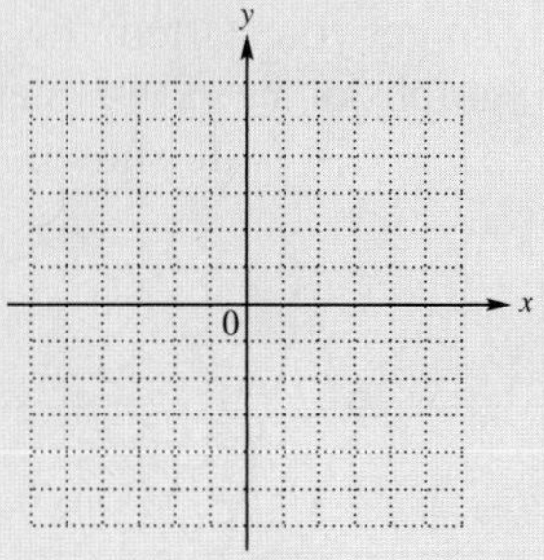

3 Graph the solution set of the system.

$$3x + 2y \leq 12$$
$$x \leq 2$$
$$y \leq 4$$

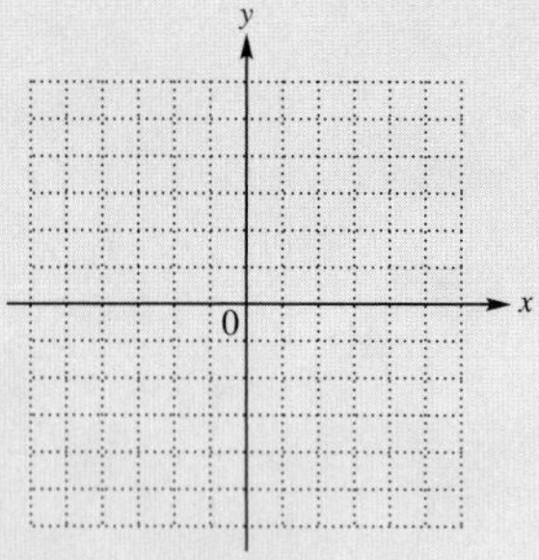

Answers

2.

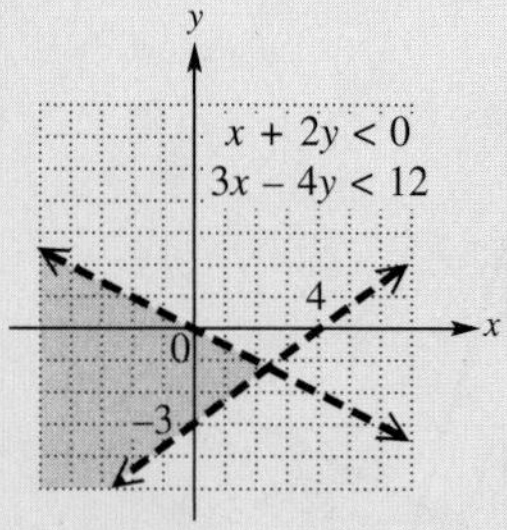

3.

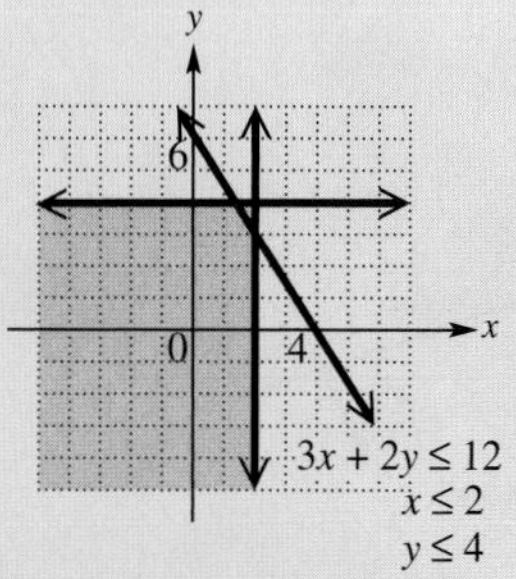

EXAMPLE 2 Solving a System of Two Linear Inequalities

Graph the solution set of the system.

$$x - y > 5$$
$$2x + y < 2$$

Figure 14 shows the graphs of both $x - y > 5$ and $2x + y < 2$. Dashed lines show that the graphs of the inequalities do not include their boundary lines. Use $(0, 0)$ as a test point to determine the region to shade for each inequality.

The solution set of the system is the region with the darkest shading. The solution set does not include either boundary line.

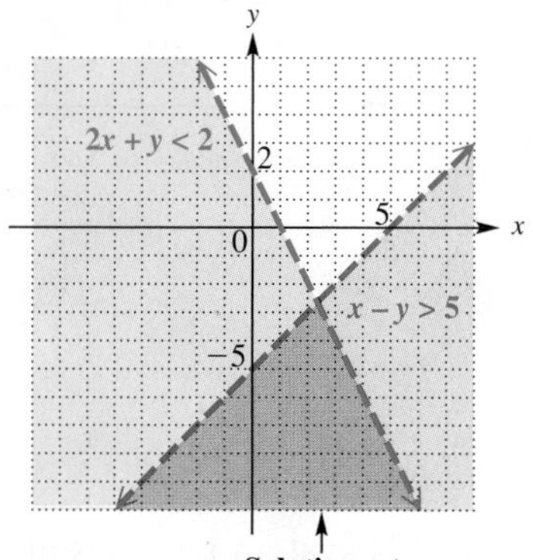

Figure 14

◀ Work Problem **2** at the Side.

EXAMPLE 3 Solving a System of Three Linear Inequalities

Graph the solution set of the system.

$$4x - 3y \leq 8$$
$$x \geq 2$$
$$y \leq 4$$

Recall that $x = 2$ is a vertical line through the point $(2, 0)$, and $y = 4$ is a horizontal line through $(0, 4)$. The graph of the solution set is the shaded region in **Figure 15,** including all boundary lines. (Here, use $(3, 2)$ as a test point to confirm that the correct region is shaded.)

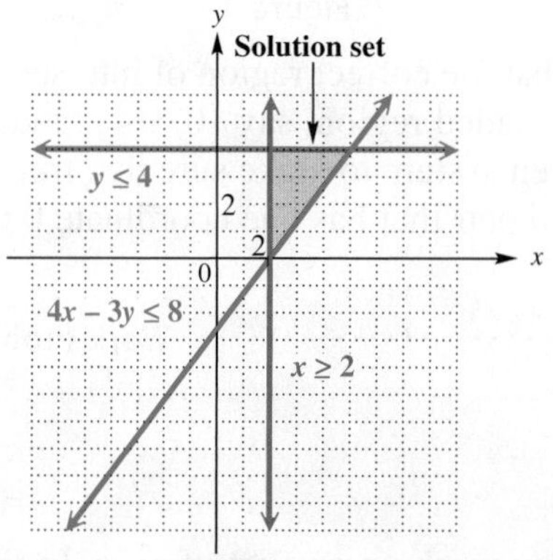

Figure 15

◀ Work Problem **3** at the Side.

15.5 Exercises

FOR EXTRA HELP

Download the MyDashBoard App

MyMathLab®

CONCEPT CHECK *Match each system of inequalities with the correct graph from choices A–D.*

1. $x \geq 5$
$y \leq -3$

2. $x \leq 5$
$y \geq -3$

3. $x > 5$
$y < -3$

4. $x < 5$
$y > -3$

A.

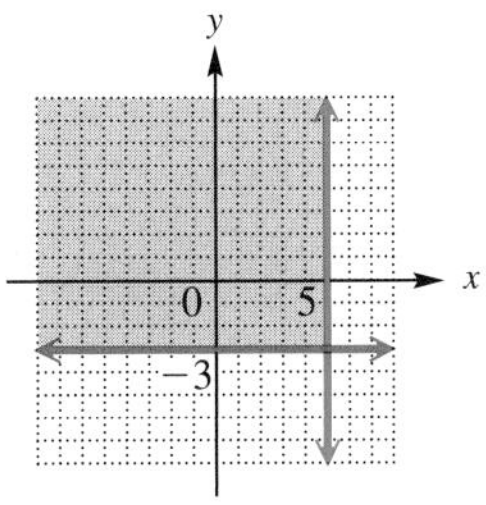

B.

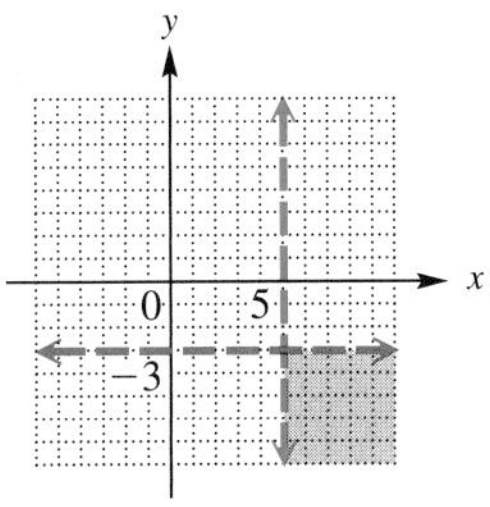

C.

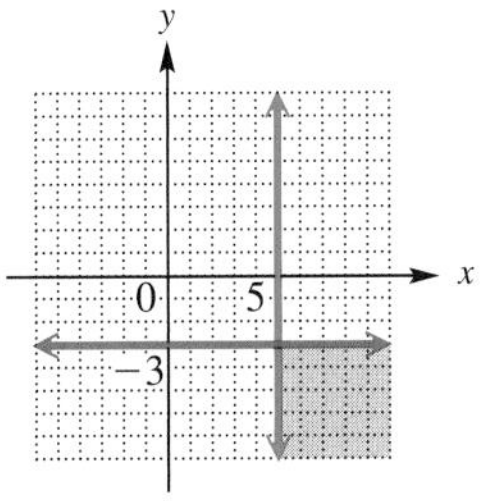

D.

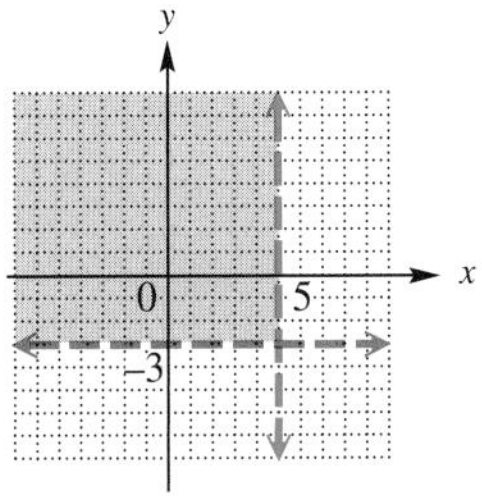

CONCEPT CHECK *In Exercises 5–8, shade the solution set of each system of inequalities. Boundary lines are already graphed.*

5. $x - 3y \leq 6$
$x \geq -4$

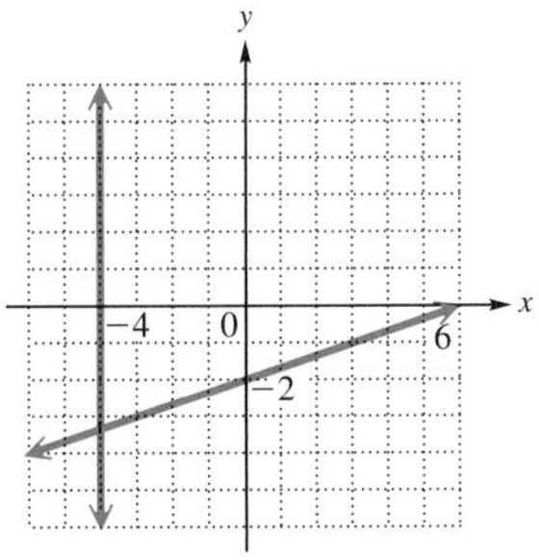

6. $x - 2y \geq 4$
$x \leq -2$

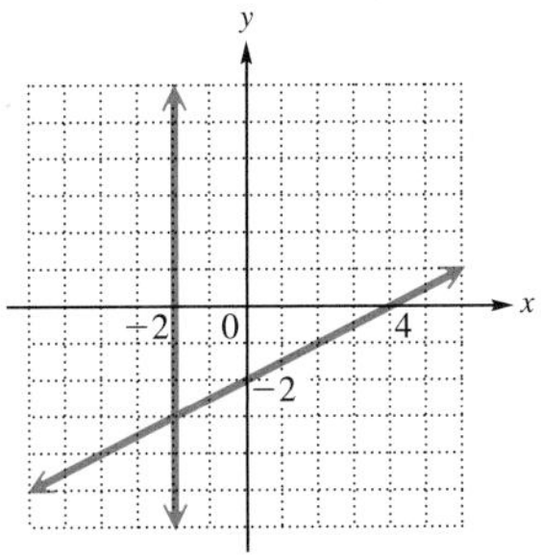

7. $x + y > 4$
$5x - 3y < 15$

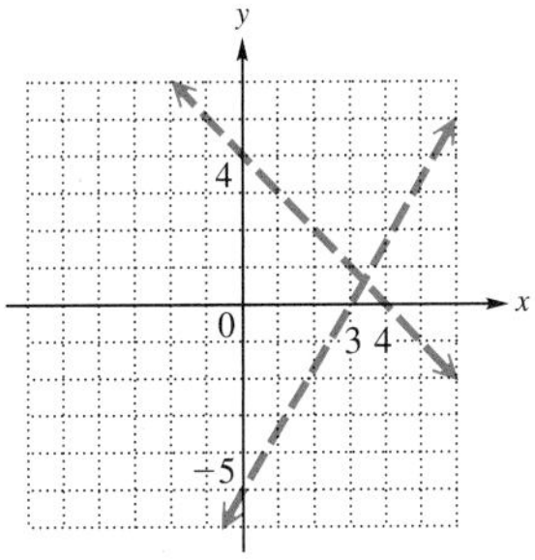

8. $3x - 2y > 12$
$4x + 3y < 12$

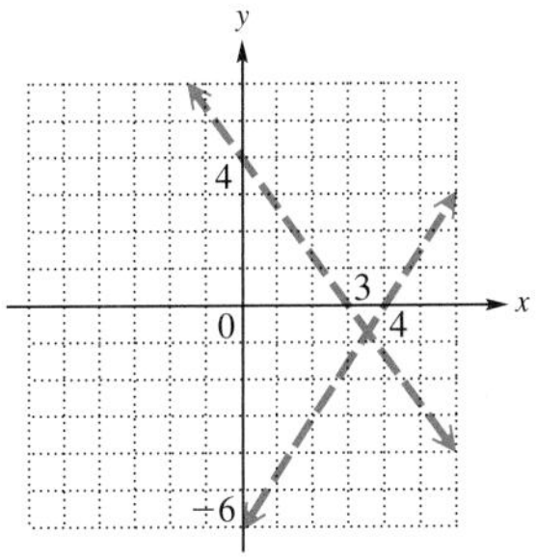

Graph the solution set of each system of linear inequalities. **See Examples 1–3.**

9. $x + y \leq 6$
$x - y \geq 1$

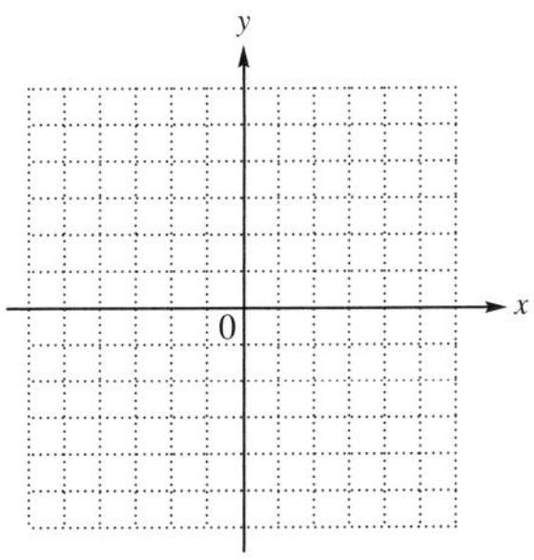

10. $x + y \leq 2$
$x - y \geq 3$

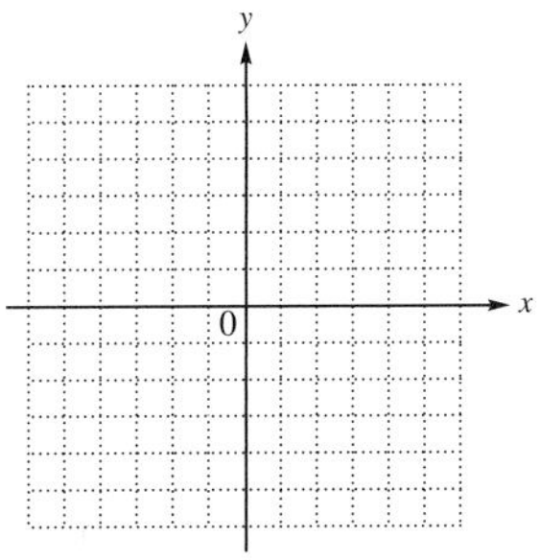

11. $4x + 5y \geq 20$
$x - 2y \leq 5$

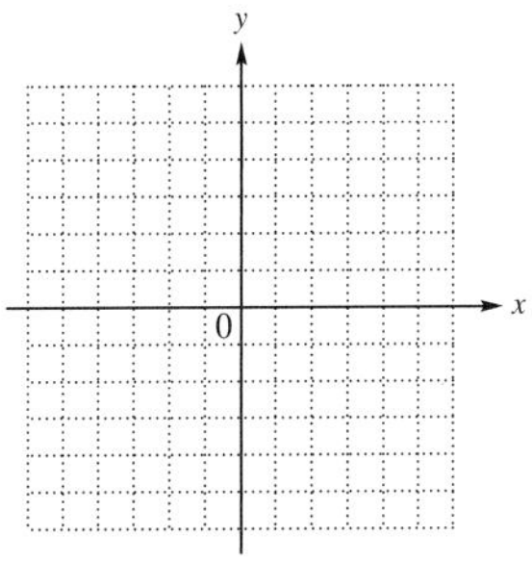

12. $x + 4y \leq 8$
$2x - y \geq 4$

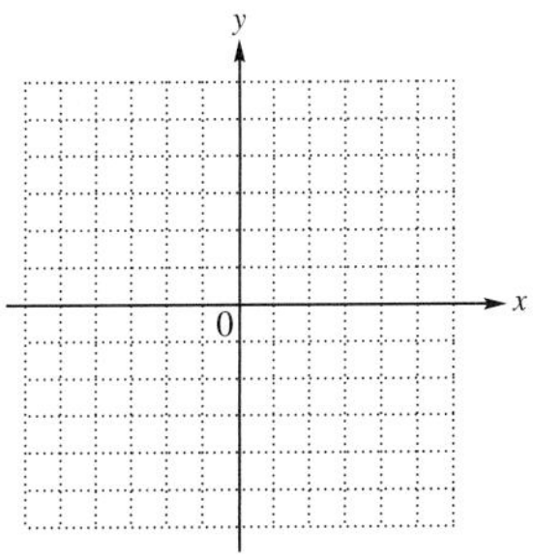

13. $2x + 3y < 6$
$x - y < 5$

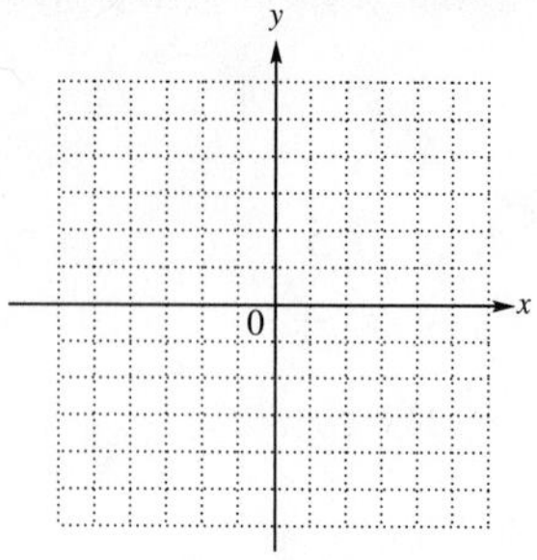

14. $x + 2y < 4$
$x - y < -1$

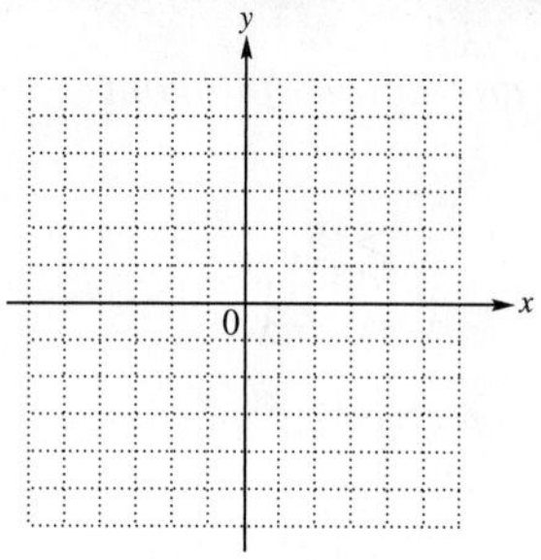

15. $y \leq 2x - 5$
$x < 3y + 2$

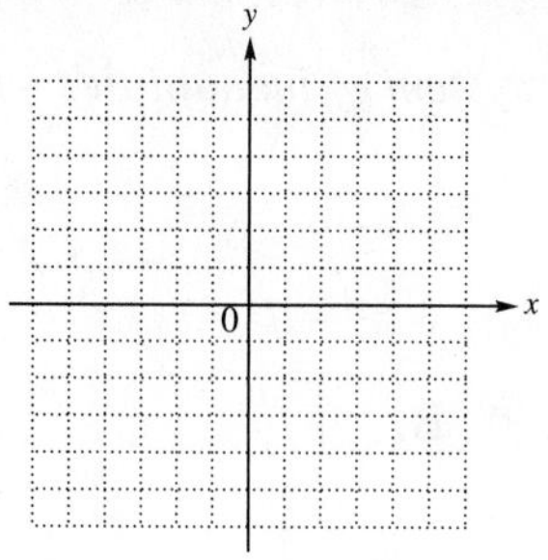

16. $x \geq 2y + 6$
$y > -2x + 4$

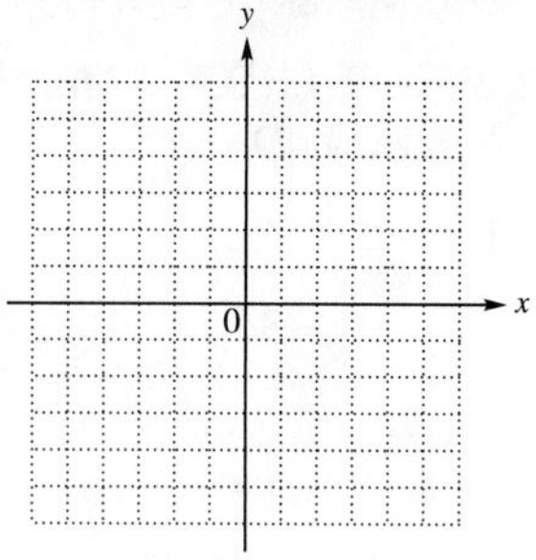

17. $4x + 3y < 6$
$x - 2y > 4$

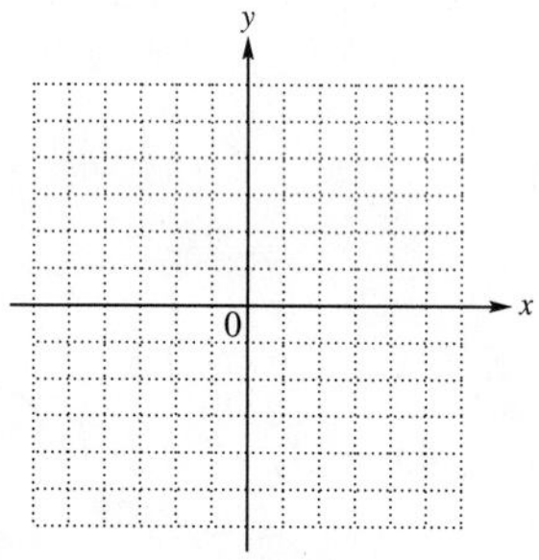

18. $3x + y > 4$
$x + 2y < 2$

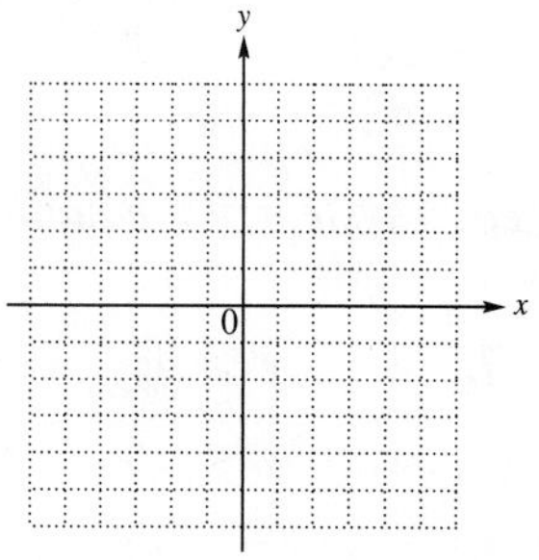

19. $x \leq 2y + 3$
$x + y < 0$

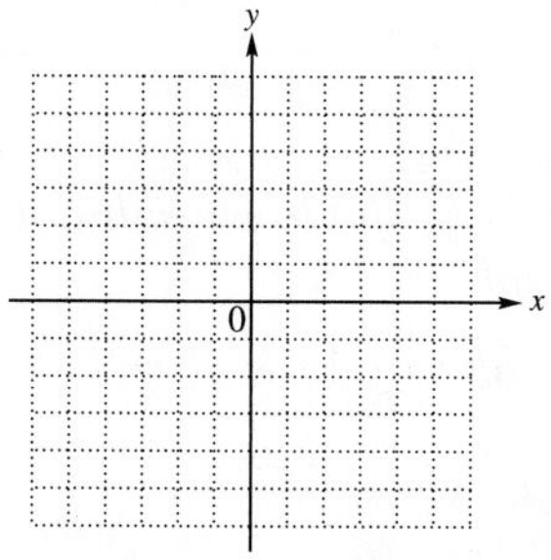

20. $x \leq 4y + 3$
$x + y > 0$

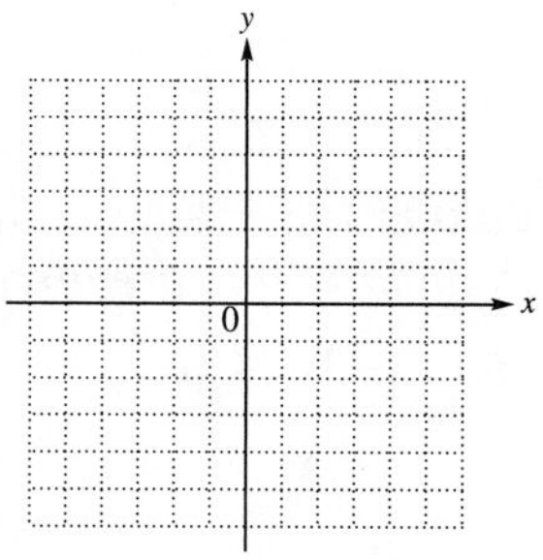

21. $4x + 5y < 8$
$y > -2$
$x > -4$

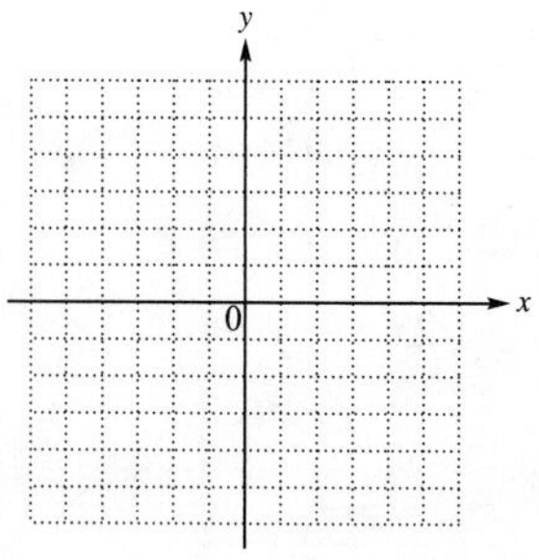

22. $x + y \geq -3$
$x - y \leq 3$
$y \leq 3$

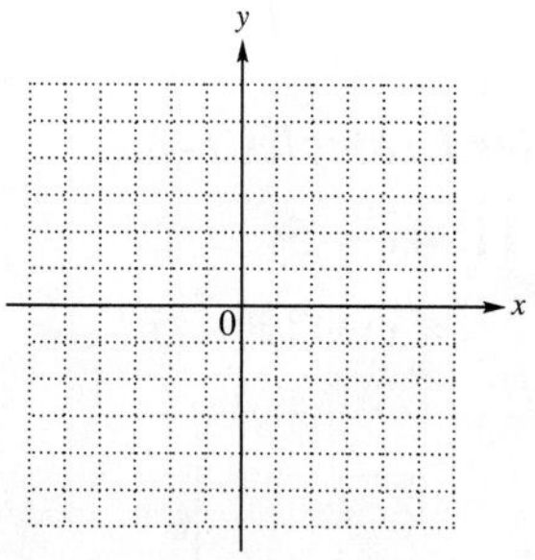

23. $3x - 2y \geq 6$
$x + y \leq 4$
$x \geq 0$
$y \geq -4$

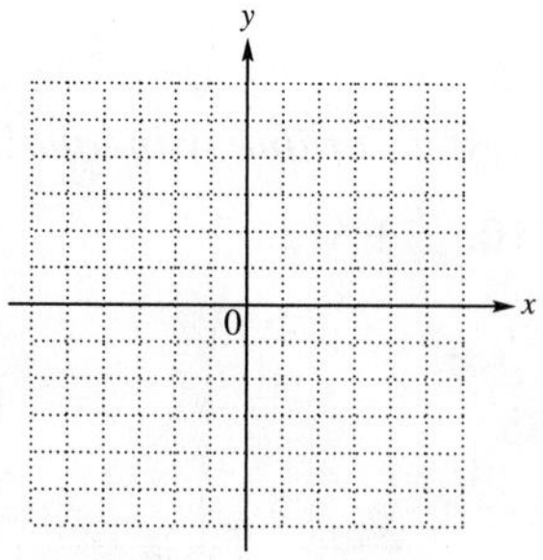

24. $2x - 3y < 6$
$x + y > 3$
$x < 4$
$y < 4$

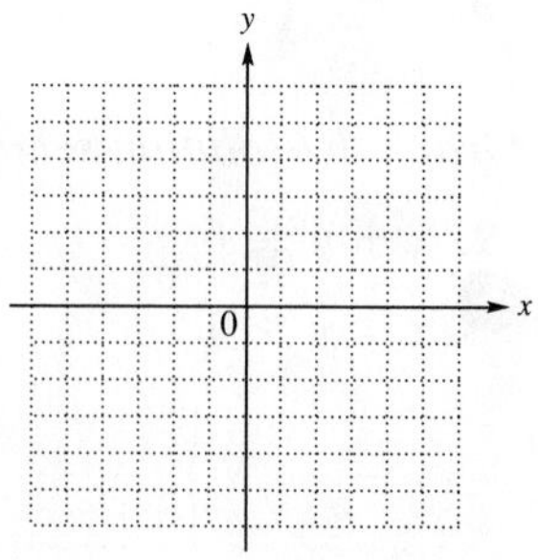

25. Every system of inequalities illustrated in the examples of this section has infinitely many solutions. Explain why this is so. Does this mean that *any* ordered pair is a solution?

Chapter 15 Summary

Key Terms

15.1

system of linear equations A system of linear equations (or **linear system**) consists of two or more linear equations with the same variables.

solution of a system A solution of a system of linear equations is an ordered pair that makes all the equations of the system true at the same time.

solution set of a system The set of all ordered pairs that are solutions of a system is its solution set.

consistent system A system of equations with at least one solution is a consistent system.

inconsistent system An inconsistent system is a system of equations with no solution.

independent equations Equations of a system that have different graphs are independent equations.

dependent equations Equations of a system that have the same graph (because they are different forms of the same equation) are dependent equations.

15.5

system of linear inequalities A system of linear inequalities contains two or more linear inequalities (and no other kinds of inequalities).

solution set of a system of linear inequalities The solution set of a system of linear inequalities includes all ordered pairs that make all inequalities of the system true at the same time.

Test Your Word Power

See how well you have learned the vocabulary in this chapter.

1. A **system of linear equations** consists of
   A. at least two linear equations with different variables
   B. two or more linear equations that have an infinite number of solutions
   C. two or more linear equations with the same variables
   D. two or more linear inequalities.

2. A **solution of a system** of linear equations is
   A. an ordered pair that makes one equation of the system true
   B. an ordered pair that makes all the equations of the system true at the same time
   C. any ordered pair that makes one or the other or both equations of the system true
   D. the set of values that make all the equations of the system false.

3. A **consistent system** is a system of equations
   A. with at least one solution
   B. with no solution
   C. with an infinite number of solutions
   D. that have the same graph.

4. An **inconsistent system** is a system of equations
   A. with one solution
   B. with no solution
   C. with an infinite number of solutions
   D. that have the same graph.

5. **Dependent equations**
   A. have different graphs
   B. have no solution
   C. have one solution
   D. are different forms of the same equation.

Answers to Test Your Word Power

1. C; *Example:* $2x + y = 7$
   $3x - y = 3$
2. B; *Example:* The ordered pair $(2, 3)$ satisfies both equations of the system in the Answer 1 example, so it is a solution of the system.
3. A; *Example:* The system in the Answer 1 example is consistent. The graphs of the equations intersect at exactly one point, in this case the solution $(2, 3)$.
4. B; *Example:* The equations of two parallel lines make up an inconsistent system. Their graphs never intersect, so there is no solution of the system.
5. D; *Example:* The equations $4x - y = 8$ and $8x - 2y = 16$ are dependent because their graphs are the same line.

Quick Review

Concepts

15.1 Solving Systems of Linear Equations by Graphing

An ordered pair is a solution of a system if it makes all equations of the system true at the same time.

To solve a linear system by graphing, follow these steps.

Step 1 Graph each equation of the system on the same axes.

Step 2 Find the coordinates of the point of intersection.

Step 3 Check. Write the solution set.

If the graphs of the equations do not intersect (that is, the lines are parallel), then the system has *no solution* and the *solution set is* $\emptyset$.

If the graphs of the equations are the same line, then the system has an *infinite number of solutions*. Use set-builder notation to write the solution set as

$$\{(x, y) \mid \_\_\_\_\_\_\_\_\},$$

where a form of the equation is written on the blank.

Examples

Is $(4, -1)$ a solution of the system $\begin{aligned} x + y &= 3 \\ 2x - y &= 9 \end{aligned}$?

Because $4 + (-1) = 3$ and $2(4) - (-1) = 9$ are both true, $(4, -1)$ is a solution.

Solve the system by graphing.

$$x + y = 5$$
$$2x - y = 4$$

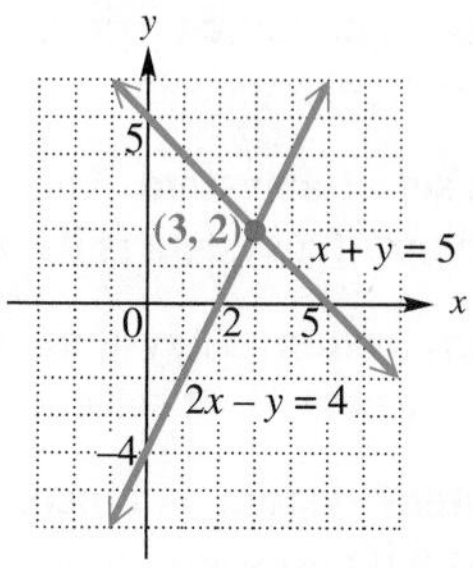

The ordered pair $(3, 2)$ satisfies both equations, so $\{(3, 2)\}$ is the solution set.

15.2 Solving Systems of Linear Equations by Substitution

Step 1 Solve one equation for either variable.

Step 2 Substitute for that variable in the other equation to get an equation in one variable.

Step 3 Solve the equation from Step 2.

Step 4 Substitute the result into the equation from Step 1 to get the value of the other variable.

Step 5 Check. Write the solution set.

Solve by substitution.

$$x + 2y = -5 \quad (1)$$
$$y = -2x - 1 \quad (2)$$

Equation (2) is already solved for y.

Substitute $-2x - 1$ for y in equation (1).

$x + 2(-2x - 1) = -5$	Let $y = -2x - 1$ in (1).
$x - 4x - 2 = -5$	Distributive property
$-3x - 2 = -5$	Combine like terms.
$-3x = -3$	Add 2.
$x = 1$	Divide by -3.

To find y, let $x = 1$ in equation (2).

$y = -2(1) - 1$	Let $x = 1$.
$y = -3$	Multiply, and then subtract.

The solution $(1, -3)$ checks, so $\{(1, -3)\}$ is the solution set.

Concepts	Examples
15.3 Solving Systems of Linear Equations by Elimination	
Step 1 Write both equations in standard form $Ax + By = C$.	Solve by elimination. $x + 3y = 7$ (1); $3x - y = 1$ (2)
Step 2 Multiply one or both equations by appropriate numbers as needed so that the sum of the coefficients of either the x- or y-terms is 0.	Multiply equation (1) by -3 to eliminate the x-terms.
Step 3 Add the equations to get an equation with only one variable (or no variable).	$-3x - 9y = -21$ Multiply (1) by -3. $3x - y = 1$ (2) $-10y = -20$ Add.
Step 4 Solve the equation from Step 3.	$y = 2$ Divide by -10.
Step 5 Substitute the solution from Step 4 into either of the original equations to find the value of the remaining variable.	Substitute to get the value of x. $x + 3(2) = 7$ Let $y = 2$ in (1). $x + 6 = 7$ Multiply. $x = 1$ Subtract 6.
Step 6 Check. Write the solution set.	Since $1 + 3(2) = 7$ and $3(1) - 2 = 1$, the solution $(1, 2)$ checks. The solution set is $\{(1, 2)\}$.
15.4 Applications of Linear Systems	
Use the modified six-step method.	
Step 1 **Read** the problem carefully.	The sum of two numbers is 30. Their difference is 6. Find the numbers.
Step 2 **Assign variables** for each unknown value. Use diagrams or tables as needed.	Let x represent one number. Let y represent the other number.
Step 3 **Write two equations** using both variables.	$x + y = 30$ (1) $x - y = 6$ (2)
Step 4 **Solve** the system.	$2x = 36$ Add. $x = 18$ Divide by 2.
Step 5 **State the answer.**	Let $x = 18$ in equation (1): $18 + y = 30$. Solve to get $y = 12$. The numbers are 18 and 12.
Step 6 **Check** the answer in the words of the original problem.	The sum of 18 and 12 is 30, and the difference between 18 and 12 is 6, so the answer checks.
15.5 Solving Systems of Linear Inequalities	
To solve a system of linear inequalities, follow these steps.	The shaded region is the solution of the following system.
Step 1 Graph each inequality on the same axes. (This was explained in **Section 11.5.**)	$2x + 4y \geq 5$ $x \geq 1$
Step 2 Choose the intersection. The solution set of the system is formed by the overlap of the regions of the two graphs.	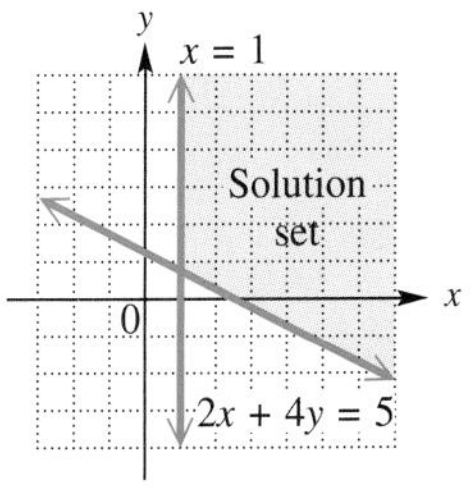

Chapter 15 Review Exercises

15.1 *Decide whether the given ordered pair is a solution of the given system.*

1. $(3, 4)$
$4x - 2y = 4$
$5x + y = 19$

2. $(-5, 2)$
$x - 4y = -13$
$2x + 3y = 4$

Solve each system by graphing.

3. $x + y = 4$
$2x - y = 5$

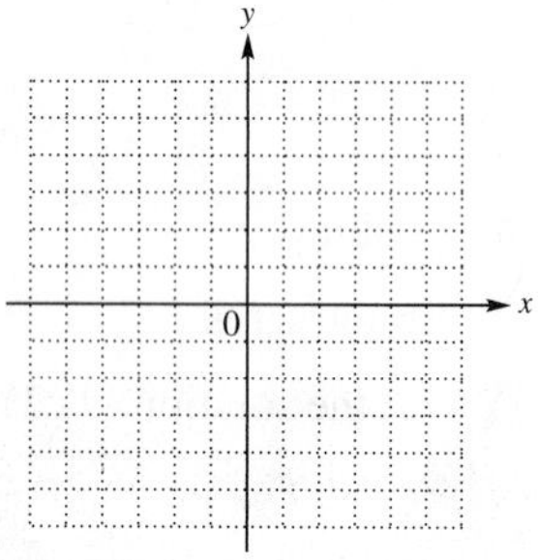

4. $x - 2y = 4$
$2x + y = -2$

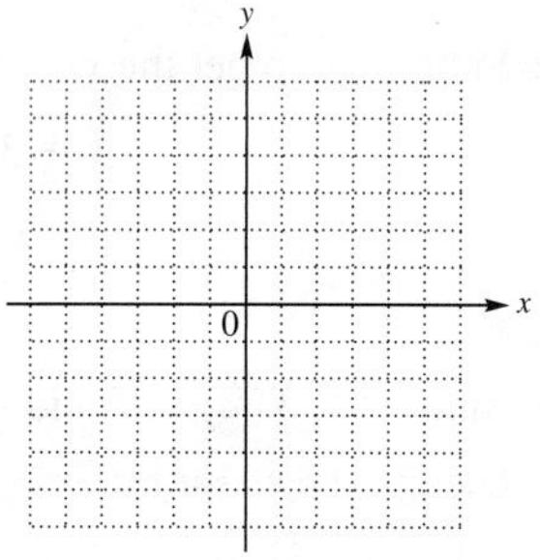

5. $x - 2 = 2y$
$2x - 4y = 4$

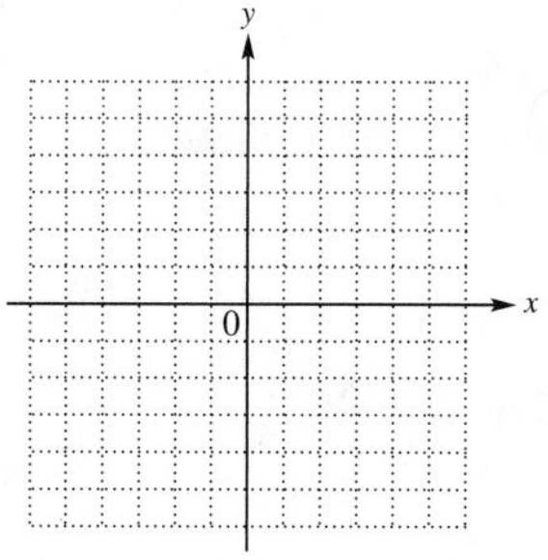

6. $2x + 4 = 2y$
$y - x = -3$

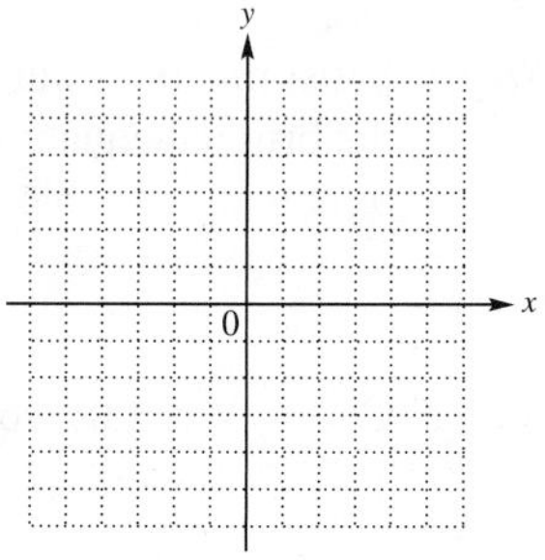

15.2 *Solve each system by the substitution method.*

7. $3x + y = 7$
$x = 2y$

8. $2x - 5y = -19$
$y = x + 2$

9. $4x + 5y = 44$
$x + 2 = 2y$

10. $5x + 15y = 3$
$x + 3y = 2$

15.3 *Solve each system by the elimination method.*

11. $2x - y = 13$
$x + y = 8$

12. $3x - y = -13$
$x - 2y = -1$

13. $-4x + 3y = 25$
$6x - 5y = -39$

14. $3x - 4y = 9$
$6x - 8y = 18$

15.1–15.3 *Solve each system by any method.*

15. $x - 2y = 5$
$y = x - 7$

16. $5x - 3y = 11$
$2y = x - 4$

17. $\frac{x}{2} + \frac{y}{3} = 7$
$\frac{x}{4} + \frac{2y}{3} = 8$

18. $\frac{3x}{4} - \frac{y}{3} = \frac{7}{6}$
$\frac{x}{2} + \frac{2y}{3} = \frac{5}{3}$

19. $0.2x + 1.2y = -1$
$0.1x + 0.3y = 0.1$

20. $0.1x + y = 1.6$
$0.6x + 0.5y = -1.4$

15.4 *For each problem, complete any tables. Then solve the problem using a system of equations.*

21. The two leading pizza chains in the United States are Pizza Hut and Domino's. In 2010, Pizza Hut had 2613 more locations than Domino's, and together the two chains had 12,471 locations. How many locations did each chain have? (*Source: PMQ Pizza Magazine.*)

22. Together, the average paid circulation for *Reader's Digest* and *People* magazines in 2011 was 9.3 million. The circulation for *People* was 2.1 million less than that of *Reader's Digest*. What were the circulation figures for each magazine? (*Source:* Audit Bureau of Circulations.)

23. Candy that sells for \$1.30 per lb is to be mixed with candy selling for \$0.90 per lb to get 100 lb of a mix that will sell for \$1 per lb. How much of each type should be used?

Number of Pounds	Cost per Pound (in dollars)	Total Value (in dollars)
______	1.30	$1.30x$
y	______	______
100	1.00	______

24. A cashier has 20 bills, all of which are \$10 or \$20 bills. The total value of the money is \$330. How many of each type does the cashier have?

Number of Bills	Denomination of Bills (in dollars)	Total Value (in dollars)
x	10	____
____	____	$20y$
____	XXXXXX	330

25. The perimeter of a rectangle is 90 m. Its length is $1\frac{1}{2}$ times its width. Find the length and width of the rectangle.

26. A certain plane flying with the wind travels 540 mi in 2 hr. Later, flying against the same wind, the plane travels 690 mi in 3 hr. Find the rate of the plane in still air and the wind speed.

27. After taxes, Ms. Cesar's game show winnings were \$18,000. She invested part of it at 3% annual simple interest and the rest at 4%. Her interest income for the first year was \$650. How much did she invest at each rate?

Amount Invested (in dollars)	Percent (as a decimal)	Interest (in dollars)
x	0.03	______
y	______	______
18,000	XXXXX	______

28. A 40% antifreeze solution is to be mixed with a 70% solution to get 90 L of a 50% solution. How many liters of the 40% and 70% solutions will be needed?

Number of Liters	Percent (as a decimal)	Amount of Pure Antifreeze
x	0.40	______
y	______	______
90	0.50	______

15.5 **CONCEPT CHECK** *Answer each question.*

29. Which system of linear inequalities is graphed in the figure?

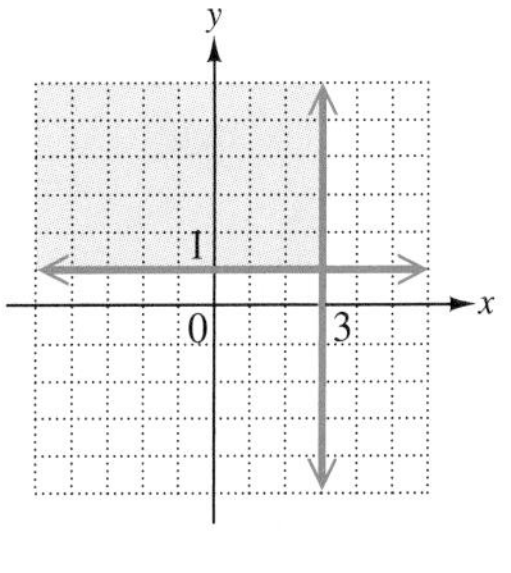

A. $x \leq 3$, $y \leq 1$ **B.** $x \leq 3$, $y \geq 1$ **C.** $x \geq 3$, $y \leq 1$ **D.** $x \geq 3$, $y \geq 1$

30. Which system of inequalities has no solution? (Do not graph.)

A. $x \geq 4$, $y \leq 3$ **B.** $x + y > 4$, $x + y < 3$ **C.** $x > 2$, $y < 1$ **D.** $x + y < 4$, $x - y < 3$

Graph the solution set of each system of linear inequalities.

31. $x + y \geq 2$
$x - y \leq 4$

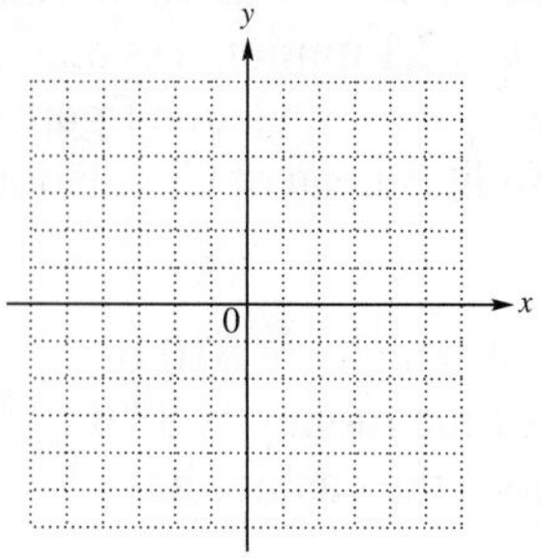

32. $y \geq 2x$
$2x + 3y \leq 6$

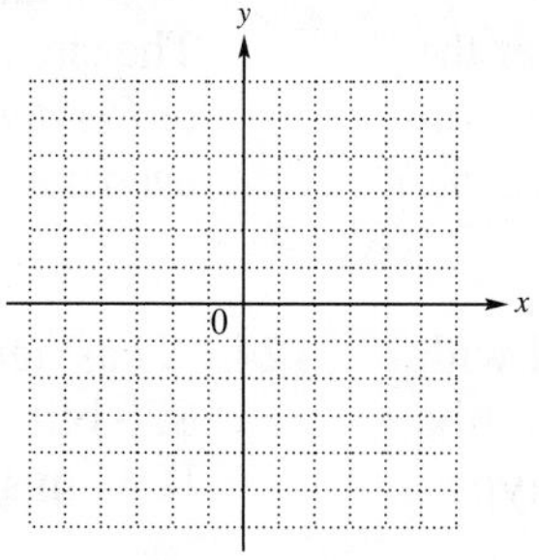

33. $x + y < 3$
$2x > y$

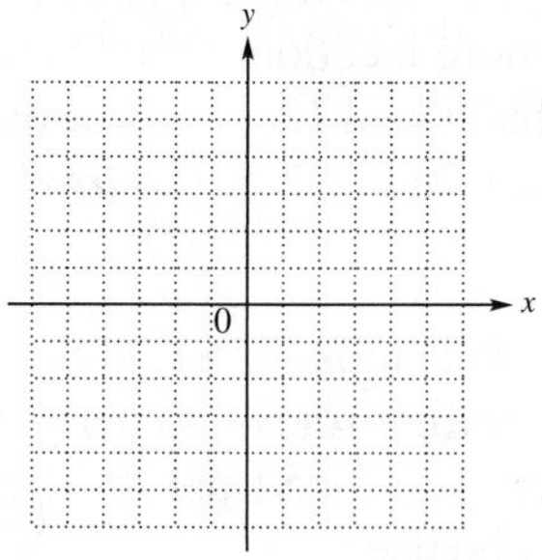

Mixed Review Exercises

Solve.

34. $3x + 4y = 6$
$4x - 5y = 8$

35. $\frac{3x}{2} + \frac{y}{5} = -3$
$4x + \frac{y}{3} = -11$

36. $x + 6y = 3$
$2x + 12y = 2$

37. $x + y < 5$
$x - y \geq 2$

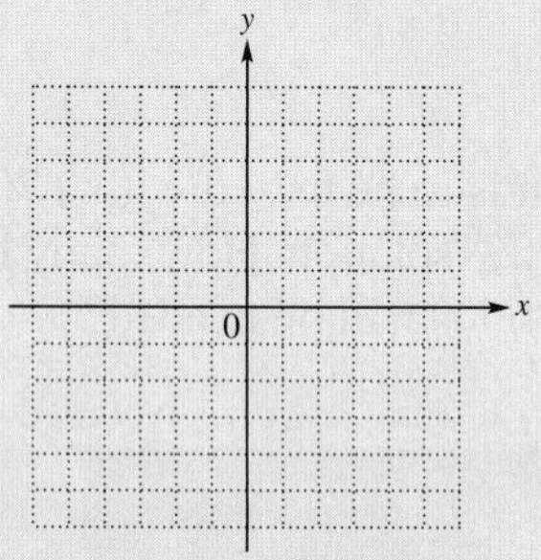

38. $y \leq 2x$
$x + 2y > 4$

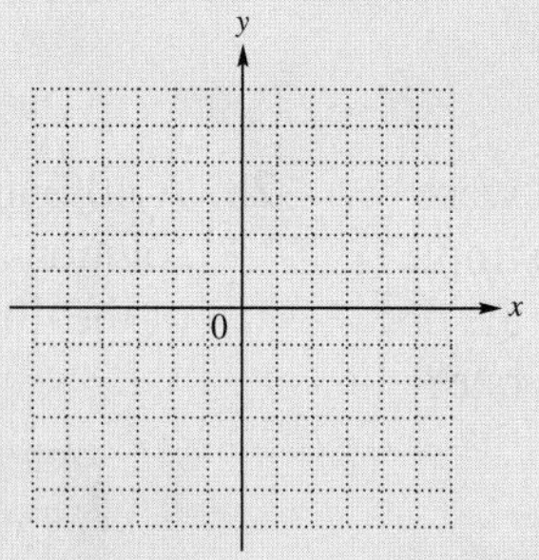

39. $y < -4x$
$y < -2$

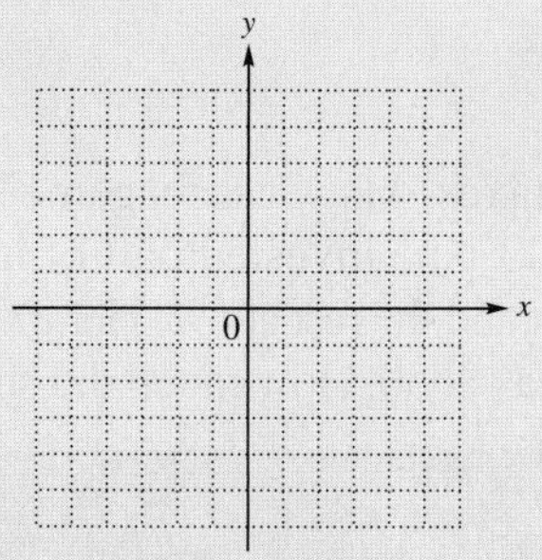

40. The perimeter of an isosceles triangle is 29 in. One side of the triangle is 5 in. longer than each of the two equal sides. Find the lengths of the sides of the triangle.

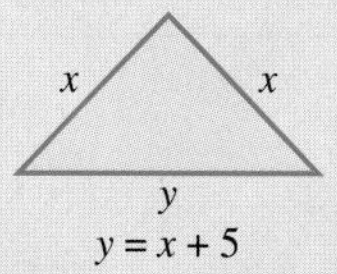

41. In Super Bowl XLVI, the New York Giants beat the New England Patriots by 4 points, and the winning score was 13 points less than twice the losing score. What was the final score of the game? (*Source:* NFL.)

42. Eboni Perkins compared the monthly payments she would incur for two types of mortgages: fixed-rate and variable-rate. Her observations led to the following graph.

(a) For which years would the monthly payment be more for the fixed-rate mortgage than for the variable-rate mortgage?

(b) In what year would the payments be the same, and what would those payments be?

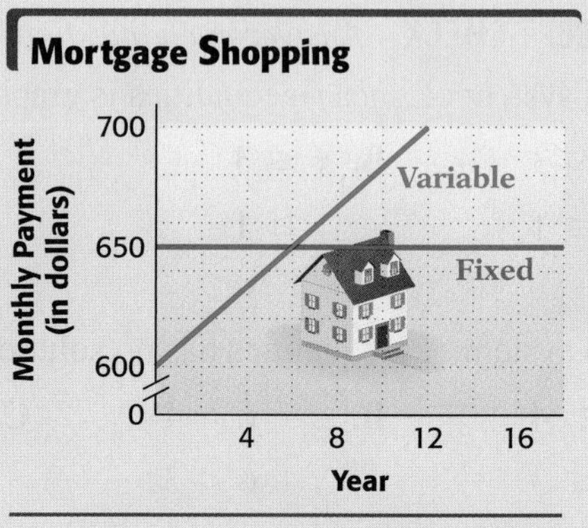

Chapter 15 Test

CHAPTER Test Prep VIDEO

The Chapter Test Prep Videos with test solutions are available on DVD, in MyMathLab, and on YouTube—search "LialDevMath" and click on "Channels."

1. Decide whether each ordered pair is a solution of the system.

$$2x + y = -3$$
$$x - y = -9$$

(a) $(1, -5)$ **(b)** $(1, 10)$ **(c)** $(-4, 5)$

2. Solve the system by graphing.

$$2x + y = 1$$
$$3x - y = 9$$

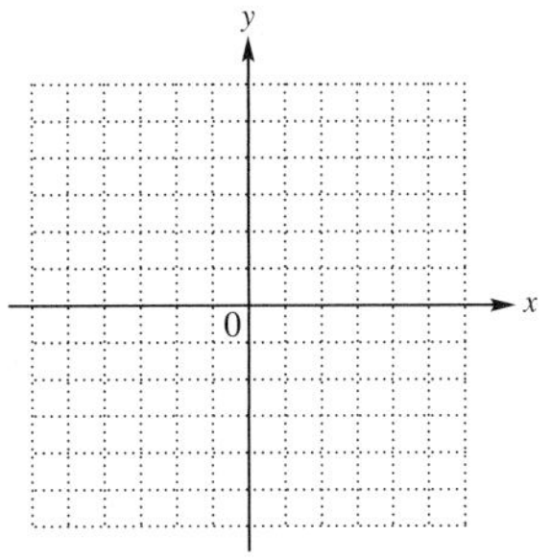

3. Suppose that the graph of a system of two linear equations consists of lines that have the same slope but different y-intercepts. How many solutions does the system have?

Solve each system by the substitution method.

4. $2x + y = -4$
$x = y + 7$

5. $4x + 3y = -35$
$x + y = 0$

Solve each system by the elimination method.

6. $2x - y = 4$
$3x + y = 21$

7. $4x + 2y = 2$
$5x + 4y = 7$

8. $3x + 4y = 9$
$2x + 5y = 13$

9. $6x - 5y = 0$
$-2x + 3y = 0$

10. $4x + 5y = 2$
$-8x - 10y = 6$

Solve each system by any method.

11. $3x = 6 + y$

$6x - 2y = 12$

12. $\frac{6}{5}x - \frac{1}{3}y = -20$

$-\frac{2}{3}x + \frac{1}{6}y = 11$

Solve each problem.

13. The distance between Memphis and Atlanta is 782 mi less than the distance between Minneapolis and Houston. Together, the two distances total 1570 mi. How far is it between Memphis and Atlanta? How far is it between Minneapolis and Houston? (*Source: Rand McNally Road Atlas.*)

14. In 2010, a total of 7.8 million people visited the Statue of Liberty and the National World War II Memorial, two popular tourist attractions. The Statue of Liberty had 0.2 million fewer visitors than the National World War II Memorial. How many visitors did each of these attractions have? (*Source:* National Park Service, Department of the Interior.)

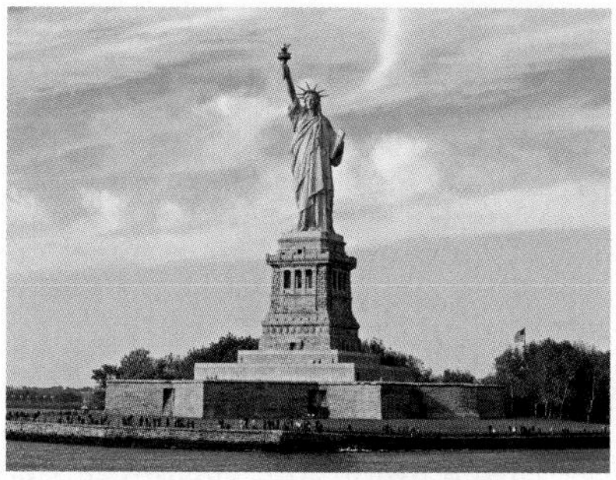

15. A 15% solution of alcohol is to be mixed with a 40% solution to get 50 L of a final mixture that is 30% alcohol. How much of each of the original solutions should be used?

Liters of Solution	Percent (as a decimal)	Liters of Pure Alcohol

16. Two cars leave from Perham, Minnesota, and travel in the same direction. One car travels $1\frac{1}{3}$ times as fast as the other. After 3 hr they are 45 mi apart. What are the rates of the cars?

	r	t	d
Faster Car			
Slower Car			

Graph the solution set of each system of inequalities.

17. $2x + 7y \leq 14$

$x - y \geq 1$

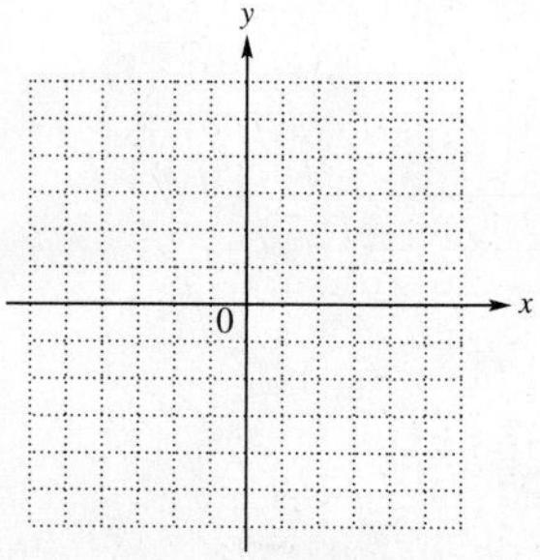

18. $2x - y > 6$

$4y + 12 \geq -3x$

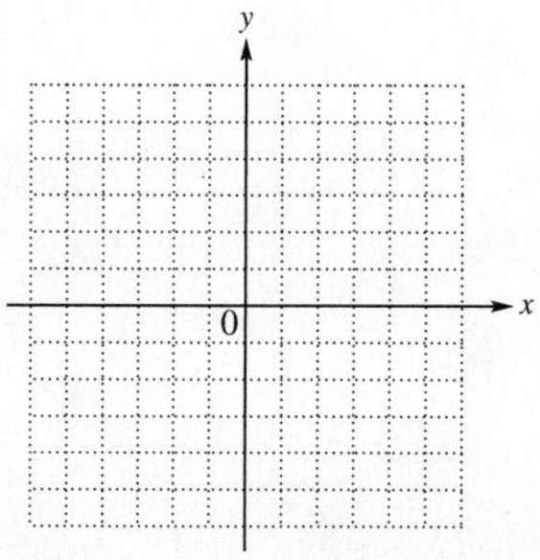

16 Roots and Radicals

Many formulas in mathematics and science, such as that for finding the time it takes for a pendulum to swing from one extreme to the other and back again, include *square roots,* the topic of this chapter.

16.1 Evaluating Roots

OBJECTIVES

1. Find square roots.
2. Decide whether a given root is rational, irrational, or not a real number.
3. Find decimal approximations for irrational square roots.
4. Use the Pythagorean theorem.
5. Find cube, fourth, and other roots.

OBJECTIVE 1 Find square roots. Recall that *squaring* a number means multiplying the number by itself.

7^2 means $7 \cdot 7$, which equals 49. The square of 7 is 49.

The opposite (inverse) of squaring a number is taking its *square root.* This is equivalent to asking

"What number when multiplied by itself equals 49?"

For the example above, one answer is 7, since $7 \cdot 7 = 49$.

This discussion can be generalized.

Square Root

A number b is a **square root** of a if $b^2 = a$.

EXAMPLE 1 Finding All Square Roots of a Number

Find all square roots of 49.

We ask, "What number when multiplied by itself equals 49?" As mentioned above, one square root is 7. Another square root of 49 is -7, because

$$(-7)(-7) = 49.$$

Thus, the number 49 has *two* square roots: 7 and -7. One square root is positive, and one is negative.

◀ Work Problem 1 at the Side.

1 Find all square roots of the given number.

GS (a) 100
Ask "What number when multiplied by itself equals 100?" There are two answers.

(b) 25 **(c)** 36 **(d)** $\frac{25}{36}$

The **positive** or **principal square root** of a number is written with the symbol $\sqrt{\ }$. For example, the positive square root of 121 is 11.

$$\sqrt{121} = 11 \qquad 11^2 = 121$$

The symbol $-\sqrt{\ }$ is used for the **negative square root** of a number. For example, the negative square root of 121 is -11.

$$-\sqrt{121} = -11 \qquad (-11)^2 = 121$$

The **radical symbol** $\sqrt{\ }$ always represents the positive square root (except that $\sqrt{0} = 0$). The number inside the radical symbol is the **radicand,** and the entire expression—radical symbol and radicand—is a **radical.**

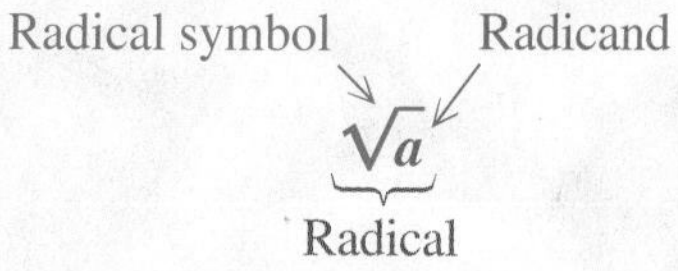

An algebraic expression containing a radical is a **radical expression.**

The radical symbol $\sqrt{\ }$ has been used since sixteenth-century Germany and was probably derived from the letter *R*. The radical symbol at the right comes from the Latin word for root, *radix*. It was first used by Leonardo of Pisa (Fibonacci) in 1220.

Early radical symbol

Leonardo of Pisa (c. 1170–1250)

Answers

1. (a) 10, −10 **(b)** 5, −5 **(c)** 6, −6 **(d)** $\frac{5}{6}, -\frac{5}{6}$

We summarize our discussion of square roots as follows.

Square Roots of a

Let a be a positive real number.

$\sqrt{a}$ is the positive or principal square root of a.

$-\sqrt{a}$ is the negative square root of a.

For nonnegative a, the following hold.

$$\sqrt{a} \cdot \sqrt{a} = (\sqrt{a})^2 = a \quad \text{and} \quad -\sqrt{a} \cdot (-\sqrt{a}) = (-\sqrt{a})^2 = a$$

Also, $\sqrt{0} = 0$.

Calculator Tip

Most calculators have a square root key, usually labeled $\sqrt{x}$, for finding the square root of a number. On some models, the square root key must be used in conjunction with the key marked INV or 2nd.

EXAMPLE 2 Finding Square Roots

Find each square root.

(a) $\sqrt{144}$

The radical $\sqrt{144}$ represents the positive or principal square root of 144. Think of a positive number whose square is 144.

$$12^2 = 144, \quad \text{so} \quad \sqrt{144} = 12.$$

(b) $-\sqrt{1024}$

This symbol represents the negative square root of 1024. A calculator with a square root key can be used to find $\sqrt{1024} = 32$. Then, $-\sqrt{1024} = -32$.

(c) $\sqrt{\frac{4}{9}} = \frac{2}{3}$ **(d)** $-\sqrt{\frac{16}{49}} = -\frac{4}{7}$ **(e)** $\sqrt{0.81} = 0.9$

Work Problem 2 at the Side. ▶

As noted above, when the square root of a positive real number is squared, the result is that positive real number. (Also, $(\sqrt{0})^2 = 0$.)

EXAMPLE 3 Squaring Radical Expressions

Find the *square* of each radical expression.

(a) $\sqrt{13}$ The square of $\sqrt{13}$ is $(\sqrt{13})^2 = 13$. Definition of square root

(b) $-\sqrt{29}$

$(-\sqrt{29})^2 = 29$

The square of a *negative* number is positive.

(c) $\sqrt{p^2 + 1}$

$(\sqrt{p^2 + 1})^2 = p^2 + 1$

Work Problem 3 at the Side. ▶

2 Find each square root.

GS **(a)** $\sqrt{16}$

$4^2 =$ ____, so

$\sqrt{16} =$ ____.

GS **(b)** $-\sqrt{169}$

$13^2 =$ ____, so

$-\sqrt{169} =$ ____.

(c) $-\sqrt{225}$ **(d)** $\sqrt{729}$

(e) $-\sqrt{\frac{36}{25}}$ **(f)** $\sqrt{0.49}$

3 Find the *square* of each radical expression.

GS **(a)** $\sqrt{41}$

$(\sqrt{41})^2 =$ ____

GS **(b)** $-\sqrt{39}$

$(-\sqrt{39})^2 =$ ____

(c) $\sqrt{120}$

(d) $\sqrt{2x^2 + 3}$

Answers

2. **(a)** 16; 4 **(b)** 169; −13 **(c)** −15 **(d)** 27 **(e)** $-\frac{6}{5}$ **(f)** 0.7

3. **(a)** 41 **(b)** 39 **(c)** 120 **(d)** $2x^2 + 3$

4 Tell whether each square root is *rational, irrational,* or *not a real number.*

(a) $\sqrt{9}$

(b) $\sqrt{7}$

(c) $\sqrt{\frac{9}{16}}$

(d) $\sqrt{72}$

(e) $\sqrt{-43}$

OBJECTIVE 2 Decide whether a given root is rational, irrational, or not a real number. Numbers with rational square roots are **perfect squares.**

$$\text{Perfect squares} \begin{cases} 25 & & \sqrt{25} = 5 \\ 144 & \text{are perfect squares since} & \sqrt{144} = 12 \\ \frac{4}{9} & & \sqrt{\frac{4}{9}} = \frac{2}{3} \end{cases} \text{Rational square roots}$$

A number that is not a perfect square has a square root that is not a rational number. For example, $\sqrt{5}$ is not a rational number because it cannot be written as the ratio of two integers. Its decimal equivalent (or approximation) neither terminates nor repeats. However, $\sqrt{5}$ is a real number and corresponds to a point on the number line.

A real number that is not rational is an **irrational number.** The number $\sqrt{5}$ is irrational. ***Many square roots of integers are irrational.***

If a is a *positive* real number that is *not* a perfect square, then

$\sqrt{a}$ is irrational.

Not every number has a real number square root. For example, there is no real number that can be squared to obtain -36. (The square of a real number can never be negative.) Because of this, $\sqrt{-36}$ ***is not a real number.***

If a is a ***negative*** real number, then $\sqrt{a}$ is *not* a real number.

CAUTION

Do not confuse $\sqrt{-36}$ and $-\sqrt{36}$. $\sqrt{-36}$ is not a real number since there is no real number that can be squared to obtain -36. However, $-\sqrt{36}$ is the negative square root of 36, which is -6.

EXAMPLE 4 Identifying Types of Square Roots

Tell whether each square root is *rational, irrational,* or *not a real number.*

(a) $\sqrt{17}$ Because 17 is not a perfect square, $\sqrt{17}$ is irrational.

(b) $\sqrt{64}$ 64 is a perfect square, 8^2, so $\sqrt{64} = 8$ is a rational number.

(c) $\sqrt{-25}$ There is no real number whose square is -25. Therefore, $\sqrt{-25}$ is not a real number.

◀ **Work Problem 4 at the Side.**

Note

Not all irrational numbers are square roots of integers. For example, π (approximately 3.14159) is an irrational number that is not a square root of any integer.

Answers

4. **(a)** rational **(b)** irrational **(c)** rational **(d)** irrational **(e)** not a real number

OBJECTIVE 3 Find decimal approximations for irrational square roots. Even if a number is irrational, a decimal that approximates the number can be found using a calculator.

EXAMPLE 5 Approximating Irrational Square Roots

Find a decimal approximation for each square root. Round answers to the nearest thousandth.

(a) $\sqrt{11}$

Using the square root key of a calculator gives $3.31662479 \approx 3.317$, where $\approx$ means **"is approximately equal to."**

(b) $\sqrt{39} \approx 6.245$ Use a calculator. **(c)** $-\sqrt{740} \approx -27.203$

Work Problem 5 at the Side. ▶

OBJECTIVE 4 Use the Pythagorean theorem. Many applications of square roots use the Pythagorean theorem. Recall from **Section 13.8** and **Figure 1** that if c is the length of the hypotenuse of a right triangle, and a and b are the lengths of the two legs, then

$$a^2 + b^2 = c^2.$$

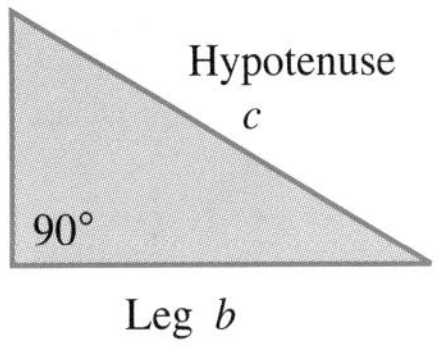

Figure 1

EXAMPLE 6 Using the Pythagorean Theorem

Find the length of the unknown side of each right triangle with sides a, b, and c, where c is the hypotenuse.

(a) $a = 3, b = 4$

$a^2 + b^2 = c^2$ Use the Pythagorean theorem.

$3^2 + 4^2 = c^2$ Let $a = 3$ and $b = 4$.

$9 + 16 = c^2$ Square.

$25 = c^2$ Add.

Since the length of a side of a triangle must be a positive number, find the positive square root of 25 to get c.

$$c = \sqrt{25} = 5$$

(b) $b = 5, c = 9$

$a^2 + b^2 = c^2$ Use the Pythagorean theorem.

$a^2 + 5^2 = 9^2$ Let $b = 5$ and $c = 9$. (Solve for a^2.)

$a^2 + 25 = 81$ Square.

$a^2 = 56$ Subtract 25.

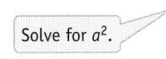

Use a calculator to find the positive square root of 56 to approximate a.

$$a = \sqrt{56} \approx 7.483$$

Work Problem 6 at the Side. ▶

5 Find a decimal approximation for each square root. Round answers to the nearest thousandth.

(a) $\sqrt{28}$ **(b)** $\sqrt{63}$

(c) $-\sqrt{190}$ **(d)** $\sqrt{1000}$

6 Find the length of the unknown side of each right triangle with sides a, b, and c, where c is the hypotenuse. Give any decimal approximations to the nearest thousandth.

GS **(a)** $a = 7, b = 24$

$\_\_\_^2 + \_\_\_^2 = \_\_\_^2$

$49 + \_\_\_ = c^2$

$\_\_\_ = c^2$

The positive square root of 625 is ____, so $c =$ ____.

(b) $b = 13, c = 15$

(c)

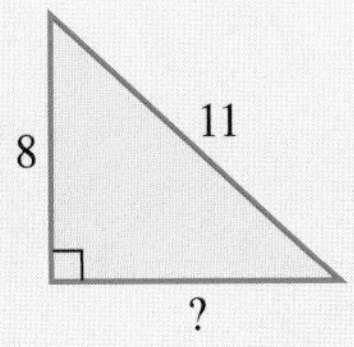

Answers

5. (a) 5.292 **(b)** 7.937 **(c)** −13.784 **(d)** 31.623

6. (a) 7; 24; c; 576; 625; 25; 25 **(b)** $\sqrt{56} \approx 7.483$ **(c)** $\sqrt{57} \approx 7.550$

7 A rectangle has dimensions 5 ft by 12 ft. Find the length of its diagonal.

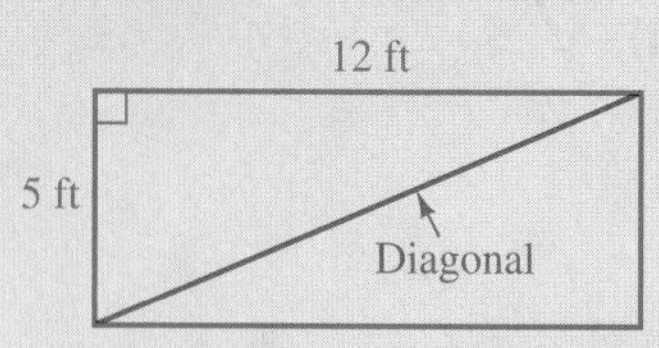

(Note that the diagonal divides the rectangle into two right triangles with itself as the hypotenuse.)

CAUTION

Be careful not to make the common mistake of thinking that $\sqrt{a^2 + b^2}$ equals $a + b$. Consider the following.

$$\sqrt{9 + 16} = \sqrt{25} = 5, \quad \text{but} \quad \sqrt{9} + \sqrt{16} = 3 + 4 = 7.$$

In general, $\sqrt{a^2 + b^2} \neq a + b$.

EXAMPLE 7 Using the Pythagorean Theorem to Solve an Application

A ladder 10 ft long leans against a wall. The foot of the ladder is 6 ft from the base of the wall. How high up the wall does the top of the ladder rest?

Step 1 **Read** the problem again.

Step 2 **Assign a variable.** As shown in **Figure 2,** a right triangle is formed with the ladder as the hypotenuse. Let a represent the height of the top of the ladder when measured straight down to the ground.

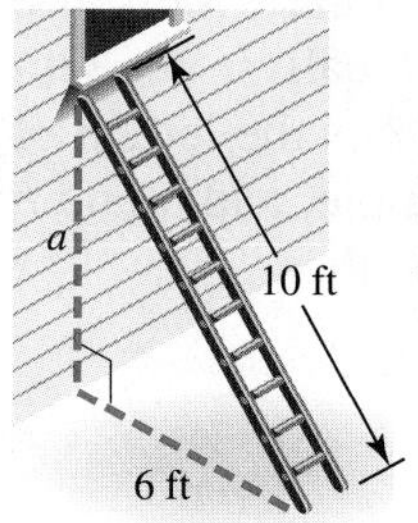

Recall that the symbol ⏋ indicates a 90° or right angle.

Figure 2

Step 3 **Write an equation** using the Pythagorean theorem.

$$a^2 + b^2 = c^2$$

Substitute carefully.

$$a^2 + 6^2 = 10^2 \qquad \text{Let } b = 6 \text{ and } c = 10.$$

Step 4 **Solve.**

$$a^2 + 36 = 100 \qquad \text{Square.}$$

$$a^2 = 64 \qquad \text{Subtract 36.}$$

$$a = \sqrt{64} \qquad \text{Solve for } a.$$

$$a = 8 \qquad \sqrt{64} = 8$$

Choose the positive square root of 64 since a represents a length.

Step 5 **State the answer.** The top of the ladder rests 8 ft up the wall.

Step 6 **Check.** From **Figure 2,** we have the following.

$$8^2 + 6^2 \stackrel{?}{=} 10^2 \qquad a^2 + b^2 = c^2$$

$$64 + 36 = 100 \checkmark \qquad \text{True}$$

The check confirms that the top of the ladder rests 8 ft up the wall.

◀ **Work Problem 7 at the Side.**

Answer

7. 13 ft

OBJECTIVE 5 Find cube, fourth, and other roots. Finding the square root of a number is the inverse (opposite) of squaring a number. There are inverses to finding the cube of a number and to finding the fourth or greater power of a number. These inverses are, respectively, the **cube root** $\sqrt[3]{a}$, and the **fourth root** $\sqrt[4]{a}$. Similar symbols are used for other roots.

$\sqrt[n]{a}$

The nth root of a, written $\sqrt[n]{a}$, is a number whose nth power equals a. That is,

$$\sqrt[n]{a} = b \quad \text{means} \quad b^n = a.$$

In $\sqrt[n]{a}$, the number n is the **index,** or **order,** of the radical.

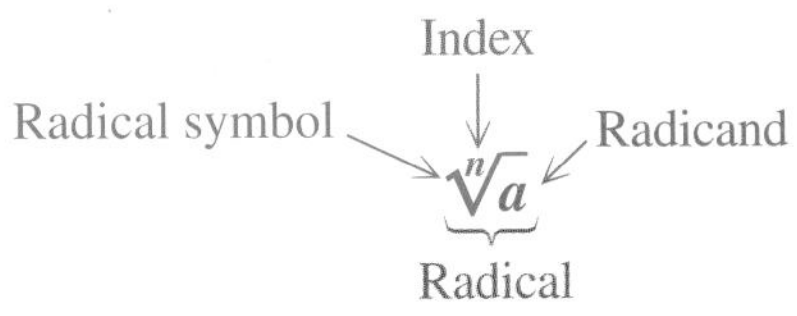

We could write $\sqrt[2]{a}$ instead of $\sqrt{a}$, but the simpler symbol $\sqrt{a}$ is customary since the square root is the most commonly used root.

Calculator Tip

A calculator that has a key marked $\sqrt[x]{y}$, x^y, or y^x (again perhaps in conjunction with the INV or 2nd key) can be used to find other roots.

When working with cube roots or fourth roots, it is helpful to memorize the first few **perfect cubes** ($1^3 = 1$, $2^3 = 8$, $3^3 = 27$, and so on) and the first few **perfect fourth powers** ($1^4 = 1$, $2^4 = 16$, $3^4 = 81$, and so on).

Work Problem 8 at the Side. ▶

EXAMPLE 8 Finding Cube Roots

Find each cube root.

(a) $\sqrt[3]{8}$

What number can be cubed to give 8? Because $2^3 = 8$, $\sqrt[3]{8} = 2$.

(b) $\sqrt[3]{-8} = -2$, because $(-2)^3 = -8$.

(c) $\sqrt[3]{216} = 6$, because $6^3 = 216$.

Work Problem 9 at the Side. ▶

Notice in **Example 8(b)** that we can find the cube root of a negative number. (Contrast this with the square root of a negative number, which is not real.) In fact, the cube root of a positive number is positive, and the cube root of a negative number is negative. ***There is only one real number cube root for each real number.***

When a radical has an ***even index*** (square root, fourth root, and so on), ***the radicand must be nonnegative*** to yield a real number root. Also, for $a > 0$,

$$\sqrt{a},\ \sqrt[4]{a},\ \sqrt[6]{a}, \textbf{ and so on are positive (principal) roots.}$$

$$-\sqrt{a},\ -\sqrt[4]{a},\ -\sqrt[6]{a}, \textbf{ and so on are negative roots.}$$

8 Complete the following list of perfect cubes and perfect fourth powers.

Perfect Cubes	Perfect Fourth Powers
$1^3 = 1$	$1^4 = 1$
$2^3 = 8$	$2^4 = 16$
$3^3 = 27$	$3^4 = 81$
$4^3 =$ ____	$4^4 =$ ____
$5^3 =$ ____	$5^4 =$ ____
$6^3 =$ ____	$6^4 =$ ____
$7^3 =$ ____	$7^4 =$ ____
$8^3 =$ ____	$8^4 =$ ____
$9^3 =$ ____	$9^4 =$ ____
$10^3 =$ ____	$10^4 =$ ____

9 Find each cube root.

GS **(a)** $\sqrt[3]{27}$

____$^3 = 27$, so

$\sqrt[3]{27} =$ ____.

(b) $\sqrt[3]{1}$

(c) $\sqrt[3]{-125}$

Answers

8. Perfect cubes: 64; 125; 216; 343; 512; 729; 1000
Perfect fourth powers: 256; 625; 1296; 2401; 4096; 6561; 10,000

9. **(a)** 3; 3 **(b)** 1 **(c)** -5

10 Find each root.

(a) $\sqrt[4]{81}$

(b) $\sqrt[4]{-81}$

(c) $-\sqrt[4]{81}$

(d) $\sqrt[5]{243}$

(e) $\sqrt[5]{-243}$

EXAMPLE 9 **Finding Other Roots**

Find each root.

(a) $\sqrt[4]{16} = 2$, because 2 is positive and $2^4 = 16$.

(b) $-\sqrt[4]{16}$

From part (a), $\sqrt[4]{16} = 2$, so the negative root is $-\sqrt[4]{16} = -2$.

(c) $\sqrt[4]{-16}$

For a real number fourth root, the radicand must be nonnegative. There is no real number that equals $\sqrt[4]{-16}$.

(d) $-\sqrt[5]{32}$

First find $\sqrt[5]{32}$. Because 2 is the number whose fifth power is 32, $\sqrt[5]{32} = 2$. Since $\sqrt[5]{32} = 2$, it follows that

$$-\sqrt[5]{32} = -2.$$

(e) $\sqrt[5]{-32} = -2$, because $(-2)^5 = -32$.

◀ **Work Problem 10 at the Side.**

Answers

10. **(a)** 3 **(b)** not a real number **(c)** -3 **(d)** 3 **(e)** -3

16.1 Exercises

FOR EXTRA HELP

Download the MyDashBoard App

MyMathLab®

CONCEPT CHECK *Decide whether each statement is* true *or* false. *If false, tell why.*

1. Every positive number has two real square roots.

2. A negative number has negative square roots.

3. Every nonnegative number has two real square roots.

4. The positive square root of a positive number is its principal square root.

5. The cube root of every real number has the same sign as the number itself.

6. Every positive number has three real cube roots.

Find all square roots of each number. **See Example 1.**

7. 9

8. 16

9. 64

10. 100

11. 169

12. 225

13. $\frac{25}{196}$

14. $\frac{81}{400}$

15. 900

16. 1600

CONCEPT CHECK *What must be true about the value of the variable a for each statement in Exercises 17–20 to be true?*

17. $\sqrt{a}$ represents a positive number.

18. $-\sqrt{a}$ represents a negative number.

19. $\sqrt{a}$ is not a real number.

20. $-\sqrt{a}$ is not a real number.

Find each square root. **See Examples 2 and 4(c).**

21. $\sqrt{64}$

GS $8^2 =$ ____, so

$\sqrt{64} =$ ____.

22. $\sqrt{9}$

GS $3^2 =$ ____, so

$\sqrt{9} =$ ____.

23. $\sqrt{1}$

24. $\sqrt{4}$

25. $\sqrt{49}$

26. $\sqrt{81}$

27. $-\sqrt{256}$

28. $-\sqrt{196}$

29. $-\sqrt{\frac{144}{121}}$

30. $-\sqrt{\frac{49}{36}}$

31. $\sqrt{0.64}$

32. $\sqrt{0.16}$

33. $\sqrt{-121}$

34. $\sqrt{-64}$

35. $-\sqrt{-49}$

36. $-\sqrt{-100}$

Find the square of each radical expression. ***See Example 3.***

37. $\sqrt{100}$ $(\sqrt{100})^2 =$ ____

38. $\sqrt{36}$ $(\sqrt{36})^2 =$ ____

39. $-\sqrt{19}$

40. $-\sqrt{99}$

41. $\sqrt{\frac{2}{3}}$

42. $\sqrt{\frac{5}{7}}$

43. $\sqrt{3x^2 + 4}$

44. $\sqrt{9y^2 + 3}$

CONCEPT CHECK *Without using a calculator, determine between which two consecutive integers each square root lies. For example,*

$\sqrt{75}$ *is between* 8 *and* 9, *because* $\sqrt{64} = 8$, $\sqrt{81} = 9$, *and* $64 < 75 < 81$.

45. $\sqrt{94}$

46. $\sqrt{43}$

47. $\sqrt{51}$

48. $\sqrt{30}$

49. $\sqrt{23.2}$

50. $\sqrt{10.3}$

Determine whether each number is rational, irrational, *or* not a real number. *If a number is rational, give its exact value. If a number is irrational, give a decimal approximation to the nearest thousandth. Use a calculator as necessary.* ***See Examples 4 and 5.***

51. $\sqrt{25}$

52. $\sqrt{169}$

53. $\sqrt{29}$

54. $\sqrt{33}$

55. $-\sqrt{64}$

56. $-\sqrt{81}$

57. $-\sqrt{300}$

58. $-\sqrt{500}$

59. $\sqrt{-29}$

60. $\sqrt{-47}$

61. $\sqrt{1200}$

62. $\sqrt{1500}$

Work Exercises 63 and 64 without using a calculator.

63. Choose the best estimate for the length and width (in meters) of this rectangle.

A. 11 by 6 **B.** 11 by 7 **C.** 10 by 7 **D.** 10 by 6

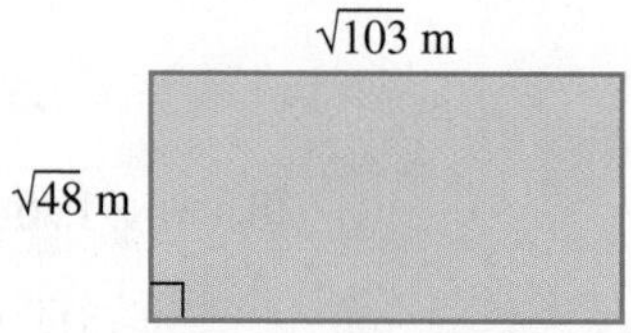

64. Choose the best estimate for the base and height (in feet) of this triangle.

A. $b = 8, h = 5$ **B.** $b = 8, h = 4$

C. $b = 9, h = 5$ **D.** $b = 9, h = 4$

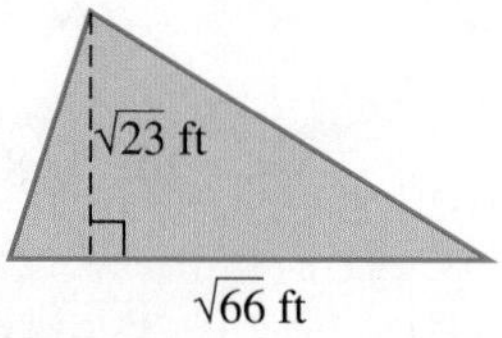

Find the length of the unknown side of each right triangle with sides a, b, and c, where c is the hypotenuse. Give any decimal approximations to the nearest thousandth. ***See* Figure 1 *and Example 6.***

65. $a = 8, b = 15$

66. $a = 24, b = 10$

67. $a = 6, c = 10$

68. $a = 5, c = 13$

69. $a = 11, b = 4$

70. $a = 13, b = 9$

Solve each problem. Give any decimal approximations to the nearest tenth.
See Example 7.

71. The diagonal of a rectangle measures 25 cm. The width of the rectangle is 7 cm. Find the length of the rectangle.

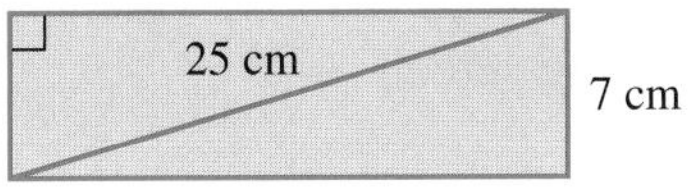

72. The length of a rectangle is 40 m, and the width is 9 m. Find the measure of the diagonal of the rectangle.

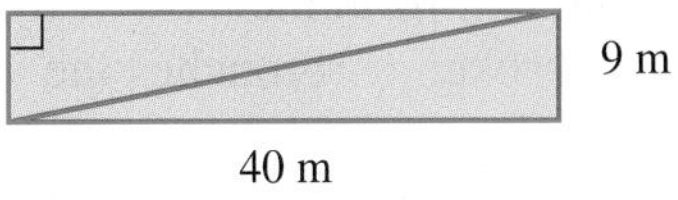

73. Tyler is flying a kite on 100 ft of string. How high is it above his hand (vertically) if the horizontal distance between Tyler and the kite is 60 ft?

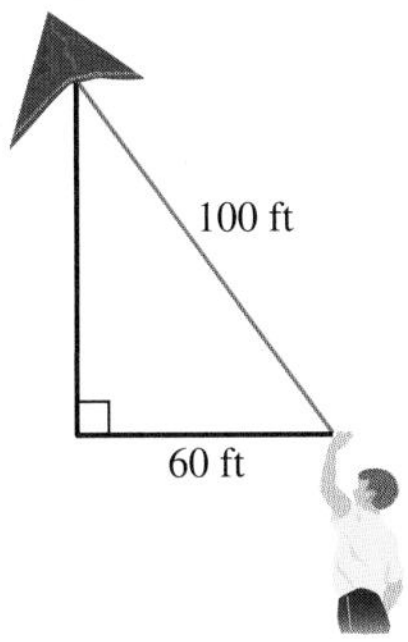

74. A guy wire is attached to the mast of a transmitting antenna. It is attached 96 ft above ground level. If the wire is staked to the ground 72 ft from the base of the mast, how long is the wire?

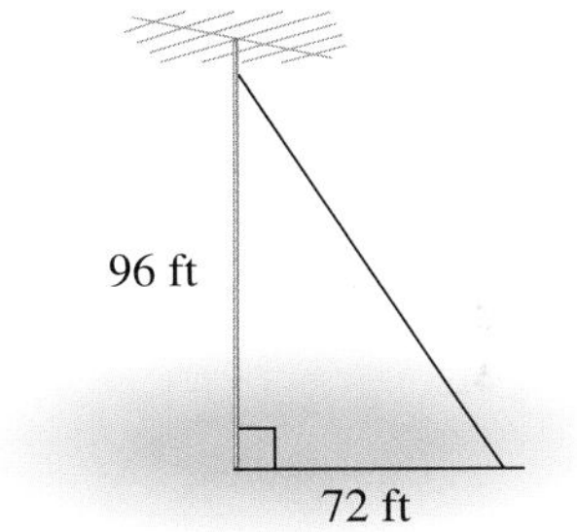

75. A surveyor measured the distances shown in the figure. Find the distance across the lake between points *R* and *S*.

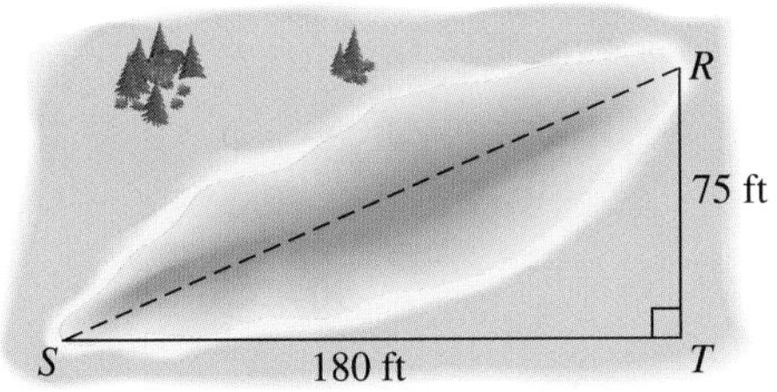

76. A boat is being pulled toward a dock with a rope attached at water level. When the boat is 24 ft from the dock, 30 ft of rope is extended. What is the height of the dock above the water?

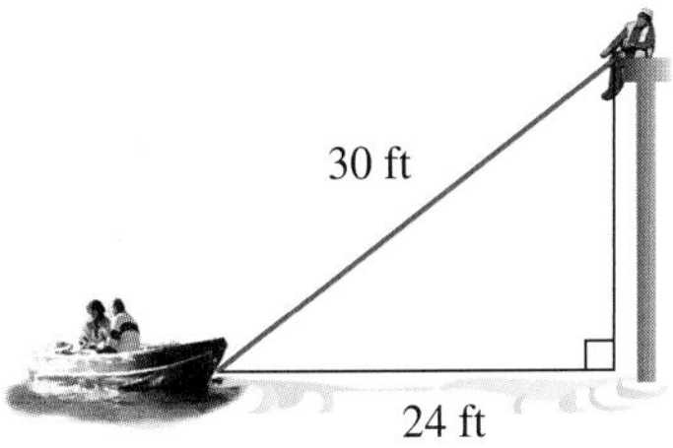

77. A surveyor wants to find the height of a building. At a point 110.0 ft from the base of the building he sights to the top of the building and finds the distance to be 193.0 ft. How high is the building?

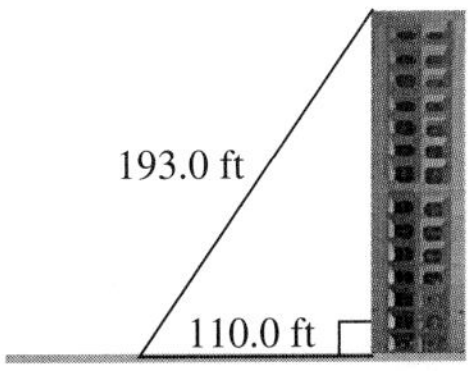

78. Two towns are separated by a dense forest. To go from Town B to Town A, it is necessary to travel due west for 19.0 mi, then turn due north and travel for 14.0 mi. How far apart are the towns?

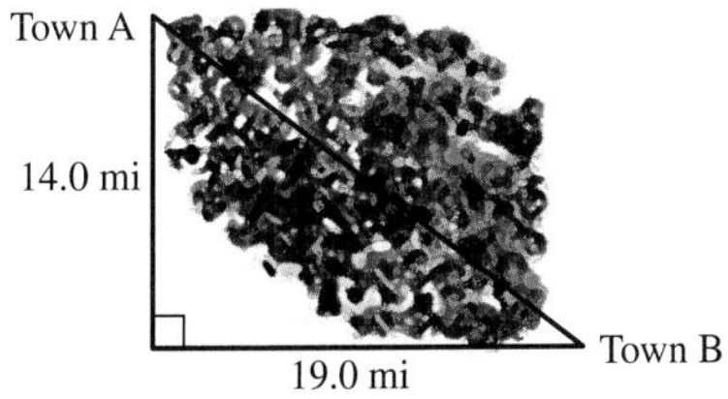

79. Following Hurricane Katrina, thousands of pine trees in southeastern Louisiana formed right triangles as shown in the photo. Suppose that, for a small such tree, the vertical distance from the base of the broken tree to the point of the break is 4.5 ft. The length of the broken part is 12.0 ft. How far along the ground is it from the base of the tree to the point where the broken part touches the ground?

80. One of the authors of this text purchased a Samsung UN46C6300 LED HDTV. A television set is "sized" according to the diagonal measurement of the viewing screen. The author purchased a 46-in. TV, so the TV measures 46 in. from one corner of the viewing screen diagonally to the other corner. The viewing screen is 40 in. wide. Find the height of the viewing screen.

81. What is the value of x (to the nearest thousandth) in the figure?

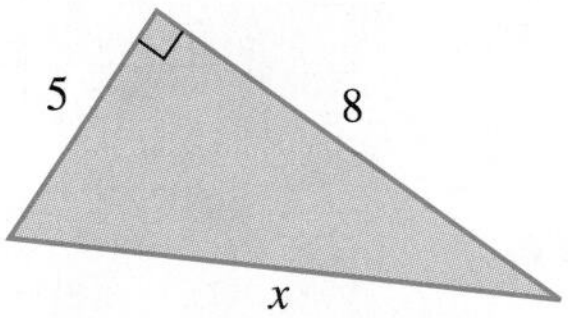

82. What is the value of y (to the nearest thousandth) in the figure?

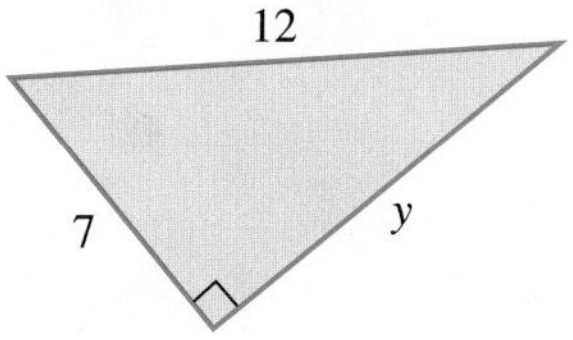

Find each root. ***See Examples 8 and 9.***

83. $\sqrt[3]{64}$

$\_\_\_\_^3 = 64$, so

$\sqrt[3]{64} = \_\_\_\_$.

84. $\sqrt[3]{343}$

$\_\_\_\_^3 = 343$, so

$\sqrt[3]{343} = \_\_\_\_$.

85. $\sqrt[3]{125}$

86. $\sqrt[3]{729}$

87. $\sqrt[3]{512}$

88. $\sqrt[3]{1000}$

89. $\sqrt[3]{-27}$

90. $\sqrt[3]{-64}$

91. $\sqrt[3]{-216}$

92. $\sqrt[3]{-343}$

93. $-\sqrt[3]{-8}$

94. $-\sqrt[3]{-216}$

95. $\sqrt[4]{256}$

96. $\sqrt[4]{625}$

97. $\sqrt[4]{1296}$

98. $\sqrt[4]{10,000}$

99. $\sqrt[4]{-1}$

100. $\sqrt[4]{-625}$

101. $-\sqrt[4]{625}$

102. $-\sqrt[4]{256}$

103. $\sqrt[5]{-1024}$

104. $\sqrt[5]{-100,000}$

16.2 Multiplying, Dividing, and Simplifying Radicals

OBJECTIVE 1 Multiply square root radicals. Consider the following.

$$\sqrt{4} \cdot \sqrt{9} = 2 \cdot 3 = 6 \quad \text{and} \quad \sqrt{4 \cdot 9} = \sqrt{36} = 6$$

This shows that $\sqrt{4} \cdot \sqrt{9} = \sqrt{4 \cdot 9}$.

The result here is a particular case of the **product rule for radicals.**

Product Rule for Radicals

If a and b are nonnegative real numbers, then the following hold.

$$\sqrt{a} \cdot \sqrt{b} = \sqrt{a \cdot b} \quad \text{and} \quad \sqrt{a \cdot b} = \sqrt{a} \cdot \sqrt{b}$$

In words, the product of two square roots is the square root of the product. The square root of a product is the product of the two square roots.

EXAMPLE 1 Using the Product Rule to Multiply Radicals

Use the product rule for radicals to find each product.

(a) $\sqrt{2} \cdot \sqrt{3}$
$= \sqrt{2 \cdot 3}$
$= \sqrt{6}$

(b) $\sqrt{7} \cdot \sqrt{5}$
$= \sqrt{35}$

(c) $\sqrt{11} \cdot \sqrt{a}$ $(a \geq 0)$
$= \sqrt{11a}$

Work Problem 1 at the Side. ▶

OBJECTIVE 2 Simplify radicals by using the product rule. ***A square root radical is simplified when no perfect square factor other than 1 remains under the radical symbol.*** This is accomplished by using the product rule.

EXAMPLE 2 Using the Product Rule to Simplify Radicals

Simplify each radical.

(a) $\sqrt{20}$ (20 has a perfect square factor of 4.)

$= \sqrt{4 \cdot 5}$ Factor; 4 is a perfect square.

$= \sqrt{4} \cdot \sqrt{5}$ Product rule

$= 2\sqrt{5}$ $\sqrt{4} = 2$

Thus, $\sqrt{20} = 2\sqrt{5}$. Because 5 has no perfect square factor (other than 1), $2\sqrt{5}$ is the **simplified form** of $\sqrt{20}$. Note that $2\sqrt{5}$ represents a product whose factors are 2 and $\sqrt{5}$.

We could factor 20 into prime factors and look for pairs of like factors.

$\sqrt{20}$

$= \sqrt{2 \cdot 2 \cdot 5}$ Each pair of like factors produces one factor outside the radical.

$= 2\sqrt{5}$

Continued on Next Page

OBJECTIVES

1. Multiply square root radicals.
2. Simplify radicals by using the product rule.
3. Simplify radicals by using the quotient rule.
4. Simplify radicals involving variables.
5. Simplify other roots.

1 Use the product rule for radicals to find each product.

GS **(a)** $\sqrt{6} \cdot \sqrt{11}$
$= \sqrt{\_\_\_ \cdot \_\_\_}$
$= \sqrt{\_\_\_}$

(b) $\sqrt{2} \cdot \sqrt{5}$

(c) $\sqrt{10} \cdot \sqrt{r}$ $(r \geq 0)$

Answers

1. (a) 6; 11; 66 **(b)** $\sqrt{10}$ **(c)** $\sqrt{10r}$

2 Simplify each radical.

GS (a) $\sqrt{8}$

$= \sqrt{\_\_\_\_ \cdot 2}$

$= \sqrt{\_\_\_\_} \cdot \sqrt{2}$

$= \_\_\_\_$

(b) $\sqrt{27}$

(c) $\sqrt{50}$

(d) $\sqrt{60}$

(e) $\sqrt{30}$

Answers

2. (a) 4; 4; $2\sqrt{2}$ (b) $3\sqrt{3}$ (c) $5\sqrt{2}$ (d) $2\sqrt{15}$ (e) cannot be simplified

(b) $\sqrt{72}$ — Look for the *greatest* perfect square factor of 72.

$= \sqrt{36 \cdot 2}$ Factor; 36 is a perfect square.

$= \sqrt{36} \cdot \sqrt{2}$ Product rule

$= 6\sqrt{2}$ $\sqrt{36} = 6$

We could factor 72 into prime factors and look for pairs of like factors.

$\sqrt{72}$

$= \sqrt{2 \cdot 2 \cdot 2 \cdot 3 \cdot 3}$ Factor into primes.

$= 2 \cdot 3 \cdot \sqrt{2}$ $\sqrt{2 \cdot 2} = 2$; $\sqrt{3 \cdot 3} = 3$

$= 6\sqrt{2}$ The result is the same.

(c) $\sqrt{300}$

$= \sqrt{100 \cdot 3}$ 100 is a perfect square.

$= \sqrt{100} \cdot \sqrt{3}$ Product rule

$= 10\sqrt{3}$ $\sqrt{100} = 10$

(d) $\sqrt{15}$ Since 15 has no perfect square factors (except 1), $\sqrt{15}$ cannot be simplified.

◀ **Work Problem 2 at the Side.**

EXAMPLE 3 Multiplying and Simplifying Radicals

Find each product and simplify.

(a) $\sqrt{9} \cdot \sqrt{75}$

$= 3\sqrt{75}$ $\sqrt{9} = 3$

$= 3\sqrt{25 \cdot 3}$ Factor; 25 is a perfect square.

$= 3\sqrt{25} \cdot \sqrt{3}$ Product rule

$= 3 \cdot 5 \cdot \sqrt{3}$ $\sqrt{25} = 5$

$= 15\sqrt{3}$ Multiply.

We could have used the product rule to get $\sqrt{9} \cdot \sqrt{75} = \sqrt{675}$, and then simplified. The method above uses smaller numbers.

(b) $\sqrt{8} \cdot \sqrt{12}$

$= \sqrt{8 \cdot 12}$ Product rule

$= \sqrt{4 \cdot 2 \cdot 4 \cdot 3}$ Factor; 4 is a perfect square.

$= \sqrt{4} \cdot \sqrt{4} \cdot \sqrt{2 \cdot 3}$ Commutative property; product rule

$= 2 \cdot 2 \cdot \sqrt{6}$ $\sqrt{4} = 2$

$= 4\sqrt{6}$ Multiply.

Continued on Next Page

(c) $2\sqrt{3} \cdot 3\sqrt{6}$

$= 2 \cdot 3 \cdot \sqrt{3 \cdot 6}$ Commutative property; product rule

$= 6\sqrt{18}$ Multiply.

$= 6\sqrt{9 \cdot 2}$ Factor; 9 is a perfect square.

$= 6\sqrt{9} \cdot \sqrt{2}$ Product rule

$= 6 \cdot 3 \cdot \sqrt{2}$ $\sqrt{9} = 3$

$= 18\sqrt{2}$ Multiply.

Work Problem 3 at the Side. ▶

Note

There is often more than one way to find a product.

$\sqrt{8} \cdot \sqrt{12}$	Example 3(b)
$= \sqrt{4 \cdot 2} \cdot \sqrt{4 \cdot 3}$	Factor.
$= 2\sqrt{2} \cdot 2\sqrt{3}$	$\sqrt{4} = 2$
$= 2 \cdot 2 \cdot \sqrt{2} \cdot \sqrt{3}$	Commutative property
Same result → $= 4\sqrt{6}$	Multiply; product rule

OBJECTIVE 3 **Simplify radicals by using the quotient rule.** The **quotient rule for radicals** is very similar to the product rule.

Quotient Rule for Radicals

If a and b are nonnegative real numbers and $b \neq 0$, then the following hold.

$$\sqrt{\frac{a}{b}} = \frac{\sqrt{a}}{\sqrt{b}} \quad \text{and} \quad \frac{\sqrt{a}}{\sqrt{b}} = \sqrt{\frac{a}{b}}$$

In words, the square root of a quotient is the quotient of the two square roots. The quotient of two square roots is the square root of the quotient.

EXAMPLE 4 Using the Quotient Rule to Simplify Radicals

Use the quotient rule to simplify each radical.

(a) $\sqrt{\frac{25}{9}}$

$= \frac{\sqrt{25}}{\sqrt{9}}$

$= \frac{5}{3}$

(b) $\frac{\sqrt{288}}{\sqrt{2}}$

$= \sqrt{\frac{288}{2}}$

$= \sqrt{144}$

$= 12$

(c) $\sqrt{\frac{3}{4}}$

$= \frac{\sqrt{3}}{\sqrt{4}}$

$= \frac{\sqrt{3}}{2}$

Work Problem 4 at the Side. ▶

3 Find each product and simplify.

GS (a) $\sqrt{3} \cdot \sqrt{15}$

$= \sqrt{3 \cdot \_\_\_\_}$

$= \sqrt{\_\_\_\_}$

$= \sqrt{\_\_\_\_ \cdot 5}$

$= \sqrt{\_\_\_\_} \cdot \sqrt{5}$

$= \_\_\_\_$

(b) $\sqrt{10} \cdot \sqrt{50}$

(c) $\sqrt{12} \cdot \sqrt{2}$

(d) $3\sqrt{5} \cdot 4\sqrt{10}$

4 Use the quotient rule to simplify each radical.

GS (a) $\sqrt{\frac{81}{16}}$

$= \frac{\sqrt{\_\_\_\_}}{\sqrt{\_\_\_\_}}$

$= \_\_\_\_$

(b) $\frac{\sqrt{192}}{\sqrt{3}}$

(c) $\sqrt{\frac{10}{49}}$

Answers

3. (a) 15; 45; 9; 9; $3\sqrt{5}$ (b) $10\sqrt{5}$ (c) $2\sqrt{6}$ (d) $60\sqrt{2}$

4. (a) 81; 16; $\frac{9}{4}$ (b) 8 (c) $\frac{\sqrt{10}}{7}$

5 Simplify $\dfrac{8\sqrt{50}}{4\sqrt{5}}$.

6 Simplify.

(a) $\sqrt{\dfrac{3}{8}} \cdot \sqrt{\dfrac{7}{2}}$

GS **(b)** $\sqrt{\dfrac{5}{6}} \cdot \sqrt{120}$

$= \sqrt{\dfrac{5}{6} \cdot \underline{\quad}}$

$= \sqrt{\underline{\quad}}$

$= \underline{\quad}$

Answers

5. $2\sqrt{10}$

6. (a) $\dfrac{\sqrt{21}}{4}$ (b) 120; 100; 10

EXAMPLE 5 Using the Quotient Rule to Divide Radicals

Simplify.

$$\frac{27\sqrt{15}}{9\sqrt{3}}$$

$$= \frac{27}{9} \cdot \frac{\sqrt{15}}{\sqrt{3}} \quad \text{Multiplication of fractions}$$

$$= \frac{27}{9} \cdot \sqrt{\frac{15}{3}} \quad \text{Quotient rule}$$

$$= 3\sqrt{5} \quad \text{Divide.}$$

◀ **Work Problem 5 at the Side.**

EXAMPLE 6 Using Both the Product and Quotient Rules

Simplify.

$$\sqrt{\frac{3}{5}} \cdot \sqrt{\frac{1}{5}}$$

$$= \sqrt{\frac{3}{5} \cdot \frac{1}{5}} \quad \text{Product rule}$$

$$= \sqrt{\frac{3}{25}} \quad \text{Multiply fractions.}$$

$$= \frac{\sqrt{3}}{\sqrt{25}}, \text{ or } \frac{\sqrt{3}}{5} \quad \text{Quotient rule; } \sqrt{25} = 5$$

◀ **Work Problem 6 at the Side.**

OBJECTIVE 4 **Simplify radicals involving variables.** Consider a radical with variable radicand, such as $\sqrt{x^2}$.

If x represents a nonnegative number, then $\sqrt{x^2} = x$.

If x represents a negative number, then $\sqrt{x^2} = -x$, the *opposite* of x (which is positive).

For example, $\sqrt{5^2} = 5$, but $\sqrt{(-5)^2} = \sqrt{25} = 5$, the *opposite* of -5.

This means that the square root of a squared number is always nonnegative. We can use absolute value to express this.

$\sqrt{a^2}$

For any real number a, the following holds.

$$\sqrt{a^2} = |a|$$

The product and quotient rules apply when variables appear under radical symbols, as long as the variables represent only *nonnegative* real numbers. ***To avoid negative radicands, we assume variables under radical symbols are nonnegative in this text unless otherwise specified.*** In such cases, absolute value bars are not necessary, since for all $x \geq 0$, $|x| = x$.

EXAMPLE 7 **Simplifying Radicals Involving Variables**

Simplify each radical. Assume that all variables represent nonnegative real numbers.

(a) $\sqrt{x^4} = x^2$, since $(x^2)^2 = x^4$.

(b) $\sqrt{25m^6}$

$= \sqrt{25} \cdot \sqrt{m^6}$ Product rule

$= 5m^3$ $(m^3)^2 = m^6$

(c) $\sqrt{8p^{10}}$

$= \sqrt{4 \cdot 2 \cdot p^{10}}$ Factor; 4 is a perfect square.

$= \sqrt{4} \cdot \sqrt{2} \cdot \sqrt{p^{10}}$ Product rule

$= 2 \cdot \sqrt{2} \cdot p^5$ $(p^5)^2 = p^{10}$

$= 2p^5\sqrt{2}$ Commutative property

(d) $\sqrt{r^9}$

$= \sqrt{r^8 \cdot r}$

$= \sqrt{r^8} \cdot \sqrt{r}$ Product rule

$= r^4\sqrt{r}$ $(r^4)^2 = r^8$

(e) $\sqrt{\frac{5}{x^2}}$

$= \frac{\sqrt{5}}{\sqrt{x^2}}$ Quotient rule

$= \frac{\sqrt{5}}{x}$ $(x \neq 0)$

Note

A quick way to find the square root of a variable raised to an even power is to divide the exponent by the index, 2.

Examples: $\sqrt{x^6} = x^3$ and $\sqrt{x^{10}} = x^5$

$6 \div 2 = 3$ $\qquad$ $10 \div 2 = 5$

Work Problem 7 at the Side. ▶

OBJECTIVE 5 **Simplify other roots.** The product and quotient rules for radicals also apply to other roots. To simplify cube roots, look for factors that are perfect cubes, as in **Section 16.1.** Recall that a perfect cube is a number with a rational cube root. For example, $\sqrt[3]{64} = 4$, and because 4 is a rational number, 64 is a perfect cube.

Other roots are handled in a similar manner.

Properties of Radicals

For all real numbers a and b where the indicated roots exist, the following hold.

$$\sqrt[n]{a} \cdot \sqrt[n]{b} = \sqrt[n]{ab} \quad \text{and} \quad \frac{\sqrt[n]{a}}{\sqrt[n]{b}} = \sqrt[n]{\frac{a}{b}} \quad (b \neq 0)$$

7 Simplify each radical. Assume that all variables represent nonnegative real numbers.

GS **(a)** $\sqrt{x^8}$

$(\_\_\_\_)^2 = x^8$, so

$\sqrt{x^8} = \_\_\_\_$.

GS **(b)** $\sqrt{36y^6}$

$= \sqrt{36} \cdot \sqrt{\_\_\_\_}$

$= \_\_\_\_$

(c) $\sqrt{100p^{12}}$

(d) $\sqrt{12z^2}$

(e) $\sqrt{a^5}$

(f) $\sqrt{\frac{10}{n^4}}$ $(n \neq 0)$

Answers

7. **(a)** x^4; x^4 **(b)** y^6; $6y^3$ **(c)** $10p^6$ **(d)** $2z\sqrt{3}$ **(e)** $a^2\sqrt{a}$ **(f)** $\frac{\sqrt{10}}{n^2}$

8 Simplify each radical.

GS **(a)** $\sqrt[3]{108}$

$= \sqrt[3]{\_\_\_\_ \cdot 4}$

$= \sqrt[3]{\_\_\_\_} \cdot \sqrt[3]{4}$

$= \_\_\_\_$

(b) $\sqrt[3]{250}$

(c) $\sqrt[4]{160}$

(d) $\sqrt[4]{\frac{16}{625}}$

9 Simplify each radical.

(a) $\sqrt[3]{z^9}$

(b) $\sqrt[3]{8x^6}$

(c) $\sqrt[3]{54t^5}$

(d) $\sqrt[3]{\frac{a^{15}}{64}}$

Answers

8. **(a)** 27; 27; $3\sqrt[3]{4}$ **(b)** $5\sqrt[3]{2}$ **(c)** $2\sqrt[4]{10}$ **(d)** $\frac{2}{5}$

9. **(a)** z^3 **(b)** $2x^2$ **(c)** $3t\sqrt[3]{2t^2}$ **(d)** $\frac{a^5}{4}$

EXAMPLE 8 Simplifying Other Roots

Simplify each radical.

(a) $\sqrt[3]{32}$

$= \sqrt[3]{8 \cdot 4}$ Factor; 8 is a perfect cube.

$= \sqrt[3]{8} \cdot \sqrt[3]{4}$ Product rule

$= 2\sqrt[3]{4}$ Take the cube root.

Remember to write the root index 3 in each radical.

(b) $\sqrt[4]{32}$

$= \sqrt[4]{16 \cdot 2}$ Factor; 16 is a perfect fourth power.

$= \sqrt[4]{16} \cdot \sqrt[4]{2}$ Product rule

$= 2\sqrt[4]{2}$ Take the fourth root.

Remember to write the root index 4 in each radical.

(c) $\sqrt[3]{\frac{27}{125}}$

$= \frac{\sqrt[3]{27}}{\sqrt[3]{125}}$ Quotient rule

$= \frac{3}{5}$ Take cube roots.

◀ **Work Problem 8 at the Side.**

Other roots of radicals involving variables can also be simplified. To simplify cube roots with variables, use the fact that for *any* real number *a*,

$$\sqrt[3]{a^3} = a.$$

EXAMPLE 9 Simplifying Cube Roots Involving Variables

Simplify each radical.

(a) $\sqrt[3]{m^6}$

$= m^2$ $(m^2)^3 = m^6$

(b) $\sqrt[3]{27x^{12}}$

$= \sqrt[3]{27} \cdot \sqrt[3]{x^{12}}$ Product rule

$= 3x^4$ $3^3 = 27$; $(x^4)^3 = x^{12}$

(c) $\sqrt[3]{32a^4}$

$= \sqrt[3]{8a^3 \cdot 4a}$ Factor; 8 is a perfect cube.

$= \sqrt[3]{8a^3} \cdot \sqrt[3]{4a}$ Product rule

$= 2a\sqrt[3]{4a}$ $(2a)^3 = 8a^3$

(d) $\sqrt[3]{\frac{y^3}{125}}$

$= \frac{\sqrt[3]{y^3}}{\sqrt[3]{125}}$ Quotient rule

$= \frac{y}{5}$ Take cube roots.

◀ **Work Problem 9 at the Side.**

16.2 Exercises

FOR EXTRA HELP

Download the MyDashBoard App

MyMathLab®

CONCEPT CHECK *Decide whether each statement is* true *or* false. *If false, show why.*

1. $\sqrt{(-6)^2} = -6$

2. $\sqrt[3]{(-6)^3} = -6$

3. The radical $2\sqrt{7}$ represents a sum.

4. In the radical $3\sqrt{11}$, the numbers 3 and $\sqrt{11}$ are factors.

Use the product rule for radicals to find each product. **See Example 1.**

5. $\sqrt{3} \cdot \sqrt{5}$

6. $\sqrt{3} \cdot \sqrt{7}$

7. $\sqrt{2} \cdot \sqrt{11}$

8. $\sqrt{2} \cdot \sqrt{15}$

9. $\sqrt{6} \cdot \sqrt{7}$

10. $\sqrt{5} \cdot \sqrt{6}$

11. $\sqrt{13} \cdot \sqrt{r}$ $(r \geq 0)$

12. $\sqrt{19} \cdot \sqrt{k}$ $(k \geq 0)$

13. CONCEPT CHECK Which one of the following radicals is simplified?

A. $\sqrt{47}$ **B.** $\sqrt{45}$ **C.** $\sqrt{48}$ **D.** $\sqrt{44}$

14. If p is a prime number, is $\sqrt{p}$ in simplified form? Explain.

Simplify each radical. **See Example 2.**

15. $\sqrt{45}$

16. $\sqrt{63}$

17. $\sqrt{24}$

18. $\sqrt{44}$

19. $\sqrt{90}$

20. $\sqrt{56}$

21. $\sqrt{75}$

22. $\sqrt{18}$

23. $\sqrt{125}$

24. $\sqrt{80}$

25. $\sqrt{145}$

26. $\sqrt{110}$

27. $\sqrt{160}$

28. $\sqrt{128}$

29. $-\sqrt{700}$

30. $-\sqrt{600}$

31. $3\sqrt{52}$

32. $9\sqrt{8}$

33. $5\sqrt{50}$

34. $6\sqrt{40}$

Use the Pythagorean theorem to find the length of the unknown side of each right triangle. Express answers as simplified radicals. **See Section 16.1.**

35.
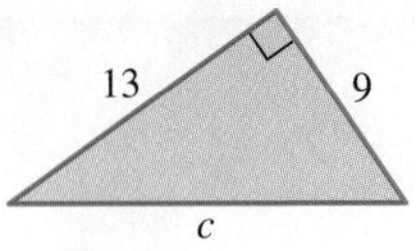

36.
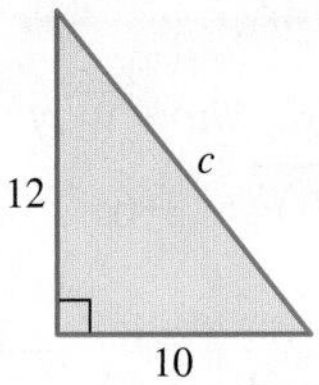

37.
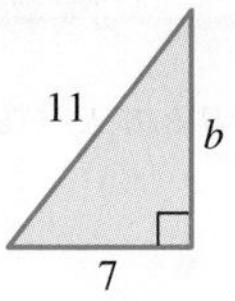

38.
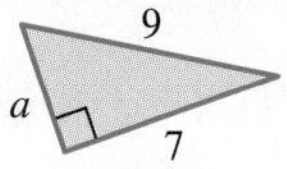

Find each product and simplify. **See Example 3.**

39. $\sqrt{3} \cdot \sqrt{18}$

40. $\sqrt{3} \cdot \sqrt{21}$

41. $\sqrt{9} \cdot \sqrt{32}$

42. $\sqrt{9} \cdot \sqrt{50}$

43. $\sqrt{12} \cdot \sqrt{48}$

44. $\sqrt{50} \cdot \sqrt{72}$

45. $\sqrt{12} \cdot \sqrt{30}$

46. $\sqrt{30} \cdot \sqrt{24}$

47. $2\sqrt{10} \cdot 3\sqrt{2}$

48. $5\sqrt{6} \cdot 2\sqrt{10}$

49. $5\sqrt{3} \cdot 2\sqrt{15}$

50. $4\sqrt{6} \cdot 3\sqrt{2}$

51. Simplify the product $\sqrt{8} \cdot \sqrt{32}$ in two ways.

Method 1: Multiply 8 by 32 and simplify the square root of this product.

Method 2: Simplify $\sqrt{8}$, simplify $\sqrt{32}$, and then multiply.

How do the answers compare? Make a conjecture (an educated guess) about using these methods when simplifying a product such as this.

52. Simplify the radical $\sqrt{288}$ in two ways.

Method 1: Factor 288 as $144 \cdot 2$ and then simplify.

Method 2: Factor 288 as $48 \cdot 6$ and then simplify.

How do the answers compare? Make a conjecture concerning the quickest way to simplify such a radical.

Simplify each radical expression. **See Examples 4–6.**

53. $\sqrt{\frac{16}{225}}$

54. $\sqrt{\frac{9}{100}}$

55. $\sqrt{\frac{7}{16}}$

56. $\sqrt{\frac{13}{25}}$

57. $\sqrt{\frac{4}{50}}$

58. $\sqrt{\frac{14}{72}}$

59. $\frac{\sqrt{75}}{\sqrt{3}}$

60. $\frac{\sqrt{200}}{\sqrt{2}}$

61. $\sqrt{\frac{5}{2}} \cdot \sqrt{\frac{125}{8}}$

62. $\sqrt{\frac{8}{3}} \cdot \sqrt{\frac{512}{27}}$

63. $\frac{30\sqrt{10}}{5\sqrt{2}}$

64. $\frac{50\sqrt{20}}{2\sqrt{10}}$

Simplify each radical. Assume that all variables represent nonnegative real numbers. ***See Example 7.***

65. $\sqrt{m^2}$ **66.** $\sqrt{k^2}$ **67.** $\sqrt{y^4}$ **68.** $\sqrt{s^4}$

69. $\sqrt{36z^2}$ **70.** $\sqrt{49n^2}$ **71.** $\sqrt{400x^6}$ **72.** $\sqrt{900y^8}$

73. $\sqrt{18x^8}$ **74.** $\sqrt{20r^{10}}$ **75.** $\sqrt{45c^{14}}$ **76.** $\sqrt{50d^{20}}$

77. $\sqrt{z^5}$ **78.** $\sqrt{y^3}$ **79.** $\sqrt{a^{13}}$ **80.** $\sqrt{p^{17}}$

81. $\sqrt{64x^7}$ **82.** $\sqrt{25t^{11}}$ **83.** $\sqrt{x^6y^{12}}$ **84.** $\sqrt{a^8b^{10}}$

85. $\sqrt{81m^4n^2}$ **86.** $\sqrt{100c^4d^6}$ **87.** $\sqrt{\frac{7}{x^{10}}}$ $(x \neq 0)$ **88.** $\sqrt{\frac{14}{z^{12}}}$ $(z \neq 0)$

89. $\sqrt{\frac{y^4}{100}}$ **90.** $\sqrt{\frac{w^8}{144}}$ **91.** $\sqrt{\frac{x^6}{y^8}}$ $(y \neq 0)$ **92.** $\sqrt{\frac{a^4}{b^6}}$ $(b \neq 0)$

Simplify each radical. ***See Example 8.***

93. $\sqrt[3]{40}$ **94.** $\sqrt[3]{48}$ **95.** $\sqrt[3]{54}$ **96.** $\sqrt[3]{135}$

97. $\sqrt[3]{128}$ **98.** $\sqrt[3]{192}$ **99.** $\sqrt[4]{80}$ **100.** $\sqrt[4]{243}$

101. $\sqrt[3]{\frac{8}{27}}$ **102.** $\sqrt[3]{\frac{64}{125}}$ **103.** $\sqrt[3]{-\frac{216}{125}}$ **104.** $\sqrt[3]{-\frac{1}{64}}$

Simplify each radical. ***See Example 9.***

105. $\sqrt[3]{p^3}$

106. $\sqrt[3]{w^3}$

107. $\sqrt[3]{x^9}$

108. $\sqrt[3]{y^{18}}$

109. $\sqrt[3]{64z^6}$

110. $\sqrt[3]{125a^{15}}$

111. $\sqrt[3]{343a^9b^3}$

112. $\sqrt[3]{216m^3n^6}$

113. $\sqrt[3]{16t^5}$

114. $\sqrt[3]{24x^4}$

115. $\sqrt[3]{\frac{m^{12}}{8}}$

116. $\sqrt[3]{\frac{n^9}{27}}$

The volume V of a cube is found with the formula $V = s^3$, where s is the length of an edge of the cube. Use this information in Exercises 117 and 118.

117. A container in the shape of a cube has a volume of 216 cm^3. What is the depth of the container?

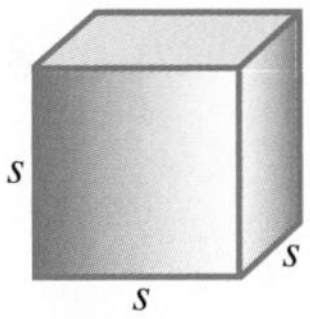

118. A cube-shaped box must be constructed to contain 128 ft^3. What should the dimensions (height, width, and length) of the box be?

The volume V of a sphere is found with the formula $V = \frac{4}{3}\pi r^3$, where r is the length of the radius of the sphere. Use this information in Exercises 119 and 120.

119. A ball in the shape of a sphere has a volume of 288π in.^{3}. What is the radius of the ball?

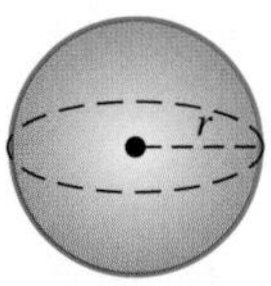

120. Suppose that the volume of the ball described in **Exercise 119** is multiplied by 8. How is the radius affected?

Work Exercises 121 and 122 without using a calculator.

121. Choose the best estimate for the area (in square inches) of this rectangle.

A. 45 **B.** 72 **C.** 80 **D.** 90

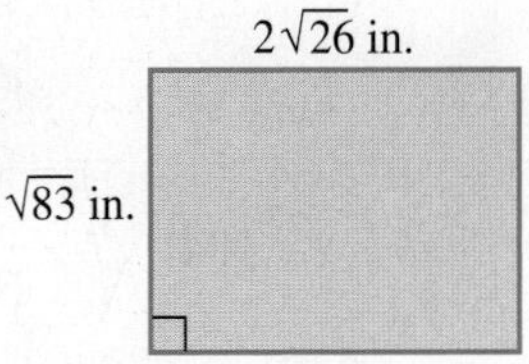

122. Choose the best estimate for the area (in square feet) of the triangle.

A. 20 **B.** 40 **C.** 60 **D.** 80

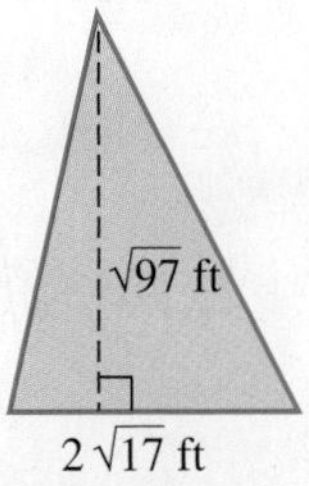

16.3 Adding and Subtracting Radicals

OBJECTIVES

1. **Add and subtract radicals.**
2. **Simplify radical sums and differences.**
3. **Simplify *more* complicated radical expressions.**

OBJECTIVE 1 Add and subtract radicals. We use the distributive property.

$8\sqrt{3} + 6\sqrt{3}$

$= (8 + 6)\sqrt{3}$ Distributive property

$= 14\sqrt{3}$ Add.

$2\sqrt{11} - 7\sqrt{11}$

$= (2 - 7)\sqrt{11}$ Distributive property

$= -5\sqrt{11}$ Subtract.

Only **like radicals**—those that are ***multiples of the same root of the same number***—can be combined in this way. Examples of **unlike radicals** are

$2\sqrt{5}$ and $2\sqrt{3}$, Radicands are different.

as well as $2\sqrt{3}$ and $2\sqrt[3]{3}$. Indexes are different.

EXAMPLE 1 Adding and Subtracting Like Radicals

Add or subtract, as indicated.

(a) $3\sqrt{6} + 5\sqrt{6}$ These are like radicals.

$= (3 + 5)\sqrt{6}$ We are factoring out $\sqrt{6}$ here.

$= 8\sqrt{6}$

(b) $5\sqrt{10} - 7\sqrt{10}$ These are like radicals.

$= (5 - 7)\sqrt{10}$

$= -2\sqrt{10}$

(c) $\sqrt{7} + 2\sqrt{7}$

$= 1\sqrt{7} + 2\sqrt{7}$

$= (1 + 2)\sqrt{7}$

$= 3\sqrt{7}$

(d) $\sqrt{5} + \sqrt{5}$

$= 1\sqrt{5} + 1\sqrt{5}$

$= (1 + 1)\sqrt{5}$

$= 2\sqrt{5}$

(e) $\sqrt{3} + \sqrt{7}$ cannot be combined. They are unlike radicals.

Work Problems 1 and 2 at the Side. ▶

OBJECTIVE 2 Simplify radical sums and differences.

EXAMPLE 2 Simplifying Radicals to Add or Subtract

Add or subtract, as indicated.

(a) $3\sqrt{2} + \sqrt{8}$

$= 3\sqrt{2} + \sqrt{4 \cdot 2}$ Factor.

$= 3\sqrt{2} + \sqrt{4} \cdot \sqrt{2}$ Product rule

$= 3\sqrt{2} + 2\sqrt{2}$ $\sqrt{4} = 2$

$= 5\sqrt{2}$ Add like radicals.

(b) $\sqrt{18} - \sqrt{27}$

$= \sqrt{9 \cdot 2} - \sqrt{9 \cdot 3}$ Factor; 9 is a perfect square.

$= \sqrt{9} \cdot \sqrt{2} - \sqrt{9} \cdot \sqrt{3}$ Product rule

$= 3\sqrt{2} - 3\sqrt{3}$ $\sqrt{9} = 3$

These are unlike radicals. They cannot be combined.

Continued on Next Page

1 Are the radicals in each pair *like* or *unlike*? Tell why.

(a) $5\sqrt{6}$ and $4\sqrt{6}$

(b) $2\sqrt{3}$ and $3\sqrt{2}$

(c) $\sqrt{10}$ and $\sqrt[3]{10}$

(d) $7\sqrt{2x}$ and $8\sqrt{2x}$

(e) $\sqrt{3y}$ and $\sqrt{6y}$

2 Add or subtract, as indicated.

GS **(a)** $8\sqrt{5} + 2\sqrt{5}$

$= (\_\_\_ + \_\_\_)\_\_\_$

$= \_\_\_$

(b) $4\sqrt{3} - 9\sqrt{3}$

(c) $4\sqrt{11} - 3\sqrt{11}$

(d) $\sqrt{15} + \sqrt{15}$

(e) $2\sqrt{7} + 2\sqrt{10}$

Answers

1. (a) like; Both are square roots and have the same radicand, 6.
(b) unlike; The radicands are different—one is 3 and one is 2.
(c) unlike; The indexes are different—one is a square root and one is a cube root.
(d) like; Both are square roots and have the same radicand, $2x$.
(e) unlike; The radicands are different—one is $3y$ and one is $6y$.

2. (a) 8; 2; $\sqrt{5}$; $10\sqrt{5}$ **(b)** $-5\sqrt{3}$
(c) $\sqrt{11}$ **(d)** $2\sqrt{15}$
(e) These are unlike radicals and cannot be combined.

3 Add or subtract, as indicated.

GS **(a)** $\sqrt{8} + 4\sqrt{2}$

$= \sqrt{\_\_\_ \cdot 2} + 4\sqrt{2}$

$= \sqrt{\_\_\_} \cdot \sqrt{2} + 4\sqrt{2}$

$= \_\_\_ + 4\sqrt{2}$

$= \_\_\_$

(b) $\sqrt{18} - \sqrt{2}$

(c) $\sqrt{27} + \sqrt{12}$

(d) $5\sqrt{200} - 6\sqrt{18}$

4 Simplify each radical expression. Assume that all variables represent nonnegative real numbers.

(a) $\sqrt{7} \cdot \sqrt{21} + 2\sqrt{27}$

(b) $\sqrt{3r} \cdot \sqrt{6} + \sqrt{8r}$

(c) $y\sqrt{72} - \sqrt{18y^2}$

(d) $\sqrt[3]{81x^4} + 5\sqrt[3]{24x^4}$

Answers

3. **(a)** 4; 4; $2\sqrt{2}$; $6\sqrt{2}$ **(b)** $2\sqrt{2}$ **(c)** $5\sqrt{3}$ **(d)** $32\sqrt{2}$

4. **(a)** $13\sqrt{3}$ **(b)** $5\sqrt{2r}$ **(c)** $3y\sqrt{2}$ **(d)** $13x\sqrt[3]{3x}$

(c) $2\sqrt{12} + 3\sqrt{75}$

$= 2(\sqrt{4} \cdot \sqrt{3}) + 3(\sqrt{25} \cdot \sqrt{3})$ Product rule

$= 2(2\sqrt{3}) + 3(5\sqrt{3})$ $\sqrt{4} = 2$; $\sqrt{25} = 5$

$= 4\sqrt{3} + 15\sqrt{3}$ Multiply.

$= 19\sqrt{3}$ Think: $(4 + 15)\sqrt{3}$ Add like radicals.

◀ **Work Problem 3 at the Side.**

OBJECTIVE 3 Simplify more complicated radical expressions.

EXAMPLE 3 Simplifying Radical Expressions

Simplify each radical expression. Assume that all variables represent nonnegative real numbers.

(a) $\sqrt{5} \cdot \sqrt{15} + 4\sqrt{3}$ (We simplify this expression by performing the operations.)

$= \sqrt{5 \cdot 15} + 4\sqrt{3}$ Product rule

$= \sqrt{75} + 4\sqrt{3}$ Multiply.

$= \sqrt{25 \cdot 3} + 4\sqrt{3}$ Factor; 25 is a perfect square.

$= \sqrt{25} \cdot \sqrt{3} + 4\sqrt{3}$ Product rule

$= 5\sqrt{3} + 4\sqrt{3}$ $\sqrt{25} = 5$

$= 9\sqrt{3}$ Add like radicals.

(b) $\sqrt{12k} + \sqrt{27k}$

$= \sqrt{4 \cdot 3k} + \sqrt{9 \cdot 3k}$ Factor.

$= \sqrt{4} \cdot \sqrt{3k} + \sqrt{9} \cdot \sqrt{3k}$ Product rule

$= 2\sqrt{3k} + 3\sqrt{3k}$ $\sqrt{4} = 2$; $\sqrt{9} = 3$

$= 5\sqrt{3k}$ Add like radicals.

(c) $3x\sqrt{50} + \sqrt{2x^2}$

$= 3x\sqrt{25 \cdot 2} + \sqrt{x^2 \cdot 2}$ Factor.

$= 3x\sqrt{25} \cdot \sqrt{2} + \sqrt{x^2} \cdot \sqrt{2}$ Product rule

$= 3x \cdot 5\sqrt{2} + x\sqrt{2}$ $\sqrt{25} = 5$; $\sqrt{x^2} = x$

$= 15x\sqrt{2} + x\sqrt{2}$ Multiply.

$= 16x\sqrt{2}$ Think: $(15x + 1x)\sqrt{2}$ Add like radicals.

(d) $2\sqrt[3]{32m^3} - \sqrt[3]{108m^3}$

$= 2\sqrt[3]{(8m^3)4} - \sqrt[3]{(27m^3)4}$ Factor.

$= 2(2m)\sqrt[3]{4} - 3m\sqrt[3]{4}$ $\sqrt[3]{8m^3} = 2m$; $\sqrt[3]{27m^3} = 3m$

$= 4m\sqrt[3]{4} - 3m\sqrt[3]{4}$ Multiply.

$= m\sqrt[3]{4}$ Subtract like radicals.

◀ **Work Problem 4 at the Side.**

16.3 Exercises

FOR EXTRA HELP

Download the MyDashBoard App

MyMathLab®

CONCEPT CHECK *Complete each statement.*

1. Like radicals have the same ____________ and the same ____________, or order. For example, $5\sqrt{2}$ and $-3\sqrt{2}$ are (*like* / *unlike*) radicals, as are $\sqrt{7}$, $-\sqrt{\_\_\_}$, and $8\sqrt{\_\_\_}$.

2. The radicals $\sqrt[4]{3xy^3}$ and $-6\sqrt[4]{3xy^3}$ are (*like* / *unlike*) radicals because both have the same root index, _____, and the same radicand, _____.

3. $\sqrt{5} + 5\sqrt{3}$ cannot be simplified because the ____________ are different. They are (*like* / *unlike*) radicals.

4. $4\sqrt[3]{2} + 3\sqrt{2}$ cannot be simplified because the ____________ are different. They are (*like* / *unlike*) radicals.

5. Simplifying the expression $5\sqrt{2} + 6\sqrt{2}$ as $(5 + 6)\sqrt{2}$, or $11\sqrt{2}$, is an application of the ____________ property.

6. The radicals $\sqrt{5x}$ and $4\sqrt{5x}$ are (*like* / *unlike*) radicals. To add them, use the identity property for multiplication to write $\sqrt{5x}$ as $\_\_\_\_\sqrt{5x}$. Thus, $\sqrt{5x} + 4\sqrt{5x}$ can be added to obtain _____.

Simplify and add or subtract wherever possible. **See Examples 1 and 2.**

7. $2\sqrt{3} + 5\sqrt{3}$

8. $6\sqrt{5} + 8\sqrt{5}$

9. $14\sqrt{7} - 19\sqrt{7}$

10. $16\sqrt{2} - 18\sqrt{2}$

11. $\sqrt{17} + 4\sqrt{17}$

12. $5\sqrt{19} + \sqrt{19}$

13. $6\sqrt{7} - \sqrt{7}$

14. $11\sqrt{14} - \sqrt{14}$

15. $\sqrt{6} + \sqrt{6}$

16. $\sqrt{11} + \sqrt{11}$

17. $\sqrt{6} + \sqrt{7}$

18. $\sqrt{14} + \sqrt{17}$

19. $5\sqrt{3} + \sqrt{12}$

20. $3\sqrt{2} + \sqrt{50}$

21. $\sqrt{45} + 4\sqrt{20}$

22. $\sqrt{24} + 6\sqrt{54}$

23. $5\sqrt{72} - 3\sqrt{50}$

24. $6\sqrt{18} - 5\sqrt{32}$

25. $-5\sqrt{32} + 2\sqrt{98}$

26. $-4\sqrt{75} + 3\sqrt{12}$

27. $5\sqrt{7} - 3\sqrt{28} + 6\sqrt{63}$

28. $3\sqrt{11} + 5\sqrt{44} - 8\sqrt{99}$

29. $2\sqrt{8} - 5\sqrt{32} - 2\sqrt{48}$

30. $5\sqrt{72} - 3\sqrt{48} + 4\sqrt{128}$

31. $4\sqrt{50} + 3\sqrt{12} - 5\sqrt{45}$

32. $6\sqrt{18} + 2\sqrt{48} + 6\sqrt{28}$

33. $\frac{1}{4}\sqrt{288} + \frac{1}{6}\sqrt{72}$

34. $\frac{2}{3}\sqrt{27} + \frac{3}{4}\sqrt{48}$

Find the perimeter of each figure.

35.

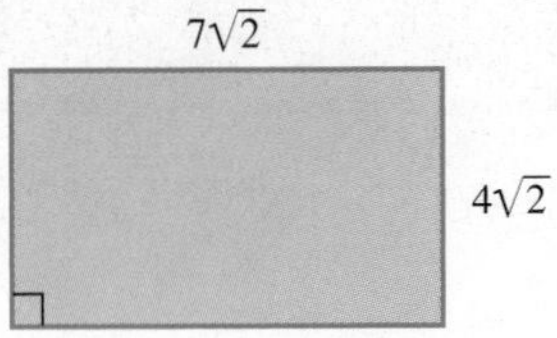

36.

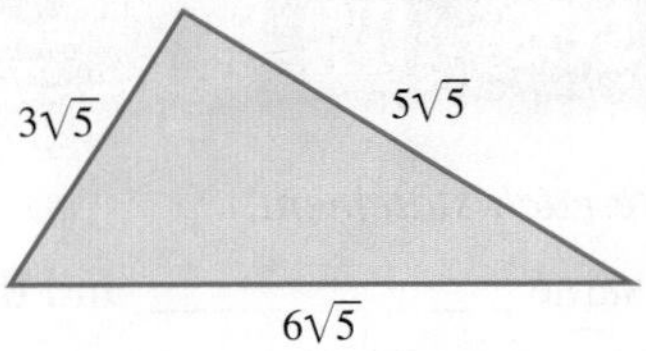

Perform the indicated operations. Assume that all variables represent nonnegative real numbers. ***See Example 3.***

37. $\sqrt{6} \cdot \sqrt{2} + 9\sqrt{3}$

38. $4\sqrt{15} \cdot \sqrt{3} + 4\sqrt{5}$

39. $\sqrt{3} \cdot \sqrt{7} + 2\sqrt{21}$

40. $\sqrt{13} \cdot \sqrt{2} + 3\sqrt{26}$

41. $\sqrt{32x} - \sqrt{18x}$

42. $\sqrt{125t} - \sqrt{80t}$

43. $\sqrt{27r} + \sqrt{48r}$

44. $\sqrt{24x} + \sqrt{54x}$

45. $\sqrt{9x} + \sqrt{49x} - \sqrt{25x}$

46. $\sqrt{4a} - \sqrt{16a} + \sqrt{100a}$

47. $\sqrt{6x^2} + x\sqrt{24}$

48. $\sqrt{75x^2} + x\sqrt{108}$

49. $3\sqrt{8x^2} - 4x\sqrt{2} - x\sqrt{8}$

50. $\sqrt{2b^2} + 3b\sqrt{18} - b\sqrt{200}$

51. $-8\sqrt{32k} + 6\sqrt{8k}$

52. $4\sqrt{12x} + 2\sqrt{27x}$

53. $2\sqrt{125x^2z} + 8x\sqrt{80z}$

54. $\sqrt{48x^2y} + 5x\sqrt{27y}$

55. $4\sqrt[3]{16} - 3\sqrt[3]{54}$

56. $5\sqrt[3]{128} + 3\sqrt[3]{250}$

57. $6\sqrt[3]{8p^2} - 2\sqrt[3]{27p^2}$

58. $8k\sqrt[3]{54k} + 6\sqrt[3]{16k^4}$

59. $5\sqrt[4]{m^3} + 8\sqrt[4]{16m^3}$

60. $5\sqrt[4]{m^5} + 3\sqrt[4]{81m^5}$

16.4 Rationalizing the Denominator

OBJECTIVES

1. Rationalize denominators with square roots.
2. Write radicals in simplified form.
3. Rationalize denominators with cube roots.

OBJECTIVE 1 **Rationalize denominators with square roots.** Although calculators now make it fairly easy to divide by a radical in an expression such as $\frac{1}{\sqrt{2}}$, it is sometimes easier to work with radical expressions if the denominators do not contain any radicals.

For example, the radical in the denominator of $\frac{1}{\sqrt{2}}$ can be eliminated by multiplying the numerator and denominator by $\sqrt{2}$, since $\sqrt{2} \cdot \sqrt{2} = \sqrt{4} = 2$.

$$\frac{1}{\sqrt{2}} = \frac{1 \cdot \sqrt{2}}{\sqrt{2} \cdot \sqrt{2}} = \frac{\sqrt{2}}{2} \quad \text{Multiply by } \frac{\sqrt{2}}{\sqrt{2}} = 1.$$

This process of changing a denominator from one with a radical to one without a radical is called **rationalizing the denominator.** ***The value of the radical expression is not changed. Only the form is changed, because the expression has been multiplied by 1 in the form of*** $\frac{\sqrt{2}}{\sqrt{2}}$.

EXAMPLE 1 Rationalizing Denominators

Rationalize each denominator.

(a) $\frac{9}{\sqrt{6}}$

$= \frac{9 \cdot \sqrt{6}}{\sqrt{6} \cdot \sqrt{6}}$ Multiply by $\frac{\sqrt{6}}{\sqrt{6}} = 1$.

$= \frac{9\sqrt{6}}{6}$ In the denominator, $\sqrt{6} \cdot \sqrt{6} = \sqrt{36} = 6$.

$= \frac{3\sqrt{6}}{2}$ Write in lowest terms.

The denominator is now a rational number.

(b) $\frac{12}{\sqrt{8}}$ While the denominator could be rationalized by multiplying by $\sqrt{8}$, simplifying the denominator first is more direct.

$\frac{12}{\sqrt{8}}$

$= \frac{12}{2\sqrt{2}}$ $\sqrt{8} = \sqrt{4} \cdot \sqrt{2} = 2\sqrt{2}$

$= \frac{12 \cdot \sqrt{2}}{2\sqrt{2} \cdot \sqrt{2}}$ Multiply by $\frac{\sqrt{2}}{\sqrt{2}} = 1$.

$= \frac{12 \cdot \sqrt{2}}{2 \cdot 2}$ $\sqrt{2} \cdot \sqrt{2} = \sqrt{4} = 2$

$= \frac{12\sqrt{2}}{4}$ Multiply.

$= 3\sqrt{2}$ Write in lowest terms; $\frac{12}{4} = 3$

Work Problem 1 at the Side. ▶

1 Rationalize each denominator.

GS **(a)** $\frac{3}{\sqrt{5}}$

$= \frac{3 \cdot \_\_\_}{\sqrt{5} \cdot \_\_\_}$

$= \_\_\_$

(b) $\frac{-6}{\sqrt{11}}$

(c) $-\frac{\sqrt{7}}{\sqrt{2}}$

(d) $\frac{20}{\sqrt{18}}$

Answers

1. **(a)** $\sqrt{5}$; $\sqrt{5}$; $\frac{3\sqrt{5}}{5}$ **(b)** $\frac{-6\sqrt{11}}{11}$

(c) $-\frac{\sqrt{14}}{2}$ **(d)** $\frac{10\sqrt{2}}{3}$

2 Simplify.

GS **(a)** $\sqrt{\dfrac{16}{11}}$

$= \dfrac{\sqrt{16}}{\_\_\_\_}$

$= \dfrac{\sqrt{16} \cdot \_\_\_\_}{\sqrt{11} \cdot \_\_\_\_}$

$= \dfrac{\_\_\_\_\sqrt{11}}{\_\_\_\_}$

(b) $\sqrt{\dfrac{5}{18}}$

(c) $\sqrt{\dfrac{8}{32}}$

Answers

2. **(a)** $\sqrt{11}$; $\sqrt{11}$; $\sqrt{11}$; 4; 11

(b) $\dfrac{\sqrt{10}}{6}$ **(c)** $\dfrac{1}{2}$

Note

In **Example 1(b),** we could also have rationalized the original denominator $\sqrt{8}$ by multiplying by $\sqrt{2}$, since $\sqrt{8} \cdot \sqrt{2} = \sqrt{16} = 4$.

$$\frac{12}{\sqrt{8}} = \frac{12 \cdot \sqrt{2}}{\sqrt{8} \cdot \sqrt{2}} = \frac{12\sqrt{2}}{\sqrt{16}} = \frac{12\sqrt{2}}{4} = 3\sqrt{2}$$

Either approach yields the same correct answer.

OBJECTIVE 2 Write radicals in simplified form. A radical is considered to be in simplified form if the following three conditions are met.

Conditions for Simplified Form of a Radical

1. The radicand contains no factor (except 1) that is a perfect square (when dealing with square roots), a perfect cube (when dealing with cube roots), and so on.
2. The radicand has no fractions.
3. No denominator contains a radical.

EXAMPLE 2 Simplifying a Radical

Simplify.

$$\sqrt{\frac{27}{5}}$$

$= \dfrac{\sqrt{27}}{\sqrt{5}}$ Quotient rule

$= \dfrac{\sqrt{27} \cdot \sqrt{5}}{\sqrt{5} \cdot \sqrt{5}}$ Rationalize the denominator.

$= \dfrac{\sqrt{27} \cdot \sqrt{5}}{5}$ $\sqrt{5} \cdot \sqrt{5} = \sqrt{25} = 5$

$= \dfrac{\sqrt{9 \cdot 3} \cdot \sqrt{5}}{5}$ Factor.

$= \dfrac{\sqrt{9} \cdot \sqrt{3} \cdot \sqrt{5}}{5}$ Product rule

$= \dfrac{3 \cdot \sqrt{3} \cdot \sqrt{5}}{5}$ $\sqrt{9} = 3$

$= \dfrac{3\sqrt{15}}{5}$ Product rule

The three conditions for a simplified radical are met.

◀ Work Problem 2 at the Side.

EXAMPLE 3 Simplifying a Product of Radicals

Simplify.

$$\sqrt{\frac{5}{8}} \cdot \sqrt{\frac{1}{6}}$$

$= \sqrt{\frac{5}{8} \cdot \frac{1}{6}}$ Product rule

$= \sqrt{\frac{5}{48}}$ Multiply fractions.

$= \frac{\sqrt{5}}{\sqrt{48}}$ Quotient rule

$= \frac{\sqrt{5}}{\sqrt{16} \cdot \sqrt{3}}$ Product rule

$= \frac{\sqrt{5}}{4\sqrt{3}}$ $\sqrt{16} = 4$

$= \frac{\sqrt{5} \cdot \sqrt{3}}{4\sqrt{3} \cdot \sqrt{3}}$ Rationalize the denominator.

$= \frac{\sqrt{15}}{4 \cdot 3}$ Product rule; $\sqrt{3} \cdot \sqrt{3} = 3$

$= \frac{\sqrt{15}}{12}$ Multiply.

Work Problem 3 at the Side. ▶

EXAMPLE 4 Simplifying Quotients Involving Radicals

Simplify. Assume that x and y represent positive real numbers.

(a) $\frac{\sqrt{4x}}{\sqrt{y}}$

$= \frac{\sqrt{4x} \cdot \sqrt{y}}{\sqrt{y} \cdot \sqrt{y}}$ Rationalize the denominator.

$= \frac{\sqrt{4xy}}{y}$ Product rule; $\sqrt{y} \cdot \sqrt{y} = y$

$= \frac{\sqrt{4} \cdot \sqrt{xy}}{y}$ Product rule

$= \frac{2\sqrt{xy}}{y}$ $\sqrt{4} = 2$

(b) $\sqrt{\frac{2x^2y}{3}}$

$= \frac{\sqrt{2x^2y}}{\sqrt{3}}$ Quotient rule

$= \frac{\sqrt{2x^2y} \cdot \sqrt{3}}{\sqrt{3} \cdot \sqrt{3}}$ Rationalize the denominator.

$= \frac{\sqrt{6x^2y}}{3}$ Product rule; $\sqrt{3} \cdot \sqrt{3} = 3$

$= \frac{\sqrt{x^2} \cdot \sqrt{6y}}{3}$ Product rule

$= \frac{x\sqrt{6y}}{3}$ $\sqrt{x^2} = x$, since $x > 0$.

Work Problem 4 at the Side. ▶

3 Simplify.

(a) $\sqrt{\frac{1}{2}} \cdot \sqrt{\frac{5}{6}}$

GS **(b)** $\sqrt{\frac{1}{10}} \cdot \sqrt{20}$

$= \sqrt{\frac{1}{10}} \cdot \sqrt{\frac{20}{\_\_\_}}$

$= \sqrt{\frac{\_\_\_}{10}}$

$= \_\_\_$

(c) $\sqrt{\frac{5}{8}} \cdot \sqrt{\frac{24}{10}}$

4 Simplify. Assume that all variables represent positive real numbers.

(a) $\frac{\sqrt{5p}}{\sqrt{q}}$

(b) $\sqrt{\frac{5r^2t^2}{7}}$

Answers

3. **(a)** $\frac{\sqrt{15}}{6}$ **(b)** 1; 20; $\sqrt{2}$ **(c)** $\frac{\sqrt{6}}{2}$

4. **(a)** $\frac{\sqrt{5pq}}{q}$ **(b)** $\frac{rt\sqrt{35}}{7}$

5 Rationalize each denominator.

GS **(a)** $\sqrt[3]{\frac{5}{7}}$

$= \frac{\sqrt[3]{5}}{\sqrt[3]{7}}$

To get a perfect cube in the denominator radicand, multiply the ________ and denominator by

$\sqrt[3]{\_\_\_ \cdot \_\_\_}$.

Complete the solution.

(b) $\frac{\sqrt[3]{5}}{\sqrt[3]{9}}$

(c) $\frac{\sqrt[3]{4}}{\sqrt[3]{25y}}$ $(y \neq 0)$

Answers

5. **(a)** numerator; 7; 7; $\frac{\sqrt[3]{245}}{7}$ **(b)** $\frac{\sqrt[3]{15}}{3}$

(c) $\frac{\sqrt[3]{20y^2}}{5y}$

OBJECTIVE 3 **Rationalize denominators with cube roots.** To rationalize a denominator with a cube root, we change the radicand in the denominator to a perfect cube.

EXAMPLE 5 Rationalizing Denominators with Cube Roots

Rationalize each denominator.

(a) $\sqrt[3]{\frac{3}{2}}$

First write the expression as a quotient of radicals. Then multiply the numerator and denominator by the appropriate number of factors of 2 to make the radicand in the denominator a perfect cube. This will eliminate the radical in the denominator. Here, multiply by $\sqrt[3]{2 \cdot 2}$, or $\sqrt[3]{2^2}$.

$$\sqrt[3]{\frac{3}{2}} = \frac{\sqrt[3]{3}}{\sqrt[3]{2}} = \frac{\sqrt[3]{3} \cdot \sqrt[3]{2 \cdot 2}}{\sqrt[3]{2} \cdot \sqrt[3]{2 \cdot 2}} = \frac{\sqrt[3]{3 \cdot 2 \cdot 2}}{\sqrt[3]{2 \cdot 2 \cdot 2}} = \frac{\sqrt[3]{12}}{2}$$

We need 3 factors of 2 in the radicand in the denominator.

$\sqrt[3]{2 \cdot 2 \cdot 2} = \sqrt[3]{2^3} = 2$
Denominator radicand is a perfect cube.

(b) $\frac{\sqrt[3]{3}}{\sqrt[3]{4}}$

Since $\sqrt[3]{4} = \sqrt[3]{2 \cdot 2}$, multiply the numerator and denominator by a sufficient number of factors of 2 to get a perfect cube in the radicand in the denominator.

$$\frac{\sqrt[3]{3}}{\sqrt[3]{4}} = \frac{\sqrt[3]{3} \cdot \sqrt[3]{2}}{\sqrt[3]{2 \cdot 2} \cdot \sqrt[3]{2}} = \frac{\sqrt[3]{6}}{\sqrt[3]{2 \cdot 2 \cdot 2}} = \frac{\sqrt[3]{6}}{2}$$

(c) $\frac{\sqrt[3]{2}}{\sqrt[3]{3x^2}}$ $(x \neq 0)$

Multiply the numerator and denominator by a sufficient number of factors of 3 and of x to get a perfect cube in the radicand in the denominator.

$$\frac{\sqrt[3]{2}}{\sqrt[3]{3x^2}} = \frac{\sqrt[3]{2} \cdot \sqrt[3]{3 \cdot 3 \cdot x}}{\sqrt[3]{3 \cdot x \cdot x} \cdot \sqrt[3]{3 \cdot 3 \cdot x}} = \frac{\sqrt[3]{18x}}{\sqrt[3]{(3x)^3}} = \frac{\sqrt[3]{18x}}{3x}$$

We need 3 factors of 3 and 3 factors of x in the radicand in the denominator.

Denominator radicand is a perfect cube.

CAUTION

A common error in a problem like the one in **Example 5(a)** is to multiply by $\sqrt[3]{2}$ instead of $\sqrt[3]{2^2}$. Doing this would give a denominator of

$$\sqrt[3]{2} \cdot \sqrt[3]{2} = \sqrt[3]{4}.$$

Because 4 is not a perfect cube, the denominator is still not rationalized.

◀ Work Problem 5 at the Side.

16.4 Exercises

FOR EXTRA HELP

Download the MyDashBoard App

MyMathLab®

1. CONCEPT CHECK When we rationalize the denominator in an expression such as $\frac{4}{\sqrt{3}}$, we multiply both the numerator and denominator by $\sqrt{3}$. By what number are we actually multiplying the given expression, and what property of real numbers justifies the fact that our result is equal to the given expression?

2. In **Example 1(a),** we show algebraically that $\frac{9}{\sqrt{6}}$ is equal to $\frac{3\sqrt{6}}{2}$. Support this result numerically by finding the decimal approximation of $\frac{9}{\sqrt{6}}$ on your calculator, and then finding the decimal approximation of $\frac{3\sqrt{6}}{2}$. What do you notice?

Rationalize each denominator. ***See Examples 1 and 2.***

3. $\frac{6}{\sqrt{5}}$ **4.** $\frac{3}{\sqrt{2}}$ **5.** $\frac{5}{\sqrt{5}}$ **6.** $\frac{15}{\sqrt{15}}$

7. $\frac{8}{\sqrt{2}}$ **8.** $\frac{12}{\sqrt{3}}$ **9.** $\frac{-\sqrt{11}}{\sqrt{3}}$ **10.** $\frac{-\sqrt{13}}{\sqrt{5}}$ **11.** $\frac{7\sqrt{3}}{\sqrt{5}}$

12. $\frac{4\sqrt{6}}{\sqrt{5}}$ **13.** $\frac{24\sqrt{10}}{16\sqrt{3}}$ **14.** $\frac{18\sqrt{15}}{12\sqrt{2}}$ **15.** $\frac{16}{\sqrt{27}}$ **16.** $\frac{24}{\sqrt{18}}$

17. $\frac{-3}{\sqrt{50}}$ **18.** $\frac{-5}{\sqrt{75}}$ **19.** $\frac{63}{\sqrt{45}}$ **20.** $\frac{27}{\sqrt{32}}$ **21.** $\frac{\sqrt{24}}{\sqrt{8}}$

22. $\frac{\sqrt{36}}{\sqrt{18}}$ **23.** $\sqrt{\frac{1}{2}}$ **24.** $\sqrt{\frac{1}{3}}$ **25.** $\sqrt{\frac{13}{5}}$ **26.** $\sqrt{\frac{17}{11}}$

Simplify each product of radicals. ***See Example 3.***

27. $\sqrt{\frac{7}{13}} \cdot \sqrt{\frac{13}{3}}$ **28.** $\sqrt{\frac{19}{20}} \cdot \sqrt{\frac{20}{3}}$ **29.** $\sqrt{\frac{21}{7}} \cdot \sqrt{\frac{21}{8}}$ **30.** $\sqrt{\frac{5}{8}} \cdot \sqrt{\frac{5}{6}}$

31. $\sqrt{\frac{1}{12}} \cdot \sqrt{\frac{1}{3}}$ **32.** $\sqrt{\frac{1}{8}} \cdot \sqrt{\frac{1}{2}}$ **33.** $\sqrt{\frac{2}{9}} \cdot \sqrt{\frac{9}{2}}$ **34.** $\sqrt{\frac{4}{3}} \cdot \sqrt{\frac{3}{4}}$

Simplify each radical. Assume that all variables represent positive real numbers. **See Example 4.**

35. $\frac{\sqrt{7}}{\sqrt{x}}$

36. $\frac{\sqrt{19}}{\sqrt{y}}$

37. $\frac{\sqrt{4x^3}}{\sqrt{y}}$

38. $\frac{\sqrt{9t^3}}{\sqrt{s}}$

39. $\sqrt{\frac{5x^3z}{6}}$

40. $\sqrt{\frac{3st^3}{5}}$

41. $\sqrt{\frac{9a^2r^5}{7t}}$

42. $\sqrt{\frac{16x^3y^2}{13z}}$

43. **CONCEPT CHECK** Which one of the following would be an appropriate choice for multiplying the numerator and denominator of $\frac{\sqrt[3]{2}}{\sqrt[3]{5}}$ by in order to rationalize the denominator?

A. $\sqrt[3]{5}$ **B.** $\sqrt[3]{25}$ **C.** $\sqrt[3]{2}$ **D.** $\sqrt[3]{4}$

44. **CONCEPT CHECK** In **Example 5(b)**, we multiplied the numerator and denominator of $\frac{\sqrt[3]{3}}{\sqrt[3]{4}}$ by $\sqrt[3]{2}$ to rationalize the denominator. Suppose we had chosen to multiply by $\sqrt[3]{16}$ instead. Would we have obtained the correct answer after all simplifications were done?

Rationalize each denominator. Assume that variables in the denominator represent nonzero real numbers. **See Example 5.**

45. $\sqrt[3]{\frac{1}{2}}$

46. $\sqrt[3]{\frac{1}{4}}$

47. $\sqrt[3]{\frac{5}{9}}$

48. $\sqrt[3]{\frac{2}{5}}$

49. $\frac{\sqrt[3]{4}}{\sqrt[3]{7}}$

50. $\frac{\sqrt[3]{5}}{\sqrt[3]{10}}$

51. $\sqrt[3]{\frac{3}{4y^2}}$

52. $\sqrt[3]{\frac{3}{25x^2}}$

53. $\frac{\sqrt[3]{7m}}{\sqrt[3]{36n}}$

54. $\frac{\sqrt[3]{11p}}{\sqrt[3]{49q}}$

In Exercises 55 and 56, **(a)** *give the answer as a simplified radical and* **(b)** *use a calculator to give the answer correct to the nearest thousandth.*

55. The period p of a pendulum is the time it takes for it to swing from one extreme to the other and back again. The value of p in seconds is given by

$$p = k \cdot \sqrt{\frac{L}{g}},$$

where L is the length of the pendulum, g is the acceleration due to gravity, and k is a constant. Find the period when $k = 6$, $L = 9$ ft, and $g = 32$ ft per sec per sec.

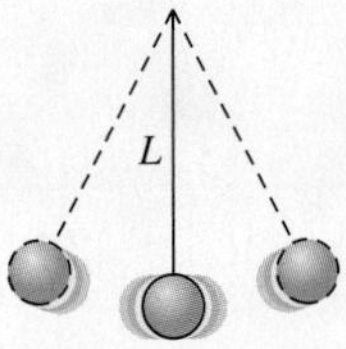

56. The velocity v in kilometers per second of a meteorite approaching Earth is given by

$$v = \frac{k}{\sqrt{d}},$$

where d is its distance from the center of Earth and k is a constant. What is the velocity of a meteorite that is 6000 km away from the center of Earth, if $k = 450$?

16.5 More Simplifying and Operations with Radicals

Apply these guidelines when simplifying radical expressions.

OBJECTIVES

1. Simplify products of radical expressions.
2. Use conjugates to rationalize denominators of radical expressions.
3. Write radical expressions with quotients in lowest terms.

Simplifying Radical Expressions

1. If a radical represents a rational number, use that rational number in place of the radical.

   Examples: $\sqrt{49} = 7$, $\sqrt{\frac{169}{9}} = \frac{13}{3}$

2. If a radical expression contains products of radicals, use the product rule for radicals, $\sqrt[n]{x} \cdot \sqrt[n]{y} = \sqrt[n]{xy}$, to get a single radical.

   Examples: $\sqrt{3} \cdot \sqrt{2} = \sqrt{6}$, $\sqrt[3]{5} \cdot \sqrt[3]{x} = \sqrt[3]{5x}$

3. If a radicand of a square root radical has a factor that is a perfect square, express the radical as the product of the positive square root of the perfect square and the remaining radicand factor. A similar statement applies to higher roots.

   Examples: $\sqrt{20} = \sqrt{4 \cdot 5} = \sqrt{4} \cdot \sqrt{5} = 2\sqrt{5}$

   $\sqrt[3]{16} = \sqrt[3]{8 \cdot 2} = \sqrt[3]{8} \cdot \sqrt[3]{2} = 2\sqrt[3]{2}$

4. If a radical expression contains sums or differences of radicals, use the distributive property to combine like radicals.

   Examples: $3\sqrt{2} + 4\sqrt{2}$ can be combined to obtain $7\sqrt{2}$.

   $3\sqrt{2} + 4\sqrt{3}$ cannot be simplified further.

5. Rationalize any denominator containing a radical.

   Examples: $\frac{5}{\sqrt{3}} = \frac{5 \cdot \sqrt{3}}{\sqrt{3} \cdot \sqrt{3}} = \frac{5\sqrt{3}}{3}$

   $\sqrt[3]{\frac{1}{4}} = \frac{\sqrt[3]{1}}{\sqrt[3]{4}} = \frac{\sqrt[3]{1} \cdot \sqrt[3]{2}}{\sqrt[3]{4} \cdot \sqrt[3]{2}} = \frac{\sqrt[3]{2}}{\sqrt[3]{8}} = \frac{\sqrt[3]{2}}{2}$

OBJECTIVE 1 Simplify products of radical expressions.

EXAMPLE 1 Multiplying Radical Expressions

Find each product, and simplify.

(a) $\sqrt{5}(\sqrt{8} - \sqrt{32})$ — Simplify inside the parentheses.

$= \sqrt{5}(2\sqrt{2} - 4\sqrt{2})$ — $\sqrt{8} = 2\sqrt{2}$; $\sqrt{32} = 4\sqrt{2}$

$= \sqrt{5}(-2\sqrt{2})$ — Subtract like radicals.

$= -2\sqrt{5 \cdot 2}$ — Product rule

$= -2\sqrt{10}$ — Multiply.

Continued on Next Page

1 Find each product, and simplify.

GS **(a)** $\sqrt{7}(\sqrt{2}+\sqrt{5})$

$= \_\_\_ \cdot \sqrt{2} + \sqrt{7} \cdot \_\_\_$

$= \_\_\_\_$

(b) $\sqrt{2}(\sqrt{8}+\sqrt{20})$

(c) $(\sqrt{2}+5\sqrt{3})(\sqrt{3}-2\sqrt{2})$

(d) $(\sqrt{2}-\sqrt{5})(\sqrt{10}+\sqrt{2})$

2 Find each product. Simplify the answers. In part (c), assume that $x \geq 0$.

GS **(a)** $(\sqrt{5}-3)^2$

$= (\_\_\_)^2 - 2(\_\_\_)(3) + \_\_\_^2$

$= \_\_\_ - \_\_\_ + 9$

$= \_\_\_\_$

(b) $(4\sqrt{2}+5)^2$

(c) $(6+\sqrt{x})^2$

Answers

1. (a) $\sqrt{7}$; $\sqrt{5}$; $\sqrt{14}+\sqrt{35}$
   (b) $4+2\sqrt{10}$ (c) $11-9\sqrt{6}$
   (d) $2\sqrt{5}+2-5\sqrt{2}-\sqrt{10}$
2. (a) $\sqrt{5}$; $\sqrt{5}$; 3; 5; $6\sqrt{5}$; $14-6\sqrt{5}$
   (b) $57+40\sqrt{2}$ (c) $36+12\sqrt{x}+x$

(b) $(\sqrt{3}+2\sqrt{5})(\sqrt{3}-4\sqrt{5})$ Use the FOIL method to multiply.

$= \underbrace{\sqrt{3}(\sqrt{3})}_{\text{First}} + \underbrace{\sqrt{3}(-4\sqrt{5})}_{\text{Outer}} + \underbrace{2\sqrt{5}(\sqrt{3})}_{\text{Inner}} + \underbrace{2\sqrt{5}(-4\sqrt{5})}_{\text{Last}}$

$= 3 - 4\sqrt{15} + 2\sqrt{15} - 8 \cdot 5$ Product rule

$= 3 - 2\sqrt{15} - 40$ Add like radicals. Multiply.

$= -37 - 2\sqrt{15}$ Combine like terms.

This does *not* equal $-39\sqrt{15}$.

(c) $(\sqrt{3}+\sqrt{21})(\sqrt{3}-\sqrt{7})$

$= \sqrt{3}(\sqrt{3}) + \sqrt{3}(-\sqrt{7}) + \sqrt{21}(\sqrt{3}) + \sqrt{21}(-\sqrt{7})$ FOIL method

$= 3 - \sqrt{21} + \sqrt{63} - \sqrt{147}$ Product rule

$= 3 - \sqrt{21} + \sqrt{9} \cdot \sqrt{7} - \sqrt{49} \cdot \sqrt{3}$ Factor; 9 and 49 are perfect squares.

$= 3 - \sqrt{21} + 3\sqrt{7} - 7\sqrt{3}$ $\sqrt{9}=3$; $\sqrt{49}=7$

Since there are no like radicals, no terms can be combined.

◀ Work Problem **1** at the Side.

Example 2 uses the following rules for the square of a binomial from **Section 12.4.**

$$(a+b)^2 = a^2 + 2ab + b^2 \quad \text{and} \quad (a-b)^2 = a^2 - 2ab + b^2$$

EXAMPLE 2 Using Special Products with Radicals

Find each product. In part (c), assume that $x \geq 0$.

(a) $(\sqrt{10}-7)^2$

$= (\sqrt{10})^2 - 2(\sqrt{10})(7) + 7^2$ $(a-b)^2 = a^2 - 2ab + b^2$. Let $a = \sqrt{10}$ and $b = 7$.

$= 10 - 14\sqrt{10} + 49$ $(\sqrt{10})^2 = 10$; $7^2 = 49$

$= 59 - 14\sqrt{10}$ Add 10 and 49.

Do *not* try to combine further here. $59 - 14\sqrt{10} \neq 45\sqrt{10}$

(b) $(2\sqrt{3}+4)^2$

$= (2\sqrt{3})^2 + 2(2\sqrt{3})(4) + 4^2$ $(a+b)^2 = a^2 + 2ab + b^2$. Let $a = 2\sqrt{3}$ and $b = 4$.

$= 12 + 16\sqrt{3} + 16$ $(2\sqrt{3})^2 = 4 \cdot 3 = 12$

$= 28 + 16\sqrt{3}$ Add 12 and 16.

Do *not* try to combine further here. $28 + 16\sqrt{3} \neq 44\sqrt{3}$

(c) $(5-\sqrt{x})^2$

$= 5^2 - 2(5)(\sqrt{x}) + (\sqrt{x})^2$ Square the binomial.

$= 25 - 10\sqrt{x} + x$ Simplify.

◀ Work Problem **2** at the Side.

Example 3 involves the product of the sum and difference of two terms.

$$(a + b)(a - b) = a^2 - b^2$$

EXAMPLE 3 Using a Special Product with Radicals

Find each product. In part (b), assume that $x \geq 0$.

(a) $(4 + \sqrt{3})(4 - \sqrt{3})$ — $(a + b)(a - b) = a^2 - b^2$; Let $a = 4$ and $b = \sqrt{3}$.

$= 4^2 - (\sqrt{3})^2$

$= 16 - 3$ — $4^2 = 16$; $(\sqrt{3})^2 = 3$

$= 13$ — Subtract.

(b) $(\sqrt{x} - \sqrt{6})(\sqrt{x} + \sqrt{6})$ — $(a + b)(a - b) = a^2 - b^2$; Let $a = \sqrt{x}$ and $b = \sqrt{6}$.

$= (\sqrt{x})^2 - (\sqrt{6})^2$

$= x - 6$ — $(\sqrt{x})^2 = x$; $(\sqrt{6})^2 = 6$

Work Problem 3 at the Side. ▶

In **Example 3,** the pair of expressions $4 + \sqrt{3}$ and $4 - \sqrt{3}$ are **conjugates** of each other, as are $\sqrt{x} - \sqrt{6}$ and $\sqrt{x} + \sqrt{6}$. Recall from **Section 12.4** that the expressions $a + b$ and $a - b$ are conjugates.

OBJECTIVE 2 Use conjugates to rationalize denominators of radical expressions. To rationalize the denominator in a quotient such as

$$\frac{2}{4 + \sqrt{3}},$$

we multiply the numerator and denominator by the conjugate of the denominator, here $4 - \sqrt{3}$.

$$\frac{2(4 - \sqrt{3})}{(4 + \sqrt{3})(4 - \sqrt{3})} \quad \text{gives} \quad \frac{2(4 - \sqrt{3})}{13}.$$

See Example 3(a).

EXAMPLE 4 Using Conjugates to Rationalize Denominators

Simplify by rationalizing each denominator. In part (c), assume that $x \geq 0$.

(a) $\dfrac{5}{3 + \sqrt{5}}$

$= \dfrac{5(3 - \sqrt{5})}{(3 + \sqrt{5})(3 - \sqrt{5})}$ — Multiply the numerator and denominator by the conjugate of the denominator.

$= \dfrac{5(3 - \sqrt{5})}{3^2 - (\sqrt{5})^2}$ — $(a + b)(a - b) = a^2 - b^2$

$= \dfrac{5(3 - \sqrt{5})}{9 - 5}$, or $\dfrac{5(3 - \sqrt{5})}{4}$ — $3^2 = 9$ and $(\sqrt{5})^2 = 5$; Subtract.

Continued on Next Page

3 Find each product. Simplify the answers. In part (d), assume that $x \geq 0$.

GS **(a)** $(3 + \sqrt{5})(3 - \sqrt{5})$

$= \_\_\_^2 - (\_\_\_)^2$

$= \_\_\_ - \_\_\_$

$= \_\_\_$

(b) $(\sqrt{3} - 2)(\sqrt{3} + 2)$

(c) $(\sqrt{5} + \sqrt{3})(\sqrt{5} - \sqrt{3})$

(d) $(\sqrt{10} - \sqrt{x})(\sqrt{10} + \sqrt{x})$

Answers

3. **(a)** 3; $\sqrt{5}$; 9; 5; 4 **(b)** -1 **(c)** 2 **(d)** $10 - x$

4 Rationalize each denominator. In part (c), assume that $x \geq 0$.

GS (a) $\dfrac{5}{4+\sqrt{2}}$

Multiply the numerator and denominator by the ________ of the denominator.

This number is ________.

Complete the solution.

(b) $\dfrac{\sqrt{5}+3}{2-\sqrt{5}}$

(c) $\dfrac{7}{5-\sqrt{x}}$

5 Write each quotient in lowest terms.

GS (a) $\dfrac{5\sqrt{3}-15}{10}$

$= \dfrac{\_\_\_(\sqrt{3}-\_\_\_)}{\_\_\_\cdot 2}$

$= \_\_\_\_\_\_$

(b) $\dfrac{12+8\sqrt{5}}{16}$

Answers

4. (a) conjugate; $4-\sqrt{2}$; $\dfrac{5(4-\sqrt{2})}{14}$

(b) $-11-5\sqrt{5}$

(c) $\dfrac{7(5+\sqrt{x})}{25-x}$ $(x \neq 25)$

5. (a) 5; 3; 5; $\dfrac{\sqrt{3}-3}{2}$ (b) $\dfrac{3+2\sqrt{5}}{4}$

(b) $\dfrac{6+\sqrt{2}}{\sqrt{2}-5}$

$= \dfrac{(6+\sqrt{2})(\sqrt{2}+5)}{(\sqrt{2}-5)(\sqrt{2}+5)}$ Multiply the numerator and denominator by the conjugate of the denominator.

$= \dfrac{6\sqrt{2}+30+2+5\sqrt{2}}{2-25}$ FOIL method; $(a+b)(a-b)=a^2-b^2$

$= \dfrac{11\sqrt{2}+32}{-23}$ Combine like terms.

$= \dfrac{-11\sqrt{2}-32}{23}$ $\quad \dfrac{a}{-b}=\dfrac{-a}{b}$

(c) $\dfrac{4}{3-\sqrt{x}}$

$= \dfrac{4(3+\sqrt{x})}{(3-\sqrt{x})(3+\sqrt{x})}$ Multiply by $\dfrac{3+\sqrt{x}}{3+\sqrt{x}}=1$.

$= \dfrac{4(3+\sqrt{x})}{9-x}$ $\quad 3^2=9$; $(\sqrt{x})^2=x$

We assume here that $x \neq 9$.

◀ **Work Problem 4 at the Side.**

OBJECTIVE 3 Write radical expressions with quotients in lowest terms.

EXAMPLE 5 Writing a Radical Quotient in Lowest Terms

Write $\dfrac{3\sqrt{3}+9}{12}$ in lowest terms.

$\dfrac{3\sqrt{3}+9}{12}$ Don't simplify yet.

$= \dfrac{3(\sqrt{3}+3)}{3(4)}$ Factor.

$= 1\cdot\dfrac{\sqrt{3}+3}{4}$ Divide out the common factor; $\frac{3}{3}=1$

$= \dfrac{\sqrt{3}+3}{4}$ Identity property; lowest terms

◀ **Work Problem 5 at the Side.**

CAUTION

An expression like the one in **Example 5** can only be simplified by factoring a common factor from the denominator and *each* term of the numerator. ***First factor, and then divide out the common factor.***

Factor $\dfrac{4+8\sqrt{5}}{4}$ as $\dfrac{4(1+2\sqrt{5})}{4}$ to obtain $1+2\sqrt{5}$.

16.5 Exercises

FOR EXTRA HELP

Download the MyDashBoard App

MyMathLab®

CONCEPT CHECK *In Exercises 1–4, perform the operations mentally, and write the answers without doing intermediate steps.*

1. $\sqrt{49} + \sqrt{36}$

2. $\sqrt{100} - \sqrt{81}$

3. $\sqrt{2} \cdot \sqrt{8}$

4. $\sqrt{8} \cdot \sqrt{8}$

5. **CONCEPT CHECK** In **Example 1(b),** a student simplified $-37 - 2\sqrt{15}$ by combining the -37 and the -2 to get $-39\sqrt{15}$, which is incorrect. **What Went Wrong?**

6. **CONCEPT CHECK** Find each product mentally.

(a) $(\sqrt{x} + \sqrt{y})(\sqrt{x} - \sqrt{y})$

(b) $(\sqrt{28} - \sqrt{14})(\sqrt{28} + \sqrt{14})$

Simplify each expression. Use the five guidelines given in this section. Assume that all variables represent nonnegative real numbers. ***See Examples 1–3.***

7. $\sqrt{5}(\sqrt{3} - \sqrt{7})$

8. $\sqrt{7}(\sqrt{10} + \sqrt{3})$

9. $2\sqrt{5}(\sqrt{2} + 3\sqrt{5})$

10. $3\sqrt{7}(2\sqrt{7} + 4\sqrt{5})$

11. $3\sqrt{14} \cdot \sqrt{2} - \sqrt{28}$

12. $7\sqrt{6} \cdot \sqrt{3} - 2\sqrt{18}$

13. $(2\sqrt{6} + 3)(3\sqrt{6} + 7)$

14. $(4\sqrt{5} - 2)(2\sqrt{5} - 4)$

15. $(5\sqrt{7} - 2\sqrt{3})(3\sqrt{7} + 4\sqrt{3})$

16. $(2\sqrt{10} + 5\sqrt{2})(3\sqrt{10} - 3\sqrt{2})$

17. $(8 - \sqrt{7})^2$

18. $(6 - \sqrt{11})^2$

19. $(2\sqrt{7} + 3)^2$

20. $(4\sqrt{5} + 5)^2$

21. $(\sqrt{a} + 1)^2$

22. $(\sqrt{y} + 4)^2$

23. $(5 - \sqrt{2})(5 + \sqrt{2})$

24. $(3 - \sqrt{5})(3 + \sqrt{5})$

25. $(\sqrt{8}-\sqrt{7})(\sqrt{8}+\sqrt{7})$ **26.** $(\sqrt{12}-\sqrt{11})(\sqrt{12}+\sqrt{11})$ **27.** $(\sqrt{y}-\sqrt{10})(\sqrt{y}+\sqrt{10})$

28. $(\sqrt{t}-\sqrt{13})(\sqrt{t}+\sqrt{13})$ **29.** $(\sqrt{2}+\sqrt{3})(\sqrt{6}-\sqrt{2})$ **30.** $(\sqrt{3}+\sqrt{5})(\sqrt{15}-\sqrt{5})$

31. $(\sqrt{10}-\sqrt{5})(\sqrt{5}+\sqrt{20})$ **32.** $(\sqrt{6}-\sqrt{3})(\sqrt{3}+\sqrt{18})$ **33.** $(\sqrt{5}+\sqrt{30})(\sqrt{6}+\sqrt{3})$

34. $(\sqrt{10}-\sqrt{20})(\sqrt{2}-\sqrt{5})$ **35.** $(\sqrt{5}-\sqrt{10})(\sqrt{x}-\sqrt{2})$ **36.** $(\sqrt{x}+\sqrt{6})(\sqrt{10}+\sqrt{3})$

37. CONCEPT CHECK Determine the expression by which you should multiply the numerator and denominator to rationalize each denominator.

(a) $\dfrac{1}{\sqrt{5}+\sqrt{3}}$ **(b)** $\dfrac{3}{\sqrt{6}-\sqrt{5}}$

38. CONCEPT CHECK A student tried to rationalize the denominator of $\frac{2}{4+\sqrt{3}}$ by multiplying by $\frac{4+\sqrt{3}}{4+\sqrt{3}}$. **What Went Wrong?** By what should he multiply?

Rationalize each denominator. Write quotients in lowest terms. Assume that all variables represent nonnegative real numbers. ***See Examples 4 and 5.***

39. $\dfrac{1}{3+\sqrt{2}}$ **40.** $\dfrac{1}{4-\sqrt{3}}$ **41.** $\dfrac{14}{2-\sqrt{11}}$ **42.** $\dfrac{19}{5-\sqrt{6}}$

43. $\dfrac{\sqrt{2}}{2-\sqrt{2}}$

44. $\dfrac{\sqrt{7}}{7-\sqrt{7}}$

45. $\dfrac{\sqrt{5}}{\sqrt{2}+\sqrt{3}}$

46. $\dfrac{\sqrt{3}}{\sqrt{2}+\sqrt{3}}$

47. $\dfrac{\sqrt{5}+2}{2-\sqrt{3}}$

48. $\dfrac{\sqrt{7}+3}{4-\sqrt{5}}$

49. $\dfrac{6-\sqrt{5}}{\sqrt{2}+2}$

50. $\dfrac{3+\sqrt{2}}{\sqrt{2}+1}$

51. $\dfrac{12}{\sqrt{x}+1}$

52. $\dfrac{10}{\sqrt{x}-4}$ $(x \neq 16)$

53. $\dfrac{3}{7-\sqrt{x}}$ $(x \neq 49)$

54. $\dfrac{1}{6-\sqrt{z}}$ $(z \neq 36)$

Write each quotient in lowest terms. ***See Example 5.***

55. $\dfrac{6\sqrt{11}-12}{6}$

56. $\dfrac{12\sqrt{5}-24}{12}$

57. $\dfrac{2\sqrt{3}+10}{16}$

58. $\dfrac{4\sqrt{6}+24}{20}$

59. $\dfrac{12-\sqrt{40}}{4}$

60. $\dfrac{9-\sqrt{72}}{12}$

61. $\dfrac{16+\sqrt{128}}{24}$

62. $\dfrac{25+\sqrt{75}}{10}$

Solve each problem.

63. The radius r of the circular top or bottom of a tin can with a surface area S and a height h is given by

$$r = \frac{-h + \sqrt{h^2 + 0.64S}}{2}.$$

(a) What radius should be used to make a can with a height of 12 in. and a surface area of 400 in.^{2}?

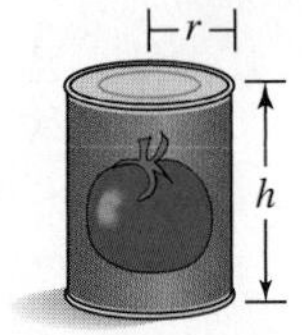

(b) What radius should be used to make a can with a height of 6 in. and a surface area of 200 in.^{2}? Round the answer to the nearest tenth.

64. If an investment of P dollars grows to A dollars in 2 yr, the annual rate of return on the investment is given by

$$r = \frac{\sqrt{A} - \sqrt{P}}{\sqrt{P}}.$$

First rationalize the denominator. Then find the annual rate of return r (as a percent) if \$50,000 increases to \$54,080.

Relating Concepts (Exercises 65–70) For Individual or Group Work

Work Exercises 65–70 in order, *to see why a common student error is indeed an error.*

65. Use the distributive property to write $6(5 + 3x)$ as a sum.

66. Your answer in **Exercise 65** should be $30 + 18x$. Why can we not combine these two terms to get $48x$?

67. Simplify $(2\sqrt{10} + 5\sqrt{2})(3\sqrt{10} - 3\sqrt{2})$. (See **Exercise 16.**)

68. Your answer in **Exercise 67** should be $30 + 18\sqrt{5}$. Many students will, in error, try to combine these terms to get $48\sqrt{5}$. Why is this wrong?

69. Write the expression similar to $30 + 18x$ that simplifies to $48x$. Then write the expression similar to $30 + 18\sqrt{5}$ that simplifies to $48\sqrt{5}$.

70. Write a short explanation of the similarities between combining like terms and combining like radicals.

16.6 Solving Equations with Radicals

A **radical equation** is an equation with a variable in a radicand.

$$\sqrt{x} = 6, \quad \sqrt{x+1} = 3, \quad \text{and} \quad 3\sqrt{x} = \sqrt{8x+9} \qquad \text{Radical equations}$$

OBJECTIVES

1. Solve radical equations having square root radicals.
2. Identify equations with no solutions.
3. Solve equations by squaring a binomial.
4. Solve problems using formulas that involve radicals.

OBJECTIVE 1 Solve radical equations having square root radicals. To solve radical equations, we use a new property, called the ***squaring property of equality.***

Squaring Property of Equality

If each side of a given equation is squared, all solutions of the original equation are *among* the solutions of the squared equation.

CAUTION

Using the squaring property can give a new equation with *more* solutions than the original equation. For example, starting with the equation $x = 4$ and squaring each side gives

$$x^2 = 4^2, \quad \text{or} \quad x^2 = 16.$$

This last equation, $x^2 = 16$, has *two* solutions, 4 or -4, while the original equation, $x = 4$, has only *one* solution, 4.

Because of this possibility, checking is more than just a guard against algebraic errors when solving an equation with radicals. It is an essential part of the solution process. ***All proposed solutions from the squared equation must be checked in the original equation.***

EXAMPLE 1 Using the Squaring Property of Equality

Solve $\sqrt{p+1} = 3$.

$$\sqrt{p+1} = 3$$

$$\left(\sqrt{p+1}\right)^2 = 3^2 \qquad \text{Use the squaring property to square each side.}$$

$$p + 1 = 9 \qquad \left(\sqrt{p+1}\right)^2 = p+1$$

Proposed solution ⟶ $\mathbf{p = 8}$ Subtract 1.

CHECK (A check is essential.)

$$\sqrt{p+1} = 3 \qquad \text{Original equation}$$

$$\sqrt{8+1} \stackrel{?}{=} 3 \qquad \text{Let } p = 8.$$

$$\sqrt{9} \stackrel{?}{=} 3 \qquad \text{Add.}$$

$$3 = 3 \checkmark \qquad \text{True}$$

Because this statement is true, $\{8\}$ is the solution set of $\sqrt{p+1} = 3$. In this case the equation obtained by squaring had just one solution, which also satisfied the original equation.

Work Problem 1 at the Side. ▶

1 Solve each equation. Be sure to check your solutions.

GS **(a)** $\sqrt{k} = 3$

$\left(\sqrt{k}\right)^{\_\_} = 3^2$

$\_\_\_\_ = 9$

CHECK $\sqrt{k} = 3$

$\sqrt{\_\_\_\_} \stackrel{?}{=} 3$

$\_\_\_\_ = 3$

(True / False)

The solution set is ____.

(b) $\sqrt{x-2} = 4$

(c) $\sqrt{9-t} = 4$

Answers

1. **(a)** 2; k; 9; 3; True; $\{9\}$ **(b)** $\{18\}$
(c) $\{-7\}$

ach equation.

$\sqrt{3x+9} = 2\sqrt{x}$

(b) $5\sqrt{x} = \sqrt{20x+5}$

3 Solve each equation.

(a) $\sqrt{x} = -4$

GS **(b)** $\sqrt{x} + 6 = 0$

$\sqrt{x} = \_\_\_\_$

$(\sqrt{x})^2 = (\_\_\_\_)^2$

$x = \_\_\_\_$

CHECK $\sqrt{x} + 6 = 0$

$\sqrt{\_\_\_\_} + 6 \stackrel{?}{=} 0$

$\_\_\_\_ = 0$

(True / False)

The solution set is ____.

Answers

2. **(a)** $\{9\}$ **(b)** $\{1\}$

3. **(a)** $\emptyset$ **(b)** -6; -6; 36; 36; 12; False; $\emptyset$

EXAMPLE 2 Using the Squaring Property with a Radical on Each Side

Solve $3\sqrt{x} = \sqrt{x+8}$.

$$3\sqrt{x} = \sqrt{x+8}$$

$$(3\sqrt{x})^2 = (\sqrt{x+8})^2 \quad \text{Squaring property}$$

$$3^2(\sqrt{x})^2 = (\sqrt{x+8})^2 \quad (ab)^2 = a^2b^2$$

Be careful here.

$$9x = x + 8 \quad (\sqrt{x})^2 = x;\ (\sqrt{x+8})^2 = x+8$$

$$8x = 8 \quad \text{Subtract } x.$$

Proposed solution → $x = 1$ Divide by 8.

CHECK

$$3\sqrt{x} = \sqrt{x+8} \quad \text{Original equation}$$

$$3\sqrt{1} \stackrel{?}{=} \sqrt{1+8} \quad \text{Let } x = 1.$$

$$3(1) \stackrel{?}{=} \sqrt{9} \quad \text{Simplify.}$$

This is *not* the solution.

$$3 = 3\ \checkmark \quad \text{True}$$

The solution set is $\{1\}$.

◀ **Work Problem 2 at the Side.**

OBJECTIVE 2 **Identify equations with no solutions.** Not all radical equations have solutions.

EXAMPLE 3 Using the Squaring Property When One Side Is Negative

Solve $\sqrt{x} = -3$.

$$\sqrt{x} = -3$$

$$(\sqrt{x})^2 = (-3)^2 \quad \text{Squaring property}$$

$$x = 9 \longleftarrow \text{Proposed solution}$$

CHECK

$$\sqrt{x} = -3 \quad \text{Original equation}$$

$$\sqrt{9} \stackrel{?}{=} -3 \quad \text{Let } x = 9.$$

$$3 = -3 \quad \text{False}$$

Because the statement $3 = -3$ is false, the number 9 is *not* a solution of the given equation. It is an **extraneous solution** and must be rejected. In fact, $\sqrt{x} = -3$ has no solution. The solution set is $\emptyset$.

Note

Because $\sqrt{x}$ represents the *principal* or *nonnegative* square root of x in **Example 3,** we might have seen immediately that there is no solution.

◀ **Work Problem 3 at the Side.**

We use the following steps when solving an equation with radicals.

Solving a Radical Equation

Step 1 **Isolate a radical.** Arrange the terms so that a radical is alone on one side of the equation.

Step 2 **Square each side.**

Step 3 **Combine like terms.**

Step 4 **Repeat Steps 1–3,** if there is still a term with a radical.

Step 5 **Solve the equation.** Find all proposed solutions.

Step 6 **Check all proposed solutions** in the original equation.

EXAMPLE 4 **Using the Squaring Property with a Quadratic Expression**

Solve $x = \sqrt{x^2 + 5x + 10}$.

Step 1 The radical is already isolated on the right side of the equation.

Step 2 Square each side.

$$x^2 = \left(\sqrt{x^2 + 5x + 10}\right)^2 \quad \text{Squaring property}$$

$$x^2 = x^2 + 5x + 10 \quad \left(\sqrt{a}\right)^2 = a$$

Step 3

$$0 = 5x + 10 \quad \text{Subtract } x^2.$$

Step 4 This step is not needed.

Step 5

$$-10 = 5x \quad \text{Subtract 10.}$$

Proposed solution → $-2 = x$ Divide by 5.

Step 6 **CHECK**

$$x = \sqrt{x^2 + 5x + 10} \quad \text{Original equation}$$

$$-2 \stackrel{?}{=} \sqrt{(-2)^2 + 5(-2) + 10} \quad \text{Let } x = -2.$$

$$-2 \stackrel{?}{=} \sqrt{4 - 10 + 10} \quad \text{Multiply.}$$

$$-2 = 2 \quad \text{False}$$

The principal square root of a quantity *cannot* be negative.

Since substituting -2 for x leads to a false result, the equation has no solution, and the solution set is $\emptyset$.

Work Problem 4 at the Side. ▶

OBJECTIVE 3 **Solve equations by squaring a binomial.** Recall the rules for squaring binomials from **Section 12.4.**

$$(a + b)^2 = a^2 + 2ab + b^2 \quad \text{and} \quad (a - b)^2 = a^2 - 2ab + b^2$$

We apply the second rule in **Example 5** on the next page with $(x - 3)^2$.

$$(x - 3)^2$$

$$= x^2 - 2x(3) + 3^2$$

$$= x^2 - 6x + 9$$

Remember the middle term when squaring.

Work Problem 5 at the Side. ▶

4 Solve $x = \sqrt{x^2 - 4x - 16}$.

5 Square each expression.

GS **(a)** $w - 5$

$(w - 5)^2$

$= \_\_\_^2 - \_\_\_ w (\_\_\_)$

$+ \_\_\_^2$

$= \_\_\_\_\_$

(b) $2k - 5$

(c) $3m - 2p$

Answers

4. $\emptyset$

5. **(a)** w; 2; 5; 5; $w^2 - 10w + 25$
(b) $4k^2 - 20k + 25$
(c) $9m^2 - 12mp + 4p^2$

6 Solve each equation.

GS **(a)** $\sqrt{6w+6} = w+1$

$(\sqrt{6w+6})^{—} = (w+1)^2$

$\_\_\_\_ = w^2 + \_\_\_\_$

$w^2 - \_\_\_\_ = 0$

$(\_\_\_\_)(\_\_\_\_) = 0$

Complete the solution.

(b) $2u - 1 = \sqrt{10u+9}$

Answers

6. **(a)** 2; $6w + 6$; $2w + 1$; $4w - 5$; $w + 1$; $w - 5$; $\{-1, 5\}$

(b) $\{4\}$

EXAMPLE 5 **Using the Squaring Property When One Side Has Two Terms**

Solve $\sqrt{2x-3} = x - 3$.

$$\sqrt{2x-3} = x - 3$$

$$(\sqrt{2x-3})^2 = (x-3)^2 \quad \text{Square each side.}$$

$$2x - 3 = x^2 - 6x + 9 \quad (x-y)^2 = x^2 - 2xy + y^2$$

This equation is quadratic because of the presence of the x^2-term. To solve it, as shown in **Section 13.7,** we must write the equation in standard form.

$$\text{Standard form} \rightarrow x^2 - 8x + 12 = 0 \quad \text{Subtract } 2x\text{, add 3, and interchange sides.}$$

$$(x-6)(x-2) = 0 \quad \text{Factor.}$$

$$x - 6 = 0 \quad \text{or} \quad x - 2 = 0 \quad \text{Zero-factor property}$$

$$x = 6 \quad \text{or} \quad x = 2 \leftarrow \text{Proposed solutions}$$

CHECK Let $x = 6$.

$$\sqrt{2x-3} = x - 3$$
$$\sqrt{2(6)-3} \stackrel{?}{=} 6 - 3$$
$$\sqrt{12-3} \stackrel{?}{=} 3$$
$$\sqrt{9} \stackrel{?}{=} 3$$
$$3 = 3 \checkmark \text{ True}$$

Let $x = 2$.

$$\sqrt{2x-3} = x - 3$$
$$\sqrt{2(2)-3} \stackrel{?}{=} 2 - 3$$
$$\sqrt{4-3} \stackrel{?}{=} -1$$
$$\sqrt{1} \stackrel{?}{=} -1$$
$$1 = -1 \quad \text{False}$$

Only 6 is a valid solution. (2 is extraneous.) The solution set is $\{6\}$.

◀ **Work Problem 6 at the Side.**

EXAMPLE 6 **Rewriting an Equation before Using the Squaring Property**

Solve $\sqrt{9x} - 1 = 2x$.

We must apply Step 1 to isolate the radical *before* squaring each side. If we skip Step 1 and begin by squaring, we obtain the following.

$$(\sqrt{9x} - 1)^2 = (2x)^2$$

$$9x - 2\sqrt{9x} + 1 = 4x^2 \quad \text{Don't do this.}$$

This equation still contains a radical. Follow the steps below.

$$\sqrt{9x} - 1 = 2x$$

This is a key step. $\sqrt{9x} = 2x + 1$ — Add 1 to isolate the radical. (Step 1)

$$(\sqrt{9x})^2 = (2x+1)^2 \quad \text{Square each side. (Step 2)}$$

No terms contain radicals. $9x = 4x^2 + 4x + 1$ — $(a+b)^2 = a^2 + 2ab + b^2$

$$4x^2 - 5x + 1 = 0 \quad \text{Standard form (Step 3)}$$

$$(4x-1)(x-1) = 0 \quad \text{Factor. (Step 5; Step 4 is not needed.)}$$

$$4x - 1 = 0 \quad \text{or} \quad x - 1 = 0 \quad \text{Zero-factor property}$$

$$x = \frac{1}{4} \quad \text{or} \quad x = 1 \leftarrow \text{Proposed solutions}$$

Continued on Next Page

CHECK Let $x = \frac{1}{4}$. (Step 6)

$$\sqrt{9x} - 1 = 2x$$

$$\sqrt{9\left(\frac{1}{4}\right)} - 1 \stackrel{?}{=} 2\left(\frac{1}{4}\right)$$

$$\frac{3}{2} - 1 \stackrel{?}{=} \frac{1}{2}$$

$$\frac{1}{2} = \frac{1}{2} \checkmark \quad \text{True}$$

Let $x = 1$. (Step 6)

$$\sqrt{9x} - 1 = 2x$$

$$\sqrt{9(1)} - 1 \stackrel{?}{=} 2(1)$$

$$3 - 1 \stackrel{?}{=} 2$$

$$2 = 2 \checkmark \quad \text{True}$$

Both proposed solutions check, so the solution set is $\left\{\frac{1}{4}, 1\right\}$.

CAUTION

When squaring each side of

$$\sqrt{9x} = 2x + 1$$

in **Example 6**, the *entire* binomial $2x + 1$ must be squared to get $4x^2 + 4x + 1$. It is incorrect to square the $2x$ and the 1 separately to get $4x^2 + 1$.

Work Problem 7 at the Side. ▶

Some radical equations require squaring twice, as in the next example.

EXAMPLE 7 Using the Squaring Property Twice

Solve $\sqrt{21 + x} = 3 + \sqrt{x}$.

$$\sqrt{21 + x} = 3 + \sqrt{x}$$

$$\left(\sqrt{21 + x}\right)^2 = \left(3 + \sqrt{x}\right)^2 \quad \text{Square each side.}$$

$$21 + x = 9 + 6\sqrt{x} + x \quad \text{Be careful here.}$$

$$12 = 6\sqrt{x} \quad \text{Subtract 9. Subtract } x.$$

$$2 = \sqrt{x} \quad \text{Divide by 6.}$$

$$2^2 = \left(\sqrt{x}\right)^2 \quad \text{Square each side again.}$$

Proposed solution ⟶ $4 = x$ Apply the exponents.

CHECK

$$\sqrt{21 + x} = 3 + \sqrt{x} \quad \text{Original equation}$$

$$\sqrt{21 + 4} \stackrel{?}{=} 3 + \sqrt{4} \quad \text{Let } x = 4.$$

$$\sqrt{25} \stackrel{?}{=} 3 + 2 \quad \text{Simplify.}$$

$$5 = 5 \checkmark \quad \text{True}$$

The solution set is $\{4\}$.

Work Problem 8 at the Side. ▶

7 Solve each equation.

GS **(a)** $\sqrt{x} - 3 = x - 15$

$$\sqrt{x} = x - \_\_\_\_$$

$$\left(\sqrt{x}\right)^{\_\_} = (x - \_\_\_\_)^2$$

$$\_\_\_\_ = x^2 - \_\_\_\_\_\_$$

$$\_\_\_\_\_\_\_\_ = 0$$

Complete the solution.

(b) $\sqrt{z + 5} + 2 = z + 5$

8 Solve each equation.

(a) $\sqrt{p + 1} - \sqrt{p - 4} = 1$

(b) $\sqrt{2x + 1} + \sqrt{x + 4} = 3$

Answers

7. **(a)** 12; 2; 12; x; $24x + 144$; $x^2 - 25x + 144$; $\{16\}$
(b) $\{-1\}$

8. **(a)** $\{8\}$ **(b)** $\{0\}$

Heron of Alexandria

First century Greek mathematician

9 Find the area of a triangle with sides of lengths 7 cm, 15 cm, and 20 cm.

OBJECTIVE 4 Solve problems using formulas that involve radicals. The most common formula for the area of a triangle is $A = \frac{1}{2}bh$, where b is the length of the base and h is the height. What if the height h is not known? **Heron's formula** allows us to calculate the area of a triangle if we know only the lengths of the sides a, b, and c. To use Heron's formula, we first find the **semiperimeter** s, which is one-half the perimeter.

$$s = \frac{1}{2}(a + b + c) \quad \text{Semiperimeter}$$

The area $\boldsymbol{A}$ is then given by the following formula.

$$A = \sqrt{s(s-a)(s-b)(s-c)} \quad \text{Heron's formula}$$

For example, a 3-4-5 right triangle has area

$$A = \frac{1}{2}(3)(4) = 6 \text{ square units,}$$

calculated using the familiar formula. We can also use Heron's formula, with $s = \frac{1}{2}(3 + 4 + 5) = 6$.

$$A = \sqrt{6(6-3)(6-4)(6-5)} \quad \text{Let } s = 6,\ a = 3,\ b = 4, \text{ and } c = 5.$$

$$A = \sqrt{6 \cdot 3 \cdot 2 \cdot 1} \quad \text{Subtract.}$$

$$A = \sqrt{36} \quad \text{Multiply.}$$

$$A = 6 \quad \text{Take the positive root.}$$

The area is again 6 square units, as expected.

EXAMPLE 8 Using Heron's Formula to Find the Area of a Triangle

The sides of a triangular garden plot have lengths 20 ft, 34 ft, and 42 ft. See **Figure 3.** Find its area.

First find the semiperimeter of the triangle.

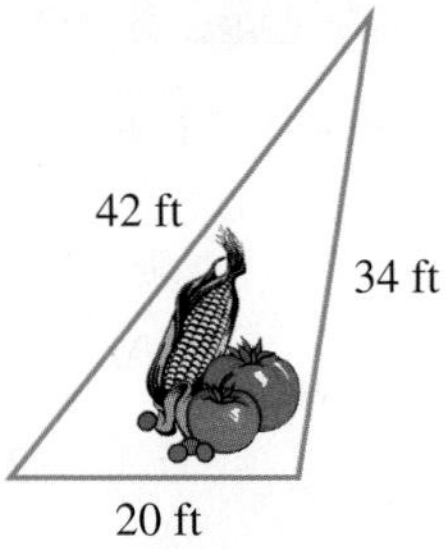

Figure 3

$$s = \frac{1}{2}(a + b + c)$$

$$s = \frac{1}{2}(20 + 34 + 42) \quad \text{Let } a = 20,\ b = 34, \text{ and } c = 42.$$

$$s = \frac{1}{2}(96), \text{ or } 48 \quad \text{Add, and then multiply.}$$

Now use Heron's formula to find the area.

$$A = \sqrt{s(s-a)(s-b)(s-c)} \quad \text{Heron's formula}$$

$$A = \sqrt{48(48-20)(48-34)(48-42)} \quad \text{Substitute.}$$

$$A = \sqrt{48(28)(14)(6)} \quad \text{Subtract.}$$

$$A = \sqrt{112{,}896} \quad \text{Multiply.}$$

$$A = 336 \quad \text{Use a calculator.}$$

The area of the garden plot is 336 ft^2.

◀ **Work Problem 9 at the Side.**

Answer

9. 42 cm^2

16.6 Exercises

FOR EXTRA HELP

Download the MyDashBoard App

MyMathLab®

CONCEPT CHECK *Fill in the blanks to complete the following.*

1. To solve an equation involving a radical, such as $\sqrt{2x-1}=5$, use the __________ property of equality. This property says that if each side of an equation is __________, all solutions of the __________ equation are among the solutions of the squared equation.

2. Solving some radical equations involves squaring a binomial using the following rules.

$(a+b)^2=$ __________

$(a-b)^2=$ __________

Solve each equation. ***See Examples 1–4.***

3. $\sqrt{x}=7$

4. $\sqrt{k}=10$

5. $\sqrt{t+2}=3$

6. $\sqrt{x+7}=5$

7. $\sqrt{r-4}=9$

8. $\sqrt{k-12}=3$

9. $\sqrt{t}=-5$

10. $\sqrt{x}=-8$

11. $\sqrt{4-t}=7$

12. $\sqrt{9-s}=5$

13. $\sqrt{2t+3}=0$

14. $\sqrt{5x-4}=0$

15. $\sqrt{3x-8}=-2$

16. $\sqrt{6x+4}=-3$

17. $\sqrt{w}-4=7$

18. $\sqrt{t}+3=10$

19. $\sqrt{10x-8}=3\sqrt{x}$

20. $\sqrt{17t-4}=4\sqrt{t}$

21. $5\sqrt{x}=\sqrt{10x+15}$

22. $4\sqrt{z}=\sqrt{20z-16}$

23. $\sqrt{3x-5}=\sqrt{2x+1}$

24. $\sqrt{5x+2}=\sqrt{3x+8}$

25. $k=\sqrt{k^2-5k-15}$

26. $s=\sqrt{s^2-2s-6}$

27. $7x=\sqrt{49x^2+2x-10}$

28. $6x=\sqrt{36x^2+5x-5}$

29. CONCEPT CHECK Consider the following incorrect "solution." **What Went Wrong?**

$$-\sqrt{x-1} = -4$$
$$-(x-1) = 16 \quad \text{Square each side.}$$
$$-x+1 = 16 \quad \text{Distributive property}$$
$$-x = 15 \quad \text{Subtract 1.}$$
$$x = -15 \quad \text{Multiply by } -1.$$

Solution set: $\{-15\}$

30. CONCEPT CHECK The first step in solving the equation

$$\sqrt{2x+1} = x - 7$$

is to square each side of the equation. Errors often occur in solving equations such as this one when the right side of the equation is squared incorrectly. What is the square of the right side?

Solve each equation. ***See Examples 5 and 6.***

31. $\sqrt{2x+1} = x - 7$

32. $\sqrt{3x+3} = x - 5$

33. $\sqrt{3k+10} + 5 = 2k$

34. $\sqrt{4t+13} + 1 = 2t$

35. $\sqrt{5x+1} - 1 = x$

36. $\sqrt{x+1} - x = 1$

37. $\sqrt{6t+7} + 3 = t + 5$

38. $\sqrt{10x+24} = x + 4$

39. $x - 4 - \sqrt{2x} = 0$

40. $x - 3 - \sqrt{4x} = 0$

41. $\sqrt{x} + 6 = 2x$

42. $\sqrt{k} + 12 = k$

Solve each equation. ***See Example 7.***

43. $\sqrt{x+1} - \sqrt{x-4} = 1$

44. $\sqrt{2x+3} + \sqrt{x+1} = 1$

45. $\sqrt{x} = \sqrt{x-5} + 1$

46. $\sqrt{2x} = \sqrt{x+7} - 1$

47. $\sqrt{3x+4} - \sqrt{2x-4} = 2$

48. $\sqrt{1-x} + \sqrt{x+9} = 4$

49. $\sqrt{2x+11} + \sqrt{x+6} = 2$

50. $\sqrt{x+9} + \sqrt{x+16} = 7$

Solve each problem.

51. The square root of the sum of a number and 4 is 5. Find the number.

52. The square root of the sum of a number and 7 is 4. Find the number.

53. Three times the square root of 2 equals the square root of the sum of some number and 10. Find the number.

54. The negative square root of a number equals that number decreased by 2. Find the number.

Use Heron's formula to solve each problem. ***See Example 8.***

55. A surveyor has measured the lengths of the sides of a triangular plot of land as 180 ft, 200 ft, and 240 ft. Find the area of the plot. (Round to the nearest whole number.)

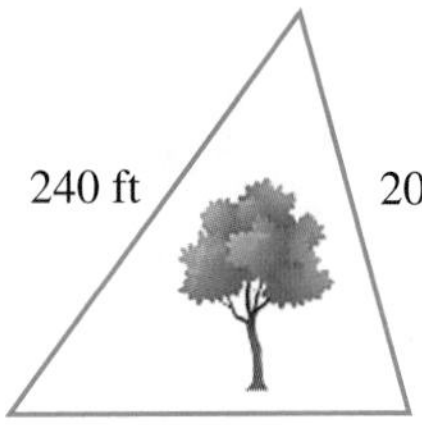

56. A triangular indoor play space in a day care center has sides of length 30 ft, 45 ft, and 65 ft. How much carpeting would be needed to cover this space? (Round to the nearest whole number.)

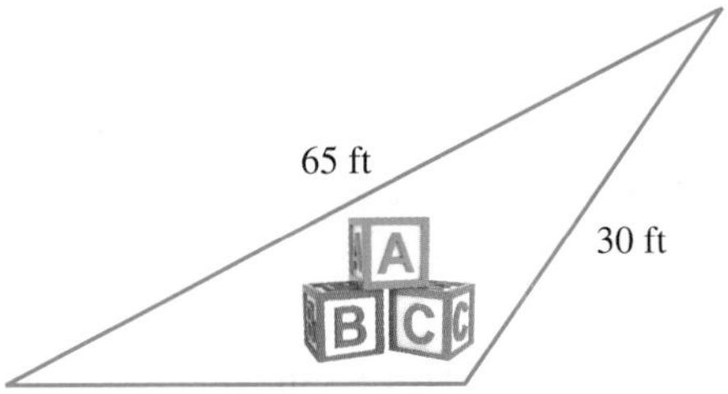

Solve each problem. Give answers to the nearest tenth.

57. To estimate the speed at which a car was traveling at the time of an accident, a police officer drives the car involved in the accident under conditions similar to those during which the accident took place and then skids to a stop. If the car is driven at 30 mph, then the speed at the time of the accident is given by

$$s = 30\sqrt{\frac{a}{p}},$$

where a is the length of the skid marks left at the time of the accident and p is the length of the skid marks in the police test. Find s for the following values of a and p.

(a) $a = 862$ ft; $p = 156$ ft

(b) $a = 382$ ft; $p = 96$ ft

(c) $a = 84$ ft; $p = 26$ ft

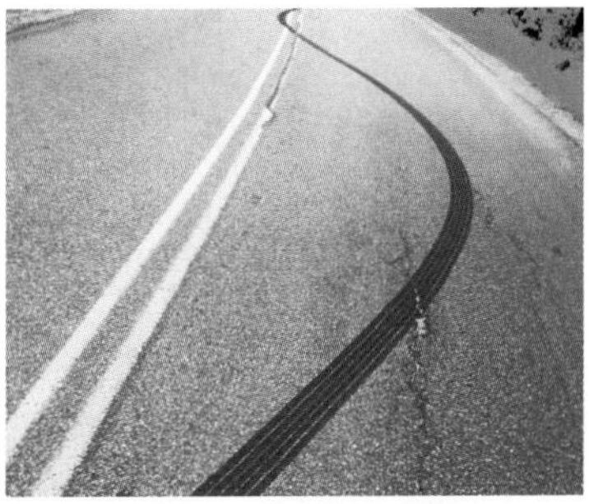

58. A formula for calculating the distance, d, one can see from an airplane to the horizon on a clear day is

$$d = 1.22\sqrt{x},$$

where x is the altitude of the plane in feet and d is given in miles.

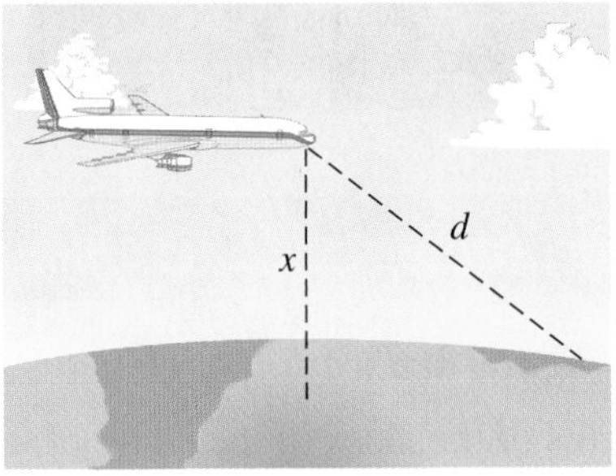

How far can one see to the horizon in a plane flying at the following altitudes?

(a) 15,000 ft

(b) 18,000 ft

(c) 24,000 ft

On a clear day, the maximum distance in kilometers that one can see from a structure is given by the formula

$$\text{sight distance} = 111.7\sqrt{\text{height of structure in kilometers}}.$$

Use this formula and the conversion equations 1 ft $\approx$ 0.3048 m *and* 1 km $\approx$ 0.621371 mi *as necessary to solve each problem. Round your answers to the nearest mile. (Source: A Sourcebook of Applications of School Mathematics,* NCTM.)

59. The London Eye is a unique structure that features 32 observation capsules. It is the fourth-tallest structure in London, with a diameter of 135 m. Does the formula justify the claim that on a clear day passengers on the London Eye can see Windsor Castle, 25 mi away? (*Source:* www.londoneye.com)

60. The Empire State Building in New York City is 1250 ft high. (The antenna reaches to 1454 ft.) The observation deck, located on the 102nd floor, is at a height of 1050 ft. How far could you see on a clear day from the observation deck? (*Source:* www.esbnyc.com)

61. The twin Petronas Towers in Kuala Lumpur, Malaysia, are 1483 ft high (including the spires). How far would one of the builders have been able to see on a clear day from the top of a spire? (*Source*: *World Almanac and Book of Facts.*)

62. The Khufu Pyramid in Giza was built in about 2566 B.C. to a height, at that time, of 481 ft. It is now only about 449 ft high. How far would one of the original builders of the pyramid have been able to see from the top of the pyramid? (*Source*: www.archaeology.com)

Relating Concepts (Exercises 63–68) For Individual or Group Work

Consider the figure, and **work Exercises 63–68 in order.**

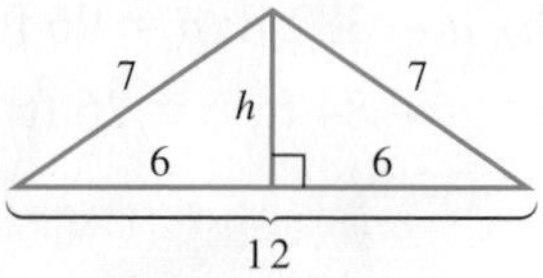

63. The lengths of the sides of the entire triangle are 7, 7, and 12. Find the semiperimeter s.

64. Now use Heron's formula to find the area of the entire triangle. Write it as a simplified radical.

65. Find the value of h by using the Pythagorean theorem.

66. Find the area of each of the congruent right triangles forming the entire triangle by using the formula $A = \frac{1}{2}bh$.

67. Double your result from **Exercise 66** to determine the area of the entire triangle.

68. How do your answers in **Exercises 64 and 67** compare?

Chapter 16 Summary

Key Terms

16.1

square root A number a is a square root of b if $a^2 = b$.

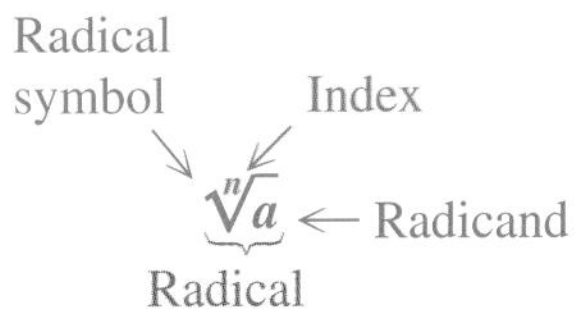

principal square root The positive square root of a number is its principal square root.

radicand The number or expression inside a radical symbol is the radicand.

radical A radical symbol with a radicand is a radical.

radical expression An algebraic expression containing a radical is a radical expression.

perfect square A number with a rational square root is a perfect square.

irrational number A real number that is not rational is an irrational number.

cube root A number a is a cube root of b if $a^3 = b$.

index (order) In a radical of the form $\sqrt[n]{a}$, the number n is the index, or order.

16.2

perfect cube A number with a rational cube root is a perfect cube.

16.3

like radicals Like radicals are multiples of the same root of the same number.

16.4

rationalizing the denominator The process of changing the denominator of a fraction from a radical (irrational number) to an expression not involving a radical is called rationalizing the denominator.

16.5

conjugate The conjugate of $a + b$ is $a - b$.

16.6

radical equation An equation with a variable in a radicand is a radical equation.

extraneous solution A proposed solution that does not satisfy the given equation is an extraneous solution.

New Symbols

$\sqrt{}$ radical symbol

$\approx$ is approximately equal to

$\sqrt[3]{a}$ cube root of a

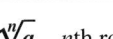

$\sqrt[n]{a}$ nth root of a

Test Your Word Power

See how well you have learned the vocabulary in this chapter.

1 A **square root** of a number is
- **A.** the number raised to the second power
- **B.** the number under a radical symbol
- **C.** a number that when multiplied by itself gives the original number
- **D.** the inverse of the number.

2 A **radicand** is
- **A.** the index of a radical
- **B.** the number or expression inside the radical symbol
- **C.** the positive root of a number
- **D.** the radical symbol.

3 A **radical** is
- **A.** a symbol that indicates the nth root
- **B.** an algebraic expression containing a square root
- **C.** the positive nth root of a number
- **D.** a radical symbol and the number or expression inside it.

4 The **principal root** of a positive number with even index n is
- **A.** the positive nth root of the number
- **B.** the negative nth root of the number
- **C.** the square root of the number
- **D.** the cube root of the number.

5 An **irrational number** is
- **A.** the quotient of two integers, with denominator not 0
- **B.** a decimal number that neither terminates nor repeats
- **C.** the principal square root of a number
- **D.** a nonreal number.

(continued)

Test Your Word Power *(continued)*

6 The **Pythagorean theorem** states that, in a right triangle,
- **A.** the sum of the measures of the angles is 180°
- **B.** the sum of the lengths of the two shorter sides equals the length of the longest side
- **C.** the longest side is opposite the right angle
- **D.** the square of the length of the longest side equals the sum of the squares of the lengths of the two shorter sides.

7 **Like radicals** are
- **A.** radicals in simplest form
- **B.** algebraic expressions containing radicals
- **C.** multiples of the same root of the same number
- **D.** radicals with the same index.

8 The **conjugate** of $a + b$ is
- **A.** $a - b$
- **B.** $a \cdot b$
- **C.** $a \div b$
- **D.** $(a + b)^2$.

9 **Rationalizing the denominator** is the process of
- **A.** eliminating fractions from a radical expression
- **B.** changing the denominator of a fraction from a radical expression to an expression not involving a radical
- **C.** clearing a radical expression of radicals
- **D.** multiplying radical expressions.

10 An **extraneous solution** is a value
- **A.** that makes an equation false and must be discarded
- **B.** that makes an equation true
- **C.** that makes an expression equal 0
- **D.** that checks in the original equation.

Answers to Test Your Word Power

1. C; *Examples:* 6 is a square root of 36 since $6^2 = 6 \cdot 6 = 36$. -6 is also a square root of 36.
2. B; *Example:* In $\sqrt{3xy}$, $3xy$ is the radicand.
3. D; *Examples:* $\sqrt{144}$, $\sqrt{4xy^2}$, and $\sqrt{4 + t^2}$
4. A; *Examples:* $\sqrt{36} = 6$, $\sqrt[4]{81} = 3$, and $\sqrt[6]{64} = 2$
5. B; *Examples:* π, $\sqrt{2}$, $-\sqrt{5}$
6. D; *Example:* In a right triangle where $a = 6$, $b = 8$, and $c = 10$, $6^2 + 8^2 = 10^2$.
7. C; *Examples:* $\sqrt{7}$ and $3\sqrt{7}$ are like radicals, as are $2\sqrt[3]{6k}$ and $5\sqrt[3]{6k}$.
8. A; *Example:* The conjugate of $\sqrt{3} + 1$ is $\sqrt{3} - 1$.
9. B; *Example:* To rationalize the denominator of $\frac{5}{\sqrt{3} + 1}$, multiply the numerator and denominator by $\sqrt{3} - 1$ (the conjugate of the denominator) to get $\frac{5(\sqrt{3} - 1)}{2}$.
10. A; *Example:* The proposed solution 2 is extraneous when $\sqrt{5q - 1} + 3 = 0$ is solved using the method of **Section 8.6.**

Quick Review

Concepts	*Examples*

16.1 Evaluating Roots

Let a be a positive real number.

$\sqrt{a}$ is the positive or principal square root of a.

$-\sqrt{a}$ is the negative square root of a. Also, $\sqrt{0} = 0$.

Examples: $\sqrt{49} = 7$ $\quad -\sqrt{81} = -9$

If a is a negative real number, then $\sqrt{a}$ is not a real number.

Example: $\sqrt{-25}$ is not a real number.

If a is a positive rational number, then $\sqrt{a}$ is rational if a is a perfect square. $\sqrt{a}$ is irrational if a is not a perfect square.

Examples: $\sqrt{\frac{4}{9}}$ and $\sqrt{16}$ are rational. $\quad \sqrt{\frac{2}{3}}$ and $\sqrt{21}$ are irrational.

Every real number has exactly one real cube root.

Examples: $\sqrt[3]{27} = 3 \quad \sqrt[3]{-8} = -2$

Pythagorean Theorem

If c is the length of the longest side (hypotenuse) of a right triangle and a and b are the lengths of the shorter sides (legs), then

$$a^2 + b^2 = c^2.$$

Example: Find a for the triangle in the figure.

$$a^2 + 10^2 = 13^2$$
$$a^2 + 100 = 169$$
$$a^2 = 69$$
$$a = \sqrt{69} \approx 8.3$$

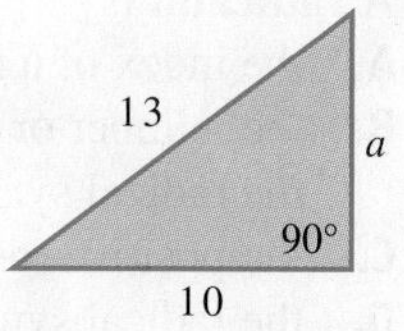

Concepts	Examples
16.2 Multiplying, Dividing, and Simplifying Radicals	
Product Rule for Radicals If a and b are nonnegative real numbers, then the following hold. $\sqrt{a}\cdot\sqrt{b}=\sqrt{ab}$ and $\sqrt{ab}=\sqrt{a}\cdot\sqrt{b}$	$\sqrt{5}\cdot\sqrt{7}=\sqrt{5\cdot 7}=\sqrt{35}$ $\sqrt{48}=\sqrt{16\cdot 3}=\sqrt{16}\cdot\sqrt{3}=4\sqrt{3}$
Quotient Rule for Radicals If a and b are nonnegative real numbers and $b\neq 0$, then the following hold. $\sqrt{\frac{a}{b}}=\frac{\sqrt{a}}{\sqrt{b}}$ and $\frac{\sqrt{a}}{\sqrt{b}}=\sqrt{\frac{a}{b}}$	$\sqrt{\frac{25}{64}}=\frac{\sqrt{25}}{\sqrt{64}}=\frac{5}{8}$ $\quad\frac{\sqrt{8}}{\sqrt{2}}=\sqrt{\frac{8}{2}}=\sqrt{4}=2$
For all real numbers a and b where the indicated roots exist, the following hold. $\sqrt[n]{a}\cdot\sqrt[n]{b}=\sqrt[n]{ab}$ and $\frac{\sqrt[n]{a}}{\sqrt[n]{b}}=\sqrt[n]{\frac{a}{b}}$ $(b\neq 0)$	$\sqrt[3]{5}\cdot\sqrt[3]{3}=\sqrt[3]{15}$ $\quad\frac{\sqrt[4]{12}}{\sqrt[4]{4}}=\sqrt[4]{\frac{12}{4}}=\sqrt[4]{3}$
16.3 Adding and Subtracting Radicals Add and subtract like radicals by using the distributive property. ***Only like radicals can be combined in this way.***	$2\sqrt{5}+4\sqrt{5}$ $=(2+4)\sqrt{5}$ $=6\sqrt{5}$ $\sqrt{8}-\sqrt{32}$ $=2\sqrt{2}-4\sqrt{2}$ $=-2\sqrt{2}$
16.4 Rationalizing the Denominator The denominator of a radical expression can be rationalized by multiplying both the numerator and denominator by a number that will eliminate the radical from the denominator.	$\frac{2}{\sqrt{3}}=\frac{2\cdot\sqrt{3}}{\sqrt{3}\cdot\sqrt{3}}=\frac{2\sqrt{3}}{3}$ $\sqrt[3]{\frac{5}{121}}=\frac{\sqrt[3]{5}\cdot\sqrt[3]{11}}{\sqrt[3]{11^2}\cdot\sqrt[3]{11}}=\frac{\sqrt[3]{55}}{11}$
16.5 More Simplifying and Operations with Radicals When appropriate, use the rules for adding and multiplying polynomials to simplify radical expressions.	$\sqrt{6}(\sqrt{5}-\sqrt{7})=\sqrt{30}-\sqrt{42}$ $(\sqrt{3}+1)(\sqrt{3}-2)$ $=3-2\sqrt{3}+\sqrt{3}-2$ FOIL method $=1-\sqrt{3}$ Combine like terms.
The formulas $(a+b)^2=a^2+2ab+b^2$, $(a-b)^2=a^2-2ab+b^2$, and $(a+b)(a-b)=a^2-b^2$ are useful when simplifying radical expressions.	$(\sqrt{13}-\sqrt{2})^2$ $=(\sqrt{13})^2-2(\sqrt{13})(\sqrt{2})+(\sqrt{2})^2$ $=13-2\sqrt{26}+2$ $=15-2\sqrt{26}$ $(\sqrt{5}+\sqrt{3})(\sqrt{5}-\sqrt{3})$ $=5-3$ $=2$

(continued)

Concepts	Examples
16.5 More Simplifying and Operations with Radicals *(continued)*	
Any denominators with radicals should be rationalized.	$\frac{3}{\sqrt{6}} = \frac{3 \cdot \sqrt{6}}{\sqrt{6} \cdot \sqrt{6}} = \frac{3\sqrt{6}}{6} = \frac{\sqrt{6}}{2}$
If a radical expression contains two terms in the denominator and at least one of those terms is a square root radical, multiply both the numerator and denominator by the conjugate of the denominator. In general, the expressions $a + b$ and $a - b$ are conjugates.	$\frac{6}{\sqrt{7} - \sqrt{2}}$ $= \frac{6(\sqrt{7} + \sqrt{2})}{(\sqrt{7} - \sqrt{2})(\sqrt{7} + \sqrt{2})}$ $\sqrt{7} + \sqrt{2}$ is the conjugate of $\sqrt{7} - \sqrt{2}$. $= \frac{6(\sqrt{7} + \sqrt{2})}{7 - 2}$ Multiply. $= \frac{6(\sqrt{7} + \sqrt{2})}{5}$ Subtract.
16.6 Solving Equations with Radicals	
Solving a Radical Equation	Solve.
Step 1 Isolate a radical.	$\sqrt{2x - 3} + x = 3$ $\sqrt{2x - 3} = 3 - x$ Isolate the radical.
Step 2 Square each side. (By the squaring property of equality, all solutions of the original equation are *among* the solutions of the squared equation.)	$(\sqrt{2x - 3})^2 = (3 - x)^2$ Square each side.
Step 3 Combine like terms.	$2x - 3 = 9 - \mathbf{6x} + x^2$ Remember the middle term when squaring $3 - x$.
Step 4 If there is still a term with a radical, repeat Steps 1–3.	$x^2 - 8x + 12 = 0$ Standard form
Step 5 Solve the equation for proposed solutions.	$(x - 2)(x - 6) = 0$ Factor. $x - 2 = 0$ or $x - 6 = 0$ Zero-factor property $x = 2$ or $x = 6$ Proposed solutions
Step 6 Check all proposed solutions from Step 5 in the original equation.	***A check is essential here.*** As shown below, 6 is extraneous. **CHECK** $\sqrt{2x - 3} + x = 3$ Original equation $\sqrt{2(6) - 3} + 6 \stackrel{?}{=} 3$ Let $x = 6$. $\sqrt{9} + 6 \stackrel{?}{=} 3$ $9 = 3$ False Verify that 2 is the only solution. The solution set is $\{2\}$.

Chapter 16 Review Exercises

16.1 *Find all square roots of each number.*

1. 49 **2.** 81 **3.** 196 **4.** 121 **5.** 256 **6.** 729

Find each root.

7. $\sqrt{16}$ **8.** $-\sqrt{0.36}$ **9.** $\sqrt[3]{-512}$ **10.** $\sqrt[4]{81}$

11. $\sqrt{-8100}$ **12.** $-\sqrt{4225}$ **13.** $\sqrt{\frac{144}{169}}$ **14.** $-\sqrt{\frac{100}{81}}$

15. Find the value of x.

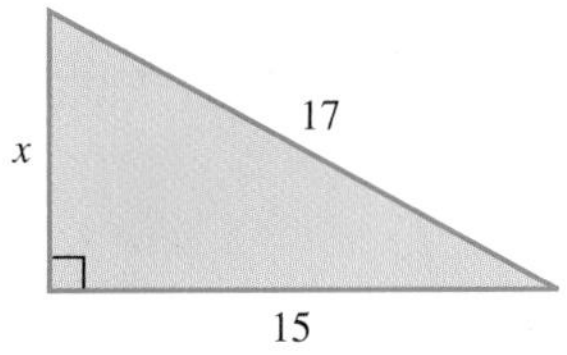

16. An HP f1905 computer monitor has viewing screen dimensions as shown in the figure. Find the diagonal measure of the viewing screen to the nearest tenth.

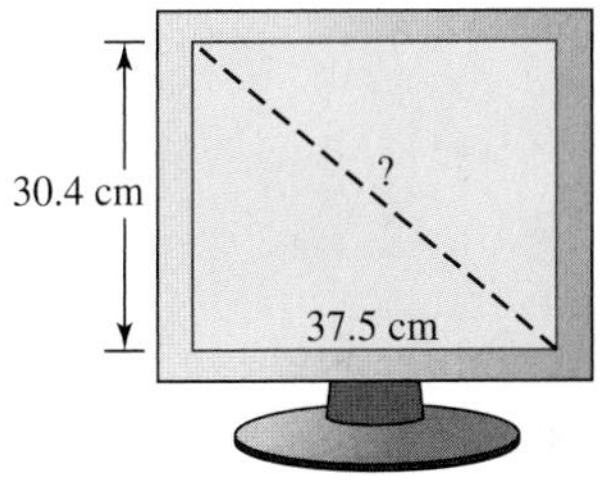

Write rational, irrational, *or* not a real number *for each number. If a number is rational, give its exact value. If a number is irrational, give a decimal approximation for the number. Round approximations to the nearest thousandth.*

17. $\sqrt{73}$ **18.** $\sqrt{169}$ **19.** $-\sqrt{625}$ **20.** $\sqrt{-19}$

16.2 *Simplify each expression.*

21. $\sqrt{48}$ **22.** $-\sqrt{288}$ **23.** $\sqrt[3]{16}$ **24.** $\sqrt[3]{375}$

25. $\sqrt{12} \cdot \sqrt{27}$ **26.** $\sqrt{32} \cdot \sqrt{48}$ **27.** $-\sqrt{\frac{121}{400}}$ **28.** $\sqrt{\frac{7}{169}}$

29. $\sqrt{\frac{1}{6}} \cdot \sqrt{\frac{5}{6}}$ **30.** $\sqrt{\frac{2}{5}} \cdot \sqrt{\frac{2}{45}}$ **31.** $\frac{3\sqrt{10}}{\sqrt{5}}$ **32.** $\frac{8\sqrt{150}}{4\sqrt{75}}$

Simplify each expression. Assume that all variables represent nonnegative real numbers.

33. $\sqrt{r^{18}}$ **34.** $\sqrt{x^{10}y^{16}}$ **35.** $\sqrt{162x^9}$ **36.** $\sqrt{\frac{36}{p^2}} \quad (p \neq 0)$

37. $\sqrt{a^{15}b^{21}}$ **38.** $\sqrt{121x^6y^{10}}$ **39.** $\sqrt[3]{y^6}$ **40.** $\sqrt[3]{216x^{15}}$

16.3 *Simplify and combine terms where possible.*

41. $7\sqrt{11} + \sqrt{11}$ **42.** $3\sqrt{2} + 6\sqrt{2}$ **43.** $3\sqrt{75} + 2\sqrt{27}$

44. $4\sqrt{12} + \sqrt{48}$ **45.** $4\sqrt{24} - 3\sqrt{54} + \sqrt{6}$ **46.** $2\sqrt{7} - 4\sqrt{28} + 3\sqrt{63}$

47. $\frac{2}{5}\sqrt{75} + \frac{3}{4}\sqrt{160}$ **48.** $\frac{1}{3}\sqrt{18} + \frac{1}{4}\sqrt{32}$ **49.** $\sqrt{15} \cdot \sqrt{2} + 5\sqrt{30}$

Simplify each expression. Assume that all variables represent nonnegative real numbers.

50. $\sqrt{4x} + \sqrt{36x} - \sqrt{9x}$ **51.** $\sqrt{16p} + 3\sqrt{p} - \sqrt{49p}$ **52.** $3k\sqrt{8k^2n} + 5k^2\sqrt{2n}$

16.4 *Perform the indicated operations, and write all answers in simplest form. Rationalize all denominators. Assume that all variables represent nonnegative real numbers.*

53. $\frac{10}{\sqrt{3}}$ **54.** $\frac{8\sqrt{2}}{\sqrt{5}}$ **55.** $\frac{12}{\sqrt{24}}$ **56.** $\sqrt{\frac{2}{5}}$

57. $\sqrt{\frac{5}{14}} \cdot \sqrt{28}$ **58.** $\sqrt{\frac{2}{7}} \cdot \sqrt{\frac{1}{3}}$ **59.** $\sqrt{\frac{r^2}{16x}} \quad (x \neq 0)$ **60.** $\sqrt[3]{\frac{1}{3}}$

16.5 *Simplify each expression.*

61. $-\sqrt{3}\left(\sqrt{5}+\sqrt{27}\right)$

62. $3\sqrt{2}\left(\sqrt{3}+2\sqrt{2}\right)$

63. $\left(2\sqrt{3}-4\right)\left(5\sqrt{3}+2\right)$

64. $\left(\sqrt{7}+2\sqrt{6}\right)\left(\sqrt{12}-\sqrt{2}\right)$

65. $\left(2\sqrt{3}+5\right)\left(2\sqrt{3}-5\right)$

66. $\left(\sqrt{x}+2\right)^2 \quad (x \geq 0)$

Rationalize each denominator. Assume that all variables represent nonnegative real numbers.

67. $\dfrac{1}{2+\sqrt{5}}$

68. $\dfrac{3}{1+\sqrt{x}}$

69. $\dfrac{\sqrt{5}-1}{\sqrt{2}+3}$

Write each quotient in lowest terms.

70. $\dfrac{15+10\sqrt{6}}{15}$

71. $\dfrac{3+9\sqrt{7}}{12}$

72. $\dfrac{6+\sqrt{192}}{2}$

16.6 *Solve each equation.*

73. $\sqrt{x}+5=0$

74. $\sqrt{k+1}=7$

75. $\sqrt{5t+4}=3\sqrt{t}$

76. $\sqrt{2p+3}=\sqrt{5p-3}$

77. $\sqrt{4x+1}=x-1$

78. $\sqrt{13+4t}=t+4$

79. $\sqrt{2-x}+3=x+7$

80. $\sqrt{x}-x+2=0$

81. $\sqrt{x+2}-\sqrt{x-3}=1$

Mixed Review Exercises

82. CONCEPT CHECK Match each radical in Column I with the equivalent choice in Column II. Choices may be used once, more than once, or not at all.

I		II	
(a) $\sqrt{64}$	**(b)** $-\sqrt{64}$	**A.** 4	**B.** 8
(c) $\sqrt{-64}$	**(d)** $\sqrt[3]{64}$	**C.** -4	**D.** Not a real number
(e) $\sqrt[3]{-64}$	**(f)** $-\sqrt[3]{-64}$	**E.** 16	**F.** -8

Simplify each expression if possible. Assume that all variables represent nonnegative real numbers.

83. $2\sqrt{27} + 3\sqrt{75} - \sqrt{300}$

84. $\dfrac{1}{5 + \sqrt{2}}$

85. $\sqrt{\dfrac{1}{3}} \cdot \sqrt{\dfrac{24}{5}}$

86. $\sqrt[3]{54a^7b^{10}}$

87. $-\sqrt{5}\left(\sqrt{2} + \sqrt{75}\right)$

88. $\sqrt{\dfrac{16r^3}{3s}} \quad (s \neq 0)$

89. $\dfrac{12 + 6\sqrt{13}}{12}$

90. $\left(\sqrt{5} - \sqrt{2}\right)^2$

91. $\left(6\sqrt{7} + 2\right)\left(4\sqrt{7} - 1\right)$

Solve each equation.

92. $\sqrt{x + 2} = x - 4$

93. $\sqrt{k} + 3 = 0$

94. $\sqrt{1 + 3t} - t = -3$

95. The diagram at the right can be used to verify the Pythagorean theorem, as follows.

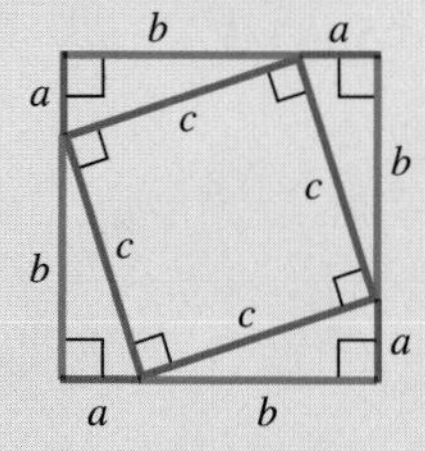

(a) Find the area of the large square.

(b) Find the sum of the areas of the smaller square and the four right triangles.

(c) Set the areas equal and simplify as much as possible.

Chapter 16 Test

CHAPTER Test Prep VIDEO

The Chapter Test Prep Videos with test solutions are available on DVD, in MyMathLab, and on YouTube—search "LialDevMath" and click on "Channels."

1. Find all square roots of 400.

2. Consider $\sqrt{142}$.

(a) Determine whether this number is rational or irrational.

(b) Find a decimal approximation to the nearest thousandth.

3. If $\sqrt{a}$ is not a real number, then what kind of number must a be?

Simplify where possible. Assume that all variables represent nonnegative real numbers.

4. $\sqrt[3]{216}$

5. $-\sqrt{54}$

6. $\sqrt{\dfrac{128}{25}}$

7. $\sqrt[3]{32}$

8. $\dfrac{20\sqrt{18}}{5\sqrt{3}}$

9. $3\sqrt{28} + \sqrt{63}$

10. $\left(\sqrt{5} + \sqrt{6}\right)^2$

11. $\sqrt{32x^2y^3}$

12. $\left(6 - \sqrt{5}\right)\left(6 + \sqrt{5}\right)$

13. $\left(2 - \sqrt{7}\right)\left(3\sqrt{2} + 1\right)$

14. $3\sqrt{27x} - 4\sqrt{48x} + 2\sqrt{3x}$

Solve each problem.

15. Find the measure of the unknown leg of this right triangle.

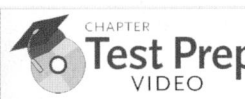

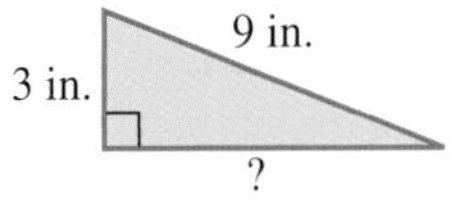

(a) Give its length in simplified radical form.

(b) Round the answer to the nearest thousandth.

16. In electronics, the impedance Z of an alternating current circuit is given by the following formula.

$$Z = \sqrt{R^2 + X^2}$$

Here, R is the resistance and X is the reactance, both in ohms. Find the value of the impedance Z if $R = 40$ ohms and $X = 30$ ohms. (*Source:* Cooke, N., and J. Orleans, *Mathematics Essential to Electricity and Radio*, McGraw-Hill.)

Rationalize each denominator.

17. $\dfrac{5\sqrt{2}}{\sqrt{7}}$

18. $\sqrt{\dfrac{2}{3x}}$ $(x > 0)$

19. $\dfrac{-2}{\sqrt[3]{4}}$

20. $\dfrac{-3}{4 - \sqrt{3}}$

21. Write $\dfrac{\sqrt{12} + 3\sqrt{128}}{6}$ in lowest terms.

Solve each equation.

22. $\sqrt{p} + 4 = 0$

23. $\sqrt{x + 1} = 5 - x$

24. $3\sqrt{x} - 2 = x$

25. $\sqrt{x + 7} - \sqrt{x} = 1$

26. Consider the following incorrect "solution." **What Went Wrong?** Give the correct solution set.

$$\sqrt{2x + 1} + 5 = 0$$

$\sqrt{2x + 1} = -5$ Subtract 5.

$2x + 1 = 25$ Square each side.

$2x = 24$ Subtract 1.

$x = 12$ Divide by 2.

The solution set is $\{12\}$.

17 Quadratic Equations

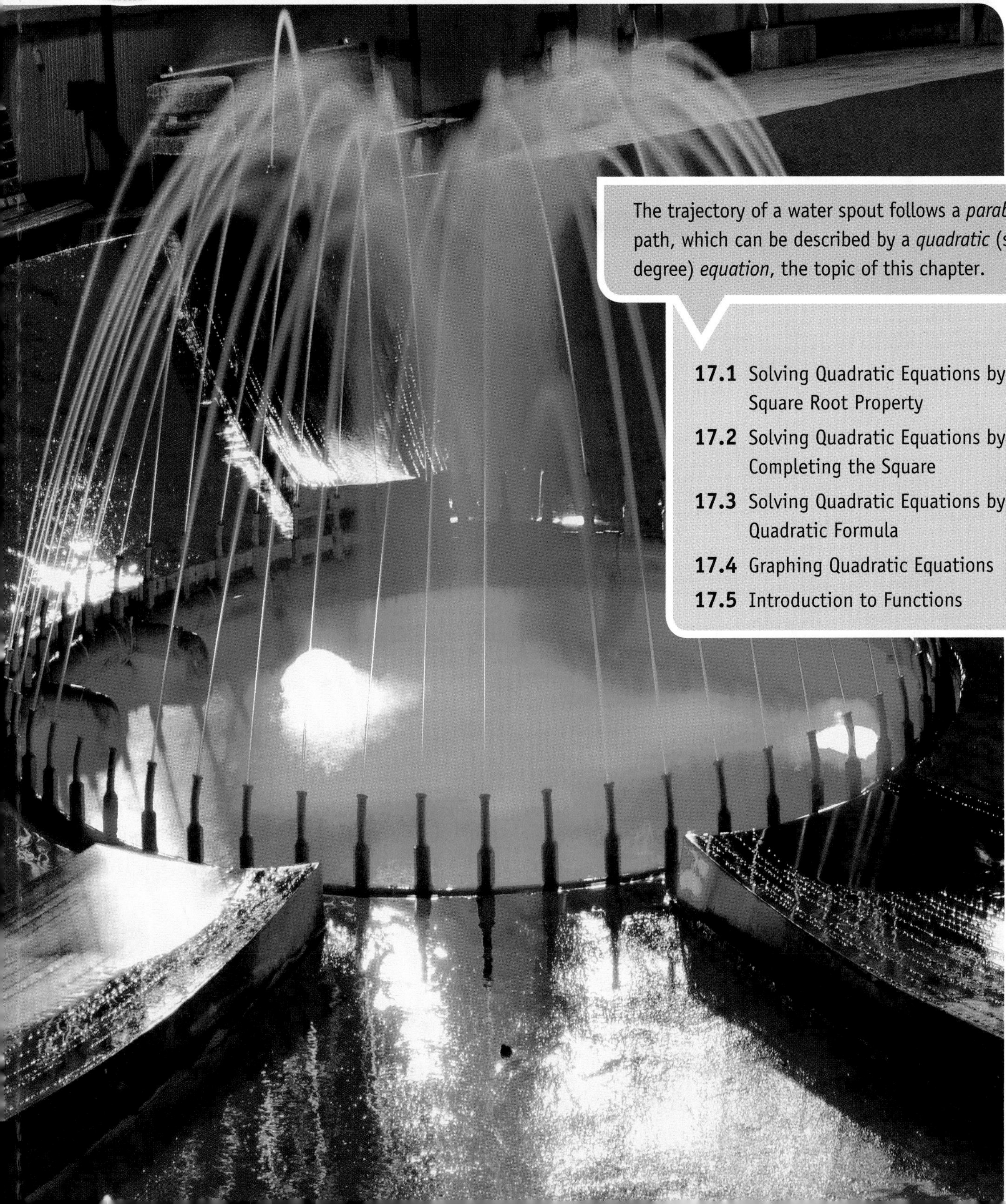

The trajectory of a water spout follows a *parabolic* path, which can be described by a *quadratic* (second-degree) *equation*, the topic of this chapter.

17.1 Solving Quadratic Equations by the Square Root Property

OBJECTIVES

1. Review the zero-factor property.
2. Solve equations of the form $x^2 = k$, where $k > 0$.
3. Solve equations of the form $(ax + b)^2 = k$, where $k > 0$.
4. Use formulas involving squared variables.

1 Solve each equation by using the zero-factor property.

GS **(a)** $x^2 - 2x - 15 = 0$

$(x + \_\_\_\_)(x - \_\_\_\_) = 0$

$x + \_\_\_\_ = 0$ or $x - \_\_\_\_ = 0$

$x = \_\_\_\_$ or $x = \_\_\_\_$

The solution set is $\{\_\_\_\_, \_\_\_\_\}$.

(b) $2x^2 - 3x + 1 = 0$

(c) $x^2 = 400$

Answers

1. (a) 3; 5; 3; 5; −3; 5; −3; 5
(b) $\left\{\frac{1}{2}, 1\right\}$ (c) $\{-20, 20\}$

OBJECTIVE 1 Review the zero-factor property. Recall that a **quadratic equation** is an equation that can be written in the form

$$ax^2 + bx + c = 0, \quad \text{Standard form}$$

for real numbers a, b, and c, with $a \neq 0$. In **Section 13.7** we solved quadratic equations by factoring, using the zero-factor property.

Zero-Factor Property

If a and b are real numbers and if $ab = 0$, then $a = 0$ or $b = 0$.

EXAMPLE 1 Solving Quadratic Equations by the Zero-Factor Property

Solve each equation by using the zero-factor property.

(a)

$x^2 + 4x + 3 = 0$

$(x + 3)(x + 1) = 0$ — Factor.

$x + 3 = 0$ or $x + 1 = 0$ — Zero-factor property

$x = -3$ or $x = -1$ — Solve each equation.

The solution set is $\{-3, -1\}$.

(b)

$x^2 = 9$

$x^2 - 9 = 0$ — Subtract 9.

$(x + 3)(x - 3) = 0$ — Factor.

$x + 3 = 0$ or $x - 3 = 0$ — Zero-factor property

$x = -3$ or $x = 3$ — Solve each equation.

The solution set is $\{-3, 3\}$.

◀ **Work Problem 1 at the Side.**

Not all quadratic equations can easily be solved by factoring, so we must develop other methods.

OBJECTIVE 2 Solve equations of the form $x^2 = k$, where $k > 0$. In **Example 1(b),** we might also have solved $x^2 = 9$ by noticing that x must be a number whose square is 9. Thus, $x = \sqrt{9} = 3$ or $x = -\sqrt{9} = -3$. This is generalized as the **square root property.**

Square Root Property

If k is a positive number and if $x^2 = k$, then

$$x = \sqrt{k} \quad \text{or} \quad x = -\sqrt{k}.$$

The solution set is $\{-\sqrt{k}, \sqrt{k}\}$, which can be written $\{\pm\sqrt{k}\}$. (The symbol $\pm$ is read "positive or negative" or "plus or minus.")

EXAMPLE 2 Solving Quadratic Equations of the Form $x^2 = k$

Solve each equation. Write radicals in simplified form.

(a) $x^2 = 16$

By the square root property, if $x^2 = 16$, then

$$x = \sqrt{16} = 4 \quad \text{or} \quad x = -\sqrt{16} = -4.$$

Check each solution by substituting it for x in the original equation. The solution set is $\{-4, 4\}$, or $\{\pm 4\}$.

This notation indicates two solutions, one positive and one negative.

(b) $z^2 = 5$

The solutions are $z = \sqrt{5}$ or $z = -\sqrt{5}$, so the solution set is $\{-\sqrt{5}, \sqrt{5}\}$, or $\{\pm\sqrt{5}\}$.

(c)

$$5m^2 - 32 = 8$$

$$5m^2 = 40 \quad \text{Add 32.}$$

$$m^2 = 8 \quad \text{Divide by 5.}$$

Don't stop here. Simplify the radicals.

$$m = \sqrt{8} \quad \text{or} \quad m = -\sqrt{8} \quad \text{Square root property}$$

$$m = 2\sqrt{2} \quad \text{or} \quad m = -2\sqrt{2} \quad \sqrt{8} = \sqrt{4} \cdot \sqrt{2} = 2\sqrt{2}$$

The solution set is $\{-2\sqrt{2}, 2\sqrt{2}\}$, or $\{\pm 2\sqrt{2}\}$.

(d) $x^2 = -4$

Because -4 is a negative number and because the square of a real number cannot be negative, ***there is no real number solution*** of this equation. (In this book, we are concerned with finding only *real number* solutions.) The solution set is $\emptyset$.

Work Problem 2 at the Side. ▶

OBJECTIVE 3 Solve equations of the form $(ax + b)^2 = k$, where $k > 0$. In each equation in **Example 2,** the exponent 2 had a single variable as its base. We can extend the square root property to solve equations in which the base is a binomial.

EXAMPLE 3 Solving Quadratic Equations of the Form $(x + b)^2 = k$

Solve each equation.

(a) Use $x - 3$ as the base.

$$(x - 3)^2 = 16$$

$$x - 3 = \sqrt{16} \quad \text{or} \quad x - 3 = -\sqrt{16} \quad \text{Square root property}$$

$$x - 3 = 4 \quad \text{or} \quad x - 3 = -4 \quad \sqrt{16} = 4$$

$$x = 7 \quad \text{or} \quad x = -1 \quad \text{Add 3.}$$

CHECK

$$(x - 3)^2 = 16$$
$$(7 - 3)^2 \stackrel{?}{=} 16 \quad \text{Let } x = 7.$$
$$4^2 \stackrel{?}{=} 16 \quad \text{Subtract.}$$
$$16 = 16 \ \checkmark \quad \text{True}$$

$$(x - 3)^2 = 16$$
$$(-1 - 3)^2 \stackrel{?}{=} 16 \quad \text{Let } x = -1.$$
$$(-4)^2 \stackrel{?}{=} 16 \quad \text{Subtract.}$$
$$16 = 16 \ \checkmark \quad \text{True}$$

The solutions are 7 and -1, and the solution set is $\{-1, 7\}$.

Continued on Next Page

2 Solve each equation. Write radicals in simplified form.

(a) $x^2 = 49$

(b) $x^2 = 11$

(c) $2x^2 + 8 = 32$

(d) $x^2 = -9$

Answers

2. (a) $\{-7, 7\}$ **(b)** $\{-\sqrt{11}, \sqrt{11}\}$ **(c)** $\{-2\sqrt{3}, 2\sqrt{3}\}$ **(d)** $\emptyset$

3 Solve each equation.

(a) $(x+2)^2 = 36$

(b) $(x-4)^2 = 3$

$x - 4 =$ ____ or $x - 4 =$ ____

$x =$ ____ or $x =$ ____

The solution set is ________.

4 Solve $(2x-5)^2 = 18$.

5 Solve each equation.

(a) $(5x+1)^2 = 7$

(b) $(7x-1)^2 = -1$

Answers

3. **(a)** $\{-8, 4\}$
(b) $\sqrt{3}; -\sqrt{3}; 4+\sqrt{3}; 4-\sqrt{3}$; $\{4+\sqrt{3}, 4-\sqrt{3}\}$

4. $\left\{\frac{5+3\sqrt{2}}{2}, \frac{5-3\sqrt{2}}{2}\right\}$

5. **(a)** $\left\{\frac{-1+\sqrt{7}}{5}, \frac{-1-\sqrt{7}}{5}\right\}$ **(b)** $\emptyset$

(b) $(x+1)^2 = 6$

$x + 1 = \sqrt{6}$ or $x + 1 = -\sqrt{6}$ Square root property

$x = -1 + \sqrt{6}$ or $x = -1 - \sqrt{6}$ Add -1.

CHECK $\left(-1+\sqrt{6}+1\right)^2 = \left(\sqrt{6}\right)^2 = 6$ ✓ Let $x = -1 + \sqrt{6}$.

$\left(-1-\sqrt{6}+1\right)^2 = \left(-\sqrt{6}\right)^2 = 6$ ✓ Let $x = -1 - \sqrt{6}$.

The solution set is $\{-1+\sqrt{6}, -1-\sqrt{6}\}$, or $\{-1 \pm \sqrt{6}\}$.

◀ Work Problem **3** at the Side.

EXAMPLE 4 Solving a Quadratic Equation of the Form $(ax+b)^2 = k$

Solve $(3r-2)^2 = 27$.

$$(3r-2)^2 = 27$$

$3r - 2 = \sqrt{27}$ or $3r - 2 = -\sqrt{27}$ Square root property

$3r - 2 = 3\sqrt{3}$ or $3r - 2 = -3\sqrt{3}$ $\sqrt{27} = \sqrt{9}\cdot\sqrt{3} = 3\sqrt{3}$

$3r = 2 + 3\sqrt{3}$ or $3r = 2 - 3\sqrt{3}$ Add 2.

$r = \frac{2+3\sqrt{3}}{3}$ or $r = \frac{2-3\sqrt{3}}{3}$ Divide by 3.

CHECK $(3r-2)^2 = 27$ Original equation

$\left(3\cdot\frac{2+3\sqrt{3}}{3} - 2\right)^2 \stackrel{?}{=} 27$ Let $r = \frac{2+3\sqrt{3}}{3}$.

$\left(2+3\sqrt{3}-2\right)^2 \stackrel{?}{=} 27$ Multiply.

$(ab)^2 = a^2b^2$ → $\left(3\sqrt{3}\right)^2 \stackrel{?}{=} 27$ Subtract.

$27 = 27$ ✓ True

The check of the other solution is similar. The solution set is

$$\left\{\frac{2+3\sqrt{3}}{3}, \frac{2-3\sqrt{3}}{3}\right\}.$$

◀ Work Problem **4** at the Side.

CAUTION

The solutions in **Example 4** are fractions that cannot be simplified. Note that 3 is *not* a common factor in the numerator.

EXAMPLE 5 Recognizing When There Is No Real Solution

Solve $(x+3)^2 = -9$.

Because the square root of -9 is not a real number, there is no real number solution for this equation. The solution set is $\emptyset$.

◀ Work Problem **5** at the Side.

OBJECTIVE 4 Use formulas involving squared variables.

EXAMPLE 6 Finding the Length of a Bass

We can approximate the weight of a bass, in pounds, given its length L and its girth (distance around) g, where both are measured in inches, using the following formula.

$$w = \frac{L^2 g}{1200}$$

Approximate the length of a bass weighing 2.20 lb and having girth 10 in. (*Source: Sacramento Bee.*)

$$w = \frac{L^2 g}{1200} \qquad \text{Given formula}$$

$$\mathbf{2.20} = \frac{L^2 \cdot \mathbf{10}}{1200} \qquad w = 2.20,\ g = 10$$

$$2640 = 10L^2 \qquad \text{Multiply by 1200.}$$

$$L^2 = 264 \qquad \text{Divide by 10. Interchange sides.}$$

$$L = \sqrt{264} \quad \text{or} \quad L = -\sqrt{264} \qquad \text{Square root property}$$

A calculator shows that $\sqrt{264} \approx 16.25$, so the length of the bass is a little more than 16 in. (We discard the negative solution $-\sqrt{264} \approx -16.25$, since L represents length.)

Work Problem 6 at the Side. ▶

6 Use the formula in **Example 6** to approximate the length to the nearest hundredth of a bass weighing 2.80 lb and having girth 11 in.

Answer

6. approximately 17.48 in.

17.1 Exercises

FOR EXTRA HELP

Download the MyDashBoard App

MyMathLab®

CONCEPT CHECK *Match each equation in Column I with the correct description of its solution in Column II.*

I	II
1. $x^2 = 12$	**A.** No real number solutions
2. $x^2 = -9$	**B.** Two integer solutions
3. $x^2 = \frac{25}{36}$	**C.** Two irrational solutions
4. $x^2 = 16$	**D.** Two rational solutions that are not integers

5. **CONCEPT CHECK** When a student was asked to solve $x^2 = 81$, she wrote $\{9\}$ as her answer. Her teacher did not give her full credit. The student argued that because $9^2 = 81$, her answer had to be correct. **What Went Wrong?** Give the correct solution set.

6. Give an example of an equation that can be solved by the square root property for equations. Then solve it.

Solve each equation by using the zero-factor property. ***See Example 1.***

7. $x^2 - x - 56 = 0$

8. $x^2 - 2x - 99 = 0$

9. $x^2 = 121$

10. $x^2 = 144$

11. $3x^2 - 13x = 30$

12. $5x^2 - 14x = 3$

Solve each equation by using the square root property. Write all radicals in simplest form. ***See Example 2.***

13. $x^2 = 81$

14. $x^2 = 121$

15. $k^2 = 14$

16. $m^2 = 22$

17. $t^2 = 48$

18. $x^2 = 54$

19. $x^2 = \frac{25}{4}$

20. $m^2 = \frac{36}{121}$

21. $x^2 = -100$

22. $x^2 = -64$

23. $z^2 = 2.25$

24. $w^2 = 56.25$

25. $r^2 - 3 = 0$

26. $x^2 - 13 = 0$

27. $7x^2 = 4$

28. $3x^2 = 10$

29. $4x^2 - 72 = 0$

30. $5z^2 - 200 = 0$

31. $3x^2 - 8 = 64$

32. $2x^2 + 7 = 61$

33. $5x^2 + 4 = 8$

34. $4x^2 - 3 = 7$

Solve each equation by using the square root property. Express all radicals in simplest form. ***See Examples 3–5.***

35. $(x - 3)^2 = 25$

36. $(x - 7)^2 = 16$

37. $(x + 5)^2 = -13$

38. $(x + 2)^2 = -17$

39. $(x - 8)^2 = 27$

40. $(x - 5)^2 = 40$

41. $(3x + 2)^2 = 49$

42. $(5x + 3)^2 = 36$

43. $(4x - 3)^2 = 9$

44. $(7x - 5)^2 = 25$

45. $(5 - 2x)^2 = 30$

46. $(3 - 2x)^2 = 70$

47. $(3x + 1)^2 = 18$

48. $(5x + 6)^2 = 75$

49. $\left(\frac{1}{2}x + 5\right)^2 = 12$

50. $\left(\frac{1}{3}x + 4\right)^2 = 27$

51. $(4x - 1)^2 - 48 = 0$

52. $(2x - 5)^2 - 180 = 0$

Solve each problem. ***See Example 6.***

53. One expert at marksmanship can hold a silver dollar at forehead level, drop it, draw his gun, and shoot the coin as it passes waist level. The distance traveled by a falling object is given by

$$d = 16t^2,$$

where d is the distance (in feet) the object falls in t seconds. If the coin falls about 4 ft, use the formula to estimate the time that elapses between the dropping of the coin and the shot.

54. The illumination produced by a light source depends on the distance from the source. For a particular light source, this relationship can be expressed as

$$d^2 = \frac{4050}{I},$$

where d is the distance from the source (in feet) and I is the amount of illumination in foot-candles. How far from the source is the illumination equal to 50 foot-candles?

55. The area A of a circle with radius r is given by the formula

$$A = \pi r^2.$$

If a circle has area 81π in.^{2}, what is its radius?

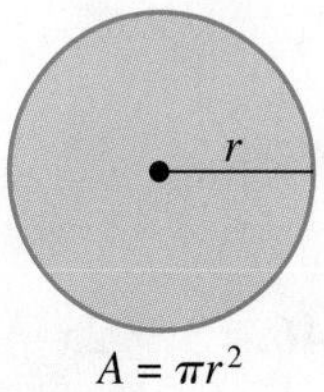

$A = \pi r^2$

56. The surface area S of a sphere with radius r is given by the formula

$$S = 4\pi r^2.$$

If a sphere has surface area 36π ft^2, what is its radius?

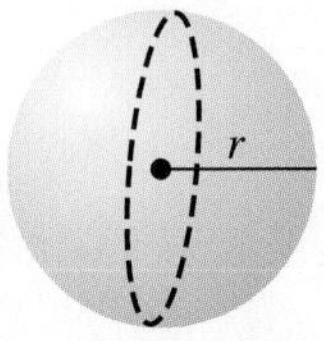

$S = 4\pi r^2$

The amount A that P dollars invested at an annual rate of interest r will grow to in 2 yr is $A = P(1 + r)^2$.

57. At what interest rate will \$100 grow to \$110.25 in 2 yr?

58. At what interest rate will \$200 grow to \$208.08 in 2 yr?

17.2 Solving Quadratic Equations by Completing the Square

OBJECTIVES

1. Solve quadratic equations by completing the square when the coefficient of the second-degree term is 1.
2. Solve quadratic equations by completing the square when the coefficient of the second-degree term is not 1.
3. Simplify the terms of an equation before solving.
4. Solve applied problems that require quadratic equations.

OBJECTIVE 1 **Solve quadratic equations by completing the square when the coefficient of the second-degree term is 1.** The methods we have studied so far are not enough to solve the equation

$$x^2 + 6x + 7 = 0.$$

If we could write the equation in the form $(x + 3)^2$ equals a constant, we could solve it with the square root property discussed in **Section 17.1.** To do that, we need to have a perfect square trinomial on one side of the equation.

Recall from **Section 13.5** that a perfect square trinomial has the form

$$x^2 + 2kx + k^2 \quad \text{or} \quad x^2 - 2kx + k^2,$$

where k represents a positive number.

EXAMPLE 1 Creating Perfect Square Trinomials

Complete each trinomial so that it is a perfect square. Then factor the trinomial.

(a) $x^2 + 8x +$ ______

The perfect square trinomial will have the form $x^2 + 2kx + k^2$. Thus, the middle term, $8x$, must equal $2kx$.

$$8x = 2kx \leftarrow \text{Solve this equation for } k.$$

$$4 = k \qquad \text{Divide each side by } 2x.$$

Therefore, $k = 4$ and $k^2 = 4^2 = 16$. The required perfect square trinomial is

$$x^2 + 8x + 16, \quad \text{which factors as} \quad (x + 4)^2.$$

(b) $x^2 - 18x +$ ______

Here the perfect square trinomial will have the form $x^2 - 2kx + k^2$. The middle term, $-18x$, must equal $-2kx$.

$$-18x = -2kx \leftarrow \text{Solve this equation for } k.$$

$$9 = k \qquad \text{Divide each side by } -2x.$$

Thus, $k = 9$ and $k^2 = 9^2 = 81$. The required perfect square trinomial is

$$x^2 - 18x + 81, \quad \text{which factors as} \quad (x - 9)^2.$$

Work Problem 1 at the Side. ▶

1 Complete each trinomial so that it is a perfect square. Then factor the trinomial.

(a) $x^2 + 12x +$ ____ factors as ________.

(b) $x^2 - 14x +$ ____ factors as ________.

(c) $x^2 - 2x +$ ____ factors as ________.

EXAMPLE 2 Rewriting an Equation to Use the Square Root Property

Solve $x^2 + 6x + 7 = 0$.

$$x^2 + 6x = -7 \qquad \text{Subtract 7 from each side.}$$

To use the square root property, the expression on the left, $x^2 + 6x$, must be written as a perfect square trinomial in the form $x^2 + 2kx + k^2$.

$$x^2 + 6x + \_\_\_\_$$ (A square must go here.)

Here, $2kx = 6x$, so $k = 3$ and $k^2 = 9$. The required perfect square trinomial is

$$x^2 + 6x + 9, \quad \text{which factors as} \quad (x + 3)^2.$$

Continued on Next Page

Answers

1. **(a)** 36; $(x + 6)^2$
   (b) 49; $(x - 7)^2$
   (c) 1; $(x - 1)^2$

Therefore, if we add 9 to *each* side of $x^2 + 6x = -7$, the equation will have a perfect square trinomial on the left side, as needed.

This is a key step.

$$x^2 + 6x + \mathbf{9} = -7 + \mathbf{9} \quad \text{Add 9.}$$

$$(x+3)^2 = 2 \quad \text{Factor on the left. Add on the right.}$$

Now use the square root property to complete the solution.

$$x + 3 = \sqrt{2} \quad \text{or} \quad x + 3 = -\sqrt{2}$$

$$x = -3 + \sqrt{2} \quad \text{or} \quad x = -3 - \sqrt{2} \quad \text{Add } -3.$$

Check by first substituting $-3 + \sqrt{2}$ and then substituting $-3 - \sqrt{2}$ for x in the original equation. The solution set is $\{-3 + \sqrt{2}, -3 - \sqrt{2}\}$.

◀ **Work Problem 2 at the Side.**

The process of changing the form of the equation in **Example 2** from

$$x^2 + 6x + 7 = 0 \quad \text{to} \quad (x+3)^2 = 2$$

is called **completing the square.** Completing the square changes only the form of the equation. To see this, multiply out the left side of $(x + 3)^2 = 2$. Then write the equation in standard form to get $x^2 + 6x + 7 = 0$.

Look again at the original equation in **Example 2.**

$$x^2 + 6x + 7 = 0$$

If we take half the coefficient of x, which is 6 here, and square it, we get 9.

$$\frac{1}{2} \cdot 6 = 3 \quad \text{and} \quad 3^2 = \mathbf{9}$$

Coefficient of x (6) — Quantity added to each side (9)

To complete the square in **Example 2,** we added **9** to each side.

EXAMPLE 3 Completing the Square to Solve a Quadratic Equation

Complete the square to solve $x^2 - 8x = 5$.

To complete the square on $x^2 - 8x$, take half the coefficient of x and square it.

$$\frac{1}{2}(-8) = -4 \quad \text{and} \quad (-4)^2 = \mathbf{16}$$

Coefficient of x (−8)

Add the result, **16**, to each side of the equation.

$$x^2 - 8x = 5 \quad \text{Given equation}$$

$$x^2 - 8x + \mathbf{16} = 5 + \mathbf{16} \quad \text{Add 16.}$$

$$(x-4)^2 = 21 \quad \text{Factor. Add.}$$

$$x - 4 = \sqrt{21} \quad \text{or} \quad x - 4 = -\sqrt{21} \quad \text{Square root property}$$

$$x = 4 + \sqrt{21} \quad \text{or} \quad x = 4 - \sqrt{21} \quad \text{Add 4.}$$

A check indicates that the solution set is $\{4 + \sqrt{21}, 4 - \sqrt{21}\}$.

◀ **Work Problem 3 at the Side.**

2 Solve $x^2 - 4x - 1 = 0$.

3 Solve each equation by completing the square.

GS **(a)** $x^2 + 4x = 1$

Take half the coefficient of x and square it.

$\frac{1}{2} \cdot$ ____ = ____,

and ____^{2} = ____.

Add ____ to each side of the equation.

$x^2 + 4x +$ ____ $= 1 +$ ____

Factor and add.

____

Complete the solution.

(b) $z^2 + 6z - 3 = 0$

Answers

2. $\{2 + \sqrt{5}, 2 - \sqrt{5}\}$

3. (a) 4; 2; 2; 4; 4; 4; 4; $(x + 2)^2 = 5$; $\{-2 + \sqrt{5}, -2 - \sqrt{5}\}$

(b) $\{-3 + 2\sqrt{3}, -3 - 2\sqrt{3}\}$

OBJECTIVE 2 Solve quadratic equations by completing the square when the coefficient of the second-degree term is not 1. If a quadratic equation has the form

$$ax^2 + bx + c = 0, \quad \text{where } a \neq 1,$$

then to obtain 1 as the coefficient of x^2, we first divide each side by a.

Solving a Quadratic Equation by Completing the Square

Step 1 **Be sure the second-degree term has coefficient 1.** If the coefficient of the second-degree term is 1, proceed to Step 2. If it is not 1, but some other nonzero number a, divide each side of the equation by a.

Step 2 **Write in correct form.** Make sure that all terms with variables are on one side of the equality symbol and that all constant terms are on the other side.

Step 3 **Complete the square.** Take half the coefficient of the first-degree term, and square it. Add the square to each side of the equation. Factor the variable side, and simplify on the other side.

Step 4 **Solve** the equation by using the square root property.

EXAMPLE 4 Solving a Quadratic Equation by Completing the Square

Solve $4x^2 + 16x = 9$.

Step 1 ***Before completing the square, the coefficient of x^2 must be 1,*** not 4. We get 1 as the coefficient of x^2 here by dividing each side by 4.

The coefficient of x^2 must be 1.

$$x^2 + 4x = \frac{9}{4} \qquad \text{Divide by 4.}$$

Step 2 The equation is already in the correct form, with the variable terms on one side and the constant on the other.

Step 3 Complete the square. Take half the coefficient of x and square it.

$$\frac{1}{2}(4) = 2 \quad \text{and} \quad 2^2 = 4$$

$$x^2 + 4x + 4 = \frac{9}{4} + 4 \qquad \text{Add 4.}$$

$$(x + 2)^2 = \frac{25}{4} \qquad \text{Factor; } \tfrac{9}{4} + 4 = \tfrac{9}{4} + \tfrac{16}{4} = \tfrac{25}{4}.$$

Step 4 Solve the equation by using the square root property.

$$x + 2 = \sqrt{\frac{25}{4}} \quad \text{or} \quad x + 2 = -\sqrt{\frac{25}{4}} \qquad \text{Square root property}$$

$$x + 2 = \frac{5}{2} \quad \text{or} \quad x + 2 = -\frac{5}{2} \qquad \text{Take square roots.}$$

$$x = \frac{1}{2} \quad \text{or} \quad x = -\frac{9}{2} \qquad \text{Subtract } 2 = \tfrac{4}{2}.$$

Continued on Next Page

4 Solve each equation by completing the square.

(a) $9x^2 + 18x = -5$

(b) $4t^2 - 24t + 11 = 0$

5 Solve each equation by completing the square.

GS **(a)** $3x^2 + 5x - 2 = 0$

Step 1
Divide each side by ____.

____

Step 2
Add ____ to each side.

____

Step 3
Take half the coefficient of x and square it. Add ____ to each side to complete the square.

$x^2 + \frac{5}{3}x + \_\_\_\_ = \frac{2}{3} + \frac{25}{36}$

Factor and add.

$(\_\_\_\_\_\_)^2 = \_\_\_\_$

Continue with Step 4 to complete the solution.

(b) $2x^2 - 4x - 1 = 0$

Answers

4. (a) $\left\{-\frac{1}{3}, -\frac{5}{3}\right\}$ **(b)** $\left\{\frac{11}{2}, \frac{1}{2}\right\}$

5. (a) 3; $x^2 + \frac{5}{3}x - \frac{2}{3} = 0$; $\frac{2}{3}$; $x^2 + \frac{5}{3}x = \frac{2}{3}$; $\frac{25}{36}$; $\frac{25}{36}$; $x + \frac{5}{6}$; $\frac{49}{36}$; $\left\{-2, \frac{1}{3}\right\}$

(b) $\left\{\frac{2+\sqrt{6}}{2}, \frac{2-\sqrt{6}}{2}\right\}$

CHECK Substitute each solution in the original equation.

$4x^2 + 16x = 9$	$4x^2 + 16x = 9$
$4\left(\frac{1}{2}\right)^2 + 16\left(\frac{1}{2}\right) \stackrel{?}{=} 9$ Let $x = \frac{1}{2}$.	$4\left(-\frac{9}{2}\right)^2 + 16\left(-\frac{9}{2}\right) \stackrel{?}{=} 9$ Let $x = -\frac{9}{2}$.
$4\left(\frac{1}{4}\right) + 8 \stackrel{?}{=} 9$	$4\left(\frac{81}{4}\right) - 72 \stackrel{?}{=} 9$
$1 + 8 \stackrel{?}{=} 9$	$81 - 72 \stackrel{?}{=} 9$
$9 = 9$ ✓ True	$9 = 9$ ✓ True

The two solutions $\frac{1}{2}$ and $-\frac{9}{2}$ check, so the solution set is $\left\{-\frac{9}{2}, \frac{1}{2}\right\}$.

◀ Work Problem **4** at the Side.

EXAMPLE 5 Solving a Quadratic Equation by Completing the Square

Solve $2x^2 - 7x - 9 = 0$.

Step 1 Transform so that 1 is the coefficient of the x^2-term.

$$x^2 - \frac{7}{2}x - \frac{9}{2} = 0 \quad \text{Divide by 2.}$$

Step 2 Add $\frac{9}{2}$ to each side to get the variable terms on the left and the constant on the right.

$$x^2 - \frac{7}{2}x = \frac{9}{2} \quad \text{Add } \tfrac{9}{2}.$$

Step 3 To complete the square, take half the coefficient of x and square it.

$$\left[\frac{1}{2}\left(-\frac{7}{2}\right)\right]^2 = \left(-\frac{7}{4}\right)^2 = \frac{49}{16}$$

Add the result, $\frac{49}{16}$, to each side of the equation.

$$x^2 - \frac{7}{2}x + \frac{49}{16} = \frac{9}{2} + \frac{49}{16}$$

Be sure to add $\frac{49}{16}$ to *each* side.

$$\left(x - \frac{7}{4}\right)^2 = \frac{121}{16} \quad \text{Factor; } \tfrac{9}{2} + \tfrac{49}{16} = \tfrac{72}{16} + \tfrac{49}{16} = \tfrac{121}{16}.$$

Step 4 Solve by using the square root property.

$$x - \frac{7}{4} = \sqrt{\frac{121}{16}} \quad \text{or} \quad x - \frac{7}{4} = -\sqrt{\frac{121}{16}} \quad \text{Square root property}$$

$$x = \frac{7}{4} + \frac{11}{4} \quad \text{or} \quad x = \frac{7}{4} - \frac{11}{4} \quad \text{Add } \tfrac{7}{4};\ \sqrt{\tfrac{121}{16}} = \tfrac{11}{4}.$$

$$x = \frac{18}{4} \quad \text{or} \quad x = -\frac{4}{4} \quad \text{Add and subtract.}$$

$$x = \frac{9}{2} \quad \text{or} \quad x = -1 \quad \text{Lowest terms}$$

A check confirms that the solution set is $\left\{-1, \frac{9}{2}\right\}$.

◀ Work Problem **5** at the Side.

EXAMPLE 6 **Solving a Quadratic Equation by Completing the Square**

Solve $4p^2 + 8p + 5 = 0$.

$$4p^2 + 8p + 5 = 0$$

The coefficient of the second-degree term must be 1.

$$p^2 + 2p + \frac{5}{4} = 0 \qquad \text{Divide by 4.}$$

$$p^2 + 2p = -\frac{5}{4} \qquad \text{Subtract } \tfrac{5}{4}.$$

The coefficient of p is 2. Take half of 2, square the result, and add it to each side.

$$p^2 + 2p + 1 = -\frac{5}{4} + 1 \qquad \left[\tfrac{1}{2}(2)\right]^2 = 1^2 = 1\text{; Add 1.}$$

$$(p + 1)^2 = -\frac{1}{4} \qquad \text{Factor. Add.}$$

We cannot use the square root property to solve this equation, because the square root of $-\frac{1}{4}$ is not a real number. This equation has no real number solution.* The solution set is $\emptyset$.

Work Problem 6 at the Side. ▶

6 Solve $5x^2 + 3x + 1 = 0$ by completing the square.

OBJECTIVE 3 **Simplify the terms of an equation before solving.**

EXAMPLE 7 **Simplifying before Completing the Square**

Solve $(x + 3)(x - 1) = 2$.

$$(x + 3)(x - 1) = 2 \qquad \text{Given equation}$$

$$x^2 + 2x - 3 = 2 \qquad \text{Multiply using the FOIL method.}$$

$$x^2 + 2x = 5 \qquad \text{Add 3.}$$

$$x^2 + 2x + 1 = 5 + 1 \qquad \text{Add } \left[\tfrac{1}{2}(2)\right]^2 = 1^2 = 1.$$

$$(x + 1)^2 = 6 \qquad \text{Factor on the left. Add on the right.}$$

$$x + 1 = \sqrt{6} \quad \text{or} \quad x + 1 = -\sqrt{6} \qquad \text{Square root property}$$

$$x = -1 + \sqrt{6} \quad \text{or} \quad x = -1 - \sqrt{6} \qquad \text{Subtract 1.}$$

The solution set is $\{-1 + \sqrt{6}, -1 - \sqrt{6}\}$.

Work Problem 7 at the Side. ▶

7 Solve each equation.

(a) $r(r - 3) = -1$

(b) $(x + 2)(x + 1) = 5$

Note

The solutions $-1 + \sqrt{6}$ and $-1 - \sqrt{6}$ given in **Example 7** are *exact.* In applications, decimal solutions are more appropriate. Using the square root key of a calculator, $\sqrt{6} \approx 2.449$, which results in the following approximate solutions.

$$x \approx 1.449 \quad \text{and} \quad x \approx -3.449$$

*The equation in **Example 6** has no solution over the *real number system.* In the **complex number system,** however, this equation does have solutions. The complex numbers include numbers whose squares are negative. These numbers are discussed in intermediate and college algebra courses.

Answers

6. $\emptyset$

7. (a) $\left\{\frac{3 + \sqrt{5}}{2}, \frac{3 - \sqrt{5}}{2}\right\}$

(b) $\left\{\frac{-3 + \sqrt{21}}{2}, \frac{-3 - \sqrt{21}}{2}\right\}$

8 Solve each problem.

(a) Suppose a ball is projected upward from ground level with an initial velocity of 128 ft per sec. Its altitude (height) s at time t (in seconds) is given by

$$s = -16t^2 + 128t,$$

where s is in feet. At what times will the ball be 48 ft above the ground? Give answers to the nearest tenth.

(b) At what times will the ball in **Example 8** be 28 ft above the ground?

Answers

8. **(a)** 0.4 sec and 7.6 sec
(b) 0.5 sec and 3.5 sec

OBJECTIVE 4 Solve applied problems that require quadratic equations. The next example illustrates an application of quadratic equations from physics.

EXAMPLE 8 Solving a Velocity Problem

If a ball is projected into the air from ground level with an initial velocity of 64 ft per sec, its altitude (height) s in feet in t seconds is given by the formula

$$s = -16t^2 + 64t.$$

At what times will the ball be 48 ft above the ground?

Since s represents the height, we substitute **48** for s in the formula and then solve this equation for time t by completing the square.

$$48 = -16t^2 + 64t \qquad \text{Let } s = 48.$$

$$-3 = t^2 - 4t \qquad \text{Divide by } -16.$$

$$t^2 - 4t = -3 \qquad \text{Interchange sides.}$$

$$t^2 - 4t + 4 = -3 + 4 \qquad \text{Add } \left[\tfrac{1}{2}(-4)\right]^2 = (-2)^2 = 4.$$

$$(t-2)^2 = 1 \qquad \text{Factor. Add.}$$

$$t - 2 = 1 \quad \text{or} \quad t - 2 = -1 \qquad \text{Square root property}$$

$$t = 3 \quad \text{or} \quad t = 1 \qquad \text{Add 2.}$$

The ball reaches a height of 48 ft twice, once on the way up and again on the way down. It takes 1 sec to reach 48 ft on the way up, and then after 3 sec, the ball reaches 48 ft again on the way down.

◀ **Work Problem 8 at the Side.**

17.2 Exercises

FOR EXTRA HELP

Download the MyDashBoard App

MyMathLab®

1. CONCEPT CHECK Which step is an appropriate way to begin solving the quadratic equation

$$2x^2 - 4x = 9$$

by completing the square?

A. Add 4 to each side of the equation.

B. Factor the left side as $2x(x-2)$.

C. Factor the left side as $x(2x-4)$.

D. Divide each side by 2.

2. CONCEPT CHECK In **Example 3** of **Section 13.7,** we solved the quadratic equation

$$4p^2 - 26p + 40 = 0$$

by factoring. If we were to solve by completing the square, would we get the same solutions, $\frac{5}{2}$ and 4?

Complete each trinomial so that it is a perfect square. Then factor the trinomial. ***See Example 1.***

3. $x^2 + 10x +$ ____

4. $x^2 + 16x +$ ____

5. $x^2 + 2x +$ ____

6. $m^2 - 2m +$ ____

7. $p^2 - 5p +$ ____

8. $x^2 + 3x +$ ____

Solve each equation by completing the square. ***See Examples 2 and 3.***

9. $x^2 - 4x = -3$

10. $x^2 - 2x = 8$

11. $x^2 + 5x + 6 = 0$

12. $x^2 + 6x + 5 = 0$

13. $x^2 + 2x - 5 = 0$

14. $x^2 + 4x + 1 = 0$

15. $x^2 - 8x = -4$

16. $m^2 - 4m = 14$

17. $t^2 + 6t + 9 = 0$

18. $k^2 - 8k + 16 = 0$

19. $x^2 + x - 1 = 0$

20. $x^2 + x - 3 = 0$

Solve each equation by completing the square. ***See Examples 4–7.***

21. $4x^2 + 4x - 3 = 0$

22. $9x^2 + 3x - 2 = 0$

23. $2x^2 - 4x = 5$

24. $2x^2 - 6x = 3$

25. $2p^2 - 2p + 3 = 0$

26. $3q^2 - 3q + 4 = 0$

27. $3k^2 + 7k = 4$

28. $2k^2 + 5k = 1$

29. $(x + 3)(x - 1) = 5$

30. $(y - 8)(y + 2) = 24$

31. $(r - 3)(r - 5) = 2$

32. $(k - 1)(k - 7) = 1$

33. $-x^2 + 2x = -5$

34. $-r^2 + 3r = -2$

Solve each problem. ***See Example 8.***

35. If an object is projected upward from ground level on Earth with an initial velocity of 96 ft per sec, its altitude (height) s in feet in t seconds is given by the formula $s = -16t^2 + 96t$. At what times will the object be 80 ft above the ground?

36. At what times will the object described in **Exercise 35** be 100 ft above the ground? Round your answers to the nearest tenth.

37. If an object is projected upward on the surface of Mars from ground level with an initial velocity of 104 ft per sec, its altitude (height) s in feet in t seconds is given by the formula $s = -13t^2 + 104t$. At what times will the object be 195 ft above the surface?

38. After how many seconds will the object in **Exercise 37** return to the surface? (*Hint:* When it returns to the surface, $s = 0$.)

39. A farmer has a rectangular cattle pen with perimeter 350 ft and area 7500 ft^2. What are the dimensions of the pen? (*Hint:* Use the figure to set up the equation.)

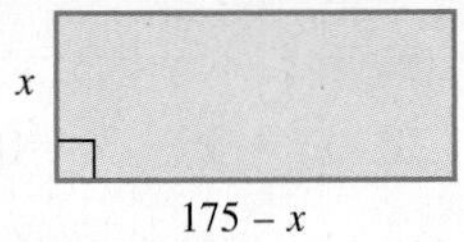

40. The base of a triangle measures 1 m more than three times the height of the triangle. Its area is 15 m^2. Find the lengths of the base and the height.

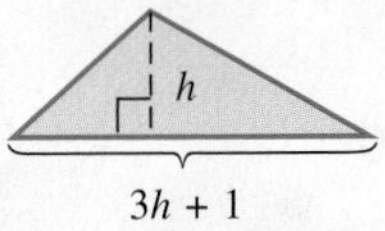

17.3 Solving Quadratic Equations by the Quadratic Formula

OBJECTIVES

1. Identify the values of a, b, and c in a quadratic equation.
2. Use the quadratic formula to solve quadratic equations.
3. Solve quadratic equations with only one solution.
4. Solve quadratic equations with fractions.

We can solve any quadratic equation by completing the square, but the method can be tedious. In this section we complete the square on the general quadratic equation

$$ax^2 + bx + c = 0, \quad \text{with} \quad a \neq 0, \qquad \text{Standard form}$$

to obtain the *quadratic formula,* which gives the solution(s) of any quadratic equation.

Note

In $ax^2 + bx + c = 0$, there is a restriction that a is not zero. If it were, the equation would be linear, not quadratic.

OBJECTIVE 1 Identify the values of *a*, *b*, and *c* in a quadratic equation. To solve a quadratic equation by this new method, we must first identify the values of a, b, and c in the standard form.

EXAMPLE 1 Identifying Values of *a*, *b*, and *c* in Quadratic Equations

Identify the values of the variables a, b, and c in each quadratic equation $ax^2 + bx + c = 0$.

(a) $2x^2 + 3x - 5 = 0$ (with a, b, c arrows above 2, 3, -5) — This equation is in standard form.

Here, $a = 2$, $b = 3$, and $c = -5$.

(b) $-x^2 + 2 = 6x$

Subtract $6x$ to write the equation in standard form $ax^2 + bx + c = 0$.

$-x^2$ means $-1x^2$.

$$-x^2 - 6x + 2 = 0 \qquad \text{Subtract } 6x.$$

Here, $a = -1$, $b = -6$, and $c = 2$.

(c) $5x^2 - 12 = 0$

The x-term is missing, so write the equation as follows.

$$5x^2 + 0x - 12 = 0$$

Then, $a = 5$, $b = 0$, and $c = -12$.

(d) $-4x^2 = -x$

In standard form, this equation is written $-4x^2 + x = 0$. The constant term c is missing, so $a = -4$, $b = 1$, and $c = 0$.

(e)

$$(2x - 7)(x + 4) = -23$$

The equation is not in standard form.

$$2x^2 + x - 28 = -23 \qquad \text{Multiply using the FOIL method.}$$

$$2x^2 + x - 5 = 0 \qquad \text{Add 23. The equation is now in standard form.}$$

Identify the required values. We see that $a = 2$, $b = 1$, and $c = -5$.

Work Problem 1 at the Side. ▶

1 Identify the values of the variables a, b, and c in each quadratic equation $ax^2 + bx + c = 0$.

(a) $5x^2 + 2x - 1 = 0$

(b) $3x^2 = x - 2$

(c) $9x^2 - 13 = 0$

(d) $-x^2 + x = 0$

(e) $(3x + 2)(x - 1) = 8$

Answers

1. (a) $a = 5, b = 2, c = -1$
(b) $a = 3, b = -1, c = 2$
(c) $a = 9, b = 0, c = -13$
(d) $a = -1, b = 1, c = 0$
(e) $a = 3, b = -1, c = -10$

OBJECTIVE 2 Use the quadratic formula to solve quadratic equations. To develop the quadratic formula, we complete the square on $ax^2 + bx + c = 0$ (where $a > 0$). For comparison, we also show the corresponding steps for solving $2x^2 + x - 5 = 0$ (from **Example 1(e)**).

Step 1 Transform so that the coefficient of x^2 is equal to 1.

$2x^2 + x - 5 = 0$	Standard form	$ax^2 + bx + c = 0$	Standard form
$x^2 + \frac{1}{2}x - \frac{5}{2} = 0$	Divide by 2.	$x^2 + \frac{b}{a}x + \frac{c}{a} = 0$	Divide by a.

Step 2 Write so that the variable terms with x are alone on the left side.

$x^2 + \frac{1}{2}x = \frac{5}{2}$	Add $\frac{5}{2}$.	$x^2 + \frac{b}{a}x = -\frac{c}{a}$	Subtract $\frac{c}{a}$.

Step 3 Add the square of half the coefficient of x to each side, factor the left side, and combine terms on the right.

$x^2 + \frac{1}{2}x + \frac{1}{16} = \frac{5}{2} + \frac{1}{16}$	Add $\frac{1}{16}$.	$x^2 + \frac{b}{a}x + \frac{b^2}{4a^2} = -\frac{c}{a} + \frac{b^2}{4a^2}$	Add $\frac{b^2}{4a^2}$.
$\left(x + \frac{1}{4}\right)^2 = \frac{41}{16}$	Factor. Add.	$\left(x + \frac{b}{2a}\right)^2 = \frac{b^2 - 4ac}{4a^2}$	Factor. Add.

Step 4 Use the square root property to complete the solution.

$x + \frac{1}{4} = \pm\sqrt{\frac{41}{16}}$	$x + \frac{b}{2a} = \pm\sqrt{\frac{b^2 - 4ac}{4a^2}}$
$x + \frac{1}{4} = \pm\frac{\sqrt{41}}{4}$	$x + \frac{b}{2a} = \pm\frac{\sqrt{b^2 - 4ac}}{2a}$
$x = -\frac{1}{4} \pm \frac{\sqrt{41}}{4}$	$x = -\frac{b}{2a} \pm \frac{\sqrt{b^2 - 4ac}}{2a}$
$x = \frac{-1 \pm \sqrt{41}}{4}$	$\boldsymbol{x = \frac{-b \pm \sqrt{b^2 - 4ac}}{2a}}$

The final result on the right is the **quadratic formula,** which is also valid for $a < 0$. ***It is a key result that should be memorized. It gives two values: one for the + sign and one for the − sign.***

Quadratic Formula

The quadratic equation $ax^2 + bx + c = 0$ (where $a \neq 0$) has solutions

$$x = \frac{-b + \sqrt{b^2 - 4ac}}{2a} \quad \text{and} \quad x = \frac{-b - \sqrt{b^2 - 4ac}}{2a},$$

or in compact form, $x = \dfrac{-b \pm \sqrt{b^2 - 4ac}}{2a}$.

CAUTION

In the quadratic formula, the fraction bar extends under $-b$ as well as the radical. ***Be sure to find the values of $-b \pm \sqrt{b^2 - 4ac}$ first, and then divide those results by the value of $2a$.***

EXAMPLE 2 **Solving a Quadratic Equation by the Quadratic Formula**

Solve $2x^2 - 7x - 9 = 0$.

$$x = \frac{-b \pm \sqrt{b^2 - 4ac}}{2a}$$ Quadratic formula

Be sure to write $-b$ in the numerator.

$$x = \frac{-(-7) \pm \sqrt{(-7)^2 - 4(2)(-9)}}{2(2)}$$ Substitute $a = 2$, $b = -7$, and $c = -9$.

$$x = \frac{7 \pm \sqrt{49 + 72}}{4}$$ Simplify.

$$x = \frac{7 \pm \sqrt{121}}{4}$$ Add.

This represents **two** solutions.

$$x = \frac{7 \pm 11}{4}$$ $\sqrt{121} = 11$

Find the two solutions by first using the plus symbol and then using the minus symbol.

$$x = \frac{7 + 11}{4} = \frac{18}{4} = \frac{9}{2} \quad \text{or} \quad x = \frac{7 - 11}{4} = \frac{-4}{4} = -1$$

Check each solution in the original equation. The solution set is $\left\{-1, \frac{9}{2}\right\}$.

Work Problem 2 at the Side. ▶

EXAMPLE 3 **Rewriting a Quadratic Equation before Solving**

Solve $x^2 = 2x + 1$.

Write the equation in standard form as $x^2 - 2x - 1 = 0$.

$$x = \frac{-b \pm \sqrt{b^2 - 4ac}}{2a}$$ Quadratic formula

Be careful substituting the negative values.

$$x = \frac{-(-2) \pm \sqrt{(-2)^2 - 4(1)(-1)}}{2(1)}$$ Substitute $a = 1$, $b = -2$, and $c = -1$.

$$x = \frac{2 \pm \sqrt{4 + 4}}{2}$$ Simplify.

$$x = \frac{2 \pm \sqrt{8}}{2}$$ Add.

$$x = \frac{2 \pm 2\sqrt{2}}{2}$$ $\sqrt{8} = \sqrt{4} \cdot \sqrt{2} = 2\sqrt{2}$

Factor first. Then divide out the common factor.

$$x = \frac{2(1 \pm \sqrt{2})}{2}$$ Factor to write in lowest terms.

$$x = 1 \pm \sqrt{2}$$ Divide out the common factor.

The solution set is $\left\{1 + \sqrt{2}, 1 - \sqrt{2}\right\}$.

Work Problem 3 at the Side. ▶

2 Solve each equation by using the quadratic formula.

(a) $2x^2 + 3x - 5 = 0$

(b) $6x^2 + x - 1 = 0$

3 Solve $x^2 + 1 = -8x$.

GS Write the equation in standard form.

____

Substitute $a =$ ____, $b =$ ____, and $c =$ ____ in the quadratic formula and complete the solution.

Answers

2. **(a)** $\left\{1, -\frac{5}{2}\right\}$ **(b)** $\left\{-\frac{1}{2}, \frac{1}{3}\right\}$

3. $x^2 + 8x + 1 = 0$; 1; 8; 1; $\left\{-4 + \sqrt{15}, -4 - \sqrt{15}\right\}$

4 Solve $9x^2 - 12x + 4 = 0$.

5 Solve.

(a) $x^2 - \frac{4}{3}x + \frac{2}{3} = 0$

GS **(b)** $x^2 - \frac{9}{5}x = \frac{2}{5}$

Multiply by the LCD, ____, to clear fractions.

Simplify and write the resulting equation in standard form.

____

Substitute $a =$ ____, $b =$ ____, and $c =$ ____ in the quadratic formula. Complete the solution.

Answers

4. $\left\{\frac{2}{3}\right\}$

5. **(a)** $\emptyset$

(b) 5; $5x^2 - 9x - 2 = 0$; 5; −9; −2; $\left\{-\frac{1}{5}, 2\right\}$

OBJECTIVE 3 **Solve quadratic equations with only one solution.**

EXAMPLE 4 **Solving a Quadratic Equation with One Solution**

Solve $4x^2 + 25 = 20x$.

Write the equation in standard form.

$$4x^2 - 20x + 25 = 0 \quad \text{Subtract } 20x.$$

Here, $a = 4$, $b = -20$, and $c = 25$. Substitute in the quadratic formula.

$$x = \frac{-(-20) \pm \sqrt{(-20)^2 - 4(4)(25)}}{2(4)} = \frac{20 \pm 0}{8} = \frac{5}{2}$$

In this case, $b^2 - 4ac = 0$, and the trinomial $4x^2 - 20x + 25$ is a perfect square. There is one distinct solution in the solution set $\left\{\frac{5}{2}\right\}$.

◀ **Work Problem 4 at the Side.**

Note

The single solution of the equation in **Example 4** is a rational number. If all solutions of a quadratic equation are rational, the equation can be solved by factoring as well.

OBJECTIVE 4 **Solve quadratic equations with fractions.**

EXAMPLE 5 **Solving a Quadratic Equation with Fractions**

Solve $\frac{1}{10}t^2 = \frac{2}{5}t - \frac{1}{2}$.

$$\frac{1}{10}t^2 = \frac{2}{5}t - \frac{1}{2}$$

$$10\left(\frac{1}{10}t^2\right) = 10\left(\frac{2}{5}t - \frac{1}{2}\right) \quad \text{Clear fractions. Multiply by the LCD, 10.}$$

$$10\left(\frac{1}{10}t^2\right) = 10\left(\frac{2}{5}t\right) - 10\left(\frac{1}{2}\right) \quad \text{Distributive property}$$

$$t^2 = 4t - 5 \quad \text{Multiply.}$$

$$t^2 - 4t + 5 = 0 \quad \text{Subtract } 4t. \text{ Add 5.}$$

Identify $a = 1$, $b = -4$, and $c = 5$.

$$t = \frac{-(-4) \pm \sqrt{(-4)^2 - 4(1)(5)}}{2(1)} \quad \text{Substitute in the quadratic formula.}$$

$$t = \frac{4 \pm \sqrt{16 - 20}}{2} \quad \text{Simplify.}$$

$$t = \frac{4 \pm \sqrt{-4}}{2} \quad \text{Subtract.}$$

Stop here.

Because $\sqrt{-4}$ does not represent a real number, the solution set is $\emptyset$.

◀ **Work Problem 5 at the Side.**

17.3 Exercises

FOR EXTRA HELP

Download the MyDashBoard App

MyMathLab®

1. CONCEPT CHECK A student writes the quadratic formula as

$$x = -b \pm \frac{\sqrt{b^2 - 4ac}}{2a}.$$

What Went Wrong? Explain the error, and give the correct formula.

2. CONCEPT CHECK If we apply the quadratic formula and find that the value of $b^2 - 4ac$ is negative, what can we conclude?

Write each equation in standard form $ax^2 + bx + c = 0$, if necessary. Then identify the values of a, b, and c. Do not actually solve the equation. ***See Example 1.***

3. $4x^2 + 5x - 9 = 0$

$a =$ ____, $b =$ ____, $c =$ ____

4. $8x^2 + 3x - 4 = 0$

$a =$ ____, $b =$ ____, $c =$ ____

5. $3x^2 = 4x + 2$

$a =$ ____, $b =$ ____, $c =$ ____

6. $5x^2 = 3x - 6$

$a =$ ____, $b =$ ____, $c =$ ____

7. $3x^2 = -7x$

$a =$ ____, $b =$ ____, $c =$ ____

8. $9x^2 = 8x$

$a =$ ____, $b =$ ____, $c =$ ____

Use the quadratic formula to solve each equation. Write all radicals in simplified form, and write all answers in lowest terms. ***See Examples 2–4.***

9. $k^2 + 12k - 13 = 0$

10. $r^2 - 8r - 9 = 0$

11. $p^2 - 4p + 4 = 0$

12. $9x^2 + 6x + 1 = 0$

13. $2x^2 = 5 + 3x$

14. $2z^2 = 30 + 7z$

15. $2x^2 + 12x = -5$

16. $5m^2 + m = 1$

17. $6x^2 + 6x = 0$

18. $4n^2 - 12n = 0$

19. $-2x^2 = -3x + 2$

20. $-x^2 = -5x + 20$

21. $3x^2 + 5x + 1 = 0$

22. $6x^2 - 6x + 1 = 0$

23. $7x^2 = 12x$

24. $9r^2 = 11r$

25. $x^2 - 24 = 0$

26. $z^2 - 96 = 0$

27. $25x^2 - 4 = 0$

28. $16x^2 - 9 = 0$

29. $3x^2 - 2x + 5 = 10x + 1$

30. $4x^2 - x + 4 = x + 7$

31. $2x^2 + x + 5 = 0$

32. $3x^2 + 2x + 8 = 0$

Use the quadratic formula to solve each equation. ***See Example 5.***

33. $\frac{3}{2}k^2 - k - \frac{4}{3} = 0$

34. $\frac{2}{5}x^2 - \frac{3}{5}x - 1 = 0$

35. $\frac{1}{2}x^2 + \frac{1}{6}x = 1$

36. $\frac{2}{3}t^2 - \frac{4}{9}t = \frac{1}{3}$

37. $\frac{3}{8}x^2 - x + \frac{17}{24} = 0$

38. $\frac{1}{3}x^2 + \frac{8}{9}x + \frac{7}{9} = 0$

39. $0.6x - 0.4x^2 = -1$

40. $0.5x^2 = x + 0.5$

41. $0.25x^2 = -1.5x - 1$

Solve each problem.

42. An astronaut on the moon throws a baseball upward. The altitude (height) h of the ball, in feet, x seconds after he throws it, is given by the equation

$$h = -2.7x^2 + 30x + 6.5.$$

At what times is the ball 12 ft above the moon's surface? Give answer(s) to the nearest tenth.

43. A frog is sitting on a stump 3 ft above the ground. He hops off the stump and lands on the ground 4 ft away. During his leap, his height h is given by the equation

$$h = -0.5x^2 + 1.25x + 3,$$

where x is the distance in feet from the base of the stump, and h is in feet. How far was the frog from the base of the stump when he was 1.25 ft above the ground?

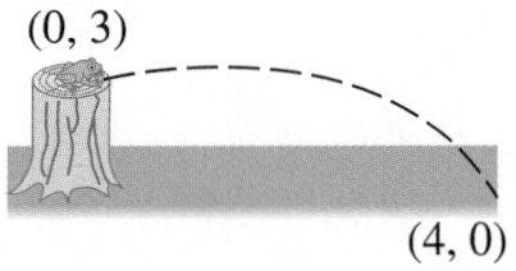

44. An old Babylonian problem asks for the length of the side of a square, given that the area of the square minus the length of a side is 870. Find the length of the side. (*Source:* Eves, H., *An Introduction to the History of Mathematics*, Saunders College Publishing.)

45. A rule for estimating the number of board feet of lumber that can be cut from a log depends on the diameter of the log. To find the diameter d required to get 9 board feet of lumber, we use the equation

$$\left(\frac{d-4}{4}\right)^2 = 9.$$

Solve this equation for d. Are both answers reasonable?

17.4 Graphing Quadratic Equations

OBJECTIVES

1. Graph quadratic equations.
2. Find the vertex of a parabola.

OBJECTIVE 1 **Graph quadratic equations.** In this section, we graph quadratic equations in two variables, of the form

$$y = ax^2 + bx + c.$$

The simplest quadratic equation, $y = x^2$ (or $y = 1x^2 + 0x + 0$), can be graphed in much the same way that straight lines were graphed in **Chapter 11,** by finding ordered pairs that satisfy the equation.

EXAMPLE 1 Graphing a Quadratic Equation

Graph $y = x^2$.

Select several values for x. Then find the corresponding y-values. For example, selecting $x = 2$ and substituting in $y = x^2$ gives

$$y = 2^2 = 4,$$

and so the point $(2, 4)$ is on the graph of $y = x^2$. ***(Recall that in an ordered pair such as (2, 4), the x-value comes first and the y-value second.)***

Work Problem 1 at the Side. ▶

If we plot the points from **Margin Problem 1** on a coordinate system and draw a smooth curve through them, we obtain the graph in **Figure 1.** A table of values is shown with the graph.

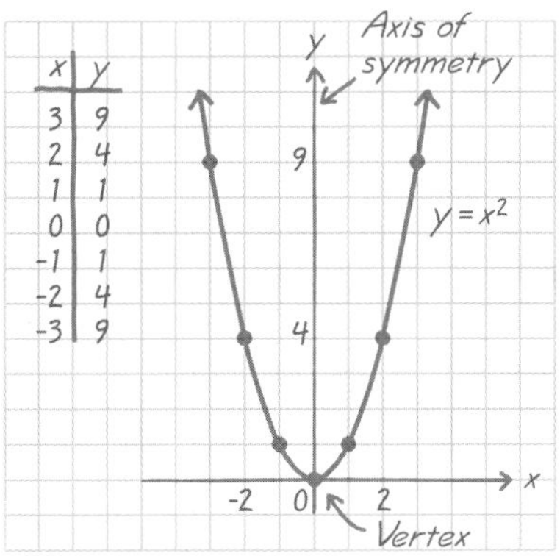

Figure 1

Work Problem 2 at the Side. ▶

The curve in **Figure 1** is a **parabola.** Every equation of the form

$$y = ax^2 + bx + c, \quad \text{with } a \neq 0,$$

has a graph that is a parabola. The point $(0, 0)$, the *lowest* point on this graph, is the **vertex** of the parabola. (If a parabola opens downward, the vertex will be the *highest* point. See **Figure 2** on the next page.)

The vertical line through the vertex of a parabola that opens upward or downward is the **axis of symmetry.** The two halves of the parabola are mirror images of each other across this line. For the parabola shown in **Figure 1,** the axis of symmetry is the y-axis.

1 Complete the table of values for $y = x^2$.

x	y
3	
2	4
1	
0	
−1	
−2	
−3	

2 Graph $y = \frac{1}{2}x^2$ by first completing the table of values.

x	y
−2	
−1	
0	
1	
2	

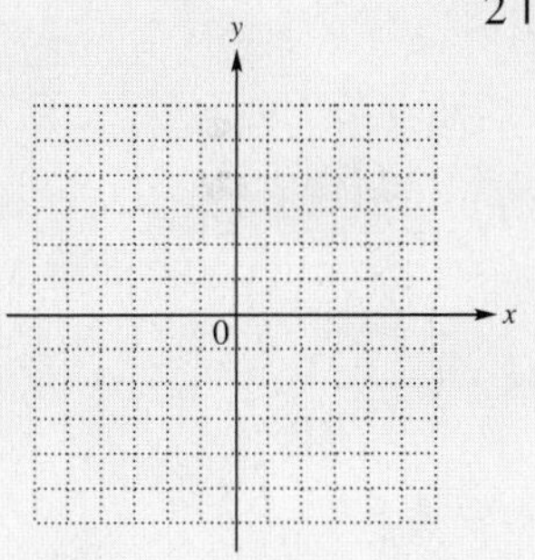

Answers

1. See the table beside the graph in **Figure 1.**
2.

x	y
−2	2
−1	$\frac{1}{2}$
0	0
1	$\frac{1}{2}$
2	2

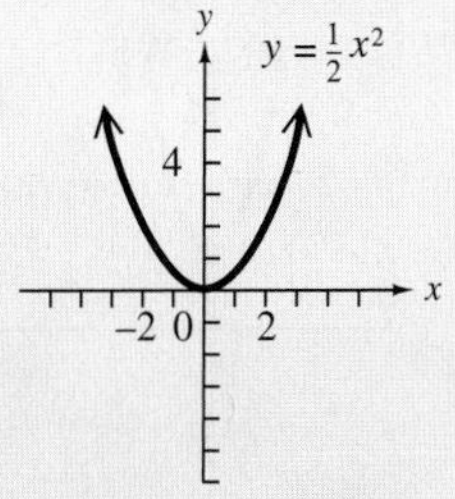

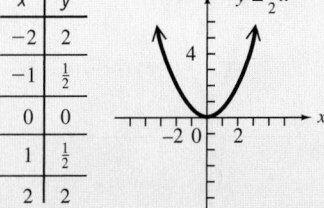

3 Complete each ordered pair for $y = -x^2 + 3$.

$(-2, \_\_)$, $(-1, \_\_)$,
$(1, \_\_)$, $(2, \_\_)$

4 Graph each equation, and identify each vertex.

(a) $y = -x^2 - 3$

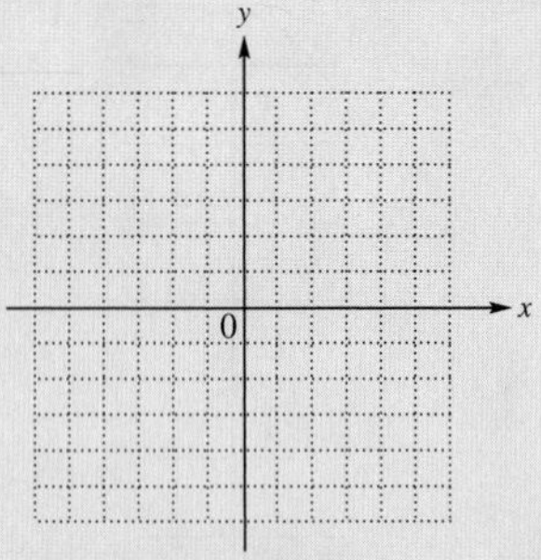

(b) $y = x^2 + 3$

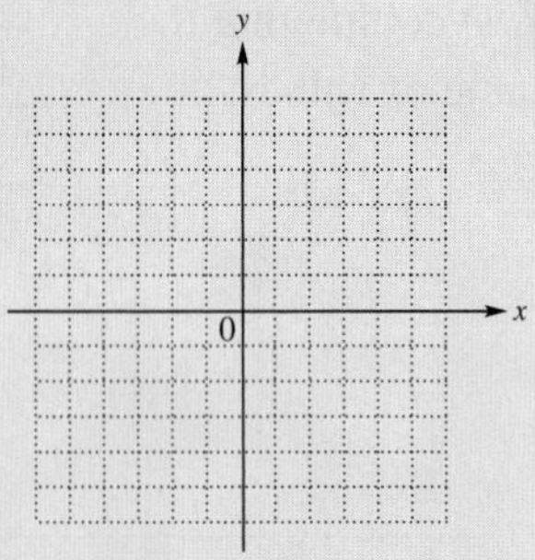

Answers

3. $(-2, -1), (-1, 2), (1, 2), (2, -1)$

4. (a)

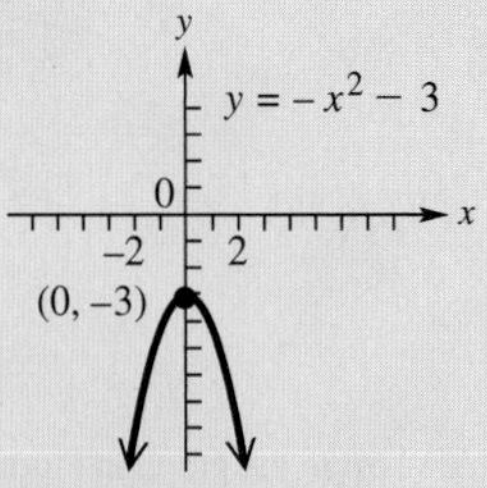

(b)

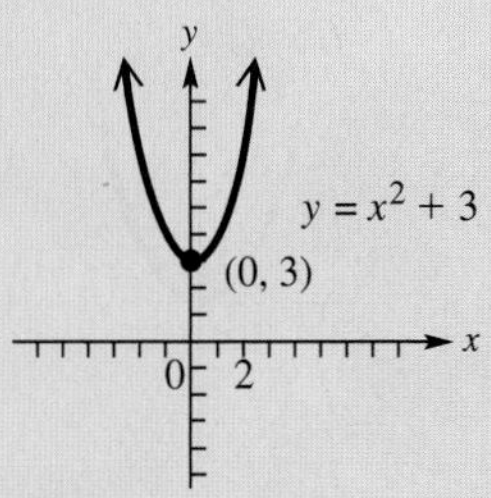

EXAMPLE 2 Graphing a Parabola by Plotting Points

Graph $y = -x^2 + 3$.

Find several ordered pairs. Let $x = 0$ to find the y-intercept.

$$y = -x^2 + 3 = -0^2 + 3 = 3$$

This gives the ordered pair $(0, 3)$.

◀ **Work Problem 3 at the Side.**

Plot the ordered pair $(0, 3)$ and the ordered pairs from **Margin Problem 3** at the side and join them with a smooth curve as shown in **Figure 2.** The vertex of this parabola is $(0, 3)$. The graph opens downward because x^2 has a negative coefficient, so the vertex is the *highest* point of the graph.

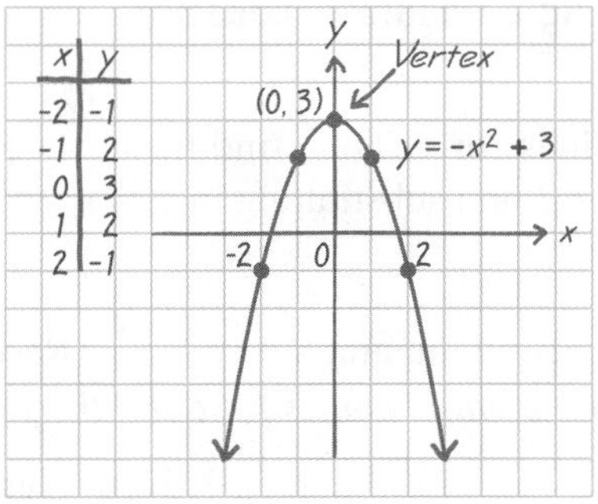

Figure 2

◀ **Work Problem 4 at the Side.**

OBJECTIVE 2 Find the vertex of a parabola. The vertex is the most important point to locate when graphing a quadratic equation.

EXAMPLE 3 Graphing a Parabola

Graph $y = x^2 - 2x - 3$.

Because of its symmetry, if a parabola has two x-intercepts, the x-value of the vertex is exactly halfway between them. Therefore, we begin by finding the x-intercepts. We let $y = 0$ in the equation, and solve for x.

$$0 = x^2 - 2x - 3 \quad \text{Let } y = 0.$$

$$0 = (x + 1)(x - 3) \quad \text{Factor.}$$

$$x + 1 = 0 \quad \text{or} \quad x - 3 = 0 \quad \text{Zero-factor property}$$

$$x = -1 \quad \text{or} \quad x = 3 \quad \text{Solve each equation.}$$

There are two x-intercepts, $(-1, 0)$ and $(3, 0)$. Since the x-value of the vertex is halfway between the x-values of the two x-intercepts, it is half their sum.

$$x = \frac{1}{2}(-1 + 3)$$

$$= 1 \leftarrow x\text{-value of the vertex}$$

We find the corresponding y-value by substituting 1 for x in the equation.

$$y = 1^2 - 2(1) - 3 \quad \text{Let } x = 1.$$

$$= -4 \leftarrow y\text{-value of the vertex}$$

The vertex is $(1, -4)$. The axis of symmetry is the vertical line $x = 1$.

Continued on Next Page

To find the y-intercept, we substitute 0 for x in the equation.

$$y = 0^2 - 2(0) - 3 \quad \text{Let } x = 0.$$
$$= -3 \quad \text{Simplify.}$$

The y-intercept is $(0, -3)$.

We plot the three intercepts and the vertex, and find additional ordered pairs as needed. For example, we let $x = 2$.

$$y = 2^2 - 2(2) - 3 \quad \text{Let } x = 2.$$
$$= -3 \quad \text{Simplify.}$$

This leads to the ordered pair $(2, -3)$. A table that includes the ordered pairs we found is shown with the graph in **Figure 3.**

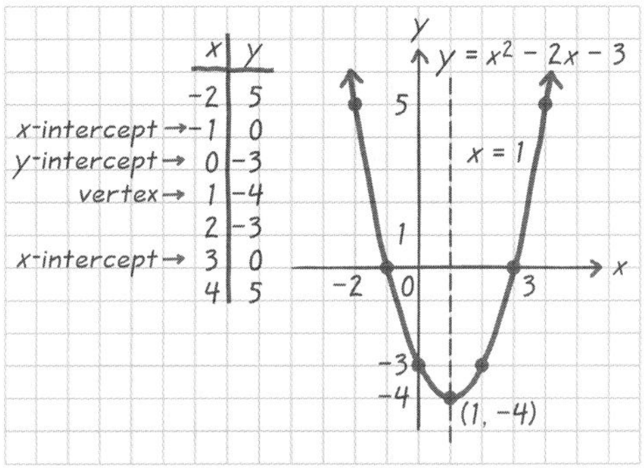

Figure 3

Work Problem 5 at the Side. ▶

We can generalize from **Example 3.** The x-coordinates of the x-intercepts for $y = ax^2 + bx + c$, by the quadratic formula, are

$$x = \frac{-b + \sqrt{b^2 - 4ac}}{2a} \quad \text{and} \quad x = \frac{-b - \sqrt{b^2 - 4ac}}{2a}.$$

Thus, the x-value of the vertex is halfway between the x-values of the x-intercepts.

$$x = \frac{1}{2}\left(\frac{-b + \sqrt{b^2 - 4ac}}{2a} + \frac{-b - \sqrt{b^2 - 4ac}}{2a}\right)$$

$$x = \frac{1}{2}\left(\frac{-b + \sqrt{b^2 - 4ac} - b - \sqrt{b^2 - 4ac}}{2a}\right)$$

$$x = \frac{1}{2}\left(\frac{-2b}{2a}\right) \quad \text{Combine like terms.}$$

$$\boldsymbol{x = -\frac{b}{2a}} \quad \text{Multiply. Write in lowest terms.}$$

For the equation in **Example 3,** $y = x^2 - 2x - 3$, we have $a = 1$ and $b = -2$. Thus, the x-value of the vertex is found as follows.

$$x = -\frac{b}{2a} = -\frac{-2}{2(1)} = 1 \leftarrow \text{This is the same } x\text{-value for the vertex found in \textbf{Example 3.}}$$

Note

The x-value of the vertex is $x = -\frac{b}{2a}$, even if the graph has no x-intercepts.

5 Graph $y = x^2 + 2x - 8$.

GS To find the x-intercepts, solve the equation.

$0 = $ ______________

$0 = (x + 4)(x - $ ____$)$

$x = $ ____ or $x = 2$

The x-intercepts are ____ and $(2, 0)$.

The x-value of the vertex is

$\frac{1}{2}(-4 + $ ____$) = $ ____.

The y-value of the vertex is ____.

The vertex is ____.

The y-intercept is ____.

The axis of symmetry is the line $x = $ ____.

Find additional ordered pairs as needed and sketch the graph.

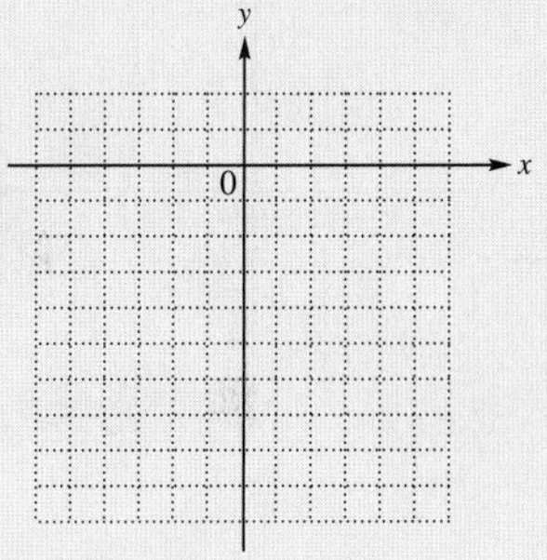

Answers

5. $x^2 + 2x - 8$; 2; -4; $(-4, 0)$; 2; -1; -9; $(-1, -9)$; $(0, -8)$; -1

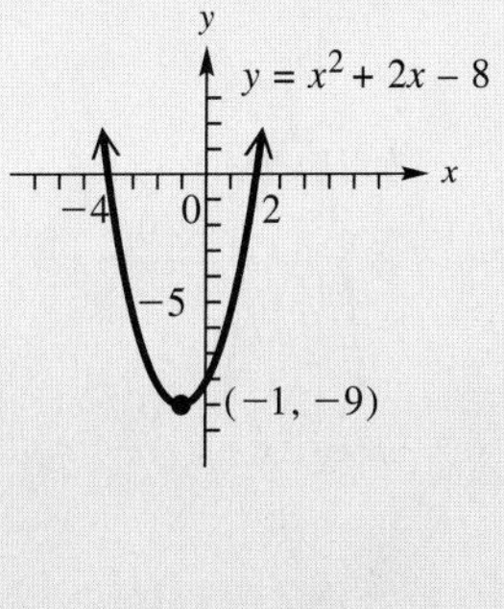

6 Complete each ordered pair for $y = x^2 - 4x + 1$.

(5, ___), (4, ___), (−1, ___)

7 Graph each parabola.

(a) $y = x^2 - 3x - 3$

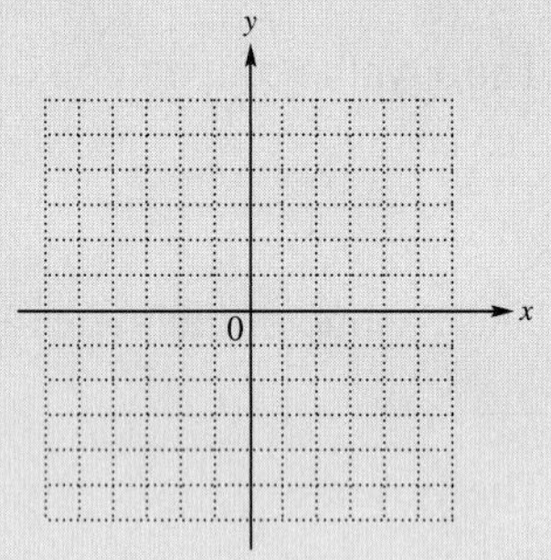

(b) $y = -x^2 + 2x + 4$

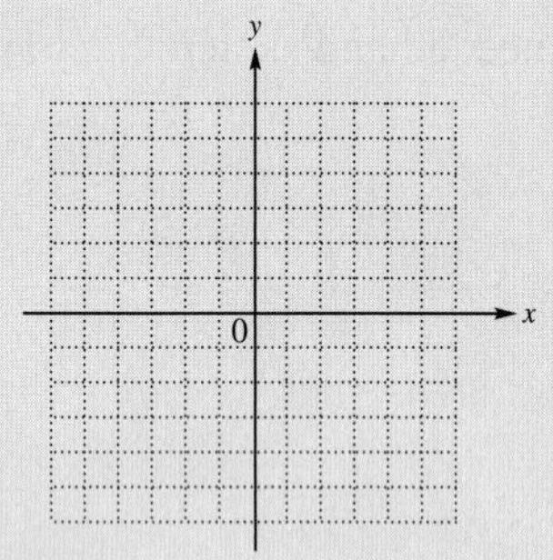

Answers

6. (5, 6), (4, 1), (−1, 6)

7. (a)

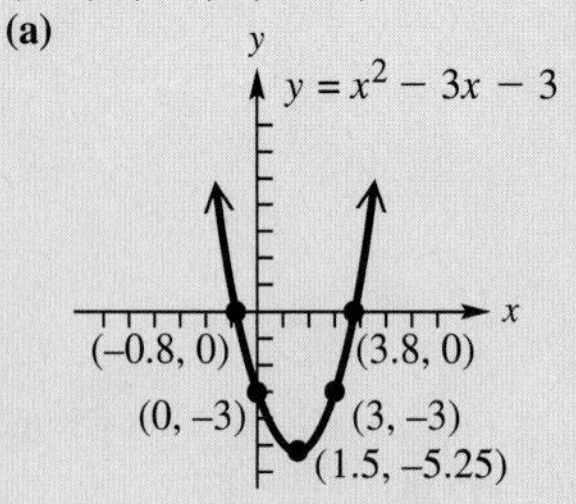

(b)

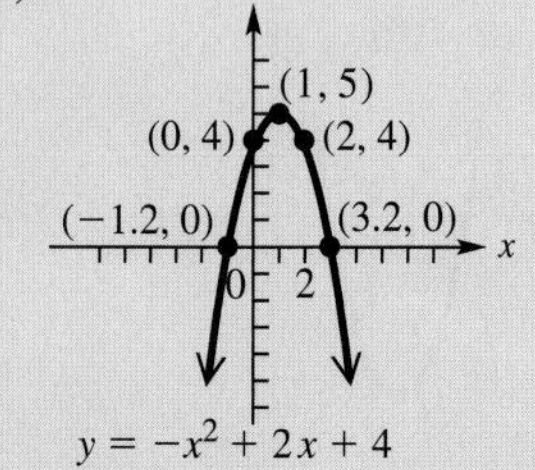

EXAMPLE 4 Graphing a Parabola

Graph $y = x^2 - 4x + 1$.

Here, $a = 1$ and $b = -4$, so we substitute to find the x-value of the vertex.

$$x = -\frac{b}{2a} = -\frac{-4}{2(1)} = 2$$

The y-value of the vertex is found as follows.

$$y = 2^2 - 4(2) + 1 \quad \text{Let } x = 2.$$

$$y = -3 \quad \text{Simplify.}$$

The vertex is $(2, -3)$. The axis is the line $x = 2$.

Now we find the intercepts. Let $x = 0$ in $y = x^2 - 4x + 1$.

$$y = 0^2 - 4(0) + 1 \quad \text{Let } x = 0.$$

$$y = 1$$

The y-intercept is $(0, 1)$. We let $y = 0$ to get the x-intercepts. If $y = 0$, the equation becomes $0 = x^2 - 4x + 1$, which cannot be solved by factoring. We use the quadratic formula to solve for x.

$$x = \frac{-(-4) \pm \sqrt{(-4)^2 - 4(1)(1)}}{2(1)} \quad \text{Let } a = 1, b = -4, c = 1 \text{ in the quadratic formula.}$$

$$x = \frac{4 \pm \sqrt{12}}{2} \quad \text{Simplify.}$$

$$x = \frac{4 \pm 2\sqrt{3}}{2} \quad \sqrt{12} = \sqrt{4} \cdot \sqrt{3} = 2\sqrt{3}$$

Factor first. Then divide out the common factor.

$$x = \frac{2(2 \pm \sqrt{3})}{2} \quad \text{Factor.}$$

$$x = 2 \pm \sqrt{3} \quad \text{Divide out 2.}$$

Using a calculator, we find that the x-intercepts are $(3.7, 0)$ and $(0.3, 0)$, to the nearest tenth.

◀ **Work Problem 6 at the Side.**

We plot the intercepts, vertex, and the points found in **Margin Problem 6,** and join these points with a smooth curve. The graph is shown in **Figure 4.**

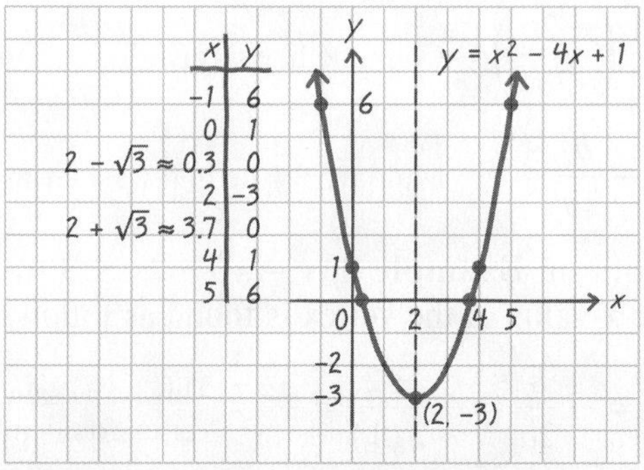

Figure 4

◀ **Work Problem 7 at the Side.**

17.4 Exercises

FOR EXTRA HELP

Download the MyDashBoard App

MyMathLab®

CONCEPT CHECK *Fill in each blank with the correct response.*

1. Every equation of the from $y = ax^2 + bx + c$, with $a \neq 0$, has a graph that is a(n) _______.

2. The _______ of a parabola is the lowest point of the graph if the parabola opens upward and the _______ point of the graph if the parabola opens downward.

3. The vertical line through the vertex of a parabola that opens upward or downward is the _____________ of the parabola. The two halves of the parabola are _______ images of each other across this line.

4. The graph of an equation of the form $y = ax^2 + bx + c$, with $a \neq 0$, has exactly _______ y-intercept(s), but may have either two, _______, or _______ x-intercepts.

Graph each equation. Give the coordinates of the vertex in each case. **See Examples 1–4.**

5. $y = 2x^2$

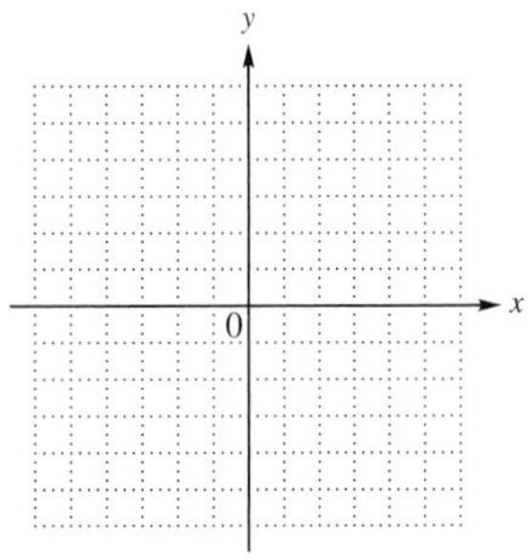

6. $y = \frac{1}{4}x^2$

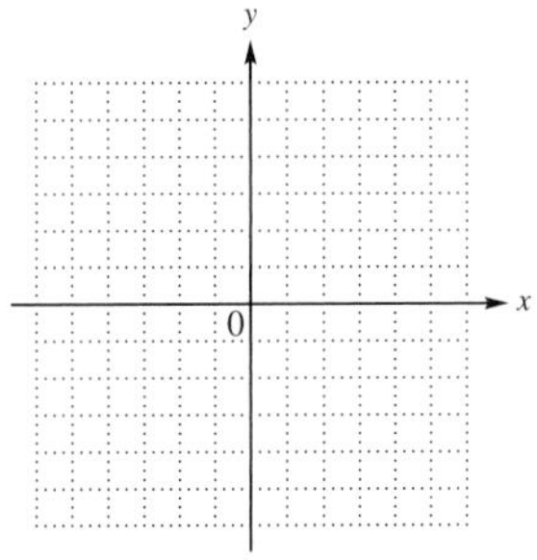

7. $y = x^2 - 4$

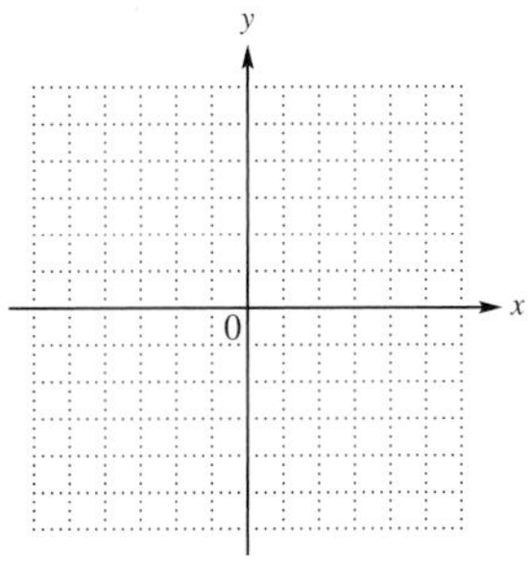

8. $y = x^2 - 2$

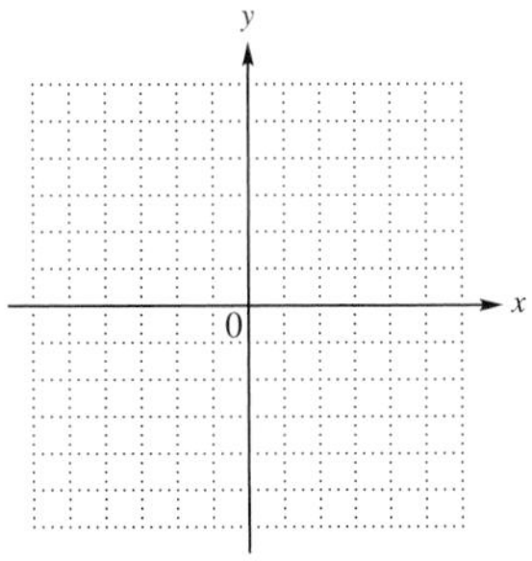

9. $y = -x^2 + 2$

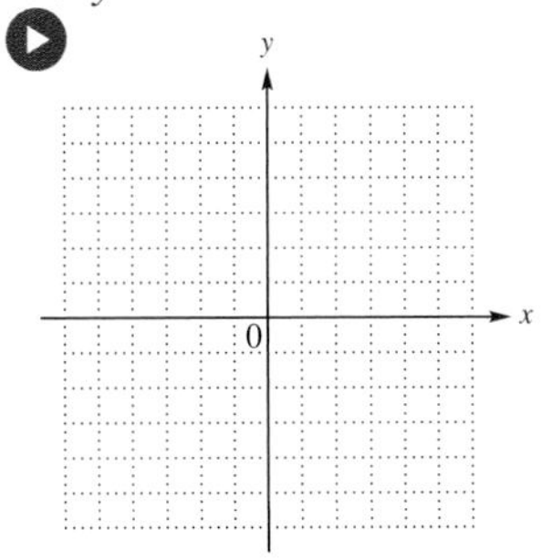

10. $y = -x^2 + 4$

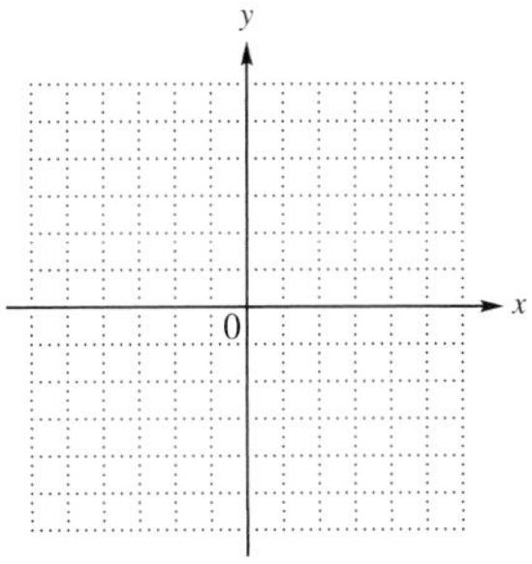

11. $y = (x + 1)^2$

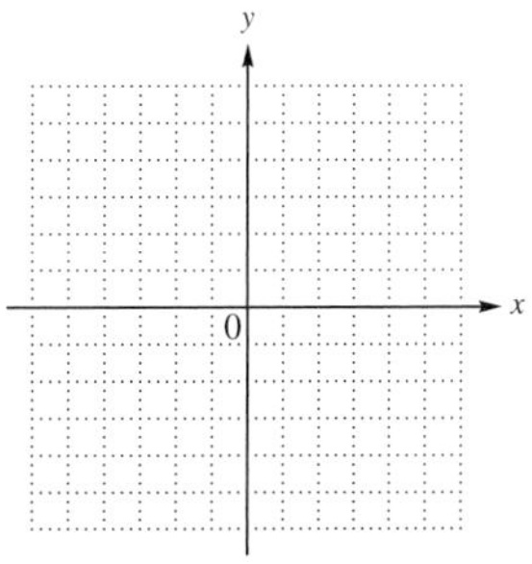

12. $y = (x - 2)^2$

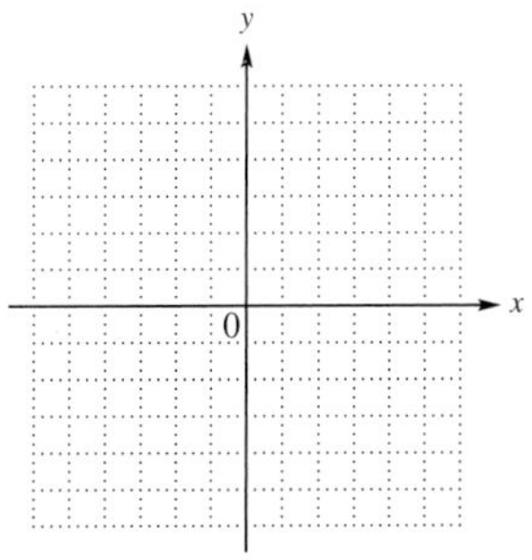

13. $y = x^2 + 2x + 3$

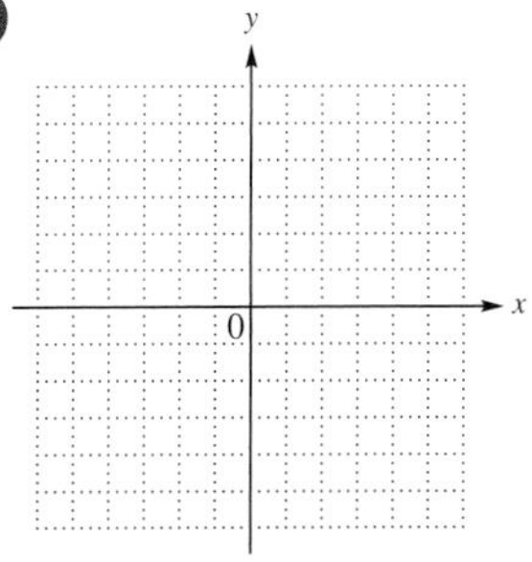

14. $y = x^2 - 4x + 3$

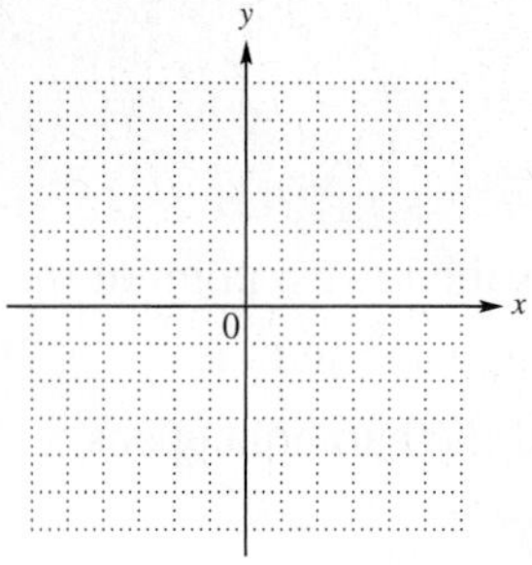

15. $y = -x^2 + 6x - 5$

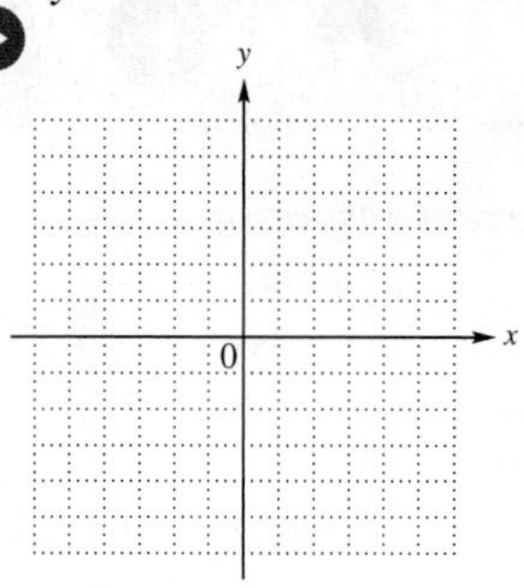

16. $y = -x^2 - 4x - 3$

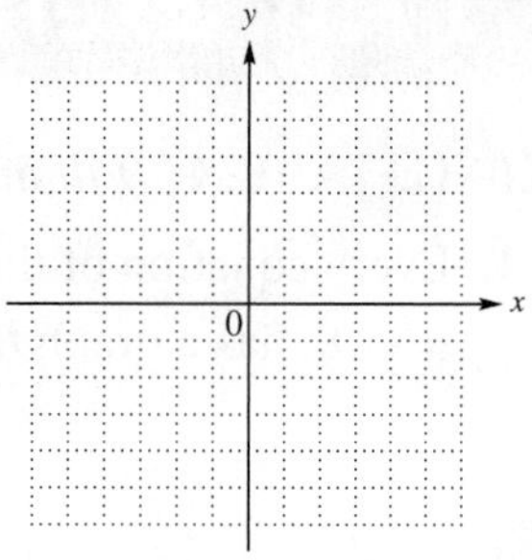

17. CONCEPT CHECK Based on your work in **Exercises 5–16**, what seems to be the direction in which the parabola

$$y = ax^2 + bx + c$$

opens if $a > 0$? If $a < 0$?

18. CONCEPT CHECK See **Exercises 12–14.** How many real solutions does a quadratic equation have if its corresponding graph has

(a) no x-intercepts

(b) one x-intercept

(c) two x-intercepts?

Solve each problem.

19. The U.S. Naval Research Laboratory designed a giant radio telescope that had a diameter of 300 ft and maximum depth of 44 ft. The graph on the right below describes a cross section of this telescope. Find the equation of this parabola. (*Source:* Mar, J. and Liebowitz, H., *Structure Technology for Large Radio and Radar Telescope Systems*, The MIT Press.)

20. Suppose the telescope in **Exercise 19** had a diameter of 400 ft and maximum depth of 50 ft. Find the equation of this parabola.

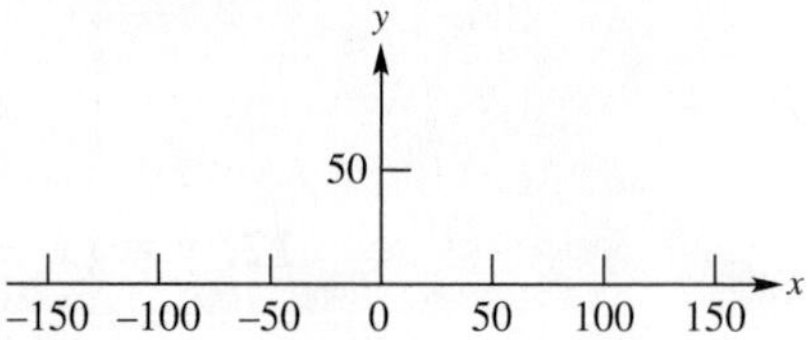

17.5 Introduction to Functions

OBJECTIVES

1. Understand the definition of a relation.
2. Understand the definition of a function.
3. Determine whether an equation represents a function.
4. Use function notation.
5. Apply the function concept in an application.

If gasoline costs \$4.00 per gal and we purchase 1 gal, then we must pay \$4.00(1) = \$4.00. For 2 gal, the cost is \$4.00(2) = \$8.00. For 3 gal, the cost is \$4.00(3) = \$12.00, and so on. Generalizing, if x represents the number of gallons, then the cost y is \$4.00x. The equation

$$y = 4.00x$$

relates the number of gallons, x, to the cost in dollars, y. The set of ordered pairs (x, y) that satisfy this equation forms a *relation.*

OBJECTIVE 1 Understand the definition of a relation. In an ordered pair (x, y), x and y are the **components** of the ordered pair. Any set of ordered pairs is a **relation.** The set of all first components in the ordered pairs of a relation is the **domain** of the relation, and the set of all second components in the ordered pairs is the **range** of the relation.

EXAMPLE 1 Identifying Domains and Ranges of Relations

Identify the domain and range of each relation.

(a) $\{(0, 1), (2, 5), (3, 8), (4, 2)\}$

This relation has

domain $\{0, 2, 3, 4\}$

and range $\{1, 5, 8, 2\}$.

The correspondence between the elements of the domain and the elements of the range is shown in **Figure 5.**

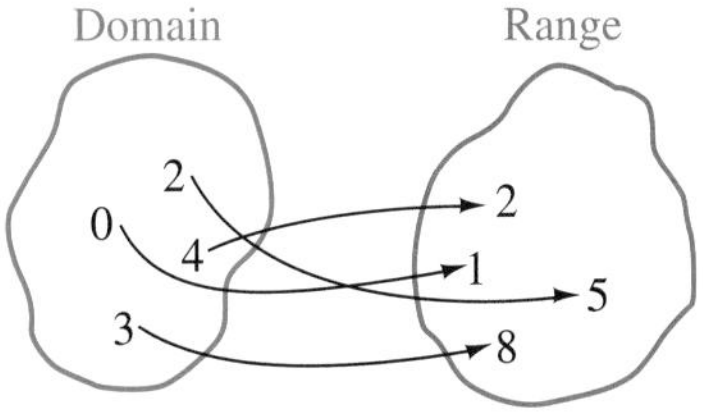

Figure 5

(b) $\{(3, 5), (3, 6), (3, 7), (3, 8)\}$

The relation has domain $\{3\}$ and range $\{5, 6, 7, 8\}$.

Work Problem 1 at the Side. ▶

1 Identify the domain and the range of each relation.

(a) $\{(5, 10), (15, 20), (25, 30), (35, 40)\}$

(b) $\{(1, 4), (2, 4), (3, 4)\}$

OBJECTIVE 2 Understand the definition of a function. We now investigate an important type of relation, called a *function.*

Function

A **function** is a set of ordered pairs in which each distinct first component corresponds to exactly one second component.

The relation in **Example 1(a)** is a function. The relation in **Example 1(b)** is *not* a function because the first component, 3, corresponds to more than one second component. If the components of the ordered pairs in **Example 1(b)** were interchanged, giving the relation

$$\{(5, 3), (6, 3), (7, 3), (8, 3)\},$$

then the relation *would* be a function. ***In that case, each domain element (first component) corresponds to exactly one range element (second component).***

Answers

1. (a) domain: $\{5, 15, 25, 35\}$; range: $\{10, 20, 30, 40\}$

(b) domain: $\{1, 2, 3\}$; range: $\{4\}$

2 Determine whether each relation is a function.

(a) $\{(-2, 8), (-1, 1), (0, 0), (1, 1), (2, 8)\}$

(b) $\{(5, 2), (5, 1), (5, 0)\}$

(c) $\{(-1, -3), (0, 2), (3, 1), (8, 1)\}$

Answers

2. **(a)** function **(b)** not a function **(c)** function

EXAMPLE 2 Determining Whether Relations Are Functions

Determine whether each relation is a function.

(a) $\{(-2, 4), (-1, 1), (0, 0), (1, 1), (2, 4)\}$

Each first component appears once and only once. The relation is a function.

(b) $\{(9, 3), (9, -3), (4, 2)\}$

The first component 9 appears in two ordered pairs and corresponds to two different second components. Therefore, this relation is not a function.

◀ **Work Problem 2 at the Side.**

Functions often have an infinite number of ordered pairs and may be represented by equations that tell how to find the second components (outputs), given the first components (inputs). Here are some everyday examples of functions.

1. The **cost** y in dollars charged by an express mail company is a function of the **weight** x in pounds determined by the equation $y = 1.5(x - 1) + 9$.
2. In Cedar Rapids, Iowa, sales tax is 7% of the price of an item. The **tax** y on a particular item is a function of the **price** x, because $y = 0.07x$.
3. The **distance** d traveled by a car moving at a constant speed of 45 mph is a function of the **time** t. Thus, $d = 45t$.

The function concept can be illustrated by an input-output "machine," as seen in **Figure 6.** The express mail company equation $y = 1.5(x - 1) + 9$ provides an output (the cost y in dollars) for a given input (the weight x in pounds).

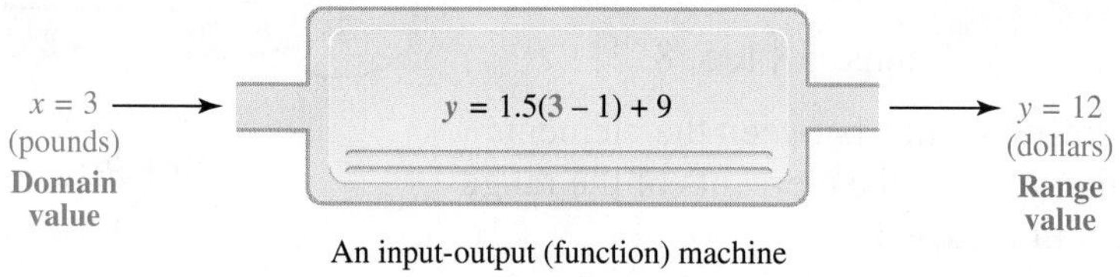

An input-output (function) machine for $y = 1.5(x - 1) + 9$

Figure 6

OBJECTIVE 3 Determine whether an equation represents a function. Given the graph of an equation, the definition of a function can be used to decide whether or not the graph represents a function. By definition, each x-value of a function must lead to exactly one y-value.

In **Figure 7(a),** the indicated x-value leads to two y-values, so this graph is not the graph of a function. A vertical line can be drawn that intersects this graph in more than one point.

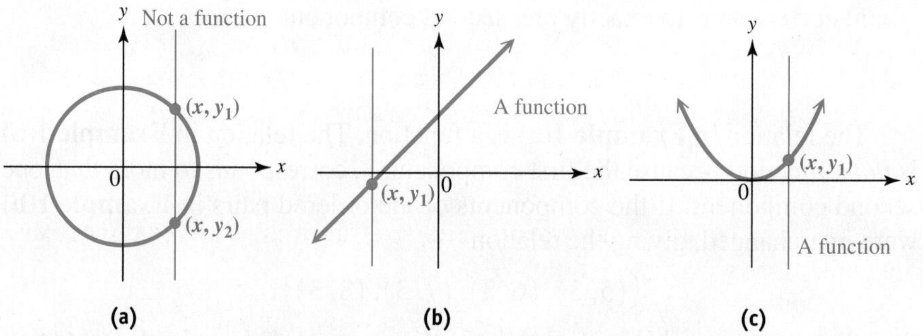

Figure 7

By contrast, in **Figure 7(b)** and **Figure 7(c)** any vertical line will intersect each graph in no more than one point, so these graphs are graphs of functions. This idea leads to the **vertical line test** for a function.

> **Vertical Line Test**
>
> If a vertical line intersects a graph in more than one point, then the graph is not the graph of a function.

As **Figure 7(b)** suggests, any nonvertical line is the graph of a function. ***Thus, any linear equation of the form $y = mx + b$ defines a function.*** (Recall that a vertical line has undefined slope.) Also, any vertical parabola, as in **Figure 7(c)**, is the graph of a function, so ***any quadratic equation of the form $y = ax^2 + bx + c$ (with $a \neq 0$) defines a function.***

EXAMPLE 3 Deciding Whether Relations Are Functions

Determine whether each relation represented by a graph or an equation is a function.

(a)

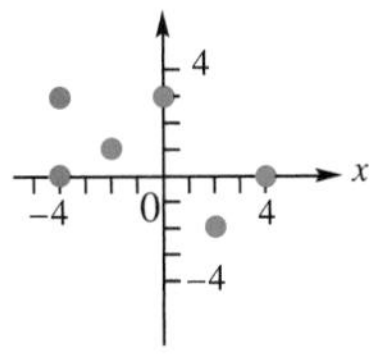

Because there are two ordered pairs with first component −4, as shown in red, this is not the graph of a function.

(b)

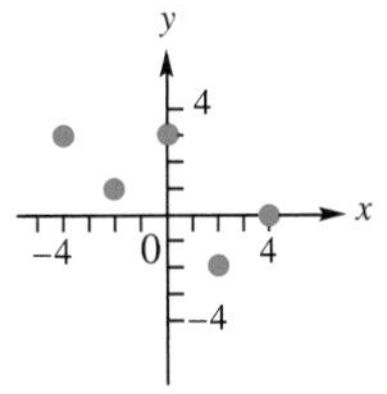

Every first component is paired with one and only one second component, and as a result, no vertical line intersects the graph in more than one point. Therefore, this is the graph of a function.

(c) $y = 2x - 9$

This linear equation is in the form $y = mx + b$. Since the graph of this equation is a line that is not vertical, the equation defines a function.

(d)

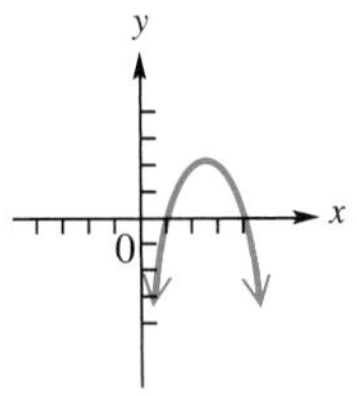

Use the vertical line test. Any vertical line intersects the graph of a vertical parabola just once, so this is the graph of a function.

(e)

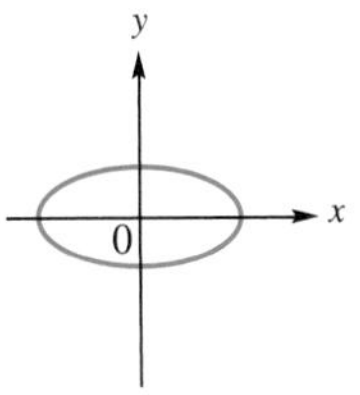

The vertical line test shows that this graph is not the graph of a function—a vertical line could intersect the graph twice.

(f) $x = 4$

The graph of $x = 4$ is a vertical line, so the equation does *not* define a function. (Every ordered pair has x-value 4.)

Work Problem 3 at the Side. ▶

3 Determine whether each relation represented by a graph or an equation is a function.

(a)

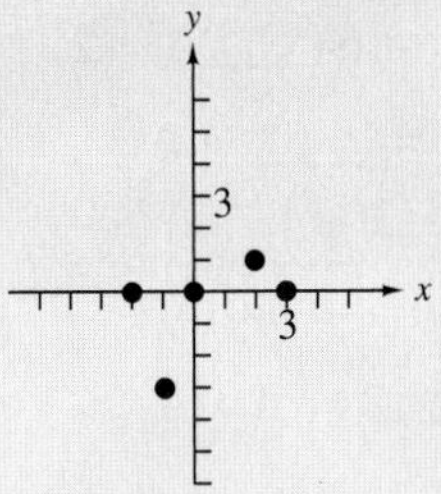

(b)

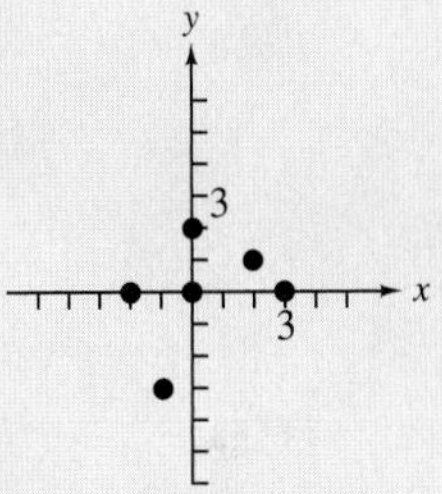

(c)

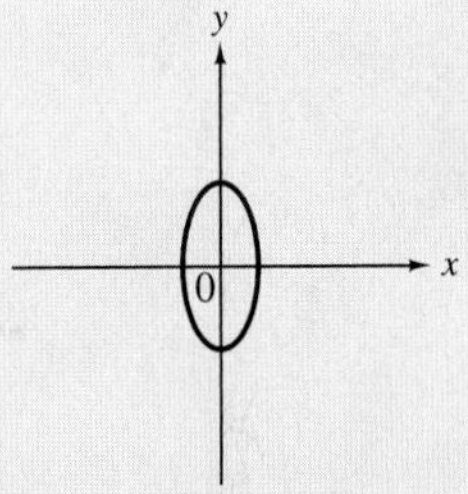

(d)

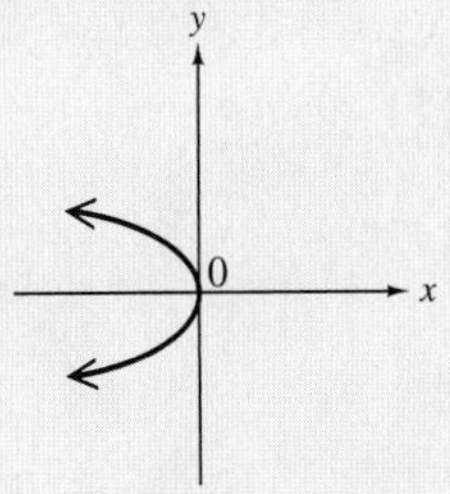

(e) $y = 3$

Answers

3. **(a)** function **(b)** not a function
(c) not a function **(d)** not a function
(e) function

4 For the function $f(x) = 6x - 2$, find each function value.

GS **(a)** $f(-1)$

$= 6(\_\_\_) - 2$

$= \_\_\_ - 2$

$= \_\_\_$

(b) $f(0)$

(c) $f(1)$

Answers

4. **(a)** -1; -6; -8
(b) -2 **(c)** 4

OBJECTIVE 4 Use function notation. The letters f, g, and h are commonly used to name functions. For example, the function $y = 3x + 5$ may be written

$$f(x) = 3x + 5,$$

where $f(x)$, which represents the value of f at x, is read "**f of x**." The notation $f(x)$ is another way of expressing the range element y for a function f. For the function $f(x) = 3x + 5$, if $x = 7$ then we find $f(7)$ as follows.

$f(7) = 3 \cdot 7 + 5$ Let $x = 7$.

$f(7) = 21 + 5$ Multiply.

$f(7) = 26$ Add.

Read this result, $f(7) = 26$, as "f of 7 equals 26." The notation $f(7)$ represents the value of y when x is 7. The statement $f(7) = 26$ says that the value of y is 26 when x is 7. It also indicates that the point $(7, 26)$ lies on the graph of f.

Similarly, to find $f(-3)$ for $f(x) = 3x + 5$, substitute -3 for x.

Use parentheses to avoid errors.

$f(-3) = 3(-3) + 5$ Let $x = -3$.

$f(-3) = -9 + 5$ Multiply.

$f(-3) = -4$ Add.

CAUTION

The notation $f(x)$ does *not* mean f times x. ***The symbol f(x) means the value of the function f when evaluated for x. It represents the y-value that corresponds to x.***

Function Notation

In the notation $f(x)$,

	f	is the name of the function,
	x	is the domain value,
and	$f(x)$	is the range value y for the domain value x.

EXAMPLE 4 Using Function Notation

For the function $f(x) = x^2 - 3$, find each function value.

(a) $f(4)$

Substitute 4 for x.

$f(x) = x^2 - 3$ Given function

$f(4) = 4^2 - 3$ Let $x = 4$.

$= 16 - 3$ Apply the exponent.

4^2 means $4 \cdot 4$.

$f(4) = 13$ Subtract.

(b) $f(0) = 0^2 - 3$ Let $x = 0$.

$f(0) = 0 - 3$

$f(0) = -3$

(c) $f(-3) = (-3)^2 - 3$ Let $x = -3$.

$f(-3) = 9 - 3$

$(-3)^2$ means $-3 \cdot (-3)$.

$f(-3) = 6$

◀ Work Problem **4** at the Side.

OBJECTIVE 5 Apply the function concept in an application. Because a function assigns to each element in its domain exactly one element in its range, the function concept is used in real-data applications in which two quantities are related.

EXAMPLE 5 Applying the Function Concept to Population

Asian-American populations (in millions) are shown in the table.

ASIAN-AMERICAN POPULATION

Year	Population (in millions)
2000	11.2
2004	13.1
2006	14.9
2010	16.7

Source: U.S. Census Bureau.

(a) Use the table to write a set of ordered pairs that defines a function f.

If we choose the years as the domain elements and the populations in millions as the range elements, the information in the table can be written as a set of four ordered pairs. In set notation, the function f is

$$f = \{(2000, 11.2), (2004, 13.1), (2006, 14.9), (2010, 16.7)\}.$$

y-values are in millions.

(b) What is the domain of f? What is the range?

The domain is the set of years, or x-values.

$$\{2000, 2004, 2006, 2010\}$$

The range is the set of populations in millions, or y-values.

$$\{11.2, 13.1, 14.9, 16.7\}$$

(c) Find $f(2000)$ and $f(2006)$.

We refer to the table or the ordered pairs in part (a) to find that

$$f(2000) = 11.2 \text{ million} \quad \text{and} \quad f(2006) = 14.9 \text{ million.}$$

(d) For what x-value does $f(x)$ equal 16.7 million? 13.1 million?

We use the table or the ordered pairs found in part (a) to determine

$$f(2010) = 16.7 \text{ million} \quad \text{and} \quad f(2004) = 13.1 \text{ million.}$$

Work Problem 5 at the Side. ▶

5 The median age at first marriage for men in the United States for selected years is given in the table.

Year	Age (in years)
2004	27.4
2006	27.5
2008	27.6
2010	28.2

Source: U.S. Census Bureau.

(a) Write a set of ordered pairs that defines a function f for the data.

(b) Give the domain and range of f.

(c) Find $f(2010)$.

(d) For what x-value does $f(x)$ equal 27.5?

Answers

5. (a) $f = \{(2004, 27.4), (2006, 27.5), (2008, 27.6), (2010, 28.2)\}$

(b) domain: $\{2004, 2006, 2008, 2010\}$; range: $\{27.4, 27.5, 27.6, 28.2\}$

(c) 28.2 **(d)** 2006

17.5 Exercises

FOR EXTRA HELP

Download the MyDashBoard App

MyMathLab®

CONCEPT CHECK *Complete the following table for the function $f(x) = x + 2$.*

	x	$x + 2$	$f(x)$	(x, y)
	0	2	2	(0, 2)
1.	1	___	___	___
2.	2	___	___	___
3.	3	___	___	___
4.	4	___	___	___

5. CONCEPT CHECK Describe the graph of function f in **Exercises 1–4** if the domain is $\{0, 1, 2, 3, 4\}$.

6. CONCEPT CHECK Describe the graph of function f in **Exercises 1–4** if the domain is the set of all real numbers.

Determine whether each relation is a function. In Exercises 7–12, give the domain and the range. ***See Examples 1–3.***

7. $\{(-4, 3), (-2, 1), (0, 5), (-2, -8)\}$

8. $\{(3, 7), (1, 4), (0, -2), (-1, -1), (-2, 5)\}$

9.

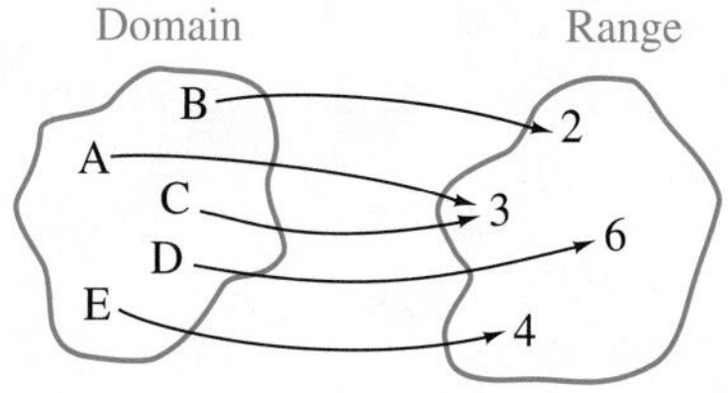

10.

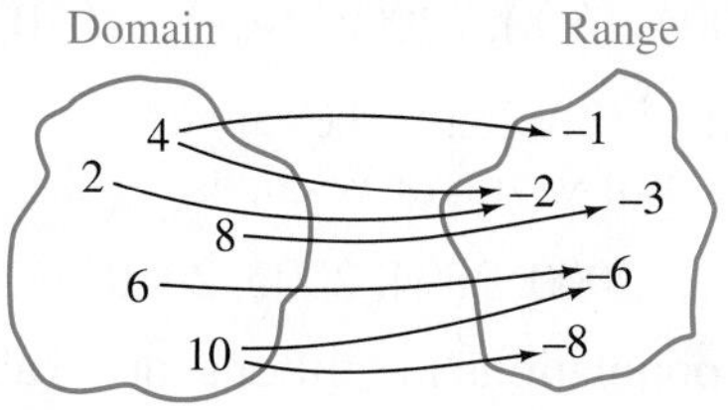

11.

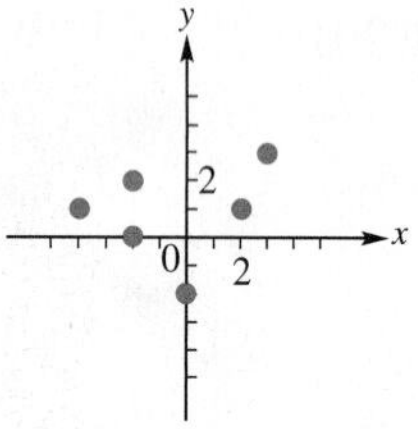

12.

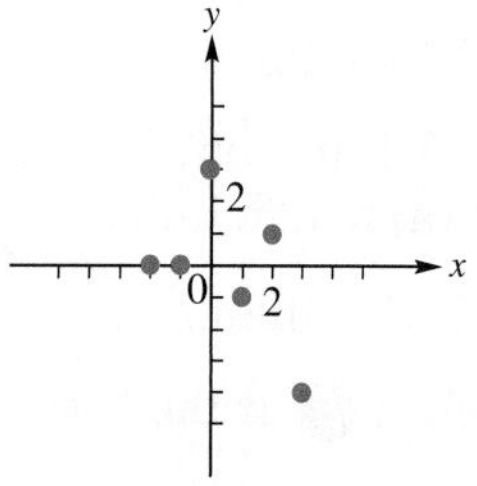

13.

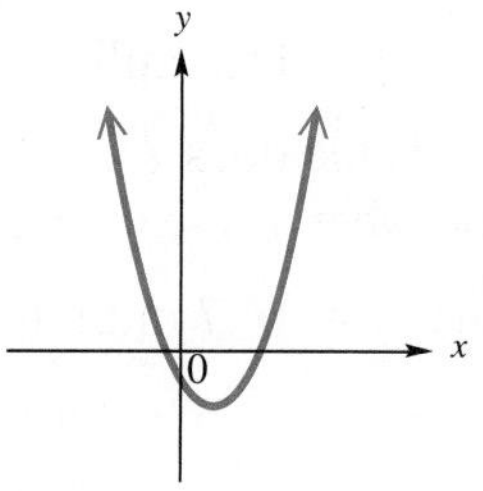

14.

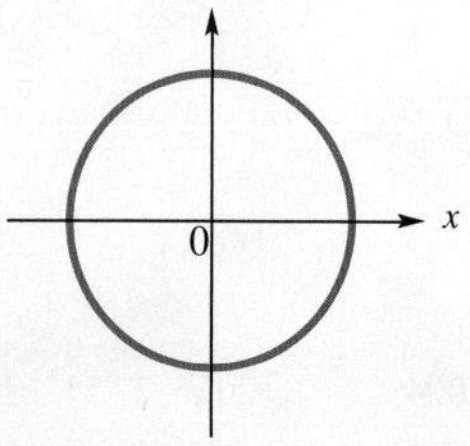

15.

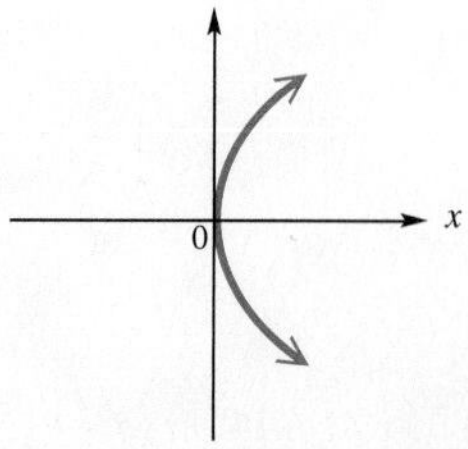

16.

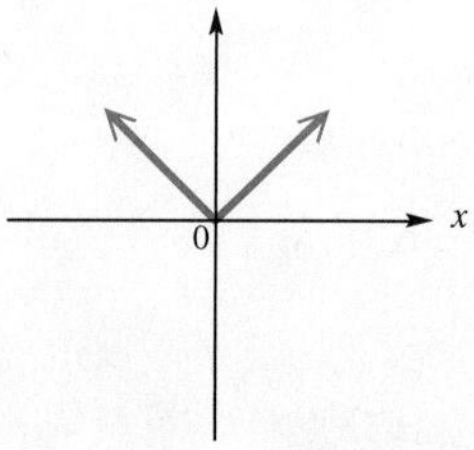

Decide whether each equation defines y as a function of x. **See Example 3.**

17. $y = 5x + 3$

18. $y = -7x + 12$

19. $y = \sqrt{x}$

20. $y = \pm\sqrt{x}$

21. $x = -7$

22. $x = 0$

23. $y = 5$

24. $y = -8$

For each function f, find **(a)** $f(2)$, **(b)** $f(0)$, *and* **(c)** $f(-3)$. **See Example 4.**

25. $f(x) = 4x + 3$

26. $f(x) = -3x + 5$

27. $f(x) = x^2 - x + 2$

28. $f(x) = x^3 + x$

29. $f(x) = |x|$

GS $|x|$ is read "the __________ of x."

(a) $f(2) = |\_\_\_|$

$f(2) = $ ____

(b) Find $f(0)$. **(c)** Find $f(-3)$.

30. $f(x) = |x + 7|$

GS $|x + 7|$ is read "the __________ of the quantity x plus ____."

(a) $f(2) = |\_\_\_ + 7|$

$f(2) = |\_\_\_|$

$f(2) = $ ____

(b) Find $f(0)$. **(c)** Find $f(-3)$.

The graph shows the number of foreign-born residents (in millions) for selected years. Use the information in the graph for Exercises 31–35. **See Example 5.**

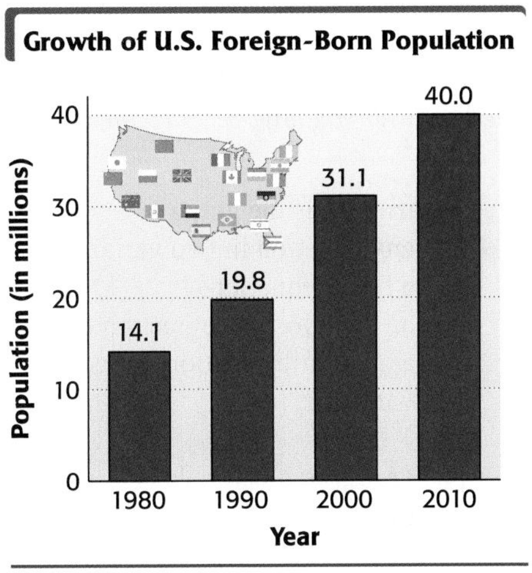

Source: U.S. Census Bureau.

31. Write the information in the graph as a set of ordered pairs. Does this set represent a function?

32. Suppose that g is the name given to this relation. Give the domain and range of g.

33. Find $g(1980)$ and $g(2000)$.

34. For what value of x does $g(x) = 40.0$ (million)?

35. Suppose $g(2008) = 38.0$ (million). What does this tell you in the context of the application?

Chapter 17 Summary

Key Terms

17.4

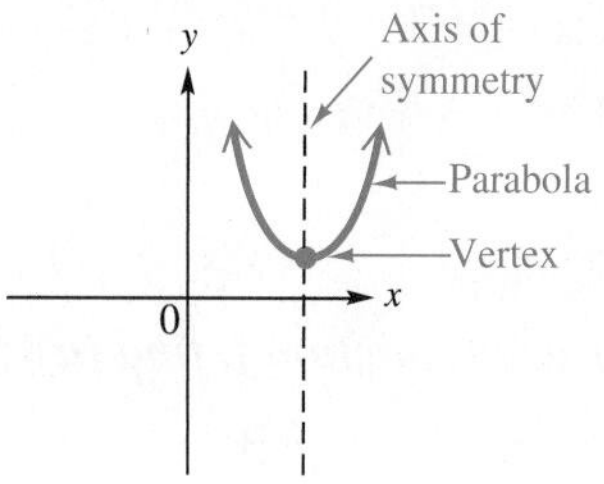

parabola The graph of the quadratic equation $y = ax^2 + bx + c$ is a parabola.

vertex The vertex of a parabola that opens upward or downward is the lowest or highest point on the graph.

axis of symmetry The axis of symmetry of a parabola that opens upward or downward is the vertical line through the vertex.

17.5

components In an ordered pair (x, y), x and y are the components.

relation Any set of ordered pairs is a relation.

domain The set of all first components in the ordered pairs of a relation is the domain of the relation.

range The set of all second components in the ordered pairs of a relation is the range of the relation.

function A function is a set of ordered pairs in which each distinct first component corresponds to exactly one second component.

New Symbols

$\pm$ positive or negative (plus or minus)

$f(x)$ function f of x

Test Your Word Power

See how well you have learned the vocabulary in this chapter.

1. A **parabola** is the graph of
   - **A.** any equation in two variables
   - **B.** a linear equation
   - **C.** an equation of degree three
   - **D.** a quadratic equation in two variables.

2. The **vertex** of a parabola is
   - **A.** the point where the graph intersects the y-axis
   - **B.** the point where the graph intersects the x-axis
   - **C.** the lowest point on a parabola that opens up or the highest point on a parabola that opens down
   - **D.** the origin.

3. A **relation** is
   - **A.** any set of ordered pairs
   - **B.** a set of ordered pairs in which each distinct first component corresponds to exactly one second component
   - **C.** two sets of ordered pairs that are related
   - **D.** a graph of ordered pairs.

4. The **domain** of a relation is
   - **A.** the set of all x- and y-values in the ordered pairs of the relation
   - **B.** the difference between the components in an ordered pair of the relation
   - **C.** the set of all first components in the ordered pairs of the relation
   - **D.** the set of all second components in the ordered pairs of the relation.

5. The **range** of a relation is
   - **A.** the set of all x- and y-values in the ordered pairs of the relation
   - **B.** the difference between the components in an ordered pair of the relation
   - **C.** the set of all first components in the ordered pairs of the relation
   - **D.** the set of all second components in the ordered pairs of the relation.

6. A **function** is
   - **A.** any set of ordered pairs
   - **B.** a set of ordered pairs in which each distinct first component corresponds to exactly one second component
   - **C.** two sets of ordered pairs that are related
   - **D.** a graph of ordered pairs.

Answers to Test Your Word Power

1. D; *Examples:* See **Figures 1–4** in **Section 17.4.**
2. C; *Example:* The graph of $y = (x + 3)^2$ has vertex $(-3, 0)$, which is the lowest point on the graph.
3. A; *Example:* $\{(0, 2), (2, 4), (3, 6), (-1, 3)\}$
4. C; *Example:* The domain in the relation given in Answer 3 is the set of x-values $\{0, 2, 3, -1\}$.
5. D; *Example*: The range of the relation given in Answer 3 is the set of y-values $\{2, 4, 6, 3\}$.
6. B; *Example*: The relation given in Answer 3 is a function since each x-value corresponds to exactly one y-value.

Quick Review

Concepts	Examples

17.1 Solving Quadratic Equations by the Square Root Property

Square Root Property of Equations
If k is positive, and if $x^2 = k$, then

$$x = \sqrt{k} \quad \text{or} \quad x = -\sqrt{k}.$$

The solution set $\{-\sqrt{k}, \sqrt{k}\}$ can also be written $\{\pm\sqrt{k}\}$.

Examples

Solve $(2x + 1)^2 = 5$.

$$2x + 1 = \sqrt{5} \quad \text{or} \quad 2x + 1 = -\sqrt{5}$$

$$2x = -1 + \sqrt{5} \quad \text{or} \quad 2x = -1 - \sqrt{5}$$

$$x = \frac{-1 + \sqrt{5}}{2} \quad \text{or} \quad x = \frac{-1 - \sqrt{5}}{2}$$

Solution set: $\left\{\frac{-1 + \sqrt{5}}{2}, \frac{-1 - \sqrt{5}}{2}\right\}$

17.2 Solving Quadratic Equations by Completing the Square

Solving a Quadratic Equation by Completing the Square

Step 1 If the coefficient of the second-degree term is 1, go to Step 2. If it is not 1, divide each side of the equation by this coefficient.

Step 2 Make sure that all variable terms are on one side of the equation and all constant terms are on the other.

Step 3 Take half the coefficient of x, square it, and add the square to each side of the equation. Factor the variable side, and combine terms on the other side.

Step 4 Use the square root property to solve the equation.

Examples

Solve $2x^2 + 4x - 1 = 0$.

$$x^2 + 2x - \frac{1}{2} = 0 \qquad \text{Divide by 2.}$$

$$x^2 + 2x = \frac{1}{2} \qquad \text{Add } \tfrac{1}{2}.$$

$$x^2 + 2x + 1 = \frac{1}{2} + 1 \qquad \left[\tfrac{1}{2}(2)\right]^2 = 1^2 = 1$$

$$(x + 1)^2 = \frac{3}{2} \qquad \text{Factor. Add.}$$

$$x + 1 = \sqrt{\frac{3}{2}} \quad \text{or} \quad x + 1 = -\sqrt{\frac{3}{2}}$$

$$x + 1 = \frac{\sqrt{3} \cdot \sqrt{2}}{\sqrt{2} \cdot \sqrt{2}} \quad \text{or} \quad x + 1 = -\frac{\sqrt{3} \cdot \sqrt{2}}{\sqrt{2} \cdot \sqrt{2}}$$

$$x + 1 = \frac{\sqrt{6}}{2} \quad \text{or} \quad x + 1 = -\frac{\sqrt{6}}{2}$$

$$x = -1 + \frac{\sqrt{6}}{2} \quad \text{or} \quad x = -1 - \frac{\sqrt{6}}{2}$$

$$x = \frac{-2}{2} + \frac{\sqrt{6}}{2} \quad \text{or} \quad x = \frac{-2}{2} - \frac{\sqrt{6}}{2}$$

$$x = \frac{-2 + \sqrt{6}}{2} \quad \text{or} \quad x = \frac{-2 - \sqrt{6}}{2}$$

Solution set: $\left\{\frac{-2 + \sqrt{6}}{2}, \frac{-2 - \sqrt{6}}{2}\right\}$

Concepts	Examples

17.3 Solving Quadratic Equations by the Quadratic Formula

Quadratic Formula

The solutions of $ax^2 + bx + c = 0$ (with $a \neq 0$) are

$$x = \frac{-b \pm \sqrt{b^2 - 4ac}}{2a}.$$

Solve $3x^2 - 4x - 2 = 0$.

$$x = \frac{-(-4) \pm \sqrt{(-4)^2 - 4(3)(-2)}}{2(3)} \qquad \text{Let } a = 3, b = -4, c = -2.$$

$$x = \frac{4 \pm \sqrt{40}}{6} \qquad \text{Simplify.}$$

$$x = \frac{4 \pm 2\sqrt{10}}{6} \qquad \sqrt{40} = \sqrt{4} \cdot \sqrt{10} = 2\sqrt{10}$$

$$x = \frac{2(2 \pm \sqrt{10})}{2(3)} \qquad \text{Factor.}$$

$$x = \frac{2 \pm \sqrt{10}}{3} \qquad \text{Divide out 2.}$$

Solution set: $\left\{\frac{2 + \sqrt{10}}{3}, \frac{2 - \sqrt{10}}{3}\right\}$

17.4 Graphing Quadratic Equations

To graph $y = ax^2 + bx + c$, follow these steps.

Step 1 Find the vertex. Use $x = -\frac{b}{2a}$ and find y by substituting this value for x in the equation.

Graph $y = 2x^2 - 5x - 3$.

$$x = -\frac{b}{2a} = -\frac{-5}{2(2)} = \frac{5}{4} \qquad a = 2, b = -5$$

$$y = 2\left(\frac{5}{4}\right)^2 - 5\left(\frac{5}{4}\right) - 3 \qquad \text{Let } x = \tfrac{5}{4}.$$

$$y = 2\left(\frac{25}{16}\right) - \frac{25}{4} - 3 \qquad \text{Apply the exponent. Multiply.}$$

$$y = \frac{25}{8} - \frac{50}{8} - \frac{24}{8} \qquad \text{Multiply. Find a common denominator.}$$

$$y = -\frac{49}{8} \qquad \text{Subtract.}$$

The vertex is $\left(\frac{5}{4}, -\frac{49}{8}\right)$.

Step 2 Let $x = 0$ to find the y-intercept.

$$y = 2(0)^2 - 5(0) - 3 \qquad \text{Let } x = 0.$$

$$y = -3 \qquad \text{Simplify.}$$

The y-intercept is $(0, -3)$.

Step 3 Let $y = 0$ and solve to find the x-intercepts (if they exist).

$$0 = 2x^2 - 5x - 3 \qquad \text{Let } y = 0.$$

$$0 = (2x + 1)(x - 3) \qquad \text{Factor.}$$

$$2x + 1 = 0 \quad \text{or} \quad x - 3 = 0 \qquad \text{Zero-factor property}$$

$$2x = -1 \quad \text{or} \quad x = 3 \qquad \text{Solve each equation.}$$

$$x = -\frac{1}{2}$$

The x-intercepts are $\left(-\frac{1}{2}, 0\right)$ and $(3, 0)$.

Concepts	Examples

Step 4 Plot the intercepts and the vertex.

Step 5 Find and plot additional ordered pairs near the vertex and intercepts as needed.

x	y
$-\frac{1}{2}$	0
0	-3
1	-6
$\frac{5}{4}$	$-\frac{49}{8}$
2	-5
3	0

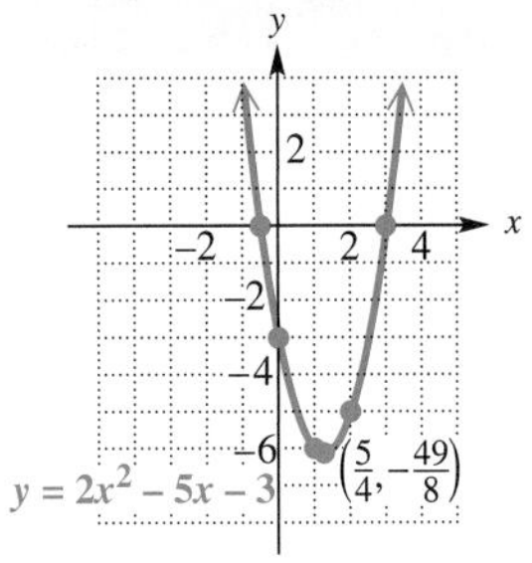

17.5 Introduction to Functions

Vertical Line Test
If a vertical line intersects a graph in more than one point, then the graph is not the graph of a function.

By the vertical line test, the graph shown is not the graph of a function.

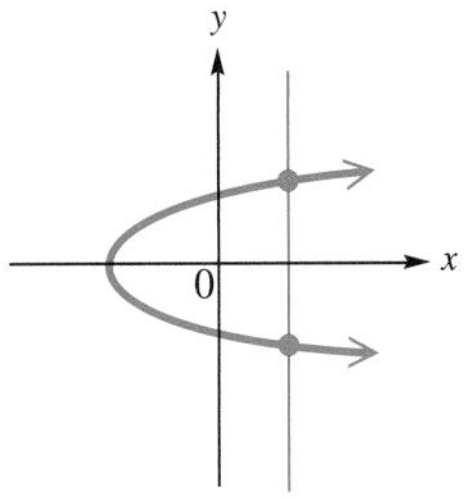

The set of all first components in the ordered pairs of a relation is the **domain** of the relation, and the set of all second components in the ordered pairs is the **range** of the relation.

The function

$$\{(10, 5), (20, 15), (30, 25)\}$$

has domain $\{10, 20, 30\}$ and range $\{5, 15, 25\}$.

To find $f(x)$ for a specific value of x, replace x by that value in the expression for the function f.

If $f(x) = 2x + 7$, then we find $f(3)$ as follows.

$f(x) = 2x + 7$

$f(3) = 2(3) + 7$ Let $x = 3$.

$f(3) = 13$ Multiply. Add.

Chapter 17 Review Exercises

17.1 *Solve each equation by using the zero-factor property.*

1. $x^2 + 3x - 28 = 0$

2. $x^2 + 14x + 45 = 0$

3. $2z^2 + 7z = 15$

4. $r^2 - 169 = 0$

Solve each equation by using the square root property. Express all radicals in simplest form.

5. $y^2 = 144$

6. $x^2 = 37$

7. $m^2 = 128$

8. $(k + 2)^2 = 25$

9. $(r - 3)^2 = 10$

10. $(2p + 1)^2 = 14$

11. $(3k + 2)^2 = -3$

12. $(3x + 5)^2 = 0$

17.2 *Solve each equation by completing the square.*

13. $m^2 + 6m + 5 = 0$

14. $p^2 + 4p = 7$

15. $-x^2 + 5 = 2x$

16. $2x^2 - 3 = -8x$

17. $4(x^2 + 7x) + 29 = -20$

18. $(4x + 1)(x - 1) = -7$

Solve each problem.

19. If an object is projected upward on Earth from a height of 50 ft, with an initial velocity of 32 ft per sec, then its height h after t seconds is given by

$$h = -16t^2 + 32t + 50,$$

where h is in feet. At what time(s) will it be at a height of 30 ft?

20. Find the lengths of the three sides of the right triangle shown.

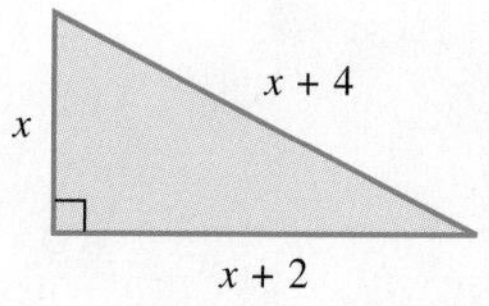

Solve each equation by using the quadratic formula.

21. $m^2 - 5m - 36 = 0$ **22.** $5r^2 = 14r$ **23.** $2x^2 - x + 3 = 0$ **24.** $4w^2 - 12w + 9 = 0$

25. $-4x^2 - 2x + 7 = 0$ **26.** $2x^2 + 8 = 4x + 11$ **27.** $x(5x - 1) = 1$

28. $\frac{1}{4}x^2 = 2 - \frac{3}{4}x$ **29.** $\frac{1}{2}x^2 + 3x = 5$ **30.** $0.2x^2 = 0.4x - 0.1$

17.4 *Sketch the graph of each equation. Identify each vertex.*

31. $y = -x^2 + 5$

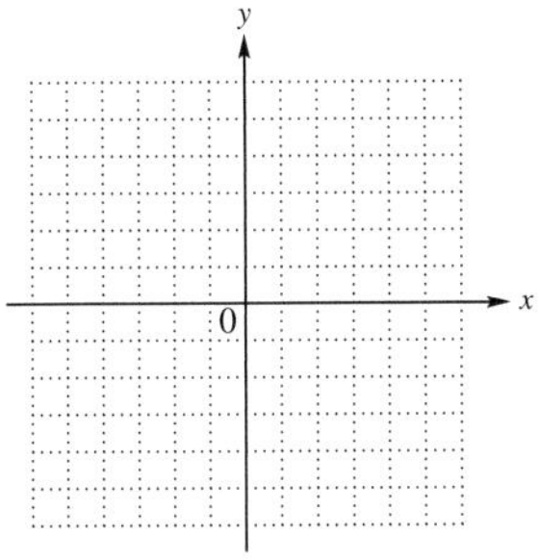

32. $y = (x - 1)^2$

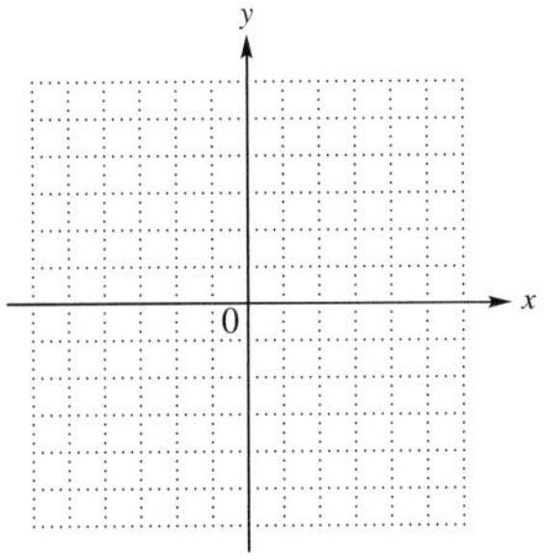

33. $y = -x^2 + 2x + 3$

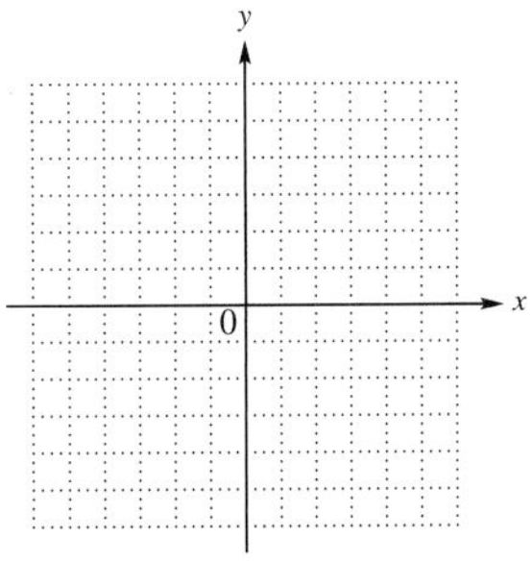

34. $y = x^2 + 4x + 2$

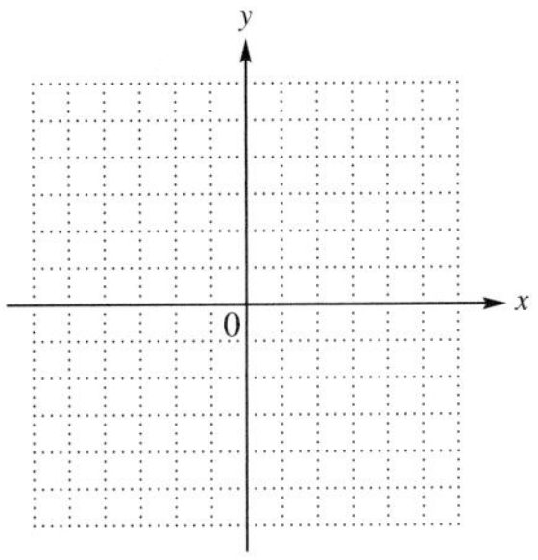

17.5 *Determine whether each relation is or is not a function. In Exercises 35 and 36, give the domain and the range.*

35. $\{(-2, 4), (0, 8), (2, 5), (2, 3)\}$ **36.** $\{(8, 3), (7, 4), (6, 5), (5, 6), (4, 7)\}$

37.

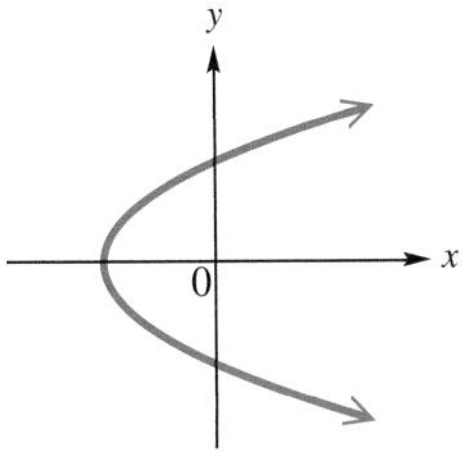

38.

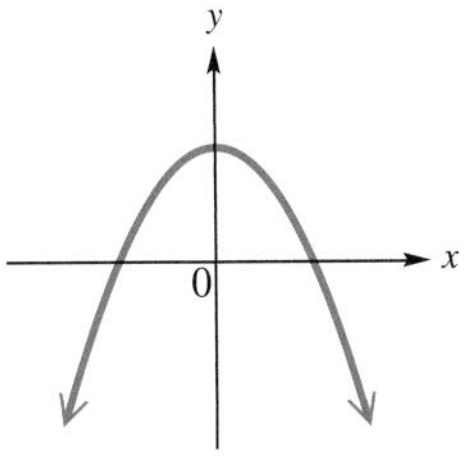

39. $2x + 3y = 12$ **40.** $y = x^2$

Find ***(a)*** $f(2)$ *and* ***(b)*** $f(-1)$.

41. $f(x) = 3x + 2$ **42.** $f(x) = 2x^2 - 1$ **43.** $f(x) = |x + 3|$ **44.** $f(x) = -x^2 + 2x - 3$

Mixed Review Exercises

Solve by any method.

45. $(2t - 1)(t + 1) = 54$

46. $(2p + 1)^2 = 100$

47. $(k + 2)(k - 1) = 3$

48. $6t^2 + 7t - 3 = 0$

49. $2x^2 + 3x + 2 = x^2 - 2x$

50. $x^2 + 2x + 5 = 7$

51. $m^2 - 4m + 10 = 0$

52. $k^2 - 9k + 10 = 0$

53. $(5x + 6)^2 = 0$

54. $\frac{1}{2}r^2 = \frac{7}{2} - r$

55. $x^2 + 4x = 1$

56. $7x^2 - 8 = 5x^2 + 8$

57. Consider the quadratic equation $x^2 - 9 = 0$.

(a) Solve the equation by factoring.

(b) Solve the equation by the square root property.

(c) Solve the equation by the quadratic formula.

(d) Compare your answers. If a quadratic equation can be solved by both factoring and the quadratic formula, should we always get the same results? Explain.

58. CONCEPT CHECK A student writes the quadratic formula as

$$x = -b \pm \sqrt{\frac{b^2 - 4ac}{2a}}.$$

What Went Wrong? Explain the error, and give the correct formula.

Chapter 17 Test

CHAPTER Test Prep VIDEO

The Chapter Test Prep Videos with test solutions are available on DVD, in MyMathLab, and on YouTube—search "LialDevMath" and click on "Channels."

Solve by using the square root property.

1. $x^2 = 39$

2. $(x + 3)^2 = 64$

3. $(4x + 3)^2 = 24$

Solve by completing the square.

4. $x^2 - 4x = 6$

5. $2x^2 + 12x - 3 = 0$

Solve by the quadratic formula.

6. $2x^2 + 5x - 3 = 0$

7. $3w^2 + 2 = 6w$

8. $4x^2 + 8x + 11 = 0$

9. $t^2 - \frac{5}{3}t + \frac{1}{3} = 0$

Solve by the method of your choice.

10. $p^2 - 2p - 1 = 0$

11. $(2x + 1)^2 = 18$

12. $(x - 5)(2x - 1) = 1$

13. $t^2 + 25 = 10t$

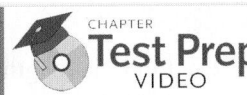

Solve each problem.

14. If an object is projected into the air from ground level with an initial velocity of 64 ft per sec, its altitude (height) s in feet after t seconds is given by the formula

$$s = -16t^2 + 64t.$$

At what time(s) will the object be at a height of 64 ft?

15. Find the lengths of the three sides of the right triangle.

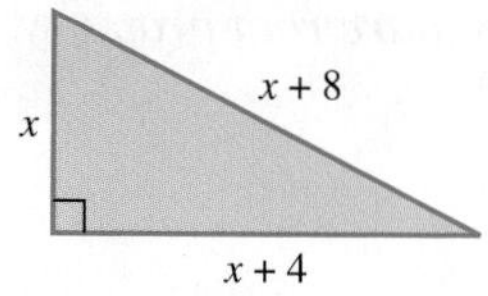

Sketch the graph of each equation. Identify each vertex.

16. $y = (x - 3)^2$

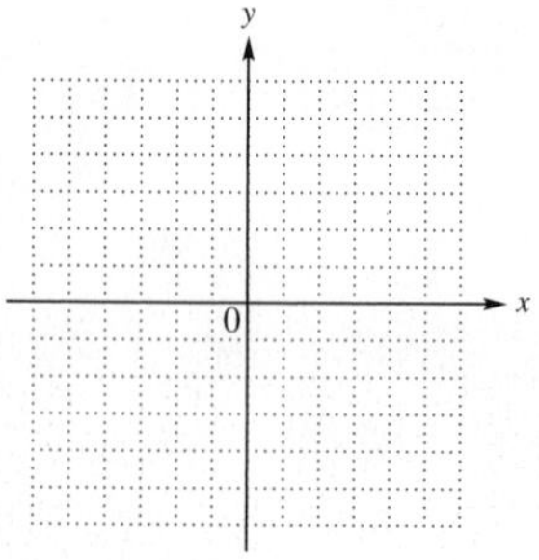

17. $y = -x^2 - 2x - 4$

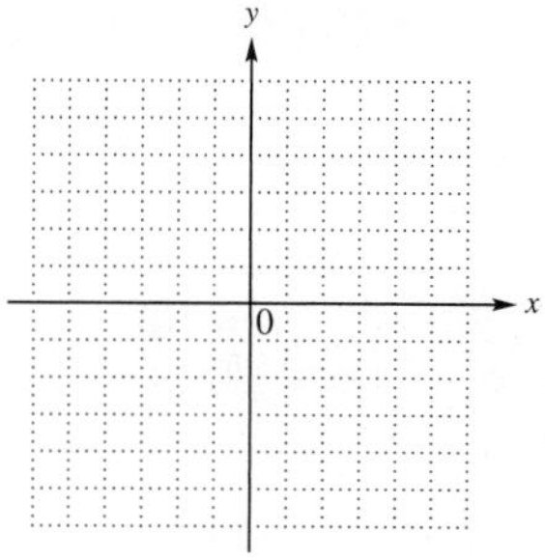

18. Determine whether each relation is a function. If it is, give the domain and the range.

(a) $\{(2, 3), (2, 4), (2, 5)\}$

(b) $\{(0, 2), (1, 2), (2, 2)\}$

19. Use the vertical line test to determine whether the graph is that of a function.

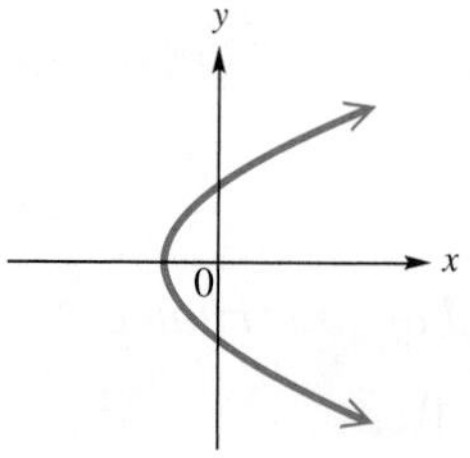

20. If $f(x) = 3x + 7$, find $f(-2)$.

Chapters 9–17 Cumulative Review Exercises

Find the value of each expression for $x = -2$ and $y = 4$.

1. $\dfrac{4x - 2y}{x + y}$

2. $x^3 - 4xy$

Perform the indicated operations.

3. $\dfrac{(-13 + 15) - (3 + 2)}{6 - 12}$

4. $-7 - 3[2 + (5 - 8)]$

Decide what property justifies each statement.

5. $(9 + 2) + 3 = 9 + (2 + 3)$

6. $-7 + 7 = 0$

7. $6(4 + 2) = 6(4) + 6(2)$

Solve each equation.

8. $2x - 7x + 8x = 30$

9. $2 - 3(t - 5) = 4 + t$

10. $2(5h + 1) = 10h + 4$

11. $d = rt$ for r

12. $\dfrac{x}{5} = \dfrac{x - 2}{7}$

13. $\dfrac{1}{3}p - \dfrac{1}{6}p = -2$

14. $0.05x + 0.15(50 - x) = 5.50$

15. $4 - (3x + 12) = (2x - 9) - (5x - 1)$

Solve each inequality. Write the solution set in interval notation and graph it.

16. $-2.5x < 6.5$

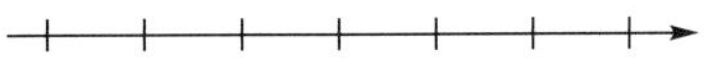

17. $4(x + 3) - 5x < 12$

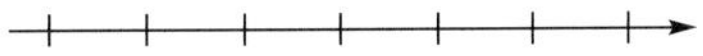

18. $\dfrac{2}{3}t - \dfrac{1}{6}t \le -2$

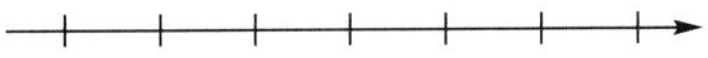

Graph the following.

19. $2x + 3y = 6$

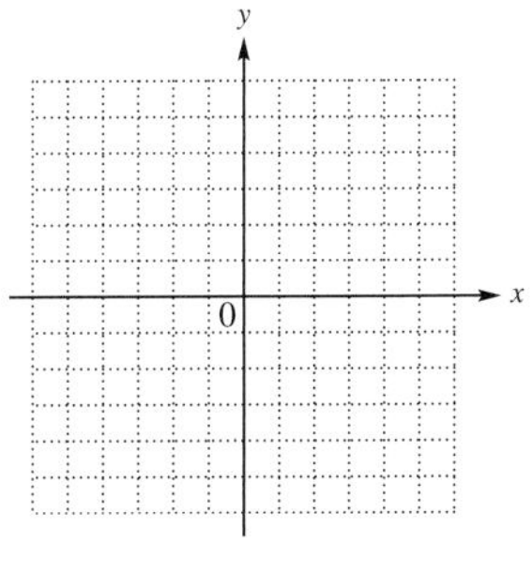

20. $y = 3$

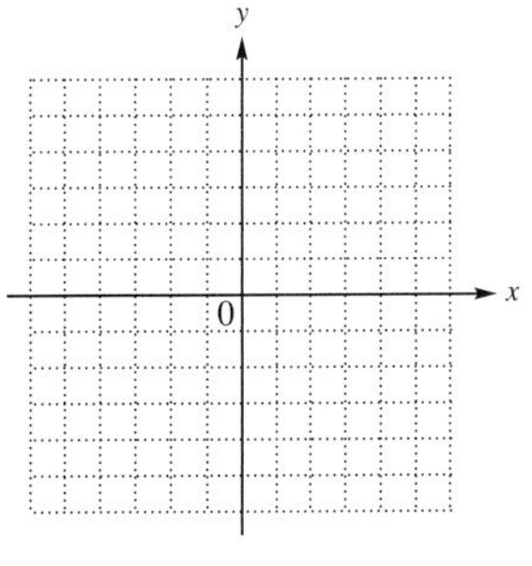

21. $2x - 5y < 10$

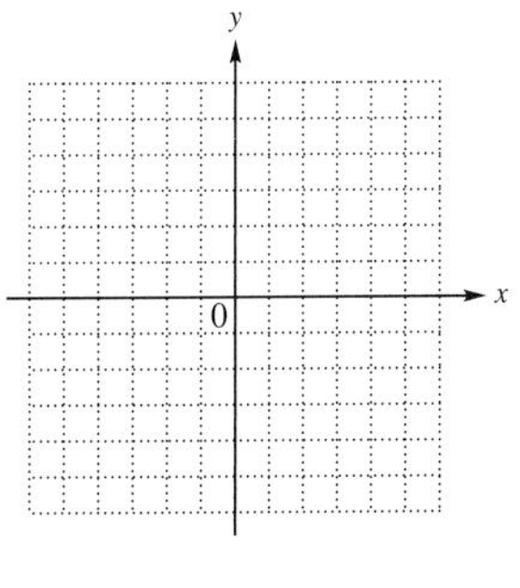

Find the slope of each line.

22. Through $(-5, 6)$ and $(1, -2)$

23. Perpendicular to the line $y = 4x - 3$

Write an equation in slope-intercept form for each line.

24. Through $(-4, 1)$ with slope $\frac{1}{2}$

25. Through the points $(1, 3)$ and $(-2, -3)$

26. **(a)** Write an equation of the vertical line through $(9, -2)$.

(b) Write an equation of the horizontal line through $(4, -1)$.

Evaluate each expression.

27. $2^{-3} \cdot 2^5$

28. $\left(\frac{3}{4}\right)^{-2}$

29. $\frac{6^5 \cdot 6^{-2}}{6^3}$

30. $\left(\frac{4^{-3} \cdot 4^4}{4^5}\right)^{-1}$

Simplify each expression and write the answer using only positive exponents. Assume no denominators are 0.

31. $\frac{(p^2)^3 p^{-4}}{(p^{-3})^{-1} p}$

32. $\frac{(m^{-2})^3 m}{m^5 m^{-4}}$

Perform the indicated operations.

33. $(2k^2 + 4k) - (5k^2 - 2) - (k^2 + 8k - 6)$

34. $(9x + 6)(5x - 3)$

35. $(3p + 2)^2$

36. $\frac{8x^4 + 12x^3 - 6x^2 + 20x}{2x}$

37. $(y^2 + 3y + 5)(3y - 1)$

38. $\frac{12p^3 + 2p^2 - 12p + 4}{2p - 2}$

Factor completely.

39. $2a^2 + 7a - 4$

40. $10m^2 + 19m + 6$

41. $8t^2 + 10tv + 3v^2$

42. $4p^2 - 12p + 9$

43. $25r^2 - 81t^2$

44. $2pq + 6p^3q + 8p^2q$

45. $2x^3 - 10x^2 + 3x - 15$

46. $8p^3 + 125$

Solve each equation.

47. $6m^2 + m - 2 = 0$

48. $8x^2 = 64x$

49. $49x^2 - 56x + 16 = 0$

50. $x^2 - 7x = -12$

51. $(x + 4)(x - 1) = -6$

52. $z^2 + 144 = 24z$

53. One of the following is equal to 1 for *all* real numbers. Which one is it?

A. $\dfrac{k^2 + 2}{k^2 + 2}$ **B.** $\dfrac{4 - m}{4 - m}$ **C.** $\dfrac{2x + 9}{2x + 9}$ **D.** $\dfrac{x^2 - 1}{x^2 - 1}$

54. Which one of the following rational expressions is *not* equivalent to $\dfrac{4 - 3x}{7}$?

A. $-\dfrac{-4 + 3x}{7}$ **B.** $-\dfrac{4 - 3x}{-7}$ **C.** $\dfrac{-4 + 3x}{-7}$ **D.** $\dfrac{-(3x + 4)}{7}$

Perform each operation, and write the answer in lowest terms.

55. $\dfrac{5}{q} - \dfrac{1}{q}$

56. $\dfrac{3}{7} + \dfrac{4}{r}$

57. $\dfrac{4}{5q - 20} - \dfrac{1}{3q - 12}$

58. $\dfrac{2}{k^2 + k} - \dfrac{3}{k^2 - k}$

59. $\dfrac{7z^2 + 49z + 70}{16z^2 + 72z - 40} \div \dfrac{3z + 6}{4z^2 - 1}$

60. $\dfrac{\dfrac{4}{a} + \dfrac{5}{2a}}{\dfrac{7}{6a} - \dfrac{1}{5a}}$

Solve each equation, and check your solutions.

61. $\dfrac{r + 2}{5} = \dfrac{r - 3}{3}$

62. $\dfrac{1}{x} = \dfrac{1}{x + 1} + \dfrac{1}{2}$

Solve each system of equations.

63. $2x - y = -8$
$x + 2y = 11$

64. $4x + 5y = -8$
$3x + 4y = -7$

65. $3x + 4y = 2$
$6x + 8y = 1$

66. $4x - y = 19$
$3x + 2y = -5$

67. $2x - y = 6$
$3y = 6x - 18$

68. Graph the solution set of the system.

$$x + 2y \le 12$$
$$2x - y \le 8$$

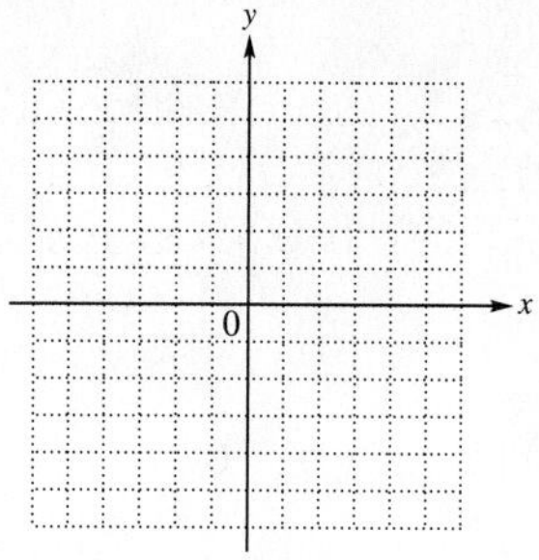

Simplify each expression.

69. $\dfrac{6\sqrt{6}}{\sqrt{5}}$

70. $3\sqrt{5} - 2\sqrt{20} + \sqrt{125}$

71. $\sqrt[3]{16a^3b^4} - \sqrt[3]{54a^3b^4}$

72. $\sqrt{27} - 2\sqrt{12} + 6\sqrt{75}$

73. $\dfrac{2}{\sqrt{3} + \sqrt{5}}$

74. $(3\sqrt{2} + 1)(4\sqrt{2} - 3)$

Solve each equation.

75. $\sqrt{x + 2} = -8$

76. $\sqrt{x} + 2 = x - 10$

Solve by any method.

77. $(5x - 6)^2 = 49$

78. $x^2 - 6x - 5 = -4$

79. $3x^2 - 2x + 1 = 0$

80. $(x + 5)(x - 2) = -4$

81. $\dfrac{1}{2}x^2 + \dfrac{1}{2}x = \dfrac{5}{4}$

82. Graph the equation $y = x^2 - 4x$. Identify the vertex. What are the x-intercepts?

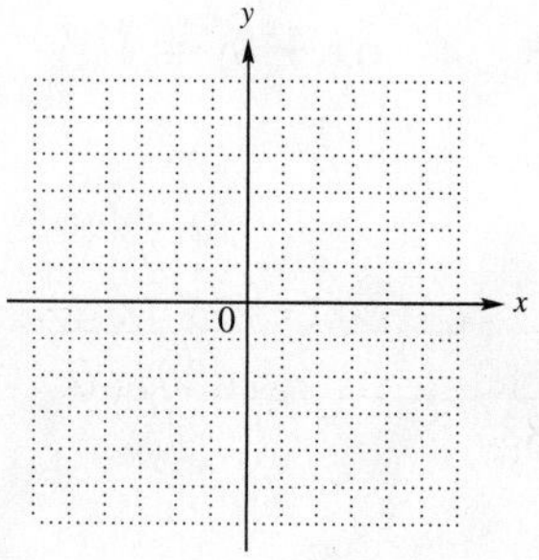

83. (a) Is $\{(0, 4), (1, 2), (3, 5)\}$ a function?

(b) Give the domain and the range of the relation in part (a).

84. If $f(x) = -2x + 7$, find $f(-2)$.

Solve each problem.

85. Des Moines and Chicago are 345 mi apart. Two cars start from these cities traveling toward each other. They meet after 3 hr. The car from Chicago has an average rate 7 mph faster than the other car. Find the average rate of each car. (*Source: State Farm Road Atlas.*)

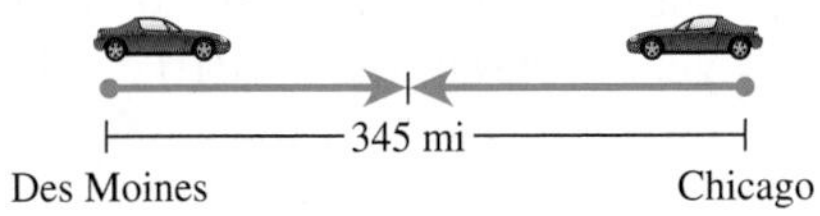

86. Trevor Brazile won the ProRodeo All-Around Championship in both 2009 and 2010. He won a total of $854,700 in these two years, winning $161,142 more in 2010 than in 2009. How much did he win each year? (*Source: World Almanac and Book of Facts.*)

87. The points on the graph show the total retail sales of prescription drugs in the United States in the years 2004–2010, along with a graph of a linear equation that models the data.

(a) Use the ordered pairs shown on the graph to find the slope of the line to the nearest whole number. Interpret the slope.

(b) Use the graph to estimate sales in the year 2008. Write your answer as an ordered pair of the form (year, sales in billions of dollars).

Retail Prescription Drug Sales

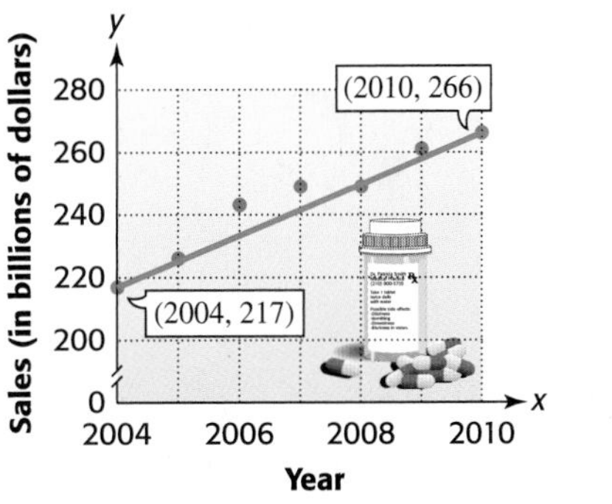

Source: National Association of Chain Drug Stores.

88. Admission prices at a high school football game were $6 for adults and $2 for children. The total value of the tickets sold was $2528, and 454 tickets were sold. How many adults and how many children attended the game?

Kind of Ticket	Number Sold	Cost of Each (in dollars)	Total Value (in dollars)
Adult	x	6	$6x$
Child	y	_____	_____
Total	454	XXXXXX	_____

89. On a business trip, Carrie Ann traveled to her destination at an average rate of 60 mph. Coming home, her average rate was 50 mph, and the trip took $\frac{1}{2}$ hr longer. How far did she travel each way?

90. Len can weed the yard in 3 hr. Bruno can weed the yard in 2 hr. How long would it take them if they worked together?

91. The winning times in seconds for the women's 1000 m speed skating event in the Winter Olympics for the years 1960 through 2010 can be closely approximated by the linear equation

$$y = -0.4336x + 94.30,$$

where x is the number of years since 1960. That is, $x = 4$ represents 1964, $x = 8$ represents 1968, and so on. (*Source: World Almanac and Book of Facts.*)

(a) Use this equation to complete the table of values. Round times to the nearest hundredth of a second.

x	y
20	
38	
42	

(b) Write ordered pairs for the data given in the table.

(c) In the context of this problem, what does the ordered pair (32, 80.42) mean?

92. At the 2010 Winter Olympics in Vancouver, Canada, the top medal winner was the United States, which won a total of 37 medals. The United States won 2 more silver medals than bronze and 4 fewer gold medals than bronze. Find the number of each type of medal won. (*Source: World Almanac and Book of Facts.*)

93. In 2010, a person with a bachelor's degree could expect to earn \$21,424 more each year than someone with a high school diploma. Together the individuals would earn \$86,528. How much could a person at each level of education expect to earn? (*Source*: Bureau of Labor Statistics.)

94. Mount Mayon in the Philippines is the most perfectly shaped conical volcano in the world. Its base is a circle with circumference 80 mi, and it has a height of about 8100 ft. (One mile is 5280 ft.) Find the radius of the circular base to the nearest mile. (*Source:* www.britannica.hk)

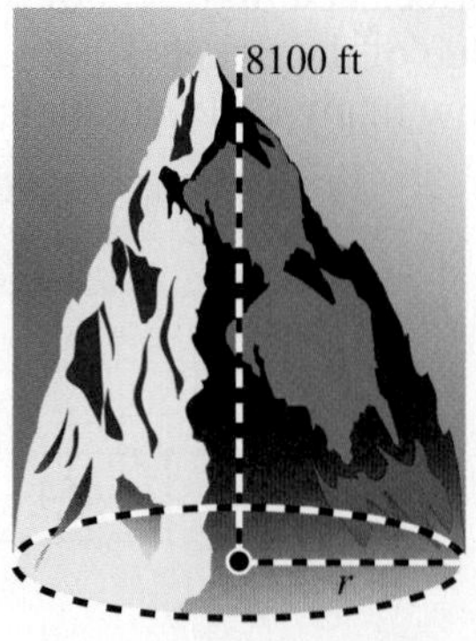

Circumference = 80 mi

Appendices

In Appendix A, you'll learn about measuring things. We need a way to measure very large things, like this meteor crater in northern Arizona that is 60 stories deep and 4150 ft across. And we need to measure tiny things, like a computer microchip that is only 5 mm long.

A Measurement

A.1 Problem Solving with U.S. Measurement Units

A.2 The Metric System—Length

A.3 The Metric System—Capacity and Weight (Mass)

A.4 Metric–U.S. Measurement Conversions and Temperature

B Inductive and Deductive Reasoning

A.1 Problem Solving with U.S. Measurement Units

OBJECTIVES

1. Learn the basic U.S. measurement units.
2. Convert among U.S. measurement units using multiplication or division.
3. Convert among U.S. measurement units using unit fractions.
4. Solve application problems using U.S. measurement units.

We measure things all the time: the distance traveled on vacation, the floor area we want to cover with carpet, the amount of milk in a recipe, the weight of the bananas we buy at the store, the number of hours we work, and many more.

In the United States, we still use the **United States (U.S.) measurement system** for many everyday activities. Examples of U.S. measurement units are inches, feet, quarts, ounces, and pounds. However, science, medicine, sports, and manufacturing use the **metric system** (meters, liters, and grams). Because the rest of the world uses *only* the metric system, U.S. businesses have been changing to the metric system in order to compete internationally.

OBJECTIVE 1 Learn the basic U.S. measurement units. Until the switch to the metric system is complete, we still need to know how to use U.S. measurement units. The table below lists the relationships you should memorize. The time relationships are used in both the U.S. and metric systems.

U.S. Measurement Unit Relationships

U.S. Length Units	U.S. Weight Units
12 inches (in.) = 1 foot (ft)	16 ounces (oz) = 1 pound (lb)
3 feet (ft) = 1 yard (yd)	2000 pounds (lb) = 1 ton
5280 feet (ft) = 1 mile (mi)	
U.S. Capacity Units	**Time (U.S. and Metric)**
8 fluid ounces (fl oz) = 1 cup (c)	60 seconds (sec) = 1 minute (min)
2 cups (c) = 1 pint (pt)	60 minutes (min) = 1 hour (hr)
2 pints (pt) = 1 quart (qt)	24 hours (hr) = 1 day
4 quarts (qt) = 1 gallon (gal)	7 days = 1 week (wk)

As you can see, there is no simple way to convert among these various measures. The units evolved over hundreds of years and were based on a variety of "standards." For example, one yard was the distance from the tip of a king's nose to his thumb when his arm was outstretched. An inch was three dried barleycorns laid end to end.

EXAMPLE 1 Knowing U.S. Measurement Units

Memorize the U.S. measurement conversions in the table above. Then fill in the blanks.

(a) 24 hr = ______ day Answer: 1 day

(b) 1 yd = ______ ft Answer: 3 ft

◀ Work Problem 1 at the Side.

OBJECTIVE 2 Convert among U.S. measurement units using multiplication or division. You often need to convert from one unit of measure to another. Two methods of converting measurements are shown here. Study each way and use the method you prefer. The first method involves deciding whether to multiply or divide.

Converting among Measurement Units

1. ***Multiply*** when converting from a larger unit to a smaller unit.
2. ***Divide*** when converting from a smaller unit to a larger unit.

1 After memorizing the U.S. measurement conversions, fill in the blanks.

(a) 1 c = ____ fl oz

(b) ____ qt = 1 gal

(c) 1 wk = ____ days

(d) ____ ft = 1 yd

(e) 1 ft = ____ in.

(f) ____ oz = 1 lb

(g) 1 ton = ____ lb

(h) ____ min = 1 hr

(i) 1 pt = ____ c

(j) ____ hr = 1 day

(k) 1 min = ____ sec

(l) 1 qt = ____ pt

(m) ____ ft = 1 mi

36IN=3Ft=

Answers

1. (a) 8 (b) 4 (c) 7 (d) 3 (e) 12 (f) 16 (g) 2000 (h) 60 (i) 2 (j) 24 (k) 60 (l) 2 (m) 5280

EXAMPLE 2 **Converting from One U.S. Unit of Measure to Another**

Convert each measurement.

(a) 7 ft to inches

You are converting from a *larger* unit to a *smaller* unit (a *foot* is longer than an *inch*), so multiply.

Because *1 ft = 12 in.*, multiply by 12.

$$7 \text{ ft} = 7 \cdot 12 = 84 \text{ in.}$$

(b) $3\frac{1}{2}$ lb to ounces

You are converting from a *larger* unit to a *smaller* unit (a *pound* is heavier than an *ounce*), so multiply.

Because *1 lb = 16 oz*, multiply by 16.

Divide 16 and 2 by their common factor of 2. 16 ÷ 2 is 8. 2 ÷ 2 is 1.

$$3\frac{1}{2}\text{ lb} = 3\frac{1}{2} \cdot 16 = \frac{7}{\cancel{2}_1} \cdot \frac{\cancel{16}^{8}}{1} = \frac{56}{1} = 56 \text{ oz}$$

(c) 20 qt to gallons

You are converting from a *smaller* unit to a *larger* unit (a *quart* is smaller than a *gallon*) so divide.

Because *4 qt = 1 gal,* divide by 4.

$$20 \text{ qt} = \frac{20}{4} = 5 \text{ gal}$$

Divide by 4.

Remember to write **gal** in your answer.

(d) 45 min to hours

You are converting from a *smaller* unit to a *larger* unit (a *minute* is less than an *hour*), so divide.

Because *60 min = 1 hr*, divide by 60 and write the fraction in lowest terms.

$$45 \text{ min} = \frac{45}{60} = \frac{45 \div 15}{60 \div 15} = \frac{3}{4}\text{ hr} \leftarrow \text{Lowest terms}$$

Divide by 60.

Work Problem 2 at the Side. ▶

OBJECTIVE 3 **Convert among U.S. measurement units using unit fractions.** If you have trouble deciding whether to multiply or divide when converting measurements, use *unit fractions* to solve the problem. You'll also use this method in science courses. A **unit fraction** is equivalent to 1. Here is an example.

$$\frac{12 \text{ in.}}{12 \text{ in.}} = \frac{\cancel{12}^{1} \cancel{\text{in.}}}{\cancel{12}_{1} \cancel{\text{in.}}} = 1$$

Use the table of measurement relationships on the previous page to find that 12 in. is the same as 1 ft. So you can substitute 1 ft for 12 in. in the numerator, or you can substitute 1 ft for 12 in. in the denominator. This makes two useful unit fractions.

$$\frac{1 \text{ ft}}{12 \text{ in.}} = 1 \quad \text{or} \quad \frac{12 \text{ in.}}{1 \text{ ft}} = 1$$

2 Convert each measurement using multiplication or division.

GS **(a)** $5\frac{1}{2}$ ft to inches

Because you are converting from a *larger* to a *smaller* unit, do you multiply or divide? ________

GS **(b)** 64 oz to pounds

Because you are converting from a *smaller* to a *larger* unit, you ________.

(c) 6 yd to feet

(d) 2 tons to pounds

(e) 35 pt to quarts

(f) 20 min to hours

Answers

2. **(a)** multiply; 66 in. **(b)** divide; 4 lb **(c)** 18 ft **(d)** 4000 lb **(e)** $17\frac{1}{2}$ qt **(f)** $\frac{1}{3}$ hr

VOCABULARY TIP

Uni- is a prefix that means "one," so a **unit fraction** is equivalent to 1.

3 First write the unit fraction needed to make each conversion. Then complete the conversion.

GS **(a)** 36 in. to feet

unit fraction $\frac{1 \text{ ft}}{12 \text{ in.}}$

$\frac{\overset{3}{\cancel{36}} \cancel{\text{in.}}}{1} \bullet \frac{1 \text{ ft}}{\underset{1}{\cancel{12}} \cancel{\text{in.}}} = \_\_\_$

(b) 14 ft to inches

unit fraction $\frac{\_\_ \text{ in.}}{\_\_ \text{ ft}}$

(c) 60 in. to feet

unit fraction ____

(d) 4 yd to feet

unit fraction ____

(e) 39 ft to yards

unit fraction ____

(f) 2 mi to feet

unit fraction ____

Answers

3. **(a)** $\frac{3 \bullet 1 \text{ ft}}{1} = 3$ ft **(b)** $\frac{12 \text{ in.}}{1 \text{ ft}}$; 168 in.

(c) $\frac{1 \text{ ft}}{12 \text{ in.}}$; 5 ft **(d)** $\frac{3 \text{ ft}}{1 \text{ yd}}$; 12 ft

(e) $\frac{1 \text{ yd}}{3 \text{ ft}}$; 13 yd **(f)** $\frac{5280 \text{ ft}}{1 \text{ mi}}$; 10,560 ft

To convert from one measurement unit to another, just multiply by the appropriate unit fraction. Remember, a unit fraction is equivalent to 1. Multiplying something by 1 does *not* change its value.

Use these guidelines to choose the correct unit fraction.

Choosing a Unit Fraction

The ***numerator*** should use the measurement unit you want in the ***answer.*** The ***denominator*** should use the measurement unit you want to ***change.***

EXAMPLE 3 Using Unit Fractions with U.S. Length Measurements

(a) Convert 60 in. to feet.

Use a unit fraction with feet (the unit for your answer) in the numerator, and inches (the unit being changed) in the denominator. Because *1 ft = 12 in.*, the necessary unit fraction is

$\frac{1 \text{ ft}}{12 \text{ in.}}$ ← Unit for your **answer** is **feet.** / ← Unit being **changed** is **inches.**

Next, multiply 60 in. times this unit fraction. Write 60 in. as the fraction $\frac{60 \text{ in.}}{1}$ and divide out common units and factors wherever possible.

$$60 \text{ in.} \bullet \frac{1 \text{ ft}}{12 \text{ in.}} = \frac{\overset{5}{\cancel{60}} \cancel{\text{in.}}}{1} \bullet \frac{1 \text{ ft}}{\underset{1}{\cancel{12}} \cancel{\text{in.}}} = \frac{5 \bullet 1 \text{ ft}}{1} = 5 \text{ ft}$$

These units should match. Divide out inches. Divide 60 and 12 by 12.

Remember to write **ft** in your answer.

(b) Convert 9 ft to inches.

Select the correct unit fraction to change 9 ft to inches.

$\frac{12 \text{ in.}}{1 \text{ ft}}$ ← Unit for your **answer** is **inches.** / ← Unit being **changed** is **feet.**

Multiply 9 ft times the unit fraction.

$$9 \text{ ft} \bullet \frac{12 \text{ in.}}{1 \text{ ft}} = \frac{9 \cancel{\text{ft}}}{1} \bullet \frac{12 \text{ in.}}{1 \cancel{\text{ft}}} = \frac{9 \bullet 12 \text{ in.}}{1} = 108 \text{ in.}$$

These units should match. Divide out feet.

CAUTION

If no units will divide out, you used the wrong unit fraction.

◄ **Work Problem 3 at the Side.**

EXAMPLE 4 Using Unit Fractions with U.S. Capacity and Weight Measurements

(a) Convert 9 pt to quarts.

First select the correct unit fraction.

$\frac{1 \text{ qt}}{2 \text{ pt}}$ ← Unit for your **answer** is **quarts.** / ← Unit being **changed** is **pints.**

Continued on Next Page

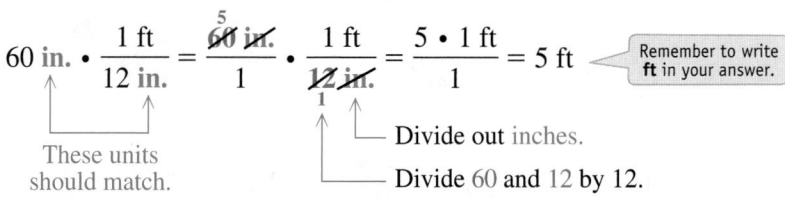

Next multiply.

Write as a mixed number.

$$9\ \text{pt} \bullet \frac{1\ \text{qt}}{2\ \text{pt}} = \frac{9\ \cancel{\text{pt}}}{1} \bullet \frac{1\ \text{qt}}{2\ \cancel{\text{pt}}} = \frac{9}{2}\ \text{qt} = 4\frac{1}{2}\ \text{qt}$$

These units should match. Divide out pints.

(b) Convert $7\frac{1}{2}$ gal to quarts.

Write as an improper fraction.

$$\frac{7\frac{1}{2}\ \cancel{\text{gal}}}{1} \bullet \frac{4\ \text{qt}}{1\ \cancel{\text{gal}}} = \frac{15}{2} \bullet \frac{4}{1}\ \text{qt}$$

Divide out gallons.

$$= \frac{15}{\cancel{2}_1} \bullet \frac{\cancel{4}^2}{1}\ \text{qt}$$

Divide 4 and 2 by their common factor of 2. $4 \div 2$ is 2. $2 \div 2$ is 1.

$$= 30\ \text{qt}$$

(c) Convert 36 oz to pounds.

Notice that **oz** divides out, leaving **lb**, the unit you want for the answer.

$$\frac{\cancel{36}^9\ \cancel{\text{oz}}}{1} \bullet \frac{1\ \text{lb}}{\cancel{16}_4\ \cancel{\text{oz}}} = \frac{9}{4}\ \text{lb} = 2\frac{1}{4}\ \text{lb}$$

Note

In **Example 4(c)** above you get $\frac{9}{4}$ lb. Recall that $\frac{9}{4}$ means $9 \div 4$. If you do $9 \div 4$ on your calculator, you get 2.25 lb. U.S. measurements usually use fractions or mixed numbers, like $2\frac{1}{4}$ lb. However, 2.25 lb is also correct and is the way grocery stores often show weights of produce, meat, and cheese.

Work Problem 4 at the Side. ▶

4 Convert using unit fractions.

GS **(a)** 16 qt to gallons

unit fraction $\frac{1\ \text{gal}}{4\ \text{qt}}$

$$\frac{16\ \cancel{\text{qt}}}{1} \bullet \frac{1\ \text{gal}}{4\ \cancel{\text{qt}}} = \frac{16}{4}\ \text{gal} = \_\_\_\_$$

GS **(b)** 3 c to pints

unit fraction $\frac{\_\_\_\ \text{pt}}{\_\_\_\ \text{c}}$

(c) $3\frac{1}{2}$ tons to pounds

(d) $1\frac{3}{4}$ lb to ounces

(e) 4 oz to pounds

EXAMPLE 5 Using Several Unit Fractions

Sometimes you may need to use two or three unit fractions to complete a conversion.

(a) Convert 63 in. to yards.

Use the unit fraction $\frac{1\ \text{ft}}{12\ \text{in.}}$ to change inches to feet and the unit fraction $\frac{1\ \text{yd}}{3\ \text{ft}}$ to change feet to yards. Notice how all the units divide out except yards, which is the unit you want in the answer.

$$\frac{63\ \cancel{\text{in.}}}{1} \bullet \frac{1\ \cancel{\text{ft}}}{12\ \cancel{\text{in.}}} \bullet \frac{1\ \text{yd}}{3\ \cancel{\text{ft}}} = \frac{63}{36}\ \text{yd} = \frac{63 \div 9}{36 \div 9}\ \text{yd} = \frac{7}{4}\ \text{yd} = 1\frac{3}{4}\ \text{yd}$$

Continued on Next Page

Answers

4. **(a)** 4 gal **(b)** $\frac{1\ \text{pt}}{2\ \text{c}}$; $1\frac{1}{2}$ pt or 1.5 pt **(c)** 7000 lb **(d)** 28 oz **(e)** $\frac{1}{4}$ lb or 0.25 lb

5 Convert using two or three unit fractions.

(a) 4 tons to ounces

$$\frac{4\text{ tons}}{1} \bullet \frac{2000\text{ lb}}{1\text{ ton}} \bullet \frac{16\text{ oz}}{1\text{ lb}} =$$

(b) 3 mi to inches

(c) 36 pt to gallons

(d) 2 wk to minutes

Answers

5. (a) 128,000 oz (b) 190,080 in.
(c) $4\frac{1}{2}$ gal or 4.5 gal (d) 20,160 min

You can also divide out common factors in the numbers.

$$\frac{\overset{7}{\cancel{\overset{21}{\cancel{63}}}}}{1} \bullet \frac{1}{\underset{4}{\cancel{12}}} \bullet \frac{1}{\underset{1}{\cancel{3}}} = \frac{7}{4} = 1\frac{3}{4}\text{ yd}$$

Instead of changing $\frac{7}{4}$ to $1\frac{3}{4}$, you can enter $7 \div 4$ on your calculator to get 1.75 yd. Both answers are correct because 1.75 is equivalent to $1\frac{3}{4}$.

(b) Convert 2 days to seconds.

Use three unit fractions. The first one changes days to hours, the next one changes hours to minutes, and the last one changes minutes to seconds. All the units divide out except seconds, which is the unit you want in your answer.

$$\frac{2\text{ days}}{1} \bullet \frac{24\text{ hr}}{1\text{ day}} \bullet \frac{60\text{ min}}{1\text{ hr}} \bullet \frac{60\text{ sec}}{1\text{ min}} = 172{,}800\text{ sec}$$

Divide out **days**.
Divide out **hr**.
Divide out **min**.

Remember to write **sec** in your answer.

◀ **Work Problem 5 at the Side.**

OBJECTIVE 4 Solve application problems using U.S. measurement units. To solve application problems, we will use the steps you learned in **Chapter 1.** Those steps are summarized here.

Step 1 **Read** the problem carefully.

Step 2 **Work out a plan.**

Step 3 **Estimate** a reasonable answer.

Step 4 **Solve** the problem.

Step 5 **State the answer.**

Step 6 **Check** your work.

EXAMPLE 6 Solving U.S. Measurement Applications

(a) A 36 oz can of coffee is on sale at Jerry's Foods for $14.39. What is the cost per pound, to the nearest cent? (*Source:* Jerry's Foods.)

Step 1 **Read** the problem. The problem asks for the cost per *pound* of coffee.

Step 2 **Work out a plan.** The weight of the coffee is given in *ounces,* but the answer must be cost *per pound.* Convert ounces to pounds. The word *per* indicates division. You need to divide the cost by the number of pounds.

Step 3 **Estimate** a reasonable answer. To estimate, round $14.39 to $14. Then, there are 16 oz in a pound, so 36 oz are a little more than 2 pounds. Finally, $\$14 \div 2 = \$7$ per pound is our estimate.

Continued on Next Page

Step 4 **Solve** the problem. Use a unit fraction to convert 36 oz to pounds.

Notice that **oz** divides out, leaving **lb** for the answer.

$$\frac{\overset{9}{\cancel{36}}\ \cancel{\text{oz}}}{1} \cdot \frac{1\text{ lb}}{\underset{4}{\cancel{16}}\ \cancel{\text{oz}}} = \frac{9}{4}\text{ lb} = 2.25\text{ lb}$$

On your calculator, $9 \div 4 = 2.25$

Then divide to find the *cost* per *pound.*

$$\begin{matrix}\text{Cost} \rightarrow \\ \text{per} \rightarrow \\ \text{pound} \rightarrow\end{matrix} \frac{\$14.39}{2.25\text{ lb}} = 6.39\overline{5} \approx 6.40 \quad \text{(rounded)}$$

Step 5 **State the answer.** The coffee costs \$6.40 per pound (to the nearest cent).

Step 6 **Check** your work. The exact answer of \$6.40 is close to our estimate of \$7.

(b) Bilal's favorite cake recipe uses $1\frac{2}{3}$ cups of milk for one cake. If he makes six cakes for a bake sale at his son's school, how many quarts of milk will he need?

Step 1 **Read** the problem. The problem asks for the number of *quarts* of milk needed for six cakes.

Step 2 **Work out a plan.** Multiply to find the number of *cups* of milk for six cakes. Then convert *cups* to *quarts* (the unit required in the answer).

Step 3 **Estimate** a reasonable answer. To estimate, round $1\frac{2}{3}$ cups to 2 cups. Then, 2 cups times 6 = 12 cups. There are 4 cups in a quart, so 12 cups ÷ 4 = 3 quarts is our estimate.

Step 4 **Solve** the problem. First multiply. Then use unit fractions to convert.

$$1\frac{2}{3} \cdot 6 = \frac{5}{\underset{1}{\cancel{3}}} \cdot \frac{\overset{2}{\cancel{6}}}{1} = \frac{10}{1} = 10\text{ cups} \quad \left\{ \text{Milk needed for six cakes} \right.$$

$$\frac{\overset{5}{\cancel{10}}\ \cancel{\text{cups}}}{1} \cdot \frac{1\ \cancel{\text{pt}}}{\underset{1}{\cancel{2}}\ \cancel{\text{cups}}} \cdot \frac{1\text{ qt}}{2\ \cancel{\text{pt}}} = \frac{5}{2}\text{ qt} = 2\frac{1}{2}\text{ qt}$$

Both **cups** and **pt** divide out, leaving **qt,** the unit you want for the answer.

Step 5 **State the answer.** Bilal needs $2\frac{1}{2}$ qt (or 2.5 qt) of milk.

Step 6 **Check** your work. The exact answer of $2\frac{1}{2}$ qt is close to our estimate of 3 qt.

Note

In Step 2 above, we *first multiplied* $1\frac{2}{3}$ cups by 6 to find the number of cups needed, then *converted* 10 cups to $2\frac{1}{2}$ quarts. It would also work to *first convert* $1\frac{2}{3}$ cups to $\frac{5}{12}$ qt, then *multiply* $\frac{5}{12}$ qt by 6 to get $2\frac{1}{2}$ qt.

Work Problem 6 at the Side. ▶

6 Solve each application problem using the six problem-solving steps.

GS **(a)** Kristin paid \$3.29 for 12 oz of extra sharp cheddar cheese. What is the price per pound, to the nearest cent?

Step 1 Problem asks for price per ________.

Step 2 So change 12 oz to what unit? ________

(b) A moving company estimates 11,000 lb of furnishings for an average 3-bedroom house. If the company made five such moves last week, how many tons of furnishings did it move? (*Source:* North American Van Lines.)

Answers

6. (a) pound; pounds; \$4.39 per pound (rounded)

(b) 27.5 or $27\frac{1}{2}$ tons

A.1 Exercises

FOR EXTRA HELP

Download the MyDashBoard App

MyMathLab®

CONCEPT CHECK *Rewrite each group of measurement units in order from smallest to largest.*

1. **(a)** foot, mile, inch, yard

(b) pint, fluid ounce, gallon, quart, cup

2. **(a)** ton, ounce, pound

(b) day, hour, week, second, minute

Fill in the blanks with the measurement relationships you have memorized. ***See Example 1.***

3. **(a)** 1 yd = ______ ft; **(b)** ______ in. = 1 ft

4. **(a)** 1 ft = ______ in.; **(b)** ______ ft = 1 mi

5. **(a)** ______ fl oz = 1 c; **(b)** 1 qt = ______ pt

6. **(a)** ______ qt = 1 gal; **(b)** 1 pt = ______ c

7. **(a)** ______ lb = 1 ton; **(b)** 1 lb = ______ oz

8. **(a)** ______ oz = 1 lb; **(b)** 1 ton = ______ lb

9. **(a)** 1 min = ______ sec; **(b)** ______ min = 1 hr

10. **(a)** 1 day = ______ hr; **(b)** ______ sec = 1 min

Convert each measurement by multiplying or dividing. ***See Example 2.***

11. **(a)** 120 sec = ______ min

(b) 4 hr = ______ min

12. **(a)** 180 min = ______ hr

(b) 5 min = ______ sec

13. **(a)** 2 qt = ______ gal

(b) $6\frac{1}{2}$ ft = ______ in.

14. **(a)** $4\frac{1}{2}$ gal = ______ qt

(b) 12 oz = ______ lb

15. An adult African elephant could weigh 7 to 8 tons. How many pounds could it weigh? (*Source: The Top 10 of Everything.*)

16. A reticulated python snake is the world's longest snake. It grows to a length of 18 to 33 feet. How many yards long can the snake be? (*Source: The Top 10 of Everything.*)

Convert each measurement in Exercises 17–38 using unit fractions. ***See Examples 3 and 4.***

17. 9 yd = ______ ft

GS unit fraction $\Big\}\ \frac{\text{ft}}{\text{yd}}$

18. 20,000 lb = ______ tons

GS unit fraction $\Big\}\ \frac{\text{ton}}{\text{lb}}$

19. 7 lb = ______ oz

20. 96 oz = ______ lb

21. 5 qt = ______ pt

22. 26 pt = ______ qt

23. 90 min = __________ hr

24. 45 sec = __________ min

25. 3 in. = __________ ft

26. 30 in. = __________ ft

27. 24 oz = __________ lb

28. 36 oz = __________ lb

29. 5 c = __________ pt

30. 15 qt = __________ gal

Use the information in the bar graph below to answer Exercises 31–32.

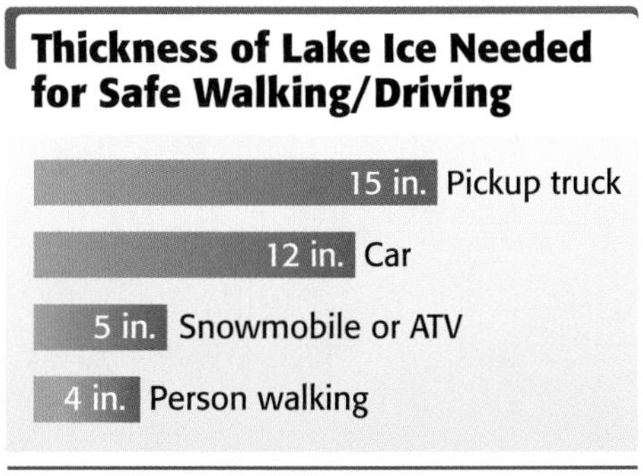

Source: Wisconsin DNR.

31. If the ice on a lake is $\frac{1}{2}$ ft thick, what will it safely support?

32. How many feet of ice are needed to safely drive a pickup truck on a lake?

33. $2\frac{1}{2}$ tons = ______ lb

34. $4\frac{1}{2}$ pt = ______ c

35. $4\frac{1}{4}$ gal = ______ qt

36. $2\frac{1}{4}$ hr = ______ min

37. After 15 years, a saguaro cactus is still only one-third to two-thirds of a foot tall, depending upon rainfall. How tall could the cactus be in inches? (*Source: Ecology of the Saguaro III.*)

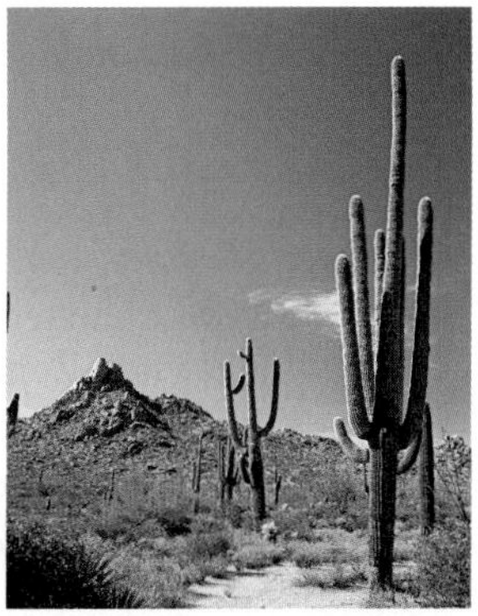

38. Kevin Durant, an NBA basketball player, is $6\frac{3}{4}$ ft tall. What is his height in inches? (*Source:* NBA.com)

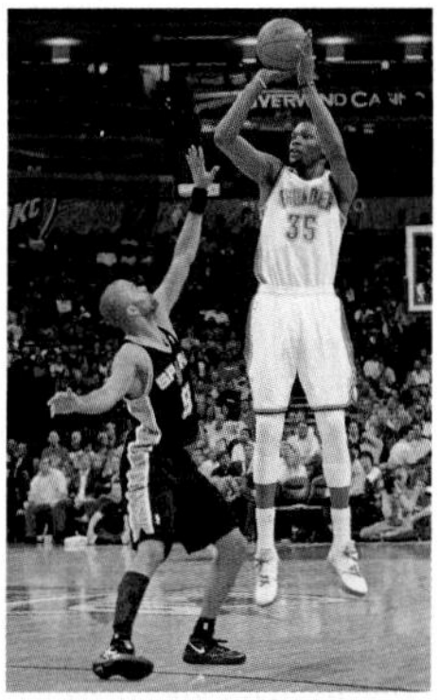

Use two or three unit fractions to make each conversion. ***See Example 5.***

39. 6 yd = ______ in.

40. 2 tons = ______ oz

41. 112 c = ______ qt

42. 336 hr = ______ wk

43. 6 days = ______ sec

44. 5 gal = ______ c

45. $1\frac{1}{2}$ tons = ______ oz

46. $3\frac{1}{3}$ yd = ______ in.

47. CONCEPT CHECK The statement 8 = 2 is *not* true. But with appropriate measurement units, it *is* true.

8 quarts = *2 gallons*

Attach measurement units to these numbers to make the statement true.

(a) 1 __________ = 16 __________

(b) 10 __________ = 20 __________

(c) 120 __________ = 2 __________

(d) 2 __________ = 24 __________

(e) 6000 __________ = 3 __________

(f) 35 __________ = 5 __________

48. CONCEPT CHECK Explain in your own words why you can add 2 feet + 12 inches to get 3 feet, but you cannot add 2 feet + 12 pounds.

Convert each measurement. ***See Example 5.***

49. $2\frac{3}{4}$ mi = __________ in.

50. $5\frac{3}{4}$ tons = __________ oz

51. $6\frac{1}{4}$ gal = __________ fl oz

52. $3\frac{1}{2}$ days = __________ sec

53. 24,000 oz = __________ ton

54. 57,024 in. = __________ mi

Solve each application problem. Show your work. ***See Example 6.***

55. Geralyn bought 20 oz of strawberries for $2.29. What was the price per pound for the strawberries, to the nearest cent?

56. Zach paid $0.90 for a 0.8 oz candy bar. What was the cost per pound?

57. Dan orders supplies for the science labs. Each of the 24 stations in the chemistry lab needs 2 ft of rubber tubing. If rubber tubing sells for $8.75 per yard, how much will it cost to equip all the stations?

58. In 2009, Marquette, Michigan, had 181.0 inches of snowfall, while Reno, Nevada, had 22.5 inches. What was the difference in snowfall between the two cities, in feet? Round to the nearest tenth. (*Source: World Almanac and Book of Facts.*)

59. Tropical cockroaches are the fastest land insects. They can run about 5 feet per second. At this rate, how long would it take the cockroach to travel one mile? (*Source: Guinness World Records.*)

Give your answer
(a) in seconds.
(b) in minutes.

60. A snail moves at an average speed of 2 feet every 3 minutes. At that rate, how long would it take the snail to travel one mile? (*Source: The Concise Nature Encyclopedia.*)

Give your answer
(a) in hours.
(b) in days.

61. At the day care center, each of the 15 toddlers drinks about $\frac{2}{3}$ cup of milk with lunch. The center is open 5 days a week.
(a) How many quarts of milk will the center need for one week of lunches?
(b) If the center buys milk in gallon containers, how many containers should be ordered for one week?

62. Farley's and Sathers Candy Company makes 995,000 pounds of gummies candy each day. (*Source:* Farley's and Sathers.)
(a) How many tons of gummies are produced during a 5-day workweek?
(b) The plant operates 24 hours per day. How many tons of gummies are produced each hour, to the nearest tenth?

Relating Concepts (Exercises 63–66) For Individual or Group Work

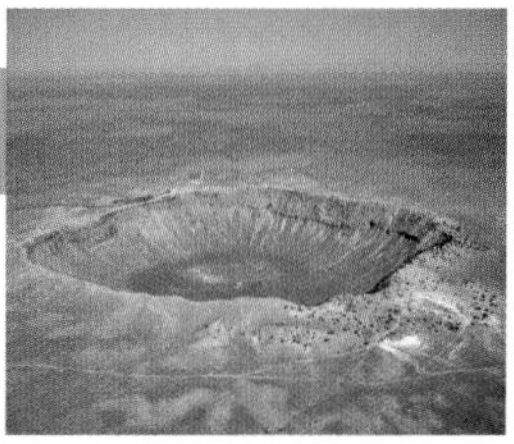

On the first page of this chapter, we said that the bowl-shaped crater in northern Arizona made by a meteor crash was sixty stories deep and 4150 ft across. Use this information as you ***work Exercises 63–66 in order.*** *(Source: The Meteor Crater Story.)*

63. **(a)** The distance across the crater is what part of a mile, to the nearest tenth?
(b) The crater is nearly circular. In a circle, the distance around the outside edge is about 3.14 times the distance across the circle. How far is it to walk around the edge of the crater, as measured in feet?
(c) How far is it to walk around the edge of the crater in miles, to the nearest tenth?

64. **(a)** The crater is 550 ft deep. The depth is how many yards, to the nearest whole number?
(b) How many inches deep is the crater?
(c) The depth of the crater is what part of a mile, to the nearest tenth?
(d) When we say that the crater is as deep as a sixty-story building, we are assuming that each story is how many feet tall, to the nearest foot?

65. On one side of the crater there are a few small juniper trees. The trees are 700 years old but only 18 inches to 30 inches tall because of the strong winds and lack of rain.
(a) How tall are the trees in feet?
(b) How many months old are the trees?
(c) At this rate of growth, how long would it take a 30-inch tree to reach a height of three feet?

66. Evidence of two huge meteor crashes has been found on the floor of the Caribbean Sea. One giant circular crater is 90 miles across and the other is 120 miles across.
(a) Using the information about circles in **Exercise 63,** what is the approximate distance around the edge of the smaller crater, to the nearest mile?
(b) Around the larger crater?

A.2 The Metric System—Length

OBJECTIVES

1. Learn the basic metric units of length.
2. Use unit fractions to convert among metric units.
3. Move the decimal point to convert among metric units.

VOCABULARY TIP

Meter means "measurement." The metric system is simply a system using certain types of measurement units. The basic unit for measuring length is the **meter**.

1 Circle the items that measure about 1 meter.

Length of a pencil

Length of a baseball bat

Height of doorknob from the floor

Height of a house

Basketball player's arm length

Length of a paper clip

Answer

1. baseball bat, height of doorknob, basketball player's arm length

Around 1790, a group of French scientists developed the metric system of measurement. It is an organized system based on multiples of 10, like our number system and our money. After you are familiar with metric units, you will see that they are easier to use than the hodgepodge of U.S. measurement relationships you used in the last section.

Note

The metric system information in this text is consistent with usage guidelines from the National Institute of Standards and Technology, www.nist.gov/metric.

OBJECTIVE 1 Learn the basic metric units of length. The basic unit of length in the metric system is the **meter** (also spelled *metre*). Use the symbol **m** for meter; do **not** put a period after it. If you put five of the pages from this textbook side by side, they would measure about 1 meter. Or, look at a yardstick—a meter is just a little longer. A yard is 36 inches long, while a meter is about 39 inches long.

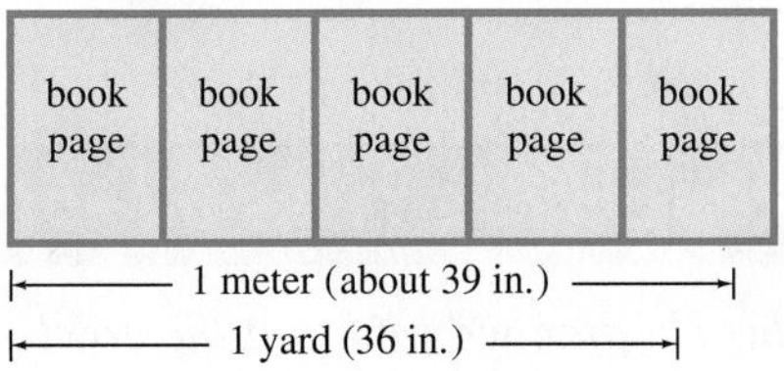

In the metric system, you use meters for things like measuring the length of your living room, talking about heights of buildings, or describing track and field athletic events.

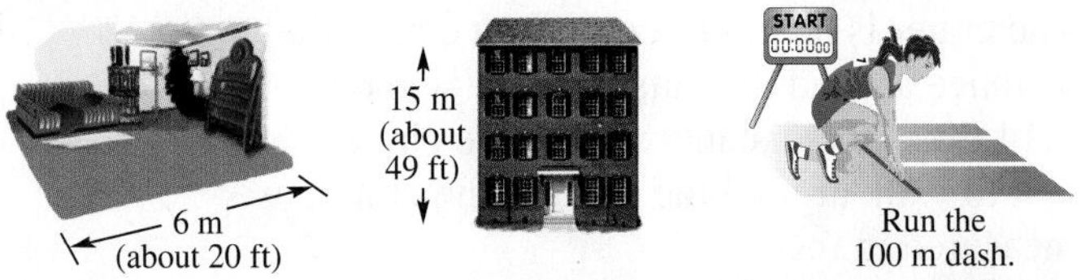

◀ **Work Problem 1 at the Side.**

To make longer or shorter length units in the metric system, **prefixes** are written in front of the word *meter*. For example, the prefix *kilo* means **1000**, so a *kilo*meter is **1000** meters. The table below shows how to use the prefixes for length measurements. It is helpful to memorize the prefixes because they are also used with weight and capacity measurements. The arrows point to the units you will use most often in daily life.

Prefix	*kilo-* meter	*hecto-* meter	*deka-* meter	meter	*deci-* meter	*centi-* meter	*milli-* meter
Meaning	1000 meters	100 meters	10 meters	1 meter	$\frac{1}{10}$ of a meter	$\frac{1}{100}$ of a meter	$\frac{1}{1000}$ of a meter
Symbol	km	hm	dam	m	dm	cm	mm

Length units that are used most often (km, m, cm, mm)

Here are some comparisons to help you get acquainted with the commonly used length units: km, m, cm, mm.

*Kilo*meters are used instead of miles. A kilometer is **1000** meters. It is about 0.6 mile (a little more than half a mile) or about 5 to 6 city blocks. If you participate in a 10 km run, you'll run about 6 miles.

A meter is divided into 100 smaller pieces called ***centi***meters. Each centimeter is $\frac{1}{100}$ of a meter. Centimeters are used instead of inches. A centimeter is a little shorter than $\frac{1}{2}$ inch. The cover of this textbook is about 21 cm wide. A nickel is about 2 cm across. Measure the width and length of your little finger on this centimeter ruler. The tip of your little finger is probably between 1 cm and 2 cm wide.

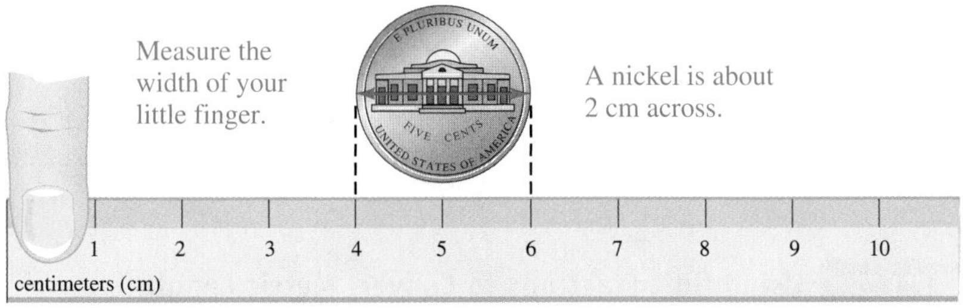

A meter is divided into 1000 smaller pieces called ***milli***meters. Each millimeter is $\frac{1}{1000}$ of a meter. It takes 10 mm to equal 1 cm, so it is a very small length. The thickness of a dime is about 1 mm. Measure the width of your pen or pencil and the width of your little finger on this millimeter ruler.

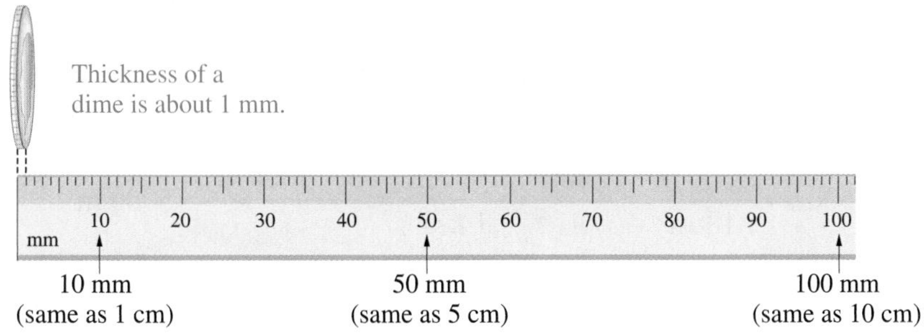

EXAMPLE 1 Using Metric Length Units

Write the most reasonable metric unit in each blank. Choose from km, m, cm, and mm.

(a) The distance from home to work is 20 ____.

20 **km** because kilometers are used instead of miles. 20 km is about 12 miles.

(b) My wedding ring is 4 ____ wide.

4 **mm** because the width of a ring is very small.

(c) The newborn baby is 50 ____ long.

50 **cm**, which is half of a meter. A meter is about 39 inches, so half a meter is around 20 inches.

Work Problem 2 at the Side. ▶

VOCABULARY TIP

Prefix **Pre-** means "before" and **fix** means "stick." So, a prefix is a word part stuck on, or attached, to the front of a word. Prefixes commonly used with **meter** include **kilo**meter (1000 m); **centi**meter ($\frac{1}{100}$ m); and **milli**meter ($\frac{1}{1000}$ m).

2 Write the most reasonable metric unit in each blank. Choose from km, m, cm, and mm.

(a) The woman's height is 168 ____.

(b) The man's waist is 90 ____ around.

(c) Louise ran the 100 ____ dash in the track meet.

(d) A postage stamp is 22 ____ wide.

(e) Michael paddled his canoe 2 ____ down the river.

(f) The pencil lead is 1 ____ thick.

(g) A stick of gum is 7 ____ long.

(h) The highway speed limit is 90 ____ per hour.

(i) The classroom was 12 ____ long.

(j) A penny is about 18 ____ across.

Answers

2. (a) cm **(b)** cm **(c)** m **(d)** mm **(e)** km **(f)** mm **(g)** cm **(h)** km **(i)** m **(j)** mm

3 First write the unit fraction needed to make each conversion. Then complete the conversion.

GS **(a)** 3.67 m to cm

unit fraction $\left.\right\}\ \dfrac{100\text{ cm}}{1\text{ m}}$

$\dfrac{3.67\ \cancel{\text{m}}}{1} \cdot \dfrac{100\text{ cm}}{1\ \cancel{\text{m}}}$

$= \dfrac{3.67 \cdot 100\text{ cm}}{1} = \underline{\qquad}$

(b) 92 cm to m

unit fraction $\left.\right\}\ \dfrac{\underline{\quad}\text{ m}}{\underline{\quad}\text{ cm}}$

(c) 432.7 cm to m

unit fraction } ______

(d) 65 mm to cm

unit fraction } ______

(e) 0.9 m to mm

unit fraction } ______

(f) 2.5 cm to mm

unit fraction } ______

Answers

3. **(a)** 367 cm **(b)** $\dfrac{1\text{ m}}{100\text{ cm}}$; 0.92 m

(c) $\dfrac{1\text{ m}}{100\text{ cm}}$; 4.327 m

(d) $\dfrac{1\text{ cm}}{10\text{ mm}}$; 6.5 cm

(e) $\dfrac{1000\text{ mm}}{1\text{ m}}$; 900 mm

(f) $\dfrac{10\text{ mm}}{1\text{ cm}}$; 25 mm

OBJECTIVE 2 **Use unit fractions to convert among metric units.** You can convert among metric length units using unit fractions. Keep these relationships in mind when setting up the unit fractions.

Metric Length Relationships

1 km = 1000 m so the unit fractions are: $\dfrac{1\text{ km}}{1000\text{ m}}$ or $\dfrac{1000\text{ m}}{1\text{ km}}$	**1 m = 1000 mm** so the unit fractions are: $\dfrac{1\text{ m}}{1000\text{ mm}}$ or $\dfrac{1000\text{ mm}}{1\text{ m}}$
1 m = 100 cm so the unit fractions are: $\dfrac{1\text{ m}}{100\text{ cm}}$ or $\dfrac{100\text{ cm}}{1\text{ m}}$	**1 cm = 10 mm** so the unit fractions are: $\dfrac{1\text{ cm}}{10\text{ mm}}$ or $\dfrac{10\text{ mm}}{1\text{ cm}}$

EXAMPLE 2 **Using Unit Fractions to Convert Metric Length Measurements**

Convert each measurement using unit fractions.

(a) 5 km to m

Put the unit for the answer (meters) in the numerator of the unit fraction; put the unit you want to change (km) in the denominator.

Unit fraction equivalent to 1 $\left\{\dfrac{1000\text{ m}}{1\text{ km}}\right.$ ← Unit for your **answer** is **m.** / ← Unit being **changed** is **km.**

Multiply. Divide out common units where possible.

$$5\text{ km} \cdot \frac{1000\text{ m}}{1\text{ km}} = \frac{5\ \cancel{\text{km}}}{1} \cdot \frac{1000\text{ m}}{1\ \cancel{\text{km}}} = \frac{5 \cdot 1000\text{ m}}{1} = 5000\text{ m}$$

These units should match.

Here, **km** divides out leaving **m**, the unit you want for your answer.

Do **not** write a period here.

5 km = 5000 m

The answer makes sense because a kilometer is much longer than a meter, so 5 km will contain many meters.

(b) 18.6 cm to m

Multiply by a unit fraction that allows you to divide out centimeters.

$$\frac{18.6\ \cancel{\text{cm}}}{1} \cdot \underbrace{\frac{1\text{ m}}{100\ \cancel{\text{cm}}}}_{\text{Unit fraction}} = \frac{18.6}{100}\text{ m} = 0.186\text{ m}$$

Do **not** write a period here.

18.6 cm = 0.186 m

There are 100 cm in a meter, so 18.6 cm will be a small part of a meter. The answer makes sense.

◀ **Work Problem 3 at the Side.**

OBJECTIVE 3 Move the decimal point to convert among metric units. By now you have probably noticed that conversions among metric units are made by multiplying or dividing by 10, by 100, or by 1000. A quick way to *multiply* by 10 is to move the decimal point one place to the *right*. Move it two places to the right to multiply by 100, three places to multiply by 1000. *Dividing* is done by moving the decimal point to the *left* in the same manner.

Work Problem 4 at the Side. ▶

An alternate conversion method to unit fractions is moving the decimal point using this **metric conversion line.**

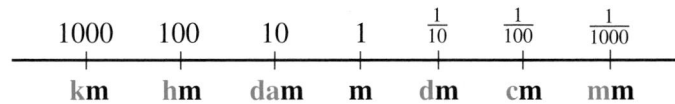

1000	100	10	1	$\frac{1}{10}$	$\frac{1}{100}$	$\frac{1}{1000}$
km	**hm**	**dam**	**m**	**dm**	**cm**	**mm**

Here are the steps for using the conversion line.

Using the Metric Conversion Line

Step 1 Find the unit you are given on the metric conversion line.

Step 2 Count the number of places to get from the unit you are given to the unit you want in the answer.

Step 3 Move the decimal point the ***same number of places*** and in the ***same direction*** as you did on the conversion line.

EXAMPLE 3 Using the Metric Conversion Line

Use the metric conversion line to make the following conversions.

(a) 5.702 km to m

Find **km** on the metric conversion line. To get to **m**, you move *three places* to the *right*. So move the decimal point in 5.702 *three places* to the *right*.

km **hm** **dam** **m** **dm** **cm** **mm**

Three places to the right

5.702

Move decimal point three places to the right.

5.702 km = 5702 m

(b) 69.5 cm to m

Find **cm** on the conversion line. To get to **m**, move *two places* to the *left*.

km **hm** **dam** **m** **dm** **cm** **mm**

Two places to the left

69.5

Move decimal point two places to the left.

69.5 cm = 0.695 m

Continued on Next Page

4 Do each multiplication or division by hand or on a calculator. Compare your answer to the one you get by moving the decimal point.

GS **(a)** (43.5) (10) = ______

43.5 gives 435.

(b) 43.5 ÷ 10 = ______

43.5 gives ______.

(c) (28) (100) = ______

28.00 gives ______.

(d) 28 ÷ 100 = ______

28. gives ______.

(e) (0.7) (1000) = ______

0.700 gives ______.

(f) 0.7 ÷ 1000 = ______

000.7 gives ______.

Answers

4. (a) 435 **(b)** 4.35; 4.35 **(c)** 2800; 2800 **(d)** 0.28; 0.28 **(e)** 700; 700 **(f)** 0.0007; 0.0007

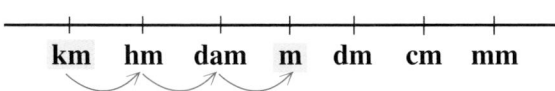

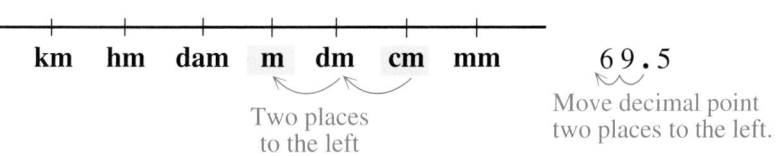

5 Convert using the metric conversion line.

GS **(a)** 12.008 km to m

km to m is three places to the right.

12.008 km = ______ m

GS **(b)** 561.4 m to km

m to km is ______ places to the ______.

561.4 m = ______ km

(c) 20.7 cm to m

(d) 20.7 cm to mm

(e) 4.66 m to cm

6 Convert using the metric conversion line.

GS **(a)** 9 m to mm 9.000

m to mm is ______ places to the ______.

9 m = ______ mm

GS **(b)** 3 cm to m

cm to m is ______ places to the ______.

3 cm = ______ m

(c) 5 mm to cm

(d) 70 m to km

(e) 0.8 m to cm

Answers

5. **(a)** 12,008 m **(b)** three; left; 0.5614 km **(c)** 0.207 m **(d)** 207 mm **(e)** 466 cm

6. **(a)** three; right; 9000 mm **(b)** two; left; 0.03 m **(c)** 0.5 cm **(d)** 0.07 km **(e)** 80 cm

(c) 8.1 cm to mm

From **cm** to **mm** is *one place* to the *right*.

km hm dam m dm cm mm 8.1

One place to the right

Move decimal point one place to the right.

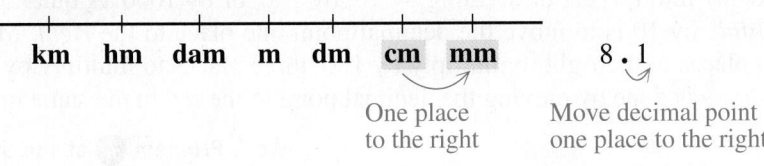

8.1 cm = 81 mm

◀ **Work Problem 5 at the Side.**

EXAMPLE 4 Practicing Length Conversions

Convert using the metric conversion line.

(a) 1.28 m to mm

Moving from **m** to **mm** is going *three places to the right*. In order to move the decimal point in 1.28 three places to the right, you must write a 0 as a placeholder.

1.28**0** Zero is written in as a placeholder.

Move decimal point three places to the right.

1.28 m = 1280 mm

The answer is a *whole number*, so you do not need to write the decimal point.

(b) 60 cm to m

From **cm** to **m** is two places to the left. The decimal point in 60 starts at the *far right side* because 60 is a whole number. Then move it two places to the left.

60. — Decimal point starts here.

60. — Move decimal point two places to the left.

60 cm = 0.60 m, and 0.60 m is equivalent to 0.6 m.

Think: $\frac{60}{100}$ in lowest terms is $\frac{6}{10}$.

(c) 8 m to km

From **m** to **km** is three places to the left. The decimal point in 8 starts at the far right side. In order to move it three places to the left, you must write two zeros as placeholders.

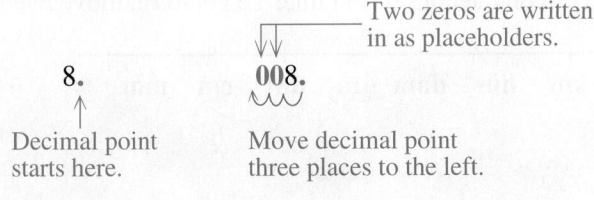

Two zeros are written in as placeholders.

8. — Decimal point starts here.

008. — Move decimal point three places to the left.

8 m = 0.008 km

◀ **Work Problem 6 at the Side.**

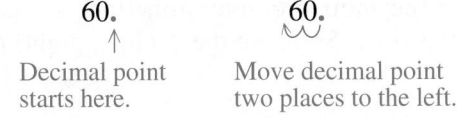

A.2 Exercises

FOR EXTRA HELP

Download the MyDashBoard App

MyMathLab®

CONCEPT CHECK *Use your knowledge of the meaning of metric prefixes to fill in the blanks.*

1. *kilo* means ___1000___, so 1 km = __________ m.

2. *deka* means ___10___, so 1 dam = __________ m.

3. *milli* means __________, so 1 mm = __________ m.

4. *deci* means __________, so 1 dm = __________ m.

5. *centi* means __________, so 1 cm = __________ m.

6. *hecto* means __________, so 1 hm = __________ m.

7. Circle the shorter length unit in each pair.
(a) cm m **(b)** m mm **(c)** km cm

8. Circle the longer length unit in each pair.
(a) m km **(b)** mm cm **(c)** km mm

Use this ruler to measure the width of your thumb and hand for Exercises 9–12.

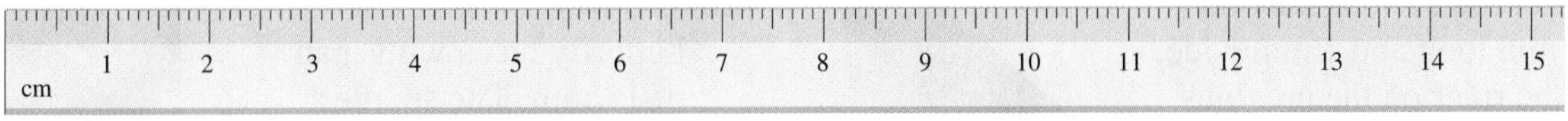

9. The width of your hand in centimeters

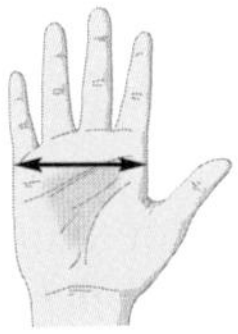

10. The width of your hand in millimeters

11. The width of your thumb in millimeters

12. The width of your thumb in centimeters

Write the most reasonable metric length unit in each blank. Choose from km, m, cm, and mm. ***See Example 1.***

13. The child was 91 _____ tall.

14. The cardboard was 3 _____ thick.

15. Ming-Na swam in the 200 _____ backstroke race.

16. The bookcase is 75 _____ wide.

17. Adriana drove 400 _____ on her vacation.

18. The door is 2 _____ high.

19. An aspirin tablet is 10 _____ across.

20. Lamard jogs 4 _____ every morning.

21. A paper clip is about 3 _____ long.

22. My pen is 145 _____ long.

23. Dave's truck is 5 _____ long.

24. Wheelchairs need doorways that are at least 80 _____ wide.

Convert each measurement. Use unit fractions or the metric conversion line.
See Examples 2–4.

25. 7 m to cm

GS $\frac{7\,\cancel{m}}{1} \cdot \frac{100\text{ cm}}{1\,\cancel{m}} = \frac{7 \cdot 100\text{ cm}}{1} =$

26. 18 m to cm

GS $\frac{18\,\cancel{m}}{1} \cdot \frac{100\text{ cm}}{1\,\cancel{m}} =$

27. 40 mm to m

28. 6 mm to m

29. 9.4 km to m

30. 0.7 km to m

31. 509 cm to m

32. 30 cm to m

33. 400 mm to cm

34. 25 mm to cm

35. 0.91 m to mm

36. 4 m to mm

37. CONCEPT CHECK Is 82 cm greater than or less than 1 m? What is the difference in the lengths?

38. CONCEPT CHECK Is 1022 m greater than or less than 1 km? What is the difference in the lengths?

39. Computer microchips may be only 5 mm long and 1 mm wide. Using the ruler on the previous page, draw a rectangle that measures 5 mm by 1 mm. Then convert each measurement to centimeters.

A greatly enlarged photo of a computer microchip

40. The world's smallest butterfly has a wingspan of 15 mm. The smallest mouse is 50 mm long. Using the ruler on the previous page, draw a line that is 15 mm long and a line 50 mm long. Then convert each measurement to centimeters. (*Source*: *Top 10 of Everything*.)

41. The Roe River near Great Falls, Montana, is the shortest river in the world, just 61 m long. How many kilometers long is the Roe River? (*Source: Guinness World Records*.)

42. There are over 100,000 km of blood vessels in the human body. How many meters of blood vessels are in the body? (*Source: The Human Body*.)

43. The median height for U.S. females who are 20 to 29 years old is about 1.64 m. Convert this height to centimeters and to millimeters. (*Source:* U.S. National Center for Health Statistics.)

44. The median height for 20- to 29-year-old males in the United States is about 177 cm. Convert this height to meters and to millimeters. (*Source:* U.S. National Center for Health Statistics.)

45. Use two unit fractions to convert 5.6 mm to km.

46. Use two unit fractions to convert 16.5 km to mm.

A.3 The Metric System—Capacity and Weight (Mass)

OBJECTIVES

1. Learn the basic metric units of capacity.
2. Convert among metric capacity units.
3. Learn the basic metric units of weight (mass)
4. Convert among metric weight (mass) units.
5. Distinguish among basic metric units of length, capacity, and weight (mass).

We use capacity units to measure liquids, such as the amount of milk in a recipe, the gasoline in our car tank, and the water in an aquarium. (The capacity units in the U.S. system are cups, pints, quarts, and gallons.) The basic metric unit for capacity is the **liter** (also spelled *litre*). The capital letter L is the symbol for liter, to avoid confusion with the numeral 1.

OBJECTIVE 1 **Learn the basic metric units of capacity.** The liter is related to metric length in this way: A box that measures 10 cm on every side holds exactly one liter. (The volume of the box is 10 cm • 10 cm • 10 cm = 1000 cubic centimeters. Volume is discussed in **Chapter 7.**) A liter is just a little more than 1 quart.

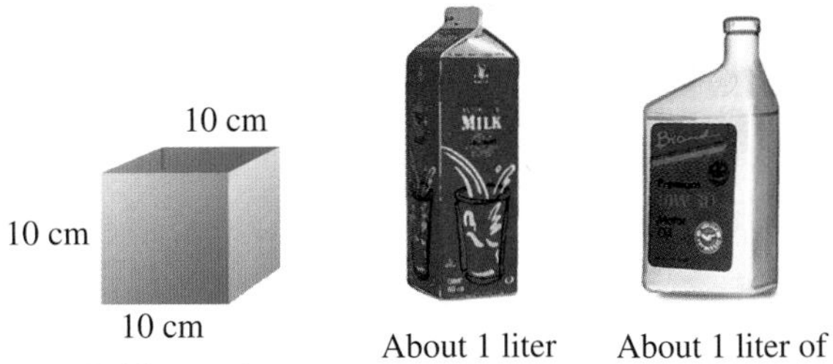

Holds exactly 1 liter (L)

About 1 liter of milk

About 1 liter of oil for your car

A liter is a little more than one quart (just $\frac{1}{4}$ cup more).

In the metric system you use liters for things like buying milk and soda at the store, filling a pail with water, and describing the size of your home aquarium.

Buy a 2 L bottle of soda.

Use a 12 L pail to wash floors.

Watch the fish in your 40 L aquarium.

Work Problem 1 at the Side. ▶

To make larger or smaller capacity units, we use the same **prefixes** as we did with length units. For example, the prefix *kilo* means 1000 so a *kilo*meter is 1000 meters. In the same way, a *kilo*liter is 1000 liters.

Prefix	*kilo-* liter	*hecto-* liter	*deka-* liter	liter	*deci-* liter	*centi-* liter	*milli-* liter
Meaning	1000 liters	100 liters	10 liters	1 liter	$\frac{1}{10}$ of a liter	$\frac{1}{100}$ of a liter	$\frac{1}{1000}$ of a liter
Symbol	kL	hL	daL	L	dL	cL	mL

Capacity units used most often (liter and milliliter)

VOCABULARY TIP

Liter Connect **liter** to a 2-liter bottle of soda. Use **liters** to measure **liquids** (things that pour).

1 Circle the things that can be measured in liters.

Amount of water in the bathtub

Length of the bathtub

Width of your car

Amount of gasoline you buy for your car

Weight of your car

Height of a pail

Amount of water in a pail

Answer

1. water in bathtub, gasoline, water in a pail

VOCABULARY TIP

L; mL Think: liter and quart; milliliter and "a drop."

2 Write the most reasonable metric unit in each blank. Choose from L and mL.

(a) I bought 8 ____ of soda at the store.

(b) The nurse gave me 10 ____ of cough syrup.

(c) This is a 100 ____ garbage can.

(d) It took 10 ____ of paint to cover the bedroom walls.

(e) My car's gas tank holds 50 ____.

(f) I added 15 ____ of oil to the pancake mix.

(g) The can of orange soda holds 350 ____.

(h) My friend gave me a 30 ____ bottle of expensive perfume.

The capacity units you will use most often in daily life are liters (L) and *milli*liters (mL). A tiny box that measures 1 cm on every side holds exactly one milliliter. (In medicine, this small amount is also called 1 cubic centimeter, or 1 cc for short.) It takes 1000 mL to make 1 L. Here are some comparisons.

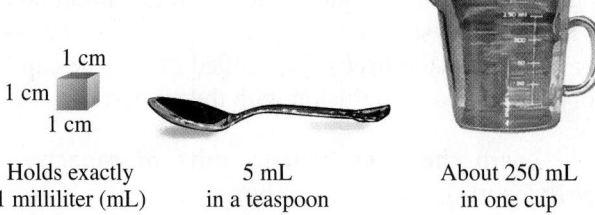

Holds exactly 1 milliliter (mL)

5 mL in a teaspoon

About 250 mL in one cup

EXAMPLE 1 Using Metric Capacity Units

Write the most reasonable metric unit in each blank. Choose from L and mL.

(a) The bottle of shampoo held 500 _______.
500 **mL** because 500 L would be about 500 quarts, which is too much.

(b) I bought a 2 _______ carton of orange juice.
2 **L** because 2 mL would be less than a teaspoon.

◀ **Work Problem 2 at the Side.**

OBJECTIVE 2 Convert among metric capacity units. Just as with length units, you can convert between milliliters and liters using unit fractions.

Metric Capacity Relationships

1 L = 1000 mL, so the unit fractions are:

$$\frac{1\text{ L}}{1000\text{ mL}} \quad \text{or} \quad \frac{1000\text{ mL}}{1\text{ L}}$$

Or you can use a metric conversion line to decide how to move the decimal point.

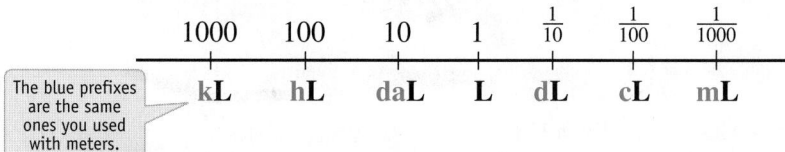

1000	100	10	1	$\frac{1}{10}$	$\frac{1}{100}$	$\frac{1}{1000}$
kL	hL	daL	L	dL	cL	mL

The blue prefixes are the same ones you used with meters.

EXAMPLE 2 Converting among Metric Capacity Units

Convert using the metric conversion line or unit fractions.

(a) 2.5 L to mL

Using the metric conversion line:

From **L** to **mL** is *three places* to the *right.*

2.500 (Write two zeros as placeholders.)

2.5 L = 2500 mL

Using unit fractions:

Multiply by a unit fraction that allows you to divide out liters.

$$\frac{2.5\text{ }\cancel{\text{L}}}{1} \cdot \frac{1000\text{ mL}}{1\text{ }\cancel{\text{L}}} = 2500\text{ mL}$$

L divides out, leaving **mL** for your answer.

Do **not** write a period here.

Continued on Next Page

Answers

2. **(a)** L **(b)** mL **(c)** L **(d)** L **(e)** L **(f)** mL **(g)** mL **(h)** mL

(b) 80 mL to L

Using the metric conversion line:

From **mL** to **L** is *three places* to the *left*.

80. (Decimal point starts here.) → 080. (Move decimal point three places to the left.)

80 mL = 0.080 L or 0.08 L

Using unit fractions:

Multiply by a unit fraction that allows you to divide out mL.

$$\frac{80\ \cancel{\text{mL}}}{1} \cdot \frac{1\ \text{L}}{1000\ \cancel{\text{mL}}}$$

mL divides out, leaving **L** for your answer.

$$= \frac{80}{1000}\ \text{L} = 0.08\ \text{L}$$

Do **not** write a period here.

Work Problem **3** at the Side. ▶

OBJECTIVE 3 Learn the basic metric units of weight (mass). The **gram** is the basic metric unit for *mass*. Although we often call it "weight," there is a difference.

Weight is a measure of the pull of gravity; the farther you are from the center of Earth, the less you weigh. In outer space you become weightless, but your mass, the amount of matter in your body, stays the same regardless of where you are.

In science courses, it will be important to distinguish between the weight of an object and its mass. But for everyday purposes, we will use the word *weight*.

The gram is related to metric length in this way: The weight of the water in a box measuring 1 cm on every side is 1 gram. This is a very tiny amount of water (1 mL) and a very small weight. One gram is also the weight of a dollar bill or a single raisin. A nickel weighs 5 grams. A plain, regular-sized hamburger (with bun) weighs from 175 to 200 grams.

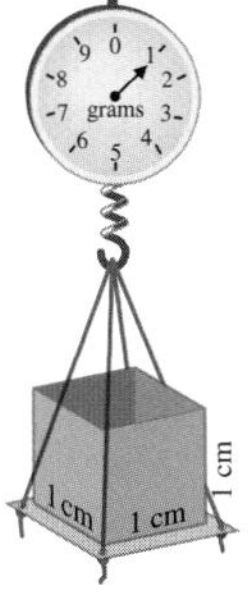

The 1 mL of water in this tiny box weighs 1 gram.

A nickel weighs 5 grams.

A dollar bill weighs 1 gram.

A plain hamburger weighs 175 to 200 grams.

Work Problem **4** at the Side. ▶

3 Convert using the metric conversion line or unit fractions.

GS **(a)** 9 L to mL

On the conversion line, L to mL is three places to the right **or** use the unit fraction $\frac{1000\ \text{mL}}{1\ \text{L}}$.

9 L = ______ mL

(b) 0.75 L to mL

(c) 500 mL to L

(d) 5 mL to L

(e) 2.07 L to mL

(f) 3275 mL to L

VOCABULARY TIP

gram The weight of a nickel (**5** cents) is **5** grams.

4 Circle the things that weigh about 1 gram.

A small paper clip

A pair of scissors

One playing card from a deck of cards

A calculator

An average-sized apple

The check you wrote to the cable company

Answers

3. **(a)** 9000 mL **(b)** 750 mL **(c)** 0.5 L **(d)** 0.005 L **(e)** 2070 mL **(f)** 3.275 L

4. paper clip, playing card, check

To make larger or smaller weight units, we use the same **prefixes** as we did with length and capacity units. For example, the prefix *kilo*- means 1000 so a *kilo*meter is 1000 meters, a *kilo*liter is 1000 liters, and a *kilo*gram is 1000 grams.

Prefix	*kilo-* gram	*hecto-* gram	*deka-* gram	gram	*deci-* gram	*centi-* gram	*milli-* gram
Meaning	1000 grams	100 grams	10 grams	1 gram	$\frac{1}{10}$ of a gram	$\frac{1}{100}$ of a gram	$\frac{1}{1000}$ of a gram
Symbol	kg	hg	dag	g	dg	cg	mg

Weight (mass) units that are used most often

The units you will use most often in daily life are kilograms (kg), grams (g), and milligrams (mg). *Kilo*grams are used instead of pounds. A kilogram is 1000 grams. It is about 2.2 pounds. Two packages of butter plus one stick of butter weigh about 1 kg. An average newborn baby weighs 3 to 4 kg, while a college football player might weigh 100 to 130 kg.

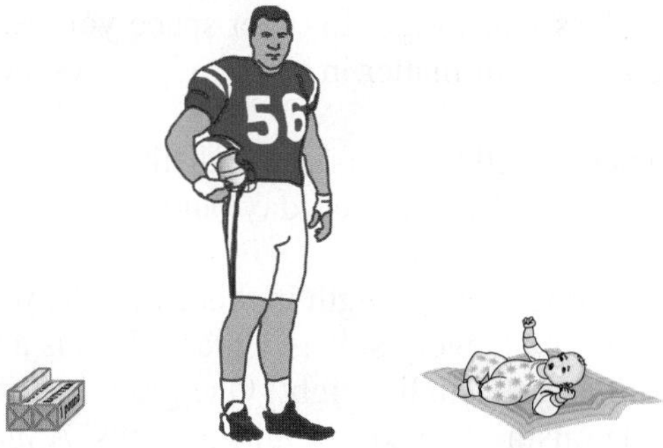

1 kilogram is about 2.2 pounds. 100 to 130 kg 3 to 4 kg

Extremely small weights are measured in *milli*grams. It takes 1000 mg to make 1 g. Recall that a dollar bill weighs about 1 g. Imagine cutting it into 1000 pieces; the weight of one tiny piece would be 1 mg. Dosages of medicine and vitamins are given in milligrams. You will also use milligrams in science classes.

Cut a dollar bill into 1000 pieces.
One tiny piece weighs 1 milligram.

EXAMPLE 3 Using Metric Weight Units

Write the most reasonable metric unit in each blank.
Choose from kg, g, and mg.

(a) Ramon's suitcase weighed 20 _____.
Write 20 **kg** because kilograms are used instead of pounds.
20 kg is about 44 pounds.

(b) LeTia took a 350 _____ aspirin tablet.
Write 350 **mg** because 350 g would be more than the weight of a hamburger, which is too much.

(c) Jenny mailed a letter that weighed 30 _____.
Write 30 **g** because 30 kg would be much too heavy and 30 mg is less than the weight of a dollar bill.

◀ **Work Problem 5 at the Side.**

VOCABULARY TIP

kg; g; mg Think: kilograms and pounds; a gram and a paper clip; a milligram and a grain of salt.

5 Write the most reasonable metric unit in each blank. Choose from kg, g, and mg.

(a) A thumbtack weighs 800 ____.

(b) A teenager weighs 50 ____.

(c) This large cast-iron frying pan weighs 1 ____.

(d) Jerry's basketball weighed 600 ____.

(e) Tamlyn takes a 500 ____ calcium tablet every morning.

(f) On his diet, Greg can eat 90 ____ of meat for lunch.

(g) One strand of hair weighs 2 ____.

(h) One banana might weigh 150 ____.

Answers

5. (a) mg **(b)** kg **(c)** kg **(d)** g **(e)** mg **(f)** g **(g)** mg **(h)** g

OBJECTIVE 4 Convert among metric weight (mass) units. As with length and capacity, you can convert among metric weight units by using unit fractions. The unit fractions you need are shown here.

Metric Weight (Mass) Relationships

1 kg = 1000 g, so the unit fractions are:

$$\frac{1\text{ kg}}{1000\text{ g}} \quad \text{or} \quad \frac{1000\text{ g}}{1\text{ kg}}$$

1 g = 1000 mg, so the unit fractions are:

$$\frac{1\text{ g}}{1000\text{ mg}} \quad \text{or} \quad \frac{1000\text{ mg}}{1\text{ g}}$$

Or you can use a metric conversion line to decide how to move the decimal point.

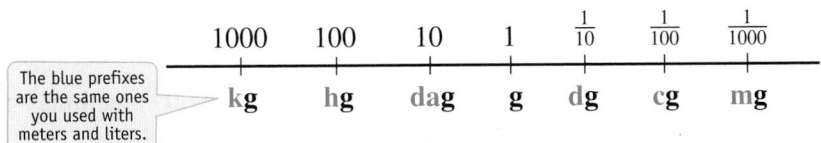

1000	100	10	1	$\frac{1}{10}$	$\frac{1}{100}$	$\frac{1}{1000}$
kg	hg	dag	g	dg	cg	mg

The blue prefixes are the same ones you used with meters and liters.

EXAMPLE 4 Converting among Metric Weight Units

Convert using the metric conversion line or unit fractions.

(a) 7 mg to g

Using the conversion line:
From **mg** to **g** is *three places* to the *left*.

7. ↑ Decimal point starts here.

.007. Move decimal point three places to the left.

7 mg = 0.007 g

Do **not** write a period here.

Using unit fractions:

Multiply by a unit fraction that allows you to divide out mg.

$$\frac{7\text{ }\cancel{\text{mg}}}{1} \cdot \frac{1\text{ g}}{1000\text{ }\cancel{\text{mg}}} = \frac{7}{1000}\text{ g}$$

$$= 0.007\text{ g}$$

Three decimal places for thousandths.

(b) 13.72 kg to g

Using the conversion line:
From **kg** to **g** is *three places* to the *right*.

13.720 Decimal point moves three places to the right.

13.72 kg = 13,720 g

This is a comma (**not** a decimal point).

Using unit fractions:

Multiply by a unit fraction that allows you to divide out kg.

$$\frac{13.72\text{ }\cancel{\text{kg}}}{1} \cdot \frac{1000\text{ g}}{1\text{ }\cancel{\text{kg}}} = 13{,}720\text{ g}$$

This is a comma (**not** a decimal point).

Work Problem 6 at the Side. ▶

6 Convert using the metric conversion line or unit fractions.

GS **(a)** 10 kg to g

On the conversion line, kg to g is _______ places to the _______ **or** use the unit fraction $\frac{1000\text{ g}}{1\text{ kg}}$.

10 kg = _______ g

(b) 45 mg to g

(c) 6.3 kg to g

(d) 0.077 g to mg

(e) 5630 g to kg

(f) 90 g to kg

Answers

6. **(a)** three; right; 10,000 g **(b)** 0.045 g **(c)** 6300 g **(d)** 77 mg **(e)** 5.63 kg **(f)** 0.09 kg

VOCABULARY TIP

Length; capacity; weight Think: meter/mile; liter/liquid; gram/gravity (weight).

7 First decide which type of units are needed: length, capacity, or weight. Then write the most appropriate unit in the blank. Choose from km, m, cm, mm, L, mL, kg, g, and mg.

(a) Gail bought a 4 ____ can of paint.

Use _______ units.

(b) The bag of chips weighed 450 ____.

Use _______ units.

(c) Give the child 5 ____ of cough syrup.

Use _______ units.

(d) The width of the window is 55 ____.

Use _______ units.

(e) Akbar drives 18 ____ to work.

Use _______ units.

(f) The laptop computer weighs 2 ____.

Use _______ units.

(g) A credit card is 55 ____ wide.

Use _______ units.

Answers

7. **(a)** L; capacity **(b)** g; weight **(c)** mL; capacity **(d)** cm; length **(e)** km; length **(f)** kg; weight **(g)** mm; length

OBJECTIVE 5 **Distinguish among basic metric units of length, capacity, and weight (mass).** As you encounter things to be measured at home, on the job, or in your classes at school, be careful to use the correct type of measurement unit.

Use ***length units*** (kilometers, meters, centimeters, millimeters) to measure:

how long	how high	how far away
how wide	how tall	how far around (perimeter)
how deep	distance	

Use ***capacity units*** (liters, milliliters) to measure liquids (things that can be poured) such as:

water	shampoo	gasoline
milk	perfume	oil
soft drinks	cough syrup	paint

Also use liters and milliliters to describe how much liquid something can hold, such as an eyedropper, measuring cup, pail, or bathtub.

Use ***weight units*** (kilograms, grams, milligrams) to measure:

the weight of something   how heavy something is

In **Chapter 7** you will use square units (such as square meters) to measure area, and cubic units (such as cubic centimeters) to measure volume.

EXAMPLE 5 **Using a Variety of Metric Units**

First decide which type of units are needed: length, capacity, or weight. Then write the most appropriate metric unit in the blank. Choose from km, m, cm, mm, L, mL, kg, g, and mg.

(a) The letter needs another stamp because it weighs 40 _______.

Use _______ units.

Use **weight** units because of the word "weighs."

The letter weighs 40 **g** because 40 mg is less than the weight of a dollar bill and 40 kg would be about 88 pounds.

(b) The swimming pool is 3 _______ deep at the deep end.

Use _______ units.

Use **length** units because of the word "deep."

The pool is 3 **m** deep because 3 cm is only about an inch and 3 km is more than a mile.

(c) This is a 340 _______ can of juice.

Use _______ units.

Use **capacity** units because juice is a liquid.

It is a 340-**mL** can because 340 liters would be more than 340 quarts.

◀ **Work Problem 7 at the Side.**

A.3 Exercises

FOR EXTRA HELP

Download the MyDashBoard App

MyMathLab®

CONCEPT CHECK *Fill in the blank with the best metric unit to measure each item. Choose from liters, milliliters, kilograms, grams, and milligrams.*

1. **(a)** For a dose of cough syrup, use ________.

(b) For a large carton of milk, use ________.

(c) For a vitamin pill, use ________.

(d) For a heavyweight wrestler, use ________.

2. **(a)** For a serving of vegetables, use ________.

(b) For a large bottle of soda, use ________.

(c) For a travel size bottle of shampoo, use ________.

(d) For a pain reliever tablet, use ________.

Write the most reasonable metric unit in each blank. Choose from L, mL, kg, g, and mg.
See Examples 1 and 3.

3. The glass held 250 ____ of water.

4. Hiromi used 12 ____ of water to wash the kitchen floor.

5. Dolores can make 10 ____ of soup in that pot.

6. Jay gave 2 ____ of vitamin drops to the baby.

7. Our yellow Labrador dog grew up to weigh 40 ____.

8. A small safety pin weighs 750 ____.

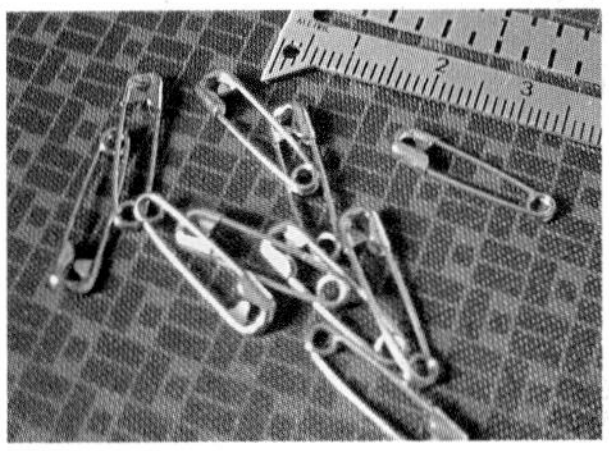

9. Lori caught a small sunfish weighing 150 ____.

10. One dime weighs 2 ____.

11. Andre donated 500 ____ of blood today.

12. Barbara bought the 2 ____ bottle of cola.

13. The patient received a 250 ____ tablet of medication each hour.

14. The 8 people on the elevator weighed a total of 500 ____.

15. The gas can for the lawn mower holds 4 ____.

16. Kevin poured 10 ____ of vanilla into the mixing bowl.

17. Pam's backpack weighs 5 ____ when it is full of books.

18. One grain of salt weighs 2 ____.

CONCEPT CHECK *Today, medical prescriptions are usually given in the metric system. But sometimes a mistake is made. Indicate whether each dose is* reasonable *or* unreasonable. *If a dose is unreasonable, indicate whether it is* too much *or* too little.

19. Drink 4.1 L of Kaopectate after each meal.

20. Drop 1 mL of solution into the eye twice a day.

21. Soak your feet in 5 kg of Epsom salts per liter of water.

22. Inject 0.5 L of insulin each morning.

23. Take 15 mL of cough syrup every four hours.

24. Take 200 mg of vitamin C each day.

25. Take 350 mg of aspirin three times a day.

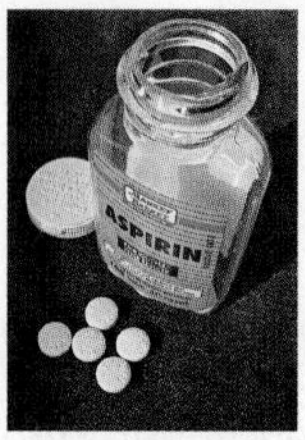

26. Buy a tube of ointment weighing 0.002 g.

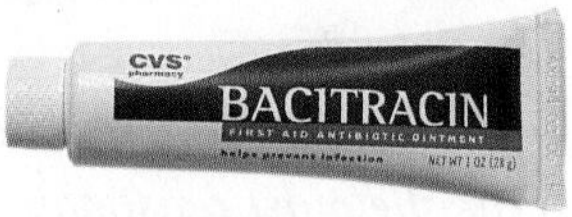

27. CONCEPT CHECK Describe how you decide which unit fraction to use when converting 6.5 kg to grams.

28. CONCEPT CHECK Write an explanation of each step you use to convert 20 mg to grams using the metric conversion line.

Convert each measurement. Use unit fractions or the metric conversion line.
See Examples 2 and 4.

29. 15 L to mL

GS

unit fraction $\left.\right\}$ $\dfrac{1000\text{ mL}}{1\text{ L}}$ **or** On conversion line, L to mL is ______ places to the ______.

15 L = ______ mL

30. 6 L to mL

GS

unit fraction $\left.\right\}$ $\dfrac{1000\text{ mL}}{1\text{ L}}$ **or** On the conversion line, L to mL is ______ places to the ______.

6 L = ______ mL

31. 3000 mL to L

32. 18,000 mL to L

33. 925 mL to L

34. 200 mL to L

35. 8 mL to L

36. 25 mL to L

37. 4.15 L to mL

38. 11.7 L to mL

39. 8000 g to kg

40. 25,000 g to kg

41. 5.2 kg to g

42. 12.42 kg to g

43. 0.85 g to mg

44. 0.2 g to mg

45. 30,000 mg to g

46. 7500 mg to g

47. 598 mg to g

48. 900 mg to g

49. 60 mL to L

50. 6.007 kg to g

51. 3 g to kg

52. 12 mg to g

53. 0.99 L to mL

54. 13,700 mL to L

Write the most appropriate metric unit in each blank. Choose from km, m, cm, mm, L, mL, kg, g, and mg. ***See Example 5.***

55. The masking tape is 19 _____ wide.

56. The roll has 55 ____ of tape on it.

57. Buy a 60 _____ jar of acrylic paint for art class.

58. One onion weighs 200 ____.

59. My waist measurement is 65 _____.

60. Add 2 ____ of windshield washer fluid to your car.

61. A single postage stamp weighs 90 _____.

62. The hallway is 10 ____ long.

Solve each application problem. Show your work. (Source: The Human Body.)

63. Human skin has about 3 million sweat glands, which release an average of 300 mL of sweat per day. How many liters of sweat are released each day?

64. In hot climates, the sweat glands in a person's skin may release up to 3.5 L of sweat in one day. How many milliliters is that?

65. The average weight of an adult human brain is 1.34 kg. An adult human's skin weighs about 5 kg. How many grams does each body part weigh?

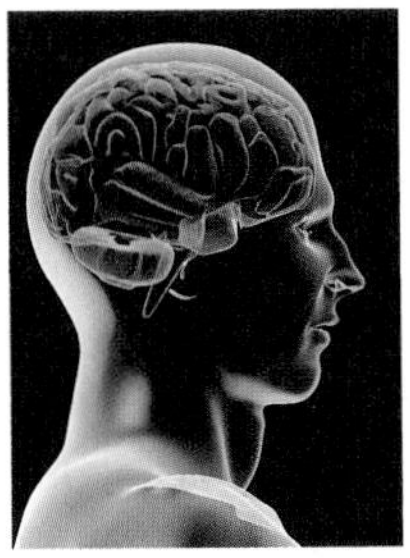

66. A healthy human heart pumps between 70 mL and 230 mL of blood per beat depending on the person's activity level. How many liters of blood does the heart pump per beat?

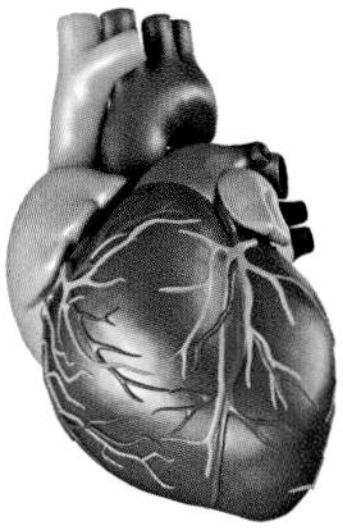

67. On average, we breathe in and out roughly 900 mL of air every 10 seconds. How many liters of air is that?

68. In the Victorian era, people believed that heavier brains meant greater intelligence. They were impressed that Otto von Bismarck's brain weighed 1907 g, which is how many kilograms?

69. A small adult cat weighs from 3000 g to 4000 g. How many kilograms is that? (*Source:* Lyndale Animal Hospital.)

70. If the letter you are mailing weighs 29 g, you must put additional postage on it. How many kilograms does the letter weigh? (*Source:* U.S. Postal Service.)

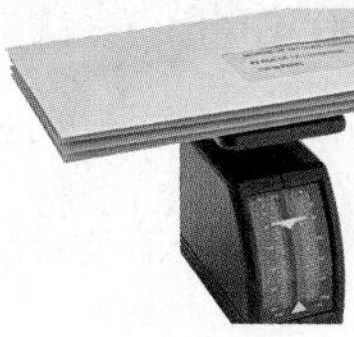

71. CONCEPT CHECK Is 1005 mg greater than or less than 1 g? What is the difference in the weights?

72. CONCEPT CHECK Is 990 mL greater than or less than 1 L? What is the difference in the amounts?

73. One nickel weighs 5 g. How many nickels are in 1 kg of nickels?

74. The ratio of the total length of all the fish to the amount of water in an aquarium can be 3 cm of fish for every 4 L of water. What is the total length of all the fish you can put in a 40 L aquarium? (*Source: Tropical Aquarium Fish.*)

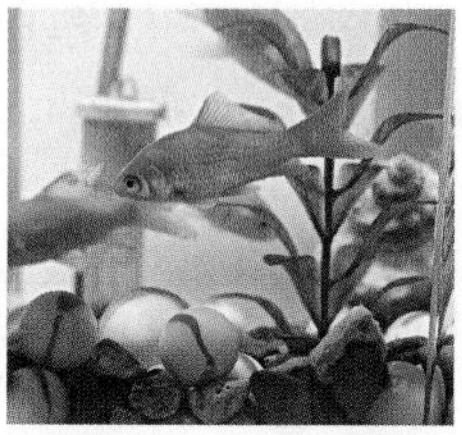

Relating Concepts (Exercises 75–78) For Individual or Group Work

Recall that the prefix ***kilo*** *means 1000, so a* ***kilometer*** *is 1000 meters. You'll learn about other prefixes for numbers greater than 1000 as you* ***work Exercises 75–78 in order.***

75. **(a)** The prefix *mega* means one million. Use the symbol M (capitalized) for *mega*. So a megameter (Mm) is how many meters?

1 Mm = __________ m

(b) Figure out a unit fraction that you can use to convert megameters to meters. Then use it to convert 3.5 Mm to meters.

76. **(a)** The prefix *giga* means one billion. Use the symbol G (capitalized) for *giga*. So a gigameter (Gm) is how many meters?

1 Gm = __________ m

(b) Figure out a unit fraction you can use to convert meters to gigameters. Then use it to convert 2500 m to gigameters.

77. **(a)** The prefix *tera* means one trillion. Use the symbol T (capitalized) for *tera*. So a *tera*meter (Tm) is how many meters?

1 Tm = ____________ m

(b) Think carefully before you fill in the blanks:

1 Tm = __________ Gm

1 Tm = __________ Mm

78. A computer's memory is measured in *bytes*. A byte can represent a single letter, a digit, or a punctuation mark. The memory for a laptop computer may be measured in megabytes (abbreviated MB) or gigabytes (abbreviated GB). Using the meanings of *mega* and *giga*, you would think that

1 MB = __________ bytes and

1 GB = __________ bytes.

However, because computers use a base 2 or binary system, 1 MB is actually 2^{20} and 1 GB is 2^{30}. Use your calculator to find the actual values.

2^{20} = __________ 2^{30} = __________

A.4 Metric–U.S. Measurement Conversions and Temperature

OBJECTIVES

1. Use unit fractions to convert between metric and U.S. measurement units.
2. Learn common temperatures on the Celsius scale.
3. Use formulas to convert between Celsius and Fahrenheit temperatures.

OBJECTIVE 1 Use unit fractions to convert between metric and U.S. measurement units. Until the United States has switched completely to the metric system, it will be necessary to make conversions between U.S. and metric units. *Approximate* conversions can be made with the help of the table below, in which the values have been rounded to the nearest hundredth or thousandth. (The only value that is exact, not rounded, is 1 inch = 2.54 cm.)

Metric to U.S. Units	U.S. to Metric Units
1 kilometer ≈ 0.62 mile	1 mile ≈ 1.61 kilometers
1 meter ≈ 1.09 yards	1 yard ≈ 0.91 meter
1 meter ≈ 3.28 feet	1 foot ≈ 0.30 meter
1 centimeter ≈ 0.39 inch	1 inch = 2.54 centimeters
1 liter ≈ 0.26 gallon	1 gallon ≈ 3.79 liters
1 liter ≈ 1.06 quarts	1 quart ≈ 0.95 liter
1 kilogram ≈ 2.20 pounds	1 pound ≈ 0.45 kilogram
1 gram ≈ 0.035 ounce	1 ounce ≈ 28.35 grams

EXAMPLE 1 Converting between Metric and U.S. Units of Length

Convert 10 m to yards using unit fractions. Round your answer to the nearest tenth if necessary.

We're changing from a *metric length* unit to a *U.S. length* unit. In the "Metric to U.S. Units" side of the table, you see that 1 meter ≈ 1.09 yards. Two unit fractions can be written using that information.

$$\frac{1\text{ m}}{1.09\text{ yd}} \quad \text{or} \quad \frac{1.09\text{ yd}}{1\text{ m}}$$

Multiply by the unit fraction that allows you to divide out meters (that is, meters is in the denominator).

$$10\text{ m} \cdot \frac{1.09\text{ yd}}{1\text{ m}} = \frac{10\ \cancel{\text{m}}}{1} \cdot \frac{1.09\text{ yd}}{1\ \cancel{\text{m}}} = \frac{(10)(1.09\text{ yd})}{1} = 10.9\text{ yd}$$

These units should match.

Meters (**m**) divide out leaving **yd**, the unit you want for the answer.

10 m ≈ 10.9 yd

Note

In **Example 1** above, you could also use the numbers from the "U.S. to Metric Units" side of the table that involve meters and yards:

$$\frac{10\ \cancel{\text{m}}}{1} \cdot \frac{1\text{ yd}}{0.91\ \cancel{\text{m}}} = \frac{10}{0.91}\text{ yd} \approx 10.99\text{ yd}$$

The answer is slightly different because the values in the table are rounded. Also, you have to divide instead of multiply, which is usually more difficult to do without a calculator. We will use the first method in this chapter.

Work Problem 1 at the Side. ▶

1 Convert using unit fractions. Round your answers to the nearest tenth.

(a) 23 m to yards (Look at the "Metric to U.S. Units" side of the table.)

(b) 40 cm to inches

(c) 5 mi to kilometers (Look at the "U.S. to Metric Units" side of the table.)

(d) 12 in. to centimeters

Answers

1. (a) 23 m ≈ 25.1 yd **(b)** 40 cm ≈ 15.6 in. **(c)** 5 mi ≈ 8.1 km **(d)** 12 in. ≈ 30.5 cm

2 Convert. Use the values from the table on the previous page to make unit fractions. Round answers to the nearest tenth.

GS **(a)** 17 kg to pounds. From the table, 1 kg ≈ 2.20 lb

$$\frac{17 \not{kg}}{1} \cdot \frac{2.20 \text{ lb}}{1 \not{kg}} =$$

(b) 5 L to quarts

(c) 90 g to ounces

(d) 3.5 gal to liters

(e) 145 lb to kilograms

(f) 8 oz to grams

Answers

2. (a) $\frac{(17)(2.20 \text{ lb})}{1} = 37.4$ lb; 17 kg ≈ 37.4 lb
(b) 5 L ≈ 5.3 qt **(c)** 90 g ≈ 3.2 oz
(d) 3.5 gal ≈ 13.3 L **(e)** 145 lb ≈ 65.3 kg
(f) 8 oz ≈ 226.8 g

EXAMPLE 2 **Converting between Metric and U.S. Units of Weight and Capacity**

Convert using unit fractions. Round your answers to the nearest tenth.

(a) 3.5 kg to pounds

Look in the "Metric to U.S. Units" side of the table on the previous page to see that 1 kilogram ≈ 2.20 pounds. Use this information to write a unit fraction that allows you to divide out kilograms.

$$\frac{3.5 \not{kg}}{1} \cdot \frac{2.20 \text{ lb}}{1 \not{kg}} = \frac{(3.5)(2.20 \text{ lb})}{1} = 7.7 \text{ lb}$$

3.5 kg ≈ 7.7 lb

The conversion value is approximate, so use the ≈ symbol in your answer.

(b) 18 gal to liters

Look in the "U.S. to Metric Units" side of the table to see that 1 gallon ≈ 3.79 liters. Write a unit fraction that allows you to divide out gallons.

gal divides out, leaving **L** for your answer.

$$\frac{18 \not{gal}}{1} \cdot \frac{3.79 \text{ L}}{1 \not{gal}} = \frac{(18)(3.79 \text{ L})}{1} = 68.22 \text{ L}$$

68.22 rounded to the nearest tenth is 68.2.
18 gal ≈ 68.2 L

(c) 300 g to ounces

In the "Metric to U.S. Units" side of the table, 1 gram ≈ 0.035 ounce.

$$\frac{300 \not{g}}{1} \cdot \frac{0.035 \text{ oz}}{1 \not{g}} = \frac{(300)(0.035 \text{ oz})}{1} = 10.5 \text{ oz}$$

g divides out, leaving **oz** for your answer.

300 g ≈ 10.5 oz

CAUTION

Because the metric and U.S. measurement systems were developed independently, almost all comparisons are approximate. Your answers should be written with the "≈" symbol to show they are approximate.

◀ **Work Problem 2 at the Side.**

OBJECTIVE 2 **Learn common temperatures on the Celsius scale.** In the metric system, temperature is measured on the **Celsius** scale. On the Celsius scale, water freezes at 0 °C and boils at 100 °C. The small raised circle stands for "degrees" and the capital **C** is for Celsius. Read the temperatures like this:

Water freezes at 0 degrees Celsius (0 °C).
Water boils at 100 degrees Celsius (100 °C).

In the metric system, °C is the symbol for "degrees Celsius."

The U.S. temperature system, used only in the United States, is measured on the **Fahrenheit** scale. On this scale:

Water freezes at 32 degrees Fahrenheit (32 °F).
Water boils at 212 degrees Fahrenheit (212 °F).

In the U.S. system, °F stands for "degrees Fahrenheit."

The thermometer below shows some typical temperatures in both Celsius and Fahrenheit. For example, comfortable room temperature is about 20 °C or 68 °F, and normal body temperature is about 37 °C or 98.6 °F.

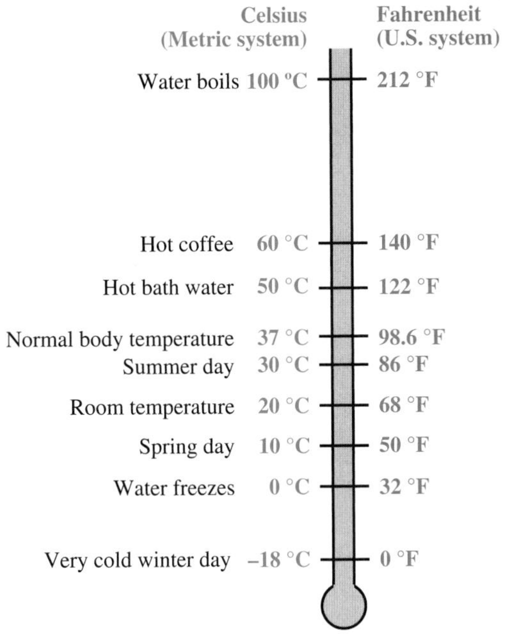

Note

The freezing and boiling temperatures are exact. The other temperatures are approximate. Even normal body temperature varies slightly from person to person.

EXAMPLE 3 **Using Celsius Temperatures**

Choose the metric temperature that is most reasonable for each situation.

(a) Warm summer day 29 °C 64 °C 90 °C

29 °C is reasonable. 64 °C and 90 °C are too hot; they're both above the temperature of hot bath water (above 122 °F).

(b) Inside a freezer −10 °C 3 °C 25 °C

−10 °C is the reasonable temperature because it is the only one below the freezing point of water (0 °C). Your frozen foods would start thawing at 3 °C or 25 °C.

Work Problem 3 at the Side. ▶

OBJECTIVE ▶ **3** **Use formulas to convert between Celsius and Fahrenheit temperatures.** You can use these formulas to convert between Celsius and Fahrenheit temperatures.

Celsius–Fahrenheit Conversion Formulas

Converting from Fahrenheit (F) to Celsius (C)	Converting from Celsius (C) to Fahrenheit (F)
$C = \frac{5(F - 32)}{9}$	$F = \frac{9 \cdot C}{5} + 32$

VOCABULARY TIP

Celsius Celsius (metric) temperatures used to be called **centi**grade temperatures, and **centi-** means hundred(th). Water boils at 100 degrees in the Celsius system.

3 Choose the metric temperature that is *most* reasonable for each situation.

GS **(a)** Set the living room thermostat at:

11 °C 21 °C 71 °C

Hint: Look at the blue side of the thermometer in the text.

(b) The baby has a fever of:

29 °C 39 °C 49 °C

(c) Wear a sweater outside because it's:

15 °C 28 °C 50 °C

(d) My iced tea is:

−5 °C 5 °C 30 °C

(e) Let's go swimming! It's:

95 °C 65 °C 35 °C

(f) Inside a refrigerator (not the freezer) it's:

−15 °C 0 °C 3 °C

(g) There's a blizzard outside. It's:

12 °C 4 °C −20 °C

(h) I need hot water to get these clothes clean. The water should be:

55 °C 105 °C 200 °C

Answers

3. **(a)** 21 °C **(b)** 39 °C **(c)** 15 °C **(d)** 5 °C **(e)** 35 °C **(f)** 3 °C **(g)** −20 °C **(h)** 55 °C

4 Convert to Celsius.

GS **(a)** 59 °F $C = \frac{5(F - 32)}{9}$

$C = \frac{5(59 - 32)}{9}$ Finish the work.

(b) 41 °F

(c) 212 °F

(d) 98.6 °F

5 Convert to Fahrenheit.

GS **(a)** 100 °C $F = \frac{9 \cdot C}{5} + 32$

$F = \frac{9 \cdot 100}{5} + 32$ Finish the work.

(b) 25 °C

(c) 80 °C

(d) 5 °C

Answers

4. (a) $C = \frac{5(\overset{3}{\cancel{27}})}{\underset{1}{\cancel{9}}} = 15\text{ °C}$ **(b)** 5 °C

(c) 100 °C **(d)** 37 °C

5. (a) $F = \frac{9 \cdot \overset{20}{\cancel{100}}}{\underset{1}{\cancel{5}}} + 32 = 180 + 32 = 212\text{ °F}$

(b) 77 °F **(c)** 176 °F **(d)** 41 °F

As you use these formulas, be sure to follow the order of operations from **Chapter 1.**

Order of Operations

1. Do all operations inside ***parentheses*** or ***other grouping symbols.***
2. Simplify any expressions with ***exponents*** and find any ***square roots.***
3. ***Multiply*** or ***divide,*** proceeding from left to right.
4. ***Add*** or ***subtract,*** proceeding from left to right.

EXAMPLE 4 Converting Fahrenheit to Celsius (U.S. to Metric)

Convert 68 °F to Celsius.

Use the formula and follow the order of operations.

$$C = \frac{5(F - 32)}{9}$$ Fahrenheit to Celsius formula

$$= \frac{5(68 - 32)}{9}$$ Work inside parentheses first: 68 − 32 is 36.

$$= \frac{5(36)}{9}$$

$$= \frac{5(\overset{4}{\cancel{36}})}{\underset{1}{\cancel{9}}}$$ Divide 36 and 9 by their common factor: 36 ÷ 9 is 4, and 9 ÷ 9 is 1. Multiply 5(4) in the numerator.

$$= 20$$

Thus, 68 °F = 20 °C.

◀ **Work Problem 4 at the Side.**

EXAMPLE 5 Converting Celsius to Fahrenheit (Metric to U.S.)

Convert 15 °C to Fahrenheit.

Use the formula and follow the order of operations.

$$F = \frac{9 \cdot C}{5} + 32$$ Celsius to Fahrenheit formula

$$= \frac{9 \cdot 15}{5} + 32$$

$$= \frac{9 \cdot \overset{3}{\cancel{15}}}{\underset{1}{\cancel{5}}} + 32$$ Divide 15 and 5 by their common factor: 15 ÷ 5 is 3, and 5 ÷ 5 is 1. Multiply 9 • 3 in the numerator.

$$= 27 + 32$$ Add.

$$= 59$$

Thus, 15 °C = 59 °F.

◀ **Work Problem 5 at the Side.**

A.4 Exercises

FOR EXTRA HELP

CONCEPT CHECK *Use the table on the first page of this section to answer Exercises 1 and 2.*

1. **(a)** Use only the "Metric to U.S." side of the table. Write **two** unit fractions using the information for converting between kilograms and pounds.

(b) When converting 5 kg to pounds, use the unit fraction from part (a) that allows you to divide out which unit? ____________ Would you use the unit fraction with kg in the *numerator* or the *denominator*?

____

2. **(a)** Use only the "U.S. to Metric" side of the table. Write **two** unit fractions using the information for converting between quarts and liters.

(b) When converting 3 qt to liters, use the unit fraction from part (a) that allows you to divide out which unit? ____________. Would you use the unit fraction with qt in the *numerator* or the *denominator*?

____

Use the table on the first page of this section and unit fractions to make approximate conversions from metric to U.S. units or U.S. to metric units. Round your answers to the nearest tenth. ***See Examples 1 and 2.***

3. 20 m to yards

$$\frac{20\ \cancel{m}}{1} \cdot \frac{1.09\ \text{yd}}{1\ \cancel{m}} =$$

4. 8 km to miles

$$\frac{8\ \cancel{km}}{1} \cdot \frac{0.62\ \text{mi}}{1\ \cancel{km}} =$$

5. 80 m to feet

6. 85 cm to inches

7. 16 ft to meters

8. 3.2 yd to meters

9. 150 g to ounces

10. 2.5 oz to grams

11. 248 lb to kilograms

12. 7.68 kg to pounds

13. 28.6 L to quarts

14. 15.75 L to gallons

15. For the 2000 Olympics, the 3M Company used 5 g of pure gold to coat Michael Johnson's track shoes. (*Source:* 3M Company.)

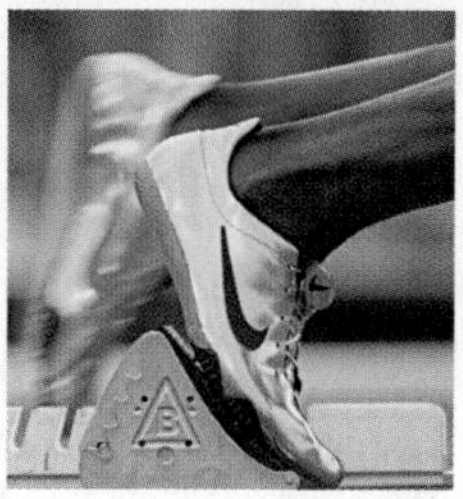

(a) How many ounces of gold were used, to the nearest tenth?

(b) Was this enough extra weight to slow him down?

16. The label on a Van Ness auto feeder for cats and dogs says it holds 1.4 kg of dry food. How many pounds of food does it hold, to the nearest tenth? (*Source:* Van Ness Plastics.)

17. The heavy-duty wash cycle in a dishwater uses 8.4 gal of water. How many liters does it use, to the nearest tenth? (*Source:* Frigidaire.)

18. The rinse-and-hold cycle in a dishwasher uses only 4.5 L of water. How many gallons does it use, to the nearest tenth? (*Source:* Frigidaire.)

19. The smallest pet fish are dwarf gobies, which are half an inch long. How many centimeters long is a dwarf gobie, to the nearest tenth? (*Hint*: First write half an inch in decimal form.)

20. The fastest nerve signals in the human body travel 120 meters per second. How many feet per second do the signals travel? (*Source: The Human Body.*)

CONCEPT CHECK *Circle the most reasonable metric temperature for each situation.* ***See Example 3.***

21. A snowy day
12 °C 28 °C −8 °C

22. Brewing coffee
80 °C 180 °C 15 °C

23. A high fever
21 °C 40 °C 103 °C

24. Swimming pool water
90 °C 78 °C 25 °C

25. Oven temperature
150 °C 50 °C 30 °C

26. Light jacket weather
0 °C 10 °C −10 °C

Use the conversion formulas from this section and the order of operations to convert Fahrenheit temperatures to Celsius, or Celsius temperatures to Fahrenheit. Round your answers to the nearest degree if necessary. ***See Examples 4 and 5.***

27. 60 °F Complete the conversion.

$$C = \frac{5(60-32)}{9} = \frac{5(28)}{9} =$$

28. 80 °F Complete the conversion.

$$C = \frac{5(80-32)}{9} = \frac{5(48)}{9} =$$

29. 104 °F

30. 36 °F

31. 8 °C

32. 18 °C

33. 35 °C

34. 0 °C

Solve each application problem. Round your answers to the nearest degree if necessary.

35. The highest temperature ever recorded on Earth was 136 °F at El Azizia, Libya, in 1922. Convert this temperature to Celsius. (*Source: World Almanac and Book of Facts.*)

36. Hummingbirds have a normal body temperature of 107 °F. But on cold nights they go into a state of torpor where their body temperature drops to 39 °F. What are these temperatures in Celsius? (*Source: Wildbird.*)

37. (a) Here is the tag on a pair of Sorel boots. In what kind of weather would you wear these boots?

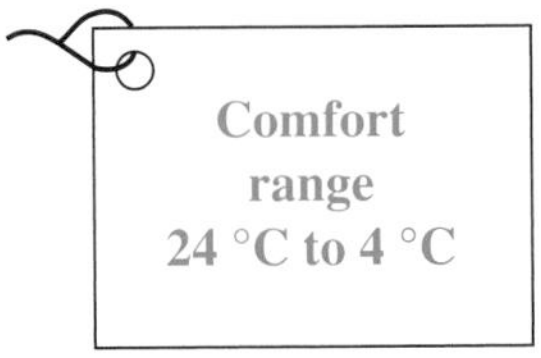

Source: Sorel.

(b) For what Fahrenheit temperatures are the boots designed?

(c) What range of metric temperatures do you have in January where you live? Would you be comfortable in these boots?

38. Sleeping bags made by Eddie Bauer are sold around the world. Each type of sleeping bag is designed for outdoor camping in certain temperatures.

Junior bag	5 °C or warmer
Removable liner bag	0 °C to 15 °C
Conversion bag	−7 °C to 0 °C

Source: Eddie Bauer.

(a) At what Fahrenheit temperatures should you use the Junior bag?

(b) What Fahrenheit temperatures is the removable liner bag designed for?

Relating Concepts (Exercises 39–46) For Individual or Group Work

The information below appeared in American newspapers. However, both Newfoundland (part of Canada) and Ireland use the metric system. Their newspapers would have reported all the measurements in metric units. Complete the conversions to metric, rounding answers to the nearest tenth.

Q. A recent news brief reported on some men who flew a model airplane from Newfoundland to Ireland. Can you provide some details of the flight?

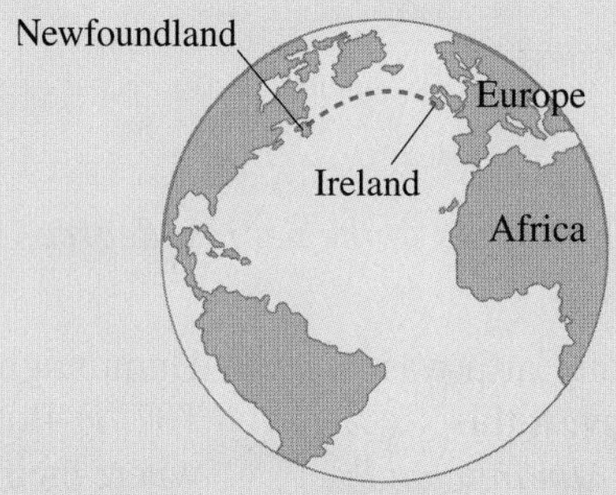

A. The model plane is 6 feet long and weighs 11 pounds. Made of balsa wood and Mylar, it crossed the Atlantic—the flight path took it 1,888.3 miles—in 38 hours, 23 minutes. It soared at a cruising altitude of 1,000 feet. The plane used a souped-up piston engine and carried less than a gallon of fuel, as mandated by rules of the Federation Aeronautique Internationale, the governing body of model airplane building. When it landed in County Galway, Ireland, it had less than 2 fluid ounces of fuel left. The plane was built by Maynard Hill of Silver Spring, Maryland.

(*Source*: nctimes.com)

39. Length of model plane

40. Weight of plane

41. Length of flight path

42. Time of flight

43. Cruising altitude

44. Fuel at the start, in milliliters

45. Fuel left after landing, in milliliters (*Hint:* First convert 2 fl oz to quarts.)

46. What *percent* of the fuel was left at the end of the flight?

B Inductive and Deductive Reasoning

OBJECTIVES

1. Use inductive reasoning to analyze patterns.
2. Use deductive reasoning to analyze arguments.
3. Use deductive reasoning to solve problems.

OBJECTIVE 1 Use inductive reasoning to analyze patterns. In many scientific experiments, conclusions are drawn from specific outcomes. After many repetitions and similar outcomes, the findings are generalized into statements that appear to be true. When general conclusions are drawn from specific observations, we are using a type of reasoning called **inductive reasoning.** The next examples illustrate this type of reasoning.

EXAMPLE 1 Using Inductive Reasoning

Find the next number in the sequence 3, 7, 11, 15, ...

To discover a pattern, calculate the difference between each pair of successive numbers.

$$7 - 3 = 4$$
$$11 - 7 = 4$$
$$15 - 11 = 4$$

Notice that the difference is always 4. Each number is 4 greater than the previous one. Thus, the next number in the pattern is 15 + 4, or 19.

Work Problem 1 at the Side. ▶

1 Find the next number in the sequence 2, 8, 14, 20, ... Describe the pattern.

EXAMPLE 2 Using Inductive Reasoning

Find the next number in this sequence.

7, 11, 8, 12, 9, 13, . . .

The pattern in this example involves addition and subtraction.

$$7 + 4 = 11$$
$$11 - 3 = 8$$
$$8 + 4 = 12$$
$$12 - 3 = 9$$
$$9 + 4 = 13$$

To get the second number, we add 4 to the first number. To get the third number, we subtract 3 from the second number. To obtain subsequent numbers, we continue the pattern. The next number is 13 − 3 = 10.

Work Problem 2 at the Side. ▶

2 Find the next number in the sequence 6, 11, 7, 12, 8, 13, ... Describe the pattern.

EXAMPLE 3 Using Inductive Reasoning

Find the next number in the sequence 1, 2, 4, 8, 16, ...

Each number after the first is obtained by multiplying the previous number by 2. So the next number would be 16 • 2 = 32.

Work Problem 3 at the Side. ▶

3 Find the next number in the sequence 2, 6, 18, 54, ... Describe the pattern.

Answers

1. 26; add 6 each time.
2. 9; add 5, subtract 4.
3. 162; multiply by 3.

4 Find the next shape in this sequence.

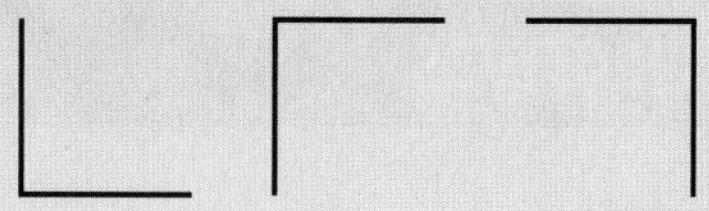

EXAMPLE 4 **Using Inductive Reasoning**

(a) Find the next geometric shape in this sequence.

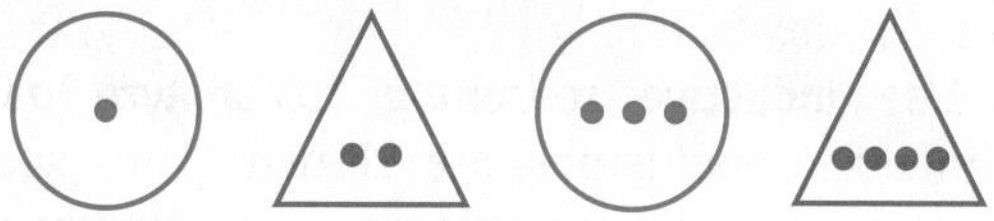

The figures alternate between a blue circle and a red triangle. Also, the number of dots increases by 1 in each subsequent figure. Thus, the next figure should be a blue circle with five dots inside it.

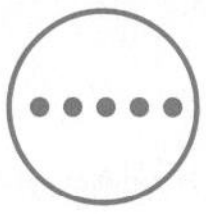

(b) Find the next geometric shape in this sequence.

The first two shapes consist of vertical lines with horizontal lines at the bottom extending first *left* and then *right.* The third shape is a vertical line with a horizontal line at the top extending to the *left.* Therefore, the next shape should be a vertical line with a horizontal line at the top extending to the *right.*

◀ **Work Problem 4 at the Side.**

OBJECTIVE 2 **Use deductive reasoning to analyze arguments.** In the previous discussion, specific cases were used to find patterns and predict the next event. There is another type of reasoning called **deductive reasoning,** which moves from general cases to specific conclusions.

EXAMPLE 5 **Using Deductive Reasoning**

Does the conclusion follow from the premises in this argument?

All Hondas are automobiles. ← Premise

All automobiles have horns. ← Premise

∴ All Hondas have horns. ← Conclusion

In this example, the first two statements are called *premises* and the third statement (below the line) is called a *conclusion.* The symbol ∴ is a mathematical symbol meaning "**therefore**." The entire set of statements is called an *argument.*

Continued on Next Page

Answer

4\.

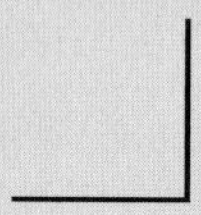

The focus of deductive reasoning is to determine whether the conclusion follows (is valid) from the premises. A set of circles called **Euler circles** is used to analyze the argument.

In **Example 5,** the statement "All Hondas are automobiles" can be represented by two circles, one for Hondas and one for automobiles. Note that the circle representing Hondas is totally inside the circle representing automobiles because the first premise states that *all* Hondas are automobiles.

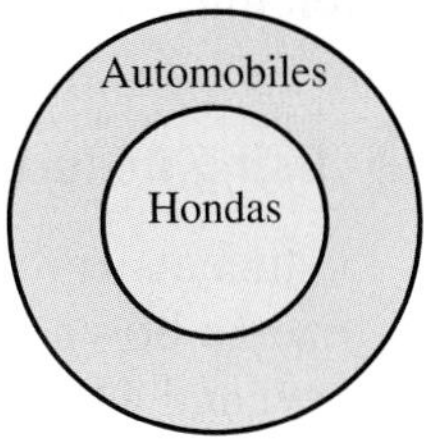

Now, a circle is added to represent the second statement, vehicles with horns. This circle must completely surround the circle representing automobiles because the second premise states that *all* automobiles have horns.

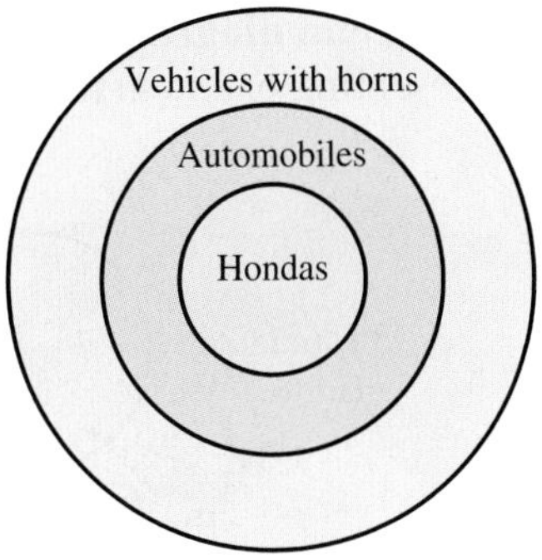

To analyze the conclusion, notice that the circle representing Hondas is *completely* inside the circle representing vehicles with horns. Therefore, it must follow that all Hondas have horns. ***The conclusion is valid.***

Work Problem 5 at the Side. ▶

5 Does the conclusion follow from the premises in the following argument?

All cars have four wheels.
All Fords are cars.
∴ All Fords have four wheels.

6 Does each conclusion follow from the premises?

(a) All animals are wild.
All cats are animals.
∴ All cats are wild.

(b) All students use math.
All adults use math.
∴ All adults are students.

EXAMPLE 6 Using Deductive Reasoning

Does the conclusion follow from the premises in this argument?

All tables are round.
All glasses are round.
∴ All glasses are tables.

Use Euler circles. Draw a circle representing tables *inside* a circle representing round objects, because the first premise states that *all* tables are round.

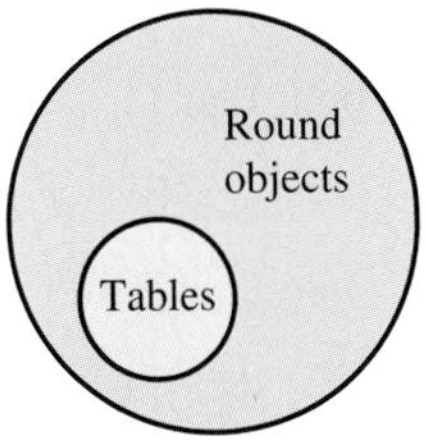

The second statement requires that a circle representing glasses must now be drawn inside the circle representing round objects, but not necessarily inside the circle representing tables. Therefore, the conclusion does ***not*** follow from the premises. This means that ***the conclusion is invalid.***

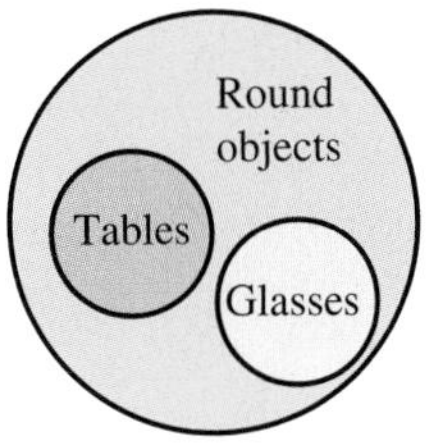

Work Problem 6 at the Side. ▶

Answers

5. The conclusion follows from the premises; it is valid.

6. (a) The conclusion follows from the premises; it is valid.
(b) The conclusion does *not* follow from the premises; it is invalid.

OBJECTIVE 3 **Use deductive reasoning to solve problems.** Another type of deductive reasoning problem occurs when a set of facts is given in a problem and a conclusion must be drawn using these facts.

EXAMPLE 7 Using Deductive Reasoning

There were 25 students enrolled in a ceramics class. During the class, 10 of the students made a bowl and 8 students made a birdbath. Three students made both a bowl and a birdbath. How many students did not make either a bowl or a birdbath?

This type of problem is best solved by organizing the data using a drawing called a **Venn diagram.** Two overlapping circles are drawn, with each circle representing one item made by the students, as shown below.

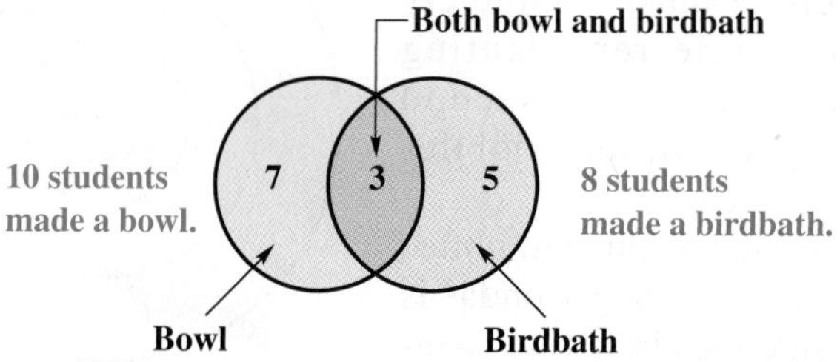

In the region where the circles overlap, write the number of students who made *both* items, namely, 3. In the remaining portion of the birdbath circle, write the number 5, which when added to 3 will give the total number of students who made a birdbath, namely, 8. In a similar manner, write 7 in the remaining portion of the bowl circle, since $7 + 3 = 10$, the total number of students who made a bowl. The total of all three numbers written in the circles is 15. Since there were 25 students in the class, this means $25 - 15$ or 10 students did not make either a birdbath or a bowl.

◄ **Work Problem 7 at the Side.**

EXAMPLE 8 Using Deductive Reasoning

Four cars in a race finish first, second, third, and fourth. The following facts are known.

(a) Car A beat Car C.

(b) Car D finished between Cars C and B.

(c) Car C beat Car B.

In which order did the cars finish?

To solve this type of problem, it is helpful to use a line diagram.

1. *Write A before C,* because Car A beat Car C (fact **a**).

   A C

2. *Write B after C,* because Car C beat Car B (fact **c**).

   A C B

3. *Write D between C and B,* because Car D finished between Car C and Car B (fact **b**).

The correct order of finish is shown below.

A C D B

◄ **Work Problem 8 at the Side.**

7 In a college class of 100 students, 35 take both math and history, 50 take history, and 40 take math. How many take neither math nor history? Draw a Venn diagram.

8 A Chevy, BMW, Cadillac, and Ford are parked side by side. The known facts are:

(a) The Ford is on the right end.

(b) The BMW is next to the Cadillac.

(c) The Chevy is between the Ford and the Cadillac.

Which car is parked on the left end?

Answers

7.

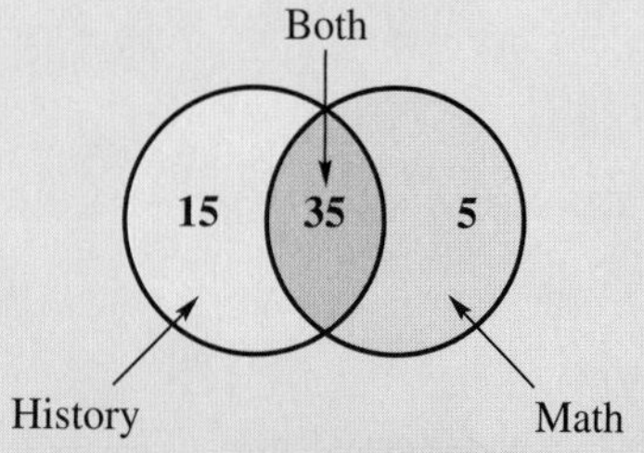

$15 + 35 + 5 = 55$

$100 - 55 = 45$

45 students take neither math nor history.

8. BMW

Appendix B Exercises

FOR EXTRA HELP

Download the MyDashBoard App

MyMathLab®

Find the next number in each sequence. Describe the pattern in each sequence. ***See Examples 1–3.***

1. 2, 9, 16, 23, 30, …

2. 5, 8, 11, 14, 17, …

3. 0, 10, 8, 18, 16, …

4. 3, 9, 7, 13, 11, …

5. 1, 2, 4, 8, …

6. 1, 4, 16, 64, …

7. 1, 3, 9, 27, 81, …

8. 3, 6, 12, 24, 48, …

9. 1, 4, 9, 16, 25, …

10. 6, 7, 9, 12, 16, …

Find the next shape in each sequence. ***See Example 4.***

11.

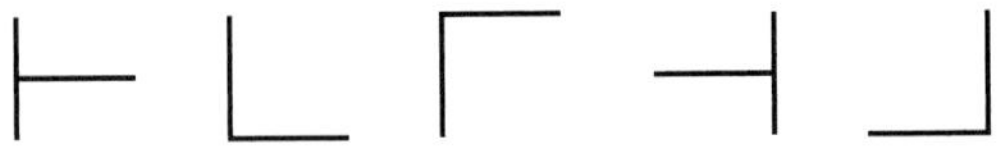

12.

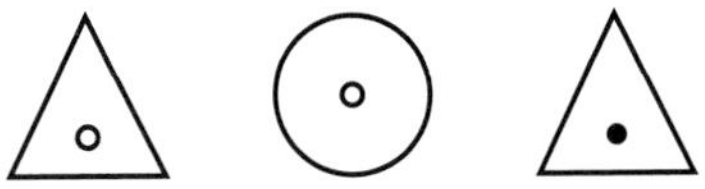

13.

14.

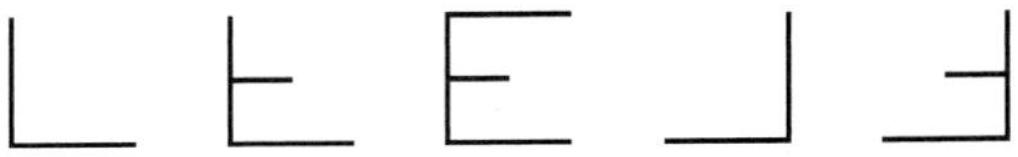

For each argument, draw Euler circles and then state whether or not the conclusion follows from the premises. ***See Examples 5 and 6.***

15. All animals are wild.
All lions are animals.
∴ All lions are wild.

16. All students are hard workers.
All business majors are students.
∴ All business majors are hard workers.

17. All teachers are serious.
All mathematicians are serious.
∴ All mathematicians are teachers.

18. All boys ride bikes.
All Americans ride bikes.
∴ All Americans are boys.

Solve each application problem. ***See Examples 7 and 8.***

19. In a given 30-day period, a husband watched television 20 days and his wife watched television 25 days. If they watched television together 18 days, how many days did neither watch television? Draw a Venn diagram.

20. In a class of 40 students, 21 students take both calculus and physics. If 30 students take calculus and 25 students take physics, how many do not take either calculus or physics? Draw a Venn diagram.

21. Tom, Dick, Mary, and Joan all work for the same company. One is a secretary, one is a computer operator, one is a receptionist, and one is a mail clerk.

(a) Tom and Joan eat dinner with the computer operator.

(b) Dick and Mary carpool with the secretary.

(c) Mary works on the same floor as the computer operator and the mail clerk.

Who is the computer operator?

22. Four cars—a Ford, a Buick, a Mercedes, and an Audi—are parked in a garage in four spaces.

(a) The Ford is in the last space.

(b) The Buick and Mercedes are next to each other.

(c) The Audi is next to the Ford but not next to the Buick.

Which car is in the first space?

Answers to Selected Exercises

In this section we provide the answers that we think most students will obtain when they work the exercises using the methods explained in the text. If your answer does not look exactly like the one given here, it is not necessarily wrong. In many cases there are equivalent forms of the answer that are correct. For example, if the answer section shows $\frac{3}{4}$ and your answer is 0.75 you have obtained the right answer but written it in a different (yet equivalent) form. Unless the directions specify otherwise, 0.75 is just as valid an answer as $\frac{3}{4}$.

In general, if your answer does not agree with the one given in the text, see whether it can be transformed into the other form. If it can, then it is the correct answer. If you still have doubts, talk with your instructor.

CHAPTER 1 Whole Numbers

SECTION 1.1 (pages 8–9)

1. C **2.** B **3.** 1; 0 **5.** 8; 2 **7.** 687 **8.** 321 **9.** 3; 561; 435 **11.** Evidence suggests that this is true. It is common to count using fingers. **12.** No doubt there is a relationship here. One answer might be that people could count using their fingers and toes and, therefore, thought of them as numbers or digits. **13.** false (no "and") **14.** true **15.** three hundred forty-six thousand, nine **17.** twenty-five million, seven hundred fifty-six thousand, six hundred sixty-five **19.** 63,163 **21.** 10,000,223 **23.** 3,200,000 parachute jumps **25.** 50,051,507 cans **27.** 54,750,000 Hot Wheels **29.** 800,000,621,020,215 **31.** public transportation; six million, sixty-nine thousand, five hundred eighty-nine **33.** seven million, eight hundred ninety-four thousand, nine hundred eleven

SECTION 1.2 (pages 18–21)

1. 97 **3.** 89 **5.** 889 **7.** 889 **9.** 7785 **11.** correct **12.** incorrect **13.** incorrect **14.** correct **15.** 78,446 **17.** 8928 **19.** 59,224 **21.** correct **22.** incorrect **23.** correct **24.** incorrect **25.** incorrect **27.** 145 **29.** 102 **31.** 1651 **33.** 1154 **35.** 413 **37.** 1771 **39.** 1410 **41.** 6391 **43.** 11,624 **45.** 17,611 **47.** 15,954 **49.** 10,648 **51.** 15,594 **53.** 11,557 **55.** 12,078 **57.** 4250 **59.** 12,268 **61.** correct **63.** incorrect; should be 769 **65.** correct **67.** incorrect; should be 11,577 **69.** correct **71.** Changing the order in which numbers are added does not change the sum. You can add from bottom to top when checking addition. **72.** Grouping the addition of numbers in any order does not change the sum. You can add numbers in any order. For example, you can add pairs of numbers that add to 10. **73.** 33 miles **75.** 38 miles **77.** $16,342 **79.** 699 people **81.** 20,157 students **83.** 970 ft **85.** 72 ft **87.** 9421 **88.** 1249 **89.** 77,762 **90.** 22,267 **91.** 9,994,433 **92.** 3,334,499 **93.** Write the largest digits on the left, using the smaller digits as you move right. **94.** Write the smallest digits on the left, using the larger digits as you move right.

SECTION 1.3 (pages 28–31)

1. 32 **2.** 4 **3.** 86 **4.** 35 **5.** 17 **7.** 213 **9.** 101 **11.** 7111 **13.** 3412 **15.** 2111 **17.** 13,160 **19.** 41,110 **21.** correct **23.** incorrect; should be 62 **25.** incorrect; should be 121 **27.** correct **29.** incorrect; should be 7222 **31.** minuend; below **32.** B and D **33.** 38 **35.** 45 **37.** 19 **39.** 281 **41.** 519 **43.** 7059 **45.** 7589 **47.** 7 **49.** 19 **51.** 2833 **53.** 7775 **55.** 503 **57.** 156 **59.** 2184 **61.** 5687 **63.** 19,038 **65.** 31,556 **67.** 6584 **68.** check **69.** subtraction **70.** addition **71.** correct **73.** correct **75.** correct **77.** correct **79.** Possible answers are: 1. $3 + 2 = 5$ could be changed to $5 - 2 = 3$ or $5 - 3 = 2$. 2. $6 - 4 = 2$ could be changed to $2 + 4 = 6$ or $4 + 2 = 6$. **80.** No, you cannot. Numbers must be subtracted in the order given. The difference found in subtraction is the result of subtracting the subtrahend from the minuend. Changing the order of the minuend and subtrahend does change the answer. **81.** 47 calories **83.** 69 more tornadoes **85.** 246 feet **87.** 3270 jobs eliminated **89.** 9539 flags **91.** $263 **93.** 141 miles **95.** 138,888 patients **97.** 310 fewer calories; 26 fewer fat grams **99.** 370 calories; 10 fat grams

SECTION 1.4 (pages 38–41)

1. the same **2.** commutative **3.** zero **4.** zeros; right **5.** 24 **7.** 0 **9.** 24 **11.** 40 **13.** Factors may be multiplied in any order to get the same answer. They are the same; you may add or multiply numbers in any order. **14.** You may shift the parentheses in a multiplication problem. Just as in addition, the different grouping results in the same answer. **15.** 210 **17.** 238 **19.** 3210 **21.** 1872 **23.** 8612 **25.** 10,084 **27.** 20,488 **29.** 258,447 **31.** 86; 172; two; 17,200 **32.** 45; 3510; three; 3,510,000 **33.** 480 **35.** 2220 **37.** 3600 **39.** 3750 **41.** 65,400 **43.** 270,000 **45.** 86,000,000 **47.** 48,500 **49.** 720,000 **51.** 1,940,000 **53.** 476 **55.** 2400 **57.** 3735 **59.** 2378 **61.** 6164 **63.** 15,792 **65.** 21,665 **67.** 15,730 **69.** 82,320 **71.** 183,996 **73.** 2,468,928 **75.** 66,005 **77.** 86,028 **79.** 19,422,180 **81.** 2,278,410 **83.** To multiply by 10, 100, or 1000, just attach one, two, or three zeros, respectively, to the answer number you are multiplying and that's your answer.

84. $\begin{array}{r} 291 \\ \times\ 307 \\ \hline 2037 \\ 000 \\ 873 \\ \hline 89{,}337 \end{array}$ $\quad$ $\begin{array}{r} 291 \\ \times\ 307 \\ \hline 2037 \\ 8730 \\ \hline 89{,}337 \end{array}$ **85.** 3000 balls **87.** $2328 **89.** 24,090 gallons

91. $600 **93.** $1560 **95.** $112,888 **97.** 50,568 **99.** 38,250 trees **101.** 7,746,712 people **103.** $14,160 **105.** (a) 452 (b) 452 **106.** commutative **107.** (a) 281 (b) 281 **108.** associative **109.** (a) 15,840 (b) 15,840 **110.** commutative **111.** (a) 6552 (b) 6552 **112.** associative **113.** No. Some examples are: $7 - 5 = 2$, but $5 - 7$ does not equal 2; $12 - 6 = 6$, but $6 - 12$ does not equal 6; $(8 - 2) - 5 = 1$, but $8 - (2 - 5)$ does not equal 1. **114.** No. Some examples are: $10 \div 2 = 5$, but $2 \div 10$ does not equal 5; $(16 \div 8) \div 2 = 1$, but $16 \div (8 \div 2)$ does not equal 1.

SECTION 1.5 (pages 51–54)

1. ×; ( )( ); • **2.** ÷; $\overline{)}$; — (fraction bar) **3.** $\begin{array}{r} 6 \\ 4\overline{)24} \end{array}$ $\frac{24}{4} = 6$ **5.** $\begin{array}{r} 5 \\ 9\overline{)45} \end{array}$ $45 \div 9 = 5$ **7.** $16 \div 2 = 8$ $\frac{16}{2} = 8$ **9.** number **10.** 0

ANSWERS

11. 1 **13.** 7 **15.** undefined **17.** 24 **19.** 0 **21.** undefined **23.** 2; 3; 5; 10 **24.** 3; 5 **25.** 2; 3 **26.** 2; 3; 5; 10 **27.** 25 **29.** 18 **31.** 304 **33.** 627 R1 **35.** 1522 R5 **37.** 309 **39.** 3005 **41.** 5006 **43.** 811 R1 **45.** 2589 R2 **47.** 2630 **49.** 12,458 R3 **51.** 10,253 R5 **53.** 18,377 R6 **55.** 1877 **57.** incorrect; should be 1908 R1 **59.** incorrect; should be 670 R2 **61.** incorrect; should be 3568 R1 **63.** correct **65.** correct **67.** incorrect; should be 9628 R3 **69.** correct **71.** Multiply the quotient by the divisor and add any remainder. The result should be the dividend. **72.** Three choices might be: A number is divisible by 2 if it ends in a 0, 2, 4, 6, or 8; A number is divisible by 5 if it ends in 0 or 5; A number is divisible by 10 if it ends in 0. **73.** 328 tables **75.** 9600 each hour **77.** \$48,500 **79.** 165 locations **81.** \$1,137,500 **83.** 862,500 berries **85.** ✓✓✓✓ **87.** ✓XXX **89.** XX✓X **91.** X✓XX **93.** ✓✓XX **95.** XXXX

SECTION 1.6 (pages 60–62)

1. 53 **2.** 34 **3.** 250 **4.** 160 **5.** 120 R7 **6.** 522 R14 **7.** 1105 R5 **8.** 9746 R1 **9.** 7134 R12 **10.** 3252 R10 **11.** 900 R100 **12.** 1101 **13.** 73 R5 **15.** 476 R15 **17.** 2407 R1 **19.** 1146 R15 **21.** 3331 R82 **23.** 850 **25.** incorrect; should be 101 R14 **27.** incorrect; should be 658 **29.** incorrect; should be 62 **31.** 117 episodes **33.** 56 floor clocks **35.** \$355 **37.** 43,200 rings **39.** **(a)** 648 each year **(b)** less than 2 each day **41.** \$0 **42.** 0 **43.** undefined **44.** impossible; if you have 6 cookies, it is not possible to divide them among 0 people. **45.** **(a)** 14 **(b)** 17 **(c)** 38 **46.** Yes. Some examples are $18 \cdot 1 = 18$; $26 \cdot 1 = 26$; $43 \cdot 1 = 43$. **47.** **(a)** 3200 **(b)** 320 **(c)** 32 **48.** Drop the same number of zeros that appear in the divisor. The result is the quotient. With the divisor 10, drop one 0; with 100, drop two zeros; with 1000, drop three zeros.

SECTION 1.7 (pages 70–73)

1. ten **2.** ten **3.** hundred **4.** hundred **5.** thousand **6.** thousand **7.** ten-thousand **8.** ten-thousand **9.** 860 **11.** 6800 **13.** 28,500 **15.** 6000 **17.** 16,000 **19.** 8,000,000,000 **21.** 600,000 **23.** 5,000,000 **25.** 4480; 4500; 4000 **27.** 3370; 3400; 3000 **29.** 6050; 6000; 6000 **31.** 5340; 5300; 5000 **33.** 19,540; 19,500; 20,000 **35.** 26,290; 26,300; 26,000 **37.** 93,710; 93,700; 94,000 **39.** 1. Locate the place to be rounded and underline it. 2. Look only at the next digit to the right. If this digit is 5 or more, increase the underlined digit by 1. 3. Change all digits to the right of the underlined place to zeros. **40.** 1. Locate the place to be rounded and underline it. 2. Look only at the next digit to the right. If this digit is 4 or less do not change the underlined digit. 3. Change all digits to the right of the underlined place to zeros. **41.** 30 60 50 80 220; 219 **43.** 80 40 40; 35 **45.** 70 30 2100; 2278 **47.** 900 700 400 800 2800; 2828 **49.** 900 400 500; 435 **51.** 800 400 320,000; 282,000 **53.** 12,760; 12,605 **55.** 700 500 200; 158 **57.** 900 30 27,000; 27,231 **59.** Perhaps the best explanation is that 3492 is closer to 3500 than 3400, but 3492 is closer to 3000 than 4000. **60.** Rounding numbers usually allows for faster calculation and results in an estimated answer prior to getting an exact answer. One example is

$$\begin{array}{r} 400 \\ -\,200 \\ \hline 200 \end{array} \qquad \begin{array}{r} 432 \\ -\,209 \\ \hline 223 \end{array}$$

Estimate: 200 *Exact:* 223

61. 80 million people; 310 million people **63.** 349,000 streets; 350,000 streets **65.** 39,840,000 tickets; 39,800,000 tickets; 40,000,000 tickets **67.** 19,266,000 players; 19,270,000 players; 19,300,000 players **69.** 71,500 **70.** 72,499 **71.** 7500 **72.** 8499 **73.** 3930; 11,240; 15,970; 17,920; 534,880; 2,788,000 **74.** 4000; 10,000; 20,000; 20,000; 500,000; 3,000,000 **75.** **(a)** When using front end rounding, all digits are 0 except the first digit. These numbers are easier to work with when estimating answers. **(b)** Sometimes when using front end rounding; the estimated answer can vary greatly from the exact answer.

SECTION 1.8 (pages 76–79)

1. exponent 2; base 3 **2.** exponent 3; base 2 **3.** exponent 2; base 5 **4.** exponent 2; base 4 **5.** 2; 8; 64 **7.** 2; 15; 225 **9.** 4 **11.** 8 **13.** 10 **15.** 12 **17.** false: 5^2 means that 5 is used as a factor 2 times, so $5^2 = 5 \cdot 5 = 25$. **18.** false: $4^2 = 4 \cdot 4 = 16$ **19.** false: 1 raised to any power is 1. In this example, $1 \cdot 1 \cdot 1 = 1$. **20.** false: a number raisecd to the first power is the number itself, so $6^1 = 6$. **21.** 36; 36; 36; 36 **23.** 625; 625 **25.** 10,000; 10,000 **27.** A perfect square is the square of a whole number. The number 25 is the square of 5 because $5 \cdot 5 = 25$. The number 50 is not a perfect square. There is no whole number that can be squared to get 50. **28.** 1. Do all operations inside parentheses or other grouping symbols. 2. Simplify any expressions with exponents and find any square roots. 3. Multiply or divide proceeding from left to right. 4. Add or subtract proceeding from left to right. **29.** true **30.** true **31.** false: $6 + 8 \div 2 = 10$. Multiplications and divisions and performed from left to right, then additions and subtractions. **32.** false: $4 + 5(6 - 4) = 4 + 5 \cdot 2 = 4 + 10 = 14$. A common error is adding 4 to 5 first and then multiplying by 2. Follow the order of operations. **33.** 12 **35.** 20 **37.** 45 **39.** 63 **41.** 118 **43.** 22 **45.** 30 **47.** 102 **49.** 9 **51.** 63 **53.** 33 **55.** 70 **57.** 7 **59.** 17 **61.** 55 **63.** 108 **65.** 26 **67.** 26 **69.** 27 **71.** 16 **73.** 16 **75.** 21 **77.** 7 **79.** 20 **81.** 14 **83.** 25 **85.** 16 **87.** 23 **89.** 233

SECTION 1.9 (pages 84–87)

1. 8; 500; $8 \cdot 500 = 4000$ **2.** $8\frac{1}{2}$; 500; $8\frac{1}{2} \cdot 500 = 4250$ **3.** 4500 stores **5.** Dollar General; about 5750 stores **7.** 1250 fewer stores **9.** 100 adults **10.** 7 career paths **11.** 9 people **13.** **(a)** Saw ad **(b)** 25 people **15.** 9 people **17.** 2014; 7000 installations **19.** 4500 installations **21.** Possible answers are: 1. shortage of units to install; 2. lack of qualified workers; 3. poor economy; 4. less demand for solar products. **22.** Possible answers are: 1. greater demand; 2. more units available; 3. many qualified workers; 4. tax incentives for solar projects. **23.** $(7 - 2) \cdot 3 - 6$ **24.** $(4 + 2) \cdot (5 + 1)$ **25.** $36 \div (3 \cdot 3) \cdot 4$ **26.** $56 \div (2 \cdot 2 \cdot 2) + \frac{0}{6}$ **27.** **(a)** $7920 + 1320 + 2640 + (5280 - 1320 - 1320) + 2640 + 1320 + 7920 + 5280$ **(b)** $31,680 \times 3 = 95,040$ ft **(c)** 18 miles

SECTION 1.10 (pages 92–95)

1. C **2.** B **3.** A **4.** B **5.** *Estimate:* $600 + 900 + 1000 + 800 + 2000 = 5300$ sandwiches; *Exact:* 5208 sandwiches **7.** *Estimate:* $\$300 - \$200 = \$100$ saved; *Exact:* \$104 saved **9.** *Estimate:* $200 \times 20 = 4000$ kits; *Exact:* 5664 kits

11. *Estimate:* 3000 ÷ 700 ≈ 4 toys; *Exact:* 4 toys **13.** *Estimate:* \$30 × 5 = \$150; *Exact:* \$170 **15.** *Estimate:* 100,000 + 300,000 = 400,000 deaths: *Exact:* 360,222 deaths **17. (a)** *Estimate:* 300,000 − 100,000 = 200,000 deaths; *Exact:* 140,082 deaths **19.** 100,000 + 300,000 + 90,000 + 200,000 = 690,000 deaths *Exact:* 618,642 deaths **21.** *Estimate:* \$2000 − \$700 − \$300 − \$400 − \$200 − \$200 = \$200; *Exact:* \$350 **23.** *Estimate:* 40,000 × 100 = 4,000,000 square feet; *Exact:* 6,011,280 square feet **25.** *Estimate:* \$300 + \$200 + \$100 + \$70 + \$400 + \$200 = \$1270; *Exact:* \$1254 **27.** *Estimate:* \$300 + \$70 + \$400 + \$200 = \$970; \$970 − \$800 = \$170; *Exact:* \$185 **29.** *Estimate:* (\$1000 × 6) + (\$900 × 20) = \$24,000; *Exact:* \$20,961 **31.** Possible answers are: Addition: more; total; gain of. Subtraction: less; loss of; decreased by. Multiplication: twice; of; product. Division: divided by; goes into; per. Equals: is; are. **32.** 1. Read the problem carefully. 2. Work out a plan. 3. Estimate a reasonable answer. 4. Solve the problem. 5. State the answer. 6. Check your work. **33.** \$20,009 **35.** 2477 pounds **37.** \$378 **39.** \$375 **41.** 20 seats

Chapter 1 REVIEW EXERCISES (pages 101–108)

1. 6; 573 **2.** 36; 215 **3.** 105; 724 **4.** 1; 768; 710; 618 **5.** seven hundred twenty-eight **6.** fifteen thousand, three hundred ten **7.** three hundred nineteen thousand, two hundred fifteen **8.** sixty-two million, five hundred thousand, five **9.** 10,008 **10.** 200,000,455 **11.** 110 **12.** 121 **13.** 5464 **14.** 15,657 **15.** 10,986 **16.** 9845 **17.** 40,602 **18.** 49,855 **19.** 36 **20.** 27 **21.** 189 **22.** 184 **23.** 6849 **24.** 4327 **25.** 224 **26.** 25,866 **27.** 49 **28.** 0 **29.** 32 **30.** 64 **31.** 45 **32.** 42 **33.** 56 **34.** 81 **35.** 40 **36.** 45 **37.** 48 **38.** 8 **39.** 0 **40.** 42 **41.** 48 **42.** 0 **43.** 84 **44.** 368 **45.** 522 **46.** 98 **47.** 5000 **48.** 2992 **49.** 5396 **50.** 45,815 **51.** 14,912 **52.** 20,160 **53.** 465,525 **54.** 174,984 **55.** 875 **56.** 2368 **57.** 1176 **58.** 5100 **59.** 15,576 **60.** 30,184 **61.** 887,169 **62.** 500,856 **63.** \$360 **64.** \$1064 **65.** \$20,352 **66.** \$684 **67.** 14,000 **68.** 23,800 **69.** 206,800 **70.** 318,500 **71.** 128,000,000 **72.** 90,300,000 **73.** 5 **74.** 7 **75.** 6 **76.** 2 **77.** 6 **78.** 4 **79.** 7 **80.** 0 **81.** undefined **82.** 0 **83.** 8 **84.** 9 **85.** 82 **86.** 98 **87.** 4422 **88.** 352 **89.** 150 R4 **90.** 124 R25 **91.** 820 **92.** 15,200 **93.** 21,000 **94.** 70,000 **95.** 3490; 3500; 3000 **96.** 20,070; 20,100; 20,000 **97.** 98,200; 98,200; 98,000 **98.** 352,120; 352,100; 352,000 **99.** 4 **100.** 7 **101.** 12 **102.** 14 **103.** 3; 7; 343 **104.** 6; 3; 729 **105.** 3; 5; 125 **106.** 5; 4; 1024 **107.** 34 **108.** 26 **109.** 9 **110.** 4 **111.** 9 **112.** 6 **113.** 8 parents **114.** 5 parents **115.** keeping bedroom clean; 25 parents **116.** hanging up wet bath towels; 3 parents **117.** *Estimate:* 40 million × 400 = 16,000 million or 16,000,000,000 checks; *Exact:* 14,600 million or 14,600,000,000 checks **118.** *Estimate:* 1000 × 60 = 60,000 revolutions; *Exact:* 84,000 revolutions **119.** *Estimate:* 40,000,000 − 30,000,000 = 10,000,000; *Exact:* 12,073,57 people **120.** *Estimate:* 700,000 − 600,000 = 100,000; *Exact:* 153,223 people **121.** *Estimate:* \$3000 + (8 × \$90) + 200 = \$3920; *Exact:* \$3583 **122.** *Estimate:* \$60 + (\$90 × \$2) = \$240; *Exact:* \$233 **123.** *Estimate:* (30 × \$2000) + (30 × \$900) = \$87,000; *Exact:* \$74,052 **124.** *Estimate:* (60 × \$20) + (20 × \$7) = \$1340; *Exact:* \$1139 **125.** *Estimate:* \$600 − \$100 = \$500; *Exact:* = \$513 **126.** *Estimate:* \$2000 − \$500 − \$400 = \$1100; *Exact:* \$1019 **127.** *Estimate:* 9000 ÷ 200 = 45 pounds; *Exact:* 50 pounds **128.** *Estimate:* 30,000 ÷ 1000 = 30 hours; *Exact:* 33 hours **129.** *Estimate:* 30,000 ÷ 600 = 50 acres; *Exact:* 52 acres **130.** *Estimate:* 6000 ÷ 200 = 30 homes; *Exact:* 32 homes **131.** 332 **132.** 448 **133.** 253 **134.** 588 **135.** 1041 **136.** 1661 **137.** 32,062 **138.** 24,947 **139.** 3 **140.** 7 **141.** 93,635 **142.** 83,178 **143.** undefined **144.** 7 **145.** 6900 **146.** 2310 **147.** 1,079,040 **148.** 130,212 **149.** 108 **150.** 207 **151.** three hundred seventy-six thousand, eight hundred fifty-three **152.** four hundred eight thousand, six hundred ten **153.** 8700 **154.** 401,000 **155.** 8 **156.** 9 **157.** \$5544 **158.** \$31,080 **159.** \$2288 **160.** \$15,782 **161.** 468 cards **162.** 52,700 pounds **163.** \$280 **164.** \$114,635 **165.** \$1905 **166.** \$12,420 **167.** 1467 ft **168.** 293 ft **169. (a)** 11,958 ft **(b)** greater than two miles by 1398 ft **170.** more than 4 football fields (a little less than 5)

Chapter 1 TEST (pages 109–110)

1. nine thousand, two hundred five **2.** twenty-five thousand, sixty-five **3.** 426,005 **4.** 8530 **5.** 112,630 **6.** 1045 **7.** 6206 **8.** 168 **9.** 171,000 **10.** 1615 **11.** 4,450,743 **12.** 7047 **13.** undefined **14.** 458 R5 **15.** 160 **16.** 6350 **17.** 76,000 **18.** 41 **19.** 28 **20.** *Estimate:* \$500 + \$500 + \$500 + \$400 − \$800 = \$1100; *Exact:* \$1140 **21.** *Estimate:* 90,000 ÷ 400 = 225 acres; *Exact:* 231 acres **22.** *Estimate:* \$2000 − \$500 − \$200 − \$200 = \$1100; *Exact:* \$948 **23.** *Estimate:* (50 × 60 × 4) + (40 × 60 × 3) = 19,200 chicks; *Exact:* 18,000 chicks **24.** 1. Locate the place to which you are rounding and underline it. 2. Look only at the next digit to the right. If this digit is a 4 or less, do not change the underlined digit. If the digit is 5 or more, increase the underlined digit by 1. 3. Change all digits to the right of the underlined place to zeros. Each person's rounding example will vary. **25.** 1. Read the problem carefully. 2. Work out a plan. 3. Estimate a reasonable answer. 4. Solve the problem. 5. State the answer. 6. Check your work.

CHAPTER 2 Multiplying and Dividing Fractions

SECTION 2.1 (pages 114–116)

1. Numerator 4; Denominator 5 **2.** Numerator 5; Denominator 6 **3.** Numerator 9; Denominator 8 **4.** Numerator 7; Denominator 5 **5.** 3; 8 **6.** 7; 16 **7.** 5; 24 **8.** 24; 32 **9.** $\frac{3}{4}; \frac{1}{4}$ **11.** $\frac{1}{3}; \frac{2}{3}$ **13.** $\frac{7}{5}; \frac{3}{5}$ **15.** $\frac{5}{6}; \frac{4}{6}; \frac{2}{6}$ **17.** $\frac{8}{25}$ **19.** $\frac{303}{520}$ **21.** Proper $\frac{1}{3}, \frac{5}{8}, \frac{7}{16}$ Improper $\frac{8}{5}, \frac{6}{6}, \frac{12}{2}$ **23.** Proper $\frac{3}{4}, \frac{9}{11}, \frac{7}{15}$ Improper $\frac{3}{2}, \frac{5}{5}, \frac{19}{18}$

25. One possibility is

$\frac{3}{4}$ ← Numerator
← Denominator

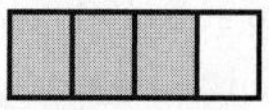

The denominator shows the number of equal parts in the whole and the numerator shows how many of the parts are being considered.
26. An example is $\frac{1}{2}$ as a proper fraction and $\frac{3}{2}$ as an improper fraction. A proper fraction has a numerator smaller than the denominator. An improper fraction has a numerator that is equal to or greater than the denominator. Drawings will vary.

SECTION 2.2 (pages 121–123)

1. true **2.** true **3.** false: Multiply 7 by 5 and add the numerator. $(7 \cdot 5) + 2 = 37; \frac{37}{5}$ **4.** false: Any mixed number can be charged to an improper fraction. The reverse is also true. **5.** false: You must add the numerator to the product of the whole number and the denominator. $6\frac{1}{2} = \frac{(6 \cdot 2) + 1}{2} = \frac{13}{2}$ **6.** true **7.** $\frac{5}{4}$ **9.** $\frac{23}{5}$ **11.** $\frac{17}{2}$ **13.** $\frac{81}{8}$ **15.** $\frac{43}{4}$ **17.** $\frac{29}{5}$ **19.** $\frac{43}{5}$ **21.** $\frac{54}{11}$ **23.** $\frac{131}{4}$ **25.** $\frac{221}{12}$ **27.** $\frac{269}{15}$ **29.** $\frac{187}{24}$

31. false: The denominator remains the same. So, $\frac{4}{3} = 1\frac{1}{3}$.
32. false: A mixed number always has a value equal to or greater than a whole number. **33.** true **34.** true **35.** $1\frac{1}{3}$ **37.** $2\frac{1}{4}$ **39.** 9 **41.** $7\frac{3}{5}$ **43.** $15\frac{3}{4}$ **45.** $5\frac{2}{9}$ **47.** $8\frac{1}{8}$ **49.** $16\frac{4}{5}$ **51.** 28 **53.** $26\frac{1}{7}$
55. Multiply the denominator by the whole number and add the numerator. The result becomes the new numerator, which is placed over the original denominator.

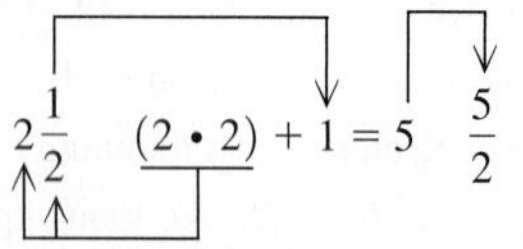

56. Divide the numerator by the denominator. The quotient is the whole number of the mixed number and the remainder is the numerator of the fraction part. The denominator is unchanged.

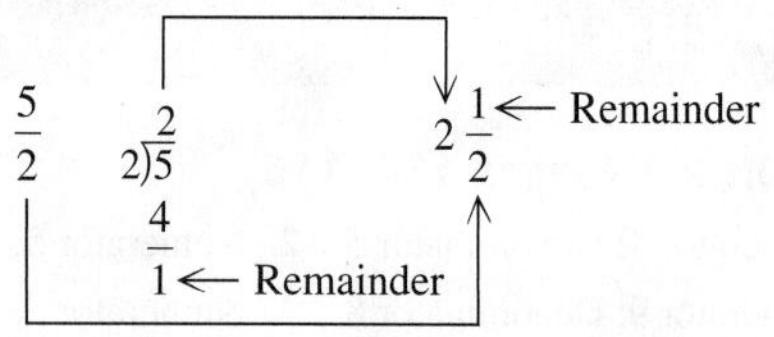

57. $\frac{501}{2}$ **59.** $\frac{1000}{3}$ **61.** $\frac{4179}{8}$ **63.** $154\frac{1}{4}$ **65.** 171 **67.** $122\frac{13}{32}$
69. $\frac{2}{3}, \frac{4}{5}, \frac{3}{4}, \frac{7}{10}$ **70. (a)** numerator; denominator

(b)

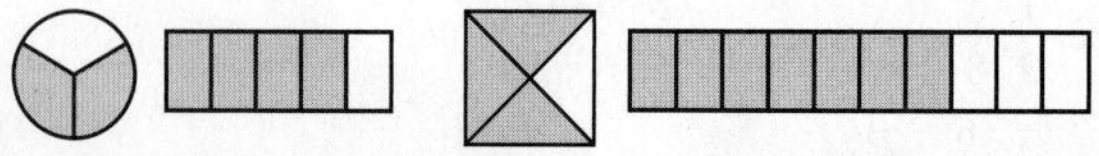

(c) less **71.** $\frac{5}{5}, \frac{10}{3}, \frac{6}{5}$

72. (a) numerator; denominator

(b)

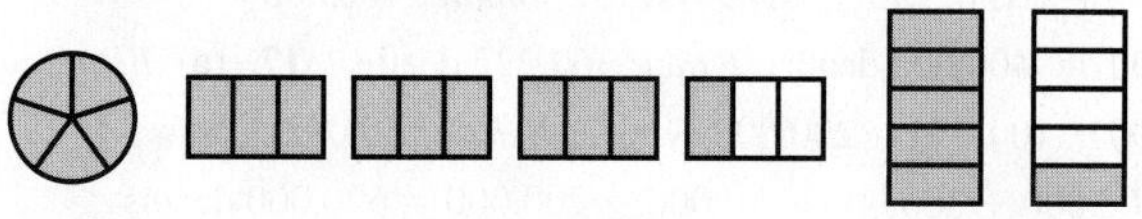

(c) greater **73.** $\frac{5}{3} = 1\frac{2}{3}; \frac{7}{7} = 1; \frac{11}{6} = 1\frac{5}{6}$ **74. (a)** improper; greater than or equal to

(b)

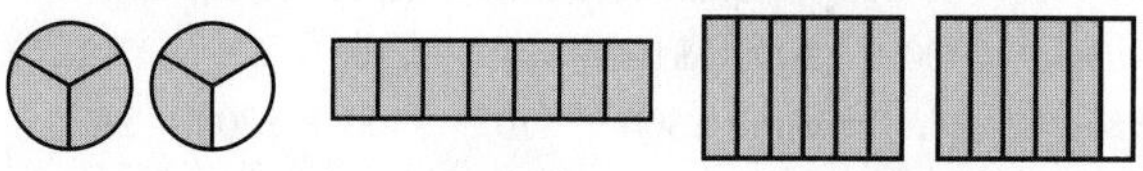

SECTION 2.3 (pages 128–129)

1. false **2.** true **3.** true **4.** false **5.** 1, 2, 3, 4, 6, 8, 12, 16, 24, 48 **7.** 1, 2, 4, 7, 8, 14, 28, 56 **9.** 1, 2, 3, 4, 6, 9, 12, 18, 36 **11.** 1, 2, 4, 5, 8, 10, 20, 40 **13.** 1, 2, 4, 8, 16, 32, 64 **15.** 1, 2, 41, 80 **17.** composite **19.** prime **21.** composite **23.** prime **25.** composite **27.** prime **29.** B **30.** C **31.** A **33.** $2 \cdot 3$ **35.** 5^2 **37.** $2^2 \cdot 17$ **39.** $2^3 \cdot 3^2$ **41.** $2^2 \cdot 11$ **43.** $2^2 \cdot 5^2$ **45.** 5^3 **47.** $2^2 \cdot 3^2 \cdot 5$ **49.** $2^6 \cdot 5$ **51.** $2^3 \cdot 3^2 \cdot 5$ **53.** A prime number is a whole number that has exactly two *different* factors, itself and 1. Examples include 2, 3, 5, 7, 11. A composite number has a factor(s) other than itself or 1. Examples include 4, 6, 8, 9, 10. The numbers 0 and 1 are neither prime nor composite. **54.** No even number other than 2 is prime because all even numbers have 2 as a factor. Many odd numbers are multiples of prime numbers and are not prime. For example, 9, 21, 33, and 45 are all multiples of 3. **55.** All the possible factors of 24 are 1, 2, 3, 4, 6, 8, 12, and 24. This list includes both prime numbers and composite numbers. The prime factors of 24 include only prime numbers. The prime factorization of 24 is $2 \cdot 2 \cdot 2 \cdot 3 = 2^3 \cdot 3$. **56.** No. The order of division does not matter. As long as you use only prime numbers, your answers will be correct. However, it does seem easier to always start with 2 and then use progressively greater prime numbers. The prime factorization of 36 is $36 = 2 \cdot 2 \cdot 3 \cdot 3 = 2^2 \cdot 3^2$. **57.** $2 \cdot 5^2 \cdot 7$ **59.** $2^6 \cdot 3 \cdot 5$ **61.** $2^3 \cdot 3 \cdot 5 \cdot 13$ **63.** $2^2 \cdot 3^2 \cdot 5 \cdot 7$ **65.** 2, 3, 5, 7, 11, 13, 17, 19, 23, 29, 31, 37, 41, 43, 47 **66.** A prime number is a whole number that is evenly divisible by itself and 1 only. **67.** No. Every other even number is divisible by 2 in addition to being divisible by itself and 1. **68.** No. A multiple of a prime number can never be prime because it will always be divisible by the prime number. **69.** $2 \cdot 2 \cdot 3 \cdot 5 \cdot 5 \cdot 7$ **70.** $2^2 \cdot 3 \cdot 5^2 \cdot 7$

SECTION 2.4 (pages 134–135)

1. even **2.** 5; 0 **3.** 0 **4.** 3 **5.** ✓✓✓✓ **7.** ✓✓XX **9.** ✓X✓✓ **11.** ✓✓XX **13.** true **14.** false **15.** false **16.** true **17.** $\frac{3}{5}$ **19.** $\frac{6}{7}$ **21.** $\frac{7}{8}$ **23.** $\frac{6}{7}$ **25.** $\frac{4}{7}$ **27.** $\frac{1}{50}$ **29.** $\frac{8}{11}$

31. $\frac{5}{9}$ **33.** $\frac{\overset{1}{\cancel{2}} \cdot \overset{1}{\cancel{3}} \cdot 3}{\underset{1}{\cancel{2}} \cdot 2 \cdot 2 \cdot \underset{1}{\cancel{3}}} = \frac{3}{4}$ **35.** $\frac{\overset{1}{\cancel{5}} \cdot 7}{2 \cdot 2 \cdot 2 \cdot \underset{1}{\cancel{5}}} = \frac{7}{8}$

37. $\frac{\overset{1}{\cancel{2}} \cdot \overset{1}{\cancel{3}} \cdot \overset{1}{\cancel{3}} \cdot \overset{1}{\cancel{5}}}{\underset{1}{\cancel{2}} \cdot 2 \cdot \underset{1}{\cancel{3}} \cdot \underset{1}{\cancel{3}} \cdot \underset{1}{\cancel{5}}} = \frac{1}{2}$ **39.** $\frac{\overset{1}{\cancel{2}} \cdot \overset{1}{\cancel{2}} \cdot \overset{1}{\cancel{3}} \cdot 3}{\underset{1}{\cancel{2}} \cdot \underset{1}{\cancel{2}} \cdot \underset{1}{\cancel{3}}} = 3$

41. $\frac{2\cdot 2\cdot 2\cdot \overset{1}{\cancel{3}}\cdot \overset{1}{\cancel{3}}}{\underset{1}{\cancel{3}}\cdot \underset{1}{\cancel{3}}\cdot 5\cdot 5}=\frac{8}{25}$ **43.** $\frac{1}{2}=\frac{1}{2}$; equivalent

45. $\frac{5}{12}\neq\frac{2}{5}$; not equivalent **47.** $\frac{5}{8}\neq\frac{35}{52}$; not equivalent

49. $\frac{7}{8}=\frac{7}{8}$; equivalent **51.** $8\neq 9$; not equivalent **53.** $\frac{5}{6}=\frac{5}{6}$; equivalent

55. A fraction is in lowest terms when the numerator and the denominator have no common factors other than 1. Some examples are $\frac{1}{2}$, $\frac{3}{8}$, and $\frac{2}{3}$.

56. Two fractions are equivalent when they represent the same portion of a whole. For example, the fractions $\frac{10}{15}$ and $\frac{8}{12}$ are equivalent.

$$\frac{10}{15}=\frac{2\cdot \overset{1}{\cancel{5}}}{3\cdot \underset{1}{\cancel{5}}}=\frac{2\cdot 1}{3\cdot 1}=\frac{2}{3}$$

$$\frac{8}{12}=\frac{\overset{1}{\cancel{2}}\cdot \overset{1}{\cancel{2}}\cdot 2}{\underset{1}{\cancel{2}}\cdot \underset{1}{\cancel{2}}\cdot 3}=\frac{1\cdot 1\cdot 2}{1\cdot 1\cdot 3}=\frac{2}{3}$$

Equivalent $\left(\frac{2}{3}=\frac{2}{3}\right)$

57. $\frac{5}{8}$ **59.** $\frac{2}{1}=2$

SECTION 2.5 (pages 144–147)

1. multiply; denominator **2.** numerator; denominator **3.** divide; denominator **4.** lowest terms **5.** $\frac{1}{4}$ **7.** $\frac{2}{35}$ **9.** $\frac{3}{4}$ **11.** $\frac{1}{4}$ **13.** $\frac{5}{12}$ **15.** $\frac{4}{5}$ **17.** $\frac{7}{16}$ **19.** $\frac{7}{48}$ **20.** true **21.** false: $\frac{4}{5}\cdot 8=\frac{4}{5}\cdot\frac{8}{1}=\frac{32}{5}=6\frac{2}{5}$

23. 15 **25.** $13\frac{1}{2}$ **27.** $31\frac{1}{2}$ **29.** 80 **31.** $94\frac{2}{3}$ **33.** 400 **35.** 810

37. $\frac{1}{4}$ square mile **39.** 9 square meters **41.** $\frac{1}{4}$ square mile

43. Multiply the numerators and multiply the denominators. An example is $\frac{3}{4}\cdot\frac{1}{2}=\frac{3\cdot 1}{4\cdot 2}=\frac{3}{8}$. **44.** You must divide a numerator and a denominator by the same number. If you do all possible divisions, your answer will be in lowest terms. One example is $\frac{3}{4}\cdot\frac{2}{3}=\frac{3\cdot 2}{4\cdot 3}=\frac{\overset{1}{\cancel{3}}\cdot \overset{1}{\cancel{2}}}{\underset{2}{\cancel{4}}\cdot \underset{1}{\cancel{3}}}=\frac{1}{2}$.

45. $1\frac{1}{2}$ square yards **47.** $3\frac{1}{2}$ square miles

49. They are both the same size: $\frac{3}{64}$ square mile

51. $4000+2000+1000+800+400+200+100+80=$ 8580 supermarkets **52.** 8912 supermarkets

53. *Estimate:* $\frac{4}{5}\cdot 2000=1600$; *Exact:* 1762 supermarkets (rounded)

54. *Estimate:* $\frac{3}{8}\cdot 200=75$; *Exact:* 61 supermarkets (rounded)

55. $\frac{4}{5}\cdot 2200=1760$ supermarkets **56.** $\frac{3}{8}\cdot 160=60$ supermarkets

SECTION 2.6 (pages 151–154)

1. times; triple; of; twice; product; twice as much **2.** Check your work. **3.** area **4.** square **5.** $\frac{1}{2}$ square foot **7.** $\frac{8}{9}$ square foot **9.** $\frac{3}{10}$ square yard

11. \$2568 **13.** \$35 **15.** **(a)** 910 women **(b)** 650 men **17.** 4 hours; 153 people **19.** $\frac{3}{4}$; 765 people **21.** Because everyone is included and fractions are given for *all* groups, the sum of the fractions must be 1 or *all* of the people. **22.** Answers will vary. Some possibilities are: 1. You made an addition error. 2. The fractions on the circle graph are incorrect. 3. The fraction errors were caused by rounding.

23. \$76,000 **25.** \$15,200 **27.** \$4750

29. The correct solution is $\frac{9}{10}\times\frac{20}{21}=\frac{\overset{3}{\cancel{9}}}{\underset{1}{\cancel{10}}}\times\frac{\overset{2}{\cancel{20}}}{\underset{7}{\cancel{21}}}=\frac{6}{7}$.

30. Yes, the statements are true. Since whole numbers are 1 or greater, when you multiply, the product will always be greater than either of the numbers multiplied. But, when you multiply two proper fractions, you are finding a fraction of a fraction, and the product will be smaller than either of the two proper fractions. **31.** \$750 **33.** 2 ft **35.** 9000 votes

37. $\frac{1}{32}$ of the estate

SECTION 2.7 (pages 160–163)

1. reciprocal **2.** 1 **3.** invert; multiplication **4.** lowest **5.** $\frac{8}{3}$ **7.** $\frac{6}{5}$ **9.** $\frac{5}{8}$ **11.** $\frac{1}{4}$ **13.** $\frac{2}{3}$ **15.** $2\frac{5}{8}$ **17.** $\frac{9}{20}$ **19.** 4 **21.** 6 **23.** $\frac{13}{16}$ **25.** $\frac{4}{5}$ **27.** 18 **29.** 24 **31.** $\frac{1}{14}$ **33.** $\frac{2}{9}$ quart **35.** 15 times

37. 88 dispensers **39.** 160 fasteners **41.** You can divide two fractions by multiplying the first fraction by the reciprocal of the divisor (the second fraction). **42.** Sometimes the answer is less and sometimes it is greater.

$$\frac{1}{4}\div\frac{7}{8}=\frac{1}{\underset{1}{\cancel{4}}}\cdot\frac{\overset{2}{\cancel{8}}}{7}=\frac{2}{7}\quad\left(\text{less than }\tfrac{7}{8}\text{ but not less than }\tfrac{1}{4}\right)$$

$$\frac{1}{2}\div\frac{1}{4}=\frac{1}{\underset{1}{\cancel{2}}}\cdot\frac{\overset{2}{\cancel{4}}}{1}=\frac{2}{1}=2\ \text{(greater)}$$

43. 12 pounds **45.** 208 homes **47.** **(a)** 124 visits **(b)** 62 visits **49.** 2432 towels **51.** double, twice, times, product, twice as much **52.** goes into, divide, per, quotient, divided by **53.** reciprocal

54. $\frac{4}{3}$; $\frac{8}{7}$; $\frac{1}{5}$; $\frac{19}{12}$ **55.** **(a)** Multiply the length of one side by 3, 4, 5, or 6. **(b)** $\frac{15}{16}\times 4=\frac{15}{\underset{4}{\cancel{16}}}\times\frac{\overset{1}{\cancel{4}}}{1}=\frac{15}{4}=3\frac{3}{4}$ in. **56.** $\frac{225}{256}$ square inch; Multiply the length by the width.

SECTION 2.8 (pages 171–175)

1. false: A reciprocal is used when dividing fractions. **2.** false: The reciprocal of 10 is $\frac{1}{10}$. **3.** true **4.** true

5. *Estimate:* $5\cdot 2=10$; *Exact:* $7\frac{7}{8}$ **7.** *Estimate:* $2\cdot 3=6$; *Exact:* $4\frac{1}{2}$

9. *Estimate:* $3\cdot 1=3$; *Exact:* 4 **11.** *Estimate:* $8\cdot 6=48$; *Exact:* 50

13. *Estimate:* $5 \cdot 2 \cdot 5 = 50$; *Exact:* $49\frac{1}{2}$ **15.** *Estimate:* $3 \cdot 2 \cdot 3 = 18$; *Exact:* 12 **17.** D **18.** A **19.** B **20.** C **21.** *Estimate:* $1 \div 4 = \frac{1}{4}$; *Exact:* $\frac{1}{3}$ **23.** *Estimate:* $3 \div 3 = 1$; *Exact:* $\frac{5}{6}$ **25.** *Estimate:* $9 \div 3 = 3$; *Exact:* $3\frac{3}{5}$ **27.** *Estimate:* $1 \div 2 = \frac{1}{2}$; *Exact:* $\frac{5}{12}$ **29.** *Estimate:* $2 \div 6 = \frac{1}{3}$; *Exact:* $\frac{3}{10}$ **31.** *Estimate:* $6 \div 6 = 1$; *Exact:* $\frac{17}{18}$

33. **(a)** *Estimate:* $1 \cdot 3 = 3$ cups; *Exact:* $1\frac{7}{8}$ cups of applesauce

(b) *Estimate:* $1 \cdot 3 = 3$ teaspoons; *Exact:* $1\frac{1}{4}$ teaspoons of salt

(c) *Estimate:* $2 \cdot 3 = 6$ cups; *Exact:* $4\frac{3}{8}$ cups of flour

35. **(a)** *Estimate:* $1 \div 2 = \frac{1}{2}$ teaspoon; *Exact:* $\frac{1}{4}$ teaspoon vanilla extract

(b) *Estimate:* $1 \div 2 = \frac{1}{2}$ cup; *Exact:* $\frac{3}{8}$ cup applesauce

(c) *Estimate:* $2 \div 2 = 1$ cup; *Exact:* $\frac{7}{8}$ cup of flour

37. *Estimate:* $329 \div 12 \approx 27$ units; *Exact:* 28 units

39. *Estimate:* $45 \cdot 20 = 900$ in.; *Exact:* $877\frac{1}{2}$ in.

41. The answer should include: *Step 1* Change mixed numbers to improper fractions. *Step 2* Multiply the fractions. *Step 3* Write the answer in lowest terms, changing to mixed or whole numbers where possible.
42. The additional step is to use the reciprocal of the divisor (the second fraction). **43.** *Estimate:* $130 \cdot 1 = \$130$ million; *Exact:* 182 million or $182,000,000 **45.** *Estimate:* $140 \cdot 1 = 140$ trips; *Exact:* 112 trips **47.** **(a)** *Estimate:* $18 \cdot 4 = 72$ in.; *Exact:* 71 in. **(b)** No, because 6 ft = 72 in. **49.** *Estimate:* $6 \cdot 6 = 36$ miles per hour; *Exact:* $33\frac{11}{16}$ miles per hour

Chapter 2 REVIEW EXERCISES (pages 184–187)

1. $\frac{1}{3}$ **2.** $\frac{5}{8}$ **3.** $\frac{2}{4}$ **4.** Proper $\frac{1}{8}, \frac{3}{4}, \frac{2}{3}$; Improper $\frac{4}{3}, \frac{5}{5}$
5. Proper $\frac{15}{16}, \frac{1}{8}$; Improper $\frac{6}{5}, \frac{16}{13}, \frac{5}{3}$ **6.** $\frac{19}{4}$ **7.** $\frac{59}{6}$ **8.** $3\frac{3}{8}$ **9.** $12\frac{3}{5}$
10. 1, 2, 3, 6 **11.** 1, 2, 3, 4, 6, 8, 12, 24 **12.** 1, 5, 11, 55
13. 1, 2, 3, 5, 6, 9, 10, 15, 18, 30, 45, 90 **14.** 3^3 **15.** $2 \cdot 3 \cdot 5^2$
16. $2^2 \cdot 3 \cdot 5 \cdot 7$ **17.** 25 **18.** 288 **19.** 1728 **20.** 2048 **21.** $\frac{24}{24} = 1$
22. $\frac{18}{24} = \frac{3}{4}$ **23.** $\frac{14}{24} = \frac{7}{12}$ **24.** $\frac{10}{24} = \frac{5}{12}$ **25.** $\frac{\overset{1}{\cancel{3}} \cdot 5}{2 \cdot 2 \cdot 3 \cdot \underset{1}{\cancel{3}}}; \frac{5}{12}$
26. $\frac{\overset{1}{\cancel{2}} \cdot \overset{1}{\cancel{2}} \cdot \overset{1}{\cancel{2}} \cdot \overset{1}{\cancel{2}} \cdot \overset{1}{\cancel{2}} \cdot 2 \cdot 2 \cdot \overset{1}{\cancel{3}}}{\underset{1}{\cancel{2}} \cdot \underset{1}{\cancel{2}} \cdot \underset{1}{\cancel{2}} \cdot \underset{1}{\cancel{2}} \cdot \underset{1}{\cancel{2}} \cdot \underset{1}{\cancel{3}}}$; 4 **27.** equivalent **28.** not equivalent
29. equivalent **30.** $\frac{3}{5}$ **31.** $\frac{3}{16}$ **32.** $\frac{1}{7}$ **33.** $\frac{4}{21}$ **34.** 15 **35.** 625
36. $1\frac{1}{3}$ **37.** $1\frac{2}{3}$ **38.** $2\frac{1}{2}$ **39.** 2 **40.** 8 **41.** 24 **42.** $\frac{5}{24}$ **43.** $\frac{2}{15}$
44. $\frac{4}{13}$ **45.** $1\frac{3}{8}$ square feet **46.** $3\frac{15}{16}$ square yards **47.** 7857 square feet
48. $5\frac{1}{2}$ square feet **49.** *Estimate:* $6 \cdot 1 = 6$; *Exact:* $6\frac{7}{8}$
50. *Estimate:* $2 \cdot 7 \cdot 1 = 14$; *Exact:* $21\frac{3}{8}$ **51.** *Estimate:* $16 \div 3 = 5\frac{1}{3}$; *Exact:* $5\frac{1}{6}$ **52.** *Estimate:* $5 \div 6 = \frac{5}{6}$; *Exact:* $\frac{3}{4}$ **53.** 512 bins
54. $\frac{3}{20}$ of the business **55.** *Estimate:* $158 \div 4 \approx 40$ pull cords; *Exact:* 36 pull cords **56.** *Estimate:* $8 \cdot 2 \cdot 50 = 800$ pounds; *Exact:* $833\frac{1}{3}$ pounds **57.** 25 pounds **58.** $186 **59.** $\frac{7}{48}$ of total raised
60. $\frac{4}{25}$ ton **61.** $\frac{3}{8}$ **62.** $\frac{2}{5}$ **63.** $31\frac{1}{4}$ **64.** $28\frac{1}{3}$ **65.** $\frac{1}{10}$ **66.** $\frac{5}{32}$
67. 30 **68.** $2\frac{3}{5}$ **69.** $1\frac{3}{5}$ **70.** $38\frac{1}{4}$ **71.** $\frac{17}{3}$ **72.** $\frac{307}{8}$
73. $\frac{\overset{1}{\cancel{2}} \cdot \overset{1}{\cancel{2}} \cdot 2}{\underset{1}{\cancel{2}} \cdot \underset{1}{\cancel{2}} \cdot 3} = \frac{2}{3}$ **74.** $\frac{\overset{1}{\cancel{2}} \cdot 2 \cdot \overset{1}{\cancel{3}} \cdot 3 \cdot 3}{\underset{1}{\cancel{2}} \cdot \underset{1}{\cancel{3}} \cdot 5 \cdot 7} = \frac{18}{35}$ **75.** $\frac{5}{6}$ **76.** $\frac{2}{3}$
77. $\frac{2}{5}$ **78.** $\frac{1}{3}$ **79.** *Estimate:* $3 \cdot 50 = 150$ ounces; *Exact:* 125 ounces
80. *Estimate:* $7 \cdot 9 = 63$ quarts; *Exact:* $67\frac{2}{3}$ quarts
81. $1\frac{17}{32}$ square inches **82.** $1\frac{31}{32}$ square yards

Chapter 2 TEST (pages 188–189)

1. $\frac{5}{6}$ **2.** $\frac{3}{8}$ **3.** $\frac{2}{3}, \frac{6}{7}, \frac{1}{4}, \frac{5}{8}$ **4.** $\frac{27}{8}$ **5.** $30\frac{3}{4}$ **6.** 1, 2, 3, 6, 9, 18
7. $3^2 \cdot 5$ **8.** $2^4 \cdot 3^2$ **9.** $2^2 \cdot 5^3$ **10.** $\frac{3}{4}$ **11.** $\frac{5}{6}$
12. Write the prime factorization of both numerator and denominator. Divide the numerator and denominator by any common factors. Multiply the remaining factors in the numerator and denominator.

$$\frac{56}{84} = \frac{\overset{1}{\cancel{2}} \cdot \overset{1}{\cancel{2}} \cdot 2 \cdot \overset{1}{\cancel{7}}}{\underset{1}{\cancel{2}} \cdot \underset{1}{\cancel{2}} \cdot 3 \cdot \underset{1}{\cancel{7}}} = \frac{2}{3}$$

13. Answers will vary. One possibility is: Multiply fractions by multiplying the numerators and multiplying the denominators. Divide two fractions by using the reciprocal of the divisor (the second fraction) and then changing division to multiplication.
14. $\frac{1}{3}$ **15.** 36 **16.** $\frac{5}{12}$ square yard **17.** 5475 seedlings **18.** $\frac{9}{10}$
19. $15\frac{3}{4}$ **20.** 24 pieces **21.** *Estimate:* $4 \cdot 4 = 16$; *Exact:* $14\frac{7}{16}$
22. *Estimate:* $2 \cdot 4 = 8$; *Exact:* $7\frac{17}{18}$ **23.** *Estimate:* $10 \div 2 = 5$; *Exact:* $4\frac{4}{15}$ **24.** *Estimate:* $9 \div 2 = 4\frac{1}{2}$; *Exact:* $4\frac{6}{7}$
25. *Estimate:* $3 \cdot 12 = 36$; *Exact:* $30\frac{5}{8}$ grams

CHAPTER 3 Adding and Subtracting Fractions

SECTION 3.1 (pages 195–196)

1. like **2.** unlike **3.** unlike **4.** like **5.** like **6.** lowest terms
7. $\frac{5}{8}$ **9.** $\frac{1}{2}$ **11.** $1\frac{1}{2}$ **13.** $\frac{5}{6}$ **15.** $1\frac{1}{2}$ **17.** $\frac{11}{27}$ **19.** $\frac{3}{8}$ **21.** $\frac{6}{11}$

ANSWERS

23. $\frac{3}{5}$ **25.** $1\frac{1}{7}$ **27.** $\frac{1}{5}$ **29.** $1\frac{1}{6}$ **31.** $1\frac{1}{10}$

33. Three steps to add like fractions are: 1. Add the numerators of the fractions to find the numerator of the sum (the answer). 2. Use the denominator of the fractions as the denominator of the sum. 3. Write the answer in lowest terms. **34.** Like fractions have the same denominator. Unlike fractions have denominators that are different.

$$\text{like: } \frac{3}{8}, \frac{1}{8}, \frac{7}{8}, \quad \text{unlike: } \frac{1}{2}, \frac{3}{4}, \frac{5}{8}$$

35. $\frac{7}{9}$ of their goal **37.** $\frac{3}{4}$ completed **39.** $\frac{1}{2}$ acre

SECTION 3.2 (pages 205–208)

1. true **2.** false; 25 is not divisible by 6. The LCM is 30. **3.** true **4.** false; 28 is not divisible by 3. The LCM is 21. **5.** 6 **7.** 15 **9.** 36 **11.** 48 **13.** 100 **15.** 40 **17.** 45 **19.** 120 **21.** 120 **23.** 216 **25.** 600 **27.** $\frac{9}{24}$ **29.** $\frac{10}{24}$ **31.** $\frac{21}{24}$ **32.** $\frac{8}{12}$ **33.** $\frac{21}{28}$ **34.** $\frac{36}{16}$ **35.** $\frac{21}{24}$ **37.** 6 **39.** 28 **41.** 12 **43.** 72 **45.** 84 **47.** 96 **49.** 27

51. It probably depends on how large the numbers are. If the numbers are small, the method using multiples of the largest number seems easiest. If the numbers are larger, or there are more than two numbers, then the factorization method or the alternative method will be better. **52.** Divide the desired denominator by the given denominator. Next, multiply both the given numerator and denominator by this number. Divide 12 by 4 to get 3. Then multiply.

$$\frac{3}{4} = \frac{3 \cdot 3}{4 \cdot 3} = \frac{9}{12}$$

Check your answer by writing $\frac{9}{12}$ in lowest terms. You should get $\frac{3}{4}$.

53. 3600 **55.** 10,584 **57.** like; unlike **58.** numerators; denominator; lowest **59.** least; smallest **60.** 40 is the least common multiple. **61.** 70 **62.** 450 **63.** 240 is a common multiple but twice as large as the least common multiple; 120 is the LCM. **64.** The least common multiple can be no smaller than the largest number in a group and the number 1760 is a multiple of 55.

SECTION 3.3 (pages 213–216)

1. like **2.** least common denominator **3.** $\frac{7}{8}$ **5.** $\frac{8}{9}$ **7.** $\frac{3}{4}$ **9.** $\frac{39}{40}$ **11.** $\frac{23}{36}$ **13.** $\frac{13}{14}$ **15.** $\frac{23}{24}$ **17.** $\frac{29}{36}$ **19.** $\frac{3}{8}$ **21.** $\frac{23}{48}$ **23.** $\frac{1}{2}$ **25.** $\frac{1}{2}$ **27.** $\frac{7}{15}$ **29.** $\frac{1}{6}$ **31.** $\frac{19}{45}$ **33.** $\frac{3}{40}$ **35.** $\frac{17}{48}$ **37.** $\frac{1}{4}$ in. **39.** $\frac{17}{40}$ for reserved seating **41.** $\frac{31}{40}$ in. **43.** $\frac{1}{24}$ of the tank

45. You cannot add or subtract until all the fractional pieces are the same size. For example, halves are larger than fourths, so you cannot add $\frac{1}{2} + \frac{1}{4}$ until you rewrite $\frac{1}{2}$ as $\frac{2}{4}$.

46. ***Step 1*** Rewrite the unlike fractions as like fractions.
Step 2 Add or subtract the like fractions.
Step 3 Simplify the answer.

47. $\frac{1}{3}$ **49.** totally honest; 400 people; $\frac{7}{12}$ of the users **51.** $\frac{3}{16}$ in.

SECTION 3.4 (pages 221–228)

1. 5 **2.** 6 **3.** 9 **4.** 13 **5.** 15 **6.** 21 **7.** 17 **8.** 3

9. *Estimate:* $6 + 3 = 9$; *Exact:* $8\frac{5}{6}$

11. *Estimate:* $7 + 4 = 11$; *Exact:* $11\frac{1}{2}$

13. *Estimate:* $1 + 4 = 5$; *Exact:* $4\frac{5}{24}$

15. *Estimate:* $25 + 19 = 44$; *Exact:* $43\frac{2}{3}$

17. *Estimate:* $34 + 19 = 53$; *Exact:* $52\frac{1}{10}$

19. *Estimate:* $23 + 15 = 38$; *Exact:* $38\frac{5}{28}$

21. *Estimate:* $13 + 19 + 15 = 47$; *Exact:* $45\frac{5}{6}$

23. *Estimate:* $15 - 12 = 3$; *Exact:* $2\frac{5}{8}$

25. *Estimate:* $13 - 1 = 12$; *Exact:* $11\frac{7}{15}$

27. *Estimate:* $28 - 6 = 22$; *Exact:* $22\frac{7}{30}$

29. *Estimate:* $17 - 7 = 10$; *Exact:* $10\frac{3}{8}$

31. *Estimate:* $19 - 6 = 13$; *Exact:* $12\frac{19}{20}$

33. *Estimate:* $20 - 12 = 8$; *Exact:* $7\frac{11}{12}$

35. $\frac{15}{4}$ **36.** $\frac{63}{8}$ **37.** $2\frac{2}{5}$ **38.** $3\frac{3}{7}$ **39.** $\frac{43}{8}$ **40.** $\frac{92}{15}$ **41.** $18\frac{2}{3}$ **42.** $3\frac{5}{8}$ **43.** $9\frac{1}{8}$ **45.** $11\frac{1}{2}$ **47.** $3\frac{5}{6}$ **49.** $6\frac{11}{12}$ **51.** $8\frac{1}{8}$ **53.** $\frac{5}{6}$ **55.** $2\frac{7}{8}$ **57.** $2\frac{7}{12}$ **59.** $5\frac{9}{20}$ **61.** $3\frac{16}{21}$ **63.** Find the least common denominator. Change the fraction parts so that they have the same denominator. Add the fraction parts. Add the whole number parts. Write the answer as a mixed number. Simplify the answer.

64. You need to regroup when the minuend (top number) is a whole number or when the fraction in the minuend is smaller than the fraction in the subtrahend (bottom number). Examples are shown below.

$$\begin{array}{r} 10 = 9\frac{3}{3} \\ -6\frac{2}{3} = 6\frac{2}{3} \\ \hline 3\frac{1}{3} \end{array} \qquad \begin{array}{r} 5\frac{1}{4} = 5\frac{1}{4} = 4\frac{5}{4} \\ -3\frac{1}{2} = 3\frac{2}{4} = 3\frac{2}{4} \\ \hline 1\frac{3}{4} \end{array}$$

65. *Estimate:* $26 - 15 = 11$ ft; *Exact:* $11\frac{1}{4}$ ft

67. *Estimate:* $3 - 3 = 0$ in.; *Exact:* $\frac{7}{16}$ in.

69. *Estimate:* $3 - 1 = 2$ in.; *Exact:* $2\frac{1}{2}$ in.

71. *Estimate:* $3 - 1 = 2$ in.; *Exact:* $2\frac{3}{16}$ in.

73. *Estimate:* $16 + 19 + 24 + 31 = 90$ ft; *Exact:* $87\frac{8}{4} = 89$ ft

75. *Estimate:* $24 + 35 + 24 + 35 = 118$ in.; *Exact:* $116\frac{1}{2}$ in.

77. *Estimate:* $100 - 10 - 14 - 9 - 19 - 12 - 10 - 14 = 12$ gallons; *Exact:* $12\frac{5}{8}$ gallons

79. *Estimate:* $527 - 108 - 151 - 139 = 129$ ft; *Exact:* 130 ft

81. *Estimate:* $59 + 24 + 17 + 29 + 58 = 187$ tons; *Exact:* $186\frac{13}{24}$ tons

83. $4\frac{11}{16}$ in. **85.** $21\frac{3}{8}$ in. **87. (a)** 30 **(b)** 28 **(c)** 25 **(d)** 264

88. least common denominator **89. (a)** $\frac{23}{24}$ **(b)** $\frac{8}{15}$ **(c)** $\frac{43}{48}$ **(d)** $\frac{4}{21}$

90. fraction parts **91.** improper; large

92. (a) $4\frac{5}{8} + 3\frac{6}{8} = 7\frac{11}{8} = 8\frac{3}{8}$; $\frac{37}{8} + \frac{30}{8} = \frac{67}{8} = 8\frac{3}{8}$

(b) $11\frac{56}{40} - 8\frac{35}{40} = 3\frac{21}{40}$; $\frac{496}{40} - \frac{355}{40} = \frac{141}{40} = 3\frac{21}{40}$

SECTION 3.5 (pages 235–238)

1.–12.

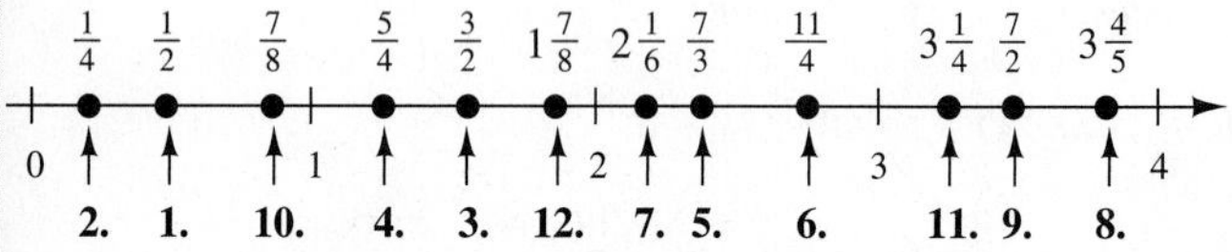

13. > **15.** < **17.** > **19.** > **21.** true; $\left(\frac{1}{2}\right)^2 = \frac{1}{2} \cdot \frac{1}{2} = \frac{1}{4}$

22. false; $\left(\frac{3}{8}\right)^2 = \frac{3}{8} \cdot \frac{3}{8} = \frac{9}{64}$ **23.** false; $\left(\frac{2}{5}\right)^3 = \frac{2}{5} \cdot \frac{2}{5} \cdot \frac{2}{5} = \frac{8}{125}$

24. true; $\left(\frac{5}{6}\right)^3 = \frac{5}{6} \cdot \frac{5}{6} \cdot \frac{5}{6} = \frac{125}{216}$ **25.** $\frac{1}{9}$ **27.** $\frac{25}{64}$ **29.** $\frac{9}{16}$ **31.** $\frac{64}{125}$

33. $\frac{81}{16} = 5\frac{1}{16}$ **35.** $\frac{81}{256}$ **37.** A number line is a horizontal line with a range of equally spaced whole numbers placed on it. The lowest number is on the left and the greatest number is on the right. It can be used to compare the size or value of numbers.

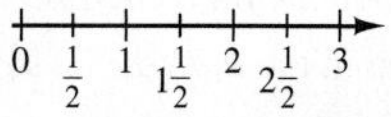

38. 1. Do all operations inside parentheses or other grouping symbols. 2. Simplify any expressions with exponents or square roots. 3. Multiply or divide, proceeding from left to right. 4. Add or subtract, proceeding from left to right. **39.** 4 **40.** 13 **41.** 10 **42.** 37 **43.** 1 **45.** $\frac{3}{16}$

47. $\frac{4}{9}$ **49.** $\frac{1}{3}$ **51.** $\frac{1}{2}$ **53.** $\frac{3}{8}$ **55.** $\frac{1}{4}$ **57.** $1\frac{1}{2}$ **59.** $\frac{1}{12}$ **61.** 3

63. $\frac{5}{16}$ **65.** $\frac{1}{4}$ **67.** $\frac{1}{32}$ **69.** $\frac{11}{50}$ in Las Vegas is greater. **71.** <; >

72. (a) like; numerators; numerator **(b)** Answers will vary.

73. $\frac{2}{45}$ **74.** $2\frac{1}{32}$

75.–80.

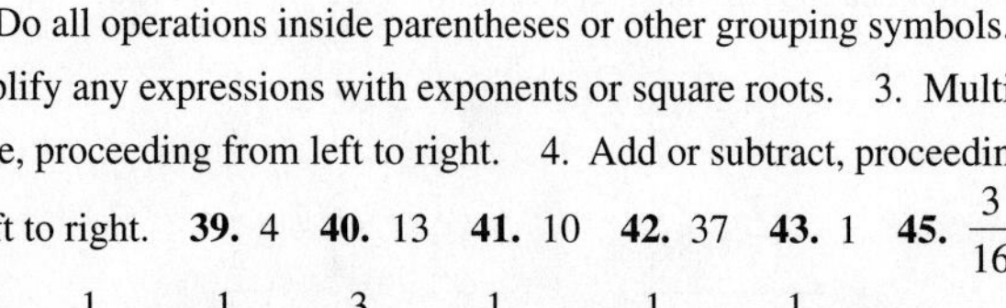

Chapter 3 REVIEW EXERCISES (pages 245–250)

1. $\frac{6}{7}$ **2.** $\frac{7}{9}$ **3.** $\frac{3}{4}$ **4.** $\frac{1}{8}$ **5.** $\frac{4}{5}$ **6.** $\frac{1}{6}$ **7.** $\frac{13}{31}$ **8.** $\frac{1}{3}$

9. $\frac{11}{12}$ of his total income **10.** $\frac{1}{4}$ more of the events **11.** 10 **12.** 12

13. 60 **14.** 24 **15.** 120 **16.** 180 **17.** 8 **18.** 21 **19.** 10

20. 45 **21.** 32 **22.** 20 **23.** $\frac{5}{6}$ **24.** $\frac{7}{8}$ **25.** $\frac{5}{8}$ **26.** $\frac{5}{12}$ **27.** $\frac{13}{24}$

28. $\frac{17}{36}$ **29.** $\frac{9}{10}$ of the students **30.** $\frac{23}{24}$ of her budget

31. *Estimate:* $19 + 14 = 33$; *Exact:* $32\frac{3}{8}$

32. *Estimate:* $23 + 15 = 38$; *Exact:* $38\frac{1}{9}$

33. *Estimate:* $13 + 9 + 10 = 32$; *Exact:* $31\frac{43}{80}$

34. *Estimate:* $32 - 15 = 17$; *Exact:* $17\frac{1}{12}$

35. *Estimate:* $34 - 16 = 18$; *Exact:* $18\frac{1}{3}$

36. *Estimate:* $215 - 136 = 79$; *Exact:* $79\frac{7}{16}$

37. $9\frac{1}{10}$ **38.** $10\frac{5}{12}$ **39.** $3\frac{1}{4}$ **40.** $1\frac{2}{3}$ **41.** $5\frac{1}{2}$ **42.** $2\frac{19}{24}$

43. *Estimate:* $19 - 6 - 7 = 6$ miles; *Exact:* $5\frac{19}{24}$ miles

44. *Estimate:* $29 + 25 = 54$ tons; *Exact:* $53\frac{5}{12}$ tons

45. *Estimate:* $8 + 3 + 5 + 3 = 19$ pounds; *Exact:* $18\frac{1}{2}$ pounds

46. *Estimate:* $1535 - 1476 = 59$ pounds; *Exact:* $59\frac{11}{16}$ pounds

47.–50.

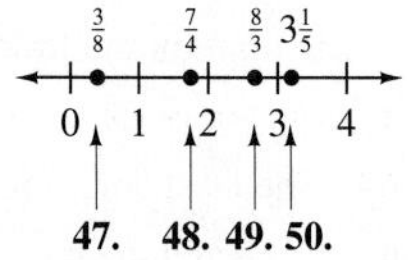

51. < **52.** < **53.** > **54.** > **55.** < **56.** > **57.** < **58.** >

59. $\frac{1}{4}$ **60.** $\frac{4}{9}$ **61.** $\frac{27}{1000}$ **62.** $\frac{81}{4096}$ **63.** $\frac{1}{2}$ **64.** $6\frac{3}{4}$ **65.** $\frac{1}{16}$

66. 1 **67.** $\frac{3}{16}$ **68.** $1\frac{25}{64}$ **69.** $\frac{3}{4}$ **70.** $\frac{2}{5}$ **71.** $\frac{19}{32}$ **72.** $\frac{11}{16}$ **73.** $2\frac{1}{6}$

74. $26\frac{1}{4}$ **75.** $5\frac{3}{8}$ **76.** $11\frac{43}{80}$ **77.** $15\frac{5}{12}$ **78.** $\frac{8}{11}$ **79.** $\frac{1}{250}$ **80.** $\frac{1}{2}$

81. $\frac{2}{9}$ **82.** $\frac{11}{27}$ **83.** > **84.** < **85.** < **86.** > **87.** 36 **88.** 120

89. 126 **90.** 18 **91.** 108 **92.** 60

93. *Estimate:* $93 - 14 - 22 = 57$ ft; *Exact:* $56\frac{7}{8}$ ft

94. *Estimate:* $4 \cdot 50 = 200$ pounds of sugar; $200 - 69 - 77 - 33 = 21$ pounds of sugar; *Exact:* $21\frac{5}{8}$ pounds of sugar

ANSWERS

Chapter 3 TEST (pages 251–252)

1. $\frac{3}{4}$ **2.** $\frac{1}{2}$ **3.** $\frac{2}{5}$ **4.** $\frac{1}{6}$ **5.** 12 **6.** 30 **7.** 108 **8.** $\frac{5}{8}$ **9.** $\frac{23}{36}$ **10.** $\frac{5}{24}$ **11.** $\frac{1}{40}$ **12.** *Estimate:* $8 + 5 = 13$; *Exact:* $12\frac{1}{2}$ **13.** *Estimate:* $16 - 12 = 4$; *Exact:* $4\frac{11}{15}$ **14.** *Estimate:* $19 + 9 + 12 = 40$; *Exact:* $40\frac{29}{60}$ **15.** *Estimate:* $24 - 18 = 6$; *Exact:* $5\frac{5}{8}$ **16.** Answers will vary. One possibility: Probably addition and subtraction of fractions are more difficult because you have to find the least common denominator and then change the fractions to the same denominator. **17.** Answers will vary. One possibility: Round mixed numbers to the nearest whole number. Then add or subtract to estimate the answer. The estimate may vary from the exact answer but it lets you know if your answer is reasonable. **18.** *Estimate:* $10 + 85 + 37 + 8 = 140$ pounds; *Exact:* $140\frac{1}{24}$ pounds **19.** *Estimate:* $148 - 69 - 37 = 42$ gallons; *Exact:* $41\frac{5}{8}$ gallons **20.** $>$ **21.** $>$ **22.** 2 **23.** $\frac{13}{48}$ **24.** $1\frac{3}{4}$ **25.** $1\frac{1}{3}$

CHAPTER 4 Decimals

SECTION 4.1 (pages 259–263)

1. 3 **2.** 4 **3.** 8 tenths **4.** 3 thousandths **5.** 7; 0; 4 **7.** 4; 7; 0 **9.** 1; 8; 9 **11.** 6; 2; 1 **13.** 410.25 **15.** 6.5432 **17.** 5406.045 **19.** $\frac{7}{10}$ **21.** $13\frac{2}{5}$ **23.** $\frac{1}{4}$ **25.** $\frac{33}{50}$ **27.** $10\frac{17}{100}$ **29.** $\frac{3}{50}$ **31.** $\frac{41}{200}$ **33.** $5\frac{1}{500}$ **35.** $\frac{343}{500}$ **37.** five tenths **39.** seventy-eight hundredths **41.** one hundred five thousandths **43.** twelve and four hundredths **45.** one and seventy-five thousandths **47.** 6.7 **49.** 0.32 **51.** 420.008 **53.** 0.0703 **55.** 75.030 **57.** Anne should not say "and" because that denotes a decimal point. **58.** Jerry used "and" twice. Only the first "and" is correct. **59.** ten thousandths inch; $\frac{10}{1000} = \frac{1}{100}$ inch **61.** 12 pounds **63.** 3-C **64.** 4-C **65.** 4-A **66.** 3-B **67.** one and six hundred two thousandths centimeters **68.** one and twenty-six thousandths centimeters **69.** millionths, ten-millionths, hundred-millionths, billionths; these match the words on the left side of the place value chart in Chapter 1 with "ths" attached. **70.** The first place to the left of the decimal point is ones, so the first place to the right could be one*ths,* like tens and ten*ths.* But anything that is 1 or more is to the *left* of the decimal point. **71.** seventy-two million four hundred thirty-six thousand nine hundred fifty-five hundred-millionths **72.** six hundred seventy-eight thousand five hundred fifty-four billionths **73.** eight thousand six and five hundred thousand one millionths **74.** twenty thousand sixty and five hundred five millionths **75.** 0.0302040 **76.** 9,876,543,210.100200300

SECTION 4.2 (pages 269–270)

1. 0 **2.** 9 **3.** Look only at the 6, which is *5 or more*. So round up by adding one thousandth to 5.709 to get 5.710. **4.** Look only at the 2, which is *4 or less*. So the part you are keeping, 10.0, stays the same. **5.** 16.9 **7.** 0.956 **9.** 0.80 **11.** 3.661 **13.** 794.0 **15.** 0.0980 **17.** 9.09 **19.** 82.0002 **21.** \$0.82 **23.** \$1.22 **25.** \$0.70 **27.** \$48,650 **29.** \$840 **31.** \$500 **33.** \$1.00 **35.** \$1000 **37. (a)** 253 miles per hour **(b)** 135 miles per hour **39. (a)** 186.0 miles per hour **(b)** 763.0 miles per hour **41.** It rounds to \$0 (zero dollars) because \$0.499 is closer to \$0 than to \$1. **42.** Rounding \$0.499 to \$0.50 would be more helpful than rounding to \$0. Round amounts less than \$1.00 to the nearest cent instead of the nearest dollar. **43.** It rounds to \$0.00 (zero cents) because \$0.0015 is closer to \$0.00 than to \$0.01. **44.** Both round to \$0.60. Rounding to nearest thousandth (tenth of a cent) would allow you to identify \$0.597 as less than \$0.601.

SECTION 4.3 (pages 275–278)

1. $\begin{array}{r} 6.420 \\ +\,10.163 \end{array}$ **2.** $\begin{array}{r} 7.000 \\ +\,9.204 \end{array}$ **3.** $\begin{array}{r} 20.0000 \\ -\,9.1263 \end{array}$ **4.** $\begin{array}{r} 137.06 \\ -\,12.00 \end{array}$ **5.** 17.48 **7.** 7.763 **9.** 77.006 **11.** 20.104 **13.** 0.109 **15.** 330.86895 **17. (a)** 24.75 in. **(b)** 3.95 in. **19. (a)** 62.27 in. **(b)** 0.39 in. **21.** 6 should be written 6.00. The sum is 46.22. **22.** The two problems are done in a different order: $8 - 2.9$ is not the same as $2.9 - 8$ because subtraction is not commutative. **23.** *Estimate:* \$20 − 7 = \$13; *Exact:* \$13.16 **25.** *Estimate:* $400 + 1 + 20 = 421$; *Exact:* 414.645 **27.** *Estimate:* $9 - 4 = 5$; *Exact:* 4.849 **29.** *Estimate:* $60 + 500 + 6 = 566$; *Exact:* 608.4363 **31.** 0.275 **33.** 6.507 **35.** 1.81 **37.** 6056.7202 **39.** *Estimate:* $80 - 30 = 50$ million people; *Exact:* 50.4 million people **41.** *Estimate:* $400 + 200 + 100 + 80 + 40 + 30 + 30 = 880$ million people; *Exact:* 940.82 million people **43.** *Estimate:* $2 + 2 + 2 = 6$ meters; *Exact:* 6.19 meters, which is less than the rhino by 0.21 meter **45.** *Estimate:* $11 - 10 = 1$ ounce; *Exact:* 0.65 ounce **47.** *Estimate:* $20 + 6 + 20 + 6 = 52$ in.; *Exact:* 52.1 in. **49.** *Estimate:* \$5 − \$5 = \$0; *Exact:* \$0.30 **51.** *Estimate:* \$19 + \$2 + \$2 + \$10 + \$2 = \$35; *Exact:* \$35.25 **53.** \$2059.36 **55.** \$103.97 **57.** \$498.22 **59.** $b = 1.39$ centimeters **61.** $q = 7.943$ ft

SECTION 4.4 (pages 281–284)

1. 3 **2.** 5 **3.** 2 **4.** 4 **5.** 0.1344 **7.** 159.10 **9.** 15.5844 **11.** \$34,500.20 **13.** 43.2 **14.** 43.2 **15.** 0.432 **16.** 0.432 **17.** 0.0432 **18.** 0.0432 **19.** 0.0000312 **21.** 0.000025 **23.** *Estimate:* $40 \times 5 = 200$; *Exact:* 190.08 **25.** *Estimate:* $40 \times 40 = 1600$; *Exact:* 1558.2 **27.** *Estimate:* $7 \times 5 = 35$; *Exact:* 30.038 **29.** *Estimate:* $3 \times 7 = 21$; *Exact:* 19.24165 **31.** unreasonable; \$289.00 **33.** reasonable **35.** unreasonable; \$4.19 **37.** unreasonable; 9.5 pounds **39.** \$945.87 (rounded) **41.** \$2.45 (rounded) **43.** \$77.10 (rounded) **45.** \$20,265 **47. (a)** Area before 1929 ≈ 23.2 in.^{2}; Area today ≈ 16.0 in.^{2} **(b)** 7.2 in.^{2} **49. (a)** 0.43 in. **(b)** 4.3 in. **51.** \$1264.04; \$2723.72 **53.** \$4.09 (rounded) **55. (a)** \$72.05 **(b)** \$27.80

57. 59.6; 32; 4.76; 803.5; 7226; 9. Multiplying by 10, decimal point moves one place to the right; by 100, two places to the right; by 1000, three places to the right. **58.** 5.96; 0.32; 0.0476; 8.035; 6.5; 52.3. Multiplying by 0.1, decimal point moves one place to the left; by 0.01, two places to the left; by 0.001, three places to the left.

SECTION 4.5 (pages 291–294)

1. C; $5\overline{)25.5}$ **2.** B; $7\overline{)42.3}$ **3.** $2.2\overline{)8.24}$ **4.** $5.1\overline{)10.5}$ **5.** 3.9 **7.** 0.47 **9.** 400.2 **11.** 36 **13.** 0.06 **14.** 0.6 **15.** 6000 **16.** 6 **17.** 25.3 **19.** 516.67 (rounded) **21.** 26.756 (rounded) **23.** 10,082.647 (rounded) **25.** unreasonable; $40 \div 8 = 5$; Correct answer is 4.725. **27.** reasonable; $50 \div 50 = 1$ **29.** unreasonable; $300 \div 5 = 60$; Correct answer is 60.2. **31.** $4.00 (rounded) **33.** $67.08 **35.** $11.92 per hour **37.** 28.0 miles per gallon (rounded) **39.** $0.03 per can (rounded) **41.** 7.37 meters (rounded) **43.** 0.08 meter **45.** 22.49 meters **47. (a)** Work inside parentheses; subtract $9.5 - 3.1$ to get 6.4. **(b)** Apply the exponent; multiply $(2.2)(2.2)$ to get 4.84. **(c)** Add 4.84 plus 6.4 to get 11.24. **49.** 14.25 **51.** 73.4 **53.** 1.205 **55.** 0.334 **57. (a)** 1,083,333 pieces (rounded) **(b)** 18,056 pieces (rounded) **(c)** 301 pieces (rounded). **59.** 100,000 box tops **61.** 2632 box tops (rounded) **63.** 0.377; 0.91; 0.0886; 3.019; 40.65; 662.57 **(a)** Dividing by 10, decimal point moves one place to the left; by 100, two places to the left; by 1000, three places to the left. **(b)** The decimal point moved to the *right* when *multiplying* by 10 or 100 or 1000. Here it moves to the *left* when *dividing* by those numbers. **64.** 402; 71; 3.39; 157.7; 460; 8730 **(a)** Dividing by 0.1, decimal point moves one place to the right; by 0.01, two places to the right; by 0.001, three places to the right. **(b)** The decimal point moved to the *left* when *multiplying* by 0.1 or 0.01 or 0.001. Here the decimal point moves to the *right* when *dividing* by those numbers.

SECTION 4.6 (pages 299–302)

1. (a) not correct; $5\overline{)2}$ **(b)** correct **(c)** correct
2. (a) correct **(b)** not correct; $3\overline{)1}$ **(c)** correct
3. $\overset{0.}{4\overline{)3.}}$ To continue dividing, write zeros in the dividend.
4. $\overset{0.}{8\overline{)1.}}$ To continue dividing, write zeros in the dividend.
5. (a) less than **(b)** greater than **(c)** less than
6. (a) less than **(b)** greater than **(c)** greater than
7. 0.5 **9.** 0.75 **11.** 0.3 **13.** 0.9 **15.** 0.6 **17.** 0.875 **19.** 2.25 **21.** 14.7 **23.** 3.625 **25.** 6.333 (rounded) **27.** 0.833 (rounded) **29.** 1.889 (rounded) **31.** $\frac{2}{5}$ **33.** $\frac{5}{8}$ **35.** $\frac{7}{20}$ **37.** 0.35 **39.** $\frac{1}{25}$ **41.** 0.2 **43.** $\frac{9}{100}$ **45.** shorter; 0.72 inch **47.** more; 0.05 inch **49.** too much; 0.005 gram **51.** 0.9991 cm, 1.0007 cm **52.** 3.0 ounces, 2.995 ounces, 3.005 ounces **53.** 0.5399, 0.54, 0.5455 **55.** 5.0079, 5.79, 5.8, 5.804 **57.** 0.6009, 0.609, 0.628, 0.62812 **59.** 2.8902, 3.88, 4.876, 5.8751 **61.** 0.006, 0.043, $\frac{1}{20}$, 0.051 **63.** 0.37, $\frac{3}{8}$, $\frac{2}{5}$, 0.4001 **65.** red box **67.** green box **69.** 1.4 in. (rounded) **71.** 0.3 in. (rounded) **73.** 0.4 in. (rounded) **75. (a)** A proper fraction like $\frac{5}{9}$ is less than 1, so it cannot be equivalent to a decimal number that is greater than 1. **(b)** $\frac{5}{9}$ means $5 \div 9$ or $9\overline{)5}$, so correct answer is 0.556 (rounded). This makes sense because both the fraction and decimal are less than 1.

76. (a) $2.035 = 2\frac{35}{1000} = 2\frac{7}{200}$, not $2\frac{7}{20}$ **(b)** Adding the whole number part gives $2 + 0.35$, which is 2.35, not 2.035. To check, $2.35 = 2\frac{35}{100} = 2\frac{7}{20}$. **77.** Just add the whole number part to 0.375. So $1\frac{3}{8} = 1.375$; $3\frac{3}{8} = 3.375$; $295\frac{3}{8} = 295.375$. **78.** It works only when the fraction part has a one-digit numerator and a denominator of 10, a two-digit numerator and a denominator of 100, and so on.

Chapter 4 REVIEW EXERCISES (pages 308–311)

1. 0; 5 **2.** 0; 6 **3.** 8; 9 **4.** 5; 9 **5.** 7; 6 **6.** $\frac{1}{2}$ **7.** $\frac{3}{4}$ **8.** $4\frac{1}{20}$ **9.** $\frac{7}{8}$ **10.** $\frac{27}{1000}$ **11.** $27\frac{4}{5}$ **12.** eight tenths **13.** four hundred and twenty-nine hundredths **14.** twelve and seven thousandths **15.** three hundred six ten-thousandths **16.** 8.3 **17.** 0.205 **18.** 70.0066 **19.** 0.30 **20.** 275.6 **21.** 72.79 **22.** 0.160 **23.** 0.091 **24.** 1.0 **25.** $15.83 **26.** $0.70 **27.** $17,625.79 **28.** $350 **29.** $130 **30.** $100 **31.** $29 **32.** *Estimate:* $6 + 400 + 20 = 426$; *Exact:* 444.86 **33.** *Estimate:* $80 + 1 + 100 + 1 + 30 = 212$; *Exact:* 233.515 **34.** *Estimate:* $300 - 20 = 280$; *Exact:* 290.7 **35.** *Estimate:* $9 - 8 = 1$; *Exact:* 1.2684 **36.** *Estimate:* 90 million − 20 million = 70 million; *Exact:* 77.0 million more cats **37.** *Estimate:* $400 − $300 − $70 = $30; *Exact:* $15.80 **38.** *Estimate:* $2 + $5 + $20 = $27; $30 − $27 = $3; *Exact:* $4.14 **39.** *Estimate:* $2 + 4 + 5 = 11$ kilometers; *Exact:* 11.55 kilometers **40.** *Estimate:* $6 \times 4 = 24$; *Exact:* 22.7106 **41.** *Estimate:* $40 \times 3 = 120$; *Exact:* 141.57 **42.** 0.0112 **43.** 0.000355 **44.** reasonable; $700 \div 10 = 70$ **45.** unreasonable; $30 \div 3 = 10$; Correct answer is 9.5. **46.** 14.467 (rounded) **47.** 1200 **48.** 0.4 **49.** $708 (rounded) **50.** $2.99 (rounded) **51.** 133 shares (rounded) **52.** $3.47 (rounded) **53.** 29.215 **54.** 10.15 **55.** 3.8 **56.** 0.64 **57.** 1.875 **58.** 0.111 (rounded) **59.** 3.6008, 3.68, 3.806 **60.** 0.209, 0.2102, 0.215, 0.22 **61.** $\frac{1}{8}$, $\frac{3}{20}$, 0.159, 0.17 **62.** 404.865 **63.** 254.8 **64.** 3583.261 (rounded) **65.** 29.0898 **66.** 0.03066 **67.** 9.4 **68.** 175.675 **69.** 9.04 **70.** 19.50 **71.** 8.19 **72.** 0.928 **73.** 35 **74.** 0.259 **75.** 0.3 **76.** $3.00 (rounded) **77.** $2.17 (rounded) **78.** $35.96 **79.** $199.71 **80.** $78.50 **81. (a)** baked potato with skin **(b)** $\frac{1}{2}$ cup green peas **(c)** 0.59 milligram **82. (a)** 2.08 milligrams **(b)** more, by 0.08 milligram

Chapter 4 TEST (pages 312–313)

1. $18\frac{2}{5}$ **2.** $\frac{3}{40}$ **3.** sixty and seven thousandths **4.** two hundred eight ten-thousandths **5.** 725.6 **6.** 0.630 **7.** $1.49 **8.** $7860 **9.** *Estimate:* $8 + 80 + 40 = 128$; *Exact:* 129.2028 **10.** *Estimate:* $80 - 4 = 76$; *Exact:* 75.498 **11.** *Estimate:* $6(1) = 6$; *Exact:* 6.948 **12.** *Estimate:* $20 \div 5 = 4$; *Exact:* 4.175 **13.** 839.762 **14.** 669.004 **15.** 0.0000483 **16.** 480 **17.** 2.625 **18.** 0.44, $\frac{9}{20}$, 0.4506, 0.451 **19.** 35.49 **20.** $1294.47 **21.** pintails, wigeons, gadwalls **22.** $5.35 (rounded) **23.** 2.8 degrees **24.** $3.79 per foot (rounded) **25.** Answers will vary.

CHAPTER 5 Ratio and Proportion

SECTION 5.1 (pages 321–324)

1. Answers will vary. One possibility is: A ratio compares two quantities with the same units. Examples will vary. **2.** You can divide out the common (same) units. **3.** 20; $\frac{4}{1}$ **4.** 5; $\frac{4}{15}$ **5.** $\frac{8}{9}$ **7.** $\frac{2}{1}$ **9.** $\frac{1}{3}$ **11.** $\frac{8}{5}$ **13.** $\frac{3}{8}$ **15.** $\frac{9}{7}$ **17.** $\frac{6}{1}$ **19.** $\frac{5}{6}$ **21.** ounces; It takes fewer steps to solve and you do not have to work with a mixed number. **22.** hours; It takes fewer steps to solve and you do not have to work with a mixed number. **23.** $\frac{8}{5}$ **25.** $\frac{1}{12}$ **27.** $\frac{5}{16}$ **29.** $\frac{4}{1}$ **31.** $\frac{1}{6}$ **33.** $\frac{15}{1}$ **35.** Answers will vary. One possibility is stocking cards of various types in the same ratios as those in the table. **36.** Answers will vary. Examples: A person may send Valentine's Day cards every year but graduation cards only once every few years; Valentine's Day may have more advertising. **37.** $\frac{6}{5}$; $\frac{36}{17}$ **39.** *White Christmas* to *It's Now or Never; White Christmas* to *I Will Always Love You; Candle in the Wind* to *I Want to Hold Your Hand.* **41.** $\frac{7}{5}$ **43.** $\frac{6}{1}$ **45.** $\frac{38}{17}$ **47.** $\frac{1}{4}$ **49.** $\frac{34}{35}$ **51.** $\frac{1}{1}$; As long as the sides all have the same length, any measurement you choose will maintain the ratio.
52. Answers will vary. Some possibilities are: $\frac{4}{5} = \frac{8}{10} = \frac{12}{15} = \frac{16}{20} = \frac{20}{25} = \frac{24}{30} = \frac{28}{35}$. **53.** It is not possible. Amelia would have to be older than her mother to have a ratio of 5 to 3.
54. Answers will vary, but a ratio of 3 to 1 means your income is 3 times your friend's income.

SECTION 5.2 (pages 329–332)

1. $\frac{5 \text{ cups}}{3 \text{ people}}$ **3.** $\frac{3 \text{ feet}}{7 \text{ seconds}}$ **5.** $\frac{18 \text{ miles}}{1 \text{ gallon}}$
7. Divide; correct set-up is $\frac{\$5.85}{3 \text{ boxes}}$ or $3\overline{)5.85}$. **8.** 5 rooms for 1 nurse
9. \$12 per hour or \$12/hour **11.** 1.25 pounds/person **13.** 325.9; 21.0 (rounded) **15.** 338.6; 20.9 (rounded) **17.** \$1.125/oz; \$0.913/oz (rounded); \$0.913/oz; best buy is 4 oz for \$3.65.
19. 14 ounces for \$2.89, about \$0.206/ounce **21.** 18 ounces for \$1.79, about \$0.099/ ounce **23.** Answers will vary. For example, you might choose Brand B because you like more chicken, so the cost per chicken chunk may actually be the same as or less than Brand A. **24.** Answers will vary. For example, if you use only half of the larger bag, you really pay \$0.30 per pound, so the smaller bag is the better buy. **25.** 1.75 pounds/week
27. \$12.26/hour **29. (a)** Penny Saver, \$0.415; Most Minutes, \$0.503 (rounded); USA Card, \$0.30 **(b)** Penny Saver, \$0.083/min; Most Minutes, \$0.101/min (rounded); USA Card, \$0.06/min. USA Card is the best buy.
31. Penny Saver total \$0.54, \$0.018/min; Most Minutes total \$0.565, \$0.019/min (rounded); USA card total \$0.55, \$0.018/min (rounded). All three unit rates are very similar. **33.** one battery for \$1.79; like getting 3 batteries so $\$1.79 \div 3 \approx \$0.597$ per battery **35.** Brand P with the 50¢ coupon is the best buy. ($\$3.39 - \$0.50 = \$2.89$; $\$2.89 \div 16.5$ ounces $\approx \$0.175$ per ounce) **37.** Plan A: \$0.09/min; Plan B: \$0.07/min; Plan B is the better buy. **38. (a)** 15 min/day **(b)** 30 min/day
39. (60 extra min) (\$0.45) = \$27 overage
$\$39.95 + \$27 \text{ overage} = \$66.95$
$$\frac{\$66.95}{450 \text{ min} + 60 \text{ min}} = \frac{\$66.95}{510 \text{ min}} \approx \$0.13 \text{ min}$$
40. (90 extra min) (\$0.40) = \$36 overage
$\$59.95 + \$36 \text{ overage} = \$95.95$
$$\frac{\$95.95}{900 \text{ min} + 90 \text{ min}} = \frac{\$95.95}{990 \text{ min}} \approx \$0.10/\text{min}$$
41. From Exercise 39: $\frac{\$69.95}{510 \text{ min}} \approx \$0.14 \text{ min}$
From Exercise 40: $\frac{\$69.95}{990 \text{ min}} \approx \$0.07 \text{ min}$

SECTION 5.3 (pages 336–337)

1. $\frac{\$9}{12 \text{ cans}} = \frac{\$18}{24 \text{ cans}}$ **3.** $\frac{200 \text{ adults}}{450 \text{ children}} = \frac{4 \text{ adults}}{9 \text{ children}}$ **5.** $\frac{120}{150} = \frac{8}{10}$
7. $\frac{3}{5} = \frac{3}{5}$; true **9.** $\frac{5}{8} = \frac{5}{8}$; true **11.** $\frac{3}{4} \neq \frac{2}{3}$; false **13.** $\frac{14}{5} = \frac{14}{5}$; true
15. $\frac{16}{9} = \frac{16}{9}$; true **17.** $\frac{7}{6} \neq \frac{9}{8}$; false **19.** Answers may vary. One example: A proportion shows that two ratios are equal. The multiplications on the proportion should show $6 \cdot 45 = 270$ and $30 \cdot 9 = 270$.
20. Answers will vary. One possibility: If the cross products are equal, the proportion is true. If the cross products are not equal, the proportion is false. **21.** 54 = 54; True **23.** $336 \neq 320$; False **25.** $2880 \neq 2970$; False **27.** 28 = 28; True **29.** $44.8 \neq 45$; False **31.** 66 = 66; True
33. $68\frac{1}{4} = 68\frac{1}{4}$; True **35.** $5\frac{2}{5} \neq 5\frac{1}{3}$; False
37. $2\frac{1}{80} \neq 2\frac{7}{100}$ or $2.0125 \neq 2.07$; False
39. $\frac{68 \text{ hits}}{200 \text{ at bats}} = \frac{153 \text{ hits}}{450 \text{ at bats}}$ $200 \cdot 153 = 30{,}600$; $68 \cdot 450 = 30{,}600$
Cross products are *equal,* so the proportion is *true.* They hit equally well.
40. The left-hand ratio compares hours to cartons, but the right-hand ratio compares cartons to hours. The correct proportion is shown.
$\frac{3.5 \text{ hours}}{91 \text{ cartons}} = \frac{5.25 \text{ hours}}{126 \text{ cartons}}$ $91 \cdot 5.25 = 477.75$; $3.5 \cdot 126 = 441$
Cross products are *not* equal, so the proportion is *false*. The men do not work equally fast.

SECTION 5.4 (pages 342–343)

1. Find the cross products. $10 \cdot 5 = 50$ and $7 \cdot x$
2. Divide both sides by 7; $\frac{\overset{1}{\cancel{7}} \cdot x}{\underset{1}{\cancel{7}}} = \frac{50}{7}$ **3.** 12; 12; $x = 4$ **5.** $x = 2$
7. $x = 88$ **9.** $x = 91$ **11.** $x = 5$ **13.** $x = 10$ **15.** $x \approx 24.44$ (rounded)
17. $x = 50.4$ **19.** $x \approx 17.64$ (rounded) **21.** $x = 1$ **23.** $x = 3\frac{1}{2}$
25. $x = 0.2$ or $x = \frac{1}{5}$ **27.** $x = 0.005$ or $x = \frac{1}{200}$
29. Find cross products: $20 \neq 30$, so the proportion is false. $\frac{6\frac{2}{3}}{4} = \frac{5}{3}$ or $\frac{10}{6} = \frac{5}{3}$ or $\frac{10}{4} = \frac{7.5}{3}$ or $\frac{10}{4} = \frac{5}{2}$

30. Find cross products: $192 \neq 180$, so the proportion is false.
$\frac{6.4}{8} = \frac{24}{30}$ or $\frac{6}{7.5} = \frac{24}{30}$ or $\frac{6}{8} = \frac{22.5}{30}$ or $\frac{6}{8} = \frac{24}{32}$

SECTION 5.5 (pages 347–350)

1. $\frac{6 \text{ potatoes}}{4 \text{ eggs}} = \frac{12 \text{ potatoes}}{x \text{ eggs}}$ **2.** $\frac{1 \text{ serving}}{15 \text{ chips}} = \frac{x \text{ serving}}{70 \text{ chips}}$ **3.** 18; x; 22.5 hours **5.** \$7.20 **7.** 42 pounds **9.** \$403.68 **11.** 10 ounces (rounded) **13.** 5 quarts **15.** 14 ft, 10 ft **17.** 14 ft, 8 ft **19.** 96 pieces chicken; 33.6 pounds lasagna; 10.8 pounds deli meats; $5\frac{3}{5}$ pounds cheese; 7.2 dozen (about 86) buns; 14.4 pounds of salad **21.** 44 students is an *unreasonable* answer because there are only 35 students in the class. **22.** 2 minutes is an *unreasonable* answer because it is less than the 30 minutes she gets in just one day. **23.** 2065 students (reasonable); about 4214 students with incorrect set-up (only 2950 students in the group) **25.** about 83 people (reasonable); about 750 people with incorrect set-up (only 250 people attended) **27.** 625 stocks **29.** 4.06 meters (rounded) **31.** 311 calories (rounded) **33.** 10.53 meters (rounded) **35.** You cannot solve this problem using a proportion because the ratio of age to weight is not constant. As Jim's age increases, his weight may decrease, stay the same, or increase. **36.** Answers will vary; Exercises 3–34 are all examples of application problems. **37.** 5610 students **39.** 120 calories and 12 grams of fiber **41.** $1\frac{3}{4}$ cups water, 3 Tbsp margarine, $\frac{3}{4}$ cup milk, 2 cups flakes **42.** $5\frac{1}{4}$ cups water, 9 Tbsp margarine, $2\frac{1}{4}$ cups milk, 6 cups flakes

Chapter 5 REVIEW EXERCISES (pages 356–359)

1. $\frac{3}{4}$ **2.** $\frac{4}{1}$ **3.** great white shark to whale shark; whale shark to blue whale **4.** $\frac{2}{1}$ **5.** $\frac{2}{3}$ **6.** $\frac{5}{2}$ **7.** $\frac{1}{6}$ **8.** $\frac{3}{1}$ **9.** $\frac{3}{8}$ **10.** $\frac{4}{3}$ **11.** $\frac{1}{9}$ **12.** $\frac{10}{7}$ **13.** $\frac{7}{5}$ **14.** $\frac{5}{6}$ **15.** $\frac{\$11}{1 \text{ dozen}}$ **16.** $\frac{12 \text{ children}}{5 \text{ families}}$ **17.** 0.2 page/minute or $\frac{1}{5}$ page/minute; 5 minutes/page **18.** \$20/hour; 0.05 hour/dollar or $\frac{1}{20}$ hour/dollar **19.** 8 ounces for \$4.98, about \$0.623/ounce **20.** 17.6 pounds for \$18.69 − \$1 coupon, about \$1.005/pound **21.** $\frac{3}{5} = \frac{3}{5}$ or $90 = 90$; true **22.** $\frac{1}{8} \neq \frac{1}{4}$ or $432 \neq 216$; false **23.** $\frac{47}{10} \neq \frac{49}{10}$ or $980 \neq 940$; false **24.** $\frac{16}{9} = \frac{16}{9}$ or $3456 = 3456$; true **25.** $4.8 = 4.8$; true **26.** $14 = 14$; true **27.** $x = 1575$ **28.** $x = 20$ **29.** $x = 400$ **30.** $x = 12.5$ **31.** $x \approx 14.67$ (rounded) **32.** $x \approx 8.17$ (rounded) **33.** $x = 50.4$ **34.** $x \approx 0.57$ (rounded) **35.** $x \approx 2.47$ (rounded) **36.** 27 cats **37.** 46 hits **38.** \$15.63 (rounded) **39.** 3299 students (rounded) **40.** 68 feet **41.** $27\frac{1}{2}$ hours or 27.5 hours **42.** 511 calories (rounded) **43.** 14.7 milligrams **44.** $x = 105$ **45.** $x = 0$ **46.** $x = 128$ **47.** $x \approx 23.08$ (rounded) **48.** $x = 6.5$ **49.** $x \approx 117.36$ (rounded) **50.** $1440 \neq 1485$; False **51.** $10.8 \neq 10.864$; False **52.** $2\frac{1}{3} = 2\frac{1}{3}$; True **53.** $\frac{8}{5}$ **54.** $\frac{33}{80}$ **55.** $\frac{15}{4}$ **56.** $\frac{4}{1}$ **57.** $\frac{4}{5}$ **58.** $\frac{37}{7}$ **59.** $\frac{3}{8}$ **60.** $\frac{1}{12}$ **61.** $\frac{45}{13}$ **62.** 24,900 fans (rounded) **63.** $\frac{8}{3}$ **64.** 75 ft for \$1.99 − \$0.50 coupon, about \$0.020/ft **65.** 21 ft long; 15 ft wide **66.** 7.5 hours or $7\frac{1}{2}$ hours **67.** $\frac{1}{2}$ teaspoon or 0.5 teaspoon **68.** 21 points (rounded) **69.** Set up the proportion to compare teaspoons to pounds on both sides.
$\frac{1.5 \text{ teaspoons}}{24 \text{ pounds}} = \frac{x \text{ teaspoons}}{8 \text{ pounds}}$
Show that cross products are equal. $(24)(x) = (1.5)(8)$
Divide both sides by 24. $\frac{\overset{1}{\cancel{24}}(x)}{\underset{1}{\cancel{24}}} = \frac{12}{24}$, so $x = \frac{1}{2}$ teaspoon or 0.5 teaspoon.
70. (a) 1400 milligrams **(b)** 100 milligrams

Chapter 5 TEST (pages 360–361)

1. $\frac{4}{5}$ **2.** $\frac{20 \text{ miles}}{1 \text{ gallon}}$ **3.** $\frac{\$1}{5 \text{ minutes}}$ **4.** $\frac{9}{2}$ **5.** $\frac{15}{4}$ **6.** Best buy is 8 in. sub, about \$0.861/inch. **7.** 16 ounces for \$1.89 − \$0.50 coupon, about \$0.087/ounce **8.** You earned less this year. Example: $\frac{\text{Last year} \rightarrow \$30{,}000}{\text{This year} \rightarrow \$20{,}000} = \frac{3}{2}$ **9.** $\frac{3}{7} \neq \frac{2}{5}$ or $252 \neq 270$; false **10.** $5.88 = 5.88$; true **11.** $x = 25$ **12.** $x \approx 2.67$ (rounded) **13.** $x = 325$ **14.** $x = 10\frac{1}{2}$ **15.** 24 orders **16.** 3.6 ounces **17.** 87 students (rounded) **18.** No, 4875 cannot be correct because there are only 650 students in the whole school. **19.** 23.8 grams (rounded) **20.** 60 ft

CHAPTER 6 Percent

SECTION 6.1 (pages 369–374)

1. %; 100 **2.** decimal point; left **3.** 0.12 **5.** 0.70 or 0.7 **7.** 0.25 **9.** 1.40 or 1.4 **11.** 0.055 **13.** 1.00 or 1 **15.** 0.005 **17.** 0.0035 **19.** 100; % **20.** decimal point; right **21.** 60% **23.** 1% **25.** 37.5% **27.** 200% **29.** 370% **31.** 3.12% **33.** 416.2% **35.** 0.28% **37.** Answers will vary. Some possibilities: No common denominators are needed with percents. The denominator is always 100 with percent, which makes comparisons easier to understand. **38.** Answers will vary. Some possible answers: using discounts on purchases, calculating sales tax, figuring interest on loans, examining investments, finding tips in restaurants, calculating interest on savings, and doing math problems in this book. **39.** 0.13 **41.** 21.8% **43.** 0.151 **45.** 17.7% **47.** 0.146 **49.** 8% **51.** 0.30 or 0.3 **53.** 12 children **54.** 500 adults **55.** 420 employees **57.** 270 chairs **59.** \$377.50 **61.** 820 commuters **63.** 26 plants **65. (a)** Since 100% means 100 parts out of 100 parts, 100% is all of the number. **(b)** Answers will vary. For example, 100% of \$72 is \$72. **66. (a)** 50% means 50 parts out of 100 parts. That's half of the number. A shortcut for finding 50% of a number is to divide the number by 2. **(b)** Answers will vary. For example, 50% of \$14 is $\$14 \div 2 = \$7$.

67. (a) Since 200% is two times a number, find 200% of the number by multiplying the number by 2 (double it). **(b)** Answers will vary. For example, 200% of \$20 is 2 • \$20 = \$40. **68. (a)** Since 300% is three times a number, find 300% of the number by multiplying the number by 3 (triple it). **(b)** Answers will vary. For example, 300% of \$10 is 3 • \$10 = \$30. **69. (a)** Since 10% means 10 parts out of 100 parts or $\frac{1}{10}$, the shortcut for finding 10% of a number is to move the decimal point in the number one place to the left. **(b)** Answers will vary. For example, 10% of \$90 is \$9. **70. (a)** Since 1% means 1 part out of 100 parts or $\frac{1}{100}$, the shortcut for finding 1% of a number is to move the decimal point in the number two places to the left. **(b)** Answers will vary. For example, 1% of \$500 is \$5. **71.** 12%; 0.12 **73. (a)** witch costume **(b)** 4.5%; 0.045 **75.** 9%; 0.09 **77. (a)** speeding **(b)** 5%; 0.05 **79.** 21%; 0.21 **81. (a)** candy **(b)** 10%; 0.10 **83.** 95% shaded; 5% unshaded **85.** 30% shaded; 70% unshaded **87.** 55% shaded; 45% unshaded **89.** 64% shaded; 36% unshaded

SECTION 6.2 (pages 380–385)

1. false **2.** true **3.** true **4.** false **5.** $\frac{17}{20}$ **7.** $\frac{5}{8}$ **9.** $\frac{1}{16}$ **11.** $\frac{1}{6}$ **13.** $\frac{1}{15}$ **15.** $\frac{1}{200}$ **17.** $1\frac{4}{5}$ **19.** $3\frac{3}{4}$ **21.** true **22.** false **23.** false **24.** true **25.** 70% **27.** 37% **29.** 62.5% **31.** 87.5% **33.** 48% **35.** 46% **37.** 35% **39.** 83.3% (rounded) **41.** 55.6% (rounded) **43.** 14.3% (rounded) **45.** B **46.** D **47.** $\frac{1}{2}$; 50% **49.** $\frac{7}{8}$; 0.875 **51.** 0.167 (rounded); 16.7% (rounded) **53.** $\frac{7}{10}$; 70% **55.** $\frac{1}{8}$; 0.125 **57.** 0.667 (rounded); 66.7% (rounded) **59.** 0.06; 6% **61.** 0.08; 8% **63.** 0.005; 0.5% **65.** $2\frac{1}{2}$; 250% **67.** 3.25; 325% **69.** There are many possible answers. Examples 2 and 3 show the steps that students should include in their answers. **70.** There are many correct answers. The table of percent equivalents shows some of the possibilities. **71.** $\frac{9}{50}$; 0.18; 18% **73.** $\frac{13}{100}$; 0.13; 13% **75.** $\frac{1}{5}$; 0.2; 20% **77. (a)** $\frac{4}{5}$; 0.80; 80% **(b)** $\frac{1}{5}$; 0.20; 20% **79.** $\frac{1}{10}$; 0.1; 10% **81.** $\frac{2}{25}$; 0.08; 8% **83.** $\frac{13}{50}$; 0.26; 26% **85.** 100; 100 **86. (a)** 765 workers **(b)** 96 letters **(c)** 21 DVDs **87.** 50; 100; half or $\frac{1}{2}$ **88.** 10; 100; 1; left **89.** 1; 100; 2; left **90. (a)** 525 homes **(b)** 37 printers **(c)** \$0.08
91. Find 10% of \$160, then add $\frac{1}{2}$ of the 10% amount.

$$\begin{array}{ccccc} 10\% & + & 5\% & = & 15\% \\ \downarrow & & \downarrow & & \downarrow \\ \$16 & + & \$8 & = & \$24 \end{array}$$

92. Find 100% of \$160, then add 50% of 160.

$$\begin{array}{ccccc} 100\% & + & 50\% & = & 150\% \\ \downarrow & & \downarrow & & \downarrow \\ \$160 & + & \$80 & = & \$240 \end{array}$$

93. From 100% of \$450, subtract 10% of \$450.

$$\begin{array}{ccccc} 100\% & - & 10\% & = & 90\% \\ \downarrow & & \downarrow & & \downarrow \\ \$450 & - & \$45 & = & \$405 \end{array}$$

94. To 100% of \$800, add 100% of \$800, and then add 10% of \$800.

$$\begin{array}{ccccccc} 100\% & + & \$100\% & + & 10\% & = & 210\% \\ \downarrow & & \downarrow & & \downarrow & & \downarrow \\ \$800 & + & \$800 & + & \$80 & = & \$1680 \end{array}$$

SECTION 6.3 (pages 390–393)

1. whole; $\frac{5}{\text{unknown}} = \frac{10}{100}$ **2.** percent; $\frac{1.5}{4.5} = \frac{\text{unknown}}{100}$ **3.** percent; $\frac{36}{24} = \frac{\text{unknown}}{100}$ **4.** part; $\frac{\text{unknown}}{72} = \frac{30}{100}$ **5.** part; $\frac{\text{unknown}}{160} = \frac{35}{100}$ **6.** whole; $\frac{20}{\text{unknown}} = \frac{25}{100}$ **7.** whole = 150 **9.** whole = 70 **11.** percent = 25% **13.** percent = 33.3% (rounded) **15.** part = 26 **17.** percent = 26.5% (rounded) **19.** 115 **21.** 0.4% **23.** 2.5% **25.** 0.3% **27.** percent; % **28.** entire or total; of **29.** whole **30.** multiply **31.** $\frac{\text{Part} \rightarrow 60}{\text{Whole} \rightarrow \text{unknown}} = \frac{10 \leftarrow \text{Percent}}{100 \leftarrow \text{Always } 100}$ **33.** $\frac{600}{800} = \frac{\text{unknown}}{100}$ **35.** $\frac{\text{unknown}}{970} = \frac{25}{100}$ **37.** $\frac{12}{\text{unknown}} = \frac{20}{100}$ **39.** $\frac{54.34}{\text{unknown}} = \frac{3.25}{100}$ **41.** $\frac{\text{unknown}}{487} = \frac{0.68}{100}$ **43.** Percent—the ratio of the part to the whole. It appears with the word *percent* or "%" after it. Whole—the entire quantity. It often appears after the word *of.* Part—the part being compared with the whole. **44.** A possible sentence is: Of the 580 cars entering the parking lot, 464 cars, or 80%, had parking stickers on their windshields. percent = 80; whole = 580; part = 464 **45.** $\frac{730}{1262} = \frac{\text{unknown}}{100}$ **47.** $\frac{86}{142} = \frac{\text{unknown}}{100}$ **49.** $\frac{\text{unknown}}{610} = \frac{23}{100}$ **51.** $\frac{4300}{8600} = \frac{\text{unknown}}{100}$ **53.** $\frac{\text{unknown}}{480} = \frac{55}{100}$ **55.** $\frac{\text{unknown}}{822} = \frac{49.5}{100}$ **57.** $\frac{168}{\text{unknown}} = \frac{12}{100}$ **59.** $\frac{\text{unknown}}{680} = \frac{45}{100}$

SECTION 6.4 (pages 401–406)

1. part = percent • whole **2.** 76 **3.** 42 test tubes **5.** 1836 military personnel **7.** 4.8 ft **9.** 315 files **11.** 819 trucks **13.** \$3.28 **15.** 1530 tables **17.** 182 cell phones **19.** \$21.60 **21.** $\frac{80}{\text{unknown}} = \frac{25}{100}$ **22.** $\frac{32}{\text{unknown}} = \frac{5}{100}$ **23.** 160 hay bales **25.** 550 students **27.** 330 mountain bikes **29.** 2800 **31.** $\frac{18}{36} = \frac{\text{unknown}}{100}$ **32.** $\frac{62}{248} = \frac{\text{unknown}}{100}$ **33.** 52% **35.** 1.5; 1.5% **37.** 18.6% (rounded) **39.** 9.2% **41.** 150% of \$30 cannot be less than \$30 because 150% is greater than 1 (100%). The answer must be greater than \$30. 25% of \$16 cannot be greater than \$16 because 25% is less than 1 (100%). The answer must be less than \$16.

42. Answers will vary. One example is: There are 600 vehicles in the parking lot and 45 of them are pickup trucks. What percent are pickup trucks? Total vehicles, 600, is the whole, and the number of pickup trucks, 45, is the part.

$$\frac{45}{600} = \frac{x}{100}$$
$$600 \cdot x = 4500$$
$$\frac{\overset{1}{\cancel{600}} \cdot x}{\underset{1}{\cancel{600}}} = \frac{4500}{600}$$
$$x = 7.5 \text{ or } 7.5\%$$

43. (a) $52.80 **(b)** $187.20 **45.** 1963 people **47.** 2144 people (rounded) **49. (a)** 0.96 million or 960,000 trips **(b)** 47.04 million or 47,040,000 trips **51.** 33% **53.** 2074 children **55.** 2% **57.** $3200 monthly earnings; $38,400 yearly earnings **59.** Breyers **61.** 76 people (rounded) **63.** 84.7% **65.** 1072 drivers **67.** 2156 products **69.** 231 customers **71.** whole; 100 **72.** whole **73.** 108 calories **74.** 300 grams **75.** 67 grams (rounded) **76.** 20 grams **77.** Yes, since they would eat 28% × 4 servings = 112% of the daily value. **78.** 17 packages (rounded). It may be possible but would result in a diet that is high in total fat, saturated fat, sodium, and total carbohydrates.

SECTION 6.5 (pages 411–414)

1. 46 percent; 780 whole **2.** 1800 whole; 14 percent **3.** 270 donors **5.** 1350 bath towels **7.** 83.2 quarts **9.** 3500 airbags **11.** 1029.2 meters **13.** $4.16 **15.** 70 percent; 476 part **16.** 270 part; 45 percent **17.** 160 patients **19.** 325 salads **21.** 1080 people **23.** 300 gallons **25.** percent; two; right; % **26.** decimal; two; left; % **27.** 76% **29.** 1.5% **31.** 250% **33.** You must first change the fraction in the percent to a decimal, then divide the percent by 100 to change it to a decimal.

$$\underbrace{2\tfrac{1}{2}\%}_{\text{Write } 2\frac{1}{2} \text{ as } 2.5.} = 2.5\% = 0.025 \leftarrow 2.5\% \text{ as a decimal}$$

34. The correct answer is $6.50. The error is in changing $\frac{1}{2}$% to a decimal.

$$\frac{1}{2}\% = 0.5\% = 0.005;$$
$$(0.005)(\$1300) = \$6.50 \quad \text{Correct}$$

Here are the incorrect answers and how your classmates got them.

$$\left.\begin{array}{ll} \frac{1}{2}\% = 0.0005; & (0.0005)(\$1300) = \$0.65 \\ \frac{1}{2}\% = 0.05; & (0.05)(\$1300) = \$65 \\ \frac{1}{2}\% = 0.5; & (0.5)(\$1300) = \$650 \end{array}\right\} \text{Incorrect}$$

35. 3.78 million or 3,780,000 office workers **37. (a)** 9625 people **(b)** 1875 people **(c)** 1000 people **39. (a)** 17,640 households **(b)** 360 households **41. (a)** 4% **(b)** 96% **43.** 36.9% (rounded) **45. (a)** relatives **(b)** 1950 families **47.** 1326 families **49.** $24,009 **51.** 478,175 Mustangs (rounded) **53.** $510,390 **55.** $439.61

SECTION 6.6 (pages 421–424)

1. whole; percent; part **2.** rate; cost of the item; sales **3.** $0.24; $6.24 **5.** 3%; $437.75 **7.** $672.60 (rounded); $12,901.60 (rounded) **9.** sales amount; commission rate; commission **10.** commission; sales amount **11.** $22.40 **13.** 20% **15.** $185.51 (rounded) **17.** whole; percent; part **18.** discount; original **19.** $20 (rounded); $179.99 (rounded) **21.** 30%; $126 **23.** $8.76; $49.64 **25.** On the basis of commission alone you would choose Company A. Other considerations might be: reputation of the company; expense allowances; other fringe benefits; travel; promotion and training. **26.** Some answers might be: calculating percent pay increases or decreases; changes in the cost of utilities, groceries, gasoline, and insurance; changes in the value of investments; the economy (inflation or deflation).
27. $203.93 (rounded) **29.** $90.62 (rounded) **31.** 5% **33.** 35.9% (rounded) **35.** 25% **37.** $134 **39.** $568.80 **41.** 4% **43.** $85.80; $304.20 **45.** 22% **47.** $18.43 (rounded) **49.** $6087.20 **51.** $14,032.12 (rounded) **53.** percent; whole **54.** rate (percent) of tax; cost of item **55.** $110.68 (rounded) **56.** Yes, the same; (10% • $95) + (6.5% • $95) = $15.68 (rounded) and (10% + 6.5%) • ($95) = $15.68 **57.** $1360.12 **58.** No. In Exercise 57 the excise tax of $15.40 cannot be added to the sales tax rate of $7\frac{3}{4}$% because the excise tax is an amount, not a percent.

SECTION 6.7 (pages 427–430)

1. 0.035 **2.** 0.075 **3.** 2.25 years **4.** 5.75 years **5.** $6 **7.** $105 **9.** $258.75 **11.** $2227.50 **13.** $\frac{3}{12}$ or $\frac{1}{4}$; 0.25 **14.** $\frac{6}{12}$ or $\frac{1}{2}$; 0.5 **15.** $\frac{9}{12}$ or $\frac{3}{4}$; 0.75 **16.** $\frac{15}{12}$ or $\frac{5}{4}$; 1.25 **17.** $6 **19.** $42.30 **21.** $16.84 (rounded) **23.** $647.06 (rounded) **25.** $210 **27.** $773.30 **29.** $2043 **31.** $3306.88 (rounded) **33.** $17,797.81 (rounded) **35.** The answer should include: Principal—This is the amount of money borrowed or loaned. Interest rate—This is the percent used to calculate the interest. Time of loan—The length of time that money is loaned or borrowed is an important factor in determining interest.
36. When time is given in months, the number of months is placed over 12 because there are 12 months in a year. This becomes a fraction of a year. Here is an example:

$$6 \text{ months} = \frac{6}{12} \text{ year} = \frac{1}{2} \text{ year} = 0.5 \text{ year}$$

37. $345.68 **39.** $26,250 **41.** $4746.88 (rounded) **43.** $277.50 **45.** $159.50 **47. (a)** $222.75 **(b)** $5622.75 **49.** $3258.50

SECTION 6.8 (pages 435–438)

1. false **2.** interest; principal **3.** $540.80 **5.** $1966.91 (rounded) **7.** $4587.79 (rounded) **9.** 1.02; 1.02; 1.02 **10.** 1.05; 1.05 **11.** $1102.50 **13.** $1873.52 (rounded) **15.** $2027.46 (rounded) **17.** $13,842.59 (rounded) **19.** 1.1941 **20.** 1.5530 **21.** 1.1717 **22.** 1.1742 **23.** $216.70 **25.** $9752; $1752 **27.** $10,976.01 (rounded); $2547.84 (rounded) **29.** Compound interest is interest paid on past interest, as well as on the principal. Many people describe compound interest as "interest on interest."

30. Compound amount is the total amount, original deposit plus interest on deposit, at the end of the compound interest period. The interest is found by subtracting the original deposit from the compound amount.
31. \$60,835 **33. (a)** \$128,402 **(b)** \$52,402
35. (a) \$87,786.23 (rounded) **(b)** \$17,786.23 **37.** principal; rate; time; p; r; t **38.** principal; interest **39.** principal; interest
40. principal; interest **41. (a)** \$254.48 **(b)** No; it has more than doubled (about five times). **(c)** Examples will vary. **42. (a)** the compound interest account; \$474.55 more **(b)** Greater amounts of interest are earned with compound interest than with simple interest because interest is earned on both principal and past interest.

Chapter 6 REVIEW EXERCISES (pages 446–451)

1. 0.35 **2.** 1.5 **3.** 0.9944 **4.** 0.00085 **5.** 315% **6.** 2% **7.** 87.5%
8. 0.2% **9.** $\frac{3}{20}$ **10.** $\frac{3}{8}$ **11.** $1\frac{3}{4}$ **12.** $\frac{1}{400}$ **13.** 75%
14. 62.5% or $62\frac{1}{2}$% **15.** 325% **16.** 0.5% **17.** 0.125 **18.** 12.5%
19. $\frac{1}{4}$ **20.** 25% **21.** $1\frac{4}{5}$ **22.** 1.8 **23.** whole = 250 **24.** part = 24
25. $\frac{\text{Part} \rightarrow 287}{\text{Whole} \rightarrow 820} = \frac{35 \leftarrow \text{Percent}}{100 \leftarrow \text{Always } 100}$ **26.** $\frac{73}{90} = \frac{\text{unknown}}{100}$
27. $\frac{\text{unknown}}{160} = \frac{14}{100}$ **28.** $\frac{418}{\text{unknown}} = \frac{16}{100}$ **29.** $\frac{3}{8} = \frac{\text{unknown}}{100}$
30. $\frac{\text{unknown}}{1280} = \frac{88}{100}$ **31.** 171 programs **32.** 870 reference books
33. 31.2 acres **34.** 2.8 kilograms **35.** 750 crates **36.** 2320 test tubes
37. 484 miles **38.** 17,000 cases **39.** 55% **40.** 5.2% (rounded)
41. 9.5% (rounded) **42.** 30.8% (rounded) **43.** 79 shark attacks (rounded)
44. 28% **45.** \$145.28 **46.** 186 trucks **47.** 0.4% **48.** 175%
49. 120 miles **50.** \$575 **51.** \$31.50; \$661.50 **52.** $7\frac{1}{2}$%; \$838.50
53. \$276 **54.** 5% **55.** \$33.75; \$78.75 **56.** 25%; \$189 **57.** \$8
58. \$67.50 **59.** \$3.50 **60.** \$152.10 **61.** \$832.50 **62.** \$1595.10
63. \$5375.60; \$1375.60 **64.** \$2187.71 (rounded); \$317.71 (rounded)
65. \$4108.32; \$508.32 **66.** \$16,337.50; \$3837.50 **67.** part = 12
68. whole = 1640 **69.** 23.28 meters **70.** 150% **71.** \$0.51
72. 1980 employees **73.** 40% **74.** 498.8 liters **75.** 0.55 **76.** 3
77. 500% **78.** 471% **79.** 0.086 **80.** 62.1% **81.** 0.00375
82. 0.06% **83.** 75% **84.** $\frac{21}{50}$ **85.** $\frac{7}{8}$ **86.** 37.5% or $37\frac{1}{2}$%
87. $\frac{13}{40}$ **88.** 60% **89.** $\frac{1}{400}$ **90.** 375% **91.** \$3331.25 **92.** \$16,520
93. (a) 96.0% (rounded) **(b)** 45.9% (rounded) **94. (a)** \$179,894.93 (rounded) **(b)** \$31,894.93 (rounded) **95.** \$6225 **96.** 18.2% (rounded)
97. \$848.40 (rounded) **98.** 1100 boaters **99.** \$13,005 **100.** 33.4% (rounded)

Chapter 6 TEST (pages 452–453)

1. 0.65 **2.** 80% **3.** 175% **4.** 87.5% or $87\frac{1}{2}$% **5.** 3.00 or 3 **6.** 0.02
7. $\frac{1}{8}$ **8.** $\frac{1}{400}$ **9.** 60% **10.** 62.5% or $62\frac{1}{2}$% **11.** 250%
12. 800 sacks **13.** 20% **14.** 125,000,000 households **15.** \$3450.60
16. 8% **17.** 25%
18. A possible answer is: Part is the increase in salary. Whole is last year's salary. Percent of increase is unknown.

$$\frac{\text{amount of increase}}{\text{last year's salary}} = \frac{p}{100}$$

19. The interest formula is $I = p \cdot r \cdot t$. If time is in months, it is expressed as a fraction with 12 as the denominator. If time is expressed in years, it is placed over 1 or shown as a decimal number.
Problems will vary. Some possibilities are:

$$I = (1000)\,(0.05)\left(\frac{9}{12}\right) = \$37.50$$

$$I = (1000)\,(0.05)\,(2.5) = \$125$$

20. \$11.52; \$84.48 **21.** \$91; \$189 **22.** \$378 **23.** \$192 **24.** \$20,880
25. (a) \$10,668 (rounded) **(b)** \$1668

CHAPTER 7 Geometry

SECTION 7.1 (pages 466–469)

1. Wording may vary slightly. A **line** is a row of points continuing in both directions forever; the drawing should have arrows on both ends. A **line segment** is a piece of a line; the drawing should have an endpoint (dot) on each end. A **ray** has one endpoint and goes on forever in one direction; the drawing should have one endpoint (dot) and an arrowhead on the other end.
2. Wording may vary slightly. An **acute angle** measures less than 90°; a **right angle** measures exactly 90° and forms a "square corner"; an **obtuse angle** measures more than 90° but less than 180°; a **straight angle** measures exactly 180° and forms a straight line. Drawings of an acute and obtuse angle may vary. **3.** line named $\overleftrightarrow{CD}$ or $\overleftrightarrow{DC}$ **5.** line segment named $\overline{GF}$ or $\overline{FG}$ **7.** ray named $\overrightarrow{PQ}$ **9.** perpendicular **11.** parallel
13. intersecting **15.** $\angle AOS$ or $\angle SOA$ **17.** $\angle AQC$ or $\angle CQA$
19. right (90°) **21.** acute **23.** straight (180°) **25.** $\angle EOD$ and $\angle COD$; $\angle AOB$ and $\angle BOC$ **27.** $\angle HNE$ and $\angle ENF$; $\angle ACB$ and $\angle KOL$
29. 50° **31.** 4° **33.** 50° **35.** 90° **37.** $\angle SON \cong \angle TOM$; $\angle TOS \cong \angle MON$ **39.** $\angle GOH$ measures 63°; $\angle EOF$ measures 37°; $\angle AOC$ and $\angle GOF$ both measure 80°. **41.** True, because $\overleftrightarrow{UQ}$ is perpendicular to $\overleftrightarrow{ST}$. **42.** True, because they form a 90° angle, as indicated by the small red square **43.** False; the angles have the same measure (both are 180°). **44.** False; $\overleftrightarrow{ST}$ and $\overleftrightarrow{PR}$ are parallel. **45.** False; $\overleftrightarrow{QU}$ and $\overleftrightarrow{TS}$ are perpendicular. **46.** True, because both angles are formed by perpendicular lines, so they both measure 90° **47.** corresponding angles: $\angle 1$ and $\angle 8$, $\angle 2$ and $\angle 5$, $\angle 3$ and $\angle 6$, $\angle 4$ and $\angle 7$; alternate interior angles: $\angle 4$ and $\angle 5$, $\angle 3$ and $\angle 8$ **49.** $\angle 2$, $\angle 4$, $\angle 6$, $\angle 8$ all measure 130°; $\angle 1$, $\angle 3$, $\angle 5$, $\angle 7$ all measure 50°. **51.** $\angle 6$, $\angle 1$, $\angle 3$, $\angle 8$ all measure 47°; $\angle 5$, $\angle 2$, $\angle 7$, $\angle 4$ all measure 133°. **53.** $\angle 6$, $\angle 8$, $\angle 4$, $\angle 2$ all measure 114°; $\angle 7$, $\angle 5$, $\angle 3$, $\angle 1$ all measure 66°. **55.** $\angle 1 \cong \angle 3$, both are 138°; $\angle 2 \cong \angle ABC$, both are 42°.

SECTION 7.2 (pages 476–479)

1. Perimeter is the distance around the outside edges; add the lengths of all the sides to find the perimeter. The area is the surface inside the figure; multiply length times width to find the area. The units for perimeter will be **yd;** the units for area will be **yd^2.**

2. A rectangle is a figure with four sides that meet to form four right angles. Each set of opposite sides is parallel and has the same length. A square has four right angles; all four sides have the same length. Drawings will vary; show your instructor.
3. $P = 16$ yd + 12 yd = 28 yd; $A = 48$ yd^2 **5.** $P = 3.6$ km; $A = 0.81$ km^2 **7.** $P = 40$ ft; $A = 100$ ft^2 **9.** $P = 196.2$ ft; $A = 1674.2$ ft^2 **11.** $P = 12$ mi; $A = 9$ mi^2 **13.** $P = 38$ m; $A = 39$ m^2 **15.** $P = 98$ m; $A = 492$ m^2 **17.** unlabeled side = 12 in.; $P = 78$ in.; $A = 234$ in.^{2} **19.** $P = 48$ m; $A = 144$ m^2 **21.** 96 in.^{2} (rounded); 77 in.^{2} (rounded); 19 in.^{2} difference **23.** \$94.81 **25.** Area of largest field is 9600 yd^2; Area of smallest field is 5000 yd^2; difference is 9600 yd^2 − 5000 yd^2 = 4600 yd^2. **27.** 53 yd **29.** $A = 6528$ ft^2

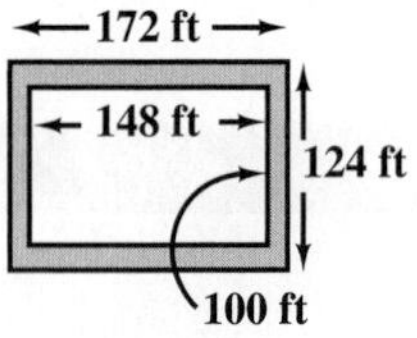

31.

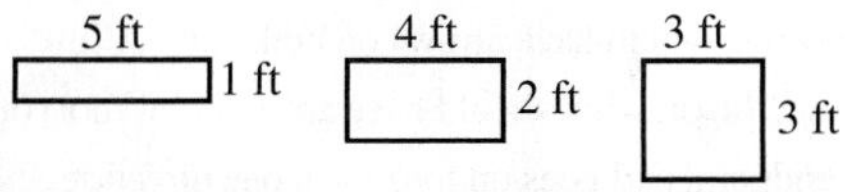

32. **(a)** 5 ft by 1 ft has area of 5 ft^2; 4 ft by 2 ft has area of 8 ft^2; 3 ft by 3 ft has area of 9 ft^2. **(b)** The square plot 3 ft by 3 ft has greatest area.
33.

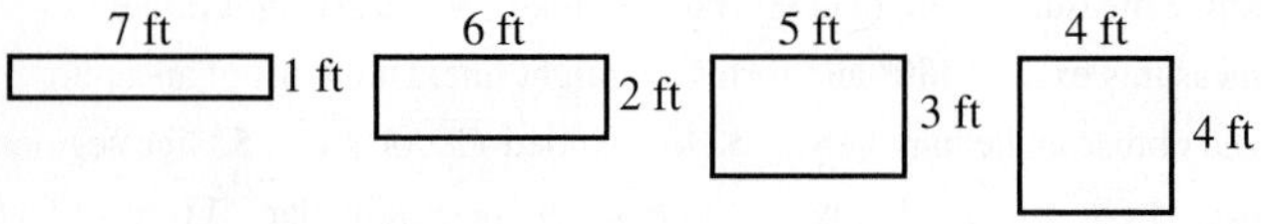

34. **(a)** 7 ft^2; 12 ft^2; 15 ft^2; 16 ft^2 **(b)** Square plots have the greatest area.
35. **(a)** $P = 10$ ft; $A = 6$ ft^2

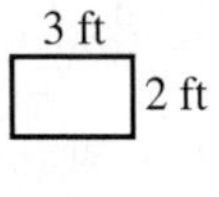

(b) $P = 20$ ft; $A = 24$ ft^2

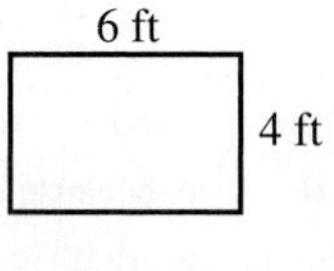

(c) Perimeter is twice the original; area is four times the original.
36. **(a)** $P = 30$ ft; $A = 54$ ft^2

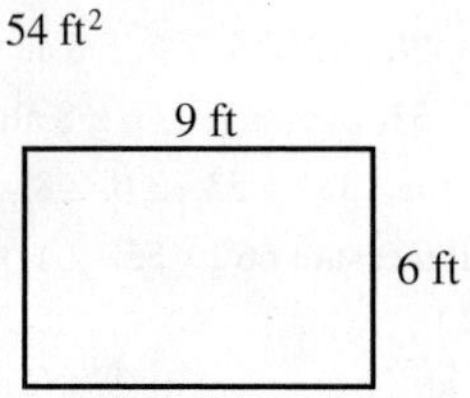

(b) Perimeter is three times the original; area is nine times the original.
(c) Perimeter will be four times the original; area will be 16 times the original.

SECTION 7.3 (pages 484–485)

1. In a parallelogram, opposite sides are parallel and have the same length. Drawings may vary; an example is shown.

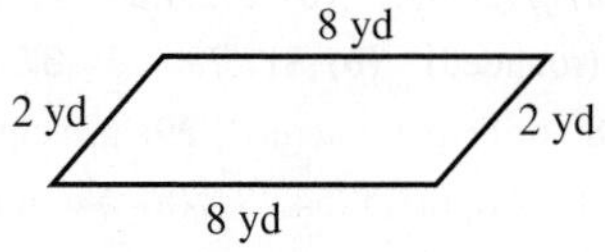

2. To find the perimeter of any figure, add the lengths of all the sides. Drawings will vary. **3.** $P = 46$ m + 46 m + 58 m + 58 m; $P = 208$ m **5.** $P = 207.2$ m **7.** $P = 6.06$ km **9.** $A = 775$ mm^2 **11.** $A = 19.25$ ft^2 **13.** $A = 3099.6$ cm^2 **15.** Height is not part of perimeter; square units are used for area, not perimeter. $P = 2.5$ cm + 2.5 cm + 2.5 cm + 2.5 cm = 10 cm **16.** The bases are the parallel lines, 22 ft and 13 ft; area is measured in square units. $A = 0.5 \cdot 11.6$ ft $\cdot$ (22 ft + 13 ft) = 203 ft^2
17. \$1410.75

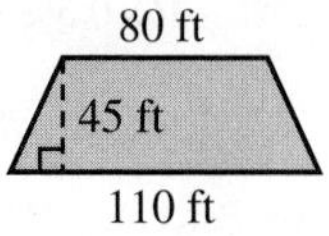

19. $A = 437.5$ in.^{2}

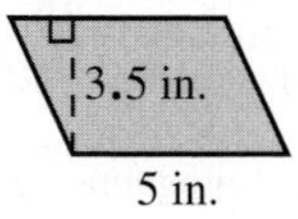

21. $A = 3.02$ m^2 **23.** $A = 25{,}344$ ft^2

SECTION 7.4 (pages 490–492)

1. Multiply $\frac{1}{2}$ times 58 m times 66 m; the units in the answers will be m^2. **2.** Add 58 m plus 72 m plus 72 m; the units in the answer will be m. **3.** $P = 202$ m; $A = 1914$ m^2 **5.** $P = 58.9$ cm; $A = 139.15$ cm^2 **7.** $P = 26\frac{1}{4}$ yd or 26.25 yd; $A = 30\frac{3}{4}$ yd^2 or 30.75 yd^2 **9.** $P = 85.2$ cm; $A = 302.46$ cm^2 **11.** Area of triangle = $0.5 \cdot 12$ m $\cdot$ 9 m = 54 m^2 Area of square = 12 m $\cdot$ 12 m = 144 m^2; Entire area = 54 m^2 + 144 m^2 = 198 m^2 **13.** $A = 1664$ m^2 **15.** 32° **17.** 48° **19.** No. Right angles are 90°, so two right angles are 180°, and the sum of all *three* angles in a triangle equals 180°. **20.** Two identical triangles form a parallelogram. The area of a parallelogram is base times height, so the area of one of the triangles is $\frac{1}{2} \cdot$ base $\cdot$ height. Sketches will vary. **21.** $7\frac{7}{8}$ ft^2 or 7.875 ft^2 **23.** **(a)** 126.8 m of curb **(b)** 672 m^2 of sod **25.** **(a)** $A = 32$ m^2 **(b)** $A = 13.41$ m^2 **27.** **(a)** 175 yd of frontage **(b)** $A = 8750$ yd^2

SECTION 7.5 (pages 499–502)

1. perimeter; ft **2.** radius; divide by 2 **3.** $d = 2 \cdot 9$ mm; $d = 18$ mm **5.** $r = 0.35$ km **7.** $C \approx 69.1$ ft; $A \approx 379.9$ ft^2 **9.** $C \approx 8.2$ m; $A \approx 5.3$ m^2 **11.** $C \approx 47.1$ cm; $A \approx 176.6$ cm^2 **13.** $C \approx 23.6$ ft; $A \approx 44.2$ ft^2 **15.** $A \approx 76.9$ in.^{2} **17.** $A \approx 57$ cm^2 **19.** π is the ratio of circumference of a circle to its diameter. If you divide the circumference of any circle by its diameter, the answer is always a little more than 3. The approximate value is 3.14, which we call π (pi). **20.** Test questions will vary. They should include finding the circumference and/or area of a circle.

21. $A \approx 7850$ yd^2 **23.** $C \approx 91.4$ in.; Bonus: 693 revolutions/mile (rounded) **25.** $A \approx 70{,}650$ mi^2

27. Watch $C \approx 3.1$ in. $A \approx 0.8$ in.^{2}

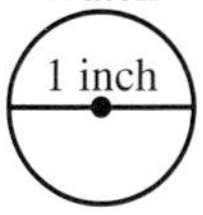

Wall Clock $C \approx 18.8$ in. $A \approx 28.3$ in.^{2}

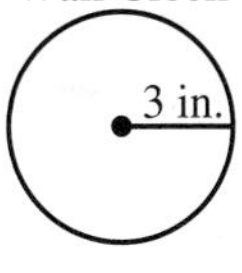

29.

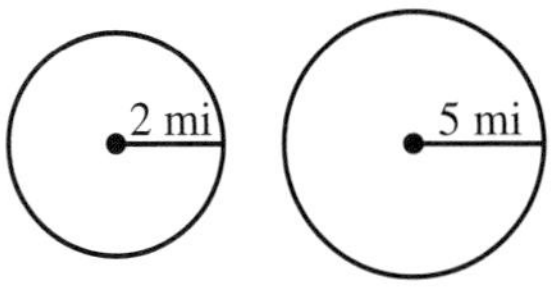

$A \approx 78.5$ mi^2 − 12.6 mi^2; Difference ≈ 65.9 mi^2

31. **(a)** $d \approx 45.9$ cm **(b)** Divide the circumference by π (3.14). **33.** \$325.83 (rounded). **35.** The prefix *rad-* tells you that radius is a ray from the center of the circle. The prefix *dia-* means the diameter goes through the circle, and the prefix *circum-* means the circumference is the distance around. **36.** The prefix *par-* means beside, so parallel lines are beside each other. (Perpendicular lines cross so that they form a right angle.) **37.** $A \approx 78.5$ in.^{2} **38.** $A \approx 113.0$ in.^{2} **39.** $A \approx 153.9$ in.^{2} **40.** small: \$0.089 (rounded); medium: \$0.080 (rounded); large: \$0.071 (rounded) Best buy **41.** small: \$0.153 (rounded); medium: \$0.124 (rounded); large: \$0.104 (rounded) Best buy **42.** small: \$0.076 (rounded) Best buy; medium: \$0.088 (rounded); large: \$0.078 (rounded)

SECTION 7.6 (pages 511–512)

1. It is not a cube because the edges have different measurements; in a cube, all edges measure the same. The units will be cm^3. **2.** Examples will vary. Some possibilities are cereal box and file cabinet for rectangular solid; water pipe and juice can for cylinder; basketball and moon for sphere. **3.** Rectangular solid; $V = 550$ cm^3; $S = 463$ cm^2 **5.** Cylinder; $V \approx 471$ ft^3; $S \approx 354.4$ ft^2 **7.** Sphere; $V \approx 44{,}579.6$ m^3 (rounded) **9.** Hemisphere; $V \approx 3617.3$ in.^{3} (rounded) **11.** Cone; $V \approx 418.7$ m^3 (rounded) **13.** Pyramid; $V = 800$ cm^3 **15.** $V = 18$ in.^{3} **17.** $V \approx 2481.5$ cm^3 (rounded) **19.** $V = 651{,}775$ m^3 **21.** $V \approx 3925$ ft^3 **23.** Student used diameter of 7 cm; should use radius of 3.5 cm in formula. Units for volume are cm^3, not cm^2. Correct answer is $V \approx 192.3$ cm^3. **24.** Both involve finding the area of a circular base and multiplying by the height. To find the volume of the cone, divide the volume of the cylinder by 3.

SECTION 7.7 (pages 517–520)

1. 4 **3.** 8 **5.** ≈ 3.317 **7.** ≈ 2.236 **9.** ≈ 8.544 **11.** ≈ 10.050 **13.** ≈ 13.784 **15.** ≈ 31.623 **17.** 30 is about halfway between 25 and 36, so $\sqrt{30}$ should be about halfway between 5 and 6, or about 5.5. Using a calculator, $\sqrt{30} \approx 5.477$. Similarly, $\sqrt{26}$ should be a little more than $\sqrt{25}$; by calculator $\sqrt{26} \approx 5.099$. And $\sqrt{35}$ should be a little less than $\sqrt{36}$; by calculator $\sqrt{35} \approx 5.916$. **18.** Squaring a number is multiplying the number times itself. Finding the square root is the opposite operation and "undoes" squaring. Examples will vary; one possibility is $7^2 = 7 \cdot 7 = 49$, so $\sqrt{49} = 7$. **19.** $\sqrt{1521} = 39$ ft **21.** $\sqrt{289} = 17$ in. **23.** $\sqrt{144} = 12$ mm **25.** $\sqrt{73} \approx 8.5$ in. **27.** $\sqrt{65} \approx 8.1$ yd **29.** $\sqrt{195} \approx 14.0$ cm **31.** $\sqrt{7.94} \approx 2.8$ m **33.** $\sqrt{65.01} \approx 8.1$ cm **35.** $\sqrt{292.32} \approx 17.1$ km

37. The student did not square the numbers correctly: 9^2 is 81 and 7^2 is 49. Also, the final answer is rounded to thousandths instead of tenths. Correct answer is $\sqrt{130} \approx 11.4$ in. **38.** The student used the formula for finding the hypotenuse, but the unknown side is a leg, so leg $= \sqrt{(20)^2 - (13)^2}$. Also, the final answer should be m, not m^2. Correct answer is $\sqrt{231} \approx 15.2$ m. **39.** hypotenuse $= \sqrt{65} \approx 8.1$ ft **41.** leg $= \sqrt{360{,}000} = 600$ m **43.** hypotenuse $= \sqrt{32.5} \approx 5.7$ ft **45.** leg $= \sqrt{135} \approx 11.6$ ft

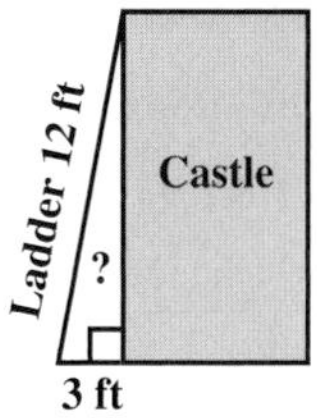

47. $\sqrt{16{,}200} \approx 127.3$ ft

48. **(a)**

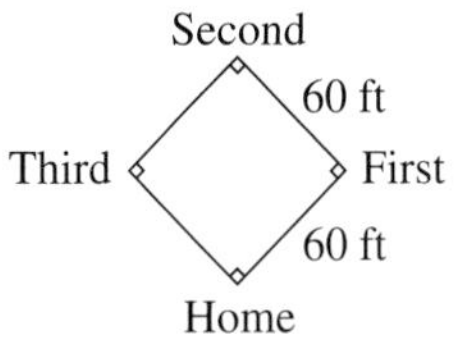

(b) $\sqrt{7200} \approx 84.9$ ft

49. The distance from third to first is the same as the distance from home to second because the baseball diamond is a square.

50. One possibility is:

$$\text{major league}\left\{\frac{90\text{ ft}}{127.3\text{ ft}} = \frac{60\text{ ft}}{x}\right\}\text{softball}$$

$x \approx 84.9$ ft

SECTION 7.8 (pages 527–530)

1. One dictionary definition is "conforming; agreeing." Examples of congruent objects include two matching chairs and two pieces of paper from the same notebook. **2.** One dictionary definition is "resembling, but not identical." Examples of similar objects are sets of different-size pots or measuring cups; small- and large-size cans of beans. **3.** $\angle 1$ and $\angle 4$; $\angle 2$ and $\angle 5$; $\angle 3$ and $\angle 6$; $\overline{AB}$ and $\overline{DE}$; $\overline{BC}$ and $\overline{EF}$; $\overline{AC}$ and $\overline{DF}$. **5.** $\angle 1$ and $\angle 6$; $\angle 2$ and $\angle 4$; $\angle 3$ and $\angle 5$; $\overline{ST}$ and $\overline{YW}$; $\overline{TU}$ and $\overline{WX}$; $\overline{SU}$ and $\overline{YX}$ **7.** $\angle 1$ and $\angle 6$; $\angle 2$ and $\angle 5$; $\angle 3$ and $\angle 4$; $\overline{LM}$ and $\overline{TS}$; $\overline{LN}$ and $\overline{TR}$; $\overline{MN}$ and $\overline{SR}$ **9.** SAS **11.** SSS **13.** ASA **15.** use SAS: $BC = CE$, $\angle ABC \cong \angle DCE$, $BA = CD$ **17.** use SAS: $PS = RS$, $m\angle QSP = m\angle QSR = 90°$, $QS = QS$ (common side) **19.** $a = 6$ cm; $\frac{7.5}{b} = \frac{6}{12}$; $b = 15$ cm **21.** $a = 5$ mm; $b = 3$ mm **23.** $a = 24$ in.; $b = 20$ in. **25.** $x = 24.8$ m; Perimeter = 72.8 m; $y = 15$ m; Perimeter = 54.6 m **27.** Perimeter = 8 cm + 8 cm + 8 cm = 24 cm; Area $\approx (0.5)(8\text{ cm})(6.9\text{ cm}) \approx 27.6$ cm^2; The area is approximate because the height was rounded to the nearest tenth. **29.** $h = 24$ ft **31.** $x = 50$ m **33.** $n = 110$ m

Chapter 7 REVIEW EXERCISES (pages 541–548)

1. line segment named $\overline{AB}$ or $\overline{BA}$ **2.** line named $\overleftrightarrow{CD}$ or $\overleftrightarrow{DC}$ **3.** ray named $\overrightarrow{OP}$ **4.** parallel **5.** perpendicular **6.** intersecting **7.** acute **8.** obtuse **9.** straight; 180° **10.** right; 90° **11.** $\angle AOB$ and $\angle BOC$; $\angle BOC$ and $\angle COD$; $\angle COD$ and $\angle DOA$; $\angle DOA$ and $\angle AOB$ **12.** $\angle ERH$ and $\angle HRG$; $\angle HRG$ and $\angle GRF$; $\angle FRG$ and $\angle FRE$; $\angle FRE$ and $\angle ERH$

13. **(a)** 10° **(b)** 45° **(c)** 83° **14.** **(a)** 25° **(b)** 90° **(c)** 147° **15.** ∠1 and ∠4 measure 30°; ∠3 and ∠6 measure 90°; ∠5 measures 60°. **16.** ∠8, ∠3, ∠6, ∠1 all measure 160°; ∠4, ∠7, ∠2, ∠5 all measure 20°. **17.** $P = 4.84$ m **18.** $P = 128$ in. **19.** 152 cm **20.** 41 ft **21.** $A = 486 \text{ mm}^2$ **22.** $A = 16.5 \text{ ft}^2$ or $16\frac{1}{2} \text{ ft}^2$ **23.** $A \approx 39.7 \text{ m}^2$ (rounded) **24.** $P = 50$ cm; $A = 140 \text{ cm}^2$ **25.** $P = 102.1$ ft; $A = 567 \text{ ft}^2$ **26.** $P = 200.2$ m; $A \approx 2206.1 \text{ m}^2$ (rounded) **27.** $P = 518$ cm; $A = 11{,}660 \text{ cm}^2$ **28.** $P = 27.1$ m; $A = 23.52 \text{ m}^2$ **29.** $P = 20\frac{1}{4}$ ft (exactly) or 20.3 ft (rounded); $A = 14 \text{ ft}^2$ **30.** 70° **31.** 24° **32.** $d = 137.8$ m **33.** $r = 1\frac{1}{2}$ in. or 1.5 in. **34.** $C \approx 6.3$ cm (rounded); $A \approx 3.1 \text{ cm}^2$ (rounded) **35.** $C \approx 109.3$ m (rounded); $A \approx 950.7 \text{ m}^2$ (rounded) **36.** $C \approx 37.7$ in. (rounded); $A \approx 113.0 \text{ in.}^2$ (rounded) **37.** $A \approx 20.3 \text{ m}^2$ (rounded) **38.** $A = 64 \text{ in.}^2$ **39.** $A = 673 \text{ km}^2$ **40.** $A = 1020 \text{ m}^2$ **41.** $A = 229 \text{ ft}^2$ **42.** $A = 132 \text{ ft}^2$ **43.** $A = 5376 \text{ cm}^2$ **44.** $A \approx 498.9 \text{ ft}^2$ (rounded) **45.** $A \approx 447.9 \text{ yd}^2$ (rounded) **46.** Rectangular solid; $V = 30 \text{ in.}^3$; $S = 59 \text{ in.}^2$ **47.** Rectangular solid; $V = 96 \text{ cm}^3$; $S = 128 \text{ cm}^2$ **48.** Rectangular solid; $V = 45{,}000 \text{ mm}^3$; $S = 8700 \text{ mm}^2$ **49.** Cylinder; $V \approx 549.5 \text{ cm}^3$; $S \approx 376.8 \text{ cm}^2$ **50.** Cylinder; $V \approx 1808.6 \text{ m}^3$ (rounded); $S \approx 1205.8 \text{ m}^2$ (rounded) **51.** Sphere; $V \approx 267.9 \text{ m}^3$ (rounded) **52.** Hemisphere; $V \approx 452.2 \text{ ft}^3$ (rounded) **53.** Cone; $V \approx 512.9 \text{ m}^3$ (rounded) **54.** Pyramid; $V = 16 \text{ yd}^3$ **55.** 7 **56.** 2.828 (rounded) **57.** 54.772 (rounded) **58.** 12 **59.** 7.616 (rounded) **60.** 25 **61.** 10.247 (rounded) **62.** 8.944 (rounded) **63.** $\sqrt{289} = 17$ in. **64.** $\sqrt{49} = 7$ cm **65.** $\sqrt{104} \approx 10.2$ cm (rounded) **66.** $\sqrt{52} \approx 7.2$ in. (rounded) **67.** $\sqrt{6.53} \approx 2.6$ m (rounded) **68.** $\sqrt{71.75} \approx 8.5$ km (rounded) **69.** SSS **70.** SAS **71.** ASA **72.** $y = 30$ ft; $x = 34$ ft; $P = 104$ ft **73.** $y = 7.5$ m; $x = 9$ m; $P = 22.5$ m **74.** $x = 12$ mm; $y = 7.5$ mm; $P = 38$ mm **75.** parallel lines **76.** line segment **77.** acute angle **78.** intersecting lines **79.** right angle; 90° **80.** ray **81.** straight angle; 180° **82.** obtuse angle **83.** perpendicular lines **84.** 81° **85.** 138° **86.** Square; $P = 18$ in.; $A \approx 20.3 \text{ in.}^2$ (rounded) or $A = 20\frac{1}{4} \text{ in.}^2$ **87.** Trapezoid; $P = 10.3$ cm; $A \approx 6.2 \text{ cm}^2$ (rounded) **88.** Circle; $C \approx 40.8$ (rounded); $A \approx 132.7 \text{ m}^2$ (rounded) **89.** Parallelogram; $P = 54$ ft; $A = 140 \text{ ft}^2$ **90.** Triangle; $P = 20$ yd; $A = 18\frac{3}{4} \text{ yd}^2$ or 18.8 yd^2 (rounded) **91.** Rectangle; $P = 7$ km; $A \approx 2.0 \text{ km}^2$ (rounded) **92.** Circle; $C \approx 53.4$ m (rounded) $A \approx 226.9 \text{ m}^2$ (rounded) **93.** Parallelogram; $P = 78$ mm; $A = 288 \text{ mm}^2$ **94.** Triangle; $P = 37.8$ mi; $A = 58.5 \text{ mi}^2$ **95.** $P = 90$ m; $A = 92 \text{ m}^2$ **96.** $P = 282$ cm; $A = 4190 \text{ cm}^2$ **97.** Cylinder; $V \approx 100.5 \text{ ft}^3$ (rounded) **98.** Rectangular solid or cube; $V \approx 3.4 \text{ in.}^3$ (rounded) or $V = 3\frac{3}{8} \text{ in.}^3$ (exactly) **99.** Rectangular solid; $V \approx 7.4 \text{ m}^3$ (rounded) **100.** Pyramid; $V = 561 \text{ cm}^3$ **101.** Cone; $V \approx 1271.7 \text{ cm}^3$ **102.** Sphere; $V \approx 1436.0 \text{ m}^3$ (rounded) **103.** $x \approx 12.6$ km (rounded) **104.** $\angle D = 72°$ **105.** $x = 12$ mm; $y = 14$ mm **106.** The prefix *dec-* in *dec*ade means 10 and the prefix *cent-* in *cent*ury means 100, so divide 200 (two centuries) by 10. The answer is 20 decades.

Chapter 7 TEST (Pages 549–550)

1. E **2.** A; 90° **3.** D **4.** G; 180° **5.** Parallel lines are lines in the same plane that never intersect. Perpendicular lines intersect to form a right angle.

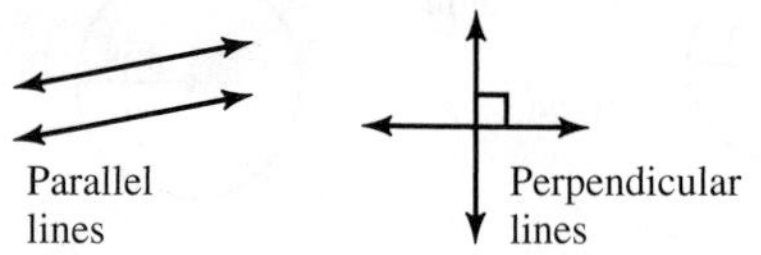

6. 9° **7.** 160° **8.** ∠1 measures 50°; ∠3 measures 90°; ∠2 and ∠4 measure 40°. **9.** Rectangle; $P = 23$ ft; $A = 30 \text{ ft}^2$ **10.** Square; $P = 72$ mm; $A = 324 \text{ mm}^2$ **11.** Parallelogram; $P = 26.2$ m; $A = 33.12 \text{ m}^2$ **12.** Trapezoid; $P = 169.4$ cm; $A = 1591 \text{ cm}^2$ **13.** $P = 32.45$ m; $A = 48 \text{ m}^2$ **14.** $P = 37.8$ yd or $37\frac{4}{5}$ yd; $A = 58.5 \text{ yd}^2$ **15.** 55° **16.** $r = 12.5$ in. or $12\frac{1}{2}$ in. **17.** $C \approx 5.7$ km (rounded) **18.** $A \approx 206.0 \text{ cm}^2$ (rounded) **19.** $A \approx 39.3 \text{ m}^2$ (rounded) **20.** Rectangular solid; $V = 6480 \text{ m}^3$; $S = 2232 \text{ m}^2$ **21.** Sphere; $V \approx 33.5 \text{ ft}^3$ (rounded) **22.** Cylinder; $V \approx 5086.8 \text{ ft}^3$; $S \approx 2599.9 \text{ ft}^2$ (rounded) **23.** $\sqrt{85} \approx 9.2$ cm (rounded) **24.** $y = 12$ cm; $z = 6$ cm **25.** Linear units like cm are used to measure perimeter, radius, diameter, and circumference. Area is measured in square units like cm^2 (squares that measure 1 cm on each side). Volume is measured in cubic units like cm^3.

CHAPTER 8 Statistics

SECTION 8.1 (pages 556–561)

1. parts **2.** protractor **3.** 360 million or 360,000,000 pets **5.** $\frac{90}{360} = \frac{1}{4}$ **7.** $\frac{90}{75} = \frac{6}{5}$ **9.** Sleeping; 8 hours 40 minutes **11.** $\frac{520}{1440} = \frac{13}{36}$ **13.** $\frac{50}{76} = \frac{25}{38}$ **15.** $\frac{210}{150} = \frac{7}{5}$ **17.** 160 people **19.** 96 people **21.** 960 people **23.** 1027 people (rounded) **25.** 121 people (rounded) **27.** 483 people (rounded) **29.** First find the percent of the total that is to be represented by each item. Next, multiply the percent by 360° to find the size of each sector. Finally, use a protractor to draw each sector. **30.** A protractor is used to measure the number of degrees in a sector. First, you must draw a line from the center of the circle to the left edge. Next, place the hole of the protractor over the center of the circle, making sure that the 0 on the protractor is on the line. Finally, make a mark at the desired number of degrees. This gives you the size of the sector. **31.** 90° **33.** 10%; 36° **35.** 54° **37.** 15% **39.** **(a)** $200,000 **(b)** 22.5°; 72°; 108°; 90°; 67.5°

(c)

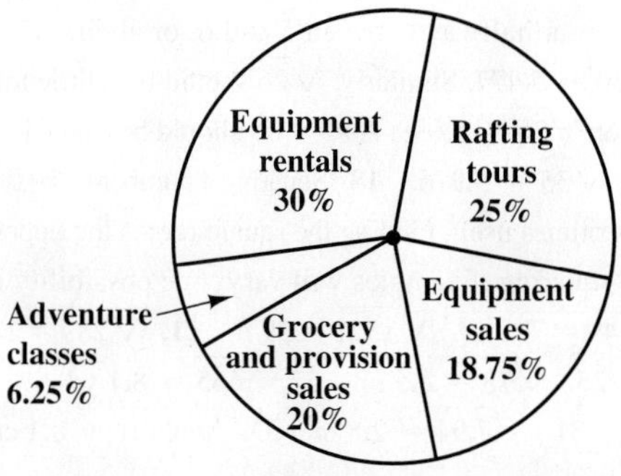

41. **(a)** 2578; 37% (rounded); 133° (rounded); 266; 4% (rounded); 14° (rounded); 1155; 16% (rounded); 58° (rounded); 1813; 26% (rounded); 94° (rounded); 749; 11% (rounded); 40° (rounded); 469; 7% (rounded); 25% (rounded)

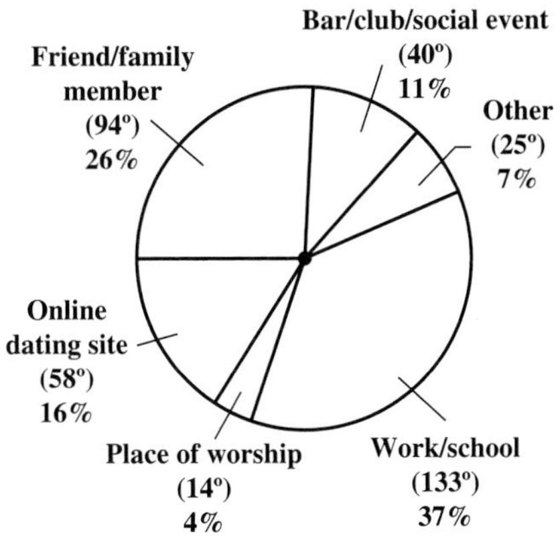

(b) No. The total is 101% due to rounding. **(c)** No. The total is 364° due to rounding.

SECTION 8.2 (pages 565–570)

1. false **2.** true **3.** India; 51.4% **5.** USA, Britain, and Australia **7.** 121 days (rounded) **9.** May; 10,000 plants **11.** 1500 plants **13.** 4000 plants **15.** 150,000 gal **17.** 2009; 250,000 gal **19.** 550,000 gal **21.** 24.1 million or 24,100,000 PCs **23.** 204.7 million or 204,700,000 PCs **25.** Answers will vary. Some possibilities are: greater demand as a result of lower price; more uses and applications for the owner; improved technology; multiple computers in each location; more laptop computers sold. **26.** Answers will vary. Some possibilities are: Fewer people will want a computer because they already own one; new technology will replace the computer with something better. **27.** 300,000 MP3 players **29.** 150,000 MP3 players **31.** 350,000 MP3 players **33.** Probably Store B with greater sales. Predicted sales might be 450,000 MP3 players to 500,000 MP3 players in 2014. **34.** Some possible answers are that Store B may have started to: do more advertising; keep longer store hours; give better training to their staff; employ more help; give better service than Store A. **35.** A single bar or a single line must be used for each set of data. To show multiple sets of data, multiple sets of bars or lines must be used. **36.** You would use a set of bars or a set of lines for each set of data. Examples will vary. **37.** $400,000 **39.** $250,000 **41.** $50,000 **43.** Answers will vary. Some possibilities are that the decrease in sales may have resulted from poor service or greater competition. The increase in sales may have been a result of more advertising or better service. **44.** As sales increase or decrease, so do profits. **45.** 23 years **46.** 20 additional flavors **47.** 214,286 rolls each day (rounded) **48.** 1.17 lb (rounded) **49.** The weight of one roll of Life Savers is much less than 1.17 lb. The answer is "correct" using the information given, but some of the data given must not be accurate. Perhaps 3 million rolls of Life Savers are produced daily. **50.** Answers will vary. Possible answers are: misprints or typographical errors; careless reporting of data; math errors.

SECTION 8.3 (pages 573–576)

1. table **2.** bar graph **3.** 18–25 years old; 40 million or 40,000,000 users **5.** 50 million or 50,000,000 users **7.** 25 million or 25,000,000 users **9.** $4100 to $5000; 16 employees **11.** 11 employees **13.** 49 employees **15.** Class intervals are the result of combining data into groupings. Class frequency is the number of data items that fit in each class interval. These are used to group data and to have multiple responses (frequency) in a class interval—this makes the data easier to interpret. **16.** If too few class intervals were used the class frequencies would be high and any differences in the data might not be observable. If too many class intervals were used, interpretation might become impossible because class frequencies would be very low or nonexistent. **17.** 𝍸 I; 6 **19.** I; 1 **21.** II; 2 **23.** IIII; 4 **25.** 𝍸; 5 **27.** III; 3 **29.** 𝍸 II; 7 **31.** III; 3 **33.** 𝍸 𝍸 𝍸 𝍸 𝍸 I; 26 **35.** 𝍸 𝍸 𝍸 II; 17 **37.** I; 1

SECTION 8.4 (pages 581–584)

1. true **2.** false **3.** grade point average **4.** most often **5.** 14 hours **7.** 2.5 in. of rain (rounded) **9.** $37,127 **11.** $72.53 **13.** 12.5 customers (rounded) **15.** 17.2 fish (rounded) **17.** $130,000 **19.** 125 guests **21.** 508 calories **23.** 21% **25.** 68 and 74 yr (bimodal) **27.** The median is a better measure of central tendency when the list contains one or more extreme values. Examples will vary. One possibility is to find the mean and the median of the following home values: $182,000; $164,000; $191,000; $115,000; $982,000; mean home value = $326,800; median home value = $182,000 **28.** The size to order is the mode. The mode is the size most worn by customers, and it would be wise to order most hats in this size. **29.** 2.42 (rounded) **31.** 2.60 **33.** 3.14 (rounded) **35.** Each made 176 sales calls. **36.** mean for Samuels: 22; mean for Stricker: 22 **37.** median for Samuels: 19.5; median for Stricker: 22 **38.** mode for Samuels: 22; mode for Stricker: 22 **39.** The mean and mode are identical for both sales representatives, and the medians are close. **40.** The number of weekly sales calls made by Samuels varies greatly from week to week, while the number of weekly sales calls made by Stricker remains fairly constant. **41.** range = 40 − 8 = 32 **42.** range = 25 − 19 = 6 **43.** No, not with any certainty. There probably are additional questions that need to be answered, such as the dollar amount of sales, number of repeat customers, and so on. **44.** Answers will vary. Some possible answers are: He works hard one week, then takes it easy the next week; the characteristics of the sales territories vary greatly; illness or personal problems may be affecting performance.

Chapter 8 REVIEW EXERCISES (pages 591–596)

1. Basketball; 18,000 teams **2.** $\frac{16{,}000}{76{,}000} = \frac{4}{19}$ **3.** $\frac{15{,}000}{76{,}000} = \frac{15}{76}$ **4.** $\frac{14{,}000}{76{,}000} = \frac{7}{38}$ **5.** $\frac{18{,}000}{16{,}000} = \frac{9}{8}$ **6.** $\frac{16{,}000}{14{,}000} = \frac{8}{7}$ **7.** 3936 companies **8.** 1728 companies **9.** 720 companies **10.** 2904 companies **11.** On-site child care and fitness centers. Answers will vary. Perhaps employers feel that they are not needed or would not be used. Or, it may be that they would be too expensive for the benefit derived. **12.** Flexible hours and casual dress. Answers will vary. Perhaps employees request them and appreciate them. Or, it may be that neither of them costs the employer anything to offer. **13.** March; 8,000,000 acre-feet

14. June; 2,000,000 acre-feet **15.** 5,000,000 acre-feet **16.** 4,000,000 acre-feet **17.** 5,000,000 acre-feet **18.** 2,000,000 acre-feet

19. $50,000,000 **20.** $20,000,000 **21.** $20,000,000 **22.** $40,000,000

23. The floor-covering sales decreased for 2 years and then moved up slightly. Answers will vary. Perhaps there is less new construction, remodeling, and home improvement in the area near Center A, or better product selection and service have reversed the decline in sales.

24. The floor-covering sales are increasing. Answers will vary. Perhaps new construction and home remodeling have increased in the area near Center B, or greater advertising has attracted more customers.

25. 28.5 digital cameras **26.** 35.2 complaints **27.** $51.05 **28.** 257.33 points (rounded) **29.** 39 forms **30.** $562 **31.** $79 **32.** 18 and 32 launchings (bimodal) **33.** 36° **34.** 35% **35.** 20%; 72°

36. 25%; 90° **37.** 10%; 36°

38.

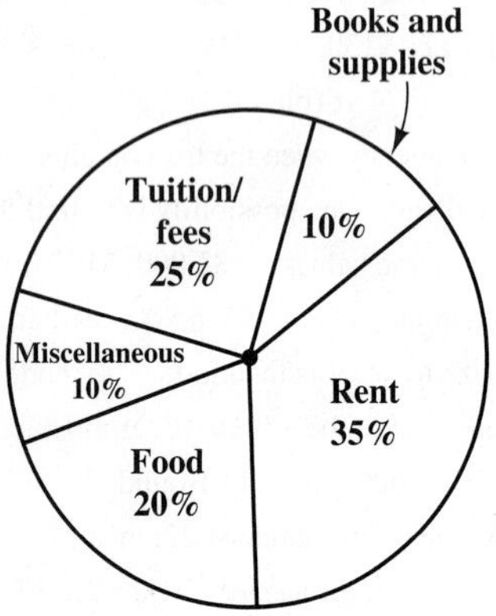

39. 100 volunteers **40.** 144.7 vaccinations (rounded) **41.** 48 applicants

42. 31 and 43 two-bedroom apartments (bimodal) **43.** 5.0 hr

44. 21 e-mails **45.** ||||; 4 **46.** |; 1 **47.** ||||| |; 6 **48.** ||; 2

49. ||||| ||||| |||; 13 **50.** ||||| ||; 7 **51.** ||||| ||; 7

52.

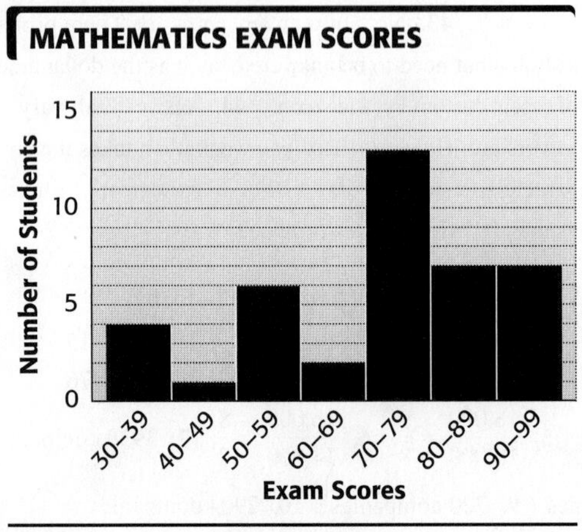

53. 64.5 (rounded) **54.** 118.8 units (rounded)

Chapter 8 TEST (pages 597–600)

1. $66.3 billion (rounded) **2.** $22.6 billion (rounded) **3.** $3.3 billion (rounded) **4.** $76.8 billion (rounded) **5.** $147.3 billion (rounded)

6. $11.8 billion (rounded) **7.** 108° **8.** 36° **9.** 72° **10.** 126° **11.** 5%

12.

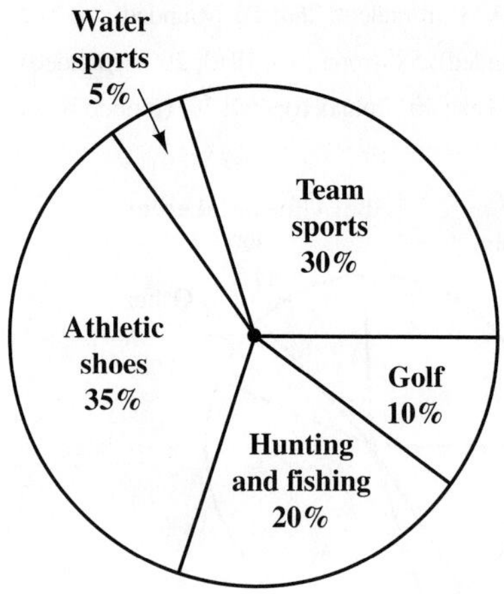

13. |||; 3 **14.** ||; 2 **15.** ||||; 4 **16.** |||; 3 **17.** |||; 3 **18.** |||||; 5

19.

ALAN'S VENDING SALES

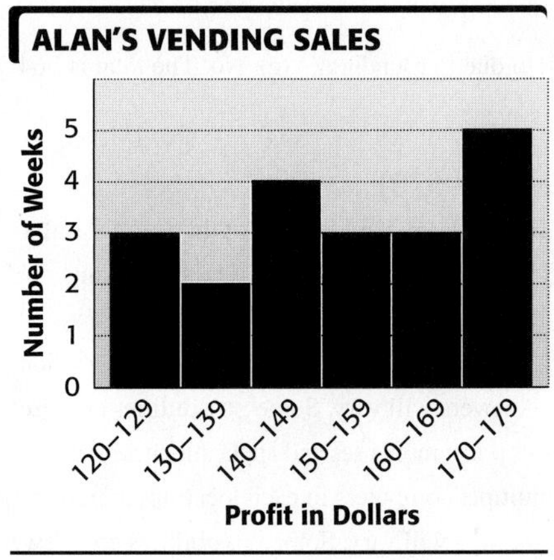

20. 75 mi **21.** 15.3 lb (rounded) **22.** 478.9 mph **23.** The weighted mean is used because different classes are worth different numbers of credits. GPA problems will vary; one possibility is shown.

Credits	Grade
3	A 3 • 4 = 12
2	C 2 • 2 = 4
4	B 4 • 3 = 12
9	**Total 28**

$$28 \div 9 \approx 3.11 \text{ (rounded)}$$

24. Arrange the values in order, from least to greatest. When there is an odd number of values in a list, the median is the middle value. Students' problems will vary; one possibility is shown.

8, 17, 23, 32, 64
 └— Median

25. $26 (rounded) **26.** 174 points (rounded) **27.** 31.5 degrees

28. 10.0 meters **29.** 52 milliliters **30.** 103 and 104 degrees (bimodal)

Chapters 1–8 CUMULATIVE REVIEW EXERCISES (pages 601–604)

1. **(a)** forty-five and two hundred three ten-thousandths **(b)** thirty billion, six hundred fifty thousand, eight **2.** **(a)** 160,000,500 **(b)** 0.075

3. 46,900 **4.** 6.20 **5.** 0.661 **6.** 10,000

7. *Estimate:* 80,000 − 50,000 = 30,000; *Exact:* 28,988
8. *Estimate:* 8 − 4 = 4; *Exact:* 4.2971
9. *Estimate:* 7000 × 700 = 4,900,000; *Exact:* 4,628,904
10. *Estimate:* 70 × 9 = 630; *Exact:* 568.11
11. *Estimate:* 40,000 ÷ 40 = 1000; *Exact:* 902
12. *Estimate:* 7 ÷ 1 = 7; *Exact:* 8.45
13. $10\frac{4}{15}$ **14.** $1\frac{11}{12}$ **15.** $44\frac{2}{5}$ **16.** $8\frac{4}{5}$ **17.** $\frac{8}{21}$ **18.** $\frac{1}{12}$ **19.** 23
20. 10 **21.** 108 **22.** $\frac{1}{4}$ **23.** 1 **24.** $\frac{1}{4}$ **25.** 0.75 **26.** 0.375
27. 0.583 (rounded) **28.** 0.55 **29.** 0.199, 0.207, 0.218, 0.22, 0.2215
30. 0.58, 0.608, $\frac{5}{8}$, 0.6319, $\frac{13}{20}$ **31.** 9.671 **32.** 93.603 **33.** 8 **34.** 45
35. $\frac{1}{8}$ **36.** $\frac{4}{1}$ **37.** $x = 6$ **38.** $x = 6$ **39.** $x = 16.2$ **40.** $x = 52$
41. 324 homes **42.** 2800 people **43.** 260% **44.** 0.03
45. 2.00 or 2 **46.** 87% **47.** 380% **48.** $\frac{2}{25}$ **49.** $\frac{5}{8}$ **50.** $1\frac{3}{4}$
51. 87.5% or $87\frac{1}{2}$% **52.** 420% **53.** 6.5% or $6\frac{1}{2}$% **54.** $117.18
55. $205.20; $250.80 **56.** $47,341.75 **57.** 45 children **58.** 255 ounces
59. Square; $P = 9$ ft; $A = 5\frac{1}{16}$ (exactly) or 5.1 ft^2 (rounded)
60. Circle; $C \approx 28.3$ mm; $A \approx 63.6\text{ mm}^2$ (both rounded)
61. Right triangle; $P = 56$ mi; $A = 84\text{ mi}^2$ **62.** Parallelogram; $P = 36$ ft; $A = 70\text{ ft}^2$ **63.** $P = 54$ ft; $A = 140\text{ ft}^2$
64. $C \approx 40.8$ m (rounded); $A \approx 132.7\text{ m}^2$ (rounded) **65.** $x \approx 12.6$ km (rounded); $A \approx 37.8\text{ km}^2$ **66.** $V \approx 3.4\text{ in.}^3$ (rounded); $S = 13.5\text{ in.}^2$
67. $P = 7$ km; $A \approx 2.0\text{ km}^2$ (rounded) **68.** $d = 17$ m; $C \approx 53.4$ m (rounded); $A \approx 226.9\text{ m}^2$ (rounded) **69.** $V \approx 339.1\text{ ft}^3$; $S \approx 282.6\text{ ft}^2$
70. Cylinder: $V \approx 549.8\text{ cm}^3$ (rounded) **71.** Rectangular solid: $V = 199.5\text{ m}^3$ **72.** 34 cm **73.** mean = 27 hookups; median = 26 hookups; mode = 19 hookups **74.** mean = 4.4 acres; median = 4.3 acres; mode = 2.85 acres **75.** 42.0% (rounded) **76. (a)** $889.80 (rounded) **(b)** $977 **77.** 38 nasal strips for $9.95, about $0.262 per strip
78. $59,225.25 **79.** $V \approx 3056.6\text{ cm}^3$ (rounded) **80.** $1\frac{1}{12}$ yd
81. 0.544 m^3 less **82.** Naomi's car; 25.2 miles per gallon (rounded)

CHAPTER 9 The Real Number System

SECTION 9.1 (pages 611–613)

1. true **2.** false; +, −, ·, and ÷ are operation symbols.
3. false; Using the rules for order of operations gives $4 + 3(8 - 2) = 4 + 3(6) = 4 + 18 = 22$. **4.** false; $3^3 = 3 \cdot 3 \cdot 3 = 27$
5. false; The correct translation is $4 = 16 - 12$. **6.** false; The correct translation is $6 = 10 - 4$. **7.** The 4 would be applied last because we work first inside the parentheses. **8.** exponents **9.** 49 **11.** 144 **13.** 64
15. 1000 **17.** 81 **19.** 1024 **21.** $\frac{16}{81}$ **23.** 0.000064
25. ② ①; 45; 58 **27.** ② ① ③; 12; 8; 13 **29.** ① ②; 32
31. ① ②; 19 **33.** $\frac{49}{30}$ **35.** 12 **37.** 36.14 **39.** 26 **41.** 4
43. 95 **45.** 12 **47.** 14 **49.** $\frac{19}{2}$ **51.** $3 \cdot (6 + 4) \cdot 2 = 60$
52. $2 \cdot (8 - 1) \cdot 3 = 42$ **53.** $10 - (7 - 3) = 6$
54. $15 - (10 - 2) = 7$ **55.** $(8 + 2)^2 = 100$ **56.** $(4 + 2)^2 = 36$
57. false **59.** true **61.** true **63.** false **65.** false **67.** true
69. $15 = 5 + 10$ **71.** $9 > 5 - 4$ **73.** $16 \neq 19$ **75.** $2 \leq 3$
77. Seven is less than nineteen; true **79.** One-third is not equal to three-tenths; true **81.** Eight is greater than or equal to eleven; false
83. $30 > 5$ **85.** $3 \leq 12$ **87.** $1.3 \leq 2.5$ **89.** $\frac{3}{4} < \frac{4}{5}$
91. (a) $14.7 - 40 \cdot 0.13$ **(b)** 9.5 **(c)** 8.075; walking (5 mph) **(d)** $14.7 - 55 \cdot 0.11$; 8.65; 7.3525, swimming **93.** Alaska, Texas, California, Idaho **95.** Alaska, Texas, California, Idaho, Missouri

SECTION 9.2 (pages 618–621)

1. B **2.** C **3.** A **4.** B, C **5.** 11 **6.** 10 **7.** $13 + x$; 16
8. expression; equation **9.** The equation would be $5x - 9 = 49$.
10. $2x^3 = 2 \cdot x \cdot x \cdot x$, while $2x \cdot 2x \cdot 2x = (2x)^3$. **11.** Answers will vary. Two such pairs are $x = 0, y = 6$ and $x = 1, y = 4$. For each pair, the value of $2x + y$ is 6. **12.** The exponent 2 applies only to the base x. The exponential must be evaluated before the product. **13. (a)** 64 **(b)** 144
15. (a) $\frac{7}{8}$ **(b)** $\frac{13}{12}$ **17. (a)** 9.569 **(b)** 14.353 **19. (a)** 52 **(b)** 114
21. (a) 12 **(b)** 33 **23. (a)** 6 **(b)** $\frac{9}{5}$ **25. (a)** $\frac{4}{3}$ **(b)** $\frac{13}{6}$
27. (a) $\frac{2}{7}$ **(b)** $\frac{16}{27}$ **29. (a)** 12 **(b)** 55 **31. (a)** 1 **(b)** $\frac{28}{17}$
33. (a) 3.684 **(b)** 8.841 **35.** $12x$ **37.** $x - 2$ **39.** $7 - \frac{1}{3}x$
41. $2x - 6$ **43.** $\frac{12}{x + 3}$ **45.** $6(x - 4)$ **47.** An equation cannot be *evaluated*. An equation that contains a variable can be *solved* to find the value or values of the variable that make it a true statement. **48.** An expression cannot be *solved*. It indicates a series of operations to be performed. An equation can be solved. **49.** no **51.** yes **53.** yes **55.** no **57.** yes
59. yes **61.** $x + 8 = 18$ **63.** $2x + 5 = 5$ **65.** $16 - \frac{3}{4}x = 13$
67. $3x = 2x + 8$ **69.** expression **71.** equation **73.** equation
75. expression **77.** 69 yr **78.** 74 yr **79.** 77 yr **80.** 79 yr

SECTION 9.3 (pages 628–630)

1. 0 **2.** integers **3.** positive **4.** right **5.** quotient; denominator
6. irrational **7.** 2,632,000 **9.** −7067

11. 

13. 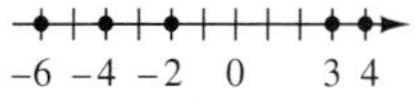

15. 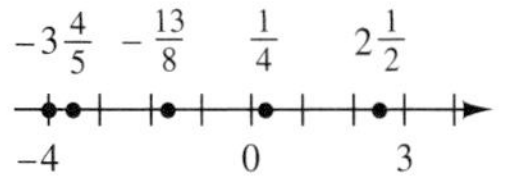

17. (a) 3, 7 **(b)** 0, 3, 7
(c) −9, 0, 3, 7 **(d)** $-9, -1\frac{1}{4}, -\frac{3}{5}, 0, 3, 5.9, 7$ **(e)** $-\sqrt{7}, \sqrt{5}$
(f) All are real numbers. **19. (a)** 11 **(b)** 0, 11 **(c)** 0, 11, −6
(d) $\frac{7}{9}, -2.\overline{3}, 0, -8\frac{3}{4}, 11, -6$ **(e)** $\sqrt{3}, \pi$ **(f)** All are real numbers.
21. 4 **22.** One example is 2.85. There are others. **23.** 0
24. One example is 4. There are others. **25.** One example is $\sqrt{13}$. There are others. **26.** 0 **27.** true **28.** false **29.** true **30.** true **31.** false
32. true *In Exercises 33–38, answers will vary.* **33.** $\frac{1}{2}, \frac{5}{8}, 1\frac{3}{4}$
34. $-1, -\frac{3}{4}, -5$ **35.** $-3\frac{1}{2}, -\frac{2}{3}, \frac{3}{7}$ **36.** $\frac{1}{2}, -\frac{2}{3}, \frac{2}{7}$

37. $\sqrt{5}, \pi, -\sqrt{3}$ **38.** $\frac{2}{3}, \frac{5}{6}, \frac{5}{2}$ **39.** -11 **41.** -21 **43.** -100 **45.** $-\frac{2}{3}$ **47.** false **49.** true **51.** (a) 2 (b) 2 **53.** (a) -6 (b) 6 **55.** (a) $\frac{3}{4}$ (b) $\frac{3}{4}$ **57.** (a) -4.95 (b) 4.95 **59.** (a) A (b) A (c) B (d) B **60.** 5; 5; 5; -5 **61.** 7 **63.** -12 **65.** $-\frac{2}{3}$ **67.** 9 **69.** false **71.** true **73.** Energy, 2008 to 2009 **75.** 2009 to 2010

SECTION 9.4 (pages 635–638)

1. negative; -5 **2.** zero (0) **3.** negative; -2

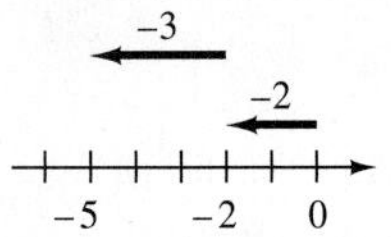

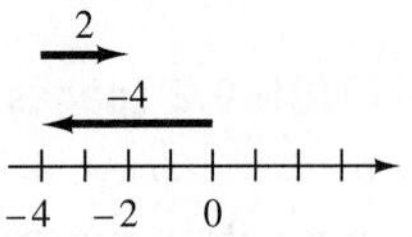

4. -3; 5 **5.** Add -2 and 5. **6.** Add -6 and 2. **7.** Add -1 and -3. **8.** Add -8 and 4. **9.** 2 **11.** -3 **13.** -10 **15.** -13 **17.** -15.9 **19.** -5; 5 **21.** 13 **23.** 0 **25.** -8 **27.** $\frac{3}{10}$ **29.** $\frac{1}{2}$ **31.** $-\frac{3}{4}$ **33.** -1.6 **35.** -8.7 **37.** -11; -14; -25 **39.** $-\frac{1}{4}$, or -0.25 **41.** true **43.** false **45.** true **47.** false **49.** true **51.** false **53.** $-5 + 12 + 6$; 13 **55.** $[-19 + (-4)] + 14$; -9 **57.** $[-4 + (-10)] + 12$; -2 **59.** $\left[\frac{5}{7} + \left(-\frac{9}{7}\right)\right] + \frac{2}{7}$; $-\frac{2}{7}$ **61.** $-\$62$ **63.** -184 m **65.** 17 **67.** 37 yd **69.** 120°F **71.** $-\$107$ **73.** -6

SECTION 9.5 (pages 643–647)

1. -8; -6; 2 **2.** 5; -4 **3.** $7 - 12$; $12 - 7$ **4.** additive; inverse; opposite **5.** -4 **6.** -22 **7.** -3; -10 **9.** -16 **11.** 4; 11 **13.** 19 **15.** -4 **17.** 5 **19.** 0 **21.** $\frac{3}{4}$ **23.** $-\frac{11}{8}$ **25.** $\frac{15}{8}$ **27.** 13.6 **29.** -11.9 **31.** 8 **33.** -2.8 **35.** -6.3 **37.** -28 **39.** $\frac{37}{12}$ **41.** -42.04 **43.** $4 - (-8)$; 12 **45.** $-2 - 8$; -10 **47.** $[9 + (-4)] - 7$; -2 **49.** $[8 - (-5)] - 12$; 1 **51.** -69°F **53.** 14,776 ft **55.** $-\$80$ **57.** -176.9°F **59.** \$1045.55 **61.** 469 B.C. **63.** \$323.83 **65.** 14 ft **67.** 40,776 ft **69.** 4.4% **70.** Americans spent more money than they earned, which means that they had to dip into savings or borrow money. **71.** \$1530 billion **73.** \$2900 **75.** $-\$20,900$ **77.** 136 ft **79.** negative **80.** positive **81.** positive **82.** negative **83.** positive **84.** negative

SECTION 9.6 (pages 656–659)

1. greater than 0 **2.** less than 0 **3.** less than 0 **4.** less than 0 **5.** greater than 0 **6.** less than 0 **7.** -28 **9.** 30 **11.** 0 **13.** $\frac{5}{6}$ **15.** -2.38 **17.** $\frac{3}{2}$ **19.** -3 **21.** -2 **23.** 16 **25.** 0 **27.** undefined **29.** $\frac{3}{2}$ **31.** C **32.** A **33.** 30; 10; 3 **35.** 7 **37.** 4 **39.** -3 **41.** -1 **43.** $\frac{7}{4}$ **45.** 68 **47.** -228 **49.** 1 **51.** 0 **53.** -6 **55.** undefined **57.** $-12 + 4(-7)$; -40 **59.** $-1 - 2(-8)(2)$; 31 **61.** $-3[3 - (-7)]$; -30 **63.** $\frac{3}{10}[-2 + (-28)]$; -9 **65.** $\frac{-20}{-8 + (-2)}$; 2 **67.** $\frac{-18 + (-6)}{2(-4)}$; 3 **69.** $\frac{-\frac{2}{3}\left(-\frac{1}{5}\right)}{\frac{1}{7}}$; $\frac{14}{15}$ **71.** $9x = -36$ **73.** $\frac{x}{4} = -1$ **75.** $x - \frac{9}{11} = 5$ **77.** $\frac{6}{x} = -3$ **79.** 29 **80.** The incorrect answer, 92, was obtained by performing all of the operations in order from left to right rather than following the rules for order of operations. The multiplications and divisions need to be done in order, before the additions and subtractions. **81.** 42 **82.** 5 **83.** $8\frac{2}{5}$ **84.** $8\frac{2}{5}$ **85.** 2 **86.** $-12\frac{1}{2}$

SECTION 9.7 (pages 666–669)

1. (a) B (b) F (c) C (d) I (e) B (f) D, F (g) B (h) A (i) G (j) H **2.** order; grouping **3.** yes **4.** yes **5.** no **6.** no **7.** no **8.** no **9.** (foreign sales) clerk; foreign (sales clerk) **10.** (defective merchandise) counter; defective (merchandise counter) **11.** -15; commutative property **13.** 3; commutative property **15.** 6; associative property **17.** 7; associative property **19.** Subtraction is not associative. **20.** Division is not associative. **21.** row 1: $-5, \frac{1}{5}$; row 2: $10, -\frac{1}{10}$; row 3: $\frac{1}{2}, -2$; row 4: $-\frac{3}{8}, \frac{8}{3}$; row 5: $-x, \frac{1}{x}$ $(x \neq 0)$; row 6: $y, -\frac{1}{y}$ $(y \neq 0)$; opposite; the same **22.** identity property **23.** commutative property **25.** associative property **27.** inverse property **29.** inverse property **31.** identity property **33.** commutative property **35.** distributive property **37.** identity property **39.** distributive property **41.** The expression following the first equality symbol should be $-3(4) - 3(-6)$. This simplifies to $-12 + 18$, which equals 6. **42.** We must multiply $\frac{3}{4}$ by 1 in the form of a fraction, $\frac{3}{3}$: $\frac{3}{4} \cdot \frac{3}{3} = \frac{9}{12}$. **43.** $7 + r$ **45.** s **47.** $-6x + (-6)7$; $-6x - 42$ **49.** $w + [5 + (-3)]$; $w + 2$ **51.** 6700 **53.** 2 **55.** 0.77 **57.** $4t + 12$ **59.** $-8r - 24$ **61.** y; -4; $-5y + 20$ **63.** $-16y - 20z$ **65.** $8(z + w)$ **67.** $5(3 + 17)$; 100 **69.** $7(2v + 5r)$ **71.** $24r + 32s - 40y$ **73.** $-24x - 9y - 12z$ **75.** $-4t - 5m$ **77.** $5c + 4d$ **79.** $3q - 5r + 8s$

SECTION 9.8 (pages 673–675)

1. B **2.** C **3.** A **4.** B **5.** $15x$ **7.** $13b$ **9.** $4r + 11$ **11.** $5 + 2x - 6y$ **13.** $-7 + 3p$ **15.** $-\frac{4}{3}y - 10$ **17.** -12 **19.** 5 **21.** 1 **23.** -1 **25.** 74 **27.** $-\frac{3}{8}$ **29.** $\frac{1}{2}$ **31.** $\frac{2}{5}$ **33.** -1.28 **35.** like **37.** unlike **39.** like **41.** unlike **43.** x; -3; $2x$; 6; $1 - 2x$ **45.** $-\frac{1}{3}t - \frac{28}{3}$ **47.** $-4.1r + 4.2$ **49.** $-2y^2 + 3y^3$ **51.** $-19p + 16$ **53.** $-2x + 4$ **55.** $-\frac{14}{3}x - \frac{22}{3}$ **57.** $-\frac{3}{2}y + 16$ **59.** $-16y + 63$ **61.** $4r + 15$ **63.** $12k - 5$ **65.** $-2k - 3$ **67.** $4x - 7$ **69.** $-23.7y - 12.6$ **71.** $(x + 3) + 5x$; $6x + 3$ **73.** $(13 + 6x) - (-7x)$; $13 + 13x$ **75.** $2(3x + 4) - (-4 + 6x)$; 12 **77.** Wording may vary. One example is "the difference between 9 times a number and the sum of the number and 2." **79.** The student made a sign error when applying the distributive property. $7x - 2(3 - 2x)$ means $7x - 2(3) - 2(-2x)$, which simplifies to $7x - 6 + 4x$, or $11x - 6$. **80.** The student incorrectly started by adding $3 + 2$. As the first step, 2 must be multiplied by $4x - 5$. Thus, $3 + 2(4x - 5)$ equals $3 + 8x - 10$, which simplifies to $8x - 7$. **81.** $1000 + 5x$ (dollars) **82.** $750 + 3y$ (dollars) **83.** $1000 + 5x + 750 + 3y$ (dollars) **84.** $1750 + 5x + 3y$ (dollars)

Chapter 9 REVIEW EXERCISES (pages 680–684)

1. 625 **2.** 0.00000081 **3.** 0.009261 **4.** $\frac{125}{8}$ **5.** 27 **6.** 200 **7.** 7 **8.** 4 **9.** $13 < 17$ **10.** $5 + 2 \neq 10$ **11.** Six is less than fifteen. **12.** Answers will vary. One example is $2 + 5 \geq \frac{16}{2}$. **13.** 30 **14.** 60 **15.** 14 **16.** 13 **17.** $x + 6$ **18.** $8 - x$ **19.** $6x - 9$ **20.** $12 + \frac{3}{5}x$ **21.** yes **22.** no **23.** $2x - 6 = 10$ **24.** $4x = 8$ **25.** equation **26.** expression

27. 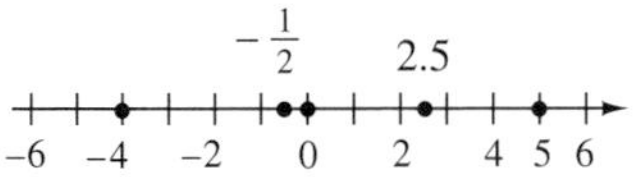

28. 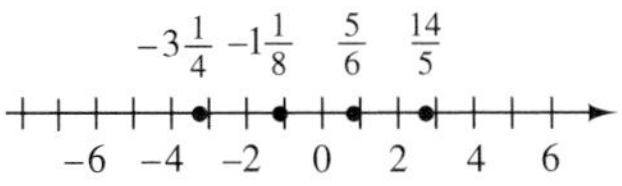

29. −10 **30.** −9 **31.** $-\frac{3}{4}$ **32.** $-|23|$ **33.** true **34.** true **35.** true **36.** false **37.** −3 **38.** −19 **39.** −7 **40.** 9 **41.** −6 **42.** −4 **43.** −17 **44.** $-\frac{29}{36}$ **45.** −10 **46.** −19 **47.** $(-31 + 12) + 19; 0$ **48.** $[-4 + (-8)] + 13; 1$ **49.** \$26.25 **50.** −10°F **51.** −11 **52.** −1 **53.** 7 **54.** $-\frac{43}{35}$ **55.** 10.31 **56.** −12 **57.** 2 **58.** 1 **59.** $-4 - (-6); 2$ **60.** $[4 + (-8)] - 5; -9$ **61.** $[18 - (-23)] - 15; 26$ **62.** $19 - (-7 - 12); 38$ **63.** 38 **64.** 12,392.69 **65.** −126 thousand **66.** 267 thousand **67.** 681 thousand **68.** −857 thousand **69.** 36 **70.** −105 **71.** $\frac{1}{2}$ **72.** 10.08 **73.** −20 **74.** −10 **75.** −24 **76.** −35 **77.** 4 **78.** −20 **79.** $-\frac{3}{4}$ **80.** 11.3 **81.** −1 **82.** undefined **83.** 1 **84.** 0 **85.** −18 **86.** −18 **87.** 125 **88.** −423 **89.** $-4(5) - 9; -29$ **90.** $\frac{5}{6}[12 + (-6)]; 5$ **91.** $\frac{12}{8 + (-4)}; 3$ **92.** $\frac{-20(12)}{15 - (-15)}; -8$ **93.** $\frac{x}{x + 5} = -2$ **94.** $8x - 3 = -7$ **95.** identity property **96.** identity property **97.** inverse property **98.** inverse property **99.** distributive property **100.** associative property **101.** associative property **102.** commutative property **103.** $(7 + 1)y; 8y$ **104.** $-12 \cdot 4 - 12(-t); -48 + 12t$ **105.** $3(2s + 4y); 6s + 12y$ **106.** $-1(-4r) + (-1)(5s); 4r - 5s$ **107.** $17p^2$ **108.** $16r^2 + 7r$ **109.** $-19k + 54$ **110.** $5s - 6$ **111.** $-45t - 23$ **112.** $-45t^2 - 23.4t$ **113.** −6 **114.** $\frac{25}{36}$ **115.** −26 **116.** $\frac{8}{3}$ **117.** $-\frac{1}{24}$ **118.** $\frac{7}{2}$ **119.** 2 **120.** 77.6 **121.** $-1\frac{1}{2}$ **122.** 11 **123.** $-\frac{28}{15}$ **124.** 24 **125.** −11 **126.** 16 **127.** −47°F **128.** 27 ft

Chapter 9 TEST (pages 685–686)

1. true **2.** false **3.** 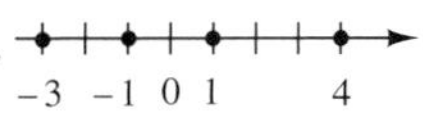

4. $-|-8|$ (or −8) **5.** −1.277 **6.** $\frac{-6}{2 + (-8)}; 1$ **7.** negative **8.** 4 **9.** $-2\frac{5}{6}$ **10.** 6 **11.** 2 **12.** 108 **13.** 11 **14.** $\frac{30}{7}$ **15.** −70 **16.** 3 **17.** 178°F **18.** 15 **19.** −\$1.42 trillion **20.** D **21.** A **22.** E **23.** B **24.** C **25.** $21x$ **26.** $-9x^2 - 6x - 8$ **27.** identity and distributive properties **28.** (a) −18 (b) −18 (c) The distributive property tells us that the two methods produce equal results.

CHAPTER 10 Equations, Inequalities, and Applications

SECTION 10.1 (pages 693–694)

1. equality; expression **2.** linear; = **3.** equivalent **4.** added to; subtracted from **5.** C **6.** A, B **7.** **(a)** expression; $x + 15$ **(b)** expression; $m + 7$ **(c)** equation; $\{-1\}$ **(d)** equation; $\{-17\}$ **8.** Replace the variable(s) in the original equation with the proposed solution. A true statement will result if the proposed solution is correct. **9.** $\{12\}$ **11.** $\{-3\}$ **13.** $\{4\}$ **15.** $\{-9\}$ **17.** $\left\{-\frac{3}{4}\right\}$ **19.** $\{6.3\}$ **21.** $\{-16.9\}$ **23.** $\{-10\}$ **25.** $\{-13\}$ **27.** $\left\{\frac{4}{15}\right\}$ **29.** $\{7\}$ **31.** $\{-3\}$ **33.** $\{-4\}$ **35.** $\{3\}$ **37.** $\{-2\}$ **39.** $\{4\}$ **41.** $\{-16\}$ **43.** $\{2\}$ **45.** $\{2\}$ **47.** $\{-4\}$ **49.** $\{4\}$ **51.** $\{0\}$ **53.** $\left\{\frac{7}{15}\right\}$ **55.** $\{7\}$ **57.** $\{-4\}$ **59.** $\{13\}$ **61.** $\{29\}$ **63.** $\{18\}$ **65.** Answers will vary. One example is $x - 6 = -8$. **66.** Answers will vary. One example is $x + \frac{1}{2} = 1$.

SECTION 10.2 (pages 699–700)

1. **(a)** and **(c)**: multiplication property of equality; **(b)** and **(d)**: addition property of equality **2.** C **3.** $\frac{3}{2}$ **4.** $\frac{5}{4}$ **5.** 10 **6.** 100 **7.** $-\frac{2}{9}$ **8.** $-\frac{3}{8}$ **9.** −1 **10.** −1 **11.** 6 **12.** 7 **13.** −4 **14.** −13 **15.** 0.12 **16.** 0.21 **17.** −1 **18.** −1 **19.** B **20.** A **21.** $\{6\}$ **23.** $\left\{\frac{15}{2}\right\}$ **25.** $\{-5\}$ **27.** $\left\{-\frac{18}{5}\right\}$ **29.** $\{12\}$ **31.** 2; 2; 0; $\{0\}$ **33.** $\{-12\}$ **35.** $\{40\}$ **37.** $\{-12.2\}$ **39.** $\{-48\}$ **41.** $\{72\}$ **43.** $\{-35\}$ **45.** $\{14\}$ **47.** $\left\{-\frac{27}{35}\right\}$ **49.** $\{3\}$ **51.** $\{-5\}$ **53.** $\{20\}$ **55.** $\{7\}$ **57.** 1; $\{-12\}$ **59.** $\{0\}$ **61.** $\{-6\}$ **63.** Answers will vary. One example is $\frac{3}{2}x = -6$. **64.** Answers will vary. One example is $100x = 17$. **65.** $-4x = 10; -\frac{5}{2}$ **67.** $\frac{x}{-5} = 2; -10$

SECTION 10.3 (pages 709–712)

1. addition; subtract **2.** left; like **3.** distributive; parentheses **4.** multiplication; $\frac{4}{3}$ **5.** fractions; 6 **6.** decimals; 10 **7.** **(a)** identity; B **(b)** conditional; A **(c)** contradiction; C **8.** D **9.** A **10.** D **11.** $\{4\}$ **13.** $\{-5\}$ **15.** $\left\{\frac{5}{2}\right\}$ **17.** $\left\{-\frac{1}{2}\right\}$ **19.** $\{5\}$ **21.** $\{1\}$ **23.** $\left\{-\frac{5}{3}\right\}$ **25.** $\{2\}$ **27.** $\left\{-\frac{5}{3}\right\}$ **29.** $\emptyset$ **31.** $\{0\}$ **33.** $\{-1\}$ **35.** $\emptyset$ **37.** {all real numbers} **39.** $\{0\}$ **41.** 14; $\{7\}$ **43.** $\{12\}$ **45.** $\{11\}$ **47.** $\{0\}$ **49.** $\left\{\frac{3}{25}\right\}$ **51.** 100; $\{60\}$ **53.** $\{4\}$ **55.** $\{5000\}$

57. $\left\{-\frac{72}{11}\right\}$ **59.** $\{0\}$ **61.** $\emptyset$ **63.** {all real numbers} **65.** $\{-6\}$ **67.** $\{15\}$ **69.** $12 - q$ **71.** $\frac{9}{z}$ **73.** $x + 29$ **75.** $a + 12$; $a - 2$ **77.** $25r$ **79.** $\frac{t}{5}$ **81.** $3x + 2y$

SECTION 10.4 (pages 721–726)

1. *Step 1:* Read the problem carefully. *Step 2:* Assign a variable to represent the unknown. *Step 3:* Write an equation. *Step 4:* Solve the equation. *Step 5:* State the answer. *Step 6:* Check the answer. **2.** Some examples are *is, are, was,* and *were.* **3.** D; There cannot be a fractional number of cars. **4.** D; A day cannot have more than 24 hr. **5.** A; Distance cannot be negative. **6.** C; Time cannot be negative. **7.** 1; 16 (or 14); −7 (or −9) **8.** odd; 2; 11 (or 15); even; 2; 10 (or 14) **9.** complementary; supplementary **10.** 90°; 180° **11.** yes, 90°; yes, 45° **12.** $x - 1$; $x - 2$ **13.** $8 \cdot (x + 6) = 104$; 7 **15.** $5x + 2 = 4x + 5$; 3 **17.** $3x - 2 = 5x + 14$; −8 **19.** $3(x - 2) = x + 6$; 6 **21.** $3x + (x + 7) = -11 - 2x$; −3 **23.** *Step 1:* We are asked to find the number of drive-in movie screens in the two states. *Step 2:* the number of screens in Pennsylvania. *Step 3:* x; $x + 3$; *Step 4:* 28; *Step 5:* 28; 28; 31; *Step 6:* 3; screens in New York; 31; 59 **25.** Democrats: 193; Republicans: 242 **27.** Bon Jovi: \$108.2 million; Roger Waters: \$89.5 million **29.** wins: 57; losses: 25 **31.** orange: 97 mg; pineapple: 25 mg **33.** 420 lb **35.** active: 225 mg; inert: 25 mg **37.** 1950 Denver nickel: \$16.00; 1944 Philadelphia nickel: \$12.00 **39.** onions: 81.3 kg; grilled steak: 536.3 kg **41.** American: 18; United: 11; Southwest: 26 **43.** $x + 5$; $x + 9$; shortest piece: 15 in.; middle piece: 20 in.; longest piece: 24 in. **45.** gold: 46; silver: 29; bronze: 29 **47.** 36 million mi **49.** *A* and *B*: 40°; *C*: 100° **51.** 68, 69 **53.** 146, 147 **55.** 10, 12 **57.** 17, 19 **59.** 10, 11 **61.** 18 **63.** 18° **65.** 20° **67.** 39° **69.** 50°

SECTION 10.5 (pages 733–738)

1. The perimeter of a plane geometric figure is the distance around the figure. **2.** The area of a plane geometric figure is the measure of the surface covered or enclosed by the figure. **3.** area **4.** area **5.** perimeter **6.** perimeter **7.** area **8.** area **9.** area **10.** area **11.** $P = 26$ **13.** $A = 64$ **15.** $b = 4$ **17.** $t = 5.6$ **19.** $h = 7$ **21.** $r = 2.6$ **23.** $A = 50.24$ **25.** $V = 150$ **27.** $V = 52$ **29.** $V = 7234.56$ **31.** $I = \$600$ **33.** $p = \$550$ **35.** 0.025; $t = 1.5$ yr **37.** length: 18 in.; width: 9 in. **39.** length: 14 m; width: 4 m **41.** shortest: 5 in.; medium: 7 in.; longest: 8 in. **43.** two equal sides: 7 m; third side: 10 m **45.** perimeter: 5.4 m; area: 1.8 m^2 **47.** 10 ft **49.** about 154,000 ft^2 **51.** 194.48 ft^2; 49.42 ft **53.** 23,800.10 ft^2 **55.** length: 36 in.; maximum volume: 11,664 in.^{3} **57.** 48°, 132° **59.** 55°, 35° **61.** 51°, 51° **63.** 105°, 105° **65.** $t = \frac{d}{r}$ **67.** $H = \frac{V}{LW}$ **69.** $b = P - a - c$ **71.** $r = \frac{C}{2\pi}$ **73.** $r = \frac{I}{pt}$ **75.** $h = \frac{2A}{b}$ **77.** $h = \frac{3V}{\pi r^2}$ **79.** $W = \frac{P - 2L}{2}$ **81.** $m = \frac{y - b}{x}$ **83.** $y = \frac{C - Ax}{B}$ **85.** $r = \frac{M - C}{C}$ **87.** $a = \frac{P - 2b}{2}$

We give one possible answer for Exercises 89–95. There are other correct forms.

89. $y = -6x + 4$ **91.** $y = 5x - 2$ **93.** $y = \frac{3}{5}x - 3$ **95.** $y = \frac{1}{3}x - 4$

SECTION 10.6 (pages 748–751)

1. < (or >); > (or <); ≤ (or ≥); ≥ (or ≤) **2.** false **3.** $(0, \infty)$ **4.** $(-\infty, \infty)$ **5.** $x > -4$ **6.** $x \geq -4$ **7.** $x \leq 4$ **8.** $x < 4$ **9.** $-1 < x \leq 2$ **10.** $-1 \leq x < 2$

11. $(-\infty, 4]$

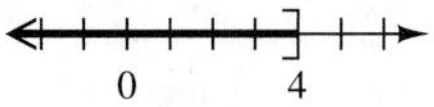

13. $(-3, \infty)$

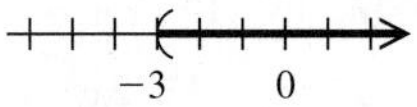

15. $[8, 10]$

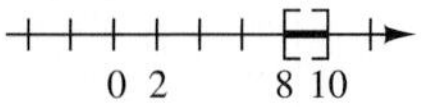

17. $(0, 10]$

19. $[1, \infty)$

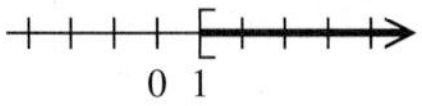

21. $[5, \infty)$

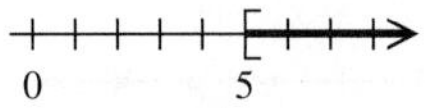

23. $(-\infty, -6)$

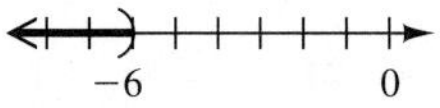

25. It must be reversed when multiplying or dividing by a negative number.

26. Divide by −5 and reverse the direction of the inequality symbol to get $x < -4$.

27. $(-\infty, 6)$

29. $[-10, \infty)$

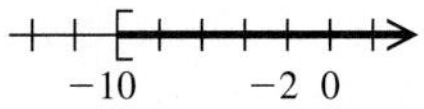

31. $(-\infty, -3)$

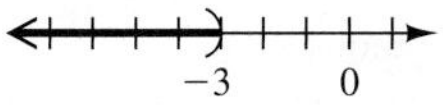

33. $(-\infty, 0]$

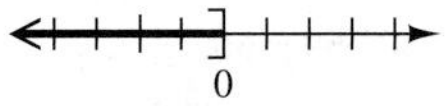

35. $(20, \infty)$

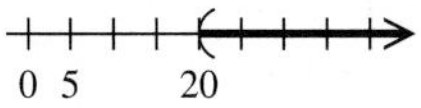

37. $[-3, \infty)$

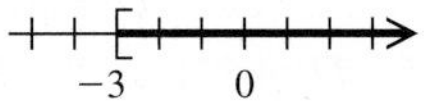

39. $(-\infty, -3]$

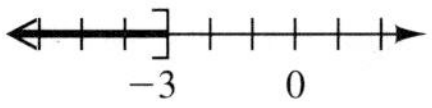

41. $(-1, \infty)$

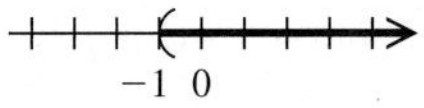

43. $[-5, \infty)$

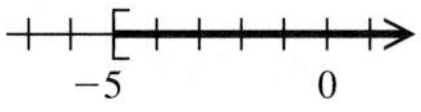

45. $(-\infty, 1)$

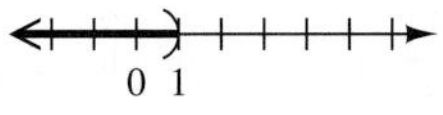

47. $(-\infty, 0]$

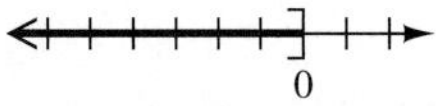

49. $\left(-\frac{1}{2}, \infty\right)$

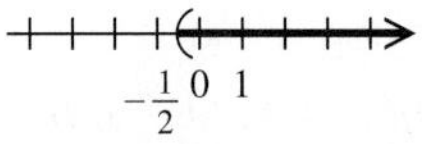

51. $[4, \infty)$

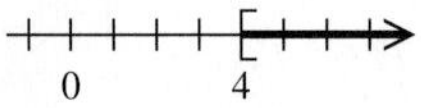

53. $(-\infty, 32)$

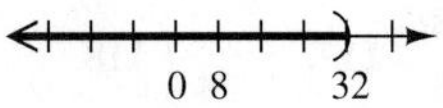

55. $\left[\frac{5}{12}, \infty\right)$

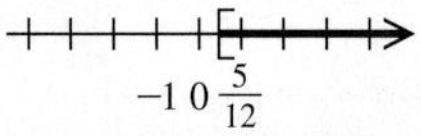

57. $(-21, \infty)$

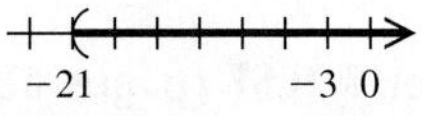

59. $x \geq 16$ **60.** $x < 1$ **61.** $x > 8$ **62.** $x \geq 12$ **63.** $x \leq 20$ **64.** $x > 40$ **65.** 88 or more **67.** 80 or more **69.** all numbers greater than 16 **71.** It has never exceeded 40°C. **73.** 32 or greater **75.** 12 min

77. $[-1, 6]$

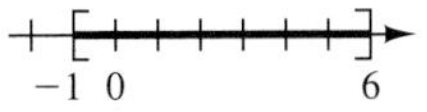

79. $(1, 3)$

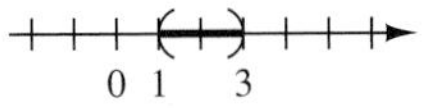

81. $\left(-\frac{11}{6}, -\frac{2}{3}\right]$

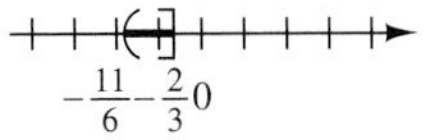

83. $[-26, 6]$

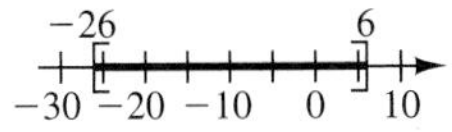

85. $[-3, 6]$

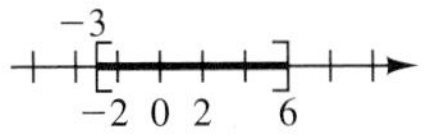

87. $\{4\}$

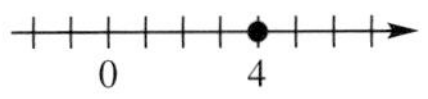

88. **(a)** $(-\infty, 4)$

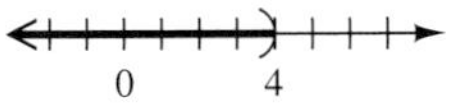

(b) $(4, \infty)$

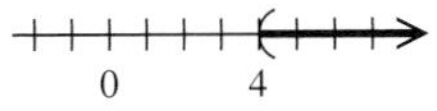

89. It is the set of all real numbers.

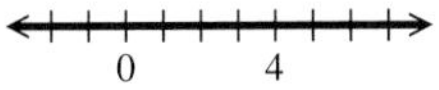

90. The graph would be the set of all real numbers; $(-\infty, \infty)$

Chapter 10 REVIEW EXERCISES (pages 756–758)

1. $\{9\}$ **2.** $\{4\}$ **3.** $\{-6\}$ **4.** $\left\{\frac{3}{2}\right\}$ **5.** $\{20\}$ **6.** $\left\{-\frac{61}{2}\right\}$ **7.** $\{15\}$ **8.** $\{0\}$ **9.** $\emptyset$ **10.** {all real numbers} **11.** $-\frac{7}{2}$ **12.** 20 **13.** Hawaii: 6425 mi^2; Rhode Island: 1212 mi^2 **14.** Seven Falls: 300 ft; Twin Falls: 120 ft **15.** 80° **16.** 11, 13 **17.** $h = 11$ **18.** $A = 28$ **19.** $r = 4.75$ **20.** $V = 3052.08$ **21.** $h = \frac{A}{b}$ **22.** $h = \frac{2A}{b + B}$

For Exercises 23 and 24, other correct forms of the answers are possible.

23. $y = -x + 11$ **24.** $y = \frac{3}{2}x - 6$ **25.** 135°, 45° **26.** 100°, 100° **27.** perimeter: 326.5 ft; area: 6538.875 ft^2 **28.** diameter: 46.78 ft; radius: 23.39 ft

29. $[-4, \infty)$

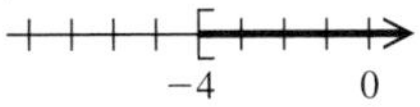

30. $(-\infty, 7)$

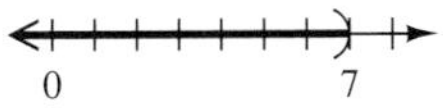

31. $[-5, 6)$

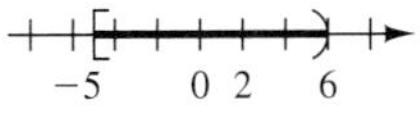

32. $\left[\frac{1}{2}, \infty\right)$

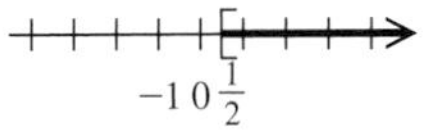

33. $[-3, \infty)$

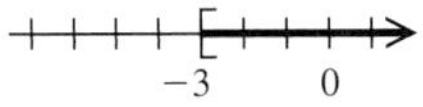

34. $(-\infty, 2)$

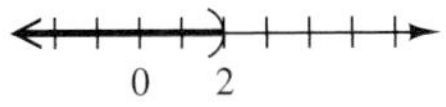

35. $[3, \infty)$

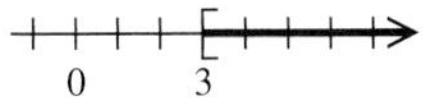

36. $[46, \infty)$

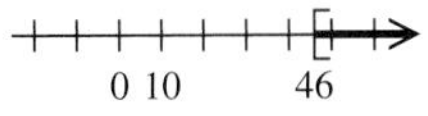

37. $(-\infty, -5)$

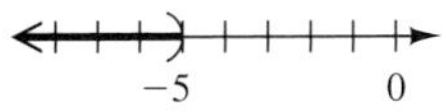

38. $(-\infty, -37)$

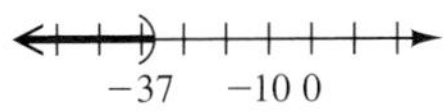

39. $\left[-2, \frac{3}{2}\right)$

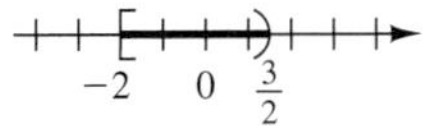

40. $(1, 5]$

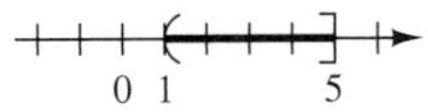

41. 88 or more **42.** all numbers less than or equal to $-\frac{1}{3}$ **43.** $\{7\}$ **44.** $r = \frac{d}{2}$ **45.** $(-\infty, 2)$ **46.** $\{-9\}$ **47.** $\{70\}$ **48.** $\left\{\frac{13}{4}\right\}$ **49.** $\emptyset$ **50.** {all real numbers} **51.** DiGiorno: \$668.7 million; Red Baron: \$268.8 million **52.** 24°, 66° **53.** gold: 14; silver: 7; bronze: 5 **54.** D: 22°; E: 44°; F: 114° **55.** 44 m **56.** 70 ft

Chapter 10 TEST (pages 759–760)

1. $\{6\}$ **2.** $\{-6\}$ **3.** $\left\{\frac{13}{4}\right\}$ **4.** $\{-10.8\}$ **5.** $\{21\}$ **6.** {all real numbers} **7.** $\{30\}$ **8.** $\emptyset$ **9.** wins: 62; losses: 20 **10.** Hawaii: 4021 mi^2; Maui: 728 mi^2; Kauai: 551 mi^2 **11.** 24, 26 **12.** 50° **13.** **(a)** $W = \frac{P - 2L}{2}$ **(b)** 18 **14.** $y = \frac{5}{4}x - 2$ (Other correct forms of the answer are possible.) **15.** 100°, 80° **16.** 75°, 75° **17.** **(a)** $x < 0$ **(b)** $-2 < x \leq 3$

18. $(-\infty, 11)$

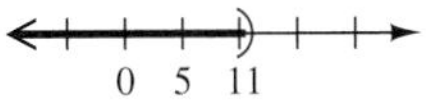

19. $[-3, \infty)$

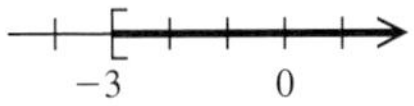

20. $(-\infty, 4]$

21. $(-2, 6]$

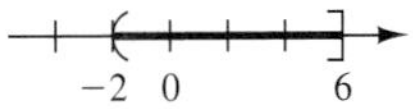

22. 83 or more

CHAPTER 11 Graphs of Linear Equations and Inequalities in Two Variables

SECTION 11.1 (pages 770–775)

1. does; do not **2.** II **3.** y **4.** 3 **5.** 6 **6.** −4 **7.** No. For two ordered pairs to be equal, their x-values must be equal and their y-values must be equal. Here we have $4 \neq -1$ and $-1 \neq 4$. **8.** Substituting $\frac{1}{3}$ for x in $y = 6x + 2$ gives $y = 6\left(\frac{1}{3}\right) + 2$. Here $6\left(\frac{1}{3}\right) = 2$, so $y = 2 + 2$, or 4. Because $6\left(\frac{1}{7}\right) = \frac{6}{7}$, calculating y for $x = \frac{1}{7}$ requires working with fractions. **9.** negative; negative **10.** negative; positive **11.** positive; negative **12.** positive; positive **13.** 2007, 2008, 2009, 2010 **15.** 2004: about 171 billion lb; 2010: about 193 billion lb **16.** U.S. milk production increased about 22 billion lb during these years, from about 171 billion lb to about 193 billion lb. **17.** 2009 to 2010; about 1.5 million **19.** The number of cars imported each year was decreasing. **21.** yes **23.** yes **25.** no **27.** yes **29.** no **31.** no **33.** 11 **35.** $-\frac{7}{2}$ **37.** −4 **39.** −5 **41.** 4, 6, −6; $(0, 4)$, $(6, 0)$, $(-6, 8)$ **43.** 3, −5, −15; $(0, 3)$, $(-5, 0)$, $(-15, -6)$ **45.** −9, −9, −9; $(-9, 6)$, $(-9, 2)$, $(-9, -3)$ **47.** −6, −6, −6; $(8, -6)$, $(4, -6)$, $(-2, -6)$ **49.** 8, 8, 8; $(8, 8)$, $(8, 3)$, $(8, 0)$ **51.** −2, −2, −2; $(9, -2)$, $(2, -2)$, $(0, -2)$ **53.** $(2, 4)$; I **55.** $(-5, 4)$; II **57.** $(3, 0)$; no quadrant **59.** If $xy < 0$, then either $x < 0$ and $y > 0$ or $x > 0$ and $y < 0$. If $x < 0$ and $y > 0$, then the point lies in quadrant II. If $x > 0$ and $y < 0$, then the point lies in quadrant IV. **60.** If $xy > 0$, then either $x > 0$ and $y > 0$ or $x < 0$ and $y < 0$. If $x > 0$ and $y > 0$, then the point lies in quadrant I. If $x < 0$ and $y < 0$, then the point lies in quadrant III.

61–72.

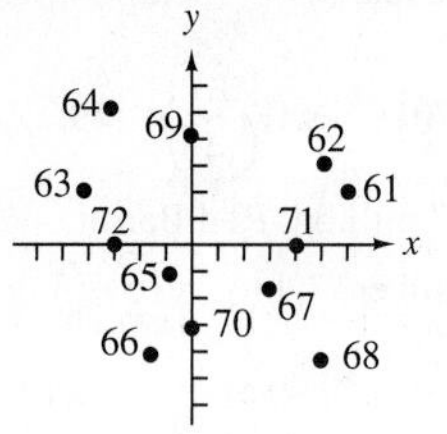

73. −3, 6, −2, 4

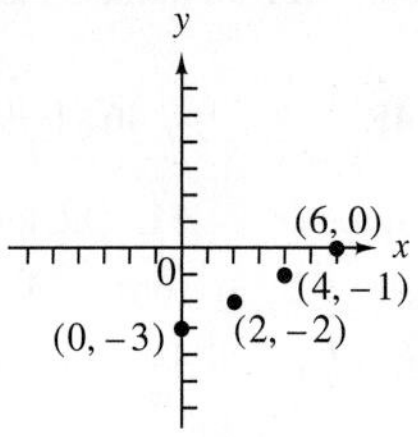

75. $-3, 4, -6, -\frac{4}{3}$

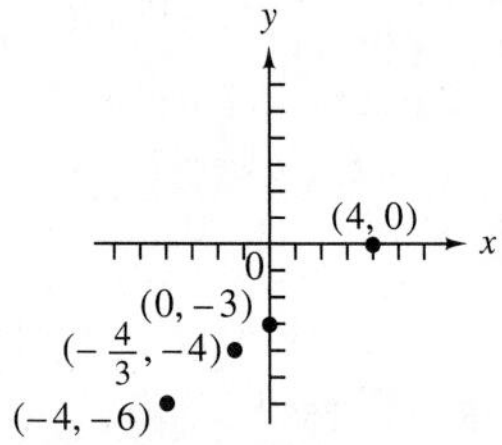

77. −4, −4, −4, −4

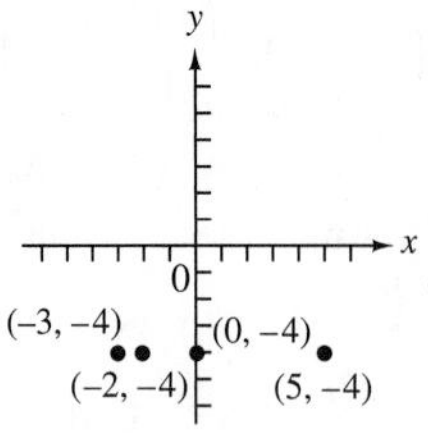

79. The points in each graph appear to lie on a straight line.
80. **(a)** They are the same (all −4). **(b)** They are the same (all 5).
81. **(a)** (5, 45) **(b)** (6, 50) **83.** **(a)** (2007, 27.1), (2008, 29.3), (2009, 28.3), (2010, 28.0), (2011, 26.9) **(b)** (2000, 32.4) means that 32.4 percent of 2-year college students in 2000 received a degree within 3 years.

(c) **2-YEAR COLLEGE STUDENTS COMPLETING A DEGREE WITHIN 3 YEARS**

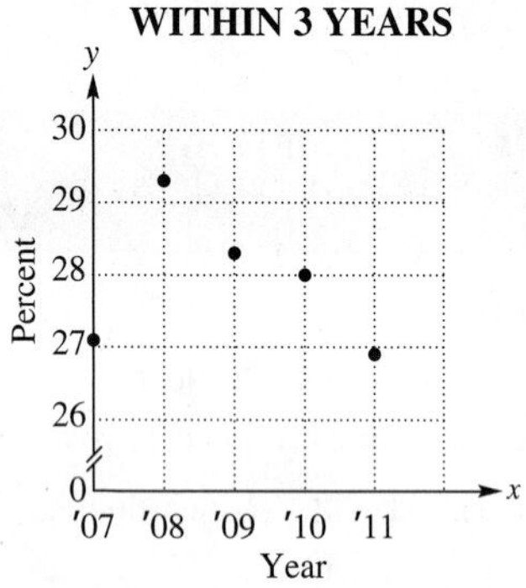

(d) With the exception of the point for 2007, the points lie approximately on a straight line. Rates at which 2-year college students complete a degree within 3 years were generally decreasing.
85. **(a)** 98, 88, 78, 68 **(b)** (20, 98), (40, 88), (60, 78), (80, 68)

(c) **TARGET HEART RATE ZONE (Lower Limit)**

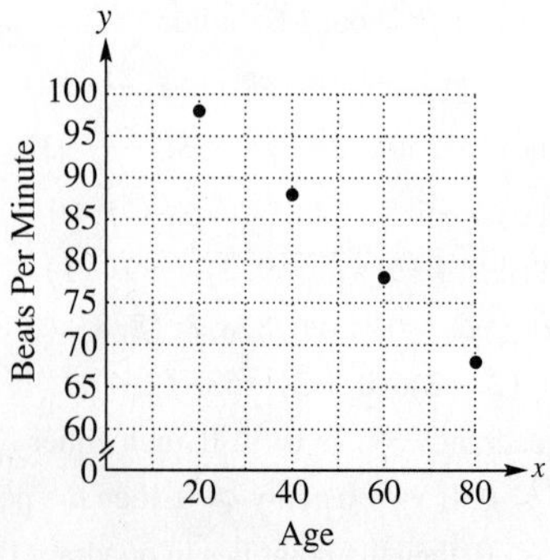

Yes, the points lie in a linear pattern.
87. between 98 and 157 beats per minute; between 88 and 141 beats per minute

SECTION 11.2 (pages 784–789)

1. *By*; *C*; 0 **2.** line; solution **3.** **(a)** A **(b)** C **(c)** D **(d)** B
4. A, C, D **5.** x-intercept: (4, 0); y-intercept: (0, −4)
6. x-intercept: (−5, 0); y-intercept: (0, 5) **7.** x-intercept: (−2, 0); y-intercept: (0, −3) **8.** x-intercept: (4, 0); y-intercept: (0, 3)

9. 5, 5, 3

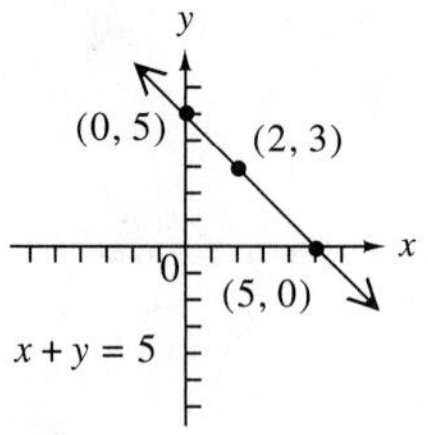

11. 1, 3, −1

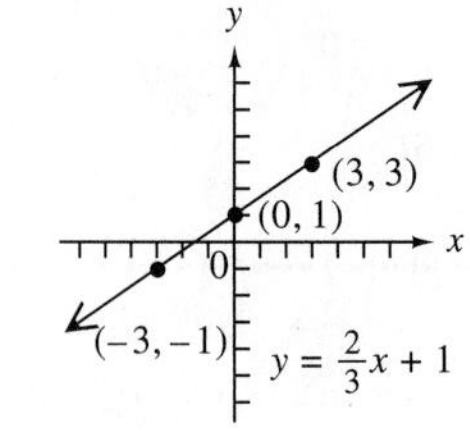

13. −6, −2, −5

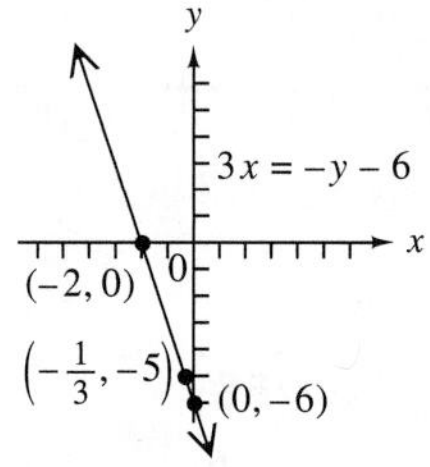

15. 0; (0, −8); 0; (8, 0)
17. (0, −8); (12, 0)
19. (0, 0); (0, 0)

21.

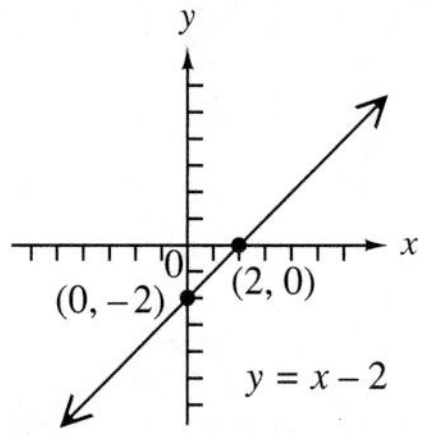

23.

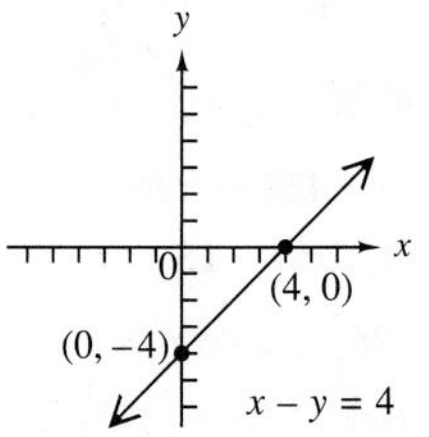

25.

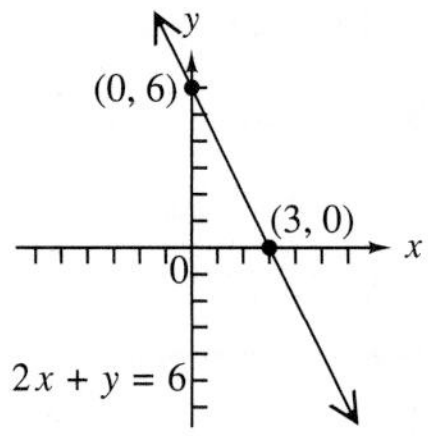

27.

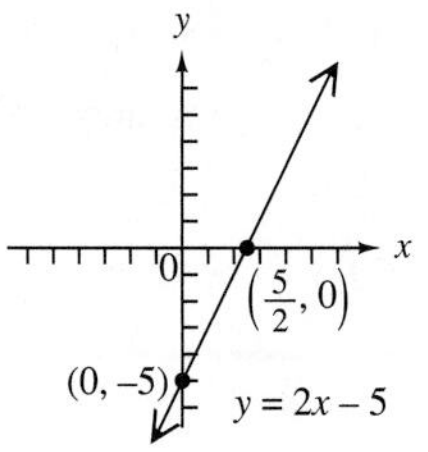

29.

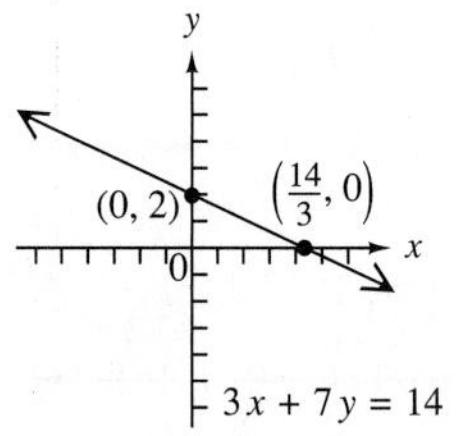

31.

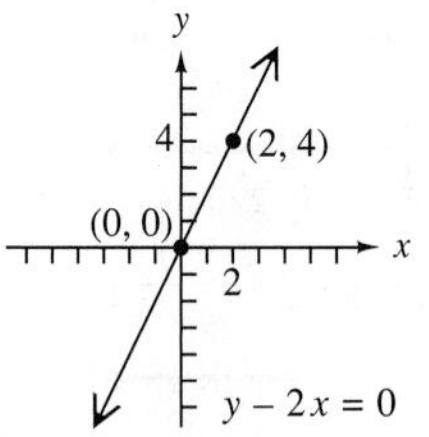

33.

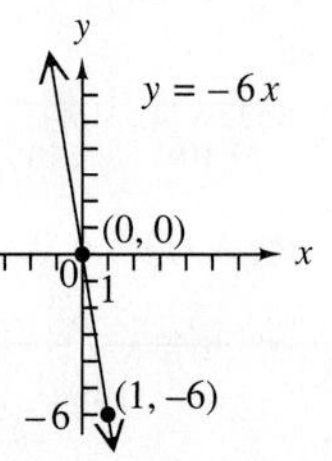

35.

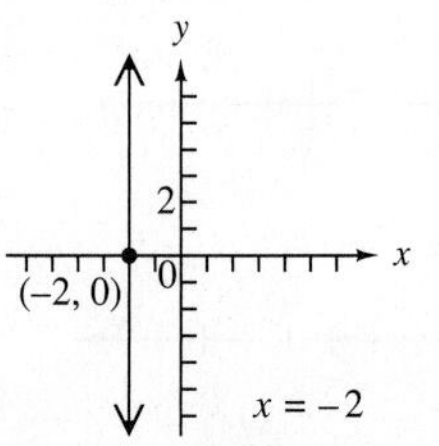

37.

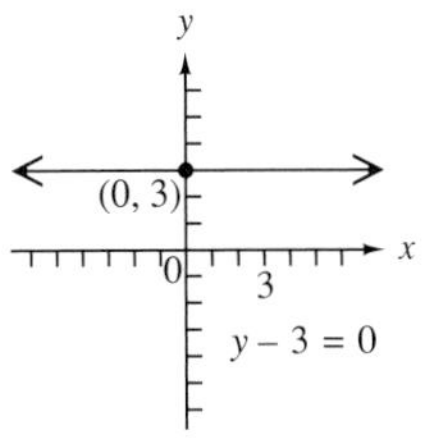

39.

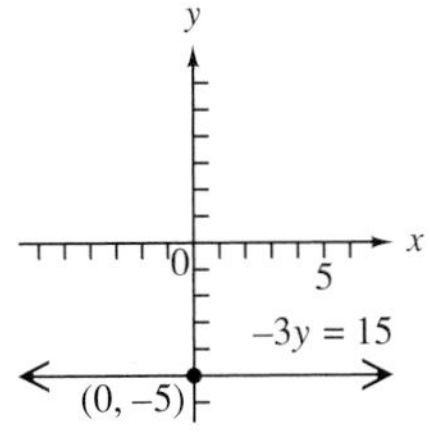

41. **(a)** D **(b)** C **(c)** B **(d)** A **42.** $y = 0$; $x = 0$

In Exercises 43–46, descriptions may vary.

43. The graph is a line with x-intercept $(-3, 0)$ and y-intercept $(0, 9)$. **44.** The graph is a vertical line with x-intercept $(11, 0)$. **45.** The graph is a horizontal line with y-intercept $(0, -2)$. **46.** The graph passes through the origin $(0, 0)$ and the points $(2, 1)$ and $(4, 2)$.

47. **(a)** 151.5 cm, 159.3 cm, 174.9 cm
(b) $(20, 151.5)$, $(22, 159.3)$, $(26, 174.9)$
(c) **HEIGHTS OF WOMEN**

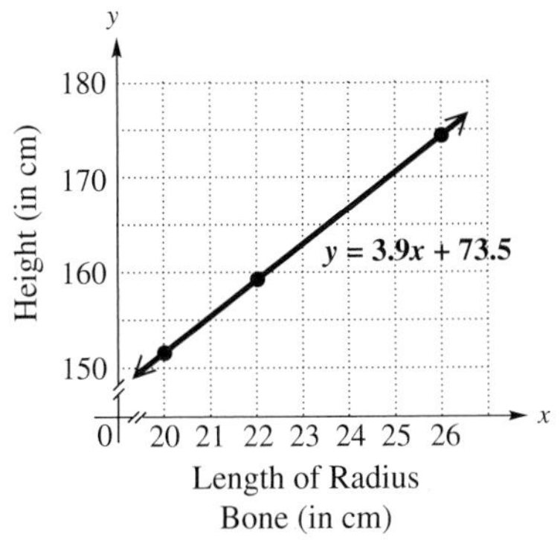

(d) 24 cm; 24 cm
49. **(a)** \$62.50; \$100 **(b)** 200 **(c)** $(50, 62.50)$, $(100, 100)$, $(200, 175)$
(d) **POSTER COSTS**

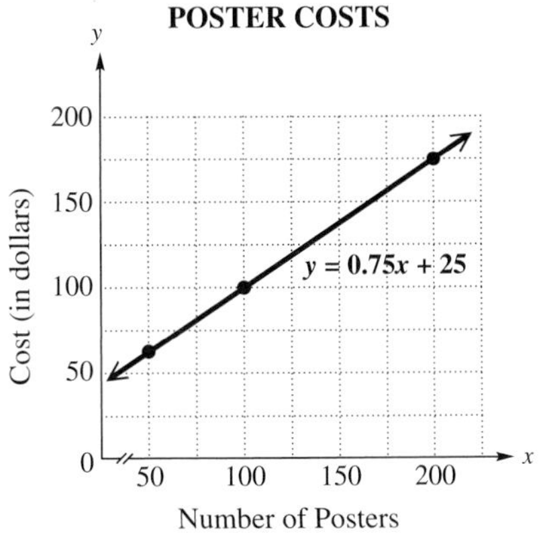

51. **(a)** \$30,000 **(b)** \$15,000 **(c)** \$5000 **(d)** After 5 yr, the SUV has a value of \$5000. **53.** **(a)** 1990: 23.8 lb; 2000: 29.0 lb; 2009: 33.8 lb **(b)** 1990: 25 lb; 2000: 30 lb; 2009: 33 lb **(c)** The values are quite close.

SECTION 11.3 (pages 798–801)

1. steepness; vertical; horizontal **2.** ratio; y; rise; x; run **3.** **(a)** 6 **(b)** 4 **(c)** $\frac{6}{4}$, or $\frac{3}{2}$; slope of the line **(d)** Yes, it doesn't matter which point we start with. The slope would be expressed as the quotient of -6 and -4, which simplifies to $\frac{3}{2}$.

4. **(a)** C **(b)** A **(c)** D **(d)** B **5.** 4 **7.** $-\frac{1}{2}$

9. Answers will vary.

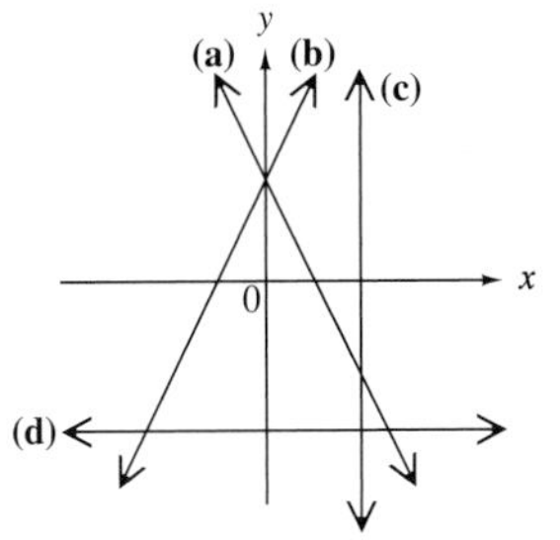

10. **(a)** falls from left to right **(b)** horizontal **(c)** vertical **(d)** rises from left to right **11.** **(a)** negative **(b)** 0 **12.** **(a)** negative **(b)** negative **13.** **(a)** positive **(b)** negative **14.** **(a)** positive **(b)** 0 **15.** **(a)** 0 **(b)** negative **16.** **(a)** 0 **(b)** positive **17.** $\frac{8}{27}$ **18.** $\frac{3}{10}$

19. $-\frac{2}{3}$ **20.** $-\frac{1}{4}$ **21.** Because the student found the difference $3 - 5 = -2$ in the numerator, he or she should have subtracted in the same order in the denominator to get $-1 - 2 = -3$. The correct slope is $\frac{-2}{-3} = \frac{2}{3}$.

22. Slope is defined as $\frac{\text{change in } y}{\text{change in } x}$, but the student found $\frac{\text{change in } x}{\text{change in } y}$. The correct slope is $\frac{-5}{8} = -\frac{5}{8}$. **23.** $\frac{5}{4}$ **25.** $\frac{3}{2}$ **27.** -3 **29.** 0

31. undefined **33.** $\frac{1}{4}$ **35.** $-\frac{1}{2}$ **37.** 5 **39.** $\frac{1}{4}$ **41.** $\frac{3}{2}$ **43.** $-\frac{3}{2}$

45. 0 **47.** undefined **49.** 1 **51.** $\frac{4}{3}$; $\frac{4}{3}$; parallel **53.** $\frac{5}{3}$; $\frac{3}{5}$; neither

55. $\frac{3}{5}$; $-\frac{5}{3}$; perpendicular **57.** 0.42 **58.** positive; increased **59.** 0.42%

60. -0.3 **61.** negative; decreased **62.** 0.3%

SECTION 11.4 (pages 809–814)

1. **(a)** D **(b)** C **(c)** B **(d)** A **2.** A, B, D **3.** **(a)** C **(b)** B **(c)** A **(d)** D **4.** **(a)** the y-axis **(b)** the x-axis **5.** slope: $\frac{5}{2}$; y-intercept: $(0, -4)$ **7.** slope: -1; y-intercept: $(0, 9)$ **9.** slope: $\frac{1}{5}$; y-intercept: $\left(0, -\frac{3}{10}\right)$ **11.** $y = 4x - 3$ **13.** $y = -x - 7$ **15.** $y = 3$

17. $x = 0$ **19.** $y = 3x - 3$ **20.** $y = 2x - 4$ **21.** $y = -x + 3$

22. $y = -x - 2$ **23.** $y = -\frac{1}{2}x + 2$ **24.** $y = \frac{3}{2}x - 3$

25.

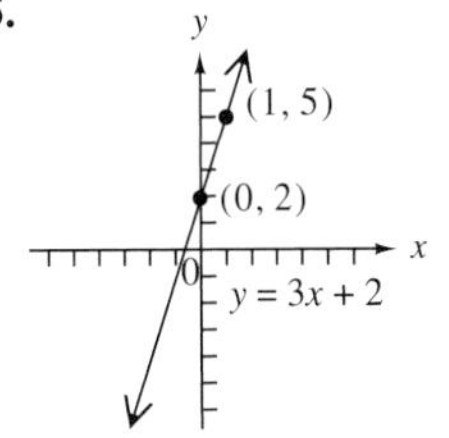

27.

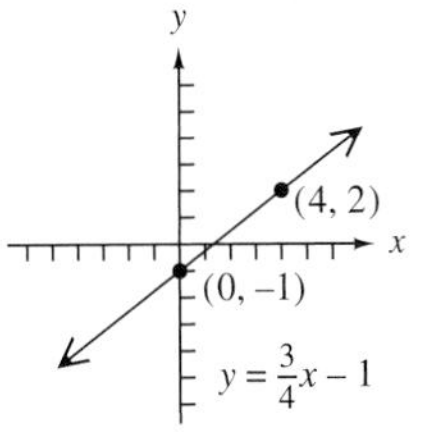

29.

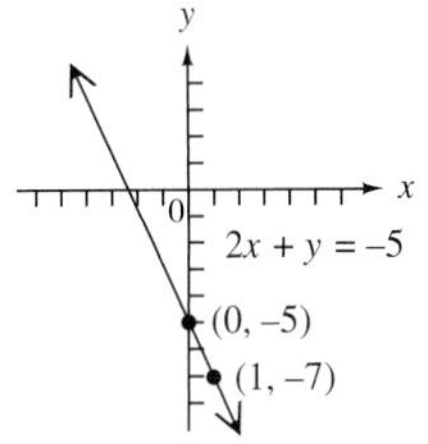

31.

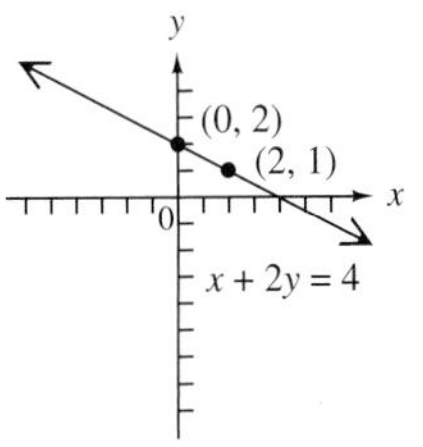

ANSWERS

33. $y = \frac{1}{2}x + 4$

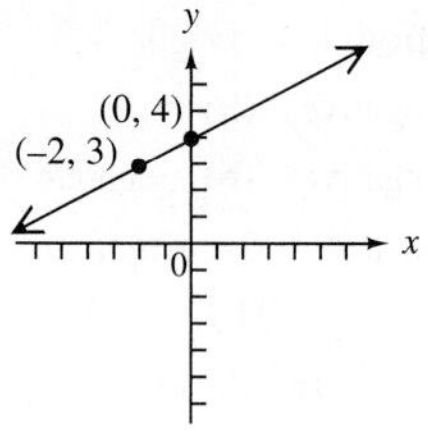

35. $y = -\frac{2}{5}x - \frac{23}{5}$

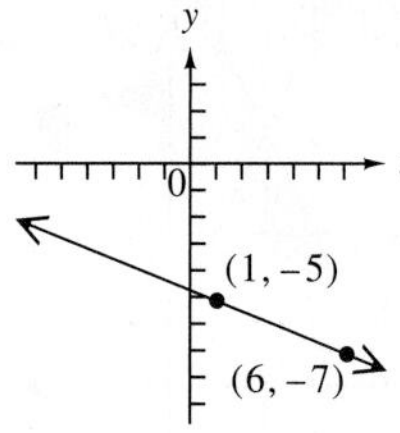

37. $y = 2$

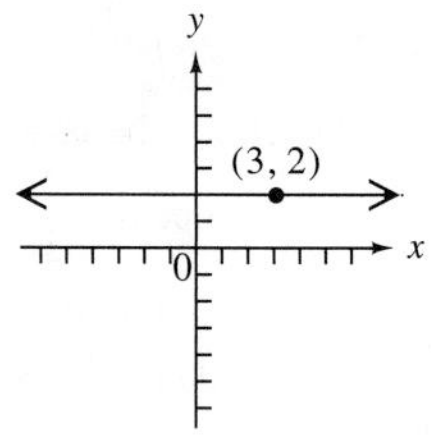

39. $x = 3$ (no slope-intercept form)

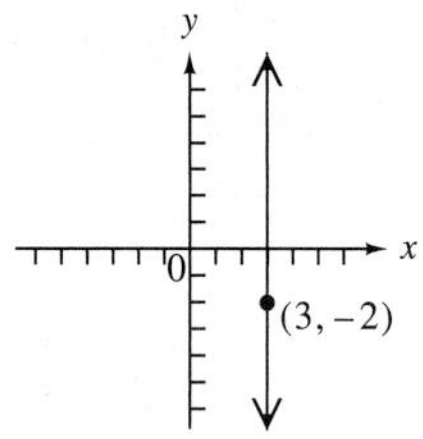

41. $y = \frac{2}{3}x$

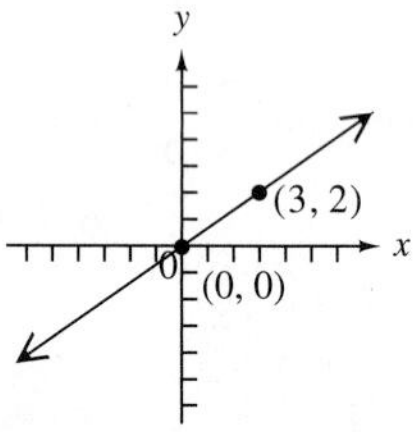

43. $y = 2x - 7$ **45.** $y = -2x - 4$ **47.** $y = \frac{2}{3}x + \frac{19}{3}$ **49.** **(a)** $y = x - 3$ **(b)** $x - y = 3$ **51.** **(a)** $y = -\frac{5}{7}x - \frac{54}{7}$ **(b)** $5x + 7y = -54$ **53.** **(a)** $y = -\frac{2}{3}x - 2$ **(b)** $2x + 3y = -6$ **55.** **(a)** $y = \frac{1}{3}x + \frac{4}{3}$ **(b)** $x - 3y = -4$ **57.** $y = -2x - 3$

59. $y = 4x - 5$ **61.** $y = \frac{3}{4}x - \frac{9}{2}$ **63.** **(a)** \$400 **(b)** \$0.25 **(c)** $y = 0.25x + 400$ **(d)** \$425 **(e)** 1500 **65.** **(a)** (1, 2294), (2, 2372), (3, 2558), (4, 2727), (5, 2963) **(b)** yes

AVERAGE ANNUAL COSTS AT 2-YEAR COLLEGES

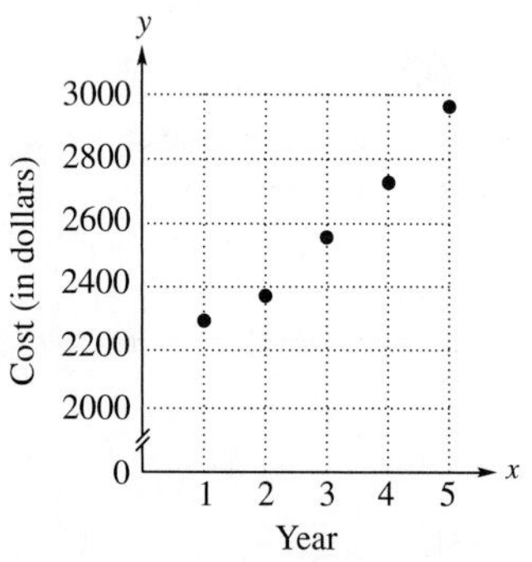

(c) $y = 197x + 1978$ **(d)** \$3160 **67.** $y = 15x - 29{,}978$

69. (0, 32); (100, 212) **70.** $\frac{9}{5}$ **71.** $F - 32 = \frac{9}{5}(C - 0)$

72. $F = \frac{9}{5}C + 32$ **73.** $C = \frac{5}{9}(F - 32)$ **74.** 86° **75.** 10° **76.** −40°

SECTION 11.5 (pages 819–821)

1. >, > **2.** < **3.** ≤ **4.** ≥ **5.** false; The point (4, 0) lies on the boundary line $3x - 4y = 12$, which is *not* part of the graph because the symbol < does not involve equality. **6.** true **7.** false; Because (0, 0) is on the boundary line $x + 4y = 0$, it cannot be used as a test point. Use a test point *off* the line.

8. false; Use a solid line for the boundary line because of the equality portion of the ≥ symbol. **9.** true **10.** true

11.

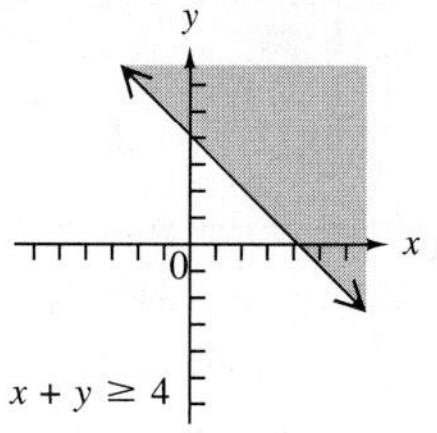

13.

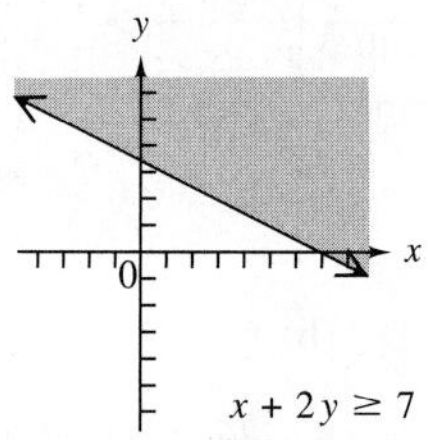

15.

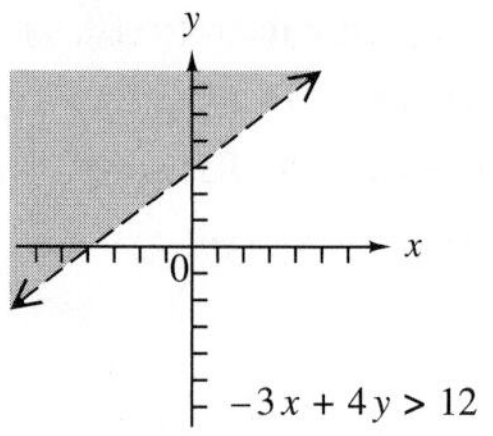

17.

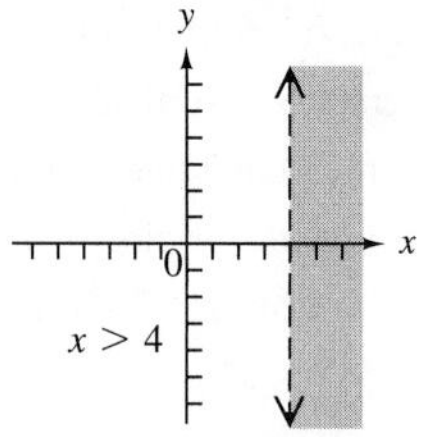

19.

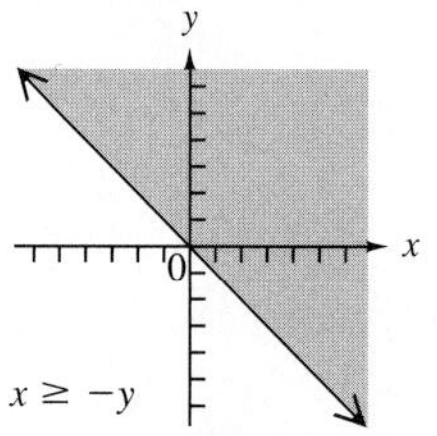

21. Use a dashed line if the symbol is < or >. Use a solid line if the symbol is ≤ or ≥.

22. A test point cannot lie on the boundary line. It must lie on one side of the boundary.

23.

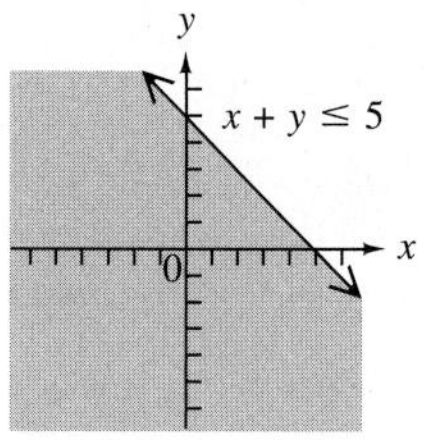

25.

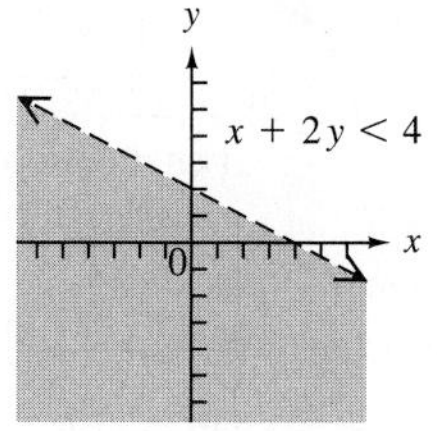

27.

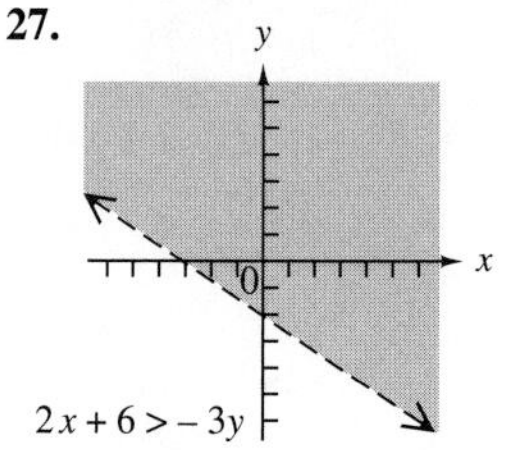

29.

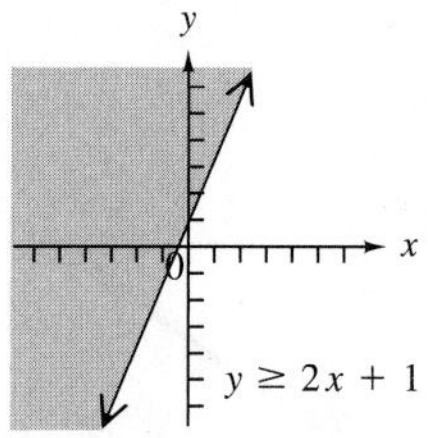

31.

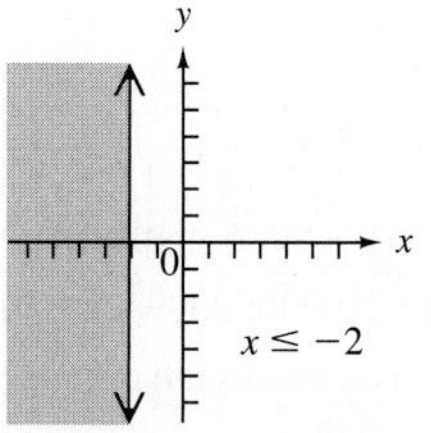

33.

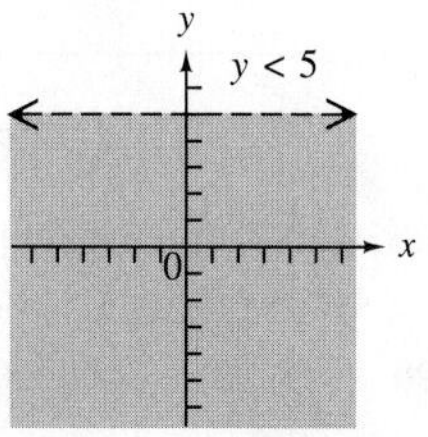

35.

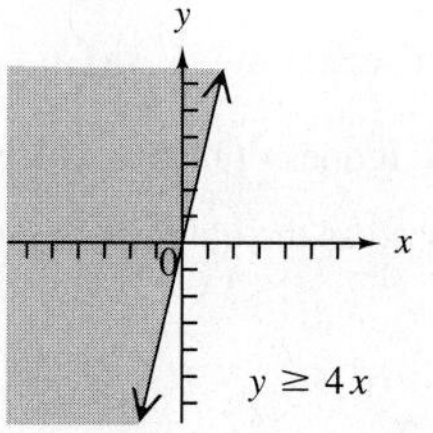

37.

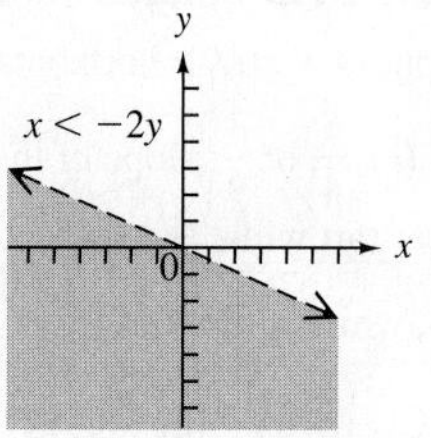

Chapter 11 REVIEW EXERCISES (pages 826–830)

1. (2006, 52.5), (2011, 55.4) **2.** In the year 2000, 52.0% of first-year college students at two-year public institutions returned for a second year. **3.** 2008 to 2009; about 53.7% **4.** 2007 to 2008; 2009 to 2010; about 2% increase each time **5.** -1; 2; 1 **6.** 2; $\frac{3}{2}$; $\frac{14}{3}$ **7.** 0; $\frac{8}{3}$; -9 **8.** 7; 7; 7 **9.** yes **10.** no **11.** yes **12.** no **13.** quadrant I **14.** quadrant II **15.** no quadrant **16.** no quadrant **17.** x is positive in quadrants I and IV; y is negative in quadrants III and IV. Thus, if x is positive and y is negative, (x, y) must lie in quadrant IV. **18.** In the ordered pair $(k, 0)$, the y-value is 0, so the point lies on the x-axis. In the ordered pair $(0, k)$, the x-value is 0, so the point lies on the y-axis.

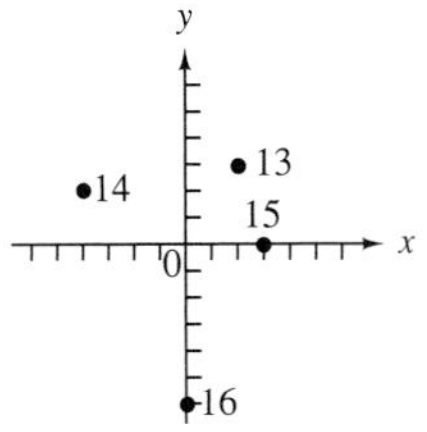

Graph for Exercises 13–16

19.
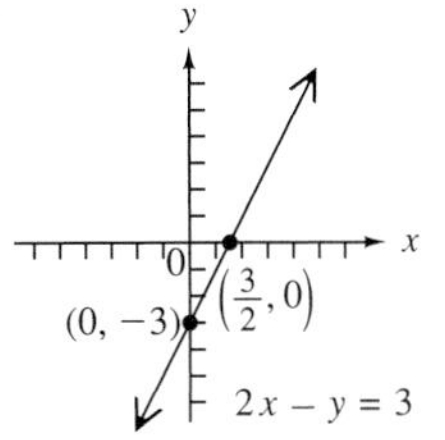

20.
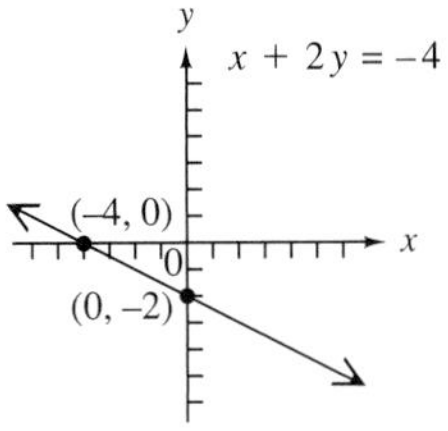

21.
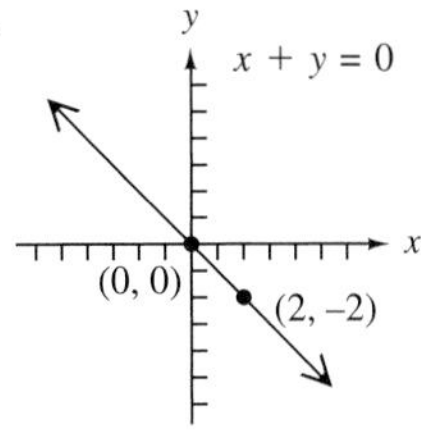

22.
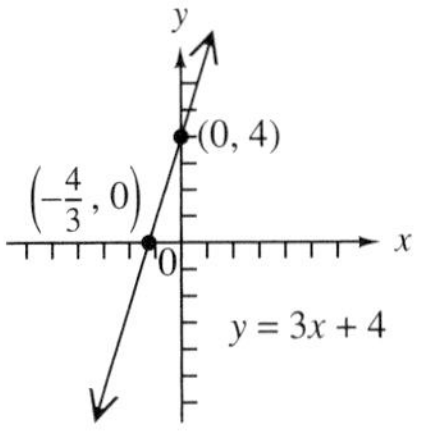

23. $-\frac{1}{2}$ **24.** $-\frac{2}{3}$ **25.** 0 **26.** undefined **27.** 3 **28.** $\frac{2}{3}$ **29.** $\frac{3}{2}$ **30.** $-\frac{1}{3}$ **31.** $\frac{3}{2}$ **32.** undefined **33.** 0 **34. (a)** 2 **(b)** $\frac{1}{3}$ **35.** parallel **36.** perpendicular **37.** neither **38.** 0 **39.** $y = -x + \frac{2}{3}$ **40.** $y = -\frac{1}{3}x + 1$ **41.** $y = x - 7$ **42.** $y = \frac{2}{3}x + \frac{14}{3}$ **43.** $y = -\frac{3}{4}x - \frac{1}{4}$ **44.** $y = -\frac{1}{4}x + \frac{3}{2}$ **45.** $y = 1$ **46.** $x = \frac{1}{3}$ **47. (a)** $y = -\frac{1}{3}x + 5$ **(b)** slope: $-\frac{1}{3}$; y-intercept: $(0, 5)$

(c)
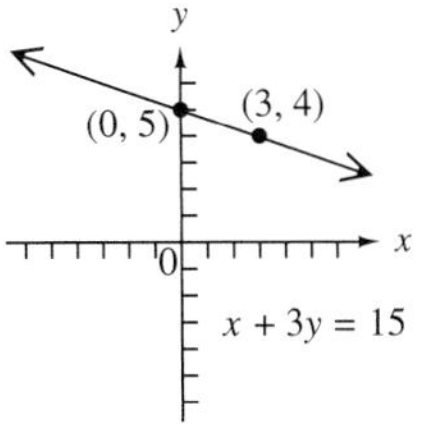

48.
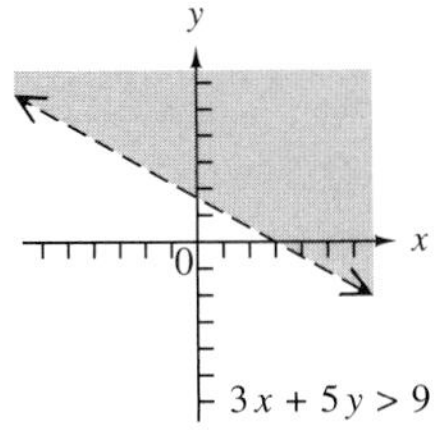

49.
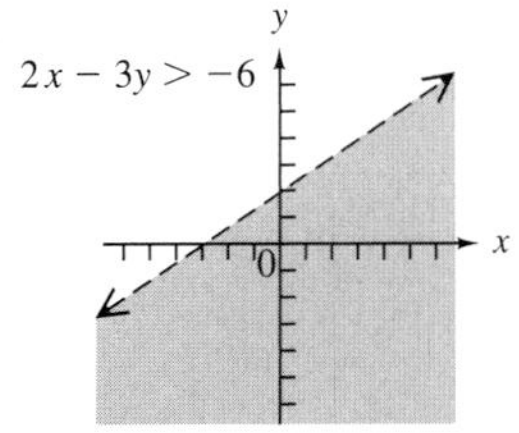

50.
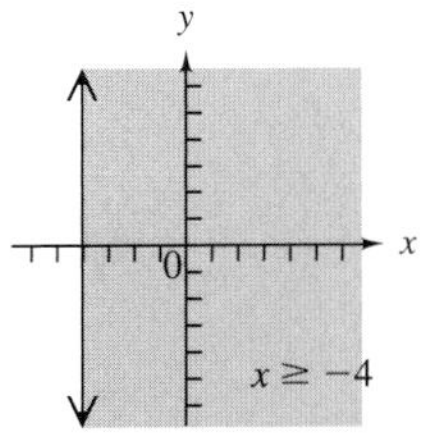

51. A **52.** C, D **53.** A, B, D **54.** D **55.** C **56.** B

57. $(0, -5)$; $\left(-\frac{5}{2}, 0\right)$; -2

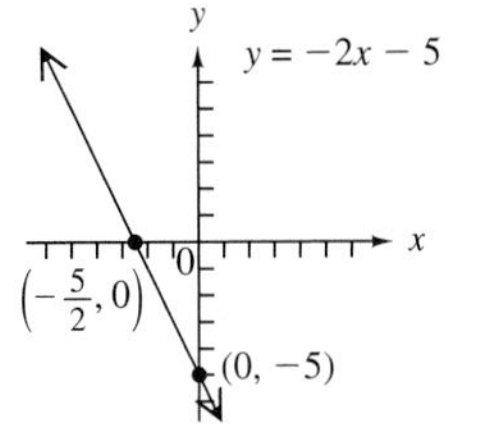

58. $(0, 0)$; $(0, 0)$; $-\frac{1}{3}$

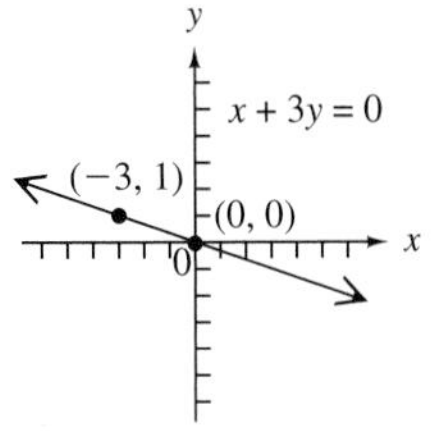

59. $(0, 5)$; none; 0

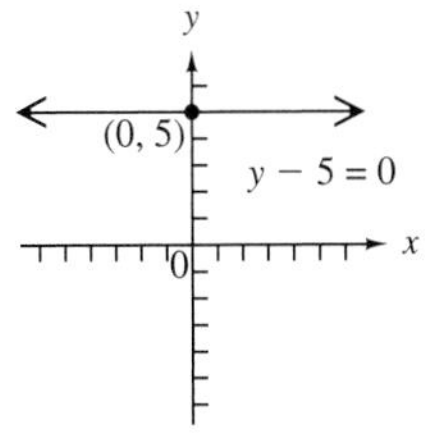

60. none; $(-1, 0)$; undefined

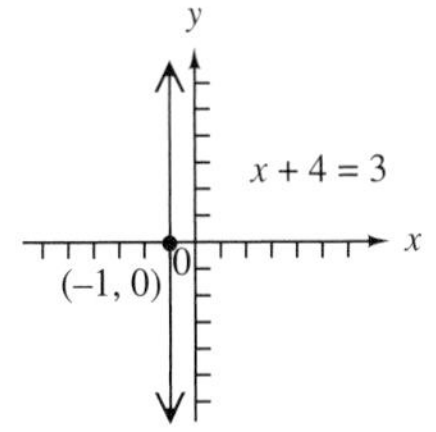

61. (a) $y = -\frac{1}{4}x - \frac{5}{4}$ **(b)** $x + 4y = -5$ **62. (a)** $y = -3x + 30$ **(b)** $3x + y = 30$ **63. (a)** $y = -\frac{4}{7}x - \frac{23}{7}$ **(b)** $4x + 7y = -23$

64.
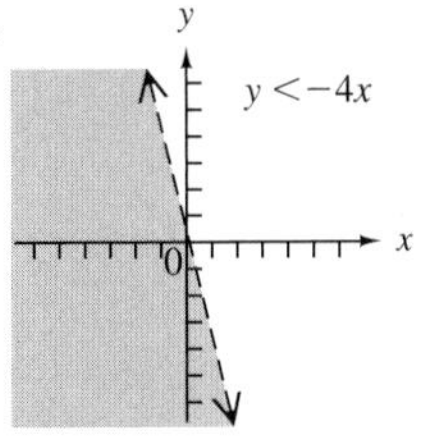

65.
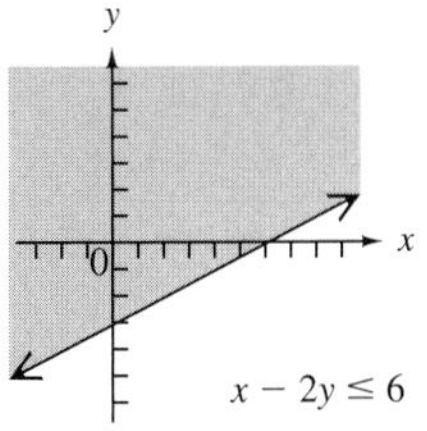

66. 1.7% **67.** Since the graph rises from left to right, the slope is positive. **68.** (2004, 42.3), (2009, 44.0) **69.** $y = 0.34x - 639.06$ **70.** 0.34; yes **71.** 42.6, 43.0, 43.3, 43.7 **72.** 44.3%; No. The equation is based on data only from 2004 through 2009.

Chapter 11 TEST (pages 831–832)

1. between 2003 and 2004, 2004 and 2005, and 2005 and 2006 **2.** The unemployment rate was increasing. **3.** 2008: 5.8%; 2009: 9.3%; increase: 3.5%

4. $(0, 6)$; $(2, 0)$

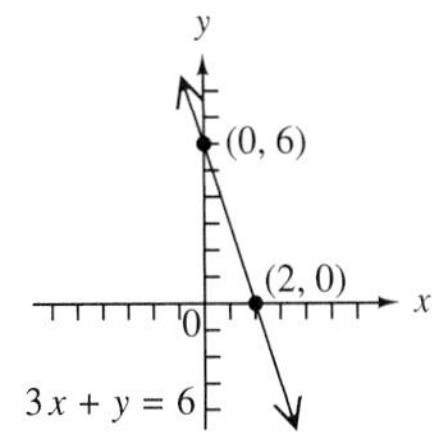

5. $(0, 0)$; $(0, 0)$

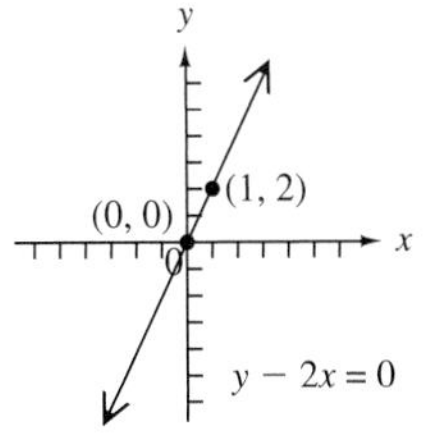

6. none; $(-3, 0)$

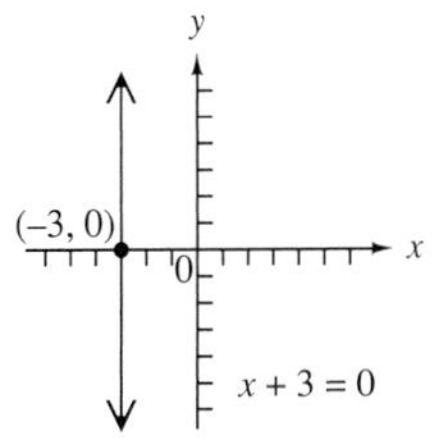

7. $(0, 1)$; none

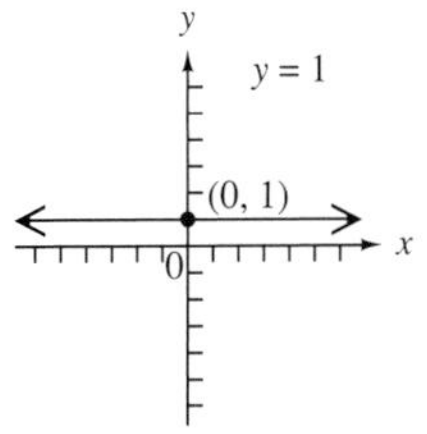

ANSWERS

8. (0, −4); (4, 0)

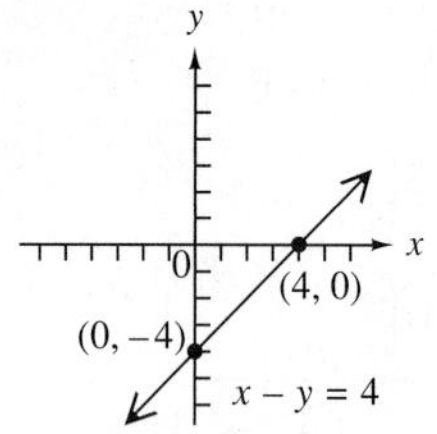

9. $-\frac{8}{3}$ **10.** −2 **11.** undefined **12.** $\frac{5}{2}$ **13.** 0 **14.** $y = 2x + 6$ **15.** $y = \frac{5}{2}x - 4$ **16.** $y = -9x + 12$ **17.** $y = -\frac{3}{2}x + \frac{9}{2}$

18.

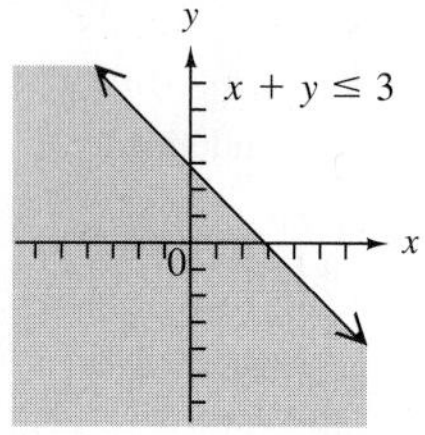

19.

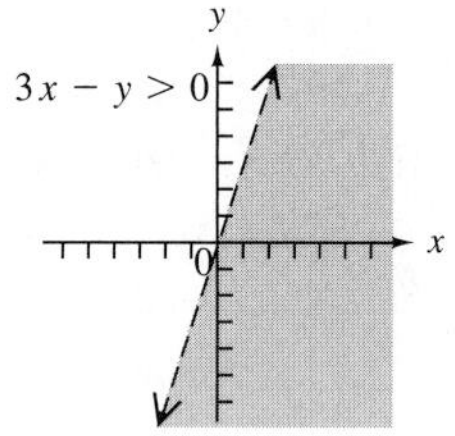

20. The slope is negative since sales are decreasing. **21.** (0, 209), (11, 123); −7.8 **22.** $y = -7.8x + 209$ **23.** 154.4 thousand; The equation gives a number that is lower than the actual sales. **24.** In 2011, worldwide snowmobile sales were 123 thousand.

CHAPTER 12 Exponents and Polynomials

SECTION 12.1 (pages 839–842)

1. 7; 5 **2.** two **3.** 8 **4.** is not **5.** 26 **6.** $5x^9$ **7.** 1; 6 **9.** 1; 1 **11.** 1; $\frac{1}{5}$ **13.** 2; −19, −1 **15.** 3; 1, −8, $\frac{2}{3}$ **17.** $2m^5$ **19.** $-r^5$ **21.** $\frac{2}{3}x^4$ **23.** cannot be simplified; $0.2m^5 - 0.5m^2$ **25.** $-5x^5$ **27.** $5p^9 + 4p^7$ **29.** $-2y^2$ **31.** already simplified; 4; binomial **33.** already simplified; $6m^5 + 5m^4 - 7m^3 - 3m^2$; 5; none of these **35.** $x^4 + \frac{1}{3}x^2 - 4$; 4; trinomial **37.** 7; 0; monomial **39.** $1.5x^2 - 0.5x$; 2; binomial **41.** (a) −1 (b) 5 **43.** (a) 19 (b) −2 **45.** (a) 36 (b) −12 **47.** (a) −124 (b) 5 **49.** $5m^2 + 3m$ **51.** $4x^4 - 4x^2$ **53.** $\frac{7}{6}x^2 - \frac{2}{15}x + \frac{5}{6}$ **55.** $12m^3 - 13m^2 + 6m + 11$ **57.** $2.9x^3 - 3.5x^2 - 1.5x - 9$ **59.** $8r^2 + 5r - 12$ **61.** $5m^2 - 14m + 6$ **63.** $4x^3 + 2x^2 + 5x$ **65.** $-18y^5 + 7y^4 + 5y^3 + 3y^2 + y$ **67.** $-2m^3 + 7m^2 + 8m - 9$ **69.** $-11x^2 - 3x - 3$ **71.** $2x^2 + 8x$ **73.** $8x^2 + 8x + 6$ **75.** $8t^2 + 8t + 13$ **77.** $13a^2b - 7a^2 - b$ **79.** $c^4d - 5c^2d^2 + d^2$ **81.** $12m^3n - 11m^2n^2 - 4mn^2$ **83.** (a) $23y + 5t$ (b) 25°, 67°, 88° **85.** 5; 175 **86.** 87 ft; (1, 87) **87.** 6; \$27 **88.** 2.5; 130

SECTION 12.2 (pages 849–850)

1. false **2.** true **3.** false **4.** true **5.** 1 **6.** three squared; five cubed; seven to the fourth power **7.** t^7 **9.** $\left(\frac{1}{2}\right)^5$ **11.** $(-8p)^2$ **13.** base: 3; exponent: 5; 243 **15.** base: −3; exponent: 5; −243 **17.** base: $-6x$; exponent: 4 **19.** base: x; exponent: 4 **21.** 5^8 **23.** 4^{12} **25.** $(-7)^9$ **27.** t^{24} **29.** $-56r^7$ **31.** $42p^{10}$ **33.** The product rule does not apply. **35.** The product rule does not apply. **37.** 4^6 **39.** t^{20} **41.** $343r^3$ **43.** 5^{12} **45.** -8^{15} **47.** $5^5x^5y^5$ **49.** $8q^3r^3$ **51.** $\frac{1}{8}$ **53.** $\frac{a^3}{b^3}$ **55.** $\frac{9^8}{5^8}$ **57.** $-8x^6y^3$ **59.** $9a^6b^4$ **61.** $\frac{5^5}{2^5}$ **63.** $\frac{9^5}{8^3}$ **65.** $2^{12}x^{12}$ **67.** -6^5p^5 **69.** $6^5x^{10}y^{15}$ **71.** x^{21} **73.** $4w^4x^{26}y^7$ **75.** $-r^{18}s^{17}$ **77.** $\frac{125a^6b^{15}}{c^{18}}$ **79.** $25m^6p^{14}q^5$ **81.** $16x^{10}y^{16}z^{10}$ **83.** Using power rule (a), simplify as follows: $(10^2)^3 = 10^{2 \cdot 3} = 10^6 = 1{,}000{,}000$. **84.** The 4 is used as an *exponent* on 3. It is *not* multiplied by 3. The correct simplification is $3^4 \cdot x^8 \cdot y^{12} = \mathbf{81}x^8y^{12}$. **85.** $30x^7$ **87.** $6p^7$

SECTION 12.3 (pages 855–857)

1. (a) B (b) D (c) A (d) C **2.** (a) C (b) A (c) B (d) D **3.** distributive **4.** binomials **5.** $15pq^2$ **7.** $-18m^3n^2$ **9.** $9y^{10}$ **11.** $-8x^{10}$ **13.** $-6m^2 - 4m$ **15.** $6p - \frac{9}{2}p^2 + 9p^4$ **17.** $10y^9 + 4y^6 + 6y^5$ **19.** $6y^6 + 4y^4 + 2y^3$ **21.** $28r^5 - 32r^4 + 36r^3$ **23.** $6a^4 - 12a^3b + 15a^2b^2$ **25.** $3m^2$; $2mn$; n^3; $21m^5n^2 + 14m^4n^3 - 7m^3n^5$ **27.** $12x^3 + 26x^2 + 10x + 1$ **29.** $6r^3 + 5r^2 - 12r + 4$ **31.** $20m^4 - m^3 - 8m^2 - 17m - 15$ **33.** $5x^4 - 13x^3 + 20x^2 + 7x + 5$ **35.** $3x^5 + 18x^4 - 2x^3 - 8x^2 + 24x$ **37.** first row: x^2; $4x$; second row: $3x$; 12; Product: $x^2 + 7x + 12$ **39.** first row: $2x^3$; $6x^2$; $4x$; second row: x^2; $3x$; 2; Product: $2x^3 + 7x^2 + 7x + 2$ **41.** $m^2 + 12m + 35$ **43.** $n^2 + n - 6$ **45.** $8r^2 - 10r - 3$ **47.** $9x^2 - 4$ **49.** $9q^2 + 6q + 1$ **51.** $15xy - 40x + 21y - 56$ **53.** $6t^2 + 23st + 20s^2$ **55.** $-0.3t^2 + 0.22t + 0.24$ **57.** $x^2 - \frac{5}{12}x - \frac{1}{6}$ **59.** $\frac{15}{16} - \frac{1}{4}r - 2r^2$ **61.** $2x^3 + x^2 - 15x$ **63.** $6y^5 - 21y^4 - 45y^3$ **65.** $-200r^7 + 32r^3$ **67.** (a) $3y^2 + 10y + 7$ (b) $8y + 16$ **69.** $30x + 60$ **70.** $30x + 60 = 600$; {18} **71.** 10 yd by 60 yd **72.** 140 yd **73.** \$2100 **74.** \$1260

SECTION 12.4 (pages 862–864)

1. square; twice; product; square **2.** positive; negative **3.** difference; squares **4.** conjugates **5.** (a) $4x^2$ (b) $12x$ (c) 9 (d) $4x^2 + 12x + 9$ **7.** $p^2 + 4p + 4$ **9.** $z^2 - 10z + 25$ **11.** $x^2 - \frac{3}{2}x + \frac{9}{16}$ **13.** $v^2 + 0.8v + 0.16$ **15.** $16x^2 - 24x + 9$ **17.** $100z^2 + 120z + 36$ **19.** $4p^2 + 20pq + 25q^2$ **21.** $0.64t^2 + 1.12ts + 0.49s^2$ **23.** $25x^2 + 4xy + \frac{4}{25}y^2$ **25.** $9t^3 - 6t^2 + t$ **27.** $48t^3 + 24t^2 + 3t$ **29.** $-16r^2 + 16r - 4$ **31.** (a) $49x^2$ (b) 0 (c) $-9y^2$ (d) $49x^2 - 9y^2$; Because 0 is the identity element for addition, it is not necessary to write "+ 0." **33.** $q^2 - 4$ **35.** $r^2 - \frac{9}{16}$ **37.** $s^2 - 6.25$ **39.** $4w^2 - 25$ **41.** $100x^2 - 9y^2$ **43.** $4x^4 - 25$ **45.** $49x^2 - \frac{9}{49}$ **47.** $9p^3 - 49p$ **49.** $m^3 - 15m^2 + 75m - 125$ **51.** $y^3 + 6y^2 + 12y + 8$ **53.** $8a^3 + 12a^2 + 6a + 1$ **55.** $81r^4 - 216r^3t + 216r^2t^2 - 96rt^3 + 16t^4$ **57.** $3x^5 - 27x^4 + 81x^3 - 81x^2$ **59.** $-8x^6y - 32x^5y^2 - 48x^4y^3 - 32x^3y^4 - 8x^2y^5$ **61.** $\frac{1}{2}m^2 - 2n^2$ **63.** $9a^2 - 4$ **65.** $\pi x^2 + 4\pi x + 4\pi$ **67.** $x^3 + 6x^2 + 12x + 8$ **69.** $(a + b)^2$ **70.** a^2 **71.** $2ab$ **72.** b^2 **73.** $a^2 + 2ab + b^2$ **74.** They both represent the area of the entire large square. **75.** 1225 **76.** $30^2 + 2(30)(5) + 5^2$ **77.** 1225 **78.** They are equal.

SECTION 12.5 (pages 872–873)

1. negative **2.** positive **3.** negative **4.** negative **5.** positive **6.** positive **7.** 0 **8.** 0 **9.** **(a)** B **(b)** C **(c)** D **(d)** B **(e)** E **(f)** B **10.** **(a)** C **(b)** F **(c)** F **(d)** B **(e)** E **(f)** B **11.** 1 **13.** 1 **15.** -1 **17.** 0 **19.** 0 **21.** 2 **23.** $\frac{1}{64}$ **25.** 16 **27.** $\frac{49}{36}$ **29.** $\frac{1}{81}$ **31.** $\frac{8}{15}$ **33.** $-\frac{7}{18}$ **35.** $\frac{1}{9}$ **37.** $\frac{1}{6^5}$, or $\frac{1}{7776}$ **39.** 216 **41.** $2r^4$ **43.** $\frac{25}{64}$ **45.** $\frac{p^5}{q^8}$ **47.** r^9 **49.** $\frac{x^5}{6}$ **51.** $3y^2$ **53.** x^3 **55.** $\frac{yz^2}{4x^3}$ **57.** $a+b$ **59.** 343 **61.** $\frac{1}{x^2}$ **63.** $\frac{64x}{9}$ **65.** $\frac{x^2z^4}{y^2}$ **67.** $6x$ **69.** $\frac{1}{m^{10}n^5}$ **71.** $\frac{5}{16x^5}$ **73.** $\frac{36q^2}{m^4p^2}$ **75.** The student attempted to use the quotient rule with unequal bases. The correct way to simplify is $\frac{16^3}{2^2} = \frac{(2^4)^3}{2^2} = \frac{2^{12}}{2^2} = 2^{12-2} = 2^{10} = 1024$.

76. The student incorrectly assumed that the negative exponent indicated a negative number. The correct way to simplify is $5^{-4} = \frac{1}{5^4} = \frac{1}{625}$.

SECTION 12.6 (pages 876–877)

1. $6x^2+8$; 2; $3x^2+4$ **2.** 0 **3.** $3x^2+4$; 2 (These may be reversed.); $6x^2+8$ **4.** is not **5.** To use the method of this section, the divisor must be a monomial. This is true of the first problem, but not the second. **6.** 8; 13; no **7.** 2 **8.** -30 **9.** $2m^2-m$ **11.** $30x^3-10x+5$ **13.** $-4m^3+2m^2-1$ **15.** $4t^4-2t^2+2t$ **17.** $a^4-a+\frac{2}{a}$ **19.** $-2x^3+\frac{2x^2}{3}-x$ **21.** $-9x^2+5x+1$ **23.** $\frac{4x^2}{3}+x-\frac{2}{3x}$ **25.** $27r^4$; $36r^3$; $6r^2$; $3r$; 2; $9r^3-12r^2-2r+1-\frac{2}{3r}$ **27.** $-m^2+3m-\frac{4}{m}$ **29.** $\frac{12}{x}-\frac{6}{x^2}+\frac{14}{x^3}-\frac{10}{x^4}$ **31.** $-4b^2+3ab-\frac{5}{a}$ **33.** $6x-2+\frac{1}{x}$ **35.** $15x^5-35x^4+35x^3$ **36.** $-72y^6+60y^5-24y^4+36y^3-84y^2$

SECTION 12.7 (pages 882–883)

1. The divisor is $2x+5$. The quotient is $2x^3-4x^2+3x+2$. **2.** Stop when the degree of the remainder is less than the degree of the divisor, or when the remainder is 0. **3.** Divide $12m^2$ by $2m$ to get $6m$. **4.** Multiply $6m$ by $2m-3$ to get $12m^2-18m$. **5.** $x+2$ **7.** $2y-5$ **9.** $p-4+\frac{44}{p+6}$ **11.** $r-5$ **13.** $2a-14+\frac{74}{2a+3}$ **15.** $4x^2-7x+3$ **17.** $3y^2-2y+2$ **19.** $2x^2-2x+3+\frac{-1}{x+1}$ **21.** $3k-4+\frac{2}{k^2-2}$ **23.** x^2+1 **25.** x^2+1 **27.** $2p^2-5p+4+\frac{6}{3p^2+1}$ **29.** x^3+6x-7 **31.** $2x^2+\frac{3}{5}x+\frac{1}{5}$ **33.** (x^2+x-3) units **35.** 33 **36.** 33 **37.** They are the same. **38.** The answers should agree.

SECTION 12.8 (pages 888–890)

1. **(a)** C **(b)** A **(c)** B **(d)** D **2.** **(a)** A **(b)** C **(c)** B **(d)** D **3.** in scientific notation **4.** in scientific notation **5.** not in scientific notation; 5.6×10^6 **6.** not in scientific notation; 3.4×10^4 **7.** not in scientific notation; 4×10^{-3} **8.** not in scientific notation; 7×10^{-4} **9.** not in scientific notation; 8×10^1 **10.** not in scientific notation; 9×10^2 **11.** A number is written in scientific notation if it is the product of a number whose absolute value is between 1 and 10 (inclusive of 1) and a power of 10. **12.** To multiply by a positive power of 10, move the decimal point to the right as many places as the exponent on 10. With a negative power of 10, move the decimal point to the left as many places as the absolute value of the exponent on 10. **13.** 5.876×10^9 **15.** 8.235×10^4 **17.** 7×10^{-6} **19.** -2.03×10^{-3} **21.** 750,000 **23.** 5,677,000,000,000 **25.** 1,000,000,000,000 **27.** -6.21 **29.** 0.00078 **31.** 0.000000005134 **33.** 6×10^{11}; 600,000,000,000 **35.** 1.5×10^7; 15,000,000 **37.** 8×10^{-3}; 0.008 **39.** 2.4×10^2; 240 **41.** 6.3×10^{-2}; 0.063 **43.** 6.426×10^4; 64,260 **45.** 3×10^{-4}; 0.0003 **47.** 4×10^1; 40 **49.** 1.3×10^{-5}; 0.000013 **51.** 5×10^2; 500 **53.** 2.6×10^{-3}; 0.0026 **55.** 7.205×10^{-6}; 0.000007205 **57.** 0.000002 **59.** 4.2×10^{42} **61.** 1.5×10^{17} mi **63.** \$3209 **65.** \$52,733 **67.** 3.59×10^2 sec, or 359 sec **69.** \$86.40

Chapter 12 REVIEW EXERCISES (pages 894–897)

1. $22m^2$; degree 2; monomial **2.** p^3-p^2+4p+2; degree 3; none of these **3.** already in descending powers; degree 5; none of these **4.** $-8y^5-7y^4+9y$; degree 5; trinomial **5.** $-5a^3+4a^2$ **6.** $2r^3-3r^2+9r$ **7.** $11y^2-10y+9$ **8.** $-13k^4-15k^2-4k-6$ **9.** $10m^3-6m^2-3$ **10.** $-y^2-4y+26$ **11.** $10p^2-3p-11$ **12.** $7r^4-4r^3-1$ **13.** 4^{11} **14.** -5^{11} **15.** $-72x^7$ **16.** $10x^{14}$ **17.** 19^5x^5 **18.** -4^7y^7 **19.** $5p^4t^4$ **20.** $\frac{7^6}{5^6}$ **21.** $27x^6y^9$ **22.** t^{42} **23.** $36x^{16}y^4z^{16}$ **24.** $\frac{8m^9n^3}{p^6}$ **25.** $125x^6$ **26.** The product rule for exponents does not apply here because we want the sum of 7^2 and 7^4, not their product. **27.** $10x^2+70x$ **28.** $-6p^5+15p^4$ **29.** $6r^3+8r^2-17r+6$ **30.** $8y^3+27$ **31.** $5p^5-2p^4-3p^3+25p^2+15p$ **32.** $x^2+3x-18$ **33.** $6k^2-9k-6$ **34.** $12p^2-48pq+21q^2$ **35.** $2m^4+5m^3-16m^2-28m+9$ **36.** $a^2+8a+16$ **37.** $9p^2-12p+4$ **38.** $4r^2+20rs+25s^2$ **39.** $r^3+6r^2+12r+8$ **40.** $8x^3-12x^2+6x-1$ **41.** $4z^2-49$ **42.** $36m^2-25$ **43.** $25a^2-36b^2$ **44.** $4x^4-25$ **45.** three; two **46.** $(a+b)^2=(a+b)(a+b)=a^2+2ab+b^2$. The term $2ab$ is not in a^2+b^2. **47.** 2 **48.** $\frac{1}{32}$ **49.** $\frac{25}{36}$ **50.** $-\frac{3}{16}$ **51.** 36 **52.** x^2 **53.** $\frac{1}{p^{12}}$ **54.** r^4 **55.** 2^8 **56.** $\frac{1}{9^6}$ **57.** 5^8 **58.** $\frac{1}{8^{12}}$ **59.** $\frac{1}{m^2}$ **60.** y^7 **61.** r^{13} **62.** $25m^6$ **63.** $\frac{y^{12}}{8}$ **64.** $\frac{1}{a^3b^5}$ **65.** $72r^5$ **66.** $\frac{8n^{10}}{3m^{13}}$ **67.** $\frac{5y^2}{3}$ **68.** $-2x^2y$ **69.** $-y^3+2y-3$ **70.** $p-3+\frac{5}{2p}$ **71.** $-x^9+2x^8-4x^3+7x$ **72.** $-2m^2n+mn^2+\frac{6n^3}{5}$ **73.** $2r+7$ **74.** $4m+3+\frac{5}{3m-5}$ **75.** $2a+1+\frac{-8a+12}{5a^2-3}$ **76.** $k^2+2k+4+\frac{-2k-12}{2k^2+1}$ **77.** 4.8×10^7 **78.** 2.8988×10^{10} **79.** 6.5×10^{-5} **80.** 8.24×10^{-8} **81.** 24,000 **82.** 78,300,000 **83.** 0.000000897 **84.** 0.00000000000995 **85.** 8×10^2; 800 **86.** 4×10^6; 4,000,000 **87.** 2.5×10^{-2}; 0.025 **88.** 1×10^{-2}; 0.01 **89.** 2.796×10^{10} calculations; 1.6776×10^{12} calculations **90.** about 3.3 **91.** 0 **92.** $\frac{243}{p^3}$ **93.** $\frac{1}{49}$ **94.** $49-28k+4k^2$

95. $y^2 + 5y + 1$ **96.** $\frac{1296r^8s^4}{625}$ **97.** $-8m^7 - 10m^6 - 6m^5$ **98.** 32 **99.** $5xy^3 - \frac{8y^2}{5} + 3x^2y$ **100.** $\frac{r^2}{6}$ **101.** $8x^3 + 12x^2y + 6xy^2 + y^3$ **102.** $\frac{3}{4}$ **103.** $a^3 - 2a^2 - 7a + 2$ **104.** $8y^3 - 9y^2 + 5$ **105.** $10r^2 + 21r - 10$ **106.** $144a^2 - 1$ **107.** $2x^2 + x - 6$; $6x - 2$ **108.** $20x^4 + 8x^2$; $25x^8 + 20x^6 + 4x^4$ **109.** The friend wrote the second term of the quotient as $-12x$ rather than $-2x$. Here is the correct method: $\frac{6x^2 - 12x}{6} = \frac{6x^2}{6} - \frac{12x}{6} = x^2 - 2x$. **110.** $2mn + 3m^4n^2 - 4n$

Chapter 12 TEST (pages 898–899)

1. $-7x^2 + 8x$; 2; binomial **2.** $4n^4 + 13n^3 - 10n^2$; 4; trinomial **3.** $4t^4 + t^3 - 6t^2 - t$ **4.** $-2y^2 - 9y + 17$ **5.** $-12t^2 + 5t + 8$ **6.** -32 **7.** $\frac{216}{m^6}$ **8.** $-27x^5 + 18x^4 - 6x^3 + 3x^2$ **9.** $2r^3 + r^2 - 16r + 15$ **10.** $t^2 - 5t - 24$ **11.** $8x^2 + 2xy - 3y^2$ **12.** $25x^2 - 20xy + 4y^2$ **13.** $100v^2 - 9w^2$ **14.** $x^3 + 3x^2 + 3x + 1$ **15.** $12x + 36$; $9x^2 + 54x + 81$ **16. (a)** positive **(b)** positive **(c)** negative **(d)** positive **(e)** zero **(f)** negative **17.** $\frac{1}{625}$ **18.** 2 **19.** $\frac{7}{12}$ **20.** 8^5 **21.** x^2y^6 **22.** $4y^2 - 3y + 2 + \frac{5}{y}$ **23.** $-3xy^2 + 2x^3y^2 + 4y^2$ **24.** $2x + 9$ **25.** $3x^2 + 6x + 11 + \frac{26}{x-2}$ **26. (a)** 3.44×10^{11} **(b)** 5.57×10^{-6} **27. (a)** 29,600,000 **(b)** 0.0000000607 **28. (a)** 1×10^3; 5.89×10^{12} **(b)** 5.89×10^{15} mi

ANSWERS

CHAPTER 13 Factoring and Applications

SECTION 13.1 (pages 909–911)

1. product; multiplying **2.** common factor; is; divides **3.** 4 **5.** 4 **7.** 6 **9.** 1 **11.** 8 **13.** $10x^3$ **15.** xy^2 **17.** 6 **19.** $6m^3n^2$ **21.** factored **22.** factored **23.** not factored **24.** not factored **25.** The correct factored form is $18x^3y^2 + 9xy = 9xy(2x^2y + 1)$. If a polynomial has two terms, the product of the factors must have two terms. $9xy(2x^2y) = 18x^3y^2$ is just one term. **26.** Verify that we have factored completely. Then we multiply the factors. The product should be the original polynomial. **27.** $3m^2$ **29.** $2z^4$ **31.** $2mn^4$ **33.** $y + 2$ **35.** $a - 2$ **37.** $2 + 3xy$ **39.** $x(x - 4)$ **41.** $3t(2t + 5)$ **43.** $m^2(m - 1)$ **45.** $-6x^2(2x + 1)$ **47.** $5y^6(13y^4 + 7)$ **49.** no common factor (except 1) **51.** $8mn^3(1 + 3m)$ **53.** $-2x(2x^2 - 5x + 3)$ **55.** $13y^2(y^6 + 2y^2 - 3)$ **57.** $9qp^3(5q^3p^2 + 4p^3 + 9q)$ **59.** $(x + 2)(c + d)$ **61.** $(2a + b)(a^2 - b)$ **63.** $(p + 4)(q - 1)$ **65.** not in factored form; $(7t + 4)(8 + x)$ **66.** not in factored form; $(5x - 1)(3r + 7)$ **67.** in factored form **68.** in factored form **69.** not in factored form **70.** not in factored form **71.** $(5 + n)(m + 4)$ **73.** $(2y - 7)(3x + 4)$ **75.** $(y + 3)(3x + 1)$ **77.** $(z + 2)(7z - a)$ **79.** $(3r + 2y)(6r - x)$ **81.** $(w + 1)(w^2 + 9)$ **83.** $(a + 2)(3a^2 - 2)$ **85.** $(4m - p^2)(4m^2 - p)$ **87.** $(y + 3)(y + x)$ **89.** $(z - 2)(2z - 3w)$ **91.** $(5 - 2p)(m + 3)$ **93.** $(3r + 2y)(6r - t)$ **95.** commutative property **96.** $2x(y - 4) - 3(y - 4)$ **97.** No, because it is not a product. It is the difference between $2x(y - 4)$ and $3(y - 4)$. **98.** $(y - 4)(2x - 3)$, or $(2x - 3)(y - 4)$; yes

SECTION 13.2 (pages 917–918)

1. a and b must have different signs. **2.** a and b must have the same sign. **3.** C **4.** Factor out the greatest common factor, $2x$. **5.** $a^2 + 13a + 36$ **6.** $y^2 - 4y - 21$ **7.** A prime polynomial is one that cannot be factored using only integers in the factors. **8.** To check a factored form, we multiply the factors. We should get the trinomial we started with. If we forget to factor out any common factors, the check will *not* indicate it. **9.** 1 and 12, -1 and -12, 2 and 6, -2 and -6, 3 and 4, -3 and -4; The pair with a sum of 7 is 3 and 4. **11.** 1 and -24, -1 and 24, 2 and -12, -2 and 12, 3 and -8, -3 and 8, 4 and -6, -4 and 6; The pair with a sum of -5 is 3 and -8. **13.** $p + 6$ **15.** $x + 11$ **17.** $x - 8$ **19.** $y - 5$ **21.** $x + 11$ **23.** $y - 9$ **25.** $(y + 8)(y + 1)$ **27.** $(b + 3)(b + 5)$ **29.** $(m + 5)(m - 4)$ **31.** $(x + 8)(x - 5)$ **33.** prime **35.** $(y - 5)(y - 3)$ **37.** $(z - 8)(z - 7)$ **39.** $(r - 6)(r + 5)$ **41.** $(a - 12)(a + 4)$ **43.** $(r + 2a)(r + a)$ **45.** $(x + y)(x + 3y)$ **47.** $(t + 2z)(t - 3z)$ **49.** $(v - 5w)(v - 6w)$ **51.** $(a + 5b)(a - 3b)$ **53.** $4(x + 5)(x - 2)$ **55.** $2t(t + 1)(t + 3)$ **57.** $-2x^4(x - 3)(x + 7)$ **59.** $a^3(a + 4b)(a - b)$ **61.** $5m^2(m^3 + 5m^2 - 8)$ **63.** $mn(m - 6n)(m - 4n)$

SECTION 13.3 (pages 921–922)

1. B **2.** D **3.** $(m + 6)(m + 2)$ **5.** $(a + 5)(a - 2)$ **7.** $(2t + 1)(5t + 2)$ **9.** $(3z - 2)(5z - 3)$ **11.** $(2s - t)(4s + 3t)$ **13.** $(3a + 2b)(5a + 4b)$ **15. (a)** 2; 12; 24; 11 **(b)** 3; 8 (Order is irrelevant.) **(c)** $3m$; $8m$ **(d)** $2m^2 + 3m + 8m + 12$ **(e)** $(2m + 3)(m + 4)$ **(f)** $(2m + 3)(m + 4) = 2m^2 + 8m + 3m + 12 = 2m^2 + 11m + 12$ **17.** $(2x + 1)(x + 3)$ **19.** $(4r - 3)(r + 1)$ **21.** $(4m + 1)(2m - 3)$ **23.** $(3m + 1)(7m + 2)$ **25.** $(2b + 1)(3b + 2)$ **27.** $(4y - 3)(3y - 1)$ **29.** $(4 + x)(4 + 3x)$, or $(x + 4)(3x + 4)$ **31.** $3(4x - 1)(2x - 3)$ **33.** $2m(m - 4)(m + 5)$ **35.** $-4z^3(z - 1)(8z + 3)$ **37.** $(3p + 4q)(4p - 3q)$ **39.** $(3a - 5b)(2a + b)$ **41.** The student stopped too soon. He needs to factor out the common factor $4x - 1$ to get $(4x - 1)(4x - 5)$ as the correct answer. **42.** The student forgot to include the common factor $3k$ in her answer. The correct answer is $3k(k - 5)(k + 1)$.

SECTION 13.4 (pages 927–928)

1. B **2.** A **3.** A **4.** B **5.** A **6.** A **7.** $2a + 5b$ **9.** $x^2 + 3x - 4$; $x + 4$, $x - 1$ **11.** $2z^2 - 5z - 3$; $2z + 1$, $z - 3$ **13.** $(4x + 4)$ cannot be a factor because its terms have a common factor of 4, but those of the polynomial do not. The correct factored form is $(4x - 3)(3x + 4)$. **14.** The student forgot to factor out the common factor 2 from the terms of the trinomial. The completely factored form is $2(2x - 1)(x + 3)$. **15.** $(3a + 7)(a + 1)$ **17.** $(2y + 3)(y + 2)$ **19.** $(3m - 1)(5m + 2)$ **21.** $(3s - 1)(4s + 5)$ **23.** $(5m - 4)(2m - 3)$ **25.** $(4w - 1)(2w - 3)$ **27.** $(4y + 1)(5y - 11)$ **29.** prime **31.** $2(5x + 3)(2x + 1)$ **33.** $-q(5m + 2)(8m - 3)$ **35.** $3n^2(5n - 3)(n - 2)$ **37.** $-y^2(5x - 4)(3x + 1)$ **39.** $(5a + 3b)(a - 2b)$ **41.** $(4s + 5t)(3s - t)$ **43.** $m^4n(3m + 2n)(2m + n)$ **45.** $-1(x + 7)(x - 3)$ **47.** $-1(3x + 4)(x - 1)$ **49.** $-1(a + 2b)(2a + b)$ **51.** $5 \cdot 7$ **52.** $(-5)(-7)$ **53.** The product of $3x - 4$ and $2x - 1$ is $6x^2 - 11x + 4$. **54.** The product of $4 - 3x$ and $1 - 2x$ is $6x^2 - 11x + 4$. **55.** The factors in **Exercise 53** are the opposites of the factors in **Exercise 54.** **56.** $(3 - 7t)(5 - 2t)$

SECTION 13.5 (pages 935–938)

1. 1; 4; 9; 16; 25; 36; 49; 64; 81; 100; 121; 144; 169; 196; 225; 256; 289; 324; 361; 400 **2.** 1; 16; 81; 256; 625 **3.** A, D **4.** The binomial $4x^2 + 16$ can be factored as $4(x^2 + 4)$. *After* any common factor is removed, a sum of squares (like $x^2 + 4$ here) *cannot* be factored. **5.** $(y + 5)(y - 5)$ **7.** $(x + 12)(x - 12)$ **9.** prime **11.** prime **13.** $(3r + 2)(3r - 2)$ **15.** $4(3x + 2)(3x - 2)$ **17.** $(14p + 15)(14p - 15)$ **19.** $(4r + 5a)(4r - 5a)$ **21.** prime **23.** $(p^2 + 7)(p^2 - 7)$ **25.** $(x^2 + 1)(x + 1)(x - 1)$ **27.** $(p^2 + 16)(p + 4)(p - 4)$ **29.** B, C **30.** No, it is not a perfect square since the middle term would have to be $30y$. **31.** $(w + 1)^2$ **33.** $(x - 4)^2$ **35.** prime **37.** $2(x + 6)^2$ **39.** $(2x + 3)^2$ **41.** $(4x - 5)^2$ **43.** $(7x - 2y)^2$ **45.** $(8x + 3y)^2$ **47.** $-2h(5h - 2y)^2$ **49.** $\left(p + \frac{1}{3}\right)\left(p - \frac{1}{3}\right)$ **51.** $\left(2m + \frac{3}{5}\right)\left(2m - \frac{3}{5}\right)$ **53.** $(x + 0.8)(x - 0.8)$ **55.** $\left(t + \frac{1}{2}\right)^2$ **57.** $(x - 0.5)^2$ **59.** 1; 8; 27; 64; 125; 216; 343; 512; 729; 1000 **60.** 3 **61.** C, D **62.** A, D **63.** $(x + 1)(x^2 - x + 1)$ **65.** $(x - 1)(x^2 + x + 1)$ **67.** $(p + q)(p^2 - pq + q^2)$ **69.** $(y - 6)(y^2 + 6y + 36)$ **71.** $(k + 10)(k^2 - 10k + 100)$ **73.** $(3x - 1)(9x^2 + 3x + 1)$ **75.** $(5x + 2)(25x^2 - 10x + 4)$ **77.** $(y - 2x)(y^2 + 2xy + 4x^2)$ **79.** $(3x - 4y)(9x^2 + 12xy + 16y^2)$ **81.** $(2p + 9q)(4p^2 - 18pq + 81q^2)$ **83.** $2(2t - 1)(4t^2 + 2t + 1)$ **85.** $5(2w + 3)(4w^2 - 6w + 9)$ **87.** $(x + y^2)(x^2 - xy^2 + y^4)$ **89.** $(5k - 2m^3)(25k^2 + 10km^3 + 4m^6)$ **91.** $(x^3 - 1)(x^3 + 1)$ **92.** $(x - 1)(x^2 + x + 1)(x + 1)(x^2 - x + 1)$ **93.** $(x^2 - 1)(x^4 + x^2 + 1)$ **94.** $(x - 1)(x + 1)(x^4 + x^2 + 1)$ **95.** The result in **Exercise 92** is completely factored. **96.** Show that $x^4 + x^2 + 1 = (x^2 + x + 1)(x^2 - x + 1)$. **97.** difference of squares **98.** $(x - 3)(x^2 + 3x + 9)(x + 3)(x^2 - 3x + 9)$

SECTION 13.6 (pages 941–942)

1. (a) B **(b)** D **(c)** A **(d)** C **(e)** A, B **2. (a)** C **(b)** E **(c)** C **(d)** B **(e)** A, B **3.** $8m^3(4m^6 + 2m^2 + 3)$ **5.** $7k(2k + 5)(k - 2)$ **7.** $(m + n)(m - 4n)$ **9.** $10nr(10nr + 3r^2 - 5n)$ **11.** $(4 + m)(5 + 3n)$ **13.** $(y^2 + 9)(y + 3)(y - 3)$ **15.** $(2p - 5)(4p^2 + 10p + 25)$ **17.** $(p - 12)^2$ **19.** $(4z - 1)^2$ **21.** prime **23.** $(p + 4)(p^2 - 4p + 16)$ **25.** $(2a + 1)(a^2 - 7)$ **27.** $(4r + 3m)^2$ **29.** prime **31.** $4(2k - 3)^2$ **33.** $(4k - 3h)(2k + h)$ **35.** $(5z - 6)(2z + 1)$ **37.** $(3y - 1)(3y + 5)$ **39.** prime **41.** $(2 - q)(2 - 3p)$ **43.** $8(5z + 4)(25z^2 - 20z + 16)$ **45.** $(10a + 9y)(10a - 9y)$ **47.** $(a + 4)^2$ **49.** prime **51.** $(5a - 7b)^2$ **53.** $(m + 3)(2m - 5n)$ **55.** $2(3m - 10)(9m^2 + 30m + 100)$

SECTION 13.7 (pages 949–951)

1. $ax^2 + bx + c$ **2.** standard **3.** factor **4.** 0; zero; factor **5. (a)** linear **(b)** quadratic **(c)** quadratic **(d)** linear **6.** Because $(x - 9)^2 = (x - 9)(x - 9)$, applying the zero-factor property leads to two solutions of 9. Thus, 9 is a double solution. **7.** Set each *variable* factor equal to 0, to get $2x = 0$ or $3x - 4 = 0$. The solution set is $\left\{0, \frac{4}{3}\right\}$. **8.** The variable x is another factor to set equal to 0, so the solution set is $\left\{0, \frac{1}{7}\right\}$. **9.** $\{-5, 2\}$ **11.** $\left\{3, \frac{7}{2}\right\}$ **13.** $\left\{-\frac{1}{2}, \frac{1}{6}\right\}$ **15.** $\left\{-\frac{5}{6}, 0\right\}$ **17.** $\left\{0, \frac{4}{3}\right\}$ **19.** $\{9\}$ **21.** $\{-2, -1\}$ **23.** $\{1, 2\}$ **25.** $\{-8, 3\}$ **27.** $\{-1, 3\}$ **29.** $\{-2, -1\}$ **31.** $\{-4\}$ **33.** $\left\{-2, \frac{1}{3}\right\}$ **35.** $\left\{-\frac{4}{3}, \frac{1}{2}\right\}$ **37.** $\left\{-\frac{2}{3}\right\}$ **39.** $\{-3, 3\}$ **41.** $\left\{-\frac{7}{4}, \frac{7}{4}\right\}$ **43.** $\{-11, 11\}$ **45.** $\{0, 7\}$ **47.** $\left\{0, \frac{1}{2}\right\}$ **49.** $\{2, 5\}$ **51.** $\left\{-4, \frac{1}{2}\right\}$ **53.** $\left\{-12, \frac{11}{2}\right\}$ **55.** $\{-2, 0, 2\}$ **57.** $\left\{-\frac{7}{3}, 0, \frac{7}{3}\right\}$ **59.** $\left\{-\frac{5}{2}, \frac{1}{3}, 5\right\}$ **61.** $\left\{-\frac{7}{2}, -3, 1\right\}$ **63.** $\{-5, 0, 4\}$ **65.** $\{-3, 0, 5\}$ **67.** $\{-1, 3\}$ **69. (a)** 64; 144; 4; 6 **(b)** No time has elapsed, so the object hasn't fallen (been released) yet. **70.** Time cannot be negative.

SECTION 13.8 (pages 958–962)

1. Read; variable; equation; Solve; answer; Check; original **2.** Only 6 is reasonable since a square cannot have a side of negative length. **3.** *Step 3:* $(2x + 1)(x + 1)$; *Step 4:* 4; $-\frac{11}{2}$; *Step 5:* 9; 5; *Step 6:* $9 \cdot 5$ **5.** *Step 3:* 192; $4x$; *Step 4:* 6; -8; *Step 5:* 8; 6; *Step 6:* $8 \cdot 6$; 192 **7.** length: 14 cm; width: 12 cm **9.** base: 12 in.; height: 5 in. **11.** length: 15 in.; width: 12 in. **13.** height: 13 in.; width: 10 in. **15.** mirror: 7 ft; painting: 9 ft **17.** 20, 21 **19.** -3, -2 or 4, 5 **21.** -3, -1 or 7, 9 **23.** -2, 0, 2 or 6, 8, 10 **25.** 7, 9, 11 **27.** 12 cm **29.** 12 mi **31.** 8 ft **33. (a)** 1 sec **(b)** $\frac{1}{2}$ sec and $1\frac{1}{2}$ sec **(c)** 3 sec **(d)** The negative solution, -1, does not make sense since t represents time, which cannot be negative. **35.** 112 ft **37.** 256 ft **39. (a)** 109 million; The result using the model is the same as the actual number from the table for 2000. **(b)** 14 **(c)** 270 million; The result is a little less than 286 million, the actual number for 2008. **(d)** 393 million

Chapter 13 REVIEW EXERCISES (pages 966–970)

1. $15(t + 3)$ **2.** $30z(2z^2 + 1)$ **3.** $11x^2(4x + 5)$ **4.** $50m^2n^2(2n - mn^2 + 3)$ **5.** $(x - 4)(2y + 3)$ **6.** $(2y + 3)(3y + 2x)$ **7.** $(x + 3)(x + 7)$ **8.** $(y - 5)(y - 8)$ **9.** $(q + 9)(q - 3)$ **10.** $(r - 8)(r + 7)$ **11.** prime **12.** $3(x^2 + 2x + 2)$ **13.** $(r + 8s)(r - 12s)$ **14.** $(p + 12q)(p - 10q)$ **15.** $-8p(p + 2)(p - 5)$ **16.** $3x^2(x + 2)(x + 8)$ **17.** $(m + 3n)(m - 6n)$ **18.** $(y - 3z)(y - 5z)$ **19.** $p^5(p - 2q)(p + q)$ **20.** $-3r^3(r + 3s)(r - 5s)$ **21.** r and $6r$, $2r$ and $3r$ **22.** Factor out z. **23.** $(2k - 1)(k - 2)$ **24.** $(3r - 1)(r + 4)$ **25.** $(3r + 2)(2r - 3)$ **26.** $(5z + 1)(2z - 1)$ **27.** prime **28.** $4x^3(3x - 1)(2x - 1)$ **29.** $-3(x + 2)(2x - 5)$ **30.** $rs(5r + 6s)(2r + s)$ **31.** $-5y(3y + 2)(2y - 1)$ **32.** prime **33.** $-mn(3m + 5)(m - 8)$ **34.** $(2a - 5b)(7a + 4b)$ **35.** B **36.** D **37.** $(n + 8)(n - 8)$ **38.** $(5b + 11)(5b - 11)$ **39.** $(7y + 5w)(7y - 5w)$ **40.** $36(2p + q)(2p - q)$ **41.** prime **42.** $(z + 5)^2$ **43.** $(3t - 7)^2$ **44.** $(4m + 5n)^2$ **45.** $(5x - 1)(25x^2 + 5x + 1)$ **46.** $(10p + 3)(100p^2 - 30p + 9)$ **47.** $\left\{-\frac{3}{4}, 1\right\}$ **48.** $\{-7, -3, 4\}$ **49.** $\left\{0, \frac{5}{2}\right\}$ **50.** $\{-3, -1\}$ **51.** $\{1, 4\}$ **52.** $\{3, 5\}$ **53.** $\left\{-\frac{4}{3}, 5\right\}$ **54.** $\left\{-\frac{8}{9}, \frac{8}{9}\right\}$ **55.** $\{0, 8\}$ **56.** $\{-1, 6\}$ **57.** $\{7\}$ **58.** $\{6\}$ **59.** $\left\{-\frac{2}{5}, -2, -1\right\}$ **60.** $\{-3, 3\}$

61. $\left\{-\frac{3}{8}, 0, \frac{3}{8}\right\}$ **62.** $\left\{-\frac{3}{4}, -\frac{1}{2}, \frac{1}{3}\right\}$ **63.** $\left\{\frac{9}{5}\right\}$ **64.** $\{-3, 10\}$ **65.** length: 10 ft; width: 4 ft **66.** 5 ft **67.** 6, 7 or $-5, -4$ **68.** $-5, -4, -3$ or 5, 6, 7 **69.** 26 mi **70.** 6 m **71.** width: 10 m; length: 17 m **72.** **(a)** 256 ft **(b)** 1024 ft **73.** 602 thousand; The result is a little higher than the 592 thousand given in the table. **74.** 882 thousand; The estimate may be unreliable because the conditions that prevailed in the years 2001–2009 may have changed, causing either a greater increase or a greater decrease predicted by the model for the number of such vehicles. **75.** D **76.** The terms of $2x + 8$ have a common factor of 2. The completely factored form is $2(x + 4)(3x - 4)$. **77.** $(3m + 4p)(5m - 4)$ **78.** $8abc(3b^2c - 7ac^2 + 9ab)$ **79.** $(z - x)(z - 10x)$ **80.** $(3k + 5)(k + 2)$ **81.** $(y^2 + 25)(y + 5)(y - 5)$ **82.** $3m(2m + 3)(m - 5)$ **83.** prime **84.** $8(z + 2y)(z^2 - 2zy + 4y^2)$ **85.** $-1(2r + 3q)(6r - 5q)$ **86.** $(10a + 3)(10a - 3)$ **87.** $(7t + 4)^2$ **88.** $\{0, 7\}$ **89.** $\{-5, 2\}$ **90.** $\left\{-\frac{2}{5}\right\}$ **91.** 15 m, 36 m, 39 m **92.** length: 6 m; width: 4 m

Chapter 13 TEST (pages 971–972)

1. D **2.** $6x(2x - 5)$ **3.** $m^2n(2mn + 3m - 5n)$ **4.** $(2x + y)(a - b)$ **5.** $(x - 7)(x - 2)$ **6.** $(2x + 3)(x - 1)$ **7.** $(3x + 1)(2x - 7)$ **8.** $3(x + 1)(x - 5)$ **9.** $(5z - 1)(2z - 3)$ **10.** prime **11.** prime **12.** $(y + 7)(y - 7)$ **13.** $(9a + 11b)(9a - 11b)$ **14.** $(x + 8)^2$ **15.** $(2x - 7y)^2$ **16.** $-2(x + 1)^2$ **17.** $4t(t + 4)^2$ **18.** $(x^2 + 9)(x + 3)(x - 3)$ **19.** $(x - 8)(x^2 + 8x + 64)$ **20.** $8(k + 2)(k^2 - 2k + 4)$ **21.** $\{-3, 9\}$ **22.** $\left\{\frac{1}{2}, 6\right\}$ **23.** $\left\{-\frac{2}{5}, \frac{2}{5}\right\}$ **24.** $\{10\}$ **25.** $\{0, 3\}$ **26.** $\left\{-8, -\frac{5}{2}, \frac{1}{3}\right\}$ **27.** 6 ft by 9 ft **28.** $-2, -1$ **29.** 17 ft **30.** 85 billion pieces

CHAPTER 14 Rational Expressions and Applications

Note: In work with rational expressions, several different equivalent forms of the answer often exist. If your answer does not look exactly like the one given here, check to see if you have written an equivalent form.

SECTION 14.1 (pages 981–982)

1. **(a)** $3; -5$ **(b)** $q; -1$ **2.** **(a)** is not **(b)** are **3.** A rational expression is a quotient of polynomials, such as $\frac{x + 3}{x^2 - 4}$. **4.** Division by 0 is undefined, so if the denominator of a rational expression equals 0, the expression is undefined. **5.** **(a)** 1 **(b)** $\frac{17}{12}$ **7.** **(a)** 0 **(b)** -1 **9.** **(a)** $\frac{9}{5}$ **(b)** undefined **11.** **(a)** $\frac{2}{7}$ **(b)** $\frac{13}{3}$ **13.** $y \neq 0$ **15.** $x \neq -6$ **17.** $x \neq \frac{5}{3}$ **19.** $m \neq -3, m \neq 2$ **21.** never undefined **23.** never undefined **25.** $3r^2$ **27.** $\frac{2}{5}$ **29.** $\frac{x - 1}{x + 1}$ **31.** $\frac{7}{5}$ **33.** $m - n$ **35.** $\frac{3(2m + 1)}{4}$ **37.** $\frac{3m}{5}$ **39.** $\frac{3r - 2s}{3}$ **41.** $\frac{x + 1}{x - 1}$ **43.** $\frac{z - 3}{z + 5}$ **45.** $\frac{a + b}{a - b}$ **47.** -1 **49.** $-(m + 1)$ **51.** -1 **53.** It is already in lowest terms.

Answers may vary in Exercises 55–59.

55. $\frac{-(x + 4)}{x - 3}, \frac{-x - 4}{x - 3}, \frac{x + 4}{-(x - 3)}, \frac{x + 4}{-x + 3}$ **57.** $\frac{-(2x - 3)}{x + 3}, \frac{-2x + 3}{x + 3}, \frac{2x - 3}{-(x + 3)}, \frac{2x - 3}{-x - 3}$ **59.** $\frac{-(3x - 1)}{5x - 6}, \frac{-3x + 1}{5x - 6}, \frac{3x - 1}{-(5x - 6)}, \frac{3x - 1}{-5x + 6}$ **61.** $x^2 + 3$

SECTION 14.2 (pages 987–988)

1. **(a)** B **(b)** D **(c)** C **(d)** A **2.** **(a)** D **(b)** C **(c)** A **(d)** B **3.** $\frac{4m}{3}$ **5.** $\frac{40y^2}{3}$ **7.** $\frac{2}{c + d}$ **9.** $4(x - y)$ **11.** $\frac{16q}{3p^3}$ **13.** $\frac{7}{r^2 + rp}$ **15.** $\frac{z^2 - 9}{z^2 + 7z + 12}$ **17.** 5 **19.** $-\frac{3}{2t^4}$ **21.** $\frac{1}{4}$ **23.** $\frac{x(x - 3)}{6}$ **25.** $x - 3; x + 3; x + 3; x - 3; \frac{10}{9}$ **27.** $-\frac{3}{4}$ **29.** -1 **31.** $\frac{9(m - 2)}{-(m + 4)}$, or $\frac{-9(m - 2)}{m + 4}$ **33.** $\frac{p + 4}{p + 2}$ **35.** $\frac{(k - 1)^2}{(k + 1)(2k - 1)}$ **37.** $\frac{4k - 1}{3k - 2}$ **39.** $\frac{m + 4p}{m + p}$ **41.** $\frac{10}{x + 10}$ **43.** $\frac{5xy^2}{4q}$

SECTION 14.3 (pages 993–994)

1. C **2.** B **3.** C **4.** A **5.** 30 **7.** x^7 **9.** $72q$ **11.** $84r^5$ **13.** $24x^3y^4$ **15.** $x^{10}y^{14}z^{23}$ **17.** $28m^2(3m - 5)$ **19.** $30(b - 2)$ **21.** $c - d$ or $d - c$ **23.** $k(k + 5)(k - 2)$ **25.** $a(a + 6)(a - 3)$ **27.** $(p + 3)(p + 5)(p - 6)$ **29.** $\frac{20}{55}$ **31.** $\frac{-45}{9k}$ **33.** $\frac{26y^2}{80y^3}$ **35.** $\frac{35t^2r^3}{42r^4}$ **37.** $\frac{20}{8(m + 3)}$ **39.** $\frac{8t}{12 - 6t}$ **41.** $\frac{14(z - 2)}{z(z - 3)(z - 2)}$ **43.** $\frac{2(b - 1)(b + 2)}{b^3 + 3b^2 + 2b}$

SECTION 14.4 (pages 1001–1004)

1. E **2.** A **3.** C **4.** H **5.** B **6.** D **7.** G **8.** F **9.** $\frac{11}{m}$ **11.** b **13.** $\frac{4}{y + 4}$ **15.** $\frac{m - 1}{m + 1}$ **17.** x **19.** $y^2 - 3y - 18; 3; 6; y - 6$ **21.** Combine the numerators and keep the same denominator. For example, $\frac{3x + 2}{x - 6} + \frac{-2x - 8}{x - 6} = \frac{x - 6}{x - 6}$. Then write in lowest terms: $\frac{x - 6}{x - 6} = 1$. **22.** Find the LCD, write each fraction as an equivalent one with this LCD, and then proceed as in **Exercise 21.** For example, $\frac{3}{x} + \frac{2}{y} = \frac{3y}{xy} + \frac{2x}{xy} = \frac{3y + 2x}{xy}$. **23.** $\frac{3z + 5}{15}$ **25.** $\frac{10 - 7r}{14}$ **27.** $\frac{-3x - 2}{4x}$ **29.** $\frac{57}{20x}$ **31.** $\frac{x + 1}{2}$ **33.** $\frac{5x + 9}{6x}$ **35.** $x(x + 3)$; $\frac{3x + 3}{x(x + 3)}$ **37.** $\frac{-k - 10}{k(k + 5)}$ **39.** $\frac{x + 4}{x + 2}$ **41.** $\frac{x^2 + 6x - 8}{(x - 2)(x + 2)}$ **43.** $\frac{3}{t}$ **45.** $m - 2$ or $2 - m$ **46.** $\frac{-4}{3 - k}$, or $-\frac{4}{3 - k}$ **47.** $\frac{-2}{x - 5}$, or $\frac{2}{5 - x}$ **49.** -4 **51.** $\frac{-5}{x - y^2}$, or $\frac{5}{y^2 - x}$ **53.** $\frac{x + y}{5x - 3y}$, or $\frac{-x - y}{3y - 5x}$ **55.** $\frac{-6}{4p - 5}$, or $\frac{6}{5 - 4p}$ **57.** $\frac{-(m + n)}{2(m - n)}$ **59.** $\frac{-x^2 + 6x + 11}{(x + 3)(x - 3)(x + 1)}$ **61.** $\frac{-5q^2 - 13q + 7}{(3q - 2)(q + 4)(2q - 3)}$ **63.** $\frac{9r + 2}{r(r + 2)(r - 1)}$

65. $\frac{2x^2+6xy+8y^2}{(x+y)(x+y)(x+3y)}$, or $\frac{2x^2+6xy+8y^2}{(x+y)^2(x+3y)}$

67. $\frac{15r^2+10ry-y^2}{(3r+2y)(6r-y)(6r+y)}$ **69. (a)** $\frac{9k^2+6k+26}{5(3k+1)}$ **(b)** $\frac{1}{4}$

SECTION 14.5 (pages 1011–1012)

1. division **2.** identity property for multiplication **3. (a)** $6; \frac{1}{6}$ **(b)** $12; \frac{3}{4}$ **(c)** $\frac{1}{6} \div \frac{3}{4}$ **(d)** $\frac{2}{9}$ **5.** -6 **7.** $\frac{1}{pq}$ **9.** $\frac{1}{xy}$ **11.** $\frac{2a^2b}{3}$ **13.** $\frac{m(m+2)}{3(m-4)}$ **15.** $\frac{2}{x}$ **17.** $\frac{8}{x}$ **19.** $\frac{a^2-5}{a^2+1}$ **21.** $\frac{3(p+2)}{2(2p+3)}$ **23.** $\frac{40-12p}{85p}$ **25.** $\frac{t(t-2)}{4}$ **27.** $\frac{-k}{2+k}$ **29.** $\frac{2x-7}{3x+1}$ **31.** $\frac{3m(m-3)}{(m-1)(m-8)}$ **33.** $\frac{6}{5}$

SECTION 14.6 (pages 1021–1024)

1. proposed; original **2.** extraneous; extraneous **3.** expression; $\frac{43}{40}x$ **5.** equation; $\left\{\frac{40}{43}\right\}$ **7.** expression; $-\frac{1}{10}y$ **9.** $\{-6\}$ **11.** $\{-15\}$ **13.** $\{7\}$ **15.** $\{-15\}$ **17.** $\{-5\}$ **19.** $\{-6\}$ **21.** $\{5\}$ **23.** $\{12\}$ **25.** $\{2\}$ **27.** $x \neq -2, x \neq 0$ **29.** $x \neq -3, x \neq 4, x \neq -\frac{1}{2}$ **31.** $x \neq -9, x \neq 1, x \neq -2, x \neq 2$ **33.** $\left\{\frac{20}{9}\right\}$ **35.** $\emptyset$ **37.** $4(x-1); \{3\}$ **39.** $\{3\}$ **41.** $\{-2, 12\}$ **43.** $\left\{-\frac{1}{5}, 3\right\}$ **45.** $\left\{-\frac{3}{5}, 3\right\}$ **47.** $\{-4\}$ **49.** $\emptyset$ **51.** $\{-1\}$ **53.** $\{-6\}$ **55.** $\left\{-6, \frac{1}{2}\right\}$ **57.** Transform the equation so that the terms with k are on one side and the remaining term is on the other. **58.** Factor out k on the left side. **59.** $F = \frac{ma}{k}$ **61.** $a = \frac{kF}{m}$ **63.** $y = mx + b$ **65.** $R = \frac{E - Ir}{I}$, or $R = \frac{E}{I} - r$ **67.** $b = \frac{2A - hB}{h}$, or $b = \frac{2A}{h} - B$ **69.** $a = \frac{2S - dnL}{dn}$, or $a = \frac{2S}{dn} - L$ **71.** $t = \frac{rs}{rs - 2s - 3r}$, or $t = \frac{-rs}{-rs + 2s + 3r}$ **73.** $c = \frac{ab}{b - a - 2ab}$, or $c = \frac{-ab}{-b + a + 2ab}$ **75.** $z = \frac{3y}{5 - 9xy}$, or $z = \frac{-3y}{9xy - 5}$

SECTION 14.7 (pages 1032–1035)

1. into a headwind: $(m-5)$ mph; with a tailwind: $(m+5)$ mph **2.** $\frac{D}{R} = \frac{d}{r}$ **3.** $\frac{1}{10}$ job per hr **4.** $\frac{2}{3}$ of the job **5. (a)** the amount **(b)** $5 + x$ **(c)** $\frac{5+x}{6} = \frac{13}{3}$ **7.** x represents the original numerator; $\frac{x+3}{(x-4)+3} = \frac{3}{2}; \frac{9}{5}$ **9.** x represents the original numerator; $\frac{x+2}{3x-2} = 1; \frac{2}{6}$ **11.** x represents the number; $\frac{1}{6}x = x + 5; -6$ **13.** x represents the quantity; $x + \frac{3}{4}x + \frac{1}{2}x + \frac{1}{3}x = 93; 36$ **15.** 18.809 min **17.** 3.565 hr **19.** 314.248 m per min **21.** table entries: (fourth column) $\frac{8}{4-x}, \frac{24}{4+x}; \frac{8}{4-x} = \frac{24}{4+x}$ **23.** 8 mph **25.** 32 mph **27.** 3 mph **29.** table entries: (second column) $\frac{1}{2}, \frac{1}{3}$; (fourth column) $\frac{1}{2}x, \frac{1}{3}x; \frac{1}{2}x + \frac{1}{3}x = 1$ **31.** $2\frac{2}{5}$ hr **33.** $4\frac{8}{19}$ hr **35.** 10 hr **37.** 36 hr **39.** 10 mL

SECTION 14.8 (pages 1038–1040)

1. direct **2.** direct **3.** inverse **4.** inverse **5.** inverse **6.** direct **7.** direct **8.** inverse **9. (a)** increases **(b)** decreases **10.** The customers in the lower-priced seats know more about the game than those in the higher-priced seats. **11.** inverse **12.** inverse **13.** direct **14.** direct **15.** 15 **17.** 300 **19.** 4 **21.** 6 **23.** 15 in.^{2} **25.** $42\frac{2}{3}$ in. **27.** 15 ft **29.** 20 lb per ft^2 **31.** 25 kg per hr **33.** 4

Chapter 14 REVIEW EXERCISES (pages 1046–1049)

1. $x \neq 3$ **2.** $x \neq 0$ **3.** $m \neq -1, m \neq 3$ **4.** $k \neq -5, k \neq -\frac{2}{3}$ **5. (a)** $-\frac{4}{7}$ **(b)** -16 **6. (a)** $\frac{11}{8}$ **(b)** $\frac{13}{22}$ **7. (a)** undefined **(b)** 1 **8. (a)** undefined **(b)** $\frac{1}{2}$ **9.** $\frac{b}{3a}$ **10.** -1 **11.** $\frac{-(2x+3)}{2}$ **12.** $\frac{2p+5q}{5p+q}$

Answers may vary in Exercises 13 and 14.

13. $\frac{-(4x-9)}{2x+3}, \frac{-4x+9}{2x+3}, \frac{4x-9}{-(2x+3)}, \frac{4x-9}{-2x-3}$ **14.** $\frac{-(8-3x)}{3-6x}, \frac{-8+3x}{3-6x}, \frac{8-3x}{-(3-6x)}, \frac{8-3x}{-3+6x}$ **15.** 2 **16.** $\frac{2}{3m^6}$ **17.** $\frac{5}{8}$ **18.** $\frac{r+4}{3}$ **19.** $\frac{3}{2}$ **20.** $\frac{y-2}{y-3}$ **21.** $\frac{p+5}{p+1}$ **22.** $\frac{3z+1}{z+3}$ **23.** 96 **24.** $108y^4$ **25.** $m(m+2)(m+5)$ **26.** $(x+3)(x+1)(x+4)$ **27.** $\frac{35}{56}$ **28.** $\frac{40}{4k}$ **29.** $\frac{15a}{10a^4}$ **30.** $\frac{-54}{18-6x}$ **31.** $\frac{15y}{50-10y}$ **32.** $\frac{4b(b+2)}{(b+3)(b-1)(b+2)}$ **33.** $\frac{15}{x}$ **34.** $-\frac{2}{p}$ **35.** $\frac{4k-45}{k(k-5)}$ **36.** $\frac{28+11y}{y(7+y)}$ **37.** $\frac{-2-3m}{6}$ **38.** $\frac{3(16-x)}{4x^2}$ **39.** $\frac{7a+6b}{(a-2b)(a+2b)}$ **40.** $\frac{-k^2-6k+3}{3(k+3)(k-3)}$ **41.** $\frac{5z-16}{z(z+6)(z-2)}$ **42.** $\frac{-13p+33}{p(p-2)(p-3)}$ **43.** $\frac{a}{b}$ **44.** $\frac{4(y-3)}{y+3}$ **45.** $\frac{6(3m+2)}{2m-5}$ **46.** $\frac{(q-p)^2}{pq}$ **47.** $\frac{xw+1}{xw-1}$ **48.** $\frac{1-r-t}{1+r+t}$ **49.** $\left\{\frac{35}{6}\right\}$ **50.** $\{-16\}$ **51.** $\{-4\}$ **52.** $\emptyset$ **53.** $\{3\}$ **54.** $t = \frac{Ry}{m}$ **55.** $y = \frac{4x+5}{3}$ **56.** $t = \frac{rs}{s-r}$ **57.** $\frac{20}{15}$ **58.** $\frac{3}{18}$ **59.** 1.843 hr **60.** 800.897 m per min **61.** $7\frac{1}{2}$ min **62.** $3\frac{1}{13}$ hr **63.** $\frac{36}{5}$ **64.** 4 cm **65.** $\frac{m+7}{(m-1)(m+1)}$ **66.** $8p^2$ **67.** $\frac{1}{6}$ **68.** 3 **69.** $\frac{z+7}{(z+1)(z-1)^2}$ **70.** $\frac{x-7}{2x+3}$ **71.** $\{4\}$ **72.** $\{-2, 3\}$ **73.** $\{2\}$ **74.** $w = v - at$ **75.** 3 **76.** $7\frac{1}{2}$ hr **77.** table entries: (third column) $x + 50$, $x - 50$; (fourth column) $\frac{400}{x+50}, \frac{200}{x-50}$; 150 km per hr **78.** 10 hr **79.** 50 amps **80.** inverse

Chapter 14 TEST (pages 1050–1051)

1. $x \neq -2, x \neq 4$ **2. (a)** $\frac{11}{6}$ **(b)** undefined **3.** (Answers may vary.) $\frac{-(6x-5)}{2x+3}, \frac{-6x+5}{2x+3}, \frac{6x-5}{-(2x+3)}, \frac{6x-5}{-2x-3}$ **4.** $-3x^2y^3$ **5.** $\frac{3a+2}{a-1}$ **6.** $\frac{25}{27}$ **7.** $\frac{3k-2}{3k+2}$ **8.** $\frac{a-1}{a+4}$ **9.** $\frac{x-5}{3-x}$ **10.** $150p^5$ **11.** $(2r+3)(r+2)(r-5)$ **12.** $\frac{240p^2}{64p^3}$ **13.** $\frac{21}{42m-84}$ **14.** 2 **15.** $\frac{-14}{5(y+2)}$ **16.** $\frac{x^2+x+1}{3-x}$, or $\frac{-x^2-x-1}{x-3}$ **17.** $\frac{-m^2+7m+2}{(2m+1)(m-5)(m-1)}$ **18.** $\frac{2k}{3p}$ **19.** $\frac{-2-x}{4+x}$ **20.** $\left\{-\frac{1}{2}, 1\right\}$ **21.** $\left\{-\frac{1}{2}, 5\right\}$ **22.** $\left\{-\frac{1}{2}\right\}$ **23.** $D = \frac{dF-k}{F}$, or $D = d - \frac{k}{F}$ **24.** -4 **25.** 3 mph **26.** $2\frac{2}{9}$ hr **27.** 27 days **28.** 27

CHAPTER 15 Systems of Linear Equations and Inequalities

SECTION 15.1 (pages 1059–1062)

1. system of linear equations; same **2.** ordered pair; true **3.** inconsistent; no; independent **4.** solution; consistent **5.** dependent; consistent; infinitely many **6.** parallel; solution **7.** It is not a solution of the system because it is not a solution of the second equation, $2x + y = 4$. **8.** $\{(x, y) \mid 3x - 2y = 4\}$ **9.** B, because the ordered pair must be in quadrant II. **10.** D, because the ordered pair must be on the y-axis, with $y < 0$. **11.** no **13.** yes **15.** yes **17.** no

We show the graphs here only for Exercises 19–23.

19. $\{(4, 2)\}$

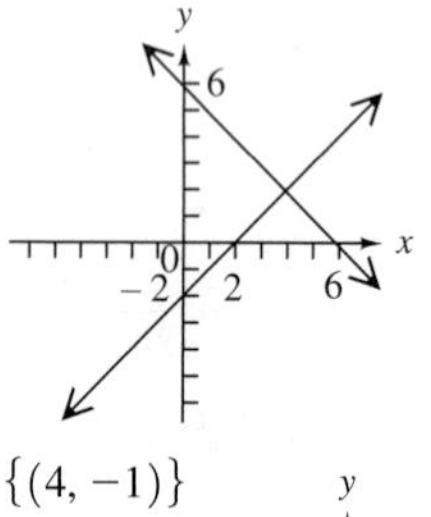

21. $\{(0, 4)\}$

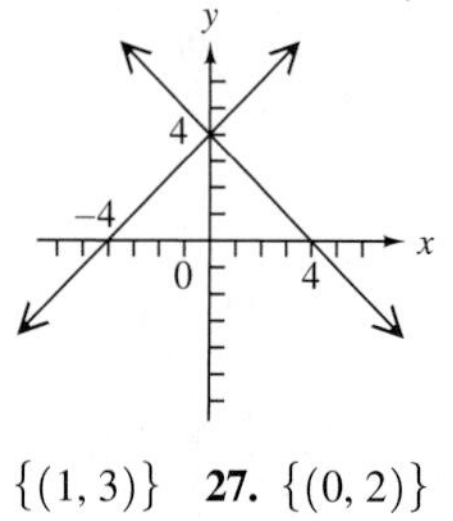

23. $\{(4, -1)\}$

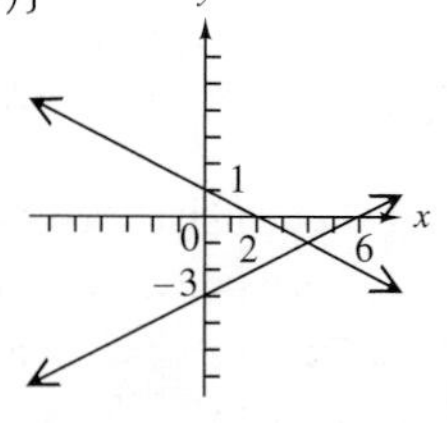

25. $\{(1, 3)\}$ **27.** $\{(0, 2)\}$ **29.** $\emptyset$ (inconsistent system) **31.** $\{(x, y) \mid 5x - 3y = 2\}$ (dependent equations) **33.** $\{(4, -3)\}$ **35.** $\{(x, y) \mid 2x - y = 4\}$ (dependent equations) **37.** $\emptyset$ (inconsistent system) **39. (a)** neither **(b)** intersecting lines **(c)** one solution **41. (a)** dependent **(b)** one line **(c)** infinite number of solutions **43. (a)** inconsistent **(b)** parallel lines **(c)** no solution **45. (a)** neither **(b)** intersecting lines **(c)** one solution **47. (a)** 40 **(b)** 30 **49.** 1980–2000 **51.** 2006; about 600 (million) units **53.** The slope would be negative. Sales of CDs were decreasing during this period. **54.** The slope would be positive. Sales of digital downloads were increasing during this period.

SECTION 15.2 (pages 1069–1070)

1. The student must find the value of y and write the solution as an ordered pair. The solution set is $\{(3, 0)\}$. **2.** The true result $0 = 0$ means that the system has an infinite number of solutions. The solution set is $\{(x, y) \mid x + y = 4\}$. **3.** A false statement, such as $0 = 3$, occurs. **4.** A true statement, such as $0 = 0$, occurs. **5.** $\{(3, 9)\}$ **7.** $\{(7, 3)\}$ **9.** $\{(-2, 4)\}$ **11.** $\{(-4, 8)\}$ **13.** $\{(3, -2)\}$ **15.** $\{(x, y) \mid 3x - y = 5\}$ **17.** $\emptyset$ **19.** $\{(x, y) \mid 2x - y = -12\}$ **21.** $\left\{\left(\frac{1}{3}, -\frac{1}{2}\right)\right\}$ **23.** $\{(2, -3)\}$ **25.** $\{(2, -4)\}$ **27.** $\{(-4, 2)\}$ **29.** $\{(7, -3)\}$ **31.** $\{(2, 3)\}$ **33.** To find the total cost, multiply the number of bicycles (x) by the cost per bicycle (400 dollars) and add the fixed cost (5000 dollars). Thus $y_1 = 400x + 5000$ gives this total cost (in dollars). **34.** $y_2 = 600x$ **35.** $y_1 = 400x + 5000$, $y_2 = 600x$; solution set: $\{(25, 15{,}000)\}$ **36.** 25; 15,000; 15,000

SECTION 15.3 (pages 1075–1077)

1. true **2.** false; Multiply by -3. **3.** true **4.** true **5.** $\{(-1, 3)\}$ **7.** $\{(-1, -3)\}$ **9.** $\{(-2, 3)\}$ **11.** $\left\{\left(\frac{1}{2}, 4\right)\right\}$ **13.** $\{(3, -6)\}$ **15.** $\{(7, 4)\}$ **17.** $\{(0, 4)\}$ **19.** $\{(-4, 0)\}$ **21.** $\{(0, 0)\}$ **23.** $\emptyset$ **25.** $\{(x, y) \mid x - 3y = -4\}$ **27.** $\{(0, 7)\}$ **29.** $\{(-6, 5)\}$ **31.** $\left\{\left(-\frac{6}{5}, \frac{4}{5}\right)\right\}$ **33.** $\left\{\left(\frac{1}{8}, -\frac{5}{6}\right)\right\}$ **35.** $\emptyset$ **37.** $\{(x, y) \mid 2x + y = 0\}$ **39.** $\{(11, 15)\}$ **41.** $\left\{\left(13, -\frac{7}{5}\right)\right\}$ **43.** $\{(6, -4)\}$ **45.** $5.66 = 2001a + b$ **46.** $7.89 = 2010a + b$ **47.** $2001a + b = 5.66$, $2010a + b = 7.89$; solution set: $\{(0.248, -490.590)\}$ **48. (a)** $y = 0.248x - 490.590$ **(b)** \$7.39; This is a bit more than the actual figure.

SECTION 15.4 (pages 1084–1089)

1. D **2.** A **3.** B **4.** B **5.** D **6.** D **7.** C **8.** C **9.** the second number; $x - y = 48$; The two numbers are 73 and 25. **11.** *The Phantom of the Opera:* 9803; *Cats:* 7485 **13.** *Harry Potter and the Deathly Hallows Part 2:* \$381 million; *Transformers: Dark of the Moon:* \$352 million **15.** Terminal Tower: 708 ft; Key Tower: 947 ft **17.** variables; width; Equation (1): $x = 38 + y$; length; twice; Equation (2): $2x + 2y = 188$; length: 66 yd; width: 28 yd **19. (a)** 45 units **(b)** Do not produce—the product will lead to a loss. **21.** table entries: (third column) $1x$ (or x), $10y$; 46 ones; 28 tens **23.** 5 DVDs of *Moneyball;* 2 Blu-ray discs of *The Concert for George* **25.** table entries: (second column) 4%, or 0.04; (third column) $0.04y$; Equation (1): $x = 2y$; Equation (2): $0.05x + 0.04y = 350$; \$2500 at 4%; \$5000 at 5% **27.** U2: \$92; Taylor Swift: \$72 **29.** table entries: (third column) $0.40x$, $0.70y$, $0.50(120)$ (or 60); 80 L of 40% solution; 40 L of 70% solution **31.** table entries: (second column) 3, 4; (third column) $6x$, $3y$, $4(90)$ (or 360); 30 lb at \$6 per lb; 60 lb at \$3 per lb **33.** nuts: 40 lb; raisins: 20 lb **35.** table entries: (third column) 4.5, 4.5; (fourth column) $4.5x$, $4.5y$; Equation (1): $4.5x + 4.5y = 495$; Equation (2): $x = 10 + y$; 60 mph; 50 mph **37.** bicycle: 13.5 mph; car: 49.3 mph **39.** car leaving Cincinnati: 55 mph; car leaving Toledo: 70 mph **41.** Roberto: 17.5 mph; Juana: 12.5 mph **43.** table entries: (third column) 3, 3; (fourth column) 24; boat: 10 mph; current: 2 mph **45.** plane: 470 mph; wind: 30 mph

SECTION 15.5 (pages 1093–1094)

1. C **2.** A **3.** B **4.** D

5.

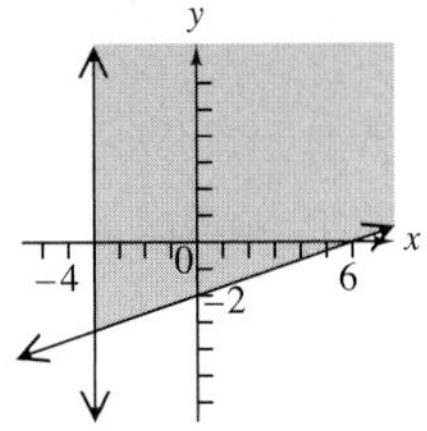

6.

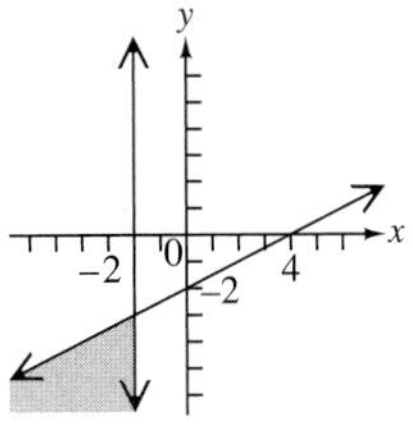

7.

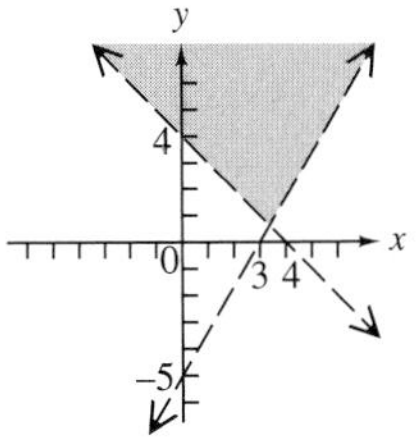

8.

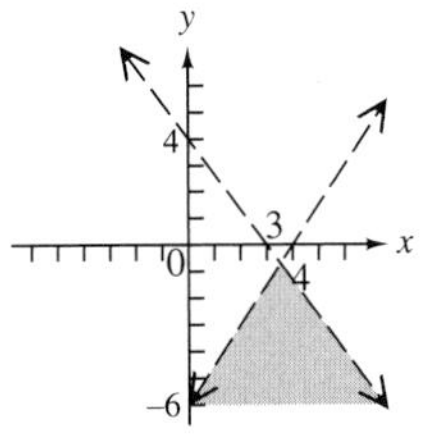

9.

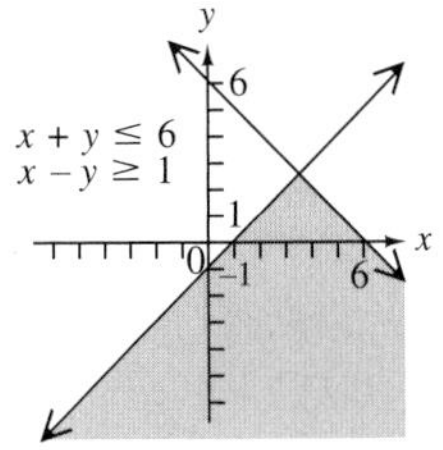

11.

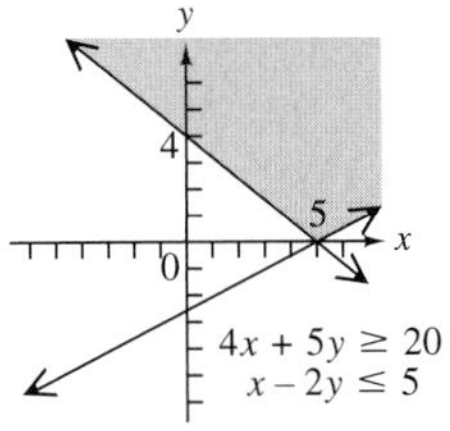

13.

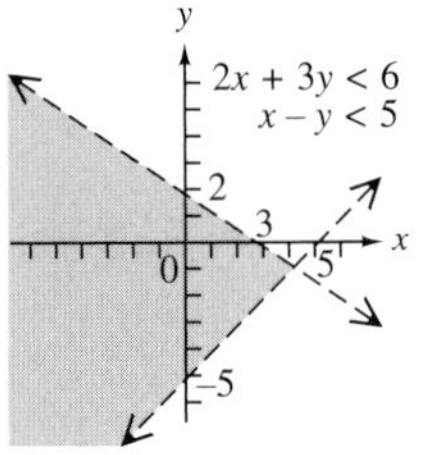

15.

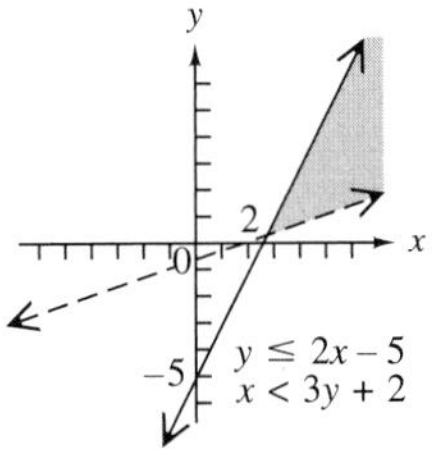

17.

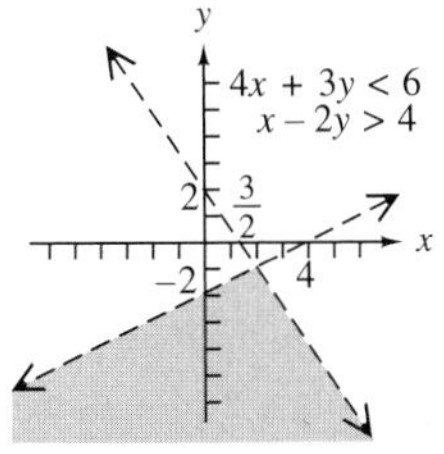

19.

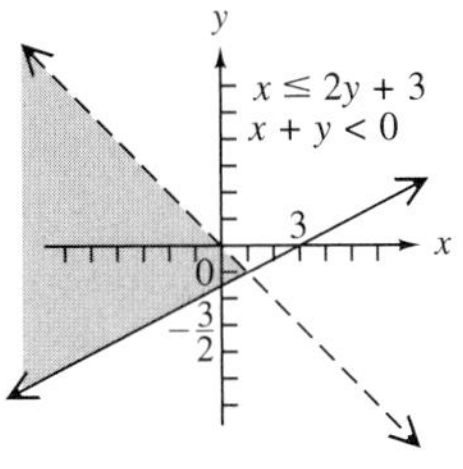

21.

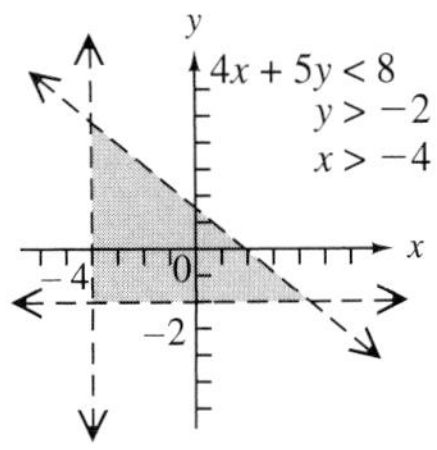

23.

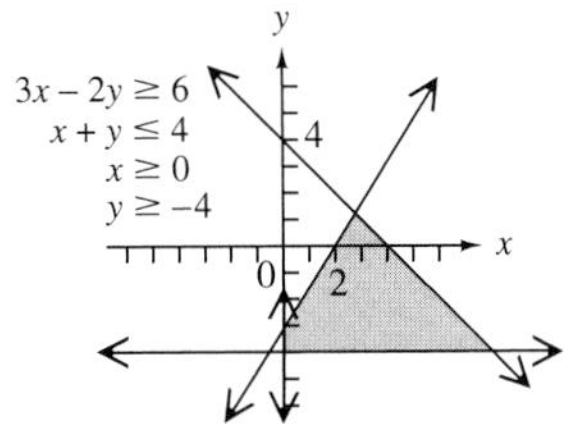

25. There are infinitely many points in any of the shaded regions. There are, however, points that are not in the shaded regions, and thus there are ordered pairs that are not solutions.

Chapter 15 REVIEW EXERCISES (pages 1098–1100)

1. yes **2.** no

We do not show the graphs for Exercises 3–6.

3. $\{(3, 1)\}$ **4.** $\{(0, -2)\}$ **5.** $\{(x, y) \mid x - 2y = 2\}$ **6.** $\emptyset$ **7.** $\{(2, 1)\}$ **8.** $\{(3, 5)\}$ **9.** $\{(6, 4)\}$ **10.** $\emptyset$ **11.** $\{(7, 1)\}$ **12.** $\{(-5, -2)\}$ **13.** $\{(-4, 3)\}$ **14.** $\{(x, y) \mid 3x - 4y = 9\}$ **15.** $\{(9, 2)\}$ **16.** $\left\{\left(\frac{10}{7}, -\frac{9}{7}\right)\right\}$ **17.** $\{(8, 9)\}$ **18.** $\{(2, 1)\}$ **19.** $\{(7, -2)\}$ **20.** $\{(-4, 2)\}$ **21.** Pizza Hut: 7542 locations; Domino's: 4929 locations **22.** *Reader's Digest*: 5.7 million; *People:* 3.6 million **23.** table entries: (first column) x; (second column) 0.90; (third column) $0.90y$, 100 (1) (or 100); 25 lb of \$1.30 candy; 75 lb of \$0.90 candy **24.** table entries: (first column) y, 20; (second column) 20; (third column) $10x$; 13 twenties; 7 tens **25.** length: 27 m; width: 18 m **26.** plane: 250 mph; wind: 20 mph **27.** table entries: (second column) 0.04; (third column) $0.03x$, $0.04y$, 650; \$7000 at 3%; \$11,000 at 4% **28.** table entries: (second column) 0.70; (third column) $0.40x$, $0.70y$, $0.50(90)$ (or 45); 60 L of 40% solution; 30 L of 70% solution **29.** B **30.** B

31.

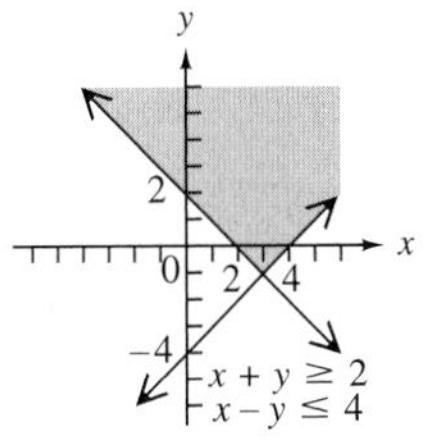

32.

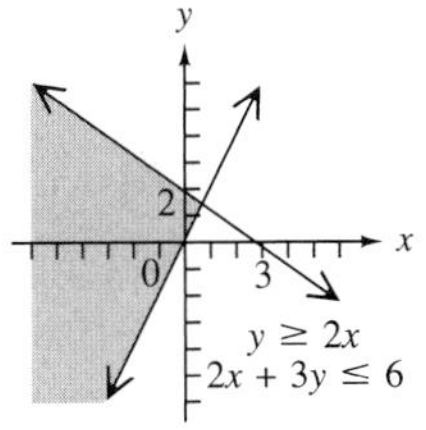

33.

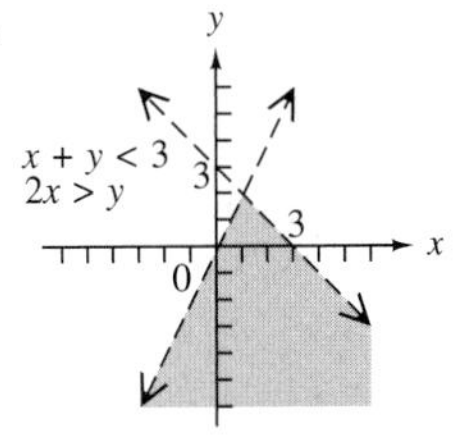

34. $\{(2, 0)\}$ **35.** $\{(-4, 15)\}$ **36.** $\emptyset$

37.

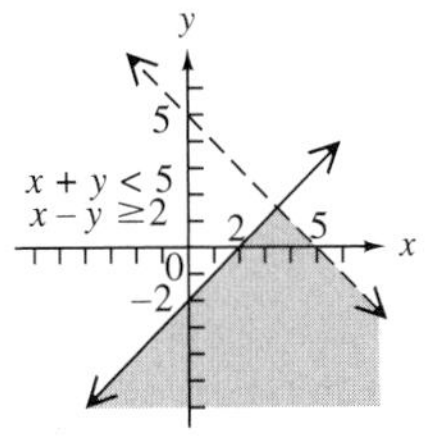

38.

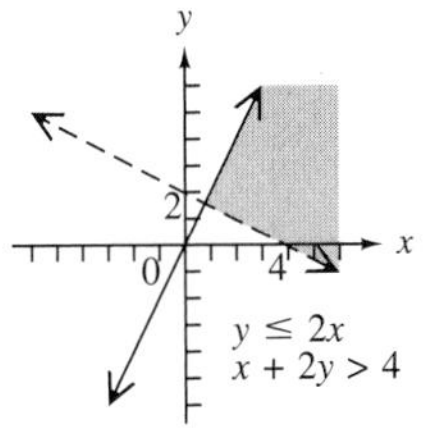

39.

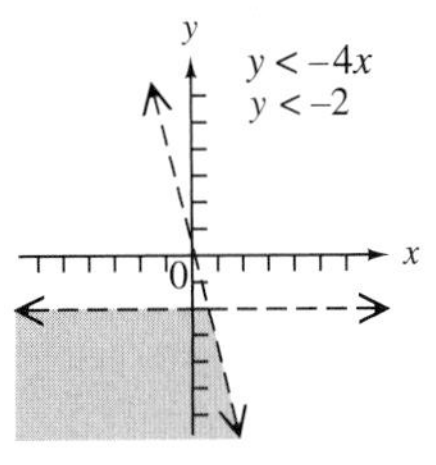

40. 8 in., 8 in., and 13 in. **41.** Giants: 21; Patriots: 17 **42.** **(a)** years 0–6 **(b)** year 6; about \$650

ANSWERS

Chapter 15 TEST (pages 1101–1102)

1. **(a)** no **(b)** no **(c)** yes **2.** $\{(2, -3)\}$ **3.** It has no solution. **4.** $\{(1, -6)\}$ **5.** $\{(-35, 35)\}$ **6.** $\{(5, 6)\}$ **7.** $\{(-1, 3)\}$ **8.** $\{(-1, 3)\}$ **9.** $\{(0, 0)\}$ **10.** $\emptyset$ **11.** $\{(x, y) \mid 3x - y = 6\}$ **12.** $\{(-15, 6)\}$ **13.** Memphis and Atlanta: 394 mi; Minneapolis and Houston: 1176 mi **14.** Statue of Liberty: 3.8 million; National World War II Memorial: 4.0 million **15.** 20 L of 15% solution; 30 L of 40% solution **16.** slower car: 45 mph; faster car: 60 mph

17.

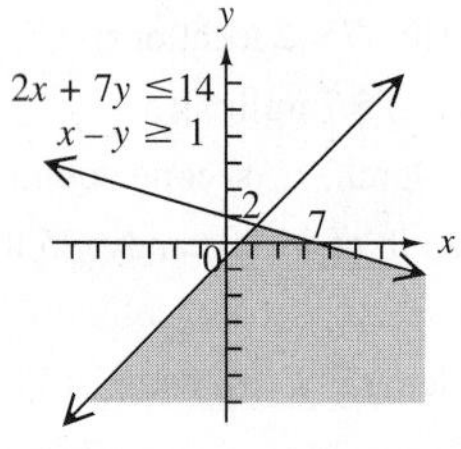

18.

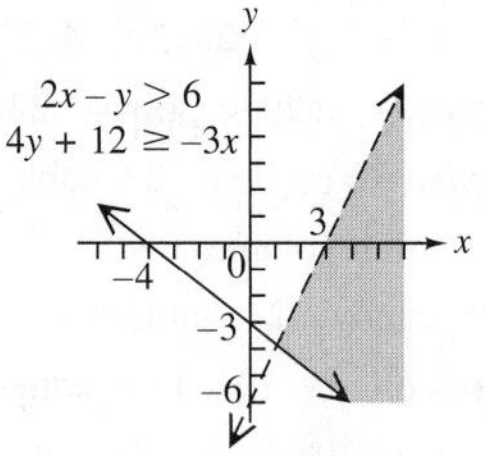

CHAPTER 16 Roots and Radicals

SECTION 16.1 (pages 1111–1114)

1. true **2.** false; A negative number has no real square roots. **3.** false; Zero has only one square root. **4.** true **5.** true **6.** false; A positive number has just one real cube root. **7.** $-3, 3$ **9.** $-8, 8$ **11.** $-13, 13$ **13.** $-\frac{5}{14}, \frac{5}{14}$ **15.** $-30, 30$ **17.** a must be positive. **18.** a must be positive. **19.** a must be negative. **20.** a must be negative. **21.** 64; 8 **23.** 1 **25.** 7 **27.** -16 **29.** $-\frac{12}{11}$ **31.** 0.8 **33.** not a real number **35.** not a real number **37.** 100 **39.** 19 **41.** $\frac{2}{3}$ **43.** $3x^2 + 4$ **45.** 9 and 10 **46.** 6 and 7 **47.** 7 and 8 **48.** 5 and 6 **49.** 4 and 5 **50.** 3 and 4 **51.** rational; 5 **53.** irrational; 5.385 **55.** rational; -8 **57.** irrational; -17.321 **59.** not a real number **61.** irrational; 34.641 **63.** C **65.** $c = 17$ **67.** $b = 8$ **69.** $c \approx 11.705$ **71.** 24 cm **73.** 80 ft **75.** 195 ft **77.** 158.6 ft **79.** 11.1 ft **81.** 9.434 **83.** 4; 4 **85.** 5 **87.** 8 **89.** -3 **91.** -6 **93.** 2 **95.** 4 **97.** 6 **99.** not a real number **101.** -5 **103.** -4

SECTION 16.2 (pages 1121–1124)

1. false; $\sqrt{(-6)^2} = \sqrt{36} = 6$ **2.** true **3.** false; $2\sqrt{7}$ represents the product of 2 and $\sqrt{7}$. **4.** true **5.** $\sqrt{15}$ **7.** $\sqrt{22}$ **9.** $\sqrt{42}$ **11.** $\sqrt{13r}$ **13.** A **14.** Yes; because a prime number cannot be factored (except as the product of itself and 1), it cannot have any perfect square factors (except 1). **15.** $3\sqrt{5}$ **17.** $2\sqrt{6}$ **19.** $3\sqrt{10}$ **21.** $5\sqrt{3}$ **23.** $5\sqrt{5}$ **25.** cannot be simplified **27.** $4\sqrt{10}$ **29.** $-10\sqrt{7}$ **31.** $6\sqrt{13}$ **33.** $25\sqrt{2}$ **35.** $5\sqrt{10}$ **37.** $6\sqrt{2}$ **39.** $3\sqrt{6}$ **41.** $12\sqrt{2}$ **43.** 24 **45.** $6\sqrt{10}$ **47.** $12\sqrt{5}$ **49.** $30\sqrt{5}$ **51.** *Method 1:* $\sqrt{8} \cdot \sqrt{32} = \sqrt{8 \cdot 32} = \sqrt{256} = 16$; *Method 2:* $\sqrt{8} = 2\sqrt{2}$ and $\sqrt{32} = 4\sqrt{2}$, so $\sqrt{8} \cdot \sqrt{32} = 2\sqrt{2} \cdot 4\sqrt{2} = 8 \cdot 2 = 16$; The same answer results. Either method can be used to obtain the correct answer. **52.** *Method 1:* $\sqrt{288} = \sqrt{144 \cdot 2} = \sqrt{144} \cdot \sqrt{2} = 12\sqrt{2}$; *Method 2:* $\sqrt{288} = \sqrt{48 \cdot 6} = \sqrt{48} \cdot \sqrt{6} = \sqrt{16 \cdot 3} \cdot \sqrt{6} = 4\sqrt{3} \cdot \sqrt{6} = 4\sqrt{18} = 4\sqrt{9 \cdot 2} = 4 \cdot 3\sqrt{2} = 12\sqrt{2}$; The same answer results. Method 1 yields the answer more quickly using the greatest perfect square factor first.

53. $\frac{4}{15}$ **55.** $\frac{\sqrt{7}}{4}$ **57.** $\frac{\sqrt{2}}{5}$ **59.** 5 **61.** $\frac{25}{4}$ **63.** $6\sqrt{5}$ **65.** m **67.** y^2 **69.** $6z$ **71.** $20x^3$ **73.** $3x^4\sqrt{2}$ **75.** $3c^7\sqrt{5}$ **77.** $z^2\sqrt{z}$ **79.** $a^6\sqrt{a}$ **81.** $8x^3\sqrt{x}$ **83.** x^3y^6 **85.** $9m^2n$ **87.** $\frac{\sqrt{7}}{x^5}$ **89.** $\frac{y^2}{10}$ **91.** $\frac{x^3}{y^4}$ **93.** $2\sqrt[3]{5}$ **95.** $3\sqrt[3]{2}$ **97.** $4\sqrt[3]{2}$ **99.** $2\sqrt[4]{5}$ **101.** $\frac{2}{3}$ **103.** $-\frac{6}{5}$ **105.** p **107.** x^3 **109.** $4z^2$ **111.** $7a^3b$ **113.** $2t\sqrt[3]{2t^2}$ **115.** $\frac{m^4}{2}$ **117.** 6 cm **119.** 6 in. **121.** D

SECTION 16.3 (pages 1127–1128)

1. radicand; index; like; 7; 7 **2.** like; 4; $3xy^3$ **3.** radicands; unlike **4.** indexes or roots; unlike **5.** distributive **6.** like; 1; $5\sqrt{5x}$ **7.** $7\sqrt{3}$ **9.** $-5\sqrt{7}$ **11.** $5\sqrt{17}$ **13.** $5\sqrt{7}$ **15.** $2\sqrt{6}$ **17.** These unlike radicals cannot be combined. **19.** $7\sqrt{3}$ **21.** $11\sqrt{5}$ **23.** $15\sqrt{2}$ **25.** $-6\sqrt{2}$ **27.** $17\sqrt{7}$ **29.** $-16\sqrt{2} - 8\sqrt{3}$ **31.** $20\sqrt{2} + 6\sqrt{3} - 15\sqrt{5}$ **33.** $4\sqrt{2}$ **35.** $22\sqrt{2}$ **37.** $11\sqrt{3}$ **39.** $3\sqrt{21}$ **41.** $\sqrt{2x}$ **43.** $7\sqrt{3r}$ **45.** $5\sqrt{x}$ **47.** $3x\sqrt{6}$ **49.** 0 **51.** $-20\sqrt{2k}$ **53.** $42x\sqrt{5z}$ **55.** $-\sqrt[3]{2}$ **57.** $6\sqrt[3]{p^2}$ **59.** $21\sqrt[4]{m^3}$

SECTION 16.4 (pages 1133–1134)

1. We are actually multiplying by 1. The identity property for multiplication justifies our result. **3.** $\frac{6\sqrt{5}}{5}$ **5.** $\sqrt{5}$ **7.** $4\sqrt{2}$ **9.** $\frac{-\sqrt{33}}{3}$ **11.** $\frac{7\sqrt{15}}{5}$ **13.** $\frac{\sqrt{30}}{2}$ **15.** $\frac{16\sqrt{3}}{9}$ **17.** $\frac{-3\sqrt{2}}{10}$ **19.** $\frac{21\sqrt{5}}{5}$ **21.** $\sqrt{3}$ **23.** $\frac{\sqrt{2}}{2}$ **25.** $\frac{\sqrt{65}}{5}$ **27.** $\frac{\sqrt{21}}{3}$ **29.** $\frac{3\sqrt{14}}{4}$ **31.** $\frac{1}{6}$ **33.** 1 **35.** $\frac{\sqrt{7x}}{x}$ **37.** $\frac{2x\sqrt{xy}}{y}$ **39.** $\frac{x\sqrt{30xz}}{6}$ **41.** $\frac{3ar^2\sqrt{7rt}}{7t}$ **43.** B **44.** yes **45.** $\frac{\sqrt[3]{4}}{2}$ **47.** $\frac{\sqrt[3]{15}}{3}$ **49.** $\frac{\sqrt[3]{196}}{7}$ **51.** $\frac{\sqrt[3]{6y}}{2y}$ **53.** $\frac{\sqrt[3]{42mn^2}}{6n}$ **55.** **(a)** $\frac{9\sqrt{2}}{4}$ sec **(b)** 3.182 sec

SECTION 16.5 (pages 1139–1142)

1. 13 **2.** 1 **3.** 4 **4.** 8 **5.** Because multiplication must be performed before addition, it is incorrect to add -37 and -2. $-2\sqrt{15}$ cannot be simplified further. The final answer is $-37 - 2\sqrt{15}$. **6.** **(a)** $x - y$ **(b)** 14 **7.** $\sqrt{15} - \sqrt{35}$ **9.** $2\sqrt{10} + 30$ **11.** $4\sqrt{7}$ **13.** $57 + 23\sqrt{6}$ **15.** $81 + 14\sqrt{21}$ **17.** $71 - 16\sqrt{7}$ **19.** $37 + 12\sqrt{7}$ **21.** $a + 2\sqrt{a} + 1$ **23.** 23 **25.** 1 **27.** $y - 10$ **29.** $2\sqrt{3} - 2 + 3\sqrt{2} - \sqrt{6}$ **31.** $15\sqrt{2} - 15$ **33.** $\sqrt{30} + \sqrt{15} + 6\sqrt{5} + 3\sqrt{10}$ **35.** $\sqrt{5x} - \sqrt{10} - \sqrt{10x} + 2\sqrt{5}$ **37.** **(a)** $\sqrt{5} - \sqrt{3}$ **(b)** $\sqrt{6} + \sqrt{5}$ **38.** There would still be a radical in the denominator. He should multiply by $\frac{4 - \sqrt{3}}{4 - \sqrt{3}}$. **39.** $\frac{3 - \sqrt{2}}{7}$ **41.** $-4 - 2\sqrt{11}$ **43.** $1 + \sqrt{2}$ **45.** $-\sqrt{10} + \sqrt{15}$ **47.** $2\sqrt{5} + \sqrt{15} + 4 + 2\sqrt{3}$ **49.** $\frac{-6\sqrt{2} + 12 + \sqrt{10} - 2\sqrt{5}}{2}$ **51.** $\frac{12(\sqrt{x} - 1)}{x - 1}$ $(x \neq 1)$ **53.** $\frac{3(7 + \sqrt{x})}{49 - x}$ **55.** $\sqrt{11} - 2$

57. $\frac{\sqrt{3}+5}{8}$ **59.** $\frac{6-\sqrt{10}}{2}$ **61.** $\frac{2+\sqrt{2}}{3}$ **63.** **(a)** 4 in. **(b)** 3.4 in. **65.** $30+18x$ **66.** They are not like terms. **67.** $30+18\sqrt{5}$ **68.** They are not like radicals. **69.** Make the first term $30x$, so that $30x+18x=48x$. Make the first term $30\sqrt{5}$, so that $30\sqrt{5}+18\sqrt{5}=48\sqrt{5}$. **70.** Both like terms and like radicals are combined by adding their numerical coefficients. The variables in like terms are replaced by radicals in like radicals.

SECTION 16.6 (pages 1149–1152)

1. squaring; squared; original **2.** $a^2+2ab+b^2$; $a^2-2ab+b^2$ **3.** $\{49\}$ **5.** $\{7\}$ **7.** $\{85\}$ **9.** $\emptyset$ **11.** $\{-45\}$ **13.** $\left\{-\frac{3}{2}\right\}$ **15.** $\emptyset$ **17.** $\{121\}$ **19.** $\{8\}$ **21.** $\{1\}$ **23.** $\{6\}$ **25.** $\emptyset$ **27.** $\{5\}$ **29.** When the left side is squared, the result should be $x-1$, not $-(x-1)$. The correct solution set is $\{17\}$. **30.** $x^2-14x+49$ **31.** $\{12\}$ **33.** $\{5\}$ **35.** $\{0,3\}$ **37.** $\{-1,3\}$ **39.** $\{8\}$ **41.** $\{4\}$ **43.** $\{8\}$ **45.** $\{9\}$ **47.** $\{4,20\}$ **49.** $\{-5\}$ **51.** 21 **53.** 8 **55.** 17,616 ft^2 **57.** **(a)** 70.5 mph **(b)** 59.8 mph **(c)** 53.9 mph **59.** yes; 26 mi **61.** 47 mi **63.** $s=13$ units **64.** $6\sqrt{13}$ sq. units **65.** $h=\sqrt{13}$ units **66.** $3\sqrt{13}$ sq. units **67.** $6\sqrt{13}$ sq. units **68.** They are both $6\sqrt{13}$.

Chapter 16 REVIEW EXERCISES (pages 1157–1160)

1. $-7, 7$ **2.** $-9, 9$ **3.** $-14, 14$ **4.** $-11, 11$ **5.** $-16, 16$ **6.** $-27, 27$ **7.** 4 **8.** -0.6 **9.** -8 **10.** 3 **11.** not a real number **12.** -65 **13.** $\frac{12}{13}$ **14.** $-\frac{10}{9}$ **15.** 8 **16.** 48.3 cm **17.** irrational; 8.544 **18.** rational; 13 **19.** rational; -25 **20.** not a real number **21.** $4\sqrt{3}$ **22.** $-12\sqrt{2}$ **23.** $2\sqrt[3]{2}$ **24.** $5\sqrt[3]{3}$ **25.** 18 **26.** $16\sqrt{6}$ **27.** $-\frac{11}{20}$ **28.** $\frac{\sqrt{7}}{13}$ **29.** $\frac{\sqrt{5}}{6}$ **30.** $\frac{2}{15}$ **31.** $3\sqrt{2}$ **32.** $2\sqrt{2}$ **33.** r^9 **34.** x^5y^8 **35.** $9x^4\sqrt{2x}$ **36.** $\frac{6}{p}$ **37.** $a^7b^{10}\sqrt{ab}$ **38.** $11x^3y^5$ **39.** y^2 **40.** $6x^5$ **41.** $8\sqrt{11}$ **42.** $9\sqrt{2}$ **43.** $21\sqrt{3}$ **44.** $12\sqrt{3}$ **45.** 0 **46.** $3\sqrt{7}$ **47.** $2\sqrt{3}+3\sqrt{10}$ **48.** $2\sqrt{2}$ **49.** $6\sqrt{30}$ **50.** $5\sqrt{x}$ **51.** 0 **52.** $11k^2\sqrt{2n}$ **53.** $\frac{10\sqrt{3}}{3}$ **54.** $\frac{8\sqrt{10}}{5}$ **55.** $\sqrt{6}$ **56.** $\frac{\sqrt{10}}{5}$ **57.** $\sqrt{10}$ **58.** $\frac{\sqrt{42}}{21}$ **59.** $\frac{r\sqrt{x}}{4x}$ **60.** $\frac{\sqrt[3]{9}}{3}$ **61.** $-\sqrt{15}-9$ **62.** $3\sqrt{6}+12$ **63.** $22-16\sqrt{3}$ **64.** $2\sqrt{21}-\sqrt{14}+12\sqrt{2}-4\sqrt{3}$ **65.** -13 **66.** $x+4\sqrt{x}+4$ **67.** $-2+\sqrt{5}$ **68.** $\frac{3(1-\sqrt{x})}{1-x}$ $(x\neq 1)$ **69.** $\frac{-\sqrt{10}+3\sqrt{5}+\sqrt{2}-3}{7}$ **70.** $\frac{3+2\sqrt{6}}{3}$ **71.** $\frac{1+3\sqrt{7}}{4}$ **72.** $3+4\sqrt{3}$ **73.** $\emptyset$ **74.** $\{48\}$ **75.** $\{1\}$ **76.** $\{2\}$ **77.** $\{6\}$ **78.** $\{-3,-1\}$ **79.** $\{-2\}$ **80.** $\{4\}$ **81.** $\{7\}$ **82.** **(a)** B **(b)** F **(c)** D **(d)** A **(e)** C **(f)** A **83.** $11\sqrt{3}$ **84.** $\frac{5-\sqrt{2}}{23}$ **85.** $\frac{2\sqrt{10}}{5}$ **86.** $3a^2b^3\sqrt[3]{2ab}$ **87.** $-\sqrt{10}-5\sqrt{15}$ **88.** $\frac{4r\sqrt{3rs}}{3s}$ **89.** $\frac{2+\sqrt{13}}{2}$ **90.** $7-2\sqrt{10}$ **91.** $166+2\sqrt{7}$ **92.** $\{7\}$ **93.** $\emptyset$ **94.** $\{8\}$ **95.** **(a)** $(a+b)^2$, or $a^2+2ab+b^2$ **(b)** c^2+2ab **(c)** Subtract $2ab$ from each side to get $a^2+b^2=c^2$.

Chapter 16 TEST (pages 1161–1162)

1. $-20, 20$ **2.** **(a)** irrational **(b)** 11.916 **3.** a must be negative. **4.** 6 **5.** $-3\sqrt{6}$ **6.** $\frac{8\sqrt{2}}{5}$ **7.** $2\sqrt[3]{4}$ **8.** $4\sqrt{6}$ **9.** $9\sqrt{7}$ **10.** $11+2\sqrt{30}$ **11.** $4xy\sqrt{2y}$ **12.** 31 **13.** $6\sqrt{2}+2-3\sqrt{14}-\sqrt{7}$ **14.** $-5\sqrt{3x}$ **15.** **(a)** $6\sqrt{2}$ in. **(b)** 8.485 in. **16.** 50 ohms **17.** $\frac{5\sqrt{14}}{7}$ **18.** $\frac{\sqrt{6x}}{3x}$ **19.** $-\sqrt[3]{2}$ **20.** $\frac{-3(4+\sqrt{3})}{13}$ **21.** $\frac{\sqrt{3}+12\sqrt{2}}{3}$ **22.** $\emptyset$ **23.** $\{3\}$ **24.** $\{1,4\}$ **25.** $\{9\}$ **26.** 12 is not a solution. A check shows that it does not satisfy the original equation. The solution set is $\emptyset$.

CHAPTER 17 Quadratic Equations

SECTION 17.1 (pages 1168–1170)

1. C **2.** A **3.** D **4.** B **5.** According to the square root property, -9 is also a solution, so her answer was not completely correct. The solution set is $\{-9,9\}$, or $\{\pm 9\}$. **6.** The equation $x^2=21$ can be solved with the square root property. Because the square of x is 21, x can be either the positive or the negative square root of 21. The solution set is $\{-\sqrt{21},\sqrt{21}\}$, or $\{\pm\sqrt{21}\}$. **7.** $\{-7,8\}$ **9.** $\{-11,11\}$ **11.** $\left\{-\frac{5}{3},6\right\}$ **13.** $\{-9,9\}$ **15.** $\{-\sqrt{14},\sqrt{14}\}$ **17.** $\{-4\sqrt{3},4\sqrt{3}\}$ **19.** $\left\{-\frac{5}{2},\frac{5}{2}\right\}$ **21.** $\emptyset$ **23.** $\{-1.5,1.5\}$ **25.** $\{-\sqrt{3},\sqrt{3}\}$ **27.** $\left\{-\frac{2\sqrt{7}}{7},\frac{2\sqrt{7}}{7}\right\}$ **29.** $\{-3\sqrt{2},3\sqrt{2}\}$ **31.** $\{-2\sqrt{6},2\sqrt{6}\}$ **33.** $\left\{-\frac{2\sqrt{5}}{5},\frac{2\sqrt{5}}{5}\right\}$ **35.** $\{-2,8\}$ **37.** $\emptyset$ **39.** $\{8+3\sqrt{3},8-3\sqrt{3}\}$ **41.** $\left\{-3,\frac{5}{3}\right\}$ **43.** $\left\{0,\frac{3}{2}\right\}$ **45.** $\left\{\frac{5+\sqrt{30}}{2},\frac{5-\sqrt{30}}{2}\right\}$ **47.** $\left\{\frac{-1+3\sqrt{2}}{3},\frac{-1-3\sqrt{2}}{3}\right\}$ **49.** $\{-10+4\sqrt{3},-10-4\sqrt{3}\}$ **51.** $\left\{\frac{1+4\sqrt{3}}{4},\frac{1-4\sqrt{3}}{4}\right\}$ **53.** about $\frac{1}{2}$ sec **55.** 9 in. **57.** 5%

SECTION 17.2 (pages 1177–1178)

1. D **2.** yes **3.** 25; $(x+5)^2$ **5.** 1; $(x+1)^2$ **7.** $\frac{25}{4}$; $\left(p-\frac{5}{2}\right)^2$ **9.** $\{1,3\}$ **11.** $\{-3,-2\}$ **13.** $\{-1+\sqrt{6},-1-\sqrt{6}\}$ **15.** $\{4+2\sqrt{3},4-2\sqrt{3}\}$ **17.** $\{-3\}$ **19.** $\left\{\frac{-1+\sqrt{5}}{2},\frac{-1-\sqrt{5}}{2}\right\}$ **21.** $\left\{-\frac{3}{2},\frac{1}{2}\right\}$ **23.** $\left\{\frac{2+\sqrt{14}}{2},\frac{2-\sqrt{14}}{2}\right\}$ **25.** $\emptyset$ **27.** $\left\{\frac{-7+\sqrt{97}}{6},\frac{-7-\sqrt{97}}{6}\right\}$ **29.** $\{-4,2\}$ **31.** $\{4+\sqrt{3},4-\sqrt{3}\}$ **33.** $\{1+\sqrt{6},1-\sqrt{6}\}$ **35.** 1 sec and 5 sec **37.** 3 sec and 5 sec **39.** 75 ft by 100 ft

SECTION 17.3 (pages 1183–1184)

1. $2a$ should be in the denominator for $-b$ as well. The correct formula is $x = \dfrac{-b \pm \sqrt{b^2 - 4ac}}{2a}$. **2.** There is no real number solution, so the solution set is $\emptyset$. **3.** 4; 5; −9 **5.** 3; −4; −2 **7.** 3; 7; 0 **9.** $\{-13, 1\}$ **11.** $\{2\}$ **13.** $\left\{-1, \frac{5}{2}\right\}$ **15.** $\left\{\frac{-6+\sqrt{26}}{2}, \frac{-6-\sqrt{26}}{2}\right\}$ **17.** $\{-1, 0\}$ **19.** $\emptyset$ **21.** $\left\{\frac{-5+\sqrt{13}}{6}, \frac{-5-\sqrt{13}}{6}\right\}$ **23.** $\left\{0, \frac{12}{7}\right\}$ **25.** $\{-2\sqrt{6}, 2\sqrt{6}\}$ **27.** $\left\{-\frac{2}{5}, \frac{2}{5}\right\}$ **29.** $\left\{\frac{6+2\sqrt{6}}{3}, \frac{6-2\sqrt{6}}{3}\right\}$ **31.** $\emptyset$ **33.** $\left\{-\frac{2}{3}, \frac{4}{3}\right\}$ **35.** $\left\{\frac{-1+\sqrt{73}}{6}, \frac{-1-\sqrt{73}}{6}\right\}$ **37.** $\emptyset$ **39.** $\left\{-1, \frac{5}{2}\right\}$ **41.** $\{-3+\sqrt{5}, -3-\sqrt{5}\}$ **43.** 3.5 ft **45.** −8, 16; Only 16 board feet is a reasonable answer.

SECTION 17.4 (pages 1189–1190)

1. parabola **2.** vertex; highest **3.** axis of symmetry; mirror **4.** one; one; no **5.** $(0, 0)$

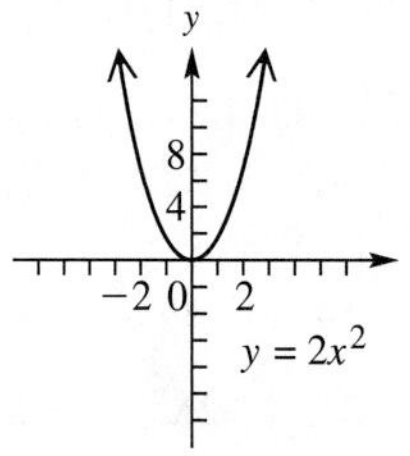

7. $(0, -4)$

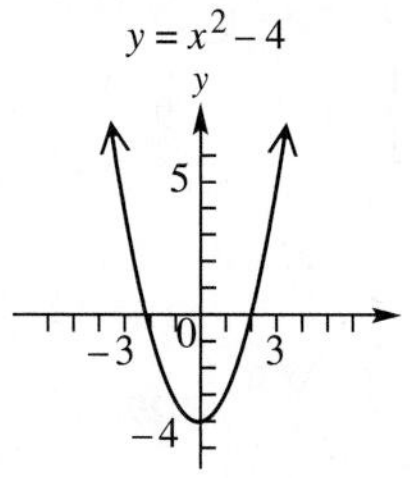

9. $(0, 2)$

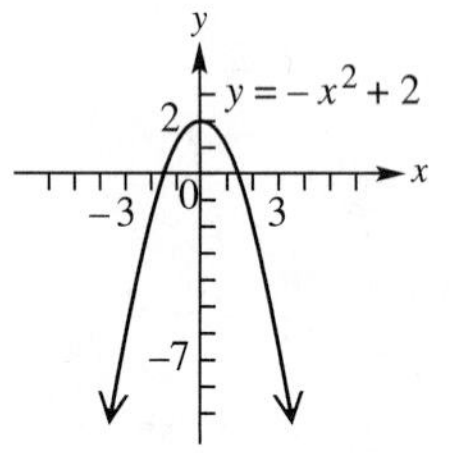

11. $(-1, 0)$

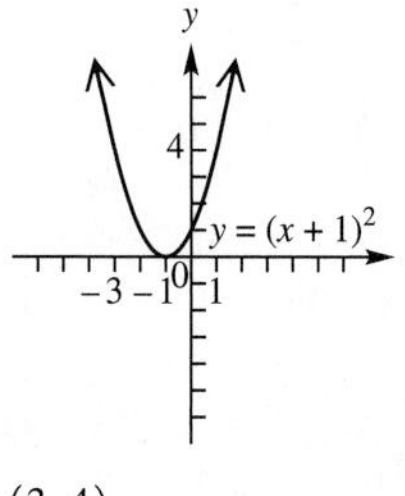

13. $(-1, 2)$

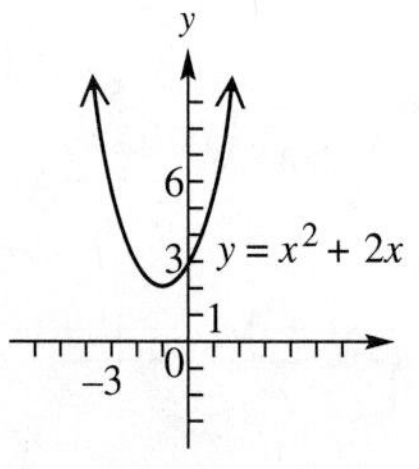

15. $(3, 4)$

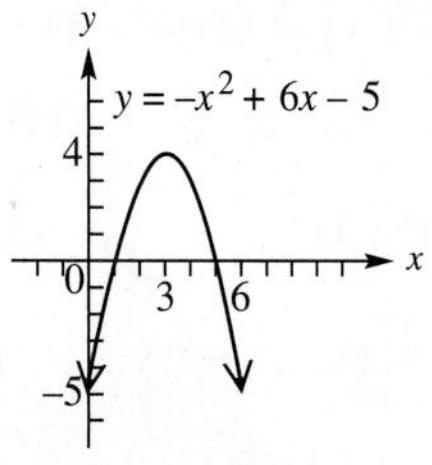

17. If $a > 0$, it opens upward, and if $a < 0$, it opens downward. **18. (a)** none **(b)** one **(c)** two **19.** $y = \frac{11}{5625}x^2$

SECTION 17.5 (pages 1196–1197)

1. 3; 3; $(1, 3)$ **2.** 4; 4; $(2, 4)$ **3.** 5; 5; $(3, 5)$ **4.** 6; 6; $(4, 6)$ **5.** The graph consists of the five points $(0, 2)$, $(1, 3)$, $(2, 4)$, $(3, 5)$, and $(4, 6)$. **6.** The graph is the line through the points $(0, 2)$, $(1, 3)$, $(2, 4)$, $(3, 5)$, and $(4, 6)$. **7.** not a function; domain: $\{-4, -2, 0\}$; range: $\{3, 1, 5, -8\}$ **9.** function; domain: {A, B, C, D, E}; range: $\{2, 3, 6, 4\}$ **11.** not a function; domain: $\{-4, -2, 0, 2, 3\}$; range: $\{-2, 0, 1, 2, 3\}$ **13.** function **15.** not a function **17.** function **19.** function **21.** not a function **23.** function **25. (a)** 11 **(b)** 3 **(c)** −9 **27. (a)** 4 **(b)** 2 **(c)** 14 **29. (a)** absolute value; 2; 2 **(b)** 0 **(c)** 3 **31.** {(1980, 14.1), (1990, 19.8), (2000, 31.1), (2010, 40.0)}; yes **33.** $g(1980) = 14.1$ (million); $g(2000) = 31.1$ (million) **35.** For the year 2008, the function gives 38.0 million foreign-born residents in the United States.

Chapter 17 REVIEW EXERCISES (pages 1202–1204)

1. $\{-7, 4\}$ **2.** $\{-9, -5\}$ **3.** $\left\{-5, \frac{3}{2}\right\}$ **4.** $\{-13, 13\}$ **5.** $\{-12, 12\}$ **6.** $\{-\sqrt{37}, \sqrt{37}\}$ **7.** $\{-8\sqrt{2}, 8\sqrt{2}\}$ **8.** $\{-7, 3\}$ **9.** $\{3+\sqrt{10}, 3-\sqrt{10}\}$ **10.** $\left\{\frac{-1+\sqrt{14}}{2}, \frac{-1-\sqrt{14}}{2}\right\}$ **11.** $\emptyset$ **12.** $\left\{-\frac{5}{3}\right\}$ **13.** $\{-5, -1\}$ **14.** $\{-2+\sqrt{11}, -2-\sqrt{11}\}$ **15.** $\{-1+\sqrt{6}, -1-\sqrt{6}\}$ **16.** $\left\{\frac{-4+\sqrt{22}}{2}, \frac{-4-\sqrt{22}}{2}\right\}$ **17.** $\left\{-\frac{7}{2}\right\}$ **18.** $\emptyset$ **19.** 2.5 sec **20.** 6, 8, 10 **21.** $\{-4, 9\}$ **22.** $\left\{0, \frac{14}{5}\right\}$ **23.** $\emptyset$ **24.** $\left\{\frac{3}{2}\right\}$ **25.** $\left\{\frac{-1+\sqrt{29}}{4}, \frac{-1-\sqrt{29}}{4}\right\}$ **26.** $\left\{\frac{2+\sqrt{10}}{2}, \frac{2-\sqrt{10}}{2}\right\}$ **27.** $\left\{\frac{1+\sqrt{21}}{10}, \frac{1-\sqrt{21}}{10}\right\}$ **28.** $\left\{\frac{-3+\sqrt{41}}{2}, \frac{-3-\sqrt{41}}{2}\right\}$ **29.** $\{-3+\sqrt{19}, -3-\sqrt{19}\}$ **30.** $\left\{\frac{2+\sqrt{2}}{2}, \frac{2-\sqrt{2}}{2}\right\}$

31. $(0, 5)$

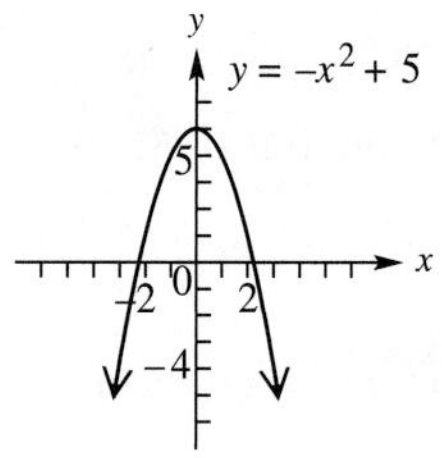

32. $(1, 0)$

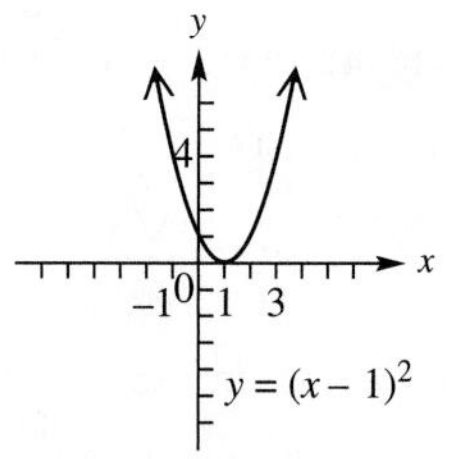

33. $(1, 4)$

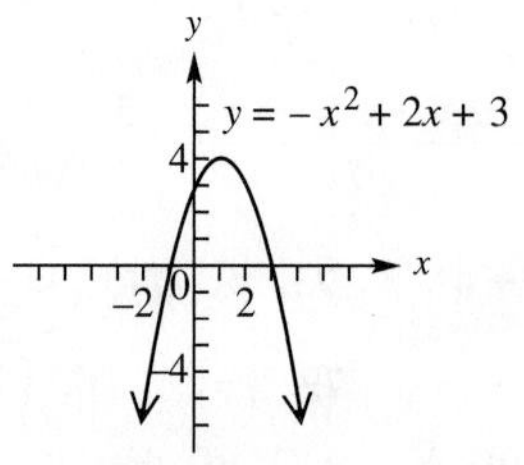

34. $(-2, -2)$

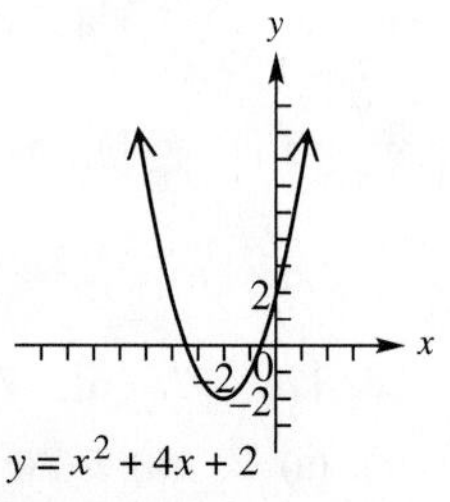

35. not a function; domain: $\{-2, 0, 2\}$; range: $\{4, 8, 5, 3\}$ **36.** function; domain: $\{8, 7, 6, 5, 4\}$; range: $\{3, 4, 5, 6, 7\}$ **37.** not a function **38.** function **39.** function **40.** function **41. (a)** 8 **(b)** −1 **42. (a)** 7 **(b)** 1 **43. (a)** 5 **(b)** 2 **44. (a)** −3 **(b)** −6 **45.** $\left\{-\frac{11}{2}, 5\right\}$ **46.** $\left\{-\frac{11}{2}, \frac{9}{2}\right\}$

47. $\left\{\frac{-1+\sqrt{21}}{2}, \frac{-1-\sqrt{21}}{2}\right\}$ **48.** $\left\{-\frac{3}{2}, \frac{1}{3}\right\}$

49. $\left\{\frac{-5+\sqrt{17}}{2}, \frac{-5-\sqrt{17}}{2}\right\}$ **50.** $\{-1+\sqrt{3}, -1-\sqrt{3}\}$

51. $\emptyset$ **52.** $\left\{\frac{9+\sqrt{41}}{2}, \frac{9-\sqrt{41}}{2}\right\}$ **53.** $\left\{-\frac{6}{5}\right\}$

54. $\{-1+2\sqrt{2}, -1-2\sqrt{2}\}$ **55.** $\{-2+\sqrt{5}, -2-\sqrt{5}\}$

56. $\{-2\sqrt{2}, 2\sqrt{2}\}$ **57.** **(a)** $\{-3, 3\}$ **(b)** $\{-3, 3\}$ **(c)** $\{-3, 3\}$ **(d)** We will always get the same results, no matter which method of solution is used. **58.** The fraction bar should be under both $-b$ and $\sqrt{b^2-4ac}$. The term $2a$ should not be in the radicand. The correct formula is $x = \frac{-b \pm \sqrt{b^2-4ac}}{2a}$.

Chapter 17 TEST (pages 1205–1206)

1. $\{-\sqrt{39}, \sqrt{39}\}$ **2.** $\{-11, 5\}$ **3.** $\left\{\frac{-3+2\sqrt{6}}{4}, \frac{-3-2\sqrt{6}}{4}\right\}$

4. $\{2+\sqrt{10}, 2-\sqrt{10}\}$ **5.** $\left\{\frac{-6+\sqrt{42}}{2}, \frac{-6-\sqrt{42}}{2}\right\}$

6. $\left\{-3, \frac{1}{2}\right\}$ **7.** $\left\{\frac{3+\sqrt{3}}{3}, \frac{3-\sqrt{3}}{3}\right\}$ **8.** $\emptyset$

9. $\left\{\frac{5+\sqrt{13}}{6}, \frac{5-\sqrt{13}}{6}\right\}$ **10.** $\{1+\sqrt{2}, 1-\sqrt{2}\}$

11. $\left\{\frac{-1+3\sqrt{2}}{2}, \frac{-1-3\sqrt{2}}{2}\right\}$ **12.** $\left\{\frac{11+\sqrt{89}}{4}, \frac{11-\sqrt{89}}{4}\right\}$

13. $\{5\}$ **14.** 2 sec **15.** 12, 16, 20

16. $(3, 0)$

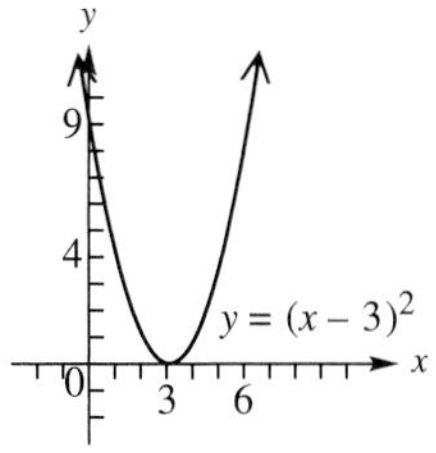

17. $(-1, -3)$

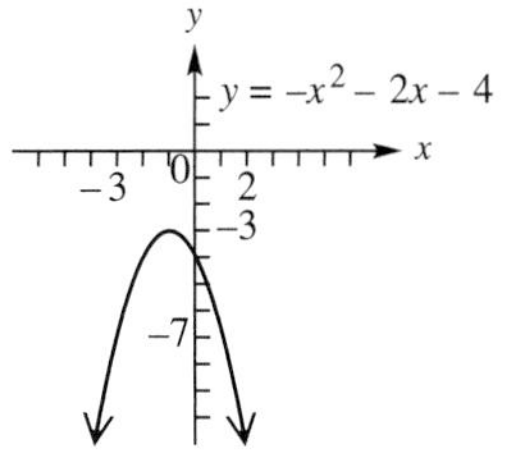

18. **(a)** not a function **(b)** function; domain: $\{0, 1, 2\}$; range: $\{2\}$ **19.** not a function **20.** 1

Chapters 9-17 CUMULATIVE REVIEW EXERCISES (pages 1207–1212)

1. -8 **2.** 24 **3.** $\frac{1}{2}$ **4.** -4 **5.** associative property **6.** inverse property **7.** distributive property **8.** $\{10\}$ **9.** $\left\{\frac{13}{4}\right\}$ **10.** $\emptyset$

11. $r = \frac{d}{t}$ **12.** $\{-5\}$ **13.** $\{-12\}$ **14.** $\{20\}$ **15.** $\{\text{all real numbers}\}$

16. $(-2.6, \infty)$

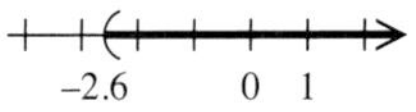

17. $(0, \infty)$

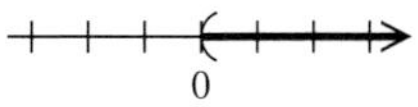

18. $(-\infty, -4]$

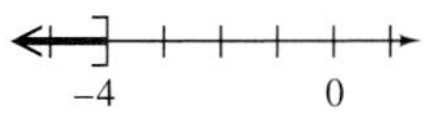

19.

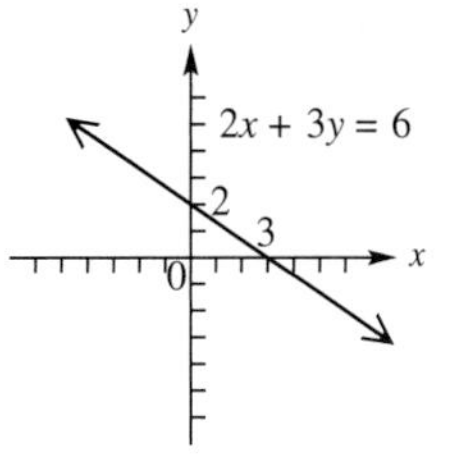

20.

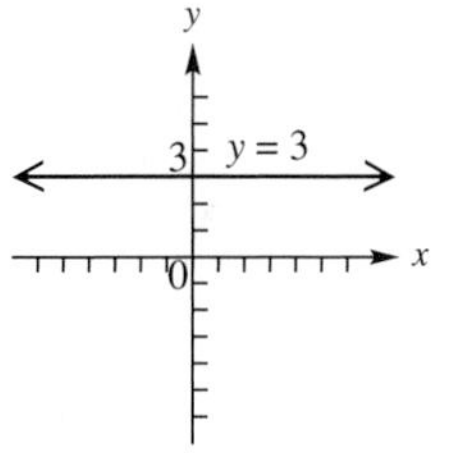

21.

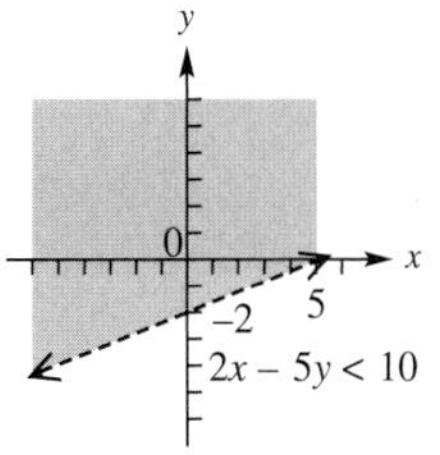

22. $-\frac{4}{3}$ **23.** $-\frac{1}{4}$

24. $y = \frac{1}{2}x + 3$ **25.** $y = 2x + 1$

26. **(a)** $x = 9$ **(b)** $y = -1$

27. 4 **28.** $\frac{16}{9}$ **29.** 1 **30.** 256

31. $\frac{1}{p^2}$ **32.** $\frac{1}{m^6}$ **33.** $-4k^2 - 4k + 8$ **34.** $45x^2 + 3x - 18$

35. $9p^2 + 12p + 4$ **36.** $4x^3 + 6x^2 - 3x + 10$

37. $3y^3 + 8y^2 + 12y - 5$ **38.** $6p^2 + 7p + 1 + \frac{3}{p-1}$

39. $(2a-1)(a+4)$ **40.** $(2m+3)(5m+2)$

41. $(4t+3v)(2t+v)$ **42.** $(2p-3)^2$ **43.** $(5r+9t)(5r-9t)$

44. $2pq(3p+1)(p+1)$ **45.** $(2x^2+3)(x-5)$

46. $(2p+5)(4p^2-10p+25)$ **47.** $\left\{-\frac{2}{3}, \frac{1}{2}\right\}$ **48.** $\{0, 8\}$ **49.** $\left\{\frac{4}{7}\right\}$

50. $\{3, 4\}$ **51.** $\{-2, -1\}$ **52.** $\{12\}$ **53.** A **54.** D **55.** $\frac{4}{q}$

56. $\frac{3r+28}{7r}$ **57.** $\frac{7}{15(q-4)}$ **58.** $\frac{-k-5}{k(k+1)(k-1)}$ **59.** $\frac{7(2z+1)}{24}$

60. $\frac{195}{29}$ **61.** $\left\{\frac{21}{2}\right\}$ **62.** $\{-2, 1\}$ **63.** $\{(-1, 6)\}$ **64.** $\{(3, -4)\}$

65. $\emptyset$ **66.** $\{(3, -7)\}$ **67.** $\{(x, y) \mid 2x - y = 6\}$

68.

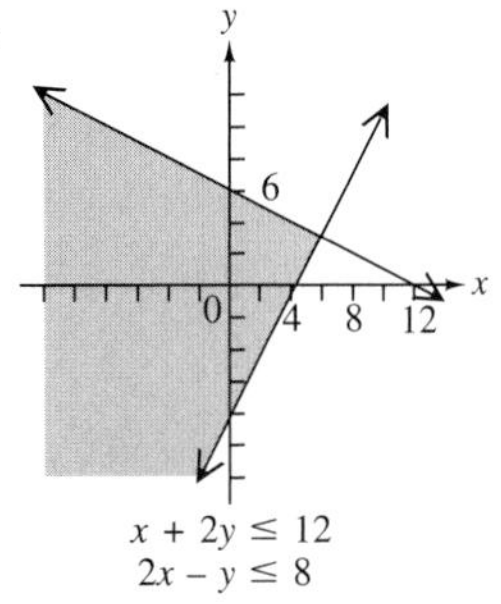

69. $\frac{6\sqrt{30}}{5}$ **70.** $4\sqrt{5}$

71. $-ab\sqrt[3]{2b}$ **72.** $29\sqrt{3}$

73. $-\sqrt{3} + \sqrt{5}$ **74.** $21 - 5\sqrt{2}$

75. $\emptyset$ **76.** $\{16\}$ **77.** $\left\{-\frac{1}{5}, \frac{13}{5}\right\}$

78. $\{3+\sqrt{10}, 3-\sqrt{10}\}$ **79.** $\emptyset$

80. $\left\{\frac{-3+\sqrt{33}}{2}, \frac{-3-\sqrt{33}}{2}\right\}$

81. $\left\{\frac{-1+\sqrt{11}}{2}, \frac{-1-\sqrt{11}}{2}\right\}$ **82.** $(2, -4)$; $(0, 0)$ and $(4, 0)$

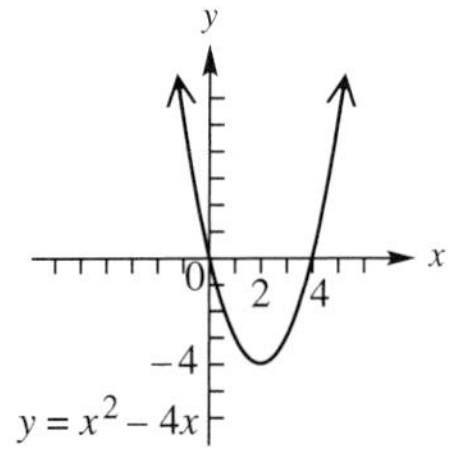

83. **(a)** yes **(b)** domain: $\{0, 1, 3\}$; range: $\{4, 2, 5\}$ **84.** 11

85. from Chicago: 61 mph; from Des Moines: 54 mph **86.** 2009: \$346,779; 2010: \$507,921 **87. (a)** 8; A slope of (approximately) 8 means that the retail sales of prescription drugs increased by about \$8 billion per year. **(b)** (2008, 250) **88.** table entries: (third column) 2; (fourth column) $2y$, 2528; 405 adults, 49 children **89.** 150 mi **90.** $1\frac{1}{5}$ hr **91. (a)** 85.63, 77.82, 76.09 **(b)** (20, 85.63), (38, 77.82), (42, 76.09) **(c)** In 1992, the winning time was approximately 80.42 sec. **92.** gold: 9; silver: 15; bronze: 13 **93.** high school diploma: \$32,552; bachelor's degree: \$53,976 **94.** 13 mi

APPENDIX A Measurement

SECTION A.1 (pages AP-8–AP-11)

1. (a) inch, foot, yard, mile **(b)** fluid ounce, cup, pint, quart, gallon **2. (a)** ounce, pound, ton **(b)** second, minute, hour, day, week **3. (a)** 3; **(b)** 12 **5. (a)** 8; **(b)** 2 **7. (a)** 2000; **(b)** 16 **9. (a)** 60; **(b)** 60 **11. (a)** 2 **(b)** 240 **13. (a)** $\frac{1}{2}$ or 0.5 **(b)** 78 **15.** 14,000 to 16,000 lb **17.** $\frac{3 \text{ ft}}{1 \text{ yd}}$; 27 **19.** 112 **21.** 10 **23.** $1\frac{1}{2}$ or 1.5 **25.** $\frac{1}{4}$ or 0.25 **27.** $1\frac{1}{2}$ or 1.5 **29.** $2\frac{1}{2}$ or 2.5 **31.** snowmobile/ATV; person walking **33.** 5000 **35.** 17 **37.** 4 to 8 in. **39.** 216 **41.** 28 **43.** 518,400 **45.** 48,000 **47. (a)** pound/ounces **(b)** quarts/pints or pints/cups **(c)** minutes/hours or seconds/minutes **(d)** feet/inches **(e)** pounds/tons **(f)** days/weeks **48.** Feet and inches both measure length, so you can add them once you've changed 12 inches into 1 ft. But pounds measure weight and cannot be added to a length measurement. **49.** 174,240 **51.** 800 **53.** 0.75 or $\frac{3}{4}$ **55.** \$1.83 (rounded) **57.** \$140 **59. (a)** 1056 sec **(b)** 17.6 min **61. (a)** $12\frac{1}{2}$ qt **(b)** 4 containers, because you can't buy part of a container **63. (a)** 0.8 mi (rounded) **(b)** 13,031 ft **(c)** 2.5 mi (rounded) **64. (a)** 183 yd (rounded) (b) 6600 in. **(c)** 0.1 mi (rounded) **(d)** 9 ft (rounded) **65. (a)** $1\frac{1}{2}$ to $2\frac{1}{2}$ ft, or 1.5 to 2.5 ft **(b)** 8400 months **(c)** 140 more years, for a total of 840 years in all **66. (a)** 283 mi (rounded) **(b)** 377 mi (rounded)

SECTION A.2 (pages AP-17–AP-18)

1. 1000 **2.** 10 **3.** $\frac{1}{1000}$ or 0.001; $\frac{1}{1000}$ or 0.001 **4.** $\frac{1}{10}$ or 0.1; $\frac{1}{10}$ or 0.1 **5.** $\frac{1}{100}$ or 0.01; $\frac{1}{100}$ or 0.01 **6.** 100; 100 **7. (a)** cm **(b)** mm **(c)** cm **9.** Answers will vary; about 8 to 10 cm. **11.** Answers will vary; about 20 to 25 mm. **13.** cm **15.** m **17.** km **19.** mm **21.** cm **23.** m **25.** 700 cm **27.** 0.040 m or 0.04 m **29.** 9400 m **31.** 5.09 m **33.** 40 cm **35.** 910 mm **37.** less; 18 cm or 0.18 m **38.** greater; 22 m or 0.022 km

39. 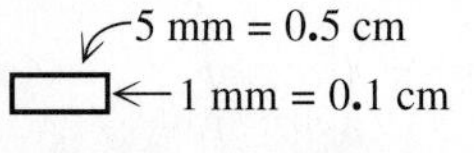

41. 0.061 km **43.** 164 cm; 1640 mm **45.** 0.0000056 km

SECTION A.3 (PAGES AP-25–AP-28)

1. (a) milliliters **(b)** liters **(c)** milligrams **(d)** kilograms **2. (a)** grams **(b)** liters **(c)** milliliters **(d)** milligrams **3.** mL **5.** L **7.** kg **9.** g **11.** mL **13.** mg **15.** L **17.** kg **19.** unreasonable; too much **20.** reasonable **21.** unreasonable; too much **22.** unreasonable; too much **23.** reasonable **24.** reasonable **25.** reasonable **26.** unreasonable; too little **27.** Unit for your answer (g) is in numerator; unit being changed (kg) is in denominator so it will divide out. The unit fraction is $\frac{1000 \text{ g}}{1 \text{ kg}}$. **28.** From mg to g is three places to the left on the metric conversion line, so move the decimal point three places to the left. 020. mg = 0.02 g **29.** three, right; 15,000 mL **31.** 3 L **33.** 0.925 L **35.** 0.008 L **37.** 4150 mL **39.** 8 kg **41.** 5200 g **43.** 850 mg **45.** 30 g **47.** 0.598 g **49.** 0.06 L **51.** 0.003 kg **53.** 990 mL **55.** mm **57.** mL **59.** cm **61.** mg **63.** 0.3 L **65.** 1340 g; 5000 g **67.** 0.9 L **69.** 3 kg to 4 kg **71.** greater; 5 mg or 0.005 g **72.** less; 10 mL or 0.01 L **73.** 200 nickels **75. (a)** 1,000,000 **(b)** $\frac{3.5 \cancel{\text{Mm}}}{1} \cdot \frac{1{,}000{,}000 \text{ m}}{1 \cancel{\text{Mm}}} = 3{,}500{,}000 \text{ m}$ **76. (a)** 1,000,000,000 **(b)** $\frac{2500 \cancel{\text{m}}}{1} \cdot \frac{1 \text{ Gm}}{1{,}000{,}000{,}000 \cancel{\text{m}}} = 0.0000025 \text{ Gm}$ **77. (a)** 1,000,000,000,000 **(b)** 1000; 1,000,000 **78.** 1,000,000; 1,000,000,000; 1,048,576; 1,073,741,824

SECTION A.4 (pages AP-33–AP-36)

1. (a) $\frac{1 \text{ kg}}{2.20 \text{ lb}}$; $\frac{2.20 \text{ lb}}{1 \text{ kg}}$ **(b)** kg; denominator **2. (a)** $\frac{1 \text{ qt}}{0.95 \text{ L}}$; $\frac{0.95 \text{ L}}{1 \text{ qt}}$ **(b)** qt; denominator **3.** 21.8 yd **5.** 262.4 ft **7.** 4.8 m **9.** 5.3 oz **11.** 111.6 kg **13.** 30.3 qt **15. (a)** about 0.2 oz **(b)** probably not **17.** about 31.8 L **19.** about 1.3 cm **21.** −8 °C **22.** 80 °C **23.** 40 °C **24.** 25 °C **25.** 150 °C **26.** 10° C **27.** 16 °C (rounded) **29.** 40 °C **31.** 46 °C (rounded) **33.** 95 °F **35.** 58 °C (rounded) **37. (a)** pleasant weather, above freezing but not hot **(b)** 75 °F to 39 °F (rounded) **(c)** Answers will vary. In Minnesota, it's 0 °C to −40 °C; in California, 24 °C to 0 °C. **39.** about 1.8 m **40.** about 5.0 kg **41.** about 3040.2 km **42.** 38 hr 23 min **43.** about 300 m **44.** less than 3790 mL **45.** about 59.4 mL **46.** about 1.6%

APPENDIX B Inductive and Deductive Reasoning

(pages AP-41–AP-42)

1. 37; add 7. **3.** 26; add 10, subtract 2. **5.** 16; multiply by 2. **7.** 243; multiply by 3. **9.** 36; add 3, add 5, add 7, etc.; or $1^2, 2^2, 3^2$, etc.

11.  **13.** 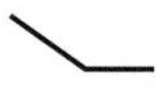

15. Conclusion follows.

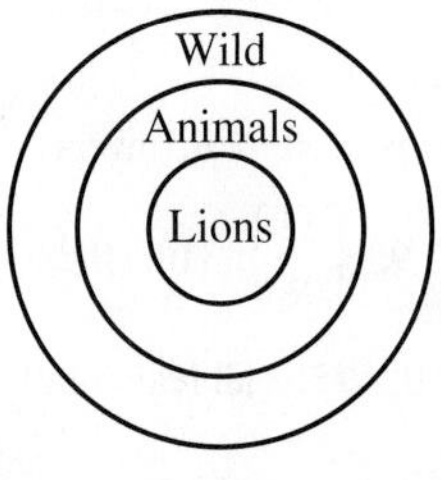

17. Conclusion does not follow.

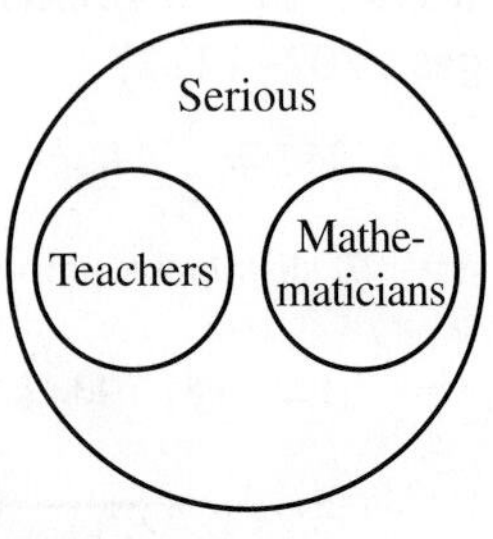

19. 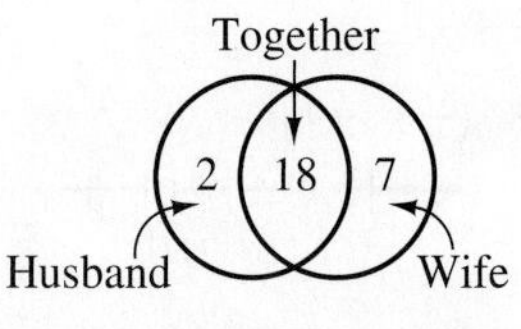

$2 + 18 + 7 = 27$

$30 - 27 = 3$

Neither watched TV on 3 days.

21. Dick is the computer operator.

Solutions to Selected Exercises

CHAPTER 1 Whole Numbers

SECTION 1.1 (pages 8–9)

9. 3,561,435; millions: 3; thousands: 561; ones: 435

15. 346,009 is three hundred forty-six thousand, nine.

31. The least-used method of transportation is public transportation. 6,069,589 in words is six million, sixty-nine thousand, five hundred eighty-nine.

SECTION 1.2 (pages 18–21)

11. Line up the numbers in columns. Then start at the right and add the ones digits. Add the ten digits next, and, finally, the hundreds digits.

$$\begin{array}{r} 932 \\ 44 \\ +\ 613 \\ \hline 1589 \end{array}$$

49. ***Step 1***

Add the digits in the ones column.

Step 2

Add the digits in the tens column, including the regrouped 2.

Step 3

Add the hundreds column, including the regrouped 1.

Step 4

Add the thousands column, including the regrouped 1.

$$\begin{array}{r} {\scriptstyle 1\,1\,2}\ \ \\ 18 \\ 708 \\ 9\,286 \\ +\ 636 \\ \hline 10{,}648 \end{array}$$

63. Add up to check addition.

$$\begin{array}{rl} 769 & \\ 179 & \\ 214 & \\ +\ 376 & \\ \hline 759 & \text{incorrect; should be 769} \end{array}$$

77. To find the total amount raised, add the amounts raised at each event.

\$3 482	flea market
+ 12,860	annual auction
\$16,342	total amount raised

SECTION 1.3 (pages 28–31)

33. In the ones column, 5 is less than 7, so in order to subtract, regroup 1 ten as 10 ones. Then subtract 7 ones from 15 ones in the ones column. Finally, subtract 3 tens from 6 tens in the tens column.

$$\begin{array}{r} {\scriptstyle 6\,15} \\ \not7\not5 \\ -37 \\ \hline 38 \end{array}$$

65.
$$\begin{array}{r} {\scriptstyle 9\ \ 9\ \ \ \ } \\ {\scriptstyle 5\ \ \not{10}\ \not{10}\ 10} \\ 6\not6,\ \not0\ \not0\ \not0 \\ -34,\ 4\ 4\ 4 \\ \hline 31,\ 5\ 5\ 6 \end{array}$$

81. To find how many fewer calories a woman burns, subtract the number of calories a woman burns from the number of calories a man burns.

187	calories man burns
−140	calories woman burns
47	fewer calories for a woman

The woman burned 47 fewer calories.

SECTION 1.4 (pages 38–41)

11. $\underbrace{(4)(5)}(2)$ or $(4)\underbrace{(5)(2)}$

$20(2) = 40$ $\qquad (4)(10) = 40$

37.
$$\begin{array}{r} 600 \\ \times\ 6 \\ \hline \end{array} \qquad \begin{array}{r} 6 \\ \times 6 \\ \hline 36 \end{array} \qquad \begin{array}{r} 600 \\ \times\ 6 \\ \hline 3600 \end{array} \quad \text{Attach 00.}$$

73. First multiply 9352 by 4. Then multiply 9352 by 6, making sure to line up the tens. Then multiply 9352 by 2, making sure to line up the hundreds. Then add the partial products.

$$\begin{array}{rl} 9\,3\,5\,2 & \\ \times\quad 2\,6\,4 & \\ \hline 3\,7\,4\,0\,8 & \leftarrow 4 \times 9352 \\ 5\,6\,1\,1\,2\ \ & \leftarrow 6 \times 9352 \\ 1\,8\,7\,0\,4\ \ \ \ & \leftarrow 2 \times 9352 \\ \hline 2{,}4\,6\,8{,}9\,2\,8 & \end{array}$$

85. *Approach* To find the number of balls purchased, multiply the number of cartons (300) by the number of balls per carton (10).

Solution

300	cartons
× 10	balls per carton
3000	balls

3000 balls were purchased.

103. Multiply the number of laptop computers purchased (12) by the price per laptop (\$970). Then multiply the number of printers purchased (8) by the price per printer (\$315).

$$\begin{array}{r} 970 \\ \times\ \ 12 \\ \hline 11{,}640 \end{array} \qquad \begin{array}{r} 315 \\ \times\ \ 8 \\ \hline 2520 \end{array}$$

The total cost is the sum of these values: \$11,640 + \$2520 + \$14,160

SECTION 1.5 (pages 51–54)

3. $24 \div 4 = 6$: $4\overline{)24}$ with quotient 6 or $\frac{24}{4} = 6$

35. $6\overline{)9\,{}^{3}1\,{}^{1}3\,{}^{1}7}$, quotient $1\ 5\ 2\ 2\ \mathbf{R}5$

Check $(6 \times 1522) + 5 = 9132 + 5$
$= 9137$

51. $7\overline{)7\,1,\,{}^{1}7\,{}^{3}7\,{}^{2}6}$, quotient $1\,0,\ 2\ 5\ 3\ \mathbf{R}5$

Check $(7 \times 10{,}253) + 5 = 71{,}771 + 5$
$= 71{,}776$

67. $9\overline{)8\,6,\,6\,5\,5}$, quotient $9\ 6\,2\,8\ \mathbf{R}7$

Check $(9 \times 9628) + 7 = 86{,}652 + 7$
$= 86{,}659$

Does not match dividend, incorrect

Rework:

$9\overline{)8\,6,\,{}^{5}6\,{}^{2}5\,{}^{7}5}$, quotient $9\ \ 6\ \ 2\ \ 8\ \mathbf{R}3$

Check $(9 \times 9628) + 3 = 86{,}652 + 3$
$= 86{,}655$

Matches dividend, correct

73. To find the number of tables that can be set, divide the number of napkins (2624) by the number of napkins it takes to set each table (8).

$8\overline{)2\,6\,{}^{2}2\,{}^{6}4}$, quotient $3\ \ 2\ \ 8$

328 tables can be set.

SECTION 1.6 (pages 60–62)

13. Use 2 as a trial divisor, since 18 is closer to 20 than to 10.

$\frac{131}{2} = 6$ with 1 left over

Because $6 \times 18 = 108$ and $131 - 108 = 23$, which is greater than the divisor, use 7 instead. To find the next digit in the quotient, use 2 as a trial divisor again.

(Continued)

SOLUTIONS

$\frac{5}{2} = 2$ with 1 left over

Because $2 \times 18 = 36$ and $59 - 36 = 23$, which is greater than the divisor, use 3 instead.

```
     7 3 R5   Check     7 3
18 )1 3 1 9           × 1 8
    1 2 6             5 8 4
      5 9             7 3
      5 4           1 3 1 4
        5         +       5
                    1 3 1 9  matches
```

25.

```
     1 0 1 R4   Check    1 0 1
35 )3 5 4 9            ×   3 5
                         5 0 5
                       3 0 3
                       3 5 3 5
                     +       4
                       3 5 3 9  incorrect
```

Rework:

```
     1 0 1 R14   Check    1 0 1
35 )3 5 4 9             ×   3 5
    3 5                   5 0 5
      0 4 9             3 0 3
        3 5             3 5 3 5
        1 4           +     1 4
                        3 5 4 9  matches
```

The correct answer is 101 **R**14.

39. **(a)** Divide the number of Big Macs eaten by 39 years.

```
       6 4 8
39 )2 5, 2 7 2
    2 3 4
      1 8 7
      1 5 6
        3 1 2
        3 1 2
            0
```

The average number of Big Macs eaten each year was 648.

(b) 2 Big Macs eaten each day would be $2 \times 365 = 730$. He has eaten less than 2 Big Macs each day.

SECTION 1.7 (pages 70–73)

9. 855 rounded to the nearest ten: 860

8$\underline{5}$5 Underline the 5 in the tens place. Next digit is 5 or more, so add one to the 5 in the tens place. All digits to the right of the underlined place change to zero.

25. **To the nearest ten:** 44$\underline{7}$6

Underline the 7 in the tens place. Next digit is 5 or more, so add 1 to the 7. All digits to the right of the underlined place are changed to zero.

4480

To the nearest hundred: 4$\underline{4}$76

Underline the 4 in the hundreds place. Next digit is 5 or more, so add 1 to the 4. All digits to the right of the underlined place are changed to zero.

4500

To the nearest thousand: $\underline{4}$476

Underline the 4 in the thousands place. Next digit is 4 or less, so leave 4 as 4 in the thousands place. All digits to the right of the underlined place are changed to zero.

4000

41.

Estimate:	*Exact:*
30	25
60	63
50	47
+ 80	+ 84
220	219

47.

Estimate:	*Exact:*
900	863
700	735
400	438
+ 800	+ 792
2800	2828

53.

Estimate:	*Exact:*
8000	8 215
60	56
700	729
+ 4000	+ 3 605
12,760	12,605

61. $\underline{7}$6,000,000 Next digit is 5 or more, so add 1 to 7. Change all digits to the right of the underlined place to zero. Add 1 to 7.

80 million people

3$\underline{1}$1,000,000 Next digit is 4 or less, so leave 1 as 1. Change all digits to the right of the underlined place to zero.

310 million people

SECTION 1.8 (pages 76–79)

5. 8^2: exponent is 2, base is 8.

$8^2 = 8 \cdot 8 = 64$

21. $6^2 = 36$, so $\sqrt{36} = 6$.

37.

$5 \cdot 3^2 + \frac{0}{8}$	Exponent
$5 \cdot 9 + \frac{0}{8}$	Multiply.
$45 + \frac{0}{8}$	Divide.
$45 + 0 = 45$	Add.

67.

$8 \cdot \sqrt{49} - 6(9 - 4)$	Parentheses
$8 \cdot \sqrt{49} - 6(5)$	Square root
$8 \cdot 7 - 6(5)$	Multiply.
$56 - 6(5)$	Multiply.
$56 - 30 = 26$	Subtract.

85.

$8 \cdot 9 \div \sqrt{36} - 4 \div 2 + (14 - 8)$	Parentheses
$8 \cdot 9 \div \sqrt{36} - 4 \div 2 + 6$	Square root
$8 \cdot 9 \div 6 - 4 \div 2 + 6$	Multiply.
$72 \div 6 - 4 \div 2 + 6$	Divide.
$12 - 4 \div 2 + 6$	Divide.
$12 - 2 + 6$	Subtract.
$10 + 6 = 16$	Add.

SECTION 1.9 (pages 84–87)

7. According to the pictograph, Family Dollar has 4500 stores. Subtract 4500 from 5750 to find the difference.

$5750 - 4500 = 1250$ fewer stores

15. According to the bar graph, $18 - 9 = 9$ more people out of 100 found their career as a result of "Studied in school" than "Luck or chance."

19. According to the line graph the increase in the number of installations from 2113 to 2014 is $7000 - 2500 = 4500$.

23. We want to insert parentheses in $7 - 2 \cdot 3 - 6$ so that it simplifies to 9. There isn't a method to use, so just use trial and error.

$(7 - 2) \cdot 3 - 6 = 5 \cdot 3 - 6$
$= 15 - 6 = 9$

SECTION 1.10 (pages 92–95)

11. ***Step 1***

Find the number of toys each child will receive.

Step 2

"Same number of toys to each" indicates division should be used.

Step 3

3000 divided by 700 gives an estimate of about 4 toys.

Step 4

$$657\overline{)2628}$$ quotient 4; 2628; remainder 0

Step 5
Each child will receive 4 toys.

Step 6
Exact answer is close to estimate. Also, $4 \times 657 = 2628$ which matches the number given in the problem.

21. *Step 1*
Find her monthly savings.

Step 2
Her monthly take home pay and expenses are given. "Remainder" indicates subtraction may be used.

Step 3
Estimate:
$\$2000 - \$700 - \$300 - \$400 - \$200 - \$200 = \$200$

Step 4

$$\begin{array}{r} \$695 \\ 340 \\ 435 \\ 240 \\ +\ 180 \\ \hline 1890 \end{array} \qquad \begin{array}{r} \$2240 \\ -\ 1890 \\ \hline \$350 \end{array}$$

Step 5
Her monthly savings are \$350.

Step 6
The exact answer seems reasonable compared to the estimate.
Check: $\$350 + \$1890 = \$2240$. Re-add the expenses to check the sum of \$1890.

29. *Step 1*
The number and cost of wheelchairs and recorder-players are given, and the total cost of all items must be found.

Step 2
Find the cost of all wheelchairs and the cost of all recorder-players. Then add these costs to get the total cost.

Step 3
The cost of the wheelchairs is about $\$1000 \times 6 = \$6000$.
The cost of the recorder-players is about $\$900 \times 20 = \$18{,}000$.
Estimate: $\$6000 + \$18{,}000 = \$24{,}000$

Step 4
cost of wheelchairs: $\begin{array}{r} \$1256 \\ \times\ \ 6 \\ \hline \$7536 \end{array}$

cost of recorder-players: $\begin{array}{r} \$895 \\ \times\ 15 \\ \hline 4475 \\ 895\ \ \\ \hline \$13{,}425 \end{array}$

Step 5
The total cost is $\$13{,}425 + \$7536 = \$20{,}961$.

Step 6
The exact answer is reasonably close to the estimate. Check by repeating Step 4.

41. *Step 1*
The total seating is given along with the information to find the number of seats on the main floor. The number of rows of seats in the balcony is given, and the number of seats in each row of the balcony must be found.

Step 2
First, the number of seats on the main floor must be found. Next, the number of seats on the main floor must be subtracted from the total number of seats to find the number of seats in the balcony. Finally, the number of seats in the balcony must be divided by the number of rows of seats in the balcony.

Step 3
Seats on main floor: $30 \times 25 = 750$
Seats in balcony: $1250 - 750 = 500$
Seats in each row in balcony:
$500 \div 25 = 20$
Since we didn't round, our estimate matches our exact answer.

Step 4
See the calculations in Step 3.

Step 5
The number of seats in each row of the balcony is 20.

Step 6
The answer matches the estimate, as expected.

Check
Seats in balcony: $20 \times 25 = 500$
Seats on main floor: $30 \times 25 = 750$
Total seats: $500 + 750 = 1250$, which matches the number given in the problem.

CHAPTER 2 Multiplying and Dividing Fractions

SECTION 2.1 (pages 114–116)

13. Each of the two figures is divided into 5 parts and 7 are shaded: $\frac{7}{5}$

Three are unshaded: $\frac{3}{5}$

21. Proper fractions: numerator *less* than denominator.

$$\frac{1}{3}, \frac{5}{8}, \frac{7}{16}$$

Improper fractions: numerator *greater than or equal to* denominator.

$$\frac{8}{5}, \frac{6}{6}, \frac{12}{2}$$

SECTION 2.2 (pages 121–123)

17. Write $5\frac{4}{5}$ as an improper fraction.

$5 \cdot 5 = 25$ Multiply 5 and 5.
$25 + 4 = 29$ Add 4. The numerator is 29.
$5\frac{4}{5} = \frac{29}{5}$ Use the same denominator.

43. $\frac{63}{4}$

Divide 63 by 4.

$$4\overline{)63}$$ quotient 15 ← Whole number part; 4; 23; 20; 3 ← Remainder

The quotient 15 is the whole number part of the mixed number. The remainder 3 is the numerator of the fraction, and the denominator remains as 4.

$$\frac{63}{4} = 15\frac{3}{4}$$

61. Write $522\frac{3}{8}$ as an improper fraction.

$522 \cdot 8 = 4176$
$4176 + 3 = 4179$ Add 3. The numerator is 4179.
$522\frac{3}{8} = \frac{4179}{8}$ Use the same denominator.

SECTION 2.3 (pages 128–129)

5. Factorizations of 48:
$1 \cdot 48 = 48$ $2 \cdot 24 = 48$ $3 \cdot 16 = 48$
$4 \cdot 12 = 48$ $6 \cdot 8 = 48$
The factors of 48 are 1, 2, 3, 4, 6, 8, 12, 16, 24, and 48.

41. 44

$11\overline{)11}$ quotient 1 — Divide 11 by 11.

$2\overline{)22}$ quotient 11 — Divide 22 by 2.

$2\overline{)44}$ quotient 22 — Divide 44 by 2.

Quotient is 1.

Because all factors (divisors) are prime, the prime factorization of 44 is $2 \cdot 2 \cdot 11 = 2^2 \cdot 11$

SECTION 2.4 (pages 134–135)

21. Because the greatest common factor of 56 and 64 is 8, divide both numerator and denominator by 8.

$$\frac{56}{64} = \frac{56 \div 8}{64 \div 8} = \frac{7}{8}$$

33. Write the prime factorization of both numerator and denominator. Then divide both numerator and denominator by any common factors, and write a 1 by each factor that has been divided. Finally, multiply the remaining factors in both numerator and denominator.

$$\frac{18}{24} = \frac{\cancel{2} \cdot \cancel{3} \cdot 3}{\cancel{2} \cdot 2 \cdot 2 \cdot \cancel{3}} = \frac{1 \cdot 1 \cdot 3}{1 \cdot 2 \cdot 2 \cdot 1} = \frac{3}{4}$$

43. $\frac{3}{6}$ and $\frac{18}{36}$ Write each fraction in lowest terms.

$$\frac{3}{6} = \frac{1 \cdot \cancel{3}}{2 \cdot \cancel{3}} = \frac{1}{2}$$

$$\frac{18}{36} = \frac{\cancel{2} \cdot \cancel{3} \cdot \cancel{3}}{\cancel{2} \cdot 2 \cdot \cancel{3} \cdot \cancel{3}} = \frac{1}{2}$$

The fractions are *equivalent* $\left(\frac{1}{2} = \frac{1}{2}\right)$.

57. Write the prime factorization of both numerator and denominator. Then divide both numerator and denominator by any common factors, and write a 1 by each factor that has been divided. Finally, multiply the remaining factors in both numerator and denominator.

$$\frac{160}{256} = \frac{\cancel{2} \cdot \cancel{2} \cdot \cancel{2} \cdot \cancel{2} \cdot \cancel{2} \cdot 5}{\cancel{2} \cdot \cancel{2} \cdot \cancel{2} \cdot \cancel{2} \cdot \cancel{2} \cdot 2 \cdot 2 \cdot 2} = \frac{5}{8}$$

SECTION 2.5 (pages 144–147)

11. Divide a numerator and denominator by the same number to use the multiplication shortcut.

$$\frac{2}{3} \cdot \frac{7}{12} \cdot \frac{9}{14} = \frac{\cancel{2}^{1}}{\cancel{3}_{1}} \cdot \frac{\cancel{7}^{1}}{\cancel{12}_{\cancel{6}\,2}} \cdot \frac{\cancel{9}^{\cancel{3}\,1}}{\cancel{14}_{2}}$$

$$= \frac{1 \cdot 1 \cdot 1}{1 \cdot 2 \cdot 2} = \frac{1}{4}$$

25. Write 36 as $\frac{36}{1}$ and multiply.

$$36 \cdot \frac{5}{8} \cdot \frac{9}{15} = \frac{\cancel{36}^{9}}{1} \cdot \frac{\cancel{5}^{1}}{\cancel{8}_{2}} \cdot \frac{\cancel{9}^{3}}{\cancel{15}_{\cancel{3}\,1}} = \frac{27}{2} = 13\frac{1}{2}$$

41. Area = length • width

$$\text{Area} = \frac{5}{6} \cdot \frac{3}{10}$$

$$= \frac{\cancel{5}^{1}}{\cancel{6}_{2}} \cdot \frac{\cancel{3}^{1}}{\cancel{10}_{2}}$$ Divide 5 and 10 by 5. Divide 3 and 6 by 3.

$$= \frac{1}{4} \text{ square mile}$$

45. Area = length • width

$$\text{Area} = 2 \cdot \frac{3}{4}$$

$$= \frac{\cancel{2}^{1}}{1} \cdot \frac{3}{\cancel{4}_{2}}$$ Divide numerator and denominator by 2.

$$= \frac{3}{2}$$ Write as a mixed number.

$$= 1\frac{1}{2}$$ Square yards

SECTION 2.6 (pages 151–154)

5. *Step 1*

The problem asks for the area of the front surface of a digital photo frame.

Step 2

To find area, multiply the length of $\frac{3}{4}$ foot by the width of $\frac{2}{3}$ foot.

Step 3

An estimate is $\frac{6}{12}$ square foot.

$$\frac{3}{4} \cdot \frac{2}{3} = \frac{\cancel{3}^{1} \cdot \cancel{2}^{1}}{\cancel{4}_{2} \cdot \cancel{3}_{1}} = \frac{1}{2}$$

Step 4

The exact value is $\frac{1}{2}$, which is the same as the estimate since we didn't round.

Step 5

The area of the digital photo frame is $\frac{1}{2}$ square foot.

Step 6

The answer, $\frac{1}{2}$ square foot, matches our estimate.

15. *Step 1*

The problem asks for the number of runners who are women.

Step 2

$\frac{7}{12}$ of the 1560 runners are women, so multiply $\frac{7}{12}$ by 1560 to find the number of runners who are women.

Step 3

Round the number of runners to 1600; then, $\frac{1}{2}$ of 1600 is 800. Since $\frac{7}{12}$ is more than $\frac{1}{2}$, our estimate is that "more than 800 runners" are women.

Step 4

$$\frac{7}{12} \cdot 1560 = \frac{7 \cdot \cancel{1560}^{130}}{\cancel{12}_{1} \cdot 1} = 910$$

Step 5

(a) 910 runners are women.

(b) 1560 total runners − 910 women = 650 men

Step 6

The exact answer, 910, fits our estimate of "more than 800."

25. From Exercise 23 the total income is \$76,000. The circle graph shows that $\frac{1}{5}$ of the income is for rent.

$$\frac{1}{5} \cdot 76{,}000 = \frac{1}{\cancel{5}_{1}} \cdot \frac{\cancel{76{,}000}^{15{,}200}}{1} = 15{,}200$$

The amount of their rent is \$15,200.

37. To find the remaining amount of the estate, subtract $\frac{7}{8}$ from 1.

$$1 - \frac{7}{8} = \frac{8}{8} - \frac{7}{8} = \frac{1}{8}$$

Multiply the remaining $\frac{1}{8}$ of the estate by the fraction going to the American Cancer Society.

$$\frac{1}{4} \cdot \frac{1}{8} = \frac{1}{32}$$

$\frac{1}{32}$ of the estate goes to the American Cancer Society.

SOLUTIONS

SECTION 2.7 (pages 160–163)

5. The reciprocal of $\frac{3}{8}$ is $\frac{8}{3}$ because

$\frac{3}{8} \cdot \frac{8}{3} = \frac{24}{24} = 1.$

29. $\dfrac{18}{\frac{3}{4}} = 18 \div \frac{3}{4}$ Rewrite using the ÷ symbol for division.

$= \frac{18}{1} \div \frac{3}{4}$ Write 18 as $\frac{18}{1}$.

$= \frac{\overset{6}{\cancel{18}}}{1} \cdot \frac{4}{\underset{1}{\cancel{3}}}$ The reciprocal of $\frac{3}{4}$ is $\frac{4}{3}$. Change "÷" to "•". Divide the numerator and denominator by 3.

$= \frac{6 \cdot 4}{1 \cdot 1}$ Multiply.

$= \frac{24}{1} = 24$

35. ***Step 1***
The problem asks for the number of times a measuring cup needs to be filled.

Step 2
Solve the problem by dividing the total number of cups (5) by the size of the measuring cup $\left(\frac{1}{3}\right)$.

Step 3
Because there are three $\frac{1}{3}$-cups in one cup, multiply 3 by 5 to get 15. So, our estimate is 15.

Step 4
$5 \div \frac{1}{3} = \frac{5}{1} \div \frac{1}{3} = \frac{5}{1} \cdot \frac{3}{1} = 15$

Step 5
They need to fill the measuring cup 15 times.

Step 6
The exact answer, 15, is the same as our estimate.

49. ***Step 1***
The problem asks for the number of towels that can be made.

Step 2
Solve the problem by dividing the 912 yards of fabric by the fraction of a yard $\left(\frac{3}{8}\right)$ needed for each dish towel.

Step 3
Round 912 yards to 900 yards. Round $\frac{3}{8}$ to $\frac{1}{2}$.

Multiply 900 by 2, the reciprocal of $\frac{1}{2}$, to get the estimate of 1800 towels.

Step 4
$912 \div \frac{3}{8} = \frac{912}{1} \cdot \frac{8}{3} = \frac{\overset{304}{\cancel{912}}}{1} \cdot \frac{8}{\underset{1}{\cancel{3}}} = 2432$

Step 5 2432 towels can be made.

Step 6
The exact answer, 2432, is close to our estimate, 1800.

SECTION 2.8 (pages 171–175)

5. $4\frac{1}{2} \cdot 1\frac{3}{4}$

Estimate: $5 \cdot 2 = 10$

To find the exact answer, change each mixed number to an improper fraction and then multiply.

Exact: $4\frac{1}{2} \cdot 1\frac{3}{4} = \frac{9}{2} \cdot \frac{7}{4} = \frac{63}{8} = 7\frac{7}{8}$

21. $1\frac{1}{4} \div 3\frac{3}{4}$

Estimate: $1 \div 4 = \frac{1}{4}$

To find the exact answer, first change each mixed number to an improper fraction. Then, use the reciprocal of the divisor (the second fraction) and multiply.

Exact:

$1\frac{1}{4} \div 3\frac{3}{4} = \frac{5}{4} \div \frac{15}{4} = \frac{\overset{1}{\cancel{5}}}{\underset{1}{\cancel{4}}} \cdot \frac{\overset{1}{\cancel{4}}}{\underset{3}{\cancel{15}}} = \frac{1}{3}$

CHAPTER 3 Adding and Subtracting Fractions

SECTION 3.1 (pages 195–196)

15. Add the numerators. Keep the denominator.

$\frac{3}{8} + \frac{7}{8} + \frac{2}{8} = \frac{3+7+2}{8} = \frac{12}{8} = 1\frac{1}{2}$

29. First, subtract the numerators and keep the denominator. Then write the fraction in lowest terms.

$\frac{47}{36} - \frac{5}{36} = \frac{47-5}{36} = \frac{42}{36}$

$= \frac{42 \div 6}{36 \div 6} = \frac{7}{6} = 1\frac{1}{6}$

39. First, add the two fractions of land that were purchased.

$\frac{9}{10} + \frac{3}{10} = \frac{9+3}{10} = \frac{12}{10}$

Next, subtract the fraction of land that was planted in carrots.

$\frac{12}{10} - \frac{7}{10} = \frac{12-7}{10} = \frac{5}{10} = \frac{1}{2}$

$\frac{1}{2}$ acre is planted in squash.

SECTION 3.2 (pages 205–208)

13. 20 and 50
Multiples of 50:
50, <u>100</u>, 150, 200, 250, . . .
100 is the first multiple of 50 that is divisible by 20. $(100 \div 20 = 5)$
The least common multiple of 20 and 50 is 100.

21. Find the prime factorization for each number.
4, 6, 8, 10

$4 = 2 \cdot 2$
$6 = 2 \cdot ③$
$8 = ② \cdot ② \cdot ②$
$10 = 2 \cdot ⑤$

LCM $= 2 \cdot 2 \cdot 2 \cdot 3 \cdot 5 = 120$
The LCM of 4, 6, 8, and 10 is 120.

37. $\frac{2}{3} = \frac{?}{9}$ Divide 9 by 3, to get 3. Now multiply both numerator and denominator of $\frac{2}{3}$ by 3.

$\frac{2}{3} = \frac{2 \cdot 3}{3 \cdot 3} = \frac{6}{9}$

41. $\frac{3}{16} = \frac{?}{64}$ Divide 64 by 16, to get 4. Now multiply both numerator and denominator of $\frac{3}{16}$ by 4.

$\frac{3}{16} = \frac{3 \cdot 4}{16 \cdot 4} = \frac{12}{64}$

55. $\frac{109}{1512}, \frac{23}{392}$

2	1512	392
2	756	196
2	378	98
3	189	~~49~~
3	63	~~49~~
3	21	~~49~~
7	7	49
7	~~1~~	7
	1	1

The LCM of the denominators is
$2 \cdot 2 \cdot 2 \cdot 3 \cdot 3 \cdot 3 \cdot 7 \cdot 7 = 10{,}584.$

61. Find the prime factorization for each number.
5, 7, 14, 10

$5 = ⑤$
$7 = ⑦$
$14 = ② \cdot 7$
$10 = 2 \cdot 5$

LCM $= 2 \cdot 5 \cdot 7 = 70$
The LCM of 5, 7, 14, and 10 is 70.

SOLUTIONS

SECTION 3.3 (pages 213–216)

3. The least common multiple of 4 and 8 is 8. Rewrite both fractions as fractions with a least common denominator of 8. Then add the numerators.

$$\frac{3}{4}+\frac{1}{8}=\frac{6}{8}+\frac{1}{8}=\frac{6+1}{8}=\frac{7}{8}$$

29. The least common multiple of 12 and 4 is 12. Rewrite both fractions as fractions with a least common denominator of 12. Then subtract the numerators and write the resulting fraction in lowest terms.

$$\frac{5}{12}-\frac{1}{4}=\frac{5}{12}-\frac{3}{12}=\frac{5-3}{12}=\frac{2}{12}=\frac{1}{6}$$

41. Add the fractions to find the total length of the screw. The least common denominator of the fractions is 40. Rewrite all three fractions with a denominator of 40. Then add the numerators.

$$\frac{1}{8}+\frac{1}{4}+\frac{2}{5}=\frac{5}{40}+\frac{10}{40}+\frac{16}{40}=\frac{5+10+16}{40}=\frac{31}{40}$$

The total length of the screw is $\frac{31}{40}$ inch.

49. One way to compare fractions accurately is to rewrite each fraction with a common denominator.

$$\frac{1}{4}=\frac{15}{60},\ \frac{1}{3}=\frac{20}{60},\ \frac{1}{5}=\frac{12}{60},\ \frac{13}{60}$$

The fraction with the largest numerator is the largest fraction. This is $\frac{20}{60}$ or $\frac{1}{3}$, which is "totally honest."

To find the number of hours, multiply.

$$\frac{1}{3}\cdot 1200=\frac{1}{\underset{1}{\cancel{3}}}\cdot\frac{\overset{400}{\cancel{1200}}}{1}=\frac{400}{1}=400$$

400 people responded "totally honest."

To find what fraction of the users responding to "totally honest" and "fib a little" add the fractions $\frac{1}{3}$ and $\frac{1}{4}$. The least common denominator is 12. Rewrite $\frac{1}{3}$ and $\frac{1}{4}$ with denominators of 12. Then add the numerators and write in lowest terms.

$$\frac{1}{3}+\frac{1}{4}=\frac{4}{12}+\frac{3}{12}=\frac{4+3}{12}=\frac{7}{12}$$

SECTION 3.4 (pages 221–228)

19. *Estimate:* *Exact:*

Estimate		Exact
23	← Rounds to	$22\frac{3}{4}=22\frac{21}{28}$
+15	← Rounds to	$+15\frac{3}{7}=15\frac{12}{28}$
38		$37\frac{33}{28}$

$$37\frac{33}{28}=37+1\frac{5}{28}=38\frac{5}{28}$$

29. *Estimate:* *Exact:*

Estimate		Exact
17	← Rounds to	17
− 7	← Rounds to	$-\ 6\frac{5}{8}$
10		

To subtract $\frac{5}{8}$, first regroup the whole number 17 into 16 + 1.

$$17=16+1=16+\frac{8}{8}$$

Now you can subtract.

$$16\frac{8}{8}-6\frac{5}{8}=10\frac{3}{8}$$

55.

$$8\frac{3}{4}=\frac{35}{4}=\frac{70}{8}$$
$$-\ 5\frac{7}{8}=\frac{47}{8}=\frac{47}{8}$$
$$\frac{23}{8}=2\frac{7}{8}$$

71. Subtract the size of the smallest hose clamp from the size of the largest hose clamp.

Estimate: 3 − 1 = 2 in.

Exact:
$$2\frac{3}{4}=\frac{11}{4}=\frac{44}{16}$$
$$-\ \frac{9}{16}=\frac{9}{16}=\frac{9}{16}$$
$$\frac{35}{16}=2\frac{3}{16}$$

The difference in size is $2\frac{3}{16}$ inches.

81. Add the weights.

Estimate:

59 + 24 + 17 + 29 + 58 = 187 tons

Exact:

$$58\frac{1}{2}=58\frac{12}{24}$$
$$23\frac{5}{8}=23\frac{15}{24}$$
$$16\frac{5}{6}=16\frac{20}{24}$$
$$29\frac{1}{4}=29\frac{6}{24}$$
$$+\ 58\frac{1}{3}=58\frac{8}{24}$$
$$184\frac{61}{24}=184+2\frac{13}{24}=186\frac{13}{24}$$

The total weight is $186\frac{13}{24}$ tons.

SECTION 3.5 (pages 235–238)

7.

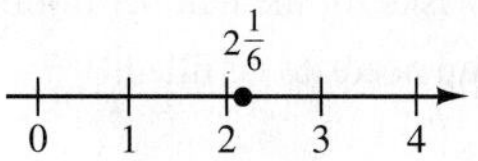

33. $\left(\frac{3}{2}\right)^4=\frac{3}{2}\cdot\frac{3}{2}\cdot\frac{3}{2}\cdot\frac{3}{2}=\frac{81}{16}=5\frac{1}{16}$

49. $6\left(\frac{2}{3}\right)^2\left(\frac{1}{2}\right)^3=6\left(\frac{2}{3}\cdot\frac{2}{3}\right)\left(\frac{1}{2}\cdot\frac{1}{2}\cdot\frac{1}{2}\right)$

Simplify the expressions with exponents.

$$=\frac{\overset{1}{\cancel{\overset{2}{\cancel{6}}}}\cdot\overset{1}{\cancel{2}}\cdot\overset{1}{\cancel{2}}\cdot 1\cdot 1\cdot 1}{\underset{1}{\cancel{3}}\cdot 3\cdot\underset{1}{\cancel{2}}\cdot\underset{1}{\cancel{2}}\cdot\underset{1}{\cancel{2}}}$$

$$=\frac{1}{3}$$

63. $\left(\frac{3}{4}\right)^2-\left(\frac{1}{2}-\frac{1}{6}\right)\div\frac{4}{3}$

$$=\left(\frac{3}{4}\right)^2-\left(\frac{3}{6}-\frac{1}{6}\right)\div\frac{4}{3}$$

Work inside parentheses first.

$$=\left(\frac{3}{4}\right)^2-\left(\frac{2}{6}\right)\cdot\frac{3}{4}$$

Simplify the expression with the exponent.

$$=\frac{3}{4}\cdot\frac{3}{4}-\left(\frac{2}{6}\right)\cdot\frac{3}{4}\quad\text{Multiply.}$$

$$=\frac{9}{16}-\frac{\overset{1}{\cancel{2}}}{\underset{2}{\cancel{6}}}\cdot\frac{\overset{1}{\cancel{3}}}{\underset{2}{\cancel{4}}}$$

$$=\frac{9}{16}-\frac{1}{4}$$ The LCD is 16. Rewrite $\frac{1}{4}$ as a fraction with a denominator of 16.

$$=\frac{9}{16}-\frac{4}{16}$$ Subtract.

$$=\frac{5}{16}$$

SOLUTIONS

CHAPTER 4 Decimals

SECTION 4.1 (pages 259–263)

5. 7 0 . 4 8 9 (tens, ones, tenths labelled above 7, 0, 4)

17. Reorder as 5 thousands, 4 hundreds, 0 tens, 6 ones, 0 tenths, 4 hundredths, 5 thousandths: 5406.045

31. $0.205 = \frac{205}{1000} = \frac{205 \div 5}{1000 \div 5} = \frac{41}{200}$

45. Three decimal places is *thousandths,* so 1.075 is one and seventy-five *thousandths*.

53. Seven hundred three ten-thousandths: $\frac{703}{10{,}000} = 0.0703$

59. 8-pound test line has a diameter of 0.010 inch. 0.010 inch is read "ten thousandths inch."

$0.010 = \frac{10}{1000} = \frac{10 \div 10}{1000 \div 10} = \frac{1}{100}$ inch

SECTION 4.2 (pages 269–270)

13. Draw a cut-off line after the tenths place: 793.9|88

The first digit cut is 8, which is *5 or more*, so round up the tenths place.

$$\begin{array}{r} 793.9 \\ +\ \ 0.1 \\ \hline 794.0 \end{array}$$

Answer: 793.988 ≈ 794.0

17. Round 9.0906 to the nearest hundredth.

Draw a cut-off line: 9.09|06

The first digit cut is 0, which is *4 or less*. The part you keep stays the same.

9.0906 ≈ 9.09

35. Round $999.73 to the nearest dollar.

Draw a cut-off line: $999|.73

The first digit cut is 7, which is *5 or more,* so round up.

$$\begin{array}{r} \$999 \\ +\quad 1 \\ \hline \$1000 \end{array}$$

Answer: $999.73 ≈ $1000

SECTION 4.3 (pages 275–278)

13. Subtract 0.291 from 0.4.

$$\begin{array}{rl} 0.400 & \leftarrow \text{Write two zeros.} \\ -\ 0.291 & \\ \hline 0.109 & \end{array}$$

(Line up decimal points.)

25.

Estimate:		*Exact:*
400	←	392.700
1	←	0.865
+ 20	←	+ 21.080
421		414.645

43. First add the three players' heights.

Estimate:		*Exact:*
2	←	2.13
2	←	1.98
+ 2	←	+ 2.08
6 meters		6.19 meters

Now subtract the players' combined height of 6.19 meters from the rhino's height of 6.4 meters.

6.4 − 6.19 = 0.21

So, the NBA stars' combined height is 0.21 meter less than the rhino's height.

49. Subtract the price of the regular fishing line, $4.84, from the price of the fluorescent fishing line, $5.14.

Estimate:		*Exact:*
$5	←	$5.14
− 5	←	− 4.84
$0		$0.30

The fluorescent fishing line costs $0.30 more than the regular fishing line.

SECTION 4.4 (pages 281–284)

15. (7.2) (0.06)

$$\begin{array}{r} 72 \\ \times 6 \\ \hline 432 \end{array} \qquad \begin{array}{rl} 7.2 & \leftarrow \text{1 decimal place} \\ \times\ 0.06 & \leftarrow \text{2 decimal places} \\ \hline 0.432 & \leftarrow \text{3 decimal places} \end{array}$$

19. (0.006) (0.0052)

$$\begin{array}{rl} 0.0052 & \leftarrow \text{4 decimal places} \\ \times\ 0.006 & \leftarrow \text{3 decimal places} \\ \hline 0.0000312 & \leftarrow \text{7 decimal places} \end{array}$$

43. Multiply the number of gallons that she pumped into her pickup truck by the price per gallon. Use a calculator.

20.510 × 3.759 = 77.09709

Round 77.09709 to the nearest cent.

Michelle paid $77.10 for the gas.

55. (a) Find the cost for three short-sleeved, solid-color shirts.

$$\begin{array}{r} \$14.75 \\ \times\quad 3 \\ \hline \$44.25 \end{array}$$

Based on this subtotal, shipping is $7.95.

Next, find the cost of the three monograms.

$$\begin{array}{r} \$4.95 \\ \times\quad 3 \\ \hline \$14.85 \end{array}$$

Add these amounts, plus $5.00 for a gift box.

$$\begin{array}{rl} \$44.25 & \leftarrow \text{Shirts} \\ 7.95 & \leftarrow \text{Shipping} \\ 14.85 & \leftarrow \text{Monograms} \\ +\ \ 5.00 & \leftarrow \text{Gift box} \\ \hline \$72.05 & \end{array}$$

The total cost is $72.05.

(b) Subtract the cost of the shirts to find the difference.

$$\begin{array}{r} \$72.05 \\ -\ 44.25 \\ \hline \$27.80 \end{array}$$

The monograms, gift box, and shipping added $27.80 to the cost of the gift.

SECTION 4.5 (pages 291–294)

11.

```
       3 6.
1.5 )5 4.0
     4 5
       9 0
       9 0
         0
```

Move decimal point in divisor and dividend 1 place; write 0 in the dividend.

13. Given: 108 ÷ 18 = 6

Find: 0.108 ÷ 1.8 by moving the decimal points.

```
      .06
1.8 )0.1 08
```

So, 0.108 ÷ 1.8 = 0.06.

19. $\frac{3.1}{0.006}$

```
         5 1 6.6 6 6
0.006 )3. 1 0 0 0 0 0
          3 0
            1 0
              6
              4 0
              3 6
                4 0     Write 0 in dividend.
                3 6
                  4 0   Write 0 in dividend.
                  3 6
                    4 0 Write 0 in dividend.
                    3 6
                      4 Stop and round answer
                        to the nearest hundredth.
```

Line up decimal points. Move decimal point in divisor and dividend three places. Write 000 in dividend.

The quotient is 516.67 (rounded).

29. $307.02 \div 5.1 = 6.2$

Estimate: $300 \div 5 = 60$

The answer 6.2 is *unreasonable* because it is so much less than 60.

$$\begin{array}{r} 6\,0.2 \\ 5.1\overline{)3\,0\,7.0\,2} \\ \underline{3\,0\,6} \\ 1\,0 \\ \underline{0} \\ 1\,0\,2 \\ \underline{1\,0\,2} \\ 0 \end{array}$$

The correct answer is 60.2, which is close to the estimate of 60.

55. $33 - 3.2(0.68 + 9) - 1.3^2$ Parentheses, Exponent

$33 - 3.2(9.68) - 1.69$ Multiply.

$33 - 30.976 - 1.69$ Subtract.

$2.024 - 1.69 = 0.334$ Subtract.

61. Divide 100,000 box tops (the answer from Exercise 59) by 38 weeks.

$$\begin{array}{r} 2\,6\,3\,1.5 \\ 38\overline{)1\,0\,0,0\,0\,0.0} \\ \underline{7\,6} \\ 2\,4\,0 \\ \underline{2\,2\,8} \\ 1\,2\,0 \\ \underline{1\,1\,4} \\ 6\,0 \\ \underline{3\,8} \\ 2\,2\,0 \\ \underline{1\,9\,0} \\ 3\,0 \end{array}$$

2631.5 rounds to 2632.

The school needs to collect 2632 box tops (rounded) during each of the 38 weeks.

SECTION 4.6 (pages 299–302)

23. $3\frac{5}{8} = \frac{29}{8} = 3.625$

$$\begin{array}{r} 3.6\,2\,5 \\ 8\overline{)2\,9.0\,0\,0} \\ \underline{2\,4} \\ 5\,0 \\ \underline{4\,8} \\ 2\,0 \\ \underline{1\,6} \\ 4\,0 \\ \underline{4\,0} \\ 0 \end{array}$$

35. $0.35 = \frac{35}{100} = \frac{35 \div 5}{100 \div 5} = \frac{7}{20}$

51. Write zeros so that all the numbers have four decimal places. The acceptable lengths must be greater than 0.9980 cm and less than 1.0020 cm.

$1.0100 > 1.0020$ *unacceptable*

$0.9991 > 0.9980$ and

$0.9991 < 1.0020$ *acceptable*

$1.0007 > 0.9980$ and

$1.0007 < 1.0020$ *acceptable*

$0.9900 < 0.9980$ *unacceptable*

The lengths of 0.9991 cm and 1.0007 cm are acceptable.

63. $\frac{3}{8}, \frac{2}{5}, 0.37, 0.4001$

$\frac{3}{8} = 0.3750$

$\frac{2}{5} = 0.4000$

$0.37 = 0.3700 \leftarrow$ least

$0.4001 = 0.4001 \leftarrow$ greatest

From least to greatest: $0.37, \frac{3}{8}, \frac{2}{5}, 0.4001$

CHAPTER 5 Ratio and Proportion

SECTION 5.1 (pages 321–324)

7. \$100 to \$50

$$\frac{\$100}{\$50} = \frac{100}{50} = \frac{100 \div 50}{50 \div 50} = \frac{2}{1}$$

19. $1\frac{1}{4}$ to $1\frac{1}{2}$

$$\frac{1\frac{1}{4}}{1\frac{1}{2}} = \frac{\frac{5}{4}}{\frac{3}{2}} = \frac{5}{4} \div \frac{3}{2} = \frac{5}{\underset{2}{\cancel{4}}} \cdot \frac{\overset{1}{\cancel{2}}}{3} = \frac{5}{6}$$

47. The increase in price is $\$12.50 - \$10 = \$2.50$.

$$\frac{\$2.50}{\$10} = \frac{2.50}{10} = \frac{2.50 \cdot 10}{10 \cdot 10} = \frac{25}{100} = \frac{25 \div 25}{100 \div 25} = \frac{1}{4}$$

The ratio of the increase in price to the original price is $\frac{1}{4}$.

SECTION 5.2 (pages 329–332)

15. Miles traveled:

$28{,}396.7 - 28{,}058.1 = 338.6$

Miles per gallon:

$$\frac{338.6 \text{ miles} \div 16.2}{16.2 \text{ gallons} \div 16.2} \approx 20.90 \text{ or } 20.9$$

19.

Size	*Cost per Unit*
12 ounces	$\frac{\$2.49}{12 \text{ ounces}} \approx \$0.208$
14 ounces	$\frac{\$2.89}{14 \text{ ounces}} \approx \$0.206$
18 ounces	$\frac{\$3.96}{18 \text{ ounces}} = \$0.22$

Because \$0.206 is the lowest cost per ounce, the best buy is 14 ounces for \$2.89.

33. One battery for \$1.79 is like getting 3 batteries, so $\$1.79 \div 3 \approx \$0.597$ per battery. An eight-pack of AA batteries for \$4.99 is $\$4.99 \div 8 \approx \$0.624$ per battery. Because \$0.597 is the lower cost per battery, the better buy is the package with one battery.

SECTION 5.3 (pages 336–337)

11. $\frac{150}{200} = \frac{200}{300}$

$\frac{150 \div 50}{200 \div 50} = \frac{3}{4}$ and $\frac{200 \div 100}{300 \div 100} = \frac{2}{3}$

Because $\frac{3}{4}$ is *not* equal to $\frac{2}{3}$, the proportion is *false*.

33. $\frac{2\frac{5}{8}}{3\frac{1}{4}} = \frac{21}{26}$

Cross products:

$2\frac{5}{8} \cdot 26 = \frac{21}{\underset{4}{\cancel{8}}} \cdot \frac{\overset{13}{\cancel{26}}}{1} = \frac{273}{4} = 68\frac{1}{4}$

$3\frac{1}{4} \cdot 21 = \frac{13}{4} \cdot \frac{21}{1} = \frac{273}{4} = 68\frac{1}{4}$

The cross products are *equal,* so the proportion is *true*.

37. $\frac{2\frac{3}{10}}{8.05} = \frac{\frac{1}{4}}{0.9}$

Cross products:

$2\frac{3}{10}(0.9) = \frac{23}{10} \cdot \frac{9}{10} = \frac{207}{100} = 2\frac{7}{100}$

or $(2.3)(0.9) = 2.07$

$(8.05)\frac{1}{4} = (8.05)(0.25) = 2.0125$

or $8\frac{5}{100} \cdot \frac{1}{4} = \frac{805}{100} \cdot \frac{1}{4} = \frac{805}{400} = 2\frac{1}{80}$

The cross products are *not* equal so the proportion is *false*.

$\left(2\frac{7}{100} \neq 2\frac{1}{80} \text{ or } 2.07 \neq 2.0125\right)$

SECTION 5.4 (pages 342–343)

9. $\dfrac{42}{x} = \dfrac{18}{39}$

$x \cdot 18 = 42 \cdot 39$ Cross products are equal.

$\dfrac{x \cdot \cancel{18}^{1}}{\cancel{18}_{1}} = \dfrac{1638}{18}$ Divide both sides by 18.

$x = 91$

Check

$42 \cdot 39 = 1638$

$91 \cdot 18 = 1638$

15. $\dfrac{99}{55} = \dfrac{44}{x}$ **OR** $\dfrac{9}{5} = \dfrac{44}{x}$

$9 \cdot x = 5 \cdot 44$ Cross products are equal.

$\dfrac{\cancel{9}^{1} \cdot x}{\cancel{9}_{1}} = \dfrac{220}{9}$ Divide both sides by 9.

$x = \dfrac{220}{9}$

$x \approx 24.44$ (rounded)

Check

$55 \cdot 44 = 2420$

$99 \cdot 24.44 = 2419.56$

Slightly different because of rounding.

23. $\dfrac{2\frac{1}{3}}{1\frac{1}{2}} = \dfrac{x}{2\frac{1}{4}}$

$1\frac{1}{2} \cdot x = 2\frac{1}{3} \cdot 2\frac{1}{4}$

$\dfrac{3}{2} \cdot x = \dfrac{7}{\cancel{3}_{1}} \cdot \dfrac{\cancel{9}^{3}}{4}$

$\dfrac{3}{2} \cdot x = \dfrac{21}{4}$

$\dfrac{\frac{3}{2} \cdot x}{\frac{3}{2}} = \dfrac{\frac{21}{4}}{\frac{3}{2}}$

$x = \dfrac{21}{4} \div \dfrac{3}{2} = \dfrac{\cancel{21}^{7}}{\cancel{4}_{2}} \cdot \dfrac{\cancel{2}^{1}}{\cancel{3}_{1}} = \dfrac{7}{2} = 3\frac{1}{2}$

27. $\dfrac{x}{\frac{3}{50}} = \dfrac{0.15}{1\frac{4}{5}}$

Change to decimals:

$\dfrac{3}{50} = 3 \div 50 = 0.06$

$1\frac{4}{5} = \dfrac{9}{5}$ and $9 \div 5 = 1.8$

$\dfrac{x}{0.06} = \dfrac{0.15}{1.8}$

$x \cdot 1.8 = (0.06)(0.15)$

$\dfrac{x \cdot \cancel{1.8}}{\cancel{1.8}} = \dfrac{0.009}{1.8}$

$x = 0.005$

Change to fractions:

$0.15 = \dfrac{15 \div 5}{100 \div 5} = \dfrac{3}{20}$

$\dfrac{x}{\frac{3}{50}} = \dfrac{\frac{3}{20}}{1\frac{4}{5}}$

$1\frac{4}{5} \cdot x = \dfrac{3}{50} \cdot \dfrac{3}{20}$

$\dfrac{1\cancel{\frac{4}{5}} \cdot x}{1\cancel{\frac{4}{5}}} = \dfrac{\frac{9}{1000}}{1\frac{4}{5}}$

$x = \dfrac{9}{1000} \div 1\frac{4}{5} = \dfrac{\cancel{9}^{1}}{\cancel{1000}_{200}} \cdot \dfrac{\cancel{5}^{1}}{\cancel{9}_{1}}$

$= \dfrac{1}{200}$

Compare answers.

$0.005 = \dfrac{5 \div 5}{1000 \div 5} = \dfrac{1}{200} \leftarrow$ Matches

SECTION 5.5 (pages 347–350)

11. $\dfrac{6 \text{ ounces}}{7 \text{ servings}} = \dfrac{x \text{ ounces}}{12 \text{ servings}}$

$7 \cdot x = 6 \cdot 12$

$\dfrac{\cancel{7}^{1} \cdot x}{\cancel{7}_{1}} = \dfrac{72}{7}$

$x \approx 10.3 \approx 10$ (rounded)

You need about 10 ounces for 12 servings.

17. The length of the dining area is the same as the length of the kitchen, which is 14 feet (from Exercise 15).

Find the width of the dining area.

4.5 inches − 2.5 inches = 2 inches on the floor plan.

$\dfrac{1 \text{ inch}}{4 \text{ feet}} = \dfrac{2 \text{ inches}}{x \text{ feet}}$

$1 \cdot x = 4 \cdot 2$

$x = 8$

The dining area is 8 feet wide.

23. $\dfrac{7 \text{ refresher}}{10 \text{ entering}} = \dfrac{x \text{ refresher}}{2950 \text{ entering}}$

$10 \cdot x = 7 \cdot 2950$

$\dfrac{\cancel{10}^{1} \cdot x \cdot}{\cancel{10}_{1}} = \dfrac{20{,}650}{10}$

$x = 2065$

2065 students will probably need a refresher course. This is a reasonable answer because it's more than half the students, but not all the students.

Incorrect set-up

$\dfrac{10 \text{ entering}}{7 \text{ refresher}} = \dfrac{x \text{ refresher}}{2950 \text{ entering}}$

$7 \cdot x = 10 \cdot 2950$

$\dfrac{\cancel{7}^{1} \cdot x}{\cancel{7}_{1}} = \dfrac{29{,}500}{7}$

$x \approx 4214$ (rounded)

The *incorrect* set-up gives an *unreasonable* estimate of 4214 entering students; there are only 2950 entering students.

33. Coretta $\left\{\dfrac{1.05 \text{ meters}}{1.68 \text{ meters}} = \dfrac{6.58 \text{ meters}}{x \text{ meters}}\right\}$ tree

$1.05 \cdot x = 1.68(6.58)$

$\dfrac{\cancel{1.05} \cdot x}{\cancel{1.05}} = \dfrac{11.0544}{1.05}$

$x = 10.528 \approx 10.53$

The height of the tree is about 10.53 meters.

39. First find the number of calories in a $\frac{1}{2}$-cup serving of bran cereal.

$\dfrac{\frac{1}{3} \text{ cup}}{80 \text{ calories}} = \dfrac{\frac{1}{2} \text{ cup}}{x \text{ calories}}$

$\dfrac{1}{3} \cdot x = 80 \cdot \dfrac{1}{2}$

$\dfrac{\cancel{\frac{1}{3}} \cdot x}{\cancel{\frac{1}{3}}} = \dfrac{\frac{40}{1}}{\frac{1}{3}}$

$x = \dfrac{40}{1} \cdot \dfrac{3}{1}$

$x = 120$ calories

Then find the number of grams of fiber in a $\frac{1}{2}$-cup serving of bran cereal.

$\dfrac{\frac{1}{3} \text{ cup}}{8 \text{ grams}} = \dfrac{\frac{1}{2} \text{ cup}}{x \text{ grams}}$

$\dfrac{1}{3} \cdot x = 8 \cdot \dfrac{1}{2}$

$\dfrac{\cancel{\frac{1}{3}} \cdot x}{\cancel{\frac{1}{3}}} = \dfrac{\frac{4}{1}}{\frac{1}{3}}$

$x = \dfrac{4}{1} \cdot \dfrac{3}{1}$

$x = 12$ grams

A $\frac{1}{2}$-cup serving of bran cereal provides 120 calories and 12 grams of fiber.

CHAPTER 6 Percent

SECTION 6.1 (pages 369–374)

15. 0.5% = 0.005

Drop the percent symbol. Attach two zeros so the decimal point can be moved two places to the left.

27. 2 = 200%

Two zeros are attached so the decimal point can be moved two places to the right. Attach a percent symbol.

53. 100% is all of the children.

So, 100% of 12 children is 12 children.

12 children are present.

79. 21% of the children eat french fries. To write 21% as a decimal, drop the percent symbol and move the decimal point two places to the left, resulting in 0.21. So, 0.21 of the children eat french fries.

SECTION 6.2 (pages 380–385)

9. First write 6.25 over 100. Then to get a whole number in the numerator, multiply the numerator and denominator by 100. Finally, write the fraction in lowest terms.

$$6.25\% = \frac{6.25}{100} = \frac{6.25(100)}{100(100)}$$

$$= \frac{625 \div 625}{10{,}000 \div 625} = \frac{1}{16}$$

41. $\frac{5}{9} = \frac{p}{100}$

$9 \cdot p = 5 \cdot 100$ Find cross products.

$\frac{\cancel{9} \cdot p}{\cancel{9}} = \frac{500}{9}$ Divide both sides by 9.

$p = \frac{500}{9}$

$p \approx 55.5\overline{5}$

Thus, $\frac{5}{9} = 55.6\%$ (rounded).

49. 87.5% = 0.875 decimal

$= \frac{875 \div 125}{1000 \div 125} = \frac{7}{8}$ fraction

63. $\frac{1}{200} = 1 \div 200 = 0.005$ decimal

$= 0.5\%$ percent

67. $3\frac{1}{4} = \frac{13}{4} = 13 \div 4 = 3.25$ decimal

$= 325\%$ percent

77. 64 out of 80 employees have cell phones.

(a) $\frac{64}{80} = \frac{64 \div 16}{80 \div 16} = \frac{4}{5}$ fraction

$\frac{4}{5} = 4 \div 5 = 0.8$ decimal

$0.8 = 0.80 = 80\%$ percent

(b) If $\frac{4}{5}$ (have cell phones)

$1 - \frac{4}{5} = \frac{1}{5}$ do not have cell phones.

$\frac{1}{5} = 1 \div 5 = 0.2$ decimal

$0.2 = 0.20 = 20\%$ percent

SECTION 6.3 (pages 390–393)

11. part = 15, whole = 60

$\frac{15}{60} = \frac{x}{100}$

$\frac{1}{4} = \frac{x}{100}$ $\frac{15}{60}$ is $\frac{1}{4}$ in lowest terms.

$4 \cdot x = 1 \cdot 100$ Find cross products.

$\frac{\cancel{4} \cdot x}{\cancel{4}} = \frac{100}{4}$ Divide both sides by 4.

$x = 25$

The percent is 25, written as 25%.

21. whole = 5000, part = 20

$\frac{20}{5000} = \frac{x}{100}$ Percent proportion

$5000 \cdot x = 20 \cdot 100$ Find cross products.

$\frac{\cancel{5000} \cdot x}{\cancel{5000}} = \frac{2000}{5000}$ Divide both sides by 5000.

$x = 0.4$

The percent is 0.4, written as 0.4%.

37. 12 injections is 20% of what number of injections?

part: 12 injections; percent: 20%; whole (unknown): what number of injections

$\frac{12}{\text{unknown}} = \frac{20}{100}$

47. 86 of 142 people is what percent?

part: 86; whole: 142 people; percent (unknown): what percent

$\frac{86}{142} = \frac{\text{unknown}}{100}$

SECTION 6.4 (pages 401–406)

15. Write 225% as a decimal, 2.25.

225% of 680 tables

(2.25)(680) = 1530

part = 1530 tables

29. part is 350; percent is 12.5

$\frac{\text{part}}{\text{whole}} = \frac{\text{percent}}{100}$

so $\frac{350}{x} = \frac{12.5}{100}$

$12.5 \cdot x = 35{,}000$ Cross products

$\frac{\cancel{12.5} \cdot x}{\cancel{12.5}} = \frac{35{,}000}{12.5}$ Divide both sides by 12.5.

$x = 2800$

$12\frac{1}{2}\%$ of 2800 is 350.

37. part is 64; whole is 344

$\frac{\text{part}}{\text{whole}} = \frac{\text{percent}}{100}$

$\frac{64}{344} = \frac{x}{100}$ **OR** $\frac{8}{43} = \frac{x}{100}$

Cross products $43 \cdot x = 8 \cdot 100$

Divide both sides by 43. $\frac{\cancel{43} \cdot x}{\cancel{43}} = \frac{800}{43}$

$x \approx 18.6$

\$64 is 18.6% (rounded) of \$344.

43. (a) part is unknown; whole is 240; percent is 22

$\frac{\text{part}}{\text{whole}} = \frac{\text{percent}}{100}$

$\frac{x}{240} = \frac{22}{100}$ **OR** $\frac{x}{240} = \frac{11}{50}$

Cross products $x \cdot 50 = 240 \cdot 11$

Divide both sides by 50. $\frac{x \cdot \cancel{50}}{\cancel{50}} = \frac{2640}{50}$

$x = 52.8$

The amount withheld is \$52.80.

(b) \$240 earnings − \$52.80 withheld = \$187.20 amount remaining. The amount remaining is \$187.20.

55. part is 960; whole is 48,000; percent is unknown

$\frac{\text{part}}{\text{whole}} = \frac{\text{percent}}{100}$

$\frac{960}{48{,}000} = \frac{x}{100}$ **OR** $\frac{1}{50} = \frac{x}{100}$

Cross products $50 \cdot x = 1 \cdot 100$

Divide both sides by 50. $\frac{\cancel{50} \cdot x}{\cancel{50}} = \frac{100}{50}$

$x = 2$

There are 2% of these jobs filled by women.

SOLUTIONS

SECTION 6.5 (pages 411–414)

3. Write 25% as the decimal 0.25. The whole is 1080. Let x represent the unknown part.

part = percent • whole

$x = (0.25)(1080)$

$x = 270$

25% of 1080 blood donors is 270 blood donors.

23. The part is 3.75 and the percent is $1\frac{1}{4}\% = 1.25\%$ or 0.0125 as a decimal. The whole is unknown.

part = percent • whole

$3.75 = (0.0125)(x)$

$$\frac{3.75}{0.0125} = \frac{\overset{1}{\cancel{(0.0125)}}(x)}{\underset{1}{\cancel{0.0125}}}$$

$300 = x$

$1\frac{1}{4}\%$ of 300 gallons is 3.75 gallons.

29. Because 160 follows *of*, the whole is 160. The part is 2.4, and the percent is unknown.

part = percent • whole

$2.4 = x \cdot 160$

$$\frac{2.4}{160} = \frac{x \cdot \overset{1}{\cancel{160}}}{\underset{1}{\cancel{160}}}$$

$0.015 = x$

0.015 is 1.5%.

1.5% of 160 liters is 2.4 liters.

43. Because 1250 follows *of*, the whole is 1250. The part is 461, and the percent is unknown.

part = percent • whole

$461 = x \cdot 1250$

$$\frac{461}{1250} = \frac{x \cdot \overset{1}{\cancel{1250}}}{\underset{1}{\cancel{1250}}}$$

$0.369 \approx x$

0.369 is 36.9%.

36.9% (rounded) of these Americans rate their health as excellent.

55. Find the sales tax on the new Polaris unit. The whole is 524. The percent is $7\frac{3}{4}\% = 7.75\%$, which is 0.0775 as a decimal.

part = percent • whole

$x = (0.0775)(524)$

$x = 40.61$

Add the sales tax to the purchase price.

$\$524 + \$40.61 = \$564.61$

Subtract the trade-in.

$\$564.61 - \$125 = \$439.61$

The total cost to the customer is $439.61.

SECTION 6.6 (pages 421–424)

5. Find the sales tax rate using the sales tax formula. The sales tax is $12.75, and the cost of the item is $425.

sales tax = rate of tax • cost of item

$\$12.75 = r \cdot \$425$

$$\frac{12.75}{425} = \frac{r \cdot \overset{1}{\cancel{425}}}{\underset{1}{\cancel{425}}}$$ Divide both sides by 425.

$0.03 = r$

0.03 is 3% Write the decimal as a percent.

The tax rate is 3% and the total cost is $\$425 + \$12.75 = \$437.75$.

11. The problem asks for the amount of commission. Use the commission formula. The rate of commission is 8%, and the sales are $280.

commission = rate of commission • sales

$= (8\%)(\$280)$

$= (0.08)(\$280)$

$= \$22.40$

The amount of commission is $22.40.

23. The problem asks for the amount of the discount and the sale price. First, find the amount of discount using the discount formula. The rate of discount is 15%, and the original price is $58.40.

amount of discount = rate of discount • original price

$= (15\%)(\$58.40)$

$= (0.15)(\$58.40)$ Write 15% as a decimal.

$= \$8.76$ Amount of discount

Now find the sale price by subtracting the amount of the discount ($8.76) from the original price: $\$58.40 - \$8.76 = \$49.64$. The amount of discount is $8.76 and the sale price is $49.64.

31. The problem asks for the rate of sales tax. Use the sales tax formula. The sales tax is $99. The cost of the door is $1980. Use r to represent the unknown rate of tax.

sales tax = rate of tax • cost of item

$\$99 = r \cdot \$1980$

$$\frac{99}{1980} = \frac{r \cdot \overset{1}{\cancel{1980}}}{\underset{1}{\cancel{1980}}}$$ Divide both sides by 1980.

$0.05 = r$

0.05 is 5%. Write the decimal as a percent.

The rate of sales tax is 5%.

47. The problem asks for the cost of the dictionary. First find the amount of discount using the discount formula. The rate of discount is 6%, and the original price is $18.50.

discount = rate of discount • original price

$= (6\%)(\$18.50)$

$= (0.06)(\$18.50)$ Write 6% as a decimal.

$= \$1.11$ Amount of discount

To find the sale price, subtract the amount of discount ($1.11) from the original price.

Sale price

= original price − amount of discount

$= \$18.50 - \$1.11 = \$17.39$ Sale price

To find the sales tax, use the sales tax formula. The rate of tax is 6% and the cost of the dictionary is $17.39.

sales tax = rate of tax • cost of item

$= (6\%)(\$17.39)$

$= (0.06)(\$17.39)$ Write 6% as a decimal.

$\approx \$1.04$ Sales tax

The total cost of the dictionary is $\$17.39 + \$1.04 = \$18.43$.

57. To find the sales tax, use the sales tax formula.

sales tax = rate of tax • cost of item

$= \left(7\frac{3}{4}\%\right)(\$1248)$

$= (0.0775)(\$1248)$ Write $7\frac{3}{4}\%$ as a decimal.

$= \$96.72$ Sales tax

The total cost is the sum of the ticket, the sales tax, and the excise tax.

$\$1248 + \$96.72 + \$15.40 = \$1360.12$

SECTION 6.7 (pages 427–430)

9. $2300 at $4\frac{1}{2}\%$ for $2\frac{1}{2}$ years

The principal (p) is $2300. The rate ($r$) is $4\frac{1}{2}\%$, or 0.045 as a decimal, and the time (t) is $2\frac{1}{2}$ or 2.5 years.

$I = p \cdot r \cdot t$

$= (2300)(0.045)(2.5)$

$= 258.75$

The interest is $258.75.

SOLUTIONS

19. \$940 at 3% for 18 months

The principal is \$940. The rate is 3% or 0.03, and the time is $\frac{18}{12}$ of a year.

$$I = p \cdot r \cdot t$$

$$= (940)(0.03)\left(\frac{18}{12}\right)$$

18 months $= \frac{18}{12}$ of a year.

$$= (28.2)(1.5) \quad \tfrac{18}{12} = (1.5)$$

$$= 42.3$$

The interest is \$42.30.

33. \$16,850 at $7\frac{1}{2}\%$ for 9 months

First find the interest. The principal is \$16,850. The rate is $7\frac{1}{2}\%$ or 0.075, and the time is $\frac{9}{12}$ of a year.

$$I = p \cdot r \cdot t$$

$$= (16{,}850)(0.075)\left(\frac{9}{12}\right)$$

$$= (1263.75)\left(\frac{3}{4}\right) \quad \text{Write } \tfrac{9}{12} \text{ in lowest terms as } \tfrac{3}{4}$$

$$= 947.81 \text{ (rounded)}$$

The interest is \$947.81.

To find the total amount due, add the principal and the interest.

amount due = *principal* + *interest*

= \$16,850 + \$947.81

= \$17,797.81

The total amount due is \$17,797.81.

43. \$14,800 at $2\frac{1}{4}\%$ for 10 months

The principal is \$14,800. The rate is $2\frac{1}{4}\%$ or 0.0225, and the time is $\frac{10}{12}$ of a year.

$$I = p \cdot r \cdot t$$

$$= (14{,}800)(0.0225)\left(\frac{10}{12}\right)$$

$$= (333)\left(\frac{5}{6}\right) \quad \text{Write } \tfrac{10}{12} \text{ in lowest terms as } \tfrac{5}{6}.$$

$$= 277.50$$

She will earn \$277.50 in interest.

SECTION 6.8 (pages 435–438)

3. \$500 at 4% for 2 years

Year	*Interest*	*Compound Amount*
1	(\$500)(0.04)(1) = \$20	
		\$500 + \$20 = \$520
2	(\$520)(0.04)(1) = \$20.80	
		\$520 + \$20.80 = \$540.80

The compound amount is \$540.80.

13. \$1400 at 6% for 5 years

Year 1 Year 2 Year 3 Year 4 Year 5

(\$1400) (1.06) (1.06)(1.06)(1.06)(1.06) ≈ \$1873.52

↑ Original deposit — 100% + 6% = 106% = 1.06 — ↑ Compound amount (rounded)

The compound amount is \$1873.52.

23. \$1000 at 4% for 5 years

Look down the column headed 4%, and across to row 5 (because 5 years = 5 time periods). At the intersection of the column and the row, read the compound amount, 1.2167.

compound amount = (\$1000)(1.2167)

= \$1216.70

Find the interest by subtracting the principal (\$1000) from the compound amount.

interest = \$1216.70 − \$1000

= \$216.70

33. (a) 6% column, row 9 from the table is 1.6895.

compound amount = (\$76,000)(1.6895)

= \$128,402

The total amount that should be repaid is \$128,402.

(b) To find the amount of interest earned, subtract the principal (\$76,000) from the compound amount.

The amount of interest earned is \$128,402 − \$76,000 = \$52,402.

CHAPTER 7 Geometry

SECTION 7.1 (pages 466–469)

7. The figure starts at point P and goes on forever in one direction, so it is a ray named $\overrightarrow{PQ}$.

9. The lines are *perpendicular* because they intersect at right angles. The small red square indicates the right angle (90°).

23. Two rays in a straight line pointing in opposite directions measure 180°. An angle that measures 180° is called a *straight angle*.

31. Find the complement of 86° by subtracting.

90° − 86° = 4° ← Complement

35. Find the supplement of 90° by subtracting.

180° − 90° = 90° ← Supplement

37. $\angle SON \cong \angle TOM$ because they are vertical angles.

$\angle TOS \cong \angle MON$ because they are vertical angles.

39. Because $\angle COE$ and $\angle GOH$ are vertical angles, they are also congruent. This means they have the same measure.

$\angle COE$ measures 63°, so $\angle GOH$ also measures 63°.

The sum of the measures of $\angle COE$, $\angle AOC$, and $\angle AOH$ equals 180°.

Therefore, $\angle AOC$ measures

180° − (63° + 37°) = 180° − 100° = 80°.

Since $\angle AOC$ and $\angle GOF$ are vertical angles, they are congruent, so $\angle GOF$ also measures 80°. Since $\angle AOH$ and $\angle EOF$ are vertical angles, they are congruent, so $\angle EOF$ measures 37°.

51. $\angle 6 \cong \angle 1$ (vertical angles), so the measure of $\angle 1$ is also 47°.

$\angle 6 \cong \angle 8$ (corresponding angles), so the measure of $\angle 8$ is also 47°.

$\angle 6 \cong \angle 3$ (alternate interior angles), so the measure of $\angle 3$ is also 47°.

Notice that the exterior sides of $\angle 6$ and $\angle 5$ form a straight angle of 180°. Therefore, $\angle 6$ and $\angle 5$ are supplementary angles and the sum of their measures is 180°. If $\angle 6$ is 47° then $\angle 5$ must be 133° because 180° − 47° = 133°. So the measure of $\angle 5$ is 133°.

$\angle 5 \cong \angle 2$ (vertical angles), so the measure of $\angle 2$ is also 133°.

$\angle 5 \cong \angle 7$ (corresponding angles), so the measure of $\angle 7$ is also 133°.

$\angle 7 \cong \angle 4$ (vertical angles), so the measure of $\angle 4$ is also 133°.

SECTION 7.2 (pages 476–479)

5. $P = 4 \cdot s$

$= 4 \cdot 0.9$ km

$= 3.6$ km

$A = s \cdot s$

$= 0.9 \text{ km} \cdot 0.9 \text{ km}$

$= 0.81 \text{ km}^2$ ← Write km² for area (**not** km).

9. A storage building that is 76.1 ft by 22 ft

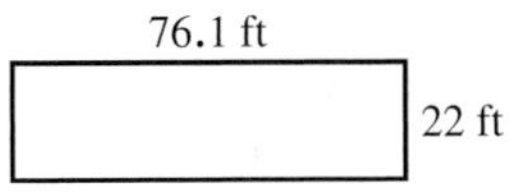

The length of this rectangle is 76.1 ft and the width is 22 ft. Use the formula $P = 2 \cdot l + 2 \cdot w$ to find the perimeter.

$$P = 2 \cdot 76.1 \text{ ft} + 2 \cdot 22 \text{ ft}$$
$$= 152.2 \text{ ft} + 44 \text{ ft}$$
$$= 196.2 \text{ ft}$$

Use the formula $A = l \cdot w$ to find the area.

$$A = 76.1 \text{ ft} \cdot 22 \text{ ft}$$
$$= 1674.2 \text{ ft}^2$$

17. The unlabeled side is 6 in. + 6 in. or 12 in.

Add the lengths of the sides to find the perimeter.

$$P = 15 \text{ in.} + 6 \text{ in.} + 6 \text{ in.} + 6 \text{ in.} + 6 \text{ in.} + 6 \text{ in.} + 9 \text{ in.} + 12 \text{ in.} + 6 \text{ in.} + 6 \text{ in.}$$
$$= 78 \text{ in.}$$

Draw two horizontal lines to break up the figure into two smaller squares and one larger rectangle. Add the areas of these squares and rectangle to find the total area.

$$A = (6 \text{ in.} \cdot 6 \text{ in.}) + (18 \text{ in.} \cdot 9 \text{ in.}) + (6 \text{ in.} \cdot 6 \text{ in.})$$
$$= 36 \text{ in.}^2 + 162 \text{ in.}^2 + 36 \text{ in.}^2$$
$$= 234 \text{ in.}^2$$

You could also draw two vertical lines to break up the figure into three rectangles.

$$A = (15 \text{ in.} \cdot 6 \text{ in.}) + (15 \text{ in.} \cdot 6 \text{ in.}) + (9 \text{ in.} \cdot 6 \text{ in.})$$
$$= 90 \text{ in.}^2 + 90 \text{ in.}^2 + 54 \text{ in.}^2$$
$$= 234 \text{ in.}^2 \leftarrow \text{Same result}$$

23. Find the perimeter of the room.

$$P = 2 \cdot 4.4 \text{ m} + 2 \cdot 5.1 \text{ m}$$
$$= 8.8 \text{ m} + 10.2 \text{ m} = 19 \text{ m}$$

She will need 19 m of the strip. Multiply by the cost per meter to find the total cost.

$$\text{Cost} = \frac{19 \cancel{\text{m}}}{1} \cdot \frac{\$4.99}{1 \cancel{\text{m}}} = \$94.81$$

Tyra will have to spend $94.81 for the strip.

27.

$$A = \text{length} \cdot \text{width}$$
$$5300 \text{ yd}^2 = 100 \text{ yd} \cdot \text{width}$$
$$\frac{5300 \text{ yd} \cdot \cancel{\text{yd}}}{100 \cancel{\text{yd}}} = \frac{\cancel{100 \text{ yd}} \cdot \text{width}}{\cancel{100 \text{ yd}}}$$
$$53 \text{ yd} = \text{width}$$

The width is 53 yd.

SECTION 7.3 (pages 484–485)

13. The figure is a trapezoid. The height (h) is 42 cm, the short base (b) is 61.4 cm, and the long base (B) is 86.2 cm.

$$A = \frac{1}{2} \cdot h \cdot (b + B)$$
$$= 0.5 \cdot 42 \text{ cm} \cdot (61.4 \text{ cm} + 86.2 \text{ cm})$$
$$= 0.5 \cdot 42 \text{ cm} \cdot (147.6 \text{ cm})$$
$$= 3099.6 \text{ cm}^2$$

19. Area of one parallelogram:

$$A = b \cdot h$$
$$= 5 \text{ in.} \cdot 3.5 \text{ in.}$$
$$= 17.5 \text{ in.}^2$$

To find the total area of the 25 parallelogram pieces, multiply the area of one parallelogram (17.5 in.^2) by the number of pieces (25).

$$\text{Total area} = 25 \cdot 17.5 \text{ in.}^2 = 437.5 \text{ in.}^2$$

21. Break the figure into two pieces, a rectangle on the left side and a parallelogram. Find the area of each piece, and then add the areas.

Area of rectangle:

$$A = \text{length} \cdot \text{width}$$
$$= 1.3 \text{ m} \cdot 0.8 \text{ m}$$
$$= 1.04 \text{ m}^2$$

Area of parallelogram:

$$A = \text{base} \cdot \text{height}$$
$$= 1.8 \text{ m} \cdot 1.1 \text{ m}$$
$$= 1.98 \text{ m}^2$$

$$\text{Total area} = 1.04 \text{ m}^2 + 1.98 \text{ m}^2$$
$$= 3.02 \text{ m}^2$$

SECTION 7.4 (pages 490–492)

9. To find the perimeter, add the lengths of the three sides.

$$P = 35.5 \text{ cm} + 21.3 \text{ cm} + 28.4 \text{ cm}$$
$$= 85.2 \text{ cm}$$

When finding the area of this right triangle, the base and height are the perpendicular sides. So the base is 28.4 cm and the height is 21.3 cm.

$$A = \frac{1}{2} \cdot b \cdot h$$
$$= 0.5 \cdot 28.4 \text{ cm} \cdot 21.3 \text{ cm}$$
$$= 302.46 \text{ cm}^2$$

13. First, find the area of the entire figure, which is a rectangle.

$$A = l \cdot w$$
$$= 52 \text{ m} \cdot 37 \text{ m}$$
$$= 1924 \text{ m}^2$$

Find the area of the unshaded triangle.

$$A = \frac{1}{2} \cdot b \cdot h$$
$$= \frac{1}{2} \cdot 52 \text{ m} \cdot 10 \text{ m}$$
$$= 260 \text{ m}^2$$

Subtract the unshaded area from the area of the entire figure to find the shaded area.

$$\text{Shaded area} = 1924 \text{ m}^2 - 260 \text{ m}^2$$
$$= 1664 \text{ m}^2$$

15. *Step 1* Add the two angles given.

$$90° + 58° = 148°$$

Step 2 Subtract the sum from 180°.

$$180° - 148° = 32°$$

The third angle measures 32°.

23. (a) To find the amount of curbing needed to go around the triangular space, find the perimeter.

$$P = 42 \text{ m} + 32 \text{ m} + 52.8 \text{ m}$$
$$= 126.8 \text{ m}$$

126.8 m of curbing will be needed.

(b) To find the amount of sod needed to cover the space, find the area. It is a right triangle, so the perpendicular sides are the base and height.

$$A = \frac{1}{2} \cdot 32 \text{ m} \cdot 42 \text{ m} = 672 \text{ m}^2$$

672 m^2 of sod will be needed.

SECTION 7.5 (pages 499–502)

13. $d = 7\frac{1}{2}$ ft or 7.5 ft

$$C = \pi \cdot d$$
$$\approx 3.14 \cdot 7.5 \text{ ft}$$
$$\approx 23.6 \text{ ft}$$

To find the area, first find the radius.

$$r = \frac{7.5 \text{ ft}}{2} = 3.75 \text{ ft}$$
$$A = \pi \cdot r \cdot r$$
$$\approx 3.14 \cdot 3.75 \text{ ft} \cdot 3.75 \text{ ft}$$
$$\approx 44.2 \text{ ft}^2$$

SOLUTIONS

17. Find the area of a whole circle with a radius of 10 cm:

$A = \pi \cdot r \cdot r$

$\approx 3.14 \cdot 10 \text{ cm} \cdot 10 \text{ cm}$

$= 314 \text{ cm}^2$

Divide the area of the whole circle by 2 to find the area of the semicircle:

$\dfrac{314 \text{ cm}^2}{2} = 157 \text{ cm}^2$

Area of the triangle:

$A = \dfrac{1}{2} \cdot b \cdot h$

$= \dfrac{1}{2} \cdot 20 \text{ cm} \cdot 10 \text{ cm}$

$= 100 \text{ cm}^2$

Subtract to find the shaded area.

$157 \text{ cm}^2 - 100 \text{ cm}^2 = 57 \text{ cm}^2$

23. A point on the tire tread moves the length of the circumference in one complete turn.

$C = \pi \cdot d$

$\approx 3.14 \cdot 29.10 \text{ in.}$

$\approx 91.4 \text{ in.}$

Bonus question:

$\dfrac{1 \text{ revolution}}{91.4 \cancel{\text{ inches}}} \cdot \dfrac{12 \cancel{\text{ inches}}}{1 \cancel{\text{ foot}}} \cdot \dfrac{5280 \cancel{\text{ feet}}}{1 \text{ mile}}$

≈ 693 revolutions/mile (rounded)

25.

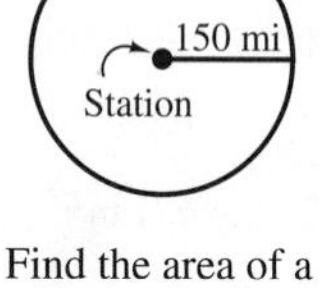

Find the area of a circle with a radius of 150 miles.

$A = \pi \cdot r \cdot r$

$\approx 3.14 \cdot 150 \text{ mi} \cdot 150 \text{ mi}$

$= 70{,}650 \text{ mi}^2$

There are about 70,650 mi^2 in the broadcast area.

31. (a) $C = 144 \text{ cm}$

$C = \pi \cdot d$

$144 \text{ cm} = \pi \cdot d$

$144 \text{ cm} \approx 3.14 \cdot d$

$\dfrac{144 \text{ cm}}{3.14} \approx \dfrac{\cancel{3.14} \cdot d}{\cancel{3.14}}$

$45.9 \text{ cm} \approx d$

The diameter is about 45.9 cm.

(b) Divide the circumference by π (3.14).

SECTION 7.6 (pages 511–512)

9. The figure is a hemisphere.

$V = \dfrac{2}{3} \cdot \pi \cdot r^3$ or $\dfrac{2 \cdot \pi \cdot r^3}{3}$

$\approx \dfrac{2 \cdot 3.14 \cdot 12 \text{ in.} \cdot 12 \text{ in.} \cdot 12 \text{ in.}}{3}$

$\approx 3617.3 \text{ in.}^3 \leftarrow$ Round to tenths.

11. The figure is a cone.

First find B, the area of the circular base.

$B = \pi \cdot r \cdot r$

$\approx 3.14 \cdot 5 \text{ m} \cdot 5 \text{ m}$

$\approx 78.5 \text{ m}^2$

Now find the volume of the cone.

$V = \dfrac{B \cdot h}{3}$

$\approx \dfrac{78.5 \text{ m}^2 \cdot 16 \text{ m}}{3}$

$\approx 418.7 \text{ m}^3 \leftarrow$ Round to tenths.

19. Use the formula for the volume of a pyramid.

First find B, the area of the square base.

$B = s \cdot s$

$= 145 \text{ m} \cdot 145 \text{ m}$

$= 21{,}025 \text{ m}^2$

Now find the volume of the pyramid.

$V = \dfrac{B \cdot h}{3}$

$= \dfrac{21{,}025 \text{ m}^2 \cdot 93 \text{ m}}{3}$

$= 651{,}775 \text{ m}^3$

The volume of the ancient stone pyramid is 651,775 m^3.

21. Use the formula for the volume of a cylinder.

First find the radius of the pipe.

$\text{radius} = \dfrac{5 \text{ ft}}{2} = 2.5 \text{ ft}$

Now find the volume.

$V = \pi \cdot r^2 \cdot h$

$\approx 3.14 \cdot 2.5 \text{ ft} \cdot 2.5 \text{ ft} \cdot 200 \text{ ft}$

$\approx 3925 \text{ ft}^3$

The volume of the city sewer pipe is about 3925 ft^3.

SECTION 7.7 (pages 517–520)

25. The unknown length is the side opposite the right angle, which is the hypotenuse. The lengths of the legs are 8 in. and 3 in.

$\text{hypotenuse} = \sqrt{(\text{leg})^2 + (\text{leg})^2}$

$= \sqrt{(8)^2 + (3)^2}$

$= \sqrt{64 + 9}$

$= \sqrt{73}$

$\approx 8.5 \text{ in.}$

35. The length of the hypotenuse is 21.6 km. The length of one of the legs is 13.2 km.

$\text{leg} = \sqrt{(\text{hypotenuse})^2 - (\text{leg})^2}$

$= \sqrt{(21.6)^2 - (13.2)^2}$

$= \sqrt{466.56 - 174.24}$

$= \sqrt{292.32}$

$\approx 17.1 \text{ km}$

43. The diagonal brace is the hypotenuse. The lengths of the legs are 4.5 ft and 3.5 ft.

$\text{hypotenuse} = \sqrt{(\text{leg})^2 + (\text{leg})^2}$

$= \sqrt{(4.5)^2 + (3.5)^2}$

$= \sqrt{20.25 + 12.25}$

$= \sqrt{32.5}$

$\approx 5.7 \text{ ft}$

The diagonal brace is about 5.7 ft long.

45.

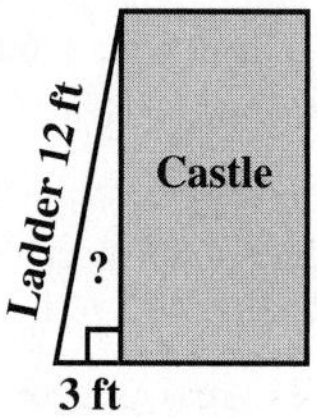

The ladder is opposite the right angle so it is the hypotenuse. Thus, the length of the hypotenuse is 12 ft and the length of one of the legs is 3 ft.

$\text{leg} = \sqrt{(\text{hypotenuse})^2 - (\text{leg})^2}$

$= \sqrt{(12)^2 - (3)^2}$

$= \sqrt{144 - 9}$

$= \sqrt{135}$

$\approx 11.6 \text{ ft}$

The ladder will reach about 11.6 ft high on the building.

SECTION 7.8 (pages 527–530)

5. Notice that *rotating* $\triangle STU$ makes it possible to slide it on top of $\triangle WXY$ so the triangles match.

The corresponding angles are:

$\angle 1$ and $\angle 6$; $\angle 2$ and $\angle 4$; $\angle 3$ and $\angle 5$.

The corresponding sides are:

$\overline{ST}$ and $\overline{YW}$; $\overline{TU}$ and $\overline{WX}$; $\overline{SU}$ and $\overline{YX}$.

13. On both triangles, two corresponding angles and the side that connects them measure the same, so the Angle–Side–Angle (ASA) method can be used to prove that the triangles are congruent.

21. You want to find the values of a and b in the smaller triangle. Notice that the side that is 6 mm long in the smaller triangle corresponds to the side that is 12 mm long in the larger triangle.

Set up a ratio of corresponding sides.

$\frac{6 \text{ mm}}{12 \text{ mm}} = \frac{6}{12} = \frac{1}{2}$ in lowest terms

Write a proportion to find a.

$\frac{a}{10} = \frac{1}{2}$

$a \cdot 2 = 10 \cdot 1$ Cross products are equal.

$\frac{a \cdot \overset{1}{\cancel{2}}}{\underset{1}{\cancel{2}}} = \frac{10}{2}$ Divide both sides by 2.

$a = 5 \text{ mm}$

Write a proportion to find b.

$\frac{b}{6} = \frac{1}{2}$

$b \cdot 2 = 6 \cdot 1$ Cross products are equal.

$\frac{b \cdot \overset{1}{\cancel{2}}}{\underset{1}{\cancel{2}}} = \frac{6}{2}$ Divide both sides by 2.

$b = 3 \text{ mm}$

27. Since triangles *CDE* and *FGH* are similar and all of the sides of *CDE* are the same length, all the sides of *FGH* are also the same length. Therefore, each missing side of triangle *FGH* is 8 cm.

Now add the lengths of all three sides to find the perimeter of triangle *FGH*.

$P = 8 \text{ cm} + 8 \text{ cm} + 8 \text{ cm}$

$P = 24 \text{ cm}$

Set up a ratio of corresponding sides to find the height (h) of triangle *FGH*.

$\frac{10.4}{12} = \frac{h}{8}$

$(12)(h) = (10.4)(8)$ Cross products are equal.

$\frac{(\overset{1}{\cancel{12}})(h)}{\underset{1}{\cancel{12}}} = \frac{83.2}{12}$ Divide both sides by 12.

$h \approx 6.9 \text{ cm}$ Rounded

Area of triangle FGH

$= 0.5 \cdot b \cdot h$

$\approx (0.5)(8 \text{ cm})(6.9 \text{ cm})$

$\approx 27.6 \text{ cm}^2$ Approximate area because height was rounded

33. The side 50 m long in the smaller triangle corresponds to the side with length n in the larger triangle.

Write a proportion to find n.

$\frac{50}{n} = \frac{100}{100 + 120} = \frac{100}{220}$

Write $\frac{100}{220}$ in lowest terms as $\frac{5}{11}$.

$\frac{50}{n} = \frac{5}{11}$

$5 \cdot n = 50 \cdot 11$ Cross products

$\frac{\overset{1}{\cancel{5}} \cdot n}{\underset{1}{\cancel{5}}} = \frac{550}{5}$ Divide both sides by 5.

$n = 110 \text{ m}$

The length of the lake is 110 m.

CHAPTER 8 Statistics

SECTION 8.1 (pages 556–561)

3. The number of pets owned in the United States is 11 million + 150 million + 90 million + 75 million + 18 million + 16 million = 360 million.

15. Time working compared to household duties/caring for others.

3 hours 30 minutes = 210 minutes

2 hours 30 minutes = 150 minutes

$\frac{210}{150} = \frac{210 \div 30}{150 \div 30} = \frac{7}{5}$

25. Need the car to go 50 miles or fewer:

$x = 4\%$ of 3021

$= (0.04)(3021)$

$= 121$ people (rounded)

33. Percent for day camp $= \frac{\$546}{\$5460}$

$= 0.10 = 10\%$

Degrees of a circle = 10% of 360°

$= (0.10)(360°)$

$= 36°$

39. **(a)** Total sales = \$12,500 + \$40,000 + \$60,000 + \$50,000 + \$37,500 = \$200,000

(b) Adventure classes = \$12,500

percent of total $= \frac{12{,}500}{200{,}000} = 0.0625$

$= 6.25\%$

number of degrees $= (0.0625)(360°)$

$= 22.5°$

Grocery and provision sales = \$40,000

percent of total $= \frac{40{,}000}{200{,}000} = 0.2 = 20\%$

number of degrees $= (0.2)(360°) = 72°$

Equipment rentals = \$60,000

percent of total $= \frac{60{,}000}{200{,}000} = 0.3 = 30\%$

number of degrees $= (0.3)(360°) = 108°$

Rafting tours = \$50,000

percent of total $= \frac{50{,}000}{200{,}000} = 0.25 = 25\%$

number of degrees $= (0.25)(360°) = 90°$

Equipment sales = \$37,500

percent of total $= \frac{37{,}500}{200{,}000} = 0.1875$

$= 18.75\%$

number of degrees $= (0.1875)(360°)$

$= 67.5°$

(c)

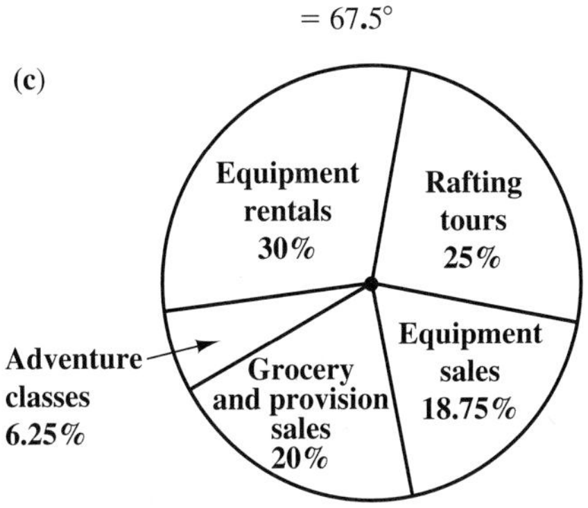

SECTION 8.2 (pages 565–570)

5. The countries in which less than 15% of household income is spent, on average, for food are USA, Britain, and Australia.

11. There are two bars for February. One is for 2012 and one is for 2013. The bar for 2013 rises to 7. Multiply 7 by 1000 because the label on the left side of the graph says *in thousands*. The bar for 2012 rises to halfway between 5 and 6. So multiply 5.5 by 1000.

Plants shipped in February of 2013 = 7000

Plants shipped in February of 2012 = 5500

7000 − 5500 = 1500

There were 1500 more plants shipped in February of 2013.

21. Above the dot for 1990 is 24.1. Looking at the left edge of the graph, notice that this number is in millions.

The number of PCs sold in 1990 was 24.1 million or 24,100,000.

31. Find the dot above 2012 on the line for store B. Use a ruler or straightedge to line up the dot with the numbers along the left edge. The dot is aligned with 350, so chain store B sold 350 • 1000 = 350,000 MP3 players in 2012.

37. Find the dot above 2013 on the line for total sales. Use a ruler or straightedge to line up the dot with the numbers along the left edge. The dot is aligned with 400. Multiply 400 by 1000 because the label on the left side of the graph says *in thousands*. The total sales in 2013 were

400 • \$1000 = \$400,000.

SECTION 8.3 (pages 573–576)

3. Because the bar associated with 18–25 is the highest, the age group of 18–25 years has the greatest number of members, which is 40,000,000.

11. The bar for 31–40 rises to 11, so there are 11 employees who earn \$3100 to \$4000.

23. 0–5; 4 tally marks and a class frequency of 4

SECTION 8.4 (pages 581–584)

5. $\text{Mean} = \dfrac{\text{sum of all values}}{\text{number of values}}$

$= \dfrac{9 + 14 + 18 + 11 + 12 + 20}{6}$

$= \dfrac{84}{6}$

$= 14$ hours

The mean (average) of the time spent playing video games was 14 hours.

13.

Customers Each Hour	Frequency	Product
8	2	(8 • 2) = 16
11	12	(11 • 12) = 132
15	5	(15 • 5) = 75
26	1	(26 • 1) = 26
Totals	**20**	**249**

weighted mean

$= \dfrac{\text{sum of products}}{\text{total number of customers}}$

$= \dfrac{249}{20} \approx 12.5$ customers (rounded)

21. Arrange the numbers in numerical order from least to greatest.

298, 346, 412, 501, 515, 521, 528, 621

The list has 8 numbers. The middle numbers are the 4th and 5th numbers, so the median is

$\dfrac{501 + 515}{2} = \dfrac{1016}{2} = 508$ calories.

25. $\underline{74}, \underline{68}, \underline{68}, \underline{68}, 75, 75, \underline{74}, \underline{74}, 70$

Because both 68 and 74 occur three times, each is a mode. This list is *bimodal*.

33.

Credits	Grade	Credits • Grade
2	A (= 4)	2 • 4 = 8
3	C (= 2)	3 • 2 = 6
4	A (= 4)	4 • 4 = 16
1	C (= 2)	1 • 2 = 2
4	B (= 3)	4 • 3 = 12
14		**44**

$\text{GPA} = \dfrac{\text{sum of (Credits • Grade)}}{\text{total number of credits}}$

$= \dfrac{44}{14} \approx 3.14$ (rounded)

CHAPTER 9 The Real Number System

SECTION 9.1 (pages 611–613)

33. $\dfrac{1}{4} \cdot \dfrac{2}{3} + \dfrac{2}{5} \cdot \dfrac{11}{3}$

$= \dfrac{1}{6} + \dfrac{22}{15}$ Multiply; $\dfrac{2}{12} = \dfrac{1}{6}$

$= \dfrac{5}{30} + \dfrac{44}{30}$ LCD = 30

$= \dfrac{49}{30}$ Add.

45. $\left(\dfrac{3}{2}\right)^2\left[\left(11 + \dfrac{1}{3}\right) - 6\right]$

$= \dfrac{9}{4}\left[\left(11 + \dfrac{1}{3}\right) - 6\right]$ Apply the exponent.

$= \dfrac{9}{4}\left[\left(\dfrac{33}{3} + \dfrac{1}{3}\right) - 6\right]$ LCD = 3

$= \dfrac{9}{4}\left(\dfrac{34}{3} - 6\right)$ Add inside parentheses.

$= \dfrac{9}{4}\left(\dfrac{34}{3} - \dfrac{18}{3}\right)$ LCD = 3

$= \dfrac{9}{4}\left(\dfrac{16}{3}\right)$ Subtract inside parentheses.

$= \dfrac{144}{12}$ Multiply.

$= 12$ Divide.

65. $\dfrac{9(7 - 1) - 8 \cdot 2}{4(6 - 1)} > 3$

$\dfrac{9(6) - 8 \cdot 2}{4(5)} > 3$ Work inside parentheses.

$\dfrac{54 - 16}{20} > 3$ Multiply.

$\dfrac{38}{20} > 3$ Subtract.

$1\dfrac{9}{10} > 3$ Write as a mixed number.

Since $1\dfrac{9}{10} < 3$, the statement is *false*.

SECTION 9.2 (pages 618–621)

31. (a) $\dfrac{3x + y^2}{2x + 3y}$

$= \dfrac{3(2) + 1^2}{2(2) + 3(1)}$ Let $x = 2$ and $y = 1$.

$= \dfrac{3(2) + 1}{4 + 3}$ Work separately above and below the fraction bar.

$= \dfrac{6 + 1}{7}$

$= \dfrac{7}{7}$, or 1

(b) $\dfrac{3x + y^2}{2x + 3y}$

$= \dfrac{3(1) + 5^2}{2(1) + 3(5)}$ Let $x = 1$ and $y = 5$.

$= \dfrac{3 + 25}{2 + 15}$ Multiply.

$= \dfrac{28}{17}$ Add.

33. (a) $0.841x^2 + 0.32y^2$

$= 0.841 \cdot 2^2 + 0.32 \cdot 1^2$ Let $x = 2$, $y = 1$.

$= 0.841 \cdot 4 + 0.32 \cdot 1$ Apply the exponents.

$= 3.364 + 0.32$ Multiply.

$= 3.684$ Add.

(b) $0.841x^2 + 0.32y^2$

$= 0.841 \cdot 1^2 + 0.32 \cdot 5^2$ Let $x = 1$, $y = 5$.

$= 0.841 \cdot 1 + 0.32 \cdot 25$

$= 0.841 + 8$

$= 8.841$

57. $\frac{z+4}{2-z}=\frac{13}{5}$

$\frac{\frac{1}{3}+4}{2-\frac{1}{3}} \stackrel{?}{=} \frac{13}{5}$ Let $z=\frac{1}{3}$.

$\frac{\frac{1}{3}+\frac{12}{3}}{\frac{6}{3}-\frac{1}{3}} \stackrel{?}{=} \frac{13}{5}$ Write with LCD = 3.

$\frac{\frac{13}{3}}{\frac{5}{3}} \stackrel{?}{=} \frac{13}{5}$ Add and subtract.

$\frac{13}{3}\cdot\frac{3}{5} \stackrel{?}{=} \frac{13}{5}$ Definition of division

$\frac{13}{5}=\frac{13}{5}$ ✓ True

The true result shows that $\frac{1}{3}$ is a solution of the equation.

SECTION 9.3 (pages 628–630)

45. In order to compare these two numbers, write them with a common denominator.

$$-\frac{2}{3}=-\frac{8}{12} \quad\text{and}\quad -\frac{1}{4}=-\frac{3}{12}$$

Since

$$\frac{8}{12}>\frac{3}{12},$$

$-\frac{2}{3}$ is farther to the left of 0 on a number line than $-\frac{1}{4}$, so $-\frac{2}{3}$ is the lesser number.

73. A decrease is represented by a negative number, and the amount of decrease is represented by its absolute value. There are three negative numbers in the table: −43.6, −16.2, and −0.6. Of these, −43.6 has the greatest absolute value, so Energy, 2008 to 2009 represents the greatest decrease.

SECTION 9.4 (pages 635–638)

39. $\left(-\frac{1}{2}+0.25\right)+\left(-\frac{3}{4}+0.75\right)$

$=\left(-\frac{1}{2}+\frac{1}{4}\right)+\left(-\frac{3}{4}+\frac{3}{4}\right)$

$=\left(-\frac{2}{4}+\frac{1}{4}\right)+0$

$=-\frac{1}{4}$, or -0.25

59. "$\frac{2}{7}$ more than the sum of $\frac{5}{7}$ and $-\frac{9}{7}$" is interpreted and evaluated as follows.

$\left[\frac{5}{7}+\left(-\frac{9}{7}\right)\right]+\frac{2}{7}$

$=-\frac{4}{7}+\frac{2}{7}$ Work inside brackets.

$=-\frac{2}{7}$ Add.

SECTION 9.5 (pages 643–647)

39. $\left(-\frac{3}{8}-\frac{2}{3}\right)-\left(-\frac{9}{8}-3\right)$

$=\left[-\frac{3}{8}+\left(-\frac{2}{3}\right)\right]-\left[-\frac{9}{8}+(-3)\right]$

$=\left[-\frac{9}{24}+\left(-\frac{16}{24}\right)\right]-\left[-\frac{9}{8}+\left(-\frac{24}{8}\right)\right]$ Get a common denominator for each pair of fractions.

$=-\frac{25}{24}-\left(-\frac{33}{8}\right)$ Add inside brackets.

$=-\frac{25}{24}-\left(-\frac{99}{24}\right)$ Get a common denominator.

$=-\frac{25}{24}+\frac{99}{24}$ Definition of subtraction

$=\frac{74}{24}$ Add.

$=\frac{37}{12}$ Lowest terms

41. $[-12.25-(8.34+3.57)]-17.88$

$=[-12.25-11.91]-17.88$

$=[-12.25+(-11.91)]-17.88$

$=-24.16-17.88$

$=-24.16+(-17.88)$

$=-42.04$

59. Sum of checks:

\$35.84 + \$26.14 + \$3.12

= \$61.98 + \$3.12

= \$65.10

Sum of deposits:

\$85.00 + \$120.76 = \$205.76

Final balance:

= Beginning balance − checks + deposits

= \$904.89 − \$65.10 + \$205.76

= \$839.79 + \$205.76

= \$1045.55

Her account balance at the end of August was \$1045.55.

SECTION 9.6 (pages 656–659)

43. $\frac{4(2^3-5)-5(-3^3+21)}{3[6-(-2)]}$

$=\frac{4(8-5)-5(-27+21)}{3[6-(-2)]}$

$=\frac{4(3)-5(-6)}{3[8]}$

$=\frac{12+30}{24}$

$=\frac{42}{24}$

$=\frac{7}{4}$

49. $\left(\frac{5}{6}x+\frac{3}{2}y\right)\left(-\frac{1}{3}a\right)$

$=\left[\frac{5}{6}(6)+\frac{3}{2}(-4)\right]\left[-\frac{1}{3}(3)\right]$ Let $x=6$, $y=-4$, and $a=3$.

$=[5+(-6)](-1)$

$=(-1)(-1)$

$=1$

69. "The product of $-\frac{2}{3}$ and $-\frac{1}{5}$, divided by $\frac{1}{7}$," is interpreted and evaluated as follows.

$\frac{-\frac{2}{3}\left(-\frac{1}{5}\right)}{\frac{1}{7}}$

$=\frac{\frac{2}{15}}{\frac{1}{7}}$ Multiply in the numerator.

$=\frac{2}{15}\cdot\frac{7}{1}$ Definition of division

$=\frac{14}{15}$ Multiply.

SECTION 9.7 (pages 666–669)

73. $-3(8x+3y+4z)$

$=-3(8x)+(-3)(3y)+(-3)(4z)$ Distributive property

$=(-3\cdot 8)x+(-3\cdot 3)y+(-3\cdot 4)z$ Associative property

$=-24x-9y-12z$ Multiply.

79. $-(-3q+5r-8s)$

$=-1(-3q+5r-8s)$

$=3q-5r+8s$

SECTION 9.8 (pages 673–675)

45. $-\frac{4}{3}+2t+\frac{1}{3}t-8-\frac{8}{3}t$

$=\left(2t+\frac{1}{3}t-\frac{8}{3}t\right)+\left(-\frac{4}{3}-8\right)$

$=\left(2+\frac{1}{3}-\frac{8}{3}\right)t+\left(-\frac{4}{3}-8\right)$

$=\left(\frac{6}{3}+\frac{1}{3}-\frac{8}{3}\right)t+\left(-\frac{4}{3}-\frac{24}{3}\right)$

$=-\frac{1}{3}t-\frac{28}{3}$

47. $-5.3r+4.9-(2r+0.7)+3.2r$

$=-5.3r+4.9-2r-0.7+3.2r$

$=(-5.3r-2r+3.2r)+(4.9-0.7)$

$=(-5.3-2+3.2)r+(4.9-0.7)$

$=-4.1r+4.2$

CHAPTER 10 Equations, Inequalities, and Applications

SECTION 10.1 (pages 693–694)

33. $10x+4=9x$

$10x+4-9x=9x-9x$ Subtract $9x$.

$1x+4=0$

$x+4-4=0-4$ Subtract 4.

$x=-4$

CHECK $10x+4=9x$

$10(-4)+4\stackrel{?}{=}9(-4)$ Let $x=-4$.

$-36=-36$ ✓ True

Solution set: $\{-4\}$

53. $\frac{5}{7}x+\frac{1}{3}=\frac{2}{5}-\frac{2}{7}x+\frac{2}{5}$

$\frac{5}{7}x+\frac{1}{3}=\frac{4}{5}-\frac{2}{7}x$

$\frac{5}{7}x+\frac{1}{3}+\frac{2}{7}x=\frac{4}{5}-\frac{2}{7}x+\frac{2}{7}x$ Add $\frac{2}{7}x$.

$\frac{7}{7}x+\frac{1}{3}=\frac{4}{5}$ Combine like terms.

$1x+\frac{1}{3}-\frac{1}{3}=\frac{4}{5}-\frac{1}{3}$ Subtract $\frac{1}{3}$.

$x=\frac{12}{15}-\frac{5}{15}$ LCD = 15

$x=\frac{7}{15}$ Subtract.

CHECK $\frac{5}{7}x+\frac{1}{3}=\frac{2}{5}-\frac{2}{7}x+\frac{2}{5}$

$\frac{5}{7}\left(\frac{7}{15}\right)+\frac{1}{3}\stackrel{?}{=}\frac{2}{5}-\frac{2}{7}\left(\frac{7}{15}\right)+\frac{2}{5}$ Let $x=\frac{7}{15}$.

$\frac{1}{3}+\frac{1}{3}\stackrel{?}{=}\frac{4}{5}-\frac{2}{15}$ Multiply.

$\frac{2}{3}\stackrel{?}{=}\frac{12}{15}-\frac{2}{15}$ LCD = 15

$\frac{2}{3}=\frac{10}{15}$, or $\frac{2}{3}$ ✓ True

Solution set: $\left\{\frac{7}{15}\right\}$

61. $10(-2x+1)=-19(x+1)$

$-20x+10=-19x-19$ Distributive property

$-20x+10+19x=-19x-19+19x$ Add $19x$.

$-x+10=-19$ Combine like terms.

$-x+10-10=-19-10$ Subtract 10.

$-x=-29$

$x=29$

A check confirms this solution.

Solution set: $\{29\}$

SECTION 10.2 (pages 699–700)

53. $\frac{2}{5}x-\frac{3}{10}x=2$

$\frac{4}{10}x-\frac{3}{10}x=2$ LCD = 10

$\frac{1}{10}x=2$ Combine like terms.

$10\cdot\frac{1}{10}x=10\cdot 2$ Multiply by 10.

$x=20$

CHECK $\frac{2}{5}x-\frac{3}{10}x=2$

$\frac{2}{5}(20)-\frac{3}{10}(20)\stackrel{?}{=}2$ Let $x=20$.

$8-6\stackrel{?}{=}2$

$2=2$ ✓ True

Solution set: $\{20\}$

61. $0.9w-0.5w+0.1w=-3$

$0.5w=-3$ Combine like terms.

$\frac{0.5w}{0.5}=\frac{-3}{0.5}$ Divide by 0.5.

$w=-6$

A check confirms this solution.

Solution set: $\{-6\}$

SECTION 10.3 (pages 709–712)

55. $0.02(5000)+0.03x=0.025(5000+x)$

Multiply both sides by 1000, *not* 100.

$1000[0.02(5000)+0.03x]$
$=1000[0.025(5000+x)]$

$20(5000)+30x=25(5000+x)$

$100{,}000+30x=125{,}000+25x$ Distributive property

$5x+100{,}000=125{,}000$ Subtract $25x$.

$5x=25{,}000$ Subtract 100,000.

$x=5000$ Divide by 5.

CHECK Substituting 5000 for x in the original equation results in a true statement, $250=250$. ✓

Solution set: $\{5000\}$

59. $-(6k-5)-(-5k+8)=-3$

$-1(6k-5)-1(-5k+8)=-3$

$-6k+5+5k-8=-3$

$-k-3=-3$

$-k=0$

$k=0$

CHECK Substituting 0 for k in the original equation results in a true statement, $-3=-3$. ✓

Solution set: $\{0\}$

SECTION 10.4 (pages 721–726)

21. ***Step 1***

Read the problem again.

Step 2

Let $x=$ the unknown number. Then $3x$ is three times the number, $x+7$ is 7 more than the number, and the sum is

$3x+(x+7)$.

$2x$ is twice the number, and

$-11-2x$

is the difference between -11 and twice the number.

Step 3

$3x+(x+7)=-11-2x$

Step 4

$4x+7=-11-2x$

$6x+7=-11$ Add $2x$.

$6x=-18$ Subtract 7.

$x=-3$ Divide by 6.

Step 5

The number is -3.

SOLUTIONS

Step 6

Check that -3 is the correct answer by substituting this result into the words of the original problem. The sum of three times this number and 7 more than the number is

$$3(-3) + (-3 + 7) = -5.$$

The difference between -11 and twice this number is

$$-11 - 2(-3) = -5.$$

The values are equal, so the number -3 is the correct answer.

49. *Step 1*

Read the problem again.

Step 2

Let $x =$ the measures of angles A and B.

$x + 60 =$ the measure of angle C.

Step 3

The sum of the measures of the angles of any triangle is 180°, so

$x + x + (x + 60) = 180.$

Step 4

$3x + 60 = 180$

$3x = 120$ Subtract 60.

$x = 40$ Divide by 3.

Step 5

Angles A and B have measures of 40°, and angle C has a measure of $40 + 60 = 100°$.

Step 6

The answer checks since

$$40 + 40 + 100 = 180.$$

SECTION 10.5 (pages 733–738)

55. Let L represent the length of the box. The girth is $4 \cdot 18 = 72$ in. Since the length plus the girth is 108, we have

$L + 72 = 108$

$L = 36$ in. Subtract 72.

The maximum volume of the box is

$V = LWH$

$V = 36(18)(18)$ Substitute.

$V = 11{,}664$ in.^{3}.

59. The two angles are complementary, so the sum of their measures is 90°.

$(8x - 1) + 5x = 90$

$13x - 1 = 90$ Combine like terms.

$13x = 91$ Add 1.

$x = 7$ Divide by 13.

Since $x = 7$, we have

$8x - 1 = 8(7) - 1 = 56 - 1 = 55$

and

$5x = 5(7) = 35.$

The angles measure 55° and 35°.

85. $M = C(1 + r)$ for r

$M = C + Cr$ Distributive property

$M - C = Cr$ Subtract C.

$\dfrac{M - C}{C} = \dfrac{Cr}{C}$ Divide by C.

$\dfrac{M - C}{C} = r$, or $r = \dfrac{M - C}{C}$

SECTION 10.6 (pages 748–751)

53. $\dfrac{2}{3}(p + 3) > \dfrac{5}{6}(p - 4)$

$6\left(\dfrac{2}{3}\right)(p + 3) > 6\left(\dfrac{5}{6}\right)(p - 4)$

Multiply by 6, the LCD.

$4(p + 3) > 5(p - 4)$

$4p + 12 > 5p - 20$

Distributive property

$-p + 12 > -20$

Subtract $5p$.

$-p > -32$

Subtract 12.

$\dfrac{-p}{-1} < \dfrac{-32}{-1}$

Divide by -1. Reverse the direction of the symbol.

$p < 32$

Solution set: $(-\infty, 32)$

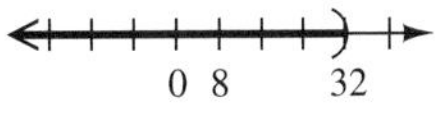

71. The Celsius temperature C must give a Fahrenheit temperature F that has *never exceeded* 104°, which translates as *is less than or equal to* 104°.

$F \le 104$

$\dfrac{9}{5}C + 32 \le 104$

$\dfrac{9}{5}C \le 72$ Subtract 32.

$\dfrac{5}{9}\left(\dfrac{9}{5}C\right) \le \dfrac{5}{9}(72)$ Multiply by $\dfrac{5}{9}$.

$C \le 40$

The temperature of Providence, Rhode Island, has never exceeded 40°C.

CHAPTER 11 Graphs of Linear Equations and Inequalities in Two Variables

SECTION 11.1 (pages 770–775)

29. Is $(5, -6)$ a solution of the equation $x = -6$?

Since y does not appear in the equation, we just substitute 5 for x.

$x = -6$

$5 \stackrel{?}{=} -6$ Let $x = 5$.

The result is false, so $(5, -6)$ is not a solution of the equation $x = -6$.

49. The given equation $x - 8 = 0$ may be written as $x = 8$. For any value of y, the value of x will always be 8. The completed table of values follows.

x	y	Ordered pairs
8	8 →	(8, 8)
8	3 →	(8, 3)
8	0 →	(8, 0)

SECTION 11.2 (pages 784–789)

39. $-3y = 15$

$y = \dfrac{15}{-3}$, or -5 Divide by -3.

For any value of x, the value of y is -5. Three ordered pairs are $(-2, -5)$, $(0, -5)$, and $(1, -5)$. Plot these points and draw a line through them. The graph is a horizontal line.

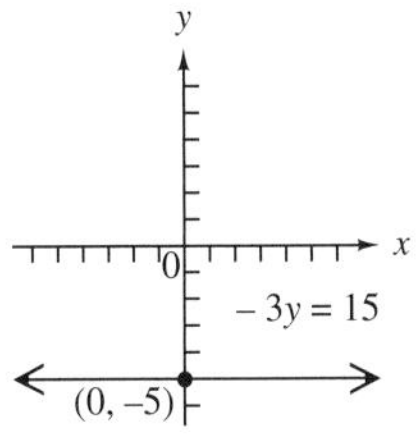

SECTION 11.3 (pages 798–801)

17. The slope (or grade) of the hill is the ratio of the rise to the run, or the ratio of the vertical change to the horizontal change. Since the rise is 32 and the run is 108, the slope is

$\dfrac{32}{108} = \dfrac{8 \cdot 4}{27 \cdot 4} = \dfrac{8}{27}.$ Lowest terms

SOLUTIONS

35. First label the given points.

$$\begin{matrix} (x_1, y_1) & & (x_2, y_2) \\ \downarrow\ \downarrow & & \downarrow\ \downarrow \\ \left(-\frac{7}{5}, \frac{3}{10}\right) & \text{and} & \left(\frac{1}{5}, -\frac{1}{2}\right) \end{matrix}$$

Then apply the slope formula.

$$\text{slope } m = \frac{\text{change in } y}{\text{change in } x}$$

$$= \frac{y_2 - y_1}{x_2 - x_1}$$

$$= \frac{-\frac{1}{2} - \frac{3}{10}}{\frac{1}{5} - \left(-\frac{7}{5}\right)} \quad \text{Substitute.}$$

$$= \frac{-\frac{5}{10} - \frac{3}{10}}{\frac{1}{5} + \frac{7}{5}} \quad \text{Get a common denominator in the numerator. Simplify the denominator.}$$

$$= \frac{-\frac{8}{10}}{\frac{8}{5}} \quad \text{Subtract in the numerator. Add in the denominator.}$$

$$= -\frac{8}{10} \div \frac{8}{5} \quad \tfrac{a}{b} = a \div b$$

$$= -\frac{8}{10} \cdot \frac{5}{8} \quad \text{Multiply by the reciprocal.}$$

$$= -\frac{1}{2} \quad \text{Multiply. Write in lowest terms.}$$

SECTION 11.4 (pages 809–814)

19. The rise is 3 and the run is 1, so the slope m is given by

$$m = \frac{\text{rise}}{\text{run}} = \frac{3}{1} = 3.$$

The y-intercept is $(0, -3)$, so $b = -3$. The equation of the line, written in slope-intercept form $y = mx + b$, is

$$y = 3x - 3. \quad m = 3, b = -3$$

55. (a) First, find the slope of the line.

$$\begin{matrix} (x_1, y_1) & & (x_2, y_2) \\ \downarrow\ \downarrow & & \downarrow\ \downarrow \\ \left(\frac{1}{2}, \frac{3}{2}\right) & \text{and} & \left(-\frac{1}{4}, \frac{5}{4}\right) \end{matrix}$$

$$m = \frac{\frac{5}{4} - \frac{3}{2}}{-\frac{1}{4} - \frac{1}{2}} = \frac{\frac{5}{4} - \frac{6}{4}}{-\frac{1}{4} - \frac{2}{4}} = \frac{-\frac{1}{4}}{-\frac{3}{4}}$$

$$= -\frac{1}{4} \div \left(-\frac{3}{4}\right) = -\frac{1}{4}\left(-\frac{4}{3}\right) = \frac{1}{3}$$

Now use the point $\left(\frac{1}{2}, \frac{3}{2}\right)$ for (x_1, y_1) and $m = \frac{1}{3}$ in the point-slope form.

$$y - y_1 = m(x - x_1)$$

$$y - \frac{3}{2} = \frac{1}{3}\left(x - \frac{1}{2}\right) \quad \text{Substitute.}$$

$$y - \frac{3}{2} = \frac{1}{3}x - \frac{1}{6} \quad \text{Distributive property}$$

$$y = \frac{1}{3}x - \frac{1}{6} + \frac{9}{6} \quad \text{Add } \tfrac{3}{2} = \tfrac{9}{6}.$$

$$y = \frac{1}{3}x + \frac{4}{3} \quad \text{Combine like terms; } \tfrac{8}{6} = \tfrac{4}{3}$$

This last equation is written in slope-intercept form $y = mx + b$.

(b) Clear the fractions in the final equation in part (a) by multiplying each term by 3.

$$y = \frac{1}{3}x + \frac{4}{3} \quad \text{From part (a)}$$

$$3y = x + 4 \quad \text{Multiply by 3.}$$

$$-x + 3y = 4 \quad \text{Subtract } x.$$

$$x - 3y = -4 \quad \text{Multiply by } -1.$$

This last equation is written in standard form $Ax + By = C$.

61. Solve the given equation for y.

$$3x = 4y + 5$$

$$3x - 4y = 5 \quad \text{Subtract } 4y.$$

$$-4y = -3x + 5 \quad \text{Subtract } 3x.$$

$$y = \frac{3}{4}x - \frac{5}{4} \quad \text{Divide by } -4.$$

The slope is $\frac{3}{4}$. A line parallel to this line has the same slope. Use the point-slope form with $m = \frac{3}{4}$ and $(x_1, y_1) = (2, -3)$.

$$y - y_1 = m(x - x_1)$$

$$y - (-3) = \frac{3}{4}(x - 2) \quad \text{Substitute.}$$

$$y + 3 = \frac{3}{4}x - \frac{3}{2}$$

$$y = \frac{3}{4}x - \frac{3}{2} - \frac{6}{2} \quad \text{Subtract } 3 = \tfrac{6}{2}.$$

$$y = \frac{3}{4}x - \frac{9}{2} \quad \text{Combine like terms.}$$

CHAPTER 12 Exponents and Polynomials

SECTION 12.1 (pages 839–842)

21. $\frac{1}{2}x^4 + \frac{1}{6}x^4$

$$= \left(\frac{1}{2} + \frac{1}{6}\right)x^4 \quad \text{Distributive property}$$

$$= \left(\frac{3}{6} + \frac{1}{6}\right)x^4 \quad \text{Write fractions with a common denominator.}$$

$$= \frac{4}{6}x^4 \quad \text{Add fractions.}$$

$$= \frac{2}{3}x^4 \quad \text{Lowest terms}$$

37. $0.8x^4 - 0.3x^4 - 0.5x^4 + 7$

$$= (0.8 - 0.3 - 0.5)x^4 + 7$$

$$= 0x^4 + 7$$

$$= 7$$

Since 7 can be written as $7x^0$, the degree of the simplified polynomial is 0. It has one term, so it is a monomial.

53. Add.

$$\begin{array}{r} \frac{2}{3}x^2 + \frac{1}{5}x + \frac{1}{6} \\ \frac{1}{2}x^2 - \frac{1}{3}x + \frac{2}{3} \end{array}$$

Rewrite so that the fractions in each column have a common denominator. Then add column by column.

$$\begin{array}{r} \frac{4}{6}x^2 + \frac{3}{15}x + \frac{1}{6} \\ \frac{3}{6}x^2 - \frac{5}{15}x + \frac{4}{6} \\ \hline \frac{7}{6}x^2 - \frac{2}{15}x + \frac{5}{6} \end{array}$$

71. Use the formula for the perimeter of a square, $P = 4s$, with $s = \frac{1}{2}x^2 + 2x$.

$$P = 4s$$

$$P = 4\left(\frac{1}{2}x^2 + 2x\right) \quad \text{Substitute.}$$

$$P = 4\left(\frac{1}{2}x^2\right) + 4(2x) \quad \text{Distributive property}$$

$$P = 2x^2 + 8x \quad \text{Multiply.}$$

SECTION 12.2 (pages 849–850)

43. $(-5^2)^6$

$\quad = (-1 \cdot 5^2)^6 \qquad$ Write the factor -1.

$\quad = (-1)^6 \cdot (5^2)^6 \qquad$ Power rule (b)

$\quad = 1 \cdot 5^{2 \cdot 6} \qquad$ Power rule (a)

$\quad = 5^{12} \qquad$ Multiply.

77. $\left(\dfrac{5a^2 b^5}{c^6} \right)^3 \quad (c \neq 0)$

$\quad = \dfrac{(5a^2 b^5)^3}{(c^6)^3} \qquad$ Power rule (c)

$\quad = \dfrac{5^3 (a^2)^3 (b^5)^3}{(c^6)^3} \qquad$ Power rule (b)

$\quad = \dfrac{125 a^6 b^{15}}{c^{18}} \qquad$ Power rule (a)

79. $(-5m^3 p^4 q)^2 (p^2 q)^3$

$\quad = (-1 \cdot 5m^3 p^4 q)^2 (p^2 q)^3 \qquad$ Write the factor -1.

$\quad = (-1)^2 \cdot 5^2 \cdot (m^3)^2 \cdot$
$\qquad (p^4)^2 \cdot q^2 \cdot (p^2)^3 \cdot q^3 \qquad$ Power rule (b)

$\quad = 1 \cdot 25 \cdot m^6 p^8 q^2 p^6 q^3 \qquad$ Power rule (a)

$\quad = 25 m^6 p^{8+6} q^{2+3} \qquad$ Product rule

$\quad = 25 m^6 p^{14} q^5$

SECTION 12.3 (pages 855–857)

57. $\left(x - \dfrac{2}{3} \right) \left(x + \dfrac{1}{4} \right)$

$\qquad$ **F** $\qquad$ **O** $\qquad$ **I** $\qquad$ **L**

$\quad = x(x) + x\left(\dfrac{1}{4} \right) + \left(-\dfrac{2}{3} \right)x + \left(-\dfrac{2}{3} \right)\dfrac{1}{4}$

$\quad = x^2 + \dfrac{1}{4}x - \dfrac{2}{3}x - \dfrac{1}{6}$

$\quad = x^2 + \left(\dfrac{3}{12}x - \dfrac{8}{12}x \right) - \dfrac{1}{6}$

$\quad = x^2 - \dfrac{5}{12}x - \dfrac{1}{6}$

SECTION 12.4 (pages 862–864)

11. $\left(x - \dfrac{3}{4} \right)^2$

$\quad = x^2 - 2(x)\left(\dfrac{3}{4} \right) + \left(\dfrac{3}{4} \right)^2$

$\qquad (a-b)^2 = a^2 - 2ab + b^2$

$\quad = x^2 - \dfrac{3}{2}x + \dfrac{9}{16}$

21. $(0.8t + 0.7s)^2$

$\quad = (0.8t)^2 + 2(0.8t)(0.7s) + (0.7s)^2$

$\qquad (a+b)^2 = a^2 + 2ab + b^2$

$\quad = 0.64t^2 + 1.12ts + 0.49s^2$

29. $-(4r - 2)^2$

First square the binomial.

$\qquad (4r - 2)^2$

$\qquad\quad = (4r)^2 - 2(4r)(2) + 2^2$

$\qquad\quad = 16r^2 - 16r + 4$

Now multiply by -1.

$\qquad -1(16r^2 - 16r + 4)$

$\qquad\quad = -16r^2 + 16r - 4$

43. $(2x^2 - 5)(2x^2 + 5)$

$\quad = (2x^2)^2 - 5^2 \qquad (a-b)(a+b)$
$\qquad\qquad\qquad\qquad = a^2 - b^2$

$\quad = 4x^4 - 25 \qquad (2x^2)^2 = 2^2 x^4 = 4x^4$

SECTION 12.5 (pages 872–873)

17. $(-2)^0 - 2^0$

$\quad = 1 - 1 \qquad a^0 = 1$

$\quad = 0 \qquad$ Subtract.

33. $-2^{-1} + 3^{-2}$

$\quad = -(2^{-1}) + 3^{-2} \qquad$ In -2^{-1}, the base is 2.

$\quad = -\dfrac{1}{2^1} + \dfrac{1}{3^2} \qquad a^{-n} = \dfrac{1}{a^n}$

$\quad = -\dfrac{1}{2} + \dfrac{1}{9} \qquad$ Apply the exponents.

$\quad = -\dfrac{9}{18} + \dfrac{2}{18} \qquad$ Write fractions with a common denominator.

$\quad = -\dfrac{7}{18} \qquad$ Add.

69. $\dfrac{(m^7 n)^{-2}}{m^{-4} n^3}$

$\quad = \dfrac{(m^7)^{-2} n^{-2}}{m^{-4} n^3} \qquad$ Power rule (b)

$\quad = \dfrac{m^{7(-2)} n^{-2}}{m^{-4} n^3} \qquad$ Power rule (a)

$\quad = \dfrac{m^{-14} n^{-2}}{m^{-4} n^3}$

$\quad = m^{-14 - (-4)} n^{-2 - 3} \qquad$ Quotient rule

$\quad = m^{-10} n^{-5}$

$\quad = \dfrac{1}{m^{10} n^5} \qquad a^{-n} = \dfrac{1}{a^n}$

SECTION 12.6 (pages 876–877)

23. $\dfrac{-3x^3 - 4x^4 + 2x}{-3x^2}$

$\quad = \dfrac{-4x^4 - 3x^3 + 2x}{-3x^2} \qquad$ Write in descending powers.

$\quad = \dfrac{-4x^4}{-3x^2} - \dfrac{3x^3}{-3x^2} + \dfrac{2x}{-3x^2} \qquad$ Divide each term by $-3x^2$.

$\quad = \dfrac{4x^2}{3} + x - \dfrac{2}{3x} \qquad$ Quotient rule; be careful with signs.

Notice how the third term is written with. in the denominator, which is **not** the same as $-\dfrac{2}{3}x$ or $-\dfrac{2x}{3}$. In $-\dfrac{2}{3x}$, we are *dividing* by x. In $-\dfrac{2}{3}x$ and $-\dfrac{2x}{3}$, we are *multiplying* by x. Applying the quotient rule to the term $\dfrac{2x}{-3x^2}$ gives the following.

$\dfrac{2x}{-3x^2} = \dfrac{2x^1}{-3x^2} = -\dfrac{2}{3}x^{1-2} = -\dfrac{2}{3}x^{-1}$

$\qquad\qquad = -\dfrac{2}{3}\left(\dfrac{1}{x} \right) = -\dfrac{2}{3x}$

$\left(\dfrac{4}{3}x^2 \text{ is an acceptable form for the first term,} \right.$

$\left. \dfrac{4x^2}{3}. \text{ Why?} \right)$

33. Use the formula for the area of a rectangle, $A = LW$, with $A = 12x^2 - 4x + 2$ and $W = 2x$.

$\qquad\qquad A = LW$

$\quad 12x^2 - 4x + 2 = L(2x) \qquad$ Substitute for A and W.

$\quad \dfrac{12x^2 - 4x + 2}{2x} = L \qquad$ Divide each side by $2x$.

$\quad \dfrac{12x^2}{2x} - \dfrac{4x}{2x} + \dfrac{2}{2x} = L \qquad$ Divide each term by $2x$.

$\quad 6x - 2 + \dfrac{1}{x} = L \qquad \dfrac{2}{2x} = \dfrac{2}{2} \cdot \dfrac{1}{x}$

$\qquad\qquad\qquad\qquad\qquad = 1 \cdot \dfrac{1}{x} = \dfrac{1}{x}$

$6x - 2 + \dfrac{1}{x}$ represents the length of the rectangle.

SECTION 12.7 (pages 882–883)

23. $(x^4 - x^2 - 2) \div (x^2 - 2)$

Use 0 as the coefficient of the missing x^3- and x-terms in the dividend and the missing x-term in the divisor.

$$
\begin{array}{r}
x^2 \qquad\qquad + 1 \\
x^2 + 0x - 2\,\overline{)\,x^4 + 0x^3 - x^2 + 0x - 2} \\
\underline{x^4 + 0x^3 - 2x^2} \qquad\qquad\quad \\
x^2 + 0x - 2 \\
\underline{x^2 + 0x - 2} \\
0
\end{array}
$$

The remainder is 0. The answer is the quotient, $x^2 + 1$.

$$\frac{\;\cdot^{4} + 11x^3 - 8x^2 - 13x + 7}{2x^2 + x - 1}$$

$$\begin{array}{r} x^3 \qquad\quad + 6x - 7 \\ 2x^2 + x - 1\overline{)2x^5 + x^4 + 11x^3 - 8x^2 - 13x + 7} \\ \underline{2x^5 + x^4 - x^3} \qquad\qquad\qquad\qquad \\ 12x^3 - 8x^2 - 13x \qquad \\ \underline{12x^3 + 6x^2 - 6x} \qquad \\ -14x^2 - 7x + 7 \\ \underline{-14x^2 - 7x + 7} \\ 0 \end{array}$$

The remainder is 0. The answer is the quotient, $x^3 + 6x - 7$.

SECTION 12.8 (pages 888–890)

51. $\dfrac{4 \times 10^5}{8 \times 10^2}$

$= \dfrac{4}{8} \times \dfrac{10^5}{10^2}$

$= 0.5 \times 10^3$

Divide; quotient rule

$= (5 \times 10^{-1}) \times 10^3$

Write 0.5 in scientific notation.

$= 5 \times (10^{-1} \times 10^3)$

Associative property

$= 5 \times 10^2$

Product rule

$= 500$

Write without exponents.

55. $\dfrac{(1.65 \times 10^8)(5.24 \times 10^{-2})}{(6 \times 10^4)(2 \times 10^7)}$

$= \dfrac{1.65 \times 5.24}{6 \times 2} \times \dfrac{10^8 \times 10^{-2}}{10^4 \times 10^7}$

Associative and commutative properties

$= 0.7205 \times \dfrac{10^6}{10^{11}}$ Product rule

$= 0.7205 \times 10^{-5}$ Quotient rule

$= (7.205 \times 10^{-1}) \times 10^{-5}$

Write 0.7205 in scientific notation.

$= 7.205 \times (10^{-1} \times 10^{-5})$

Associative property

$= 7.205 \times 10^{-6}$ Product rule

$= 0.000007205$

Write without exponents.

CHAPTER 13 Factoring and Applications

SECTION 13.1 (pages 909–911)

79. $18r^2 + 12ry - 3xr - 2xy$

$= (18r^2 + 12ry) + (-3xr - 2xy)$

Group the terms.

$= 6r(3r + 2y) - x(3r + 2y)$

Factor each group.

$= (3r + 2y)(6r - x)$

Factor out the common factor, $3r + 2y$.

87. $y^2 + 3x + 3y + xy$

$= y^2 + 3y + xy + 3x$

Rearrange terms.

$= (y^2 + 3y) + (xy + 3x)$

Group the terms.

$= y(y + 3) + x(y + 3)$

Factor each group.

$= (y + 3)(y + x)$

Factor out the common factor, $y + 3$.

SECTION 13.2 (pages 917–918)

41. $a^2 - 8a - 48$

Find two integers whose product is -48 and whose sum is -8. Since c is negative, one integer must be positive and one must be negative.

Factors of -48	Sums of Factors
$-1, 48$	47
$1, -48$	-47
$-2, 24$	22
$2, -24$	-22
$-3, 16$	13
$3, -16$	-13
$-4, 12$	8
$4, -12$	-8 ←
$-6, 8$	2
$6, -8$	-2

Thus,

$a^2 - 8a - 48$ factors as $(a + 4)(a - 12)$.

57. $-2x^6 - 8x^5 + 42x^4$

First, factor out the negative common factor, $-2x^4$.

$-2x^6 - 8x^5 + 42x^4$

$= -2x^4(x^2 + 4x - 21)$

Now factor $x^2 + 4x - 21$.

Factors of -21	Sums of Factors
$1, -21$	-20
$-1, 21$	20
$3, -7$	-4
$-3, 7$	4 ←

Thus,

$x^2 + 4x - 21$ factors as $(x - 3)(x + 7)$.

The completely factored form is

$-2x^6 - 8x^5 + 42x^4$

$= -2x^4(x - 3)(x + 7)$.

61. $5m^5 + 25m^4 - 40m^2$

First, factor out the GCF, $5m^2$.

$5m^5 + 25m^4 - 40m^2$

$= 5m^2(m^3 + 5m^2 - 8)$

The trinomial $m^3 + 5m^2 - 8$ cannot be factored further, so the polynomial is in completely factored form.

63. $m^3n - 10m^2n^2 + 24mn^3$

First, factor out the GCF, mn.

$m^3n - 10m^2n^2 + 24mn^3$

$= mn(m^2 - 10mn + 24n^2)$

The expressions $-6n$ and $-4n$ have a product of $24n^2$ and a sum of $-10n$. The completely factored form is

$m^3n - 10m^2n^2 + 24mn^3$

$= mn(m - 6n)(m - 4n)$.

SECTION 13.3 (pages 921–922)

29. $16 + 16x + 3x^2$

Write the trinomial in descending powers.

$3x^2 + 16x + 16$

To factor, find two integers whose product is $3(16) = 48$ and whose sum is 16. The integers are 12 and 4.

$3x^2 + 16x + 16$

$= 3x^2 + 12x + 4x + 16$

$= (3x^2 + 12x) + (4x + 16)$

Group the terms.

$= 3x(x + 4) + 4(x + 4)$

Factor each group.

$= (x + 4)(3x + 4)$

Factor out the common factor, $x + 4$.

35. $-32z^5 + 20z^4 + 12z^3$

First factor out the negative common factor, $-4z^3$.

$-32z^5 + 20z^4 + 12z^3$

$= -4z^3(8z^2 - 5z - 3)$

To factor $8z^2 - 5z - 3$, find two integers whose product is $8(-3) = -24$ and whose sum is -5. These integers are -8 and 3. Now rewrite the given trinomial and factor it.

$-32z^5 + 20z^4 + 12z^3$

$= -4z^3(8z^2 - 5z - 3)$

$= -4z^3(8z^2 - 8z + 3z - 3)$

$-5z = -8z + 3z$

$= -4z^3[(8z^2 - 8z) + (3z - 3)]$

Group the terms.

$= -4z^3[8z(z - 1) + 3(z - 1)]$

Factor each group.

$= -4z^3[(z - 1)(8z + 3)]$

Factor out the common factor, $z - 1$.

$= -4z^3(z - 1)(8z + 3)$

SECTION 13.4 (pages 927–928)

43. $6m^6n + 7m^5n^2 + 2m^4n^3$

Factor out the GCF, m^4n.

$$6m^6n + 7m^5n^2 + 2m^4n^3 = m^4n(6m^2 + 7mn + 2n^2)$$

Now factor $6m^2 + 7mn + 2n^2$ by trial and error. Possible factors of $6m^2$ are $6m$ and m or $3m$ and $2m$. Possible factors of $2n^2$ are $2n$ and n.

$$(3m + 2n)(2m + n) = 6m^2 + 7mn + 2n^2 \quad \text{Correct}$$

The completely factored form is

$$6m^6n + 7m^5n^2 + 2m^4n^3 = m^4n(3m + 2n)(2m + n).$$

SECTION 13.5 (pages 935–938)

51. $4m^2 - \frac{9}{25}$

Because $4m^2 = (2m)^2$ and $\frac{9}{25} = \left(\frac{3}{5}\right)^2$, $4m^2 - \frac{9}{25}$ is a difference of squares.

$$4m^2 - \frac{9}{25} = (2m)^2 - \left(\frac{3}{5}\right)^2 = \left(2m + \frac{3}{5}\right)\left(2m - \frac{3}{5}\right)$$

57. $x^2 - 1.0x + 0.25$

The first and last terms are perfect squares, x^2 and $(-0.5)^2$. The trinomial is a perfect square, since the middle term is

$$2 \cdot x \cdot (-0.5) = -1.0x.$$

$$x^2 - 1.0x + 0.25 = (x)^2 - 2(x)(0.5) + (0.5)^2 = (x - 0.5)^2$$

SECTION 13.7 (pages 949–951)

57. $9y^3 - 49y = 0$

To factor the polynomial, begin by factoring out the greatest common factor.

$$y(9y^2 - 49) = 0$$

Now factor $9y^2 - 49$ as the difference of two squares.

$$y(3y + 7)(3y - 7) = 0$$

Set each of the three factors equal to 0 and solve.

$y = 0$ or $3y + 7 = 0$ or $3y - 7 = 0$

$3y = -7$ or $3y = 7$

$y = -\frac{7}{3}$ or $y = \frac{7}{3}$

Solution set: $\left\{-\frac{7}{3}, 0, \frac{7}{3}\right\}$

59. $(2r + 5)(3r^2 - 16r + 5) = 0$

$(2r + 5)(3r - 1)(r - 5) = 0$ Factor $3r^2 - 16r + 5$.

Set each factor equal to 0 and solve.

$2r + 5 = 0$ or $3r - 1 = 0$ or $r - 5 = 0$

$2r = -5$ or $3r = 1$ or $r = 5$

$r = -\frac{5}{2}$ or $r = \frac{1}{3}$

Solution set: $\left\{-\frac{5}{2}, \frac{1}{3}, 5\right\}$

65. $r^4 = 2r^3 + 15r^2$

Write with 0 on one side of the equation.

$r^4 - 2r^3 - 15r^2 = 0$ Subtract $2r^3$ and $15r^2$.

$r^2(r^2 - 2r - 15) = 0$ GCF $= r^2$

$r \cdot r \cdot (r + 3)(r - 5) = 0$ Factor.

Set each of the factors equal to 0, and solve the resulting equations. (Since r is a factor twice, it gives a double solution.)

$r = 0$ or $r + 3 = 0$ or $r - 5 = 0$

$r = -3$ or $r = 5$

Solution set: $\{-3, 0, 5\}$

SECTION 13.8 (pages 958–962)

15. Let x = the length of a side of the square painting.

Then $x - 2$ = the length of a side of the square mirror.

Since the formula for the area of a square is $A = s^2$, the area of the painting is x^2, and the area of the mirror is $(x - 2)^2$. The difference between their areas is 32.

$$x^2 - (x - 2)^2 = 32$$
$$x^2 - (x^2 - 4x + 4) = 32$$
$$x^2 - x^2 + 4x - 4 = 32$$
$$4x - 4 = 32$$
$$4x = 36$$
$$x = 9$$

The length of a side of the painting is 9 ft. The length of a side of the mirror is

$$9 - 2 = 7 \text{ ft}.$$

As a check,

$9^2 - 7^2 = 81 - 49 = 32,$ as required.

23. Let x = the least even integer.

Then $x + 2$ = the next even integer and

$x + 4$ = the third even integer.

$x^2 + (x + 2)^2 = (x + 4)^2$

$x^2 + x^2 + 4x + 4 = x^2 + 8x + 16$ Square the binomials.

$2x^2 + 4x + 4 = x^2 + 8x + 16$ Combine like terms.

$x^2 - 4x - 12 = 0$ Standard form

$(x + 2)(x - 6) = 0$ Factor.

$x + 2 = 0$ or $x - 6 = 0$ Zero-factor property

$x = -2$ or $x = 6$

If $x = -2$, then $x + 2 = 0$ and $x + 4 = 2$.
If $x = 6$, then $x + 2 = 8$ and $x + 4 = 10$.
The integers are -2, 0, and 2 or 6, 8, and 10.

CHECK

$(-2)^2 + 0^2 \stackrel{?}{=} 2^2$	$6^2 + 8^2 \stackrel{?}{=} 10^2$
$4 + 0 \stackrel{?}{=} 4$	$36 + 64 \stackrel{?}{=} 100$
$4 = 4$ ✓	$100 = 100$ ✓
True	True

CHAPTER 14 Rational Expressions and Applications

SECTION 14.1 (pages 981–982)

9. (a) $\frac{(-3x)^2}{4x + 12}$

$= \frac{(-3 \cdot 2)^2}{4(2) + 12}$ Let $x = 2$.

$= \frac{(-6)^2}{8 + 12} = \frac{36}{20} = \frac{9}{5}$

(b) $\frac{(-3x)^2}{4x + 12}$

$= \frac{[-3(-3)]^2}{4(-3) + 12}$ Let $x = -3$.

$= \frac{9^2}{-12 + 12},$ or $\frac{81}{0}$

Since substituting -3 for x makes the denominator 0, the given rational expression is *undefined* for $x = -3$.

35. $\frac{12m^2 - 3}{8m - 4}$

$= \frac{3(4m^2 - 1)}{4(2m - 1)}$ Factor out the common factors.

$= \frac{3(2m + 1)(2m - 1)}{4(2m - 1)}$ Factor again in the numerator.

$= \frac{3(2m + 1)}{4}$ Divide out the common factor.

SOLUTIONS

43. We factor the numerator and denominator by grouping in this rational expression.

$$\frac{zw+4z-3w-12}{zw+4z+5w+20}$$

$=\frac{(zw+4z)-(3w+12)}{(zw+4z)+(5w+20)}$ Group the terms.

$=\frac{z(w+4)-3(w+4)}{z(w+4)+5(w+4)}$ Factor each group.

$=\frac{(w+4)(z-3)}{(w+4)(z+5)}$ Factor by grouping.

$=\frac{z-3}{z+5}$ Divide out the common factor.

61. $LW=A$ Formula for the area of a rectangle

$W=\frac{A}{L}$ Divide by L.

$W=\frac{x^4+10x^2+21}{x^2+7}$ Substitute for A and L.

$W=\frac{(x^2+7)(x^2+3)}{x^2+7}$ Factor.

$W=x^2+3$ Divide out the common factor.

Note: If it is not apparent that A,

$$x^4+10x^2+21,$$

factors as $(x^2+7)(x^2+3)$, we can use "long division" to find the quotient $\frac{A}{L}$. Remember to insert zeros for the coefficients of the missing terms.

$$\begin{array}{r} x^2 \qquad\quad + 3 \\ x^2+0x+7\overline{)x^4+0x^3+10x^2+0x+21} \\ \underline{x^4+0x^3+\ \ 7x^2} \qquad\qquad \\ 3x^2+0x+21 \\ \underline{3x^2+0x+21} \\ 0 \end{array}$$

The width of the rectangle is x^2+3.

SECTION 14.2 (pages 987–988)

31. $\frac{6(m-2)^2}{5(m+4)^2}\cdot\frac{15(m+4)}{2(2-m)}$

$=\frac{6\cdot 15(m-2)^2(m+4)}{5\cdot 2(m+4)^2(2-m)}$ Multiply.

$=\frac{2\cdot 3\cdot 3\cdot 5(m-2)(m-2)(m+4)}{5\cdot 2(m+4)(m+4)(-1)(m-2)}$ Factor.

$=\frac{3\cdot 3(m-2)}{(m+4)(-1)}$ Divide out the common factors.

$=\frac{9(m-2)}{-(m+4)}$, or $\frac{-9(m-2)}{m+4}$

39. $\frac{m^2+2mp-3p^2}{m^2-3mp+2p^2}\div\frac{m^2+4mp+3p^2}{m^2+2mp-8p^2}$

$=\frac{m^2+2mp-3p^2}{m^2-3mp+2p^2}\cdot\frac{m^2+2mp-8p^2}{m^2+4mp+3p^2}$

Multiply by the reciprocal.

$=\frac{(m+3p)(m-p)(m+4p)(m-2p)}{(m-2p)(m-p)(m+3p)(m+p)}$

Multiply numerators and multiply denominators. Factor.

$=\frac{m+4p}{m+p}$

Divide out the common factors.

43. The formula for the area of a rectangle is $A=LW$. Solving for W gives $W=\frac{A}{L}$, which can be written as $W=A\div L$, since the fraction bar indicates division. Substituting the given expressions for A and L gives the width.

$$\overset{A}{\downarrow}\qquad\quad\overset{L}{\downarrow}$$

$$\frac{5x^2y^3}{2pq}\div\frac{2xy}{p}$$

$=\frac{5x^2y^3}{2pq}\cdot\frac{p}{2xy}$ Multiply by the reciprocal.

$=\frac{5xy^2}{4q}$ Multiply. Divide out the common factors.

SECTION 14.3 (pages 993–994)

39. $\frac{-4t}{3t-6}=\frac{?}{12-6t}$

Factor each denominator. (On the right, rewrite $12-6t$ as $-6t+12$, and factor.)

$$\frac{-4t}{3(t-2)}=\frac{?}{-6(t-2)}$$

The missing factor is -2, so multiply the fraction on the left by $\frac{-2}{-2}$.

$$\frac{-4t}{3(t-2)}\cdot\frac{-2}{-2}=\frac{8t}{-6(t-2)},\ \text{or}\ \frac{8t}{12-6t}$$

SECTION 14.4 (pages 1001–1004)

31. $\frac{x+1}{6}+\frac{3x+3}{9}$

Begin by writing the second fraction in lowest terms.

$$\frac{3x+3}{9}$$

$$=\frac{3(x+1)}{9},\ \text{or}\ \frac{x+1}{3}$$

The sum becomes

$$\frac{x+1}{6}+\frac{x+1}{3}.$$

The LCD is 6.

$\frac{x+1}{6}+\frac{x+1}{3}$

$=\frac{x+1}{6}+\frac{x+1}{3}\cdot\frac{2}{2}$ Get a common denominator.

$=\frac{x+1}{6}+\frac{2(x+1)}{6}$

$=\frac{x+1+2x+2}{6}$ Add numerators; distributive property

$=\frac{3x+3}{6}$ Combine like terms.

$=\frac{3(x+1)}{3\cdot 2}$ Factor.

$=\frac{x+1}{2}$ Lowest terms

43. $\frac{t}{t+2}+\frac{5-t}{t}-\frac{4}{t^2+2t}$

$=\frac{t}{t+2}+\frac{5-t}{t}-\frac{4}{t(t+2)}$ Factor.

$=\frac{t}{t+2}\cdot\frac{t}{t}+\frac{5-t}{t}\cdot\frac{t+2}{t+2}-\frac{4}{t(t+2)}$

LCD $=t(t+2)$

$=\frac{t\cdot t+(5-t)(t+2)-4}{t(t+2)}$

Write using the LCD.

$=\frac{t^2+5t+10-t^2-2t-4}{t(t+2)}$

Multiply in the numerator.

$=\frac{3t+6}{t(t+2)}$ Combine like terms.

$=\frac{3(t+2)}{t(t+2)}$ Factor.

$=\frac{3}{t}$ Lowest terms

61. $\frac{2q+1}{3q^2+10q-8}-\frac{3q+5}{2q^2+5q-12}$

$=\frac{2q+1}{(3q-2)(q+4)}-\frac{3q+5}{(2q-3)(q+4)}$

Factor.

$=\frac{(2q+1)(2q-3)}{(3q-2)(q+4)(2q-3)}$

$-\frac{(3q+5)(3q-2)}{(2q-3)(q+4)(3q-2)}$

LCD $=(3q-2)(q+4)(2q-3)$

$=\frac{(2q+1)(2q-3)-(3q+5)(3q-2)}{(3q-2)(q+4)(2q-3)}$

Subtract numerators.

$=\frac{(4q^2-4q-3)-(9q^2+9q-10)}{(3q-2)(q+4)(2q-3)}$

Multiply in the numerator.

$=\frac{4q^2-4q-3-9q^2-9q+10}{(3q-2)(q+4)(2q-3)}$

Distributive property

$=\frac{-5q^2-13q+7}{(3q-2)(q+4)(2q-3)}$

Combine like terms.

SOLUTIONS

65. $\dfrac{x+3y}{x^2+2xy+y^2}+\dfrac{x-y}{x^2+4xy+3y^2}$

$=\dfrac{x+3y}{(x+y)(x+y)}+\dfrac{x-y}{(x+3y)(x+y)}$

$=\dfrac{(x+3y)(x+3y)}{(x+y)(x+y)(x+3y)}$

$+\dfrac{(x-y)(x+y)}{(x+3y)(x+y)(x+y)}$

LCD $=(x+y)(x+y)(x+3y)$

$=\dfrac{(x^2+6xy+9y^2)+(x^2-y^2)}{(x+y)(x+y)(x+3y)}$

$=\dfrac{2x^2+6xy+8y^2}{(x+y)(x+y)(x+3y)}$,

or $\dfrac{2x^2+6xy+8y^2}{(x+y)^2(x+3y)}$

There is no need to factor out 2 in the numerator, since the denominator does not have 2 as a factor. The answer is in lowest terms.

SECTION 14.5 (pages 1011–1012)

17. $\dfrac{\frac{1}{x}+x}{\frac{x^2+1}{8}}$ Use Method 2.

$=\dfrac{8x\left(\frac{1}{x}+x\right)}{8x\left(\frac{x^2+1}{8}\right)}$ LCD $=8x$

$=\dfrac{8+8x^2}{x(x^2+1)}$ Distributive property

$=\dfrac{8(1+x^2)}{x(x^2+1)}$ Factor.

$=\dfrac{8}{x}$ Lowest terms

31. $\dfrac{\frac{1}{m-1}+\frac{2}{m+2}}{\frac{2}{m+2}-\frac{1}{m-3}}$ Use Method 2.

$=\dfrac{(m-1)(m+2)(m-3)\left(\frac{1}{m-1}+\frac{2}{m+2}\right)}{(m-1)(m+2)(m-3)\left(\frac{2}{m+2}-\frac{1}{m-3}\right)}$

LCD $=(m-1)(m+2)(m-3)$

$=\dfrac{(m+2)(m-3)+2(m-1)(m-3)}{2(m-1)(m-3)-(m-1)(m+2)}$

Distributive property

$=\dfrac{(m-3)[(m+2)+2(m-1)]}{(m-1)[2(m-3)-(m+2)]}$

Factor out $m-3$ in the numerator and $m-1$ in the denominator.

$=\dfrac{(m-3)[m+2+2m-2]}{(m-1)[2m-6-m-2]}$

Distributive property

$=\dfrac{3m(m-3)}{(m-1)(m-8)}$

Combine like terms.

33. $2-\dfrac{2}{2+\frac{2}{2+2}}$

$=2-\dfrac{2}{2+\frac{2}{4}}$ Simplify the denominator.

$=2-\dfrac{2}{\frac{5}{2}}$ $2+\frac{2}{4}=\frac{4}{2}+\frac{1}{2}=\frac{5}{2}$

$=2-\left(2\cdot\frac{2}{5}\right)$ $\frac{2}{\frac{5}{2}}=2\div\frac{5}{2}=2\cdot\frac{2}{5}$

$=2-\frac{4}{5}$ Multiply.

$=\frac{10}{5}-\frac{4}{5}$ $2=\frac{10}{5}$

$=\frac{6}{5}$ Subtract.

SECTION 14.6 (pages 1021–1024)

11. $\frac{3x}{5}-6=x$

$5\left(\frac{3x}{5}-6\right)=5(x)$ Multiply by the LCD, 5.

$5\left(\frac{3x}{5}\right)-5(6)=5x$ Distributive property

$3x-30=5x$ Multiply.

$-30=2x$ Subtract $3x$.

$-15=x$ Divide by 2.

Check by substituting -15 for x in the original equation.

Solution set: $\{-15\}$

21. $\frac{q+2}{3}+\frac{q-5}{5}=\frac{7}{3}$

$15\left(\frac{q+2}{3}+\frac{q-5}{5}\right)=15\left(\frac{7}{3}\right)$

Multiply by the LCD, 15.

$15\left(\frac{q+2}{3}\right)+15\left(\frac{q-5}{5}\right)=5\cdot 7$

$5(q+2)+3(q-5)=35$

$5q+10+3q-15=35$

$8q-5=35$

$8q=40$

$q=5$

A check confirms that 5 is the solution.

Solution set: $\{5\}$

41. $\frac{2}{m}=\frac{m}{5m+12}$

$m(5m+12)\left(\frac{2}{m}\right)=m(5m+12)\left(\frac{m}{5m+12}\right)$

Multiply by the LCD, $m(5m+12)$.

$(5m+12)(2)=m(m)$ Divide out the common factors.

$10m+24=m^2$ Multiply.

$m^2-10m-24=0$ Standard form

$(m+2)(m-12)=0$ Factor.

$m+2=0$ or $m-12=0$

Zero-factor property

$m=-2$ or $m=12$

The proposed solutions are -2 and 12.

CHECK $\frac{2}{-2}\stackrel{?}{=}\frac{-2}{5(-2)+12}$ Let $m=-2$.

$-1=\frac{-2}{2}$ ✓ True

CHECK $\frac{2}{12}\stackrel{?}{=}\frac{12}{5(12)+12}$ Let $m=12$.

$\frac{1}{6}=\frac{12}{72}$ ✓ True

Solution set: $\{-2, 12\}$

69. $d=\dfrac{2S}{n(a+L)}$ for a

We need to isolate a on one side of the equation.

$d=\dfrac{2S}{n(a+L)}$

$d\cdot n(a+L)=\dfrac{2S}{n(a+L)}\cdot n(a+L)$

Multiply by $n(a+L)$.

$dn(a+L)=2S$

$dna+dnL=2S$ Distributive property

$dna=2S-dnL$ Subtract dnL.

$a=\dfrac{2S-dnL}{dn}$, Divide by dn.

or $a=\dfrac{2S}{dn}-L$

75. $9x+\frac{3}{z}=\frac{5}{y}$ for z

$yz\left(9x+\frac{3}{z}\right)=yz\left(\frac{5}{y}\right)$ Multiply by the LCD, yz.

$yz(9x)+yz\left(\frac{3}{z}\right)=yz\left(\frac{5}{y}\right)$ Distributive property

$9xyz+3y=5z$

$9xyz-5z=-3y$ Get the z-terms on one side.

$z(9xy-5)=-3y$ Factor out z.

$z=\dfrac{-3y}{9xy-5}$ Divide by $9xy-5$.

or $z=\dfrac{3y}{5-9xy}$

SECTION 14.7 (pages 1032–1035)

27. Let $x =$ the rate of the current of the river. Then $(12 - x)$ is the rate upstream (against the current) and $(12 + x)$ is the rate downstream (with the current).

Use $t = \frac{d}{r}$ to complete the table.

	d	r	t
Upstream	6	$12 - x$	$\frac{6}{12 - x}$
Downstream	10	$12 + x$	$\frac{10}{12 + x}$

Since the times are equal, we get the following equation.

$$\frac{6}{12 - x} = \frac{10}{12 + x}$$

$$(12 + x)(12 - x)\frac{6}{12 - x} = (12 + x)(12 - x)\frac{10}{12 + x}$$

Multiply by the LCD, $(12 + x)(12 - x)$.

$$6(12 + x) = 10(12 - x)$$
$$72 + 6x = 120 - 10x$$
$$16x = 48$$
$$x = 3$$

The rate of the current of the river is 3 mph.

35. Let $x =$ the number of hours it would take Brenda to paint the room by herself.

	Rate	Time Working Together	Fractional Part of the Job Done When Working Together
Hilda	$\frac{1}{6}$	$3\frac{3}{4} = \frac{15}{4}$	$\frac{1}{6} \cdot \frac{15}{4} = \frac{5}{8}$
Brenda	$\frac{1}{x}$	$3\frac{3}{4} = \frac{15}{4}$	$\frac{1}{x} \cdot \frac{15}{4} = \frac{15}{4x}$

Since together Hilda and Brenda complete 1 whole job, we must add their individual fractional parts and set the sum equal to 1.

$$\frac{5}{8} + \frac{15}{4x} = 1$$

$$8x\left(\frac{5}{8} + \frac{15}{4x}\right) = 8x(1) \quad \text{Multiply by the LCD, } 8x.$$

$$8x\left(\frac{5}{8}\right) + 8x\left(\frac{15}{4x}\right) = 8x$$
$$5x + 30 = 8x$$
$$30 = 3x$$
$$10 = x$$

It will take Brenda 10 hr to paint the room by herself.

37. Let $x =$ the number of hours to fill the pool with both pipes left open.

	Rate	Time to Fill the Pool	Part Done by Each Pipe
Inlet Pipe	$\frac{1}{9}$	x	$\frac{1}{9}x$
Outlet Pipe	$\frac{1}{12}$	x	$\frac{1}{12}x$

Part done by inlet pipe − Part done by outlet pipe = Full pool

$$\frac{1}{9}x - \frac{1}{12}x = 1$$

$$36\left(\frac{1}{9}x - \frac{1}{12}x\right) = 36(1)$$

Multiply by the LCD, 36.

$$36\left(\frac{1}{9}x\right) - 36\left(\frac{1}{12}x\right) = 36$$
$$4x - 3x = 36$$
$$x = 36$$

It will take 36 hr to fill the pool.

SECTION 14.8 (pages 1038–1040)

1. For a constant time of 3 hr, if the rate of the pickup truck *increases,* then the distance traveled *increases*. Thus, the variation between the quantities is *direct.*

23. For a given base, the area A of a triangle varies directly as its height h, so there is a constant k such that $A = kh$. Find the value of k.

$$10 = k(4) \quad \text{Let } A = 10, h = 4.$$

$$k = \frac{10}{4} = 2.5 \quad \text{Solve for } k.$$

For $k = 2.5$, $A = kh$ becomes

$$A = 2.5h.$$

Now find A for $h = 6$.

$$A = 2.5(6) = 15$$

When the height of the triangle is 6 in., the area of the triangle is 15 in.^{2}

CHAPTER 15 Systems of Linear Equations and Inequalities

SECTION 15.1 (pages 1059–1062)

27. $2x - 3y = -6$

$y = -3x + 2$

To graph $2x - 3y = -6$, find the intercepts.

$$2x - 3(0) = -6 \quad \text{Let } y = 0.$$
$$2x = -6$$
$$x = -3$$

The x-intercept is $(-3, 0)$.

$$2(0) - 3y = -6 \quad \text{Let } x = 0.$$
$$-3y = -6$$
$$y = 2$$

The y-intercept is $(0, 2)$.

Plot the intercepts, $(-3, 0)$ and $(0, 2)$, and draw the line through them.

To graph the second line, which is in slope-intercept form, start by plotting the y-intercept, $(0, 2)$. From this point, move 3 units down and 1 unit to the right (because the slope is -3) to reach the point $(1, -1)$. Draw the line through $(0, 2)$ and $(1, -1)$.

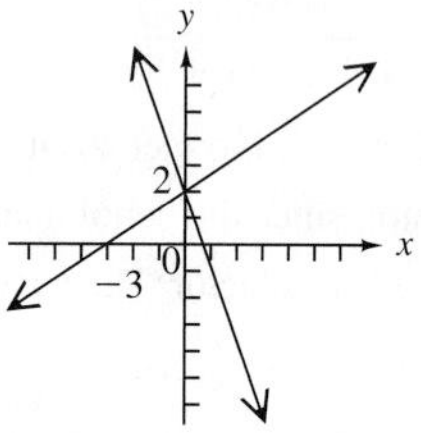

The lines intersect at their common y-intercept, $(0, 2)$.

Solution set: $\{(0, 2)\}$

35. $4x - 2y = 8 \quad (1)$

$2x = y + 4 \quad (2)$

Graph the line $4x - 2y = 8$ using its intercepts, $(2, 0)$ and $(0, -4)$.

Graph $2x = y + 4$ using its intercepts, $(2, 0)$ and $(0, -4)$. Since both equations have the same intercepts, they are equations of the same line.

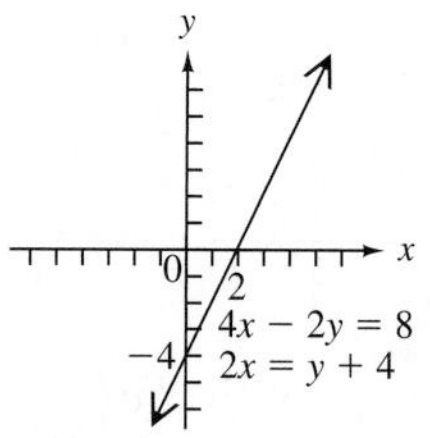

There are an infinite number of solutions, so we use set-builder notation to write the solution set. Rewrite equation (2) in standard form.

$$2x - y = 4$$

The solution set is

$$\{(x, y) \mid 2x - y = 4\}.$$

The given system consists of dependent equations.

SECTION 15.2 (pages 1069–1070)

21. $6x - 8y = 6$ (1)

$2y = -2 + 3x$ (2)

Solve equation (2) for y.

$$2y = -2 + 3x$$

$$y = \frac{3x - 2}{2} \quad (3)$$

Substitute $\frac{3x-2}{2}$ for y in equation (1).

$$6x - 8y = 6 \quad (1)$$

$$6x - 8\left(\frac{3x - 2}{2}\right) = 6$$

$$6x - 4(3x - 2) = 6$$

$$6x - 12x + 8 = 6$$

$$-6x + 8 = 6$$

$$-6x = -2$$

$$x = \frac{-2}{-6}, \quad \text{or} \quad \frac{1}{3}$$

To find y, let $x = \frac{1}{3}$ in equation (3).

$$y = \frac{3x - 2}{2} = \frac{3\left(\frac{1}{3}\right) - 2}{2} = \frac{1 - 2}{2} = -\frac{1}{2}$$

Solution set: $\left\{\left(\frac{1}{3}, -\frac{1}{2}\right)\right\}$

27. $\frac{x}{5} + 2y = \frac{16}{5}$ (1)

$\frac{3x}{5} + \frac{y}{2} = -\frac{7}{5}$ (2)

Multiply each side of equation (1) by 5.

$$5\left(\frac{x}{5} + 2y\right) = 5\left(\frac{16}{5}\right)$$

$$x + 10y = 16 \quad (3)$$

Multiply each side of equation (2) by 10.

$$10\left(\frac{3x}{5} + \frac{y}{2}\right) = 10\left(-\frac{7}{5}\right)$$

$$6x + 5y = -14 \quad (4)$$

We now have an equivalent system that does not include fractions.

$$x + 10y = 16 \quad (3)$$

$$6x + 5y = -14 \quad (4)$$

Solve equation (3) for x.

$$x = 16 - 10y \quad (5)$$

Substitute $16 - 10y$ for x in equation (4).

$$6x + 5y = -14 \quad (4)$$

$$6(16 - 10y) + 5y = -14$$

$$96 - 60y + 5y = -14$$

$$-55y = -110$$

$$y = 2$$

To find x, let $y = 2$ in equation (5).

$$x = 16 - 10y \quad (5)$$

$$x = 16 - 10(2)$$

$$x = -4$$

Solution set: $\{(-4, 2)\}$

SECTION 15.3 (pages 1075–1077)

25. $-x + 3y = 4$ (1)

$-2x + 6y = 8$ (2)

Multiply equation (1) by -2 and add the result to equation (2).

$$2x - 6y = -8 \quad (3)$$

$$-2x + 6y = 8 \quad (2)$$

$$0 = 0 \quad \text{True}$$

The true result, $0 = 0$, indicates that the equations of the original system are dependent. To obtain an equation in standard form $Ax + By = C$, with $A > 0$, multiply equation (1) by -1.

$$x - 3y = -4$$

Solution set: $\{(x, y) \mid x - 3y = -4\}$

37. $6x + 3y = 0$ (1)

$-18x - 9y = 0$ (2)

Multiply equation (1) by 3 and add the result to equation (2).

$$18x + 9y = 0 \quad (3)$$

$$-18x - 9y = 0 \quad (2)$$

$$0 = 0 \quad \text{True}$$

The true result, $0 = 0$, means that the system has an infinite number of solutions. To obtain an equation for the solution set in which the coefficients have greatest common factor 1, divide both sides of equation (1) by 3 to obtain the equivalent equation

$$2x + y = 0.$$

Solution set: $\{(x, y) \mid 2x + y = 0\}$

SECTION 15.4 (pages 1084–1089)

19. $C = 85x + 900$ (1)

$R = 105x$ (2)

No more than 38 units can be sold.

(a) Since $C = R$, substitute $105x$ for C and solve for x.

$$105x = 85x + 900$$

$$20x = 900 \quad \text{Subtract } 85x.$$

$$x = 45 \quad \text{Divide by 20.}$$

The break-even quantity is 45 units.

(b) Since no more than 38 units can be sold and the break-even quantity is 45, do not produce the product $(45 > 38)$. Doing so will lead to a loss.

37. ***Step 1***

Read the problem again.

Step 2

Assign variables.

Let x = the average rate of the bicycle;

y = the average rate of the car.

	r	t	d
Bicycle	x	7.5	$7.5x$
Car	y	7.5	$7.5y$

Use the formula $d = rt$.

Step 3

The total distance traveled by the bicycle and the car is 471 mi.

$$7.5x + 7.5y = 471 \quad (1)$$

The car traveled 35.8 mi faster than the bicycle.

$$y = x + 35.8 \quad (2)$$

These equations form the following system.

$$7.5x + 7.5y = 471 \quad (1)$$

$$y = x + 35.8 \quad (2)$$

Step 4

Because equation (2) is already solved for y, the substitution method is a good choice. Substitute $x + 35.8$ for y in equation (1).

$$7.5x + 7.5y = 471 \quad (1)$$

$$7.5x + 7.5(x + 35.8) = 471 \quad \text{Let } y = x + 35.8.$$

$$7.5x + 7.5x + 268.5 = 471 \quad \text{Distributive property}$$

$$15x + 268.5 = 471 \quad \text{Combine like terms.}$$

$$15x = 202.5 \quad \text{Subtract 268.5.}$$

$$x = 13.5 \quad \text{Divide by 15.}$$

To find the value of y substitute 13.5 for x in equation (2).

$$y = x + 35.8 \quad (2)$$

$$y = 13.5 + 35.8 \quad \text{Let } x = 13.5.$$

$$y = 49.3$$

Step 5

The average rate of the bicycle is 13.5 mph, and the average rate of the car is 49.3 mph.

Step 6

In 7.5 hr, the total distance traveled by the bicycle and the car is

$$7.5(13.5) + 7.5(49.3) = 471 \text{ mi}.$$

Since $49.3 - 13.5 = 35.8$, the car traveled 35.8 mph faster than the bicycle, as stated.

SECTION 15.5 (pages 1093–1094)

19. $x \le 2y + 3$

$x + y < 0$

Graph $x = 2y + 3$ as a solid line through $(3, 0)$ and $(5, 1)$. Using $(0, 0)$ as a test point results in the true statement $0 \le 3$, so shade the region containing the origin. Graph $x + y = 0$ as a dashed line through $(0, 0)$ and $(1, -1)$. We cannot use $(0, 0)$ as a test point because it is on the boundary line. Using $(1, 0)$ results in the false statement $1 < 0$, so shade the region *not* containing $(1, 0)$.

The solution set of the system is the intersection of the two shaded regions. It includes the portion of the line $x = 2y + 3$ that bounds the region but not the portion of the line $x + y = 0$.

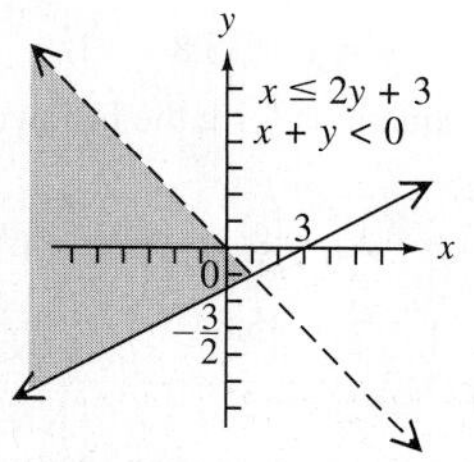

23. $3x - 2y \ge 6$

$x + y \le 4$

$x \ge 0$

$y \ge -4$

Graph $3x - 2y = 6$, $x + y = 4$, $x = 0$, and $y = -4$ as solid lines. All four inequalities are true for $(2, -2)$. Shade the region bounded by the four lines, which contains the test point $(2, -2)$. The solution set is the shaded region.

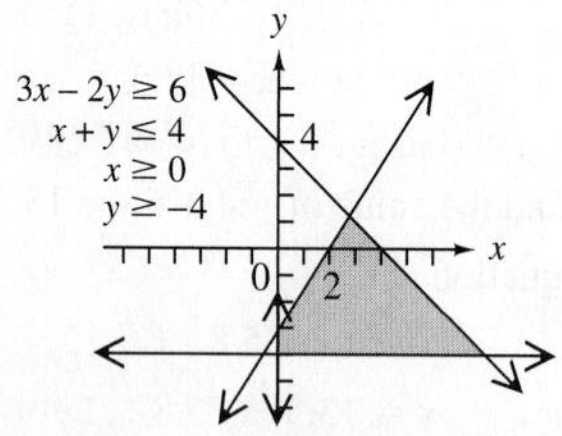

CHAPTER 16 Roots and Radicals

SECTION 16.1 (pages 1111–1114)

63. $\sqrt{103} \approx \sqrt{100} = 10$ Length

$\sqrt{48} \approx \sqrt{49} = 7$ Width

The best estimate for the length and width, in meters, of the rectangle is 10 by 7, choice C.

79. Let $a = 4.5$ and $c = 12.0$. Use the Pythagorean theorem.

$c^2 = a^2 + b^2$

$(12.0)^2 = (4.5)^2 + b^2$ Substitute carefully.

$144 = 20.25 + b^2$ Square.

$123.75 = b^2$ Subtract 20.25.

$b = \sqrt{123.75}$ Solve for b.

$b \approx 11.1$ Use a calculator.

The distance from the base of the tree to the point where the broken part touches the ground is 11.1 ft (to the nearest tenth).

93. $-\sqrt[3]{-8} = -(-2) = 2$

SECTION 16.2 (pages 1121–1124)

31. $3\sqrt{52}$

$= 3 \cdot \sqrt{4 \cdot 13}$ Factor; 4 is a perfect square.

$= 3\sqrt{4} \cdot \sqrt{13}$ Product rule

$= 3 \cdot 2 \cdot \sqrt{13}$ $\sqrt{4} = 2$

$= 6\sqrt{13}$ Multiply.

49. $5\sqrt{3} \cdot 2\sqrt{15}$

$= 5 \cdot 2 \cdot \sqrt{3 \cdot 15}$ Commutative property; product rule

$= 10\sqrt{45}$ Multiply.

$= 10\sqrt{9 \cdot 5}$ Factor; 9 is a perfect square.

$= 10\sqrt{9} \cdot \sqrt{5}$ Product rule

$= 10 \cdot 3 \cdot \sqrt{5}$ $\sqrt{9} = 3$

$= 30\sqrt{5}$ Multiply.

57. $\sqrt{\dfrac{4}{50}}$

$= \sqrt{\dfrac{2}{25}}$ Write the fraction in lowest terms.

$= \dfrac{\sqrt{2}}{\sqrt{25}}$ Quotient rule

$= \dfrac{\sqrt{2}}{5}$ $\sqrt{25} = 5$

119. Use the formula for the volume of a sphere.

$$V = \frac{4}{3}\pi r^3$$

Let $V = 288\pi$ and solve for r.

$288\pi = \frac{4}{3}\pi r^3$ $V = 288\pi$

$\frac{3}{4}(288\pi) = \frac{3}{4}\left(\frac{4}{3}\pi r^3\right)$ Multiply by $\frac{3}{4}$.

$216\pi = \pi r^3$ Simplify.

$216 = r^3$ Divide by π.

$\sqrt[3]{216} = r$ Take cube roots.

$6 = r$

The radius is 6 in.

121. $2\sqrt{26} \approx 2\sqrt{25} = 2 \cdot 5 = 10$ Length

$\sqrt{83} \approx \sqrt{81} = 9$ Width

Using 10 and 9 as estimates for the length and the width of the rectangle gives us

$$10 \cdot 9 = 90$$

as an estimate for the area, in square inches. Choice D is the best estimate.

SECTION 16.3 (pages 1127–1128)

29. $2\sqrt{8} - 5\sqrt{32} - 2\sqrt{48}$

$= 2(\sqrt{4} \cdot \sqrt{2}) - 5(\sqrt{16} \cdot \sqrt{2}) - 2(\sqrt{16} \cdot \sqrt{3})$ Product rule

$= 2(2\sqrt{2}) - 5(4\sqrt{2}) - 2(4\sqrt{3})$ $\sqrt{4} = 2; \sqrt{16} = 4$

$= 4\sqrt{2} - 20\sqrt{2} - 8\sqrt{3}$ Multiply.

$= -16\sqrt{2} - 8\sqrt{3}$ Subtract like radicals.

33. $\frac{1}{4}\sqrt{288} + \frac{1}{6}\sqrt{72}$

$= \frac{1}{4}\sqrt{144 \cdot 2} + \frac{1}{6}\sqrt{36 \cdot 2}$

$= \frac{1}{4}(\sqrt{144} \cdot \sqrt{2}) + \frac{1}{6}(\sqrt{36} \cdot \sqrt{2})$ Product rule

$= \frac{1}{4}(12\sqrt{2}) + \frac{1}{6}(6\sqrt{2})$ $\sqrt{144} = 12; \sqrt{36} = 6$

$= 3\sqrt{2} + 1\sqrt{2}$ Multiply.

$= 4\sqrt{2}$ Add like radicals.

35. Use the formula for the perimeter of a rectangle.

$P = 2L + 2W$

$P = 2(7\sqrt{2}) + 2(4\sqrt{2})$ Let $L = 7\sqrt{2}$ and $W = 4\sqrt{2}$.

$P = 14\sqrt{2} + 8\sqrt{2}$ Multiply.

$P = 22\sqrt{2}$ Add like radicals.

59. $5\sqrt[4]{m^3} + 8\sqrt[4]{16m^3}$

$= 5\sqrt[4]{m^3} + 8\sqrt[4]{16 \cdot m^3}$

$= 5\sqrt[4]{m^3} + 8\sqrt[4]{16} \cdot \sqrt[4]{m^3}$ Product rule

$= 5\sqrt[4]{m^3} + 8 \cdot 2 \cdot \sqrt[4]{m^3}$ $\sqrt[4]{16} = 2$

$= 5\sqrt[4]{m^3} + 16\sqrt[4]{m^3}$ Multiply.

$= 21\sqrt[4]{m^3}$ Add like radicals.

SECTION 16.4 (pages 1133–1134)

41. $\sqrt{\frac{9a^2r^5}{7t}}$

$= \frac{\sqrt{9a^2r^5}}{\sqrt{7t}}$ Quotient rule

$= \frac{\sqrt{9a^2r^4 \cdot r}}{\sqrt{7t}}$ Factor.

$= \frac{\sqrt{9a^2r^4} \cdot \sqrt{r}}{\sqrt{7t}}$ Product rule

$= \frac{3ar^2\sqrt{r}}{\sqrt{7t}}$ $\sqrt{9} = 3$; $\sqrt{a^2} = a$, since $a > 0$; $\sqrt{r^4} = r^2$

$= \frac{3ar^2\sqrt{r} \cdot \sqrt{7t}}{\sqrt{7t} \cdot \sqrt{7t}}$ Rationalize the denominator.

$= \frac{3ar^2\sqrt{7rt}}{7t}$ Product rule; $\sqrt{7t} \cdot \sqrt{7t} = 7t$

47. $\sqrt[3]{\frac{5}{9}}$

$= \frac{\sqrt[3]{5}}{\sqrt[3]{9}}$ Quotient rule

$= \frac{\sqrt[3]{5} \cdot \sqrt[3]{3}}{\sqrt[3]{9} \cdot \sqrt[3]{3}}$ Multiply numerator and denominator by $\sqrt[3]{3}$ to make the radicand in the denominator a perfect cube.

$= \frac{\sqrt[3]{15}}{\sqrt[3]{27}}$ Product rule

$= \frac{\sqrt[3]{15}}{3}$ $\sqrt[3]{27} = 3$

55. (a) $p = k \cdot \sqrt{\frac{L}{g}}$

$p = 6 \cdot \sqrt{\frac{9}{32}}$ Let $k = 6$, $L = 9$, $g = 32$.

$p = \frac{6\sqrt{9}}{\sqrt{32}}$ Quotient rule; multiply.

$p = \frac{6 \cdot 3}{\sqrt{16 \cdot 2}}$ $\sqrt{9} = 3$; Factor.

$p = \frac{18}{4\sqrt{2}}$ Multiply; product rule

$p = \frac{9 \cdot 2}{2 \cdot 2\sqrt{2}}$ Factor.

$p = \frac{9}{2\sqrt{2}}$ Divide out 2.

$p = \frac{9 \cdot \sqrt{2}}{2\sqrt{2} \cdot \sqrt{2}}$ Rationalize the denominator.

$p = \frac{9\sqrt{2}}{4}$ Multiply.

The period of the pendulum is $\frac{9\sqrt{2}}{4}$ sec.

(b) Using a calculator, $\frac{9\sqrt{2}}{4} \approx 3.182$ sec.

SECTION 16.5 (pages 1139–1142)

15. $(5\sqrt{7} - 2\sqrt{3})(3\sqrt{7} + 4\sqrt{3})$

$= 5\sqrt{7}(3\sqrt{7}) + 5\sqrt{7}(4\sqrt{3}) - 2\sqrt{3}(3\sqrt{7}) - 2\sqrt{3}(4\sqrt{3})$ FOIL method

$= 15 \cdot 7 + 20\sqrt{21} - 6\sqrt{21} - 8 \cdot 3$ Multiply.

$= 105 + 14\sqrt{21} - 24$ Multiply numbers. Add like radicals.

$= 81 + 14\sqrt{21}$ Subtract.

45. $\frac{\sqrt{5}}{\sqrt{2} + \sqrt{3}}$

$= \frac{\sqrt{5}(\sqrt{2} - \sqrt{3})}{(\sqrt{2} + \sqrt{3})(\sqrt{2} - \sqrt{3})}$

Multiply numerator and denominator by the conjugate of the denominator.

$= \frac{\sqrt{5} \cdot \sqrt{2} - \sqrt{5} \cdot \sqrt{3}}{(\sqrt{2})^2 - (\sqrt{3})^2}$

Distributive property; $(a + b)(a - b) = a^2 - b^2$

$= \frac{\sqrt{10} - \sqrt{15}}{2 - 3}$ Product rule; $(\sqrt{a})^2 = a$

$= \frac{\sqrt{10} - \sqrt{15}}{-1}$ Subtract.

$= -\sqrt{10} + \sqrt{15}$

59. $\frac{12 - \sqrt{40}}{4}$

$= \frac{12 - \sqrt{4} \cdot \sqrt{10}}{4}$ Product rule

$= \frac{12 - 2\sqrt{10}}{4}$ $\sqrt{4} = 2$

$= \frac{2(6 - \sqrt{10})}{2(2)}$ Factor numerator and denominator.

$= \frac{6 - \sqrt{10}}{2}$ Lowest terms

SECTION 16.6 (pages 1149–1152)

23. $\sqrt{3x - 5} = \sqrt{2x + 1}$

$(\sqrt{3x - 5})^2 = (\sqrt{2x + 1})^2$ Squaring property

$3x - 5 = 2x + 1$ $(\sqrt{a})^2 = a$

$x = 6$ Subtract $2x$. Add 5.

CHECK $\sqrt{3x - 5} = \sqrt{2x + 1}$

$\sqrt{3(6) - 5} \stackrel{?}{=} \sqrt{2(6) + 1}$ Let $x = 6$.

$\sqrt{13} = \sqrt{13}$ ✓ True

Solution set: $\{6\}$

33. $\sqrt{3k + 10} + 5 = 2k$

$\sqrt{3k + 10} = 2k - 5$ Subtract 5.

$(\sqrt{3k + 10})^2 = (2k - 5)^2$ Squaring property

$3k + 10 = 4k^2 - 20k + 25$ $(\sqrt{a})^2 = a$; Square the binomial.

$4k^2 - 23k + 15 = 0$ Standard form

$(4k - 3)(k - 5) = 0$ Factor.

$k = \frac{3}{4}$ or $k = 5$ Zero-factor property

Checking both proposed solutions in the original equation will show that 5 is a solution, but $\frac{3}{4}$ is not.

Solution set: $\{5\}$

49. $\sqrt{2x + 11} + \sqrt{x + 6} = 2$

$\sqrt{2x + 11} = 2 - \sqrt{x + 6}$ Isolate a radical.

$(\sqrt{2x + 11})^2 = (2 - \sqrt{x + 6})^2$ Square both sides.

$2x + 11 = 4 - 4\sqrt{x + 6} + x + 6$ $(\sqrt{a})^2 = a^2$; Square the binomial.

$x + 1 = -4\sqrt{x + 6}$ Isolate the radical.

$(x + 1)^2 = (-4\sqrt{x + 6})^2$ Square both sides again.

$x^2 + 2x + 1 = 16(x + 6)$ Square the binomial; $(a\sqrt{b})^2 = a^2b$

$x^2 + 2x + 1 = 16x + 96$ Distributive property

$x^2 - 14x - 95 = 0$ Standard form

$(x + 5)(x - 19) = 0$ Factor.

$x = -5$ or $x = 19$ Zero-factor property

Checking both proposed solutions in the original equation will show that -5 is a solution, but 19 is not.

Solution set: $\{-5\}$

53. Let $x =$ the unknown number.

Write an equation.

$3\sqrt{2} = \sqrt{x + 10}$

To solve this equation, square both sides and then solve the resulting equation.

$(3\sqrt{2})^2 = (\sqrt{x + 10})^2$ Square both sides.

$18 = x + 10$ $(3\sqrt{2})^2 = 9 \cdot 2 = 18$

$8 = x$ Subtract 10.

Since $\sqrt{8 + 10} = \sqrt{18}$, and $\sqrt{18} = 3\sqrt{2}$, the number is 8.

CHAPTER 17 Quadratic Equations

SECTION 17.1 (pages 1168–1170)

27. $7x^2 = 4$

$x^2 = \frac{4}{7}$ Divide by 7.

$x = \sqrt{\frac{4}{7}}$ or $x = -\sqrt{\frac{4}{7}}$

Square root property

$x = \frac{\sqrt{4}}{\sqrt{7}} \cdot \frac{\sqrt{7}}{\sqrt{7}}$ or $x = -\frac{\sqrt{4}}{\sqrt{7}} \cdot \frac{\sqrt{7}}{\sqrt{7}}$

Rationalize denominators.

$x = \frac{2\sqrt{7}}{7}$ or $x = -\frac{2\sqrt{7}}{7}$

Solution set: $\left\{-\frac{2\sqrt{7}}{7}, \frac{2\sqrt{7}}{7}\right\}$

49. $\left(\frac{1}{2}x + 5\right)^2 = 12$

$\frac{1}{2}x + 5 = \sqrt{12}$ or $\frac{1}{2}x + 5 = -\sqrt{12}$

$\frac{1}{2}x = -5 + \sqrt{12}$ or $\frac{1}{2}x = -5 - \sqrt{12}$

Subtract 5.

$\frac{1}{2}x = -5 + 2\sqrt{3}$ or $\frac{1}{2}x = -5 - 2\sqrt{3}$

$\sqrt{12} = \sqrt{4} \cdot \sqrt{3} = 2\sqrt{3}$

$x = -10 + 4\sqrt{3}$ or

$x = -10 - 4\sqrt{3}$

Multiply each term by 2.

Solution set:

$\{-10 + 4\sqrt{3}, -10 - 4\sqrt{3}\}$

57. $A = P(1 + r)^2$

$110.25 = 100(1 + r)^2$

Let $A = 110.25$, $P = 100$.

$1.1025 = (1 + r)^2$

Divide by 100.

$(1 + r)^2 = 1.1025$

Interchange sides.

$1 + r = \sqrt{1.1025}$ or

$1 + r = -\sqrt{1.1025}$

Square root property

$r = -1 + 1.05$ or $r = -1 - 1.05$

Subtract 1. Use a calculator.

$r = 0.05$ or $r = -2.05$

Reject the solution -2.05 since the rate of interest cannot be negative. The rate is 0.05, or 5%.

SECTION 17.2 (pages 1177–1178)

17. $t^2 + 6t + 9 = 0$

The left side of this equation is already a perfect square.

$(t + 3)^2 = 0$ Factor.

$t + 3 = 0$ Square root property

$t = -3$ Subtract 3.

A check verifies that the solution is -3.

Solution set: $\{-3\}$

19. $x^2 + x - 1 = 0$

$x^2 + x = 1$ Add 1.

Take half of 1, the coefficient of x, square it, and add the result to each side.

$x^2 + x + \frac{1}{4} = 1 + \frac{1}{4}$

Add $\left[\frac{1}{2}(1)\right]^2 = \frac{1}{4}$.

$\left(x + \frac{1}{2}\right)^2 = \frac{5}{4}$

Factor and add.

$x + \frac{1}{2} = \sqrt{\frac{5}{4}}$ or $x + \frac{1}{2} = -\sqrt{\frac{5}{4}}$

$x + \frac{1}{2} = \frac{\sqrt{5}}{2}$ or $x + \frac{1}{2} = -\frac{\sqrt{5}}{2}$

$x = -\frac{1}{2} + \frac{\sqrt{5}}{2}$ or $x = -\frac{1}{2} - \frac{\sqrt{5}}{2}$

$x = \frac{-1 + \sqrt{5}}{2}$ or $x = \frac{-1 - \sqrt{5}}{2}$

Solution set: $\left\{\frac{-1 + \sqrt{5}}{2}, \frac{-1 - \sqrt{5}}{2}\right\}$

23. $2x^2 - 4x = 5$

$x^2 - 2x = \frac{5}{2}$ Divide by 2.

$x^2 - 2x + 1 = \frac{5}{2} + 1$

Add $\left[\frac{1}{2}(-2)\right]^2 = 1$.

$(x - 1)^2 = \frac{7}{2}$ Factor and add.

$x - 1 = \sqrt{\frac{7}{2}}$ or $x - 1 = -\sqrt{\frac{7}{2}}$

$x - 1 = \frac{\sqrt{7}}{\sqrt{2}}$ or $x - 1 = -\frac{\sqrt{7}}{\sqrt{2}}$

$x - 1 = \frac{\sqrt{14}}{2}$ or $x - 1 = -\frac{\sqrt{14}}{2}$

Rationalize denominators by multiplying by $\frac{\sqrt{2}}{\sqrt{2}}$.

$x = 1 + \frac{\sqrt{14}}{2}$ or $x = 1 - \frac{\sqrt{14}}{2}$

$x = \frac{2}{2} + \frac{\sqrt{14}}{2}$ or $x = \frac{2}{2} - \frac{\sqrt{14}}{2}$

$x = \frac{2 + \sqrt{14}}{2}$ or $x = \frac{2 - \sqrt{14}}{2}$

Solution set: $\left\{\frac{2 + \sqrt{14}}{2}, \frac{2 - \sqrt{14}}{2}\right\}$

SECTION 17.3 (pages 1183–1184)

23. $7x^2 = 12x$

Write the equation in standard form.

$7x^2 - 12x = 0$

Substitute $a = 7$, $b = -12$, and $c = 0$ into the quadratic formula.

$x = \frac{-b \pm \sqrt{b^2 - 4ac}}{2a}$

$x = \frac{-(-12) \pm \sqrt{(-12)^2 - 4(7)(0)}}{2(7)}$

$x = \frac{12 \pm \sqrt{144 - 0}}{14}$

$x = \frac{12 \pm 12}{14}$

$x = \frac{12 + 12}{14} = \frac{24}{14} = \frac{12}{7}$ or

$x = \frac{12 - 12}{14} = \frac{0}{14} = 0$

Solution set: $\left\{0, \frac{12}{7}\right\}$

29. $3x^2 - 2x + 5 = 10x + 1$

Write the equation in standard form.

$3x^2 - 12x + 4 = 0$

Substitute $a = 3$, $b = -12$, and $c = 4$ into the quadratic formula.

$x = \frac{-b \pm \sqrt{b^2 - 4ac}}{2a}$

$x = \frac{-(-12) \pm \sqrt{(-12)^2 - 4(3)(4)}}{2(3)}$

$x = \frac{12 \pm \sqrt{144 - 48}}{6}$

$x = \frac{12 \pm \sqrt{96}}{6}$

$x = \frac{12 \pm 4\sqrt{6}}{6}$

$\sqrt{96} = \sqrt{16} \cdot \sqrt{6} = 4\sqrt{6}$

$x = \frac{2(6 \pm 2\sqrt{6})}{2 \cdot 3}$ Factor.

$x = \frac{6 \pm 2\sqrt{6}}{3}$ Divide out 2.

Solution set: $\left\{\frac{6 + 2\sqrt{6}}{3}, \frac{6 - 2\sqrt{6}}{3}\right\}$

39. $0.6x - 0.4x^2 = -1$

To eliminate the decimals, multiply each side by 10.

$$6x - 4x^2 = -10$$

Write this equation in standard form.

$$4x^2 - 6x - 10 = 0$$

Divide each side by 2 so that we can work with smaller coefficients in the quadratic formula.

$$2x^2 - 3x - 5 = 0$$

Use the quadratic formula with $a = 2$, $b = -3$, and $c = -5$.

$$x = \frac{-(-3) \pm \sqrt{(-3)^2 - 4(2)(-5)}}{2(2)}$$

$$x = \frac{3 \pm \sqrt{9 + 40}}{4}$$

$$x = \frac{3 \pm \sqrt{49}}{4}$$

$$x = \frac{3 \pm 7}{4}$$

$$x = \frac{3 + 7}{4} = \frac{10}{4} = \frac{5}{2} \quad \text{or}$$

$$x = \frac{3 - 7}{4} = \frac{-4}{4} = -1$$

Solution set: $\left\{-1, \frac{5}{2}\right\}$

SECTION 17.4 (pages 1189–1190)

19. Because the vertex is at the origin, an equation of the parabola is of the form

$$y = ax^2.$$

As shown in the figure, one point on the graph has coordinates $(150, 44)$.

$y = ax^2$ General equation

$44 = a(150)^2$ Let $x = 150$, $y = 44$.

$44 = 22{,}500a$ Apply the exponent.

$a = \frac{44}{22{,}500}$ Solve for a.

$a = \frac{4 \cdot 11}{4 \cdot 5625}$, or $\frac{11}{5625}$ Lowest terms

Thus, an equation of the parabola is

$$y = \frac{11}{5625}x^2.$$

Photo Credits

1 Marcito/Fotolia; **4** Yuri Arcurs/Fotolia; **9** (top left) Nicola Sutton/Life File/Photodisc/Getty Images, (top right) DKatz/Fotolia, (bottom left) Alexander Crispin/Getty Images, (bottom right) Monart Design/Fotolia; **15** Bew111/Dreamstime; **20** (left) Andres Rodriguez/Fotolia, (right) Emiliau/Fotolia; **30** (top left) Victor Zastol'skiy/Fotolia, (top right) Robhainer/Fotolia, (bottom left) Kenneth Graff/Shutterstock, (bottom right) Stan Salzman; **40** (top) Jonathan Alcorn/ZUMA Press/Newscom, (bottom) John A. Rizzo/Photodisc/Getty Images; **53** (top left) Anette Linnea Rasmussen/Shutterstock, (top right) Beth Anderson/Pearson Education, Inc., (bottom left) Sueharper/Dreamstime, (bottom right) Stan Salzman; **61** (left) Edwin Remsberg/Alamy, (right) Stan Salzman; **62** (left) Spflaum/Dreamstime, (right) Todd Taulman/Fotolia; **63** Graeme Dawes/Shutterstock; **82** A.F. Archive/Alamy; **85** Stan Salzman; **86** Elenathewise/Fotolia; **90** Stan Salzman; **93** Lightpainter/Dreamstime; **94** Svlumagraphica/Dreamstime; **95** (left) Beth Anderson/Pearson Education, Inc./SUBWAY® is a registered trademark of Doctor's Associates Inc. © Doctor's Associates Inc. All Rights Reserved, (right) Vacclav/Fotolia; **106** (left) Sascha Burkard/Fotolia, (right) Vladimir Caplinskij/Shutterstock; **107** (left) Jason Stitt/Shutterstock, (right) Lily/Fotolia; **108** Karl Weatherly/Photodisc/Getty Images; **111** (both) Jelly Belly Candy Company. Used with permission; **115** (left) Photodisc/Getty Images, (right) U.S. Mint; **146** (left) Sascha Burkard/Shutterstock, (top right) Pakhnyushchyy/Fotolia, (bottom right) Kotafoty/Fotolia; **147** Iofoto/Fotolia; **151** (left) Antares614/Dreamstime, (right) Mark Breck/Shutterstock; **152** (left) Lynn Watson/Shutterstock, (right) Brodogg1313/Dreamstime; **153** Stan Salzman; **154** (left) Slocummedia/Fotolia, (right) Arim44/Dreamstime; **161** (left) V. J. Matthew/Shutterstock, (right) Ulrich Willmünder/Shutterstock; **162** (left) Stephen Coburn/Fotolia, (right) Oleg Zabielin/Fotolia; **163** Collection of Stan Salzman; **173** Jelly Belly Candy Company. Used with permission; **184** Giuseppe Porzani/Fotolia; **187** (left) Sylvana Rega/Fotolia, (right) Kropic/Fotolia; **190** Olga Nayashkova/Fotolia; **191** ZUMA Press/Newscom; **196** (both) Stan Salzman; **215** (left) Chris Hill/Shutterstock, (right) Stan Salzman; **216** Elenathewise/Fotolia; **224** (left) ZUMA Press/Newscom, (right) Imageegami/Fotolia; **226** (left) Pavel Losevsky/Fotolia, (right) Stan Salzman; **227** (left) Gudellaphoto/Fotolia, (right) ZUMA Press/Newscom; **237** (left) Konstantin Sutyagin/Shutterstock, (right) Stan Salzman; **245** (left) Kim Steele/Digital Vision/Getty Images, (right) Stan Salzman; **246** (left) Amy Myers/Shutterstock, (right) Stan Salzman; **248** (both) Stan Salzman; **250** (left) Alisonhancock/Fotolia, (right) Lise F. Young/Shutterstock; **253** Pressmaster/Fotolia; **262** Pressmaster/Fotolia; **263** Carlosseller/Fotolia; **269** (both) Beth Anderson/Pearson Education, Inc.; **270** (top) Perry Mastrovito/Corbis, (bottom) Charles Smith/Corbis; **293** U.K. History/Alamy; **294** (left) Feng Yu/Fotolia, (top right) National Motor Museum/Motoring Picture Library/Alamy, (bottom right) Used with permission of General Mills Marketing, Inc. (GMMI); **300** (left) Diane Macdonald/Stockbyte/Getty Images, (right) Flying Colours Ltd/Photodisc/Getty Images; **301** Neo Edmund/Fotolia; **309** (left) Stockbyte/Getty Images, (right) Beth Anderson/Pearson Education, Inc.; **311** Beth Anderson/Pearson Education, Inc.; **314** Twixx/Fotolia; **315** Edbockstock/Fotolia; **322** Picturequest/JupiterImages; **324** Alan Pappe/Photodisc/Getty Images; **331** Graca Victoria/Shutterstock; **332** Edbockstock/Fotolia; **337** Sly/Fotolia; **348** Ryan McVay/Photodisc/Getty Images; **350** Jjava/Fotolia; **358** (left) Doug Menuez/Photodisc/Getty Images, (right) Jack Hollingsworth/Photodisc/Getty Images; **359** (left) Comstock/Thinkstock, (right) David Buffington/Photodisc/Getty Images; **362** Steve Byland/Fotolia; **363** Pavel Losevsky/Fotolia; **370** (left) Stan Salzman, (right) Digital Vision/Getty Images; **371** (left) Getty Images/Digital Vision, (right) Yuri Arcurs/Fotolia; **372** Annette Shaff/Shutterstock; **383** (left) Stockbyte/Getty Images, (right) Beth Anderson/Pearson Education, Inc.; **384** (left) Regien Paassen/Shutterstock, (right) Siri Stafford/Digital Vision/Getty Images; **392** (top left) Yuri Arcurs/Fotolia, (top right, bottom left) Beth Anderson/Pearson Education, Inc., (bottom right) Lisa F. Young/Fotolia; **393** (left) Beth Anderson/Pearson Education, Inc., (right) Actionpics/Fotolia; **396** Pete Saloutos/Fotolia; **404** (left) Alexander/Fotolia, (right) Stan Salzman; **405** Okea/Fotolia; **406** Givaga/Fotolia; **413** Stan Salzman; **414** (left) Vac/Fotolia, (right) Sean Locke/Photodisc/Getty Images; **423** (left) PA Saez/Alamy, (top right) Michael Shake/Fotolia, (bottom right) Marc Romanelli/Photodisc/Getty Images; **430** (left) Stan Salzman, (right) Andresr/Shutterstock; **437** (left) Roza/Fotolia, (right) Gabriel Blaj/Fotolia; **447** (left) Peternile/Fotolia, (right) Kratuanoiy/Fotolia; **451** (top left) Jj2theb/Fotolia, (top right) Stan Salzman, (bottom left) Halfdark/PhotoAlto/Getty Images, (bottom right) Jack Hollingsworth/ Digital Vision/Getty Images; **454** Michael Flippo/Fotolia; **455** National Oceanic and Atmospheric Administration (NOAA); **477** (left) Mark Karrass/Spirit/Corbis, (right) Shock/Fotolia; **478** Tom Gannam/AP Images; **500** (left) Beth Anderson/Pearson Education, Inc., (right) Jacques Descloitres, MODIS Rapid Response Team/GSFC/NASA; **501** (left) Stephen Ausmus/U.S. Department of Agriculture, (right) Joe Sohm/Visions of America/Digital Vision/Getty Images; **502** Brian Tolbert/Spirit/Corbis; **512** (left) Neil Beer/Photodisc/Getty Images, (right) Motorlka/Fotolia; **519** Earl Orf; **551** Yuri Arcurs/Fotolia; **554** Beth Anderson/Pearson Education, Inc.; **557** Kurhan/ Fotolia; **559** Hwahl3/Fotolia; **560** (left) Karl Weatherly/Photodisc/Getty Images, (right) Sheer Photo, Inc/Stockbyte/Getty Images; **568** Noel Hendrickson/Blend Images/Getty Images; **569** Vibe Images/Fotolia; **570** Beth Anderson/Pearson Education, Inc.; **573** Diego Cervo/Fotolia; **574** Maksym Yemelyanov/Fotolia; **575** Andresr/Shutterstock; **576** Pressmaster/Fotolia; **578** Stan Salzman; **583** Monkey Business Images/Shutterstock; **584** Evgenyb/Fotolia; **587** Michael Shake/ Shutterstock; **591** Rena Schild/Shutterstock; **593** Beth Anderson/Pearson Education, Inc.; **595** Robert Kneschke/Fotolia; **596** Corbis; **597** Kristin Smith/Shutterstock; **604** Sean Justice/Corbis; **605** Michael Flippo/Fotolia; **613** Darren Baker/Fotolia; **621** Monkey Business/Fotolia; **634** Sirena Designs/Fotolia; **638** (left) Bochkarev Photography/Shutterstock, (top right) UryadnikovS/Fotolia, (bottom right) Vlad G/Fotolia; **646** (left) Jesse Kunerth/Fotolia, (top right) John Hornsby, (bottom right) Igor Kushnir/Fotolia; **647** Horticulture/Fotolia; **659** Rob Byron/Fotolia; **661** Jason Stitt/ Fotolia; **682** Paul Maguire/Fotolia; **684** Elainekay/Fotolia; **687** Huaxiadragon/Fotolia; **714** Dimitar Dilkoff/AFP/Getty Images/Newscom; **723** (top) Emkaplin/Fotolia, (bottom) DenisNata/Fotolia; **725** Tmass/Fotolia; **735** Roger Cracknell 01/classic/Alamy; **745** Lightpoet/Shutterstock; **758** (left) Monkey Business/Fotolia, (right) Michaeljung/Fotolia; **759** Sababa66/Shutterstock; **761** Eppic/ Fotolia; **762** Konstantin Yuganov/Fotolia; **763** Dianapaz/Fotolia; **767** KingPhoto/Fotolia; **769** Sanjay Goswami/Fotolia; **775** Igor Mojzes/Fotolia; **776** Rhona Wise/EPA/Newscom; **783** Zentilia/ Fotolia; **808** Karen Roach/Fotolia; **814** Rebecca Fisher/Fotolia; **833** (top) Jezper/Fotolia, (bottom) Sara Piaseczynski; **889** Johan Swanepoel/Shutterstock; **890** (left) Carlos Santa Maria/Fotolia, (right) Ivan Cholakov/Shutterstock; **899** JPL-Caltech/STScI/NASA; **901** World History Archive/ Alamy; **943** Nickolae/Fotolia; **951** OnFilm/iStockphoto; **955** Offscreen/Shutterstock; **957** Gino Santa Maria/Fotolia; **969** Giordano Aita/Fotolia; **972** Mikesch112/Fotolia; **973** Matthew O'Haren/ Icon SMI 394/Newscom; **1030** Ernest Prim/Fotolia; **1033** (left) Jason Roberts/EPA/Newscom, (right) Gero Breloer/EPA/Newscom; **1034** Epantha/Fotolia; **1036** John Hornsby; **1039** Mary Durden/Fotolia; **1052** Everett Collection, Inc.; **1053** Geewhiz/Fotolia; **1070** Vadim Ponomarenko/ Fotolia; **1077** Sashkin/Fotolia; **1078** Julijah/Fotolia; **1085** (left) Scott Rothstein/Shutterstock, (top right) AbleStock/Thinkstock, (bottom right) Sashkin/Fotolia; **1087** (left) Everett Collection Inc./ Alamy, (right) Everett Collection Inc./Alamy; **1088** Eric Gevaert/Fotolia; **1089** Ints/Fotolia; **1102** (left) Iophoto/Shutterstock, (right) Marty/Fotolia; **1103** Danny E Hooks/Shutterstock; **1104** Beth Anderson/Pearson Education, Inc.; **1114** (left) John Hornsby, (right) Patric Martel/ Fotolia; **1134** Apollo 17 Crew, NASA; **1142** Serhiy Kobyakov/Fotolia; **1148** Betacam-SP/Fotolia; **1151** Timothy Large/Shutterstock; **1152** (top left) Lotharingia/Fotolia, (top right) Beboy/Fotolia, (bottom left) Tmax/Fotolia, (bottom right) Sculpies/Fotolia; **1162** Comstock/Thinkstock; **1163** Badmanproduction/Fotolia; **1167** Sablin/Fotolia; **1184** NASA Headquarters; **1192** Lisa F. Young/Shutterstock; **1211** Tyler Olson/Fotolia; **1212** Zimmytws/Fotolia; **AP-1** Ricknoll/Fotolia; **AP-8** (left) Paul Banton/Shutterstock, (right) Sharyn Young/Shutterstock; **AP-9** (top left) Denis Pepin/Fotolia, (bottom left) Morey Milbradt/Brand X Pictures/JupiterImages, (right) EPA European Pressphoto Agency B.V./Alamy; **AP-11** (left) Valueline/PunchStock, (top right) Stockbyte/Getty Images, (bottom right) Ricknoll/Fotolia; **AP-18** (left) Sergey Yarochkin/Fotolia, (right) Vladimir Sazonov/Fotolia; **AP-25** (top left) Tinka/Fotolia, (top, middle, and bottom right) Beth Anderson/Pearson Education, Inc., (bottom left) Doug Menuez/Photodisc/Getty Images; **AP-26** (both) Beth Anderson/Pearson Education, Inc.; **AP-27** (top left) Beth Anderson/Pearson Education, Inc., (top right) Comugnero Silvana/Fotolia, (bottom left) Sebastian Kaulitzki/ Shutterstock, (bottom right) Max Delson Martins Santos/iStockphoto; **AP-28** (top left) GK Hart/Vikki Hart/Photodisc/Getty Images, (top right) Ryan McVay/Photodisc/Getty Images, (bottom left) Beth Anderson/Pearson Education, Inc., (bottom right) David Frazier/Corbis; **AP-34** Bongarts/SportsChrome/Newscom; **AP-35** (left) Ryan McVay/Photodisc/Getty Images, (right) Mark Herreid/Fotolia

Index

D

Q

R

Geometry Formulas

Square
Perimeter: $P = 4s$
Area: $A = s^2$

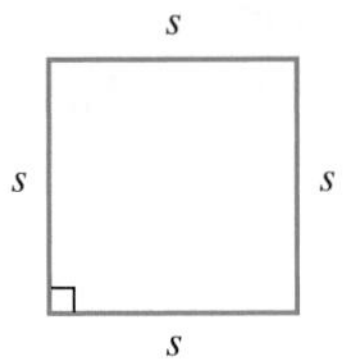

Rectangular Solid
Volume: $V = LWH$
Surface area: $A = 2HW + 2LW + 2LH$

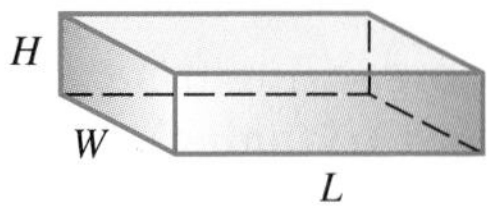

Rectangle
Perimeter: $P = 2L + 2W$
Area: $A = LW$

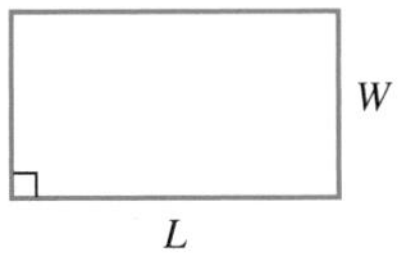

Cube
Volume: $V = e^3$
Surface area: $S = 6e^2$

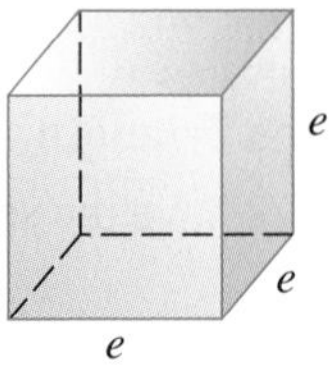

Triangle
Perimeter: $P = a + b + c$
Area: $A = \frac{1}{2}bh$

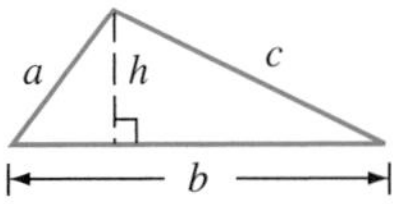

Right Circular Cylinder
Volume: $V = \pi r^2 h$
Surface area: $S = 2\pi rh + 2\pi r^2$
(Includes both circular bases)

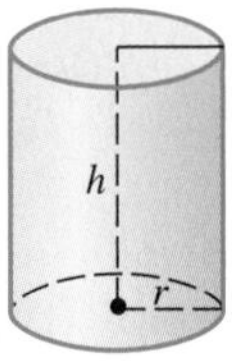

Parallelogram
Perimeter: $P = 2a + 2b$
Area: $A = bh$

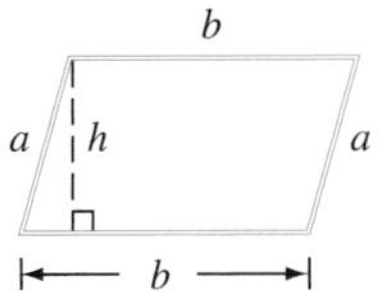

Cone
Volume: $V = \frac{1}{3}\pi r^2 h$

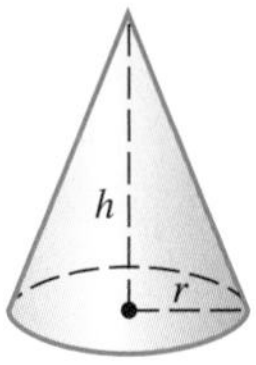

Trapezoid
Perimeter: $P = a + b + c + B$
Area: $A = \frac{1}{2}h(b + B)$

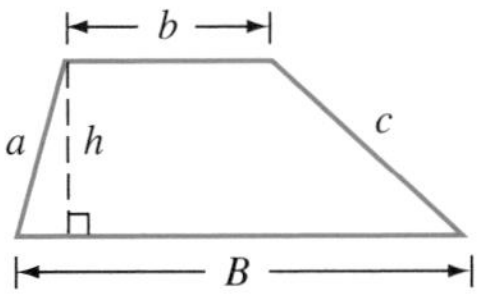

Right Pyramid
Volume: $V = \frac{1}{3}Bh$
B = area of the base

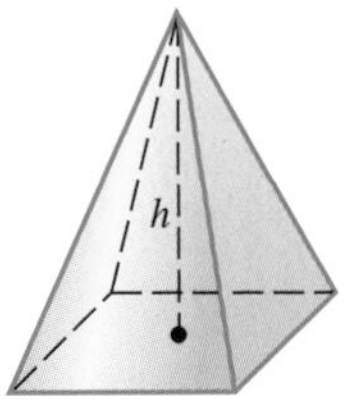

Circle
Diameter: $d = 2r$
Circumference: $C = 2\pi r$
or $C = \pi d$
Area: $A = \pi r^2$

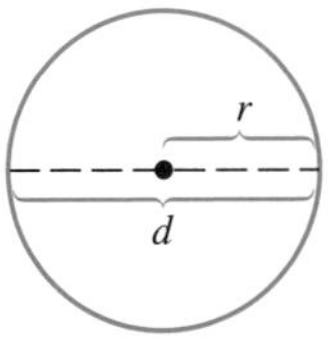

Sphere
Volume: $V = \frac{4}{3}\pi r^3$

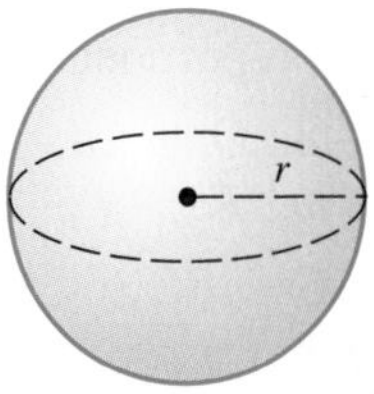

Triangles and Angles

Right Triangle
Triangle has one 90° (right) angle.

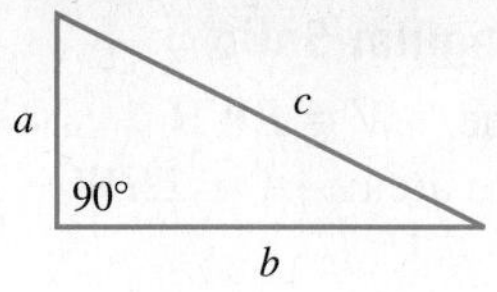

Pythagorean Theorem (for right triangles)
$a^2 + b^2 = c^2$

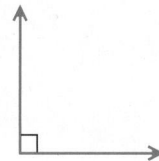

Isosceles Triangle
Two sides have the same length.

$AB = BC$

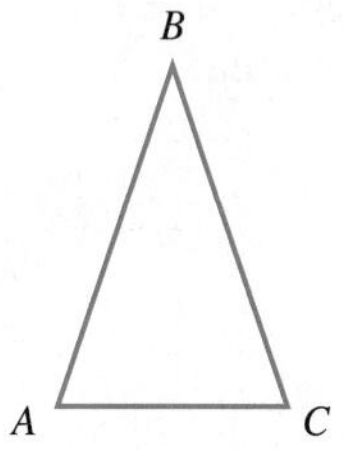

Equilateral Triangle
All sides have the same length.

$AB = BC = CA$

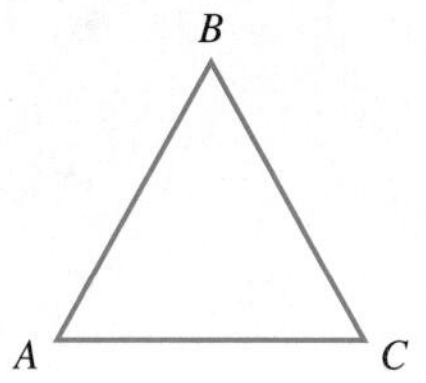

Sum of the Angles of Any Triangle
$m\angle A + m\angle B + m\angle C = 180°$

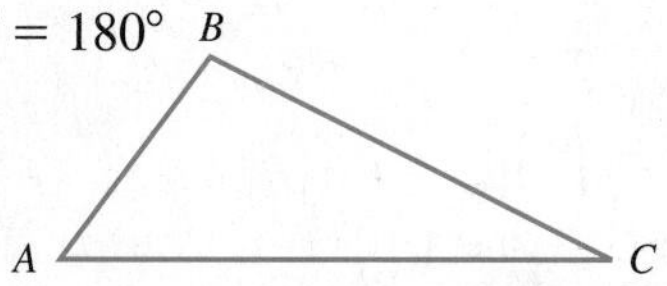

Similar Triangles
Corresponding angles have equal measures. Corresponding sides are proportional.

$m\angle A = m\angle D,\ m\angle B = m\angle E,\ m\angle C = m\angle F$

$\frac{AB}{DE} = \frac{AC}{DF} = \frac{BC}{EF}$

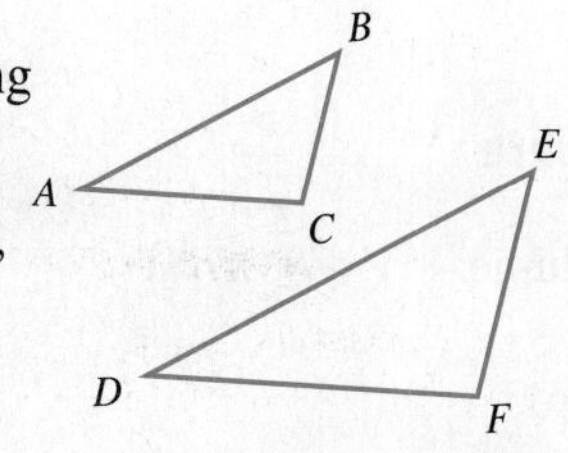

Acute Angle
Measure is less than 90°.

Right Angle
Measure is 90°.

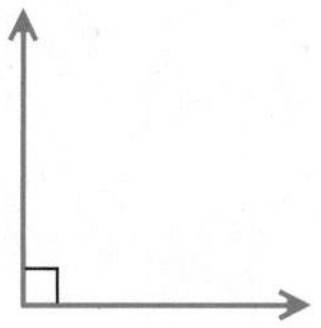

Obtuse Angle
Measure is more than 90° but less than 180°.

Straight Angle
Measure is 180°.

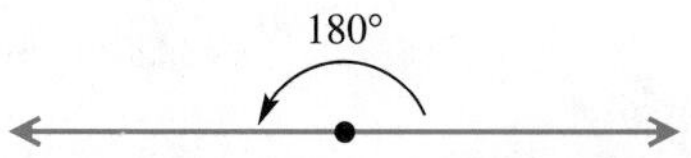

Complementary Angles
The sum of the measures of two complementary angles is 90°.

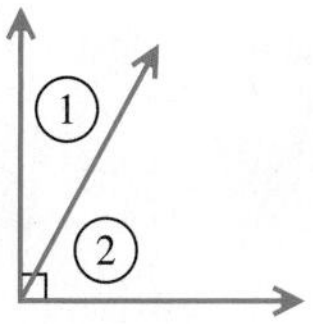

Angles ① and ② are complementary.

Supplementary Angles
The sum of the measures of two supplementary angles is 180°.

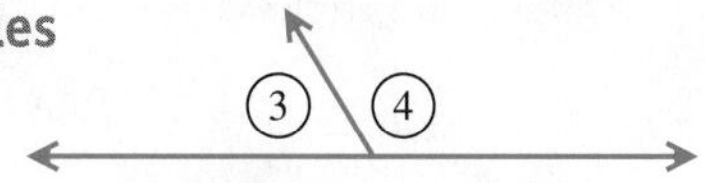

Angles ③ and ④ are supplementary.

Vertical Angles
Vertical angles have equal measures.

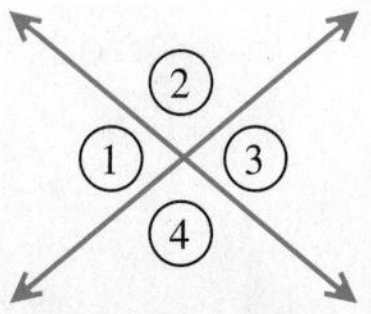

Angle ① = Angle ③

Angle ② = Angle ④